Handbook of Lung Targeted Drug Delivery Systems

Handbook of Lung Targeted Drug Delivery Systems

Recent Trends and Clinical Evidences

Edited by
Yashwant Pathak and Nazrul Islam

CRC Press
Taylor & Francis Group
Boca Raton London New York

CRC Press is an imprint of the
Taylor & Francis Group, an **informa** business
A CHAPMAN & HALL BOOK

First edition published 2022
by CRC Press
6000 Broken Sound Parkway NW, Suite 300, Boca Raton, FL 33487-2742

and by CRC Press
2 Park Square, Milton Park, Abingdon, Oxon, OX14 4RN

CRC Press is an imprint of Taylor & Francis Group, LLC

Library of Congress Cataloguing-in-Publication Data
Names: Pathak, Yashwant, editor. | Islam, Nazrul, editor.

Title: Handbook of lung targeted drug delivery systems : recent trends and clinical evidences / edited by Yashwant Pathak and Nazrul Islam.
Description: First edition. | Boca Raton : CRC Press, 2021. | Includes bibliographical references and index.
Identifiers: LCCN 2021021768 (print) | LCCN 2021021769 (ebook) | ISBN 9780367490676 (hardback) | ISBN 9781003046547 (ebook)
Subjects: LCSH: Respiratory organs--Diseases--Treatment. | Drug delivery systems.
Classification: LCC RC731 .H36 2021 (print) | LCC RC731 (ebook) | DDC 616.2/00461--dc23
LC record available at https://lccn.loc.gov/2021021768
LC ebook record available at https://lccn.loc.gov/2021021769

ISBN: 978-0-367-49067-6 (hbk)
ISBN: 978-0-367-49552-7 (pbk)
ISBN: 978-1-003-04654-7 (ebk)

DOI: 10.1201/9781003046547

Typeset in Times New Roman
by MPS Limited, Dehradun

To the loving memories of my parents and Dr. Keshav Baliram Hedgewar, who gave proper direction to my life; to my beloved wife, Seema, who gave positive meaning, and my son, Sarvadaman, who gave a golden lining to my life.

Yashwant Pathak

In gratitude to my beloved parents, who were my inspiration, and to my wife, Dr Nelufa Begum and two sons Tahseen and Taneem for their wholehearted support.

Nazrul Islam

Contents

Editors

Dr Yashwant Pathak completed Ph.D. in Pharmaceutical Technology from Nagpur University, India and EMBA & MS in Conflict Management from Sullivan University. He is Professor and Associate Dean for Faculty Affairs at College of Pharmacy, University of South Florida. Tampa, Florida. With extensive experience in academia as well as industry, he has more than 100 publications and 2 patent and 2 patent application, 20 edited books published, including 7 books in Nanotechnology and 6 in Nutraceuticals and drug delivery systems. Dr Yashwant Pathak is also a Professor of Global health with College of Public health at USF, and recently his student completed PhD in Global health on Homeopathy medicine and its efficacy, a survey of 1274 patients in Pune in India.

Dr. Nazrul Islam completed his PhD from Monash University (Australia), focusing on the development of dry powder inhaler (DPI) formulation for deep lung delivery of drugs. He worked at Mayne Pharma (Australia) as a formulation scientist. Currently, he is working in the Pharmacy Discipline at the Faculty of Health, Queensland University of Technology (QUT), Australia. As a senior lecturer, he is engaged in teaching and research in drug delivery. Dr. Islam has made key contributions to understanding the mechanism of micronized drug dispersion from dry powder agglomerates, which stipulate the formulation strategies to increase drug delivery efficiency from the formulations. He has published more than 50 research articles and 2 book chapters. He is a member of the American Association for the Pharmaceutical Scientist (AAPS) and the Australian Pharmaceutical Science Association (APSA). His key research areas are pulmonary drug delivery, nanotechnology, biodegradable polymer nanoparticles, DPI formulation, aerosol delivery, drug formulation and delivery, and stability studies of drug and drug products.

List of Contributors

Samson A. Adeyemi
Wits Advanced Drug Delivery Platform Research Unit
Department of Pharmacy and Pharmacology
School of Therapeutic Science
Faculty of Health Sciences, University of the Witwatersrand
Johannesburg, South Africa

Amelia Alberts
Graduate & Postdoctoral Affairs
Department of Molecular Pharmacology & Physiology
Morsani College of Medicine
University of South Florida
Tampa, Florida, USA

Rex Frimpong Anane
CAS Center for Excellence in Animal Evolution and Genetics
Kunming Institute of Zoology
Chinese Academy of Sciences
Kunming, Yunnan, China

Mary Ann Archer
Department of Pharmaceutics
School of Pharmacy and Pharmaceutical Sciences
University of Cape Coast
Cape Coast, Ghana

Christina Osei Asare
Department of Pharmaceutical Sciences
School of Applied Sciences, Central University
Miotso, Ghana

Seth Kwabena Amponsah
Department of Medical Pharmacology
University of Ghana Medical School
Accra, Ghana

Mudassir Ansari
Shobhaben Pratapbhai Patel School of Pharmacy and Technology Management
SVKM's NMIMS University
Mumbai, India

Shivani Baliyan
Amity Institute of Pharmacy
Amity University Uttar Pradesh
Noida, India

Colin J. Barrow
School of Life and Environmental Sciences
Centre for Chemistry and Biotechnology
Deakin University
Victoria, Australia

Urvashi Bhati
Reliance Life Sciences
Dhirubhai Ambani Life Sciences Center
Maharashtra, India

Sai HS. Boddu
Department of Pharmaceutical Sciences
College of Pharmacy and Health Sciences, Ajman University
Ajman, UAE

Nayanmoni Boruah
Department of Pharmaceutical Sciences
Faculty of Science and Engineering
Dibrugarh University
Assam, India

Unnati D. Chanpura
Graduate School of Pharmacy
Gujarat Technological University
Gandhinagar Campus
Gujarat, India

Soumalya Chakraborty
National Institute of Pharmaceutical Education and Research (NIPER)
Punjab, India

Pronobesh Chattopadhyay
Pharmaceutical Technology Division Defence Research Laboratory
Assam, India

Sunita Chaudhary
Arihant School of Pharmacy and Bio Research Institute
Gujarat, India

Yahya E. Choonara
Wits Advanced Drug Delivery Platform Research Unit
Department of Pharmacy and Pharmacology
School of Therapeutic Science
Faculty of Health Sciences
University of the Witwatersrand
Johannesburg, South Africa

Jeffrey Cruz
Graduate & Postdoctoral Affairs
University of South Florida Morsani College of Medicine
Tampa, Florida, USA

Ofosua Adi-Dako
Department of Pharmaceutics & Microbiology
School of Pharmacy, University of Ghana
Accra, Ghana

Aparoop Das
Department of Pharmaceutical Sciences, Dibrugarh University
Dibrugarh, India

Sanghita Das
Pharmaceutical & Fine Chemical Division
Department of Chemical Technology
University of Calcutta
Kolkata, India

Rajiv Dahiya
School of Pharmacy
Faculty of Medical Sciences
The University of the West Indies, St. Augustine
Trinidad & Tobago, West Indies

Sunita Dahiya
Department of Pharmaceutical Sciences
School of Pharmacy
Medical Sciences Campus
University of Puerto Rico
San Juan, USA

Raj Das
School of Engineering
RMIT University
Melbourne, Australia

Sanjoy Kumar Das
Institute of Pharmacy
West Bengal, India

Herita R. Desai
Bombay College of Pharmacy
Kalina, Santacruz
Maharashtra, India

Akash Dey
National Institute of Pharmaceutical Education and Research (NIPER)
Punjab, India

Ahmed S. Fahad
Department of Pharmaceutical and Policy Sciences
The University of Toledo
Health Science Campus
Toledo, Ohio, USA

Anusmriti Ghosh
School of Mechanical
Medical and Process Engineering
Queensland University of Technology
Brisbane, Australia

Urvashee Gogoi
Department of Pharmaceutical Sciences
Dibrugarh University
Assam, India

Mohammad Rahimi-Gorji
Faculty of Medicine and Health Sciences
Ghent University
Belgium

Y.T. Gu
School of Mechanical
Medical and Process Engineering
Science and Engineering Faculty
Queensland University of Technology
Brisbane, Australia

Swati Gupta
Amity Institute of Pharmacy
Amity University Uttar Pradesh
Noida, India

Nana Ama Mireku-Gyimah
Department of Pharmacognosy and Herbal Medicine
School of Pharmacy
University of Ghana
Accra, Ghana

Louis Hamenu
Department of Chemistry
School of Physical and Mathematical Sciences
University of Ghana
Accra, Ghana

Hemanga Hazarika
Department of Pharmaceutical Sciences
Dibrugarh University
Assam, India

Fen Huang
School of Energy and Environment
Southeast University
Nanjing, China

Johirul Islam
Department of Pharmaceutical Sciences
Dibrugarh University
Assam, India

Mohammad S. Islam
School of Mechanical and Mechatronic Engineering
University of Technology Sydney (UTS)
Ultimo, Australia

Nazrul Islam
Queensland University of Technology
Pharmacy Discipline
School of Clinical Sciences
Faculty of Health
Brisbane, QLD, Australia

Abhay Ittadwar
Professor and Head
Gurunanak College of Pharmacy
Nagpur, India

Anuj Jain
Department of Applied Mechanics
Motilal Nehru National Institute of Technology Allahabad
Prayagraj, India

Deepti Kaushalkumar Jani
Associate Professor and Head
Department of Pharmacology
Babaria Institute of Pharmacy, BITS Edu Campus
Vadodara, India

Vinita Kale
Professor
Department of Pharmaceutics
Gurunanak College of Pharmacy
Nagpur, India

Jagat R. Kanwar
School of Medicine
Faculty of Health
Institute for Innovation in Mental and Physical Health and Clinical Translation (IMPACT)
Deakin University
Victoria, Australia

Rupinder K. Kanwar
Department of Translational Medicine Centre
All India Institute of Medical Sciences (AIIMS)
Madhya Pradesh, India

Arnab Kapat
Reliance Life Sciences
Dhirubhai Ambani Life Sciences Center
Maharashtra, India

Sheshanka Kesani
Taneja College of Pharmacy
University of South Florida
Tampa, Florida, USA

Shraddha Khairnar
Shobhaben Pratapbhai Patel School of Pharmacy and Technology Management
SVKM's NMIMS University
Mumbai, India

Pradeep Kumar
Wits Advanced Drug Delivery Platform Research Unit
Department of Pharmacy and Pharmacology
School of Therapeutic Science
Faculty of Health Sciences
University of the Witwatersrand
Johannesburg, South Africa

Sakshi Kumar
Amity Institute of Pharmacy
Amity University Uttar Pradesh
Noida, India

Doris Kumadoh
Centre for Plant Medicine Research
Mampong-Akuapem, Ghana

Puchanee Larpruenrudee
School of Mechanical and Mechatronic Engineering
University of Technology Sydney (UTS)
Ultimo, NSW, Australia

Amy Le
Taneja College of Pharmacy
University of South Florida
Tampa, Florida, USA

Renjie Li
Department of Chemical Engineering
Monash University
Victoria, Australia

Bhaskar Mazumder
Department of Pharmaceutical Sciences
Dibrugarh University
Dibrugarh, India

Caline McCarthy
University of South Florida Morsani College of Medicine
Department of Molecular Pharmacology & Physiology
Tampa, Florida, USA

Tejal Mehta
Institute of Pharmacy
Nirma University
Ahmedabad, India

Humera Memon
Institute of Pharmacy
Nirma University
Ahmedabad, India

Hao Miao
Department of Chemical Engineering
Monash University
Clayton, Victoria, Australia

Srijan Mishra
Department of Pharmacology
Institute of Pharmacy
Nirma University
Ahmedabad, India

Keshav Moharir
Assistant Professor
Gurunanak College of Pharmacy
Nagpur, India

Samyak Nag
Amity Institute of Pharmacy
Amity University Uttar Pradesh
Noida, India

Jerry Nesamony
Department of Pharmacy Practice
The University of Toledo
Health Science Campus
Toledo, Ohio, USA

Esther Eshun Oppong
Department of Pharmaceutics
School of Pharmacy and Pharmaceutical Sciences
University of Cape Coast
Cape Coast, Ghana

Yaa Asantewaa Osei
Department of Pharmaceutics
Faculty of Pharmacy and Pharmaceutical Sciences
Kwame Nkrumah University of Science Technology
Cape Coast, Ghana

Komal Parmar
ROFEL, Shri G.M. Bilakhia College of Pharmacy
Gujarat, India

Urvashi K. Parmar
SVKM's Bhanuben Nanavati College of Pharmacy
Mumbai, India

Anita Patel
Faculty of Pharmacy
Nootan Pharmacy College
Sankalchand Patel University
Gujarat, India

Dasharath M. Patel
Graduate School of Pharmacy
Gujarat Technological University
Gandhinagar Campus
Gujarat, India

Dhaval Patel
Saraswati Institute of Pharmaceutical Sciences
Gujarat, India

Jasmin Patel
MBBS, SMIMER
Surat, Gujarat, India

Jayvadan Patel
Nootan Pharmacy College
Faculty of Pharmacy
Sankalchand Patel University
Gujarat, India

Veer J. Patel
Graduate School of Pharmacy
Gujarat Technological University
Gandhinagar Campus
Gujarat, India

Vimal Patel
Institute of Pharmacy
Nirma University
Gujarat, India

Manash Pratim Pathak
Pratiksha Institute of Pharmaceutical Sciences
Assam, India

Manash K. Paul
Scientist, Principal investigator
Division: Pulmonary and Critical Care Medicine
David Geffen School of Medicine
University of California Los Angeles
California, USA

Kalyani Pathak
Department of Pharmaceutical Sciences
Dibrugarh University
Assam, India

Yashwant Pathak
Professor and Associate Dean for Faculty Affairs
USF Health Taneja College of Pharmacy
University of South Florida
Adjunct Professor
Faculty of Pharmacy
Airlangga University
Surabaya, Indonesia

Pompy Patowary
Department of Pharmaceutical Technology
Defense Research Laboratory
Assam, India

Atul Pathak
HBT Labs Inc.
Vanguard Way Brea
California, USA

Akshoy R. Paul
Department of Applied Mechanics
Motilal Nehru National Institute of Technology
Uttar Pradesh, India

Gunther Paul
James Cook University
Australian Institute of Tropical Health and Medicine
Townsville, Australia

Viness Pillay
Wits Advanced Drug Delivery Platform Research Unit
Department of Pharmacy and Pharmacology
School of Therapeutic Science
Faculty of Health Sciences
University of the Witwatersrand
Johannesburg, South Africa

Charles Preuss
University of South Florida Morsani College of Medicine
Department of Molecular Pharmacology & Physiology
Tampa, Florida, USA

Venkata Ramana
Reliance Life Sciences
Dhirubhai Ambani Life Sciences Center
Maharashtra, India

Abdur Rashid
Department of Pharmaceutics
College of Pharmacy
King Khalid University
Abha, Saudi Arabia

Raja Reddy Bomma Reddy
Senior Scientists
Tergus Pharma LLC
North Carolina, USA

Subhabrata Ray
Dr. B. C. Roy College of Pharmacy and Allied Health Sciences
Durgapur, India

Sudipta Roy
Bengal School of Technology (A College of Pharmacy)
West Bengal, India

Prabin K. Roy
Department of Pharmaceutical Sciences
Dibrugarh University,
Assam, India

Riya Saikia
Department of Pharmaceutical Sciences
Dibrugarh University
Assam, India

Surovi Saikia
Natural Products Chemistry Group
CSIR North East Institute of Science & Technology, Jorhat
Assam, India

Dhananjoy Saha
Directorate of Technical Education, Bikash Bhavan
Kolkata, India

Suvash C. Saha
School of Mechanical and Mechatronic Engineering
University of Technology Sydney (UTS)
Sydney, Australia

Somchai Sawatdee
Walailak University
Nakhon Si Thammarat, Thailand

Hiral Shah
Arihant School of Pharmacy & BRI
Gujarat, India

Jigna Shah
Professor
Department of Pharmacology
Institute of Pharmacy
Nirma University
Ahmedabad, Gujarat, India

Vandit Shah
Post-graduate scholar
Department of Pharmacology
L. M. College of Pharmacy
Ahmedabad, Gujarat, India

Hemanta Kumar Sharma
Department of Pharmaceutical Sciences
Faculty of Science and Engineering
Dibrugarh University
Assam, India

Himangshu Sharma
Department of Pharmaceutical Sciences
Faculty of Science and Engineering
Dibrugarh University
Guwahati, Assam, India

Ujwala Shinde
Department of Pharmaceutics
Bombay College of Pharmacy
Mumbai, Maharashtra, India

Kavita Singh
Shobhaben Pratapbhai Patel School of Pharmacy and Technology Management
SVKM's NMIMS University
Mumbai, India

Agnivesh Shrivastava
Gattefosse India Pvt. Limited
Mumbai, Maharashtra, India

Kianna Samuel
Taneja College of Pharmacy
University of South Florida
Tampa, Florida, USA

Teerapol Srichana
Department of Pharmaceutical Technology
Prince of Songkla University
Hat Yai, Songkla, Thailand

Tan Suwandecha
Prince of Songkla University
Hat Yai, Songkla, Thailand

Anurag Tiwari
Department of Applied Mechanics
Motilal Nehru National Institute of Technology Allahabad
Prayagraj, India

Truong Tran
College of Pharmacy
University of South Florida
Tampa, Florida, USA

Zhenbo Tong
School of Energy and Environment
Southeast University
Nanjing, China

Khushali Vashi
MPH, BDS, CPH
University of South Florida
Tempa, FL, USA

Deepa U. Warrier
Bombay College of Pharmacy, Kalina, Santacruz
Mumbai, India

Hui Wang
Livzon Microsphere Technology Co. Ltd
Zhuhai, Guangdong, China

Haoqin Yang
Institute for Process Modelling and Optimization
JITRI, Suzhou, China

Jiaqi Yu
Institute for Process Modelling and Optimization
JITRI, Suzhou, China

Xudong Zhou
Department of Chemical Engineering
Monash University
Clayton, Victoria, Australia

Kamaruz Zaman
Associate Professor
Department of Pharmaceutical Sciences
Dibrugarh University, Dibrugarh
Assam, India

Foreword

For centuries, people have been inhaling drugs for both recreational and medicinal purposes. The human lung is a unique organ, well suited for drug delivery. It provides a large surface area, a thin alveolar epithelium layer with an exceedingly vast network of capillaries ensuring an excellent blood supply. These factors enhance lung drug absorption, resulting in increased bioavailability and rapid onset of action. The respiratory tract has evolved several defense mechanisms to prevent the inhaled materials out of the lungs and to remove or inactivate them once they get deposited. Despite these barriers, interest in pulmonary drug delivery is ever increasing and has attracted scientists around the world to work in the field of lung drug delivery systems and formulations because of the tremendous opportunities it offers.

Currently, most of the focus is on the direct delivery of drugs into the lungs using pressurized metered dose inhalers or dry powder inhalers. The advent of nanotechnology has triggered significant interest in nano-drug delivery systems targeted toward the lungs. Nanoparticles provide enormous surface area. In addition, the large absorption area of the lungs helps in the absorption of large numbers of nanoparticles in a short time.

Lung delivery of drugs offers a non-invasive administration route for systemic and local delivery of drugs. Lung drug delivery is also drawing attraction due to the fact that the respiratory epithelial cells have a prominent role in regulation of air passage tone and airway lining fluid production

Progress in genomics and molecular biology has helped in better understanding of the lung morphology and physiology. This new knowledge is being applied to develop better drug delivery systems via the lungs.

Interest in drug delivery to the lungs has resulted in the development of new environmentally friendly propellants, non-aqueous inhalers, user-friendly dry powder inhalers, nebulizers, and other related devices.

The recent Covid-19 pandemic has shown how important it is to maintain healthy lungs and how not doing so can lead to many health complications. A number of research groups, both in universities and industry, are working on the development of a nasal vaccine against the Covid-19 virus. Success of a nasal vaccine for influenza, approved by the FDA, has stimulated great interest in this area.

This book is focused on targeted drug delivery systems for lungs and covers most recent trends in the research and provides clinical evidence for the usefulness of lung delivery systems. The handbook includes 48 chapters covering various aspects of lung drug delivery systems. The leading scientists and academicians in this field have written the majority of these chapters.

This is a very timely and wonderful reference book for academicians and industry people who are working in this area. I would like to congratulate the editors, Yashwant Pathak and Nazrul Islam, for their sincere hard work to put this book together and to all the chapter authors who have contributed to this volume.

I am sure the readers will find it of great benefit, and this handbook will contribute to the world of science in the field of lung targeted drug delivery systems.

M.N.A. Rao, M.Pharm., Ph.D.,
General Manager, R&D, Divis Laboratories,
Hyderabad, India

Preface

The lungs are the first organ affected by Covid-19. In the early days of infection, the virus enters the lung cells and starts showing the effects that can lead to severe failure of the lungs. As a result of the number of unprecedented fatalities due to the Covid-19 pandemic, systems that deliver drugs to the lungs are currently receiving significant attention, not simply for local drug delivery, but because they can also be used as a route of extremely fast absorption for systemic effects of the drugs. Lung drug delivery systems have shown great advantages, including high permeability, large surface area—approximately 100 square meters—and the rapid onset of drug action, comparable to intravenous drug delivery systems.

With the advent of nanotechnology, smaller particles are found to be the most suitable for lung delivery systems. Particle size and size distribution can be a major factor to improve the absorption from lungs. Lung drug delivery is advantageous in the treatment of respiratory diseases when compared with other routes of administration. Inhalation therapy can provide the direct application of a drug within the lungs. The local lung deposition and delivery of the administered drug facilitates a targeted treatment of respiratory diseases, such as pulmonary arterial hypertension, without the need for high dose exposure to drugs by other routes.

This handbook of lung-targeted drug delivery systems, including recent trends and clinical evidence, is a collection of 47 chapters written by well-known scientists who have a lot of experience and expertise in this field. The book is divided into 10 subthemes:

1. Understanding Physiology and Pharmacology of Lungs
2. Modeling for Lung delivery
3. Polymeric Nanoparticles for Lung Delivery
4. Lipid Nanoparticles for Lung Delivery
5. Metallic Nanoparticles
6. Specific Formulations Targeted to Pulmonary Systems
7. Devices for Lung Drug Delivery
8. Regulatory Requirements for Lung Drug Delivery Systems
9. Recent Trends in Clinical Applications
10. Future Directions and Challenges

For each of the subthemes we have included a few chapters, which has covered relevant aspects of lung-targeted drug delivery systems.

We believe this handbook will be an excellent reference book for the scientists and industry people who are working in this field. We express our sincere thanks to all the authors who contributed to this book, without their help, we would not have achieved this enormous task.

Our sincere gratitude to Ms Hillary Lafoe, Jessica Poile, Rachel Panthier, and other CRC press, all their employees, Mr Manmohan Negi and all his team, the printing press employees who helped to make this book see the daylight.

It took a lot of our time from our families; we wish to recognize their sacrifices for this book.

If you find any errors in this book, kindly send us the details and we will update in the second edition. We are aware this was a tremendously big task on our part and there might be some lacunas left in the book, we express our sincere apologies for the same.

We would appreciate constructive comments from readers that will help us on our future projects.

Yashwant Pathak and Nazrul Islam

1

Introduction to Lung Physiology from a Drug Delivery Perspective

Aparoop Das[1,#], Manash Pratim Pathak[1,2], Pompy Patowary[1,3], and Sanghita Das[3,4]
[1]Department of Pharmaceutical Sciences, Dibrugarh University, Dibrugarh, India
[2]Pratiksha Institute of Pharmaceutical Sciences, Assam, India
[3]Division of Pharmaceutical Technology, Defence Research Laboratory, Assam, India
[4]Pharmaceutical & Fine Chemical Division, Department of Chemical Technology, University of Calcutta, Kolkata, India

1.1 Introduction

1.1.1 Brief Introduction to Nanotechnology and Nanoparticle Mediated Drug Delivery

Nanoparticles (NPs) are nanosized materials used to embed drugs, imaging agents and genes intended for targeted drug delivery by covalent conjugation or noncovalent attachment (1). As defined by the International Organization for Standardization, NPs are those having at least one dimension less than 100 nm. The American Food and Drug Administration (FDA) has cited another broader definition of NPs as 'engineered to exhibit properties or phenomena attributable to dimensions up to 1000 nm', which is typically adopted in academic research (2). To deliver the appropriate amount of the desired drug precisely to the target organ without causing any side effects while also taking care of the induction of drug resistance is a daunting task; however, it is an important requirement in a targeted drug delivery system (DDS) (3). NP drug carriers can modulate drug distribution via passive and active targeting. Passive targeting is the process by which nanoscaled particles accumulate in tumors/sites of inflammation merely due to their size, whereas active targeting works through the attachment of biochemical moieties which facilitate delivery to diseased tissues expressing biomarkers distinguishing them from the surrounding healthy tissue (4). The term *nanotechnology* involves manipulating and controlling nanoscale (1–100 nm) objects. For over 28 years, the potential benefit of nanotechnology has been appreciated by most researchers and has revolutionized the field of drug delivery and drug targeting with the hope of transforming Ehrlich's hypothetical concept of a "magic bullet" into clinical reality.

The new interdisciplinary platform of nanotechnology-based DDS shows a great deal of promise with several advantages such as improved solubility and bioavailability of hydrophobic drugs, rendering them suitable for parenteral administration encapsulation efficacy which enhances the drug release profiles, high drug payload, extended drug half-life, improved therapeutic index of peptides, oligonucleotides, etc., controlled release along with reduced immunogenicity, and toxicity (5). They can further be used for drug delivery to the central nervous system (CNS) owing to their smaller size and higher barrier permeability.

1.1.2 The Inhalation Route

Drug delivery via the inhalation route has been known since the distant past, when the Egyptians inhaled black henbane vapour to treat various diseases (6). The lung is taken into high consideration as an attractive target among a myriad of routes of administration due to its several advantages (2) over conventional oral administration. Drug delivery to the lungs by inhalation offers a targeted drug therapy for a wide range of respiratory diseases including lung cancer, chronic obstructive pulmonary disease (COPD), asthma, cystic fibrosis (CF), tuberculosis (TB), pneumonia and pulmonary hypertension (7), among which COPD, lower respiratory infections and lung cancer are the third, fourth and sixth causes of death worldwide respectively. Other than pulmonary disease, inhaled medications are also beneficial for treatment of anxiety, hypertensive crisis, certain seizures, arrhythmias, spasms, nausea and myriad forms of pain (8).

The inhalation route has received widespread attention within the drug delivery field as it possesses many characteristics ideal for drug transportation and has many advantages over other methods, since the lungs have a large surface area (100 m^2) and a very thin epithelial layer (0·2–0·7 μm) for high blood perfusion, elevated blood flow (5 L/min), extensive vascularization to promote systemic delivery, high solute permeability, avoidance of first pass metabolism, and sustained release of the drug and thus reduced dosing frequency (9–11). In addition, drug-metabolizing enzymes are present in much lower concentrations in the lungs than the gastrointestinal tract and liver, thereby limiting enzymatic and proteolytic interaction with the inhaled molecules (8). Moreover, the lung is capable of absorbing

First Author and Primary Contact

DOI: 10.1201/9781003046547-1

pharmaceuticals either for local deposition or for systemic delivery by targeting the drug delivery carriers to the alveolar region with uniform distribution. Additionally, in inhalation drug delivery there is no risk of needle injuries and health-care staff are not needed, resulting in better patient compliance. The pulmonary tissue has notably high levels of systemic bioavailability for macromolecules, making it the best non-invasive absorption route. Although drug molecules are absorbed more efficiently from the lung than from any other non-invasive routes, however, the therapeutic efficacy of inhaled drugs is limited by their rapid clearance in the lungs (12).

The practice of respiratory medicine has entered into a brave new era through the development of new therapeutic strategies and improvement of current therapies, with some NPs already developed into commercial products. Although inhalation devices and aerosols containing various drugs have been used since the early 19th century, currently there are three main delivery devices used for pulmonary delivery of drugs, viz. nebulizers, pressurized metered-dose inhalers and dry powder inhalers (DPIs) to deliver solutions, suspensions and dry particles respectively (13, 14) and an array of carrier systems, which hold great potential for treating diseases that require direct lung delivery. An ideal delivery device has to generate an aerosol in the range of 0.5–5 μm and provide reproducible drug dosing without affecting the physical and chemical stability of the drug formulation (15). Moreover, the ideal inhalation system must be a simple, convenient, inexpensive and portable device. Although efficient and reproducible pulmonary deposition of aerosol medicines is now possible with current technology, however, the mechanism of the action of particles after pulmonary deposition is highly complex due to the various barriers and clearance mechanisms in the respiratory tract, which pose significant challenges in formulation development (6). The fate of particle–lung interactions are still being researched and have become indispensable for pharmaceutical scientists to develop more effective inhalable formulations. Moreover, toxicity and regulatory concerns limit approval of nanotechnology-based medicinal products (16).

The primary focus of this chapter is to provide an insight into the physiological and efficacy aspects of the mechanism of pulmonary DDS with respect to the lung structure and characteristics, and also the research progress on various parameters to be considered during formulation development and how molecular properties affect rate and extent of pulmonary drug absorption, clearance and metabolic mechanisms, etc. Herein, we have also highlighted the future opportunities for nanotechnology focusing on the treatment of lung diseases.

1.2 Anatomy and Physiology of Lungs

A. Anatomy of lungs/gross tissue organization:

i. The respiratory system:

Energy is produced in the form of adenosine triphosphate (ATP) in the body through cellular respiration. For cellular respiration, cells need oxygen (O_2) as a reactant, and carbon dioxide (CO_2) is produced as a waste product. As excessive CO_2 may become toxic, it is eliminated from the body by the joint cooperation of the respiratory and cardiovascular systems. However, intake of O_2 and expulsion of CO_2 is performed solely by the respiratory system, and the cardiovascular system supports the entire system by aiding in the transportation of blood containing the gases between the body cells and lungs. Apart from the regular gas exchange, the respiratory system also participates in a repertoire of functions like blood pH regulation, aiding in smelling, and production of the voice, and it helps the body gets rid of heat and water in exhaled air (17, 18).

Structurally, the respiratory system consists of two main parts: (a) upper respiratory system and (b) lower respiratory system. The nose, nasal cavity, and pharynx and its associated structures comprise the upper respiratory system and the larynx, trachea, bronchi and bronchioles comprise the lower respiratory system (17). Functionally, also, the respiratory system is divided into two parts, i.e. the conducting zone and the respiratory zone. The conducting zone is not directly involved in gas exchange which includesthe nose, nasal cavity, pharynx, larynx, trachea, bronchi, bronchioles and terminal bronchioles. These components filter, moisten or warm the air and facilitate conduction of the air. The respiratory zone is the site where gaseous exchange takes place and this zone includes respiratory bronchioles, alveolar ducts, alveolar sacs and alveoli, and here the gaseous exchange takes place between air and blood (18).

ii Upper respiratory system:

a. Nose and nasal cavity:

The nose is the external and visible component of the respiratory system and is followed by the internal component inside the skull called the nasal cavity or internal nose. The nose is made up of bone and hyaline cartilage which is covered by muscle and skin and is lined by mucous-secreting cells called goblet cells. The nasal cavity is a two-way irregular passage for the entry of air and is divided by a septum. The posterior part of the septum is made up of ethmoid bone and vomer, and anteriorly made up of hyaline cartilage. The nasal cavity is lined up with the vascular ciliated columnar epithelium containing goblet cells that secretes mucous which coats the nasal hairs and make them sticky. Bones in the face and cranium that are filled with air are called paranasal sinuses and are present in the nasal cavity. Some of the sinuses are maxillary sinuses in the lateral wall, frontal and sphenoidal sinuses in the roof and ethmoidal sinuses in the upper portion of the lateral wall of the nasal cavity (17, 19).

Inspired air passes through the nose, which is the first respiratory passage, and then the air enters the nasal cavity where air is filtered, moistened or humidified, and warmed. Sticky hairs in the nares of the nasal cavity filter the air from any particles or microbes from the external environment. As the filtered air passes through the mucosa, which are in moist condition, the air becomes moistened. There is a huge vascularity surrounding the mucosa that makes the passing air warm inside the nasal cavity. Other than the regular respiratory functions, the nose and nasal cavity perform some non-respiratory functions like aiding in olfactory stimuli and modifying voice through speech vibrations.

b. Pharynx and its associated structures:

The pharynx is the continuation of the nasal cavity and is a tube-like structure formed of skeletal muscle which is lined by mucous membrane. The associated structures of the pharynx are basically its three major regions, namely the nasopharynx, the oropharynx and the laryngopharynx.

The nasopharynx is the superior portion of the pharynx that is posterior to the nasal cavity and extends up to the soft palate. It acts only as an airway and houses the pharyngeal tonsils in its top. Auditory tubes (or eustachian tubes), which connect to the middle ear cavity, directly open into the nasopharynx. The oropharynx is the intermediate part of the pharynx. The position of the oropharynx lies posterior to the oral cavity and extends towards the hyoid bone from the soft palate. It acts as the passageway for both air and food. There is a distinct change observed in the epithelium lining where the pseudostratified ciliated columnar epithelium changes to stratified squamous epithelium during the transformation of nasopharynx to oropharynx. Finally, the laryngopharynx forms the inferior portion of the pharynx, posterior to the larynx, which begins from the level of the hyoid bone. It continues as the passageway for air and ingested material where the respiratory and the digestive systems move away from its inferior part. Finally, the laryngopharynx ends in the larynx and enters the esophagus (18).

iii Lower respiratory system:

a. Larynx

The larynx or "voice box" is the connecting point between the laryngopharynx and the trachea. The size of the larynx differs slightly in males and females. It is usually larger in males than females, which is why males have a larger "Adam's apple" and deeper voices. The Adam's apple, or thyroid cartilage, is made up of two hyaline cartilages that are fused to form the anterior wall of the larynx, which makes its shape triangular. The thyroid cartilage (anterior), epiglottis (superior) and cricoid cartilage (inferior) make up the structure of the larynx.

The epiglottis is a leaf-shaped fibro-elastic cartilage attached to the thyroid cartilage and covers the opening of the trachea. It lies obliquely behind the tongue and behind the hyoid bone, and forms a lid over the 'glottis. The epiglottis is lined up by squamous epithelium. It closes the larynx during the swallowing of food materials and thus protects the lower airway passage from food particles.

The vocal cords are stretched across the opening of the larynx and are made up of non-keratinized stratified squamous epithelium, goblet cells and basal cells. Tension in the vocal cords produced by the surrounding muscles determines the pitch of the voice. Vocal cords produce low-pitched sounds when the muscles around the vocal cords are relaxed followed by opening of the vocal cords; this phenomenon is called *abducted* (open). When the surrounding muscles contracts, the vocal cords are stretched out and produce a high-pitched voice; this phenomenon is called *adducted* (closed) (19).

b. Trachea:

The trachea or windpipe is the continuation of the larynx, located anterior to the esophagus and divide into left and right primary bronchi at a junction known as the *carina*. The layers of the wall of trachea, starting from superficial to deep are (1) adventia, (2) hyaline cartilage, (3) submucosa, and (4) mucosa. The function of the trachea is quite similar to the nasal cavity, i.e. same protection against dust with the help of a mucous membrane lining consisting of ciliated pseudostratified columnar epithelium.

c. Bronchi and bronchioles:

Bronchi are separated to left and right at the carina. The carina is surrounded by nervous tissue that detects foreign particles and induces coughing to expel them. Cartilage rings give mechanical support to the bronchioles and prevent them from collapse. Pseudostratified ciliated columnar epithelium lines up the bronchus that contains goblet cells that produces mucus. Bronchi act as a passageway for air and trap foreign particles in and out of the lungs. Bronchioles, which are 1 mm in diameter, start from the tertiary bronchi and branch further to terminal bronchioles that facilitates gaseous exchange through the alveolr sacs.

d. Alveoli:

With the respiratory bronchioles begins the respiratory zone. When respiratory bronchioles reach deeper, the epithelial lining changes from simple cuboidal to simple squamous epithelium. The respiratory bronchioles are further subdivided into alveolar ducts and the terminal parts are called alveolar sacs, which are analogous to a cluster of grapes. Alveoli basically participate in the gas exchange. Two types of cells are found in the lining of the alveoli: type I alveolar cells and type II alveolar cells. Ninety-seven percent of the alveolar surface area is made up of type I alveolar cells which are the main site for gaseous exchange. Type II alveolar cells secrete alveolar fluids that help in maintaining moisture between the cells and the air, thus reducing the surface tension of the alveoli. The alveolar wall is surrounded by alveolar macrophages that phagocytes foreign particles and pathogens that reach the alveoli (17, 18).

B. Particle deposition:

The deposition of particles in the lungs takes place four different ways (20). They are interception, impaction, sedimentation and diffusion.

i. Interception: When a particle travels very close to the epithelial surface, the particle tends to be intercepted; however, the size and length of the particle determines the interception. Fibres having the dimension of 1 micrometer (μm) and length of 200 μm suffer interception.
ii. Impaction: Suspended particles in air tend to travel in their own pathways; however, when there is a band along the path, the particle tends to stick or impact on

the surface of their path. This type of impaction depends on the velocity of the air and mass of the particle. Particles greater than 10 μm are deposited in the nose or throat and are unable to reach the lower respiratory system.

iii. Sedimentation: This type of deposition happens in the case of particles having size greater than 0.5 μm and takes place in the bronchi and bronchioles. When particle size is very small, the buoyancy of the particle is overcome by gravitational force and air resistance, and as a result, the particle is deposited on the surface.

iv. Diffusion: This type of deposition happens in the case of particles of sizes smaller than 0.5 μm and takes place in small airways and alveoli. Particles with such smaller sizes act like gas molecules and follow the *Brownian motion*. Due to the smaller size of the particles, their movement is vigorous and they are deposited mostly in the lower respiratory region; however, the may also deposit in the upper respiratory region.

C. Pulmonary surfactant:

Pulmonary surfactant lines the surface of the alveoli, lowering surface tension and preventing atelectasis (complete or partial collapse of the entire lung or some area of the lung) during breathing. Surfactant consists of a phospholipid, dipalmitoylphosphatidylcholine, and four surfactant-associated proteins, SP-A, SP-B, SP-C, and SP-D. Out of the four, SP-A and SP-D are hydrophilic and participate in the pulmonary host defence. SP-B and SP-C are hydrophobic and assist in reducing the surface tension of the surfactant. The cuboidal type-II cells in the epithelium produce pulmonary surfactant. Initially, a thin layer of any inhaled drug will be deposited in the pulmonary surfactant, followed by deposition and absorption in other respiratory regions (21, 22). The surface tension reducing property of the surfactant prevents the drug particle from forming drug droplet aggregation, and so, when designing a inhaled drug formulation, care should be taken to select drugs or associated compositions that do not disturb or damage the pulmonary surfactant.

D. Mechanism of drug absorption and clearance:

Study of pulmonary absorption and elimination is still in the nascent stage. The complex structure of the lungs and the presence of surfactant have made the study difficult. Soluble macromolecules are absorbed from the lungs to the body essentially by two mechanisms: absorptive transcytosis (passage through the cells) or transcellular transport and paracellular transport (passage between the cells).

Absorptive transcytosis may be independent of any receptor or transporters, like passive diffusion, and may involve receptor or transporter mediated binding to facilitate absorption. For a drug molecule to be absorbed, following passive diffusion, from the lung epithelial lining into the blood, the drug has to pass through various barriers such as the surfactant, epithelium, basement membrane and the vascular endothelium. However, the process of absorption may become more complex if the drug follows active transport and utilizes receptors or transporters (transport molecules). Transporters are surface molecules that facilitate the movement of drug molecules inside (influx transporter) or outside (efflux transporter) of the cell through the ATP dependent mechanism. Most of the drugs, mostly hydrophobic, are absorbed through passive diffusion. There are mainly two different families of transporters. They are the solute carrier transporters superfamily (SLC) and ABC. OCT1, OCT2 and OCT3 are some significant transporters belonging to the SLC family, and P-gp, MRP1, MRP2, MRP3, MRP4, MRP5 belong to the ABC family.

When transportation of molecules takes place in the junctional complex between two cells, then it is called paracellular transport. Paracellular transport may also occur between three cells when there is a specific spot on the circumference between endothelial and epithelial cells. In another type of paracellular transport, a large "pore" is formed in the basement membrane as a result of the death of a cell. This pore is eventually filled up by new cells and these new cells seal the monolayer in the basement membrane. Small hydrophilic and small peptide molecules can be absorbed by paracellular transport. Sometimes, absorption enhancers like chitosan may disrupt the tight paracellular junction, which may result in intercellular transport of the drug (23, 24).

Drugs are cleared from the respiratory system mainly by mucociliary clearance, enzymatic degradation and clearance by alveolar macrophage. All these clearance types depend upon the size, chemical nature and site of deposition of the drug particles.

i. Mucociliary clearance:

Mucociliary clearance is instrumental in removal of foreign particles like inhaled pollutants and pathogens as well as secretions. The mucociliary apparatus is made up of cilia, a gel-sol layer consisting of a *gel* phase that contains mucins and other glycoproteins overlaying the watery *sol* phase or periciliary fluid where cilia can beat without resistance. The mucus is produced by goblet cells in the epithelium and the submucosal glands. The ciliated cells transport the mucus containing the foreign particles in a proximal direction to get the mucus swallowed or expectorated. Particles larger than 6 μm in diameter are eliminated from the airways by mucociliary clearance; however, smaller particles penetrate the mucus and escape the mucociliary clearance (23, 25).

ii. Enzymatic degradation:

Similar to the liver, some metabolizing enzymes like CYP2S, CYP2F and CYP4B1 from the cytochrome P450 family are also found in the lungs. Some phase-II enzymes like UDP UGT, SULT, GST, estrase and peptidase are also reported to be expressed to some extent in the lungs. Apart from the cytochrome P450 enzymes, proteases also play an important role in the stability and pharmacokinetics of inhaled drugs and may contribute to the pathology of many respiratory diseases. Increased enzyme concentration in the lungs due to diseased condition plays as a barrier for delivery of drugs to treat respiratory diseases (26).

iii. Clearance by alveolar macrophage:

Any drug that tends to dissolve slowly is eliminated by the alveolar macrophages by means of phagocytosis. Macrophages transport the insoluble drug to the mucociliary escalator along the alveolar surface or translocate to the tracheobronchial lymph, or may be disposed of by enzymatic degradation internally. Phagocytosis is size dependent, and is optimal for particles having sizes of 1.5–3 μm (23, 25). So the particle size should be kept in mind to avoid clearance by the alveolar macrophages. Many engineered particles have successfully avoided alveolar clearance by increasing the porosity or by PEGylation (27, 28).

1.3 Nanoparticle-based Systems for Pulmonary Application

Nano-based drug delivery technology gains new horizons as unfavourable physical properties of the particles have been modified, including enhancing drug solubility and encapsulation efficiency, and improving the drug release profile in order to obtain better pharmacokinetic and therapeutic efficacy. In the 1990s, conventional colloidal carriers like emulsions, liposomes and polymeric nanoparticles were replaced by solid lipid nanoparticles (SLN), aqueous nanoscale suspensions made up of phospholipids and triglycerides because of the advantages such as improved stability at room as well as body temperature, lipophilic and hydrophilic drugs can be loaded, targeted drug delivery, higher drug content capacity, water-based technology, easy to sterilize, biodegradable and biocompatible (5, 29–31). SLN provide affordable and patient-friendly drug delivery through various routes, and hydrophilic SLN has shown promising effects for various ailments such as cancer and tuberculosis (32). Most importantly, SLN compositions exhibit physiological resemblance that allow them to exert the least toxic effects and make them ideal for pulmonary drug delivery. Benefits from administration of SLN encapsulated drugs through pulmonary routes include controlled release characteristics and rapid degradation. Also, SLN exerts physiological tolerability and can be drug targeted to pulmonary macrophages which makes SLN a useful tool in treating lung infections (33). Some core lung regions exhibit the predominant appearance of phospholipids which play a pivotal role in maintaining essential respiratory mechanisms, e.g. alveolar phospholipid-based surfactant proteins essential in regulating optimal surface tension and decreasing friction in the lung tissue (32, 34). Drug delivery through the pulmonary route opens up new horizons for drug administration to the lungs directly and is one of the most important non-invasive techniques because of its higher bioavailability as lower proteolytic activity and a highly vascularized, thin epithelial barrier, as well as a large alveolar surface area, enables more efficient drug absorption, reduced dosing frequency, and includes very few side effects along with bypassing first pass metabolism and the appropriate size desired for pulmonary delivery promotes transepithelial transport by avoiding alveolar macrophage clearance (32, 35, 36). Nebulization of SLN has gained increased interest in current research as it has been observed that aerosolized drugs showed quick absorption because of the large surface area of the alveoli and the presence of thin walls of the alveolar airspaces, leading to augmented blood flow and permitting entry to systemic circulation (31, 32). SLN can be chosen as a carrier to improve bioavailability of drugs in treating lung cancer, where nebulization of SLN of antitubercular drugs reduced the dosing frequency and also imparted improved drug bioavailability in the treatment of pulmonary tuberculosis. Other studies revealed that biodistribution of radio-labelled SLN has shown significant uptake of the radio-labelled SLN into the lymphatics after inhalation (37).

A. Solid lipid microparticles (SLM)

Solid lipid microparticles (SLM) are the least exploited particulate delivery system, as their suitability in comparison to submicron-sized particles are limited. Despite being the least popular, the main interest about lipid microparticles (LM) is due to their simpler production by high pressure homogenization, lucid characterization techniques, increase in size leading to extended release properties (38, 39) and reduced dose of microparticles as loading capacity is much higher. Solid lipid microparticles (SLN) are almost same as they have the same composition, with biocompatible lipid and solid particle matrixes and characteristic control release properties. Differences observed among SLN and SLM are in the size ranges, as SLN has a submicron size of 50–1000 nm and SLM is 1–50 μm. Interestingly, LMs display higher physicochemical stability than LNs. Lyophilization can be performed for LMs without any changing of particle characteristics, but aggregation of LNs occurs if lyophilization is carried out without cryoprotectors (40). Reduced surface area makes LM take up less surfactant than LNs to stabilize as maximizing surfactant ratios is prone to exhibit adverse effects (41). Particles in a dry powder inhaler tends to aggregate in lungs and undergo macrophagic clearance rapidly with geometric sizes between 1 and 3 μm and mass density near 1 g/cm^3 (42). Particle size is an important parameter as particles of <1 μm are easily exhaled before reaching the target and those smaller than 1–0.5 μm are deposited in the alveolar region where particles of >5 μm are deposited at the oropharynx and upper respiratory tract, and for appropriate lung deposition through inhalation drug delivery, particles should have an aerodynamic size range of 1–6 μm. SLM can be a suitable option for targeting drug in lungs through pulmonary delivery as anatomically the lungs curb the efficiency of inhaled aerosols depending on the particle size. ,The optimized particle size of 1–5 μm has been considered to reach and retain in the targeted airway region for stipulated period of time to exert the desired effect as SLMs of 1–50 μm in sizes express excellent flow properties and large geometric size helps in avoiding phagocytosis by macrophages in lung and attains the sustained release of formulations attainable (31). Some research on SLM has evidenced the promising drug delivery of quercetin to exert anti-oxidant and anti-inflammatory properties for treating ailments like asthma (43).

B. Polymers

Polymers are covalently bonded macromolecules and can be used for better delivery, and a drug's effects can be extended as it attaches to drugs by an encapsulation technique. A polymeric drug delivery system enables delivery of drug to the site of action with improved safety and efficacy due to benefits that include improved surface area; higher, non-degradable encapsulation efficiency of the drug; prolonged drug delivery and a long shelf life. The putative role of polymers is to encapsulate nanoparticles, microparticles and large porous particles (LLPs) for fabricating formulations. Natural (albumin, chitosan, hyaluronic acid, etc.) as well as synthetic (poly-lactic-co-glycolic acid, poly-lactic acid, polyethylene glycol) polymers are most commonly used for drug delivery systems. Therapeutically poly (lactic acid) (PLA), poly (lactic-*co*-glycolic acid) (PLGA), poly(ε-caprolactone) (PCL), alginate, chitosan and gelatine bases are used frequently (44). Polymer selection for encapsulation of drugs in nano, micro, or other polymeric forms needs great attention in order to make a formulation with all the essential characteristics, such as size, charge, biocompatibility, rate of drug release, dosing frequency, drug accumulation in the site of action and rate of polymer degradation. (45). Polymers can act in a controlled release manner, which is of utmost importance in pulmonary drug delivery. Mechanisms behind the drug release behaviors of polymers are drug diffusion, polymer swelling and polymer degradation (46). Poly lactic-co-glycolic acid (PLGA) describes a broad category of drug encapsulation and is the mostly explored polymer for pulmonary drug delivery in a sustained release manner (47). Biocompatibility and biodegradability properties of PLGA exerted nontoxic potential in several in vitrostudies of human airway cell lines, and safety profiling was confirmed by in vivo pulmonary drug drlivery in mice model (48). PLGA is used in pulmonary drug delivery, and it has been noticed that various natural and synthetic polymers affect the viability of cells and hampered bioaccumulation potential. Augmented lung inflammation, more non-viable cells, neutrophil infiltration and hemorrhagic conditions were noted upon use of PLA and PLGA polymers formulation, and these studies showed that PLGA exerts higher cytotoxic effects than sodium hyaluronate and chitosan (49). Besides this, because of their mucoadhesive property, chitosan encapsulated PLGA nanoparticles have shown higher residence time in the lungs (50). The chitosan based PLGA encapsulated palmitic acid-conjugated exendin-4 antidiabetic nano drugs showed to be released in 3 days and hypoglycemia was induced in vivo in mice within 4 days. Thease outcomes suggested the discontinuation of the use of chitosan for pulmonary delivery as it might cause edema in lungs because of the widening of pulmonary epithelial junctions (51).

C. Liposomes

Liposomes can be described as one or more spherical concentric lipid bilayers separated by aqueous buffer compartments (52). Lipid based drug delivery systems adopting sustained release properties making them release over longer periods of time with maximum drug effects are a very tempting option for pulmonary delivery as the basic material is phospholipids, which are intrinsically present in lungs. Other than microparticles, among nanoparticle-based advanced drug delivery strategies, liposomes hold an advantage as an enormous amount of the molecules, proteins and peptides can be incorporated in liposomes, and a very attractive thing about liposomes in pulmonary delivery is that they have reached the clinical development stage (53) and carry advantages like the ability to solubilize poorly soluble drugs, the capacity to provide a reservoir for sustained release, the ability to prolong local and systemic therapeutic levels, their facilitation of intracellular delivery of drugs, especially to alveolar macrophages, the avoidance of local irritation of lung tissue, the ability to target specific cell populations using surface-bound ligands or antibodies, and the potential to be absorbed across the epithelium intact to reach systemic circulation (54).

Classification of liposomes depends on their size and morphology into multilamellar vesicles (MLVs; 0.1–20 μm), large unilamellar vesicles (LUVs; 0.1–1 μm) and small unilamellar vesicles (SUVs; 25–100 nm) (55). Approximately 20% of human pulmonary surfactants in lung tissue by dry weight constitute lipids, and others are phospholipids, which are chemically the same so liposome composition makes liposome an ideal choice for pulmonary drug delivery. Certain parameters like particle size, which is a critical factor in depositing drug into deep lungs should be taken into account in developing inhalable liposome particles. Stability assessment of liposomes in bronchoalveolar lavage fluid depends on the increased transition temperature of the lipid mixture, which is determined by an important variable — rigidity — as heat produced during nebulization promulgates impact on the leakage of liposome encapsulated drugs because the temperature is higher than the phase transition temperature (Tc) of the lipid mixture (56). In the case of inhalational liposomal dry powder formulations, leakage of encapsulated drug occurs during lyophilization and jet milling. Therapeutically, liposomes were shown to be efficiently effective in treating pulmonary diseases, In pulmonary tuberculosis the use of liposomes provides an improved strategy for delivery of antitubercular drugs in lungs. Citing lack of effectiveness from broad spectrum antibiotics in oral and IV routes, Conley et al. reported treatment of *Francisellatularensis* infection in mice using ciprofloxacin encapsulated liposomes administered through jet nebulization (57). In this area of research, liposomes have been suggested as surfactant replacement therapy in patients with respiratory distress syndrome. Respiratory distress syndrome in neonates has recently been prevented using lung surfactants based on mixtures of phospholipids, e.g. Survanta (58). Many studies have also shown that drugs entrapped in liposomes are safe for pulmonary delivery since liposomal drugs work in a controlled release manner, which means doses causing huge adverse effects can be reduced to produce a therapeutic effect. Liposomal drug delivery systems have been shown to modify the delivery of chemotherapeutic agents in the management of lung cancer and prophylaxis of metastasis when compared to the IV route of administration. An example of a chemotherapeutic agent, cisplatin, used in lung cancer treatments has several disadvantages, like dose-limiting toxicity (DLT) in systemic administration, followed by nephrotoxicity, peripheral neuropathy and ototoxicity, but clinical Phase I trials showed pharmacokinetics, safety and efficacy with aerosolized liposomal cisplatin at maximum tolerated doses in metastatic lung carcinoma (59).

1.4 Treatment of Chronic Diseases Through the Pulmonary Route

A. Tuberculosis (TB)

Since the 1950s, the choice of drug delivery through the pulmonary route has gained interest as the occurrence of pulmonary tuberculosis (TB) arises from the portal entry of a pathogen, such as mycobacterium, that causes TB (60). In pulmonary delivery the absorption of drugs into systemic circulation takes place from the large surface of the lung mucosa and maintains high drug concentration in lung tissue along with bypassing the hepatic first pass effect (61). In pulmonary TB management, advantages of pulmonary delivery lie in the fact that lungs play a pivotal role as, starting from site of drug absorption, drug delivery to the primary site of action takes place in the lung. Also, phagocytosis of substances like macromolecular drugs, particulates or vesicular drug delivery in lungs are carried out by alveolar macrophages (AM) that house TB bacilli. It also makes possible the delivery of such agents that may change the rationality of host–pathogen (62). Possibly available aerosol delivery methods in treating TB are advanced technologies, along with conventional nebulization, that have been in clinical practice for a long time (63–65). The readily available and known measures for delivering medicaments to pulmonary regions like the lungs and airways are pressurized, metered-dose inhalers (pMDI) or metered dose inhalers (MDI). MDIs act by regulating the optimized dose to be administered at the site of action by using a dose-metering valve and propellant aerosols (66, 67). Delivery of a drug under positive pressure can be done by nebulization using dry powder inhalers (DPI), which depend on the indrawn breath of the patient to pull in a dry powder (68). Intratracheal delivery of drugs has been a choice of pulmonary delivery in in vivo animal experiments in treating TB. The most widely used aerosolized strategy for drug delivery of liquid or dry powders through the lungs in animal models is accomplished with the PennCentury Micro Sprayer, which quantifies the amount delivered accurately as it ignores the nose and throat and is directly inserted at the bronchial bifurcation down the trachea of the unconscious animal (62). Moreover, distribution studies of formulations can be achieved using dyes or fluorescent dyes. Despite the conventional oral or injectables routes, drug delivery through inhalational enables administration of anti-TB agents in the lungs. Sukhanov et al. described the advantages of resolving lung lesions in human patients by the administration of cycloferonact as an immunomodulator through the inhalation route for 5 weeks (69). Also, delivery of surfactants through inhalation can be achieved for TB therapies. Efficacy of treatment through inhalational drug delivery in infected animal lungs has been studied elsewhere. Aerosol formulations of a variety of drugs, including rifampicin, rifabutin, rifapentine, isoniazid, ethambutol, ethionamide and capreomycin have been investigated to treat TB and found to reduce bacterial contamination in lungs (62).

B. Lung cancer

About 1.3 million annually reported cases of cancer and deaths all over the world result from a highly prevalent ailment: lung cancer (70). Treatment of lung cancer by pulmonary drug delivery has gained notice as the disadvantages of other routes of administration can be avoided. Oral and intravenous (IV) injection routes face the difficulty of maintaining optimum concentration of chemotherapeutic drugs, as very low concentrations can reach the lung (71, 72). The inhalational route of drug delivery has received much attention in recent days as human insulin powder of recombinant DNA for inhalational delivery has been approved, for example, Exubera® (later withdrawn by Pfizer citing poor acceptance by patients and clinicians) and Afrezza® (manufactured by MannKind Corporation, Danbury, CT, USA, and approved by the US Food and Drug Administration in 2014). Micro- and nanosized particles encapsulating chemotherapeutic drugs have been investigated for delivery to the lungs through the inhalational route (73). Gill et al. prepared micelles made up of polyethylene glycol $(PEG)_{5000}$–distearoylphosphatidylethanolamine (PEG_{5000}–DSPE) caryingpaclitaxel showed well, maintaining paclitaxel concentrations in lungs for long durations after intra-tracheal delivery (74). Small lipid nanoparticles of paclitaxel delivered by inhalation were found to be more efficient in reducing hyperplasia and hypertrophy in comparison to IV delivery. A study also explored and reported the benefits of DOX-conjugated dendrimer in lung cancer therapy through pulmonary delivery (73).

C. Diabetes

Diabetes is a chronic metabolic disorder characterized by inadequate insulin secretion with an increase in glucose level (75). Since 1920, insulins have been in use for treating diabetes via various routes of delivery such as oral insulin, implantable peritoneal insulin pumps, subcutaneous insulin, etc. In 1924, pulmonary insulin delivery was first examined (76, 77) and inhaled insulin treated individuals with type 2 diabetes mellitus as an adjuvant therapy. Most excitingly, it has been seen that a patient's compliance with and acceptance of inhaled insulin is greater than for subcutaneous insulin (78). Inhalation of insulin showed a decrease in the blood glucose level (77, 79). The efficiency of insulin inhalation depends on the breathing activity, as slow inhalation facilitates the best penetration to the alveolar spaces and attains target, where rapid inhalation leads to inappropriate or loss of drug concentration in the target site of the lung (80, 81). Some examples of inhaled insulin systems extensively studied in clinical protocols are Exubera, a drug approved by the Food and Drug Administration (FDA) and the European Medicines Agency (EMA) for type 1 diabetes mellitus and type 2 diabetes mellitus therapy, which was developed in 2006 as a collaboration between Nektar Therapeutics and Pfizer. The device delivers dry powder formulation packed in blister packets containing 1 mg or 3 mg of regular human insulin. Aradigm Corporation and Novo Nordisk together developed the AERx insulin diabetes management system (AERx® iDMS), which delivers insulin in aerosol droplets along with an electronic control device that helps the patient to inhale the insulin in a reproducible pattern. The Technosphere system introduced a placebo formulation of dry powder recombinant human insulin microspheres for patients with type 2 diabetes mellitus. This system was developed by Mannkind Corp. with the MedTone inhaler (Pharmaceutical Discovery Corp.) and this system currently has reached phase III clinical trials (82).

D. Pulmonary arterial hypertension

Pulmonary arterial hypertension (PAH) is a chronic vascular disorder characterized by elevated levels of pulmonary vascular resistance (PVR) resulting from imbalanced pulmonary vascular remodeling. Vasodilation of pulmonary arteries and blood circulation serves as the lifesaving cure in the pathological condition of PAH (83). Despite other modes of therapy and routes of administration, inhaled vasodilators find their way to provide effective and specific pulmonary vasodilation without hampering systemic pressure (84–88). Advantages of inhalation therapy regarding PAH treatment open up new horizons for treating such conditions as it ranges from the local delivery of drugs and exerts enhanced efficacy as a higher concentration of drugs is retained at the site of action. Interestingly, pulmonary delivery minimizes drug concentration in systemic circulation, which in turn lowers the chances of having systemic adverse effects like systemic hypotension caused by the vasodilation activity of medicaments. This delivery system offers effective therapy in a cost-effective way along with minimizing frequency and drug dose as well as achieving the rapid onset of action and drug absorption (57–60).

Although the development of drug delivery systems is just emerging, these systems show a promising future. Lungs offer a vast variety of advantages over conventional oral drug delivery, making pulmonary drug delivery rapidly gain importance. The enormous gas-exchange surface of the lungs represents a versatile, highly promising and little-exploited route for drug delivery in the treatment of chronic diseases. Furthermore, such a route for drug administration may also be manipulated for systemic drug delivery. The interaction of inhaled particles with the lung has been dynamically developed and has been the subject of intense research in recent years. In this chapter, we have reviewed efficient pulmonary drug delivery systems by understanding the characteristic features of pulmonary drug transport, the deposition mechanics, the nature of therapeutic agents, properties of delivery systems, the molecular basis of pulmonary diseases and barriers to drug delivery. Pulmonary drug delivery of nanoparticles is a non-invasive method that can be designed to target specific cells or organs while sustaining the release of the therapeutics locally or systemically. However, the future of pulmonary drug delivery research requires more reasonable regulation and better understanding of the properties of nanoparticles in achieving different biological effects along with development of more safety excipients and better manufacturing technologies. Moreover, international standards of guidelines of the Food and Drug Administration (FDA) and the European Medicines Agency (EMA) need to be defined for in vitro and in vivo models in order to allow valid comparisons between studies and to define the safety and efficacy of proposed treatments.

Again, the ligand-based carrier system for pulmonary targeting should be evaluated for stability and effectiveness to overcome regulatory restrictions. Addressing the toxicity of inhaled therapeutic nanocarriers is matter of inescapable importance. Scientists and researchers have given many and varied answers in an attempt to address all the rising issues, find alternatives and engineer adequate systems to fulfill requirements and needs. Because of the high degree of heterogeneity and complexity in diseased lungs, there is still a large, underdeveloped area of research regarding how inhaled nanomedicines will act under these circumstances. With the growing global trend to look for more precise medicines and diagnosis, application of nanotechnology to biologic therapeutics for developing the next generation delivery devices for better management of pulmonary disease/disorders looks bright. In conclusion, it can be stated that the presence of nanotechnology opens up new vistas for developing novel nanoparticle-based drug formulations to achieve safer pulmonary drug delivery in the treatment of chronic diseases, however thorough physicochemical and nanotoxicological analysis for possible human application is needed.

REFERENCES

1. McNeil SE. Unique Benefits of Nanotechnology to Drug Delivery and Diagnostics. In: McNeil S. (ed) Characterization of Nanoparticles Intended for Drug Delivery. Methods in Molecular Biology (Methods and Protocols), vol. 697. Humana Press, Totowa, New Jersey. 2011.
2. Pontes JF and Grenha A. Multifunctional nanocarriers for lung drug delivery. *Nanomaterials*. 2020; 10(2): 183.
3. Dhand C, Prabhakaran MP, Beuerman RW, et al. Role of size of drug delivery carriers for pulmonary and intravenous administration with emphasis on cancer therapeutics and lung-targeted drug delivery. *RSC Advances*. 2014; 4(62): 32673–32689.
4. Sahoo SK and Labhasetwar V. Nanotech approaches to drug delivery and imaging. *Drug Discov Today*. 2003; 8: 1112–1120.
5. Paranjpe M and Müller-Goymann CC. Nanoparticle-mediated pulmonary drug delivery: A review. *International Journal of Molecular Sciences*. 2014; 15(4): 5852–5873.
6. Chow SF, Weng J, Xuan B, et al. How can the challenges faced by nanoparticle-based pulmonary drug formulations be overcome. *Therapeutic Delivery*. 2019; 10 (2): 87–89.
7. Kurmi BD, Kayat J, Gajbhiye V. et al. Micro- and nanocarrier-mediated lung targeting. *Expert Opinion on Drug Delivery*. 2010; 7(7): 781–794.
8. Patton JS and Byron PR. Inhaling medicines: delivering drugs to the body through the lungs. *Nature Reviews Drug Discovery*. 2007; 6(1): 67–74.
9. Ruge CA, Kirch J, and Lehr CM. Pulmonary drug delivery: From generating aerosols to overcoming biological barriers-therapeutic possibilities and technological challenges. *The Lancet Respiratory Medicine*. 2013; 1(5): 402–413.
10. Willis L, Hayes D, and Mansour HM. Therapeutic liposomal dry powder inhalation aerosols for targeted lung delivery. *Lung*. 2012; 190(3): 251–262.
11. Gill S, Löbenberg R, Ku T, et al. Nanoparticles: Characteristics, mechanisms of action, and toxicity in pulmonary drug delivery-A review. *Journal of Biomedical Nanotechnology*. 2007; 3(2): 107–119.
12. Todoroff J and Vanbever R. Fate of nanomedicines in the lungs. *Current Opinion in Colloid & Interface Science*. 2011; 16; 246–254.
13. Mansour HM, Rhee YS, and Wu X. Nanomedicine in pulmonary delivery. *International Journal of Nanomedicine*. 2009; 4: 299–319.
14. Hickey AJ and Mansour HM. Chapter 5: Delivery of drugs by the pulmonary route. In: Florence AT and Siepmann J (eds) *Modern Pharmaceutics*, vol. 2, 5th edn. Taylor and Francis, New York. 2009; 191–219.
15. Pilcer G and Amighi K. Formulation strategy and use of excipients in pulmonary drug delivery. *International Journal of Pharmaceutics*. 2010; 392(1–2): 1–19.
16. Joshi M, Nagarsenkar M, and Prabhakar B. Albumin nanocarriers for pulmonary drug delivery: An attractive approach. *Journal of Drug Delivery Science and Technology*. 2020; 56, 101529.
17. Tortora GJ and Derrickson BH. *Principles of Anatomy and Physiology*. John Wiley & Sons, Hoboken, New Jersey. 2018; 851–867.
18. Betts JG Anatomy & Physiology. 2013; 787–846. ISBN 978-1-938168-13-0. Retrieved 05 July 2020.
19. Waugh A. and Grant A. Ross & Wilson anatomy and physiology in health and illness E-book. *Elsevier Health Sciences*. 12th edition. 2014; 243–254.
20. How Do Particulates Enter the Respiratory System? www.ccohs.ca/oshaanswers/chemicals/how_do.html
21. Mobley C and Hochhaus G. Methods used to assess pulmonary deposition and absorption of drugs. *Drug Discovery Today*. 2011; 6(7): 367–375.
22. Han S and Mallampalli RK. The role of surfactant in lung disease and host defense against pulmonary infections. *Annals of the American Thoracic Society*. 2015; 12(5): 765–774.
23. Ibrahim M and Garcia-Contreras L. Mechanisms of absorption and elimination of drugs administered by inhalation. *Therapeutic Delivery*. 2013; 4(8): 1027–1045.
24. Patton JS. Mechanisms of macromolecule absorption by the lungs. *Advanced Drug Delivery Reviews*. 1996; 19(1), 3–36.
25. Olsson B. et al. Pulmonary Drug Metabolism, Clearance, and Absorption. In: Smyth H and Hickey A. (eds) Controlled Pulmonary Drug Delivery. Advances in Delivery Science and Technology. Springer, New York, NY. 2011.
26. Woods A, Andrian T, Sharp G., et al. Development of new in vitro models of lung protease activity for investigating stability of inhaled biological therapies and drug delivery systems. *European Journal of Pharmaceutics and Biopharmaceutics*. 2020; 146, 64–72.
27. Bur M, Huwer H, Lehr CM et al. Assessment of transport rates of proteins and peptides across primary human alveolar epithelial cell monolayers. *European Journal of Pharmaceutical Sciences*. 2006; 28(3), 196–203.
28. Kreyling WG, Semmler M, Erbe F. et al. Translocation of ultrafine insoluble iridium particles from lung epithelium to extrapulmonary organs is size dependent but very low. *Journal of Toxicology and Environmental Health, Part A*. 2002; 65(20): 1513–1530.
29. Garud A, Singh D, and Garud N. Solid lipid nanoparticles (SLN): Method, characterization and applications. *International Current Pharmaceutical Journal*. 2012; 1(11): 384–393.
30. Qi J, Lu Y, and Wu, W Absorption, disposition and pharmacokinetics of solid lipid nanoparticles. *Current Drug Metabolism*. 2012; 13(4): 418–428.
31. Loira-Pastoriza C, Todoroff J and Vanbever R. Delivery strategies for sustained drug release in the lungs. *Advanced Drug Delivery Reviews*. 2014; 75: 81–91.
32. Üner M. and Yener G. Importance of solid lipid nanoparticles (SLN) in various administration routes and future perspectives. *International Journal of Nanomedicine*. 2007; 2(3): 289.
33. Müller RH, Mäder K, and Gohla S. Solid lipid nanoparticles (SLN) for controlled drug delivery–A review of the state of the art. *European Journal of Pharmaceutics and Biopharmaceutics*. 2000; 50(1): 161–177.
34. Beck-Broichsitter M, Ruppert C, Schmehl T, et al. Biophysical investigation of pulmonary surfactant surface properties upon contact with polymeric nanoparticles in vitro. *Nanomedicine: Nanotechnology, Biology and Medicine*. 2011; 7(3): 341–350.

35. Rytting E, Nguyen J, Wang X, et al. Biodegradable polymeric nanocarriers for pulmonary drug delivery. *Expert Opinion on Drug Delivery*. 2008; 5(6): 629–639.
36. Pison U, Welte T, Giersig M, et al. Nanomedicine for respiratory diseases. *European Journal of Pharmacology*. 2006; 533: 341–350.
37. Videira MA, Botelho MF, Santos AC, et al. Lymphatic uptake of pulmonary delivered radiolabelled solid lipid nanoparticles. *Journal of Drug Targeting*. 2002; 10(8): 607–613.
38. Jaspart S, Piel G, Delattre L, et al. Solid lipid microparticles: Formulation, preparation, characterisation, drug release and applications. *Expert Opinion on Drug Delivery*. 2005; 2(1): 75–87.
39. Brannon-Peppas, L. Recent advances on the use of biodegradable microparticles and nanoparticles in controlled drug delivery. *International Journal of Pharmaceutics*. 1995; 116(1): 1–9.
40. Scalia S, Young PM, and Traini D. Solid lipid microparticles as an approach to drug delivery. *Expert Opinion on Drug Delivery*. 2015; 12(4): 583–599.
41. Liwarska-Bizukojc E, Miksch K, Malachowska-Jutsz A, et al. Acute toxicity and genotoxicity of five selected anionic and nonionic surfactants. *Chemosphere*. 2005; 58(9): 1249–1253.
42. Sznitowska M, Wolska E, Baranska H, et al. The effect of a lipid composition and a surfactant on the characteristics of the solid lipid microspheres and nanospheres (SLM and SLN). *European Journal of Pharmaceutics and Biopharmaceutics*. 2017; 110: 24–30.
43. Silva LFC, Kasten G, de Campos CEM, et al. Preparation and characterization of quercetin-loaded solid lipid microparticles for pulmonary delivery. *Powder Technology*. 2013; 239: 183–192.
44. Louey MD and Garcia-Contreras L. Controlled release products for respiratory delivery. *American Pharmaceutical Review*. 2004; 7: 82–87.
45. Sheth P and Myrdal PB. Polymers for pulmonary drug delivery. In: Smyth HDC and Hickey AJ (eds.) *Controlled Pulmonary Drug Delivery*. Springer, New York, NY. 2011; 265–282.
46. Alhusban FA and Seville PC. Carbomer-modified spray-dried respirable powders for pulmonary delivery of salbutamol sulphate. *Journal of Microencapsulation*. 2009; 26(5): 444–455.
47. Cook RO, Pannu RK, and Kellaway IW. Novel sustained release microspheres for pulmonary drug delivery. *Journal of Controlled Release*. 2005; 104(1): 79–90.
48. Tomoda K, Ohkoshi T, Hirota K, et al. Preparation and properties of inhalable nanocomposite particles for treatment of lung cancer. *Colloids and Surfaces B: Biointerfaces*. 2009; 71(2): 177–182.
49. Salama R, Traini D, Chan HK, et al.Recent advances in controlled release pulmonary therapy. *Current Drug Delivery*. 2009; 6(4): 404–414.
50. Tahara K, Sakai T, Yamamoto H, et al. Improved cellular uptake of chitosan-modified PLGA nanospheres by A549 cells. *International Journal of Pharmaceutics*. 2009; 382(1–2): 198–204.
51. Kim H, Lee J, Kim TH, et al. Albumin-coated porous hollow poly (lactic-co-glycolic acid) microparticles bound with palmityl-acylated exendin-4 as a long-acting inhalation delivery system for the treatment of diabetes. *Pharmaceutical Research*. 2011; 28(8): 2008–2019.
52. Rawat M, Singh D, Saraf S, et al. Lipid carriers: A versatile delivery vehicle for proteins and peptides. *YakugakuZasshi*. 2008; 128(2): 269–280.
53. Weiner N, Martin F, and Riaz M. Liposomes as a drug delivery system. *Drug Development and Industrial Pharmacy*. 1989; 15(10): 1523–1554.
54. Gaspar MM, Bakowsky U, and Ehrhardt C. Inhaled liposomes-current strategies and future challenges. *Journal of Biomedical Nanotechnology*. 2008; 4: 1–13
55. Rudokas M, Najlah M, Alhnan MA, et al. Liposome delivery systems for inhalation: A critical review highlighting formulation issues and anticancer applications. *Medical Principles and Practice*. 2016; 25(2): 60–72.
56. Swaminathan J and Ehrhardt C. Liposomes for pulmonary drug delivery. In: SmythHDC and Hickey AJ (eds.).Controlled Pulmonary Drug Delivery.Springer, New York, NY. 2011; 313–334.
57. Conley J, Yang H, Wilson T, et al. Aerosol delivery of liposome-encapsulated ciprofloxacin: Aerosol characterisation and efficacy against Francisellatularensis infection in mice. *Antimicrobial Agents and Chemotherapy*. 1997; 41: 1288–1292
58. Paul S, Rao S, Kohan R, et al. Poractant alfa versus beractant for respiratory distress syndrome in preterm infants: A retrospective cohort study. *Journal of Paediatrics and Child Health*. 2013; 49: 839–844.
59. Wittgen BP, Kunst PW, van der Born K, et al. Phase I study of aerosolized slit cisplatin in the treatment of patients with carcinoma of the lung. *Clinical Cancer Research*. 2007; 13: 2414–2421
60. Berishvili TA. PAS inhalation therapy of respiratory tuberculosis in the sanatorium LIBANI; preliminary communication. *Problemy Tuberkuleza*. 1954; 4: 69.
61. Mathias NR and Hussain MA. Non-invasive systemic drug delivery: Developability considerations for alternate routes of administration. *Journal of Pharmaceutical Sciences*. 2010; 99: 1e20.
62. Misra A, Hicke, AJ, Rossi C, et al. Inhaled drug therapy for treatment of tuberculosis. *Tuberculosis*. 2011; 91(1): 71–81.
63. Pilipchuk NS and Protsiuk RG. Effect of inhalation of aerosols of antitubercular preparations on pulmonary surfactant. *Vrachebnoe Delo*. 1986; (7): 21–25.
64. Protsiuk RG. Effect of the inhalation of tuberculostatic aerosols on the dynamics of bacterial excretion in pulmonary tuberculosis patients. *Problemy Tuberkuleza*. 1983; (5): 38–42.
65. Semenova EV. Basis for the administration of streptomycin and isoniazid in ultrasonic aerosols in the treatment of intrathoracic tuberculosis. *Antibiotiki*. 1977; 22(5): 469–471.
66. Berger W. Aerosol devices and asthma therapy. *Current Drug Delivery*. 2009; 6:38e49.
67. Fink JB. Metered-dose inhalers, dry powder inhalers, and transitions. *Respiratory Care*. 2000; 45: 623e35.
68. Onoue S, Misaka S, Kawabata Y, et al. New treatments for chronic obstructive pulmonary disease and viable formulation/device options for inhalation therapy. *Expert Opinion on Drug Delivery*. 2009; 6: 793e811.

69. Sukhanov DS, Ivanov AK, Kovalenko AL, et al. Pathogenetic therapy of tuberculosis of respiratory organs during sanatorium-and-spa treatment. *VoprKurortolFizioter Lech FizKult*. 2009; 6: 34–37.
70. Lemjabbar-Alaoui H, Hassan OU, Yang YW, et al. Lung cancer: Biology and treatment options. *Biochimica et Biophysica Acta (BBA)-Reviews on Cancer*. 2015; 1856(2): 189–210.
71. Karathanasis E, Ayyagari AL, Bhavane R, et al. Preparation of in vivo cleavable agglomerated liposomes suitable for modulated pulmonary drug delivery. *Journal of Controlled Release*. 2005; 103: 159–175.
72. Cipolla DC and Gonda I. Formulation technology to repurpose drugs for inhalation delivery. *Drug Discovery Today: Therapeutic Strategies*. 2011; 8: 123–130.
73. Lee WH, Loo CY, Ghadiri M, et al. The potential to treat lung cancer via inhalation of repurposed drugs. *Advanced Drug Delivery Reviews*. 2018; 133: 107–130.
74. Mangal S, Gao W, Li T, et al. Pulmonary delivery of nanoparticle chemotherapy for the treatment of lung cancers: Challenges and opportunities. *Acta Pharmaceutica Sinica B*. 2017; 38: 782–797.
75. Bliss M. The Discovery of Insulin. University of Chicago Press, Chicago, IL. 1982.
76. Laqueur E and Grevenstuk A. Uber die wirkunkintratrachealerzuführung von insulin. *KlinWochenschr*. 1924; 3: 1273–1274.
77. Gänsslen M. Uber inhalation von insulin. *KlinWochenschr*. 1925; 4: 71.
78. Elliott RB, Edgar BW, Pilcher CC, et al. Parenteral absorption of insulin from the lung in diabetic children. *Australian Paediatric Journal*. 1987; 23(5): 293–297.
79. Heubner W, de Jongh S and Laquer E. Uber inhalation von insulin. *KlinWochenschr*. 1924; 3: 2342–2343
80. Katz IM, Schroeter JD and Martonen TB. Factors affecting the deposition of aerosolized insulin. *Diabetes Technology & Therapeutics*. 2001; 3(3): 387–397.
81. Farr SJ, McElduff A, Mather LE, et al. Pulmonary insulin administration using the AERx system: Physiological and physiochemical factors influencing insulin effectiveness in healthy fasting subjects. *Diabetes Technology & Therapeutics*. 2000; 2(2): 185–197.
82. Mastrandrea LD. Inhaled insulin: Overview of a novel route of insulin administration. *Vascular Health and Risk Management*. 2010; 6: 47.
83. Badesch DB, Champion HC, Sanchez MA, et al. Diagnosis and assessment of pulmonary arterial hypertension. *Journal of the American College of Cardiology*. 2009; 30 (54): S55–S66.
84. Preston IR, Sagliani KD, Roberts KE, et al. Comparison of acute hemodynamic effects of inhaled nitric oxide and inhaled epoprostenol in patients with pulmonary hypertension. *Pulmonary Circulation*. 2013; 3(1): 68–73.
85. Labiris NR andDolovich MB. Pulmonary drug delivery. Part I: Physiological factors affecting therapeutic effectiveness of aerosolized medications. *British Journal of Clinical Pharmacology*. 2003; 56(6): 588–599.
86. Tayab ZR and Hochhaus G. Pharmacokinetic/pharmacodynamic evaluation of inhalation drugs: Application to targeted pulmonary delivery systems. *Expert Opinion on Drug Delivery*. 2005; 2(3): 519–532.
87. Olschewski H. Inhaled iloprost for the treatment of pulmonary hypertension. *European Respiratory Review*. 2009; 18(111): 29–34.
88. Groneberg DA, Witt C, Wagner U, et al. Fundamentals of pulmonary drug delivery. *Respiratory Medicine*. 2003; 97(4): 382–387.

2

Introduction to Pharmacology of the Lung from a Drug Delivery Perspective

Caline McCarthy[1] **and Charles Preuss**[2]
[1]*University of South Florida Morsani College of Medicine, Department of Molecular Pharmacology & Physiology, Tampa, Florida, USA*
[2]*University of South Florida Morsani College of Medicine, Department of Molecular Pharmacology & Physiology, Tampa, Florida, USA*

2.1 The Respiratory Tract: An Overview

The respiratory system can be divided into an upper and lower region. The **upper region** comprises the structures that allow air to enter the body and carry air to the trachea. Air enters through the mouth and/or nose, then travels through the **pharynx**, which is a joint structure for the respiratory and digestive system where food and air both travel. At the bottom of the pharynx, the pathways divide into two distinct structures: the esophagus (which carries food into the stomach) and the **larynx**. A small flap called the **epiglottis** covers the larynx when food is swallowed, making sure that food does not enter the airways and lungs. The larynx contains the vocal cords. The vocal cords are two bands of smooth muscle. When air passes over the cords, they vibrate, producing sounds that allow us to vocalize, speak and sing. The **lower region** of the respiratory system includes all the structures we classically label as the lobes of the lung. The lower region includes all respiratory structures found past the larynx and vocal folds, including the trachea (which carries air into the lungs), bronchi, bronchioles, and alveoli.

The respiratory system includes the lungs and the major airways that deliver air to the lungs. The airways can be further divided into two main divisions: the **conducting zone** and the **respiratory zone**.

2.1.1 The Conducting Zone

The conducting zone consists of the airways that carry air into and out of the respiratory zone for gas exchange. The conducting zone includes the nose, nasopharynx, larynx, trachea, primary bronchi, bronchioles, and terminal bronchioles. The trachea branches first into two bronchi, each leading into a lung. Branching continues further into bronchioles and then further into terminal bronchioles. Each time the structures branch from the trachea down to the bronchioles, the diameter of the airway becomes increasingly smaller as the sheer number of airways increases geometrically. Because of this exponential increase, the *total cross-sectional area* actually increases with each branching. While the trachea is our largest airway, the total cross-sectional area of the trachea is smaller than the total cross-sectional area of the terminal bronchioles, whose *individual* surface area is a fraction of the size of the trachea.

The conducting zone also functions to humidify, warm, and filter air prior to gas exchange. Alveoli in the respiratory zone are very delicate structures and susceptible to damage under drastic temperature changes. It is important that air from the environment is a consistent temperature before reaching the respiratory zone. Air is warmed to the body's heat and humidified by water evaporating from the mucosal lining of the airways. Air is also filtered in the trachea and bronchi. Both airway structures have epithelium with cilia that clear foreign particles and fluid up in an upward sweeping motion out of the airways (1).

The walls of the conducting zone contain smooth muscle which can dilate or constrict in response to autonomic stimulation and is responsive to both sympathetic and parasympathetic control. The sympathetic adrenergic neurons activate the β_2 adrenergic receptors on the bronchiole smooth muscle, leading to bronchodilation of the airways. These receptors can also be activated through circulating β_2 adrenergic agonists, normally released from the adrenal medulla. These receptors serve as an ideal target for synthetic β_2 adrenergic agonist drugs (like albuterol and salmeterol) in individuals with diseases that cause constriction of the upper airways, like asthma. A change in the diameter of the airway influences airway resistance, which impacts the work of breathing and airflow. Targeting the upper airways with a β_2 adrenergic agonist causes bronchodilation, which decreases airway resistance and the work of breathing, allowing airflow to return to normal (2).

2.1.2 The Respiratory Zone

The respiratory zone is the airways lined with alveoli, including the **respiratory bronchioles, alveolar ducts, and alveolar sacs**. The respiratory bronchioles are the transition structure between the conducting and respiratory zones; they contain cilia embedded in the smooth muscle of the airways to clear away debris (like the bronchioles in the conducting zone), but also contain occasional alveoli. Alveoli are the microscopic structures responsible for gas exchange, therefore the respiratory zone includes all the airways

DOI: 10.1201/9781003046547-2

that participate in gas exchange. The respiratory zone structures branch off in the same fashion as the branching that occurs in the conducting zone; as the diameter of each structure decreases, the total surface area increases. **Alveoli** are air sacs that are one cell thick. Each lung has approximately 300–400 million alveoli, each of their diameters is ~200 μm, for a total cross-sectional area of 50–70 m^2. Such a substantial total cross-sectional area helps alveoli function in their role of facilitating gas exchange of O_2 and CO_2 with the bordering pulmonary capillaries. Diffusion can occur quickly and efficiently due to the thin natural of the alveolar walls and sizable surface area. (1).

There are two main conventions to use the respiratory system as a route of drug delivery. The first is for the management and treatment of respiratory diseases, such as asthma and chronic obstructive pulmonary disease. The majority of drugs delivered through the respiratory route aim to bind receptors located *within* the respiratory system in order to manage pulmonary related diseases. However, the treatment of pulmonary diseases is not the only incentive when considering inhalation as a route to drug delivery. Pulmonary circulation holds an intimate relationship with systemic blood. The lungs are the only organ, besides the heart, that interact with total systemic blood volume. Before blood is pumped from the right ventricle into systemic circulation, blood travels through pulmonary circulation to the lungs to participate in gas exchange. At the alveolar–capillary membrane, blood hemoglobin picks up oxygen and delivers carbon dioxide. It is at the capillary–alveolar interface that inhaled aerosols have the opportunity to travel to distal tissue binding sites. The lungs function as a portal of entry into the body, distributing drugs to tissues *outside* of the respiratory system. In order to make it to the capillary–alveolar interface, the therapeutic aerosol has to be the proper size, inhaled at the proper velocity, and evade inactivating protease encountered in the respiratory system (Figure 2.1). (3)

Particles or droplets that fail to reach the lung and deposit in the mouth or throat are subsequently swallowed, whereas particles that reach the lung but fail to deposit are exhaled (5). Drug particles that successfully clear pulmonary enzymes and dissolve in the fluid lining the epithelial tissue are now able to be absorbed into the lung tissue, interact with cell receptors, and transmit their therapeutic effect.

2.2 Respiratory Pathology

Respiratory pathology arises when the respiratory system is not functioning as it should. This next section covers the four

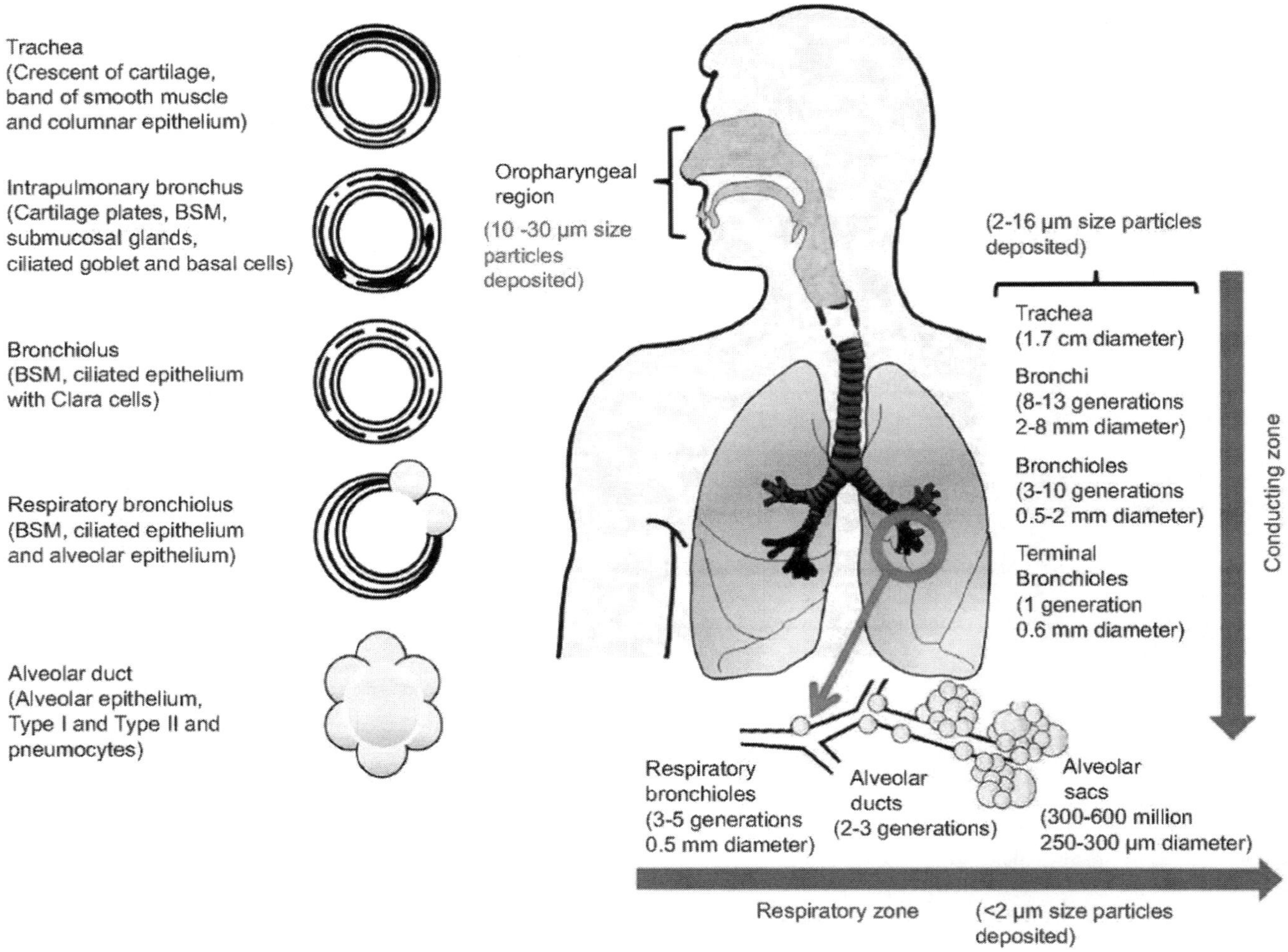

FIGURE 2.1 The Respiratory System (Pleasants and Hess (4))

common respiratory pathologies requiring aerosol drug delivery: asthma, chronic bronchitis, chronic obstructive pulmonary disease (COPD), and cystic fibrosis. In addition to introducing each disease and its underlying pathophysiology, this section will address the overarching goals of disease-specific treatments and disease management.

2.2.1 Asthma Overview

Asthma is a disease characterized by intermittent reversible bronchoconstriction of hyperreactive airways. These episodes can be stimulated by various endogenous and/or exogenous stimuli, such as allergens, exercise, and air temperature changes. Underlying chronic inflammation of the respiratory airways is also associated with asthma. The severity of patients' asthma will indicate the treatment they should receive. The severity classifications range from intermittent to severe, taking into account factors such as symptoms, nighttime awakenings, lung function (forced expiratory volume and forced vital capacity), and symptoms' interference with normal activity (6).

2.2.1.1 Classifications and Goals of Treatment

Asthma is generally classified based on two main characteristics, **phenotype** and **severity of symptoms**.

Phenotypes include allergic asthma, adult/late onset asthma, eosinophilic asthma, and non-eosinophilic asthma, amongst others. The risk factors for each recognized phenotype include **genetic**, **environmental**, and unique **situational** contributors. There have been over 100 genes and 18 genomic regions found to be associated with allergy and asthma that have been identified in 11 different populations (7). There is strong supporting evidence from family and twin studies confirming the connection between genetics and the development of asthma and allergies. Allergic asthma is more frequently diagnosed in children, whereas nonallergic asthma is found to occur more frequently in adults. Both indoor and outdoor environments also pose a significant contribution to risk. Inhaled agents such as allergens, pollutants, and viruses have the ability to provoke a hyperactive immune response, exacerbating asthmatic symptoms such as bronchoconstriction, mucous production, and hyperreactivity (8). Depending on the time of exposure in someone's life, environmental triggers pose different degrees of risk. For example, the annual epidemic of asthma exacerbations in young children in the northern hemisphere each September has been found to be primarily driven by the seasonal rhinovirus infection (9). Lastly, factors specific to each person's lifestyle and upbringing may contribute to the development of asthma. Risk factors in the prenatal period are complex, including maternal nutrition and diet, tobacco use, stress, mode of delivery, and antibiotic use during pregnancy. After birth factors such as gender, socioeconomic status, access to healthcare, family size, exposure to tobacco and animals, and occupational setting are unique factors which can contribute to the development of asthma across the lifespan.

Asthma symptoms are classified as either intermittent, mild, moderate persistent, or severe persistent. These classifications are determined through lung function tests as well as the frequency of symptoms. Both the phenotype and severity of a person's symptoms will contribute to the treatment regimen. According to the guidelines from the National Asthma Education and Prevention Program, there are two main goals of long-term asthma management. The first is to reduce impairment, such as the frequency and intensity of symptoms and functional limitations posed by asthma. The second is to reduce risk by mitigating future asthma attacks, the progressive decline in lung function, and any possible side effects of medication (10).

2.2.2 Chronic Bronchitis Overview

Chronic bronchitis is commonly defined as a productive cough for at least 3 months occurring over a span of 2 years and is commonly secondary to chronic obstructive pulmonary disease (COPD) (11). Goblet cells normally secrete mucus to keep the respiratory tract well lubricated and protect epithelial cells from foreign material. The inhalation of irritants such as cigarette smoke, smog, toxic chemicals, and pollutants into the respiratory tract stimulate goblet cells, which are scattered throughout the respiratory system. In the case of chronic bronchitis, goblet cells over-secrete mucus. Additionally, alveolar epithelial cells of the airways release local inflammatory mediators such as IL-8 and GM-CSF, which attract neutrophils and monocytes into the lung. Cigarette smoke is also known to impair bronchial mucociliary function of the lung, decelerating the lungs' ability to remove foreign irritants and mucus (12). Consequently, in smaller airways the inflammatory response, as well as buildup of mucus and debris, impedes airflow and further increases debris accumulation in the airway, causing additional airway irritation. Eventually, this leads to the characteristic bronchitis cough. In more than 50% of patients, the cough is also accompanied by sputum (13). Acute bronchitis is also a common respiratory disease but is frequently due to a lower respiratory viral infection lasting only a few weeks.

2.2.2.1 Chronic Bronchitis Goals of Treatment

While there are currently no cures for chronic bronchitis, pharmacological intervention and lifestyle modifications can slow the progression of the disease, while relieving symptoms and helping to prevent further complications. The primary goals of therapy are to lower the cough by reducing mucus production and inflammation generally through pharmacological intervention and cessation of smoking.

2.2.3 Chronic Obstructive Pulmonary Disease (COPD) Overview

COPD has been classified as a chronic respiratory disease characterized by the joint presentation of chronic obstructive bronchitis and emphysema. Chronic obstructive bronchitis presents as chronic bronchitis (see Section 2.2.2 for a description) with the addition of airway obstruction, which can be identified through pulmonary function tests and is characterized by incompletely reversible expiratory airflow limitation. Emphysema is pathologically defined as loss of lung parenchyma, eventually leading to the loss of alveolar recoil and diminished radial airway traction. These two events increase the airways' susceptibility to collapse. Airflow limitation caused by airway collapse, mucus hypersecretion, loss of alveolar traction, and destroyed alveolar septa increases the work associated with breathing (14, 15). Symptomatically, COPD presents as a productive cough as the initial symptom with progressive dyspnea. Prolonged inhalation exposure to cigarette smoke is the primary cause of COPD.

2.2.3.1 COPD Goals of Treatment

COPD is a disease for which there is no current cure. The current treatments focus on symptom management and mitigating disease progression. The progressive lung tissue loss associated with COPD is non-reversible because lung tissue does not regenerate. Therefore, ideal treatments involve the reduction of inflammation and decreasing further tissue loss. One of the essential steps in COPD treatment is quitting all smoking. For patients with a mild form of COPD, stopping smoking may be the extent of treatment necessary to make breathing easier and stop the progression of lung tissue damage. For patients with more advanced stages of the disease, there are extensive medications and lung therapy options to treat the symptoms, improve quality of life, and slow the progression of COPD. Medications include short and long acting bronchodilators, inhaled steroids, combination steroids, oral steroids, theophylline, and anti-microbials. Pulmonary rehabilitation, including strength and endurance training, and educational, nutritional, and psychosocial support, improves symptoms and exercise tolerance, but is frequently under-implemented. Supplemental oxygen for patients with resting hypoxemia (defined as Spo2 <89%) improves survival (16).

2.2.4 Cystic Fibrosis Overview

Cystic Fibrosis is an inherited condition in an autosomal recessive pattern causing mutations in the *CTFR* gene. The initial indicators of CF may come as early as the first 48 hours of life, presenting as the failure to pass a meconium stool (17). Mutation of the *CTFR* gene disrupts the gene's ability to regulate the flow of chloride ions and water across cell membranes. Such dysregulation leads to the buildup of a thick, sticky mucus, primarily in the respiratory, gastrointestinal, and reproductive systems. Not only does the mucus clog the airways and make it difficult for individuals to breath, individuals with CF are also more susceptible to opportunistic infections in the airway due to the lack of normal particle movement throughout the airways. Additionally, CF patients commonly experience complications with the respiratory and digestive systems, fertility, osteoporosis, and electrolyte imbalances.

2.2.4.1 Cystic Fibrosis Goals of Treatment

Due to the multifaceted complications associated with CF, there are multiple goals of treatment. Treatment targeting the respiratory system aims to prevent and control lung infections, keep airway inflammation at bay, and dilate the airways, as well as diminish mucus viscosity and remove it from the airways (18). In addition, there are treatments focusing on improving gastrointestinal mobility, adequate nutrition and preventing dehydration.

2.3 Respiratory Drug Delivery Mechanisms

The next segment will cover the different mechanisms of aerosol drug delivery, including the physical methods of administering inhaled medications as well as inhaled and oral drug therapies related to the lung.

2.3.1 Therapeutic Administrations of Inhaled Medications

There are numerous devices used to dispense drugs into the lungs, each demonstrating unique advantages and limitations. Drug delivery by inhalation is the principal tactic in treating obstructive lung diseases. Medications are inhaled by patients and then dispersed into the airways through aerosol droplets, mist, or power. The deposition of aerosol in the airway is a function of the particles' size, shape, and density, along with human-dependent factors including efficiency and force of inhalation, inhaled volume, and flow. The location where the aerosol deposits influences its ability to work effectively. For example, in asthma and COPD, all airways (large, medium, and small) experience pathological structure changes, therefore deposition throughout the airways is beneficial (4). The widespread availability of aerosol delivery devices has brought independence to patients in their ability to self-administer medication. However, optimal delivery is often compromised due to a lack of proper physician–patient education surrounding device usage, as well as other factors.

While the location and extent of deposition are critical components of drug optimization and efficiency, particle size is another key factor. The terminal and optimal site of deposition in the respiratory tract depends on the pathology being treated. Deposition is commonly measured using mass median aerodynamic diameter (MMAD) and the geometric standard deviation (GSD). The MMAD signifies the aerodynamic diameter at which half of the aerosolized drug mass lies below the indicated diameter. The optimal MMAD for obstructive lung disease is 1–5 μm. Particles with an aerodynamic diameter of approximately 0.5 to 5 μm have the highest probability of depositing in the lung, with the smaller particles having a greater probability of penetrating into the deep lung and larger particles more likely to deposit in the mouth and upper airways (19). The high relative humidity of the airway affects particle deposition due to hygroscopic growth, contributing to depositioning of drug on the respiratory epithelium (3).

Most commonly, the droplets act on local receptors in the respiratory system, but they are also capable of taking a systemic effect elsewhere in the body. For the application of treating respiratory illnesses, the drug will take effect in the respiratory system. For the application of treating illnesses external to the respiratory system, the drug will access systemic blood after inhalation, and bind to receptors external to the respiratory system. In either case, the drug must be inhaled

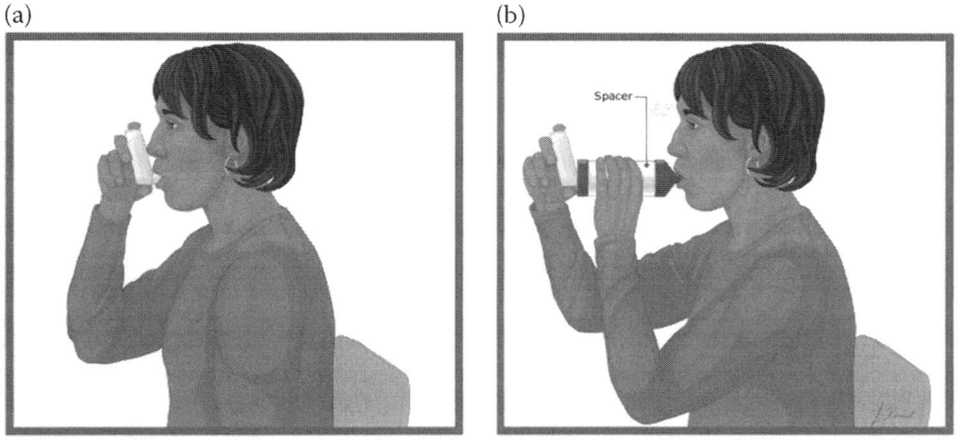

FIGURE 2.2 Adult Using a Pressurized Metered Dose Inhaler Without and With Spacer (UpToDate, 2020)

through a dispensary device. The principle types of devices used to deliver aerosol particles into the lung are inhalers and nebulizers. Within those two groups, pressurized metered dose inhalers (pDMI), dry powder, and soft mist inhalers, as well as jet, ultrasonic, and mesh nebulizers will be discussed. The use of spacers and valve holding chambers that assist users in their delivery coordination and efficacy will also be addressed.

2.3.2 Inhaler Devices

2.3.2.1 Pressurized Metered Dose Inhaler

The pressurized metered dose inhaler, or pDMI, was the first modern-day inhaler to be commercially used (in the 1950s). The pDMI is a pressurized multi-dose aluminum canister that delivers single doses through a metered valve and mouthpiece. A button on the canister, when pressed, releases a single reliable dose of medication. The key constituents of the canister include solution, suspension, or co-suspension, and other components such as propellant, suspending agents, co-solvents, and surfactants. The canister interior is coated to decrease surface adhesion and promote expulsion (4). The pressurized drug turns into small aerosol droplets when the drug reaches air through evaporation due to the difference in boiling points between room temperature air and the drug formulation.

Pressurized metered dose inhalers are commonplace in the delivery of drugs for treating obstructive pulmonary diseases such as asthma and COPD. In fact, all inhaled drug classes for obstructive lung diseases are commercially available in pDMI form, whether alone or in a combination formula (4). However, users require coordination between their inspiration and dispersement of the drug. This leads to many potential technical difficulties. In addition, with the use of pDMIs there is a high rate of oropharyngeal deposition. This leads to less aerosol making it past the oral cavity and upper airways and less aerosol deposited in the small lower airways, so pDMI are frequently used in conjunction with a **valved holding chamber (VHC)** or **spacer** to mitigate the high oropharyngeal deposition in addition to the difficulty in coordinating an immediate inhalation after activating the button. VHCs have a one-way valve situated behind the mouthpiece and connect to the mouthpiece of the pDMI. When a drug is dispensed, the VHC holds the aerosols deposited by the pDMI until the user is ready and able to inhale. This is intended for children, the elderly, and people with compromised motor and/or cognitive coordination. The spacer decelerates the aerosols after their release, and the deposited particle cloud becomes static prior to inhalation. By slowing down the particles and retaining large particles within the VHC, oropharyngeal deposition decreases with their use by 80–90% (20). In addition, this provides an evenly distributed delivery of aerosols throughout inhalation, rather than a highly concentrated surge (Figure 2.2).

2.3.2.2 Dry Powder Inhaler

A dry poweder inhaler (DPI) consists of an active pharmaceutical ingredient (API) of suitable aerodynamic size (usually 1–5 μm) for inhalation, contained within a device which, upon inhalation, provides sufficient deagglomeration of particles to deliver a therapeutic dose to the lungs. DPIs commonly contain micronized drug particles which, because of their strong predisposition to aggregate, have poor aerosolization performance. To mitigate this issue, carrier proteins are added. The carrier proteins function to improve flowability of drug particles to facilitate filling the DPI, increasing spreading of drug particles during emission, and to dilute the drug to improve accurate dose delivery (21).

Drug dispersal from a DPI is dependent on the inspiratory force of the patient. Once the device is activated and the patient inhales, airflow introduced in the devices creates turbulence. As air is introduced into the powder bed, the powder becomes aerated and enters the airways. At this point, the drug and carrier proteins separate and the drug travels into the airways while the carrier protein prematurely deposits in the oropharynx. Each DPI device has an internal structural resistance and threshold inspiratory force necessary to overcome such resistance. Overcoming such resistance is necessary in order to deagglomerate the drug/carrier protein combination and disperse the powder formation within the respiratory system to achieve a useful therapeutic response. One advantage of utilizing a patient's inspiratory airflow as the main source of energy is that such devices are breath actuated; this inherently avoids the need to synchronize the actuation and inspiration maneuver by the patient. The downside of this approach is that devices currently available show a device-specific airflow resistance, and this often demands a relatively high inspiratory effort (22) which might be a hurdle for patient populations suffering from obstructive airway diseases such as asthma or COPD, the elderly, or the very young (23).

There are a few main types of DPIs that can be categorized as either single dose or multi-dose. The single-use dose inhaler requires the patient to load a single gelatin capsule containing the drug powder formation before each use. There are also single use pre-metered devices that are discarded after each use. As for the multiple dose units, devices deliver individual doses from pre-metered replaceable blisters, disks, dimples, or tubes. Multiple-dose reservoir inhalers contain a bulk amount of drug powder in the device with a built-in mechanism to meter a single dose, and individual doses are delivered with each actuation (24, 25).

2.3.2.3 Respimat Soft Mist Inhaler

The soft mist inhaler was developed in response to the need for a pocket-sized device that can generate a single-breath, inhalable aerosol from a drug solution using a patient-independent, reproducible, and environmentally friendly energy supply. The mist comes out much slower and has a longer aerosol duration than when delivered by a pDMI, allowing for less drug to deposit in the mouth and more drug to make its way to the lungs (26). The medication in the SMI is stored inside the cartridge. When the clear base is turned, a spring inside the SMI is pressed. When the dose release button is pressed, the energy from the spring presses the medication through the nozzle, releasing the fine mist that patients breathe in. There are no parts to fix, so once the cartridge is properly inserted, individual medication doses will be given each time the base is rotated half a full turn and the dose-release button pressed. When the last dose is given, the base can no longer be rotated (27). When compared to a pDMI, scintigraphy studies have shown that lung deposition is doubled and oropharyngeal deposition is reduced when using the Respimat SMI (Table 2.1) (28).

2.3.2.4 Nebulizers

A nebulizer is a machine that delivers a steady stream of medicine in a fine mist. Nebulizers transform liquid formulations and suspension into a medical aerosol by converting a bulk liquid solution or suspension formulation into small droplets. The drug released is contained in particles 1–5 μm in diameter. The atomization process can be done through compressed air or ultrasonic energy/sound vibrations. Unlike the DPI and pDMI, patients using a nebulizer must wear a mouthpiece covering their mouth and nose during the drug delivery process. Although it is better to use a mouthpiece, nebulizers are often used with facemasks when the patient is sick or uncooperative (20). After connecting the tubing from the compressor (where the drug is initially stored) to the mouthpiece, patients turn on the compressor; patients must inhale the drug for 10–15 minutes, until all the drug is gone.

The three main types of nebulizers are the jet, ultrasonic, and mesh nebulizers. Table 2.2 details the major advantages and limitations of the inhaler types covered in this section, including the three main types of nebulizers. A major benefit of nebulizers is that they do not require patient synchronization between inhalation and actuation, thus making them useful for pediatric, elderly, ventilated, and unconscious patients, or those who are unable to use pMDIs or DPIs. Nebulization time is important for patient adherence to therapy in the out-patient setting and clinician supervision for hospitalized patients. A limitation to nebulizers is the longer administration times necessary for nebulizer administration compared to drug delivery from other aerosol devices. The benefit of an elongated administration time is that the nebulizer is capable of delivering larger doses compared to other aerosol devices. A shorter nebulization time that still delivers an effective dose is optimal (20).

2.3.4 Patient Education

A significant barrier in optimizing the efficacy of the previously discussed inhalers is proper healthcare provider and patient education surrounding correct technique. Educating patients regarding proper inhaler use has been shown to not only improve their technique, but also their clinical outcomes. Proper use of the inhaler devices is of crucial benefit to patients because it decreases their risk of hospitalizations, emergency room visits, courses of anti-microbials, and oral steroids (31). In a large meta-analysis, research showed that in the last 40 years, the high prevalence of poor technique with the use of pDMIs and DPIs has not improved. With error rates as high as 49% in testable technique categories such as MDI coordination, and an overall prevalence of correct technique seen in just 31% of patients, it is evident that many patients are not receiving the maximum benefits from their inhaler devices (32).

TABLE 2.1
Characteristics of an Ideal Inhaler

Drug Delivery	Patient Use	Pharmaceutical Concerns
High lung deposition	Simple to use	Absence of propellants
Aerosol generation independent of inspiration	Portable and pocket-sized	Uniformity of dose
Prolonged actuation time	Multi-dose (>50 actuations)	Resistant to contamination
High fine particle dose of aerosol	Dose counter	Resistance to humidity
Slow velocity aerosol	Reliability	Drug-drug interactions

(adapted from (29)).

TABLE 2.2

Comparison of Inhalers

Device	Advantage	Limitations	
Pressurized metered dose inhaler	Portable	Requires coordination of inspiration and actuation	Not suitable for young children (without use of a valve holding chamber)
	Pressurized metered dose inhaler	Compact	
	Multi-dose device	High oropharyngeal deposition (without use of a valve holding chamber)	
	Dose delivered and particle side relatively independent of inhalation maneuver	Propellant required	
	Quick and easy to use for many patients	Need to shake prior to use	
	Available for many formulations	Need to prime if not used recently	
	Not breath-actuated	Not all medications available	
Dry powder inhaler	Portable	Moderate to high inspiratory flow required for most devices	
	Compact	Not suitable for young children; some devices are single dose	
	Breath-actuated	Many not be suitable for emergencies	
	Less coordination needed	Some devices susceptible to environmental humidity	
	Short treatment time	Unable to use with valved holding chamber	
	Readily available for many formulations	Not all medications available	
Soft mist inhaler	Portable	Device needs to be assembled initially	Lack of data for adults regarding delivery through a spacer or valved holding chamber (data available for children)
	Multi-dose device	Not breath actuated	
	Less dependence on inspiratory flow	Need to prime if not used recently	
	Soft mist inhaler	Slow velocity aerosol	
	Relatively high lung deposition	Not all medications are available	
	Long aerosol cloud duration		
	Less coordination needed		
	Ease of administration		
	No propellant/Does not require spacer		
	Suitable for children		
Jet nebulizer	Less coordination needed	Cost	Pressurized gas source required
	Effective with tidal breathing	Less portable	
	Jet nebulizer	High doses can be easily administered	
	Dose modification possible	Lengthy treatment time	
	Combination therapy if drugs are compatible	Contamination possible	
	Some are breath actuated	Device preparation	
		Not all medications available	
Ultrasonic nebulizer	Less coordination needed	Cost	Potential for airway irritations
	High doses can be given	Need for electrical power	
	Small dead volume	Contamination possible	
	Quiet	Prone to malfunction	
	Faster delivery than jet nebulizers	Possible drug degradation	
	Less drug loss during exhalation	Device preparation	
		Some are breath actuated	
		Not all medications available	
Mesh nebulizer	Less coordination needed	Cost	
	Effective with tidal breathing	Contamination possible	
	High doses easily given	Device preparation	
	Dose modification possible	Not all medication available	
	Some are breath actuated		
	Small dead volume		
	Quiet		
	Faster delivery than jet nebulizers		
	Less drug loss during exhalation		
	Portable and compact		
	High dose reproducibility		

(Modified from Pleasants and Hess (4, 30)).

With as many as 25% of patients never receiving verbal instruction in how to use their device, there is vast room for improvement (33). In addition, verbal instruction is not enough. A comprehensive hands-on, step-by-step demonstration with emphasis on why the patient needs the inhaler and the importance of its regular, correct use, has been found to be educational and motivational for the patient.

There has been evidence that training patients on proper device usage should be performed in a repetitive fashion by periodic reassessment and re-education, all while allowing enough time for the patient to replicate and demonstrate the learned technique (34). This can be challenging if healthcare professionals are constrained by the amount of time they have with patients or they themselves have limited knowledge of device use. In a recent study, only 14% of over 1,500 surveyed physicians reported having adequate knowledge of inhaled therapy (35). The skilled training of physicians, physiotherapists, and respiratory nurses (amongst others) in inhaled therapy education is arguably just as important as the education of their patients. In a real world setting, all inhalers have their flaws and it is important that they are not viewed as equivalents; instead, providers should take into account each patient's strengths and limitations before administering an inhaler. Being mindful to take into account the patient's peak inspiratory flow, their cognitive and physical status, as well as inhaler preference, will contribute to improved technique and retention.

2.4 Overview of Inhaled Therapies

The abundance of cell receptors in the respiratory tract which are prime targets in the treatment of airway diseases make inhaled drug therapy an efficient mode of drug delivery. Pulmonary drug delivery bypasses the first pass effect of the gastrointestinal tract, reaching key target cells within the respiratory tract at higher levels of bioavailability with a low risk of systemic side effects (36). Classes of drugs commonly used in the management of airway diseases included **β_2** adrenergic agonists, muscarinic receptor antagonists (anticholinergics), and corticosteroids.

2.4.1 β_2 Adrenergic Agonists

Compared to all the asthma therapeutics available, β_2 adrenergic receptor agonists (e.g. beta agonists) are the most commonly prescribed therapeutic agents for the management of asthma and other obstructive pulmonary diseases (37). Adrenergic receptors are found throughout the body in a variety of organ systems such as the central nervous, cardiovascular, gastrointestinal, and musculoskeletal systems. There are α and β receptor subtypes. The α receptor subtype exhibits primarily excitatory function and the β receptor subtype exhibits primarily inhibitory function, with the exception of the myocardium. The β_2 adrenergic receptor subtype is found in the uterine and vascular smooth muscle, and most importantly the bronchioles. For this reason, inhaled delivery is essential in minimizing systemic side effects. The primary and clinically relevant effect of inhaled β_2 agonists on airway smooth muscle cell receptors is bronchodilation. The concentration of β_2 receptors remain constant throughout the airway, so that the number of receptors in the large airways is comparable to those in the small airways (38). The β_2 receptor is a G protein coupled receptor, exhibiting 7 transmembrane domains with an extracellular N-terminus and intracellular C-terminus.12

Following the binding of a β_2 agonist to its adrenergic receptor in the airway, a transmembrane cascade is initiated stimulating the release of effector molecule, adenylyl cyclase. Adenylyl cyclase then increases intracellular cAMP via the hydrolysis of ATP. The elevated cAMP concentration serves to activate cAMP-dependent protein kinase A (PKA). PKA can phosphorylate intracellular substrates, which modulate various effects within the cell. In airway smooth muscle, PKA functions to phosphorylate Gq-coupled receptors leading to a cascade of intracellular signals which have been proposed to reduce intracellular Ca2+ or decrease the sensitivity of Ca2+ (39). The change in Ca2+ results in the inhibition of myosin light chain phosphorylation, blocking airway smooth muscle contraction.

2.4.1.1 β_2 Adrenergic Agonists Subtypes: SABA and LABA

β_2 adrenergic agonists are further categorized based on the onset of action and duration of effect. There are two principle subtypes of the β_2 adrenergic receptor agonist, they are short acting (SABA) and long acting (LABA). After being administered, a short acting beta-agonist (SABA) begins to take effect and relieve symptoms within minutes and will continue to promote bronchodilatory effects and maintain open airways for 3–6 hours. The short half life makes SABAs ideal in emergency situations for the treatment of acute symptoms, for example, in response to sudden or severe asthmatic symptoms such as coughing, wheezing, chest tightness, or shortness of breath. The short half life also poses a challenge for the use of SABAs in long term disease management. Because of this, SABAs are often used in conjunction with inhaled corticosteroids, LABAs, or long acting muscarinic agonists, to make sure patients' symptoms are consistently managed. Common SABAs include salbutamol (albuterol), terbutaline, levalbuterol, and pirbuterol. SABAs are typically administered through inhalatory mediums such as pressurized metered dose inhalers or dry powder inhalation.

LABAs are often used in conjunction with inhaled corticosteroid therapy for long term symptom management in patients with COPD and asthma. The typical duration of LABA onset is >5 minutes with a duration of action of at least 12 hours. The extended effectiveness may in part be attributed to the presence of large side chains on the molecular structure, increasing the lipophilicity of the molecule. This generally increases retention of the molecule in the lipid bilayer of the cell membrane, effectively extending half life. LABAs also exhibit a stronger affinity for the β_2 adrenergic receptor than SABAs. For maintenance (daily) management treatment, LABAs should be used every 12 hours, delivered through inhalation. A few common LABAs include salmeterol, formoterol, and arformoterol (40).

2.4.1.2 β_2 Adrenergic Agonists Side Effects

The body contains abundant subtypes of the adrenergic receptor. In order to increase selectivity for the respiratory β_2 receptor and minimize systemic side effects, chemical manipulation of the side chain of the adrenaline and isoprenaline molecules are made. Increasing the size of the molecule attached to the amine has been shown to increase receptor selectivity, and modifications to the aromatic ring have been shown to prolong the duration of action. It should be recognized that all SABA and LABAs currently on the market for asthmatics have at least some selectivity for the β_2 adrenergic receptor, but the degree is variable. The adverse effects associated with bronchodilators are due to activation of the sympathetic response. Common side effects include trembling, nervousness, sudden, noticeable heart palpitations, and muscle cramps (38).

2.4.2 Corticosteroids

The term *corticosteroids* is an umbrella term used to describe agents that exhibit glucocorticoid effects. The main endogenous glucocorticoid is cortisol. Cortisol is produced in the adrenal gland, derived from cholesterol, and is a potent inhibitor of the inflammatory process. Inflammatory cells such as eosinophils, T-lymphocytes, mast cells, and dendritic cells are commonly found to be elevated in asthmatic airways along with an increased production of chemotactic mediators, mucus, and adhesion molecules. Because glucocorticoids are derived from cholesterol and demonstrate lipophilic properties, they are able to diffuse across the cell membrane and bind to glucocorticoid receptors (GR) in the cytoplasm and translocate into the nucleus. From the nucleus, corticosteroids are able to activate and suppress many genes relevant to inflammation, including the upregulation of the β_2 adrenergic receptor (Table 2.3). Nuclear GR may also interact with coactivator molecules, such as the CREB-binding protein (CBP), which is typically activated by proinflammatory transcription factors, such as nuclear factor-κB (NF-κB). This inhibitory interaction switches off the inflammatory genes that are activated by these transcription factors, decreasing their production (41).

The increase in gene transcription of β_2 adrenergic receptors due to corticosteroid effects results in increased expression of β_2 adrenergic cell surface receptors. In this way, corticosteroids protect against the downregulation of β_2 adrenergic receptors following their long term administration. For this reason, it is beneficial to co-administer these two classes of drugs. Common ICS include fluticasone, beclomethasone, budesonide, and mometasone.

2.4.2.1 Corticosteroids Side Effects

The most common local side effect of inhaled corticosteroids (ICS) are oral candidiasis (thrush) and dysphonia. Both are the result of oropharyngeal deposition. The risk for oral candidiasis can be decreased with the addition of a spacer and rinsing of the mouth after ICS use, reducing the amount of ICS deposited in the mouth. The risk of developing candidiasis is dose dependent, with a higher dose increasing the risk for local side effects. Dysphonia, or hoarseness of the voice, may occur in over 50% of patients using MDI. Dysphonia is not appreciably reduced by using spacers, but may be less with dry powder devices.

Systemic side effects are multifaceted and can affect the hypothalamic–pituitary–adrenal axis, bone metabolism, and hematologic function. The potential for adverse effects is significantly less than those associated with systemic corticosteroids, and the benefits of their use still is more significant than the side effects. Children and COPD patients should be monitored closely. Children may experience transient periods of decreased growth velocity while using ICS, with an average decrease in adult height of 1.20 cm (42). While using ICS, COPD patients are at an increased risk for pneumonia. Studies have shown that budesonide and fluticasone, delivered alone or in combination with a LABA, are associated with increased risk of serious adverse pneumonia events. However, neither significantly affected mortality compared with controls (43). As with local adverse effects, systemic adverse effects have been found to be dose dependent, with a higher dose leading to an increased risk of developing adverse symptoms.

2.4.3 Anti-cholinergic Agents

Anti-cholinergic agents target the airway muscarinic receptors. Acetylcholine (ACH), the predominant parasympathetic neurotransmitter in the airway, is the main neurotransmitter binding to

TABLE 2.3

Transcriptional Effects of Corticosteroids

Increased Transcription	Decreased Transcription
Lipocortin-1	Inflammatory cytokines (*IL-2, IL-3, IL-4, IL-5, IL-6, IL-11, IL-13, IL-15, TNFα, GM-CSF, SCF*)
β_2 Adrenergic receptor	Chemokines (*IL-8, RANTES, MIP-1α, eotaxin*)
Secretory leukocyte inhibitory protein	Inflammatory peptides (Endothelin-1)
IκB-α (inhibitor of NF-κB)	Inflammatory enzymes (*Inducible nitric oxide synthase (iNOS), inducible cyclo-oxygenase (COX-2), inducible phospholipase A2 (cPLA2)*)
Anti-inflammatory or inhibitory cytokines (*IL-10, IL-12, IL-1 receptor antagonist*)	
Mitogen-activated protein kinase phosphatase-1 (MKP-1, inhibits MAP kinase pathways)	

Modified from Barnes (41).

the muscarinic receptors. The effect of ACH-muscarinic receptor binding is an array of asthma and COPD pathophysiology, including smooth muscle contraction, mucus secretion, inflammation, and airway remodeling. There are five muscarinic receptor subtypes that belong to the G-coupled protein receptor family: M_1, M_2, and M_3 are all located on airway epithelial cells, smooth muscle, and submucosal glands and are shown to play significant roles in airway physiology. M_1 and M_3 are excitatory and promote ACH release and coupling through Gq/G11 to activate phospholipase C leading to calcium release, causing an increase in intracellular calcium and receptor activation. M_2 receptors inhibit adenylyl cyclase activity through G-protein (Gi), which results in prolonged opening of ion channels and flow of calcium and potassium. Potassium channel activation leads to hyperpolarization of the cell membrane. In addition, ACH activation of M_2 receptors reduces ACH release from the vesicle. The summative effect of muscarinic receptor antagonists is decreased airway tone with improvement in expiratory air flow, which can benefit asthmatics and individuals with COPD (44).

Anti-cholinergics are reversible competitive inhibitors of the M_1, M_2, and M_3 receptors. The duration of action of each drug is dependent on the amount of time spent at the muscarinic receptor. The duration of drug effect is highly variable, and drugs can exhibit affinity and selectivity for the different M receptors, this leads to different outcomes as antagonism at each receptor forges different outcomes. For example, antagonism at the M_3 receptor appears to be the most clinically relevant for bronchodilation and for decreasing mucin hypersecretion driven by neutrophil elastase. A few common anti-cholinergics include ipratropium, tiotropium, and glycopyrronium. Ipratropium is short-acting and nonselective; it blocks all 3 muscarinic receptors (M_1, M_2, M_3). Ipratropium's duration of effect is 3–6 hours, with maximum effect after 15–90 minutes (45). Tiotropium, a long-acting agent, selectively blocks M_1 and M_3 receptors. Tiotropium's duration of effect is at least 24 hours, making it a once daily medication, with an onset of 30 minutes and maximum effect on FEV1 1.5–5 hours post inhalation.

2.4.3.1 *Anti-cholinergic Side Effects*

The most common side effect associated with anti-cholinergics is xerostomia (dry mouth). In 1-year studies of 906 patients who received tiotropium, approximately 14% of patients experienced dry mouth (46). Onset of dry mouth generally begins 3–4 weeks into therapy, and resolves itself with continued treatment.

2.5 Overview of Oral Therapies

Although inhaled therapies are the preferred maintenance treatment for obstructive lung diseases, oral therapies can contribute additional pharmacological effects and offer an easy route of administration that may reach deeper regions of the lung that pose difficulties when treating with inhaled therapy. Unlike inhaled therapies, oral drugs must pass through the gastrointestinal tract, the portal vein, and the liver prior to absorption into the systemic circulation. Therefore, the rate and extent of absorption is initially influenced by the dissolution kinetics and the solubility in gastrointestinal fluid (47). The majority of oral therapies induce immunomodulatory or anti-inflammatory effects, and have shown to be most effective in decreasing COPD exacerbations. This section will provide an overview of mucolytics/expectorants, phosphodiesterase (PDE) inhibitors, macrolides, and leukotriene modifiers.

2.5.1 Mucolytics/Expectorants

Mucolytics are medications taken orally to loosen sputum, making it easier to cough up and clear from the lungs. Chronic bronchitis can cause the lungs of a person with COPD to produce more mucus than usual, making these drugs of particular importance for COPD patients. Mucoactive drugs can induce cough or increase the volume of secretions (expectorants), mucolytics reduce the mucus viscosity, and mucokinetic drugs such as β_2 agonists increase mucus mobility and transportability. Expectorants improve the ability to discharge purulent secretions without altering ciliary beat frequency or mucociliary clearance (48). A commonly administered mucolytic is guaifenesin. Guaifenesin peak blood levels occur in <1 hour, and the drug exhibits a short-half half-life of 0.8 hours over the range of 600–1200 mg. N-acetylcysteine is a common expectorant. After an oral dose of 600 mg with a peak after 1 hour, and it is quickly metabolized to cysteine. The plasma half-life has been reported to be 2.5 hours.

2.5.1.1 *Mucolytics/Expectorants Side Effects*

There are minimal risks associated with mucolytics or expectorants at normal dosages. At high doses, guaifenesin is emetogenic and some data from rat studies suggest that, as an antioxidant, N-acetylcysteine may increase cancer risk (49). There is also the potential that mucolytics or expectorants can make the mucus too thin, making expectoration more challenging.

2.5.2 Phosphodiesterase (PDE) Inhibitors

Oral phosphodiesterase inhibitors are currently shown to reduce exacerbations in COPD patients alongside first line inhaled maintenance therapy. Oral PDE inhibitors are not currently recommended for asthmatic patients. PDE normally metabolizes the intracellular second messenger molecule, cyclic adenosine monophosphate (cAMP), therefore, PDE inhibitors increase the intracellular cAMP by blocking PDE activity. An increased intracellular cAMP concentration promotes airway smooth muscle relaxation and decreases the release of inflammatory cytokines and immune-stimulatory cells. Roflumilast and cilomilast are commonly prescribed narrow-spectrum PDE-4 inhibitors for COPD. After an oral dose, the half-lives of roflumilast and its active metabolite N-oxide are approximately 17 hours and 30 hours, respectively, which means that steady-state plasma concentrations are reached in less than 1 week.

Theophylline has broad spectrum effects, including inhibition of PDE 3 and 4. Theophylline has anti-inflammatory, immunomodulatory, and bronchopretective effects that potentially contribute to its efficacy as a prophylactic anti-asthma drug and for COPD. The bronchodilatory effects are attributed to PDE3 inhibition, and the anti-inflammatory effect to PDE4 inhibition and activation of histone deacetylases, which are generally reduced in severe asthma and COPD (50). The support for the use of theophylline for asthmatics is not strong. Global Initiative for Asthma (GINA) guidelines recommend theophylline as an alternative to inhaled corticosteroids and other bronchodilators due to its weak efficacy and strong adverse effects (10).

2.5.2.1 Phosphodiesterase (PDE) Inhibitor Side Effects

Roflumilast interacts with inducers of CYP3A4 (rifampicin, phenobarbital, and carbamazepine) as well as with inhibitors of CYP3A4 and CYPA12 (erythromycin, ketoconazole, and cimetidine). The multitude of possible drug interactions warrants careful dosing. GI and neurological side effects are common with related side effects including diarrhea, nausea, abdominal pain, sleeplessness, and headache.

2.5.3 Macrolides

Of the macrolides, azithromycin is associated with the lowest interactions and adverse effects. Due to its positive outcomes, it is also the most investigated. The primary role of this drug is to reduce exacerbations seen in COPD patients. Azithromyocin offers immunomodulatory and anti-inflammatory effects by accumulating substantially in neutrophils and macrophages. This reduces the build-up of pro-inflammatory mediators, influences the release of cytokines such as tumor necrosis factor and IL-8, and reduces the production of endogenous chemotactic factors which normally attract leukocytes (48). Macrolides have also been shown to relax constricted airway smooth muscle, however current GINA guidelines only make recommendations regarding macrolides for adult patients with persistent symptomatic asthma despite moderate high dose ICS and LABA therapy (10). Most clinical studies have used either 250 mg daily or 500 mg 3 times/week (51). Azithromycin has a substantial half life of >70 hours, allowing for less frequent administration for long term use and short administration periods for acute use.

2.5.3.1 Macrolide Side Effects

Macrolides are prokinetic agents, increasing motilin release and promoting diarrhea. As an antibiotic, macrolides are also capable of influencing the gut microbiome. Macrolides demonstrate dose-dependent sensorineural effects that should be considered when prescribing to patients who experience other risk factors for hearing loss. Although the hearing loss is normally mild and reversible, sometimes it can be irreversible and patients at risk should be monitored. In addition, macrolides can cause a prolonged QTc which may be associated with increased risk of ventricular arrhythmia. This is a heavily researched and debated adverse effect of macrolides with few case reports, but the risks of prolonged QTc should still be considered by providers.

2.5.4 Leukotriene Modifiers

Leukotrienes, such as cysteinyl-LTs (LTC4, LTD4, and LTE4), are potent biological lipid mediators derived from arachidonic acid through the 5-lipoxygenase (5-LO) pathway and are powerful bronchoconstrictors. The cysteinyl-LTs are likely to contribute to airway remodeling that characterizes persistent asthma (52). Leukotriene modifiers, specifically montelukast, zafirlukast, and zileuton, have been cleared for asthmatics for 20 years as alternative or add-on therapies to other therapies such as ICS and LABA. They are effective in preventing asthmatic responses related to exercise or aspirin, or those that are allergen induced. Montelukast and zafirlukast are cysteinyl leukotriene type 1 (CysLT1) receptor antagonists, and zileuton inhibits 5-lipo-oxygenase, the enzyme responsible for the production of inflammatory leukotrienes. The cysteinyl leukotriene type 1 (CysLT1) receptor is localized in the human airways and synthesized by a variety of cells including mast cells, eo- sinophils, and basophils. All leukotrienes are derived from 5-lipo-oxygenase activity, forming LTA4 from arachidonic acid. Blocking 5-lipoxygenase ceases production of all leukotriene active products, leading to modest bronchodilation and an improved FEV1 greater than that of receptor antagonists. CysLT1 receptor antagonists improve lung symptoms, reduce exacerbations, and decrease airway and blood eosinophilia in adults and children with asthma of different severities (53).

Montelukast given once daily at a dose of 10 mg protected against exercise-induced broncho-constriction over a 12-week period in adults with asthma. Zileuton is dosed twice daily as a controlled release tablet. Zafirlukast is dosed twice daily, 1 hour before or 2 hours post food consumption because its oral bioavailability is significantly impacted by food (Figure 2.3).

2.5.4.1 Leukotriene Modifier Side Effects

Adverse effects are uncommon and mild, most commonly presenting as headache, gastrointestinal disorders, fatigue, pharyngitis, upper respiratory tract infection, and rash (55). Montelukast has the fewest drug interactions, whereas zafirlukast and zileuton are more inclined to drug interactions. Steady state zafirlukast coadministered with warfarin results in prolonged prothrombin time in healthy individuals (56).

2.6 Recommended Therapies for Asthma and COPD Management

This next section covers recommended therapies for the two most prevalent obstructive lung diseases, asthma and COPD.

2.6.1 Asthma

The treatment of asthma requires a multidisciplinary approach to address patient education, compliance/adherence to medications, identification and control of triggers, and consistent monitoring of patients once an optimal medication regimen has been identified. Asthma severity is classified to help healthcare

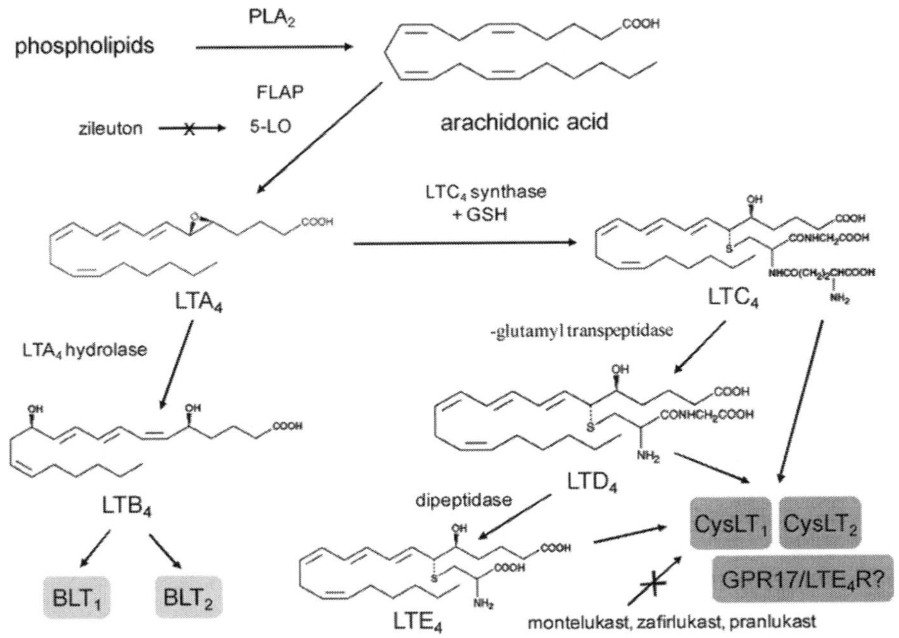

Abbreviations: CysLT = cysteinyl-leukotrienes; FLAP = five-lipoxygenase activating protein; GSH = gluthathione; 5-LO = 5-lipoxygenase; PLA$_2$ = phopsholipase A$_2$.

FIGURE 2.3 Leukotriene Production Schema with Modifiers (Montuschi et al. (54))

TABLE 2.4

Asthma Classifications

Asthma Classification	Signs and Symptoms
Mild intermittent	Mild symptoms up to two days a week and up to two nights a month
Mild persistent	Symptoms more than twice a week, but no more than once in a single day
Moderate persistent	Symptoms once a day and more than one night a week
Severe persistent	Symptoms throughout the day on most days and frequently at night

Modified from (38).

providers choose the best course of treatment (Table 2.4). Patients with mild intermittent symptoms have historically been treated with a SABA taken for relief of symptoms. GINA recommends an additional low-dose glucocorticoid (GC) whenever a SABA is used. A current novel recommended by GINA is use of a combination inhaler containing low-dose GC and a fast-acting LABA, formoterol. When a symptomatic trigger is anticipated (i.e. exercise, allergen), pretreatment with a SABA or formoterol and GC approximately 5 to 20 minutes prior to known exposure is recommended (37).

Patients with mild persistent asthma are recommended to take low-dose GC or a combination GC-LABA inhaler daily. This decreases the need for SABA rescue treatment and decreases risk of serious exacerbations. Leukotriene modifiers may be used if a patient wishes to avoid GCs, but they are considered to be less efficacious. Patients should continue with SABA therapy as needed for relief of symptoms and prior to known exposure. Patients with moderate persistent asthma are recommended to start on a combination low dose inhaled GC–LABA. The addition of an inhaled long-acting muscarinic agonist (LAMA) such as tiotropium to an inhaled GC has proven to be equally effective as GC–LABA. The inhaled GC–LAMA combination is a good alternative for patients who demonstrate intolerance or contraindications to LABAs (37).

Patients with severe persistent asthma generally require two or more controller agents to achieve improved control of their symptoms. Initial controller agents include oral and inhaled GCs and a LABA. Patients who are not on any controller medication and are consistent with national asthma education and prevention program (NAEPP) criteria for severe persistent asthma will need an initial brief course of oral GCs to bring symptoms under control. Therapy is adjusted in subsequent follow-up visits based upon assessment of asthma control and adverse reactions to the medication. NAEPP guidelines suggest trial of leukotriene modifiers as add-on therapy in patients with late onset disease, particularly those with aspirin sensitivity (58) or those who are not controlled on high-dose inhaled GC–LABA (59).

2.6.2 COPD

COPD pharmacotherapy is initiated based on the assessment of level of symptoms and risk of exacerbation. COPD therapies fall under two main categories: long-term management and acute exacerbation therapy. For all patients with COPD, short-acting bronchodilators are recommended for as-needed relief for intermittent increases in dyspnea. For minimally symptomatic patients with low risk of exacerbation, a short-acting bronchodilator is recommended. Either a SABA or SAMA is shown to rapidly improve lung function and symptoms. A short-acting bronchodilator is usually the only therapy necessary for minimally symptomatic patients with low risk of exacerbation (60).

For more symptomatic patients with low risk of exacerbation, a LABA or LAMA is recommended in addition to a short-acting bronchodilator. For patients prescribed a LABA, a SABA or combination SAMA–SABA is recommended for rescue use. For patients prescribed a LAMA, a SABA is recommended for rescue use. The efficacy and safety of LABAs and LAMAs appear comparable, and while both reduce exacerbation, LAMAs have a greater effect (61).

For minimally symptomatic patients with high risk of exacerbation, a LAMA is recommended to reduce the exacerbation rate. For symptomatic patients with high risk of exacerbations, a LAMA is the initial therapy of choice, as a LAMA alone will reduce dyspnea and exacerbations in most patients. A LAMA–LABA combination is recommended in patients with severe breathlessness and persistent exacerbations. For patients with frequent exacerbations on LABA/LAMA therapy and blood eosinophils ≥300 cells/microL, GOLD strategy suggests a LABA–ICS combination as an alternative therapy. For patients with frequent exacerbations on LABA/LAMA therapy and blood eosinophils ≦300 cells/microL, GOLD strategy suggests adding a macrolide such as azithromycin or a PDE inhibitor such as roflumilast (60).

2.7 Systemic Drug Delivery via the Lungs

This section will cover the use of the lungs for systemic drug delivery. Although inhaled therapeutics are generally targeting the treatment of pulmonary diseases, the lungs are a portal to the body. The lungs allow easy access to systemic circulation. With a rapid absorption, high permeability, and a reduced amount of drug altering enzymes compared to the GI tract, the lungs offer significant potential to influence the way that systemic drugs are delivered. In comparison to oral delivery (another non-invasive route) the lungs have just a fraction of the drug-metabolizing and efflux transporter activity of the gut and liver. Bypassing first pass metabolism leads to a higher bioavailability and fewer metabolic byproducts to take into consideration. All aerosol dosage forms pose as possible methods for administration of drugs to the lungs for systemic effect.

Proteins and peptides are of particular interest for delivery via the lungs as they are intensely subject to peptidase through other non-invasive routes. The bioavailability of peptides and proteins is 10 to 200 times greater by the pulmonary route as compared with other non-invasive routes (62). A notable outcome of the continual research in this field is the creation of an inhaled insulin for diabetics. In 2014, the FDA approved Afrezza, an inhalable, needle-less, short-acting insulin (63). Afrezza lowers blood sugar rapidly, reaching peak activity 35 to 45 minutes after using, and stops remediating glucose levels after 1.5 to 3 hours, mimicking the action of endogenous insulin. Afrezza is approved for type 1 and type 2 diabetics; however, type 1 diabetics must still take a long-acting insulin alongside Afrezza (64). When compared to subcutaneous injection, insulin is absorbed more rapidly after inhalation and provides a more physiological response to a meal (65).

REFERENCES

1. Costanzo LS. Physiology. Saunders Elsevier, Philadelphia, PA. 2014.
2. Barrett KE and Ganong WF. Ganong's Review of Medical Physiology. McGraw-Hill Medical, New York, NY. 2010.
3. Labiris NR and Dolovich MB. Pulmonary drug delivery. Part I: Physiological factors affecting therapeutic effectiveness of aerosolized medications. *British Journal of Clinical Pharmacology*. 2003; 56(6): 588–599. doi:10.1046/j.1365-2125.2003.01892.x
4. Pleasants RA and Hess DR. Aerosol delivery devices for obstructive lung diseases. *Respiratory Care*. 2018; 63(6): 708–733. doi:10.4187/respcare.06290
5. Chrystyn, H. Methods to identify drug deposition in the lungs following inhalation. *British Journal of Clinical Pharmacology*. 2001; 51(4): 289–299.
6. National Heart, Lung, and Blood Institute. (2012). *Asthma Care Quick Reference: Diagnosing and Managing Asthma*. National Heart, Lung, and Blood Institute, Bethesda, MD.
7. Subbarao P, Mandhane PJ, and Sears, MR. Asthma: Epidemiology, etiology and risk factors. *Canadian Medical Association Journal*. 2009; 181(9): 181–190. doi:10.1503/cmaj.080612
8. Diette GB, McCormack MC, Hansel NN, Breysse PN, and Matsui EC. Environmental issues in managing asthma. *Respiratory Care*. 2008; 53(5): 602–617.
9. Sears MR and Johnston NW. Understanding the September asthma epidemic. *The Journal of Allergy and Clinical Immunology*. 2007; 120: 526–529.
10. Global Initiative for Asthma. Global Strategy for Asthma Management and Prevention. 2020. Available at: http://ginasthma.org
11. Kemp WL, Burns DK, and Brown, TG. *Pathology: The Big Picture*. Vol. 1. McGraw-Hill Medical, New York, NY. 2008. Available at: https://www.mheducation.com/
12. McGrath J and Stampfli MR. The immune system as a victim and aggressor in chronic obstructive pulmonary disease. *Journal of Thoracic Disease*. 2018; 10(Suppl 17): S2011–S2017. doi:10.21037/jtd.2018.05.63
13. Widysanto A and Mathew G. Chronic Bronchitis. [Updated 2019 Aug 3] StatPearls [Internet] StatPearls Publishing, Treasure Island, FL. 2020, https://www.ncbi.nlm.nih.gov/books/NBK482437/
14. Wise RA, By, Wise RA, & Professional Manuals. Topic Page. Last Revision Datel Content last modified Nov 2018. (2018, November). Chronic Obstructive Pulmonary Disease (COPD) - Pulmonary Disorders. Retrieved June 1, 2020, from https://www.merckmanuals.com/professional/

pulmonary-disorders/chronic-obstructive-pulmonary-disease-and-related-disorders/chronic-obstructive-pulmonary-disease-copd
15. Lozano R, et al. Global and regional mortality from 235 causes of death for 20 age groups in 1990 and 2010: A systematic analysis for the Global Burden of Disease Study 2010. *The Lancet*. 2012; 380: 2095–2128.
16. Riley CM and Sciurba FC. Diagnosis and outpatient management of chronic obstructive pulmonary disease: A review. *JAMA*. 2019; 321(8): 786–797. doi:10.1001/jama.2019.0131
17. Gardner J. What you need to know about cystic fibrosis. *Nursing2007*. July 2007; 37(7): 52–55 doi:10.1097/01.NURSE.0000279437.30155.1e.
18. Rafeeq MM and Murad HAS. Cystic fibrosis: Current therapeutic targets and future approaches. *Journal of Translational Medicine*. 2017; 15: 84. doi:10.1186/s12967-017-1193-9
19. Sheth, P., Stein, SW, and Myrdal, PB. Factors influencing aerodynamic particle size distribution of suspension pressurized metered dose inhalers. *AAPS PharmSciTech*. 2015; 16(1): 192–201. doi: 10.1208/s12249-014-0210-z
20. Hess DR. Aerosol delivery devices in the treatment of asthma. *Respiratory Care*. 2008; 53(6): 699–723.
21. Kaialy W, Ticehurst M, and Nokhodchi A. Dry powder inhalers: Mechanistic evaluation of lactose formulations containing salbutamol sulphate. *International Journal of Pharmaceutics*. 2012; 423: 184–194
22. Clark AR and Hollingworth AM. The relationship between powder inhaler resistance and peak inspiratory conditions in healthy volunteers - Implications for in vitro testing. *Journal of Aerosol Medicine*. 1993; 6(2): 99–110.
23. Tiddens HA, Geller DE, Challoner P, et al. Effect of dry powder inhaler resistance on the inspiratory flow rates and volumes of cystic fibrosis patients of six years and older. *Journal of Aerosol Medicine*. 2006; 19(4): 456–465.
24. Lavorini F, Pistolesi M, and Usmani OS. Recent advances in capsule-based dry powder inhaler technology. *Multidisciplinary Respiratory Medicine*. 2017; 12: 11. doi:10.1186/s40248-017-0092-5
25. Finlay, WH. (2001). The mechanics of inhaled pharmaceutical aerosols: An introduction. Academic Press, San Diego, CA.
26. Hochrainer D, Holz H, Kreher C, Scaffidi L, Spallek M, and Wachtel H. Comparison of the aerosol velocity and spray duration of Respimat Soft Mist inhaler and pressurized metered dose inhalers. *Journal of Aerosol Medicine*. 2005; 18(3): 273–282.
27. Gardenhire DS, Hess DR, Myers TR, and Rau JL. A guide to aerosol delivery devices for respiratory therapists. 3rd edition. American Association for Respiratory Care, 2013.
28. Kunkel G, Magnussen H, Bergmann K, Juergens UR, de Mey C, Freund E, et al. Respimat (a new soft mist inhaler) delivering fenoterol plus ipratropium bromide provides equivalent bronchodilation at half the cumulative dose compared with a conventional metered dose inhaler in asthmatic patients. *Respiration*. 2000; 67(3): 306–314.
29. Wachtel H, Kattenbeck S, Dunne S, and Disse B. The Respimat® development story: Patient-centered innovation. *Pulmonary Therapy*. 2017; 3(1): 19–30. doi:10.1007/s41030-017-0040-8.
30. "Adult Using a Metered Dose Inhaler." *Uptodate.com*, UpToDate, Inc., 2020, www.uptodate.com/contents/image?imageKey=PI%2F114257&topicKey=PI%2F369&source=see_link.
31. Melani AS, Bonavia M, Cilenti V, et al. Inhaler mishandling remains common in real life and is associated with reduced disease control. *Respiratory Medicine*. 2011; 105: 930–938. doi:10.1016/j.rmed.2011.01.005
32. Sanchis J, Gich I, and Pefersen S. Systematic review of errors in inhaler use: Has patient technique improved over time? *Chest*. 2016; 128(5): 3198–3204. doi:10.1016/j.chest.2016.03.041
33. Lavorini F, Magnan A, Dubus JC, et al. Effect of incorrect use of dry powder inhalers on management of patients with asthma and COPD. *Respiratory Medicine*. 2008; 102: 593–604. doi:10.1016/j.rmed.2007.11.003
34. Lavorini F, Levy ML, Corrigan C, and Crompton G, ADMIT Working Group. The ADMIT series – Issues in inhalation therapy. Training tools for inhalation devices. *Primary Care Respiratory Journal*. 2010; 19: 335–341.
35. Plaza V, Sanchis J, Roura P, Molina J, Calle M, Quirce S, et al. Physicians' knowledge of inhaler devices and inhalation techniques remains poor in Spain. *Journal of Aerosol Medicine and Pulmonary Drug Delivery*. 2012; 25(1): 16–22.
36. Barnes PJ. Distribution of receptor targets in the lung. *Proceedings of the American Thoracic Society*. 2004; 1(4): 345–351. doi:10.1513/pats.200409-045ms
37. GINA 2006 National Heart Lung and Blood Institute; National Institute of Health. Global Strategy for Asthma Management and Prevention: Global Initiative for Asthma (GINA). 2006 Update. Available at http://www.ginasthma.org.
38. Khurana S, and Jarjour NN. Systematic approach to Asthma of varying severity. *Clinics in Chest Medicine*. 2019; 40(1): 59–70.
39. Billington CK, Ojo OO, Penn RB, and Ito S. cAMP regulation of airway smooth muscle function. *Pulmonary Pharmacology & Therapeutics*. 2013 Feb; 26(1): 112–120
40. Billington CK, Penn RB, and Hall IP. 2 agonists. *Handbook of Experimental Pharmacology*. 2017; 237: 23–40.
41. Williams DW and Rubin, BK. Clinical pharmacology of bronchodilator medications. *Respiratory Care*. 2018; 63(6): 641–651. doi:10.4187/respcare.06051
42. Barnes PJ. Inhaled corticosteroids. *Pharmaceuticals (Basel, Switzerland)*. 2010; 3(3): 514–540. doi:10.3390/ph3030514
43. Loke YK, Blanco P, Thavarajah M, and Wilson AM. Impact of inhaled corticosteroids on growth in children with asthma: Systematic review and meta-analysis. *PLoS ONE*. 2015; 10(7): e0133428. Published 2015 Jul 20. doi:10.1371/journal.pone.0133428
44. Kew KM and Seniukovich A. Inhaled steroids and risk of pneumonia for chronic obstructive pulmonary disease. *Cochrane Database of Systematic Reviews*. 2014; (3):CD010115. Published 2014 Mar 10. doi:10.1002/14651858.CD010115.pub2.
45. Ferrando M, Bagnasco D, Braido F, et al. Umeclidinium for the treatment of uncontrolled asthma. *Expert Opinion on Investigational Drugs*. 2017; 26(6): 761–766.
46. Rebuck AS, Chapman KR, Abboud R, Pare PD, Kreisman H, Wolkove N, and Vickerson F *The American Journal of Medicine*. 1987 Jan; 82(1): 59–64

47. Karner C, Chong J, and Poole P. Tiotropium versus placebo for chronic obstructive pulmonary disease. *Cochrane Database of Systematic Reviews*. 2012; 7: CD009285.
48. Borghardt JM, Kloft C, and Sharma, A. Inhaled therapy in respiratory disease: The complex interplay of pulmonary kinetic processes. *Canadian Respiratory Journal*. 2018; 2018: 2732017. doi:10.1155/2018/2732017
49. Pleasants RA and Hess DR. Review of *aerosol delivery devices for obstructive lung diseases. Respiratory Care*. 1 June 2018; 63(6), rc.rcjournal.com/content/63/6/708.
50. Sayin VI, Ibrahim MX, Larsson E, Nilsson JA, Lindahl P, and Bergo MO. Antioxidants accelerate lung cancer progression in mice. *Science Translational Medicine*. 2014; 6(221): ra15.
51. Barnes PJ. Theophylline use in asthma. *American Journal of Respiratory and Critical Care Medicine*. 2013 Oct; 188(8): 901–906.
52. Wedzicha J, Calverley PMA, Albert RK, et al. Prevention of COPD exacerbations: A European Respiratory Society/ American Thoracic. Society guideline. *European Respiratory Journal*. 2017; 50(3): 1602265.
53. Holgate ST, Peters-Golden M., Panettieri RA, and Henderson WR. Roles of cysteinyl leukotrienes in airway inflammation, smooth muscle function, and remodeling. *The Journal of Allergy and Clinical Immunology*. 2003; 111: S18–S36
54. Peters-Golden M, and Henderson WR Jr. Leukotrienes. *The New England Journal of Medicine*. 2007 Nov 1; 357(18): 1841–1854.
55. Montuschi P, Sala A, Dahlén SE, and Folco G. Pharmacological modulation of the leukotriene pathway in allergic airway disease. *Drug Discovery Today*. 2007; 12: 404–412
56. Calapai G, Casciaro M, Miroddi M, Calapai F, Navarra M, and Gan-gemi S. Montelukast-Induced adverse drug reactions: A review of case reports in the literature. *Pharmacology* 2014;94(1-2):60–70.
57. Accolate package insert. Astra Zeneca Pharmaceuticals, Wilming- ton. *Delaware*. Jul 2009. Accessed Nov 10, 2017.
58. Dahlén B, Nizankowska E, Szczeklik A, et al. Benefits from adding 5-lipoxygenase inhibitor Zileuton to conventional therapy in aspirin-intolerant asthmatics. *American Journal of Respiratory and Critical Care Medicine*. 1998; 157 (4 Pt 1):1187.
59. National Asthma Education and Prevention Program: Expert panel report III: Guidelines for the diagnosis and management of asthma. Bethesda, MD: National Heart, Lung, and Blood Institute, 2007. (NIH publication no. 08–4051) www.nhlbi.nih.gov/guidelines/asthma/asthgdln.htm
60. Global Initiative for Chronic Obstructive Lung Disease (GOLD). Global Strategy for the Diagnosis, Management and Prevention of Chronic Obstructive Pulmonary Disease: 2020 Report. www.goldcopd.org (Accessed on September 04, 2020).
61. Donohue JF, van Noord JA, Bateman ED, et al. A 6-month, placebo-controlled study comparing lung function and health status changes in COPD patients treated with tiotropium or salmeterol. *Chest*. 2002; 122: 47.
62. Mortensen NP and Hickey AJ. Targeting inhaled therapy beyond the lungs. *Respiration*. 2014; 88: 353–364. doi:10.1159/000367852
63. FDA: FDA approves Afrezza to treat diabetes. 2014. http://www.fda.gov/newsevents/newsroom/pressannouncements/ucm403122.htm.
64. Afrezza [package insert]. MannKind Corporation; 2018.
65. Patton JS, Bukar J, and Nagarajan S. Inhaled insulin. *Advanced Drug Delivery Reviews*. 1999; 35: 235–247.

3

Mechanism and Ways of Pulmonary Drug Administration

Ahmed S. Fahad[1], Sai HS. Boddu[2], and Jerry Nesamony[1]
[1]Department of Pharmaceutical and Policy Sciences, The University of Toledo, Health Science Campus, Toledo, Ohio, USA
[2]Department of Pharmaceutical Sciences, College of Pharmacy and Health Sciences, Ajman University, Ajman, UAE

3.1 Introduction

Medications and other substances have been administered through the pulmonary route for hundreds, if not thousands of years. An excellent historical overview of the development and use of inhalation therapies written by Mark Sanders in 2007 provides details about the origins of delivery of drugs and other active ingredients into the respiratory tract (1). The author has created a digital repository of media compilation at www.inhalatorium.com that shows examples of historical articles, advertisements, devices, patents, and other resources related to the history of inhalation formulations and devices. The inhalation method for delivering materials into the respiratory tract has been in use since ancient times and the earliest historical and archeological evidence for therapeutic inhalation was obtained from Egypt and dates back to 1554 BC (1). Similarly, the accounts of recreational and therapeutic inhalations of opium, tobacco, and herbal materials/medicines can be found in ancient China, India, Rome, Greece, and the Middle East. A more modern predecessor of a new era of the inhalation route of drug delivery is the inhalation device designed and developed in 1654 by the English physician Christopher Bennet. The historic evidence suggests that the earliest methods mostly used smoke from burned or burning material to deliver substances into the lungs. Over 100 years after Christopher Bennet's inhaler, in the late 1700s devices that generated mists referred to as *vapors* were introduced (2). In the mid to late 1800s devices that generated nebulized mists were used. This time period also saw the invention of the first dry powder inhaler. The first metered dose inhaler that used chlorofluorohydrocarbons (freons) as the propellant was launched in 1956. Since then the three primary categories of inhalation therapies, namely mist-based nebulized dosage forms, aerosol-particles-based metered dose inhalations, and dry powder inhalations have undergone several advancements with respect to devices and formulations including excipients (3).

Currently, the market size of global inhalable drugs is valued over US$25.0 billion and is estimated to reach US$41.5 billion by 2026 (4,5). Approximately 1.21% of marketed pharmaceutical products are delivered via the inhalation route (3). The pulmonary route of drug administration allows a quick onset of action, high drug concentration at the site with a low dose (10–20% of the amount given orally), improves local drug release at the disease site, and bypasses hepatic first-pass metabolism (6). For many years inhaled drugs have been used for treatment of common respiratory diseases such as asthma, chronic obstructive pulmonary disease (COPD), and chronic bronchitis. Certain drugs such as corticosteroids are preferred via the pulmonary route for treating lung diseases due to their systemic side effects. The pulmonary route offers greatest bioavailability when compared to all other routes of non-invasive drug delivery. Additionally the inhalation route offers the fastest onset of biological action when compared to all other non-invasive routes for a wide range of drugs and molecules. The advantages of inhaled drugs for treatment of respiratory and systemic diseases are highlighted in Table 3.1(7).

Considering these advantages, pulmonary administration of protein therapeutics and antibodies has gained a lot of popularity in recent times. Efforts to deliver such drugs as insulin, thyroid-stimulating hormone [TSH], calcitonin, follicle-stimulating hormone [FSH], growth hormones, immunoglobulins, cyclosporine A, recombinant-methionyl human granulocyte colony-stimulating factor, and pancreatic islet autoantigen are underway (12). When compared to various non-invasive routes of delivery, proteins and polypeptides are 10–200 times more bioavailable when administered through the pulmonary route. An inhaled form of insulin (Afrezza®) for systemic absorption of insulin is currently available in the market. In the future, inhalable medications may be available for gene therapy and to deliver various therapeutic proteins and polypeptides. In this chapter, our focus is mainly on the mechanisms of drug permeation in the lungs along with particle deposition and clearance from the lungs. In addition, the ways of pulmonary drug administration are also emphasized.

3.2 Mechanisms of Drug Permeation into the Lungs

The respiratory system is divided into three major areas: the oropharynx, the nasopharynx, and tracheobronchial pulmonary region. The airway circulation starts with the nasal cavity and

DOI: 10.1201/9781003046547-3

TABLE 3.1

Advantages of Pulmonary Drug Delivery for Treatment of Respiratory and Systemic Diseases

Respiratory Diseases	Systemic Diseases
Reduce risk of systemic side-effects	A non-invasive 'needle-free system
Rapid onset of action	Compatibility with a wide range of substances ranging from small to very large molecules (8,9)
Avoid harsh gastrointestinal environment and hepatic first-pass metabolism	Enormous absorptive surface area for absorption and highly preamble membrane in the alveolar area (10)
Ability to deliver a high drug concentration to the site of disease	Less harsh environment, low enzymatic activity, and bypass of hepatic first-pass metabolism
Achieve therapeutic effect at a fraction of the systemic dose	Prolong the residence time due to slow mucociliary clearance mechanism (11)
Sustained release effect	Reproducible absorption kinetic profile (9)

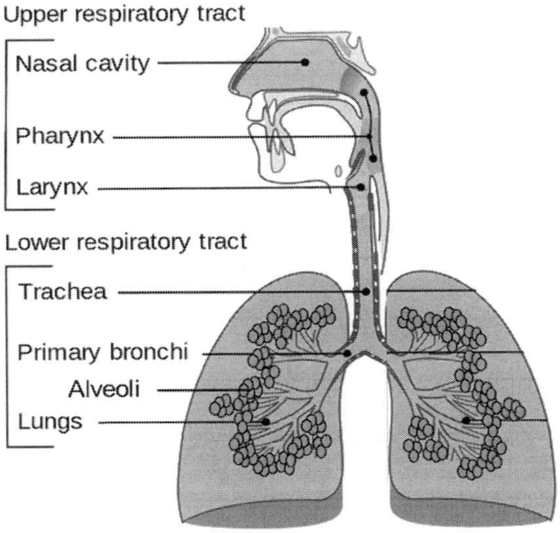

FIGURE 3.1 Anatomy of Human Respiratory System. Modified from https://en.wikipedia.org/wiki/Lung#/media/File:Illu_conducting_passages.svg

sinuses, and then proceeds to the nasopharynx, oropharynx, larynx, trachea, bronchi, and bronchioles, alveolar ducts, and alveolar sacs (Figure 3.1). The rate and extent of drug absorption varies in different regions of the lung. For instance, the conducting airways offer ~2 m^2 for drug absorption, while alveolar surfaces offer ~140 m^2. In addition, the epithelial thickness and cell population are different in the airways and alveolar region (13). The absorptive area in the lung is mainly the alveolar epithelium, which basically includes type I pneumocytes.

The pulmonary epithelium acts as a barrier to systemic absorption of drugs delivered via inhalation. However, its permeability is not uniform throughout the respiratory tract. The thickness of the epithelium is about 66 μm in the bronchi, and about 13 μm in the terminal bronchioles (14, 15). As one goes deeper into the lungs and into the alveoli where gas exchange occurs, the epithelium is extremely thin and is about 0.1–0.2 μm. The total area through which absorption can occur also undergoes drastic changes from the airways to the alveoli. The total airways surface area is about a few square meters; however, the alveolar surface area is very high and varies with respiration state. The alveolar surface area is reported to be about 35 m^2 after a deep expiration and 100 m^2 after a deep inspiration (14). Although good conclusions regarding the aerosol size beneficial for pulmonary drug delivery has been established, the ideal deposition site where optimal drug absorption occurs upon pulmonary delivery is not known. High permeability to water, gases, and to lipophilic materials is a characteristic feature of both alveolar epithelium and capillary endothelium.

Generally, inhaled drugs are absorbed either via paracellular or transcellular pathways (Figure 3.2). In the paracellular pathway, drugs pass through the tight junctions (integral proteins of claudins and occludins) present between the lung epithelial cells. The epithelial electrical resistance is highest in the upper airways, decreases to a minimum in the distal airways, and again becomes a maximum in the alveoli. This implies that the paracellular absorption occurs in the distal bronchioles. For example, small molecular weight hydrophilic drugs such as insulin (Mol wt: 5808 Da) was reported to undergo absorption via paracellular transport in the lungs. Furthermore, the alveolar type I cells limit the entry of molecules of size less than 1.2 nm diameter due to the presence of tight junctions, while endothelial junctions have larger gaps ranging from 4 to 6 nm (16). The rate of diffusion of drug molecules through the intercellular pores is found to be inversely proportional to the molecular radius (17). The use of permeation enhancers to modulate the tight junctions is considered to be a promising strategy as it enhances the delivery of hydrophilic

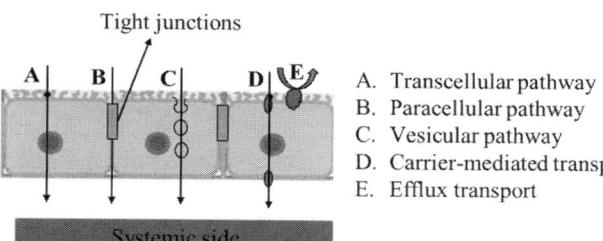

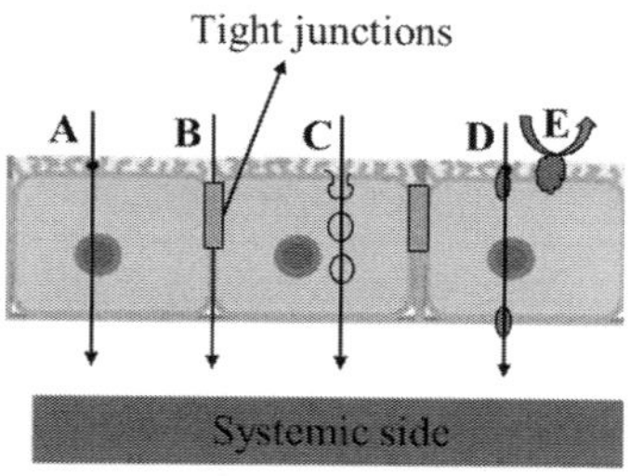

A. Transcellular pathway
B. Paracellular pathway
C. Vesicular pathway
D. Carrier-mediated transport
E. Efflux transport

FIGURE 3.2 Transport Mechanisms across Pulmonary Epithelium

compounds and protein drugs. Paracellular transport of drugs can be increased by using chitosan, hypertonic saline, sodium caprate (sodium salt of medium chain fatty acid), and oleic acid (fatty acid), which decreases the tightness between the paracellular junctions in a reversible manner (18, 19). In a recent study, sodium decanoate was shown to significantly increase the permeation of Flu-Na (paracellular marker) and PXS25 (anti-fibrotic compound) through the reversible opening of tight junction modulator in Calu-3 lung epithelial cells (20).

Low molecular weight hydrophobic drugs are absorbed through the lungs via transcellular pathways, wherein drugs passively diffuse through the cell membrane. Studies have also demonstrated that lipophilic molecules (log $P > 0$) are readily absorbed in about 1 minute and absorption of hydrophilic molecules (log $P < 0$) takes about 1 hour. A study conducted by Schanker et al. concluded that the compounds administered to lungs absorbed by passive diffusion had a rate of absorption that increased with lipophilicity of compounds (partition coefficients, chloroform/buffer pH 7.4, ranging from –3 to 2) (21). However, in the case of hydrophilic compounds, the absorption rate was inversely related to the molecular weight within the range 60–75000 Da (17). Large intracellular gaps within microvascular epithelium are more permeable and they allow proteins and all molecular sizes to enter systemic circulation. Apart from passive diffusion, transcellular pathway also occurs via transporter molecules expressed on the surface of cell membranes. In comparison to the passive diffusion, drug transport via transporters and receptors is very limited.

Two classes of transporters are expressed in lung cells, namely the solute carrier (SLC) and ATP binding cassette (ABC) transporters. The SLC family includes both organic cationic transporters (OCT) and organic anionic transports (OAT) such as OCT1, OCT2, OCTN1, OAT2, OAT3, OAT4, PEPT2, OATP1A2, OATP1B3, OATP2B1, and PGT/OATP2A1 as confirmed through QTAP LCMS/MS analysis in human lung tissue. The expression of ABC transporters such as MDR1, MRP1, MRP3, MRP4, MRP5, MRP6, MRP8, and BCRP are also reported. Of all these, the highest expression was observed in OCTN1, MRP1, BCRP, and PEPT2 proteins (22). For example, salbutamol (albuterol) undergoes absorption via OCT1, OCT3, OCTN1, and OCTN2 as it is positively charged at the physiologic pH in the lungs. An increased apical to the basolateral transport of albumin, transferrin, and immunoglobulin G was observed in rat alveolar monolayers via adsorptive endocytotic and/or receptor-mediated processes (23). Further details on drug transporters in the lungs and their impact on distribution of pulmonary administered drugs are highlighted in a recent proceedings of the workshop on Drug Transporters in the Lungs. (22)

Studies have also shown the presence of membrane vesicles within the epithelial type I cells and alveolar endothelial cells. These vesicles are non-coated or smooth-coated vesicles recognized as caveolae (24). The number of caveolae-like structures were found to be more in the endothelium compared with the alveolar type I epithelium. Membrane vesicles were shown to be involved in the transcytosis or the vesicular movement of macromolecules across endothelial cells (25). For example, caveolae-mediated transport of albumin was observed in the rat pulmonary endothelium. It has also been scientifically established that regular smoking and the presence of pulmonary disease increases lung permeability (16).

3.3 Deposition of Aerosol Particles in the Respiratory Airways

3.3.1 Mechanisms of Particle Deposition in the Respiratory Airways

Three major mechanisms are responsible for particle deposition in the lungs: inertial impaction, sedimentation, and Brownian diffusion (16). The deposition mechanism directly correlates with the particle diameter and determines the deposition of the particles in a particular area of the respiratory airways (26).

3.3.1.1 Inertial Impaction

This is the most important mechanism of aerosol particles deposition with Mass Medium Aerodynamic Diameter (MMAD) of more than 5 µm. When velocity and mass of particles lead to an impact on the airway track, as in the case of a bifurcation or further subdivision of the airway tube, the particles tend to settle in the upper respiratory airways. Partial blockade of the respiratory airways and changes in the direction of inspired air improve the chances for deposition through this mechanism. Thus, as the airflow changes direction when the airway tube branches out, the aerosols tend to retain their existing path due to the particle momentum (27). This will cause the aerosol to eventually collide with the respiratory airway walls and deposit at the site of impaction. The larger and denser the aerosol particle, the greater will be its momentum and the greater will be the probability for it to deposit through inertial impaction. The probability of aerosol particles to deposit through inertial impaction can be expressed as a function of the Stokes number (Stk). The Stokes number is defined as

$$\text{Stokesnumber} = \text{Stk} = \rho_p.\ d_p^2.\ u)/(18\mu d)$$

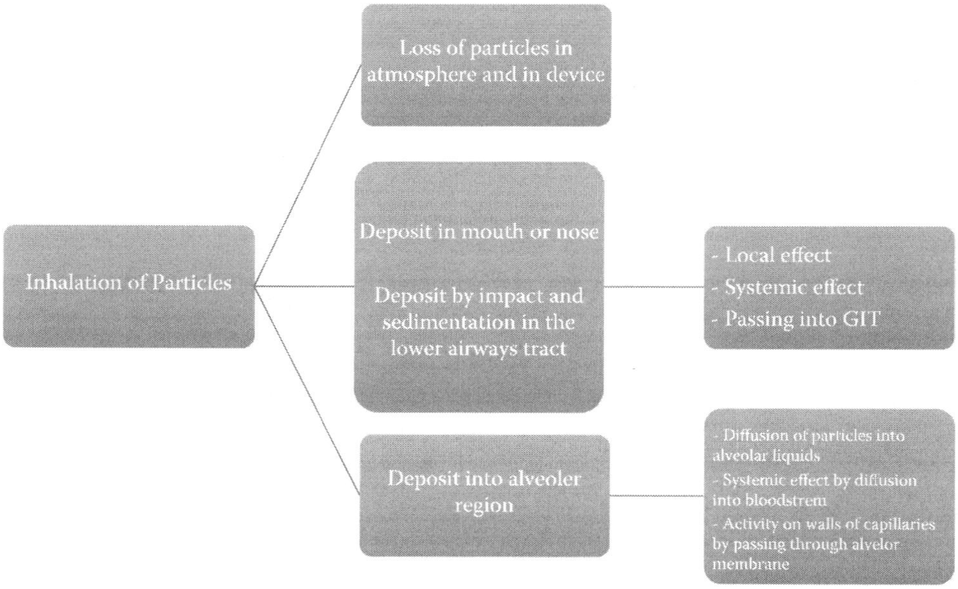

FIGURE 3.3 Pathway of Aerosolized Drug Particles in the Body (16)

where ρ_p is the particle density, d_p is the particle diameter, u is the mean velocity, μ is the dynamic viscosity of the carrier gas, and d is the airway tube diameter. Hyperventilation, a condition in which overbreathing patterns develop due to various reasons, may significantly affect particle deposition via impaction (16).

3.3.1.2 Sedimentation

Sedimentation occurs in the peripheral airways and involves aerosols with an MMAD from 1 to 5 µm. This mechanism occurs due to the action of gravitational forces on particles and occurs primarily in small airways and alveolar cavities. Particle motion is not considered a factor that has an effect on this type of particle deposition because gravitational settling occurs predominantly when the distance for particle deposition is very small (28). However, breath holding has an impact on particle sedimentation and can improve deposition. The terminal settling velocity of aerosols V_s is determined using the equation

$$V_s = \frac{\rho_p \cdot d_p^2}{18\mu} \cdot g,$$

where g is the acceleration due to gravity.

3.3.1.3 Brownian Diffusion

This is the predominant deposition mechanism for particles with an MMAD of less than or equal to approximately 0.5 µm. These particles move arbitrarily with gas molecules and collide against the airway walls. This is the mechanism through which respiratory gas exchange happens within the functional units, acini, of the lung. Hence, it is hypothesized that aerosol particles depositing via Brownian motion deposit in the acinus part of the human respiratory apparatus. A contrasting characteristic of Brownian deposition is that as particle size decreases, deposition increases. Particle deposition via Brownian diffusion is directly proportional to the diffusion constant DB

$$DB = (ckT)/(3\pi\mu d_p),$$

where c is the Cunningham constant that factors the influence of air resistance on the extreme small size of aerosols, k is the Boltzmann constant and T is the absolute temperature. Around 80% of particles with an MMAD of less than or equal to 0.5 µm are removed out of the respiratory tract during the exhalation process (29). The pattern of deposition of aerosol particles in the body is shown in Figure 3.3.

3.3.2 Factors Affecting Particle Deposition

Various physicochemical properties, physiological, and anatomical factors affect the deposition of aerosol particles in the bronchial tree. Among the physicochemical factors, aerodynamic characteristics such as particle size, particle shape, density, and aerosol velocity are perhaps the most important determinants of aerosol deposition. Lung physiology can also impact aerosol deposition and the depth of penetration into the respiratory tract. Pulmonary function tests (PFT) are used to evaluate the impact and progression of pulmonary diseases on lung physiology. PFTs are used to distinguish between asthma and COPD, and are also used to categorize/stage the severity of the condition. The severity of the pulmonary disease determines the therapies recommended to treat the condition and when it is necessary to adopt changes in treatment regimens. Four general parameters should be considered to evaluate the size and morphology of an aerosol particle:

1. Mass Median Diameter (MMD) is the diameter of the particles of which 50% w/w of particles have lower diameter and 50% w/w have a higher diameter.

2. Percentage of weight of particles with a geometrical diameter of less than 5 µm.
3. Geometric Standard Deviation (GSD) is the ratio of the diameters of particles from aerosols corresponding to 84% and 50% of the cumulative distribution curve of the weights of the particles.
4. Mass Medium Aerodynamic Diameter (MMAD) describes the size and morphology of the aerosol's particles by considering their geometrical diameter, shape, and the density: MMAD = MMD x Density^{½}(16).

3.3.3 Effect of Particle Size

The particle size of an aerosol has a crucial impact on its mechanism and site of deposition in the lung. Large-size particles with diameters of (>10 µm) that come in contact with the upper airway tract are removed by mucociliary clearance. Aerosols with a particle diameter range from 0.5 to 5 µm deposit through sedimentation mechanism (16). Aerosol particles that are intended to penetrate the lung were determined to be in the size range around 2–3 µm (30). Particles with very small diameter may be exhaled before depositing in the lung; however, holding the breath can prevent this. Extremely small diameter particles (<0.1 µm) are not easy to manufacture, though they efficiently settle by Brownian diffusion mechanism. Human studies performed using particles in the size range of between 0.005 and 15 µm demonstrated that minimum deposition was seen at a particle size of 0.5 µm. For particles greater than 0.5 µm the deposition increased due to an increase of inertial and gravitational deposition, whereas deposition increased as the particles were progressively smaller than 0.5 µm due to an increase in diffusion transport. It has also been observed that at a particular particle size, higher deposition is seen with an increase in tidal volume, a clear indication that lung physiology and depth of inhalation can have an impact on aerosol deposition. Nevertheless, researchers have not been able to confirm an exact geometrical diameter that results in deposition of inhaled particles because even large particles that have a porous internal structure can penetrate and deposit in the lungs (31).

3.1 Respiratory Clearance of Inhaled Particles

Mucociliary clearance or a combination of mucociliary and alveolar clearance mechanisms are responsible for removing settled particles that have not entered the lung epithelium and other unwanted particles from entering the respiratory system (16). The clearance mechanisms offers an important challenge that has to be overcome when formulating aerosol products (28).

3.4.1 Mucociliary Clearance

This is the respiratory system's unique mechanism against materials that enter the respiratory tract from the outside environment during breathing. The mucociliary clearance is an efficient self-respiratory cleaning besides other clearance mechanisms such as cough and alveolar clearance (16). From the trachea to the terminal bronchioles, the ciliated epithelium is extended and covered by a double-layered mucus blanket: a low-viscosity periciliary sol layer covered by a high-viscosity gel layer. The mucus is secreted by airway epithelial goblet cells and submucosal glands. The upward movement of the mucus clears the trapped insoluble particles toward the pharynx from where it can go into the gastrointestinal tract (28). The efficiency of mucociliary clearance is significantly diminished in respiratory diseases like asthma and cystic fibrosis (32).

3.4.2 Alveolar Clearance

Absorptive and non-absorptive clearance mechanisms clear deposited particles from the terminal airways (33). The absorptive mechanism involves either direct penetration into the epithelium cells or uptake and clearance by alveolar, interstitial, intravascular, and airway microphages. The alveolar epithelial surface where respiratory gas exchange occurs is populated by alveolar macrophages that engulf and remove undissolved particles that deposit in the alveoli. The uptake by macrophages can severely limit the systemic bioavailability of drugs contained in such particles due to enzymatic degradation inside the macrophages. The non-absorptive mechanism carries particles from the alveoli to the ciliated area from which the mucociliary clearance process further removes the particles from the conducting airways (16).

3.5 Ways of Pulmonary Drug Administration

There are numerous devices and formulations that are commercially available to deliver drugs and pharmaceuticals into the respiratory tract. All of these methods and devices in the market can be placed under one of the three broad categories that includes: 1. pressurized metered dose inhalers that are also referred to as metered dose inhalers (pMDIs), 2. dry powder inhalers (DPIs), and 3. nebulizers (34). All three different types of pulmonary drug delivery methods utilize different techniques and technologies in formulations, product components, packaging and use. An important aspect that determines the efficacy of an inhalation product is the technique used when administering the product. Thus, each product is accompanied with detailed instructional material, virtual animated, and real instructional videos about proper methods of administration. Healthcare providers, including pharmacists play a very important role in delivering patient education related to proper use of inhalation devices.

3.5.1 Pressurized Metered Dose Inhalers

The pMDI revolutionized pulmonary drug delivery. The pMDI consists of a pressurized canister that contains the drug in the form of a solution or suspension along with a suitable propellant. Initially chlorofluorocarbons (CFCs) were used as the propellant in pMDIs. But due to the deleterious impact on the ozone layer and its subsequent banning, CFCs are no longer used in pMDIs. Currently, hydrofluoroalkanes (HFA) are used as the propellant in marketed formulations. A dose metering valve is a part of the sealed end of the canister that inserts into the actuator. As the actuator is depressed, it presses the valve stem assembly down through an attached spring mechanism. As the valve stem moves down it disconnects the metering chamber, allowing the propellant and formulation to eventually emanate through the valve orifice. This releases the pressurized propellant along with the drug solution or suspension into the actuator. The unique design features of the actuator extend out as the mouthpiece delivers the high-velocity aerosol spray. The mouthpiece has a removable cover that prevents dust and particle contamination. The basic aspects of this system have not undergone considerable change; however, the methods that are used in formulating the drug suspension have undergone some changes. MDIs have to be primed before using, and when the formulation is a suspension, the system has to be shaken well prior to use (35).

Numerous studies have been done to evaluate how much of the emitted dose eventually deposits in the lungs. It has been established that the lung deposition of the emitted dose from a pMDI is between 10 and 20%. The low lung deposition is partly due to oropharyngeal deposition of about 50–80% of the emitted dose (28). Additionally, efficient delivery of the aerosol from the pMDI into the lungs significantly depends on the use of proper technique when self-administering the dose. Some common mistakes that were found among patients are errors related to the breathing pattern and inspiratory air flow. A considerable number of patients were found to hold their breath after emitting the dose from the device leading to loss of drug in the oropharyngeal region. When inspiring the emitted dose from a pMDI, the patient is expected to breathe in slowly and deeply. If the patient breathes in rapidly this alone will cause the aerosol particles to deposit mostly in the oropharyngeal region. After proper breathing in of the aerosol, most products require the patient to hold the breath for 10 seconds, which in itself may not be always possible due to the decline in respiratory anatomy and physiology in advanced disease states (36). Numerous patients have difficulty with the hand–mouth coordination that is necessary while using the product.

Accessories that can be attached to the pMDI mouthpiece, such as spacers and valved holding chambers, have helped to alleviate the problems associated with hand–mouth coordination. These devices function as an aerosol reservoir into which the dose is actuated and which then can be inhaled, reducing the problems related to hand–mouth coordination. Another innovation that was developed in pMDIs to address hand–mouth coordination is the breath actuated pMDI (BApMDI). BApMDIs use an inspiratory airflow triggered system that releases the dose when the patient inhales after the device mouthpiece is placed appropriately (37). The inhalation causes the firing of the device and no hand-based actuation is needed. Despite all these disadvantages, pMDIs are still the most popular method for pulmonary drug delivery due to the small size, convenience of handling, and its sheer presence and optimization in the pulmonary drug delivery realm for over 65 years.

Table 3.2 shows various pMDIs currently in the US market along with certain characteristics associated with each product. As is evident from the table, the two most commonly used propellants are HFA-134a and HFA-227. Most products have minimal amounts of other excipients when present in the formulation. All the products except ciclesonide (Alvesco®) are suspensions. Ciclesonide is a solution and the product does not have to be shaken prior to use. The Bevespi Aerosphere® product uses a co-suspension of the active ingredients in porous phospholipid particles prepared from 1,2-distearoyl-sn-glycero-3-phosphocholine (DSPC) stabilized by calcium chloride. The aerosphere technology coupled with the co-suspension delivery

TABLE 3.2

Examples of pMDIs in the US Market

Active Ingredient	Product Name	Propellant	Other Excipients (if Applicable)
Albuterol	Ventolin® HFA	HFA-134a (1,1,1,2-tetrafluoroethane)	No excipients, microcrystalline albuterol sulfate in propellant
	Proventil® HFA	HFA-134a	Ethanol, oleic acid
	ProAir® HFA	HFA-134a	Ethanol
Levalbuterol	Xopenex® HFA	HFA-134a	Dehydrated alcohol, USP, Oleic acid
Beclomethasone	QVAR Redihaler (breath actuated)	HFA-134a	Ethanol
Mometasone	Asmanex® HFA	HFA-227 (1,1,1,2,3,3,3-heptafluoropropane)	Ethanol, oleic acid
Fluticasone	Flovent® HFA	HFA-134a	No other excipients
Ciclesonide	Alvesco®	HFA-134a	Ethanol
Salmeterol + Fluticasone	Advair® HFA	HFA-134a	No other excipients
Formoterol + budesonide	Symbicort®	HFA-227	Povidone K25, polyethylene glycol 1000
Formoterol + mometasone	Dulera®	HFA-227	Anhydrous alcohol, oleic acid
Ipratropium	Atrovent® HFA	HFA-134a	Sterile water, dehydrated alcohol, anhydrous citric acid
Formoterol + glycopyrrolate	Bevespi Aerosphere®	HFA-134a	Co-suspension of drug microcrystals in porous particles comprising of 1,2-Distearoyl-sn-glycero-3-phosphocholine (DSPC) and calcium chloride

technique is a notable advancement among the pMDI product platforms. The aerosphere (also referred to as pulmosphere) technique is used to first manufacture highly porous phospholipid microparticles (17). An emulsion of DSPC in perfluorooctyl bromide (perflubron) and water stabilized by calcium chloride is atomized and sprayed into a dryer. As the drying proceeds, the perflubron evaporates, leaving behind nano- and micrometer sized pores on the dried DSPC microparticles. The DSPC microparticles can then be blended/mixed with one or more micronized active ingredients within a single product. The drug crystals interact strongly with the DSPC particles allowing for formulating a combination product with no drug–drug interactions that may destabilize the product. This technique also allows for delivering a higher than conventionally possible dose of the active ingredients due to the unique particulate microstructure, physicochemical stability, and suspension uniformity. The aerosphere-co-suspension technology has at least partially addressed some of the challenges traditionally attributed to pMDIs, such as low lung penetration, dose inconsistency, poor suspension stability, etc. The highly porous DSPC particles enter the lungs with ease through the inspiratory airflow where they absorb moisture and collapse into a mucinous mass that readily assimilates into the pulmonary membrane. This causes the active ingredient(s) to be released locally where it produces its biological action.

3.5.2 Dry Powder Inhalers (DPIs)

DPIs were first introduced in the market in the 1970s as single dose systems. The medication was present in gelatin capsules that were placed in the device and then inhaled. One important distinction between DPIs and pMDIs related to product usage is that the patient is expected to inhale forcefully and deeply when using the DPI. DPIs do not need any propellants and use inspiratory air to deliver the aerosol into the lungs. Thus, high-velocity inspiratory air will enable efficient entry of the particles into the lungs. Additionally, forceful inspiration will produce a turbulent stream of air that is capable of breaking down solid agglomerates propelling the aerosol into the lungs, thereby minimizing oropharyngeal deposition (38). Approximately 12–40% of the emitted dose from a DPI deposits in the lungs. A significant problem associated with DPIs is the loss of about 25% of the emitted dose within the device. Issues related to poor lung deposition from DPIs are partly due to the inability of the inspired air to de-aggregate large solid particulates in the formulation (39). The design of the device helps to address this issue to a certain extent, but the aerodynamics of the inspiratory air flow, humidity in the air flow, and temperature changes within the respiratory tract can determine how well large particulate agglomerates break down when a DPI is being used. The device itself can be an impediment to generating an optimal inspiratory air flow. Research on approaches such as the use of compressed air and powered propellers to disperse the aerosol powder have not yet lead to commercialization. The DPI products currently in the market are all inhalation driven and are available in different designs. Table 3.3 lists most of the currently available DPIs in the United States. One of the first modern DPIs approved in early 2000s was the Advair Diskus®. This product was one of the most significant innovations among pulmonary drug delivery systems to obtain widespread approval and market penetration since the first MDI was introduced in 1956.

Soon after the Diskus® entered the market, the Turbuhaler® design, which was a reservoir-type multi-dose non-reusable device, became available. Additionally, one of the first inhalation driven, single-dose reusable types of DPI designs, the Handihaler® device, was introduced in the market at the same time. Based on the nature of the DPI devices and their design, the currently available products can be categorized as non-reusable or reusable types. And within the non-reusable type there are two categories: the ones that contain a premeasured dose of medications placed in blister strips and those that have a dry powder reservoir that can be metered to deliver the appropriate dose. The Diskus® device contains a blister strip of medication and the Turbuhaler® and Respiclick® designs are examples of the reservoir type. The reusable DPIs consist of a device into which a capsule or cartridge containing the dry aerosol powder is placed. The capsule is then pierced using needles connected to hand activated buttons on the device, after which the powder is inhaled (40).

An important improvement in DPI design was the introduction of the Ellipta® design. This device contains two blister strips delivering two or more different medications that cannot be combined in a single blend due to incompatibility or other technical difficulties. Another notable development among DPI products was the introduction of Diskhaler® technology (41, 42). This is a breath-actuated multi-dose reusable type product. An important and distinct characteristic is that the medication is placed in a disc shaped strip with each strip containing four blisters containing the antiviral medication zanamivir mixed with lactose. The disk is loaded into the device prior to use, punctured using the piercing mechanism in the device, and the patient inhales the powdered aerosol through the mouthpiece. Most DPIs use lactose monohydrate as the sole carrier excipient and some formulations contain magnesium stearate as a lubricant (43). The tobramycin Podhaler® uses DSPC microparticles as the carrier (44). A remarkable breakthrough among DPIs came through the introduction of fumaryl diketopiperazine (FDKP) microparticles. FDKP (bis-3,6(4-fumarylaminobutyl)-2,5-diketopiperazine) particles are manufactured using the patented Technosphere® technology that leads to the formation of highly porous and homogenous microparticles with an aerodynamic diameter between 2 and 2.5 μm. Over 90% of FDKP particles prepared in this manner are in the respirable size range (0.5–5.8 μm). FDKP, a fumaramide derivative of diketopiperazine at first self-assembles into planar nanocrystals in an acidic medium (pH < 5.2). A small amount of Tween 80 is also added in the liquid medium, and residual Tween 80 is present in the final product. The planar nanocrystals then further aggregate to form three-dimensional spherical aggregates with high porosity. The drug can be mixed into the initial excipient solution which gets adsorbed and incorporated in the final porous crystalline microparticles. The highly biocompatible preparation method led to the first successful inhaled insulin product, Afrezza® (an earlier inhaled insulin product, Exubera®, was withdrawn in 2007). Afrezza® is a tremendous technological achievement because of the non-invasive pulmonary method

TABLE 3.3

Examples of DPIs in the US Market

Active Ingredient	Product Name	Product Type	Excipients
Albuterol	ProAir Respiclick®	Inhalation driven, multi-dose, non-reusable	Alpha-lactose monohydrate
Mometasone	Asmanex Twsithaler®	Inhalation driven, multi-dose, non-reusable	Lactose monohydrate
Fluticasone	Flovent Diskus®	Inhalation driven, multi-dose, non-reusable	Lactose monohydrate
	Arnuity Ellipta®	Inhalation driven, multi-dose, non-reusable	Lactose monohydrate
	ArmonAir Respiclick®	Inhalation driven, multi-dose, non-reusable	Lactose monohydrate
Salmeterol	Serevent Diskus®	Inhalation driven, multi-dose, non-reusable	Lactose monohydrate
Salmeterol + fluticasone	Advair Diskus®	Inhalation driven, multi-dose, non-reusable	Lactose monohydrate
	Airduo Respiclick®	Inhalation driven, multi-dose, non-reusable	Lactose monohydrate
Tiotropium	Spiriva Handihaler®	Inhalation driven, single-dose, reusable	Lactose monohydrate
Umeclidinium	Incruse Ellipta®	Inhalation driven, multi-dose, non-reusable	Lactose monohydrate, magnesium stearate
Aclidinium	Tudorza Pressair®	Inhalation driven, multi-dose, non-reusable	Lactose monohydrate
Glycopyrrolate	Seebri Neohaler®	Inhalation driven, single-dose, reusable	Lactose monohydrate, magnesium stearate
Indacaterol	Arcapta Neohaler®	Inhalation driven, single-dose, reusable	Lactose monohydrate
Vilanterol + fluticasone	Breo Ellipta®	Inhalation driven, multi-dose, non-reusable	Lactose monohydrate, magnesium stearate
Vilanterol + umeclidinium	Anoro Ellipta®	Inhalation driven, multi-dose, non-reusable	Lactose monohydrate, magnesium stearate
Indacaterol + glycopyrrolate	Utibron neohaler®	Inhalation driven, single-dose, reusable	Lactose monohydrate, magnesium stearate
Vilanterol + umeclidinium	Anoro Ellipta®	Inhalation driven, multi-dose, non-reusable	Lactose monohydrate, magnesium stearate
Umeclidinium + vilanterol + fluticasone	Trelegy Ellipta®	Inhalation driven, multi-dose, non-reusable	Lactose monohydrate, magnesium stearate
Insulin	Afrezza®	Inhalation driven, single-dose, reusable	Fumaryl diketopiperazine (FDKP) and polysorbate 80
Tobramycin	TOBI® podhaler	Inhalation driven, single-dose, reusable	Porous particles made from 1,2-distearoyl-sn-glycero-3-phosphocholine (DSPC), calcium chloride, and sulfuric acid
Zanamavir	Relenza Diskhaler®	Inhalation driven, multi-dose, reusable	Lactose

used to deliver a fast-acting insulin. The highly porous and homogenous FDKP particles containing insulin are capable of penetrating deep into the lungs where they readily dissolve at pH close to neutral, releasing the loaded insulin (45, 46).

DPIs are popular and very effective means of delivery inhaled therapeutics for systemic and local activity. Despite its unique advantages, there are several aspects related to the DPIs that have to be addressed. One particular issue is the loss of aerosol particles within the device. Because of the need for forceful inhalation during their use, DPIs are not recommended for children under 5 years of age. Even among older and adult patients significant errors related to proper use of DPIs have been reported. Some of the most commonly seen errors include incorrect breathing pattern during usage and wrong method of loading the inhaler. The DPIs also suffer from large inter-individual variabilities related to deposited dose, since the optimal delivery is mostly dependent on the patient's respiratory pattern during product use.

3.5.3 Nebulizers

Modern nebulizers use external energy to produce a fine mist that is inhaled by the patient. Most of the formulations for nebulization are sterile solutions, except for Pulmicort Respules®, which is a sterile suspension. Table 3.4 lists various nebulizer formulations currently available in the US. Traditionally nebulized formulations

TABLE 3.4

Examples of Nebulizers in the US Market

Active Ingredient	Product Name	Product Type	Excipients
Albuterol	Albuterol 0.5% nebulization solution	Solution in unit dose vials	Sulfuric acid to adjust pH, water for injection
Levalbuterol	Xopenex®	Solution in unit dose vials	Sodium chloride to adjust tonicity, ulfuric acid to adjust pH
Budesonide	Pulmicort Respules®	Micronized budesonide suspension	Disodium edetate, sodium chloride, sodium citrate, citric acid, polysorbate 80, water for injection
Ipratropium + albuterol	Combivent® Respimat®	Sterile solution and soft mist inhaler	Water for injection, benzalkonium chloride, edetate disodium, hydrochloric acid
	Duoneb® (nebulized)	Solution in unit dose vials	Water for injection, sodium chloride, hydrochloric acid to adjust to pH 4, and edetate disodium
Tiotropium	Spiriva® Respimat®	Sterile solution and soft mist inhaler	Water for injection, edetate disodium, benzalkonium chloride, and hydrochloric acid
Glycopyrrolate	Lonhala Magnair®	Solution in unit dose vials	Water for injection, citric acid and sodium hydroxide
Revefenacin	Yupelri®	Solution in unit dose vials	Sodium chloride, citric acid, sodium citrate, and water for injection
Olodaterol	Striverdi® Respimat®	Sterile solution and soft mist inhaler	Water for injection, benzalkonium chloride, edetate disodium, and anhydrous citric acid
Formoterol	Perforomist®	Solution in unit dose vials	Sodium chloride, citric acid, sodium citrate, and water for injection
Arformoterol	Brovana®	Solution in unit dose vials	Sodium chloride, citric acid, sodium citrate, and water for injection
Olodaterol + tiotropium	Stiolto™ Respimat®	Sterile solution and soft mist inhaler	Water for injection, benzalkonium chloride, edetate disodium, and hydrochloric acid
Loxapine	Adasuve®	Thermally generated mist	No excipients

are administered using jet nebulizers or ultrasonic mist nebulizers (47). Jet nebulizers use compressed air that is passed using tubing into the nebulizing device chamber in which the solution or suspension formulation is placed. The gas is made to pass through a narrow opening, creating a huge pressure differential, increasing the air velocity. At the point where the high-velocity air exits the opening of an inner tube in the nebulizing chamber, a thin film of the liquid formulation is provided through a capillary effect operated feed. At the narrow orifice a negative pressure is produced, and the liquid film gets drawn into the path of the air where it collapses into fine mist droplets due to liquid surface tension. A significant portion of the mist developed consists of very large and non-respirable 15–50 µm droplets. A suitably placed baffle removes such large droplets, feeding the liquid back into the nebulizing chamber. Smaller respirable droplets are allowed to ensue out of the nebulizer into an appropriate mouthpiece or a mask that can be placed on a critically ill patient. This method of delivering inhalation mist is extensively used in institutional settings where compressed air outlets are standard fixtures by patient bedsides. Jet nebulizers are popular in home settings and are used in conjunction with a portable air compressor.

Jet nebulizers are available in the reservoir, breath-enhanced, and breath-actuated types. The reservoir type employs continuous mist generation and cause over 80% of the aerosol to be lost to the environment. The breath-enhanced nebulizer uses two valves to minimize loss of the generated mist when the patient exhales. When in use, as the patient inhales, one valve opens on the vent through which air flows into the nebulizer and is carried through the mist and the mouthpiece. As the patient exhales, a valve present in the mouthpiece expels the exhaled air and mist out, but the vent valve closes, minimizing the loss of nebulized mist to the environment. Breath-actuated nebulizers have a valve that opens and aerosol generation is activated only when the patient inhales. Ultrasonic nebulizers use a piezo-electric plate that vibrates at ultrasonic frequency when electric current flows through it. The ultrasonic energy is transmitted to a mesh that vibrates when activated by the sound energy. The drug solution is in contact with the piezo-electric unit and the vibrating mesh. As the solution is vibrated, as well, cavitation causes water to break apart into small mist droplets as it extrudes out through the vibrating mesh. Some disadvantages related to nebulizers have been the need to clean the parts such as the reservoir regularly to prevent microbial contamination. Additionally, these devices also suffer from high loss of generated mist out into the environment (48).

The soft mist inhaler (SMI) Respimat® uses mechanical energy that is generated as a spring contained in the device is twisted and released to generate the mist. A cartridge containing the drug solution is loaded into the device. Following this step, the bottom of the device is twisted 180°, enabling the spring to be compressed to the desired tension. The twisting action also delivers an accurately metered volume of the drug solution into a pump cylinder. Next the patient presses the dose release button that causes the tension in the spring to be released and the energy from the spring forces the metered solution through one of the most crucial components in the system, called a uniblock (49). The uniblock is a precision engineered microchannel system comprising a silicon wafer bonded to a borosilicate glass plate. The silicon wafer has inlet, outlet, and filter channels etched on it using technology similar to that used to engineer microchips. Through rigorous and meticulous engineering, the channels have been produced to generate two fine jets of the drug solution that converge at a preset angle to generate a slow-moving cloud of mist ($1/10^{th}$ the speed of aerosol emitted from an MDI). The metered dose generated from a Respimat® device is expelled over ~1.2 s compared to the 0.1 s from an MDI. Although the breathing technique when using the Respimat® is similar to that of pMDIs, the lung deposition of various medications were found to be, in some instances, more than twice that obtained from pMDIs (50). The Loxapine (Adasuve®) oral inhalation product listed in Table 3.4 is unique in many respects. It is a single dose, breath-actuated, non-reusable mist system. After the device is properly activated by pulling and removing a tab, the device mouthpiece is placed in the mouth. The patient inhales through the device and the inspiratory air activates a drug-coated heat source that vaporizes the drug in less than 1 second. The drug vapor condenses into an aerosol mist with an MMAD between 1 and 3 μm (51).

3.6 Conclusion

Although modern techniques and devices for pulmonary drug delivery were introduced in the mid–20th century, methods of administering medications into the lungs have been employed variously for over 2000 years. Devices used to deliver medications into the respiratory tract can be categorized into pMDIs, DPIs, and nebulizers. Over the past decade, numerous advancements in excipients, particle production, and device engineering have led to progress in terms of the diversity of active ingredients that can be formulated into an inhalation product. Additionally, the innovations and developments have addressed some of the challenges associated with delivering therapeutic aerosols into the lungs. The area of pulmonary drug delivery is going through a phase of renewed interest and rigorous research. The introduction of an inhaled rapid-acting insulin product has opened the possibilities for more biological macromolecules to enter the market in the near future.

REFERENCES

1. Sanders M. Inhalation therapy: An historical review. *Primary Care Resplratory Journal*. 2007; 16(2): 71–81.
2. Stein SW and Thiel CG. The history of therapeutic aerosols: A chronological review. *Journal of Aerosol Medicine and Pulmonary Drug Delivery*. 2017; 30(1): 20–41.
3. Zhong, H., et al. A comprehensive map of FDA-approved pharmaceutical products. *Pharmaceutics*. 2018; 10(4): 263.
4. Yıldız-Peköz A and Ehrhardt C. Advances in Pulmonary Drug Delivery. *Pharmaceutics*.2020; 12(10): 911. doi:10.3390/pharmaceutics12100911.
5. Research GV. Inhalable Drugs Market Size, Share & Trends Analysis Report By Product (Aerosol, Dry Powder Formulation, Spray), By Application (Respiratory & Non-Respiratory Diseases), And Segment Forecasts, 2019–2026. 2019. Global Industry Report, Grand View Research, San Fransisco.
6. Zheng Y., et al., Pulmonary delivery of a dopamine D-1 agonist, ABT-431, in dogs and humans. *International Journal of Pharmaceutics*. 1999; 191(2): 131–140.
7. Byron PR. Drug delivery devices: issues in drug development. *Proceedings of the American Thoracic Society*. 2004; 1(4): 321–328.
8. Wolff R., Safety of inhaled proteins for therapeutic use. *Journal of Aerosol Medicine*. 1998; 11(4): 197–219.
9. Byron PR and Patton JS. Drug delivery via the respiratory tract. *Journal of Aerosol Medicine*. 1994; 7(1): 49–75.
10. Patton JS. Mechanisms of macromolecule absorption by the lungs. *Advanced Drug Delivery Reviews*. 1996; 19(1): 3–36.
11. Barnes P. and Thomson N. Drug-induced asthma. Lippincott-Raven, Asthma. Philadelphia. 1997: 1245–1249.
12. Uchenna Agu R., et al., The lung as a route for systemic delivery of therapeutic proteins and peptides. *Respiratory Research*. 2001; 2(4): 198.
13. Gonda I., The ascent of pulmonary drug delivery. *Journal of Pharmaceutical Sciences*. 2000; 89(7): 940–945.
14. Ibrahim, M. and Garcia-Contreras L. Mechanisms of absorption and elimination of drugs administered by inhalation. *Therapeutic Delivery*. 2013. **4**(8): 1027–1045.
15. Fröhlich E., et al., Measurements of deposition, lung surface area and lung fluid for simulation of inhaled compounds. *Frontiers in Pharmacology*. 2016; 7: 181.
16. Courrier H, Butz N, and Vandamme TF, Pulmonary drug delivery systems: Recent developments and prospects. *Critical Reviews™ in Therapeutic Drug Carrier Systems*. 2002; 19(4-5): 425–498.
17. Tronde A. Pulmonary drug absorption: In vitro and in vivo investigations of drug absorption across the lung barrier and its relation to drug physicochemical properties. Acta Universitatis Upsaliensis. 2002.
18. Verma, RK, Ibrahim M, and Garcia-Contreras L. Lung anatomy and physiology and their implications for pulmonary drug delivery. *Pulmonary Drug Delivery*. 2015: 1–18.
19. Ghadiri M., et al. Modulating tight junctions on airway epithelial cells to enhance paracellular transport of antifibrotic drugs. *Journal of Aerosol Medicine and Pulmonary Drug Delivery*. 2016; 29(3): A11–A12.
20. Ghadiri M., et al., The effect of non-specific tight junction modulators on the transepithelial transport of poorly permeable drugs across airway epithelial cells. *Journal of Drug Targeting*. 2017; 25(4): 342–349.
21. Brown R. and L. Schanker. Absorption of aerosolized drugs from the rat lung. *Drug Metabolism and Disposition*. 1983; 11(4): 355–360.
22. Ehrhardt C., et al., Current progress toward a better understanding of drug disposition within the lungs: summary proceedings of the first workshop on drug transporters in the lungs. *Journal of Pharmaceutical Sciences*. 2017; 106(9): 2234–2244.
23. Matsukawa Y., et al., Rates of protein transport across rat alveolar epithelial cell monolayers. *Journal of Drug Targeting*. 1999; 7(5): 335–342.
24. Gumbleton M. Caveolae as potential macromolecule trafficking compartments within alveolar epithelium. *Advanced Drug Delivery Reviews*. 2001; 49(3): 281–300.

25. Crandall, ED and Matthay MA. Alveolar epithelial transport: basic science to clinical medicine. *American Journal of Respiratory and Critical Care Medicine*. 2001; 163(4): 1021–1029.
26. Washington N, Washington C, and Wilson C. Physiological pharmaceutics: barriers to drug absorption, Taylor & Francis, United Kingdom.2000
27. Darquenne C. Aerosol deposition in the human lung in reduced gravity. *Journal of Aerosol Medicine and Pulmonary Drug Delivery*. 2014; 27(3): 170–177.
28. Labiris N. and Dolovich M. Pulmonary drug delivery. Part I: Physiological factors affecting therapeutic effectiveness of aerosolized medications. *British Journal of Clinical Pharmacology*. 2003; 56(6): 588–599.
29. Tsuda A, Henry FS, and Butler JP. Particle transport and deposition: Basic physics of particle kinetics. *Comprehensive Physiology*. 2013; 3(4): 1437–1471.
30. Gonda I. A semi-empirical model of aerosol deposition in the human respiratory tract for mouth inhalation. *Journal of Pharmacy and Pharmacology*. 1981; 33(1): 692–696.
31. Klas P. et al. Does lung retention of inhaled particles depend on their geometric diameter? *Experimental Lung Research*. 2000; 26(6): 437–455.
32. Houtmeyers E., et al.. Regulation of mucociliary clearance in health and disease. *European Respiratory Journal*. 1999; 13(5): 1177–1188.
33. Pavia, D. *Aerosols and the Lung: Clinical and Experimental Aspects*. 2015. Elsevier Science, United Kingdom.
34. Ibrahim M, Verma R, and Garcia-Contreras L. Inhalation drug delivery devices: Technology update. *Medical Devices (Auckland, NZ)*. 2015; 8: 131.
36. Marshik, P., et al.. A novel breath actuated device (Autohaler™) consistently actuates during the early phase of inspiration. *Journal of Aerosol Medicine*. 1995; 8(2): 187–195.
35. Roche N. and Dekhuijzen PR. The evolution of pressurized metered-dose inhalers from early to modern devices. *Journal of Aerosol Medicine and Pulmonary Drug Delivery*. 2016; 29(4): 311–327.
37. Giraud V and Allaert F. Improved asthma control with breath-actuated pressurized metered dose inhaler (pMDI):The SYSTER survey. *European Review for Medical and Pharmacological Sciences*. 2009; 13(5): 323–330.
38. Islam N and Gladki E. Dry powder inhalers (DPIs)—A review of device reliability and innovation. *International Journal of Pharmaceutics*. 2008; 360(1-2): 1–11.
39. Labiris NR and Dolovich MB. Pulmonary drug delivery. Part II: the role of inhalant delivery devices and drug formulations in therapeutic effectiveness of aerosolized medications. *British Journal of Clinical Pharmacology*. 2003; 56(6): 600–612.
40. Tarsin WY, et al. Emitted dose estimates from Seretide® Diskus® And Symbicort® Turbuhaler® following inhalation by severe asthmatics. *International Journal of Pharmaceutics*. 2006; 316(1-2): 131–137.
41. Jones TL, Neville DM, and Chauhan AJ. The Ellipta® in asthma and chronic obstructive pulmonary disease: Device characteristics and patient acceptability. *Therapeutic Delivery*. 2018; 9(3): 169–176.
42. Boulet L., et al. Comparison of DiskusTM inhaler, a new multidose powder inhaler, with DiskhalerTM inhaler for the delivery of Salmeterol to asthmatic patients. *Journal of Asthma*. 1995; 32(6): 429–436.
43. Lee HJ, et al. The role of lactose carrier on the powder behavior and aerodynamic performance of bosentan microparticles for dry powder inhalation. *European Journal of Pharmaceutical Sciences*. 2018; 117: 279–289.
44. Buttini F., et al.. Dose administration maneuvers and patient care in tobramycin dry powder inhalation therapy. *International Journal of Pharmaceutics*. 2018; 548(1): 182–191.
45. Gelber C and Rousseau K. Cell transport compositions and uses thereof. Google Patents. 2009.
46. Potocka E., et al.. Pharmacokinetic characterization of the novel pulmonary delivery excipient fumaryl diketopiperazine. *Journal of Diabetes Science and Technology*. 2010; 4(5): 1164–1173.
47. Thorat S and Meshram S. Formulation and product development of pressurised metered dose inhaler: An overview. *PharmaTutor*. 2015; 3(9): 53–64.
48. Vecellio L, et al. Deposition of aerosols delivered by nasal route with jet and mesh nebulizers. *International Journal of Pharmaceutics*. 2011; 407(1-2): 87–94.
49. Mehri R, et al. Respimat soft mist inhaler (SMI) in-vitro aerosol delivery with the ODAPT adapter and facemask. *Canadian Journal of Respiratory, Critical Care, and Sleep Medicine*. 2019; 5: 1–10.
50. Lavorini F, Buttini F, and Usmani OS. 100 years of drug delivery to the lungs. *Concepts and Principles of Pharmacology: 100 Years of the Handbook of Experimental Pharmacology*. 2019; 260: 143–159.
51. Zeller S., et al. Response to inhaled loxapine in patients with schizophrenia or bipolar I disorder: PANSS-EC responder analyses. *BJPsych Open*. 2017; 3(6): 285–290.

4

Transepithelial Route of Drug Delivery through the Pulmonary System

Sakshi Kumar and Swati Gupta*
Amity Institute of Pharmacy, Amity University Uttar Pradesh, Noida, India

4.1 Introduction

The pulmonary route is one of the non-invasive routes of drug delivery to achieve both systemic and local effects through inhalation. It has been used to deliver drugs targeting different tissues and organs, such as CNS, the heart, etc. A wide variety of drugs can be delivered through this route because of the presence of a different type of epithelia that enables ease of absorption; lungs are a highly perfused organ, hence more degree of absorption is seen, especially at the alveolar site, and this also permits distribution to different organs via a pulmonary vein.

The inhalation route of delivery has gained importance in recent years, since various chronic diseases are not targetable if treated by various conventional routes because of biological barriers like the blood–brain barrier, or other factors for example limiting bioavailiability of the drug while the pulmonary drug delivery can improve the systemic concentration. Also, it provides low side effects with high efficacy. Chronic diseases like asthma and chronic obstructive pulmonary diseases are targeted mainly by the inhalational route because of their ability to provide systemic delivery of a drug. It also helps in overcoming local side effects, like pulmonary alleviation, and pulmonary inflammation and constriction.

However, despite high absorption and various other advantages of targeting the lungs, there are various barriers that reduce drug motility inside the respiratory tract. These barriers can be physiological, mechanical, due to drug properties, etc. Pulmonary clearance is the major drawback seen in this type of delivery system. Such problems can be overcome by selecting the appropriate system, such as dry powders or metered-dose inhalers. Nowadays nano-drug delivery systems are being explored, including liposomal systems and inhaled solid nanoparticles. Many formulations are being developed, including inhalable insulin, which has been extensively used by the pharmaceutical industry (1).

4.2 Macrostructure of Lungs

Lungs are responsible for the exchange of gases that distributes oxygenated blood to various parts of the body, since the lung is a highly perfused organ, it can have a vascular supply which allows it to be a site of interest for transmission of various drugs for targeted delivery of a drug. The respiratory epithelium is lined by mucociliary lining, which plays a vital role in mucociliary clearance (2).

The respiratory tree is broadly differentiated into:

1. Airways
2. Blood circulation

This defines the air passage into the lungs. There are two types of zones found in the respiratory tree:

a. Conducting zone
b. Respiratory zone

The conducting portion is composed of trachea, bronchi, bronchioles, larynx, pharynx, sinuses, and nasal cavity. This zone starts from the nose or mouth and bifurcates 17 times until it reaches the respiratory or gas exchange zone. Every part has a varying epithelial lining, for example, from the nasal cavity to bronchi are pseudostratified columnar ciliated epithelium, and bronchioles are lined by ciliated columnar epithelium. From the larynx to the periphery the branching increases dichotomously. There are two daughter bronchi for each parent bronchiole. As branching increases, the cross-section area increases exponentially with the decrease in the diameter of the passage. The conducting part terminates at the bronchioles, which mark the beginning of the acinus that is the beginning of the respiratory part, which includes the alveolar region.

The primary function of this zone is to transport the gas from the nose/mouth to the gas exchange zone. If the air inspired is cool and dry, and absorbed in the same condition, it can drop the core body temperature, and the expiration of humidified air can cause unnecessary loss of water from the body. Hence, this is another role of the conducting zone. Proper humidification of inhaled air by this zone plays an important role in maintaining respiratory health, since inhalation of dry air can cause damage, such as (3)

- Cilia destruction
- Disorganization of columnar and cuboidal epithelia

DOI: 10.1201/9781003046547-4

- Disorganization of the basement membrane
- Cellular degeneration
- Mucosal ulceration

Furthermore, hyper-humidification of inspired air can lead to water intoxication and bronchoconstriction, which can also cause serious damage to the conducting passage. These factors can lead to impaired drug delivery and also affect mucociliary clearance. Hyper humidification can also be a reason for an infectious condition.

The conducting tract can be divided into two parts:

- Extrapulmonary air conduits, which define the outer part of the lungs starting from the nose, pharynx, and larynx. The trachea is the bridge between the larynx and primary bronchi. It is covered with 16–20 hyaline rings which are the cartilage rings to support the trachea.
- Intrapulmonary air conduits begin from intralobular bronchi to terminal bronchioles which are surrounded by cartilage surrounded by cylindrical musculature airway tubes.

a. Respiratory or gas exchange zone
   This is composed of distal to terminal bronchioles which are part of the conducting zone but contribute to the respiratory zone, alveoli and alveoli, and alveolar sacs, which are lined with a thin, simple squamous epithelial lining. This zone is suitable for a gaseous exchange due to its physiological characteristics, like the presence of a thin diffusive layer and alveolar ducts that have a significantly high surface area of about 102 m^2, hence providing more area for gas diffusion through this zone.

   The outer epithelia of alveoli contain the following cells, Type 1 pneumocyte cells line the alveolar walls which are surrounded by the capillaries, surfactants secreting cells secreting dipalmitoylphosphatidylcholine. Type 2 pneumocyte cells are found in between the type 1 cells, they are high in phospholipid which is the precursor to surfactant production which plays an important role in maintaining alveolar surface tension to prevent collapse.

   The blood–air barrier is formed with alveolar epithelial cells and vessels with a thin basement membrane (0.1–0.5 micro m) contributing efficient gaseous exchange from lungs' alveoli to blood. This promotes drug delivery through pulmonary nasal epithelia
b. Blood circulation
   Lungs are highly perfused organs and are supplied by two different circulatory systems: pulmonary and respiratory. The pulmonary circulation conditions the inspired air and provides nutrients to the same, and the bronchial circulation is a part of systemic circulation which receives a low cardiac output of about 1%, mainly supply from the trachea to terminal bronchioles. Pulmonary circulation is under high cardiac output and covers the alveoli to obtain an effective gaseous exchange. The anastomosis can be seen in the medium-sized bronchi–bronchioles region.

4.3 Drug Targeting: Anatomical Sites

As is clear from the above discussion, the major absorption takes place from the alveolar region due to its physiology, hence drug delivery should be fashioned in such a way that it ensures the delivery of the promised dose to the alveoli that enables drug delivery to deep in the lungs. Due to the physiology of the conducting zone, there is no such absorption seen from that region, whereas non-respiratory bronchioles exhibit poor drug absorption ability and hence are not preferred as a site to target drugs. Many factors play an important role in drug delivery to the respiratory region which can greatly be affected by various factors as mentioned in Table 4.1, including the pathophysiology of the lungs and the type of dosage form used; for example, the metered dosage form delivers 20% of the dose to the alveolar region, while spacers and patient parameters like actuation can lead to the delivery of more than 20% of the dose to the target site. Drug deposition and cilia clearance are major factors affecting the drug flow in the lungs. These factors can be termed as anatomical barriers that resistat the drug to the target site (4).

4.3.1 Anatomical Barriers to Drug Flow

Inhaled air contains certain pollutants which are treated as a foreign bodies by the body, in response to which various cells like mucus-secreting cells, macrophage, and lymphocyte, produce a certain immunological response which can be

TABLE 4.1

Factors Affecting Pulmonary Drug Delivery

Factors Affecting Pulmonary Drug Delivery **Drug-related Factors**	**Formulation-related Factors**	
Lipophilicity	**Physicochemical properties**	**Dosage form**
Polymorphism	pH and mucosal irritation	Aerosolised form
Chemical nature	Viscosity	Metered dosage form
Molecular weight		
pKa and partition coefficient	Toxicity	Dry Powder dosage form
Dissolution and solubility		

considered as barriers to the drug flow. There can be many such factors, as follows:

- Epithelial lining – The epithelia of conducting airways is primarily formed of goblet cells, mucus-forming cells which allow the formation of a protective layer over the surface of conducting tubes and hence forms a barrier to gaseous absorption. Also, the presence of cilia causes clearance of the tract which acts as a major barrier in drug flow towards the absorption site.
- Endothelia – Pulmonary endothelial lining is primarily formed of capillary endothelium; the alveolar endothelial contain organelle free domains contributing as a barrier to gaseous absorption. The presence of a high number of endocytotic vesicles makes them when there is an increase in hydrostatic pressure.
- Alveolar macrophages – The presence of macrophages in the alveolar region restricts entry of any foreign particle into the systemic circulation which is further cleared by fluid lining and mucociliary clearance.
- Interstitium and basement membrane – The interstitium membrane is the space between two cells which generally contains cells like monocytes, leukocytes, fibroblasts, etc. The major function of this layer is to form a tight junction between two cells which are further drained of interstitial fluids present in the basement membrane, forming a barrier to gaseous exchange.
- Epithelial fluid lining – As seen in gastric mucosa, respiratory epithelia contain a fluid lining which wets the drug, allowing efficient absorption. The major composition of the mucin layer is phospholipid and protein which needs to be cleared with time, hence, it increases the ciliary beats and mucociliary clearance.
- Surfactant – The surfactants help increase surface area by reducing surface tension. Hence, in the lungs, they help prevent collapse and increase effective surface area for gaseous exchange. They are produced by pneumocyte type II cells in the alveolar region and also have bacteriostatic bacteriocidal action. As drug delivery is considered, this may increase the absorption of certain drugs and reduce the same for others, depending upon the physicochemical properties of the drug and the surfactant.
- Mucociliary clearance and drug retention time. The upper respiratory tract is lined by cilia which does not adhere to the airway epithelia by mucus secretion. The ciliary movement, also called ciliary beating, causes upward movement of mucus. The drug particles are foreign particles, in response to which mucus is secreted. This mucociliary layer reduces in thickness in the peripheral region; this enhances drug deposition in the central respiratory tract. Drug absorption enhances retention time.

Pulmonary retention of drugs is affected by the physicochemical property of the drugs, their particle size, and solubility; more lipophilic drugs have a higher retainability in the lungs. Hepatic enzymes like CYP1A1 and CYP2EA have metabolizing capacity and reduce retention time. Various xenobiotic compounds like serotonin by phase 1 and 2 enzymes are located in epithelial cells. The CYP enzymes are largely present in Clara and alveolar type 2 cells. Due to high perfusion and high cardiac rate, the pulmonary metabolism is very high. In the tracheobronchial region, the perfusion rate is low, hence there is more equilibrium and higher retention (5).

Lungs' histology varies from nasal to terminal bronchioles and alveolar regions. The airways consist of many barriers, as discussed, which interfere with drug delivery and absorption. Major drug absorption is seen in the alveolar region, which contributes to the barrier in absorption. Due to this, a different route of absorption is seen via the pulmonary route as shown in Figure 4.1. The process of absorption depends upon drug properties like pKa, lipophilicity, and the partition coefficient of the drug. Membrane properties also affect the absorption; the dissolution process is necessary before absorption which requires an adequate amount of fluid for the wetting of the drug particle (6). Absorption can happen by passive diffusion of particles, that is, direct absorption of particles through the membrane without any requirement of energy. Mostly lipophilic drugs having a particle size less than 50 Da can absorb through this mechanism. The bigger particles are absorbed through carrier-mediated transport where membrane-bound proteins are present which helps in translocating the particles. This can be facilitated diffusion in which no energy is required; it can be seen for the larger size hydrophobic particles and small-sized hydrophilic drugs. Active transport of drugs can be seen for the particles which are large—mostly for the hydrophilic particle. In lungs, alveoli, which are highly perfused, are the absorption site of drugs. Also, the presence of surfactants reduces the surface tension, thereby reducing the particle size and enhancing particle absorption.

The major mechanisms of drug delivery are as follows:

- **Passive diffusion**

  It is in the process of absorption where the flow of components is seen in the direction of the concentration

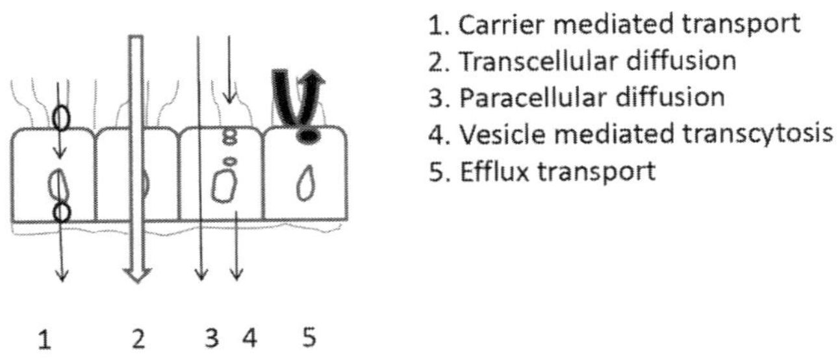

FIGURE 4.1 Mechanisms Involved in Drug Absorption

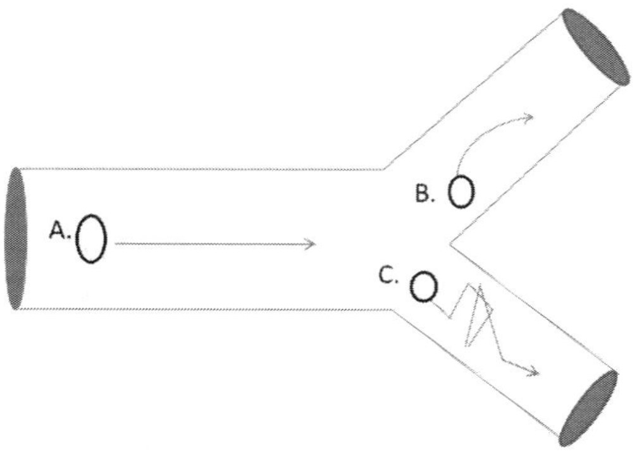

A. Impaction
B. Sedimenation
C. Diffusion

FIGURE 4.2 Mechanism of Drug Deposition

gradient. Lipophilic component absorption takes place by membrane diffusion, as the partition coefficient absorption increases. Whereas the hydrophilic compounds move through intercellular junction pores and are absorbed through passive diffusion. The absorption of hydrophilic components is inversely related to their molecular weight. Most of the exogenous molecules with molecular weight less than 40 kDa are absorbed through air spaces present between tight junctions.

- **Efflux and transporter-mediated absorption**
  Transport mediators like PEPT2, which is a high-affinity peptide transporter, are present in various sites of pulmonary epithelia, especially in the apical region of the trachea, type II cells, etc. Receptor-mediated or endocytotic movement of molecules like albumin and immunoglobulin is possible through the alveolar region.
  Efflux proteins, like MDR1-P-glycoprotein (P-GP), are associated with multiple drug resistance.
- **Vesicle-mediated transport**
  In the lungs' epithelia, many non-coated vesicles called caveolae are present. Their main role may be to transport the drug molecule. Vesicle-mediated transport is said to be a minor pathway for the transport of materials like protein across the alveolar epithelium, where the air–blood barrier plays an important restricting factor.
- **Non-specific particle trapping**

Nanoparticles with particle size less than 100 nm are taken up by alveolar cell type 1 by endocytosis. To reach circulatory system circulation, they are drained into the lymphatic system.

4.4 Drug Deposition: Mechanism

Particles given through the pulmonary route are very fine and pre-aerosolized, with a fine radius. This enables particles to travel along the pathway and reach the absorption site, that is the alveoli. The respiratory tree gets narrow as we move towards the periphery; the radius of bronchioles reduces, increasing the pressure of air, thereby increasing the velocity. With an increase in surface area, the particles are more exposed to the walls of the airway. This phenomenon imparts resistance to the flow by allowing particles to stick to the walls. This is called drug deposition. Drug deposition starts in the device itself and the remaining fraction occurs in the respiratory system, either in the conducting airways or in the alveoli region. This in turn increases the residence time, which promotes drug deposition by various mechanisms, either by inertial impaction sedimentation or diffusion as shown in Figure 4.2. Inertial impaction occurs predominantly in the extrathoracic airways and in the tracheobronchial tree, where the airflow velocity is high and rapid changes in airflow direction occur. For small-sized particles, sedimentation occurs, and diffusional transport can be seen for an ultrafine particle which shows Brownian motion.

4.4.1 Physiological Factors Affecting Deposition

1. Mode of inhalation
   The method of inhalation by a patient strongly influences the degree and extent of molecular throat deposition, which will be swallowed and contributes to a non-therapeutic dose. Irrespective of the mode of drug delivery, this factor plays a crucial role in drug deposition, since it will indirectly explain the therapeutic dose available which will ultimately reach the targeted site of action.
   Drug deposition in mouth and throat can lead to insufficient drug efficacy and can vary amongst the population. Factors like proper breathing time and inspirational frequency can increase lung deposition, which can be done by using devices like pMDI which increases the velocity of aerosolized particles, hence can reach up to deeper portions of the lungs ultimately increasing lungs deposition.
   Thus the criteria for selecting the mode of drug delivery should be based on the fact that more drug deposition is seen in the respiratory zone and less in the conducting zone.
2. Oropharyngeal deposition
   As discussed earlier, lungs have a complex histology where there is a high level of physiological variability and this factor plays a major role in influencing drug deposition. Since the pathophysiological conditions vary from one person to another, the therapeutic dose also varies. The particle size is modulated in such a way

that it can reach up to the respiratory zone easily since the conducting airways are narrower and can restrict the particle flow. We will see how the diseased state affects the deposition within the lungs in the following section.

4.4.1.1 Effect of Diseased State on Drug Deposition

Diseases can be infectious or obstructive when we talk about a pulmonary diseased state. In the case of obstructive diseases like chronic pulmonary obstructive diseases, the movement of a drug to the absorption site becomes negligible. Obstructive diseases either increase the pulmonary ciliary clearance, due to which the drug gets cleared off the lungs as seen in COPD, or the bronchiole epithelial thickens with hypersecretions of mucus, in the case of diseases like chronic bronchitis. In the case of emphysema, the destruction of alveolar walls result in permanent enlargement of the gas exchange zone. The obstruction causes deposition of particles on the bronchial airways. Infectious diseases like bacterial exposure are commonly seen in the upper respiratory tract;

Insufficient inhalation ability is observed in the case of zygomycosis, a pseudomonas infection which interferes with tidal volume

Pulmonary clearance

Drug retention time is explained by the anatomical and physiological condition of airways, which is directly influenced by the drug clearance rate, which can be explained as the time taken by the airways to completely clear the deposited particles from their surface. This is of great importance as chronic immune responses can be undone and also resist the entry of foreign particles. But despite this, it also interferes with drug absorption and therapeutic efficacy of the drug, whch influences the dose.

4.5 Levels of Clearance

- Mucociliary clearance
  The epithelia of the conducting airway are composed of a "clearance escalator" which is composed of ciliated epithelial, and goblet or mucus-producing cells are responsible for the upward beating. This allows the removal of the deposited particle in the range of 24 hours.
  In turn, the dosage form has to be selected in such a way that particles reach the target site before depositing on the edges of the conducting airway (7). This can be done by ensuring the aerodynamic nature of the particle and use of an efficient dosage form which can minimize the retention time and accelerate the particle movement.
- Phagocytosis or alveolar macrophage clearance
  This has been discussed in this chapter. Alveoli are composed of cells like macrophages which ensure that entry of any particle which can lead to the induction of systemic immune response is restricted. Phagocytosis is one of the mechanisms which restricts the movement of particles until the alveolar region and particles are engulfed by macrophages. This type of clearance can be called *alveolar macrophage clearance*. The dosing regimen can again be affected by the clearance of the drug in the alveolar region.
- Alveolar and epithelial absorption
  Lungs have a large surface area with a high surface to volume ratio and good airway epithelial permeability which are some of the prerequisites for the absorption of particles, hence drug particles can be absorbed when deposited and also can be cleared rapidly by mucociliary clearance from the epithelial lining. Hence, different approaches have to be followed to reduce absorption at the non-targeted regions of airways by identifying the physio-chemical nature of the drug as well as the targeted site. For example, rapid absorption can be seen in the case of hydrophobic molecules with less molecular weight; if they have to reach the deeper airways for their absorption, they must be treated in such a way that they can resist absorption in the upper airways by altering their physicochemical properties.
- Metabolic degradation
  Lungs secrete various enzymes, either through the epithelial endothelial layer that is a membrane-associated enzyme or from the alveolar region, such as alveolar macrophages which will affect enzymatic degradation of the drug. This can directly affect the drug dose, which can influence the therapeutic activity of the drug (8).

4.5.1 The Mechanism to Overcome Pulmonary Clearance

Various methods are being adapted to overcome clearance of deposited aerosolized particles, like the controlled release dosage form where particles are deposited and the release is controlled by the use of novel mechanisms; in this, we can control the amount of drug release and the targeted delivery of the drug (9). Pulmonary clearance is affected by various factors, which are shown in Figure 4.3 and these factors can be altered in order to alter the pulmonary drug delivery. Summary of all the methods adapted to avoid pulmonary clearance are mentioned in Table 4.2.

1. Mucociliary clearance

The major amount of drug clearance is seen by mucociliary clearance, hence it is present in the maximum area of conducting airways, which are the primary contact surfaces for the drug. Mucociliary clearance is one of the host defensive mechanisms which protects against foreign particles inhaled with air, hence drug particles are cleared by this mechanism.

The upward beating of cilia along with mucus produced by mucus-secreting cells efficiently removes the particles and creates a pathogen-free environment. This is called a mucociliary escalator, which can be avoided using various mechanisms.

- Mechanisms to avoid mucociliary escalator
  Particle residence time or particle deposition time plays an important role in drug release. It can be increased to about a day or more by altering the aerodynamic

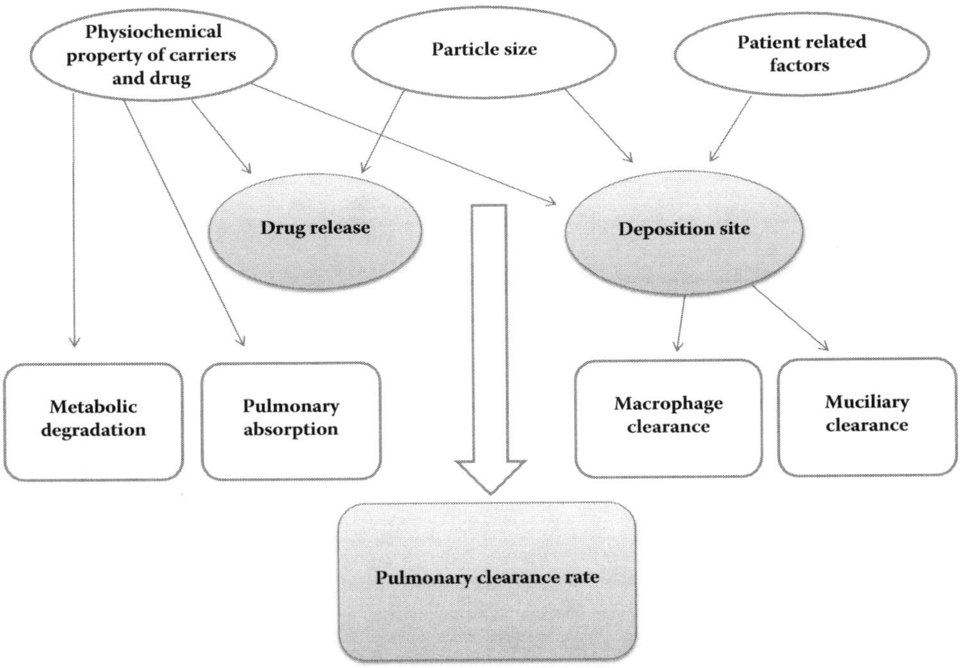

FIGURE 4.3 Factors Affecting Pulmonary Clearance

TABLE 4.2

Summary of Types of Pulmonary Clearance with the Mechanisms to Avoid Them

Pulmonary Clearance	Mechanisms to Avoid Pulmonary Clearance
Mucociliary clearance	Aerodynamic property
	Polymers like chitosan, hyaluronan
Rapid drug absorption	Drug structure modifications
	Microencapsulation
Macrophage clearance	Particle size
	Stealth characterstics

property of a drug particle. It ensures negligible deposition in the conducting airways, hence more drugs can reach the deeper airways for absorption. Aerodynamic diameter (d_a) is the diameter of a particle with a density of 1 g/cm^3 and its settling velocity is the same as that of a nanosphere with some arbitrary density. Deposition mechanisms are defined by the aerodynamic diameter and flow dynamics of the inhaled particle; by altering these we can achieve suitable particle flow according to our requirement.

Physical mechanisms of deposition are explained in the previous segment, which includes Brownian motion, particle sedimentation, inertial force causing particle impaction, and electrostatic deposition. Particle deposition and flow mechanisms of particles are directly dependent on d_a. For example, particles with $d_a > 0.5$ micrometers will be deposited by gravity, while particles with $d_a < 0.5$ are exhaled since their time of interaction with the lung lumen is less and the retention time is low. This can be altered by increasing breath-holding which will increase the retention of particles, hence the diffusion pattern is changed for the ultrafine particles.

Particles of more than 5 micrometers are deposited in the proximal lung regions like larynx, mouth, etc. For them, impaction is the major mechanism for deposition. Particles of size range varying from 1–5 micrometers are said to be optimal for efficient drug delivery to the alveolar region of the lungs. They are least deposited in the upper region of the pulmonary tract; mucociliary clearance does not interfere with this size range of particles. Also they are generally used for the aerosol delivery of a drug; for instance, terbutaline is used by the asthmatic patient as a bronchodilator where particle size ranges from 1.8 to 2 micrometers.

- Avoiding extracellular barriers
The airways are covered with airway surface liquid (ASL) which functions as a primary defense mechanism. ASL is composed of two types of fluids. One is the preciliary liquid layer (PCL), the water layer. The other is the mucus layer formed by goblet cells over epithelia

on top of PCL. These two are found to be associated with cilia; the upward flow of cilia causes the flow of the fluid lining. This is termed *mucociliary clearance* because it resists any retention of drug particles or any foreign material on the epithelia. Hence, it can be avoided by increasing the diffusion rate of particles through the mucociliary lining.

Despite comprising highly complex biochemistry, mucus shows one additional property of mucoadhesion. The particles adhere to the mucus surface through the forces of interaction, like hydrophobic interactions, Van der Waals forces of attraction and surface characteristics, providing effective mucus absorption. Diffusion takes place as a polymer chain diffuses into the mucus membrane. As retention of particles increases on the mucus surface, diffusion also increases, which will increase the absorption of drugs, thus improving their bioavailability.

Increased contact time or mucoadhesion can be achieved by the use of polymers comprised of hydrophobic domains, like hyaluronan, chitosan, mucin, tween 80, etc. Physiochemical properties of the formulations can also affect the mucoadhesion, including particle size, molecular weight, concentration density, chemical structure, etc. Agents like S-carboxy methylcystein, a mucolytic agent, break the glycoproteins present in mucus and weaken the mucus membrane allowing easy passage of particles. Other mechanisms include the use of polymers like lecithin; when linked to the dosage form, they increase the particle internalization through the membrane.

2. Rapid drug absorption

The small sized particles which have high chances for rapid absorption through airways, reducing the pulmonary local concentration, increases dosing hence reducing patient compliance. Mechanisms discussed further can be adapted to avoid the rapid absorption of the drug.

- Modification in drug structure
  Chemical modifications can allow retention of the drug molecule by different mechanisms such as extended drug release. These modifications should not alter the therapeutic value of the drug. The overall motive behind this idea is to increase the retention time on pulmonary epithelia. Such an alteration can cause variations, including increasing hydrophobicity, which can increase drug retention on the tissues, and thus increases the local effect inside the pulmonary airway system.

  Another approach can be the addition of a positive charge on the lipophilic drug molecule, increasing its molecule binding capacity. Such approaches can increase the therapeutic affinity of the drug.
- Microencapsulation
  Controlled drug release can be achieved by controlling or extending the diffusion rate. Extending the release can be done by using polymers, which work as a carrier or vehicle promoting slow diffusion of the drug. The major step is the selection of carrier, which will influence important parameters like which site to target, the release pattern of drug, diffusion rate, tissue binding capacity retention time, etc. Dosage form-related factors are also influenced by the choice of the vehicle incuding shelf life, stability, toxicity, etc.

  Various examples of the controlled release dosage form are liposomes, microparticles, and PEGylation.

3. Macrophage clearance

As discussed in the previous section of this chapter, macrophage phagocytosis is the process in which macrophages engulf the foreign particles as an immune response. Macrophage clearance can be avoided by changing some physical factors of the dosage form, like particle shape and particle size (10).

- Particle size
  Particles are developed in such a way that they can invade deeper airways without any resistance provided by mucociliary escalator and thus deposition. This is done by altering a particle's aerodynamics, hence particles having an aerodynamic diameter in the range of 0.5–1 micrometers reached the deeper airways with negligible deposition. But particles with this diameter are more attacked by macrophages, so it is a challenge to develop a formulation that can invade with the least deposition and also can avoid macrophage clearance.

  Large porous particles (LLPs) offer one approach with particles having a geometric diameter of more than 5 micrometers, but their aerodynamic diameter is less than 5 micrometers. This is because of their low density; they show better flowability as compared to conventional forms. Due to their large size, they can avoid macrophages. They can deposit homogenously on the cell surface and show no toxicity.

  Geometric diameter plays an important role when we talk about avoiding macrophages. Another approach is reducing geometric diameter to nanometers; this way they can remain on the surface and can resist ciliary as well as macrophage clearance. These nanoparticles, NPs, can extend drug release, hence can be used for controlled drug release. They offer advantages over conventional dosage forms, but some disadvantages restrict their use. For example, their minute size makes them easy to expel from lungs before they reach deeper airways—this can be avoided by forced inhalation—or they can be incorporated in microparticles and hence can be delivered efficiently.

  Low-density micron-scale particles, such as Trojan particles, are produced by using polymer-like polystyrene, hence polystyrene nanoparticles (PS-NP), are assembled into hollow micron-scale particles while spray drying. Trojan microparticles are formed during spray drying; assembly is driven by van der Waals forces. These low-density hollow particles offer good flowability and dispersibility characteristics, and as they are delivered to deeper airways they are able to produce sustained/controlled drug release.

- Stealth characteristics
  This is one of the promising ways of avoiding macrophage clearance. This can be altered by attaching or coating the main therapeutic ingredient with a material that provides stealthiness to the agent. Altering stealth characteristics can provide controlled drug delivery in the pulmonary airway.
  Agents like PEG can increase the circulatory time in the airway and also reduce the clearance. Microparticles are coated with this material, which increases their circulatory time.
  Other ingredients like hyaluronic acid, known for their mucoadhesive properties, are used in many pulmonary therapeutics. Also, they can be used for the controlled delivery of drugs.
- Drug release within or from macrophages
  When the drug-loaded vehicles undergo phagocytosis, macrophages act as a reservoir for the drug release. This can serve in maintaining the dose inside the body which helps to reduce dosing frequency and also improves patient compliance. Some of the recent approaches to increasing drug retention inside the lungs are to use of nanocarriers coated with specific ligands that have an affinity towards MR or mannose receptors which are found on macrophages. Mannose acts as an indicator between self and non-self as they are not present at the terminal ending of mammalian glycoproteins but are present on the surface of pathogenic glycoproteins. This promotes receptor-based cellular uptake of various mannosylated formulations like liposomes coated with specific ligands with an affinity for MRs. Mannose derivatives include O-palmitoyl mannan (OPM) and p-aminophenylmannopyranoside (PAM).

4.6 Airway Cells, Pulmonary Circulation, and Receptors: Importance and Function

4.6.1 Airway Cells

Drug absorption is observed in the respiratory region, which is marked by a thinner epithelial, responsible for the diffusion of gases and drug delivery. The alveolar region is a site of interest for pharmaceutical scientists seeking to provide sufficient therapeutic effect. Hence, a detailed study of the alveolar region is necessary. The alveolar region is composed of two types of cells, alveolar type 1 cells (AT1) and alveolar type 2 cells (AT2) as discussed in the earlier portion of this chapter. AT1 covers over 95% of the area, whereas AT2 cells are more numerous, in the ratio of 2:1 (AT2:AT1).

AT1 is highly water permeable and is responsible for the transport of macromolecules because of the presence of a high number of vesicles and caveolae, thus reduce cell thickening (11). AT2 cells are responsible for the secretion of surfactants, and are also responsible for the recycling of the same. They either proliferate into more AT2 cells or differentiate into AT1 cells. They are embedded in the alveolar epithelia in such a way that they form a highly tight layer, preventing leakage.

4.6.2 Airway Receptors

Understanding of pulmonary receptors is of great importance as it increases the targeted delivery efficacy. Pulmonary epithelia comprise of many membrane-bound receptors especially in the smooth muscle and epithelia of airways. Pulmonary epithelia are composed of parasympathetic receptors which can be of two types, adrenergic and muscarinic receptors; both are G protein-coupled receptors and are targeted to relax the smooth muscle, which can cause bronchodilation.

Amongst all the most commonly explored receptors are adrenergic beta 2 receptors, BAR. They are most prominently present on alveolar cells of both AT1 and AT2 cells. beta1 adrenergic receptors are majorly present on alveolar epithelia. Beta 2 adrenergic receptors can be seen on the mast cell, epithelia, and endothelia. Targeted delivery of beta-receptor agonists can result in a response like the relaxation of smooth muscle bronchodilation and hence are used in COPD. Overexposure results in resistance, which can cause sensitization of receptors against agonistic action. All the actions are summarized in Figure 4.4.

Activation of G-protein coupled receptors seems to have more therapeutic use because of their higher specificity for COPD than asthma. A subtype of the muscarinic receptor, M_3, can lead to the release of acetylcholine, causing bronchoconstriction by smooth muscle constriction, which is contradictory to the beta 2 receptor bronchodilation action (12).

4.6.3 Effect of Blood Circulation on Drug Delivery

Bronchial blood circulation can play a major role in clearing out foreign particulate material, balancing the temperature of the pulmonary environment as an immunological precursor which is due to the presence of various cells like mast cells, lymphocytes, etc.

In the human lungs, the pulmonary artery carries blood from the right ventricle, and then deoxygenated blood is carried to the lungs, where gaseous exchange takes place in the alveolar region and the oxygenated blood is carried by pulmonary vein to the left atrium, where oxygenated blood is then pumped to the different parts of the body.

Different pathophysiological conditions can result in physiological changes like an increase in mucosal vasculature and dilation of arteries. This condition can be seen in COPD, where dilation results in edema and bronchial constriction. Such physiological changes interfere with various factors like drug delivery, therapeutic dose; lung hyperinflation can result in constriction of blood vessels like an arterial vessel. Drug absorption is affected by blood flow; more blood flow results in high absorption. However, if the drug is intended for local systemic effect, due to high mucociliary clearance, high blood flow can result in certain adverse effects (13).

Pulmonary airways are lined by two major cells which are responsible for secretions found in pulmonary airways: serous and mucous cells. They are found beneath the epithelial lining of bronchial airways which are responsible for immunological and anti-microbial activity that is by certain secretions like anti-microbial peptide, lysozyme, and antibody-like IgA. Types of secretions with their secretory cells are listed in Table 4.3.

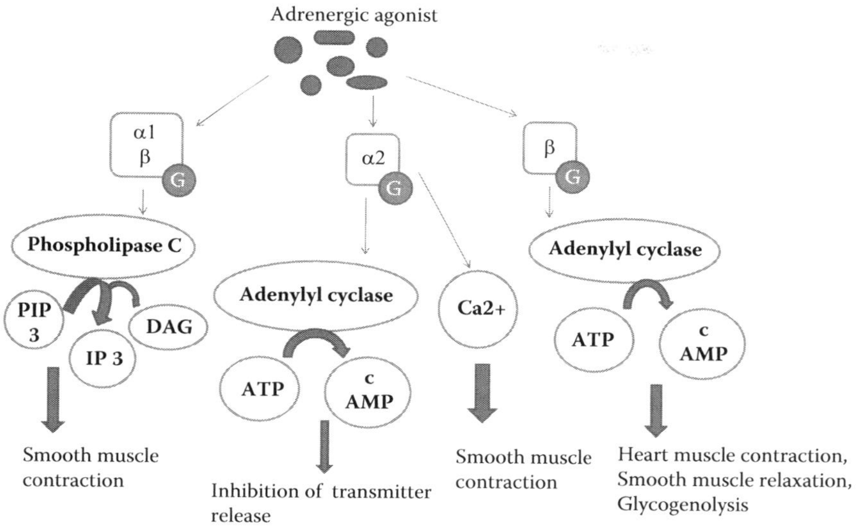

FIGURE 4.4 Adrenergic Receptor Mechanism of Action

TABLE 4.3

Various Types of Secretory Cells and their Secretions Found in the Pulmonary Airway

Mucous Cell	Serous Cell
• Mucin • Anti-microbial peptide	• Mucin • Lysozyme • Proteoglycan • IgA • Lactoferrin • Anti-microbial peptide • Albumin

The extent of mucus secretion depends on the patient's pathological condition and disease state if any. Mucus is secreted by the interference of muscarinic receptors (M_{1-3}), which is cholinergic stimulation. Cholinergic innervations can also result in the synthesis of macromolecules like albumin and epithelial changes by interfering with active ion transport and passive water efflux.

Drug delivery is significantly affected by mucus production. If a drug is hydrophilic, then molecules eventually undergo dissolution and absorption as soon as they come in contact with the mucus membrane. Whereas, if the drug molecules have different properties, then drug particles will show low dissolution and absorption.

4.7 Pulmonary Drug Delivery: Dissolution, Metabolism, Absorption, and Clearance

4.7.1 Pulmonary Dissolution

Particles are dissolved as they enter the body before absorption in the lungs. This process is enhanced by the presence of fluid lining the airways (14). The airways are lined by bi-phasic-phospholipid and mucus lining in the bronchi-bronchioles region and by surfactants in the alveolar region. They both affect the dissolution in their own ways. The process is affected by the physicochemical properties of drugs and their solubility in fluid. Drugs are restrained by mucus, and dissolution is enhanced at the alveolar region by the presence of surfactants.

4.7.2 Pulmonary Absorption

The process of absorption depends upon drug properties like pKa, lipophilicity, and the partition coefficient. Membrane properties also affect the absorption; the dissolution process is necessary before absorption, which requires an adequate amount of fluid for the wetting of the drug particle. Absorption can happen by passive diffusion of particles, that is, direct absorption of particles through the membrane without any requirement of energy. Mostly lipophilic drugs having a particle size less than 50 Da can absorb through this mechanism. The bigger particle is absorbed through carrier-mediated transport where membrane-bound proteins are present which helps in translocating the particles. This can be facilitated diffusion in which no energy is required, it can be seen for the larger size hydrophobic particles and small-sized hydrophilic drug. Active transport of drugs can be seen for the particles, which are large, mostly for the hydrophilic particle. In lungs, alveoli are the absorption site of drugs because they are highly perfused. Also, the presence of surfactants reduces the surface tension, thereby reducing the particle size and enhancing particle absorption.

4.7.3 Mucociliary Clearance

The upper respiratory tract is lined by cilia which does not adhere to the airway epithelia by mucus secretion. The ciliary movement, also called ciliary beating, causes the upward movement movement of mucus. The drug particle is the foreign particles in response to which mucus is secreted. This mucociliary layer reduces in thickness in the peripheral region this enhances drug deposition in the central respiratory tract. Drug absorption enhances retention time. The alveolar region is lined by lymph nodes and cells like macrophages cause phagocytic clearance of the drug by macrophage clearance. This process is slow, and since they are highly perfused, the absorption is carried out more quickly.

4.7.4 Pulmonary Retention

Pulmonary retention of drugs is affected by the physicochemical properties of the drugs, such as their particle size and solubility. More lipophilic drugs have a high retainability in the lungs. Hepatic enzymes like CYP1A1 and CYP2EA have metabolizing capacity and reduce retention time. Various xenobiotic compounds like serotonin are located in epithelial cells. The CYP enzymes are significantly present in Clara and alveolar type 2 cells. Due to high perfusion and high cardiac rate the pulmonary metabolism is very high. In the tracheobronchial region, the perfusion rate is low, hence, more equilibrium and high retention.

4.8 Affect of Lung Physiology and Pathophysiology on Drug Absorption

Many factors like high altitude pulmonary edema, deep breathing, hyperventilation, etc. can cause variation in breathing patterns from one person to another by the expansion of alveolar epithelia, as well as increased permeability of transcellular pores across a membrane; smokers have increased permeability of the blood–brain barrier which can cause increased uptake of a molecule.

A pathophysiological condition which causes bronchoconstriction can affect the absorption capacity of lungs, like COPD or cystic fibrosis in which hypersecretion of epithelia causes airways constriction. They play a large role in the selection of drug delivery mechanisms. The main motive is to target the site where high absorption is found with the desired local effect. Hence, pathophysiological factors play an important role in the delivery of drugs to target sites where it can overcome the effect of the blood–air barrier with a high degree of absorption.

To get the desired therapeutic effect, sometimes overlooked is multiple drug therapy where one drug targets the primary disorder and the other reduces the resistance interfering with the action of the primary drug. For example, vasoconstrictors are taken up by the patient with bronchodilators where vasoconstrictors constrict the blood vessel reducing their vasicular uptake and increasing the pulmonary drug concentration.

4.8.1 Pharmacokinetics of Nasal Drug Delivery

Drug absorption through the intranasal route is expressed in the terms of absolute absorption, or Ae. Ae can be expressed by the following equation

$$Ae = \frac{(AUC)i.\ n(Dose)i.\ v}{(AUC)i.\ v(Dose)i.\ n}$$

where (AUC)in is the area under the curve by intranasal administration, and (AUC)iv is the area under the curve by intravenous; (dose)in and (dose)iv are the intranasal and intravenous doses respectively.

The other way of calculating absolute availability is through urinary excretion data before intranasal and intravenous drug delivery that is expressed by unmetabolized drug excreted in the urine form, or Au^{∞}.

$$Ae = \frac{(A\infty)i.\ n(Dose)i.\ v}{(A\infty)i.\ v(Dose)i.\ n}$$

Absolute absorption through urinary data collection can be applicable only when the fraction of drug absorbed and excreted in urine is the same as in intranasal delivery.

4.8.2 Pharmacokinetic Processes of Oral, Intravenous, and Inhalation Administration

Oral drugs first interact with the gastrointerstitial tract where they are absorbed into systemic circulation after reaching the liver through the portal vein. Here, the drug-release kinetics are influenced by dissolution and solubility in the gastrointestinal environment. Here, the first-pass metabolism takes place, which decreases bioavailability through hepatic clearance.

After that, the drug reaches the intestinal environment where drug crosses the mucosa through membrane permeability transporter affinity and drug solubility and other physicochemical properties. When a drug is taken via an intravenous route, the drug can resist the gastrointestinal absorption and hepatic clearance (15).

To have high pulmonary clearance but low systemic bioavailability, the drug should have a low systemic affinity. Also, the systemic clearance should be high to increase pulmonary retention or selectivity.

Pulmonary retention may also be increased by the high protein binding affinity of drugs which reduces free drug concentration in plasma. In turn, this reduces bioavailability.

The PK or airway selectivity is the approach by which efficient delivery in the pulmonary targeted delivery is seen. Airway selectivity can be optimized by varying particle size, charge, etc.

4.9 Pulmonary Drug Delivery: Different Molecular Size

4.9.1 Smaller Molecules Used to Deliver Drugs through the Pulmonary Route

The systemically acting molecules, when given via inhalation, can show a high rate of systemic absorption. They can be classified based on the molecule's size; particles with a molecular weight less then 1000 Da are said to be small molecules. The bigger molecules are proteins, peptides, and vaccines.

Drugs with smaller weight are being widely used in conventional form, due to their lower molecular weight, and their pharmacokinetic and bioavailability parameters. Inhalation is being widely used for drugs given by inhalation, due to rapid absorption.

As lungs provide absorption for both lipid-soluble and water-soluble drugs by the cellular membrane and aqueous porin channels respectively, smaller molecules are easily absorbed through pulmonary epithelia. Various example of nano peptide molecules given as drugs via pulmonary transepithelial route include treprostinil.

This has an antiplatelet and vasodilatory action when given either by a systemic route or by pulmonary inhalation. Treprostinil is given by subcutaneous infusion because it is less cumbersome and does not require a central catheter for its application. Even though the drug has a higher half-life, it has to be administered by subcutaneous infusion, and it causes complications like infection at the site of action. It also causes discomfort because of the needle-based delivery of the drug. For this reason it is given via the pulmonary route.

4.9.2 Large Microporous Molecules

Large porous molecules have been used to treat many pulmonary diseases. The major problem in pulmonary drug delivery is ciliary clearance. The molecules having a mean geometric radius range between 1 and 5 microns will have high, deep pulmonary retention. Deep pulmonary retention reduces clearance with high systemic effect and to increase the retention molecules have to remain in the above range.

Microspheres are the type of delivery system in which the drug is uniformly distributed in the polymer matrix. Microspheres can be macro- (3 micro m), meso- (1–3 micro m), and nanoporous (200–500 nm). Porosity provides pulmonary retention, hence, macroporous microspheres have sustainable action. Aerodynamic radius is affected by the diameter range of the particle.

These large, highly porous molecules tend to resist accumulation. The molecules in the presence of alveolar macrophages undergo phagocytosis.

4.9.3 Pulmonary Delivery of Large Peptides and High Molecular Weight Drugs

Drugs like insulin and low molecular weight heparins are used extensively by the patients with lifelong persisting diseases, hence much research has been done to make the delivery system non-invasive. Pulmonary delivery provides a scope for lower invasiveness, so this route is explored for such drugs.

4.9.3.1 Insulin

Inhalable insulin is short-acting insulin for patients with diabetes type I or II. In the case of diabetes I, a combination of long-acting insulin with inhalable insulin is recommended. The recombinant human insulin is given by a thumb-size inhaler with a powder of fumaryl diketopiperazine. The microparticles are then absorbed into the circulation as soon as they come in contact with the lung's surface. The sudden absorption of insulin makes them fast-acting. This is known as *technosphere technology*. By this, the rotating dose of 2.5 micro g of insulin can be taken before every meal. Many disadvantages of the inhalable insulin include decreased diffusing lung capacity for carbon monoxide as compared to subcutaneous delivery of insulin. This route is not convenient for all patients, such as those who smoke or those with diabetes ketoacidosis.

4.9.3.2 Low Molecular Weight Heparins (LMWH)

These drugs come under the category of blood thinners, given in conditions like deep venous thrombosis or DVT. The nature of low molecular weight heparins is anionic, hence when given by the oral route it causes inadequate absorption through the gastric mucus membrane and even through the pulmonary route. LMWH carry an anionic charge because of the presence of carboxylic acid and sulfate groups at the glucosaminoglycan region.

Using drug delivery carriers like dendrimers, which have a positive charge, can create an overall non-ionic environment, causing ease to absorption by the pulmonary route.

4.10 Nanocarriers in Pulmonary Delivery of Drugs

- Liposomal delivery system
  The liposomal nanocarriers can be utilized in aerosol as they are highly aqueous compatible. For this reason, the retention seen in hydrophobic compounds is not seen in drug–liposomal complexes, hence reduced toxicity is another advantage of this system. It can be used to deliver drugs via the intracellular route, so they are good carriers if the drug needs to be delivered to the alveolar macrophages. They are advantageous because they can provide action for systemic as well as local irritation.

  Another factor that makes liposomes a great tool for drug delivery is the depot effect. This is the effect in which carriers mimic the immune system by sustained release of an antibody. Aerosols equipped with jet nebulizers are used to deliver liposomal formulations for treating many respiratory problems. Alveofact was the first liposomal formulation given via the pulmonary route, to treat diseases like respiratory distress syndrome (RDS). These formulations are given in liquid form through aerosols.
- Solid lipid nanoparticle in pulmonary delivery
  They contain solid lipids which remain solid at room temperature, water, and a surfactant. These SLNs are being used widely because they provide a sustain and controlled release pattern when given through the pulmonary route. When compared to other polymer-based nanoparticles, these physiological lipid-based nanoparticles show low pulmonary tolerability and high toxicological properties (16–25). They are generally given as

dry powder by aerosol. A few of the examples of SLN-based drug delivery are isoniazid and rifampicin, given with controlled and sustained delivery for treatment of tuberculosis through the pulmonary route.

- Dendrimer based pulmonary delivery
  These are nanoparticles which have a star-shaped structure. They are highly branched macromolecules. They are characterized by layers between each cascade point known as generations. The structure can be divided into three portions: inner central core, inner branches, and outer core. Their high branching and entrapping efficiency are used in delivering drugs of varying molecular weights and hydrophilic entities by host–guest interaction, and hydrophobic entities by covalent bonding. The major advantage provided by their structure is that they provide wide scope for functionality, increasing the chances of structural modification of by conjugation of different drugs and functional groups which amplifies the functionality. They have a higher surface to molecular functional group ratio. Hence, they are widely used as a vector in gene and drug delivery. Several disadvantages of dendrimers can be high toxicity, short half-life, and low aqueous solubility. The dendrimer form is generally used in gene delivery to the pulmonary region.
- Micelle
  These are the colloidal system, liquid crystal nanodispersion containing small particles ranging from 10–400 nm in diameter. The system contains a vesicle, in which the drug can be entrapped and delivered in higher concentrations greater than their water solubility. The other advantage offered by this system is that it can be modified based on the requirement; the outer shell of the micelles can be modified with certain functional groups.

REFERENCES

1. Patil JS and Sarasija S. Pulmonary drug delivery strategies: A concise, systematic review. *Lung India*. 2012; 29(1): 44–49.
2. Tronde A. Pulmonary drug absorption: In vitro and in vivo investigations of drug absorption across the lung barrier and its relation to drug physicochemical properties. *Acta universitatis upsaliensis. Comprehensive summaries of uppsala dissertations from the faculty of pharmacy*. 275: 86 Uppsala. ISBN 91-554-5373-2. 2002
3. Kevin O'Donnell P and Smyth HDC. Macro- and Microstructure of the Airways for Drug DeliVery. Pharmaceutical research. Springer. 2011. 1–16.
4. Newman SP. Drug delivery to the lungs: Challenges and opportunities. *Therapeutic Delivery*. 2017; 8(8): 647–661.
5. Mansour HM, Rhee YS, and Wu X. Nanomedicine in pulmonary delivery. *International Journal of Nanomedicine*. 2009; 4: 299–319.
6. Barnes PJ. Airway receptors. *Postgraduate Medical Journal*. 1989; 65: 532– 542.
7. Ibrahim M, Sherbiny E, Villanueva DG, Herrera D, and Smyth HDC. Overcoming lung clearance mechanisms for controlled release drug delivery. *Controlled Release Society*. 2011: 101–121.
8. Vyas SP and Khar RK. Nasopulmonary drug delivery. *Authors and Vallabh Prakashan*. 2012: 301–368.
9. Santos D, Mauricio AC, Senacadas V, Santos JD, Fernandes MH and Gomes PS. Spray drying: An overview. *IntechOpen*. 2017: 10–31.
10. Buels KS and Fryer AD. Muscarinic receptor antagonists: Effects on pulmonary function. *Springer-Verlag Berlin Heidelberg*. 2012. 208: 317–341.
11. Markus J, Borghardt C. Kloft, and Sharma A. Inhaled Therapy in Respiratory Disease: The Complex Interplay of Pulmonary Kinetic Processes. *Canadian Respiratory Journal*. 2018: 1–8.
12. Thorat S. Formulation and product development of nebuliser inhaler: An overview. *International Journal of Pharmaceutical Science and Research*. 2016; 1(5): 30–35.
13. Narang AS and Mahato RI. Targeted delivery of small and macromolecular drugs. CRC Press Taylor and Francis. 2010, 372–385.
14. Mohanty RR and Das S. Inhaled insulin – Current direction of insulin Research. *Journal of Clinical and Diagnostic Research*. 2017; 11(4): 1–2.
15. Paranjpe M and Goymann CCM. Nanoparticle mediated pulmonary drug delivery a review. *International Journal of Molecular Sciences*. 2014; 15: 5852–5873.
16. Thorat SR and Meshram SM. Formulation and product development of pressurized metered dose inhaler. *PharmaTutor*. 2015; 3(9): 53–64.
17. Yurdasiper A, Arıcı M, and Ozyazici M. Nanopharmaceuticals: Application in Inhaler Systems. In *The Design, Applications, and Toxicology of Nanopharmaceuticals and Nanovaccines Micro and Nano Technologies Micro and Nano Technologies*. 2018: 165–201.
18. Smola M, Vandamme T, and Sokolowski A. Nanocarriers as pulmonary drug delivery systems to treat and diagnose respiratory and non-respiratory diseases. *International Journal of Nanomedicine*. 2008; 3(1): 1–19.
19. Shajia J and Shaikh M. Current development in the evaluation methods of pulmonary drug delivery system. *The Indian Journal of Pharmacy*. 2016.
20. Gupta V and Thomas C. Principle and Practice of Pulomonary Drug Delivery. CRC Group, Handb Exp Pharmacol, 2010. 371–412.
21. Labiris Nr And Dolovich Mb. Pulmonary Drug Delivery Part I: Physiological Factors Affecting Therapeutic Effectiveness of Aerosolized Medications. *British Journal of Clinical Pharmacology*. 2003; 56(6): 588–599.
22. Widdicombe JG. Pulmonary and respiratory tract receptors. *The Journal of Experimental Biology*. 1982; ioo: 41–57.
23. Olsson B, Bondesson E, Borgström L, Edsbäcker S, Eirefelt S, Ekelund K, Gustavsson L, and Myrbäck TH. Pulmonary drug metabolism, clearance, and absorption. *Controlled Release Society*. 2011: 21–42.
24. Xu Z and Hickey AJ. The physics of aerosol droplet and particle generation from inhalers. *Controlled Release Society*. 2011: 75–96.
25. Sakagami M and Gumbleton M. Targeted drug delivery through the respiratory system: Molecular control on lung absorption and disposition. *Controlled Release Society*. 2011: 127–128.

5

Understanding the Pharmacokinetics and Pharmacodynamics of Lung and Lung Drug Delivery

Seth Kwabena Amponsah PhD
Department of Medical Pharmacology, University of Ghana Medical School, Accra, Ghana

5.1 Introduction

The propensity of a drug to reach its target and elicit an effect is partly dependent on drug formulation and route of drug administration. Furthermore, the time course for the movement of a drug through and out of the body (pharmacokinetics) and the biochemical and physiological effects of the drug at the target site (pharmacodynamics) are key determinants during the process of drug development. Therefore, understanding the pharmacokinetics (PK) and pharmacodynamics (PD) of a drug is essential. Drugs can sometimes be formulated to aid delivery to specific tissues or organs, and in their review, Dawidczyk, Russell, and Searson (1), highlight the relevance of understanding the PK and PD of drug delivery.

Over the last few decades there have been studies that have explored using the lungs as a means of drug delivery. The lungs or pulmonary route is non-invasive and can be used in the administration of drugs for both local and systemic effects. This route can be used to deliver drugs to patients with chronic obstructive pulmonary diseases (COPD) such as asthma. Also, due to its high permeability and large absorptive surface area, the lungs are ideal for drugs intended for systemic effects. The epithelium of the alveoli has been found to be an absorption site for a number of therapeutic agents (2, 3). Furthermore, the pulmonary route has an advantage over enteral routes of drug administrations because of minimal first-pass metabolism.

In this chapter, we delve into the PK and PD of the lung and lung drug delivery with emphasis on recent trends and clinical evidence.

5.2 Pharmacokinetics of the Lung and Lung Drug Delivery

5.2.1 Absorption of Drugs in the Lungs

Drug delivery to the lungs may have two main purposes: local or systemic effects. For drugs administered via the lungs or pulmonary route for systemic effects, drugs have to be absorbed into the bloodstream across a first barrier: the alveolar blood (air–blood) barrier. The alveolar blood barrier is made up of an epithelial cell, basement membrane, and endothelial cell. While this anatomical arrangement facilitates movement of gases or volatile liquids across the lungs, it can be a major barrier to other large molecules. Before gases or volatile liquids are absorbed into systemic circulation, they must move across epithelial lining fluid. The epithelial lining fluid is usually at the corners of the alveoli and also has a layer of surfactant. The surfactant is made up of phospholipids such as phosphatidylcholine and phosphatidylglycerol. The surfactant also contains key apoproteins (4). Functions of the surfactant are diverse, two of which include maintaining fluid homeostasis in the alveoli and eliciting a number of defense mechanisms. A number of studies have suggested that the surfactant may impair diffusion of molecules (drugs) out of the alveoli (5, 6).

There are a number of mechanisms by which drugs are known to be absorbed from the lungs. It has been hypothesized that drug absorption occurs via transcytosis (adsorptive or receptor mediated), through large transitory pores in the epithelium or paracellular transport between bi-junctions or tri-junctions (7). It is noteworthy that often the high bioavailability of drugs administered into lungs may be due to the lungs' large absorptive surface area, thin diffusion layer, and relatively slow elimination (7).

There are conflicting results on the effect of pathophysiological changes (e.g. inflammation) on drug absorption from the lungs. Some have reported that permeability of the lungs is increased during inflammation (8, 9). Some researchers have demonstrated that absorption is not affected, while others suggest a decrease in absorption (10, 11). Those who have reported that absorption may not be affected have explained this as restitution after epithelial injury (10, 11).

5.2.2 Elimination of Drugs in the Lungs

Elimination of xenobiotics from the body occurs via metabolism and excretion. Metabolism refers to biochemical transformation of xenobiotics catalyzed by specialized enzymes usually located in the liver, intestines, and lungs. Generally, metabolism converts lipophilic xenobiotics into more hydrophilic products that facilitate their excretion from the body.

DOI: 10.1201/9781003046547-5

Xenobiotic metabolism may involve different biotransformation reactions that lead to the generation of water soluble and sometimes reactive metabolites (Phase I reactions). Phase I reactions may be followed by conjugation of metabolites with endogenous molecules such as glucuronic acid, acetate, sulfate, and glutathione (Phase II reactions). Phase I reactions may be oxidation reactions catalyzed by monooxygenase enzymes, i.e. cytochrome P450 (CYP) monooxygenases and flavin monooxygenases (FMN). Phase I reactions could also be reductive reactions catalyzed by CYP reductases.

CYP enzymes are responsible for oxidative metabolism of xenobiotics such as drugs, environmental pollutants, and carcinogens. CYP enzymes also metabolize endogenous compounds such as hormones. In humans, the CYP enzyme system is made up of about 57 isoforms that show differences in catalytic activity. CYPs act by using one atom of molecular oxygen to oxidize xenobiotics. The CYPs usually require NADPH-cytochrome P450 reductase to donate electrons for the reduction of the second oxygen, and this yields water. Functionality of the NADPH-cytochrome P450 reductase is relevant for xenobiotic oxidation (12).

Metabolism of xenobiotics in the lungs differs significantly from what pertains in the liver and intestines. Expression levels of enzymes in the lungs are generally low, and the patterns of drug-metabolizing enzymes differ. In the lungs, CYP3A5, CYP2B6, CYP1B1, and CYP1A1 and CYP2E1 appear to be the most common CYP enzymes (13–17). The most abundant CYP enzyme in the liver, CYP3A4, has a relatively low expression in the lungs. The isoform CYP3A5 is by far the most important CYP enzyme in pulmonary tissue (18–20). Data is scanty on the role of Phase II enzymes in the lungs (21). It is noteworthy, however, that the expression and catalytic activity of sulphotransferase enzymes in the lungs appears to be about the same in the liver. Also, peptidase activity has been found to be high in the lung, as found elsewhere in the body (22). As already indicated, activity and level of expression of xenobiotic metabolizing enzymes in the lung is low, and this may account for disparate data reported in the literature.

Drug-metabolizing potential of the lungs is generally lower than that of the liver, and there is little evidence to suggest that the lungs play a major role in systemic clearance of xenobiotics. Xenobiotics after administration into the lungs have near complete bioavailability because of the low metabolic activity. Nonetheless, some drugs (salmeterol, theophylline, etc.) still undergo some extent of metabolism in the lungs (23–27). There is the likelihood of formation of drug metabolites in the lungs, hence, in the process of drug development there is the need to assess risk of metabolic interactions and toxicity. Metabolism of drugs in the lungs could also aid the activation of certain prodrugs. For example, beclomethasone dipropionate is metabolized by esterases to 17-beclamethasone monopropionate, an active metabolite (28).

The lungs also have the capacity to remove secretions and foreign agents that are deposited in the airway. There are ciliated cells that move mucus and any foreign agent out of the lungs in a proximal direction (29). At the end, the mucus and foreign agents are expectorated and/or swallowed. This process is termed *mucociliary clearance* (30–32). Inhaled drugs that may contain insoluble particles (diameter > 6 mm) are eliminated from the airway via mucociliary clearance (33, 34). Drugs administered into the lungs that have particle sizes with diameter less than 6 mm can move across the mucus layer and enter the bronchial epithelium (35). Additionally, drugs administered into the lungs that have small particle sizes may be deposited in the alveoli and eventually dissolve in them (37,35).

The human anatomy also makes possible mucociliary clearance occurring from the peripheral toward the central airways (36). Mucociliary clearance can be affected by age and airway disease (41, 38, 44). Studies have shown that pharmacotherapy can be used to improve mucociliary clearance in patients with COPD (39, 47, 40, 49).

Drugs administered into the lungs can also undergo alveolar clearance aided by macrophages (42). Drugs that dissolve in the alveoli may be phagocytosed by alveolar macrophages (43–45). To improve bioavailability via the pulmonary route, drugs are now being designed to escape alveolar clearance (55, 35, 43). Current pharmacological research suggests that alveoli macrophages may play beneficial roles in the management of diseases such as leishmaniasis and tuberculosis (46 ,44).

5.3 Pharmacodynamics of Lung Drug Delivery: Recent Trends and Clinical Evidence

5.3.1 Inhaled Antibiotics

Research into delivery of antibiotics into the lungs is still at the early stage. At present, a number of studies have focused on inhalers formulated into dried powders that deliver antibiotics directly into airways where there may be infectious microorganisms. Some commercially available inhaled antibiotics are shown in Table 5.1.

TABLE 5.1

Current Commercially Available Inhaled Antibiotics

Antibiotic	Indication	Approving Agency	Reference(s)
Aztreonam (inhalation solution)	To improve respiratory symptoms in cystic fibrosis patients with *Pseudomonas aeruginosa* infection aged ≥ 7 years	US FDA	(45)
Tobramycin (inhalation solution and powder)	Management of cystic fibrosis patients with *Pseudomonas aeruginosa* infection aged ≥ 6 years	US FDA	(46)
Colistimethate (dry powder for inhalation)	Management of chronic pulmonary infections due to *Pseudomonas aeruginosa* in patients with cystic fibrosis aged ≥ 6 years	European Medicines Agency	(47)

US FDA – United States Food and Drug Administration

TABLE 5.2
Inhaled Pulmonary Aerosols Undergoing Clinical Trials

Drug	Phase	Type	Potential Clinical Use	Company	References
Amikacin (Arikace™)	Phase I I II	Liposomal formulation for nebulization	Cystic fibrosis	Transave Inc.	(57)
Amikacin (BAY41–6551)	Phase III	Solution for inhalation	Pneumonia	Bayer	(58)
Aztreonam	Approved for marketing	Lyophilized drug for nebulization	Cystic fibrosis	Gilead Sciences	(59)
Ciprofloxacin	Phase III	Dry powder for inhalation	Pseudomonas infection	Bayer	(60)
Levofloxacin	Phase III	Solution for inhalation	Cystic fibrosis	Mpex Pharmaceuticals	(61)
Tobramycin	Phase III	Dry powder for inhalation	Cystic fibrosis	Novartis	(62)

Delivery of antibiotics with inhalers into the lungs could result in high drug concentrations at the infection site. This will eventually decrease the risk of systemic side effects. By comparison, the use of oral or intravenous antibiotics often leads to low drug concentration in the lungs. Inhaled aerosols of antibiotics have been shown to have significant clinical benefit in the management of pulmonary infections (53-55). Additionally, several clinical trials of inhalation antibiotics are ongoing; examples of a few are listed in Table 5.2.

5.3.2 Hormones Administered through the Pulmonary Route

5.3.2.1 Inhaled Insulins

The use of exogenous insulin has greatly improved diabetes mellitus management. Despite this, pharmacotherapy with exogenous insulin remains suboptimal in many people with diabetes mellitus. Additionally, available insulins administered parenterally have metabolic actions that are delayed and with prolonged period of effect. The aforementioned challenges could lead to postprandial hyperglycemia and late postprandial hypoglycemia. Scientists have investigated other possible routes for insulin administration. Among these other routes, lung delivery of insulins has shown promise.

Available insulin preparations on the market are usually administered via parenteral routes. However, the lungs provide an alternative for administration of some polypeptides such as insulin. The lung is well vascularized, has a large surface area, good solute exchange, and an alveolar epithelium that can facilitate systemic delivery of insulin. A number of clinical trials have shown the possibility of insulin therapy via lung delivery.

Inhalable insulin is usually formulated into a powder and delivered to the lungs with an inhaler. This approach is considered as a paradigm shift in insulin therapy, as it differs in diverse ways from the traditional. Earlier work with inhalable insulin concluded that although this formulation appears to be equally effective, it is not superior to injected insulin (56–61). Additionally, inhalable insulin may not be cost effective (62).

Two rapid-acting insulins administered by inhalation, Exubera and Afrezza (Technosphere® Inhaled Insulin), have been an innovation milestone in the search for alternative routes of insulin administration. Inhaled insulin most likely would have advantages for diabetic patients because some of these patients have needle phobia and also have challenges with correctly administering insulin via the parenteral route. It is noteworthy, however, that in 2007, Exubera was withdrawn from the market. This withdrawal was as a result of poor sales and difficulty in handling inhalers by some patients (63).

In a study to assess the efficacy of a newly developed inhaled insulin among 25 patients with Type 1 diabetes mellitus, the baseline-corrected glucose infusion rate (GIR) after administration of 8 U of Technosphere® rose faster and declined sooner than in patients who received 10 U of subcutaneous insulin lispro (64). Peak GIR after 8 U of Technosphere® Inhaled Insulin was reached 30 min after administration, whereas subcutaneous administration of insulin lispro lead to a GIR peak of approximately 150 min. In the same study, time to 50% of maximal effect (T50% C_{max}) was 19 min for Technosphere® Inhaled Insulin and 50 min for insulin lispro. After 2 hours, Technosphere® Inhaled Insulin had delivered 60% of the total glucose lowering effect compared to 33% for insulin lispro. The GIR for Technosphere® Inhaled Insulin returned to baseline at approximately 3 hours, whiles that of insulin lispro was 5 hours (64).

In another study among 12 patients with Type 1 diabetes, peak GIR was achieved at 53 min with Technosphere® Inhaled Insulin administered using the Gen2 inhaler compared with 108 min for insulin lispro (65).

Additionally, inhaled insulin was used to manage subcutaneous insulin resistance syndrome, a rare condition that occurs as a result of rapid degradation of insulin in subcutaneous tissue (66). Data from "The Medical Letter on Drugs and Therapeutics" that compared inhaled and conventional insulin treatments showed that glycemic control as assessed by mean decrease in hemoglobin A1c from baseline to end point did not differ significantly (67). Furthermore, adverse events associated with inhaled insulin appear to be comparable to subcutaneous administration, although cough may occur with the former, which tends to decrease with continued use (64).

5.3.2.2 Inhaled Growth Hormone

Since the advent of growth hormone replacement therapy among children in the 1950s (68), use of injections has been the only route of administration. Growth hormone administration by injection is associated with non-compliance and early termination of treatment (69–73). Despite advances in drug formulation and delivery methods, there have been few alternatives to the administration of human growth hormone aside from parenteral routes. Since compliance with injections is a significant challenge with drug delivery among children, the development of an inhaled formulation as an alternative to daily injections is imperative.

Currently, drug delivery via aerosol technology has improved with the use of drugs that have larger particle size, low densities and a tendency to agglomerate. The aforementioned improvements tend to increase efficiency of deep lung delivery and also improve systemic absorption. Application of this has enabled development of somatropin inhalation powder (SIP), an inhalation formulation of human growth hormone.

Walvoord and colleagues in 2009 reported that inhaled SIP administered for 7 days to growth hormone deficient children was well tolerated and resulted in a dose-dependent increase in the blood levels of insulin-like growth factor 1 (IGF-I) and growth hormone. Of the expected drug bioavailability based on previous adult data, SIP was found to be only 50% (73). The limiting factor in the development of SIP has been its bioavailability data. Future studies are thus required to improve the low bioavailability of SIP.

5.3.3 Inhaled Corticosteroids

The effects of corticosteroids are mediated via the glucocorticoid receptor. Generally, glucocorticoid receptors are ubiquitous in the body, hence, corticosteroids may act on these receptors in several tissues. There are two known glucocorticoid receptors: Type I and Type II. Most ligands appear to bind to the Type II receptor. The therapeutic and adverse effects of inhaled corticosteroids are mediated by the Type II receptor (72, 73). Stimulation of gene transcription by corticosteroids is known to be associated with several adverse effects, whereas repression of transcription factors, such as activator protein-1 and nuclear factor-B, appear to lead to the relevant anti-inflammatory effects of corticosteroids (72, 73).

Receptor binding affinities of corticosteroids are usually estimated in comparison with standard dexamethasone which has an affinity of 100. Mometasone furoate is reported to have the highest affinity of 2300, followed by fluticasone propionate (1800) and beclomethasone monopropionate (1345) (75). The new corticosteroid ciclesonide, which has an active metabolite desciclesonide, also shows a high receptor binding affinity of 1200 (76, 77). Reports suggest that receptor binding affinity differences among drugs can be compensated for by administering dose equivalents (78).

In a recent systematic review, Halpin and colleagues concluded that there is no evidence that suggests that the use of inhaled corticosteroids may lead to adverse or beneficial outcomes in acute respiratory infections due to coronavirus (79). They further suggested that randomized controlled trials are needed to assess the benefits of inhaled corticosteroids in the management of coronavirus disease 2019 (COVID-19) in patients with or without chronic respiratory disease (79).

REFERENCES

1. Dawidczyk CM, Russell LM, and Searson PC. Nanomedicines for cancer therapy: State-of-the-art and limitations to pre-clinical studies that hinder future developments. *Frontiers in Chemistry*. 2014. https://doi.org/10.3389/fchem.2014.00069
2. Thorley AJ, Ruenraroengsak P, Potter TE, and Tetley TD. Critical determinants of uptake and translocation of nanoparticles by the human pulmonary alveolar epithelium. *ACS Nano*. 2014. https://doi.org/10.1021/nn505399e
3. Weber S, Zimmer A, and Pardeike J. Solid lipid nanoparticles (SLN) and nanostructured lipid carriers (NLC) for pulmonary application: A review of the state of the art. *European Journal of Pharmaceutics and Biopharmaceutics*. 2014. https://doi.org/10.1016/j.ejpb.2013.08.013
4. Han SH and Mallampalli RK. The role of surfactant in lung disease and host defense against pulmonary infections. *Annals of the American Thoracic Society*. 2015. https://doi.org/10.1513/AnnalsATS.201411-507FR
5. Akella A and Deshpande SB. Pulmonary surfactants and their role in pathophysiology of lung disorders. *Indian Journal of Experimental Biology*. 2013; 51(1): 5–22.
6. Morales JO, Peters JI, and Williams RO. Surfactants: Their critical role in enhancing drug delivery to the lungs *Therapeutic Delivery*. 2011. https://doi.org/10.4155/tde.11.15
7. Labiris NR and Dolovich MB. Pulmonary drug delivery. Part I: Physiological factors affecting therapeutic effectiveness of aerosolized medications. *British Journal of Clinical Pharmacology*. 2003. https://doi.org/10.1046/j.1365-2125.2003.01892.x
8. Folkesson HG, Westrom BR, Pierzynowski SG, and Karlsson BW. Lung to blood passage of different-sized molecules during lung inflammation in the rat. *Journal of Applied Physiology*. 1991. https://doi.org/10.1152/jappl.1991.71.3.1106
9. Ilowite JS, Bennett WD, Sheetz MS, Groth ML, and Nierman DM. Permeability of the bronchial mucosa to 99mTc-DTPA in asthma. *American Review of Respiratory Disease*. 1989. https://doi.org/10.1164/ajrccm/139.5.1139
10. Erjefält JS and Persson CGA. Airway epithelial repair: Breathtakingly quick and multipotentially pathogenic. *Thorax*. 1997. https://doi.org/10.1136/thx.52.11.1010
11. Greiff L, Andersson M, Svensson J, Wollmer P, Stefan L, and Persson CGA. Absorption across the nasal airway mucosa in house dust mite perennial allergic rhinitis. *Clinical Physiology and Functional Imaging*. 2002. https://doi.org/10.1046/j.1475-097x.2002.00401.x
12. Phang-Lyn S and Llerena VA. Biochemistry, biotransformation - StatPearls - NCBI Bookshelf. StatPearls. 2019.
13. Badyal DK and Dadhich AP. Cytochrome P450 and drug interactions. *Indian Journal of Pharmacology*. 2001.
14. Klotz U. Pharmacokinetics and drug metabolism in the elderly. *Drug Metabolism Reviews*. 2009. https://doi.org/10.1080/03602530902722679
15. Manikandan P and Nagini S. Cytochrome P450 structure, function and clinical significance: A review. *Current Drug Targets*. 2017. https://doi.org/10.2174/1389450118666170125144557
16. Preissner S, Kroll K, Dunkel M, Senger C, Goldsobel G, Kuzman D, Guenther S, Winnenburg R, Schroeder M, and Preissner R. SuperCYP: A comprehensive database on cytochrome P450 enzymes including a tool for analysis of CYP-drug interactions. *Nucleic Acids Research*. 2009. https://doi.org/10.1093/nar/gkp970
17. Zanger UM and Schwab M. Cytochrome P450 enzymes in drug metabolism: Regulation of gene expression, enzyme activities, and impact of genetic variation. *Pharmacology and*

Therapeutics. 2013. https://doi.org/10.1016/j.pharmthera.2012.12.007
18. Anzenbacher P and Zanger UM. *Metabolism of Drugs and Other Xenobiotics*. 2012. https://doi.org/10.1002/9783527630905
19. Meyer MR and Maurer HH. Absorption, distribution, metabolism and excretion pharmacogenomics of drugs of abuse. 2011. *Pharmacogenomics*. https://doi.org/10.2217/pgs.10.171
20. Zanger UM Introduction to drug metabolism. *Metabolism of Drugs and Other Xenobiotics*. 2012. https://doi.org/10.1002/9783527630905.ch10
21. Hodges RE and Minich DM. Modulation of metabolic detoxification pathways using foods and food-derived components: a scientific review with clinical application. *Journal of Nutrition and Metabolism*. 2015. https://doi.org/10.1155/2015/760689
22. Vliegen G, Raju TK, Adriaensen D, Lambeir AM, and de Meester I. The expression of proline-specific enzymes in the human lung. *Annals of Translational Medicine*. 2017. https://doi.org/10.21037/atm.2017.03.36
23. Bellmann R and Smuszkiewicz P. Pharmacokinetics of antifungal drugs: Practical implications for optimized treatment of patients. *Infection*. 2017. https://doi.org/10.1007/s15010-017-1042-z
24. Conaway C, Yang Ym, and Chung FL. Isothiocyanates as cancer chemopreventive agents: Their biological activities and metabolism in rodents and humans. *Current Drug Metabolism*. 2005. https://doi.org/10.2174/1389200023337496
25. Khairul I, Wang QQ, Jiang YH, Wang C, and Naranmandura H. Metabolism, toxicity and anticancer activities of arsenic compounds. *Oncotarget*. 2017. https://doi.org/10.18632/oncotarget.14733
26. Metabolism of Human Diseases. *Metabolism of Human Diseases*. 2014. https://doi.org/10.1007/978-3-7091-0715-7
27. Scott LJ, McKeage K, Keam SJ, and Plosker GL. Tacrolimus: A further update of its use in the management of organ transplantation. *Drugs*. 2003. https://doi.org/10.2165/00003495-200363120-00006
28. Würthwein G and Rohdewald P. Activation of beclomethasone dipropionate by hydrolysis to beclomethasone-17-monopropionate. *Biopharmaceutics & Drug Disposition*. 1990. https://doi.org/10.1002/bdd.2510110503
29. Duncan D. Chronic obstructive pulmonary disease: An overview. *British Journal of Nursing*. 2016. https://doi.org/10.12968/bjon.2016.25.7.360
30. Chu KK, Unglert C, Ford TN, Cui D, Carruth RW, Singh K, Liu L, et al. In vivo imaging of airway cilia and mucus clearance with micro-optical coherence tomography. *Biomedical Optics Express*. 2016. https://doi.org/10.1364/boe.7.002494
31. Trindade SHK, Mello JFD, Mion ODG, Lorenzi-Filho G, Macchione M, Guimarães ET, and Saldiva PHN. Methods for studying mucociliary transport. *Brazilian Journal of Otorhinolaryngology*. 2007. https://doi.org/10.1016/S1808-8694(15)30133-6
32. Woods A, Patel A, Spina D, Riffo-Vasquez Y, Babin-Morgan A, De Rosales TM, Sunassee K, et al. In vivo biocompatibility, clearance, and biodistribution of albumin vehicles for pulmonary drug delivery.*Journal of Controlled Release*. 2015. https://doi.org/10.1016/j.jconrel.2015.05.269
33. Olsson AC, Gustavsson P, Kromhout H, Peters S, Vermeulen R, Brüske I, Pesch B, Siemiatycki J, Pintos J, Brüning T, Cassidy A Wichmann HE, et al. Exposure to diesel motor exhaust and lung cancer risk in a pooled analysis from case-control studies in Europe and Canada. *American Journal of Respiratory and Critical Care Medicine*. 2011;183(7): 941–948.
34. Holsbeke CV, Marshall J, Backer JD, and Vos W. Median mass aerodynamic diameter (MMAD) and fine particle fraction (FPF): Influence on lung deposition? *European Respiratory Journal*.2014; 44: P910.
35. Dhand C, Prabhakaran MP, Beuerman RW, Lakshminarayanan R, Dwivedi N, and Ramakrishna S. Role of size of drug delivery carriers for pulmonary and intravenous administration with emphasis on cancer therapeutics and lung-targeted drug delivery. *RSC Advances*. 2014; 4(62): 32673–32689. https://doi.org/10.1039/c4ra02861a
36. Button BM and Button B. Structure and function of the mucus clearance system of the lung. *Cold Spring Harbor Perspectives in Medicine*. 2013. https://doi.org/10.1101/cshperspect.a009720
37. Darquenne C. Aerosol deposition in health and disease. *Journal of Aerosol Medicine and Pulmonary Drug Delivery*. 2012; 25(3)140–147.
38. Edsbäcker S, Wollmer P, Selroos O, Borgström L, Olsson B, and Ingelf J. Do airway clearance mechanisms influence the local and systemic effects of inhaled corticosteroids? *Pulmonary Pharmacology and Therapeutics*. 2008. https://doi.org/10.1016/j.pupt.2007.08.005
39. Melloni B and Germouty J. The influence of a new beta agonist: Formoterol on mucociliary function. *Influence Sur La Fonction Muco-Ciliaire d'un Nouveau Beta-Agoniste: Formoterol*. 1992.
40. Kim V and Criner GJ. Chronic bronchitis and chronic obstructive pulmonary disease. *American Journal of Respiratory and Critical Care Medicine*. 2013. https://doi.org/10.1164/rccm.201210-1843CI
41. Hogg JC, Chu F, Utokaparch S, Woods R, Elliott WM, Buzatu L, Cherniack RM, Rogers RM, Sciurba FC, Coxson HO, and Paré PD. The nature of small-airway obstruction in chronic obstructive pulmonary disease.*The New England Journal of Medicine*. 2004; 350(26):2645–2653.
42. Joshi N, Walter JM, and Misharin AV. Alveolar macrophages. *Cellular Immunology*. 2018. https://doi.org/10.1016/j.cellimm.2018.01.005
43. Garg T, Goyal AK, Rath G, and Murthy RSR. Spray-dried particles as pulmonary delivery system of anti-tubercular drugs: design, optimization, in vitro and in vivo evaluation. *Pharmaceutical Development and Technology*. 2016. https://doi.org/10.3109/10837450.2015.1081613
44. Bhowmik A, Chahal K, Austin G, and Chakravorty I.Improving mucociliary clearance in chronic obstructive pulmonary disease.*Respiratory Medicine*. 2009;103(4); 496–502.
45. Shiehzadeh F and M Tafaghodi. Dry powder form of polymeric nanoparticles for pulmonary drug delivery. *Current Pharmaceutical Design*. 2016. https://doi.org/10.2174/1381612822666160128150449

46. Truzzi E, Nascimento TL, V Iannuccelli, Costantino L, Lima EM, Leo E, Siligardi C, Gualtieri ML, and Maretti E. In vivo biodistribution of respirable solid lipid nanoparticles surface-decorated with a mannose-based surfactant: A promising tool for pulmonary tuberculosis treatment? *Nanomaterials*. 2020. https://doi.org/10.3390/nano10030568
47. Dalby C, Polanowski T, Larsson T, Borgström L, Edsbäcker S, and Harrison TW. The bioavailability and airway clearance of the steroid component of budesonide/formoterol and salmeterol/fluticasone after inhaled administration in patients with COPD and healthy subjects: a randomized controlled trial. *Respiratory Research*. 2009; 10(1):104.
48. Patel B, Gupta N, and Ahsan F. Particle engineering to enhance or lessen particle uptake by alveolar macrophages and to influence the therapeutic outcome. *European Journal of Pharmaceutics and Biopharmaceutics*. 2015. https://doi.org/10.1016/j.ejpb.2014.12.001
49. Daley-Yates PT. Inhaled corticosteroids: potency, dose equivalence and therapeutic index. *British Journal of Clinical Pharmacology*. 2015; 80(3):372–380.
50. Das S, Asim Halder PR, and Mukherjee A. Biogenic gold nanoparticles against wild and resistant type visceral leishmaniasis. In *Materials Today: Proceedings*. 2018; 5: 2912–2920. Elsevier Ltd. https://doi.org/10.1016/j.matpr.2018.01.086
51. FDA approves inhaled aztreonam for cystic fibrosis. n.d. Accessed October 1, 2020. https://www.medscape.com/viewarticle/717553
52. Hagerman JK, Knechtel SA, and Klepser ME. Tobramycin solution for inhalation in cystic fibrosis patients: A review of the literature. *Expert Opinion on Pharmacotherapy*. 2007. https://doi.org/10.1517/14656566.8.4.467
53. Antoniu SA and Cojocaru I. Inhaled colistin for lower respiratory tract infections. *Expert Opinion on Drug Delivery*. 2012. https://doi.org/10.1517/17425247.2012.660480
54. Wenzler E, Fraidenburg DR, Scardina T, and Danziger LH. Inhaled antibiotics for gram-negative respiratory infections. *Clinical Microbiology Reviews*. 2016. https://doi.org/10.1128/CMR.00101-15
55. Chono S, Tanino T, and Morimoto K. Influence of particle size on drug delivery to rat alveolar macrophages following pulmonary administration of ciprofloxacin incorporated into liposomes. *Drug Target*. 2006; 14(8): 557–566.
56. Okusanya OO, Bhavnani SM, Hammel JP, Forrest A, Bulik CC, Ambrose PG, and Gupta R. Evaluation of the pharmacokinetics and pharmacodynamics of liposomal amikacin for inhalation in cystic fibrosis patients with chronic pseudomonal infections using data from two phase 2 clinical studies. *Antimicrobial Agents and Chemotherapy*. 2014; 58 (9): 5005–5015. https://doi.org/10.1128/AAC.02421-13
57. Niederman MS, Alder J, Bassetti M, Boateng F, Cao B, Corkery K, Dhand R, et al. Inhaled amikacin adjunctive to intravenous standard-of-care antibiotics in mechanically ventilated patients with gram-negative pneumonia (INHALE): A double-blind, randomised, placebo-controlled, phase 3, superiority trial. *The Lancet. Infectious Diseases*. 2020; 20(3): 330–340. https://doi.org/10.1016/S1473-3099(19)30574-2
58. Elson EC, Mermis J, Polineni D, and Oermann CM. Aztreonam lysine inhalation solution in cystic fibrosis. Clinical Medicine Insights: Circulatory, Respiratory and Pulmonary Medicine. SAGE Publications Ltd. 2019. https://doi.org/10.1177/1179548419842822
59. Chorepsima S, Kechagias KS, Kalimeris G, Triarides NA, and Falagas ME. Spotlight on inhaled ciprofloxacin and its potential in the treatment of non-cystic fibrosis bronchiectasis. Drug Design, Development and Therapy. 2018; 12: 4059–4066. https://doi.org/10.2147/DDDT.S168014
60. Inhaled Levofloxacin (Quinsair^{TM}) | CFF Clinical Trials Tool." n.d. Accessed October 1, 2020. https://www.cff.org/Trials/Pipeline/details/9/Inhaled-Levofloxacin
61. Safety of tobramycin inhalation powder (TIP) vs tobramycin solution for inhalation in patients with cystic fibrosis - full text view - ClinicalTrials.Gov." n.d. Accessed October 1, 2020. https://clinicaltrials.gov/ct2/show/NCT00388505
62. Black C, Cummins E, Royle P, Philip S, and Waugh N. The clinical effectiveness and cost-effectiveness of inhaled insulin in diabetes mellitus: A systematic review and economic evaluation. *Health Technology Assessment*. 2007. https://doi.org/10.3310/hta11330
63. Bailey CJ and Barnett AH. Why is exubera being withdrawn? *BMJ*. 2007. https://doi.org/10.1136/bmj.39409.507662.94
64. Heinemann L, Baughman R, Boss A, and Hompesch M. Pharmacokinetic and pharmacodynamic properties of a novel inhaled insulin. *Journal of Diabetes Science and Technology*. 2017. https://doi.org/10.1177/1932296816658055
65. Baughman RA, Amin N, Watkins E, Chung PC, Nezamis JP, and Boss. A phase 1, open-label, randomized dose proportionality study of technosphere insulin inhalation powder (TI) doses up to 80 U administered with the Gen2 inhaler in healthy subjects. *Diabetes*. 2013. https://doi.org/http://dx.doi.org/10.2337/db13-859-1394
66. van Alfen-van der Velden AAEM, Noordam C, de Galan BE, Hoorweg-Nijman JJG, Voorhoeve PG, and Westerlaken C. Successful treatment of severe subcutaneous insulin resistance with inhaled insulin therapy. *Pediatric Diabetes*. 2010. https://doi.org/10.1111/j.1399-5448.2009.00597.x
67. An inhaled insulin (Afrezza). *Medical Letter on Drugs and Therapeutics*. 2015.
68. Ayyar VS. History of growth hormone therapy. *Indian Journal of Endocrinology and Metabolism*. 2011. https://doi.org/10.4103/2230-8210.84852
69. Cutfield WS, Derraik JGB, Gunn AJ, Reid K, Delany T, Robinson E, and Hofman PL. Non-compliance with growth hormone treatment in children is common and impairs linear growth. *PLoS ONE*. 2011. https://doi.org/10.1371/journal.pone.0016223
70. Deal CL, Tony M, Hoÿbye C, Allen DB, Tauber M, Christiansen JS, Ambler GR, et al. Growth hormone research society workshop summary: Consensus guidelines for recombinant human growth hormone therapy in prader-willi syndrome. *Journal of Clinical Endocrinology and Metabolism*. 2013. https://doi.org/10.1210/jc.2012-3888
71. Rosenfeld RG and Bakker B. Compliance and persistence in pediatric and adult patients receiving growth hormone therapy. *Endocrine Practice*. 2008. https://doi.org/10.4158/EP.14.2.143
72. Bosscher KD and Haegeman G. Minireview: Latest perspectives on antiinflammatory actions of glucocorticoids.

Molecular Endocrinology. 2009. https://doi.org/10.1210/me.2008-0283

73. Walvoord EC, de la Peña A, Park S, Silverman B, Cuttler L, Rose SR, Cutler G, Drop S, and Chipman JJ. Inhaled growth hormone (GH) compared with subcutaneous GH in children with GH deficiency: Pharmacokinetics, pharmacodynamics, and safety. *The Journal of Clinical Endocrinology and Metabolism*. 2009; 94(6): 2052–2059. https://doi.org/10.1210/jc.2008-1897
74. King EM, Chivers JE, Rider CF, Minnich A, Giembycz MA, and Newton R. Glucocorticoid repression of inflammatory gene expression shows differential responsiveness by transactivation- and transrepression-dependent mechanisms. *PLoS ONE*. 2013. https://doi.org/10.1371/journal.pone.0053936
75. Valotis A and Högger P. Human receptor kinetics and lung tissue retention of the enhanced-affinity glucocorticoid fluticasone furoate. *Respiratory Research*. 2007. https://doi.org/10.1186/1465-9921-8-54
76. Druzgala P, Hochhaus G, and Bodor N. Soft drugs-10. Blanching activity and receptor binding affinity of a new type of glucocorticoid: Loteprednol etabonate. *Journal of Steroid Biochemistry and Molecular Biology*. 1991. https://doi.org/10.1016/0960-0760(91)90120-T
77. Rohatagi S, Arya V, Zech K, Nave R, Hochhaus G, Jensen BK, and Barrett JS. Population pharmacokinetics and pharmacodynamics of ciclesonide. *Journal of Clinical Pharmacology*. 2003. https://doi.org/10.1177/0091270002250998
78. Allen DB, Bielory L ,Derendorf Hartmut, Robert Dluhy, Colice GL, and Szefler SJ. Inhaled corticosteroids: Past lessons and future issues . *Journal of Allergy and Clinical Immunology*. 2003. https://doi.org/10.1016/S0091-6749(03)01859-1
79. Halpin DMG, Singh D, and Hadfield RM. Inhaled corticosteroids and COVID-19: A systematic review and clinical perspective. *European Respiratory Journal*. European Respiratory Society. 2020. https://doi.org/10.1183/13993003.01009-2020

6

Chronic Lung Diseases: Treatment, Challenges, and Solutions

Aparoop Das[1], Riya Saikia[1], Kalyani Pathak[1], Urvashee Gogoi[1], and Surovi Saikia[2]
[1]Department of Pharmaceutical Sciences, Dibrugarh University, Assam, India
[2]Natural Products Chemistry Group, CSIR North East Institute of Science & Technology, Jorhat, Assam, India

6.1 Introduction

Lung diseases over the years have become one of the leading causes of death and disability across the world. Chronic obstructive pulmonary disease (COPD) affects about 65 million people and kills 3 million annually, making it the third leading cause of death worldwide. Around 334 million people suffer from asthma, the most common chronic disease amongst children, affecting 14% of them globally. Respiratory ailments alone account for substantial morbidity and mortality, but factors like poverty, environmental exposures, crowding and generally poor living conditions increase vulnerability to such ailments. The highest mortality due to chronic respiratory diseases comes from Asia while the lowest is from sub-Saharan Africa (1).

Chronic lung disease (CLD) is the term for a wide variety of persistent lung disorders. It usually develops slowly and may get worse over time. CLDs may be caused by smoking tobacco or by breathing in second-hand tobacco smoke, chemical fumes, dust or other forms of air pollution and frequent lower respiratory infections during childhood. CLDs are generally not curable, but the symptoms can be managed by use of certain medications that help dilate major air passages and improve shortness of breath.

6.1.1 Types of Chronic Lung Diseases

CLDs impose an immense worldwide health burden. The following constitute the majority of the global cases of severe illness and death.

6.1.1.1 Asthma

Asthma affects both children and adults, and is one of the most common chronic ailments encountered in any clinical setting. Asthma is a chronic inflammatory disorder of the airways, usually associated with airway hyper-responsiveness and variable airflow obstruction, that is often reversible spontaneously or under treatment (2). Asthma sets in quite early and can attack all age groups of people. It is often characterized by repeated attacks of breathlessness and wheezing, which differ in severity and frequency from person to person. This happens due to airway inflammation in the lungs and affects the sensitivity of the nerve endings of the airways, making them easily irritated. In an attack, the lining of the passages swell, causing broncho spasm and reducing the flow of air across the lungs. A combination of factors encompassing specific genes, prenatal maternal smoking, diet, and environmental pollution are found to increase the risk of asthma in individuals (3).

6.1.1.1.1 Pathophysiology

Cellular inflammation: A characteristic feature of asthma is congestion of the airway lumen by a tenacious mucus plug which is composed of exuded plasma proteins and mucus glycoproteins secreted from surface epithelial cells (4). Multiple cellular mediators are involved in the inflammatory process, viz. T lymphocytes, mast cells, macrophages, eosinophils, neutrophils, and epithelial cells (5).

Tissue remodeling: Structural airway changes occur chiefly in the mucosal and submucosal tissues of the airways. Pathological changes in the mucosa include epithelial hyperplasia and goblet cell metaplasia with increase in mucus production. Submucosal changes include smooth muscle hypertrophy, collagen deposition, and enlarged mucous glands, causing the airways to narrow and increasing mucus production during an asthma attack (6).

Bronchoconstriction and airway hyper-responsiveness: Seen during acute exacerbations of asthma, a bronchoconstrictor stimulus occurs rapidly to constrict the airways in response to inhalation of an allergen or exposure to occupational sensitizers and irritants (7). Multiple mechanisms, including inflammation, dysfunctional neuroregulation, and structural changes, influence airway hyper-responsiveness.

6.1.1.2 Chronic Bronchitis

Clinically, chronic bronchitis (CB) is defined as production of cough and sputum for at least 3 months a year for two consecutive years (8). About one third of patients with COPD suffer from CB, however it has also been seen in individuals who are otherwise healthy, with prevalence estimates ranging widely both in population-based studies (2.6–16%) (9) and

DOI: 10.1201/9781003046547-6

among COPD patients (7.4–53%) (10). Chronic bronchitis results from an array of insults to the lung over time. These predominantly include cigarette smoking, respiratory infections, and environmental pollutants and irritants. Viruses cause upto 50% of acute exacerbations of chronic bronchitis and several causative agents have been implicated so far, viz. influenza virus, RSV, rhinovirus, and parainfluenza virus. The other 50% of acute exacerbations are bacterial in nature, with the most common pathogens being *Haemophilus influenzae, Streptococcus pneumoniae,* and *Moraxella catarrhalis*(11).

6.1.1.2.1 Pathophysiology

The pathological hallmark of chronic bronchitis is mucous metaplasia, a process where mucus is overproduced following an inflammatory signal. The dominant mechanisms responsible for copious amounts of mucus in CB are overproduction and hypersecretion of mucus by goblet cells and decreased elimination of mucus. The heightened mucus secretion may develop as a result of a variety of factors that include cigarette smoke exposure (12), acute and chronic viral infections (13), bacterial infections (14), or activation of inflammatory mediators for mucin gene transcription via activation of the epidermal growth factor receptor. This instigates overproduction and hypersecretion of mucus from increased degranulation by neutrophil-mediated elastase. All of these factors team up to cause difficulty in clearing secretions because of poor ciliary function, distal airway occlusion, and ineffective cough secondary to respiratory muscle weakness and reduced peak expiratory flow (15).

6.1.1.3 Chronic Obstructive Pulmonary Disease (COPD)

COPD is a leading cause of death globally. Apart from increasing healthcare expenditures (16), it imposes a significant burden with respect to disability and impaired life quality (17). COPD is an umbrella term for several conditions that block the flow of air in the bronchi and trachea. COPD has been defined as "a diseased state characterized by airflow limitation that is not fully reversible. The airflow limitation is usually both progressive and associated with an abnormal inflammatory response of the lungs to noxious particles or gases" (18),(19). The technical definition of COPD by the Global Initiative for Obstructive Lung Diseases (GOLD) is based on spiromeric criteria and uses the post-bronchodilator forced expiratory volume in one second (FEV1) and its ratio tothe forced vital capacity (FVC). COPD is characterized by a FEV1/FVC ratio <70% (20). Due to its heterogeneity, a variety of different definitions of COPD have become popular that include limitation of airflow, components of destruction of the lung parenchyma (emphysema), development of hypoxemia, and chronic sputum production (bronchitis). However, the terms *chronic bronchitis* and *emphysema* have now been omitted from the formal COPD definition (21).

6.1.1.3.1 Risk Factors

Tobacco smoke remains the prime determining factor leading to the development of COPD. Tobacco smoke causes destruction of lung tissue (emphysema) and obstruction of the small airways with inflammation and mucus (chronic bronchitis), producing the cardinal symptoms of COPD, namely shortness of breath and cough. Other environmental and occupational exposures, air pollution (both indoor and outdoor), genetic abnormalities (such as 1-antitrypsin deficiency), childhood respiratory infections (such as TB, pneumonia, or chronic asthma) may also contribute to the development of COPD. Any factor that affects the lung growth during gestation and childhood has also been found to have the potential of increasing an individual's risk of developing COPD (22). Pulmonary hypertension is a common sequel of chronic airflow obstruction (23). Diet and nutrition may not directly lead to chronic respiratory diseases, however, obesity has been found to coordinate with dyspnea and thus aid in producing symptoms of chronic respiratory diseases. Age-induced physiological lung-function decline can also predispose individuals to COPD.

6.1.1.3.2 Pathophysiology

Mucus hypersecretion and ciliary dysfunction: Airway mucus hypersecretion causes chronic productive cough. This hypersecretion is as a result of squamous metaplasia, increase in the number of goblet cells, and increased size of submucosal glands of the bronchi in response to chronic irritation by harmful particles and gases. Metaplasia of the squamous epithelial cells leads to ciliary dysfunction which is seen in the form of an abnormal mucociliary escalation and difficulty in expectorating (24).

Airflow obstruction and hyperinflation or air trapping: The primary sites of airflow obstruction are the small conducting airways that are < 2 mm in diameter. This is due to inflammation together with narrowing (airway remodeling) and presence of inflammatory exudates in the small airways. In addition, loss of lung elastic recoil and destruction of alveolar support also contribute to airflow obstruction.

Gas exchange abnormalities: Such abnormalities are seen in the advanced stages and can be identified in the form of arterial hypoxemia with or without hypercapnia.

Pulmonary hypertension: This also develops in the later stages of COPD together with severe gas exchange abnormalities. Pulmonary arterial constriction (as a result of hypoxia), endothelial dysfunction, remodeling of the pulmonary arteries (smooth muscle hypertrophy and hyperplasia), and destruction of the pulmonary capillary bed can trigger this condition.

Systemic effects: Systemic manifestations in the form of inflammation and skeletal muscle wasting restricts the exercise capacity of patients and worsens the prognosis, regardless of the degree of airflow obstruction.

6.2 Different Treatment Strategies of Chronic Lung Diseases along with Their Pharmacology

CLD is a group of disorders that affect the lungs and other parts of the respiratory system. CLD may be caused by smoking tobacco or by breathing in second-hand tobacco smoke, chemical fumes, dust, or other forms of air pollution.

CLDs include asthma, chronic bronchitis, pulmonary fibrosis, COPD, asbestosis, pneumonitis, and other lung conditions. These types of lung diseases may affect the airways, lung tissues, or circulation of blood in and out of the lungs (25).

6.2.1 Current Treatment Strategy for Lung Diseases

6.2.1.1 Treatment of Chronic Asthma (26–28)

Asthma is a CLD that inflames and narrows airways in the lungs and causes difficulties in breathing. Asthma occurs in people of all ages and generally progresses during childhood. It may cause wheezing or coughing, or a feeling of tightness in the chest. Treatment can only ease symptoms and prevent the complications. Long-term control medications are the only treatment for severe asthma. They help in preventing asthma symptoms and complications. These include:

a. Inhaled corticosteroids (ICS): ICSs are one of the effective controllers of asthma. They reduce inflammation by inhibiting multiple activated inflammatory genes through reversing histone acetylation (HDAC2), e.g. fluticasone, budesonid, mometasone, beclomethasone, and ciclesonide.
b. Inhaled long-acting beta agonists: Inhaled long-acting beta agonists open the airways by relaxing the smooth muscles around the airways, e.g. salmeterol, formoterol.
c. Inhaled long-acting anticholinergics: These relax the airways in the lungs, making breathing easier for the patients, e.g. aclidinium, umeclidinium, glycopyrronium, iIpratropium, tiotropium.
d. Leukotriene modifiers: Leukotriene modifiers are also used to prevent asthma. Leukotriene modifiers decrease the body's production of the leukotrienes that worsen both asthma and allergic reactions, e.g. montelukast sodium, zafirlukast, and zileuton.
e. Cromolyn sodium: Cromolyn sodium prevents the bronchial hyper-reactivity induced by chronic allergen exposure. Cromolyn is found to be effective in treatment of mild to moderate symptoms of chronic asthma in 60% of patients.
f. Theophylline: Theophylline is mainly used for treatment of lung diseases like asthma, bronchitis, emphysema, etc. It should be used on regular basis to prevent wheezing and breathing problems. It acts by relaxing the smooth muscles around the airways of lung.
g. Oral corticosteroids (OCS): OCS are commonly used for treatment of asthma symptoms and complications. It reduces inflammation and swelling in the airways and relaxes the airways. OCS has been shown to reduce emergency room visits and hospitalizations for asthma. Some examples are cortisone, prednisone, dexamethasone, prednisolone, betamethasone, and hydrocortisone.
h. Quick-relief medications: They can also relieve asthma attacks. These include:
    i. Inhaled short-acting beta-agonists: They provide quick relief of asthma symptoms, albuterol and levalbuterol.
    ii. Inhaled short-acting anticholinergics: They act by widening the airways by blocking the cholinergic nerves which release chemicals that can cause the muscles lining the airways to tighten. They can be used for patients with moderate symptomatic asthma, e.g. tiotropium and ipratropium.
    iii. Combination of an inhaled short-acting anticholinergic and inhaled short-acting beta agonist: This therapy is the most effective treatment of asthma in adults who are treated in emergency departments. It is very effective in controlling symptoms and complications compared with beta agonists alone.
i. Biologics

   Biologic drugs improve the immune system to treat asthma. They inhibit the activity of chemicals that swell up the airways. These drugs can prevent asthma attacks or make the attacks much milder. Four monoclonal antibodies are currently approved to treat severe asthma:

   1. Reslizumab (Cinqair): Used to treat severe asthma which is a specific type of white blood cell present in blood.
   2. Mepolizumab (Nucala): Also used to treat severe asthma caused by eosinophil (eosinophilic asthma).
   3. Omalizumab (Xolair): Used to treat severe asthma that is triggered by allergies.
   4. Benralizumab (Fasenra): Used to treat severe asthma caused eosinophil (eosinophilic asthma).

6.2.1.2 Current Treatment Strategy of Bronchitis (29,30)

The treatment for bronchitis based on its types whether is acute or chronic. In general, acute bronchitis does not need any treatment. Over-the-counter drugs can break up mucus and treat fever or pain. Treatment of chronic bronchitis is different than acute bronchitis. Chronic bronchitis is often considered as not curable. Only Symptoms can be treated by using drugs, oxygen therapy, pulmonary rehabilitation, combination therapy, and surgery.

Medications used to treat chronic bronchitis

a. Antibiotics: They are used to treat worsening coughs, breathlessness, and mucus production caused by infections in chronic bronchitis. For example, Amoxicillin, Augmentin.
b. Anti-inflammatory drugs: They are used for reducing swelling and mucus output, such as corticosteroids (also called steroids). Steroids help in relieve from symptoms in upper respiratory tract infections. They reduce the inflammation of the lining of the nose and throat. They also help in improving the symptoms of the bronchitis. Steroids can have several side effects like swelling in feet and hands, mood changes, increased appetite and weight gain, etc. Some important corticosteroids are cortisone, prednisone, prednisolone, methylprednisolone, dexamethasone, etc.
c. Bronchodilators: They relax the smooth muscles around the airways to make the airways stay open. There are

long-acting and short-acting bronchodilators. Short-acting drugs are most commonly known as rescue drugs as they act quickly. The three most widely used bronchodilators are:

1. Beta-2 agonists, such as salbutamol, vilanterol, salmeterol, and formoterol.
2. Anticholinergics like ipratropium, aclidinium,tiotropium, and glycopyrronium.

d. Combination drugs: They contain a mix of steroids and long or short-acting bronchodilators. This therapy is combined with an inhaled corticosteroid and two long-acting bronchodilators. It may be used for severe bronchitis to overcome the breathing problem. Combinations of two long-acting bronchodilators such as aclidinium/formoterol (Duaklir) and glycopyrrolate/ formoterol (Bevespi Aerosphere)

6.2.1.3 *Current Treatment Strategy of COPD (31–34)*

COPD is characterized by trouble breathing, cough, wheezing, and pain in the chest. COPD is often caused by smoking but is also caused by inhaling toxins from the environment. There is no cure for COPD. The damage caused in the lungs and airways is often permanent. Several drugs can help in reducing the inflammation and open the airways to help make breathing easier. These include:

a. Short-acting bronchodilators

Bronchodilators open the airways to make breathing easier. Short-acting bronchodilators are often used for emergency condition for quick relief. It can be taken by using an inhaler or nebulizer. Examples of short-acting bronchodilators are mainly-albuterol, levalbuterol, ipratropium, etc.

b. Corticosteroids

In COPD, airways of lungs can be inflamed, swollen and irritated. Corticosteroids are the drugs which reduce inflammation in the body, making the air flow easier to lungs. They are usually prescribed in combination with a long-acting COPD drug. These forms can be used on a short-term basis when COPD suddenly gets worse. The corticosteroids often prescribe for COPD are:

1. Fluticasone (Flovent): This is used as an inhaler twice daily. Side effects may occur, like headache, sore throat, voice changes, nausea, cold-like symptoms, and thrush.

2. Budesonide (Pulmicort): This drug is used as a handheld inhaler or in a nebulizer.

3. Prednisolone: This drug comes in form of a pill and liquid. It is only given for urgent rescue treatment. Side effects may be headache, muscle weakness, upset stomach, and weight gain.

c. Methylxanthines

This drug is given to those patients who have severe COPD and whom the typical first-line treatments, such as fast-acting bronchodilators and corticosteroids, cannot help. When this happens, a drug called theophylline, along with a bronchodilator, can be used. Theophylline is a anti-inflammatory drug and it works by relaxing the muscles in the airways.

d. Long-acting bronchodilators

Long-acting bronchodilators are used to treat COPD for a longer period of time. They are taken once or twice daily using inhalers or nebulizers. They gradually help to ease breathing. The long-acting bronchodilators available today are aclidinium, arformoterol, formoterol, glycopyrrolate, etc.

e. Combination drugs

Several COPD drugs come as combination medications. These are the combinations of two long-acting bronchodilators or an inhaled corticosteroid and a long-acting bronchodilator. Triple therapy is also recommended, which is a combination of an inhaled corticosteroid and two long-acting bronchodilators for treatment of severe COPD. Combinations of two long-acting bronchodilators include aclidinium, glycopyrrolate, etc. Combinations of an inhaled corticosteroid and a long-acting bronchodilator include budesonide, fluticasone etc.

f. Roflumilast

Roflumilast (Daliresp) is a drug that falls under the class of a phosphodiesterase-4 inhibitor. It can be orally administered as a pill. Roflumilast helps relieve inflammation, which can improve air flow to the lungs.

g. Mucoactive drugs

COPD complications can cause increased levels of mucus in the lungs. Mucoactive drugs reduce mucus secretion and help in easy breathing. They often come in form of pill and include drugs like carbocysteine, erdosteine, and N-acetylcysteine.

h. Vaccines

It is important for patients with COPD to get a yearly flu vaccine. Pneumococcal vaccine is usually recommended for COPD patients. Vaccines can reduce the risk of severe sickness and can help to avoid infections and other complications related to COPD.

6.2.2 Bioactive Compounds

Plants have played an important role in the drug discovery process. The lead or bioactive compounds isolated from herbs can be used for treatment of chronic asthma, chronic bronchitis, and COPD (35). They have fewerside effects than modern allopathic medicines.

6.2.2.1 *Bioactive Compounds for Treatment of Asthma (36,37)*

Several bioactive compounds derived from plant sources are currently used in the treatment of chronic asthma. Some of the important compounds are

1. Flavonoids

Flavonoids are natural bioactive compounds obtained from leaves, nuts, and fruits. Some of the important flavonoids used in treatment of asthma are

a. Chrysin: A flavone found in *Passiflora caerulea* and *Passiflora incarnate* flowers. It is able to suppress the proliferation of airway around smooth muscles cells to promote a reduction in the levels of pro-inflammatory cytokines which is a key factor in asthma treatment.
b. Baicalin: A bio-metabolite mainly found in leaves and bark of several species of the *Scutellaria* genus. Studies performed on baicalin confirmed that it possesses anti-inflammatory properties, decreasing the inflammatory cell infiltration and the levels of TNF-α in the broncho alveolar lavage fluids (BALF).
c. Luteolin: A flavonoid compound which had also been demonstrated for anti-asthmatic activity. It is widely distributed in aromatic flowering plants and green vegetables.
d. Oroxylin: A flavone obtained from the extracts of *Scutellaria baicalensis Georgi* and *Oroxylum indicum* trees was found to be effective in lowering the airway hyperactivity in an OVA-induced asthma murine model and decreasing the levels of IL-4, IL-5, IL-13, and OVA-specific IgE in BALF.
e. Quercetin: It is a flavonol compound mainly found in citrus fruits, onions, apples, broccoli, cereals, grapes, tea, and wine, and has been known as the main active compound of plants known for their widespread use in traditional medicine for the treatment of inflammatory, allergic, and viral diseases. The studies using this compound as an anti-asthmatic show its high efficacy in reducing inflammatory processes. The anti-inflammatory mechanism of quercetin is attributed to the lipoxygenase and PDE4 inhibition, which promote a reduction in the pro-inflammatory cytokine formation.

2. Resveratrol

Resveratrol is a natural stilbenoid compound obtained from the bark of red fruits. Scientific reports demonstrated that resveratrol was found to be effective against the asthmatic mouse model by significantly lowering the activity of pro-inflammatory cytokines and reducing the airway hyper-responsiveness.

6.2.2.2 Bioactive Compounds for Treatment of Chronic Bronchitis (38, 39)

Several bioactive compounds originating from plants had been reported to be very useful in curing the complications of chronic bronchitis. Some of the reported potent bioactive compounds used against chronic bronchitis are

a. Terpenoid compounds

   Sesquiterpenoid compounds such as farnesol and caryophyllene; flavonoids like apigenin; monoterpenes like linalool, cineole, and fenchone; and triterpenoids such as ursolic acid have been reported to inhibit various viral infections of DNA and RNA viruses. They can be effectively used against viral pathogens involved in bronchitis.
b. Phenolic compounds

   Bioactive components isolated from *Plantago major* contain biologically active compounds like aucubin, baicalein, baicalin luteolin, caffeic acid, chlorogenic acid, ferulic acid, *p*-coumaric acid, etc. In vitro antiviral activity of these compounds has been sucessfully demonstrated against three types of human adenovirus at different concentrations. They are found to be effective in treatment of chronic bronchitis infection.
c. Bioactive compounds of *Zingiber officinale*

   The active compounds isolated from *Z. officinale*, such as β-sesquiphellandrene, α-zingiberene, and β-bisabolene and some flavonoids have been investigated for their antiviral activity. All these compounds showed 50% inhibition against viral pathogens at different concentrations.

6.2.2.3 Bioactive Compounds for Treatment of COPD (40, 41)

COPD is a disease characterized by progressive airflow limitation associated with abnormal inflammatory response of the lung to noxious particles. The inhibition of inflammation by lowering the levels of inflammatory mediators in the COPD pathology such as TNFα, LTB4, and IL-8 by some active compounds from natural sources were investigated and found to be effective in treatment of COPD. They include

1. Austrasulfone, an important bioactive compound obtained from the soft coral *Cladiellaaustralis* has been successfully used for treatment of COPD. It mediated the regulation of the Akt and heme oxygenase (HO)-1 signaling pathways.
2. 1-O-(Myristoyl) glycerol (MG), obtained from the head of the fish *Ilishaelongate*, induces 42% of the neurite outgrowth of rat PC12 cells through the activation of element-binding protein (CREB) and PI3K signaling pathways.
3. Sargaquinoic acid, isolated from a marine alga, *Sargassum macrocarpum*, enhances neuritere generation and protected rat PC12D cells from oxidative stress through PI3K signaling pathways.
4. Bafilomycin(s), a family of toxic macrolide antibiotics obtained from marine *Streptomyces griseus*, can inhibit autophagy by preventing fusion of autophagosomes with lysosomes.
5. Austrasulfone, an important bioactive compound isolated from the soft coral *Cladiella australis*, has shown potent anti-apoptotic activity on neuronal cells SH-SY5Y mediated through the regulation of the Akt and heme oxygenase (HO)-1 signaling pathways.

6.3 Conventional Drug Delivery Systems for Mitigating Chronic Lung Diseases

The drug delivery mechanism for treating lung diseases has increased due to the potential for localized therapeutic options. Some CLDs are irreversible and fatal and there exist few effective treatments for reversing the lung function (42). Pharmacotherapy traditionally available for CLD is recently classified into types based on the forms of therapeutic agents used. These include a variety of antibodies, peptides, chemical drugs, and genetic molecules like SiRNA, shRNA, and miRNA (43–45). The following sections show the different types of therapeutics used as CLD therapeutics.

6.3.1 Material Based

6.3.1.1 Multifunctional Nanocarriers

A wide range of materials are used in the preparation of multifunctional nanocarriers which have the benefit of acting as carriers providing specific effects with improved efficacies and performances, either in the bound or adsorbed state. These varieties currently meet the recent challenges and address the various needs of diseases which are discussed in the following sections.

Delivery of protein and related materials:

Advances in biotechnological research recently saw a huge surge in protein-based drugs. Oral delivery of such drugs is difficult due to the degrading effect of proteases. As a result they are solely injection-based with much therapeutic noncompliance (46). These protein-based drugs also have applications in treatment of local lung diseases, as the first candidate to reach market as an inhalable protein was recombinant human DNase (rhDnase) for treating cystic fibrosis. Nanoparticles with aerodynamic properties suitable to reach the alveolar zone of the lungs was made in the form of a nano micro-system using spray drying (to encapsulate the particles with mannitol). This helps the microspheres of nanoparticles to be released post dissolving in the lining fluid of the lungs (47). Solid lipid nanoparticles (SLN) also provide homogenous distribution on lungs upon nebulization, as found from studies carried out on diabetic rats (48). However, it has been shown that the contact of nanoparticles with the surfactant of the alveolar region promotes coating of nanocarriers via a corona (bimolecular) mainly composed of lipids and proteins (49). Hydrophobicity of nanoparticles is affected by this corona and thereby enhances the biorecognition, showing interactions with other cells and biological entities with no effect on therapeutic levels.

Antibiotic materials:

Antibiotic delivery to the lungs is a very reasonable therapeutic approach in combating lung infections and the most common routes for its administration remain parenternal and oral. Some inhaled antibiotics are available in the market, like colistin, tobramycin, and aztreonan, which are used in cystic fibrosis (50). Other applications of such aerosolized antibiotics are in pneumonia (hospital acquired) (51). All the above formulations are based on nanoparticles as many published works in the literature show that nanoparticles can improve the efficacy of the drugs including side effects and improved kinetic profiles.

6.3.1.2 Hydrogels

Based on the polymeric source used, hydrogels may be homopolymer or copolymer, which have the ability to swell in contact with water (52). They are nontoxic, biodegradable, and biocompatible with high absorption capacity; due to this they have a wide range of therapeutic applications in tissue engineering, drug delivery, and wound healing. Further, temperature sensitive and pH responsive smart hydrogels also help in site-specific targeted drug delivery.

6.3.1.3 Micelle

They are formed through self assembly, due to dispersion of amphiphilic molecules comprising both hydrophobic and hydrophilic components in solution. Due to their minimal toxicity and high stability they are of high importance for sustained and controlled drug delivery (53).

6.3.1.4 Dendrimer

These are hyperbranched macromolecules with many functional groups which have emerged has an ideal drug delivery vehicle. Drugs conjugated to such vehicles are of high stability, increased bioavailability, and long self-life (54).

6.3.1.5 Liposomes

Biodegradable, biocompatible drug delivery vehicles with low toxicity used for site specific drug delivery. They can encapsulate both hydrophobic and hydrophilic drug molecules, preventing the rapid decomposition of the drug released at the specific site (55).

6.4.2 Administration Based

6.4.2.1 Pulmonary Drug Administration

Alveolar ducts, respiratory bronchioles, and alveolar sacs comprise the respiratory region of humans. Methods of transporting drugs along the respiratory transepithelial layer differ a lot and are limited to the upper airways. The administration of drugs via the pulmonary route is mainly by two methods: intranasal and oral inhalation. The former has anatomical limitations, such as narrow lumen, and the later stands to be a better therapeutic option for delivering drugs of very small particles, and loss of concentration is 20% as compared to that of 85% via the nasal route. Oral administration is divided into intra-tracheal instillation and intra-tracheal inhalation (56).

6.4.2.2 Inhalation Delivery

This procedure is non-invasive, and easy to carry and use with improved pharmacokinetics (57). The required doses needed for enhanced treatment of lung diseases are fewer and off-target effects are avoided (58). Lymphatic circulation redistributes drugs into the peripheral airways and prevents first pass metabolism with low enzymatic activity on lungs (59).

6.4.2.3 *Systematic Delivery*

A quick administration procedure in lung diseases with immediate dissolution to blood stream via IV injection. It is also easily controlled with adjustable drug dosage and patients are more compliant with injections and oral administration. However, higher doses are required for therapeutic implications and molecules have shorter life in bloodstreams (57).

6.4.3 Different Drug Delivery Systems for Different Categories of Patients

6.4.3.1 *Elderly Patients*

Local tissues and compartments which are not accessible in elderly patients through oral administration are done via the parenteral route (58). It is considered the emerging route for delivery of peptides and proteins with optimal drug release (59). The intravenous (IV) route is commonly used in hospitalized old patients avoiding fast mass metabolism of the drug and remains to be the first choice during emergencies (60). Intramuscular injection (IM), on the other hand, has the potential advatnages of self-administration and earlier hospital discharge (61). Subcutaneous (SC) delivery is preferred for supplying fluid therapy in elderly patients (dehydrated) and is also considered cost effective, can be made for self-injection, and is less invasive with few side effects as compared to the IV route (62, 63). The oral route is the generally preferred route for drug delivery in elderly patients, including oral films, sublingual tablets, medicated gums, orally disintegrating tablets, buccal patches, and sprays that deliver the drug via the mucosal surfaces of the mouth (64, 65).

6.4.3.2 *Pregnant Patients*

The use of nanomedicine during pregnancy has shown increased bioavailability in the target tissues and minimized concerns regarding side effects to fetus and the transplacental passage (66). An increase in molecular size for restricting transplacental movement can be achieved by attaching the drug to macromolecular carriers like cyclodextrins, or using liposomes or dendrimers (67,68). Fetal growth restriction can be cured by selective drug delivery to a poor-functioning placenta (69).

6.4.3.3 *Obese Patients*

Oral administration is the most commonly used route for administering drugs, and obesity causes some important metabolic changes which include higher cardiac output, increased gastrointestinal perfusion of blood and splanchnic blood flow, high gastric emptying, and disturbances in enterohepatic recirculation. These change the rate and extent of drug absorption (70). Highly obese individuals may need a higher dosage of drug due to higher plasma volume and delayed gastrointestinal absorption (71). Other administration routes such as transdermal, subcutaneous, and intramuscular administration are affected due to the deposition of subcutaneous fat (70). Also, the intrinsic physiological properties of a drug influence its distribution preferences in obese individuals (72).

6.4 Recent Development in Targeted Drug Delivery Systems for Mitigating Chronic Lung Diseases

Developments in this area include delivery devices and drug formulations which include microspheres, liposomes, and nanoformulations for intranasal delivery. AERx pulmonary technology developed Aradigm, which helps in transferring morphine and insulin to the pulmonary route. Improved inhalation of patients on ventilators are now in the form of a baby mask (42). An advanced strategy for targeted drug delivery in CLD is nanocarriers, which can improve the pharmacokinetics of loaded drugs based on their individual physical properties. Adverse effects of drugs can be minimized based on the current progress in nanoparticles, to target diverse motifs for selective delivery (73). Gene delivery attempts using nanoparticles also are promising therapeutics for CLD, as most of them are related to the chronic failure of defense mechanisms (pulmonary) due to genetic disorders (74). A new class of drug such as mepolizumab (75), (CXCR)-2 antagonist (76), and phosphoinositide 3-kinase (PI3K) inhibitors has (77) gained prominence as a new class of therapeutics for the efficient treatment of CLD.

6.5 Challenges and Solutions

6.5.1 Asthma

Any condition in which an individual's airway becomes inflammed, resulting in a narrow passage that produces extra mucus and causes difficulty in breathing, is termed *asthma*. The effects of asthma can be observed both in children and adults, and it has a variable course in children younger than 5 years of age. And it is this variability that produces challenges in the treatment of asthma. For many years, the SABA bronchodilator was considered the primary treatment for asthma instead of the use of anti-inflammatory agents like ICS (78–80). However, when treatment with a Short Acting Beta Agonist (SABA) was unable to control the effects associated with asthma, and there were occurrences of asthma exacerbation, then ICS was introduced.

Now, the use of ICS has created other challenges, because it was recommended as a maintenance therapy, and the dosage set by physicians was to be adjusted with the clinical response. But the dose often remained unchanged over long time periods. Moreover, the shift to regular therapy of ICS along with SABA was not adopted very well by parents, young patients, adult patients, and caregivers. Later on, to overcome the problems of ICS therapy combined with SABA, Long Acting Beta Agonist (LABA) therapy combined with ICS was introduced (81, 82), which led to further confusion and safety concerns. The Treating Children to Prevent Exacerbations of Asthma (TREXA) trial found that low doses of beclomethasone combined with albuterol decreased the exacerbation frequency to a great extent as compared to albuterol alone. Hence, ICS along with albuterol was used as a stepdown rescue therapy for children who have controlled asthma (83). The same therapy was also found to be effective in adults in that it could lower the frequency of exacerbations. Use of the drug budesonide in combination with formoterol was also an effective maintenance and reliever therapy that could control exacerbations in adults as compared to the combined therapy of budesonide, formoterol, and SABA (84, 85). Further investigation by the researchers revealed that, taking into account the needs of patients, the combined use of ICS along with a specific dose of fast-acting β2-agonist could potentially overcome the difficulties related to the continuous and dependent use of ICS and SABA monotherapy (86–88).

6.5.2 Bronchial Disorders

The diseased state that causes inflammation of the bronchial tube linings that are responsible for carrying the air from and into the lungs can be termed *bronchial disorders*. Because of the characteristic structure exhibited by the airways, there can be subtle changes in the physiology and clinical aspects related to bronchiolitis (89). The diagnosis of bronchiolitis becomes challenging only when there are greater abnormalities indicated in the physiologic test results of the patient (90). The test patterns of patients with bronchiolitis include trapping of air, and obstructive and mixed obstructive patterns (91). Bronchiolitis may also be evident in those patients who have normal PFTs. Most of the time, the results obtained from chest radiograph reports are normal and hence the small airways need to be subjected to a High Resolution Computed Tomography (HRCT) scan. HRCT cannot scan the normal bronchioles but can identify the diseased bronchioles, either by indirect or direct signs. Ways of managing bronchiolitis may differ depending upon the clinical features and the severity of the diseased state. But the general therapies include treatment of already existing infections in airways and lungs, gastroesophageal reflux, and omitting potential exposures. Use of macrolide antibiotics will not only help to cure different forms of bronchiolitis but will also impact the remodeling of airways in an effective manner (92–94).

6.5.3 COPD

A diseased state whereby various lung diseases block the flow of air and cause difficulty in breathing can be termed *COPD*. The reports of qualitative findings have stated the ways people suffering from severe COPD fight with the various changes in their condition (95). However, in a wide-ranging sense, self-care serves as the primary motivator for all those individuals who have been facing the long term effects of COPD (96). Recognizing and responding to various credences like inappropriate symptom attribution, perceptions related to emotion and low sense of control, should be considered the central basis of the support of self-care and self-management. Self-management cannot be defined as a single phenomenon; rather, to maintain the changes in behavior, continuous support is of utmost importance. Thus, it can be defined as a continuous process. The main challenge with self-management of COPD is that half of the patients have multiple morbidities and most of them have comorbidity. Common morbidities include anxiety, depression, coronary diseases, and pain (97). However, there are no specific self-management guidelines for patients with multiple morbidities.

There is no clear guidance on self-management appropriate for patients with multiple morbidities. Therefore, to help patients with multiple morbidities, counseling and advice from professionals are highly advised. This will help the patients analyse, interpret, and positively respond to the worsening status so that they can survive without undue suffering. Hence, a an effective implementation of a whole system approach is required that can support self-management (98). The approach aims ato address patients with multiple morbidities, ensuring easy and flexible access to the advice of professionals and a continuous care process (99, 100). Currently, self-management is aided within pulmonary rehabilitation programs, but it is expected that in the coming days it will be supported with primary care systems that would facilitate and improve the ongoing support enhancing the efficiency of the complete process (101).

6.6 Future Prospects

The main aim of the respiratory care unit is to look after the coronary care, respiratory failure, and stroke care units. Along with various developments, the medicines used for the intensive care of these patients have given rise to new challenges for the physicians associated with respiratory systems (102). Therefore, it is important to employ various educational programs that can address the problems related to rising requirements and growing challenges. Even though the connection between intensive care and respiratory medicine is not yet clear, they hold certain connections in the domiciliary ventilation area. Respiratory medicine is only utilized in weaning centers, respiratory intermediate units, and home ventilation management in some countries. In the same manner, the subject of respiratory medicine is not even incorporated into the curriculum of respiratory physicians in many countries across the world. Also, it can be seen that conventional intensive care units inadequately represent respiratory medicine. Patient-centered acute care training is initiated by the European Society of Intensive Care Medicine (ESICM) for carrying out an up-to-date intensive care curriculum. The objective behind bringing out such a program is to achieve a harmonized and advanced medicine practice and training that can enhance the quality of acute and critical care. The Pain and Intensive Care Curriculum of the Medicine and Anesthesiology branch was developed by the European Board of Anesthesiology. Further, a diploma course in Anesthesiology and Intensive Care was also introduced by the European Society of Anesthesiology. Also the European Diploma in Intensive Care was developed by ESICM. The physicians of respiratory units should look after respiratory intermediate units, home ventilation units, intensive care units, and weaning centers. Respiratory medicine should also set standards for all other professional units that are associated with intensive care medicine. They should also hold the potential to define the educational standards of weaning units, intermediate respiratory care units, and units of home ventilation (103–108).

6.7 Conclusion

Treatment of patients with CLDs have become a major challenge over the past few decades. Due to a combination of various factors, both past and present, include smoking habits and an ageing population; it is the only major chronic disease that is still associated with rising mortality. However, the last decade has witnessed the rise of technology in almost all fields of healthcare. New advances in the field of treatment strategies and technologies were also introduced and accordingly adapted to overcome various obstacles and challenges in the treatment of CLDs. The data on recent treatment strategies have shown that surgery to reduce lung volume is a promising intervention for COPD; leukotriene blocking agents have gained much attention for treatment of chronic asthma as they have both bronchodilator and anti-inflammatory properties in asthma. Further, positron emission tomography is proved to be a highly sensitive and specific diagnostic tool for diagnosis of CLDs. The monoclonal antibody against immunoglobulin E plays a pivotal role in atopic disease and also has putative immunomodulatory properties, because of which it seems to be very close to being marketed. Lowering the doses of inhaled steroids have resulted in improvements in symptoms. Also, the interleukin 4 receptor antagonist can inactivate naturally occurring interleukin 4 and is considered to be an important pro-inflammatory mediator in asthma. Even though there are some additional obstacles in the treatment of CLDs and controlling the mortality rate, further research must still be carried out so that the mortality rate associated with different CLDs can be controlled to a great extent utilizing various aspects of technology.

Acknowledgments

The authors would like to thank Dibrugarh University for providing the necessary infrastructure and technical support in proceeding with this work. The authors are also grateful to the All India Council for Technical Education (AICTE), Department of Biotechnology (DBT), Indian Council of Medical Research (ICMR), University Grants Commission (UGC), and Ministry of Tribal Affairs (MoTA), New Delhi, Govt. of India, for providing financial support under AICTE-RPS to AD and research fellowships to RS, KP, UG, and SS.

Conflicts of Interest

There are no conflicts of interest among the authors to report with respect to this chapter.

REFERENCES

1. Soriano JB, Kendrick PJ, Paulson KR, Gupta V, Abrams EM, Adedoyin RA, Adhikari TB, Advani SM, Agrawal A, Ahmadian E, and Alahdab F. Prevalence and attributable health burden of chronic respiratory diseases, 1990–2017: A systematic analysis for the global burden of disease study 2017. *The Lancet Respiratory Medicine*. 2020 Jun 1; 8(6): 585–596.
2. Bateman ED, Hurd SS, Barnes PJ, Bousquet J, Drazen JM, FitzGerald M, Gibson P, Ohta K, O'Byrne P, Pedersen SE, and Pizzichini E. Global strategy for asthma management and prevention: GINA executive summary. *European Respiratory Journal*. 2008 Jan 1; 31(1): 143–178.
3. Heederik D and von Mutius E. Does diversity of environmental microbial exposure matter for the occurrence of allergy and asthma? *Journal of Allergy and Clinical Immunology*. 2012 Jul 1; 130(1): 44–50.
4. Bousquet J, Jeffery PK, Busse WW, Johnson M, and Vignola AM. Asthma: From bronchoconstriction to airways inflammation and remodeling. American Journal of Respiratory and Critical Care Medicine. 2000 May 1;161(5): 1720–1745.
5. Busse WW, and Lemanske RF. Expert panel report 3: Moving forward to improve asthma care. *Journal of Allergy and Clinical Immunology*. 2007;120(5): 1012–1014.
6. Fahy JV. Type 2 inflammation in asthma – present in most, absent in many. *Nature Reviews Immunology*. 2015 Jan; 15(1): 57–65.
7. Busse W, Corren J, Lanier BQ, McAlary M, Fowler-Taylor A, Della Cioppa G, van As A, and Gupta N. Omalizumab, anti-IgE recombinant humanized monoclonal antibody, for the treatment of severe allergic asthma. *Journal of Allergy and Clinical Immunology*. 2001 Aug 1; 108(2): 184–190.
8. Stuart-Harris CH, Crofton J, Gilson JC, Gough J, Holland W, Knowelden J, et al. Definition and classification of chronic bronchitis. *Lancet*. 1965; 1: 775–779.
9. Cerveri I, Accordini S, Verlato G, Corsico A, Zoia MC, Casali L, Burney P, and de Marco R. Variations in the prevalence across countries of chronic bronchitis and smoking habits in young adults. *European Respiratory Journal*. 2001 Jul 1; 18(1): 85–92.
10. Kim V, Han MK, Vance GB, Make BJ, Newell JD, Hokanson JE, Hersh CP, Stinson D, Silverman EK, and Criner GJ, COPDGene investigators. The chronic bronchitic phenotype of COPD: An analysis of the COPDGene Study. *Chest*. 2011 Sep 1;140(3): 626–633.
11. Mandell LA and Robert C.Read, Infections of the lower respiratory tract. In: Finch RG, Greenwood D, Norrby SR, and Whitley RJ (eds). Antibiotic and Chemotherapy E-Book. 9th ed. United Kingdom, PA, Elsevier Health Sciences UK/Saunders 2010; 574–588.
12. Ebert RV and Terracio MJ. The bronchiolar epithelium in cigarette smokers: Observations with the scanning electron microscope. *American Review of Respiratory Disease*. 1975 Jan; 111(1): 4–11.
13. Holtzman MJ, Tyner JW, Kim EY, Lo MS, Patel AC, Shornick LP, Agapov E , and Zhang Y. Acute and chronic airway responses to viral infection: Implications for asthma and chronic obstructive pulmonary disease. *Proceedings of the American Thoracic Society*. 2005 Aug; 2(2): 132–140.
14. Burgel PR and Nadel JA. Roles of epidermal growth factor receptor activation in epithelial cell repair and mucin production in airway epithelium. *Thorax*. 2004 Nov 1; 59(11): 992–996.
15. Hogg JC, Chu F, Utokaparch S, Woods R, Elliott WM, Buzatu L, Cherniack RM, Rogers RM, Sciurba FC, Coxson HO, and Paré PD. The nature of small-airway obstruction in chronic obstructive pulmonary disease. *New England Journal of Medicine*. 2004 Jun 24; 350(26): 2645–2653.

16. Sullivan SD, Ramsey SD, and Lee TA. The economic burden of COPD. *Chest*. 2000 Feb 1; 117(2): 5S–9S.
17. Ferrer M, Alonso J, Morera J, Marrades RM, Khalaf A, Aguar MC, Plaza V, Prieto L, and Anto JM. Chronic obstructive pulmonary disease stage and health-related quality of life. *Annals of Internal Medicine*. 1997 Dec 15; 127(12): 1072–1079.
18. Crapo RO. Guidelines for methacholine and exercise challenge testing-1999. This official statement of the American Thoracic Society was adopted by the ATS Board of Directors, July 1999. *American Journal of Respiratory Critical Care Medicine*. 2000; 161: 309–329.
19. Rabe KF, Hurd S, Anzueto A, Barnes PJ, Buist SA, Calverley P, Fukuchi Y, Jenkins C, Rodriguez-Roisin R, Van Weel C, and Zielinski J. Global strategy for the diagnosis, management, and prevention of chronic obstructive pulmonary disease: GOLD executive summary. *American Journal of Respiratory and Critical Care Medicine*. 2007 Sep 15; 176(6): 532–555.
20. Pauwels RA. GOLD scientific committee. Global strategy for the diagnosis, management, and prevention of chronic obstructive pulmonary disease. NHLBI/WHO global initiative for chronic obstructive lung disease (GOLD) workshop summary. *Respiratoty Critical Care Medicine*. 2001; 163: 1256–1276.
21. Celli BR, MacNee WA, Agusti AA, Anzueto A, Berg B, Buist AS, Calverley PM, Chavannes N, Dillard T, Fahy B, and Fein A. Standards for the diagnosis and treatment of patients with COPD: A summary of the ATS/ERS position paper. *European Respiratory Journal*. 2004 Jun 1; 23(6): 932–946.
22. Eisner MD, Anthonisen N, Coultas D, Kuenzli N, Perez-Padilla R, Postma D, Romieu I, Silverman EK, and Balmes JR. An official American Thoracic Society public policy statement: Novel risk factors and the global burden of chronic obstructive pulmonary disease. *American Journal of Respiratory and Critical Care Medicine*. 2010 Sep 1; 182(5): 693–718.
23. Turato G, Zuin R, and Saetta M. Pathogenesis and pathology of COPD. *Respiration*. 2001; 68(2): 117–128.
24. Mac Nee W. Pathology, pathogenesis, and pathophysiology. *BMJ*. 2006 May 20; 332(7551): 1202–1204.
25. Barnes PJ. Mediators of chronic obstructive pulmonary disease. *Pharmacological Reviews*. 2004b; 56(4): 515–548.
26. Barnes PJ. Drugs for asthma. *British Journal of Pharmacology*. 2006; 147(1): 297–303.
27. Adams BK and Cydulka RK. Asthma evaluation and management. *Emergency Medicine Clinics of North America*. 2003; 21(2): 315–330.
28. Kaufman G. Asthma: Pathophysiology, diagnosis and management. *Nursing Standard*. 2011; 26(5): 48–57, 2011.
29. Perotin JM, Launois C, Dewolf M, et al. Managing patients with chronic cough: Challenges and solutions. *Therapeutics and Clinical Risk Management*. 2018; 14: 1041–1051.
30. Smith DRM, Dolk FCK, Pouwels KB, Christie M, Robotham JV, and Smieszek T. Defining the appropriateness and inappropriateness of antibiotic prescribing in primary care. *Journal of Antimicrobial Chemotherapy*. 2018; 73(Suppl. 2): ii11–ii18.
31. Aaron SD, Vandemheen K, Fergusson D, et al. The Canadian optimal therapy for COPD trial: Design, organization and patient recruitment. *Canadian Respiratory Journal* 2004; 11: 581–585.
32. Barnes PJ. Effect of β-agonists on inflammatory cells. *The Journal of Allergy and Clinical Immunology Journal of Allergy and Clinical Immunology*. 1999; 104: S10–S17.
33. Anthonisen NR, Manfreda J, Warren CP, et al. Antibiotic therapy in exacerbations of chronic obstructive pulmonary disease. *Annals of Internal Medicine*. 1987; 106: 196–204.
34. Celli BR. Current thoughts regarding treatment of chronic obstructive pulmonary disease. *Medical Clinics of North America*. 1996; 80(3): 589–609.
35. Mainardi T, Kapoor S, and Bielory L. Complementary and alternative medicine: Herbs, phytochemicals and vitamins and their immunologic effects. *Journal of Allergy and Clinical Immunology*. 2009; 123(2): 283–294.
36. Bielory L. Complementary and alternative interventions in asthma, allergy, and immunology. *Annals of Allergy, Asthma & Immunology*. 2004; 93: 45–54.
37. Huntley A and Ernst E. Herbal medicines for asthma: A systematic review. *Thorax*. 2000; 55(11): 925–929.
38. Amber R, Adnan M, Tariq A, and Mussarat S. A review on antiviral activity of the Himalayan medicinal plants traditionally used to treat bronchitis and related symptoms. *Journal of Pharmacy and Pharmacology*. 2017; 69(2): 109–122.
39. Peera C and Efferth T. Antiviral medicinal herbs and phytochemicals. *Journal of Pharmacognosy*. 2012; 3: 1106–1118.
40. Gao Z, Li FS, Yang J, Xu D, and Yang CH. Investigation on the literature concerning the clinically used traditional Chinese medicines for the treatment of chronic obstructive pulmonary disease in the past ten years. *Chinese Journal of Experimental Traditional Medical Formulae*. 2010; 16: 286–288.
41. Zhou M, Wei X, Li A, et al. Screening of traditional Chinese medicines with therapeutic potential on chronic obstructive pulmonary disease through inhibiting oxidative stress and inflammatory response. *BMC Complementary Medicine and Therapies*. 2016; 16: 360.
42. Bhavna K, Deepika S, and Kartik G. Recent approaches for novel treatment for pulmonary diseases. *International Journal of Pulmonary & Respiratory Sciences*. 2018; 2(4): 555–593.
43. Ruppert C, Schmidt R, Grimminger F, et al. Chemical coupling of a monoclonal antisurfactant protein-B antibody to human urokinase for targeting surfactant-incorporating alveolar fibrin. *Bioconjugate Chemistry*. 2002; 13: 804–811.
44. Durham AL, Caramori G, Chung KF, and Adcock IM. Targeted anti-inflammatory therapeutics in asthma and chronic obstructive lung disease. *Translational Research*. 2016; 167(1): 192–230.
45. Fujita Y, Takeshita F, Kuwano K, and Ochiya T. RNAi therapeutic platforms for lung diseases. Pharmaceuticals. 2013; 6(2): 223–250.
46. Pacheco RM. (Ed.) Tratado de Tecnología Farmacéutica, Volumen III: Formas de Dosificación; Editorial Síntesis, S.A., Madrid, Spain, 2017; 458.

47. Grenha A, Seijo B, and Remuñán-Lopez C. Microencapsulated chitosan nanoparticles for lung protein delivery. *European Journal of Pharmaceutial Sciences*. 2005; 25: 427–437.
48. Liu J, Gong T, Fu H, Wang C, et al. Solid lipid nanoparticles for pulmonary delivery of insulin. *International Journal Pharmaceutics*. 2008; 356: 333–344.
49. Hu Q, Bai X, Hu G, and Zuo YY. Unveiling the molecular structure of pulmonary surfactant corona on nanoparticles. *ACS Nano*. 2017; 11: 6832–6842.
50. Woods A and Rahman KM. Antimicrobial molecules in the lung: Formulation challenges and future directions for innovation. *Future Medicinal Chemistry*. 2018; 10: 575–604.
51. Luyt CE, Hékimian G, Bréchot N, and Chastre J. Aerosol therapy for pneumonia in the intensive care unit. *Clinics in Chest Medicine*. 2018; 39: 823–836.
52. Zhao W, Jin X, Cong Y, Liu Y, and Fu J. Degradable natural polymer hydrogels for articular cartilage tissue engineering. *Journal of Chemical Technology & Biotechnology*. 2013; 88: 327–339.
53. Gaucher G, Dufresne Marie-Hélène, Sant VP, Kang N, Maysinger D, and Leroux Jean-Christopheet. Block copolymer micelles: Preparation, characterization and application in drug delivery. *Journal of Controlled Release*. 2005; 109: 169–188.
54. Gillies ER and Frechet JM. Dendrimers and dendritic polymers in drug delivery. *Drug Discovery Today*. 2005; 10: 35–43.
55. Akbarzadeh A, Rezaei-Sadabady R, Davaran S, *et al*. Liposome: Classification, preparation, and applications. *Nanoscale Reserach Letters*. 2013; 8: 102.
56. Lizio R, Klenner T, Borchard G, Romeis P, Sarlikiotis AW, Reissmann T, and Leher CM. Delivery of the GnRH antagonist centrolix by intratracheal instillation in Anesthetized rats. *European Journal of Pharmaceutial Sciences*. 2000; 9: 253–258.
57. Reed MD, Tellez CS, Grimes MJ, et al. Aerosolised 5-azacytidine suppresses tumour growth and reprogrammes the epigenome in an orthotopic lung cancer model. *British Journal of Cancer*. 2013; 109: 1775–1781.
58. Koshkina NV, Waldrep JC, Roberts LE, Golunski E, Melton S, and Knight V. Paclitaxel liposome aerosol treatment induces inhibition of pulmonary metastases in murine renal carcinoma model. *Clinical Cancer Research*. 2001; 7 (10): 3258–3262.
59. Blank F, Stumbles PA, Seydoux E, *et al*. Size-dependent uptake of particles by pulmonary antigen-presenting cell populations and trafficking to regional lymph nodes. *American Journal of Respiratory Cell and Molecular Biology*. 2013; 49(1): 67–77.
60. Gordon LF and Roberts MS. Physical and Biophysical Foundation of Pharmacy Practice: Issue in Drug Delivery. Michigan Publishing, Michigan USA, 2015.
61. Bose S. Parenteral Drug Delivery for Older Patients. In: Stegemann S. (ed.) Developing Drug Products in an Aging Society: From Concept to Prescribing. Springer International Publishing, Cham. 2016; 291–329.
62. Waitt C, Waitt M, and Pirmohamed M. Intravenous therapy. *Postgraduate Medical Journal*. 2004; 80: 1–6.
63. Milkovich G and Piazza CJ. Considerations in comparing intravenous and intramuscular antibiotics, *Chemotherapy*. 1991; 37(2): 1–13.
64. Stoner KL, Harder H, Fallowfield LJ, and Jenkins VA. Intravenous versus subcutaneous drug administration. Which Do Patients Prefer? A Systematic Review. *Patient*. 2015; 8: 145–153.
65. Logrippo S, Ricci G, Sestili M, *et al*. Oral drug therapy in elderly with dysphagia: Between a rock and a hard place. *Clinical Intervention in Aging*. 2017; 12: 241–251.
66. Keelan JA. Nanotoxicology: Nanoparticles versus the placenta. *Nature Nanotechnology*. 2011; 6(5): 263–264.
67. Andaluz A, Santos L, Garcia F, Ferrer RI, Fresno L, and Moll X. Maternal and foetal cardiovascular effects of the anaesthetic alfaxalone in 2-hydroxypropyl-cyclodextrin in the pregnant ewe. *Scientific World Journal*. 2013; 6.
68. Menjoge AR, Rinderknecht AL, Navath RS, et al. Transfer of PAMAM dendrimers across human placenta: Prospects of its use as drug carrier during pregnancy. *Journal of Controlles Release*. 2011; 150(3); 326–338.
69. Harris LK. Could peptide-decorated nanoparticles provide an improved approach for treating pregnancy complications? *Nanomedicine (Lond.)*. 2016; 11(17): 2235–2238.
70. Smit C, De Hoogd S, Brüggemann RJM, Knibbe CAJ. Obesity and drug pharmacology: A review of the influence of obesity on pharmacokinetic and pharmacodynamic parameters. *Expert Opinion in Drug Metabolism Toxicology*. 2018; 14(3): 275–285.
71. Michalaki MA, Gkotsina MI, Mamali I, et al. Impaired pharmacokinetics of levothyroxine in severely obese volunteers. *Thyroid*. 2011; 21(5):477–481.
72. Shank BR and Zimmerman DE. Demystifying drug dosing in obese patients. *American Society of Health System Pharmacists*. 2015; E4825; ISBN: 978-1-58528-482-5.
73. Yhee JY, Im J, and Nho RS. Advanced therapeutic strategies for chronic lung disease using nanoparticle-based drug delivery. *Journal of Clinical Medicine*. 2016; 5: 82.
74. Kolb M, Martin G, Medina M, Ask K, and Gauldie J. Gene therapy for pulmonary diseases. *Chest*. 2006; 130: 879–884.
75. Pavord ID, Korn S. Howarth P, et al. Mepolizumab for severe eosinophilic asthma (DREAM): A multicentre, double-blind, placebo-controlled trial. *Lancet*. 2012; 380: 651–659.
76. Boppana NB, Devarajan A, Gopal K, et al. Blockade of CXCR2 signalling: A potential therapeutic target for preventing neutrophil-mediated inflammatory diseases. *Experimental Biology and Medicine*. 2014; 239: 509–518.
77. Horiguchi M, Oiso Y, Sakai H, MotomuraT, and Yamashita C. Pulmonary administration ofphosphoinositide 3-kinase inhibitor is a curative treatment for chronic obstructive pulmonary disease by alveolar regeneration. *Journale of Controlled Release*. 2015; 213: 112–119.
78. . GINA. Global Initiative for Asthma 2017; 2017. http://ginasthma.org/. Accessed February 2018.
79. Education NA, Program P. Expert panel report 3 (EPR-3): guidelines for the diagnosis and management of asthma-summary report 2007. *The Journal of allergy and clinical immunology*. 2007 Nov;120(5 Suppl):S94–138.

80. Beasley R, Weatherall M, Shirtcliffe P, et al. Combination corticosteroid/βagonist inhaler as reliever therapy: A solution for intermittent and mild asthma? *Journal of Allergy and Clinical Immunology*. 2014;133: 39–41.
81. Dissanayake SB. Safety of beta2-agonistsin asthma: Linking mechanisms, meta analyses and regulatory practice. *AAPS Journal*. 2015; 17: 754–757.
82. Spitzer WO, Suissa S, Ernst P, et al. The use of beta-agonists and the risk of death and near death from asthma. *New England Journal of Medicine*. 1992; 326: 501–506.
83. Martinez FD, Chinchilli VM, Morgan WJ, et al. Use of beclomethasone dipropionate as rescue treatment for children with mild persistent asthma (TREXA): A randomised, double-blind, placebo-controlled trial. *Lancet*. 2011; 377: 650–657.
84. Bisgaard H, Le Roux P, Bjamer D, et al. Budesonide/formoterol maintenance plus reliever therapy: A new strategy in pediatric asthma. *Chest*. 2006; 130: 1733–1743.
85. Papi A, Canonica GW, Maestrelli P, et al. Rescue use of beclomethasone and albuterol in a single inhaler for mild asthma. *New England Journal Medicine*. 2007; 356: 2040–2052.
86. Papi A, Caramori G, Adcock IM, et al. Rescue treatment in asthma. More than as-needed bronchodilation. *Chest*. 2009; 135: 1628–1633.
87. Hancox RJ, Le Souef PN, Anderson GP, et al. Asthma: Time to confront some inconvenient truths. *Respirology*. 2010; 15: 194–201.
88. Price D, Fletcher M, and vande r Molen T. Asthma control and management in 8,000 European patients: The recognise asthma and link to symptoms and experience (REALISE) survey. *NPJ Primary Care Respiratory Medicine*. 2014; 24: 14009.
89. Barker, AF, et al. Obliterative bronchiolitis. *New England Journal Medicine*, 2014. 370(19): 1820–1828.
90. King, MS, et al. Constrictive bronchiolitis in soldiers returning from Iraq and Afghanistan. *New England Journal of Medicine*. 2011; 365(3): 222–230.
91. Ryu JH, Myers JL, and Swensen SJ. Bronchiolar disorders. *American Journal of Respiratory Critical Care Medicine*. 2003; 168(11): 277–292.
92. Akpinar-Elci M, et al. Bronchiolitis obliterans syndrome in popcorn production plant workers. *Euroepan Respiratory Journal*. 2004; 24(2): 298–302.
93. Corris PA, Ryan VA, Small T, Lordan J, Fisher AJ, Meachery G, Johnson G, and Ward C. A randomised controlled trial of azithromycin therapy in bronchiolitis obliterans syndrome (BOS) post lung transplantation. *Thorax*. 2015; 70(5): 442–450.
94. Ruttens D, Verleden SE, Vandermeulen E, Bellon H, Vanaudenaerde BM, Somers J, Schoonis A, Schaevers V, Van Raemdonck DE, Neyrinck A, and Dupont LJ. Prophylactic azithromycin therapy after lung transplantation: post hoc analysis of a randomized controlled trial. *American Journal of Transplantation*. 2016; 16(1): 254–261.
95. Pinnock H, Kendall M, Murray S, et al. Living and dying with severe chronic obstructive pulmonary disease: Multi-perspective longitudinal qualitative study. *BMJ*. 2011; 342: d142.
96. Kaptein AA, Sharloo M, Fischer MJ, et al. Illness perceptions and COPD: An emerging field for COPD patient management. *Journal of Asthma* 2008; 45: 625–629.
97. Barnett K, Mercer SW, Norbury M, et al. Epidemiology of multimorbidity and implications for health care, research, and medical education: A cross-sectional study. *Lancet* 2012; 380: 37–43.
98. Taylor SJC, Pinnock H, Epiphaniou E, et al. A rapid synthesis of the evidence on interventions supporting self-management for people with long-term conditions: PRISMS – Practical systematic review of self-management support for long-term conditions. *Health Services Delivery Research*. 2014; 2: 53.
99. Kielmann T, Huby G, Powell A, et al. From support to boundary: A qualitative study of the border between self-care and professional care. *Patient Educationa and Counseling*. 2010; 79: 55–61.
100. Fairbrother P, Pinnock H, Hanley J, et al. Continuity, but at what cost? The impact of telemonitoring COPD on continuities of care: A qualitative study. *Prim Care Respiratory Journal*. 2012; 21: 322–328.
101. Taylor SJ, Sohanpal R, Bremner SA, et al. Self-management support for moderate-to-severe chronic obstructive pulmonary disease: A pilot randomised controlled trial. *British Journal of General Practise*. 2012; 62: e687–e695.
102. Van Aken H, Mellin-Olsen J, and Pelosi P. Intensive care medicine: A multidisciplinary approach. *European Jorunal of Anaesthesiology*. 2011; 28: 313–315.
103. Ambrosino N, Venturelli E, Vagheggini G, et al. Rehabilitation, weaning and physical therapy strategies in chronic critically ill patients. *European Respiratory Journal*. 2012; 39: 487–492.
104. Corrado A, Roussos C, Ambrosino N, et al. Respiratory intermediate care units: A European survey. *European Respiratory Journal*. 2002; 20: 1343–1350.
105. Farre R, Lloyd-Owen SJ, Ambrosino N, et al. Quality control of equipment in home mechanical ventilation: A European survey. *European Respiratory Journal*. 2005; 26: 86–94.
106. Artigas A, Pelosi P, Dellweg D, et al. Respiratory critical care HERMES syllabus: Defining competencies for respiratory doctors. *European Respiratory Journal*. 2012; 39: 1294–1297.
107. Artigas A, Vassilakopoulos T, Brochard L, et al. Respiratory Critical Care HERMES: A European core syllabus in respiratory critical care medicine. *Breathe*. 2012; 8: 217–229.
108. European Society of Intensive Care Medicine. CoBaTrICE: In international competency based training programme in intensive care medicine. www.cobatrice.org/en/index.asp.

7

Understanding of Lung Diseases with a Focus on Applications of Nano-particulate Drug Delivery Systems

Shraddha Khairnar[1], Mudassir Ansari[1], Ujwala Shinde[2], Agnivesh Shrivastava[3], and Kavita Singh[1]
[1]*Shobhaben Pratapbhai Patel School of Pharmacy and Technology Management, SVKM's NMIMS University, Mumbai, India*
[2]*Department of Pharmaceutics, Bombay College of Pharmacy, Mumbai, Maharashtra, India*
[3]*Gattefosse India Pvt. Limited, Mumbai, Maharashtra, India*

Abbreviations

IPF	Idiopathic pneumonic fibrosis
COPD	Chronic obstructive pulmonary disease
WHO	World Health Organization
TB	Tuberculosis
CF	Cystic fibrosis
EMT	Epithelial-mesenchymal transition
ACIF	Airway-centered interstitial fibrosis
LHCH	Langerhans cell histiocytosis
RB	Respiratory bronchiolitis
ARDS	Adult respiratory distress syndrome
BPD	Broncho pulmonary dysplasia
GOLD	Global initiative for chronic obstructive lung disease
GINA	Global initiative for asthma
SABA	Short-acting β2-agonists
MRA	Muscarinic receptor antagonist
SAMA	Short-acting muscarinic receptor antagonist
SDMI	Spring-driven mist inhaler
ICS	Inhaled corticosteroids
NAC	N-acetylcysteine
LAMA	Long-acting muscarinic antagonist
PDE4	Phosphodiesterase-4
NPPV	Non-invasive positive-pressure ventilation
RCT	Randomized controlled trial
CFC	Chlorofluorocarbon
MMAD	Mass median aerodynamic diameter
PNAPs	Porous nanoparticle aggregate particles
MALT	Mucosa-associated lymphoid tissue
SLN	Solid lipid nanoparticles
PLGA	Polylactic-co-glycolic acid
LVRS	Lung volume reduction surgery
FEV	Forced expiratory volume
HLA	Human leukocyte antigen

7.1 Introduction

Lungs are the key organ of the respiratory system and considered a vital and vulnerable part of the breathing system which plays an essential function in oxygen exchange in the body (1). Because of its daily exposure to pollutants, irritants, and infectious agents, the respiratory system is more susceptible to foreign invaders and thus inflammation of and damage to the whole system. It is estimated that 1 billion people inhale polluted outdoor air and are exposed to tobacco smoke. Moreover, poor living conditions and exposure to environmental toxins increases vulnerability leading to respiratory impairment causing disability and death in all regions of the world (2). Respiratory diseases impose an immense worldwide health burden with lung ailments as the main source of mortality, killing around 4,000,000 individuals every year on the planet (1,3).

Lung diseases range from acute conditions to chronic illness, genetic and congenital anomalies to acquired ailments, allergic to nonallergic, and curable to controllable disorders. Among them are asthma, chronic obstructive pulmonary disease (COPD), and bronchitis which have been extensively explored for understanding of their pathogenesis and treatment strategies; however, there are some, like interstitial lung diseases (ILD) which still needed deeper understanding and more research. Today, nanotechnology is playing a major role in addressing the challenges for treating lung diseases, especially in administration of drugs to the respiratory tissues with a special emphasis on lung tissue. Some of the nano-delivery methods such as nanoparticles, liposomes, micelles, dendrimers, etc. are designed to be delivered to the specific site, avoiding undesirable distribution. Moreover, degradation, elimination from lung tissue, and changes in the lung physiology on administering drugs through nano-systems is well studied and understood.

Advancement in science and medicine has justifiably addressed this area; however, there is still a lacuna in the area where scientists need to do more work. Therefore, this chapter

DOI: 10.1201/9781003046547-7

highlights various lungs diseases, their pathologies, and the application of nano delivery systems in addressing those lung diseases or disorders.

7.2 Lung Diseases: A Brief Insight

Any health condition that prevents the lungs from working properly can be termed *lung disease* . Diseases of the lungs affect various regions of the organ that forms the basis of their classification (Figure 7.1), leading to difficulty in gaseous exchange and thus breathing. The diseases affecting the lungs differ in their origins, including the disorders hampering the airways, ailments of the interstitium and the pleural cavity, and defects in the vasculature of the pulmonary system. The four previously mentioned lung illnesses share qualities such as incessant, dynamic, diminished lung capacity, and irritation (1,4,5). On the other hand, smoking, air contamination, co-morbid conditions, and hereditary parameters are among the hazard factors that influence the development of these illnesses (6,7). Moreover, cough, dyspnea, chest tightness, shortness of breath, and mucus production are known as some of the mutual clinical indications of disease conditions (5,7,8).

7.2.1 Obstructive Lung Disease (OLD)

Obstructive lung disease (OLD) is characterized by difficulty in exhaling the air present in the airways and alveoli, thus causing reduction in airflow and shortness of breath in exhalation. The primary cause for this type of pulmonary disease is inflammation that ultimately leads to constrictions of airways, disallowing ease in exchange of gases. Inflammatory

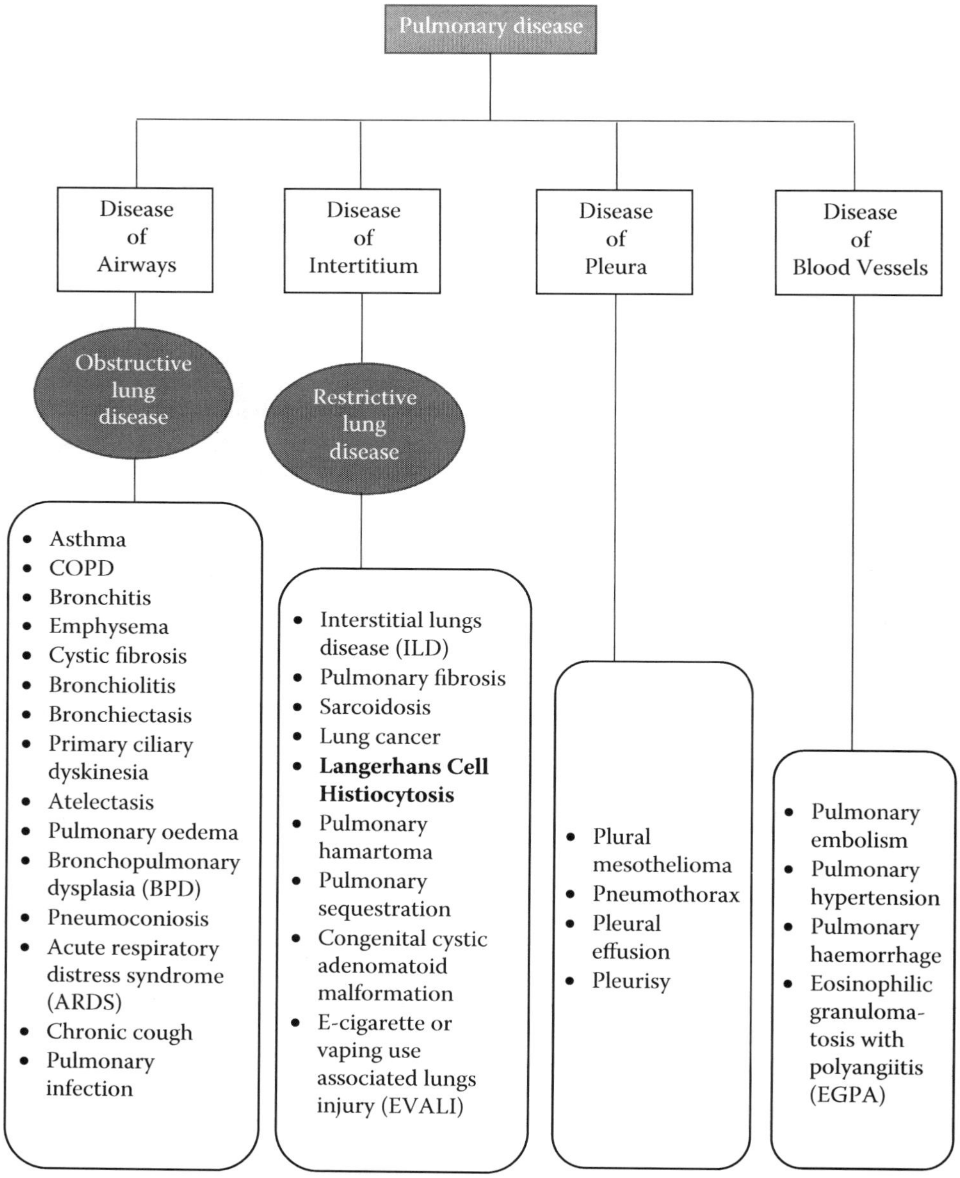

FIGURE 7.1 Classification of lung diseases

lung diseases can be divided into chronic and acute diseases. Acute inflammatory lung diseases are caused by environmental stimuli without the involvement of genetics. Airway disorders with chronic inflammation, such as asthma, COPD, and cystic fibrosis (CF), are categorized as chronic disorders which are affected by a combination of environmental, epigenetic, and genetic factors (9).

Asthma causes reversible constriction of airways (10) due to inflammation that leads to hyperactivity and tightness of the airway's smooth muscles with mucus accumulation in its lumen, making it narrower, thus difficult for air passage. Clinically it is characterized by shortness of breath, wheezing, and chest tightness. Asthma is a chronic inflammation caused by persistent infiltration of mast cells and eosinophils associated with the poor response of Th2 and associated cytokines. Th2 inflammatory mediators are all involved in airway hypersensitivity, increased mucus secretion, and high levels of IgE. The global incidence of asthma is reported to be 300 million people, constituting 21% of adults. Moreover, the incidence of asthma is increasing, mainly in children of less than 10 years (11). Both genetic and environmental factors play roles in the pathogenesis of the disease. Traditionally, asthma was divided into allergic (intrinsic) and nonallergic (extrinsic) asthma, but in recent years, within nonallergic asthma several so-called endotypes have been identified. Therefore, asthma is no longer regarded as a single disease but rather a syndrome (12). These endotypes differ concerning genetic susceptibility, environmental risk factors, age of onset, clinical presentation, prognosis, and response to treatment (13).

The next most common disease that causes deaths due to lung disorders is COPD, which has variable causes, from genetic alpha-1-antitrypsin deficiency to environmental factors such as dust and chemicals. Nearly 329 million people (approximately 5% of the world population) are struggling with COPD. The incidence rate from 1990 to 2015 has indicated an increase of 44.2% (1). Additionally, COPD is listed as the leading cause of morbidity by the World Health Organization (WHO) (6). In 2015, about 3.2 million people died of COPD and the death rate had increased by 11.6% compared to 1990 (1). Increased levels of inflammatory markers and neutrophils in blood circulation are characteristic features of COPD and are associated with oxidative stress (14). COPD is a chronic, uncurable condition which comprises two lung ailments: emphysema and chronic bronchitis. Emphysema is defined as an enlargement of alveolar spaces combined with the destruction and remodeling of the alveolar septa, usually resulting in the numerical loss of alveoli (15). Emphysema destroys the alveoli and hampers their function by making them stiff, thus preventing the air from getting out, whereas chronic bronchitis causes irritated, red, swollen airways with mucus overproduction due to inflammation causing cough with phlegm and shortness of breath (1,7).

Another deadly and fatal lung disease is *cystic fibrosis* (CF), a genetic disease that forms thick and sticky mucus that blocks the lung airways which precipitates into recurrent lung infections and breathing complications that ultimately damage lungs and cause patient death. Mutations in the CFTR gene are responsible for the disease having an autosomal recessive pattern (16). Cystic fibrosis mostly has its incidence in Caucasians where it was found to be 1:3600 in North America and Europe. Despite advances in treatment regimens, CF still has no cure and more than 90% of deaths occur as a result of lung failure (17).

Bronchiolitis is often associated with either bronchitis, such as in asthma, or with pneumonia, such as organizing pneumonia. However, there are two reasons to discuss bronchiolitis separately. Bronchiolitis is the underlying pathology of small airways disease, and furthermore, it does occur sometimes as an isolated disease. At present we best classify bronchiolitis into acute bronchiolitis, chronic bronchiolitis, COPD-associated bronchiolitis, and distinct forms of bronchiolitis, however there are several more types as shown in Table 7.1.

Bronchiectasis is a incurable disease that affects the cilia of the airways; moreover, it causes smooth muscle stretch in the airways creating a pocket that acts as a substratum of mucus and foreign particles. Since the ciliary functions are hampered, clearance of mucus and foreign particles is compromised, causing frequent pulmonary infections damaging to the airway.

TABLE 7.1

Types of Bronchiolitis

Sr. No.	Type of Bronchiolitis	Etiology
1	Bronchiolitis obliterans	Graft versus host disease, vascular collagen diseases, cardiac graft rejection, idiopathic disease
2	Organizing pneumonia	Unresolved bacterial pneumonia, inhalation of poisonous smoke, inhalation of insecticides / pesticides, inhalation of gastric juice, autoimmune disorders, toxicity of medications, idiopathic disease
3	Constrictive bronchiolitis	Graft versus host disease, vascular collagen disorders, cardiac transplantation rejection, drug reaction
4	Respiratory bronchiolitis and RB combined interstitial lung disease	Smoking cigarettes, rare idiopathic medication
5	Follicular bronchiolitis	Recurrent viral infection, immunodeficiency, autoimmune diseases, idiopathic immune defects (T cells or NK); HP / EAAA portion
6	Diffuse pan bronchiolitis	HLA system-related immune defect
7	Airway-centered interstitial fibrosis (ACIF)	Hypersensitivity pneumonia, vascular collagen infections, hazardous material inhalation, idiopathic

Primary ciliary dyskinesia is a genetic form of bronchiectasis affecting infants and children.

Atelectasis is defined as an alveolar collapse due to a lack of air filling. In newborns there exists a condition of primary atelectasis; however, normally the lung extends with the first inspiration and the alveoli are filled with air. In rare cases, this inspiration does not happen, mainly as the result of severe cerebral malformations. In other cases, primary lung injury, such as meconium aspiration, sepsis, or persistent pulmonary hypertension, can also cause severe or partial atelectasis (18). Secondary atelectasis can occur at any age after birth. The causes of atelectasis in childhood are infantile myofibromatosis (19), infantile bronchial obstruction or atresia (20,21), or compression by cysts, as in congenital adenomatoid pulmonary malformation (22).

Pulmonary edema happens due to the accumulation of fluid in the alveoli and its surrounding area due to the leakage of lung capillaries. Fluid enters the peripheral lung from the circulation via the interstitium into alveoli. Among the variable causes, the most common is congestion of the pulmonary circulation, most often caused by heart failure due to infarction, valvular diseases, and the like. In these cases, the venous flow into the left atrium is reduced, resistance in the venous part of the circulation increases, and leakage of the pulmonary veins increases. The gaps between the endothelial cells increase in size and serum gets into the interstitium and causes interstitial edema. Another important cause includes lung infections such as pneumonia and tuberculosis.

Bronchopulmonary dysplasia (BPD) is a chronic disease that occurs in newborns and infants causing scarring of the lungs, including alveoli and bronchi. It occurs in babies born prematurely when the lungs are underdeveloped, requiring the use of ventilators. Due to undeveloped lungs, the oxygen in the alveoli causes overstretching of the sac leading to damage of the airway lining; moreover, it affects the blood vessels surrounding them, leading to pulmonary hypertension. Infection and inflammation are major contributors to the pathogenesis of BPD, which is often initiated by a respiratory distress response, and exacerbated by mechanical ventilation and exposure to supplemental oxygen (23). Similar to Wilson-Mikity syndrome, infectious organisms such as cytomegalovirus (CMV) have been reported to cause BPD (24). Other risk factors include maternal smoking, preeclampsia, drug use, etc.

Pneumoconiosis, for example, asbestosis and coal worker disease, are lung disorders caused by the inhalation of a substance which damages the alveoli and causes scarring and stiffness, making it difficult to transport oxygen into systemic circulation.

Acute respiratory distress syndrome (ARDS) is a condition in whch the lungs suffer sudden depression and slow their function due to the existing condition, including severe lung infections like COVID-19 such that the person requires ventilators until recovery (25).

Chronic cough is more of a symptom than a disease that can last for more than 8 weeks due to underlying conditions such as tuberculosis, COPD, severe pneumonia, etc.

As far as *respiratory infections* are concerned, discussion of it is beyond the limit of this chapter, since they include a long list of bacterial, viral, and fungal diseases which includes COVID-19, aspergillosis, influenza, coccidioidomycosis, common cold, hantavirus, croup, pertussis, pneumonia, tuberculosis, respiratory syncytial virus, SARS, MERS, histoplasmosis, human metapneumovirus, Legionnaires' disease, mycobacterium avian complex (MAC) disease, and non-tuberculosis mycobacterium (NLM) lung disease.

7.2.2 Restrictive Lung Disease (RLD)

Restrictive lung disease (RLD) are characterized by the inability of the lungs to inhale, thus making the lungs unable to expand. The primary reason for this is the inflammation and fibrosis of the lung tissue, causing it to stiffen. One of the most prevalent RLDs is *interstitial lung disease* (ILD), a class of irreversible lung diseases causing lung scarring, thus preventing lungs from transporting oxygen into bloodstream. ILDs consist of various subclass of lung ailments such as idiopathic pulmonary fibrosis and sarcoidosis. ILDs are mostly due to inhalation of hazardous materials, like cigarette smoke, or through medications such as chemotherapy; the cause can also genetic. Shortness of breath is the most common symptom in addition to chest discomfort, dry cough, fatigue, and weight loss. Since the disease is progressive, the treatment involves managing the symptoms and improving quality of life.

Idiopathic pulmonary fibrosis (IPF) is a chronic disease characterized by excessive accumulation of extracellular proteins, such as collagen, in parenchymal tissue, fibroblast proliferation, and scarring of lung epithelial (26). The incidence of IPF is found to be approximately 23 in every 100,000 people. Parenchymal fibrosis is probably a product of activated alveolar epithelial cells that causes epithelial–mesenchymal transition (EMT), extracellular matrix construction, and accumulation of fibroblasts and myofibroblasts (10,27). Whereas *sarcoidosis* is an autoimmune inflammatory disorder causing granulomas in the lung and other body parts, which at its most severity causes heart failure.

Despite advances in diagnosis and treatment of *lung cancer*, this disease has the highest rate of cancer-related deaths worldwide. Though most lung cancer starts in the lungs, some cases start in other parts of the body and spread to the lungs. The two main types of lung cancer are small cell and non small cell, which grow and spread in different ways and hence each type may be treated differently. Like COPD, non small cell lung cancer (NSCLC) is thought to be generated due to long term exposure to cigarette smoke. Moreover, breathing second-hand smoke can also increase a person's chance of developing the disease.

Langerhans cell histiocytosis (LHCH, histiocytosis X, eosinophilic granuloma) is a very rare interstitial lung disease caused by excessive inhalation of tobacco smoke; tobacco plant antigens present within the tobacco smoke cause accumulation and proliferation of Langerhans cells (28,29). The continuous exposure of Langerhans cells (antigen presenting reticulum cell population) to plant proteins causes proliferation of these cells to keep up with the increasing number of antigens to be processed. Inhaled antigens are presented to LH cells, which are taken up and processed by specific mechanisms involving toll receptors and langerin, a molecule with a C-type lectin domain (30,31). Patients suffering from this disease experience acute respiratory failure and asphyxia.

Pulmonary hamartoma is a common benign tumor of lungs made up of connective tissue, cartilage, fat, muscle, and bone. *Pulmonary sequestration* is the outgrowth of tissues from lungs, also known as accessory lungs, but it is not connected with the bronchial tree or pulmonary artery. *Congenital cystic adenomatoid malformation* is a rare congenital form of pulmonary sequestration which appears before birth and hampers the normal breathing pattern of a fetus. *E-cigarette or vaping use associated lung injury (EVALI)* is a unique lung disorder identified in 2019 whereby the lung tissue is injured due to vitamin E acetate present in e-cigarettes.

7.2.3 Pleural Lung Disease

Disease of the pleura involves *pleural mesothelioma*, a rare form of pleural cancer that is usually due to inflammation of the pleural lining due to the inhalation of hazardous substances like asbestos fibres. Another common ailment is *pneumothorax* , also known as collapsed lungs, which occurs when there is an air leakage between the lungs and the chest wall cavity, making it impossible for the lungs to expand and work efficiently. *Pleural effusion* is a form of pneumothorax in which the cavity between lungs and chest wall is filled with fluid that can lead to heart failure. *Pleurisy*, also known as pleuritis, is an inflammation of both the pleural membranes, making them swollen and creating a gap between the two that leads to infection if filled with a fluid.

7.2.4 Vascular Lung Disease

Pulmonary embolism and *pulmonary hypertension* are the two most common diseases affecting the pulmonary blood vessels. Other important but less discussed diseases include *pulmonary hemorrhage* and *eosinophilic granulomatosis with polyangiitis* (EGPA), an inflammation of the tiny blood vessels of the lungs that damages the organ.

7.3 Management and Treatment of Lung Diseases

The treatment of pulmonary disease involves both pharmacological and non-pharmacological interventions since certain lung ailments are incurable and require supportive and lifelong interventions to improve the quality of life (Figure 7.2). Pharmacological measures involve the use of conventional and some newly approved therapies that include bronchodilators, corticosteroids, antimalarials, TNF inhibitors, etc., whereas non-pharmacological therapy includes oxygen support, ventilation, pulmonary rehabilitation, etc. Different lung diseases differ in their characteristic features, which means clinicians must make a wise selection of a particular therapy and a step-by-step strategy for disease management. This led to the formation the of Global Initiative for Asthma (GINA) and Global Initiative for Chronic Obstructive Lung Disease (GOLD) guidelines for the treatment of asthma and COPD, respectively, and other guidelines, too, for the management of other pulmonary ailments (32). Below is detailed information regarding the therapy used in different lung diseases with a special emphasis on asthma and COPD.

7.3.1 Pharmacological Approaches to Treating Lung Diseases

7.3.1.1 Bronchodilators

Bronchodilators are the mainstay agents for the management of lung disease associated with the airways, such as asthma, COPD,

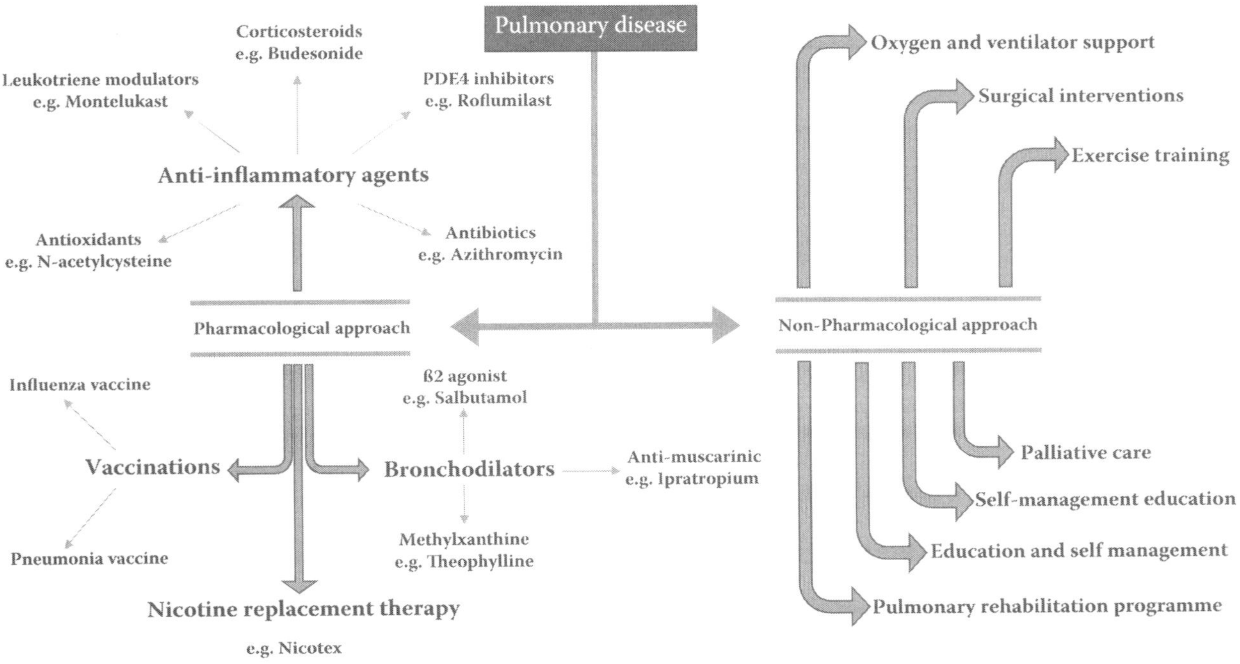

FIGURE 7.2 Treatment of lung diseases

bronchitis, atelectasis, emphysema, BPD, etc. Bronchodilators dilate, the airway smooth muscles thus increasing the force of the expiratory volume (FEV1), and reducing dynamic hyperinflation at rest and during exercise, thus improving exercise performance (33). These medications are usually given regularly to prevent or reduce the symptoms of bronchoconstriction associated with the underlying disease.

7.3.1.1.1 Beta-2 Agonists

Beta-2 agonists, including short-acting (SABA) and long-acting (LABA) agents, relax airway smooth muscle. The primary short-acting inhaled B2AR (beta-2 adrenergic receptor) agonist (SABA) used for bronchodilation is albuterol (known as salbutamol in Europe). Levalbuterol, the R-enantiomer of albuterol is also available for inhalation. The use of long-acting beta-2 agonists (LABA), includimg formoterol and salmeterol, has evolved over the last several years in the treatment of both asthma and COPD. Stimulation of beta-2 adrenergic receptors has been observed to produce resting sinus tachycardia and precipitate cardiac rhythm disturbances in susceptible patients. Moreover, exaggerated somatic tremor occurs in some patients treated with higher doses of beta-2 agonists.

7.3.1.1.2 Anti-muscarinic Drugs

Anticholinergic drugs, specifically anti-muscarinic agents, act on bronchi and inhibit the contraction of smooth muscle by acetylcholine. It was observed that the short-acting muscarinic antagonist (SAMA), provides small benefits over the short-acting beta-2 agonist in terms of lung function, health status, and the requirement for oral steroids. Ipratropium bromide is the major SAMA used in the US and Europe, and is available in nebulizer and spring-driven mist inhaler (SDMI) products. Unlike atropine, ipratropium bromide in the airway is poorly systemically absorbed and tends not to cross into the central nervous system, and as a result has limited systemic adverse effects. In addition to that, an unexpected small increase in cardiovascular events was reported in COPD patients regularly treated with ipratropium bromide. However, a large trial reported no differences in mortality, cardiovascular morbidity, or exacerbation rates when using tiotropium (34). As far as long-acting muscarinic antagonists (LAMA) such as tiotropium are concerned, clinical trials have shown a greater effect on exacerbation rates when compared to LABA treatment (35). Long-acting muscarinic antagonist (LAMA) treatment improves symptoms and health status (36), improves the effectiveness of pulmonary rehabilitation, and reduces exacerbations and related hospitalizations.

7.3.1.1.3 Methylxanthines

Theophylline exerts a modest bronchodilator effect in stable COPD, and improves FEV1 and breathlessness when added to salmeterol. There is limited and contradictory evidence regarding the effect of low-dose theophylline on exacerbation rates. Toxicity of theophylline is dose-related, which is a problem as most of the benefit occurs when near-toxic doses are given (37).

7.3.1.1.4 Combination Bronchodilator Therapy

Combining bronchodilators with different mechanisms and durations of action may increase the degree of bronchodilation with a lower risk of side effects compared to increasing the dose of a single bronchodilator. There are numerous combinations of long-acting beta agonists (LABA) and long-acting muscarinic antagonists (LAMA) in a single inhaler available. These combinations improve lung function compared to placebo and have a greater impact on patient-reported outcomes compared to monotherapies. It was found that LABA plus LAMA improves symptoms and health status in COPD patients (38) and is more effective than long-acting bronchodilator monotherapy for preventing exacerbations. Moreover, it was found that this combination decreases exacerbations to a greater extent than an inhaled corticosteroid (ICS)/LABA combination (39). As far as Asthma–COPD Overlap Syndrome (ACOS) is concerned, a combination of inhaled products containing the SABA albuterol or salbutamol with the SAMA agent ipratropium can be used as a relief medication or as a maintenance medication if taken every 6 hours. These combined inhaled products are available by nebulization (NEB) or by spring-driven mist inhaler (SDMI).

7.3.1.2 Anti-Inflammatory Agents

Anti-inflammatory drugs are the first line therapy for the management and treatment of both obstructive lung disease and restrictive lung disease, since inflammation is the primary reason for the emergence of pulmonary disorder. The disorders where these agents are used include asthma, COPD, chronic bronchitis, BPD, cystic fibrosis, EVALI, emphysema, EGPA, IPF, ILD, sarcoidosis, pleurisy, etc. (40). These agents act on various targets, hence are classified accordingly, such as corticosteroids, phosphodiesterase-4-inhibitors, antibiotics, leukotriene modulators, and antioxidants, the details of which are discussed below.

7.3.1.2.1 Corticosteroids

Inhaled corticosteroids (ICS), for example, budesonide and fluticasone as initial maintenance therapy, is advocated for the treatment of both adult and childhood asthma (41). A recent set of trials (TRIMARIN and TRIGGER) in uncontrolled asthma patients found that the triple fixed dose combination (FDC) inhaler (ICS + LABA + LAMA) was superior to a double FDC inhaler (ICS + LABA) in reducing asthma exacerbations and improving lung function (42). Short courses of less than a week of systemic corticosteroids (oral or parenteral) remains the mainstay therapy for exacerbations of COPD (43,44). However, long term daily treatment of COPD with oral glucocorticoids lacks benefit due to a high rate of systemic complications. It was reported in clinics that an ICS combined with a LABA is more effective than the individual components in improving lung function and health status, and reduces exacerbations in patients with moderate to very severe COPD. Additionally, triple inhaled therapy of ICS/LAMA/LABA improves lung functions, symptoms and health status, and reduces exacerbations compared to ICS/LABA or LAMA monotherapy. Nevertheless, survival is not affected by combination therapy (45).

7.3.1.2.2 Phosphodiesterase-4-Inhibitors (PDE4 Inhibitors)

PDE4 inhibitors, such as roflumilast have been studied and proposed in asthma patients because of their anti-inflammatory mechanisms (46). Roflumilast, use in patients with asthma, is associated with reduced allergen-induced acute bronchoconstriction as measured by both forced expiratory volume (FEV) in asthma patients and modulation of inflammatory biomarkers (47). Roflumilast reduces moderate and severe exacerbations treated with systemic corticosteroids in patients with chronic bronchitis and severe to very severe COPD. However, phosphodiesterase-4 (PDE4) inhibitors have more adverse effects than inhaled medications for COPD. The most frequent are diarrhea, nausea, reduced appetite, weight loss, abdominal pain, sleep disturbance, and headache. Furthermore, roflumilast should be avoided in underweight patients and used with caution in patients with depression (48,49).

7.3.1.2.3 Antibiotics

Azithromycin (250 mg/day or 500 mg three times per week) or erythromycin (500 mg two times per day) for 1 year reduces the risk of exacerbation in patients suffering from COPD. Since bacteria too resides in the patient airways causing purulent sputum that eventually causes pneumonia. Azithromycin also showed a reduction in exacerbation rate in former smokers but is associated with an increased incidence of bacterial resistance. However, pulse moxifloxacin therapy in patients with chronic bronchitis and frequent exacerbations does not reduce exacerbation rate.

7.3.1.2.4 Leukotriene Modulators

Leukotriene pathway modifiers, including both leukotriene synthesis inhibitors (e.g. meclofenamate sodium) and receptor antagonists (e.g. montelukast) are widely available and used in asthma and other atopic diseases, however, they have not been adequately tested in COPD (32,50).

7.3.1.2.5 Antioxidants

Inhaled carbocysteine and N-acetylcysteine (NAC) are classified as mucolytics since they reduce sputum viscosity and elasticity, improving mucociliary clearance. Additionally, they also possess anti-inflammatory properties and are considered antioxidants. Regular treatment with them reduces exacerbations and modestly improves health status in patients not receiving ICS (49). NAC is used orally and intravenously as a glutathione replacement in the prevention and treatment of acetaminophen-induced hepatitis. The innate antioxidant protection in airway tissue is thought to be in part from its glutathione-like characteristics such that a transformation of its thiol group from reduced to oxidize state provides a potent antioxidant effect (51). In addition to these, macrolide antibiotics such as azithromycin have an immuno-modulatory function in addition to their anti-bacterial effects (52).

7.3.1.3 Vaccinations

Influenza vaccination reduces serious illness, death, the risk of ischemic heart disease (53), and the total number of exacerbations (54). Influenza vaccination containing either killed or live inactivated viruses is recommended for all patients with COPD as they are more effective in elderly patients with COPD. Pneumococcal vaccinations, PCV13, and PPSV23 are recommended for all patients > 65 years of age. The PPSV23 is also recommended for younger COPD patients with significant comorbid conditions, including chronic heart or lung disease (55).

7.3.1.4 Nicotine Replacement Therapy

Nicotine replacement therapy involves the use of nicotine in a form (e.g. gums, sprays) other than tobacco, mainly for smoking cessation to combat the withdrawal symptoms so as to prevent and treat diseases associated with it. The therapy is more beneficial and better than the other placebo effects in treating smoking-related lung disease (56–58).

7.3.2 Non-pharmacological Approach

7.3.2.1 Oxygen Therapy and Ventilatory Support

7.3.2.1.1 Oxygen Therapy

The long-term administration of oxygen (>15 hours per day) to patients with chronic respiratory failure increases survival in patients with severe resting hypoxemia. Long-term oxygen therapy provides sustained benefit for any of the measured outcomes in patients with stable COPD and resting or exercise-induced moderate arterial oxygen desaturation (59).

7.3.2.1.2 Ventilatory Support

Whether to use non-invasive positive-pressure ventilation (NPPV) chronically at home to treat patients with acute or chronic respiratory failure following hospitalization remains undetermined. Randomized controlled trials (RCTs) have yielded conflicting data on the use of home NPPV on survival and re-hospitalization in chronic hypercapnic COPD. In patients with both COPD and obstructive sleep apnea, continuous positive airway pressure improves survival and avoids hospitalization (60).

7.3.2.2 Surgical Interventions

7.3.2.2.1 Lung Volume Reduction Surgery

An RCT confirmed that COPD patients with upper lobe emphysema and low post-rehabilitation exercise capacity experienced improved survival when treated with lung volume reduction surgery (LVRS) compared to medical treatment. In patients with high post-pulmonary rehabilitation exercise capacity, no difference in survival was noted after LVRS, although health status and exercise capacity were improved. LVRS has been demonstrated to result in higher mortality than medical management in severe emphysema patients with a forced expiratory volume (FEV1) ≤ 20% (61)).

7.3.2.2.2 Bullectomy

In selected patients with relatively preserved underlying lung, bullectomy (surgical removal of lung parenchyma filled with air with a size of more than 1 cm) is associated with decreased dyspnea, and improved lung function and exercise tolerance.

7.3.2.2.3 Lung Transplantation

In selected patients, lung transplantation has been shown to improve health status and functional capacity but not has been seen to prolong survival. However, bilateral lung transplantation has been reported to result in longer survival than single lung transplantation in COPD patients, especially those < 60 years of age (62). In selected patients with very severe COPD and without relevant contraindications, lung transplantation may be considered (63).

7.3.2.2.4 Bronchoscopic Interventions

Less invasive bronchoscopic approaches to lung reduction have been developed. In selected patients with heterogeneous or homogenous emphysema and significant hyperinflation refractory to optimized medical care, surgical or bronchoscopic modes of lung volume reduction (e.g. endobronchial one-way valves or lung coils) may be considered. Prospective studies have shown that the use of bronchial stents is not effective while the use of lung sealant caused significant morbidity and mortality (64).

7.3.2.3 Education and Self-Management

An individual patient's evaluation and risk assessment (e.g. exacerbations, patient's needs, preferences, and personal goals) should aid the design of personalized self-management.

7.3.2.4 Pulmonary Rehabilitation Programs

Patients with high symptom burden and risk of exacerbations should take part in a full rehabilitation program that considers the individual's characteristics and comorbidities (65).

7.3.2.5 Exercise Training

A combination of constant load or interval training with strength training provides better outcomes than either method alone. Adding strength training to aerobic training is effective in improving strength, but does not improve health status or exercise tolerance. Upper extremity exercise training improves arm strength and endurance, and improves capacity for upper extremity activities (66).

7.3.2.6 Self-Management Education

An educational program should include smoking cessation, basic information about COPD, aspects of medical treatment (respiratory medications and inhalation devices), strategies to minimize dyspnea, advice about when to seek help, and possibly a discussion of advance directives and end-of-life issues.

7.3.2.7 Palliative Care

Patients should be informed that should they become critically ill, they or their family members may need to decide whether a course of intensive care is likely to achieve their personal goals for care. Simple, structured conversations about these possible scenarios should be discussed while patients are in a stable state (67).

7.4 Application of Nano-Drug Delivery Systems in Lung Diseases

The use of nanocarriers in drug delivery has seen a growing interest in recent times for respiratory products (68). Nanocarriers are classified, in particular, as drug carriers for controlled release and guided distribution. After the NIH launch of the National Nano Initiative in 2000, work on nanocarriers in the field of drug distribution is growing. ISO describes *nanoscale* as a size range from approximately 1 to 100 nm, and *particles*, as described by the European Commission, are known to be a *minute piece of matter with specified physical limits* . A nanomaterial as described in this recommendation should consist of 50% or more of particles having a size between 1 nm and 100 nm. Nanoparticles may be characterized as particulate drug carrier networks that contain the drug and consist of excipient material—sometimes a polymer—to a significant degree that covers, protects, or functionalizes the particulate matter or a nanoscale substance consisting (mostly) of the drug, such as nanocrystals (69).

7.4.1 Characteristics of Pulmonary Nano-Drug Delivery

7.4.1.1 Pulmonary Distribution of Drug

Nanoparticles can achieve a fairly uniform drug delivery across alveoli. It is especially valid if nanoparticles may be distributed to produce an ultra-fine aerosol; it is well known that this results in the more peripheral deposition, hence reduce dosage will be required for therapeutic benefit. Specific deposition trends have been observed for pulmonary nanoparticles, whereby smaller particles enter the peripheral lung and larger specimens are preferentially deposited in the central region. Lung distribution usually needs particle sizes between 1 and 5 µm (aerodynamic size)—which ensures that nanoparticles will only be distributed and successfully stored in the lung if larger particles or droplets are transported inside (70).

7.4.1.2 Improved Solubility/Dissolution Rate

Drug nanocrystals (nanoscale active pharmaceutical ingredient, API) can exhibit improved solubility/dissolution rate of the product and thus increase diffusion and permeation. It has been shown to boost the bioavailability of improperly water-soluble medications delivered orally, which may theoretically even be used to enhance the potency of medication after administration to the lungs (71).

7.4.1.3 Sustained Release Properties

Nanocarriers may act as a medication reservoir and thus can be used to regulate the rate of drug release in the lungs continuously and in a controlled manner. This may also result in decreased dosing duration.

7.4.1.4 Delivery of Macromolecules

Nanocarriers may be used as a sensitive device for their API load, thus enabling the delivery of macromolecules which are highly susceptible to uncertainty without additional security.

7.4.1.5 Internalization by Cells

Due to their small size, nanocarriers can enter as untouched, undissolved particles (72) by phagocytosis/macro pinocytosis. This provides the capacity for transmission of intracellular drugs after cell internalization. Furthermore, if nanocarriers are functionalized with targeting moieties, this may allow for cell-specific targeting and delivery. There are mainly pulmonary macrophages and dendritic cells in the respiratory system which are part of the pulmonary mucosal immune system. Nanocarriers, especially nanoparticles, can therefore be an alternative for the supply of immune-active APIs (e.g. vaccination). Seeing that macrophages play a significant role in the development of lung tuberculosis disease, targeting macrophages for (sustained) intracellular antibiotic transmission is another fascinating application.

7.4.2 Fate of Nanocarriers in the Lungs

Once the nano formulation reaches the lung, deposits and nanocarriers are made available from the dosage form; they will be subjected to lung clearance processes such as mucociliary clearance and macrophage clearance (73). In the conducting airways, the lung epithelium is lined with ciliated cells and goblet cells that create a mucus coating which offers efficient mechanisms to remove contaminants from the lung. The pulmonary mucus could be outlined as a tightly packed network of highly heterogeneous pore sizes from 100 nm to several micrometers allowing multipurpose small-affinity interactions via negatively charged mucin and uncharged regions side chains of glycan. It also shows some size-filtering effect and electrostatic/hydrophobic interactions. Furthermore, phagocytose particulate content is found in macrophages and pulmonary dendritic cells (DCs). Because macrophages do not move to local lymph nodes, content that ends up there is often lost unless the target inside the macrophages is pathogens. Dendritic cells, on the other side, may absorb and present materials to the immune system via the lymphatic system, rendering this route important for an immune distribution (74). Clearance is much slower in the lung periphery (in the alveolar region), due to the absence of mucociliary clearance. The primary approach here is the incorporation of contaminants into the phagocytotic cells. Furthermore, alveolar dendritic cells are said to be more effective in the presentation of antigen relative to DCs found in the conducting airways (75).

7.4.3 Strategies to Overcome Clearance of Nanocarriers

Different strategies are addressed to prevent lung clearing processes. In addition to the scale and location-specific uptake or avoidance of uptake, mucoadhesive and extremely penetrative nanoparticles were investigated, respectively. Mucoadhesion was historically used as a way of reducing clearance. On the other hand, experiments have shown that tiny particles having an overall negative charge on the hydrophilic surface (virus-like particles) penetrate the mucus and thus prevent mucociliary clearance improving contact of particles with the underlying cells. PEG-coating is also used to allow further penetration, as it provides the particles with a clear, hydrophilic surface. Reports observe that particles of PEG-coated polystyrene had improved diffusion in human mucus in vitro relative to those that were COOH-modified. Moreover, Schneider et al. (76) showed that PEG-coated PLGA particles had enhanced lung retention in vivo in a mouse model compared to free poly (lactic-co-glycolic acid) (PLGA) particles.

7.4.4 Nano-Formulation for Drug Delivery to Lungs

Nanomedicine is emerging very rapidly in two areas where the impacts are likely to be most significant: the diagnostic field and the medicine and new treatment area (77). Indeed, various nanocarriers are in clinical use or under development to develop systems for imaging and drug delivery (Table 7.2). Nano vectors may allow overcoming several challenges in drug delivery including protect them from degradation, a controlled release of therapeutics enabling persistent drug delivery and treatments at diminished doses, targeted drug delivery which allows reducing administration doses and thus lowering side-effects, toxicity, and exposure of non-target organs. However, certain drugs are difficult to be efficiently transported through physiological barriers and are administered systemically, which results in possible metabolism and clearance in the liver. Moreover, higher dosing is necessary to achieve therapeutic effects with possible target and non-target organ toxicity. An important field of nanomedicine research is consequently the development of new treatment strategies for a controlled, selective and efficient transport of drugs through biological barriers such as the lungs. However, the safety of nanomedicine is a principal issue of concern as shown by in vivo and in vitro studies which demonstrated induction of toxicity in multiple organ systems and thus have adverse health effects that have been poorly studied so far (94).

7.4.4.1.1 Polymeric Nanoparticles

Sustained-release products with greater intrinsic durability are polymeric stable particles. Due to its biodegradability and nontoxicity, polylactic-co-glycolic acid (PLGA) nanoparticles are of prime importance. Particulate PLGA has a robust shape with a uniform particle size and morphology (75). Degradation times may be adjusted to match the option of polymer and processing method (95). In one of the reports prolonged-release inhaled PLGA nanoparticle was formulated for the cell-specific treatment of tuberculosis (96). Rifampicin is a common antibiotic in TB care, which is administered orally for months. The explanation behind this repetitive administration is mycobacterium tuberculosis which exists within the pulmonary macrophages where an appropriate antibiotic delivery

TABLE 7.2

Application of Nanocarriers for Diagnosis and Therapies of Pulmonary Diseases

Sr. No.	Nanocarriers	Route	Description	Use	Reference
1	Respirocytes	Hypodermal injection	Nanodevices which act as red blood cells, but with greater effectiveness	Oxygen supply to tissues	(78)
2	Quantum dots	Intra-abdominal organs space	Nanocrystals which are produced to fluoresce; Stimulated by illumination	Lung cancer imaging	(79)
3	Fullerenes	Intra-tracheal bolus	Water-soluble C60 fullerenes	Blocking allergic reaction	(80)
4	INGN401	Intravenous	Tumor suppression gene nano particulate formulation FUS1	Lung cancer	(81)
5	ABRAXANE®	Intravenous infusion	Taxane albumin-bound particles	Non small cell lung cancer	(81)
6	Liposomes	Intravenous	Uni-multilamellar spherical nanoparticles made of lipid bilayer membranes	Cancer chemotherapy/ gene therapy	(81)
7	Poly PLA homopolymers conjugated with PEG	Intravenous	Betamethasone encapsulated by poly PLA homopolymers	Asthma	(82)
8	PEG-PLGA	Subcutaneous injection	Nanoparticle compacted with NF-κB	Pulmonary arterial hypertension	(83)
9	PLGA and VS(72)-10		Salbutamol-loaded polymeric nanoparticle	Respiratory diseases	(84)
10	Gelatin nanoparticles	Oral	Natural polymer encapsulated with rifampicin	Tuberculosis	(85)
11	Poly(L-aspartic acidco-lactic acid)/DPPE co-polymer NPs	Intraperitoneal injection	Amphiphilic biodegradable poly(L-aspartic acid-co-lactic acid)/DPPE co-polymer NPs loaded with doxorubicin (DOX)	Lung cancer	(86)
12	PEG-dendritic block telodendrimer	Intravenous injection	Self-assembling nanoparticles containing Dex	Allergic Asthma	(87)
13	pDNA nanoparticles (NPs)	Intranasal	Chitosan/IFN-gamma pDNA NPs (CIN)	Allergic Asthma	(88)
14	poly (DL-lactidecoglycolide) NPs	Inhalation	poly (DL-lactideco-glycolide) loaded with ATDs	Tuberculosis	(89)
15	Polybutyl cyanoacrylate NPs	Intravenous injection	DOX-loaded NPs were incorporated into inhalable effervescent and non-effervescent carrier particles using a spray-freeze drying technique	Lung cancer	(90)
16	Poly (beta-amino ester) (PBAE) polymers	Intratumoral injection	Biodegradable PBAE polymers that self-assemble with DNA	Lung cancer	(91)
17	LPH (liposomepolycation-hyaluronic acid) nanoparticles	Intravenous injection	LPH nanoparticle formulation modified with tumor-targeting single-chain antibody fragment for systemic delivery of siRNA and micro RNA efficiently downregulated the target genes(c-Myc/MDM2/VEGF)	Cancer lung metastasis	(92)
18	Nanodiamond	Subcutaneously into the right flank area	The carbon nanomaterial nanodiamond (ND) is nontoxic and biocompatible because cytotoxicity is not caused in lung cells	Labeling and tracking of cancer cells and lung cancer therapy	(93)

is very challenging (97). However, when the medication is delivered through inhalation, local concentration rises, but rifampicin is quickly extracted from the lungs upon dissolution. But when rifampicin is manufactured in PLGA nanoparticles [e.g. as porous nanoparticle aggregate particles' (PNAPs)] and delivered by the inhaled pathway, particles are ingested in lung macrophages at the effector site substantially increasing local drug concentration (98).

7.4.4.1.2 Antigenic Nanoparticles

Nanoparticles are also tested for the transmission of macromolecules to the lungs (99), where they have proven viable for the distribution of siRNA (100) including micro RNA. A unique feature of macromolecule distribution throughout the process of mucosal vaccination is the transmission of antigens into the lungs. The goal here is to promote local antigen storage and presentation by the mucosal immune system (dendritic cells and also the mucosa-associated lymphoid tissue (MALT). This provides a special immune system that is distinguished by a very controlled humoral or cellular response and a specifically secreted antibody, sIgA (101). This form of immune response is particularly beneficial in defending against infectious diseases, intracellular bacteria, infections with parasites, etc. The availability of a particulate antigen is an essential requirement for local production, which may resemble a natural pathogen and therefore has an adjuvant operation. However, a soluble antigen migrates into the lymph nodes and does not cause immunity; therefore, it is a common goal for formulating nanoparticles that contains antigens. Moreover, compared to parenteral vaccines, vaccination via the respiratory tract further has the beauty of non-invasiveness, which does not require medical personnel and is not associated with needle-stick injuries or infections (102). Besides, no sterile liquid formulations are required with this expanding formulation option (e.g. dry powder) which may further offer increased storage stability. As such, it is a perfect way of vaccination for remote areas of the world. Additionally, to further increase specificity in delivery, the size of nanoparticles can be used for the avoidance of macrophage phagocytosis or specific uptake into macrophages, respectively. It is known that dendritic cells take up particles from the low nanometer range up to the micrometer scale, whereas macrophages prefer particles larger than 500 nm. The latter will hardly get into the lung unless they are large porous particles' having a low density and thus small aerodynamic size despite a larger geometric diameter. In turn, macrophage targeting can be achieved with particles between 500 nm and 3 μm, whereas for preferential DC uptake in the course of mucosal vaccination, particles smaller than 500 nm should be used (103).

7.4.4.1.3 Solid Lipid Nanoparticles

The advantage of local delivery to the target site with the reduction of systemic side effects is especially attractive for the therapies with chemotherapeutics, therefore, lung cancer is a natural target for local respiratory therapy. However, so far chemotherapeutics are mostly administered systemically by intravenous infusion, possibly due to difficulties of water-insoluble materials to formulate them for respiratory delivery. This problem can be overcome by formulation of nanoparticles which not only increases the therapeutic index but also controls the release of a chemotherapeutic agents. This would allow the delivery of a chemotherapeutic to its site of action, which means efficient local doses and low side effects. Numerous attempts in different stages have been made to use nanoparticles for the treatment of lung cancer, and these are excellently reviewed elsewhere (104,105). Interestingly, systemic side effects were diminished, while local side effects (e.g. pharyngitis) remained pronounced, especially for liposome-based therapies with hydrophilic drugs in clinical studies indicating that this system might be too fluid to efficiently inhibit drug leakage and, with this, side effects in non-targeted areas of the respiratory tract. Delivery by nebulization, as mostly used in studies, will probably further increase liposomal membrane instabilities and add to the problem (106). To overcome these, solid nanocarriers for chemotherapeutics might be an alternative with a range of different polymeric, inorganic, and lipid-based systems. One promising lipid-based system showing increased stability compared to liposomes along with maintaining the physiologic-alike lipid composition, was described for paclitaxel by being incorporated in solid lipid nanoparticles (SLN). These are prone to very low side effects and showed remarkable retention times for the drug in vivo in rodents (twice weekly administration) which was attributed to increased cell uptake of those 100 nm SLNs. When nanocarriers are combined with targeting moieties such as ligands, antibodies, or antibody fragments, and are delivered locally to the lungs, preferential uptake into cancer cells was observed.

7.4.4.1.4 Depots

If nanoparticles are to be used for continuous release to prevent repeated dosing, the formulation must be capable of creating depots to remain at the release/action site (107). However, in this type of formulation, usually a greater dosage of the API is required to enable product delivery for a longer period. Moreover, due to their small size and large surface, nanoparticles cannot be deemed an ideal preparation for a depot, and fast diffusion speed and gradual dissolution remain other blocks. Therefore, to overcome these, suitable excipients (e.g. insoluble polymers or lipids forming a diffusion barrier or biodegradable polymers) are required to produce a nanoparticle with sustained-release properties, which often reduces the total mass of material to be applied for continued release (108).

7.4.4.1.5 Pressure Sensitive Metered Dose Inhalers (pMDIs)

With the reformulation of CFC-containing pMDIs after the 1987 prohibition on chlorofluorocarbon (CFC) propellants in the Montreal Agreement, it has been observed that the precipitation of API from pMDI formulations was altered due to the specific properties of hydrofluoroalkane (HFA) propellants (109). Many APIs which had been formulated in the old CFC-pMDIs as a suspension had a higher solubility in HFA which enabled them to be formulated as a solution rather than suspensions. Besides, the overall aerodynamic scale was calculated not by the particle size of micronized API in the formulation but by the droplet size and evaporation behavior Salinity at the exit of the nozzle is faster due to the higher vapor pressure, which contributes to smaller droplets and smaller solidified particles. Those, in effect, are retained in the oropharynx to a smaller degree and are dispersed more peripherally in the lung. Such formulations have a mass median aerodynamic diameter (MMAD) about 1 μm suggesting that the nanometer size is a large proportion of particles. Those drugs are potentially the first nano-particulate formulations for respiratory distribution. Beclomethasone propionate (QVAR®) is an indicator of a drug comprising such "ultrafine fragments" and it has been seen that lung deposition rises from 8% to 56% (110).

7.4.4.2 Vesicular Nano-Drug Delivery to Lungs

7.4.4.2.1 Liposomes

Throughout the years the FDA and other authorities have licensed growing amounts of nano-drug carriers (111), with Doxil® (liposomal doxorubicin) becoming the first nano-pharmaceutical to be licensed in 1995 (112). Liposomes are auto-assembled vesicles in an aqueous media, formed from phospholipids and cholesterol. These consist of a lipid bilayer structure and have an inner aqueous nucleus (113,114). They can be used in parenteral supply to raise the therapeutic index and the same is applicable to the respiratory system. An indication of this is Curosurf®, a liposomal defensive alfa preparation for acute respiratory distress syndrome therapy. It is delivered by instillation and makes improved distribution with a limited amount, thus showing decreased toxicity relative to non-liposomal preparation (115). Furthermore, the controlled release of drug can be obtained with antibiotic liposomal preparations for the local management of lung infections. An example of which includes a colloidal formulation composed of hydrogenated soy phosphatidylcholine (HSPC), unilamellar small liposomes, and 50 mg/ml cholesterol-containing ciprofloxacin (Lipoquin®) for nebulized delivery (116).

Due to the very limited size of the liposomes (50–100 nm), the formed droplets are dispersed very uniformly during nebulization and stay intact, while larger liposomes are considered to be (partially) disturbed when nebulized. Clinical research has demonstrated that this approach helps the dosage level to be decreased to one-day care while retaining a strong local concentration. A further illustration in this axis is liposomal amikacin (Insmed's Arikayce®) (117). This formulation was well received and demonstrated an improved concentration and clinical enhancement of intracellular products. A liposomal amphotericin B (Ambisome®) is being applied to the lung via nebulisation in phase 2 clinical trials to combat allergic bronchopulmonary aspergillosis. Nonetheless, liposomes as vesicles based on phospholipids exhibit some fluidity that make them inherently unstable. Furthermore, they are prepared from dispersion and thus colloidal stability (storage) is an essential parameter to be considered. To improve storage flexibility, liposomes might be converted either by freeze-drying or spray-drying to a dry powder composition. Spray-drying has the benefit of producing individual microparticles, the scale of which can be modified by modifying the surface features suited to the inhalation criteria (118).

7.4.4.2.2 Nanoemulsions

Nanoemulsions were also examined for the delivery of drugs by nebulisation to the lungs (119). They can be particularly useful when administering low melting point drugs (e.g. CoQ10) as a formulation technique. These can lead to improved therapies with decreased occurrence of side effects and better patient compliance with the above-mentioned benefits of nanoparticles.

7.4.4.3 Advantage of Particulate Drug Delivery to Lungs

Nanoparticles in aqueous dispersion will not be stable but tend to agglomerate, which ultimately results in non-redispersible aggregates. Thus, an important aspect of the formulation of nanoparticles is the physical stabilization of individual nanoparticles. This can be achieved by the embedding of nanoparticles into a solid matrix by spray drying. By this approach, microparticles can be used to tailor size and surface characteristics to optimize aerodynamic properties and ensure lung deposition at the desired target. Other approaches include the formation of "Trojan particles" in which nanoparticles form the shell of a hollow dry microparticle or the assembly of "porous nanoparticle aggregates particles" (PNAP). These approaches are especially feasible for "solid" polymeric or lipid nanoparticles, whereas fluidic vesicular systems such as liposomes and micelles are more difficult to transfer to a dry powder while maintaining their particulate nature (120,121).

7.5 Conclusion

Asthma, COPD, acute lower respiratory tract infections, TB, and lung cancer are the most severe cause of mortality in the world. Understanding the pathogenesis of pulmonary disease is essential for its proper management and treatment. The emergence of nanocarriers in the field of medicine has played a major role in addressing the challenges faced by current clinical therapy or diagnosis. Thus, understanding various aspects of nano-drug delivery from its preparation to its fate in vitro and in vivo is essential for the drug to reach the target site to render its efficacy while improving its safety. However, the intervention of nano-drug delivery in treating lung disorders in clinics is still lacking, although the demands for the same is very high, perhaps due to the high cost of its production. Some products in the market have shown success in treating lung diseases and many more are in the pipeline. Thus, further research in this field would be very much helpful in understanding both the pathogenesis and a nano-drug delivery system for the management and treatment of lung diseases.

REFERENCES

1. Soriano JB, Abajobir AA, Abate KH, Abera SF, Agrawal A, Ahmed MB, Aichour AN, Aichour I, Aichour MTE, and Alam K. Global, regional, and national deaths, prevalence, disability-adjusted life years, and years lived with disability for chronic obstructive pulmonary disease and asthma, 1990–2015: A systematic analysis for the Global Burden of Disease Study 2015. *The Lancet Respiratory Medicine*. 2017; 5(9): 691–706.
2. Wang H, Naghavi M, Allen C, Barber RM, Bhutta ZA, Carter A, Casey DC, Charlson FJ, Chen AZ, and Coates MM. Global, regional, and national life expectancy, all-cause mortality, and cause-specific mortality for 249 causes of death, 1980–2015: A systematic analysis for the Global Burden of Disease Study 2015. *The Lancet*. 2016; 388(10053): 1459–1544.
3. Chuchalin AG, Khaltaev N, Antonov NS, Galkin DV, Manakov LG, Antonini P, Murphy M, Solodovnikov AG, Bousquet J, and Pereira MHS. Chronic respiratory diseases and risk factors in 12 regions of the Russian Federation. *International Journal of Chronic Obstructive Pulmonary Disease*. 2014; 9: 963.
4. García-Sancho C, Buendía-Roldán I, Fernández-Plata MR, Navarro C, Pérez-Padilla R, Vargas MH, Loyd JE, and Selman M. Familial pulmonary fibrosis is the strongest risk factor for idiopathic pulmonary fibrosis. *Respiratory Medicine*. 2011; 105(12): 1902–1907.
5. Vestbo J, Anderson J, Brook RD, Calverley PMA, Celli BR, Crim C, Haumann B, Martinez FJ, Yates J, and Newby DE. The Study to Understand Mortality and Morbidity in COPD (SUMMIT) study protocol. *European Respiratory Society*. 2013; 41(5): 1017–1022
6. Obeidat M, Nie Y, Chen V, Shannon CP, Andiappan AK, Lee B, Rotzschke O, Castaldi PJ, Hersh CP, and Fishbane N. Network-based analysis reveals novel gene signatures in peripheral blood of patients with chronic obstructive pulmonary disease. *Respiratory Research*. 2017; 18(1): 72.

7. Postma DS and Rabe KF. The asthma-COPD overlap syndrome. *New England Journal of Medicine*. 2015; 373(13): 1241–1249.
8. Reddel HK, Bateman ED, Becker A, Boulet L-P, Cruz AA, Drazen JM, Haahtela T, Hurd SS, Inoue H, and De Jongste JC. A summary of the new GINA strategy: A roadmap to asthma control. *European Respiratory Journal*. 2015; 46(3): 622–639.
9. Soriano JB and Rodríguez-Roisin R. Chronic obstructive pulmonary disease overview: epidemiology, risk factors, and clinical presentation. *Proceedings of the American Thoracic Society*. 2011; 8(4): 363–367.
10. Duck A, Pigram L, Errhalt P, Ahmed D, and N Chaudhuri. IPF Care: A support program for patients with idiopathic pulmonary fibrosis treated with pirfenidone in Europe. *Advances in Therapy*. 2015; 32(2): 87–107.
11. Peters SP, Ferguson G, Deniz Y, and Reisner C. Uncontrolled asthma: A review of the prevalence, disease burden and options for treatment. *Respiratory Medicine*. 2006; 100(7): 1139–1151.
12. Kontakioti E, Domvri K, Papakosta D, and Daniilidis M. HLA and asthma phenotypes/endotypes: A review. *Human Immunology*. 2014; 75(8): 930–939.
13. Anderson GP. Endotyping asthma: New insights into key pathogenic mechanisms in a complex, heterogeneous disease. *The Lancet*. 2008; 372(9643): 1107–1119.
14. Decramer M, Rennard S, Troosters T, Mapel DW, Giardino N, Mannino D, Wouters E, Sethi S, and Cooper CB. COPD as a lung disease with systemic consequences–Clinical impact, mechanisms, and potential for early intervention. *COPD: Journal of Chronic Obstructive Pulmonary Disease*. 2008; 5(4): 235–256.
15. Matsuba K and Thurlbeck WM. The number and dimensions of small airways in emphysematous lungs. *The American Journal of Pathology*. 1972; 67(2): 265.
16. Ebrahimi A and Sadroddiny E. MicroRNAs in lung diseases: Recent findings and their pathophysiological implications. *Pulmonary Pharmacology & Therapeutics*. 2015; 34: 55–63.
17. Gibson RL, Burns JL, and Ramsey BW. Pathophysiology and management of pulmonary infections in cystic fibrosis. *American Journal of Respiratory and Critical Care Medicine*. 2003; 168(8): 918–951.
18. Guarnieri M and Balmes JR. Outdoor air pollution and asthma. *The Lancet*. 2014; 383(9928): 1581–1592.
19. Acharya KR and Ackerman SJ. Eosinophil granule proteins: Form and function. *Journal of Biological Chemistry*. 2014; 289(25): 17406–17415.
20. Fusonie D and Molnar W. Anomalous pulmonary venous return, pulmonary sequestration, bronchial atresia, aplastic right upper lobe, pericardial defect and intrathoracic kidney: An unusual complex of congenital anomalies in one patient. *American Journal of Roentgenology*. 1966; 97(2): 350–354.
21. Riedlinger WFJ, Vargas SO, Jennings RW, Estroff JA, Barnewolt CE, Lillehei CW, Wilson JM, Colin AA, Reid LM, and Kozakewich HPW. Bronchial atresia is common to extralobar sequestration, intralobar sequestration, congenital cystic adenomatoid malformation, and lobar emphysema. *Pediatric and Developmental Pathology*. 2006; 9(5): 361–373.
22. Zylak CJ, Eyler WR, Spizarny DL, and Stone CH. 2002. Developmental lung anomalies in the adult: Radiologic-pathologic correlation. *Radiographics* 22 (Suppl. 1):S25–S43.
23. Bose CL, Dammann CEL, and Laughon MM. Bronchopulmonary dysplasia and inflammatory biomarkers in the premature neonate. *Archives of Disease in Childhood-Fetal and Neonatal Edition*. 2008; 93(6): F455–F461.
24. Coclite E, Natale CD, and Nigro G. Congenital and perinatal cytomegalovirus lung infection. *The Journal of Maternal-Fetal & Neonatal Medicine*. 2013; 26(17): 1671–1675.
25. Donahoe M. Acute respiratory distress syndrome: A clinical review. *Pulmonary Circulation*. 2011; 1(2): 192–211.
26. Rafii R, Juarez MM, Albertson TE, and Chan AL. A review of current and novel therapies for idiopathic pulmonary fibrosis. *Journal of Thoracic Disease*. 2013; 5(1): 48.
27. Maghsoudloo M, Jamalkandi SA, Najafi A, and Masoudi-Nejad A. Identification of biomarkers in common chronic lung diseases by co-expression networks and drug-target interactions analysis. *Molecular Medicine*. 2020; 26(1): 1–19.
28. Koethe SM, Kuhnmuench JR, and Becker CG. Neutrophil priming by cigarette smoke condensate and a tobacco anti-idiotypic antibody. *The American Journal of Pathology*. 2000; 157(5): 1735–1743.
29. Youkeles LH, Grizzanti JN, Liao Z, Chang CJ, and Rosenstreich DL. Decreased tobacco-glycoprotein-induced lymphocyte proliferation in vitro in pulmonary eosinophilic granuloma. *American Journal of Respiratory and Critical Care Medicine*. 1995; 151(1): 145–150.
30. Chabrol E, Nurisso A, Daina A, Vassal-Stermann E, Thepaut M, Girard E, Vivès RR, and Fieschi F. Glycosaminoglycans are interactants of Langerin: Comparison with gp120 highlights an unexpected calcium-independent binding mode. *PLoS One*. 2012; 7(11): e50722.
31. Lenormand C, Spiegelhalter C, Cinquin B, Bardin S, Bausinger H, Angénieux C, Eckly A, Proamer F, Wall D, and Lich B. Birbeck granule-like "organized smooth endoplasmic reticulum" resulting from the expression of a cytoplasmic YFP-tagged langerin. *PLoS ONE*. 2013; 8(4): e60813.
32. Albertson TE, Chenoweth JA, Pearson SJ, and Murin S. The pharmacological management of asthma-chronic obstructive pulmonary disease overlap syndrome (ACOS). *Expert Opinion on Pharmacotherapy*. 2020; 21(2): 213–231.
33. Thomas M, Decramer M, and O'Donnell DE. No room to breathe: The importance of lung hyperinflation in COPD. *Primary Care Respiratory Journal*. 2013; 22(1): 101–111.
34. Wise RA, Anzueto A, Cotton D, Dahl R, Devins T, Disse B, Dusser D, Joseph E, Kattenbeck S, and Koenen-Bergmann M. Tiotropium respimat inhaler and the risk of death in COPD. *New England Journal of Medicine*. 2013; 369: 1491–1501.
35. Vogelmeier C, Hederer B, Glaab T, Schmidt H, Mölken MPMH R-v, Beeh KM, Rabe KF, and Fabbri LM. Tiotropium versus salmeterol for the prevention of exacerbations of COPD. *New England Journal of Medicine*. 2011; 364(12): 1093–1103.
36. Karner C, Chong J, and Poole P. Tiotropium versus placebo for chronic obstructive pulmonary disease. *Cochrane Database of Systematic Reviews*. 2014; 21(7): CD009285

37. Cosío BG, Shafiek H, Iglesias A, Yanez A, Córdova R, Palou A, Rodriguez-Roisin R, Peces-Barba G, Pascual S, and Gea J. Oral low-dose theophylline on top of inhaled fluticasone-salmeterol does not reduce exacerbations in patients with severe COPD: A pilot clinical trial. *Chest*. 2016; 150(1): 123–130.
38. Bateman ED, Chapman KR, Singh D, D'Urzo AD, Molins E, Leselbaum A, and Gil EG. Aclidinium bromide and formoterol fumarate as a fixed-dose combination in COPD: Pooled analysis of symptoms and exacerbations from two six-month, multicentre, randomised studies (ACLIFORM and AUGMENT). *Respiratory Research*. 2015; 16(1): 92.
39. Mahler DA, Kerwin E, Ayers T, FowlerTaylor A, Maitra S, Thach C, Lloyd M, Patalano F, and Banerji D. FLIGHT1 and FLIGHT2: Efficacy and safety of QVA149 (indacaterol/glycopyrrolate) versus its monocomponents and placebo in patients with chronic obstructive pulmonary disease. *American Journal of Respiratory and Critical Care Medicine*. 2015; 192(9): 1068–1079.
40. Anzueto AR, Kostikas K, Mezzi K, Shen S, Larbig M, Patalano F, Fogel R, Banerji D, and Wedzicha JA. Indacaterol/glycopyrronium versus salmeterol/fluticasone in the prevention of clinically important deterioration in COPD: Results from the FLAME study. *Respiratory Research*. 2018; 19(1): 121.
41. Chung LP and JY Paton. Two sides of the same coin?—Treatment of chronic asthma in children and adults. *Frontiers in Pediatrics*. 2019; 7: 62.
42. Virchow JC, Kuna P, Paggiaro P, Papi A, Singh D, Corre S, Zuccaro F, Vele A, Kots M, and Georges G. Single inhaler extrafine triple therapy in uncontrolled asthma (TRIMARAN and TRIGGER): Two double-blind, parallel-group, randomised, controlled phase 3 trials. *The Lancet*. 2019; 394(10210): 1737–1749.
43. Celli BR . Pharmacological therapy of COPD: Reasons for optimism . *Chest*. 2018; 154(6): 1404–1415.
44. Walters JAE, Tan DJ, White CJ, Gibson PG, Wood-Baker R, and Walters EH. Systemic corticosteroids for acute exacerbations of chronic obstructive pulmonary disease. *Cochrane Database of Systematic Reviews*. 2014; 1 (9). CD001288
45. Dransfield MT and Singh D. Predicting pneumonia in COPD: Have we unraveled the network of risks? *American Journal of Respiratory and Critical Care Medicine*. 2020; 9: 1026–1027.
46. Al-Sajee D, Yin X, and Gauvreau GM. An evaluation of roflumilast and PDE4 inhibitors with a focus on the treatment of asthma. *Expert Opinion on Pharmacotherapy*. 2019; 20(5): 609–620.
47. Bardin P, Kanniess F, Gauvreau G, Bredenbröker D, and Rabe KF. Roflumilast for asthma: Efficacy findings in mechanism of action studies. *Pulmonary Pharmacology & Therapeutics*. 2015; 35: S4–S10.
48. Chong J, Leung B, and Poole P Phosphodiesterase 4 inhibitors for chronic obstructive pulmonary disease. *Cochrane Database of Systematic Reviews*. 2017; 9 (9): CD002309
49. Poole P, Sathananthan K, and Fortescue R. Mucolytic agents versus placebo for chronic bronchitis or chronic obstructive pulmonary disease. *Cochrane Database of Systematic Reviews*. 2019; 29 (5): CD001287
50. Ingebrigtsen TS, Marrot JL, Nordestgaard BG, Lange P, Hallas J, and Vestbo J. Statin use and exacerbations in individuals with chronic obstructive pulmonary disease. *Thorax*. 2015; 701 : 33–40.
51. Sanguinetti CM . N-acetylcysteine in COPD: Why, how, and when? *Multidisciplinary Respiratory Medicine*. 2015; 11(1): 1–11.
52. Cramer CL, Patterson A, Alchakaki A, and Soubani AO. Immunomodulatory indications of azithromycin in respiratory disease: A concise review for the clinician. *Postgraduate Medicine*. 2017; 129(5): 493–499.
53. Huang C-L, Nguyen PA, Kuo P-l, Iqbal U, Hsu YHE, and Jian WSn. Influenza vaccination and reduction in risk of ischemic heart disease among chronic obstructive pulmonary elderly. *Computer Methods and Programs in Biomedicine*. 2013; 111(2): 507–511.
54. Poole Phillippa, Chacko EE, Wood-Baker R, and Cates CJ. Influenza vaccine for patients with chronic obstructive pulmonary disease. *Cochrane Database of Systematic Reviews*. 2006; (1): CD002733
55. Kobayashi M, Bennett NM, Gierke R, Almendares O, Moore MR, Whitney CG, and Pilishvili T. Intervals between PCV13 and PPSV23 vaccines: Recommendations of the Advisory Committee on Immunization Practices (ACIP). *Morbidity and Mortality Weekly Report*. 2015; 64(34): 944–947.
56. Beard E, West R, Michie S, and Brown J. Association between electronic cigarette use and changes in quit attempts, success of quit attempts, use of smoking cessation pharmacotherapy, and use of stop smoking services in England: Time series analysis of population trends. *BMJ*. 2016; 354: i4645.
57. McNeil A, Brose LS, Calder R, Hitchman SC, Hajek P, and McRobbie H. E-cigarettes: An evidence update. A report commissioned by Public Health England. *Public Health England*. 2015: 1–113.
58. van Eerd E, Rvd Meer, Ov Schayck, and D Kotz. 2017. Smoking cessation for people with chronic obstructive pulmonary disease: A systematic review. *Challenges in Smoking Cessation for People With Chronic Obstructive Pulmonary Disease*: 15.
59. Long-Term Oxygen Treatment Trial Research Group . A randomized trial of long-term oxygen for COPD with moderate desaturation. *New England Journal of Medicine*. 2016; 375(17): 1617–1627.
60. Struik FM, Sprooten RTM, Kerstjens HAM, Bladder Gerrie, Zijnen M, Asin J, Cobben NAM, Vonk JM, and Wijkstra PJ. Nocturnal non-invasive ventilation in COPD patients with prolonged hypercapnia after ventilatory support for acute respiratory failure: A randomised, controlled, parallel-group study. *Thorax*. 2014; 69(9): 826–834.
61. National Emphysema Treatment Trial Research Group . A randomized trial comparing lung-volume-reduction surgery with medical therapy for severe emphysema. *New England Journal of Medicine*. 2003; 348(21): 2059–2073.
62. Marchetti N and Criner GJ. 2015. Surgical approaches to treating emphysema: Lung volume reduction surgery, bullectomy, and lung transplantation. *Seminars in Respiratory and Critical Care Medicine*.

63. Laube BL, Carson KA, Sharpless G, Paulin LM, and Hansel NN. Mucociliary clearance in former tobacco smokers with both chronic obstructive pulmonary disease and chronic bronchitis and the effect of roflumilast. *Journal of Aerosol Medicine and Pulmonary Drug Delivery*. 2019; 32(4): 189–199.
64. Come CE, Kramer MR, Dransfield MT, Abu-Hijleh M, Berkowitz D, Bezzi M, Bhatt SP, Boyd MB, Cases E, and Chen AC. A randomised trial of lung sealant versus medical therapy for advanced emphysema. *European Respiratory Journal*. 2015; 46(3): 651–662.
65. Garvey C, Bayles MP, Hamm LF, Hill K, Holland A, Limberg TM, and Spruit MA. Pulmonary rehabilitation exercise prescription in chronic obstructive pulmonary disease: Review of selected guidelines. *Journal of Cardiopulmonary Rehabilitation and Prevention*. 2016; 36(2): 75–83.
66. Velloso M, Nascimento NHd, Gazzotti MR, and Jardim JR. Evaluation of effects of shoulder girdle training on strength and performance of activities of daily living in patients with chronic obstructive pulmonary disease. *International Journal of Chronic Obstructive Pulmonary Disease*. 2013; 8: 187.
67. Au DH, Udris EM, Engelberg RA, Diehr PH, Bryson CL, Reinke LF, and Curtis JR. A randomized trial to improve communication about end-of-life care among patients with COPD. *Chest*. 2012; 141(3): 726–735.
68. Yang W, Peters JI, and Williams RO III. Inhaled nanoparticles—A current review. *International Journal of Pharmaceutics*. 2008; 356(1-2): 239–247.
69. Koopaei NN and Abdollahi M. Opportunities and Obstacles to the Development of Nanopharmaceuticals for Human Use. DARU Journal of Pharmaceutical Science. 2016; 24: 23
70. Cipolla DC and Gonda I. Formulation technology to repurpose drugs for inhalation delivery. *Drug Discovery Today: Therapeutic Strategies*. 2011; 8(3-4): 123–130.
71. Scherließ R and Etschmann C. DPI formulations for high dose applications–Challenges and opportunities. *International Journal of Pharmaceutics*. 2018; 548(1): 49–53.
72. Yameen B, Choi W Il, Vilos C, Swami A, Shi J, and Farokhzad OC. Insight into nanoparticle cellular uptake and intracellular targeting. *Journal of Controlled Release*. 2014; 190: 485–499.
73. Blank F, Fytianos K, Seydoux E, Rodriguez-Lorenzo L, Petri-Fink A, Garnier CV, and Rothen-Rutishauser B. Interaction of biomedical nanoparticles with the pulmonary immune system. *Journal of Nanobiotechnology*. 2017; 15(1): 1–9.
74. Murgia X, Carvalho C d S, and Lehr C-M. Overcoming the pulmonary barrier: New insights to improve the efficiency of inhaled therapeutics. *European Journal of Nanomedicine*. 2014; 6(3): 157–169.
75. Ungaro F, d'Angelo I, Miro A, Rotonda ML, and Quaglia F. Engineered PLGA nano-and micro-carriers for pulmonary delivery: Challenges and promises. *Journal of Pharmacy and Pharmacology*. 2012; 64(9): 1217–1235.
76. Schneider CS, Xu Q, Boylan NJ, Chisholm J, Tang BC, Schuster BS, Henning A, Ensign LM, Lee E, and Adstamongkonkul P. Nanoparticles that do not adhere to mucus provide uniform and long-lasting drug delivery to airways following inhalation. *Science Advances*. 2017; 3(4): e1601556.
77. Azarmi S, Roa WH, and Löbenberg R. Targeted delivery of nanoparticles for the treatment of lung diseases. *Advanced Drug Delivery Reviews*. 2008; 60(8): 863–875.
78. Frcitas RA. Exploratory design in medical nanotechnology: A mechanical artificial red cell. *Artificial Cells, Blood Substitutes, and Biotechnology*. 1998; 26(4): 411–430.
79. Iga AM, Robertson JHP, Winslet MC, and Seifalian AM. 2007. Clinical potential of quantum dots. *BioMed Research International*.2007; 10: 76087
80. Roursgaard M, Poulsen S, Kepley C, Hammer M, Nielsen G, and Larsen S. Polyhydroxylated C60 fullerene (fullerenol) attenuates neutrophilic lung inflammation in mice. *Basic & Clinical Pharmacology & Toxicology*. 2008; 103(4): 386.
81. Surendiran A, Sandhiya S, Pradhan SC, and Adithan C. Novel applications of nanotechnology in medicine. *Indian Journal of Medical Research*. 2009; 130(6): 689–701.
82. Matsuo Y, Ishihara T, Ishizaki J, Miyamoto KI, Higaki M, and Yamashita N. Effect of betamethasone phosphate loaded polymeric nanoparticles on a murine asthma model. *Cellular Immunology*. 2009; 260(1): 33–38.
83. Kimura S, Egashira K, Chen L, Nakano K, Iwata E, Miyagawa M, Tsujimoto H, Hara K, Morishita R, and Sueishi K. Nanoparticle-mediated delivery of nuclear factor κB decoy into lungs ameliorates monocrotaline-induced pulmonary arterial hypertension. *Hypertension*. 2009; 53 (5): 877–883.
84. Beck-Broichsitter M, Gauss J, Gessler T, Seeger W, Kissel T, and Schmehl T. Pulmonary targeting with biodegradable salbutamol-loaded nanoparticles. *Journal of Aerosol Medicine and Pulmonary Drug Delivery*. 2010; 23(1): 47–57.
85. Saraogi GK, Gupta P, Gupta UD, Jain NK, and Agrawal GP. Gelatin nanocarriers as potential vectors for effective management of tuberculosis. *International Journal of Pharmaceutics*. 2010; 385(1–2): 143–149.
86. Han S, Liu Y, Nie X, Xu Q, Jiao F, Li W, Zhao Y, Wu Y, and Chen C. Efficient delivery of antitumor drug to the nuclei of tumor cells by amphiphilic biodegradable poly (L-aspartic acid-co-lactic acid)/DPPE co-polymer nanoparticles. *Small*. 2012; 8(10): 1596–1606.
87. Kenyon NJ, Bratt JM, Lee J, Luo J, Franzi LM, Zeki AA, and Lam KS. Self-assembling nanoparticles containing dexamethasone as a novel therapy in allergic airways inflammation. *PLoS ONE*. 2013; 8(10): e77730.
88. Kumar M, Kong X, Behera AK, Hellermann GR, Lockey RF, and Mohapatra SS. Chitosan IFN-γ-pDNA nanoparticle (CIN) therapy for allergic asthma. *Genetic Vaccines and Therapy*. 2003; 1(1): 3.
89. Pandey R, Sharma A, Zahoor A, Sharma S, Khuller GK, and Prasad B. Poly (DL-lactide-co-glycolide) nanoparticle- based inhalable sustained drug delivery system for experimental tuberculosis. *Journal of Antimicrobial Chemotherapy*. 2003; 52(6): 981–986.
90. Roa WH, Azarmi S, Kamal Al-Hallak MHD, Finlay WH, Magliocco AM, and Löbenberg R. Inhalable nanoparticles, a non-invasive approach to treat lung cancer in a mouse model. *Journal of Controlled Release*. 2011; 150(1): 49–55.

91. Kamat CD, Shmueli RB, Connis N, Rudin CM, Green JJ, and Hann CL. Poly (β-amino ester) nanoparticle delivery of TP53 has activity against small cell lung cancer in vitro and in vivo. *Molecular Cancer Therapeutics*. 2013; 12(4): 405–415.
92. Chen Y, Zhu X, Zhang X, Liu B, and Huang L. Nanoparticles modified with tumor-targeting scFv deliver siRNA and miRNA for cancer therapy. *Molecular Therapy*. 2010; 18(9): 1650–1656.
93. Liu K-K, Zheng W-W, Wang C-C, Chiu Y-C, Cheng C-L, Lo Y-S, Chen C, and Chao J-I. Covalent linkage of nanodiamond-paclitaxel for drug delivery and cancer therapy. *Nanotechnology*. 2010; 21(31): 315106.
94. Boland S, Guadagnini R, Baeza-Squiban A, Hussain S, and Marano F. Nanoparticles used in medical applications for the lung: Hopes for nanomedicine and fears for nanotoxicity. Journal of Physics: Conference Series. 2011.
95. Janke J and Scherließ R. Vaccination via the respiratory tract-porous PLGA-nanoparticles for a faster release. 9th World Meeting on Pharmaceutics, Biopharmaceutics and Pharmaceutical Technology, 2014, Lisbon.
96. Parumasivam T, Chang RYK, Abdelghany S, Ye TT, Britton WJ, and Chan H-K. Dry powder inhalable formulations for anti-tubercular therapy. *Advanced Drug Delivery Reviews*. 2016; 102: 83–101.
97. Pham D-D, Fattal E, and Tsapis N. Pulmonary drug delivery systems for tuberculosis treatment. *International Journal of Pharmaceutics*. 2015; 478(2): 517–529.
98. Sung JC, Padilla DJ, Garcia-Contreras L, VerBerkmoes JL, Durbin D, Peloquin CA, Elbert KJ, Hickey AJ, and Edwards DA. Formulation and pharmacokinetics of self-assembled rifampicin nanoparticle systems for pulmonary delivery. *Pharmaceutical Research*. 2009; 26(8): 1847–1855.
99. Paranjpe M and Müller-Goymann CC. Nanoparticle-mediated pulmonary drug delivery: A review. *International Journal of Molecular Sciences*. 2014; 15(4): 5852–5873.
100. Youngren-Ortiz SR, Gandhi NS, España-Serrano L, and Chougule MB. Aerosol delivery of siRNA to the lungs. Part 2: Nanocarrier-based delivery systems. *KONA Powder and Particle Journal*. 2017; 34: 44–69
101. Lycke N. Recent progress in mucosal vaccine development: Potential and limitations. *Nature Reviews Immunology*. 2012; 12(8): 592–605.
102. Sullivan VJ, Mikszta JA, Laurent P, Huang J, and Ford B. Noninvasive delivery technologies: Respiratory delivery of vaccines. *Expert Opinion on Drug Delivery*. 2006; 3(1): 87–95.
103. Fifis T, Gamvrellis A, Crimeen-Irwin B, Pietersz GA, Li J, Mottram PL, McKenzie IFC, and Plebanski M. Size-dependent immunogenicity: Therapeutic and protective properties of nano-vaccines against tumors. *The Journal of Immunology*. 2004; 173(5): 3148–3154.
104. Garbuzenko OB, Mainelis G, Taratula O, and Minko T. Inhalation treatment of lung cancer: The influence of composition, size and shape of nanocarriers on their lung accumulation and retention. *Cancer Biology & Medicine*. 2014; 11(1): 44.
105. Lee W-H, Loo C-Y, Traini D, and Young PM. Inhalation of nanoparticle-based drug for lung cancer treatment: Advantages and challenges. *Asian Journal of Pharmaceutical Sciences*. 2015; 10(6): 481–489.
106. Rudokas M, Najlah M, Alhnan MA, and Elhissi A. Liposome delivery systems for inhalation: A critical review highlighting formulation issues and anticancer applications. *Medical Principles and Practice*. 2016; 25(Suppl. 2): 60–72.
107. Aragao-Santiago L, Bohr A, Delaval M, Dalla-Bona AC, Gessler T, Seeger W, and Beck-Broichsitter M. Innovative formulations for controlled drug delivery to the lungs and the technical and toxicological challenges to overcome. *Current Pharmaceutical Design*. 2016; 22(9): 1147–1160.
108. Raula J, Eerikäinen H, Peltonen L, Hirvonen J, and Kauppinen E. Aerosol-processed polymeric drug nanoparticles for sustained and triggered drug release. *Journal of Controlled Release: Official Journal of the Controlled Release Society*. 2010; 148(1): e52.
109. Carvalho TC, Peters JI, and Williams RO III. Influence of particle size on regional lung deposition–What evidence is there? *International Journal of Pharmaceutics*. 2011; 406(1-2): 1–10.
110. Leach C, Colice GL, and Luskin A. Particle size of inhaled corticosteroids: Does it matter? *Journal of Allergy and Clinical Immunology*. 2009; 124(6): S88–S93.
111. Ventola CL. Progress in nanomedicine: Approved and investigational nanodrugs. *Pharmacy and Therapeutics*. 2017; 42(12): 742.
112. Barenholz YC. Doxil®—The first FDA-approved nanodrug: Lessons learned. *Journal of Controlled Release*. 2012; 160(2): 117–134.
113. Heneweer C, Gendy SEM, and Peñate-Medina O. Liposomes and inorganic nanoparticles for drug delivery and cancer imaging. *Therapeutic Delivery*. 2012; 3(5): 645–656.
114. Hofer N, Jank K, Resch E, Urlesberger B, Reiterer F, and Resch B. Meconium aspiration syndrome – A 21-years' experience from a tertiary care center and analysis of risk factors for predicting disease severity. *Klinische Pädiatrie*. 2013; 225(07): 383–388.
115. Speer CP, Gefeller O, Groneck P, Laufkötter E, Roll C, Hanssler L, Harms K, Herting E, Boenisch H, and Windeler J. Randomised clinical trial of two treatment regimens of natural surfactant preparations in neonatal respiratory distress syndrome. *Archives of Disease in Childhood-Fetal and Neonatal Edition*. 1995; 72(1): F8–F13.
116. Cipolla D, Blanchard J, and Gonda I. Development of liposomal ciprofloxacin to treat lung infections. *Pharmaceutics*. 2016; 8(1): 6.
117. Olivier KN, Griffith DE, Eagle G, McGinnis JP, Micioni L, Liu K, Daley CL, Winthrop KL, Ruoss S, and Addrizzo-Harris DJ. Randomized trial of liposomal amikacin for inhalation in nontuberculous mycobacterial lung disease. *American Journal of Respiratory and Critical Care Medicine*. 2017; 195(6): 814–823.
118. Mönckedieck, M, Kamplade J, Littringer EM, Mescher A, Gopireddy S, Hertel M, Gutheil E, Walzel P, Urbanetz NA, and Köster M. Spray Drying Tailored Mannitol Carrier Particles for Dry Powder Inhalation with Differently Shaped Active Pharmaceutical Ingredients. In: Process-S pray: Functional Particles Produced in Spray Processes. UdoFritsching Ed. P.hD., 517–566. Springer International Publishing. 2016.

119. Amani A, York P, Chrystyn H, and Clark BJ. Evaluation of a nanoemulsion-based formulation for respiratory delivery of budesonide by nebulizers. *AAPS PharmSciTech*. 2010; 11(3): 1147–1151.
120. Abdelaziz HM, Gaber M , Abd-Elwakil MM, Mabrouk MT, Elgohary M, Kamel NM, Kabary DM, Freag MS, Samaha MW, Mortada SM , Elkhodairy KA, Fang J-Y, and Elzoghby AO. Inhalable particulate drug delivery systems for lung cancer therapy: Nanoparticles, microparticles, nanocomposites and nanoaggregates. *Journal of Controlled Release*. 2018; 269: 374–392.
121. Yang L, Luo J, Shi S, Zhang Q, Sun X, Zhang Z, and Gong T. Development of a pulmonary peptide delivery system using porous nanoparticle-aggregate particles for systemic application. *International Journal of Pharmaceutics*. 2013; 451(1-2): 104–111.

8

Model for Pharmaceutical aerosol transport through stenosis airway

Puchanee Larpruenrudee[1], Mohammad S. Islam[1], Gunther Paul[2], Akshoy R. Paul[3], Y.T. Gu[4], and Suvash C. Saha[1]
[1]School of Mechanical and Mechatronic Engineering, University of Technology Sydney (UTS), Sydney, Australia, Australia
[2]James Cook University, Australian Institute of Tropical Health and Medicine, Townsville, Australia
[3]Department of Applied Mechanics, Motilal Nehru National Institute of Technology Allahabad, Uttar Pradesh, India
[4]School of Mechanical and Mechatronic Engineering University of Technology Sydney (UTS), Ultimo, Australia

8.1 Introduction

Air pollution is one of the critical factors that has threatened human health for a long time. In the early 20th century, people worldwide were aware of the health effects that were caused by air pollution. Especially in London in 1952, the "killer fog" caused the deaths of over 12,000 people (1). According to the World Health Organization (2), over 7 million deaths were caused by air pollution effects in terms of both ambient and household air pollutions. The effects of this pollution can cause many diseases, particularly in respiratory diseases.

Although the respiratory system has many defense mechanisms to protect internal organs from foreign substances, air pollutants can still pass through those mechanisms. The main factors that cause the penetrative processes are the size and chemical nature of the air pollutants (3). Several recent articles (4–9) have examined the effects of air pollution on the risk of respiratory diseases such as asthma, respiratory infections, and chronic obstructive pulmonary disease (COPD). They point out that this pollution can increase the risk of respiratory diseases for both adults and children. Chronic asthma and COPD are two of the respiratory diseases that have significantly affected human health worldwide since 1990. Based on the reports from the (10), the majority of deaths were caused by COPD. From the statistics of deaths in 2017, over 7500 deaths were caused by COPD among people aged over 55 years, whereas 400 deaths were caused by asthma from among people aged 5 to 34 years. For worldwide deaths from diseases, both asthma and COPD are the leading cause of death and injury in clinical treatment (11). The evidence from the clinical treatment of Ahmed and Athar (12), is that the mortality and morbidity from these diseases often occur because of intrinsic positive end-expiratory pressure (PEEPi) and dynamic hyperinflation (DH). By this point, it can create overwork of breathing (WOB).

In some cases, mechanical ventilation is used for patients who cannot breathe by themselves because of exhaustion of the respiratory muscles (11,12). Some recent articles (13–16) and have studied the characteristics of airborne particulate matter which causes respiratory diseases. They have found that having accurate data on the estimation of aerosol deposition within the human respiratory system is significant to investigating and treating the disease's symptoms. Furthermore, it is essential to estimate drug delivery in the case of mechanical ventilation patients (17–19). However, there are some difficult situations in terms of using mechanical ventilation because there are different airway characteristics. Normally, the respiratory system can completely grow and develop by 18–20 years of age (20,21). In contrast, the ability of the respiratory system will decrease among the elderly population. In this case, the lung structure will be changed with aging, which generates alveolar dead space in the lung airways (22). The thickness of tissue within the lung airways tends to be based on aging. The diameter of the lung airways for the elderly between 50 and 80 years of age has been found to be smaller than other ages, reduced by approximately 7% (23). In terms of other ages, the diameter of the lung airways will decrease around 1% per year at 30 years of age and later (23,24). With this smaller diameter condition, the lung will need more pressure in order to change the volume in the lung. Therefore, the patient will have more difficulty in breathing (25).

Computational fluid dynamics (CFD) has been immensely used in biomedical science and engineering, preferentially in the pharmaceutical industries. This is due to the simulation abilities, which provide predictability and optimization in terms of inhaled therapies. It is also used to simulate the drug delivery process, which includes particulate, aerosol, and gaseous drug types by starting from the entire process through lung airways (26,27).

Studies in the last two decades have analyzed the airflow and particle deposition based on symmetric and asymmetric lung bifurcations from both realistic and non-realistic models. Most realistic models have been generated from CT scans, whereas non-realistic models have been created and developed from Weibel's model. Balashazy and Hoffman (28) have examined

DOI: 10.1201/9781003046547-8

the deposition of aerosols in both symmetric and asymmetric bifurcations and have found that the aerosols are mostly deposited firmly at the wall of a bifurcation. Nowak, Kakade, and Annapragada (29) have simulated airflow and aerosol deposition by comparing the non-realistic lung model from Weibel (30) and realistic lung model from a CT scan. For this simulation, the lung model from G0 to G3 has been considered under the steady-state conditions for inhalation and exhalation. They state that the deposition fractions at bifurcations are totally the same for both models. Lee and Lee (31) have generated the lung model based on a (30) model with the first four generations to study the aerosol dispersion, whereas Liu et al. (32) have studied the airflow characteristics of the asymmetric lung airway at the G5 to G11 (30) model. The results of their studies indicate that the airways, having different sizes, cannot affect human breathing because they have the same airflow rate for both the medial branches and their mother branches. In the same year, Kleinstreuer and Zhang (33) analyzed the target of aerosol deposition of the drug delivery within the symmetrically bifurcating at G3 to G6 under different hemispherical tumor conditions based on Weibel's (30) model. The results of this study showed that the airflow pattern is based on the tumor size and inlet flow rate, whereas the particle deposition is based on the tumor size and upstream flow.

In some cases, the nanoparticle deposition has been considered based on Weibel's lung model. It has focused on the mouth to the trachea through G3 to G5 and specific lung regions with a magnetic method (34,35,36). The authors concluded that the deposition efficiency is influenced by the nanoparticle size and inlet Reynolds numbers. Nanoparticles usually deposit at the bifurcation and inside the wall around the bifurcation. Moreover, nanoparticle deposition has an insignificant effect on turbulent fluctuations of upper airways. Cebral and Summers (37) have studied the airflow through stenosis areas by using the realistic lung models and only considering the upper airways. The results from this study indicate that the pressure would decrease, whereas the shear stress would increase in the stenosis area. Van Ertbruggen et al. (27) evaluated the gas flow and particle deposition by using a realistic model that starts from the trachea to G7 (the segmental bronchi). The authors summarized that an increase of inspiratory flow causes an increase of particle inertia that leads to greater deposition percentage. The main factors of deposition percentage also depend on the length, angles, and diameters of branches.

The literature survey on aerosol deposition from Borgstrom et al. (38) reported that the deposition at the throat region is a significant factor for lung deposition. Gemci et al. (39) studied the airflow in a 17-generation realistic model. The results demonstrate that the airflow characteristics within asymmetric airways are depended on the branching conduits. Ma and Lutchen (16) simulated the aerosol deposition in the upper airways based on a realistic model of the healthy lung. Different particle sizes from 1 to 30 μm have been used for this simulation. The results have shown that the deposition efficiency could be affected by particle size rather than the flow rate. They also point out that the turbulent flow will be generated only from the upper airway. Therefore, the model without the upper airway could not provide a realistic airflow within the lung. In contrast, there are some studies that consider focusing on the model starting at the trachea region. In terms of drug delivery for the treatment of respiratory diseases, drug delivery efficiency can be improved by an increase the dosage of the drug that transports to the disease area (40,41). To improve this efficiency, the trachea region is an important area where it should be prioritized to be a first step for the drug injection (42,43).

The articles in this last decade examined the airflow and deposition patterns of the aerosol particles under various conditions, such as particle sizes, velocity, and the airway's characteristics (44–48). Other than that, some authors have analyzed airflow and particle deposition based on the combination between helium and oxygen (49,50). Saber and Heydari (51) analyzed the flow patterns and deposition fraction of particles between 0.1 μm to 10 μm at the upper airways from trachea to the third generations based on the healthy lung airway from the lung model of Weibel (30). They reported that the number of deposition fraction is significantly based on the Stokes and Reynolds numbers. Srivastav et al. (52) found a significant impact of the cartilaginous rings on airflow and particle transport through human airways. Srivastav et al. (53) in another study, investigated the effects of particle deposition in glomous tumor obstructed diseased human airways. They concluded from both the studies that the magnitude and location of maximum wall shear stress is an indication of probable wall injury.

Sul et al. (54) used the CFD to study the airflow characteristics in normal and obstructed airways. They only considered the lower airways between G8 to G14 for both symmetric and asymmetric models. The results from this study provide a clear understanding of shear stresses that are independent on the respiratory rate, but the shear stresses are dependent on the distribution of the obstructed areas within the lung. Augusto et al. (55,56) studied the aerosol deposition under three respiratory conditions: inhalation, exhalation and breath-holding. In this case, four generations (G3 to G6 of Weibel's model) and three bifurcations model with 5 μm of particle size are considered to investigate the particle transport. The result shows that the number of aerosols deposition on the wall is higher for the breath-holding. The articles from Islam et al. (57) and Islam et al. (58) have examined particle transport and particle deposition in a large scale 17th generation based on the realistic lung model. It has found that the majority of smaller particles could escape from the 17th generation and seem to continue flowing to the 23rd generation. Zhang et al. (59), have analyzed the particle deposition in realistic lung airways from CT scans. This article used the specific lung model for the COPD patient to compare with the healthy lung model. For the stenosis location of COPD models, the results from this study indicated that the deposition efficiency and deposition fraction are affected by the stenosis areas. The deposition efficiency will increase in the same region of stenosis location, whereas the deposition fraction from the upper and lower lobes will decrease as a result of stenosis at the main bronchus.

Islam et al. (60) have studied and evaluated the particle transport and deposition in the lung airways from trachea to third generation of human airways. They have used three different lung models to compare the particle characteristics between the realistic and non-realistic lung models. For the non-realistic models, the first model is based on a symmetric lung model from Weibel's model, while the second model was developed to become an asymmetric lung model. The results showed that the airflow rate distribution of the realistic model is higher than the non-realistic models. Larger particle sizes have an essential effect on the turbulence deposition. Farghadan et al. (61) have analyzed the particle transport in human lung airways in order to predict the particle source and destination for both inhalation and exhalation. They used a realistic lung model, defining the trachea region to be an injection area for 1 μm diameter.

The result of their simulations indicates that the number of particles at the airway wall has a greater number than other regions. Islam et al. (62) reviewed the airflow characteristics and particle deposition from the respiratory anatomical development. The results have found that both turbulence dispersion and flow rates, which are lower than 25 L/min, could not have a significant effect on deposition patterns of nanoparticle. Singh et al. (63) studied the airflow and particle transport through stenosis airways by employing the realistic model from CT scans based on the healthy human lung. The models have been reconstructed and created the random stenosis areas from first generation to third generation. They have compared the stenosis area with the different cases that are stenosis at the left and right side of the main lung. Two main variables, particle size and initial velocity, have been applied to evaluate the airflow and particle patterns. The authors report the results that stenosis airway could generate a complex velocity. The deposition efficiency could increase, which is influenced by the flow rate and particle size.

The study from Koullapis et al. (64) has examined the effect of the inlet velocity profile, inhalation flow rate and electrostatic charge for particle deposition in the respiratory tract. The realistic model, which involves the upper airways through the lower airways from first generation to the seventh generation, has been considered in this study. The authors state that the initial velocities have a significant effect on particle deposition for the large sizes at the tongue, whereas the small particle sizes are often diffusible in the same area. The case study of Lalas et al. (65) has particularly analyzed the airflow and inhaled particles through a comparison between three different lung airways that start from oral to fifth generation by three different cases involving the normal lung airway, the abnormal lung airway with narrows in random areas, and the asymmetric lung model with narrow areas at the right side. In this case, the particle diameters from 1 to 30 μm are considered with only one flow rate of inlet velocity that is 12.6 L/min. For the results, the authors point out that the particles are often deposited in the opposite lung side from the inflammation areas. In terms of deposition fraction, particle deposition is based on the particle sizes and particle densities.

From the recent articles, most of them mainly focus on the realistic model with stenosis area at different generations, while the non-realistic model with stenosis areas are less considered. In cases of the non-realistic model, these articles only consider the non-realistically symmetric model. There are only a few articles that focus on the asymmetric model. Based on recent studies that use the realistic model from Lalas et al. (65), although this case study is useful for the medication, it does not consider the narrow airway at the first generation and only considers one inlet velocity that is appropriate to transport particle sizes from 1 to 10 μm. In contrast, Koullapis et al. (64) state that the initial velocities have an essential effect on the particle deposition for both larger and smaller sizes. In order to enhance the understanding of the respiratory tract, the objective of this study is to analyze the airflow and mainly consider particle transport of asymmetric lung models with stenosis sections at different generations from first generation to fourth generation. Moreover, two main variables, initial velocity and particle diameter, will be applied to evaluate the effect that influences the airflow and particle transport. In terms of the breathing condition, an inhaled breathing condition will be considered in this chapter.

In case of the effects of aging on human lung airways, Phalen and Oldham (66) and Isaacs and Martonen (67) have studied the particle deposition within the lung under the function of age: children and young adults aged 2 to 18, based on particle sizes 0.1 μm to 10 μm. There are few studies that have investigated these effects by comparing the normal lung to the emphysematous lungs, known as senile lung, in order to analyze the flow characteristics and the shear stress (68,69). Kim et al. (25) have studied the characteristics of some mechanical properties based on a comparison of the cases of people between 50 and 80 years of age. The maximum pressure drop and shear stress were found in case of 80 years' old. The authors report that one of the key aspects that could influence the change of mechanical properties is having different lung diameters. However, these studies only focus on the analysis of a group of people and do not consider the effect of flow rates. Therefore, this chapter will focus on understanding adult lung and elderly lung conditions that affect the airflow patterns and particle transport. Two different lungs which have different diameters will be analyzed, based on the comparison between the normal asymmetric lung and the lungs that having smaller diameters.

8.2 Numerical Method

The conservation of mass and momentum assumptions were used to solve the fluid flow and particle transport:

$$\frac{\partial \rho}{\partial t} + \nabla \cdot (\rho \vec{v}) = S_m \tag{8.1}$$

Where, S_m is the mass source, and

$$\frac{\partial}{\partial t}(\rho \vec{v}) + \nabla \cdot (\rho \vec{v}\vec{v}) = -\nabla p + \nabla \cdot \left(\mu \left[\left(\nabla \vec{v} + \nabla \vec{v}^{T} \right) - \frac{2}{3} \nabla \cdot \vec{v} \mathrm{I} \right] \right) + \rho \vec{g} + \vec{F} \tag{8.2}$$

Where p is fluid static pressure

$\rho \vec{g}$ is a body force of gravity

$\vec{F}$ is a body force of external force (particle – fluid interaction)

In terms of flow characteristic, turbulent flow, which based on $k - \varepsilon$ model, was considered in this study. For the calculation of turbulent kinetic energy:

$$\frac{\partial(\rho k)}{\partial t} + \frac{\partial(\rho k u_i)}{\partial x_i} = \frac{\partial}{\partial x_j}\left[\frac{\mu}{\sigma_k}\frac{\partial_k}{\partial x_j}\right] + 2\mu_t E_{ij} E_{ij} - \rho \quad (8.3)$$

And for dissipation rate of an inertial frame:

$$\frac{\partial(\rho\varepsilon)}{\partial t} + \frac{\partial(\rho\varepsilon u_i)}{\partial x_i} = \frac{\partial}{\partial x_j}\left[\frac{\mu_t}{\sigma_\varepsilon}\frac{\partial_\varepsilon}{\partial x_j}\right] + \varepsilon_{1\varepsilon}\frac{\varepsilon}{k}2\mu_t E_{ij} E_{ij} - \varepsilon_{2\varepsilon}\rho\frac{\varepsilon^2}{k} \quad (8.4)$$

Where, u_i is the velocity component in the corresponding direction.

E_{ij} is the component of the rate of deformation,

u_t is eddy viscosity.

Based on Singh et al. (63), there are some constants that were collected during numerous iterations of appropriated data for the turbulent flows:

$C_\mu = 0.09$, $\sigma_k = 1.00$, $\sigma_u = 1.30$, $C_{1\varepsilon} = 1.44$, $C_{2\varepsilon} = 1.92$

There are two variables of fluid flow within the lung airway. The main component is air, which has the density of 1.225 kg/m^3 and viscosity at 1.7893x10^{5}kg/ms. The secondary component is aerosol which has the density at 1100 kg/m^3 under the discrete phase model. In terms of the method, particles were injected through the inlet surface at trachea under the surface injection method. The particle transport was calculated by:

$$\vec{F}_{D,i} = \frac{1}{2} C_D \frac{\pi d_{p,i}^2}{4} \rho(\vec{v}_{p,i} - \vec{v})|\vec{v}_{p,i} - \vec{v}| \quad (8.5)$$

Where C_D *is* drag coefficient

d_p *is particle diameter*

$\vec{v}_p$ *is the particle velocity*

The second law of Newton is employed for a single-particle motion *i*:

$$m_{p,i}\frac{\partial \vec{v}_{p,i}}{\partial t} = \vec{F}_{D,i} + m_{p,i}\vec{g} \quad (8.6)$$

The various particle sizes, which include 1 μm, 2.5 μm, 5 μm, and 10 μm, were injected with three different flow rates that involve 7.5 lpm, 15 lpm, and 30 lpm. The boundary conditions in this report were taken as the inlet velocity at trachea and outlet outflow at the end of fourth generation. In terms of wall conditions, stationary walls and no slip walls were used, while the boundary of wall conditions were based on the discrete phase model and a heat flux thermal. Under the discrete phase model, the wall was set as "trapped" condition which means if any particle particle touches the wall, will be deposited. Since the wall is covered by the mucus layer which is very sticky, this condition is proved to be accurate. See the summarize of properties as following (Tables 8.1 and 8.2).

8.3 Geometrical Development

To enhance the understanding of the airflow and particle patterns of the asymmetrical lung model, this study considers two different cases of asymmetric lung airways. In this section, all of the 3D models were created based on the smooth surface condition by using SolidWorks 2019. These models have been reconstructed from an asymmetric lung model from Islam et al. (60) which was based on a symmetric lung model from Weibel's lung model. The normal asymmetric lung model will be the basis of other lungs that have different airway conditions. The diameter of branches at the right side were smaller than the left side, around 25%.

The first case is the asymmetric lung model, having aging effects that cause a smaller diameter for the whole lung model.

TABLE 8.1

Air Properties

Main Component: It Is Considered as Continuous Phase	
Density (kg/m^3)	1.225
Viscosity (kg/ms)	1.7893x10^5

TABLE 8.2

Aerosol Properties

Secondary Component: It Is Considered as Discrete Phrase	
Density (kg/m^3)	1100
Particle size (μm)	1, 2.5, 5, 10
Inlet flow rate (lpm)	7.5, 15, 30

The diameter for this case was reduced by 10% and 20% from the normal asymmetric lung model. Seeg more information in Table 8.3 to Table 8.5.

The second condition is an asymmetric lung model with stenosis area. There were five different cases that were generated with the stenosis region in the different generations. All of these models used the same dimension except at the obstructive areas. The diameter in this area was reduced by 50% from the normal diameter of the normal airways. Moreover, the narrow airways were located in the random positions which start from first generation to fourth generation. See more detail following Table 8.6.

Figure 8.1a represents asymmetric lung airways without stenosis area. Figure 8.1b represents the stenosis section on the right side of the first generation. Figure 8.1c and Figure 8.1d indicate the stenosis regions at the left and right sides of second generation while Figure 8.1e and Figure 8.1f show the stenosis areas at the right side of third generation and fourth generation respectively.

8.4 Grid Generation and Validation

Grid generation is an essential feature of the the model to obtain an accurate result for CFD simulation. In order to get an appropriate grid refinement, considering the mesh quality is the first aspect that should be prioritized. The ANSYS mesh module was used to generate the mesh for the whole geometry by using adapted sizing. However, there were some features that need to be used to control the flow characteristics inside the domain. Therefore, the inflation layer would be used as a boundary to control the change of velocity and the particle movement inside the domain. In this study, the mesh generation for all of the models was generated based on the

TABLE 8.3

The Dimension of Human Lung Airways for Normal Asymmetric Lung

Number of Generations	Location	Diameter				Length (cm)	Number
		Left Lung (cm)	Total Cross-Sectional Area (cm^2)	Right Lung (cm)	Total Cross-Sectional Area (cm^2)		
0	Trachea		1.80 cm		2.54 cm^2	12.0	1
1	Bronchi	1.22	1.17	0.92	0.66	4.8	2
2	Bronchi	0.83	0.54	0.62	0.30	1.9	4
3	Bronchi	0.56	0.24	0.42	0.14	0.8	8
4	Bronchioles	0.45	0.16	0.34	0.09	1.3	16

TABLE 8.4

The Dimension of Human Lung Airways for Asymmetric Lung Case with Aging Effects: With the Reduction of 10%

Number of Generations	Location	Diameter				Length (cm)	Number
		Left Lung (cm)	Total Cross-Sectional Area (cm^2)	Right Lung (cm)	Total Cross-Sectional Area (cm^2)		
0	Trachea		1.62 cm		2.06 cm^2	12.0	1
1	Bronchi	1.09	0.93	0.83	0.54	4.8	2
2	Bronchi	0.75	0.44	0.56	0.25	1.9	4
3	Bronchi	0.50	0.19	0.38	0.11	0.8	8
4	Bronchioles	0.41	0.13	0.31	0.08	1.3	16

TABLE 8.5

The Dimension of Human Lung Airways for Asymmetric Lung Case with Aging Effects: With the Reduction of 20%

Number of Generations	Location	Diameter				Length (cm)	Number
		Left Lung (cm)	Total Cross-Sectional Area (cm^2)	Right Lung (cm)	Total Cross-Sectional Area (cm^2)		
0	Trachea		1.44 cm		1.63 cm^2	12.0	1
1	Bronchi	0.98	0.75	0.74	0.43	4.8	2
2	Bronchi	0.66	0.34	0.49	0.19	1.9	4
3	Bronchi	0.45	0.16	0.34	0.09	0.8	8
4	Bronchioles	0.36	0.10	0.27	0.06	1.3	16

TABLE 8.6

The Dimension of Human Lung Airways for Asymmetric Lung Case with Stenosis Areas

Number of Generations	Location	Diameter				Length (cm)	Number	Diameter of Stenosis Area (cm)
		Left Lung (cm)	Total Cross-Sectional Area (cm^2)	Right Lung (cm)	Total Cross-Sectional Area (cm^2)			
0	Trachea		1.80 cm		2.54 cm^2	12.0	1	–
1	Bronchi	1.22	1.17	0.92	0.66	4.8	2	0.46
2	Bronchi	0.83	0.54	0.62	0.30	1.9	4	0.31
3	Bronchi	0.56	0.24	0.42	0.14	0.8	8	0.21
4	Bronchioles	0.45	0.16	0.34	0.09	1.3	16	0.17

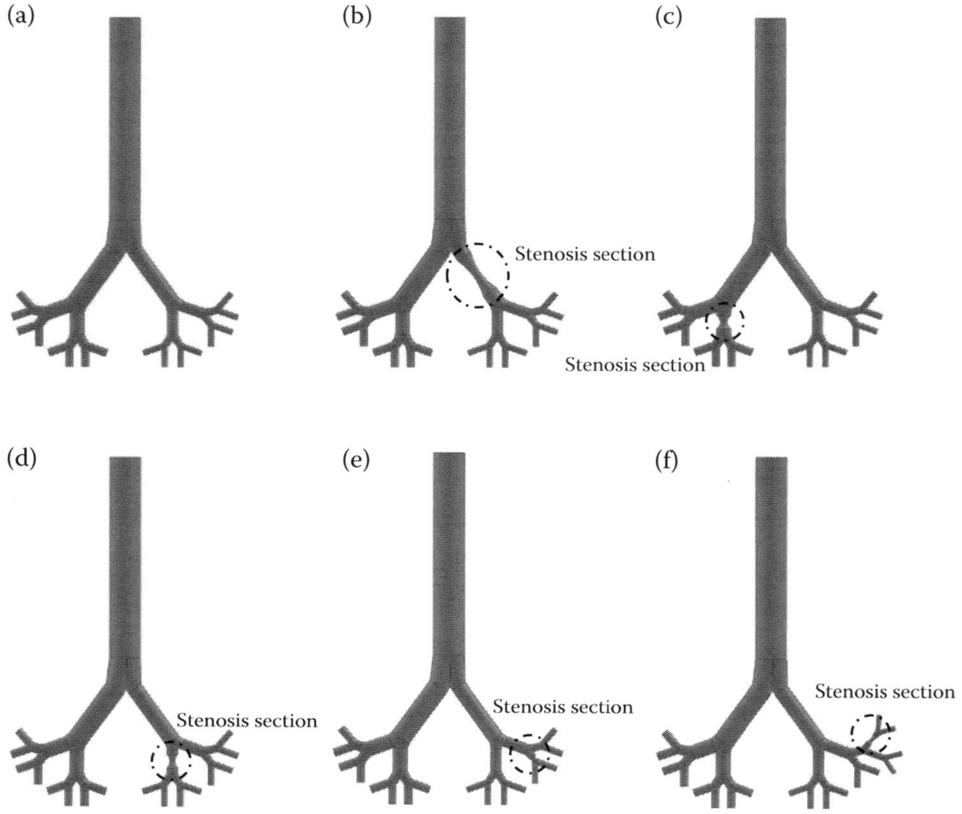

FIGURE 8.1 Reconstructed Models of Asymmetric Lung Airways: (a) Normal Asymmetric Lung, (b) Stenosis-right at First Generation, (c) Stenosis-left at Second Generation, (d) Stenosis-right at Second Generation, (e) Stenosis-right at Third Generation, and (f) Stenosis-right at Fourth Generation

unstructured elements (Figure 8.2a). Figure 8.2b shows a boundary of inflation layer for the inside of the model. Figure 8.2c and Figure 8.2d present the mesh at the entry trachea and outlet mesh of third generation. Figure 8.2e and Figure 8.2f represent the mesh at the stenosis area and mesh at a bifurcating branch, respectively.

Grid refinement was generated based on flow condition and appropriate *y*+ value was selected as outlined in Srivastav Paul and Jain (70). An orthogonal quality was applied to be the criteria to determine the mesh quality. The patch conforming method was also applied for the whole geometry. The model with the different number of elements was recalculated several times to obtain the maximum velocity each time. After recalculation, the maximum velocities became stable when the number of elements reached around 1.4 million (Figure 8.3a) with an average of minimum orthogonal quality around 0.13. Figure 8.3 presents the result for grid-independence test.

The numerical results were compared with the available benchmark for experimental and computational measurements (71,73–75). A mouth–throat model was used for validation as no similar stenosis model was found in the literature. Figure 8.3b shows the deposition fraction comparison at the mouth–throat section for different flow rates. The numerical results were in the range of the experimental measurement and showed a good agreement with the published literature.

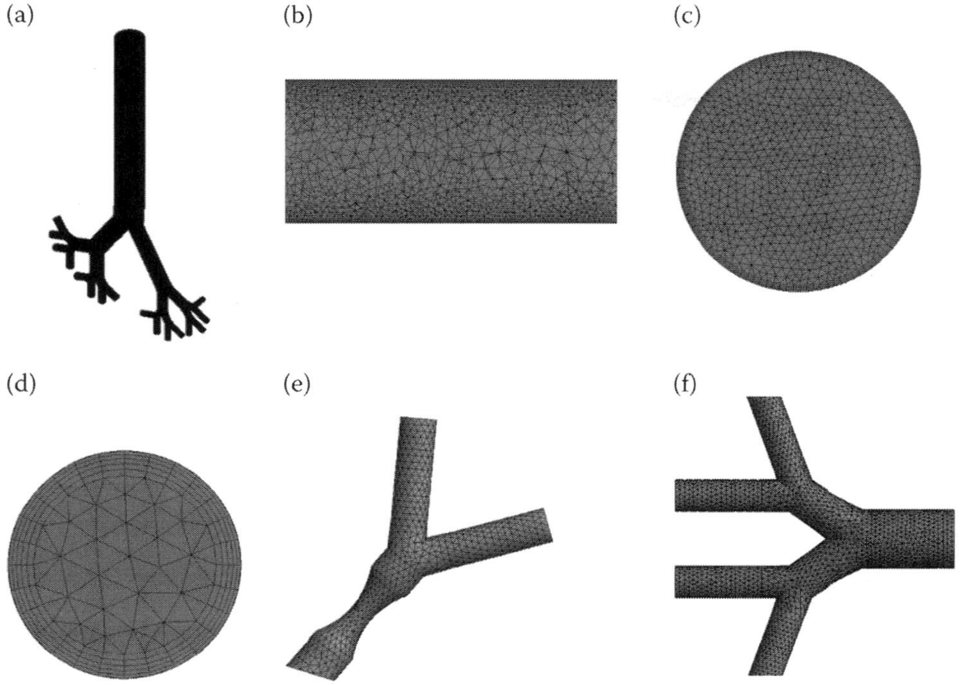

FIGURE 8.2 Unstructured Mesh for the Model: (a) The Complete Computational Model, (b) Inflation Layer Mesh Near to the Wall, (c) Inlet Mesh, (d) Outlet Mesh of Third Generation, (e) Mesh at Stenosis Area, and (f) Mesh at a Bifurcating Branch

8.5 Results and Discussion

All models with both aging effect and stenosis cases were used to study the airflow patterns and particle transport based on two variable conditions that are inlet velocities and particle sizes. In terms of initial velocities, this report will focus on three different inlet velocities: 0.4915 m/s, 0.9829 m/s, and 1.9658 m/s, as flow rate 7.5 lpm, 15 lpm, and 30 lpm respectively. In terms of particle sizes, this report will employ four different sizes: 1 μm, 2.5 μm, 5 μm, and 10 μm.

8.5.1 Effects of Aging Cases

8.5.1.1 Airflow Analysis

8.5.1.1.1 Velocity Profiles

The velocity profiles will be located at seven different cross-sections on three different asymmetric lung models under the particle size of 10 μm. The comparison will be based on three different flow rates that are 7.5 lpm, 15 lpm, and 30 lpm. Figure 8.4 shows the velocity profiles for normal asymmetric lung whereas Figure 8.5 and Figure 8.6 represent the velocity profiles for the reduction of diameter by 10% and 20% respectively.

- *Velocity profiles for asymmetric lung model*

Overall, in Figure 8.4 the velocities are continually increasing from the inlet and continually decreasing after through the airway at third generation. The maximum velocity magnitude is at Line 4 where it is the airway from the left lung.

- *Velocity profiles for asymmetric lung model with diameter reduction of 10%*

From Figure 8.5, the velocities tend to continually increase from the inlet at trachea area to second generation of the lung airways. However, after second generation, the velocities tend to decrease then increase again at third and fourth generations, respectively. The maximum velocity magnitude is found at the airway from the left lung (Line 4) where it has a larger lung scale.

- *Velocity profiles for asymmetric lung model with a diameter reduction of 20%*

From Figure 8.6, the velocities tend to continually increase, beginning from the inlet at trachea through first and second generations. After that, the velocities seem to decrease at third generation and increase at fourth generation. The maximum velocity magnitude is seen at the left lung (Line 4) that has a larger diameter.

Overall, of velocity profiles from three different cases that have various lung volumes (Figures 8.4 to 8.6), the diameter of the whole lung airway could have a significant effect on the velocity patterns. The velocity tends to continually increase from the inlet at trachea to second generation. After that, velocity decreases at third generation and becomes increasingly again at fourth generation. In addition, the higher velocity usually occurs at the left lung where it has a larger diameter. Based on the studies from Islam et al. (58) and Singh et al. (63), due to the airflow resistance of smaller airways, it can flow the air to the opposite lung side. In order to support this theory, the velocity magnitude from the reducing diameter

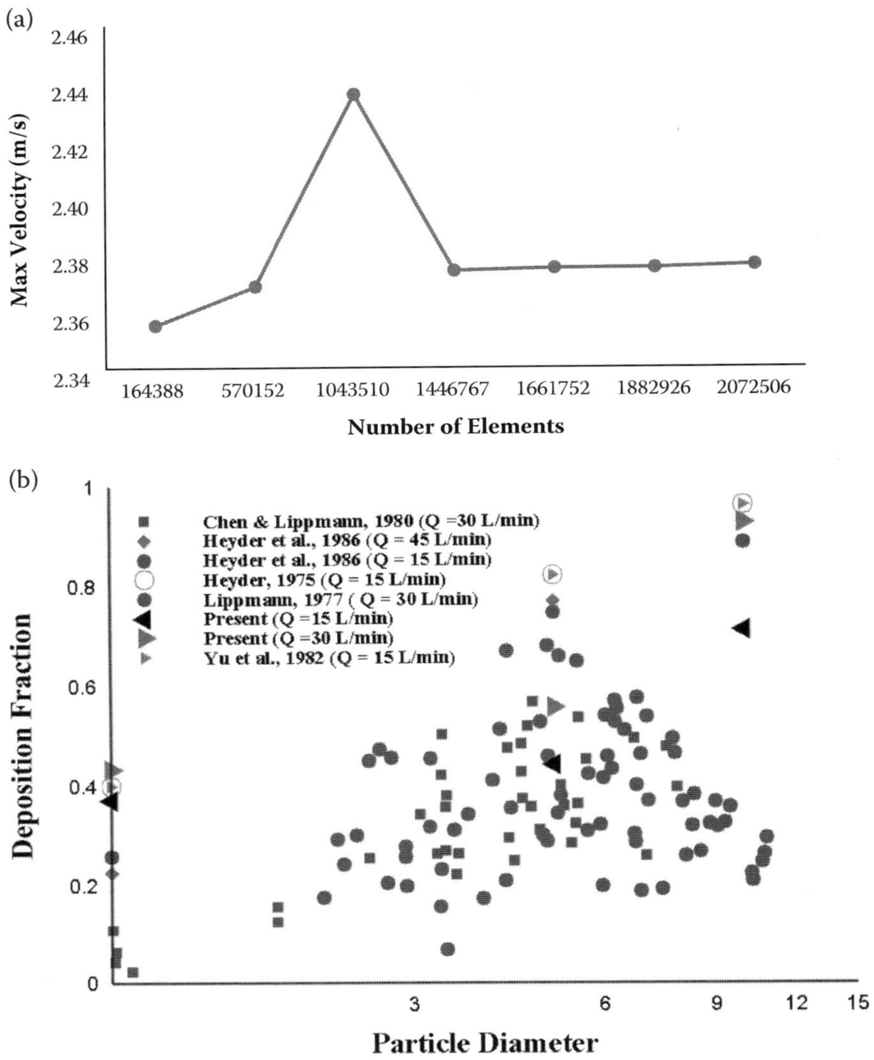

FIGURE 8.3 (a) Grid-Independent for the Maximum Velocity: Under the Inlet Condition at 30 lpm and (b) Deposition Fraction Comparison with Published Literatures (70–74)

cases have higher values as compared to the normal lung case. By the diameter reduction of 20% we observed maximum velocity in the airways. As the reduction percentage of the diameter is decreased the velocity is also reduced.

8.5.1.1.2 Velocity Contours

Two types of velocity contours will be presented in this report. The first type is the velocity contour for the whole domain on the XY plane, whereas the second type inlcudes the contours at the cross-section planes. Figure 8.7 is the velocity contours for the asymmetric lung having different diameters. Figure 8.7a is for normal asymmetric lung, whereas Figure 8.7b and c show the reduction of diameter by 10% and 20% respectively.

Similar to the comparison from velocity profiles, the airflow velocity continually increases from the inlet at the trachea area. From Figure 8.7, all three cases have similar velocity patterns that have a higher velocity magnitude at the left lung, where it is bigger than the right lung. However, the maximum velocity magnitude tends to be found in the case that has 20% of the diameter reduction (Figure 8.7c), whereas the normal lung case (Figure 8.7a) has the lowest value.

The selected planes are located at the random position which begins at trachea (0 generation) to fourth generation based on cross-sectional contours. The objective of these contours is to compare the levels of velocity between the normal asymmetric lung and the lung that has smaller diameter with 10% and 20% of the diameter reduction. Figure 8.8 shows the location of selected planes at different generations. Figure 8.9 is the velocity contours for all three cases at different locations.

From Figure 8.9, the velocity contours at trachea area (Plane 1) are found to be similar for all three cases. Similarly, the velocity contours at first generation (Planes 2 and 3) and other generations (Plane 4 to Plane 9) also tend to be similar for all three cases. However, the higher and highest levels are usually found at the left-lung airways that have larger diameter (Plane 4, 6, 8). According to Singh et al. (63), the velocity pattern of velocity contour can be changed by the shape of airway

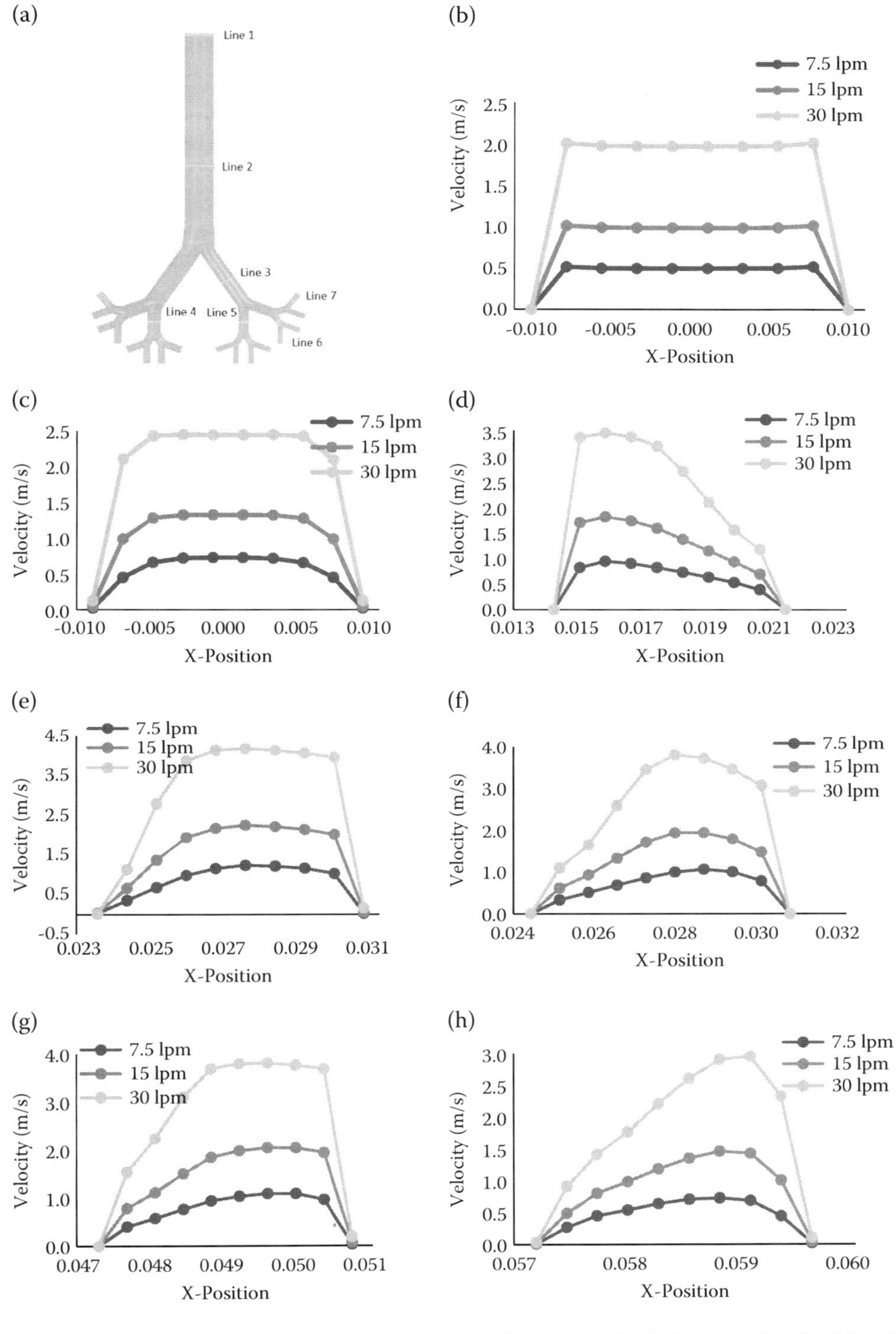

FIGURE 8.4 Selected Line for Velocity Profiles of Asymmetric Lung Model: (a) Selected Location (b) Line 1, (c) Line 2, (d) Line 3, (e) Line 4, (f) Line 5, (g) Line 6, and (h) Line 7

conditions. Therefore, the maximum velocity magnitude for each plane belongs to the smallest airway (Figure 8.9c).

8.5.1.2 Pressure Drop

The pressure drop and pressure contours are presented in this section.

8.5.1.2.1 Pressure Distribution

The randomly selected planes are firstly located in the middle of the trachea through the whole lung to fourth generation. The calculation for pressure drop is based on 10 μm of particle size and three different flow rates that are 7.5 lpm, 15 lpm, and 30 lpm. Figure 8.10 presents the pressure drop at nine selected planes for three cases.

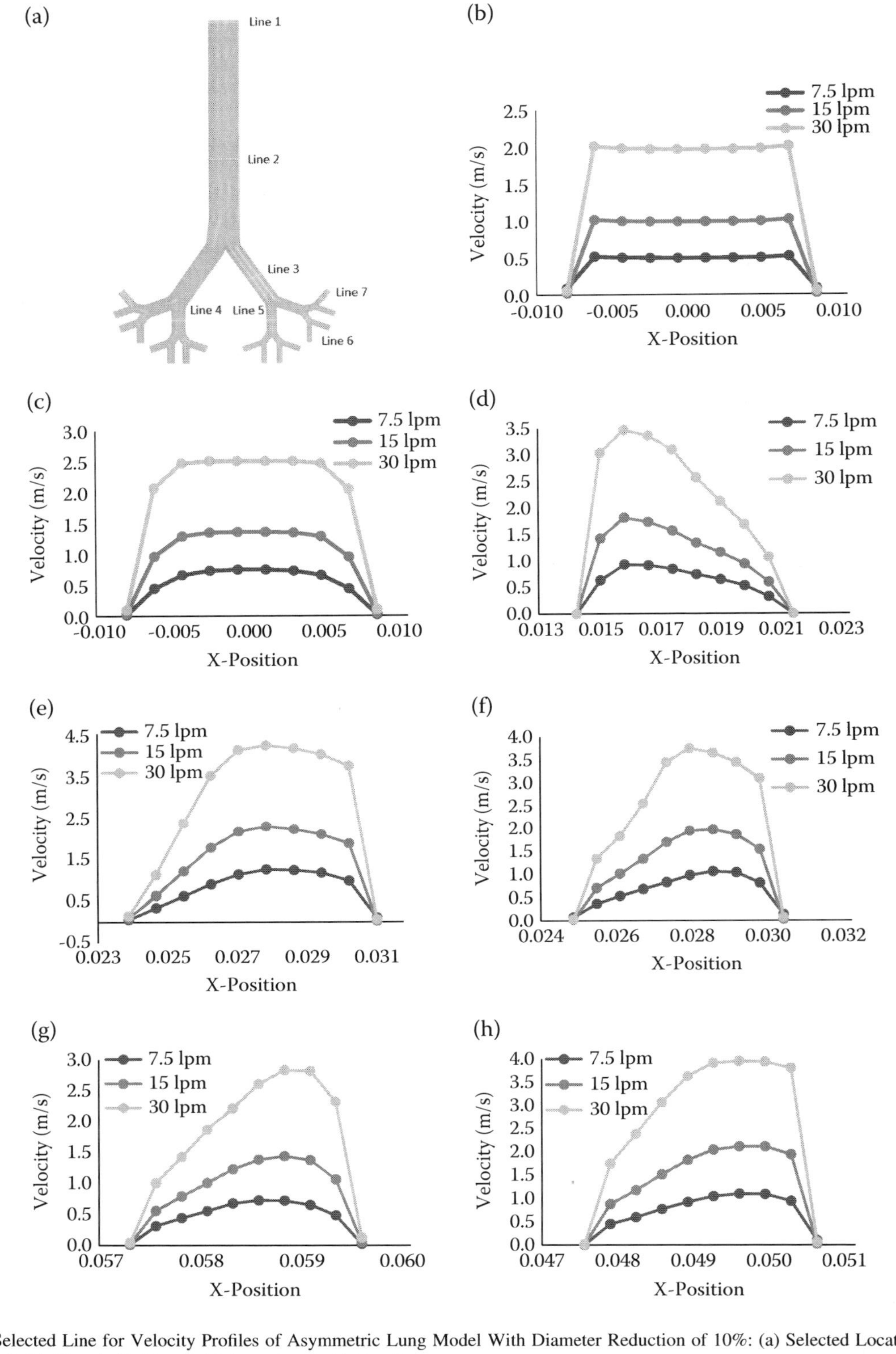

FIGURE 8.5 Selected Line for Velocity Profiles of Asymmetric Lung Model With Diameter Reduction of 10%: (a) Selected Location (b) Line 1, (c) Line 2, (d) Line 3, (e) Line 4, (f) Line 5, (g) Line 6, and (h) Line 7

The results from Figure 8.10 indicate that the pressure continually decreases from the inlet because of the air direction that always flows from the higher-pressure area to the lower-pressure area. The maximum pressure drop is found at the lowest branch at fourth generation (Plane 9) for the highest flow rate at 30 lpm (Figure 8.10c). In terms of maximum pressure, the case that has the smallest diameter usually has maximum pressure at the inlet as compared to the other two cases.

To summarize, the level of pressure drop tends to be based on the flow rate. A higher flow rate can cause higher pressure

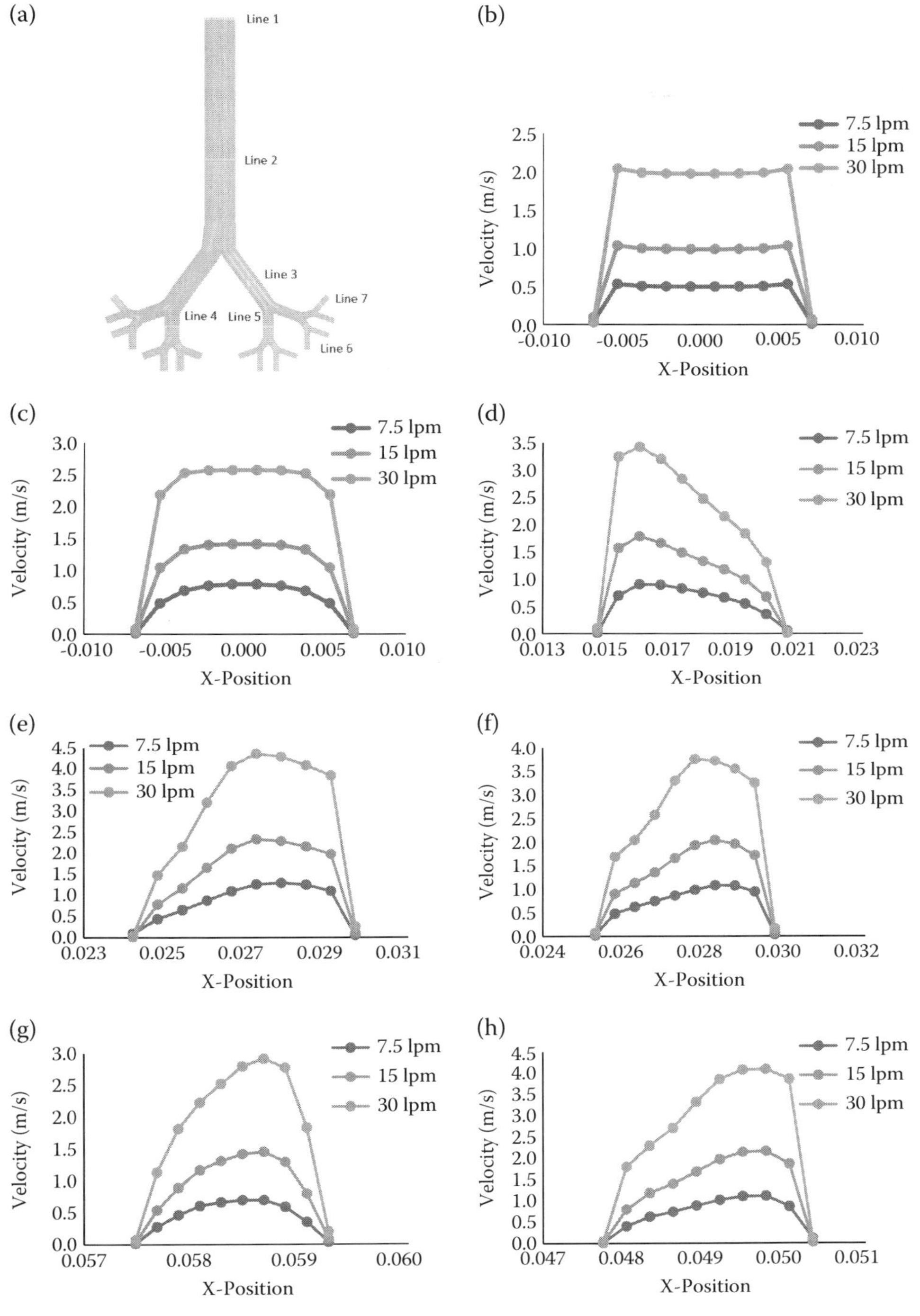

FIGURE 8.6 Selected Line for Velocity Profiles of Asymmetric Lung Model With Diameter Reduction of 10%: (a) Selected Location (b) Line 1, (c) Line 2, (d) Line 3, (e) Line 4, (f) Line 5, (g) Line 6, and (h) Line 7

and a higher pressure drop. In terms of the effect of the diameters for the whole lung, the case that has a diameter reduction of 10% has a similar pressure level to the normal lung case. On the other hand, the 20% of diameter reduction case has a higher pressure level than the other two cases.

8.5.1.2.2 Pressure Contours

The overall of pressure for the whole lung model is provided based on the XY plane of the contours feature. Highest flow rates (30 lpm) and largest particle size (10 μm) were used for the simulation. Figure 8.11 indicates the pressure contour at the XY plane for three different diameters.

From Figure 8.11, all three cases have similar patterns in terms of the level of the pressure. The pressure tends to continually decrease from the inlet at trachea to other generations. However, a higher pressure always occurs at the bifurcation for all generations. Furthermore, the highest pressure is generally located at the bifurcation of the first generation for all three

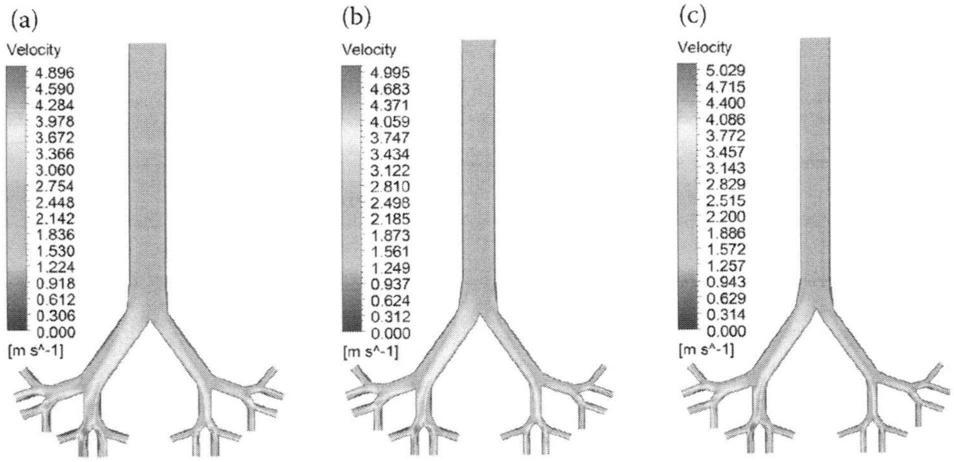

FIGURE 8.7 Velocity Contours at XY Plane for Three Lung Models: (a) Normal Asymmetric Lung, (b) Reduction of Diameter of 10%, (c) Reduction of Diameter of 20%

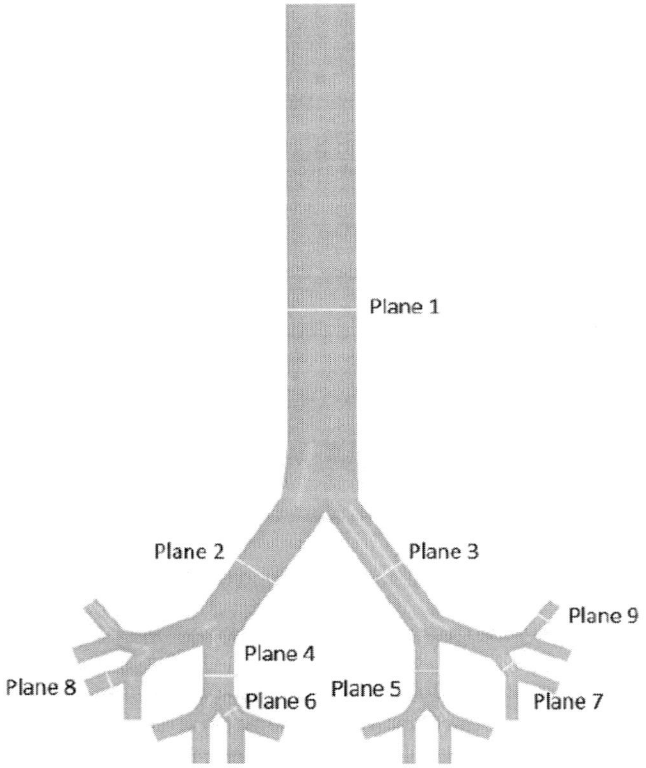

FIGURE 8.8 Selected Planes for Different Positions of Asymmetric Lung Models

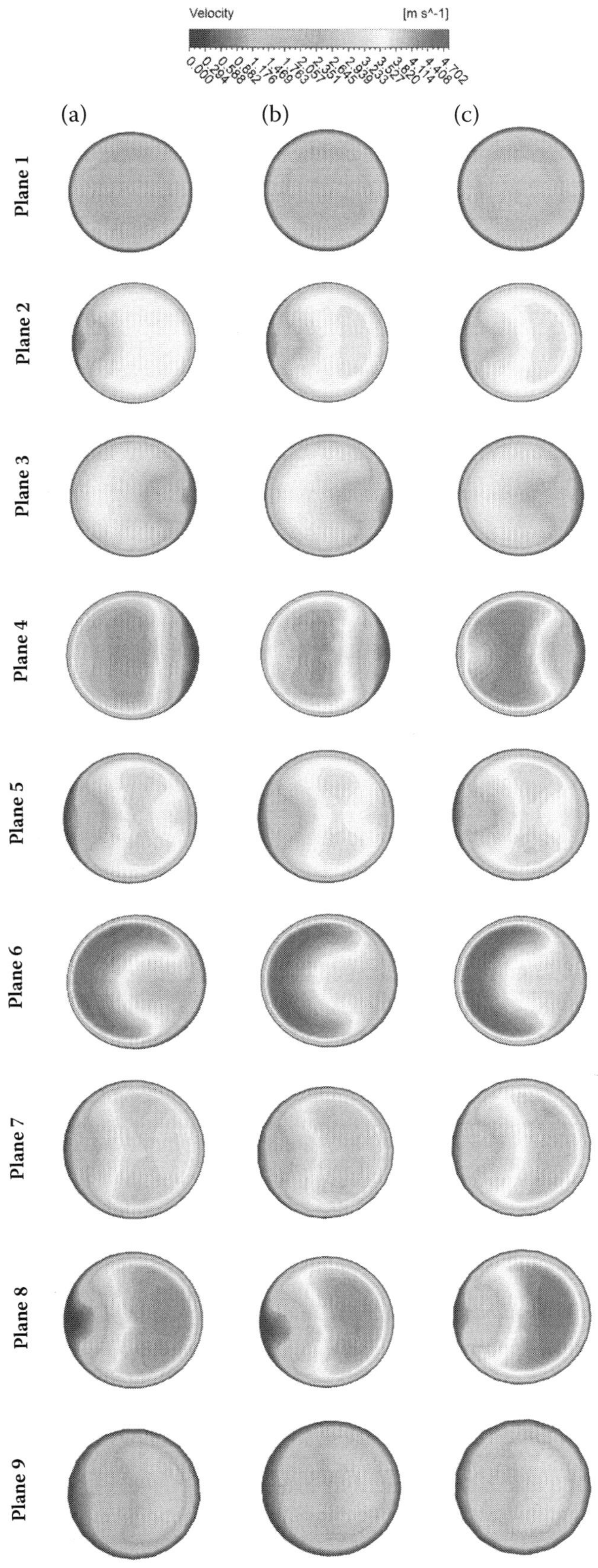

FIGURE 8.9 Velocity Contours at Different Positions of Asymmetric Lung Models at 30 lpm: (a) Normal Asymmetric Lung, (b) Reduction of Diameter for 10%, (c) Reduction of Diameter for 20%

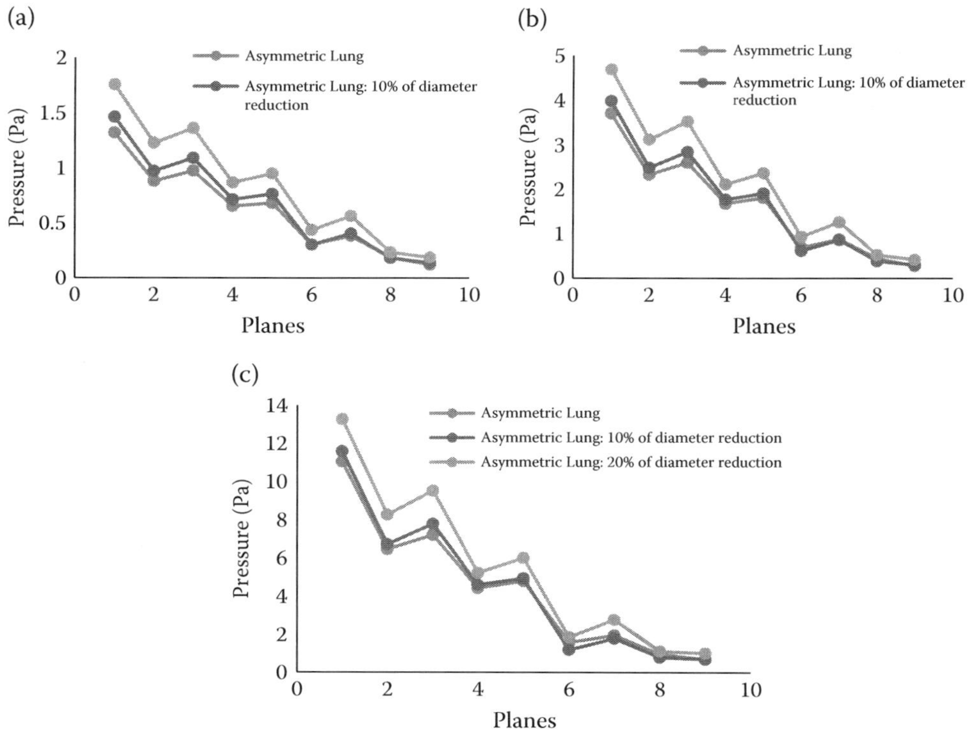

FIGURE 8.10 Pressure Drop at Different Plane Positions for Asymmetric Lung with Different Diameters: (a) 7.5 lpm, (b) 15 lpm, and (c) 30 lpm

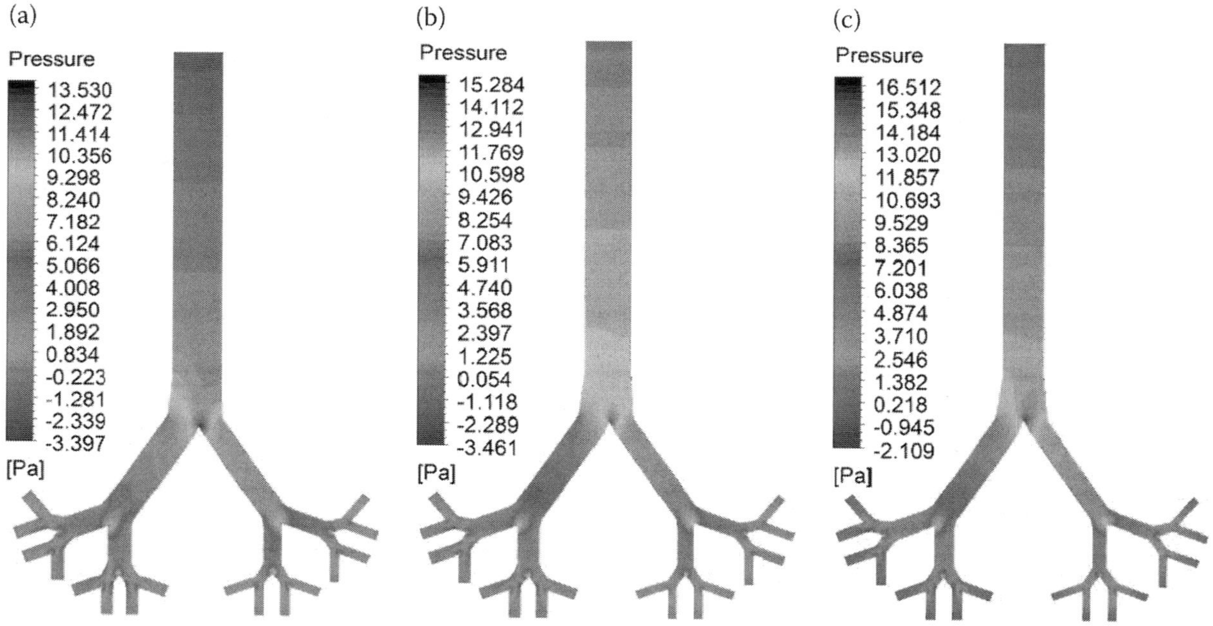

FIGURE 8.11 Pressure Contours at XY Plane for Three Lung Models: (a) Normal Asymmetric Lung, (b) Reduction of Diameter of 10%, (c) Reduction of Diameter of 20%

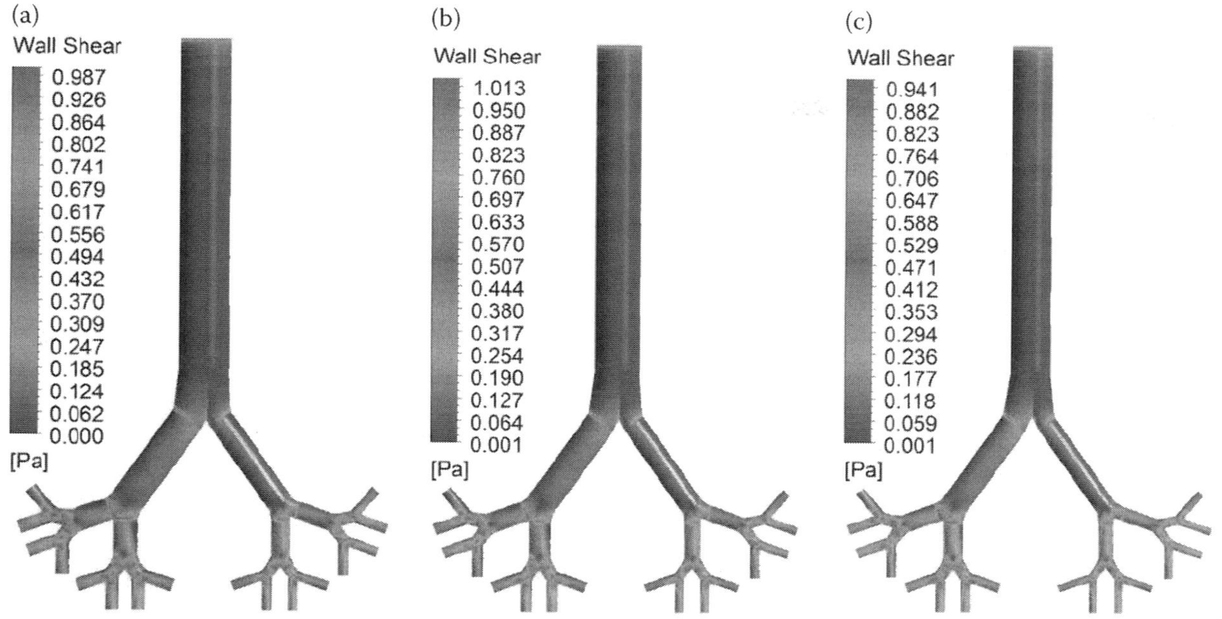

FIGURE 8.12 Wall Shear for Three Lung Models at 30 lpm: (a) Normal Asymmetric Lung, (b) Reduction of Diameter of 10%, (c) Reduction of Diameter of 20%

cases. It can be noticed that the case that has the smallest diameter has the highest range of pressure (Figure 8.11c). Referring to Nowak, Kakade, and Annapragada (29), lung volume can lead to different pressure levels in bifurcation. It can be summarized that a smaller diameter will lead to a higher pressure level.

8.5.1.3 Wall Shear

The shear stress at the wall is obtained based on the calculation with the flow rate of 30 lpm and particle diameter of 10 μm. Figure 8.12 shows the shear stress at the wall for normal asymmetric lung and two of diameter reductions.

Figure 8.12 indicates that the shear stress at the wall is higher at the inlet area for all three cases. Then it decreases and begins to increase again after passing first generation of the airway. The higher shear stress is always found in the bifurcation area for all cases. However, the highest shear stress is found at the case which has 10% reduction of diameter (Figure 8.12b), whereas the second-highest shear stress belongs to the normal lung case (Figure 8.12a). Consequently, the level of shear stress at the wall is slightly influenced by the airway volume and has a greater influence from the bifurcation area.

8.5.1.4 Turbulent Intensity

In this section, the comparison of the level of turbulent intensity for three different cases is provided based on the 30 lpm flow rate and 10 μm particle size. Figure 8.13 presents the turbulent contour for the whole lung with three different cases.

From Figure 8.13, turbulent flow is observed at the inlet region and at the bifurcation area for each generation. It can be seen that turbulence intensity occurs at the same location for both normal lung case and the 20% reduction of diameter case (Figure 8.13a,c), whereas the second case (Figure 8.13b) has differently turbulence occurring. This case only has turbulence intensity at fourth bifurcation of the left lung. To compare the three cases that have different diameters, the highest turbulence intensity is located in the first case (Figure 8.13a) that has a normal diameter. However, there is only a slight variation of turbulence intensity between all three cases. Therefore, it can be summarized that the diameter could have an insignificant effect on the level of turbulence intensity.

Turbulent intensity at a different positions is calculated based on nine selected planes that involve zero generation to fourth generation. The flow rate at 30 lpm is used for the calculation of three different lung models, as shown in Figure 8.14.

From the graph of Figure 8.14, it can be seen that the normal lung airway has a greater amount of turbulence intensity. The highest turbulence fluctuation also occurs at Plane 6 where it is located at third generation. Both cases that have the diameter reducing conditions have identical trends. However, the airflow becomes stable at the fourth generation and tends to be continually stable for all lower generations.

8.5.1.5 Particle Transport

8.5.1.5.1 Deposition Efficiency

The comparison of particle deposition efficiency for asymmetric lung models with different sizes is presented based on three different flow rates: 7.5 lpm, 15 lpm, and 30 lpm (Figure 8.15).

From Figure 8.15, it is clear that the airway volume can significantly affect deposition efficiency. Smallest lung airways have a greater amount of particle deposition for all flow rates whereas the other two lungs, which have bigger diameters, have similar levels. One of the key factors that could influence the level of particle efficiency is the flow rate. If

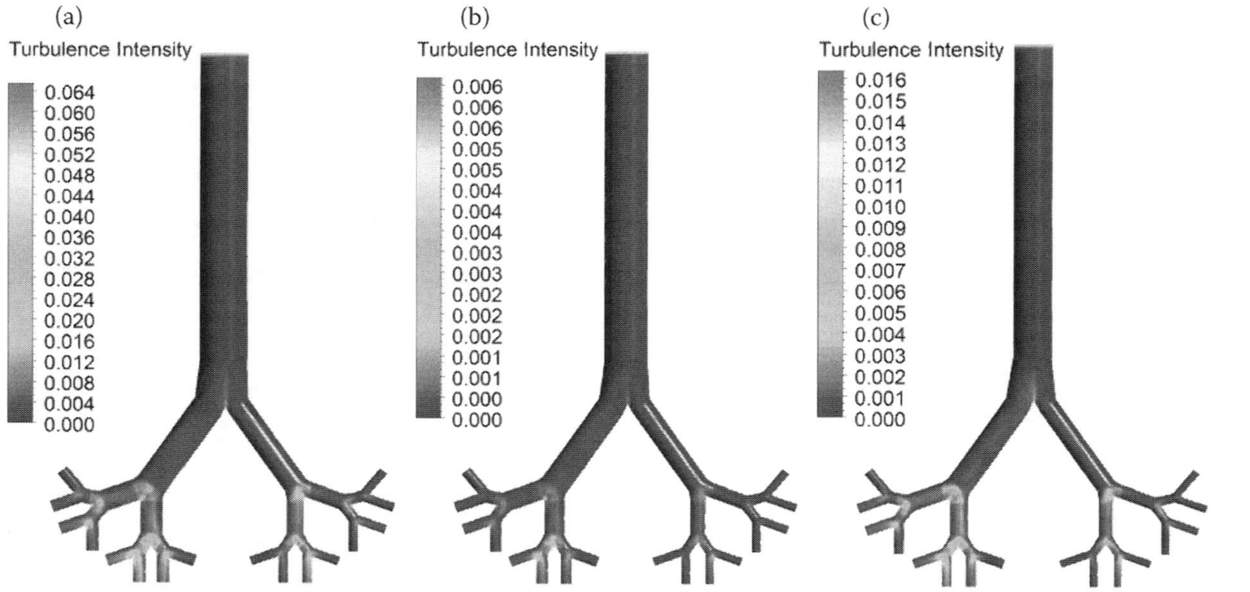

FIGURE 8.13 Turbulence Intensity Contour for Three Lung Models at 30 lpm: (a) Normal Asymmetric Lung, (b) Reduction of Diameter of 10%, (c) Reduction of Diameter of 20%

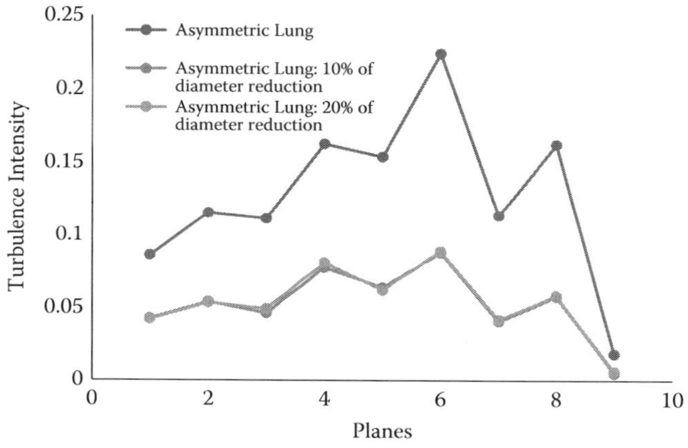

FIGURE 8.14 Turbulence Intensity at Different Plane Positions for Three Lung Models at 30 lpm

compared to all flow rates from above figure, the highest flow rate that is 30 lpm (Figure 8.15c) has the highest deposition efficiency which reaches around 48% in terms of 10 μm of particle size, whereas the flow rates at 7.5 lpm and 15 lpm (Figure 8.15a,b) could reach around 6% and 17% respectively.

It can be summarized that a higher particle inertia, which is caused by larger particle size and higher flow rate, could increase the deposition efficiency [van Ertbruggen et al. (27, 16, 62)]. In addition, the diameter of the lung airways also affects the deposition efficiency rate, especially in the case of a 20% diameter reduction.

8.5.1.5.2 Deposition Fraction

The calculation of deposition fraction for different particle diameters is shown in Figure 8.16.

From Figure 8.16, the particle sizes of 1 μm and 2.5 μm (Figure 8.16a,b) have similar deposition fraction rates that are around 2% to 5% for all three flow rates, while the particle size as 5 μm (Figure 8.16c) has the range between 2% to 8% for all three flow rates. However, the largest particle size as 10 μm (Figure 8.16d) has the highest range between 6% to 48%. The highest deposition fraction is found at the highest flow rate, at 30 lpm. For the overall deposition fraction, in the case of having three different diameters of the lung airways, it can be seen that the smallest lung airway, with 20% diameter reduction, usually has a higher deposition rate for all particle sizes, whereas the other two lung airways have slightly different deposition rates.

To summarize, the deposition fraction is generally based on the particle size and flow rate, including density (63,65,76). Larger particle size and higher flow rates can increase the deposition fraction. The airway volume can also influence the deposition fraction. However, this is only for the case of a 20% diameter reduction; the case of a 10% diameter reduction has similar ranges to normal airways.

Four different particle sizes are injected from the inlet at trachea area under the different diameters of three asymmetric

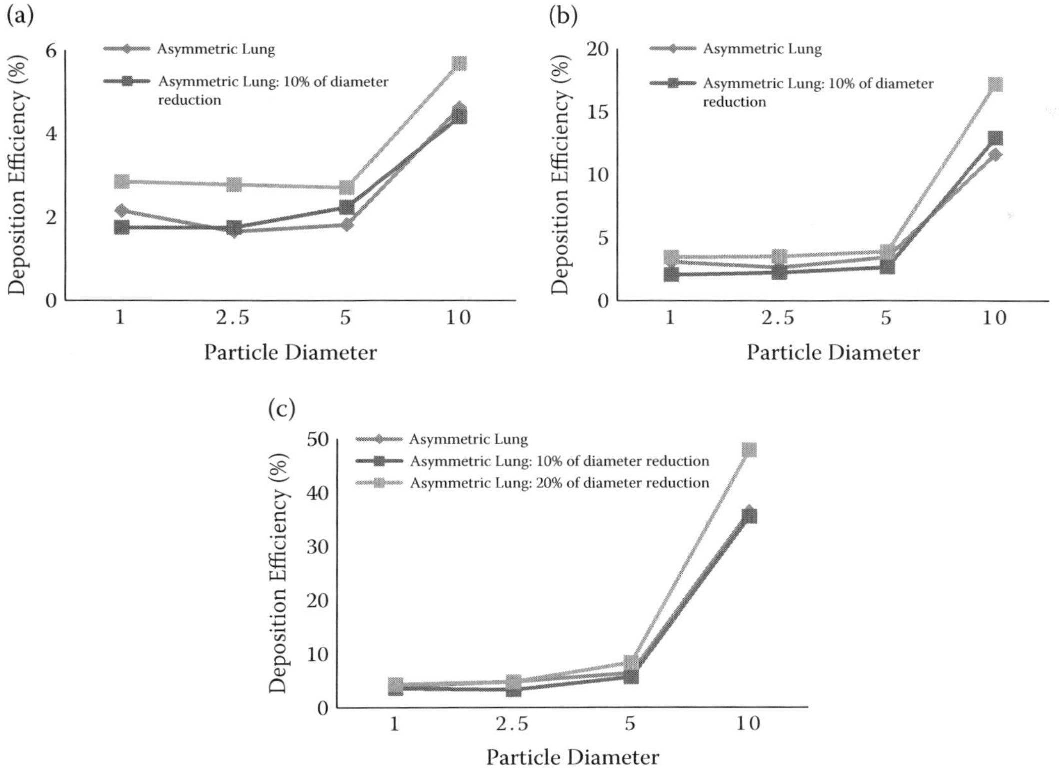

FIGURE 8.15 Deposition Efficiency in Different Regions of Asymmetric Lung for Three Different Diameters: (a) 7.5 lpm, (b) 15 lpm, and (c) 30 lpm

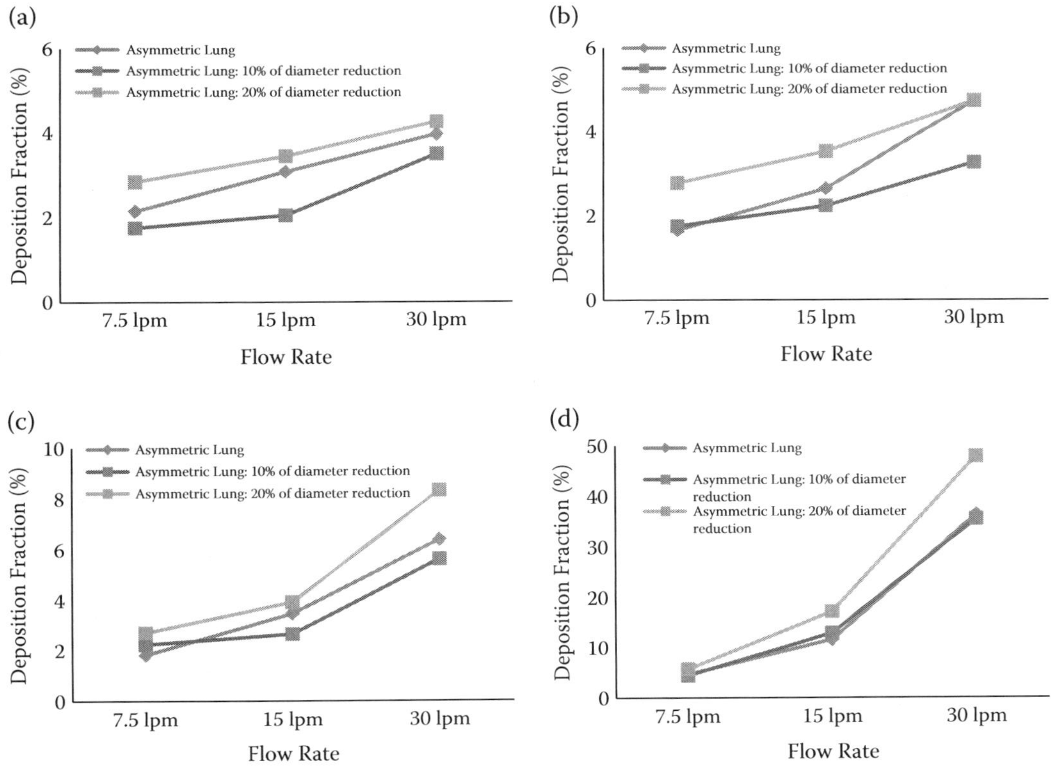

FIGURE 8.16 Deposition Fraction of Different-Diameter Particles in Asymmetric Lung for Three Different Diameters: (a) 1 μm diameter, (b) 2.5 μm diameter, and (c) 5 μm diameter, and (d) 10 μm diameter

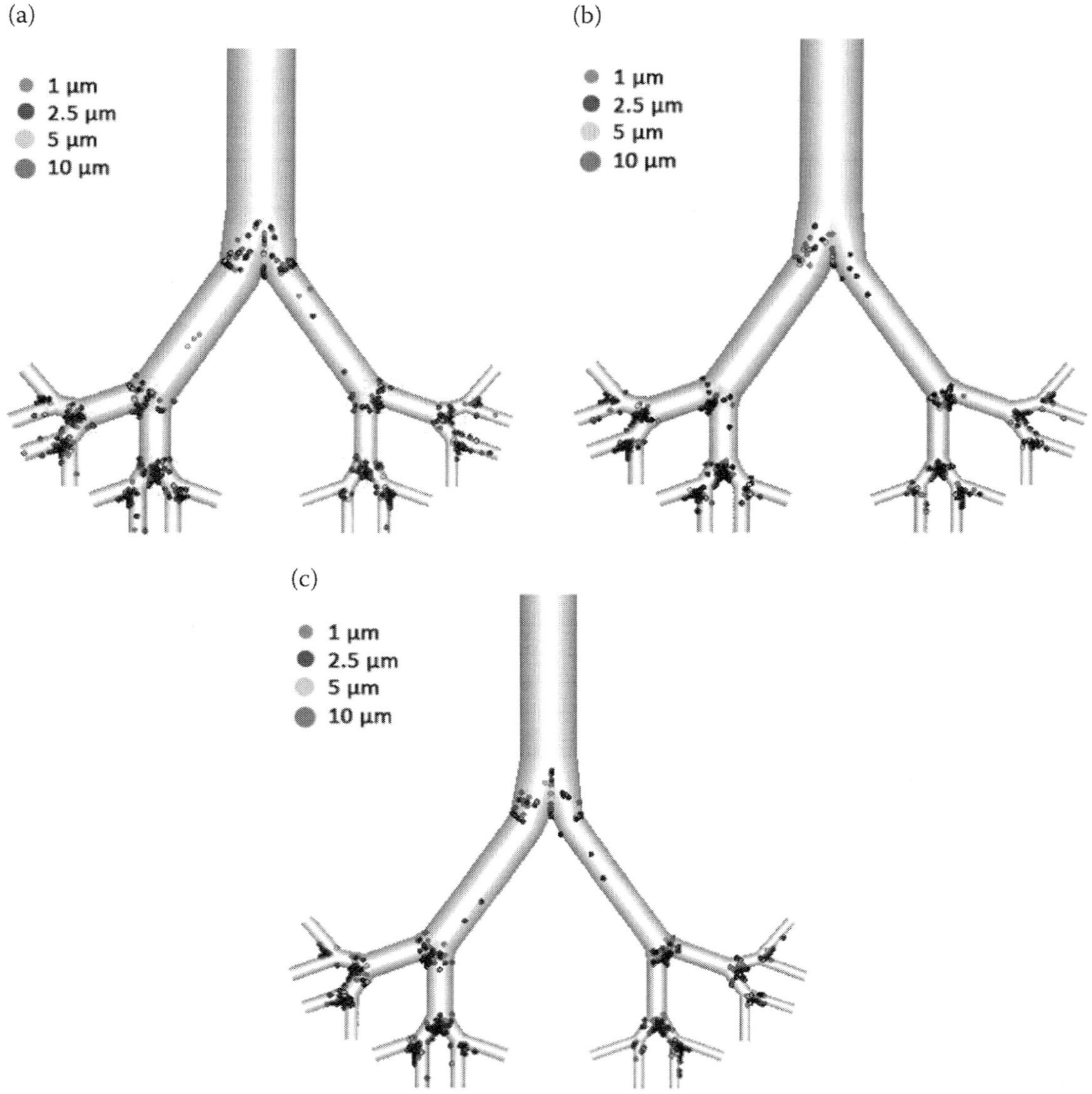

FIGURE 8.17 Particle Deposition Scenario at 30 lpm for Three Lung Models: (a) Normal Asymmetric Lung, (b) Reduction of Diameter of 10%, (c) Reduction of Diameter of 20%

lungs with the flow rate at 30 lpm. Figure 8.17 shows the particle deposition with three lung conditions.

Over all three cases, particles are usually trapped at the left lung rather than the right lung because it has a higher lung volume. Most particles are trapped at the wall of a bifurcation for all generations [(28); Farghadan et al. (61)]. It can be seen that after the first generation, the particles of 10 μm seem to be deposited more than other particle sizes. If compared to the deposition for all generations, particle size of 10 μm is likely trapped at the bifurcation of the second generation, whereas other particle sizes are slightly trapped at this generation. It can be summarized that the majority of particle sizes (10 μm) can be trapped in the lung airways from the first to fourth generations.

8.5.2 Stenosis Cases

8.5.2.1 Airflow Analysis

In order to understand the airflow characteristics, this report will provide a clear understanding of airflow patterns by considering two features: velocity profiles and velocity contours.

8.5.2.1.1 Velocity Profiles

Velocity profiles can be classified as the variation in velocity on a line at right angles to the original flow's direction. It could be representative of the velocity magnitude and change of the flow direction caused by the various shapes within the domain. In other words, velocity profiles can represent flow's behavior during transportation through the domain.

In this section, the velocity profiles will be plotted at seven different cross-sections on the five different cases in an asymmetric model with stenosis sections using 10 μm as the particle size. Figure 8.18 will be for the lung with stenosis area at first generation. Figures 8.19 and 8.20 will be for the airways with stenosis areas at second generation in left and right lungs. Figures 8.21 and 8.22 will be for the geometry with stenosis areas at third and fourth generations. In details, it can be easily described that Line 1will be the representative of the inlet, Line 2 will be the middle of the trachea, Line 3 will be for the right lung at first generation, and Lines 4 and 5 will be for the left lung and right lung at second generation. In contrast, Lines 6 and 7 will be for the right lung at third and fourth generation respectively.

- *Velocity profiles for asymmetric lung model with right-stenosis area at firstgeneration*

Overall, in Figure 8.18, the velocities are continually increasing from the inlet and becoming lower after the stenosis region at first generation. The lowest velocity magnitude occurs in Line 7 where it is located in the fourth generation. The maximum velocity magnitude is at Line 3, which is the stenosis area.

- *Velocity profiles for asymmetric lung model with left-stenosis area at secondgeneration*

Overall, in Figure 8.19, the velocities are continually increasing from the inlet and continually decreasing after through the airway at third generation. The maximum velocity magnitude, in this case, occurs in the stenosis area (Line 4). It can be summarized that the velocity at the left lung is higher than the right lung.

- *Velocity profiles for asymmetric lung model with right-stenosis area at second generation*

Overall, in Figure 8.20, the velocities are continually increasing from the inlet and decreasing from third generation. The maximum velocity magnitude occurs in the stenosis area (Line 5), where it is in the second generation.

- *Velocity profiles for asymmetric lung model with right-stenosis area at thirdgeneration*

In Figure 8.21, the velocities are continually increasing from the inlet and becoming lower again after through the stenosis region at third generation (Line 6). The maximum velocity can be found in the stenosis area (Line 6).

- *Velocity profiles for asymmetric lung model with right-stenosis area at fourthgeneration*

In Figure 8.22, the velocities are continually increasing from the inlet and becoming stable from the second generation. The maximum velocity magnitude for this case is at the left lung (Line 4).

The results of velocity profiles from five cases (Figures 8.18 to 8.22) and from a normal asymmetric lung case (Figure 8.4) could significantly indicate the relationship of the stenosis region, the dimension of the airway, and velocity. The airflow velocity is repetitively increasing from the inlet for all flow rates. If compared to all cases, the velocity profiles at the inlet and the middle of the trachea have a similar velocity magnitude. The maximum velocity always occurs at the stenosis area, and after that, velocity continually decreases. A higher velocity always occurs at the left lung, where it has a bigger diameter than the right lung. In contrast, a lower velocity always initiates at the fourth generation compared with the velocity at first to third generations. Besides, the stenosis region at the fourth generation could have only a few effects on the velocity because it is lower than the velocity at the left lung, which has a larger diameter. According to Islam et al. (58) and Singh et al. (62), higher airflow velocity at smaller airways or stenosis areas can be caused by airflow distribution within the lungs. With these conditions, air will flow to the opposite side because of the airflow resistance in these areas. The results in this report could support these authors because the maximum velocity for asymmetric lung without any narrowing could be found at the left lung (Figure 8.4d) where it is larger than the right lung. If compared to other stenosis cases, the velocity at the left lung (Line 4) will be lower than the velocity at stenosis areas.

8.5.2.1.2 Velocity Contours

Velocity contours represent fluid flow which discharges at the velocity point that is considered. It can be converted and represented as the average cross-sectional for the flow velocity at a selected point.

This report will provide two types of velocity contours with 30 lpm for the flow rates. The first type is a velocity contour for the whole domain that will be generated at the XY plane. Another type is the velocity contour at nine selected planes. Figure 8.23 shows the velocity contour at XY plane in different lung conditions. Figure 8.23a is for normal asymmetric lung, while Figure 8.23b is for an asymmetric lung with stenosis area at generation 1. Figure 8.23c,d is the geometry for the stenosis area at second generation of the left and right lungs, respectively. Figure 8.23e,f represents stenosis areas at third and fourth generations.

The results from this section are similar to the results from the velocity profiles section. The airflow velocity is continuing to increase from the inlet for all cases. For normal asymmetric lung (Figure 8.23a), velocity is significantly increasing at the left lung that has bigger airways, especially at the second branch. For the stenosis cases (Figure 8.23b–e), maximum airflow velocity occurs at stenosis areas, but it is only for the cases that have stenosis areas at first and third generations. The stenosis area at fourth generation (Figure 8.23f) does not have a considerable effect on airflow velocity. However, it can be seen that the bifurcation after the stenosis region has a lower airflow velocity if compared with normal asymmetric lung.

The randomly selected plans are generated at various locations from trachea (0 generation) to fourth generation by considering the comparison between the normal asymmetric lung

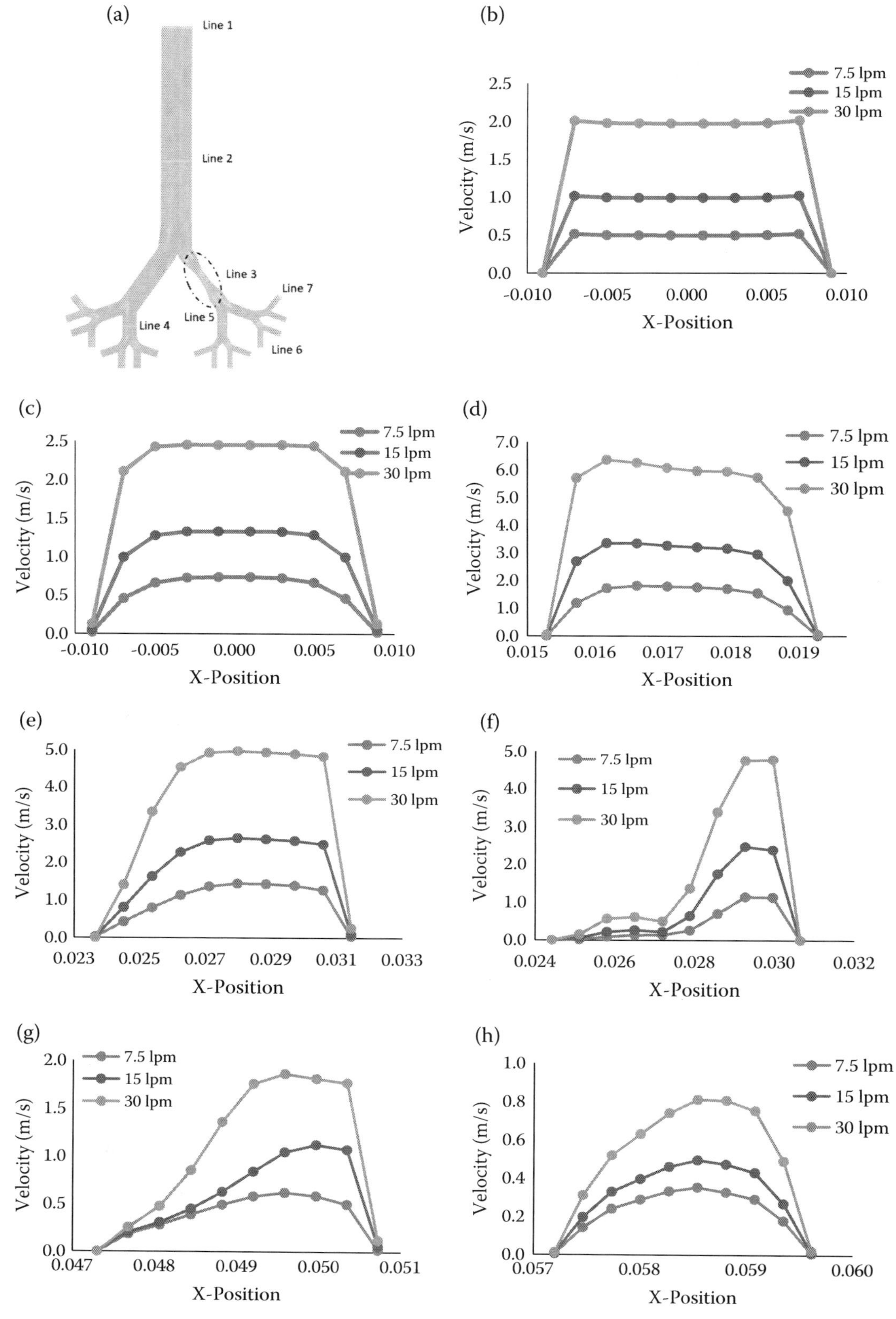

FIGURE 8.18 Selected Line for Velocity Profiles of Asymmetric Lung Model with Stenosis Area at First Generation: (a) Selected Location (b) Line 1, (c) Line 2, (d) Line 3 at Stenosis Area, (e) Line 4, (f) Line 5, (g) Line 6, and (h) Line 7

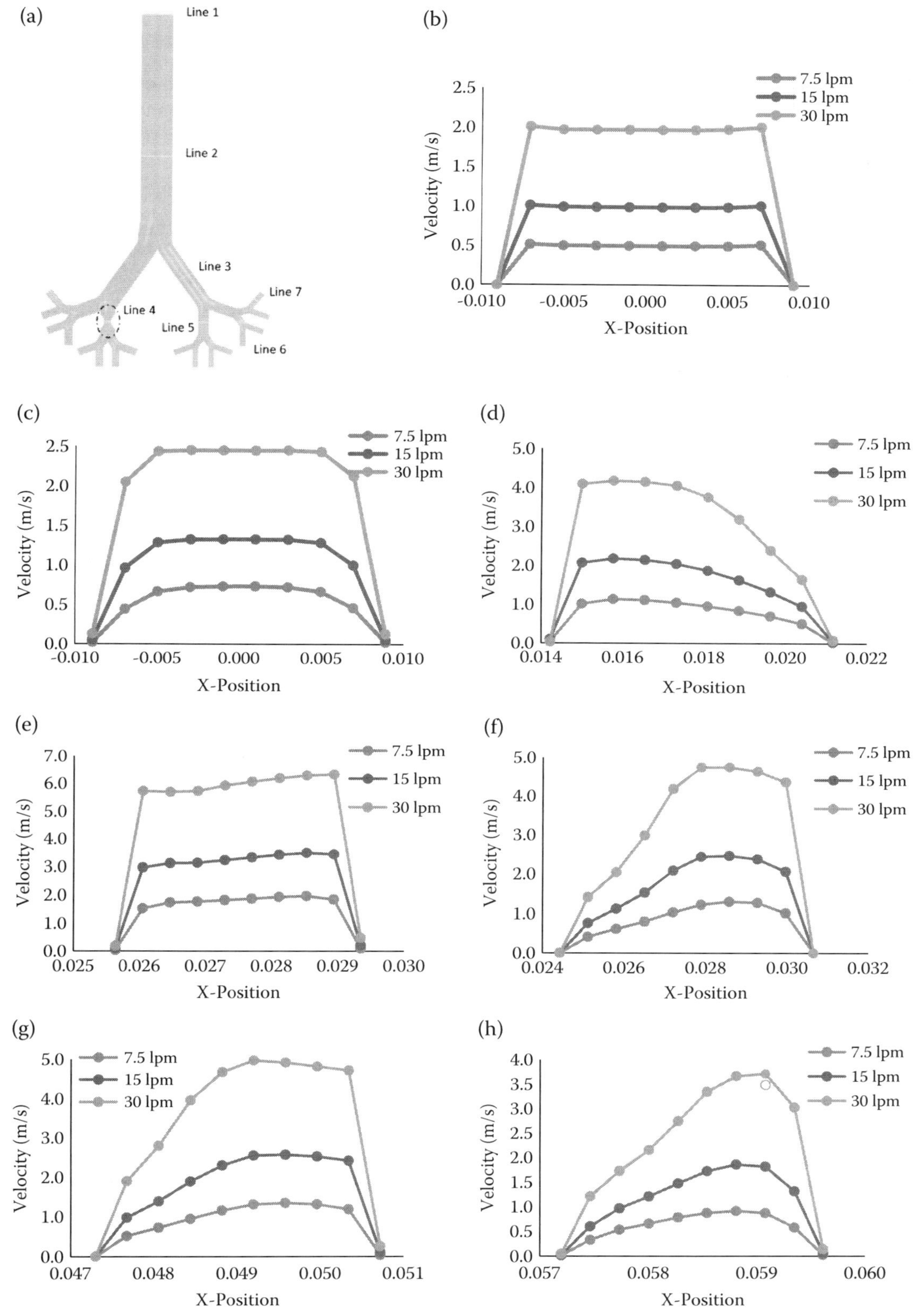

FIGURE 8.19 Selected Line for Velocity Profiles of Asymmetric Lung Model with Left-Stenosis area at Second Generation: (a) Selected Location (b) Line 1, (c) Line 2, (d) Line 3, (e) Line 4 at Stenosis Area, (f) Line 5, (g) Line 6, and (h) Line 7

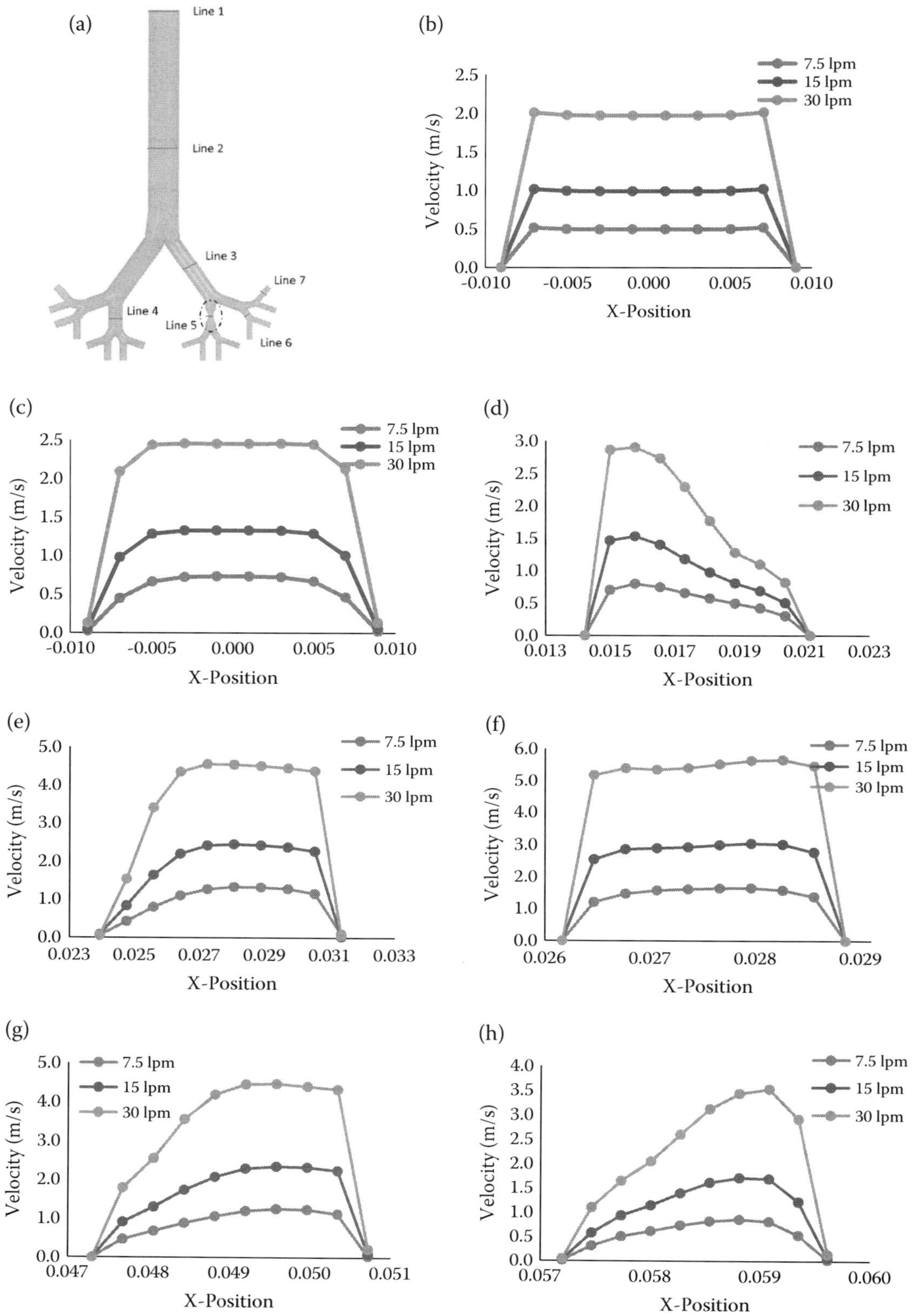

FIGURE 8.20 Selected Line for Velocity Profiles of Asymmetric Lung Model with Right-Stenosis Area at Second Generation: (a) Selected Location (b) Line 1, (c) Line 2, (d) Line 3, (e) Line 4, (f) Line 5 at Stenosis Area, (g) Line 6, and (h) Line 7

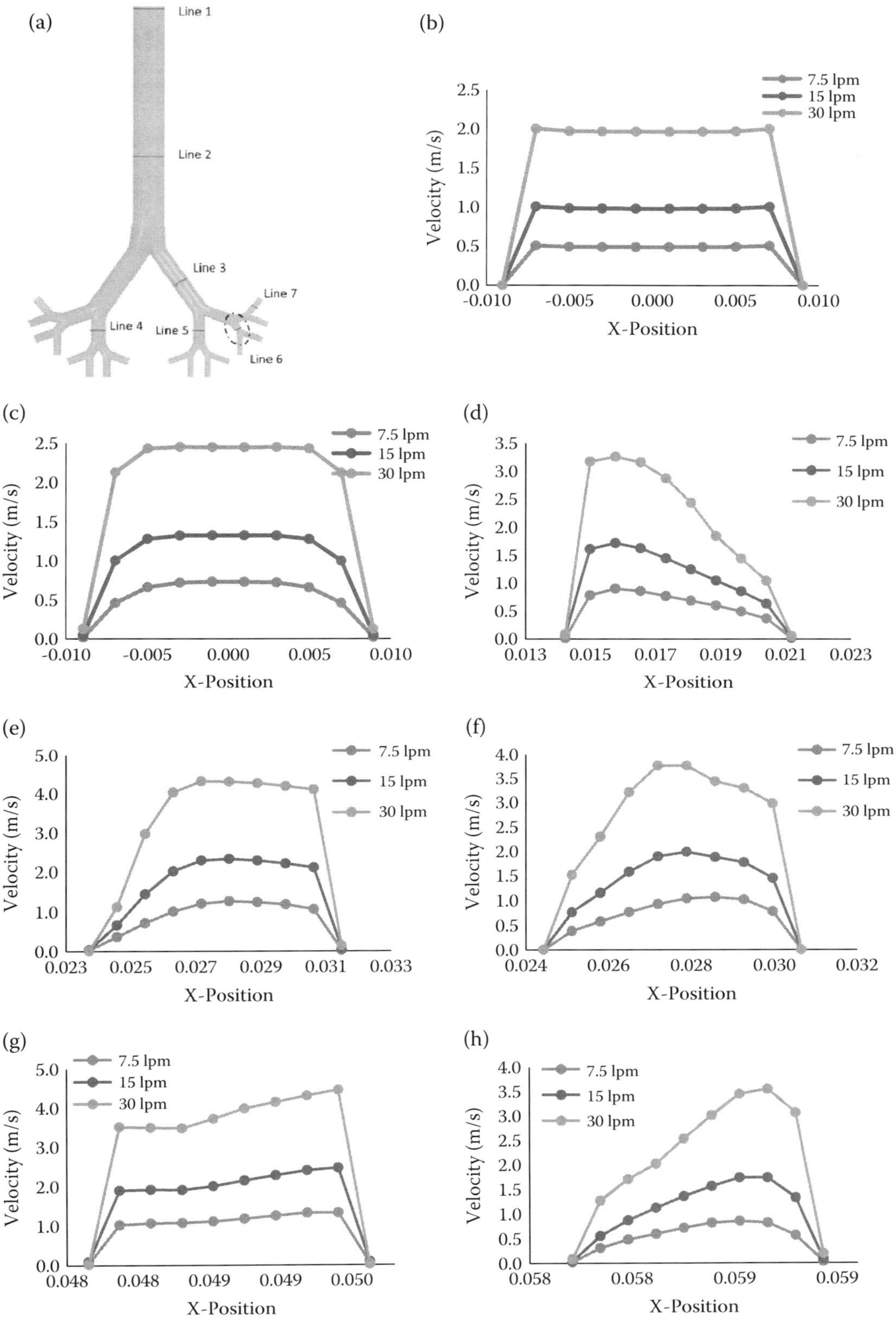

FIGURE 8.21 Selected Line for Velocity Profiles of Asymmetric Lung Model with Right-Stenosis Area at Third Generation: (a) Selected Location (b) Line 1, (c) Line 2, (d) Line 3, (e) Line 4, (f) Line 5, (g) Line 6 at Stenosis Area, and (h) Line 7

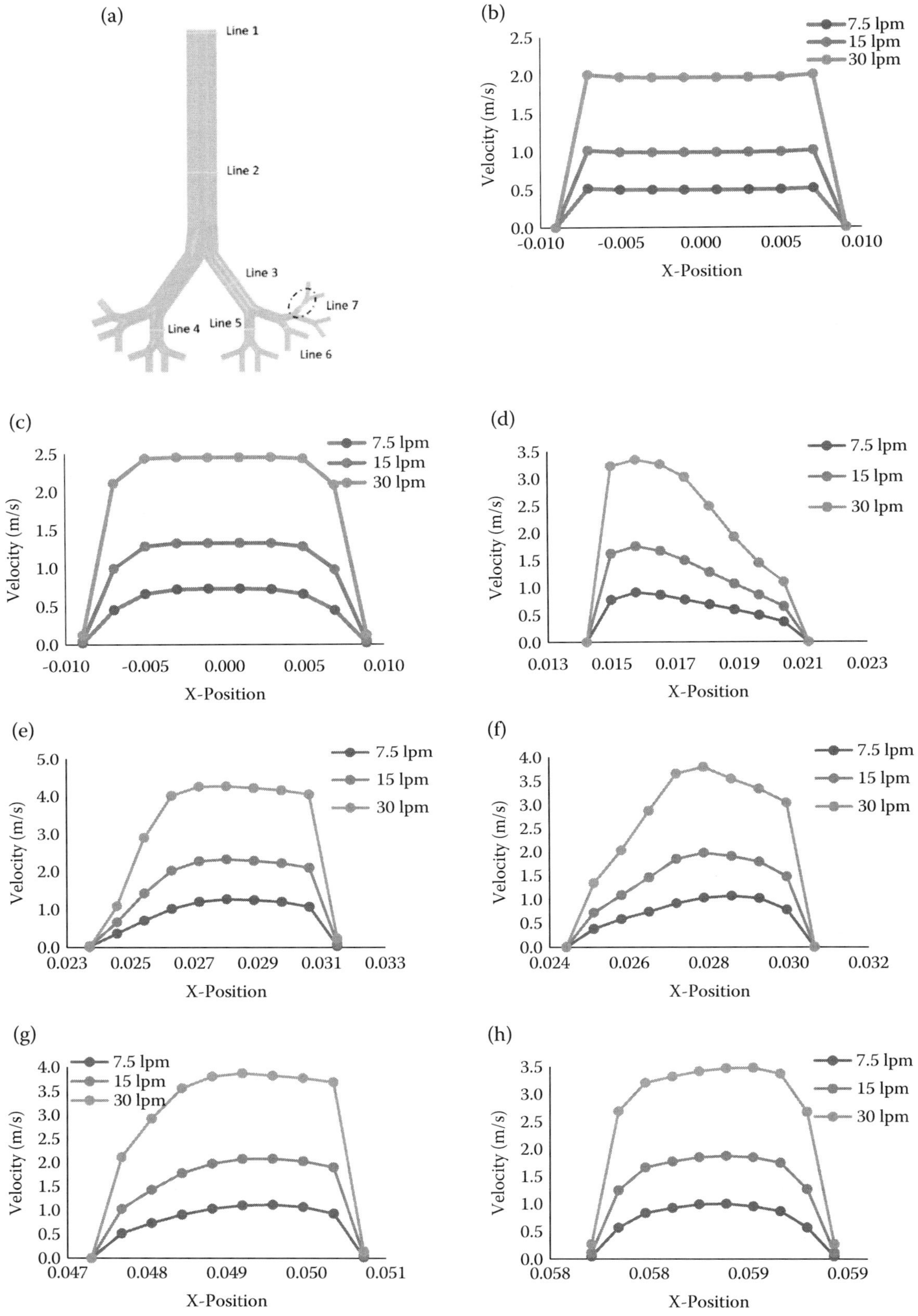

FIGURE 8.22 Selected Line for Velocity Profiles of Asymmetric Lung Model with Right-Stenosis Area at Fourth Generation: (a) Selected Location (b) Line 1, (c) Line 2, (d) Line 3, (e) Line 4, (f) Line 5, (g) Line 6, and (h) Line 7 at Stenosis Area

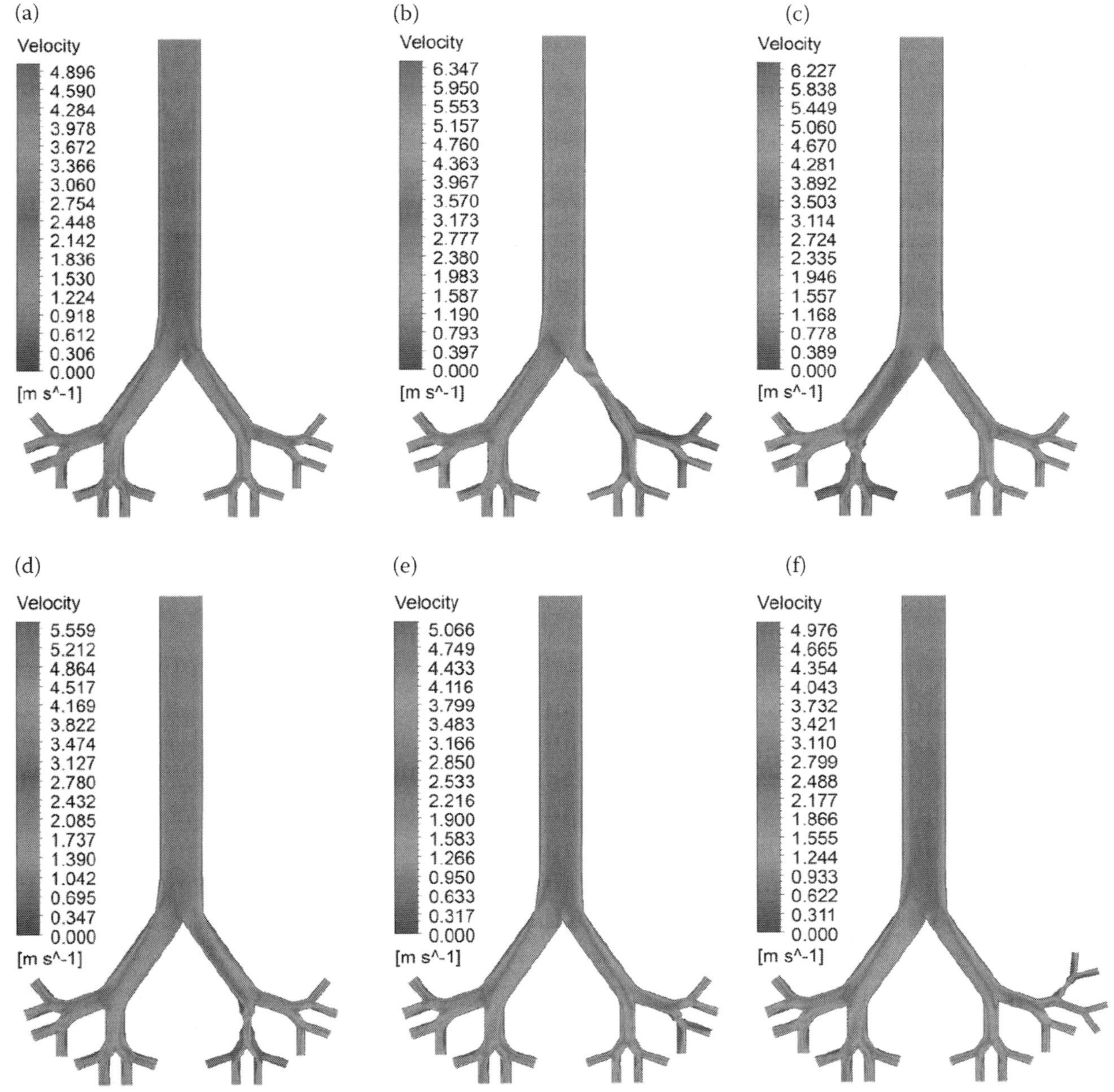

FIGURE 8.23 Velocity Contours at XY Plane for Six Lung Models: (a) Normal Asymmetric Lung, (b) Stenosis-Right at First Generation, (c) Stenosis-Left at Second Generation, (d) Stenosis-Right at Second Generation, (e) Stenosis-Right at Third generation, and (f) Stenosis-Right at Fourth Generation

and asymmetric lung with obstructive regions. Figure 8.24 shows the location of selected planes from six different cases. Figure 8.24a is an asymmetrical case without any narrowing. Figure 8.24b is for the case that has narrowing at first generation, while Figure 8.24c, d is for the cases that narrow at second generation in the left and right lungs. Figure 8.24e,f is for the models that have narrowing at third and fourth generations. Figure 8.24 represents the comparison of velocity contours between six different lung airways.

Figure 8.25 indicates that the airflow velocity in the middle of the trachea (Plane 1) for all cases has a similar pattern. For the first generation (Planes 2 and 3), the flow patterns are still similar except for the flow pattern at the stenosis region (Figure 8.25b). Similarly, the flow patterns at the second generation (Planes 4 and 5) are also similar except at stenosis areas (Figure 8.25c,d) that are located on different sides of the lung. Moreover, the area that is under the stenosis area also has a different flow pattern (Figure 8.25b of Plane 5). Planes 6 and 7are generated to see the airflow patterns at third generation. It can be seen that the airflow pattern in the stenosis area (Figure 8.25e of Plane 7) does not differ from other areas. However, it is obvious that the areas that connect to the stenosis area have different airflow patterns. For deep details of changed airflow patterns, Plane 6 of Figure 8.25c and Plane 7 of Figure 8.25b vary from others because of the connection between stenosis areas for generations 1 and 2

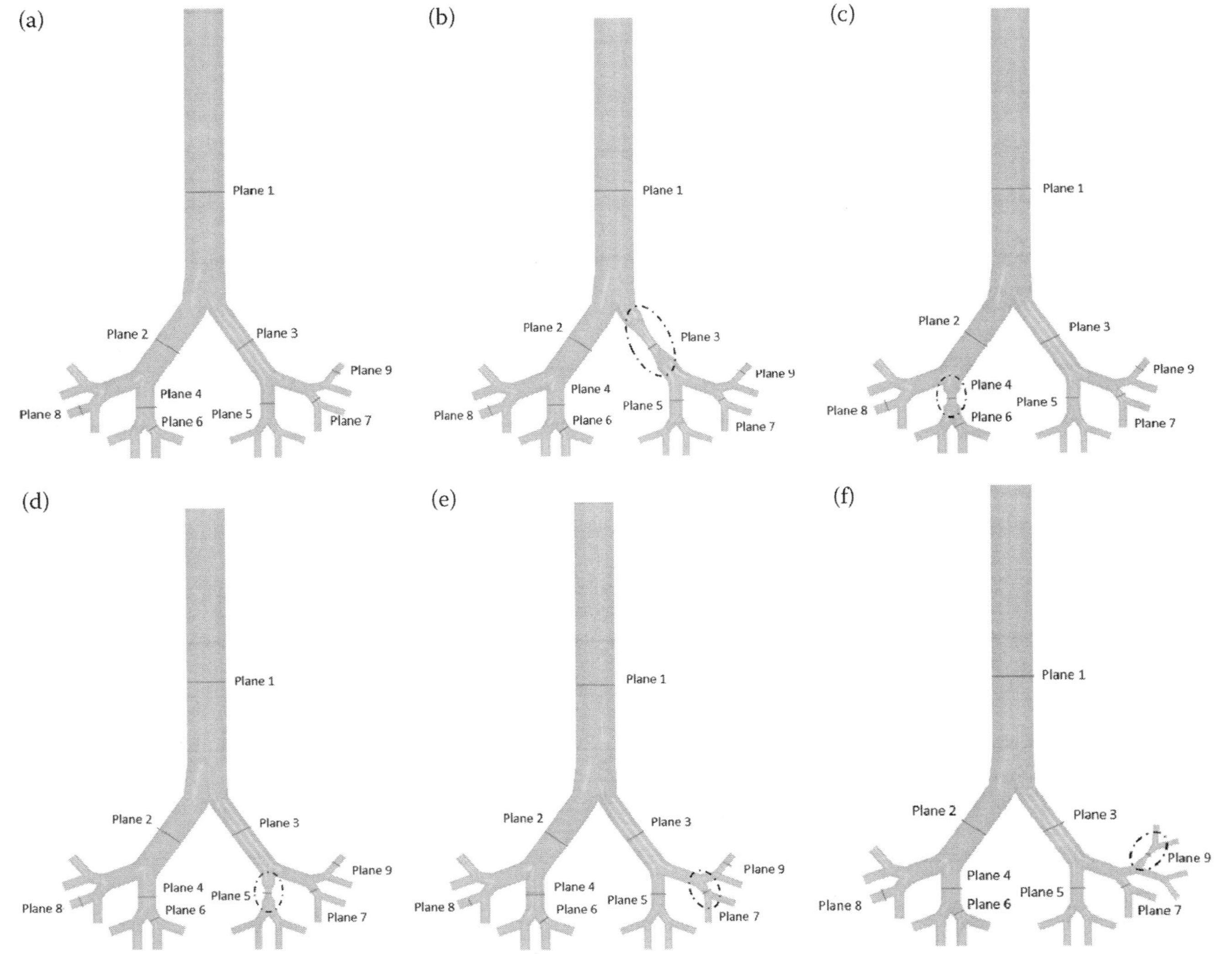

FIGURE 8.24 Selected Planes for Different Positions of Asymmetric Lung Model and Asymmetric Lung Model with Stenosis Areas: (a) Normal Asymmetric Lung, (b) Stenosis-Right at First Generation, (c) Stenosis-Left at Second Generation, (d) Stenosis-Right at Second Generation, (e) Stenosis-Right at Third Generation, and (f) Stenosis-Right at Fourth Generation

respectively. In terms of the airflow pattern at fourth generation (Planes 8 and 9), it is also similar to third generations that means stenosis area does not make a change in airflow patterns (Plane 9 of Figure 8.25f for stenosis area at fourth generation). Furthermore, only Plane9 of Figure 8.25b has a different pattern because it directly connects to the stenosis area from first generation.

Overall, it can be summarized that the airflow pattern can change due to the obstructive areas. And then, other flow patterns, which are under obstructive airways, usually have different airflow patterns if compared with other cases without obstruction. Based on the study from Singh et al. (63), the airflow velocity is significantly affected by several factors such as turbulence fluctuation at the narrow section, the shapes of airways in different conditions, and the force that is pressure driven. However, obstructions at third and fourth generations could not have significant effects on airflow patterns.

8.5.2.2 *Pressure Drop*

Pressure drop is significantly related to airway resistance. Airway resistance is defined as the thing that resists flowing, which is caused by friction forces. The degree of resistance to flow within the airways usually depends on the flow characteristics that are laminar or turbulent, the viscosity of the gas, and the dimensions of the airways (76). This section contains two main parts that are related to airway pressure: pressure drop and pressure contours.

8.5.2.2.1 Pressure Drop

The randomly selected planes are generated from the middle of the trachea to fourth generation for various asymmetric lung models. The pressure drop is calculated based on 10 μm for the particle size. Three different flow rates are considered in this part in order to evaluate the changes in pressure drop. Figure 8.26 represents the pressure drop at different selected

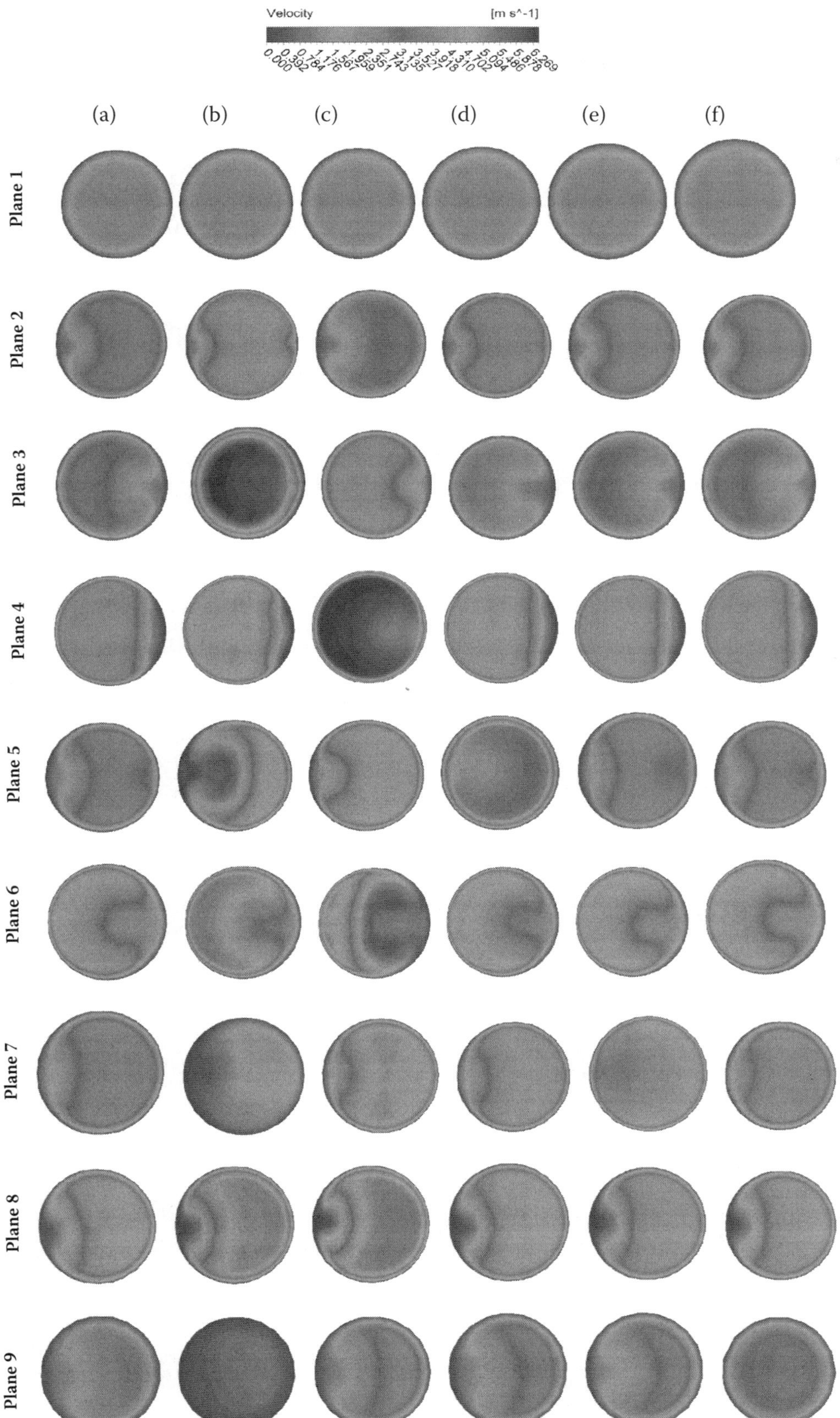

FIGURE 8.25 Velocity Contours at Different Positions of Asymmetric and Stenosis Airways at 30 lpm: (a) Normal Asymmetric Lung, (b) Stenosis-Right at First Generation, (c) Stenosis-Left at Second Generation, (d) Stenosis-Right at Second Generation, (e) Stenosis-Right at Third Generation, and (f) Stenosis-Right at Fourth Generation

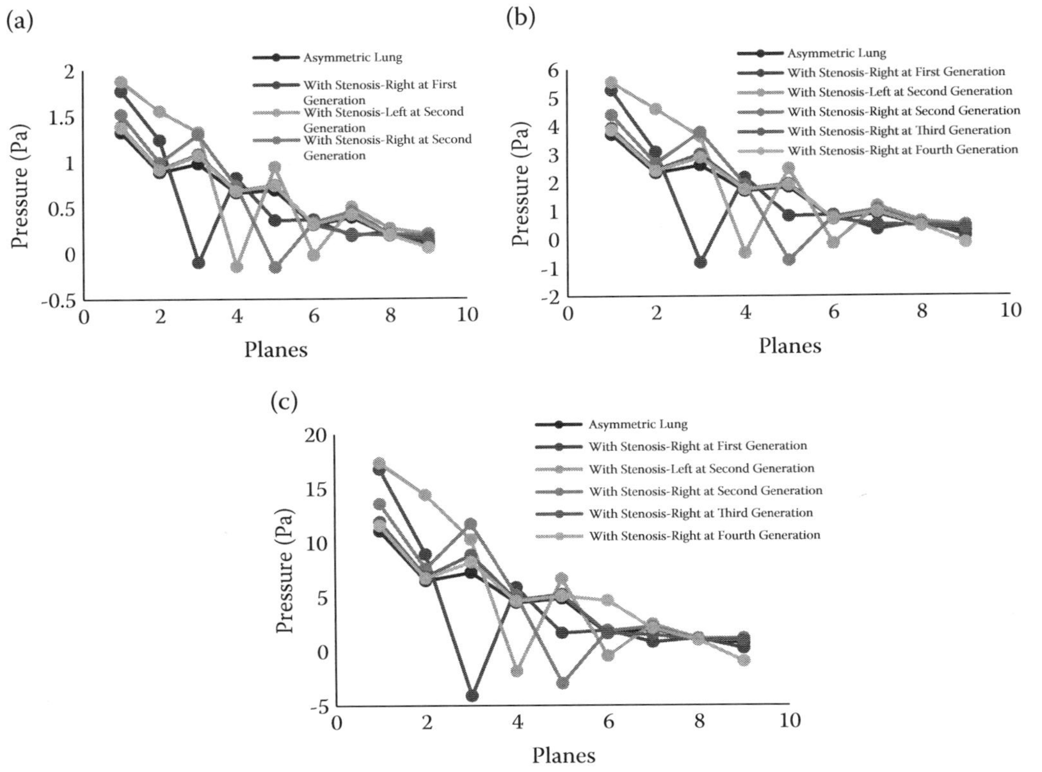

FIGURE 8.26 Pressure Drop at Different Plane Positions for Asymmetric Lung and Asymmetric Lung with Stenosis Airways: (a) 7.5 lpm, (b) 15 lpm, and (c) 30 lpm

planes for all six cases; Figure 8.26a is for flow rate at 7.5 lpm, Figure 8.26b is for 15 lpm, and Figure 8.26c is for 30 lpm.

According to Singh et al. (63), atmospheric pressure must be higher than the airway pressure because of the direction of the air, which always flows from the higher pressure zone to the lower pressure zone. From the results of Figure 8.26, it is obvious that other selected planes from all six cases with three different flow rates have a lower pressure than the initial pressure on Plane 1. Moreover, the highest pressure drop always occurs in an area that has an obstruction. For the lowest flow rate at 7.5 lpm (Figure 8.26a), the maximum pressure drop is located in the stenosis area at the right lung of generation 2 (Plane 5). However, both flow rates at 15 lpm and 30 lpm (Figure 8.26b,c) have the highest pressure drop at the stenosis area of the right lung at generation 1 (Plane 3). In contrast, the highest pressure is also located at Plane 1 of stenosis-left at the second generation, which has 30 lpm for the flow rate (Figure 8.26c). For a normal asymmetric lung without any stenosis section, initial pressure is lower than other cases that consist of stenosis areas. More details for stenosis cases, it can be seen that the area underneath the stonosis still has a higher pressure drop.

To summarize, it can be noticed that the flow rate can influence the level of pressure drop. Comparing all three flow rates, the highest pressure drop at the stenosis area is located at the highest flow rate, that is, 30 lpm. In addition, stenosis condition also has an essential effect on pressure drop. Referring to Singh et al. (63), airway pressure could influence human breathing. A higher pressure drop in the stenosis region could increase airway resistance. As a result, it causes difficulty in inhaled conditions.

8.5.2.2.2 Pressure Contours

Pressure contours present the pressure variation within the domain. It could be changed based on the fluid flows during the domain. The XY plane is created at the center of all six cases with 30 lpm for flow rate and 10 μm for particle size. Figure 8.27 is representative of pressure contours at the XY plane for different lung conditions. Figure 8.27a is for an asymmetrical lung model, whereas Figure 8.27b involves the stenosis area at first generation. Figure 8.27c,d is for left and right stenosis cases at second generation. Figure 8.27e,f is for stenosis conditions at third and fourth generations, respectively.

To compare all six cases, the highest pressure is found at the first bifurcation, and then the pressure continually decreases along the airways. However, the pressure increases again in the region where it is near the bifurcation and has a significant increase in the area before the obstruction. In contrast, the areas before stenosis regions at third and fourth generations do not have any higher pressure due to the length and dimension of the airways. Moreover, the lowest pressure is usually found at the center of the obstructive area. After that, it increases again. For asymmetric lung without stenosis condition (Figure 8.27a), the overall pressure in this case is similar to other cases except for the branches that are related to the stenosis regions.

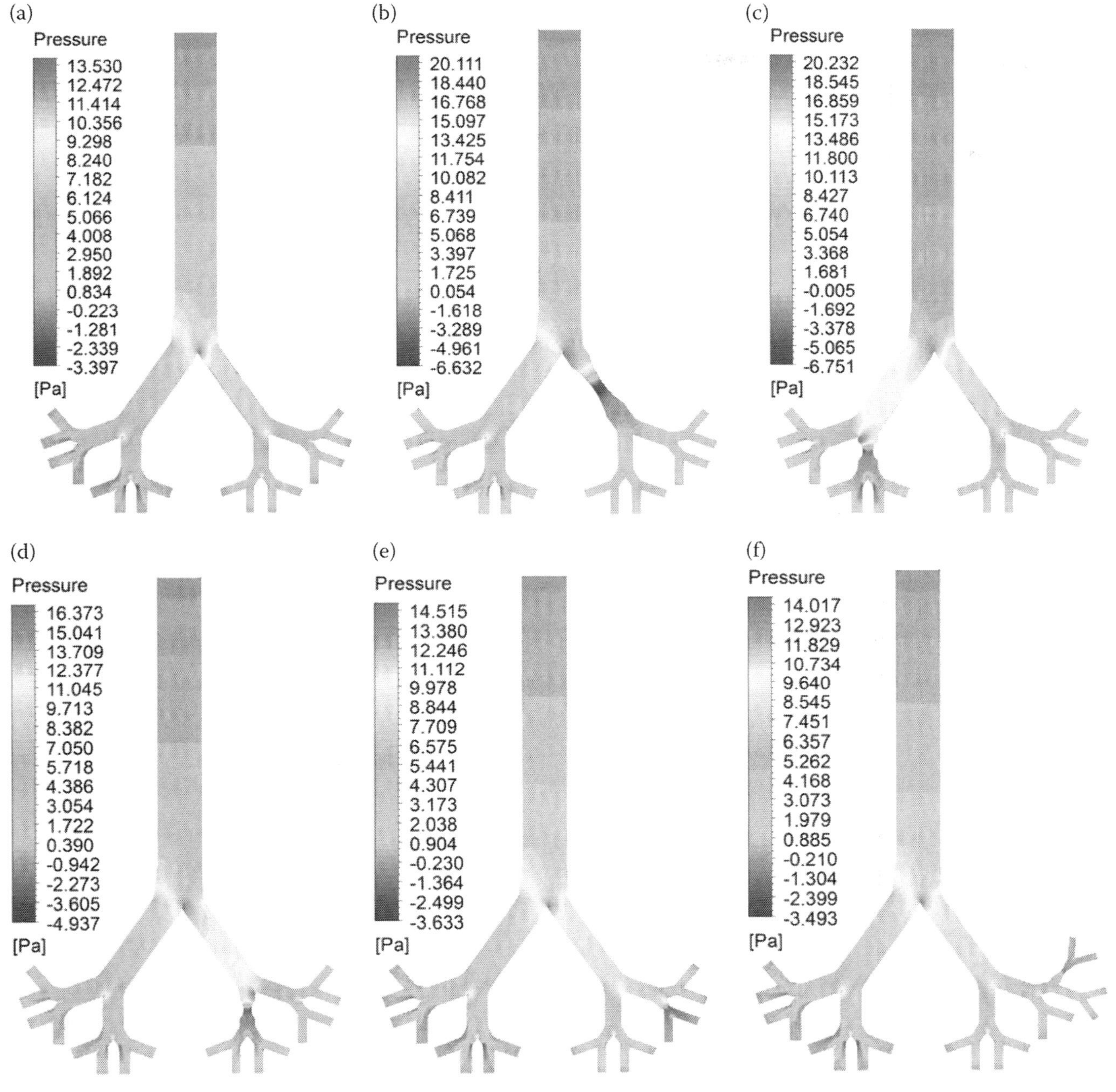

FIGURE 8.27 Pressure Contours at XY Plane for Six Lung Models: (a) Normal Asymmetric Lung, (b) Stenosis-Right at First Generation, (c) Stenosis-Left at Second Generation, (d) Stenosis-Right at Second Generation, (e) Stenosis-Right at Third Generation, and (f) Stenosis-Right at Fourth Generation

Consequently, there is higher pressure at a bifurcation and the nearly obstructed region. The lowest pressure is always found at the center of the obstructive region. Based on the study of Nowak, Kakade, and Annapragada (29), pressure in bifurcating airways will be different based on the airway resistance, which is caused by the lung volume as the dimension of the airway and overall flow rate. The pressure in all six cases can vary because of various dimensions. The overall pressure at right lung is higher than left lung because of its smaller dimension.

8.5.2.3 Wall Shear

Shear stress is classified as one of six types of stresses. However, wall shear stress is normally known as the *shear stress of fluid*, which is always occurring in the layer next to the wall of a domain. It has been considered for the clinical field because of the fluid characteristic. Normally, fluid flows fastest at the center of the domain and slowest close to the wall. Moreover, wall shear stress can be generated based on the flow friction that is related to velocity magnitude.

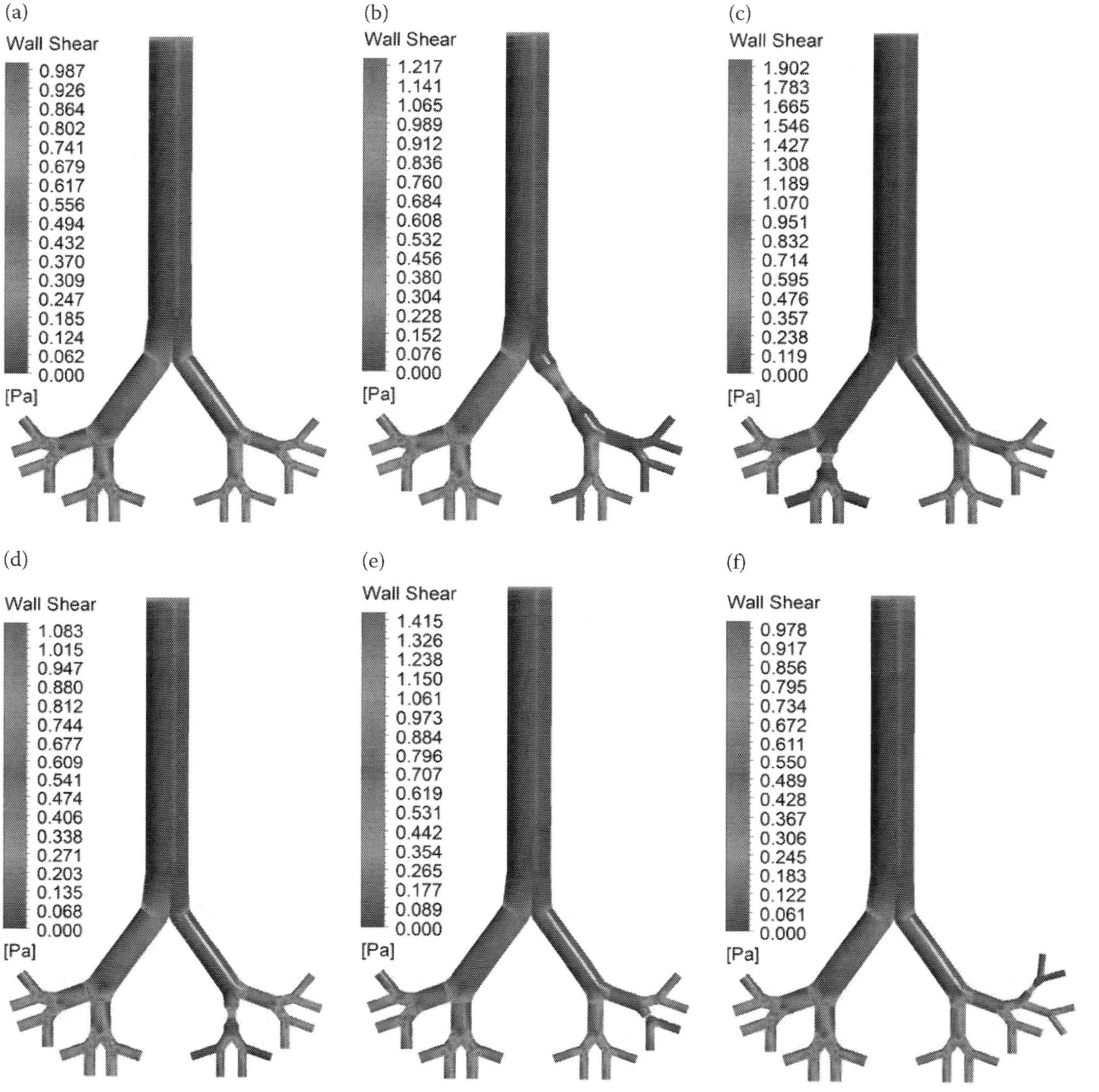

FIGURE 8.28 Wall Shear for Six Lung Models at 30 lpm: (a) Normal Asymmetric Lung, (b) Stenosis-Right at First Generation, (c) Stenosis-Left at Second Generation, (d) Stenosis-Right at Second Generation, (e) Stenosis-Right at Third Generation, and (f) Stenosis-Right at Fourth Generation

In this section, all six lung airway models are simulated with 30 lpm for flow rate and 10 μm for the particle size. Figure 8.28 represents the wall shear contour for six different cases. Figure 8.28a is for a normal asymmetric lung airway, whereas Figure 8.28b is for an asymmetric lung with stenosis area at first generation. Figure 8.28c,d is for the lung with stenosis areas at the left and right lungs at second generation. Figure 8.28e,f is for the lung with stenosis areas at third and fourth generations, respectively.

From these figures, wall shear stress is higher at the initial airway, becomes more stable, and starts to be higher again at the first bifurcation area. It can be seen that most of the bifurcations have a higher wall shear stress excepting the branch, which directly connects to the stenosis area. In terms of the stenosis area, it can be found that the stenosis area has an essential effect on wall shear stress. Most stenosis areas have a higher wall shear stress in the middle of itself, and lower branches will have a lower wall shear stress.

The results in this report are similar to the study of Cebral and Summers (37). The pressure and shear stress are always opposite to each other. From the previous discussion on pressure contours, the pressure is found lower in the stenosis area, whereas shear stress is higher in this area. Furthermore, wall shear stress at the left lung has a higher magnitude if compared with the right lung. In a nutshell, airway dimension and bifurcation can influence the wall shear stress.

8.5.2.4 Turbulence Intensity

Turbulence intensity is representative of the level of turbulence, which uses the intensity measurement of the flow fluctuations. The level of the turbulence intensity is generated by the contour of the whole domain. The flow rate at 30 lpm and particle size at 10 μm are used as the input data for all lung models. Figure 8.29 shows the turbulence intensity contour for six different airway conditions. Figure 8.29a is an asymmetric airway without any narrow, whereas Figure 8.29b and others are the airways with stenosis areas at different generations, which are from first to fourth generations.

From Figure 8.29, turbulence intensity often occurs in the inlet region and most of the bifurcations. It can be seen that the level of turbulence intensity from a normal asymmetric lung (Figure 8.29a) is quite different from other lungs that have stenosis areas (Figure 8.29b–f). The lobes, which are under the stenosis airway, obviously have a higher turbulence intensity compared to the normal case (Figure 8.29a). From Figure 8.29b and 8.29c, the lung, which is opposite stenosis lung, also has lower turbulence intensity if compared to normal airways (Figure 8.29a). However, stenosis areas at third and fourth generations (Figure 8.29e,f) has few effects on the turbulence flow. Ma and Lutchen (16) claimed that only from the upper airway can turbulent flow be generated. In contrast, the results from this chapter and Singh et al. (63) have similar points, that the stenosis area could influence the flow pattern, which can increase turbulence intensity.

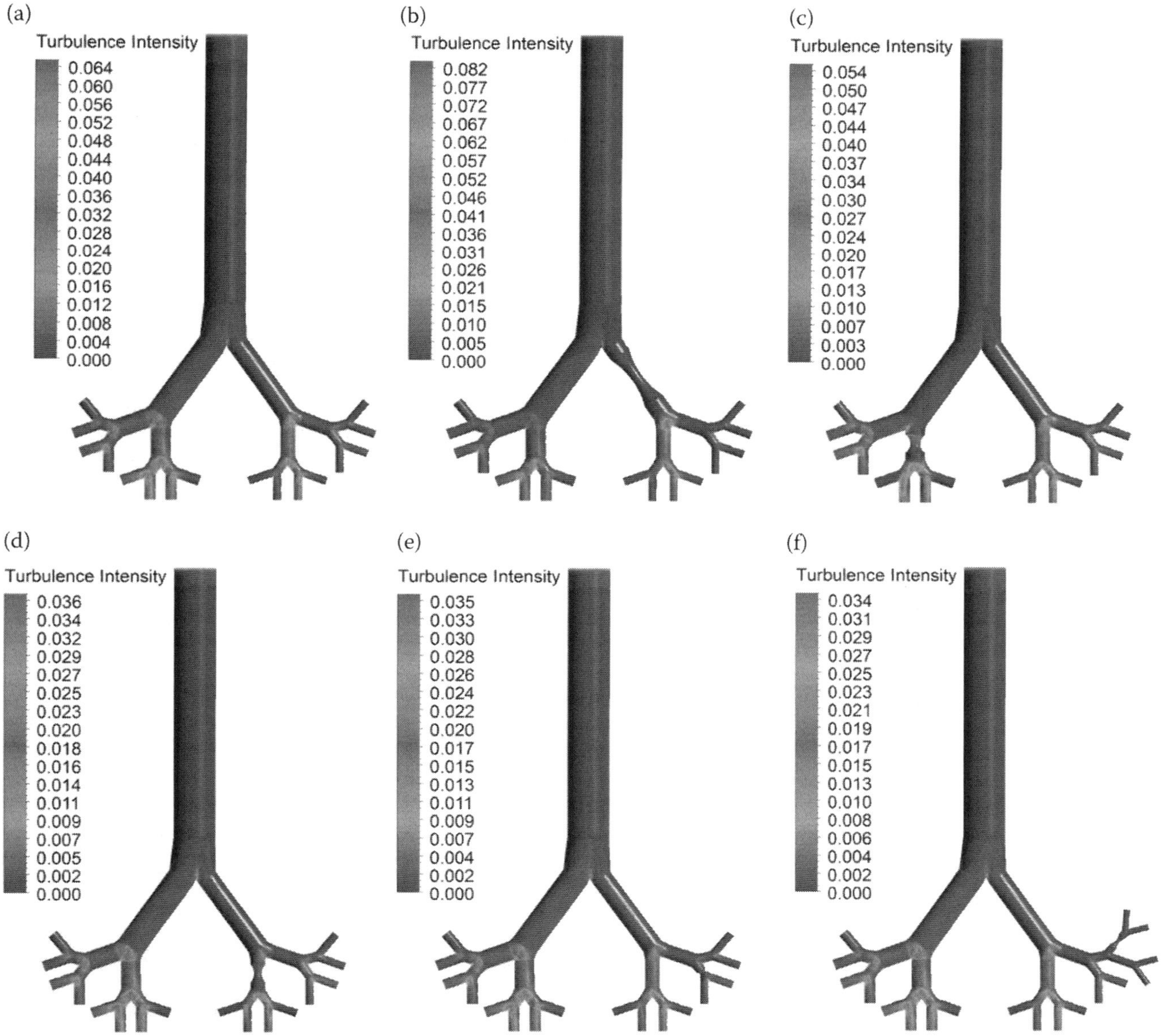

FIGURE 8.29 Turbulence Intensity Contour for Six Lung Models at 30 lpm: (a) Normal Asymmetric Lung, (b) Stenosis-Right at First Generation, (c) Stenosis-Left at Second Generation, (d) Stenosis-Right at Second Generation, (e) Stenosis-Right at Third Generation, and (f) Stenosis-Right at Fourth Generation

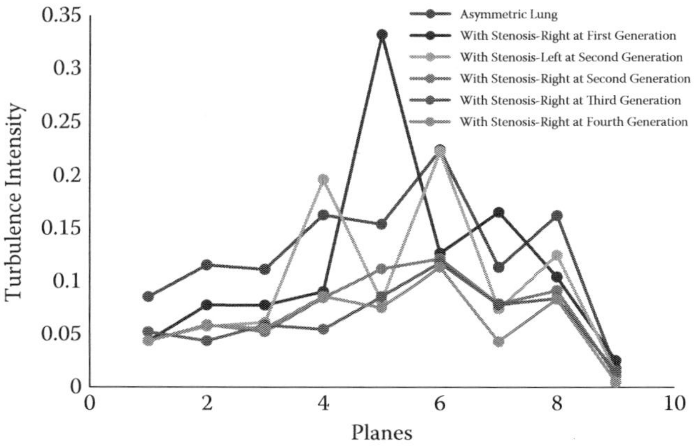

FIGURE 8.30 Turbulence Intensity at Different Plane Positions for Six Lung Models at 30 lpm

There are nine selected planes that are created to calculate the turbulence intensity values in order to make the comparisons between six models. This calculation is based on the flow rate at 30 lpm. Figure 8.30 shows the turbulence intensity values at randomly selected planes with different airway conditions.

From Figure 8.30, the airway with the stenosis area at first generation has the highest turbulence fluctuation at Plane 5, which is under the stenosis branch. The second highest turbulence values are found from two cases at Plane 6, which are the cases that have stenosis area at left lung at second generation and the case that has no stenosis area. However, all six cases have similar lowest turbulence values at Plane 9 where it locates in fourth generation. It can be summarized that having a stenosis area can increase the turbulence fluctuation, especially at first generation. On the other hand, having a stenosis area at fourth generation could have an insignificant effect on turbulence values.

To conclude, looking at the overall turbulence intensity from both contours and plotting graph, turbulence intensity can be influenced by the stenosis area but only for first generation and second generation. Stenosis areas at third and fourth generations could have a minor effect on flow patterns and turbulence intensity.

8.5.2.5 Particle Transport

Particle transport in lung airways plays a vital role in drug delivery. At the same time, it can also induce respiratory diseases due to inhalation of harmful substances such as dust and gaseous pollutants. There are various types of these substances in the environment that are classified based on shapes and sizes ranging from smaller sizes to larger sizes [(78); Farghadan et al. (61)].

In order to enhance the understanding of particle transport within asymmetric lung airways with stenosis area at different generations of the lung, this report will investigate and analyze the particle deposition under various conditions. Then this report will provide a comparison based on two variables from deposition efficiency and deposition fraction.

8.5.2.5.1 Deposition Efficiency

Deposition efficiency is the ratio of the number of particles that have various sizes and the amount of each particle size that is deposited at a given area. The particle deposition efficiency for six lung conditions is calculated based on three different flow rates and provided as Figure 8.31.

Obstruction has an extreme effect on deposition efficiency. From the three flow rates (Figure 8.31), it is obvious that the stenosis area at second generation has the highest deposition efficiency. Moreover, the stenosis area at first generation has the second-highest deposition efficiency compared to the asymmetric lung without any stenosis area. Another important thing is that the flow rate could influence the deposition efficiency. Flow rate at 30 lpm (Figure 8.31c), could cause the highest deposition or level of trapped particles: 49% in the case of 10 μm particle size. In contrast, other flow rates at 7.5 lpm and 15 lpm have deposition rates of 6% and 20%, respectively.

A higher flow rate, especially with larger particle size, can increase the deposition efficiency rate due to greater particle inertia [van Ertbruggen et al. (27); (16,63)]. It can be concluded that the flow rate and particle size could influence deposition efficiency. However, based on the above discussion, the stenosis area could increase greater deposition efficiency, especially the stenosis area at first and second generations.

8.5.2.5.2 Deposition Fraction

Deposition fraction is referred as the number of particles, focusing on only one size and the amount of that particle size deposited at a given area. Figure 8.32 shows the deposition fraction of different-diameter particles in six different cases.

From Figure 8.32, it can be seen that particle sizes of 1 μm and 2.5 μm (Figure 8.32a,b) have a similar deposition fraction, approximately 5% at 30 lpm, while the flow rate at 7.5 lpm has a deposition fraction around 2%. In the case of particle size 5 μm (Figure 8.32c), stenosis case at second generation has a higher deposition fraction of around 10% compared to other stenosis cases which have deposition fractions around 4% to 7%. In terms of the particle size of 10 μm (Figure 8.32d), a higher deposition fraction is around 49% at 30 lpm of the flow rate. This case is still the case of the stenosis area in the second generation. Furthermore, it is also the highest value if compared to other conditions.

In conclusion, the particle deposition of each deposition fraction is usually based on the particle diameter, flow rate, and density (63,65,74). For example, particles of greater diameter will have a higher particle deposition rate. In the same way, a higher flow rate will have a higher particle deposition rate. In

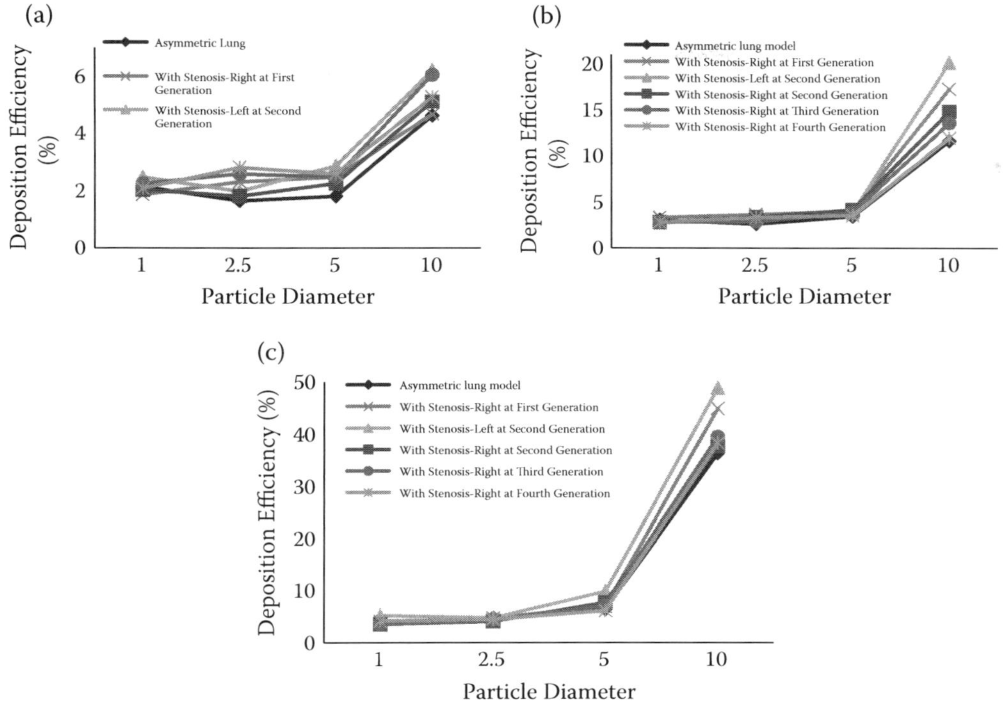

FIGURE 8.31 Deposition Efficiency in Different Regions of Asymmetric Lung and Asymmetric Lung with Stenosis Areas: (a) 7.5 lpm, (b) 15 lpm, and (c) 30 lpm

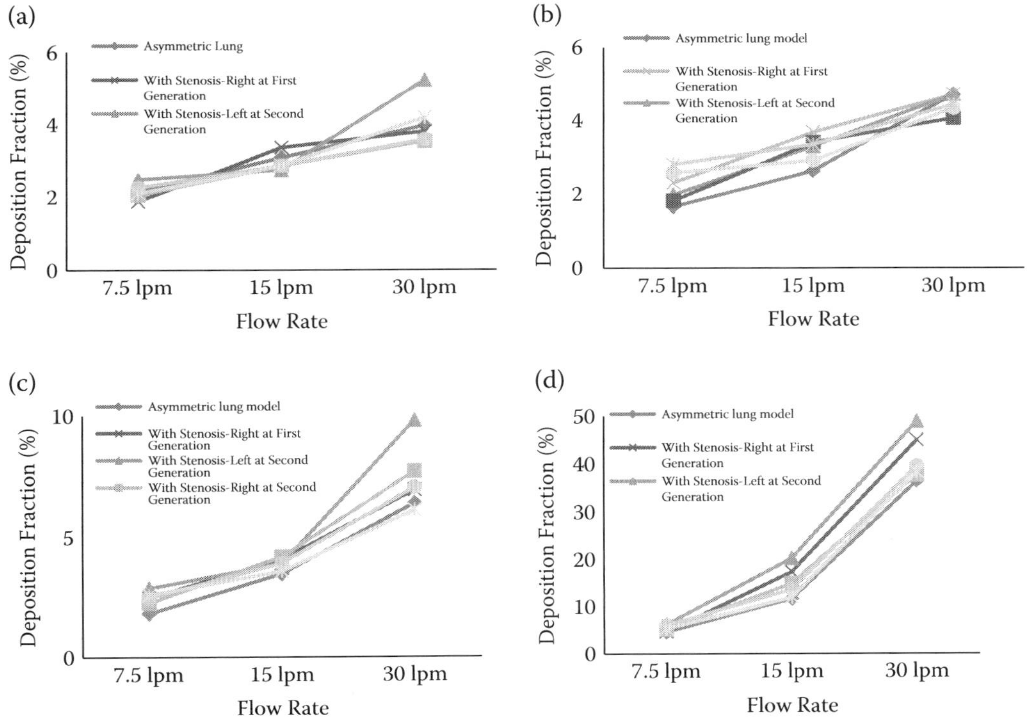

FIGURE 8.32 Deposition Fraction of Different-Diameter Particles in Asymmetric Lung and Asymmetric Lung with Stenosis Airways: (a) 1 µm Diameter, (b) 2.5 µm Diameter, (c) 5 µm Diameter, and (d) 10 µm Diameter

terms of obstructive cases, the smaller airway volume can increase the particle deposition (63). From this chapter, obstructive areas in the first and second generations have a particular effect on increasing the deposition fraction rate.

The various particle sizes have been injected into all six lungs under stenosis conditions in different areas with 30 lpm of flow rate. Figure 8.33 shows the particle deposition scenario with six different lung conditions.

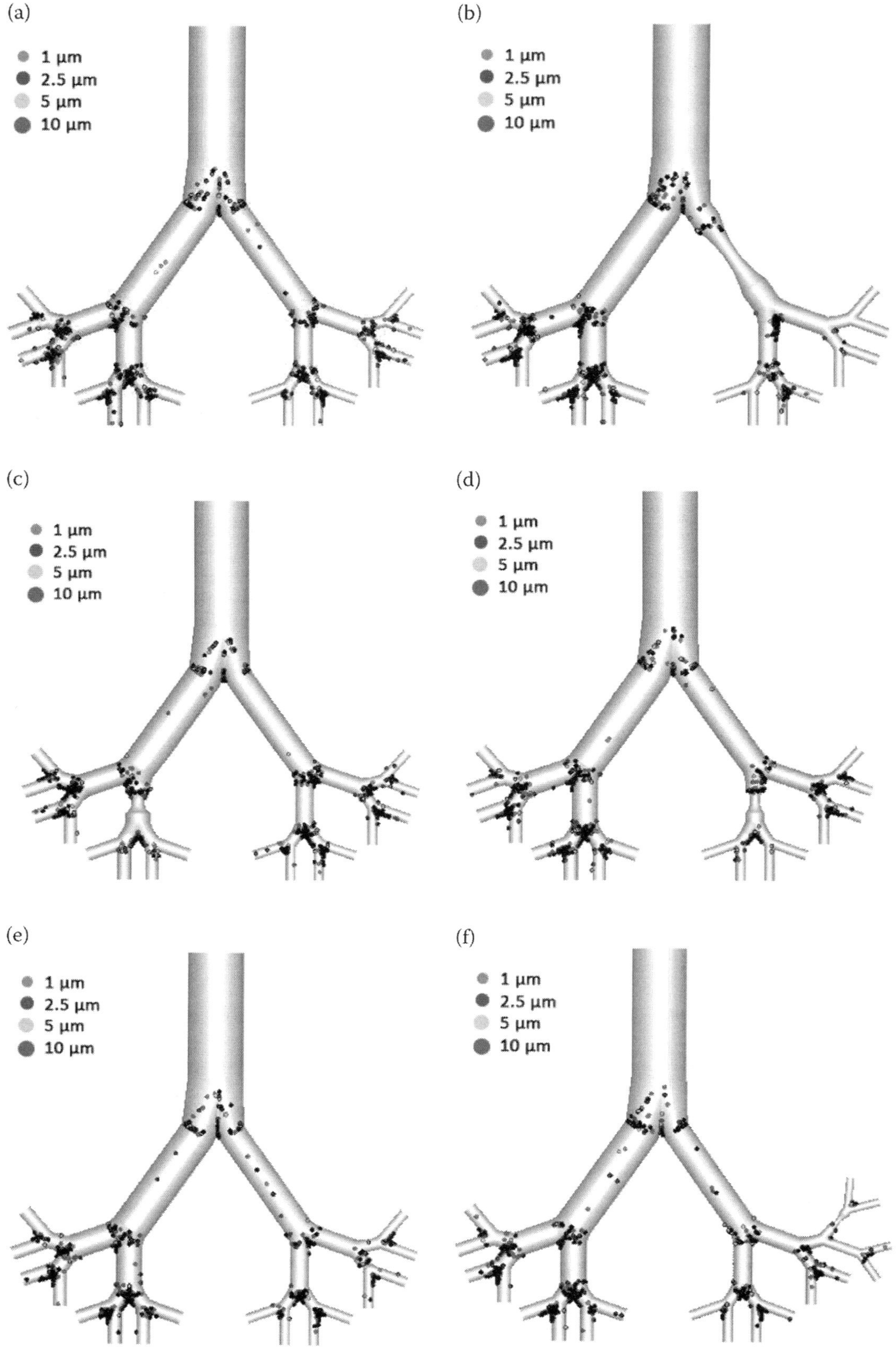

FIGURE 8.33 Particle Deposition Scenario at 30 lpm for Six Lung Models: (a) Normal Asymmetric Lung, (b) Stenosis-Right at First Generation, (c) Stenosis-Left at Second Generation, (d) Stenosis-Right at Second Generation, (e) Stenosis-Right at Third Generation, and (f) Stenosis-Right at Fourth Generation

Overall, in six cases, the majority of particles are trapped at the airway, which is opposite the stenosis area, especially for the largest particle size of 10 μm. All of these particles are trapped at the wall of bifurcation for each generation. The lobe of second generation has more particle deposition if compared to other areas. In cases of asymmetric lung models without a stenosis area (Figure 8.33a), the left lung has higher particle deposition than the right lung because it has a larger diameter.

To summarize, the majority of particles are usually trapped at the wall of a bifurcation [(28); Farghadan et al. (61)]. The results of this study support this idea. An increase in velocity at the stenosis area could influence the particle pattern, which is significantly deposited at the bifurcation area (63). In addition, it can be summarized that the largest particle size, at 10 μm, is usually trapped in the bifurcation of second generation.

8.6 Limitation of the Study

This chapter has used zero gauge pressure as the outlet boundary for all lung airway models. but in real life, some unknown pressures should be inside the lung in the third and fourth generations. In terms of wall condition, it has been considered as a stationary wall condition which has no dynamic wall motion. Moreover, this report only considers inhaled breathing for both airflow and particle transport. This report focuses on the abnormal lung having an asymmetric condition with stenosis area, not considering airway inflammation cases.

8.7 Conclusions

In this chapter, airflow and particle transport have been investigated under an asymmetric lung airways condition. A normal asymmetric lung has been compared to two asymmetric lungs with aging effects that result in smaller diameters. Moreover, this study also compares a normal asymmetric lung to other asymmetric lungs with stenosis sections in the different generations from the first generation to the fourth generation. Calculations have been performed based on two variables: flow rates and particle sizes. The following information will conclude this study.

- *Airflow velocity*

For effects of aging cases: The airflow velocity magnitude is influenced by the volume in the lung airways. A smaller diameter could generate higher velocity magnitude, especially for the left-lung airways.

For stenosis cases: The airflow velocity could be influenced by the stenosis section. The velocity in this area is higher than in other areas compared to the normal asymmetric lung. However, it seems that the airflow velocity can be increased by stenosis conditions from first generation to the second generation, while the stenosis areas at third and fourth generations could not have much effect on velocity magnitude.

- *Pressure*

For effects of aging cases: The airway volume and the flow rate can affect pressure levels. A smaller airway could lead pressure to be higher. Among the three cases, the maximum pressure is located at the trachea for all cases while the minimum pressure has been found at the fourth generation.

For stenosis cases: The level of the pressure is dependent on the airway volume and the flow rate. The overall pressure at the right lung is higher than the left lung because it has a smaller dimension. The maximum pressure has been found in the trachea region, while the lowest pressure is located at the stenosis area where it is at the right lung of the first generation.

- *Wall shear*

For effects of aging cases: The bifurcation can have a greater effect on wall shear stress, whereas the volume of the lung airways can have a minor effect on shear stress at the wall.

For stenosis cases: Airway dimension and bifurcation can have a significant effect on the wall shear stress. Wall shear is found to be higher at the bifurcation and in the middle of the stenosis area.

- *Turbulence intensity*

For effects of aging cases: Normal asymmetric lungs can generate more turbulence intensity while smaller airways have a lower turbulence intensity and also have similar levels. The airflow begins to be stable from the fourth generation of the airways, and it tends to be continually stable for all lower generations.

For stenosis cases: Overall turbulence intensity of the whole lung is similar for all six cases. Local turbulence intensity has been found to be higher at the bifurcation and the stenosis area where it is only for first and second generations. On the other hand, stenosis areas at the third and fourth generations could not have a significant effect on turbulence intensity.

- *Particle deposition*

For effects of aging cases: Both deposition efficiency and deposition fraction are based on the particle size, flow rate, and airway volume. The smaller diameter of a 20% reduction usually has higher deposition efficiency and deposition fraction in different flow rates and particle sizes. On the bifurcation area (in the carinal angle region) for all generations can deposit more particles, especially of 10 μm size.

For stenosis cases: The trend of deposition efficiency and deposition fraction is based on the stenosis area, particle size, and flow rate. The deposition efficiency and deposition fraction for stenosis areas at first and second generations tend to be higher than the stenosis area at other generations. The particles are usually trapped at the wall of a bifurcation for each generation. The majority of particles sized 10 μm is deposited in the bifurcation of second-generation.

This study has presented a numerical simulation of the airflow and particle transport in asymmetric human lungs with stenosis and the effects of aging conditions. Eight different lung models have been used to analyze the velocity profiles, pressure drop, wall shear stress, turbulence intensity, and particle deposition. The model has been developed based on the symmetric lung from Weibel's model by reducing the diameter 25% for the whole right lung and 50% for stenosis areas. In terms of the effects of aging cases, the diameter for the whole model of the asymmetric lung has been reduced by 10% and 20% and remained the same lengths for the whole model. This study enhances the understanding of the effects of the obstruction and the aging on the airflow patterns and particle movements within abnormal human lungs. The understanding of pressure analysis could be useful for clinical treatment, especially with the severe cases coming with the usage of mechanical ventilation. The knowledge of particle transport could improve the understanding of the drug–aerosol delivery passing stenosis regions. Respiratory diseases, such as asthma and COPD, may have variant stenosis conditions as well as the effect of aging on the respiratory system. The comprehensive study of the asymmetric lung with stenosis areas and the effects of aging should be examined with more variables. In a further study, specific lung disease should be considered.

REFERENCES

1. Davis DL, Bell ML, and Fletcher T. A look back at the London smog of 1952 and the half century since. *Environmental Health Perspectives*. 2002; 110(12): A734–A735.
2. World Health Organization. Monitoring health for the sustainable development goals. *World Health Statistics*. 2020. https://www.who.int/gho/publications/world_health_statistics/en/
3. D'Amato G, Cecchi L, D'Amato M, and Liccardi G. Urban air pollution and climate change as environmental risk factors of respiratory allergy: An update. *Journal of Investigational Allergology and Clinical Immunology*. 2010; 20(2): 95–102.
4. Guan WJ, Zheng XY, Chung KF, and Zhong NS. Impact of air pollution on the burden of chronic respiratory diseases in China: Time for urgent action. *The Lancent*. 2016; 388(10054): 15–21.
5. Hu G, Zhong N, and Ran P. Air pollution and COPD in China. *Journal of Thoracic Disease*. 2014; 7(1): 59–66.
6. Schiavoni G, D'Amato G, and Afferni C. The dangerous liaison between pollens and pollution in respiratory allergy. *Annals of Allergy Asthma & Immunology*. 2017; 118(3): 269–275.
7. Kim D, Chen Z, Zhou LF, and Huang SX. Air pollutants and early origins of respiratory diseases. *Chronic Diseases and Translational Medicine*. 2018; 4(2): 75–94.
8. Kurt OK, Zhang J, and Pinkerton KE. Pulmonary health effects of air pollution. *Current Opinion in Pulmonary Medicine*. 2016; 22(2): 138–143.
9. Yao H and Rahman I. Current concepts on oxidative/carbonyl stress, inflammation and epigenetics in pathogenesis of chronic obstructive pulmonary disease. *Toxicology and Applied Pharmacology*. 2011; 254(2): 72–85.
10. Australian Institute of Health and Welfare. Chronic Obstructive Pulmonary Disease (COPD). Cat. no. ACM 35. Canberra: AIHW. 2019. https://www.aihw.gov.au/reports/chronic-respiratory-conditions/copd
11. Vicente E, Almengor JC, Caballero LA, and Campo JC. Invasive mechanical ventilation in COPD and asthma. *Medicina Intensiva*. 2011; 35(5): 288–298.
12. Ahmed SM and Athar M. Mechanical ventilation in patients with chronic obstructive pulmonary disease and bronchial asthma. *Indian Journal of Anaesthesia*. 2015; 59(9): 589–598.
13. Dominici F, Peng RD, Bell ML, et al. Fine particulate air pollution and hospital admission for cardiovascular and respiratory diseases. *JAMA*. 2006; 295(10): 1127–1134.
14. Gauderman WJ, Avol E, Gilliland F, et al. The effect of air pollution on lung development from 10 to 18 years of age. *The New England Journal of Medicine*. 2004; 352: 1057–1067.
15. Moss OR and Oldham MJ. Dosimetry counts: Molecular hypersensitivity may not drive pulmonary hyperresponsiveness. *Journal of Aerosol Medicine*. 2006; 19(4): 555–564.
16. Ma B and Lutchen KR. CFD simulation of aerosol deposition in an anatomically based human large-medium airway model. *Annals of Biomedical Engineering*. 2009; 37: 271–285.
17. Dhand R. Aerosol delivery during mechanical ventilation: From basic techniques to new devices. *Journal of Aerosol Medicine and Pulmonary Drug Delivery*. 2008; 21(1): 45–60.
18. Dhand R. How should aerosols be delivered during invasive mechanical ventilation. *Respiratory Care*. 2017; 62(10): 1343–1367.
19. Dugernier J, Ehrmann S, Sottiaux T, et al. Aerosol delivery during invasive mechanical ventilation: A systematic review. *Critical Care*. 2017; 21(1): 264 (11 pages)
20. Mauderly JL and Hahn FF. The effects of age on lung function and structure of adult animals. *Advances in Veterinary Science and Comparative Medicine*. 1982; 26: 35–77.
21. Polger G and Weng TR. The functional development of the respiratory system: From the period of gestation to adulthood. *American Review of Respiratory Disease*. 1979; 120: 625–695.
22. Sharma G and Goodwin J. Effect of aging on respiratory system physiology and immunology. *Clinical Interventions in Aging*. 2006; 1(3): 253–260.
23. Lai-Fook SJ and Hyatt RE. Effects of age on elastic moduli of human lungs. *Journal of Applied Physiology*. 2000; 89(1): 163–168.
24. Faria ACD, Lopes AJ, Jansen JM, et al. Influence of the ageing process on the resistive and reactive properties of the respiratory system. *Clinics*. 2009; 64(11): 1065–1073.
25. Kim JW, Heise RL, Reynolds AM, et al. Aging effects on airflow dynamics and lung function in human bronchioles. *PLoS ONE* 2017; 12(8). e0183654.
26. Tu J, Yeoh GH, and Liu C. Computational Fluid Dynamics: A Practical Approach. 2nd edn. Elsevier, USA. 2012.
27. Van Ertbruggen C, Hirsch C, and Paiva M. Anatomically based three-dimensional model of airways to simulate flow and particle transport using computational fluid dynamics. *Journal of Applied Physiology (1985)*. 2005; 98(3): 970–980.
28. Balashazy I and Hoffman W. Deposition of aerosols in asymmetric airway bifurcations. *Journal of Aerosol Science*. 1995; 26(2): 273–292.
29. Nowak N, Kakade PP, and Annapragada AV. Computational fluid dynamics simulation of airflow and aerosol deposition in

human lungs. *Annals of Biomedical Engineering*. 2003; 31(5): 374–390.

30. Weibel ER. 1963. Morphometry of the human lung. Academic Press, ISBN978-1-4832-0076-7.
31. Lee DY and Lee JW. Dispersion of aerosol bolus during one respiration cycle in a model of lung airways. *Journal of Aerosol Science*. 2002; 33(9): 1219–1234.
32. Liu Y, So RMC, and Zhang CH. Modeling the bifurcating flow in an asymmetric human lung airway. *Journal of Biomechanics*. 2003; 36(7): 951–959.
33. Kleinstreuer C and Zhang Z. Targeted drug aerosol deposition analysis for a four-generation pulmonary airway model with hemispherical tumors. *Journal of Biomechanical Engineering*. 2003; 125(2): 197–206.
34. Shi H, Kleinstreuer C, and Zhang Z. Nanoparticle transport and deposition in bifurcating tubes with different inlet conditions. *Physics of Fluids*. 2004; 16(7): 2199–2213.
35. Zhang Z and Kleinstreure C. Airflow structures and nanoparticle deposition in a human upper airway model. *Journal of Computational Physics*. 2004; 198(1): 178–210.
36. Ghosh A, Islam MS, and Saha SC. Targeted drug delivery of magnetic nano-particle in the specific lung region. *Computation*. 2020; 8(1): 10.
37. Cebral JR and Summers SM. Tracheal and central bronchial aerodynamics using virtual bronchoscopy and computational fluid dynamics. *IEEE Transactions on Medical Imaging*. 2004; 23(8): 1021–1033.
38. Borgstrom L, Olsson B and Thorsson L. Degree of throat deposition can explain the variability in lung deposition of inhaled drugs. *Journal of Aerosol Medicine*. 2006; 19(4): 473–483.
39. Gemci T, Ponyavin V, Chen Y, Chen H, and Collins R. CFD simulation of airflow in a 17-generation digital reference model of the human bronchial tree. *Series on Biomechanics*. 2007; 23(1): 5–18
40. Hess DR, Anderson P, Dhand R, et al. Device selection and outcomes of aerosol therapy: Evidence-based guidelines. *Chest*. 2005; 127(1): 335–371.
41. Vestbo J, Hurd SS, Agusti AG, et al. Global strategy for the diagnosis, management, and prevention of chronic obstructive pulmonary disease: GOLD executive summary. *American Journal of Respiratory and Critical Care Medicine*. 2013; 187(4): 347–365.
42. Tsuda A, Henry FS, and Butler JP. Particle transport and deposition: Basic physics of particle kinetics. *Comprehensive Physiology*. 2011; 3(4): 1437–1471.
43. Zhang Z and Kleinstresuer C. Effect of particle Inlet distributions on deposition in a triple bifurcation lung airway model. *Journal of Aerosol Medicine*. 2001; 14(1): 13–29.
44. Gu Q, Qi S, Yue Y, et al. Structural and functional alterations of the tracheobronchial tree after left upper pulmonary lobectomy for lung cancer. *Biomedical Engineering Online*. 2019; 18(1): 1–18.
45. Islam MS, Saha SC, Sauret E, et al. Numerical investigation of diesel exhaust particle and deposition in the CT-scan based lung airway. *AIP Conference Proceedings*. 2017c; 1851(1): 020092.
46. Islam MS, Saha SC, Gemci T, et al. Polydisperse microparticle transport and deposition to the terminal bronchioles in a heterogeneous vasculature tree. *Scientific Reports*. 2018a; 8(1): 1–9.
47. Islam MS, Saha SC, Sauret E, et al. Euler-Lagrange approach to investigate respiratory anatomical shape effects on aerosol particle transport and deposition. *Toxicology Research and Application*. 2019b; 3: 1–15.
48. Saha SC, Islam MS, Rahimi-Gorji M, et al. Aerosol particle transport and deposition in a CT-scan based mouth-throat model. *AIP Conference Proceedings 2121*. 2019; 1: 040011.
49. Islam MS, Saha SC, and Young PM. Aerosol particle transport and deposition in a CT-based lung airway for helium-oxygen mixture. *Proceedings of the 21st Australasian Fluid Mechanics Conference Adelaide Australia*, Adelaide, Australia, 2018b: 10–13.
50. Islam MS, Gu YT, Farkas A, et al. Helium-oxygen mixture model for particle transport in CT-based upper airways. *International Journal of Environmental Research and Public Health*. 2020a; 17(10): 3574.
51. Saber EM and Heydari G. Flow patterns and deposition fraction of particles in the range of 0.1-10 μm at trachea and the first third generations under different breathing conditions. *Computers in Biology and Medicine*. 2012; 42(5): 631–638.
52. Srivastav VK, Paul AR, and Jain A. Effects of cartilaginous rings on airflow and particle transport through simplified and realistic models of human upper respiratory tracts. *Acta Mechanica Sinica*. 2013; 29(6): 883–892. DOI: 10.1007/s10409-013-0086-2
53. Srivastav VK, Kumar A, Shukla S, Paul AR, Bhatt AD, and Jain A. Airflow and aerosol-drug delivery in a CT scan based human respiratory tract with tumor using CFD. *Journal of Applied Fluid Mechanics*. 2014; 7(2): 345–356.
54. Sul B, Wallqvist A, Morris MJ, et al. Computational study of the respiratory airflow characteristics in normal and obstructed human airways. *Computers in Biology and Medicine*. 2014; 52: 130–143.
55. Augusto LLX, Lopes GC, and Goncalves JAS. A CFD study of deposition of pharmaceutical aerosols under different respiratory conditions. *Brazilian Journal of Chemical Engineering*. 2015; 33(3): 549–558.
56. Australian Institute of Health and Welfare. Asthma. Cat. no. ACM 33. AIHW, Canberra 2019. https://www.aihw.gov.au/reports/chronic-respiratory-conditions/asthma
57. Islam MS, Saha SC, Sauret E, et al. Ultrafine particle transport and deposition in a large scale 17-generationlung model. *Journal of Biomechanics*. 2017a; 64: 16–25.
58. Islam MS, Saha SC, Sauret E, et al. Pulmonary aerosol transport and deposition analysis in upper 17 generations of the human respiratory tract. *Journal of Aerosol Science*. 2017b; 108: 29–43.
59. Zhang B, Qi S, Yue Y, et al. Particle disposition in the realistic airway tree models of subjects with tracheal bronchus and COPD. *BioMed Research International*. 2018; 2018: 1–15.
60. Islam MS, Saha SC, Gemci T, et al. Euler-Lagrange prediction of diesel-exhaust polydisperse particle transport and deposition in lung: Anatomy and turbulence effects. *Scientific Reports*. 2019a; 9(1): 1–16.
61. Farghadan A, Coletti F, Arzani A . Topological analysis of particle transport in lung airways: Predicting particle source and destination. *Computers in Biology and Medicine*. 2019; 115: 103497.

62. Islam MS, Paul G, Ong HX, et al. A review of respiratory anatomical development, air flow characterization and particle deposition. *International Journal of Environmental Research and Public Health*. 2020b; 17(2): 380.
63. Singh P, Raghav V, Padhmashali V, et al. Airflow and particle transport prediction through stenosis airways. *International Journal of Environmental Research and Public Health*. 2020; 17(3): 1119 (19 pages).
64. Koullapis PG, Kassinos SC, Bivolarova MP, et al. Particle deposition in a realistic geometry of the human conducting airways: Effects of inlet velocity profile, inhalation flowrate and electrostatic charge. *Journal of Biomechanics*. 2016; 49(11): 2201–2212.
65. Lalas A, Nousias S, Kikidis D, et al. Substance deposition assessment in obstructed pulmonary system through numerical characterization of airflow and inhaled particles attributes. *BMC Medical Informatics and Decision Making*. 2017; 17: 173 (66 pages).
66. Phalen RF and Oldham MJ. Methods for modeling particle deposition as a function of age. *Respiration Physiology*. 2001; 128(1): 119–130.
67. Isaacs KK and Martonen TB. Particle deposition in children's lungs: Theory and experiment. *Journal of Aerosol Medicine*. 2005; 18(3): 337–353.
68. Verbeken EK, Cauberghs M, Mertens I, et al. The senile lung: Comparison with normal and emphysematous lungs. *Structural Aspects Chest*. 1992; 101(3): 793–799.
69. Xia G, Tawhai MH, Hoffman EA, et al. Airway wall stiffening increases peak wall shear stress: A fluid-structure interaction study in rigid and compliant airways. *Annals of Biomedical Engineering*. 2010; 38(5): 1836–1853.
70. Srivastav VK, Paul AR, and Jain A. Capturing the wall turbulence in CFD simulation of human respiratory tract. *Mathematics and Computers in Simulation*. 2019; 160(6): 23–38. DOI: 10.1016/j.matcom.2018.11.019
71. Chan TL and Lippmann M. Experimental measurements and empirical modelling of the regional deposition of inhaled particles in humans. *The American Industrial Hygiene Association Journal*. 1980; 41: 399–409.
72. Cheng KH and Swift DL. Calculation of total deposition fraction of ultrafine aerosols in human extrathoracic and intrathoracic regions. *Aerosol Science and Technology*. 1995; 22(2): 194–201.
73. Heyder J, Gebhart J, Rudolf G, et al. Deposition of particles in the human respiratory tract in the size range 0.005-15 µm. *Journal of Aerosol Science*. 1986; 17: 811–825.
74. Lippmann M. Regional deposition of particles in the human respiratory tract. *Comprehensive Physiology*. 2011; Supliment 26: Handbook of Physiology, Reactions to Environmental Agents. https://doi.org/10.1002/cphy.cp090114
75. Yu C and Diu C. A comparative study of aerosol deposition in different lung models. *The American Industrial Hygiene Association Journal*. 1982; 43: 54–65.
76. Islam MS, Saha SC, Sauret E, et al. Numerical investigation of aerosol particle transport and deposition in realistic lung airway. *Proceeding of the 6th International Conference on Computational Methods,* 2015 2: 1–9.
77. Esther PH, Gordon GP, and Lawrence DL. Mathematical simulation of pulmonary O_2 and CO_2 exchange. *American Journal of Physiology*. 1973; 224(4): 904–917.
78. Asgharian B, Price OT, Oldham M, et al. Computational modeling of nanoscale and microscale particle deposition, retention and dosimetry in the mouse respiratory tract. *Journal of Inhalation Toxicology*. 2014; 26(14): 829–842.

9

Design of Efficient Dry Powder Inhalers

Anurag Tiwari[1], Akshoy Ranjan Paul[1], Anuj Jain[1], and Suvash C. Saha[2]
[1]Department of Applied Mechanics, Motilal Nehru National Institute of Technology Allahabad, Prayagraj, India
[1]School of Mechanical and Mechatronic Engineering, University of Technology Sydney (UTS), Sydney, Australia

9.1 Introduction

Rapid urbanization and industrialization in developing countries has resulted in a sharp increase in air pollution. Global air pollution has reached alarming levels in recent times, and 97% of low- and middle-income cities with populations of more than 100,000 do not meet the air quality guidelines of the World Health Organization (1–5). More than 90% of the population in China, the world's most populous country, are breathing air loaded with three times the recommended limits of particulate matter (PM), while 40% of the Indian population is exposed to five times the recommended limit of PM in the air.

Air pollution may adversely affect human health in more than one way, ranging from biochemical to physiological changes causing simple to serious problems. Prolonged exposure to air pollution leads to cough, breathing difficulties, and cardiac and even renal disorders. According to research (1–9), PM in the air decreases life span and even causes human infertility. To enable the transportation of oxygen, a large amount of air (400 million liters on average over a lifetime) is absorbed by the blood stream through the lungs. Therefore, the internal surface of the respiratory system is predominantly exposed to the air pollutants. The air pollution mainly effects the pulmonary and cardiovascular systems. Some common diseases such as hay fever, asthma, laryngitis, bronchitis, pneumonia, tuberculosis, emphysema, lung cancer, and respiratory distress syndrome (RDS) that are affecting human population are triggered from continuous exposure to air pollution. The World Health Organization (WHO) estimates that there are 334 million asthma sufferers worldwide, as well as 230 million people with COPD (10). Unfortunately, the essential drug list of the WHO for treatment, inhaled corticosteroids and inhaled bronchodilators, are either unavailable or unaffordable (11–13). Moreover, drug delivery through inhalation is a potential drug administration route for many systemic diseases because human lungs offer a large surface area for the absorption of medication and provide higher bioavailability as compared to medicines administered orally due to bypass of the hepatic first pass movement (14–18). Therefore, development and refinement of inhalation treatments are key concerns for the healthcare community.

Inhalation delivery systems need intensive optimization for physiological variations and limitations of human respiratory system. The human respiratory tract comprises several branched tubings of airways starting from trachea and ending at the alveolar sacs. The diameter of the airways are gradually decreased wherever bifurcation occurs. The larger particles are deposited in upper airways and finer particles reach to the alveolar region.

Inhalation therapies have been used since ancient times, with the evidence of people smoking *dhatura* based plants and other infusions in India 4000 years ago. Clay inhalers became popular for inhalation of these infusions in the late 18th and in the 19th century. However, the respiratory medication conveyance innovation has progressed past traditional use. Subsequently, more sophisticated devices, such as atomizers and nebulizers, came into use in France in the mid 18th century. The chemical derivatives, such as stramonium-based cigarettes, were prescribed for asthma and other respiratory troubles early in the 20th century. Riker Laboratories launched the first commercial pressurized metered-dose inhaler (pMDI) in 1956 containing epinephrine and isoproterenol drugs. However, in 1987, the use of chlorofluorocarbon propellants in pMDIs was banned under the Montreal Protocol, and dry powder inhaler (DPI) technology started becoming popular. Nebulizers, pMDI, and DPI are the main three types of pulmonary delivery systems that are currently being used and are discussed in brief in the following subsections.

The nebulizer generates small liquid droplets by atomization and this droplet-laden air is inhaled by the patient. Two types of nebulizers are found — the air jet type and ultrasonic type. Air jet nebulizers work on the aerosolization of liquid using a stream of compressed air forced into a small diameter tubing located above the surface of the nebulizing liquid. The liquid is drawn up by the Venturi phenomenon and broken up into droplets. The nebulizer has baffles to make sure that the particles are generated in the respirable range of mean median aerodynamic diameter (MMAD) of 1–5 μm. The working of an ultrasonic nebulizer is based on the vibrations of a piezoelectric crystal driven by an alternating electrical field. When the vibration intensity is sufficient, cavitation occurs and droplets are generated. Ventilation enables airflow to cross the nebulizer and be inhaled as aerosol droplets. Both types of nebulizers require a power source for operation, need considerable time to deliver drugs, and are comparatively costly. More recent medical advice does not support the use of nebulizers for severe acute respiratory syndrome (SARS)

DOI: 10.1201/9781003046547-9

patients because its use could spread the disease, hence the reliance on DPI technology. Design progressions for nebulizers and secondary devices can be accomplished using the analytical design approach laced with simple physics or empirical correlations, CFD, and experimental trial and error. Analytic principles are available for the development of nebulizer devices, as well as correlations for aerosol deposition in the mouth and throat (oropharynx) regions, including the effect of a mouthpiece.

9.1.2 Pressurized Metered Dose Inhaler (pMDI)

The essential components of a metered dose inhaler (MDI) are an actuator, a pressurized canister containing drug suspension or solution with pressurized propellant and surfactant, and a metering chamber. The pressure inside the canister drives the liquid formulation through a nozzle and atomizes it. The device is small, easy to handle, portable, and inexpensive. However, it has some limitations, including poor coordination between actuation and breathing, especially for children and older patients, leading to poor drug delivery, and the device injects the particles initially with large droplet size and high velocity, leading to high oropharyngeal deposition and low deposition in pulmonary airways (18, 21).

pMDI using chroloflurocarbon (CFC) gases as propellant were widely used as inhalation devices. The Montreal Protocol in 1987 prohibited the use of CFC propellants by 1 January 1996. Since then, hydrofluoroalkanes (HFA) are being used as a propellant, but its performance is poor as compared to CFCs and may also be banned in future as it is also a greenhouse gas. Moreover, pMDIs can't be used with all kinds of drug formulations. Substituting CFC-driven MDI by the DPI was decided by the signatory countries to attain the goal.

9.1.3 Dry Powder Inhaler (DPI)

Among the types of pulmonary drug delivery devices, DPIs seem to be the most promising for use in future (19). The market value of the world's pulmonary inhaler devices is anticipated to rise to US$43,214 million by the end of 2025 and a major share of this will be owned by DPIs. But more than its commercial value, it offers the greater social value of making a healthy society.

The drug in a DPI is kept as fine powder with or without the combination of a carrier substance. The significant features of DPI are that it is portable, free from propellant, easy to operate and cost-effective, with better stability of the drug formulation because the drug is kept in dry state (22, 23). The device is breath actuated. Spinhaler® was the first commercial DPI device launched in the market in 1970 for pulmonary drug delivery. The DPI became popular among the people of the European countries as an inhalation device (24). Around 40% of patients use the DPI device for the treatment of asthma and COPD (25, 26). DPI devices are also used for delivery of effective drugs such as antibiotics (27–29), insulin (30), drugs for neurological disorders like Parkinson's disease (31), anti-tuberculosis drugs (33), anticoagulant heparin (33, 35), drugs for sexual dysfunction (37), anti-hypertensive nifedipine (39), opioids and fentanyl for the management of cancer pain (41–44,) and delivery of the atropine sulfate nanoparticle as an antidote for organo-phosphorus poisoning with enhanced bioavailability (45) through the pulmonary route. DPIs can help reduce various physiological complications by means of an improved pulmonary drug delivery either for local or general purposes (22). The particles, which are sufficiently small in size and inhalable in nature, are subjected to forces like accumulation and cohesion, resulting in non-uniform and poor dispersion in the flow (46, 47). A DPI device is preferred for pulmonary drug delivery because of its potential ability to produce high, fine particle fractions (FPF) enabling a reasonably high drug deposition in the lungs, quick and easy operation by the patients, and most importantly, due to its breath-actuated use and portability. A significant feature of a DPI is its clinical efficacy. Because of these advantages, it has captured 68% of the market share among all pulmonary drug delivery devices available. Still, this technology is not without challenges and there is scope to improve the deposition efficiency of drug.

9.2 Classification of DPI

DPI devices can be divided into three types: single unit dose, multiple-unit dose and multiple dose (48, 49) as shown in Figure 9.1. In a single unit dose DPI device, each capsule/blister pack is specifically filled with the drug formulation in the form of powder (10). The powdered formulation is filled in a wheel, blister pack, or strip in multiple-unit dose DPI devices (49). The device can be in rotary planar configuration as shown in Figure 9.1(b).

DPI is used as a means of an alternative but potential drug delivery device through the pulmonary route, for successful local treatment [asthma, COPD or cystic fibrosis (CF)], for the complete delivery of drugs with irreproducible and poor to no bioavailability when delivered via the gastrointestinal passage (e.g. peptides and smaller proteins able to pass through the alveolar membrane, like insulin) and for the drugs which require quick onset reaction (e.g. morphine). The later applications have posed a challenge for pulmonologists and drug makers in the last decades, as they attempt to accomplish a superiorly controlled and more reproducible peripheral (deep) drug deposition in the human lungs.

The CF patients are given treatment in the form of inhaled antibiotics, usually in high doses, probably ≥ 100 mg of powder in each dose. This task can be accomplished as new inhalers are able to handle such a high dose. Such amounts would certainly need a unique de-agglomeration technique to produce aerosol ofsuitable diameter. On the otehr hand, several other high potency, new drugs with highly lipophilic materials are administered at a low dose. The reproducible administration of the extremely low doses must be reproducible and hence need special doses as well as de-agglomeration. In order to meet such diverse needs, a variety of designs of DPI devices are available in the market.

In general, DPI comprises a drug formulation in the form of powder, a dose mechanism, a powder de-agglomeration principle dispersing the powder into the inhaled air stream, and the inhaler's mouthpiece. Numerous secondary inhaler parts are also used along with the DPI device to accomplish a large variety of

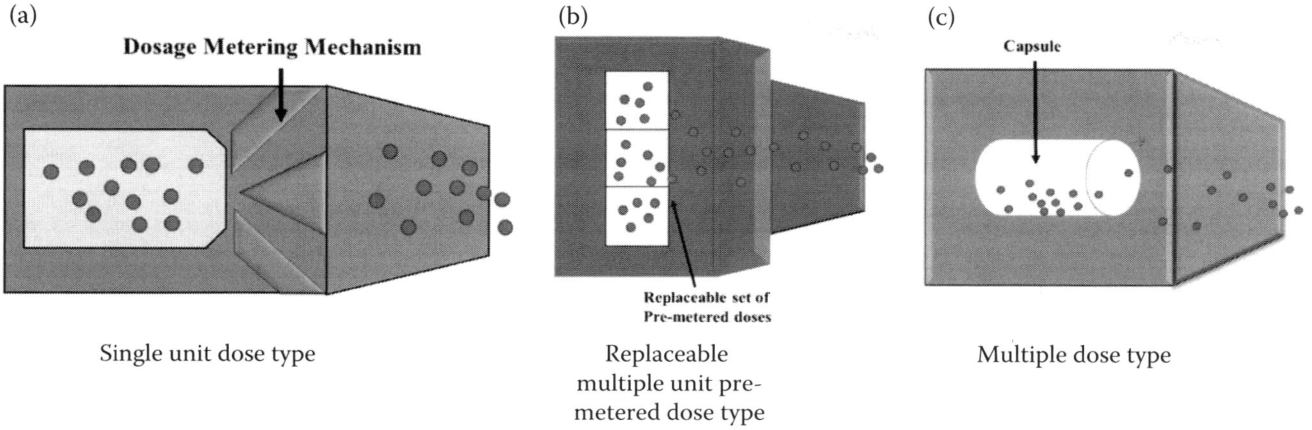

FIGURE 9.1 Types of Dry Power Inhaler

functions, like safety, ease of operation, and it also protects the powdered drug from moisture. Commonly, multi-dose DPI devices need flowability and homogeneity in the drug formulation. A DPI has a relatively complex delivery system and its performance is determined by a number of parameters with complex interrelationships (Figure 9.2).

The maneuvering of inspiratory flow and the design of the inhalation device are two important factors for any DPI design. The flow maneuver is governed by the flow resistance of the DPI device. The flow resistance of a DPI device is usually pretty high as compared to other pulmonary drug administering devices, like nebulizers and MDI. This is due to the design, which requires restriction in the flow passage leading to an increase in the kinetic energy of the air flow through the inhaler, causing de-agglomeration for ample entrainment of powdered drugs. The de-agglomeration principle is the key mechanism in a DPI determining the deposition efficiency of the drug deep into the lungs. The de-agglomeration breaks up pellets, which are generally spherical in shape, into primary drug or separate drug particles from the carrier crystals in adhesive mixtures or nucleus agglomerates during inhalation (Figure 8). Its purpose is to produce an aerosol containing more drug particles in the MMAD of 1–5 μm suitable to penetrate the targeted region for better deposition. An ideal DPI device must produce FPF adjusted to the inhalation flow rate of each patient. The magnitudes of adhesive and cohesive forces in the drug powder with the de-agglomeration forces are important in the force balance among them, and also influence the underlying method of de-agglomeration.

The de-agglomeration can be increased by increasing the turbulence. The air flow pattern (turbulence) can be changed by modifying the geometry/structure of the DPI device leading to enhanced DPI performance. The turbulence level for various inhalers available in the market are different and a considerable improvement can be made by modifying the design of DPIs and adopting a robust de-agglomeration principle.

A study shows that replacement cost of salbutamol HFA MDI with a DPI device is around US$1.5 billion a year, on a recurring basis, which is considered a business incentive to develop a novel DPI. Most DPI devices use a micronized active pharmaceutical ingredient (API) combined with coarse carrier particles, normally lactose, in the drug formulation. This makes the metering and delivery of the drug more reproducible. Shear force and/or impaction of particle/particle or particle surface are crucial for de-agglomeration of the powder formulation in DPI. These forces usually result in de-agglomeration, and the drug is released from the DPI device. Larger carrier particles and any API that is still agglomerated

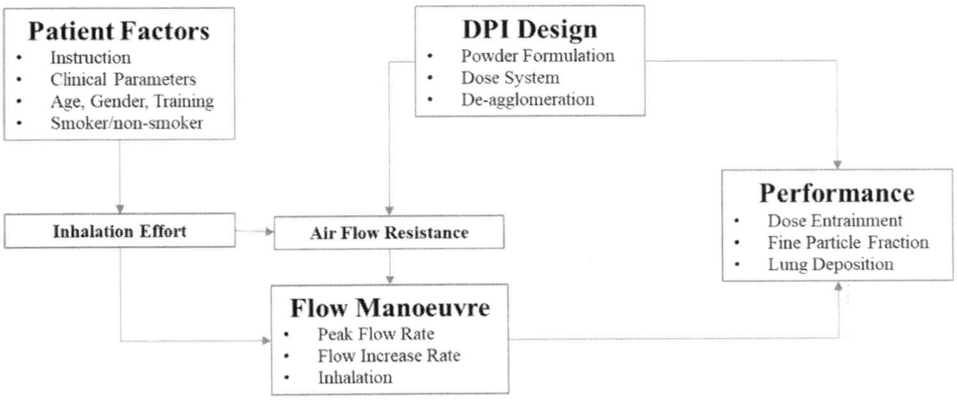

FIGURE 9.2 Scheme of the Major Variables and Interactions in DPI Performance

consequently impact the patient's throat, while the smaller API particles further reach to the lungs.

9.3 DPI Design

Figure 9.3 represents the factors associated with patients, formulations and the device design that need to be addressed to achieve optimal drug delivery from the DPI inhalers. The geometrical design and structure of an inhaler strongly influences the deposition of drug particles in the human lungs. An ideal system must deliver drugs in accurate and consistent doses in the lungs while maintaining the stability of the drug formulation. It is also expected that the drug delivery devices are simple and handy and can be comfortably used by the patients. DPI devices are gaining popularity because of their easy use and better drug powder stability. Pressurized MDI devices, on the other hand, are still facing challenges from the formulation and design points of view. Besides, researchers are remodeling nebulizers to widen their applicability. Actually, there is no single device that satisfies the multitude of requirements in order to get the drugs of various physicochemical properties in required concentration to desired locations in the lungs. Medical practitioners need to understand the functioning of different inhalers available in the market and match the capabilities of each inhaler with the needs of the patient, according to their health condition, to accomplish the best therapeutic results. The features of various commercial DPI inhalers are furnished in Table 9.1.

9.3.1 Basic Principle of DPI Design

The factor that govern the improvement of new DPI designs are listed below.

Pharmaceutical factors

- Regulatory and pharmacopoeial obligations.
- Delivery systems for new chemical entities (NCEs) including non-respiratory drugs.
- Optimized design for performance and efficiency.

Clinical factors

- Power-assisted dispersion: an alternative to pMDI devices, which have patient coordination and reformulation problems.
- Devices that are free from flow-rate dependence.
- Development of device equivalent pMDIs, nebulizers in terms of efficacy.
- Device enabled for targeted drug delivery in particular areas in the lungs.
- New devices should exhibit clinical benefits over the existing DPI devices and/or be cheaper.

Commercial factors

- Generic drugs require innovative DPI designs to influence markets.
- Launch of a viable alternative to CFC propellant driven pMDI device.
- Newer design should account for manufacturing costs, environmental impact, ergonomics and clinical compliance.
- Patent extensions and protection of intellectual property.

The creation of newer anti-inflammatory drug formulation for the cure of asthma, such as corticosteroids, mometasone furoate etc., have also provoked newer DPI designs for effective drug delivery (46). In this context, a DPI device under brand name Twisthaler™ (Kenilworth, NJ, USA) marketed by Schering-Plough (USA) has been developed, which is similar to Turbohaler.

9.3.2 Improving Performance of DPI

The development of DPI devices, such as Turbohaler and Diskus, are based on their ability to deliver optimum dosing of drug by increasing FPF and the consistency of dosing to individual patients. Debate on the DPI designs centres around a reservoir-based design (like Turbohaler) versus the factory-set metering (like Diskus). The later is preferred because of its

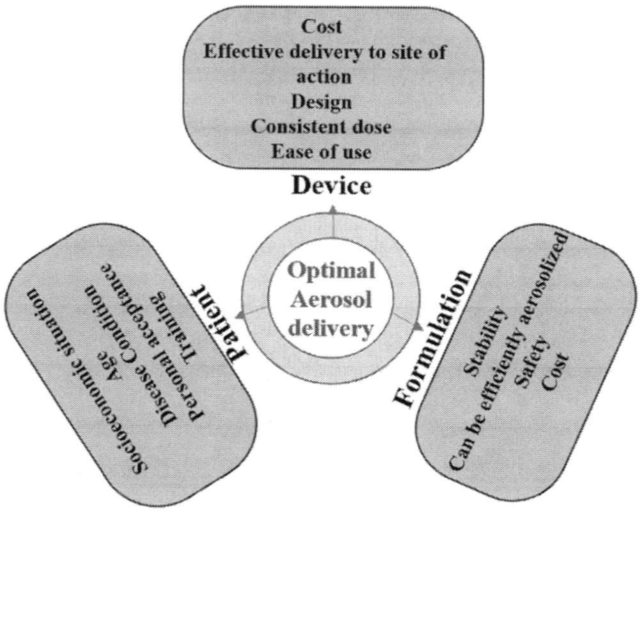

FIGURE 9.3 Device Variables Affecting Optimum Distribution of Aerosol from Inhaler

TABLE 9.1

The Features of Various Commercial Dry Powder Inhalers

Name of DPI Device	Manufacturer (Year of Launch)	Type of Device (Metering)	Method of Delivery	Resistance of Device	Inhalation (l/min)	Inhalation Resistance (kPa0.5 l/min)	Drug Deposition (%)	Remarks (Reusable, Diseases)
Breezhaler	Novartis	Single dose	Capsule	Low	111	0.017		Yes, COPD
Cyclohaler/ Aerolizer	Novartis	Unit dose	Capsule	Low	102	0.019		Yes, Asthma, COPD
Ellipta	GSK, 2013	Multi-dose	Blister Strip	Low	74	0.027		Asthma and COPD
Novolizer	Meda Pharmaceuticals Ltd.	Multi-dose	Reservoir	Medium	72	0.027		Asthma
Genuair	Almirall,2012	Multi-dose	Cartridge	Medium	64	0.031		COPD
Diskus	(Accuhaler)	GlaxoSmith-Kline	Multi-unit dose	Blister Strip	Medium	72	0.027	
Asthma/ COPD								
Nexthaler	Chiesi, 2012	Multi-dose	Reservoir	Medium	54	0.036		Ashtma
Turbuhaler	AstraZeneca	Multi-dose		Medium	54	0.039		
Handihaler	Boehringer Ingelheim	Unit dose	Capsule	High	37	0.058		COPD
Podhaler	Novartis, 2013	Multi-unit dose	Reservoir	Medium	78	0.025		Cystic fibrosis
Conix	3M, 2009	Multi-dose		Medium	---	---		Asthma, COPD
TwinCaps	Daiichi Sankyo, 2010	Multi-unit dose	Capsule	High	---	0.057		Viral Infection
Staccato	Teva select brands,2012	Single unit dose	Thin drug film	Medium		0.025		CNS disorder
Dreamboat	Mankind,2014	Multi-unit dose	Cartridge	Medium	---	0.093		Diabetes
Spinhaler	Aventis	Unit dose	Capsule	Low				Asthma, COPD
Diskhaler	GlaxoSmithKline	Multi-unit dose	Blister pack	Low				Asthma/ Influenza
Turbospin	PH&T	Unit dose	Capsule	Medium				Lung infection/ CF
Novolizer	ASTA	Multi-dose	Cartridge	Medium				Asthma
Easyhaler	Orion	Multi-dose	Reservoir	High				Asthma/COPD
Clickhaler	ML Labs	Multi-dose	Reservoir	High				Asthma
Pulvinal	Chiesi Ltd.	Multi-dose	Reservoir	High				Asthma
Prohaler	Aptar Pharma	Multi-unit dose	Blisters	High				Asthma/COPD
Orbital	Pharmaxis	Multi-unit dose	Reservoir	High				Asthma
Eclipse	Aventis Pharma	Unit dose	Blister	High				Asthma
Certihaler	SkyePharma	Multi-dose	Reservoir	N/A				Asthma

consistency in doses for wide range of flow rates, whereas the former has the ability to generate higher FPF mass. However, patients using a DPI device should inhale the powdered drug with sufficient inhalation flow rate for effective aerosolization as it does not usually contain propellant. The turbulence must be created in the inhaler passage to effectively create the aerosolized particle cloud. Hence, a threshold airflow rate with efficient DPI design is necessary for the creation of required turbulence shear forces. The physical properties of powder formulation and DPI design are therefore closely related (50).

Some DPI devices require a threshold flow rate value (around 30 l/min or even more) to effectively start producing the drug in the form of powdered aerosol. But this might not be effective for children and aged patients having lower inspiratory rates. Therefore, there is a growing interest in including a source of internal energy in the DPI design helping to aerosolize the dose, which is independent of the patient's ability. For example, a battery-operated motor is used in the Spiros™ device (San Diego, USA) to produce an aerosol cloud even for a very low inhalation rate and with higher FPF mass. Other energy sources used for aerosolization include a spring-loaded hammer and compressed air.

9.3.3 DPI Performance Independent of Inhalation Flow Rate

The device resistance (often measured by pressure drop across a DPI) and the flow rate are crucial in developing low resistance DPI. DPI devices usually require a minimum flow rate to create FPF mass apt for inspiration by children and elderly patients. Everard (51) depicted that a rise in flow rate influenced the distribution of particle size in aerosolized drugs considerably. Hence, attainment of an inspiratory flow rate around 30 l/min before ingesting 150 ml of air through the DPI device may result in an increase of MMAD of the dispersed drug cloud leading to a drop in drug deposition efficiency of the DPI device. In in vitro studies, pre-programmed inhalation profiles recorded from the breathing profiles of the patients, are fed to the DPI device for a range of flow rates and flow accelerations. The generated drug clouds are analyzed in terms of dosing, FPF, and MMAD using an impactor/impinger. This study has helped to understand the intricate relationship of various parameters influencing the DPI performance and hence has become a standard for testing of any new DPI device.

9.3.4 Generic Drugs

Generic drugs are now given their due importance everywhere, especially in developing countries. With the end of exclusive patent rights on various pulmonary drugs (e.g. salbutamol), the newer, highly efficient DPI devices can incorporate these off-patent drugs for penetration in the drug market. However, Hill et al. adopted an alternative move by using dose-targeting in a DPI and a therapeutically equivalent with an MDI device. The in vitro and in vivo tests for the DPI device were conducted using scintigraphy and this move looks promising for fully engineered products to come in the near future, where the design of the DPI will be largely controlled by a certain dosing of drug.

The breath-actuated DPI device is classified in various categories. In one category, drugs to be inhaled use gelatin capsules or blisters as dose compartments. Inspiratory flows are extended to these compartments and the drug particles are entrained with the air flow due to turbulent shear or collision forces (52,53). However, this mechanism cannot produce high FPF dosing, because the residence time for the disruptive force is very small. In other types of capsule based DPI devices, the capsules are given a desired motion (linear or spiral) to release and disperse the drug powders.

Drug deposition in lungs is assisted by higher flow resistance in a DPI as it decreases aerosol velocity in the pulmonary passage, thus enhancing the residence time for penetration into deep lung regions. It is assumed that higher inspiratory effort is required while operating a high resistance DPI device. However, it is not true, as the patient's effort is not directly related to DPI design.

In the case of a DPI device, aerosolization is triggered by the inspiration of the patients. Hence, unlike in pMDI, less time coordination between device actuation and inspiration is needed for drug delivery. It is complex yet important that the manufacturing process must ensure the aerosolization characteristics for a DPI device are more complex (54). This is very important as performance of a DPI device is much dependent on synchronization of powder formulation and the design of the inhaler.

The DPI design is crucial to deliver its desired performance. The design of DPI devices like Clickhaler, Multihaler, and Diskus delivers high airflow responsible for de-agglomerating the drug particles by splitting. Turbuhaler and Spinhaler devices, on the other hand, creates impaction of drug particles in the device surface for de-agglomeration. In earlier DPI devices, high airflow ensured higher FPF mass, but the drug particles are more likely deposited in oral cavities, not deep in the lungs. Hence, a proper balance between the flow rate and device resistance is necessary.

The updated drug formulation and capsule filling equipment eventually made DPI based drug delivery affordable. However, the amount of drug delivered through DPI device at a required MMAD has not yet reached to the desired quantity as revealed by both in vitro and *in* vivo investigations. Many researchers (55–57) studied drug-to-drug and drug-to-carrier interactions, but the results failed to register higher drug deposition at the desired location in lungs. As the lactose or particulate based carriers possess some morphological difficulties, innovative mechanical means, like creating vibration using a tape in a taper DPI device (3M) or isostatic pressure (Jethaler, MAGhaler) are used to control carrier morphology (58–60). Engineering techniques are also used for drug particles, like spray drying using improved atomization (61, 62), fluid crystallization (63, 64), dry mechano-fusion (65), and fluidized bed design (66) for demonstrating better dispersion property during release of drug formulation. A few additives, called force control agents, are also useful in lowering the inter-particulate forces of the drug powder, and exhibit better dispersion and deposition even in narrow airways (67–69). de Boer et al. (70) opined that the research on DPI are not about controlling a couple of parameters, but understanding the interrelation of the parameters influencing the DPI performance and acting upon it. Therefore, the current design modification of DPI devices must be seen as an integrated improvement formulation–device combination.

The fine particle doses delivered in in vitrocondition for various DPI devices available in the market are shown in Figure 9.4 as a percentage of label claim (FPF <5 μm). The DPI devices were tested for both high (4 kPa) and low flow resistance (2 kPa) and for 40–75 l/min of flow rate using a next-generation impactor (71). The actual dose may be lower than the label claims because of losses in the DPI device and drug deposition in oropharynx region, often as high as 50% of the delivered drug (72,73). Several studies (74–81) showed that DPI design is an important determinant in ensuring higher

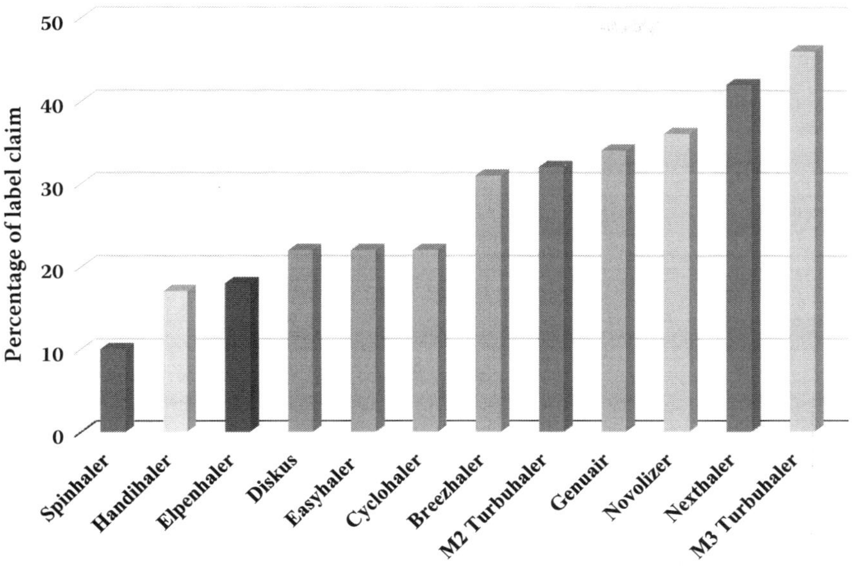

FIGURE 9.4 FPF Doses Delivered In Vitro Condition for Various DPI Devices (95)

drug deposition in lungs. In the absence of a complete bronchial tree, lung deposition is assessed using computational fluid dynamics (CFD) modeling, pharmacokinetic techniques (82–86), and scintigraphy (87). CFD modeling includes both simplistic (88) or diseased (89) lung geometry. However, the results often differ due to non-mimicking of biological conditions that are often referred to as boundary conditions in CFD. The pulmonary passage has long been considered an effective alternative route for vaccination, but developing a DPI for the same is still considered a challenging task, as the vaccine delivery through DPI in the pulmonary passage is one-off process with no immediate, measurable therapeutic response (90). Hence the DPI inhaler for administering vaccines should be consistent over a wide range of flow rates, and effective use of the device, preferred location, choice of adjuvants, mouthpiece design, training of vaccine recipients by health workers, and special attention to pediatric needs are crucial issues (91–94).

9.4 Computational Modeling of DPI

With the rapid advancement of computation and availability of various types of software, bio-fluid dynamics and drug delivery increasingly involve computational modeling to find out the niche areas of drug delivery, which is otherwise impossible with in vivo or in vitro experimentation. A major portion of this chapter is dedicated to exploring the importance of using powerful computational modeling in the development of efficient DPI devices.

9.4.1 Numerical Techniques used in DPI

DPI is the popular device used to deliver drug particles into lungs. This device has been used since 1971 (96,97) for the treatment of different types of respiratory diseases like COPD, asthma, cystic fibrosis (CF), bronchitis, etc. (98, 100). In the DPI device, drugs are used in dry powder form, which are chemically more stable, and due to its small size it is portable and easy to handle (101).

Recently designed DPIs having poor aerosol delivery performance were less than 30% able to reach into the respiratory tract with particle size 1–5 μm (104). Numerical modeling plays an important role in understanding the air and particle flow in DPI.

CFD modeling is an important tool in determining the multiphase flow (air and drug particles) behavior of DPIs. Chan and his co-workers have used CFD modeling to predict the deagglomeration of air and drug particles into DPI (102–108).

The discrete nature of drug particles and their interactions cannot be predicted by CFD models. Therefore, it is necessary to use a CFD–DEM model for explaining and calculating interparticle forces, like particle–particle and particle–wall interactions (109). The CFD–DEM model explains flow and force properties of each particle, which exactly predicts the dispersion behavior of drug particles. This model is very useful to overcome the limitations of existing DPI devices and helpful in designing efficient inhalers.

9.4.2 Development of CFD Model

CFD is an important tool which is used to analyze physical phenomena by numerical calculations. It has broad scope in different field of engineering technology. In biomedical engineering it can be used to design devices for the human respiratory system and cardiovascular system, and in designing medical instruments. It provides reproducible and complete simulated results of real-life problems which are not possible by experimentation. The motion of fluid has been characterized by different governing equations such as conservation of mass, momentum, and energy in the forms of partial differential equations (110, 111) as furnished below.

9.4.2.1 Mass Conservation

This law states that the mass of the system must be constant over time. Mass of a system does not change until some mass is added or removed. It is often referred to as a continuity equation in fluid dynamics and is represented as

$$\frac{\partial \rho}{\partial t} + \frac{\partial(\rho u_x)}{\partial x} + \frac{\partial(\rho u_y)}{\partial y}\frac{\partial(\rho u_z)}{\partial z} = 0$$

where ρ is density (kg/m^3); t is time (s); u_x, u_y and u_z are the velocity components in x, y and z directions.

9.4.2.2 Conservation of Momentum

The total external force on an element is equal to the change in momentum for a certain fluid element; this equation is called the Navier-Stokes equation

$$\frac{\partial(\rho_f \varepsilon \bar{u})}{\partial t} + \nabla\cdot(\rho_f \varepsilon \bar{u}\bar{u}) = -\nabla P - F_{fp} + \nabla\cdot(\varepsilon\tau) + \rho_f \varepsilon g + \nabla \cdot(-\rho\overline{u'u'})$$

where, ε is porosity; ρ_f refers to fluid density; P is pressure; $\bar{u}$ is fluid mean velocity; u' is turbulent velocity fluctuation; and τ indicates viscous stress tensor of the fluids.

In the past few decades, many researchers have studied the dispersion process and performance of fine particles in the field of pharmaceuticals. but experimentation alone cannot exactly demonstrate the dispersion mechanism of drug particles and their quantitative analysis. On account of this, CFD simulation techniques are very robust and simpler due to the availability of high-quality commercial software for analysis of turbulence and particle models.

9.4.2.3 Turbulence Models

Mainly two types of flow exist in the inhaler: swirling flow and axial flow. In swirling flow, the flow structure is very complex and highly turbulent in nature (112,113). But these behaviors have been simulated by using Reynolds-averaged Navier-Stokes (RANS) equations and the different turbulence modeling. When the k-ε turbulence model fails to determine the complexity of the flow, use of two or more than two equation models, like the shear stress transport (SST) k-ω model (114–116), are used by many researchers to simulate the turbulence behavior of fluid flow (117). For the anisotropy nature of turbulence flow, a 6-equation Reynold stress model (RSM) is used to predict the turbulence behavior.

A few researchers used Ansys-CFX for a swirling flow in the inhaler and reported a very poor convergence (118). While in the same inhaler some researchers (119) used RSM model to give fast convergence due to its solution algorithms. In general, the RSM model is computationally efficient (i.e. consuming less time and storage) and also quite sensitive to the design and generation of the grid. Many comparisons are done by the researchers by using different turbulence models like standard k-ε, k-ω models, SST k-ω, and renormalization group (RNG) k-ε models and found that the SST k-ω model gives the correct prediction for wall bounded flows (120).

While large eddy simulation (LES) requires much more computational resources for the unsteady and anisotropic nature of flow, the sub-grid-scale (SGS) model is used to predict the small-scale turbulence in the flow. This is especially suitable for wall bounded flows and uses a blending function to connect the wall-bounded region with the free-stream flow. Despite using large computational resources, LES is considered a potential numerical approach to investigate the unsteady flows with anisotropic turbulence, as the technique can resolve only bigger eddies and hence more detailing of flow structure is revealed based on the mathematical filter and grid spacing used (121). Table 9.2 summarizes the use of CFD in inhaler-based pulmonary drug delivery.

9.4.3 Coupling of CFD-DEM Model

The discrete element method (DEM) is used to numerically analyze the bulk behavior of many small-scale entities. DEM mainly classified into two approach first one is the hard sphere and second one soft sphere. Cundall and Strack (113) developed the soft-sphere technique. Particles are allowed to undergo minute deformations in this approach. Elastic, plastic, and frictional forces between particles are measured using deformations. The main advantages of DEM are the defined size and density of the drug particles in the simulation and coefficient of restitution, Young modulus, Poisons ratio and the particle–particle interactions. Nevertheless, due to the fine particle size of the drug, it is not possible to apply DEM properly (148).

The rapid development of CFD-DEM as an interdisciplinary research tool has been made possible with the development of the high speed computation and powder technology (149). The first CFD–DEM method has been introduced by Tsuji et al. (114) and later on followed by others. The discrete nature of the drug particle used in DPIs like particle–particle and particle wall interactions has been determined by DEM applying Newton's second law of motion. Whereas the traditional CFD has been used to investigate the fluid flow behavior at continuum level.

The granular flow of particles exhibits in the form of rotational and translational motion. In the governing equations, the mass of the particle is m_i and inertia moment of inertia, I_i, is represented as:

$$m_i\frac{dv_i}{dt} = \sum_j F_{ij}^c + \sum_k F_{ik}^{nc} + F_i^f + F_j^g$$

$$I_i\frac{dw_i}{dt} = \sum_j M_{ij}$$

In the above equation v_i and $w_i$represent the translational and angular velocities the particles; M_{ij} is the torque acting on the drug particle i by particle j; $F_{ij}{}^{c}$, $F_{ikc}{}^{n}$, $F_i{}^{f}$ and $F_j{}^{g}$ are the contact force, non-contact force, particle–fluid interaction force, and gravitational force respectively. Figure 9.5 represents the force and torque acting on the particle i from the particles j and k.

TABLE 9.2

Summary of the Use of CFD in Inhaler Based Pulmonary Drug Delivery

Authors (Year)	Numerical Method and Turbulence Model	Types of Inhaler	Flow Rate [l/min]	Particle Tracking Approach	Diameter of Drug and Carrier Particle	Major Findings
(122)	RANS/k-ω-SST	Aerolizer	30, 45 and 60	DPM (D, TD)	Mono, 3.2 μm	Influence of grid type and mouthpiece length on wall collisions and drug retention
(123)	RANS/k-ω-SST	Aerolizer	30 to 120	DPM (D, TD)	Mono, 3.2 μm	Influence of flow rate on powder retention in different regions, FPF, wall impact number and velocity
(124)	RANS/k-ω-SST	Aerolizer	30, 45 and 60	DPM (D, TD)	Mono, 3.2 μm	Resolving rotating capsule and influence on FPF
(125)	RANS/k-ω-SST	Aerolizer	30, 45 and 60	DPM (D, TD)	Mono, 3.2 μm	influence of inlet cross-section on wall collisions and FPF
(126)	RANS/k-ω-SST	Aerolizer	30, 45 and 60	DPM (D, TD)	Mono, 3.2 μm	Influence of mouthpiece geometry, different expansion ratios
(127)	RANS/RSM (Reynolds stress turbulence model)	Aerolizer/ cyclohaler	55–165	DEM (D, G??)	Compact agglomerates diameter 50 μm, primary particles: 2.6–4.1 μm	Single compact agglomerate tracking, analysis of agglomerate dispersion and FPF through flow stresses and wall impacts, effect of agglomerate composition
(128)	RANS/RSM (Reynolds stress turbulence model)	Impaction throats	60–150	DEM (D, G??)	Compact agglomerates, primary particle size distribution 1–8 μm	Breakage of agglomerates in a knee-type impaction, throat and primary particle deposition, determination of fragment size and FPF
(129)	RANS/k-ω-SST	Aerolizer, Handyhaler	60	DPM (SD)	Mono, 32, 108 and 275 μm	Determination of wall collision rates for two inhalers and different carriers
(130)	RANS/RSM (Reynolds stress turbulence model)	Aerolizer	28–138	DEM (D, G??)	Compact agglomerates, primary particle size distribution 1–8 μm	Analysis of agglomerate breakage, dispersion and particle deposition in an Aerolizer with rotating capsule, determination of FPF
(131, 132)	RANS/k-ε, k-ε-RNG, k-ωSST and LES	Turbohaler	30 to 70	DPM (D, G, B, TD??)	Poly: 1–20 μm (mean 2.2 μm) and mono 1–5 μm	Performance of different turbulence models, analysis of drug particle wall deposition based on critical impact velocity
(133, 134)	RANS/low-Re-k-ω	Handyhaler,	straight flow passages45, 60 and 75	DPM (D/C, G, B, TD)	Mono: 1 μm	Analysis of parameters affecting de-agglomeration and aerosolization and influence of different inserts in flow passage
(135)	RANS/k-ω-SST	Turbohaler	Flow rate increase to peak values of 30 to 60	DPM (D, G, B, TD??)	Poly: 1–20 μm, mean 2.2 μm	Influence of dynamic flow rate increase (low and fast) on powder deposition and comparison with measurements
(136)	RANS/low-Re-k-ω	Straight and knee-type inhaler	45	DPM (D, G, B, TD)	Mono, 1 μm	Influence of capsule chamber orientation and rod array in mouthpiece on retentions and FPF
(137– 139)	Navier-Stokes, no turbulence	NA	NA	DEM (D, G, P)	Cluster (carrier: 50–106 μm plus drugs: 5 μm)	Flow-induced detachment of drug particles from carrier particles, very low velocities
(140)	No flow	NA	NA	DEM (–)	Cluster (carrier 87.5 μm, drugs 5 μm)	Analysis of drug detachment from carrier at different impact angle and velocity, extremely low impact velocity, no cluster rotation

(Continued)

TABLE 9.2 (Continued)

Summary of the Use of CFD in Inhaler Based Pulmonary Drug Delivery

Authors (Year)	Numerical Method and Turbulence Model	Types of Inhaler	Flow Rate [l/min]	Particle Tracking Approach	Diameter of Drug and Carrier Particle	Major Findings
(141)	RANS/RSM (Reynolds stress turbulence model)	Aerolizer	83–138	DPM for inhaler and DEM for single clusters collisions	Carrier: 70–130 μm, drug particles: 2 μm	Evaluation of wall collision-induced drug detachment by CFD-DEM and derivation of empirical correlations, Lagrangian simulations of carrier motion through inhaler and determination of FPF from collision statistics
(142, 143)	RANS/RNG-k-ε	Cyclohaler	60	DPM (??)	Small drug particles, no size given	Inhaler optimization, influence of mouthpiece geometry on strain rate.
(144)	RANS/k-ω-SST	Cyclohaler	70 and 100	DPM (D, G, SL, RL, TD)	Mono: 50, 110, 500 μm, poly: 500 μm	Influence of flow rate, carrier particle size and fluid forces on flow stresses acting on particles and wall collision statistics
(145)	LES with SGS turbulence model	Knee-type inhaler	70 and 100 μm	DEM for carrier motion & single clusters Carrier	Carrier: 70 and 100 μm, drug particles: 2 μm	Evaluation of drug detachment by DEM and derivation of empirical correlations, DEM simulations of carrier motion through inhaler and determination of FPF from derived correlations, deposition accounted for.
(146)	RANS/k-ω-SST	Turbohaler	30 to 70	DPM(D, G, B, TD??)	Poly: 1–20 μm, mean 2.2 μm	Optimization of inhaler geometry (roof of inlet chamber) with respect to reduced deposition.
(147)	No flow	NA	NA	DEM (–)	Carrier: 100 μm, drug particles 5 μm	Analysis of drug detachment from carriers by wall impact, different velocities and angles as well as rotation

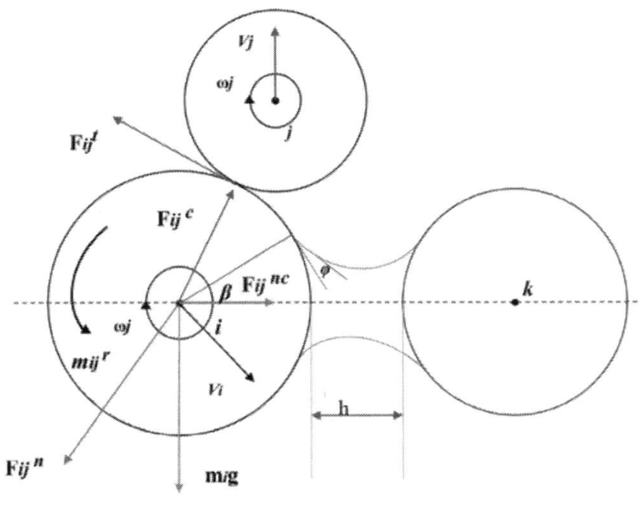

FIGURE 9.5 Force and Torque Acting on the Particle *i* from the Particles *j* and *k* (150)

DEM modeling has been used for the solid flow for a particular particle level, while CFD modeling is used for the fluid flow at computation cell level. At each time step the velocity, position, and forces of individual particles are determined by the DEM; the volumetric drag force and porosity are calculated after that by using these parameters to determine the interaction (drag force) from the fluid to individual particles. For prediction of motion of individual particles for the next step, it is required to pass the combined resulting force into DEM (151). Figure 9.6 represents the

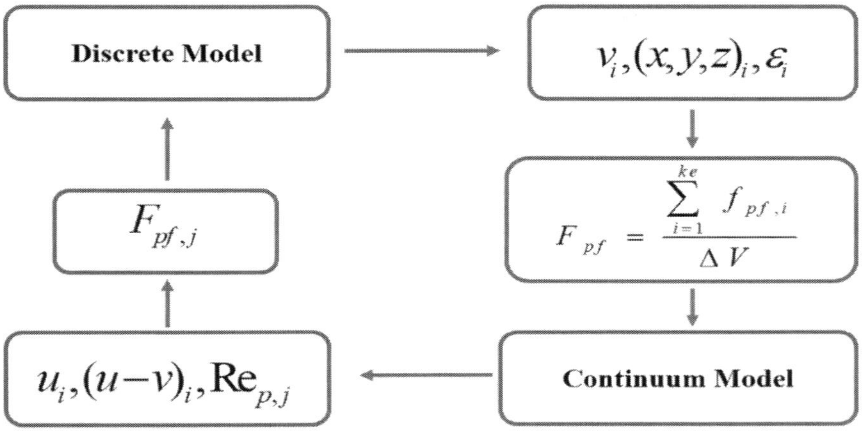

FIGURE 9.6 Flowchart for Coupling of DEM and CFD (Continuum) Models

TABLE 9.3

Major Forces Applied in CFD–DEM Modeling

	Force	Nature of Force	Force Representation	Formulation
Particle–Particle and particle–wall forces	Normal force (154–158, 159)	Contact	$F_{cn,ij}$	$-\frac{E}{3(1-\nu^2)}\sqrt{2R_i}\,\delta_n{\frac{3}{2}}n$
		Damping	$F_{dn,ij}$	$-C_n\left(\frac{3m_iE}{\sqrt{2}(1-\nu^2)}\sqrt{R\delta_n}\right)^{\frac{1}{2}}V_{n,ij}$
	Tangential force (160–162)	Contact	$F_{ct,ij}$	$-\left\{\frac{\mu_s F_{cn,ij}}{\lvert\delta_t\rvert}\right\}\left[1-\left(\frac{1-\min\left\{\left\lvert\delta_t,\delta_{t,\max}\right\rvert\right\}}{\delta_{t,\max}}\right)^{\frac{3}{2}}\right]\delta_t$
		Damping	$F_{dt,ij}$	$-C_t\left(6m_i\mu_s\left\lvert F_{cn,ij}\right\rvert\sqrt{1-\frac{\delta_t}{\delta_{t,\max}}/\delta_{t,\max}}\right)^{\frac{1}{2}}V_{t,ij}$
Particle–fluid interactions	Drag force (163,164)		$F_{pf,i}$	$V_{p,i}\nabla P$

flowchart for coupling of DEM and CFD (continuum) models. In the process engineering, the fluid flow of the fine particles is handled by the combined CFD and DEM model (152,153). Major forces applied in CFD–DEM modeling are furnished in Table 9.3.

9.4.4 Discrete Phase Modeling (DPM)

DPM is one of the popular multiphase models used by the researchers investigating drug deposition in human lungs. DPM is employed to compute the particle motion in the human pulmonary model where air-phase equations will remain the same as described earlier and particle-phase equations will be formulated based on the Lagrangian concept as follows. In this equation, the inertial effect of particles is equated with the sum of drag force and gravitational forces acting on the particles and is written as

$$\frac{\partial u_p}{\partial t} = F_D(u - u_p) + \frac{g(\rho_p - \rho)}{\rho_p}$$

where, F_D $(u\text{-}u_p)$ is the drag force per unit particle mass, while the second term of R.H.S. denotes gravitational force.

F_D is defined as

$$F_D = \frac{18\mu}{\rho_p d_p^2}\frac{C_D\,\mathrm{Re}}{24}$$

In the above equation, the velocity of air is denoted as u, velocity of particle is u_p, dynamic viscosity of air is μ, air density is ρ, particle density is ρ_p, and particle diameter is d_p.

Relative Reynolds number Re is defined as

$$\mathrm{Re} = \frac{\rho d_p \lvert u_p - u\rvert}{\mu}$$

The drag coefficient (C_D) is computed from the relation

$$C_D = a_1 + \frac{a_2}{\mathrm{Re}} + \frac{a_3}{\mathrm{Re}^2}$$

where a_1, a_2 and a_3 are constants applicable to smooth spherical particles for a wide range of Reynolds numbers given by Morsi and Alexander (248).

9.5 Pharmacological Aspects of DPI Design

Hickey, Concessio, Oort, and Platz (69) described the properties of irregular particle morphological features and demonstrated the application in the DPI device by characterization of powder and its performance level. Zeng et al. (73) investigated the similar particle size with different morphological feature affected the drug delivery while using the DPIs. Zeng et al. (73) also discussed the effects of carrier morphological features on the dispersion and de-agglomeration of the drug. The morphological features for inhalation aerosols and variation in such factors may be one of the main causes of the batch-to-batch variation in drug delivery. Secondly, the engineering of carrier particles to produce precisely designed shapes uses this strategy to improve the efficient delivery of drug. Increasing either the surface smoothness or the elongation ratio of lactose crystals will increase the potentially respirable fraction of SS from dry powder formulations for inhalation. Larhrib et al. (165) studied the properties of aerosols, like particle density and size distribution, and demonstrated the optimum lung deposition for targeting and systemic delivery. Crowder et al. (166) found that the Stokes law effectively predicted the aerosolization behavior of a drug particle and optimized drug delivery by considering various morphological parameters of the particle. Larhrib et al. (165) studied the crystallization of salbutamol from aqueous solution, engineering smooth crystals of lactose with different elongation ratios and deposition profiles. Replacing micronized salbutamol sulfate with needle-shaped drug further enhanced deposition. The highest deposition of salbutamol sulfate was obtained by combining needle-shaped lactose (NL) with needle-shaped drug. Crowder et al. (166) in another study, investigated the effect of the input of vibrational energy into a powder on aerosol entrainment. Dispersion demonstrated that the input of vibrational energy increased the dose emission while the input of frequencies specific to the powders improved the reproducibility. An ability to tailor energy input to match the flow properties of a given powder formulation may significantly improve reproducibility of dose delivery. Hassan et al. (167) examined the drug delivery performance of pollen-shaped hydroxyapatite (HA) carriers and compared the results with conventional lactose (LA) carriers. Its morphology allows a high drug mixing ratio without sacrificing the delivery efficiency compared to traditional lactose (LA) carriers, especially at lower inhalation flow rates. Suitability of pollen-shape carriers, especially at low flow rates and high drug mixing ratios.

Later, Peng et al. (168) worked on different factors affecting the performance of DPI, such as particle morphology (shape, size, and density), and the physical properties of carrier particles and their surface roughness. Benke et al. (169) studied the specific factors in the case of a carrier-based DPI system and how it affected the lung deposition by introducing surface modification and surface properties. In the case of a carrier-based formulation, the surface modifications are achieved by recrystallization of the carriers and their surface dissolution, fluidized-bed coating, and by mechanical dry coating. Development of physical mixing and optimization parameters such as interparticulate interactions may further help improve pulmonary deposition. Recently, Ali et al. (151) used differently shaped drug microparticles in airflow and calculated the drag force. The best shape of drug particles with the ordinarily used DPI device was found to be the triangular shape with aspect ratio 2:1, as it offered the highest flowability and lowest drag force. This triangular shape is expected to have a flowability two times higher than the control shape (3 mm diameter circle). Properties and parameters of pharmaceutical drug particles used in DPI are furnished in Table 9.4.

9.5.1 Methods for Preparation of the Pharmaceutical Aerosol

There are many strategies available for the formulation of inhalation powders, which vary in value, scalability, compatibility, and among other matters, APIs and the way they allow engineered particle technology. Some formulation techniques are illustrated in Figure 9.7 and are described below.

Milling: This is a drop-down approach for obtaining the desireable range of particle size but lacks of the information regarding the particle breakup mechanism (165–168). Consequently, milling is not always applicable to the rational engineering of

TABLE 9.4

Properties and Parameters of Pharmaceutical Drug Particles used in DPI

Properties	Parameters
Aerosol properties	• Mass median aerodynamic diameter (MMAD) • Geometric standard deviation (GSD) • Fine particle fraction (FPF) • Air/particle velocity
Particle properties	• Aerodynamic diameter • Bulk density • Tap density • Shape • Charge
Physiochemical properties	• Solubility • Hygroscopicity

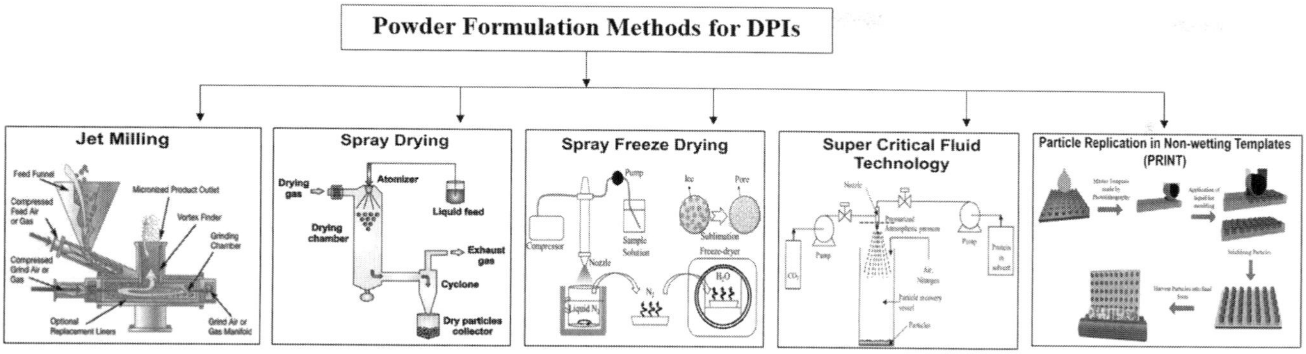

FIGURE 9.7 Powder Formulation Methods for DPI

particles with respect to their size, density, or surface nature. However, milling can change the particle size and the nature of the surface (170–172). It affects the hydrophobic and particle flow characteristics of the powder (173–175). The same applies for high-dose powders that do not depend on coarse excipient for high flow capacity (176). A comparison of wet and dry milling using succinic acid and sucrose found that wet milling has comparatively greater surface energy due to sensitivity to mechanical stress.

***Spray Drying*:** This is a bottom-up approach to formulation for engineered particles. Vehring et al. (164) reviewed spray-drying extensively (177). The nozzle is used for spraying the solution of the dissolved drug and the droplets into heat chamber for drying. In this process, small size particles are obtained by controlling the droplet size. Literature also validates that the drugs are more chemically stable than in other forms (178), and the nature of the spray-dried particle is amorphous (179). The performance of the spray-dried particle can be enhanced by using the modified enhancer (180) and drug release devices (181) for respiratory drug delivery.

***Supercritical Fluid Technique*:** This is a comparatively updated method for the production of micronized drug particles. Most importantly, their morphological features and API remain unchanged even at high temperatures, but excessive pressure may affect the activity of the drug aerosol (182). Even now, since viral inactivation is reflected by this method, it may not be a good technique for vaccine production (183).

***Particle Replication in Non-wetting Template Technique*:** This method has become popular to make efficient drug particles based on their size, shape, and flowability. This is also called PRINT in the particle engineering technology and used for "micro molding" the particle (184). The micro-mold is used to press the API or API with an excipient to form unique particle shapes and sizes. The particles generated from this method have an MMAD 3 μm suitable for respiratory delivery (185).

9.5.2 Factors Affecting the Aerosol Performance

Formation of aerosol depends on the suspension of the liquid or solids in the gaseous medium (186). The droplet size of the liquid is in a circular form (187) but the solid particles vary, depending on the processing technique. Particles that are trapped in a flow is expected to follow the course of that flow. The laminar flow regime is characterized by high speed flow with low circulation. These are the effects of particle forces like inertial, electrostatic, gravitational, and thermal, where the path of the particles is deviated from the streamline (188).

9.5.3 Aerodynamic Nature of Pharmaceutical Aerosol

***Slip*:** Any particle is subjected to slip on its surface when particle diameter closes down to the mean free path' of the flow. This would lead to including the Cunningham slip correction factor in the Stokes number, if drug particles are less than 1 μm.

Particle Density: Particle density is an important physical parameter to calculate the aerodynamic diameter of a particle. The particle aerodynamic diameter is equal to a sphere, whose density is equal to water, and that settles in still air at the same velocity as the particle in question (189), where, d_{ae} is represented as follows:

$$d_{ae} = \sqrt{\rho_p}\, d_g$$

where ρ_g and d_g are the density and geometrical diameter of the particles. It is necessary to use low density and a large geometric diameter of the particle to improve the delivery performance of drug particles into the respiratory tract.

***Hygroscopic Properties of Particles*:** Hygroscopic aerosols contain a certain amount of moisture that depends on some specific state of temperature and relative humidity. It has much more significance for agglomeration and dispersion of the drug particles and for aerodynamic behaviors of particle flowability into airways (190). Hygroscopic accumulation is a common phenomenon in pharmacological aerosols (191).

***Aerosol Agglomeration and De-agglomeration Process*:** The attractive forces, which are applied for the formulation of dry powder at the molecular level are the same between the particles. These forces are classified as Van der Waals forces, capillary force, electrostatic force, and mechanical interlocking forces (192). The process of agglomeration and de-agglomeration of drug and carrier particles is illustrated in Figure 9.8.

***Van der Waals Forces*:** These arise due to the interaction effect of the pharmaceutical particles. The particle considered as rigid and homogeneous, and it describes the origin adhesive/cohesive forces which are proportional to their particle size (193).

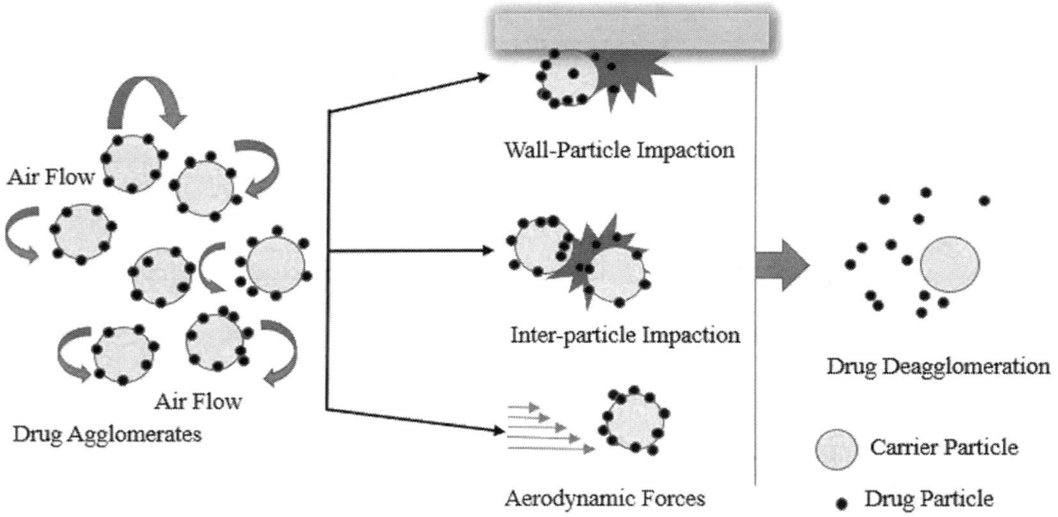

FIGURE 9.8 Process Agglomeration and De-aglomeration in Drug and Carrier Particles

For two similar particles, R_1=R_2=R

$F_{VdW} = \frac{AR}{6h^2}\frac{R_1R_2}{R_1+R_2} = \frac{AR}{6h^2}$ For particle and surface R_2>>>>R_1

$$F_{VdW} = \frac{AR}{6h^2}$$

where A, R and h are Hamakerconstant (A ~10^{-19} J), particle radius (m) and separation distance respectively.

The Van der Waals force depends much more on geometrical and dimensional parameters like particle morphology, size, and roughness. Due to the heterogeneous and deformable nature of the pharmaceutical particles, it is difficult to obtain the exact value of the Hamaker constant, so the exact determination of the Van der Waals force between the particles is not possible.

Capillary Force: The interaction of water with solid particles is possible in three ways: adsorption, capillary condensation, and deliquescence, a process which is based on the particles' properties and relative humidity (194). An increased value of the relative humidity decreases the dipersion performance of the aerosol by increasing the particle interaction force (195). The capillary force can be represented as

$$F_{cap} = 4\pi R^* y_L \cos\theta + 4\pi R^* y_{SL}$$

where R^*, θ, y_L, and Y_{SL} represents the hormonic mean particle radius, contact angle, water surface tension, and interfacial surface tension between solid and liquid.

Electrostatic Force: The nature of the electrostatic forces is both attractive and repulsive, and weaker than the Van der Waals forces for the pharmaceutical particles. The pharmaceutical particles have poor conductivity. So it is necessary to consider the electrostatic forces generated from contact charging, coulombic interaction, and tribo-electrification (196). The contact potential force can be represented as

$$F_C = \frac{2\pi q^2}{A}$$

where charge q is generated from the detachment process from substrate, and A is the contact area of particle and substrate.

Coulombic force is generated by the integrations of the opposite charge of the particles (197)

$$F_q = k_e \frac{q^2}{d^2}$$

where q, k_e and d are respresented as the charge on the particles, Coulomb constant, and diameter of the particle respectively.

Due to frictional charging, the tribo-electrification forces are generated (198,199)

$$F_e = q^2\left[1 - \frac{h}{(R^2 + h^2)^{0.5}}\right]\frac{1}{16\pi\varepsilon_0 h^2}$$

where, R, q, h and ε_0 represent the particle radius, charge, separation between two particles, and permittivity of vacuum respectively.

Lift and Drag Forces: Two aerodynamic forces, lift and drag, play an important role in the dispersion of powdered drug. Inertial forces generated from dispersion are responsible for effective separation of powdered drug, hence the DPI performance is less dependent on the mixture and properties of drugs (200). However, the dispersion dependent on lift and drag forces of drug particles can be proved much more effective if combined with mixtures having carrier particles (201,202).

9.5.4 Mechanism of Drug Deposition

Inertial Impaction: Inertial particle transport is a physical process that primarily affects the deposition of larger particles. When the dose is emitted from the inhaler, the airflow (in the mouth) is quick. The particles in the released dose are applied to the airflow down into the airways. In the mouth and down to the larynx, and further down the trachea, the airflow is quick. Particles greater than 5 µm will be deposited in the oropharyngeal zone. These particles will be swallowed, and the drug will be absorbed into the systemic circulation following gastrointestinal absorption. Particles <5 μm enter the lungs through the right and left bronchus and into the broad airways. Here the airflow is still relatively quick and thus the particles just below 5 μm are deposited by impaction at the bifurcation of these larger airways (203).

Sedimentation: The second main process is sedimentation, i.e. gravitational attraction, which is responsible for the accumulation of smaller particles < 5 μm to 1 μm. As the inhaled air stream flows down the narrowing airways, the flow is slower and slower. In the smaller airways of the lungs, the particles are trapped in a very slow-moving airflow. These particles would be deposited due to gravity. This sedimentation mechanism is proportional to the aerodynamic size of the particle and the time during which the drug particles reside in the lungs (204–207). As a result, the risk of sedimentation rises with the period of residency in these more peripheral airways (157). That is why breathing maneuvers (i.e. breath holding) after inhalation is necessary (208–210).

Diffusion: The third mechanism is the Brownian diffusion, which is the primary transport mechanism for very small particles, particularly those between 1 μm and 0.1–1 μm in diameter. Particles are suspended in a very slow-moving air stream and travel by colliding with each other and, if they collide with the airway wall, the particles deposit otherwise they will be exhaled (211). This mechanism is inversely proportional to the size of the particle and is directly proportional to the length of the lung stay (212). It is necessary to hold the breath again. Strong and slow airflows have a profound effect on the total accumulation of inhaled drug particles in the lung. Increased inspirational flow would improve deposition by inertial effect both in the upper airways and in the oro-pharynx (213). Increased inhalation volumes would also increase the penetration of particles deeper into the lung and therefore increase deposition in the alveolar region (214).

In addition, the breathing duration increases the particle deposition by gravitational sedimentation and diffusion, because it increases the time that the particles remain in the lung (215,216). Patient training the correct inhalation technique improves the delivery of drugs to the lungs (49), especially in terms of the inhalation rate and duration of breathing (217–219). The most important consideration for pulmonary deposition is the patient's inhalation technique. This is also discussed thoroughly later. In addition, significant patient dependent factors play a role in the delivery of pulmonary drugs. These involve oro-pharynx and larynx morphology and the patient's inspiring volume and flow rate. The calibration of the airways also affects the amount of drug stored in the airways (220).

9.5.5 Empirical Relationships for Drug Deposition

There are several formulae and empirical relationships available to predict drug deposition in the human lungs. Reynolds number of flow (Re_f) is computed as $Re_f = \rho uD/\mu$, while particle Reynolds number is defined as $Re_p = \rho v_s d_p/\mu$. Characteristic dimension of the geometry carrying the flow is denoted as D, while particle diameter is denoted as d_p. It is worthwhile to mention that the settling velocity of the particle is expressed as $v_s = C_c\rho_p g d_p{}^2/18\mu$, where C_c is the Cunningham slip factor, which is dependent on the mean free path of molecules in the air and is the particle diameter (d).

For particle Reynolds number Re_p <<1, Stokes number (St_k) decides whether the drug particle will undergo inertial impaction on the human airway walls and is defined as

$$St_k = \frac{v_s/g}{D/u} = \frac{u\rho_p d_p^2 C_c}{18\mu D}$$

The Stokes numbers for all particle sizes and all flow rates are less than unity (221).

Froude number (Fr) is a useful parameter to compare between impaction and sedimentation of drug particles. It is defined as

$$Fr = \frac{u^2}{gD}$$

A higher value of the Froude number indicates that the sedimentation is negligible as compared to inertial impaction.

Chan and Lippman (58) suggested an empirical relationship to predict the probability of impaction (P_i) which is given as

$$P_i = 1.606St_k + 0.0023$$

The accuracy of the above equation was confirmed by Borojeni et al. (59) by conducting rigorous experiments with high-resolution CT-scan imaging-based human pulmonary models of a few adult subjects.

9.5.6 Formulation Challenge for Pulmonary Delivery of High Powder Dose

DPI is a device that has been used to treat respiratory disease using the active ingredient of pharmaceutical aerosols for 50 years. Their use is mainly aimed at treating respiratory disease, where the dose of the drug (formoterol and fluticasone propionate) dose varies from 6 to –500 μg (222–224).

Conventional powder mixtures contain micronized drug particles ranging from 1 to 5 μm mixed with an inactive, large 40 μm stimulant like mannitol, sucrose, lactose, sorbitol, or glucose (225). Particle engineering technology is an important tool to produce particles using various main materials and currently only pure drug is used to produce micronized drug particles (226). Due to the cohesive nature of the drug particle, mainly Van der Waals force's influence of unoptimized powder causing particle agglomeration, makes the powder difficult to disperse (227, 228)

9.6 Various Factors Affecting DPI Performance

Two forces mainly have an effect on the efficiency of DPIs. The first is the patient-created inspiratory flow and the second is the turbulence developed within the inhaler, which is uniquely dependent on its basic technical design (229–232). These are the two factors are capable of greatly influencing the disaggregation of the dosage of the powdered substance, inhaled particles diameter, consistency, and dose variability. Micro dispersion of powder is achieved by the patient's inspiratory flow. The patient's inspiratory flow depends on airway and lung conditions, and partially on the intrinsic resistance of the device. So these balanced two forces represent the critical factor which resolves the efficacy of the coupled molecule–device.

Only half of the total powdered drug can be converted into aerosols; the rest leave the DPI device in pallet form. The dispersion of powder has increased by increasing the airflow and it increases the impaction losses in the proximal airways, therefore, less of the drug reaches into the lungs (233–235). On the other hand, when the airflow decreased, deposition of the drug powder into lungs increased, even if the lower airflow can limit deposition by affecting drug powder disaggregation and dispersion. Obviously, changes in both these forces can only be made by changing the characteristics of airflow or the original DPI design.

9.6.1 Device-related Factors

***Pressure Drop and Airflow Resistance*:** The DPI device has different airflow resistance (R), differentiated on the basis of the proportional change in its pressure drop (ΔP) and flow rate (Q).

$$\sqrt{\Delta P} = Q * R$$

The above equation at zero resistance maximum inspiratory flow rate is obtained. Highest pressure drop and minimum inspiratory pressure drop are obtained at infinite resistance.

Capsule Piercing Performance: In DPI devices, hard capsules are used because of their many effective advantages. The powder contained within the capsule requires release of the powder during aerosolization from the pierced hole of the capsule (66). During inhalation, spin occurs in the capsule; as the capsule starts to spin, centrifugal force becomes generated and the powder starts coming out from the hole of the capsule. The rotational speed of the capsule is directly depending on the air flow (66). It often enhances the centrifugal force and thus enables the release of the drug powder. In fact, the FPF also increases as the airflow rate increases (236, 237). Consequently, a large difference in drug retained in the capsule can be reported at a higher flow rate than the flow rate of relative <30 l/min.

***Fluid Dynamics*:** The performance of a DPI can be improved by making small modifications to the design, such as changing the airflow field generated within the device. CFD, a comparatively convenient and cost-effective method, is used to visualize the flow behaviors of a fluid in the inhaler (238). Advanced experimental diagnostic tools, like several non-intrusive optical instrument systems (namely, LDV, PIV, etc.) can also be employed to conduct the in vitro experimentation on DPI and lung models.

9.6.2 Powder-Related Factors

***Engineered Drug Formulation*:** Along with device design, drug formulation is an important factor in effective drug delivery from the inhaler. The drug delivery mechanism and physiochemical properties are controlled by the effective drug formulation with proper composition (69,239). Some important properties are responsible for the effective dispersion of the drug particle, like density (71), surface area (240,241), size distribution (242), size of particle (74), morphological properties (75), (76), and cohesive and adhesive forces (243, 244).

***Morphology of Particles*:** In the formulation of DPI, the surface morphological properties of the drug particle affected the cohesive/adhesive forces and the agglomeration and de-agglomeration processes. The surface roughness of the powder is likely to affect the interaction forces between drug particles and the contact area. It also affects the FPF in the formulation of drug particle with carrier particles (245–247). Surface coating is the advanced technique used for the modification of surface morphology and delivered FPF (248).

***Carrier Particles*:** The formulation of drug particles has occurred, either agglomerating active micronized drugs or as a mixture containing an inactive carrier. Generally, lactose is used as a carrier particle for the effective formulation. Some carrier particles, the alternative of sugars like sorbitol, mannitol, glucose, trehalose, xylitol, etc., are used in the formulation of the drug particle (80, 249).

***Formulations with Nanoparticles*:** Nanoparticles are increasingly used in pharmaceuticals for fast absorption in the epithelium layers of human lungs due to strong interparticle forces and for averting the mucociliary clearance (250). Hence, nanoparticles help the drug to reach deep into the lungs even if the adhesive forces are weakened. Many drug formulations with soluble excipients exhibit quick dispersion in nanoparticles, indicative of better therapeutic performance. Nanoparticles are even considered a promising factor for the future delivery of vaccines through pulmonary routes.

9.6.3 Patient-Related Factors

The inhalers, in all senses, are meant for patient use. DPI is no exception. Hence, six E's, as outlined below, must be ensured while designing a DPI device.

Effectiveness: The device must generate enough FPF mass for wide inhalation rates.

Efficiency: The device must deliver high deposition in the lungs throughout.

Engagement: The device must be patient-friendly.

Error-Free: The device must be robust and suitable for daily usage.

Easy to Learn: The correct use of device must be easily learnt by the patients and by the healthcare professionals.

Easy to Change: The device must be simple in design to use for an easy switch-over.

9.7 Future of DPI

A plethora of publications on DPIs shows that the almost half of the total published research is related to drug formulation and pharmacology of DPI, while only a small number of papers discussed DPI design and its consequences. Among the topics addressed by the present researchers working on various aspects of DPI are use of nanosystems to increase residence

TABLE 9.5

Challenges Associated with Dry Powder Inhaler Devices

Patient compliant device	• Easy yet robust device design • Patient friendly device • Device needs minimum training for patients • Universal inhaler handling all medication
Drug formulation and dosing	• Can handle high dosing per inhalation • Free from excipients • Capsule-free dosing • Stable and cost-effective formulation
Improved DPI design	• Compact, cost-effective design • Use of bio-compatible, anti-bacterial material in device • Design having less dependency on interparticulate forces • Design assisting high FPF mass at low flow rates • Higher lung deposition efficiency
DPI for special use	• Disposable device for delivering hygroscopic drug • Delivering vaccines as one-off dose • Delivering antibiotics with localized concentration • Patient specific design

time and bioavailability, however lower deposition efficiency and poor drug stability are matters of concern (83). Experts summarized that an optimal formulation integrated with a robust aerosol release system is required for an DPI device to succeed in producing consistent dosing. A number of attachments and/or small parts like swirlers, grids, and lacunas are used with DPI devices to attain effective aerosol dispersion, obviously a a cost of increasing flow resistance of different magnitudes (251–259). Treating infection in the lungs using DPI-delivered aerosolized antibiotics has recently gained attention. Localized concentrated drug delivery in certain areas of lungs where infection and/or damage occurs eventually reduces the drug loss and side effects. Table 9.5 summarizes the challenges associated with the DPI devices.

Development in computer aided design (CAD) and additive manufacturing technology (like three-dimensional printing) in the past couple of decades have boosted experimental and computational research of DPI design using sophisticated instrumentation (non-intrusive optical technique), CFD and DEM. Optimization of flow and aerosol movement from the DPI device to deep into lungs are predicted using the computational technique and later validated using in vitro experimentation based on CT-scan based realistic DPI and lung models (85). With the increase of computational power in recent decades, simulation of flow behavior from DPI to lower branches of bronchi are carried out using multi- and mesoscale modeling techniques in spite of high time and cost.

REFERENCES

1. Conforti A, Mascia M, Cioffi G, Angelis C, Coppola G, Rosa PD, Pivonello R, Alviggi C, and Placido GD. Air pollution and female fertility: A systematic review of literature. *Reproduction Biology Endocrinology*. 2018 Dec 30. doi: 10.1186/s12958-018-0433-z
2. Chauhan AJ and Johnston SL. Air pollution and infection in respiratory illness. *British Medical Bulletin*. 2003; 68: 95–112. doi: 10.1093/bmb/ldg022.
3. Yang RY, AB Yu, SK Choi, MS Coates and HK Chan. Agglomeration of fine particles subjected to centripetal compaction. *Powder Technology*. 2008; 184(1): 122–129.
4. Brilliantov NV, Spahn F, Hertzsch JM, and Pöschel T. Model for collisions in granular gases. *Physical Review E*. 1996; 53: 5382–5392.
5. Langston PA, Tüzün U, and Heyes DM. Continuous potential discrete particle simulation of stress and velocity fields in hoppers: Transition from fluid to granular flow. *Chemical Engineering Science*. 1994; 49: 1259.
6. Xu BH and Yu AB. Numerical simulation of the gas-solid flow in a fluidized bed by combining discrete particle method with computational fluid dynamics. *Chemical Engineering Science*. 1997; 52: 2786–2809.
7. Zhu HP, Zhou ZY, Yang RY, and Yu AB. Discrete particle simulation of particulate systems: Theoretical developments. *Chemical Engineering Science*. 2007; 62: 3378–3396.
8. Bosquillon C, Préat V, and Vanbever R. Pulmonary delivery of growth hormone using dry powders and visualization of its local fate in rats. *Journal of Controlled Release*. 2004; 96: 233–244
9. Crowder TM, Rosati JA, Schroeter JD, Hickey AJ, and Ted BM. Effects of particle morphology on lung delivery. *Pharmaceutical Research*. March 2002; 19(3).
10. Daniher DI and Zhu J. Dry powder platform for pulmonary drug delivery. *Particuology*. 2008; 6: 225–238. 10.1016/j.partic.2008.04.004.
11. Islam N and Cleary MJ. Developing an efficient and reliable dry powder inhaler for pulmonary drug delivery – A review for multidisciplinary researchers. *Medical Engineering & Physics*. 2012; 34: 409–427.
12. Vidgrén MT, Vidgrén PA, and Paronen TP. Comparison of physical and inhalation properties of spray-dried and mechanically micronized disodium cromoglycate. *International Journal of Pharmaceutics*. 1987; 35: 139–144.
13. Petros RA and Desimone JM. Strategies in the design of nanoparticles for therapeutic applications. *Nature Reviews Drug Discovery*. 2010; 9: 615–627.

14. Grant M and Leone-Bay A Peptide therapeutics: it's all in the delivery. *Therapeutic Delivery*. 2012; 3: 981–99610.4155/tde.12.79.
15. Hooton JC, German CS, Allen S, Davies MC, Roberts CJ, Tendler SJ and Williams PM. An atomic force microscopy study of the effect of nanoscale contact geometry and surface chemistry on the adhesion of pharmaceutical particles. *Pharmaceutical Research*. 2004; 21: 953–961. https://doi.org/10.1023/B:PHAM.0000029283.47643
16. Kandasam R and Chandrasekaran K. Sustained release aerosol for pulmonary drug delivery system: A review. *International Journal of Pharmacy and Pharmaceutical Sciences*. 2013; 5:126–130.
17. Israelachvili JN. Intermolecular and Surface Forces. Academic Press, London, UK, 1992.
18. Ramsey JM, Hibbitts A, Barlow J, Kelly C, Sivadas N, and Cryan SA. 'Smart' non-viral delivery systems for targeted delivery of RNAi to the lungs. *Therapeutic Delivery*. 2013; 4: 59–76. 10.4155/tde.12.133.
19. Hamishehkar H, Rahimpour Y, and Javadzadeh Y. The Role of Carrier in Dry Powder Inhaler, In: Sezer AD (ed). *Recent Advances in Novel Drug Carrier Systems*. 2012. InTech Open. ISBN: 978-953-51-0810-8.
20. Traini D, Young PM, Thielmann F, and Acharya M. The influence of lactose pseudopolymorphic form on salbutamol sulfate-lactose interactions in dry powder inhaler formulations. *Drug Development and Industrial Pharmacy*. 2008; 34: 992–1001.
21. Yang Y, Tsifansky M, Wu CJ, Yang HI, Schmidt G, Yeo Y. Inhalable antibiotic deliveryusing a dry powder codelivering recombinant deoxyribonuclease and ciprofloxacin fortreatment of cystic fibrosis. *Pharmaceutical Research*. 2010; 27: 151–160.
22. Prime D. Review of dry powder inhalers. *Advanced Drug Delivery Reviews*. 1997; 26: 51–58. 10.1016/s0169-409x(97)00510-3.
23. Adi H, Traini D, Chan H-K, and Young PM. The influence of drug morphology on the aerosolisation efficiency of dry powder inhaler formulations. *Journal of Pharmaceutical Sciences*. 2008; 97: 2780–2788.
24. Marriott C and Frijlink HW. Lactose as a carrier for inhalation products: breathing new life into an old carrier. *Advanced Drug Delivery Reviews*. 2012; 64: 217–219. 10.1016/j.addr.2011.11.003.
25. Adi S, Tong Z, Chan H-K, Yang R, and Yu A. Impact angles as an alternative way to improve aerosolisation of powders for inhalation? *European Journal of Pharmaceutical Sciences*. 2010; 41(2): 320–327.
26. Atkins PJ. Dry powder inhalers: an overview. *Respiratory Care*. 2005; 50(10): 1304–1312.
27. Ahlneck C and Zografi G. The molecular basis of moisture effects on the physical and chemical stability of drugs in the solid state. *International Journal of Pharmaceutics*.1990;62(2–3): 87–95.
28. Le Brun PPH, de Boer AH, Frijlink HW, and Heijerman HGM. A review of the technical aspects of drug nebulization. *Pharmacy World & Science*. 2000; 22: 75–81.
29. Euliss LE, DuPont JA, Gratton S, et al. Imparting size, shape, and composition control of materials for nanomedicine. *Chemical Society Reviews*. 2006; 35: 1095–1104.
30. Rolland JP, Maynor BW, Euliss LE, et al. Direct fabrication and harvesting of monodisperse, shape-specific nanobiomaterials. *Journal of the American Chemical Society*. 2005; 127: 10096–10100.
31. Stoessl J. Potential therapeutic targets for Parkinson's disease. *Expert Opinion on Therapeutic Targets*.2008; 12: 425–436. 10.1517/14728222.12.4.425.
32. Christoph K, Goniva C, Aichinger G, and Pirker S. Comprehensive DEM-DPM-CFD Simulations - Model Synthesis. Experimental Validation and Scalability. Seventh International Conference on CFD in the Minerals and Process Industries CSIRO, Melbourne, Australia 9–11 December 2009.
33. Anon. Method of and apparatus for effecting delivery of fine powders. *IP Com J*. 2008; 8(18): 13.
34. Brocklebank D, Ram F, Wright J, et al. Comparison of the effectiveness of inhaler devices in asthma and chronic obstructive airways disease: a systematic review of the literature. Health Technology Assessment. 2001; 5(1-149): 142.
35. Rawat A, Majumder QH, and Ahsan F. Inhalable large porous microspheres of low molecular weight heparin: In vitro and in vivo evaluation. *Journal of Controlled Release*. 2008;128: 224–232. 10.1016/j.jconrel.2008.03.013.
36. Brown BAS, Rasmussen J, Becker D, and Friend DR. A piezo-electronic inhaler for local & systemic applications. ed. *Drug Development and Delivery*. 2004; 62(10): 1413–1414.
37. Cheatham WW, Leone-Bay A, Grant M, Fog PB, and Diamond DC. Pulmonary delivery of inhibitors of phosphodiesterase type 5. Application. Mannkind Corporation. 2006; 23.
38. Anderson PJ. History of aerosol therapy: liquid nebulization to MDIs to DPIs. *Respiratory Care*. 2005; 50: 1139–1149.
39. Plumley C, Gorman EM, El-Gendy N, Bybee CR, Munson EJ, and Berkland C. Nifedipine nanoparticle agglomeration as a dry powder aerosol formulation strategy. *International Journal of Pharmaceutics*. 2008; 369: 136–143. 10.1016/j.ijpharm.2008.10.016.
40. Minne A, Boireau H, Horta MJ, and Vanbever R. Optimization of the aerosolization properties of an inhalation dry powder based on selection of excipients. *European Journal of Pharmaceutics and Biopharmaceutics*. 2008; 70: 839–844.
41. Farr SJ and Otulana BA. Pulmonary delivery of opioids as pain therapeutics. *Advanced Drug Delivery Reviews*. 2006; 58: 1076–1088. 10.1016/j.addr.2006.07.013.
42. Burden of asthma and chronic obstructive pulmonary disease and access to essential medicines in low-income and middle-income countries. *The Lancet Respiratory Medicine*. 2015; 3: 159–170. Available from: http://www.thelancet.com/journals/lanres/article/PIIS2213-2600(15)00004-1/abstract
43. Kleinstreuer C, Zhang Z, and Donohue JF. Targeted drug-aerosol delivery in the human respiratory system. *Annual Review of Biomedical Engineering*. 2008; 10:195–220. 10.1146/annurev.bioeng.10.061807.160544.
44. Fleischer W, Reimer K, and Leyendecker P. Opioids for the treatment of the chronic obstructive pulmonary disease. 2005. Euro-Celtique SA.

45. Gratton SEA, Williams SS, Napier ME, et al. The pursuit of a scalable nanofabrication platform for use in material and life science applications. *Accounts of Chemical Research*. 2008; 41: 1685–1695.
46. Surendrakumar K, Martyn GP, Hodgers ECM, Jansen M, and Blair JA. Sustained release of insulin from sodium hyaluronate based dry powder formulations after pulmonary delivery to beagle dogs. *Journal of Controlled Release*. 2003; 91: 385–394.
47. Hillery AM, Lloyd AW and Swarbrick J. Pulmonary Drug Delivery. In: Drug Delivery and Targeting for Pharmacists and Pharmaceutical Scientists. CRC Press, London. 2001. ISBN-13: 978-8123916194, ISBN-10: 9780415271981.
48. Usmani OS, Biddiscombe MF, and Barnes PJ. Regional lung deposition and bronchodilator response as a function of beta2-agonist particle size. *American Journal of Respiratory and Critical Care Medicine*. 2005; 172: 1497–1504.
49. Newman SP, Weisz AW, Talaee N. and Clarke SW. Improvement of drug delivery with a breath actuated pressurized aerosol for patients with poor inhaler technique. *Thorax*. 1991b; 46: 712–716.
50. Overhoff KA, Clayborough R , and Crowley M. Review of the TAIFUN® Multidose Dry Powder Inhaler Technology. *Drug Development and Industrial Pharmacy*. 2008; 34; 960–965. 10.1080/03639040802154897.
51. Everard, ML. Role of inhaler competence and contrivance in "difficult asthma". *Paediatric Respiratory Reviews*. 2003; 4: 135–142.
52. Pavia D, Thomson ML, Clarke SW and Shannon HS. Effect of lung function and mode of inhalation on penetration of aerosol into the human lung. *Thorax*. 1977; 32: 194–197.
53. Dolovich MB and Dhand R. Aerosol drug delivery: Developments in device design and clinical use. *Lancet*. 2011; 377: 1032–1045
54. Ibrahim M, Garcia-Contreras L, and Verma R. Inhalation drug delivery devices: technology update. *Medical Devices: Evidence and Research*. 2015; 8: 131–139. 10.2147/mder.s48888.
55. Newman SP, Pavia D, and Clarke SW. Simple instructions for using pressurized aerosol bronchodilators. *Journal of the Royal Society of Medicine*. 1980; 73: 776–779.
56. Lipworth BJ and Clark DJ. Effects of airway calibre on lung delivery of nebulised salbutamol. *Thorax*. 1997; 52: 1036–1039.
57. Finlay WH. The Mechanics of Inhaled Pharmaceutical Aerosols: An Introduction. Academic Press. London, 2019. ISBN: 978-0-08-102749-3.
58. Chan TL and Lippmann M. Experimental measurement and empirical modelling of the regional deposition of inhaled particles in humans. *American Industrial Hygiene Association Journal*. 1980; 41: 399–409.
59. Borojeni AAT, Noga ML, Vehring R, and Finlay WH. Measurements of total aerosol deposition in intrathoracic conducting airway replicas of children. *Journal of Aerosol Science*. 2014; 73, 39–47.
60. Smith IJ and Parry-Billings M. The inhalers of the future? A review of dry powder devices on the market today. *Pulmonary Pharmacology & Therapeutics*. 2003; 16: 79–95. http://dx.doi.org/ 10.1016/S1094-5539(02)00147-5
61. Lippmann M, Yeates DB, and Albert RE. Deposition, retention, and clearance of inhaled particles. *The British Journal of Industrial Medicine*. 1980; 37(4): 337–362.
62. Chan JG, Duke CC, Ong HX, et al. A novel inhalable form of rifapentine. *Journal of Pharmaceutical Sciences*. 2014; 103(5): 1411–1421.
63. Labiris NR and Dolovich MB. Pulmonary drug delivery. Part I: Physiological factors affecting therapeutic effectiveness of aerosolized medications. *British Journal of Clinical Pharmacology*. 2003; 56(6): 588–599.
64. Virchow JC. Guidelines versus clinical practice – which therapy and which device. *Respiratory Medicine*. 2004; 98(Suppl B): S28–S34.
65. Kruger P, Ehrlein, Zier M, and Greguletz R. Inspiratory flow resistance of marketed dry powder inhalers. Munich, Germany: Oral presentation at the European Respiratory Society Annual Meeting, September 6th–10th, 2014, proceedings.
66. Coates MS, Chan HK, Fletcher DF, and Raper JA. Influence of air flow on the performance of a dry powder inhaler using computational and experimental analyses. *Pharmaceutical Research*. 2005; 22: 1445–1453.
67. Azouz W and Chrystyn H. Clarifying the dilemmas about inhalation techniques for dry powder inhalers: Integrating science with clinical practice. *Primary Care Respiratory Journal*. 2012; 21: 208–213.
68. Bass K, Farkas D, and Longest W. Optimizing aerosolization using computational fluid dynamics in a pediatric air-jet dry powder inhaler. *AAPS PharmSciTech*. 2019 Nov 1; 20(8): 329. doi: 10.1208/s12249-019-1535-4.
69. Hickey AJ, Concessio NM, Oort VM, and Platz RM. Factors influencing the dispersion of dry powders as aerosols. *Pharmaceutical Technology*. 1994; 18(8): 58–84.
70. de Boer AH, Hagedoorn P, Woolhouse R, and Wynn E. Computational fluid dynamics (CFD) assisted performance evaluation of the Twincer™ disposable high-dose dry powder inhaler. *Journal of Pharmacy and Pharmacology*. 2012; 64: 1316–1325
71. Edwards DA, Hanes J, Caponetti G, Hrkach J, Ben-Jebria A, Eskew ML, et al. Large porous particles for pulmonary drug delivery. *Science*. 1997; 276: 1868–1871.
72. Young PM, Cocconi D, Colombo P, Bettini R, Price R, Steele DF, et al. Characterization of a surface modified dry powder inhalation carrier prepared by particle smoothing. *Journal of Pharmacy and Pharmacology*. 2002; 54: 1339–1344.
73. Zeng XM, Martin GP, Marriott C, and Pritchard J. The effects of carrier size and morphology on the dispersion of salbutamol sulfate after aerosolization at different flow rates. *Journal of Pharmacy and Pharmacology*. 2000; 52: 1211–1221.
74. Islam N, Stewart P, Larson I, et al. Effect of carrier size on the dispersion of salmeterol xinafoate from interactive mixtures. *Journal of Pharmaceutical Sciences*. 2004; 93: 1030–1038.
75. Martinac A, Filipovic-Grcic J, Voinovich D, Perissutti B, and Franceschinis E. Development and bioadhesive properties of chitosan-ethylcellulose microspheres for nasal delivery. *International Journal of Pharmaceutics*. 2005; 291: 69–77.

76. Donovan MJ, Kim SH, Raman V, and Smyth HD. Dry powder inhaler device influence on carrier particle performance. *Journal of Pharmaceutical Sciences*. 2012; 101: 1097–1107
77. Louey MD and Stewart PJ. Particle interactions involved in aerosol dispersion of ternary interactive mixtures. *Pharmaceutical Research*. 2002; 19: 1524–1531.
78. Islam N, Stewart P, Larson I, and Hartley P. Surface roughness contribution to the adhesion force distribution of salmeterol xinafoate on lactose carriers by atomic force microscopy. *Journal of Pharmaceutical Sciences*. 2005; 94: 1500–1511.
79. Kawashima Y, Serigano T, Hino T, Yamamoto H, and Takeuchi H. Effect of surface morphology of carrier lactose on dry powder inhalation property of pranlukast hydrate. *International Journal of Pharmaceutics*. 1998; 172: 179–188.
80. Steckel H and Bolzen N. Alternative sugars as potential carriers for dry powder inhalations. *International Journal of Pharmaceutics*. 2004; 270: 297–306.
81. Amaro MI, Tajber L, Corrigan OI, and Healy AM. Design of experiment to study the effect of spray dryer operating variables on sugar powders intended for inhalation. *Journal of Pharmacy and Pharmacology*. 2010; 62: 1413–1414.
82. Huang M, Ma Z, Khor E, and Lim L-Y. Uptake of FITC-Chitosan nanoparticles byA549 cells. *Pharmaceutical Research*. 2002; 19: 1488–1494.
83. Ali ME and Lamprecht A. Spray freeze drying for dry powder inhalation of nanoparticles. *European Journal of Pharmaceutics and Biopharmaceutics*. 2014; 87: 510–517.
84. Levy ML, Carroll W, Jose´ L, Alonso I, Keller C, Lavorini F, and Lehtimaki L. Understanding dry powder inhalers: Key technical and patient preference attributes. *Advances in Therapy*. 2019; 36: 2547–2557. https://doi.org/10.1007/s12325-019-01066-6
85. Suwandecha T, Wongpoowarak K, and Schrichana T. Computer-aided design of dry powder inhalers using computational fluid dynamics to assess performance. *Pharmaceutical Development and Technology*. 2016; 21: 54–60.
86. Islam N and Gladki E. Dry powder inhalers (DPIs) - A review of device reliability and innovation. *International Journal of Pharmaceutics*. 2008; 360(1–2): 1–11.
87. Van Holsbeke CS, Leemans G, Vos WG, et al. Functional respiratory imaging as a tool to personalize respiratory treatment in patients with unilateral diaphragmic paralysis. *Respiratory Care*. 2014; 59: e127–e131.
88. Köbrich R, Rudolf G, and Stahlhofen W. A mathematical model of mass deposition in man. *Annals of Occupational Hygiene*. 1994; 38: 15–23.
89. Patton JS and Byron PR. Inhaling medicines: Delivering drugs to the body through the lungs. *Nature Reviews Drug Discovery*. 2007; 6: 67–74.
90. Shah UV, Karde V, Ghoroi C, and Heng JYY. Influence of particle properties on powder bulk behaviour and processability. *International Journal of Pharmaceutics*. 2017; 518: 138–154. http://dx.doi.org/10.1016/j.ijpharm.2016.12.045.
91. Chawla A, Taylor KMG, Newton JM, and Johnson MCR. Production of spray dried salbutamol sulphate for use in dry powder aerosol formulation. *International Journal of Pharmaceutics*. 1994; 108: 233–240.
92. Frijlink HW and De Boer AH. Dry powder inhalers for pulmonary drug delivery. *Expert Opinion on Drug Delivery*. 2004; 1(1): 67–86.
93. Garcia-Contreras L, Fiegel J, Telko MJ, et al. Inhaled large porous particles of capreomycin for treatment of tuberculosis in a guinea pig model. *Antimicrob Agents Chemother*. 2007; 51: 2830–2836.
94. Cheng, YS. Mechanisms of pharmaceutical aerosol deposition in the respiratory tract. *AAPS PharmSciTech*. 2014; 15(3), June 2014.
95. Newman S, Malik S, Hirst P, et al. Lung deposition of salbutamol in healthy human subjects from the MAGhaler dry powder inhaler. *Respiratory Medicine*. 2002; 96: 1026–1032.
96. Padrela L, Rodrigues MA, Tiago J, et al. Tuning physiochemical properties of theofylline by cocrystallization using the super critical fluid enhanced atomization technique. *Powder Technology*. 2014; 86: 129–136.
97. Tong HY, Shekunov BY, York P, et al. Characterization of two polymorphs of salmeterol xinafoate crystallized from supercritical fluids. *Pharmaceutical Research*. 2001; 18: 852–858.
98. Begat P, Morton DA, Shur J, et al. The role of force control agents in high-dose dry powder inhaler formulations. *Journal of Pharmaceutical Sciences*. 2009; 98: 2770–2783.
99. Farkas DR, Hindle M, and Longest PW. Characterization of a new high dose dry powder inhaler (DPI) based on fluid bed design. *Annals of Biomedical Engineering*. 2015; 43: 2804–2815.
100. Buttini F, Brambilla G, Copelli D, et al. Effect of flow rate on in vitro aerodynamic performance of NEXThaler in comparison with diskus and turbuhaler dry powder inhalers. *Journal of Aerosol Medicine and Pulmonary Drug Delivery*. 2015; 28: 1–12.
101. de Boer A, Hagedoorn P, Hoppentocht M, Buttini F, Grasmeijer F, and Frijlink HW. Dry powder inhalation: past, present and future. *Expert Opinion on Drug Delivery*. 2017; 14(4): 499–512. https://doi.org/10.1080/17425247.2016.1224846
102. Demoly P, Hagedoorn P, de Boer AH, et al. The clinical relevance of dry powder inhaler performance for drug delivery. *Respiratory Medicine*. 2014; 108: 1195–1203.
103. DeHaan WH and Finlay WH. Predicting extrathoracic deposition from dry powder inhalers. *Journal of Aerosol Science*. 2004; 35: 309–331.
104. Hoppentocht M, Akkerman OW, Hagedoorn P, et al. Tolerability and pharmacokinetic evaluation of inhaled dry powder tobramycin free base in non-cystic fibrosis bronchiectasis patients. *PLoS ONE*. 2016; 11: e0149768. doi:10.1371/journal.pone.0149768.
105. Weibel ER and Gomez DM. Architecture of the human lung. Use of quantitative methods establishes fundamental relations between size and number of lung structures. *Science*. 1962; 137: 577–585.
106. Srivastav VK, Kumar A, Shukla SK, Paul AR, Bhatt AD and Jain A. Airflow and aerosol-drug delivery in a CT scan based human respiratory tract with tumor using CFD. *Journal of Applied Fluid Mechanics (SCI)*, April 2014; 7(2): 345–356. doi: 10.36884/jafm.7.02.20282

107. Lexmond AJ, Kruizinga TJ, Hagedoorn P, et al. Effect of inhaler design variables on pediatric use of dry powder inhalers. *PLoS ONE*. 2014; 19(6). doi:10.1371/journal.pone.0099304.
108. Bell JH, Hartley PS, and Cox JSG. Dry powder aerosols. 1. New powder inhalation device. *Journal of Pharmaceutical Sciences*. 1971; 60: 1559.
109. Frijlink HW and de Boer AH. Dry powder inhalers for pulmonary drug delivery. *Expert Opinion on Drug Delivery*. 2004; 1: 67–86.
110. Tong Z, Yu A, Chan H-K and Rang Y. Modelling of powder dispersion in dry powder inhalers - A brief review. *Current Pharmaceutical Design*. 2015; 21: 3966–3973.
111. Chan HK. Dry powder aerosol drug delivery—Opportunities for colloid and surface scientists. *Colloids and Surfaces A: Physicochemical and Engineering Aspect*. 2006; 284-285: 50–55.
112. Coates MS, Chan HK, Fletcher DF, and Chiou H. Influence of mouthpiece geometry on the aerosol delivery performance of a dry powder inhaler. *Pharmaceutical Research*. 2007; 24: 1450–1456.
113. Cundall PA and Strack ODL. A discrete numerical model for granular assemblies. Géotechnique. 1979; 29: 47.
114. Tsuji Y, Kawaguchi T, and Tanaka T. Discrete particle simulation of two-dimensional fluidized bed. *Powder Technology*. 1993; 77: 79–87.
115. Sommerfeld M and Schmalfus S. Numerical analysis of carrier particle motion in dry powder inhaler. *Journal of Fluids Engineering*. 2016; 138(No. 041308-1 to 12, April).
116. Menter FR. Two-equation eddy-viscosity turbulence models for engineering applications. *AIAA Journal*. 1994; 32: 269–289.
117. Coates MS, Fletcher DF, Chan H-K, and Raper JA. Effect of design on the performance of a dry powder inhaler using computational fluid dynamics. Part 1: Grid structure and mouthpiece length. *Journal of Pharmaceutical Sciences*. 2004; 93(11): 2863–2876.
118. Tong ZB, Yang RY, Chu KW, Yu AB, Adi S, and Chan HK. Numerical study of the effects of particle size and polydispersity on the agglomerate dispersion in a cyclonic flow. *Chemical Engineering Journal*. 2010; 164: 432–441.
119. Milenkovic JM. Airflow and Particle Deposition in a Dry Powder Inhaler: A CFD and Particle Computational Approach. Ph.D. Thesis. Centre for Research and Technology Hellas, Laboratory of Polymer Reaction Engineering, Thessaloniki, Greece, 2015
120. Coates MS, Fletcher DF, Chan HK, and Raper JA. Effect of design on the performance of a dry powder inhaler using computational fluid dynamics. Part 1: Grid structure and mouthpiece length. *Journal of Pharmaceutical Sciences*. 2004; 93: 2863–2876.
121. Coates MS, Chan HK, Fletcher DF, and Raper JA. Effect of design on the performance of a dry powder inhaler using computational fluid dynamics. Part 2: Air inlet size. *Journal of Pharmaceutical Sciences*. 2006; 95: 1382–1392.
122. Tong ZB, Adi S, Yang RY, Chan HK, and Yu AB. Numerical investigation of the de-agglomeration mechanisms of fine powders on mechanical impaction. *Journal of Aerosol Science*. 2011; 42: 811–819.
123. Tong ZB, Zheng B, Yang RY, Yu AB, Chan HK. CFD-DEM investigations of the dispersion mechanisms in commercial dry powder inhalers. *Powder Technology*. 2013; 240: 19–24.
124. Milenkovic J, Alexopoulos AH, and Kiparissides C. Flow and particle deposition in the Turbohaler: A CFD simulation. *International Journal of Pharmaceutics*. 2013; 448: 205–213.
125. Longest PW, Son Y-J, Holbrook L, and Hindle M. Aerodynamic factors responsible for the deaggregation of carrier-free drug powders to form micrometer and submicrometer aerosols. *Pharmaceutical Research*. 2013; 30: 1608–1627.
126. Behara SRB, Longest PW, Farkas DR, and Hindle, M. Development and comparison of new high-efficiency dry powder inhalers for carrier-free formulations. *Journal of Pharmaceutical Sciences*. 2014; 103: 465–477.
127. Yang J, Wu C-Y, and Adams M. Three-dimensional DEM-CFD analysis of air flowinduced detachment of API particles from carrier particles in dry powder inhalers. *Acta Pharmaceutica Sinica B*. 2014; 4: 52–59
128. Yang J, Wu C-Y, and Adams M. DEM analysis of the effect of electrostatic interaction on particle mixing for carrier-based dry powder inhaler formulations. *Particuology*. 2015; 23: 25–30.
129. Tong Z, Yu A, Chan H-K, and Yang R. Discrete modelling of powder dispersion in dry powder inhalers – A brief review. Current Pharmaceutical Design. 2015b; 21: 3966–3973.
130. van Wachem B, Thalberg K, Remmelgas J., and Niklasson-Bjorn I. Simulation of dry powder inhalers: Combining micro-scale, meso-scale and macro-scale modeling. *AIChE Journal*. 2017; 63: 501–516.
131. Milenkovic J, Alexopoulos AH, and Kiparissides C. Optimization of a DPI inhaler: A computational approach. *Journal of Pharmaceutical Sciences*. 2017; 106: 850–858.
132. Ariane M, Sommerfeld M, and Alexiadis A. Wall collision and drug-carrier detachment in dry powder inhalers: Using DEM to devise a sub-scale model for CFD calculations. *Powder Technology*. 2018; 334: 65–75.
133. Johnson KL. Contact Mechanics. Cambridge University Press, Cambridge, 1985.
134. Wong W, Fletcher DF, Traini D, Chan H-K, and Young PM. The use of computational approaches in inhaler development. *Advanced Drug Delivery Reviews*. 2012; 64: 312–322.
135. Laube Beth L. The expanding role of aerosols in systemic drug delivery, gene therapy and vaccination: an update.*Translational Respiratory Medicine*. 2014; 2: 3.
136. Mindlin RD and Deresiewicz H. Elastic spheres in contact under varying oblique forces. Journal of Applied Mechanics. 1953; 20: 327–344.
137. World Health Organization. Global surveillance, prevention and control of chronic respiratory diseases. A comprehensive approach. WHO, 2007. Available from: http://www.who.int/gard/publications/ GARD_Manual/en/
138. Yang J, Wu CY, and Adams M. DEM analysis of the effect of particle wall impact on the dispersion performance in carrier-based dry powder inhalers. *International Journal of Pharmaceutics*. 2015; 487(1-2): 32–38.

139. Kou X and Cao X. Review of dry powder inhaler devices. *American Pharmaceutical Review*. 2016.
140. Moskal A and Sosnowski TR. Computational fluid dynamics (CFD) and direct visualization studies of aerosol release from two cyclohaler-type dry powder inhalers. *Journal of Drug Delivery Science and Technology*. 2012 Jan 1; 22(2): 161–165.
141. Labiris NR and Dolovich M. Pulmonary drug delivery. Part I: Physiological factors affecting therapeutic effectiveness of aerosolized medications Br. *The Journal of Clinical Pharmacology*. 2003 Dec; 56(6): 588–599. doi: 10.1046/j.1365-2125.2003.01892.x
142. Yang L, Li C, and Tang X. The impact of $PM_{2.5}$ on the host defense of respiratory system . *Cell Development Biology*. 2020; 8: 1–9. https://doi.org/10.3389/fcell.2020.00091.
143. Longest P and Holbrook LT. In silico models of aerosol delivery to the respiratory tract - Development and applications. *Advanced Drug Delivery Reviews*. 2012; 64(4): 296–311.
144. Zeng XM, Martin GP, Tee S-K, Ghoush AA, and Marriott C. Effects of particle size and adding sequence of fine lactose on the deposition of salbutamol sulfate from a dry powder formulation. *International Journal of Pharmaceutics*. 1999; 182: 133–144.
145. Price Y, Edge S, and Staniforth JN. The influence of relative humidity on particulate interactions in carrier-based dry powder inhaler formulations. *International Journal of Pharmaceutics*. 2002; 246: 47–59. https://doi.org/10.1016/0378-5173(90)90221-OR
146. Podczeck F. Particle–Particle Adhesion in Pharmaceutical Powder Handling. Imperial College Press, London, UK, 1998.
147. Hickey AJ and Mansour HM. Delivery of Drugs by the Pulmonary Route. In: Modern Pharmaceutics, Florence AT and Siepmann J (eds), Vol. 2. Taylor & Francis, New York. 2009; 191–219
148. Young PM, Price R, Tobyn MJ, Buttrum M, and Dey F. Investigation into the effect of humidity on drug-drug interactions using the atomic force microscope. *Journal of Pharmaceutical Sciences*. 2003; 92: 815–822. http://dx.doi.org/10.1002/jps.10250.
149. Young PM, Price R, Tobyn MJ, Buttrum M, and Dey F. *Journal of Pharmaceutical Sciences*. 2004; 93: 753–761.
150. Byron PR and Patton JS. Drug delivery respiratory. *Journal of Aerosol Medicine*. 1994; 7(1); 49–75,
151. Ali AM, Ahmed S, Dena A, Yacoub MH and El-Sherbiny IM. Exploring the influence of particle shape and air velocity on the flowability in the respiratory tract: A computational fluid dynamics approach. *Drug Development and Industrial Pharmacy*. 2019. DOI: 10.1080/03639045.2019.1600534
152. TM Crowder, AJ Hickey, MD Louey and N. Orr. A Guide to Pharmaceutical Particulate Science. Interpharm Press/CRC, Boca Raton, FL, 2003.
153. Hickey AJ. Back to the future: Inhaled drug products. *Journal of Pharmaceutical Sciences*. 2013; 102: 1165–1172.
154. Crowder TM, Rosati JA, Schroeter JD, Hickey AJ, and Ted BM. Effects of particle morphology on lung delivery. *Pharmaceutical Research*. March 2002; 19(3), (© 2002).
155. Claus S, Weiler C, Schiewe J, and Friess W. How can we bring high drug doses to the lung?. *European Journal of Pharmaceutics and Biopharmaceutics*. 2013; 86: 1–6
156. Sommerfeld M, Cui Y, and Schmalfus S. Potential and constraints for the application of CFD combined with Lagrangian particle tracking to dry powder inhalers. *European Journal of Pharmaceutical Sciences*, 2019.
157. Mack P, Horvath K, Tully J, and Maynor B. Particle engineering for inhalation formulation and delivery of biotherapeutics. *Inhalation*. 2012; 6: 16–20.
158. Coates M, Chan H-K, Fletcher D, and Chiou H. Influence of mouthpiece geometry on the aerosol delivery performance of a dry powder inhaler. *Pharmaceutical Research*. 2007; 24(8): 1450–1456.
159. Frijlink HW and de Boer AH. Trends in the technology-driven development of new inhalation devices. *Drug Discovery Today: Technologies*. 2005; 2(1): 47–57.
160. de Boer AH, Chan HK, and Price R. A critical view on lactose-based drug formulation and device studies for powder inhalation; which are relevant and what interactions to expect. *Advanced Drug Delivery Reviews*. 2012; 64: 257–274.
161. Grasmeijer F, Grasmeijer N, Hagedoorn P, et al. Recent advances in the fundamental understanding of adhesive mixtures for inhalation. *Current Pharmaceutical Design*. 2015; 21: 5900–5914.
162. Berard V, Lesniewska E, Andres C, Pertuy D, Laroche C, and Pourcelot Y. Dry powder inhaler: influence of humidity on topology and adhesion studied by AFM. *International Journal of Pharmaceutics*. 2002; 232: 213–224.
163. van der Palen J, Ginko T, Kroker A, van der Valk P, Goosens M, Padullés L, Seoane B, Rekeda L, and Garcia Gil E. Preference, satisfaction and errors with two dry powder inhalers in patients with COPD, *Expert Opinion on Drug Delivery*. 2013; 10: 1023–1031.
164. Vehring R. Pharmaceutical particle engineering via spray drying. *Pharmaceutical Research*. 2008; 25: 999–1022. http://dx.doi.org/10.1007/s11095-007-9475-1.
165. Larhrib H, Martin GP, Marriott C, and Prime D. The influence of carrier and drug morphology on drug delivery from dry powder formulations. *International Journal of Pharmaceutics*. 2003; 12(1-2): 283–296.
166. Crowder T and Hickey A. Powder specific active dispersion for generation of pharmaceutical aerosols. *International Journal of Pharmaceutics*. 2006; 327(1-2):65–72.
167. Hassan MS and LauRFeasibility study of pollen-shape drug carriers in dry powder inhalation. *Journal of Pharmaceutical Sciences*, 2010; 99; 1309–1321 10.1002/jps.21913.
168. Peng T, Lin S, Niu B, Wang X, Huang Y, Zhang X, Li G, Pan X, and Wu C. Influence of physical properties of carrier on the performance of dry powder inhalers. *Acta Pharmaceutica Sinica B*. 2016; 6: 308–318 10.1016/j.apsb.2016.03.011.
169. Benke E, Szabó-Révész P, Hopp B, and Ambrus R. Characterization and development opportunities of carrier-based dry powder inhaler systems. *Acta Pharmaceutica Hungarica*. 2017; 87: 59–68.
170. Islam N, Stewart P, Larson I, and Hartley P. Effect of carrier size on the dispersion of salmeterol xinafoate from

interactive mixtures. *Journal of Pharmaceutical Sciences*. 2004; 93: 1030–1038.

171. Sitz R. Current innovations in dry powder inhalers. *Ondrugdelivery*. 2010: 10–12. Available from: www.ondrugdelivery.com
172. Laube BL, Janssens HM, De Jongh FH, Devadason SG, Dhand R, Diot P, Everard ML, Horvath I, Navalesi P, Voshaar T, and Chrystyn H. What the pulmonary specialist should know about the new inhalation therapies. *European Respiratory Journal*. 2011; 37: 1308–1417.
173. Lavorini F and GA Fontana. Inhaler technique and patient's preference for dry powder inhaler devices. *Expert Opinion on Drug Delivery*. 2013; 11(1): 1–3. doi: 10.1517/17425247.2014.846907.
174. Tong Z, Zheng B, Yang R, Yu A, and Chan H. CFD-DEM investigation of the dispersion mechanisms in commercial dry powder inhalers. *Powder Technology*. 2013; 240: 19–24.
175. Kohler D. The Novolizer®: Overcoming inherent problems of dry powder inhalers. *Respiratory Medicine 98*. 2004; 98(Supplement 1) : S17–S21.
176. Parisini I, Cheng SJ, Symons DD, and Murnane D. Potential of a cyclone prototype spacer to improve in vitro dry powder delivery. *Pharmaceutical Research*. 2013; 31(5): 1133–1145.
177. Shur J, Lee S, Adams W, Lionberger R, Tibbatts J, and Price R. Effect of device design on the in vitro performance and comparability for capsule-based dry powder inhalers. *AAPS Journal*. 2012; 14: 667–676.
178. Heng JYY, Thielmann F, and Williams DR. The effects of milling on the surface properties of form I paracetamol crystals. *Pharmaceutical Research*. 2006; 23: 1918–1927. http://dx.doi.org/10.1007/s11095-006-9042-1.
179. Steckel H, and Brandes HG. A novel spray-drying technique to produce low density particles for pulmonary delivery. *International Journal of Pharmaceutics*. 2004; 278: 187–195.
180. Velaga SP, Berger R, and Carlfors J. Supercritical fluids crystallization of budesonide and flunisolide. *Pharmaceutical Research*. 2002; 19: 1564–1571.
181. Chen L, Heng R-L, Delele MA, Cai J, Du D-Z, and Opara UL. Investigation of dry powder aerosolization mechanisms in different channel designs.*International Journal of Pharmaceutics*. 2013; 457: 143–149
182. Castellanos A. The relationship between attractive interparticle forces and bulk behaviour in dry and uncharged fine powders. *Advances in Physics*. 2005; 54(4): 263–376, DOI: 10.1080/17461390500402657
183. Chan H-K, Young PM, Traini D, and Coates M. Dry powder inhalers: Challenges and goals for next generation therapies. *Pharmaceutical Technology Europe*. 2007; 19(4): 19–24.
184. Dhand R and Fink JB. Dry powder inhalers. *Respiratory Care*. 1999; 44: 940–951.
185. Telko Martin J and Hickey Anthony J. Dry powder inhaler formulation. *Respiratory Care*. 2005; 50: 1209–1227.
186. Lindert S, Below A, and Breitkreutz J. Performance of dry powder inhalers with single dosed capsules in preschool children and adults with improved upper airway models. *Pharmaceutics*. 2014; 6: 36–51.
187. Najafabadi A, Gilani K, Barghi M, and Rafiee-Tehrani M. The effect of vehicle on physical properties and aerosolisationbehaviour of disodium cromoglycate microparticles spray dried alone or with L-leucine. *International Journal of Pharmaceutics*. 2004; 285: 97–108.
188. Napier ME and DeSimone JM. Nanoparticle drug delivery platform. *Polymer Reviews*. 2007; 47: 321–327.
189. Negro RWD. Dry powder inhalers and the right things to remember: a concept review. *Multidisciplinary Respiratory Medicine*. 2015; 10: 13.
190. Zijlstra GS, Hinrichs WLJ, Boer AHd, and Frijlink HW. The role of particle engineering in relation to formulation and de-agglomeration principle in the development of a dry powder formulation for inhalation of cetrorelix. *European Journal of Pharmaceutical Sciences*. 2004; 23: 139–149.
191. Srivastav VK, Paul AR, and Jain A. Effects of cartilaginous rings on airflow and particle transport through simplified and realistic human upper respiratory tracts. *Acta Mechanica Sinica*. 2013; 29(6): 883–892. DOI 10.1007/s10409-013-0086-2
192. Stank K and Steckel H. Physico-chemical characterisation of surface modified particles for inhalation. *International Journal of Pharmaceutics*. 2013; 448: 9–18.
193. Zhou Y, Sun J, and Cheng Y-S. Comparison of deposition in the USP and physical mouth-throat models with solid and liquid particles. *Journal of Aerosol Medicine and Pulmonary Drug Delivery*. 2011; 24: 277–284.
194. David L-B, Chatan C, Cheryl LMS, et al. Trileucine improves aerosol performance and stability of spray-dried powders for inhalation. *Journal of Pharmaceutical Sciences*. 2008; 97: 287–302.
195. Seville PC, Learoyd TP, Li HY, Williamson IJ, and Birchall JC. Amino acid-modified spraydried powders with enhanced aerosolisation properties for pulmonary drug delivery. *Powder Technology*. 2007; 178: 40–50.
196. Zhou Q, Tong Z, Tang P, Citterio M, Yang R, and Chan H-K. Effect of device design on the aerosolization of a carrier-based dry powder inhaler — A case study on Aerolizer® Foradile®. *The AAPS Journal*. 2013; 15: 511–522.
197. Gerrity TR. Pathophysiological and disease constraints on aerosol deposition. In: Byron PR. (ed) Respiratory Drug Delivery. CRC Press, Inc., Boca Raton, FL, 1990; 1–38.
198. Mulvaney MD and Stewart PJ. Characterization of adhesional properties of lactose carriers using atomic force microscopy. *Journal of Pharmaceutical and Biomedical Analysis*. 2001; 25: 559–567.
199. Hoppentocht M, Hagedoorn P, Frijlink HW, and de Boer AH. Technological and practical challenges of dry powder inhalers and formulations. *Advanced Drug Delivery Reviews*. 2014; 75: 18–31.
200. Hoppentocht M, Hagedoorn P, Frijlink HW, and de Boer AH. Developments and strategies for inhaled antibiotic drugs in tuberculosis therapy: A critical evaluation. *European Journal of Pharmaceutics and Biopharmaceutics*. 2013; 86: 23–30.
201. Podczeck F, Newton JM, and James MB. The adhesion force of micronized salmeterol xinafoate particles to pharmaceutically relevant surface materials. *Journal of Physics D*. 1996; 29: 1878–1884

202. https://www.disabled-world.com/health/respiratory/air-pollutants.php
203. https://www.disabled-world.com/health/respiratory/pollution-lungs.php
204. Learoyd TP, Burrows JL, French E, and Seville PC. Chitosan-based spray-dried respirable powders for sustained delivery of terbutaline sulfate. *European Journal of Pharmaceutics and Biopharmaceutics*. 2008; 68: 224–234
205. Lexmond, AJ, Hagedoorn P., van der Wiel E., ten Hacken NHT, Frijlink HW, and de Boer AH. Adenosine dry powder inhalation for bronchial challenge testing, part 1: Inhaler and formulation development and in vitro performance testing. *European Journal of Pharmaceutics and Biopharmaceutics*. 2013; 86: 105–114.
206. Tonnis WF, Kersten GF, Frijlink HW, et al. Pulmonary vaccine delivery: A realistic approach? *Journal of Aerosol Medicine and Pulmonary Drug Delivery*. 2012; 25: 1–12.
207. Ruzycki CA, Javaheri E, and Finlay WH. The use of computational fluid dynamics in inhaler design. *Expert Opinion on Drug Delivery*. 2013; 10: 307–323.
208. Suarez S and Hickey AJ. Drug properties affecting aerosol behavior. *Respiratory Care*. 2000; 45(6): 652–666.
209. Sanders M. Inhalation therapy: An historical review. *Primary Care Respiratory Journal*. 2007; 16: 71–81.
210. Tonnis WF, AJ Lexmond, HW Frijlink, AH de Boer, and WL Hinrichs. Devices and formulations for pulmonary vaccination. *Expert Opinion on Drug Delivery*. 2013; 10: 1383–1397.
211. Da G, Witt C, Wagner U, Chung K, and Fischer A. Fundamentals of pulmonary drug delivery. *Respiratory Medicine*. 2003; 97: 382–387.
212. Hashmi A, Soomro JA, Memon A, and Soomro TK. Incorrect inhaler technique compromising quality of life of asthmatic patients. *Journal of Medicine*. 2012; 13: 16–21
213. Beran D, Zar HJ, Perrin C, Menezes AM, and Burney P. Burden of asthma and chronic obstructive pulmonary disease and access to essential medicines in low-income and middle-income countries.*The Lancet Respiratory Medicine*. 2015; 3(2): 159–170. doi: 10.1016/S2213-2600(15)00004-1.
214. Gonda I. Physicochemical Principles in Aerosol Delivery. Topics in Pharmaceutical Sciences 1991. Medpharm Scientific Publishers, Stuttgart, Germany, 1992; 95–115.
215. Zhou Q, Tang P, Leung SSY, Chan JGY, and Chan H-K. Emerging inhalation aerosol devices and strategies: Where are we headed? *Advanced Drug Delivery Reviews*. 2014; 75: 3–17.
216. Coates MS, Chan H-K, Fletcher DF, and Raper JA. Effect of design on the performance of a dry powder inhaler using computational fluid dynamics. Part 2: Air inlet size. *Journal of Pharmaceutical Sciences* 2006; 95(6): 1382–1392.
217. Donovan MJ and Smyth HDC. Influence of size and surface roughness of large lactose carrier particles in dry powder inhaler formulations. *International Journal of Pharmaceutics*. 2010; 402: 1–9.
218. Newman SP, Weisz AW, Talaee N, and Clarke SW. Improvement of drug delivery with a breath actuated pressurized aerosol for patients with poor inhaler technique. Thorax. 1991b; 46: 712–716.
219. Ngoc NTQ, Chang L, Jia X, and Lau R. Experimental investigation of design parameters on dry powder inhaler performance. *International Journal of Pharmaceutics*. 2013; 457: 92–100.
220. Usmani OS, Biddiscombe MF and Barnes PJ. Regional lung deposition and bronchodilator response as a function of beta2-agonist particle size. *American Journal of Respiratory and Critical Care Medicine*. 2005; 172: 1497–1504.
221. Sprittles JE and Shikhmurzaev YD. The dynamics of liquid drops and their interaction with solids of varying wettabilities. *Physics of Fluids*. 2012; 24: 1–17.
222. Menter FR, Kuntz M, and Langtry R. Ten Years of Industrial Experience with the SST Turbulence Model. In: Hanjalic K and Tummers M. (eds). *Turbulence, Heat and Mass Transfer*. Begell House, Inc., West Redding. 2003.
223. Smutney CC, Grant M, and Kinsey PS. Device factors affecting pulmonary delivery of dry powders.*Therapeutic Delivery*. 2013; 4: 939–949.
224. Mezhericher M, Brosh T, and Levy A. Modeling of particle pneumatic conveying using DEM and DPM methods. *Particulate Science and Technology: An International Journal*. 2011; 29(2): 197–208.
225. Visser J. Particle adhesion and removal: A review. *Particulate Science and Technology*. 1995; 13(3-4): 169–196, DOI: 10.1080/02726359508906677
226. Finlay WH. The Mechanics of Inhaled Pharmaceutical Aerosols: An Introduction. Academic Press, London, UK, 2001
227. Milenkovic J, Alexopoulos AH, and Kiparissides C. Deposition and fine particle production during dynamic flow in a dry powder inhaler: A CFD simulation. *International Journal of Pharmaceutics*. 2014; 461: 129–136.
228. Virchow JC, Crompton GK, Dal Negro RW, Pedersen S, Magnan A, Seidemberg J, et al. Importance of inhaler devices in the management of airway diseases. *Respiratory Medicine*. 2008; 102: 10–19.
229. Longest PW, Son Y-J, Holbrook L, and Hindle M. Aerodynamic factors responsible for the deaggregation of carrier-free drug powders to form micrometer and submicrometer aerosols. *Pharmaceutical Research*. 2013; 30: 1608–1627.
230. Bailey MML and Berkland CJ. Nanoparticle formulations in pulmonary drug delivery. *Medicinal Research Reviews*. 2009 Jan; 29(1): 196–2120.
231. Barnes PJ. Distribution of receptor targets in the lung. *Proceedings of the American Thoracic Society*. 2004; 1: 345–351
232. Yang J., Wu C-Y, and Adams M. DEM analysis of the effect of particle-wall impact on the dispersion performance in carrier-based dry powder inhalers. *International Journal of Pharmaceutics*. 2015; 487: 32–38.
233. Louey MD, Razia S, and Stewart PJ. Influence of physico-chemical carrier propertieson the in vitro aerosol deposition from interactive mixtures. *International Journal of Pharmaceutics*. 2003; 252: 87–98.
234. Yang J, Wu C-Y, and Adams MJ. Numerical modelling of agglomeration and deagglomeration in dry powder inhalers: A review. *Current Pharmaceutical Design*. October 2015; 21(999).
235. Zeng X-M, Martin GP, Marriott C, and Pritchard J. Lactose as a carrier in dry powder formulations: The influence of surface characteristics on drug delivery. *Journal of Pharmaceutical Sciences*. 2001; 90: 1424–1434.

236. Martonen TB, Bell K, Phalen R, Wilson A, and Ho A. Growth rate measurements and deposition modeling of hygroscopic aerosols in human tracheobronchial models. *Annals of Occupational Hygiene*. 1982; 26: 93–108.
237. Yang RY, Zou RP, and Yu AB. Computer simulation of the packing of fine particles. *Physical Review E*. 2000; 62: 3900–3908.
238. Rabbani NR and Seville PC. The influence of formulation components on the aerosolisation properties of spray-dried powders. *Journal of Controlled Release*. 2005; 110: 130–140.
239. Learoyd TP, Burrows JL, French E, Seville PC. Chitosan-based spray-dried respirable powders for sustained delivery of terbutaline sulfate. *European Journal of Pharmaceutics and Biopharmaceutics*. 2008; 68: 224–234.
240. Gilani K, RouholaminiNajafabadi A, Barghi M, and Rafiee-Tehrani M. Aerosolisation of beclomethasone dipropionate using spray dried lactose/polyethylene glycol carriers. *European Journal of Pharmaceutics and Biopharmaceutics*. 2004; 58: 595–606.
241. Arakawa T, Tsumoto K, Kita Y, Chang B, and Ejima D. Biotechnology applications of amino acids in protein purification and formulations. *Amino Acids*. 2007; 33: 587–605.
242. Gavini E, Hegge AB, Rassu G, et al. Nasal administration of carbamazepine using chitosan microspheres: In vitro/in vivo studies. *International Journal of Pharmaceutics*. 2006; 307: 9–15.
243. Cook RO, Pannu RK, and Kellaway IW. Novel sustained release microspheres for pulmonary drug delivery. *Journal of Controlled Release*. 2005; 104: 79–90.
244. WHO Global Ambient Air Quality Database. World Health Organization, Geneva, 2018. https://www.who.int/airpollution/data/AAP_database_methods_2018_final.pdf
245. Geller DE, Konstan MW, Smith J, Noonberg SB, and Conrad C. Novel tobramycin inhalation powder in cystic fibrosis subjects: Pharmacokinetics and safety. *Pediatric Pulmonology*. 2007; 42: 307–313 10.1002/ppul.20594.
246. Briottet X. Radiometry in the optical domain. In: Baghdadi N, and Zribi M, eds. Optical Remote Sensing of Land Surface. Elsevier, Oxford, UK, 2016; 1–56.
247. Beran D, Zar HJ, Perrin C, Menezes AM, Burney P, Forum of International Respiratory Societies working group. Burden of asthma and chronic obstructive pulmonary disease and access to essential medicines in low-income and middle-income countries. *The Lancet Respiratory Medicine*. 2015; 3: 159–170.
248. Morsi S and Alexander A. An investigation of particle trajectories in two-phase flow systems. *Journal of Fluid Mechanics*. 1972; 55(2): 193–208, doi:10.1017/S0022112072001806
249. Ali AM, Ahmed S, Dena A, Yacoub MH, and El-Sherbiny IM. Exploring the influence of particle shape and air velocity on the flowability in the respiratory tract: A computational fluid dynamics approach. *Drug Development and Industrial Pharmacy*. 2019. DOI: 10.1080/03639045.2019.1600534
250. Tsuda A, Henry FS, and Butler JP. Particle transport and deposition: Basic physics of particle kinetics. *Comprehensive Physiology*. 2013 Oct; 3(4): 1437–1471. doi: 10.1002/cphy.c100085
251. Podczeck F. Particle–Particle Adhesion in Pharmaceutical Powder Handling. Imperial College Press, London, UK, 1998.
252. Morrow PE. Factors determining hygroscopic aerosol deposition in airways. *Physiological Reviews*. 1986; 66: 330–376
253. Martonen TB. Analytical model of hygroscopic particle behavior in human airways. *Bulletin of Mathematical Biology*. 1982; 44: 425–442.
254. Hickey AJ and Martonen TB. Behavior of hygroscopic pharmaceutical aerosols and the influence of hydrophobic additives. *Pharmaceutical Research*. 1993; 10: 1–7.
255. Ashurst I, Malton A, Prime D, and Sumby B. Latest advances in the development of dry powder inhalers. *Pharmaceutical Science & Technology Today*. 2000; 3; 246–256. 10.1016/s1461-5347(00)00275-3.
256. Luner PE, Zhang Y, Abramov YA, and Carvajal MT. Evaluation of milling method on the surface energetics of molecular crystals using inverse gas chromatography. *Crystal Growth & Design*. 2012; 12: 5271–5282. http://dx.doi.org/10.1021/cg300785z
257. Hickey AJ, Mansour HM, Telko MJ, Xu Z, Smyth HDC, Mulder T, McLean R, Langridge J, and Papadopoulos D. Physical characterization of component particles included in dry powder inhalers. I. Strategy review and static characteristics. *Journal of Pharmaceutical Sciences*. 2007; 96: 1282–1301.
258. Van Holsbeke CS, Leemans G, Vos WG, De Backer JW, Vinchurkar SC, Geldof M, Verdonck PR, Parizel PM, Van Schil PE, and De Backer WA. Functional respiratory imaging as a tool to personalize respiratory treatment in patients with unilateral diaphragmatic paralysis. *Respiratory Care*.2013; 59, e127–e131 10.4187/respcare.02756.
259. Sjoholm P, Ingham DB, Lehtimaki M, Perttu-roiha I, Goodfellow H, and Torvela H. Gas-cleaning technology. In Goodfellow H and Tähti E (eds). *Industrial Ventilation Design Guidebook*. Academic Press, 2001; 1197–1316. 10.1016/B978-012289676-7/50016-3.

10

In Vivo Animal Models for Lung Targeted Drug Delivery Systems

Khushali Vashi MPH BDS CPH[1], Jasmin Patel MBBS[2], and Yashwant Pathak PhD.[3]
[1]University of South Florida, Tampa, FL, USA
[2]SMIMER, Surat, Gujarat, India
[3]USF Health Taneja College of Pharmacy, University of South Florida and Adjunct Professor Faculty of Pharmacy, Airlangga University, Surabaya, Indonesia

10.1 Introduction

In the modern era of persisting chronic diseases, the efficacy of a treatment is highly reliant on the mode of transportation by which the drug is delivered with the most advantageous concentration of the drug (1). The gaps in this efficacy of the treatment of severe diseases has revealed a pressing need for a multidisciplinary approach for drug delivery to the target organ (1). Lung targeted drug delivery systems have become of tremendous scientific interest in the field of research because the organ is capable of absorbing pharmaceutical drugs either for local deposition or for systemic delivery. There have been several studies considering different drug delivery systems using newer mathematical models. The use of computational fluid dynamics (CFD) and physiologically based pharmacokinetic (PBPK) modeling are some of the newer concepts that have shown excellent in vivo results and several other modeling techniques are showing promising results (2). There are studies focusing on each and every aspect of the drug delivery systems while using newer techniques such as using aerosolized drug forms rather than the conventional oral forms which in turn increase patient compliance, encourage ideal drug dosage, and decrease toxicity.

10.2 Coupled In Silico Platform: Computational Fluid Dynamics (CFD) and Physiologically Based Pharmacokinetic (PBPK) Modeling

There are a number of factors that influence the formulation of inhaled drugs, these factors such as quality of the drug, efficacy of the drug, and physiological parameters in the population make it challenging to get a final product(2). To overcome such challenges, in silico predictive tools can be used to facilitate design of inhaled pharmaceutics. The complexity of physiological processes and interactions within it can be studied through computational models and advanced simulation tools which can further facilitate the development of inhalation medicines (2). Technological advancements, like cheaper computer power and ease in availability of powerful software, have led to an increase in the use of computer-based simulations (3). CFD is used in determining airflow patterns and turbulence levels for a device design and also to provide an accurate simulation of the particles and droplets by subjecting them to several forces, turbulence, and wall interactions (3). Particle interaction cannot be considered when using the CFD technique which is the main disadvantage (4). There are several other models available, like the discrete phase model (DPM), two-fluid model, mixture model, dense dispersed phase model (DDPM), and the discrete element method (DEM) (5). Dry powder inhalers (DPIs) are used to deliver drugs in a dry powder form without the need of a propellant. This form of drug delivery produces aerosol particles that might be advantageous to deliver the drug faster to the lungs (6). The only disadvantage is that the efficiency is less, so <30% of the dose reaches the lungs (7). The major factor affecting the deposition profile of DPIs are patient-related factors and physical properties of dry powder formulations (8). A study by Zhou et al. (9) in 2013 researched the device design and its effect on commercially used DPIs and other kinds of commercially available products. A study by Jiang et al. (10) in 2012 researched using a similar method investigated the effect of powder residence time on the performance of the Aerolizer DPI. There have been several other studies that have focused on different aspects of the DPIs and their interactions with human lungs. Computer-based analysis provided information that can help us understand dispersion mechanisms of different drug delivery models better. This knowledge can further be helpful in enhancing inhaler design in order to create more efficient drug delivery device (2).

10.3 Modeling and Simulation of Biopharmaceutical Performance in Lung Drug Delivery

Biopharmaceutical modeling has gained much popularity in the field of pharmacology. *Biopharmaceutical performance* refers

DOI: 10.1201/9781003046547-10

to the influence of pharmaceutical formulation variables on in vivo performance (11). A 2019 study by Sou et al. researched antibiotic resistance of *Pseudomonas aeruginosa* (PA) lung infections using quorum sensing inhibitors (QSIs) that are meant for inhibiting biofilm formation and sensitize PA to antibiotic treatments. For such respiratory conditions, it is essential to have targeted lung drug delivery systems available for which proper models need to be developed. This study has demonstrated that pharmacometrics modeling can be an essential tool to facilitate drug selection, maintain drug dosage, and develop future development of the treatment (12,13). This kind of developed model can also be used in association with an infection model for studying drug exposure-response relationships to improve the understanding of the pharmacokinetic–pharmacodynamic relationships of QSIs (12). Thus, this is an important concept to understand and there is a need for development of bio-models that can facilitate pulmonary drug absorption while maintaining patient safety.

10.4 Clinically Relevant Test Methods to Establish In Vitro Equivalence for Spacers and Valved Holding Chambers Used with Pressurized Metered Dose Inhalers (pMDIs)

Bioequivalence (BE) is defined as the absence of a significant difference in the rate and extent to which the active ingredient or active moiety in a pharmaceutical equivalent becomes available at the site of drug action when administered at the same molar dose under similar conditions in an appropriately designed study (Code of Federal Regulations Title 21, 320.1). Spacers (S) and valved holding chambers (VHCs) are pressurized metered-dose inhaler (pMDI) accessory devices, designed to overcome commonly faced problems by patients when administering aerosol via a pMDI. The development of spacers/VHCs (S/VHC) was in response to the frequently encountered patient-related issues with pMDI techniques involving poor coordination between actuation and inhalation, and local side effects arising from oropharyngeal deposition (14,15). Moreover, S/VHCs substantially modify the aerodynamic particle size distribution (APSD) of the inhaled medication, and potentially the spatial distribution of the mass of active pharmaceutical ingredient(s) [API(s)] depositing in the respiratory tract (16). Clinically relevant in vitro BE testing is not yet much emphasized by the standard authorities, however future consideration of such testings would help better understand the clinical outcomes, performance predictions, similarities, and differences between reference and test products (16). Currently, a three-part strategy is proposed whereby in vitro testing for BE can simulate more clinically relevant conditions than in the current compendial procedures: (i) The inclusion of a short delay between inhaler actuation and sampling onset is appropriate when determining APSD at flow rate(s) suitable for the intended patient population. (ii) Assessment of total emitted mass ex S/VHC by simulating tidal breathing pattern(s) appropriate for intended use. (iii) Incorporation of appropriate face model(s), representative of the intended patient age range(s), into test procedures for S/VHCs with facemasks, enabling clinically appropriate dead space and fit-to-face to be simulated (16).

10.5 Lipid Nanoparticles' Biocompatibility and Cellular Uptake in a 3D Human Lung Model

In the last couple of decades, researchers have explored the use of nanoparticle-based drug delivery systems to carry, protect, and to deliver drugs directly to the site of infection, leading to reduction in the amount and frequency of dosage of the drug and thereby preventing the toxicities related to the therapy while improving patient compliance (17). The physical and chemical properties of nanoparticles (NP), including particle size, surface, and morphology, are determinant factors that influence their transport and deposition within the respiratory tract (18,19). After inhalation, the NPs are usually deposited in the alveolar region of the lungs (20). Having a large epithelial surface of 150 m^2, the respiratory tract has a dense network of immune cells consisting of macrophages and dendritic cells (DC) (21). These cells play a crucial role as a protective barrier for the inhaled drug particles (22). Other factors contributing to the deposition of drug particles include the type of device for drug delivery and the magnetic field (23).

Nanostructured lipid carriers (NLC) are another alternative that can be used as a lung drug delivery system because of their biocompatibility, high drug loading capacity, modifications to size and morphology, target specificity, and stability (20). The disadvantages for NLC are their potential toxicity and biodistribution in lungs (20). The study by Magalhaes et al., focuses on studying these potential toxicities on monocultures and 2D co-cultures. This study focused on testing on 3D lung models, developing non-mannosylated NLCs and mannosylated NCLS (M-NLC) where they were characterized on the basis of size, morphology, polydispersity (PDI), and zeta (ζ)-potential. Their cellular uptake was labelled using a fluorophore and these NLCs were submerged or used through a pseudo air–liquid interface (ALI) approach in vivo. A post 24-hour exposure check was performed to assess the biocompatibility and targeting efficiency of NLCs and M-NLCs using techniques like cell viability, pro-inflammatory assays, and visualization of the lung tissue with confocal laser scanning microscopy. The applications for nano-based systems for lung drug delivery have been extensively researched to improve the treatment and drug delivery for infectious diseases targeting the lungs (20). The study by Magalhaes et al., has proven that exposure to all the tested NLCs formulations does not affect or alter the cell membrane integrity and cellular morphology in the lungs, nor do they elicit any cytotoxic or pro-inflammatory responses after the 24-hour exposure window. These drug delivery systems can thus be good candidates for lung drug delivery.

10.6 Designing In Vitro Bioequivalence Studies for Pressurized Metered Dose Inhalers with Spacer or Valved Holding Chamber Add-On Devices

Bioequivalence (BE) is defined as the absence of a significant difference in the rate and extent to which the active ingredient or active moiety in a pharmaceutical equivalent becomes available at the site of drug action when administered at the same molar dose under similar conditions in an appropriately designed study (Code of Federal Regulations Title 21, 320.1). A study by Sandell and Mitchell (24) researched the analytical test methods and associated statistical considerations by considering the laboratory testing of pressurized metered dose inhaler–spacer/valved holding chamber (pMDI-S/VHC) combinations for in vitro bioequivalence (IVBE) and concluded that, for these add-on devices, information provided in developing correlations were in support of IVBE. The methods used for this study were presented using four different scenarios for comparing TEST ("second entry" or "generic") versus REF ("innovator"): (i) innovator and second entry product pMDI alone without any S/VHC (baseline comparison); (ii) innovator and second entry pMDI product with the same S/VHC; (iii) innovator pMDI product with existing S/VHC and second entry product with a different S/VHC; and (iv) introduction of a second, different S/VHC to be used with a given innovator pMDI product (24). The important aspects to be considered include delayed inhalation, utilizing age-appropriate flow rates, and using anatomically appropriate face models for evaluation of devices with a facemask (24). The results showed that a proper statistical design of experiment was provided for each scenario, while also using alternative approaches for calculation of confidence intervals for the mean of TEST/REF relationship. It was thus concluded that precise calculations can be obtained using different statistical approaches (24).

10.7 International Guidelines for Bioequivalence of Locally Acting Orally Inhaled Drug Products: Similarities and Differences

Bioequivalence (BE) is defined as the absence of a significant difference in the rate and extent to which the active ingredient or active moiety in a pharmaceutical equivalent becomes available at the site of drug action when administered at the same molar dose under similar conditions in an appropriately designed study (Code of Federal Regulations Title 21, 320.1). BE is vital to infer therapeutic equivalence (TE) between the generic and corresponding reference drug product in the abbreviated new drug applications (ANDAs). It also supports the formulation modifications during the new drug product development phase, as well as for post-approval changes in drug applications (25). Currently, to determine BE for systemically acting drug products, several regions make use of pharmacokinetic studies (26). These drugs reach their action sites through systemic circulation. However, such an approach of locally acting drugs may not be enough in all cases to establish BE since their intended actions and deliveries do not rely on systemic circulation. Therefore, this presents a need for an alternative approach for BE of locally acting drugs (25). Orally inhaled drug products (OIDPs) includes products such as the dry powder inhaler (DPI), metered-dose inhaler (MDI), and nebulization products. Most inhalation products are designed to act as a local agent for the lungs, and their drug delivery does not necessarily rely on systemic circulation (25). Another challenge is to find a biomarker that is sensitive enough to clinically detect the potential differences in local drug delivery. Thus, establishment of proper BE standards for locally acting OIDPs has been an ongoing challenge for regulatory agencies across the globe (25). Currently, only limited international jurisdictions have regulatory guidelines on BE standards with varying recommended BE standards; details are available in Table 10.1 (25).

10.8 In Vitro Evaluation of Aerosol Delivery by Different Nebulization Modes in Pediatric and Adult Mechanical Ventilators

A nebulizer is a small machine that is used to convert liquid medicine into aerosolized form. These aerosolized forms of medications are mainly used to treat mechanically ventilated patients for the administration of bronchodilators (36). A study conducted by Wan et al., 2014 in (36) studied the delivery of aerosolized drug administered through three modes of nebulization for mechanical ventilation considering the influence of the type of aerosol generator, pattern of nebulization, and a patient's breathing pattern. This study also compared the efficiency of pneumatic nebulization modes provided by a ventilator using the pediatric and adult in vitro lung models. The results of the study showed that the percentage of total dose in both lung models was 5.1–7.5%, without any statistical significance among the three modes. Median nebulization times for inspiratory intermittent [IIM], continuous [CM], and expiratory intermittent [EIM] were 38.9, 14.3, and 17.7 minutes, respectively, and nebulization time for the three modes significantly differed ($P < 0.001$). The inhaled drug mass for the three nebulization modes with the adult lung model was similar to that of the pediatric lung model (7.39 ± 0.76 vs $6.27 \pm 0.69\%$, $P = 0.77$). The conclusion of the study was that aerosol drug delivery with a jet nebulizer placed proximal to the ventilator was not dependent on nebulization mode during simulated adult or the pediatric conventional mechanical ventilation, and that the use of EIM and CM modes of nebulization should be considered to decrease the treatment time (36).

Another study by Ke et al., 2020 in (37) studied soft mist inhalers that can generate aerosols with a smaller particle size as compared to the pressurized metered-dose inhalers. This study aimed to measure the particle size distribution of the soft mist inhalers when they are coupled with an inhalation aid, as well as to measure the efficacy of the delivery system in different actuation timings and circuit positions during mechanical ventilation with a pressurized metered-dose inhaler used as a suitable comparison. The results concluded that soft mist inhalers generated a smaller mass medium aerodynamic diameter than the pressurized metered-dose inhalers. During mechanical ventilation, the optimum way to deliver the both the inhalations was at 15 cm from the Y-piece and actuated at the end of expiration and the onset of inspiration, respectively. The conclusions from the study were that soft mist inhalers, when used with an inhalation aid, showed marginal improvement on the particle size distribution. Other factors that played an important role were the inhaler type, actuation timing, and position within the circuit (37).

TABLE 10.1

Similarities and Differences in the BE Guidelines for Orally Inhaled Drugs from Four Organizations: Australia – Therapeutic Goods Administration (TGA), Canada – Health Canada (HC), European Union (EU) – European Medicines Association (EMA), and the United States of America – Food and Drug Administration (FDA), with the Focus on the OIDPs Intended for Local Sites for Action

International Regulatory Authority	Agency	BE Guidelines Referenced	Scope	Date Posted/ Effective
Australia	Therapeutic Goods Administration	Guidance 19: Inhalation and nasal medicines/19.2 Generic MDI (27,28)	Support applications to register new inhalation and nasal medicine; applications to register generic inhalation and nasal medicines; requests to vary registered inhalation and nasal medicines. Include medicines for treating asthma; chronic obstructive pulmonary disease (COPD), other conditions of the lungs	August 9, 2013
Canada	Health Canada	Release of the Draft Guidance Document Data Requirements for Safety and Effectiveness of Subsequent Market Entry Inhaled Corticosteroid Products for Use in the Treatment of Asthma for Industry (26)	Support Abbreviated New Drug Submissions (ANDS); Supplemental New Drug Submissions (SNDS). Include medicines for: corticosteroid for asthma treatment (not COPD)	September 19, 2011
Canada	Health Canada	Guidance to Establish Equivalence or Relative Potency of Safety and Efficacy of a Second Entry Short-Acting Beta2-Agonist Metered-Dose Inhaler (29)	Short-acting beta-2 agonist metered-dose inhaler only	February 1, 1999
Canada	Health Canada	Guidance for Industry Pharmaceutical Quality of Inhalation and Nasal Products (section Regional information) (30)	Addresses quality aspects of new marketing authorization applications (including for generic products) and does not outline expected quality aspects related to changes in existing inhalation and nasal productsIncludes products for administration of the drug substance to the lungs, such as pressurized MDI, DPI, products for nebulization, and non-pressurized MDIs, as well as pressurized metered-dose nasal sprays, nasal powders, and nasal liquids	October 1, 2006
European Union	European Medicines Agency	Doc. Ref. CPMP/EWP/4151/00 Rev. 1: Guideline on The Requirements for Clinical Documentation for Orally Inhaled Products (OIP) Including the Requirements for Demonstration of Therapeutic Equivalence Between Two Inhaled Products for Use in the Treatment of Asthma And Chronic Obstructive Pulmonary Disease (COPD) in Adults and for Use in the Treatment of Asthma in Children and Adolescents(31,32)	Recommendations for demonstration of therapeutic equivalence between two inhaled products, in the context of abridged applications or variations/ extensions to a marketing authorization, used in the management and treatment of adult patients with asthma and/or COPD and children and adolescents with asthma. Include dosage forms: pressurized MDI; breath-operated MDI; pressurized MDI with spacers or holding chambers; non-pressurized MDI; solution and suspensions for nebulization; DPIs	January 22, 2009
United States of America	Food and Drug Administration	Draft Bioequivalence Recommendations for Specific Product: Fluticasone Propionate/ Salmeterol Xinafoate Dry Powder Inhaler (FP-SX DPI) (33,34)	For FP-SX DPI	September 2013
United States of America	Food and Drug Administration	Draft Bioequivalence Recommendations for Specific Product: Nebulized Budesonide Inhalation Suspension (35)	For nebulized budesonide inhalation suspension	September 2012
United States of America	Food and Drug Administration	Draft Bioequivalence Recommendations for Specific Product: Albuterol Sulfate Metered-Dose Inhaler (33)	For albuterol sulfate MDI	June 2013

10.9 Conclusion

Pulmonary drug delivery is a complex process with potentially promising future prospects to serve a number of considerations for physical and biological merits and demerits. With supportive lung physiological parameters and favorable *ex vivo* models study outcomes, lung drug delivery could potentially outweigh the conventional drug delivery, having its own limitations of high dosing frequency, toxicity, and low efficacy. In the last 20 years, CFD models have been exceptional and made significant progress in understanding the physics behind pulmonary drug delivery (38). Newer mathematical models have made it easier to compute and find effective drug delivery systems while considering vital factors such as minimizing dosage and time, individual lung profile, and patient compliance and safety (38). The applications for nano-based systems (NP, NLC, M-NLC) for lung drug delivery have been extensively researched using 3D lung models to improve the treatment and drug delivery targeting the lungs (20). Availability of limited information on BE regarding orally inhaled drugs results in a few international jurisdictions having regulatory guidelines on BE standards, with varying recommendations also a potential problem. A study by Fok et al. 1996 and Goralski and Davis (39) addresses challenges with pediatric aerosol drug delivery and issues with current drug delivery systems. There have been several in vivo study models which have shown promising results in theory but there needs to be more research in to accurately design models that will work efficiently for lung drug delivery.

REFERENCES

1. Patil JS and Sarasija S. Pulmonary drug delivery strategies: A concise, systematic review. *Lung India: Official Organ of Indian Chest Society*. 2012; 29(1): 44–49. https://doi.org/10.4103/0970-2113.92361
2. Vulović A, Šušteršič T, Cvijić S, Ibrić S, and Filipović N. Coupled in silico platform: Computational fluid dynamics (CFD) and physiologically-based pharmacokinetic (PBPK) modelling. *European Journal of Pharmaceutical Sciences: Official Journal of the European Federation for Pharmaceutical Sciences*. 2018; 113: 171–184. https://doi.org/10.1016/j.ejps.2017.10.022
3. Wong W, Fletcher DF, Traini D, Chan HK, and Young PM. The use of computational approaches in inhaler development. *Advanced Drug Delivery Reviews*. 2012; 64(4): 312–322. https://doi.org/10.1016/j.addr.2011.10.004
4. Longest PW and Holbrook LT. In silico models of aerosol delivery to the respiratory tract - Development and applications. *Advanced Drug Delivery Reviews*. 2012; 64(4): 296–311. https://doi.org/10.1016/j.addr.2011.05.009
5. Kleinstreuer C, Feng Y, and Childress E. Drug-targeting methodologies with applications: A review. *World Journal of Clinical Cases*. 2014; 2(12), 742–756. https://doi.org/10.12998/wjcc.v2.i12.742
6. Chan HK. Dry powder aerosol delivery systems: Current and future research directions. *Journal of Aerosol Medicine*. 2006; 19(1): 21–27. https://doi.org/10.1089/jam.2006.19.21
7. Islam N and Gladki E. Dry powder inhalers (DPIs) - A review of device reliability and innovation. *International Journal of Pharmacology*. 2008; 360(1–2): 1–11. https://doi.org/10.1016/j.ijpharm.2008.04.044
8. Mossaad DMR. Drug Delivery to the Respiratory Tract Using Dry Powder Inhalers. 2014. Electronic Thesis and Dissertation Repository. 2036. Retrieved from https://ir.lib.uwo.ca/etd/2036/
9. Zhou Q, Tong Z, Tang P, Yang R, and Chan HK. CFD analysis of the aerosolization of carrier-based dry powder inhaler formulations. *AIP Conference Proceedings*. 2013; 1542(1): 1146–1149. https://doi.org/10.1063/1.4812139
10. Jiang L, Tang Y, Zhang H, Lu X, Chen X, and Zhu J. Importance of powder residence time for the aerosol delivery performance of a commercial dry powder inhaler Aerolizer®. *Journal of Aerosol Medicine for Pulmonary Drug Delivery*. 2012; 25(5): 265–279. https://doi.org/10.1089/jamp.2011.0908
11. Lionberg RA and Zhang X. Modeling and simulation of biopharmaceutical performance. *Intelligent Pharmaceuticals*. 2014; 95(5): 480–482. https://doi.org/10.1038/clpt.2014.40
12. Sou T, Kukavica-Ibrulj I, Soukarieh F, Halliday N, Levesque RC, Williams P, Stocks M, Cámara M, Friberg LE, and Bergström C. Model-based drug development in pulmonary delivery: Pharmacokinetic analysis of novel drug candidates for treatment of Pseudomonas aeruginosa lung infection. *Journal of Pharmaceutical Sciences*. 2019; 108(1): 630–640. https://doi.org/10.1016/j.xphs.2018.09.017
13. Sundaram S, Trivedi R, Durairaj C, Ramesh R, Ambati BK, and Kompella UB. Targeted drug and gene delivery systems for lung cancer therapy. *Clinical Cancer Research: An Official Journal of the American Association for Cancer Research*. 2009; 15(23): 7299–7308. https://doi.org/10.1158/1078-0432.CCR-09-1745
14. Nikander K, Nicholls C, Denyer J, and Pritchard J. The evolution of spacers and valved holding chambers. *Journal of Aerosol Medicine and Pulmonary Drug Delivery*. 2014; 27(S1): S4–S23. doi:10.1089/jamp.2013.1076
15. Slator L, von Hollen D, Sandell D, and Hatley RH. In vitro comparison of the effect of inhalation delay and flow rate on the emitted dose from three valved holding chambers. *Journal of Aerosol Medicine and Pulmonary Drug Delivery*. 2014; 4(27): 37–43. doi:10.1089/jamp.2013.1061
16. Mitchell J and Dolovich MB. Clinically relevant test methods to establish in vitro equivalence for spacers and valved holding chambers used with pressurized metered dose inhalers (pMDIs). *Journal of Aerosol Medicine and Pulmonary Drug Delivery*. 2012; 25(4): 217–242. doi:10.1089/jamp.2011.0933
17. Zazo H, Colino CI, and Lanao JM. Current applications of nanoparticles in infectious diseases. *Journal of Controlled Release*. 2016; 224: 86–102. https://doi.org/10.1016/j.jconrel.2016.01.008
18. Magalhães J, Vieira A, Pinto S, Pinheiro S, Granja A, Santos S, Pinheiro M, and Reis S. New approaches from nanomedicine and pulmonary drug delivery for the treatment of tuberculosis. In: Neves AR and Reis S (eds). *Nanoparticles in the Life Sciences and Biomedicine*. Taylor and Francis, Boca Raton, FL. 2018; 197–234.

Retrieved from https://www.taylorfrancis.com/books/e/9781351207355/chapters/10.1201/9781351207355-8
19. Rothen-Rutishauser B, Blank F, Mühlfeld C, and Gehr P. In vitro models of the human epithelial airway barrier to study the toxic potential of particulate matter. *Expert Opinion on Drug Metabolism & Toxicology*. 2008; 4(8): 1075–1089. doi: 10.1517/17425255.4.8.1075
20. Magalhaes J, Pinheiro M, Drasler B, Septiadi D, Petri-Fink A, Santos SG, Rothen-Rutishauser B, & Reis S. Lipid nanoparticles biocompatibility and cellular uptake in a 3D human lung model. *Nanomedicine*. 2019; 15(3): 259–271. https://doi.org/10.2217/nnm-2019-0256
21. Gehr P, Bachofen M, and Weibel ER. The normal human lung: Ultrastructure and morphometric estimation of diffusion capacity. *Respiratory Physiology*. 1978; 32(2): 121–140. https://doi.org/10.1016/0034-5687(78)90104-4
22. Lieber M, Smith B, Szakal A, Nelson-Rees W, and Todaro G. A continuous tumor-cell line from a human lung carcinoma with properties of type II alveolar epithelial cells. *International Journal of Cancer*. 1976; 17(1): 62–70. https://doi.org/10.1002/ijc.2910170110
23. Mohammadian M and Pourmehran O. CFPD simulation of magnetic drug delivery to a human lung using an SAW nebulizer. *Biomechanics and Modeling in Mechanobiology*. 2019; 18: 547–562. https://doi.org/10.1007/s10237-018-1101-0
24. Sandell D and Mitchell JP. Considerations for designing in vitro bioequivalence (IVBE) studies for pressurized metered dose inhalers (pMDIs) with spacer or valved holding chamber (S/VHC) add-on devices. *Journal of Aerosol Medicine and Pulmonary Drug Delivery*. 2015; 28(3): 156–181. https://doi.org/10.1089/jamp.2014.1150
25. Lu D, Lee SL, Lionberger RA, Choi S, Adams W, Caramenico HN, Chowdhury BA, Conner DP, Katial R, Limb S, Peters JR, Yu L, Seymour S, and Li BV. International Guidelines for Bioequivalence of Locally Acting Orally Inhaled Drug Products: Similarities and differences. *The AAPS Journal*. 2015; 17(3): 546–557. https://doi.org/10.1208/s12248-015-9733-9
26. Davit B, Braddy AC, Conner DP, and Yu LX. International Guidelines for Bioequivalence of Systemically Available Orally Administered Generic Drug Products: A survey of similarities and differences. *The AAPS Journal*. 2013; 15(4): 974–990. https://doi.org/10.1208/s12248-013-9499-x
27. Australia. Australian Guidance 19: Inhalation and Nasal Medicines. Section 19.2.1: Selecting a Reference Medicine. 2013. Retrieved from http://www.tga.gov.au/industry/pm-argpm-guidance-19-02.htm
28. Code of Federal Regulations Title 21, 320.1. Retrieved from https://www.accessdata.fda.gov/scripts/cdrh/cfdocs/cfcfr/CFRSearch.cfm?fr=320.1
29. Health-Canada. Guidance to Establish Equivalence or Relative Potency of Safety and Efficacy of a Second Entry Short-acting Beta2-Agonist Metered Dose Inhaler. 1999. http://www.hc-sc.gc.ca/dhp-mps/alt_formats/hpfb-dgpsa/pdf/prodpharma/mdi_bad-eng.pdf
30. Health-Canada. Guidance for Industry: Pharmaceutical Quality of Inhalation and Nasal Products. 2006. http://www.hc-sc.gc.ca/dhp-mps/alt_formats/hpfb-dgpsa/pdf/prodpharma/inhalationnas-eng.pdf
31. EMA. Guideline on the Requirements for Clinical Documentation for Orally Inhaled Products (OIP) Including the Requirements for Demonstration of Therapeutic Equivalence Between Two Inhaled Products for Use in the Treatment of Asthma and Chronic Obstructive Pulmonary Disease (COPD) in Adults and for Use in the Treatment of Asthma in Children and Adolescents. 2009. Retrieved from http://www.ema.europa.eu/docs/en_GB/document_library/Scientific_guideline/2009/09/WC500003504.pdf
32. Fok TF, Monkman S, Dolovich M, Gray S, Coates G, Paes B, Rashid F, Newhouse M, and Kirpalani H. Efficiency of aerosol medication delivery from a metered dose inhaler versus jet nebulizer in infants with bronchopulmonary dysplasia. *Pediatric Pulmonology*. 1996; 21(5): 301–309. https://doi.org/10.1002/(SICI)1099-0496(199605)21:5<301::AID-PPUL5>3.0.CO;2-P
33. US-FDA. Draft Guidance on Albuterol Sulfate. 2013. Retrieved from http://www.fda.gov/downloads/Drugs/GuidanceComplianceRegulatoryInformation/Guidances/UCM346985.pdf
34. US-FDA. Draft Guidance on Fluticasone Propionate; Salmeterol Xinafoate. 2013. Retrieved from http://www.fda.gov/downloads/Drugs/GuidanceComplianceRegulatoryInformation/Guidances/UCM367643.pdf
35. US-FDA. Draft Guidance on Budesonide. 2012. Retrieved from http://www.fda.gov/downloads/Drugs/GuidanceComplianceRegulatoryInformation/Guidances/UCM319977.pdf
36. Wan G, Lin H, Fink JB, Chen Y, Wang W, Chiu Y, Kao YY, and Liu C. In vitro evaluation of aerosol delivery by different nebulization modes in pediatric and adult mechanical ventilators. *Respiratory Care*. 2014; 59(10): 1494–1500. doi: https://doi.org/10.4187/respcare.02999
37. Ke W, Wang W, Lin T, Wu C, Huang S-H, Wu H-D, and Chen C. In vitro evaluation of aerosol performance and delivery efficiency during mechanical ventilation between soft mist inhaler and pressurized metered-dose inhaler. *Respiratory Care*. 2020; 65(7): 1001–1010. https://doi.org/10.4187/respcare.06993
38. Longest PW, Bass K, Dutta R, Rani V, Thomas ML, El-Achwah A, and Hindle M. Use of computational fluid dynamics deposition modeling in respiratory drug delivery. *Expert Opinion on Drug Delivery*. 2019; 16(1): 7–26. https://doi.org/10.1080/17425247.2019.1551875
39. Goralski JL and Davis SD. Breathing easier: Addressing the challenges of aerosolizing medications to infants and preschoolers. *Respiratory Medicine*. 2014; 108(8): 1069–1074. https://doi.org/10.1016/j.rmed.2014.06.004

11

Advances in In Silico Study of Generic Orally Inhaled Drug Products

Hao Miao[1], Renjie Li[1], Xudong Zhou[1], Jiaqi Yu[2], Fen Huang[1,3], Haoqin Yang[2], and Zhenbo Tong[3]
[1]Department of Chemical Engineering, Monash University, Clayton, VIC, Australia
[2]Institute for Process Modelling and Optimization, JITRI, Suzhou, China
[3]School of Energy and Environment, Southeast University, Nanjing, China

11.1 Introduction

Orally inhaled drug products (OIDPs) refer to a drug-device system in which drugs in the form of aerosols, powders, solutions, or suspensions are delivered into the lung via oral inhalation (1). An OIDP usually consists of two components, the drug formulation and an inhaler device which allows the administration of the drug and the deposition of active ingredients in the targeted site within patients. OIDPs show great capability of therapeutic effectiveness to treat pulmonary diseases, such as chronic obstructive pulmonary disease (COPD), cystic fibrosis (CF), and asthma (2), as well as systemic diseases, such as diabetes (3), schizophrenia (4), and Parkinson's disease (5). Based on inhalation method, OIDPs are categorized into pressurized metered-dose inhalers (pMDIs), dry powder inhalers (DPIs) and the newly developed soft mist inhalers (SMIs).

There is a huge market for generic OIDPs but they need to be bioequivalent to the reference listed drug (RLD) (in vitro and in vivo study, drug and device) according to regulatory guidance in different countries to ensure their safety, effectiveness, and quality (1)6. In vitro studies include particle size distribution (PSD) measurement, dose uniformity test, aerodynamic particle size distribution (APSD), and inhalation angle study according to the specific requirement for each type of OIDPs. In vivo studies include pharmacokinetic (PK) and clinical pharmacodynamic (PD) modeling or a comparative end-point BE study. A generic OIDP needs to overcome three technical barriers, i.e. drug formulation, inhaler design, and human factors (8). For example, dry powder formulation usually involves powder mixing of a large portion of lactose carrier and a few micrograms of active pharmaceutical ingredient (API) particles. Generic inhalers should also have different designs from the original ones due to patent limitations and design space (9). Issues of patient adherence or patients misusing the inhaler may lead to unexpected clinical results, thus a generic inhaler is expected to substitute for the referred device without extra training or intervention from healthcare (6,10).

In silicomodeling approaches can save time and cost for development of generic OIDPs (6). Computational fluid dynamics (CFD) including laminar, transitional, and turbulent flow modeling can be combined with the discrete phase method (DPM) or discrete element method (DEM) to predict the dispersion process and drug delivery. These methods have been proven to be useful in exploring the underlying mechanism of the aerosolization process and particle deposition (11). The modeling of the airflow and particle delivery can also be applied for device optimization. Application of CFD in constructed drug-specific physiologically based pharmacokinetic (PBPK) models can be useful to estimate the pulmonary drug absorption based on the deposition prediction (12). This chapter will discuss in silico methods that have been applied in generic OIDPs development.

11.2 Basic Consideration for Generic OIDPs

The major challenge for developing a generic OIDP is to understand the critical quality attributes (CQAs) of drugs and devices that could affect the aerosolization performance of the product. This can be facilitated with the aid of computational tools.

11.2.1 Formulation Considerations

For generic OIDPs development, both qualitative and quantitative factors need to be considered. *Qualitatively the same* (Q1) refers to the sameness of active and inactive ingredients between the generic and reference products. *Quantitatively the same* (Q2) means the same amount of components, however, quantitative difference allows the same inactive ingredients but different concentrations within the levels applied in other approved OIDPs (13). *Physicochemical attributes of a specific dosage form sameness* (Q3) refers to the same physicochemical properties. For some orally inhaled products, FDA draft guidance recommends in vitro BE comparative studies including polymorphic form and crystalline habit (shape) of the drug substance (Q3 study). In general, Q1, Q2, and Q3 need to meet the regulatory requirements (14).

CQAs of drugs need to be considered before BE studies. *CQAs* refer to physical, chemical, and biological functionalities that must be in a proper range to achieve targeted product

DOI: 10.1201/9781003046547-11

quality (15). The first order CQAs of OIDPs formulation, i.e. electrostatic force, surface energy, and flowability, are determined by the properties of particles, such as texture and crystallinity. They have further impact on the second order CQAs such as mass median aerodynamic diameter (MMAD) and fine particle fraction (FPF).

11.2.2 Device Considerations

In general, there are three types of inhalation devices. pMDIs are widely used because of their portability and low cost. pMDIs can accurately deliver specific and consistent volume of drug suspension or solution with propellent via a metering device (10,16). However, the existence of propellant is not environmentally friendly, and the demand of hand-to-breath coordination may impede the correct use of such devices in patient groups (10). DPIs have gained increasing interest after the Montreal Protocol in 1987 (8). A typical DPI contains powder formulation of fine API mixed with carrier excipients (17). Particle engineering technologies such as spraying drying, spray freeze drying, etc., have broadened the applications of DPI in several new fields. DPIs are portable and do not require patient's coordination. But the delivery efficiency of DPI products relies on the inhalation maneuvers of the patient (18). SMIs generate aerosol mist by the mechanical force from a spring (19). SMIs can deliver high content of drugs with long duration and slow speed compared to pMDIs, thus facilitating the coordination between actuation and inhalation (20).

In terms of patient compliance, it is suggested that the targeted patient population use the test device properly and effectively with no significant change in their inspiratory maneuvers (13). The following requirements between the generic and the reference DPI device units are proposed by the Office of Generic Drugs of the FDA: same energy source, such as passive or active; same metering mechanism, such as pre-metered single-unit-dose system (e.g. capsule-based Handihaler or Breezhaler), pre-metered multi-unit-dose system (e.g. blister strip-based Diskus or Ellipta), or device metered multi-dose system (e.g. reservoir-based Turbuhaler); same dose counting principle and same indicator of doses; similar size and shape (13). Possible internal design differences, such as geometry or dimension of air channels, are required to be assessed by BE studies (13).

11.2.3 Requirement of BE for Generic OIDPs

According to the FDA, BE refers to the equivalent rate and level of APIs at the site of drug action under the same administration dose and conditions (21). The requirements of BE for generic OIDPs are complex, including the similarity of PK study, in vitro test, formulation and device, and PD or clinical endpoint studies. There are several key questions that need to be considered before a BE study: Is the dose available to the lung equivalent? Is the regional or spatial distribution equivalent (central versus peripheral: c/p ratio)? Do the deposited drugs stay in the airways for an equivalent time (22)? These questions are well connected to the pulmonary drug deposition, because the drug absorption in the lungs is well connected to the deposition data. When the dosage form is powder, the inhaled particles have various fates including dissolution, blood circulation, local action, translocation, transcytosis, systemic or sensory-nerves uptake, and lymphatic uptake and clearance (macrophages and mucociliary) (23). The fates depend on the solubility, landing site, and size of particles. Soluble particles can act locally or pass into the systemic system, while insoluble particles will be translocated out of the respiratory tract (23–26). As to the solution or suspension type, the process of dispersion tends to be complicated due to the secondary breakup of liquid.

PK in the lungs is difficult to predict compared with other routes of administration due to the complicated pulmonary delivery and deposition process. After the respiratory inhalation of OIDPs, most of the drugs will be deposited in the mouth and pharynx (40–90% of the ED) before being swallowed, while a small portion (10–60% of the ED) will be deposited in the airway (22). The swallowed part will go through the absorption of the gastrointestinal (GI) tract and first-pass inactivation of liver to the systematic bloodstream, whereas the pulmonary-deposited fraction will be absorbed completely or be cleared by mucosa cilia (22). The prediction of pulmonary drug absorption based on the drug deposition data is vital before clinical trials, as the direct measurement of dissolved drug in interstitial fluid of lung tissue by a microdialysis technique during open chest surgery is impractical for a BE study (27–29). Without direct measurements, PK modeling is promising for understanding the differences in OIDPs according to the prediction of lung tissue drug concentrations (27). As a result, mathematical models to assess PK in clinical drug therapeutic use have attracted increasing interests recently (29) to depict the correlations between the plasma or tissue concentration of drug and time (30). The PK models can be categorized into empirical PK models (PK parameters are estimated according to clinical data) and PBPK models (PK parameters are based on the formulation properties and physiological characteristics) (29).

There were two different compartmental models, both using in vitro experimental data as inputs for predicting plasma concentrations after administration of OIDPs (31). In the modeling process, aerosol deposition, dissolution, and mucociliary clearance and dissipation need to be considered (27,32). A commercial PBPK software, GastroPlus™ (Simulations Plus, Rochester, NY, USA), includes a single path model providing lung deposition data, with limited ability to calculate local drug absorption (33,34). The software can be combined with other in silico deposition models to acquire the plasma profiles (33,35,36). For example, the in silico plasma profiles of surface modified API can be predicted by coupling the software with deposition models (33). The model is more accurate when using realistic lung deposition data together with experimental peripheral permeability data (37). Furthermore, simulations based on the model indicated that local concentration of drugs did not correspond with systemic exposure of drugs in some cases (32).

11.3.1 CFD Modeling Techniques

CFD models can predict the flow properties by solving the Navier-Stokes (NS) equations according to the mass and momentum conservation laws. It should be noticed that the properties of the air flow from devices to the parafacial lungs are varying. For example, turbulence flow plays an important role in dispersing agglomerated drug particles in DPIs. The turbulence flow is presented in the extrathoracic (ET) region while laminar flow is dominant in the tracheobronchial (TB) and the alveolar regions (38). As a result, predicting the air flow properties and selecting the proper turbulence models are vital for the prediction of pulmonary deposition. The most accurate turbulence model is direct numerical simulation (DNS) with extremely high computational cost. Other turbulence models comprising Reynolds-averaged Navier-Stokes (RANS) models and large eddy simulation (LES) should be carefully selected according to the situation; the common turbulence models in CFD were summarized by Sommerfeld et al. in 2019 (39).

CFD models can be coupled with DPM for aerosol delivery, assuming that the discrete phase has no influence on the airflow and the particle interaction is negligible (40). Particles are highly likely to be trapped once contacting the airway, while either reflection or trap will happen in DPIs according to the angle and velocity of particles. The drag force and gravitational force are considered in most studies (41–58). When the size of particles is large, the gravity may not be ignored. CFD models can also be coupled with DEM models where particles are regarded as discrete elements and moving based on Newton's second law. DEM models consider the particle–particle and particle–wall forces, which are influenced by the particle intrinsic properties (59). In CFD-DEM, particle–fluid interactions are also considered, and the major forces considered are listed in Table 11.1. With the combination of DEM and CFD simulation, experimental trends regarding drug substance adhesion, flow rate, and device geometry can be correctly predicted (60). CFD-DEM can also study how particles of different diameters leave inhalation devices by considering particle-particle collisions (61).

The relevance of CFD and its proof of verification is usually considered when evaluating in silico models, as it is vital to verify the credibility of CFD models by in vivo and in vivo and in vitro data. Several approaches of validation have been proposed by researchers. In vivo data collected from imaging methods are often used for validating spatial deposition of CFD models, as the distribution of inhaled drugs in the lung has an essential influence on the effectiveness, especially for drugs requiring targeted administration (63). The imaging approaches such as gamma scintigraphy, positron emission tomography (PET), and single-photon emission computed tomography (SPECT) provide a direct visualization of radio-labeled aerosol distribution within the airway models (64). The data from the imaging approaches can demonstrate the reliability of CFD models. Several studies have validated the models via data collected from imaging methods (65–68). Longest et al. compared a whole lung airway model predictions on regional deposition of DPI aerosol with published in vivo data (69). The imaging methods would be highly useful for development of OIDPs as it can help researchers track local and total deposition in the lungs.

In vitro experiments based on replicas of computational airway geometry are used to validate predictions of total and regional deposition of CFD models. However, data that can be applied for validation of highly localized CFD predictions are still limited due to the inconsistency of the conditions using CFD models and *in* vitro experiments. To improve the relevance between CFD models and in vitro data, Holbrook et al. summarized the following criteria: (1) The resolution between in vitro data should resemble that of regional CFD predictions. (2) Computational models should be fully described and provide the geometric model. (3) It should describe experimental and simulated conditions in detail (70). Several researchers validated CFD models with in vitro experimental data (46,70–73). For example, Huang et al. investigated three mouth–throat models using CFD methods with validation of in vitro experiments data (75). However, in vivo and in vitro data are not easily accessible due to the limitations of modeling techniques, medical ethics, and funds (74), which lead to the challenges in evaluating the credibility of the CFD models.

11.3.2 In Silico Modeling of pMDIs

pMDIs contain liquefied hydrofluoroalkane (HFA) as propellants, with API in the dosage form of suspension or solution (75). When pMDIs are actuated, the pressurized drug and propellant mixture with metered dose are emitted from the canister of the inhaler (76). The actuation process requires hand–mouth coordination and propellant that is not environmentally friendly (71,78–81). Besides this, the plume temperature of the drug emitted from a pMDI may cause patient discomfort and inconsistent dose delivery (81).

Simulation of flow is difficult in pMDIs as the properties of emitted aerosol are transient, unsteady, and turbulent (76). The computational model of pMDIs is different from the stochastic

TABLE 11.1
Interactions Considered in CFD-DEM (62)

Interactions	
Particle–particle/wall forces	Normal contact force
	Tangential contact force
	Van der Waals force
	Capillary force
	Electrostatic force
Particle–fluid interactions	Drag force
	Saffman force
	Magnus force

particle deposition model of DPIs. The empirical formulas of DPIs assume that particles have the same entrance velocity as the airflow (82). However, the formulas don't work in the model of pMDIs, as the velocity of aerosol is higher than the flow speed (82). The internal flow model (IFM) was developed based on the homogeneous frozen model (HFM) (83,84) for predicting the variables of flow in pMDIs (85). Furthermore, Gavtash et al. found that the aerosol size of pMDIs is largely influenced by the viscosity and surface tension of solution instead of ethanol function by numerical investigation (75).

11.3.3 In Silico Modeling of DPIs

DPIs should be carefully designed, as the number of residual particles in devices cannot be ignored and the velocity of outlet flow will determine the particle deposition of ET region. Most DPIs are passive, where the delivery of particles depends on the inhalation maneuvers. The passive DPIs are suitable for people who have normal lung function due to the superior patient compliance, while active DPIs are necessary when the patients have low inspiratory capacity and pulmonary pressure.

According to in silico investigation, a slight change of device such as variations of grid or mouthpiece geometry can influence the aerosol performance significantly (62,86). For example, using CFD-assisted methods, simple modifications to Cyclohaler® and Rotahaler® can make inhalers perform better (87). The drug emission could be independent of inhalation maneuvers after the optimization of the entrainment of DPIs, which was conducted by CFD (88). CFD predictions on DPI entrainment geometries also proved that the Euler-Euler (EE) method could predict the entrainment amount of formulations in DPIs accurately (89). Similarly, the mechanisms of grid design on the atomization effect of different DPI dosage forms was studied using Aerolizer® (90). Furthermore, the influence of size of the air bypass was calculated by CFD simulation (91).

The difference of formulation types and properties should also be considered in the evaluation of device performance. In carrier-based DPIs, where fine respirable drug particles are blended with a coarser carrier (α-lactose monohydrate in most cases) (92,93), the influence of the physical properties (size, shape) of carrier particles (anhydrous, granular lactose carrier, and budesonide) on dispersion and delivery of particles (Aerolizer® and Handihaler®) can be simulated (94). Other CFD-DEM results show that increasing the mass ratio of carriers to drugs can improve the aerosol performance, as shown in Figure 11.1 (96), and the detachment process of API from two different carriers, TS-80 (polystyrene beads with the size of 82.8 μm) and TS-230 (polystyrene beads with the size of 277.5 μm), is shown in Figure 11.2 (97).

The influence of inhalation flow rate on the particle dispersion and delivery should be carefully considered, as the pulmonary function of patients are various. The inhalation flow rate is determined by the resistance of inhalers and the pressure drop. The inhalation resistance of commercial devices has been summarized by Hira et al. in 2018 (96), while the pressure drop via orifice is related to the inhalation maneuvers. In general, the optimal inhalation flow rate needs to be weighed, as the airflow with excessively low rate struggles to disperse the agglomeration of drugs, while the flow with high speed will increase the drug deposition in the mouth–throat.

11.3.4 In Silico Modeling of SMIs

SMIs can achieve high pulmonary deposition with low doses (97), but it is not breath actuated and not favoured in most countries due to the extremely high cost (98). In Respimat® SMI, the soft mist is generated by the collision of two fluids through a subtle nozzle system (99). The nozzle system, called Uniblock, includes a filter structure and two fluid channels, meaning two liquid jets are ejected from the nozzle at a fixed angle and merge, causing the drug solution to collide and decompose into an inhalable, slow-moving soft mist aerosol.

The scale of the channels in the nozzle is micron sized while the space of the emitted aerosols is millimeter level, making the simulation of soft mist inhalers difficult. Results of a simulation study indicated that optimization of mouthpiece and control of patient inhalation maneuvers are highly likely to reduce drug deposition in the device and mouth–throat (100).

11.4 In Silico Modeling of Lung Deposition

11.4.1 Airway Geometry

Airway geometry varies for inter-subjects and intra-subjects, which can influence the lung deposition, making the prediction of pulmonary deposition difficult (101). The human airways can be divided into the ET region, the TB region, and the alveolar region (102). The respiratory tract can be divided into 23 generations from conducting airways (G0–15) including trachea, bronchi, bronchioles, and terminal bronchioles to acinar airways (G15–23) covering respiratory bronchioles, alveolar ducts, and alveolar sacs (103–105).

Most of the inhaled drug tends to be deposited in the ET region with the proportion of around 55–60%, while a small fraction of drugs (5–10%) is high likely to be deposited in the G0–G15, and the rest (about 35%) will deposited in G16–23. As a result, it is vital to figure out the deposition mechanism of the ET region.

11.4.2 CFD Modeling of Lung Deposition

There have been various HRT models for the prediction of lung deposition. In terms of ET region, both idealized models including US pharmacopeia throat (USP), idealized mouth–throat (IMT), and realistic models such as mouth-throat (RMT) can be applied. The USP which is a 90° bended model without physiological features was widely used because of its low computation cost (40,106–111). IMT is more comparable with in vivo data as it retains the key anatomical features (111). RMT tends to be the most accurate model, but considering the inter-subject and intra-subject variation, the realistic models may not be representative.

A case in point is the comparison between in vitroand in silico data of USP and IMT and in silico data of RMT, as shown in Figure 11.3 (111). According to Figure 11.3, both

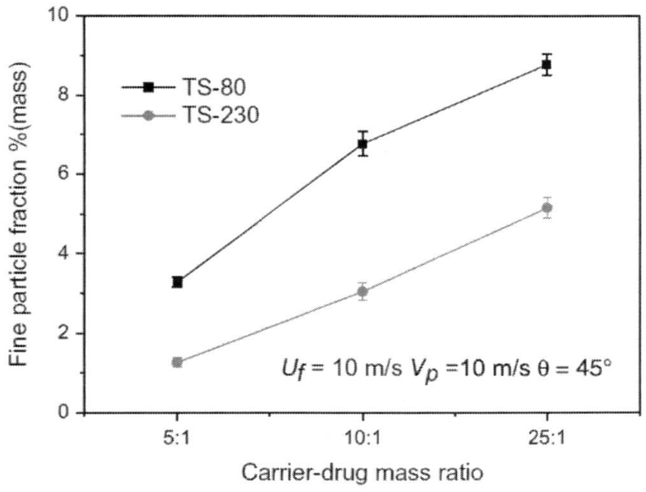

FIGURE 11.1 FPF as Carrier-Drug Mass Ratio (95)

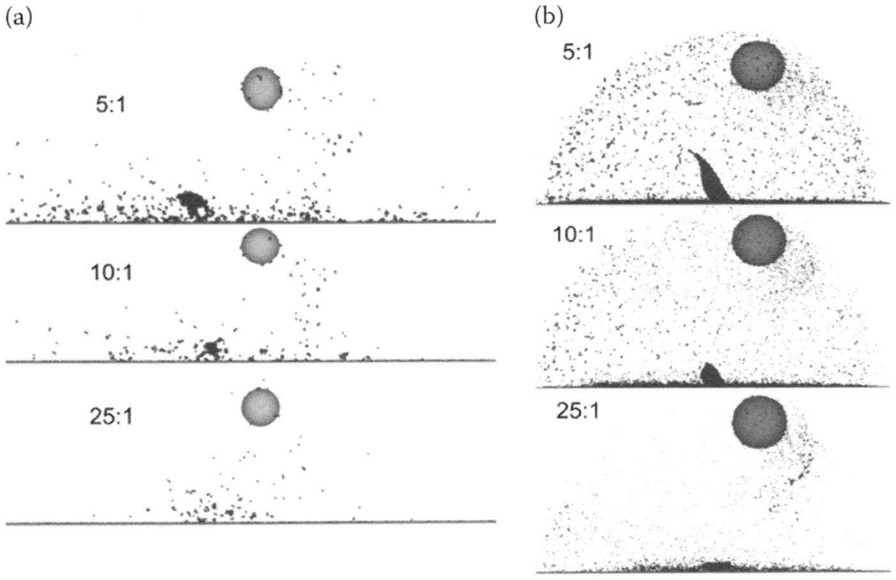

FIGURE 11.2 (a) Detachment of API from TS-80. (b) Detachment of API from the TS-230 (95)

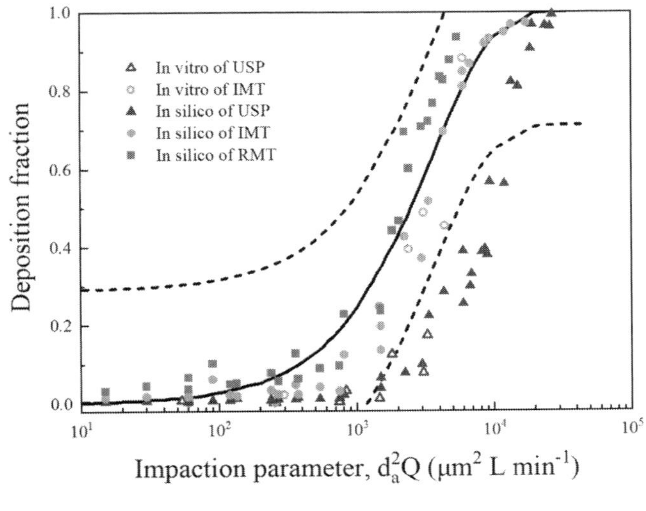

FIGURE 11.3 Comparison Between In Vitro and In Silico Data of USP and IMT and In Silico Data of RMT (111)

IMT and RMT models fit the in vivo data better than USP models. RMT is highly likely to be superior to IMT due to the more realistic anatomical structure. As a result, idealized models are indispensable, which can be used as a standard of OIDPs horizontal comparison, and the realistic models should be integrated to form a database.

11.4.3 CFD-PBPK Modeling

PBPK model coupled with CFD is a promising method to predict the inhaled drug bio-performance (112). Though current models have shown promising results, there is still room for improvements in the future. Most models for the pulmonary drug absorption are zero-dimensional compartmental models (113), which generally divide lung airways into two sections (airways and alveolar) for the ease of simulation (114). However, the credibility of the solutions delivered by such block models are questioned, and the spatial drug concentration in lungs is not obtainable due to the limitation on model dimensions. Three-dimensional CFD methods can customize the lung geometries, but it is hard to achieve the balance between the high computational cost and accuracy. It may take the order of days to simulate a few breathing cycles, which is undesirable when dealing with huge amounts of data.

In order to solve the computational cost versus model performance dilemma, Kannal et al. (115) proposed a compartment-Quasi3D (Q3D) multiscale method to simulate the transport and absorption of the inhaled drug. The Q3D multiscale model was a combination of original Q3D models and compartmental models, where the spatially resolved section (i.e. Quasi-3D) ended at G8. The rest of the airways, together with the alveoli and other organs, were using a compartmental model proposed by Yu et al. (114,115). The simulation speed of this method was approximately 25,000 times faster than that of CFD, while the 3D lung geometry and the accuracy of the solutions were preserved (115). The Q3D method can simulate the mucociliary transport of the undissolved and dissolved drug in the mucus lining with much higher precision. This Q3D framework may have broader range of applications in the future, for example, to detect and quantify the constriction of regional diseased lung by spirometry data (116), and to build a transit and absorption model for oral inhaled drug delivery (117).

11.5 Summary and Perspectives

Numerical models are vital for the development of generic OIDPs as it is impractical to measure the deposition and absorption of inhaled drugs directly. The main difficulty of in silico models is the prediction of drug delivery and deposition. In simulations, there are various factors which need to be considered, including formulation properties, device geometry, and inhalation maneuvers. The validation based on experimental and in vivo data is crucial, as it can increase the credibility and facilitate the development of generic OIDPs. To investigate the drug delivery precisely by numerical methods, the models should comprise particle, device, and human respiratory tract, as each one of them is indispensable. Moreover, the idealized models need to be optimized to improve the in vivo/in vitro correlations (IVIVC) while a large number of realistic models need to be integrated to form a database.

Some novel statistical approaches are promising for predicting the pulmonary deposition of aerosol particles. Artificial neural network (ANN) is a statistical model based on functionally related nodes (i.e. artificial neurons) which behave like neurons in a biological brain (118–120). An ANN model usually consists of one input layer, several hidden layers, and one output layer. All the neurons in each layer are connected to the neurons in the adjacent layers by synaptic weights, which act as signal multipliers (119–120). ANN has shown satisfying results in predictions based on massive datasets, thus it may be possible to correlate several in vitro data with the performances of OIDPs by this method. De Matas et al. (121) built a three-layered ANN to evaluate the IVIVCs for DPI delivery. For pharmaceutical aerosols, Muddle et al. (122) applied 31 ANN architectures to evaluate the feasibility of predicting FPF based on formulation and device variables. An optimal architecture which included API, formulation, and device inputs demonstrated its potential to predict the effect of these factors on DPIs. At this stage, it is hard to develop and train the model to fulfill the requirement of ANN, owing to the limited sample size of experimental datasets. But the solid performance of ANN has proven its potential to be useful for OIDP development.

REFERENCES

1. Pharmacopeia U. USP 39 NF 34. 2015.
2. Pilcer G and Amighi K. Formulation strategy and use of excipients in pulmonary drug delivery. *International Journal of Pharmaceutics*. 2010; 392(1–2): 1–19.
3. De Galan B, Simsek S, Tack C, and Heine R. Efficacy and safety of inhaled insulin in the treatment of diabetes mellitus. *Netherlands Journal of Medicine*. 2006; 64(9): 319–325.
4. Lee H-G, Kim D-W, and Park C-W. Dry powder inhaler for pulmonary drug delivery: Human respiratory system, approved products and therapeutic equivalence guideline. *Journal of Pharmaceutical Investigation*. 2018; 48(6): 603–616.
5. Koranne M, Monsuur F, and Quadflieg J. APV Focus Group Drug Delivery. https://www.apv-mainz.de/fileadmin/dateiablage/apv-mainz/Newsletter_Drug_Delivery/APV-DD-Newsletter_2019-01_final.pdf.
6. Walenga RL, Babiskin AH, and Zhao L. In silico methods for development of generic drug-device combination orally inhaled drug products. *CPT: Pharmacometrics and Systems Pharmacology*. 2019; 8(6): 359–370.
7. Lee SL, Saluja B, Garciaarieta A, et al. Regulatory considerations for approval of generic inhalation drug products in the US, EU, Brazil, China, and India. *AAPS Journal*. 2015; 17(5): 1285–1304.
8. Hoppentocht M, Hagedoorn P, Frijlink HW, and De Boer AH. Technological and practical challenges of dry powder inhalers and formulations. *Advanced Drug Delivery Reviews*. 2014; 75(2): 18–31.
9. Hickey AJ. Inhaled Pharmaceutical Product Development Perspectives || Regulatory Strategy. Elsevier, Amsterdam. 2018; 73–84.

10. Nelson HS. Inhalation devices, delivery systems, and patient technique. *Annals of Allergy, Asthma & Immunology*. 2016; 117(6): 606–612.
11. Koullapis PG, Ollson B, Kassinos SC, and Sznitman J. Multiscale in silico lung modeling strategies for aerosol inhalation therapy and drug delivery. *Current Opinion in Biomedical Engineering*. 2019; 11: 130–136.
12. Vulovic A, Sustersic T, Cvijic S, Ibric S, and Filipovic N. Coupled in silico platform: Computational fluid dynamics (CFD) and physiologically-based pharmacokinetic (PBPK) modelling. *European Journal of Pharmaceutical Sciences*. 2017; 113: 171–184.
13. Lee SL. Scientific and Regulatory Considerations for Bioequivalence (BE) of Dry Powder Inhalers (DPIs). 2011 GPhA/FDA Fall Technical Conference, October 4, 2011.
14. Choi S, Wang Y, Conti DS, et al. Generic drug device combination products: Regulatory and scientific considerations. *International Journal of Pharmaceutics*. 2017; 544(2): 443–454.
15. Zeng XM, Martin GP, Tee S-K, Ghoush AA, and Marriott C. Effects of particle size and adding sequence of fine lactose on the deposition of salbutamol sulphate from a dry powder formulation. *International Journal of Pharmaceutics*. 1999; 182(2): 133–144.
16. Haidl P, Heindl S, Siemon K, Bernacka M, and Cloes RM. Inhalation device requirements for patients' inhalation maneuvers. *Respiratory Medicine*. 2016; 118: 65–75.
17. Healy AM, Amaro MI, Paluch KJ, and Tajber L. Dry powders for oral inhalation free of lactose carrier particles. *Advanced Drug Delivery Reviews*. 2014; 75: 32–52.
18. Ashurst II MA, Prime D, and Sumby B. Latest advances in the development of dry powder inhalers. *Pharmscitechnoltoday*. 2000; 3(7): 246–256.
19. Mehri R, Alatrash A, Matida EA, and Fiorenza F. Respimat soft mist inhaler (SMI) in-vitro aerosol delivery with the ODAPT adapter and facemask. *Canadian Journal of Respiratory, Critical Care, and Sleep Medicine*. 2019; 5(1):3–12.
20. Pitcairn G, Reader S, Pavia D, and Newman S. Deposition of corticosteroid aerosol in the human lung by Respimat Soft Mist inhaler compared to deposition by metered dose inhaler or by Turbuhaler dry powder inhaler. *Journal of Aerosol Medicine the Official Journal of the International Society for Aerosols in Medicine*. 2005; 18(3): 264.
21. Food and Drug Administration. Guidance for Industry: Bioavailability and Bioequivalence Studies for Orally Administered Drug Products—General Considerations. Food and Drug Administration, Washington, DC. 2003.
22. Hochhaus G, Horhota S, Hendeles L, Suarez S, And Rebello J. Pharmacokinetics Of Orally Inhaled Drug Products. Springer. 2015.
23. Colombo P, Traini D, and Buttini F. Inhalation Drug Delivery -Techniques and Products. John Wiley & Sons, Ltd, UK. 2013.
24. Martonen T. On the fate of inhaled particles in the human: A comparison of experimental data with theoretical computations based on a symmetric and asymmetric lung. *Bulletin of Mathematical Biology*. 1983; 45(3): 409–424.
25. Blank F, Rothen-Rutishauser BM, Schurch S, and Gehr P. An optimized in vitro model of the respiratory tract wall to study particle cell interactions. *Journal of Aerosol Medicine*. 2006; 19(3): 392–405.
26. Finlay WH. The Mechanics of Inhaled Pharmaceutical Aerosols: An Introduction. Academic Press. 2001.
27. Bäckman P, Arora S, Couet W, Forbes B, de Kruijf W, and Paudel A. Advances in experimental and mechanistic computational models to understand pulmonary exposure to inhaled drugs. *European Journal of Pharmaceutical Sciences*. 2018; 113: 41–52.
28. Marchand S, Chauzy A, Dahyot-Fizelier C, and Couet W. Microdialysis as a way to measure antibiotics concentration in tissues. *Pharmacological Research*. 2016; 111: 201–207.
29. Borghardt JM, Weber B, Staab A, and Kloft C. Pharmacometric models for characterizing the pharmacokinetics of orally inhaled drugs. *The AAPS Journal*. 2015; 17(4): 853–870.
30. Jones H and Rowland-Yeo K. Basic concepts in physiologically based pharmacokinetic modeling in drug discovery and development. *CPT: Pharmacometrics & Systems Pharmacology*. 2013; 2(8): 1–12.
31. Bhagwat S, Schilling U, Chen M-J, et al. Predicting pulmonary pharmacokinetics from in vitro properties of dry powder inhalers. *Pharmaceutical Research*. 2017; 34(12): 2541–2556.
32. Bäckman P, Tehler U, and Olsson B. Predicting exposure after oral inhalation of the selective glucocorticoid receptor modulator, AZD5423, based on dose, deposition pattern, and mechanistic modeling of pulmonary disposition. *Journal of Aerosol Medicine and Pulmonary Drug Delivery*. 2017; 30(2): 108–117.
33. Wu S, Zellnitz S, Mercuri A, Salar-Behzadi S, Bresciani M, and Fröhlich E. An in vitro and in silico study of the impact of engineered surface modifications on drug detachment from model carriers. *International Journal of Pharmaceutics*. 2016; 513(1-2): 109–117.
34. ICoRP. Human respiratory tract model for radiological protection. A report of a Task Group of the International Commission on Radiological Protection. *Annals of the ICRP*. 1994; 24(1–3): 1–482.
35. Chaudhuri S, Lukacova V, Bolger M, Woltosz W, eds. Modeling Regional Lung Deposition and Disposition (ADME-PK) Behavior of Aerosolized Fentanyl Following Inhaled Administration in Humans. 29th Annual Conference of the American Association for Aerosol Research, Portland, OR. 2010.
36. Chaudhuri SR, Lukacova V, Woltosz W, eds. Simulating the Disposition of Budesonide from Dry Powder Inhalers (DPIs) and Nebulizer. Proceedings of AAPS Annual Meeting, San Antonio. 2013.
37. Salar-Behzadi S, Wu S, Mercuri A, Meindl C, Stranzinger S, and Fröhlich E. Effect of the pulmonary deposition and in vitro permeability on the prediction of plasma levels of inhaled budesonide formulation. *International Journal of Pharmaceutics*. 2017; 532(1): 337–344.
38. Longest PW and Vinchurkar S. Validating CFD predictions of respiratory aerosol deposition: Effects of upstream transition and turbulence. *Journal of Biomechanics*. 2007; 40(2): 305–316.
39. Sommerfeld M, Cui Y, and Schmalfuß S. Potential and constraints for the application of CFD combined with

Lagrangian particle tracking to dry powder inhalers. *European Journal of Pharmaceutical Sciences*. 2019; 128: 299–324.
40. Dehbi A. A CFD model for particle dispersion in turbulent boundary layer flows. *Nuclear Engineering and Design*. 2008; 238(3): 707–715.
41. Jayaraju ST, Brouns M, Verbanck S, and Lacor C. Fluid flow and particle deposition analysis in a realistic extrathoracic airway model using unstructured grids. *Journal of Aerosol Science*. 2007; 38(5): 494–508.
42. Li Z. Particle deposition in oral-tracheal airway models with very low inhalation profiles. *Journal of Bionic Engineering*. 2012; 9(2): 252–261.
43. Ilie M, Matida EA, and Finlay WH. Asymmetrical aerosol deposition in an idealized mouth with a DPI mouthpiece inlet. *Aerosol Science and Technology*. 2008; 42(1): 10–17.
44. Jayaraju ST, Brouns M, Lacor C, Belkassem B, and Verbanck S. Large eddy and detached eddy simulations of fluid flow and particle deposition in a human mouth–throat. *Journal of Aerosol Science*. 2008; 39(10): 862–875.
45. Koullapis PG, Hofemeier P, Sznitman J, and Kassinos SC. An efficient computational fluid-particle dynamics method to predict deposition in a simplified approximation of the deep lung. *European Journal of Pharmaceutical Sciences*. 2018; 113: 132–144.
46. Kolanjiyil AV and Kleinstreuer C. Computationally efficient analysis of particle transport and deposition in a human whole-lung-airway model. Part I: Theory and model validation. *Computers in Biology and Medicine*. 2016; 79: 193–204.
47. Kolanjiyil AV and Kleinstreuer C. Computational analysis of aerosol-dynamics in a human whole-lung airway model. *Journal of Aerosol Science*. 2017; 114: 301–316.
48. Longest PW, Hindle M, Das Choudhuri S, and Byron PR. Numerical simulations of capillary aerosol generation: CFD model development and comparisons with experimental data. *Aerosol Science and Technology*. 2007; 41(10): 952–973.
49. Jin HH, Fan JR, Zeng MJ, and Cen KF. Large eddy simulation of inhaled particle deposition within the human upper respiratory tract. *Journal of Aerosol Science*. 2007; 38(3): 257–268.
50. Koullapis P, Kassinos SC, Muela J, et al. Regional aerosol deposition in the human airways: The SimInhale benchmark case and a critical assessment of in silico methods. *European Journal of Pharmaceutical Sciences*. 2018; 113: 77–94.
51. Koullapis PG, Kassinos SC, Bivolarova MP, and Melikov AK. Particle deposition in a realistic geometry of the human conducting, airways: Effects of inlet velocity profile, inhalation flowrate and electrostatic charge. *Journal of Biomechanics*. 2016; 49(11): 2201–2212.
52. Lambert AR, O'Shaughnessy P, Tawhai MH, Hoffman EA, and Lin CL. Regional deposition of particles in an image-based airway model: Large-eddy simulation and left-right lung ventilation asymmetry. *Aerosol Science and Technology*. 2011; 45(1): 11–25.
53. Longest PW, Hindle M, Das Choudhuri S, and Xi J. Comparison of ambient and spray aerosol deposition in a standard induction port and more realistic mouth–throat geometry. *Journal of Aerosol Science*. 2008; 39(7): 572–591.
54. Longest PW, Hindle M, and Das Choudhuri S. Effects of generation time on spray aerosol transport and deposition in models of the mouth-throat geometry. *Journal of Aerosol Medicine and Pulmonary Drug Delivery*. 2009; 22(2): 67–83.
55. Liu ZJ, Li AG, Xu XX, and Gao R. Computational fluid dynamics simulation of airflow patterns and particle deposition characteristics in children upper respiratory tracts. *Engineering Applications of Computational Fluid Mechanics*. 2012; 6(4): 556–571.
56. Feng Y, Zhao J, Kleinstreuer C, et al. An in silico intersubject variability study of extra-thoracic morphology effects on inhaled particle transport and deposition. *Journal of Aerosol Science*. 2018; 123: 185–207.
57. Chen X, Zhong W, Zhou X, Jin B, and Sun B. CFD–DEM simulation of particle transport and deposition in pulmonary airway. *Powder Technology*. 2012; 228: 309–318.
58. Chen XL, Zhong WQ, Sun BB, Jin BS, and Zhou XG. Study on gas/solid flow in an obstructed pulmonary airway with transient flow based on CFD-DPM approach. *Powder Technology*. 2012; 217: 252–260.
59. van Wachem B, Thalberg K, Remmelgas J, and Niklasson-Björn I. Simulation of dry powder inhalers: Combining micro-scale, meso-scale and macro-scale modeling. *AIChE Journal*. 2017; 63(2): 501–516.
60. Nguyen D, Remmelgas J, Björn IN, van Wachem B, and Thalberg K. Towards quantitative prediction of the performance of dry powder inhalers by multi-scale simulations and experiments. *Internation Journal of Pharmaceutics*. 2018; 547(1–2): 31–43.
61. Remmelgas J, Thalberg K, Björn IN, and van Wachem B. Simulation of the flow of cohesive particles in a model inhaler using a CFD/DEM model. *Procedia Engineering*. 2015; 102: 1526–1530.
62. Tong Z, Yu A, Chan H-K, and Yang R. Discrete modelling of powder dispersion in dry powder inhalers-a brief review. *Current Pharmaceutical Design*. 2015; 21(27): 3966–3973.
63. Nokhodchi A and Martin GP. Pulmonary Drug Delivery: Advances and Challenges. John Wiley & Sons, Ltd, Chichester, West Sussex, United Kingdom. 2015.
64. Conway J. Lung imaging - two dimensional gamma scintigraphy, SPECT, CT and PET. *Advanced Drug Delivery Reviews*. 2012; 64(4): 357–368.
65. Newman S, Salmon A, Nave R, and Drollmann A. High lung deposition of 99mTc-labeled ciclesonide administered via HFA-MDI to patients with asthma. *Respiratory Medicine*. 2006; 100(3): 375–384.
66. Newman SP, Pitcairn GR, Hirst PH, et al. Scintigraphic comparison of budesonide deposition from two dry powder inhalers. *European Respiratory Journal*. 2000; 16(1): 178–183.
67. De Backer JW, Vos WG, Vinchurkar SC, et al. Validation of computational fluid dynamics in CT-based airway models with SPECT/CT. *Radiology*. 2010; 257(3): 854–862.
68. Tian G, Hindle M, Lee S, and Longest PW. Validating CFD predictions of pharmaceutical aerosol deposition

with in vivo data. *Pharmaceutical Research*. 2015; 32(10): 3170–3187.
69. Longest PW, Tian G, Khajeh-Hosseini-Dalasm N, and Hindle M. Validating whole-airway CFD predictions of DPI aerosol deposition at multiple flow rates. *Journal of Aerosol Med icine and Pulmonary Drug Delivery*. 2016; 29(6): 461–481.
70. Holbrook LT and Longest PW. Validating CFD predictions of highly localized aerosol deposition in airway models: In vitro data and effects of surface properties. *Journal of Aerosol Science*. 2013; 59: 6–21.
71. Longest PW, Tian G, Walenga RL, and Hindle M. Comparing MDI and DPI aerosol deposition using in vitro experiments and a new stochastic individual path (SIP) model of the conducting airways. *Pharmaceutical Research*. 2012; 29(6): 1670–1688.
72. Huang F, Zhang Y, Tong ZB, Chen XL, Yang RY, and Yu AB. Numerical investigation of deposition mechanism in three mouth–throat models. *Powder Technology*. 2021; 378: 724–735.
73. Milenkovic J, Alexopoulos A, and Kiparissides C. Deposition and fine particle production during dynamic flow in a dry powder inhaler: A CFD approach. *International Journal of Pharmaceutics*. 2014; 461(1–2): 129–136.
74. Pathmanathan P, Gray RA, Romero VJ, and Morrison TM. Applicability analysis of validation evidence for biomedical computational models. *Journal of Verification, Validation and Uncertainty Quantification*. 2017; 2(2): 021005.
75. Gavtash B, Versteeg HK, Hargrave G, et al. A model of transient internal flow and atomization of propellant/ethanol mixtures in pressurized metered dose inhalers (pMDI). *Aerosol Science and Technology*. 2018; 52(5): 494–504.
76. Crosland BM, Johnson MR, and Matida EA. Characterization of the spray velocities from a pressurized metered-dose inhaler. *Journal of Aerosol Medicine and Pulmonary Drug Delivery*. 2009; 22(2): 85–98.
77. Crompton GK. Dry powder inhalers: Advantages and limitations. *Journal of Aerosol Medicine*. 1991; 4(3): 151–156.
78. Perez P and Reyes J. An integrated neural network model for PM10 forecasting. *Atmospheric Environment*. 2006; 40(16): 2845–2851.
79. Newman S and Busse W. Evolution of dry powder inhaler design, formulation, and performance. *Respiratory Medicine*. 2002; 96(5): 293–304.
80. Lee W-H, Loo C-Y, Traini D, and Young PM. Inhalation of nanoparticle-based drug for lung cancer treatment: Advantages and challenges. *Asian Journal of Pharmaceutical Sciences*. 2015; 10(6): 481–489.
81. Brambilla G, Church T, Lewis D, and Meakin B. Plume temperature emitted from metered dose inhalers. *International Journal of Pharmaceutics*. 2011; 405(1–2): 9–15.
82. Farkas Á, Horváth A, Kerekes A, et al. Effect of delayed pMDI actuation on the lung deposition of a fixed-dose combination aerosol drug. *International Journal of Pharmaceutics*. 2018; 547(1–2): 480–488.
83. Fletcher G. Factors Affecting the Atomization of Saturated Liquids. Loughborough University. 1975.
84. Clark AR. Metered Atomisation for Respiratory Drug Delivery. Loughborough University. 1991.
85. Gavtash B, Versteeg H, Hargrave G, et al. CFD simulation of pMDI aerosols in confined geometry of USP-IP using predictive spray source. *Journal of Aerosol Medicine and Pulmonary Drug Delivery*. 2015; 26.
86. Milenkovic J, Alexopoulos AH, and Kiparissides C. Optimization of a DPI inhaler: A computational approach. *Journal of Pharmaceutical Sciences*. 2017; 106(3): 850–858.
87. Suwandecha T, Wongpoowarak W, and Srichana T. Computer-aided design of dry powder inhalers using computational fluid dynamics to assess performance. *Pharmaceutical Development and Technology*. 2016; 21(1): 54–60.
88. Kopsch T, Murnane D, and Symons D. Optimizing the entrainment geometry of a dry powder inhaler: Methodology and preliminary results. *Pharmaceutical Research*. 2016; 33(11): 2668–2679.
89. Kopsch T, Murnane D, and Symons D. Computational modelling and experimental validation of drug entrainment in a dry powder inhaler. *International Journal of Pharmaceutics*. 2018; 553(1–2): 37–46.
90. Leung CMS, Tong Z, Zhou QT, et al. Understanding the different effects of inhaler design on the aerosol performance of drug-only and carrier-based DPI formulations. Part 1: Grid structure. *The AAPS Journal*. 2016; 18(5): 1159–1167.
91. Kopsch T, Murnane D, and Symons D. A personalized medicine approach to the design of dry powder inhalers: Selecting the optimal amount of bypass. *International Journal of Pharmaceutics*. 2017; 529(1–2): 589–596.
92. Malcolmson RJ and Embleton JK. Dry powder formulations for pulmonary delivery. *Pharmaceutical Science & Technology Today*. 1998; 1(9): 394–398.
93. Telko MJ and Hickey AJ. Dry powder inhaler formulation. *Respiratory Care*. 2005; 50(9): 1209–1227.
94. Donovan MJ, Kim SH, Raman V, and Smyth HD. Dry powder inhaler device influence on carrier particle performance. *Journal of Pharmaceutical Sciences*. 2012; 101(3): 1097–1107.
95. Tong Z, Yang R, and Yu A. CFD-DEM study of the aerosolisation mechanism of carrier-based formulations with high drug loadings. *Powder Technology*. 2017; 314: 620–626.
96. Hira D, Koide H, Nakamura S, et al. Assessment of inhalation flow patterns of soft mist inhaler co-prescribed with dry powder inhaler using inspiratory flow meter for multi inhalation devices. *PLoS One*. 2018; 13(2): e0193082.
97. Hodder R and Price D. Patient preferences for inhaler devices in chronic obstructive pulmonary disease: Experience with Respimat® Soft Mist™ Inhaler. *International journal of chronic obstructive pulmonary disease*. 2009; 4: 381.
98. Lavorini F. The challenge of delivering therapeutic aerosols to asthma patients. *ISRN Allergy*. 2013; 102418.
99. Dalby R, Spallek M, and Voshaar T. A review of the development of Respimat® Soft Mist™ Inhaler. *International Journal of Pharmaceutics*. 2004; 283(1–2): 1–9.
100. Longest PW and Hindle M. Evaluation of the Respimat Soft Mist inhaler using a concurrent CFD and in vitro

approach. *Journal of Aerosol Medicine and Pulmonary Drug Delivery*. 2009; 22(2): 99–112.

101. Grgic B, Finlay W, Burnell P, and Heenan A. In vitro intersubject and intrasubject deposition measurements in realistic mouth–throat geometries. *Journal of Aerosol Science*. 2004; 35(8): 1025–1040.
102. Finlay WH. Chapter 5 - Introduction to the respiratory tract. In: Finlay WH, ed. The Mechanics of Inhaled Pharmaceutical Aerosols (Second Edition). Academic Press, London. 2019; 103–116.
103. Agnihotri V, Ghorbaniasl G, Verbanck S, and Lacor C. On the multiple LES frozen field approach for the prediction of particle deposition in the human upper respiratory tract. *Journal of Aerosol Science*. 2014; 68: 58–72.
104. Weibel ER, Sapoval B, and Filoche M. Design of peripheral airways for efficient gas exchange. *Respiratory Physiology & Neurobiology*. 2005; 148(1–2): 3–21.
105. Ahookhosh K, Pourmehran O, Aminfar H, Mohammadpourfard M, Sarafraz MM, and Hamishehkar H. Development of human respiratory airway models: A review. *European Journal of Pharmaceutical Sciences*. 2020; 145: 105233.
106. Breuer M, Baytekin HT, and Matida EA. Prediction of aerosol deposition in 90° bends using LES and an efficient Lagrangian tracking method. *Journal of Aerosol Science*. 2006; 37(11): 1407–1428.
107. Longest PW and Xi JX. Effectiveness of direct Lagrangian tracking models for simulating nanoparticle deposition in the upper airways. *Aerosol Science and Technology*. 2007; 41(4): 380–397.
108. Bass K and Worth Longest P. Recommendations for simulating microparticle deposition at conditions similar to the upper airways with two-equation turbulence models. *Journal of Aerosol Science*. 2018; 119: 31–50.
109. Zhang Y, Finlay WH, and Matida EA. Particle deposition measurements and numerical simulation in a highly idealized mouth-throat. *Journal of Aerosol Science*. 2004; 35(7): 789–803.
110. Zhang Y, Chia TL, and Finlay WH. Experimental measurement and numerical study of particle deposition in highly idealized mouth-throat models. *Aerosol Science and Technology*. 2006; 40(5): 361–372.
111. Huang F, Zhu Q, Zhou X, et al. Role of CFD based in silico modelling in establishing an in vitro-in vivo correlation of aerosol deposition in the respiratory tract. *Advanced Drug Delivery Reviews*. 2021; 170: 369–385.
112. Vulović A, Šušteršič T, Cvijić S, Ibrić S, and Filipović N. Coupled in silico platform: Computational fluid dynamics (CFD) and physiologically-based pharmacokinetic (PBPK) modelling. *European Journal of Pharmaceutical Science*. 2018; 113: 171–184.
113. Cabal A, Jajamovich G, Mehta K, Guo P, and Przekwas A. In-silico lung modeling platform for inhaled drug delivery. *Proceedings of the Drug Delivery to the Lungs*. 2016; 27: 82–86.
114. Yu JY and Rosania GR. Cell-based multiscale computational modeling of small molecule absorption and retention in the lungs. *Pharmaceutical Research*. 2010; 27(3): 457–467.
115. Kannan RR, Singh N, and Przekwas A. A compartment-quasi-3D multiscale approach for drug absorption, transport, and retention in the human lungs. *International Journal for Numerical Methods in Biomedical Engineering*. 2018; 34(5): e2955.
116. Kannan RR, Singh N, and Przekwas A. A quasi-3D compartmental multi-scale approach to detect and quantify diseased regional lung constriction using spirometry data. *International Journal for Numerical Methods in Biomedical Engineering*. 2018; 34(5): e2973.
117. Kannan R and Przekwas A. A multiscale absorption and transit model for oral drug delivery: Formulation and applications during fasting conditions. *International Journal for Numerical Methods in Biomedical Engineering*. 2020; 36(3): e3317.
118. Bui VKH, Moon J-Y, Chae M, Park D, and Lee Y-C. Prediction of aerosol deposition in the human respiratory tract via computational models: A review with recent updates. *Atmosphere*. 2020; 11(2): 137.
119. Biancofiore F, Busilacchio M, Verdecchia M, et al. Recursive neural network model for analysis and forecast of PM10 and PM2. 5. *Atmospheric Pollution Research*. 2017; 8(4): 652–659.
120. Grivas G and Chaloulakou A. Artificial neural network models for prediction of PM10 hourly concentrations, in the Greater Area of Athens, Greece. *Atmospheric Environment*. 2006; 40(7): 1216–1229.
121. de Matas M, Shao Q, Richardson CH, and Chrystyn H. Evaluation of in vitro in vivo correlations for dry powder inhaler delivery using artificial neural networks. *European Journal of Pharmaceutical Science*. 2008; 33(1): 80–90.
122. Muddle J, Kirton SB, Parisini I, et al. Predicting the fine particle fraction of dry powder inhalers using artificial neural networks. *Journal of Pharmaceutical Sciences*. 2017; 106(1): 313–321.

12

Effect of Aerosol Devices and Administration Techniques on Drug Delivery in a Simulated Spontaneously Breathing Pediatric Tracheostomy Model

Amelia Alberts[1] and Charles Preuss[2]
[1]Graduate & Postdoctoral Affairs, University of South Florida Morsani College of Medicine, Tampa, Florida, USA
[2]Department of Molecular Pharmacology & Physiology, Morsani College of Medicine, University of South Florida, Tampa, Florida, USA

12.1 Introduction

The effectiveness of aerosol drug delivery in pediatric patients is contingent on several factors: the anatomy of the developing lungs, the physiology of the disease, the pharmacology of the drug being dispensed, and the mode and technique of drug administration. From this, it is clear that optimization of aerosol drug delivery requires clinical understanding, as well as knowledge of proper devices and administration techniques. This chapter focuses on aerosol drug delivery devices and modes of administration in pediatric patients, supported by recent research using simulated, spontaneously breathing pediatric tracheostomy models. Spontaneously breathing tracheostomy models allow researchers to mimic the respiratory patterns of pediatric patients. For most of the research discussed, spontaneous breathing was simulated using tracheostomy tubes, a training lung with a lift bar, and a connected ventilator. The ventilator cycles, filling up the trigger chamber, and the lift bar simulates inspiration by lifting the test chamber. Expiration is passive and resistance levels can be set. Breathing parameters varied between studies called upon during this chapter. In general, when mimicking pediatric patients, studies had an average respiratory rate of 25 breaths/min, tidal volume of 150 ml, inspiratory time of 0.8 seconds, and peak inspiratory flow of 20 l/min (1). From this, recommendations on delivery will be provided, with special considerations that should be taken for pediatric patients.

There are many potential benefits of using inhalation drug therapy rather than oral or intravenous. Aerosol delivery is less invasive, smaller doses can be used to achieve the same therapeutic effect, and the onset of the effect can be more rapid. In addition, delivery into the lungs exposes the drug to a rich supply of blood. Lung delivery avoids an initial first pass through the liver or stomach, potentially avoiding adverse reactions or metabolic inactivation of drug factors. Not only is aerosol inhalation the predominant portal of entry of drugs for pulmonary diseases, but lung delivery can also be considered to treat some systematic issues (2). Optimizing the delivery of aerosol drugs will contribute to long-term artificial airway treatments in patients with tracheostomies as well as the acute and chronic treatment of respiratory diseases in spontaneously breathing children.

12.2 Anatomical and Physiological Factors

Initially, aerosol drugs can only reach regions that encounter gas flow. This means that the anatomy and physiology of the lungs have a strong effect on the delivery of an inhaled drug. The branching of the respiratory tract increases the surface area available for absorption of the drug, from the air into pulmonary epithelium. This enhances the delivery efficiency of drugs inhaled into the pharynx, larynx, bronchi, bronchioles, alveoli, and finally capillaries, which then transport the drug to its target tissue.

Deposition is when the gaseous drug converts into a solid form. The goal of treatment is for this solid to then be absorbed into the respiratory tissue. The specific region that deposition occurs in can affect the absorption and thus therapeutic effectiveness of the treatment. Deposition can occur in the epithelium of the mouth, nose, pharynx, trachea, bronchi, and alveoli. When drugs are deposited in the upper respiratory tract much is lost or exhaled and only a fraction of drug will reach target tissue. Drug delivery devices are constantly being innovated to optimize this by depositing drugs further down the respiratory tract. Pressure, temperature, and moisture can affect drug deposition by directly modifying the active pharmaceutical ingredients or the ability of the respiratory cells to absorb the dissolved drug (3). This means that the type of device and drug chosen will greatly influence the deposition rate that occurs in target tissue. Factors that affect drug deposition must be considered in order to optimize delivery of the drug. Another factor that will affect drug deposition is the physiology of breathing.

DOI: 10.1201/9781003046547-12

Measurement of the movement of incoming air or administered drugs is termed *flow rate*. Peak inspiratory flow rate can be used to measure the deposition abilities of certain devices, such as dry powder inhalers and pressurized metered-dose inhalers. Air flow resistance is the force that opposes the movement of air in or out of the lungs. The work of breathing is the work a patient must exert to breathe spontaneously. In other words, the physiologic force of air needed to oppose the resistance of air flow. When choosing an aerosol drug delivery device, consider one that decreases resistance and thus imposed work of breathing, while also promoting high levels of deposition into the desired respiratory tissue.

Physiologic work can be measured using esophageal balloons and a Campbell diagram to integrate work from the balloon's pressure measurements (4). The imposed work of breathing is an important measurement when dealing with inhalation drug therapy because it describes the work a patient must exert to breathe spontaneously through a machine, such as an inhaler or ventilator. This is the work needed to oppose the resistance of the respiratory apparatus itself. The total work of breathing is the summation of physiologic and imposed work. Imposed work exceeding physiologic work may lead to an increase in respiratory muscle load, which can lead to muscle fatigue, especially in pediatric patients.

The issue of muscle fatigue can be avoided by minimization and monitoring of imposed work. Transducers, polygraph recorders, X-Y plotters, and other technology can be used to measure imposed work, but operating these devices requires training (4). This is why computerized pediatric respiratory monitors, such as the CP-100 Pediatric, Bicore Monitoring Systems, are used. Validation studies have found these devices to accurately measure imposed work, tidal volume, flow rate, and tracheal airway pressure during spontaneous ventilation. The device that is chosen needs to be set in order to support ventilation and keep imposed work as close to zero as possible. Some ways to decrease imposed load include larger diameter endotracheal tubes, humidifiers, positive airway pressures, and expiratory valves (4).

12.3 Devices

Nasal sprays are a simple way to deliver drugs to the upper respiratory system or sinuses (5). Although they are simple and can be effective for some drugs, these devices are restricted as to the formulations they can deliver. Antibiotics, mucolytics, liposomal, and recombinant drugs are difficult to safely and efficiently delivery with inhalers or sprays. Examples of respiratory devices for drug delivery are shown in Table 12.1.

Another common therapeutic aerosol delivery system is the inhaler. Different types of inhalers include pressurized metered-dose, dry powder, and soft mist. Pressurized metered-dose inhalers (pMDI) administer the drug through a mouthpiece, using a pressurized propellant. pMDI are compact, quiet, fast, and inexpensive. The drug itself is typically sealed and protected from the environment and the dosing is easily reproducible. That being said, the types of drugs pMDI can deliver are more limited than other delivery modalities. There is also a high level of oropharynx drug deposition when using this technique, which is sometimes associated with less efficient absorption. Finally, it may be difficult to synchronize patient breathing with the device, especially with very young patients. Using spacers and valve holding chambers can mitigate these downfalls. When using a pMDI, patients can either use an open-mouth technique, closed-mouth technique, or a spacer. A spacer captures aerosol to decrease oropharynx deposition. The valve holding chamber manually synchronizes actuation of the drug with the patient's inhalation (1).

Dry-powder inhalers (DPI) deliver drug in a powdered form, typically paired with an actuated dosing system. DPI are portable and breath actuated, meaning they do not use propellants like pMDI, but administer medication using the patient's inspiratory flow. DPI are easy to use and do not require synchronization of breathing and thus have no need for spacers. On the other hand, DPI require more force and effort from the patient which can lead to muscle fatigue, especially in younger patients. The harder and deeper patients are able to inhale, the more medication that can be delivered. This means that dosing can be inconsistent, and the drug is susceptible to environmental effects like humidity and temperature (9).

The final type of inhaler this chapter will discuss is soft mist inhalers (SMI). SMI create clouds of medication with a higher concentration of particles than MDI or DPI. No propellant is needed, and lower doses can be used because the mist is produced at a slower speed, reducing oropharynx deposition. Some research has linked SMI administration of tiotropium with a higher risk of death in people with chronic obstructive pulmonary disease (COPD) (23).

The development of nebulizers revolutionized the delivery of aerosol drugs by allowing quick and efficient transformation of liquid drug into an inhalant. Small-volume nebulizers (SVN) are typically categorized as jet, ultrasonic, or mesh. SVN are powered by compressed air, oxygen, a compressor, and an electricity source to generate aerosol from a liquid drug (9).

Jet nebulizers (JN) are a common choice for the treatment of many pulmonary diseases but can be inconvenient for patients because they are not optimally portable and require a power source to function. JN use compressed gas to shoot a drug-containing aqueous solution into a stream of droplets via a mouthpiece or mask. This makes it an option for children of any age group and cooperation level, because there is no advanced technique required to use the device. Many different types of drugs can be delivered with JN, at considerably high doses. On the other hand, JN are sometimes avoided because they can be bulky, susceptible to contamination, and can be wasteful due to low delivery efficiency (1).

Ultrasonic nebulizers are more discreet and efficient, but are restricted as to what kind of drugs can be used because the heat produced during administration can denature protein-based drugs. Mesh nebulizers use micropump technology and low-frequency waves that avoid this heat and are thus able to deliver more sensitive, protein-containing suspensions (6). Vibrating mesh nebulizers (VMN) contain a plate that vibrates to generate aerosol particles. VMN must be handled with care and cleaned regularly to avoid blockage. VMN are discreet, portable, and have fast nebulizer times, meaning more effectiveness at converting liquid medicine into the fine spray necessary for absorption (1).

TABLE 12.1

Respiratory Devices for Drug Delivery

Device	Typical Appearance	Description
Pressurized metered dose inhaler (pMDI)	A handheld, portable device with a mouthpiece on the bottom and a formulation container on the top. The mouthpiece is covered by a lid that must be removed prior to use.	Administers the drug through a mouthpiece, using a pressurized propellant. pMDI are quiet, fast, and inexpensive. Sometimes known as a "puffer".
Soft mist inhaler (SMI)	Similar to the pMDI, the SMI has a mouthpiece, but it may be on the top or bottom. On the opposite end of the mouthpiece is a drug-containing cartridge. There is a dose release button and capillary tube with a spring in order to deliver formulations.	Create clouds of medication with a higher concentration of particles than MDI or DPI. No propellant is needed and lower doses can be used.
Dry-powder inhaler (DPI)	DPI come in several different designs, but typically are either round or straight with a capped mouthpiece, protective screen leading to a capsule chamber on the inside, and button for aerosol release. DPI come in single and multi-dose designs.	Delivers drug in a powdered form, typically paired with an actuated dosing system. DPI do not use propellants like pMDI, but administer medication using the patient's inspiratory flow.
Jet nebulizer (JN)	JN come with a mouthpiece or mask, connected to a drug-containing chamber, attached to a tube, which is connected to and powered by an air compressor.	Uses compressed gas to shoot a drug-containing aqueous solution into a stream of droplets.
Ultrasonic nebulizer	These nebulizers are smaller and more portable than the JN. Ultrasonic nebulizers also use mouthpieces or masks, but instead of tubing they are directly connected to the power source and formulation chamber.	Discreet and efficient, but are restricted in what kind of drug can be used because the heat produced during administration can denature protein-based drugs.
Vibrating mesh nebulizer (VMN)	VMN visually resemble ultrasonic mesh nebulizers but contain a membrane with many small holes on top of the liquid reservoir to disperse the formulation droplets.	Uses micropump technology and low-frequency waves that avoid heat and are thus able to deliver more sensitive, protein-containing suspensions.
Breath-enhanced nebulizer (BEN)	BEN have a chamber with a mouthpiece and liquid formulation container. This chamber is connected to tubing which connects to a power source.	Continues to deliver aerosol during exhalation as well as inhalation.
Breath-actuated nebulizer (BAN)	BAN look similar to BEN, except BAN contain a vent with a valve to allow air flow upon inspiration, but not expiration.	Delivers medication during inhalation only.

Breath-actuated nebulizers are part of a new, potentially more efficient generation of jet nebulizers that expel medication only during inspiration (7). The difference between breath-enhanced and breath-actuated nebulizers is that breath-actuated delivers medication during inhalation and breath-enhanced continues to deliver aerosol during exhalation as well. The administration of medication during exhalation can result in waste, because the patient is not inhaling the drug into respiratory tissue that the drug can optimally be deposited and absorbed into. A 2014 study using adult breathing models found that breath-actuated nebulizers delivered a more consistent dose, which allowed more confidence in titrating to lower effective doses. The same study also found that there was a reduced risk of under-dosing during disease progression with breath-actuated nebulizers (8,9).

VMN deliver higher inhaled mass percentage in adult models, and jet nebulizers (JN) deliver a lesser mass percentage in pediatric models (1). One study comparing JN, breath-enhanced nebulizers (BEN), breath-actuated nebulizers (BAN), manually triggered nebulizers (MTN), and vibrating mesh nebulizers (VMN) found that the VMN delivered the greatest inhaled drug dose (35.5%) and lowest residual dose (2.8%), and the JN and BENs delivering the lowest inhaled dose (15.0% and 17.7% respectively) and highest residual (62.3% and 66.2%). MTN and BAN had the lowest exhaled doses (1.7% and 2.7%), JN had the highest exhaled dose (15%), and VMNs resulted in relatively high exhaled dose as well (11.1%). This study was done using an in vitro and ex vivo adult model, so the lack of pediatric considerations in this data must also be noted (10).

As previously mentioned, humidification and temperature are two major factors affecting deposition and therefore absorption of the drug. Normally, the upper airway (mouth, pharynx, larynx) contributes moisture. This is bypassed in the long-term aerosol drug delivery in patients with tracheostomies. Special considerations, such as combining nebulizer use with humidification therapy, are considered for these patients. Three major types of humidification devices are typically used: heated, unheated, and heat-and-moisture exchangers (HME). A 2016 study on aerosol drug delivery and humidification therapy found that unheated humidifiers provided for more efficient drug delivery than heated, but that HME demonstrated the most efficiency. Specifically, the best drug delivery was seen in HME with passive exhaled humidification. This study also compared MTN and JN, finding MTN more effective (11).

12.4 Techniques

There is variation in the performance and convenience of the different drug delivery devices, depending on the patient and drug. Patients with artificial airways, like tracheostomies and endotracheal tubes, often require the administration of aerosol drugs. This may have an effect on drug deposition, and minimization of absorption barriers requires additional considerations.

When using a jet nebulizer, a T-connector and tracheostomy collar can be used. One study comparing the albuterol delivery of different types of nebulizers with a spontaneously breathing pediatric tracheotomy model found that using a jet nebulizer with a T-piece and resuscitation bag resulted in higher deposition than without. Other studies have shown T-pieces are more efficient than tracheostomy masks (1).

Aerosol drugs can be administered assisted or unassisted. In other words, via a manual resuscitation bag or directly. A study by Berlinski (12) showed that using a manual resuscitation bag and T-piece with a tracheostomy tube results in the highest inhaled dose. They compared albuterol delivery with a JN alone, with a corrugated tube, and with a corrugated tube and resuscitation bag. The study also compared T-pieces versus tracheostomy masks. The highest drug delivery was found with JN in conjugation with resuscitation bags and T-pieces. Positioning of the T-piece may also result in more or less efficient deposition of drug, but more research on this is needed (12).

In another study, pMDI was found to be more efficient when the T-piece was placed proximal to the spacer. This study simulated the unassisted technique by connecting the T-piece of a nebulizer to another T-piece, which was connected to a tracheostomy tube (TT). Corrugated tubing was used to connect the T-pieces and a pMDI canister was placed into the nozzle inlet of the spacer, connecting it to the TT. The assisted set-up involved jet and vibrating nebulizers attached to 450 ml pediatric manual resuscitation bags via T-piece adapters and corrugated tubing. Inhaled mass percent of nominal and emitted dose was used to determine efficiency. Nominal dose is the amount of drug that is contained within the reservoir of the nebulizer. Emitted dose is the amount that actually leaves the nebulizer as aerosol (1).

Frequent and long-term administration of pMDIs directly into patients' tracheostomy is not recommended unless an AeroChamber is used, as well. Without the chamber addition, that method has been shown to lead to hemoptysis and granulation tissue formation in bronchi. It is possible to adapt a volumetric spacer to fit a pediatric tracheostomy tube by cutting the barrel of a standard bladder irrigation syringe and attaching it to the outflow end of the spacer and the tube (1).

A study comparing aerosol delivery with a soft mist inhaler paired with a non-electrostatic valved holding chamber and deadspace volume mask reported transnasal delivery was more efficient compared to nasal and oral modalities for 5- and 14-month-olds (12). Another study compared transnasal albuterol delivery between VMN and continuous-output JN for different breathing parameters, set to mimic 7-month and 5-year-old children for tidal volumes of 25, 50, and 155 ml. Breathing frequencies were set to 40, 30, and 25 breaths/min, respectively. Inspiratory to expiratory ratios (I:E) of 1:3 and 1:2 were used. This study found that lung doses were larger for both age groups simulated when the JN was used transnasally. Lung dose is the portion of the nominal dose that is actually delivered to lung tissue. The JN produced 0.51%, 1.05%, and 0.97% lung doses on average, when delivered transnasally with the 7-month old model. The JN reached 0.44% and 1.14% for 50 and 155 ml tidal volumes, respectively, for the 5-month old model. As with the 5-month old JN model, VMN was not able to produce a measurably significant lung dose with the 25 ml tidal volume. With VMN, the 7-month-old model recorded 0.13% and 0.87% lung dose for 50 and 155 ml, and the 5-month-old model reached only 0% and 0.42%. The same study also showed that using a tightly sealed loading masking increased JN delivery significantly at lower tidal volumes (12).

A cascade impactor is used to measure distributed particulate size, which is important in determining the effective diameter of the nozzle of an administration device. Researchers can use this technology to determine what route of administration is best for certain particulate sizes. In general, the smaller the particulate size, the higher the lung dose recorded. One study found that the nasal route reported a higher percentage of deposition for particles smaller than 5 μm (97% vs 90%) and between 1 and 3 μm (49% vs 38%) when compared to the oronasal route (12).

Tidal volume is the amount of air that is displaced during non-forced inhalation and exhalation. For children, this is between 5 and 8 ml/kg of their body weight (13). One study found that increasing tidal volume during inhalation resulted in decreased lung dose. The study used budesonide administered via a pMDI with an 9-month-old model. The same study also showed lung dose was significantly higher with hydrofluoroalkane beclomethasone compared to chlorofluorocarbon beclomethasone (12).

Placement and size of nasal cannulas have also been shown to have an effect on flow and subsequent delivery of aerosolized drug. The cannula is a thin tube used to administer medication from the reservoir to the mouthpiece for patient inhalation. A study compared inspired dose percentages between three typical cannula sizes: infant, pediatric, and adult. The study found that the smaller the cannula, the lower the inspired dose. The study also found that increasing the flow rate from 3 and 5 to 10 l/min significantly decreased inhaled dose to almost nothing. While size of the cannula cannot always be modified, flow rate can and should be considered (12).

12.5 Pathological Considerations

In general, it is important to consider a child's ability to dependably operate and consistently use their medical inhalation device. A simple, easily transportable design is ideal, but optimization of drug delivery must be considered as well. The intricacies of the child's specific disease, unique living environment, and mental capacity should be taken into consideration before prescribing a treatment. First- or secondhand tobacco smoke, pollutants, infection exposure, nutrition, and other lifestyle attributes may play a role in the prognosis of chronic respiratory diseases and should also direct the treatment planning.

Lung disease can be categorized as either obstructive or restrictive. Obstructive lung diseases like asthma, emphysema, COPD, and cystic fibrosis make it difficult for patients to exhale air from their lungs. Medications that reduce the inflammation and narrowing of the airways can be used to treat symptoms. Inhalation medications that reduce narrowing by relaxing the smooth muscle of the respiratory tract are known as bronchodilators and include albuterol, ipratropium, formoterol, salmeterol, and tiotropium. Medications that target inflammation of the airways are typically corticosteroids, e.g. beclomethasone (14).

Restrictive pulmonary diseases are caused by a reduced ability to expand the lungs, such as with interstitial lung disease, sarcoidosis, some neuromuscular diseases, and some cases of scoliosis. Stiffness of the lungs or chest wall, muscle weakness, and damaged nerves can cause these diseases. Medications vary with the central cause, but therapies often involve anti-inflammatories. Both obstructive and restrictive lung diseases are associated with shortness of breath upon exertion and coughing (14).

With children, one of the most common forms of obstructive lung disease is asthma. The condition is chronic and often manifests in childhood, with symptoms that range from mild to severe. When approaching the treatment of asthma, medications are either for short-term relief or control. Short-term relief medications are used to prevent or relieve the symptoms of an asthma attack and include short-acting beta-2 agonsits (SABA), corticosteroids, or short-acting anti-cholinergics. Albuterol, a SABA, is one of the most commonly prescribed rescue medications. The delivery of albuterol using nebulizers with high gas flow has been found to be inefficient by some studies (1). Control medications may also be prescribed to these patients, intended for daily intake to help prevent symptoms even when an asthma attack is not imminent. Control medications include corticosteroids, monoclonal antibodies, mast cell stabilizers, inhaled long-acting bronchodilators (LABAs), and subcutaneous immunotherapy (14).

Cystic fibrosis is an inherited disease with variable penetrance that leads to thickening of mucus. Individuals with this disease may have issues breathing and a higher susceptibility to pulmonary infection resulting from the defective clearance of respiratory mucus. The ability to deliver anti-microbials quickly and effectively is of utmost importance with these patients. *Pseudomonas aeruginosa* is a frequent infection with these patients and can be treated with aztreonam in children older than 7 and tobramycin in children older than 6 (15).

Mucolytics are sometimes used in patients with cystic fibrosis or others that have mucus hypersecretion and retention. This pathology can also result from ciliary disease or from mucus-producing tumors that cause over-proliferation of goblet cells and submucosal glands. Mucolytics are sometimes called mucoactive agents and degrade polymer gels to promote clearance of sputum and a decrease of mucus hypersecretion. Aerosolized N-acetylcysteine uses its free sulfhydryl group to hydrolyze disulfide bonds in mucin and other proteins. *N*-acetylcysteine is a commonly use mucolytic but has a couple of issues, such as being inactivated at the airway surface and pre-systemically metabolized, lowering the concentration absorbed. It is also associated with a foul odor and airway irritation. Peptide mucolytics like dornase alfa can degrade polymerized collections of DNA and actin that may form during prolonged airway inflammation. Dornase has only shown to be effective in treating cystic fibrosis patients. Bland aerosols and bicarbonate can be used to produce an effective cough, but this is by irritating airway tissue and thus is not recommended (16).

Expectorants or secretagogues trigger coughing and the clearance of mucus by increasing water concentration in the airway. One example is hyperosmolar 7% saline and mannitol, an inexpensive solution that has been shown effective in patients with cystic fibrosis. It works due to mannitol's ability to attract water and secretions into the airway, and can sometimes trigger bronchospasms. P2Y2 purinergic pathway agonists, like denufosol, can promote chloride transport through functional membrane channels, also helpful with patients with cystic fibrosis. The use of an epithelial sodium channel inhibitor like amiloride is not recommended because it has not been shown effective and has been connected with exacerbated lung function (16).

Hospital-acquired and ventilator-associated pneumonia can occur with pediatric patients who are already hospitalized, undergoing ventilation therapy, and contract a respiratory bacterial infection. Inhaled antibiotics are recommended for patients not responding to intravenous medication or with gram-negative bacilli, bacteria that are only susceptible to aminoglycosides or polymyxins. Food and Drug Administration (FDA)-approved and commercially available inhalation antibiotics for this form of pneumonia include aztreonam, tobramycin solution, and tobramycin powder (15).

Other common causes of respiratory infection in children include general lower respiratory infections, bronchiolitis, influenza, respiratory syncytial viruses, the common cold, croup, streptococcal pharyngitis (also known as strep throat), and viral pneumonia. Four FDA-approved antivirals include oseltamivir phosphate, zanamivir, peramivir, and baloxavir marboxil. Croup, strep, and many cases of lower respiratory infections are bacterial and are often only treated with antibiotic inhalation therapy for severe cases (17).

The lungs are one of the last organs to mature in developing children. One issue in premature babies may be a collapse of alveoli due to a lack of surfactant. Surfactant keeps the alveoli open, allowing for gas exchange to occur. Surfactant replacement therapy can be used to prevent respiratory distress in neonates, when paired with supplementation of oxygen or ventilation. Surfactant replacement is a mixture of phospholipids and proteins that coat the alveoli to reduce surface tension and prevent atelectasis. It can be given prophylactically or as a rescue treatment and has been shown to reduce infant mortality and respiratory morbidity. Surfactant treatment may also be considered with neonates suffering from severe meconium aspiration syndrome, pulmonary hemorrhage, or severe respiratory syncytial virus-induced failure (18).

Aerosolized drugs have also been manufactured to deliver gene and peptide therapies. There are several different mechanisms these drugs work by. Some inhaled agents of gene therapy contain nucleic acids that can block abnormal gene production. Inhalation gene therapies are currently being investigated to treat patients with lung cancer, *Mycobacterium tuberculosis* infection, alpha-1 antitrypsin deficiency, and cystic fibrosis. Other medications may contain peptides that can be delivered directly to lung tissues, in order to treat pulmonary disease such as asthma, sarcoidosis, pulmonary hypertension, and cystic fibrosis. Administration of peptide drugs that target systemic circulation, such as insulin and calcitonin, are also being considered for inhalation therapies. Viral vectors, plasmids, cationic molecules, and interfering RNAs are among some of the formulations used to deliver those categories of drugs (19).

Some children with hemodynamic or circulatory instability require treatments that increase cardiac output and vascular resistance. Systemic vasoactive medication taken orally or intravenously may have minimal selectiveness and therefore cannot target pulmonary epithelium unless delivered via aerosol. If blood pressure issues persist after the use of fluid resuscitation, inhalation therapy may be considered. Vasopressor drugs may help treat these patients by increasing contractility and heart rate or vasoconstricting peripherally. Vasodilators can also help by controlling vascular resistance. For instance, the drugs typically used to treat pediatric pulmonary hypertension target pulmonary vascular resistance and can result in either vasoconstriction or vasodilation. Vasopressors, like endothelin-1 receptor A agonists and hypoventilation, may be beneficial, depending on the patient's specific disease. Vasodilators such as nitric oxide, phosphodiesterase type V inhibitors, endothelin-1 receptor B agonists, prostanoids, rho-kinase inhibitors, and serotonin antagonists may also be used in inhalation therapy as well (20).

The use of a helium–oxygen mixture called Heliox can be used as a treatment for pediatric patients with upper and lower airway obstructions. The reduced density of the helium and oxygen mixture provides laminar flow within the airway and reduces the work of breathing. It has been used as an adjunctive treatment in children with bronchiolitis, asthma, and croup. One study found that using Heliox: 80% helium and 20% oxygen improved delivered dose at high flows when compared to oxygen alone. This study used a non-anatomically correct model with pediatric patterns of 100 ml tidal volume, 20 breaths/min frequency, and a ventilation with an I:E ratio of 1:2 (12). That being said, recent reviews show that the use of Heliox correlates with no significant reduction in rate of intubation, no reduced rate of discharge, and no decrease in length of treatment except with infants who initially started with nasal continuous positive airway pressure (NCPAP) due to severe respiratory distress. The same report suggests that the addition of Heliox therapy is recommended only in infants with severe RSV bronchiolitis, during the first hour of treatment (21).

12.6 Pharmalogical Considerations

Preparation of the drug, such as drug particle size and the speed of delivery, can affect deposition and thus drug delivery. Altering these properties of the drug can affect the delivery system by changing inertial impaction, gravitational sedimentation, and diffusion. The smaller the particle, the farther it can travel down the respiratory tract and the more likely it is to participate in diffusion in the lower airway. Larger particles (>3 μm) are more likely to deposit via inertial impaction in the conducting zone, such as the oropharynx. The mass median diameter quantifies particle size in polydisperse aerosols, or medications with a mix of particle sizes (9).

Larger doses and smaller particles are typically delivered with more efficiency. That being said, because each inhalation device operates differently, there will be a variation in optimum particle size, respirable dose, and deposition. One drug may have several different formulations and can be delivered with different devices. Manufacturers of these drugs do not always provide recommendations on the specific formulation and device to use.

Examples of some medications available for inhalation therapy are listed in Tables 12.2–12.5.

Some common drug formulations include lactose carrier systems, liposomes, porous particles, and biodegradable polymers. These formulations are molecules designed to facilitate better delivery and absorption of the drug into target tissue. Lactose carrier systems involve mixing the drug with a sugar carrier molecule in order to improve the flow of drug from the administration device. Flow of drugs may be hindered when the surface electric forces of the powder exceed the gravitational forces that aid in deposition and absorption. The carrier molecule contains an active site that binds the drug particles but is also able to separate when the drug is administered, to allow the drug to be released into target tissue. Potential side effects may result if the lactose does not separate from the drug upon aerosol generation. This may result in unintentional deposition into the oropharyngeal region, which can cause adverse reactions when using corticosteroids. Dry powder inhalers often contain either the drug alone or in a mixture with a bulky carrier molecule, such as α-lactose monohydrate. Lactose, glucose, and mannitol are sugars approved for use as drug carriers by the FDA. For use in drug formulations, these carrier molecules can be generated as either fine or coarse particles, or a mix of both. While drug particles will bind to both fine and coarse particles, there are some differences in the outcome when using them. Therapeutic

TABLE 12.2

COPD Drugs That Can Be Delivered by MDI (22,23)

Steroids	Bronchodilators	Combination Steroid/Dilator
Beclomethasone	Albuterol	Budesonide-formoterol
Ciclesonide	Levalbuterol	Fluticasone-salmeterol
Fluticasone	Ipratropium	Mometasone-formoterol

TABLE 12.3

COPD Drugs That Can Be Delivered by DPI (22)

Steroids	Bronchodilators	Combination Steroid/Dilator
Budesonide	Albuterol	Fluticasone-vilanterol
Fluticasone	Salmeterol	Fluticasone-salmeterol
Mometasone	Tiotropium	

TABLE 12.4

Corticosteroids by DPI, MDI, or SVN (9)

Drug	Device
Beclomethasone	pMDI (≥5 YO)
Budesonide	SVN or DPI (≥6 YO)
Ciclesonide	pMDI (≥12 YO)
Flunisolide	pMDI
Fluticasone propionate	DPI or pMDI
Fluticasone furoate	DPI
Mometasone furoate	pMDI or DPI

TABLE 12.5

Other Common Drugs, Devices, and Uses (9)

Drug	Device	Potential Use
Zanamivir	DPI	Influenza A and B (≥5 YO)
Ribavirin	SPAG	Respiratory syncytial virus
Tobramycin	SVN or DPI	*P. aeruginosa* in cystic fibrosis (≥6 YO)
Aztreonam	SVN	*P. aeruginosa* in cystic fibrosis (≥7 YO)
Cromolyn sodium	SVN	Bronchial asthma prophylaxis (≥2 YO)
Mannitol	DP	Diagnostic bronchoconstrictor
Dornase Alpha, N-Acetylcysteine	SVN	Mucoactivity in cystic fibrosis
Hyperosmolar saline	SVN	Mucoactive

*SPAG is an abbreviation for *small-particle aerosol generator* and is a specialized large-volume nebulizer for the delivery of ribavirin, YO = years old

aerosol mixtures with higher ratios of fine lactose particles have been found to improve disaggregation. This is because fine particles have less surface roughness, allowing the drug to separate more readily. This leads to higher respirable fractions, meaning more efficient drug delivery. Flowability of the drug is higher when using coarse lactose particles. The addition of ternary agents to lactose formulations, such as l-leucine, is hypothesized to increase dispersibility of aerosolized drugs. L-leucine may be able to occupy some of the drugs' high-energy binding sites on lactose, leaving only low-energy binding sites for the drug itself. This leads to an overall decreased strength of interaction between the drug and lactose, and thus higher respirable fractions (24).

Liposomes can be used to enhance the sustained release of drugs in the airways or alveoli, rather than allowing rapid systemic absorption. The use of liposomes as pulmonary drug vehicles can minimize the risks associated with drugs that cause side-effects when absorbed systemically. Airway-targeted deposition using drug vehicles like liposomes can increase the absorbed dose of drug therapies. This means it may be appropriate to decrease dosing frequency, which can lead to higher levels of drug-use compliance by the patient. Historically, liposomes have been used to improve phospholipid delivery to alveolar tissue of neonates undergoing respiratory distress syndrome. Lately, liposomes are being studied for their usefulness in prolonging the release of drugs used for gene therapy and to treat lung disease. Once liposomes reach the alveoli they are cleared by macrophages. This process is different for liposome-bound drugs than for other inhaled drugs. Liposomal processing occurs in a similar manner as endogenous surfactant. When designing liposome formulations, it is important to consider polymer surface coatings that protect the liposomes from the patient's immune system in order to prolong circulation (24).

Drugs that are lipophilic will intercalate within the lipid bilayers, while lipophobic ones will only contact the interface. Drugs with intermediate solubility are not transported well by liposomes, so are not typically used in these formulations. Drugs that are weak acids or bases can be manipulated to gather in the interior of the liposome, facilitating higher levels of drug retention. The usefulness of liposome formulations in delivering antibiotics to infected lungs is being investigated. Animal studies have shown sustained release and higher susceptibility of bacteria when liposome–drug formulations were used. Liposomal formulations of tobramycin were detected in the lungs of *Pseudomonas aeruginosa* infected mice 16 hours after administration. Free tobramycin only was detectable for 15 minutes. In a similarly designed study, tobramycin was shown to reduce the colony forming units (CFU) in infected rats treated with liposomal tobramycin. The number of bacteria went from 1.4×10^6 to 4.3×10^5 CFU/lung (24). When comparing this to the increase in CFU that was found in rats treated with free tobramycin, it seems clear liposomal formulations have the potential to enhance pulmonary drug delivery and efficacy.

Other animal studies, investigating the effects of liposomal formulations of anti-asthma drugs, also found more sustained-release and fewer systemic effects when using liposomes during delivery. In human trials, liposomes were shown to improve penetration and slow clearance of the anti-asthma drug beclomethasone dipropionate. Because they are lipophilic, one would assume corticosteroids are easily made into liposomal formulations. In actuality, using liposomes with some of these drugs, such as triamcinolone acetonide, has shown to result in no sustained release or improvement in target tissue absorption. Versions of the same drug, such as with hydrophilic pro-drug triamcinolone acetonide phosphate, do show sustained release and longer occupancy in the receptor. Considering the therapeutic value of enhancing drug delivery and sustained release, more research is needed to establish guidelines about which drugs should and should not be incorporated into liposomal formulations (24).

Liposomes can be made from endogenous pulmonary phospholipids submerged in aqueous solution. The aqueous solution becomes entrapped by either one or multiple lipid bilayers, known as unilamellar or multilamellar vesicles, respectively. Unilamellar vesicles can either be large or small. There is another type of liposome which are known as long circulating liposomes, and they typically utilize a hydrophilic polymer. Vesicular properties such as size, fluidity, charge, and method of preparation can influence the chemical behavior of a drug. Liposome size influences circulating half-life and degree of drug encapsulation. Smaller vesicles have a longer half-life and slower release rate because they are opsonized at a lower rate, compared to large vesicles. When designing liposomes for clinical use, a suggested size of 50–200 nm in diameter is recommended in order to avoid phagocytosis. The fluidity of a liposome depends on the specific phase transition temperature (T_C), which depends on the type of fatty acids chains that make up the lipid bilayer. When a bilayer is in an environment below the T_C the lipids are considered to be in the gel phase. Gel phase lipids are organized and rigid. Above the T_C, the lipids are known to be in the fluid phase or with liquid-crystalline state lipids. With this in mind, it logically follows that lipids with a high T_C (above 37 °C) make the liposome more gel-like, less fluid and therefore less leaky. Cholesterol adds to the stability of lipid bilayers and, when used at high concentrations, can make liposomes less leaky as well (24).

Porous particles are a relatively new type of formulation which use Pulmospheres^{TM} to prolong drug circulation. Pulmospheres readily disperse and are made of phosphatidylcholine. Production of these particles is by solvent evaporation and spray-drying technique, using either polymeric or nonpolymeric excipients. Pulmospheres are large and have a low density, allowing for less aggregation and higher respirable fractions. The particles' large size protects the drug from macrophage engulfment, allowing longer time spent in the alveolar region. Using a rat model and cromolyn, only between 8% and 12.5% of macrophages contained Pulmospheres, whereas between 30% and 39% of macrophages had consumed the nonporous form of the drug. Pulmospheres have also been shown to decrease side effects by reducing oropharyngeal deposition. Cromolyn Pulmospheres formulations resulted in a respirable fraction of 68%, compared with 24% when using micronized cromolyn particles. With more research, porous particles may be used to minimize drug dose and maximize therapeutic effect. Porous particles have also been suggested to increase systemic circulation of insulin and testosterone (24).

Biodegradable polymer microspheres are also being investigated as potential drug formulations to promote prolonged sustained release. Polylactic acid (PLA) is used clinically, however, it may not be appropriate for pulmonary inhalation therapy because of its long half-life. Oligolactic acid is an oligomer of lactic acid and may be a better choice because of its shorter half-life. Hydroxypropyl cellulose (HPC) is a mucoadhesive polymer and may be used to reduce mucociliary clearance of beclomethasone dipropionate (BDP). One experiment, using powder aerosol administration on guinea pigs, showed that 180 minutes after drug administration 86% of crystalline BDP with HPC microspheres remained, whereas less than 20% remained of crystalline BDP alone. In this same experiment, eosinophil inhibition was used to measure BDP's activity. The outcomes show BDP/HPC microspheres were active for 24 hours, compared to 1–6 hours with BDP alone. Nanocarriers, like polymer microspheres, show great promise in increasing the sustained release and efficacy of corticosteroids, though research must be done on the potential toxicities of such drug carriers (24).

As you can see, there are many choices when it comes to drug class, formulation, dosing, and administration device. The right combination of choices can lead to a treatment plan that optimizes drug bioavailability and minimizes premature clearance. Differences can exist between drug output, efficiency, and particle size for different nebulizers. One study found a 10-fold difference between delivery systems' efficiency. Another study found nebulizer efficiency ranged anywhere from 30% to <5% of initial dose depending on the nebulizer used. Using a device without understanding its level of efficiency can lead to underdosing. As the quality of device technology, drug design, and delivery formulation improves, so does the quality of therapy. Selecting the proper drug and optimal device is critical in ensuring the appropriate therapeutic dose is delivered to lung tissue (24).

12.7 Pediatric Considerations

Pediatric patients require special consideration due to the wide range of age, size, breathing patterns, and cooperation level (1). Diaphragmatic breathing requires breathing muscles that do not develop until later in infancy. Sometimes it is difficult to administer aerosolized medication using the necessary devices to pediatric patients. For instance, to use a pMDI, the patient must remove the inhaler's cap, shake it with the mouthpiece facing down, breathe out, press the inhaler, and at the same time take a deep breath through the mouth for a few seconds, hold breath for several seconds to absorb the medicine into airways, relax, and repeat for as many puffs as indicated. If the patient inhales the medication too quickly it will hit the back of his/her throat and will not reach the lungs. Synchronizing a child's inhalations to the device can pose a challenge. There is still a need for more research comparing the different types of nebulizers and determining which delivers the drug most efficiently to pediatric patients, depending on the specific disease at hand.

Inhaled medications are at the forefront of pediatric pulmonary disease treatment. As mentioned prior, delivering aerosolized medicine to children is compounded by the rapid anatomical and physiological changes associated with normal growth. Children younger than 18 months are obligate nasal breathers, decreasing the effectiveness of inhaled drugs. One study used computational fluid dynamic methods to compare the parameters between difference age groups and found significant differences in regional deposition when comparing 10-day, 7-month, 3-year, 5-year, and 63-year-old parameters. Knowing the patient's demographics, and choosing the right device and technique can optimize inhaled drug deposition. Patients with asthma, under respiratory distress requiring invasive mechanical or non-invasive ventilation support, and those requiring transnasal support are only a few subsets that can benefit from improved delivery of inhalation drugs (12). Most pediatric studies are done in vitro with a simulated apparatus because radiolabeled deposition and pharmacokinetic studies can be challenging in children (12). Many studies are still done only in adults and should be repeated with parameters set for children. In addition, the studies that are being done lack a consistency between parameters dictated for the same age group.

Some inhaled drugs are used in ways not directly stated as proper usage in order to make them more therapeutically beneficial for different age groups or support devices (12). More research should be done on pharmaceutical guidelines and the optimization of aerosol drug particle size in order to enhance drug deposition in pediatric patients. When reviewing data on drug deposition it is important to account for differences between loading doses or look at absolute doses and percentages for accurate comparison. A final consideration that should be made with pediatric patients is the crucial role education plays for both patients and practitioners. Patients need to know how to deliver the medication themselves or have a legal guardian that will understand and be able to administer the medication when needed. Health care providers need to be aware of what devices exist, and how to use the chosen device properly, so they can educate patients as a part of the treatment plan. Providers should also know the options available and limitations that may exist for their patients that make one device a better choice over another. Some of these limitations include drug and device availability, cost, ease of use, patient cooperation, and acceptability by family (12).

12.8 Conclusion

This chapter aimed to clarify differences between aerosol devices and drug administration techniques for pediatric patients. Several studies using spontaneously breathing tracheostomy models were discussed in order to understand the best way to treat both pulmonary and systemic illnesses using inhalation drug therapy. The proper drug, device, and delivery method depends on the unique way in which the patient's disease presents. This chapter outlines some of options that are available, as supported in current research.

When treating pediatric patients using aerosol drug delivery, the anatomical and physiological differences between children and adults should be noted. Some of the research discussed in this chapter suggests that because children have less developed respiratory systems and muscles, different breathing patterns and mechanisms, as well as unique pharmacological considerations, there are special considerations that should be taken into account when it comes to aerosol drug delivery.

There are many different aerosolized drug delivery devices available, as well as variations in the technique of delivery, which was discussed to help familiarize readers with some of the many choices at hand. Some of the devices covered include pressurized metered-dose inhalers, soft mist inhalers, dry-powder inhalers, jet nebulizers, ultrasonic nebulizers, vibrating mesh nebulizers, breath-enhanced nebulizers, and breath-actuated nebulizers.

At some ages, children may not be able to operate these devices optimally. Some of the research discussed showed that these patients may benefit from the use of assistance in the form of T-pieces paired with resuscitation bags. AeroChambers, holding chambers, volume masks, and cascade impactors, as well as cannula design are other technical features of aerosol drug delivery that were discussed in this chapter, that can be modified to enhance the therapy of pediatric patients.

Several diseases that are sometimes treated using inhalation drug therapy were summarized because it is essential to understand what kind of therapy would best benefit the patient depending on the pathology of the disease. The most common lung disease inhalation therapy is used in children with asthma. Cystic fibrosis is also a major focus of research for inhalation drug therapies. Pneumonia, hemodynamic instability, generalized respiratory distress, and airway obstruction are some of the other conditions discussed in this chapter.

Another important factor to consider is the drug formulation being delivered. While smaller drug particle sizes are typically preferred, molecular drug carrier systems exist and are being researched in order to optimize the delivery of the drug and thus efficiency of therapy. Lactose carriers, liposomes, porous particles, and polymer microspheres were the carrier systems discussed. In general, the studies discussed indicate that carrier systems increase the bioavailability of drugs and thus greater efficiency of drug delivery to the lungs, but more research is needed on these mechanisms and what therapies drug carrier systems would be most useful for.

Overall, it is clear that more research should be done on the usefulness of inhalation drug therapy in terms of drug carrier formulation, device, and technique in order to optimize the way aerosolized drugs are being administered to pediatric patients, as well as to assess the usefulness of this specialized mode of delivery in treating systemic diseases.

REFERENCES

1. Alhamad BR, Fink JB, Harwood RJ, Sheard MM, and Ari A. The effect of aerosol devices and admin tech on drug delivery in a simulated spont breathing pediatric model with a tracheostomy. *Respiratory Care.* 2015 July; 60(7): 1026–1032. http://rc.rcjournal.com/content/60/7/1026.
2. Darquenne C. Aerosol deposition in health and disease. *Journal of Aerosol Medicine and Pulmonary Drug Delivery.* 2012 Jun; 25(3): 140–147. https://www.ncbi.nlm.nih.gov/pmc/articles/PMC3417302/.
3. Ali M. Engineered aerosol medicine and drug delivery methods for optimal respiratory therapy. *Respiratory Care.* 2014 Oct; 59(10): 1608–1610. http://rc.rcjournal.com/content/59/10/1608.

4. Berman LS, Banner MJ, Blanch PB, and Widner LR. A new pediatric respiratory monitor that accurately measures imposed work of breathing: A validation study. *Journal of Clinical Monitoring*. 1995; 11(1): 14–17. https://www.ncbi.nlm.nih.gov/pubmed/7745447.
5. Cheng YS. Mechanisms of pharmaceutical aerosol deposition in the respiratory tract. *AAPS PharmSciTech*. 2014 Jun; 15(3): 630–640. https://www.ncbi.nlm.nih.gov/pmc/articles/PMC4037474/.
6. Ari A. Jet, ultrasonic, and mesh nebulizers: An evaluation of nebulizers for better clinical outcome. *Eurasian Journal of Pulmonology*. 2014; 16: 1–6. https://scholarworks.gsu.edu/cgi/viewcontent.cgi?referer=&httpsredir=1&article=1001&context=rt_facpub.
7. Leung K, Louca E, and Coates AL. Comparison of breath-enhanced to breath-actuated nebulizers for rate, consistency, and efficiency. *Chest*. 2004; 126(5): 1619–1627. https://pubmed.ncbi.nlm.nih.gov/15539736/.
8. Suggett J, Nagel M, Doyle C, Ali R, and Mitchell J. Delivery of medication by breath-actuated nebulizer (BAN) is similar when used with differing inhalation/exhalation ratios: A contrast to breath enhanced nebulizer (BEN) behavior. *European Respiratory Journal*. 2014; 44: 3819. https://erj.ersjournals.com/content/44/Suppl_58/P3819.
9. Gardenhire DS, Burnett D, Strickland S, and Myers TR. A guide to aerosol delivery devices for respiratory therapists. *AARC*. 2017; 4: 1–42. https://www.aarc.org/wp-content/uploads/2015/04/aerosol_guide_rt.pdf.
10. Lin HL, Fang TP, Cho HS, Wan GH, Hsieh MJ, and Fink JB. Aerosol delivery during spont breathing with different types of nebulizers. *Pulmonary Pharmacology and Therapeutics*. 2018; 48: 225–231. https://www.ncbi.nlm.nih.gov/pubmed/29277689.
11. Ari A, Harwood R, Sheard M, Alquaimi MM, Alhamad B, and Fink JB. Quantifying aerosol delivery in simulated spontaneously breathing patients with tracheostomy using different humidification systems with or without exhaled humidity. *Respiratory Care*. 2016 May; 61(5): 600–606. http://rc.rcjournal.com/content/61/5/600
12. Berlinski A. Pediatric aerosol therapy. *Respiratory Care*. 2017 June; 62(6): 662–677. http://rc.rcjournal.com/content/62/6/662#:~:text=Many%20aerosol%20delivery%20devices%20are,are%20available%20on%20the%20market.
13. Moore RH. Delivery of inhaled medication in children. *UpToDate*. 2020. https://www.uptodate.com/contents/delivery-of-inhaled-medication-in-children.
14. National Heart Lung and Blood Institute. Health Topics on COPD, Asthma, and Childhood Interstitial Lung Disease. NCBI. 2020 May. https://www.nhlbi.nih.gov/health-topics/copd.
15. Daniels LM, Juliano J, Marx A, and Weber DJ. Inhaled antibiotics for hospital-acquired and ventilator-associated pneumonia. *Clinical Infectious Diseases*. 2017 Feb; 64(3): 386–387. https://academic.oup.com/cid/article/64/3/386/2736669
16. Rubin BK. Aerosol medications for treatment of mucus clearance disorders. *Respiratory Care*. 2015 Jun; 60(6): 825–832. http://rc.rcjournal.com/content/60/6/825#T1
17. CDC. What You Should Know About Flu Antiviral Drugs. National Center for Immunization and Respiratory Diseases (NCIRD). 2019 Apr. https://www.cdc.gov/flu/treatment/whatyoushould.htm
18. Daco S. Surfactant Administration in the NICU. Nursing Clinical Effectiveness Committee of The Royal Children's Hospital Melbourne. 2018 Jan. https://www.rch.org.au/rchcpg/hospital_clinical_guideline_index/Surfactant_Administration_in_the_NICU/
19. Laube B. Aerosolized medications for gene and peptide therapy. *Respiratory Care*. 2015 June; 60(6): 806–824. http://rc.rcjournal.com/content/60/6/806.
20. Hawkins A and Tulloh R. Treatment of pediatric pulmonary hypertension. *Vascular Health and Risk Management*. 2009; 5(2): 509–524. doi:10.2147/vhrm.s4171. https://www.ncbi.nlm.nih.gov/pmc/articles/PMC2697585/.
21. Liet JM, Ducruet T, Gupta V, and Cambonie G. Heliox inhalation therapy for bronchiolitis in infants. *Cochrane Database of Systematic Reviews*. 2015; (9): CD006915. https://pubmed.ncbi.nlm.nih.gov/26384333/.
22. Watson S and Carter A. Inhalers for COPD: Types, instructions, pros, and cons. *Healthline*. 2018 Aug. https://www.healthline.com/health/copd/inhaler-nebulizer#metereddose-inhaler.
23. Singh S, Loke YK, Enright PL, and Furberg CD. Mortality associated with tiotropium mist inhaler in pts with COPD. *BMJ*. 2011 April; 342: d3215. https://www.bmj.com/content/342/bmj.d3215.
24. Labiris NR and Dolovich MB. Pulmonary drug delivery Part II: The role of inhalant delivery devices and drug formulations in therapeutic effectiveness of aerosolized medications. *British Journal of Clinical Pharmacology*. 2003; 56(6): 600–612. https://www.ncbi.nlm.nih.gov/pmc/articles/PMC1884297/.

13

Pulmonary Drug Delivery: The Role of Polymeric Nanoparticles

Ofosua Adi-Dako[1], Doris Kumadoh[2], Esther Eshun Oppong[3], Christina Osei Asare[4], and Mary Ann Archer[2]
[1]Department of Pharmaceutics and Microbiology, School of Pharmacy, University of Ghana, Accra, Ghana
[2]Centre for Plant Medicine Research, Mampong-Akuapem, Ghana
[3]Department of Pharmaceutics, School of Pharmacy and Pharmaceutical Sciences, University of Cape Coast, Cape Coast, Ghana
[4]Department of Pharmaceutical Sciences, School of Applied Sciences, Central University, Miotso, Ghana

13.1 Introduction

13.1.1 Pulmonary diseases

Generally pulmonary diseases or disorders could result from inhaled irritants, or viral infections, subsequently lowering the host's resistance and predisposing the patient to secondary infection or inflammation of the respiratory tract. The infections could be fatal if the airways become blocked by discharge and inflammatory swelling, or if the infection spreads through the lungs to other organs (1). For instance, the global pandemic, Covid-19, a pulmonary infection caused by the novel coronavirus, SARS-CoV-2 pathogen, has greatly affected the lives of people around the world. The disease is transmitted by the respiratory virus with droplets from infected persons which are inhaled into the lungs causing pneumonia and acute respiratory syndrome (2). Several complications from this lung infection, include coagulopathy, lung injury, and multiple organ failure (3–5) have resulted in morbidity and mortality worldwide.

Pulmonary disorders could be classified as infectious diseases, e.g. tuberculosis and pneumonia; obstructive conditions, e.g. asthma; restrictive conditions, e.g. fibrosis; and vascular diseases, e.g. pulmonary hypertension (1,6,7).

The nasal inhalation of nanoparticles for pulmonary therapy can be exploited for the treatment of disease. Recent reports indicate that the pulmonary route of drug delivery has generated immense research interest and investigations into the use of both local and systemic drug delivery systems. The focus of interest in pulmonary drug delivery is the high permeability and surface area of the lung, which is essential in the delivery of drugs for the treatment of pulmonary diseases (1).

Pulmonary drug delivery to the lungs is attractive and advantageous as there is relatively higher drug bioavailability due to the larger surface area of the lungs and a fast onset of action. There is improved patient compliance, as the formulation is self-administered, and the route of administration is non-invasive high drug permeability coupled with minimal drug degradation. The pulmonary route has been used for the delivery of vaccines, chemotherapeutics, antibiotics, proteins, peptides, protease inhibitors, and interferons (8).

Conventional dosage forms have associated limitations, such as frequency of administration due to a shorter half-life, patient non-compliance, and issues with peak and valley plasma concentration. These properties of conventional dosage forms are unable to achieve site specificity or targeted delivery of drugs. Formulation of modified release and targeted delivery systems, such as polymeric nanoparticles for pulmonary disease, is of immense benefit due to the ease of preparation, the control of the size distribution, good retention, and protection of the drug (9,10). The key areas to be considered in achieving efficient pulmonary drug delivery are the delicate balance between the patient, the design of the drug formulation, and the inhalation device employed (1).

Recent reports indicate that polymers have gained considerable attention as pulmonary drug delivery systems. They are well suited for drug delivery as they have high drug encapsulation efficiency and protect the drug from degradation, and exhibit modified and sustained drug delivery with a long shelf life. Polymers such as chitosan, gelatin, alginate, poly(lactic acid) (PLA), poly(lactic-co-glycolic acid) (PLGA), and poly(ε-caprolactone) (PCL) are used for therapeutic purposes (8,11).

Functionalized polymer nano-particulate drug delivery systems have made a remarkable impact in inhalational drug delivery. Such polymer drug delivery systems are tailored to suit the prevention and treatment of lung diseases and disorders (9). Polymeric nano-based drug delivery systems are well suited to be developed into useful therapeutic strategies for emerging diseases such as Covid-19, and existing disorders of the lung which would require targeted drug delivery for optimal therapeutic outcomes (12).

An evaluation of the current strategies, approaches, advances, and future prospects of targeted polymeric nanoparticle drug delivery systems in the field of pulmonary drug delivery are essential for the prevention, treatment, and management of prevailing and emerging lung diseases and disorders.

DOI: 10.1201/9781003046547-13

13.1.2 Anatomy and Physiology of the Respiratory System

The upper respiratory system consists of the nose, larynx, and pharynx, and the lower respiratory tract is composed of the trachea, bronchi, and lungs (1). The alveoli in the lungs play a key role in the respiratory system. The exchange of gases occurs at the alveoli, after which there is diffusion into the arterioles. The respiratory tract consists of the nose, oropharynx, larynx, trachea, bronchi and bronchioles. The lungs ensure the uptake and exchange of gases such as oxygen, which is absorbed through the alveoli, capillaries, and the arteries, and perfuses the tissue, after which carbon dioxide, the metabolic product, is exhaled (1,13,14).

13.1.3 Pulmonary Drug Delivery

13.1.3.1 Pulmonary Drug Delivery and the Treatment of Pulmonary Disorders

Pulmonary drug delivery has been employed for so many years for the local treatment of diseases like asthma, cystic fibrosis, chronic obstructive pulmonary disease (COPD), and infections. Developments and advances in pulmonary delivery have led to the systemic treatment of cancer, infections, autoimmune diseases, diabetes, and immune deficiencies. Generally the oral and nasal inhalation routes are used for pulmonary drug delivery. However a higher drug deposition has been associated with oral inhalation (15,16).

Drugs administered as aerosols are suitable for use in a variety of inhaler devices and delivery systems. Formulations that can be aerosolized are essential for inhalation drug delivery. Moreover, the delivery systems should deliver an aerosol with particles of appropriate size or droplets for deep lung deposition or at the peripheral airways for an optimal therapeutic outcome (16). Advantages associated with pulmonary drug delivery include the large surface area of absorption in the lung, the local or systemic effect that is produced, the fast onset of action, and the avoidance of the influence of pH, food and first-pass metabolism, the minimal effective dose associated with a reduction in side effects, and a non-invasive drug delivery approach. Usually the inhalation devices used are tamper resistant (14,16,17).

The efficiency of inhalation delivery depends on the site of drug deposition in the lung. This is crucial as the sophistication of deposition of drug in the lung relies heavily on the anatomy and physiology of the lung, physicochemical characteristics of the inhaled drug, drug formulation properties, and the kind of delivery system employed. The flow and deposition of the aerosolized drug depends on the nature and dimensions of the airways. The drug particles of a mean size (1–5 μm) and shape can be targeted at the alveoli, where there is maximal absorption.

Pulmonary infections can produce changes in the airways leading to a disrupted flow and deposition of inhaled drug. There are mathematical models available that predict the drug deposition and distribution of the inhaled drug (14).

Pulmonary drug delivery can be improved by attaching site specific ligands which improves targeting and site specificity, avoids the exposure of healthy cells to the drug, and allows dose reduction and less toxicity. In addition, delivery of proteins and genes could be achieved through the pulmonary route. This enhances drug stability as such drugs could be degraded by metabolic enzymes in the gastrointestinal tract and the liver if administered orally.

The application of nanotechnology in the design and formulation of drug delivery systems has enhanced targeted delivery of drugs specifically to diseased lung tissue, which is coupled with the reduction of side effects (18–20).

Pulmonary drug delivery is a convenient and effective approach for drug administration that is associated with a fast onset of drug action, is devoid of first-pass metabolism, has no potential lung toxicity, is non-invasive, and provides a simple approach for treatment.

The approach employed involves the use of the intranasal and oral inhalation approach. The intranasal route is convenient, safe, and enhances patient compliance. The nasal mucosa consists of the nasal epithelium, which is readily accessible for the absorption of drugs.

However, the oral inhalation route is usually preferred, as the intranasal route presents with narrow airway lumen as a limitation of that route (21).

Oral inhalation enhances the diffusion of drug into the peripheral areas and the alveoli of the lungs, thereby improving drug distribution with a better therapeutic outcome. Administration by the nasal route could be associated with a drug concentration loss of 85% as compared to that of 20% when administered by oral inhalation (22,23).

13.1.3.2 Challenges in Pulmonary Drug Delivery

The respiratory airway has ciliated epithelial cells which move mucus and alveolar fluids to the airways above. This mode of mucociliary clearance is useful for the removal of undesirable remains of degraded substances, unwanted secretions, and inhaled particles. In this way, the mucus undergoes a cycle of production and disposal. In addition, the alveolar macrophages serve as the host's defense mechanism by engulfing inhaled undesirable particles. The action of mucociliary clearance and macrophages could hinder the deposition and higher residence time of inhaled drug particles. Hence the appropriate drug delivery strategies should be able to overcome these limitations so the drug reaches the intended target in the lungs (20,24).

Usually the inhalation and deposition of drug in the lungs is influenced by processes such as inertial impaction, gravitational sedimentation, and diffusion. The drug particles size distribution in the aerosol is important. One challenge in the deposition of drug occurs in the highly branched airways, which have a narrow lumen between the trachea and the alveolar sacs. This causes the particles to only collide with the airway wall upon administration and become deposited deep or on the periphery of the lungs instead of flowing through the airways.

Pulmonary diseases such as acute respiratory distress syndrome in infants, pneumonia, cystic fibrosis, pulmonary hypertension, lung cancer, and lung infections can change the normal anatomy of the lung. This results in narrowing of the airway, thickening of the mucous, poor blood flow, which ultimately adversely affects the deposition of drug. In pulmonary disorders, the constriction in the airway rather paves the way for the deposition of drug in the upper airways by the process of impaction. The air flow rate and the inhaled and exhaled volume affects the length of the residence time of the particles inhaled (20,24).

13.1.4 Pulmonary Drug Deposition

Pulmonary drug delivery could be achieved by the use of nebulizers, dry powder inhalers (DPIs), metered-dose inhalers (MDI), and aerosols, usually containing preparations such as nano-formulations, and biodegradable nanoparticles (15,25).

The use of aerosols is efficient in delivering drugs into the airways. Aerosols are pressurized systems that release a metered dose of fine mist spray after the activation of an associated valve system. A pharmaceutical aerosol contains the drug or therapeutic agent dispersed in a propellant in a suitable pressurized container and is administered for a local or topical effect. They can be administered to the lungs as inhalation aerosols or to the mouth as nasal aerosols. Aerosols usually target the delivery of a small quantity of the drug particles to a specific site for absorption. This produces a fast response (22).

Generally aerosols are introduced via nasal or oral inhalation. Usually the oral inhalation route is preferred as the nasal route is accompanied by constraints such a narrow airway lumen. Three main mechanisms associated with drug deposition after pulmonary administration follow.

13.1.4.1 Inertial Impaction

This is the movement of large particles that develop inertia, compelling them to move out of the direction of the main stream of particles. Subsequently these large particles move in a straight path and eventually impact the walls of the airways. Impaction in this way is useful for the treatment of asthma, tracheobronchitis, and COPD.

13.1.4.2 Sedimentation

Air velocity decreases when particles move deeper into the lung. The probability that deposition would be by impaction is decreased. Hence the gravitational force facilitates deposition by sedimentation.

13.1.4.3 Diffusion

Deposition of smaller particles sizes is influenced by diffusion due to their random movement or Brownian motion of the particles (1,16).

There are also other associated factors with lung deposition.

13.1.4.4 The Physicochemical Properties of the Drug and Formulation

Drug deposition in the lungs is affected by the diameter or particle size, density, particle shape, hygroscopicity of the particle, and electrical charge, and by the type of formulation, e.g. suspension, solution, or powder.

13.1.4.5 The Type of Delivery Device

Generally the devices for inhalation produce aerosols with varied particle sizes, even though particles of similar sizes would be preferred (1,22,25).

13.1.5 Pulmonary Drug Delivery Devices

Currently a lot of modern devices are used to target drugs to the lungs in the prevention and treatment of lung diseases, e.g. asthma and chronic obstructive pulmonary disease. The design of these pulmonary drug delivery systems is influenced by factors such as

1. Physicochemical properties of the drug
2. The patients to be treated, e.g. children, elderly
3. The clinical goals to be achieved
4. Regulatory requirements and legislation

Categories of pulmonary drug devices employed are

1. Dry powder Inhalers (DPIs): These are devices that deliver an aerosol of dry powder of the pure drug, or a combination of drugs with a drug carrier. The particles of drug are deposited in the upper airways or the deep lung. The velocity during deposition and the particle size are crucial for drug deposition.
2. Metered-dose inhalers (MDIs): The device is used to deliver drug in a fixed dose to the airways in patients with asthma, emphysema, bronchitis, and chronic lung disease. The device consists of the container, which is the canister, and the formulation, which includes the drug, excipients, surfactants and propellant, a metering valve, and an actuator or mouthpiece.
3. Nebulizers: These are designed to deliver suspensions or solutions of drug in an aerosol form into the respiratory tract. They can be useful for children with cough or respiratory and inhalation problems, and delivering local anaethesia in the trachea.

13.1.6 Methods of Formulation

Compatibility of excipients with active pharmaceutical ingredients are essential for formulation. Excipients found in polymeric pulmonary drug delivery systems play a role in drug absorption. The compatibility of these pulmonary drug delivery excipients with the drugs for treatment should be established before formulation. Long-term administration, especially of pulmonary formulations, requires that the potential for lung toxicity is eliminated.

13.1.6.1 Preparation of Particulate Matter

There are quite a number of techniques employed in the preparation of suitable drug particle sizes for pulmonary drug delivery. These techniques include spray drying, super-critical fluid technology, crystallization, double emulsion/solvent evaporation, and particle replication with non-wetting template methods.

13.1.6.1.1 Spray Drying Technique

Spray drying is a useful technique for the formulation of fine particles for pulmonary drug delivery, usually for the preparation of dry powder inhalers (DPIs) as it's a rapid process resulting in a product with unique properties. During the spray drying process, fine droplets, which are atomized from solutions or slurries, are passed into a stream of hot air, with subsequent evaporation of the moisture resulting in dry spherical particles. Spray drying produces free-flowing particles, with easily controllable sizes suitable for production on a large scale (25).

13.1.6.1.2 Supercritical Fluid Technology

Supercritical fluids exhibit the properties of both liquids and gases. These fluids behave as liquids, with properties such as solvency, flow, and polarity when above their critical temperature and pressure. The technique involves the controlled crystallization of drugs which is conducted from a dispersion in supercritical fluids, usually achieved with gases, e.g. carbon dioxide, nitrous oxide. The production of pulmonary drug delivery systems, e.g. nanoparticles, proteins, peptides, etc., can be carried out with this technique, which also has the potential of improving the formulation characteristics of the drug for pulmonary administration.

13.1.6.1.3 Crystallization

This method involves the production of a supersaturated solution, which subsequently encourages the formation of crystals. Even though the pace of the process is slow, there is minimal aggregation, defined crystal lattice, purity of the crystals formed, and a maximum yield.

13.1.6.1.4 Double Emulsion/Solvent Evaporation Technique

This method involves the preparation of a double oil-in-water or triple water-in-oil-in-water emulsion. The oil phase is eliminated by subjecting it to either non-solvent extraction or solvent dilution or evaporation. In the process the diffusion and evaporation of the organic solvent results in drug-loaded polymer particles.

13.1.6.1.5 Particle Replication

This process allows the formulation of similar sized nanoparticles, where the shape, size and constituents, and surface modification can be monitored and controlled (1).

13.2 Polymeric Nanoparticle Drug Delivery Systems

Nanoparticles are usually colloidal particles within the size range of 1–100 nm and could be classified as metal, lipid-based, ceramic, and polymeric nanoparticles based on the physicochemical properties. Advances and developments in the formulation and application of polymeric nanoparticles has generated immense research interest. Polymers employed for such drug delivery systems are of natural, synthetic, or semi-synthetic origin. The biodegradable forms are preferred due to chronic toxicity associated with the non-biodegradable forms. Polymeric nanoparticles could contain a drug, protein, and DNA material intended to target a cell or particular organ. The drug could be made soluble, attached, encapsulated, or entrapped in the matrix of the nanoparticle. Polymeric nanoparticles are regarded as smart polymers, due to their target specificity, reduction in side effects and rate controlling properties. Formulation of polymeric nanoparticles is less complicated than for regular nanoparticles and therefore more investigated. Polymeric nanoparticles have a larger surface area that allows them to exhibit a high number of surface functional groups such as ligands. The advantage of a smaller size is that it enables the polymeric particles to enter small capillaries, thereby targeting the cell of interest. This type of nanoparticle shows a good control of size and size distribution, and has a longer clearance time as compared to other nanoparticles, which is indicative of the fact that small quantities of drug are able to elicit better therapeutic effects with less side effects. Polymeric particles can be tailored to suit therapeutic needs, rate controlling with a high drug loading capacity. Limitations of this group of nanoparticles are linked to toxicity concerns and inability to halt therapy in an emergency. Scaling up for industrial purposes would require sophisticated equipment and high costs coupled with a strenuous process (9,26).

13.2.1 Types of Polymeric Nanoparticles

13.2.1.1 Lipid Polymer Hybrid Nanoparticles

Solid lipid nanoparticles are carriers, which contain lipids, and are useful for the delivery of therapeutic agents. This formulation is termed a lipid-polymer hybrid nanoparticles have the benefits of containing both solid and lipid phases and exhibit controlled release, protection of the drug load, and good tolerability with both lipophilic and hydrophilic properties. Due to the hydrophilic and lipophilic nature of the lipid-polymer hybrid nanoparticles, they are amenable to various formulations encapsulating different kinds of drugs (9,27,28).

13.2.1.2 Solid-Lipid Polymer Hybrid Nanoparticles

Development of solid lipid polymeric nanoparticles has yielded solid lipid polymeric nanoparticles with a unique formulation design and formulated with a core-shell, in contrast to solid lipid nanoparticles which are rather coated with a polymer shell. Solid lipid polymer nanoparticles have the advantage of the high encapsulation efficiency of hydrophilic drugs. Their drug delivery systems can be formulated with biopolymers like pectin and bovine serum albumin, and are biocompatible. The polymeric coating maintains the stability and colloidal integrity, and the solid lipid core enhances encapsulation efficiency (9,29).

13.2.1.3 Functionalized Polymeric Nanoparticles

Modification of the polymer in polymeric nanoparticles results in a functionalized polymeric nanoparticle. Functionalized polymeric nanoparticles have enhanced distribution and are better protected from phagocytosis by the reticuloendothelial system. Consequently, since the nanoparticles are protected from degradation, there is an increased amount of drug in the blood circulation.

Functionalized polymeric nanoparticles can be categorized as long stealth nanoparticles, lectin-based polymerized nanoparticles, polysaccharide-based nanoparticles, and ligand-based nanoparticles (9).

13.2.1.4 Polysaccharide Conjugated Polymeric Nanoparticles

Polysaccharide conjugated polymeric nanoparticles are functionalized polymeric nanoparticles which usually consist of adsorbed polysaccharides on the surface of the nanoparticles. For instance hydrophobic polyesters such as PLGA and PLA are covalently incorporated with a polysaccharide made up dextran, chitosan, and hyaluronic acid. This drug delivery system has the advantage of being biocompatible (9,30).

13.2.1.5 Ligand-Based Polymeric Nanoparticles

Recent advances in polymeric nano-formulations have resulted in the formulation of ligand-based polymeric nanoparticles. They are designed to deliver therapeutic agents at targeted sites. For instance, for the sensing of cancerous cells, the green fluorescent protein is used. Slight modifications in the ligands enhances the cell affinity. Ligand-based nano-particulate systems are suitable for the diagnosis of disease and intracellular drug delivery (9,31).

13.2.1.6 Fluorescence Polymeric Nanoparticles

Luminescent polymeric nanoparticles have been developed with the use of fluorescent compounds such as luminescent polymers, fluorescent substances that exhibit coordination with metals. Aggregation induced emissions (AIE) are able to strongly radiate fluorescence due to the aggregation of fluorophores. Aggregation induced emission substances can are suitable for bio or chemo sensors. Fluorescent organic nanoparticles in water have good emission characteristics. The conjugation of aggregation-induced emission dyes in combination with polymers and block co-polymers are useful as agents for bio-imaging and the treatment of disease (9,32,33).

13.3 Applications of Polymer Nanoparticles in Pulmonary Drug Delivery

Research in the area of polymeric nanoparticles has attracted heightened interest as there is increased therapeutic efficacy with less toxicity, a longer clearance period, and good size control. Different forms of polymeric nanoparticles exist as ligand-based nanoparticles, polymeric micelles, and dendrimers. Polymeric nanoparticles are carriers of therapeutic agents such as proteins, drugs, and DNA material for targeted delivery to an organ or cell (9).

13.3.1 Modifications of Polymer Nanoparticle Pulmonary Drug Delivery Systems and Safety Evaluations for Better Performance

Strategies recently adopted to improve hydrophilicity of nanoparticles which could subsequently enhance mucopenetration in pulmonary delivery, and avoid attack by opsonins, were considered and evaluated. This involved addition of polymers to the surface of nanoparticles. Polymers to be considered in this regard were methoxy polyethylene glycol (MPEG), polyethylene glycol (PEG), 1,2-dilauroyl-sn-glycero-3-phosphocholine (DLPC), and vitamin E, etc. The nanoparticles are either encapsulated in polymer carriers, dispersed in the polymer matrix, or loaded onto the surface of the polymer. A combination of polymers such as polyelectrolyte complexes, use oppositely charged polymers to entrap drugs into the polymer matrix of nanoparticles. Subsequently the drug is released via diffusion or polymer degradation.

Efficient lung deposition through the formulation of nanocomposites, was also investigated. Nanocomposites are produced by a combination of nanoparticle aggregates and an excipient such as a polymer, e.g. poly-DL-lactide-co-glycolide acid (PLGA). PLGA, which is safe to use in nanotechnology, has good controlled release characteristics and stability. Other applications of nanocomposites for efficient lung deposition and fast release of therapeutics such as salmon-calcitonon that have been studied involved the adsorption of salmon-calcitonin on PLGA, which when used as a coating on a lactose carrier is effective in the formulation of a nanocomposite.

Attempts have been made to exploit the electrostatic aggregation of nanoparticles as an approach to modify the carrier surface to include both cationic and anionic nanoparticle adsorption, in the field of the formulation of a nanocomposite for the development of a dry powder inhaler. The findings suggest that cationic and anionic poly lactic-co-glycolic acid/phosphatidylcholine (PLGA/PC) lipid and polymer hybrid nanoparticles could be adsorbed onto the surface of chitosan carrier nanoparticles. This hybrid polymer system has become of interest and consists of a polymer nanoparticle core and a liposomal layer (8).

13.3.2 Progress in Safety Evaluations

PLGA nanoparticles are extensively used in formulations delivered through the inhalation route. In view of its effectiveness, Haque et al., analysed the kinetics and clearance of polylactide-co-glycolide (PLGA) nanoparticles from the lungs to provide more information on the safety of these nanoparticles after extended use and provide a basis for clinical studies. The findings showed that lung kinetics and lung retention was significantly affected by particle size and lung clearance was affected by particle charge. There were temporary inflammatory changes observed after a single dose administration, influencing lung retention times. The study highlighted the significant insight into the role of the particle size and charge in the evaluation of the kinetics and the processes involved in the pulmonary delivery and clearance of PLGA particles (34).

13.4 Strategies, Approaches and Applications of Polymeric Nanoparticles in the Prevention and Treatment of Infectious Diseases

13.4.1 Pulmonary Infections

13.4.1.1 Advances in the Treatment of Pulmonary Viral Infections

Pulmonary disorders could be classified as infectious diseases, e.g. tuberculosis and pneumonia; obstructive conditions, e.g. asthma; restrictive conditions, e.g. fibrosis; and vascular diseases, e.g. pulmonary hypertension (1,6,7).

Several antivirals and immunomodulating drugs have ben hypothesized to be of immense benefit or investigated for their efficacy against the novel Covid-19 coronavirus that is in progress at the time of writing. The approach in this regard is to target the inhibition of the activity of proteases in the host cell and block the entry of the virus. Other approaches adopted are the development of nucleoside analogues which would target RNA polymerase, resulting in the inhibition of RNA synthesis, e.g. remdesivir, decreasing the production of pro-inflammatory cytokines or activation of CD8-T cells. Recent reports indicate the association of these drugs with adverse effects at the approved doses and also at higher doses. A formulation that could control the release of drug and maintain the minimum effective concentration, as well as reducing the side effects, could be recommended in this situation.

Polymeric nano-based systems have been studied in this regard. Mehta et al., hypothesized that advanced drug delivery systems that could enhance drug absorption and intracellular drug delivery, and maintain the concentration of drug in the lungs and systemic circulation with reduction of side effects would hold great promise as nanocarriers in the treatment of Covid-19.

Aerosol based drug delivery systems were usually employed in the treatment of pulmonary disorders. However, there is the likelihood of the undesirable transmission of viruses due to fugitive emissions from aerosol therapy, which could compromise the safety of healthcare workers treating Covid-19 patients. Novochizol, a chitosan biodegradable nanoparticle-based aerosol system holds great promise, as it adheres to the mucous membranes in the lung epithelium and provides sustained drug release. Novochizol is easily formulated. This polymer nanosystem is able to ensure the maintenance of optimal drug concentrations in the lungs and mitigate unwanted systemic distribution of drug (12).

There are remarkably promising approaches in nanotechnology that could be suitable for Covid-19 drug delivery and treatment. Noteworthy properties of nanoparticles useful for drug delivery include their small size improving targeted delivery, increased surface to volume ratio, thereby enhancing drug loading, and improved penetration of negatively charged mucosal membranes, due to the surface charge modification. Mechanisms of nanoparticle deposition in the respiratory tract include impaction and sedimentation for large particles and diffusion for smaller particles. Consequently, a combination approach of both nano- and microparticles has been investigated by Gartner et al. to possibly reduce mucociliary clearance and improve deposition in pulmonary delivery for the treatment of Covid-19 (35). Relevant strategies adopted to enhance Covid-19 therapeutic systems are the encapsulation of the therapeutic agent in the core of the nanoparticle to increase stability, improved targeting, and polymer performance to decrease the amount of required drug in the nano-based drug delivery system- (36).

Pulmonary viral infections, e.g. influenza viruses A and B, and SARS-Cov 2 and other microbial infections could be life threatening in patients with underlying pulmonary comorbidities and immuno-compromised patients. Oral or parenteral anti-infectives and anti-inflammatory agents administered in such situations could be associated with insufficient therapeutic levels of drug concentration at the site of interest in the lung. Localized inhalational drug delivery is well suited for drug administration in such situations. Chitosan is employed in such drug delivery systems for its mucoadhesive properties, binding action to microbial DNA with inhibition of mRNA, and protein synthesis as chitosan infiltrates the nuclei of microbes. Chitosan inhibits microbial growth by binding to the nutrients of the micro-organisms. It is reported that low-molecular-weight chitosan has greater anti-microbial activity (20).

13.4.1.2 Advances in the Treatment of Pulmonary Bacterial Infections

Tuberculosis is caused by *Mycobacterium tuberculosis* resulting in an infection primarily of the lungs. The bacteria multiply within the granuloma, which are complex structures, which need to be accessible by the anti-tubercular drugs. Drugs to be systemically administered need to be administered in high doses for effective treatment of pulmonary tuberculosis (37). The pulmonary route is beneficial for tuberculosis treatment as it allows a higher bioavailability, with a larger lung surface area, and with higher perfusion, which targets the site of infection and avoids the first-pass effect. A study of chitosan nanoparticles showed a higher lung retention and longer residence time of administered anti-bacterials.

Research conducted by encapsulating isoniazid in chitosan nanoparticles enhanced the efficacy of isoniazid in *in* vitro and in vivo models. Nanoparticles are able to better target the phagocytic cells where *M. tuberculosis* usually replicates.

Such results with promising outcomes could be evaluated in clinical studies as a suitable alternative for drug-resistant tuberculosis (38,39).

The current treatment of tuberculosis involves the administration of solid dosage forms such as tablets and capsules for four to six months. This dosage regimen could lead to side effects. Development of a chitosan nanoparticle-based dry powder formulation of rifampicin was successful in achieving dose reduction and frequency of administration as compared to the pure rifampicin powder.

Pulmonary delivery of the nanoparticles achieved sustained release as compared to the pure powder inhalation and showed an optimal pharmacokinetic profile. This study provided a useful therapeutic approach for the treatment of tuberculosis of the lung (40).

A formulation of ethionamide-loaded chitosan alginate nanoparticles was stabilized with different quantities of carageenan.

The findings indicate that in this study carageenan improved the stability in the processing as well as the entrapment efficiency of ethionamide. The nanoparticles exhibited controlled release over 96 hours, which reduced with increasing amounts of carageenan, with no drug excipient interactions. The formulated nanoparticles exhibited a significant activity against the *Mycobacterium* strain H37RA, and held great promise for inhalation therapy. The study also reiterates the immense potential of polymer nanoparticles in the pulmonary delivery of anti-tubercular drugs, with associated reduction in undesirable systemic effects. Polymeric nanoparticles have demonstrated their great potential in enhancing the efficacy of therapeutic agents and the ability to reduce their off target adverse effects (41).

Pneumonia is an infection that could adversely affect the alveoli in the lung causing respiratory failure. Development of antibiotic resistance to conventional treatment hampers the therapeutic outcomes. An anti-microbial peptide, BP 100, was studied as a suitable antibiotic alternative for treatment of pneumonia.

The antibacterial activity of BP 100 is optimized when carriers such as gold nanoparticles and polymers as coatings are used. With the use of a lung surfactant model, the effect of BP 100 transposition with polymeric carriers, e.g. polyethylene glycol (PEG) and polystyrene (PS), was evaluated. The results showed that the polymer PEG works by a mechanism of ligand competition for protection on the polymeric gold nanoparticle and BP 100 system. The findings highlighted the propensity of the use of PEG with gold nanoparticle carrier systems for anti-microbial peptides as a potential candidates for pulmonary disease (42).

13.4.2 Advances in the Treatment of Pulmonary Diseases and Disorders

13.4.2.1 Lung Cancer

Chitosan has attracted considerable interest in anti-cancer interventions, due to its mucoadhesive, penetration enhancement, and cell targeting properties for cancer treatment. The aerodynamic behavior of nanoparticles necessary for pulmonary inhalation could be augmented by microencapsulation of nanoparticles in pulmonary drug delivery.

Chitosan and its derivatives are useful materials in nano and micro carriers in pulmonary delivery of anti-cancer agents. It is biodegradable, biocompatible, anti-proliferative, and anti-microbial. The potential of chitosan as a backbone in lung cancer treatment and a carrier in pulmonary drug delivery has been examined.

Chitosan-based pulmonary delivery systems exhibit a reduction in systemic toxicity in anti-cancer therapy, a higher drug absorption in the lungs, and overall improved efficacy of anti-proliferative drug in lung cancer. Targeting with a ligand has shown better cell uptake and cancer cell apoptosis. Most studies have been conducted in vitro and ex vivo, and, as such, an in vivo evaluation for aerodynamic properties for inhalation in humans is necessary to establish their potential in lung cancer therapy (20).

The pulmonary delivery of therapeutics, e.g. peptides, is effective for the systemic absorption of therapeutic agents with problems of poor absorption. Moreover, pulmonary delivery is an attractive approach for the systemic absorption of anti-cancer agents, rather than the use of chemotherapy and radiotherapy in cancer treatment. Consideration for the use of phosphorylcholine, consisting of the copolymer MPC-DPA, is appealing as it is biocompatible and self-assembling to form diverse nanosystems (18).

The formulation of nanoparticles consisting of poly (2-methacryloyloxyethyl phosphorylcholine) and *b* poly (2-(diisopropylamino) ethyl methacrylate) (MPC-DPA) as carriers with a coating of n-trimethyl chitosan chloride which improves delivery for optimal concentrations in the cell, has exhibited remarkable applications in nanoformulation.

A number of nanosystems were analysed for their potential as suitably optimized MPC-DPA nanoparticles loaded with different quantities of an emerging therapeutical agent, curcumin, for their pulmonary delivery applications in the treatment of lung cancer. The necessary strategies to be adopted included the prediction, controlled and consistent production of particle sizes, drug loading capacity, effective airway travel, and permeation of lung tissue with suitable anti-proliferative indices. The nanosystems developed were effective and are recommended for further investigation in inhalation delivery and in vitro evaluation for pulmonary delivery for the treatment of lung cancer. This was a novel and promising study in the evaluation of anti-cancer curcumin-loaded nanoformulations of MPC-DPA for pulmonary delivery (43).

Ahmad et al., investigated novel chitosan-coated PLGA nanoparticles of catechin hydrate to assess the pharmacokinetics and probable improvement of the bioavailability after pulmonary delivery through the nose to the lungs, with the use of H1299 lung cancer cells. The particle size of the chitosan-coated nanoparticles were ~150 nm, with a polydispersity index of ~0.306. A high entrapment efficiency was achieved as the release was seen to be targeted at the cancer cells. A high apoptosis of cancer cells as well as a remarkable mucoadhesion was observed. The design of the chitosan-coated nanoparticles enhanced the safe delivery of catechin in rat lungs.

The findings showed a higher bioavailability after administration of the chitosan-coated PLGA nanoparticles of catechin hydrate in the lungs of the rat and provided a successful outcome for further investigations in the treatment of lung cancer (26).

In many ways, other polymeric nanoparticle applications have shown great promise in the treatment of cancer. The anti-cancer applications of gelatin-based nanoparticles were investigated. Anti-cancer drugs can be encapsulated in polymeric nano carrier systems. A formulation of cisplatin-loaded gelatin-based nanoparticles showed significant anti-proliferative activity against lung adenocarcinoma cells. The droplets of the nebulized aerosol were found to be suitable for deep lung drug delivery in vivo (44).

Doxorubicin-loaded poly-isobutylcyanoacrylate nanoparticles also exhibited cytotoxicity via alveolar macrophages. Alveolar macrophages show anti-cancer activity via the mechanism of the phagocytosis of the polymeric nanoparticles.

Doxorubicin can be conjugated with polymeric nano carrier systems for lung delivery, but the large particle sizes formed created a problem for pulmonary delivery. To overcome this, an aerosol of liquid droplets was formulated for the deep lung delivery of doxorubicin. Conjugation of hyaluronan-cisplatin with nanoparticles was done and administered by lung instillation in vivo. There was an approximately five-fold enhanced anti-cancer activity than when the intravenous route was used. With regard to the delivery of cisplatin, both kidney and brain toxicities were reduced with the use of polymeric nanoparticles for drug delivery (45–48).

13.4.3 Advances in the Treatment of Asthma and Chronic Obstructive Pulmonary Disease

An optimal therapeutic outcome in the treatment of asthma and chronic obstructive pulmonary disease (COPD) is achieved when the drug is targeted at the site of action.

Therapeutic agents encapsulated in biodegradable polymeric nanoparticles are useful in pulmonary delivery. It is hypothesized that a combination of theophylline and budesonide improve the therapeutic outcome in the treatment of respiratory disease. A combination of budesonide and theophylline was encapsulated in polylactic acid (PLA) nanoparticles and investigated for their potential in pulmonary drug delivery. The nanoparticles were evaluated for their particle size, drug loading, zeta potential, and *in vitro* deposition characteristics during a nebulization procedure. The formulation produced a sustained drug release over 24 hours. The fine particle proportion of nebulization was in the ratio 75% of theophylline and 48% of budesonide. The study findings indicate that polymeric nanoparticles loaded with a combination of theophylline and budesonide are highly recommended in the treatment of asthma and COPD (49,50).

Budesonide has been extensively used in the treatment of asthma. However after inhalational delivery, it suffers the limitation of low absorption in the lungs. Ahmad et al., developed chitosan-coated budesonide nanoparticles, intended to improve the bioavailability, dispersion of the aerosol particles and the lung deposition, and evaluate the pharmacokinetic profile of budesonide. The particle size obtained after formulation was ~196 nm with a spherical shape and zeta potential of ~11.8. Pulmonary delivery of budesonide using the chitosan approach exhibited a higher bioavailability and lung deposition in the animal model, up to three times higher, than when administered orally, and twice as high when administered intravenously. There was no observed toxicity with this approach. The study is indicative of the potential of the chitosan-coated nanoparticles in enhancing the pulmonary delivery of budesonide (51).

Pulmonary targeting in respiratory diseases such as COPD and asthma improves drug efficacy and reduces the side effects (18). The particle size in inhalation therapy is crucial for drug delivery, drug deposition, and drug uptake. Usually the challenges to be overcome by the drug delivery system are the resistance posed by the secreted mucus, and the effects of the reticuloendothelial system on the nanoparticles. Other strategies useful for disruption of the mucosal barrier are the formulation of nanoparticles with magnetic properties, the use of mucolytic agents, and hydrolyzing enzymes. Moreover, biodegradable polymers are used to overcome the major barriers such as macrophage clearance and mucociliary barriers in drug delivery. Such polymers, e.g. dextran, gelatin, chitosan, poly (lactic-co-glycolic acid) (PGLA), poly(ε-caprolactone) (PCL), and alginate produce a longer half-life, higher diffusion ability, and a prolonged therapeutic effect (52). PEGylated dendrimers are suitable for inhalation delivery and for the treatment of disorders of the endothelium and to prevent inflammation. Muralidharan et al. studied the treatment of pulmonary inflammation with the use of inhalable dry powders in combination with dimethyl fumarate (53,54).

Poly-ethylene imines (PEIs), dendrimers, chitosan, and poly (lactic-co-glycolic acid) (PLGA) nanoparticles have useful applications in gene delivery (55). There are reports indicating the suitability of chitosan interferon (IFN)-γ-pDNA (CIN) in reducing hyper-responsiveness of airways in methacholine and ovalbumin-induced asthma in vivo. CIN has the ability to reduce to reduce inflammatory cytokine, and CD8+ T lymphocytes. A formulation of poly-l-lysine and polyethylene glycol nanoparticles linked by a cysteine residue was useful in the delivery of thymulin, exhibited anti-inflammatory, collagen deposition, and anti-fibrotic activity in an animal model. This nanoparticle drug delivery system was reported to be safe and immune compatible for the lungs of humans. In addition, biodegradable particles loaded with cytosine-phosphate-guanine adjuvant were found to be a promising therapy in the treatment of allergies caused by dust and were able to inhibit the Th2 asthmatic response. There have been concerns expressed over the safety and toxicity of inhalable nanoparticles. However, biodegradable nanoparticles with the use of polymers and lipids have been found to lower adverse inflammatory responses as compared to other non-biodegradable nano-drug delivery systems (56).

13.4.4 Advances in the Treatment of Lung Fibrosis

There has recently been a heightened interest in albumin for the formulation of nanoparticles as it is biocompatible, biodegradable, and less toxic. Albumin nanoparticles are able to exhibit targeted and controlled drug release which makes it suitable for inhalational delivery in infectious disease, lung cancer, and asthma. Targeted delivery with albumin nanoparticles can be achieved by ligand binding. Pulmonary delivery of albumin nanoparticles has met with some physical, physiological, and immunological limitations. These limitations could be addressed by monitoring the particle properties to obtain optimal lung deposition and targeted delivery of drug. Moreover, there is an associated improvement in patient compliance and therapeutic outcome with the inhalation route as compared to the parenteral route. Joshi et al., formulated albumin nanoparticles with tacrolimus, which is an immunosuppressant drug useful for the treatment of fibrosis. The particle size of the formulation was ~182 nm and had a zeta potential of -34.5 mV. Inhalational delivery of the tacrolimus albumin nanoparticles showed a 24-hour slow release, and exhibited a higher anti-fibrotic effect than when administered by the peritoneal route. This formulation approach for tacrolimus albumin nanoparticles that was adopted, coupled with the simple method of preparation, enables the ease of scale-up for industrial purposes (57).

Leal et al. sought to address unmet needs in the formulation of drug delivery systems for gene therapy relating to the intracellular delivery of the therapeutic agent. There was the need to overcome both mucus and cellular penetration challenges in the pulmonary delivery of therapeutic agents in the treatment of cystic fibrosis. A peptide coated PEGylated nanoparticle drug delivery system enhanced the lung delivery of therapeutics over 600-fold in mice. This investigated intervention holds great promise in improving the absorption and efficacy of therapeutic agents in the treatment of cystic fibrosis (1)58.

13.4.5 Advances in the Treatment of Pulmonary Hypertension

Developments in targeted pulmonary delivery for the treatment of pulmonary hypertension have been insightful.

The activation of the mTOR pathway has been found to play a role in the progression of pulmonary hypertension in vascular modeling. Segurra-Ibarra et al. hypothesized that rapamycin-loaded polyethylene glycol- poly(ε-caprolactone) nanoparticles would be useful for targeted lung delivery in the treatment of pulmonary hypertension. The findings after the study showed that the use of rapamycin-loaded nanoparticles was useful in inhibiting the activated mTOR pathway, and effective in treating pulmonary arterial hypertension with fewer side effects. The findings indicate the potential of targeted lung delivery of the nanoparticles and further encourage the exploratory studies for inhalational delivery in the treatment of pulmonary hypertension (59).

Researchers such as Makled et al. investigated the pulmonary delivery of inhaled or nebulized sildenafil to the lungs. Before the study, it was observed that sildenafil was administered through the oral and parenteral route. The physicochemical properties, drug load, drug release, toxicity, and stability after sterilization and nebulization of formulated solid lipid nanoparticles loaded with sildenafil was conducted. The findings indicated a high encapsulation efficiency of above 80%, sustained drug release over 24 hours. Sterilization and nebulization did not affect the stability and drug load of the nanoparticles. The effects of the emulsifying agents in the coating on the mucin secretions was encouraging. It is envisaged that more interesting information would be derived from clinical studies (60).

It has been suggested that activation of the mTOR pathway plays a role in the progression of pulmonary hypertension. The efficacy of formulation of conjugated polyethylene glycol–distearoyl-phosphoethanolamine micelles loaded with fasudil was studied for the potential treatment of pulmonary arterial hypertension. The controlled release nanoparticle formulation exhibited an entrapment efficiency of 58%, enhanced the cell uptake, improved the half-life (five times higher than the control), and accumulated in the pulmonary vessels, suitable for reduction of the pulmonary arterial pressure (61).

13.5 Future Perspectives for Polymeric Nanoparticle Pulmonary Delivery of Therapeutic Agents in Emerging and Existing Lung Diseases

The inhalation of therapeutic agents in the form of aerosols is a convenient route for the pulmonary delivery which also enhances patient compliance. The use of polymeric nanoparticles among other nano-particulate drug delivery systems holds great promise for non-invasive targeted delivery for efficient cellular uptake of drugs to treat lung disease and assist lung repair. The prospects for polymeric nanoparticles, including its combination with cytokines, growth factors, small molecules, and stem cells for lung disease treatment are highly favourable (62).

Current trends indicate that chitosan and its modified forms have been extensively studied as the backbone of pulmonary targeted nanocarriers. They are useful for the delivery of therapeutic agents in cancer and infections. Chitosan has the advantage of being mucoadhesive, biodegradable, biocompatible, anti-proliferative, and anti-microbial. Developments in this area of research show a considerable interest in the use of chitosan over conventional carriers such as lactose. It is envisaged that further investigations in *in vitro* aerodynamic parameters, pharmacokinetics, and clinical studies would provide valuable information for therapeutics in this regard (20).

With regard to emerging antiviral medications, polysaccharides or polymers are of considerable research interest in the quest for anti-coronavirus interventions due to their peculiar antiviral properties and mechanisms of action. Such antiviral polymers are able to disrupt the life cycle of the infecting virus or improve upon immunity of the host. Carageenan, chitosan, and some traditional Chinese medicine polymers, of marine origin, have exhibited anti-coronavirus properties with a myriad of mechanisms of action potentially useful in the treatment of the Covid-19 infection. Hybridized polymeric nanoparticles have been investigated for their applications in the diagnosis and treatment of the influenza virus, human adenovirus, and HIV (63).

Recent studies with high prospects include the use of carageenan nasal spray with enormous antiviral efficacy against the influenza A virus, human coronavirus.

Further investigations with these findings in the intense search for novel ideas with the use of polymeric nanoparticles for pulmonary drug delivery are being birthed and should be pursued with much vigor in this era in the search for solutions for the treatment of Covid-19 (64,65).

13.5.1 Prospects for Emerging Disease Therapy with Polymeric Nanoparticles

The Covid-19 global pandemic has brought to light the importance of evolving drug development approaches and strategies in targeted pulmonary nanoparticle drug delivery systems with controlled drug release. Success in this area of research involves research collaborations and a concerted effort leading to the prevention of infection together with efficacious therapeutic drug delivery systems that would reduce mortality and improve control and recovery rates, and the response to epidemics (2,12).

Even though the changing nature of Covid-19 has created uncertainties in effective interventions, there are numerous promising therapies for the coronavirus infection with polymeric nanoparticles via the inhalational route, which addresses the critical issue of lung failure. Major developmental steps are being taken in research in this regard with potential successful outcomes (66).

Research into polymer drug conjugate-based nanoparticles is receiving much attention due to its immense potential in therapy, e.g. cancer chemotherapy. Polymer drug conjugate nanoparticles are designed to target tumor cells and tissues. Drug release can be triggered in response to certain stimuli. The strategy employed is to reduce systemic toxicity associated with the cancer drugs and optimize the therapeutic efficacy. Inhalational delivery of these polymer drug conjugate nanoparticles is attractive, as this approach is non-invasive (67). Polymer drug conjugates after inhalational delivery have the ability to improve the pharmacokinetics of the drug that is loaded, as well as optimize the controlled release of the drug as compared to the oral delivery or inhalation of the drug only. Polymer drug conjugates have a great potential if developed as pulmonary delivery systems. There is the need to further investigate and develop the physicochemical properties to enhance the effect of the drug conjugation on the polymer surface, the rate of drug release, therapeutic efficacy, required for preclinical studies necessary for rigorous clinical studies in cancer nanomedicine. It is envisaged that if exhaustive investigations are carried out in this regard, there would be remarkable advances necessitating clinical trials in the future (68).

Inhalable nanoparticle powder formulations play a key role in targeted pulmonary delivery. Safety of inhalable nanoparticle powders can be established through aggressive research. The use of polymers is essential for formulation development and the safety of inhalable drug delivery systems. Vibrant research investigations have been witnessed in the area of inhalable nanoparticulate powders with respect to vaccines, systemic drug delivery, surface modification in nanoparticle formulations, stability, and deep lung deposition. Formulation techniques in the processing of inhalational formulations have been improved with regard to freeze-drying, advanced spray drying, supercritical fluid extraction, condensation aerosol growth, and thermal condensation (8).

Strategies and developments in the area of polymer nanoparticles for pulmonary delivery with associated fields of interest have been evaluated. Challenges with this new era of research have been highlighted with potential solutions. There are interesting emerging opportunities for clinical studies and future development of promising therapies for lung disease (62).

REFERENCES

1. Deshmukh R, Bandyopadhyay N, Abed SN, Bandopadhyay S, Pal Y, and Deb PK. Chapter 3 - Strategies for pulmonary delivery of drugs. In: Tekade RK. (ed.) Drug Delivery Systems. Academic Press. 2020.
2. Ari A. Practical strategies for a safe and effective delivery of aerosolized medications to patients with COVID-19. *Respiratory Medicine*. 2020; 167: 105987.
3. Dariya B and Nagaraju GP. Understanding novel COVID-19: Its impact on organ failure and risk assessment for diabetic and cancer patients. *Cytokine & Growth Factor Reviews*. 2020; 53: 43–52.
4. Latimer G, Corriveau C, Debiasi RL, Jantausch B, Delaney M, Jacquot C, Bell M, and Dean T. Cardiac dysfunction and thrombocytopenia-associated multiple organ failure inflammation phenotype in a severe paediatric case of COVID-19. *The Lancet Child & Adolescent Health*. 2020; 4: 552–554.
5. Loganathan S, Kuppusamy M, Wankhar W, Gurugubelli KR, Mahadevappa VH, Lepcha L, and Choudhary AK. Angiotensin-converting enzyme 2 (ACE2): COVID 19 gate way to multiple organ failure syndromes. *Respiratory Physiology & Neurobiology*. 2021; 283: 103548.
6. Kovacs G, Dumitrescu D, Barner A, Greiner S, Grünig E, Hager A, Köhler T, Kozlik-Feldmann R, Kruck I, Lammers AE, et al. Definition, clinical classification and initial diagnosis of pulmonary hypertension: Updated recommendations from the Cologne Consensus Conference 2018. *International Journal of Cardiology*. 2018; 272: 11–19.
7. Sajjadi S, Akbari Rad M, Hejazi S, Firoozi A, Akbari Rad F, Azami G, Afrazeh M, and Khodashahi R. The relationship between diabetes mellitus and pulmonary diseases: A systematic review. *Journal of Cardio Thoracic Medicine*. 2018; 6: 274–281.
8. Muralidharan P, Malapit M, Mallory E, Hayes Jr D, and Mansour HM. Inhalable nanoparticulate powders for respiratory delivery. *Nanomedicine: Nanotechnology, Biology and Medicine*. 2015; 11: 1189–1199.
9. Sur S, Rathore A, Dave V, Reddy KR, Chouhan RS, and Sadhu V. Recent developments in functionalized polymer nanoparticles for efficient drug delivery system. *Nano-Structures & Nano-Objects*. 2019; 20: 100397.
10. Pirtarighat S, Ghannadnia M, and Baghshahi S. Green synthesis of silver nanoparticles using the plant extract of Salvia spinosa grown in vitro and their antibacterial activity assessment. *Journal of Nanostructure in Chemistry*. 2019; 9: 1–9.
11. Paranjpe M and Müller-Goymann CC. Nanoparticle-mediated pulmonary drug delivery: A review. *International Journal of Molecular Sciences*. 2014; 15: 5852–5873.
12. Mehta M, Prasher P, Sharma M, Shastri MD, Khurana N, Vyas M, Dureja H, Gupta G, Anand K, Satija S, Chellappan DK, and Dua K. Advanced drug delivery systems can assist in targeting coronavirus disease (COVID-19): A hypothesis. *Medical Hypotheses*. 2020; 144: 110254.
13. Molgat-Seon Y, Peters CM, and Sheel AW. Sex-differences in the human respiratory system and their impact on resting pulmonary function and the integrative response to exercise. *Current Opinion in Physiology*. 2018; 6: 21–27.
14. Verma RK, Ibrahim M, and Garcia-Contreras L. Lung anatomy and physiology and their implications for pulmonary drug delivery. In: Nokhodchi A , and Martin GP (eds). *Pulmonary Drug Delivery: Advances and Challenges*. Wiley On-line Library. 2015: 1–18.
15. Lexmond A and Forbes B. Drug delivery devices for inhaled medicines. In: Page C, and Barnes P (eds.). *Pharmacology and Therapeutics of Asthma and COPD*. Springer. 2016. https://doi.org/10.1007/164_2016_6
16. Rangaraj N, Pailla SR, and Sampathi S. Insight into pulmonary drug delivery: Mechanism of drug deposition to device characterization and regulatory requirements. *Pulmonary Pharmacology & Therapeutics*. 2019; 54: 1–21.
17. Dhanani J, Fraser JF, Chan H-K, Rello J, Cohen J, and Roberts JA. Fundamentals of aerosol therapy in critical care. *Critical Care*. 2016; 20: 1–16.

18. Kuzmov A and Minko T. Nanotechnology approaches for inhalation treatment of lung diseases. *Journal of Controlled Release*. 2015; 219: 500–518.
19. Majumder J, Taratula O, and Minko, T. Nanocarrier-based systems for targeted and site specific therapeutic delivery. *Advanced Drug Delivery Reviews*. 2019; 144: 57–77.
20. Rasul RM, Tamilarasi Muniandy M, Zakaria Z, Shah K, Chee CF, Dabbagh A, Rahman NA, and Wong TW. A review on chitosan and its development as pulmonary particulate anti-infective and anti-cancer drug carriers. *Carbohydrate Polymers*. 2020; 250: 116800.
21. Newman SP. Drug delivery to the lungs: Challenges and opportunities. *Therapeutic Delivery*. 2017; 8: 647–661.
22. Deb PK, Abed SN, Maher H, Al-Aboudi A, Paradkar A, Bandopadhyay S, and Tekade RK. Aerosols in pharmaceutical product development. In: Tekade RK (ed.). *Drug Delivery Systems*. Elsevier. 2020.
23. Nelson HS. Inhalation devices, delivery systems, and patient technique. *Annals of Allergy, Asthma Immunology*. 2016; 117: 606–612.
24. Fröhlich, E. Biological obstacles for identifying in vitro-in vivo correlations of orally inhaled formulations. *Pharmaceutics*. 2019; 11: 316.
25. Lin Y-W, Wong J, Qu L, Chan H-K, and Zhou QT. Powder production and particle engineering for dry powder inhaler formulations. *Current Pharmaceutical Design*. 2015; 21: 3902–3916.
26. Ahmad N, Ahmad R, Alrasheed RA, Almatar HMA, Al-Ramadan AS, Buheazah TM, Alhomoud HS, Al-Nasif HA, and Alam MA. A chitosan-PLGA based catechin hydrate nanoparticles used in targeting of lungs and cancer treatment. *Saudi Journal of Biological Sciences*. 2020; 27: 2344–2357.
27. Dave V, Kushwaha K, Yadav RB, and Agrawal, U. Hybrid nanoparticles for the topical delivery of norfloxacin for the effective treatment of bacterial infection produced after burn. *Journal of Microencapsulation*. 2017; 34: 351–365.
28. Wang T, Bae M, Lee J-Y, and Luo Y. Solid lipid-polymer hybrid nanoparticles prepared with natural biomaterials: A new platform for oral delivery of lipophilic bioactives. *Food Hydrocolloids*. 2018; 84: 581–592.
29. Zheng X, Bian Q, Ye C, and Wang G. Visible light-, pH-, and cyclodextrin-responsive azobenzene functionalized polymeric nanoparticles. *Dyes and Pigments*. 2019; 162: 599–605.
30. Srinivasarao M, Galliford CV, and Low PS. Principles in the design of ligand-targeted cancer therapeutics and imaging agents. *Nature Reviews Drug Discovery*. 2015; 14: 203–219.
31. Chen J, Lin L, Guo Z, Xu C, Li Y, Tian H, Tang Z, He C, and Chen X. N-Isopropylacrylamide modified polyethylenimines as effective siRNA carriers for cancer therapy. *Journal of Nanoscience and Nanotechnology*. 2016; 16: 5464–5469.
32. Mao L, Liu Y, Yang S, Li Y, Zhang X, and Wei, Y. Recent advances and progress of fluorescent bio-/chemosensors based on aggregation-induced emission molecules. *Dyes and Pigments*. 2019; 162: 611–623.
33. Wan Q, Huang Q, Liu M, Xu D, Huang H, Zhang X, and Wei Y. Aggregation-induced emission active luminescent polymeric nanoparticles: Non-covalent fabrication methodologies and biomedical applications. *Applied Materials Today*. 2017; 9: 145–160.
34. Haque S, Pouton CW, Mcintosh MP, Ascher DB, Keizer DW, Whittaker M , and Kaminskas LM. The impact of size and charge on the pulmonary pharmacokinetics and immunological response of the lungs to PLGA nanoparticles after intratracheal administration to rats. *Nanomedicine: Nanotechnology, Biology and Medicine*. 2020;30: 102291.
35. Gartner III TE and Jayaraman A. Modeling and simulations of polymers: A roadmap. *Macromolecules*. 2019; 52: 755–786.
36. Bhavana V, Thakor P, Singh SB, and Mehra NK. COVID-19: Pathophysiology, treatment options, nanotechnology approaches, and research agenda to combating the SARS-CoV2 pandemic. *Life Sciences*. 2020; 261: 118336.
37. Hickey A, Durham P, Dharmadhikari A, and Nardell E. Inhaled drug treatment for tuberculosis: Past progress and future prospects. *Journal of Controlled Release*. 2016; 240: 127–134.
38. Costa-Gouveia J, Aínsa JA, Brodin P, and Lucía A. How can nanoparticles contribute to antituberculosis therapy? *Drug Discovery Today*. 2017; 22: 600–607.
39. Garg T, Rath G, and Goyal AK. Inhalable chitosan nanoparticles as antitubercular drug carriers for an effective treatment of tuberculosis. *Artificial Cells, Nanomedicine, & Biotechnology*. 2016; 44: 997–1001.
40. Rawal T, Parmar R, Tyagi RK, and Butani S. Rifampicin loaded chitosan nanoparticle dry powder presents an improved therapeutic approach for alveolar tuberculosis. *Colloids and Surfaces B: Biointerfaces*. 2017; 154: 321–330.
41. Abdelghany S, Alkhawaldeh M, and Alkhatib HS. Carrageenan-stabilized chitosan alginate nanoparticles loaded with ethionamide for the treatment of tuberculosis. *Journal of Drug Delivery Science and Technology*. 2017; 39: 442–449.
42. Souza F, Fornasier F, Carvalho A, Silva B, Lima M, and Pimentel A. Polymer-coated gold nanoparticles and polymeric nanoparticles as nanocarrier of the BP100 antimicrobial peptide through a lung surfactant model. *Journal of Molecular Liquids*. 2020; 314: 113661.
43. Elzhry Elyafi AK, Standen G, Meikle ST, Lewis AL, and Salvage JP. Development of MPC-DPA polymeric nanoparticle systems for inhalation drug delivery applications. *European Journal of Pharmaceutical Sciences*. 2017; 106: 362–380.
44. Ahmed E. Editorial (Thematic Issue: Nanocarriers Based on Natural Polymers as Platforms for Drug and Gene Delivery Applications). *Current Pharmaceutical Design*. 2016; 22: 3303–3304.
45. Elzoghby AO, Mostafa SK, Helmy MW, Eldemellawy MA, and Sheweita SA. Superiority of aromatase inhibitor and cyclooxygenase-2 inhibitor combined delivery: Hyaluronate-targeted versus PEGylated protamine nanocapsules for breast cancer therapy. *International Journal of Pharmaceutics*. 2017; 529: 178–192.
46. Freag MS. Hyaluronate-lipid nanohybrids: Fruitful harmony in cancer targeting. *Current Pharmaceutical Design*. 2017; 23: 5283–5291.
47. Freag MS, Elnaggar YS, Abdelmonsif DA, and Abdallah OY. Layer-by-layer-coated lyotropic liquid crystalline nanoparticles for active tumor targeting of rapamycin. *Nanomedicine*. 2016; 11: 2975–2996.

48. Ray L. Chapter 4 - Polymeric nanoparticle-based drug/gene delivery for lung cancer. In: Kesharwani P. (ed.) Nanotechnology-Based Targeted Drug Delivery Systems for Lung Cancer. Academic Press. 2019.
49. Buhecha MD, Lansley AB, Somavarapu S, and Pannala AS. Development and characterization of PLA nanoparticles for pulmonary drug delivery: Co-encapsulation of theophylline and budesonide, a hydrophilic and lipophilic drug. *Journal of Drug Delivery Science and Technology.* 2019; 53: 101128.
50. Yeh H-W and Chen D-R. In vitro release profiles of PLGA core-shell composite particles loaded with theophylline and budesonide. *International Journal of Pharmaceutics.* 2017; 528: 637–645.
51. Ahmad N, Ahmad R, Almakhamel MZ, Ansari K, Amir M, Ahmad W, Ali A, and Ahmad FJ. A comparative pulmonary pharmacokinetic study of budesonide using polymeric nanoparticles targeted to the lungs in treatment of asthma. *Artificial Cells, Nanomedicine, and Biotechnology.* 2020; 48: 749–762.
52. Dua K, Bebawy M, Awasthi R, Tekade RK, Tekade M, Gupta G, De Jesus Andreoli Pinto T, and Hansbro PM. Application of chitosan and its derivatives in nanocarrier based pulmonary drug delivery systems. *Pharmaceutical Nanotechnology.* 2017; 5: 243–249.
53. Khan OF, Zaia EW, Jhunjhunwala S, Xue W, Cai W, Yun DS, Barnes CM, Dahlman JE, Dong Y, Pelet, JM, et al. Dendrimer-inspired nanomaterials for the in vivo delivery of siRNA to lung vasculature. *Nano Letters.* 2015; 15: 3008–3016.
54. Muralidharan P, Hayes D, Black SM, and Mansour HM. Microparticulate/nanoparticulate powders of a novel Nrf2 activator and an aerosol performance enhancer for pulmonary delivery targeting the lung Nrf2/Keap-1 pathway. *Molecular Systems Design & Engineering.* 2016; 1: 48–65.
55. Da Silva AL, Cruz FF, Rocco PRM, and Morales, MM. New perspectives in nanotherapeutics for chronic respiratory diseases. *Biophysical Reviews.* 2017; 9: 793–803.
56. Wadhwa R, Aggarwal T, Thapliyal N, Chellappan DK, Gupta G, Gulati M, Collet T, Oliver B, Williams K, Hansbro PM, Dua K, and Maurya PK. Chapter 5 - Nanoparticle-based drug delivery for chronic obstructive pulmonary disorder and asthma: Progress and challenges. In: Maurya PK and Singh S. (eds.) Nanotechnology in Modern Animal Biotechnology. Elsevier. 2019.
57. Joshi M, Nagarsenkar M, and Prabhakar, B. Albumin nanocarriers for pulmonary drug delivery: An attractive approach. *Journal of Drug Delivery Science and Technology.* 2020; 56: 101529.
58. Leal J, Peng X, Liu X, Arasappan D, Wylie DC, Schwartz SH, Fullmer JJ, Mcwilliams BC, Smyth HDC, and Ghosh D. Peptides as surface coatings of nanoparticles that penetrate human cystic fibrosis sputum and uniformly distribute in vivo following pulmonary delivery. *Journal of Controlled Release.* 2020; 322: 457–469.
59. Segura-Ibarra V, Amione-Guerra J, Cruz-Solbes AS, Cara FE, Iruegas-Nunez DA, Wu S, Youker KA, Bhimaraj A, Torre-Amione G, Ferrari M, Karmouty-Quintana H, Guha A, and Blanco E. Rapamycin nanoparticles localize in diseased lung vasculature and prevent pulmonary arterial hypertension. *International Journal of Pharmaceutics.* 2017; 524: 257–267.
60. Makled S, Nafee N, and Boraie N. Nebulized solid lipid nanoparticles for the potential treatment of pulmonary hypertension via targeted delivery of phosphodiesterase-5-inhibitor. *International Journal of Pharmaceutics.* 2017; 517: 312–321.
61. Gupta N, Ibrahim HM, and Ahsan F. Peptide–micelle hybrids containing fasudil for targeted delivery to the pulmonary arteries and arterioles to treat pulmonary arterial hypertension. *Journal of Pharmaceutical Sciences.* 2014; 103: 3743–3753.
62. Iyer R, Hsia CCW, and Nguyen KT. Nano-therapeutics for the lung: State-of-the-art and future perspectives. *Current Pharmaceutical Design.* 2015; 21: 5233–5244.
63. Mukherjee S, Mazumder P, Joshi M, Joshi C, Dalvi SV, and Kumar M. Biomedical application, drug delivery and metabolic pathway of antiviral nanotherapeutics for combating viral pandemic: A review. *Environmental Research.* 2020; 191: 110119.
64. Chen X, Han W, Wang G, and Zhao X. Application prospect of polysaccharides in the development of anti-novel coronavirus drugs and vaccines. *International Journal of Biological Macromolecules.* 2020; 164: 331–343.
65. Kerry RG, Malik S, Redda YT, Sahoo S, Patra JK, and Majhi S. Nano-based approach to combat emerging viral (NIPAH virus) infection. *Nanomedicine: Nanotechnology, Biology and Medicine.* 2019; 18: 196–220.
66. Uludağ H, Parent K, Aliabadi HM, and Haddadi A. Prospects for RNAi therapy of COVID-19. *Frontiers in Bioengineering and Biotechnology.* 2020; 8: 916.
67. Feng Q and Tong R. Anticancer nanoparticulate polymer-drug conjugate. *Bioengineering & Translational Medicine.* 2016; 1: 277–296.
68. Marasini N, Haque S, and Kaminskas LM. Polymer-drug conjugates as inhalable drug delivery systems: A review. *Current Opinion in Colloid & Interface Science.* 2017; 31: 18–29.

14

Polymeric Nanoparticle-Based Drug–Gene Delivery for Lung Cancer

Keshav Moharir[1], Vinita Kale[2], Abhay Ittadwar[3], and Manash K. Paul Dr. PhD[4]
[1]Assistant Professor, Gurunanak College of Pharmacy, Nagpur, India
[2]Professor Dept. of Pharmaceutics, Gurunanak College of Pharmacy, Nagpur, India
[3]Professor and Head, Gurunanak College of Pharmacy, Nagpur, India
[4]Scientist, Principal investigator, Division: Pulmonary and Critical Care Medicine, David Geffen School of Medicine, University of California Los Angeles, California, USA

14.1 Introduction and Background

Cancer is the second leading cause of death globally, with lung cancer securing first spot in terms of mortality rate for data of men and women combined, according to the Global Cancer Statistics 2020: GLOBOCAN reports. The American Cancer Society estimates about 235,760 new cases of lung cancer (119,100 in men and 116,660 in women) and about 131,880 deaths from lung cancer (69,410 in men and 62,470 in women), approximately 25% of cancer-related deaths in the United States for the year 2021. Based on the microscopic manifestation of cancerous cells, lung cancers are classified as non-small cell lung cancer (NSCLC) and small cell lung cancer (SCLC). Among all lung cancers, approximately 80–85% are NSCLC while remaining 10–15% are SCLC. Adenocarcinoma, large cell carcinoma, and squamous cell carcinoma are grouped as subtypes of NSCLC as they have similar clinical features, outcomes and courses of treatment (1). Lung cancer is an important public health concern due to the high mortality rate and[1] relatively low 5-year survival rate ("US National Institute of Health, National Cancer Institute, SEER Cancer Statistics Report 2010-2016" Accessed from 25/12/2020).

As early-stage detection of lung cancer is difficult due to asymptomatic conditions and lack of robust detection techniques, sizeable numbers of patients are at an advanced stage with metastasis to tissues and organs away from the lungs. Chemotherapy, radiotherapy, immunotherapy, or a combination of these are possible treatment options apart from surgical removal of localized lung portion, that too depending on stage, malignancy, and metastasis at the time of diagnosis (2). Conventional treatment regimens include chemotherapy with drug delivery systems that often lack specificity and therapeutic efficacy toward target tumor cells. The non-selectivity of chemotherapeutics toward target cells necessitates higher doses, resulting in adverse effects on normal cells and tissues. In addition, subtherapeutic bioavailability leads to chances of multiple drug resistance and hence relapse of cancer (3,4). Thus, it becomes imperative to overcome these limitations and develop drug delivery systems that direct the drugs in a controlled manner at cancer cells only.

Novel drug delivery systems incorporate nanotechnology-enabled carriers or smart systems, which get rid of problems associated with conventional therapeutic strategies (5). Nanomedicine exhibits improved site-specific targeting, increased drug bioavailability, reduced toxicity, and agility of loading multiple components having multi-functionality for therapy as well as diagnosis (6). These nanomaterials include but are not limited to polymeric NPs, carbon nanotubes (CNT), dendrimers, ethosomes, polymeric micelles, liposomes, solid lipid nanoparticles (SLN), exosomes, metal-based NPs , pH-responsive smart polymeric materials, niosomes, quantum dots, and mesoporous silica substrates (7–9). Similarly, polymeric nanocarriers also have been explored as non-viral vectors that facilitate to safely direct genetic materials at desired target sites (10,11). Nanocarriers are also being investigated to carry tumor-targeting moieties like antibodies, peptides, proteins, siRNA, and aptamers (single strand RNA or DNA selectively binding with precise targets) which bind covalently to the external surfaces of these carriers (12–14).

This chapter attempts to provide updates on polymeric NPs as carriers for drugs and genetic therapeutic agents for lung cancer treatment. Also discussed are the mechanism of drug release, examples, and future prospects. Lung cancer pathophysiology is not discussed, as it is covered elsewhere in this book.

14.2 Nanoparticles for Delivering Drugs and Genetic Materials in Cancer: Rationale and Need

Polymeric NPs are often colloidal, typically of a size ranging below 1000 nm. The need for nanocarriers arises

1 Replace 20.5% by 21.7% and 2010-2016 by 2011-2017 accessed 25th June 2021. https://seer.cancer.gov/statfacts/html/lungb.html

DOI: 10.1201/9781003046547-14

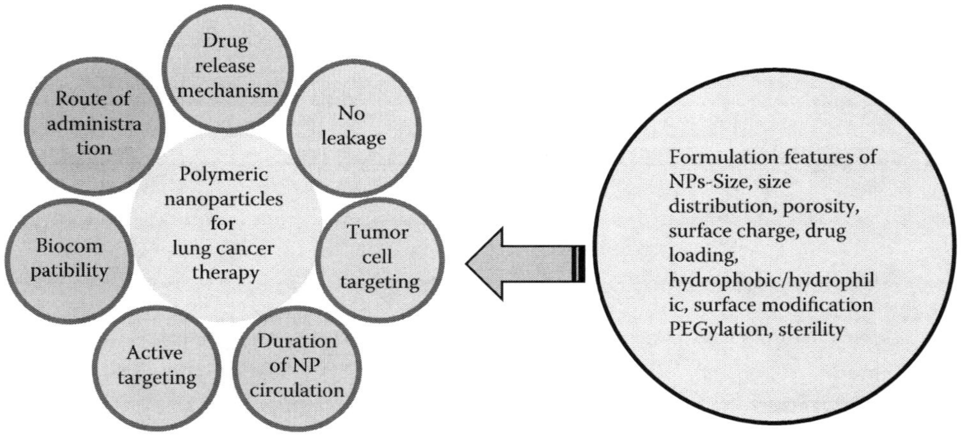

FIGURE 14.1 Requirements of Ideal NPs for Cancer Therapies. Required Properties in Turn Depend Upon Formulation Features of NPs and the Nature of Ingredients

mainly out of limitations, drawbacks, and disadvantages associated with conventional drug delivery systems as discussed previously. NPs as carriers of anti-cancer agents have several convincing points. They can be available in different sizes, shapes, geometries, structures, architecture, and chemical properties (15). Reasons for developing polymeric NPs include

1. Overcoming solubility and stability associated with hydrophobic anti-cancer drug molecules. Paclitaxel is a classical example of a hydrophobic drug used in NSCLC. The incorporation of paclitaxel in albumin-stabilized NPs improved solubility and clinical response (16).
2. NPs improve biodistribution and retention time of the anti-cancer agents (like peptides, proteins, nucleic acids) in the tumor microenvironments. Stability and half-life of paclitaxel improved when conjugated with polyethylene glycol (PEG) (17,18).
3. Enhancement in targeting of tumor cells specificity by "stealth mechanism." PEGylation (conjugating with PEG) results in reduced detection by the host immune system with lower uptake by the reticuloendothelial system (RES) (19).
4. NPs can be fabricated to release an anti-cancer drug in response to a stimulus or trigger. Doxorubicin has been formulated into a dual pH-responsive nanoparticulate system that uses tumor extracellular pH for cell internalization of drug-loaded NPs and drug release at acidic endosomal pH (20).
5. Nanoparticulate systems can reduce drug resistance by tumor cells. NPs can circumvent the drug resistance of tumor cells by activation of apoptosis and manipulating drug efflux (21).
6. Nanocarriers can make it feasible for selective accumulation of systemically administered anti-cancer agents at tumor sites via enhanced permeability and retention (EPR) effect of the poor tumor vasculature and underdeveloped lymphatic drainage (22).

14.3 Types of Nanoparticles Employed in Cancer Therapeutics

Several types of nanocarriers have been explored that can deliver anti-cancer drugs at tumor sites. The specific receptors that are highly expressed in tumors can be targeted by NPs. They are effectively used not only in treatments but also in the early detection of cancers. The ability of NPs for improved bioavailability and target will depend on size, shape, surface charge and many more physicochemical characteristics as shown in Figure 14.1. Irrespective of their type, NPs should have ideal properties that will ensure efficient, targeted anti-cancer drug delivery at required cells with longer circulation/retention time with minimum toxic effects. These required properties are mentioned in Figure 14.1.

Depending on their composition and structure, NPs can be broadly classified as (a) polymer-based such as polymeric micelles, polymeric NPs (23), PEGylated and peptide dendrimers, polymerosomes, polymer conjugates, and nanocapsules; (b) metal-based nanocarriers, for example, gold, silver, iron oxide and platinum NPs (24); (c) lipoidal nanocarriers, like liposomes (25), nanoemulsions (26), solid lipid nanoparticles, nanostructured lipid nanocarriers and lipoplexe s (d) carbon-based nanosystems (27); (e) viral vector as nanosystem (28); (f) combinations of various nanoparticle systems as hybrids (29); (g) silica-based nanocarriers (30); and (h) miscellaneous like [2] and quantum dots (31,32). These structurally divergent nanocarrier systems are depicted in Figure 14.2.

14.4 Polymeric Nanoparticles for Cancer Therapy

Polymeric nanoparticles (PNPs) are described as colloidal or solid particles in size range of 10–1000 nm. PNPs are widely explored as nanocarriers in sustained and controlled drug release and delivery. They may be in the shape of nanospheres or

2 Replace 'dendrimers' with ' PEGylated and peptide dendrimers ' , keeping the same references.

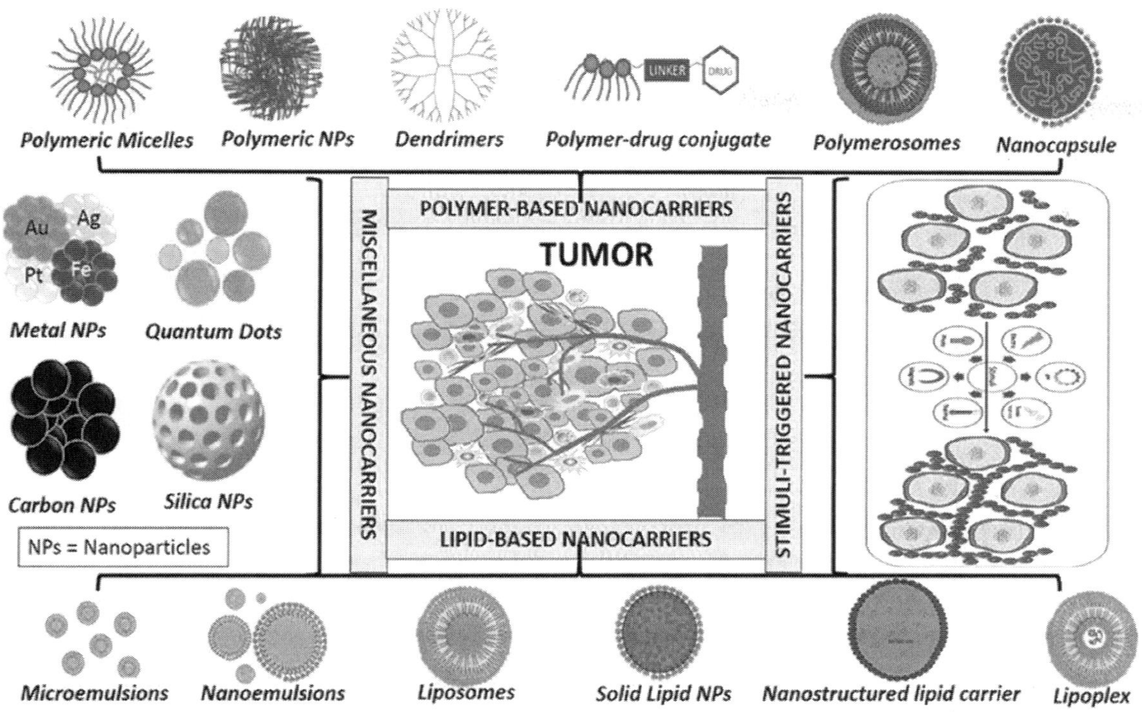

FIGURE 14.2 Comprehensive Overview of Types of Nanoparticles for Lung Cancer Therapy. Adapted from (33)

nanocapsules. While nanospheres are spherical, solid nanoparticles, where drug molecules are either layered on them or inside them as matrix, nanocapsules are solid shells with an inner cavity where drug molecules reside as core material.

Different types of polymers are used, namely, natural or synthetic for preparing polymeric nanoparticles. The last decade has seen exploration of biodegradable polymers such as albumin, poly (lactic-co-glycolic) acid (PLGA), chitosan, poly (lactic acid) (PLA), *N*-(2-Hydroxypropyl) methacrylamide (HPMA) copolymers gelatin, polyalkyl-cyanoacrylates, and polycaprolactone due to their properties suitable for controlled and sustained release as well as biocompatibility. Oral drug delivery has not been practicable for lung cancer treatment, as a minimum therapeutic drug level is not achieved at the lung tumor site for a longer time. In such scenarios, polymeric nanoparticles are promising and a ray of hope as newer drug delivery strategies are evolving (34). The NPs have further enhanced their efficacy by incorporating RNA interference (RNAi)-involving technology such as siRNA (small interfering RNA), shRNA (short-hairpin RNA), and miRNA (micro RNA), that selectively knock down the genes of interest. Thus, a combination of this technology with chemotherapeutic agent–loaded NPs could be a potentially advantageous strategy for cancer therapy (35). Polymeric NPs are synthesized by self-assembly of copolymers or block copolymers that contain more than two polymer chains with different solubility characteristics. Physicochemical properties of polymers can be modulated to achieve desired drug dissolution and release rates from the polymeric NPs. The targets usually selected for targeting lung cancer cells are several of the specific overexpressed receptors such as EGFR (epidermal growth factor receptor), FRA (folate receptor - α), transferrin, integrin, CD44, IL22 (interleukin), VEGF (vascular endothelial growth factor), ALK (anaplastic lymphoma kinase), ROS (receptor tyrosine kinase) and many more (36). We further discuss different polymeric nanoparticles for cancer therapy.

14.4.1 Dendrimers

Dendrimers, also known as "starburst polymers" are polymeric structures with core that expand with successive branches. They are spherical and appear as layered structures, each layer of branches being called *generations*. The branch closest to center G0 is called G1, while the successive branches are called G2, G3 , and so on. They are about 2–10 nm in size with three-dimensional structures, hyperbranched, and can be tailor-made to the need. Their typical features make them suitable for easy structural modifications and conjugation with nucleic acids, i.e. RNA and DNA (37). They can be flanked with different functional groups on the periphery thus modifying the entrapment capacity. They have distinctive properties like good polydispersity index, water solubility, modifiable structure, and space available in the internal structure. The most commonly studied dendrimers are polyamidoamine (PAMAM) (38). PAMAM dendrimers are biocompatible, nonimmunogenic and water soluble, with peripheral amine functional groups that are easy to modify for targeted drug delivery.

Dendrimers have shown the capability to target specific cells and tissues. Conjugates of dendrimers-antibody have proven better than free antibodies in terms of efficiency (39). PEGylated dendrimers have shown improved solubility, higher bioavailability , and lower toxicity (40). Dendrimers modified by 2,2-bis (hydroxymethyl) propanoic acid when conjugated with doxorubicin exhibited all these advantages in vivo (41). It has also been observed that surface modification of dendrimers by peptides, folic acid, and sugar groups improve efficacy of the bound or conjugated drug as these agents accommodate anti-cancer drugs in a better manner (42).

14.4.2 Polymeric Magnetic Nanoparticles

The magnetic nanoparticles exhibit super paramagnetism. Thus drug-loaded magnetic NPs can be moved to the exact tumor location by applying an external magnetic field. They improve the therapeutic efficacy of the drug and reduce associated side effects. The polymeric super-paramagnetic iron oxide nanoparticles (SPIONs) are suitable for coupling with micelles or polymers to act as carriers of drugs (43). In one study, platinum, cisplatin, carboplatin, or oxaliplatin with holmium-165 as a radionuclide agent were incorporated in magnetic nanoparticles. The therapy was selectively targeted toward NSCLC. Results indicated that neutron-activated holmium-drug complex (^{166}HoIG-cisplatin) shows more cytotoxicity toward NSCLC cell line A549 than the free drug. In the same study, external magnetic field application achieved a higher concentration of radium as compared to normal cells (44). SPION — folic acid nanoparticles coated with PEI (polyethylenimine) and PEG polymers loaded with doxorubicin — are described by Huang et al. Dual targeted drug delivery was achieved as drug release was magnetically delivered, as well as dependant on pH change in the tumor microenvironment, with enhanced anti-cancer activity (45).

14.4.3 Polymeric Micelles

These are 5–100 nm size colloidal particles, aggregates of surfactants, usually in the aqueous medium. The structure shows a polar head and non-polar tail wherein the inner core cavity of a micelle is hydrophobic and the hydrophilic region faces solvent. Micelles improve the solubility of poorly water-soluble anti-cancer agents, prolong circulation time and induce passive targeting to tumor location by the EPR (enhanced permeation and retention) process (46). The poorly water - soluble hydrophobic drug camptothecin was embedded in chitosan self-assembling micelles which were conjugated with o-nitrobenzyl succinate, a light-sensitive agent. The dual impact of pH and UV radiation resulted in targeted drug release in a controlled manner with quick internalization in MCF-7 tumor cells (47). Similarly, Karabasz et al. reported hydrophobic curcumin loaded sodium alginate micelles which exhibited micellar stabilization of curcumin for a longer time. Strong cytotoxic activity was observed for tumor cells–mouse endothelial cell line while normal cells were unaffected (48). Another micellar formulation was designed with hydrazine as a pH sensitive linker. Jin et al. formulated a dextran hydrazine conjugate with doxorubicin and deoxycholic acid. When inside a tumor cell, the hydrazine linker is hydrolyzed at endosomal pH since the conjugate is acid sensitive. Doxorubucin release was then quick, inhibiting growth of MCF-7 cancer cells *in vitro* with low systemic toxicity (49).

14.4.4 Carbon Nanotubes (CNTs)

These are cylindrical structure made of graphene with diameter in the range of nanometers. Their unique property is the length-to-diameter ratio greater than 1,000,000. CNTs are classified as single-walled carbon nanotubes (SWNTs) or multi-walled carbon nanotubes (MWNTs) depending on layers of graphene from which they are formed. They are recognized as a remarkable drug carrier due to high surface area with optical and electronic properties. Their walls can accommodate more than one drug compound (50). The last two decades have witnessed tremendous research on CNTs for various applications in healthcare, like imaging, drug delivery, biosensors, cancer therapies and tissue scaffold fabrication (50). Since SWNTs act as high absorbing materials with absorption in the wavelength range of near infrared and Raman spectroscopy, in vivo photothermal treatment for cancer was useful in animal models (51). Liu et al. bonded anti-cancer drug to the CNT wall covalently and reported that blood circulation time of PEG conjugated paclitaxel, bonded with SWNTs via ester linkage, was notably higher as compared to free paclitaxel. This showed enhanced drug efficacy when tested for 4T1 murine cancer in mice (52).

14.4.5 Quantum Dots (QDs)

QDs were first reported in the 1980s, but their use in biomedical research started later owing to specific advantages, such as low toxicity as compared to other inorganic nanomaterials, improved aqueous solubility and strong fluorescence. QDs have shown good reports with respect to cancer diagnostics by way of imaging studies. When given intravenously, they highlight the tumor region by exhibiting strong fluorescence in the presence of markers (53). A controlled drug delivery system of anti-cancer drugs was demonstrated by zinc oxide QDs, where pH sensitive element was incorporated with doxorubicin. PEG was applied over QDs, while doxorubicin was bound through Zn ions. Doxorubicin released in surrounding tumor pH and killed cancer cells; here it was proposed that zinc oxide–loaded doxorubicin QDs improved the apoptosis process (54).

14.4.6 Polymeric Nanoparticles Based on Natural Polymers–Drug Conjugates

Natural polymers have an added advantage of biocompatibility. They can be modified as a whole entity or on the surface to tune their properties as desired. Conjugation is possible with suitable ligands, making them targetable toward selected tissues. They can also give controlled and sustained drug delivery when desired. Following are some examples of nanoparticles of natural polymers investigated for cancer therapy.

Chitosan nanoparticles prepared by ionotropic gelation and loaded with ascorbic acid exhibited antioxidant property. *In vitro* studies indicated lower survival rate of HeLa cancer cells while normal cells were unaffected (55). In yet another study, Nascimento et al. reported targeted chitosan nanoparticles for epidermal growth factor receptor (EGFR), which delivered siRNA (small interfering RNA) to treat non small cell lung cancer (NSCLC). Mad2 siRNA was complexed with PEG 2000 conjugated chitosan to form NPs. The selective uptake by NSCLC was higher as compared to non-conjugated NPs. Apoptosis of cancer cells was observed as siRNA eliminated Mad2 (a mitotic checkpoint component) (56).

NPs of **hyaluronic acid** (HA) are known to coat siRNA and smaller molecules for drug delivery. Anti-tumor combination therapy by HA-based NPs has been reported by Ganesh et al. They encapsulated cisplatin for drug delivery with siRNA and near IR (NIR) light region absorbing dye indocyanine green (ICG) for tumor imaging. Since siRNA and HA are negatively charged, HA must be loaded with cationic moieties like amines to encapsulate siRNA. Such an siRNA in a study showed selective uptake by solid tumor cells in mice models (57).

Thiolated **sodium alginate** NPs conjugated to docetaxel showed selective and strong toxic effects toward HT-29 cells, excluding normal cells. Since sodium is stable towards enzymatic degradation and hydrolytic reactions, its NPs are stable at lower pH also (58). **Dextran** is another abundant natural polymer that can self-assemble to NPs and incorporate anticancer drugs in a targeted manner. For example, dextran with curcumin forms NPs in water by self-assembly. Methotrexate was released at a sustained rate from this complex with fast internalization inside the tumor cells. Curcumin and methotrexate showed synergistic action, reinforcing significance of drug–conjugate significance in drug delivery for cancer (60).

Albumins are highly water soluble with low immunity and no toxicity. Drugs can form covalent bonds with albumin due to the presence of many binding sites on albumin. Albumin was found to improve the solubility and bioavailability of curcumin when encapsulated with albumin nanoparticles. Curcumin showed higher anti-tumor activity with better pharmacokinetics *in vivo*(59). Therapeutic efficacy of gelatin NPs encapsulated resveratrol was investigated by Karthikeyan et al. Drug release was in sustained manner with rapid uptake by NSCLC tumor cells. Stronger anti-tumor activity of resveratrol NPs was observed when compared to free drug (61).

14.4.7 Polymeric Nanoparticles from Synthetic Polymers

Nanocapsulation of the drugs is beneficial for cancer therapy in many ways, especially when compared with conventional drug delivery systems. They help drug targeting without affecting normal cells. Solubility as well as absorption rate of the drug is improved. Biodegradable polymers like poly (lactic acid), poly (lactic-co-glycolic) acid (PLGA), modified chitosan and albumin, (PLA), polyalkyl- cyanoacrylates, HPMA (hydroxyl propyl methyl acrylate) , and polycaprolactone are nowadays preferred over non-biodegradable alternatives because the former have better biocompatibility and control over drug release. Following are few examples of polymeric NPs for cancer therapy.

Polylactic acid and polyethylene glycol (PLA-PEG) NPs with taxane as anti-cancer drug improved chemoradiotherapeutic activity *in vivo* and *in vitro*(62). Recently, Kim et al. developed cisplatin and paclitaxel NPs with PLA-PEG polymers for advanced NSCLC, and the same polymeric NP system was used to deliver gemcitabine for treating metastatic lung cancer (63). An oral formulation of anti-tumor drug was reported by Zhao et al. PCL and self-synthesized TPGS-*b*-(PCL-*ran*-PGA) diblock copolymer; mucoadhesive nature of the polymer resulted in longer stay of formulation at tumor site followed by higher drug uptake compared to free drug (64).

Paclitaxel nanocapsulated with mPEG-PDLLA [monomethoxy-poly (ethylene glycol)-block-poly(D,L-lactide)], popularly known as Genexol-PM is indicated for advanced NSCLC. It has completed phase II clinical trials. Similar phase II clinical trials for Genexol-PM and gemcitabine formulation are also completed as reported by clinical trials registry (https://clinicaltrials.gov/ct2/show/NCT01770795).

TABLE 14.1

Nanoparticles Undergoing Clinical Trials or Approved for Use in Lung Cancer Therapy

Type of Nanoparticle	Drug	Specific Target	Stage of Clinical Trial/Approved	Brand Name	Reference
Polymeric micelles	Paclitaxel	NSCLC	Phase III	Genexol-PM	(117)
Polymeric nanoparticles	Albumin particle bound paclitaxel	Advanced NSCLC	Approved	Abraxane	(118)
Encapsulated liposome	FUS1 (TUSC2)	Lung cancer	Phase I/II	Oncoprex	
PEGylated liposome	188Re-N,N-bis (2-mercaptoethyl)-N0,N0-diethylethylenediamine	Advanced solid tumors	--	188Re-BMEDA-liposome	
Bacteria derived nanoparticles	Anti-EGFR bispecific antibody minicell with microRNA	NSCLC	Phase I completed	TargomiRs	
Cyclodextrin based nanoparticle conjugates	Camptothecin	Small cell lung cancer	Phase II completed	CRLX101	
PEG polymer coated silica-gold nanoshells for thermal ablation	Photothermal ablation of target lesions	Metastatic lung cancer	1 Completed and 1 terminated	AuroLase	
Polymeric micelles	Epirubicin	Solid tumors	Phase 1 completed	K-912/NC-6300	(119)
Polymeric micelles	7-Ethyl-10-hydroxycamptothecin	Advanced solid tumor	Phase II	NK012 (Nippon Kayaku)	(66)
PEGylated docetaxel nanoparticles	Docetaxel	Various solid tumors	Phase I	NKTR-105 (Nektar)	
Polymeric micelles	Docetaxel	Solid tumors	Phase I	CriPec	
Liposomes	Cisplatin	Various advanced cancers	Phase II	LiPlaCis	

Gelatin NPs were nanocapsulated with biotin embedded epithelial growth factor (EGF) tumor–specific ligand for targeting lung cancer. Increased cellular uptake of cisplatin on lung adenocarcinoma A549 cells *in vitro* and *in vivo* was demonstrated by aerosol administration to lung tumor cells in a mouse model (65). Table 14.1 illustrates nanoparticles undergoing clinical trials or approved for use in lung cancer therapy.

14.4.8 Polymersomes

Polymersomes are polymeric self-assembling vesicles synthesized from amphiphilic block copolymers. The polymers used are biodegradable, incorporating hydrophobic and hydrophilic blocks. As compared to liposomes, they have improved stability and longer circulation time with ability to control the rate of drug release (67). Vesicle membrane of polymersomes is thicker than that of liposomes, owing to differences in molecular weights of polymers used in their formulation. Therefore, polymersomes are more stable, dissociate slowly, show low drug loss in circulation and can withstand higher fluid pressure (68). These are one of the few vesicular nanoparticles that can incorporate both hydrophilic and hydrophobic drugs.

Recently, Hou et al. loaded doxorubicin hydrochloride and Nile red dye into the polymersome and drug release was induced by UV radiation. Polymersomes were prepared with poly (N,N'-dimethylacrylamide)(PDMA) as the hydrophilic block and poly (o-nitrobenzyl acrylate) (PNBA) as the hydrophobic block by reversible addition-fragmentation chain transfer (RAFT)polymerization method. It was reported and concluded that release of hydrophobic drug can be regulated at specific wavelengths due to fluorescence quenching by this technique (69).

14.4.9 PEG-Drug Conjugates

PEG has achieved important place in drug delivery systems involving nanoparticles because they (a) are not immunogenic and antigenic, (b) have excellent solubility in the majority of solvents, and (c) are FDA approved. In addition, the PEGylation process involving surface coating of polymeric NPs with PEG helps them escape uptake by immune cells and avoids phagocytosis. The RES (reticuloendothelial system) fails to recognize PEGylated NPs as foreign objects, resulting in maximum drug bioavailability (66).

14.5 Inhalable and Pulmonary Nanoparticles for Lung Cancer

Because it is non-invasive, the systemic circulation of drugs is mostly avoided when drugs are delivered to lungs by the inhalable route. The drug reaches the target site having minimum interference with other organs. However, there are barriers like ciliated columnar epithelial cells that flush out external objects from upper airways, while deep at the alveoli level macrophages are highly functional at removing foreign particles. In recent years, surface modified, functional nanoparticles have been investigated, which overcome these barriers. These nanocarriers are capable of bearing the force of ejection from nebulizers and aerosolized packages (70).

14.5.1 Polymeric NPs with Chemotherapeutic Agents by Inhalation

Biodegradable polymeric NPs can hold chemotherapeutics and deliver them to lungs efficiently. They should be in the size range of 0.5–5 µm in order to reach and stay at the desired site. Further, DPIs (dry powder inhalers) prepared by spray drying or freeze drying are preferred, as they remain stable for longer duration due to their dry nature (71–73).

PLGA NPs containing TAS-103 anti-cancer agent when spray dried with trehalose, on inhalation achieved about 13 times higher drug concentration in rat lungs when compared with free drug administered by the intravenous route (74). In another study, human serum albumin-doxorubicin (HSA-DOX) NPs with average particle size of 341.6 nm were formulated, wherein HAS was prior treated with octyl aldehyde to make self-assembled NPs hydrophobic. As the particle size was very small and unsuitable for direct delivery to lungs, they were aerosolized to deliver drug deep inside lungs (74). Nephrotoxicity and neurotoxicity of cisplatin was reduced when it was conjugated with hyaluronan to form NPs and delivered to lungs via pulmonary route. At the same time, cisplatin concentration showed about five times higher cytotoxic activity for lung cancer as compared to free drug given intravenously (75). All these examples confirm efficacy of the inhalable route of drug delivery for lung cancer, but little is said about improvement in stability, reproducibility in formulation and feasibility to scale up these techniques to a mass scale.

14.5.2 Pulmonary Gene Delivery by NPs

Gene delivery can be accomplished either by viral or nonviral vectors. While viral vectors may cause serious unwanted effects like inflammation, insertion into normal genes, activation of proto oncogenes and immune response, polymeric nanoparticles are superior for gene delivery in terms of safety, production, stability and handling (76). Polyethyleneimine (PIE) is the polymer of choice as it bonds with nucleic acid easily and strongly. Transfection of nucleic acid from this polymer is easy as PEI has positive charges that interact with negatively charged moieties in RNA and DNA (77).

A study by Jia et al. reported that nanoparticles of PEI-IL-12 (interleukin 12) complex achieved 87% reduction in nodules of lung metastasis by 0.5 mm when compared with free IL-12 in a mice model for metastatic lung cancer. Further, no adverse effects of IL-12 on other organs were observed when given for more than a month, twice in a week, as an aerosol (78). In yet another study, the toxicity associated with amines in PEI was reduced by binding with chitosan to form co-polymeric NPs. These chitosan–PEI NPs were loaded with shRNA (short RNA) in a mouse with lung cancer. Probably, chitosan shielded amines of PEI, reducing its toxicity and improving efficacy in A549 and HeLa cell lines (79). One more study used a co-delivery concept where the chemotherapeutic agent as well as RNA were loaded onto suitable NP and delivered to the lung cancer site

TABLE 14.2

Overview of Polymeric Nanoparticle Formulations for Gene Delivery in Lung Cancer Therapy

Genetic Material Used	Polymer	Type of Model for Lung Cancer	References
siRNA	Polymeric micelles	A549	(120)
siRNA	Chitosan-graft-PEI	A549	(121)
shRNA	PLL-alkyl-PEI	--	(122)
Plasmid DNA	PEG - PEI	A549	(123)
Plasmid DNA	PLGA	A549	(124)
Oligonucleotides/siRNA	PLGA/chitosan	A549	(125)
siRNA	PEI/AH-PEG	A549	(15)
2'-O-methyl-RNA	Chitosan	A549	(15)
siRNA	PLGA	A549	(82)

successfully by the pulmonary route. Doxorubicin–PEI complex was conjugated with siRNA while pH responsive butyl carbazate was used as the linker. More than 60% apoptosis rate was observed when the siRNA-doxorubicin were given as polyplexes in combination, while individual doxorubicin and siRNA expressed around 18% and 36% apoptosis respectively (80).

Gene delivery to lungs by inhalation or pulmonary route is an attractive and promising drug delivery system to treat lung cancer. Multifunctional nanoparticles are the first choice for this, as multiple studies show that polyplexes or nanoparticles given by this route are more efficacious as compared to drugs given without complex/conjugation in free form (81). Table 14.2 gives an overview of polymeric nanoparticles that engage genes or nucleic acids for the treatment of lung cancer.

14.6 New Gene Delivery Platforms–Cell Penetrating Peptides (CPP) and CRISPR-Cas9 for Lung Cancer Therapy

Gene delivery using peptide based vectors is an efficacious gene transport medium because it provides several advantages. They can bind with plasmid DNA electrostatically, cross cells by endocytosis easily, having quick translocation and avoid enzymatic degradation of DNA by carrying them toward specific receptors. Cell penetrating peptides (CPP) are nonviral peptides with up to 30–35 amino acids sequence. Net charge on CPPs is positive due to arginine presence that helps their conjugation with negatively charged cellular membrane contents electrostatically (83,84). To quote an example, improved cytotoxic activity of angiotensin II type 2 (AT2R) in a xenograft mouse model for lung carcinoma was reported when delivered as an aerosol spray. The cytotoxicity for tumor cells was observed when AT2R was complexed with a CPP called dimerized TAT and pDNA (plasmid DNA) in the presence of condensing calcium ions. CPP helped gene delivery by direct penetration, endocytosis and translocation (85).

Clustered Regularly Interspaced Short Palindromic Repeats/ CRISPR associated protein 9 (CRISPR/Cas9) system is a versatile and flexible gene editing tool that gives the ability to add or delete DNA in the genome in a sequence-specific manner. The CRISPR/Cas9 tool consists of two parts, endonuclease and single stranded guide RNA (sgRNA). Endonuclease cuts both DNA strands in a specific manner as directed by sgRNA. The genome repair takes place by double-stranded break (DSB) repair mechanism. In translational research, CRISPR/Cas9 is used currently in studying prevention-treatment-prognosis of all types of cancers. Many studies are being envisaged for use of CRISPR/Cas9 in lung cancer treatment. In one study, cancer metastasis-associated gene screening was carried out to study how mutagenesis takes place and what genetic changes actually occur in this process. Non-metastatic lung cancer cells with sgRNA library were injected into mice. Formed cells from these mice were transplanted to mice with weak immunity. These mice exhibited metastatic lung cancer, from which enriched sgRNA were extracted for sequencing. This helped to identify and build a gene pool associated with lung cancer metastasis (86). The CRISPR/Cas9 system due to its high molecular weight and complex nature, is difficult to deliver correctly. Viral, nonviral and physical vectors have been employed for its delivery. Nonviral vectors such as nanoparticles show low toxicity and immune reaction as compared to viral vectors. NPs containing cationic polymers can be used safely as CRISPR/Cas9 are negatively charged and can form electrostatic complex with them (87). These NPs can cross cellular membranes by endocytosis or phagocytosis. Investigative studies also proved that polymeric NPs enhance target cancer genome editing with endonuclease leading to successful CRISPR/Cas9 system delivery (88).

14.7 An Overview of Targets for Polymeric NP in Lung Cancer Therapeutics

Targeting of anti-cancer drug-containing nanoparticles is necessary to improve the uptake of therapeutic agents by cancer cells instead of normal cells. There are two key modes through which anti-cancer agents carrying polymeric nanoparticle can target lung cancer cells. Passive targeting goes after enhanced permeation and retention (EPR–a mechanism by which non-targeted drugs and prodrugs concentrate in tissues with greater vascular permeability, as in cancerous cells, when compared with normal cells) effect in surrounding tumor microenvironment. Active targeting is the process by which an interaction between nanoparticle ligand and

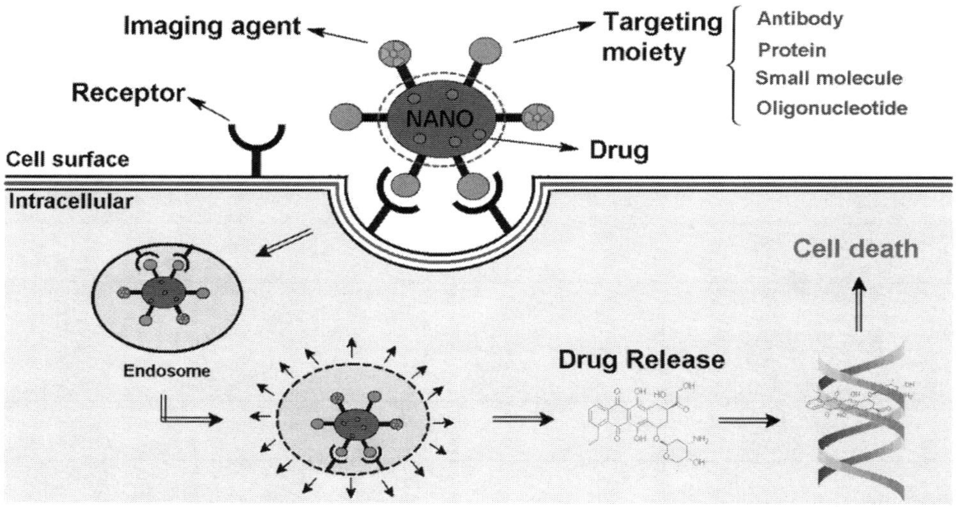

FIGURE 14.3 Schematic Representation of the Mechanism of Action of Targeted Nanoparticle (NP). Adapted from (95). Initial Step Involves Binding of NP to Tumor Cell Specific Receptor Followed by Internalization through Endocytosis. NP Breakdown Releases Anti-cancer Agent That Kills Tumor Cells Selectively

tumor specific interaction takes place, mainly through endocytosis (89). The effectiveness with which active targeting works is governed by surface charge, ligand-receptor binding affinity, stability and molecular size of NPs, and ligand concentration at receptor sites (90). Similarly, the course of action of nanoparticles is also influenced by tumor related biological changes such as angiogenesis — formation of new blood vessels due to proliferation of cells, tumor pericytes — accumulate cancer stem cells, tumor fibroblasts, tumor extracellular matrix, tumor assisted macrophages, microenvironment pressure change, hypoxia and pH change. These changes are sometimes utilized as opportunities to deliver drugs or genes to tumor sites specifically (33).

14.7.1 Passive Targeting

Passive targeting follows EPR, meaning pathological changes and tumor microenvironments are utilized for drug delivery to tumors. The leaking endothelial cells in tumor with gaps are used for drug delivery. Passive drug delivery systems using this mechanism are called first generation nanomedicines (91). No receptor specific ligand is attached to the nanoparticles, and drug release is by a natural, simple diffusion mechanism. In general, passive targeting with diffusion process is not full proof in terms of cytotoxicity of the drugs and targeting tumor cells (92). Further, pathophysiological changes associated with tumors cause limitations in drug reaching the tumor. This has certainly raised concerns and the need for development of second generation nanocarriers for cancer treatment (93). Thus, active targeting has evolved with expectations of better clinical outcomes.

14.7.2 Active Targeting

Active targeting involves binding or attaching ligand with high affinity to the nanocarriers. The targeted delivery is confirmed by selective binding of ligand to the receptor on target tumor cells. Many types of ligands have been explored, such as peptides, carbohydrates, proteins, folic acid, aptamers, etc. These ligands bind specifically to receptors on target tumor cells, the most commonly explored being folate receptors, transferrin, glycoproteins, vascular endothelial growth factor (VEGF) receptors and the epidermal growth factor receptor (EGFR). Selection of ligand should be such that it will bind to tumor cells with high affinity and with minimum affinity to healthy cells, and should avoid leakage of the drug and drug degradation while circulation (94). The basic mechanism by which nanoparticles release anti-cancer agents include an initial step that involves binding of NP to tumor cell specific receptor, followed by internalization through endocytosis. NPs break down and release anti-cancer agent that kills tumor cells selectively (95). This can be pictorially explained by the illustration given in Figure 14.3.

14.7.2.1 Folate Receptors

Cells of lung cancer overexpress glycoproteins and folate receptors alpha (FRA) as they are derived from epithelial lineage. FRA has a role in controlling cell growth by combining with folic acid and methotrexate (96). Folate-chitosan grafted PEI copolymer NPs loaded with shRNA and siRNA targeted to folate receptors revealed endocytosis with transfection almost double than that of non-conjugated polymer (97). In yet another study, temozolomide was delivered into lung cancer cells as inhalable dry powder. Nanoparticles were prepared by dextran-PEG-folate combination. Decrease in tumor formation rate was observed in A549 lung carcinoma cells (98).

14.7.2.2 Transferrin Receptors

Transferrin receptor (TfR; CD71) is a type II transmembrane glycoprotein receptor that plays an important role in regulating intercellular iron transport as well as cell growth (99). Expression of this receptor is very low in healthy cells as compared to overexpression in tumor cells, thus making it a target for ligands in NPs for cancer therapy. Nanoparticles of paclitaxel and plasmid DNA targeted for transferrin displayed augmented anti-cancer activity for lung cancer (100).

14.7.2.3 Luteinizing Hormone Releasing Hormone (LHRH) Receptors

LHRH is one of the receptors overexpressed by lung cancer cells, as observed in other cancers. Nanoparticles of nanostructured lipid nanocarriers were loaded with siRNA (BCL2-MRP1) and paclitaxel to target LHRH receptors. The nanoparticles delivered by inhalation achieved approximately 16 times higher anti-cancer effect when compared with free drug. Selectivity for target was confirmed by imaging techniques indicating accumulation of NPs in timorous cells (101). Enhanced anti-cancer activity was also reported on targeting LHRH receptors when cisplatin and doxorubicin were loaded to mesoporous silica nanoparticles along with siRNA (MRP1 and BCL2) (102).

14.7.2.4 Epidermal Growth Factor Receptors (EGFRs)

A tyrosine kinase protein-based receptor which regulates cell growth, division and proliferation is epidermal growth factor receptors (EGFR). On an average, 50–60% of lung cancer cells express EGFR receptors (103). EGFR-targeted inhalable formulation of SPIONs exhibited cognizable magnetic hyperthermia led tumor growth inhibition within the mouse model. Significant tumor inhibition in cell lines A549 and A549-luc was observed when compared with non-targeted SPION nanoparticles (104). Similarly, chitosan conjugated ligand targeted to EGRF receptors of lung tumors and loaded with siRNA (Mad2) resulted in tumor cell death in NSCLC. Here, EGRF targeted NPs with siRNA exhibited complete gene silencing as compared to non-targeted NPs (56).

14.7.2.5 CD44 Receptors

CD44 is a cell surface associated glycoprotein concerned with cell proliferation, cell migration, cell differentiation, and presentation of cytokines, angiogenesis and growth factors to the concerned receptors. It acts as a hyaluronan receptor involved in drug resistance in cancer and metastasis (105). CD44 interacts with glycan, a cell matrix forming agent, and helps in migration and adhesion of cell. Hyaluronic acid specifically targets CD44 receptor proteins on the cell surface of cancer cells, and this can be achieved by conjugating hyaluronic acid on the surface of polymeric nanocarriers (106). PEG-coated PEI NPs conjugated with hyaluronic acid could successfully deliver siRNA to lung cancer receptors. Increased cellular update was observed with in vivo studies revealing significant gene silencing (57).

14.7.2.6 Integrin Receptors

Integrins are receptors of extracellular matrix (ECM), a kind of adhesion molecules. They facilitate cells to cells and cells to ECM interactions. The cross communication between tumor cells and microenvironment results in significant signaling activities and encourage the malignant cancer proliferation. Lung cancer subtype decides expression of integrins (82). Integrin subtype αVβ3 is a clinically important integrin, targeting to this yields better results and outcomes. This αVβ3 can be targeted by RGD peptide that can be easily surface bound to polymeric nanoparticles. Surface modified dextran nanoparticles loaded with paclitaxel and RGD peptide attached were fabricated by a spray drying method. Nanoparticles were used in a dry powder inhaler for treating NSCLC. Paclitaxel NP improved cytotoxicity in comparison with therapeutic activity of the pure drug (107).

14.7.2.7 Vascular Endothelial Growth Factor (VEGF) Receptor

The vascular endothelial growth factors (VEGFs) are responsible for tumor angiogenesis and neovascularization as they can bind and activate the VEGF receptor signaling chain. Tumor hypoxia and oncogenes increase VEGF levels in the cancer cells, resulting in overexpression of VEGFR-1 (fms-like tyrosine kinase) and VEGFR-2 (fetal liver kinase-1) on endothelial cells of the tumor. Polymeric nanoparticles can have ligand on their surface which can specifically bind to VEGF receptors on lung cancer cells, resulting in prevention of angiogenesis (108).

14.8 Overcoming Drug Resistance by Tumor Cells

Apart from these overexpressed receptors on cancer cell surface as targets, polymeric nanoparticles-loaded anti-cancer agents have also been investigated for overcoming multi-drug resistance by cancer cells. Tumor drug resistance works on the mechanisms that include physiological and cell related factors, such as overexpression of ATP binding cassette (ABC) transporters like efflux transporter, failed apoptosis and increased interstitial fluid pressure with hypoxic and acidic conditions in tumor microenvironment (109). Polymeric NPs can control the drug release, incorporate pH responsive polymers, and employ redox sensitive and temperature responsive agents to overcome drug resistance by tumor cells. Combination therapy with multiple anti-cancer agents is also an alternative. In current research, the ways to overcome drug resistance of cancer cells include (a) targeting efflux transporters, (b) targeting apoptotic pathway, (c) targeting hypoxia, and (d) PEGylation to escape RES and deliver maximum drug at the site of action (110).

14.9 Polymeric Nanoparticle Aspects in Drug–Gene Delivery to Lung Cancer

Physicochemical properties of polymeric NPs do influence the extent of drug solubility and rate of drug release. Also, an excessive immune response to NPs may result in inflammation. Size and size distribution is responsible for the extent to which NPs are going into tumor cells. NPs with size <100nm are likely to visit other organs through systemic circulation (111). Inflammation in the lungs is evident in cases of non-biodegradable polymers. Polymers that undergo fast biodegradation are suitable for lung cancer (112). Different structure and shape of NPs prepared from the same polymer can influence their efficacy as change in dimensions can influence drug loading and adherence of NPs to cancer cell surfaces. For drug and gene delivery, the charge on NPs plays a vital role as negative surface charge bearing NPs are more tolerable (113).

14.10 Merits and Limitations of Polymeric NPs

With modern techniques, NPs are not difficult to formulate and optimize for their efficacy. They have high surface area due to small particle size, thus higher drug loading capacity. They aid in controlled drug delivery, can carry imaging agents, and can be surface modified and loaded with genes, antibody fragments, peptides and proteins. NPs with PEGylation can escape host immune system with longer circulation time in body fluids. However, low reproducibility in formulation and difficulty in scale up are drawbacks with NPs (38,114). The d evelopment of inhalable NPs for treating lung cancers is a critical process and many technical issues need to be taken care of, like size distribution of NPs, biodegradation, surface charge, structure and shape, as well as interactions with alveolar surfactants (115,116).

14.11 Conclusion and Future Perspectives

Lung cancers have the highest death rate among all types of cancers. Conventional drug delivery systems have their own set of disadvantages which can be overcome by novel therapeutic systems. Advances in nanotechnology have opened doors to vast possibilities in this area. Undoubtedly, multifunctional polymeric nanoparticles have achieved success in terms of improvement in selective targeting to the tumor cells, sparing the healthy cells and tissues. Polymeric nanoparticles can be modified and tuned to the desired properties. They do score over traditional systems of drug delivery for improved therapeutic efficiencies, specific receptor targeting, nominal side effects and delayed drug release.

On the other hand, certain issues need to be addressed, such as toxicity of nanoparticles, reproducibility in formulation, insufficient drug loading, scale-up for mass production and overcoming drug resistance by tumor cells. Translational research is still away from bench to bedside for the benefit of cancer patients. A deeper understanding and knowledge is necessary to overcome these concerns. Investigating applications and utility of polymeric nanoparticles in a genome editing tool like CRISPR-Cas9 must be explored. Nevertheless, nanotechnological advances have progressed tremendously, but still, there is need for further research to have maximum gain in the diagnosis and treatment of lung cancer.

REFERENCES

1. American Cancer Society Medical and Editorial Content Team. What Is Lung Cancer? About and Key Statistics. 2021. Accessed January 1. https://www.cancer.org/cancer/lung-cancer.html
2. Mangal S, Gao W, Li T, and Zhou QT. Pulmonary delivery of nanoparticle chemotherapy for the treatment of lung cancers: Challenges and opportunities. *Acta Pharmacologica Sinica*. 2017; 38(6): 782–797. doi:10.1038/aps.2017.34
3. Senapati S, Mahanta AK, Kumar S, and Maiti P. Controlled drug delivery vehicles for cancer treatment and their performance. *Signal Transduction and Targeted Therapy*. 2018; 3(March): 7. Nature Publishing Group UK. doi:10.1038/s41392-017-0004-3
4. Sriraman SK, Aryasomayajula B, and Torchilin VP. Barriers to drug delivery in solid tumors. *Tissue Barriers*. 2014; 2(3): e29528. Taylor & Francis.
5. Babu A, Templeton AK, Munshi A, and Ramesh R. Nanodrug delivery systems: A promising technology for detection, diagnosis, and treatment of cancer. *AAPS PharmSciTech*. 2014 15 (3): 709–721. Springer.
6. Menon JU, Kuriakose A, Iyer R, Hernandez E, Gandee L, Zhang S, Takahashi M, Zhang Z, Saha D, and Nguyen KT. Dual-drug containing core-shell nanoparticles for lung cancer therapy. *Scientific Reports*. 2017; 7(1): 13249. doi:10.1038/s41598-017-13320-4
7. Pérez-Herrero E and Fernández-Medarde A. Advanced targeted therapies in cancer: Drug nanocarriers, the future of chemotherapy. *European Journal of Pharmaceutics and Biopharmaceutics*. 2015; 93: 52–79. doi:https://doi.org/10.1016/j.ejpb.2015.03.018
8. Sharma P, Mehta M, Dhanjal DS, Kaur S, Gupta G, Singh H, Thangavelu L, et al. Emerging trends in the novel drug delivery approaches for the treatment of lung cancer. *Chemico-Biological Interactions*. 2019; 309: 108720. doi:https://doi.org/10.1016/j.cbi.2019.06.033
9. Yu H-P, Aljuffali IA, and Fang J-Y. Injectable drug-loaded nanocarriers for lung cancer treatments. *Current Pharmaceutical Design*. 2017; 23 (3): 481 – 494. doi: http://dx.doi.org/10.2174/1381612822666161027113654
10. Gopalan B, Ito I, Branch CD, Stephens C, Roth JA, and Ramesh R. Nanoparticle based systemic gene therapy for lung cancer: Molecular mechanisms and strategies to suppress nanoparticle-mediated inflammatory response. *Technology in Cancer Research & Treatment*. 2004; 3(6): 647–657. Sage Publications Inc. doi: 10.1177/153303460400300615. :
11. Ramesh R. Nanoparticle-mediated gene delivery to the lung. *Gene Therapy Protocols. Methods in Molecular Biology™*. 2008; 433: 301–332. Humana Press. doi: https://doi.org/10.1007/978-1-59745-237-3_19
12. Codony-Servat J, García-Roman S, Molina-Vila MÁ, Bertran-Alamillo J, Giménez-Capitán A, Viteri S, Cardona AF, D'Hondt E, Karachaliou N, and Rosell R. Anti-epidermal growth factor vaccine antibodies enhance the efficacy of tyrosine kinase inhibitors and delay the emergence of resistance in <em>EGFR</Em> mutant lung

cancer cells. *Journal of Thoracic Oncology*. 2018; 13(9): 1324–1337. Elsevier. doi:10.1016/j.jtho.2018.04.030

13. Liu J, Wei T, Zhao J, Huang Y, Deng H, Kumar A, Wang C, Liang Z, Ma X, and Liang X-J. Multifunctional aptamer-based nanoparticles for targeted drug delivery to circumvent cancer resistance. *Biomaterials*. 2016; 91: 44–56. doi:https://doi.org/10.1016/j.biomaterials.2016.03.013
14. Mukherjee, A, Waters AK, Babic I, Nurmemmedov E, Glassy MC, Kesari S, and Yenugonda VM. Antibody drug conjugates: Progress, pitfalls, and promises. *Human Antibodies*. 2019; 27: 53–62. IOS Press. doi: 10.3233/HAB-180348
15. Amreddy N, Babu A, Muralidharan R, Munshi A, and Ramesh R. Polymeric nanoparticle-mediated gene delivery for lung cancer treatment. *Topics in Current Chemistry*. 2017; 375(2): 1–23. Springer International Publishing. doi: 10.1007/s41061-017-0128-5
16. Williams HD, Trevaskis NL, Charman SA, Shanker RM, Charman WN, Pouton CW, and Porter CJH. Strategies to address low drug solubility in discovery and development. *Pharmacological Reviews*. 2013; 65(1): 315 – 499. Drug Delivery, Disposition and Dynamics, Monash Institute of Pharmaceutical Sciences, Monash University, Parkville, Victoria, Australia. doi:10.1124/pr.112.005660
17. Luo T, Magnusson J, Préat V, Frédérick R, Alexander C, Bosquillon C, and Vanbever R. Synthesis and in vitro evaluation of polyethylene glycol-paclitaxel conjugates for lung cancer therapy. *Pharmaceutical Research*. 2016; 33(7): 1671–1681. doi:10.1007/s11095-016-1908-2
18. Conde J, Doria G, and Baptista P. Noble metal nanoparticles applications in cancer. *Journal of Drug Delivery*. 2012; 2012 : 1–12. doi:10.1155/2012/751075
19. Detampel P, Witzigmann D, Krähenbühl S, and Huwyler J. Hepatocyte targeting using pegylated asialofetuin-conjugated liposomes. *Journal of Drug Targeting*. 2014; 22(3): 232–241. Taylor & Francis. doi: 10.3109/1061186X.2013.860982
20. Du J-Z, Du X-J, Mao C-Q, and Wang J. Tailor-made dual PH-sensitive polymer–doxorubicin nanoparticles for efficient anticancer drug delivery. *Journal of the American Chemical Society*. 2011; 133(44): 17560–17563). American Chemical Society. doi:10.1021/ja207150n
21. Conde J, de la Fuente JM, and Baptista PV. Nanomaterials for reversion of multidrug resistance in cancer: A new hope for an old idea? *Frontiers in Pharmacology*. 2013; 4(October): 1–5. doi:10.3389/fphar.2013.00134
22. Gerlowski LE and Jain RK. Microvascular permeability of normal and neoplastic tissues. *Microvascular Research*. 1986; 31(3): 288–305. doi:https://doi.org/10.1016/0026-2862(86)90018-X
23. Shen J, Sun H, Xu P, Yin Q, Zhang Z, Wang S, Yu H, and Li Y. Simultaneous inhibition of metastasis and growth of breast cancer by co-delivery of twist ShRNA and paclitaxel using pluronic P85-PEI/TPGS complex nanoparticles. *Biomaterials*. 2013; 34(5): 1581–1590. doi:https://doi.org/10.1016/j.biomaterials.2012.10.057
24. Connor EE, Mwamuka J, Gole A, Murphy CJ, and Wyatt MD. Gold nanoparticles are taken up by human cells but do not cause acute cytotoxicity. *Small*. 2005; 1(3): 325–327. doi:https://doi.org/10.1002/smll.200400093
25. Zhang, X, Liu Y, Kim YJ, Mac J, Zhuang R, and Wang P. Co-delivery of carboplatin and paclitaxel via cross-linked multilamellar liposomes for ovarian cancer treatment. *RSC Advances*. 2017; 7(32): 19685–19693. The Royal Society of Chemistry. doi: 10.1039/C7RA01100H
26. Izadiyan Z, Basri M, Masoumi HRF, Karjiban RA, Salim N, and Shameli K. Modeling and optimization of nanoemulsion containing sorafenib for cancer treatment by response surface methodology. *Chemistry Central Journal*. 2017; 11(1): 21. doi:10.1186/s13065-017-0248-6
27. Sanginario A, Miccoli B, and Demarchi D. Carbon nanotubes as an effective opportunity for cancer diagnosis and treatment. *Biosensors*. 2017; 7(1): 9. MDPI. doi: 10.3390/bios7010009.
28. Ling C, Lu Y, Cheng B, McGoogan KE, Gee SWY, Ma W, Li B, Aslanidi GV, and Srivastava A. High-efficiency transduction of liver cancer cells by recombinant adeno-associated virus serotype 3 vectors. *Journal of Visualized Experiments* . 2011; (49): e2538. MyJoVE Corp. doi: 10.3791/2538
29. Shi J, Kantoff PW, Wooster R, and Farokhzad OC. Cancer nanomedicine: Progress, challenges and opportunities. *Nature Reviews Cancer*. 2017; 17(1): 20–37. doi:10.1038/nrc.2016.108
30. Yu M, Jambhrunkar S, Thorn P, Chen J, Gu W, and Yu C. Hyaluronic acid modified mesoporous silica nanoparticles for targeted drug delivery to CD44-overexpressing cancer cells. *Nanoscale*. 2013; 5(1): 178–183. The Royal Society of Chemistry. doi: 10.1039/C2NR32145A
31. Mendes LP, Pan J, and Torchilin VP. Dendrimers as nanocarriers for nucleic acid and drug delivery in cancer therapy. *Molecules*. 2017; 22(9): 1–21. doi:10.3390/molecules22091401
32. Riyaz B, Sudhakar K, and Mishra V. Chapter 13 - Quantum dot-based drug delivery for lung cancer . In: Kesharwani P. (ed.) Nanotechnology-Based Targeted Drug Delivery Systems for Lung Cancer. Academic Press. 2019; 311–326. doi:https://doi.org/10.1016/B978-0-12-815720-6.00013-7
33. Fernandes C, Suares D, and Yergeri MC. Tumor microenvironment targeted nanotherapy. *Frontiers in Pharmacology* 2018; 9:(October): 1–25. doi:10.3389/fphar.2018.01230
34. Blasiak B, Van Veggel FC JM, and Tomanek B. Applications of nanoparticles for MRI cancer diagnosis and therapy. *Journal of Nanomaterials*. 2013; 2013: 1–17. doi:10.1155/2013/148578
35. Hiss DC, Gabriels GA, and Folb PI. Combination of tunicamycin with anticancer drugs synergistically enhances their toxicity in multidrug-resistant human ovarian cystadenocarcinoma cells. *Cancer Cell International*. 2007; 7 (1): 5. doi:10.1186/1475-2867-7-5
36. Griffin R and Ramirez RA. Molecular targets in non-small cell lung cancer. *The Ochsner Journal*. 2017; 17(4): 388–392. The Academic Division of Ochsner Clinic Foundation. https://pubmed.ncbi.nlm.nih.gov/29230123
37. Palmerston Mendes L, Pan J, and Torchilin VP. Dendrimers as nanocarriers for nucleic acid and drug delivery in cancer therapy. *Molecules*. 2017; 22 (9): 1. doi: 10.3390/molecules22091401.
38. Ali I, Alsehli M, Scotti L, Scotti MT, Tsai ST, Yu RS, Hsieh MF, and Chen JC. Progress in polymeric nano-

medicines for theranostic cancer treatment. *Polymers*. 2020; 12(3): 1–32. doi:10.3390/polym12030598.
39. Patri AK, Myc A, Beals J, Thomas TP, Bander NH, and Baker JR. Synthesis and in vitro testing of J591 antibody–dendrimer conjugates for targeted prostate cancer therapy. *Bioconjugate Chemistry*. 2004; 15(6): 1174–1181. American Chemical Society. doi: 10.1021/bc0499127
40. Kurmi BD, Gajbhiye V, Kayat J, and Jain NK. Lactoferrin-conjugated dendritic nanoconstructs for lung targeting of methotrexate. *Journal of Pharmaceutical Sciences*. 2011; 100(6): 2311–2320. Elsevier. doi: 10.1002/jps.22469
41. Padilla De Jesús OL, Ihre HR, Gagne L, Fréchet JMJ, and Szoka FC. Polyester dendritic systems for drug delivery applications: In vitro and in vivo evaluation. *Bioconjugate Chemistry*. 2002; 13(3): 453–461. American Chemical Society. doi: 10.1021/bc010103m
42. Kesharwani P, Jain K, and Jain NK. Dendrimer as nanocarrier for drug delivery . *Progress in Polymer Science*. 2014; 39(2): 268–307. Elsevier.
43. Ulbrich K, Holá K, Šubr V, Bakandritsos A, Tuček J, and Zbořil R. Targeted drug delivery with polymers and magnetic nanoparticles: Covalent and noncovalent approaches, release control, and clinical studies . *Chemical Reviews*. 2016; 116(9): 5338–5431. American Chemical Society. doi: 10.1021/acs.chemrev.5b00589
44. Munaweera I, Shi Y, Koneru B, Saez R, Aliev A, Di Pasqua AJ, and Balkus KJ. Chemoradiotherapeutic magnetic nanoparticles for targeted treatment of nonsmall cell lung cancer . *Molecular Pharmaceutics*. 2015; 12(10): 3588–3596. American Chemical Society. doi: 10.1021/acs.molpharmaceut.5b00304
45. Huang Y, Mao K, Zhang B, and Zhao Y. Superparamagnetic iron oxide nanoparticles conjugated with folic acid for dual target-specific drug delivery and MRI in cancer theranostics. *Materials Science and Engineering: C*. 2017; 70: 763–771. doi:https://doi.org/10.1016/j.msec.2016.09.052
46. Zhang Y, Huang Y, and Li S. Polymeric micelles: Nanocarriers for cancer-targeted drug delivery. *AAPS PharmSciTech*. 2014; 15(4): 862–871. doi:10.1208/s12249-014-0113-z
47. Meng L, Huang W, Wang D, Huang X, Zhu X, and Yan D. Chitosan-based nanocarriers with PH and light dual response for anticancer drug delivery. *Biomacromolecules*. 2013; 14 (8): 2601–2610. American Chemical Society. doi: 10.1021/bm400451v
48. Karabasz A, Lachowicz D, Karewicz A, Mezyk-Kopec R, Stalińska K, Werner E, Cierniak A, Dyduch G, Bereta J, and Bzowska M. Analysis of toxicity and anticancer activity of micelles of sodium alginate-curcumin. *International Journal of Nanomedicine*. 2019; 14: 7249–7262. doi:10.2147/IJN.S213942
49. Jin R, Guo X, Dong L, Xie E, and Cao A. Amphipathic dextran-doxorubicin prodrug micelles for solid tumor therapy. *Colloids and Surfaces B: Biointerfaces*. 2017; 158: 47–56. doi:https://doi.org/10.1016/j.colsurfb.2017.06.023
50. Wang X and Liu Z. Carbon nanotubes in biology and medicine: An overview. *Chinese Science Bulletin*. 2012; 57(2): 167–180. doi:10.1007/s11434-011-4845-9.
51. Robinson JT, Welsher K, Tabakman SM, Sherlock SP, Wang H, Luong R, and Dai Hongjie. High performance in vivo near-IR (>1 μm) imaging and photothermal cancer therapy with carbon nanotubes. *Nano Research*. 2010; 3 (11): 779–793. doi:10.1007/s12274-010-0045-1
52. Liu Z, Chen K, Davis C, Sherlock S, Cao Q, Chen X, and Dai H. Drug delivery with carbon nanotubes for in vivo cancer treatment. *Cancer Research*. 2008; 68(16): 6652 – 6660. doi:10.1158/0008-5472.CAN-08-1468.
53. Cai W, Shin D-W, Chen K, Gheysens O, Cao Q, Wang SX, Gambhir SS, and Chen X. Peptide-labeled near-infrared quantum dots for imaging tumor vasculature in living subjects. *Nano Letters*. 2006; 6(4): 669–676. American Chemical Society. doi: 10.1021/nl052405t
54. Cai X, Luo Y, Zhang W, Du D, and Lin Y. PH-sensitive ZnO quantum dots–doxorubicin nanoparticles for lung cancer targeted drug delivery. *ACS Applied Materials & Interfaces*. 2016; 8(34): 22442–22450. American Chemical Society. doi: 10.1021/acsami.6b04933
55. Sekar V, Rajendran K, Vallinayagam S, Deepak V, and Mahadevan S. Synthesis and characterization of chitosan ascorbate nanoparticles for therapeutic inhibition for cervical cancer and their in silico modeling. *Journal of Industrial and Engineering Chemistry*. 2018; 62: 239–249. doi:https://doi.org/10.1016/j.jiec.2018.01.001.
56. Nascimento AV, Singh A, Bousbaa H, Ferreira D, Sarmento B, and Amiji MM. Mad2 checkpoint gene silencing using epidermal growth factor receptor-targeted chitosan nanoparticles in non-small cell lung cancer model. *Molecular Pharmaceutics*. 2014; 11(10): 3515–3527. American Chemical Society. doi: 10.1021/mp5002894
57. Ganesh S, Iyer AK, Morrissey DV, and Amiji MM. Hyaluronic acid based self-assembling nanosystems for CD44 target mediated SiRNA delivery to solid tumors. *Biomaterials*. 2013; 34(13): 3489–3502. doi:https://doi.org/10.1016/j.biomaterials.2013.01.077
58. Chiu HI, Ayub AD, Yusuf SNM, Yahaya N, Abd Kadir E, and Lim V. Docetaxel-loaded disulfide cross-linked nanoparticles derived from thiolated sodium alginate for colon cancer drug delivery. *Pharmaceutics*. 2020; 12(1): 1–25. doi:10.3390/pharmaceutics12010038
59. J ithan AV, Madhavi K, Madhavi M, and Prabhakar K. Preparation and characterization of albumin nanoparticles encapsulating curcumin intended for the treatment of breast cancer. *International Journal of Pharmaceutical Investigation*. 2011; 1(2): 119–125. http://www.jpionline.org/index.php/ijpi/article/view/59
60. Curcio M, Cirillo G, Tucci P, Farfalla A, Bevacqua E, Vittorio O, Iemma F, and Nicoletta FP. Dextran-curcumin nanoparticles as a methotrexate delivery vehicle: A step forward in breast cancer combination therapy. *Pharmaceuticals*. 2020 . doi: 10.3390/ph13010002
61. Karthikeyan S, Rajendra Prasad N, Ganamani A, and Balamurugan E. Anticancer activity of resveratrol-loaded gelatin nanoparticles on NCI-H460 non-small cell lung cancer cells. *Biomedicine & Preventive Nutrition*. 2013; 3(1): 64–73. doi:https://doi.org/10.1016/j.bionut.2012.10.009.
62. Jung J, Park S-J, Chung HK, Kang H-W, Lee S-W, Seo MH, Park HJ, Song SY, Jeong S-Y, and Choi EK. Polymeric nanoparticles containing taxanes enhance chemoradiotherapeutic efficacy in non-small cell lung cancer. *International Journal of Radiation Oncology, Biology, Physics*. 2012; 84(1): e77–e83. Elsevier. doi: 10.1016/j.ijrobp.2012.02.030

63. Kim D-W, Kim S-Y, Kim H-K, Kim S-W, Shin SW, Kim JS, Park K, Lee MY, and Heo DS. Multicenter phase II trial of genexol-PM, a novel cremophor-free, polymeric micelle formulation of paclitaxel, with cisplatin in patients with advanced non-small-cell lung cancer. *Annals of Oncology*. 2007; 18 (12): 2009–2014. Elsevier. doi: 10.1093/annonc/mdm374.
64. Zhao T, Chen H, Yang L, Jin HAI, Zhigang L, Han LIN, Fanglin L, and Zhiyun X. DDAB-modified TPGS-b-(PCL-Ran-PGA) nanoparticles as oral anticancer drug carrier for lung cancer chemotherapy. *Nano*. 2013; 08(02): 1350014. World Scientific Publishing Co. doi: 10.1142/S1793292013500148
65. Tseng C-L, Su W-Y, Yen K-C, Yang K-C, and Lin F-H. The use of biotinylated-EGF-modified gelatin nanoparticle carrier to enhance cisplatin accumulation in cancerous lungs via inhalation. *Biomaterials*. 2009; 30(20): 3476–3485. doi:https://doi.org/10.1016/j.biomaterials.2009.03.010.
66. Jin KT, Lu ZB, Chen JY, Liu YY, Lan HR, Dong HY, Yang F, Zhao YY, and Chen XY. Recent trends in nanocarrier-based targeted chemotherapy: Selective delivery of anticancer drugs for effective lung, colon, cervical, and breast cancer treatment. *Journal of Nanomaterials*. 2020; 2020. doi:10.1155/2020/9184284
67. Li Z, Pan T, Jin W, Song B, Gao Y, Yuan M, Ren H, Zhang T, and Mu Y. Preparation of functionalized polymersomes and the in vivo imaging. *International Journal of Polymeric Materials and Polymeric Biomaterials*. 2015; 64(3): 117–124. Taylor & Francis. doi: 10.1080/00914037.2014.909422
68. Cho HK, Cheong IW, Lee JM, and Kim JH. Polymeric nanoparticles, micelles and polymersomes from amphiphilic block copolymer. *Korean Journal of Chemical Engineering*. 2010; 27(3): 731–740. doi:10.1007/s11814-010-0216-5
69. Hou W, Liu R, Bi S, He Q, Wang H, and Gu J. P hotoresponsive polymersomes as drug delivery.*Molecules*. 2020; 25(21): 542–552.
70. Abdelaziz HM, Gaber M, Abd-Elwakil MM, Mabrouk MT, Elgohary MM, Kamel NM, Kabary DM, et al. Inhalable particulate drug delivery systems for lung cancer therapy: Nanoparticles, microparticles, nanocomposites and nanoaggregates. *Journal of Controlled Release*. 2018; 269: 374–392. doi:https://doi.org/10.1016/j.jconrel.2017.11.036.
71. Abdelwahed W, Degobert G, Stainmesse S, and Fessi H. Freeze-drying of nanoparticles: Formulation, process and storage considerations. *Advanced Drug Delivery Reviews*. 2006; 58(15): 1688–1713. doi:https://doi.org/10.1016/j.addr.2006.09.017.
72. Ali ES, Sharker SMd, Islam MT, Khan IN, Shaw S, Rahman MdA, Uddin SJ, et al. Targeting cancer cells with nanotherapeutics and nanodiagnostics: Current status and future perspectives. *Seminars in Cancer Biology*. 2020; 69: 52–68. doi:10.1016/j.semcancer.2020.01.011.
73. Yang W, Peters JI, and Williams RO. Inhaled nanoparticles—A current review. *International Journal of Pharmaceutics*. 2008; 356(1): 239–247. doi:https://doi.org/10.1016/j.ijpharm.2008.02.011.
74. Tomoda K, Ohkoshi T, Hirota K, Sonavane GS, Nakajima T, Terada H, Komuro M, Kitazato K, and Makino K. Preparation and properties of inhalable nanocomposite particles for treatment of lung cancer. *Colloids and Surfaces B: Biointerfaces*. 2009; 71(2): 177–182. doi:https://doi.org/10.1016/j.colsurfb.2009.02.001.
75. Elzoghby AO, Mostafa SK, Helmy MW, ElDemellawy MA, and Sheweita SA. Superiority of aromatase inhibitor and cyclooxygenase-2 inhibitor combined delivery: Hyaluronate-targeted versus PEGylated protamine nanocapsules for breast cancer therapy. *International Journal of Pharmaceutics*. 2017; 529(1): 178–192. doi:https://doi.org/10.1016/j.ijpharm.2017.06.077.
76. Nayerossadat N, Maedeh T, and Ali P. Viral and nonviral delivery systems for gene delivery PT - REVI. *Advanced Biomedical Research.* 2012; 1(2): 27–37.
77. Hong, S-H, Park S-J, Lee S, Cho CS, and Cho M-H. Aerosol gene delivery using viral vectors and cationic carriers for in vivo lung cancer therapy. *Expert Opinion on Drug Delivery*. 2015; 12(6): 977–991. Taylor & Francis. doi: 10.1517/17425247.2015.986454
78. Jia S-F, Worth LL, Densmore CL, Xu B, Duan X, and Kleinerman ES. Aerosol gene therapy with PEI. *Clinical Cancer Research*. 2003; 9(9): 3462 – 3468. http://clincancerres.aacrjournals.org/content/9/9/3462.abstract
79. Jiang H-L, Xu C-X, Kim Y-K, Arote R, Jere D, Lim H-T, Cho M-H, and Cho C-S. The suppression of lung tumorigenesis by aerosol-delivered folate–chitosan-graft-polyethylenimine/Akt1 ShRNA complexes through the Akt signaling pathway. *Biomaterials*. 2009; 30(29): 5844–5852. doi:https://doi.org/10.1016/j.biomaterials.2009.07.017
80. Xu C, Wang P, Zhang J, Tian H, Park K, and Chen X. Pulmonary codelivery of doxorubicin and SiRNA by pH-sensitive nanoparticles for therapy of metastatic lung cancer. *Small*. 2015; 11(34): 4321–4333. John Wiley & Sons, Ltd. doi: https://doi.org/10.1002/smll.201501034
81. Lee AY, Cho MH, and Kim S. Recent advances in aerosol gene delivery systems using non-viral vectors for lung cancer therapy. *Expert Opinion on Drug Delivery*. 2019; 16(7): 757–772. Taylor & Francis. doi: 10.1080/17425247.2019.1641083
82. Su C-y, Li J-q, Zhang L-l, Wang H, Wang F-h, Tao Y-w, Wang Y-q, et al. The biological functions and clinical applications of integrins in cancers. *Frontiers in Pharmacology*. 2020. https://www.frontiersin.org/article/10.3389/fphar.2020.579068
83. Kuşcu L and Sezer AD. Future prospects for gene delivery systems . *Expert Opinion on Drug Delivery*. 2017; 14(10): 1205–1215. Taylor & Francis. doi: 10.1080/17425247.2017.1292248
84. Zhang D, Wang J, and Xu D. Cell-penetrating peptides as noninvasive transmembrane vectors for the development of novel multifunctional drug-delivery systems. *Journal of Controlled Release*. 2016; 229: 130–139. doi:https://doi.org/10.1016/j.jconrel.2016.03.020
85. Tesauro D, Accardo A, Diaferia C, Milano V, Guillon J, Ronga L, and Rossi F. Peptide-based drug-delivery systems in biotechnological applications: Recent advances and perspectives. *Molecules*. 2019; 24(2): 1–27. doi:10.3390/molecules24020351
86. Chen S, Sanjana NE, Zheng K, Shalem O, Lee K, Shi X, Scott DA, et al. Genome-wide CRISPR screen in a mouse model of tumor growth and metastasis. *Cell*. 2015; 160(6): 1246–1260. Elsevier. doi: 10.1016/j.cell.2015.02.038

87. Wang M, Glass ZA, and Xu Q. Non-viral delivery of genome-editing nucleases for gene therapy. *Gene Therapy*. 2017; 24(3): 144–150. doi:10.1038/gt.2016.72
88. Ryu N, Kim M-A, Park D, Lee B, Kim Y-R, Kim K-H, Baek J-I, Kim WJ, Lee K-Y, and Kim U-K. Effective PEI-mediated delivery of CRISPR-Cas9 Complex for targeted gene therapy. *Nanomedicine: Nanotechnology, Biology and Medicine*. 2018; 14(7): 2095–2102. doi:https://doi.org/10.1016/j.nano.2018.06.009
89. Srinivasarao M, Galliford CV, and Low PS. Principles in the design of ligand-targeted cancer therapeutics and imaging agents. *Nature Reviews Drug Discovery*. 2015; 14(3): 203–219. doi:10.1038/nrd4519
90. Prabhu RH, Patravale VB, and Joshi MD. Polymeric nanoparticles for targeted treatment in oncology: Current insights. *International Journal of Nanomedicine*. 2015; 10: 1001–1018. doi:10.2147/IJN.S56932
91. Golden PL, Huwyler J, and Pardridge WM. Treatment of large solid tumors in mice with daunomycin-loaded sterically stabilized liposomes. *Drug Delivery*. 1998; 5(3): 207–212. Department of Medicine, UCLA School of Medicine, Los Angeles, CA 90095, USA. doi: 10.3109/10717549809052036
92. Hobbs SK, Monsky WL, Yuan F, Roberts WG, Griffith L, Torchilin VP, and Jain RK. Regulation of transport pathways in tumor vessels: Role of tumor type and microenvironment. *Proceedings of the National Academy of Sciences*. 1998; 95(8): 4607 – 4612. doi:10.1073/pnas.95.8.4607
93. Jain RK and Stylianopoulos T. Delivering nanomedicine to solid tumors. *Nature Reviews Clinical Oncology*. 2010; 7(11): 653–664. doi:10.1038/nrclinonc.2010.139.
94. Wicki A, Witzigmann D, Balasubramanian V, and Huwyler J. Nanomedicine in cancer therapy: Challenges, opportunities, and clinical applications. *Journal of Controlled Release*. 2015; 200 : 138–157. Elsevier B.V. doi: 10.1016/j.jconrel.2014.12.030
95. Srinivasan M, Rajabi M, and Mousa S A. Multifunctional nanomaterials and their applications in drug delivery and cancer therapy. *Nanomaterials*. 2015; 5(4): 1690–1703. doi:10.3390/nano5041690
96. Wibowo AS, Singh M, Reeder KM, Carter JJ, Kovach AR, Meng W, Ratnam M, Zhang F, and Dann CE. Structures of human folate receptors reveal biological trafficking states and diversity in folate and antifolate recognition. *Proceedings of the National Academy of Sciences*. 2013; 110(38): 15180 – 15188. doi:10.1073/pnas.1308827110
97. Jiang H-L, Kim Y-K, Arote R, Nah J-W, Cho M-H, Choi Y-J, Akaike T, and Cho C-S. Chitosan-graft-polyethylenimine as a gene carrier. *Journal of Controlled Release*. 2007; 117(2): 273–280. doi:https://doi.org/10.1016/j.jconrel.2006.10.025
98. Rosière R, Gelbcke M, Mathieu V, Antwerpen PV, Amighi K, and Wauthoz N. New dry powders for inhalation containing temozolomide-based nanomicelles for improved lung cancer therapy. *International Journal of Oncology*. 2015; 47 (3): 1131–1142. doi:10.3892/ijo.2015.3092
99. Whitney JF, Clark JM, Griffin TW, Gautam S, and Leslie KO. Transferrin receptor expression in nonsmall cell lung cancer. Histopathologic and clinical correlates. *Cancer*. 1995; 76(1): 20–25. John Wiley & Sons, Ltd. doi: https://doi.org/10.1002/1097-0142(19950701)76:1<20::AID-CNCR2820760104>3.0.CO;2-3
100. Daniels TR, Bernabeu E, Rodríguez JA, Patel S, Kozman M, Chiappetta DA, Holler E, Ljubimova JY, Helguera G, and Penichet ML. The transferrin receptor and the targeted delivery of therapeutic agents against cancer. *Biochimica et Biophysica Acta (BBA) - General Subjects*. 2012; 1820(3): 291–317. doi:https://doi.org/10.1016/j.bbagen.2011.07.016
101. Taratula O, Kuzmov A, Shah M, Garbuzenko OB, and Minko T. Nanostructured lipid carriers as multifunctional nanomedicine platform for pulmonary co-delivery of anticancer drugs and SiRNA. *Journal of Controlled Release: Official Journal of the Controlled Release Society*. 2013; 171(3): 349–357. doi:10.1016/j.jconrel.2013.04.018
102. Taratula O, Garbuzenko OB, Chen AM, and Minko T. Innovative strategy for treatment of lung cancer: Targeted nanotechnology-based inhalation co-delivery of anticancer drugs and SiRNA. *Journal of Drug Targeting*. 2011; 19 (10): 900–914. Taylor & Francis. doi: 10.3109/1061186X.2011.622404
103. Han W and Lo H-W. Landscape of EGFR signaling network in human cancers: Biology and therapeutic response in relation to receptor subcellular locations. *Cancer Letters*. 2012; 318(2): 124–134. doi:https://doi.org/10.1016/j.canlet.2012.01.011
104. Sadhukha T, Wiedmann TS, and Panyam J. Inhalable magnetic nanoparticles for targeted hyperthermia in lung cancer therapy. *Biomaterials*. 2013; 34(21): 5163–5171. doi:https://doi.org/10.1016/j.biomaterials.2013.03.061
105. Misra S, Hascall VC, Markwald RR, and Ghatak S. Interactions between hyaluronan and its receptors (CD44, RHAMM) regulate the activities of inflammation and cancer . *Frontiers in Immunology*. 2015; 6(May): 1–31. doi:10.3389/fimmu.2015.00201
106. Leung EL-H, Fiscus RR, Tung JW, Tin VP-C, Cheng LC, Sihoe AD-L, Fink LM, Ma Y, and Wong MP. Non-small cell lung cancer cells expressing CD44 are enriched for stem cell-like properties. *PLOS One*. 2010; 5(11): e14062. Public Library of Science. https://doi.org/10.1371/journal.pone.0014062
107. Torrico Guzmán EA, Sun Q, and Meenach SA. Development and evaluation of paclitaxel-loaded aerosol nanocomposite microparticles and their efficacy against air-grown lung cancer tumor spheroids . *ACS Biomaterials Science & Engineering*. 2019; 5(12): 6570–6580. American Chemical Society. doi: 10.1021/acsbiomaterials.9b00947
108. Devery AM, Wadekar R, Bokobza SM, Weber AM, Jiang Y, and Ryan AJ. Vascular endothelial growth factor directly stimulates tumour cell proliferation in non-small cell lung cancer. *International Journal of Oncology*. 2015; 47(3): 849–856. doi:10.3892/ijo.2015.3082
109. Scagliotti GV, Novelle S, and Selvaggi G. Multidrug resistance in non-small-cell lung cancer. *Annals of Oncology*. 1999; 10(Suppl 5): S83–S86. Elsevier Masson SAS. doi: 10.1093/annonc/10.suppl_5.S83
110. Yao Y, Zhou Y, Liu L, Xu Y, Chen Q, Wang Y, Wu S, Deng Y, Zhang J, and Shao A. Nanoparticle-based drug delivery in cancer therapy and its role in overcoming drug

resistance. *Frontiers in Molecular Biosciences*. 2020;7: 1–14. https://www.frontiersin.org/article/10.3389/fmolb.2020.00193

111. Medina C, Santos-Martinez MJ, Radomski A, Corrigan OI, and Radomski MW. Nanoparticles: Pharmacological and toxicological significance. *British Journal of Pharma-cology*. 2007; 150(5): 552–558. John Wiley & Sons, Ltd. doi: https://doi.org/10.1038/sj.bjp.0707130
112. Dailey LA, Wittmar M, and Kissel T. The role of branched polyesters and their modifications in the development of modern drug delivery vehicles. *Journal of Controlled Release*. 2005; 101(1): 137–149. doi:https://doi.org/10.1016/j.jconrel.2004.09.003
113. Harush-Frenkel O, Bivas-Benita M, Nassar T, Springer C, Sherman Y, Avital A, Altschuler Y, Borlak J, and Benita S. A safety and tolerability study of differently-charged nanoparticles for local pulmonary drug delivery. *Toxicology and Applied Pharmacology*. 2010; 246(1): 83–90. doi:https://doi.org/10.1016/j.taap.2010.04.011
114. Rezvantalab S, Drude NI, Moraveji MK, Güvener N, Koons EK, Shi Y, Lammers T, and Kiessling F. PLGA-based nanoparticles in cancer treatment. *Frontiers in Pharmacology*. 2018; 9(Nov): 1–19. doi:10.3389/fphar.2018.01260
115. Bahadar H, Maqbool F, Niaz K, and Abdollahi M. Toxicity of nanoparticles and an overview of current experimental models. *Iranian Biomedical Journal*. 2016; 20(1): 1–11. Pasteur Institute of Iran. doi: 10.7508/ibj.2016.01.001
116. Khattab SN, Abdel Naim SE, El-Sayed M, El Bardan AA, Elzoghby AO, Bekhit AA, and El-Faham A. Design and synthesis of new S-triazine polymers and their application as nanoparticulate drug delivery systems. *New Journal of Chemistry*. 2016; 40(11): 9565–9578. The Royal Society of Chemistry. doi: 10.1039/C6NJ02539K
117. Jabir NRet al. An overview on the current status of cancer nanomedicines. *Current Medical Research and Opinion*. 2018; 34: 911–921.
118. Anselmo AC and Mitragotri S.Nanoparticles in the clinic: An update. *Bioengineering & Translational Medicine*. 2019; 4(3): e10143. https://doi.org/10.1002/btm2.10143.
119. Mukai H, Kogawa T, Matsubara Net al. A first-in-human Phase 1 study of epirubicin-conjugated polymer micelles (K-912/NC-6300) in patients with advanced or recurrent solid tumors. *Invest New Drugs*. 2017; 35: 307–314. https://doi.org/10.1007/s10637-016-0422-z.
120. Tangsangasaksri Met al. siRNA-loaded polyion complex micelle decorated with charge-conversional polymer tuned to undergo stepwise response to intra-tumoral and intra-endosomal pHs for exerting enhanced RNAi efficacy. *Biomacromolecules*. 2016;17(1): 246–255.
121. Jere Det al. Chitosan-graft-polyethylenimine for Akt1 siRNA delivery to lung cancer cells. *International Journal of Pharmaceutics*. 2009; 378(1-2): 194–200.
122. Askarian Set al. Cellular delivery of shRNA using aptamer-conjugated PLL-alkyl-PEI nanoparticles. *Colloids and Surfaces B: Biointerfaces*.2015; 136(1):355–364.
123. Kleemann Eet al. Nano-carriers for DNA delivery to the lung based upon a TAT-derived peptide covalently coupled to PEG-PEI. *Journal of Controlled Release*. 2005; 109: 299–316.
124. Zhang B, ZhangY, and Yu D., Lung cancer gene therapy: Transferrin and hyaluronic acid dual ligand-decorated novel lipid carriers for targeted gene delivery. *Oncology Reports*. 2017;37(2): 937–944. https://doi.org/10.3892/or.2016.5298.
125. Babu A, Amreddy N, Muralidharan R, et al. Chemodrug delivery using integrin-targeted PLGA-Chitosan nanoparticle for lung cancer therapy. *Scientific Reports*. 2017; 7: 14674. https://doi.org/10.1038/s41598-017-15012-5.

15

Inhalable Polymeric Nano-particulate Powders for Respiratory Delivery

Anita Patel and Jayvadan Patel
Faculty of Pharmacy, Nootan Pharmacy College, Sankalchand Patel University, Gujarat, India

15.1 Introduction

Inhalation therapy has a long and beneficial history in the management of different respiratory diseases using a variety of natural inhalation remedies, for example, leaves from plants and vapors from aromatic plants, balsams, and myrrh. Demand for manufacturing tailor-made inhalable drug formulations, coupled with progress in biology and engineering, have led to a more optimized therapeutic effectiveness (1).

The pulmonary route of administration has achieved a great deal of interest since the early 1990s as a substitute for the parenteral route. From the beginning of the 19th century, liquid nebulizers have been urbanized and employed as acceptable inhalable pharmaceutical treatments. Examples of nebulizer drugs that had been produced and explored at this time were adrenaline, porcine insulin, penicillin, and steroids. The discovery of liquid nebulizers paved the way for the development of different types of inhaler devices. In 1956, the pressured metered dose inhaler (pMDI) was launched and turned out to be the main therapy for asthma. Regardless of progress in drug formulation technology, most of the inhalable aerosol therapeutics put up with limitations such as short half-life and low bioavailability, resulting in the need for increasingly repeated dosing. As a result, there was an urgent requirement for an effectual inhalation therapy that could defeat these limitations and give a sustained therapeutic effect. To develop an effective inhalation therapy, the anatomy of the respiratory system, lung deposition mechanisms, and lung defense mechanisms should be completely understood (2,3).

The respiratory system is divided into two main parts: the upper respiratory tract, comprising nose, nasal cavity, and pharynx, and the lower respiratory tract comprising larynx, trachea, bronchi, alveoli, and lungs. Lungs are in charge of gas exchange all over the body. Healthy lungs breathe in 1 pint (0.47 l) of air about 12–15 times every minute. The lungs are made of five lobes; the right lung has three lobes, while the left lung has two lobes. The interior of the lung consists of bronchi, alveoli, blood vessels, and lymph nodes. The bronchi are divided into bronchioles which branch in the lung, forming passageways for air, and end with the alveoli, which is accountable for gas exchange (4,5).

There are more than 300 million alveoli in the lung, and every alveolus is lined with pulmonary capillaries forming a giant network containing over 280 billion capillaries, which present a large surface area of approximately 70 m^2 existing as a blood-gas barrier. The alveolar gas exchange generally takes place at the interface made from the alveolar epithelium, endothelium, and interstitial cell layers, where the distance between the capillaries and alveolar is very small, around 0.5 μm, and so facilitates gas exchange via diffusion.

The alveoli are covered with a layer of fluids and mucus, mostly made up of phospholipid and surface proteins, which decrease the surface tension and are vital for the proper functioning of gas exchange. Lower respiratory passages are covered by a thin layer of connective tissue, surrounded with different cells, namely fibroblasts, nerves, macrophages, and lymph vessels (5,6).

The respiratory epithelial cells have a major function in the regulation of airway tone and the formation of airway lining fluid. From this perspective, growing consideration has been given to the potential of a pulmonary system as a non-invasive route of administration for systemic and local delivery of therapeutics, because of the high permeability, large absorptive surface area, and high vascularity of lungs (7–9). The alveolar epithelium of the distal lung has been shown to be an absorption site for the majority of the therapeutics and a variety of macromolecules (10–13). Further, benefits over peroral applications are the relatively low enzymatic activity, rapid absorption of the drug, and the capability for overcoming the first-pass metabolism. It has been previously reported that the local respiratory disorders and a number of systemic diseases can be well treated by delivering the drugs through the pulmonary route. Drugs can be administering locally for action in the lungs for the therapy of respiratory diseases, for instance, asthma or cystic fibrosis, local infectious diseases, pulmonary hypertension, the systemic use of insulin, human growth hormones, and oxytocin (14–21). This targeted delivery can possibly result in reducing the overall dose and the side effects that develop from high levels of systemic drug exposure. On the other hand, systemic drug delivery can be attained by targeting delivery to the alveolar region where the drug can be absorbed through the thin layer of epithelial cells and into the systemic circulation (22,23). This leads to improving permeability, rapid onset of action, and evasion of first-pass metabolism. In addition, the lungs can be targeted for delivery to specific lung cells, like alveolar macrophages (24), for the

DOI: 10.1201/9781003046547-15

treatment of diseases like tuberculosis. Furthermore, the latest progress demonstrates the enormous potential for well-organized pulmonary delivery of proteins and growth hormones, that cannot be delivered by the oral route and require parenteral delivery (25).

15.2 Challenges for Pulmonary Drug Delivery

The predominant part of present inhalation therapies suffers from a short half-life and low drug bioavailability at the targeted site, resulting in a suboptimal therapeutic effect and serious side effects. The short half-life and low drug bioavailability of inhaled drugs are as a result of three chief clearance mechanisms (26):

1. Pulmonary clearance consists of mucociliary clearance and alveolar macrophages,
2. Enzymatic degradation, and
3. Rapid systemic absorption

15.2.1 Pulmonary Clearance Mechanisms

The most important function of the respiratory defense mechanism is to put off foreign particles from entering the respiratory system and to keep it healthy and sterile. As with foreign particles, when aerosol particles are administered, the respiratory system removes the aerosol particles to avoid their interaction with the lung cells, which leads to treatment failure. The clearance mechanism of inhaled particles is maintained by the deposition site within the lungs. For instance, particles deposited in the tracheobronchial tree are quickly eliminated by mucociliary escalator, while particles deposited in the lower alveolar region are eliminated by macrophages (27).

15.2.1.1 Mucociliary Clearance

Mucociliary clearance is the main defense mechanism in the upper respiratory tract. The upper airways are lined with epithelial cells composed of two layers: ciliated cells and goblet (mucus-producing) cells, which both are referred to as a mucociliary escalator. The ciliated cells are covered with airway surface liquid and are made up of two layers: mucus layer and periciliary layer. This periciliary layer supplies a desired liquid environment that facilitates the cilia displacement toward mucus clearance from the lung to the mouth. The principle of mucociliary escalator contains entrapment of the foreign/inhalable particles in the mucus layer before moving to the lower respiratory regions, which after that are propelled along with mucus out of the trachea, either by coughing or swallowing (26,28). The mucociliary escalator removes most of the inhaled particles of sizes more than 6 μm. In contrast, smaller particles escape the mucociliary escalator because they quickly reach the epithelium and preferably deposit in the alveolar region, where they dissolve or hold for a long time span in the lungs. The mucociliary escalator gives an efficient clearance owing to its ability to balance between the function of ciliated cells and mucus-producing cells. However, during lung infection or inflammation, the balance of mucociliary escalator upsets, resulting in accelerating clearance of the inhaled drug and consecutively lessening its retention time and efficiency (29).

15.2.1.2 Alveolar Macrophage Clearance

Besides the mucociliary escalator clearance, in the deep lung, particularly in the alveoli, there is one more influential clearance mechanism than alveolar macrophages. The alveolar macrophages are phagocyte cells derived from monocytes and are present in large numbers in the lungs. Each alveolus is characteristically cleaned by 12–14 alveolar macrophages to keep it free from any foreign particles. Because of the presence of alveolar macrophages, the half-life of inhalable drugs within the alveoli cannot go beyond a few hours, which in turn results in a rising dose frequency. It was also reported that alveolar macrophages swallow up particles of sizes ranging between 1.5 and 3 μm (27).

This size-discriminating characteristic has been employed as a base for developing inhalable drugs, which can escape the alveolar macrophages and give a controlled drug release in the deep lung. Although the alveolar macrophage clearance seems to be understood, the precise mechanism behind particle uptake, transport, and clearance in the alveolar epithelium is still unidentified (29–31).

15.2.2 Enzymatic Degradation

Inhaled drugs are also highly vulnerable to enzymatic degradation in the lung, resultant in suboptimal therapy. The primary detoxification enzyme in the lung is the cytochrome P450 (CYP) families, which provide a line of defense against ingested or inhaled xenobiotics. There are several CYP isoforms expressed in the lungs (32), which are capable of degrading a broad spectrum of chemically-different inhalable drugs, pollutants, toxicants, etc. Numerous inhaled drugs like budesonide, salmeterol, and theophylline are enzymatically degraded in the lung. As well, peptide/protein drugs like insulin are highly susceptible to peptidase and proteases enzymes present in the lung. Even though the lung has low metabolic activity, in comparison to other organs such as the liver, enzymatic degradation considerably influences a drug's bioavailability at the lung and so should be cautiously assessed during drug formulation (4,29).

15.2.3 Rapid Systemic Absorption

Another significant challenge facing inhalation therapeutics is their fast systemic absorption from the lung. The rapid systemic absorption arises from the lung's large surface area, good epithelial permeability, and high vascularity, as well as the highly dispersed nature of therapeutic aerosols. The most advantageous absorption of inhalable aerosol relies on its site of action, either locally or systemically. So, to achieve an ideal local effect, an inhaled drug must be absorbed and terminated in the lung, whilst any systemic absorption results in speedy elimination of the drug and unfavorable side effects. Alternatively, the systemic effect is

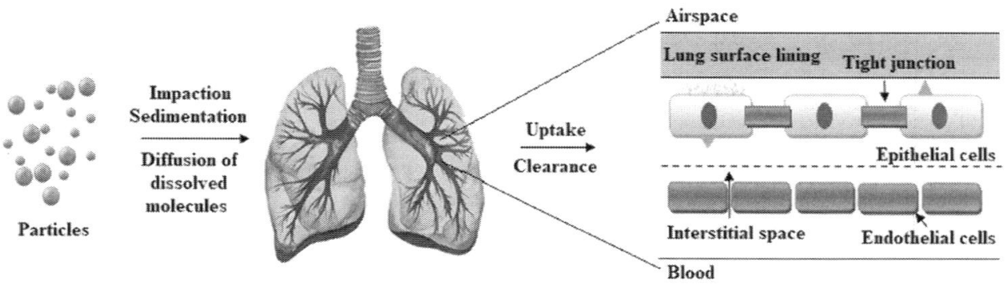

FIGURE 15.1 Deposition Mechanism and Uptake of Particles in the Lungs

attained when the inhaled drug is systemically absorbed from the lung into the bloodstream. The air-to-blood transfer of inhalable drugs often begins with the interaction between the drugs, and the surfactant following deposition on the mucosa of tracheobronchial airways or alveolar region. This interaction is very much influenced by the drug's nature and finds out whether the drug is absorbed or eliminated. For example, the contact between peptide drugs and lung surfactant causes particle aggregations, which in turn compromise their dissolution and speed up their clearance via alveolar macrophage (5,29) (Figure 15.1).

Quite the opposite, the contact between small lipophilic drugs like glucocorticosteroids and lung surfactant enhances their solubility and augments the rate and extent of absorption (30,31). Instantaneously below the lung surfactant layer, there is a 0.01–10 μm thick lining layer where the drug can diffuse to the epithelium, followed by interstitium, and finally diffuse to the bloodstream (33). The mechanisms of systemic drug absorption comprise passive and active transport mechanisms like paracellular or transcellular transport pore formation, vesicular transport, and lymphatic drainage. The mechanisms of drug absorption across the epithelium is greatly reliant on the drug nature and molecular weight, and the targeted site (4,23).

Investigations reported that small lipophilic and hydrophobic drugs are comprehensively absorbed within 12 minutes from the lung into the systemic circulation via passive diffusion, while small hydrophilic drugs are absorbed within 65 minutes through the tight junction. Studies also found that hydrophilic and highly cationic small molecules exhibit a prolonged absorption (4,23,34–36). Beyond these facts, the systemic absorption of some drugs like peptide is still unclear. But a few hypothesized mechanisms stated that the absorption of peptide drugs takes place either via transcytosis through caveoli or paracellularly through the tight junctions. Most of the researchers propose that absorption through the tight junction is the predominant mechanism for peptide drugs. As a result, these findings claim that drug absorption site and mechanism should definitely be considered while designing inhalation therapy to attain optimum lung-tissue retention and permeability (5,27,35–37).

15.3 Significance of Particulate-based Pulmonary Drug Delivery

The inhaled drugs show very low bioavailability at the lungs owing to the strong pulmonary clearance mechanisms in addition to fast systemic absorption. The low bioavailability of the drug at the lung stands for the main obstruction toward formulating inhaled drugs with high therapeutic effectiveness and sustained drug release. Drug bioavailability at the targeted site (lung) is addressed as the key factor for the best possible therapy because it finds out whether a drug induces a complete treatment or a partial treatment with high toxicity.

Quite a few approaches have been developed to conquer the faster drug absorption and extend its half-life (4). Among various approaches, particulate-based drug delivery systems that depend on using carriers, which encapsulate the inhaled drug, seem to be valuable for pulmonary delivery over other approaches. Particulate-based drug delivery has several benefits: (a) protects the drug from enzymatic degradation, (b) circumvents pulmonary clearance, (c) slows down the drug absorption, (d) delivers the drug to the targeted site at the lungs, (e) facilitates a controlled drug release, (f) decreases dose frequency, (g) maximizes the therapeutic efficiency, and (h) lessens unpleasant side effects. Particulate-based pulmonary drug delivery systems present great possibilities to develop local as well as systemic targeted therapy for a range of diseases like respiratory diseases, diabetes, and cancer therapy. A number of carriers, for example, liposomes, solid lipid nanoparticles, polymeric nanoparticles, micelles, etc., have been employed for developing different particulate-based pulmonary drug delivery systems aiming to optimize drug loading, residence half-life, drug release, and toxicity, and at the same time conquer various lung clearance mechanisms, enzymatic degradation, and rapid systemic absorption (38,39).

From this different particulate-based pulmonary drug delivery system, the exploitation of polymeric nanoparticles has earned much importance, as they symbolize a main class of nanotherapeutics. Polymeric nanocarriers are extensively employed as drug delivery systems and administered via different routes, counting the inhalation route (40). Polymeric nanoparticles also present potential benefits over other carriers like liposomes, because of their high stability, high drug loading capability, slower drug release mechanism, and longer pharmacological activity of payloads. Additionally, polymeric nanoparticles demonstrate great flexibility: their morphological as well as surface properties, like size, shape, and zeta potential, can be effortlessly adjusted by selecting different polymers lengths and kinds of surfactants and solvents for the production. Among synthetic polymers, poly-lactic acid (PLA) and poly-lactic-co-glycolic acid (PLGA), both Food and Drug Administration (FDA) approved, are the most commonly employed because of their great biocompatibility along with biodegradability (41). Specifically, PLGA tenders the possibility to modify its biodegradation rate by altering its compositional ratio. To fulfill the prerequisite of pulmonary delivery, their properties can be tailored by the addition of specific functional groups, drug moieties, and target ligands on their surfaces (42). Polymeric nanoparticles have been mostly utilized to evade side effects and avert the early degradation of drug molecules, and also to improve the pharmacokinetics of drugs with limited bioavailability (43). Thus, the assortment of drug carriers is of considerable importance, but some other factors like physicochemical properties of the drug, the used inhaled device, targeted site, disease condition, and the nature and safety of the carrier should also be taken into consideration.

15.4 Transepithelial Transport of Drugs

The formulation of drug delivery systems for pulmonary application necessitates an in-depth understanding of the lung in its healthy state, as well as different diseased states. A lung is made up of more than 40 different cells (44). The human respiratory system is a multifaceted organ system having a close structure–function relationship. This system generally consists of two fundamental regions: the conducting airways and the respiratory region. The transepithelial transfer of drugs along the respiratory epithelium from these two regions is distinguished by large quantitative differences. Transport of the drug in upper airways is restricted because of smaller surface area and lesser regional blood flow. Moreover, this region acquires a high filtering capability and eliminates up to 90% of delivered drug particles. Further, inhaled substances stick on the mucus layer, which covers the walls of the conducting airways. Mucus is secreted by goblet plus submucosal gland cells and develops a gel-like film composed of mucin as the main constituent. Ciliated cells are also present in this region; they induce propulsion of mucus upward and out of the lung. As a result, the lung will be cleared of foreign materials (45). On the other hand, the smaller airway and alveolar space account for more than 95% of the lung's surface area and are straightforwardly connected to the systemic circulation via the pulmonary circulation. Besides this, the morphology of the main alveolar epithelial cells, the pulmonary blood-gas barrier system, the size of pores, and tight junction depth of alveolar and endothelial cells are the most probable basis that manages the transepithelial drug transport (46).

15.5 Lung Compatibility of Formulation Excipients/Polymers

The important consideration to be given in the progress of a pulmonary drug delivery system is the compatibility of polymers employed in the design of particulate carriers. The safety profiles of these polymers are required to be first determined and their compatibility with lung fluid is a big issue. The polymers employed to extend the release rate for chronic exploit may build up in the lung, particularly in the lung periphery, which is not used by mucociliary clearance. Chronic inhalation of carrier particles has been shown to provoke the depletion of surfactant with succeeding recruitment of phagocytic cells (47). The chance of the presence of residual solvent in the final product brings about pulmonary toxicity. For that reason, manufacturing techniques and formulation constituents must be carefully screened to circumvent the toxic effects. Carriers utilized in the development of dry powder inhalation formulations, like cyclodextrin and sugars, can induce bronchoconstriction in numerous hypersensitive persons (48). Chronic use of proteins and other carriers, like absorption enhancers and enzyme inhibitors, can generate immunogenicity, local irritation, and toxicity. Augmented permeability may also permit the transport of other toxins and antigens across the epithelial barrier (49). These are a few fundamental concerns, which can be appropriately rectified through suitable models.

15.6 Fundamental Properties of Active Pharmaceutical Ingredient (API) Used by the Inhalation Route

To be successful vehicles for inhaled drugs, powder particles must be de-agglomerated in the airflow to form micrometer-sized particles as schematically shown in Figure 15.2.

The result of the de-agglomeration process depends on the inhaler design, applied flow rate, and properties of the powder. This is why the pharmaceutical powder has to be suitably formulated prior to its being employed as a drug carrier. The most important cohesive interactions which act against powder re-dispersion are caused by Van der Waals forces, electrostatic and capillary forces, by hydrogen bonds, and by mechanical interlocking attributable to surface roughness (50). These interactions depend not only on the particle properties and their formulation method, but also on ambient conditions like air humidity.

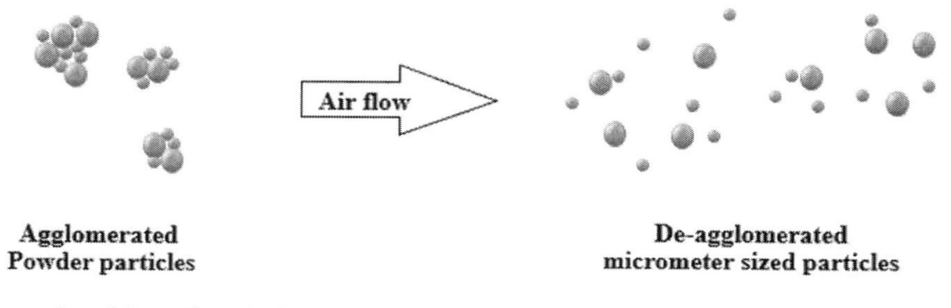

FIGURE 15.2 De-agglomeration of Power Drug Particles in the Airflow

Approaches that are employed to formulate powder particles with the properties that assist their de-agglomeration comprise (a) controlling the surface smoothness of the particles (51,52), (b) growing interparticle distance, e.g. by adding nanoparticles (53), and (c) altering the properties of the particle surface, e.g. by coating with leucine, magnesium stearate, or surfactants as force-control agents (54,55). It can also be advantageous to exploit micrometer-size nanostructured partly porous particles instead of compact ones with an identical size (56,57). All these approaches can be employed both for drugs prepared as interactive mixtures (API-excipient) and for excipient-free formulations. Inhalable particles can also be manufactured with API embedded in a polymer matrix with PLGA being the most familiar matrix designer (58). The pulmonary drug delivery system has the supplementary benefit of sustained drug release in the target area. A novel conception of inhalation powders was also mentioned by investigators regarding the application of solid lipid nanoparticles (SLN) and nanostructured lipid carriers (NLC) forming a dry lipid matrix which can be loaded with the API and delivered to the lungs from dry powder inhalers (DPIs) without added excipients (59,60).

Spray drying, spray freeze drying, and enhanced crystallization/precipitation (ultrasonic-assisted or utilizing supercritical fluid technology) are applied to achieve engineered powder drug particles (61,62). Due to the technological viability and effortless scale-up, spray drying is probably the most suitable method of manufacturing inhalable powders. It also permits control of the particle structure, size, and aerosolization properties (63,64). The selection of the right composition and concentration of the precursor (i.e. liquid solution or suspension), as well as of the drying parameters, are influential in achieving powders with the preferred particle size and shape (65,66).

15.7 Modes of Pulmonary Drug Administration

During the past decade, the systemic absorption of a broad range of therapeutic agents after the pulmonary application has been confirmed in animals as well as humans. The drug can be administered by two primary modes via the pulmonary route: first, by intranasal administration, which has an anatomical restriction, like a narrow airway lumen; second, by oral inhalational administration. In oral inhalative administration, considerably superior consequences can be predictable as it permits administering very small particles with a concentration loss of only 20% in comparison to 85% by the nasal route. Oral inhalation administration can again be categorized as intra-tracheal instillation and intra-tracheal inhalation. The most popular technique used in the laboratory is the intra-tracheal instillation. In the intra-tracheal instillation, a small quantity of drug solution or dispersion is administered into the lungs by a special syringe. This is a rapid and scientific technique of drug delivery to the lungs. The localized drug deposition is attained with a fairly small absorptive area. As a result, the instillation procedure is very simple, not expensive, and has uniform drug distribution. In a preclinical animal study, intra-tracheal instillation has normally been utilized to evaluate the pulmonary absorption as well as systemic bioavailability, predominantly in relation to the accurate dosing and efficiency relatable with this procedure (67,68). Alternatively, intra-tracheal instillation is not a physiological route for application, and outcomes achieved from these studies may not be transferable to aerosol applications in humans. In contrast, the inhalation method uses the aerosol techniques by which we can get more standardized distribution with great penetration. But this method is pricey and it is difficult to determine the exact dose in the lungs (68).

15.8 Mechanism of Deposition of a Particle in the Lungs

The deposition of drugs by inhalation administration in the pulmonary airway is mostly carried out by three mechanisms: gravitational sedimentation, inertial impaction, and diffusion. If the drug particle size is relatively large, then deposition takes place by the first two mechanisms, where either sedimentation comes about as a result of gravitational force or inertial impaction takes place as a result of hyperventilation. As the particle size is smaller they deposit mostly by diffusion mechanism that, in turn, is based on the Brownian motion (68). Normally, there are no mathematical equations that set up a relationship between particle velocity and its deposition behavior. Therefore, determining the particle trajectories, inertia, and diffusion under the synchronized conditions is crucial to predict the distribution pattern of inhaled particles (69).

15.8.1 Gravitational Sedimentation

The settlement of the particle under the action of gravity addresses the gravitational sedimentation mechanism for the deposition of the particle in the lungs. Particles sized between 0.5 and 5 μm have been settled by this mechanism. An effectual deposition by this mechanism takes place in the small airways and the alveoli. However, the deposition also takes place in the upper respiratory tract. The terminal settling velocity for the spherical particles is expressed by

$$Vs = \frac{(\varrho_p d^2{}_p g}{18\mu} \tag{15.1}$$

where,

1. d_p and ϱ_p are particle diameter and density
2. g is the gravitational acceleration
3. μ is the dynamic viscosity of carrier gas

The likelihood of the particle deposition by gravitational sedimentation is proportional to the terminal settling velocity. The probability rises with increases in particle size, density, and rising residence time (70).

15.8.2 Inertial Impaction

Inertial impaction presents a prospective mechanism of deposition for the particles between 2 and 5 μm. Inertial impaction delivers the drug in the complex geometry of the upper respiratory tract and conducting airway bifurcations. The Stokes number clarifies the possibility of the particle to diverge from streamline of carrier gas:

$$S_{tk} = \frac{(\varrho_p d^2{}_p u}{(18\mu d} \quad (15.2)$$

where,

1. u = μ are the linear velocity and dynamic viscosity of carrier gas
2. d is the diameter of the air space

More S_{tk} refers to an efficient deposition of the particle by inertial impaction (70).

15.8.3 Brownian Diffusion

Deposition by Brownian diffusion takes place by the collisions of the particle with gas molecules. It is a potential mechanism of the deposition for the particles having less than 0.5 μm diameters, and in the acinar region of the lungs. The particles below 0.01 μm are effective in the nose, mouth, and pharyngeal airways. The diffusion rate is proportional to the Brownian diffusion coefficient (70)

$$Db = \frac{(ckT)}{(3\pi\mu d_p)} \quad (15.3)$$

where

1. k is Boltzmann's constant
2. T is the absolute temperature
3. c is the Cunningham Correction factor

15.8.4 Electrostatic Precipitation

Electrostatic precipitation occurs when charged particles get close to the opposite charge surfaces in airways (70).

15.8.5 Interception

Interception serves as a feasible mechanism of the particle deposition in smaller airways and alveoli. Interception entails the deposition of an elongated particle in the gas space while one of its ends touches the alveolar wall under the effect of gravity (70).

Besides the pulmonary morphological features and ventilatory parameters, size of the particles or droplets and the geometry is fairly essential. The size of a particle or droplet with regard to diameter, along with the surface electrical charges, the shape of the particulate matter if it is a fiber, and hygroscopy also provide insight into drug deposition via the pulmonary route (68,71).

15.9 Particle Clearance

Elimination of inhaled particulate compounds deposited in the respiratory tract is essential for clearance. Clearance mechanisms are different in different regions of the respiratory tract and also rely on the physicochemical properties of the deposited material. The particle clearance from the respiratory tract comprises the clearance from nasopharyngeal, tracheobronchial, and pulmonary compartments. Clearance from the nasopharyngeal regions consists of the mucociliary and mechanical clearance with absorption into the circulation. The tracheobronchial clearance includes mucociliary clearance, endocytosis into the peribronchial region, and absorption into circulation. Clearance from the pulmonary compartment takes place by alveolar macrophage-mediated clearance, endocytosis by lung epithelial cells into the interstitium, and absorption into circulation. Factors such as age, exercise, influenza, and pneumonia have an effect on the clearance rate of deposited particles. The rate of clearance declines with age while exercise augments the rate of clearance. Clearance may be delayed up to 3 months to 1 year in cases of influenza and pneumonia (70).

15.10 Factors Affecting Drug Bioavailability via the Pulmonary Route

Bioavailability of administered drugs via the pulmonary route rely on numerous factors. Different physicochemical, biological, and formulation factors have played a significant role in the absorption of drugs through the lungs (Figure 15.3).

Diffusion and absorption of drugs from the lungs' interface is dependent on the lipid solubility of the molecule, its size, degree of ionization, and the area of the absorptive surface. It is expected that lipid-soluble drugs will diffuse more quickly because of the lipoidal nature of the cell membrane. Smaller molecules have a propensity to penetrate membranes more quickly than larger molecules. The majority of drugs are weak organic acids or bases, existing in ionized as well as un-ionized forms in an aqueous environment. The ionized form has lower lipid solubility and high electrical resistances and so cannot penetrate cell membranes easily. An un-ionized form is generally lipid-soluble and diffuses readily across cell membranes. The proportion of the un-ionized form present, and accordingly the drug's ability to cross a membrane, is measured by the pH and the drug's pKa. At the physiological pH 7.4, drugs with lower pH have more uncharged particles, which are free to cross the alveolar mucosal membranes and produce faster absorption than with higher pKa drugs (72–75).

The alveolar mucosal membrane is highly vascular, allowing the drug to go into the systemic circulation quickly and to arrive at its target organ with minimum delay. Drugs may also be delivered directly into the target organ, resulting in the site-specific therapeutic action of drugs. This is imperative for the delivery of anticancer agents, anti-inflammatory agents, antiviral agents, bronchodilators, mucolytics, and phospholipid-protein mixtures in serious and chronic conditions like asthma,

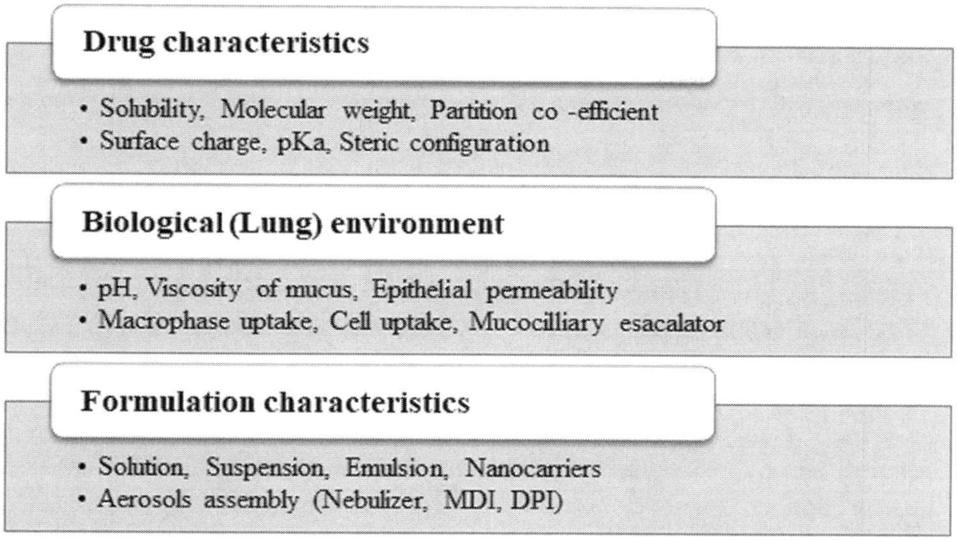

FIGURE 15.3 Factors Affecting Drug Clearance and Drug Bioavailability

cystic fibrosis, tuberculosis, and others. Additionally, the particle size of drugs has an effect on the absorption, dissolution, and deposition of drugs in the epithelial lining fluid. Small nanoparticles will demonstrate better bioavailability, faster dissolution, and different deposition patterns when compared with larger particles (73,75). Numerous investigations have revealed that 0.5 and 5 μm mass median aerodynamic diameters (MMAD) have resourcefully deposited particles inside the lungs. Furthermore, nanoparticles, slow inhalation, method, and condition during administration will be influential features for the deepest deposition of particles in the lungs (76). In recent times, 5-fluorouracil nanoparticles have been produced by a supercritical antisolvent method and showed regular shape with 248 nm size, smooth surfaces with a high surface area, and better aerodynamic performance in presence of lactose (77). Zeman et al. have stated that the deposition of inhaled drugs relies on the particles' aerodynamic size as well as the patient's breathing pattern. Outcomes showed that the conducting airway deposition of large particles (9.5 μm MMAD) and smaller particles (5 μm MMAD) were 35 ± 8% and 27 ± 11%, respectively (78). Likewise, high azithromycin loading powders for inhalation have been manufactured by spray-drying methods. Results showed that optimized powder had improved flowability, aerodynamic diameter (3.82 μm), and in vitro deposition (51%) and also enhanced drug targeting index via intra-tracheal route than intravenous injection (79).

A range of physiological factors can have an effect on the absorption and deposition of drugs, for example, airway caliber, breathing pattern, lung morphology, lung surface, lung surfactant, mucus, inhalation time, and disease status. The lung has a huge, highly perfused surface area that is greater than the surface of a person's skin. Inhalation of highly diffusive agents might cause undesirable systemic spillover, bioavailability, as well as pharmacokinetics, which basically reflects the intravenous infusion in such cases. Additional studies have shown resourceful absorption and lower metabolism of smaller sized drugs through lungs as compared with oral delivery (75,80).

A variety of physiological mechanisms have established the absorption of drugs via lung-like active transport, passive transport, paracellular, transcellular, receptor-mediated endocytosis, and transcytosis pathways. Studies confirmed that sugar, proteins, and peptides follow paracellular, transcellular, endocytosis, and transcytosis pathways, whereas active drugs and lipids follow receptor-mediated endocytosis and pinocytosis pathways. Furthermore, studies revealed that for smaller peptides (i.e. 40 kDa) paracellular transport may dominate, whereas for larger proteins (i.e. 40 kDa) transcytosis may be more significant. In general, larger macromolecules are non-specifically absorbed by airspace using transcytosis as well as endocytosis transport (33,75).

Mucociliary clearance has also been the most important restraint for the pulmonary absorption of drugs. The volumes in addition to the composition of epithelial lining fluid of alveolar epithelial cells are also contributing features for the absorption of drug molecules. This fluid is, in turn, wrapped by a monolayer of lung surfactant. The epithelial cell layer represents the main barrier to the absorption of drugs. Alveolar epithelium plays a vital role in the absorption of inhaled drugs within the lungs. The absorption of drugs from mucus barriers is largely reliant on particle charge, lipophilicity, solubility, and size (81,82). The airway caliber through which the aerosol passes to arrive at the more distal parts of the lungs is very important. Airway caliber may change lung deposition as well as bioavailability of inhaled drugs because narrowed airways decrease peripheral drug delivery (70). Airway obstacle is a respiratory dilemma that decreases the quantity of drug absorption because the deposition is greater in the central large airways than in peripheral regions comprising small airways and alveoli. Under regular breathing conditions, sub-micron particles are breathed out. On the other hand, particles of ultrafine sizes around 0.01 μm can be deposited through diffusion. Holding of breath or slowing down of respiration rate raises the dwell time, diminishes turbulent flow, and allows more time for these particles to deposit by sedimentation as well as diffusion (83).

TABLE 15.1
Advantages and Disadvantages of Delivery Devices for Pulmonary Administration

Type of Device	Advantages	Disadvantages
Nebulizers	Easy to use. Can be employed in mechanically ventilated patients.	Require water soluble drug. Incompetent delivery. Considerable drug wasted. Many factors can influence delivery. Device inconsistency. Not convenient, bulky, expensive, and time consuming. Requires electrical power supply.Nebulizers should be kept clean. Potential for breakdown of drug.
Metered dose inhalers	Require smallest time for treatment. Small and handy. Can be employed in mechanically ventilated patients.	Require momentous coordination. Complexity in determining number of doses left behind. Many excipients.
Dry powder inhalers	Breath actuated, needs minimum coordination. Improves pulmonary drug deposition. Some are single dose units. Need assembly.	Young children and actually obstructed patients may not create sufficient inspiratory flow to actuate. Wants change in inhalation method. Augmented oropharyngeal drug deposition and systemic absorption add to risk for systemic adverse effects. Potential for drug to aggregate. Potential to cause cough. Cannot be utilized in mechanically ventilated patients.

Types of drug formulation and method of administration also have a significant effect on the deposition of inhaled aerosol particles. Droplet size, the shape of particles, the velocity of particles, and the type of particulate systems have also influenced bioavailability along with the distribution of drugs within the lungs. Furthermore, novel colloidal carriers can be utilized to direct the duration of local or systemic drug activity by modulating the solute-release features. Pulmonary delivery of pharmaceuticals can be attained by powder formulation using aqueous formulations for nebulizers, dry powder inhalers, or solvent formulations for metered dose inhalers (75,84). These devices require mechanical energy for the production of aerosol particulates, like micronization of powder in DPIs, breakdown of a liquid droplet in nebulizers, and evaporation of dispersing solvent in MDIs (85). The most important challenges in pulmonary drug delivery are the poor effectiveness of the conventional devices to deliver dry powder of small molecules, peptide, and protein to the lower regions of the respiratory tract. Low inertia elements get breathed out and larger macromolecular drugs may give up poor bioavailability as they get deposited on the oropharynx and larynx. Pulmonary drug delivery demonstrates inappropriate dosing reproducibility, and it is often complicated to regulate doses as per patient need (86).

15.11 Devices for Pulmonary Delivery

The delivery device in the pulmonary administration of drugs plays a key role in the victory of this system. In recent times, great strides have been made in the growth of sophisticated devices. However, devices are far less explored than powder formulations (87). There is a broad range of passive breath-driven and active power-driven single/multiple dose DPIs existing in the market. The assortment of the device for the administration of drugs to the lungs is a significant feature in the formulation design and choice of device depends on the physicochemical and dosing demands related to the drug substances to be breathed in. If the drug is intended for a specific part of the lungs, then the preferred device must be able to produce and delive particles/droplets of explicit aerodynamic diameter. The devices most widely employed for respiratory delivery are nebulizers, MDIs, and DPIs (68).

Appropriate devices are employed to deliver the drug to the lungs successfully and their advantages and disadvantages for pulmonary delivery are presented in Table 15.1.

A range of technology is being assessed to conquer difficulties associated with a conventional aerosol device designed for dry powder inhalation systems and liquid inhalation systems. The aerosol devices employed to administer the drug substance to the lung are as follows.

15.11.1 Jet Nebulizers

Several types of nebulizers are available, namely jet nebulizers, ultrasonic nebulizers, and vibrating mesh nebulizers. Jet nebulizers are driven by compressed air. The ultrasonic nebulizers utilize a piezo-electric transducer to generate droplets from an open liquid reservoir. Vibrating mesh nebulizers make use of perforated membranes actuated by an annular piezo element to vibrate in a resonant bending mode. The openings in the membrane have a big cross-section size on the liquid supply side and a thin cross-section size on the side where the droplets come out. Depending on the therapeutic application, the hole sizes, as well as the number of holes, can be attuned (68). Jet nebulizers are an increasingly used inhalation device. These are the division of the most common twin fluid atomizers. The working process comprises a pressurized air source that provides high-pressure air which flows via a nozzle where the air acceleration occurs. The pressure close to the nozzle is designed so that high-velocity air flows over a short section of the liquid surface supplied by the liquid feed tube. This is the place where the generation of liquid droplets takes place. The leading features are geometric parameters such as nozzle shape, dynamic non-dimensional parameters such as Mach number in gas jet, gas Weber number, Reynolds number in the gas jet, Reynolds number in the liquid surrounding the gas jet, viscosity ratio, and density ratio. Key determinants of both effectiveness and output rate are the ability of the nebulizer to return droplets to the nebulizer reservoir for re-nebulization. The most important determinant of the output rate in a jet nebulizer is the rate at which primary droplets are formed, which increases with a rise in the in the flow rate of liquid into droplet production site with all other variables unaffected (69).

The choice of an appropriate device is influenced by parameters, like the nature of the drug and its formulation, the site of action, and pathophysiology of the lung. Aqueous solutions and suspensions are nebulized efficiently. Aerosols with reference to mechanically formed vibration mesh technologies also have been exploited productively to administer proteins to lungs and are right now being employed in the clinical trials of protein as well as peptide-based pharmaceuticals (88–90).

15.11.2 Dry Powder Inhalers (DPI)

DPI systems have the possible benefits of storing drugs in a dry state, which can present long-term stability as well as sterility. The main principle of DPI is to introduce a prescribed dose of powder aerosol into air inhaled by the patient during a single breath at the same time as the powder is averted from being exposed to ambient air by the time the patient is prepared to breathe in. Coughing may be provoked by inhalation of a larger amount of powder; total amounts of inhaled powder are generally lower than 10–20 mg. The powder to be inhaled from a DPI has to be carried away by the air being inhaled, a procedure referred to as "fluidization." The mechanics of DPIs are complicated for irregular shapes and coarse particles.

Most DPI formulations are made up of micronized drug blended with larger sized carrier particles which augment flow, diminish aggregation, and assist in dispersion. The combination of fundamental physicochemical characteristics, area, size, shape, and morphology have an impact on forces of interaction as well as aerodynamic properties which decide dispersion and fluidization, along with delivery to lungs and deposition in peripheral airways. Upon actuating, the formulation gets fluidized and goes into the airway. The drug parts from the carrier and gets carried deep into the lungs. The larger carrier particles collide on the oropharyngeal surfaces, and are cleared. If consistent forces are powerful, shear does not break up carrier from the drug, which results in lower deposition effectiveness (69).

The first form of DPI has been used in the patient's breathing to disperse and deliver powdered micron-sized particles. In addition, progress has concentrated equally on the improvement of devices that disperse powder with compressed air or electric motors and on the manufacturing of particles that are effortlessly respirable (91). The larger geometric particle size strengthens the dispersion properties of these particles and, thus, resulted in an ordinary inhaler for powder delivery to the lungs with greater lung deposition of the delivered dose (equal to 59%) (92). On account of their minute aerodynamic diameter, these particles would deposit in the alveolar area and stay away from phagocytosis as they were a little large to be engulfed by alveolar macrophages. Porous particles are commonly fabricated with GRAS (generally recognized as safe) inert pharmaceutical materials through spray drying, a cost-effective and single-step pharmaceutical procedure (93,94).

DPIs are increasingly being used devices to deliver drugs, especially proteins, to the lungs. Several commercial-grade DPIs take this into consideration: Spinhaler (Fisons Pharmaceuticals, Rochester, NY) and Rotahaler (GSK, RTP, NC) (68). Despite that, the latest applications of computational fluid dynamics have been supportive of the design and implementation of DPI devices and recognize the influence of airflow changes and de-agglomeration in the inhaler system (95,96).

15.11.3 Metered Dose Inhalers (MDI)

The practical aspects of MDI entail a transitory, cavitation turbulent fluid that flashes into speedily evaporating droplets. MDIs continue to be a standard aerosolized delivery system which can generate an aerodynamic particle size of below 5 μm with hydrofluoroalkane (HFA) propellants. HFA-propelled MDIs are simply portable, tamper-proof, and multi-dose. They shield the remaining product from light, moisture, and oxidation as well as providing a trouble-free and cost-effective technology with precise liquid actuation by volume. The aerosol drug dose exits the MDI mouthpiece as a fast-moving large droplet cloud. At about the distance of the back of the throat, the droplet width is condensed as a result of propellant evaporation, and a sensible proportion of the "poly-dispersed" aerosol cloud is at this moment little enough to penetrate the lung. A proportion of each metered dose is lost in the actuator mouthpiece, and a further ratio is lost in the oropharynx, attributable to inertial impaction of the "ballistic portion" of the spray (97).

15.12 Targeting Inhalable Polymeric Nanoparticles for Local Lung Delivery

The pulmonary delivery of nanoparticles also has the prospective to increase and sustain local lung drug concentration for the treatment of respiratory diseases. Targeting the site of action possibly reduces the required body dose, thus reducing systemic side effects. In addition, oral dosing variability resulting from changing gastric conditions could be circumvented.

Zatta and colleagues developed 5-fluorouracil nanoparticles by the piezo-electric atomization technique, for the treatment of metastatic melanoma by pulmonary administration, aiming at local and systemic action; 5-fluorouracil nanoparticles exhibited appropriate aerodynamic properties and dose uniformity for efficient pulmonary delivery. The formulations were tested for their cytotoxic action on melanoma cancer cells (A2058 and A375) and nano-particulate formulation of 5-fluorouracil presented a cytotoxic effect 1.7 times greater than the drug 5-fluorouracil when pure. The results showed that 5-fluorouracil nanoparticles have favorable complementary properties for lung delivery. If combined in a unique therapeutic system, the powders presenting different granulometrics could be administered by means of a DPI with a satisfactory drug distribution along the respiratory tract (98).

Researchers developed chitosan nanoparticles-based DPI formulations of prothionamide, which can sustain release for several hours in the lungs. Chitosan, a biodegradable polymer, was employed to coat prothionamide and further freeze-dried to prepare DPI with the aerodynamic particle size of 1.76 μm. In vitro release study showed an initial-burst release followed by sustained release up to 96.91% in 24 hours. Prepared DPI maintained the prothionamide concentration above minimum inhibitory concentration for greater than 12 hours after single-dose administration and increased the prothionamide residence time in the lung's tissue more than 24 hours. An animal study also revealed the reduction of dose in the administration, which will improve the management of tuberculosis (99).

DPIs comprising tadalafil-loaded PLGA nanoparticles were prepared for the treatment of life-threatening pulmonary arterial hypertension by an emulsion solvent evaporation technique. Tadalafil-loaded polymeric PLGA nanoparticles showed sustained release behavior over 24 hours. The results revealed that encapsulation efficiency, zeta potential, and particle size of PLGA nanoparticles were more affected by the aqueous/organic phase ratio, while release efficiency was more affected by the polymer/drug ratio. A good in vitro aerosolization performance of developing DPI containing tadalafil–PLGA nanoparticles recommends these powders would be predominately deposited in deep regions of the lung following inhalation. The results indicated that prepared dry powders containing tadalafil-loaded PLGA nanoparticles were appropriate for inhalation and have the potential for the treatment of pulmonary arterial hypertension (100).

Simvastatin nanoparticles encapsulated with PLGA have been fruitfully formulated using the solvent and anti-solvent precipitation technique for inhalation using the polymeric nanoparticle technologies and characterized in vitro. This study has demonstrated that the formulation was stable up to 9 months in a freeze-dried form and can be correctly reconstituted for nebulization into the deep lungs. Additionally, it was shown to acquire controlled release properties that could be advantageous for sustained anti-inflammatory properties, reducing absorption and permitting higher simvastatin concentrations to remain in the airways for therapeutic effectiveness and reduced mucus production. This study proposes that simvastatin nanoparticle nebulization could potentially be used for the treatment of chronic pulmonary diseases (101).

Doxorubicin-loaded nanoparticles can be integrated into inhalable effervescent as well as non-effervescent carrier particles with a spray-freeze drying method. The prepared inhalable powders were tested in a tumor-bearing BALB/c mouse model (an albino, laboratory-bred strain of the house mouse). They observed that the animals treated with effervescent nanoparticle carriers showed longer survival times than animals treated with non-effervescent nanoparticle carriers, and also the lungs of animals treated with inhalable effervescent doxorubicin nanoparticles showed fewer and much smaller tumors compared to the control groups as visualized by magnetic resonance imaging. The study demonstrated that inhalable effervescent doxorubicin nanoparticles are an effective way to treat lung cancer (102).

In recent times, simultaneously manufactured nano-in-macro (SIMANIM) particles have been produced and functional for the pulmonary administration of antibodies. These SIMANIM particles were fabricated by spray drying of a double emulsion containing IgG antibodies. The in vitro release of antibodies at pH 2.5 was observed for 35 days, and the action of the released antibodies was tested with gel electrophoresis and enzyme-linked immunosorbent assay. Both assays established stability and the effectiveness of the released antibodies even after 35 days (103).

Researchers formulated chitosan nanoparticles encapsulated into mannitol microspheres and in vitro tested on respiratory epithelium cells. The in vitro study revealed that chitosan nanoparticle-encapsulated microspheres have superior aerodynamic properties, higher drug loading efficiency, and show high biocompatibility, which proves their potential for formulating a resourceful pulmonary drug delivery system (104).

Azarmi and colleagues developed a dry powder formulation of doxorubicin encapsulated in nanoparticles. This vector was formed with a polymerization/emulsion method and then dried. After being re-dissolved in deionized water, the particles had an average size of 173 ± 43 nm. The doxorubicin-loaded nanoparticles demonstrated superior cytotoxicity compared to free doxorubicin (105).

Researchers explored the possibility of local delivery of itraconazole nanoparticles for the treatment of *Aspergillus fumigatus* fungal infections, which classically enter the body through the lungs and circulate through the lymph system. Itraconazole nanoparticles prepared with polysorbate 80 and poloxamer 407, and delivered to mice via the pulmonary route, provided considerably greater lung tissue concentrations in comparison to oral gavage of solutions of either drug or nanoparticles. Additionally, the body dose essential to attain therapeutic lung and serum levels was notably lesser with pulmonary delivery than that needed for oral delivery, consequently lowering the potential for systemic side effects (106).

Yamamoto and colleagues fabricated PLGA nanoparticles surface-modified with chitosan and encapsulating the peptide elcatonin. Nanoparticles administered to the lungs of guinea pigs generated momentous diminutions in blood calcium levels with respect to the initial calcium concentrations, with prolonged effects up to 24 hours. This was appreciably longer than the results achieved with unmodified nanoparticles. The chitosan-modified nanoparticles were demonstrated to be eliminated more slowly than the unmodified version, which suggests that sustained effects were a result of the retention of nanoparticles (107).

In other instances, targeting certain cell types is desired in addition to sustaining drug levels in the lungs. Pulmonary tuberculosis (TB), which is caused by the bacterium *Mycobacterium tuberculosis*, requires targeted drug delivery to alveolar macrophages in which the bacterium resides (108). A series of studies evaluated the strategy of encapsulating three TB antibiotics (rifampicin, isoniazid, and pyrazinamide) together in different types of nanoparticles with pharmacokinetic and pharmacodynamic testing in a guinea pig model of TB. PLGA nanoparticles delivered to the lungs of guinea pigs maintained therapeutic drug concentrations in the plasma for 6–8 days and in the lungs for 9–11 days (109). These sustained levels of drugs in the lungs, the site of drug action, lower the total body dose of drugs essential for treatment. One more possible approach to transform nanoparticles entailed PLGA nanoparticles that were surface-functionalized with wheat germ agglutinin. The incentive for this approach was that agglutinin, a lectin with bioadhesive properties, can potentially interact with lectin receptors on the alveolar epithelium, in that way sustaining further drug levels in the lungs (110). These antibiotic-containing nanoparticles elevated drug levels a little longer than non-functionalized particles, that is to say, 6–14 days in the plasma and 15 days in the lungs. The final approach investigated the exploiting of the natural polymer sodium alginate formulated along with chitosan into nanoparticles encapsulating the three anti-tubercular drugs (111). Drug levels in guinea pigs were measured in the plasma for 10–14 days and in the lungs for up to 15 days. Effectiveness of the three different types of nanoparticle formulations was examined in guinea pigs infected intramuscularly with *M. tuberculosis*. The animals were exposed to the aerosolized nanoparticles every 10–15 days for a total of three to five doses or were given oral drugs every day. At the end of these investigations, no tubercle bacilli were measurable in the lungs of animals treated with aerosol nanoparticles or oral drugs, whereas tubercle bacilli were measured in the lungs of untreated control animals. These outcomes signified that 3 to 5 pulmonary treatments were satisfactory to substitute for the otherwise necessary 45–46 daily doses of oral drug.

Researchers have developed inhalable alginate nanoparticles as anti-tubercular drug carriers against experimental tuberculosis. The relative bioavailability of all drugs from the formulation has found considerably higher relative to oral free drugs when tested in guinea pigs (111).

15.13 Technical Issues of Nanoparticle Applications

Nanoparticles are, in most cases, delivered to the lungs by nebulization of colloidal solutions. However, nanoparticles stored in an aqueous medium will, over a period, result in polymer hydrolysis and loss of drug. Solution instability is an additional problem on account of particle agglomeration and settling (112), a result of the small size and strong particle–particle interactions of nanoparticles, which could bring about poor functionality of the nebulizer. To formulate nanoparticles for ultimate commercial use, the lyophilization of nanoparticles has been explored as a means to give a storage form that can be rehydrated to deliver nanoparticles in solution (113,114). Resuspension is complicated and the retention, post-hydration, of the same nanoparticle size necessitates the use of large quantities of stabilizers during the freeze-drying method. These cryoprotectant sugars and surfactants remain in the reconstituted solution and are delivered along with the particles. In addition, individual nanoparticles do not deposit proficiently in the lungs by diffusion, sedimentation, or impaction, which results in the exhalation of a majority of the inhaled dose (25).

15.14 Conclusion

Pulmonary drug delivery of nanoparticles is a non-invasive method that can be designed to target specific cells or organs at the same time sustaining the release of the therapeutics locally or systemically. Fabricating nanoparticles into dry-powder aerosols of nanosized particles offers stability, comfort of handling, and delivery from simple inhalers. Inhaled nano-particulate therapy maintains the immense potential for treating diseases that have a need for direct lung delivery, like tuberculosis, with reduced drug dosage and dosing frequency, leading to fewer systemic side effects and better patient compliance. The application of pulmonary delivery to biotherapeutics, like insulin, is an important one to evade the need for needle injections. Associating proteins or peptides with nanoparticles provides an additional advantage of sustaining systemic effects, potentially reducing dosing frequency. An issue for nano-particulate drug delivery to the lungs is to understand the fate of particles and their interactions with biological systems. Quick particle clearance lessens sustained delivery of the drug, and particle translocation might bring nanoparticles to undesired areas of the body. A more detailed study into the effect of particle physicochemical properties (e.g. nanoparticle size and material) on extending particle persistence in the lungs and their influence on particle fate is essential to aid the design of improved systems. Pulmonary delivery of inhalable nanoparticles has created scientific curiosity recently, predominantly in vaccine applications, systemic drug delivery in the treatment of pain, and non-invasive brain targeting. Further coordination with environmental health research is key to understanding the implications that particle fate and toxicology might have on drug delivery.

REFERENCES

1. Weers JG, Bell J, Chan H-K, Cipolla D, Dunbar C, Hickey AJ, and Smith IJ. Pulmonary formulations: What remains to be done? *Journal of Aerosol Medicine and Pulmonary Drug Delivery*. 2010; 23(S2): S5–S23.
2. Courrier H, Butz N, and Vandamme TF. Pulmonary drug delivery systems: Recent developments and prospects. *Critical Reviews in Therapeutic Drug Carrier Systems*. 2002; 19(4–5): 425–498.
3. Ungaro F, d'Angelo I, Miro A, La Rotonda MI, and Quaglia F. Engineered PLGA nano-and micro-carriers for pulmonary delivery: Challenges and promises. *Journal of Pharmacy and Pharmacology*. 2012; 64(9): 1217–1235.
4. Smola M, Vandamme T, and Sokolowski A. Nanocarriers as pulmonary drug delivery systems to treat and to diagnose respiratory and non respiratory diseases. *International Journal of Nanomedicine*. 2008; 3(1): 1–19.
5. Paranjpe M and Muller-Goymann CC. Nanoparticle-mediated pulmonary drug delivery: A review. *International Journal of Molecular Science*. 2014; 15(4): 5852–5873.
6. McCorry LK. Essentials of Human Physiology for Pharmacy. CRC Press, Boca Raton, FL. 2004.
7. Groneberg DA, Nickolaus M, Spinger J, Doring F, Daniel H, and Fischer A. Localization of peptide transporter PEPT2 in the lung: Implications of pulmonary oligopeptide uptake. *American Journal of Pathology*. 2001; 158: 707–714.
8. Groneberg DA, Eynott PR, Doring F, Thai D, Oates T, Barnes P, Chung K, Daniel H, and Fischer A. Distribution and function of the peptide transporter PEPT2 in normal and cystic fibrosis human lung. *Thorax*. 2002; 57: 55–60.
9. Groneberg DA, Witt C, Wagner U, Chung KF, and Fischer A. Fundamentals of pulmonary drug delivery. *Respiratory Medicine*. 2003; 97: 382–387.
10. Tuncer DI and Nevin C. Controlled delivery of peptides and proteins. *Current Pharmaceutical Design*. 2007; 13: 99–117.

11. Sangwan S, Agosti JM, Bauer LA, Otulana BA, Morishige RJ, Cipolla DC, Blanchard JD, and Smaldone GC. Aerozolized protein delivery in asthma: Gamma camera analysis of regional deposition and perfusion. *Journal of Aerosol Medicine*. 2001; 14: 185–195.
12. Scheuch G and Siekmeier R. Novel approaches to enhance pulmonary delivery of proteins and peptide. *Journal of Physiology and Pharmacology*. 2007; 58: 615–625.
13. Siekmeier R and Scheuch G. Systemic treatment by inhalation of macromolecules: Principles, problems and examples. *Journal of Physiology and Pharmacology*. 2008; 59: 53–79.
14. Flume P and Klepser ME. The rationale for aerosolized antibiotics. *Pharmacotherapy*. 2002; 22: 71S–79S.
15. Mastrandrea LD and Quattrin T. Clinical evaluation of inhaled insulin. *Advanced Drug Delivery Reviews*. 2006; 58: 1061–1075.
16. Patton JS, Bukar JG, and Eldon MA. Clinical pharmacokinetics and pharmacodynamics of inhaled insulin. *Clinical Pharmacokinetics*. 2004; 43: 781–801.
17. Guntur VP and Dhand R. Inhaled insulin: Extending the horizons of inhalation therapy. *Respiratory Care*. 2007; 52: 911–922.
18. Codrons V, Vanderbist F, and Verbeeck RK. Systemic delivery of parathyroid hormone (1-34) using inhalation dry powders in rats. *Journal of Pharmaceutical Sciences*. 2003; 92: 938–950.
19. Gessler T, Schmehl T, Olschewski H, Grimminger F, and Seeger W. Aerosolized vasodilators in pulmonary hypertension. *Journal of Aerosol Medicine*. 2002; 15: 117–122.
20. Gessler T, Seeger W, and Schmehl T. Inhaled prostanoids in the therapy of pulmonaryhypertension. *Journal of Aerosol Medicine*. 2008; 21: 1–12.
21. Sermet-Gaudelus I, Cocguic YL, Ferroni A, Clairicia M, Barthe J, Delaunay JP, Brousse V, and Lenoir G. Nebulized antibiotics in cystic fibrosis. *Pediatric Drugs*. 2002; 4(7): 455–467.
22. Patton JS, Fishburn CS, and Weers JG. The lungs as a portal of entry for systemic drug delivery. *Proceedings of the American Thoracic Society*. 2004; 1: 338–344.
23. Patton JS and Byron PR. Inhaling medicines: Delivering drugs to the body through the lungs. *Nature Reviews Drug Discovery*. 2007; 6: 67–74
24. Chellat F, Merhi Y, Moreau A, and Yahia L. Therapeutic potential of nanoparticulate systems for macrophage targeting. *Biomaterial*. 2005; 26: 7260–7275.
25. Sung JC, Pulliam BL, and Edwards DA. Nanoparticles for drug delivery to the lungs. *TRENDS in Biotechnology*. 2007; 25: 563–570.
26. Patton JS, Brain JD, Davies LA, Fiegel J, Gumbleton M, Kim KJ, Sakagami M, Vanbever R, and Ehrhardt C. The particle has landed-characterizing the fate of inhaled pharmaceuticals. *Journal of Aerosol Medicine and Pulmonary Drug Delivery*. 2010; 23(S2): S71–S87.
27. El-Sherbiny IM, Villanueva DG, Herrera D, and Smyth HD. Overcoming lung clearance mechanisms for controlled release drug delivery. In: Smyth HDC and Hickey JA. (eds.) Controlled Pulmonary Drug Delivery. Springer, New York, NY. 2011; 101–126.
28. Hogg J. Response of the lung to inhaled particles. *Medical Journal of Australia*. 1985; 142(13): 675–678.
29. Olsson B, Bondesson E, Borgstrom L, et al. Pulmonary drug metabolism, clearance, and absorption. In: Smyth HDC and Hickey JA. (eds.) Controlled Pulmonary Drug Delivery. Springer, New York, NY. 2011; 21–50.
30. Oberdorster, G. Lung clearance of inhaled insoluble and soluble particles. *Journal of Aerosol Medicine*. 1988; 1(4): 289–330.
31. Parkinson A. Biotransformation of xenobiotics. In: Klaassen C. (ed.) Casarett and Doull's Toxicology: The Basic Science of Poisons. McGraw-Hill, New York. 2001; 133–224.
32. Wiedmann T, Bhatia R, and Wattenberg L. Drug solubilization in lung surfactant. *Journal of Controlled Release*. 2000; 65(1): 43–47.
33. Patton JS. Mechanisms of macromolecule absorption by the lungs. *Advanced Drug Delivery Reviews*. 1996; 19(1): 3–36.
34. Liao X and Wiedmann TS. Solubilization of cationic drugs in lung surfactant. *Pharmaceutical Research*. 2003; 20(11): 1858–1863.
35. Mansour HM, Rhee YS, and Wu X. Nanomedicine in pulmonary delivery. *International Journal of Nanomedicine*. 2009; 4: 299–319.
36. Beck-Broichsitter M, Gauss J, Packhaeuser CB, Lahnstein K, Schmehl T, Seeger W, Kissel T, and Gessler T. Pulmonary drug delivery with aerosolizable nanoparticles in an *ex-vivo* lung model. *International Journal of Pharmaceutics*. 2009; 367(1): 169–178.
37. Labiris N and Dolovich M. Pulmonary drug delivery. Part I: Physiological factors affecting therapeutic effectiveness of aerosolized medications. *British Journal of Clin ical Pharmacology*. 2003; 56(6): 588–599.
38. Thulasiramaraju TV, Tejeswar KB, and Nikilesh BM. Pulmonary drug delivery systems - An overview. *Asian Journal of Research in Pharmaceutical Sciences and Biotechnology*. 2013; 1(1): 16–34.
39. Jaafar-Maalej C, Elaissari A, and Fessi H. Lipid-based carriers: Manufacturing and applications for pulmonary route. *Expert Opinion on Drug Delivery*. 2012; 9(9): 1111–1127.
40. Kuzmov A and Minko T. Nanotechnology approaches for inhalation treatment of lung diseases. *Journal of Controlled Release*. 2015; 219: 500–518.
41. Cryan SA. Carrier-based strategies for targeting protein and peptide drugs to the lungs. *AAPS Journal*. 2005; 7: E20–E41.
42. Feng SS and Chien S. Chemotherapeutic engineering: Application and further development of chemical engineering principles for chemotherapy of cancer and other diseases. *Chemical Engineering Science*. 2003; 58(18): 4087–4114.
43. Velino C, Carella F, Adamiano A, Sanguinetti M, Vitali A, Catalucci D, Bugli F, and Iafisco M. Nanomedicine approaches for the pulmonary treatment of cystic fibrosis. *Frontiers in Bioengineering and Biotechnology*. 2019; 7: Article 406.
44. Fores B and Ehrhardt C. Human respiratory epithelial cell culture for drug delivery applications. *European Journal of Pharmaceutics and Biopharmaceutics*. 2005; 60: 193–205.
45. Evans CM and Koo JAS. Airway mucus: The good, the bad, the sticky. *Pharmacology & Therapeutics*. 2009; 121(3): 332–348.

46. Palecanda A and Kobzik L. Receptors for unopsonized particles: The role of alveolar macrophages scavenger receptors. *Current Molecular Medicine*. 2001; 1: 589–595.
47. Perez-Gil J. Structure of pulmonary surfactant membranes and films: The role of proteins and lipid-protein interactions. *Biochimica et Biophysics Acta*. 2008; 1778: 1676–1695.
48. Hickey AJ and Garcia-Contreras L. Immunological and toxicological implications of short-term studies in animals of pharmaceutical aerosol delivery to the lungs: Relevance to humans. *Critical Reviews in Therapeutic Drug Carrier Systems*. 2001; 18: 387–431.
49. Heinemann L, Klappoth W, Rave K, Hompesch B, Linkeschowa R, and Heise T. Intra-individual variability of the metabolic effect of inhaled insulin together with an absorption enhancer. *Diabetes Care*. 2000; 23: 1343–1347.
50. Ramachandran V, Murnane D, Hammond RB, et al. Formulation pre-screening of inhalation powders using computational atom–atom systematic search method. *Molecular Pharmacology*. 2015; 12: 18–33.
51. Chew NYK and Chan HK. Use of solid corrugated particles to enhance powder aerosol performance. *Pharmaceutical Research*. 2001; 18: 1570–1577.
52. Geller DE, Weers J, and Heuerding S. Development of an inhaled dry-powder formulation of tobramycin using Pulmo-Sphere™ technology. *Journal of Aerosol Medicine and Pulmonary Drug Delivery*. 2011; 24: 175–182.
53. Begat P, Price R, Harris H, Morton DAV, and Staniforth JN. The influence of force control agents on the cohesive-adhesive balance in dry powder inhaler formulations. *Kona Powder and Particle Journal*. 2005; 23: 109–121.
54. Begat, P, Morton, DA, Shur, J, Kippax, P, Staniforth, JN, and R Price. The role of force control agents in high-dose dry powder inhaler formulations. *Journal of Pharmaceutical Sciences*. 2009; 98:2770–2783.
55. Sosnowski TR and Gradon L. Modification of inhalable powders by pulmonary surfactant components adsorbed on droplets during spray-drying process. *Colloids and Surfaces A: Physicochemical and Engineering Aspects*. 2010; 365: 56–61.
56. Tsapis N, Bennett D, Jackson B, Weitz DA, and Edwards DA. Trojan particles: large porous carriers of nanoparticles for drug delivery. *Proceedings of the National Academy of Sciences USA*. 2002; 99: 12001–12005.
57. Jablczynska K, Janczewska M, Kulikowska A, and Sosnowski TR. Preparation and characterization of biocompatible polymer particles as potential nanocarriers for inhalation therapy. *International Journal of Polymer Science*. 2015; 2015: Article ID 763020.
58. Liang Z, Ni R, Zhou J, and Mao S. Recent advances in controlled pulmonary drug delivery. *Drug Discovery Today*. 2014; 20: 380–389.
59. Weber S, Zimmer A, and Pardeike J. Solid lipid nanoparticles (SLN) and nanostructured lipid carriers (NLC) for pulmonary application: A review of the state of the art. *European Journal of Pharmaceutics and Biopharmaceutics*. 2014; 86: 7–22.
60. Li YZ, Sun X, Gong T, Liu J, Zuo J, and Zhang ZR. Inhalable microparticles as carriers for pulmonary delivery of thymopentin-loaded solid lipid nanoparticles. *Pharmaceutical Research*. 2010; 27: 1977–1986.
61. Gradon L and Sosnowski TR. Formation of particles for dry powder inhalers. *Advanced Powder Technology*. 2014; 25: 43–55.
62. Kaialy W and Nokhodchi A. Particle engineering for improved pulmonary drug delivery through dry powder inhalers. In: Nokhodhi A and Martin GP. (eds.) Pulmonary Drug Delivery. Advances and Challenges (Advances in Pharmaceutical Technology). John Wiley & Sons, Chichester, UK. 2015; 171–197.
63. Vehring R. Expert review: Pharmaceutical particle engineering via spray drying. *Pharmaceutical Research*. 2008; 25: 999–1022.
64. Weers JG and Miller DP. Formulation design of dry powders for inhalation. *Journal of Pharmaceutical Sciences*. 2015; 104: 3259–3288.
65. Nandiyanto ABD and Okuyama K. Progress in developing spray-drying methods for the production of controlled morphology particles: From the nanometer to submicrometer size ranges. *Advanced Powder Technology*. 2011; 22: 1–19.
66. Kramek-Romanowska K, Odziomek M, Sosnowski TR, and Gradon L. Effects of process variables on the properties of spray-dried mannitol and mannitol/disodium cromoglycate powders suitable for drug delivery by inhalation. *Industrial and Engineering Chemistry Research*. 2011; 50: 13922–13931.
67. Lizio R, Klenner T, Borchard G, Romeis P, Sarlikiotis AW, Reissmann T, and Lehr CM. Systemic delivery of the GnRH antagonist centrolix by intratracheal instillation in anesthetized rats. *European Journal of Pharmaceutical Sciences*. 2000; 9(3): 253–258.
68. Patil JS and Sarasija S. Pulmonary drug delivery strategies: A concise, systematic review. *Lung India*. 2012; 29(1): 44–49.
69. Finlay WH. The Mechanics of Inhaled Pharmaceutical Aerosols: An Introduction, 2nd edition. Academic Press, London. 2019; 103–182.
70. Kaur G, Narang RK, Rath G, and Goyal AK. Advances in pulmonary delivery of nanoparticles. *Artificial Cells, Blood Substitutes, and Biotechnology*. 2012; 40: 75–96.
71. Chono S, Tanino T, Seki T, and Morimoto K. Influence of particle size on drug delivery to rat alveolar macrophages following pulmonary administration of ciprofloxacin incorporated into liposomes. *Journal of Drug Targeting*. 2006; 14: 557–566.
72. Chaudhuri SR and Lukacova V. Simulating Delivery of Pulmonary (and Intranasal) Aerosolised Drugs. Simulations Plus, Inc, Lanchester. 2010.
73. Desai A and Lee M. Gibaldi's Drug Delivery Systems in Pharmaceutical Care. American Society of Health-System Pharmacist, Bethesda, MD. 2007.
74. Labiris NR and Dolovich MB. Pulmonary drug delivery. Part I: Physiological factors affecting therapeutic effectiveness of aerosolized medications. *British Journal of Clinical Pharmacology*. 2003; 56: 588–599.
75. Vyas SP and Khar RK. Targeted & Controlled Drug Delivery, Novel Carrier Systems. CBS Publishers, New Delhi. 2002.
76. Rubin BK. Air and soul: The science and application of aerosol therapy. *Respiratory Care*. 2010; 55: 911–921.

77. Kalantarian P, Najafabadi AR, Haririan I, Vatanara A, Yamini Y, Darabi M, and Gilani K. Preparation of 5-fluorouracil nanoparticles by supercritical antisolvents for pulmonary delivery. *International Journal of Nanomedicine*. 2010; 5: 763–770.
78. Zeman KL, Wu J, and Bennett WD. Targeting aerosolized drugs to the conducting airways using very large particles and extremely slow inhalations. *Journal of Aerosol Medicine and Pulmonary Drug Delivery*. 2010; 23: 363–369.
79. Zhang Y, Wang X, Lin X, Liu X, Tian B, and Tang X. High azithromycin loading powders for inhalation and their *in-vivo* evaluation in rats. *International Journal of Pharmaceutics*. 2010; 395: 205–214.
80. Tronde A, Norden B, Marchner H, Wendel AK, Lennernas H, and Bengtsson UH. Pulmonary absorption rate and bioavailability of drugs *in-vivo* in rats: Structure-absorption relationships and physicochemical profiling of inhaled drugs. *Journal of Pharmaceutical Sciences*. 2003; 92: 1216–1233.
81. Tyrrell J and Tarran R. Gaining the upper hand on pulmonary drug delivery. *Journal of Pharmacovigilance*. 2014; 2(1): 118.
82. El-Sherbiny IM, El-Baz NM, and Yacoub MH. Inhaled nano- and microparticles for drug delivery. *Global Cardiology Science & Practice*. 2015; 2015(1): 1–14.
83. Tronde A, Bosquillon C, and Forbes B. The isolated perfused lungs for drug absorption studies. In: Ehrhardt C and Kwang-Jin K. (eds.) Drug Absorption Studies: In-Situ, In-Vitro and In-Silico Models. Springer Sciences, New York. 2008; 135–154.
84. Dandekar P, Venkataraman C, and Mehra A. Pulmonary targeting of nanoparticle drug matrices. *Journal of Aerosol Medicine and Pulmonary Drug Delivery*. 2010; 23: 343–353.
85. Noymer PD, Myers DJ, Cassella JV, and Timmons R. Assessing the temperature of thermally generated inhalation aerosols. *Journal of Aerosol Medicine and Pulmonary Drug Delivery*. 2011; 24: 11–15.
86. Shaikh S, Nazim S, Khan T, Shaikh A, Zameeruddin M, and Quazi A. Recent advances in pulmonary drug delivery system: A review. *International Journal of Applied Pharmaceutics*. 2010; 2: 27–31.
87. Chan HK. Inhalation drug delivery devices and emerging technologies. *Expert Opinion on Therapeutic Patents*. 2005; 13(9): 1333–1343.
88. Geller, DE. New liquid aerosol generation devices: System that force pressurized liquids through nozzles. *Respiratory Care*. 2002; 47: 1392–1404.
89. Geller DE, Thippawong J, and Otulana B. Bolus inhalation of rhDNase with the AERx system in subjects with cystic fibrosis. *Journal of Aerosol Medicine*. 2003; 16: 175–182.
90. Henry RR, Mudaliar SR, Howland III WC, Chu N, Kim D, An B, and Reinhardt RR. Inhaled insulin using the AERx insulin diabetes management system in healthy and asthmatic subjects. *Diabetes Care*. 2003; 26: 764–769.
91. Atkins PJ. Dry powder inhalers: An overview. *Respiratory Care*. 2005; 50: 1304–1312.
92. Dunbar C, Scheuch G, Sommerer K, DeLong M, Verma A, and Batycky R. *In-vitro* and *in-vivo* dose delivery characteristics of large porous particles for inhalation. *International Journal of Pharmaceutics*. 2002; 245(1–2): 179–189.
93. Gupta R, Vanbever R, Mintzes J, Nice J, Chen D, Batycky R, Langer R, and Edwards DA. Physical characterization of large porous particles for inhalation. *Pharmaceutical Research*. 2000; 17: 1437–1438.
94. Bosquillon C, Lombry C, Preat V, and Vanbever R. Influence of formulation excipients and physical characteristics of inhalation dry powders on their aerosolization performance. *Journal of Controlled Release*. 2001; 70(3): 329–339.
95. Voss AP and Finley WH. De-agglomeration of dry powder pharmaceutical aerosols. *International Journal of Pharmaceutics*. 2003; 248: 39–50.
96. Coates MS, Fletcher DF, Chan HK, and Raper JA. Effect of design on the performance of a dry powder inhaler using computational fluid dynamics. Part I: Grid structure and mouthpiece length. *Journal of Pharmaceutical Sciences*. 2004; 93(11): 2863–2876.
97. Byron PR. Drug delivery devices: Issues in drug development. *Proceedings of the American Thoracic Society*. 2004; 1: 321–328.
98. Zatta KC, Frank LA, Reolon LA, Amaral-Machado L, Egito EST, Gremiao MPD, Pohlmann AR, and Guterres SS. An inhalable powder formulation based on microand nanoparticles containing 5-fluorouracil for the treatment of metastatic melanoma. *Nanomaterials*. 2018; 8(2): 75.
99. Debnath SK, Saisivam S, Debanth M, and Omri A. Development and evaluation of chitosan nanoparticles based dry powder inhalation formulations of prothionamide. *PLoS ONE*. 2018; 13(1): e0190976.
100. Varshosaz J, Taymouri S, Hamishehkar HH, Vatankhah R, and Yaghubi S. Development of dry powder inhaler containing tadalafil-loaded PLGA nanoparticles. *Research in Pharmaceutical Sciences*. 2017; 12(3): 222–232.
101. Tulbah AS, Pisano E, Scalia S, Young PM, Traini D, and Ong HX. Inhaled simvastatin nanoparticles for inflammatory lung disease. *Nanomedicine*. 2017; 12(20): 2471–2485.
102. Roa WH, Azarmi S, Al-Hallak MH, Finlay WH, Magliocco AM, and Lobenberg R. Inhalable nanoparticles, a non-invasive approach to treat lung cancer in a mouse model. *Journal of Controlled Release*. 2011; 150: 49–55.
103. Kaye RS, Purewal TS, and Alpar HO. Simultaneously manufactured nano-in-micro (SIMANIM) particles for dry-powder modified-release delivery of antibodies. *Journal of Pharmaceutical Sciences*. 2009; 98(11): 4055–4068.
104. Grenha A, Grainger CI, Dailey LA, Seijo B, Martin GP, Remunan-Lopez C, and Forbes B. Chitosan nanoparticles are compatible with respiratory epithelial cells in vitro. *European Journal of Pharmaceutical Sciences*. 2007; 31(2): 73–84.
105. Azarmi S, Tao X, Chen H, Wang Z, Finlay WH, Lobenberg R, and Roa WH. Formulation and cytotoxicity of doxorubicin nanoparticles carried by dry powder aerosol particles. *International Journal of Pharmaceutics*. 2006; 319: 155–161.
106. Vaughn JM, McConville JT, Burgess D, Peters JI, Johnston KP, Talbert RL, and Williams III RO. Single dose and multiple dose studies of itraconazole nanoparticles. *European Journal of Pharmaceutics and Biopharmaceutics*. 2006; 63(2): 95–102.

107. Yamamoto H, Kuno Y, Sugimoto S, Takeuchi H, and Kawashima Y. Surface-modified PLGA nanosphere with chitosan improved pulmonary delivery of calcitonin by mucoadhesion and opening of the intercellular tight junctions. *Journal of Controlled Release*. 2005; 102(2): 373–381.
108. Pandey R and Khuller GK. Antitubercular inhaled therapy: Opportunities, progress and challenges. *Journal of Antimicrobial Chemotherapy*. 2005; 55: 430–435.
109. Pandey R, Sharma A, Zahoor A, Sharma S, Khuller GK, and Prasad B. Poly (DL-lactide-co-glycolide) nanoparticle-based inhalable sustained drug delivery system for experimental tuberculosis. *Journal of Antimicrobial Chemotherapy*. 2003; 52(6): 981–986.
110. Sharma A, Sharma S, and Khuller GK. Lectin-functionalized poly (lactide-co-glycolide) nanoparticles as oral/aerosolized antitubercular drug carriers for treatment of tuberculosis. *Journal of Antimicrobial Chemotherapy*. 2004; 54(4): 761–766.
111. Ahmad Z, Sharma S, and Khuller GK. Inhalable alginate nanoparticles as antitubercular drug carriers against experimental tuberculosis. *International Journal of Antimicrobial Agents*. 2005; 26(4): 298–303.
112. Dailey LA, Schmehl T, Gessler T, Wittmar M, Grimminger F, Seeger W, and Kissel T. Nebulization of biodegradable nanoparticles: Impact of nebulizer technology and nanoparticle characteristics on aerosol features. *Journal of Controlled Release*. 2003; 86(1): 131–144.
113. Wendorf J, Singh M, Chesko J, Kazzaz J, Soewanan E, Ugozzoli M, and O'Hagan D. A practical approach to the use of nanoparticles for vaccine delivery. *Journal of Pharmaceutical Sciences*. 2006; 95(12): 2738–2750.
114. Saez A, Guzman M, Molpeceres J, and Aberturas MR. Freeze-drying of polycaprolactone and poly(-lactic glycolic) nanoparticles induce minor particle size changes affecting the oral pharmacokinetics of loaded drugs. *European Journal of Pharmaceutics and Biopharmaceutics*. 2000; 50(3) :379–387.

16

Recent Trends in Applications of Nano-Drug Delivery Systems

Doris Kumadoh[1], Ofosua Adi-Dako[2], Yaa Asantewaa Osei[3], and Mary Ann Archer[1]
[1]Centre for Plant Medicine Research, Mampong-Akuapem, Ghana
[2]Department of Pharmaceutics & Microbiology, School of Pharmacy, University of Ghana, Accra, Ghana
[3]Department of Pharmaceutics, Faculty of Pharmacy and Pharmaceutical Sciences, Kwame Nkrumah University of Science Technology, Cape Coast, Ghana

16.0 Recent Trends in Applications of Nano-Drug Delivery Systems

16.1 Nano-Drug Delivery Systems

Nanomedicine is basically the coupling of medicines with nanotechnology tools and techniques for advanced therapy with the aid of molecular knowledge. Nanoscale drug delivery systems help to improve the pharmacokinetics and increase the bio-distribution of therapeutic agents to target organs, thereby resulting in improved efficacy while limiting drug toxicity. Nanoparticles (NPs) can be defined as complex drug carrier systems with materials having at least one diameter measuring 100 nm or less (1) which incorporate and protect a certain drug or particle. Nanoparticles can be administered through various routes including intravenous injection, oral administration, and pulmonary inhalation (2). NPs may be classified as organic (including liposomes, nano crystals, dendrimers, polymeric NPs) and inorganic (including metals and silica) (3). Various nanoparticles are applied in lung disease management. Figure 16.1 indicates some types used.

16.1.1 Solid–Lipid Particulate Delivery System

Solid lipid nanoparticles (SLNs) are novel lipid based nano-carriers with sizes ranging between 10 and 1000 nm. The main advantage of SLNs is the fact that they are prepared with physiologically well-tolerated lipids. SLNs were introduced to overcome problems of polymeric nanoparticles, lipid emulsion, and liposomes (4,5). SLNs also have the advantage of good compatibility and biodegradability, and serve as a good carrier for lipophilic drugs. SLNs can be used for various routes of drug administration such as topical, parenteral, dermal, and pulmonary. Disadvantages include poor drug loading capacity, a relatively high water content of the dispersions (70–99.9%), drug expulsion after polymeric transition during storage, and the low capacity to load hydrophilic drugs due to partitioning effects during the production process (4). Formulations of SLNs include three main constituents comprising solid lipids, surfactants, and water. Lipids comprise triglycerides (for example, tristearin), incomplete glycerides, fatty acids (for example, stearic acid), hormones (such as cholesterol) and waxes (such as cetyl palmitate) (4,5).

16.1.2 Nano Structured Lipid Carriers

Nanostructured lipid carriers (NLCs) are lipid-based formulations that have a solid matrix at room temperature. NLCs are considered superior to many other traditional lipid-based nanocarriers such as nano-emulsions, liposomes and solid lipid nanoparticles (SLNs) due to their enhanced physical stability, improved drug loading capacity, and biocompatibility (48). NLCs are known to give protection to sensitive compounds and increase feasibility, bioavailability, and solubility. NLCs can enhance drug distribution to target organs, change the pharmacokinetic characteristics of drug carriers to enhance the therapeutic effect, and reduce adverse side effects (6,7).

The formulation of NLCs involves the nano-emulsification of a lipohilic phase composed of a mixture of liquid lipid and solid lipid (e.g. tripalmitin and squalene, respectively) in an aqueous solution of water-soluble surfactants/emulsifiers (7).

16.1.3 Nano-emulsions

Nano-emulsions are diverse systems consisting of dispersion of two immiscible liquids such as water and oil; they are completely stable by use of surfactant molecules in various proportions. The droplet size of a nano-emulsion typically falls in the range of 20–200 nm (8). The method of preparing the nano-emulsions is emulsification, thus nanoparticles use a polymerizable monomer as the dispersed phase where nano-emulsion droplets act as nanoreactor. An example of a nano-emulsion delivery system is phospholipid-based nano-emulsion system (8,9).

DOI: 10.1201/9781003046547-16

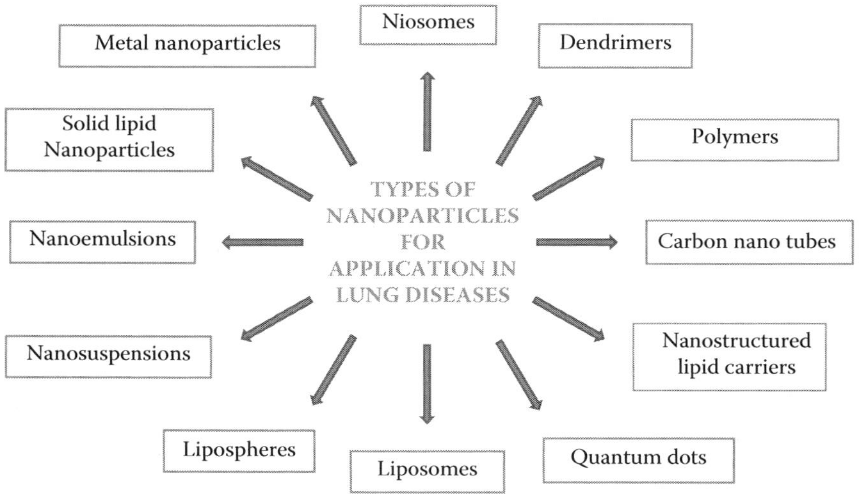

FIGURE 16.1 Types of Nanoparticles Applied for Lung Diseases

16.1.4 Niosomes

Niosomes are non-ionic surfactant vesicles having a two-layer structure formed by self-assembly of hydrated surfactant monomers and cholesterol in an aqueous phase. Due to the biodegradable, biocompatible, and nonimmunogenic structure of niosomes, they are promising drug carriers for delivering drugs directly to the body part where the therapeutic effect is required. This reduces the dose frequency to achieve the desired effects, which subsequently decreases the side effects (10). Examples of surfactants used in the formulation of niosomes include alkyl ethers, alkyl glyceryl ethers, sorbitan fatty acid esters, and polyoxyethylene fatty acid esters (11).

16.1.5 Liposomes

Liposomes are small spherical vesicles (single or bilayer) in which one or more aqueous compartments are completely enclosed by molecules that have hydrophilic and hydrophobic functionality. Liposomes as amphiphilic molecules can carry both hydrophobic and hydrophilic drugs. The shape, surface, charge, size, and functional groups in liposomes can be easily changed according to the drug and the target site. Liposomes can be encapsulated with various drugs such as anti-cancer drugs, neurotransmitters, antibiotics, and anti-inflammatory drugs (2,12,13). Liposomes are also effective in reducing systemic toxicity and preventing early degradation of the encapsulated drug after administration. They can be covered with polymers, such as polyethylene glycol (pegylated or stealth liposomes), to exhibit prolonged half-life in blood circulation. Liposomes can also be conjugated to antibodies or ligands in order to enhance target specificity (12,13).

16.1.6 Lipospheres

Lipospheres are lipid-based dispersion systems where the drug is dissolved or dispersed in a lipidic core, the surface of which is embedded with an emulsifier layer. Lipospheres consist of two distinct layers, the inner solid particle that contains the entrapped drug with a phospholipid outer layer. Lipospheres are among the promising particulate drug delivery systems for improving dissolution rate of water insoluble drugs. Lipospheres are attractive nano-drug delivery systems for topical delivery due to characteristics such as film forming ability, occlusive properties, controlled release from solid lipid matrix resulting in prolonged release of drug, and retarded systemic absorption of drugs (14,15).

16.1.7 Polymeric Nanoparticles

Polymeric nanoparticles are nanoparticles manufactured from various kinds of polymers, organic solvents, and surfactants. The use of polymeric NPs generally offers precise control over particle composition, size, and structure. This control contributes to improved pharmacokinetics, enhanced aqueous solubility of insoluble hydrophobic drugs, and stabilization of therapeutic agents against possible hydrolytic and/or enzymatic degradation. The ability to incorporate surface decorating moieties (e.g. cell-targeting peptides) onto the polymeric nanoparticle, mediation of sustained and/or stimuli-responsive release of the payloads, and mimicking of natural nanosystems (e.g. viruses, lipoproteins, and proteins) for cell-specific targeted delivery with increased bioavailability, is an added advantage (16).

16.1.8 Nanosuspension

Nanosuspensions can be defined as submicron colloidal dispersions of nano-sized drug particles (10–1000 nm) stabilized by surfactants. Nanosuspensions help to increase drug saturation solubility and dissolution rates, reduce drug administration doses, side effects, and cost of therapy. One type of nanosuspension is PEGylated nanoparticles, which are used for passive targeting (17).

16.1.9 Dendrimers

Dendrimers are synthetic nano-material modern day polymers with well-defined size and structural branched chains and a monodisperse character. Drugs can be incorporated into either the internal surface/core or surface by covalent bonds. Dendrimers are unimolecular miceller in nature and this enhances the solubility of poorly soluble drugs and compatibility with DNA, heparin, and polyanions to make them more versatile (18,19).

16.1.10 Carbon Nanotubes

Carbon nanotubes (CNTs) are known as nano-architectured allotropes of carbon, having graphene sheets that are wrapped or rolled in different ways forming a cylindrical shape to make them either metals or narrow-band semiconductors. CNTs exhibit some unusual properties, like a high degree of stiffness, a large length-to-diameter ratio, and exceptional resilience, which gives them the potential to be used in a variety of drug delivery applications (20).

16.1.11 Quantum Dots

Quantum dots are semi-conductor particles (nanocrystals) having size ranging from 1 to 10 nm. They are made to fluorescent in different colours depending on their size. And they can show strong light absorbance, bright and narrow symmetric emission band. Quantum dots (QDs) have the simultaneous potential of targeting and imaging. Quantum dots are therefore used in drug delivery in pharmaceutical and biomedical applications, especially in clinical diagnostics (21).

16.1.12 Metal Nanoparticles

Metal nanoparticles (MNP) such as gold, silver, iron, platinum, and ceramic, are used due to their optical, magnetic, electrical properties, and size, which leads to less solvent contamination and uniform distribution. Most metal-based NPs possess unique physicochemical properties. For instance, the unique optical properties of plasmon-resonant noble metals (i.e. Au, Ag, Pt, and Pd) and luminescent semiconductor nanocrystals [quantum dots (QDs)] make them useful as markers for biological systems imaging. Gold nanoparticles (Au NPs) may be used in biological applications, such as cancer therapy, cell labeling, drug delivery, and diagnostics. The magnetic features of Fe_2O_3 can be exploited to drive loaded NPs to specific target tissues by applied magnetic fields. Silver NPs are also used in common household products as additives with anti-microbial activity against more than 650 different types of disease-causing organisms, including viruses and cancer therapy (22).

16.2 Applications of Nano-Drug Delivery Systems in the Management of Lung Diseases

16.2.1 Lung Diseases

Lung diseases normally refer to several types of diseases that prevent the lungs from working properly. Lung diseases can affect respiratory function, or the ability to breathe properly, and pulmonary function (23,24).

The lung is the internal organ most vulnerable to infection and injury from the external environment because of its constant exposure to particles, chemicals, and infectious organisms in ambient air (24,25).

Lung diseases affect infants, young, and old worldwide. Chronic obstructive pulmonary disease (COPD), asthma, acute lower respiratory tract infections, tuberculosis, and lung cancer which are normally known as the "big five" diseases of the respiratory system, are among the most common causes of severe illness and death worldwide (25).

It is estimated that more 384 million people have COPD; 65 million endure moderate-to-severe COPD and 3 million die from it each year, making it the third leading cause of death worldwide (26,27). Globally, asthma is the most common chronic disease of childhood, affecting 14% of children; 334 million people suffer from asthma (28), while 10.3 million people develop tuberculosis, and 1.3 million die from it each year, making it the most common fatal infectious disease (29). It is estimated that more than 1.76 million people die from lung cancer each year (30), acute lower respiratory infections have been among the top three cause of death among children and adults, and 4 million people die from lower respiratory tract infections and pneumonia each year (25). Every minute, two children less than 5 years old die from pneumonia; 80% of pneumonia deaths are in children under 2 years (31). Lung diseases may be categorized into three main types which are airway, tissue, and circulatory disease.

16.2.1.1 Lung Airway Diseases

These diseases affect the tubes (airways) that take oxygen and other gases into and out of the lungs. They normally cause a narrowing or blockage of the airways and include asthma, COPD (chronic bronchitis and emphysema), and bronchiectasis (24,32).

16.2.1.2 Lung Tissue Diseases

These diseases affect the structure of the lung tissue. Scarring or inflammation of the tissue makes the lungs unable to expand fully (restrictive lung disease). This makes it hard for the lungs to take in oxygen and release carbon dioxide. A feeling of "wearing a too-tight sweater or vest" is normally experienced by people suffering from lung tissue diseases. This results in an inability to breathe deeply. Pulmonary fibrosis, interstitial lung disease, and pulmonary sarcoidosis are examples of lung tissue disease (24,25).

16.2.1.3 Lung Circulation Diseases

These diseases are normally caused by clotting, scarring, or inflammation of the blood vessels in the lungs. Lung circulation diseases primarily affect the ability of the lungs to take up oxygen and release carbon dioxide. They may also affect the normal functioning of the heart. Examples of lung circulation diseases are pulmonary hypertension, embolism, and edema (24) (Figure 16.2).

The use of nano-drug delivery systems in the lungs represents an attractive alternative to the intravenous or oral routes, due to

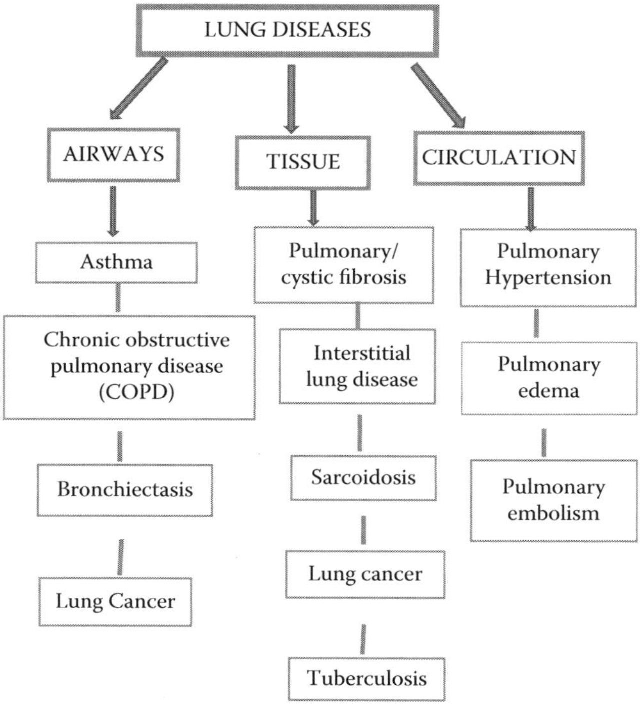

FIGURE 16.2 Categorization of Lung Diseases Based on Affected Areas of the Lung

the unique anatomical and physiological features of the lungs and the minimal interactions between the targeted site and other organs. Some of the widely used nanocarriers for the treatment of chronic pulmonary diseases, via pulmonary route, include polymeric nanoparticles, liposomal nanocarriers, solid lipid nanoparticles, and submicron emulsions (33,34). The application of nanomedicines is increasing rapidly due to the targeted and more efficient nature of drug delivery in comparison to most conventional therapy. Disadvantages of conventional therapy are normally addressed, such as enhancement of site-specific drug delivery, reduced side effects, and better treatment outcomes (25,36). Targeted delivery of medicines through the respiratory system is a potential approach to improve drug accumulation and efficacy in the management of lung diseases while decreasing their negative side effects (35–37). Nanocarrier systems also provide the advantage of sustained drug release in the lung tissue resulting in reduced dosing frequency and improved patient compliance (38). The application of nanoparticle-based delivery in the management of lung diseases helps to minimize pulmonary clearance mechanisms, enhance drug therapeutic efficacy, and control the release behavior of some drugs. Influential parameters for lung drug delivery using nanoparticle delivery systems includes the nature of drug carrier, inhalation technology, and health status of the patient's respiratory system (36–38). Various applications of nanoparticle drug delivery systems are currently being investigated with some being employed for the management of varied lung diseases including COPD, asthma, pneumonia, lung cancer, and tuberculosis (33–40).

16.2.2.1 Asthma

This occurs when the airways are constantly inflamed and may spasm, causing wheezing, coughing, chest tightness, and shortness of breath. Air pollution, infections, and allergens can activate asthma symptoms (24,25). It is the most common chronic disease in children and is more severe in children in developing countries (25). Some causes of asthma include genetic predisposition, exposure to environmental allergens, air pollution, dietary factors, and abnormal immunological responses (25,41,42). Management of asthma involves the use of corticosteroids, long acting beta-2 agonists, short acting beta-2 agonists, long-acting anti-cholinergics, and other preventive strategies such as avoiding known allergens (42).

The use of NPs may help facilitate drug delivery directly to the target tissues, improve the lung drug deposition rate and the therapeutic effects of anti-asthmatic drugs while attenuating adverse side effects such as immunosuppression (35). Nanomaterials used for anti-asthmatic studies may escape from several physiological barriers, such as airway mucus, to cross the endothelium and deliver expected pharmacological effects. It has been reported that steroids encapsulated in NPs produced more sustained therapeutic effects at the site of airway inflammation compared to free steroids (35,43,44).

Dendrimers have been studied for application in the treatment of asthma. To reduce the side effects of long-term high-dose hormone inhalation and further improve the bioavailability of hormones, PEGylated Poly(amidoamine) (PAMAM) dendrimer could be used as a carrier for beclomethasone dipropionate and other insoluble drugs. This could improve drug solubility and increase lung accumulation capacity, thereby improving the bioavailability of the drug, and reducing dosage, dosing frequency, and toxic side effects. This nanocarrier could also be used to deliver hydrophobic drugs like dexamethasone directly into the lungs, leading to a decrease in allergic pulmonary inflammation and decreasing the eosinophils and inflammatory cytokines (43).

Liposomes have also been studied and seen to prolong the retention of salbutamol sulfate in the lungs while maintaining the effective drug concentration for more than 10 hours. When compared with the micronized salbutamol sulfate, the nanoparticles loaded with the same drug were less affected in the human oropharynx and had higher peripheral deposition (1,35,43,45).

Solid lipid nanoparticles of circumin, which is an adjunct therapy in asthma therapy, improved bioavailability and therapeutic efficacy of this anti-inflammatory compound. Inhalable montelukast-loaded nanostructured lipid carriers improved systemic bioavailability (12-fold) and eventual therapeutic outcomes compared to those of montelukast aqueous solutions (35).

Despite the inability of the nanocarrier-based gene delivery system to achieve sustained and high levels in gene-expression in vivo, it provides a highly adjustable platform for therapeutic gene delivery and therefore requires more studies for effective application in asthma management (43). Studies on nanocarriers for anti-asthmatic drugs include polyethyleneimine (PEI), one of the most important polymeric gene carriers, chitosan, polyamidoamine dendrimers, and biodegradable poly(lactic-co-glycolic) acid (PLGA) in the delivery of nucleic acids to lungs. Chitosan-IFN-γ pDNA nanoparticles (CIN) can significantly lower airway hyper-responsiveness and lung histopathology in BALB/c mice with allergic asthma induced by ovalbumin. Thiolated chitosan nanoparticles and nanoparticles containing chitosan and cyclodextrin, which prolong drug retention in the respiratory tract and enhance their anti-inflammatory effects, have also been used as lung delivery systems (43,46).

Ca^{2+}/calmodulin-dependent protein kinase II (CaMKII) inhibition has been shown to decrease disease phenotypes of allergic asthma, providing the rationale for targeting CaMKII as a potential therapeutic approach for asthma. Studies in mice models have indicated that CaMKII inhibitor peptide (CaMKIIN)-loaded NPs using nano-sized chitosan mitigated the severity of allergic asthma (35).

Lanza et al. (44) has studied the anti-angiogenic prodrug therapy delivered by a micelle therapy in a house dust mite allergic asthma model. Angiogenesis is known to be a feature of asthmatic inflammatory responses. Results indicated a significantly reduced microvascularity, bronchial remodeling, and airway hyper-responsiveness in a rat model of allergic asthma suggesting that direct anti-neovascular therapy could contribute significantly to asthma management (35,44,47).

16.2.2.2 Chronic Obstructive Pulmonary Disease (COPD)

This disorder normally occurs as chronic obstructive bronchitis (brings a long-term wet cough) and emphysema which subsequently limits inhalation, making normal breathing difficult. COPD affects the bronchial tubes (airways) and causes permanent inflammation which can lead to the production of excess mucus. The excess mucus generally results in a tenacious cough and has the possibility of leaving the patient more susceptible to infections. The associated emphysema affects the air sacs at the end of the bronchial tube, causing less oxygen to reach the bloodstream, resulting in symptoms such as shortness of breath, coughing, and wheezing (24,42).

Although the most important cause of COPD is tobacco smoking, which normally results in emphysema and bronchitis, several other causes includes environmental exposure to hazardous agents and genetic syndromes such as α1-antitrypsin deficiency (48,49). Smoke exposure in childhood may predispose to the development of chronic lung disease in adult life (50,51).

COPD may be prevented by avoiding exposure to smoke and hazardous agents and air pollutants. Management of COPD includes use of medication, vaccination, pulmonary rehabilitation, oxygen therapy, and surgery (49,51).

Nano-drug delivery systems normally provide sustained release of drugs and help in overcoming airway defenses while simultaneously targeting diseased cells in chronic inflammatory lung diseases (52–54). The major challenge in using nanocarriers for treatment is their penetration across the mucus layer due to the thicker and highly viscoelastic mucus layer in COPD. The particle size, shape, and surface charge of nanocarriers used in the nano-drug delivery system is vital in developing and determining the effectiveness of the system (52–54,55,56).

Nano-drug delivery systems that have been studied for use in COPD management include stealth liposome encapsulating budesonide, a potent corticosteroid which was able to control the release of the drug and enable the therapeutic concentrations to be retained for longer periods in rat lungs while reducing the systemic toxicity (52). Studies of liposomes comprised of two drugs, namely beclomethasone (glucocorticoid steroid) and formoterol (beta-2-selective receptor agonist), revealed that beclomethasone retained its effect for longer times and formoterol improved lung function and enhanced peripheral lung deposition. The mucociliary escalator was not affected (52,57).

Theranostics (combination of diagnosis with therapy) opens new opportunities for effective lung healthcare. Nano-based theranostics offer many advantages for COPD patients such as real-time diagnosis of the state of lung inflammation and treatment for individual patients with reduced adverse effects of drugs. The use of multifunctional airway-targeting nanocarriers designed to carry therapeutic agents coupled with imaging moieties will fulfil the dual functions of therapy and diagnostics. Liposomes and amphiphilic block copolymers are potential materials for the fabrication of theranostic hybrids. Multifunctional probe and drug-loaded polymeric vesicles normally synthesized by blending of two synthetic polymers such as poly (ethylene glycol)-poly(lactic-co-glycolic acid) PLGA-PEG are used for delivery of COPD drugs (prednisolone, corticosteroid, and/or anti-inflammatory bronchodilator, theophylline) and diagnosis. Hydrophilic polymers such as PEG, PEI, PEO, PVA, and biodegradable and biocompatible co-polymers have been used in therapeutic devices (57). When compared to free inhaled steroids, positively charged nano formulations of polymeric micelle of steroid drugs such as budesonide may accumulate at the site of airway inflammation with higher anti-inflammatory benefits in COPD (52). Liposomes that have been conjugated to cell penetrating peptides may also show better cellular uptake to airway cells in gene therapy, and this can be of much advantage in the management of COPD (52,57,58). Adenoviral vector-based gene therapy has been regarded as a convenient strategy for lung diseases. Studies have shown potential of viral liposome complexes carrying the AAT (α1 anti-trypsin) gene (51,57,59).

Lipid based carriers that have been used for the formulation of approved drugs that may be used in the management of COPD include lipid microparticles, liposomes, and co-suspension formulations of salbutamol, formoterol, glycopyrronium, budesonide, gentamicin, tobramycin, and ciprofloxacin. Advantages of lipid-based nanocarriers for drug delivery include the possibility of decrease in frequency of administration and adverse events, and improvement in compliance and therapeutic outcomes (57,60).

The use of hygroscopic excipients such as sodium chloride or citric acid in DPI formulations may produce submicron aerosol particles which increase in size due to the humid environment of the respiratory tract and ensure deposition in the lungs (54,61).

16.2.2.3 Bronchioectasis

Bronchiectasis is the irreversible, pathological dilatation of the small and medium-sized bronchi normally resulting from a cycle of inflammation, bacterial colonisation, and infection of the lungs. Bronchiectasis occurs in some pathologic processes that are associated with recurrent or acute infection and inflammation, such as cystic fibrosis, chronic obstructive pulmonary disease, and asthma, leading to permanent structural changes in the airways such as permanent lung scarring, impaired lung function, respiratory failure, and in very severe cases, death in early adulthood (62,63).

Symptoms of bronchiectasis may include a chronic productive cough, overproduction of mucus and the dilation of the bronchus resulting in the potential for recurrent infections (62–64). The goal of management of bronchiectasis is to suppress the inflammation and prevent infections, reduce the sputum volume, viscosity, and purulence, and lessen the cough, dyspnoea, and fever in instances of excerbations (63–65). Treatments include therapy to optimize airway clearance using supplemental oxygen where prescribed in addition to long-term antibiotic therapy (oral or nebulized). Physiotherapy and transplant surgery may also be used in some cases (62,65).

In a study to decrease the rapid clearance of ciprofloxacin hydrochloride from the lungs, two forms of the drug have been tested for sustaining levels of ciprofloxacin in the lungs. These involved a liposomal formulation where ciprofloxacin is encapsulated in small unilamellar vesicles, and a dry powder formulation of the practically insoluble zwitterionic form of the drug. In a large multicenter, phase 3 clinical studies included more than 1000 bronchiectasis patients. Results obtained from the study indicates both formulations were generally well tolerated with most adverse events found to be mild to moderate in intensity. The liposomal nanoparticle formulation of ciprofloxacin may have potential for application in the treatment of bronchiectasis (68).

16.2.2.4 Acute Lower Respiratory Infections

Lower respiratory tract infections are normally caused by viruses (67). Acute lower respiratory infections include pneumonia (caused by viruses, bacteria, or fungi) (68) as well as infections affecting the airways such as acute bronchitis (normally caused by viruses such as rhinovirus, enterovirus, influenza A and B, parainfluenza, coronavirus, human metapneumovirus, and respiratory syncytial virus) (69) and bronchiolitis, influenza, and whooping cough. Acute respiratory infections are a leading cause of illness and death in children and adults across the world (67,69). Some acute respiratory infections such as pneumonia can be prevented by immunization, adequate nutrition, and by addressing environmental factors, while those caused by bacteria can be treated with antibiotics (68,69).

Nanoparticle forms of the antibiotics ciprofloxacin (liposomal form), amikacin (liposomal form), tobramycin (PLGA/PVA/chitosan/alginate forms), and levofloxacin (PLGA/PCL form) have been shown to produce better antibiotic effects compared to their respective free drugs (70,71).

Silver nanoparticle formulations have also been suggested for delivery in the treatment of respiratory infections (72)

16.2.2.5 Lung Cancer

Lung cancer occurs when abnormal cells proliferate uncontrolled in one or both lungs. Besides interfering with how the lung functions, cancer cells can spread from the tumor into the bloodstream or lymphatic system where they can spread to other organs. In other cases, the lung cancer may be the result of spread of cancer from other organs of the body to the lungs (23,73). Symptoms of lung cancer include a persistent cough, shortness of breath, wheezing, coughing up blood, chest pain, and recurring pneumonia (73,74). Tobacco or cigarette smoking is the most common cause of lung cancer. Those who are passively exposed to tobacco or cigarette smoke may also develop lung cancer. Other causes of lung cancer include diesel air pollution, radon gas exposure, exposure to occupational chemicals like benzene, formaldehyde, asbestos, and other environmental and workplace elements (73–75). The choice of treatment and diagnosis is normally dependent on the specific type of tumor and includes surgery, combining targeted therapy, immunotherapy, radiotherapy, and chemotherapy (77).

Conventional methods of treatments of lung cancer such as surgery, radiotherapy, and chemotherapy are normally associated with severe side effects. Current lung cancer treatment research now focuses on developing methods or drugs which are specifically targeted and actively absorbed by tumor cells only, leaving healthy tissues unharmed while producing limited side effects (37,77,78). The use of nanoparticle drug delivery systems in the management of lung cancer may provide the needed resource for this targeted therapy.

Photodynamic therapy (PDT) of lung cancer

PDT functions using three main parameters: photosensitive (PS) drugs, light, and oxygen. Upon the administration of a PS drug to a patient, it can either passively or actively accumulate within a tumor site, and once exposed to a specific wavelength of light, it is stimulated to produce reactive oxygen species (ROS), resulting in tumor destruction. The efficacy of ROS generation for tumor destruction is highly dependent on the accumulation of the PS in tumor cells, which is the case of non-targeted drug delivery mechanisms; only a small amount of PS is able to passively accumulate in tumor sites due to the enhanced permeability and retention (EPR) effect, and the remainder distributes into healthy tissues, causing side effects (79,80). Improving the efficacy of PDT requires the development of specific receptor-based photosynthetic nanocarrier drugs, which promote the active uptake and absorption of PS drugs in tumor sites only and prevent unwanted side effects, is crucial (79).

The use of NPs helps to enhance the selectivity and stability of PSs, reduce unwanted side effects, limit dark toxicity, and enhance the solubility of the PSs by making them hydrophilic (37,78,79). NP drug delivery systems can protect therapeutic drugs from immune system barriers by mimicking biological molecules and improving drug circulation, tissue distribution, and cellular absorption rates (78,79).

NPs are also found to accumulate in tumor cells more predominantly, due to the passive EPR uptake effect which allows NP PS drug carriers to enter between the spaces of tumor cells, suppresses lymphatic filtration, and increases drug uptake in tumor cells, while reducing the overall toxicity in healthy cells. Optimizing NPs as carriers for PS drug delivery is essential as the NP-drug diameter and the tumor location, type, and size can affect the EPR PS accumulation in target cells. Further research focuses on the functionalization of NPs into active PS drug delivery systems (third generation PSs) through the attachment of molecules, which are specifically compatible with only targeted tumor cells (79).

Nano-drug carrier platforms and PDT third generation PS targeting strategies for lung cancer PDT PS NP platforms consist of various types of organic and inorganic compounds, such as liposomes, quantum dots, gold NPs, polymers, micelles, magnetic NPs, dendrimers, and carbon-based NPs (79,80). This platform may either be passive or active. In the passive NP PS drug delivery systems, PSs can only accumulate in lung tumor cells and surrounding environments due to EPR uptake. NP composition and size can affect this passive uptake (79). Examples of passive PDT NP PS drug delivery systems include micelles and liposomes, polymeric particles, dendrimers, metal oxide, ceramic, silica, and alumina organic based NPs (79–81). Some studies have shown that hypocrellin B and zinc phthalocyanine (ZnPc) are promising PSs for the treatment of lung cancer. Silver, gold, and ceramide, as well as polymeric NPs have been shown to significantly enhance passive drug delivery in lung cancer cells (80,81) and so requires further research. The disadvantage of passive PSs carrying NPs may be that the system cannot exclusively differentiate between cancerous versus normal cells and so occasionally distributes in healthy tissues (79).

In the active PDT NP drug carriage, a molecular recognition process is used to deliver the conjugated PS to a specific targeted tumor site (79,81). Active NP PS drug delivery platforms are normally inorganic (quantum dots, solid lipids, self-illuminating nanocrystals, theranostic, hydrogels, immune-conjugates, metal-oxide based or up-converted NPs). A promising study has shown that over 95% of in vitro lung cancer cells were dead after PDT laser irradiation photo-toxicity assays with a pheophorbide-A (PheoA) PS that was conjugated to gold nanoparticles (AuNPs), which were further linked to a cancer-targeting agent known as hyaluronic acid (HA) (79).

Liposomes have been applied in the management of lung cancer. Liposomes containing surface-grafted lipid derivatives conjugated with polyethylene glycol (PEG) have been found to have long circulation times in the blood as a consequence of reduced uptake by the reticuloendothelial system. Some chemotherapeutic agents, such as doxorubicin and vincristine, have been encapsulated in PEGylated liposomes for investigation and validation. PEGylated liposomes achieve a higher drug load in tumors due to a passive targeting process, which exploits the "enhanced permeability and retention effect" (EPR), resulting from increased vascular permeability inherent in many solid tumors (82,83).

A study involving the repurposing of pirfenidone (PFD), an anti-fibrotic drug, for non small cell lung cancer (NSCLC) treatment by encapsulation in a cationic liposomal carrier, the therapeutic potential against NSCLC cell lines in vitro and ex vivo was carried out. Results indicate ex vivo studies using 3D tumor spheroid models revealed superior efficacy of PFD-loaded liposomes against NSCLC, as compared to plain PFD (84). Liposomal formulations of anti-cancer drugs such as cisplatin, placlitaxel and lipoplatin have also been studied for application in the treatment of lung cancer (82–84).

Polymeric nanoparticles are also of immense importance in the treatment of lung cancer. Abraxane, an FDA-approved albumin-based nanoparticle carrying paclitaxel, is indicated for first-line treatment of locally advanced or metastatic NSCLC in combination with carboplatin in non-curative surgery or radiation therapy patients. Other drugs developed through research using polymeric nanoparticles include cremophor, a free nano-formulation of paclitaxel and cisplatin using block copolymers of PEG and polylactic acid for the treatment of lung cancer (83).

Metal nanoparticles are also being studied in the delivery of lung cancer medications. Gold nanoparticle conjugates of methotrexate (MTX), a drug with high water solubility and low tumor retention, have shown high tumor retention and enhanced therapeutic efficacy in a Lewis lung carcinoma mouse model (83,85). Multifunctional mesoporous silica nanoparticles (MSN) have been used in intracellular labeling and animal magnetic resonance imaging studies. MSN-based drug delivery systems capable of delivering doxorubicin and cisplatin combined with two types of siRNA targeted to MRP1 and BCL2 mRNA for suppression of pump and non-pump cellular resistance in NSLC have been developed for inhalation treatment of lung cancer (83).

16.2.2.6 Tuberculosis (TB)

Tuberculosis is an airborne bacterial infection caused by the microorganism *Mycobacterium tuberculosis.* TB primarily affects the lungs, although other organs and tissues may be involved. Tuberculosis is one of the ten top causes of death worldwide (86). It is spread from infected person to another by expelling the mycobacteria into air. The emergence and spread of mult-idrug-resistant strains of TB (MDR-TB) is currently a great concern. Management for MDR-TB patients is complicated and involves adequately planned management model balancing, available specialized resources, and prevalence of disease. Treatments are normally based on proven efficacy of the available medications normally in specialized centers. Examples of medicines currently recommended and used include appropriate combinations of amikacin, moxifloxacin, ethionamide, clofazimine, pyrazinamide, high dose isoniazid, ethambutol, and bedaquiline (86–88).

Tuberculosis (TB) due to the continuous emergence of resistance challenges the global effort to control TB cases across the world. Newer nanotechnology-based drug-delivery approaches would provide more advantages over conventional systems of treatment. Controlled and sustained release of drugs is one of the advantages of nanoparticle-based anti-tuberculosis drugs over free drug. It also reduces the dosage frequency and resolves the difficulty of poor compliance due to the ability to encapsulate multiple drugs in the matrix as well as decreased side effects (89–91).

The evaluation of efficacy of anti-TB drugs after oral administration of three leading anti-tubercular drugs, namely, rifampin, isoniazid, and pyrazinamide encapsulated by PLG NPs has been conducted. Drug levels of the drug-loaded PLG NPs were maintained above the minimum inhibitory concentration in mice after a single oral administration for 6–9 days in the plasma. Free drugs were, however, absent from plasma within 12–24 hours following the oral administration. In addition, complete bacterial clearance from the organs was observed when Mtb infected mice were treated with the nanoparticle-bound drugs (five oral doses every 10th day). Only after administration of 46 doses were free drugs able to generate the same results (90).

Another study involving the oral/aerosol administration of wheat germ agglutinin-coated poly(DLlactide-co-glycolide (PLG) nanoparticles in mice showed prolonged plasma levels of 6–7 days for RIF and 13–14 days for INH and PYZ as compared to uncoated PLGNPs (4–6 days for RIF and 8–9 days for INH and PYZ). Three oral/nebulized doses of these lectin-coated nanoparticles every 14 days (versus 45 daily doses of free drugs) resulted in complete bacterial clearance. The presence of the three nanoparticle formulations were found in lungs, liver, and spleen for 15 days (90,91). A similar study also showed a single subcutaneous injection of PLG nanoparticles loaded with RMP, INH, and PZA resulted in sustained therapeutic drug levels in plasma for 32 days and in lungs or spleen for 36 days. This produced complete sterilization of organs of Mtb infected mice and demonstrated better therapeutic efficacy as compared with 35 doses of the daily oral free drugs (90).

Liposomes have been evaluated for the constant delivery of anti-TB drugs. PEGylated liposomes help to prevent quick removal of drug from the body and broaden circulation times. Evaluation of liposome incorporated gentamicin for antimicrobial activity compared to free drug in a mouse model of dispersed *M. avium* complex infection showed the encapsulated drug considerably reduced the bacterial load in liver and spleen though sterilization was not found. A similar outcome has been obtained with diverse liposome-entrapped second-line antibiotics. A liposomes-in-hydrogel phospholipids combination form of isoniazide prolonged in vitro/in vivo drug release behavior and high biocompatibility for in vivo applications during clinical studies against *M. tuberculosis* has been successfully formulated for localized treatment of bone TB (90,92). Liposomes encapsulated drugs of rifampicin and isoniazid administered alone were more potent in clearing mycobacterial infection when compared to the free drugs with the effective dose decreased to one weekly administration for 6 weeks upon coadministration of the liposome forms of the two drugs with no hepatotoxicity. Similar results have also been obtained using stealth liposomes (90).

Niosomes can be used as orally controlled and sustained release systems for hydrophilic oral anti-tubercular drugs such as isoniazid and pyrazinamide, as well as rifampicin and gatifloxacin to be developed as low dose effective treatments with better patient compliance. Oral forms of alginate nanoparticles of rifampicin, isoniazid, pyrazinamide, and ethambutol may also have similar or even better potential due to the ability of alginate nanoparticles to load more drugs with lower consumption of polymer (90,91).

Alginate supplemented with chitosan such as poly(L-lysine)-chitosan for loading of rifampicin, isoniazid, pyrazinamide, and ethambutol may result in improved sustained release characteristic, pharmacokinetics, and therapeutic efficacy. The hydrophilic character of alginate prevents fast clearance by the mononuclear phagocyte system when alginate nanoparticle drug formulations are administered intravenously, giving them an advantage over neutral polymers or liposomes (90–92).

Biotinylated oligonucleotides probes have been used for recognition and detection of specific mycobacterial target DNA through a sandwich-like hybridization, which has been used for the detection of *Mycobacterium tuberculosis* and *Mycobacterium avium* subsp. *paratuberculosis.* To produce the fluorescent signal, cadmium selenite quantum dots were used in conjugation with streptavidin and specific-specific probes, and for the isolation of DNA targets, the magnetic beads in conjugation with streptavidin and genus-specific probes were used. This nanotechnology-based method is much more reliable than conventional methods because it reduces the drawback of PCR-based diagnostic assay, where there are chances of generation of false negative results, especially in the case of clinical samples such as feces (93).

16.2.2.7 Cystic Fibrosis

Cystic fibrosis (CF) is a lifelong, autosomal hereditary disease stemming from the mutation of the cystic fibrosis transmembrane conductance regulator (CFTR). The disease causes thick, sticky mucus to form in the lungs, pancreas, and other organs. In the lungs, this mucus blocks the airways, causing lung damage, making it hard to breathe and leading to serious lung infections (24,94,95). Treatment of CF includes the use of nutritional support, pulmonary therapy airway clearance, and pharmacological treatment involving the use of medications such as inhaled hypertonic saline, dornase alfa, aztreonam, tobramycin, azithromycin, ivaclafor, lumacaftor, and ibuprofen (94,95).

The main objective on application of nanotechnology in the management of CF is to decrease the overproduction of the mucus secreted in the lungs, along with its altered composition and consistency, which results in airway obstruction, making the lungs susceptible to recurrent and persistent bacterial infections and endobronchial chronic inflammation. This can be done using nanotechnologies to improve and enhance drug delivery or gene therapies (96–100). Overcoming the CF mucus barrier, which is a major obstacle, may be done by the coating of the NP's surface with a dense layer of low-molecular-weight polyethylene glycol (PEG) to eliminate both electrostatic and hydrophobic interactions. Inhaled NPs must reach the deep airways and penetrate the mucus covering the alveoli for efficacy. NPs must be small enough to penetrate the mucus but big enough to avoid rapid exhalation. Nebulization can be used to overcome some NP-to-lung delivery issues (96,98,99)

Administration of NPs with DPIs may consist in the formation of nano-embedded microparticles (NEMs) which are able to transport and efficaciously release NPs within the lungs after the dissolution of the microparticle matrix, allowing NP penetration within the mucus barrier. NP suspensions are normally changed into stable, inhalable microparticles by spray drying. Materials used for NEMs include biocompatible and biodegradable polymers such as PLGA, chitosan, gelatin, and polyacrylate, which are usually coupled to mannitol, lactose, or cyclodextrins as excipients (96,98).

Liposomes, polymeric NPs, solid lipid NPs, and dendrimers represent the main types of NPs for inhalation therapy that have been investigated for CF treatment. Antibiotic-loaded liposomes that have been extensively studied to improve their pharmacokinetics and bioavailability, decrease their toxicity, and achieve target selectivity include fluidosome-loaded tobramycin and gentamicin (98–99).

Phosphatidylcholines (PC), phosphatidylglycerols (PG), phosphatidylethanolamines (PE), phosphatidylinositols (PI), dipalmitoyl-phosphatidylcholine, (DPPC), distearoylphosphatidylcholine (DSPC), dimyristoylphosphatidylglycerol, (DMPG), dimyristoyl-phosphatidylcholine, (DMPC), and dioleoyl-phosphatidylethanolamine (DOPE) are examples of some phospholipids used in the nano-drug delivery system (96,98).

N-acetyl cysteine, a mucolytic used in cystic fibrosis management in combination with a PEG coating on NP carriers, upon investigation was found to enhance mucus penetration across cystic fibrosis mucus (100).

Solid lipid nanoparticles of Chol/lecithin-tobramycin and DSPC/albumin/gelatin/chitosan-ciprofloxacin have also been studied for delivery in CF. Solid lipid nanoparticles of the anti-inflammatory agent budesonide formulated using GMS have been studied and shown to have good drug dispersion within the matrix (96,98,99).

Current research in CFTR gene therapies focuses on treating CF by restoring the mutated CFTR gene. The use of polylactic-*co*-glycolic acid (PLGA) NPs, which are biodegradable, and can be aerosolized and loaded with therapeutic proteins, has shown evidence of cellular uptake and transport to the cytoplasm and Golgi, and are easily functionalized with targeting moieties, and PEG is widely researched (100).

Dendrimer-siRNA nanocomplexes for gene therapy have been developed, which is an attractive approach for the treatment of several pulmonary diseases characterized by an over-expression of genes. A more efficient cellular uptake and a good gene silencing capability was obtained when the dendrimer drug complex was compared to other types of carriers (97).

16.2.2.8 Pulmonary Fibrosis

Idiopathic pulmonary fibrosis is a chronic disease characterized by the aberrant accumulation of fibrotic tissue in the lung's parenchyma. Causes may include viruses and exposure to tobacco smoke as well as hereditary considerations. Symptoms include shortness of breath, a persistent dry cough, tiredness, loss of appetite, weight loss, and rounded and swollen fingertips (101,102). Current treatments involve the use of medicines such as pirfenidone and nintedanib, and the use of pulmonary rehabilitation and supplemental oxygen in cases of hypoxemia (101). Lung transplant may also be necessary in some cases. In addition, living a healthy lifestyle may help to slow the progression of the disease (102).

A study involving the inhalational delivery of liposomal form of PGE2 in mice IPF model found that local delivery of PGE2 has a high therapeutic potential, though not all targeted proteins were effectively suppressed and some signs of IPF (most notably interstitial lung edema, inflammation, and excessive collagen production) were still present. The study suggested a combinatorial local lung delivery of PGE2 and suppressors of proteins responsible for inflammation, extracellular matrix degradation, and hypoxic damage in a bid to improve therapeutic outcomes (103).

Nanostructured lipid carriers have been studied as an inhalational delivery system for PGE2 with and without siRNAs targeted to MMP3, CCL12, and HIF1Alpha mRNAs. NLC-PGE2 in combination with three siRNAs delivered locally to the lungs gave promising results for use in the management of idiopathic pulmonary fibrosis. Mouse body mass, lung hydoxyproline content, and lung tissue damage were significantly decreased, and animal mortality was prevented (104).

16.2.2.9 Pulmonary Sarcoidosis

Sarcoidosis is a systemic disease that can affect any organ of the body; however, the lungs and intrathoracic lymph nodes are the most affected sites (105). The disease causes a presence of noncaseating granuloma, a cluster of macrophages, epitheloid cells, mononuclear cells, and $CD4^+$ T cells, and a few $CD8^+$ T cells in the peripheral zone. Causes are still not clearly understood; however, it is generally thought that genetic predisposition and environmental factors may be major factors for consideration. Symptoms include dyspnea, cough, and chest tightness. Generally, patients with disabling symptoms and progressive organ damage are considered for treatment as some forms of the disease may be self-limiting. Corticosteroids are normally the first line of drug treatment whilst glucocorticoid-sparing drugs and biologic agents are used in recurrent disease (105,106). Some studies aimed at improving the uniform tissue distribution of free glucocorticoids and their short half-life in biological fluids have involved the use of new delivery nanoparticle vehicles including PEGylated liposomes, polymeric micelles, polymer-drug conjugates, inorganic scaffolds, and hybrid nanoparticles (107).

16.2.2.10 Pulmonary Hypertension

Pulmonary hypertension may be defined as a mean pulmonary arterial pressure (mPAP) greater than 20 mm Hg at rest (108).

Pulmonary hypertension due to lung disease may be caused by obstructive pulmonary diseases, restrictive pulmonary diseases, other lung disease with mixed restrictive/obstructive pattern, and developmental lung disorders (108,109). Treatments include the use of supportive therapy and medications such as PDE-5i and guanylate cyclase stimulators (sildenafil, tadalafil and vardenafil) and prostacyclin analogs and prostacyclin receptor agonists (epoprostinol, iloprost, and treprostinil) (109).

NP-based drug delivery strategies in PAH involve concentrating on improvements in vascular remodeling and hemodynamics. Various nano forms of drugs used in the treatment of pulmonary hypertension have been developed and studied in animal models for subsequent application in humans. These include liposomal forms of iloprost combining POPC, DOPTAP, PVP, SA, DPPE-PEG2000, and CH, which were found to increase vasodilation. Various liposomal forms of fasudil were also found to decrease arterial alsowall thickness and increase vasodilatory effects duration. Liposomal forms of nitric oxide, hydrogel polymer forms of nitric oxide, PEG-PLA NP forms of beraprost also resulted in a decrease in effective dose needed. PLGA forms of pitavastatin, PLGA-NP forms of the imatinib PEG-PCL NP form of rapamycin, and the PEG-PLGA NP form of NF-κB decoy oligodeoxynucleotide have also been studied for improvement in various defects leading to the development of pulmonary hypertension (110,111).

Inhalable, dry powder pharmaceutical formulations, which are able to swell on administration to the lungs of a patient, consisting of polymeric nanoparticles encapsulated within crosslinked polymeric hydrogel microparticles carrying therapeutic agents such as prostacyclin synthetic analogs, PPAR β agonists and NO donors loaded within them have been patented for application in the management of pulmonary hypertension (112).

16.2.2.11 Noncardiogenic Pulmonary Edema

Noncardiogenic pulmonary edema is caused by lung injury with a resultant increase in pulmonary vascular permeability leading to the movement of fluid to the alveolar and interstitial compartments. Acute lung injury with severe hypoxemia is normally referred to as acute respiratory distress syndrome (ARDS) and is seen in various conditions directly affecting the lungs. Progressively worsening dyspnea, tachypnea, and rales (or crackles) on examination with associated hypoxia are some characteristics experienced with the edema (113). The edema may be caused by other lung diseases such as pneumonia (113,114).

Treatments for pulmonary edema include the use of supportive therapy in addition to drugs such as glyceryl trinitrate, frusemide, morphine, and inotropes (114).

Novel sterically stabilized phospholipid nanomicelles (SSM), which are long-acting, biocompatible, and biodegradable phospholipid-based nano-drug delivery vehicles, have been used in combination with TREM1 peptide and GLP-1 and 17-AAG, a selective Hps90 inhibitor in the animal model experimental treatment of acute lung injury. These nanoparticles can be introduced by systemic administration using passive delivery or directly into the lung through inhalation, intranasal or oropharyngeal aspiration. These studies have shown some promising results with a relative decrease in lung inflammation being one of the effects seen. Studies involving DNA nanoparticles targeted to the β_2-adrenergic receptor have shown that treatment with a polyethylenimine/β_2-adrenergic receptor improved survival of test animals with lung injury. Magnetic nanoparticles modified with a synthetic ligand, zinc-coordinated bis(dipicolylamine), have been developed for utilization for highly selective and rapid separation of bacteria and their toxins from whole blood using a magnetic microfluidic device in lung inflammation and injury (115).

16.2.2.12 Pulmonary Embolism

Pulmonary embolism is the blockage of a branch of the pulmonary artery by a substance that has travelled from elsewhere in the body through the bloodstream. This condition can be immediately life threatening. The cause of the disease is normally the blockage of an artery in the lungs, usually by a clot. There are several factors that affect the occurrence of pulmonary embolism, such as age over 40 years, major surgery, fractures, and prolonged bed stay (more than 5 days). The treatment of pulmonary embolism may be surgical or by the administration of anticoagulants such as warfarin and heparin, general support circulation, antibiotics, thrombolytics, and analgesics (116,117).

Some studies have been done in animal models on the application of nano-drug delivery systems in the treatment of embolism with promising results. Novel self-assembly antiplatelet aggregation peptides containing L-arginine and L-aspartic acid complexed with Cu(II) to form stable nanoparticles achieved significantly higher anti-thrombotic activity comparable to aspirin and equally effective in reducing the thrombus weight. Aspirin conjugated to the Arg-Gly-Asp-Val (RGDV) tetrapeptide, resulting in a nano-assembly targeting glycoprotein IIb/IIIa, the receptor for RGD peptide on the surface of the activated platelets, has exhibited very strong anti-thrombotic effect compared to free aspirin. Heparin-functionalized carbon nanocapsules produced superior anti-thrombotic activity compared to free heparin. Lipid nanoparticles of hirulog, an analogue of natural thrombin inhibitor containing a fibrin-binding peptide, has also been reported to produce significantly higher levels of anti-thrombin activity compared with non-targeted particles. Liposomes and perfluorocarbon nanoparticles bound to a potent thrombin inhibitor, d-phenylalanyl-l-prolyl-l-arginyl-chloromethyl ketone, upon investigation produced promising results. Magneto fluorescent crosslinked dextran-coated iron oxide nanoparticles conjugated to tPA have been studied and found to present with better activity compared to tPA Magnetite, gelatin and zinc bound nanoparticles. uPA-coated, self-assembled chitosan and tripolyphosphate nanoparticles produced a significant improvement in the thrombolytic effect compared with free uPA upon administration. This can be applied in the treatment of pulmonary embolism, which is a form of deep vein thrombosis (116).

16.3 Future Perspective

Some nanoparticle drug delivery systems have been approved for use in the management of some lung diseases, more studies on the application of proposed nanoparticle drug delivery systems in the management of lung diseases in human subjects will go a long way to enhance treatment outcomes while reducing adverse side effects encountered with the use of most conventional medications for treatment of various lung diseases. This application will provide the much-needed specific targeting of affected areas of the lung, for enhanced and sustained release effects of medications. Toxicity of these nanoparticle drug delivery systems should also be continuously studied to optimize their usage.

REFERENCES

1. Tuxana SR. The potential role of nanomedicine in lung diseases. *Medical Research Archives*. 2018; 6(5):1–9. ISSN 2375-1924. Available at: https://journals.ke-i.org/mra/article/view/1723.
2. Kuzmov A and Minko T. Nanotechnology approaches for inhalation treatment of lung diseases. *Journal of Controlled Release: Official Journal of the Controlled Release Society*. 2015; 219: 500-518 doi:10.1016/j.jconrel.2015.07.024.
3. Witharana C and Wanigasekara J. Drug delivery systems: A new frontier in nano-technology. *International Journal of Medical Research & Health Sciences*. 2017; 6(9): 11–14.
4. Singh J, Garg T, Rath G and Goyal AK. Advances in nanotechnology-based carrier systems for targeted delivery of bioactive drug molecules with special emphasis on immunotherapy in drug resistant tuberculosis – A critical review. *Drug Delivery*. 2016; 23(5): 1676–1698. doi:10.3109/10717544.2015.1074765
5. Uddhav BS, Pisal VV, Solanki NV and Karnavat A. Current status of solid lipid nanoparticles: A review. *Modern Applications of Bioequivalence and Bioavailability*. 2018; 3(4): 1-9: MABB.MS.ID.555617.
6. Alsaad A, Hussien A, and Gareeb M. Solid lipid nanoparticles (SLN) as a novel drug delivery system: A theoretical review. *Systematic Reviews in Pharmacy*. 2020; 11(5): 259–273. doi:10.31838/srp.2020.5.39
7. Sharma A and Baldi A. Nanostructured lipid carriers: A review. *Journal of Developing Drugs*. 2018; 7: 191. doi:10.4172/2329-6631.1000191
8. Haider M, Abdin SM , Kamal L, and Orive G. Nanostructured lipid carriers for delivery of chemotherapeutics: A review. *Pharmaceutics*. 2020; 12: 288. doi:10.3390/pharmaceutics12030288
9. Nikam TH, Patil MP, Patil SS, Vadnere GP, and Lodhi S. Nanoemulsion: A brief review on development and application in parenteral drug delivery. *Advance Pharmaceutical Journal*. 2018; 3(2): 43–54.
10. Gurpreet K and Singh SK. Review of nanoemulsion formulation and characterization techniques. *Indian Journal of Pharmaceutical Sciences*. 2018; 80(5): 781–789.
11. Yeo P, Lim C, Chye S, Kiong Ling A, and Koh R. Niosomes: A review of their structure, properties, methods of preparation, and medical applications. *Asian Biomedicine*. 2017; 11(4): 301–314. doi: https://doi.org/10.1515/abm-2018-0002
12. Karami N, Moghimipour E, and Salimi A. Liposomes as a novel drug delivery system: Fundamental and pharmaceutical application definition and history. *Asian Journal of Pharmaceutics*. 2018; 12(1): 31–41. doi:10.22377/ajp.v12i01.2037
13. Hea H, Lua Y, Qia J, Zhub Q, Chenb Z, and Wua W. Adapting liposomes for oral drug delivery. *Acta Pharmaceutical Sinica B*. 2019; 9(1): 36–48.
14. Dixit N, Gupta A, Asija R, and Sharma M. Liposphere: A novel approach of drug delivery. *World Journal of Pharmaceutical and Medical Research*. 2017; 3(6): 124–130.
15. Rathore M, Gupta DMK, and Sharma V. A review on oral liposphere of candesartan cilexetil. *International Journal of Pharmaceutical and Biological Science Archive*. 2018; 6(01): 1-4 Retrieved from http://www.ijpba.in/index.php/ijpba/article/view/77
17. Rudd R, Vasilev K, and Simovic S. Nanosuspension technologies for delivery of poorly soluble drugs. *Journal of Nanomaterials*. 2015; 2015: 1–13. 10.1155/2015/216375
16. Lim YH, Tiemann KM, Hunstad DA, Elsabahy M, and Wooley KL. Polymeric nanoparticles in development for treatment of pulmonary infectious diseases. *WIREs Nanomedicine and Nanobiotechnology*. 2016; 8: 842–871. doi: 10.1002/wnan.1401
18. Shahi S, Kulkarni M, Karva G, Giram P, and Gugulkar R. Review article: Dendrimers. *International Journal of Pharmaceutical Sciences and Research*. 2015; 33: 187–198.
19. Dhakar RC, Prajapati SK, Maurya S, Das M, Tilak V, and Verma K. Potential application of dendrimers in drug delivery: A concise review and update. *Journal of Drug Delivery and Therapeutics*. 2016; 6: 71–88. doi:10.22270/jddt.v6i2.1195
20. Rahman G, Najaf Z, Mehmood A, Bilal S, Shah AHA, Mian SA, and Ali G. An overview of the recent progress in the synthesis and applications of carbon nanotubes. *C—Journal of Carbon Research*. 2019; 5(1): 3.
21. Bajwa N, Mehra NK, Jain K, and Jain NK. Pharmaceutical and biomedical applications of quantum dots . *Artificial Cells, Nanomedicine, and Biotechnology*. 2016; 44: 758–768.
22. Gatto F and Bardi G. Metallic nanoparticles: General research approaches to immunological characterization. *Nanomaterials* 2018; 8: 753. doi:10.3390/nano8100753
23. National Institute for Environmental Health Sciences. Lung Diseases. National Institute for Health, US department of Health and Human Services. 2019; pp. 1, 2.
24. https://medlineplus.gov/ency/article/000066.htm. Date accessed 20/08/2020.
25. Forum of International Respiratory Societies. The Global Impact of Respiratory Disease, Second Edition. European Respiratory Society, Sheffield. 2017.
26. The Global Strategy for the Diagnosis, Management, and Prevention of Chronic Obstructive Pulmonary Disease. 2019 Report.
27. World Health Organization. Chronic Respiratory Diseases, Burden of COPD. Date accessed 21/08/2020. www.who.int/respiratory/copd/burden/en/index.html

28. World Health Organization. Chronic Respiratory Disease, Asthma. Date accessed 21/08/2020. www.who.int/respiratory/asthma/en/
29. World Health Organization. Global Tuberculosis Report 2019. www.who.int/tb/publications/global_report/en/. Date accessed 21/08/2020.
30. World Health Organization. 2018 Cancer Report.
31. Fighting for Breath: A Call to Action on Childhood Pneumonia. Save the Children. 2017.
32. https://www.webmd.com/lung/lung-diseases-overview#1-2. Date accessed 23/08/2020.
33. Witharana C and Wanigasekara J. Drug delivery systems: A new frontier in nanotechnology. *International Journal of Medical Research &Health Sciences*. 2017; 6(9): 11–14.
34. Hossein Bahmanpour A, Ghaffari M, Ashraf S, and Mozafari M. Nanoengineered Biomaterials for Advanced Drug Delivery Nanotechnology for Pulmonary and Nasal Drug Delivery. Woodhead Publishing Series in Biomaterials. 2020; pp. 561–579. https://doi.org/10.1016/B978-0-08-102985-5.00023-1
35. Singh AP, Biswas A, Shuklal A, and Maiti P. Targeted therapy in chronic diseases using nanomaterial-based drug delivery vehicles. *Signal Transduction and Targeted Therapy*. 2019; 4: 33. https://doi.org/10.1038/s41392-019-0068-3
36. Dabbagh A, Kasim NHA, Yeong CH, Wong TW, and Rahman NA. Critical parameters for particle-based pulmonary delivery of chemotherapeutics. *Journal of Aerosol Medicine and Pulmonary Drug Delivery*. 2018; 31(3): 139–154.
37. Anderson CF, Grimmett ME, Domalewski CJ, and Cui H Inhalable nanotherapeutics to improve treatment efficacy for common lung diseases. *WIREs Nanomedicine and Nanobiotechnology*. 2020; 12: e1586. https://doi.org/10.1002/wnan.1586
38. Alexescu TG, Tarmure S, Negrean V, Cosnarovici M, Ruta VM, Popovici I, Para I, Perne MG, Orasan OH, and Todea DA. Nanoparticles in the treatment of chronic lung diseases. *Journal of Mind and Medical Sciences*. 2019; 6(2): Article 7. doi:10.22543/7674.62.P224231. Available at: https://scholar.valpo.edu/jmms/vol6/iss2/7
39. Mehta M, Chellappan DK, Wich P, Hansbro N, Hansbro P, and Dua K. miRNA nanotherapeutics: Potential and challenges in respiratory disorders. *Future Medicinal Chemistry*. 2020; 12: 987–990. doi: 10.4155/fmc-2020-0066
40. El-Sherbiny IM , Elbaz NM , Sedki M, Elgammal A, and Yacoub MH. Magnetic nanoparticles-based drug and gene delivery systems for the treatment of pulmonary diseases. *Nanomedicine (London)*. 2017; 12(4): 387–402.
41. Health Care Guideline Diagnosis and Management of Asthma, 11th Edition. Institute for Clinical Systems Improvement (ICSI). 2016.
42. Henderson W. 8 of the Most Common Lung Diseases in Women; Pulmonary Fibrosis. 2017.
43. Wang L, Feng M, Li Q, Qiu C, and Chen R. Advances in nanotechnology and asthma. *Annals of Translational Medicine*. 2019; 7(8): 20. doi:10.21037/atm.2019.04.62
44. Lanza GM, Jenkins J, Schmieder AH, Moldobaeva A, Cui G, Zhang H, Yang X, Zhong Q, Keupp J, Sergin I, Paranandi KS, Eldridge L, Allen JS, Williams T, Scott MJ, Razani B, and Wagner EM. Anti-angiogenic nanotherapy inhibits airway remodeling and hyper-responsiveness of dust mite triggered asthma in the Brown Norway rat. *Theranostics*. 2017; 7(2): 377–389.
45. Kolanjiyil AV, Kleinstreuer C, and Sadikot RT. Computationally efficient analysis of particle transport and deposition in a human whole-lung-airway model. Part II: Dry powder inhaler application. *Computers in Biology and Medicine*. 2017; 247–253. [PubMed: 27836120]
46. Dua K, Bebawy M, Awasthi R, Tekade RK, Tekade M, Gupta G, De Jesus Andreoli Pinto T, and Hansbro PM. Application of Chitosan and its derivatives in nanocarrier based pulmonary drug delivery systems. *Pharmaceutical Nanotechnology*.2017; 5: 243–249. doi: 10.2174/2211738505666170808095258
47. Upadhyay S, Ganguly K, and Palmberg L. Wonders of nanotechnology in the treatment for chronic lung diseases. *Journal of Nanomedicine and Nanotechnology*. 2015; 6: 337. doi:10.4172/2157-7439.1000337
48. Pocket Guide to COPD Disease Management. A Guide for Health Care Professionals. Global Initiative for Chronic Obstructive Lung Diseases. 2018.
49. ATS Patient Education Series. Chronic obstructive pulmonary disease. *American Journal of Respiratory and Critical Care Medicine*. 2019; 199: P1–P2.
50. Global Initiative for Chronic Obstructive lung Disease. Global Strategy for the Diagnosis Management and Prevention of Chronic Obstructive Lung Disease, 2020 Report. 2020.
51. WHO. Chronic Obstructive Pulmonary Disease. 2017; Date accessed 25/08/2020. https://www.who.int/news-room/fact-sheets
52. Passi M, Shahid S, Chockalingam S, Sundar IK, and Packirisamy G. Conventional and nanotechnology based approaches to combat chronic obstructive pulmonary disease: Implications for chronic airway diseases. *International Journal of Nanomedicine*. 2020; 15: 3803–3826. https://doi.org/10.2147/IJN.S242516
53. Muralidharan P, Hayes D, and Mansour HM. Dry powder inhalers in COPD, lung inflammation and pulmonary infections. *Expert Opinion on Drug Delivery*. 2015; 12(6): 947–962. doi: 10.1517/17425247.2015.977783
54. Rogliani P, Calzetta L, Coppola A, Cavalli F, Ora J, Puxeddu E, Matera M, and Cazzola M. Optimizing drug delivery in COPD: The role of inhaler devices. *Respiratory Medicine*. 2017; 124:6–14. doi: 10.1016/j.rmed.2017.01.006
55. Velino C, Carella F, Adamiano A, Sanguinetti M, Vitali A, Catalucci D, Bugli F, and Iafisco M. Nanomedicine approaches for the pulmonary treatment of cystic fibrosis. *Frontiers in Bioengineering and Biotechnology*. 2019; 7: 406. doi: 10.3389/fbioe.2019.00406
56. Chenthamara D, Subramaniam S, and Ramakrishnan SG. Therapeutic efficacy of nanoparticles and routes of administration. *Biomaterials Research*. 2019; 23: 20. https://doi.org/10.1186/s40824-019-0166-x
57. Suer H and Bayram H. Liposomes as potential nanocarriers for theranostic applications in chronic inflammatory lung diseases. *Biomedical and Biotechnology Research Journal*. 2017; 1: 1–8. doi:10.4103/bbrj.bbrj_54_17
58. Kim N, Duncan G, Hanes J, and Suk J. Barriers to inhaled gene therapy of obstructive lung diseases: A review. *Journal of Controlled Release*. 2016; 240: 1–23. doi:10.1016/j.jconrel.2016.05.031

59. Chiuchiolo MJ and Crystal RG. Gene therapy for alpha-1 antitrypsin deficiency lung disease. *Annals of the American Thoracic Society*. 2016; 13(Suppl 4): S352–S369. doi:10.1513/AnnalsATS.201506-344KV
60. Newman S. Drug delivery to the lungs: Challenges and opportunities. *Therapeutic Delivery*. 2017; 8: 647–661. doi:10.4155/tde-2017-0037
61. Haddrel AE, Lewis D, Church T, Vehring R, Murnane D, and Reid JP. Pulmonary aerosol delivery and the importance of growth dynamics. *Therapeutic Delivery*. 2017; 8(12): 1051–1061. https://doi.org/10.4155/tde-2017-0093
62. Cope G. Bronchiectasis—A growing respiratory problem. *Nurse Prescribing*. 2017; 15(9): 1–6.
63. bpac.org.nz/2020/bronchiectasis.aspx. Preventing and managing bronchiectasis in high-risk paediatric populations. Date accessed 28/08/2020.
64. Abo-Leyah H and Chalmers JD. New therapies for the prevention and treatment of exacerbations of bronchiectasis. *Current Opinion in Pulmonary Medicine*. 2017; 23(3): 218–224. Medline doi:10.1097/MCP.0000000000000368
65. Polverino E, Goeminne PC, McDonnell MJ, Aliberti S, Marshall SE, Loebinger MR, Murris M, Cantón R, Torres A, Dimakou K, Soyza AD, Hill AT, Haworth CS, Vendrell M, Ringshausen FC, Subotic D, Wilson R, Vilaró J, Stallberg B, Welte T, Rohde G, Blasi F, Elborn S, Almagro M, Timothy A, Ruddy T, Tonia T, Rigau D, Chalmers JD. European Respiratory Society Guidelines for the Management of Adult Bronchiectasis. *European Respiratory Journal*. 2017; 50(3): 1700629. doi: 10.1183/13993003.00629-2017
66. Weers J. Comparison of phospholipid-based particles for sustained release of ciprofloxacin following pulmonary administration to bronchiectasis patients. *Pulmonary Therapy*. 2019; 5: 127–150. https://doi.org/10.1007/s41030-019-00104-6
67. Pham HT, Nguyen PTT, Tran ST, and Phung TTB. Clinical and pathogenic characteristics of lower respiratory tract infection treated at the Vietnam National Children's Hospital. *Canadian Journal of Infectious Diseases and Medical Microbiology*. 2020; Article ID 7931950: 6 pages. https://doi.org/10.1155/2020/7931950
68. WHO. 2019. Date accessed 28/08/2020. https://www.who.int/news-room/fact-sheets/detail/pneumonia
69. Kinkade S and Long NA. Acute bronchitis. *American Family Physician*. 2016; 94(7): 560–565.
70. Prat Aymerich C and Lacoma A. Bacteria in the respiratory tract—How to treat? Or do not treat? *International Journal of Infectious Diseases*. 2016; 51: 113–122. doi:10.1016/j.ijid.2016.09.005.
71. Csoka I, Karimi K, Mukhtar M, and Ambrus R. Pulmonary drug delivery: Role of antibiotic formulations for treatment of respiratory tract infections. *Acta Pharmaceutica Hungarica*2019; 89: 43–62.
72. Zachar O. Nanomedicine Formulations for Respiratory Infections by Inhalation Delivery - Covid-19 and Beyond. 2020. doi: 10.31219/osf.io/adnyb
73. Murtaza Mustafa AR, Jamalul Azizi EL, IIIzam A, Nazirah AM, and Sharifa SAA. Lung cancer: Risk factors, management, and prognosis. *IOSR Journal of Dental and Medical Sciences (IOSR-JDMS)*. 2017; 15(10, Ver. IV): 94–101.
74. WHO. WHO Report on Cancer Setting Priorities, Investing Wisely and Providing Care For All. World Health Organization. 2020.
75. Latimer KM and Mott TF. Lung cancer: Diagnosis, treatment principles, and screening. *American Family Physician*. 2015; 91(4): 250–256.
76. Yuan M, Huang L, and Chen J. The emerging treatment landscape of targeted therapy in non-small-cell lung cancer. *Signal Transduction and Targeted Therapy*. 2019; 4: 61. https://doi.org/10.1038/s41392-019-0099-9
77. Pontes JF and Grenha A. Multifunctional nanocarriers for lung drug delivery. *Nanomaterials*. 2020; 10: 183. doi:10.3390/nano10020183
78. Sarkara S, Osama K, Jamal QMS, Amjad Kamal M, Sayeed U, Khan MKA, Haris Siddiqui M, and Akhtar S. Advances and implications in nanotechnology for lung cancer management. *Current Drug Metabolism*. 2017; 18(1): 31.
79. Mokwena MG, Kruger CA, Ivan M-T, and Heidi A. A review of nanoparticle photosensitizer drug delivery uptake systems for photodynamic treatment of lung cancer. *Photodiagnosis and Photodynamic Therapy*. 2018; 22: 147–154.
80. Crous A and Heidi A. Targeted Photodynamic therapy for Improved Lung cancer treatment (In Alba Fabiola Coster Torres (eds.) Lung Cancer- Strategies for Diagnosis and Treatment). 2018; 153–172.
81. Kruger CA and Abrahamse H. Utilisation of targeted nanoparticle photosensitiser drug delivery systems for the enhancement of photodynamic therapy. *Molecules*. 2018; 23: 2628. doi:10.3390/molecules23102628
82. Hussain S. Nanomedicine for treatment of lung cancer. In: Ahmad A., Gadgeel S. (ed.). *Lung Cancer and Personalized Medicine: Novel Therapies and Clinical Management. Advances in Experimental Medicine and Biology*. Springer, Cham. 2016; 890. doi:10.1007/978-3-319-24932-2_8
83. Bandyopadhyay A, Das T, and Yeasmin S. Nanoparticles in Lung Cancer Therapy – Recent Trends. Springer Science Publishers, New Delhi. 2015.
84. Parvathaneni V, Kulkarni NS, Shukla SK, Farrales PT, Kunda NK, Muth A, and Gupta V. Systematic development and optimization of inhalable pirfenidone liposomes for non-small cell lung cancer treatment. *Pharmaceutics*. 2020; 12: 206.
85. Mou X, Ali Z, Song L, and He N. Applications of magnetic nanoparticles in targeted drug delivery system. *Journal of Nanoscience and Nanotechnology*. 2015; 15: 54–62.
86. WHO. Global Tuberculosis Report. WHO. 2019.
87. Sulis G, Centis R, and Sotgiu G. Recent developments in the diagnosis and management of tuberculosis. *npj Primary Care Respiratory Medicine*. 2016; 26: 16078. https://doi.org/10.1038/npjpcrm.2016.78
88. MDR/XDR-TB Management of Patients and Contacts: Challenges Facing the New Decade. The 2020 Clinical Update by the Global Tuberculosis Network.
89. Singh J, Garg T, Rath G, and Goyal AK. Advances in nanotechnology-based carrier systems for targeted delivery of bioactive drug molecules with special emphasis on immunotherapy in drug resistant tuberculosis – A critical

review. *Drug Delivery*. 2016; 23(5): 1676–1698. doi: 10.3109/10717544.2015.1074765
90. Nasiruddin M, Kausar Neyaz Md., and Das S. Nanotechnology-based approach in tuberculosis treatment. *Tuberculosis Research and Treatment*. 2017; 2017: Article ID 4920209, 12 pages.
91. da Silva PB, de Freitas ES, Bernegossi J, Gonçalez ML, Sato MR, Leite CQF, Pavan FR, and Chorilli M. Nanotechnology-based drug delivery systems for treatment of tuberculosis-A review. *Journal of Biomedical Nanotechnology*. 2016; 12: 241–260.
92. Eleraky NE, Allam A, Hassan SB, and Omar MM. Nanomedicine fight against antibacterial resistance: An overview of the recent pharmaceutical innovations. *Pharmaceutics*. 2020; 12: 142. doi:10.3390/pharmaceutics12020142
93. Stephen BJ, Singh SV, Datta M, Jain N, Jayaraman S, Chaubey KK, Gupta S, Singh M, Aseri GK, Khare N, Yadav P, Dhama K, and Sohal JS. Nanotechnological approaches for the detection of mycobacteria with special references to *Mycobacterium avium* subspecies *Paratuberculosis* (MAP). *Asian Journal of Animal and Veterinary Advances*. 2015; 10: 518–526.
94. Carlos M and Moss RB. Recent advancement in cystic fibrosis. *Current Opinion in Paediatrics*. 2015; 27(3): 317–324. doi: 10.1097/MOP.0000000000000226
95. DeSimone E, Tilleman J, and Giles ME. Cystic fibrosis: Update on treatment guidelines and new recommendations. *US Pharmacist*. 2018; 43(5): 16–21.
96. Velino C, Carella F, Adamiano A, Sanguinetti M, Vitali A, Catalucci D, Bugli F, and Iafisco M. Nanomedicine approaches for the pulmonary treatment of cystic fibrosis. *Frontiers in Bioengineering and Biotechnology*. 2019; 7: 406. doi: 10.3389/fbioe.2019.00406
97. Brockman SM, Bodas M, Silverberg D, Sharma A, and Vij N. Dendrimer-based selective autophagy-induction rescues dF508-CFTR and inhibits Pseudomonas aeruginosa infection in cystic fibrosis. *PLoS ONE*. 2017; 12: e0184793. doi:10.1371/journal.pone.0184793
98. Trandafir LM, Leon MM, Frasinariu O, Baciu G, Dodi G, and Cojocaru E. Current practices and potential nanotechnology perspectives for pain related to cystic fibrosis. *Journal of Clinical Medicine*. 2019; 8(7): 1023. https://doi.org/10.3390/jcm8071023
99. Ong V, Mei V, Cao L, Lee K, and Chung E. Nanomedicine for cystic fibrosis. *SLAS TECHNOLOGY: Translating Life Sciences Innovation*. 2019; 24: 247263031882433. doi:10.1177/2472630318824334
100. Strug L, Stephenson A, Panjwani N, and Harris A. Recent advances in developing therapeutics for cystic fibrosis. *Human Molecular Genetics*. 2018; 27: R173–R186. doi: 10.1093/hmg/ddy188
101. Idiopathic pulmonary fibrosis; Date accessed 1/09/2020. www.nhs.uk
102. Sgalla G, Lovene B, and Richeldi L. Idiopathic pulmonary fibrosis: Pathogenesis and management. *Respiratory Research*. 2018; 19 (32): 1–18.
103. Minko T. Nanotechnology Approach for Inhalation Treatment of Pulmonary Fibrosis; Thesis Abstract. Rutgers University, New Brunswick, NJ. 2017.
104. Garbuzenko O, Ivanova V, Kholodovych V, Reimer D, Reuhl K, Yurkow E, Adler D, and Minko T. Combinatorial treatment of idiopathic pulmonary fibrosis using nanoparticles with prostaglandin E and siRNA(s). *Nanomedicine: Nanotechnology, Biology and Medicine*. 2017;13: 1983–1992. doi:10.1016/j.nano.2017.04.005
105. Ungprasent P and Ryu JH. Clinical Manifestations, Diagnosis and Treatment of Sarcoidosis. 2019; ncbi.nim.nih.gov
106. American Thoracic Society Patient. Treatment of sarcoidosis, education series. *American Journal of Respiratory and Critical Care Medicine*. 2018; 197: 9–10.
107. Lühder F and Reichardt HM. Novel drug delivery systems tailored for improved administration of glucocorticoids. *International Journal of Molecular Sciences*. 2017; 18: 1836.
108. Simonneau G, Montani D, Celermajer DS, et al. Haemodynamic definitions and updated clinical classification of pulmonary hypertension. *European Respiratory Journal*. 2019; 53: 1801913. doi:10.1183/13993003.01913-2018
109. Yaghi S, Novikov A, and Trandafirescu T. Clinical update on pulmonary hypertension. *Journal of Investigative Medicine*. 2020; 68(4): 821–827.
110. Segura-Ibarra V, Wu S, Hassan N, Moran-Guerrero JA, Ferrari M, Guha A, Karmouty-Quintana H, and Blanco E. Nanotherapeutics for treatment of pulmonary arterial hypertension. *Frontiers in Physiology*. 2018; 9: 890. https://doi.org/10.3389/fphys.2018.00890
111. Nakamura K, Matsubara H, Akagi S, Sarashina T, Ejiri K, Kawakita N, Yoshida M, Miyoshi T, Watanabe A, Nishii N, and Ito H. Nanoparticle-mediated drug delivery system for pulmonary arterial hypertension. *Journal of Clinical Medicine*. 2017; 6(5): 48. https://doi.org/10.3390/jcm6050048
112. European Patent Office. European Patent Application EP 2 893 922 A1, Publication date: 15.07.2015; Bulletin 2015/29; 2015.
113. Malek R, Soufi S. Pulmonary edema. [Updated 2020 Apr 29]. In: StatPearls [Internet]. StatPearls Publishing, Treasure Island, FL. 2020; Available from https://www.ncbi.nlm.nih.gov/books/NBK557611
114. Bird A. Acute pulmonary oedema management in general practice. *Australian Family Physician*. 2010; 39(12): 910–914.
115. Sadikot RT, Kolanjiyil AV, Kleinstreuer C, and Rubinstein I. Nanomedicine for treatment of acute lung injury and acute respiratory distress syndrome. *Biomedicine Hub*. 2017; 2: 1–12. doi:10.1159/000477086
116. Cicha I. Thrombosis: Novel nanomedical concepts of diagnosis and treatment. *World Journal of Cardiology*. 2015; 7(8): 434–441. https://doi.org/10.4330/wjc.v7.i8.434
117. Sanchez O, Planquette B, and Meyer G. Update on acute pulmonary embolism. *European Respiratory Review: An Official Journal of the European Respiratory Society*. 2009; 18: 137–147. doi:10.1183/09059180.000046

17

Nanocarrier Systems for Lung Drug Delivery

Sai HS. Boddu[1] **and Jerry Nesamony**[2]
[1]Department of Pharmaceutical Sciences, College of Pharmacy and Health Sciences, Ajman University, Ajman, UAE
[2]Department of Pharmacy Practice, The University of Toledo, Health Science Campus, USA

17.1 Introduction

Pulmonary drug delivery has always been an important route for pharmaceutical and biomedical researchers since the lung is capable of absorbing pharmaceuticals either for local action or for systemic absorption. Since the early 1990s, pulmonary route has been used as an alternative to the parenteral route due to its non-invasive nature (1–4). For example, an aqueous solution of morphine was investigated for the possibility of systemic administration by pulmonary administration using the AERx pulmonary drug delivery system (5). The result of aerosolized morphine delivery was approximately 100% bioavailability compared to intravenous infusion. Pulmonary delivery can result in higher bioavailability of drugs used for treating central nervous system (CNS) disorders leading to reduced dose and costs necessary to achieve therapeutic effect (6). There are several drugs used for treating CNS disorders that are currently under investigation for pulmonary delivery, such as anxiety (alprazolam), Parkinson's disease (apomorphine), analgesia (morphine, fentanyl), and migraine (loxapine, prochlorperazine) (7). Additionally, in 2012, the inhaled dibenzodiazepine loxapine (Adasuve, Alexza Pharmaceuticals, Mountain View, CA, USA) was approved by the Food and Drug Administration (FDA) for treatment of agitation associated with schizophrenia or bipolar I disorder in adults (6). Nicotine can be formulated and optimized for pulmonary delivery to treat patients with nicotine addiction, because the commercialized products for smoking cessation that are currently marketed to deliver nicotine or its substitutes (such as bupropion or varenicline) are very slow compared to inhalation of cigarette smoke (8).

Inhalation route is traditionally used to deliver drug molecules such as mucolytics, beta-agonists, anti-infectives, anti-cholinergic drugs, and corticosteroids (9). However, lately more research is being conducted to deliver opioids, hormones, anti-cancer agents, peptides, antibodies, vaccines, and genetic molecules (e.g. SiRNA, shRNA, and miRNA) through this route (9,10). It is reported that macromolecules have substantially high bioavailability across the pulmonary epithelium, which can be up to 200 times higher than via any other non-invasive route into the body (11). Insulin is one among many macromolecules that has been investigated for decades to systemically deliver it through the pulmonary route. In a study, insulin was administered by the pulmonary route in healthy individuals (12). When compared to subcutaneous insulin administration that showed onset in 50–60 minutes and hypoglycemic effect over 10–120 minutes, pulmonary insulin resulted in faster absorption within 7–20 minutes with rapid significant hypoglycemic effect in 60–70 minutes. In 2015, the US FDA approved the Afrezza Inhalation System for use in both type 1 and type 2 diabetes patients. The Afrezza Inhalation System is a drug/delivery device combination product consisting of Technosphere® Insulin inhalation powder (TI) (fumaryl diketopiperazine-based porous micron-sized carrier particles with recombinant human insulin; MannKind Corporation, Valencia, CA, USA) prefilled into single-use cartridges and the Afrezza inhaler. However, cough has been reported in more than 25% of patients and this product is currently not being used in smokers and chronic obstructive pulmonary disease (COPD) patients (13). Besides insulin, heparin (anticoagulant), calcitonin and parathyroid hormone (for treatment of osteoporosis), human growth hormone (growth hormone deficiency therapy), and erythropoietin (used in anemia) are being evaluated in clinical or preclinical trials for feasibility in pulmonary delivery (14,15).

Lung cancer is a common form of cancer worldwide and has a poor survival rate usually associated with late diagnosis (16). Lung cancer is usually treated with chemotherapy, surgery, radiotherapy, or a combination of more than one method. Most chemotherapeutics marketed for treatment of lung cancer are available as an intravenous injection or in infusion dosage forms, with a few options that can be given orally (17). Systemic administration of chemotherapy is not site specific. It results in a low concentration at the site of lung tumor compared with the initial dose given, which makes systemic adverse effects more aggressive and noticeable (6). Furthermore, chemotherapeutic agents are expensive and used for long-term treatment. Therefore, administration of anti-cancer drugs for local effect by pulmonary drug delivery could have real benefits, such as better patient adherence and higher treatment efficiency (17). Some of the advantages of the pulmonary route include its non-invasive nature, high permeability of the respiratory mucosa, large absorptive surface area (~70–140 m^2 in adult humans), thin alveolar epithelium (0.1–0.2 μm), low enzymatic activity, circumvention of first-pass hepatic metabolism, reduced dose administered to the patients, rapid absorption of the drug, and active blood supply to the lungs (18).

DOI: 10.1201/9781003046547-17

17.2 Barriers to Drug Absorption via Pulmonary Route

Despite these advantages, the pulmonary route is not feasible for delivering many drug molecules. The drug molecule's solubility and permeability affect its rate and extent of pulmonary absorption. In addition, formulation design strongly impacts the pharmacokinetic profile of drug concentration and elimination half-life (19). Highly soluble drug molecules such as albuterol can dissolve when they come in contact with lung fluids, while for poorly soluble drug molecules, dissolution can affect the absorption rate through the pulmonary mucosa (6). The availability of pulmonary fluid at the dissolution site is an important factor. An estimation of total volume of lung fluid is 15–17 mL in humans (20). However, the proportion of the fluid volume that an aerosol particle is exposed to is unpredictable (21), and the thickness of the lining layer along with the volume of the lung lining fluid vary between the central and peripheral parts of the lungs. It is reported that particles that landed in the upper part of the lung dissolve faster than those that deposited in the alveoli due to the presence of a large solid-liquid interface in the upper respiratory system (22). Phospholipids present in the pulmonary surfactant, to some extent, can improve the solubility of drug molecules within the lungs (23,24). Pulmonary surfactants are a mixture of lipids (90%) and proteins (10%) that are secreted into the alveolar space by the type II epithelial cells (25). The surfactant helps lower the surface tension at the air/liquid interface within the alveoli of the lung. However, pulmonary surfactants can compromise the stability of active biopharmaceuticals such as peptides and proteins (11). To summarize, the rate and extent of the dissolution process is the first challenge for aerosol particles to be absorbed by the lung tissue.

In general, reduced frequency of administration and duration of therapy can enhance patient adherence. Physiological clearance such as mucociliary and macrophage clearance are the two main hurdles to maintain sustained drug levels at the site of deposition (6). Pulmonary mucus is a thick, viscoelastic hydrogel layer up to 30 µm thick, and it mainly contains water and glycoproteins (mucin) (26,27). The mucociliary clearance mechanism is highly effective and not specific, with 80–90% of the inhaled material being cleared from the upper and central lung within 24 hours of inhalation (28,29). One way to improve sustained drug levels at the site of action is to minimize clearance. Very little is known about the nature of the interaction between surfactant components and inhaled particles at the molecular level (6). Several researchers have reported on particle–surfactant interactions and evaluated the clearance of particles into the surfactant layer (30) and the impact of such interaction on the biophysical functionality of the surfactant film (31).

The surface properties of inhaled particles affect the occurrence as well as the intensity of interactions with pulmonary surfactant components, in particular, alveolar macrophage clearance (32,33), which shows size-dependent uptake that is considered most effective for particles with a geometric diameter range of 0.5–3 µm (34). Alveolar macrophages, which are in close proximity to the pulmonary surfactant, have the ability to internalize and clear biologics before reaching the systemic circulation. The uptake by alveolar macrophages is increased by immunoglobulin G (IgG) antibodies present in the both pulmonary surfactant and pulmonary mucus (35). When developing a formulation that can overcome or minimize the lung clearance mechanisms on droplet aerosols, methods have to be devised to reduce pulmonary clearance mechanisms, improve bioavailability, and maintain sustained concentration of drugs given by inhalation route.

17.3 Conventional Strategies to Enhance Drug Absorption via the Pulmonary Route

For successful pulmonary drug delivery, the aerosol particles should be efficiently deposited in the lungs. To achieve this goal, both the inhalation device and drug formulation have to be optimized. Several advancements in inhalation formulations and device technologies have been proposed (36,37). Dry powder inhalation is a very popular formulation used in pulmonary drug delivery due to the fact that it enhances stability of pharmaceuticals and utilizes established manufacturing techniques such as spray drying (38). Prodrug is a pharmaceutical approach used in pulmonary dosage forms to improve the drug solubility. In the prodrug approach, a drug molecule is structurally modified to enhance solubility and dissolution (39). For example, a prodrug of treprostinil (hexadecyl-treprostinil) (C16TR) in lipid nanoparticles effectively inhibited bleomycin-induced pulmonary fibrosis in a rat model (40). Selective formation of a particular polymorph of higher solubility using crystal engineering and the formation of amorphous forms has also been investigated in inhalation dosage forms (21,22).

For macromolecules such as proteins and peptides, solubility is less challenging, but they suffer from enzymatic degradation in the lung fluid. However, the enzymatic degradation of these macromolecules can be minimized by chemical modifications that block peptidases, or via linkage to create ring-shaped biomolecules (11). For example, protease inhibitors (aprotinin, bacitracin, and soybean trypsin) along with absorption enhancers (sodium glycocholate, linoleic acid-surfactant mixed micelles, and N-lauryl-β-d-maltopyranoside) were used for absorption of insulin and calcitonin (41,42). The nature of the excipients used in a formulation also plays a significant role in the absorption of drugs given through pulmonary administration (6). Among the excipients that have been approved and commonly used for pulmonary delivery are lactose, mannitol, and glucose (38,43). Phospholipids, such as dipalmitoyl phosphatidylcholine, can be used to improve solubility of poorly water-soluble drug molecules (38). On the other hand, a fusion of exogenous and endogenous surfactants might hinder the free diffusion of drug molecules, leading to reduced dissolution and absorption (6). Yet at times, a combination of different solubility enhancers could be a valid option to overcome toxicity limitations that are encountered during formulation development. Nevertheless, the integrity of pulmonary air–blood barrier is a major concern during evaluation of pulmonary excipients (6,38). Conventional strategies generally improve the pulmonary absorption of drugs administered as a solution or dry powder, which are beneficial for the treatment of some respiratory diseases, such as asthma.

However, for treating conditions such as lung cancer, infections, and fibrosis, a more sustained exposure of drugs to the lungs is a better strategy (44). This requirement has generated significant interest in the development of inhalable nanocarrier systems, which can provide sustained drug exposure in the lungs and limit side effects related to systemic exposure of highly toxic or irritant drugs. Both microcarriers and nanocarriers have been reported to be viable approaches in pulmonary delivery; however, the final selection of size should be dependent on the drug under investigation. The scope of microcarriers has been highlighted the in literature (45,46), but is out of the purview of this chapter. In the subsequent sections, this chapter will focus on the potential application of nanocarrier systems in pulmonary delivery.

17.4 Deposition and Clearance of Inhaled Particles

17.4.1 Deposition of Inhaled Particles

The deposition of inhaled particles in the lungs is very complex. It mainly depends on the particle size, which is usually expressed in terms of its mass median aerodynamic diameter (MMAD), shape (47), surface charge (48), and composition (49,50). Size plays an important role in achieving deep-lung deposition of inhaled nanocarrier systems. Inhaled particles with an aerodynamic diameter ranging between 0.5 and 5 μm have the greatest potential to be deposited in the lung. Three different mechanisms of drug deposition are mentioned in the literature: impaction (3–6 μm), sedimentation (1–3 μm), and diffusion (<1 μm) (51). The transport of microparticles is mainly explained through inertial impaction and sedimentation mechanisms, while nanoparticle transport relies on the diffusion mechanism (51). Impaction involves the passage of aerosol particles through the oropharynx and upper respiratory at a high velocity. The centrifugal force associated with particles makes them collide with the respiratory wall, which is followed by deposition in the oropharynx regions. Generally, DPIs and MDIs with particle sizes greater than 5 μm shows this type of mechanism. In sedimentation, particles of sufficient mass and sizes (1–3 μm) deposit in smaller airways and bronchioles predominantly due to gravitational forces. Sedimentation is partly influenced by the breathing pattern; slow breathing provides a sufficient time span for sedimentation. Deposition of particles (<1 μm) in the deeper alveolar areas of the lungs via diffusion mechanism is explained by the Brownian motion. The random motion of particles occurs like gas molecules in the air, which results in deposition of particles in the alveolar region (the lower lung area) (Figure 17.1).

Nanocarriers coated with chemicals naturally present in the alveolar epithelial lining (albumin or phospholipids) can help translocate the particles across the alveolus–capillary barrier (52). It is widely accepted that particles with MMAD less than 5 μm exhibit efficient lung deposition. However, the deposition of particles with MMAD lower than 1 μm appears to be controversial. Some reports suggest that particles lower than 1 μm are mainly exhaled without eliciting any therapeutic activity within the lungs due to their extremely low settling velocity. Others argue that particles ~100 nm are able to deposit efficiently throughout the alveolar region due to the Brownian motion (53,54). Nevertheless, the amount of particles lost due to exhalation could be reduced through appropriate inhalation procedure followed by an adequate breath-holding (54).

17.4.2 Clearance of Inhaled Particles

Mucociliary clearance plays a major role in the defense mechanism of lungs and clearance of inhaled particles. The surface of the upper airways from the trachea to the tertiary bronchi is lined by a thick mucus film, which acts as a protective layer to entrap and remove the foreign particles. The mucociliary clearance dispels the particles through either coughing or swallowing, before they could move toward the lower areas of the lung. The mucociliary clearance eliminates a majority of the inhaled particles of sizes greater than 6 μm, while smaller particles escape the mucociliary clearance with preferential deposition in the deeper areas of the lungs such as the alveolar region, where they dissolve/stay for a longer time (55). In the alveolar region, the transport mechanism is more complex as the lining consists of a variety of lipids and proteins, which act as a barrier for the transport of the molecules.

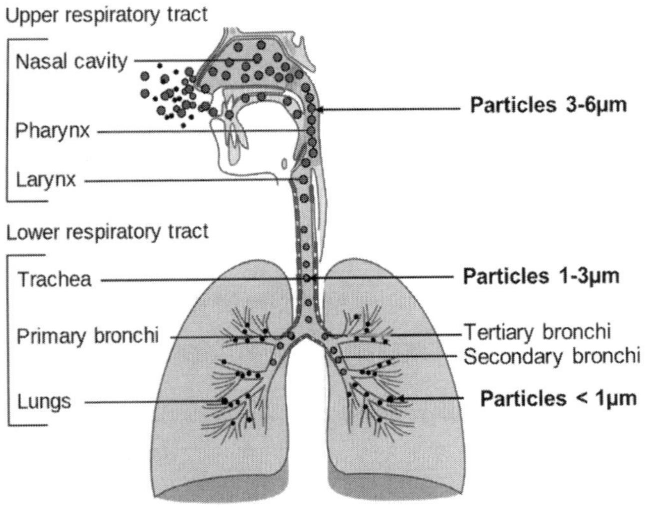

FIGURE 17.1 Size-Dependent Deposition of Particles in the Lung

The tight junctions present on the epithelial cell serve as the primary barrier to the transport along the alveolar lining. However, depending on the nature and structure drug molecules, the transporter proteins help in their transport via active absorption or passive diffusion (56). Drug molecules that cross the epithelial cells are either phagocytized by the alveolar macrophages or absorbed into the systemic circulation (21). The alveolar phagocytosis in rats was found to be dependent on the particle size, with particles in the size range of ~2–3 µm showing maximal phagocytosis and attachment (57).

Makino et al. observed a similar trend in the alveolar macrophage uptake of polystyrene particles with higher uptake of 1 µm particles compared to those with diameters <500 nm or >6 µm. The engulfing of polystyrene particles of 1 µm in diameter by alveolar macrophages is also influenced by their surface properties. Particles having primary amine groups on the surface are effectively trapped when compared to those with carboxyl, sulfate, or hydroxyl groups. Based on this information, more effective uptake and therapeutic action of nanoparticles might be achieved with nanoparticle size <500 nm in diameter to avoid the alveolar macrophages, unless intended (58). The relation between size and shape of particles in phagocytosis by alveolar macrophages was studied by Mitragotri et al. using six different types of polystyrene particles (spheres, oblate ellipsoids, prolate ellipsoids, elliptical disks, rectangular disks, and UFO-shaped particles) (59). Elliptical disks showed rapid internalization (<6 minutes) upon perpendicular attachment to cell membrane along the long axis, while attachment of the short-axis side of the particle to cell membrane did not show noticeable phagocytic activity. This study concluded that particle shape at the point of initial contact with cells plays an important role in phagocytosis of the particle.

From these findings, it is apparent that spheres or oblate ellipsoids may be more suitable for targeting macrophages, especially in intracellular infections such as tuberculosis, via pulmonary administration. Similar findings were reported in a more recent study for inhalable particles of size ranging between 1 and 5 µm. The effect of particle properties such as shape on the biological action of inhaled particles were investigated. Salbutamol sulphate (a highly water soluble drug) and budesonide (a poorly water soluble drug) were compared in terms of dissolution, permeation, and preferential uptake by epithelial cells in comparison with macrophages after spray drying and jet milling. Spray drying of salbutamol sulphate resulted in spherical shaped particles (mostly amorphous), whereas jet milling resulted in irregular shaped particles (mostly crystalline). The spray dried form of both drugs resulted in lower respirable fractions and higher permeability/cell uptake rates compared to the needle shaped particles. Spray dried particles were preferentially taken up by macrophages, suggesting the possibility of targeted delivery with processing (60).

In spite of advances in the formulation development, there is still some lack of information with respect to the exact uptake, transport and clearance of nanocarrier systems in the alveolar epithelium and how drug molecules reach the systemic circulation. The literature suggests that regardless of chemical composition and shape, nanoparticles with an approximate size of 34 nm are rapidly translocated from the alveolar luminal surface into lymph nodes and bloodstream. Charge on the surface also plays an important role in the biodistribution of these nanoparticles.

For optimal pulmonary delivery, non-cationic nanoparticles with an approximate size range between 6 and 34 nm would have the highest potential. Although several *in vitro* models have been established for studying the uptake and permeation of drug molecules in the pulmonary epithelium (air–liquid interface models), there are still open questions with respect to the behavior of the cells in a diseased condition. A majority of studies published in the literature looked at the clearance of drugs loaded inside nanocarrier systems, rather than clearance of nanocarrier systems from the lungs. These studies do not reveal the kinetics of nanocarrier systems. Further research into the clearance of nanocarriers is necessary to understand the accumulation behavior of nanocarriers following multiple dose administration to prevent "nanoparticle overload" and local inflammatory responses (61).

17.5 Nanocarrier Systems for Pulmonary Delivery

Nanocarrier systems have further increased the scope of the pulmonary route as they are considered to be superior to conventional medicines in terms of ability to show enhanced drug solubility, uniform distribution of drugs in the alveoli, sustained drug release, and improved pharmacokinetic/pharmacodynamic profiles of drugs (62,63). It is also possible to alter the biological properties of nanocarriers through changes on the particle surface. Nanocarriers with size <200 nm could theoretically escape detection by alveolar macrophages, leading to more effective uptake and action (64). Drugs encapsulated in nanocarriers were used for local action in the treatment of respiratory diseases such as asthma, COPD, cystic fibrosis, tuberculosis, and lung cancer (65). For example, COPD patients are characterized by bronchial hypersecretion, wherein lipophilic corticosteroids are hindered from reaching their receptors localized within the cytoplasm of bronchial epithelial cells. However, drugs such as beclomethasone dipropionate showed enhanced penetration through the mucus layer and provided localized therapy when loaded in non-phospholipid vesicles such as niosomes (66). Nanocarrier systems are heavily exploited for therapeutic purpose to carry the drug in the body in a controlled manner from the site of administration to the target area. This implies the passage of the drug molecules and drug delivery system across numerous physiological barriers, which represent the most challenging goal in drug delivery. Broadly, nanocarriers are categorized into metallic/inorganic and organic nanoparticles. Inorganic nanoparticles, both ceramic and metallic, have been used for pulmonary administration in magnetic resonance imaging and stimuli-responsive therapeutic and/or diagnostic delivery. However, limited surface chemical availability, instability, and poor biocompatibility are considered as serious drawbacks of inorganic nanoparticles, limiting their use mostly as an imaging agent (67). On the contrary, organic nanocarrier systems such as polymeric nanoparticles, liposomes, micelles, and nanoemulsions have shown extremely useful properties from pharmacological and

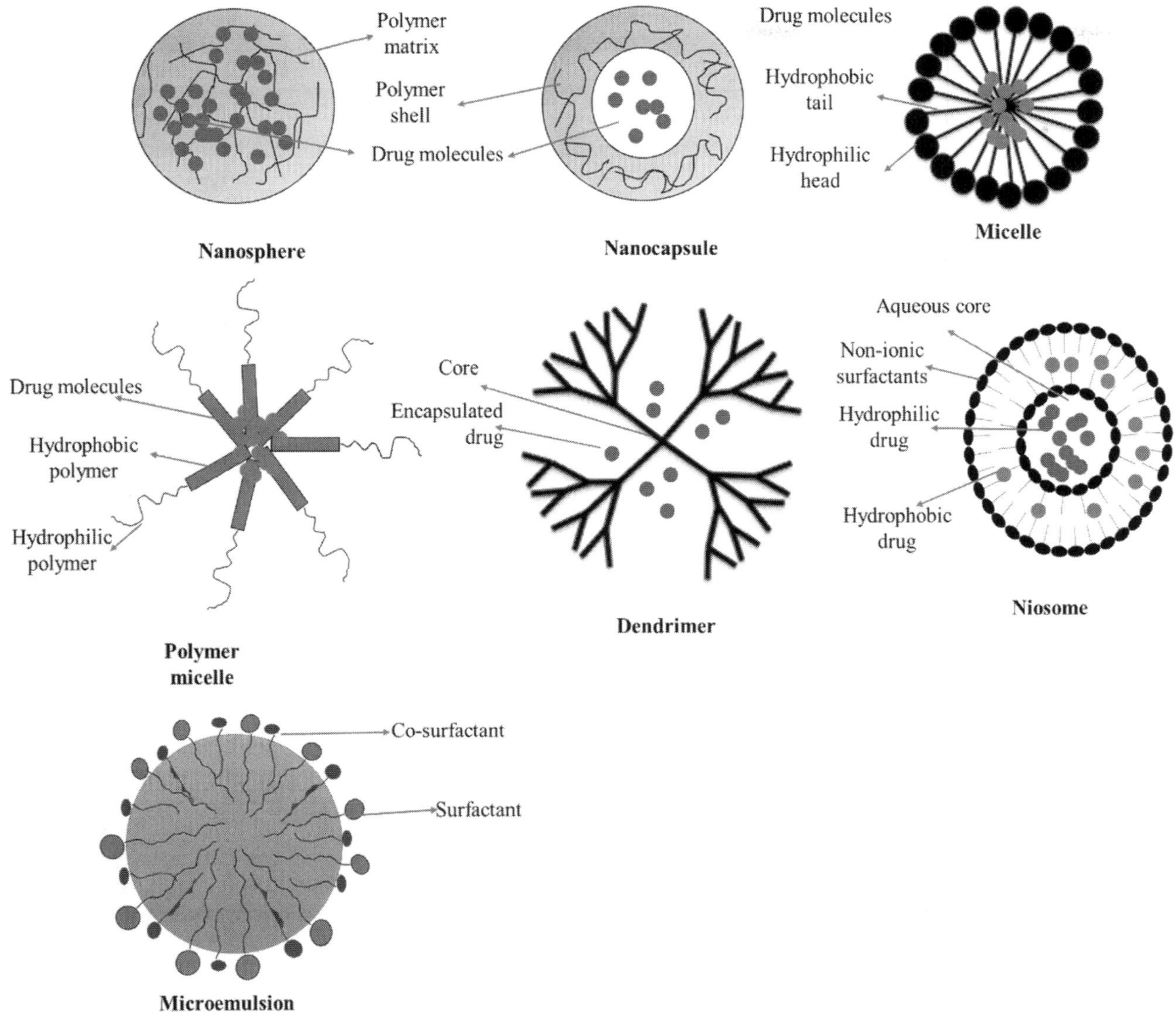

FIGURE 17.2 Various Types of Organic Nanocarrier Systems

therapeutic standpoints in treating various conditions such as asthma and COPD following pulmonary administration. In this chapter, our focus will be mainly on organic nanocarrier systems in pulmonary drug delivery. In organic nanocarrier systems, drugs are either retained with different covalent and electrostatic interactions or captured inside the particle using different techniques and materials such as lipids (liposomes, solid lipid nanoparticles, niosomes, microemulsions, lipidic micelles) or polymers (polymeric nanoparticles, polymer micelles, dendrimers) (Figure 17.2) (68,69).

17.5.1 Nanosuspensions

Nanosuspensions contain nano-sized drug particles dispersed and stabilized in an aqueous solution. Nanosuspensions are suitable for drugs that are poorly soluble in aqueous and organic solvents (70). They are drug particles with size ranging between 200 and 600 nm and are stabilized using surfactant/polymer(s). Unlike the other nanocarriers such as nanoparticles and liposomes, no matrix material is added to the system. For pulmonary delivery, aqueous nanosuspensions are nebulized using mechanical or ultrasonic nebulizers for lung delivery. The small size of drug particles allows uniform distribution, rapid diffusion/dissolution, and prolonged residence of drugs in the lungs (71). For example, budesonide nanosuspensions showed outstanding deposition in a modified multistage liquid collision method when compared with normal and micronized particles. Following inhalation of nanosuspension, budesonide concentration after 1 hour in the unit mass of rat lung tissue was significantly higher compared with normal particles ($p < 0.01$) and micronized particles ($p < 0.05$) (72). Nanosuspensions show decreased mucociliary clearance upon pulmonary administration. In a study by Fu et al., intra-tracheally dosed nanosuspensions of fluticasone propionate in rats showed decreased mucociliary clearance and extended pulmonary absorption in comparison to a microsuspension (73). Other drugs, such as buparvaquone (74) and coenzyme Q10 (75), were also tested for pulmonary delivery (76). Research into nanosuspension technology for delivery of poorly water soluble drugs are progressing at a rapid pace; however, more efforts should be invested in understanding the potential toxicological effects and related implications.

17.5.2 Liposomes

Liposomes are the most extensively investigated carriers for pulmonary delivery because they can be fabricated from pulmonary surfactants and endogenous phospholipids, and as a result they are biocompatible, biodegradable, and relatively nontoxic (53). Liposomes are produced in a broad size range, and both hydrophilic as well as hydrophobic drugs can be incorporated. These drugs include cytotoxic agents, antimicrobial and antiviral drugs, asthma medications, and drugs intended for systemic absorption (77). Researchers have been investigating liposomes as drug carriers for over 30 years, and many therapeutic agents have been incorporated into liposomes as a strategy to develop pulmonary drug delivery leading to animal and human studies (78). Liposomal aerosols have several advantages over other traditional pulmonary aerosol carriers such as extended duration of release, no significant local irritation, minimal toxicity, and improved drug stability. For example, pulmonary distribution and clearance of 99mTc-labeled beclomethasone dipropionate liposomes in 11 healthy volunteers showed sustained drug release and drug action in the lower airways (79).

Successful development of an inhalable liposomal product depends mainly on formulation composition and nebulizer design, which play a major role in enhancing the deposition of vesicles in the deep lung and alleviating the influence of shearing on liposome stability. A liposomal formulation is delivered to the airways by direct instillation, microspray, or by jet/ultrasonic/vibrating-mesh nebulization. Generally, the shear generated during the nebulization process could result in physical stress on liposomal bilayers, causing a loss in the entrapped drug (80). The loss/leakage of the drug could be prevented by using a vibrating-mesh nebulizer customized with large mesh apertures. Vibrating-mesh nebulizers that use lower-frequency waves are considered suitable for delivery of liposomes and nucleic acids (81,82). In one study, the impact of nebulization on liposomes (conventional liposomes, PEGylated liposomes and cationic polymer coated liposomes) was assessed with three commercially available inhaler systems (air-jet, ultrasonic, and vibrating-mesh). This study concluded that conventional liposomes were more stable as compared to polymer-coated and cationic polymer-coated liposomes in terms of aggregation and drug leakage, while the release rates of a model hydrophilic drug were highest with the vibrating-mesh nebulizers regardless of the surface characteristics of the liposomes (83).

Though PEGylation of liposomes prevents their uptake by the alveolar macrophages and increases the duration of action (84), it might adversely affect the stability of liposomes during nebulization and promote drug leakage (85). The stability of liposomes during nebulization is also improved through incorporation of cholesterol and high phase transition phospholipids such as distearoylglycerophosphatidyl-choline in the lipid bilayer. For example, Arikace™, a liposomal inhalation amikacin used in the treatment of non-tuberculosis mycobacterial infections, is made using cholesterol-enriched dipalmitoylphosphatidylcholine (high-phase transition phospholipid) and delivered using a Pari e-Flow vibrating mesh nebulizer. Arikayce is considered an orphan drug by the FDA and the European Medicines Agency for the management of *Pseudomonas aeruginosa* infections in cystic fibrosis patients. Liposomal technology is by far the most successful strategy for delivering drugs via inhalation route among other nanocarrier systems. Several liposomal inhalation products are currently under investigation in clinical trials for treating various pulmonary diseases such as cystic fibrosis, pulmonary aspergillosis, pain management, and lung cancer (44) (Table 17.1). Lipoquin™ (liposomal ciprofloxacin for inhalation) has recently completed phase 3 clinical trials and is in the process obtaining approval from the FDA for treating non–cystic fibrosis bronchiectasis patients (80). Pulmaquin™ (Aradigm Corp., USA) is a variant of Lipoquin that contains both liposomal ciprofloxacin and free ciprofloxacin. The free ciprofloxacin in Pulmaquin is meant to provide an initially high peak concentration of the drug in the lungs (52). The success of Linhaliq™ would encourage pharmaceutical companies to invest more in the development of inhalable liposomal products, especially in the delivery of macromolecules such as DNA.

17.5.3 Polymeric Nanoparticles

They have a particle size between 10 and 1000 nm, and can be prepared using a variety of natural and synthetic polymers (86). Several polymers have been used in fabricating nanoparticles: poly lactic acid (PLA), albumin, poly glycolic co-lactic acid (PLGA), poly(ε-caprolactone) (PCL), alginate, chitosan, and gelatin base (87). Drug molecules are either being uniformly distributed within the matrix surface or loaded inside the nanocarrier core, utilizing both natural or synthetic biodegradable polymers (88). Hydrophobic and hydrophilic drug molecules can be incorporated inside nanoparticles in which they improve physiochemical stability of drugs by protecting from enzymatic degradation and providing sustained drug release of the incorporated molecule. In a study, pirfenidone-loaded PLGA nanoparticles showed enhanced anti-fibrotic efficacy and sustained drug levels in the lungs and bronchoalveolar lavage after 24 hours and 1 week following intra-tracheal instillation. Studies also indicate that nanoparticles facilitate transport of drugs to the epithelium, while resisting mucociliary clearance (89). Mucoadhesive polymers can be utilized in fabricating nanoparticles to enhance bioavailability by increasing the retention time of nanoparticles within pulmonary mucosa and minimizing the action of mucociliary clearance mechanism (53). Chitosan-modified paclitaxel-loaded PLGA nanoparticles with a mean diameter of 200–300 nm exhibited enhanced uptake across the endothelial cells of the lung tumor capillary due to electrical interaction (90). Similarly, PEGylated polyaspartamide–polylactide fluorescent nanoparticles containing ibuprofen exhibited better mucus-penetrating properties in artificial mucus compared to the un-PEGylated particles (Figure 17.3) (91).

TABLE 17.1

List of Liposomal Drugs Completed or Undergoing Clinical Trials. *Indicates Approval by the US FDA

Study	Device/Technique	Status	ClinicalTrials.gov Identifier
A phase 4, randomized, controlled study of nebulized amikacin in patients with acute exacerbation of non-cystic fibrosis bronchiectasis.	Nebulization	Completed	NCT02081963
A phase 3 pharmacokinetic study of inhaled liposomal amphotericin B (AmBisome®) in lung transplant recipients.	Nebulization	Completed	NCT00177710
A phase 2 study to assess the efficacy, safety and tolerability of once daily dosing of liposomal-amikacin (Arikayce™) for inhalation for the treatment of *Mycobacterium abscessus* lung disease.	Inhalation	Completed	NCT03038178
A phase 1/2 open label safety and efficacy study designed to determine the maximum tolerated dose of inhaled cisplatin liposomal in patients with relapsed/progressive osteosarcoma metastatic to the lung.	Inhalation	Completed	NCT00102531
A phase 2 study to assess the efficacy, safety and tolerability of liposomal-amikacin Arikayce) in patients with treatment refractory non-tuberculous mycobacteria lung disease.	Inhalation	Completed	NCT01315236
A phase 3 study to evaluate the effectiveness of inhaled liposomal amikacin in patients with non-tuberculous mycobacterial lung infection (CONVERT).	Inhalation	Completed*	NCT02344004
A phase 3 study to evaluate the long term safety and tolerability of inhalation liposomal amikacin in patients with chronic infection due to pseudomonas aeruginosa.	PARI Investigational eFlow® Nebulizer System (eFlow®)	Completed	NCT01316276
A phase 1/2 to assess the safety and tolerability of a multi-dose study of liposomal amikacin for inhalation (Arikayce) in cystic fibrosis subjects.	PARI Investigational eFlow® Nebulizer System (eFlow®)	Completed	NCT00777296
A phase 3 randomized, controlled clinical trial of liposomal cyclosporine-A for the treatment of bronchiolitis obliterans syndrome in adults following a single lung transplant.	PARI Investigational eFlow® Nebulizer System (eFlow®)	Recruiting	NCT03657342
A phase 3 randomized, controlled clinical trial of liposomal cyclosporine-A for the treatment of bronchiolitis obliterans syndrome.	PARI Investigational eFlow® Nebulizer System (eFlow®)	Recruiting	NCT03656926
A phase 3, open-label safety extension study to assess the safety and tolerability of liposomal amikacin for inhalation in participants with non-tuberculous mycobacterial lung infections due to *Mycobacterium avium* complex.	Inhalation	Completed	NCT02628600
A phase 3 study to determine whether Arikayce is effective in treating chronic lung infections caused by chronic *Pseudomonas aeruginosa* (Pa) infections and further comparison with Tobramycin TOBI®, an inhalation antibiotic already available for use.	PARI Investigational eFlow® Nebulizer System (eFlow®)	Completed	NCT01315678
A phase 1 study to evaluate the pharmacokinetics, safety and tolerability of liposomal Treprostinil (L606) inhalation solution in healthy volunteers.	Inhalation	Recruiting	NCT04041648
A phase 2 study to evaluate the longer term safety, tolerability and efficacy of 560 mg once daily dose of Arikayce administered for 6 cycles over 18 months.	Inhalation	Completed	NCT03905642
A phase 2/3 randomized double-blind study comparing the safety and the efficacy of nebulized AmBisome to prevent invasive pulmonary aspergillosis in neutropenic hemato-oncologic patients.	Inhalation	Completed	NCT00263315
A phase 3 study to compare the therapeutic (clinical and radiological) efficacy of a 6-month treatment by itraconazole and nebulized Zmbisome® (liposomal amphotericin B = LAmB) versus treatment by itraconazole alone, in immunocompromised patients affected by chronic pulmonary aspergillosis.	Inhalation	Recruiting	NCT03656081
A phase 3 study (ARD-3150-1201, ORBIT-3) to evaluate the safety and efficacy of inhaled pulmaquin (ciprofloxacin dispersion for inhalation) compared to inhaled placebo in subjects who have a confirmed diagnosis of non-cystic fibrosis (non-CF) bronchiectasis with a history of pulmonary exacerbations and chronic *P. aeruginosa* infections.	Inhalation	Completed	NCT01515007
To assess the clinical data of patients with non-tuberculous mycobacteria *M. abscessus* infections and treated with liposomal amikacin inhalation.	Inhalation	Recruiting	NCT04163601
To determine the concentration of 9-nitrocamptothecin in the alveolar fluid over time following inhalation in patients with metastatic or recurrent cancer of the endometrium or the lung.	Inhalation	Completed	NCT00277082

(Continued)

TABLE 17.1 (Continued)

List of Liposomal Drugs Completed or Undergoing Clinical Trials. *Indicates Approval by the US FDA

Study	Device/Technique	Status	ClinicalTrials.gov Identifier
A phase 3 multicenter open-label, extension clinical trial to demonstrate the effectiveness and safety of liposomal cyclosporine A inhalation solution in the treatment of bronchiolitis obliterans syndrome in patients after lung transplantation.	PARI Investigational eFlow® Device	Enrolling by invitation	NCT04039347
A phase 2 study to determine the safety and tolerability of Arikayce versus placebo in patients who have bronchiectasis and chronic infection due to pseudomonas infection.	PARI eFlow® nebulizer	Completed	NCT00775138
A phase 1/2 study to evaluate the efficacy of liposomal cyclosporine A in the treatment of chronic rejection in lung transplant recipients with bronchiolitis obliterans syndrome.	Inhalation	Completed	NCT01650545
A phase 1b/2a multi-dose safety and tolerability study of liposomal amikacin for inhalation (Arikayce) in cystic fibrosis patients with chronic infections due to *Pseudomonas aeruginosa*.	PARI eFlow nebulizer	Completed	NCT00558844
A phase 1 trial to study the effectiveness of inhaled doxorubicin in treating patients with primary lung cancer or lung metastases.	Inhalation	Completed	NCT00004930
A phase 1 trial to study the effectiveness of inhaled doxorubicin in treating patients with advanced solid tumors affecting the lungs	Inhalation	Completed	NCT00020124
A phase 2 trial to evaluate the overall tolerability and the efficacy of aerosolized amphotericin B as a lipid complex (Abelcet®) for primary prophylaxis of invasive pulmonary aspergillosis (IPA) in pediatric patients with acute leukemia undergoing intensive chemotherapy.	Inhalation	Completed	NCT01615809
A phase 1 study to assess the tolerance analysis, safety and pharmacokinetic and distribution studies of nebulized amphotericin, but when administered to patients it also carries out the preliminary assessment of the safety and efficacy of the treatment.	Inhalation	Recruiting	NCT04267497

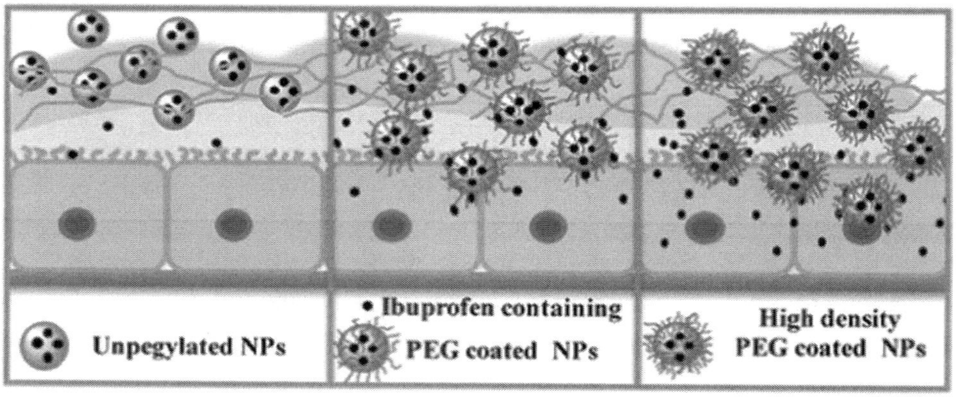

FIGURE 17.3 Mucus-Penetrating Nanoparticles of Ibuprofen for Pulmonary Administration. Reproduced with Permission from Craparo et al. (91)

Polymeric nanoparticles are required to have excellent biocompatibility and fast degradation rates when frequent administration is necessary. No substantial alteration in the biophysical properties of pulmonary surfactant was observed upon inhalation of PLGA nanoparticles (92). In one study, the potential of albumin nanoparticles for controlled release following pulmonary drug delivery was studied in mice using SPECT/CT imaging. The rate of clearance of bovine serum albumin nanoparticles (t½ = 86.6 hours) was significantly higher than the same concentration of albumin solution (t½ = 28.9 hours). Further, the same group also demonstrated the digestion of albumin nanoparticles in 48 hours suggesting their potential for use as a controlled release system with simultaneous biodegradation alleviating the concerns of accumulation and related toxicity in the lung (93). Based on the experimental evidence from literature, polymeric nanoparticles appear an enticing approach for inhalation delivery. However, depositing the exact dose of drugs into the lungs in a reliable and reproducible manner using dry powder inhalers is challenging in reality. Further efforts to scale up the manufacturing procedure are required to produce batches of nanoparticles with similar size, shape, density, and surface properties.

17.5.4 Solid Lipid Nanoparticles

Solid lipid nanoparticles (SLNs) offer an alternative to conventional colloidal systems such as liposomes and polymeric micro- and nanoparticles (94). In fact, SLNs have advantages such as high safety and ability to scale up production (94). Drug loading capacity of SLNs depends on factors like miscibility and solubility of the drug in the SLN lipid, physiochemical properties of the lipid solid matrix, and the polymorphic state of the lipid component within the SLNs (53). A controlled release profile can be achieved using SLNs as a function of the ratio of lipid/surfactant concentration, and through modifying production parameters. The release of drugs can be modulated to achieve prolonged release up to 5–7 weeks (53). SLNs have demonstrated an improved local bioavailability following pulmonary administration, which is dependent on the particle size. For example, in a recent study, Yuxingcao essential oil (YEO)-loaded SLNs of various sizes (200, 400 and 800 nm) were prepared using Compritol 888 ATO as lipid and polyvinyl alcohol as an emulsifier. SLNs showed prolonged pulmonary residence time and increased area under the curve values (15.4, 18.2 and 26.3 μg/g h for SLN-200, SLN-400 and SLN-800, respectively) in rats following intra-tracheal administration. SLNs of YEO showed 4.5–7.7-fold increase in the area under the curve values compared to the intra-tracheal dosing of YEO solution and a 257–438-fold increase compared to the intravenous dosing of YEO solution, respectively. These results demonstrated the potential of SLNs in improving the local bioavailability upon inhalation (95).

The lipid matrix used in the preparation of SLNs should be nontoxic and biocompatible. SLN dispersion can be nebulized or it can be used as a solid powder in a DPI device for pulmonary administration (53). SLNs showed clearance similar to liposomes in the lung tissues. In a recent study, the pulmonary pharmacokinetic behavior and patterns of lung clearance of inhalable nanocarriers (anionic ~150 nm liposomes and solid lipid nanoparticles) made out of 3H-labeled structural lipids (phosphatidylcholine and tristearin) were assessed in rats. This study reported similar clearance rates for both SLNs and liposomes from the lungs, despite deposition of SLNs in the upper respiratory track after intra-tracheal instillation. Further lipids also exhibit prolonged lung exposure and association with the lung tissue for over 2 weeks as compared to bronchoalveolar lavage fluid (44). SLNs are considered a good platform for delivering lipophilic drugs; however, issues such as low drug-loading and initial burst release of drugs due to the rearrangement of lipids to form a perfect crystal lattice should be resolved (96,97). To overcome these disadvantages, nanostructured lipid carriers (NLCs), which contain appropriate proportions of liquid lipids and solid lipids, have gained momentum. NLCs have a reduced degree of matrix crystallinity resulting in improved storage stability, encapsulation efficiency, and release profiles.

NLCs are capable of accommodating more drug between the fatty acid chains of glycerides compared to SLNs (98). In a recent comparative study, SLNs of levofloxacin in the presence of DNase as mucolytic enzyme were prepared using myristyl myristate. Further, NLCs of levofloxacin were also developed to improve the drug encapsulation and storage. This study reported that the encapsulation efficiency of levofloxacin was higher in NLCs (55.9 ± 1.6%) when compared with SLNs (20.1 ± 1.4%). NLCs of levofloxacin exhibited a more controlled release profile compared to SLNs; however, both showed a biphasic drug release pattern with an initial burst release in the first 5 hours followed by a prolonged release (99). Multifunctional NLCs of an anti-cancer drug (doxorubicin or paclitaxel) and siRNA were developed and characterized for delivery into the lungs by inhalation. NLCs were found to effectively deliver the drugs into lung cancer cells and significantly decreased the exposure of drugs to healthy organs when compared with intravenous injection (100). Overall, NLCs can be considered a better alternative compared to SLNs and can be easily manufactured on a large scale. Further studies identifying more lipids compatible with pulmonary fluids should be conducted.

17.5.5 Nanomicelles

Nanomicelles are colloidal carrier systems with size ranging from 5 to 200 nm. They are simple aggregates of either amphiphilic surfactant molecules (surfactant nanomicelles) or block copolymers (polymeric micelles). Based on the orientation of amphiphilic molecules, nanomicelles are categorized into normal or reverse micelles. Normal micelles are formed when the hydrophobic moieties form a cluster in the core and the hydrophilic part aligns outward in water, while an opposite alignment results in aggregates of reverse micelles. Normal micelles are used in the delivery of hydrophobic drugs, whereas reverse nanomicelles are used to deliver hydrophilic drugs (101). Cyclosporine A pharmacokinetics were examined by the respiratory or intravenous route in adult and young rats (102). In this study, intra-tracheal administration of saline suspension of cyclosporine A resulted in a bioavailability of 78.1 ± 6.9% with a peak plasma level at 30 minutes in adult rats. Whereas, the intra-tracheal instillation of cyclosporine A with surfactant micelles of Cremophor® EL resulted in a bioavailability of 77.4 ± 7.2% and 66.3 ± 4.3% in adult and young rats, respectively. The peak plasma level was attained after 5 minutes with the micellar formulation. In addition, ethanol solution of cyclosporine A delivered by aerosol showed a bioavailability of 80.1 ± 4.1% and the peak plasma level at 20 minutes. From the data it was evident that the micelle formulation was the fastest among all formulations evaluated, and the effect is attributed to the permeability of micelles into the lung tissues. In conclusion, pulmonary delivery can be utilized as an efficient route for delivery of cyclosporine A to limit autoimmune diseases and allergic reactions in transplantation procedure (102).

Polymeric micelles such as Pluronic® or poloxamer micelles have the ability to efficiently deliver a great variety of lipophilic drugs. They have a small size (<50 nm) with modifiable surface properties, which makes them suitable for drug delivery purposes. Recent studies have demonstrated the usefulness of Pluronic micelles in increasing the lung absorption of insulin and other poorly-water soluble drugs such as budesonide (103,104). Modification of the micellar surface with poly (ethylene glycol) (PEG) facilitates drug diffusion through the

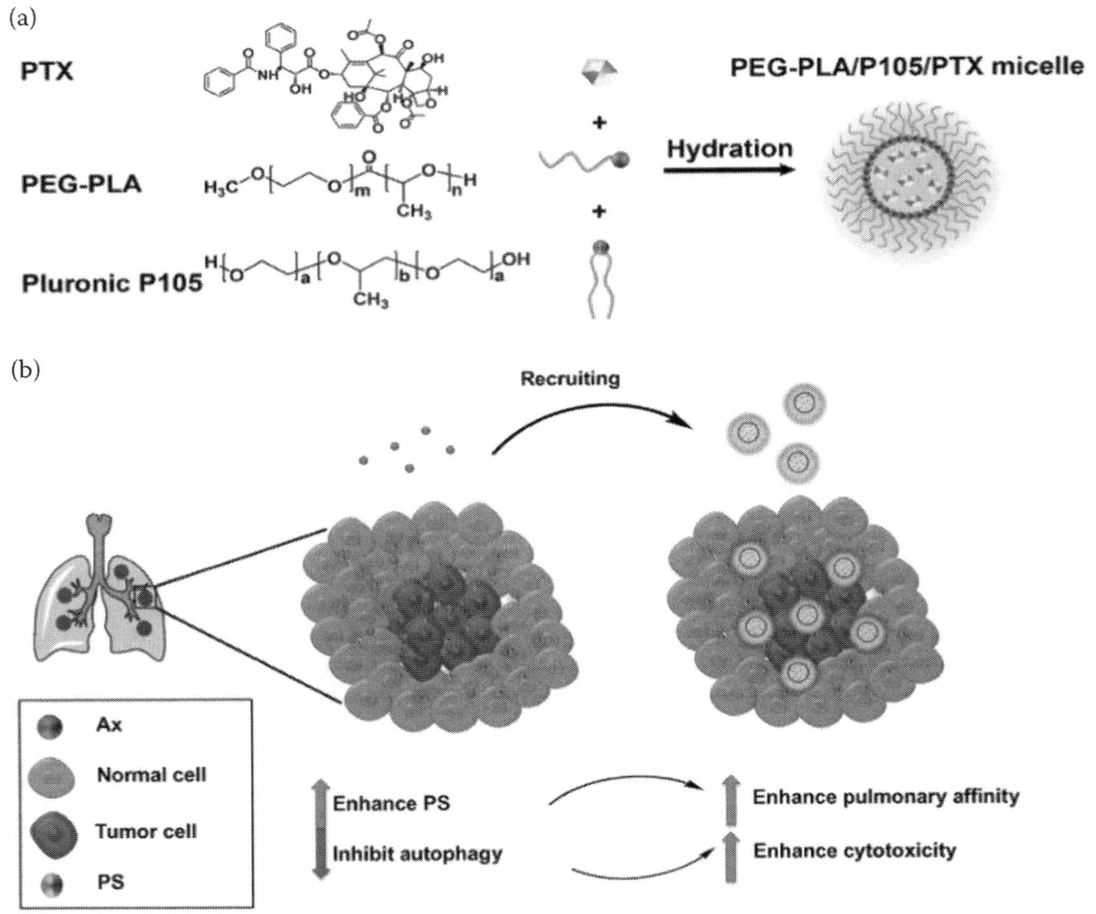

FIGURE 17.4 Image of Pulmonary-tendency of Paclitaxel Polymer Micelles in Response to Biological Functions of Ambroxol for Synergistic Lung Cancer Therapy. (A) Preparation of PEG-PLA/P105/Paclitaxel Micelles. (B) Ambroxol Can Up-regulate the Secretion of Pulmonary Surfactant and Inhibit Tumor Cell Autophagy; Subsequently, the PEG-PLA/P105/Paclitaxel Micelles Respond to the Biological Functions of Ambroxol with Stronger Lung Tendency and Cytotoxicity. Abbreviations: Ax, Ambroxol; PTX, Paclitaxel; PEG-PLA, Polyethyleneglycol-Polylactic Acid; PS, Pulmonary Surfactant. Reproduced with Permission from He et al. (106)

human airway mucus. For example, curcumin acetate-loaded in PEG-PLGA micelles showed prolonged pulmonary residence time and sustained drug release locally through efficient translocation across the air-blood (105). In a recent study, paclitaxel was entrapped in nanomicelles of polyethyleneglycol-polylactic acid (PEG-PLA) and Pluronic P105 and was tested for therapeutic efficacy in lung cancer. Ambroxol was delivered simultaneously to alter the microenvironment by up-regulating the pulmonary surfactant secretion and inhibiting autophagy of the lung, thereby promoting the lung accumulation and increasing cell-killing sensitivity of the micelles. The in vivo anti-tumor efficacy indicated higher cell-killing sensitivity in an orthotopic A549 lung tumor model (Figure 17.4) (106).

17.5.6 Microemulsion

Microemulsions are transparent, isotropic, thermodynamically stable systems made up of water, oil, and amphiphiles (surfactant and co-surfactant). Microemulsion possess a low surface tension with high spreading coefficient. Pulmonary drug delivery using emulsion or microemulsion formulations has been investigated in a few studies (107).

Microemulsion is considered as an aggregation colloid with droplet size ~100 nm. An aerosolized form of reverse water-in-CFC micelles stabilized by lecithin and loaded with peptides was studied (108). This system was stable and efficient to deliver peptides and proteins to the respiratory tract. However, its use is limited because CFCs were banned from commercial products. These formulations have the potential to be modified to utilize ozone friendly propellants such as HFA, hydrocarbons, or fluorocarbons in MDI applications (53). Reverse water-in-fluorocarbon emulsions of neat fluorocarbons were investigated for pulmonary delivery of drugs, genes, and hydrophilic bioactive materials to the lungs. A series of perfluoroalkylated amphiphiles with a dimorpholinophosphate polar head group were used to emulsify water-in-fluorocarbons. F8H11DMP was used in preparing both stable water-in-fluorocarbon emulsions and microemulsions. Water-in-fluorocarbon microemulsions (size ~12 nm) were formed at F8H11DMP concentrations higher than 5%. However, this study did not report the toxicity of the microemulsion formulation (109,110). In another study, lecithin inverse microemulsions containing dimethylethyleneglycol and hexane as models for dimethyl ether and propane were prepared and characterized for pulmonary delivery. The iodine solubilization method indicated microemulsion formation at ~10^{-4} to 10^{-5} molal lecithin. Dimethyl ether/lecithin demonstrated microemulsion characteristics similar to those in the model propellant. These microemulsion based formulations were reported to be stable for more than 4 weeks at room temperature with an internal aqueous diameter of around 3 µm and respirable fraction of approximately 36%. This study was the first that used lecithin reverse microemulsion for lung delivery of polar drug molecules (111). In conclusion, using reverse microemulsion (versus micelles) might enhance solubility and increase drug loading of a wide range of polar drug molecules. Aerosolized microemulsions delivered using MDIs are still under extensive evaluation.

17.5.7 Dendrimers

Interest in dendrimer-based pulmonary delivery is a rapidly growing domain due to their unique structural/physiological properties and biological attributes. Dendrimers are nanostructures of hyper-branched polymers with a well-defined, homogeneous, and monodisperse structure. A dendrimer consists of a symmetric core, an inner shell, and an outer shell. The unique hyper-branching nature helps in entrapping both hydrophilic and lipophilic drugs into their structure (112,113). Polyamidoamine (PAMAM) is the most notable dendrimer due to its high aqueous solubility and amenable surface functional groups. Studies indicate the safety of dendrimer formulations on the lung tissues with enhanced drug uptake. PEGylated polylysine dendrimers exhibited size-dependent absorption from the lungs with an inverse correlation to their molecular weight. PEGylated dendrimers with higher molecular weight (78 kDa) resulted in only 2% systemic absorption, while dendrimers with smaller molecular weight (<22 kDa) showed up to 20–30% absorption. Increasing the dendrimer molecular weight resulted in slower absorption and prolonged retention in the bronchoalveolar lavage fluid and lung tissue (114). This study concluded that small-molecular-weight PEGylated dendrimers deliver drugs to the blood following inhalation, while high-molecular-weight PEGylated dendrimers provide controlled delivery of medications to the blood or lung tissue. The surface charge of dendrimers also plays an important role in the pulmonary delivery. For example, enoxaparin is a low-molecular-weight heparin used to treat vascular thromboembolism. A study by Bai et al. in a rodent model concluded that positively charged PAMAM increased the relative bioavailability of enoxaparin by 40% without affecting the mucociliary transport rate, while a negatively charged dendrimer had no effect (115). In a recent study, the absorption-enhancing effects of PAMAM dendrimers were also established using fluorescein isothiocyanate-labeled dextrans of various molecular weights (116). Many studies have shown the potential of dendrimers in pulmonary delivery; however, these studies were performed in *in vitro* experimental models or animal models and cannot be extrapolated to humans (117). Further studies are needed to understand the biocompatibility and dendrimer-associated toxicity.

17.6 Enhancing Pulmonary Deposition of Nanocarrier Systems

17.6.1 Nanoparticle-in-Microparticles Formulations

Approximately 80% of nanoparticles of size less than 1 µm are easily exhaled from the lungs after inspiration exhaled due to the lack of inertial and gravitational forces needed for deposition (118,119). To overcome this problem, a group of researchers has developed large porous particles, known as Trojan particles. These are hybrid porous particles in the micrometer range composed of nanoparticles that could improve particle deposition in the lung (120). These porous particles once deposited in the lungs dissociate to yield nanoparticles for drug absorption. PLGA nanoparticles of tobramycin (size ~250 nm) were embedded in an inert microcarrier made of lactose (size ~11–12 µm) and used for pulmonary delivery. Further, the surface of nanoparticles was modified with hydrophilic polymers such as poly(vinyl alcohol) (PVA) and chitosan (CS). Biodistribution studies revealed that chitosan-modified particles were found in the upper airways, lining lung epithelial surfaces, while PVA-modified nanoparticles reached the deep lung (121). A study conducted by a group of researchers concluded that large porous particles have appropriate aerodynamic properties for alveolar deposition without being internalized by alveolar microphages (122). Moreover, nanoparticles encapsulated within micron-sized carriers that disintegrate on deposition could help to circumvent clearance of nanoparticles by alveolar macrophages (120).

17.6.2 Use of Mucoadhesive Polymers

Sustained release particles can be permanently removed from the respiratory system by an active clearance mechanism in the lungs. Therefore, many approaches have been evaluated to overcome this issue. One of the approaches was based on the use of a mucoadhesive formulation, but mucus turnover can impair the efficacy of this approach (123). However, mucoadhesive formulations using polymers such as glycol chitosan (GCS) and polyethylene glycol were investigated with significant improvements in drug bioavailability in pulmonary delivery (124,125). In addition to mucoadhesion properties, polymers such as chitosan can improve permeability across the epithelial membranes (126–128).

17.6.3 Use of Active Targeting Approach

The concepts of active or passive targeting is also used in pulmonary drug delivery (36). In passive targeting, the particle deposition within the lungs is strongly dependent on the aerodynamic diameter of aerosol particles as well as the patient and device factors like breathing or dose released (36,129). A variation of these parameters determines the site of delivery to either the alveoli or the airways in which the selectivity of the right and left sides is not applicable (6). An active targeting approach, such as magnetic targeting, utilizes nanomagnetosols (aerosol droplets containing super paramagnetic iron oxide nanoparticles), which can be guided by an external magnetic field to the desired location inside the lungs (130). Therefore, this approach can be utilized in targeting chemotherapeutic agents for which localized therapy would be highly favorable. In addition, several approaches have been investigated to improve alveolar macrophage targeting to increase intracellular bioavailability during treatment of tuberculosis (131,132). Targeting epithelial cells using gene-based therapy for cystic fibrosis, asthma (133) and alpha-1-antitrypsin deficiency is also being investigated (10,134,135).

17.7 Toxicity Concerns with Inhalational Nanomedicines

Inhalation of nanocarrier systems results in a local response due to increased macrophages as a result of the insoluble nature of the carrier. This is often considered as a non-alarming physiological response; however, long-term exposure of nanocarriers increases the deposition that may lead to intense inflammatory responses such as irritation, edema, cellular injury, phagocytosis impairment, and breakdown in defense mechanisms (136,137). The safety and compatibility of polymers (used in the preparation of nanocarriers) should be determined with lung fluids. Further, the processing techniques, formulation components, and residual solvents in the final product should be screened thoroughly in order to avoid toxic consequences (137). For example, carriers used in DPIs, such as sugars and cyclodextrins, are shown to cause broncho-constriction in hypersensitive individuals (138). Chronic use of formulation components such as absorption enhancers and enzyme inhibitors is known to produce immunogenicity, local irritation, and toxicity. Increased permeability from the use of absorption enhancers may also lead to the transport of other toxins and antigens across the epithelial barrier. The use of chemicals that are endogenous to lungs, such as dipalmitoylphosphatidylcholine, can significantly reduce the formulation-related toxicity.

Studies also support toxicity due to migration of inhaled nanoparticles from the lung epithelium into the systemic circulation and diffusion into organs such as the bone marrow, lymph nodes, spleen, heart, and to the central nervous system and ganglia (136). Toxicity of nanocarriers depends on the particle size. Ultrafine particles are found to be more toxic compared to larger particles due to their huge surface area, resulting in cytotoxicity, allergic reaction, or inflammation (139). In a recent study, radioactive iridium-192 nanoparticle aerosols (nominal sizes: 10, 15, 35, and 75 nm) were administered in female Wistar-Kyoto rats via nose inhalation. The content of iridium-192 in the lung, head, gastrointestinal tract, and various other organs/tissues was measured. This study reported that the majority of iridium-192 was in the lung and GIT immediately post-exposure. For the smallest particle size (10 nm), the levels of iridium-192 were similar in both the lung and GIT, while for others the lung concentration was higher. The concentration of iridium-192 in the head and other tissues was significantly lower (140). Particles of size 0.5–3 μm are predominantly cleared from the alveoli by macrophages, while smaller particles of size less than 260 nm eludes the phagocytosis and are not easily cleared from the pulmonary region (141). Such nanocarriers can remain in the lungs, mainly in the alveolar cells or in the interstitium, for up to 700 days causing toxicity issues. Hence, it is important to understand the fate of inhaled nanoparticles in the alveoli, to establish the dose-response relationship of both drug-loaded nanoparticles and environmental toxins.

17.8 Conclusion and Future Directions

Pulmonary drug delivery has received more attention as an alternative and non-invasive means for local treatment of major lung diseases. However, this route of administration poses some challenges for medicinal and formulation chemists mainly in terms of drug absorption, control of particle size, suitable toxicology models, and patient compliance. Using nanotechnology, it is possible to overcome the challenges of conventional drug delivery systems. Nanotechnology aids in the preparation of drug carriers with better properties that maximize the drug delivery to the target site. Nanocarrier properties such as material, size, and surface modification can be tailored to stimulate/inhibit specific immune reactions for treating immune diseases such as allergic asthma. However, this could only be achieved through collaborative and interdisciplinary research between biologists, physicists, chemists, and material scientists. Moreover, thorough characterization and control of the nanocarrier synthesis procedure is required for assuring high-quality nanocarrier batches with good reproducibility.

Particle size of nanocarriers must be carefully optimized as it affects both dose deposition and distribution of aerosol particles in the lungs. Nanosized particles distribute more in peripheral airways with less drug deposited in a unit surface area, while micron-sized particles deposit predominantly in the central airways with more drug deposited in a unit surface area. Dose deposition and distribution of aerosol particles affect the overall therapeutic efficacy of drugs. In addition, the control of particle size is also an important safety consideration as humid environment (~99.5% RH) in the respiratory system promotes aggregation of drug particles, potentially causing irritancy issues in addition to reducing efficacy. The aggregation behavior mostly affects the settling velocity and the location of the deposition of hygroscopic particles. Suitable animal models are required to understand the aggregation behavior of particles with reasonable accuracy. The development of animal models in which formulations can be administered intra-nasally or intra-tracheally requires a lot of planning by technical experts. It is also necessary to ensure that animals breathe in during the drug administration while monitoring other factors such as dosage timing, volume, and the depth of anesthesia. These studies can provide any early indication of any possible signs of lung irritation that can guide developmental decisions for minimizing the potential of unexpected and costly failures at the later stages.

The outcome of pulmonary delivery using an inhalation device also depends on the user, the device, and the drug. Most drugs applied by this route are meant to be inhaled deeply into the lungs with a proper aerosolization of the drug. Alterations in either drug formulation or device design or variation in the way patients handle and use the devices, can result in dose variability and affect drug bioavailability. Inhalation devices must be engineered to precisely deliver an accurate dose from the first use to the last. The drug must be formulated to remain stable across a range of environments, such as variations in temperature and humidity, and resist significant physical stresses. Currently, researchers are investigating and testing more sophisticated devices and aerosolized particles to optimize pulmonary drug delivery, minimize clearance, and improve drug targeting within the lungs. Overall, these developments will likely attract new investments which help tailor nanocarrier systems for translating technological advances into clinical benefits.

Acknowledgments

The authors would like to acknowledge Mr. Ahmed Fahad for his contribution.

REFERENCES

1. Groneberg D, et al. Distribution and function of the peptide transporter PEPT2 in normal and cystic fibrosis human lung. *Thorax*. 2002; 57(1): 55–60.
2. Groneberg D, et al. Fundamentals of pulmonary drug delivery. *Respiratory Medicine*. 2003; 97(4): 382–387.
3. Groneberg DA, et al. Localization of the peptide transporter PEPT2 in the lung: implications for pulmonary oligopeptide uptake. *The American Journal of Pathology*. 2001; 158(2): 707–714.
4. Zheng Y, et al. Pulmonary delivery of a dopamine D-1 agonist, ABT-431, in dogs and humans. *International Journal of Pharmaceutics*. 1999; 191(2): 131–140.
5. Ward ME, et al. Morphine pharmacokinetics after pulmonary administration from a novel aerosol delivery system. *Clinical Pharmacology & Therapeutics*. 1997; 62(6): 596–609.
6. Ruge CA, Kirch J, and Lehr C-M. Pulmonary drug delivery: From generating aerosols to overcoming biological barriers—Therapeutic possibilities and technological challenges. *The Lancet Respiratory Medicine*. 2013; 1(5): 402–413.
7. Noymer P, et al. Pulmonary delivery of therapeutic compounds for treating CNS disorders. *Therapeutic Delivery*. 2011; 2(9): 1125–1140.
8. Islam N and Rahman S. Improved treatment of nicotine addiction and emerging pulmonary drug delivery. *Drug Discoveries & Therapeutics*. 2012; 6(3): 123–132.
9. Lalan M, et al. Inhalation drug therapy: Emerging trends in nasal and pulmonary drug delivery. In: Misra A and Shahiwala A. (eds.) Novel Drug Delivery Technologies: Innovative Strategies for Drug Re-positioning. Springer Singapore, Singapore. 2019; 291–333.
10. Xi J and Talaat M. Nanoparticle deposition in rhythmically moving acinar models with interalveolar septal apertures. *Nanomaterials*. 2019; 9(8): 1126.
11. Patton JS, Fishburn CS, and Weers JG. The lungs as a portal of entry for systemic drug delivery. *Proceedings of the American Thoracic Society*. 2004; 1(4): 338–344.
12. Farr SJ, et al. Pulmonary insulin administration using the AERx® system: Physiological and physicochemical factors influencing insulin effectiveness in healthy fasting subjects. *Diabetes Technology & Therapeutics*. 2000; 2(2): 185–197.
13. Heinemann L and Parkin CG. Rethinking the viability and utility of inhaled insulin in clinical practice. *Journal of Diabetes Research*. 2018; .
14. Siekmeier R and Scheuch G. Treatment of systemic diseases by inhalation of biomolecule aerosols. *Journal of Physiology and Pharmacology*. 2009; 60(suppl 5): 15–26.
15. Tuinman PR, et al. Nebulized anticoagulants for acute lung injury-a systematic review of preclinical and clinical investigations. *Critical Care*. 2012; 16(2): R70.
16. Jemal A, et al. Global cancer statistics. *CA: A Cancer Journal for Clinicians*. 2011; 61(2): 69–90.
17. Carvalho TC, Carvalho SR, and McConville JT. Formulations for pulmonary administration of anticancer agents to treat lung malignancies. *Journal of Aerosol Medicine and Pulmonary Drug Delivery*. 2011; 24(2): 61–80.
18. Borghardt JM, Kloft C, and Sharma A. Inhaled therapy in respiratory disease: The complex interplay of pulmonary kinetic processes. *Canadian Respiratory Journal*. 2018; 2018: 2732017.
19. Forbes B, et al. Challenges in inhaled product development and opportunities for open innovation. *Advanced Drug Delivery Reviews*. 2011; 63(1): 69–87.
20. Patton JS. Mechanisms of macromolecule absorption by the lungs. *Advanced Drug Delivery Reviews*. 1996; 19(1): 3–36.
21. Patton JS, et al. The particle has landed—Characterizing the fate of inhaled pharmaceuticals. *Journal of Aerosol Medicine and Pulmonary Drug Delivery*. 2010; 23(S2): S-71–S-87.
22. Olsson B, et al. Pulmonary drug metabolism, clearance, and absorption. In: Smyth, H and Hickey, A (eds.) Controlled Pulmonary Drug Delivery. Springer, New York, USA. 2011; 21–50.
23. Morales, JO, Peters JI, and Williams RO. Surfactants: Their critical role in enhancing drug delivery to the lungs. *Therapeutic Delivery*. 2011; 2(5): 623–641.
24. Wiedmann T, Bhatia R, and Wattenberg L. Drug solubilization in lung surfactant. *Journal of Controlled Release*. 2000; 65(1): 43–47.
25. Pérez-Gil J. Structure of pulmonary surfactant membranes and films: the role of proteins and lipid–protein interactions. *Biochimica et Biophysica Acta (BBA)-Biomembranes*. 2008; 1778(7): 1676–1695.
26. Antunes MB, Gudis DA, and Cohen NA. Epithelium, cilia, and mucus: Their importance in chronic rhinosinusitis. *Immunology and Allergy Clinics of North America*. 2009; 29(4): 631–643.
27. Sanders N, et al. Extracellular barriers in respiratory gene therapy. *Advanced Drug Delivery Reviews*. 2009; 61(2): 115–127.
28. Geiser M, et al. Assessment of particle retention and clearance in the intrapulmonary conducting airways of hamster lungs with the fractionator. *Journal of Microscopy*. 1990; 160(1): 75–88.
29. Geiser M and Kreyling WG. Deposition and biokinetics

of inhaled nanoparticles. *Particle and Fibre Toxicology*. 2010; 7(1): 1.
30. Schürch S, et al. Surfactant displaces particles toward the epithelium in airways and alveoli. *Respiration Physiology*. 1990; 80(1): 17–32.
31. Harishchandra RK, Saleem M, and Galla H-J. Nanoparticle interaction with model lung surfactant monolayers. *Journal of The Royal Society Interface*. 2010; 7(Suppl 1): S15–S26.
32. Chroneos ZC, Sever-Chroneos Z, and Shepherd VL. Pulmonary surfactant: An immunological perspective. *Cellular Physiology and Biochemistry*. 2009; 25(1): 13–26.
33. Ruge CA, et al. Uptake of nanoparticles by alveolar macrophages is triggered by surfactant protein A. *Nanomedicine: Nanotechnology, Biology and Medicine*. 2011; 7(6): 690–693.
34. Chono S, et al. Influence of particle size on drug delivery to rat alveolar macrophages following pulmonary administration of ciprofloxacin incorporated into liposomes. *Journal of Drug Targeting*. 2006; 14(8): 557–566.
35. Anselmo AC, Gokarn Y, and Mitragotri S. Non-invasive delivery strategies for biologics. *Nature Reviews Drug Discovery*. 2018; 18: 19–40.
36. Dolovich MB and Dhand R. Aerosol drug delivery: Developments in device design and clinical use. *The Lancet*. 2011; 377(9770): 1032–1045.
37. Labiris NR and Dolovich MB. Pulmonary drug delivery. Part II: the role of inhalant delivery devices and drug formulations in therapeutic effectiveness of aerosolized medications. *British Journal of Clinical Pharmacology*. 2003; 56(6): 600–612.
38. Pilcer G and Amighi K. Formulation strategy and use of excipients in pulmonary drug delivery. *International Journal of Pharmaceutics*. 2010; 392(1): 1–19.
39. Patton JS and Byron PR. Inhaling medicines: Delivering drugs to the body through the lungs. *Nature Reviews Drug Discovery*. 2007; 6(1): 67–74.
40. Corboz MR, et al. Therapeutic administration of inhaled INS1009, a treprostinil prodrug formulation, inhibits bleomycin-induced pulmonary fibrosis in rats. *Pulmonary Pharmacology & Therapeutics*. 2018; 49: 95–103.
41. Yamamoto A, Umemori S, and Muranishi S. Absorption enhancement of intrapulmonary administered insulin by various absorption enhancers and protease inhibitors in rats. *Journal of Pharmacy and Pharmacology*. 1994; 46(1): 14–18.
42. Yamamoto A, Fujita T, and Muranishi S. Pulmonary absorption enhancement of peptides by absorption enhancers and protease inhibitors. *Journal of Controlled Release*. 1996; 41(1–2): 57–67.
43. Islam N and Cleary MJ. Developing an efficient and reliable dry powder inhaler for pulmonary drug delivery–A review for multidisciplinary researchers. *Medical Engineering & Physics*. 2012; 34(4): 409–427.
44. Haque S, et al. A comparison of the lung clearance kinetics of solid lipid nanoparticles and liposomes by following the 3H-labelled structural lipids after pulmonary delivery in rats. *European Journal of Pharmaceutics and Biopharmaceutics*. 2018; 125: 1–12.
45. Eleftheriadis GK, et al. Polymer–Lipid microparticles for pulmonary delivery. *Langmuir*. 2018; 34(11): 3438–3448.
46. Takahashi T and Kubo H. The role of microparticles in chronic obstructive pulmonary disease. *International Journal of Chronic Obstructive Pulmonary Disease*. 2014; 9: 303.
47. Hirn S, et al. Particle size-dependent and surface charge-dependent biodistribution of gold nanoparticles after intravenous administration. *European Journal of Pharmaceutics and Biopharmaceutics*. 2011; 77(3): 407–416.
48. Gessner A, et al. Nanoparticles with decreasing surface hydrophobicities: influence on plasma protein adsorption. *International Journal of Pharmaceutics*. 2000; 196(2): 245–249.
49. Lim YH, et al. Polymeric nanoparticles in development for treatment of pulmonary infectious diseases. *Wiley Interdisciplinary Reviews: Nanomedicine and Nanobiotechnology*. 2016; 8(6): 842–871.
50. Card JW, et al. Pulmonary applications and toxicity of engineered nanoparticles. *American Journal of -Physiology-Lung Cellular and Molecular Physiology*. 2008; 295(3): L400–L411.
51. Tena AF and Clarà PC. Deposition of inhaled particles in the lungs. *Archivos de Bronconeumología (English Edition)*. 2012; 48(7): 240–246.
52. Kato T, et al. Evidence that exogenous substances can be phagocytized by alveolar epithelial cells and transported into blood capillaries. *Cell and Tissue Research*. 2003; 311(1): 47–51.
53. Courrier H, Butz N, and Vandamme TF. Pulmonary drug delivery systems: recent developments and prospects. *Critical Reviews™ in Therapeutic Drug Carrier Systems*. 2002; 19(4–5).
54. Jabbal S, Poli G, and Lipworth B. Does size really matter?: Relationship of particle size to lung deposition and exhaled fraction. *Journal of Allergy and Clinical Immunology*. 2017; 139(6): 2013–2014.e1.
55. El-Sherbiny IM, El-Baz NM, and Yacoub MH. Inhaled nano- and microparticles for drug delivery. *Global Cardiology Science & Practice*. 2015; 2015: 2.
56. Paranjpe M and Müller-Goymann CC. Nanoparticle-mediated pulmonary drug delivery: A review. *International Journal of Molecular Sciences*. 2014; 15(4): 5852–5873.
57. Champion JA, Walker A, and Mitragotri S. Role of particle size in phagocytosis of polymeric microspheres. *Pharmaceutical Research*. 2008; 25(8): 1815–1821.
58. Makino K, et al. Phagocytic uptake of polystyrene microspheres by alveolar macrophages: Effects of the size and surface properties of the microspheres. *Colloids and Surfaces B: Biointerfaces*. 2003; 27(1): 33–39.
59. Champion JA and Mitragotri S. Role of target geometry in phagocytosis. *Proceedings of the National Academy of Sciences*. 2006; 103(13): 4930–4934.
60. Zellnitz S, et al. Impact of drug particle shape on permeability and cellular uptake in the lung. *European Journal of Pharmaceutical Sciences*. 2019; 139: 105065.
61. Oberdorster G. Lung particle overload: Implications for occupational exposures to particles. *Regulatory Toxicology and Pharmacology*. 1995; 21(1): 123–135.
62. Choi YH and Han H-K. Nanomedicines: Current status and future perspectives in aspect of drug delivery and pharmacokinetics. *Journal of Pharmaceutical Investigation*. 2018; 48(1): 43–60.

63. Mansour HM, Rhee Y-S, and Wu X. Nanomedicine in pulmonary delivery. *International Journal of Nanomedicine*. 2009; 4: 299.
64. Dandekar P, Venkataraman C, and Mehra A. Pulmonary targeting of nanoparticle drug matrices. *Journal of Aerosol Medicine and Pulmonary Drug Delivery*. 2010; 23(6): 343–353.
65. Trapani A, et al. Nanocarriers for respiratory diseases treatment: Recent advances and current challenges. *Current Topics in Medicinal Chemistry*. 2014; 14(9): 1133–1147.
66. Terzano C, et al. Non-phospholipid vesicles for pulmonary glucocorticoid delivery. *European Journal of Pharmaceutics and Biopharmaceutics*. 2005; 59(1): 57–62.
67. Lim YH, et al. Polymeric nanoparticles in development for treatment of pulmonary infectious diseases. *Wiley Interdisciplinary Reviews. Nanomedicine and Nanobiotechnology*. 2016; 8(6): 842–871.
68. Smola M, Vandamme T, and Sokolowski A. Nanocarriers as pulmonary drug delivery systems to treat and to diagnose respiratory and non respiratory diseases. *International Journal of Nanomedicine*. 2008; 3(1): 1–19.
69. van Rijt SH, Bein T, and Meiners S. Medical nanoparticles for next generation drug delivery to the lungs. *European Respiratory Journal*. 2014.
70. Yadollahi R, Vasilev K, and Simovic S. Nanosuspension technologies for delivery of poorly soluble drugs. *Journal of Nanomaterials*. 2015; 44: 765–774
71. Jacob S, Nair AB, and Shah J. Emerging role of nanosuspensions in drug delivery systems. *Biomaterials Research*. 2020; 24(1): 3.
72. Zhang Y and Zhang J. Preparation of budesonide nanosuspensions for pulmonary delivery: Characterization, in vitro release and in vivo lung distribution studies. *Artificial Cells, Nanomedicine, and Biotechnology*. 2016; 44(1): 285–289.
73. Fu T-T, et al. Fluticasone propionate nanosuspensions for sustained nebulization delivery: An in vitro and in vivo evaluation. *International Journal of Pharmaceutics*. 2019; 572: 118839.
74. Hernández-Trejo N, et al. Characterization of nebulized buparvaquone nanosuspensions—Effect of nebulization technology. *Journal of Drug Targeting*. 2005; 13(8–9): 499–507.
75. Rossi I, et al. Nebulized coenzyme Q10 nanosuspensions: A versatile approach for pulmonary antioxidant therapy. *European Journal of Pharmaceutical Sciences*. 2018; 113: 159–170.
76. Patel VR and Agrawal YK. Nanosuspension: An approach to enhance solubility of drugs. *Journal of Advanced Pharmaceutical Technology & Research*. 2011; 2(2): 81–87.
77. Zeng XM, Martin GP, and Marriott C. The controlled delivery of drugs to the lung. *International Journal of Pharmaceutics*. 1995; 124(2): 149–164.
78. Kellaway IW and Farr SJ. Liposomes as drug delivery systems to the lung. *Advanced Drug Delivery Reviews*. 1990; 5(1–2): 149–161.
79. Saari M, et al. Pulmonary distribution and clearance of two beclomethasone liposome formulations in healthy volunteers. *International Journal of Pharmaceutics*. 1999; 181(1): 1–9.
80. Elhissi A. Liposomes for pulmonary drug delivery: The role of formulation and inhalation device design. *Current Pharmaceutical Design*. 2017; 23(3): 362–372.
81. Ari A. Jet, ultrasonic, and mesh nebulizers: an evaluation of nebulizers for better clinical outcomes. *Eurasian Journal Pulmonol.*2014;16: 1–7.
82. Elhissi AMA and Taylor KMG. Delivery of liposomes generated from proliposomes using air-jet, ultrasonic, and vibrating-mesh nebulisers. *Journal of Drug Delivery Science and Technology*. 2005; 15(4): 261–265.
83. Lehofer B, et al. Impact of atomization technique on the stability and transport efficiency of nebulized liposomes harboring different surface characteristics. *European Journal of Pharmaceutics and Biopharmaceutics: Official Journal of Arbeitsgemeinschaft fur Pharmazeutische Verfahrenstechnik e.V.* 2014; 88(3): 1076–1085.
84. Konduri KS, et al. The use of sterically stabilized liposomes to treat asthma. In: Methods in Enzymology. Elsevier. 2005; 413–427.
85. Rudokas M, et al. Liposome delivery systems for inhalation: A critical review highlighting formulation issues and anticancer applications. *Medical Principles and Practice*. 2016; 25(Suppl. 2): 60–72.
86. Bachu RD, et al. Ocular drug delivery barriers—Role of nanocarriers in the treatment of anterior segment ocular diseases. *Pharmaceutics*. 2018; 10(1): 28.
87. Menon JU, et al. Polymeric nanoparticles for pulmonary protein and DNA delivery. *Acta Biomaterialia*. 2014; 10(6): 2643–2652.
88. Speiser P. Poorly soluble drugs: A challenge in drug delivery. In: Muller , RH, Benita S, and Bohm BHL (eds.) Emulsions and Nanosuspensions for the Formulation of Poorly Soluble Drugs. Medpharm Scientific Publishers, Stuttgart. 1998; 15–28.
89. Sham JOH, et al. Formulation and characterization of spray-dried powders containing nanoparticles for aerosol delivery to the lung. *International Journal of Pharmaceutics*. 2004; 269(2): 457–467.
90. Yang R, et al. Lung-specific delivery of paclitaxel by chitosan-modified PLGA nanoparticles via transient formation of microaggregates. *Journal of Pharmaceutical Science*. 2009; 98(3): 970–984.
91. Craparo EF, et al. Pegylated polyaspartamide–Polylactide-based nanoparticles penetrating cystic fibrosis artificial mucus. *Biomacromolecules*. 2016; 17(3): 767–777.
92. Beck-Broichsitter M, et al. Biophysical investigation of pulmonary surfactant surface properties upon contact with polymeric nanoparticles in vitro. *Nanomedicine*. 2011; 7(3): 341–350.
93. Woods A, et al. Albumin nanoparticles for drug delivery to the lungs-In vitro investigation of biodegradation as a mechanism of clearance. *Journal of Aerosol Medicine and Pulmonary Drug Delivery*. 2014; 27(4): A15.
94. Muèller RH, Maèder K, and Gohla S. Solid lipid nanoparticles (SLN) for controlled drug delivery–a review of the state of the art. *European Journal of Pharmaceutics and Biopharmaceutics*. 2000; 50(1): 161–177.
95. Zhao Y, et al. Solid lipid nanoparticles for sustained pulmonary delivery of Yuxingcao essential oil: Preparation, characterization and in vivo evaluation. *International Journal of Pharmaceutics*. 2017; 516(1–2): 364–371.

96. Kaur IP, Rana C, and Singh H. Development of effective ocular preparations of antifungal agents. *Journal of Ocular Pharmacology and Therapeutics*. 2008; 24(5): 481–494.
97. Mishra V, et al. Solid lipid nanoparticles: Emerging colloidal nano drug delivery systems. *Pharmaceutics*. 2018; 10(4): 191.
98. Khan S, et al. Nanostructured lipid carriers: An emerging platform for improving oral bioavailability of lipophilic drugs. *International Journal of Pharmaceutical Investigation*. 2015; 5(4): 182–191.
99. Islan GA, et al. Smart lipid nanoparticles containing levofloxacin and DNase for lung delivery. Design and characterization. *Colloids and Surfaces B: Biointerfaces*. 2016; 143: 168–176.
100. Taratula O, et al. Nanostructured lipid carriers as multifunctional nanomedicine platform for pulmonary co-delivery of anticancer drugs and siRNA. *Journal of Controlled Release*. 2013; 171(3): 349–357.
101. Bachu RD, et al. Ocular drug delivery barriers-Role of nanocarriers in the treatment of anterior segment ocular diseases. *Pharmaceutics*. 2018; 10(1): 28.
102. Taljanski W, et al. Pulmonary delivery of intratracheally instilled and aerosolized cyclosporine A to young and adult rats. *Drug Metabolism and Disposition*. 1997; 25(8): 917–920.
103. Pellosi DS, et al. In vitro/in vivo investigation on the potential of Pluronic® mixed micelles for pulmonary drug delivery. *European Journal of Pharmaceutics and Biopharmaceutics*. 2018; 130: 30–38.
104. Andrade F, et al. Pharmacological and toxicological assessment of innovative self-assembled polymeric micelles as powders for insulin pulmonary delivery. *Nanomedicine*. 2016; 11(17): 2305–2317.
105. Wijagkanalan W, et al. Efficient targeting to alveolar macrophages by intratracheal administration of mannosylated liposomes in rats. *Journal of Controlled Release*. 2008; 125(2): 121–130.
106. He W, et al. Pulmonary-affinity paclitaxel polymer micelles in response to biological functions of ambroxol enhance therapeutic effect on lung cancer. *International Journal of Nanomedicine*. 2020; 15: 779.
107. Lawrence MJ and Rees GD. Microemulsion-based media as novel drug delivery systems. *Advanced Drug Delivery Reviews*. 2012; 64: 175–193.
108. Evans RM and SJ Farr. Aerosol formulations including proteins and peptides solubilized in reverse micelles and process for making the aerosol formulations. 1993. Google Patents.Patent: 5,230,884.
109. Patel N, Marlow M, and Lawrence M. Microemulsions: A novel pMDI formulation. In: Morton D (ed.) Drug Delivery to the Lungs IX. The Aerosol Society, Bristol, London. 1998; 160–163.
110. Courrier HM, Vandamme TF, and Krafft MP. Reverse water-in-fluorocarbon emulsions and microemulsions obtained with a fluorinated surfactant. *Colloids and Surfaces A: Physicochemical and Engineering Aspects*. 2004; 244(1): 141–148.
111. Sommerville ML, et al. Lecithin inverse microemulsions for the pulmonary delivery of polar compounds utilizing dimethylether and propane as propellants. *Pharmaceutical Development and Technology*. 2000; 5(2): 219–230.
112. Gaudana R, et al. Recent perspectives in ocular drug delivery. *Pharmaceutical Research*. 2009; 26(5): 1197.
113. Nasr M, et al. PAMAM dendrimers as aerosol drug nanocarriers for pulmonary delivery via nebulization. *International Journal of Pharmaceutics*. 2014; 461(1–2): 242–250.
114. Ryan GM, et al. Pulmonary administration of PEGylated polylysine dendrimers: Absorption from the lung versus retention within the lung is highly size-dependent. *Molecular Pharmaceutics*. 2013; 10(8): 2986–2995.
115. Bai S, Thomas C, and Ahsan F. Dendrimers as a carrier for pulmonary delivery of enoxaparin, a low-molecular weight heparin. *Journal of Pharmaceutical Sciences*. 2007; 96(8): 2090–2106.
116. Lu J, et al. The effect of absorption-enhancement and the mechanism of the PAMAM dendrimer on poorly absorbable drugs. *Molecules (Basel, Switzerland)*. 2018; 23(8): 2001.
117. Santos A, Veiga F, and Figueiras A. Dendrimers as pharmaceutical excipients: synthesis, properties, toxicity and biomedical applications. *Materials*. 2020; 13(1): 65.
118. Heyder J, et al., Deposition of particles in the human respiratory tract in the size range 0.005–15 μm. *Journal of Aerosol Science*. 1986; 17(5): 811–825.
119. Heyder J and Rudolf G. Mathematical models of particle deposition in the human respiratory tract. *Journal of Aerosol Science*. 1984; 15(6): 697–707.
120. Tsapis N, et al., Trojan particles: Large porous carriers of nanoparticles for drug delivery. *Proceedings of the National Academy of Sciences*. 2002; 99(19): 12001–12005.
121. Ungaro F, et al. Dry powders based on PLGA nanoparticles for pulmonary delivery of antibiotics: Modulation of encapsulation efficiency, release rate and lung deposition pattern by hydrophilic polymers. *Journal of Controlled Release*. 2012; 157(1): 149–159.
122. Edwards DA, Ben-Jebria A, and Langer R. Recent advances in pulmonary drug delivery using large, porous inhaled particles. *Journal of Applied Physiology*. 1998; 85(2): 379–385.
123. Lehr C-M, et al. An estimate of turnover time of intestinal mucus gel layer in the rat in situ loop. *International Journal of Pharmaceutics*. 1991; 70(3): 235–240.
124. Makhlof A, et al. Nanoparticles of glycol chitosan and its thiolated derivative significantly improved the pulmonary delivery of calcitonin. *International Journal of Pharmaceutics*. 2010; 397(1): 92–95.
125. Wang YY, et al. Addressing the PEG mucoadhesivity paradox to engineer nanoparticles that "slip" through the human mucus barrier. *Angewandte Chemie International Edition*. 2008; 47(50): 9726–9729.
126. Andrews GP, Laverty TP, and Jones DS. Mucoadhesive polymeric platforms for controlled drug delivery. *European Journal of Pharmaceutics and Biopharmaceutics*. 2009; 71(3): 505–518.
127. Bernkop-Schnürch A. Thiomers: A new generation of mucoadhesive polymers. *Advanced Drug Delivery Reviews*. 2005; 57(11): 1569–1582.
128. Thanou M, Verhoef J, and Junginger H. Oral drug absorption enhancement by chitosan and its derivatives. *Advanced Drug Delivery Reviews*. 2001; 52(2): 117–126.
129. Carvalho TC, Peters JI, and Williams RO. Influence of particle size on regional lung deposition–What evidence is there? *International Journal of Pharmaceutics*. 2011; 406(1): 1–10.

130. Dames P, et al. Targeted delivery of magnetic aerosol droplets to the lung. *Nature Nanotechnology*. 2007; 2(8): 495–499.
131. Griffiths G, et al. Nanobead-based interventions for the treatment and prevention of tuberculosis. *Nature Reviews Microbiology*. 2010; 8(11): 827–834.
132. Lawlor C, et al. Cellular targeting and trafficking of drug delivery systems for the prevention and treatment of MTb. *Tuberculosis*. 2011; 91(1): 93–97.
133. Mathieu M and Demoly P. Gene therapy for asthma. In: Factor P (ed.)Gene Therapy for Acute and Acquired Diseases. Springer. 2001; 107–126.
134. Aneja MK, et al. Targeted gene delivery to the lung. *Expert Opinion on Drug Delivery*. 2009; 6(6): 567–583.
135. Gill D, et al. The development of gene therapy for diseases of the lung. *Cellular and Molecular Life Sciences*. 2004; 61(3): 355–368.
136. Ahmad J, et al. Nanotechnology-based inhalation treatments for lung cancer: State of the art. *Nanotechnology, Science and Applications*. 2015; 8: 55.
137. Patil JS and Sarasija S. Pulmonary drug delivery strategies: A concise, systematic review. *Lung India: Official Organ of Indian Chest Society*. 2012; 29(1): 44–49.
138. Hickey AJ and Garcia-Contreras L. Immunological and toxicological implications of short-term studies in animals of pharmaceutical aerosol delivery to the lungs: Relevance to humans. *Critical Reviews™ in Therapeutic Drug Carrier Systems*. 2001; 18(4): 387–431.
139. Lanone S and Boczkowski J. Biomedical applications and potential health risks of nanomaterials: Molecular mechanisms. *Current Molecular Medicine*. 2006; 6(6): 651–663.
140. Buckley A, et al. Size-dependent deposition of inhaled nanoparticles in the rat respiratory tract using a new nose-only exposure system. *Aerosol Science and Technology*. 2016; 50(1): 1–10.
141. Yang, W, Peters JI, and Williams III RO. Inhaled nanoparticles—A current review. *International Journal of Pharmaceutics*. 2008; 356(1-2): 239–247.

18

Nanomedicine for the Management of Pulmonary Disorders

Pompy Patowary[1,2], Manash Pratim Pathak[1,2], Pronobesh Chattopadhyay[2], and Kamaruz Zaman[1]
[1]Department of Pharmaceutical Sciences, Dibrugarh University, Dibrugarh, Assam, India
[2]Pharmaceutical Technology Division, Defence Research Laboratory, Assam, India

18.1 Introduction

The National Nanotechnology Initiative defines nanotechnology as the control and understanding of matter at dimensions between approximately 1 and 100 nm, where unique phenomena enable novel applications. The emerging field of nanotechnology includes a wide variety of sciences and is fundamentally based on nano-sized engineered substances with all the physicochemical properties of that specific element in improvised form (1,2). The miniaturization of systems differs greatly from its bulk-sized counterparts with many advantages, viz. new material properties, increased conductivity, enlarged surface area to volume ratio, decreased transport time of molecules, high linear flow rate and optical properties (3). Other than its extensive use in various fields of science and technology, one of the most revolutionary applications of nanotechnology is nanomedicine (4). Since the early 2000s, nanomedicine has emerged as a fascinating interdisciplinary field of research based on the use of nanoscale materials or devices for diagnosis and drug delivery as well as therapeutic methods for the development of nanopharmaceuticals. Driven by the massive progress in the nanomedicine field, solutions to clinical problems have emerged in the form of novel drugs and reformulation of already existing drugs with improved biopharmaceutical features, increased efficacy and minimized drug toxicity (5). Although nanomedicine is presently in its nascent stage of development, various nanotechnology-based pharmaceutical products have already hit the market. Ranging from quantum dots for molecular imaging and diagnostics (6) to therapy using nanocarriers and integrated medical nanosystems (7), nanomedical developments may perform multifaceted repair actions at the cellular level inside the body. Moreover, the permeability of nanoparticles (NPs) through cell membranes, including the blood–brain barrier (8), make them easily accessible to the cells and translocation via blood and lymph (9), thus making them an attractive tool for drug-delivery treatment and nutraceuticals. Applied nanotechnology can be harnessed for generating new concepts for dealing with challenging diseases like cancer, diabetes, lung and cardiovascular problems, inflammatory or infectious diseases, and neurological disorders.

Respiratory diseases, ranging from self-limiting ailments from the common cold to life-threatening chronic inflammatory lung diseases, have imposed an immense worldwide health burden, making them the most common lethal neoplasm in the world. Moreover, respiratory diseases make up 5 of the 30 most common causes of death and the numbers are likely to increase in the near future. The World Health Organisation (WHO) estimates that by the year 2030, the potentially fatal respiratory diseases, viz. tuberculosis (TB), pneumonia, chronic obstructive pulmonary disorder (COPD) and lung cancer, will account for about one in five deaths worldwide. It is noteworthy to mention that prevention, control and cure of these diseases and promotion of respiratory health must be a top priority in global decision-making in the health sector.

In recent years, nanomedicine has gained immense popularity for the targeted delivery of therapeutic and diagnostic compounds to the lung (10), mostly because it is non-invasive nature and bypasses first-pass metabolism (11). But the delivery of inhaled medicines to the lungs has proven to be highly inefficient due to large depositional losses and aerosols with a size range of about 40 nm, which lack effective deposition in the lungs and are exhaled out (12). Also, nanomedicine delivery to lungs can be challenging because of low predetermined lung–site deposition efficiencies. Again, the effectiveness of pulmonary delivery of protein drugs is affected by biobarriers like macrophages, mucus, ciliated cells and proteases in the epithelium of lungs, thus reducing their overall bioavailability, and also by the barrier between alveolar air and capillary blood (2). Furthermore, hypersecretion of mucus is the most common symptom in pulmonary disorders, which is a primary obstacle to circumvent targeted delivery of drugs. Therefore, direct drug delivery to the required lung sites has become an enticing research area. The primary advantage of NPs is their ability to evade clearance by macrophages, thereby entering the respiratory epithelium more easily than larger-sized particles. Among the various drug delivery systems available for pulmonary application, the use of biodegradable polymeric NPs represents several advantages for the treatment of respiratory diseases (13). Nanomedicines, like liposomes, dendrimers, micelles, nanotubes and nanocarriers, present the possibility to increase bioavailability and favor intracellular penetration of specific drugs into the lung tissue

DOI: 10.1201/9781003046547-18

(14,15). In recent years, a number of respiratory diseases like COPD, TB, cystic fibrosis, asthma, cancer, etc. have been approached using NPs, since nanocarrier systems can be administered to the airways easily. Local delivery of medications to the pulmonary system represent an unmet medical need and is highly desirable where conventional therapy proves to be ineffective in maintaining the desired drug concentration in the blood plasma for a prolonged period (16).

In this chapter, we will summarize and discuss recent trends and developments in nanotechnology for the detection and therapy of various pulmonary disorders, focusing on NPs.

18.2 Physicochemical Characterization for Nanoparticle-Based Systems

The in vivo distribution and behavior of nanometric drug carriers depend mostly on the physicochemical characteristics. Thus, it is necessary to categorize suitable and robust techniques that can be used for this purpose. Again, the characterization of conventional nanomedicines is based on the assessment of physicochemical properties such as molecular weight, composition, identity, purity, solubility and stability. Routine techniques applied for characterization of conventional pharmaceuticals can be used for the characterization of nanomedicines (17), however, numerous specific characteristics of nanomaterials, viz. size, surface composition/energy/charge and shape are crucially important and need to be well investigated. Addressed below are brief descriptions of some of the most commonly used methods to evaluate the specific physicochemical properties of nanomaterials, and a tabular representation has been given along with their advantages (Table 18.1).

18.2.1 Dynamic Light Scattering

Dynamic light scattering (DLS), also known as photon correlation spectroscopy or quasi-elastic light scattering, is one of the most popular light scattering techniques which allows particle sizing (<1 nm diameter) in solutions and the performance of size-distribution studies. DLS is typically used for studing the stability of formulations according to time and/or variations of temperature, for identifying the presence of aggregates in formulations prepared by different procedures, and for rapid determination of the particle size of monodisperse samples (18). Characteristic applications are emulsions, micelles, polymers, colloids, proteins and nanoparticles. The basic principle of DLS is that the sample is illuminated by a laser beam and the fluctuations of the scattered light are detected at a known scattering angle, θ, by a fast photon detector.

Simple DLS instruments that measure at a fixed angle can determine the mean particle size in a limited size range. More elaborate multi-angle instruments can determine the full particle size distribution. From a microscopic point of view, the particles scatter the light and thereby imprint information about their motion. Analysis of the fluctuation of the scattered light thus yields information about the particles. Nanomaterials tend to aggregate in water, changing their size and surface properties, thereby leading to different interactions with the water molecules that surround the nanoparticles, which is a major drawback of DLS (19). As a result, the size obtained from DLS may be overestimated and size distribution may be altered due to environmental dependence.

18.2.2 X-ray Diffraction

X-ray diffraction (XRD) is a primary tool for analyzing the tertiary structures of crystal or polycrystal structures at the atomic scale in the range of 1–100 nm (20) and crystallinity mapping can be done along with DSC. For analyzing lipid-based formulations, both techniques can be used simultaneously. X-ray diffractograms of nanomaterials provide information from phase composition to crystallite size and from lattice strain to crystallographic orientation (21). In XRD, the sample is exposed to a collimated beams of X-rays, with detection of the type and intensity of scattering by stacked

TABLE 18.1

Commonly Used Analytical Techniques for Evaluation of the Physicochemical Characteristics of Nanomaterials

Characterization Technique	Physicochemical Characteristics Analyzed	Advantages
Dynamic light scattering (DLS)	Particle size and hydrodynamic size distribution	Wide time range, cheap and simple experimental setup
X-ray diffraction	Size, degree and orientation, phase, and chemical composition of crystalline materials	Least expensive, convenient and the best method for phase analysis
Scanning electron microscopy (SEM)	Size and size distribution, structure/shape, stability, identification of elemental composition, shape aggregation	Generates detailed three-dimensional and topographical images in digital form; fast and requires minimal preparation actions
Transmission electron microscopy (TEM)	Particle size, size distribution, structure/shape, stability, shape heterogeneity, aggregation, dispersion	Offers very powerful resolution and magnification, radiation resistant, good signal-to-noise ratio, simple sample preparation on grid
Zeta potential	Surface charge, surface chemistry and reactivity	Highest resolution as compared to other methods, small sample volume, rapid measurement
Ultraviolet-visible spectroscopy (UV-Vis)	Particle size and size distribution	High sensitivity, linearity over wide concentration ranges, small sample volume required

parallel crystalline atomic planes of the examined specimen, according to Bragg's law: $2d \sin\theta = n\lambda$, where n is an integer, λ is the wavelength, θ is the scattering angle, and d is the interplanar distance (22). XRD is non-contact and non-destructive, which makes it ideal for in situ studies and characterization of polymer-layered silicate nanocomposites.

18.2.3 Scanning Electron Microscopy

Scanning electron microscopy (SEM) uses beams of accelerated electrons and electromagnetic lenses for imaging, in contrast to light microscopy that uses visible light and glass lenses, with the main improvements including greater depth of field and higher magnification (>100,000×) (23). Consequently, the incident electron beam is scanned in a raster pattern across the surface of the sample, and the emitted electrons are detected by an electron detector for each position in the scanned area. Among these emissions, detection of the secondary electrons is the most common mode in SEM and can achieve resolution smaller than 1 nm (24). The sample electron emission can include either elastic or inelastic scattering events. This electron–sample interaction is used to extrapolate information about the exterior of the particles, including their morphology, orientation and composition. The size, distribution and shape of nanomaterials can be directly acquired from SEM, however, it can only be used for certain biological materials, due to degradation caused by the electron beam. Additionally, while scanned by an electron beam, nonconductive biomolecule samples may acquire charge and inadequately deflect the electron beam, leading to imaging artifacts (25). Therefore, coating an ultrathin layer of electrically conducting material onto the biomolecules is often required for sample preparation.

18.2.4 Transmission Electron Microscopy

Transmission electron microscopy (TEM) is one of the most frequently used techniques for characterizing nanomaterials at spatial scales ranging from <1 to 100 nm and up to the micrometer level, making it effective for novel applications. In TEM, a beam of electrons passes through an ultrathin sample specimen and the crystalline sample interacts with the electron beam by the process of diffraction generating an image that can be magnified/focused by an objective lens. Apart from particle size measurement, TEM images can be used to evaluate whether good dispersion has been achieved or whether agglomeration is present in the system. One important criterion of TEM is that the samples must be prepared as a thin foil (not more than 1 μm in diameter) for the electron beams to penetrate. Again, the samples must be held at liquid nitrogen temperatures post embedding to withstand the high vacuum inside the instrument. TEM is mostly used for metallurgy and biological sciences (26–28).

18.2.5 Zeta Potential Measurements

Zeta potential measurements are essential for predicting the stability of particles that can be measured using light-scattering techniques. In an ionic solution, the surface of a charged particle is surrounded by an electrical double layer of of a thin liquid layer named the Stern layer and an outer diffuse layer consisting of loosely associated ions known as the diffuse layer. In an ionic solution, an electrical double layer surrounds the surface of a charged particle (29). Given the tangential motion driven by an external force or Brownian motion of the charged particle, the movement of the charged particle shears ions migrating with the charge particle in the diffuse layer from ions staying with the bulk dispersant outside the layer. The zeta potential is the potential difference between the dispersion medium and the stationary layer of fluid attached to the dispersed particle, determined by evaluation of the velocity of the charged species moving toward the electrode, in the presence of an external electric field across the sample solution. The higher the zeta potential, the higher is the repulsion between the particles and zeta potential with a value of ± 30 mV usually chosen to infer particle stability (30). Electrophoresis light scattering (ELS) also known as laser Doppler micro-electrophoresis, is currently used for zeta-potential determination. It is worth noting that zeta potential is a property sensitive to environmental changes, including alteration of pH and ionic strength (31).

18.3 Criteria of Nano-formulations Intended for Pulmonary Delivery

Large surface area of approximately 70–140 m^2 coupled with the property of fast absorption owing to an efficient vascularization system makes adult human lungs an ideal candidate as a drug delivery system for various therapeutic agents. Moreover, the thin blood–alveolar barrier, rapid onset of action, high therapeutic ratio, lower administered dose, increased selectivity and bypassing of first pass metabolism in the lungs proved to be an added advantage for pulmonary targeted drug development (32–34). A majority of the pulmonary delivery systems found in the market basically fall under three categories: dry powder inhalation (DPI), metered dose inhalators (MDI) and nebulizers. However, most of the marketed pulmonary targeted products are short-acting, where drug concentration peaks initially, followed by prompt decline, and may lead to adverse systemic effects. Moreover, patients require inhaling several times a day to get the desired effect, which affects patient compliance (33). To overcome all these shortcomings, controlled release drug delivery was developed, but this system comes with formidable airway clearance mechanisms such as mucociliary clearance, alveolar or macrophage clearance, systemic absorption and metabolic degradation. To overcome these shortcomings, different formulations come into play, of which nano-formulations are the most prominent.

Different forms of nano-formulations are used for delivery in the pulmonary system, of which solid lipid nanoparticles (SLN) are well-known. Particle size of the drug delivery system plays an important role in deciding the targeted site, and depending upon the particle size, the administered drug gets deposited in various regions of the respiratory system. Upper airways, smaller airways and bronchioles and alveoli are the three regions where administered particles get deposited depending upon their sizes. Inertial impaction, gravitational sedimentation and Brownian diffusion are the three widely accepted mechanisms of drug deposition which depends upon the particle size of the drug delivered (32). Apart from the drug delivery system, breathing parameters, specific disease types and cellular aspects of the transportation of pulmonary drugs are some of the vital parameters needing to be taken care of during development of drugs for pulmonary routes. Apart from all those criteria, choice of patient specific inhalation device in nano-based drug delivery for pulmonary administration is also of paramount importance. Thus, this section of the book chapter will focus on various aspects such as nano-formulation types, particle size, site and mechanism of deposition, cellular aspects of the transportation of pulmonary drug and inhaler types in specific disease conditions that play a crucial role in the case of nanoparticles intended for pulmonary delivery

18.3.1 Pulmonary Directed Nano-Drug Delivery Formulations

Till date various types of nano-formulations, viz. liposomes, micelles, SLN, dendrimers and polymers, are being developed to be used for drug delivery. Liposomes are basically lipid bilayer fabricated from cholesterol and natural nontoxic phospholipids. Their hydrophobic as well as hydrophilic characters and high stability make them unique among other nano-formulation types and are being utilized as carriers for some delicate drugs such as genetic materials and vaccines as well as steroids (35,36). SLN are an alternative to polymeric NPS, having solid lipid as a matrix material for drug delivery (37). Lipid analogous to endogenous lipids and phosoholipids like cholesterol, which are well tolerated by the body, are utilized for fabrication of SLN. Mobility of drug in solid lipid is low in comparison to liquid oils, so SLN are regarded as an ideal formulation to achieve controlled drug release (38).

Micelles are made from amphiphilic copolymers where the hydrophobic blocks of the copolymers form the micelle core, whereas the hydrophilic blocks are present in the outer core, thus enabling drugs to travel through hydrophilic channels, allowing the drugs to stay in the systemic circulation thus delaying the excretion rate (39,40). Apart from a wide range of drug carriers, dendrimers are developed to be a carrier for genetic materials acting as nonviral vectors. Amine groups found in dendrimers condense the nucleic acid into NPs by ionic interaction and guard the nucleic acid from enzymatic degradation (41). Lipid-based NPs like liposomes have their demerits, too. They lack encapsulation efficiency; leakage of water-soluble drugs occurs in the presence of blood components and they have storage stability issues. But polymeric NPs are devoid of such issues whereas they possess potential controlled release properties equipped with increased stability of drugs (42). Again, nanospheres and nano capsules are two types of widely fabricated polymeric NPs. A drug candidate is uniformly distributed in a matrix system in the case of nanospheres, whereas drugs are concentrated to a cavity having a suitable polymeric membrane in the case of nano capsules (43).

Incorporation and adsorption are the two methods by which drugs are loaded into polymeric NPs, but incorporation gives maximum drug loading as compared to the latter (44). Drugs administration via pulmonary route is a challenging task, owing to the unique physicochemical, physiological, biochemical as well as anatomical parameters of the respiratory tract. Deviation of agglomerated particles from their path due to the humid environment of the respiratory tract, abrupt pulmonary clearance in the form of mucociliary or alveolar clearance of the aerosolized drug, as well as particle accession by macrophages are various challenges that need to be dealt with while designing drugs for pulmonary administration (22,45). However, lipid-based NPs are advantageous over other polymeric based counterparts in pulmonary drug delivery due to the lipophilic environment of the respiratory system. Alveolar fluids and mucus are mainly composed of phospholipids, cholesterol as well as surface proteins that reduce the surface tension required for proper functioning of gas exchanges and removal of foreign particles by cilia of the epithelial cells (32).

Out of all lipid-based NPs, SLN holds promise to be the best formulation in terms of acceptability in the respiratory system. SLN are mainly made up of phospholipids and triglycerides, both endogenous to respiratory system, due to which there is high probability of acceptance and less toxicity (46). SLN nebulization in mice yields no pro-inflammatory cytokines or chemokines, which is an indication of little or no toxicity upon administration of SLN (47).

18.3.2 Particle Size and Materials

The site of deposition of a drug depends upon its particle size. Sizes above 5 μm deposit in the oropharyngeal region, sizes from 1–5 μm in the smaller airways as well as in the bronchial area, and sizes less than 0.5 μm deposit in the peripheral alveoli. Basically, NPS deposits in the peripheral alveolar region through the sedimentation mechanism, owing to their particle sizes. Many times, NPs become aggregated after releasing from the aerosol system and form micrometer size ranges, settle in the bronchial region due to increasing mass and thereby the desired effect is achieved (32). SLN, being a favorable NPs formulation, has recently grabbed attention worldwide due to their higher success rate during nebulization. Because SLN is a colloidal formulation and due to its nano-size, it is an ideal candidate for lymphatic interstitium penetration (48). When SLN was inhaled by a group of adult Wistar rats, the study showed significant uptake of a radio-labeled SLN into the lymphatics (49). A similar study in a model of non small cell lung cancer (NSCLC) in human alveolar adenocarcinoma epithelial A549 cells showed promising results of an erlotinib (ETB)-loaded SLN based formulation of dry powder inhaler (ETB-SLN DPI) having size below 4 μm; 3-(4,5-Dimethylthiazol-2-yl)-2,5-diphenyltetrazolium bromidefor (MTT) assay and 4',6-Diamidino-2-Phenylindole (DAPI) staining revealed enhanced cytotoxicity compared to free ETB (50).

Lipid based nano-formulation offers the best compatibility with the pulmonary delivery system compared to other material-based nano-formulations due to the enhanced dissolution rate, uniform drug distribution and because it is analogous to endogenous surfactant found in the entire respiratory system (51). Cholesterol, lecithin and phospholipids are different types of pulmonary surfactants which are utilized synthetically in the fabrication of SLN.

18.3.3 Particle Size and Deposition in Various Regions of the Respiratory System

The respiratory system is subdivided into upper airways, smaller airways and bronchioles, and finally peripheral alveoli in terms of particle deposition. The pattern in which the particles are deposited in various regions of the lungs follows three widely accepted mechanisms: inertial impaction, gravitational sedimentation and Brownian diffusion. The deposition mechanism in various region of the respiratory system has been depicted in Figure 18.1. Deposition of particle size greater than 5 μm follows the mechanism of impaction. Dry powder inhalation (DPI) and metered dose inhalators (MDI) whose particle sizes are above 5 μm undergo the mechanism of impaction. On release of the particles from the inhalers, the high-speed particles collide in the respiratory wall with high impact under centrifugal force and get deposited in the oropharynx region. The mechanism of sedimentation mainly follows the law of gravitation where particle sizes ranging from 1–5 μm and having a sufficiently large mass are deposited in the smaller airways and bronchioles. Particle sizes less than 0.5 μm are deposited in the peripheral alveoli with the help of Brownian movement of the lung surfactant and smaller particles exhaled out from the system (32,45).

18.3.4 Peptide Based Nano-formulation for Transportation of Pulmonary Drugs

Two major cell types are found in pulmonary epithelium: Type I and Type II pneumocytes. Type I pneumocytes cover ~95% of all the alveolar epithelial surface, whereas Type II pneumocytes cover only 5%. Type I pneumocytes have endocytotic vesicles

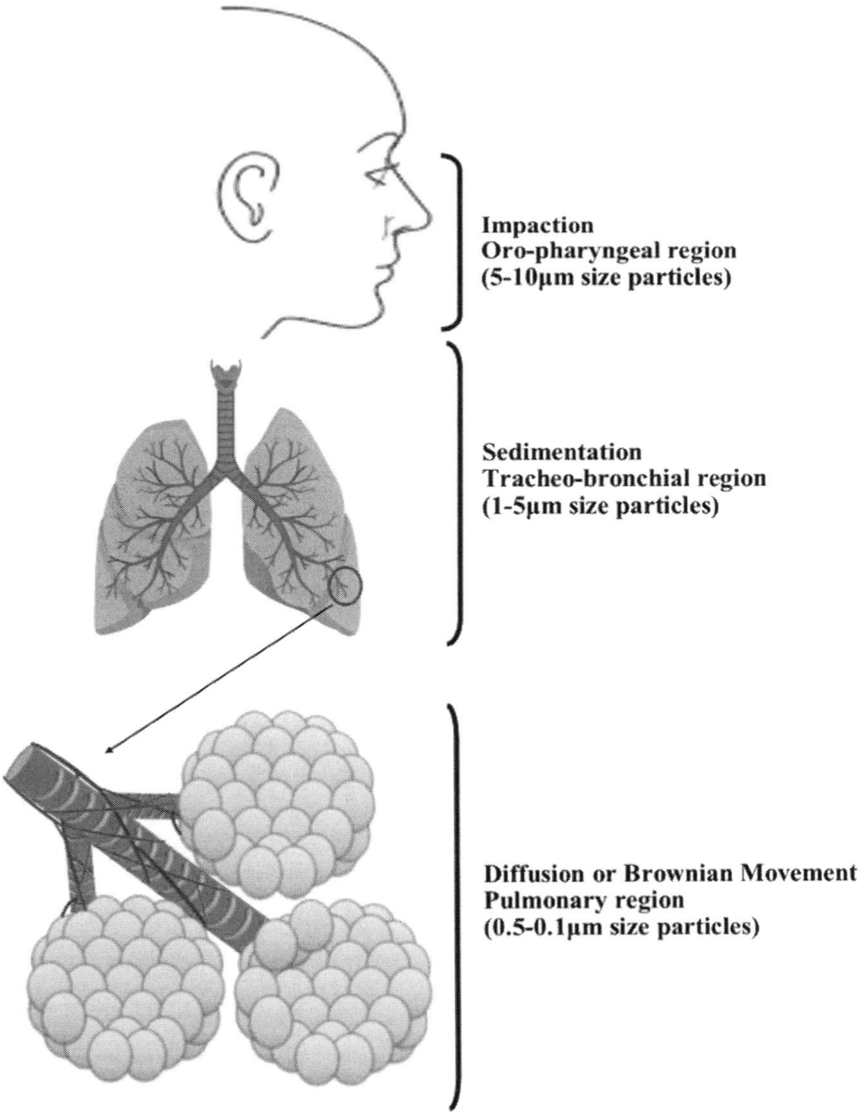

FIGURE 18.1 The Deposition Mechanism in Various Region of the Respiratory System Along with Varying Particle Size

which may function as a carrier of large proteins (34,52), and Type II pneumocytes produce surfactant protein and differentiate into Type I cells after injury in the epithelial barrier (53,54). Peptide transporter (Pept) 1 and Pept2 are two low-affinity and high-capacity transporters, mainly expressed in the small intestine, are also found in human lung tissue where Pept2 is abundantly found in Type-II cells responsible for oligopeptide transport as reported in a study involving rat lung epithelial cells (55). So, with the help of a Pept2-like cell transporter, recently many new technologies like peptide conjugation and peptidomimetic strategies have been developed. Peptide based NPs for pulmonary drug delivery is relatively a new and emerging field for systemic as well as local treatment in the lungs. Peptides are preferred for nonviral vector therapy, especially for their cell penetrating capacity (56). Recently, a peptide-based NP conjugate study demonstrated promising results in lung cancer (57). In the study intratrachealstudy, intra-tracheal (i.t.) administration of a dimerized transactivator of transcription (TAT) peptide (dTAT)-based nonviral gene delivery technique attenuated an acutely growing mouse Lewis lung carcinoma allografts in mouse lungs. The dTAT-based nonviral gene delivery technique contains an apoptosis-inducer gene, angiotensin II type 2 receptor (AT2R) alongwith along with a complex of Ca2+ (dTATpAT2R-Ca2þ).

18.3.5 Choice of Inhalation Device Depending Upon Diseased Condition

A successful pulmonary administration of aerosolized drugs relies on three factors: drug formulation, inhalation device and the disease condition the patient is suffering from. Often the administered drug in the form of an aerosol does not reach the desired site due to diseased condition or improper administration of drug arising from the inhaler. There are four different types of inhalers found in the market, viz., metered dose inhalers (MDIs), DPIs, soft mist inhalers (SMIs) and nebulizers. The aerodynamic particle size distribution of almost all the inhalers falls within the range of 1–5 µm. Drug administration via MDIs and DPIs as well as SMIs require inspiratory flow from the patient. But in severe conditions of diseases like asthma, COPD and pulmonary hypertension, owing to a congested condition, the patient is unable to inspire the aerosolized drug, leading to deposition of drug in the upper respiratory tract and resulting in inadequate effect at the site of action (58). This condition worsens in the case of infant, elderly and comatose patients where drug cannot be administered with the help of the above-mentioned inhalers. In such cases, a nebulizer comes to the rescue. A repertoire of nebulizer designs is found in the market, i.e. jet nebulizer, breath enhanced nebulizer, manual actuated nebulizer, breath actuated nebulizer, ultrasonic nebulizer and vibrating mesh nebulizer. Performance of a nebulizer is highly dependent on its specific design. A recently concluded study identified the vibrating mesh nebulizer as an efficient design in terms of inhaled and residual dose when tested in a spontaneous breathing lung model as compared to other contemporary designs. Many recent studies have reported the potential of NPs when administered via nebulizers. In two recently concluded studies, the researchers reported the potential of SLN loaded sildenafil citrate (SC) in pulmonary hypertension. In both the studies, the researchers developed a nano-formulation of SLN loaded SC and studied its efficacy in an in vivo model of the Wistar rat by nebulizing with a jet nebulizer. The researchers found the colloidal stability or the drug entrapment was intact with minimum occurrence of bleeding in lung parenchyma (59,60). These studies indicated that nebulizers are the most preferred inhalation devices over other devices nowadays when it comes to the administration of nano-formulations. Administration of nano-formulation in the pulmonary route is a challenging task, due to various challenges the formulation has to go through. Particles must be of specific sizes for the desired site of deposition in the respiratory system where nano-sized particles generally get deposited in the alveolar region.

Again, lipid-based nano-formulation is the widely accepted nano-formulation for pulmonary administration, of which SLN is a promising lipid-based nano-formulation due to the presence of physiological components in its composition. Peptide based nano-formulations are a relatively new technology that may be used for pulmonary administration due to their cell penetrating capacity and transportation of oligopeptides. The inhalation device also plays a crucial role in transporting drug to its desired site. Among all inhalation devices, the nebulizer has the advantage in administering drug to elderly, infant and comatose patients without any inspiration support from the patients. Most recently, researchers are utilizing jet and vibrating mesh nebulizers to nebulize their fabricated nano-formulations to the desired site. Although there is much hype surrounding the lipid-based drug delivery for pulmonary administration, there is paucity of information about the exact mechanism of transportation of drugs across pulmonary cells and these formulations have yet to testify clinically. Apart from the above discussed criteria of NPS intended for pulmonary delivery, there are other aspects, such as nanotoxicology, flow rate and volume administered as well as acceptability of these formulations in humans, etc., these must be addressed thoroughly before possible clinical trials. Some of the widely used nanocarriers and their advantages for pulmonary drug delivery are listed in Table 18.2.

18.4 Pulmonary Infectious Diseases and Nanomedicinal Approach

Worldwide, pulmonary diseases are among the primary causes of mortality and morbidity, however, a few nanomedicine based life-prolonging therapies and targeted drug deliveries have been developed for these maladies. With the revolution of nanotechnology many therapeutic strategies are being investigated for drug delivery to the respiratory system which includes small molecules, protein and peptide drugs, or siRNA targeted to specific genes. Most of these studies were generally focused on localized application of drugs via nanocarriers for obtaining the best possible therapeutic outcomes for chronic lung diseases (61). Lipid-based carrier systems, including liposomes (mostly lung surfactants and synthetic lipids) and nanoliposomes, are the most promising encapsulation technologies employed in the fast developing field of nanotechnology for controlled drug delivery to the lungs (11). Since a variety of nanocarriers have been

TABLE 18.2

Widely Used Nanocarriers for Pulmonary Drugs

Nanocarrier	Size	Advantages
Polymeric nanoparticles	10–1000 nm	Nontoxic, non-inflammatory, and non-immunogenic
Liposomal nano-carriers	0.5 μm–100 nm	Selective targeting to tumor tissue, enhanced pharmacokinetic effect and therapeutic index, decreased toxicity
Solid lipid nano-carriers	50–1000 nm	High drug loading and stability, incorporation of both hydrophilic and hydrophobic drugs, nontoxic, minimization of organic solvent use
Submicron emulsions	10–500 nm	High kinetic stability and solubilizing capacity, tiny globule size
Dendrimers		Polyvalent, self-assembling, electrostatic interactions, chemically stable, less cytotoxic and highly soluble
Nanocrystals	less than 500 nm	Enhanced solubility, zero-dimensional
Inorganic nanoparticles	3–6 nm	Nontoxic, hydrophilic, biocompatible and highly stable

investigated that can be transferred easily to the airways for the delivery of macromolecules to the respiratory tract, the ability of a nano-structured system may be approached for treating a number of respiratory diseases including COPD, genetic disorders affecting the airways and infectious diseases including TB and cancer.

18.4.1 Asthma

According to reports of the WHO, more than 300 million individuals globally suffer from asthma (62) and its socioeconomic burden exceeds the sum of human adenovirus/acquired immune deficiency syndrome (HIV/AIDS) and TB. Little has been done in the field of asthma therapy during the last 20 years, and with the exception of a very few drugs like tiotropium, indacaterol and omalizumab, no revolutionary changes have occurred (63). Treatment with corticosteroids and the combinations of steroids with bronchodilators is the first-line control strategy for asthma (64) that can manage asthma symptoms on a day-to-day basis but it has severe adverse effects, including hypertension, cataracts, osteoporosis in elderly patients and stunted growth in children (65). Novel therapeutic agents with effective routes of administration and prolonged action may be an alternative to corticosteroids for management of asthma. PEGylated poly(amidoamine) (PAMAM) dendrimer, a typical dendrimer, has been widely studied and applied (66) which can perk up drug solubility and increase lung accumulation capacity for insoluble drugs like beclomethasone dipropionate (BDP) (67). Recently, bilirubin-based nanoparticles (BRNPSs) composed of freely water dispersible, entirely PEGylated bilirubin were developed based on the report that asthmatic symptoms are rapidly relieved during jaundice, indicating a positive role of augmented bilirubin levels in serum (68). The effects of BRNPSs on T-helper-2-type (Th2) immune responses were investigated both in vivo and in vitro showing potent anti-inflammatory effects in an *A. oryzae* protease allergen-induced asthma model. It has been reported that nanocarrier-encapsulated steroids achieved better and long-term therapeutic effects in airway inflammation sites as compared with that of free steroids. Again, a study reported that in the lungs, liposomes delayed the retention of salbutamol sulfate maintaining an effective drug concentration for more than 10 hours, thereby achieving significantly higher efficacy than the free drug solution (69). Moreover, the selection and construction of apposite nanocarriers are cutting-edge topics in the gene therapy of asthma. According to a study reported by Kumar et al., chitosan-IFN-γ pDNA NPS (CIN) can significantly reduce airway hyper-responsiveness and lung histopathology scoring in BALB/c mice with ovalbumin (OVA) induced allergic asthma (70).

Since angiogenesis is a feature of asthmatic inflammatory responses, anti-angiogenesis nano-therapy could offer a new therapeutic approach to this serious disease. Recently, the results of study done by Lanza et al. indicated that lipase labile phospholipid prodrug forms of fumagillin (Fum-PD) or docetaxel (Dxtl-PD) incorporated into lipid-based αvβ3-integrin targeted micelles for drug delivery ameliorated asthma in the house dust-mite-triggered Brown Norway rat model (71). Again, SLN formulation of curcumin was capable of overcoming its poor solubility and rapid metabolization, which was the major drawback of curcumin affecting its clinical efficacy as a potent anti-asthmatic. In another recent study, it was reported that curcumin loaded in SLNs suppressed airway hyper-responsiveness and infiltration of inflammatory cells by significantly inhibiting the expression of Th2 cytokines in a murine model of OVA induced asthma (72). Furthermore, the bioavailability and efficacy of andrographolide (AG), a labdane diterpene lactone was improved over free drug by encapsulating it in PLGA nanoparticles, thereby, resulting in a promising therapeutic agent against OVA induced asthma in mice (73).

18.4.2 COPD

COPD, the culmination of chronic bronchitis, small airways disease and emphysema (existing separately or in combination) is associated with structural changes and inflammatory response contributing to bronchial hyper-responsiveness and reduced lung function. Again, increased number of T-lymphocytes, particularly CD8+ T cells, macrophages and neutrophils is a characteristic of COPD (5). COPD is considered to be among the greatest global health hazards in terms of mortality and morbidity, and approximately 210 million people in the world are currently estimated to suffer from COPD, with a large disease burden in developing countries (74) and it is estimated that COPD will be ranked the third cause of death by 2020. Moreover, it has been reported as the second leading cause of disability and third leading cause of mortality in the United States (US) (75). Unfortunately, a suitable model demonstrating the role of nanomedicine in COPD is lacking. Additionally, COPD is basically steroid insensitive and is therefore unresponsive to traditional inhaled anti-inflammatory steroid therapies (1). There are few reports where lipospheres and proliposomes have been used as drugs for dry powder inhalation aerosol delivery (76). Various preclinical studies with murine models have proven nanoparticle-based therapy to show promising results against lung inflammation (77). Morever, inhaled NPs, owing to their small size, can easily traverse to distal airspaces, where they are deposited exclusively by diffusion (78). Properly designed nanoparticles, like methotrexate-loaded albumin NP and doxorubicin-loaded SLN, are specialized for the purpose of pulmonary delivery and have proven to be highly distributed in the lungs. Polyethylene glycol (PEG) is a polymer that can rapidly navigate through the mucus barrier (79). Therefore, if NPs are adequately coated with such muco-alert polymers, it can easily gain access to the underlying epithelia. Furthermore, special emphasis has been put on the inhibition of neutrophil infiltration, the hallmark feature in the pathogenesis of COPD.

8081 In COPD, the common pathogenic mechanisms is found to be progressive inflammatory response involving a significant increase in number of neutrophils and macrophages. In a recent study, neutrophil-targeted polylactide-coglycolide (PLGA)-PEG NPs (PNPs) for ibuprofen delivery has been demonstrated as an effective drug-delivery nanosystem for treating COPD (80). In this study, PNPs is found to exert its ability to control neutrophil mediated proteas anti-protease imbalance in Pa-LPS/cigarette-smoke induced COPD. Since cytokines and chemokines possess a critical role in the orchestration of COPD, therefore, the highly selective monoclonal antibodies are valuable treatment options for COPD. Clinical trials of anti-interleukin (IL)-8 antibodies abgenix, anti-tumor necrosis factor (TNF)-α antibody infliximab has been conducted, however the outcome was not totally fruitful in improving lung function (82). Thus, the use of nano delivery system (NDS) may allow administration by inhalation, increasing local concentrations and limiting possible side effects. Another such monoclonal antibody is cetuximab, which could be a valuable option for the treatment of COPD.

Again, alveolar macrophages, the most abundant, versatile cells in the respiratory tract, play a major role in the pathogenesis of COPD as a key source of mediators (83). A recent study with biocompatible PEGylated dextran-coated magnetic iron oxide NPs has been reported to specifically target one subpopulation of macrophages in the lung of a COPD animal model by maintaining a balance in the number of M1 and M2 macrophages (84). Lipid NPs, viz. SLN and nanostructured lipid carriers (NLC), result in a longer dosing interval and better patient compliance and have been investigated as a possibility to improve therapy of COPD.

18.4.3 Tuberculosis

TB is another global public health problem due to its high prevalence in many countries, and is closely linked to socio-economic conditions, making it the second leading cause of death from an infectious disease after HIV. According to WHO fact sheets an estimated 1.7 million, 10.4 million and 1.6 million cases of death due to TB were recorded in the years 2009, 2016 and 2017, respectively (85–87). Although TB is practically curable in 100% of the cases, a majority of cases occur in developing countries where diagnosing TB is difficult due to technical limitations and lack of resources.

The first-line drugs or multiple anti-tubercular drugs (ATDs) must be administered continuously for a minimum of 6 months, which leads to patient noncompliance with prescribed treatment regimens, treatment-related adverse effects and misuse, which may cause emergence of mult-drug-resistant TB (MDR-TB). Again, the options for treatment of MDR-TB are the second line ATDs which are comparatively more toxic, less effective and not easily available due to their high cost (88). Misuse or indiscriminate use of second-line ATDs further results in the development of extensively drug resistant TB (XDR-TB) (89). Moreover, the situation is worsened by the fact that many countries lack proper infrastructure to accurately diagnose MDR-TB and XDR-TB, which increases TB associated mortality. Besides these limitations, the current therapies available for TB have inadequate ability to penetrate granulomas and also have reduced effects on dormant bacilli (90). In this context, improvement of the therapeutic index of existing ATDs, by their encapsulation into NDs, should be emphasized.

Recent advances in nanotechnology have enabled the development of novel strategies and new diagnostic platforms for TB, which are aimed at more sensitive and faster pathogen detection and effective parenteral or mucosal TB vaccine delivery (91,92). As reviewed previously, inhalable nanocarriers like nanocrystals and NPs as well as aerosols offer a latent value in local and passive delivery of anti-TB therapy (93). Pulmonary delivery of vaccines may prove to be an effective alternative to intradermal (i.d.) or subcutaneous (s.c.) routes, since the lung is the primary site of infection for *Mycobacterium tuberculosis* (Mtb), the causative strain for TB (94). Drug carriers and radiopharmaceuticals like 99Tc-EMB, graphene oxide, etc., that are nontoxic to humans, can encapsulate potent ATDs to efficiently deliver in the target site (95). Furthermore, liposomes are attractive nanocarriers for their versatile and unique structures and are one of the successful drug delivery systems for increasing the effectiveness of antibiotics (96).

PEGylated, phosphatidylcholine, dimyristoylphosphatidylcholine (DMPC), DPPC, dimyristoylphosphatidylglycerol (DMPG) and distearoylphosphatidylcholine (DSPC)-based systems are some of the examples of liposomes having remarkable potential as direct drug delivery systems for frontline ATDs to lungs (97–99). However, one disadvantage of liposomes is that they are vulnerable to intestinal lipases, therefore oral route of administration cannot be performed (100). Moreover, the preferential accumulation of liposomes with ATDs after i.v administration occurs at the liver and spleen, which can be controlled by the addition of ligands to liposomes to accomplish lung specificity (101). In contrast to the potential problems and poor chemical stability (102) with liposomal and biodegradable polymer delivery systems, Clemens et al. reported mesoporous silica (MS) NPs as a promising delivery platform for ATDs in effectively killing intracellular Mtb as compared to that of free drug. In an attempt to incorporate ATDs in controlled release proniosome derived niosomes, high encapsulation efficiency was obtained and might prove to be adantageous to solve the problem of MDR-TB (103).

18.4.4 Lung Cancer

Worldwide, lung cancer is the second most common cancer in both men and women and is the leading cause of 23% of total cancer deaths, with a dismal 5-year survival rate of only 15% (104). Lung cancer is manifested in one of two forms: small cell lung carcinoma (SCLC), which represents 13% of the total lung cancer cases, and non small cell lung carcinoma (NSCLC), which represents 80–85% and is considered significantly more aggressive (105). Further, NSCLC is categorized as epidermoid, large cell, bronchoalveolar, adenocarcinoma, and squamous cell carcinoma (104). Presently, chemotherapy, radiotherapy and surgery are the available therapies for lung cancer treatment. However, multi-drug resistance to cytotoxic agents, nonspecific toxicity to healthy cells, deficiency in early-stage diagnostics, lack of therapeutic efficiency and intratumoral genetic heterogeneity (106) paved the path for the need for efficient and less toxic treatment alternatives. It is noteworthy to mention that for effective drug targeting, parameters like physicochemical characteristics of the chemotherapeutics, carrier properties and type of the inhalation device, the tumor type and location as well as targeting strategy must be considered. Active targeting via NP-based ligand or monoclonal antibodies may hold a stronger potential than passive targeting based on the well-known enhanced permeability and retention (EPR) effect. Active targeting strategies are advantageous as this ideally allows for cell-specific detection and killing of both primary tumor cells and metastatically spread circulating cancer cells (107). Furthermore, the main route of application in recent preclinical studies is based on I.V. delivery of NPs (108) which remained with partial success due to poor site-specific drug availability (109). Therefore, methods to improve tumor-targeted delivery of chemotherapeutics that will result in increased drug, improved pharmacological properties and least toxicity to normal tissues retain their precedence in cancer therapy.

NPs can optimize the biodistribution profile, enhance transport properties and target the drug to the tumor regions, which increases therapeutic efficiency and reduces nonspecific toxicity of anti-cancer drugs (110). Recently, the US Food and Drug Administration (FDA) approved an albumin-bound formulation (Abraxane®) of cremophor EL and paclitaxel for NSCLC in combination with carboplatin, in an attempt to reduce the toxicity of cremophor EL and to improve the overall efficacy of paclitaxel (111). Furthermore, nanotechnology-based therapeutic delivery with pulmonary/inhalational route of administration has received much attention in the recent decade for lung cancer treatment. The inhalational delivery of doxorubicin (DOX)-loaded NPSs in a cancer-bearing mouse model (BALB/c model) (112) and 5-fluorouracil (5-FU) NPS produced by a supercritical antisolvent process in lung cancer were successfully established (113). Again, numerous proteins have been targeted by small interfering ribonucleic acid (siRNA) embedded into nonviral delivery systems in cancer pathologies based on cell cycle, proliferation, apoptosis and angiogenesis pathway studies. The encapsulation of hydrophilic drugs such as nucleic acids and proteins has been already developed for numerous formulation processes (114,115). Yet again, perfect sequence homology has been observed between the strands of endogen micro RNA (miRNA) and messenger RNA (mRNA), referred to as the RNAi concept. In a murine mouse model of lung cancer, a lipid/polymer nanoparticle delivering miRNA slowed tumor growth (116). This plethora of evidence confirms that polymeric NPs have been extensively explored using various highly used pulmonary drugs for the treatment of asthma, COPD, TB and lung cancer.

However, the biodegradability and toxicity of the polymers for long term use should be closely examined in the formulation of polymeric NPs for pulmonary delivery, to avoid accretion of polymer carriers following repeated doses.

18.5 Regulation and Guidelines for Marketing

Nanotechnology is viewed by the European Union (EU) and the Scientific Committee on Emerging and Newly Identified Health Risks (SCENIHR) as one of the major technological drivers of innovation (117). Since 1995, more than 50 nanopharmaceuticals have been approved by regulatory authorities, including liposomes, non-liposomal lipid formulations, peptides, PEGylated proteins, nanocrystals, aptamers, protein-drug conjugates, polymer-based nanoformulations, surfactant-based nanoformulations, virosomes and metal-based nanoformulations (118) and are presently available for clinical use (119). Among those, 18 are marketed nanopharmaceuticals and other products for imaging and diagnostic applications. Moreover, more than 70 nanomedical products are undergoing clinical trials.

The approval process for nanomedicines in humans is regulated by the FDA in the US (120). Both the FDA and the EU have already set up robust schemes for the approval and legislation of new nanopharmaceuticals and provide comprehensive pharmacological and toxicological profiles of novel nanopharmaceuticals in maintaining a balance between innovation and safety (121). Following the discovery or invention, the material undergoes a preclinical phase of testing, which usually involves animal studies for demonstration of efficacy, safety, toxicity profile and identification of appropriate dose ranges (122).

In June 2014, the FDA issued one draft and three final guidance documents for the industry regarding the use of nanotechnology in FDA-regulated products, including nano drugs (123). Data demonstrating the physicochemical properties, efficacy and toxicity of a nano drug are then compiled into an Investigational New Drug (IND) application for FDA review and approval. Moreover, researchers can send nanomaterials to the Nanotechnology Characterization Laboratory (NCL) to have them tested and validated according to a series of emerging protocols (124). Following FDA approval, clinical trials are undertaken with human volunteers to determine safety and efficacy. Further, the fourth phase of the clinical trial, i.e. post-marketing surveillance, is initiated at the request of the FDA or health care professionals to assess the risks associated with nano drugs. The entire process is estimated to cost $1 billion per new drug and take approximately 10–15 years. Some of the US FDA-approved nanomedicines for pulmonary disorders have been listed in Table 18.3.

TABLE 18.3

Nanomedicines for Pulmonary Disorders in Clinical Trials or US FDA-Approved (111,119,140, 141, 142)

Name	Description	Medical Indications	Current Status
Arikayce	Inhaled liposomal formulation of amikacin	Serious chronic lung infections	Marketed
BLP25	Liposome vaccine	Lung neoplasms non small cell lung carcinoma	Phase II completed
CRLX101	A drug–conjugate formulation of camptothecin and a cyclodextran-PEG polymer	Lung cancers (SCLC and NSCLC)	Phase 1 and 2 clinical trials ongoing
Curosurf	Phospholipid fraction from porcine lung	Respiratory distress syndrome	Marketed
Pulmaquin	Combination of liposomal and aqueous-phase ciprofloxacin	Cystic fibrosis (CF) or non-CF bronchiectasis	Completed company-sponsored phase 2 studies
SN-38	An active metabolite of the topoisomerase inhibitor irinotecan	Solid tumors, NSCLC	Two phase 1 trials have been completed
UMIN000014940	PLGA NP- pitavastatin	Pulmonary artery hypertension	Completed investigator-initiated phase I clinical trial

18.6 Safety Assessment of NPs

Although nanomedicines have undoubtedly shown potential for the treatment of chronic lung diseases, possible organ toxicity is a major concern in pulmonary medicine. With the rapid expansion of new applications of nanomedicine and use of nanotechnologies in the 21st century for innovative treatment strategies, a balance between therapeutic efficacy and safety of the NDS with the aim to increase the benefit-to-risk ratio is the need of the hour. With regard to toxicological response, it is noteworthy to mention that the lung is the primary and probably the most important target during inhalation of NPs and a secondary target after I.V. injection due to the high blood irrigation of this organ and the small space between alveolar epithelial cells and the blood capillaries. Furthermore, a wide range of NP-based strategies have been reported to cause respiratory disorders (2,82). Lung inflammation has been observed following the administration of carbon nanotubes via i.t. installation. However, the type of response, whether innate or inflammatory, varies between molecular and nano-sized forms (125), surface area and charge (126), long-term toxicity verses subacute and acute toxicity (127), exposure level (128), type of particles (biological or non-biological) used, etc. Again, the probabilty for adverse health effects may occur due to direct exposure to intentionally produced nanomaterials and/or byproducts allied with their applications. The major cellular responses caused by nanotoxicology in the lungs include release of reactive oxygen (ROS) and nitrogen species (RNS), proinflammatory/inflammation-associated proteins as well as injury of nuclear DNA (129).

Furthermore, functional disturbances could be observed after NP exposure, including airway hyper-reactivity (AHR), tissue injury and effects on existing pulmonary inflammation. Therefore, reliable and reproducible screening protocols are needed to understand the novel physicochemical properties of nanomaterials and their interactions at the nano/bio interface responsible for biological hazards (130).

But before proceeding with the nanomaterial toxicity testing, some of the discussion points become crucial to answer, such as which toxicological end points to screen for, the comprehensiveness of the screening effort, the correct balance of in vitro versus in vivo testing, the cost of the effort, whether current regulation, testing and classification protocols are suitable and who should be responsible for screening and safety assessment of nanomaterials (131). It is also a matter of concern that generally the toxicity of the whole formulation is analyzed before marketing a product, but results of the toxicity profile of NPs are often ignored. Consequently, a specific emphasis on the toxicity of the non-drug loaded particles, specially for the non-biodegradable and inorganic ones, should be given since biodegradable NPs becom degraded by metabolic pathways (118). From a safety perspective, the need to connect several major themes, viz. safety, regulation, research quality and innovation of current respiratory nanotoxicology is apparent, including:

a. Comprehensive nanoparticle characterization in the relevant delivery vehicle and physiological matrices (132).
b. Predictive in vitro studies with the self-propagating human lung cell lines A549 or BEAS-2B can help to predict the hazard potential of a series of ambient particles based on an established mechanistic pathway at the molecular, cellular, organismal and ecosystem levels, and an in vivo outcome (133). However, analysis of in vitro studies has to be cautiously performed since they are usually carried out at high doses.
c. Cell culture studies are the hallmark of nanotoxicological analysis; however, data obtained from in vitro experiments could be misleading for varied reasons as in the case of a report produced by Monteiro-Riviere et al. where the dye-based MTT assay produced invalid results with some NPs. Hence, more than a single assay may be performed when determining NP toxicity for risk assessment (134).
d. Nanomaterials may interfere with the read-out systems of commonly used assays for cell viability and/or mitochondrial function, generating false results. In spite of performing in vitro studies, the toxicological status of NPs should further be verified from in vivo experiments by examining molecular markers of oxidative stress and/or inflammation measured within the bronchoalveolar lavage fluid (BALF) and histology (135) in both diseased and healthy lungs.

e. The adverse health effects of particulate matter (PM) are measurable as exacerbations of respiratory disease and deaths as well as hospitalizations and deaths from respiratory and cardiovascular disease (136), inflammation being the common factor that binds these adverse effects.
f. Preclinical testing of novel nano-therapies for lung cancer with a detailed and stratified analysis in mice including, first, toxicological and pharmacokinetic analysis of the nanoengineered formulations in healthy animals, and second, theranostic application of the nanoparticles with simultaneous imaging of particle biodistribution, drug release and testing of the therapeutic efficacy in appropriate lung cancer models (137). The molecular heterogeneity and histopathological features of lung tumors need to be defined in detail to permit personalized approaches of nanomedical therapies. Comprehensive identification of differential receptor expression of tumor versus normal cells needs to be defined by the omics technologies that have fostered systems-based analysis of different tumors (138, 139).
g. Tailoring the surface of NPs by functionalization or tuning parameters like surface charge, spacer length, hydrophobicity, aggregation propensity, etc., is a scheduled process to improve the behavior of NPs for nanomedicine application. Moreover, to assess the role of surface charge on the toxicity and inflammogenicity of NPs in the lung environment, well-engineered, nontoxic and chemically stable model NPs are preferred (140).

18.7 Conclusion

It has become essential to bring material and pharmaceutical scientists together to find new avenues through multidisciplinary approaches that can speed up the development of new diagnostic and therapeutic solutions. Nanopharmaceuticals have massive potential to address the failures of traditional drugs that could not be effectively formulated due to factors such as poor water solubility or a lack of target specificity. Although studies on NPs for pulmonary application are still in an initial phase, studies performed so far suggest that NPs are an interesting choice in the systemic or local treatment of respiratory diseases. Again, some novel herbal NDs together with NPs are under exploration for establishment for operative delivery of phytomedicine resulting in reduced toxic total side effects and enhanced patient compliance.

REFERENCES

1. Thorley AJ and Tetley TD. New perspectives in nanomedicine. *Pharmacology & Therapeutics* 2013; 140(2):176–185.
2. Pison U, Welte T, Giersig M, et al. Nanomedicine for respiratory diseases. *European Journal of Pharmacology*. 2006; 533(1–3): 341–350.
3. Bhaskar S, Tian F, Stoeger T, et al. Multifunctional nanocarriers for diagnostics, drug delivery and targeted treatment across blood-brain barrier: Perspectives on tracking and neuroimaging. *Particle and Fibre Toxicology*. 2010; 7(1): 3.
4. Rosi NL and Mirkin CA. Nanostructures in biodiagnostics. *Chemical Reviews*. 2005; 105(4): 1547–1562.
5. Roy I and Vij N. Nanodelivery in airway diseases: Challenges and therapeutic applications. *Nanomedicine-Nanotechnology, Biology and Medicine*. 2010; 6(2): 237–244.
6. Akerman ME, Chan WC, Laakkonen P, et al. Nanocrystal targeting in vivo. *Proceedings of the National Academy of Sciences*. 2002; 99(20): 12617–12621.
7. Moghimi SM, Hunter AC, andMurray JC. Nanomedicine: Current status and future prospects. *FASEB Journal*. 2005; 19(3): 311–330.
8. Rao KS, Ghorpade A, and Labhasetwar V. Targeting anti-HIV drugs to the CNS. *Expert Opinion on Drug Delivery*. 2009; 6(8): 771–784.
9. Buzea C, Pacheco II, and Robbie K. Nanomaterials and nanoparticles: Sources and toxicity. *Biointerphases*. 2007; 2(4): MR17–MR71.
10. Foldvari M and Elsabahy M. Nanotechnology enables superior medical therapies. *Current Drug Delivery*. 2011; 8(3): 225–226.
11. Mansour HM, Rhee YS, and Wu X. Nanomedicine in pulmonary delivery. *International Journal of Nanomedicine*. 2009; 4: 299–319.
12. Jaques PA and Kim CS. Measurement of total lung deposition of inhaled ultrafine particles in healthy men and women. *Inhalation Toxicology*. 2000; 12(8): 715–731.
13. Sadikot RT, Kolanjiyil AV, Kleinstreuer C, et al. Nanomedicine for treatment of acute lung injury and acute respiratory distress syndrome. *Biomedicine Hub*. 2017; 2(2): 1–12.
14. Resnier P, Mottais A, Sibiril Y, et al. Challenges and successes using nanomedicines for aerosol delivery to the airways. *Current Gene Therapy*. 2016; 16(1): 34–46.
15. van Rijt SH, Bein T, and Meiners S. Medical nanoparticles for next generation drug delivery to the lungs. *European Respiratory Journal*. 2014; 44: 765–774.
16. Codrons V, Vanderbist F, Verbeeck RK, et al. Systemic delivery of parathyroid hormone (1–34) using inhalation dry powders in rats. *Journal of Pharmaceutical Sciences*. 2003; 92(5): 938–950.
17. Patri A, Dobrovolskaia M, Stern S, et al. Preclinical characterization of engineered nanoparticles intended for cancer therapeutics. In: Mansoor M. Amiji (ed.) Nanotechnology for Cancer Therapy. CRC Press. 2006; 105–138.
18. Brar SK and Verma M. Measurement of nanoparticles by light-scattering techniques. *Trends in Analytical Chemistry*. 2011; 30(1): 4–17.
19. Hall JB, Dobrovolskaia MA, Patri AK, et al. Characterization of nanoparticles for therapeutics. *Nanomedicine (London)*. 2007; 2(6): 789–803.
20. Sapsford KE, Tyner KM, Dair BJ, et al. Analyzing nanomaterial bioconjugates: A review of current and emerging purification and characterization techniques. *Analytical Chemistry*. 2011; 83: 4453–4488.
21. Joshi M, Bhattacharyya A, and Ali SW. Characterization techniques for nanotechnology applications in textiles. *Indian Journal of Fibre & Textile Research*. 2008; 33: 304–317.

22. Cantor CR and Schimmel PR. Techniques for the Study of Biological Structure and Function. W. H. Freeman, San Francisco, CA. 1980.
23. Manaia EB, Abuçafy MP, Chiari-Andréo BG, et al. Physicochemical characterization of drug nanocarriers. *International Journal of Nanomedicine*. 2017; 12: 4991.
24. Johal MS. Understanding Nanomaterials. CRC Press, Boca Raton, FL. 2011.
25. Lin PC, Lin S, Wang PC, et al. Techniques for physicochemical characterization of nanomaterials. *Biotechnology Advances*. 2014; 32(4): 711–726.
26. Anjum DH. Characterization of nanomaterials with transmission electron microscopy. *IOP Conference Series: Materials Science and Engineering*. 2016; 146(1): 012001.
27. Akbari B, Tavandashti MP, and Zandrahimi M. Particle size characterization of nanoparticles-A practical approach. *Iranian Journal of Materials Science and Engineering*. 2011; 8(2): 48–56.
28. Smith DJ. Characterization of nanomaterials using transmission electron microscopy. In: Nanocharacterisation. 2, 2015; 1–29.
29. Clogston J and Patri A. Zeta potential measurement. In: McNeil SE. (ed.) Characterization of Nanoparticles Intended for Drug Delivery. Humana Press. 2011; 63–70.
30. Mueller R. Solid lipid nanoparticles (SLN) for controlled drug delivery-A review of the state of the art. *European Journal of Pharmaceutics and Biopharmaceutics*. 2000; 50: 161–177.
31. Xu R. Progress in nanoparticles characterization: Sizing and zeta potential measurement. *Particuology*. 2008; 6: 112–115.
32. Paranjpe M and Müller-Goymann C. Nanoparticle-mediated pulmonary drug delivery: A review. *International Journal of Molecular Science*. 2014; 15(4): 5852–5873.
33. Liang Z, Ni R, Zhou J, et al. Recent advances in controlled pulmonary drug delivery. *Drug Discovery Today*. 2015; 20(3): 380–389.
34. Groneberg DA, Witt C, Wagner U, et al. Fundamentals of pulmonary drug delivery. *Respiratory Medicine*. 2003; 97(4): 382–387.
35. Jeevanandam J, San Chan Y, Danquah MK. Nanoformulations of drugs: Recent developments, impact and challenges. *Biochimie*. 2016; 128: 99–112.
36. Akbarzadeh A, Rezaei-Sadabady R, Davaran S, et al. Liposome: Classification, preparation, and applications. *Nanoscale Research Letters*. 2013;8 (1): 102.
37. Müller RH, Mäder K, Gohla S. Solid lipid nanoparticles (SLN) for controlled drug delivery-A review of the state of the art. *European Journal of Pharmaceutics and Biopharmaceutics*. 2000; 50(1): 161–177.
38. Mehnert W and Mäder K. Solid lipid nanoparticles: Production, characterization and applications. *Advanced Drug Delivery Reviews*. 2012; 64: 83–101.
39. Sen S, Han Y, Rehak P, et al. Computational studies of micellar and nanoparticle nanomedicines. *Chemical Society Reviews*. 2018; 47(11): 3849–3860.
40. Kwon GS, Yokoyama M, Okano T, et al. Biodistribution of micelle-forming polymer–drug conjugates. *Pharmaceutical Research*. 1993; 10(7): 970–974.
41. Eichman JD, Bielinska AU, Kukowska-Latallo JF, et al. The use of PAMAM dendrimers in the efficient transfer of genetic material into cells. *Pharmaceutical Science and Technology Today*. 2000; 3(7): 232–245.
42. Soppimath KS, Aminabhavi TM, Kulkarni AR, et al. Biodegradable polymeric nanoparticles as drug delivery devices. *Journal of Controlled Release*. 2001; 70(1-2): 1–20.
43. Langer R. Biomaterials in drug delivery and tissue engineering: One laboratory's experience. *Accounts of Chemical Research*. 2000; 33(2): 94–101.
44. Ueda M, Iwara A, and Kreuter J. Influence of the preparation methods on the drug release behaviour of loperamide-loaded nanoparticles. *Journal of Microencapsulation*. 1998; 15(3): 361–372.
45. Cheng YS. Mechanisms of pharmaceutical aerosol deposition in the respiratory tract. *AAPS PharmSciTech*. 2014; 15(3): 630–640.
46. Beck-Broichsitter M, Ruppert C, Schmehl T, et al. Biophysical investigation of pulmonary surfactant surface properties upon contact with polymeric nanoparticles in vitro. *Nanomedicine: Nanotechnology, Biology and Medicine*. 2011; 7(3): 341–350.
47. Nassimi M, Schleh C, Lauenstein HD, et al. A toxicological evaluation of inhaled solid lipid nanoparticles used as a potential drug delivery system for the lung. *European Journal of Pharmaceutics and Biopharmaceutics*. 2010; 75(2): 107–116.
48. Reddy LH, Sharma RK, Chuttani K, et al. Influence of administration route on tumor uptake and biodistribution of etoposide loaded solid lipid nanoparticles in Dalton's lymphoma tumor bearing mice. *Journal of Controlled Release*. 2005; 105(3): 185–198.
49. Videira MA, Botelho MF, Santos AC, et al. Lymphatic uptake of pulmonary delivered radiolabelled solid lipid nanoparticles. *Journal of Drug Targeting*. 2002; 10(8): 607–613.
50. Bakhtiary Z, Barar J, Aghanejad A, et al. Microparticles containing erlotinib-loaded solid lipid nanoparticles for treatment of non-small cell lung cancer. *Drug Development and Industrial Pharmacy*. 2017; 43(8): 1244–1253.
51. Khadka P, Ro J, Kim H, et al. Pharmaceutical particle technologies: An approach to improve drug solubility, dissolution and bioavailability. *Asian Journal of Pharmaceutical Sciences*. 2014; 9(6): 304–316.
52. Beck-Broichsitter M, Merkel OM, and Kissel T. Controlled pulmonary drug and gene delivery using polymeric nano-carriers. *Journal of Controlled Release*. 2012; 161(2): 214–224.
53. Groneberg DA, Eynott PR, Doring F, et al. Distribution and function of the peptide transporter PEPT2 in normal and cystic fibrosis human lung. *Thorax*. 2002; 57: 55–60.
54. Groneberg DA, Nickolaus M, Springer J, et al. Localization of the peptide transporter PEPT2 in the lung: Implications for pulmonary oligopeptide uptake. *American Journal of Pathology*. 2001; 158: 707–714.
55. Gukasyan HJ, Uchiyama T, Kim KJ, et al. Oligopeptide transport in rat lung alveolar epithelial cells is mediated by Pept2. *Pharmaceutical Research*. 2017; 34(12): 2488–2497.
56. Sebbage V. Cell-penetrating peptides and their therapeutic applications. *Bioscience Horizons*. 2009; 2(1): 64–72.

57. Ishiguro S, Alhakamy NA, Uppalapati D, et al. Combined local pulmonary and systemic delivery of AT2R gene by modified TAT peptide nanoparticles attenuates both murine and human lung carcinoma xenografts in Mice. *Journal of Pharmaceutical Sciences*. 2017; 106(1): 385–394.
58. Prajapati S, Saha S, Kumar CD, et al. Nebulized drug delivery: An overview. *International Journal of Pharmaceutical Sciences and Research*. 2019; 10(8): 3575–3582.
59. Makled S, Nafee N, and Boraie N. Nebulized solid lipid nanoparticles for the potential treatment of pulmonary hypertension via targeted delivery of phosphodiesterase-5-inhibitor. *International Journal Pharmaceutics*. 2017; 517(1–2): 312–321.
60. Nafee N, Makled S, and Boraie N. Nanostructured lipid carriers versus solid lipid nanoparticles for the potential treatment of pulmonary hypertension via nebulization. *European Journal of Pharmaceutical Sciences*. 2018; 125: 151–162.
61. Upadhyay S, Ganguly K, and Palmberg L. Wonders of nanotechnology in the treatment for chronic lung diseases. *Journal of Nanomedicine and Nanotechnology*. 2015; 6(6): 1.
62. Yin LM, Jiang GH, Wang Y, et al. Use of serial analysis of gene expression to reveal the specific regulation of gene expression profile in asthmatic rats treated by acupuncture. *Journal of Biomedical Science*. 2009; 16(46): 1–13.
63. Szelenyi I. Nanomedicine: Evolutionary and revolutionary developments in the treatment of certain inflammatory diseases. *Inflammation Research*. 2012; 61(1): 1–9.
64. Tan LD, Bratt JM, Godor D, et al. Benralizumab: A unique IL-5 inhibitor for severe asthma. *Journal of Asthma and Allergy*. 2016; 9: 71–81.
65. Inapagolla R, Guru BR, Kurtoglu YE, et al. In vivo efficacy of dendrimer-methylprednisolone conjugate formulation for the treatment of lung inflammation. *International Journal of Pharmaceutics*. 2010; 399: 140–147.
66. Wang L, Feng M, Li Q, et al. Advances in nanotechnology and asthma. *Annals of Translational Medicine*. 2019; 7(8): 180.
67. Nasr M, Najlah M, D'Emanuele A, et al. PAMAM dendrimers as aerosol drug nanocarriers for pulmonary delivery via nebulization. *International Journal of Pharmaceutics*. 2014; 461: 242–250.
68. Kim DE, Lee Y, Kim M, et al. Bilirubin nanoparticles ameliorate allergic lung inflammation in a mouse model of asthma. *Biomaterials*. 2017; 140: 37–44.
69. Chen X, Huang W, Wong BC, et al. Liposomes prolong the therapeutic effect of anti-asthmatic medication via pulmonary delivery. *International Journal of Nanomedicine*. 2012; 7: 1139–1148.
70. Kumar M, Kong X, Behera AK, et al. Chitosan IFN-gamma-pDNA nanoparticle (CIN) therapy for allergic asthma. *Genetic Vaccines and Therapy*. 2003; 1: 3.
71. Lanza GM, Jenkins J, Schmieder AH, et al. Anti-angiogenic nanotherapy inhibits airway remodeling and hyper-responsiveness of dust mite triggered asthma in the Brown Norway rat. *Theranostics*. 2017; 7(2): 377.
72. Wang W, Chen T, Xu H, et al. Curcumin-loaded solid lipid nanoparticles enhanced anticancer efficiency in breast cancer. *Molecules*. 2018; 23(7): 1578.
73. Chakraborty S, Ehsan I, Mukherjee B, et al. Therapeutic potential of andrographolide-loaded nanoparticles on a murine asthma model. *Nanomedicine: Nanotechnology, Biology and Medicine*. 2019;20:102006
74. Adeloye D, Chua S, Lee C, et al. Global Reference Health Epidemiology Group. Global and regional estimates of COPD prevalence: Systematic review and meta-analysis. *Journal of Global Health*. 2015; 5(2): 020415.
75. Mercado N, Ito K, and Barnes PJ. Accelerated ageing of the lung in COPD: New concepts. *Thorax*. 2015; 70(5): 482–489.
76. Sadikot RT. The potential role of nanomedicine in lung diseases. *Medical Research Archives*. 2018; 6(5): 1–9.
77. Ghosh N, Singh B, and Chaudhury K. Recent advances in nanomedicine for respiratory diseases: A systematic review. *Anti-Cancer Drugs*. 2014; 18: 23.
78. Verbanck S and Paiva M. Gas mixing in the airways and airspaces. *Comprehensive Physiology*. 2011; 1(2): 809–834.
79. Lai SK, O'Hanlon DE, Harrold S, et al. Rapid transport of large polymeric nanoparticles in fresh undiluted human mucus. *Proceedings of the National Academy of Science*. 2007; 104(5): 1482–1487.
80. Vij N, Min T, Bodas M, et al. Neutrophil targeted nano-drug delivery system for chronic obstructive lung diseases. *Nanomedicine: Nanotechnology, Biology and Medicine*. 2016; 12(8): 2415–2427.
81. Hogg JC. Pathophysiology of airflow limitation in chronic obstructive pulmonary disease. *Lancet*. 2004; 364(9435): 709–721.
82. Boland S, Guadagnini R, Baeza-Squiban A, et al. Nanoparticles used in medical applications for the lung: Hopes for nanomedicine and fears for nanotoxicity. *Journal of Physics: Conference Series*. 2011; 304: 012031.
83. Alber A, Howie SE, Wallace WA, and Hirani N. The role of macrophages in healing the wounded lung. *International Journal of Experimental Pathology*. 2012; 93(4): 243–251.
84. Al Faraj A, Shaik AS, Afzal S, Al Sayed B, and Halwani R. MR imaging and targeting of a specific alveolar macrophage subpopulation in LPS-induced COPD animal model using antibody-conjugated magnetic nanoparticles. *International Journal of Nanomedicine*. 2014; 9: 1491.
85. Ballester M, Nembrini C, Dhar N, et al. Nanoparticle conjugation and pulmonary delivery enhance the protective efficacy of Ag85B and CpG against tuberculosis. *Vaccine*. 2011; 29(40): 6959–6966.
86. Gordillo-Marroquín C, Gómez-Velasco A, Sánchez-Pérez H, et al. Magnetic nanoparticle-based biosensing assay quantitatively enhances acid-fast bacilli count in paucibacillary pulmonary tuberculosis. *Biosensors*. 2018; 8(4): 128.
87. Bothamley G. The tuberculosis network European trials group (TBNET): New directions in the management of tuberculosis. *Breathe*. 2017; 13(3): 65–71.
88. Costa A, Pinheiro M, Magalhães J, et al. The formulation of nanomedicines for treating tuberculosis. *Advanced Drug Delivery Reviews*. 2016; 102: 102–115.
89. Sun Z, Zhang J, Song H, et al. Concomitant increases in spectrum and level of drug resistance in Mycobacterium tuberculosis isolates. *International Journal of Tuberculosis and Lung Disease*. 2010; 14(11): 1436–1441.
90. Wallis RS and Hafner R. Advancing host-directed therapy for tuberculosis. *Nature Reviews Immunology*. 2015; 15(4): 255.

91. Gao L, He X, Ju L, et al. A label-free method for the detection of specific DNA sequencesusing gold nanoparticles bifunctionalized with a chemiluminescent reagent and a catalyst as signal reporters. *Analytical and Bioanalytical Chemistry*. 2016; 408(30): 8747–8754.
92. Hawkridge T and Mahomed H. Prospects for a new, safer and more effective TB vaccine. *Paediatric Respiratory Reviews*. 2011; 12(1): 46–51.
93. Mehanna MM, Mohyeldin SM, and Elgindy NA. Respirable nanocarriers as a promising strategy for antitubercular drug delivery. *Journal of Controlled Release*. 2014; 187: 183–197.
94. Källenius G, Pawlowski A, Brandtzaeg P, et al. Should a new tuberculosis vaccine be administered intranasally? *Tuberculosis (Edinburgh)*. 2007; 87(4): 257–266.
95. Saifullah B, Chrzastek A, Maitra A, et al. Novel antituberculosis nanodelivery formulation of ethambutol with graphene oxide. *Molecules*. 2017; 22(10): 1560.
96. Sosnik A, Carcaboso AM, Glisoni RJ, Moretton MA, and Chiappetta D. New old challenges in tuberculosis: Potentially effective nanotechnologies in drug delivery. *Advanced Drug Delivery Reviews*. 2010; 62(4–5): 547–559.
97. Anabousi S, Kleemann E, Bakowsky U, et al. Effect of PEGylation on the stability of liposomes during nebulisation and in lung surfactant. *Journal of Nanoscience and Nanotechnology*. 2006; 6(9–10): 3010–3016.
98. Finlay WH and Wong JP. Regional lung deposition of nebulized liposome-encapsulated ciprofloxacin. *International Journal of Pharmaceutics*. 1998;167(1–2): 121–127.
99. Pandey R and Khuller GK. Nanotechnology based drug delivery system(s) for the management of tuberculosis. *Indian Journal of Experimental Biology*. 2006; 44: 357–366.
100. Deol P, Khuller G, and Joshi K. Therapeutic efficacies of isoniazid and rifampin encapsulated in lung-specific stealth liposomes against Mycobacterium tuberculosis infection induced in mice. *Antimicrobial Agents and Chemotherapy*. 1997; 41(6): 1211–1214.
101. Foradada M, Pujol MD, Bermúdez J, and Estelrich J. Chemical degradation of liposomes by serum components detected by NMR. *Chemistry and Physics of Lipids*. 2000; 104(2): 133–148.
102. Clemens DL, Lee BY, Xue M, et al. Targeted intracellular delivery of antituberculosis drugs to Mycobacterium tuberculosis-infected macrophages via functionalized mesoporous silica nanoparticles. *Antimicrobial Agents and Chemotherapy*. 2012; 56(5): 2535–2545.
103. Jemal A, Bray F, Center MM, Ferlay J, Ward E, and Forman D. Global cancer statistics. *CA Cancer Journal for Clinicians*. 2011; 61(2): 69–90.
104. Gennatas S, Noble J, Stanway S, et al. Patterns of relapse in extrapulmonary small cell carcinoma: Retrospective analysis of outcomes from two cancer centres. *BMJ Open*. 2015; 5(1): e006440.
105. Burrell RA, McGranahan N, Bartek J, and Swanton C. The causes and consequences of genetic heterogeneity in cancer evolution. *Nature*. 2013; 501(7467): 338–345.
106. Peer D, Karp JM, Hong S, Farokhzad OC, Margalit R, and Langer R. Nanocarriers as an emerging platform for cancer therapy. *Nature Nanotechnology*. 2007; 2(12): 751–760.
107. Choi SH, Byeon HJ, Choi JS, et al. Inhalable self-assembled albumin nanoparticles for treating drug-resistant lung cancer. *Journal of Controlled Release*. 2015; 197: 199–207.
108. Yuan D, Lv Y, Yao Y, et al. Efficacy and safety of Abraxane in treatment of progressive and recurrent non-small cell lung cancer patients: A retrospective clinical study. *Thoracic Cancer*. 2012; 3(4): 341–347.
109. Akhter S, Ahmad I, Ahmad MZ, et al. Nanomedicines as cancer therapeutics: Current status. *Current Cancer Drug Targets*. 2013; 13(4): 362–378.
110. Landesman-Milo D, Ramishetti S, and Peer D. Nanomedicine as an emerging platform for metastatic lung cancer therapy. *Cancer and Metastasis Reviews*. 2015; 34(2): 291–301.
111. Roa WH, Azarmi S, Al-Hallak MH, et al. Inhalable nanoparticles, a non-invasive approach to treat lung cancer in a mouse model. *Journal of Controlled Release*. 2011; 150: 49–55.
112. Kalantarian P, Najafabadi AR, Haririan I, et al. Preparation of 5-fluorouracil nanoparticles by supercritical antisolvents for pulmonary delivery. *International Journal of Nanomedicine*. 2010; 5: 763–770.
113. Wu SY, Singhania A, Burgess M, Putral LN, Kirkpatrick C, Davies NM, et al. Systemic delivery of E6/7 siRNA using novel lipidic particles and its application with cisplatin in cervical cancer mouse models. *Gene Therapy*. 2011; 18: 14e22.
114. David S, Resnier P, Guillot A, Pitard B, Benoit JP, and Passirani C. siRNA LNCs e a novel platform of lipid nanocapsules for systemic siRNA administration. *European Journal of Pharmaceutics and Biopharmaceutics*. 2012; 81: 448e52.
115. Xia W, Wang J, Xu Y, Jiang F, and Xu, L. L-BLP25 as a peptide vaccine therapy in non-small cell lung cancer: A review. *Journal of Thoracic Disease*. 2014; 6(10): 1513.
116. Savolainen K, Backman U, Brouwer D, et al. Nanosafety in Europe 2015-2025: Towards Safe and Sustainable Nanomaterials and Nanotechnology Innovations. Finnish Institute of Occupational Health. 2013.
117. Cockburn A, Bradford R, Buck N, et al. Approaches to the safety assessment of engineered nanomaterials (ENM) in food. *Food and Chemical Toxicology*. 2012; 50(6): 2224–2242.
118. Caster JM, Patel AN, Zhang T, and Wang A. Investigational nanomedicines in 2016: A review of nanotherapeutics currently undergoing clinical trials. *Wiley Interdisciplinary Reviews Nanomedicine and Nanobiotechnology*. 2017; 9(1): e1416.
119. Eifler AC and Thaxton CS. Nanoparticle therapeutics: FDA approval, clinical trials, regulatory pathways, and case study. In: Hurst SJ. (ed.) Biomedical Nanoetechnology. Humana Press, US. 2011; 325–338.
120. Kim J, Chankeshwara SV, Thielbeer F, et al. Surface charge determines the lung inflammogenicity: A study with polystyrene nanoparticles. *Nanotoxicology*. 2016; 10(1): 94–101.
121. Bobo D, Robinson KJ, Islam J, Thurecht KJ, and Corrie SR. Nanoparticle-based medicines: A review of FDA-approved materials and clinical trials to date. *Pharmaceutical Research*. 2016; 33(10): 2373–2387.

122. Wolfram J, Zhu M, Yang Y, et al. Safety of nanoparticles in medicine. *Current Drug Targets*. 2015;1 6(14): 1671–1681.
123. Ventola CL. Progress in nanomedicine: Approved and investigational nanodrugs. *Pharmacology & Therapeutics*. 2017; 42(12): 742.
124. Ferreira AJ, Cemlyn-Jones J, and Cordeiro CR. Nanoparticles, nanotechnology and pulmonary nanotoxicology. *Revista Portuguesa de Pneumologia (English Edition)*. 2013; 19(1): 28–37.
125. Vinardell MP. Toxicological highlight: In vitro cytotoxicity of nanoparticles in mammalian germ-line stem cell. *Toxicological Sciences*. 2005; 88(2): 285–286.
126. Chan VS. Nanomedicine: An unresolved regulatory issue. *Regulatory Toxicology and Pharmacology*. 2006; 46(3): 218–224.
127. Oberdörster G. Safety assessment for nanotechnology and nanomedicine: Concepts of nanotoxicology. *Journal of Internal Medicine*. 2010; 267(1): 89–105.
128. Muhlfeld C, Rothen-Rutishauser B, Blank F, Vanhecke D, Ochs M, and Gehr P. Interactions of nanoparticles with pulmonary structures and cellular responses. *American Journal of Physiology-Lung Cellular and Molecular Physiology*. 2008; 294(5): L817–L829.
129. Nel AE, Mäedler L, Velegol D, et al. Understanding biophysicochemical interactions at the nano-bio interface. *Nature Materials*. 2009; 8: 543–557.
130. Donaldson K, Borm P, Castranova V, and Gulumian M. The limits of testing particle-mediated oxidative stress in vitro in predicting diverse pathologies; Relevance for testing of nanoparticles. *Particle and Fibre Toxicology*. 2009; 6: 13.
131. Kumar A, Dailey LA, and Forbes B. Lost in translation: What is stopping inhaled nanomedicines from realizing their potential? *Therapeutic Delivery*. 2014; 5(7): 757–761.
132. Nel A, Xia T, Madler L, et al. Toxic potential of materials at the nanolevel. *Science*. 2006; 311: 622–627.
133. Monteiro-Riviere NA, Inman AO, and Zhang LW. Limitations and relative utility of screening assays to assess engineered nanoparticle toxicity in a human cell line. *Toxicology and Applied Pharmacology*. 2009; 234(2): 222–235.
134. Maurer-Jones MA and Haynes CL. Toward correlation in in vivo and in vitro nanotoxicology studies. *The Journal of Law, Medicine & Ethics*. 2012; 40(4): 795–801.
135. Brook RD, Franklin B, Cascio W, et al. Air pollution and vascular disease. A statement for healthcare professionals from the Expert Panel on Population and Prevention Science of the American Heart Association. *Circulation*. 2004; 109: 2655–2671.
136. Bölükbas DA and Meiners S. Lung cancer nanomedicine: Potentials and pitfalls. *Nanomedicine*. 2015; 10(21): 3203–3212.
137. Suzuki H, Aoki K, Chiba K, et al. Mutational landscape and clonal architecture in grade II and III gliomas. *Nature Genetics*. 2015; 47(5): 505–511.
138. Rizvi NA, Hellmann MD, Snyder A, et al. Mutational landscape determines sensitivity to PD-1 blockade in non-small cell lung cancer. *Science*. 2015; 348(6230): 124–128.
139. Kim J, Chankeshwara SV, Thielbeer F, et al. Surface charge determines the lung inflammogenicity: A study with polystyrene nanoparticles. *Nanotoxicology*. 2016; 10(1): 94–101.
140. Explore 320,846 Research Studies in All 50 States and in 209 Countries. https://clinicaltrials.gov
141. https://www.medicines.org.uk/emc/product/6450/smpc
142. Nakamura K, Matsubara H, Akagi S, et al. Nanoparticle-mediated drug delivery system for pulmonary arterial hypertension. *Journal of Clinical Medicine*. 2017; 6(5): 48.

19

Recent Approaches in Dendrimer-Based Pulmonary Drug Delivery

Rajiv Dahiya[1] **and Sunita Dahiya**[2]
[1]School of Pharmacy, Faculty of Medical Sciences, The University of the West Indies, St. Augustine, Trinidad & Tobago, West Indies
[2]Department of Pharmaceutical Sciences, School of Pharmacy, Medical Sciences Campus, University of Puerto Rico, San Juan, USA

19.1 Introduction

The term *dendrimer* was derived from two Greek words: *dendri* means *tree-like* and *meros* means *part of*(1). Dendrimers are also known as *arborols* from the Latin word *arbor* which means *series of branches* or *cascade* molecules (2) which is accepted as the polymer of 21st century as a fourth class of polymers after three traditional polymer types: linear, crosslinked, and branched (3). Since their commencement in the mid-1980s, dendrimers have progressed as distinguished synthetic macromolecules with noticeable biomedical applications. Dendrimers possess an exceedingly branched three-dimensional unique architecture whose structural components can be tailored to affect its internal as well as surface properties. Dendrimer properties are quite different from conventional polymers as they show combined features of molecular chemistry, as they are synthesized by stepwise controlled processes, as well as of polymer chemistry as it is composed of monomers (4,5). Vogtle et al. (1978) first synthesized a dendrimer-like compound, polypropylenimine (PPI) (2), however it formed a low generation compound as a result of difficulties encountered during the synthesis. In the mid- and post-1980s, some researchers synthesized well-defined higher generation dendrimers. Since then, more than 100 varieties of dendrimer structures have been synthesized (6–11). Dendrimer-based delivery systems emerged as promising nanomedicines due to their distinguished structural, physicochemical, and biological properties ,which provide excellent platforms for targeting, diagnostic and theranostic applications (4,5).

The higher surface area of lungs (80–120 m^2) and alveolar epithelium (0.2–0.7 μm), high permeability, and rich vasculature promote the drug's administration, allowing high and rapid drug absorption and slow enzymatic degradation (12). These properties are advantageous for lipophilic molecules that have low oral absorption since such compounds can be absorbed at higher rates after pulmonary delivery. Moreover, lung anatomy possesses favorable properties which permit a fast onset of action of inhaled drugs if the drug permeates through the associated biological barriers. The pulmonary tissue can be an important non-invasive absorption route for certain molecules like peptides and proteins, which show stability and bioavailability issues by oral route. For treatment of lung diseases, pressurized metered-dose inhalers (pMDIs), nebulizers, and dry powder inhalers (DPIs) are the most commonly used dosage forms (13,14). Based on the disease's pathophysiology, aerosol type, and formulation type, these systems have their own merits and demerits. For the successful inhalation aerosol formulation, particle size distribution of formulation, flow rate of inhalation and the capacity of the aerosol device to achieve dispersion are critical factors (15,16).

Due to the progressing nanomedicine field, advanced treatment options for fatal pulmonary disorders like respiratory tract infections, chronic obstructive pulmonary disease (COPD), and lung cancer are feasible. The advanced treatment includes different respiratory formulations like liposomes; solid lipid nanoparticles; polymeric nanoparticles; polymeric micelles; microspheres and microparticles that have considerable merits including stability, in vivo drug deposition, tissue distribution, and bioavailability as compared to conventional delivery. Due to their small sizes, these novel systems have demonstrated sufficient promise in drug targeting with increased safety and decreased toxicity (13,17,18). In recent times, multifunctional bioactive platforms including dendrimers, polymer–drug conjugates, hybrid nanoparticles, inorganic nanocarriers like quantum dots, etc., have drawn the attention of formulation scientists. Among these nanocarriers, dendrimers have been comprehensively studied for diagnostic and treatment applications (19).

19.2 Dendrimer Concept

19.2.1 General Features and Applicabilities

The well-defined nano-sized dendrimer structures has the "central core" from which branches originate during synthesis (20–22). The distinguished properties such as small size, globular shape, highly specific architecture, internal cavities, and surface functionality are unique for variety of biomedical applications (23–26). Dendrimers differ from traditional polymers due to three unique structural constituents: a central core, internal branches,

DOI: 10.1201/9781003046547-19

and functional terminal groups (20,27–30). The atomistical structure of dendrimers has enabled syntheses of more than fifty dendrimer families having unique properties. Dendrimers' core, internal cavities (interior void spaces), and surface (terminal surface functional groups) can be modified to achieve diversified applications which expand their scope in the biomedical field (31–33). The applicability of dendrimer nanostructures has been investigated as promising pharmaceutical excipients (33). Being highly branched with functional surfaces, dendrimers can combine polymeric excipients like polyethylene glycol (PEG) to achieve controlled drug release or to develop novel pharmaceutical formulations (34). Certain other properties such as nanosize, precise molecular weight, presence of interior voids, biocompatibility, and absence of immunogenicity make dendrimers promising nanocarriers for drug and gene delivery (35,36).

19.2.2 Structural Aspects

A typical dendrimer structure (37) with its functional abilities is presented in Figure 19.1. The structural arrangement inside the dendrimer creates internal cavities which can accommodate drug to provide increased solubility and stability (38,39). The dendrimer core, comprising of an atom or group of atoms, allows addition of branches through synthesis reaction producing a globular dendritic structure.

A dendritic structure without a core is named a *dendron*. During stepwise synthesis of dendrimers, each consecutive reaction step adds a generation or branch, whereas the number of steps is referred to as the dendrimer *generation* (G). If a dendrimer has G < 4, it is a low-generation dedrimer, whereas a dendrimer with G > 4 is a high-generation dendrimer (40). The dendrimer core is denoted as generation zero (G0) (41). Dendrimer's physicochemical properties are highly influenced by the branching units and the terminal functional groups. This is because the surface functional groups interact with the external molecules to provoke different physicochemical properties (42). The physical and chemical properties of a dendrimer can be controlled during its synthesis by controlling the core groups, the generation numbers, and the type and number of surface functional groups. The controllable dendrimer synthesis allows the attainment of desired properties to meet formulation or clinical needs and enhances their scope as versatile nanomedicine (43).

19.3 Dendrimer Syntheses

19.3.1 Classical Approaches

19.3.1.1 Divergent Approach

Tomalia proposed dendrimer synthesis via divergent growth approach, which is currently the most widely used approach for dendrimer synthesis (44). In this process, the dendrimer starts assembling from the central core to the periphery indicating that the globular dendrimer structure grows in a cascade from the first generation (G1) to the second generation (G1 + 1 = G2), and so on via sequential reactive steps. In the first step of divergent growth method, the core is activated or modified to couple with the first monomer to create the first generation of the dendrimer. In the second step, the first generation (G1) is deprotected or activated to couple the second generation (G2), and accordingly the generations are added (45,46). The divergent process requires the previous chemical reaction to complete before addition of the new generation to avoid formation of deficient dendrimer branches (47). The dendrimer surface may be readily modified at each step to obtain desired dendrimer properties at the end of the synthesis. Although the divergent approach results in highly symmetric dendrimer structure, heterogeneously functionalized dendrimers with different types of functional groups attached to the dendrimer surfaces have been studied (45,47).

19.3.1.2 Convergent Approach

As an alternate to the divergent method, the convergent approach was first proposed by Fréchet and Hawker in 1989–1990 (47). Unlike the divergent method, the convergent process synthesizes dendrimers from the periphery, not from the central core. In the first step of the convergent method, generally two surface groups are coupled with monomer to give the dendritic segment which is known as dendron or

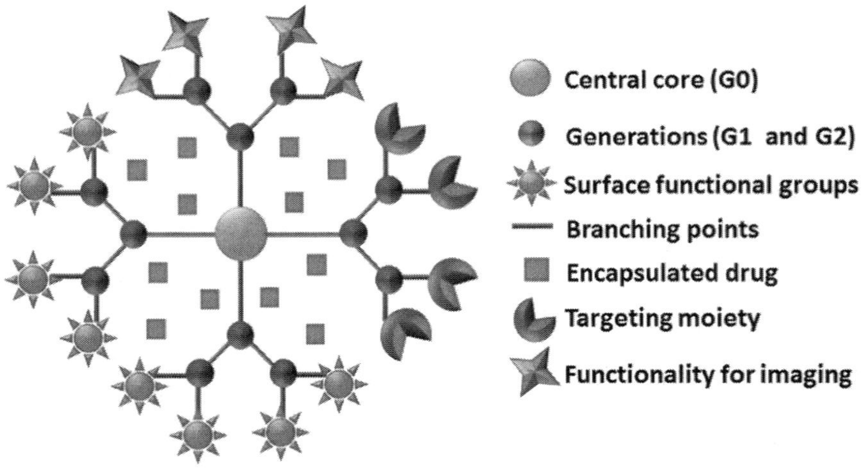

FIGURE 19.1 General Dendrimer Structure with Functional Abilities

generation zero (G0). In the second step of convergent process, this dendritic segment is activated to couple with other monomers to create a first generation dendritic block. This procedure can be repeated until the desired generation dendrimers are created, which can be coupled to give a multifunctional core. The convergent synthesis completes at the core where two or more dendrons are joined, constructing the final dendrimer structure during the final step. Since the coupling occurs at the central point of the developing dendron, the preparation of large generation dendrimers (above G6) is difficult due to steric inhibition resulting in reduced yields (44,48). The convergent synthesis is advantageous over divergent processes due to a smaller number of coupling reactions involved at each step, thereby producing dendrimers with greater structural control and high purity. Convergent approach can couple two different segments together, enabling synthesis of asymmetric dendrimers with heterogeneous morphologies. These heterogeneous structures extend opportunities to incorporate several active sites in one dendrimer resulting in a multifunctional dendrimer (26).

19.3.2 Accelerated Approaches

Classical synthesis approaches show certain drawbacks due to the high number of reaction steps involved, such as increased chemical waste and increased possibility of structural defects, as well as difficulty in assuring substitution of all the reactive groups in high generation dendrimers. These limitations make the dendrimer synthesis process expensive and tedious with limited scope of commercialization. To overcome these drawbacks, the accelerated approaches were developed, which not only reduced the number of reaction steps and time, but also decreased the amount of raw materials and the total expenditure on dendrimer synthesis (48).

19.3.2.1 Double Exponential Growth Technique

Moore (1995) introduced the double exponential growth technique, which emerged as the breakthrough in the dendrimer synthesis. This technique utilized a combination of the divergent and the convergent methods to form two types of monomers from a single starting material by the convergent and divergent growth methods (49). This technique is very versatile due to the possibility of classical, supramolecular, or asymmetrical dendrimer preparation (48).

19.3.2.2 Double-Stage Convergent Method

This method is derived from the classical convergent technique and also known as the *hypercore approach*(48). It uses two different monomers in the dendron synthesis to form hypercore which permits the dendrimer formation to have chemically differentiated external and internal branches. The hypercore decreases the steric hindrance and aids to attain higher generations monodisperse dendrimers as compared to the classical convergent technique (48,50).

19.3.2.3 Branched Monomer Approach

This method is also known as hypermonomer approach, as it employs the monomer containing a larger number of functional groups than traditional monomers, which allows the synthesis of higher generation dendrimers with a high number of functional groups in fewer reaction steps (51).

19.4 Dendrimer Characteristics

Dendrimers are unique macromolecules with specific characteristics which make them potential candidates in the field of biomedicine. These characteristic properties are summarized in Table 19.1 and discussed in this section.

19.4.1 Nanodimension

Dendrimers possess many unique properties as compared to linear polymers, making them suitable for various biomedical applications. Nanoscale size and protein-like characteristics impart biomimetic properties to dendrimers, which allows their use as artificial proteins (52). In addition, the protein-like

TABLE 19.1

Dendrimer Properties and Associated Benefits

Dendrimer Characteristics	Benefit
Unique molecular architecture	Enhances physical and chemical properties as compared to traditional linear polymers
Monodispersity	Controls size and molecular mass during synthesis procedures
Low viscosity	Promotes dendrimer–drug complex formation and release of the drugs
Globular structure and availability of internal cavities	Facilitates encapsulation of guest molecules in the interior void spaces
Nanoscale size and protein-like shape	Imparts biomimetic properties and assists non-recognition of dendrimers as artificial proteins by the RES, minimizing the early clearance
Polyvalent surface	Favors interaction of surface groups with biological targets; Allows surface modifications for tailoring properties
High solubilization potential	Provides more hydrophobic internal cavity for encapsulation of poorly water soluble compounds
Wide loading capability	Allows loading of wide range of therapeutics via electrostatic surface interactions, covalent surface conjugation, or encapsulation into the internal cavities

globular shape of dendrimer can also be modulated to resemble enzymes, antibodies, and globular proteins closely using molecular engineering techniques. For instance, dimensions of insulin (3 nm), cytochrome C (4 nm), and hemoglobin (5.5 nm) are similar to the G3, G4, and G5 PAMAM dendrimers with ammonia cores, respectively (53). Moreover, nano-size helps dendrimers to pass through biological barriers like the blood–brain barrier and tumor cell membranes while reducing their early clearance through the liver or spleen. These properties make dendrimer an excellent nanocarrier for drug delivery (54).

19.4.2 Monodispersity

Exceedingly branched and well-defined molecular architecture of dendrimers exhibit high degrees of monodispersibility due to their controllable synthesis (55). Dendrimer monodispersibility allows prediction of their pharmacokinetic performance in a biological organism, providing useful information for drug selection based on specific pathology and administration route to achieve maximum therapeutic response and lowest toxicity (56).

19.4.3 Polyvalent Surface

The presence of many reactive terminal groups at dendrimer surface (polyvalency) accounts for its solubility and reactivity. Polyvalency favors interaction of dendrimers with biological receptor sites like proteins, viruses, and polymer cells whereas dendrimer's surface modification is constructed to imitate biological receptors, substrates, inhibitors, and cofactors. The formation of free surface group can complex or conjugate with drug or ligands when crosslinking agents are used. This sort of reversible reaction enhances receptor–ligand affinity and can be used to amplify signal transduction (57,58). With increase in dendrimer generation, surface functional groups also increase, which favors the interaction of surface groups with the biological targets and demonstrates greater therapeutic response (59).

19.4.4 High Solubilization Power

In general, dendrimers are soluble (60). The solubilization power of dendrimers is attributed to ionic interactions, hydrogen bonding, and hydrophobic interactions which enable them to improve the solubility, biodistribution, and efficacy of drugs (57). One of the attribute for high solubilization and reactivity of dendrimers is the availability of many terminal surface groups. The solubility of higher generation dendrimers mainly depends on the characteristics of their surface groups. For instance, polar solvents solubilize dendrimers with hydrophilic surface groups, whereas nonpolar solvents solubilize dendrimers with hydrophobic surface groups. Besides the surface functional groups, other parameters like the type of the repetition units, the generation number, and the core contribute to the dendrimer solubility. Also, the nature of medium in which dendrimer is found is an important factor that influences the solubility and encapsulation possibility of drug molecule in the dendrimer. For this reason, poorly soluble drug molecules will be favored to encapsulate into the dendrimer providing a more hydrophobic environment (61,62).

19.4.5 Low Viscosity

The linear polymers show increased intrinsic viscosity with increased polymers in the molecular mass. The opposite behavior is shown by the dendrimer macromolecules in solution, showing considerably smaller viscosity values than linear polymers. In general, the viscosity of dendrimer increases with the number of monomers up to less than G4 dendrimers. However, from G4 and higher generations, the dendrimer viscosity decreases with increasing functional groups. The low viscosity facilitates the dendrimer–drug complexation for the immediate release of drugs (60,63).

19.4.6 Wide Loading Capability

A broad variety of inorganic or organic therapeutics can be loaded in dendrimer structure by three mechanisms: (i) surface adsorption through electrostatic interactions, (ii) surface conjugation through covalent bonding, or (iii) encapsulation inside dendrimer cavities. As the number of surface functional groups on dendrimer molecule increases, their loading capabilities also increases. Addition of each one generation doubles the number of surface terminal groups which is also available for drug interactions. The presence of ionizable groups such as the amine and carboxyl groups favors the electrostatic attachment of ionizable drugs whereas dendrimer surface groups show strong covalent attachment through hydrolysable or biodegradable linkages that can provide controlled drug release opportunities. Penicillin V conjugates with G2.5 dendrimer via amide linkage, whereas it conjugated to G3 PAMAM dendrimer through ester linkages. The conjugation by the amide linkage provided higher stability whereas the ester linkage demonstrated increased drug circulation time compared to liposome-based drug-delivery systems (64,65).

The encapsulation of a drug inside a dendrimer cavity occurs by interactions with specific structures via physical entrapment (66). The empty hydrophobic internal cavities of the dendrimer allows interactions with poorly soluble drugs, whereas internal cavities with the presence of oxygen and nitrogen atoms allow the drug to interact by hydrogen bond formation (67). The low molecular weight bioactive molecules, if carried to the dendrimer surface, can induce undesired immunogenicity (68).

19.4.7 Altered Conformational Behavior

Dendrimers respond to the adjacent chemical environment such as pH, salts, and solvent, which alters their conformational behavior to exert specific types of effects. The protonated or deprotonated state of the dendrimer largely influence the solubilization potential of dendrimers. The basic terminal groups and basic interior containing tertiary amine are present in amino-terminated PPI and PAMAM dendrimers, for which the pH less than 4 usually leads to extended conformations. At this pH, with increased generations, the dendrimer interior becomes "hollow" by reason of repulsion between the cationic amine surface groups and the tertiary amines in the interior. At neutral pH, the positively charged amines at the surface and the tertiary amines in the interior are back folded. At pH higher than 10, the charge of the dendrimer molecule becomes neutral while minimum repulsion exists among the dendrimer branches and the surface groups which leads the dendrimer to contract and acquire globular structure (69). The presence of excess salt provides high ionic strength which strongly affects the charged dendrimers, such as PPI. In addition, the excess salt concentration leads to a higher level of back folding, as it is seen in case of high pH, favoring contracted dendrimer conformation. When the salt concentration is low, repulsive forces exist between the charged dendrimer portions, resulting in an elongated conformation to decrease charge repulsion on the structure (70). A higher degree of back folding is seen with decreased solvation in all generations of dendrimers; the highest back folding tendency is, exhibited by the low generation dendrimers because of poor solvation, than that of higher generation dendrimers. Nonpolar solvent such as benzene can poorly solvate the dendrimer and favors intramolecular interactions among dendrimer fragments causing the back-folding (71). A weakly acidic solvent like chloroform donates hydrogen to the interior amines of a basic dendrimer like PPI, resulting in hydrogen bond formation between the dendrimer's amine and the solvent, causing an elongated dendrimer conformation.

19.5 Dendrimer Applications in Medicine

The advancements in dendrimer chemistry have given enormous opportunities for development and delivery of dendrimer-based medicines through different routes such as oral, injectable, ocular, pulmonary, transdermal, and rectal. The applications of dendrimers in medicine are mainly attributed to their abilities in modulating the physicochemical and pharmacokinetic properties of the therapeutic agent. In the present time, researchers are combining dendrimers and dendrimer-based nanomedicines in treatment and prevention of various diseases for better treatment options. Dendrimers also find applications in the imaging, diagnosis, or monitoring of several deadly clinical conditions. Targeted biomedicine using dendrimers are able to target and deliver medicines very precisely to the diseased area. Dendrimer-based biomedicines are also employed as contrast agents for use in molecular imaging, fluorescent imaging, and magnetic resonance imaging in identification of diseased cells. Targeted dendrimers act as flexible carriers to achieve greater therapeutic effects using conventional drugs while minimizing the adverse effects (72–74).

19.6 Pulmonary Drug Delivery Devices

The anatomy and physiology of the pulmonary airways enable the use of almost 4000-year-old inhalation therapy in modern remedies (75,76). Inhalation therapy is capable of treating different intra- and extra-pulmonary diseases like asthma, pulmonary hypertension, pneumonia, tuberculosis, chronic obstructive pulmonary disease (COPD), and diabetes (75,76). As inhalation is a low-invasive therapy, it shows improved patient compliance with the treatment. Eventually, availability of easy-to-use inhalation formulations and devices have made inhalation therapy a more patient friendly treatment (77). In present times, promising interest is seen for design and development of various types of nanoparticulate, macromolecular, and/or polymer conjugate systems for pulmonary delivery applications of various therapeutics. Among these, some systems are designed and prepared to be taken up by macrophages for direct drug delivery to bacteria (78,79), which are useful to treat pulmonary tuberculosis (75,80). Also, these systems can achieve enhanced drug accumulation inside the tumors via passive or active targeting, thus are effective for airway cancer therapy (76,80). In spite of many advantages, easy exhalation from the lungs after pulmonary delivery is a major drawback of pulmonary delivery systems (80,81). Therefore, strategies which form nanoparticles with suitable aerodynamic diameters, such as nano-agglomeration and nanoparticle-rooted microparticles have been employed to overcome this limitation (75,79,80).

Pulmonary route has the capacity to deliver a wide range of molecules. However, poor water solubility of nonpolar therapeutic agents retards their delivery to pulmonary airways. The non-polar therapeutic agents can be formulated as nanoparticulates using functional or biocompatible polymers and fabricated as polymeric nanoparticles, polymeric micelles or core–shell nanoparticles. Advanced particle engineering techniques like wet bead milling, high-pressure homogenization, emulsification diffusion, and supercritical fluid technology can also be used to design and develop nano-sized particles for pressurized metered dose inhaler (pMDI). The typical pMDI contains a pressurized metal container to hold a drug and a surfactant, where a metering valve is placed in a plastic sheath with an attached mouthpiece (82). The pMDI formulation possesses the drug particles dispersion in a hydrofluoroalkane (HFA) propellant with surfactant(s) to stabilize the suspended drug particles as well as to control their aerodynamic performance. Some other excipients like antioxidants (e.g. ascorbic acid), flavoring agents (e.g. saccharin or menthol), and valve lubricants (e.g. polyvinylpyrrolidone, oleic acid, polyethylene glycol) are included in pMDI formulations. When delivered by pulmonary route, nano-sized particles demonstrate enhanced dispersibility along with enhanced total delivered drug dose and enhanced capacity to bypass macrophage clearance in the alveolar region (82,83). Therefore, development of inhalable dendrimer pMDI formulations with enhanced aerodynamic performance and physical stability is one of greatest research interests.

Nebulizers are widely used pulmonary delivery devices. Unlike pMDIs, nebulizers do not need coordination of actuation and inhalation (79). In nebulizers, drug dose is given by the usual tidal breathing or by unusual breathing during exacerbation in very sick, old, or young patients. The treatment of various pulmonary diseases such as asthma, cystic fibrosis, bronchiectasis, pulmonary hypertension, pneumonia, COPD, and tuberculosis have been investigated using nebulizers. For the pulmonary delivery, three major techniques such as air-blast atomization, high frequency vibration, and colliding liquid jets are available for dispersing a drug-containing formulation into vapor droplets. Modern simple, portable, and battery-operated nebulizers are available for a wide range of inhalable drugs (84–86). However, nebulizers are limited to stationary use only. Dry powder inhalers (DPIs) can resolve limitations of nebulizers and pMDIs to some extent. Major advantages of DPIs include: improved physicochemical stability compared to solution- or suspension-based inhaled formulations; no need of refrigeration, cold chain storage, or reconstitution; and dispersibility capacity of respirable powder as fine particles by the inspiratory breathing of the patient (75).

A DPI system is composed of micronized drug, carrier molecules, and an inhaler device. The carrier molecule builds formulation bulk facilitating the dispensing and the actuation of the micronized substance with the ease of handling. The most commonly used carrier in DPI systems is alpha-lactose monohydrate. The actuation is highly important for low dose DPIs for which precise dose up to 50–500 mg per actuation is needed (75–77). The DPIs have been used for inhalation therapy for the delivery of wide variety of actives including steroids, bronchodilators, anti-cholinergics, or some combination. The factors such as particle size distribution, particle shape, particle surface, inhalation flow rate, and dispersion power from the DPI device are crucial to the effective DPI formulation (75,76). In recent times, novel drug delivery-based DPIs containing microparticles, polymeric nanoparticles, polymeric micelles, lipid vesicles, and polymer–drug conjugates have shown promise to resolve the carrier-associated problems, proffering improved flow properties and drug dispersion and superior physicochemical stability, as well as enhanced tissue distribution and bioavailability (76–79).

19.7 Dendrimers in Pulmonary Drug Delivery Applications

The distinguished dendrimer characteristics are beneficial from the synthesis, formulation, and pharmacological perspectives in development of dendrimer-based pulmonary drug delivery (80,81). Several researchers have studied dendrimer macromolecules for their potential in pulmonary delivery of various therapeutics.

19.7.1 PAMAM Dendrimers

PAMAM is the most extensively investigated dendrimer for pulmonary drug delivery. The endocytic uptake of anionic PAMAM dendrimers in lung tissues was compared with that of dextran probes which had identical molecular sizes as PAMAM. Both dendrimer and dextran exhibited passive absorption in the intact lung, however, the dendrimer demonstrated slower absorption, lung biocompatibility, rapid uptake into the pulmonary epithelium, and extended lung transportation as compared to dextran supporting the use of inhaled PAMAM-based delivery for controlled drug release to the lungs (87,88).

The inhalable G4-amineterminated PAMAM dendrimer-siRNA complexes (dendriplexes) were developed by spray drying using mannitol, as opposed to emulsification diffusion using chitosan-g-lactic acid. The microparticles were dispersed in propellant HFA-227 for pMDI formulation in which single-puff mannitol and chitosan-g-lactic acid were found to contain 9.5 and 26 ng of engineered microparticles, respectively, whereas propellant did not interrupt the gene silencing activity of siRNA while maintaining the efficiency of targeted lung epithelial cells (89). The G3 PAMAM dendrimers nanocarriers were synthesized using carboxy terminated poly(d,l-lactide-co-glycolide) (PLGA) with/without functionalization with fluorescein isothiocyanate to form various nanoblends. The functionalized PLGA nanoparticles exhibited smooth surfaces and satisfactory encapsulation efficiency, and were found to show better cellular internalization than dendrimer alone, indicating nanoblends as a dynamic approach for improving cellular uptake and transport to achieve systemic or local actions of drugs (90). The improved aerodynamic behavior of siRNA in a pMDI formulation was achieved using triphenylphosphonium (TPP)-decorated G4 PAMAM dendrimers (dendriplexes). The dendriplexes containing 12 TPP molecules with an N/P ratio of 30 exhibited improved fine particle parameters and in vitro gene knockdown efficiency as compared to DPI, indicating suitability of TPP-decorated dendrimer for pulmonary delivery of siRNA using a portable inhaler device (91). Anionic PAMAM dendrimer-based nanocarriers formulated for pMDI exhibited size-dependent passive transport across pulmonary airways and demonstrated rapid respiratory uptake with improved lung biocompatibility and extended lung transport kinetics (92). The higher generation G3 PAMAM dendrimer showed the smallest particle size and significantly improved plasma drug concentration, whereas low generation G1 showed improved solid state properties such as bulk density, porosity, and electrical charges as well as sustained drug release up to 72 hours (93). The 3D printing was utilized to prepare siRNA-dendrimer nanocomplexes which were further processed to obtain spray-dried microparticles using a mixture of inulin and trehalose for DPI formulation. The spray-dried microparticles showed better stabilization, cellular uptake, and gene silencing capacity as compared to crystalline mannitol carrier (94).

The pH-sensitive and inhalable doxorubicin–PAMAM conjugates spray-dried with mannitol exhibited quick redispersion in aqueous medium to disperse as nanoscale conjugates that could enter alveolar phagocytosis. In addition, the microparticles showed improved stability at pH 7.4 with higher sustained drug release over 120 hours in phosphate buffer at pH 5.0, indicating microparticles as an effective DPI strategy for direct delivery of large drug doses to lung tumors (95). The higher generation G4 PAMAM dendrimer improved the solubility of beclometasone dipropionate (BMDP) when delivered as BMDP-PAMAM conjugate, and displayed sustained drug release upto 8 hours, revealing that the aerodynamic performance of nebulizers was significantly influenced by the formulation properties and the device geometry rather than dendrimer generation (96). The PPI–siRNA complex nanoparticles showed highest lung distribution, whereas lowest distribution was observed in the kidney, liver, and spleen (97). The positively charged G3 PAMAM dendrimer formed complex with enoxaparin (low molecular weight heparin) due to electrostatic interactions between the anionic sulfate/carboxylic acid groups of heparin and the cationic amino groups of the PAMAM dendrimers, which showed 2.5-fold higher bioavailability of enoxaparin. The enhanced bioavailability was attributed to improved pulmonary absorption due to decreased negative surface charge density of heparin, indicating the potential of cationic dendrimer for delivery of heparin in the treatment of deep vein thrombosis after intra-tracheal instillation (98). The doxorubicin was covalently attached to PAMAM to form conjugate (PAMAM-DXR) via acid–labile hydrazone bonds. The covalent conjugate showed 80% drug release at pH 7.4, whereas only 4% of doxorubicin was released at pH 7.4 indicating the role of acid–labile hydrazone linkage in pH-dependent release of doxorubicin. Moreover, the particle sizes (9.7 nm) and zeta potential (+13.8 mV) of conjugates were found to be favorable for cellular internalization in pulmonary epithelia with sustained doxorubicin release at lysozomal pH 5 and greater stability at extracellular physiological pH 7.4. The absence of cytotoxicity below 30 µm concentrations as well as reduction in lung metastasis upon pharyngeal aspiration of dendrimers compared to its I.V. injection at an equivalent dose of 1 mg doxorubicin per kg were another potential result of dendrimer conjugates, indicating the potential of intracellular labile bonds for effective pulmonary delivery of anti-cancer agents (99). A modified PAMAM dendrimer (polyplex) with 10-bromodecanoic acid showed improved efficiency against A549 cells for effective delivery of RNAi-based genes in treatment of lung cancer (100).

19.7.2 PEGylated Dendrimers

Certain components such as amphiphilic lipids, heavy metal ions, vitamins, bile salts, systemic proteins, and genetic materials can non-specifically bind to the free amino group of the PAMAM dendrimer surfaces which results in loss of activity of these essential elements causing dendrimer-based systemic toxicities (101). PEG conjugation or PEGylation has been widely employed technique to minimize the toxicity of the dendrimer's active surface groups due to inertness, water solubility, non-immunogenicity, and non-antigenicity properties of PEG. In addition, other problems of dendrimer delivery including drug leakage, hemolysis, immunogenicity, and systemic cytotoxicity that can also be reduced via PEGylated PAMAM-based drug delivery. Moreover, it is possible to achieve enhanced solubilization of hydrophobic drug along with DNA transfection, siRNA delivery, and tumor targeting through PEGylation of dendrimers (102). The cellular biodistribution in lungs achieved peak plasma concentration in short time using positive or negatively charged PEGylated G3 PAMAM after pulmonary delivery. These results were encouraging to investigate more dendrimer-based pulmonary drug deliveries using PEGylation approach (87). The other approach to mask the free amino group on dendrimer surfaces is acetylation of the free amino groups to overcome certain limitations of dendrimers in drug delivery. However, the PEGylation dendrimer nanocarriers provide additional advantages, like their escape from reticuloendothelial system (RES), which enhances the retention and blood circulation time of the formulation within the body (103,104). The enhanced retention promotes nanocarriers' transportation across porous cancer cells resulting in enhanced permeability and retention (EPR) effect of the nanocarriers leading to increased tumor accumulation which facilitates passive tumor targeting of the dendrimer nanocarriers (105,106).

Among other dendrimers of its class, PAMAM is the most efficient for site-specific delivery of peptides and anticancer agents as it possesses highly controllable size, polyvalent surface, low systemic toxicity, and easy market availability (107–110) in addition to its hydrophilicity, biocompatibility, non-immunogenicity, bioavailability enhancement, and reduced dosing abilities (111). However, the major drawbacks of PAMAM such as cationic toxicities and related biological implications can be minimized through several surface modification approaches such as PEGylation. The surface modification of the dendrimers with target-specific ligands find promising applications for active targeting of anticancer agents to specific tumor tissues in treatment of various cancers (106,112).

The PEGylation can gradually modify the degradation pattern of the dendrimers. The PEGylated G4 dendrimer exhibited poor cellular internalization but efficient transportation through epithelial cell monolayers as compared to G3 polyester dendrimers. The PEGylation can improve the aerodynamic performance resulting in good dispersibility and deposition patterns of the aerosol particles via enhanced propellant solvation effect in pMDIs, along with improvement in the cellular transportation of both G3 and G4 hydroxyl-terminated polyester dendrimers (113). The PEGylation of PAMAM dendrimer enhanced the transport of doxorubicin across pulmonary epithelium employing a cosolvent-free doxorubicin-PAMAM conjugate in pMDI, indicating PEGylation as effective delivery strategy for anticancer agents via pulmonary route (114). The increased surface hydrophobicity of dendrimers such as G5 polylysine significantly impacted absorption, metabolism mucociliary elimination, and pulmonary pharmacokinetics of anticancer agents like methotrexate following intra-tracheal administration via aerosol microsprayer. The increased surface hydrophobicity which was achieved by substituting 50% surface PEG groups using hexapeptide linkages exhibited improved pulmonary bioavailability (115). The bioavailability and clearance of PEGylated dendrimer was influenced by size of the species. The PEGylated G4 dendrimer showed similar bioavailability in sheep and rats, but exhibited faster clearance from the sheep lungs as compared to rat lungs (116). The pulmonary delivery of PEGylated polylysine dendrimers can be better tolerated over intravenous delivery. The doxorubicin was conjugated to PEGylated G5 polylysine dendrimer using acid-labile benzoic acid linkers for the treatment of lung metastasis. When administered twice weekly via intra-tracheal instillation, a two-fold decrease in lung tumor burden was observed after 2 weeks compared to intravenous administration of doxorubicin solution (117).

19.8 Dendrimer Toxicities and Countermeasure Strategies

Although dendrimers have potential biomedical and pharmaceutical applications, these macromolecules must be addressed for their toxicities. The biological interactions of dendrimers with essential bodily components such as iron and zinc via complex formation is mainly attributed to the small dendrimer sizes (1–100 nm) affecting biological functions of hemoglobin and kidneys respectively. This is due to the tendency of the dendrimers to interact with cellular components of closely similar

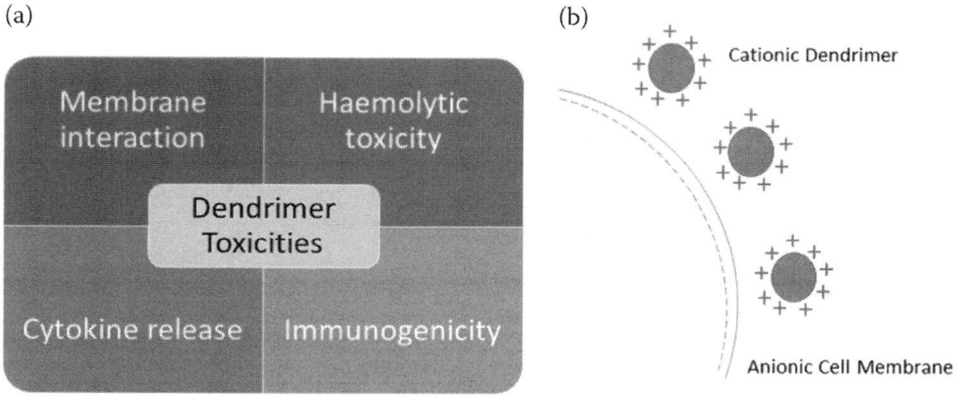

FIGURE 19.2 (a) Dendrimers' Toxicities. (b) Illustration of Dendrimer-Cell Membrane Electrostatic Interaction

size range. Another potential reason behind dendrimer-associated toxicity is surface charge, which affects the in vivo biological fate of dendrimers by manipulating their biodistribution and pharmacokinetics (118,119). The major toxicity issues associated with dendrimers are depicted in Figure 19.2(a).

The terminal groups present on the dendrimer surfaces are highly responsible for cytotoxicity. The cationic dendrimer shows electronic interactions with the negative charges of the phosphate groups of the lipid bilayer generating nanopores which result in reduced stability and increased cellular permeability as illustrated in Figure 19.2(b). The increase in dendrimer generation, molecular weight and concentration lead to more prominent ruptures of cell membrane due to increased nanopores formation. In addition, more efficient in drug transport observed with globular dendrimer structures as compared to linear polymers due to increased membrane permeability, although increased membrane permeability is one of the reason for cellular lysis (120). However, the dendrimer-induced altered permeability effects were not permanent, therefore the leaked cytoplasmic proteins return to their normal levels upon removal of dendrimers (121). The hemolytic toxicity has been observed due to interactions of cationic surface groups of the dendrimer with the red blood cells. The G4 and G5 amine-terminated PPI exhibited hemolytic effect, which was found to be reduced when the PPI surfaces were coated with galactose (122). On the other hand, absence of hemolytic effect was observed with the anionic dendrimers, indicating that the cationic dendrimer with higher generations contributes higher cationic charge, leading to increased hemolysis (119).

The release of the ROS (reactive oxygen species) can be modulated by the dendrimers which further increases the cytokine production. The PAMAM dendrimer exhibit increased intracellular ROS production and cytokine-induced cytotoxicity leading to cell death as the generation number and concentrations of dendrimer increases indicating that the toxicity of PAMAM dendrimer directly relates to the number of primary amino groups on dendrimer surface (123). Moreover, the dendrimer's generation, terminal surface groups and surface charge significantly influences their anti-inflammatory activity. The conjugation of PAMAM dendrimer surfaces with terminal amine (-NH_2), hydroxyl (-OH) or carboxylic (-COOH) groups of glucosamine inhibit the cytokine release suggesting application of dendrimers as a potential tool for various therapies like rheumatoid arthritis (124,125). In addition, the dendrimer size and surface terminal groups can contribute differently to the dendrimer immunogenicity in terms of no response, weak response, or no induction of specific antibody production. The G5 PPI produced detectable humoral immune response indicating that the body does not recognize the dendrimer nanocarrier as a foreign particle. This feature is useful to render the dendrimer suitable for drug transportation across the body (126).

Several strategies have been developed to countermeasure the dendrimer toxicities (38,121). Table 19.2 summarizes strategies used to reduce the toxicities associated with dendrimers. The most effective strategy to combat dendrimer toxicity is development of biodegradable surface functionality. However, the central core or branches can also be made biocompatible to address dendrimer toxicity. When the surface modified dendrimers are administered, they undergo degradation into nontoxic products followed by their elimination, and functions as excipients with no toxic or immunological effects (143). Surface engineering is an attractive approach to solve dendrimers' toxicities in which the modification of the dendrimer surface groups is involved. For instance, the surface modification protects cationic amino group while using neutral or anionic molecules which in turn prevents their electrostatic interactions with the cell membrane so that the cytotoxicity of the cationic surface groups can be avoided. The additional benefits of surface engineering approach include improved encapsulation, increased solubility; controllable drug release, improved biodistribution and pharmacokinetis; target specificity, and improved stability as well as increased antiviral or antibacterial activity of drugs (121).

19.9 Major Challenges

Due to ongoing interest in dendrimer macromolecules, several attempts have been made to design and develop novel dosage forms employing these molecules. Although various researchers have expanded great efforts for design, synthesis, and development synthesis of respirable dendrimer-based formulations for pulmonary delivery, its full potential has

TABLE 19.2

Strategies to Counter Dendrimer Toxicities

Strategy	Technique	Effects	Reference
Development of biocompatible or Biodegradable dendrimer	Use of polyether dendrimer	Absence of cellular toxicity	127
	Use of peptide dendrimer	Absence of hemolytic toxicities	128
	Use of polyester dendrimer	Absence of cellular toxicity and reduced drug toxicity	129
	Use of polyether imine dendrimer	Reduces hemolytic toxicities	130
	Use of phosphate dendrimer	Absence of hemolytic toxicity and cytotoxicity	131
	Use of melamine dendrimer	Reduces hepatotoxicity	132
	Use of triazine dendrimer	Absence of all toxicities	133
Development of surface-engineered dendrimer	PEGylation	Reduces hemolytic toxicity of PAMAM dendrimer with improved drug loading	134
	Carbohydrate-conjugated dendrimer	Reduces hemolytic toxicity of PPI dendrimers	135
	Acetylation	Reduces toxicity of PAMAM dendrimer with high transepithelial permeability	136
	Peptide-conjugated dendrimer	Reduces cytotoxicity of the cationic PAMAM dendrimer	137
	Antibody-conjugated dendrimer	Reduces systemic toxicity and enhances cellular internalization of the conjugate	138
	Tuftsin-conjugated dendrimer	Lowers cytotoxicity of conjugated complex dendrimer than uncomplexed PPI dendrimer	139
	Folic acid-conjugated dendrimer	Reduces hemolysis of folic acid-dendrimer conjugate compared to the PAMAM dendrimer	140
	Drug-conjugated dendrimer	Reduced hemolytic toxicity of the conjugate compared to the PAMAM dendrimer	141
	Half generation dendrimer	Reduces cytotoxicity of the PAMAM dendrimer	142

still not been explored. Although literature shows several reports which discuss particle size, generation number, surface charge, aerodynamic performanc, pulmonary biodistribution, and pharmacokinetics on in vivo fate, there are other challenges, such as achievement of sufficiently high pulmonary drug concentration and delivery of drugs to pulmonary airways, which impact the performance of dendrimer-based pulmonary delivery. Respiratory diseases and lung morphology are complex where other important parameters such as severity of disease, and the patient's age and breathing pattern have significant roles in determining the success of pulmonary therapy (144). In addition, factors such as the formulation, pulmonary device design, and structure play very important roles (75,76). Most researchers have investigated synthesis techniques, drug loading methods, formulation development strategies, and pulmonary pharmacokinetic issues, however, research efforts in the area of aerosol device-related issues are still inadequate. As the pulmonary device architecture is complex and must be focused on device design, function, and metering system, it requires strong collaborative efforts among formulation scientists and engineers to utilize the full potential of multifunctional dendrimer macromolecules for successful pulmonary drug delivery.

19.10 Current Clinical Status

Although preclinical studies in animals have studied fundamental aspects about dendrimer safety, clinical trials are required to realize interactions of dendrimers with the human body. The clinical trials related to fordendrimers can be searched in the European portal at the EU Clinical Trials Register (https://www.clinicaltrialsregister.eu) and USFDA portal (https://clinicaltrials.gov). The clinical trials have been performed with dendrimers for non small cell lung cancer, for primary hepatocellular cancer, and for understanding the pharmacokinetics, safety, and tolerability profiles (145).

For the development of dendrimer-based pharmaceutical and biomedical products, Starpharma Ltd. (Melbourne, Australia) is a world leader which presently holds more than 100 patents under its own proprietary dendrimer technology. The Starpharma's VivaGel BV employs sulfonate-terminated carbosilane dendrimers and the product is indicated for bacterial vaginosis. Recently, a novel delivery technology platform has been developed by Starpharma Ltd., known as DEP (Dendrimer Enhanced Products). Presently, DEP® product DEP® docetaxel (DTX-SPL8783, intravenous dose 200 mg) is in Starpharma's development pipeline undergoing Phase 2 clinical trials for non small cell lung cancer, whereas DEP® cabazitaxel and DEP® irinotecan are Starpharma's other clinical candidates. Dr Tomalia created a new group named Nano-Synthons in 2010, which is committed to design and develop superior quality dendrimers and dendrimer-based products (146). Some commercially available dendrimers from various suppliers are Starburst (PAMAM dendrimers), Astramol (PPI dendrimers), and Priostar (polyetherhydroxylamine PEHAM dendrimers). In spite of a myriad of research efforts, commercialization of dendrimer-based products is still very limited (145,146).

19.11 Conclusion

Dendrimers possess potential for pulmonary drug delivery applications for both respiratory and non-respiratory diseases. However, more strategic efforts are still required to synthesize, design, and skilfully authenticate dendrimer-based inhalable formulations. In this context, it is crucial to identify superior dendrimer carriers to develop best pulmonary formulation targeting their clinical translation. To achieve this goal, future research efforts should consider the current knowledge to bridge the gaps in dendrimer research instead focusing only on designing and developing novel dendrimers.

Conflicts of Interest

The authors report no conflicts of interest regarding the present work.

REFERENCES

1. Tomalia DA, Baker H, Dewald J, et al. A new class of polymers: Starburst-dendritic macromolecules. *Polymer Journal (Tokyo)*. 1985; 17: 117–132.
2. Buhleier E, Wehner W, and Vogtle F. Cascade and nonskid-chain-like syntheses of molecular cavity topologies. *Synthesis*. 1978; 2: 155–158.
3. Tomalia DA. Birth of a new macromolecular architecture: Dendrimers as quantized building blocks for nanoscale synthetic polymer chemistry. *Progress in Polymer Science*. 2005; 30: 294–324.
4. Caminade AM, Laurent R, and Majorel JP. Characterization of dendrimers. *Advanced Drug Delivery Reviews*. 2005; 57: 2130–2146.
5. Malik A, Chaudhary S, Garg G, and Tomar A. Dendrimers: A tool for drug delivery. *Advances in Biological Research*. 2012; 6(4): 165–169.
6. Newkome GR, Yao Z, Baker GR, and Gupta VK. Micelles. Part 1. Cascade molecules: A new approach to micelles. A [27]-arborol. *Journal of Organic Chemistry*. 1985; 50: 2003–2004.
7. Denkewalter RG, Kolc J, and Lukasavage WJ. US Pat. 4289872. 1981.
8. Newkome GR, Moorefield CN, and Vogtle F. Dendrimers and Dendrons: Concepts, Syntheses, Applications. Wiley-VCH, Weinheim. 2001; p. 623.
9. Frechet JMJ and Tomalia DA. Dendrimers and Other Dendritic Polymers. John Wiley & Sons, Ltd. 2002.
10. Tomalia DA, Naylor AM, and Goddard WA. Starburst dendrimers: Molecular-level control of size, shape, surface chemistry, topology, and flexibility from atoms to macroscopic matter. *Angewandte Chemie International Edition*. 1990; 29: 138–175.
11. Fischer M and Vogtle F. Dendrimers: From design to application – A progress report. *Angewandte Chemie International Edition*. 1999; 38: 884–905.
12. Weber S, Zimmer A, and Pardeike J. Solid lipid nanoparticles (SLN) and nanostructured lipid carriers (NLC) for pulmonary application: A review of the state of the art. *European Journal of Pharmaceutics and Biopharmaceutics*. 2014; 86(1): 7–22.
13. Mehta PP. Dry powder inhalers: Upcoming platform technologies for formulation development. *Therapeutic Delivery*. 2019; 10(9): 551–554.
14. Newman SP. Drug delivery to the lungs: Challenges and opportunities. *Therapeutic Delivery*. 2017; 8(8): 647–661.
15. Garcia A, Mack P, Williams S, et al. Microfabricated engineered particle systems for respiratory drug delivery and other pharmaceutical applications. *Journal of Drug Delivery*. 2012; 2012: 941243.
16. Carvalho TC and McConville JT. The function and performance of aqueous aerosol devices for inhalation therapy. *Journal of Pharmacy and Pharmacology*. 2016; 68(5): 556–578.
17. Pawar A, Rajalakshmi S, Mehta P, Shaikh K, and Bothiraja C. Strategies for formulation development of andrographolide. *RSC Advances*. 2016; 6; 69282–69300.
18. Mehta P and Pawar V. Electrospun Nanofiber Scaffolds: Technology and Applications, Applications of Nanocomposite Materials in Drug Delivery. Woodhead Publishing Series in Biomaterials. 2018; 509–573.
19. Dutta T, Jain NK, McMillan NA, and Parekh HS. Dendrimer nanocarriers as versatile vectors in gene delivery. *Nanomedicine*. 2010; 6(1): 25–34.
20. Tomalia DA, Baker H, Dewald J, et al. Dendritic macromolecules: Synthesis of starburst dendrimers. *Macromolecules*. 1986; 19: 2466–2468.
21. Tomalia DA and Frechet JMJ. Discovery of dendrimers and dendritic polymers: A brief historical perspective. *Journal of Polymer Science Part A: Polymer Chemistry*. 2002; 40: 2719–2728.
22. Dahiya R and Dahiya S. Advanced Drug De livery Applications of Self-assembled Nanostructures and Polymeric Nanoparticles. In: Anand K, Saravanan M, Chandrasekaran B, Kanchi S, Panchu SJ, and Chen QS (eds.). *Handbook on Nano-Biomaterials for Therapeutics and Diagnostic Applications*. Elsevier, Netherlands. 2021; 297–339.
23. Bryszewska M and Klajnert B. Dendrimers in biomedical applications. *Current Medicinal Chemistry*. 2012; 19(29): 4895.
24. Mostafavi E, Babaei A, and Ataie A. Synthesis of nanostructured La0.6Sr0.4Co0.2Fe0.8O3 perovskite by co-precipitation method. *Journal of Ultrafine Grained and Nanostructured Materials*. 2015; 48: 45–52.
25. Frechet JM. Functional polymers and dendrimers: Reactivity, molecular architecture, and interfacial energy. *Science*. 1994; 263: 1710–1715.
26. Dufes C, Uchegbu IF, and Schatzlein AG. Dendrimers in gene delivery. *Advanced Drug Delivery Reviews*. 2005; 57: 2177–2202.
27. Tomalia DA, Naylor AM, and Goddard WA. Starburst dendrimers: Molecular-level control of size, shape, surface chemistry, topology, and flexibility from atoms to macroscopic matter. *Angewandte Chemie, International Edition in English*. 1990; 29: 138–175.
28. Tomalia DA, Berry V, Hall M, and Hedstrand DM. Starburst dendrimers. 4. Covalently fixed unimolecular assemblages reminiscent of spheroidal micelles. *Macromolecules*. 1987; 20: 1164–1167.

29. Tomalia DA, Hall M, and Hedstrand DM. Starburst dendrimers. III. The importance of branch junction symmetry in the development of topological shell molecules. *Journal of the American Chemical Society*. 1987; 109: 1601–1603.
30. Naylor AM, Goddard WA, Kiefer GE, and Tomalia DA. Starburst dendrimers. 5. Molecular shape control. *Journal of the American Chemical Society*. 1989; 111(6): 2339–2341.
31. Araújo RV, Santos SDS, Igne Ferreira E, and Giarolla J. New advances in general biomedical applications of PAMAM dendrimers. *Molecules*. 2018; 23(11): 2849.
32. Fischer M and Vogtle F. Dendrimers: From design to application – A progress report. *Angewandte Chemie, International Edition*. 1999; 38: 884–905.
33. Roy U, Rodríguez J, Barber P, Das NJ, Sarmento B, and Nair M. The potential of HIV-1 nanotherapeutics: From in vitro studies to clinical trials. *Nanomedicine*. 2015; 10: 3597–3609.
34. Siepmann J, Faham A, Clas SD, et al. Lipids and polymers in pharmaceutical technology: Lifelong companions. *International Journal of Pharmaceutics*. 2018; 558: 128–148.
35. Dufès C, Uchegbu IF, and Schätzlein AG. Dendrimers in gene delivery. *Advanced Drug Delivery Reviews*. 2005; 57(15): 2177–2202.
36. Kim, Y, Park EJ, and Na DH. Recent progress in dendrimer-based nanomedicine development. *Archives of Pharmacal Research*. 2018; 41: 571–582.
37. Abd AS, Aziz E, and Agatemor C. Emerging opportunities in the biomedical applications of dendrimers. *Journal of Inorganic and Organometallic Polymers and Materials*. 2018; 28: 369–382.
38. Santos A, Veiga F, and Figueiras A. Dendrimers as pharmaceutical excipients: Synthesis, properties, toxicity and biomedical applications. *Materials (Basel)*. 2019; 13(1): 65.
39. Otto DP and de Villiers MM. Poly(amidoamine) dendrimers as a pharmaceutical excipient. Are we there yet? *Journal of Pharmaceutical Sciences*. 2018; 107(1): 75–83.
40. Fox LJ, Richardson RM, and Briscoe WH. PAMAM dendrimer-Cell membrane interactions. *Advances in Colloid and Interface Science*. 2018; 257: 1–18.
41. Boas U, Christensen JB, and Heegaard PMH. Dendrimers: Design, synthesis and chemical properties. In: Dendrimers in Medicine and Biotechnology. New Molecular Tools. The Royal Society of Chemistry, Cambridge, UK. 2006; 1–27.
42. Sherje AP, Jadhav M, Dravyakar BR, and Kadam D. Dendrimers: A versatile nanocarrier for drug delivery and targeting. *International Journal of Pharmaceutics*. 2018; 48(1): 707–720.
43. Parry ZA and Pandey R. The concept of dendrimers. In: Dendrimers in Medical Science. Taylor & Francis Group, Ed.; Apple Academic Press, Oakvile, Canada. 2015; 9–16.
44. Grayson SM and Fréchet JM. Convergent dendrons and dendrimers: From synthesis to applications. *Chemical Reviews*. 2001; 101(12): 3819–3868.
45. Araújo RV, Santos SDS, Igne Ferreira E, and Giarolla J. New advances in general biomedical applications of PAMAM dendrimers. *Molecules*. 2018; 23(11): 2849.
46. Caminade AM, Turrin CO, Laurent R, Ouali A, and Delavaux-Nicot B. Syntheses of dendrimers and dendrons. In: Dendrimers: Towards Catalytic, Material and Biomedical Uses. John Wiley & Sons, Chichester, West Sussex, UK. 2011; 1–33.
47. Abbasi E, Aval SF, Akbarzadeh A, et al. Dendrimers: Synthesis, applications, and properties. *Nanoscale Research Letters*. 2014; 9(1): 247.
48. Parata A and Felder-Flescha D. General introduction on dendrimers, classical versus accelerated syntheses and characterizations. In: Dendrimers in Nanomedicine. Jenny Stanford Publishing, New York, NY, USA. 2016; 1–22.
49. Kawaguchi T, Walker KL, Wilkins CL, and Moore JS. Double exponential dendrimer growth. *Journal of the American Chemical Society*. 1995; 117(8): 2159–2165.
50. Wooley KL, Hawker CJ, and Frechet JMJ. Hyperbranched macromolecules via a novel double-stage convergent growth approach. *Journal of the American Chemical Society*. 1991; 113: 4252–4261.
51. Maraval V, Caminade AM, Majoral JP, and Blais JC. Dendrimer design: How to circumvent the dilemma of a reduction of steps or an increase of function multiplicity? *Angewandte Chemie International Edition*. 2003; 42: 1822–1826.
52. Tomalia DA and Fréchet JMJ. Discovery of dendrimers and dendritic polymers: A brief historical perspective. *Journal of Polymer Science Part A: Polymer Chemistry*. 2002; 40: 2719–2728.
53. Hecht S and Fréchet JMJ. Dendritic encapsulation of function: Applying nature's site isolation principle from biomimetics to materials science. *Angewandte Chemie International Edition in English*. 2001; 40: 74–91.
54. Chauhan AS, Jain NK, Diwan PV, and Khopade AJ. Solubility enhancement of indomethacin with poly(amidoamine) dendrimers and targeting to inflammatory regions of arthritic rats. *Journal of Drug Targeting*. 2004; 12(9–10): 575–583.
55. Kumar PP, Meena K, Kumar P, Choudhary C, Singh Thakur D, and Bajpayee P. Dendrimer: A novel polymer for drug delivery. *Journal of Innovative Trends in Pharmaceutical Sciences*. 2010; 1: 252–269.
56. Gillies ER and Fréchet JMJ. Dendrimers and dendritic polymers in drug delivery. *Drug Discovery Today*. 2005; 10: 35–43.
57. Gestwicki JE, Cairo CW, Strong LE, Oetjen KA, and Kiessling LL. Influencing receptor-ligand binding mechanisms with multivalent ligand architecture. *Journal of the American Chemical Society*. 2002; 124(50): 14922–14933.
58. Mulder A, Huskens J, and Reinhoudt DN. Multivalency in supramolecular chemistry and nanofabrication. *Organic & Biomolecular Chemistry*. 2004; 2: 3409–3424.
59. Devarakonda B, Hill RA, and de Villiers MM. The effect of pamam dendrimer generation size and surface functional group on the aqueous solubility of nifedipine. *International Journal of Pharmaceutics*. 2004; 284(1–2): 133–140.
60. Inoue K. Functional dendrimers, hyperbranched and star polymers. *Progress in Polymer Science*. 2000; 25: 453–571.
61. Svenson S and Chauhan AS. Dendrimers for enhanced drug solubilization. *Nanomedicine (London)*. 2008; 3(5): 679–702.
62. Maiti PK, Çagin T, Lin ST, and Goddard WA. Effect of solvent and pH on the structure of PAMAM dendrimers. *Macromolecules*. 2005; 38: 979–991.

63. Lu Y, An L, and Wang ZG. Intrinsic viscosity of polymers: General theory based on a partially permeable sphere model. *Macromolecules*. 2013; 46: 5731–5740.
64. Mane PP, Jadhav VS, Humbe VP, Hakke SC, and Kale BB. Dendrimer: A novel polymer and tool for drug delivery. *American Journal of Pharmaceutical Research*. 2014; 4: 3851–3862.
65. Yang H and Lopina ST. Penicillin V-conjugated PEG-PAMAM star polymers. *Journal of Biomaterials Science, Polymer Edition*. 2003; 14: 1043–1056.
66. Bugno J, Hsu HJ, and Hong S. Tweaking dendrimers and dendritic nanoparticles for controlled nano-bio interactions: Potential nanocarriers for improved cancer targeting. *Journal of Drug Targeting*. 2015; 23(7–8): 642–650.
67. Cheng Y, Xu Z, Ma M, and Xu T. Dendrimers as drug carriers: Applications in different routes of drug administration. *Journal of Pharmaceutical Sciences*. 2008; 97(1): 123–143.
68. Sampathkumar S and Yarema KJ. Dendrimers in cancer treatment and diagnosis. In: Kumar CSSR. (ed.) Nanomaterial for Cancer Diagnosis. Wiley VCH, Weinheim, Germany. 2007; 7: 1–43.
69. Gupta U, Agashe HB, and Jain NK. Polypropylene imine dendrimer mediated solubility enhancement: Effect of pH and functional groups of hydrophobes. *Journal of Pharmacy & Pharmaceutical Sciences*. 2007; 10: 358–367.
70. Ramzi A, Scherrenberg R, Joosten J, Lemstra P, and Mortensen K. Structure-property relations in dendritic polyelectrolyte solutions at different ionic strength. *Macromolecules*. 2002; 35: 827–833.
71. Chai M, Niu Y, Youngs WJ, and Rinaldi PL. Structure and conformation of DAB dendrimers in solution via multidimensional NMR techniques. *Journal of the American Chemical Society*. 2001; 123: 4670–4678.
72. Le NTT, Nguyen TNQ, Cao VD, Hoang DT, Ngo VC, and Hoang Thi TT. Recent progress and advances of multi-stimuli-responsive dendrimers in drug delivery for cancer treatment. *Pharmaceutics*. 2019; 11(11): 591.
73. Najlah M and D'Emanuele A. Synthesis of dendrimers and drug-dendrimer conjugates for drug delivery. *Current Opinion in Drug Discovery & Development*. 2007; 10(6): 756–767.
74. Baker JR Jr. Dendrimer-based nanoparticles for cancer therapy. *Hematology: American Society of Hematology Education Program*. 2009;2009(1): 708–719.
75. Mehta P, Bothiraja C, Kadam S, and Pawar A. Potential of dry powder inhalers for tuberculosis therapy: Facts, fidelity and future. *Artificial Cells, Nanomedicine, and Biotechnology*. 2018; 46(Suppl 3): S791–S806.
76. Mehta P, Bothiraja C, Mahadik K, Kadam S, and Pawar A. Phytoconstituent based dry powder inhalers as biomedicine for the management of pulmonary diseases. *Biomedicine & Pharmacotherapy*. 2018; 108: 828–837.
77. Fröhlich E. Biological obstacles for identifying in vitro-in vivo correlations of orally inhaled formulations. *Pharmaceutics*. 2019; 11(7): 316.
78. Shadab M, Haque S, Sheshala R, Meng LW, Meka VS, and Ali J. Recent advances in non-invasive delivery of macromolecules using nanoparticulate carriers system. *Current Pharmaceutical Design*. 2017; 23(3): 440–453.
79. Yetisgin AA, Cetinel S, Zuvin M, Kosar A, and Kutlu O. Therapeutic nanoparticles and their targeted delivery applications. *Molecules*. 2020; 25(9): E2193.
80. Patil K, Bagade S, Bonde S, Sharma S, and Saraogi G. Recent therapeutic approaches for the management of tuberculosis: Challenges and opportunities. *Biomedicine & Pharmacotherapy*. 2018; 99: 735–745.
81. Labiris NR and Dolovich MB. Pulmonary drug delivery. Part I: Physiological factors affecting therapeutic effectiveness of aerosolized medications. *British Journal of Clinical Pharmacology*. 2003; 56(6): 588–599.
82. El-Sherbiny IM, El-Baz NM, and Yacoub MH. Inhaled nano- and microparticles for drug delivery. *Global Cardiology Science & Practice*. 2015; 2015: 2.
83. Myrdal PB, Sheth P, and Stein SW. Advances in metered dose inhaler technology: Formulation development. *AAPS PharmSciTech*. 2014; 15(2): 434–455.
84. Ari A and Fink JB. Recent advances in aerosol devices for the delivery of inhaled medications. *Expert Opinion on Drug Delivery*. 2020; 17(2): 133–144.
85. Martin AR and Finlay WH. Nebulizers for drug delivery to the lungs. *Expert Opinion on Drug Delivery*. 2015; 12(6): 889–900.
86. Pritchard JN. Nebulized drug delivery in respiratory medicine: What does the future hold? *Therapeutic Delivery*. 2017; 8(6): 391–399.
87. Zhong Q, Merkel OM, Reineke JJ, and da Rocha SR. Effect of the route of administration and PEGylation of poly(amidoamine) dendrimers on their systemic and lung cellular biodistribution. *Molecular Pharmaceutics*. 2016; 13(6): 1866–1878.
88. Oddone N, Lecot N, Fernández M, Rodriguez-Haralambides A, Cabral P, Cerecetto H, and Benech JC. In vitro and in vivo uptake studies of Pamam G4.5 dendrimers in breast cancer. *Journal of Nanobiotechnology*. 2016; 14(1): 45.
89. Conti DS, Brewer D, Grashik J, Avasarala S, and da Rocha SR. Poly(amidoamine) dendrimer nanocarriers and their aerosol formulations for Sirna delivery to the lung epithelium. *Molecular Pharmaceutics*. 2014; 11(6): 1808–1822.
90. Bharatwaj B, Dimovski R, Conti DS, and da Rocha SR. Polymeric nanocarriers for transport modulation across the pulmonary epithelium: Dendrimers, polymeric nanoparticles, and their nanoblends. *AAPS Journal*. 2014; 16(3): 522–538.
91. Bielski E, Zhong Q, Mirza H, Brown M, Molla A, Carvajal T, and da Rocha SRP. TPP-dendrimer nanocarriers for Sirna delivery to the pulmonary epithelium and their dry powder and metered-dose inhaler formulations. *International Journal of Pharmaceutics*. 2017; 527(1–2): 171–183.
92. Morris CJ, Aljayyoussi G, Mansour O, Griffiths P, and Gumbleton M. Endocytic uptake, transport and macromolecular interactions of anionic Pamam dendrimers within lung tissue. *Pharmaceutical Research*. 2017; 34: 2517–2531.
93. Rajabnezhad S, Casettari L, Lam J, Nomani AM, et al. Pulmonary delivery of rifampicin microspheres using lower generation polyamidoamine dendrimers as a carrier. *Powder Technology*. 2016; 291: 366–374.

94. Agnoletti M, Bohr A, Thanki K, Wan F, Zeng X, Boetker JP, Yang M, and Foged C. Inhalable siRNA-loaded nano-embedded microparticles engineered using microfluidics and spray drying. *European Journal of Pharmaceutics and Biopharmaceutics*. 2017; 120: 9–21.
95. Zhong Q. Co-spray dried mannitol/poly(amidoamine)-doxorubicin dry-powder inhaler formulations for lung adenocarcinoma: Morphology, in vitro evaluation, and aerodynamic performance. *AAPS PharmSciTech*. 2018; 19(2): 531–540.
96. Nasr M, Najlah M, D'Emanuele A, and Elhissi A. PAMAM dendrimers as aerosol drug nanocarriers for pulmonary delivery via nebulization. *International Journal of Pharmaceutics*. 2014; 461(1–2): 242–250.
97. Garbuzenko OB, Mainelis G, Taratula O, and Minko T. Inhalation treatment of lung cancer: The influence of composition, size and shape of nanocarriers on their lung accumulation and retention. *Cancer Biology & Medicine*. 2014; 11(1): 44–55.
98. Bai S, Thomas C, and Ahsan F. Dendrimers as a carrier for pulmonary delivery of enoxaparin, a low-molecular weight heparin. *Journal of Pharmaceutical Sciences*. 2007; 96(8): 2090–2106.
99. Zhong Q, Bielski ER, Rodrigues LS, Brown MR, Reineke JJ, and da Rocha SR. Conjugation to poly(amidoamine) dendrimers and pulmonary delivery reduce cardiac accumulation and enhance antitumor activity of doxorubicin in lung metastasis. *Molecular Pharmaceutics*. 2016; 13(7): 2363–2375.
100. Ayatollahi S, Salmasi Z, Hashemi M, Askarian S, Oskuee RK, Abnous K, and Ramezani M. Aptamer-targeted delivery of Bcl-xL shRNA using alkyl modified PAMAM dendrimers into lung cancer cells. *International Journal of Biochemistry and Cell Biology*. 2017; 92: 210–217.
101. Zeng Y, Kurokawa Y, Win-Shwe TT, Zeng Q, Hirano S, Zhang Z, and Sone H. Effects of PAMAM dendrimers with various surface functional groups and multiple generations on cytotoxicity and neuronal differentiation using human neural progenitor cells. *Journal of Toxicological Sciences*. 2016; 41(3): 351–370.
102. Luong D, Kesharwani P, Deshmukh R, Mohd Amin MCI, Gupta U, Greish K, and Iyer AK. PEGylated PAMAM dendrimers: Enhancing efficacy and mitigating toxicity for effective anticancer drug and gene delivery. *Acta Biomaterialia*. 2016; 43: 14–29.
103. Wang H, Zheng L, Guo R, Peng C, Shen M, Shi X, and Zhang G. Dendrimer-entrapped gold nanoparticles as potential CT contrast agents for blood pool imaging. *Nanoscale Research Letters*. 2012; 7(1): 190.
104. Paolucci V, Mejlsøe SL, Ficker M, Vosch T, and Christensen JB. Photophysical properties of fluorescent core dendrimers controlled by size. *Journal of Physical Chemistry B*. 2016; 120(36): 9576–9580.
105. Gorain B, Choudhury H, Pandey M, et al. Dendrimers as effective carriers for the treatment of brain tumor. In: Nanotechnology-Based Targeted Drug Delivery Systems for Brain Tumors. Elsevier. 2018; pp. 267–305.
106. Sharma AK, Gothwal A, Kesharwani P, Alsaab H, Iyer AK, and Gupta U. Dendrimer nanoarchitectures for cancer diagnosis and anticancer drug delivery. *Drug Discovery Today*. 2017; 22(2): 314–326.
107. Pettit MW, Griffiths P, Ferruti P, and Richardson SC. Poly(amidoamine) polymers: Soluble linear amphiphilic drug-delivery systems for genes, proteins and oligonucleotides. *Therapeutic Delivery*. 2011; 2(7): 907–917.
108. Qin W, Yang K, Tang H, Tan L, Xie Q, Ma M, Zhang Y, and Yao S. Improved GFP gene transfection mediated by polyamidoamine dendrimer-functionalized multi-walled carbon nanotubes with high biocompatibility. *Colloids and Surfaces B: Biointerfaces*. 2011; 84(1): 206–213.
109. Bharatwaj B, Mohammad AK, Dimovski R, Cassio FL, Bazito RC, Conti D, Fu Q, Reineke J, and da Rocha SR. Dendrimer nanocarriers for transport modulation across models of the pulmonary epithelium. *Molecular Pharmaceutics*. 2015; 12(3): 826–838.
110. Duncan R and Izzo L. Dendrimer biocompatibility and toxicity. *Advanced Drug Delivery Reviews*. 2005; 57(15): 2215–2237.
111. Palmerston Mendes L, Pan J, and Torchilin VP. Dendrimers as nanocarriers for nucleic acid and drug delivery in cancer therapy. *Molecules*. 2017; 22(9): 1401.
112. Satija J, Gupta U, and Jain NK. Pharmaceutical and biomedical potential of surface engineered dendrimers. *Critical Reviews in Therapeutic Drug Carrier Systems*. 2007; 24(3): 257–306.
113. Heyder RS, Zhong Q, Bazito RC, and da Rocha SRP. Cellular internalization and transport of biodegradable polyester dendrimers on a model of the pulmonary epithelium and their formulation in pressurized metered-dose inhalers. *International Journal of Pharmaceutics*. 2017; 520(1–2): 181–194.
114. Zhong Q, Humia BV, Punjabi AR, Padilha FF, and da Rocha SRP. The interaction of dendrimer-doxorubicin conjugates with a model pulmonary epithelium and their cosolvent-free, pseudo-solution formulations in pressurized metered-dose inhalers. *European Journal of Pharmaceutical Sciences*. 2017; 109: 86–95.
115. Haque S, McLeod VM, Jones S, et al. Effect of increased surface hydrophobicity via drug conjugation on the clearance of inhaled pegylated polylysine dendrimers. *European Journal of Pharmaceutics and Biopharmaceutics*. 2017; 119: 408–418.
116. Ryan GM, Bischof RJ, Enkhbaatar P, et al. A comparison of the pharmacokinetics and pulmonary lymphatic exposure of a generation 4 PEGylated dendrimer following intravenous and aerosol administration to rats and sheep. *Pharmaceutical Research*. 2016; 33(2): 510–525.
117. Kaminskas LM, McLeod VM, Ryan GM, et al. Pulmonary administration of a doxorubicin-conjugated dendrimer enhances drug exposure to lung metastases and improves cancer therapy. *Journal of Controlled Release*. 2014; 183: 18–26.
118. Maysinger D. Nanoparticles and cells: Good companions and doomed partnerships. *Organic & Biomolecular Chemistry*. 2007; 5(15): 2335–2342.
119. Yellepeddi VK, Kumar A, and Palakurthi S. Surface modified poly(amido)amine dendrimers as diverse nanomolecules for biomedical applications. *Expert Opinion on Drug Delivery*. 2009; 6(8): 835–850.
120. Lee H and Larson RG. Lipid bilayer curvature and pore formation induced by charged linear polymers and dendrimers: the effect of molecular shape. *Journal of Physical Chemistry B*. 2008; 112(39): 12279–12285.

121. Jain K, Kesharwani P, Gupta U, and Jain NK. Dendrimer toxicity: Let's meet the challenge. *International Journal of Pharmaceutics*. 2010; 394(1–2): 122–142.
122. Bhadra D, Yadav AK, Bhadra S, and Jain NK. Glycodendrimeric nanoparticulate carriers of primaquine phosphate for liver targeting. *International Journal of Pharmaceutics*. 2005; 295(1–2): 221–233.
123. Naha PC, Davoren M, Lyng FM, and Byrne HJ. Reactive oxygen species (ROS) induced cytokine production and cytotoxicity of PAMAM dendrimers in J774A.1 cells. *Toxicology and Applied Pharmacology*. 2010; 246(1–2): 91–99.
124. Avti PK and Kakkar A. Dendrimers as anti-inflammatory agents. *Brazilian Journal of Pharmaceutical Sciences*. 2013; 49: 57–65.
125. Roberts JC, Bhalgat MK, and Zera RT. Preliminary biological evaluation of polyamidoamine (PAMAM) starburst dendrimers. *Journal of Biomedical Materials Research*. 1996; 30(1): 53–65.
126. Agashe HB, Dutta T, Garg M, and Jain NK. Investigations on the toxicological profile of functionalized fifth-generation poly (propylene imine) dendrimer. *Journal of Pharmacy and Pharmacology*. 2006; 58(11): 1491–1498.
127. Malik N, Wiwattanapatapee R, Klopsch R, et al. Dendrimers: Relationship between structure and biocompatibility in vitro, and preliminary studies on the biodistribution of 125I-labelled polyamidoamine dendrimers in vivo. *Journal of Controlled Release*. 2000; 65(1–2): 133–148.
128. Agrawal P, Gupta U, and Jain NK. Glycoconjugated peptide dendrimers-based nanoparticulate system for the delivery of chloroquine phosphate. *Biomaterials*. 2007; 28(22): 3349–3359.
129. Twibanire JK and Grindley TB. Polyester dendrimers: Smart carriers for drug delivery. *Polymers*. 2014; 6(1): 179–213.
130. Rama Krishna T, Jain S, Tatu U, and Jayaraman N. Synthesis and biological evaluation of 3-amino-propan-1-ol based poly(ether imine) dendrimers. *Tetrahedron*. 2005; 61: 4281–4288.
131. Domański DM, Bryszewska M, and Salamończyk G. Preliminary evaluation of the behavior of fifth-generation thiophosphate dendrimer in biological systems. *Biomacromolecules*. 2004; 5(5): 2007–2012.
132. Neerman MF, Chen HT, Parrish AR, and Simanek EE. Reduction of drug toxicity using dendrimers based on melamine. *Molecular Pharmaceutics*. 2004; 1(5): 390–393.
133. Lim J, Guan B, Nham K, Hao G, Sun X, and Simanek EE. Tumor uptake of triazine dendrimers decorated with four, sixteen, and sixty-four PSMA-targeted ligands: Passive versus active tumor targeting. *Biomolecules*. 2019; 9(9): 421.
134. Bhadra D, Bhadra S, Jain S, and Jain NK. A PEGylated dendritic nanoparticulate carrier of fluorouracil. *International Journal of Pharmaceutics*. 2003; 257(1–2): 111–124.
135. Klajnert B, Appelhans D, Komber H, et al. The Influence of densely organized maltose shells on the biological properties of poly(propylene imine) dendrimers: New effects dependent on hydrogen bonding. *Chemistry*. 2008; 14(23): 7030–7041.
136. Kolhatkar RB, Kitchens KM, Swaan PW, and Ghandehari H. Surface acetylation of polyamidoamine (Pamam) dendrimers decreases cytotoxicity while maintaining membrane permeability. *Bioconjugate Chemistry*. 2007; 18(6): 2054–2060.
137. Yang H and Kao WJ. Synthesis and characterization of nanoscale dendritic RGD clusters for potential applications in tissue engineering and drug delivery. *International Journal of Nanomedicine*. 2007; 2(1): 89–99.
138. Shukla R, Thomas TP, Peters JL, et al. HER2 specific tumor targeting with dendrimer conjugated anti-HER2 mAb. *Bioconjugate Chemistry*. 2006; 17(5): 1109–1115.
139. Dutta T, Garg M, and Jain NK. Targeting of efavirenz loaded tuftsin conjugated poly(propyleneimine) dendrimers to HIV infected macrophages in vitro. *European Journal of Pharmaceutical Sciences*. 2008; 34(2–3): 181–189.
140. Singh P, Gupta U, Asthana A, and Jain NK . Folate and folate-PEG-PAMAM dendrimers: Synthesis, characterization, and targeted anticancer drug delivery potential in tumor bearing mice. *Bioconjugate Chemistry*. 2008; 19(11): 2239–2252.
141. Asthana A, Chauhan AS, Diwan PV, and Jain NK. Poly (amidoamine) (PAMAM) dendritic nanostructures for controlled site-specific delivery of acidic anti-inflammatory active ingredient. *AAPS PharmSciTech*. 2005; 6(3): E536–E542.
142. Jevprasesphant R, Penny J, Jalal R, Attwood D, McKeown NB, and D'Emanuele A. The influence of surface modification on the cytotoxicity of PAMAM dendrimers. *International Journal of Pharmaceutics*. 2003; 252(1–2):263–266.
143. Mishra V, Gupta U, and Jain NK. Surface-engineered dendrimers: A solution for toxicity issues. *Journal of Biomaterials Science Polymer Edition*. 2009; 20(2): 141–166.
144. Giovagnoli S, Schoubben A, and Ricci M. The long and winding road to inhaled TB therapy: Not only the bug's fault. *Drug Development and Industrial Pharmacy*. 2017; 43(3): 347–363.
145. Mehta P, Kadam S, Pawar A, and Bothiraja C. Dendrimers for pulmonary delivery: Current perspectives and future challenges. *New Journal of Chemistry*. 2019; 43: 8396–8409.
146. Starpharma.Dendrimer drug delivery. Available from: https://www.starpharma.com/drug_delivery

20

Hybrid Lipid/Polymer Nanoparticles for Pulmonary Delivery of siRNA: Development and Fate Upon In Vitro Deposition on the Human Epithelial Airway Barrier

Samyak Nag[1], Shivani Baliyan[1], Atul Pathak[2], and Swati Gupta*[1]
[1]*Amity Institute of Pharmacy, Amity University Uttar Pradesh, Noida, India*
[2]*HBT Labs Inc., Vanguard Way Brea, CA, USA*

20.1 Introduction

The pulmonary delivery of drugs is the most striking target and of incredible scientific and biomedical attention in the research of health care because the lung is able to absorb pharmaceuticals for limited deposition or also for systemic delivery (1). The cells of epithelium for respiration play a vital role in regulating the tone of an airway system and for producing fluid that lines the airway (2). So rising consideration has been focused on the potential of route of pulmonary for both the local as well as systemic delivery of therapeutic agents as a non-invasive administration, due to the higher permeability and large absorption surface of lungs, and a good supply of blood (3). Further it has several advantages over applications of peroral that have lower enzyme activity in comparison, fast drug absorption and potential to overcome first-pass metabolism.

20.1.1 Mechanism for Administration of Pulmonary siRNA Delivery

In pulmonary route, administration of drug is done by two primary modes: (i) intranasal administration, having some structural limitations, like narrow air-pathway lumen, (ii) oral inhalation administration (4). Orally, much better outcomes are observed because it controls the administration of very smaller particles with a loss of concentration of 20% only, while comparing nasal route up to 85%. It can be further sub-classified as intra-tracheal instillation and intra-tracheal inhalation (5).

The widely used method for laboratory purpose is the intra-tracheal instillation. In this, a little amount of solution of drug or dispersion is released in the lungs via special syringe. This is the fast and assessable procedure for delivering the drugs into the lungs. The local deposition of drug is attained with a relatively smaller absorption area (6). So the process of instillation is very simple, inexpensive, and has irregular distribution of drug. On the other hand, method of inhalation utilizes a technique of aerosol through which we could obtain more uniformly distributing pattern with greater penetration. However, this procedure is costly and problematic for the measurement of the precise dose in the lungs.

Figure 20.1 illustrates the mechanism of delivery of siRNA into the lungs. RNA intrusion states that the repression of gene assertion was done by small RNA molecules of double strands (also known as siRNA) that direct mRNA technology for the degradation of a particular type of mRNA. The siRNA therapeutic aspects possess the simulated potential for the specific silencing of any gene, primarily uneven intricate pathologies. In a precise manner, chronic diseases like cancer of lung or cystic fibrosis (CF), currently epitomize an actual interest in the field of various applications for the therapies that are based on siRNA. The stripped siRNA possesses a half-life less than an hour in the plasma of humans and is also incapable of penetrating the cellular membranes (7). Considering siRNA as an inhaler for local delivery, the existence of higher mucin, deoxyribonucleic acid, and actin absorptions made the air route mucus a multifaceted blockade that is difficult to defeat. So, the ideal formulations for inhalation purpose are essentia: for the stability enhancement of enhanced siRNA and to defeat the barriers of non-cellular and cellular lung, and also to rise obtain ability of siRNA on the targeted portion.

The encasing of siRNA hooked on transporters of colloids has been regarded as a fascinating method for the inhalation treatment of severe lung disease. Here in this study, the ecological poly(lactic-co-glycolic) acid (PLGA) nanoparticles (NPs) are attaining significant attention as they can also offers a great fortification or the beneficial consignment from the enzymatic breakdown, extended delivery profile (i.e. decreasing the amount of introductions), and enhanced holding within the lungs. Detailed in vitro analytical studies are permitted to evaluate the providence and the outcome of hNPs over the sprinkle on a tripartite cell co-culture model (TCCC) grown at the interface of air and liquid, which mimics the airway for human epithelial barricade. Ultimately, the properties of in vitro for hNPs that are loaded with siRNA over the countenance of aENaC and bENaC units were characterized on a human model for the cellular linings of epithelial of lungs.

DOI: 10.1201/9781003046547-20

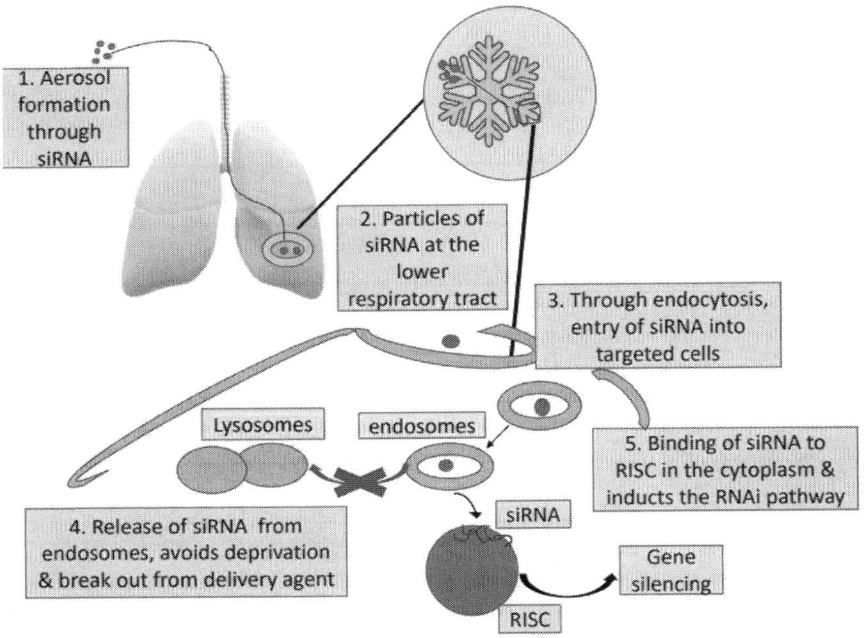

FIGURE 20.1 Schematic Demonstrations Implied in Delivery of siRNA into the Lungs

20.1.2 siRNA Delivery Barricades

siRNA clutches the potential for therapeutically active silencing of gene, but still a number of barriers are present to achieving operative and skillful in vivo distribution.

Figure 20.2 will show that the very first organic barriers come across with the ease for the ordered siRNA is the activity of nuclease within the tissues and plasma (8). The main nuclease in the claret is the 30 exonuclease, though slicing of inter-nucleotide ties could also be considered. The testified half-life of basic siRNA within the serum varies from minutes up to 1 hour. With the accumulation of mingling nuclease deprivation and kidney clearance, a key barrier for the in vivo distribution of siRNA is uptaken by the reticulo-endothelial system (RES) (9).

This RES is comprised of cells which are phagocytic in nature; as well as mixing monocytes and also tissue macrophages, their physical function is to eliminate external plague and to eradicate dead remains of cells and abate cells (10,11).

Table 20.1 illustrates the issues with bare siRNAs for clinical implementations.

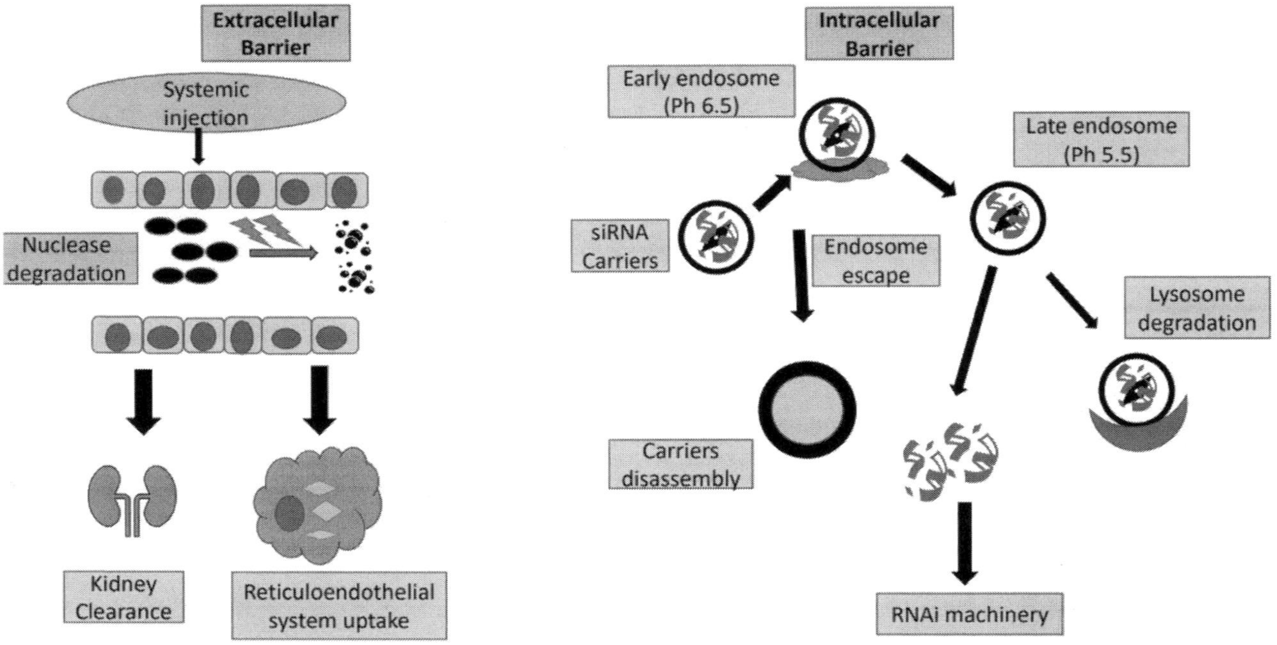

FIGURE 20.2 Barricades Encountered by Naked siRNA

TABLE 20.1

Issues with Bare siRNA for Clinical Implementations

Problems with Naked siRNA	Reason	Approach	References
Short half-life	Dilapidation through serum nucleases, fast clearance by RES and filtration through kidney, phagocytic uptake	Chemically modified siRNA, i.e. phosphorothioate, sugar, or oligo-ribonucleotides base as stabilizers, nanoparticle carriers	(12)
Lowered cellular uptake	Anionic in nature, the size is too big to pass through membrane	Complexation of siRNA with cationic polymeric nanoparticle, cell penetrating peptide (CPP) that facilitates internalization	(13)
Toxicological effects inundation of RNAi machinery, stimulation of immune response	Approachability to miRNA was poor, TLR activation and Type 1 IFN response	Correct dose and definite binding to the targeted cell, chemically modified	(14)
Off-target effect	Non-TLR facilitated inborn immune response, miRNA offers off-target silencing	In vitro effect through 2'OH methylation-screening of siRNA	(15)

Other encounter for therapy of siRNA is *stimulation of immunity*, that is, the identification of duplex of siRNA with the help of inborn resistant power or system. It is investigated that if insertion of many siRNA in numbers is done, then it will give the outcomes for the stimulation of inborn immune reactions or expressions (16).

To combat such problems, the expansion of an innocuous and active system of in vivo release is required.

20.1.3 Structural Components of Lipid Polymers Hybrid Nanoparticles and Their Arrangement Mechanism

As shown in Figure 20.3 below, the central and the outer cover constituents are comprised of diverse elements of polymers, oils, metal oxides, and biological and inert composite from natural and artificial bases that are magnificently active for the production of the nanoparticles (15,17). This type of system is

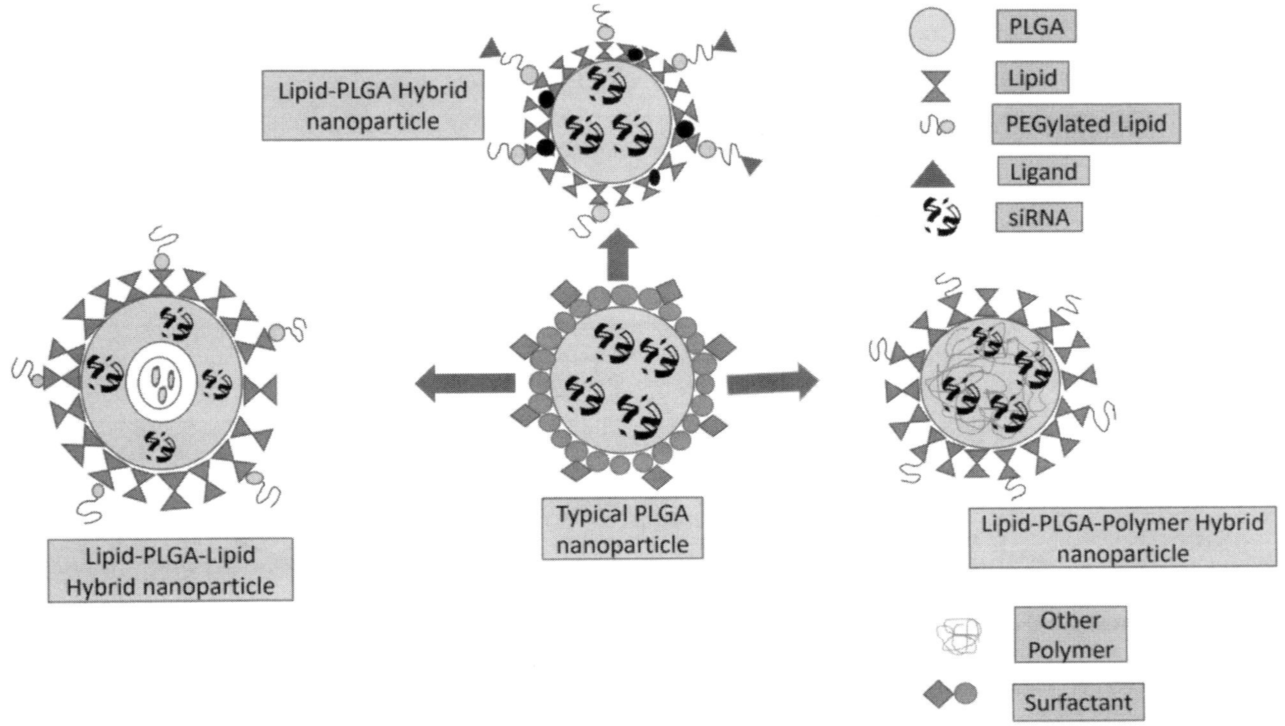

FIGURE 20.3 An Outline of Major Hybrid Poly (lactic-co-glycolic acid) (PLGA) nanoparticles. (A) Labeled Sketch Representation of the Lipid–Poly (lactic-co-glycolic acid) (PLGA)–Lipid Hybrid Nanostructure, (B) Illustrative Transmission Electron Microscopy and (C) Confocal Laser Scanning Fluorescence Image of the Hybrid Microparticles Revealing the Reality of an Outer Lipid–PEG Layer (Blue) and Inner Lipid Layer (green), Separated by a PLGA Layer (Pink). Abbreviations: PEG, Polyethylene Glycol; siRNA, Short Interfering RNA.

comprised of subsequent chief deposits and apparatuse, specified as follows:

- The internal film comprised of diverse polymers andis biological and inert constituents, which in turn are denoted as fundamental material. These fundamental constituents could be layered with some other mediators or also able to formulate the matrix assembly which is then performed and evaluated by the help of various directing moieties (18,19). Thus, sheaths of the nanoparticles could capture the drugs which are hydrophilic/hydrophobic in nature.
- The next film of such hybrid nanoparticles is capsulated within the original or nature derivative lipids which imparts various characteristics of pharmacokinetics that are necessary for drug delivery system (14). The subsequent film encases the vital principal of polymers and increases the camaraderie with that of the organic system. Also, it proceeds as a barrier for controlling penetrability which edges the delivery of encumbered therapeutically representatives for the purpose of water perforation.
- A third film is comprised of lipids and might be a conjugation of various polymers which aids for the operational and superficial embellishment of nanoparticles to offer the therapeutically interests for evaluating the effects of pharmacokinetics by means of targeting precise delivery and enhanced retaining course of the nanoparticles into the biotic system (20).

However, the machinery for the preparation in diverse films and further assembling within each other would require additional evaluation.

20.2 RNA Interference Mechanism (RNAi)

20.2.1 Therapeutically Active Silencing of Gene with the Help of siRNA

RNA interference (RNAi) is a vital path that originates in several eukaryotes as well as animals (21,22). Figure 20.4 will represent that the triggering of RNAi because of the existence of extended RNA possessing double strands that is also fragmented and is identified as siRNA which is comprised of long nucleotides about 21 to 23 in counts, by the help of endonuclease speculator. siRNA was encumbered on top of RNA-induced silencing complex (RISC) (Farheen et al. 2014). This RISC encompasses a protein known as Argonaute (Ago-2) which is capable of slicing and eliminating the nearside strands of siRNA folds. The singlet strand of RNA gives the directions for connotation with the complex of RISC based protein which further guides the preciseness for targeting the identification of mRNA by means of corresponding base combination (23,24). The protein Argonaute-2 damages that particular mRNA which corresponds to that of the reverse strand and endonucleolytic divisions, which arise between 10th and 11th base compared to the 50 ends of the strand of reverse siRNA, thus instigating splicing of genes and deterioration of messenger RNA.

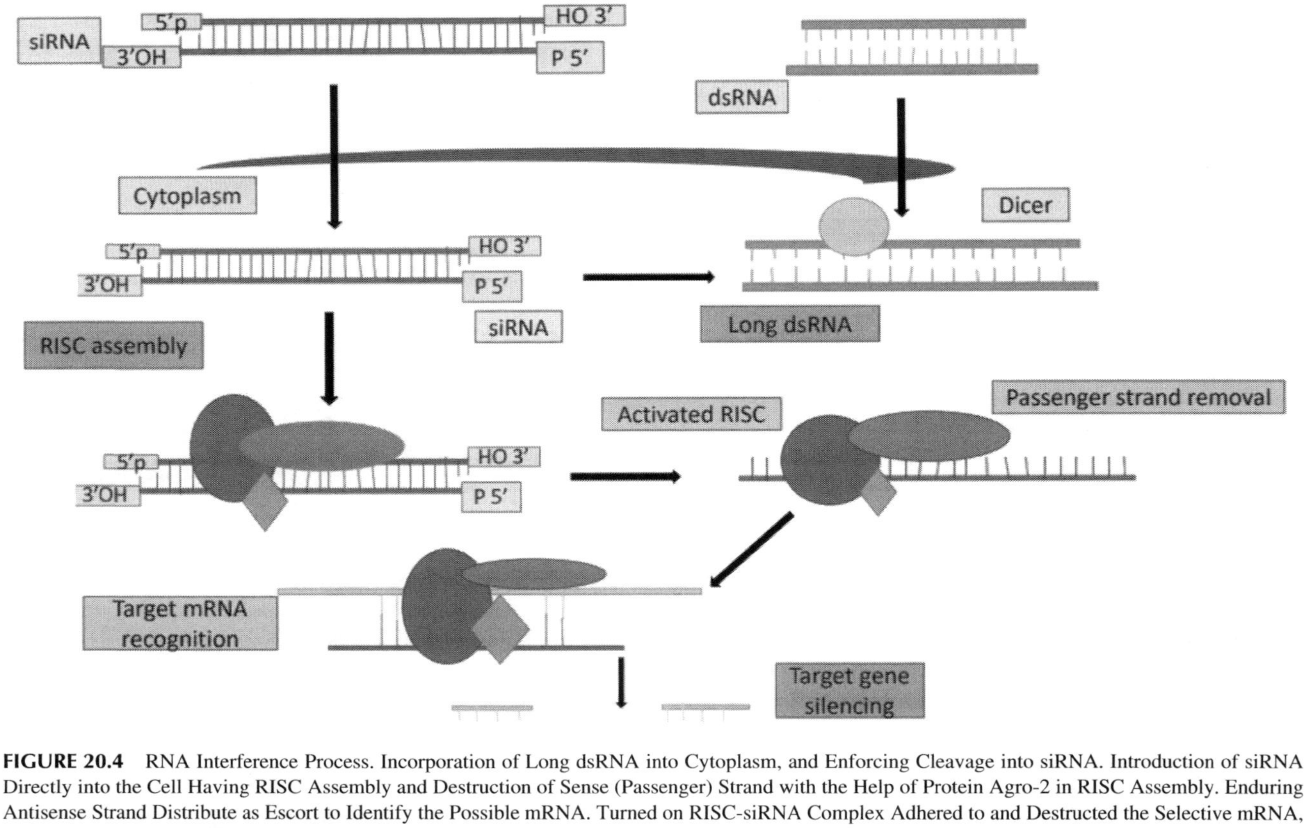

FIGURE 20.4 RNA Interference Process. Incorporation of Long dsRNA into Cytoplasm, and Enforcing Cleavage into siRNA. Introduction of siRNA Directly into the Cell Having RISC Assembly and Destruction of Sense (Passenger) Strand with the Help of Protein Agro-2 in RISC Assembly. Enduring Antisense Strand Distribute as Escort to Identify the Possible mRNA. Turned on RISC-siRNA Complex Adhered to and Destructed the Selective mRNA, and Enforcing the Silencing of Selective Gene.

20.3 General Method of Preparation for LPHNPs

20.3.1 Two-Step Method

20.3.1.1 Conventional Method

Throughout the initiation of the study done by Zeng et al., 2019 (25), a distinctive two-step method was often implied for the formation of LPHNPs, in which already-prepared polymeric NPs were thoroughly mixed with the vesicles of lipid that had been formed already and were surface integrated against the prior one, driven by electrostatic forces of propellers. This preparation of polymeric nanoparticles is usually done by the nano-precipitation method, emulsification–solvent-evaporation (ESE), or by homogenization using high pressure.

Thus, according to the study of Liu et al., 201026, it can be further divided into two categories: (i) already existing preformed polymeric nanoparticles were added directly to the film of lipid which was dried earlier, or (ii) preformed nanoparticles were added to existing vesicles of lipid, which was prepared by preliminary hydration of the lipid thin film. Moreover, according to Li et al., 2014 27 in both the cases, the hybrids are gathered by supplying the energy from outside through vertexing/ultrasonication of the suspension and providing heat at a certain temperature far apart from the temperature of phase transition of the constituents of lipid. In the process of purification, separation of free lipid and LPHNPs is done accordingly by differential centrifugation. For obtaining a homogeneous nanoparticle suspension with single modal distribution of particles, the hybrid nanoparticles undergo progressive extrusion/homogenization after the preparation. At the time of sample extrudation (done at laboratory scale), the nanoparticle suspension is downscaled by passing through a porous membrane.

20.3.1.2 Non-conventional Method

Apart from the above elaborated methods, there are other methodologies also, that have been applied for manufacturing LPHNPs as done by Zhang et al., 2009). For instance, they made polymeric nanoparticles (like polyglutamic acid) of usual size ranges from 400 to 500 nm, formed by spray-drying method, by dispersing them in DCM that was composed of lipids (tripalmitin, tristearin, etc.). After that, the suspension was again further spray dried for the preparation of LPHNPs up to the range of 0.9–1.2 μm in size with a spray dryer, which was not appropriate for the formation of nanoparticles.

20.3.1.2.1 Formulation Parameters Optimization

According to the work done by Zhang et al., 2010 29, the physical characterization of LPHNPs was done by the two-step method measured by the described formulating parameters: (i) prior prepared lipid vesicle size and polydispersity index, (ii) potential of the surface for the external shell of lipid, (iii) ionic measure of strength of the aqueous phase, and (iv) ratio of lipid with respect to polymer. Generally, particles that are produced by extrusion are much smaller and more homogeneous in comparison to those produced by direct hydration of the film of lipid within the solution of polymer (14). Moreover, the homogeneousness of LPHNPs is abruptly dependent on the lipid vesicles charge. Although minimum aggregation of particle with the higher stability of colloids and slight PDI range was achieved by the use of only a single type of lipid for the formation of vesicles, i.e. the zwitterionic lipid-DPPC or the cationic-lipid DPTAP.

The studies done by Zhang et al., 2010 (30,31), proved that, similar to liposomes, LPHNPs possess a drawback, i.e. their weaker stability for the colloids with respect to that of the strength of ionic solution: the electrostatic force of attraction can't stabilize the particles of hybrid in aqueous medium containing >10 mm NaCl of ionic strength. Similarly, like liposomes, it is alleviated by incorporation of PEG-conjugated lipids into the mixture of lipid (DPPC/DPTAP) to convey PEG chains that have been stabilized over the surface of the LPHNPs. Along with average size, PDI, and potential of the surface (measurement of stability of colloids), the other significant parameters for the formulation are percentage of drug loading (%DL), percentage of encapsulation efficiency (%EE), and in vitro release drug profile from the LPHNPs (Kandil et al. 2019).

20.3.2 One-Step Method

In consonance with Jaglal et al., 2021 (32), there is one drawback for the two-step method, that is, the preparation of polymeric nanoparticles and vesicles of lipid distinctly make the procedure ineffective with regard to energy and the time expenditure. So the one-step method is used that proves to be more efficient. The vesicles of lipid that have been preformed and polymeric nanoparticles are not the rudiments for the one-step method. The method only entails mixing of solutions of polymer and lipids that later on tend to self-assemble for the formation of LPHNPs. Mostly the methods used for this purpose are ESE/or nano-precipitation; both of them are applied for the formation of nonhybrid-polymeric nanoparticles. The lipids/PEG–lipids utilized here as stabilizers done by Khalili et al., 2021 (29) for production of hybrid, whereas surfactants which are ionic or nonionic in nature, like PVA, DMAB, poloxamer are commonly used as stabilizing agents in the formulation of regular, nonhybrid polymeric nanoparticles.

20.3.3 Nano-precipitation

According to the study by Sengel-Turk et al., 2020 (33), the method of nano-precipitation demands the drug and the polymer should dissolve collectively in an organic solvent miscibility with water like acetone or EtOH, and the lipid/lipid–PEG mixed in aqueous medium, i.e. water. It is compulsory to heat the solution of lipid/lipid–PEG until it gets converted from gel to the liquid state at a transitional temperature for achieving a homogeneous dispersed liquid-crystalline phase. Further, in consonance with Abadi et al., 2021 (29), it was proceeded by adding the polymer in a dropwise manner into the lipid aqueous dispersion with constant stirring. This activates the polymer to coil further into nanoparticles with simultaneous self-assembling of the lipids that surround the polymer due to hydrophobic interactions, wherever tails that are hydrophobic in nature for the lipids are focused toward the internal nanoparticles and the hydrophilic-head part expressed outward to the exterior aqueous solution. The tails of hydrophobic lipid of the lipid–PEG fuse into the interior shell of lipid, whereas its PEG chains, expressed out to the aqueous environment, are responsible for the stabilization of the hybrid. Then evaporation of organic media is done, which leads to the formation of the LPHNPs and further completion by centrifugation.

20.3.3.1 Optimization of Formulation Parameters in Nano-precipitation

According to the study done by Elkateb et al., 2020 (34), the proposed ratio of lipid/polymer was optimized upto 15% (w/w) for optimum casing of the polymer, where the polymer used was PLGA and lipid was lecithin plus DSPE–PEG, for the purpose of generating stabilized mono-disperse LPHNPs (60–80 nm). This criteria for optimization was vital because the higher ratio of L/P (w/w) will lead to greater concentration of lipid as compare to CMC of the polymer (critical micelle concentration) that would be responsible for forming liposomes with addition to the hybrids, whereas lowering the ratio of L/P (w/w) will cause agglomeration. Similarly, in accordance with Pukale et al., 2020 (35), LPHNPs, which are cationic in nature, having a size range of ~65 nm approximately were formulized by single-step nano-precipitation for the delivery of siRNA systematically, including a lipid of cationic nature (BHEM-Chol) and amphiphilic polymer mPEG–PLA, where the ideal lipid/polymer ratio was stated to be 10% (w/w).

Pandita et al. (36) presented an outstanding characteristic of the coating of lipid (considered as ratio of lipid/polymer): for the protection of the core of the polymer. This controls %EE and release of drug, indirectly. Whereas, for the nanoparticles that are nonhybrid polymeric in nature, the physicochemical property of the polymer is the solitary effect on %EE and the release profile.

The stability of colloidal properties of LPHNPs (designed either by one-step or two-step method) done by Bose et al. (37), is dependent on the lipid–PEG existiung in the formulation of lipid. By increasing the PEGylated lipid in the formulation of lipid, the hybrid nanoparticles thus formed are more stable, whereas the ideal amount of lipid–PEG was obtained at ~25% (w/w). So, alterations in the concentration of PEG at a certain higher level on the surface of the particle did not change the release profile kinetics or %EE. The fraction of PEG exists in the preparation will lead to effect the average size of LPH nanoparticles.

The other formulating parameters, like molecular weight and the amount of polymer, and ratio of water to that of solvent, have analogous properties on the size of both nonhybrid as well as hybrid polymeric nanoparticles.

20.3.4 Emulsification–Solvent Evaporation

This method of evaporation can be subclassified as the methods of single and double emulsification (38). A single ESE method is viable for those drugs which are soluble in solvents of hydrophobic nature (i.e. oil phase). In this process, an oil-in-water (o/w) emulsion is obtained during mixing of the drug and an oil phase of polymer which is immiscible in water along with an aqueous phase holding dissolved lipid by means of ultra-sonication or continuous stirring. The core of the polymer is formed by the process of evaporation of the biological media and the lipids were assembled concurrently at the surrounding of polymer core.

According to the work done by Dave et al. (39), the double ESE method, i.e. (w/o/w), is useful for those drugs that are soluble in water. Firstly, emulsification is done by mixing drug aqueous solution in an organic solvent (i.e. oil phase) which contains polymer and lipid to obtain a mixture of w/o. A w/o/w emulsion is produced when this mixture is run into the emulsification process again in an aqueous phase comprising the lipid–PEG, further followed by consequent evaporation of oil phase, to produce the LPH nanoparticles.

20.3.4.1 Optimization in ESE

Similar to nano-precipitation, the ratio of lipid to polymer is the most influential within all the parameters used in the formulation to direct ESE method. It was found that use of PLGA, TPGS, and PC, leads to a decrease in size with rise in the ratio L/P and accomplished an ideal yield of production in terms of (w/w).

The ratio of L/P was showed an impact on %EE. The ideal concentration for lipid in aspect of small size and higher %EE was found to be 0.04% (w/v). Amazingly, the ESE method characteristically resulted in high %EE as compare to the nano-precipitation because of its production of large-sized particles.

20.3.5 Manufacturing of DPPC/PLGA Hybrid NPs

d'angelo et al. (39) approached by emulsion or solvent diffusion for manufacturing DPPC/PLGA hybrid NPs. Water in oil emulsion (w/o) was attained by incorporating 100 μl of water into 1 ml of methylene blue accommodating PLGA with quantity 10 mg and 1% w/v and DPPC at distinct DPPC/PLGA weight ratios containing 1:20, then 1:50, then 1:100, and 1:150 with the help of vortex mixing (40).

They incorporated PEI into internal water phase and keeping consideration of loading of 0.016 mg per 100 mg of PGLA. When the mixing was finished, the w/o emulsion was incorporated into 12.5 ml of ethanol 96% (v/v) with the help of moderate magnetic stirring and enforced to instantaneous nanoparticle precipitation.

They attenuated the formulation with the help of 12.5 ml of Milli-Q water by enforcing stirring for 10 minutes and rotary evaporation was enforced to evaporate residue organic solvent under vacuum at 30 °C.

They extracted hNPs from the produced hNPs colloidal dispersion (5 ml) with the help of centrifugation at 7000 rcf at time interval of 20 minutes at 4 °C and incorporated in Milli-Q water.

According to Pukale et al., 2020 (40), siRNA accommodated hNPs were manufactured in optimized formulation situation and considered the loading of 1 nmol/100 mg of PLGA and incorporating siRNA to the internal water phase.

TABLE 20.2

Structural Components and Applied Method for Preparation of Lipid Polymer Hybrid Nanoparticles as a Novel Outlook for siRNA Delivery System

Structural Components and siRNA	Inferences	Applied Method	Applications	References
PLGA, dioleoyltrimethylammonium propane (DOTAP)	Gene silencing, amplified intracellular consumption via receptor mediated endocytosis	Double emulsion solvent evaporation method	Chronic obstructive pulmonary diseases (COPD)	(41)
Polycaprolactone (PCL), hydrogenated soyphosphatidylcholine, 1,2-distearoyl- sn-glycero-3-phosphoethanolamine-N-methoxy (polyethylene glycol)-2000 (DSPE-PEG2000)	Determined targeting in lung cancer, inhibition of tumor growth, lowered toxicity	Single-step sonication method	Small cell lung cancer and solid tumors	(42)

Table 20.2 illustrates structural components and applied methodd for manufacturing of LPHNPs. Fluorescently designated hNPs enforcing DPPC/PLGA and DPPC/PLGA were manufactured with the help of rhodamine designated PLGA incorporating in organic phase at total concentration of 10% w/w.,

According to Yalcin et al., 2020 (40) Calculated SAXS profile informed the disperse radiation strength as a reference of momentum shift, and with noticeable mingle of 1 and 6 m sample to detector distance spectra. They acquired solvent subtraction by calculating water-filled capillaries and blank capillaries.

20.4 Evaluation of Lipid Polymer Hybrid NPs Based siRNA Delivery

20.4.1 Evaluation by Size and Zeta Potential

According to the study done by d'Angelo et al. (39) Enforcement of hydrodynamic diameter, then polydispersity index and zeta potential of hNPs were regulated by DLS, with the help of Zetasizer Nano ZS.

20.4.2 Evaluation by Incorporating siRNA to the Interior of hNPs

Calculating siRNA actual incorporated indirectly by quantifying the content of unenclosed siRNA. hNPs were obtained by centrifugation at 6000 rcf for 25 minutes at 5 °C and supernatant was examined for siRNA content was evaluated by Sakurai et al. (43). Spectrofluorimetry was enforced for quantitative examination at lambda 480 nm.

Outcome was reported as actual incorporation as nmol of enclosed siRNA per mg of yielded hNPs and effectiveness of encapsulation was obtained from three different batches as per studies done by d'Angelo et al. (39).

20.4.3 Evaluation by SAXS Spectroscopy

Calculation by Synchotron SAXS were enforced at ID02 high brilliance beam line with area of 200 × 400 μm wavelength at lambda 0.1 nm, by enforcing two distinct sample calculating distances.

Samples were incorporated in plastic capillaries with interior diameter of 2 mm and thickness wall of 0.05 mm and surrounded with polyethylene caps.

Capillaries were incorporated horizontally onto the sample case. The subjection time of each measurement was 0.1 seconds and spectra was verified for radiation destruction.

20.4.4 Evaluation by In Vitro Interaction with Mucin

According to the study of Chen et al., 2020 (44), the dispersing light of aqueous suspension of hNPs involving or without mucin was analyzed by spectrophotometry. Type II mucin with saturated solution was manufactured by centrifugation at 6000 rcf at interval of 20 minutes by dispersing mucin in water and stirring overnight.

Dispersing hNPs in mucin solution was done by taking concentration of 1 mg/ml and enforcing the vortexing for 1 minute. Dispersing light of mucin or hNPs blend was calculated by spectrophotometric assessment at 650 nm at time 0 and finishing incubation for 30 and 60 minutes at room temperature. Reference absorbances of mucin and of 1 mg/ml hNPs dispersion in water were also evaluated

They obtained triplicate reading and indicating the outcome at 650 nm ± SD over time. By the studies of d'Angelo et al. (39) the inter-reactivity of hNPs with mucin are examined by DLS. hNP solution in mucin were analyzed using Zetasizer Nano ZS. Triplicate experiments were carried out and outcomes are obtained as typical size distribution by strength of hNPs in mucin versus hNPs in water.

20.4.5 Evaluation by Particle Stability in Artificial Mucus

The stability of siRNA loaded hNPs and their ability to diffuse inside airway mucus has been studied by Sakurai at el. (43) and SAXS analysis were enforced on suspension of hNP in the presence of artificial mucus.

AM was manufactured by incorporating 25 μl of sterile egg yolk dispersion along with 25 mg of mucin and then 20 mg of DNA and also 30 μl of aqueous DTPA, also NaCl 25 mg, then 11 mg of KCl and 100 μl of RPMI 1640 incorporated into 5 ml of water.

By the studies of Peng et al. (45,46), stirring of dispersion was carried out for a length of time prior to homogenous mixture being acquired. Aqueous solution of siRNA-incorporated hNPs were blended with AM in 1:3 volume ratio for including total volume of sample measuring 40 µl. Equilibration at intervals of 1 hour was enforced. After 12 hours enforcing incubation with AM the structure of particle proved by replicating SAXS calculations as done by d'Angelo et al. (39).

20.4.6 Evaluation by In Vitro Aerosol Performance

Enforcement of characteristics of aerosolization to fluorescently designated DPPC or PLGA and PEI or DPPC or PLGA hNP suspension were scrutinized in vitro post-delivery from the marketed nebulizer (for instance, Aeroneb Go Nebulizer).

In essence, the suspension of 1 ml of hNP was conveyed to nebulizer at the reservoir part, and attached to Next Generation Impactor having part of induction port. Working of nebulizer was enforced at 10 l/minutes and aerosol drawn through impactor for 5 minutes after achieving dryness.

As reported by Tümay et al., 2016 (45), the content of fluorescent designated hNPs endure interior of nebulizer assembly which was accumulated on the seven NGI accumulation bowls and retrieved quantitatively of the induction port in NaOH 0.5 M. After vortexing at room temperature at an interval of 1 hour, the content of fluorescent labeled hNPs in the prepared solutions was assessed by spectrofluorimetric analysis at lambda max 553 nm. Make sure to prepare calibrationcCurves and obtained from DPPC or PLGA or PEI or DPPC or PLGA standard solutions manufactured from a collection of hNPs deteriorated in NaOH 0.5 M.

Calculation of emitted dose as the subtracting between total content of hNPs initially kept and content left in nebulizer assembly.

Calculation of MMAD, i.e. mass median aerodynamic diameter, along with GSD, i.e. geometric SD, in compliance with pharmacopoeia standards which expressed a plot of cumulative mass of particle maintained in each accumulation bowl against cut off diameter of sequential stages as studied by d'Angelo et al. (39).

20.4.7 Evaluation by In Vitro Studies on Airway Triple Cell Cocultures

Faber et al. (47) impersonated the human epithelial barrier, all in vitro subjection experiments were enforced with three-dimensional triple cell cocultures, i.e. TCCC. Impersonating human epithelial barrier which consist of human bronchial epithelial cells, i.e. 16HBE14o- and along with human blood monocyte-obtained macrophages, i.e. MDM and also dendritic cells, i.e. MDDC.

For emerging TCCC in vitro model, 16HBE14o-cell cultures were augmented with MDM on the portion of apical and with MDDC at the basolateral portion obtained from human blood. When the fifth day arrived, detached the medium from the assembly at both upper and lower portion, and inserts go around upside trunk below and the lower part was eroded watchfully with enforcing cell scraper.

According to d'Angelo et al. (39), incubation of inserts was enforced with 65 µl medium comprising 80 × 10000 MDDC/ml at the basal portion at an interval of 2 hours.

20.4.8 Evaluation by Nanoparticle Accumulation Effectiveness

As studied by Nyugen et al. (48,49) assembly for cell culture inserts accommodating glass coverslips (24 × 24 mm) and subjected to distinct contents of aerosolized rhodamine designated DPPC or PLGA and PEI or DPPC or PLGA hNPs. When subjection was finished, coverslips were detached and washed with NaOH 0.5 M at 500 µl. They enforced vortexing at room temperature at an interval of 1 hour, the total content of fluorescent designated hNPs in the dispersion was analyzed by quantitatively enforcing spectrofluorimetry similar to in vitro aerosol performance.

Outcomes were designated as microgram of hNPs accumulated per cm square ± numbers of SD obtained from analysis in triplicate manner as per the studies of d'Angelo et al. (39).

20.4.9 Evaluation by Cell Uptake Analysis

Subjection of TCCC to distinct contents of aerosolized rhodamine designated DPPC or PLGA and then with PEI or DPPC or PLGA hNPs. Post subjection finished at 24 hours, attached the TCCC at an interval of 15 minutes with enforcing 3% formaldehyde in PBS at room temperature and rinsed the cells three times in PBS. Cytoskeleton of F-actin and all cells having DNA were labeled with phalloidin and 4′, 6-diamidin-2-phenylindol (DAPI) at a ratio of 1:100 and 1:50 attenuated in 0.2% Triton X-100 in PBS.

As reported in the study of d'Angelo et al. (39) for histoscopy, compartments were implanted in GlycerGel. Executed sample assessment with an inverted Zeiss confocal laser scanning histoscope 710. Emerging images were documented and considering three freedom fields by enforcing 3D rebuilding software.

20.4.9.1 Evaluation by Cytotoxicity Assay

According to the study done by Bardoliwala et al. (48) subjection of TCCC to distinct contents of aerosolized siRNA loaded DPPC or PLGA and siRNA loaded DPPC or PLGA and referring blank hNPs. Post subjection interval of 24 hours was finished, liberation of lactate dehydrogenase (LDH) by TCCC, which is referred to cell membrane detrimental, was analyzed by LDH cytotoxicity observation kit.

Executed of test in triplicates and judged by comparing with negative controls, which are TCCC subjected to aqueous solution of 0.5 mM sodium chloride or mannitol at same content of hNPs dispersions. Controlling positive, TCCC were remedied with 100 µl of 0.3% Triton X-100 in PBS on the portion of apical and enforcing incubation at interval of 24 hours at 37 °C and considered 5% CO2 as stated by d'Angelo et al. (39).

20.4.9.2 Evaluation by In Vitro Gene Silencing Effect

In consonance with d'Angelo et al. (39) cell cytotoxicity, A549 cells were coated on the portion of 96-well plates and enforcing incubation with hNPs at interval of 24 and 48–72 hours. They judged the cell viability by MTT assay.

20.4.9.3 Evaluation by Western Blotting

According to the studies done by Ehrmann et al. (50) macerated entire proteins form the cells and splitted on 10% sodium dodecyl sulphate-polyacrylamide gel electrophoresis and conveyed electrically on polyvinylidene fluoride (PVDF). Enforcing the incubation of membrane for 2 hours by taking 5% nonfat milk to obstruct nonspecific attaching sites and enforcing incubation at interval of 2 hours at room temperature with enforcing antibody, i.e. anti-betaENaC and anti-beta-actin and enforcing analysis by densitometry as reported by d'Angelo et al. (39).

20.5 Fate Upon In Vitro Deposition on Human Epithelial Barrier

20.5.1 Fate and Cytotoxicity of hNPs Aerosolized in TCCC

According to the studies done by d'Angelo et al. (39) the potential of aerosolized hNPs to puncture airway epithelial barrier was investigated in vitro on TCCC. In the course of nebulization process, the stability of hNPs was assured by transmission electron microscopy (TEM) which analyzed hNPs loaded with siRNA that is freeze dried in existence of mannitol in ratio of 1:25 NP/mannitol prior to and after nebulization.

The accumulation effectiveness was also enacted for both experimented formulations, spotlight in each case a uniform dispersion of hNPs placed on coverslips by entering into exposure chamber.

To be precise 0.0053 and 0.0127 of hNPs were accumulated on coverslips by taking low and high concentrations post nebulization of DPPC/PLGA as stated by d'Angelo et al. (39).

By identical value of GSD, PEI/DPPC/PLGA hNPs designated minimal MMAD of experimented formulation and bigger FPF and RF values in contrast to DPPC/PLGA.

As reported by the studies done by d'Angelo et al. (39) by analyzing cell uptake, hNPs which are fluorescently labeled were aerosolized on TCCC. Results designated that hNPs with red labeled comprising DPPC/PLGA and PEI/DPPC/PLGA can productively puncture into cells and are restricted intracellularly which are displayed by CLSM.

Proinflammatory properties of empty aerosolized and hNPs which are siRNA loaded on TCCC was also probed by analyzing secretion of TNF-alfa at duration of 24 hours post exposure.

hNPs that are loaded with siRNA, at all concentrations designated a huge making of TNF-alfa after comparison with negative control.

20.6 Applications of Lipid Polymer Hybrid Nanoparticles of siRNA Delivery in Various Clinical Conditions

20.6.1 Lipid-Polymer Hybrid Nanoparticles for Sustained siRNA Delivery and Gene Silencing in Treatment of Lung Cancer

The hybrid lipid-polymer nanoparticles generally presents excellent silencing efficacy, and siRNA release from nanoparticles by temporally is prolonged for exceeding 1 month. Shi et al. (2014) evaluated on luciferase-expressed Hela cells and A549 lung carcinoma cells post transfection for short period of time, the siRNA LPHNPs presented better sustained silencing function than lipofectamine 2000-siRNA complexes. Sustained silencing of prohibitin 1 (PHB1) presents excellent inhibition of tumor cell growth in vitro and in vivo than the lipofectamine complexes.

Sivakumar et al. (59) initiated safe and effective delivery of siRNA to tumor cells and they developed some cationic lipid-polymer based nanoparticle and focused into clinical studies in patients. They hypothesized that siRNA delivery in sustained manner could emerge to extended term, knockdown of target genes in effective manner and preventing from repeated administration of therapeutic siRNA to regulate the silencing action and lowering systemic side effects.

Sivakumar et al. (59) prepared hybrid lipid-polymer nanoparticles by modified double emulsion solvent evaporation technique and self assembly method. They have synthesized a new cationic lipid-like compound (G0-C_{14}) by interacting 1,2-epoxytetradecane with PAMAM dendrimer (generation 0), and applied for formulating siRNA nanoparticle. They generated the release profile of siRNA by calculating the dye-labeled siRNA persisting in the nanoparticle or lipofectamine 2000 at separate time points. They have calculated luciferase-expression change in luciferase expressing HeLa cells and transfected with nanoparticle or lipofectamine for duration of 6 hours. Similarly, they have also tested PHB1 expression level in A549 lung cancer cells by immunoblotting for duration of 14–24 days, post treatment siRNA LPHNPs. They monitored in vitro proliferation of A549 cell by AlamarBlue assay for period of 12 days. They studied A549 xenograft tumor growth by applying 6 week old BALB/C nude mice.

The hybrid lipid polymer nanoparticle comprised of an aqueous siRNA core which is stabilized by positively charged lipid like compound G0-C_{14}, containing hydrophobic PLGA polymer shell at middle and comprising neutral-charge lipid-PEG surface layer Sivakumar et al., 2019 (59). They have extended half-release time of siRNA to approx 9 days, in comparison to approx 8 hours for lipofectamine 2000. They have lowered the signal of luciferase at day 2 less than 10% in siRNA LPHNPs and lipofectamine 2000-siRNA and still remained same at day 4, but it is recuperated to over 30% in Lipofectamine 2000-siRNA.So, they resulted in the recuperation of luciferase expression at approx 30% and 60% at day 7 and day 10 respectively, although they observed much lesser than those in Lipofectamine 2000-siRNA transfected cells.

TABLE 20.3

Combination of Anti-cancer Drugs and siRNA Delivery System for Pulmonary Targeting Designed into Lipid Polymer Hybrid Nanoparticles

Delivery System	Carriers	Target Genes	Co-treated Drugs	References
siRNA delivery	Lipid-polymer hybrid nanoparticles	MCL1 Survivin	Suberoylanilidehydroxamic acid Gadolinium-linked lipid, rhodamine-labeled lipid for imaging	(28)
siRNA delivery	Lipid-polymer hybrid nanoparticles	EphA2RhoAMCLFLI1-EWSHER-2STAT3	Paclitaxel___	(33)
siRNA coupling	Lipid polymer hybrid nanoparticles Aptamer-siRNA chimera EpCAM aptamer-siRNA chimera	TWIST1PSMAPLK1	___	(51)

Shi et al. (2014) observed effective PHB1 silencing and was regulated by siRNA LPHNPs over 2 weeks and expression of PHB1 was recuperated at 24 days.

20.6.1.1 Combinatorial of siRNA Delivery and Anti-cancer Drugs Delivery

Table 20.3 illustrates combination of anti-cancer drugs and siRNA pulmonary delivery. There are numerous kinds of nanocarriers that have been studied for the conduction of siRNA and anti-cancer drugs with an approach to improve anti-cancer possessions through disabling multiple drug resistant characteristic or administrating diverse caspase-based facilitated cell necrosis routes. Thoughtfully considered as multiple functions of developing drug/siRNA delivery, nanocarriers will significantly raise the in vivo growth of tumor through active and passive capabilities for targeting the tumor/lumps. So, versatile nanomedicines provide countless advantages in overpowering the disadvantages of existing behavior, together with chemotherapy.

To release nucleic acids, these particles functionalized with positive charged quaternary ammonium or may be layered with a lipid bi-layer of cations is been investigated.

20.6.2 Clinical Trials on Aerosolized siRNA Delivery Based Lipid Polymer Hybrid Nanoparticle

Table 20.4 illustrates ongoing clinical trials on siRNA along with nano-therapeutic drugs designed into LPHNPs. Excellair™, manufactured by ZaBeCor Pharmaceuticals, is an inhaled siRNA for nursing asthma. Spleen tyrosine kinase (Syk) was generally inhibited by siRNA and transcription was inhibited by which is regulated by SyK. Syk regulates the signaling from B-cell receptor and initiates the downstream signaling cascades that result in the activation of many pro-inflammatory transcription factors. Excellair lessens the inflammation persisting in asthma by specific inhibition of Syk. During Phase I study, asthma patients received inhaled siRNA therapeutic for continuous 21 days. More than 75% of patients treated showed enhancement of breathing or minimized rescue inhaler use, but placebo patients showed no such enhancement. These results initiated Excellair into phase II clinical trials in 2009

Alnylam Pharmaceuticals designede a nebulizer and nasal spray to deliver siRNA therapeutic. It was designed for the treatment of respiratory syncytial virus (RSV) which targets the nucleocapsid protein and recognized for prophylactic treatment against RSV infections in healthy patients and for nursing RSV infection in lung transplant patients. The ALN-RSV01 siRNA comprised of RNA duplex having double stranded containing 19 base pairs of complementary and 2-nt dT overhangs at pair of 3′ ends.

20.6.3 Delivery of siRNA as Nonviral Gene Therapy to the Lungs for Reducing Airway Inflammation

siRNA cannot transverse biological membranes to approach at their specific sites because of highly negative charged, hydrophilicity, and with bigger size (approx. 13.8 kDa). The perfect siRNA delivery system should:

1. Deliquesce siRNA into particles of nano-size
2. Shield siRNA from enzymatic destruction
3. Ease cellular uptake
4. Encourage endosomal diversion to liberate siRNA to the cytoplasm where location of RNA-induced silencing complex (RISC) can be tracked
5. Having insignificant effects on gene silencing activity or specificity
6. Having insignificant toxicity

Nonviral delivery comprises naked siRNA and vectors for delivery are considered as lipid polymer hybrid nanoparticles.

siRNA which are naked or have not formulated require the siRNA delivery by not utilizing delivery vehicle or carrier. Superiority of this strategy comprised the good facility of preparation and ease of delivery by different routes like inhalation, intra-tracheal, or intranasal routes. Crucial limitation of this strategy is that siRNA which is delivered are allowed for poor cell approach and uptake, and during the pathway it gets degraded in airways.

TABLE 20.4

Ongoing siRNA Along with Nano-therapeutic Drugs Clinical Trials Designed into Lipid Polymer Hybrid Nanoparticle in Clinical Therapy of Tumors and Carcinomas

siRNA in Nano-therapeutic Drugs	Carrier	Target Gene	Disease	Delivery Route	Phase	References
CALAA-01	Transferrin receptor-targeted Cyclodextrin (CD) and lipid hybrid nanoparticle	RRM2	Solid tumor on lungs, non small cell carcinoma including adenocarcinoma	IV	II	(34)
ALN-VSP02	Solid lipid nanoparticle and biodegradable polymer hybrid	VEGF and KSP	Solid tumor on lungs, squamous cell carcinoma	IV	I	(35)
Atu027	Lipoplex-biodegradable polymer hybrid siRNA	PKN-3	Advanced solid tumor on lungs	IV	I	(36)
siG12D LODER siG12D LODER + chemotherapy	Miniature biodegradable lipoplex-siRNA miniature biodegradable polymeric matrix	KRASG12D	Pancreatic cancer	EUS biopsy needle	III	(37)
TKM 080301	Polymer hybrid-cyclodextrin NPs	PLK-1	Alveolar sarcoma, alveolar rhabdomyosarcoma	IV	I	(38)
TKM 080301	Solid lipid nanoparticle and biodegradable polymer hybrid	PLK-1	Advanced solid tumor	IV	I	(39)
siRNA-EphA2-	Cationic lipid holymer-Hybrid nanoparticle(G0-$C_{14)}$	EphA2	Advanced cancer	IV	I	(52)
DOPCEZN-2968	liposomeLocked nucleic acid and polymer cyclodextrin hybrid NPs	HIF-1	Tracheal cancer, bronchioloalveolar carcinoma	IV	I	(53–56)

Crucial barriers for siRNA delivery to the lung involved the existence of mucus, alveolar fluid, alveolar macrophages, and clearance by mucociliary route. Adverse physicochemical properties of siRNAs having negative charge with high molecular weight and uncertainty in plasma ($t1/2 = 10$ minutes) siRNA in the lysosomes by the process of endocytosis (57,58).

Functions of siRNA therapeutics gets declined due to deteriorate effects of siRNA in the lysosomes. Enzymatic deterioration can occur in siRNA which are not modified, therefore new technology included like chemical modification of siRNA have been initiated to enhance stability. Specificity and potency are also improved by chemical modification and lower down the immune response and effects which are off target.

siRNA which are instilled directly into the lungs through routes like intranasal or intra-tracheal appeared to be in direct connection with epithelial cells present in the lungs. siRNA which are not modified leads to enhancement of activation of immune system by nonspecific way by binding to toll-like receptor 7 pathway (59).

Structure of RNA duplex which is enforced to be processed under chemical modification by increasing biological stability without any unfortunate gene-silencing process and avert activation of nonspecific immune system. Technology in modifications comprised like assimilating 2′-O-methyl into sugar backbone of chosen nucleotides inside both sense and antisense strands.

Boosting the delivery of siRNA by increasing therapeutic efficacy enforcing coupling of small molecules or peptides to siRNA having sense strand.

In new strategy, antagomir-122 was derived from hydroxyproline-linked cholesterol solid support and 2′-O-methyl phosphoramidites. Noticeable lowering in endogenous miR-122 levels in the liver was found after administration. Strategies have been implemented which protect siRNA against deterioration by endonucleases which assist them to approach at site of action.

Three main procedures enforced for the manufacturing of siRNA in vitro (59)which comprised (i) chemical derivation, (ii) small RNAs transcription by in vitro, and (iii) transcription of long RNAs by in vitro and dissolving of long dsRNAs with the help of family enzyme RNase III (for instance, Dicer).

Design of siRNA segment is required before concoction of siRNA by these three procedures. Merging T7 RNA polymerase and RNA-dependent RNA polymerase by in vitro system of bacteriophage f6 to produce siRNA molecules. siRNA was emerged by in vivo RNA replication system (60).

siRNA which are modified or not modified can be emerged by enforcing reconstitution in simple way by taking normal saline or 5% solution of dextrose for delivery through inhalation route. Supervising the aerosolized siRNA solution can be enforced using a nebulizer.

With a new technology, the latest group of naked siRNA was developed designated Ribophorin II, i.e. PnkRNA and nkRNA. Emerged as naked or RNAi agent which are not modified, for instance RPN2-PnkRNA, which has been handpicked as targeted delivery as hybrid lipid polymer nanocarrier for asthma by inhalation appeared to be systematic inhibition of airway inflammation unaccompanied by any noteworthy toxicity.

For example, mice which are expressed with luciferase were nursed by delivery through intra-tracheal route by administering 10 nmol siRNA duplex or phosphorothioate having 10 nmol which is secured in nucleic acids without anti-sense oligonucleotide.

Imaging of oligonucleotides by IVIS stamped with Cy5 merged with confocal microscopy was enforced to identify their biodistribution. Oligonucleotides enforced to be under quick and systemic distribution by transcytosis and renal clearance after transcytosis and renal clearance once administered.

Phosphorothioate secured with nucleic acid anti-sense oligonucleotides, which are up taken by kidney and liver, enforced to cause gene expression in these organs. Intra-tracheal delivery of free TNF-alfa siRNA post hemorrhage was not effective in lowering the symptoms associated with septic shock model of acute asthma attack.

20.6.4 Delivery of Nebulized LPNHPs into Epithelial Cells of Bronchiole Possessing a Therapeutically Active siRNA Approach for Anti-inflammatory Properties and Fate Upon In Vitro Deposition

The localized tenderness of a tissue of lung is an ordinary functional retort to the respirational (61) insult by any foreign substance and airborne pathogens. Although, if the inflammation gets more chronic, wide incursion by the cells of immune system, remodeling of lung tissue by the time and contraction of the air pathways will ultimately cause the obstruction of lung and damage to the function of lung. Those disruptive provocative diseases of lung include chronic obstructive pulmonary disease (COPD), cystic fibrosis (CF), respiratory diseases, and enduring communicable lung diseases. All these can exhibit higher starting inflammation interrupted by exacerbations.

The first contour of defensiveness against airborne encounters to the lung is the epithelium of airway. The epithelium having linings of bronchial possess a vital role (62) in the arrangement of the lung inflammatory chute because of its capability for regulating inborn and adaptive response of immune system.

The microparticles and nanoparticles that are inhalable are being emerging as auspicious carriers for delivering active pharmaceutical ingredients (APIs) straight to epithelium of the lung and also towards its fundamental layers. These active pharmaceutical ingredients include, apart from others, anti-sense oligonucleotides which exercise its therapeutic result through RNA-interference (RNAi) path by obstructing the countenance of definite proteins that included in pathways of disease.

Most of the focus of efforts have been emphasizing particle-regulated distribution of small interfering RNAs (siRNAs) to the epithelium of airway, using siRNA acts as consignments could considered as substantial beneficial potential. Artificial siRNA mimics (63) have often been working for purpose of research activities, although their probable practice in medication treatments is an additional current investigation.

In this succeeding review, we examine the capability of nebulized-lipid–polymer hybrid-nanoparticles (LPNs) for delivering siR-17 to that of BECs and to giveaway expression of IL-8. For this approach, we arranged LPNs loaded with siR-17 comprised of decomposable polymer, i.e. poly- (DL-lactic-co-glycolic acid) (PLGA), and also the lipid 1, 2-dioleoyloxy-3-(trimethylammonium) propane (DOTAP) that is cationic in nature.

Here in this experimental proof, it's been shown that nebulization enables the aerosolization of PLGA–LPN suspensions modified by DOTAP, which produces aerosol dewdrops that are ideal for delivering cargo to the epithelium of bronchioles.

The technique used for encapsulation of siR-17 mimics was solvent evaporation of double emulsion, NC#1 and double stranded DNA-oligonucleotide with that of LPNs modified within DOTAP.

A spectroscopy named as photon-correlation (64) was used to measure polydispersity index and z-average by the help of light scattering (dynamic) method. Determination of zeta potential was done by the help of laser-doppler microelectrophoresis.

Lyophilized LPN or trehalose cakes were reformed in free water of RNAse in the existence/absence of 0.9% saline for the duration of 10 minutes at the room temperature further by mild vortexing for 3 minutes. The nebulization was directed with that of Aerogens single vibrational mesh-nebulizer. The dispersions of LNP were nebulized in a centrifuging tube of 15 ml that is attached to the Aerogen and collected subsequently to the nebulization for duration of 10 seconds at $<500\times g$. The PDI, z-average and zeta-potential were determined prior and after that of the nebulization.

The mass median aerodynamic diameter (MMAD) and the geometric standard deviation (GSD) of the dispersion of nebulized LPN were determined by the next-generation impactor (65) at a drift rate of 15 ± 0.5 l/min. Calculation of the number of the nanoparticles was achieved by the help of the Nanosight NS300. The volume median diameter (VMD) of the nebulized dispersed LPNs and 0.9% saline was determined by the use of the Spraytec-lazer diffraction system.

The NuLi-1 BEC-line of human was cultivated in a plasticware of coated-collagen in a growth medium of eronchial Epithelial. While for several experiments, the stimulation of the cells was done over a period of 24 hours with 10 μg·ml/1 LPS of *Pseudomonas aeruginosa*.

The concentrations of IL-8 and IL-6 were determined in a medium that is removed from the cultures of cells with sandwiching of ELISA and centrifugation is done up to 5 minutes at 2000×g at 4 °C.

Constant variables were marked as the mean including standard deviations. The quantitative outcomes were equated through paired or unpaired t-test, when required suitably. For the analysis of regression, both the linear and nonlinear (goodness-of-fit) representations were characterized.

20.6.5 Potential of siRNA Delivery by Gene Therapy for Treating Cystic Fibrosis through LPNHPs

Cystic fibrosis is marked by immense secretion of thick mucus (66) and emerge due to faulty CFTR protein. It emerged as airway obstruction which construct the lungs vulnerable to repeated and constant bacterial infections and endobronchial chronic inflammation which is the main reason for development of bronchiectasis, respiratory failure and eventually patient death.

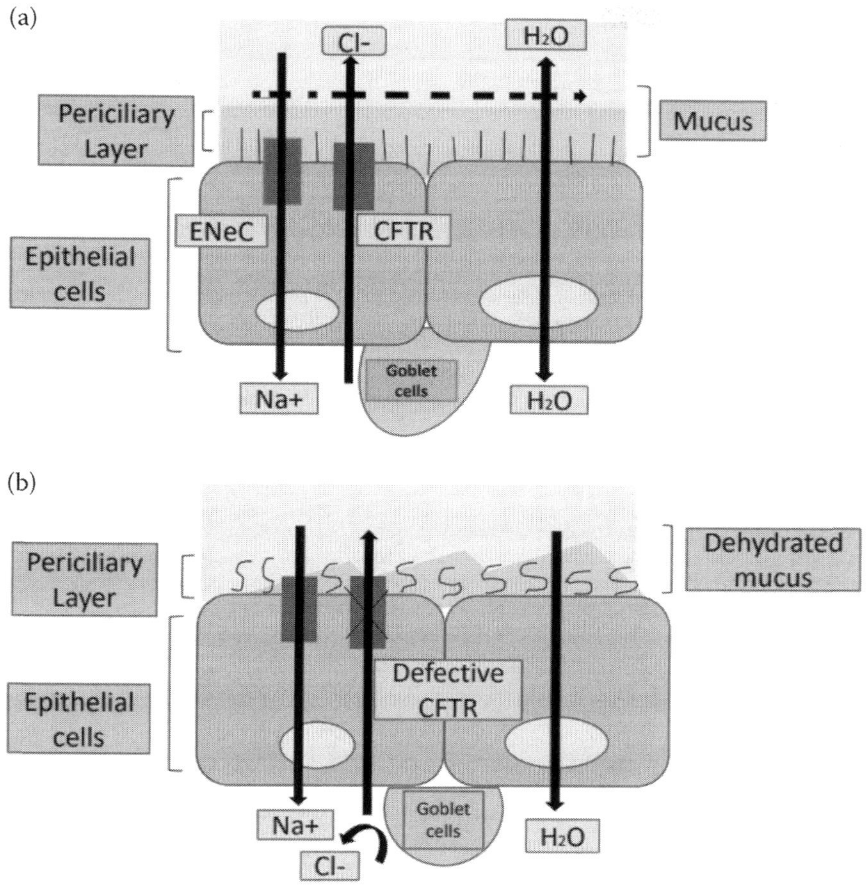

FIGURE 20.5 Suggested Mechanisms for Design and Development of CF Lung Disease. In Normal Passage of Airways (A), Hydration of Lung Lining Fluids Is Regulated by Na+ Absorption and Cl– Secretion. In CF Airways (B), the Non-existence of CFTR Leads to Uncontrolled Na+ Absorption and Related Dehydration of the Airway Surface Fluid Layer with Inevitably Poor Mucociliary Clearance.

Emerging technology of gene therapy established new techniques to treat cystic fibrosis.

Etiology of this disease by gene which undergoes mutation on chromosome 7 which conceals cystic fibrosis transmembrane conductance regulator (CFTR) protein (67).

Figure 20.5 will represent that CFTR protein emerged as chloride channel on epithelial cells having apical membrane and it is accountable for adjustment of secretion of chloride ions and undergoes reabsorption of sodium ions

Mutations have been categorized in five major groups by influential effect on CFTR function:

1. Mutation obstructing with protein derivation
2. Mutation influencing protein ontogenesis
3. Mutation adjusting channel adjustment
4. Mutation influencing chloride delivery conductance
5. Mutation decreasing the extent of normally performing at apical membrane

The emerged mutations on CFTR leads to defect in chloride ions secretion and tremendous absorption of sodium ions across epithelia. Habitation of microorganisms inside static mucus provides a suitable atmosphere to settle and prevention from target by immune response and antimicrobial drugs.

Pathogen which is responsible for this is *Pseudomonas aeruginosa* (PA) a gram-negative organism which is marked with production of mucoid coating comprised of alginate and shields the bacteria against antibiotics and phagocytosis. Therefore, it emerged as impoverished lung penetration of antibiotics and results in clinical irritation and inadequacy of therapies.

The goal of nursing CF mainly focused on down regulation effects of faulty function in CFTR. Novel technology to focused reinstitute or replace CFTR gene in spite of nursing the defects of pathologies (68).

Process of gene therapy involved delivery of nucleic acids into a cell and reinstitute genes or functions biologically. Diverse vectors have been screened and viral systems are most widespread in approach but nonviral vectors have evidence of safety, easier and economical options compared to viral vectors.

The technology of CRISPR/Cas9 has emerged as promising method for genome study in widespread organisms.

Approach is focused on rectifying transport defect of chloride ion which is emerged as influential on nasal epithelium by employing E1-deficient adenovirus and exposure of CFTR protein in airway epithelial cells.

Evidence of results achieved by screening the therapy in three patients achieved effective rectification of faulty in chloride transport without proof of severe side effects.

Regulation of transduction in most tissues was compensated by adverse effects, for instance responses of inflammatory reaction in the host organism. Adeno-associated viral Vectors have emerged as an alternate option for gene delivery.

Ideal characteristics of this gene delivery are high transduction regulation and wide serotype-contingent tissue tropism (69). They designed AAV2 in aerosolized form as repeated dose comprising complementary DNA (cDNA) in CFTR. AAVs showed some limitations emerged mainly by employing viruses as vectors. Current research work focused to mark and ameliorate limitations and discovering formulation as safer and novel approach.

20.7 Conclusion

Design and emerged novel hNPs constituting PLGA neighbored by DPPC shell, designating size and surface properties which are optimized for siRNA transfer and pulmonary delivery. Optimized formulations are enforced for enclosing of mixture of siRNA against alfa and beta subunits of ENaC channel with maximum efficiency and optimized by including PEI into the formulation. Freeze drying procedure with mannitol enforced to manufacture stable hNP-based dry powders with mannitol, an osmotic agent for nursing cystic fibrosis. Cell consumption studies upon in vitro aerosolization on TCCC assured the potential of hNPs which are fluorescently labeled to be restrained inside human airway epithelial barrier. Emerged concept of design does not exercise cytotoxic or acute proinflammatory effect with regard to cell components of TCCC model. In the end, in vitro gene silencing studies spotlighted that potential of released siRNA minimize ENaC protein expression in human lung epithelial cells. All inclusively, results showed that great ability of DPPC/PLGA hNPs as carriers for siRNA pulmonary delivery deposited on human epithelial airway barrier, encouraged toward high technological studies of their therapeutic effectiveness in serious lung diseases.

REFERENCES

1. Dheer D, Arora D, Jaglan S, Rawal RK, and Shankar R Polysaccharides based nanomaterials for targeted anticancer drug delivery. *Journal of Drug Targeting*. 2017; 25(1): 1–6.
2. Ojiaku CA, Yoo EJ, and Panettieri Jr RA Transforming growth factor β1 function in airway remodeling and hyperresponsiveness. The missing link? *American Journal of Respiratory Cell and Molecular Biology*. 2017; 56(4): 432–442.
3. Emami F, Yazdi SJ, and Na DH. Poly (lactic acid) poly (lactic-co-glycolic acid) particulate carriers for pulmonary drug delivery. *Journal of Pharmaceutical Investigation*. 2019; 49(4): 427–442.
4. Gandhi J, Desai N, Golwala P, and Shah P Medication conveyance through nose: Factors affecting and novel applications. *Drug Delivery Letters*. 2018; 8(3): 169–183.
5. Parumasivam T, Chang RY, Abdelghany S, Ye TT, Britton WJ, and Chan HK. Dry powder inhalable formulations for anti-tubercular therapy. *Advanced Drug Delivery Reviews*. 2016; 102: 83–101.
6. Katz MG, Fargnoli AS, Gubara SM, Fish K, Weber T, Bridges CR, Hajjar RJ, and Ishikawa K. Targeted gene delivery through the respiratory system: Rationale for intratracheal gene transfer. *Journal of Cardiovascular Development and Disease*. 2019; 6(1): 8.
7. James R, Manoukian OS, and Kumbar SG. Poly (lactic acid) for delivery of bioactive macromolecules. *Advanced Drug Delivery Reviews*. 2016; 107: 277–288.
8. McMillan NA. Lowering the siRNA delivery barrier: Alginate scaffolds and immune. Nanotechnology for the delivery of therapeutic. *Nucleic Acids*. 2013; 4: 193.
9. Kashaw SK, Sahu P, Rajoriya V, Jana P, Kashaw V, Sau S, and Iyer AK. Exploring siRNA umpired nanogels: A tale of barrier combating carrier. *Current Pharmaceutical Design*. 2020; 26(27): 3234–3250.
10. Ingram JP. The Role of the Innate Immune System in Programmed Cell Death. Temple University, Philadelphia; 2018.
11. Swart OL. Analysis and Effect of Small-molecules Targeting Pre-microRNA Structures and Synthetic Efforts Toward a Novel Scaffold for RNA Targeting. PhD Thesis. University of Rochester; 2019.
12. Xu CF and Wang J. Delivery systems for siRNA drug development in cancer therapy. *Asian Journal of Pharmaceutical Sciences*. 2015; 10(1): 1–2.
13. Mukherjee A, Waters AK, Kalyan P, Achrol AS, Kesari S, and Yenugonda VM. Lipid–polymer hybrid nanoparticles as a next-generation drug delivery platform: State of the art, emerging technologies, and perspectives. *International Journal of Nanomedicine*. 2019; 14: 1937.
14. Tahir N, Haseeb MT, Madni A, Parveen F, Khan MM, Khan S, Jan N, and Khan A. Lipid polymer hybrid nanoparticles: A novel approach for drug delivery. In: Role of Novel Drug Delivery Vehicles in Nanobiomedicine. IntechOpen. 2019.
15. Khan MM, Madni A, Torchilin V, Filipczak N, Pan J, Tahir N, and Shah H. Lipid-chitosan hybrid nanoparticles for controlled delivery of cisplatin. *Drug Delivery*. 2019; 26(1): 765–772.
16. Hsu KJ and Turvey SE. Functional analysis of the impact of ORMDL3 expression on inflammation and activation of the unfolded protein response in human airway epithelial cells. *Allergy, Asthma & Clinical Immunology*. 2013; 9(1): 4.
17. Tahir N, Madni A, Correia A, Rehman M, Balasubramanian V, Khan MM, and Santos HA. Lipid-polymer hybrid nanoparticles for controlled delivery of hydrophilic and lipophilic doxorubicin for breast cancer therapy. *International Journal of Nanomedicine*. 2019; 14: 4961.
18. Ishak RA, Mostafa NM, and Kamel AO. Stealth lipid polymer hybrid nanoparticles loaded with rutin for effective brain delivery–Comparative study with the gold standard (Tween 80): Ooptimization, characterization and biodistribution. *Drug Delivery*. 2017; 24(1): 1874–1890.
19. Mandal B, Bhattacharjee H, Mittal N, Sah H, Balabathula P, Thoma LA, and Wood GC. Core–shell-type lipid–polymer hybrid nanoparticles as a drug delivery platform. *Nanomedicine: Nanotechnology, Biology and Medicine*. 2013; 9(4): 474–491.
20. Fellmann C and Lowe SW. Stable RNA interference rules for silencing. *Nature Cell Biology*. 2014; 16(1): 10–18.

21. Farheen Badrealam K, Zubair S, and Owais M. SiRNA nanotherapeutics – The panacea of diseases? *Current Gene Therapy*. 2015; 15(2): 201–214.
22. Michaels YS, Barnkob MB, Barbosa H, Baeumler TA, Thompson MK, Andre V, Colin-York H, Fritzsche M, Gileadi U, Sheppard HM, and Knapp DJ. Precise tuning of gene expression levels in mammalian cells. *Nature Communications*. 2019; 10(1): 1–2.
23. Vishnu PG, Bhattacharya TK, Bhushan B, Kumar P, Chatterjee RN, Paswan C, Dushyanth K, Divya D, and Prasad AR. In silico prediction of short hairpin RNA and in vitro silencing of activin receptor type IIB in chicken embryo fibroblasts by RNA interference. *Molecular Biology Reports*. 2019; 46(3): 2947–2959.
24. Ramasamy T, Tran TH, Choi JY, Cho HJ, Kim JH, Yong CS, Choi HG, and Kim JO. Layer-by-layer coated lipid–polymer hybrid nanoparticles designed for use in anticancer drug delivery. *Carbohydrate Polymers*. 2014; 102: 653–661.
25. Krishnamurthy S, Vaiyapuri R, Zhang L, and Chan JM. Lipid-coated polymeric nanoparticles for cancer drug delivery. *Biomaterials Science*. 2015; 3(7): 923–936.
26. De A, Bose R, Kumar A, and Mozumdar S. Targeted Delivery of Pesticides Using Biodegradable Polymeric Nanoparticles. Springer India, New Delhi. 2014.
27. Singh A, Trivedi P, and Jain NK. Advances in siRNA delivery in cancer therapy. *Artificial Cells, Nanomedicine, and Biotechnology*. 2018; 46(2): 274–283.
28. Lee SJ, Kim MJ, Kwon IC, and Roberts TM. Delivery strategies and potential targets for siRNA in major cancer types. *Advanced Drug Delivery Reviews*. 2016; 104: 2–15.
29. Singh AP, Biswas A, Shukla A, and Maiti P. Targeted therapy in chronic diseases using nanomaterial-based drug delivery vehicles. *Signal Transduction and Targeted Therapy*. 2019; 4(1): 1–21.
30. Kandil R and Merkel OM. Pulmonary delivery of siRNA as a novel treatment for lung diseases.*Therapeutic Delivery*. 2019; 10(4): 203–206.
31. Patil JS and Sarasija S. Pulmonary drug delivery strategies: A concise, systematic review. *Lung India: Official Organ of Indian Chest Society*. 2012; 29(1): 44.
32. Lam JK, Liang W, and Chan HK. Pulmonary delivery of therapeutic siRNA. *Advanced Drug Delivery Reviews*. 2012; 64(1): 1–5.
33. Merkel OM, Rubinstein I, and Kissel T. siRNA delivery to the lung: What's new? *Advanced Drug Delivery Reviews*. 2014; 75: 112–128.
34. d'Angelo I, Conte C, La Rotonda MI, Miro A, Quaglia F, and Ungaro F. Improving the efficacy of inhaled drugs in cystic fibrosis: Challenges and emerging drug delivery strategies. *Advanced Drug Delivery Reviews*. 2014; 75: 92–111.
35. Pandita D, Kumar S, and Lather V. Hybrid poly (lactic-co-glycolic acid) nanoparticles: Design and delivery prospectives. *Drug Discovery Today*. 2015; 20(1): 95–104.
36. Bose RJ, Lee SH, and Park H. Lipid-based surface engineering of PLGA nanoparticles for drug and gene delivery applications. *Biomaterials Research*. 2016; 20(1): 34.
37. Hadinoto K, Sundaresan A, and Cheow WS. Lipid–polymer hybrid nanoparticles as a new generation therapeutic delivery platform: A review. *European Journal of Pharmaceutics and Biopharmaceutics*. 2013; 85(3): 427–443.
38. Dave V, Tak K, Sohgaura A, Gupta A, Sadhu V, and Reddy KR. Lipid-polymer hybrid nanoparticles: Synthesis strategies and biomedical applications.*Journal of Microbiological Methods*. 2019; 160: 130–142.
39. d'Angelo I, Costabile G, Durantie E, Brocca P, Rondelli V, Russo A, Russo G, Miro A, Quaglia F, Petri-Fink A, and Rothen-Rutishauser B. Hybrid lipid/polymer nanoparticles for pulmonary delivery of siRNA: Development and fate upon in vitro deposition on the human epithelial airway barrier. *Journal of Aerosol Medicine and Pulmonary Drug Delivery*. 2018; 31(3): 170–181.
40. Rahman M, Alharbi KS, Alruwaili NK, Anfinan N, Almalki WH, Padhy I, Sambamoorthy U, Swain S, and Beg S. Nucleic acid-loaded lipid-polymer nanohybrids as novel nanotherapeutics in anticancer therapy. *Expert Opinion on Drug Delivery*. 2020; 17(6): 805–816.
41. Foldvari M, Chen DW, Nafissi N, Calderon D, Narsineni L, and Rafiee A. Non-viral gene therapy: Gains and challenges of non-invasive administration methods. *Journal of Controlled Release*. 2016; 240: 165–190.
42. Sakurai Y, Mizumura W, Ito K, Iwasaki K, Katoh T, Goto Y, Suga H, and Harashima H. Improved stability of siRNA-loaded lipid nanoparticles prepared with a PEG-monoacyl fatty acid facilitates ligand-mediated siRNA delivery. *Molecular Pharmaceutics*. 2020; 17(4): 1397–1404.
43. Kakavandi A and Akbari M. Experimental investigation of thermal conductivity of nanofluids containing of hybrid nanoparticles suspended in binary base fluids and propose a new correlation. *International Journal of Heat and Mass Transfer*. 2018; 124: 742–751.
44. Peng H, Yang H, Song L, Zhou Z, Sun J, Du Y, Lu K, Li T, Yin A, Xu J, and Wei S. Sustained delivery of siRNA/PEI complex from in situ forming hydrogels potently inhibits the proliferation of gastric cancer. *Journal of Experimental & Clinical Cancer Research*. 2016; 35(1): 57.
45. Tümay SO. A novel selective "turn-on" fluorescent sensor for Hg2+ and its utility for spectrofluorimetric analysis of real samples. *Journal of the Turkish Chemical Society Section A: Chemistry*. 2020; 7(2): 505–516.
46. Faber SC, McNabb NA, Ariel P, Aungst ER, and McCullough SD. Exposure effects beyond the epithelial barrier: Trans-epithelial induction of oxidative stress by diesel exhaust particulates in lung fibroblasts in an organotypic human airway model. *Toxicological Sciences*. 2020; 177(1): 140–155.
47. Nguyen TV, Edington CD, Suter EC, Carrier RL, Trumper DL, and Griffith LG, Inventors; Northeastern University China, Massachusetts Institute of Technology, assignee. Device for Controlled Apical Flow in Cell Culture Inserts. United States patent US 10,323,221. 2019.
48. Bardoliwala D, Patel V, Javia A, Ghosh S, Patel A, and Misra A. Nanocarriers in effective pulmonary delivery of siRNA: Current approaches and challenges. *Therapeutic Delivery*. 2019; 10(5): 311–332.
49. Ehrmann I, Crichton JH, Gazzara MR, James K, Liu Y, Grellscheid SN, Curk T, de Rooij D, Steyn JS, Cockell S, and Adams IR. An ancient germ cell-specific RNA-binding protein protects the germline from cryptic splice site poisoning. *Elife*. 2019; 8: e39304.
50. Lee SH, Kang YY, Jang HE, and Mok H. Current preclinical small interfering RNA (siRNA)-based conjugate

systems for RNA therapeutics. *Advanced Drug Delivery Reviews*. 2016; 104: 78–92.
51. Hallan SS, Kaur P, Kaur V, Mishra N, and Vaidya B. Lipid polymer hybrid as emerging tool in nanocarriers for oral drug delivery. *Artificial Cells, Nanomedicine, and Biotechnology*. 2016; 44(1): 334–349.
52. Silva EJ, Souza LG, Silva LA, Taveira SF, Guilger RC, Liao LM, Queiroz Junior LH, Santana MJ, and Marreto RN. A novel polymer-lipid hybrid nanoparticle for the improvement of topotecan hydrochloride physicochemical properties. *Current Drug Delivery*. 2018; 15(7): 979–986.
53. Sarett SM, Nelson CE, and Duvall CL. Technologies for controlled, local delivery of siRNA. *Journal of Controlled Release*. 2015; 218: 94–113.
54. Chaudhary Z, Ahmed N. Rehman A, and Khan GM. Lipid polymer hybrid carrier systems for cancer targeting: A review. *International Journal of Polymeric Materials and Polymeric Biomaterials*. 2018; 67(2): 86–100.
55. Madni A, Batool A, Noreen S, Maqbool I, Rehman F, Kashif PM, Tahir N, and Raza A. Novel nanoparticulate systems for lung cancer therapy: An updated review. *Journal of Drug Targeting*. 2017; 25(6): 499–512.
56. de Carvalho Vicentini FT, Borgheti-Cardoso LN, Depieri LV, de Macedo Mano D, Abelha TF, Petrilli R, and Bentley MV. Delivery systems and local administration routes for therapeutic siRNA. *Pharmaceutical Research*. 2013; 30(4): 915–931.
57. Ballarín-González B, Thomsen TB, and Howard KA. Clinical translation of RNAi-based treatments for respiratory diseases. *Drug Delivery and Translational Research*. 2013; 3(1): 84–99.
58. Tai W and Gao X. Ribonucleoprotein: A biomimetic platform for targeted siRNA delivery. *Advanced Functional Materials*. 2019 (35): 1902221.
59. Sivakumar P, Kim S, Kang HC, and Shim MS. Targeted siRNA delivery using aptamer-siRNA chimeras and aptamer-conjugated nanoparticles. *Wiley Interdisciplinary Reviews: Nanomedicine and Nanobiotechnology*. 2019 (3): e1543.
60. Shrestha J, RazaviBazaz S, Aboulkheyr Es H, Yaghobian Azari D, Thierry B, Ebrahimi Warkiani M, and Ghadiri M. Lung-on-a-chip: The future of respiratory disease models and pharmacological studies. *Critical Reviews in Biotechnology*. 2020; 40(2): 213–230.
61. Zhou X, Wei T, Cox CW, Jiang Y, Roche WR, and Walls AF. Mast cell chymase impairs bronchial epithelium integrity by degrading cell junction molecules of epithelial cells. *Allergy*. 2019(7): 1266–1276.
62. Nogimori T, Furutachi K, Ogami K, Hosoda N, Hoshino SI. A novel method for stabilizing microRNA mimics. Biochemical and biophysical research communications. 2019; 511(2): 422–426.
63. Sandy AR, Zhang Q, and Lurio LB. Hard x-ray photon correlation spectroscopy methods for materials studies. *Annual Review of Materials Research*. 2018; 48: 167–190.
64. Yoshida H, Kuwana A, Shibata H, Izutsu KI, and Goda Y. Comparison of aerodynamic particle size distribution between a next generation impactor and a cascade impactor at a range of flow rates. *AAPS PharmSciTech*. 2017; 18(3): 646–653.
65. Rossi GA, Morelli P, Galietta LJ, and Colin AA. Airway microenvironment alterations and pathogen growth in cystic fibrosis. *Pediatric Pulmonology*. 2019; 54(4): 497–506.
66. Li H, Salomon JJ, Sheppard DN, Mall MA, and Galietta LJ. Bypassing CFTR dysfunction in cystic fibrosis with alternative pathways for anion transport. *Current Opinion in Pharmacology*. 2017; 34: 91–97.
67. Bell SC, Mall MA, Gutierrez H, Macek M, Madge S, Davies JC, Burgel PR, Tullis E, Castaños C, Castellani C, and Byrnes CA. The future of cystic fibrosis care: A global perspective. *The Lancet Respiratory Medicine*. 2020; 8(1): 65–124.
68. Lee CS, Bishop ES, Zhang R, Yu X, Farina EM, Yan S, Zhao C, Zeng Z, Shu Y, Wu X, and Lei J. Adenovirus-mediated gene delivery: Potential applications for gene and cell-based therapies in the new era of personalized medicine. *Genes & Diseases*. 2017; 4(2): 43–63.

21

Solid Lipid Nanoparticle-Based Drug Delivery for Lung Cancer

Deepa U. Warrier[1], Harita R. Desai[1], Kavita H. Singh[2], and Ujwala A. Shinde[1]
[1]Bombay College of Pharmacy, Kalina, Santacruz, Mumbai, India
[2]Shobhaben Pratapbhai Patel School of Pharmacy and Technology Management, SVKM's NMIMS University, Mumbai, India

21.1 Introduction

21.1.1 Epidemiology

Lung cancer is also known as lung carcinoma and is a malignant form of lung tumor characterized by uninhibited cell growth in lung tissues (https://www.msdmanuals.com/professional/pulmonary-disorders/tumors-of-the-lungs/lung-carcinoma). Lung cancer is also known as bronchogenic cancer, i.e. uncontrolled growth of the epithelial cells of the lungs (1,2).

Lung cancer is classified into two major histological categories, viz. small cell lung carcinoma (SCLC) representing 15% of all lung cancers and non small cell lung cancer (NSCLC) representing 85% of all lung cancers. Adenocarcinoma, squamous cell carcinoma (SqCC), and large cell carcinoma are the subdivisions of NSCLC (Figure 21.1).

In 2015, the WHO revised and published the recent histopathological categorization of lung cancer. The WHO classification (Figure 21.2 and Figure 21.3) was revised, built on molecular reports and targetable genetic variations which were newly identified in lung cancer (3).

The second most prevalent cancer across both males and females is lung cancer (small cell and non small cell). Approximately 228,820 additional cases of lung cancer (116,300 in males and 112,520 in females) and approximately 135,720 deaths from lung cancer (72,500 in males and 63,220 in females) are predicted by the American Cancer Society for the year 2020 ("Lung Carcinoma–Pulmonary Disorders–MSD Manual Professional Edition").

21.1.2 RisK Factors Leading to Lung Cancer

Internationally, lung cancer remains the main cause of cancer-related mortality in females and males (4). Cigarette smoking is the most known risk factor in the development of lung cancer. Tobacco smoking by cigars and pipes is also linked to high risk of developing cancer of lung (5). Other risks to the development of lung cancer include second-hand smoke contact; unprocessed biomass fuels; nonsmokers with a record of chronic bronchitis, pneumonia, or tuberculosis; asbestos exposure among workers in asbestos mining; ship building; construction; textiles etc.; and other organic and metal exposures which include beryllium, cadmium, chromium, silica, formaldehyde, etc. (6).

21.2 Current Management

21.2.1 Diagnosis

Apart from cessation of smoking, perhaps the highest reduction in death rates for lung cancer has to do with early diagnosis supported by surgical resection. Although chest X-rays and sputum cytology screening have shown no benefit, the USPSTF now endorses early detection through low-dose chest tomography (CT) (7). CT screening is, however, associated with increased rates of interesting results and can lead to the identification of some low-aggressive lung cancers (8). It includes improved risk stratification using lung cancer risk prediction models or biomarkers, and thus a detailed knowledge of the biological features of aggressive cancers to optimize the value of screening. Research has focused on the discovery of biomarkers of lung cancer risk, aggressive behaviour, and prognosis in early cancer. The neoplastic cells, the tumor microenvironment, or the hosts themselves may produce all such biomarkers. A range of methods have been tested to identify biosignatures using tissue and biofluid-based assays, including GWAS, epigenetics, microRNA, and proteomics (9). Many single nucleotide polymorphisms (SNPs) were detected using GWAS on different chromosomal loci, such as the locus 15q25, associated with tobacco use and lung cancer (10). A few studies have evaluated biomarkers that circulate with microRNA. Early-stage NSCLC patients with 80% accuracy were reported to have a 34-microRNA signature (11) whereas others appeared to foretell recurring disease (12,13) in plasma, surgically biopsied samples, and in tiny biopsies (14). Eventually, serum proteomic signatures were integrated with CT imagery to assess diagnosis of lung cancer in people with indeterminate lung nodules (15). It has been shown that a signature of seven auto-antibodies has high sensitivity and specificity for lung cancer related antigens for early detection of lung cancer in a high-risk population, as well as for the differentiation between benign and malignant diseases in CT-detected lung nodules for which prospective research is underway (16). Following significant progress in the discovery of biomarkers, the

DOI: 10.1201/9781003046547-21

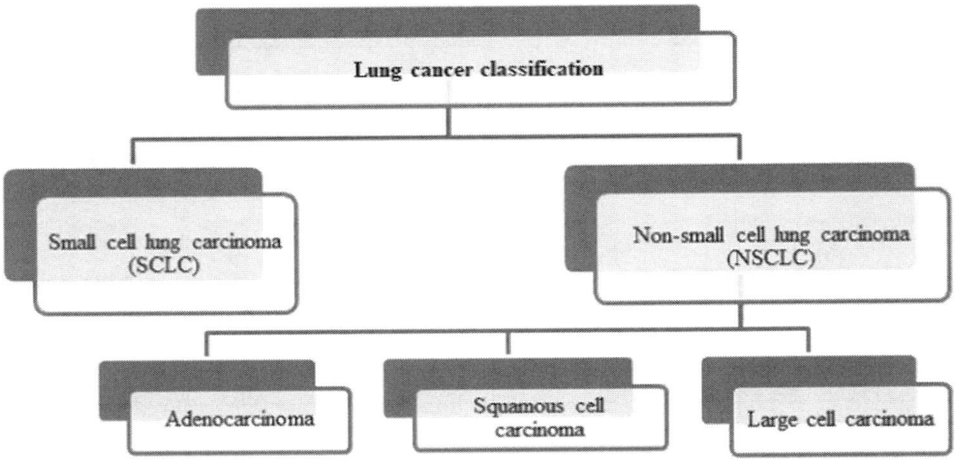

FIGURE 21.1 Initial Classification of Lung Cancer

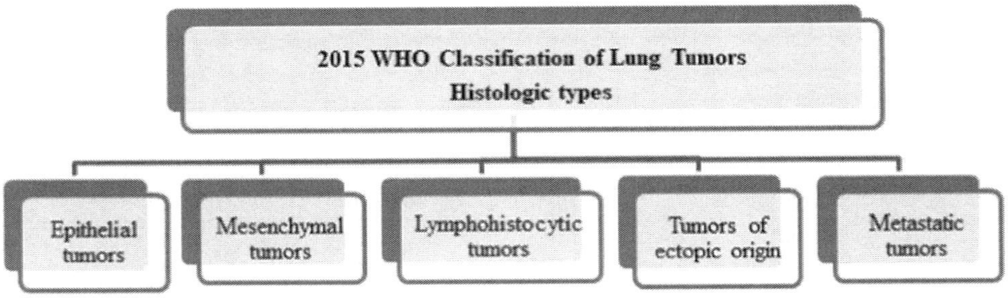

FIGURE 21.2 2015 WHO Classification of Lung Tumors

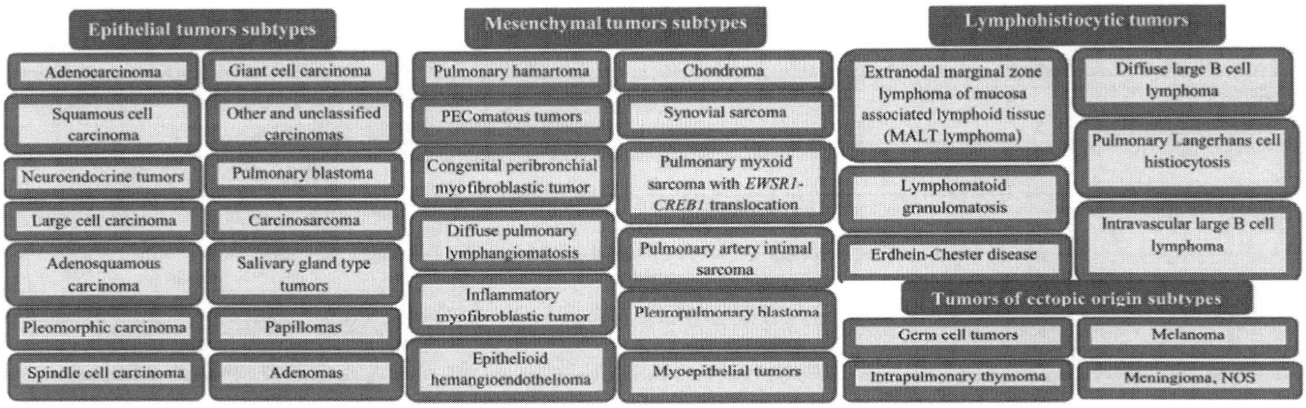

FIGURE 21.3 Classification of Lung Tumor According to Subtypes

remaining challenges include the choice of suitable candidate signatures based on tumor specificity and high-throughput methods, tumor genetic heterogeneity, and reproducibility in relevant validation studies (17,18).

21.2.2 Standard of Care

Present treatment depends on the type of carcinoma and the time of diagnosis and requires the involvement of chemotherapy, radiation therapy, surgery, photodynamic therapy, and target specific therapy (19,20) (Table 21.1)

21.3 Role of Nanocarriers in Treating Lung Cancer

Nanomaterials of size range 10–200 nanometers in diameter are called nanocarriers and they are believed to be a possible vehicle as a drug delivery system. Nanocarriers have reduced the cytotoxicity of anti-tumor drugs and experiments have been performed which show improved therapeutic efficacy of drugs used in the treatment of lung cancer (26,27). Several nanocarriers are frequently used to deliver biologically active chemotherapeutic drugs

TABLE 21.1

USFDA Approved Drugs (June 23, 2020) Used for the Treatment of Lung Cancer

Drugs	Mechanism of Action	Reference
Non Small Cell Lung Cancer		
Abraxane (Paclitaxel albumin-stabilized nanoparticle formulation)	130 nm albumin-bound paclitaxel–the gp60 protein-albumin compound causes the internalization of plasma constituents via the vascular endothelium by caveolin-1-mediated pathways. Paclitaxel attaches to tubulin and prevents microtubule disassembly resulting in a cell division inhibition.	(21)
Avastin (Bevacizumab)	It works by binding with VEGF and inhibits the binding of VEGF receptors, consequently restricting tumor blood vessels from growing and expanding.	(22)
Afinitor (Everolimus)	Remains bound to the immunophiline FK Binding Protein-12 (FKBP-12) to create an immunosuppressive complex which binds and restricts the stimulation of the mammalian Rapamycin Target (mTOR), a critical regulatory kinase. Suppression of mTOR stimulation results in inhibitory activity of T lymphocyte stimulation and multiplication associated with antigen and cytokine (IL-2, IL-4, and IL-15), and retardation of the production of antibodies.	(https://www.cancer.gov/publications/dictionaries/cancer-drug/def/everolimus?redirect=true)
Carboplatin	Carboplatin comprises a platinum atom combined with two groups of ammonia and a remnant of cyclobutane-dicarboxyl. This compound is intracellularly stimulated to form reactive platinum clusters that adhere to DNA nucleophilic units like GC-rich sites, causing cross-links of intrastrand and interstrand DNA, along with cross-links of DNA-proteins. These effects of carboplatin induced DNA and protein lead to cell death and a control of cell growth.	(23)
Docetaxel	Docetaxel exhibits powerful and strong antitumor characteristics. It combines with and stabilizes tubulin, thus hindering microtubular disassembly resulting in cell cycle arrest in the G2/M process and cell death process.	(24)
Gefitinib	Works by inhibiting EGFR tyrosine kinase site.	(https://pubmed.ncbi.nlm.nih.gov/16736978)
Small Cell Lung Cancer		
Atezolizumab (Tecentriq)	A humanized, Fc-optimized, monoclonal antibody targeted against protein ligand PD-L1 (programmed cell death-1 ligand 1), with apparent inhibitory immune control point and anti-tumor action. Atezolizumab has the ability to bind to PD-L1, obstructing its binding to and activation of its receptor programmed death 1 (PD-1) displayed on activated T-cells, which can significantly improve the T-cell-mediated immune response to neoplasms and inactivate reverse T-cells.	(https://www.cancer.gov/publications/dictionaries/cancer-drug/def/atezolizumab?redirect=true)
Topotecan Hydrochloride (Hycamtin)	Topotecan selectively helps to stabilize topoisomerase I-DNA covalent compounds during the S phase of the cell cycle, hindering topoisomerase I-mediated single-strand DNA breaks, and generating extremely lethal double-strand DNA breaks when DNA replication machine encounters the complexes.	(https://www.cancer.gov/publications/dictionaries/cancer-drug/def/topotecan-hydrochloride?redirect=true)
Doxorubicin	Doxorubicin binds with DNA via the intercalation and macromolecular biosynthesis inhibition.	(25)
Durvalumab (Imfinzi)	A Fc-optimized monoclonal antibody targeting programmed cell death-1 ligand 1 (PD-L1; B7 homolog 1; B7H1), with possible immune modulation and antineoplastic activity. After intravenous administration, durvalumab associates with PD-L1, hence preventing its attachment to and activation of its receptor programmed death 1 (PD-1), found on activated T cells. This however may reverse the inactivation of T-cells and enable the immune system to impose a cytotoxic T-lymphocyte (CTL) response to tumor cells that express PD-L1.	(https://www.cancer.gov/publications/dictionaries/cancer-drug/def/durvalumab)
Methotrexate	Acts by inhibiting DHFR, an enzyme that helps in synthesizing tetrahydro folate.	(https://pubmed.ncbi.nlm.nih.gov/17922664)

Vascular endothelial growth factor – VEGF, B-cell lymphoma – bcl-2, Epidermal growth factor receptor – EGFR, Mammalian Target of Rapamycin – mTOR, dihydrofolate reductase– DHFR.

for treating lung cancer. They can be categorized into nanocarriers based on lipids (SLN, micelles, liposomes, and lipid nano capsules) and non-lipid-based nano-systems (polymeric nanoparticles, mesoporous nanoparticles, dendrimers, and metallic or inorganic nanoparticles) (28). Nanoparticles have brought more attention to nano-biomedical technology because of the smaller size which has an immense amount of energy, thus helping the particles to adsorb and transport both hydrophilic and hydrophobic macromolecules to specified locations (29). Hydrophobic and hydrophilic drugs can be loaded in nanostructures and offers controlled drug release. The technique of nanocarrier is used for its precise targeting which is very important in targeted chemotherapy. Two essential processes for site-specific targeting, that is, active and passive mechanisms, are pursued (30).

The significant advantages of nanoparticles are displayed in Figure 21.4. Thus, drug delivery based on nanocarriers can be tailored to the needs of tumor regions by selectively delivering drugs and thereby reducing the risk of unspecific transmission to healthy tissues. This further helps to reduce side-effects of the antitumor therapeutic agents. Hence nano-drug delivery is the solution to summarize the novel options for a lung cancer cure (26).

21.4 Solid Lipid Nanoparticles as a Drug Delivery System for Treating Lung Cancer

One of the encouraging nano systems for the controlled delivery of drugs is solid lipid nanoparticles (SLNs), formerly known as lipospheres (31,32). SLNs are a class of colloidal particles comprising of physiologically safe and biodegradable solid lipids, i.e. lipids that remain solid at ambient and body temperatures, dispersed in an aqueous surfactant phase (33). The researchers have been using the notion that the use of solid lipids alternatively to liquid oils can provide controlled drug release, since the drug's fluidity in a solid lipid matrix is significantly lower, unlike liquid oil. The mean diameter of SLNs ranges within 50 and 1000 nm in the submicron scale (34,35). The impressive feature of SLNs is that they can bear a range of therapies including small drug molecules, large biomacromolecules (polysaccharides), genetic material (DNA/siRNA), and even vaccine antigens (36,37). SLNs have been reported to be effective for the treatment of a wide range of tumors. Lung tumors are of interest because they can be cured by inhalation using SLNs which are administered directly into the lungs. Targeted drug delivery systems based on nanotechnology can deliver anticancer drugs to the cancer cells selectively. In addition to site-specific delivery, SLNs also improve plasma solubility and bioavailability, enhance stability, and provide controlled release of drugs. Each of these aspects lead to a reduction in the total dose requirements which eventually leads to a reduction in side effects and improves patient compliance (38,39).

21.5 Role of Solid Lipid Nanoparticles in Drug Delivery for Treatment of Lung Cancer

Paclitaxel (PTX) is a taxane plant alkaloid, a diterpenoid resulting from the bark and needles of the Pacific yew tree. It

FIGURE 21.4 Advantages of SLNs as Drug Reservoirs

prevents replication of cells by stabilizing the microtubule cytoskeleton by inhibiting the cells in the late G2-mitotic cell cycle process and thus acts as a potent antineoplastic agent (40). Being a poorly water-soluble molecule and thus hydrophobic in nature, 1:1 mixture of cremophor EL (a polyethoxylated-castor oil) and ethanol is used to solubilise PTX. Yet cremophor EL has a variety of side effects, including hypersensitivity, nephrotoxicity, and neurotoxicity (41). Using cremophor EL as a vehicle also tends to change the biochemical properties of lipoproteins, like high density lipoprotein. It has also been shown to partially mediate PTX's cytotoxic activity in primary tumor cell cultures from patients (42).

As an alternative carrier, Da-Bing Chen et al. studied the inclusion of PTX into SLNs using stearic acid as the solid lipid carrier. The objective of this research was to prepare surface-modified long circulating stearic acid nanoparticles containing PTX, and to compare two formulations of PEGylated nanoparticles in terms of rate of drug release and pharmacokinetics. Being a type of native physiological molecule, stearic acid is bio-acceptable and biodegradable in the human body and therefore is a good carrier. The team prepared two formulations; the first being PTX loaded SLNs using Brij 78 (polyoxyethylene-20 stearyl-ether) as a non-ionic surfactant, and the second, PTX loaded SLNs using poloxamer F68 as the non-ionic surfactant. Brij78 has the PEG cain as hydrophilic end and stearyl alcohol is its hydrophobic end. A phospholipid substituted by PEG is PEG-DSPE. PTX-loaded SLNs were formulated by emulsion evaporation method followed by low temperature solidification. The *in vitro* release studies displayed slow release kinetics with the release of PTX from F 68-SLNS being linear. The release of PTX from the Brij 78-SLN followed the Weibull distribution, which means that drug encapsulation is matrix or reservoir type and the release pattern is affected by the SLN surface properties. The results of the study also suggest that PTX is distributed in the lipid matrix which is responsible for sustaining the release of PTX from SLNs. After injection of PTX formulations in Cremophor EL or in Brij 78-SLN and F68-SLN, pharmacokinetics was assessed in KM mice. PTX encapsulation in both SLN formulations has shown marked variations in comparison to the free pharmacokinetics of PTX. From this research, the authors were able to conclude that PTX loaded stearic acid — F68 and Brij 78 — SLNs were long circulating with half-lives of 10.06 and 4.88 hours respectively as compared to PTX injection which had a half-life of 1.36 hours (43).

Cancer cells can develop drug resistance mechanisms and escape chemotherapy (44). Multi-drug resistance (MDR) phenotype, primarily due to the expression of the encoding of the MDR gene family for the membrane proteins P-glycoprotein (P-gp), is a major problem in chemotherapy (45).

An intracellular reduction of the drug concentration can be obtained using the ability of P-gp proteins which can extrude various positively charged xenobiotics along with few anti-cancer therapeutic agents (46). Previous studies have shown that nano-particulate drug delivery systems can integrate PTX or doxorubicin (DOX) into cells and increase the drug's intracellular concentration (47). Researchers have also revealed through studies that intracellular entry of drug-laden nanocarrier delivery systems can be through endocytosis, followed by release of the entrapped drug into cytoplasm. This is an alternative drug entry route which allows to bypass or inhibit the P-gp-mediated efflux (48). This enabled the SLN-loaded anti-cancer therapeutics to enter the cells via an endocytic pathway, leading to an increased concentration of intracellular drug and drug cytotoxicity, thus evading P-gp dependent efflux and reversing MDR activity in MDR cancer cells. PTX and DOX were chosen as hydrophobic cytotoxic drugs to prove this. The goal was to formulate drug loaded (PTX and DOX) SLN and evaluate and analyse their drug resistance reversal behaviour in MDR cells (PTX and DOX resistant) in comparison to free drug solutions. The solvent dispersion method was used to prepare drug/s loaded SLN using monostearin as the solid lipid. The SLNs so obtained were re-dispersed by probe sonication with 20 sonic bursts 200W, active every 2 s in the aqueous solution of 0.1% poloxamer 188 (w/v). The resulting dispersion was then easily frozen in a deep freezer under −64°C for 5 hours and the sample was then transferred to the freeze-dryer. The drying period was monitored for 72 hours and the SLN powders were collected for analysis of *in vitro* release. The *in vitro* release profiles showed maximum drug release of 40% for PTX and 65% for DOX over a period of 24 hours. These findings suggested that PTX and DOX can be released gradually from the SLNs and the SLNs can hold a constant drug/s concentration for a long time. This may therefore reduce the length and frequency of administration which is beneficial for clinical application. In this study, two human cancer cells (human breast cancer cells MCR-7 and human ovarian cancer cells SKOV3) and their multi-drug-resistant variants were used to further evaluate the cytotoxicity of the drug loaded SLN. The cytotoxicity studies showed that the drug loaded in SLN not only improves cytotoxicity against sensitive cancer cells, which was also shown in previous *in vitro* (49) and in vivo tests (50), but also improves cytotoxicity against resistant cancer cells as compared to free drug solution. Consequently, MDR reversal action in both multi-drug resistant cell lines (MCF-7/ADR and SKOV3-TR30) were exhibited by drug/s loaded SLNs. By this study, the authors concluded that transportation of standard anti-cancer drugs, viz. DOX or PTX, into the tumor cells can be enhanced by SLNs as drug delivery system and boost their cytotoxicity against vulnerable cancer cells and multi-drug resistant modified cells in comparison to free drug solutions. In other words, lipid matrix in the SLNs will shield the encapsulated drug from the cell's P-gp efflux mechanism and overcome multi-drug resistance and reveal a possible application of this reversal mechanism of drug resistance in drug-resistant human cancer cells (51).

The use of curcumin (CUR) has been documented in literature to efficiently hinder tumor cell proliferation, migration, and invasion in malignant tumors of brain (52). Other research found that curcumin had the function of overcoming drug resistance for anti-carcinoma medications (53).

Curcumin's clinical use was, however, limited due to its enormously low aqueous solubility, rapid metabolism in vivo and limited bioavailability (54). In the study investigated by Wang et al., CUR loaded SLNs were prepared for treatment for

TABLE 21.2
Examples of Drug Loaded SLNs for Lung Cancer Treatment

Drug	Lipid	Method of SLN Preparation	Significance
Docetaxel	Compritol 888 ATO	Melt – High energy	SLN-DTX blocks tumor cell growth and inhibits lung metastasis by lowering serum IL-6 development, triggering tumor cell apoptosis, and reducing tumor cell proliferation and BCL-2 expression (Da (57)).
Artemether	GMS:Compritol (50:50)	High pressure homogenization technique	The ART – SLNs were stabilized by $MPEG_{2000}$-DSPE which made them long circulating and the ART-SLNs showed promise for oral anticancer drug delivery (58).
Curcumin	Stearic acid and lecithin	Sol – gel method	CUR – SLNs in the ratio of 4:1 enhanced CUR solubility, bioavailability, and improved accumulation of CUR in tumor and lungs of mice bearing A549 xenograft (56).
Sclareol	Precirol	Hot homogenization method	Higher geno-cytotoxicity exhibited by Sclareol loaded SLNs against A549 cancer cell lines (59).
5-fluorouracil	Monostearin	Solvent dispersion	5-FU SLNs modified with chitosan improved encapsulation efficiency and cytotoxic effect (60).
Myricetin	Compritol 888 ATO, Gelucire 44/14, oleic acid	Ultrasonication – high speed homogenization	Myricetin loaded SLNs increased cytotoxic effects of myricetin by stimulating necroptosis in A549 cells (61).
Camptothecin (CPT)	Hydrogenated soybean phosphotidylcholine (HSPC), triglycerol myristate ™, Stearylamine (SA), gellucire 53/10	Hot homogenization - ultrasonication	The CPT loaded PEGylated SLns increased intracellular uptake and prolonged circulation time by evading nonspecific RES uptake mechanism (35).

non small cell lung cancer. The anti-tumor activity of SLN-curcumin was investigated *in vitro* as well as in vivo in the anticipation of evaluating a reasonably effective and safe drug potential for treating lung cancer. The CUR-SLN in the ratio 2:1 and 4:1 was prepared by solvent-dispersion technique using stearic acid as the solid lipid and Myrj 53 as the surfactant. The CUR-SLN was successfully prepared with a circular shape ranging from 20 to 80 nm, which enables the nanoparticle to pass vascular wall cell clearance. SLN's lipid matrix helps the drug encapsulated in nanoparticles to penetrate carcinoma cells more effectively than plain drug (55). Curcumin stability in PBS was also evaluated, and the results showed that free curcumin in PBS deteriorated rapidly and that only 50% remained after the 6 hours incubation compared to the introduced sample. Nevertheless, in the same conditions, the nanoparticulate CUR was up to 80% stable. An increased entrapment efficiency of 75% was displayed by CUR-SLN in the 4:1 ratio as compared to the 2:1 ratio, which represented an entrapment efficiency of just 62%. The IC_{50} of CUR-SLN (4:1) was 9 μM for NCL-H1299 and 4 μM for A549, 1/20 and 1/19 of that of the original curcumin. The CUR plasmid concentration was increased in mice after intraperitoneal injection of CUR loaded SLN. In addition, it was also observed that CUR-SLN improved curcumin targeting to the lung tumors which eventually increased curcumin's inhibition efficiency from 19.5% to 69.3%. Analysis of flow cytometry (FCM) and immune staining indicated that the effect of inhibition was due to apoptosis rather than from necrosis. CUR-SLN's tumor targeting and deep tumor inhibition effect suggested its medical application for the treatment of lung cancer, and offered a novel tool for the production of new anti-cancer agents (56) (Table 21.2).

21.6 Ligand–Anchored Solid Lipid Nanoparticles

The drug delivery field has seen increasing potential of SLNs to deliver targeted anti-cancer systems by reducing the uptake by reticuloendothelial system. Effective delivery of bioactives can be accomplished by anchoring the surface of the SLNs using a targeting ligand.

Lactoferrin (Lf), often recognized as lactotransferrin, is a transferrin family multifunctional protein. Lactoferrin is a globular glycoprotein with a molecular mass of approximately 80 kDa, frequently represented in diverse secretory fluids such as saliva, milk, nasal secretion, and tears. It is linked to a broad array of biologically essential processes which include host defense, regulation of cell proliferation, and cell differentiation (62). Lactoferrin boosts Fas expression and cell death in the colon mucous membrane of rats treated with azoxymethane and may also be chimeric aptamers in the delivery of drugs targeted at cancer cells (63). It is a type of glycoprotein which binds to iron and has had antioxidant, antimicrobial, and antineoplastic activity to display. It also has the innate ability to aggravate cell death and impede the proliferation of cancer cells. Recent research indicates that Lf induces apoptosis and interferes with cancer cell proliferation. This also facilitates the regeneration after chemotherapy of the leucocytes and erythrocytes (64). The Lf's biological properties are mediated on the surface of target cells by specific receptors. The Lf receptors are expressed in bronchial epithelial cells on the apical surface. Such expressed Lf receptors can be used for the transport of other nanocarrier systems or Lf-conjugated drug. Lf conjugation thus triggers the anticancer function of drug-SLN complex by precise delivery to the lungs. The study performed by Pandey et al. reveals Lf-coupled SLNs targeting ability for site-specific lung delivery of PTX. The SLNs were prepared by solvent injection method using tristearin and hydrogenated soya phosphatidylcholine (HSPC) in the ratio 60:40 and 1 w/w % stearic acid. Tween 80 was used as the non-ionic surfactant. After the formulation of PTX loaded SLNs, conjugation of Lf with SLNs was done by cardbodiimide reaction where the carboxylate group of Lf was conjugated with the amino group present of SLNs using EDC as coupling agent. Better anticancer activity was displayed by Lf-coupled SLNs as compared to plain SLNs and free PTX when ex vivo cytotoxicity studies were performed on human bronchial epithelial cell lines, BEAS-2B. Studies on in vivo biodistribution displayed greater concentrations of PTX accumulated through Lf-coupled SLNs in the lungs unlike plain SLNs and free PTX. These findings demonstrate that Lf-coupled PTX-SLNs can be used as a prospective targeting carrier for the delivery of anticancer drug to the lungs with least side effects (65).

Protein interaction–carbohydrates has been well studied, and cell interaction and binding by it is well facilitated. These interactions are mediated by the lectin receptors expressed on the neoplastic cell surface. These lectin receptors have very high affinity for carbohydrate molecules. Hence, glycosylation plays a significant role in lectin receptor mediated targeting (66). A plethora of research well documents the specific entry, enhanced activity, and reduced toxicity of mannose (67). The enhanced activity is due to the favoured uptake by receptor-mediated internalization via lectin receptors at tumor cell surface (68). Thus, mannose anchored drug delivery vehicles have emerged as a forthcoming strategy for site specific delivery of anti-cancer agents to the tumor tissues (69). One such study was performed with the aim to synthesize mannosylated SLNs (M-SLNs) encapsulated with PTX for parenteral administration. PTX SLNs were prepared by solvent injection method using tristearin as solid lipid (core) and distearoyl-phosphatidylethanolamine (DSPE) as a stabilizer. The ring opening reaction of aldehyde group of mannose with the free amine group of stearylamine was the mechanism involved in mannosylation. The efficiency of PTX-SLNs and M-SLNs for drug trapping was found to be 84.5 % and 79.4% correspondingly. Due to dissolution and consequent loss of surface adsorbed drug in sodium acetate buffer (pH 4.0), which was used as a medium for mannosylation of SLNs, the material trapping capacity of M-SLNs was found to diminish. A549 human epithelium lung cancer cell line was used to analyse the *in vitro* cytotoxicity of M-SLNs. In addition, haemolytic toxicity was also performed for the study of biocompatibility and biodistribution to determine the preferred position of the drug at the target. Analysis of cytotoxicity on A549 cell lines and drug biodistribution showed that M-SLNs deliver a higher concentration of PTX compared to PTX-SLNs at an alveolar cell site. Results showed that PTX drug solution has less cytotoxicity at a dose level of 40 mg/ml with 42.8% cell growth inhibition. This may be attributed to least drug uptake after incubation with cell for 48 hours. In addition, PTX-SLNs showed cell growth inhibition of 54.4% which could be due to entry of PTX-SLNs into the cells most likely via passive diffusion. M-SLNs showed higher cell control growth inhibition of 66.8% up to 48 hours, which may be attributed to receptor mediated endocytosis. Conjugation of mannose with SLNs, drastically decreased RBC hemolysis possibly due to prevention of RBC interaction with charged amine groups associated with unconjugated SLNs. M-SLNs decreased drug build-up in liver. This can be because mannose receptors are over-expressed on the lung cancer cell. This illustrates M-SLN's selective targeting to lung tissue. The authors suggest that results indicate M-SLNs encapsulated with PTX have promising potential as alternative parenteral formulations for successful treatment of lung cancer (69).

Etoposide (ETPS) is a plant alkaloid, a podophyllotoxin semisynthetic derivative that possesses anti-tumor activity. ETPS forms a complex with topoisomerase II and DNA and thus blocks DNA synthesis. This complex causes breaks in dual stranded DNA and inhibits repair by binding topoisomerase II. Cumulative defects in DNA prohibit cell division entry into the mitotic process and contribute to death of the cells. ETPS mainly operates in the G2 and S cell cycle phases. Transferrin (Tf) is a polypeptide glycoprotein consisting of about 679 amino acids. The iron transport in the human body is an important process controlled by Tf. For their rapid growth and proliferation, cancer cells require more iron which results in the up-regulation of Tf-receptors in many other malignant tumors including lung, brain, breast, colorectal, and prostate (70–74). Tf has been recognized and investigated as a possible targeting ligand for drug delivery to brain and breast cancer (74–84). However, to investigate its role in lung cancer, Pooja D et al. conducted a study on Tf conjugated SLNs to improve delivery of ETPS to lung cancer. The ETPS loaded SLNs were prepared by single emulsification and solvent evaporation method. Glyceryl monostearate (GMS) and stearic acid were employed as solid lipid, soya lecithin was used as a stabilizer, and 1.5 % w/v tween 80 was used as the non-ionic surfactant. Surface conjugation of transferrin to the lyophilized ESN SLNs were performed by reaction with EDC/NHS and incubation for 8 hours. Carboxylic groups present on the 85 ESN surface were activated in the first step by addition of EDC and stabilized by NHS. Activated carboxylic groups were allowed to interact with the free amine groups present on Tf in the second step. The conjugation efficiency was found to be 71.8±4.6% when determined by Bradford assay. An antiproliferation assay was used to examine the *in vitro* anti-cancer efficacy of three ETPS formulations against A549 human lung cancer cells. The results suggested that plain ETPS had low anti-cancer activity against the cell line. Nevertheless, after encapsulation into SLNs, its performance was increased up to 6.6 times. However, the lower viability of Tf-ESN-treated cells than ESN-treated cells

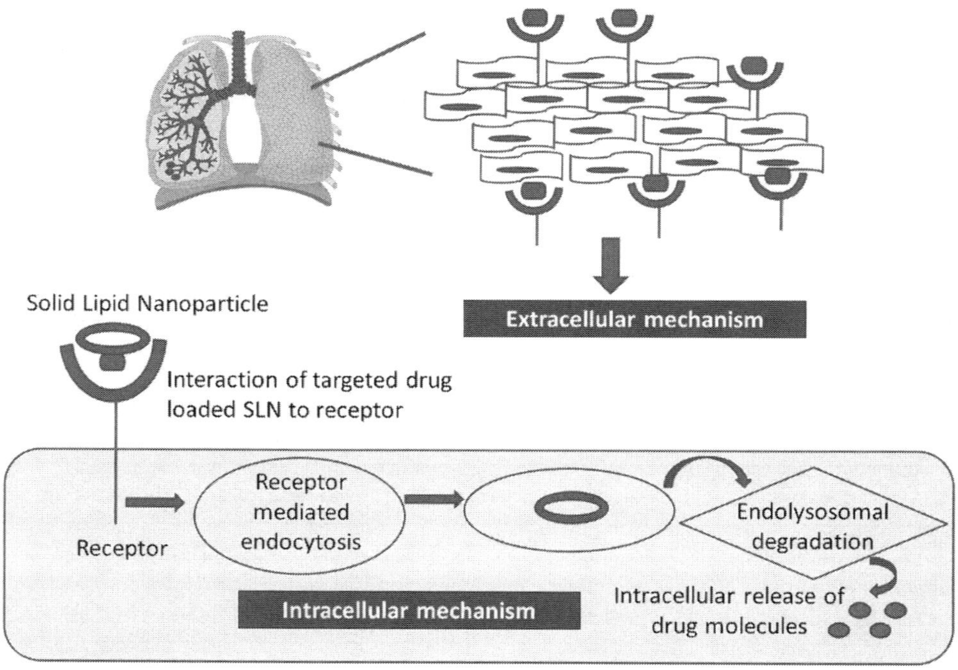

FIGURE 21.5 Receptor Mediated Drug Delivery of Solid Lipid Nanoparticles to Cancer Cells of the Lungs

suggested that Tf's conjugation on the SLN's surface enhanced the cytotoxicity of ETPS nanoparticles. The increased uptake of nanoparticles by the A549 cells was related to the active targeting of Tf-ETPS-SLNs by Tf-receptors. Tf-ESN had higher plasma concentration, longer blood circulation, and reduced clearance of encapsulated ETPS than marketed Etosid®, an etoposide formulation. The authors concluded that the study gave promising results indicating that targeting nanomedicines to Tf receptors could increase the therapeutic efficacy of non small cell lung cancer treatment (85) (Figure 21.5).

As targets for NSCLC therapy, various pathways have been explored, like the epidermal growth factor (EGF) pathway, which is specific of the oncogenic receptor, tyrosin kinase (TK) pathway. EGF receptor (EGFR) inhibitors are currently being used in a subset of patients with advanced stages of the disease as first-line therapy choices. Due to intrinsic resistance, 20–30% of NSCLC patients with activating EGFR mutations do not respond to treatment with EGFR-TK inhibitors and nearly all patients with an early positive response begin developing acquired resistance within a short time span of treatment. EGF receptor activation (EGFR1, EGFR2, EGFR3, EGFR4) in response to ligands (e.g. EGF, TGFα, and others) triggers several downstream signaling paths through three main signaling branches.: signal transducers and activators of transcription (STAT), rat sarcoma (RAS), and phosphatidylinositol 3′-kinase (PI3K). Every member of the EGFR family has been involved in driving initiation and development of NSCLC.

To minimize and conquer these limitations, the authors, Garbuzenko OB et al., proposed a unique five-step treatment which includes gene therapy using a pool of small interfering RNA (siRNAs) to suppress four types of EGFR-TKs; using an anti-tumor drug, PTX, to induce cell death; employing passive targeting approach by delivering therapeutic agents directly to lungs through inhalation delivery system; employ active targeting, i.e. using synthetic luteinizing hormone-releasing hormone (LHRH) peptide for receptor mediated targeting to the cancer cells; and increase stability and improve the penetration of siRNA and PTX using nanostructured lipid carriers (NLCs) as a means of nanomedicine drug delivery. NLCs of PTX were prepared by modified melted ultrasonic method using Precirol ATO 5 as the solid lipid, squalene as liquid lipid, and soyabean phosphatidylcholine as lipophilic emulsifier. The aqueous phase consisted of Tween 80 as the surfactant, and 1,2-dioleoyl-3-trimethylammonium propane (DOTAP) as cationic lipid. PEG coating of the NLCs was performed by adding DSPE-PEG 2000 to the aqueous phase. The LHRH peptide for active targeting was also added to the aqueous phase. The siRNA-NLC complexes was prepared by adding the siRNA solution to the NLCs prepared in the previous step and the mixture was vortexed and incubated at room temperature. The LHRH-PEG-PTX-siRNA NLCs hence synthesized, were further characterized, and assessed *in vitro* using human lung cancer cells with different sensitivities to gefitinib, which is an inhibitor of EGFR. The cellular internalization of the novel NLCs was studied in human A549 NSCLC cells using a confocal microscope. Figure 21.6 shows the fluorescent images of the cells treated with the novel NLCs. The authors found that the NLCs along with PTX and siRNA were successfully internalized (86).

The scientists also developed an orthotopic mouse model of human lung cancer in their laboratory. Figure 21.7 shows the instillation for inhalation delivery of drug and siRNA NLC formulations.

The biodistribution of the novel NLCs given by inhalation route was compared with the non-targeted NLCs which were delivered intravenously and by inhalation. After inhalation delivery, it was found that targeting the NLCs explicitly to tumor cells by the

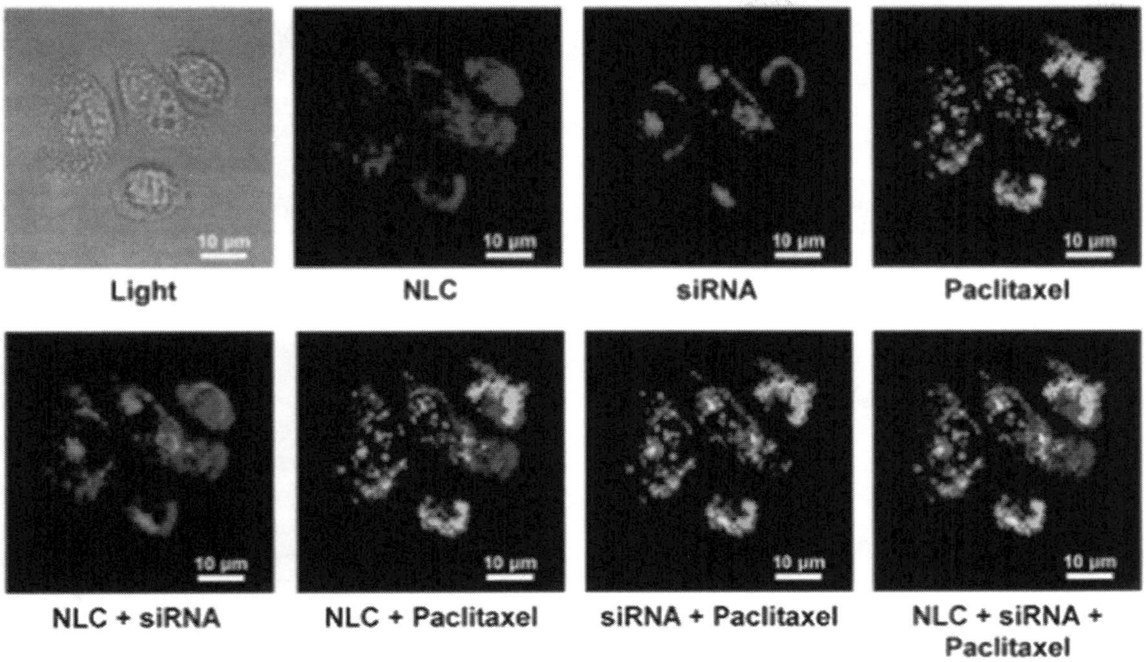

FIGURE 21.6 Cellular Internalization of Nanostructured Lipid Carriers (NLC). siRNA and PTX Delivered by NLC (confocal microscope Leica G-STED SP8). A549 Adenocarcinomic Human Basal Epithelial (Alveolar Type II Pneumocytes) Non-Small Cell Lung Cancer (NSCLC) Cells were Incubated for 18 hours with NLC (Blue Color Represents Near-infrared Fluorescence) Containing siRNA (Red Fluorescence) and PTX (Green Fluorescence). Superimposition of Red and blue Colors Gives Pink Color; Superimposition of Blue and Green Gives Cyan Color; Superimposition of Red and Green Colors Gives Yellow Color; Superimposition of Red, Green and Blue Colors Gives White Color. (Reprinted with permission from Olga B. Garbuzenko et al., Strategy to Enhance Lung Cancer Treatment by Five Essential Elements: Inhalation Delivery, Nanotechnology, Tumor-Receptor Targeting, Chemo- and Gene Therapy (Theranostics, 2019, 9 (26): 8362–76) (86))

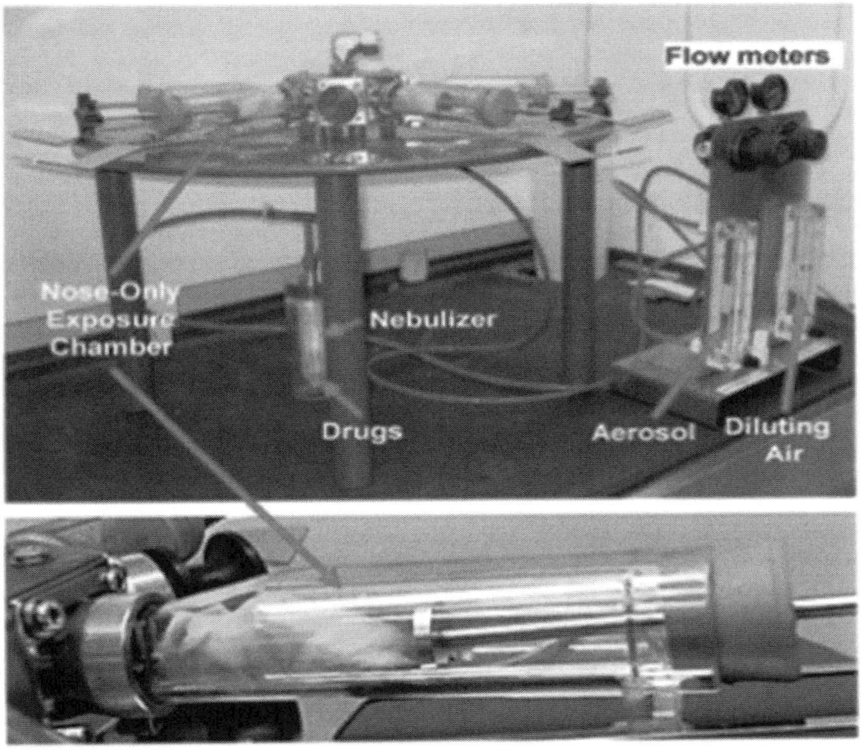

FIGURE 21.7 Instillation for Inhalation Delivery of Drug and siRNA NLC-Formulations. (Reprinted with permission from Olga B. Garbuzenko et al., Strategy to Enhance Lung Cancer Treatment by Five Essential Elements: Inhalation Delivery, Nanotechnology, Tumor-Receptor Targeting, Chemo- and Gene Therapy (Theranostics, 2019, 9 (26): 8362–76) (Garbuzenko et al. 2019)

LHRH peptide greatly increased accumulation of LHRH-NLC nanoparticles in the lungs with tumor. After inhalation, the accumulation of tumor-targeted nanoparticles (LHRH-NLC) was greater compared not just with intravenous administration of these particles but also to the non-targeted NLCs which were also administered by inhalation. The effectiveness of inhalation treatment of lung cancer using the novel targeted NLCs was assessed by observing suppression efficiency of all four variants of EGFR-TK receptors, induction of cell death in tumor cells and inhibition of tumor development. Finally, the authors observed that combined targeted inhalation of siRNAs with PTX (a blend of gene- and chemotherapy) culminated in the death of NSCLC cells, restricting and suppressing tumor growth with an efficacy that cannot be accomplished if either therapy is administered alone or separately (86).

21.7 Solid Lipid Nanoparticles Administration by Pulmonary Drug Delivery System

Aerosol drug delivery system has emerged with vast field of information with an advantage of treating lung cancer. Pulmonary administration of antineoplastic agents allows superior efficiency of drug build-up in lung tumors. The pulmonary route is a non-invasive route of administration for both systemic and local delivery of drugs. In addition, the pulmonary route has many advantages over the oral route such as rapid absorption due to high surface area and high vascularization which helps to circumvent the first pass effect (87). Drug encapsulation in SLNs can further enhance drug accumulation and reduce drug-associated toxicity. However, maintaining the significant physicochemical parameters for improved inhalation is the challenge for developing a colloidal system for pulmonary drug administration (88). Pulmonary drug delivery systems to provide therapeutic agents into the lungs consist of dry powder inhalers (DPIs), nebulizers, pressurized metered dose inhalers (pMDIs), and soft mist inhalers. Nebulizers, being propellant-free, produce a large amount of aerosolized drug droplets from suspension or drug solution with increased time duration and thus are suitable for children below the age of two years and geriatric patients with chronic disorders. This makes nebulizers more advantageous over pMDIs.

Videira et al. conducted a study on new formulation of PTX loaded in SLN and used for lung cancer treatment an experimental mouse mammary carcinoma. The team assessed the efficacy of treatment with SLN loaded with PTX (SLN-PTX) on minimizing the number and size of lung metastases compared to the treatment with intravenous administration of the same medication utilizing conventional formulation. PTX-SLN were prepared by melted homogenization using 3% w/v glyceryl palmitostearate and 50% w/w polysorbate 80. The PTX-SLN cytotoxicity was studied using rat peritoneal macrophage cells and the PTX–SLN *in vitro* efficacy was studied in murine mammary adenocarcinoma – MTX- B2 cell line. The macrophage evaluation showed that cell viability decreased for the highest concentration of NPs (37.5 mg/ml) which appears to be correlated with medium turbidity and lipid concentrations. In addition, cell permeability can promote deposition of NPs in cytosol, resulting in increased lysosomal activity and potential loss of cell viability via endocytic pathway saturation. The Taxol displayed a 20-fold lower cytotoxic effect than the one detected with SLN-PTX. The PTX-SLN therapeutic efficacy study was performed to compare the effectiveness of inhaled SLNs with that of intravenously administered Taxol. The researchers used B6D2F1 female mice (6 weeks, 18–20 g) which were randomly divided into properly defined groups of six animals per cage for the control group and four animals per cage in remaining groups. The animals developed experimental lung metastases after MXT-B2 cell inoculation (50,000 cells/100 μl) in the tail-vein, as confirmed by the control group, at day 30. The findings suggested that SLN-PTX inhalation was more successful in suppressing experimental pulmonary metastases than Taxol's intravenous administration. It was observed that the SLN-PTX inhalation strategy of pulmonary administration exhibits a high therapeutic efficacy which resulted in a significant reduction in the 2-week treatment and a complete removal of experimental metastases in the lung surfaces when the number of administrations (long-term treatment) was extended. Hence, inhaled SLN-PTX may represent a potential method for regional delivery to the lung area, capable of addressing current clinical strategy issues. In fact, this correlates to a new concept that allows targeted delivery to the deeper lung regions, also capable of reaching the lymphatic system, which has proved critical in breast and lung adenocarcinoma, particularly during the metastatic progression. These tests, taken together, show that SLN-PTX efficiently suppressed the progression of lung metastases (89) as shown in Figure 21.8.

The therapeutic performance of pulmonary drug delivery system is influenced by many factors which include aerodynamic diameter, surface area, mechanical properties, solid state drug properties, mucoadesiveness, polymer properties, and surface dissolution kinetics (90).

In a recent study conducted by Kaur et al., PTC and DOX co-loaded SLNs have been formulated by the solvent evaporation method using various surfactants like Cremophor EL, Tween 80, and Tween 20, followed by spray drying. As previously discussed in this chapter regarding the role of P-gp efflux mechanism causing multi-drug resistance, it has been proposed that due to the high permeability, increasing solubility, and diverse molecular weight of surfactants, we can overcome the challenges of P-gp mediated drug efflux (91). Based on this assumption, the authors have conducted in vivo (lung deposition, bio-distribution and lungs histology) to select appropriate formulations for improved lung cancer targeting. It was observed that DPIs with PTX-DOX co-loaded SLNs prepared using Cremophor EL demonstrated higher drug distribution in the lungs compared to the plain drug. The cellular uptake was performed on A-549 highly resistant cancer cell lines expressing P-gp. *In vitro* findings suggested that in lung adenocarcinoma A549 cells, which have increased P-gp, optimized DPIs with Cremophor EL, improve the cell uptake and transfection efficacy. Studies in vivo showed that optimized formulation shows improved retention and drug accumulation in the lung without any sign of tissue abnormality, thus helping to minimize the toxic effects in non-target tissues. This recent study supports the vast research literature that inhalable drug loaded SLNs offer a potential chemotherapeutic strategy for lung cancer (92).

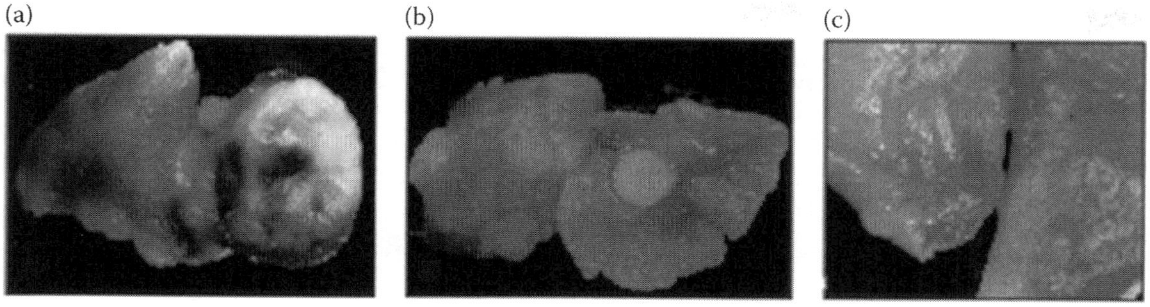

FIGURE 21.8 Therapeutic Efficacy of SLN-PTX was Compared With the IV Administration of the Drug. Photographs of Lung Surface Invasion and Tumor Growth Obtained With a Camera Attached to a Magnifying Lens (40x) respectively from: (a) Lung Surface of Animals not Treated (group A) Presenting 100 % of Lung Injury; (b) Treated with Taxol (120 μg/0.1 mL PBS) Twice a Week (group E); (C) Pulmonary Administration of SLN-PTX (100 μg/2 ml) Twice a Week (group F). A Decrease in the Number and Volume of Lung Metastases can be Observed in the Group Treated During the Invasion and Metastases Growth, the Therapeutic Stage, where Only a Few Small Metastases were Observed (group D), Contrasting With the Metastatic Surface Presented in the Lung Collected from an Animal in Group E. (Reprinted with permission from Mafalda Videira et al., Preclinical Evaluation of a Pulmonary Delivered Paclitaxel-Loaded Lipid Nanocarrier Antitumor Effect, Nanomedicine: Nanotechnology, Biology, and Medicine, 2012, 8 (7): 1208–15) (89)

21.7.1 Characterization of Dry Powders for Inhalation

Different principles of particle engineering and technological advancements in the design of inhalers have made DPIs an attractive drug delivery system for use in pulmonary delivery. However, the physical properties and fate of the nanoparticles are very important to achieve the desired efficacy.

21.7.1.1 In Vitro Aerodynamic Properties

The aerodynamic properties are crucial factors to achieve fine lung deposition. The drug accumulation site is determined by particle size, whereas the deposition pattern across the human respiratory tract is determined by aerodynamic particle size. The nanoparticles, due to their small size, inherently tend to penetrate and accumulate within the leaky tumor vasculature when the drug is administered via systemic route. This is known as enhanced permeation and retention effect (EPR) (93–96). However, when nanoparticles are administered by inhalation, the EPR effect may not play a role in tumor deposition. But the drug's delivery directly to the lungs allows for passive targeting of the lung tumour. In addition, nanocarriers are taken via endocytosis into the cancer cell, which usually does not occur in the case of solubilized drugs (97,98). Nanoparticles can thus improve the penetration and deposition of inhaled drugs in tumor tissues and cells, resulting in greater anti-tumor activity compared to the free drug (99–102). A problem with NP pulmonary delivery is that its size is not appropriate for deep deposition nanoparticles in the lungs.

For deep lung delivery a carrier system like lactose microparticles is required. Important requirements like appropriate mass median aerodynamic diameter (MMAD) and a suitable fine particle fraction (FPF) should be met by the carrier particles (103–105). The aerodynamic particle diameter is one fundamental characteristic that defines whether drug is deposited in the mouth-throat area or in the airways (106–108). Drug particles with an aerodynamic diameter of about 0.5–5 μm have the greatest potential to accumulate in the lung. Thus, smaller particles accumulate more peripherally in the lung, for example in the alveolar space, while larger particles deposit more centrally in the large conducting airways. Particles greater than 5 μm tend to accumulate in the region of the mouth-throat (109) and thus minimize the lung dose. The optimal size for deposition in the lung for aerosol powders is considered to be particle diameters ranging from about 1 to 5 μm (90). Delivery of drug-laden nanoparticles to the lungs integrates the targeted delivery principles with the benefits of using nanoparticles in lung cancer therapy. The airborne particles inhaled by means of DPIs may be carried through the mouth to the systemic circulation. The particle's ability to enter the head airway based on its aerodynamic diameter is known as inhalability. With particles smaller than 5 μm, inhalability was found to be close to 100%. The size is necessary because of the morphological size of the alveolar sacs. With an increase in particle size, the inhalability decreases.

When the drug is inhaled, the airflow undergoes varied changes that can lead to various particle deposition mechanisms. For local effect, the particle size of the mixture should not exceed 5 μm and for the systemic effect to be achieved no more than 2 μm. Such systems have demonstrated deep pulmonary penetration by amalgamating product output and powdered formulation (110). Roa et al. optimized effervescent and non-effervescent carrier particles for local lung delivery of doxorubicin loaded nanoparticles. The tests were performed in lung cancer bearing mice. Spray–freeze drying was used to prepare the carrier particles with a desired MMAD, rather than traditional spray drying or freeze drying (112). Utilizing spray–freeze drying, the drug is not subjected to high temperatures as it may be with spray drying, so the NPs are naturally formulated with a uniform distribution throughout the carrier particle instead of collecting at the evaporation front as may happen with spray drying (111). The active release mechanism used in effervescent formulations scatters the NPs and inhibits aggregation when the carrier matrix is dissolved and released as stated previously (104). The prepared NPs had an average size of 137.22±1.53 nm. The doxorubicin-loaded nanoparticles had a particle size of was 145±20 nm after redissolution of the effervescent. The mass aerodynamic diameters of carrier particles loaded with blank and doxorubicin NPs were respectively 3.45±0.11 and 3.41±0.22 μm (n=6). The authors found that tumor-infected mice treated with an effervescent nanoparticle carrier displayed prolonged survival time than the non-effervescent nanoparticle carrier treated animals (104). The survival time of tumor infected mice treated with inhalable DOX nanoparticles was also more as compared to same drug dose intravenously administered (112).

21.7.1.2 Simulated Lung Deposition

Several in vivo methods, based on the use of radioactively labeled aerosols, can assess particle deposition in the lung. Technically the method is complex, requires different tracers, and is costly. In addition, specific knowledge is provided by existing methods such as single photon emission computed tomography (SPECT), positron emission tomography (PET), and π-scintigraphy. Data are indicated mainly as total lung deposition, consisting of deposition in the conducting and peripheral airways. The penetration index (PI; the ratio between conductive deposition and peripheral airways) provides information as to the degree to which the particles penetrate the alveoli, where absorption occurs primarily. For example, researchers have developed liposomal curcumin dry powder inhaler (LCD) for inhalation treatment of primary lung cancer. The *in vitro* simulated lung deposition of LCDs was determined using the Next Generation Impactor (NGI, Copley, Nottingham, UK). The FPF (5 μm) was calculated with equation. The fine particle fraction (FPF) of LCDs was found to be 46.7175.23%.

$$FPF = \frac{\text{Drugs in stages } 2-7}{\text{Drugs in all stages}} \, x \, 100\%$$

Various receptors are over-expressed by lung tumor cells, like folate receptor, epidermal growth factor receptor (EGF receptor), and luteinizing hormone-releasing hormone (LHRH receptors) (113–115). Tseng et al. demonstrated that EGF-targeted biotinylated gelatin nanoparticles were more specifically deposited in cancer cells due to receptor-mediated ingestion and caused no lung injury (121,122). EGF-targeted inhalable magnetic nanoparticles showed improved absorption in cancer cells relative to non-targeted particles and improved anti-tumor activity (123,124).

LHRH peptide-coated mesoporous silica nanoparticles (MSNs) were developed by Taratula et al. for the delivery of anti-cancer drugs, viz. doxorubicin and cisplatin, and antisense oligonucleotides targeted to MRP1 and BCL-2 against resistant lung cancer. Inhalation therefore allowed higher concentrations of drug/siRNA to be accumulated in the lungs relative to intravenous administration. The targeted nanoparticles were effectively internalized into human lung cancer cells and exhibited an improved activity against cancer (125). Taratula et al. also demonstrated that lipid nanoparticles targeted at LHRH receptors can promote the selective accumulation of siRNA and DOX in cells of lung tumor and decrease accumulation in healthy tissues of lung (126).

Models of aerosol deposition have been in progress for decades to help understand the complex method of delivering respiratory drugs. Computational fluid dynamics (CFD) is a scientific simulation technique which can provide spatial and temporal solved predictions of several aspects involved in the delivery of respiratory drugs from initial aerosol formation to the absorption of drugs by the respiratory cells. CFD respiratory drug delivery models have some benefits compared to semi-empirical and 1D whole-lung methods. CFD simulations are based on the underlying transport equations that can directly account for factors such as transient flow, turbulence and turbulent particle dispersion, changes in hygroscopic particle size, and interactions between fluid and wall in complex geometries. Realistic airway geometries are employed with CFD simulations, which are important to account for accumulation in complex structures such as the larynx, bifurcations of airways and restricted airways. CFD simulations can directly predict the effects of jet and spray momentum on an aerosol, as it is emitted from an inhaler into the mouth-throat and upper tracheobronchial airways, considering pharmaceutical aerosols. CFD models' limitations include uncertainty in capturing physics related to the production and distribution of pharmaceutical aerosols, challenge of resolving flow dynamics in the large space of the bifurcating airways, and computational cost. As a consequence, CFD models are usually restricted to parts of the respiratory tract, such as from the oral cavity to around the sixth generation of airways (127).

21.8 Solid Lipid Nanoparticles for Gene Delivery

Gene therapy is an alternative strategy which shows promise in the treatment of lung cancer. Viral vectors and non-viral vectors are the two main groups of vectors which gene therapy comprises of. Viral vectors have immunogenicity and toxicity concerns leading to increasing research interest on non-viral systems for gene delivery. The nanoparticle gene delivery system is classified into anionic and cationic nanoparticles. Cationic nanoparticles include ionic interaction between cationic polymers and anionic plasmid DNA leading to formation of stable polymer-lipid-DNA complex. Research has shown that treating early endobronchial cancer using p53 gene-cationic lipid complexes is a substitute for viral vectors (128).

However, there are certain cellular obstacles which serve as challenges for achieving improved levels of transfection activity. These include cell surface binding, cellular internalization, escape from endolysosomal compartment, and translocation through the nuclear envelope (129). These barriers can be overcome by surface functionalization of SLN surface using moieties such as chitosan, folate chitosan, cetyl trimethyl ammonium bromide (CTAB), and TAT peptides (130).

In a study conducted by Shao et al in 2015, transferrin (Tf) conjugated NLCs were formulated for the co-delivery of PTX and enhanced green fluorescence protein plasmid (DNA). The aim was to achieve a stable co-loaded formulation, maximize the DNA loading capacity and target the NCl-H460 cells using Tf as a ligand for the Tf receptors. The PTX-DNA co-loaded NLCs were prepared by microemulsion technique using a mixture of GMS (solid lipid), oleic acid (liquid lipid) and soya lecithin (stabilizer). The aqueous phase consisted of DNA, N-[1-(2,3-dioleyloxy) propyl]-N, N,N-trimethyl-ammonium chloride

(DOTMA) and Tween 80. Tf conjugated ligands were synthesized using PEG of three different molecular weight. Tf and PEG were reacted in the presence of sodium cyanoborohydride in acetate buffer. Following this, the Tf-PEG and phosphatidylethanolamine (PE) were reacted in the presence of 1-ethyl-3 -[3-dimethylaminopropyl]-carbodiimide hydrochloride (EDC HCl). To coat the co-loaded PTX-DNA NLCS with Tf-PEG-PE, the NLC mixture was slowing added to Tf-PEG-PE solution. The co-loaded NLCS displayed a particle size of 79 nm with a high zets potential which was due to the cationic surfactant (DOTMA). The DNA loading capacity in the NLCs were around 90%. The Tf-PTX-DNA NLCs exhibited 70% accumulated release in 72 hours, whereas the non-Tf-PTX-DNA NLCs achieved 80% release after 24 hours. The sustained release for the Tf-PTX-DNA NLCs was due to the long PEG chains. High gene transfection efficiency and low cytotoxicity were displayed by Tf-PTX-DNA NLCs. The anti-tumor efficacy of Tf-PTX-DNA NLCs were assessed and the tumor inhibition rate of the novel Tf-decorated NLCs was found to be greater than the non-Tf conjugated NLCs. The Tf decoration also enhanced the active targeting ability of the carriers to NCl-H460 cells. The researchers concluded that nanostructured lipid carriers as a means of nanomedicine delivery system is a promising strategy for combined drug and gene delivery for tumor therapy (131).

In a study conducted by Choi et al. in 2008, *p53*-null H1299 lung cancer cells were transfected in SLN-carrier p53. The team was able to prove the efficient expression of p53 protein in comparison to commercially available Lipofectin, portraying that SLNs could be used as highly efficient gene therapy vehicles in lung cancer (132).

Researchers have also efficiently loaded SLNs with PTX and Bcl-2 siRNA for synergistic combination therapy as well as coencapsulated CdSe/ZnS quantum dots to bestow optical traceability (133).

21.9 Role of Solid Lipid Nanoparticles in Combinatorial Drug Delivery

A combination of multiple agents with specific properties and therapeutic efficiencies are often required for effective cancer treatment. To reduce multiple drug resistance, cytotoxicity associated with chemotherapeutic agents and small molecule inhibitors combinatorial drug delivery systems have emerged (134). One such approach of combinatorial delivery is encapsulation of gene (nucleic acid) and a chemotherapeutic agent in SLNs and further functionalize the surface of these SLNs with an appropriate ligand to enhance targeting ability.

Yiqun Han et al. synthesized transferrin-polyethylene glycol (PEG)-phosphatidyl-ethanolamine (PEG-PE) and modified the surface of gene and drug loaded SLN. To improve the cancer therapy effectiveness, they co-loaded DOX and green fluorescence protein plasmid (pEGFP) into SLNs to establish a novel multi-functional delivery system that can target lung tumor cells. These SLNs were formulated independently and then combined to form co-loaded SLN (SLN/DE). The efficiency of modified vectors *in vitro* transfection was assessed using a human alveolar adenocarcinoma cell line (A549 cells) and the efficiency of in vivo transfection was assessed using mice carrying A549 tumors. The Tf-modified co-encapsulated DOX and pEGFP SLN (T-SLN/DE) had a particle size of 267 nm with a surface charge of 42 mV. T-SLN/DE's *in vitro* cytotoxicity was poor (cell viability in comparison with controls was between 80 and 100%). In addition, Tf-SLN/DE demonstrated best transfection efficiency at 48 and 72 hours in A549 lung cancer cells. T-SLN / DE also had an excellent therapeutic effect on drug delivery as well as gene therapy (135).

Lipid–polymer hybrid nanoparticles (LPHNPs) are core–shell nanostructures of the next generation, derived functionally from both liposome and polymeric nanoparticles (NPs), where a polymer center remains enveloped by a lipid layer. This hybrid model could be a robust rostrum of drug delivery with well-defined release kinetics, high encapsulation efficiency, well tolerated serum stability, and well-triggered tissue, molecular, and cellular targeting properties (136).

In a similar manner, multi-compartmental lipid-protein nanohybrids (MLPNs) have been successfully utilized as delivery systems for combined delivery of anti-cancer drugs (137). Protein-lipid nanohybrids consist of the drug encapsulated in a protein shell which are consequently coated with lipids. They can be formulated by functionalizing the surface of lipid nanocarriers with proteins through various mechanisms which include covalent conjugation, desolvation, high pressure homogenization, emulsification, or electrostatic coating. Protein nanocarriers have also been encapsulated within liposomes by film hydration, modified desolvation, and solvent injection methods. Their advantages include reduced rate of clearance by the RES and hence increasing their blood circulation time, improving targeting efficacy, improving oral bioavailability, and increasing the solubility and intestinal permeability of poorly water soluble and low permeable drugs (138).

Several MLPNs have been developed by Kamel et al. (142) in their laboratory. One such research conducted by the team includes utilizing the hydrophobic and biodegradable protein of plant origin: zein. Zein has been used as a coating for SLNs due to its low absorption of moisture, high thermal resistance, and good mechanical barrier properties. The aim of this study was to utilise the ability of zein to form MLNPs for combinatorial drug delivery, i.e. genistein (GEN) tretinoin (TRE) for lung cancer therapy. Previous research has reported that clinical application of GEN is hindered due to its low plasma half-life, low solubility, and low bioavailability. While TRE has a powerful therapeutic impact in the treatment of lung cancer, there is strong resistance to the TRE impact from non small cell lung cancer. When TRE and GEN are combined, cancer cells arrest the cell cycle in both G2-M and G0–G1 stages (139,140). Tyrosine kinase inhibitors were found to enhance TRE's anti-cancer activity. Therefore, GEN was expected to exhibit synergy with TRE-induced apoptosis of cancer cells through the deactivation of extracellular signal-regulated protein kinase due to its anti-PTK effect (141).

TRE also has the power to upregulate VEGF expression on another path. High levels of VEGF can contribute to angiogenesis and cancer-related progression, which in turn can inhibit the anti-cancer effect of TRE. Thus, in this study, GEN-SLNs were prepared by solvent injection technique using GMS as the lipid and Tween 80 as the surfactant. Following this, zein dissolved in ethanol was injected into the GEN-SLN dispersion which lead to the precipitation of zein around the GEN-SLNs. To prepare the TRE-GEN co-loaded MLPNs, stearylamine (SA) was used as an ion-pairing reagent. SA and TRE were dissolved in ethanol and a specified concentration of zein was added to this mixture. This SA-TRE-zein ethanolic solution was then injected into the GEN-SLNs aqueous phase with homogenization to form TRE-GEN-MLPNs. The authors found that there was an increase in drug entrapment due to ion-pairing with SA. The *in vitro* release studies demonstrated an initial burst release due to the surface adsorbed drug and the zein further contributed to the sustained delivery. TRE/GEN-MLPNs demonstrated higher cytotoxicity to lung cancer cells against A549 compared to free drugs. These findings revealed the synergistic cytotoxicity of combined GEN and TRE delivery through MLPNs. The powerful anti-tumor efficacy of TRE-GEN MLPNs were revealed after the in vivo study on urethane-induced lung cancer bearing mice. The anti-tumor effect of the co-loaded MLPNs was compared to the free TRE/GEN solution through evaluating biomarkers such as VEGF, caspase-3 as well as histopathological and immunohistochemical analysis. The authors concluded that TRE-GEN MLPNs can be administered parenterally for the effective treatment of lung cancer (142).

21.10 Stimuli Responsive Solid Lipid Nanoparticles

Currently researchers are focusing on design smart SLN which give targeted delivery and can circumvent the RES clearance and opsonization mechanisms. These smart nanoparticles can be synthesized by modifying their surface by anchoring ligands which are responsive to different stimuli. The stimuli may include pH, enzymes, or exogenous factors like magnetic fields, ultrasound, and light (143). Certain lipids and polymers can undergo changes in their physicochemical properties as a response to minute changes in temperature or pH. These properties of lipids and polymers can be used to design smart SLNs (144). Doxorubicin (DOX) is an antineoplastic drug used widely in lung cancer. However, there are certain disadvantages associated with it such as low penetration and limited distribution in solid tumors (145). The extracellular environment of tumor exhibits low pH, due to which weakly basic drugs like DOX get protonated and show reduced cellular uptake (146). Mussi et al. loaded DOX in the SLNs along with docosahexaenoic acid (DHA), a polyunsaturated fatty acid, to evaluate the anti-cancer enhancing property of DHA. SLNs were prepared by the hot melt homogenization method using an emulsification ultrasound. Compritol was used as the solid lipid carrier and Tween 80 was used as the surfactant since the formulation was intended for intravenous delivery. The effect of ion pair formation between DOX which is a cationic amphiphilic drug and the anionic lipophilic DHA was investigated. There was a significant increase in encapsulation of DOX (almost 100%) in the SLNs as the concentration of DHA was increased. The researchers proposed that lipophilicity of DOX increased due to ion-pair formation with DHA leading to an increase in encapsulation efficiency. The *in vitro* release studies showed that DOX release from SLN with DHA was slower than that observed for SLN without DHA. A release of 38 % after 1 hour was observed for SLN with DHA. The cytotoxicity studies were conducted on the human lung adenocarcinoma epithelial cell line 410 (A549) and the results revealed that DOX-DHA coloaded SLNs exhibit higher cytotoxicity in comparison to free DOX. This study represents that lipophilic agents like DHA can effectively be used to enhance the anticancer activity of DOX, which improves treatment of lung cancer (147).

21.11 Solid Lipid Microparticles

Nanoparticles comprising microparticles may serve as a promising DDS for enhancing molecular targeted therapy of lung cancer. Erlotinib (ETB) is a tyrosine kinase inhibitor which has been commonly used as the second line therapeutic agent in metastatic non small cell lung cancer. Zahra Bakhtiary et al. developed an ETB loaded SLN based dry powder inhaler (ETB-SLN DPI microparticles) using compritol as the solid lipid carrier and poloxamer 407 as the stabilizer. ETB loaded SLNs were prepared by combining homogenization and sonication method. The optimized SLN formulation was spray dried with mannitol to obtain inhaler powder in the 1–5 μm size range. The *in vitro* drug release studies revealed 13% ETB release from the SLNs in 24 hours. The cytotoxicity study was performed using A549 cell line. The results showed that SLNs were able to decrease the IC_{50} of ETB from 48.09 μM to 34.11 μM after 24 hours of treatment, while blank SLNs were biocompatible and did not elicit significant cytotoxic impact (148). This study helps us to understand that size is an important factor that influences the deposition of particles in the lungs. Owing to the size of the SLNs, they can be converted to microparticles for efficient pulmonary drug delivery and enhanced treatment of lung cancer.

In another study conducted by Meenach et al. the role of biocompatible phospholipids was explained using surfactants which simulate those present in the lungs such as dipalmitoylphosphatidylethanolamine methoxy (polyethylene glycol) (DPPE-PEG) and dipalmitoylphosphatidylcholine (DPPC). These surfactants perform a key role in lowering the particle surface tension, enabling the particles to migrate into the deep lung regions. In this study, the researchers used a special phospholipid core lipopolymer and PEGylated shell to formulate promising inhalable dry powder of PTX in which a diluted organic solution of DPPC and DPPE-PEG along with PTX MPs were cospray-dried in a closed-mode method. Results showed that aerosol performance was significantly affected by PEG chain due to which longer PEG 5k resulted in poor aerosolization with MMAD (6.8 mm) an, d FPF (55.4%), whereas shorter PEG 2k significantly improved inhalable MP aerosolization with MMAD (4 mm) and FPF (64%) in addition to the reduced moisture content (149).

21.12 Toxicological Evaluation of Solid Lipid Nanoparticles

Toxicological risk assessment of the SLNs needs to be performed for their clinical use. M. Nassimi et al. used *in vitro*, *ex vivo*, and *in vivo* approaches to examine the toxicological and inflammatory potential of the SLNs. The SLNs comprised of phospholipid, phospholipon P90G, and triglyceride Softisan S154 as the solid lipid matrix. Solutol HS 15 was used as the stabilizer. The SLNs were prepared by high pressure homogenization method. The human type II pneumocyte-like cells (A549 cells) were exposed to various doses of nanoparticle suspension for estimating the toxic dose *in vitro*. The MTT and NRU assays were used to test nanosuspension cytotoxicity, and the inflammatory potential was calculated by measuring the IL-8 concentration in the supernatant. For the ex vivo method, WST assay and live/dead staining for confocal microscopy using precision-cut lung slices (PCLS) determined the cytotoxicity of nanoparticles. The inflammatory response in the supernatants was evaluated by measuring chemokine KC and TNF-a material. The researchers conducted a 16-day inhalation toxicity test to determine the in vivo condition. Experimental studies of the invitro and ex vivo dose showed toxic effects rising at levels of about 500 lg/ml. For the in vivo experiments, therefore, the scientists used 1–200 lg doses of deposits/animal. SLNs did not cause any major signs of inflammation, even after 16 days of test with a 200-lg deposit dose. These findings indicate that repeated exposure to formulated SLNs at concentrations below a 200-lg deposit dose in a murine inhalation model was safe (150–152).

21.13 Conclusion

Cancer holds its status as one of the leading causes of mortality, lung cancer being a major contributor to it. Nevertheless, due to the massive research efforts over the last 20 years and a surge in the technological advancement, there has been an improvement in the health of the patients. Various nano-drug delivery methods have been performed to improve the conventional methods of lung tumor treatment. This chapter envisages the epidemiology and the vast SLN approaches to enhance the treatment of lung cancer. Most of the drugs have physicochemical and bioavailability related disadvantages which have been conquered through tremendous research efforts by solid lipid nanoparticle-based drug delivery. The newer strategies of SLNs for gene delivery and targeted combinatorial drug delivery methods are leading the way for effective treatment of lung cancer.

REFERENCES

1. Anti-EGFR mechanism of action: antitumor effect and underlying cause of adverse events - PubMed. n.d. Accessed August30, 2020. https://pubmed.ncbi.nlm.nih.gov/16736978/.
2. Armitage P and Doll R. The age distribution of cancer and a multi-stage theory of carcinogenesis. *British Journal of Cancer.* 1954; 8 (1): 1–12. https://doi.org/10.1038/bjc.1954.1.
3. Inamura, K. Lung cancer: understanding its molecular pathology and the 2015 WHO classification. *Frontiers in Oncology* 2017; 7 (AUG): 193. https://doi.org/10.3389/fonc.2017.00193.
4. Fitzmaurice C, Dicker D, Pain A, Hamavid H, Moradi-Lakeh M, MacIntyre MF, Allen C, et al. The global burden of cancer 2013. *JAMA Oncology* 2015; 1 (4): 505–527. https://doi.org/10.1001/jamaoncol.2015.0735.
5. Boffetta P, Pershagen G, Jöckel KH, Forastiere F, Gaborieau V, Heinrich J, Jahn I, et al. Cigar and pipe smoking and lung cancer risk: a multicenter study from Europe. *Journal of the National Cancer Institute* 1999; 91 (8): 697–701. https://doi.org/10.1093/jnci/91.8.697.
6. Barta JA, Powell CA, and Wisnivesky JP. Global epidemiology of lung cancer. *Annals of Global Health.* 2019 Ubiquity Press. https://doi.org/10.5334/aogh.2419.
7. Bach PB, Mirkin JN, Oliver TK, Azzoli CG, Berry DA, Brawley OW, Byers T, et al. Benefits and harms of CT screening for lung cancer: a systematic review. *Journal of the American Medical Association.* 2012 JAMA. https://doi.org/10.1001/jama.2012.5521.
8. Reich JM A critical appraisal of overdiagnosis: estimates of its magnitude and implications for lung cancer screening. *Thorax.* 2008. https://doi.org/10.1136/thx.2007.079673.
9. Powell CA, Halmos B, and Nana-Sinkam SP. Update in lung cancer and mesothelioma 2012. *American Journal of Respiratory and Critical Care Medicine.* 2013; 188 (2): 157–166. https://doi.org/10.1164/rccm.201304-0716UP.
10. Buczkowski K, Sieminska A, Linkowska K, Czachowski S, Przybylski G, Jassem E, and Grzybowski T. Association between genetic variants on chromosome 15q25 locus and several nicotine dependence traits in polish population: a case-control study. *BioMed Research International.* 2015 2015 (January). https://doi.org/10.1155/2015/350348.
11. Bianchi F, Nicassio F, Marzi M, Belloni E, Dall'Olio V, Bernard L, Pelosi G, Maisonneuve P, Veronesi G, and Fiore PPD. A serum circulating mirna diagnostic test to identify asymptomatic high-risk individuals with early stage lung cancer. *EMBO Molecular Medicine.* 2011; 3 (8): 495–503. https://doi.org/10.1002/emmm.201100154.
12. Lu Y, Govindan R, Wang L, Liu PY, Goodgame B, Wen W, Sezhiyan A, et al. MicroRNA profiling and prediction of recurrence/relapse-free survival in stage I lung cancer. *Carcinogenesis.* 2012; 33 (5): 1046–1054. https://doi.org/10.1093/carcin/bgs100.
13. Lung carcinoma - pulmonary disorders - MSd manual professional edition. n.d. Accessed August30, 2020. https://www.msdmanuals.com/professional/pulmonary-disorders/tumors-of-the-lungs/lung-carcinoma.
14. Huang W, Hu J, Yang DW, Fan XT, Jin Y, Hou YY, Wang JP, et al. Two MicroRNA panels to discriminate three subtypes of lung carcinoma in bronchial brushing specimens. *American Journal of Respiratory and Critical Care Medicine.* 2012; 186 (11): 1160–1167. https://doi.org/10.1164/rccm.201203-0534OC.
15. Pecot CV, Li M, Zhang XJ, Rajanbabu R, Calitri C, Bungum A, Jett JR, et al. Added value of a serum proteomic signature in the diagnostic evaluation of lung nodules. *Cancer Epidemiology Biomarkers and Prevention.* 2012; 21 (5): 786–792. https://doi.org/10.1158/1055-9965.EPI-11-0932.
16. Jett JR, Peek LJ, Fredericks L, Jewell W, Pingleton WW, and Robertson JF. Audit of the autoantibody test, EarlyCDT®-Lung, in 1600 patients: an evaluation of its performance in routine clinical practice. *Lung Cancer* 2014; 83: 51–55. https://doi.org/10.1016/j.lungcan.2013.10.008.

17. Hassanein Mohamed, Callison JC, Callaway-Lane C, Aldrich MC, Grogan EL, and Massion PP. The state of molecular biomarkers for the early detection of lung cancer. *Cancer Prevention Research.* 2012 American Association for Cancer Research. https://doi.org/10.1158/1940-6207.CAPR-11-0441.
18. Mazzone PJ, Sears CR, Arenberg DA, Gaga M, Gould MK, Massion PP, Nair Vish S, et al. American Thoracic Society documents Evaluating molecular biomarkers for the early detection of lung cancer: when is a biomarker ready for clinical use? an official american thoracic society policy statement. 2017 https://doi.org/10.1164/rccm.201708–1678ST.
19. Babu A, Templeton AK, Munshi A, and A Rajagopal . Nanoparticle-based drug delivery for therapy of lung cancer: progress and challenges. *Journal of Nanomaterials.* 2013 https://doi.org/10.1155/2013/863951.
20. Sivarajakumar R, Mallukaraj D, Kadavakollu M, Neelakandan N, Chandran S, Bhojaraj S, and Reddy Karri VVS. Nanoparticles for the treatment of lung cancers. *Journal of Young Pharmacists.* 2018; 10 (3): 276–281. https://doi.org/10.5530/jyp.2018.10.62.
21. Miele E, Spinelli GP, Miele E, Tomao F, and Tomao S. Albumin-bound formulation of paclitaxel (Abraxane® ABI-007) in the treatment of breast cancer. *International Journal of Nanomedicine.* Dove Press. 2009 4: 99–105.
22. Mukherji SK. Bevacizumab (Avastin). *American Journal of Neuroradiology.*2010 31(2): 235–236. https://doi.org/10.3174/ajnr.A1987
23. Sousa GFd, SR Wlodarczyk, and Monteiro G. Carboplatin: Molecular Mechanisms of Action Associated with Chemoresistance.*Brazilian Journal of Pharmaceutical Sciences.* 2014;50: 693–701.
24. Herbst RS and Khuri FR. Mode of action of docetaxel - a basis for combination with novel anticancer agents. *Cancer Treatment Reviews.* 2003; 29 (5): 407–415. https://doi.org/10.1016/S0305-7372(03)00097-5.
25. Jackson TL. Intracellular accumulation and mechanism of action of doxorubicin in a spatio-temporal tumor model. *Journal of Theoretical Biology.* 2003; 220 (2): 201–213. https://doi.org/10.1006/jtbi.2003.3156.
26. Ruman U, S Fakurazi, MJ Masarudin, and MZ Hussein. Nanocarrier-based therapeutics and theranostics drug delivery systems for next generation of liver cancer nanodrug modalities. *International Journal of Nanomedicine.* 2020; 15 (March): 1437–1456. https://doi.org/10.2147/IJN.S236927.
27. Rusch V, Klimstra D, Venkatraman E, Pisters PWT, Langenfeld J, Dmitrovsky E, Biostatistics Service, and [E Dl. n.d. Overexpression of the epidermal growth factor receptor and its ligand transforming growth factor a is frequent in resectable non-small cell lung cancer but does not predict tumor progression. Vol. 3. Accessed August30, 2020. https://bloodcancerdiscov.aacrjournals.org.
28. Abdelaziz HM, Gaber M, Abd-Elwakil MM, Mabrouk MT, Elgohary MM, Kamel NM, Kabary DM, et al. Inhalable particulate drug delivery systems for lung cancer therapy: nanoparticles, microparticles, nanocomposites and nanoaggregates. *Journal of Controlled Release.* 2018 269: 374–392. https://doi.org/10.1016/j.jconrel.2017.11.036.
29. Alivisatos AP. Semiconductor clusters, nanocrystals, and quantum dots. *Science.* 1996; 271 (5251): 933–937. https://doi.org/10.1126/science.271.5251.933.
30. Feng B, Hong RY, Wang LS, Guo L, Li HZ, Ding J, Zheng Y, and Wei DG. Synthesis of Fe3O4/APTES/PEG Diacid functionalized magnetic nanoparticles for MR imaging. *Colloids and Surfaces A: Physicochemical and Engineering Aspects.* 2008; 328 (1–3): 52–59. https://doi.org/10.1016/j.colsurfa.2008.06.024.
31. Cortesi R, Esposito E, Luca G, and Nastruzzi C. Production of lipospheres as carriers for bioactive compounds. *Biomaterials.* 2002; 23 (11): 2283–2294. https://doi.org/10.1016/S0142-9612(01)00362-3.
32. Paliwal R, Rai S, Vaidya B, Khatri K, Goyal AK, Mishra N, Mehta A, and Vyas SP. Effect of lipid core material on characteristics of solid lipid nanoparticles designed for oral lymphatic delivery. *Nanomedicine: Nanotechnology, Biology, and Medicine.* 2009; 5 (2): 184–191. https://doi.org/10.1016/j.nano.2008.08.003.
33. Kumar M, Kakkar V, Mishra AK, Chuttani K, and Kaur IP.Intranasal delivery of streptomycin sulfate (STRS) loaded solid lipid nanoparticles to brain and blood. *International Journal of Pharmaceutics.* 2014; 461 (1–2): 223–233. https://doi.org/10.1016/j.ijpharm.2013.11.038.
34. Gastaldi L, Battaglia L, Peira E, Chirio D, Muntoni E, Solazzi I, Gallarate M, and Dosio F. solid lipid nanoparticles as vehicles of drugs to the brain: current state of the art. *European Journal of Pharmaceutics and Biopharmaceutics.* 2014; 87(3): 433–444. https://doi.org/10.1016/j.ejpb.2014.05.004
35. Weber S, Zimmer A, and Pardeike J. Solid lipid nanoparticles (SLN) and nanostructured lipid carriers (NLC) for pulmonary application: a review of the state of the art. *European Journal of Pharmaceutics and Biopharmaceutics.* 2014; 86(1): 7–22. https://doi.org/10.1016/j.ejpb.2013.08.013
36. Paliwal R, Paliwal SR, Agrawal GP, and Vyas SP. Biomimetic solid lipid nanoparticles for oral bioavailability enhancement of low molecular weight heparin and its lipid conjugates: in vitro and in vivo evaluation. *Molecular Pharmaceutics.* 2011; 8 (4): 1314–1321. https://doi.org/10.1021/mp200109m.
37. Vyas S, Rai S, Paliwal R, Gupta P, Khatri K, Goyal A, and Vaidya B. Solid lipid nanoparticles (SLNs) as a rising tool in drug delivery science: one step up in nanotechnology. *Current Nanoscience.* 2008; 4 (1): 30–44. https://doi.org/10.2174/157341308783591816.
38. Harde H, Das M, and Jain S. Solid lipid nanoparticles: an oral bioavailability enhancer vehicle. *Expert Opinion on Drug Delivery.* 2011 Taylor & Francis. https://doi.org/10.1517/17425247.2011.604311.
39. Sanvicens Nuria and Pilar Marco M. Multifunctional nanoparticles - properties and prospects for their use in human medicine. *Trends in Biotechnology.* 2008 Trends Biotechnol. https://doi.org/10.1016/j.tibtech.2008.04.005.
40. Kohler DR and Goldspiel BR Paclitaxel (Taxol). *Pharmacotherapy: The Journal of Human Pharmacology and Drug Therapy.* 1994; 14 (1): 3–34. https://doi.org/10.1002/j.1875-9114.1994.tb02785.x.
41. Allwood MC and Martin H. The extraction of diethylhexylphthalate (DEHP) from polyvinyl chloride components of

intravenous infusion containers and administration sets by paclitaxel injection. *International Journal of Pharmaceutics.* 1996; 127 (1): 65–71. https://doi.org/10.1016/0378-5173(95)04128-1.
42. Fjällskog ML, Frii L, and Bergh J. Is cremophor EL, solvent for paclitaxel, cytotoxic? *The Lancet*. Lancet. 1993 https://doi.org/10.1016/0140-6736(93)92735-C.
43. Chen DB, Yang TZ, Lu WL, and Zhang Q. In vitro and in vivo study of two types of long-circulating solid lipid nanoparticles containing paclitaxel. *Chemical and Pharmaceutical Bulletin.* 2001; 49 (11): 1444–1447. https://doi.org/10.1248/cpb.49.1444.
44. Krishna R and Lawrence DM. Multidrug resistance (mdr) in cancermechanisms, reversal using modulators of MDR and the role of MDR modulators in influencing the pharmacokinetics of anticancer drugs. *European Journal of Pharmaceutical Sciences.* 2000; 11(4): 265–283. https://doi.org/10.1016/S0928-0987(00)00114-7.
45. Ambudkar SV, Dey S, Hrycyna CA, Ramachandra M, Pastan I, and Gottesman MM. Biochemical, cellular, and pharmacological aspects of the multidrug transporter. *Annual Review of Pharmacology and Toxicology.* 1999 https://doi.org/10.1146/annurev.pharmtox.39.1.361.
46. Zhou C, Shen P, and Cheng Y. Quantitative study of the drug efflux kinetics from sensitive and MDR human breast cancer cells. *Biochimica et Biophysica Acta - General Subjects.* 2007; 1770 (7): 1011–1020. https://doi.org/10.1016/j.bbagen.2007.02.011.
47. Miglietta A, Cavalli R, Bocca C, Gabriel L, and Gasco MR. Cellular uptake and cytotoxicity of solid lipid nanospheres (SLN) incorporating doxorubicin or paclitaxel. *International Journal of Pharmaceutics* 2000; 210 (1–2): 61–67. https://doi.org/10.1016/S0378-5173(00)00562-7.
48. Goren D, Horowitz AT, Tzemach D, Tarshish M, Zalipsky S, and Gabizon A. Nuclear delivery of doxorubicin via folate-targeted liposomes with bypass of multidrug-resistance efflux pump. *Clinical cancer research: an official journal of the American Association for Cancer Research.* 2000; 6(5): 1949–1957. doi: 10.3389/fphar.2014.00159
49. Serpe L, Catalano MG, Cavalli R, Ugazio E, Bosco O, Canaparo R, Muntoni E, et al. Cytotoxicity of anticancer drugs incorporated in solid lipid nanoparticles on HT-29 colorectal cancer cell line. *European Journal of Pharmaceutics and Biopharmaceutics.* 2004; 58 (3): 673–680. https://doi.org/10.1016/j.ejpb.2004.03.026.
50. Barraud L, Merle P, Soma E, Lefrançois L, Guerret S, Chevallier M, Dubernet C, Couvreur P, Trépo C, and Vitvitski L. Increase of doxorubicin sensitivity by doxorubicin-loading into nanoparticles for hepatocellular carcinoma cells in vitro and in vivo. *Journal of Hepatology.* 2005; 42 (5): 736–743. https://doi.org/10.1016/j.jhep.2004.12.035.
51. Miao J, Du YZ, Yuan H, Zhang XG, and Hu FQ.Drug resistance reversal activity of anticancer drug loaded solid lipid nanoparticles in multi-drug resistant cancer cells. *Colloids and Surfaces B: Biointerfaces.* 2013; 110 (October): 74–80. https://doi.org/10.1016/j.colsurfb.2013.03.037.
52. Lim KJ, Bisht S, Bar EE, Maitra A, and Eberhart CG. A polymeric nanoparticle formulation of curcumin inhibits growth, clonogenicity and stem-like fraction in malignant brain tumors. *Cancer Biology and Therapy.* 2011; 11 (5): 464–473. https://doi.org/10.4161/cbt.11.5.14410.
53. Ye MX, Zhao YL, Li Y, Miao Q, Li ZK, XL Ren, LQ Song, H Yin, and Zhang J. Curcumin reverses CIS-platin resistance and promotes human lung adenocarcinoma a549/ddp cell apoptosis through hif-1α and caspase-3 mechanisms. *Phytomedicine.* 2012; 19 (8–9): 779–787. https://doi.org/10.1016/j.phymed.2012.03.005.
54. Tan Q, Wu J, Li Y, Mei H, Zhao C, and Zhang J. A supermolecular curcumin for enhanced antiproliferative and proapoptotic activities: molecular characteristics, computer modeling and in vivo pharmacokinetics. *Nanotechnology.* 2013; 24 (3): 035102. https://doi.org/10.1088/0957-4484/24/3/035102.
55. Neves AR, Lucio M, Lima JLC, and Reis S. Resveratrol in medicinal chemistry: a critical review of its pharmacokinetics, drug-delivery, and membrane interactions. *Current Medicinal Chemistry.* 2012; 19 (11): 1663–1681. https://doi.org/10.2174/092986712799945085.
56. Wang P, Zhang L, Peng H, Li Y, Xiong J, and Xu Z. The formulation and delivery of curcumin with solid lipid nanoparticles for the treatment of on non-small cell lung cancer both in vitro and in vivo. *Materials Science and Engineering C.* 2013; 33 (8): 4802–4808. https://doi.org/10.1016/j.msec.2013.07.047.
57. Rocha MCOD, Silva PBD, Radicchi MA, Andrade BYG, Oliveira JVD, Venus T, Merker C, Estrela-Lopis I, Longo JPF, and Báo SN. Docetaxel-loaded solid lipid nanoparticles prevent tumor growth and lung metastasis of 4t1 murine mammary carcinoma cells. *Journal of Nanobiotechnology.* 2020; 18 (1): 43. https://doi.org/10.1186/s12951-020-00604-7.
58. Khatri Hiren, Chokshi N, Rawal S, and Patel MM. Fabrication, characterization and optimization of artemether loaded pegylated solid lipid nanoparticles for the treatment of lung cancer. *Materials Research Express.* 2019; 6 (4): 045014. https://doi.org/10.1088/2053-1591/aaf8a3.
59. Hamishehkar, H, Bahadori MB, Vandghanooni S, Eskandani M, Nakhlband A, and Eskandani M. Preparation, characterization and anti-proliferative effects of sclareol-loaded solid lipid nanoparticles on a549 human lung epithelial cancer cells. *Journal of Drug Delivery Science and Technology.* 2018; 45 (June): 272–280. https://doi.org/10.1016/j.jddst.2018.02.017.
60. Zielińska W, Nawrocka A, Rydzkowski M, Kokocha A, Matulewicz K, Hałas-Wiśniewska M, and Izdebska Ma. Possibilities in the application of solid lipid nanoparticles in combination with 5-fluorouracil to overcome the drugresistance of non-small cell lung cancer cell line A549. *Medical Research Journal.* 2018. October; 5(1): 1–8. https://doi.org/10.5603/mrj.a2019.0037
61. Khorsandi L, Mansouri E, Rashno M, Karami MA, and Ashtari Atefeh. Myricetin Loaded Solid Lipid Nanoparticles Upregulate MLKL and RIPK3 in Human Lung Adenocarcinoma. *International Journal of Peptide Research and Therapeutics.* 2020; 26 (2): 899–910. https://doi.org/10.1007/s10989-019-09895-3.
62. Fujita KI, Matsuda E, Sekine K, Iigo M, and Tsuda H. Lactoferrin enhances fas expression and apoptosis in the

colon mucosa of azoxymethane-treated rats. *Carcinogenesis.* 2004; 25 (10): 1961–1966. https://doi.org/10.1093/carcin/bgh205.
63. Liao Y, R Jiang, and B Lönnerdal. Biochemical and molecular impacts of lactoferrin on small intestinal growth and development during early life. *Biochemistry and Cell Biology.* 2012 https://doi.org/10.1139/o11-075.
64. Gibbons JA Lactoferrin and cancer in different cancer models. *Frontiers in Bioscience.* 2011; S3 (1): 1080. https://doi.org/10.2741/212.
65. Pandey Vikas, Gajbhiye KR, and Soni V. Lactoferrin-appended solid lipid nanoparticles of paclitaxel for effective management of bronchogenic carcinoma. *Drug Delivery.* 2015; 22 (2): 199–205. https://doi.org/10.3109/10717544.2013.877100.
66. Dennis JW, Granovsky M, and Warren CE. Glycoprotein glycosylation and cancer progression. *Biochimica et Biophysica Acta - General Subjects.* 1999 Biochimica et Biophysica Acta. https://doi.org/10.1016/S0304-4165(99)00167-1.
67. Budzynska R, Nevozhay D, Kanska U, Jagiello M, Opolski A, Wietrzyk J, and Boratynski J. Antitumor activity of mannan-methotrexate conjugate in vitro and in vivo. *Oncology Research.* 2007; 16 (9): 415–421. https://doi.org/10.3727/000000007783980837.
68. Nag Alo and Ghosh PC Assessment of targeting potential of galactosylated and mannosylated sterically stabilized liposomes to different cell types of mouse liver. *Journal of Drug Targeting.* 1999; 6 (6): 427–438. https://doi.org/10.3109/10611869908996849.
69. Sahu PK, DK Mishra, N Jain, V Rajoriya, and AK Jain. Mannosylated solid lipid nanoparticles for lung-targeted delivery of paclitaxel. *Drug Development and Industrial Pharmacy.* 2015; 41 (4): 640–649. https://doi.org/10.3109/03639045.2014.891130.
70. Kukulj S, Jaganjac M, Boranic M, Krizanac S, Santic Z, and Poljak-Blazi M Altered iron metabolism, inflammation, transferrin receptors, and ferritin expression in non-small-cell lung cancer. *Medical Oncology.* 2010; 27 (2): 268–277. https://doi.org/10.1007/s12032-009-9203-2.
71. Prost AC, Ménégaux F, Langlois P, Vidal JM, Koulibaly M, Jost JL, Duron JJ, et al. Differential transferrin receptor density in human colorectal cancer: a potential probe for diagnosis and therapy. *International Journal of Oncology.* 1998; 13 (4): 871–875. https://doi.org/10.3892/ijo.13.4.871.
72. Whitney JF, Clark JM, Griffin TW, Gautam Shiva, and Leslie KO. Transferrin receptor expression in nonsmall cell lung cancer. histopathologic and clinical correlates. *Cancer.* 1995; 76 (1): 20–25. https://doi.org/10.1002/1097-0142(19950701)76:1<20::AID-CNCR2820760104>3.0.CO;2-3.
73. Widera A, Norouziyan F, and Shen WC. Mechanisms of TfR-mediated transcytosis and sorting in epithelial cells and applications toward drug delivery. *Advanced Drug Delivery Reviews.* 2003; 55 (11): 1439–1466. https://doi.org/10.1016/j.addr.2003.07.004.
74. Zheng Y, Yu B, Weecharangsan W, Piao L, Darby M, Mao Y, Koynova R, et al. Transferrin-conjugated lipid-coated PLGA nanoparticles for targeted delivery of aromatase inhibitor 7α-APTADD to breast cancer cells. *International Journal of Pharmaceutics.* 2010; 390 (2): 234–241. https://doi.org/10.1016/j.ijpharm.2010.02.008.
75. Cui Y, Xu Q, Chow PKH, Wang D, and Wang CH. Transferrin-conjugated magnetic silica PLGA nanoparticles loaded with doxorubicin and paclitaxel for brain glioma treatment. *Biomaterials* 2013; 34 (33): 8511–8520. https://doi.org/10.1016/j.biomaterials.2013.07.075.
76. Definition of Atezolizumab - NCI Drug Dictionary - National Cancer Institute. n.d. Accessed August30, 2020. https://www.cancer.gov/publications/dictionaries/cancer-drug/def/atezolizumab?redirect=true.
77. Definition of Durvalumab - NCI Drug Dictionary - National Cancer Institute. n.d. Accessed August30, 2020. https://www.cancer.gov/publications/dictionaries/cancer-drug/def/durvalumab.
78. Definition of Etoposide Phosphate - NCI Drug Dictionary - National Cancer Institute. n.d. Accessed August30, 2020. https://www.cancer.gov/publications/dictionaries/cancer-drug/def/etoposide-phosphate?redirect=true.
79. Definition of Everolimus - NCI Drug Dictionary - National Cancer Institute. n.d. Accessed August30, 2020. https://www.cancer.gov/publications/dictionaries/cancer-drug/def/everolimus?redirect=true.
80. Definition of Topotecan Hydrochloride - NCI Drug Dictionary - National Cancer Institute. n.d. Accessed August30, 2020. https://www.cancer.gov/publications/dictionaries/cancer-drug/def/topotecan-hydrochloride?redirect=true.
81. Gan CW and Feng SS. Transferrin-conjugated nanoparticles of poly(lactide)-d-α-tocopheryl polyethylene glycol succinate diblock copolymer for targeted drug delivery across the blood-brain barrier. *Biomaterials.* 2010; 31 (30): 7748–7757. https://doi.org/10.1016/j.biomaterials.2010.06.053.
82. Jain TK, Morales MA, Sahoo SK, Leslie-Pelecky DL, and Labhasetwar V. Iron oxide nanoparticles for sustained delivery of anticancer agents. *Molecular Pharmaceutics.* 2005; 2 (3): 194–205. https://doi.org/10.1021/mp0500014.
83. Porru M, Zappavigna S, Salzano G, Luce A, Stoppacciaro A, Balestrieri ML, Artuso S, et al. Medical treatment of orthotopic glioblastoma with transferrin conjugated nanoparticles encapsulating zoledronic acid. *Oncotarget.* 2014; 5 (21): 10446–10459. https://doi.org/10.18632/oncotarget.2182.
84. Yan F, Wang Y, He S, Ku S, Gu W, and L Ye. Transferrin-conjugated, fluorescein-loaded magnetic nanoparticles for targeted delivery across the blood-brain barrier. *Journal of Materials Science: Materials in Medicine* 2013; 24 (10): 2371–2379. https://doi.org/10.1007/s10856-013-4993-3.
85. Pooja D, Kulhari H, Tunki L, Chinde S, Kuncha M, Grover Paramjit, Rachamalla SS, and R Sistla. Nanomedicines for targeted delivery of etoposide to non-small cell lung cancer using transferrin functionalized nanoparticles. *RSC Advances* 2015; 5 (61): 49122–49131. https://doi.org/10.1039/c5ra03316k.
86. Garbuzenko OB, Kuzmov A, Taratula O, Pine SR, and Minko T. Strategy to enhance lung cancer treatment by five essential elements: inhalation delivery, nanotechnology, tumor-receptor targeting, chemo- and gene therapy.

Theranostics 2019; 9 (26): 8362–8376. https://doi.org/10.7150/thno.39816.
87. Sung JC, Pulliam BL, and Edwards DA. Nanoparticles for drug delivery to the lungs. *Trends in Biotechnology*. Elsevier Current Trends. 2007. https://doi.org/10.1016/j.tibtech.2007.09.005.
88. Paranjpe M and Müller-Goymann CC. Nanoparticle-mediated pulmonary drug delivery: a review. *International Journal of Molecular Sciences*. 2014; 15(4): 5852–5873. https://doi.org/10.3390/ijms15045852
89. Videira M, AJ Almeida, and A Fabra. Preclinical evaluation of a pulmonary delivered paclitaxel-loaded lipid nanocarrier antitumor effect. *Nanomedicine: Nanotechnology, Biology, and Medicine*. 2012; 8 (7): 1208–1215. https://doi.org/10.1016/j.nano.2011.12.007.
90. Finlay WH. Introduction. In *The mechanics of inhaled pharmaceutical aerosols*, 1–2. Elsevier. 2001 https://doi.org/10.1016/b978-012256971-5/50002-x.
91. Rege BD, Kao JPY, and Polli JE. Effects of nonionic surfactants on membrane transporters in Caco-2 cell monolayers. *European Journal of Pharmaceutical Sciences*. 2002; 16 (4–5): 237–246. https://doi.org/10.1016/S0928-0987(02)00055-6.
92. Kaur P, Mishra V, Shunmugaperumal T, Goyal AK, Ghosh G, and Rath G. Inhalable spray dried lipidnanoparticles for the co-delivery of paclitaxel and doxorubicin in lung cancer. *Journal of Drug Delivery Science and Technology*. 2020; 56 (April): 101502. https://doi.org/10.1016/j.jddst.2020.101502.
93. Matsumura Y and Maeda H. A new concept for macromolecular therapeutics in cancer chemotherapy: mechanism of tumoritropic accumulation of proteins and the antitumor agent smancs. *Cancer Research*. 1986; 46 (12 Part 1).
94. Maeda H. The enhanced permeability and retention (EPR) effect in tumor vasculature: the key role of tumor-selective macromolecular drug targeting. *Advances in Enzyme Regulation*. 2001; 41 (1): 189–207. https://doi.org/10.1016/S0065-2571(00)00013-3.
95. Hobbs SK, Monsky WL, Yuan F, Roberts WG, Griffith L, Torchilin VP, and Jain RK. Regulation of transport pathways in tumor vessels: role of tumor type and microenvironment. *Proceedings of the National Academy of Sciences of the United States of America*. 1998; 95 (8): 4607–4612. https://doi.org/10.1073/pnas.95.8.4607.
96. Maeda H, Wu J, Sawa T, Matsumura Y, and Hori K. Tumor vascular permeability and the epr effect in macromolecular therapeutics: a review. *Journal of Controlled Release*. 2000; 65 (1–2): 271–284. https://doi.org/10.1016/S0168-3659(99)00248-5.
97. Markman JL, Rekechenetskiy A, Holler E, and Ljubimova JY. NanOmedicine Therapeutic Approaches To Overcome Cancer Drug Resistance. *Advanced Drug Delivery Reviews*. Adv Drug Deliv Rev. 2013 https://doi.org/10.1016/j.addr.2013.09.019.
98. Patel NR, Pattni BS, Abouzeid AH, and Torchilin VP. Nanopreparations to overcome multidrug resistance in cancer. *Advanced Drug Delivery Reviews*. NIH Public Access. 2013 https://doi.org/10.1016/j.addr.2013.08.004.
99. Garbuzenko OB, Mainelis G, Taratula O, and Minko T. Inhalation treatment of lung cancer: the influence of composition, size and shape of nanocarriers on their lung accumulation and retention. *Cancer Biology and Medicine*. 2014; 11 (1): 44–55. https://doi.org/10.7497/j.issn.2095-3941.2014.01.004.
100. Rosière R, Woensel MV, Mathieu V, Langer I, Mathivet T, Vermeersch M, Amighi K, and Wauthoz N. Development and evaluation of well-tolerated and tumor-penetrating polymeric micelle-based dry powders for inhaled anti-cancer chemotherapy. *International Journal of Pharmaceutics* 2016; 501 (1–2): 148–159. https://doi.org/10.1016/j.ijpharm.2016.01.073.
101. Tomoda K, Ohkoshi T, Hirota K, Sonavane GS, Nakajima T, Terada H, Komuro M, Kitazato K, and Makino K. Preparation and properties of inhalable nanocomposite particles for treatment of lung cancer. *Colloids and Surfaces B: Biointerfaces*. 2009; 71 (2): 177–182. https://doi.org/10.1016/j.colsurfb.2009.02.001.
102. Zarogoulidis Paul, E Chatzaki, K Porpodis, K Domvri, W Hohenforst-Schmidt, EP Goldberg, N Karamanos, and K Zarogoulidis. Inhaled chemotherapy in lung cancer: future concept of nanomedicine.*International Journal of Nanomedicine*. 2012; 7;1551–1572. https://doi.org/10.2147/IJN.S29997
103. Azarmi S, Löbenberg R, Roa WH, Tai S, and Finlay WH. Formulation and in vivo evaluation of effervescent inhalable carrier particles for pulmonary delivery of nanoparticles. *Drug Development and Industrial Pharmacy*. 2008; 34 (9): 943–947. https://doi.org/10.1080/03639040802149079.
104. Ely L, W Roa, WH Finlay, and Löbenberg R. Effervescent dry powder for respiratory drug delivery. *European Journal of Pharmaceutics and Biopharmaceutics*. 2007; 65 (3): 346–353. https://doi.org/10.1016/j.ejpb.2006.10.021.
105. Finlay WH, TR Desai, Wong JP, and Hancock REW. A novel approach to the pulmonary delivery of liposomes in dry powder form to eliminate the deleterious effects of milling. *Journal of Pharmaceutical Sciences*. 2002; 91 (2): 482–491. https://doi.org/10.1002/jps.10021.
106. Edsbäcker S and CJ Johansson. Airway selectivity: an update of pharmacokinetic factors affecting local and systemic disposition of inhaled steroids. *Basic and Clinical Pharmacology and Toxicology*. 2006. Basic Clin Pharmacol Toxicol. https://doi.org/10.1111/j.1742-7843.2006.pto_355.x.
107. Tsuda Akira, Henry FS, and Butler JP. Particle transport and deposition: basic physics of particle kinetics. *Comprehensive Physiology*. 2013; 3 (4): 1437–1471. https://doi.org/10.1002/cphy.c100085.
108. Understanding the mechanisms of action of methotrexate: implications for the treatment of rheumatoid arthritis - PubMed. n.d. Accessed August30, 2020. https://pubmed.ncbi.nlm.nih.gov/17922664/.
109. Sheth P, SW Stein, and PB Myrdal. Factors influencing aerodynamic particle size distribution of suspension pressurized metered dose inhalers. *AAPS PharmSciTech*. 2014; 16 (1): 192–201. https://doi.org/10.1208/s12249-014-0210-z.
110. Deshkar, SS, and Vas AS. Recent updates on dry powder for inhalation for pulmonary drug delivery systems. *International Journal of Research in Pharmaceutical*

Sciences. J. K. Welfare and Pharmascope Foundation. 2019; 10(4): 2944–2959. https://doi.org/10.26452/ijrps.v10i4.1575.

111. Tsapis N, Bennett D, Jackson B, Weitz DA, and Edwards DA. Trojan particles: large porous carriers of nanoparticles for drug delivery. *Proceedings of the National Academy of Sciences of the United States of America.* 2002; 99 (19): 12001–12005. https://doi.org/10.1073/pnas.182233999.
112. Roa WH, Azarmi S, Al-Hallak MHDK, Finlay WH, Magliocco AM, and Löbenberg R. Inhalable nanoparticles, a non-invasive approach to treat lung cancer in a mouse model. *Journal of Controlled Release.* 2011; 150 (1): 49–55. https://doi.org/10.1016/j.jconrel.2010.10.035.
113. Bremer RE, Scoggin TS, Somers EB, O'Shannessy DJ, and Tacha DE. Interobserver agreement and assay reproducibility of folate receptor α expression in lung adenocarcinoma: a prognostic marker and potential therapeutic target. *Archives of Pathology and Laboratory Medicine.* 2013; 137 (12): 1747–1752. https://doi.org/10.5858/arpa.2013-0039-OA.
114. Leamon CP and JA Reddy. Folate-targeted chemotherapy. *Advanced Drug Delivery Reviews.* 2004; 56 (8): 1127–1141. https://doi.org/10.1016/j.addr.2004.01.008.
115. Sudimack Jennifer, and Lee RJ. Targeted drug delivery via the folate receptor. *Advanced Drug Delivery Reviews.* Elsevier Science Publishers B.V. 2000. https://doi.org/10.1016/S0169-409X(99)00062-9.
116. Cagle PT, QJ Zhai, L Murphy, and PS Low. Folate receptor in adenocarcinoma and squamous cell carcinoma of the lung: potential target for folate-linked therapeutic agents. *Archives of Pathology and Laboratory Medicine.* 2013; 137 (2): 241–244. https://doi.org/10.5858/arpa.2012-0176-OA.
117. Chavanpatil MD, Y Patil, and J Panyam. Susceptibility of nanoparticle-encapsulated paclitaxel to P-glycoprotein-mediated drug efflux. *International Journal of Pharmaceutics.* 2006; 320 (1–2): 150–156. https://doi.org/10.1016/j.ijpharm.2006.03.045.
118. Dharap SS, Wang Y, Chandna P, Khandare JJ, Qiu B, Gunaseelan S, Sinko PJ, Stein S, Farmanfarmaian A, and Minko T. Tumor-specific targeting of an anticancer drug delivery system by LHRH peptide. *Proceedings of the National Academy of Sciences of the United States of America.* 2005; 102 (36): 12962–12967. https://doi.org/10.1073/pnas.0504274102.
119. O'Shannessy DJ, Yu G, Smale R, Fu YS, Singhal S, Thiel RP, Somers EB, and Vachani A. Folate receptor alpha expression in lung cancer: diagnostic and prognostic significance.*Oncotarget.* 2012; 3 (4): 414–425. https://doi.org/10.18632/ONCOTARGET.489.
120. Khandare JJ, Chandna P, Wang Y, Pozharov VP, and Minko T. Novel polymeric prodrug with multivalent components for cancer therapy. *Journal of Pharmacology and Experimental Therapeutics.* 2006; 317 (3): 929–937. https://doi.org/10.1124/jpet.105.098855.
121. Tseng CL, Wang TW, Dong GC, Wu SYH, Young TH, Shieh MJ, Lou PJ, and Lin FH. Development of gelatin nanoparticles with biotinylated EGF conjugation for lung cancer targeting. *Biomaterials.* 2007; 28 (27): 3996–4005. https://doi.org/10.1016/j.biomaterials.2007.05.006.
122. Tseng CLWu SYH,Wang WH,Peng CL,Lin FHLin CCYoung THShieh MJ. Targeting efficiency and biodistribution of biotinylated-EGF-conjugated gelatin nanoparticles administered via aerosol delivery in nude mice with lung cancer. *Biomaterials.* 2008; 29 (20): 3014–3022. https://doi.org/10.1016/j.biomaterials.2008.03.033.
123. Sadhukha T, Wiedmann TS, and Panyam J. Inhalable magnetic nanoparticles for targeted hyperthermia in lung cancer therapy. *Biomaterials.* 2013; 34 (21): 5163–5171. https://doi.org/10.1016/j.biomaterials.2013.03.061.
124. Tseng CL, WY Su, KC Yen, KC Yang, and FH Lin. The use of biotinylated-EGF-modified gelatin nanoparticle carrier to enhance cisplatin accumulation in cancerous lungs via inhalation. *Biomaterials.* 2009; 30 (20): 3476–3485. https://doi.org/10.1016/j.biomaterials.2009.03.010.
125. Taratula O, OB Garbuzenko, AM Chen, and T Minko. Innovative strategy for treatment of lung cancer: targeted nanotechnology-based inhalation co-delivery of anticancer drugs and SiRNA. *Journal of Drug Targeting.* 2011; 19 (10): 900–914. https://doi.org/10.3109/1061186X.2011.622404.
126. Taratula O, Kuzmov A, Shah M, Garbuzenko OB, and Minko T. Nanostructured lipid carriers as multifunctional nanomedicine platform for pulmonary co-delivery of anticancer drugs and SiRNA. *Journal of Controlled Release.* 2013; 171 (3): 349–357. https://doi.org/10.1016/j.jconrel.2013.04.018.
127. Longest PW, Bass K, Dutta R, Rani V, Thomas ML, El-Achwah A, and Hindle M. Use of computational fluid dynamics deposition modeling in respiratory drug delivery. *Expert Opinion on Drug Delivery.* Taylor and Francis Ltd. 2019 https://doi.org/10.1080/17425247.2019.1551875.
128. Mathur V, Mathur V, Satrawala Y, Rajput MS, Kumar P, Shrivastava P, and Vishvkarma A. Solid lipid nanoparticles in cancer therapy. *International Journal of Drug Delivery.* 2011; 2 (3): 192–199. https://www.arjournals.org/index.php/ijdd/article/view/43.
129. Vighi E, Leo E, Montanari M, Mucci A, Hanuskova M, and Iannuccelli V. Structural investigation and intracellular trafficking of a novel multicomposite cationic solid lipid nanoparticle platform as a PDNA carrier. *Therapeutic Delivery.* 2011; 2 (11): 1419–1435. https://doi.org/10.4155/tde.11.118.
130. Vighi E, Ruozi B, Montanari M, Battini R, and Leo E. PDNA condensation capacity and in vitro gene delivery properties of cationic solid lipid nanoparticles. *International Journal of Pharmaceutics.* 2010; 389 (1–2): 254–261. https://doi.org/10.1016/j.ijpharm.2010.01.030.
131. Shao Z, Shao J, Tan B, Guan S, Liu Z, Zhao Z, He F, and J Zhao. Targeted lung cancer therapy: preparation and optimization of transferrin-decorated nanostructured lipid carriers as novel nanomedicine for co-delivery of anticancer drugs and DNA. *International Journal of Nanomedicine* 2015; 10 (February): 1223–1233. https://doi.org/10.2147/IJN.S77837.
132. Choi SH, Jin SE, Lee MK, Lim SJ, Park JS, Kim BG, Ahn WS, and Kim CK. Novel cationic solid lipid nanoparticles enhanced p53 gene transfer to lung cancer cells. *European Journal of Pharmaceutics and Biopharmaceutics.* 2008; 68 (3): 545–554. https://doi.org/10.1016/j.ejpb.2007.07.011.

133. Bae KH, Lee JY, Lee SH, Park TG, and YS Nam. Optically traceable solid lipid nanoparticles loaded with sirna and paclitaxel for synergistic chemotherapy with in situ imaging. *Advanced Healthcare Materials.* 2013; 2 (4): 576–584. https://doi.org/10.1002/adhm.201200338.
134. Ahmed R, N Amreddy, A Babu, A Munshi, and R Ramesh. Combinatorial nanoparticle delivery of SiRNA and antineoplastics for lung cancer treatment. In *Methods in Molecular Biology.* 2019; 1974: 265–290. Humana Press Inc. https://doi.org/10.1007/978-1-4939-9220-1_20.
135. Han Y, Zhang P, Chen Y, Sun J, and Kong F. Co-delivery of plasmid DNA and doxorubicin by solid lipid nanoparticles for lung cancer therapy. *International Journal of Molecular Medicine.* 2014; 34 (1): 191–196. https://doi.org/10.3892/ijmm.2014.1770.
136. Mukherjee A, Waters AK, Kalyan P, Achrol AS, Kesari S, and Yenugonda VM Lipid-polymer hybrid nanoparticles as a nextgeneration drug delivery platform: state of the art, emerging technologies, and perspectives. *International Journal of Nanomedicine.* Dove Medical Press Ltd. 2019; 14: 1937–1952. https://doi.org/10.2147/IJN.S198353.
137. Mandal B, Mittal NK, Balabathula P, Thoma LA, and Wood GC Development and in vitro evaluation of core-shell type lipid-polymer hybrid nanoparticles for the delivery of erlotinib in non-small cell lung cancer. *European Journal of Pharmaceutical Sciences.* 2016; 81 (January): 162–171. https://doi.org/10.1016/j.ejps.2015.10.021.
138. Gaber M, Medhat W, Hany M, Saher N, Fang JY, and Elzoghby A. Protein-lipid nanohybrids as emerging platforms for drug and gene delivery: challenges and outcomes. *Journal of Controlled Release.* Elsevier BV. 2017; 254:75–91. https://doi.org/10.1016/j.jconrel.2017.03.392.
139. Schultze Eduarda, Ourique A, Yurgel VC, Begnini KR, Thurow H, Leon PMMD, Campos VF, et al. Encapsulation in lipid-core nanocapsules overcomes lung cancer cell resistance to tretinoin. *European Journal of Pharmaceutics and Biopharmaceutics.* 2014; 87 (1): 55–63. https://doi.org/10.1016/j.ejpb.2014.02.003.
140. Mukherjee Sumita, Acharya BR, Bhattacharyya B, and Chakrabarti G.Genistein arrests cell cycle progression of A549 cells at the G_2 /M phase and depolymerizes interphase microtubules through binding to a unique site of tubulin. *Biochemistry.* 2010; 49 (8): 1702–1712. https://doi.org/10.1021/bi901760d.
141. Zhang J, Su H, Li Q, Li J, and Zhao Q. Genistein decreases A549 cell viability via Inhibition of the PI3K/AKT/HIF-1α/VEGF and NF-KB/COX-2 Signaling Pathways. *Molecular Medicine Reports.* 2017; 15 (4): 2296–2302. https://doi.org/10.3892/mmr.2017.6260.
142. Kamel NM, Helmy MW, Samaha MW, Ragab D, and Elzoghby AO. Multicompartmental lipid-protein nanohybrids for combined tretinoin/herbal lung cancer therapy. *Nanomedicine.* 2019; 14 (18): 2461–2479. https://doi.org/10.2217/nnm-2019-0090.
143. Karimi Mahdi, Hamed Mirshekari, Masoumeh Aliakbari, Parham Sahandi-Zangabad, and Hamblin MR Smart mesoporous silica nanoparticles for controlled-release drug delivery. *Nanotechnology Reviews.* Walter de Gruyter GmbH. 2016. https://doi.org/10.1515/ntrev-2015-0057.
144. Torchilin VP CHAPTER 1. Fundamentals of stimuli-responsive drug and gene delivery systems. In, 1–32. Royal Society of Chemistry. 2018. https://doi.org/10.1039/9781788013536-00001.
145. Primeau AJ, Rendon A, Hedley D, Lilge L, and Tannock IF. The distribution of the anticancer drug doxorubicin in relation to blood vessels in solid tumors. *Clinical Cancer Research.* 2005; 11 (24): 8782–8788. https://doi.org/10.1158/1078-0432.CCR-05-1664.
146. Trédan O, CM Galmarini, K Patel, and IF Tannock. Drug resistance and the solid tumor microenvironment. *Journal of the National Cancer Institute.* 2007;99(19): 1441–1454.
147. Mussi SV, Silva RC, Oliveira MCD, Lucci CM, Azevedo RBD, and Antônio L, Ferreira M New approach to improve encapsulation and antitumor activity of doxorubicin loaded in solid lipid nanoparticles. *European Journal of Pharmaceutical Sciences.* 2013; 48 (1–2): 282–290. https://doi.org/10.1016/j.ejps.2012.10.025.
148. Bakhtiary Z, Barar J, Aghanejad A, Saei AA, Nemati E, Dolatabadi JEN, and Omidi Y. Microparticles containing erlotinib-loaded solid lipid nanoparticles for treatment of non-small cell lung cancer. *Drug Development and Industrial Pharmacy.* 2017; 43 (8): 1244–1253. https://doi.org/10.1080/03639045.2017.1310223.
149. Meenach SA, Vogt FG, Anderson KW, Hilt JZ, McGarry RC, and Mansour HM Design, physicochemical characterization, and optimization of organic solution advanced spray-dried inhalable dipalmitoylphosphatidylcholine (DPPC) and dipalmitoylphosphatidylethanolamine poly (ethylene glycol) (DPPE-PEG) microparticles and nanoparticles for targeted respiratory nanomedicine delivery as dry powder inhalation aerosols. *International Journal of Nanomedicine.* 2013; 8 (January): 275–293. https://doi.org/10.2147/IJN.S30724.
150. Nassimi M, Schleh C, Lauenstein HD, Hussein R, Hoymann HG, Koch W, and G Pohlmann, et al. A Toxicological evaluation of inhaled solid lipid nanoparticles used as a potential drug delivery system for the lung. *European Journal of Pharmaceutics and Biopharmaceutics.* 2010; 75 (2): 107–116. https://doi.org/10.1016/j.ejpb.2010.02.014.
151. Non-small cell lung cancer treatment (PDQ®)–patient version - national cancer institute. n.d. Accessed August 30, 2020. https://www.cancer.gov/types/lung/patient/non-small-cell-lung-treatment-pdq.
152. Nuclear delivery of doxorubicin via folate-targeted liposomes with bypass of multidrug-resistance efflux pump | Clinical Cancer Research. n.d. Accessed August30, 2020. https://clincancerres.aacrjournals.org/content/6/5/1949.

22

Lipid-Based Pulmonary Delivery System: A Review and Future Considerations of Formulation Strategies and Limitations

Komal Parmar[1] **and Jayvadan Patel**[2]
[1]*ROFEL, Shri G.M. Bilakhia College of Pharmacy, Gujarat, India*
[2]*Nootan Pharmacy College, Faculty of Pharmacy, Sankalchand Patel University, Gujarat, India*

22.1 Introduction

Leading unconventional drug delivery systems are constantly designed to provide improvised therapeutic efficacy. Error in specificity might cause unanticipated drug release resulting in dose dumping and adverse effects. Moreover, higher dose is compensation for drug distribution and phagocytic uptake which may lead to overdosing, further resulting in cytotoxicity. In addition, curbing the dose to the specific amount required to produce the therapeutic effect will reduce the excessive drug wastage and ultimately lessen the cost of dosage. Novel drug delivery systems have been utilized to prevent the dose wastage by providing controlled release of drug. But non-specific distribution of drug may lead to rapid clearance and poor pharmacokinetics (1,2). Thus, various routes of administration are adopted to deliver the drug in order to achieve acceptable therapeutic efficacy with minimum adverse effects. Various drug dosage forms are investigated and administered via various alternative routes other than oral course.

Lately, pulmonary route of administration has drawn attention to assuage the drawbacks of the conventional route of administration. Historical evidences of practice of pulmonary drug delivery could be dated back to the 16th century, where an inhaler instrument was used to produce opium vapor for the treatment of cough (3). Indian history elaborates usage of therapeutic aerosols of herbal compositions for the treatment of asthma (4). Other respiratory diseases cured using conventional pulmonary drug delivery included tuberculosis, chronic pulmonary ailment, and cystic fibrosis (5–7).

Pulmonary drug delivery system provides an non-invasive route for drug delivery via nose or mouth using the physiology of the respiratory system. Lungs possess the advantage of a large surface area of 100 m^2 and thin alveolar epithelium with excellent permeability (8). This provides ease in systemic bioavailability and rapid onset of action of drug with direct exposure of drug at the desired region. The route of administration also bypasses the hepatic metabolism thereby preventing the excessive metabolism of drug by the liver. In addition, patient compliance can be improved by averting the adverse effects caused by the conventional dose of drug, thereby reducing the waste of drug in the treatment. While approachability of drug might differ in the conventional inhaled powder form, lipid-based pulmonary drug delivery systems grant a uniform dispersion of drug in the lung alveoli with enhanced dissolution rate due to small droplet size and increased contact surface area (9,10). Further they provide uniform and controlled targeted drug release in the lung region, ultimately improving the overall bioavailability of drug in the lung (11–14).

The present chapter review is intended to discuss the various lipid-based drug delivery systems that can be utilized for pulmonary drug delivery. Concepts of various lipid-based formulations, strategies to delivery via pulmonary route, and limitations of the formulations will be essayed.

22.2 Deposition in Pulmonary Circulatory System

In an adult human lung can be deposited a varied size range of particles categorized as coarse (>5 μm), fine (0.5–5 μm), and ultra-fine (<0.5 μm). Of these, coarse particles are found to remain on the upper part of trachea, while the fine particles can reach to alveoli and ultra-fine particles are unable to settle in the lung area as most of them are exhaled due to failure of minimum size requirement to settle. Thus, the aerodynamic diameter of the particle is responsible for the deposition process in the pulmonary system. Mechanisms involved in the deposition process might involve one of more of the mentioned: interception, inertial impaction, Brownian diffusion, gravity settling, electrostatic forces, and to a less extent, turbulent flow (Figure 22.1). Each deposition mechanism is substantial with certain material and insubstantial with others.

Interception involves passage of particles in such a way that their edges encounter the airway wall. This is due to the shape and size of the particles and largely associated with elongated materials such as fibers. Particles with spherical structure rarely intercept owing to the uniformity in shape. In one study, polypropylene fibers of diameters ranging 5–9 μm and length of 60–130 μm were found in the peribronchial region of lungs (15). The fibers with small diameter will flow in the airway

DOI: 10.1201/9781003046547-22

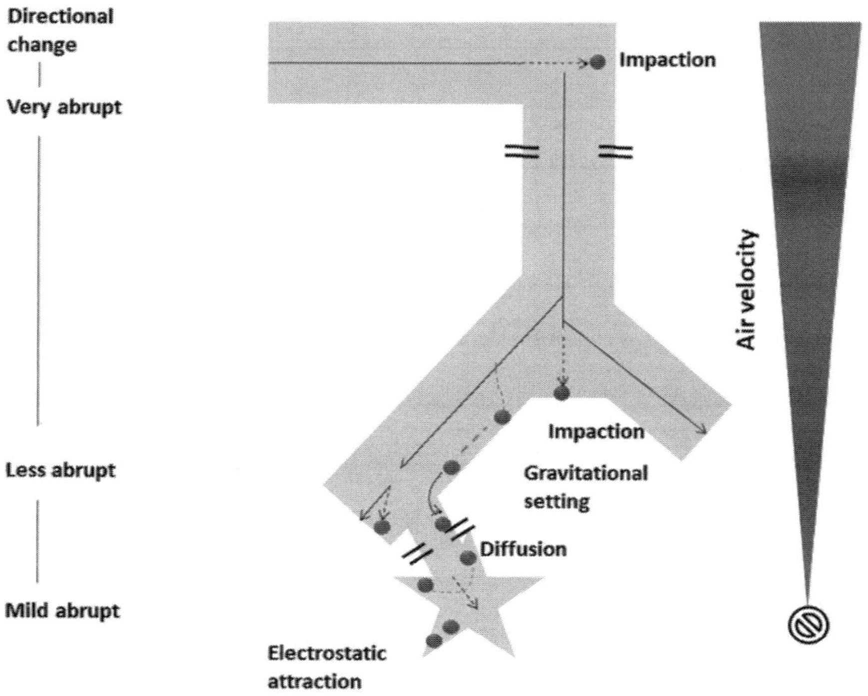

FIGURE 22.1 Different Mechanisms of Particle Deposition in Lung Airways

passage, intercepting the wall and thereby may enter the adjacent smaller daughter tubes (16). Thus, the chances of interception increase with decrease in the diameter of the airway passage. Inertial impaction is observed with the coarse particles with size greater than 5 μm. The inertia associated with such big particles will not allow them to deviate from their existing trajectory even due to sudden change in air stream, which then results in collision with the airway walls and thereby their deposition (17). This tendency of the coarse particle to deviate from the airflow path and thereby deposit on the airway walls increases with increase in the particle size, air flow rate, and carrier mass (18). This behavior of deposition of larger particles (> 8 μm) is typically observed in the upper respiratory tract with high velocity and turbulent airflow (19).

Deposition of ultra-fine particles follows the Brownian diffusion mechanism which is due to inherent movement of the particle resulted on collision with the surrounding air molecules. This movement is found to be maximum with low airflow velocity and decrease in particle size (20). Predominant diffusion of fine particles occurs through Brownian diffusion in the alveolar region and smaller airways. which is responsible for extended residence time (21). Deposition mechanism follows the mechanism of diffusion from higher concentration of particles in lung to lower concentration in small airways (22). Gravitational settling is a time dependent mechanism and generally affects particles of size range of 1–8 μm depositing in the smaller alveoli and airways (23). Here, the larger particles follow inertial impaction due to increase in particle density, size, and air flow rate, wherein the smaller particles deposit via Brownian diffusion (24). Gravitational forces and air resistance will ultimately overcome the buoyancy of the particles in the respiratory tract particularly observed in bronchi and bronchioles. Electrostatic charges are found on the drug carriers produced by aerosol devices (25). This charging of particles might be due to contact or friction between the aerosol particles and between particles and the aerosol container component during dispersal. Overall formulation composition, inhaler container components, relative humidity, and solid-state properties influence the generation of particle charge. It is observed that with increased charge level, particles tend to deposit at the upper airway before loss of charge due to humidity (26). Coulomb's law states the inverse relationship between electrostatic charge and square of the distance between the charged particles. Thus, electrostatic deposition of particles is particularly applicable to smaller airways and alveoli. Electrostatic charge comes from space charge and image charge force. Space charge force refers to repulsion between charged particles in the particle cloud, and image charge refers to induced charge force on the particle surface. Image charges are believed to be induced on the surface in the small airways in peripheral lung (27). Rarely turbulent flow deposition affects particle settling in the upper respiratory tract and large airways (28). It refers to the irregular flow of the particles due to fluctuation in the fluid speed thereby resulting in a change in direction and magnitude of the particle flow and eventually deposition on the airway walls. High respiratory flow is considered responsible for in aerosol deposition in upper respiratory tract by turbulent mixing (29). Turbulence flow in the respiratory tract depends on density of gas and lowering the density will convert some or all turbulent flow into laminar flow.

22.3 Formulation Strategies

On administering the pulmonary formulation, the particles are going to travel through the human respiratory airway with gradual reduction in airway diameter from larger at the upper

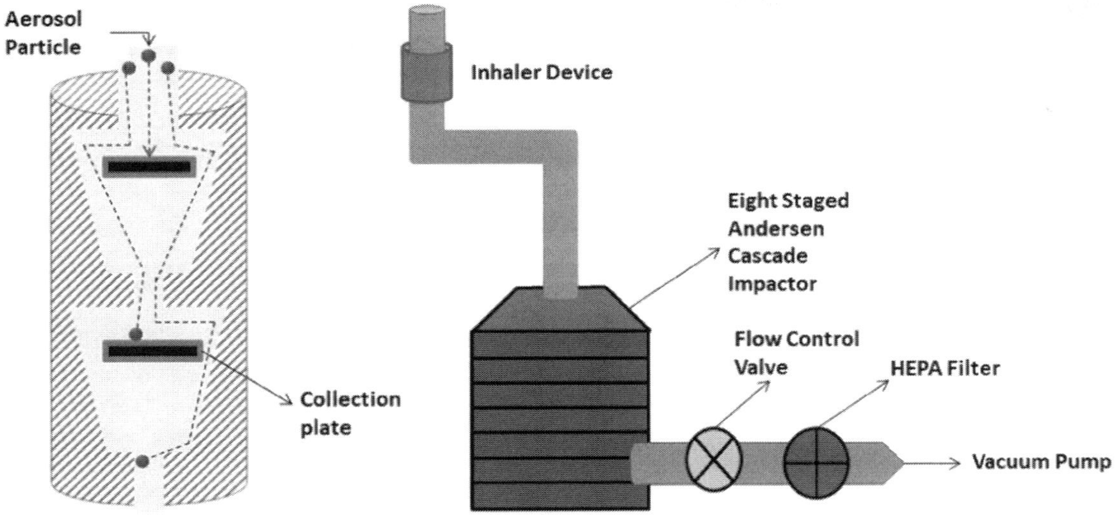

FIGURE 22.2 Schematic Diagram of a Cascade Impactor

portion to smaller size at the lower. Less than 20% of drug deposition in lung is estimated following pulmonary administration on account of various factors (30). This suggests that limitations and challenges associated with aerosol formulation strategies need to be understood.

Pulmonary drug delivery system is delivered via suitable inhaler in aerosolized form. Commonly metered dose inhalers or nebulizers are used, which delivers dispersion or solution of the formulation. A dry powder inhaler is used to deliver powder formulation via inhalation. Inhalers provide a form of drug targeting, whether locally at the site of action in the lungs or at the site of absorption for drugs acting systemically. Formulation strategies for the successful absorption of the drug particles into the lung could be modulating the aerodynamic performance of the lipid-particulate system. Performance of aerosol is estimated by measuring fine particle fraction (FPF) and mass median aerodynamic diameter (MMAD) using a cascade impactor. Figure 22.2 demonstrates the schematic diagram of a cascade impactor. FPF is the measure of mass of particles with an aerodynamic diameter less than 5 μm (31). MMAD is median of aerodynamic particle size distribution. Aerodynamic diameter of an arbitrary particle is equal to the diameter of the spherical particle with a density of 1 g/ml, with similar inertial properties. Aerodynamic diameter describes the diameter of particles for which deposition is mainly governed by sedimentation and impaction. The aerodynamic performance depends on aerodynamic diameter (D_{aero}), geometric diameter (D_{geo}), particle density (p_p/p_o), and shape factor (χ) cumulatively, as shown in equation (22.1). (Dynamic shape factor is the ratio of drag force experienced by irregular particle to the drag force experienced by spherical particle of equivalent volume traveling at same velocity and medium). It is observed that a small dense particle will travel at the similar settling velocity as a large particle with less density.

$$D_{aero} = D_{geo}\left(\sqrt{\frac{P_p}{P_o \cdot \chi}}\right) \tag{22.1}$$

On inhalation, the aerodynamic diameter of globules or particles greater than 5 μm get deposited in the upper airways following inertial impaction mechanism, while between 1 and 5 μm deposit in the lower airways following sedimentation. Droplets or particles smaller than 1 μm could be deposited following the Brownian diffusion mechanism, while ultra-fine particles may get exhaled during exhalation (32). Depending on wettability and solubility properties of drug particles, they further get absorbed following deposition. While undissolved particles will be cleared by the clearance mechanism of the respiratory system involving alveolar macrophages (33). Further, on inhalation nanocarriers will distribute with fractional deposition in different stages of airway. Geometric standard deviation defines the efficiency of the particle distribution. It is calculated as the ratio of the square root of particle diameter at 84.1 and 15.9% of total cumulative number (10).

In vitro assessment of deposition efficiency of the spray droplets from the inhaler is been reported by various investigators (34,35). Volume mean diameter refers to the diameter measured at midpoint of the volume sprayed, of which half is in smaller droplets and the other half is larger than the median. However, uniformity of droplet size distribution is demonstrated by parameter named, relative span factor (RSF). The closer the value to zero, the more uniform is the droplet size distribution. This dimensionless factor is given by equation (22.2)

$$RSF = \frac{D_{90} - D_{10}}{D_{50}} \tag{22.2}$$

Where, D_{10}, D_{50}, and D_{90} represent the cumulative globule size distribution of 10%, 50%, and 90% globules, respectively (36).

It is evident that the performance of the liquid spray will depend on its surface tension and viscosity characteristics. The small droplet size in low surface tension fluids at the same spatial location yield significant centrifugal dispersion because of lower inertia resulting into a wider spray cone angle (37). High viscosity liquids will lead to a dispersion of droplets move toward the periphery leading to increased cone angle (38). Thus apparently, higher viscosity and surface tension will cause increase in aerodynamic diameter and compromise the performance of aerosol. Use of mixture of surfactants with lipid carriers will facilitate reduction in surface tension and viscosity of the formulation, thereby also improve the drug solubility. Surfactants such as polysorbates and lecithin are widely used for the said purpose in various formulations (39,40).

In addition, pH of the liquid inhaler is considerably important parameter for the formulation. Apparently physiological pH is considered to be biocompatible, however, European Pharmacopoeia demonstrates pH value of 8 for liquid formulations stated for pulmonary utilization (41). Any deviation from the stated value might result in irritation of lung mucosa or compromise drug stability.

22.4 Device Selection

Choice of inhaler depends on the patient's compatibility and preference; it also should deliver multiple drugs in several doses. Further, the device should protect the drug from the outer environmental and inner factors, thereby retaining physical and chemical stability of the drug. Inhalers are made to deliver the formulation in a narrow particle size range of 0.5–5 μm, thus should deliver the particles without fluctuation in performance. Therefore, an ideal inhaler device should be convenient for thge patient, simple to use, and inexpensive. However, various novel inhaler sprays are developed for enhanced efficiency of spray performance. Mesh nebulizers, owing to increased portability, convenience, and energy efficiency, have gained popularity in nebulized pharmaceutical products. Mesh nebulizesr use a mesh at the dispenser part to deliver particles of consistent size (42,43). The Medspray® inhaler comprises a spray nozzle in combination with a specifically structured pump system. The solution is spread into droplets through the nozzle with mechanical mode. A special mouth piece is available that mixes the dispersed droplets with air inspired by the patient (44). Table 22.1 illustrates the various inhalers available in the market.

22.5 Patient Education

Loss of drug can be attributed to improper use of the inhaler device by the untrained patients, particularly children, elderly, and physically challenged users (45). Poor understanding of the device utilization poses a potential challenge to the success of a pulmonary drug delivery system. Practical demonstrations to patients, along with written instructions regarding use of the inhaler device, are to be given initially and regularly as reminders.

22.6 Lipid-Based Drug Delivery

Role of lipidic excipients in novel lipid-based drug delivery has been reformed in recent years. These lipid excipients with appropriate safety and regulatory standards have proved to be successful in improving the bioavailability of drugs (46). Pulmonary therapy of lipid-based carriers for drugs can prove to be a superior drug delivery in direction of the targeted and effective treatment of respiratory diseases. Characteristics of lipid-based carriers like versatility, biodegradability, and biocompatibility provide the specific means for targeted and controlled drug delivery of drugs with varied properties. Lipid-based drug delivery can be altered in several ways to meet a wide range of formulation requirements. Micro- and nano-emulsions made up of oils are reported to improve the bioavailability of drug (47,48). Lipid carriers in the form of micro- and nano-emulsions are safe and effective for the delivery of vaccines (49,50), diagnostic agents (51,52), and bioactives (53,54). Due to the biocompatibility of the oils used in the preparation of emulsions, they are less toxic in nature (55). Lipidic micelles are self-assembled wheel-like structured nano aggregates in solution. They improve drug stability and can be modified on the surface for efficient controlled and targeted drug release (56,57). Liposomes are the spherical sacs of phospholipid that encapsulate the drug and nutrients to provide targeted release in the specific tissues (58,59). Solid lipid nanoparticles and nano-structured lipid carriers are known to improve the stability and bioavailability of drugs (60). Different lipid structures and their recent applications are demonstrated in Figure 22.3 and Table 22.2.

22.6.1 Micro and Nano-Emulsions

Micro- or nano-emulsions are emulsions having micro/nano-sized droplets respectively, manufactured to improve the delivery of bioactive or active pharmaceutical ingredient. Such emulsions are thermodynamically stable isotropic mixtures of two immiscible liquids forming one homogeneous phase with

TABLE 22.1

Marketed Inhaler Products

Inhaler Product	Drug	Therapy Indication	Aerosol Device
Asmanex®	Mometasone furoate	Asthma	Twisthaler® dry powder inhaler
Relenza®	Zanamivir	Antiviral	Diskhaler® dry powder inhaler
TOBI®	Tobramycin	Anti-bacterial	PARI LC PLUS™ nebulizer
Arcapta Neohaler®	Indacaterol	Chronic obstructive pulmonary disease (COPD)	Neohaler® dry powder Inhaler
PULMICORT RESPULES®	Budesonide	Anti-inflammatory	Jet nebulizer
PARI VELOX®	Salbutamol	Asthma	Vibrating membrane nebulizer®
Seebri Breezhaler(®)	Glycopyrronium	COPD	Breezhaler(®) dry powder inhaler
VENTOLIN HFA®	Albuterol sulfate	Asthma	Metered dose inhaler
ProAir® Digihaler®	Albuterol sulfate	Asthma	Digihaler®

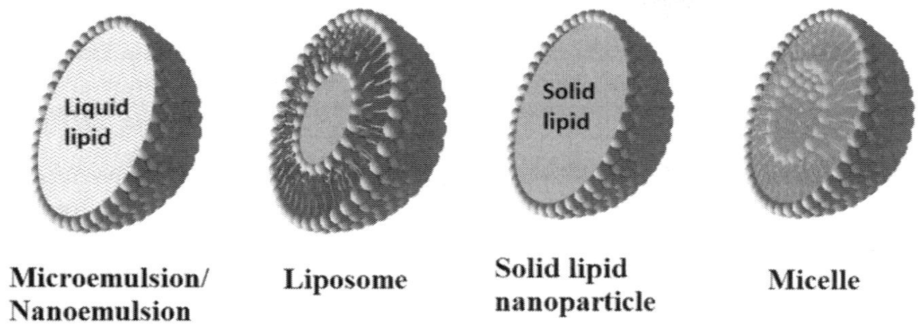

FIGURE 22.3 Various types of Lipid Based Nanocarriers

TABLE 22.2
Recent Applications of Lipid Carrier Drug Delivery Systems

Drug	Formulation	Ingredients Used	Purpose	Reference
Zataria multiflora essential oil	Nano-emulsion	Zataria multiflora essential oil, Tween 80,	Antihydatid and scolicial activity	(61)
Adapalene	Nano-emulsion	Tea tree oil, dimethyl sulfoxide, Tween 20, 60, and 80, and Span 80, ethanol	Topical anti-bacterial activity	(62)
Curcumin	Nano-emulsion	Rice bran oil, Tween 20, sodium caseinate	Bioavailability enhancement	(63)
Amphotericin B	Nano-emulsion	Isopropyl myristate, Kolliphor® HS15, Brij® 52	Bioavailability enhancement	(64,65)
Zotepine	Solid lipid nanoparticles	Dynasan®118, Dynasan®114, Compritol® 888 ATO, soyalecithin, Captex®355 and Captex®200	Bioavailability enhancement	(66)
Atorvastatin calcium	Solid lipid nanoparticle	Poloxamer 188, Compritol® 888 ATO, Phospholipon 90 H, polyethylene glycol 400	Treatment of age related macular degeneration	(67)
Duloxetine	Solid lipid nanoparticle	Stearyl alcohol, poloxamer 188, Tween 80	Treatment of depression	(68)
Paclitaxel	Solid lipid nanoparticle	Glyceryl monostearate, lyceryl tripalmitate, glyceryl trimyristate, glyceryl tristearate, stearic acid, lectin, d-α-tocopheryl polyethylene glycol 1000 succinate, Tween®80, poloxamer 188, and poly(vinyl)Alcohol	Bioavailability enhancement	(69)
Rhodamine B, metronidazole, doxorubicin hydrochloride	Liposomes	1,2-dipalmitoyl-sn-glycero-3-phosphocholine, n-heptyltriphenylphosphonium bromide, n-nonyltriphenylphosphonium bromide, n-Decyltriphenylphosphonium bromide, n-dodecyltriphenylphosphonium bromide, n-tetradecyltriphenylphosphonium bromide	Targeted drug delivery	(70)
Calcein	Liposomes	1,2-Distearoyl-sn-glycero-3-phosphoethanolamineN-[amino (polyethylene glycol)-2000, 1,2-dipalmitoyl-sn-glycero-3-phosphocholine, cholesterol, human serum albumin, arginylglycylaspartic acid, holotransferrin human	Targeted drug delivery	(71)
Vitamin K	Lipid micelles	Egg phosphatidylcholine, 1,2-distearoyl-sn-glycero-3-phosphoethanolamine-(polyethylene glycol)-2000 and glycocholic acid	Improvement in stability and bioavailability	(72)
Rituximab	Lipid micelles	1,2-Distearoyl-sn-glycero-3-phosphoethanolamine-N-[succinimidyl (polyethylene glycol)], Cy3- conjugated methoxy poly(ethylene oxide)-b-poly(ε-caprolactone)-b-poly(α-propargylcarboxylate-ε-caprolactone), Cy3-conjugated methoxy poly(ethylene oxide)-b-poly(α-benzylcarboxylate-εcaprolactone)-b-poly(α-propargylcarboxylate-ε-caprolactone)	Targeting cancer cell	(73)
Ergosterol	Lipid nanostructures	Glyceryl monostearate, stearic acid, decanoyl/octanoyl-glycerides, oleic acid, ethylis oleas, olive oil, poloxamer 188, and Tween 80	Enhancing oral bioavailability	(74)
Simvastatin	Lipid nanostructures	Glyceryl dibehenate, glyceryl behenate, macrogol 6 glycerol caprylocaprate, poloxamer 188, Span 80, polyvinyl alcohol, polyvinyl pyrrolidone,	Enhancing bioavailability	(75)

the help of surfactant molecules located at the interface of dispersed droplets (76,77). High energy emulsification techniques are investigated for the fabrication of mico/nano-emulsions which are responsible for the diminution of the droplets to micro/nano range (78). Micro/nano-emulsions are having advantage of solubilizing lipophilic drugs in oil system. Micro/nano-emulsions possess potential in pulmonary drug delivery as the droplets are small in size to make them retain in the respiratory tract (79). However little research work is reported demonstrating application of small-sized emulsions for pulmonary drug delivery.

In a study, ibuprofen loaded oil in water nano-emulsion mist was prepared using Tween 80 and Cremophor RH 40 as surfactants; Transcutol P, Capryol 90, and PEG 400 as cosurfactants; and Labrafac Lipophile Wl 1349 (a medium-chain triglyceride) as an oil (80). Average globule size was found to be 100 nm; a cell toxicity study in NIH 3T3 cells suggested biocompatibility. The nebulized mist suggested characteristics ideal for pulmonary drug delivery. Another investigation demonstrated nano-emulsion nebulized curcuminoid lung delivery using limonene and oleic acid as oil phase. The average FPF and MMAD of the nebulized nano-emulsion formulations prepared with limonene oil ranged from 50% and 4.6 μm to 45% and 5.6 μm, for 1000 μg/ml and 100 μg/ml, respectively, and with oleic acid ranged from 46% and 4.9 μm to 44% and 5.6 μm, for 1000 μg/ml and 100 μg/ml respectively. Genotoxicity studies showed biocompatibility with lymphocyte cells supporting positive performance of the nebulized formulation for lung delivery (81). Quercetin loaded nano-emulsion based aerosol was developed using palm oil ester/ricinoleic acid as oil phase for lung cancer therapy. The system exhibited an acceptable median mass aerodynamic diameter of 3.09 ± 0.05 μm and geometric standard deviation of 1.77 ± 0.03 with high FPF (90.52 ± 0.10%). Percent dispersed and percent inhaled for deposition in deep lung were observed to be 83.12 ± 1.29% and 81.26 ± 1.28% respectively. Cytotoxicity studies using A549 lung cancer cells illustrated positive results (82). For tuberculosis therapy, rifampicin oleic acid first generation and chitosan- and chitosan-folate conjugate-decorated second- and third-generation nano-emulsions based nebulizers were optimized. Aerodynamic assessment data revealed aerosol output and aerosol output rate enhancement by nano-emulsion fabrication. The internalization of drug loaded nano-emulsion was achieved by the alveolar macrophages, suggesting successful targeting. In addition, pharmacokinetic data supported targeting of drug with high lung-drug content (83).

A lecithin/D-a-tocopheryl polyethylene glycol succinate based spray-dried self-micro-emulsifying pulmonary drug delivery system was reported of a poorly soluble drug. In vitro aerodynamic studies suggested significant deposition with micro-sized droplets. In vitro cytotoxicity illustrated safety of the formulation (39). A lecithin micro-emulsion based metered dose inhaler was investigated using water in oil propellant: dimethyl ether and propane. The MMAD and FPF of the resulting aerosol was observed to be 3.1 μm and 59% respectively for dimethyl ether, with lecithin content 3% and water content 2.5% (84). In another study, aerosolized dioctyl sodium sulfosuccinate micro-emulsion-based oral drug delivery was reported. A pseudo ternary system was prepared using medium-chain triglycerides with oleic acid/glycerol monooleate and water. Droplet size measurements with result of 42 nm revealed efficiency of micro-emulsion based system (86).

22.6.2 Solid Lipid Nanoparticles

Solid lipid nanoparticles are considered the second generation of nanoparticles, similar to emulsions composed of solid lipids (mostly natural lipids) as a substitute of oil phase dispersed in aqueous solution consisting of surfactant (85). The presence of solid lipid increases the stability of the active ingredients. Surfactants impart stability by reducing the interfacial tension between the two liquids. High energy techniques exist for the fabrication of nano-sized solid lipid nanoparticles (87). Due to their small size, prolonged drug release, stability, and low toxicity, solid lipid nanoparticles are considered suitable for pulmonary route of administration (88). In one study, budesonide solid lipid nanoparticles were fabricated using an ultrasonication technique. Hydrogenated palm oil and soyabean lecithin were employed as lipids and poloxamer 188 and sorbitol as emulsifiers. In vitro aerosolization performance of optimized solid lipid nanoparticles indicated enhancement in FPF value with 17.50 % as compared to commercial budenoside suspension with value of 14.16 % (89,90). In another study, microencapsulate solid lipid nanoparticle hybrid dry powder inhaler was developed for rifabutin pulmonary drug delivery for the treatment of tuberculosis. Aerosolization studies suggested deep lung deposition of the multiparticulated system. In vivo biodistribution demonstrated antibiotic availability at the tested organ, whereas mycobacterial activity illustrated enhancement of anti-tuberculosis activity against the microbe (91). Ethambutol loaded solid lipid nanoparticles using Compritol 888 ATO as solid lipid were fabricated as dry powder inhalable formulation for tuberculosis treatment. A hot homogenization process and ultrasonication were employed for the preparation of nanoparticles. The aerodynamic behavior of the prepared dry powder was assessed and the values of Emitted Dose %, Fine Particle Fraction %, Mass Median Aerodynamic Diameter, and Geometric Standard Deviation were 91.76 ± 0.83, 23.98 ± 0.38, 5.629 ± 0.15 and 2.65 ± 0.09 respectively. Biocompatibility and lack of toxicity of prepared nanoparticles was established through MTT assay (92). In another study, rifampicin inhalable solid lipid nanoparticles were developed employing palmitic acid and cholesteryl myristate as lipid using a melt emulsification technique. Solid lipid nanoparticles were functionalized with a synthesized mannosylated surfactant to achieve active targeting to mannose receptors for intramacrophagic transport. Aerodynamic diameters were found to be in the range of 210–676 nm. Cytotoxicity studies reported biocompatibility, determined using MH-S cell line suggesting it is safe and nontoxic (93).

22.6.3 Liposomes

Liposomes refers to a vesicular drug delivery made up of at least one lipid bilayer enclosing a water core. Resemblance of liposomes to bio-vesicles make their uptake easy by the cell and tissues (94). Due to biocompatibility, biodegradability, and nano-size, liposomes are extensively investigated for targeting of several drugs (95). Enhanced permeability and retention effect owing to nano-size offers great potential for pulmonary drug delivery of liposomes (96). In an investigation, rifampicin loaded liposomes were prepared and administered in dry powder form for tuberculosis therapy. Soya lecithin and cholesterol were employed as lipid carriers in the fabrication of the liposomes. An in vitro drug deposition study was carried out after intra-tracheal administration in rats and the results suggested a 9.16-fold increase in AUC values of the optimized lyophilized liposomal powder of rifampicin (97). In other research, salbutamol sulfate liposome dry powder was fabricated using soyaphosphatidyl choline and cholesterol for the treatment of asthma. An optimized spray-dried liposomal formulation with lactose exhibited an average size of about 167.2 ± 0.170 nm, and in vitro pulmonary deposition studies using a cascade impactor illustrated results of fine particle fraction: MMAD and geometric standard deviation as 64.01 ± 0.43, 3.49 ± 0.12 and 1.08 ± 0.04 respectively (98). Budesonide and colchicine liposomal dry powder inhaler was fabricated using 1, 2-Dipalmitoyl-snglycero-3-phosphoglycerolsodium, hydrogenated soyaphosphotidylcholine, soyaphosphatidylcholine, dipalmitoylphosphatidylcholine, and 1,2-dioleoyl-sn-glycero-3-ethylphosphocholine as lipids. Intra-tracheal administration of optimized dry powder inhaler significantly reduced the neutrophils count previously augmented using bleomycin. The reduction in count was observed to be from 58.56 ± 4.5 (× 103 15 /ml) to 2.4 ± 0.6 (×103 16 /ml) after a 14-day treatment suggesting successful therapy of the formulation (99). Ciprofloxacin liposomal nanocrystal formulation was developed for controlled release in respiratory tract. Hydrogenated soy phosphatidylchocline and cholesterol were employed as lipids in the preparation produced by extrusion method. Aerosol performance of ciprofloxacin nanocrystal liposomal dry powders containing 2:1 ratio of sucrose to lipid illustrated values of fine particle fraction and MMAD were 60.3 and 2.88 respectively (100).

22.6.4 Micelle

Lipid micelle are self-assembling nano-sized colloidal spherical structures in aqueous solution. Formation of micelle occurs spontaneously due to its amphipathic nature of fatty acids, i.e. have hydrophilic heads and hydrophobic tails. Lipid micelles show enhanced permeability and high retention, and the nanoparticle size makes it a potential candidate for targeted drug delivery (101). In a study, cholesterol conjugated polyamidoamine micelle was synthesized and used as a carrier for resveratrol and heme oxygenase-1 gene delivery to lungs via inhalation. The lipid carrier delivered the drug and gene efficiently by inhalation and induced therapeutic effects on the animal model (102). In another study, spray-dried inhalable powder containing polymeric micelles was fabricated for pulmonary delivery of paclitaxel for lung cancer treatment. Lipids employed were tocopheryl succinate-polyethylene glycol 1000 and 5000. Cytotoxicity demonstrated enhanced cytotoxic activity of the formulation as compared to free drug. In vitro deposition results of the inhalable micellar powder illustrated high fine-particle fraction of 60%, suggesting a potential platform for the lung cancer therapy (103). Inhalable itraconazole based stearic acid grafted chitosan polymeric micelles for pulmonary drug delivery. Hydrophobically modified chitosan were synthesized by conjugation of stearic acid to the hydrophilic depolymerized chitosan. The critical micelle concentration of stearic acid grafted chitosan was found to be 1.58×10^{-2} mg/ml. The nebulization efficiency of the nebulizer was up to 89% and the fine particle fraction varied from 38 to 47% (104).

Fate of the drug particles in the lung can be justified by the selection of permissible/clinically approved carrier for the pulmonary route of administration. Predominant factors affecting the passage of the lipid-based drug delivery system depend on the aerodynamic diameter and size distribution of the particles. Selection of lipid carrier depends on many factors such as viscosity, osmolality, surface tension, pH, and hygroscopicity. Nature-based lipid polymers are preferred for the pulmonary administration.

22.7 Future Perspective

To have an optimize formulation for pulmonary administration, proper strategy should be applied. Numerous factors should be considered while preparation drug delivery for pulmonary route. Appropriate in vivo models should be developed to help to generate in vitro–in vivo study correlation regarding efficiency of distribution of inhaled drug particles. Prolonged cytotoxic studies should be carried out for lipid-based carriers administered through pulmonary mode. Lipid based carriers are biodegradable and biocompatible, which provides an excellent advantage for the pulmonary route of administration. Various lipid-based nanoparticles have been successfully investigated for the purpose of pulmonary drug delivery. To generate safety and efficiency data consideration should be given for extensive evaluation of the developed lipid carrier drug delivery for pulmonary route of administration.

REFERENCES

1. Kim CS, Duncan B, Creran B, and Rotello VM. Triggered nanoparticles as therapeutics. *Nano Today*. 2013; 8(4): 439–447.
2. Li SD and Huang L. Pharmacokinetics and biodistribution of nanoparticles. *Molecular Pharmaceutics*. 2008; 5(4): 496–504.
3. Sanders M. Inhalation therapy: An historical review. *Primary Care Respiratory Journal*. 2007; 16 (2): 71–81.
4. Stein SW and Thiel CG. The history of therapeutic aerosols: A chronological review. *Journal of Aerosol Medicine and Pulmonary Drug Delivery*. 2017; 30 (1): 20–41. https://doi.org/10.1089/jamp.2016.1297.
5. Magnussen H. Inhalation therapy for bronchial asthma: Strategies and targets. *Current Opinion in Pulmonary Medicine*. 2003; 9: S3–S7.
6. Muttil P, Wang C, and Hickey AJ. Inhaled drug delivery for tuberculosis therapy. *Pharmaceutical Research*. 2009; 26 (11): 2401–2416.
7. Price O, Sarkar C, and Konda S. An update on the use of inhaled therapy in COPD. *Clinical Medicine (London)*. 2018; 18(5): 387–390.

8. Knudsen L and Ochs M. The micromechanics of lung alveoli: Structure and function of surfactant and tissue components. *Histochemistry and Cell Biology*. 2018; 150(6): 661–676. doi:10.1007/s00418-018-1747-9
9. Al-Kassas R, Bansal M, and Shaw J. Nanosizing techniques for improving bioavailability of drugs. *Journal of Controlled Release*. 2017; 260: 202–212.
10. Laouini A, Andrieu V, Vecellio L, Fessi H, and Charcosset C. Characterization of different vitamin E carriers intended for pulmonary drug delivery. *International Journal of Pharmaceutics*. 2014; 471(1–2): 385–390.
11. Mathur P, Sharma S, Rawal S, Patel B, and Patel MM. Fabrication, optimization, and in vitro evaluation of docetaxel-loaded nanostructured lipid carriers for improved anticancer activity. *Journal of Liposome Research*. 2020; 30(2): 182–196.
12. Moreno-Sastre M, Pastor M, Esquisabel A, Sans E, Viñas M, Fleischer A, Palomino E, Bachiller D, and Pedraz JL. Pulmonary delivery of tobramycin-loaded nanostructured lipid carriers for *Pseudomonas aeruginosa* infections associated with cystic fibrosis. *International Journal of Pharmaceutics*. 2016; 498(1–2): 263–273.
13. Patel AR, Chougule MB, Ian T, Patlolla R, Wang G, and Singh M. Efficacy of aerosolized celecoxib encapsulated nanostructured lipid carrier in non-small cell lung cancer in combination with docetaxel. *Pharmaceutical Research*. 2013; 30(5): 1435–1446.
14. Taymouri S, Alem M, Varshosaz J, Rostami M, Akbari V, and Firoozpour L. Biotin decorated sunitinib loaded nanostructured lipid carriers for tumor targeted chemotherapy of lung cancer. *Journal of Drug Delivery Science and Technology*. 2019; 50: 237–247.
15. Atis S, Tutluoglu B, Levent E, Ozturk C, Tunaci A, Sahin K, Saral A, Oktay I, Kanik A, and Nemery B. The respiratory effects of occupational polypropylene flock exposure. *European Respiratory Journal*. 2005; 25: 110–117; DOI: 10.1183/09031936.04.00138403.
16. Chen X, Zhong W, Zhou X, Jin B, and Sun B. CFD–DEM simulation of particle transport and deposition in pulmonary airway. *Powder Technology*. 2012; 228: 309–318.
17. Darquenne C. Aerosol deposition in health and disease. *Journal of Aerosol Medicine and Pulmonary Drug Delivery*. 2012; 25(3): 140–147.
18. Ibrahim M, Verma R, and Garcia-Contreras L. Inhalation drug delivery devices: Technology update. *Medical Devices (Auckl)*. 2015; 8: 131–139.
19. Thomas RJ. Particle size and pathogenicity in the respiratory tract. *Virulence*. 2013; 4(8): 847–858.
20. Augusto LLX, Goncalves JAS, and Lopes GC. CFD evaluation of the influence of physical mechanisms, particle size, and breathing condition on the deposition of particulates in a triple bifurcation airway. *Water, Air, Soil and Pollution*. 2016; 227: 56. https://doi.org/10.1007/s11270-016-2753-y.
21. Hussain M, Madl P, and Khan A. Lung deposition predictions of airborne particles and the emergence of contemporary diseases. Part-I. *The Health*. 2011; 2(2): 51–59.
22. Tsuda A, Henry FS, and Butler JP. Particle transport and deposition: Basic physics of particle kinetics. *Comprehensive Physiology*. 2013; 3(4): 437–471.
23. Darquenne C. Deposition mechanisms. *Journal of Aerosol Medicine and Pulmonary Drug Delivery*. 2020; 33 (4): 1–5.
24. Heyder J. Deposition of inhaled particles in the human respiratory tract and consequences for regional targeting in respiratory drug delivery. *Proceedings of the American Thoracic Society*. 2004; 1: 315–320.
25. Kwak PCL and Chan HK. Electrostatics of pharmaceutical inhalation aerosols. *Journal of Pharmacy and Pharmacology*. 2009; 61: 1587–1599.
26. Xi J, Si X, and Longest W. Electrostatic charge effects on pharmaceutical aerosol deposition in human nasal-laryngeal airways. *Pharmaceutics*. 2014; 6(1): 26–35.
27. Majid H, Madl P, Hofmann W, and Alam K. Implementation of charged particles deposition in stochastic lung model and calculation of enhanced deposition. *Aerosol Science and Technology*. 2012; 46 (5): 547–554.
28. Peterson JB, Prisk GK, and Darquenne C. Aerosol deposition in the human lung periphery is increased by reduced-density gas breathing. *Journal of Aerosol Medicine and Pulmonary Drug Delivery*. 2008; 21(2): 159–168.
29. Darquenne C and Prisk GK. Aerosol deposition in the human respiratory tract breathing air and 80:20 heliox. *Journal of Aerosol Medicine*. 2004; 17(3): 278–285.
30. Newman P. Drug delivery to the lungs: Challenges and opportunities. *Therapeutic Delivery*. 2017; 8(8): 647–661.
31. Pacławski A, Szlęk J, Lau R, Jachowicz R, and Mendyk A. Empirical modeling of the fine particle fraction for carrier-based pulmonary delivery formulations. *International Journal of Nanomedicine*. 2015; 10: 801–810.
32. Labiris NR and Dolovich MB. Pulmonary drug delivery. Part I: Physiological factors affecting therapeutic effectiveness of aerosolized medications. *Brazilian Journal of Clinical Pharmacology*. 2003; 56: 588–599.
33. Asgharian B, Hofmann W, and Miller FJ. Mucociliary clearance of insoluble particles from the tracheobronchial airways of the human lung. *Journal of Aerosol Science*. 2001; 32(6): 817–832.
34. Bennett G, Joyce M, Sweeney L, and MacLoughlin R. In vitro determination of the main effects in the design of high-flow nasal therapy systems with respect to aerosol performance. *Pulmonary Therapy*. 2018; 4: 73–86.
35. Carvalho TC and McConville JT. The function and performance of aqueous aerosol devices for inhalation therapy. *Journal of Pharmacy and Pharmacology*. 2016; 68: 556–578.
36. Piacentini E, Drioli E, and Giorno L. Membrane emulsification technology: Twenty-five years of inventions and research through patent survey. *Journal of Membrane Science*. 2014; 468: 410–422.
37. Butler Ellis MC, Tuck CR, and Miller PCH. How surface tension of surfactant solutions influences the characteristics of sprays produced by hydraulic nozzles for pesticide application. *Colloids Surface A Physicochemistry Engineering Aspects*. 2001; 180: 267–276.
38. Davanlou A, Lee JD, Basu S, and Kumar R. Effect of viscosity and surface tension on breakup and coalescence of bicomponent sprays. *Chemical Engineering Science*. 2015; 131: 243–255.
39. Ishak RAH and Osman R. Lecithin/TPGS-based spray-dried self-microemulsifying drug delivery systems: In vitro

pulmonary deposition and cytotoxicity. *International Journal of Pharmaceutics*. 2015; 485(1–2): 249–260.
40. Kaur P, Garg T, Rath G, Murthy RSR, and Goyal AK. Development, optimization and evaluation of surfactant-based pulmonary nanolipid carrier system of paclitaxel for the management of drug resistance lung cancer using Box-Behnken design. *Drug Development and Industrial Pharmacy*. 2016; 23: 1912–1925.
41. Council of Europe. European Pharmacopeia 8.0. Strabourgs: Council of Europe: European Directorate for the Quality of Medicines and Healthcare; 2014. pp. 363–365.
42. Pritchard JN, Hatley RH, Denyer J, and Hollen DV. Mesh nebulizers have become the first choice for new nebulized pharmaceutical drug developments. *Therapeutic Delivery*. 2018; 9(2): 121–136.
43. Sweeney L, McCloskey AP, Higgins G, Ramsey JM, Cryan SA, and MacLoughlin R. Effective nebulization of interferon-γ using a novel vibrating mesh. *Respiratory Research*. 2019; 20: 66. https://doi.org/10.1186/s12931-019-1030-1
44. de Boer AH, Wissink J, Hagedoorn P, Heskamp I, de Kruijf W, Bunder R, Zanen P, Munnik P, van Rijn C, and Frijlink HW. In vitro performance testing of the novel medspray® wet aerosol inhaler based on the principle of rayleigh break-up. *Pharmaceutical Research*. 2008; 25(5): 1186–1192.
45. Al-Jahdali H, Ahmed A, Al-Harbi A, Khan M, Baharoon S, Bin Salih S, Halwani R, and Al-Muhsen S. Improper inhaler technique is associated with poor asthma control and frequent emergency department visits. *Allergy, Asthma and Clinical Immunology*. 2013; 9(1): 8. doi: 10.1186/1710-1492-9-8.
46. Savla R, Browne J, Plassat V, Wasan KM, and Wasan EK. Review and analysis of FDA approved drugs using lipid-based formulations. *Drug Development and Industrial Pharmacy*. 2017; 43 (11): 1743–1758.
47. Guo R, Guo X, Hu X, Abbasi AM, Zhou L, Li T, Fu X, and Liu RH. Fabrication and optimization of self-microemulsions to improve the oral bioavailability of total flavones of *Hippophaë rhamnoides* L. *Journal of Food Science*. 2017; 82 (12): 2901–2909.
48. Li YJ, Hu XB, Lu XL, Liao DH, Tang TT, Wu JY, and Xiang DX. Nanoemulsion-based delivery system for enhanced oral bioavailability and Caco-2 cell monolayers permeability of berberine hydrochloride. *Drug Delivery*. 2017; 24 (1): 1868–1873.
49. Yang Y, Chen L, Sun HW, Guo H, Song Z, You Y, Yang LY, Tong YN, Gao JN, Zeng H, Yang WC, and Zou QM. Epitope-loaded nanoemulsion delivery system with ability of extending antigen release elicits potent Th1 response for intranasal vaccine against *Helicobacter pylori*. *Journal of Nanobiotechnology*. 2019; 17(1): 6. doi: 10.1186/s12951-019-0441-y.
50. Makidon PE, Nigavekar SS, Bielinska AU, Mank N, Shetty AM, Suman J, Knowlton J, Myc A, Rook T, and Baker Jr JR. Characterization of stability and nasal delivery systems for immunization with nanoemulsion based vaccines. *Journal of Aerosol Medicine and Pulmonary Drug Delivery*. 2010; 23(2): 77–89.
51. Janjic JM and Ahrens ET. Fluorine-containing nanoemulsions for MRI cell tracking. *Wiley Interdisciplinary Reviews: Nanomedicine and Nanobiotechnology*. 2009; 1(5): 492–501.
52. Kadakia RT, Xie D, Guo H, Bouley B, Yu M, and Que EL. Responsive fluorinated nanoemulsions for 19F magnetic resonance detection of cellular hypoxia. *Dalton Transactions*. 2020; https://doi.org/10.1039/D0DT01182G
53. Kumar DHL and Sarkar P. Encapsulation of bioactive compounds using nanoemulsions. *Environmental Chemistry Letters*. 2018; 16: 59–70.
54. Espitia PP, Fuenmayor CA, and Otoni CG. Nanoemulsions: Synthesis, characterization, and application in bio-based active food packaging. *Comprehensive Reviews in Food Science and Food Safety*. 2019; 18(1): 264-285.
55. Wooster TJ, Moore SC, Chen W, Andrews H, Addepalli R, Seymour RB, and Osborne SA. Biological fate of food nanoemulsions and the nutrients they carry – internalisation, transport and cytotoxicity of edible nanoemulsions in Caco-2 intestinal cells. *RSC Advances*. 2017; 7: 40053–40066.
56. Zhou XX, Jin L, Qi RQ, and Ma T. pH-responsive polymeric micelles self-assembled from amphiphilic copolymer modified with lipid used as doxorubicin delivery carriers. *Royal Society of Open Science*. 2018; 5(3): 171654. http://dx.doi.org/10.1098/rsos.171654.
57. Hayama A, Yamamoto T, Yokoyama M, Kawano K, Hattori Y, and Maitani Y. Polymeric micelles modified by folate-PEG-lipid for targeted drug delivery to cancer cells in vitro. *Journal of Nanoscience and Nanotechnology*. 2008; 8(6): 3085–3090.
58. Marwah M, Perrie Y, Badhan RKS and Lowry D. Intracellular uptake of EGCG-loaded deformable controlled release liposomes for skin cancer. *Journal of Liposome Research*. 2020; 30(2): 136–149.
59. Sakpakdeejaroen I, Somani S, Laskar P, Mullin M, and Dufes C. Transferrin-bearing liposomes entrapping plumbagin for targeted cancer therapy. *Journal of Interdisciplinary Medicine*. 2019; 4(2): 54–71.
60. Shirodkar RK, Kumar L, Mutalik S, and Lewis S. Solid lipid nanoparticles and nanostructured lipid carriers: Emerging lipid based drug delivery systems. *Pharmaceutical Chemistry Journal*. 2020; 53(5): 440–453.
61. Moazeni M, Borji H, Darbandi MS, and Saharkhiz MJ. In vitro and in vivo antihydatid activity of a nano emulsion of Zataria multiflora essential oil. *Research in Veterinary Science*. 2017; 114: 308–312.
62. Najafi-Taher R, Ghaemi B, and Amani A. Delivery of adapalene using a novel topical gel based on tea tree oil nanoemulsion: Permeation, antibacterial and safety assessments. *European Journal of Pharmaceutical Sciences*. 2018; 120: 142–151.
63. Cuomo F, Perugini L, Marconi E, Messia MC, and Lopez F. Enhanced curcumin bioavailability through nonionic surfactant/caseinate mixed nanoemulsions. *Journal of Food Science*. 2019; 84(9): 2584–2591.
64. Araujo GMF, Barros ARA, Oshiro-Junior JA, Soares LF, da Rocha LG, de Lima AAN, da Silva JA, Converti A, and de Lima Damasceno BPG. Nanoemulsions loaded with amphotericin B: Development, characterization and leishmanicidal activity. *Current Pharmaceutical Design*. 2019; 25(14): 1616–1622.

65. Arbain NH, Salim N, Masoumi HRF, Wong TW, Basri M, and Rahman MBA. In vitro evaluation of the inhalable quercetin loaded nanoemulsion for pulmonary delivery. *Drug Delivery and Translational Research*. 2019; 9: 497–507.
66. Nagraj B, Tirumalesh C, Dinesh S, and Narendar D. Zotepine loaded lipid nanoparticles for oral delivery: Development, characterization, and in vivo pharmacokinetic studies. *Future Journal of Pharmaceutical Sciences*. 2020; 37. https://doi.org/10.1186/s43094-020-00051-z
67. Yadav M, Schiavone N, Guzman-Aranguez A, Giansanti F, Papucci L, Perez de Lara MJ, Singh M, and Kaur IP. Atorvastatin-loaded solid lipid nanoparticles as eye drops: Proposed treatment option for age-related macular degeneration (AMD). *Drug Delivery and Translational Research*. 2020; 10: 919–944.
68. Rana I, Khan N, Ansari MM, Shah FA, ud Din F, Sarwar S, Imran M, Qureshi OS, Choi HI, Lee CH, Kim JK, and Zeb A. Solid lipid nanoparticles-mediated enhanced antidepressant activity of duloxetine in lipopolysaccharide-induced depressive model. *Colloids and Surfaces B: Biointerfaces*. 2020; 194: 111209.
69. Pooja D, Kulhari H, Kuncha M, Rachamalla SS, Adams DJ, Bansal V, and Sistla R. mproving efficacy, oral bioavailability, and delivery of paclitaxel using protein-grafted solid lipid nanoparticles. *Molecular Pharmaceutics*. 2016; 13 (11): 3903–3912.
70. Kuznetsova DA, Gaynanova GA, Vasileva LA, Sibgatullina GV, Samigullin DV, Sapunova AS, Voloshina AD, Galkina IV, Petrov KA, and Zakharova LY. Mitochondria-targeted cationic liposomes modified with alkyltriphenylphosphonium bromides loaded with hydrophilic drugs: Preparation, cytotoxicity and colocalization assay. *Journal of Materials Chemistry B*. 2019; 7: 7351–7362.
71. Awad NS, Paul V, Mahmoud MS, Sawaftah NMA, Kawak PS, Al Sayah MH, and Husseini GA. Effect of pegylation and targeting moieties on the ultrasound-mediated drug release from liposomes. *ACS Biomaterials Science and Engineering*. 2020; 6 (1): 48–57.
72. Sun F, Jaspers TCC, van Hasselt PM, Hennink WE, and van Nostrum CF. A mixed micelle formulation for oral delivery of vitamin K. *Pharmaceutical Research*. 2016; 33: 2168–2179.
73. Saqr A, Vakili MR, Huang YH, Lai R, and Lavasanifar A. Development of traceable rituximab-modified PEO-polyester micelles by postinsertion of PEG-phospholipids for targeting of B-cell lymphoma. *ACS Omega*. 2019; 4: 18867–18879.
74. Dong Z, Iqbal S, and Zhao Z. Preparation of ergosterol-loaded nanostructured lipid carriers for enhancing oral bioavailability and antidiabetic nephropathy effects. *AAPS PharmSciTech*. 2020; 21: 64. DOI: 10.1208/s12249-019-1597-3.
75. Raj SB, Chandrasekhar KB, and Reddy KB. Formulation, in vitro and in vivo pharmacokinetic evaluation of simvastatin nanostructured lipid carrier loaded transdermal drug delivery system. *Future Journal of Pharmaceutical Sciences*. 2019; 5: 9. https://doi.org/10.1186/s43094-019-0008-7
76. Aswathanarayan J and Vittal R. Nanoemulsions and their potential applications in food industry. *Frontiers in Sustainable Food Systems*. 2019; 3: 95. doi: 10.3389/fsufs.2019.00095.
77. Sharma A, Garg T, Goyal AK, and Rath G. Role of microemuslsions in advanced drug delivery. *Artificial Cells, Nanomedicine, and Biotechnology*. 2016; 44(4): 1177–1185.
78. Cappellani MR, Perinelli DR, Pescosolido L, Schoubben A, Cespi M, Cossi R, and Blasi P. Injectable nanoemulsions prepared by high pressure homogenization: Processing, sterilization, and size evolution. *Applied Nanoscience*. 2018; 8: 1483–1491.
79. Kamali H, Abbasi S, Amini MA, and Amani A. Investigation of factors affecting aerodynamic performance of nebulized nanoemulsion. *Iranian Journal of Pharmaceutical Research*. 2016; 15(4): 687–693.
80. Nesamony J, Shah IS, Kalra A, and Jung R. Nebulized oil-in-water nanoemulsion mists for pulmonary delivery: Development, physico-chemical characterization and in vitro evaluation. *Drug Development and Industrial Pharmacy*. 2014; 40(9): 1253–1263.
81. Al ayoub Y, Gopalan RC, Najafzadeh M, Mohammad MA, Anderson D, Paradkar A, and Assi KH. Development and evaluation of nanoemulsion and microsuspension formulations of curcuminoids for lung delivery with a novel approach to understanding the aerosol performance of nanoparticles. *International Journal of Pharmaceutics*. 2019; 557: 254–263.
82. Arbain NH, Salim N, Masoumi HRF, Wong TW, Basri M, and Abdul Rahman MB. In vitro evaluation of the inhalable quercetin loaded nanoemulsion for pulmonary delivery. *Drug Delivery and Translational Research*. 2019; 9(2): 497–507. doi: 10.1007/s13346-018-0509-5.
83. Shah K, Chan LW, and Wong TW. Critical physicochemical and biological attributes of nanoemulsions for pulmonary delivery of rifampicin by nebulization technique in tuberculosis treatment. *Drug Delivery*. 2017; 24(1): 1631–1647.
84. Sommerville ML and Hicky AJ. Aerosol generation by metered-dose inhalers containing dimethyl ether/propane inverse microemulsions. *AAPS PharmSciTech*. 2003; 4: 455–461.
85. Paliwal R, Paliwal SR, Kenwat R, Kurmi BD, and Sahu MK. Solid lipid nanoparticles: A review on recent perspectives and patents. *Expert Opinion on Therapeutic Patents*. 2020; 30(3): 179–194.
86. El-Laithy HM. Preparation and physiochemical characterization of dioctyl sodium sulfosuccinate (aerosol OT) microemulsion for oral drug delivery. AAPsPharmSciTech 2003; 4 (1) Article 11. https://doi.org/10.1208/pt040111.
87. Duan Y, Dhar A, Patel C, Khimani M, Neogi S, Sharma P, Kumar ND, and Vekariya RL. A brief review on solid lipid nanoparticles: Part and parcel of contemporary drug delivery systems. *RSC Advances*. 2020; 10(45): 26777–26791.
88. Weber S, Zimmer A, and Pardeike J. Solid lipid nanoparticles (sln) and nanostructured lipid carriers (NLC) for pulmonary application: A review of the state of the art. *European Journal of Pharmaceutics and Biopharmaceutics*. 2014; 86(1): 7–22.

89. Esmaeili M, Aghajani M, Abbasalipourkabir R, and Amani A. Budesonide-loaded solid lipid nanoparticles for pulmonary delivery: Preparation, optimization, and aerodynamic behavior. *Artificial Cells, Nanomedicine and Biotechnology*. 2016; 44(8): 1964–1971.
90. Espitia PJP, Fuenmayor CA, and Otoni CG. Nanoemulsions: Synthesis, characterization, and application in bio-based active food packaging. *Comprehensive Reviews in Food Science and Food Safety*. 2018; 18 (1): 264–285.
91. Gaspar DP, Gaspar MM, Eleuterio CV, Grenha A, Blanco M, Goncalves LMD, Taboada P, Almeida AJ, and Remunan-Lopez C. Microencapsulated solid lipid nanoparticles as a hybrid platform for pulmonary antibiotic delivery. *Molecular Pharmaceutics*. 2017; 14(9): 2977–2990.
92. Nemati E, Mokhtarzadeh A, Panahi-Azar V, Mohammadi A, Hamishehkar H, Mesgari-Abbasi M, Ezzati Nazhad Dolatabadi J, and de la Guardia M. Ethambutol-loaded solid lipid nanoparticles as dry powder inhalable formulation for tuberculosis therapy. *AAPS PharmSciTech*. 2019; 20(3): 120. doi: 10.1208/s12249-019-1334-y.
93. Maretti E, Rustichelli C, Lassinantti Gualtieri M, Costantino L, Siligardi C, Miselli P, Buttini F, Montecchi M, Leo E, Truzzi E, and Lannuccelli V. The impact of lipid corona on rifampicin intramacrophagic transport using inhaled solid lipid nanoparticles surface-decorated with a mannosylated surfactant. *Pharmaceutics*. 2019; 11(10): 508. doi: 10.3390/pharmaceutics11100508
94. Akbarzadeh A, Rezaei-Sadabady R, Davaran S, Joo SW, Zarghami N, Hanifehpour Y, Samiei M, Kouhi M, and Nejati-Koshki K. Liposome: classification, preparation, and applications. *Nanoscale Research Letters*. 2013; 8(1): 102. doi: 10.1186/1556-276X-8-102.
95. Daraee H, Etemadi A, Kouhi M, Alimirzalu S, and Akbarzadeh A. Application of liposomes in medicine and drug delivery. *Artificial Cells, Nanomedicine and Biotechnology*. 2016; 44(1): 381–391.
96. Rudokas M, Najlah M, Alhnan MA, and Elhissi A. Liposome delivery systems for inhalation: A critical review highlighting formulation issues and anticancer applications. *Medical Principles and Practice*. 2016; 25(2): 60–72.
97. Patil JS, Devi VK, Devi K, and Sarasija S. A novel approach for lung delivery of rifampicin-loaded liposomes in dry powder form for the treatment of tuberculosis. *Lung India*. 2015; 32(4): 331–338.
98. Honmane S, Hajare A, More H, Osmani RAM, and Salunkhe S. Lung delivery of nanoliposomal salbutamol sulfate dry powder inhalation for facilitated asthma therapy. *Journal of Liposome Research*. 2019; 29(4): 332–342.
99. Chennakesavulu S, Mishra A, Sudheer A, Sowmya C, Reddy CS, and Bhargav E. Pulmonary delivery of liposomal dry powder inhaler formulation for effective treatment of idiopathic pulmonary fibrosis. *Asian Journal of Pharmaceutical Sciences*. 2018; 13(1): 91–100.
100. Khatib I, Tang P, Ruan J, Cipolla D, Dayton F, Blanchard JD, and Chan HK. Formation of ciprofloxacin nanocrystals within liposomes by spray drying for controlled release via inhalation. *International Journal of Pharmaceutics*. 2020; 578: 119045.
101. Torchilin VP. Lipid-core micelles for targeted drug delivery. *Current Drug Delivery*. 2005; 2(4): 319–327.
102. Kim G, Piao C, Oh J, and Lee M. Self-assembled polymeric micelles for combined delivery of anti-inflammatory gene and drug to the lungs by inhalation. *Nanoscale*. 2018; 10: 8503–8514.
103. Rezazadeh M, Davatsaz Z, Emami J, Hasanzadeh F, and Jahanian-Najafabadi A. Preparation and characterization of spray-dried inhalable powders containing polymeric micelles for pulmonary delivery of paclitaxel in lung cancer. *Journal of Pharmacy and Pharmaceutical Sciences*. 2018; 21(1s), 200s–214s.
104. Moazeni E, Gilani K, Najafabadi AR, Rouini MR, Mohajel N, Amini M, and Barghi MA. Preparation and evaluation of inhalable itraconazole chitosan based polymeric micelles. *DARU Journal of Pharmaceutical Sciences*. 2012; 20: 85. DOI: 10.1186/2008-2231-20-85.

23

Respirable Controlled Release Polymeric Colloidal Nanoparticles

Nazrul Islam[1,2] and Abdur Rashid[3]
[1]Queensland University of Technology, Pharmacy Discipline, School of Clinical Sciences, Faculty of Health, Brisbane, Queensland, Australia
[2]Tier 2 Research Centre, Centre for Immunology and Infection, Queensland University of Technology, Brisbane, Queensland, Australia
[3]Department of Pharmaceutics, College of Pharmacy, King Khalid University, Abha, Saudi Arabia

23.1 Introduction

The nano-particulate colloidal aerosols by inhalation are used to improve the health of human beings for many years. A wide variety of medicinal colloids has been administered to the lungs by oral inhalation for the treatment of diverse disease states especially for the management of asthma and obstructive airway diseases. Direct delivery of drugs into the pulmonary regions of the lung enables lower doses with an equivalent therapeutic action compared to oral or parenteral routes because of the large surface area (~100 m^2) of the lungs. Biodegradable polymeric nanoparticles are promising for direct delivery of drugs to the lungs for sustained/ controlled drug release and produce the desired therapeutic actions against various diseases. This delivery technology is beneficial as it overcomes the limitations of the conventional drug formulations. Delivery of drugs into the lungs ensures low systemic side-effects compared with oral or intravenous administration and no first-pass metabolism; which is a common problem for orally administered drugs. Direct delivery of drugs into the lungs also enables lower doses with an equivalent therapeutic action compared to oral or parenteral routes (1). The nebulized vancomycin resulted in higher lung tissue concentration than that of the I.V. route in a ventilated piglets (2).

The controlled drug release to the deep lungs, although in infancy, is an emerging drug delivery research, which offers new opportunities to enrich the drug delivery research and develop the inhaled products for the management of both local and systemic diseases (3–9). The physicochemical properties (size, shape) of NPs ensures the particles ability to accumulate into the lung cells confirming the access to the target cells. Owing to the limited clearance of drugs or metabolites from the lungs, the delivery of drugs via controlled-release fashion is very challenging. The polymeric NPs that are used to design the controlled release of drugs for lung delivery is still not well established as the toxicity of the polymers and their degraded products are not known. The applications of polymeric nanocarriers for the delivery of various drugs into lungs with particular interest that exhibit multifunctional properties i.e. target specificity and preferential accumulation in the target cells with improved performance are essential. The applications of nanocarriers with multifunctional properties for the delivery of various drugs such as proteins and peptides, antibiotics and anti-cancer drugs are extensively reviewed (10). Although a large number of reports demonstrated the biodegradability and biocompatibility of polymers into lungs, however, no comprehensive research outcomes are available to ensure the applicability of polymers for controlled release lung drug delivery. This chapter details the recent advances in the development of drug-loaded respirable polymeric colloidal particles for controlled drug release for the treatment of various lung diseases.

23.2 Effect of Particle Size on Distribution, Diffusion and Clearance after Lung Delivery

The drug deposition, distribution and clearance of particles following inhalation are size-dependent. It is well established that the aerodynamic particle size for respiratory delivery should be <5 μm, for proper deposition in the airways and absorption. The aerodynamic diameter (D_{ae}) is defined as the diameter of a spherical particle of unit density having a settling velocity from an air stream to the particle in question (11).

$$D_{ae} = d\sqrt{\rho/\rho_0}$$

where d is the diameter of the sphere, ρ is the density of the sphere and ρ_0 is the unit density (i.e. 1g/cm^3). From the equation, it is clear that a particle having higher than unit density will show the actual diameter smaller than its D_{ae} and particle with smaller than unit density will show the diameter larger than its $D_{ae.}$ The size distributions of aerosol particles may be characterized as monodisperse (uniform sizes, the geometric standard deviation of <1.2) or polydisperse (non-uniform sizes, the geometric standard deviation of ≥ 1.2) (12).

DOI: 10.1201/9781003046547-23

Although the particle size and their deposition into the different regions of pulmonary airways has been explained in terms of breathing pattern, deposition of particles of 0.005–15 μm diameter in the respiratory tract has also been reported (13) and aerosol particles below 0.5 μm penetrated the lung deeply but showed a greater tendency to exhale without deposition because of their low mass and momentum (14).

In the design of appropriate carriers for the effective lung delivery of drugs is dependent on the optimal particle size. There are conflicting requirements for the optimal particle size for deep lung deposition (500 nm–5 μm), efficient mucus penetration and intercellular transportation (~200 nm), and surface chemistry to avoid clearance by alveolar macrophases (15,16). In a study, the particle size range between 1.0 nm – 1.0 μm, are reported to have the capability of escaping from the macrophage uptake (17).

The distribution of polymeric NPs across a thick mucus layer depends on the particle size of around 30–500 nm (18,19), and 100 nm sized particles diffused perfectly to the mucus layer (20). The mucociliary clearance of particles deposited into the different parts of the lungs is depended on the particle size, shape and surface charge; however, the particles trapped in the mucus layer is independent of the size (21–23).

23.3 Surface Properties and Diffusion of the Drug from Nanoparticles

The NPs surface properties are an important factor for the preferential penetration of the NPs into the target cells, release and distribution of the coated drugs into the lung cells. The particle surface is often modified into the charged and or hydrophilic or hydrophobic characteristics to improve the therapeutic benefits of the deposited drugs. Steps have been investigated such as PEGylation for avoiding macrophase uptake, cationic surface enhances the cellular uptake, or to change the hydrophobicity and hydrophilicity for drug dispersion and release form the NPs. D'Angelo et al. (24) modified the surface of cationic protein colistin loaded CS NPs with PVA, which promoted efficient entrapment of the NPs into the bacterial biofilm, diffusion through the mucus and extended the antibiofilm activity of the drug for the treatment of CF patients (24). The authors prepared the drug-loaded PLGA NPs and modified their surfaces by either CS or PVA shell to get the drug into the biofilm and sustained release the drug for 15 days. Here the positively charged surface due to CS coating and negatively charged surface owing to PVA helped get the drug accumulated in the mucus and release the drug with prolonged antibiofilm activity. In another study, Shukla et al. (25) demonstrated the sorafenib loaded inhalable PLGA NPs coated with a cationic polymer PEI, which enhanced preferential cellular accumulation of the drug in various cancer cells and continued drug release occurred for 72 h. Ungaro and co-workers (6) also found the preferential accumulation of tobramycin coated in PLGA NPs modified by positively charged CS in mucin and the cationic surface of the NPs was the dominant factor in facilitating the interaction with the mucin. The cationic surface of nicotine-loaded CS NPs demonstrated higher dispersion of NPs compared to the drug alone particles (26). In another study, L-leucine conjugated with CS showed better dispersion and drug release because of the surface hydrophobicity and hydrophilicity, respectively from the diltiazem-loaded CS NPs (9). The surface hydrophobicity of the NPs achieved by the hydrophobic domain of leucine of the conjugated CS NPs. The conjugated CS showed higher dispersibility of the drug compared to the non-conjugated CS NPs. The authors claimed that the higher dispersibility was attributed to the amphiphilic environment of the L-leucine conjugate and hydrophobic cross-links. It was concluded that the conjugated CS NPs could be useful as an alternative carrier for lung drug delivery with enhanced aerosolization and sustained drug release from nano-particulate DPI formulations. Thus the surface properties of polymer NPs have a significant role in affecting the drug target cell interaction and prolonged drug release from the NPs.

23.3.1 Polymer Nanoparticles in Controlled Release Pulmonary Drug Delivery System

A lot of biodegradable polymeric nanoparticles have been extensively investigated for lung delivery of various drugs (27,28). The inhaled biodegradable polymeric nanoparticles hold great promise to improve controlled and targeted drug delivery to the lungs. Various drug-loaded polymer nanoparticles for lung drug delivery are currently under investigation. Among the natural biodegradable polymers, chitosan, PLA, PLGA, PCL, alginate and some other polymers are widely considered as a nontoxic and biocompatible polymer for lung drug delivery (28–31). A comprehensive list of some promising polymer-based nanoparticulate lung drug delivery is presented in Table 23.1. The following subsections are dedicated to the lung delivery of various polymer NPs containing various therapeutic agents with promising outcomes.

23.3.2 PLGA/PLA for Lung Drug Delivery

The PLGA and PLA are most commonly investigated synthetic polymers used in drug delivery research owing to their biodegradation and biocompatible degraded products (glycolic acid and lactic acid), which are eliminated from the body via the citric acid cycle. Among the polymers, the PLA/PLGA particles presented high structural integrity with enhanced stability, higher drug loading, and prolonged drug release (51). Tobramycin encapsulated in PLGA and di-block of PEG-PLGA NPs produced promising outcome against *Pseudomonas aeruginosa*. The NPs (225–231 nm) showed effectiveness against the biofilm barriers of the organism (52). The polymer-based drug showed anti-bacterial effects at a very low concentration (0.77 mg/L); whereas, the free drug showed similar effects at a very high concentration (1,000mg/L). This is a promising outcome of tobramycin loaded PLGA/PEG NPs against lung infection.

Guendy et al. (41) prepared ciprofloxacin-loaded PLGA NPs (190 nm) against *Pseudomonas aeruginosa* infections in cystic fibrosis (CF). The controlled release of ciprofloxacin was achieved within 8 h without burst effect followed by a slow

TABLE 23.1
Various Drug Encapsulated Polymer NPs Investigated for Lung Delivery

Polymer	Drugs	Formulation/ Delivery	Main Outcomes	Reference
CS	Insulin	Microencapsulated insulin loaded CS NP; rat model	Improved absorption for systemic action	(32,33)
GCS + PLGA	Palmitic acid modified Extendin 4	Nebulization of NPs in rats	Sustained drug release; higher efficacy than that of free drug	(34)
CS + liposome	N-acetylcysteine	N-acetylcysteine -loaded-liposome NPs coated with CS	Prolonged drug release for 19 hours with preferential accumulation into lung cells.	(35)
CS + liposome	Elcatonin	CS-oligosaccharide and PVA with a hydrophobic surface modifier	Sustained drug release; the modified liposomes adhered to lung tissues and caused opening of tight junctions and enhanced absorption;	(36)
CS +PLGA	Calcitonin (hormone)	CS-modified PLGA NPs loaded with drug	Sustained drug release with the enhanced absorption owing to the opening of tight junctions	(37,38)
CS	Isoniazid	Drug-loaded CS NPs spray dried with lactose, mannitol and maltodextrin alone or with leucine.	CS based drug decreased activity against *M. avium;* leucine increased dispersibility (FPF 45%)	(39)
CS	Prothionamide	Drug-loaded CS NPs	Sustained release of the drug (97%) and increased lung drug residency for 24h	(40)
PLAG, PEG, Liposome	Levofloxacin	In vitro lung deposition from DPI, release and anti-microbial activities	Prolonged release over 72 hours; improved activity against P. aerugenosa	
PLGA	Ciprofloxacin	Drug-loaded PLGA NPs (190 nm)	Controlled release of the drug against P. aeruginosa in CF patients	(41)
CS + guar gum	Rifampicin + isoniazid	CS coated and guar gum coated CS particles, spray dried with mannitol and leucine	Improved activity against tuberculosis with reduced cytotoxicity	(42)
Phospholipid	Vancomycin and Clarithromycin	Intra-tracheal administration in rats for tuberculosis	The NPs of dry powder aerosols showed controlled release of antibiotics for 12 hours.	(43)
CS + PLGA+ Alginate	Tobramycin	Intra-tracheal administration in rats for tuberculosis	Alginate allowed efficient drug entrapment within NPs and its release for one month; modified surface improved aerosolization and deposition	(6)
CS + PLGA	Voriconazole	Lung delivery of drug loaded PLGA NPs coated with CS in mice	Sustained release of drug from coated NPs with improved bioavailability	(44)
CS + fucoidan	Gentamicin	Drug-loaded NPs against *K. pneumoniae;* intra-tracheal rat model	Sustained drug release profile of drug up to 99% in 72 hours; reduced toxicity	(45)
CS derivative + Mannose	Etofylline	Mannose-anchored N,N,N-trimethyl CS NPs	Sustained drug release for 12 hours, target specific	(46)
CS	Salbutamol sulfate	Spray-dried CS microspheres of drug and lactose based dry powder	Delayed drug release	(47)
CM-CS + β-cyclodextrin	Theophylline	Spray-dried microspheres for lung delivery	Biocompatible; potential carrier for lung drug delivery	(48)
CS + leucine	Terbutaline sulfate + Beclomethasone dipropionate	In vitro dispersibility and release studies of spray dried drug	Sustained release of drug with improved dispersibility for leucine	(49)
CS + PEVA	Carvedilol	Drug-loaded PEVA NPs coated with CS and spray dried with lactose/ mannitol	CS caused mucoadhesive and controlled drug release; higher FPF (58.8%)	(50)
CS	Nicotine; Diltiazem HCl	Drug-loaded CS NP	Promising controlled drug release for 144 hours; improved efficacy	(7–9)

Note: DX: Dextran; XG: Xanthan gum; GCS: glycol CS; CM CS: carboxymethyl CS; PEVA: poly(ethylene-co-vinyl acetate); TMCS: *N*-trimethyl chitosan.

release of the drug (up to 90.5%) for 14 days. The prepared PLGA NPs were nontoxic and enhanced anti-bacterial effect when applied as nanoparticles. The colloidal stability of NPs in mucus was promising. The authors emphasized that the ciprofloxacin-loaded PLGA NPs could be used against *P. aeruginosa* infections in CF patients. The cationic peptides, colistin loaded PLGA NPs (267 nm) showed promising sustained drug release activity against the lung infection caused by multi-drug resistant *Pseudomonas aeruginosa* commonly experienced by the patients suffering from CF (28). Additionally, the modified NPs with CS and PVA allowed modulating surface properties of the NPs to improve the transport of drug through the CF mucus with prolonged efficiency in eradicating the biofilm compared to that of free colistin. This outcome is enormous for the treatment of lung infection using the surface-modified colistin loaded PLGA NPs.

Very recently, the combined inhaled gefitinib and aminolevulinic acid (5-ALA) loaded PLGA NPs showed promising anti-cancer effects on A549 cells (53). The combined therapy produced remarkable synergistic effect against lung cancer. The authors demonstrated that the typical lung cancer markers such as CD31, VEGF, NF-κB p65 and Bcl-2, were significantly reduced upon intra-tracheal administration of a rat model. However, no controlled release properties of drugs from NPs have been demonstrated. This outcome has opened a new era of managing lung cancers by the PLGA coated combined drug therapy. In another study, sorafenib loaded inhalable PLGA NPs coated with a cationic polymer PEI for the treatment of NSCLC has been demonstrated (25). The authors also used PLA to impart a positive charge on the surface of NPs. The prepared cationically modified NPs (<200 nm) exhibited the enhanced cellular internalization and cytotoxicity (~5-fold IC50 reduction vs SF) in various lung cancer cells with 100% drug release in 72 h. Moreover, the localized delivery of drug-loaded NPs showed higher anti-tumor activity compared to that of sorafenib alone.

The colloidal nanoparticles of bosentan, an endothelin receptor antagonist used for the management of pulmonary arterial hypertension (PAH) has been found effective, safe and sustained drug release upon pulmonary delivery(54). The estimated mean plasma concentration (Cmax) after oral and intra-tracheal administration of drug-loaded PLGA NPs in a rat model was 105.1±18.12 ng/ml and, 1264.8±323.68 ng/ml, respectively (Figure 23.1), which proved that the drug encapsulated PLGA nanoparticles (420 nm) showed 12-fold higher bioavailability and sustained vasodilation effect for more than 12 hours compared to the oral administration. The authors concluded that the inhaled nanoparticles could be an advanced drug delivery technology for the management of PAH. Sildenafil citrate, a promising drug for the treatment of PAH, has been investigated for the controlled release profile upon lung delivery. The solid lipid nanoparticles (100–250 nm) loaded with Sildenafil citrate ensured high stability, encapsulation efficiency (88–100%) and sustained release of the drug for 12 hours. The prepared NPs could be potential for pulmonary delivery of this drug against PH with high safety, better performance and patient compliance (55).

Craparo et al. (56) demonstrated the sustained ibuprofen release and uptake capacity within bronchial epithelial cells in the presence of CF artificial mucus. They prepared the fluorescent derivatives of α,β-poly(N-2-hydroxyethyl)-D,L-aspartamide by chemical synthesis with rhodamine, polylactide, and PLGA, to achieve polyaspartamide-polylactide derivatives. The PEGylated

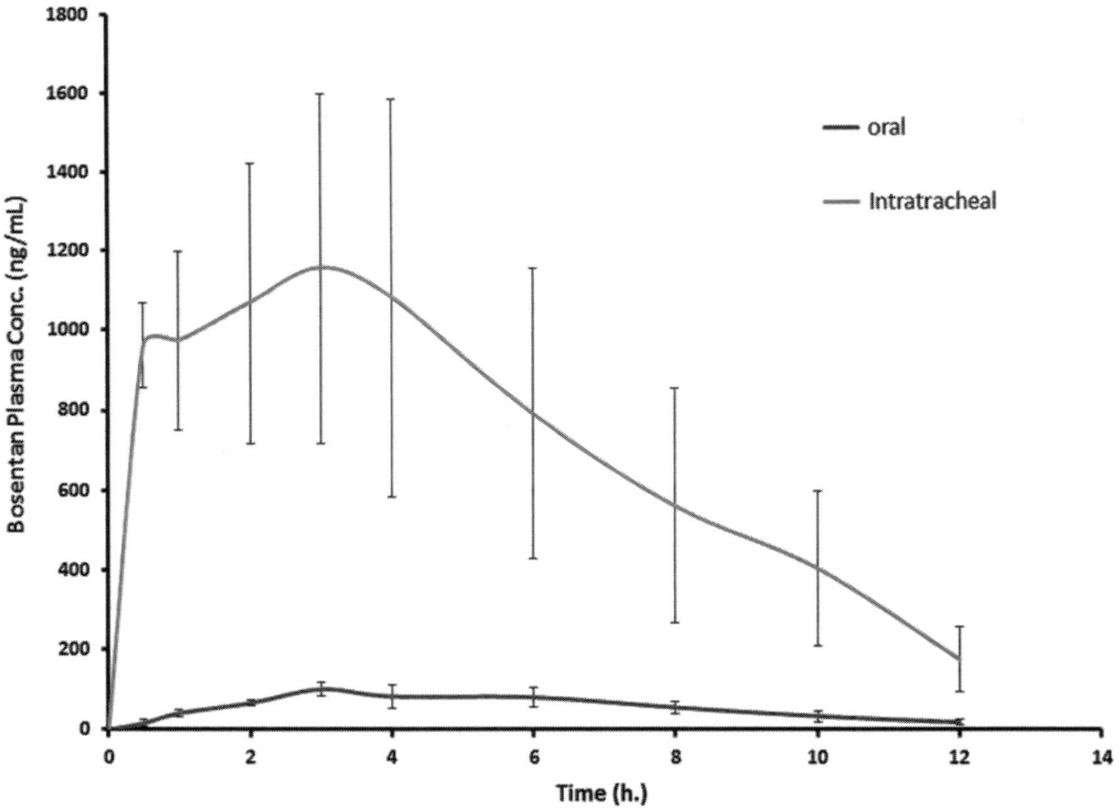

FIGURE 23.1 The Mean Bosentan Concentrations in Plasma of Rats After Intra-tracheal Delivery of the Optimized Bosentan Colloidal NPs Compared to the Oral Administration of Bosentan Suspension. Adapted from (54)

NPs (126–187 nm) containing ibuprofen showed better mucus-penetrating properties and rapidly diffused through the mucus barrier and sustained release Ibuprofen in the site of disease. This significant outcome could be an opportunity for the treatment of lung inflammation in patients with CF.

23.3.3 Chitosan Polymer

Chitosan (CS), obtained by deacetylation of naturally occurring chitin, is a linear polysaccharide composed of β-(1→4)-linked glucosamine (2-amino-2-deoxy-β-D-glucopyranose). It is insoluble in aqueous solutions; however, soluble in aqueous acidic media (pH<5). The cationic charge of CS provides mucoadhesive properties (57) that enable it to adhere to the mucosa of lung epithelial cells and contribute to the prolonged release of encapsulated drugs (8,34,42). The positively charged CS has been found to improve drug absorption by opening the intercellular tight junctions of the lung epithelium (36,37).

The controlled delivery of CS coated insulin (58) and terbutaline sulfate (59), nicotine (8), diltiazem hydrochloride (9) from DPI formulations have been investigated. CS-based DPI formulations of beclomethasone dipropionate (47), clindamycin (60) and salbutamol sulfate (60) showed promising controlled release profiles. In another study, the CS-based DPI formulation of terbutaline sulfate and beclomethasone dipropionate, produced promising sustained release delivery of both drugs (61). Pulmonary delivery of proteins (32,58) tobramycin (6) and anti-cancer drug methotrexate (62) loaded CS NPs from DPI formulations have produced controlled release profiles of those drugs. These studies demonstrated the potential of CS for lung delivery of various drugs.

Wang et al. investigated CS coated nicotine NPs for lung delivery with controlled-release profile (7,8). The dispersibility of the prepared NPs was promising with a fine particle doses (FPD) of NHT ranging between 1.7 and 3.2 mg from DPI formulations. The invitro nicotine release showed initial burst release in the first 8 hours and the release was continued for 140 hours (8). The nicotine release rate from the NPs was dependent on the concentration of the loaded nicotine in the Cs NPs as demonstrated in Figure 23.2. The dispersibility of the nicotine loaded CS NPs was increased due to the surface positive charge of the CS (26). Using male C57BL/6 naïve mice model, the locomotion activity was found to be dependent on the dose of NPs. A 75 mg nicotine loaded CS NPs produced similar locomotion effects as 50 mg of nicotine with controlled-release profile (7). The short term exposure (24 hours) of NPs to the lung cells revealed very minimal inflammatory cell infiltration around the alveolar ducts and adjacent vasculature at the dose of nicotine (50 mg) and nicotine NPs (75 mg) as presented in (Figure 23.3). This promising outcome suggested that the prepared nicotine loaded CS NPs is useful to develop DPI formulation for lung delivery, which is a potential therapeutic strategy for better management of nicotine dependence as a kind of nicotine replacement therapy with controlled release profile. Before this study, Muhsin et al. (9) developed the diltiazem loaded CS and l-leucine conjugated CS NPs for pulmonary delivery with controlled release of the drug for 800 hours.

Inhaled delivery of pirfenidone to the lungs of patients with idiopathic pulmonary fibrosis showed promising efficacy (64). Using a ship model, a 49 mg lung-deposited dose produced Cmax of 62 ± 23 mg/L, and plasma Cmax of 3.1 ± 1.7 mg/L. The authors claimed that the plasma pirfenidone concentration reached faster and at a higher concentration than in lymph. In a

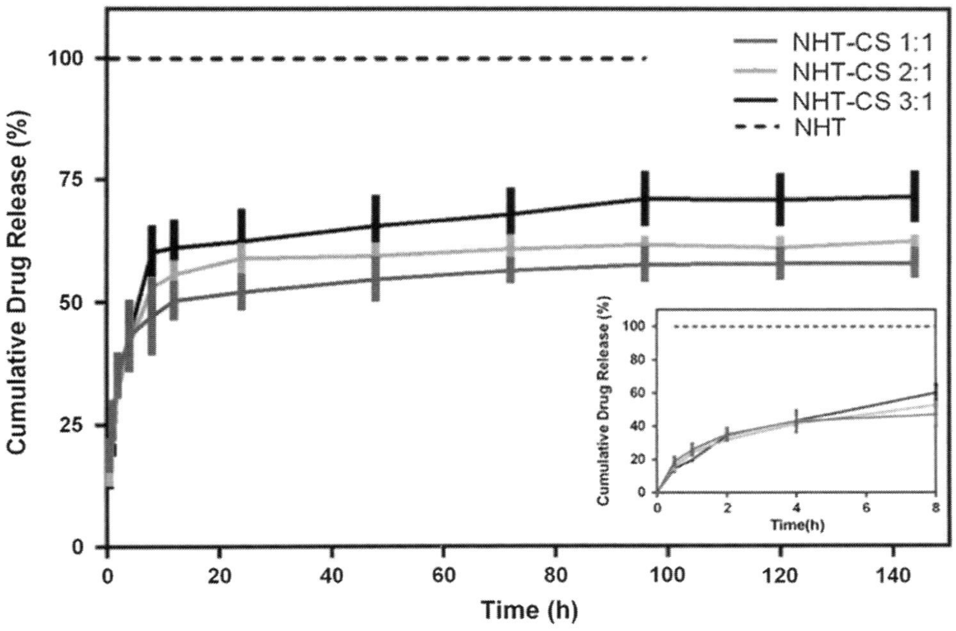

FIGURE 23.2 Cumulative NHT Release from CS Coated NPs Over a Period of 6 Days and Initial Release in the first 8 h Presented in Right Bottom; Data presented as mean ± S.D., n = 3). Adapted from Wang et al. (8)

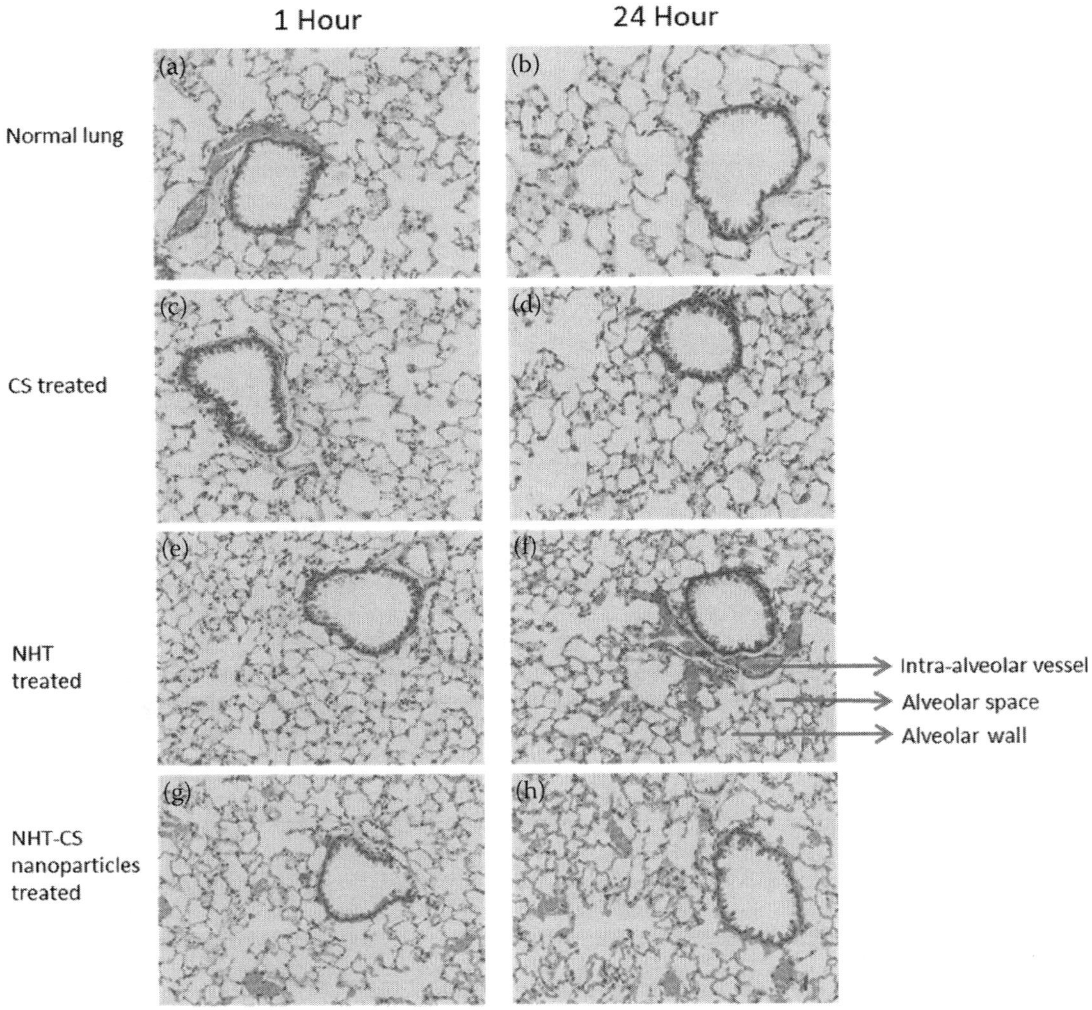

FIGURE 23.3 Histological Changes of Lung Tissues from Different Groups (10×40, n=6). (a-b) Healthy Mice Lungs; (c) CS Exposure Mouse in One Hour; (d) CS Exposure Mouse in 24 h; (e) Nicotine Exposure (50 mg) Mouse in 1 h; (f) Nicotine Exposure (50 mg) Mouse in 24 h; (g) Nicotine-CS Nanoparticles Exposure (75 mg) Mouse in 1 h; (h) Nicotine-CS Nanoparticles Exposure (75 mg) Mouse in 24 h. Adapted from (7)

previous study, the lung delivery of pirfenidone-loaded CS microspheres produced the sustained release of the drug with a biphasic release pattern (65). The relative bioavailability produced promising pulmonary absorption of this drug in a rat model upon pulmonary administration. The histopathological data showed changes to the rat lung after the administration of drug-loaded chitosan particles.

Both the polymers such as CS and PLGA have different physicochemical properties and have investigated for the lung delivery of various medications. Recently, pulmonary delivery of anti-fungal drug voriconazole showed sustained release of the drug-loaded -PLGA- coated with chitosan (44). The drug released from the chitosan-coated NPs was faster up to 24 hours followed by a slower rate up to 8 days. The bioavailability of the drug was higher form the coated NPs; however, the Cmax was greater for uncoated drug NPs in both the plasma and lung. This could be due to the slow drug release. The authors concluded that the chitosan coating on PLGA NPs loaded with voriconazole was promising for the improved bioavailability and effective pulmonary delivery for efficacious treatment of recurrent lung-fungal infection. In a study, the NPs of PLGA coated with CS produced higher biodistribution and low toxicity compared with those of non-coated PLGA NPs (66). Additionally, CS coated NPs induced no inflammatory responses as evidenced by the low levels of cytokine release after intra-tracheal administration. Furthermore, the surface positive charge of NPs due to CS influenced the transporting of NPs through the mucus layer especially for the CF patients. These findings support the application of the combined polymer NPs as pulmonary drug delivery carriers owing to their promising efficiency, biocompatibility, and low toxicity.

23.3.4 CS Nanoparticles and Anti-tubercular/Anti-bacterial Drugs

The DPI formulation of ethambutol dihydrochloride (EDH)-loaded CS NPs produced lower toxicity, higher potency and better permeation than the pure drug particles (67). The CS coated EDH showed a higher MIC (<1 μg/ml) compared with that of a pure drug (MIC 2 μg/ml) against *Mycobacterium bovis*. The CS coated NP showed higher (71%) permeability across lipid bilayer than that of the pure drug (48%). In another study, inhalable powders isoniazid-loaded CS NPs showed sustained delivery of the drug to the lung for 144 hours (39). The rate of drug release from NPs was dependent on the amount of CS used in the coating process and the drug released was decreased by increasing the CS concentration. Garg et al. (68) demonstrated the application of the combined isoniazid and rifampicin loaded CS NPs for controlled delivery for more than 24 hours.

Goyal et al. (42) investigated the release profile of rifampicin and isoniazid loaded spray dried nano-embedded microparticles against tuberculosis and found a biphasic (initial burst and sustained release) of isoniazid and rifampicin. The rifampicin and isoniazid loaded various formulations (chitosan, guar gum, mannose and guar gum coated chitosan) were prepared by spray drying technique. The in vitro test revealed that the preferential uptake of drugs encapsulated in guar gum occurred due to the selective uptake of mannose moiety to the specific cell surface of macrophages. In vivo lung delivery study showed that guar gum coated chitosan formulations showed the prolonged residence of the drug at the target site and produced a 5-fold reduction of bacteria compared to the pure drugs. This study suggested that the guar gum coated chitosan NPs could be a promising carrier for selective delivery of drugs for effective treatment of tuberculosis. A similar outcome was also demonstrated by Garg et al. who used nanoparticles (230 nm) of CS coated isoniazid and rifampicin (68). Ungaro and co-workers (6) investigated the lung delivery of tobramycin coated in PLGA NPs modified by positively charged CS, which showed higher interaction with mucin than those of unmodified NPs. Therefore, the positively charged surface of CS was the dominant factor in facilitating the interaction mechanism of NPs. Recently, lung delivery of CS-based gentamicin NPs showed anti-microbial and antioxidant activities (45). They used fucoidan with CS to produce gentamicin encapsulated NPs and found the sustained release of the drug for 72 hours. The authors suggested that the developed NPs are the potential carriers in pulmonary delivery of gentamicin for the management of pneumonia.

23.3.5 Polymer Nanoparticles and Anti-Asthmatic Drugs

The pulmonary delivery of various anti-asthmatic drugs is well established with clinical significance. The polymer-based formulations of drugs are yet to be approved for lung delivery; however, some currently available drugs such as salbutamol sulfate (47,60), theophylline (48), terbutaline sulfate and beclomethasone dipropionate (61) have been reported for sustained release delivery of drugs. Using CS-based salbutamol sulfate, the inhaled delivery of drug showed sustained release of the drug delivery (47). Kamble et al. used the same salbutamol sulfate-loaded CS microspheres for sustained release (94%) of the drug for 120 min (60). The inhaled delivery of polymer CS-based beclomethasone dipropionate (59) and terbutaline sulfate (61) for sustained release profile have also been reported. The time for 100% drug release was observed with an increasing molecular weight of CS. These investigators also studied the sustained release of the combination of hydrophilic (terbutaline sulfate) and hydrophobic (beclomethasone dipropionate) drugs from CS-based microparticles and a similar pattern of sustained drug released was demonstrated. The improved aerosolization efficiency of PLGA and soy lecithin NPs from the DPI formulation containing large porous particles of CS NPs with improved aerosolization was observed (69).

23.3.6 Anti-cancer Drugs

Lung cancer is the most causative factors of the death of a large number of populations around the world. The lung delivery of anti-cancer drugs have been found to be an effective route of drug delivery owing to the targeted delivery of therapeutics into the cancer cells; however, the presence of some barriers such as mucus, ciliated cells and tissue macrophages caused to produce a significant impact on the diffusion and absorption of the inhaled drugs. Some critical particle parameters such as particle size, morphology, surface characteristics and the targeting moieties are important to get the drug to the target sites to achieve the clinical outcomes. A detailed particle parameter for the targeted lung delivery of anti-cancer drugs have been reported (70). A comprehensive summary of the inhaled anti-cancer drugs with various polymers is presented in Table 23.2. The CS and CS coated with other polymers against various lung cancers with promising outcomes have been reported (30). The pulmonary delivery of sunitinib loaded poly (3-hydroxybutyrate-co-3-hydroxyvalerate acid) (PHBV) NPs (168 nm) showed sustained drug release efficiency of 56% for 24 hours (73). The FPF of the drug from DPI formulation was 61%. MTT assay confirmed the maximum efficacy of this drug was observed while it was loaded in the PHBV NPs. The authors concluded that the developed DPI formulation might be promising against lung cancer. The docetaxel-loaded CS particles produced to the maximum release of the drug into lungs with the sustained release profile (i.e. 70% drug released in 19 days) (82). The CS-coated PLGA NPs containing 2-methoxyestradiol (2-ME) showed a sustained release of drug for 60 hours (62). The authors demonstrated that the NPs significantly enhanced the cytotoxic effect of 2-ME without inflammation in rat lungs. This is a promising outcome of inhaled CS-coated 2-ME NPs for targeted and effective treatment of lung cancer. The CS coated cisplatin micro/NPs with the smallest size showed more activity than that of uncoated cisplatin on A549 human cells (71) with sustained release of the drugs. Using a rat model, the lung delivery of the CS-modified PLGA NPs containing paclitaxel showed significantly increased uptake and cytotoxicity against lung cancer (A549) (83). The modified CS derivatives have been investigated to pulmonary drug delivery (84). Additionally, the derivatives of CS and PEG showed the applicability of them for lung delivery of drugs with controlled drug release up to 4 days (84). Very recently, the etofylline encapsulated mannose-anchored N, N, N-tri-Me chitosan NPs showed targeted delivery, high penetrability, controlled release of the drug, and biocompatibility of the NPs (46). The prepared NPs were completely redispersible even after 6 months. Furthermore, the conjugation of CS with mannose NPs (223 nm) provided selective macrophage targeting via receptor-mediated endocytosis. The initial burst release was observed in 1 hour and

TABLE 23.2

Pulmonary Delivery of Various Anti-cancer Agents from DPI Formulations Studied for Lung Cancer

Polymer	Active Drugs	Formulations and Delivery	Findings	Reference
PEG	Cisplatin	Drug loaded PEG NPs delivered into mice trachea	Controlled release of cisplatin with promising dispersion (FPF 37–52%)	(4)
CS	Cisplatin	Drug loaded chitosan nanoparticles for cytotoxic activity	Sustained drug release, accumulated in lung cancer cells with promising activities on A549 and A2780 human cells	(71,72)
PHBV	Sunitinib	Drug loaded polymer nanoparticles	Sustained drug release for 24 hours	(73)
Dendrimer	Doxorubicin	Lung delivery of drug conjugated dendrimer particles	Enhanced accumulation of drug in lung cells and anti-tumor activity in lung metastasis compared to I.V. route	(74)
PPI, liposome and PEG	Doxorubicin	Drug loaded liposome, PPI dendrimer and PEG NPs delivered into mice lungs	Higher cancer cell death compared to I.V. route	(75)
PEG	Paclitaxel	Spray dried PEGylated drug NPs; DPPE-PEG with drug NPs delivered into mice lungs	Controlled release of paclitaxel with enhanced cytotoxicity on A459 cancer cells	(76,77)
Gelatine	Cisplatin	Drug loaded gelatine NPs	Sustained release; no animal studies done	(78)
Gelatine	Doxorubicin	Drug loaded gelatine mesospheres	Sustained release; gelatin-mannosylated-DOX-mesospheres showed higher cytotoxic compared to that of free DOX and gelatin-DOX-mesospheres.	(79)
Gelatine	Gemcitabine hydrochloride	Gemcitabine (Gem)-loaded gelatin nanocarriers (GNCs) cross-linked with genipin	Effective sustained-release with FPF of 75%	(80)
PLGA	Sorafenib	Drug loaded inhalable PLGA NPs coated with PLA and PEI	NPs showed enhanced penetration into the cancer cells, sustained drug release for 72 hours.	(25)
PLGA	Celecoxib	Drug encapsulated PLGA NPs	Prolonged drug release (30 h); promising FPF (53–70%); improved cytotoxicity against A549 cells	(81)
PLGA	Gefitinib and 5-aminolevulinic acid	Drug loaded PLGA NPs	High anti-lung cancer effects with the reduction of some typical lung cancer markers, such as CD31, VEGF, NF-κB p65 and Bcl-2; The combined therapy could be a promising strategy for treating lung cancers.	(53)

PHBV: poly (3-hydroxybutyrate-co-3-hydroxyvalerate acid.

the complete drug release was continued until 12 hours, which indicated the degradation of the NPs in the system. The mannosylated NPs exhibited 42-fold higher drug Cmax compared to that of the drug only in a rat model. The authors suggested that a mannosylation is a promising approach for the development of drug-loaded CS NPs for efficient therapy of airway diseases.

Gelatin, another widely used polymer has been demonstrated for the lung delivery of anti-cancer drug methotrexate (85). The drug-loaded gelatine NPs demonstrated a preferential accumulation of the drug in the lung cancer cells with a 4–5 fold reduction in drug IC50. The prepared NPs showed good aerosolization properties and provided a remarkable avenue for the treatment of lung cancer. Very recently, the surface engineered DOX-loaded gelatin mesospheres showed sustained drug release of 79% for 120 hours. The prepared mesospheres exhibited minimal hemolytic toxicity. The gelatin-mannosylated-DOX-mesospheres were showed higher lung accumulation and cytotoxic effect compared to that of the free DOX. The authors concluded that the gelatin-based DOX NPs could be useful to target lung cancer with reduced adverse effects (79). In another study (80), gemcitabine-loaded gelatin NPs (178 nm) cross-linked with genipin showed effective against A549 and H460 cells with the controlled release of the drug for 72 hours. The prepared NPs produced a fine particle fraction (FPF) of 75% and the formulation reduced the complex viscosity of the lung mucus layer that indicated the increased mobility of nanoparticles in the mucus. Rao et al. (78) studied the cisplatin loaded gelatin NPs (314 nm) as colloidal drug carriers for lung delivery against cancer. They mixed the prepared NPs with inhalable large carriers lactoses to further improve the flow property. The prepared formulation showed higher FPF of 31.4% compared to that of the conventional dry powder formulation; however, no controlled release profile of cisplatin has been demonstrated. Tseng et al. (86) studied the cisplatin loaded- gelatin NPs decorated with epidermal growth factor (EGF) tumor-specific ligand and found that the developed NPs was more potent (IC50 1.2 μg/ml that inhibited A549 cell growth) than that of the cisplatin alone. In another study, the same group of investigators developed gelatin NPs modified with biotinylated EGF, which was mainly accumulated in the lung cancer cells. The authors revealed that the EGF targeted gelatin NPs could be beneficial carriers for the treatment of lung cancer (87).

Using p-aminobenzylcarbonyl (PABC) as a spacer, and valine-citrulline (VC) as a substrate of cathepsin B, linked with PEG and PTX, Liang et al. prepared paclitaxel NPs (100–200 nm), which showed controlled drug release for 75 hours (>50% released in 24 hours) and exhibited promising anti-tumor and anti-angiogenic activities in an in vivo model (88). Additionally, the PTX conjugates showed a thousand-time or more water solubility compared to the PTX. The authors used the Pegylated PTX (PEG-PTX) and Taxol formulations as control, which produced lesser action compared to the nano-formulation with PABC.

The liposomal levofloxacin NP for the treatment of CF showed anti-bacterial activity against *P. aeruginosa* with controlled release of drug for 72 hours. The anionic liposomal NP showed better safety on A549 epithelial cells than that of cationic NPs. It has been concluded that the inhalation of levofloxacin-loaded anionic liposomes could be the preferred method for the treatment of *P. aeruginosa* in CF patients (89).

Very recently, Hamedinasab et al. (35) developed N-acetylcysteine-loaded chitosan-coated liposome for lung delivery with prolonged release of the coated drug. The release of N-acetylcysteine from the chitosan-coated NPs was much lower (38%) than that of uncoated particles (51%) in 9 hours and this outcome confirmed the prolonged release of the drug from the chitosan-coated liposome. Furthermore, good deposition and accumulation of drug into lungs (cellular uptake) from chitosan CH-coated liposome particles even after 24 hours. The reason behind this was due to mucoadhesion property of the cationic chitosan. This outcome further justifies the use of chitosan coating in the prolonged release of drugs form nanoparticles upon lung delivery.

23.3.7 Pulmonary Delivery of Proteins/Peptides/Genes with Different Polymeric Nanoparticles

The pulmonary delivery of genetic therapeutics for the treatment of lung cancer is an emerging avenue of drug delivery research. The inhaled peptides/genes to target the lung as an alternative route of therapeutic administration for systemic action. The aerosolized gene therapies to treat lung cancer, *M. tuberculosis*, alpha-1 antitrypsin deficiency, and CF. Additionally, the inhaled peptides to treat lung diseases such as asthma, pulmonary hypertension, and CF are progressing (90,91). Although the future of lung protein/peptides delivery is in infancy, it is expected that the pulmonary delivery of peptides/gene delivery will provide promising outcomes for cancer treatments. The design of efficient pulmonary gene delivery with controlled release of therapeutics is challenging; however, the researchers are investigating the ways to find out the best option. Using a mouse model, the PEGylated DNA particles demonstrated superior biodistribution, improved safety profiles and efficient gene transfer compared to the non-PEGylated NPs (92). The authors claimed their work as an innovative approach for enhanced transfection of the gene with PEG coating for efficacious lung delivery for CF patients. Cellular penetration of drug is a major challenge in the pulmonary delivery of drugs for patients with CF. Lung delivery of peptide coated PEGylated nanoparticles showed ~600-fold better penetration into the human CF mucus than that of uncoated control (93). The improved uptake into lung epithelial cells compared to uncoated or PEGylated nanoparticles. The prepared nanoparticles showed 65–75% retention in the airways after 24 hours post-administration. The NPs also exhibited enhanced uniform distribution and retention in the mouse lung airways. The authors concluded that these peptides may efficiently serve as surface modifications to increase the amount of drug and drug carriers delivered to the human mucus and improve clinical outcomes of the CF patients. The positively charged dendrimers complexed with enoxaparin, a low molecular weight heparin, delivered into the lungs of anesthetized rats was found to produce significant outcomes in preventing vein thrombosis (94). These complexes enhanced the pulmonary absorption of the drug by reducing the negative surface charge density of the drug molecule. These research outcomes have extended lung drug delivery technology at a high level.

Pulmonary delivery of siRNA targeting A549 lung cancer cells, both in vitro and in vivo, produced promising outcomes. The aerosol delivery of PE4K-A13-0.33C6 and PE4K-A13-0.33C10 NPs to implanted orthotopic lung tumors, showed preferential accumulation in the lung cancer cells within 4 hours. The PEGylated particles reduced the surface charge of NPs, resulting in an increase of serum stability, which resulted in significant gene silencing in the A549 cancer cells (95). No controlled release of the drug was investigated. The inhaled CS-based siRNA NPs showed promising cellular uptake and gene silencing (96). In another study (97), inhalable CS-based siRNA powder containing chitosan and mannitol, achieved a prolonged release of the siRNA from the CS NPs on the lung epithelial surface and produced effective gene silencing against the lung tumor cells in mice. Using polyethyleneimine based RNAi showed effectively decreased B16F10 lung metastases (98).

Menon et al. (63) investigated the pulmonary delivery of plasmid DNA encapsulated in six different polymers such as gelatin, chitosan, alginate, the combined PLGA-chitosan and PLGA-PEG. The authors found PLGA-based NPs and natural polymer NPs exhibited the highest cytocompatibility by human alveolar epithelial cells. The overall results support both PLGA and gelatin NPs as promising carriers for pulmonary protein/DNA delivery. Using the BSA as a model protein, they studied the sustained release of BSA. All NPs showed a bi-phasic release consisting of a burst release of the protein for the first 2 days following by sustained release for 3 weeks. Chitosan, gelatin and PLGA Nos showed an initial burst release for more than 40% loaded protein within 4 days.

The chitosan containing formulations were found to enhance the level of reporter gene expression in human lung carcinoma cells (99). Using a mouse model, CS-pDNA complexes produced higher transfection potency than the aqueous solutions containing the same amount of DNA (100, 101) and the CS showed improved pDNA stability of the prepared formulations (102). Using a mouse model, the CS-β-interferon gene complex (pCMV-Muβ) showed higher expression than the intravenous or intra-tracheal solution formulations (103). Dong et al demonstrated (104) that higher cellular OMR uptake of the CS-coated PLGA based nanoplexes of antisense 2'-O-Me RNA (OMR) compared with the product without CS. These outcomes are encouraging in the treatment of lung cancer by inhaled gene therapy using polymer nano-particulate technology.

Pulmonary delivery of the combined drugs and genes therapy showed very effective anti-tumor effects on lung cancer cells (105). Using a highly positively charged polyethylenimine (PEI) as the carrier, the combined doxorubicin and Bcl2 siRNA for lung cancer showed high cellular uptake and inhibited cell proliferation. In a mouse model, the prepared DOX and siRNA via pulmonary administration produced a higher accumulation of the drugs into the lungs compared to those of the free drugs. Therefore, inhalable PEI-based drug therapy could be useful for the treatment of lung cancers. The lung delivery of doxorubicin, paclitaxel and siRNA encapsulated in the nanostructured lipid carriers (NLC) have demonstrated (106). The inhaled route showed preferential accumulation of drugs into cancer cells and enhanced efficiency compared to the IV route. The pulmonary delivery of PEI and polyamidoamine dendrimers complexed with DNA for the treatment of chronic lung diseases showed promising outcomes. The PEI/DNA complexes showed a higher level of luciferase gene expression in the lung compared to that of DNA/polyamidoamine dendrimers complex (107).

D-cycloserine-loaded alginate-chitosan NPs (339–349 nm) showed sustained release of drug from the NPs for 24 hours (108) with limited lung inflammatory effect. This outcome proved the importance of drug encapsulated alginate NPs as a potential alternative to the existing conventional therapy in multi-drug-resistant tuberculosis. MDR-TB. In an earlier study, aerosolized alginate NPS (235 nm) encapsulating isoniazid (INH), rifampicin (RIF) and pyrazinamide (PZA) showed higher relative bioavailability of drugs compared to the drug alone (109). The release of drugs from the NPs were detected in lungs, liver and spleen above the minimum inhibitory concentration until 15 days after the nebulization. The nebulized free drugs remained in the organs only for 1 day. The authors concluded that the prepared drug-loaded alginate NPs can be considered as a potential carrier for the controlled release of anti-tubercular drugs. Despite the enormous hurdles to peptide/gene therapy via the pulmonary route, a lot of studies are to be conducted to ensure the treatment options available in clinics. To date, much progress has been made in the protein/ gene therapy to achieve the specific cell transfer, minimize immune responses, and create new genes for the management of various diseases. In future, the world will see the successful gene therapy treatment for genetic lung diseases and that would lead a useful mechanism for the management of other genetic disorders in other organs.

23.3.8 Pulmonary Delivery of Vaccines With Polymeric Nanoparticles

Pulmonary delivery of vaccines elicits both mucosal and systemic immunity and reduces the antigen dose required to induce protective immunity (26, 110–112). Pulmonary delivery of dry powder formulations of hepatitis B (113), measles (114), tuberculosis (115) and influenza (116) vaccines have been investigated in various animal models with significant success. The inhaled vaccination has been studied for a while; however, no vaccine products are commercially available yet. The inhaled vaccines induce not only the mucosal immune response but also the systemic immune response that is obtained by IV administration. A significant number of nano-particulate colloidal vaccine formulations for lung delivery have been studied (117–119). CS is one of the most widely used polymers that has been used as an adjuvant due to its mucoadhesive properties (120). Chitosan's positive charge has been found to activate the immune system (121), enhance dispersion of particles (26) and improve drug absorption by opening the intercellular tight junctions of the lung epithelium (37). Formulation of chitosan-based influenza vaccine showed dose sparing with the enhanced humoral and cellular immune response against influenza virus (122). Very recently, nasal vaccination with the mast cell activator C48/80 with chitosan nanoparticles showed promising activity against hepatitis B virus (123) showing the effectiveness of chitosan for vaccine delivery. Lung delivery of inhalable tuberculosis (TB) antigens to target the site of infection (lungs) have been investigated for the development of subunit vaccines against TB (124). The PLGA although investigated for vaccine delivery; the particles of H56 TB vaccine loaded in the copolymers of PLGA with leucine and sucrose showed promising immunogenicity. This study developed the foundation of the inhaled vaccine delivery against TB. Bivas-Benita et al. (125) investigated the lung delivery of DNA vaccines against tuberculosis, and the CS-based vaccine NPs increased immunogenicity with increased levels of IFN-γ secretion compared with pulmonary delivery of plasmid in solution for intramuscular immunization. In another study, the inhaled CS complexed with *M. tuberculosis* DNA vaccine enhanced the mucosal immunity (126) and better protection was observed compared with that of the vaccine without CS. The pulmonary delivery of CS-based DNA NPs was demonstrated to induce antigen-specific immunity (127). The CS-DNA NPs significantly enhanced the Th1 associated immune response due to elevated IL-8 and TNF-α levels. The overall results were promising for pulmonary DNA vaccination against intracellular pathogens (*M. tuberculosis* or influenza virus). Amidi et al. (128) investigated the potential of *N*-trimethyl derivative of CS for lung delivery of diphtheria toxoid (DT) and found promising outcomes against.

23.4 Conclusion and Future Directions

The aerosol delivery is expanding with the advancement of science and technology especially in the development of drug-loaded polymeric NPs to target the systemic disorders. The controlled release of drugs from respirable polymeric colloidal NPs is an advanced level of drug delivery research with a bright future. Researchers around the world investigated the application of various drug loaded-polymeric colloidal NPs for the management of different diseases such as asthma, COPD, CF, lung cancers, pneumonia etc. Among the polymers, chitosan, PEG, PLGA, alginate and gelatine are the most widely investigated polymers for lung drug delivery. A large number of drugs such as anti-asthmatic, anti-bacterials, anti-cancer, proteins and peptides have been successfully investigated for lung delivery. Despite the overwhelming understanding of the future prospect of inhaled drug delivery system, the real impact of polymers and their NP on the lung cells are still unknown. The degradation of polymer or polymeric colloidal particles in the lungs is very limited (129, 130). The *Ex vivo* models applied to the determination of the degradation of polymeric nano-particulate are not well established. The release rate of drugs form inhaled nanoparticles, the kinetics of drug release and the effect of drugs in lung cells are still not clear. Although, the isolated, perfused, and ventilated lung models are promising tools to evaluate the controlled release of drugs encapsulated in polymeric colloidal NPs intended for pulmonary delivery; there are no promising data that can address the requirement of in vivo experiments. More exclusive studies for pharmacokinetics of inhaled polymeric nano-particulate drugs in human are required. The inhaled nano-particulate colloidal drugs for the treatment of specific diseases with very expensive medicines is inevitable. The lung delivery research of large molecules like proteins and peptides for various chronic diseases is advancing rapidly. The pulmonary delivery is the most beneficial route for potent drugs as a small amount of dose will require to get therapeutic benefits. It is anticipated that the future of controlled release lung drug delivery is very bright for the management of a wide range of chronic diseases with more patient compliance. Therefore, research in controlled release lung delivery may contribute to the effective delivery of various drugs to combat many life-threatening diseases in future.

REFERENCES

1. Valle MJdJ, Gonzalez Lopez F, and Sanchez Navarro A. Pulmonary versus systemic delivery of levofloxacin. *Pulmonary Pharmacology & Therapeutics*. 2008; 21(2): 298–303.
2. Morais CLdM, Nascimento JWL, Ribeiro AC, Cortinez LI, Carmona MJC, Maia DRR, et al. Nebulization of vancomycin provides higher lung tissue concentrations than intravenous administration in ventilated female piglets with healthy lungs. *Anesthesiology*. 2020; 132(6): 1516–1527.
3. Beck-Broichsitter M, Merkel OM, and Kissel T. Controlled pulmonary drug and gene delivery using polymeric nano-carriers. *Journal of Controlled Release*. 2012; 161(2): 214–224.
4. Levet V, Rosiere R, Merlos R, Fusaro L, Berger G, Amighi K, et al. Development of controlled-release cisplatin dry powders for inhalation against lung cancers. *International Journal of Pharmaceutics*. 2016; 515(1–2): 209–220.
5. Shoyele SA. Controlling the release of proteins/peptides via the pulmonary route. Methods in Molecular Biology (Totowa, NJ, United States). 2008; 437 (Drug Delivery Systems): 141–148.
6. Ungaro F, d'Angelo I, Coletta C, d'Emmanuele di Villa Bianca R, Sorrentino R, Perfetto B, et al. Dry powders based on PLGA nanoparticles for pulmonary delivery of antibiotics: modulation of encapsulation efficiency, release rate and lung deposition pattern by hydrophilic polymers. *Journal of Controlled Release*. 2012; 157(1): 149–159.
7. Wang H, Holgate J, Bartlett S, and Islam N. Assessment of nicotine release from nicotine-loaded chitosan nanoparticles dry powder inhaler formulations via locomotor activity of C57BL/6 mice. *European Journal of Pharmaceutics and Biopharmaceutics*. 2020; 154: 175–185.
8. Wang H, George G, Bartlett S, Gao C, and Islam N. Nicotine hydrogen tartrate loaded chitosan nanoparticles: Formulation, characterization and in vitro delivery from dry powder inhaler formulation. *European Journal of Pharmaceutics and Biopharmaceutics*. 2017; 113: 118–131.
9. Muhsin MDA, George G, Beagley K, Ferro V, Wang H, and Islam N. Effects of chemical conjugation of L-leucine to chitosan on dispersibility and controlled release of drug from a nanoparticulate dry powder inhaler formulation. *Molecular Pharmaceutic*. 2016; 13(5): 1455–1466.
10. Pontes JF and Grenha A. Multifunctional nanocarriers for lung drug delivery. *Nanomaterials*. 10(2): 183.
11. Gonda I. Aerosols for delivery of therapeutic and diagnostic agents to the respiratory tract. *Critical Reviews in Therapeutic Drug Carrier Systems*. 1990; 6(4): 273–313.
12. Suarez S and Hickey AJ. Drug properties affecting aerosol behavior. *Respiratory Care*. 2000; 45(6): 652–666.
13. Heyder J, Gebhart J, Rudolf G, Schiller CF, and Stahlhofen W. Deposition of particles in the human respiratory tract in the size range 0.005–15 µm. *Journal of Aerosol Science*. 1986; 17(5): 811–825.
14. Byron PR. Prediction of drug residence times in regions of the human respiratory tract following aerosol inhalation. *Journal of Pharmaceutical Sciences*. 1986; 75(5): 433–438.
15. Ibricevic A, Guntsen SP, Zhang K, Shrestha R, Liu Y, Sun JY, et al. PEGylation of cationic, shell-crosslinked-knedel-like nanoparticles modulates inflammation and enhances cellular uptake in the lung. *Nanomedicine*. 2013; 9(7): 912–922.
16. Sung JC, Pulliam BL, and Edwards DA. Nanoparticles for drug delivery to the lungs. *Trends in Biotechnology*. 2007; 25(12): 563–570.
17. Nahar K, Absar S, Patel B, and Ahsan F. Starch-coated magnetic liposomes as an inhalable carrier for accumulation of fasudil in the pulmonary vasculature. *International Journal of Pharmaceutics (Amsterdam, Netherlands)*. 2014; 464(1–2): 185–195.
18. Sanders NN, De SSC, Van RE, Simoens P, De BF, and Demeester J. Cystic fibrosis sputum: A barrier to the transport of nanospheres. *American Journal of Respiratory and Critical Care Medicine*. 2000; 162(5): 1905–1911.
19. Suk JS, Lai SK, Boylan NJ, Dawson MR, Boyle MP, and Hanes J. Rapid transport of muco-inert nanoparticles in cystic fibrosis sputum treated with N-acetyl cysteine. *Nanomedicine (London)*. 2011; 6(2): 365–375.
20. Ensign LM, Henning A, Schneider CS, Maisel K, Wang Y-Y, Porosoff MD, et al. Ex vivo characterization of particle transport in mucus secretions coating freshly excised mucosal tissues. *Molecular Pharmaceutics*. 2013; 10(6): 2176–2182.
21. Kirch J, Guenther M, Doshi N, Schaefer UF, Schneider M, Mitragotri S, et al. Mucociliary clearance of micro- and nanoparticles is independent of size, shape and charge-an ex vivo and in silico approach. *Journal of Controlled Release*. 2012; 159(1): 128–134.

22. Henning A, Schneider M, Nafee N, Muijs L, Rytting E, Wang X, et al. Influence of particle size and material properties on mucociliary clearance from the airways. *Journal of Aerosol Medicine and Pulmonary Drug Delivery*. 2010; 23(4):233–241.
23. Lim YH, Tiemann KM, Hunstad DA, Elsabahy M, and Wooley KL. Polymeric nanoparticles in development for treatment of pulmonary infectious diseases. *Wiley Interdisciplinary Reviews - Nanomedicine and Nanobiotechnology*. 2016; 8(6): 842–871.
24. d'Angelo I, Casciaro B, Miro A, Quaglia F, Mangoni ML, and Ungaro F. Overcoming barriers in Pseudomonas aeruginosa lung infections: Engineered nanoparticles for local delivery of a cationic antimicrobial peptide. *Colloids and Surfaces B: Biointerfaces*. 2015; 135: 717–725.
25. Shukla SK, Kulkarni NS, Farrales P, Kanabar DD, Parvathaneni V, Kunda NK, et al. Sorafenib loaded inhalable polymeric nanocarriers against non-small cell lung cancer. *Pharmaceutical Research*. 2020; 37(3): 67.
26. Wang H, George G, and Islam N. Nicotine-loaded chitosan nanoparticles for dry powder inhaler (DPI) formulations - Impact of nanoparticle surface charge on powder aerosolization. *Advanced Powder Technology*. 2018; 29(12): 3079–3086.
27. Islam N and Richard D. Inhaled micro/nanoparticulate anticancer drug formulations: An emerging targeted drug delivery strategy for lung cancers. *Current Cancer Drug Targets*. 2019; 19(3): 162–178.
28. d'Angelo I, Conte C, Miro A, Quaglia F, and Ungaro F. Pulmonary drug delivery: A role for polymeric nanoparticles? *Current Topics in Medicinal Chemistry (Sharjah, United Arab Emirates)*. 2015; 15(4): 386–400.
29. Islam N and Ferro V. Recent advances in chitosan-based nanoparticulate pulmonary drug delivery. *Nanoscale*. 2016; 8: 14341–14358.
30. Tuli RA, George GA, Dargaville TR , and Islam N. Studies on the effect of the size of polycaprolactone microspheres for the dispersion of salbutamol sulfate from dry powder inhaler formulations. *Pharmaceutical Research*. 2012; 29(9): 2445–2455.
31. Abdelaziz HM, Gaber M, Abd-Elwakil MM, Mabrouk MT, Elgohary MM, Kamel NM, et al. Inhalable particulate drug delivery systems for lung cancer therapy: Nanoparticles, microparticles, nanocomposites and nanoaggregates. *Journal of Controlled Release*. 2018; 269: 374–392.
32. Al-Qadi S, Grenha A, Carrion-Recio D, Seijo B, and Remunan-Lopez C. Microencapsulated chitosan nanoparticles for pulmonary protein delivery: In vivo evaluation of insulin-loaded formulations. *Journal of Controlled Release*. 2012; 157(3): 383–390.
33. Yan C, Wang J, Gu J, Hou D, Lei L, Jing H, et al. The influence of molecular parameters of chitosan on pulmonary absorption of insulin loaded chitosan nanoparticles. *Latin American Journal of Pharmacy*. 2013; 32(6): 860–868.
34. Lee C, Choi JS, Kim I, Oh KT, Lee ES, Park E-S, et al. Long-acting inhalable chitosan-coated poly(lactic-co-glycolic acid) nanoparticles containing hydrophobically modified exendin-4 for treating type 2 diabetes. *International Journal of Nanomedicine*. 2012; 8: 2975–2983.
35. Hamedinasab H, Rezayan AH, Mellat M, Mashreghi M, and Jaafari MR. Development of chitosan-coated liposome for pulmonary delivery of N-acetylcysteine. *International Journal of Biological Macromolecules*. 2020; 156: 1455–1463.
36. Murata M, Nakano K, Tahara K, Tozuka Y, and Takeuchi H. Pulmonary delivery of elcatonin using surface-modified liposomes to improve systemic absorption: Polyvinyl alcohol with a hydrophobic anchor and chitosan oligosaccharide as effective surface modifiers. *European Journal of Pharmaceutics and Biopharmaceutics*. 2012; 80(2): 340–346.
37. Yamamoto H, Kuno Y, Sugimoto S, Takeuchi H, and Kawashima Y. Surface-modified PLGA nanosphere with chitosan improved pulmonary delivery of calcitonin by mucoadhesion and opening of the intercellular tight junctions. *Journal of Controlled Release*. 2005; 102(2): 373–381.
38. Yang M, Yamamoto H, Kurashima H, Takeuchi H, Yokoyama T, Tsujimoto H, et al. Design and evaluation of inhalable chitosan-modified poly (DL-lactic-co-glycolic acid) nanocomposite particles. *European Journal of Pharmaceutical Sciences*. 2012; 47(1): 235–243.
39. Pourshahab PS, Gilani K, Moazeni E, Eslahi H, Fazeli MR, and Jamalifar H. Preparation and characterization of spray dried inhalable powders containing chitosan nanoparticles for pulmonary delivery of isoniazid. *Journal of Microencapsulation*. 2011; 28(7): 605–613.
40. Debnath SK, Saisivam S, Debanth M, and Omri A. Development and evaluation of Chitosan nanoparticles based dry powder inhalation formulations of Prothionamide. *PLoS ONE*. 2018; 13(1): e0190976/1-e/12.
41. Guenday Tuereli N, Torge A, Juntke J, Schwarz BC, Schneider-Daum N, Tuereli AE, et al. Ciprofloxacin-loaded PLGA nanoparticles against cystic fibrosis P. aeruginosa lung infections. *European Journal of Pharmaceutics and Biopharmaceutics*. 2017; 117: 363–371.
42. Goyal AK, Garg T, Rath G, Gupta UD, and Gupta P. Development and characterization of nanoembedded microparticles for pulmonary delivery of antitubercular drugs against experimental tuberculosis. *Molecular Pharmaceutics*. 2015; 12(11): 3839–3850.
43. Park C-W, Li X, Vogt FG, Hayes D, Jr., Zwischenberger JB, Park E-S, et al. Advanced spray-dried design, physicochemical characterization, and aerosol dispersion performance of vancomycin and clarithromycin multifunctional controlled release particles for targeted respiratory delivery as dry powder inhalation aerosols. *International Journal of Pharmaceutics*. 2013; 455(1–2): 374–392.
44. Paul P, Sengupta S, Mukherjee B, Shaw TK, Gaonkar RH, and Debnath MC. Chitosan-coated nanoparticles enhanced lung pharmacokinetic profile of voriconazole upon pulmonary delivery in mice. *Nanomedicine (London, UK)*. 2018; 13(5): 501–520.
45. Huang Y-C, Li R-Y, Chen J-Y, and Chen J-K. Biphasic release of gentamicin from chitosan/fucoidan nanoparticles for pulmonary delivery. *Carbohydrate Polymers*. 2016; 138: 114–122.
46. Pardeshi CV, Agnihotri VV, Patil KY, Pardeshi SR, and Surana SJ. Mannose-anchored N,N,N-trimethyl chitosan

nanoparticles for pulmonary administration of etofylline. *International Journal of Biological Macromolecules*. 2020; 165(Part A): 445–459.

47. Xu E-y, Yang M, Guo J, Song S, Jiang J-f, Xu Y, et al. Multifunctional dry powder formulations for pulmonary drug delivery. *Zhongguo Xinyao Zazhi*. 2013; 22(23): 2822–2826.
48. Zhang WF, Zhao XT, Zhao QS, Zha SH, Liu DM, Zheng ZJ, et al. Biocompatibility and characteristics of theophylline/carboxymethyl chitosan microspheres for pulmonary drug delivery. *Polymer International*. 2014; 63(6): 1035–1040.
49. Learoyd Tristan P, Burrows Jane L, French E, Seville Peter C. Sustained delivery by leucine-modified chitosan spray-dried respirable powders. *International Journal of Pharmaceutics*. 2009; 372(1–2): 97–104.
50. Varshosaz J, Taymouri S, and Hamishehkar H. Fabrication of polymeric nanoparticles of poly(ethylene-co-vinyl acetate) coated with chitosan for pulmonary delivery of carvedilol. *Journal of Applied Polymer Science*. 2014; 131(1): 39694/1-/8.
51. Emami F, Mostafavi Yazdi SJ, and Na DH. Poly(lactic acid)/poly(lactic-co-glycolic acid) particulate carriers for pulmonary drug delivery. *Journal of Pharmaceutical Investigation*. 2019; 49(4): 427–442.
52. Ernst J, Klinger-Strobel M, Arnold K, Thamm J, Hartung A, Pletz MW, et al. Polyester-based particles to overcome the obstacles of mucus and biofilms in the lung for tobramycin application under static and dynamic fluidic conditions. *European Journal of Pharmaceutics and Biopharmaceutics*. 2018; 131: 120–129.
53. Zhang T, Bao J, Zhang M, Ge Y, Wei J, Li Y, et al. Chemo-photodynamic therapy by pulmonary delivery of gefitinib nanoparticles and 5-aminolevulinic acid for treatment of primary lung cancer of rats. *Photodiagnosis and Photodynamic Therapy*. 2020; 31: 101807.
54. Hanna LA, Basalious EB, and El Gazayerly ON. Respirable controlled release polymeric colloid (RCRPC) of bosentan for the management of pulmonary hypertension: In vitro aerosolization, histological examination and in vivo pulmonary absorption. *Drug Delivery*. 2017; 24(1): 188–198.
55. Makled S, Nafee N, and Boraie N. Nebulized solid lipid nanoparticles for the potential treatment of pulmonary hypertension via targeted delivery of phosphodiesterase-5-inhibitor. *International Journal of Pharmaceutics (Amsterdam, Netherlands)*. 2017; 517(1-2): 312–321.
56. Craparo EF, Porsio B, Sardo C, Giammona G, and Cavallaro G. Pegylated polyaspartamide-polylactide-based nanoparticles penetrating cystic fibrosis artificial mucus. *Biomacromolecules*. 2016; 17(3): 767–777.
57. Yamamoto H, Kuno Y, Sugimoto S, Takeuchi H, and Kawashima Y. Surface-modified PLGA nanosphere with chitosan improved pulmonary delivery of calcitonin by mucoadhesion and opening of the intercellular tight junctions. *Journal of Controlled Release*. 2005; 102(2): 373–381.
58. Grenha A, Seijo B, and Remunan-Lopez C. Microencapsulated chitosan nanoparticles for lung protein delivery. *European Journal of Pharmaceutical Sciences*. 2005; 25(4–5): 427–437.
59. Learoyd TP, Burrows JL, French E, and Seville PC. Modified release of beclometasone dipropionate from chitosan-based spray-dried respirable powders. *Powder Technology*. 2008; 187(3): 231–238.
60. Kamble MS, Mane OR, Borwandkar VG, Mane SS, and Chaudhari PD.Formulation and evaluation of Clindamycin HCl - Chitosan microspheres for dry powder inhaler formulation. *Drug Invention Today*. 2012;4(10):527–530.
61. Learoyd TP, Burrows JL, French E, and Seville PC. Sustained delivery by leucine-modified chitosan spray-dried respirable powders. *International Journal of Pharmaceutics*. 2009; 372(1–2): 97–104.
62. Guo XH, Zhang XX, Ye L, Zhang Y, Ding R, Hao YW, et al. Inhalable microspheres embedding chitosan-coated PLGA nanoparticles for 2-methoxyestradiol. *Journal of Drug Targeting*. 2014; 22(5): 421–427.
63. Menon JU, Ravikumar P, Pise A, Gyawali D, Hsia CCW, and Nguyen KT. Polymeric nanoparticles for pulmonary protein and DNA delivery. *Acta Biomaterialia*. 2014; 10(6): 2643–2652.
64. Kaminskas LM, Landersdorfer CB, Bischof RJ, Leong N, Ibrahim J, Davies AN, et al. Aerosol pirfenidone pharmacokinetics after inhaled delivery in sheep: A viable approach to treating idiopathic pulmonary fibrosis. *Pharmaceutical Research*. 2020; 37(1): 3.
65. Li D and Gong L. Preparation of novel pirfenidone microspheres for lung-targeted delivery: In vitro and in vivo study. *Drug Design, Development and Therapy*. 2016; 10: 2815–2821.
66. Aragao-Santiago L, Hillaireau H, Grabowski N, Mura S, Nascimento Thais L, Tsapis N, et al. Compared in vivo toxicity in mice of lung delivered biodegradable and non-biodegradable nanoparticles. *Nanotoxicology*. 2015: 1–11.
67. Ahmad MI, Nakpheng T, and Srichana T. The safety of ethambutol dihydrochloride dry powder formulations containing chitosan for the possibility of treating lung tuberculosis. *Inhalation Toxicology*. 2014; 26(14): 908–917.
68. Garg T, Rath G, and Goyal Amit K. Inhalable chitosan nanoparticles as antitubercular drug carriers for an effective treatment of tuberculosis. *Artificial Cells, Nanomedicine, and Biotechnology*. 2015: 1–5.
69. Yang Y, Cheow WS, and Hadinoto K. Dry powder inhaler formulation of lipid-polymer hybrid nanoparticles via electrostatically-driven nanoparticle assembly onto microscale carrier particles. *International Journal of Pharmaceutics (Amsterdam, Netherlands)*. 2012; 434(1-2): 49–58.
70. Dabbagh A, Abu KNH, Yeong CH, Wong TW, Abdul RN, and Abdul RN. Critical parameters for particle-based pulmonary delivery of chemotherapeutics. *Journal of Aerosol Medicine and Pulmonary Drug Delivery*. 2018; 31(3): 139–154.
71. Cafaggi S, Russo E, Stefani R, Leardi R, Caviglioli G, Parodi B, et al. Preparation and evaluation of nanoparticles made of chitosan or N-trimethyl chitosan and a cisplatin-alginate complex. *Journal of Controlled Release*. 2007; 121(1-2): 110–123.
72. Singh DJ, Lohade AA, Parmar JJ, Hegde DD, Soni P, Samad A, et al. Development of chitosan-based dry powder inhalation system of cisplatin for lung cancer. *Indian Journal of Pharmaceutical Sciences*. 2012; 74(6): 521–526.

73. Otroj M, Taymouri S, Varshosaz J, and Mirian M. Preparation and characterization of dry powder containing sunitinib loaded PHBV nanoparticles for enhanced pulmonary delivery. *Journal of Drug Delivery Science and Technology*. 2020; 56(Part_A): 101570.
74. Zhong Q, Bielski ER, Rodrigues LS, Brown MR, Reineke JJ, and da Rocha SRP. Conjugation to poly(amidoamine) dendrimers and pulmonary delivery reduce cardiac accumulation and enhance antitumor activity of doxorubicin in lung metastasis. *Molecular Pharmaceutics*. 2016; 13(7): 2363–2375.
75. Garbuzenko OB, Mainelis G, Taratula O, and Minko T. Inhalation treatment of lung cancer: The influence of composition, size and shape of nanocarriers on their lung accumulation and retention. *Cancer Biology & Medicine*. 2014; 11(1): 44–55.
76. Meenach SA, Anderson KW, Hilt JZ, McGarry RC, and Mansour HM. High-performing dry powder inhalers of paclitaxel DPPC/DPPG lung surfactant-mimic multifunctional particles in lung cancer: Physicochemical characterization, in vitro aerosol dispersion, and cellular studies. *AAPS PharmSciTech*. 2014; 15(6): 1574–1587.
77. Meenach SA, Anderson KW, Zach Hilt J, McGarry RC, Mansour HM. Characterization and aerosol dispersion performance of advanced spray-dried chemotherapeutic PEGylated phospholipid particles for dry powder inhalation delivery in lung cancer. *European Journal of Pharmaceutical Sciences*. 2013; 49(4): 699–711.
78. Rao AK, Shrikhande S, and Bajaj A. Development of cisplatin nanoparticles as dry powder inhalers for lung cancer. *Current Nanoscience*. 2013; 9(4): 447–450.
79. Gautam L, Sharma R, Shrivastava P, Vyas S, and Vyas SP. Development and characterization of biocompatible mannose functionalized mesospheres: An effective chemotherapeutic approach for lung cancer targeting. *AAPS PharmSciTech*. 2020; 21(5): 190.
80. Youngren-Ortiz SR, Hill DB, Hoffmann PR, Morris KR, Barrett EG, Forest MG, et al. Development of optimized, inhalable, gemcitabine-loaded gelatin nanocarriers for lung cancer. *Journal of Aerosol Medicine and Pulmonary Drug Delivery*. 2017; 30(5): 299–321.
81. Emami J, Pourmashhadi A, Sadeghi H, Varshosaz J, and Hamishehkar H. Formulation and optimization of celecoxib-loaded PLGA nanoparticles by the Taguchi design and their in vitro cytotoxicity for lung cancer therapy. *Pharmaceutical Development and Technology*. 2015; 20(7): 791–800.
82. Wang H, Xu Y, and Zhou X. Docetaxel-loaded chitosan microspheres as a lung targeted drug delivery system: In vitro and in vivo evaluation. *International Journal of Molecular Sciences*. 2014; 15(3): 3519–3532.
83. Yang R, Yang S-G, Shim W-S, Cui F, Cheng G, Kim I-W, et al. Lung-specific delivery of paclitaxel by chitosan-modified PLGA nanoparticles via transient formation of microaggregates. *Journal of Pharmaceutical Sciences*. 2009; 98(3): 970–984.
84. El-Sherbiny IM and Smyth HDC. Biodegradable nano-micro carrier systems for sustained pulmonary drug delivery: (I) Self-assembled nanoparticles encapsulated in respirable/swellable semi-IPN microspheres. *International Journal of Pharmaceutics*. 2010; 395(1–2): 132–141.
85. Abdelrady H, Hathout RM, Osman R, Saleem I, and Mortada ND. Exploiting gelatin nanocarriers in the pulmonary delivery of methotrexate for lung cancer therapy. *European Journal of Pharmaceutical Sciences*. 2019; 133: 115–126.
86. Tseng C-L, Su W-Y, Yen K-C, Yang K-C, and Lin F-H. The use of biotinylated-EGF-modified gelatin nanoparticle carrier to enhance cisplatin accumulation in cancerous lungs via inhalation. *Biomaterials*. 2009; 30(20): 3476–3485.
87. Tseng C-L, Wu SYH, Wang W-H, Peng C-L, Lin F-H, Lin C-C, et al. Targeting efficiency and biodistribution of biotinylated-EGF-conjugated gelatin nanoparticles administered via aerosol delivery in nude mice with lung cancer. *Biomaterials*. 2008; 29(20): 3014–3022.
88. Liang L, Lin S-W, Dai W, Lu J-K, Yang T-Y, Xiang Y, et al. Novel cathepsin B-sensitive paclitaxel conjugate: Higher water solubility, better efficacy and lower toxicity. *Journal of Controlled Release*. 2012; 160(3): 618–629.
89. Derbali RM, Aoun V, Moussa G, Frei G, Tehrani SF, Del'Orto JC, et al. Tailored nanocarriers for the pulmonary delivery of levofloxacin against *Pseudomonas aeruginosa*: A comparative study. *Molecular Pharmaceutics*. 2019; 16(5): 1906–1916.
90. Laube Beth L. The expanding role of aerosols in systemic drug delivery, gene therapy, and vaccination. *Respiratory Care*. 2005; 50(9): 1161–1176.
91. Katz MG, Fargnoli AS, Gubara SM, Fish K, Weber T, Bridges CR, et al. Targeted gene delivery through the respiratory system: Rationale for intratracheal gene transfer. *Journal of Cardiovascular Development and Disease*. 2019; 6(1): 8 pp.
92. Osman G, Rodriguez J, Chan SY, Chisholm J, Duncan G, Kim N, et al. PEGylated enhanced cell penetrating peptide nanoparticles for lung gene therapy. *Journal of Controlled Release*. 2018; 285: 35–45.
93. Leal J, Peng X, Liu X, Arasappan D, Wylie DC, Schwartz SH, et al. Peptides as surface coatings of nanoparticles that penetrate human cystic fibrosis sputum and uniformly distribute in vivo following pulmonary delivery. *Journal of Controlled Release*. 2020; 322: 457–469.
94. Bai S, Thomas C, and Ahsan F. Dendrimers as a carrier for pulmonary delivery of enoxaparin, a low-molecular weight heparin. *Journal of Pharmaceutical Sciences*. 2007; 96(8): 2090–2106.
95. Yan Y, Zhou K, Xiong H, Miller JB, Motea EA, Boothman DA, et al. Aerosol delivery of stabilized polyester-siRNA nanoparticles to silence gene expression in orthotopic lung tumors. *Biomaterials*. 2017; 118: 84–93.
96. Jeong EJ, Choi M, Lee J, Rhim T, and Lee KY. The spacer arm length in cell-penetrating peptides influences chitosan/siRNA nanoparticle delivery for pulmonary inflammation treatment. *Nanoscale*. 2015; 7(47): 20095–20104.
97. Okuda T, Kito D, Oiwa A, Fukushima M, Hira D, and Okamoto H. Gene silencing in a mouse lung metastasis model by an inhalable dry small interfering RNA powder prepared using the supercritical carbon dioxide technique.

Biological and Pharmaceutical Bulletin. 2013; 36(7): 1183–1191.
98. Zamora-Avila DE, Zapata-Benavides P, Franco-Molina MA, Saavedra-Alonso S, Trejo-Avila LM, Resendez-Perez D, et al. WT1 gene silencing by aerosol delivery of PEI-RNAi complexes inhibits B16-F10 lung metastases growth. *Cancer Gene Therapy*. 2009; 16(12): 892–899.
99. Li H-Y and Birchall J. Chitosan-modified dry powder formulations for pulmonary gene delivery. *Pharmaceutical Research*. 2006; 23(5): 941–950.
100. Okamoto H and Danjo K. Local and systemic delivery of high-molecular weight drugs by powder inhalation. *Yakugaku Zasshi*. 2007; 127(4): 643–653.
101. Okamoto H, Nishida S, Todo H, Sakakura Y, Iida K, and Danjo K. Pulmonary gene delivery by chitosan-pDNA complex powder prepared by a supercritical carbon dioxide process. *Journal of Pharmaceutical Sciences*. 2003; 92(2): 371–380.
102. Okamoto H, Sakakura Y, Shiraki K, Oka K, Nishida S, Todo H, et al. Stability of chitosan-pDNA complex powder prepared by supercritical carbon dioxide process. *International Journal of Pharmaceutics*. 2005; 290(1–2): 73–81.
103. Okamoto H, Shiraki K, Yasuda R, Danjo K, and Watanabe Y. Chitosan-interferon-β gene complex powder for inhalation treatment of lung metastasis in mice. *Journal of Controlled Release*. 2011; 150(2): 187–195.
104. Dong M, Muerdter TE, Philippi C, Loretz B, Schaefer UF, Lehr CM, et al. Pulmonary delivery and tissue distribution of aerosolized antisense 2'-O-Methyl RNA containing nanoplexes in the isolated perfused and ventilated rat lung. *European Journal of Pharmaceutics and Biopharmaceutics*. 2012; 81(3): 478–485.
105. Xu C-N, Tian H-Y, Wang Y-B, Du Y, Chen J, Lin L, et al. Anti-tumor effects of combined doxorubicin and siRNA for pulmonary delivery. *Chinese Chemical Letters*. 2017; 28(4): 807–812.
106. Taratula O, Kuzmov A, Shah M, Garbuzenko OB, and Minko T. Nanostructured lipid carriers as multifunctional nanomedicine platform for pulmonary co-delivery of anticancer drugs and siRNA. *Journal of Controlled Release*. 2013; 171(3): 349–357.
107. Rudolph C, Lausier J, Naundorf S, Muller RH, and Rosenecker J. In vivo gene delivery to the lung using polyethylenimine and fractured polyamidoamine dendrimers. *The Journal of Gene Medicine*. 2000; 2(4): 269–278.
108. Shaji J and Shaikh M. Formulation, optimization, and characterization of biocompatible inhalable D-cycloserine-loaded alginate-chitosan nanoparticles for pulmonary drug delivery. *Asian Journal of Pharmaceutical and Clinical Research*. 2016; 9(Suppl. 2): 82–95.
109. Zahoor A, Sharma S, and Khuller GK. Inhalable alginate nanoparticles as antitubercular drug carriers against experimental tuberculosis. *International Journal of Antimicrobial Agents*. 2005; 26(4): 298–303.
110. Jeyanathan M, Shao Z, Yu X, Harkness R, Jiang R, Li J, et al. AdHu5Ag85A respiratory mucosal boost immunization enhances protection against pulmonary tuberculosis in BCG-primed non-human primates. *PLoS ONE*. 2015; 10(8): e0135009/1-e/20.
111. Neuhaus V, Chichester JA, Ebensen T, Schwarz K, Hartman CE, Shoji Y, et al. A new adjuvanted nanoparticle-based H1N1 influenza vaccine induced antigen-specific local mucosal and systemic immune responses after administration into the lung. *Vaccine*. 2014; 32(26): 3216–3222.
112. White AD, Sarfas C, West K, Sibley LS, Wareham AS, Clark S, et al. Evaluation of the immunogenicity of Mycobacterium bovis BCG delivered by aerosol to the lungs of macaques. *Clinical and Vaccine Immunology*. 2015; 22(9): 992–1003.
113. Muttil P, Prego C, Garcia-Contreras L, Pulliam B, Fallon JK, Wang C, et al. Immunization of guinea pigs with novel hepatitis B antigen as nanoparticle aggregate powders administered by the pulmonary route. *AAPS Journal*. 2010; 12(3): 330–337.
114. Kisich KO, Higgins MP, Park I, Cape SP, Lindsay L, Bennett DJ, et al. Dry powder measles vaccine: Particle deposition, virus replication, and immune response in cotton rats following inhalation. *Vaccine*. 2011; 29(5): 905–912.
115. Ashhurst AS, Parumasivam T, Chan JGY, Lin LCW, Florido M, West NP, et al. PLGA particulate subunit tuberculosis vaccines promote humoral and Th17 responses but do not enhance control of *Mycobacterium tuberculosis* infection. *PLoS ONE*. 2018; 13(3): e0194620/1-e/19.
116. Sou T, Morton DAV, Williamson M, Meeusen EN, Kaminskas LM, and McIntosh MP. Spray-dried influenza antigen with trehalose and leucine produces an aerosolizable powder vaccine formulation that induces strong systemic and mucosal immunity after pulmonary administration. *Journal of Aerosol Medicine and Pulmonary Drug Delivery*. 2015; 28(5): 361–371.
117. Johnson LM, Mecham JB, Quinn F, and Hickey AJ. Nanoparticle technology for respiratory tract mucosal vaccine delivery. *KONA Powder Part Journal*. 2020; 37: 97–113.
118. Imam SS, Aqil M, and Gupta H. Pulmonary vaccine delivery systems: A novel approach for immunization. *Current Drug Therapy*. 2014; 9(3): 166–172.
119. Jabbal-Gill I, Watts P, and Smith A. Chitosan-based delivery systems for mucosal vaccines. *Expert Opinion on Drug Delivery*. 2012; 9(9): 1051–1067.
120. Huang J, Garmise RJ, Crowder TM, Mar K, Hwang CR, Hickey AJ, et al. A novel dry powder influenza vaccine and intranasal delivery technology: Induction of systemic and mucosal immune responses in rats. *Vaccine*. 2004; 23(6): 794–801.
121. Gendon YZ, Markushin SG, and Krivtsov GG. Akopova II. Chitosan as an adjuvant for parenteral inactivated influenza vaccines. *Vopr Virusol*. 2008; 53(5): 14–19.
122. Sadati Seyed F, Sadati Seyed F, Jamali A, Kheiri Masoumeh T, Abdoli A, Soleymani S, et al. Simultaneous formulation of influenza vaccine and chitosan nanoparticles within CpG oligodesoxi nucleotides lead to dose sparing and protect against lethal challenge in the mouse model. *Pathogens and Disease*. 2018; 76(8).
123. Bento D, Jesus S, Lebre F, Goncalves T, Borges O, Bento D, et al. Chitosan plus compound 48/80: Formulation and preliminary evaluation as a hepatitis B vaccine adjuvant. *Pharmaceutics*. 2019; 11(2).

124. Roces CB, Hussain MT, Schmidt ST, Christensen D, and Perrie Y. Investigating prime-pull vaccination through a combination of parenteral vaccination and intranasal boosting. *Vaccines (Basel, Switz)*. 2020; 8(1): 10.
125. Bivas-Benita M, van Meijgaarden KE, Franken KLMC, Junginger HE, Borchard G, Ottenhoff THM, et al. Pulmonary delivery of chitosan-DNA nanoparticles enhances the immunogenicity of a DNA vaccine encoding HLA-A*0201-restricted T-cell epitopes of *Mycobacterium tuberculosis*. *Vaccine*. 2004; 22(13–14): 1609–1615.
126. Ai W-q, Luo M-x, Cao Y-x, Xiong S-d, and Xu W. The mucosal immunity induced by chitosan-delivered gene vaccine and its significance in protection against tuberculosis. *Xiandai Mianyixue*. 2011; 31(2): 93–100.
127. Heuking S, Rothen-Rutishauser B, Raemy DO, Gehr P, and Borchard G. Fate of TLR-1/TLR-2 agonist functionalised pDNA nanoparticles upon deposition at the human bronchial epithelium in vitro. *Journal of Nanobiotechnology*. 2013; 11: 29/1-/10.
128. Amidi M, Pellikaan HC, Hirschberg H, de Boer AH, Crommelin DJA, Hennink WE, et al. Diphtheria toxoid-containing microparticulate powder formulations for pulmonary vaccination: Preparation, characterization and evaluation in guinea pigs. *Vaccine*. 2007; 25(37–38): 6818–6829.
129. Islam N, Islam N, Dmour I, and Taha Mutasem O. Degradability of chitosan micro/nanoparticles for pulmonary drug delivery. *Heliyon*. 2019; 5(5): e01684.
130. Islam N, Wang H, Islam N, Maqbool F, Ferro V, and Ferro V. In vitro enzymatic digestibility of glutaraldehy de-crosslinked chitosan nanoparticles in lysozyme solution and their applicability in pulmonary drug delivery. *Molecules*. 2019; 24(7), Art. no. 1271.

24

Lung Clearance Kinetics of Liposomes and Solid Lipid Nanoparticles

Soumalya Chakraborty[1], Sudipta Roy[2], Akash Dey[1], Sanjoy Kumar Das[3], Dhananjoy Saha[4], Subhabrata Ray[5], and Bhaskar Mazumder[6]
[1]National Institute of Pharmaceutical Education and Research (NIPER), Punjab, India
[2]Bengal School of Technology (A College of Pharmacy) Sugandha, West Bengal, India
[3]Institute of Pharmacy, West Bengal, India
[4]Directorate of Technical Education, Bikash Bhavan, Kolkata, India
[5]Dr. B. C. Roy College of Pharmacy and Allied Health Sciences, Durgapur, India
[6]Department of Pharmaceutical Sciences, Dibrugarh University, Dibrugarh, India

24.1 Introduction

Currently, the pulmonary route has obtained massive research attention as a parallel non-invasive way for drug delivery to both local tissue and systemic circulation due to many advantageous attributes like the extensive area for absorption, more vasculature, and membranes with high permeability and low extracellular and intracellular enzyme activity (1). Respiratory diseases represent a vast spectrum of clinical conditions with versatile therapeutic challenges, ranging from acute to chronic infections, and there is an unmet need to develop new therapeutic approaches for optimizing treatment in difficult-to-treat infections, cystic fibrosis (CF), non-cystic fibrosis bronchiectasis (NCFB), and immune-compromised and mechanically ventilated patients (2).

Pulmonary drug delivery acts to deliver a high concentration of bioactive at the main site of action with reduced systemic toxicity (3). In the case of drug delivery through pulmonary route, there are three principal barriers that retard the accumulation of the aerosols in the deep lung portions (i.e. respiratory bronchioles and the alveolar region), and these are (a) the anatomic barrier — the tracheobronchial tree structure of the pulmonary system plays the main protective role against foreign particles and pollutants; (b) the pathological barrier — the pathological condition may impact the consistency of the mucous membrane surrounding the respiratory tract epithelium, and (c) the immunological barrier — alveolar macrophages are involved in the defense mechanism and hence particles that get deposited in the alveolar part can be engulfed and moved to the upper respiratory tract where the mucociliary escalator can eliminate the particles (4).

The efficient delivery of drugs through pulmonary routes is associated with several development challenges. One of the mostly challenging questions is regarding the development of inhalable drugs that are sufficiently stable and of suitable size (1). Different particle and patient-related attributes affect optimal drug deposition. Particle deposition patterns bear critical dependence on size (ideally particles in the range between 1 and 5 μm) and shape, and viscosity has an inverse relationship with the rate of aerosolization. Patient-related factors include the diameter of the large airways (relative to age), the quality of the inhalation technique in spontaneously breathing and mechanically ventilated patients, the existence of abnormalities on the airway structure (including mucus of the airways), and the lungs' ability (5). It is worth remembering that to obtain therapeutic utility from inhaled drug particles, they need to be in "fine particle fraction" (FPF; i.e. capable of reaching the bronchioles and alveoli). To achieve it, the aerodynamic diameter of the particles should not exceed 5 or 6 μm, with particles smaller than 2 μm is the most appropriate for alveolar deposition (4).

A plethora of available inhalation devices are in the market, but none of them possess the ability to meet the requirements of optimal pulmonary drug delivery. Currently, metered-dose inhalers (MDI) and dry powder inhalers (DPI) are the most popular for chronic diseases such as chronic obstructive pulmonary disease (COPD). But these devices generate particles with extensively large size. The inertial impaction in the upper respiratory tract hinders the efficient deposition of these particles with a large diameter (1). The tendency for rapid absorption of the pulmonary administered drugs (e.g. bronchodilators and corticosteroids) from the lungs and subsequent needs for a frequent dosing regimen, represents another limiting aspect of drug delivery through pulmonary route (6). The inhaled delivery of drugs as a solution or dry powder gives a prompt therapeutic response within the lungs, which is useful for treating a few respiratory maladies, such as asthma. Be that as it may, enhancing the duration of exposure of the lungs to drugs may be a superior way for treating ailments such as lung cancer, infections, and fibrosis. This has incited noteworthy intrigued within the advancement of inhalable nanomedicines which can impart prolonged and controlled drug exposure within the lungs,

DOI: 10.1201/9781003046547-24

constrain systematic presentation and related side effects, as well as retard drug-related side effects as a consequence of the inhaled administration of exceedingly harmful or irritant drugs (7).

Different nanometric drug carriers such as liposomes and nanoparticles show some well-characterized and delicate qualities, which have made an alluring and productive methodology for pulmonary delivery of bioactives (1). First found and reported by Bangham et al. in 1960 (8), liposomes have gained consideration for conveying therapeutics because of their high flexibility joined with their high biocompatible nature (9). Prior investigations have exhibited that liposomes are stable within the presence of sputum and inhibitory elements, with the capacity to shield the drugs from inactivation, even though the sputum may go about as a boundary to bigger liposomes (10). Liposomes also possess the ability to deposit drugs in alveolar macrophages through phagocytosis, causing improvement in the treatment of intracellular infections (although many chronic lung infections reside in the conducting airways of the lung lumen) (2). Four kinds of devices are utilized for the pulmonary delivery of liposomes: pressurized MDI, DPI, soft mist inhalers, and medical nebulizers. Liposome nebulization can prompt changes in their physical properties and testimony of liposomes inside alveolar macrophages may influence their activities, even though there is no proof of changes in the lipidic content of pulmonary surfactant or lung immune response (4).

SLNs are a recently developed alternative to the conventional colloidal drug delivery framework. SLNs, first created by Müller et al. in 1991, consolidate the benefits of polymeric nanoparticles and liposomes while dodging acute and chronic toxicity (11,12). SLNs' safety attribute was credited to component lipids that are highly biocompatible and exceptionally tolerable by the lungs and body (1). Additionally, after being delivered to the lungs, SLNs show prompt absorption and degradation, coupled with sustained-release (13). Regardless of the potential benefits of inhalable nanomedicines, we have a fragmented comprehension of how the drug carriers and the individual parts of the carriers (as opposed to the loaded drugs) are disposed of from the lungs. In this way, it becomes needful to clarify regarding the kinetics of the carriers in the lungs after administering through pulmonary route, since the deposition of nanoparticles in the lungs over multiple dosing regimens can prompt "nanoparticle over-burden" and inflammatory impacts to the neighboring tissue (14). The last is somewhat because of the natural capacity of all nano-sized materials (counting proteins) to initiate mild lung reactions, the degree of which is generally reliant on their lung residence time and lung surface exposure (7). This chapter aims to give a clear understanding regarding the mechanism and pharmacokinetic attributes of the clearance of liposomes and SLNs from the lungs after pulmonary administration. Different important aspects like the anatomy and physiology of lungs, general lung clearance mechanisms, as well as a comparative discussion between the lung clearance attributes of SLNs and liposomes, have been discussed in an elaborated and lucid way.

24.2 Anatomy and Physiology of Lungs

The human lungs are paired organs having a pyramid-like shape and are bordered by a diaphragm. It is joined to the trachea by the left and right bronchi and is enclosed by pleurae attached to the mediastinum. The pleural space evokes a mesodermal cavity during embryogenesis and the lungs get eventually developed. Pleura adjacent to the lungs get transformed into the visceral pleura and finally to the parietal pleura after coming in contact with the mediastinum, diaphragm, and the chest wall (15). After complete development, 0.5 to 2.0 ml of pleural fluid is present. The mesothelial layer present in the inner side of the pleural space is active metabolically and is involved in maintaining cellular and humoral immunity. This layer also maintains the production and reabsorption of the pleural fluid. The lungs are thus surrounded by the pleura which is a serous membrane and the right and left lungs are enclosed within the right and left pleurae respectively (16). Figure 24.1 exhibits the anatomical features of the pleural cavity. The pleural fluid generation is driven by Starling forces that are defined as the gradient of oncotic and hydrostatic pressures present between the pleural space and the opposing pleural capillary beds. In addition to Starling forces, solute flux, hydraulic conductivity, total surface area, solute reflection coefficient, and the diffusive permeability of pleural capillary endothelium and interstitium contributes to the formation of the pleural fluid (15,17). The superior region of the lung is the apex while the base located near the diaphragm serves as the opposite region. The coastal surface borders the ribs and the mediastinal faces the midline (16).

Each of the pair of lungs consists of smaller units called lobes. The lobes are separated from each other by fissures. Three lobes of the right lung are the superior, middle, and inferior lobes, while the two of the left lung are the superior and inferior lobes. The right lung is shorter and wider as compared to the left lung, and the volume of left lung is lower than the right. Each division of a lobe is called a bronchopulmonary segment and each segment further receives air from its tertiary bronchus supplied by its arterial blood (18). Bronchioles are branches of bronchi having subdivisions called pulmonary lobules. Further, each of the lobules receives its large bronchiole having multiple branches and an interlobular septum separates the lobules from each other. The epithelial lining of respiratory bronchioles, upon deeper penetration into the lungs, gets converted to from simple squamous from simple cuboidal. These bronchioles further subdivide into 2–11 alveolar ducts having around 25 orders of branching. First-order branching refers to the branching from the trachea into primary bronchi, whereas second-order branching extends to the secondary bronchi and the alveolar ducts from the primary bronchi (18,19). Numerous alveoli and alveolar sacs surround the alveoli. An alveolus is a cup-shaped pouch lined by simple squamous epithelium, supported by a thin elastic basement membrane. An alveolar sac contains two or more alveoli with a common opening. Alveolar macrophages, also called dust cells, are associated with the alveolar wall and are responsible for phagocytosis, thereby removing dust particles and other

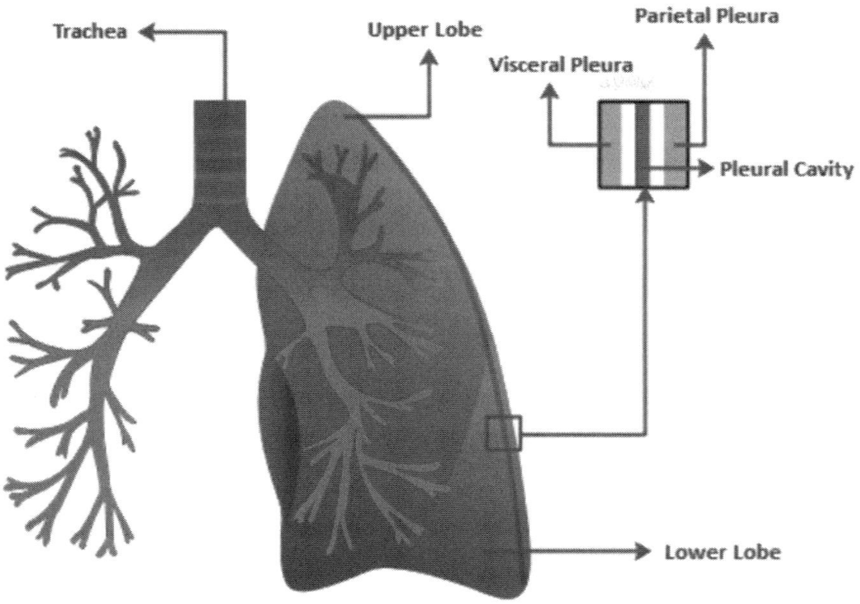

FIGURE 24.1 Anatomy of the Pleural Cavity

debris from the alveolar spaces (20). The exchange of oxygen and carbon dioxide between the air spaces in the lungs and the blood takes place by diffusion across the alveolar and capillary walls, which together form the respiratory membrane (19,20).

Blood supply to the lungs occurs through the pulmonary circulation and helps in gaseous exchange by the lungs. Deoxygenated blood travels to the lungs and picks up oxygen for transportation to the tissues of the body. The pulmonary artery follows the bronchi and branches multiple times with progressive reductions in its diameter. One arteriole along with venule supplies and drains one pulmonary lobule. Reaching near to the alveoli, the pulmonary capillary network is formed consisting of tiny vessels with thin walls. These wall of the capillaries thus formed meets the walls of alveoli forming the respiratory membrane. Oxygenated blood is drained off from the alveoli by pulmonary veins which finally exit the lungs through the hilum (17).

Bronchoconstriction and bronchodilation of the airway passage are achieved through the innervations of parasympathetic and sympathetic nervous systems, respectively. In addition to these, the autonomic nervous system also regulates coughing as well as oxygen and carbon dioxide levels of the lungs. Further, nitric oxide acts also as a neurotransmitter of the inhibitory nonadrenergic non-cholinergic nerves (21,22).

The process of gaseous exchange in the human body, called respiration, has three steps:

- Pulmonary ventilation or breathing: This is defined as the inhalation or inflow and exhalation or outflow of air and deals with the exchange of air between the alveoli of the lungs and the atmosphere.
- External (pulmonary) respiration: This is the exchange of gases between the alveoli of the lungs and the blood in pulmonary capillaries across the respiratory membrane. In this process, pulmonary capillary blood gains O_2 and loses CO_2.
- Internal (tissue) respiration: This is the exchange of gases between blood in systemic capillaries and tissue cells. In this step, the blood loses O_2 and gains CO_2(23).

Inspiration is the process of breathing. Prior to each inhalation the air pressure inside the lungs becomes the same as the air pressure of the atmosphere, which at sea level is about 760 millimeters of mercury (mmHg) or 1 atmosphere (atm). For flow of air to the lungs, the atmospheric pressure must be greater than the alveolar air pressure. Enhancement of lung size leads to achievement of this condition. Expiration is breathing out and occurs due to a pressure difference between the lungs and atmosphere, but in an opposite direction in comparison to inspiration. Due to the lack of involvement of muscular contractions, normal exhalation during quiet breathing, unlike inhalation, is a passive process (23,24).

24.2.1 Lung Volumes and Capacities

During rest, an adult breathes 12 times a minute, during which about 500 ml of air moves into and out of the lungs with each inhalation and exhalation (25). The total volume of one breath is known as the tidal volume (TV). The minute ventilation (MV), the total volume of air inhaled and exhaled each minute, is respiratory rate multiplied by tidal volume (23,25).

A spirometer or respirometer apparatus gives an account of the exchanged air volume during breathing and the respiratory rate, and the record is called a spirogram. An upward deflection indicates inhalation, and the downward deflection indicates exhalation (23,25).

The alveolar ventilation rate is defined as the volume of air that reaches the respiratory zone per minute. After a very deep breath, one can inhale more than 500 ml and this excess inhaled air is called the inspiratory reserve volume. Similar is the expiratory reserve volume that is the volume of air exhaled by forceful expiration in addition to 500 ml of tidal volume (25). Some amount of air has been found remaining in the lungs even after forceful expiration due to subatmospheric intrapleural pressure. This volume is called the residual volume and spirometry cannot give a measurement of this (23,25).

24.2.2 Mechanism of Breathing

During tidal inspiration, intrapleural pressure drops from −5 to −8 cm H_2O, leading to the drop of intra-alveolar pressure to 1 cm H_2O below atmospheric pressure (PAtm), thereby resulting in airflow into alveoli. The decreased intrapleural pressure, in turn, diminishes the airway resistance by dilating the small airways causing increased airflow. Tidal expiration is the reverse of this sequence. Intrapleural pressure increases from −8 back to −5 cm H_2O and intra-alveolar pressure increases 1 cm H_2O above atmospheric pressure when inspiratory muscles relax, leading to the decreased dimensions of the thoracic cage. This leads to the flow of air outside the alveoli following the pressure gradient and is hence regarded as a passive process because no muscle contraction is required to carry out expiration. Accessory muscles of inspiration have to be activated in case inspiration above the tidal limit is required. Similarly, while undergoing a forceful expiration, the thoracic cage is compressed to the maximum. Both intrapleural pressure and intra-alveolar pressure rises above the atmospheric pressure. Intra-airway pressure also exhibits a gradual decline from the area next to the alveoli upwards (24,26,27). Figure 24.2 depicts Intrapleural pressure and intra-alveolar pressure during inspiration (a) and expiration (b).

24.3 General Clearance Mechanism from Lungs

Mucociliary clearance of the lungs is considered to be the basic innate defense mechanism of the organ against all types of inhaled particles. It is made up of a layer of mucus and ciliated epithelium. In addition to these, ciliated columnar cells, airway surface liquid layer, and tracheobronchial epithelium consisting of the goblet cells help in the defense (28). Defense mechanisms of the lung also include various anatomical barriers, cough, aerodynamic changes, and different immune mechanisms. Tracheobronchial (TB) clearance is regarded as the fast-clearance phase and the mechanical clearance from the alveolar region is regarded as the slow-clearance phase. Mucociliary clearance in the tracheobronchial region occurs during the first 24 hours of exposure followed by the slow bronchial clearance phase (29). Hence, the proper ciliary function is essential for effective mucociliary clearance.

Cilia are an organelle responsible for transporting foreign materials in the respiratory tract toward the mouth where they can be either swallowed or expectorated. Each cilium size ranges about 6 μm long with a diameter of 250 nm. The number of cilia present in the airways is around 10^9 cilia per cm^2 usually being longer and more densely packed in the larger respiratory airways than in the bronchioles (30). To achieve this, the cilia

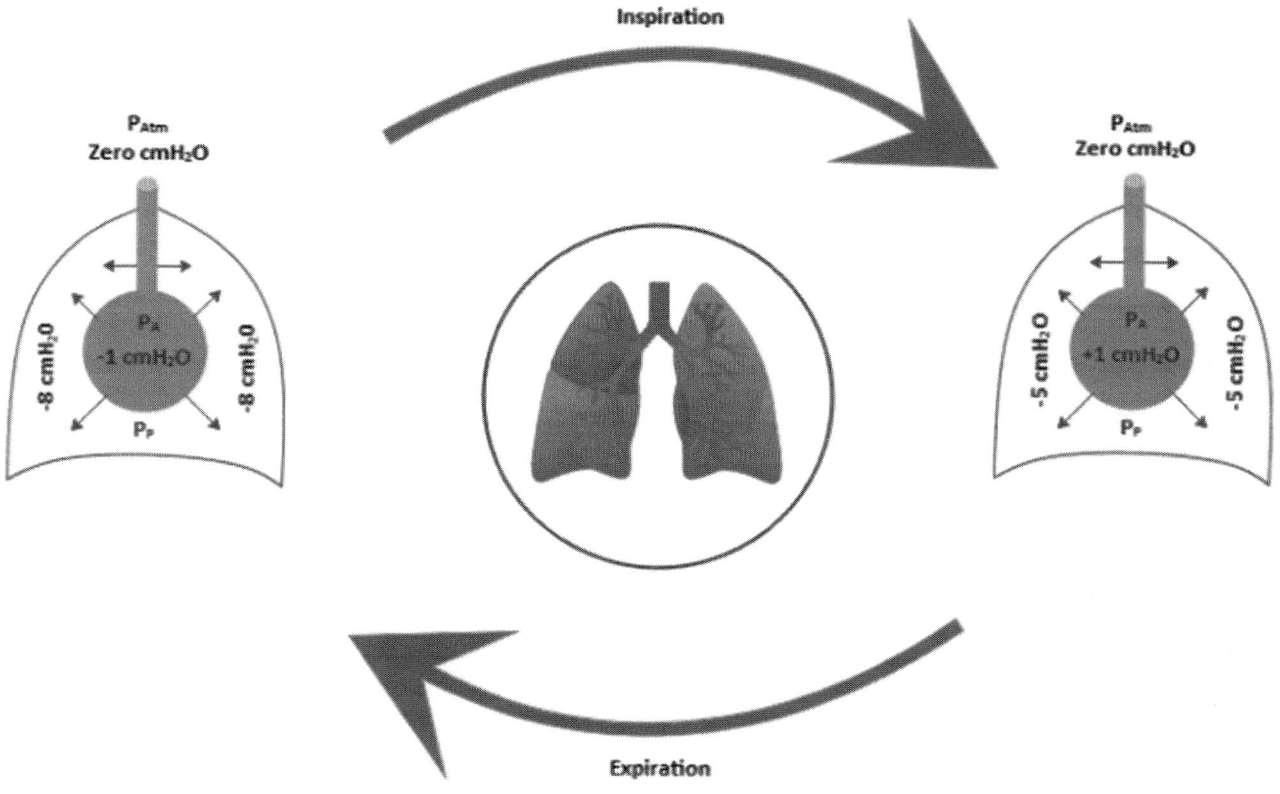

FIGURE 24.2 Intrapleural Pressure and Intra-alveolar Pressure During Inspiration (a) and Expiration (b) (P_{Atm} is Atmospheric Pressure, P_A is Intra-Alveolar Pressure, P_P is Intra-Pleural Pressure)

beat in coordinated metachronal waves at a beat frequency that has multiple physiological regulators (31). The airway mucus consists of the sole phase, a periciliary liquid, approximately 5–10 mm deep, allowing the cilia to beat, and a gel phase on the surface of the cilia of 2–20 mm thickness. Another defense mechanism, cough, leads to the expulsion of debris or infected mucus and is generated by forced expiration (32). Mucociliary clearance could be disrupted by hindrance in movements of the cilia due to genetic factors or temporary dysfunction of cilia due to environmental influences, and dehydration of the mucus may lead to ineffective clearance (30). Various approaches have been designed for controlling the volume of airway mucus and they involve the mucociliary transport, mass exchange between the mucus and the epithelium or the submucosal glands, interaction between lungs, airflow and the mucus, and water exchange between mucus and inhaled or exhaled air (33). Different other processes have also followed that lead to the ventilation of the obstructed regions of the lung, like collateral ventilation, different positioning techniques for airway clearance to enhance ventilation. Collateral ventilation occurs between the adjacent lung segments through collateral channels, which allow air to move through these pathways due to the pressure differences between adjacent lung units and thus function to minimize the collapse of lung units. There are three types of collateral connections, namely, channels of Lambert, pores of Kohn, and pathways/channels of Martin (34).

Figure 24.3 depicts the optimal positioning for airway clearance techniques to enhance ventilation to obstructed regions of the lung.

Activation of airway epithelial sodium channels is responsible for postnatal lung fluid resorption and thus has a significant biological role in the prevention of respiratory

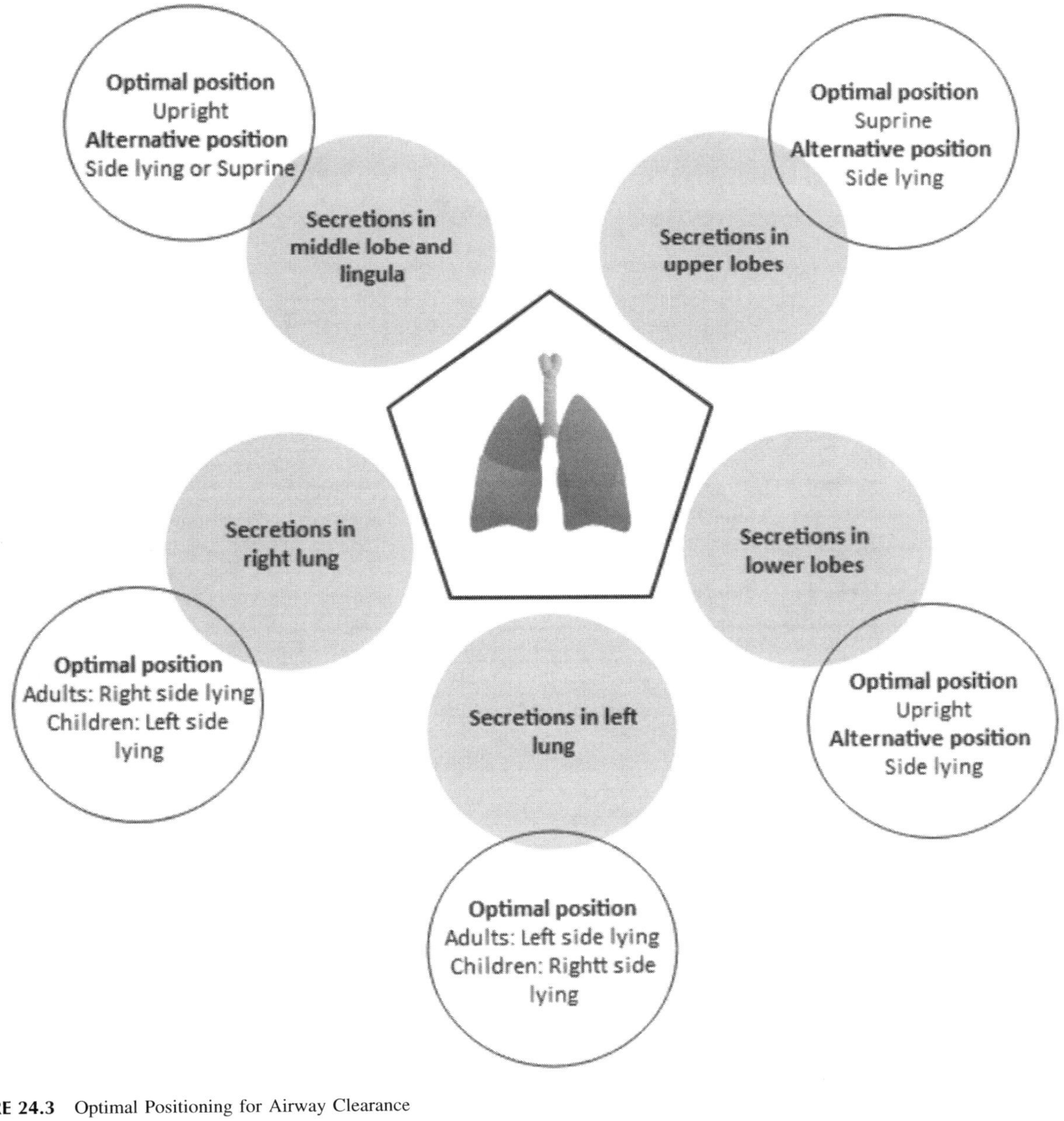

FIGURE 24.3 Optimal Positioning for Airway Clearance

distress in pediatric patients (35). Treatment of the pediatric respiratory population involves respiratory physiotherapy that focuses on facilitating the expectoration of excessive mucus. This helps to overcome chronic obstructive respiratory diseases. Airway clearance techniques (ACTs) aims at attaining mucociliary clearance by enhancing mucus mobilization, reducing airway resistance, and improvement of ventilation and gaseous exchange, thus reducing work of breathing (36).

Several mechanisms have also been proposed for the withdrawal of protein from the alveoli during the slow clearance phase and involve clearance by the mucociliary escalator, passive diffusion between cells in the epithelial barrier, intra-alveolar catabolism, phagocytosis by macrophages, and endocytic transport across the epithelial cells in vesicles, a process known as transcytosis (37). Thus the general clearance occurs through the mucociliary clearance, the removal of alveolar protein, and the alveolar degradation with the help of macrophages.

24.4 Clearance Kinetics of Liposome from Lungs

Through many complex clearance mechanisms, deposited particles are actively purged from the lungs. Swallowing, expectoration, and cough are the first line of clearance pathways involved in the nasal/oropharynx and tracheobronchial tree. *Mucociliary escalator* is an important system of clearance for inhaled aerosols, including liposomes (6). In the work of Farr et al. (38) the patterns of temporal clearance that are consistent with the function of the mucociliary escalator have been shown in the aerosolized radiolabeled liposomes delivered into central areas of the tracheobronchial. The primary clearance process in the alveolar regions of the lung is the uptake by pulmonary alveolar macrophages (38). That endocytic uptake of aerosolized liposomes by pulmonary alveolar macrophages occurs in vivo and has been reported by Forsgren et al. (39) and by Myers et al. (40). Contrary to other inhaled particles that reach the alveoli, liposomes are also absorbed into the surfactant phospholipid reservoir where the phospholipids are collected, processed, and recycled by the alveolar type II cell (41,42).

The uptake of radiolabeled dipalmitoylphosphatidylcholine (DPPC) in phosphate-buffered saline, administered as an aerosol in rats, has been investigated by the work of Geiger et al. (43). Type I and II cells have been found to take up rapidly with a slower aggregation in alveolar macrophages. The radiolabel was quickly emitted to circulation from the cells of type I, while for the cells of type II retention was maintained for several hours. Only a small fraction of the label was subjected to airway clearance, which states that an alveolar deposition was primarily the result of recorded droplet size of 1.5 μm. Later in the work of Hallman et al. (44) ^{32}p label surfactant was added to the lungs of rabbits and a clearance rate of 3.5–6.0 % h^{-1} was observed with a significant proportion of a label associated with an endogenous surfactant pool. Oyarzun et al. (41) measured the lung clearance of DPPC from spontaneously respirating anesthetized rabbits after administering radiolabeled SUVs (L-DPPC & "Mixed" L-DPPC and L-PG vesicle) into the left lung (20%). In the case of enhancement in the ventilation of 100% with respect to the control (normal breathing) [^{14}C] L-DPPC alveolar clearance was increased from 7.8% h^{-1} to 13.3% h^{-1}. The fusogenic lipid PG and regular breathing alveolar clearance was 13%.

The clearance between animals controlled and experimental, or between isomers in the lung as a whole, was no different, with a mean of 3.1% h^{-1}(41). Haque et al. (45) studied the impact of pulmonary inflammation on lung distribution and clearance kinetics. The authors studied the pulmonary pharmacokinetics of a ^{3}H-labeled PEGylated liposome loaded with a model drug (ciprofloxacin) after intra-tracheal administration to healthy rats and compared the outcomes with rats with bleomycin-induced lung inflammation after the same experimental protocol. The initial clearance of the liposomes from inflamed lungs was more rapid than the healthy lungs but became similar after 3 days. In spite of higher retention of mucus in inflamed lungs in comparison to the healthy lungs, the mucociliary clearance was more efficient from healthy lungs. The pharmacokinetic profile of ciprofloxacin was not different between rats with healthy or inflamed lungs after pulmonary administration. However, the plasma pharmacokinetics of 3H-phosphatidylcholine evidenced higher liposome bioavailability and more prolonged absorption from inflamed lungs (45).

Claypool et al. (1984) isolated ethanol/ether soluble apoproteins from rat lung surfactant and examined its impact on the absorption of synthetic liposomal phospholipid by type II cells. After 24 hours incubation, an estimated 60% rise in apoproteins was achieved in comparison to the controls. These apoproteins have also increased the involvement of alveolar macrophage-mediated uptake and were thus suggested as the possible physiological regulators for lung surfactant phospholipid clearance (46). A publication by Pettenazzo et al. (47) showed the clearance of phosphatidylcholine and cholesterol from metaproterenol containing liposomes from the lungs of adult rabbits. After tracheal administration of the liposomes with [^{14}C] radiolabeling on cholesterol and phosphatidylcholine the pharmacokinetic events were observed for 24 hours. The drug did not impact cholesterol or phosphatidylcholine clearance rates. The cholesterol clearance was similar (20–30% per 24 hours) for the three preparations and less than the phosphatidylcholine clearance calculated (35–56% per 24 hours). The liposomal lipid clearance was not altered by metaproterenol (47). Chougule et al. (48) formulated tacrolimus encapsulated liposomes by a thin film hydration method followed by spray drying. In vivo studies showed maximal residence of liposome-encapsulated tacrolimus within the lungs of 24 hours, suggesting delayed clearance from the lungs in comparison to free tacrolimus (48).

24.5 Clearance Kinetics of Solid Lipid Nanoparticles from Lungs

Still, now limited studies have been conducted investigating the lung clearance kinetics of SLN. Usually, the retention time of the drug in the lung in the form of SLN after pulmonary administration is more in comparison to the lung targeted delivery of the same drug through other routes. It is supported by the work of Varshosaz et al. (3) where 99mTc labelled amikacin-loaded cholesterol SLNs were administered through pulmonary and intravenous routes to male rats and biodistribution studies were conducted. The outcomes were compared with the same study with free drug. It was observed that, followed by I.V. administration, the SLNs were distributed in almost all organs, but after pulmonary administration, the drug was accumulated more in the lungs, even after 6 hours of administration (3). In the case of pulmonary administration, the lung retention time of drug-loaded SLN also proved to be greater in comparison to the solution of the same drug. Zhao et al. (49) developed SLNs for sustained pulmonary delivery of Yuxingcao essential oil (YEO). Enhancement in pulmonary retention (up to 24 hours) and in AUC values (4.5-7.7–folds) were observed after intra-tracheal administration to rats, in comparison to the intra-tracheally dosed YEO solution. When the dose of YEO solution administered intratracheally to rats was 1.0 mg/kg, the lung concentration of 2-undecanone was only detected within 6 hours. The tendency of hydrophobic small molecules to undergo rapid pulmonary absorption clearance was responsible for the rapid elimination of 2-undecanone (49). Shilpi and Dayal (50) developed lactoferrin (Lf)-coupled SLNs bearing rifampicin as a target-oriented drug delivery system for the lungs. An in vivo biodistribution study showed 3.05 times enhancement of drug uptake by the lungs in comparison with uncoupled SLNs. The fluroscence study also showed enhanced uptake of Lf-coupled SLNs in the lung (50).

24.6 Conclusion

Liposomes and SLNs are considered an emerging surrogate for the traditional inhalable drug delivery strategies with several possible advantages over them, including controlled and prolonged delivery as well as a more satisfying safety profile. The convincing safety profile associated with these nanocarriers lies in the utilization of biocompatible lipids in their fabrication that are highly tolerable to the lungs and the body. As reviewed herein, many investigations on the clearance mechanism of loaded drugs in liposomes and SLNs after pulmonary administration have been performed. But in comparison to that, the works addressing the clearance mechanism of different parts of these nanocarriers (except loaded drugs) are limited. It ultimately results in a fragmented understanding of this issue. The rate and mechanism of the clearance of the nanocarrier components (including loaded drugs) may also be modified under certain physiological as well as pathological conditions. The importance of research regarding the mechanism and the molecular insight of clearance and disposition of these nanocarrier components after pulmonary administration is very relevant and obvious. Further study is required to enlighten this matter and to predict the in vivo disposition pattern of these nanocarriers after pulmonary administration.

REFERENCES

1. Liu J, Gong T, Fu H, Wang C, Wang X, Chen Q, Zhang Q, He Q, and Zhang Z. Solid lipid nanoparticles for pulmonary delivery of insulin. *International Journal of Pharmaceutics*. 2008; 356: 333–344.
2. Clancy J, Dupont L, Konstan M, Billings J, Fustik S, Goss C, Lymp J, Minic P, Quittner A, and Rubenstein R. Phase II studies of nebulised Arikace in CF patients with *Pseudomonas aeruginosa* infection. *Thorax*. 2013; 68: 818–825.
3. Varshosaz J, Ghaffari S, Mirshojaei SF, Jafarian A, Atyabi F, Kobarfard F, and Azarmi S. Biodistribution of amikacin solid lipid nanoparticles after pulmonary delivery. *BioMed Research International*. 2013; 2013: 1–8.
4. Rudokas M, Najlah M, Alhnan MA, and Elhissi A. Liposome delivery systems for inhalation: A critical review highlighting formulation issues and anticancer applications. *Medical Principles and Practice*. 2016; 25: 60–72.
5. Bassetti M, Vena A, Russo A, and Peghin M. Inhaled liposomal antimicrobial delivery in lung infections. *Drugs*. 2020; 10: 1309–1318.
6. Schreier H, Gonzalez-rothib RJ, and Stecenkoc AA . Pulmonary delivery of liposomes. *Journal of Controlled Release*. 1993; 24: 209–223.
7. Haque S, Whittaker M, Mcintosh MP, Pouton CW, Phipps S, and Kaminskas LM. A comparison of the lung clearance kinetics of Solid lipid nanoparticles and Liposomes by following the 3H-labelled structural lipids after pulmonary delivery in rats. *European Journal of Pharmaceutics and Biopharmaceutics*. 2018; 62: 210–232.
8. Skupin-mrugalska P. Liposome-Based Drug Delivery for Lung Cancer. In: Kesharwani P (ed.), Nanotechnology-Based Targeted Drug Delivery Systems for Lung Cancer. Elsevier Inc. 2019; pp. 123–160.
9. Allen T and Cullis P. Liposomal drug delivery systems: From concept to clinical applications. *Advanced Drug Delivery Reviews*. 2013; 65: 210–231.
10. Alipour M, Suntres ZE, Halwani M, Azghani AO, and Omri A. Activity and interactions of liposomal antibiotics in presence of polyanions and sputum of patients with cystic fibrosis. *PLoS ONE*. 2009; 4: 20–41.
11. Dolatabadi J, Valizadeh H, and Hamishehkar H. Solid lipid nanoparticles as efficient drug and gene delivery systems: Recent breakthroughs. *Advanced Pharmaceutical Bulletin*. 2015; 5: 211–224.
12. Müller R, Mäder K, and Gohla S. Solid lipid nanoparticles (SLN) for controlled drug delivery – a review of the state of the art. *European Journal of Pharmaceutics and Biopharmaceutics*. 2000; 50: 161–177.
13. Ji P, Yu T, Liu Y, Jiang J, Xu J, Zhao Y, Hao Y, Qio Y, Zhao W, and Wu C. Naringenin-loaded solid lipid nanoparticles: Preparation, controlled delivery, cellular uptake, and pulmonary pharmacokinetics. *Drug Design, Development and Therapy*. 2016; 10: 911–925.
14. Oberdorster G. Lung particle overload: Implications for occupational exposures to particles. *Regulatory Toxicology and Pharmacology*. 1995; 21: 123–135.
15. Akulian J, Lonny Y, and David F. The evaluation and clinical application of pleural physiology. *Clinics in Chest Medicine*. 2013; 34: 11–19.
16. Charalampidis C, Andrianna Y, George L, Sofia B, Ioannis M, Vasilis K, and Loannis K. Pleura space anatomy. *Journal of Thoracic Disease*. 2015; 7: S27–S34.
17. Anatomy & Physiology. Houston, Texas: OpenStax College, Rice University, 2013; pp. 973–1008. https://openstax.org/details/books/anatomy-and-physiology

18. Patwa A., and Shah A. Anatomy and physiology of respiratory system relevant to anaesthesia. *Indianesia*. 2015; 59: 533–541.
19. Mahto K, Janna S, Khatun T, and Sznitman J. Respiratory physiology on a chip. *Scientifica*. 2012; 8: 212–234.
20. Hyde DM, Qutayba H, and Charles GI. Anatomy, pathology, and physiology of the tracheobronchial tree: emphasis on the distal airways. *Journal of Allergy and Clinical Immunology*. 2009; 124: S72–S77.
21. Delaunois L. Anatomy and physiology of collateral respiratory pathways. *European Respiratory Journal*. 1989; 2: 893–904.
22. Ricciardolo FLM, Sterk PJ, Gaston B, and Folkerts G Nitric Oxide in health and disease of the respiratory system. *Physiological Reviews*. 2004; 84: 731–765.
23. Tortora GJ and Derrickson B. Principles of Anatomy and Physiology. John Wiley & Sons, Inc. United States of America. 2009.
24. Papa S, Giuseppe F, Giulia MP and Riccardo P. Asthma and respiratory physiology: Putting lung function into perspective. *Respirology*. 2014; 19: 960–969.
25. Ince LM, Marie P, and Julie EG. Lung physiology and defense. *Current Opinion in Physiology*. 2018; 5: 9–15.
26. Plantier L Cazes A, Dinh-Xuan AT Bancal C Marchand-Adam S, and Crestani B. Physiology of the lung in idiopathic pulmonary fibrosis. European Respiratory Review. 2018; 27: 111–118.
27. Lutfi MF. The physiological basis and clinical significance of lung volume measurements. *Multidisciplinary Respiratory Medicine*. 2017; 12: 1–12.
28. Clarke SW and Pavia D. Lung mucus production and mucociliary clearance: Methods of assessment. *British Journal of Clinical Pharmacology*. 1980; 9: 537–546.
29. Hofmann W and Bahman A. The effect of lung structure on mucociliary clearance and particle retention in human and rat lungs. *Toxicological Sciences*. 2003; 73: 448–456.
30. Munkhon M and Mortensen J. Mucociliary clearance: Pathophysiological aspects. *Clinical Physiology and Functional Imaging*. 2014; 34: 171–177.
31. Ximena M and Ostrowski E. Cilia and Mucociliary Clearance. Cold Spring Harbor. 2017; 9.
32. Nicod LP. Lung defences: An overview. *European Respiratory Society*. 2005; 14: 45–50.
33. Karamaoun C, Benjamin S, Benjamin M, Alain VM, and Benoît H. New insights into the mechanisms controlling the bronchial mucus balance. *PLoS one*. 2018; 13: 1–17.
34. Maggie M, Judy Bradley J, Stuart E, and Fidelma M. Personalising airway clearance in chronic lung disease. *European Respiratory Review*. 2017; 14: 224–231.
35. Katz C, Bentur L, and Elias N. Clinical implication of lung fluid balance in the perinatal period. *Journal of Perinatology*. 2011; 31: 230–235.
36. Lauwers E, Kris I, Kim VH, and Stijn V. Outcome measures for airway clearance techniques in children with chronic obstructive lung diseases: A systematic review. *Respiratory Research*. 2020; 21: 1–16.
37. Hastings RH, Hans GF, and Michael AM. Mechanisms of alveolar protein clearance in the intact lung. *American Journal of Physiology - Lung Cellular and Molecular Physiology*. 2004; 286: L679–L689.
38. Farr SJ, Kellaway IW, Parry-Jones DR, and Woolfrey SG. ^{99m}Technetium as a marker of liposomal deposition and clearance in the human lung. *International Journal of Pharmaceutics*. 1985; 26: 303–316.
39. Forsgren PE, Modig JA, Dahlback CMO, and Axelsson BI. Prophylactic treatment with an aerosolized corticosteroid liposome in a porcine model of early ARDS induced by endotoxaemia. *Acta Orthopaedica Scandinavica*. 1990; 156: 423–431.
40. Myers MA, Niven RW, Strauh LE, Wichert BM, Schreier H, Hood CI, and Gonzalez-Rothi RJ. Pulmonary effects of chronic exposure to liposome aerosols in mice. *Experimental Lung Research*. 1993; 19: 1–19.
41. Oyarzun MJ and Clements JA, and Baritusso A. Ventilation enhances pulmonary alveolar clearance of radioactive dipalmitoyl phosphatidylcholine in liposomes. *The American Review of Respiratory Disease*. 1980; 121: 709–772.
42. Wright JR and Clements JA. Metabolism and turnover of lung surfactant. *The American Review of Respiratory Disease*. 1987; 135: 426–444.
43. Geiger K, Gallagher ML, and Hedley-White J. Cellular distribution and clearance of aerosolized dipalmitoyl lecithin. *Journal of Applied Physiology*. 1975; 39: 759–766.
44. Hallman M, Epstein BL, and Gluck L. Analysis of labelling and clearance of lung surfactant phospholipids in rabbit. *Journal of Clinical Investigation*. 1981; 68: 742–751.
45. Haque S, Feeney O, Meeusen E, Boyd BJ, Mcintosh MP, Pouton CW, Whittaker M, and Kaminskas LM. Local inflammation alters the lung disposition of a drug loaded pegylated liposome after pulmonary dosing to rats. *Journal of Controlled Release*. 2019; 307: 32–43.
46. Claypool WD, Wang DL, Chander A, and Fisher AB. An ethanol/ether soluble apoprotein from rat lung surfactant augments liposome uptake by isolated granular pneumocytes. *Journal of Clinical Investigation*. 1984; 74: 677–684.
47. Pettenazzo A, Jobc A, lkcgamL M, Abra R, Hogue E, and Mihalko P. Clearance of phosphatidylcholine and cholesterol from liposomes, liposomes loaded with mctaproterenol and rabbit surfactant from adult rabbit lungs. *The American Review of Respiratory Disease*. 1989; 139: 752–758.
48. Chougule M, Padhi B, and Mishra A. Nano-liposomal dry powder inhaler of tacrolimus: Preparation, characterization, and pulmonary pharmacokinetics. *International Journal of Nanomedicine*. 2007; 2: 675–688.
49. Zhao Y, Chang Y, Hu X, Liu C, Quan L, and Liao Y. Solid lipid nanoparticles for sustained pulmonary delivery of Yuxingcao essential oil: Preparation, characterization and in vivo evaluation. *International Journal of Pharmaceutics*. 2016; 516: 364–371.
50. Shilpi S and Dayal V. Assessment of lactoferrin-conjugated solid lipid nanoparticles for efficient targeting to the lung. *Progress in Biomaterials*. 2015; 4: 55–63.

25

Solid Lipid Nanoparticles for Sustained Pulmonary Delivery of Herbal Drugs for Lung Delivery: Preparation, Characterization, and In Vivo Evaluation

Nayanmoni Boruah[1], Himangshu Sharma[2], and Hemanta Kumar Sharma[1]
[1]Department of Pharmaceutical Sciences, Faculty of Science and Engineering University, Dibrugarh, Assam, India
[2]Life Sciences Division, Institute of Advanced Study in Science and Technology, Guwahati, Assam, India

25.1 Introduction

Solid lipid nanoparticles (SLN) are a practically feasible, emerging novel drug delivery system for targeted and sustained drug delivery. SLN have many advantages over the conventional drug delivery system as well as other novel drug delivery systems. Biocompatbility, less toxicity, sustained release, increase in drug stability, drug targeting, scale-up possibility, and the ability to incorporate both lipophilic and hydrophilic drugs are the significant advantages of SLN. Due to the capability to incorporate both lipophilic and hydrophilic drugs, SLN are emerging as potential drug delivery carriers for herbal drugs. Because of tremedous research in the field of herbal drugs, new molecules have been derived from herbal sources, especially anti-cancer molecules. Herbal drugs and extracts have the potential to treat different diseases. However, the stability of the herbal drug formulation and drug concentration at the target site are the major obstacles in using herbal drugs. Many herbal drugs have shown promising activity for respiratory diseases. To achieve the desired concentration of the drug at the target site of the lung, there is need of a novel drug delivery system. SLN have been demonstrated as a potential carrier for delivery of drugs to the lungs to get the desired effect of the drugs. The capability of SLN to graft with ligand molecules enhances therapeutic index and site specific drug delivery. SLN can be formulated into aerosol products for pulmonary delivery, which is one of the most requisite parameters for respiratory delivery of SLN. Therefore, SLN are promising carriers for herbal drugs to treat asthma, pneumonia, cystic fibrosis, tuberculosis, lung cancer, etc. In this chapter, we will discuss the various aspects of SLN for pulmonary delivery of herbal drugs and their evaluation.

25.2 The Respiratory System

The lungs consist of two functional parts: the tracheobronchial region and the alveolar region. The tracheobronchial region ranges from the larynx to the terminal bronchioles, and the alveolar region consist of the respiratory bronchioles and alveoli. The respiratory tract is highly bifurcated and there are 23 bifurcations and therefore, 23 generations of airways. The tracheobronchial region consists of the nasal cavity, the nasopharynx, the trachea, the bronchi, and the bronchioles. It is the pathway through which the air moves into and out of the lung. The tracheobronchial region is lined by the respiratory epithelium. The respiratory epithelium is covered with thick mucus and ciliated in nature (1). The respiratory epithelium consists of the basal cells, the ciliated cells, the goblet cells, and the brush cells to cellular monolayer. The thickness of the epithelium decreases from the bronchiole to the alveoli progressively to a final thickness of 0.1–0.4 μm (1,2). About 90% of the total surface area of the lungs is composed of the alveolar area (3). It is estimated that there are about 300 million alveoli in the human lung through which gas exchange occurs by diffusion (4). The alveolar epithelium consists of type I and type II alveolar epithelial cells. The lung parenchyma is protected by a connective tissue fiber network and the surfactant system (3). The subepithelial tissue under the basement membrane of alveoli consists of blood and lymphatic vessels (2). The lymphatic vessels terminate in the hilar and mediastinal lymph nodes (5).

25.3 Natural Compounds in Respiratory Diseases

25.3.1 Antiviral Natural Compounds

Many naturally derived compounds have shown potent activity against different types of respiratory disease causing viruses. Chebulagic acid and punicalagin were isolated from *Terminalia chebula.* They inactivated free virus particles and inhibited early viral entry including attachment and penetration phase, but did not affect cell to cell transmission (6). Another compound, cimicifugin, inhibited viral attachment and internalization steps stimulating IFN-β secretion (7). Uncinoside A and B are two natural compounds found in *Selaginella uncinata.* They exhibited antiviral acitivity against respiratory syncytial virus

DOI: 10.1201/9781003046547-25

(RSV) (8). In addition, genkwanol B, genkwanol C, and stelleranol exhibited antiviral activity against RSV. These compounds were isolated from *Radix Wikstroemiae* (9). Resveratrol was also found to reduce virus-induced airway inflammation via down-regulation of IFN-γ levels during RSV infection (10).

25.3.2 Natural Compounds Against Coronavirus

Coronavirus is a group of viruses. Severe acute respiratory syndronme (SARS) coronavirus is the first coronaviruses that infected a large number of people in 2003. Later, Middle East coronavirus (MERS) emerged in 2015. It has been infecting people across the world. Recently, a novel coronavirus (SARS-CoV-2) emerged from Wuhan Province in China, and it became a pandemic within a very short period of time. This virus took thousands of lives by infecting millions of people. People with weak immune systems and other diseases are at high risk of severe SARS-CoV-2 infection. Due to the unavailability of a disease specific drug molecule as well as specific vaccine for life long immunity, tremendous effort has been applied for repurposing of drugs and development of vaccines. Humanity has defeated deadly diseases in the past to reach the present state. The natural sources have a crucial role in the fight against various diseases. Various plant sources have shown activity against deadly virus diseases, including coronavirus. The natural compounds of herbal sources for antiviral activity could help us in tackling a pandemic situation. Phenolic compounds of *Isatis indigotica* have been isolated and evaluated against SARS-CoV 3CL protease. These compounds exhibited inhibitory activity on the SARS-CoV 3CL protease (11). Amentoflavone, a compound derived from *Torreya nucifera,* exhibited inhibitory activity against SARS-CoV 3CL protease (12). Scutellarein and myricetin exhibited potent inhibitory activity on the SARS-CoV helicase protein (13). Significant inhibitory activity of *Houttuynia cordata* water extract against SARS-CoV 3CL protease and viral polymerase was demonstrated by Lau et al.(14).

25.3.3 Natural Compounds Against the Influenza Virus

Homoisoflavonoids were isolated from *Caesalpinia sappan.* These natural products exhibited antiviral activity against IFA by inhibiting viral neuraminidases inhibition (15). Similarly, flavonoid rich extract from the roots of *Scutellaria baicalensis* exhibited protective activity against influenza A virus (IAV)-induced acute lung injury by inhibition of complement system. The activity on complement system was attributed to the metabolites of the compounds of the root extract (16). Other classes of compounds, such as xanthones from *Polygala karensium* inhibited neuraminidases from IAVs (17). Similarly, Chalcones are the novel inhibitors of influenza A (H1N1) neuraminidase. They were isolated from Glycyrrhiza inflate (18). Spirooliganone B, a unique compound, was isolated from the roots of *Illicium oligandrum.* It exhibited antiviral activity against influenza A (19). Aqueous extract from *Taraxacum officinale* inhibited NP RNA levels and polymerase activity (20). Root extract of *Pelargonium sidoides* exhibited antiviral activity against influenza A. It inhibited viral entry and release, and inhibited viral hemagglutination and NA activity (21).

25.3.4 Anti-inflammatory and Anti-asthmatic Compounds

Chung et al. evaluated kaempferol-3-O-rhamnoside and kaempferol in lipopolysaccharide induced acute lung injury in mice. It attenuated lung injury in mice by activation of MAPK and NF-κB pathways and causes reduction of inflammatory cells (22). Using the same pathway, asperuloside decreased acute lung injury in a murine model pretreated with asperuloside. It decreased the levels of TNF-α, IL-1β and IL-6 (23). Similarly, mitraphylline inhibited lipopolysaccharide-mediated activation of primary human neutrophils by reduction of IL-1α, IL-1β, IL-17, TNF-α, IL-6, and IL-8 (24,25). Mitraphylline is isolated from *Uncaria tomentosa* bark (24). Naringin is a natural product isolated from *Citrus grandis*. Naringin exhibited reduction in airway hyper-responsiveness and cough exacerbation. It exerted inhibitory action on increase in leukocytes, interleukins in a model of asthma using guinea pigs (26). Moreover, Guihua et al. reported inhibitory activity of naringin on increase airway resistance and eosinophil infiltration induced by ovalbumin. It causes reduction in the levels of IL-4 and INF-δ (27).

Curcumin, one of the important natural compounds, exhibited anti-inflammation on the airway of lungs. Curcumin causes down-regulation of Notch signaling pathway (Notch1/2) receptors and transcription factor GATA-3 in ovalbumin-sensitized mice (28). Protective effects quercetin was observed on lipopolysaccharide-induced acute lung injury (29,30). Quercetin exerted inhibitory activity on inflammatory cells, thereby a protective effect on lipopolysaccharide-induced acute lung injury in mice was observed (30). Moreover, Rogerio et al. reported the activity of isoquercitrin and quercetin in allergic asthma. These two compounds exhibited anti-inflammatory activity in an experimental murine allergic asthma model. It caused reduction of influx of iNOS expression, HMGB1, and p65NF-κB. IL-10 secretion was increased (31). Another compound, eriodictyol, exhibited inhibitory activity on lipopolysaccharide-induced acute lung injury in mice. Eriodictyol exerts its effects on nuclear factor erythroid-2-related factor 2 (Nrf2) pathway and inhibits the production of inflammatory cytokines in macrophages (32). Jung et al. isolated kuwanon G from the root bark of *Morus alba.* Kuwanon G decreased the levels of IgE in ovalbumin-induced allergic asthma in mice. Reduced levels of interleukins including IL-4, IL-5, and IL-13 was observed in the sera and bronchoalveolar lavage of asthma-induced mice (33).

Another natural compound, apigenin, exhibited protective activity against ovalbumin-induced asthma through the regulation of Th17 cells. Reduction in infiltration of eosinophil in lung tissue as well as a decrease in levels of interleukins was observed during treatment with apigenin (34). Intranasal delivery of curcumin exhibited a protective effect on the airways of lung, and it also exhibited remodeling activity including airway smooth muscle thickening and mucus secretion in ovalbumin-induced chronic asthma in a murine model (35). *Siegesbeckia glabrescens* is a traditional medicinal plant in Korea. Treatment with extracts of *Siegesbeckia glabrescens* resulted in a reduction of mucus overproduction in airways in asthma murine model. The expression of iNOS and COX-2, level of interleukins and number of inflammatory cells decreases during treatment with the extract (36). Quercetin flavonoid derivatives, including

3,4-O-dimethylquercetin, 3,7-O-dimethylquercetin, and 3-O-methylquercetin are found in *Siegesbeckia glabrescens* (37). Sakuranetin is a flavonoid isolated from *Baccharis retusa*. The levels of IL-5 and eotaxin decreases on treatment with sakuranetin in an asthma induced murine model. The reduction in the number of inflammatory cells in the lung was observed on treatment with sakuranetin. Inhibition of NF-κB in lung is presumed as the mode of action of sakuranetin (38). Taguchi et al. reported the prevention of elastase-induced emphysema in mice by flavanone from *Baccharis retusa*. It prevented alveolar destruction by regulating NF-κB, MMP-9 and MMP-12 (39). Extracts of *Astragalus membranaceus*, a traditional chinese herb, consist of various pharmacologically active molecules including huangqiyiesaponin C, quercetin, and kaempferol. The extract of this plant was reported to modulate Th1/2 immune balance and activate PPARγ in a murine asthma model (40).

Luteolin, a flavonoid compound, inhibited release of cytokines content in the bronchoalveolar lavage fluid (BALF). The decrease in activity of catalase and superoxide dismutase activity resulted in the control of oxidative damage and lipid peroxidation. The mechanism of action of luteolin involves inhibition of NF-κB and the MAPK activity (41). Rungsung et al. reported that luteolin attenuated acute lung injury in experimental mouse model of sepsis through suppression in ICAM-1, NF-kappa B, oxidative stress, and partially iNOS pathways (42). Luteolin inhibited HMGB1 and protects ALI in endotoxin mice. Luteolin increased HO-1 expression through ERK1/2 signaling in a time- and concentration-dependent manner. Additionally, luteolin inhibited pro-inflammatory mediators (HMGB1, iNOS/NO, COX-2, and NF-κB activity) in lipopolysaccharide-activated RAW264.7 cells (43). Luteolin alleviated lipopolysaccharide-induced bronchopneumonia injury in vitro and in vivo by down-regulating microRNA-132 expression (44). Interestingly, luteolin exhibited anti-metastatic effects. Luteolin exhibited inhibitory activity on tumor nodules of lung metastasis induced by intravenous administration of CT-26 cells. Luteolin suppressed Raf activity and exhibited inhibitory activity on PI3K (45).

The dried root of *Paeonia lactiflora* has been used in traditional Chinese medicine. There are more than 15 components present in the water/ethanol extract of the dried *Paeonia lactiflora*. A few compounds in the extract are paeoniflorin, albiflorin, oxypaeoniflorin, and benzoylpaeoniflorin. He and Dai reported that glucosides of *Paeonia lactiflora* markedly suppressed lipopolysaccharide-induced nitric oxide production. These glucosides also suppressed the expression of inducible nitric oxide synthase (iNOS) in peritoneal macrophages (46). The production of nitric oxide and PGE2 induced by lipopolysaccharide was inhibited by paeoniflorin in stimulated macrophages (47). Further experiments demonstrated that paeoniflorin inhibited the release of IL-1β and TNF-α stimulated by lipopolysaccharide (48).

Picroside II is known as a major glycoside constituent of *Picrorhiza scrophulariiflora* (49). The rhizome of *Picrorhiza scrophulariiflora* has significant importance in traditional systems of medicine. It has been used for the treatment of a broad range of diseases including tumors in the traditional system (50). Noh et al. demonstrated that Picroside II suppressed neutrophilic lung inflammation in RAW 264.7 cells and an in vivo model of lipopolysaccharide-induced ALI (51). The anti-inflammatory effect of Picroside II might be associated with TGF-β signaling. The suppression of redox-sensitive inflammation was observed on treatment with ethanol extract of *P. scrophulariiflora* (52), whereas attenuation of the classical pathway of complement activation, ROS production, and T lymphocytes proliferation was observed on treatment with crude extract of *P. scrophulariiflora* (53). Extracts of various parts of *Morinda citrifolia* have been used for the treatment of different pulmonary diseases (54,55). The extracts of *Morinda citrifolia* are enriched with the presence of several antioxidant natural constituents such as iridoid glycosides, flavonoids, etc. (56,57). In addition, Song et al. reported that Picroside C suppressed airway inflammation induced by cigarette smoke. Picroside C attenuatedthe neutrophil influx, production of ROS production, IL-6, and TNF-α as well as the activity of elastase. The mechanism of action of Picroside C is hypothesized to be the inhibition of the NF-κB pathway (58).

25.3.5 Anti-cancer Compounds

Green tea (*Camellia sinensis* leaves) contains different types of polyphenol derivatives. These polyphenol derivatives have exhibited potent chemopreventive activity against lung cancer. A few polyphenol compounds present in *Camellia sinensis* are catechin, epicatechin, epigallocatechin, epicatechin gallate, and epigallocatechin gallate. The mechanism of action of the constituents of *Camellia sinensis* against cancer involves induction of apoptosis, inhibition of TNF-α expression, induction of phase II enzymes, and auto-oxidation (59,60). Another natural compound, resveratrol, exhibited inhibitory activity on lung cancer cells. The mechanism of action of resveratrol is induction of premature senescence through ROS-mediated DNA damage. Red wine is a major source of resveratrol and many other bioactive compounds (61,62). Wogonin is a flavonoid compound isolated from the plant *Scutellariae radix*. It exhibited cytotoxicity against A549 and A427 cells without affecting normal cells at a concentration of 50 μM. Increase in the levels of caspases and generation of ROS was associated with the cell apoptosis and inhibition of tumor growth (63).

Alkaloidal compounds were found to exert potent anti-cancer activity against lung cancer. A tambjamine analog exhibited potent inhibitory action of survivin by repressing gene expression through the Janus kinase/Signal Transducer and Transcription-3 (JAK/STAT3)/survivin signaling pathway (64). Lee et al. reported that the anti-cancer effect of luteolin is mediated by down regulation of TAM receptor tyrosine kinases in non small cell lung cancer cells (65). Luteolin could act as a radio-sensitizer in non small cell lung cancer cells by enhancing apoptotic cell death through activation of a p38/ROS/caspase cascade (66). Many protein molecules are highly expressed in cancer cells in lungs. Claudin-2 is one of the highly expressed protein molecules in human lung adenocarcinoma tissues. Knockdown of claudin-2 decreases cell proliferation, therefore it could be a novel target for cancer chemotherapy. Luteolin was found to decrease claudin-2 expression in lung adenocarcinoma A549 cells in a concentration-dependent manner (67). Other anti-cancer mechanisms include pro-apoptotic effect and anti-migration effects on A549 lung adenocarcinoma cells through the activation of MEK/ERK signaling pathway (68). Many studies have reported anti-cancer activities of luteolin involving different mechanisms (69–75). Many natural compounds have exhibited potent activity against lung cancer.

25.3.6 Saikosaponins in Lung Diseases

Saikosaponins are naturally occurring triterpene glycosides isolated from plants. They are found in *Bupleurum* spp., *Heteromorpha* spp., and *Scrophularia scorodonia.* Saikosaponins (A, B_2, C and D) were reported to exert antiviral activity against HCoV-229E. Saikosaponin B_2 exhibited potent antiviral activity against HCoV-229E (IC_{50} = 1.7 ± 0.1 μmol/l). Interestingly, no cytotoxic effects on target cells was observed on treatment with Saikosaponin B_2 at concentrations that achieved antiviral activity (CC_{50} = 383.3 ± 0.2 μmol/l; SI = 221.9). They inhibited viral attachment and penetration stages against HCoV-22E9 (76). Saikosaponin A is a triterpene saponin isolated from *Radix bupleuri.* Zhu et al. reported that Saikosaponin A exhibited anti-inflammatory activity in lipopolysaccharide-stimulated RAW 264.7 cells through its activity on NF-κB and MAPK pathways (77). Saikosaponin A inhibited lipopolysaccharide-induced TNF-α and IL-1β production in primary mouse macrophages (78). Saikosaponin A has been reported to protect against experimental sepsis via inhibition of NF-κB activation in vivo (79). Saikosaponin A exerted anti-inflammatory activity on lipopolysaccharide-induced ALI through suppression of lipopolysaccharide-induced NF-κB activation and NLRP3 inflammasome expression (80). Similarly, Saikosaponin A inhibited cigarette smoke-induced oxidant stress and inflammatory responses. In cigarette smoke-induced lung inflammation, Saikosaponin A significantly inhibited inflammatory cell infiltration and production of nitric oxide, TNF-α, and IL-1β (81). Saikosaponin D exhibits potential activity in anti-inflammation, cytoprotection of normal cells, liver disease, and systemic lupus erythematosus through different signaling pathways (82,83). Hsu et al. reported that Saikosaponin D inhibited the cell growth in human lung cancer cell line. Induction of apoptosis and blocking cell cycle progression in the G1 phase are the mechanisms of action through which Saikosaponin D exerts its anti-cancer activity. Significant increase in the expression of p53 and p21/WAF1 protein was observed on treatment with the compound which contributes to cell cycle arrest (84). Saikosaponin D exhibited initiation of apoptosis and inhibition of proliferation in A549 cells. Saikosaponin D prevented the transition of subG1 to S phase and subG2 to M phase. It arrested the cell cycle at both G1 and G2 phases with the concomitant accumulation of subG1 and subG2 cells. Due to the incomplete previous phase and activation of the next one, the accumulated subG1 and subG2 cells could promote early apoptosis and inhibit late cell proliferation, respectively (85). Fnag et al. reported isolation of five novel Saikosaponins from the roots of *Bupleurum marginatum.* Three saikosaponin compounds (nepasaikosaponin k, saikosaponin n and saikosaponin h) exhibited more potent antiviral activity than ribavirin against the influenza A virus A/WSN/33 (H1N1) in 293T-Gluc cells (86).

25.4 Solid Lipid Nanoparticles for Pulmonary Deliver

Pulmonary delivery of drugs is an emerging route of drug delivery to lungs to achieve high bioavailability of drugs at target site, sustained drug concentration, and reduced systemic toxicities. There are several advantages of drug delivery to lung, including a large surface area for absorption, thin alveolar epithelium, absence of first-pass metabolism, and high bioavailability. With the invention of nebulizer therapy, pulmonary therapy has been used to treat many diseases, mainly asthma and chronic pulmonary diseases. The efficacy and safety of the inhalation depend on the drug molecules, formulation, and nebulizer system. With the advancement in drug delivery system, novel formulations are gaining importance for inhalation therapy because of their potential properties in drug delivery. Liposomes, SLN, magnetic nanoparticles, and dendrimers are the primary nano-sized drug carriers for targeted drug delivery. Due to favorable properties for targeted drug delivery, SLN are potential carriers for different lung diseases. SLN have been widely studied for anti-cancer and anti-tuberculosis drug delivery to lungs due to their biocompatible and biodegradable nature. Physical stability, biocompatibility, drug stability, controlled release, and low cytotoxicity are the favorable properties of the SLN (87). Moreover, SLN can be formulated without organic solvent (88). Due to their size, nanoparticles can be easily aerosolized into droplets with aerodynamically suitable properties for sufficiently deep lung deposition. Furthermore, nanoparticles can be grafted to various ligands to enhance the adherence to the mucosal surface of the lung for a longer period of time (89). Biodegradation of excipients does not cause any significant toxicity (90).

Various components are used for preparation of solid lipid nanoparticles. They are made up of solid lipid, emulsifier, water, and solvent. Triglycerides, partial glycerides, fatty acids, steroids, and waxes are used as lipid in preparation of solid lipid nanoparticles. Emulsifiers are used to stabilize the lipid dispersion. They are used alone or in combination.

25.5 Methods for Preparation of Solid Lipid Nanoparticles

Formulation of solid lipid nanoparticles is influenced by many factors, including physicochemical properties of the drug, desired particle characteristics, stability of the formulation, route of administration, and availability of equipment.

25.5.1 High Shear Homogenization

The high shear homogenization method involves melt emulsification. The particle size and zeta potential is influenced by emulsification time, rate of stirring, and cooling condition. High shear homogenization includes two general methods: hot homogenization and cold homogenization (91). These methods are easy to handle and used frequently. However, the presence of microparticles compromises the dispersion quality of the product (92).

25.5.2 Hot Homogenization

Hot homogenization is a well-known technique for preparation of SLN. In the hot homogenization technique, the drug is incorporated into melted lipid. A coarse o/w emulsion is formed by dispersing the lipid phase into a hot, aqueous surfactant solution under continuous stirring. It is then homogenized at the temperature above the melting point of the lipid using high pressure homogenizer to form o/w nano-emulsion, which is cooled to room temperature for solidification and formation of solid lipid nanoparticles. In exceptional cases, cooling to refrigeration temperatures may be required to produce SLN (93). Due to the high processing temperature, decrease in viscosity of the lipidic phase occurs, which results in lower particle sizes. However, a high processing temperature might accelerate degradation of drug molecules and other excipients. SLN particles are obtained after three to five passes through a high pressure homogenizer. Good quality SLN are formed, usually after three to five homogenization cycles at 500–1500 bar. Particle coalescence may occur due to increasing homogenization. Therefore, homogenization should be minimized to reduce high kinetic energy of the particles.

25.5.3 Cold Homogenization

Many problems associated with the hot homogenization process, such as temperature, accelerated drug degradation, and degradation of carrier lipid can be overcome by using a cold homogenization process (91). The first preparation step is similar to the preparation of SLN by hot homogenization techniques. The drug is dissolved or dispersed in the molten lipid. A homogenous drug distribution in the lipid matrix is achieved by cooling the molten lipid rapidly. The drug containing the solid lipid is pulverized into fine powder of microparticles by ball milling. The typical sizes are attained in the range of 50 to 100 μm. The microparticles are dispersed into a cold surfactant solution, followed by high pressure homogenization to generate SLN. The temperature is controlled at room or below room temperature with appropriate temperature control to prevent temperature rise during high pressure processing. However, the homogenization process requires high energy, which affects the homogenization process. Thus, SLN produced by hot homogenization are monodispersed as compared to cold homogenization.

25.5.4 Solid Lipid Nanoparticles Prepared by Solvent Evaporation

In this method, the lipid material and drug are dissolved in immiscible organic solvent. The lipid solution is emulsified in an aqueous phase under continuous stirring. Nanoparticle dispersion is formed by the precipitation of the lipid in the aqueous medium due to evaporation of the organic solvent under reduced pressure. This process results in small size SLN. Chloroform, ethyl acetate, and cyclohexane are commonly used for preparing SLN by solvent emulsification.

25.5.5 Microemulsion Based Technique

A microemulsion is an optically isotropic, thermodynamically stable formulation containing oil and water, stabilized by surfactant. This method involves stirring a composition consisting of lipid, emulsifiers, and water. The drug is solubilized in the molten lipid, followed by emulsification of the molten lipid in the hot surfactant solution under continuous stirring. The resultant hot microemulsion is finally dispersed in cold water under continuous stirring. The dilution ratio of hot to cold water should be between 1:25 and 1:50 for production of a good nanoparticle suspension. The determination of the dilution ratio should be based on the composition of the microemulsion. The particle size of the nanoparticles is influenced by the velocity of distribution of the lipid phase into the aqueous phase. More lipophilic solvents result in higher particle size.

25.5.6 Microwave-Assisted Microemulsion Technique

The microwave-assisted microemulsion technique consists of two steps. The first step involves the preparation of microemulsion containing lipid and drug, followed by dispersing the microemulsion into cold water. The hot microemulsion is produced under constant stirring with the application of microwave heating. The hot microemulsion is dispersed in 2–4 °C cold water under constant stirring to get the particles. Microwave heating results in desired particle characteristics than thermal heating. It provides improved loading capacity & encapsulation efficiency (94).

25.5.7 Ultrasonication

Ultrasonication is a dispersing technique used for preparation of SLN. For preparation of a stable formulation, ultrasonication and high speed stirring are used in combination and the process is carried out at high temperature. Lipid nanoparticle dispersions are obtained by dispersing the melted lipid in the warm aqueous phase containing surfactants by high sheer homogenization followed by ultrasonication. Initially, lipid is melted at 5–10 °C above its melting point. Drug is added to the molten lipid. The emulsion is formed by dispersing the lipid melt into an aqueous surfactant solution under high-speed stirring at the same temperature. The emulsion is subjected to ultrasonication to reduce the droplet size of the emulsion. After that, the emulsion is subjected to gradual cooling of the emulsion below the crystallization temperature of the lipid. The process results in nanoparticle dispersion. The major problem of this method is border particle size distribution and physical instability.

25.5.8 Double Emulsion Method

In this method, drug is dissolved in water to prepare an aqueous solution along with stabilizers such as poloxamer or gelatin. The aqueous solution is emulsified in molten lipid to produce w/o emulsion. The primary emulsion is dispersed in another aqueous solution of stabilizer to produce a w/o/w double emulsion under constant stirring. After stirring for a long period, SLN are precipitated out.

25.5.9 Solvent Diffusion Method

In this technique, the water-miscible organic solvent is diffused into the aqueous phase to produce SLN. Initially, the solvent is saturated with water to ensure thermodynamic equilibrium. Lipid is solubilized in the water-miscible solvents such as benzyl alcohol, butyl lactate, isobutyric acid, isovaleric acid, and tetrahydrofuran. Drug is dissolved into the organic solvent. The organic phase containing lipid and drug is emulsified with aqueous phase containing emulsifier, followed by addition of water to the primary emulsion to extract the organic solvent into the water phase. Extraction of organic solvent into aqueous phase results in precipitation of SLN. The emulsion–water ratio should be 1:10.

25.5.10 Using Supercritical Fluid Method

In this technique, SLN are prepared by rapid expansion of supercritical carbon dioxide solution. No residual solvent is left in the nanoparticle after the process. Several variations have been developed in this technique to prepare nanoparticle formulations. Carbon dioxide is a good solvent for this method. In supercritical fluid extraction, lipid and drug are solubilized in organic solvent with addition of surfactant under heating and stirring. The emulsion is formed by dispersing the organic solution into an aqueous solution with or without a co-surfactant. The mixture is subjected to high pressure homogenization to form an o/w emulsion. The o/w emulsion and supercritical fluid are allowed to move counter currently at a constant flow rate to extract the organic solvent and results in SLN.

25.5.11 Membrane Contactor Method

This technique is based on membrane emulsification. The main instrument involved in the technique is membrane contactor consisting of membranes with pores. The lipid is melted and pressed through the pores of the membrane. Aqueous phase flows tangentially to the membrane surface of the membrane contactor. While passing through the pores of the membrane, small droplets of lipid melt are formed. The aqueous phase carries the droplets to form SLN. The size of the nanoparticles is influenced by many factors including flow velocity and temperature of the aqueous phase, volume of the phase, and area of the membrane.

25.5.12 Coacervation Technique

Coacervation is one of the most widely used techniques of microencapsulation. Coacervation is a phase separation technique. In case of microencapsulation, coacervation is three steps (simple coacervation) or four steps (complex coacervation). For prepraration of SLN, the coacervation technique involves acidification of alkaline salts of fatty acids (95). In this technique, an aqueous solution of polymeric stabilizer is prepared by heating. A sodium salt of the lipid is dispersed in the aqueous solution under continuous stirring above the Krafft point of the salt form of the lipid. Under constant and continuous stirring, a clear solution is formed. The drug is solubilized in a water miscible organic solvent. The drug solution is added to the clear solution under contant stirring. Then the coacervating solution is added to the solution and causes formation of drug containing SLN.

25.5.13 Phase Inversion Temperature Technique

Phase inversion is the process of conversion of o/w to w/o or vice versa. Phase inversion can be induced changing volume of phases, temperature, and emulsifier affinity (96). Phase inversion occurs at a temperature known as phase inversion temperature. Above phase inversion temperature, rapid cooling of an emulsion yields a very fine droplet and stable emulsion (97). The phase inversion occurs when the curvature of the oil–water interface passes through a zero spontaneous curvature at the inversion point and minimum interfacial tension. This is associated with a shift in the surfactant nature from water-soluble to oil-soluble, or vice versa. At the inversion point, the surfactant has a similar affinity toward both phases. Hasan Ali & Sandeep Kumar Singh reported formulation of SLN using this technique. The lipid and drug is mixed and heated together above 10 °C of melting point of the lipid to obtain a clear mixture. Hot aqueous phase is added to the lipid mixture with continuous stirring on a magnetic stirrer. The temperature of the emulsion is allowed to cool to get a colloidal suspension of SLN (98). Other versions of phase inversion temperature technique have been reported by different authors (99).

25.5.14 Spray Drying Method

It is low costly industrially viable method for preparation for SLN preparation. Spray drying method is an alternative method for lyophilization of an aqueous SLN dispersion into a drug product. High processing temperature, partial melting and shear forces results in particle aggregation. Lipid with melting point >70 °C results in better nanoparticles in the spray drying process.

25.6 Critical Parameters of Pulmonary Delivery

The pulmonary delivery of novel formulations is a challenging area for pharmaceutical scientists. Various factors influence the delivery of novel formulations to the desired site of the lung. An updated understanding of cellular interaction, fluid dynamics of particles, pharmacokinetic behavior, and other important parameters is necessary for designing of an effective formulation for pulmonary delivery. As we are discussing solid lipid nanoparticles, so the particle properties have a significant role in the chemical components of the particles, their size and shape, particle density, dispersion state, viscosity, rate of entrapment, targeting moieties, etc.

25.6.1 Structural Composition

Biodegradable materials are of interest for development of formulation for pulmonary delivery. SLN and NLC have been employed for pulmonary delivery of phenethyl isothiocyanate (100), paclitaxel (101), and other drugs. They can be targeted to lower lung regions by balancing the aerodynamic parameters.

After inhalation, SLN could reach the lymphatic nodes, subsequently deposited in periaortic, axillar and inguinal lymph nodes (102). Lipid composition (103,104) and particle size (105) of the SLN are crucial factors for pulmonary delivery. Other factors such as drug–lipid ratio (104), process parameters (106), and delivery method are important for achieving local drug concentration. Many excipients are used for enhancement of stability, dispersability, and flowability of SLN. Mannitol is found to be a good excipient for dispersibility of dry powder inhaler (107). The addition of these excipients can reduce the aggregation of particles to prevent rapid excretion (108).

26.6.2 Biocompatibility

For pulmonary applications, tolerability of the SLN product is an essential factor. Tolerability of SLN can be achieved by using biocompatible and biodegradable lipids and other excipients.

25.6.3 Particle Size and Shape

The mass median aerodynamic diameter (MMAD) significantly impacts distribution patterns and deposition sites of the particles after inhalation (109,110). The particles are deposited on the different sites of lungs by impaction, interception, sedimentation, or diffusion mechanisms (111). The particles largely accumulate at the upper airways, eliminated by mucociliary clearance in the epithelia tissue for particle size larger than 6 μm. When the foreign particles are trapped in mucus, cilia beat in a coordinated direction (pharynx) to remove the freight by either coughing or swallowing (112). Aerosol particles with d_{ae} values of 1–5 μm tend to deposit predominantly in the smaller bronchiolar (lower) airways down to the bronchioalveolar region, whereas those aerosol particles with d_{ae} values in the range 5–10 μm deposit in the middle and larger bronchial (upper) airways (113,114). Particle shape influences the macrophage clearance rate in lungs. Polystyrene particles with different shapes were cleared differently by the phagocytosis process. It indicates that the particle shape influences the cellular interaction of particles with macrophages (115,116). Spherical particles are easily engulfed by macrophages due to their low curvature shape (115).

25.6.4 Surface Properties of the Particles

The surface properties of the particles influence the interaction of the particle with mucus of the respiratory tract and cells. Mucoadhesiveness of particles accelerates the rate of mucociliary clearance. Surface charge and hydrophilicity are the major factors that influence the particle interaction with cells. The stability of the formulation containing particles is improved significantly for zeta potential above ±30 Mv. Positively charged particles can easily be engulfed by cells due to their interaction with cells (89,117). Interaction of particles with mucus causes rapid clearance. Particle charge should be controlled by proper selection of component materials as well as their ratio. Otherwise, rapid mucociliary clearance may result in product failure. Minimization of absorption, degradation, and entrapment of SLN can be achieved by manipulation of surface charge. Conjugation of lipids or the hydration layer on particles can reduce the elimination rate by minimizing phagocytosis (118,119).

25.6.5 Isotonicity and pH Value

Aerosolized solutions can induce cough at very low and high values of pH and tonicity (43). Therefore, it is necessary to adjust pH and osmolarity of the product to minimize the exhalation of inhaled particles. SLN formulation can be isotonized by using glycerol or other isotonizing agents. Ionic isotonization agents such as sodium chloride or other salts can be used, but they might exhibit an undesirable effect on the stability of SLN. (42,44,45). The pH of the formulated SLN value of SLN or NLC formulations can be adjusted to desired values using buffers with low ion concentrations, acids, or bases (44).

25.6.6 Ligands for Targeting of Solid Lipid Nanoparticles

Ligands are used to target specific drug delivery. They increase the concentration of drugs in a particular class of target cells. Various small molecules, monoclonal antibodies, peptides, and aptamers are used for site specific delivery of SLN. Targeting ligands enhances the availability of drugs at target site. Conjugation of tumor necrosis factor-related apoptosis-inducing ligand (Apo2L/TRAIL) with lipids resulted in binding to the death receptors 4 and 5 (DR4 and DR5), and subsequently internalization of the particles. It enhances the availability of drugs at specific sites (120).

25.6.7 Pulmonary Delivery Tools

Nebulizers and dry powder inhalers are the most useful tools for pulmonary delivery of drugs at the current stage of pulmonary drug delivery. Nebulizers deliver formulations in aerosol form, and are able to delivery large amount of dosage over a period of time (121). Dry power inhalers were developed to enhance the stability of the formulation. Dry powder inhalers are able to deliver specific doses to the lung. Due to their stability, safety, and cost-effectiveness, these products are getting increasing response in the pharmaceutical field (122).

25.6.8 Patient Related Factors

Various patient related factors influence the delivery of drugs to desired sites. Cancer and other lung diseases change the physiological parameters of the lung. In support of the above statement, it has been observed that overproduction of a highly viscous mucus occurs in patients who suffer from cystic fibrosis. It drastically reduces the drug bioavailability in the tumor site due to rapid clearance. The mucus production was found to be more than a 10-fold increase in chronic bronchitis patients compared with healthy subjects. The disease state can significantly alter the delivery of the formulation, as well as the availability of drug at local site (main). Depending on the patient, the dosage requirement may vary. Alternation of lung structure and obstruction of lung pathway due to various diseases results in change in cross-sectional area and air flow in the lung. Due to airway obstruction, particles may move to healthy regions of the lung. This will reduce the effectiveness of the formulation as well as increase adverse effects (123). Liposomal 9-nitrocamptothecin (9NC) with a droplet size of 1–3 μm was found to tolerable at a dose of 13.3 μg/kg per day with minimal adverse effects, including for nausea, cough, anemia, vomiting, wheezing, fatigue, anemia, peribuccal rash, and neutropenia. But higher doses resulted in serious side effects. Therefore, disease condition must be considered while designing a formulation (124).

25.7 Characterization

Quality control is a fundamental aspect of pharmaceutical processes. As SLN are a novel drug delivery, many challenges are associated with developing stable formulations. The characterization of SLN is a challenging approach due to its complexity and the dynamic nature of the drug delivery system. Various parameters must be evaluated for a SLN preparation. Important parameters are surface morphology, particle size, size distribution kinetics, coexistence of additional colloidal structures, degree of crystallinity, lipid modification, drug content, in vitro drug release, and shelf life of the product.

25.7.1 Surface Morphology, Particle Size, and Size Distribution

Nanoparticles are defined as particles up to 100 nm in size. Particle size of drug delivery systems influences the physicochemical properties and fate of the particles. Various formulation parameters including raw material, method of prepration, processing time, equipment, lyophilization, and sterilization influence the particle size. Transmission electron microscope (TEM), scanning electron microscope (SEM), atomic force microscopy (AFM) and other advanced microscopy techniques such as scanning tunneling microscopy (STM) and freeze fracture electron microscopy (FFEM) allow the study of the surface morphology of the particle.

The particle size and size distribution of particles can be evaluated using Zetasizer. Dynamic light scattering is used as the technique for particle size measurement in a Zetasizer. Measurement range varies from 0.3 nm to 10 μm. Zeta potential of a formulation is determined using the principle of electrophoretic light scattering (125). Many other techniques are also used for determination of particle size and molecular size. Laser diffraction (LD) and dynamic correlation spectroscopy are routinely used for measurement of paticle size of nanoparticles.

25.7.2 Transmission Electron Microscopy

This technique is commonly used to study the morphology of the particles in colloidal carrier systems. Freeze-fracture, negative staining, and vitrification by plunge freezing are the procedures for preparation of samples for transmission electron microscopy. For staining methods, sample preparation is easy and quick, and it does not require additional equipment. However, low resolution micrographs are obtained by staining methods and may not provide true characteristics of the sample. Freeze-fracture transmission electron microscopy is a useful method for determination of size and shape of nanoparticles in a colloidal carrier system. It can distinguish liposomes, SLN, emulsions, etc. Polymorphic transition can be studied with this method (126).

25.7.3 Scanning Electron Microscopy

This method has been used to investigate the morphology of SLN. The particles are freeze dried and coated with a conducting substance (gold) before carrying out SEM. Application of vacuum may change the characteristics of the particles. However, solid nanoparticles in a dispersed state can be observed in their true condition using cryo-field emission SEM (127).

25.7.4 Dynamic Correlation Spectroscopy

This method is used for determination of particle size of SLN in a dispersion. In this particular technique, size range from a few nanometers to about 3 μm can be studied. In DCS, particles suspended in the dispersion medium are irradiated with a laser beam of a particular wavelength. DCS measures variation of the intensity of the scattered light caused by the Brownian movement of particles as a function of time. Small particles cause high intensity fluctuations because of their high diffusion coefficient. Large particles cause lower fluctuations because of their low diffusion coefficient. This variation is quantified by compilation of an autocorrelation function. The advantages of the method are sensitivity, quick analysis, and no requirement for calibration. However, it is an indirect method of particle size determination and depends on diffusion coefficient. Particle size measurements may be influenced by the presence of large particles. It is not a suitable method for measuring particle size in the upper nanometer range or anisometric particles.

25.7.5 Laser Diffraction Spectroscopy

This is a technology that utilizes diffraction patterns of a laser beam passed through any object ranging from nanometers to millimeters in size to quickly measure geometrical dimensions of a particle. Laser diffraction analysis is based on the Fraunhofer diffraction theory, stating that the intensity of light scattered by a particle is directly proportional to the particle size. The angle of the laser beam and particle size have an inversely proportional relationship, where the laser beam angle increases as particle size decreases and vice versa. At high angles, intense scattering is caused by smaller particles compared to the larger partricles. Size range of particles from a nanometer to a few millimeters can be studied by laser diffraction spectroscopy. Moreover, a laser diffractometer is suitable for determination of polydispersity of nanoparticles. When particles of several populations of variable sizes are present in the sample, this technique is unsuitable for estimation. A combination of laser diffraction spectroscopy with polarization intensity differential scattering provides more accurate information on particle size distribution due to increased sensitivity of smaller particles (126).

25.7.6 Acoustic Methods

Acoustic spectroscopy works by transmitting sound waves through a sample and measuring attenuation over a wide range of frequencies of sound waves through the fitting of physically relevant equations. The particle size is calculated from the measured spectra using software that models the physical basis of sound attenuation in concentrated systems. It claims the advantage over light based systems of being able to measure particle size in highly concentrated samples without requiring dilution. Information on surface charge can be determined from the oscillating electric field generated due to movement of the charged particles under acoustic analysis.

25.7.7 Surface Charge

Surface charge of the particles influences the stability and pharmacological behavior of the particles. Electrostatic and steric repulsion are two main factors influencing the stability of the formulation. Zeta potential is an important parameter for determination of surface charge of the nanoparticles. Zeta potential is the electrical potential at slipping plane of the interfacial double layer. It is the potential difference between the stationary layer around the particle and dispersion medium of the colloidal system. Zeta potential is the key indicating factor of stability of a colloidal system. Higher zeta potential of a colloidal system indicates a stable system, whereas low zeta potential indicates unstable system which tends to flocculate or coagulate. The zeta potential is influenced by characteristics of the formulation. Ionic strength and pH are the main factors that influence the zeta potential of the formulation. Direct measurement of zeta potential is not possible, but it is measured based on electrokinetic phenomena and electroacoustic phenomena. In electrophoresis, an electric field is applied across the colloidal system. On application of electric field, particles migrate toward the electrode of opposite charge with a velocity proportional to the zeta potential. A laser Doppler anemometer measures the velocity of the particles. Particle mobility is converted into zeta potential using the Helmholtz-Smoluchowski equation by inputting dielectric permittivity and dispersant viscosity.

25.7.8 Nuclear Magnetic Resonance

Nujclear Magnetic Resonance (NMR) spectroscopy can provide a familiar means of looking at the molecular structure of molecules bound to surfaces of nanomaterials as well as a method to determine the size of nanoparticles in solution. NMR can be used for identification of degraded components in a formulation.

25.7.9 Crystallinity and Polymorphism

The physical stability of aqueous dispersions of SLN is usually more than three years. The crystallinity index of SLN preparation increases slowly during storage and unstable modifications lead to crystallization (128). Polymorphic transitions cause change in other characteristics of the formulation, such as change in particle size. These changes decrease the shelf-life of the formulation. Differential scanning calorimetry (DSC) and X-ray diffraction (XRD) are frequently used for determination of crystallinity index and polymorphism. DSC is a useful technique for study of polymorphism of SLN, whereas XRD provides us with detail structural data. Dilution or freeze drying of SLN dispersion may alter particle size. It may change crystallinity and polymorphism of the particles. The nature of the final product should be considered while carrying out crystallinity and polymorphism studies. Drug crystals may appear in freeze-dried SLN.

25.7.10 Differential Scanning Calorimetry

Differential scanning calorimetry is performed to identify drug-excipient interactions and stability of the drug in the formulation. The output of a DSC experiment is expressed as a curve of heat flux versus temperature or time. DSC can be used to determine glass transition temperature, crystallization temperature, and melting temperature of polymers. The solid state of lipid nanoparticles can be confirmed through a melting transition in a DSC curve. Recrystallization index and changes in physical properties can be studied using DSC. It can detect and quantify polymorphic transitions and changes in crystanallity. The crystallinity of drug in SLN formulation can be characterized by interpreting the DSC curve (94). The presence of melting peak of the drug in the DSC curve suggests the presence of crystalline drugs in the formulation (129). In general, the desired state of SLN in colloidal system should be solid or at least partially solid. However, SLN prepared through melt homogenization may not be necessarily present in the solid state after processing (130). These nanoparticles are present in a supercooled dispersed state. The crystallization is retarded and occurs at a lower termperature than bulk lipids. Even a difference between the melting maximum and the onset of crystallization in colloidal dispersed systems may be up to 50 °C. Changes in particle sizes of SLN can be interpreted from DSC thermograms. The Gibbs–Thomson equation is the basis for the effect of particle size on melting point. Polymorphic changes with temperature and time can be studied using DSC. The solid state of SLN in semi-solid formulation can be confirmed by DSC (131).

25.7.11 X-ray Diffraction

X-ray diffraction has been most commonly used in the characterization of SLN to confirm the solid, crystalline nature of particles and to identify the polymorphic form of the lipid matrix after preparation. It is an indispensable tool for deriving information on the various properities of SLN. Two different techniques for elastic scattering involve X-ray and neutron small angle scattering, and X-ray and neutron wide angle scattering also known as diffraction. All these dispersions are characterized by the presence (in the case of crystalline or liquid crystalline particles) or absence (supercooled melts) of characteristic thermal transitions in the DSC heating curve allowing us to draw conclusions on their physical state. These conclusions may be confirmed by additional X-ray investigations which are particularly helpful in the case of liquid crystalline nanoparticles, the thermal transitions of which can be very small (132). XRD can differentiate polymorphs by determining the distance between the single molecular layers (133,134). The polymorphic transitions in SLN formulation can be interpreted using X-ray diffractograms. Polymorphic changes occur more rapidly in SLN as compared to bluk lipid (95). XRD can provide data to identify the effect of drugs or model substances on the polymorphic changes in SLN (135,136). The effect of drug loading on SLN and the state of the drug in SLN can be studied with XRD. The relative intensity and position of the signals of the component and carrier matrix influence the detection limit in XRD. Superimposed signals of drug and carrier matrix make it difficult to interpret the diffractograms. Jenning et al. used XRD to confirm the presence of β′-form in glyceryl behenate nanoparticles after incorporation into an oil/water cream (137).

25.7.12 Determination of Process Yield and Moisture Content

The lyophilized powder for the prepared SLN flocculates can be calculated using the following expression:

$$\text{Process yield (\%)} = \frac{\textit{Recovered mass}}{\textit{Total amount of raw materials}} \times 100\%$$

The moisture content of the lyophilized solid lipid nanoparticle content can be assessed in triplicate by a Karl Fischer titrator.

25.7.13 Flow Properties of Solid Lipid Nanoparticles

In the case of dried formulation, it is necessary to evaluate the flow characteristics of SLN formulations.

25.7.13.1 Angle of Repose

The pile is carefully built up by dropping the material through a funnel until it reaches the tip of the funnel (usual height of 2 cm). The angle of repose is calculated by inversing tangentially the ratio of height and radius of the formed pile.

25.7.13.2 Bulk and Tapped Density

Bulk density is determined by filling the powder in a 10 ml measuring cylinder, followed by 500 taps (tap sufficient to obtain the plateau condition) using a tap density tester to determine tapped density. Carr's compressibility index and Hausner's ratio are determined by the following:

$$Carr's\ index = \frac{Tapped\ density - Bulk\ density}{Tapped\ density} \times 100\%$$

$$Hausner's ratio = \frac{Tapped desnity}{Bulk density}$$

25.7.14 Drug Content and Entrapment Efficiency

Development of a validated assay protocol is necessary for accurate quantification of herbal components. Different methods such as ultraviolet (UV) spectroscopy, high performance liquid chromatography (HPLC), and high performance thin layer chromatography (HPTLC) can be utilized for development of a validated assay to determine the drug content and entrapment efficiency.

25.7.14.1 UV Spectroscopy

For analysis of herbal drugs containing SLN formulation using UV spectroscopy, the compounds must be extracted and isolated, otherwise other compounds will interfere with the quantification of the desired compounds.

25.7.14.2 High Performance Liquid Chromatography

Validated HPLC protocols for quantification of herbal compounds is the sophisticated method. This technique provides accurate measurement of drugs as well as simultaneous determination of multiple components within a short period of time. A formulation can be directly processed for quantification without going for additional extraction or isolation procedures.

25.7.14.3 High Performance Thin Layer Chromatography

This is the sophisticated form of thin layer chromatography (TLC). It is not only qualitative but also allows quantitative analysis of components in formulation. It is a highly useful technique for quantification and identification of multiple components in herbal formulations. Piezoelectric devices and inkjet printers have improved the sample application technique significantly. Simultaneous analysis of multiple samples or formulations at the time is the advantage of HPTLC over HPLC.

25.7.15 In Vitro Aerosol Dispersion Performance

The aerosolization efficacy is determined by the emitted fraction (EF %) and respirable fraction (RF %). EF % is defined as the percent of total loaded powder mass exiting the capsule, which can be determined gravimetrically. The respirable dose (RD) is the quantity of the particles per capsule deposited in the lower stage of the TSI after aerosolization, while RF % is the ratio of RD to total loaded dose. High EF % and RF % result in efficient aerosolization of the powder. Median mass aerodynamic diameter (MMAD), geometric standard deviation (GSD), and fine particle fraction (FPF) are measured using equipment such as the Westech 8-stage Non-Viable Cascade Impactor, a next-generation impactor. The aerosolized parameters, including MMAD, FPF, and emitted dose (ED) are examined at a predefined flow rate such as 30 and 60 l/min for a predefined time period. Cutoff diameters of eight stages at a flow rate of 30 l/min are used. Each plate of the impactor is precoated with 1% of silicone oil in octanol to prevent particle bounce and re-entrainment. Powder is aerosolized using a DPI device such as Rotahaler or HandiHaler. The amount of drug recovered from the induction port, pre-separator and different impactor stages is regarded as ED. The quantity of drugs can be estimated using chromatographic techniques or gravimetric methods, depending on the type of equipment (138).

25.7.16 In Vitro Activity Study

In vitro studies for pharmacological studies can be carried out using different cell lines depending on the investigated activities of the formulation containing the herbal drugs. Various cell lines are available for the investigation of lung inflammation in cystic fibrosis. CFTE29o–[Tracheal (F508del homozygous)], CFBE41o–[Bronchial (F508del homozygous)], CuFi-1 [Bronchial] are human based cystic fibrosis origin cell lines used for the study of lung inflammation in cystic fibrosis (139). Various lung cancer cell lines are used based on mutated genes for specific studies. The common somatic mutations are LRP1B, KRAS, TP53, KMT2C, KRAS, KEAP1, FAT4, EGFR, CDKN2A, and FAT2.

Lung Cancer Cell Line	Mutated Gene
A549	KRAS, KEAP1, KMT2C, FAT4
PC-9	CDKN2A, EGRF, TP53
CALU 1	LRP1B
NCI-H358	FAT2
SHP-77	TP53, LRP1B, KRAS, KEAP1
COR-L47	FAT4, KMT2C, LRP1B
DMS 153, DMS 273	TP53, LRP1B
NCIH358	KRAS, KMT2C

Drug response studies, metastasis studies, and toxicological studies can be carried out using the cancer cell lines. The frequency of non small cell lung cancer (NSCLC) is significantly higher than small lung cancer cell. PC9 cancer cell line is frequently used for study of NSCLC. Respiratory infections are major cause of lung diseases and fatality due to lung diseases. Mycobacterium is one of the major causative agent responsible for fatality. In vitro activity of herbal nanoparticles can be evaluated against various infection causing agents using a well designed protocol in a laboratory setup.

Classes of Infection Causing Agents	Different Infection Causing Agents
Virus	*Rhinoviruses, Adenoviruses, Respiratory syncytial virus, Influenza viruses, Parainfluenza viruses, Coronaviruses*
Bacteria	*Streptococcus pneumonia, Streptococcus pyrogens, Klebsiella pneumonia, Haemophilus influenza, Escherichia coli, Mycoplasma pneumonia, Mycobacterium tuberculosis and other Mycobaterium species, Staphylococcus aureus, Chlamydia trachomatis, Chlamydia pneumonia, Corynebacterium diphtheria, Mycoplasma hominis*
Fungi	*Histoplasma capsulatum, Blastomyces dermitidis, Coccidicides immitis, Candida albicans, Aspergillus fumigates, Aspergillus species, Paracoccidiodes brasiliensis*

25.8 Pulmonary Delivery Strategies

The major focus of lung delivery is to achieve site specific delivery and controlled release of drugs. The targeted lung delivery could reduce the risk of systemic toxicity of drugs. The active targeting of the nanoparticles using various targeting agents is effective to enhance the passive targeting. By combining the passive and active targeting, nanoparticles can be targeted to specific cells, which could be beneficial to different diseases, including cancer. Passive targeting promotes enhanced permeation and retention. It relies on the formulation characteristic for example size, and disease conditions such as vascularity and leakiness in cancer tissues. Inhalation of formulation is a passive strategy to achieve local high concentration of drug at lungs. Passive targeting strategies have a few limitations in targeted drug delivery. They may deliver drugs off target and may not be able to exert a strong EPR effect. However, active targeting of drugs can overcome the above limitations. Active targeting directly targets specific cells through antibodies, peptides, aptamers, proteins, and small molecules. Active targeting results in better interaction of nanoparticles with cells, thereby enhancing cellular uptake of the drug and achieving high concentration of the drug at the targeted cells, for example cancer cells.

Various properties of chitosan, such as high mucoadhesion, cationic nature, low toxicity, and high bioavailability, have resulted in more researchers using this polysaccharide as a coating for SLN to improve the delivery of bioactive compounds (140,141). Chitosan-coated SLN have been used for delivery of drugs through other delivery routes, including nasal, vaginal, and skin (142). As chitosan has a positive charge, chitosan conjugated SLN can be taken up by cells easily. It will localize concentration of the drugs in lungs. However, there may be rapid elimination due to interaction with mucus. Similarly, caffeic acid-loaded SLN coated with alginate chitosan showed higher antioxidant activity and sustained release (143).

HACC-modified SLN is another modified chitosan-loaded SLN that was recently developed to improve the stability in the gastrointestinal environment for sustained release (144). Shi et al. reported the formulation of HACC modified SLN particles. The SLN particles were evaluated for oral bioavailablity of docetaxel. These SLN particles exhibited good stability in simulated gastric and intestinal fluids, and increased the area under curve by 2.45 as compared to docetaxel. It indicates that HACC-modified SLN could have higher bioavailability of drug as compared to conventional formulation of the drug (145). A recent study reported that HACC-modified chitosan exhibited a higher drug bioavailability via various absorption mechanisms, including transcellular, paracellular, and M cell uptake, when administered orally (146). In addition, the toxicity of the HACC-modified SLN was studied in Caco-2 cells, and the results showed no toxic effect and no irritation in the mucosa of the rats (145).

25.9 Pharmacokinetics of Pulmonary Delivery

The earliest use of inhalation as means of drug administration can be traced to 1,554 BCE, when Egyptian physicians used the vapor produced when black henbane was placed onto hot bricks to treat patients with breathing problems (147). Since then, delivery of drugs by inhalation has demonstrated several advantages for local diseases such asthma, including the rapid delivery of the drug to the site of action and achieving higher local concentrations compared to systemic delivery (148). For drugs that may have adverse systemic effects, this approach can limit the systemic concentrations and has the potential to increase the therapeutic index of the drug (149). This is particularly useful for drugs that have limited oral bioavailability due to poor intestinal absorption or drugs that otherwise would be degraded or metabolized in the gastrointestinal tract (GI) (150).

Drugs can be delivered to the lungs either by a nebulizer or soft mist inhaler (SMI), pressurized metered dose inhaler (pMDI), or a dry powder inhaler (DPI). Nebulizers, pMDI, and SMIs are liquid-based pulmonary inhalation aerosol delivery systems. Of the liquid-based aerosol systems, pMDI require the use of a propellant whereas nebulizers and SMIs do not contain propellants (151). DPI were first developed to treat asthma and chronic obstructive pulmonary disease (COPD), with the first capsule-based DPI (Spinhaler®, Fisons) commercialized in 1972 (147,152). The first multi-dose DPI was developed by AstraZeneca (Turbuhaler®) containing 200 doses in a supply chamber (153). The factors that influence the deposition of liquid or dry powder aerosols are particle shape, size and distribution, hygroscopicity, static charge, anatomy of the airways, and breathing patterns, such as frequency and tidal volume.

It should be noted that the anatomy of the airways and breathing patterns are significantly different among laboratory animals, healthy humans, and patients at different stages of disease (154). After deposition in the different regions of the lung, droplets/particles are either cleared from the lungs or absorbed into blood/lymphatic circulation. Depending on the site of droplet/particle deposition, the clearance mechanisms include mechanical clearance, mucociliary clearance, uptake by alveolar macrophages and enzymatic degradation (155). Therapeutic aerosols can be mechanically cleared by coughing, sneezing, or swallowing when they are deposited in the upper region of the respiratory tract. For large inhaled doses, coughing may be spontaneously provoked, particularly if aerosolized particles are 10 μm; thus, it is important that particles are ≤10 μm to improve the efficiency of treatment. An aerodynamic particle size of 1–3 μm is required for optimal delivery to the lung. A combination of lower MMAD and higher FPF are required for efficient delivery of formulation to the lung. A FPF of approximately 70% is sufficiently high to deliver the drug efficiently deeper in the lung (156).

Several particle properties influence the aerosol performance including the aerodynamic diameter (d_{ae}), particle size distribution, dispersibility, particle morphology, surface roughness, and interfacial interparticulate interactions (157,158). Human serum albumin (HSA) was hydrophobically modified with both octyl aldehyde and DOX to impart amphiphilic character, thus inducing self-assembly. The self-assembled nanoparticles (NPs) (341.6 nm) were not suitable for inhalation treatment of lung cancer due to their low inertia resulting from small density which makes them carried by the air stream to exhalation. Using an aerosolizer that produces micronized liquid droplets, deep lung deposition of albumin NPs was successfully achieved (159,160). Therefore particle size is an important criterion for development of aerosol formulation of solid lipid nanoparticles.

If the inhaled particles are formed of slowly dissolving or insoluble materials, they may be removed by mucociliary clearance, with most particles deposited in the trachea-bronchial region being cleared within 24 hours in healthy subjects (161). Alveolar macrophages could act like Kupffer cells in exocytosing NP fragments after initial phagocytosis (162). Particles of sizes 1.5–3 μm are known to be more favorably uptaken by alveolar macrophages (161). It is important to note that there are large interspecies differences in the lung clearance of inhaled particles (163). Gamma scintigraphy analysis showed that solid lipid nanoparticle administered through I.V. distributed in almost all organs, but the drug was concentrated more in the lungs, even after 6 hours of pulmonary administration of solid lipid nanoparticle loaded with the drug (164).

25.10 Pulmonary Disposition

A series of studies have been carried out to analyze the lung deposition of particles. One of the factors influencing the efficacy of pulmonary drug delivery is that the dose is able to reach targets in the lung. The most important mechanisms of particle deposition in the respiratory tract are inertial impaction, gravitational sedimentation, and diffusion (Brownian motion). The particles in the size range 1–5 μm deposit in the small airways and alveolar ducts via sedimentation. The high inspiratory flow rates during breathing increase deposition of particles by inertial impaction, whereas gravitational sedimentation and Brownian diffusion decreases disposition (165). Particles of 0.5–5 μm desposit in the small airways and alveoli. Particles having a size of <0.5 μm, as a consequence of Brownian diffusion, are generally not deposited and are expelled upon exhalation (166). Fabian et al. reported 87% particles in exhaled air were under the size of 1 μm (167). However, nanoparticles of 100 nm are able to deposit in the alveolar region in acceptable amounts (168–171). Drug NPs usually deposit by sedimentation after being released from the aerosol device due to an agglomeration process in the lung. These agglomerated NPs are able to sediment for longer periods in the tracheobronchial section, thereby improving the biological activity of the delivered therapeutic agent.

25.11 Elimination of Solid Lipid Nanoparticles

Lungs is continuously undergo aerosol treatment due to suspended particles in the air. Many particles deposit on the airway and alveolar surfaces. External particles are mostly expelled from the lungs by coughing, sneezing, mucociliary elimination, or solubilization in the lung lining fluid. Lung phagocytes remove particles by phagocytosis to maintain a sterile and clean environment in lungs. Macrophages are the predominant lung phagocytes. They have been observed in airways and aveoli of lungs, lung stroma, pulmonary capillaries, and the pleural space. Dendritic cells are another class of lung phagocytes (172).

Epithelial cells can also take up the deposited particles. Sipos et al. reported the uptake of PNP particles into alveolar epithelial cells in a time- and concentration-dependent manner (173). The retention time of particles in the lung depends on the deposition site and their interaction cells (174). In rodents, microparticles in airways are usually cleared within 24–48 hours, whereas prolonged retention was observed for micro- and nanoparticles in humans (175).

The human respiratory tract model of the International Commission on Radiological Protection provides deposition data of inhaled particles from 1 nm to 10 μm in healthy adult female and male human subjects, at different breathing patterns and physiological activities (176). Lymphatic drainage is an important pathway for elimination of the particles. The pathway involves translocation of particles from alveoli into interstitial space, subsequent uptake by macrophages, followed by migration from lymphatics to lymph node (5). Videira et al. reported that significant uptake of radiolabeled solid lipid nanoparticles into lymphatics after inhalation. A rapid deposition of particles was observed within the few minutes. Solid lipid nanoparticles (200 nm) were rapidly cleared from the lungs into lymph in rats within the first hour post-administration. Higher rates of distribution was observed in axillar, periaortic, and inguinal lymph nodes. The study indicated the biodistribution of the particles through lymphatic routes. However, lung showed 23.3 ± 7.6% activity per gram tissue at 4 hours post-inhalation. This indicates a certain degree of SLN retention in the lung (177).

In contrast, Liu et al. reported that polyacrylamide nanoparticles (20–40 nm) were shown to remain primarily in the lungs over 24 hours (178). Choi et al. found that albumin nanoparticles of approximately 200 nm remained in the lungs for more than 48 hours post-administration (159). Langenback et al. reported the failure of inhaled polystyrene microparticles of 2.85 μm to penetrate interstitium in adult sheep, cleared through mucociliary clearance. The elimination process showed three phases of elimination. The slow phases of elimination might be due to sequestration of SLN particles by macrophage (179). Mucociliary clearance is a prominent pathway by which 3H-lipids from both SLN and liposomes are cleared from the lungs. This clearance mechanism constitutes a dominant lung clearance pathway for large biodegradable macromolecules as well as non-biodegradable nanoparticles (180,181). In agreement with the above statement, Haque et al. demonstrated approximately 50% of the 3H-labeled SLN cleared by mucociliary clearance within the first 3 days after pulmonary dosing by intra-tracheal administration. The bioavailability of 3-H radiolabeled lipids from the SLN was approximately 70% as compared with 32% for liposomes over 7 days. After the first phase of rapid clearance, the rate of lung clearance decreased. Approximately 30% and 10–15% of the pulmonary dose was found in the lungs after 7 and 14 days, respectively. Confocal images showed that the SLN were engulfed by alveolar macrophages in the form of 3–4 μm aggregates. Distinct SLN particles were observed in human lung epithelial A549 cells.

The data from the study indicate that SLN are more stable in the endosomes of lung cells as compared to liposome (182). The rate of mucociliary clearance is critically dependent on the mucoadhesiveness of the nanoparticle. Increase in mucoadhesiveness correlates with increased mucociliary clearance of the particles (183). High mucoadhesive strength is associated with the patchy and inhomogeneous lung deposition pattern of the SLN. It increases initial rapid mucociliary clearance (184). The distribution patterns of the SLN and liposomes in the lungs suggested that the SLN were localized mainly in the upper respiratory region at early time points, whereas the liposomes showed better access to deeper regions of the lung, likely as a result of the presence of SLN aggregates in the dosing solution (185). Many parameters of the particles are crucial to the elimination process, including physiochemical characteristics of structural components of the particles and particle size. While mucociliary clearance is the major mechanism by which nanomaterials are cleared from the upper respiratory tract, alveolar macrophages have a more important role in the clearance of inhaled particles from the alveolar regions of the lungs where there is an absence of significant ciliated epithelia and mucus (186).

Interestingly, the retention time of drugs (loaded into liposomes or SLN) in BALF and lung tissue have been found to be significantly shorter than the nanocarrier itself (187). Bondesson et al. determined significant differences in intrapulmonary deposition patterns by varying the size and the flow rate of the aerosol. A larger aerosol is inhaled rapidly and deposition occurs in central lung, whereas a smaller aerosol is inhaled slowly and deposition occurs in the peripheral lung. The study found significant differences in the pulmonary elimination half-life of ^{99m}Tc-Nanocoll (8 hours and 45 min) as compared to that of ^{99m}Tc-DTPA (<2 hours). After deposition, hydrophilic ^{99m}Tc-DTPA was mainly eliminated through the transepithelial route (188). The endocytosis and exocytosis of positively charged FITC-labeled maltodextrin NPs (60 nm) in human bronchial epithelial cell line was studied by fluorescence confocal microscopy. The NPs were rapidly endocytosed after as little as 3 min of incubation, and endocytosis appeared faster than binding since most of the NPs were found in the middle of the cells around the nuclei. No further increase was observed after 40 min, probably due to equilibrium between endocytosis and exocytosis. The endocytosis was dramatically reduced at 4 °C compared to 37 °C and by sodium azide treatment, suggesting an energy dependent process. Protamine pretreatment of the cells inhibited NP uptake, suggesting that the NPs are endocytosed via the clathrin pathway. Cholesterol depletion caused a 3-fold increase in NP uptake blocking of exocytosis (189). Lung retention data also showed that with the exception of initial rapid clearance from the lungs of rats, the nanomaterials showed, in general, prolonged lung residence (>10% dose remaining in the lungs after 14 days).

25.12 Sterilization of Solid Lipid Nanoparticle Formulations

SLN must be sterile for intravenous and ocular delivery. During autoclaving, the high processing temperature presumably causes formation of a hot o/w microemulsion which may result in a change in the size of the nanodroplets. During slow cooling, nanodroplets may coalesce to produce particles larger than the particles before autoclaving. The quantity of surfactant and cosurfactant present in the preparation of SLN during autoclaving is smaller because SLN are washed before sterilization. The low quantity of surfactant may cause instability of the product. Noha et al. reported better physical stability of solid lipid nanoparticles loaded with brimonidine after autoclaving at 121 °C for 15 min. The enhancement of physical stability is associated with an increase in zeta potential values. The autoclaving process causes decrease in crystallinity index, which results in less ordered lattice. Less ordered lattice increases accommodation of drug molecules, thereby increases entrapment efficiency (190). Steam sterilization results in the formation of oil in water emulsion because the lipid melts, which again converts into solid particles after recrystallization (88). A fewstudies reported no change in particle size following autoclaving of SLN (191–193). However, Gokce et al. reported change in particle size from nano-size to micron size following autoclaving (194).

The chemical constituents play an important role in the particle size due to steam sterilization. Cavalliet al. reported the change of the particle size of SLN composed of behenic acid. A similar result had been observed for SLN composed of stearic acid. However, the steam sterilization did not change the average particle size of SLN composed of monostearate monocitrate diglyceride (192). Heiati et al. reported no change in the particle size of azidothymidine palmitate loaded SLN particles stabilized with trilaurin and phospholipid (191). The selection of the emulsifier is a significant parameter for the stability of the sample. Schwarz reported lecithin as a suitable surfactant for steam sterilization. However, the particle size of poloxamer stabilized SLN was changed after steam sterilization. The increase in temperature caused dehydration of the ethylene glycol. It caused a decrease in steric stabilization of this type of emulsifier (195). The decrease in temperature can reduce the temperature-induced particle aggregation to a large extent; γ-irradiation was studied as an alternative method to steam sterilization. The change in particle size was low in γ-irradiation as compared to steam sterilization (196,197).

25.13 Storage Challenges For Solid Lipid Nanoparticle Products

Lipid modification determines the stability of the SLN in dispersion. With increased lipid modification, the stability of solid lipid nanoparticles decreases. Presence of the other colloidal structures also influences the stability of the formulation. The major problems are encountered in particle size increase, change in zeta potential, dose dumping, and gelation phenomenon. Lyophilization is a process to increase the stability of SLN over an extended period of time. It involves the transformation of aqueous dispersion to powder through the process of freezing of the aqueous dispersion, followed by the evaporation of water under vacuum. Redispersion of the lyophilized SLN is the second transformation before the use of SLN. Although it is expected that lyophilization will increase the physical and chemical stability of the formulation, freezing of the SLN might cause changes in osmolarity and pH. The surfactants have a protective role in the stability of SLN. Lyophilization can affect the protective role of the surfactant. Because of this, the addition of cryoprotectors is necessary to prevent or reduce aggregation of SLN and obtain a better product on redispersion.

During lyophilization, the temperature drops; the water starts to freeze. It leaves behind a slush of concentrated solutes. In order to dilute the concentrated solutes, the water rushes out of the SLN because of osmotic activity. The cryoprotectors minimize osmotic activity and favor the glassy state of the frozen sample (88). The commonly used cryoprotectors are sorbitol, mannose, trehalose, glucose, and polyvinylpyrrolidone. Lyophilization may increase the particle size. Cavalli reported an increase in particle size after lyophilization. The concentration of cryoprotectants plays an significant role in the stability of the SLN after reconstitution. Suitable concentration is necessary to stabilize the system and get particles of uniform size after reconstitution (198). Trehalose, glucose, lactose, and mannitol were compared to find out their effect on the particle size of SLN on reconstitution of lyophilized SLN samples. Trehalose was found to be have the most protective cryoprotective effect on the SLN. The ratio of sugar and lipid should be 2.6–3.9 to have a good cryoprotective effect (191).

During lyophilization, the crystal structure has an important role in the stability and the particle size of the SLN. It affects the sublimation speed as well as reconstitution of the SLN. Schwarz recommended that SLN preparation should be subjected to rapid freezing using liquid nitrogen (192). However, Zimmermann reported that the best result was obtained with slow freezing process (199). Slow freezing of SLN preparation results in formation of large crystals and unrestricted sublimation. However, slow freezing can cause the destabilization of the preparation. Spray drying is considered an alternative procedure to lyophilization. Freitas reported conversion of dispersion of solid lipid nanoparticles using a spray drying technique. Spray drying of the prepared SLN resulted in dry and reconstitutable powders. Size distribution of the SLN preparation was affected by the process parameters, including the temperature, lipid concentration, and sugar concentration. Optimization of the parameters is necessary to get an uniform size distribution of SLN particles.

The increase in sugar concentration with the decrease in temperature and in lipid concentration resulted in uniform size distribution. Increase in sugar concentration built a protective layer around the particles during spray drying and prevented coalescence of SLN particles. However, the yield of the prepared formulation was reduced with an increase in the concentration of sugar due to adherence of granulate in the glass of the dryer. A high power respirator can be used to prevent adherence to the glass equipment. However, it may cause loss of small particles (200). The nature of the lipid is a key factor for preparations of spray drying. The melting point of lipid is a crucial factor for spray drying of solid lipid nanoparticles. The particles achieve a temperature below 15–20 °C of the outlet temperature during spraying in a co-current temperature (201). If the particles achieve a temperature load of 40–50 °C, then the melting point of the particles should be above this range. A minimum melting point of 65 °C is recommended when selecting lipids for spray drying (200). The higher melting point of the lipid will prevent the lipid from melting during the process, thereby enhancing the stability of the particles. The decrease in the amount of lipid will reduce the numbers of lipid particles. It will cause decrease in particle contact and subsequent aggregation.

25.14 In Vivo Models for the Study of Lung Diseases

25.14.1 Animal Models of Asthma

Asthma is a chronic inflammatory disease of the airway with extensive airway remodeling. Asthma is characterized by reversible airway obstruction, increased airway responsiveness, enhanced mucus production, airway inflammation, and remodeling of the airways (202). Remodeling of the airways in asthma includes epithelial fibrosis, goblet cell metaplasia and hyperplasia, and mucus hypersecretion, as well as hypertrophy and hyperplasia of airway smooth muscles, which depends on repeated exposure to the allergen as well as airway inflammation severity and duration (203). The development of animal models of asthma involves a process of sensitizing an animal to an antigen followed by sneezing and nose rubbing effects on the airways in order to present allergic responses. The physiological and immunological airway responses could be different between species, based on the method of sensitization and the antigen used (204). The purpose of development of asthma is to mimic the disease characteristics in order to understand the underlying pathophysiology and aid in the development of new drugs.

Inducing Agent	Animals	Methods	References
Ovalbumin (OVA)	**Rat**	IP 1 mg OVA+20 mg $Al(OH)_3$ on days 0, 7 and 14, IT 1.1% OVA on day 21	(205)
	Guinea pig	IP and SC 100 mg (1 ml) OVA on day 1, IP 10 mg OVA on day 8, expose to aerosol 4% OVA from day 14, for 18 days, 4 min/day	(206)
	Rabbit	IP 0.1 mg OVA+10 mg $Al(OH)_3$, days 1 & 14, expose to OVA (10 mg/ml) aerosol, days 28–30, 10 min/day	(207)
	Sheep	SC OVA (100 mg/1 ml saline+50 µl $Al(OH)_3$ 3 times at 2-week intervals, boost SC after 4 weeks	(208)
House Dust Mite (HDM)	**Mice**	IV 25 µg/10 µl saline, 5 consecutive days/week, 7 weeks	(209)
	Sheep	Intramammary 1 mg solubilized HDM every 2 weeks for 3–4 weeks	(210)
	Monkey	IP (3.6 ml/kg) + IM (0.4 ml/kg) 3 times weekly, IM 0.4 ml/kg + 5mg/50mg/ml $Al(OH)_3$, 5weeks later	(211)
Lipopolysaccharide	**Rat**	Expose to 1 mg/ml lipopolysaccharide/PBS (pH 7.4), 15 min, turning off the nebulizer, plugged the chamber inlet and outlets, rats remained in the chamber for further 15 min to breathe the remaining aerosolized solution	(212)

Many other agents are also used for developing asthma models, including recombinant *Blattella germanica*, allergens including extracts of house dust mite, ragweed, *Ascaris suum*, latex, cockroach extracts, *Dermatophagoides farinae*, etc.

The ovalbumin (OVA)-induced asthma model in the Brown Norway rat is a suitable model to assess in vivo efficacy of anti-asthma drugs. This animal model features many similarities to human allergic asthma, including the presence of eosinophilic lung inflammation, the release of inflammatory mediators and cytokines, and the presence of airway hyperresponsiveness after the antigen challenge. The evaluation of various parameters includes flexivent lung function, bronchoalveolar lavage (BAL) with leukocyte differential counts, biomarker analysis (gene, protein expression patterns), cytokine analysis, and PK/PD blood analysis.

25.14.2 Bacterial Lung Infection Models

Many bacterial infections cause severe damage to the lung leading to pneumonia. Bacterial infections casuses more fatalities in patients with cystic fibrosis and immunocompromised conditions. Due to the development of resistance against antibiotics, evaluation of new candidate molecules is necessary against microorganisms involved in lung infection. The candidate molecules can be assessed in various models of murine lung infection including *Staphylococcus aureus, Pseudomonas aeruginosa, Klebsiella pneumonia, Legionella pneumophila*, and *Streptococcus pneumonia*. Experimental pulmonary infections were conducted on a variety of animal species, such as mice, rats, guinea pigs, rabbits, ferrets, and hamsters. Mouse models are valuable for host defense studies, whereas rat models are less susceptible to many infectious human pathogens. However, unilateral pneumonia cannot be achieved in mice properly. Rats are suitable for production of unilateral pneumonia (213). Various pathogens, including bacteria and fungus, caused pneumonia in rat models in a reproducible manner (214,215). The selection of the animal species depends on the susceptibility of the animal to human pathogens. Infection of BALB/c mice with *Coccidioides immitis* resulted in 100% mortality, whereas infection of DBA/2 mice with *Coccidioides immitis* showed an initial increase in bacterial cell count followed by a decrease in bacterial cell count and resulted in survival of the animals (216).

The route of administration of infectious agents includes intranasal inoculation, aerosol administration, and direct instillation. In intranasal inoculation, a suspension of pathogens is applied to the nose, keeping the animals in vertical position. Variable deposition may occur in different animals due to variability of respiratory frequency. Significant amount of ingestion may also occur. Anesthesia is essential prior to treatment with pathogens to minimize variability of lung deposition of pathogens. Aerosol administration involves inhalation of pathogens via nose by animals. Exprosure time generally ranges from 30 to 60 min (213). This technique is suitable for infecting a large number of animals at a time (217). Direct instillation technique involves intrabronchial or intra-tracheal inoculation of pathogens. Direct instillation allows inoculation of a definite number of pathogens. Unilateral pneumonia can be produced using the direct inoculation method. However, surgical skill is essential to carry out the direct instillation technique.

The efficacy of drug molecules can be determined using various parameters including clinical signs, bacterial load in the lung, levels of cytokines and cellular infiltrate, mortality rate, cellular infiltrate, and cytokines in the BAL fluid. Gross pathology and histopathology can be carried out to assess the lung in these models.

Bacterial infection models (gram positive bacteria and gram negative bacteria), mycotic infection models, viral infection models, and parasitic infection models have been described in different literatures. Animals with intact host defense or impaired host defense mechanisms have been used in the animal models. Pneumonia-induced acute lung infections were produced in mice (218), guinea pigs, and rabbits (219). *Staphylococcus aureus* was also used to produce acute or chronic pulmonary infections in mice (220), rats, or rabbits (221). In the immunocompromised animals, mortality up to 100% was observed. Also, methicillin-resistant *S. aureus* (MRSA) was used to establish pneumonia. Other bacterial pulmonary infection models include *Pseudomonas aeruginosa, Haemophilus influenza, Klebsiella pneumonia, Mycoplasma pneumonia, Chlamydia pneumonia, Escherichia coli, Legionella pneumophila, Acinetobacter baumannii, Enterobacter cloacae*, and *Pasteurella haemolytica* (213). Various models for mycotic pulmonary infections have been described.

25.14.3 Influenza Models

The pandemic of swine flu, which arose from Mexico, and the constant threat of new strains such as H5N1, have highlighted the potential risk to human health that influenza poses globally. Considering the potential risk due to new strains of influenza, the development of novel vaccines and anti-influenza drugs is essential for protection against the new and existing strains of Influenza. Mouse influenza and and ferret influenza models are used for evaluation of the effects of new drug molecules and vaccines. Evaluation of various parameters include viral load in BAL fluid or nasal washes, levels of cellular infiltrate and cytokines, change in body weight, and temperature and rate of survival.

25.14.4 Mouse Influenza Model

The mouse influenza model is suitable for lead identification and vaccine development due to the specifically characterized nature of the murine immune system, and are also suitable for evaluation of immunological parameters in vaccine development.

25.14.5 Ferret Influenza Model

The ferret influenza model is a specific model for the study of influenza because they are sensitive to human influenza strains. Ferrets exhibit symptoms of influenza infections like those of humans, therefore ferret models are suitable for the study of new candidate drugs and vaccines. Drugs can be evaluated against different strains of human influenza viruses in ferret influenza models.

25.14.6 Lipopolysaccharide-Induced Pulmonary Neutrophilia Model

Acute respiratory distress syndrome (ARDS), cystic fibrosis (CF), and COPD are characterized by neutrophil recruitment and inflammatory mediator release. Acute pulmonary neutrophilia models can be used to investigate the underlying mechanisms of these diseases. The mouse model and rat model are used for lipopolysaccharide-induced pulmonary neutrophilia. Lipopolysaccharide-induced neutrophilia was also studied in hamsters. Lipopolysaccharide treatment causes marked changes in neutrophils, total cell counts, level of elastase, and lymphocytes in BALB/c mice. Although neutrophils account for 90–95% of the total cells in the BAL fluid, a significant change in lymphocyte count is also observed. The selection of strain of mouse and rat is a crucial factor. BALB/c mice exhibited reproducible and significant inflammation after lipopolysaccharide treatment; however, a small change in neutrophil infiltration was observed in C57BL/6 (222). C57BL/6 mice are less sensitive to lipopolysaccharide as compared to BALB/c mice because of inbred genetic variations between different strains (223).

25.14.7 Mouse Model

In the mouse model, mice are challenged with lipopolysaccharide via the intranasal route. It induces a marked neutrophilia in BAL fluid, and also increases other mediators such as TNF. Neutrophils invade the airway lumen of the lung in a dose-dependent and time-dependent manner after lipopolysaccharide administration. These models are useful for study of steroidal drugs. The increase in elastase and myeloperoxidase activities in BAL fluid is a result of activation of neutrophil. Increase in neutrophil proteases and lung hemorrhage are the markers of lung damage. For reasonable measurement of activation of neutrophils, a time period of 24 hours is needed post challenge with lipopolysaccharide. The neutrophil level returns to baseline after 5 days.

25.14.8 Rat Model

In rat models, the lung neutrophilia model can be induced by administration of lipopolysaccharide via intra-tracheal route or aerosol delivery. Sprague/Dawley rats and Wistar rats can be used to study lipopolysaccharide induced lung neutrophilia because of genetic susceptibility. Candidate molecules can be assessed for potency and efficacy through multiple routes. Marked neutrophil recruitment and cytokines can be observed in this model after lipopolysaccharide challenge. The evaluation of the models include BAL fluid analysis, levels of inflammatory mediators, lung histology, and immunohistochemistry (224).

25.14.9 Pulmonary Fibrosis Model

25.14.9.1 Mouse and Rat Model of Bleomycin-Induced Lung Fibrosis

Bleomycin-induced pulmonary fibrosis is the most common experimental study model of human lung fibrosis. Bleomycin belongs to a sub-family of complex glycopeptides with antineoplastic properties, initially isolated from a strain of the fungus *Streptomyces verticillus* (225). Its primary clinical use is as an antitumor antibiotic for various carcinomas and lymphomas (226). Bleomycin-induced toxicity occurs predominantly in the organs of the lung, skin, and mucous membranes due to the lack of the bleomycin-inactivating enzyme, bleomycin hydrolase, in those tissues (227,228). Delivery of bleomycin to the lung causes pulmonary injury, inflammation, and subsequent fibrosis (229). Bleomycin is administered through transoral instillation, endotracheal injection, and intravenous and intraperitoneal routes. Choosing the appropriate endpoint in the bleomycin model depends on the specific parameter of the injury and fibrotic process that need to be evaluated. Analysis during the first 3 days will reflect the acute injury and inflammatory response to bleomycin. To evaluate established fibrosis in lung tissue, 3–4 weeks after drug treatment should be considered because the peak of collagen deposition appears at this time point. Evaluation consists of:

- Histopathology and qualification via a modified Ashcroft score
- Collagen quantification (directly or via image analysis of lung sections)
- BALF analysis with leukocyte differential counts
- Biomarker analysis (gene, protein expression patterns)

In mouse models, fibrosis is assessed by hydroxyproline content and lung pathology. Basically it has been assessed over a 28-day time period; end time-point is determined on the basis of experimental aim, usually day 21. BALF analysis of cellularity can be assessed using these models. Parameters including inflammatory mediator content and in-life lung function can be analyzed. In the rat model, fibrosis has been assessed over a 28-day time period. The typical end point of the study is day 28.

25.14.10 Pulmonary Cancer Model

A number of the mutations found in human NSCLC have been introduced in the mouse. These include, among others, Kras, Braf, Egfr, Lkb1, Rac1, NFkB, and p53. Kras mutation, EGRF mutation, ALK fusion gene, and pathway addiction are involved in lung cancer development.

25.14.11 Mouse Models for Non Small Cell Lung Cancer

Mouse Mutant	Tumor Induction	Reference
LSL-KrasG12D; P53lox/lox	Ad5-CMV-Cre	(230)
LSL-KrasG12D endogenous control	Sporadic infection of lung cells with Adeno-Cre virus	(231)
LL-BrafV600E	Ad5 CMV Cre	(232)
Tet-op-PIK3CA H1047R;CCSP-rtTA	Tumor induction by doxycyclin	(233)
LSL-KrasG12D; P53frt/frt	Ad5-CMV-Cre	(234)
LL-BrafV600ELSL-KrasG12D	Ad5 CMV Cre	(235)
LSL-KrasG12Vgeo	Cre-ERT2 (RERT-ert)	(236)

25.14.12 Mouse Models for Squamous Cell Carcinomas

Squamous cell carcinomas (SCC) constitute an important fraction of the NSCLC in humans (237). Overexpression of genes is found in squamous cell carcinoma, such as SOX2 (238). Loss of Lkb1 has been shown to give tumors with squamous makers (239). Chemically induced as well as transgenic mouse models have been developed to study squamous cell carcinomas (237).

25.14.13 Mouse Models for Small Cell Lung Cancer

Genetically engineered mouse models are the tumor models in which mutant mice develop autochthonous tumors upon targeted alterations in cancer genes. Patient-derived xenografts and mouse-derived allografts can be transplanted in mice. Using allograft models, mouse tumors can be transplanted and expanded in recipient mice. SCLC patient-derived xenograft (PDX) models have been developed in immunocompromised mice (240). The SCLC PDX model can be used for investigating the response of human SCLC to chemotherapy or targeted therapies (241). SCLC patients have more circulating tumor cells (CTC) than other types of cancer. It allows for the generation of CTC-derived explant (CDX) models, i.e. xenografts derived from CTCs.

25.15 Conclusion

Extensive research in the field of SLN and pulmonary drug delivery provides the necessary information for rapid designing of a SLN based delivery system for herbal drugs. Although pulmonary delivery of herbal drugs and extracts is a challenging task due to the thermal stability aspects, this obstacle can be over by proper selection and desiging of SLN formulation technique. While formulating herbal drug loaded SLN products, the effect of herbal drugs on the composition of lipid matrix, its components, surface properties of SLN, and stability of the product must be evaluated carefully. The dissolution/solubilization of herbal drugs at the target site is one of the important citeria for successful development of the formulation. Sterility of herbal drug-loaded SLN is another challenging task for researchers. However, the field of herbal drug delivery is expanding rapidly and pulmonary drug delivery of herbal drugs emerging as a potential pathway for existing as well as novel diseases.

Conflict of Interest

Nil

REFERENCES

1. Yang W, Peters JI, and Williams RO. Inhaled nanoparticles — a current review. *International Journal of Pharmaceutics*. 2008; 356(1-2): 239–247.
2. Patton JS. Mechanisms of macromolecule absorption by the lungs. *Advanced Drug Delivery Reviews*. 1996; 19(1): 3–36.
3. Knudsen L and Ochs M. The micromechanics of lung alveoli: structure and function of surfactant and tissue

components. *Histochemistry and Cell Biology*. 2018; 150(6): 661–676.
4. Khan YS and Lynch DT. Histology, Lung. In: StatPearls [Internet]. Treasure Island (FL): StatPearls Publishing; 2020.
5. Corry D, Kulkarni P, and Lipscomb MF. The migration of bronchoalveolar macrophages into hilar lymph nodes. *The American Journal of Pathology*. 1984; 115(3): 321–328.
6. Lin LT, Chen TY, Lin SC, et al. Broad-spectrum antiviral activity of chebulagic acid and punicalagin against viruses that use glycosaminoglycans for entry. *BMC Microbiology*. 2013; 13: 187.
7. Wang KC, Chang JS, Lin LT, Chiang LC, and Lin CC. Antiviral effect of cimicifugin from *Cimicifuga foetida* against human respiratory syncytial virus. *The American Journal of Chinese Medicine*. 2012; 40(5): 1033–1045.
8. Ma LY, Ma SC, Wei F, et al. Uncinoside A and B, two new antiviral chromone glycosides from Selaginella uncinata. *Chemical and Pharmaceutical Bulletin*. 2003; 51(11): 1264–1267.
9. Huang W, Zhang X, Wang Y, et al. Antiviral biflavonoids from Radix Wikstroemiae (Liaogewanggen). *Chinese Medicine*. 2010; 5: 23.
10. Zang N, Xie X, Deng Y, et al. Resveratrol-mediated gamma interferon reduction prevents airway inflammation and airway hyperresponsiveness in respiratory syncytial virus-infected immunocompromised mice. *Journal of Virology*. 2011; 85(24): 13061–13068.
11. Lin CW, Tsai FJ, Tsai CH, et al. Anti-SARS coronavirus 3C-like protease effects of Isatis indigotica root and plant-derived phenolic compounds. *Antiviral Research*. 2005; 68(1): 36–42.
12. Ryu YB, Jeong HJ, Kim JH, et al. Biflavonoids from *Torreya nucifera* displaying SARS-CoV 3CL(pro) inhibition. *Bioorganic & Medicinal Chemistry*. 2010; 18(22): 7940–7947.
13. Yu MS, Lee J, Lee JM, et al. Identification of myricetin and scutellarein as novel chemical inhibitors of the SARS coronavirus helicase, nsP13. *Bioorganic & Medicinal Chemistry Letters*. 2012; 22(12): 4049–4054.
14. Lau KM, Lee KM, Koon CM, et al. Immunomodulatory and anti-SARS activities of *Houttuynia cordata*. *Journal of Ethnopharmacology*. 2008;118(1): 79–85.
15. Jeong HJ, Kim YM, Kim JH, et al. Homoisoflavonoids from *Caesalpinia sappan* displaying viral neuraminidases inhibition. *Biological and Pharmaceutical Bulletin*. 2012; 35(5): 786–790.
16. Zhi H, Jin X, Zhu H, Li H, Zhang Y, Lu Y, and Chen D. Exploring the effective materials of flavonoids-enriched extract from *Scutellaria baicalensis* roots based on the metabolic activation in influenza A virus induced acute lung injury. *Journal of Pharmaceutical and Biomedical Analysis*. 2020; 177: 112876.
17. Dao TT, Dang TT, Nguyen PH, Kim E, Thuong PT, and Oh WK. Xanthones from *Polygala karensium* inhibit neuraminidases from influenza A viruses. *Bioorganic & Medicinal Chemistry Letters* . 2012; 22(11): 3688–3692.
18. Dao TT, Nguyen PH, Lee HS, et al. Chalcones as novel influenza A (H1N1) neuraminidase inhibitors from *Glycyrrhiza inflata*. *Bioorganic & Medicinal Chemistry Letters*. 2011; 21(1): 294–298.
19. Ma SG, Gao RM, Li YH, et al. Antiviral spirooliganones A and B with unprecedented skeletons from the roots of Illicium oligandrum. *Organic Letters*. 2013; 15(17): 4450–4453.
20. He W, Han H, Wang W, and Gao B. Anti-influenza virus effect of aqueous extracts from dandelion. *Virology Journal*. 2011; 8: 538.
21. Theisen LL, and Muller CP. EPs® 7630 (Umckaloabo®), an extract from *Pelargonium sidoides* roots, exerts anti-influenza virus activity in vitro and in vivo. *Antiviral Research*. 2012; 94(2): 147–156.
22. Chen X, Yang X, Liu T, et al. Kaempferol regulates MAPKs and NF-κB signaling pathways to attenuate LPS-induced acute lung injury in mice. *International Immunopharmacology*. 2012; 14(2): 209–216.
23. Qiu J, Chi G, Wu Q, Ren Y, Chen C, and Feng H. Pretreatment with the compound asperuloside decreases acute lung injury via inhibiting MAPK and NF-κB signaling in a murine model. *International Immunopharmacology*. 2016; 31: 109–115.
24. Rojas-Duran R, González-Aspajo G, Ruiz-Martel C, et al. Anti-inflammatory activity of Mitraphylline isolated from Uncaria tomentosa bark. *Journal of Ethnopharmacology*. 2012; 143(3): 801–804.
25. Montserrat-de la Paz S, Fernandez-Arche A, de la Puerta R, et al. Mitraphylline inhibits lipopolysaccharide-mediated activation of primary human neutrophils. *Phytomedicine*. 2016; 23(2): 141–148.
26. Jiao H, Su W, Li P, et al. Therapeutic effects of naringin in a guinea pig model of ovalbumin-induced cough-variant asthma. *Pulmonary Pharmacology & Therapeutics*. 2015; 33: 59–65.
27. Guihua X, Shuyin L, Jinliang G, and Wang S. Naringin protects ovalbumin-induced airway inflammation in a mouse model of asthma. *Inflammation*. 2016; 39(2):891–899.
28. Chong L, Zhang W, Nie Y, et al. Protective effect of curcumin on acute airway inflammation of allergic asthma in mice through Notch1-GATA3 signaling pathway. *Inflammation*. 2014; 37(5): 1476–1485.
29. Takashima K, Matsushima M, Hashimoto K, et al. Protective effects of intratracheally administered quercetin on lipopolysaccharide-induced acute lung injury. *Respiratory Research*. 2014; 15(1): 150.
30. Wang L, Chen J, Wang B, et al. Protective effect of quercetin on lipopolysaccharide-induced acute lung injury in mice by inhibiting inflammatory cell influx. *Experimental Biology and Medicine*. 2014; 239(12): 1653–1662.
31. Rogerio AP, Kanashiro A, Fontanari C, et al. Anti-inflammatory activity of quercetin and isoquercitrin in experimental murine allergic asthma. *Journal of the European Histamine Research Society*. 2007; 56(10): 402–408.
32. Zhu GF, Guo HJ, Huang Y, Wu CT, and Zhang XT. Eriodictyol, a plant flavonoid, attenuates LPS-induced acute lung injury through its antioxidative and anti-inflammatory activity. *Experimental and Therapeutic Medicine*. 2015; 10(6): 2259–2266.
33. Jung HW, Kang SY, Kang JS, Kim AR, Woo ER, and Park YK. Effect of Kuwanon G isolated from the root bark of Morus alba on ovalbumin-induced allergic response in a mouse model of asthma. *Phytotherapy Research*. 2014; 28(11): 1713–1719.

34. Li J and Zhang B. Apigenin protects ovalbumin-induced asthma through the regulation of Th17 cells. *Fitoterapia*. 2013; 91: 298–304.
35. Chauhan PS, Subhashini, Dash D, and Singh R. Intranasal curcumin attenuates airway remodeling in murine model of chronic asthma. *International Immunopharmacology*. 2014; 21(1): 63–75.
36. Jeon CM, Shin IS, Shin NR, et al. Siegesbeckia glabrescens attenuates allergic airway inflammation in LPS-stimulated RAW 264.7 cells and OVA induced asthma murine model. *International Immunopharmacology*. 2014; 22(2):414–419.
37. Kim JY, Lim HJ, and Ryu JH. In vitro anti-inflammatory activity of 3-O-methyl-flavones isolated from Siegesbeckia glabrescens. *Bioorganic & Medicinal Chemistry Letters*. 2008; 18(4): 1511–1514.
38. Toledo AC, Sakoda CPP, Perini A, et al. Flavonone treatment reverses airway inflammation and remodelling in an asthma murine model. *British Journal of Pharmacology*. 2013; 168(7): 1736–1749.
39. Taguchi L, Pinheiro NM, Olivo CR, et al. A flavanone from *Baccharis retusa* (Asteraceae) prevents elastase-induced emphysema in mice by regulating NF-κB, oxidative stress and metalloproteinases. *Respiratory Research*. 2015; 16(1): 79.
40. Chen SM, Tsai YS, Lee SW, et al. Astragalus membranaceus modulates Th1/2 immune balance and activates PPARγ in a murine asthma model. *Biochemistry and Cell Biology*. 2014; 92(5): 397–405.
41. Kuo MY, Liao MF, Chen FL, et al. Luteolin attenuates the pulmonary inflammatory response involves abilities of antioxidation and inhibition of MAPK and NFκB pathways in mice with endotoxin-induced acute lung injury. *Food and Chemical Toxicology*. 2011; 49(10): 2660–2666.
42. Rungsung S, Singh TU, Rabha DJ, et al. Luteolin attenuates acute lung injury in experimental mouse model of sepsis. *Cytokine*. 2018; 110: 333–343.
43. Park EJ, Kim YM, Kim HJ, and Chang KC. Luteolin activates ERK1/2- and Ca(2+)-dependent HO-1 induction that reduces LPS-induced HMGB1, iNOS/NO, and COX-2 expression in RAW264.7 cells and mitigates acute lung injury of endotoxin mice. *Inflammation Research*. 2018; 67(5): 445–453.
44. Liu X and Meng J. Luteolin alleviates LPS-induced bronchopneumonia injury in vitro and in vivo by down-regulating microRNA-132 expression. *Biomedicine & Pharmacotherapy*. 2018; 106: 1641–1649.
45. Kim HY, Jung SK, Byun S, et al. Raf and PI3K are the molecular targets for the anti-metastatic effect of luteolin. *Phytotherapy Research*. 2013; 27(10): 1481–1488.
46. He DY and Dai SM. Anti-inflammatory and immunomodulatory effects of *Paeonia lactiflora* pall., a traditional chinese herbal medicine. *Frontiers in Pharmacology*. 2011; 2: 10.
47. Kim ID and Ha BJ. Paeoniflorin protects RAW 264.7 macrophages from LPS-induced cytotoxicity and genotoxicity. *Toxicology in Vitro*. 2009; 23(6): 1014–1019.
48. Cao W, Zhang W, Liu J, et al. Paeoniflorin improves survival in LPS-challenged mice through the suppression of TNF-α and IL-1β release and augmentation of IL-10 production. *International Immunopharmacology*. 2011; 11(2): 172–178.
49. Azaizeh H, Saad B, Khalil K, and Said O. The state of the art of traditional arab herbal medicine in the eastern region of the mediterranean: A review. *Evidence-Based Complementary and Alternative Medicine*. 2006; 3(2): 229–235.
50. An N, Wang D, Zhu T, et al. Effects of scrocaffeside A from *Picrorhiza Scrophulariiflora* on immunocyte function in vitro. *Immunopharmacol Immunotoxicol*. 2009; 31(3): 451–458.
51. Noh S, Ahn KS, Oh SR, Kim KH, and Joo M. Neutrophilic lung inflammation suppressed by picroside ii is associated with TGF-β signaling. *Evidence-Based Complementary and Alternative Medicine*. 2015; 2015: 897272.
52. He LJ, Liang M, Hou FF, Guo ZJ, Xie D, and Zhang X. Ethanol extraction of *Picrorhiza scrophulariiflora* prevents renal injury in experimental diabetes via anti-inflammation action. *Journal of Endocrinology*. 2009; 200(3): 347–355.
53. Smith LJ, Kalhan R, Wise RA, et al. Effect of a soy isoflavone supplement on lung function and clinical outcomes in patients with poorly controlled asthma: A randomized clinical trial. *JAMA*. 2015; 313(20): 2033–2043.
54. Huang HL, Liu CT, Chou MC, Ko CH, and Wang CK. Noni (*Morinda citrifolia* L.) fruit extracts improve colon microflora and exert anti-inflammatory activities in Caco-2 cells. *Journal of Medicinal Food*. 2015; 18(6): 663–676.
55. Murata K, Abe Y, Futamura-Masudaa M, Uwaya A, Isami F, and Matsuda H. Activation of cell-mediated immunity by *Morinda citrifolia* fruit extract and its constituents. *Natural Product Communications*. 2014; 9(4): 445–450.
56. Ghosh S, Banerjee S, and Sil PC. The beneficial role of curcumin on inflammation, diabetes and neurodegenerative disease: A recent update. *Food and Chemical Toxicology*. 2015; 83: 111–124.
57. He Y, Yue Y, Zheng X, Zhang K, Chen S, and Du Z. Curcumin, inflammation, and chronic diseases: how are they linked? *Molecules*. 2015; 20(5): 9183–9213.
58. Song HH, Shin IS, Woo SY, et al. Piscroside C, a novel iridoid glycoside isolated from *Pseudolysimachion rotundum* var. *subinegrum* suppresses airway inflammation induced by cigarette smoke. *Journal of* Ethnopharmacology. 2015; 170: 20–27.
59. Khan N and Mukhtar H. Tea polyphenols in promotion of human health. *Nutrients*. 2018; 11(1): 39.
60. Clark J and You M. Chemoprevention of lung cancer by tea. *Molecular Nutrition & Food Research*. 2006; 50(2): 144–151.
61. Chen KY, Chen CC, Chang YC, and Chang MC. Resveratrol induced premature senescence and inhibited epithelial-mesenchymal transition of cancer cells via induction of tumor suppressor Rad9. *PLoS ONE*. 2019; 14(7): e0219317.
62. de B. Oliveira AL, Monteiro VVS, Navegantes-Lima KC, et al. Resveratrol role in autoimmune disease-A mini-review. *Nutrients*. 2017; 9(12): 1306.
63. Wang C and Cui C. Inhibition of lung cancer proliferation by wogonin is associated with activation of apoptosis and

generation of reactive oxygen species. *Balkan Medical Journal*. 2019; 37(1): 29–33.
64. Martínez-García D, Pérez-Hernández M, and Korrodi-Gregório L. The natural-based antitumor compound T21 decreases survivin levels through potent STAT3 inhibition in lung cancer models. *Biomolecules*. 2019; 9(8): 361.
65. Lee YJ, Lim T, Han MS, et al. Anticancer effect of luteolin is mediated by downregulation of TAM receptor tyrosine kinases, but not interleukin-8, in non-small cell lung cancer cells. *Oncology Reports*. 2017; 37(2): 1219–1226.
66. Cho HJ, Ahn KC, Choi JY, et al. Luteolin acts as a radiosensitizer in non-small cell lung cancer cells by enhancing apoptotic cell death through activation of a p38/ROS/caspase cascade. *International Journal of Oncology*. 2015; 46(3): 1149–1158.
67. Sonoki H, Tanimae A, Endo S, et al. Kaempherol and luteolin decrease claudin-2 expression mediated by inhibition of STAT3 in lung adenocarcinoma A549 cells. *Nutrients*. 2017; 9(6): 597.
68. Meng G, Chai K, Li X, Zhu Y, and Huang W. Luteolin exerts pro-apoptotic effect and anti-migration effects on A549 lung adenocarcinoma cells through the activation of MEK/ERK signaling pathway. *Chemico-Biological Interactions*. 2016; 257: 26–34.
69. Yu Q, Zhang M, Ying Q, et al. Decrease of AIM2 mediated by luteolin contributes to non-small cell lung cancer treatment. *Cell Death & Disease*. 2019; 10(3): 218.
70. Chen KC, Chen CY, Lin CR, et al. Luteolin attenuates TGF-β1-induced epithelial-mesenchymal transition of lung cancer cells by interfering in the PI3K/Akt-NF-κB-Snail pathway. *Life Science*. 2013; 93(24): 924–933.
71. Kasala ER, Bodduluru LN, Barua CC, and Gogoi R. Antioxidant and antitumor efficacy of Luteolin, a dietary flavone on benzo(a)pyrene-induced experimental lung carcinogenesis. *Biomedicine & Pharmacotherapy*. 2016; 82: 568–577.
72. Yan J, Wang Q, Zheng X, et al. Luteolin enhances TNF-related apoptosis-inducing ligand's anticancer activity in a lung cancer xenograft mouse model. *Biochemical and Biophysical Research Communications*. 2012; 417(2): 842–846.
73. Lee HZ, Yang WH, Bao BY, and Lo PL. Proteomic analysis reveals ATP-dependent steps and chaperones involvement in luteolin-induced lung cancer CH27 cell apoptosis. *European Journal of Pharmacology*. 2010; 642(1-3): 19–27.
74. Ma L, Peng H, Li K, et al. Luteolin exerts an anticancer effect on NCI-H460 human non-small cell lung cancer cells through the induction of Sirt1-mediated apoptosis. *Molecular Medicine Report*. 2015; 12(3): 4196–4202.
75. Bai L, Xu X, Wang Q, et al. A superoxide-mediated mitogen-activated protein kinase phosphatase-1 degradation and c-Jun NH(2)-terminal kinase activation pathway for luteolin-induced lung cancer cytotoxicity. *Molecular Pharmacology*. 2012; 81(4): 549–555.
76. Cheng PW, Ng LT, Chiang LC, and Lin CC. Antiviral effects of saikosaponins on human coronavirus 229E in vitro. *Clinical and Experimental Pharmacology and Physiology*. 2006; 33(7): 612–616.
77. Zhu J, Luo C, Wang P, He Q, Zhou J, and Peng H. Saikosaponin A mediates the inflammatory response by inhibiting the MAPK and NF-κB pathways in LPS-stimulated RAW 264.7 cells. *Experimental and Therapeutic Medicine*. 2013; 5: 1345–1350.
78. Wei Z, Wang J, Shi M, Liu W, Yang Z, and Fu Y. Saikosaponin a inhibits LPS-induced inflammatory response by inducing liver X receptor alpha activation in primary mouse macrophages. *Oncotarget*. 2016; 7(31): 48995–49007.
79. Zhao H, Li S, Zhang H, Wang G, Xu G, and Zhang H. Saikosaponin A protects against experimental sepsis via inhibition of NOD2-mediated NF-κB activation. *Experimental and Therapeutic Medicine*. 2015; 10(2): 823–827.
80. Du ZA, Sun MN, and Hu ZS. Saikosaponin a ameliorates LPS-Saikosaponin a ameliorates LPS-induced acute lung injury in mice. *Inflammation*. 2018; 41(1): 193–198.
81. Chen RJ, Guo XY, Cheng BH, Gong YQ, Ying BY, and Lin MX. Saikosaponin a inhibits cigarette smoke-induced oxidant stress and inflammatory responses by activation of Nrf2. *Inflammation*. 2018; 41(4): 1297–1303.
82. Liu A, Tanaka N, Sun L, et al. Saikosaponin d protects against acetaminophen-induced hepatotoxicity by inhibiting NF-κB and STAT3 signaling. *Chemico-Biological Interactions*. 2014; 223: 80–86.
83. Leung CY, Liu L, Wong RNS, Zeng YY, Li M, and Zhou H. Saikosaponin-d inhibits T cell activation through the modulation of PKCtheta, JNK, and NF-kappaB transcription factor. *Biochemical and Biophysical Research Communications*. 2005; 338(4): 1920–1927.
84. Hsu YL, Kuo PL, and Lin CC. The proliferative inhibition and apoptotic mechanism of Saikosaponin D in human non-small cell lung cancer A549 cells. *Life Sciences*. 2004; 75(10): 1231–1242.
85. Chen X, Liu C, Zhao R, et al. Synergetic and antagonistic molecular effects mediated by the feedback loop of p53 and JNK between Saikosaponin D and SP600125 on lung cancer A549 Cells. *Molecular Pharmaceutics*. 2018; 15(11): 4974–4984.
86. Fang W, Yang YJ, GuoBL, and Cen S. Anti-influenza triterpenoid saponins (saikosaponins) from the roots of *Bupleurum marginatum* var. stenophyllum. *Bioorganic & Medicinal Chemistry Letters*. 2017; 27(8): 1654–1659.
87. Wissing SA, Kayser O, and Müller RH. Solid lipid nanoparticles for parenteral drug delivery. *Advanced Drug Delivery Reviews*. 2004; 56(9): 1257–1272.
88. Mehnert W and Mäder K. Solid lipid nanoparticles: production, characterization and applications. *Advanced Drug Delivery Reviews*. 2001; 47(2-3): 165–196.
89. Jacobs C and Müller RH. Production and characterization of a budesonide nanosuspension for pulmonary administration. *Pharmaceutical Research*. 2002; 19(2): 189–194.
90. Pilcer G and Amighi K. Formulation strategy and use of excipients in pulmonary drug delivery. *International Journal of Pharmaceutics*. 2010; 392(1-2): 1–19.
91. Gupta S, Kesarla R, Chotai N, Misra A, and Omri A. Systematic approach for the formulation and optimization of solid lipid nanoparticles of efavirenz by high pressure homogenization using design of experiments for brain targeting and enhanced bioavailability. *BioMed Research International*. 2017; 2017: 5984014.
92. Ahlin P, Kristl J, and Smid-Korbar J. Optimization of procedure parameters and physical stability of solid lipid

nanoparticles in dispersions. *Acta Pharmaceutica*. 1998; 48: 259–267.
93. Lim SJ and Kim CK. Formulation parameters determining the physicochemical characteristics of solid lipid nanoparticles loaded with all-trans retinoic acid. *International Journal of Pharmaceutics*. 2002; 243(1-2): 135–146.
94. Shah RM, Malherbe F, Eldridge D, Palombo EA, and Harding IH. Physicochemical characterization of solid lipid nanoparticles (SLNs) prepared by a novel microemulsion technique. *Journal of Colloid and Interface Science*. 2014; 428: 286–294.
95. Chirio D, Gallarate M, Peira E, Battaglia L, Serpe L, and Trotta M. Formulation of curcumin-loaded solid lipid nanoparticles produced by fatty acids coacervation technique. *Journal of Microencapsulation*. 2011; 28(6): 537–548.
96. Kumar A, Li S, Cheng CM, and Lee D. Recent developments in phase inversion emulsification. *Industrial & Engineering Chemistry Research*. 2015; 54(34): 8375–8396.
97. Förster TH, Schambil F, and Tesmann H. Emulsification by the phase inversion temperature method: the role of self-bodying agents and the influence of oil polarity. *International Journal of Cosmetic Science*. 1990; 12(5): 217–227.
98. Ali H and Singh SK. Preparation and characterization of solid lipid nanoparticles of furosemide using quality by design. *Particulate Science and Technology*. 2018; 36(6): 695–709.
99. Heurtault B, Saulnier P, Pech B, Proust JE, and Benoit JP. A novel phase inversion-based process for the preparation of lipid nanocarriers. *Pharmaceutical Research*. 2002; 19(6): 875–880.
100. Dharmala K, Yoo JW, and Lee CH. Development of chitosan-SLN microparticles for chemotherapy: in vitro approach through efflux-transporter modulation. *Journal of Controlled Release*. 2008; 131(3): 190–197.
101. Videira M, Almeida AJ, and Fabra A. Preclinical evaluation of a pulmonary delivered paclitaxel-loaded lipid nanocarrier antitumor effect. *Nanomedicine*. 2012; 8(7): 1208–1215.
102. Videira MA, Gano L, Santos C, Neves M, and Almeida AJ. Lymphatic uptake of lipid nanoparticles following endotracheal administration. *Journal of Microencapsulation*. 2006; 23(8): 855–862.
103. Hong SS, Kim SH, and Lim SJ. Effects of triglycerides on the hydrophobic drug loading capacity of saturated phosphatidylcholine-based liposomes. *International Journal of Pharmaceutics*. 2015; 483(1-2): 142–150.
104. Kannan V, Balabathula P, Divi MK, Thoma LA, and Wood GC. Optimization of drug loading to improve physical stability of paclitaxel-loaded long-circulating liposomes. *Journal of Liposome Research*. 2015; 25(4): 308–315.
105. Mayer LD, Tai LC, Ko DS, et al. Influence of vesicle size, lipid composition, and drug-to-lipid ratio on the biological activity of liposomal doxorubicin in mice. *Cancer Research*. 1989; 49(21): 5922–5930.
106. Kulkarni SB, Betageri GV, and Singh M. Factors affecting microencapsulation of drugs in liposomes. *Journal of Microencapsulation*. 1995; 12(3): 229–246.
107. Steckel H and Bolzen N. Alternative sugars as potential carriers for dry powder inhalations. *International Journal of Pharmaceutics*. 2004; 270(1-2): 297–306.
108. Kawashima Y, Serigano T, Hino T, Yamamoto H, and Takeuchi H. A new powder design method to improve inhalation efficiency of pranlukast hydrate dry powder aerosols by surface modification with hydroxypropylmethylcellulose phthalate nanospheres. *Pharmaceutical Research*. 1998; 15(11): 1748–1752.
109. Bosquillon C, Lombry C, Préat V, and Vanbever R. Influence of formulation excipients and physical characteristics of inhalation dry powders on their aerosolization performance. *Journal of Controlled Release*. 2001; 70(3): 329–339.
110. Minne A, Boireau H, Horta MJ, and Vanbever R. Optimization of the aerosolization properties of an inhalation dry powder based on selection of excipients. *European Journal of Pharmaceutics and Biopharmaceutics*. 2008;70(3): 839–844.
111. Smola M, Vandamme T, and Sokolowski A. Nanocarriers as pulmonary drug delivery systems to treat and to diagnose respiratory and non respiratory diseases. *International Journal of Nanomedicine*. 2008; 3(1): 1–19.
112. Lee WH, Loo CY, Traini D, and Young PM. Inhalation of nanoparticle-based drug for lung cancer treatment: Advantages and challenges. *Asian Journal of Pharmaceutical Science*. 2015; 10(6): 481–489.
113. Suarez S and Hickey AJ. Drug properties affecting aerosol behavior. *Respiratory Care*. 2000; 45(6): 652–666.
114. Vehring R, Foss WR, and Lechuga-Ballesteros D. Particle formation in spray drying. *Journal of Aerosol Science*. 2007; 38(7): 728–746.
115. Champion JA and Mitragotri S. Shape induced inhibition of phagocytosis of polymer particles. *Pharmaceutical Research*. 2009; 26(1): 244–249.
116. Champion JA and Mitragotri S. Role of target geometry in phagocytosis. *Proceedings of the National Academy of Sciences of the United States of America*. 2006; 103(13): 4930–4934.
117. Honary S and Zahir F. Effect of zeta potential on the properties of nano-drug delivery systems - a review (Part 1). Tropical Journal of Pharmaceutical Research. 2013; 12(2): 255–264.
118. Surendrakumar K, Martyn GP, Hodgers ECM, Jansen M, and Blair JA. Sustained release of insulin from sodium hyaluronate based dry powder formulations after pulmonary delivery to beagle dogs. *Journal of Controlled Release*. 2003; 91(3): 385–394.
119. Muralidharan P, Mallory E, Malapit M, Hayes DJ, and Mansour HM. Inhalable PEGylated phospholipid nanocarriers and PEGylated therapeutics for respiratory delivery as aerosolized colloidal dispersions and dry powder inhalers. *Pharmaceutics*. 2014; 6(2): 333–353.
120. De Miguel D, Gallego-Lleyda A, Ayuso JM, et al. TRAIL-coated lipid-nanoparticles overcome resistance to soluble recombinant TRAIL in non-small cell lung cancer cells. *Nanotechnology*. 2016; 27(18): 185101.
121. Le VNP, Leterme P, Gayot A, and Flament MP. Aerosolization potential of cyclodextrins-influence of the operating conditions. *PDA Journal of Pharmaceutical Science and Technology*. 2006; 60(5): 314–322.

122. Kaialy W and Nokhodchi A. Dry powder inhalers: Physicochemical and aerosolization properties of several size-fractions of a promising alterative carrier, freeze-dried mannitol. *European Journal of Pharmaceutical Sciences*. 2015; 68: 56–67.
123. Zarogoulidis P, Chatzaki E, Porpodis K, et al. Inhaled chemotherapy in lung cancer: Future concept of nanomedicine. *International Journal of Nanomedicine*. 2012; 7: 1551–1572.
124. Verschraegen CF, Gilbert BE, Loyer E, Huaringa A, Walsh G, Newman RA, and Knight V. Clinical evaluation of the delivery and safety of aerosolized liposomal 9-nitro-20(s)-camptothecin in patients with advanced pulmonary malignancies. *Clinical Cancer Research*. 2004; 10(7): 2319–2326.
125. Aljamali NM. Zetasizer technique in biochemistry. *Analytical Biochemistry*. 2015; 4(2):1–4.
126. Jores K, Mehnert W, Drechsler M, Bunjes H, Johann C, and Mäder K. Investigations on the structure of solid lipid nanoparticles (SLN) and oil-loaded solid lipid nanoparticles by photon correlation spectroscopy, field-flow fractionation and transmission electron microscopy. *Journal of Controlled Release*. 2004; 95(2): 217–227.
127. Saupe A, Gordon KC, and Rades T. Structural investigations on nanoemulsions, solid lipid nanoparticles and nanostructured lipid carriers by cryo-field emission scanning electron microscopy and Raman spectroscopy. *International Journal of Pharmaceutics*. 2006; 314(1): 56–62.
128. Freitas C and Müller RH. Correlation between long-term stability of solid lipid nanoparticles (SLN™) and crystallinity of the lipid phase. *European Journal of Pharmaceutics and Biopharmaceutics*. 1999; 47(2): 125–132.
129. Frederiksen HK, Kristensen HG, and Pedersen M. Solid lipid microparticle formulations of the pyrethroid gamma-cyhalothrin-incompatibility of the lipid and the pyrethroid and biological properties of the formulations. *Journal of Controlled Release*. 2003; 86(2-3): 243–252.
130. Domb AJ. Long acting injectable oxytetracycline-liposphere formulations. *International Journal of Pharmaceutics*. 1995; 124: 271–278.
131. Wissing S, Lippacher A, and Müller R. Investigations on the occlusive properties of solid lipid nanoparticles (SLN). *Journal of Cosmetic Science*. 2001; 52(5): 313–324.
132. Bunjes H and Unruh T. Characterization of lipid nanoparticles by differential scanning calorimetry, X-ray and neutron scattering. *Advanced Drug Delivery Reviews*. 2007; 59(6): 379–402.
133. Aquilano D, Cavalli R, and Gasco MR. Solid lipospheres obtained from hot microemulsions in the presence of different concentrations of cosurfactant: The crystallization of stearic acid polymorphs. *Thermochimica Acta*. 1993; 230: 29–37.
134. Garti N, Sarig S, and Wellner E. Determination of the composition of mixtures of fatty acid polymorphs by DTA. *Thermochimica Acta*. 1980; 37(2): 131–136.
135. Jendrzejewska I, Zajdel P, Pietrasik E, Barsova Z, and Goryczka T. Application of X-ray powder diffraction and differential scanning calorimetry for identification of counterfeit drugs. *Monatshefte Fur Chemie*. 2018; 149(5): 977–985.
136. Cavalli R, Caputo O, Marengo E, Pattarino F, and Gasco M. The effect of the components of microemulsions on both size and crystalline structure of solid lipid nanoparticles (SLN) containing a series of model molecules. *Pharmazie*. 1998; 53: 392–396.
137. Jenning V, Schäfer-Korting M, and Gohla S. Vitamin A-loaded solid lipid nanoparticles for topical use: Drug release properties. *Journal of Controlled Release*. 2000; 66(2-3): 115–126.
138. Li X, Vogt FG, Hayes DJ, and Mansour HM. Design, characterization, and aerosol dispersion performance modeling of advanced spray-dried microparticulate/nanoparticulate mannitol powders for targeted pulmonary delivery as dry powder inhalers. *Journal of Aerosol Medicine and Pulmonary Drug Delivery*. 2014; 27(2): 81–93.
139. Castellani S, Di Gioia S, di Toma L, and Conese M. Human cellular models for the investigation of lung inflammation and mucus production in cystic fibrosis. *Analytical Cellular Pathology (Amsterdam)*. 2018; 2018: 3839803.
140. Sandri G, Motta S, Bonferoni MC, et al. Chitosan-coupled solid lipid nanoparticles: Tuning nanostructure and mucoadhesion. *European Journal of Pharmaceutics and Biopharmaceutics*. 2017; 110: 13–18.
141. Campos J, Varas-Godoy M, and Haidar ZS. Physicochemical characterization of chitosan-hyaluronan-coated solid lipid nanoparticles for the targeted delivery of paclitaxel: A proof-of-concept study in breast cancer cells. *Nanomedicine*. 2017; 12(5): 473–490.
142. Vijayakumar A, Baskaran R, Jang YS, Oh SH, and Yoo BK. Quercetin-loaded solid lipid nanoparticle dispersion with improved physicochemical properties and cellular uptake. *AAPS PharmSciTech*. 2017; 18(3): 875–883.
143. Fathi M, Mirlohi M, Varshosaz J, and Madani G. Novel caffeic acid nanocarrier: Production, characterization, and release modeling. *Journal of Nanomaterials*. 2013; 2013: 434632.
144. Cai J, Dang Q, Liu C, et al. Preparation, characterization and antibacterial activity of O-acetyl-chitosan-N-2-hydroxypropyl trimethyl ammonium chloride. *International Journal of Biological Macromolecules*. 2015; 80: 8–15.
145. Shi LL, Lu J, Cao Y, et al. Gastrointestinal stability, physicochemical characterization and oral bioavailability of chitosan or its derivative-modified solid lipid nanoparticles loading docetaxel. *Drug Development and Industrial Pharmacy*. 2017; 43(5): 839–846.
146. Shi LL, Xie H, Lu J, et al. Positively charged surface-modified solid lipid nanoparticles promote the intestinal transport of docetaxel through multifunctional mechanisms in rats. *Molecular Pharmaceutics*. 2016; 13(8): 2667–2676.
147. Sanders M. Inhalation therapy: An historical review. Primary Care Respiratory Journal. 2007; 16(2): 71–81.
148. Boger E, Evans N, Chappell M, et al. Systems pharmacology approach for prediction of pulmonary and systemic pharmacokinetics and receptor occupancy of inhaled drugs. *CPT: Pharmacometrics & Systems Pharmacology*. 2016; 5(4): 201–210.
149. Bondesson E, Borgström L, Edsbäcker S, et al. Pulmonary drug metabolism, clearance, and absorption. In: Smyth HDC, Hickey AJ ed., Control. Pulm. Drug Deliv. New York, NY: Springer 2011; 21–50.

150. Winkler J, Hochhaus G, and Derendorf H. How the lung handles drugs: Pharmacokinetics and pharmacodynamics of inhaled corticosteroids. *Proceedings of the American Thoracic Society*. 2004; 1(4): 356–363.
151. Hickey AJ and Mansour HM. Delivery of Drugs by the Pulmonary Route. In Florence AT, Siepmann J ed., Modern Pharmaceutics: Applications and Advances. 5th ed. Vol. 2. New York, NY: Informa Healthcare 2009; 191–219.
152. Chrystyn H. The Diskus: A review of its position among dry powder inhaler devices. *International Journal of Clinical Practice*. 2007; 61(6): 1022–1036.
153. Wetterlin K. Turbuhaler: A new powder inhaler for administration of drugs to the airways. *Pharmaceutical Research*. 1988; 5(8): 506–508.
154. Kukut Hatipoglu M, Hickey AJ, and Garcia-Contreras L. Pharmacokinetics and pharmacodynamics of high doses of inhaled dry powder drugs. *International Journal of Pharmaceutics*. 2018; 549(1-2): 306–316.
155. Houtmeyers E, Gosselink R, Gayan-Ramirez G, and Decramer M. Regulation of mucociliary clearance in health and disease. *European Respiratory Journal*. 1999; 13(5): 1177–1188.
156. Patil-Gadhe A, Kyadarkunte A, Patole M, and Pokharkar V. Montelukast-loaded nanostructured lipid carriers: Part II pulmonary drug delivery and in vitro-in vivo aerosol performance. *European Journal of Pharmaceutics and Biopharmaceutics*. 2014; 88(1): 169–177.
157. Chow AHL, Tong HHY, Chattopadhyay P, and Shekunov BY. Particle engineering for pulmonary drug delivery. *Pharmaceutical Research*. 2007; 24(3): 411–437.
158. Hickey AJ, Mansour HM, Telko MJ, et al. Physical characterization of component particles included in dry powder inhalers. I. Strategy review and static characteristics. *Journal of Pharmaceutical Sciences*. 2007; 96(5): 1282–1301.
159. Choi SH, Byeon HJ, Choi JS, et al. Inhalable self-assembled albumin nanoparticles for treating drug-resistant lung cancer. *Journal of Controlled Release*. 2015; 197: 199–207.
160. Sabra S, Abdelmoneem M, Abdelwakil M, et al. Self-Assembled nanocarriers based on amphiphilic natural polymers for anti-cancer drug delivery applications. *Current Pharmaceutical Design*. 2017; 23(35): 5213–5229.
161. Oberdörster G. Pulmonary deposition, clearance and effects of inhaled soluble and insoluble cadmium compounds. *IARC Scientific Publications*. 1992; 118: 189–204.
162. Abdelaziz HM, Gaber M, Abd-Elwakil MM, et al. Inhalable particulate drug delivery systems for lung cancer therapy: Nanoparticles, microparticles, nanocomposites and nanoaggregates. *Journal of Controlled Release*. 2018; 269: 374–392.
163. Bailey MR, Kreyling WG, Andre S, et al. An interspecies comparison of the lung clearance of inhaled monodisperse cobalt oxide particles—Part I: Objectives and summary of results. *Journal of Aerosol Science*. 1989; 20(2): 169–188.
164. Varshosaz J, Ghaffari S, Mirshojaei SF, et al. Biodistribution of amikacin solid lipid nanoparticles after pulmonary delivery. *BioMed Research International*. 2013; 2013: 136859.
165. Darquenne C and Prisk GK. Aerosol deposition in the human respiratory tract breathing air and 80:20 heliox. *Journal of Aerosol Medicine and Pulmonary Drug Delivery*. 2004; 17(3): 278–285.
166. Tena AF and Clarà PC. Deposition of Inhaled Particles in the Lungs. *Archivos de Bronconeumología*. 2012; 48(7): 240–246.
167. Fabian P, Brain J, Houseman EA, Gern J, and Milton DK. Origin of exhaled breath particles from healthy and human rhinovirus-infected subjects. *Journal of Aerosol Medicine and Pulmonary Drug Delivery*. 2011; 24(3): 137–147.
168. Courrier HM, Butz N, and Vandamme TF. Pulmonary drug delivery systems: Recent developments and prospects. *Critical Reviews™ in Therapeutic Drug Carrier*. 2002; 19(4-5): 425–498.
169. Bhavna, AFJ, Mittal G, et al. Nano-salbutamol dry powder inhalation: A new approach for treating broncho-constrictive conditions. *European Journal of Pharmaceutics and Biopharmaceutics*. 2009;71(2): 282–291.
170. Rogueda PG and Traini D. The nanoscale in pulmonary delivery. Part 1: Deposition, fate, toxicology and effects. *Expert Opinion on Drug Delivery*. 2007; 4(6): 595–606.
171. Wu L, Wen X, Wang X, et al. Local intratracheal delivery of perfluorocarbon nanoparticles to lung cancer demonstrated with magnetic resonance multimodal imaging. *Theranostics*. 2018; 8(2): 563–574.
172. Geiser M. Morphological aspects of particle uptake by lung phagocytes. *Microscopy Research and Technique*. 2002; 57(6): 512–522.
173. Sipos A, Kim KJ, Chow RH, Flodby P, Borok Z, and Crandall ED. Alveolar epithelial cell processing of nanoparticles activates autophagy and lysosomal exocytosis. *American Journal of Physiology Lung Cellular and Molecular Physiology*. 2018; 315(2): L286–L300.
174. Geiser M and Kreyling WG. Deposition and biokinetics of inhaled nanoparticles. *Particle and Fibre Toxicology*. 2010; 7: 2.
175. Schulz H. Particle-lung interactions. *Journal of Aerosol Science*. 1999; 30: S585–S588.
176. Human respiratory tract model for radiological protection. A report of a Task Group of the International Commission on Radiological Protection. *Annals of the ICRP*. 1994; 24(1-3): 1–482.
177. Videira MA, Botelho MF, Santos AC, Gouveia LF, de Lima JJP, and Almeida AJ. Lymphatic uptake of pulmonary delivered radiolabelled solid lipid nanoparticles. *Journal of Drug Targeting*. 2002; 10(8): 607–613.
178. Liu Y, Ibricevic A, Cohen JA, et al. Impact of hydrogel nanoparticle size and functionalization on in vivo behavior for lung imaging and therapeutics. *Molecular Pharmaceutics*. 2009; 6(6): 1891–1902.
179. Langenback EG, Bergofsky EH, Halpern JG, and Foster WM. Supramicron-sized particle clearance from alveoli: Route and kinetics. *Journal of Applied Physiology*. 1990; 69(4): 1302–1308.
180. Ryan GM, Kaminskas LM, Kelly BD, Owen DJ, McIntosh MP, and Porter CJH. Pulmonary administration of PEGylated polylysine dendrimers: Absorption from the lung versus retention within the lung is highly size-dependent. *Molecular Pharmaceutics*. 2013; 10(8): 2986–2995.

181. Olbrich C and Müller RH. Enzymatic degradation of SLN-effect of surfactant and surfactant mixtures. *International Journal of Pharmaceutics*. 1999; 180(1): 31–39.
182. Haque S, Whittaker M, McIntosh MP, Pouton CW, Phipps S, and Kaminskas LM. A comparison of the lung clearance kinetics of solid lipid nanoparticles and liposomes by following the 3H-labelled structural lipids after pulmonary delivery in rats. *European Journal of Pharmaceutics and Biopharmaceutics*. 2018; 125: 1–12.
183. Henning A, Schneider M, Nafee N, et al. Influence of particle size and material properties on mucociliary clearance from the airways. *Journal of Aerosol Medicine and Pulmonary Drug Delivery*. 2010; 23(4): 233–241.
184. Luo Y, Teng Z, Li Y, and Wang Q. Solid lipid nanoparticles for oral drug delivery: Chitosan coating improves stability, controlled delivery, mucoadhesion and cellular uptake. *Carbohydrate Polymers*. 2015; 122: 221–229.
185. Lauweryns JM and Baert JH. Alveolar clearance and the role of the pulmonary lymphatics. *The American Review of Respiratory Disease*. 1977; 115(4): 625–683.
186. Patton JS and Byron PR. Inhaling medicines: Delivering drugs to the body through the lungs. *Nature Reviews Drug Discovery*. 2007; 6(1): 67–74.
187. Chougule M, Padhi B, and Misra A. Nano-liposomal dry powder inhaler of tacrolimus: Preparation, characterization, and pulmonary pharmacokinetics. *International Journal of Nanomedicine*. 2007; 2(4): 675–688.
188. Bondesson E, Bengtsson T, Nilsson LE, and Wollmer P. Site of deposition and absorption of an inhaled hydrophilic solute. *British Journal of Clinical Pharmacology*. 2007; 63(6): 722–731.
189. Dombu CY, Kroubi M, Zibouche R, Matran R, and Betbeder D. Characterization of endocytosis and exocytosis of cationic nanoparticles in airway epithelium cells. *Nanotechnology*. 2010; 21(35): 355102.
190. El-Salamouni NS, Farid RM, El-Kamel AH, and El-Gamal SS. Effect of sterilization on the physical stability of brimonidine-loaded solid lipid nanoparticles and nanostructured lipid carriers. *International Journal of Pharmaceutics*. 2015; 496(2): 976–983.
191. Heiati H, Tawashi R, and Phillips NC. Drug retention and stability of solid lipid nanoparticles containing azidothymidine palmitate after autoclaving, storage and lyophilization. 1998; *Journal of Microencapsulation*. 15(2): 173–184.
192. Schwarz C and Mehnert W. Freeze-drying of drug-free and drug-loaded solid lipid nanoparticles (SLN). *International Journal of Pharmaceutics*. 1997; 157(2): 171–179.
193. Nayak AP, Tiyaboonchai W, Patankar S, Madhusudhan B, and Souto EB. Curcuminoids-loaded lipid nanoparticles: Novel approach towards malaria treatment. *Colloids and Surfaces B: Biointerfaces*. 2010; 81(1): 263–273.
194. Gokce EH, Sandri G, Bonferoni MC, et al. Cyclosporine A loaded SLNs: Evaluation of cellular uptake and corneal cytotoxicity. *International Journal of Pharmaceutics*. 2008;364(1): 76–86.
195. Liedtke S, Zimmermann E, Müller RH, and Mäder K. Physical characterisation of solid lipid nanoparticles (SLNTM). in: Proc Int Symp Control Release Bioact Mater. 1999; 26: 595–596.
196. Schwarz C and Mehnert W. Sterilization of drug-free and tetracaine-loaded solid lipid nanoparticles (SLN). Proc 1st World Meet. APGI/APV. 1995; 485–486.
197. Schwarz C, Freitas C, Mehnert W, and Muller RH. Sterilization and physical stability of drug-free and etomidate-loaded solid lipid nanoparticles. *International Symposium on Control Release Bioacta Material*. 1995; 22: 766–767.
198. Cavalli R, Caputo O, Carlotti ME, Trotta M, Scarnecchia C, and Gasco MR. Sterilization and freeze-drying of drug-free and drug-loaded solid lipid nanoparticles. *International Journal of Pharmaceutics*. 1997; 148(1): 47–54.
199. Zimmermann E, Müller RH, and Mäder K. Influence of different parameters on reconstitution of lyophilized SLN. *International Journal* of Pharmaceutics. 2000; 196(2): 211–213.
200. Freitas C and Müllerä RH. Spray-drying of solid lipid nanoparticles (SLN TM). *European Journal of Pharmaceutics and Biopharmaceutics*. 1998;46(2): 145–151.
201. Broadhead J, Edmond Rouan SK, and Rhodes CT. The spray drying of pharmaceuticals. *Drug Development and Industrial Pharmacy*. 1992; 18(11-12): 1169–1206.
202. James AL, Bai TR, Mauad T, et al. Airway smooth muscle thickness in asthma is related to severity but not duration of asthma. *European Respiratory Journal*. 2009; 34(5): 1040–1045.
203. Shinagawa K and Kojima M. Mouse model of airway remodeling: Strain differences. *American Journal of Respiratory and Critical Care Medicine*. 2003; 168(8): 959–967.
204. Al Suleimani M, Ying D, and Walker MJA. A comprehensive model of allergic rhinitis in guinea pigs. *Journal of Pharmacological and Toxicological Methods*. 2007; 55(2): 127–134.
205. Dong F, Wang C, Duan J, Zhang W, Xiang D, and Li M. Puerarin attenuates ovalbumin-induced lung inflammation and hemostatic unbalance in rat asthma model. *Evidence-Based Complementary and Alternative Medicine*. 2014; 2014: 726740.
206. Keyhanmanesh R, Boskabady MH, Eslamizadeh MJ, Khamneh S, and Ebrahimi MA. The effect of thymoquinone, the main constituent of *Nigella sativa* on tracheal responsiveness and white blood cell count in lung lavage of sensitized guinea pigs. *Planta Medica*. 2010; 76(3): 218–222.
207. Kamaruzaman NA, Sulaiman SA, Kaur G, and Yahaya B. Inhalation of honey reduces airway inflammation and histopathological changes in a rabbit model of ovalbumin-induced chronic asthma. *BMC Complementary Medicine and Therapies*. 2014; 14: 176.
208. Van Gramberg JL, de Veer MJ, O'Hehir RE, Meeusen ENT, and Bischof RJ. Induction of allergic responses to peanut allergen in sheep. *PLoS ONE*. 2012; 7(12): e51386.
209. Le DD, Rochlitzer S, Fischer A, et al. Allergic airway inflammation induces the migration of dendritic cells into airway sensory ganglia. *Respiratory Research*. 2014; 15(1): 73.
210. Dunphy JL, Barcham GJ, Bischof RJ, Young AR, Nash A, and Meeusen ENT. Isolation and characterization of a novel eosinophil-specific galectin released into the lungs in response to allergen challenge. *Journal of Biological Chemistry*. 2002; 277(17): 14916–14924.

211. Iwashita K, Kawasaki H, Sawada M, In M, Mataki Y, and Kuwabara T. Shortening of the induction period of allergic asthma in cynomolgus monkeys by *Ascaris suum* and house dust mite. *Journal of Pharmacological Sciences*. 2008; 106(1): 92–99.
212. Chiang PC, Hu Y, Thurston A, et al. Pharmacokinetic and pharmacodynamic evaluation of the suitability of using fluticasone and an acute rat lung inflammation model to differentiate lung versus systemic efficacy. *Journal of Pharmaceutical Sciences*. 2009; 98(11): 4354–4364.
213. Bakker-Woudenberg IAJM. Experimental models of pulmonary infection. *Journal of Microbiological Methods*. 2003; 54(3): 295–313.
214. Bakker-Woudenberg IAJM, ten Kate MT, Guo L, Working P, and Mouton JW. Ciprofloxacin in polyethylene glycol-coated liposomes: Efficacy in rat models of acute or chronic *Pseudomonas aeruginosa* infection. *Antimicrobial Agents and Chemotherapy*. 2002; 46(8): 2575–2581.
215. Leenders AC, de Marie S, ten Kate MT, Bakker-Woudenberg IA, and Verbrugh HA. Liposomal amphotericin B (AmBisome) reduces dissemination of infection as compared with amphotericin B deoxycholate (Fungizone) in a rate model of pulmonary aspergillosis. *Journal of Antimicrobial Chemotherapy*. 1996; 38(2): 215–225.
216. Cox RA, Pavey EF, and Mead CG. Course of coccidioidomycosis in intratracheally infected guinea pigs. *Infection and Immunity*. 1981; 31(2): 679–686.
217. Cook CH, Zhang Y, McGuinness BJ, Lahm MC, Sedmak DD, and Ferguson RM. Intra-abdominal bacterial infection reactivates latent pulmonary cytomegalovirus in immunocompetent mice. *The Journal of Infectious Diseases*. 2002; 185(10): 1395–1400.
218. Guirado E, Rajaram MV, Chawla A, et al. Deletion of PPARγ in lung macrophages provides an immunoprotective response against *M. tuberculosis* infection in mice. *Tuberculosis (Edinburgh)*. 2018; 111: 170–177.
219. Song J, Wang G, Hoenerhoff MJ, et al. Bacterial and pneumocystis infections in the lungs of gene-knockout rabbits with severe combined immunodeficiency. *Frontiers in Immunology*. 2018; 9: 429.
220. Yu B, Qiao J, Shen Y, and Li L. Protective effects of tenuigenin on *Staphylococcus aureus*-induced pneumonia in mice. *Microbial Pathogenesis*. 2017; 110: 385–389.
221. Kong D, Liu X, Li X, et al. Mesenchymal stem cells significantly improved treatment effects of Linezolid on severe pneumonia in a rabbit model. *Bioscience Reports*. 2019; 39(9): BSR20182455.
222. Corteling R, Wyss D, and Trifilieff A. In vivo models of lung neutrophil activation. Comparison of mice and hamsters. *BMC Pharmacology*. 2002; 2: 1.
223. Wesselkamper SC, Prows DR, Biswas P, Willeke K, Bingham E, and Leikauf GD. Genetic susceptibility to irritant-induced acute lung injury in mice. *American Journal of Physiology-Lung Cellular and Molecular Physiology*. 2000; 279(3): L575–L582.
224. Bosnar M, Čužić S, Bošnjak B, et al. Azithromycin inhibits macrophage interleukin-1β production through inhibition of activator protein-1 in lipopolysaccharide-induced murine pulmonary neutrophilia. *International Immunopharmacology*. 2011; 11(4): 424–434.
225. Kong J, Yi L, Xiong Y, et al. The discovery and development of microbial bleomycin analogues. *Applied Microbiology and Biotechnology*. 2018; 102(16): 6791–6798.
226. Yu Z, Yan B, Gao L, et al. Targeted delivery of bleomycin: A comprehensive anticancer review. *Current Cancer Drug Targets*. 2016; 16(6): 509–521.
227. Rudders RA and Hensley GT. Bleomycin pulmonary toxicity. *Chest*. 1973; 63(4): 627–628.
228. Ge V, Banakh I, Tiruvoipati R, and Haji K. Bleomycin-induced pulmonary toxicity and treatment with infliximab: A case report. *Clinical Case Reports*. 2018; 6(10): 2011–2014.
229. Walters DM and Kleeberger SR. Mouse models of bleomycin-induced pulmonary fibrosis. *Current Protocols in Pharmacology*. 2008; 5: 5.46.
230. Chen Z, Cheng K, Walton Z, et al. A murine lung cancer co-clinical trial identifies genetic modifiers of therapeutic response. *Nature*. 2012; 483(7391): 613–617.
231. Jackson EL, Willis N, Mercer K, et al. Analysis of lung tumor initiation and progression using conditional expression of oncogenic K-ras. *Genes & Development*. 2001; 15(24): 3243–3248.
232. Dankort D, Filenova E, Collado M, Serrano M, Jones K, and McMahon M. A new mouse model to explore the initiation, progression, and therapy of BRAFV600E-induced lung tumors. *Genes & Development*. 2007; 21(4): 379–384.
233. Engelman JA, Chen L, Tan X, et al. Effective use of PI3K and MEK inhibitors to treat mutant Kras G12D and PIK3CA H1047R murine lung cancers. *Nature Medicine*. 2008; 14(12): 1351–1356.
234. Singh M, Lima A, Molina R, et al. Assessing therapeutic responses in Kras mutant cancers using genetically engineered mouse models. Nature *Biotechnology*. 2010; 28(6): 585–593.
235. Trejo CL, Juan J, Vicent S, Sweet-Cordero A, and McMahon M. MEK1/2 inhibition elicits regression of autochthonous lung tumors induced by KRASG12D or BRAFV600E. *Cancer Research* . 2012; 72(12): 3048–3059.
236. Puyol M, Martín A, Dubus P, Mulero F, et al. A synthetic lethal interaction between K-Ras oncogenes and Cdk4 unveils a therapeutic strategy for non-small cell lung carcinoma. Cancer *Cell*. 2010; 18(1): 63–73.
237. You MS, Rouggly LC, You M, and Wang Y. Mouse models of lung squamous cell carcinomas. *Cancer and Metastasis Reviews*. 2013; 32(1-2): 77–82.
238. Lu Y, Futtner C, Rock JR, et al. Evidence that SOX2 overexpression is oncogenic in the lung. *PLoS ONE*. 2010; 5(6): e11022.
239. Ji H, Ramsey MR, Hayes DN, et al. LKB1 modulates lung cancer differentiation and metastasis. *Nature*. 2007; 448(7155): 807–810.
240. Morton JJ, Bird G, Refaeli Y, and Jimeno A. Humanized mouse xenograft models: Narrowing the tumor-microenvironment gap. *Cancer Research*. 2016; 76(2): 6153–6158.
241. Gardner EE, Lok BH, Schneeberger VE, et al. Chemosensitive relapse in small cell lung cancer proceeds through an EZH2-SLFN11 axis. *Cancer Cell*. 2017; 31(2): 286–299.

26

Improved Solid Lipid Nano-formulations for Pulmonary Delivery of Paclitaxel for Lung Cancer Therapy

Samson A. Adeyemi, Pradeep Kumar, Viness Pillay, and Yahya E. Choonara
Wits Advanced Drug Delivery Platform Research Unit, Department of Pharmacy and Pharmacology, School of Therapeutic Science, Faculty of Health Sciences, University of the Witwatersrand, Johannesburg, South Africa

26.1 Introduction: Lung Cancer

Cancer is the second cause of death globally after cardiovascular disease and was responsible for an estimated 9.6 million deaths in 2018. On a global scale, about one in six deaths is as a result of cancer disease (1). More people die of cancer disease every year than from AIDS, tuberculosis, and malaria combined (2). Meanwhile, cancer pandemic account for majority of deaths in developing countries as more than 60% of the cases, and 70% of cancer deaths, occur in Africa, Asia, Central and South America (3). In 2020, should the present trends continue, about 16 million new cancer cases will occur per annum with more than 10 million death recorded (2). By 2030, it is postulated that there will be 27 million cases of cancer, 17 million cancer deaths per annum, and 75 million people living with cancer within 5 years of diagnosis (4). The 5-year survival rate of patients with lung cancer is approximately 13–15% (5). The two main lung cancer types are small cell lung carcinoma (SCLC) and non small cell lung carcinoma (NSCLC). Approximately 80% of lung cancer cases are NSCLC, which has diverse molecular-biological features and clinical course forms of the disease (5).

Over the years, chemo, radiation, and surgery have been the prominent treatment options for patients diagnosed with NSCLC. Until recently, with the advent of cutting-edge approaches using nanotechnology, chemotherapeutics employed in the management of NSCLC were hampered with diverse challenges such as suboptimal dosage, cytotoxicity to normal cells due to non and targeted delivery, short circulation time as well as multiple resistance due to the reticulo-endothelial system (RES) (4,6). However, the improvements seen in symptom relief, survival, and enhanced life quality due to recent developments in the discovery of a new generation of drugs have led to cautious optimism in the fight against cancer. Recent advances in the development of new-generation drugs including paclitaxel, vinorelbine, docetaxel, and gemcitabine have increased the hope that they can foster sufficient benefits to suppress the fear and horror surrounding lung cancer (7).

Just like other biological systems, the human lung consists of a network of tissues and organs that coordinate the respiration process (8). Functionally, the lung is made up of conducting air passages and a respiratory section. The airway passage is made of the nasal cavity, oral cavity, and the associated sinuses, nasopharynx, oropharynx, larynx, trachea, bronchi, and bronchioles. Similarly, the respiratory section is subdivided into the respiratory bronchioles, alveolar ducts, alveolar sacs, and alveoli. The airway passage region conditions and regulates the air from the atmosphere, while the respiratory section facilitates the exchange of oxygen and carbon dioxide (9). The large unique surface area of the lung, as well as its remarkably thin interface, ranging between 0.2 μm and 0.7 μm between the alveolar and capillary lumens, enhances a quick and efficient physiological respiration (9,10). The alveolar epithelium possesses both type and II alveolar cells. The type I cells of the alveolar epithelium are thin squamous cells that serve as a bridge between air and internal components of the alveolar wall, which envelopes 95% of the circumference of the alveolar. Alveolar type II cells help in the regeneration of both the type I and II cells, generate lung surfactant, and occupy 5% of the alveolar surfaces (11). Brush cells and macrophages are relatively present on the circumference of the alveolar (10) (Figure 26.1).

26.2 Pulmonary Drug Delivery for Lung Cancer by Inhalation

Lung cancer remains one of the most deadly cancers and the second most frequent cancer among males and females (13). Patients treated using conventional methods such as surgical resection, chemotherapy. and radiology have a very low long-term survival rate (14). Drug delivery through the systemic approach has a limited success in targeting the lung tumor because only small amount of the chemotherapeutic drugs reach the tumor sites regardless of the size of the dosage. The majority of the conventional drugs affect normal cells, reducing their growth, which in turn makes the patient extremely weak and fragile, and in most cases leads to death (15).

DOI: 10.1201/9781003046547-26

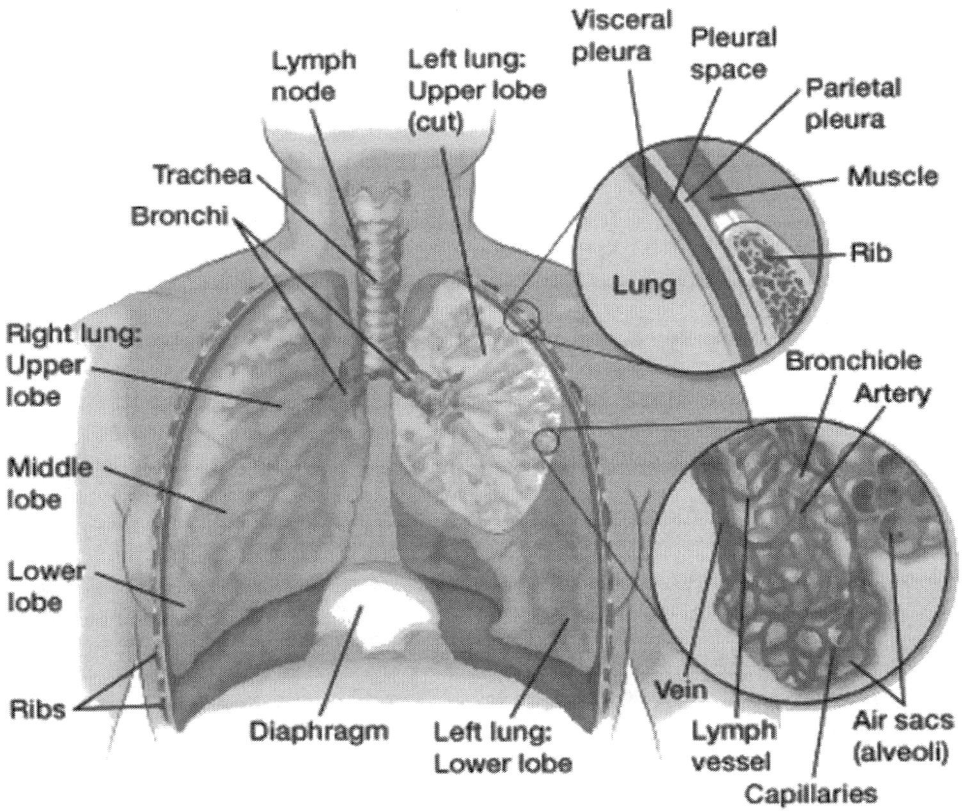

FIGURE 26.1 Normal Structure and Function of the Lungs (Figure Adapted from (12))

Therefore, improving the drug delivery approach has a great potential in the fight against lung cancer by delivering anti-cancer agents precisely to the tumor site in the lung, thereby reducing the drugs' exposure to systemic circulation.

In recent times, there has been an increased interest in the formulation of drugs explored for pulmonary delivery. One reason is the potential of the lung to serve as a portal for the passage of drugs, including peptides and proteins, into the body. The unique and excellent large surface area (about 100 m^2) available for absorption, the thin absorption membrane (0.1–0.2 μm), and the increased blood flow (5 l/min) which enhances the distribution of molecules throughout the body, establishes the lungs as an efficient entry port for drugs into the bloodstream. Most importantly, the lungs show apparently low local metabolic activity as opposed to the oral route of drug administration. In this way, first-pass metabolism is bypassed in pulmonary inhalation (16). A number of pulmonary diseases including asthma, microbial infections, and systemic diseases such as diabetes have been treated locally using pulmonary delivery. Meanwhile, pulmonary delivery also possess a huge potential application for gene delivery and its usage in cancer nanomedicines has gained enormous attention and results (17). Research has shown that the administration of oncolytics through inhalation in primary or metastatic lung cancer could enhance the exposure of lung tumor to the drug while reducing the side effects due to systemic circulation (18).

Inhalation can be define as a non-invasive, organ-specific approach of delivering the active agents to the lungs to treat lung diseases (19). In addition, cell and organ targeted pulmonary drug delivery can be formulated by attaching specific targeting ligands onto the surface of the drug carrier (20). In this way, healthy cells and systemic organs are protected from the therapeutics because it enables a reduced drug dose and prevents undue negative effects of drugs (21). A favorable approach for the specific delivery of protein- and gene-based therapeutics is the pulmonary drug delivery system. This is due to the absence or limited number of intracellular and extracellular drug-metabolizing enzymes in the lung, unlike in the gastrointestinal tract and liver (22,23). Until now, four types of inhalation devices, including pressurized metered dose inhalers, dry powder inhalers, nebulizers, and soft mist inhalers, have been employed for the delivery of pulmonary therapeutics through the inhalation route (24). A number of factors coordinate the inhalation and deposition pattern of particles in the lung. These include inertial impaction, gravitational sedimentation, and diffusion processes primarily as a function of the aerodynamic particle size distribution of the aerosolized medicines (25,26). A major limiting factor in pulmonary drug delivery is the high degree of branching of the airways that have diverse lengths and diameters (27). These branching airways, which and extend narrow downward from the carina of the trachea all though to the alveolar sac, influence the inhaled particles to strike the airway walls instead of gaining penetration into the lung due to an increased velocity of the particle (28).

As such, the inhaled particles are unable to flow in the direction of the airway branching and go along the airstream to reach the deep and peripheral lung regions (26). The ciliated epithelial cells lined up the respiratory airway, and are responsible for the transport of mucus and alveolar fluids to the upper section of the airway (29,30). The debris, excessive secretions and unwanted inhaled particles are cleared by the mucociliary clearance, which is a physiological function (27,31). In order to protect the lung epithelial cells from deleterious particles and substances, this glycosylated mucus, which is rich in protein, is constantly secreted, spread and replaced (31–33). The respiratory system makes us of its alveolar macrophages as defense cells to engulf foreign particles. These two combined mechanisms – mucus action and ciliary clearance with macrophage activity, decrease the residence time and the chances of particles to interact with the intended lung cell population (34). This however is opposed to the drug delivery aim of the inhaled particles especially when the lung cells, rather than the macrophages, are the target site of the drug action.

26.3 Solid Lipid Nanoparticles for Lung Cancer Therapy

26.3.1 Lipid-Based Nanocarriers

Lipid-based nanocarries used for the delivery of drugs and genes are mainly classified into three types: liposomes, micelles (oil dispersions), and lipid nanoparticles. Hydrophilic and hydrophobic drugs are usually delivered using liposomes, which are double-layed phospholipid vesicles, by either incorporating the therapeutics into the lipid bilayer or encapsulation in the inner aqueous matrix, respectively. When the number of lipid bilayers are reduced, nano-sized liposomes (nanoliposomes) are produced which enable the encapsulated drugs to exhibit a prolonged circulation time and tumor localization properties (35). Recently, there have been increased applications of liposomes as delivery vehicles for anti-cancer therapeutics because of their efficient biocompatibility characteristics. Over the last 10 years, liposomal research has gained enormous attention and applications in the production of new liposomal formulations such as cationic liposomes (36), temperature-responsive liposomes (37), virosomes (38) and archaeosomes (39). Regardless of these mammoth advances in liposomal research at the bench, only two liposomal formulations have been approved by the Food and Drug Administration (FDA) including DOXIL, a doxorubicin liposome injection for use for ovarian cancer, and Marqibo, an injection containing liposomal vincristine sulfate for lymphoblastic leukemia.

In lung cancer treatment, liposome-based delivery systems for drugs and genes could be a promising approach. Cisplatin as a model drug for the treatment of NSCLC for the last two decades is implicated in the development of nephrotoxicity in one out of every five (20%) patients receiving high doses (40). Lipoplastin, a cisplatin loaded liposome, was formulated by Boulikas in 2004 and was employed to reduce systemic toxicity of cisplatin (41). Using a multiple rat tumor model, the scientist further showed that lipoplatin injection significantly decrease nephrotoxicity to negligible levels when compared to standard therapy (42). Meanwhile, the phase III clinical trial testing of lipoplatin was completed in 2014 (43,44). The use of paclitaxel for the treatment of lung cancer has been widely reported. It was historically encapsulated using Cremophore EL in order to increase its solubility in physiological fluids. Nonetheless, its systemic delivery was complicated as the formulation resulted in hypersensitivity reactions. Meanwhile, a phase I clinical investigation in NSCLC patients with malignant pleural effusions in 2010 showed that in all the cases studied, the liposome-based paclitaxel formulation displayed an enhanced therapeutic efficacy (45). Moreover, liposomal–paclitaxel formulation has been shown in a recent preclinical report to decrease the incidence of drug resistance in targeted lung cancer cells (46). Specifically, the liposomal surface was modified with a mitochondrial targeting molecule, d-α-tocopheryl polyethylene glycol 1000 succinatetriphenylphosphine conjugate (TPGS1000-TPP). These targeted paclitaxel formulations could increase their cellular uptake and induce mitochondria-mediated apoptotic cell death in human A549 lung cancer cells. Presently, a paclitaxel–liposome-based formulation, Lipusu, is available commercially with many other formulations under clinical investigation (47). Table 26.1 shows the some of the current examples of liposomal formulations in clinical trials proposed for cancer treatment.

In another randomized phase III multicenter trial, a combination therapy containing liposomal formulation of cisplatin and paclitaxel attained an effective therapeutic efficacy with reduced nephrotoxicity in NSCLC patients (48). Interestingly, both the primary tumor as well as the metastasis were effectively targeted by this liposomal drug combination therapy. Similarly, cancer vaccines for the treatment of exiting cancers have been delivered using liposomes. A number of studies have shown the efficiency of therapeutic vaccine Biomira Liposomal Protein 25 (BLP25) in the treatment of advanced NSCLC (49). Using BLP25, tumor-associated antigen MUC1 is targeted through the liposomal carrier in order to prevent tumor growth. A low dose of pretreatment with cyclophosphamide followed by two cycles of liposome BLP25 treatment has been reported to drastically decrease the number of tumor foci during a preclinical study in a human MUC1 transgenic lung cancer mouse model (Hmucl. Tg) (50). Interestingly, phase III clinical studies using the liposomal formulation of BLP25 are in progress (51).

Lu and colleagues (2012) evaluated and reported the effectiveness of liposomal-based nano-vectors for gene delivery in mouse cancer models. A number of tumor suppressor genes, including p53, TUSC2/FUS1, or mda-7/IL-24, were loaded into a nanoliposome comprised of 1,2-dioleoyl-3- trimethylammonium propane (DOTAP):cholesterol (Chol.) in a preclinical investigation (52) which efficicently delivered these therapeutic genes to metastatic tumor sites with enhanced therapeutic efficacy and improved animal survival. Furthermore, the safe DOTAP:Chol nanoparticulate system upon its preclinical efficiency was employed in clinical testing for the delivery of a TUSC2/FUS1 tumor suppressor gene in NSCLC patients. Interestingly, the phase I clinical trial results proved its safety and compatibility with no treatment-related toxicity upon its intravenous administration. In addition, results from the study demonstrated the efficient internalization of the nanoparticles by both the primary and metastatic tumors, transgene and gene products were expressed, and specific alterations in TUSC2-regulated signaling pathways were observed (52). Subsequently, further phase II study of lung cancer was proposed and additional

TABLE 26.1

Examples of Some Liposomal Formulations in Clinical Trials Proposed for Cancer Treatment

Trade Name/Composition	Indication	Phase	Stage
PEGylated Liposomal Doxorubicin	AIDS-associated non-Hodgkin's lymphoma	I	Completed
Doxil Liposomal Noxorubicin	Resistant solid malignancies	I	Completed
Liposomal Entrapped Paclitaxel Easy to Use (LEP-ETU)	Advanced cancer	I	Completed
Liposomal Encapsulated Docetaxel (LE-DT)	Advanced solid tumors	I	Completed
BLP25 Liposome Vaccine	Lung neoplasms, non small cell lung carcinoma	II	Completed
Liposomal Daunorubicin	Hematologic cancer, chronic myelomonocytic leukemia, previously treated Myelodysplastic syndromes, and recurrent adult acute myeloid leukemia	II	Completed
Liposomal LE-SN38	Advanced cancer	I	Completed
Liposome Encapsulated Mitoxantrone (LEM)	Advanced cancer	I	Completed
Liposomal Encapsulated Docetaxel (LE-DT)	Advanced solid tumors	I	Completed

phase I trials testing the use of DOTAP-Chol-based nanoparticle treatment for a number of cancers, including ovarian, breast, and pancreatic cancers, were investigated. It was envisaged that a specific therapeutic gene will be delivered to a specific cancer type (44).

26.3.2 Solid Lipid Nano-Vectors for Lung Cancer

The emergence of combination therapy in which multiple drugs are synergistically co-encapsulated has emerged as a promising therapeutic option for cancer treatment due to its potential to foster an effective approach to overcome the major limitations of conventional chemotherapy. The therapeutic effect of cancer chemotherapeutics such as paclitaxel and doxorubicin is drastically reduced by cancer chemo-resistance, which is developed often during prolonged treatment. This chemo-resistance of tumors has been shown to be associated with the suppression of apoptosis and increased regulation of drug-efflux pumps or detoxification enzymes. Meanwhile, once tumors develop resistance, increased doses of anti-cancer drugs must be administered in order to destroy tumors, resulting in deleterious side effects in healthy cells.

To circumvent this, a promising modality is to simultaneously deliver multiple drugs that can reverse chemo-resistance through synergistic efficacies of various therapeutic mechanisms. For instance, different types of cationic polymer nanoparticles have been fabricated for simultaneous delivery of anti-cancer chemotherapeutics with small interfering RNA (siRNA), which has the potential to inhibit the expression of cancer-related genes such as polo-like kinase-1 (Plk-1) and B-cell lymphoma-2 (Bcl-2) (53–57). Suppressing and silencing Plk-1 results in impairment of mitosis machinery and induction of apoptosis in cancer cells without affecting normal cells (58). On the other hand, Bcl-2 protein is responsible for blocking the release of cytochrome C, resulting in fodrin cleavage and apoptotic nuclear morphology, and this a principal regulator of anti-apoptotic cellular defense (54,57,59,60). These investigations showed that targeting these genes with the administration of anti-cancer drugs with siRNA can yield more effective cancer therapy by mobilizing cancer cells in a gene-specific manner. In a recent study, siRNA-mediated knockdown of drug-efflux pumps, such as P-glycoprotein and multi-drug resistance-associated protein (MRP), showed the potential to improve the therapeutic effect of coadministered anti-cancer chemotherapeutics by enhancing their aggregation within the cell (61,62).

In order to fully harness the above-mentioned benefits of combination therapy, an efficient, nontoxic and cytocompatible nanocarrier, which can simultaneously deliver multiple types of therapeutics, is needed. Solid lipid nanoparticles (SLNs) are a suitable candidate for this purpose due to their enhanced physical stability that can withstand increase drug loading and improved pharmacokinetic properties (63,64).

SLNs are a group of vectors for the delivery of drugs and genes. Unlike their lipid counterparts, SLNs possess and exhibit higher stability, increased drug loading, enhanced biocompatibility, and ease of large-scale industrial production. In a transfection experiment, Chio and colleagues transfected p53-null H1299 lung cancer cells with SLN-carrier p53 (65). Their results showed the capability of SLNs to successfully express p53 protein compared to lipofectin, which is a commercially available lipid-based nano-vector, therefore presenting SLNs as potentially efficient gene therapy vectors in lung cancer. In another recent study, SLNs were successfully loaded with Bcl-2 siRNA and paclitaxel for an enhanced synergistic combination therapeutic effect. Similarly, CdSe/ZnS quantum dots were co-encapsulated within the gene-loaded SLNs for optical imaging (66). SLNs can also be employed to encapsulate inorganic nanocrystals for imaging applications (67,68). A new class of cationic SLNs were fabricated by Kim and colleagues (2008) which can be coupled with siRNA by mimicking naturally occurring apolipoprotein E-free low-density lipoproteins (LDL) (69). These newly formed SLNs displayed high efficiency in delivering siRNA to cancer cells in addition to their ability to inhibit targeted gene expression with minimal cytotoxicity. In a subsequent experiment, the same researchers showed that the systemic administration of c-Met-specific siRNA/SLN complexes suppress tumor growth in a mouse brain along with tumor-tropic localization, signifying their ability to tackle and inhibit glioblastoma in the clinic (70). Overall, the unique and excellent characteristics of SLNs are suitable for combined chemo and/or gene therapy and molecular imaging of cancer theragnostics.

26.4 Paclitaxel for Lung Cancer Therapy

26.4.1 Paclitaxel, Its Mechanisms and Drug Resistance

The Pacific yew tree, *Taxux brevifolia*, produces a natural product that is extracted from its bark, known as taxol of paclitaxel. Paclitaxel has been shown to exhibit a wide range of applications as an anti-cancer agent for the treatment of a number of tumors such as lung, breast, ovarian, head, and neck cancers, and Kaposi's sarcoma (71–75). Paclitaxel as a diterpene alkaloid is an active chemotherapeutic agent characterized by different pharmacological properties (Table 26.2), including the destabilization of microtubules by disassembling and blocking the cell cycle, initiating apoptosis, and causing cell death, autophagy, and anti-angiogenesis (76). In addition to its function as a cytoskeletal drug, paclitaxel also functions as a radio-sensitizer that targets the tubulin (77,78). The treatment of KB cells with paclitaxel demonstrated a downregulation of the expression of polycomb repressive complex 1 and cyclin B2, thereby causing microtubule polymer stabilization and ultimately hindering it from disassembly. Furthermore, it could play a role in the formation of mitotic spindle, chromosome segregation, and cell division (76). Similarly, paclitaxel could cause a mitotic arrest by blocking the G2/M phase of the cell cycle and modulating the radio-responsiveness of tumor cells (72). A number of pathways, including activation of the cell receptor pathway, cysteine aspartic-specific protease (caspase) cascades, and mitochondria pathways, are employed by paclitaxel to act against several diseases by activating different cancer deaths via apoptosis (72). Also, paclitaxel causes the upregulation of the proapoptotic proteins Bax (79) and Bak (74,80), as well as the downregulation and inactivation of the antiapoptotic protein, Bvl-2 (81,82).

The expression levels of Becclin-1 and autophagy protein 5, which are needed for autophagosome formation, are increased by paclitaxel in order to induce autophagy to prevent tumorigenesis. In this way, the expression of p53 and LC3B (necessary for the autophagosome), are enhanced, thereby regulating the initiation of autophagy (83). Autophagy that is induced by paclitaxel is of great importance in modulating caspase-independent cancer cell death. Additionally, both microvessel density and VEGF production were reported to be significantly reduced with paclitaxel in an in vivo experiment. Meanwhile, the preferential accumulation of paclitaxel in the endothelial cells might be responsible for its anti-angiogenic efficacies (84). Interestingly, a number of survival pathways including Raf/mitogen-activated protein kinsase (MEK), extracellular signal-regulated kinase pathway, and P13K/AKT signaling pathway to antiapoptosis are activated by paclitaxel (85). Also, the expression of survival factors including apoptosis inhibitor survivin (86), as well as the cyclin-dependent kinase inhibitor protein, p21/WAF1/Cip1 (87) are enhanced by paclitaxel, by continuously inducing the activation of cyclin B-cdk1, which results in the phosphorylation and stabilization of surviving proteins (88). As such, combining paclitaxel with MEK inhibitors or the synergetic application of paclitaxel with P13K inhibitors might produce promising approaches for chemotherapy.

Despite the significant anti-cancer efficacies of paclitaxel, its cytotoxicity on healthy cells and drug resistance remain great hurdles to its optimum efficiency. A major dose-limiting adverse side effect of paclitaxel is distal axonal polyneuropathy (mostly sensory polyneuropat(89–91). Up until now, the mechanism of paclitaxel resistance was not fully comprehended due to the complexities that exist between tumor and drug emanating from multiple genes and processes. Nonetheless, recent investigations have shown that alterations of the A-tubulin gene and its associated protein might result in the resistance of paclitaxel in ovarian cancer patients (92). Similarly, multi-drug resistance protein 1 (MDR1) might also be a contributing factor to paclitaxel resistance (76,93,94). A number of other factors, such as the deletion of multiple apoptotic factors through the overexpression of autophagic pathway, changes in cell death response, and collateral sensitivity to platinum have been suggested to account for paclitaxel resistance (95). Additionally, the activation of the MAPK and P13K/AKT signaling pathways in human melanoma A375B cells has been reported to produce resistance to paclitaxel at a concentration range of 0.001 μmol/l to 0.1 μmol/l) (95).

26.4.2 Improved Anti-angiogenic Efficacy of Paclitaxel through Newly Established Pharmaceutical Formulations

Over the last 10 years, a number of attempts have been made toward the removal of Cremophor EL from use in order to reduce its hypersensitive side effects, or to enhance the pharmacokinetic properties of paclitaxel (120). To this end, various nano-vectors including liposomes, pegylated liposomes, protein nanoparticles, and polymeric nanoparticles have been investigated for the delivery of anti-cancer agents such as paclitaxel (121). The use of liposomes as an established drug delivery system for lipophilic substances, and their ability to enhance the pharmacokinetics and therapeutic index of anti-cancer chemotherapeutics are well documented (122). Indeed, paclitaxel-loaded liposomes demonstrated enhanced solubility and the same in vitro cytotoxicity against different cancer cell lines in comparison with the results obtained from Cremophor EL paclitaxel formulation (123). Nonetheless, a major limitation in the clinical application of convectional liposome is due to its on-the-spot clearance by the reticuloendothelial system upon its systemic administration, as well as its low bioavailability (124).

Conversely, positively charged cationic liposomes have been formulated to bind preferentially to and be engulfed by negatively charged angiogenic endothelial cells that are expressed on tumors and in areas of chronic inflammation (125). Based on these modalities, Thurston and colleagues reported that actively growing angiogenic endothelial cells showed a preferential uptake through endocytosis for positively charged

TABLE 26.2

Selected Pharmacological Efficacies of Paclitaxel on Angiogenesis

Models	Pharmacological Efficacies at Cellular Level	Pharmacological Efficacies at Molecular Level	References
In vitro			
HUVEC	Inhibition of migration	Increase of the levels of acetylated tubulin; increase of forkhead box O3a translocation into the nucleus	(96)
HUVEC	Reduction of the capillary network formation in matrigel		(97)
HUVEC	Strong anti-proliferative activity		(98)
HUVEC, rat aortic ring explants	Inhibition of cell proliferation, migration, and tube formation at one-tenth the concentration needed to achieve a similar effect on tumor cell lines		(99)
HMVEC-d, HUVEC	Selective inhibition of cell proliferation and induction of apoptosis at low concentrations for prolonged periods of time		(100)
Human ovarian cancer cell lines		Decrease of survival factors such as Ang-1 and VEGF	(101)
HUVEC	Inhibition of cell proliferation, motility and invasiveness in a concentration-dependent manner		(102)
Human leukemia cell lines		VEGF downregulation in vitro (even in drug resistant cells)	(103)
HUVEC, HMEC-1	Initiation, without completion, of the mitochondrial apoptotic pathway leading to a slowing down of the cell cycle	Cytotoxic effects mediated by microtubule network disturbance, G2-M arrest, increase in Bax/Bcl-2 ratio, and mitochondria permeabilization	(104)
HUVEC, HMEC-1		Increase of interphase microtubule dynamics in vitro	(105)
HUVEC		Induction of gene and protein expression of TSP-1 at metronomic concentrations	(106)
Rat bone marrow (BM)-derived endothelial progenitor cells (EPC) cell line (TR-BME)	Inhibition of tube formation and migration of cells at low concentrations		(107)
In vivo			
Cornea assay	Inhibition of FGF-2 and VEGF-induced neovascularization		(108)
Transgenic murine Met-1 breast cancer model	Reduced intratumor angiogenesis	VEGF downregulation	(103)
Nude mice bearing murine breast cancer		VEGF downregulation	(109)
Human oral squamous cell carcinoma, lung tumor	Inhibition of tumor angiogenesis	Reduced the immunohistochemical expression of CD31, VEGF and VEGF mRNA	(110,111)
Melanoma spontaneous metastases model	Inhibition of angiogenesis in melanoma tissue lesions	Reduction of VEGF-A expression	(112)
Chick chorioallantoic membrane	Inhibition of neovascularization		(113)
Matrigel pellet in mice	Inhibition of neovascularization		(114)
Chick chorioallantoic membrane	Inhibition of neovascularization at low concentrations		(115)
4T1 metastatic breast cancer	Strong anti-angiogenic and anti-lymphangiogenic activities of low doses		(116)
Rats bearing syngeneic prostate cancer (Dunning AT-1) not expressing TSP-1		Re-induction of TSP-1 expression in tumors	(117)
HT-29 colon cancer model; 4T1 metastatic breast cancer		Upregulation of TSP-1 expression	(116,118)
Ovarian carcinoma xenograft model		Downregulation of VEGF-B, -D and -A; upregulation of Tie-1, Tie-2 and VEGFR-2	(119)

TABLE 26.3

Selected In Vitro and In Vivo Anti-angiogenic Effects of Novel Paclitaxel Formulations

Paclitaxel Formulations	Models	Pharmacological Efficacies	References
Paclitaxel encapsulated in cationic liposomes	In vivo metastatic melanoma model	Inhibition of newly blood vessels, prevention of melanoma growth and invasiveness, improvement of mice survival	(127)
Paclitaxel encapsulated in cationic liposomes	In vivo amelanotic hamster melanoma A-Mel-3 model	Increased endothelial deposition of paclitaxel in tumor vessels; a remarkable retardation of tumor growth and appearance of regional lymph node metastases	(126)
Sterically stabilized liposomes containing paclitaxel (SSL-PTX)	HUVEC proliferation and migration in vitro; MDA-MB-231 breast cancer xenograft models	Inhibition of cell proliferation and migration at low concentrations; decrease of microvascular density of tumors treated with low doses of paclitaxel	(131)
Paclitaxel entrapped in emulsifying wax nanoparticles (PTX NPs)	Colon cancer xenograft models	Increase of anti-angiogenic effect in the colon cancer xenograft models	(132)
Polymeric nanospheres loaded with paclitaxel	In vitro HUVEC culture and ex vivo rat aortic rings	Inhibition of proliferation and inhibition of endothelial sprouts	(133)
ABI-007, a Cremophor EL-free, albumin-bound, 130-nm form of paclitaxel	Rat aortic rings, human endothelial cell proliferation and tube formation.Tumor xenografts	Inhibition of rat aortic microvessel outgrowth, human endothelial cell proliferation, and tube formation; inhibition of tumor growth	(138)
Hyaluronic acid conjugates of paclitaxel (HA-PTX)	Female nude mice bearing ovarian cancer cells	Anti-tumor and anti-angiogenic effects with a marked increase of TSP-1	(139)
PEG-VC-PABC-PTX; paclitaxel (PTX) conjugated with p-aminobenzylcarbonyl (PABC), valine-citrulline (VC), and polyethylene glycol (PEG)	MCF-7 tumor xenografts	Anti-tumor and anti-angiogenic effects in vivo; decrease of microvessel density	(140)

cationic liposomes (125). The work of Schmitt-Sody and colleagues, in which paclitaxel was encapsulated in cationic liposomes, showed novel neovascular targeting on tumor selectivity and anti-tumoral effects when compared to paclitaxel free drug in an in vivo experiment (126) as presented in Table 26.3.

In comparison to Cremophor EL paclitaxel formulation, there is a significant increase of endothelial deposition of paclitaxel in tumor vessels during the infusion time, and there was a remarkable delay in both the subcutaneous tumor growth and the appearance of regional lymph node metastases in an amelanotic hamster melanoma model (126). In another report, Kunstfeld and colleagues reported the ability of paclitaxel-loaded cationic liposomes to specifically target blood vessels, and hinder melanoma growth and invasive potential, which ultimately increased the survival of the tumor-bearing mice (127). More importantly, the paclitaxel-loaded liposomes decrease the interfacial vessel density between the tumor and the dermis, and decrease mitosis in the endothelial cell. Conversely, no reduction in the mitotic index of the endothelium in vivo was observed, and minute effects were seen on tumor growth upon treatment with equimolar concentrations of paclitaxel solubilized in Cremophor EL (127). Interestingly, theses results corroborate the findings in an animal model of prostate cancer. As such, the encapsulation of paclitaxel in liposomes (EndoTAG-1) exhibited a drastic and viable reduction in microvessel density in tumors when compared to paclitaxel alone. This in turn, validates the inhibition of angiogenesis when compared to the conventional treatment (128) (Table 26.3).

Surface modification of liposomes (PEGylation) using inert and biocompatible polymers, such as polyethylene glycol (PEG), presents another alternative for the formulation of liposome with prolonged half-life (129). In this way, liposomes are shielded from recognition by opsonins and their clearance by the reticuloendothelial system is reduced (130). Using this approach, the surfaced modified liposomes (PEGylated liposomes) are often referred to as sterically stabilized liposomes (SSL). Huang and colleagues reported the anti-angiogenic activity of sterically stabilized liposomes loaded with paclitaxel (SSL-PTX). Both endothelial cell proliferation and migration were effectively inhibited by SSL-PTX in a concentration-dependent manner. Upon its metronomic administration, a significant reduction in tumor growth was observed in the MDA-MB-231 xenograft model through an anti-angiogenic mechanism (reduced microvessel density), in comparison to the administration of paclitaxel solubilized in Cremophor EL (131). As presented in Table 26.3, new nanoparticles have been produced, and evaluated for in vitro and in vivo characteristics so as to enhance the activity of low concentrations of Cremophor EL-free paclitaxel, on endothelial cell proliferation, tube formation, and motility, as well as facilitate delivery to blood vessels. Drug delivery to tumors using colloidal carriers have been shown to improve tumor therapy as a result of their ability to increase permeability and retention of the encapsulated drug. In this regard, paclitaxel encapsulated in emulsifying wax nanoparticles (PTX NPs) displayed improved therapeutic response against colon cancer xenograft models, subdue paclitaxel resistance, and aggravate its anti-angiogenic efficacy (132). Additionally, spherical nanoparticles (polymeric nanospheres), which have the capacity to specifically target the induced vascular endothelium and deliver a combination and synagestic efficacies of paclitaxel encapsulated with other anti-angiogenic agents have been synthesized (133). Paclitaxel-loaded nanospheres have been successfully evaluated for their anti-angiogenic effect in HUVECs in vitro, as well as in rat aorta rings ex vivo. The results showed an improved efficacy than those seen with free paclitaxel (134).

The use of albumin has been reported as a very promising vector for the delivery of antineoplastic drugs (135). It has been reported that albumin enhances endothelial transcytosis of protein-bound and unbound plasma constituents, usually through a cell-surafce glycoprotein receptor (gp60). Caveolin-1, which is an intracellular protein, binds to the gp60 protein and subsequently produces transcytotic vesicles. Similarly, gp60 binds to osteonectin. Interestingly, both proteins (caveolin-1 and osteonectin), are usually found in certain neoplasms (e.g. breast, lung, and prostate cancer), which could facilitate the aggregation of albumin in some tumors. In this way, it becomes easier to facilitate the intratumoral accumulation of albumin-bound drugs (135). In this regard, albumin bound paclitaxel ABI-007 is a novel, Cremophor EL-free, 130 nm nanoparticle formulation of paclitaxel (136). To produce this formulation, a high-pressure homogenization of paclitaxel with serum albumin is performed, which yields a colloidal suspension (137).

ABI-007 has been shown to exhibit a higher penetration into tumor cells, with enhanced anti-tumoral activity, in comparison to the same equimolar dose of standard paclitaxel, using breast cancer xenografts in preclinical studies (136). Interestingly, in a pivotal study by Ng and colleagues, the anti-angiogenetic effect of ABI-007 was reported using a therapeutic metronomic strategy (138). A recent result has demonstrated the application of clinically relevant concentrations of drug carriers such as Cremophor EL and polysorbate 80 disproved the anti-angiogenic efficacy of taxanes (66). Indeed, such results could hinder the application of paclitaxel metronomic dosage. To circumvent this anomaly, Ng and colleagues examined ABI-007 in both in vitro and in vivo models. It was discovered that ABI-007 reduced the growth of rat aortic microvessels, human endothelial cell proliferation, and tube formation in a significant manner (138). In addition, tumor growth was drastically supressed in xenograft models with a minimal toxicity profile of metronomic regimen of ABI-007.

A metronomic dose of paclitaxel-conjugated hyaluronic acid (HA-PTX) was recently shown to exhibit potent anti-tumor efficacy, having an enhanced anti-angiogenic potential compared with those obtained with maximum tolerated dose (MTD) administration or when treated with free PTX in female nude mice with ovarian cancer cells xenograft (139). Interestingly, there was a drastic decrease in tumor weight in the metronomic HA-PTX treated groups while there were no significant changes observed in the MTD group. Meanwhile, metronomic HA-PTX administration yielded a significant increase in TSP-1, validating the anti-angiogenic mechanism of action of this novel formulation (139). In another novel experiment, Liang and colleagues reported a new paclitaxel nano system. Using p-aminobenzylcarbonyl (PABC), a spacer, and valine-citrulline (VC), a substrate of cathepsin B, to link polyethylene glycol (PEG) and PTX, the novel paclitaxel formulation (PEG-VC-PABC-PXT) was fabricated. The newly formed formulation exhibited marked anti-tumor and anti-angiogenic efficacies in an in vivo experiment when compared against the control conjugate with a lesser effect (140). The need to develop an oral formulation of paclitaxel is crucial for the clinical application of metronomic chemotherapy. In a research study by Moes and colleagues, the oral bioavailability of paclitaxel was enhanced through a combination therapy with ritonavir in a solid dispersion formulation called ModraPac001 (141). This ModraPac001 formulation is presently in first clinical trial with oral metronomic administration of paclitaxel (141).

In summary, ABI-007 remains the most successful in clinical trials out of all the newly formulated paclitaxel-based therapeutics with anti-angiogenic properties. After its initial approval by the US FDA in the year 2005, and application for treatment with metastatic breast cancer patients, ABI-007 is presently approved for use in 42 countries around the world, including Canada, Japan, India, and the European Union (137). Based on the global approval and usage, laudable and remarkable improvements have been reported in a very short time in our understanding of the mechanism of action of paclitaxel as an anti-angiogenic agent (142). However, more clinical research and laboratory investigations are still needed to fully maximize the advantages of the benefits of the anti-angiogenic characteristics of paclitaxel for the management of patients with different cancer types. Based on this modality, insights and research on the metronomic chemotherapy based on the different newly developed paclitaxel formulations, including ABI-007, HA-PTX, and ModraPac001, form the most interesting approaches employed in recent times for enhancing the clinical control of angiogenesis, both in primary and metastatic tumor (143).

In addition, such treatment approaches should be anchored on strong and reliable outputs of randomized phase III clinical experiments, and not solely on their safety profiles or on their effect on the quality of life and on the preferences of patients. Notwithstanding, the future of cancer care will be based on tolerability and compliance, as the most necessary factors in accordance with the overriding emerging importance of quality of life (144). Oncologists now have several other effective treatment alternatives for the management of patients who already display resistance to standard doses of chemotherapy, due to the availability of new drugs for metronomic chemotherapeutic approach (i.e. ABI-007, HA-PTX, and ModraPac001), either applying them singly or in combination with other targeted therapeutics (143). Indeed, preclinical (145) and clinical reports (146,147) have proposed that metronomic chemotherapy could be the effective strategy to stabilize tumors that are already resistant to the same chemotherapeutics' maximum tolerated dose (MTD) treatments. Most importantly, paclitaxel's potential to inhibit diverse pathways in the process of angiogenesis may be an added advantage in the treatment of tumors that have exhibited resistance to drugs that regulate angiogenesis with a specific/single targeting potential (e.g. bevacizumab and sunitinib).

26.5 Chitosan and Its Derivatives in Lung Cancer Therapy

Chitin has been referred to as one of the principal sources of nitrogen accessible to numerous living terrestrial and aquatic organisms (148). It constitutes the main exoskeletal anatomy of several arthropods such as insects, spiders, and crustaceans, and forms the internal structures of other invertebrates (149). Chitin is made of β (1→4) linked residues of N-acetyl-2-amino-2-deoxy-Dglucose and 2-amino-2-deoxy-D-glucose (148). The hydrogen bond between the acetamido moieties accounts for its high crystallinity thereby characterized its extreme poor solubility in aqueous solution (150,151). Chitin is soluble in water when its degree of deacetylation is partial (152,153). A deacetylation of less than 10% is possible with chitin having a high molecular weight of $1–2.5 \times 10^6$ Da as opposed to a degree of polymerization of approximately 5000–10,000 monomeric residues (154). The most widely used derivative of chitin is chitosan produced by N-deacetylation of chitin using alkali under heat (155). In this regard, chitosan with the degree of deacetylation ranging from 40 to 98% and having a molecular weight between 5×10^4 Da and 2×10^6 Da can be synthesized (154–157). Upon the deacetylation of the chitin backbone, the free amino functional groups of chitosan are accessible for protonation in order to activate polymer solubilization or react chemically and grafted to generate new chitosan derivatives with unique physiochemical and biological characteristics.

Chitosan is a polysaccharide with multiple functional groups including NeH and OHe that influence its flexibility and allow for its modification for targeted pharmaceutical or medical applications (158). In forming polyelectrolyte complexes, the cationic amino moieties at the carbon 2 (C2) geometry of its repeating units freely undergo an electrostatic interaction with the anionic groups of other polyanions (159). A number of natural polyanions including alginate, chondroitin sulfate, pectin, carrageenan, xanthan gum, carboxymethyl cellulose, dextran sulfate, hyaluronic acid, or other synthetic polyanions such as poly (acrylic acid), polyphosphoric acid, and poly (L-lactide) have been complexed with chitosan to facilitate specific drug delivery characteristics (160–163). Chitosan can be manipulated and changed into nano-vectors with specified sizes and surface charges, by changing its molecular weight and degree of deacetylation (164). Similarly, its plastic property can be molded into amorphous forms (165). Processing parameters and formulation functions through electrostatic interaction, hydrogen bonding, and Van der Waals forces, can be employed to campact its molecular chains for nanoscale application (166).

The mucoadhesive property of chitosan as a polysaccharide and viscous characteristics have important applications in drug encapsulation, drug release, and the modulation of drug absorption kinetics (163). Based on this essential property, chitosan as a cationic polyelectrolyte exhibits high affinity for the negatively charged mucosal interface which possesses robust sialic acid and O-sulfosaccharides (166). Another unique potential of chitosan is its ability to decrease the electrical resistance across epithelial membranes by disrupting tight junctions between the epithelial cells (167). Para- and transcellular drug transport is facilitated and enhanced by chitosan through the protein kinase C pathway, by loosening the tight junction of mucosa through the translocation of tight junction proteins from the membrane to the cytoskeleton (168–170).

The biodegradable and biocompatible nature of chitosan has been well explored as a therapeutic polymer with characteristic anti-cancer (171), anti-fungal (172), anti-bacterial (173), anti-oxidative (174), anti-diabetic (174,175), and anti-inflammatory activities (171). Based on this excellent variety of applications, the use of chitosan in the production of pulmonary therapeutics is regarded more highly than that of lactose. In cancer nanomedicines, the physicochemical properties of chitosan-based nano-vectors, including shape, size, surface morphology, and charge, the nature and availability of targeting moieties are discovered to play crucial roles in influencing effective cell targeting, cellular internalization, and anti-cancer efficacy of cancer therapeutics (176,177).

Positively charged spherical and small-sized nano-vectors with decreased surface area to disrupt the cancer cell membrane and increased propensity for the negatively charged cancer cell surfaces enhance cellular uptake and internalization (178). A major limitation in the use of targeting ligand is the sublevel or an excessive fraction application in targeted drug delivery. This, in turn, will result in the inability of the ligand-functionalized nano-vector to interact with the cancer cell surface proteins and poor intracellular therapeutic availability owing to insufficient ligand or steric hindrance of overcrowded ligands at the binding site. The same results were reported for infection nanomedicines. For instance, in infectious diseases like pulmonary tuberculosis, drug targeting the alveolar macrophages is crucial because tubercle bacilli accumulate at these sites (179). Nanocarriers with hydrophobic surfaces display a higher degree of opsonization by macrophages. Chitosan nanovectors, made with high molecular weight or viscosity, can inhibit transmembrane drug transport into the macrophages (180). In this way, the nano-vector can interact with the extracellular membrane of macrophages and form a viscous obstruction to drug diffusion or endocytosis.

26.5.1 Chitosan-Based Delivery Systems in Lung Cancer Therapy

The use of chemotherapeutics such as cisplatin, paclitaxel, docetaxel, and vinorelbine remains one of the first alternatives for systemic administration for anti-cancer therapy (181). However, these anti-cancer agents present deleterious side effects such as pain, nerve deterioration, and skin allergic reactions (22). For instance, considering the intravenous paclitaxel, the solubilizing mixture of Cremophor EL and the dehydrated alcohol has been reported as the reason for neurotoxicity and drug hypersensitivity reaction (182). Conversely, on-site administration of anti-cancer chemotherapeutics by inhalation can minimize the adverse effects of systemic administration, with optimum drug accumulation at the active site.

The first clinical studies on pulmonary administration of 5-fluorouracil solution as an anti-cancer chemotherapeutic for the treatment of NSLC has been reported (183). Only minimal amounts of 5-fluorouracil were found in the plasma without local or systemic negative effects. Interestingly, the tissues of the lung tumor exhibit an higher, up to 5- to 15-fold, drug concentrations than the normal/healthy lung tissues. In another experiment, intravenous docetaxel coupled with celecoxib solution was successfully nebulized to treat lung cancer. The report showed the therapeutic efficacy of the aerosolized celecoxib as compared to oral dosage form, with a lower dose requirement and tendency to develop negative side effects (183). Meanwhile, despite the laudable benefits of pulmonary delivery, the effectiveness of inhaled tumor chemotherapeutics can be militated by the branching pattern of the respiratory tract and the biological barriers present within the respiratory passage systems, including mucus, ciliated cells, and resident macrophages inhibiting the localization, penetration, and adsorption of chemotherapeutics in the lung (184).

Chitosan-based particulate vectors of cancer chemotherapeutics are fabricated though ionic gelation, emulsification, and/or spray-drying techniques. In addition, both chemical vapor deposition and nano-precipitation methods are employed to co-synthesize chitosan nanoparticles with lipid or carbon-based composition. The aerodynamic property needed for pulmonary inhalation is considered in the design and fabrication of microparticles or microencapsulated nanoparticles (185–187). In order to enhance the affinity of the nano-vector with cancer targeting and cellular internalization, targeting ligands, including folic and hyaluronic acids, as well as permeation enhancers like gallic acid, are conjugated to the chitosan backbone (186,188). Meanwhile, quercetin can be employed to inhibit p-glycoprotein in addition to the drug to decrease drug efflux from the cancer cells (185). The average particle sizes of vector for targeting lung tumor range between 50 nm and 9.6 μm (178). A major influencing parameter on the cytotoxicity and cellular uptake of inhaled chitosan-based nanoparticles is particle size.

Quite often, chitosan-based nanoparticles possess positive surface charges (zeta potential values close to +30 mV), which in turn enhances physical stability by decreasing particle aggregation through electrical repulsion (22). Meanwhile, the endothelial cells of the tumor vasculature are characterized by the overexpression of negatively charged surface molecules such as glycoproteins, proteoglycans, and anionic phospholipids (189,190). Interestingly, accumulation of cationic chitosan particles within the lung capillaries can be enhanced through the electrical interaction between the positively charged chitosan particles and the negatively charged endothelial cells of the tumor vasculature (191,192). The oligosaccharide derivatives of chitosan, in particular, show anti-cancer activity (193–196). While its mechanism of action is not fully understood, it is assumed to be closely related to its electrostatic charges, changes in membrane permeability of cancer cells, regulating the expression of tumor factors including vascular endothelial growth factor and/or metalloproteinase-9, and pro-apoptotic effects.

The role and importance of chitosan's overall surface charge in the inhibition of cancer growth is perfectly modeled by quaternized chitosan derivatives which showed enhanced selective and active efficacy as a cancer therapeutic compared with the native chitosan (193). Chitosan cytotoxic efficacies are well defined by apoptosis through the induction of caspase 3 and caspase 8 (188,197). The fabrication and application of chitosan as a pulmonary drug delivery vehicle and as conjugate backbone for lung cancer treatment has gained wide recognition and has been widely investigated with successful results (177,197–202). Pulmonary drug delivery vectors based on chitosan formulations have been reported to decrease systemic toxicity of anti-cancer chemotherapeutics anf enhance drug bioavailability in lungs as well as increase the anti-cancer effects of the drugs. By modifying the surfaces of particulate systems using targeting ligands, engineered particles have been shown to improve cellular uptake and internalization which, in turn, increases the propensity of cancer cell apoptosis (178).

However, a majority of the studies conducted so far are based on in vitro and/or ex vivo experiments. Meanwhile, more research into the aerodynamic evaluation and in vivo characterization of the properties of these particles via natural inhalation is needed to show their actual therapeutic potentials. Usually, these nanoparticles are susceptible to exhalation and exposed to mucocilliary transport (22,23). When their diameter is greater than 200 nm, bigger particles have been reported to initiate non-specific scavenging by monocytes and clearance by the reticulo-endothelial system (177) with possible phagocytosis by the alveolar macrophages (203). Thus, the smaller the particles, the better their chances for survival and prolonged retention in the systemic circulation. Converting nanoparticles into microparticles with average aerodynamic diameter greater than 6 μm means they will be deposited at the upper airways and stand the chance of possible clearance by mucocilliary transport in the epithelial tissue (204). Meanwhile, those with average aerodynamic diameters between 1 μm and 5 μm could be deposited in the narrower airways and thereby escape mucocilliary transport, but their smaller fractions with average diameters between 2 μm and 3 μm are susceptible to recognition and clearance by the resident alveolar macrophages (22). Thus, one way to overcome these hurdles in order to obtain an effective drug delivery to lung cancer cells is to encapsulate nano-vectors with particle sizes below 200 nm into the microparticles having average aerodynamic diameters between 3 μm and 5 μm (204).

26.6 Folate-Decorated Nano Systems for Lung Cancer Therapy

Functionalizing the surface of nanocarriers remains a strategic approach to circumvent the limitations accrued by anti-cancer drugs and obtain an optimal targeted delivery (205). In this regard, the overexpressed unique proteins on cancer cells surfaces can be explored for the design of targeted delivery systems in cancer therapy. Interestingly, folic acid receptors are uniquely overexpressed as molecular signatures on the cellular membrane of various cancer cells, including lung cancer (206,207). There is a high binding between folate and folate receptors overexpressed on lung cancer cell surface, which could be internalized via receptor-mediated endocytosis (Figure 26.2) (208,209). The folate-receptor is a 38 kD glycosyl-phosphatidylinositol conjugated glycoprotein. Folate is important for the synthesis of nucleotide bases, i.e. purines and pyrimidines. Meanwhile, normal cells transport folate only in reduced form such as 5-methyl-tetrahydrofolate and do not transport folate conjugates across their membrane.

Over the past few decades, folate receptor-mediated drug delivery systems have gained wider attention in design, fabrication, and applications. Usually, folate receptors are oversecreted together with glycosyl phosphatidylinositol. As such, there is a

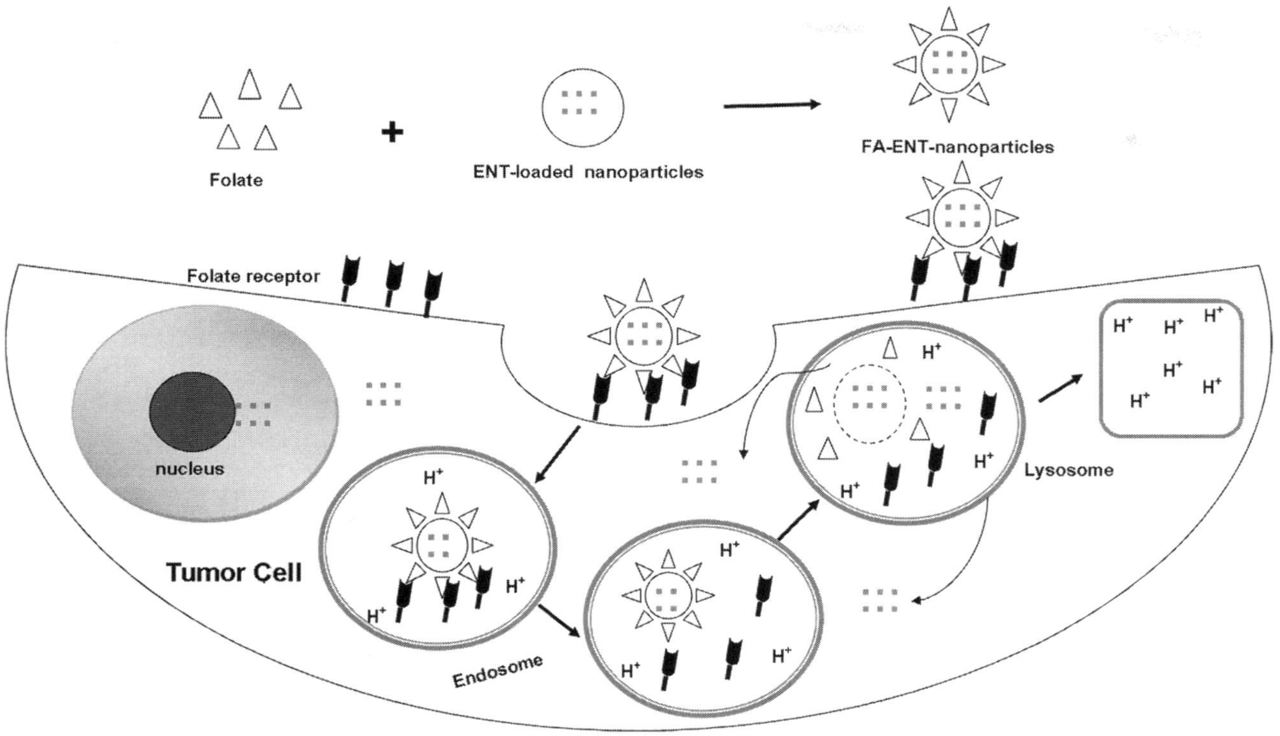

FIGURE 26.2 Graphical Illustration of the Binding Potential of Folate to Folate-Receptors Overexpressed on Cancer Surface and Their Internalization through Receptor-Mediated Endocytosis

membrane glycoprotein linkage in malignant tumors and in several cancer types, including pulmonary cancers, when compared to normal cells (210). Among other ligands, folic acid stands unique owing to its high affinity for the folate receptor. A number of folate and folate receptor conjugates have been reported to exhibit viable enhanced delivery to folate receptor-positive tumor cells (211). Upon their binding to the folate receptor that is present on the surface of cancer cells, folate or its conjugates are internalized into the cellular compartments to produce endosomes (212). Reports have shown that the desired specific targeting and the needed drug accumulation does not occur in cancer cells as long as the folate receptor does not come in contact with circulating folate-receptor-mediated drug (213).

Few unique characteristics have endeared folate as an attractive choice for targeted molecular treatment of tumors. The expression of folate receptors are very limited in normal cells as opposed their overexpression on the surface of human cancer cells (210,214). In addition, the small size of folate enhances its eligibility for specific cell targeting. Furthermore, convectional covalent bonding of small molecules (215) does not influence the binding of folate to its receptor and its subsequent internalization through endocytosis. A number of investigations have been reported in which folate-receptors on the surface of tumor cells have been exploited for targeted delivery of anti-cancer drugs, genes, and radiopharmaceuticals through folate-receptor endocytosis (216,217). Similarly, folate has also been applied as a ligand with cationic liposomes (218) and other polymers such as chitosan (209,219,220), polyethyleneimine (PEI) (209,221), and poly (L-lysine) (222,223). Enhanced transfection was observed through nanoparticle endocytosis facilitated by folate–folate receptor cellular interaction (219). In other reports, conjugated folate-PEG modified PEI in vitro and in vivo (224–227) enhanced targeted-specific gene delivery with better performance when compared to native PEI (224,226). The work of Benns and colleagues reported a remarkable anti-tumor efficacy through intro-tumor administration of therapeutic genes, delivered by folate-PEG-PEI nano-cargo. Folate-chitosan nano system, was employed for conjugation with carbon nanotubes for enhanced efficacy as a drug delivery vector in lung cancer based on its nontoxicity, biocompatibility, and availability for further modification of functional groups (228–231).

26.7 Conclusion, Recommendation, and Future Prospects

Nanoparticle-based therapy possesses enormous potential with novel applications being fabricated for use in cancer diagnosis, detention, imaging, and treatment. These nano systems are already contributing immensely in proffering solutions to cogent limitations with convectional anti-cancer drugs, including non-specific targeting, adverse side effects, reduced therapeutic efficacies, and drug resistance, as well as exhibiting convincing significance that exceeded their predecessors with the capacity for early detention in metastatic tumors. The unique potential of nanoparticles that allow for their ease of manipulation for a personalized medicine approach set them apart as excellent cargoes for the treatment of lung cancer. A number of nanoparticle-based investigations as therapeutics for lung cancer make use of a combinatorial strategy that strike the balance with design, targeting, moieties tracking and anti-cancer drugs. On a general note, nanoparticles with multifaceted structures enable easy design flexibility in drug delivery of chemotherapeutic agents with poor solubility in water, enhancing the ability to by-pass bio-systemic obstacles, and target preferred sites within the body selectively.

A number of unique and excellent characteristics, including biodegradability, biocompatibility, mucoadhesiveness, and antimicrobial and anti-cancer properties as well as enhanced permeability, present chitosan and its derivatives as an excellent skeletal framework for nano- and microparticulate vehicles for pulmonary delivery of anti-cancer therapeutics. Unlike lactose as a major conventional drug delivery carrier, chitosan and its derivatives have gained enormous attention in recent years for investigation, and their suitable applications as the pulmonary particulate vector has been well documented. Meanwhile, the use of chitosan and its derivatives as delivery cargoes are nascent and therefore need thorough analysis, including in vitro aerodynamic characterization and in vivo examination of the pharmacokinetics and functionality of these particles through natural inhalation to correlate to their actual therapeutic efficacy.

The coronavirus disease caused by SARS-CoV-2 infection, which results in the acute respiratory distress syndrome, predisposes cancer patients as potential targets to be diagnosed with Covid-19 (232). Interestingly, the latest innovative strategies around chitosan have shown the application of this biopolymer to tackle coronavirus-associated diseases as well as cancer. Chitosan derivative N-(2-hydroxypropyl)-3-trimethylammonium chloride, having an average molecular weight of 250 kDa and degree of deacetylation of between 50 and 80% have been reported to exhibit a potent interaction with the recombinant ectodomain of the S protein of HCoV, thereby inhibiting S protein–host cellular interaction and virus infection (233). The oligosaccharide derivative of chitosan, prepared by an electrical discharge plasma degradation from chitosan powder (having a molecular weight of 1000–12000 Da, degree of deacetylation = 65–70 %), showed apoptotic death in MCF-7 breast cancer cells (234). Meanwhile, the lower molecular weight of these chitosan derivatives has been documented to exhibit efficient efficacy against PC3 (prostate cancer cells), A549 (lung cancer cells), and HepG2 (hepatoma cells) than their counterparts with larger molecular weights.

Summarily, these recent discoveries in chitosan and its derivatives as therapeutic agents to treat infection and cancer are expected to generate enormous research interests proffering synergistic approaches that harness both the drug delivery and drug discovery characteristics of these excellent biomaterials. In this regard, harnessing the relationship between their structural and bioactivity properties in association with coronavirus disease as well as lung cancer treatment, is of great interest from the drug delivery and therapeutic efficacy standpoints at the molecular levels.

REFERENCES

1. Cancer. https://www.who.int/news-room/fact-sheets/detail/cancer (accessed September 10, 2019).
2. Global Cancer Statistics 2020. CANSA – Cancer Associa. *South Africa*. https://cansa.org.za/global-cancer-statistics/ (accessed September16, 2020).
3. Siegel RL, Miller KD, and Jemal A. Cancer statistics, 2016. *Cancer Journal Clinics*. 2016; 66: 7–30. https://doi.org/10.3322/caac.21332.
4. Adebowale AS, Choonara YE, Kumar P, du Toit LC, and Pillay V. Functionalized nanocarriers for enhanced bioactive delivery to squamous cell carcinomas: targeting approaches and related biopharmaceutical aspects. *Current Pharmaceutical Design*. 2015; 21(22): 3167–3180. doi: 10.2174/1381612821666150531165331.
5. Zamay TN, Zamay GS, Kolovskaya OS, Zukov RA, Petrova MM, Gargaun A, Berezovski MV, and Kichkailo AS. Current and prospective protein biomarkers of lung cancer. *Cancers*. 2017; 9: https://doi.org/10.3390/cancers9110155.
6. Zappa C and Mousa SA. Non-small cell lung cancer: current treatment and future advances. *Translational Lung Cancer Research*. 2016; 5: 288–300. https://doi.org/10.21037/tlcr.2016.06.07.
7. Clegg A, Scott D, Hewitson P, Sidhu M, and Waugh N. Clinical and cost effectiveness of paclitaxel, docetaxel, gemcitabine, and vinorelbine in non-small cell lung cancer: a systematic review. *Thorax*. 2002; 57: 20–28. https://doi.org/10.1136/thorax.57.1.20.
8. Leslie KO and Wick MR. 1 - Lung Anatomy, in: KO Leslie and MR Wick (Eds.), Practical Pulmonary Pathology: A Diagnostic Approach 3rd Ed., Elsevier, 2018, 1–14.e2. https://doi.org/10.1016/B978-0-323-44284-8.00001-6.
9. Ruigrok MJR, Frijlink HW, and Hinrichs WLJ. Pulmonary administration of small interfering RNA: The route to go? *Journal of Controlled Release*. 2016; 235: 14–23. https://doi.org/10.1016/j.jconrel.2016.05.054.
10. Rasul RM, Muniandy MT, Zakaria Z, Shah K, Chee CF, Dabbagh A, Rahman NA, and Wong TW. A review on chitosan and its development as pulmonary particulate anti-infective and anti-cancer drug carriers. *Carbohydrate Polymers*. 2020; 250: 116800. https://doi.org/10.1016/j.carbpol.2020.116800.
11. Chang L-Y, Crapo JD, Gehr P, Rothen-Rutishauser B, Mühfeld C, and Blank F. 8.04 - Alveolar Epithelium in Lung Toxicology*, in: McQueen CA (Ed.), Comprehensive Toxicolog 2nd Ed., Elsevier, Oxford, 2010: 59–91. https://doi.org/10.1016/B978-0-08-046884-6.00904-0.
12. American Cancer Society. What Is Lung Cancer? Types of Lung Cancer, https://www.cancer.org/cancer/lung-cancer/about/what-is.html (accessed September 9, 2020).
13. Goel A, Baboota S, Sahni JK, and Ali J. Exploring targeted pulmonary delivery for treatment of lung cancer. *International Journal of Pharmaceutical Investigation*. 2013; 3: 8–14. https://doi.org/10.4103/2230-973X.108959.
14. Roa WH, Azarmi S, Al-Hallak MHDK, Finlay WH, Magliocco AM, and Löbenberg R. Inhalable nanoparticles, a non-invasive approach to treat lung cancer in a mouse model. *Journal of Controlled Release*. 2011; 150: 49–55. https://doi.org/10.1016/j.jconrel.2010.10.035.
15. Tseng C-L, Wu SYH, Wang W-H, Peng C-L, Lin F-H, Lin C-C, Young T-H, and Shieh M-J. Targeting efficiency and biodistribution of biotinylated-EGF-conjugated gelatin nanoparticles administered via aerosol delivery in nude mice with lung cancer. *Biomaterials*. 2008; 29: 3014–3022. https://doi.org/10.1016/j.biomaterials.2008.03.033.

16. Pilcer G and Amighi K. Formulation strategy and use of excipients in pulmonary drug delivery. *International Journal of Pharmaceutics*. 2010; 392: 1–19. https://doi.org/10.1016/j.ijpharm.2010.03.017.
17. Zhang L-J, Xing B, Wu J, Xu B, and Fang X-L. Biodistribution in mice and severity of damage in rat lungs following pulmonary delivery of 9-nitrocamptothecin liposomes. *Pulmonary Pharmacology & Therapeutics*. 2008; 21: 239–246. https://doi.org/10.1016/j.pupt.2007.04.002.
18. Alipour S, Montaseri H, and Tafaghodi M. Preparation and characterization of biodegradable paclitaxel loaded alginate microparticles for pulmonary delivery. *Colloids and Surfaces B: Biointerfaces*. 2010; 81: 521–529. https://doi.org/10.1016/j.colsurfb.2010.07.050.
19. Kuzmov A and Minko T. Nanotechnology approaches for inhalation treatment of lung diseases. *Journal of Controlled Release*. 2015; 219: 500–518. https://doi.org/10.1016/j.jconrel.2015.07.024.
20. Majumder J, O Taratula, and Minko T. Nanocarrier-based systems for targeted and site specific therapeutic delivery. *Advanced Drug Delivery Reviews*. 2019; 144: 57–77. https://doi.org/10.1016/j.addr.2019.07.010.
21. Senapati S, Mahanta AK, Kumar S, and Maiti P. Controlled drug delivery vehicles for cancer treatment and their performance.*Signal Transduction and Targeted Therapy*. 2018; 3: 1–19. https://doi.org/10.1038/s41392-017-0004-3.
22. Lee W-H, Loo C-Y, Traini D, and Young PM. Inhalation of nanoparticle-based drug for lung cancer treatment: Advantages and challenges *Journal of Pharmaceutical Sciences*. 2015; 10: 481–489. https://doi.org/10.1016/j.ajps.2015.08.009.
23. Lee W-H, C-Y Loo, D Traini, and PM Young. Nano- and micro-based inhaled drug delivery systems for targeting alveolar macrophages. *Expert Opin. Drug Deliv*. 2015; 12: 1009–1026. https://doi.org/10.1517/17425247.2015.1039509.
24. Ruigrok MJR, Frijlink HW, and Hinrichs WLJ. Pulmonary administration of small interfering RNA: The route to go?. *Journal of Controlled Release*. 2016; 235: 14–23. https://doi.org/10.1016/j.jconrel.2016.05.054.
25. Alhajj N, Zakaria Z, Naharudin I, Ahsan F, Li W, and Wong TW. Critical physicochemical attributes of chitosan nanoparticles admixed lactose-PEG 3000 microparticles in pulmonary inhalation. *Asian Journal of Pharmaceutical Science*. 2020; 15374–384. https://doi.org/10.1016/j.ajps.2019.02.001.
26. Youngren-Ortiz SR, Gandhi NS, España-Serrano L, and Chougule MB. Aerosol delivery of siRNA to the lungs. Part 1: rationale for gene delivery systems. *KONA Powder and Particle Journal*. 2016; 33: 63–85. https://doi.org/10.14356/kona.2016014.
27. Fröhlich E. Biological obstacles for identifying in vitro-in vivo correlations of orally inhaled formulations. *Pharmaceutics*. 2019; 11: 316. https://doi.org/10.3390/pharmaceutics11070316.
28. Lam JKW, W Liang, and H-K Chan. Pulmonary delivery of therapeutic siRNA. *Advanced Drug Delivery Reviews*. 2012; 64: 1–15. https://doi.org/10.1016/j.addr.2011.02.006.
29. Roy I and Vij N. Nanodelivery in airway diseases: challenges and therapeutic applications, Nanomedicine *Nanomedicine: Nanotechnology, Biology and Medicine*. 2010; 6:237–244. https://doi.org/10.1016/j.nano.2009.07.001.
30. Sanders N, Rudolph C, Braeckmans K, De Smedt SC, and Demeester J. Extracellular barriers in respiratory gene therapy. *Advanced Drug Delivery Reviews*. 2009; 61: 115–127. https://doi.org/10.1016/j.addr.2008.09.011.
31. Yang MY, Chan JGY, and Chan H-K. Pulmonary drug delivery by powder aerosols. *Journal of Controlled Release*. 2014; 193: 228–240. https://doi.org/10.1016/j.jconrel.2014.04.055.
32. Lai SK, Wang Y-Y, and Hanes J. Mucus-penetrating nanoparticles for drug and gene delivery to mucosal tissues. *Advanced Drug Delivery Reviews*. 2009; 61: 158–171. https://doi.org/10.1016/j.addr.2008.11.002.
33. Qiu Y, Lam JKW, Leung SWS, and Liang W. Delivery of RNAi therapeutics to the airways—from bench to bedside. *Molecules*. 2016; 21: 1249. https://doi.org/10.3390/molecules21091249.
34. CA Ruge, J Kirch, and C-M Lehr. Pulmonary drug delivery: from generating aerosols to overcoming biological barriers-therapeutic possibilities and technological challenges. *Lancet Respir. Med*. 2013; 1: 402–413. https://doi.org/10.1016/S2213-2600(13)70072-9.
35. Schiffelers RM, Metselaar JM, Fens MHAM, Janssen APCA, Molema G, and Storm G. Liposome-encapsulated prednisolone phosphate inhibits growth of established tumors in mice. *Neoplasia*. 2005; 7: 118–127. https://doi.org/10.1593/neo.04340.
36. Simões S, Filipe A, Faneca H, Mano M, Penacho N, Düzgünes N, and de Lima MP. Cationic liposomes for gene delivery. *Expert Opinion on Drug Delivery*. 2005; 2: 237–254. https://doi.org/10.1517/17425247.2.2.237.
37. Lindner LH, ME Eichhorn, H Eibl, N Teichert, M Schmitt-Sody, RD Issels, and M Dellian. Novel temperature-sensitive liposomes with prolonged circulation time. *Clinical Cancer Research*. 2004; 10: 2168–2178. https://doi.org/10.1158/1078-0432.CCR-03-0035.
38. Adamina M, Guller U, Bracci L, Heberer M, Spagnoli GC, and Schumacher R. Clinical applications of virosomes in cancer immunotherapy. *Expert Opinion on Biological Therapy*. 2006; 6: 113–1121. https://doi.org/10.1517/14712598.6.11.1113.
39. Krishnan L, Deschatelets L, Stark FC, Gurnani K, and Sprott GD. Archaeosome adjuvant overcomes tolerance to tumor-associated melanoma antigens inducing protective CD8+ T cell responses. *Clinical and Developmental Immunology*. 2010; 2010(2011): e578432. https://doi.org/10.1155/2010/578432.
40. Yao X, Panichpisal K, Kurtzman N, and Nugent K. Cisplatin nephrotoxicity: a review. *International Journal of Medical Sciences*. 2007; 334: 115–124. https://doi.org/10.1097/MAJ.0b013e31812dfe1e.
41. Boulikas T. Low toxicity and anticancer activity of a novel liposomal cisplatin (Lipoplatin) in mouse xenografts. *Oncology Report*. 2004; 12: 3–12. https://doi.org/10.3892/or.12.1.3.
42. Devarajan P, Tarabishi R, Mishra J, Ma Q, Kourvetaris A, Vougiouka M, and Boulikas T. Low renal toxicity of lipoplatin compared to cisplatin in animals. *Anticancer Research*. 2004; 24: 2193–2200.

43. Liposomal cisplatin (Nanoplatin) for advanced non-small cell lung cancer – first line | Innovation Observatory, (n.d.). http://www.io.nihr.ac.uk/report/liposomal-cisplatin-nanoplatin-for-advanced-non-small-cell-lung-cancer-first-line/ (accessed September 18, 2020).
44. Babu A, Templeton AK, Munshi A, and Ramesh R. Nanoparticle-based drug delivery for therapy of lung cancer: progress and challenges. *Journal of Nanomaterials*. 2013; 2013: e863951. https://doi.org/10.1155/2013/863951.
45. Wang X, Zhou J, Wang Y, Zhu Z, Lu Y, Wei Y, and Chen L. A phase I clinical and pharmacokinetic study of paclitaxel liposome infused in non-small cell lung cancer patients with malignant pleural effusions.*European Journal of Cancer*. 2010; 46: 1474–1480. https://doi.org/10.1016/j.ejca.2010.02.002.
46. Zhou J, Zhao W-Y, Ma X, Ju R-J, Li X-Y, N Li, Sun M-G, Shi J-F, Zhang C-X, and Lu W-L. The anticancer efficacy of paclitaxel liposomes modified with mitochondrial targeting conjugate in resistant lung cancer. *Biomaterials*. 2013; 34: 3626–3638. https://doi.org/10.1016/j.biomaterials.2013.01.078.
47. Koudelka Š and Turánek J. Liposomal paclitaxel formulations. *Journal of Controlled Release*. 2012; 163: 322–334. https://doi.org/10.1016/j.jconrel.2012.09.006.
48. Stathopoulos GP, Antoniou D, Dimitroulis J, Michalopoulou P, Bastas A, Marosis K, Stathopoulos J, Provata A, Yiamboudakis P, Veldekis D, Lolis N, Georgatou N, Toubis M, Pappas. Ch, and Tsoukalas G. Liposomal cisplatin combined with paclitaxel versus cisplatin and paclitaxel in non-small-cell lung cancer: a randomized phase III multicenter trial. *Annals of Oncology*. 2010; 21: 2227–2232. https://doi.org/10.1093/annonc/mdq234.
49. North S, and Butts C. Vaccination with BLP25 liposome vaccine to treat non-small cell lung and prostate cancers. *Expert Review of Vaccines*. 2005; 4: 249–257. https://doi.org/10.1586/14760584.4.3.249.
50. Wurz GT, Gutierrez AM, Greenberg BE, Vang DP, Griffey SM, Kao C-J, Wolf M, and DeGregorio MW. Antitumor effects of L-BLP25 antigen-specific tumor immunotherapy in a novel human MUC1 transgenic lung cancer mouse model. *Journal of Translational Medicine*. 2013; 11: 64. https://doi.org/10.1186/1479-5876-11-64.
51. Wu Y-L, Park K, Soo RA, Sun Y, Tyroller K, Wages D, Ely G, Yang JCH, and Mok T. INSPIRE: A phase III study of the BLP25 liposome vaccine (L-BLP25) in Asian patients with unresectable stage III non-small cell lung cancer. *BMC Cancer*. 2011; 11: 430. https://doi.org/10.1186/1471-2407-11-430.
52. Lu C, Stewart DJ, Lee JJ, Ji L, Ramesh R, Jayachandran G, Nunez MI, Wistuba II, Erasmus JJ, Hicks ME, Grimm EA, Reuben JM, Baladandayuthapani V, Templeton NS, McMannis JD, and Roth JA. Phase I clinical trial of systemically administered TUSC2(FUS1)-nanoparticles mediating functional gene transfer in humans. *PLoS ONE*. 2012; 7: e34833. https://doi.org/10.1371/journal.pone.0034833.
53. Wang Y, Gao S, Ye W-H, Yoon HS, and Yang Y-Y. Co-delivery of drugs and DNA from cationic core-shell nanoparticles self-assembled from a biodegradable copolymer. *Nature Materials*. 2006; 5: 791–796. https://doi.org/10.1038/nmat1737.
54. Beh CW, Seow WY, Wang Y, Zhang Y, Ong ZY, Ee PLR, and Yang Y-Y. Efficient delivery of Bcl-2-targeted siRNA using cationic polymer nanoparticles: downregulating mRNA expression level and sensitizing cancer cells to anticancer drug. *Biomacromolecules*. 2009; 10: 41–48. https://doi.org/10.1021/bm801109g.
55. Zhu C, Jung S, Luo S, Meng F, Zhu X, Park TG, and Zhong Z. Co-delivery of siRNA and paclitaxel into cancer cells by biodegradable cationic micelles based on PDMAEMA-PCL-PDMAEMA triblock copolymers. *Biomaterials*. 2010; 31: 2408–2416. https://doi.org/10.1016/j.biomaterials.2009.11.077.
56. Sun T-M, Du J-Z, Yao Y-D, Mao C-Q, Dou S, Huang S-Y, Zhang P-Z, Leong KW, Song E-W, and Wang J. Simultaneous delivery of siRNA and paclitaxel via a "two-in-one" micelleplex promotes synergistic tumor suppression. *ACS Nano*. 2011; 5: 1483–1494. https://doi.org/10.1021/nn103349h.
57. Walensky LD. BCL-2 in the crosshairs: tipping the balance of life and death. *Cell Death & Differentiation*. 2006; 13: 1339–1350. https://doi.org/10.1038/sj.cdd.4401992.
58. Bu Y, Yang Z, Li Q, and Song F. Silencing of polo-like kinase (Plk) 1 via siRNA causes inhibition of growth and induction of apoptosis in human esophageal cancer cells. *Oncology*. 2008; 74: 198–206. https://doi.org/10.1159/000151367.
59. Cao N, Cheng D, Zou S, Ai H, Gao J, and Shuai X. The synergistic effect of hierarchical assemblies of siRNA and chemotherapeutic drugs co-delivered into hepatic cancer cells. *Biomaterials*. 2011; 32: 2222–2232. https://doi.org/10.1016/j.biomaterials.2010.11.061.
60. Kluck RM, Bossy-Wetzel E, Green DR, and Newmeyer DD. The release of cytochrome c from mitochondria: a primary site for Bcl-2 regulation of apoptosis. *Science*. 1997; 275: 1132–1136. https://doi.org/10.1126/science.275.5303.1132.
61. Saad M, Garbuzenko OB, and Minko T. Co-delivery of siRNA and an anticancer drug for treatment of multidrug-resistant cancer. *Nanomed*. 2008; 3: 761–776. https://doi.org/10.2217/17435889.3.6.761.
62. Patil YB, Swaminathan SK, Sadhukha T, Ma L, and Panyam J. The use of nanoparticle-mediated targeted gene silencing and drug delivery to overcome tumor drug resistance. *Biomaterials*. 2010; 31: 358–365. https://doi.org/10.1016/j.biomaterials.2009.09.048.
63. Müller RH, Mäder K, and Gohla S. Solid lipid nanoparticles (SLN) for controlled drug delivery - a review of the state of the art . *European Journal of Pharmaceutics and Biopharmaceutics*. 2000; 50: 161–177. https://doi.org/10.1016/s0939-6411(00)00087-4.
64. Mehnert W and Mäder K. Solid lipid nanoparticles: production, characterization and applications.*Advanced Drug Delivery Reviews*. 2001; 47: 165–196. https://doi.org/10.1016/s0169-409x(01)00105-3.
65. Choi SH, Jin S-E, Lee M-K, Lim S-J, Park J-S, Kim B-G, Ahn WS, and Kim C-K. Novel cationic solid lipid nanoparticles enhanced p53 gene transfer to lung cancer cells. *European Journal of Pharmaceutics and Biopharmaceutics*. 2008; 68: 545–554. https://doi.org/10.1016/j.ejpb.2007.07.011.

66. Bae KH, JY Lee, Lee SH, Park TG, and Nam YS. Optically traceable solid lipid nanoparticles loaded with siRNA and paclitaxel for synergistic chemotherapy with in situ imaging. *Advanced Healthcare Materials*. 2013; 2: 576–584. https://doi.org/10.1002/adhm.201200338.
67. Schroeder JE, Shweky I, Shmeeda H, Banin U, and Gabizon A. Folate-mediated tumor cell uptake of quantum dots entrapped in lipid nanoparticles. *Journal of Controlled Release*. 2007; 124: 28–34. https://doi.org/10.1016/j.jconrel.2007.08.028.
68. Hsu M-H and Su Y-C. Iron-oxide embedded solid lipid nanoparticles for magnetically controlled heating and drug delivery. *Biomed. Microdevices*. 2008; 10: 785. https://doi.org/10.1007/s10544-008-9192-5.
69. Hyun Ryong K, Ik K, Kh B, Sh L, and Tg. LYP. Cationic solid lipid nanoparticles reconstituted from low density lipoprotein components for delivery of siRNA. *Molecular Pharmaceutics*. 2008; 5: 622–631. https://doi.org/10.1021/mp8000233.
70. Jin J, Bae KH, Yang H, Lee SJ, Kim H, Kim Y, Joo KM, Seo SW, Park TG, and Nam D-H. In vivo specific delivery of c-Met siRNA to glioblastoma using cationic solid lipid nanoparticles. *Bioconjugate Chemistry*. 2011; 22: 2568–2572. https://doi.org/10.1021/bc200406n.
71. Wang H, Liu B, Zhang C, Peng G, Liu M, Li D, Gu F, Chen Q, Dong J-T, Fu L, and Zhou J. Parkin regulates paclitaxel sensitivity in breast cancer via a microtubule-dependent mechanism. *The Journal of Pathology*. 2009; 218: 76–85. https://doi.org/10.1002/path.2512.
72. Hsiao J-R, Leu S-F, and Huang B-M. Apoptotic mechanism of paclitaxel-induced cell death in human head and neck tumor cell lines. *Journal of Oral Pathology & Medicine*. 2009; 38: 188–197. https://doi.org/10.1111/j.1600-0714.2008.00732.x.
73. Honore S, Kamath K, Braguer D, Horwitz SB, Wilson L, Briand C, and Jordan MA. Synergistic suppression of microtubule dynamics by discodermolide and paclitaxel in non-small cell lung carcinoma cells. *Cancer Research*. 2004; 64: 4957–4964. https://doi.org/10.1158/0008-5472.CAN-04-0693.
74. Li G, Zhao J, Peng X, Liang J, Deng X, and Chen Y. Radiation/paclitaxel treatment of p53-abnormal non-small cell lung cancer xenograft tumor and associated mechanism. *Cancer Biotherapy and Radiopharmaceuticals*. 2012; 27: 227–233. https://doi.org/10.1089/cbr.2011.1154.
75. Collins TS, Lee LF, and Ting JP. Paclitaxel up-regulates interleukin-8 synthesis in human lung carcinoma through an NF-kappaB- and AP-1-dependent mechanism. *Cancer Immunology, Immunotherapy*. 2000; 49: 78–84. https://doi.org/10.1007/s002620050605.
76. Zhang X, Chen L-X, Ouyang L, Cheng Y, and Liu B. Plant natural compounds: targeting pathways of autophagy as anticancer therapeutic agents. *Cell Proliferation*. 2012; 45: 466–476. https://doi.org/10.1111/j.1365-2184.2012.00833.x.
77. Mollinedo F and Gajate C. Microtubules, microtubule-interfering agents and apoptosis *Apoptosis*. 2003; 8: 413–450. https://doi.org/10.1023/a:1025513106330.
78. Abal M, JM Andreu, and Barasoain I. Taxanes: microtubule and centrosome targets, and cell cycle dependent mechanisms of action. Current Cancer Drug Targets. 2003; 3: 193–203. https://doi.org/10.2174/1568009033481967.
79. Liu S-Y, Song S-X, Lin L, and Liu X. Molecular mechanism of cell apoptosis by paclitaxel and pirarubicin in a human osteosarcoma cell line. *Chemotherapy*. 2010; 56: 101–107. https://doi.org/10.1159/000305257.
80. Li J, Wang WL, Yang XK, Yu XX, Hou YD, and Zhang J. Inducible overexpression of Bak sensitizes HCC-9204 cells to apoptosis induced by doxorubicin. *Acta Pharmaceutica Sinica B*. 2000; 21: 769–776.
81. Ferlini C, Raspaglio G, Mozzetti S, Distefano M, Filippetti F, Martinelli E, Ferrandina G, Gallo D, Ranelletti FO, and Scambia G. Bcl-2 down-regulation is a novel mechanism of paclitaxel resistance. *Molecular Pharmacology*. 2003; 64: 51–58. https://doi.org/10.1124/mol.64.1.51.
82. Sgadari C, Toschi E, Palladino C, Barillari G, Carlei D, Cereseto A, Ciccolella C, Yarchoan R, Monini P, Stürzl M, and Ensoli B. Mechanism of paclitaxel activity in Kaposi's sarcoma. The Journal of Immunology. 2000; 165: 509–517. https://doi.org/10.4049/jimmunol.165.1.509.
83. Lanni JS, Lowe SW, Licitra EJ, Liu JO, and Jacks T. p53-independent apoptosis induced by paclitaxel through an indirect mechanism. *Proceedings of the National Academy of Sciences of the United States of America*. 1997; 94: 9679–9683. https://doi.org/10.1073/pnas.94.18.9679.
84. Merchan JR, Jayaram DR, Supko JG, He X, Bubley GJ, and Sukhatme VP. Increased endothelial uptake of paclitaxel as a potential mechanism for its antiangiogenic effects: potentiation by Cox-2 inhibition. *International Journal of Cancer*. 2005; 113: 490–498. https://doi.org/10.1002/ijc.20595.
85. MacKeigan JP, Taxman DJ, Hunter D, Earp HS, Graves LM, and Ting JPY. Inactivation of the antiapoptotic phosphatidylinositol 3-kinase-Akt pathway by the combined treatment of taxol and mitogen-activated protein kinase kinase inhibition. *Clinical Cancer Research*. 2002; 8: 2091–2099.
86. Fortugno P, Wall NR, Giodini A, O'Connor DS, Plescia J, Padgett KM, Tognin S, Marchisio PC, and Altieri DC. Survivin exists in immunochemically distinct subcellular pools and is involved in spindle microtubule function. *Journal of Cell Science*. 2002; 115: 575–585.
87. Mitsuuchi Y, Johnson SW, Selvakumaran M, Williams SJ, Hamilton TC, and Testa JR. The phosphatidylinositol 3-kinase/AKT signal transduction pathway plays a critical role in the expression of p21WAF1/CIP1/SDI1 induced by cisplatin and paclitaxel. *Cancer Research*. 2000; 60: 5390–5394.
88. Greenberg VL and Zimmer SG. Paclitaxel induces the phosphorylation of the eukaryotic translation initiation factor 4E-binding protein 1 through a Cdk1-dependent mechanism. *Oncogene*. 2005; 24: 4851–4860. https://doi.org/10.1038/sj.onc.1208624.
89. Yang IH, Siddique R, Hosmane S, Thakor N, and Höke A. Compartmentalized microfluidic culture platform to study mechanism of paclitaxel-induced axonal degeneration. *Experimental Neurology*. 2009; 218: 124–128. https://doi.org/10.1016/j.expneurol.2009.04.017.
90. Bennett GJ, Liu GK, Xiao WH, Jin HW, and Siau C. Terminal arbor degeneration - a novel lesion produced by the antineoplastic agent paclitaxel. *European Journal of Neuroscience*. 2011; 33: 1667–1676. https://doi.org/10.1111/j.1460-9568.2011.07652.x.

91. Melli G, Jack C, Lambrinos GL, Ringkamp M, and A Höke. Erythropoietin protects sensory axons against paclitaxel-induced distal degeneration. *Neurobiology of Disease*. 2006; 24: 525–530. https://doi.org/10.1016/j.nbd.2006.08.014.
92. Mozzetti S, Ferlini C, Concolino P, Filippetti F, Raspaglio G, Prislei S, Gallo D, Martinelli E, Ranelletti FO, Ferrandina G, and Scambia G. Class III B-tubulin overexpression is a prominent mechanism of paclitaxel resistance in ovarian cancer patients. *Clinical Cancer Research*. 2005; 11: 298–305.
93. Zhang J, Zhao J, Zhang W, Liu G, Yin D, Li J, Zhang S, and Li H. Establishment of paclitaxel-resistant cell line and the underlying mechanism on drug resistance. *International Journal of Gynecological Cancer*. 2012; 22: 1450–1456. https://doi.org/10.1097/IGC.0b013e31826e2382.
94. Luo S, Deng W, Wang X, Lü H, Han L, Chen B, Chen X, and Li N. Molecular mechanism of indirubin-3'-monoxime and Matrine in the reversal of paclitaxel resistance in NCI-H520/TAX25 cell line. *Chinese Medical Journal*. 2013; 126: 925–929.
95. Zhang X, Zhang L, Liu Y, Xu H, Sun P, Song J, and Luo Y. Molecular mechanism of chemosensitization to paclitaxel in human melanoma cells induced by targeting the EGFR signaling pathway. *Zhonghua Zhong Liu Za Zhi*. 2013; 35: 181–186. https://doi.org/10.3760/cma.j.issn.0253-3766.2013.03.005.
96. Bonezzi K, Belotti D, North BJ, Ghilardi C, Borsotti P, Resovi A, Ubezio P, Riva A, Giavazzi R, Verdin E, and Taraboletti G. Inhibition of SIRT2 potentiates the anti-motility activity of taxanes: implications for anti-neoplastic combination therapies. *Neoplasia N.Y. N*. 2012; 14: 846–854. https://doi.org/10.1593/neo.12728.
97. Hayot C, Farinelle S, De Decker R, Decaestecker C, Darro F, Kiss R, and Van Damme M. In vitro pharmacological characterizations of the anti-angiogenic and anti-tumor cell migration properties mediated by microtubule-affecting drugs, with special emphasis on the organization of the actin cytoskeleton. *International Journal of Oncology*. 2002; 21: 417–425.
98. Iwahana M, Utoguchi N, Mayumi T, Goryo M, and Okada K. Drug resistance and P-glycoprotein expression in endothelial cells of newly formed capillaries induced by tumors. *Anticancer Research*. 1998; 18: 2977–2980.
99. Dicker AP, Williams TL, Iliakis G, and Grant DS. Targeting angiogenic processes by combination low-dose paclitaxel and radiation therapy. *American Journal of Clinical Oncology*. 2003; 26: e45–e53. https://doi.org/10.1097/01.COC.0000072504.22544.3C.
100. Bocci G, Nicolaou KC, and Kerbel RS. Protracted low-dose effects on human endothelial cell proliferation and survival in vitro reveal a selective antiangiogenic window for various chemotherapeutic drugs. *Cancer Research*. 2002; 62: 6938–6943.
101. Hata K, Osaki M, Dhar DK, Nakayama K, Fujiwaki R, Ito H, Nagasue N, and Miyazaki K. Evaluation of the antiangiogenic effect of Taxol in a human epithelial ovarian carcinoma cell line. *Cancer Chemotherapy and Pharmacology*. 2004; 53: 68–74. https://doi.org/10.1007/s00280-003-0693-x.
102. Belotti D, Vergani V, Drudis T, Borsotti P, Pitelli MR, Viale G, Giavazzi R, and Taraboletti G. The microtubule-affecting drug paclitaxel has antiangiogenic activity. *Clinical Cancer Research*. 1996; 2: 1843–1849.
103. Lau DH, Xue L, Young LJ, Burke PA, and Cheung AT. Paclitaxel (Taxol): an inhibitor of angiogenesis in a highly vascularized transgenic breast cancer. *Cancer Biotherapy and Radiopharmaceuticals*. 1999; 14: 31–36. https://doi.org/10.1089/cbr.1999.14.31.
104. Pasquier E, Carré M, Pourroy B, Camoin L, Rebaï O, Briand C, and D Braguer. Antiangiogenic activity of paclitaxel is associated with its cytostatic effect, mediated by the initiation but not completion of a mitochondrial apoptotic signaling pathway. *Molecular Cancer Therapeutics*. 2004; 3: 1301–1310.
105. Pasquier E, Honore S, Pourroy B, Jordan MA, Lehmann M, Briand C, and Braguer D. Antiangiogenic concentrations of paclitaxel induce an increase in microtubule dynamics in endothelial cells but not in cancer cells. *Cancer Research*. 2005; 65: 2433–2440. https://doi.org/10.1158/0008-5472.CAN-04-2624.
106. Bocci G, Francia G, Man S, Lawler J, and Kerbel RS. Thrombospondin 1, a mediator of the antiangiogenic effects of low-dose metronomic chemotherapy. *Proceedings of the National Academy of Sciences of the United States of America*. 2003; 100: 12917–12922. https://doi.org/10.1073/pnas.2135406100.
107. Muta M, Yanagawa T, Sai Y, Saji S, Suzuki E, Aruga T, Kuroi K, Matsumoto G, Toi M, and Nakashima E. Effect of low-dose Paclitaxel and docetaxel on endothelial progenitor cells. *Oncology*. 2009; 77: 182–191. https://doi.org/10.1159/000236016.
108. Klauber N, Parangi S, Flynn E, Hamel E, and D'Amato RJ. Inhibition of angiogenesis and breast cancer in mice by the microtubule inhibitors 2-methoxyestradiol and taxol. *Cancer Research*. 1997; 57: 81–86.
109. Lau DH, Xue L, Young LJ, Burke PA, and Cheung AT. Paclitaxel (Taxol): an inhibitor of angiogenesis in a highly vascularized transgenic breast cancer. *Cancer Biotherapy and Radiopharmaceuticals*. 1999; 14: 31–36. https://doi.org/10.1089/cbr.1999.14.31.
110. Guo L, Burke P, Lo SH, Gandour-Edwards R, and Lau D. Quantitative analysis of angiogenesis using confocal laser scanning microscopy. *Angiogenesis*. 2001; 4: 187–191. https://doi.org/10.1023/a:1014010801754.
111. Myoung H, Hong SD, Kim YY, Hong SP, and Kim MJ. Evaluation of the anti-tumor and anti-angiogenic effect of paclitaxel and thalidomide on the xenotransplanted oral squamous cell carcinoma. *Cancer Letters*. 2001; 163: 191–200. https://doi.org/10.1016/s0304-3835(00)00701-1.
112. Wang F, Cao Y, Liu H-Y, Xu S-F, and Han R. Anti-invasion and anti-angiogenesis effect of taxol and camptothecin on melanoma cells. *Journal of Asian Natural Products Research*. 2003; 5: 121–129. https://doi.org/10.1080/1028602021000054973.
113. Dordunoo SK, Jackson JK, Arsenault LA, Oktaba AM, Hunter WL, and Burt HM. Taxol encapsulation in poly (epsilon-caprolactone) microspheres. *Cancer Chemotherapy and Pharmacology*. 1995; 36: 279–282. https://doi.org/10.1007/BF00689043.

114. Belotti D, Vergani V, Drudis T, Borsotti P, Pitelli MR, Viale G, Giavazzi R, and Taraboletti G. The microtubule-affecting drug paclitaxel has antiangiogenic activity. *Clinical Cancer Research*. 1996; 2: 1843–1849.
115. Vacca A, Ribatti D, Iurlaro M, Merchionne F, Nico B, Ria R, and Dammacco F. Docetaxel versus paclitaxel for antiangiogenesis. *Journal of Hematotherapy & Stem Cell Research*. 2002; 11: 103–118. https://doi.org/10.1089/152581602753448577.
116. Jiang H, Tao W, Zhang M, Pan S, Kanwar JR, and Sun X. Low-dose metronomic paclitaxel chemotherapy suppresses breast tumors and metastases in mice. *Cancer Investigation*. 2010; 28: 74–84. https://doi.org/10.3109/07357900902744510.
117. Damber J-E, Vallbo C, Albertsson P, Lennernäs B, and Norrby K. The anti-tumour effect of low-dose continuous chemotherapy may partly be mediated by thrombospondin. *Cancer Chemotherapy and Pharmacology*. 2006; 58: 354–360. https://doi.org/10.1007/s00280-005-0163-8.
118. Zhang M, Tao W, Pan S, Sun X, and Jiang H. Low-dose metronomic chemotherapy of paclitaxel synergizes with cetuximab to suppress human colon cancer xenografts. *Anticancer Drugs*. 2009; 20: 355–363. https://doi.org/10.1097/CAD.0b013e3283299f36.
119. Thijssen VLJL, Brandwijk RJMGE, Dings RPM, and Griffioen AW. Angiogenesis gene expression profiling in xenograft models to study cellular interactions. *Experimental Cell Research*. 2004; 299: 286–293. https://doi.org/10.1016/j.yexcr.2004.06.014.
120. Bocci G, Di Paolo A, and Danesi R. The pharmacological bases of the antiangiogenic activity of paclitaxel. *Angiogenesis*. 2013; 16: 481–492. https://doi.org/10.1007/s10456-013-9334-0.
121. Byrne JD, Betancourt T, and Brannon-Peppas L. Active targeting schemes for nanoparticle systems in cancer therapeutics. *Advanced Drug Delivery Reviews*. 2008; 60: 1615–1626. https://doi.org/10.1016/j.addr.2008.08.005.
122. Straubinger RM, Arnold RD, Zhou R, Mazurchuk R, and Slack JE. Antivascular and antitumor activities of liposome-associated drugs. *Anticancer Research*. 2004; 24: 397–404.
123. Sharma A, Sharma US, and Straubinger RM. Paclitaxel-liposomes for intracavitary therapy of intraperitoneal P388 leukemia. *Cancer Letter*. 1996; 107: 265–272. https://doi.org/10.1016/0304-3835(96)04380-7.
124. Poste G, Bucana C, Raz A, Bugelski P, Kirsh R, and Fidler IJ. Analysis of the fate of systemically administered liposomes and implications for their use in drug delivery. *Cancer Research*. 1982; 42: 1412–1422.
125. Thurston G, McLean JW, Rizen M, Baluk P, Haskell A, Murphy TJ, Hanahan D, and McDonald DM. Cationic liposomes target angiogenic endothelial cells in tumors and chronic inflammation in mice. *Journal of Clinical Investigation*. 1998; 101: 1401–1413. https://doi.org/10.1172/JCI965.
126. Schmitt-Sody M, Strieth S, Krasnici S, Sauer B, Schulze B, Teifel M, Michaelis U, Naujoks K, and Dellian M. Neovascular targeting therapy: paclitaxel encapsulated in cationic liposomes improves antitumoral efficacy. *Clincal Cancer Research*. 2003; 9: 2335–2341.
127. Kunstfeld R, Wickenhauser G, Michaelis U, Teifel M, Umek W, Naujoks K, Wolff K, and Petzelbauer P. Paclitaxel encapsulated in cationic liposomes diminishes tumor angiogenesis and melanoma growth in a "humanized" scid mouse model. *Journal of Investigative Dermatology*. 2003; 120: 476–482. https://doi.org/10.1046/j.1523-1747.2003.12057.x.
128. Bode C, Trojan L, Weiss C, Kraenzlin B, Michaelis U, Teifel M, Alken P, and Michel MS. Paclitaxel encapsulated in cationic liposomes: a new option for neovascular targeting for the treatment of prostate cancer. *Oncology Reports*. 2009; 22: 321–326.
129. Adeyemi SA, Kumar P, Choonara YE, and Pillay V. Stealth properties of nanoparticles against cancer: surface modification of NPs for passive targeting to human cancer tissue in zebrafish embryos, in: YV Pathak (Ed.), Surface Modification of Nanoparticles for Targeted Drug Delivery, Springer International Publishing, Cham, 2019, 99–124. https://doi.org/10.1007/978-3-030-06115-9_5.
130. Moghimi SM and Patel HM. Opsonophagocytosis of liposomes by peritoneal macrophages and bone marrow reticuloendothelial cells. *Biochimica et Biophysica Acta*. 1992; 1135: 269–274. https://doi.org/10.1016/0167-4889(92)90230-9.
131. Huang Y, Chen X-M, Zhao B-X, Ke X-Y, Zhao B-J, Zhao X, Wang Y, Zhang X, and Zhang Q. Antiangiogenic activity of sterically stabilized liposomes containing paclitaxel (SSL-PTX): in vitro and in vivo. *AAPS PharmSciTech*. 2010; 11: 752–759. https://doi.org/10.1208/s12249-010-9430-z.
132. Koziara JM, Whisman TR, Tseng MT, and Mumper RJ. In vivo efficacy of novel paclitaxel nanoparticles in paclitaxel-resistant human colorectal tumors. *Journal of Controlled Release*. 2006; 112: 312–319. https://doi.org/10.1016/j.jconrel.2006.03.001.
133. Hammady T, Rabanel J-M, Dhanikula RS, Leclair G, and Hildgen P. Functionalized nanospheres loaded with anti-angiogenic drugs: cellular uptake and angiosuppressive efficacy. *Journal of Arbeitsgemeinschaft fur Pharmazeutische Verfahrenstechnik*. 2009; 72: 418–427. https://doi.org/10.1016/j.ejpb.2009.01.007.
134. Hammady T, Rabanel J-M, Dhanikula RS, Leclair G, and Hildgen P. Functionalized nanospheres loaded with anti-angiogenic drugs: cellular uptake and angiosuppressive efficacy. *Recent Pat Anticancer Drug Discovery*. 2009; 72: 418–427. https://doi.org/10.1016/j.ejpb.2009.01.007.
135. Fu Q, Sun J, Zhang W, Sui X, Yan Z, and He Z. Nanoparticle albumin-bound (NAB) technology is a promising method for anti-cancer drug delivery. *Recent Patents Anticancer Drug Discovery*. 2009; 4: 262–272. https://doi.org/10.2174/157489209789206869.
136. Miele E, Spinelli GP, Miele E, Tomao F, and Tomao S. Albumin-bound formulation of paclitaxel (Abraxane ABI-007) in the treatment of breast cancer. *International Journal of Nanomedicine*. 2009; 4: 99–105. https://doi.org/10.2147/ijn.s3061.
137. Yamamoto Y, Kawano I, and Iwase H. Nab-paclitaxel for the treatment of breast cancer: efficacy, safety, and approval. *OncoTargets and Therapy*. 2011; 4: 123–136. https://doi.org/10.2147/OTT.S13836.
138. Ng SSW, Sparreboom A, Shaked Y, Lee C, Man S, Desai N, Soon-Shiong P, Figg WD, and Kerbel RS. Influence of formulation vehicle on metronomic taxane chemotherapy: albumin-bound versus cremophor EL-based paclitaxel.

Clinical Cancer Research. 2006; 12: 4331–4338. https://doi.org/10.1158/1078-0432.CCR-05-2762.
139. Lee SJ, Ghosh SC, Han HD, Stone RL, Bottsford-Miller J, Shen DY, Auzenne EJ, Lopez-Araujo A, Lu C, Nishimura M, Pecot CV, Zand B, Thanapprapasr D, Jennings NB, Kang Y, Huang J, Hu W, Klostergaard J, and Sood AK. Metronomic activity of CD44-targeted hyaluronic acid-paclitaxel in ovarian carcinoma. *Clinical Cancer Research*. 2012; 18: 4114–4121. https://doi.org/10.1158/1078-0432.CCR-11-3250.
140. Liang L, Lin S-W, Dai W, Lu J-K, Yang T-Y, Xiang Y, Zhang Y, Li R-T, and Zhang Q. Novel cathepsin B-sensitive paclitaxel conjugate: Higher water solubility, better efficacy and lower toxicity. *Journal of Controlled Release*. 2012; 160: 618–629. https://doi.org/10.1016/j.jconrel.2012.02.020.
141. Moes J, Koolen S, Huitema A, Schellens J, Beijnen J, and Nuijen B. Development of an oral solid dispersion formulation for use in low-dose metronomic chemotherapy of paclitaxel. *European Journal of Pharmaceutics and Biopharmaceutics*. 2013; 83: 87–94. https://doi.org/10.1016/j.ejpb.2012.09.016.
142. Ai B, Bie Z, Zhang S, and Li A. Paclitaxel targets VEGF-mediated angiogenesis in ovarian cancer treatment. *American Journal of Cancer Research*. 2016; 6: 1624–1635.
143. Kerbel RS. Strategies for improving the clinical benefit of antiangiogenic drug based therapies for breast cancer. *Journal of Mammary Gland Biology and Neoplasia*. 2012; 17: 229–239. https://doi.org/10.1007/s10911-012-9266-0.
144. Rodrigues G and Sanatani M. Age and comorbidity considerations related to radiotherapy and chemotherapy administration. *Seminars in Radiation Oncology*. 2012; 22: 277–283. https://doi.org/10.1016/j.semradonc.2012.05.004.
145. Emmenegger U, Francia G, Chow A, Shaked Y, Kouri A, Man S, and Kerbel RS. Tumors that acquire resistance to low-dose metronomic cyclophosphamide retain sensitivity to maximum tolerated dose cyclophosphamide. *Neoplasia New York, NY*. 2011; 13: 40–48. https://doi.org/10.1593/neo.101174.
146. Allegrini G, Di Desidero T, Barletta MT, Fioravanti A, Orlandi P, Canu B, Chericoni S, Loupakis F, Di Paolo A, Masi G, Fontana A, Lucchesi S, Arrighi G, Giusiani M, Ciarlo A, Brandi G, Danesi R, Kerbel RS, Falcone A, and Bocci G. Clinical, pharmacokinetic and pharmacodynamic evaluations of metronomic UFT and cyclophosphamide plus celecoxib in patients with advanced refractory gastrointestinal cancers. *Angiogenesis*. 2012; 15: 275–286. https://doi.org/10.1007/s10456-012-9260-6.
147. Allegrini G, Falcone A, Fioravanti A, Barletta MT, Orlandi P, Loupakis F, Cerri E, Masi G, Di Paolo A, Kerbel RS, Danesi R, Del Tacca M, and Bocci G. A pharmacokinetic and pharmacodynamic study on metronomic irinotecan in metastatic colorectal cancer patients. *British Journal of Cancer*. 2008; 98: 1312–1319. https://doi.org/10.1038/sj.bjc.6604311.
148. Daniel EAK and Hamblin MR. Chitin and chitosan: production and application of versatile biomedical nanomaterials. *International Journal of Advanced Research*. 2016; 4 (3): 411-427.
149. Brigham C. Chapter 3.22 - Biopolymers: Biodegradable Alternatives to Traditional Plastics, In: Török B and Dransfield T (Eds.), Elsevier, Green Chem, 2018, 753–770. https://doi.org/10.1016/B978-0-12-809270-5.00027-3.
150. Uto T, Idenoue S, Yamamoto K, and Kadokawa J-I. Understanding dissolution process of chitin crystal in ionic liquids: Theoretical study. *Physical Chemistry Chemical Physics*. 2018; 20: 20669–20677. https://doi.org/10.1039/c8cp02749h.
151. Zhu H, Luo W, Ciesielski PN, Fang Z, Zhu JY, Henriksson G, Himmel ME, and Hu L. Wood-derived materials for green electronics, biological devices, and energy applications. *Chemical Reviews*. 2016; 116: 9305–9374. https://doi.org/10.1021/acs.chemrev.6b00225.
152. Muzzarelli RAA, Muzzarelli C, Cosani A, and Terbojevich M. 6-Oxychitins, novel hyaluronan-like regiospecifically carboxylated chitins. *Carbohydrate Polymer*. 1999; 39: 361–367. https://doi.org/10.1016/S0144-8617(99)00027-2.
153. Ying G, Xiong W, Wang H, Sun Y, and Liu H. Preparation, water solubility and antioxidant activity of branched-chain chitosan derivatives. *Carbohydrate Polymer*. 2011; 83: 1787–1796. https://doi.org/10.1016/j.carbpol.2010.10.037.
154. Chawla SP, Kanatt SR, and Sharma AK. Chitosan, in: Ramawat KG, Mérillon J-M (Eds.), Polysaccharides - Bioactivity and Biotechnology, Springer International Publishing, Cham, 2015, 219–246. https://doi.org/10.1007/978-3-319-16298-0_13.
155. de Queiroz Antonino RSCM, Lia Fook BRP, de Oliveira Lima VA, de Farias Rached RÍ, Lima EPN, da Silva Lima RJ, Peniche Covas CA, and Lia Fook MV. Preparation and characterization of chitosan obtained from shells of shrimp (*Litopenaeus vannamei* Boone). *Marine Drugs*. 2017; 15: https://doi.org/10.3390/md15050141.
156. Hejazi R and Amiji M. Chitosan-based gastrointestinal delivery systems. *Journal of Controlled Release*. 2003; 89: 151–165. https://doi.org/10.1016/s0168-3659(03)00126-3.
157. Mourya VK and Inamdar NN. Chitosan-modifications and applications: opportunities galore. *Reactive & Functional Polymers*. 2008; 68: 1013–1051. https://doi.org/10.1016/j.reactfunctpolym.2008.03.002.
158. Kyzas GZ and Bikiaris DN. Recent modifications of chitosan for adsorption applications: a critical and systematic review. *Marine Drugs*. 2015; 13: 312–337. https://doi.org/10.3390/md13010312.
159. Kulig D, Zimoch-Korzycka A, Jarmoluk A, and Marycz K. Study on alginate–chitosan complex formed with different polymers ratio. *Polymers*. 2016; 8: https://doi.org/10.3390/polym8050167.
160. Arriagada R, Bergman B, Dunant A, Le Chevalier T, Pignon J-P, and Vansteenkiste J. International adjuvant lung cancer trial collaborative group, cisplatin-based adjuvant chemotherapy in patients with completely resected non-small-cell lung cancer. *The New England Journal of Medicine*. 2004; 350: 351–360. https://doi.org/10.1056/NEJMoa031644.
161. Gnavi S, Barwig C, Freier T, Haastert-Talini K, Grothe C, and Geuna S. Chapter One - The Use of Chitosan-Based Scaffolds to Enhance Regeneration in the Nervous System, in: S Geuna, Perroteau I, Tos P, Battiston B (Eds.), International Review of Neurobiology, Academic Press, 2013, 1–62. https://doi.org/10.1016/B978-0-12-420045-6.00001-8.
162. Hamman JH. Chitosan based polyelectrolyte complexes as potential carrier materials in drug delivery systems.

Marine Drugs. 2010; 8: 1305–1322. https://doi.org/10.3390/md8041305.
163. Quiñones JP, Peniche H, and Peniche C. Chitosan based self-assembled nanoparticles in drug delivery. *Polymers*. 2018; 10: https://doi.org/10.3390/polym10030235.
164. Mao S, Sun W, and Kissel T. Chitosan-based formulations for delivery of DNA and siRNA. *Advanced Drug Delivery Reviews*. 2010; 62: 12–27. https://doi.org/10.1016/j.addr.2009.08.004.
165. Dang NTT, Chau TTL, Duong HV, Le HT, Tran TTV, Le TQ, Vu TP, Nguyen CD, Nguyen LV, and Nguyen T-D. Water-soluble chitosan-derived sustainable materials: towards filaments, aerogels, microspheres, and plastics. *Soft Matter*. 2017; 13: 7292–7299. https://doi.org/10.1039/c7sm01292f.
166. Zhang Y, Chan JW, Moretti A, and Uhrich KE. Designing polymers with sugar-based advantages for bioactive delivery applications. *Journal of Controlled Release*. 2015; 219: 355–368. https://doi.org/10.1016/j.jconrel.2015.09.053.
167. Yeh T-H, Hsu L-W, Tseng MT, Lee P-L, Sonjae K, Ho Y-C, and Sung H-W. Mechanism and consequence of chitosan-mediated reversible epithelial tight junction opening. *Biomaterials*. 2011; 32: 6164–6173. https://doi.org/10.1016/j.biomaterials.2011.03.056.
168. Rosenthal R, Günzel D, Finger C, Krug SM, Richter JF, Schulzke J-D, Fromm M, and Amasheh S. The effect of chitosan on transcellular and paracellular mechanisms in the intestinal epithelial barrier. *Biomaterials*. 2012; 33: 2791–2800. https://doi.org/10.1016/j.biomaterials.2011.12.034.
169. Smith JM, Dornish M, and Wood EJ. Involvement of protein kinase C in chitosan glutamate-mediated tight junction disruption. *Biomaterials*. 2005; 26: 3269–3276. https://doi.org/10.1016/j.biomaterials.2004.06.020.
170. Smith J, Wood E, and Dornish M. Effect of chitosan on epithelial cell tight junctions. *Pharmaceutical Research*. 2004; 21: 43–49. https://doi.org/10.1023/b:pham.0000012150.60180.e3.
171. Azuma K, Osaki T, Minami S, and Okamoto Y. Anticancer and anti-inflammatory properties of chitin and chitosan oligosaccharides. *Journal of Functional Biomaterial*. 2015; 6: 33–49. https://doi.org/10.3390/jfb6010033.
172. Verlee A, Mincke S, and Stevens CV. Recent developments in antibacterial and antifungal chitosan and its derivatives. *Carbohydrate Polymers*. 2017; 164: 268–283. https://doi.org/10.1016/j.carbpol.2017.02.001.
173. Kara F, Aksoy EA, Yuksekdag Z, Hasirci N, and Aksoy S. Synthesis and surface modification of polyurethanes with chitosan for antibacterial properties. *Carbohydrate Polymers*. 2014; 112: 39–47. https://doi.org/10.1016/j.carbpol.2014.05.019.
174. Ngo D-H, Vo T-S, Ngo D-N, Kang K-H, Je J-Y, Pham HND, Byun H-G, and Kim S-K. Biological effects of chitosan and its derivatives. *Food Hydrocolloids*. 2015; 51: 200–216. https://doi.org/10.1016/j.foodhyd.2015.05.023.
175. Liu S-H, Chang Y-H, and Chiang M-T. Chitosan reduces gluconeogenesis and increases glucose uptake in skeletal muscle in streptozotocin-induced diabetic rats. *Journal of Agricultural and Food Chemistry*. 2010; 58: 5795–5800. https://doi.org/10.1021/jf100662r.
176. Musalli AH, Talukdar PD, Roy P, Kumar P, and Wong TW. Folate-induced nanostructural changes of oligochitosan nanoparticles and their fate of cellular internalization by melanoma. *Carbohydrate Polymers*. 2020; 244: 116488. https://doi.org/10.1016/j.carbpol.2020.116488.
177. Wang F, Wang Y, Q Ma, Cao Y, and Yu B. Development and characterization of folic acid-conjugated chitosan nanoparticles for targeted and controlled delivery of gemcitabinein lung cancer therapeutics. *Artificial Cells, Nanomedicine, and Biotechnology*. 2017; 45: 1530–1538. https://doi.org/10.1080/21691401.2016.1260578.
178. Rasul RM, Tamilarasi Muniandy M, Zakaria Z, Shah K, Chee CF, Dabbagh A, Rahman NA, and Wong TW. A review on chitosan and its development as pulmonary particulate anti-infective and anti-cancer drug carriers. *Carbohydrate Polymers*. 2020; 250: 116800. https://doi.org/10.1016/j.carbpol.2020.116800.
179. Shah K, Chan LW, and Wong TW. Critical physicochemical and biological attributes of nanoemulsions for pulmonary delivery of rifampicin by nebulization technique in tuberculosis treatment. *Drug Delivery*. 2017; 24: 1631–1647. https://doi.org/10.1080/10717544.2017.1384298.
180. Chachuli SHM, Nawaz A, Shah K, Naharudin I, and Wong TW. In vitro investigation of influences of chitosan nanoparticles on fluorescein permeation into alveolar macrophages. *Pharmaceutical Research*. 2016; 33: 1497–1508. https://doi.org/10.1007/s11095-016-1893-5.
181. Paumier A and Le Péchoux C. Radiotherapy in small-cell lung cancer: where should it go?. *Lung Cancer Amsterdam. Netherlands*. 2010; 69: 133–140. https://doi.org/10.1016/j.lungcan.2010.04.019.
182. Gelderblom H, Verweij J, Nooter K, and Sparreboom A. Cremophor EL: the drawbacks and advantages of vehicle selection for drug formulation. *European Journal of Cancer*. 1990; 37(2001): 1590–1598. https://doi.org/10.1016/s0959-8049(01)00171-x.
183. Nakano J, Huang C, Liu D, Masuya D, Nakashima T, Yokomise H, Ueno M, Wada H, and Fukushima M, Mortada SM. Evaluations of biomarkers associated with 5-FU sensitivity for non-small-cell lung cancer patients postoperatively treated with UFT. British Journal of Cancer. 2006; 95(5): 607–615. https://doi.org/10.1038/sj.bjc.6603297.
184. Darquenne C. Aerosol deposition in health and disease. *Journal of Aerosol Medicine and Pulmonary Drug Delivery*. 2012; 25: 140–147. https://doi.org/10.1089/jamp.2011.0916.
185. Liu K, Chen W, Yang T, Wen B, Ding D, Keidar M, Tang J, and Zhang W. Paclitaxel and quercetin nanoparticles co-loaded in microspheres to prolong retention time for pulmonary drug delivery, *International Journal of Nanomedicine*. 2017; 12: 8239–8255. https://doi.org/10.2147/IJN.S147028.
186. Rosière R, Woensel MV, Gelbcke M, Mathieu V, Hecq J, Mathivet T, Vermeersch M, Antwerpen PV, Amighi K, and Wauthoz N. New folate-grafted chitosan derivative to improve delivery of paclitaxel-loaded solid lipid nanoparticles for lung tumor therapy by inhalation. *Molecular Pharmaceutics*. 2018; 15(3): 899–910. https://doi.org/10.1021/acs.molpharmaceut.7b00846.

187. Silva MC, Silva AS, Fernandez-Lodeiro J, Casimiro T, Lodeiro C, and Aguiar-Ricardo A. Supercritical CO_2-assisted spray drying of strawberry-like gold-coated magnetite nanocomposites in chitosan powders for inhalation. *Mater. Basel Switzerland*. 2017; 10: https://doi.org/10.3390/ma10010074.
188. Almutairi FM, Abd-Rabou AA, and Mohamed MS. Raloxifene-encapsulated hyaluronic acid-decorated chitosan nanoparticles selectively induce apoptosis in lung cancer cells. *Bioorganic & Medicinal Chemistry*. 2019; 27: 1629–1638. https://doi.org/10.1016/j.bmc.2019.03.004.
189. Iozzo RV and Schaefer L. Proteoglycan form and function: A comprehensive nomenclature of proteoglycans. *International Society for Matrix Biology*. 2015; 42: 11–55. https://doi.org/10.1016/j.matbio.2015.02.003.
190. Ran S, Downes A, and Thorpe PE. Increased exposure of anionic phospholipids on the surface of tumor blood vessels. *Cancer Research*. 2002; 62: 6132–6140.
191. Honary S and Zahir F. Effect of zeta potential on the properties of nano-drug delivery systems - a review (part 2), trop. *Journal of Pharmacy Research*. 2013; 12: 265–273. https://doi.org/10.4314/tjpr.v12i2.20.
192. Adeyemi SA, Choonara YE, P Kumar, du Toit LC, and Pillay V. Synthesis and in vitro characterization of a pH-responsive chitosan- polyethylenimine nanosystem for the delivery of therapeutic proteins. *Journal of Drug Delivery Science and Technology*. 2017; 39: 266–276. https://doi.org/10.1016/j.jddst.2017.03.022.
193. Chokradjaroen C, Rujiravanit R, Watthanaphanit A, Theeramunkong S, N Saito, Yamashita K, and Arakawa R. Enhanced degradation of chitosan by applying plasma treatment in combination with oxidizing agents for potential use as an anticancer agent. *Carbohydrate Polymers*. 2017; 167: 1–11. https://doi.org/10.1016/j.carbpol.2017.03.006.
194. Gibot L, Chabaud S, Bouhout S, Bolduc S, Auger FA, and Moulin VJ. Anticancer properties of chitosan on human melanoma are cell line dependent. *International Journal of Biological Macromolecules*. 2015; 72: 370–379. https://doi.org/10.1016/j.ijbiomac.2014.08.033.
195. Oberemko A, Salaberria AM, Saule R, Saulis G, Kaya M, Labidi J, and Baublys V. Physicochemical and in vitro cytotoxic properties of chitosan from mushroom species (*Boletus bovinus and Laccaria laccata*). *Carbohydrate Polymers*. 2019; 221: 1–9. https://doi.org/10.1016/j.carbpol.2019.05.073.
196. Dario Rafael O-H, Luis Fernándo Z-G, Abraham P-T, Pedro Alberto V-L, Guadalupe G-S, and Pablo PJ. Production of chitosan-oligosaccharides by the chitin-hydrolytic system of *Trichoderma harzianum* and their antimicrobial and anticancer effects. *Carbohydrate Research*. 2019; 486: 107836. https://doi.org/10.1016/j.carres.2019.107836.
197. Bharathi D, Ranjithkumar R, Chandarshekar B, and Bhuvaneshwari V. Bio-inspired synthesis of chitosan/copper oxide nanocomposite using rutin and their anti-proliferative activity in human lung cancer cells. *International Journal of Biological Macromolecules*. 2019; 141: 476–483. https://doi.org/10.1016/j.ijbiomac.2019.08.235.
198. Chi J, Jiang Z, Qiao J, Peng Y, Liu W, and Han B. Synthesis and anti-metastasis activities of norcantharidin-conjugated carboxymethyl chitosan as a novel drug delivery system. *Carbohydrate Polymers*. 2019; 214: 80–89. https://doi.org/10.1016/j.carbpol.2019.03.026.
199. Cirillo G, Vittorio O, Kunhardt D, Valli E, Voli F, Farfalla A, Curcio M, Spizzirri UG, and Hampel S. Combining carbon nanotubes and chitosan for the vectorization of methotrexate to lung cancer cells. *Mater. Basel Switzreland* 2019; 12: https://doi.org/10.3390/ma12182889.
200. Li W, Hu X, Wang S, Xing Y, Wang H, Nie Y, Liu T, and Song K. Multiple comparisons of three different sources of biomaterials in the application of tumor tissue engineering in vitro and in vivo. *International Journal of Biological Macromolecules*. 2019; 130: 166–176. https://doi.org/10.1016/j.ijbiomac.2019.02.136.
201. Pandey P, Dua K, and Dureja H. Erlotinib loaded chitosan nanoparticles: formulation, physicochemical characterization and cytotoxic potential. *International Journal of Biological Macromolecules*. 2019; 139: 1304–1316. https://doi.org/10.1016/j.ijbiomac.2019.08.084.
202. Tao L, Jiang J, Gao Y, Wu C, and Liu Y. Biodegradable alginate-chitosan hollow nanospheres for codelivery of doxorubicin and paclitaxel for the effect of human lung cancer A549 cells. *BioMed Research International*. 2018; 2018: 4607945. https://doi.org/10.1155/2018/4607945.
203. Patel B, Gupta N, and Ahsan F. Particle engineering to enhance or lessen particle uptake by alveolar macrophages and to influence the therapeutic outcome. *European Journal of Pharmaceutics and Biopharmaceutics*. 2015; 89: 163–174. https://doi.org/10.1016/j.ejpb.2014.12.001.
204. A Dabbagh, NH Abu Kasim, CH Yeong, TW Wong, and N Abdul Rahman. Critical parameters for particle-based pulmonary delivery of chemotherapeutics. *Journal of Aerosol Medicine and Pulmonary Drug Delivery*. 2018; 31: 139–154. https://doi.org/10.1089/jamp.2017.1382.
205. A Jain, A Jain, NK Garg, RK Tyagi, B Singh, OP Katare, TJ Webster, and V Soni. Surface engineered polymeric nanocarriers mediate the delivery of transferrin–methotrexate conjugates for an improved understanding of brain cancer. *Acta Biomaterialia*. 2015; 24: 140–151. https://doi.org/10.1016/j.actbio.2015.06.027.
206. Tie Y, Zheng H, He Z, Yang J, Shao B, Liu L, Luo M, Yuan X, Liu Y, Zhang X, Li H, Wu M, and Wei X. Targeting folate receptor β positive tumor-associated macrophages in lung cancer with a folate-modified liposomal complex. *Signal Transduction and Targeted Therapy*. 2020; 5: 1–15. https://doi.org/10.1038/s41392-020-0115-0.
207. O'Shannessy DJ, Yu G, Smale R, Fu Y-S, Singhal S, Thiel RP, Somers EB, and Vachani A. Folate receptor alpha expression in lung cancer: diagnostic and prognostic significance. *Oncotarget*. 2012; 3: 414–425. https://doi.org/10.18632/oncotarget.489.
208. Das M and Sahoo SK. Folate decorated dual drug loaded nanoparticle: role of curcumin in enhancing therapeutic potential of nutlin-3a by reversing multidrug resistance. *PLoS ONE*. 2012; 7: e32920. https://doi.org/10.1371/journal.pone.0032920.
209. Adeyemi SA, Choonara YE, Kumar P, du Toit LC, Marimuthu T, Kondiah PPD, and Pillay V. Folate-decorated, endostatin-loaded, nanoparticles for anti-proliferative chemotherapy in esophaegeal squamous cell

carcinoma. *Biomedicine Pharmacotherapy*. 2019; 119: 109450. https://doi.org/10.1016/j.biopha.2019.109450.

210. Weitman SD, Lark RH, Coney LR, Fort DW, Frasca V, Zurawski VR, and Kamen BA. Distribution of the folate receptor GP38 in normal and malignant cell lines and tissues. *Cancer Research*. 1992; 52: 3396–3401.
211. Li K, Liang N, Yang H, Liu H, and Li S. Temozolomide encapsulated and folic acid decorated chitosan nanoparticles for lung tumor targeting: improving therapeutic efficacy both in vitro and in vivo.*Oncotarget*. 2017; 8: 111318. https://doi.org/10.18632/oncotarget.22791.
212. Zhang Y, Guo L, Roeske RW, Antony AC, and Jayaram HN. Pteroyl-gamma-glutamate-cysteine synthesis and its application in folate receptor-mediated cancer cell targeting using folate-tethered liposomes. *Analytical Biochemistry*. 2004; 332: 168–177. https://doi.org/10.1016/j.ab.2004.05.034.
213. Antony AC. The biological chemistry of folate receptors. *Blood*. 1992; 79: 2807–2820.
214. Antony AC. Folate receptors. *Annual Review of Nutrition*. 1996; 16: 501–521. https://doi.org/10.1146/annurev.nu.16.070196.002441.
215. Lee RJ and Low PS. Delivery of liposomes into cultured KB cells via folate receptor-mediated endocytosis. *Journal of Biological Chemistry*. 1994; 269: 3198–3204.
216. Lee RJ and Low PS. Folate-mediated tumor cell targeting of liposome-entrapped doxorubicin in vitro. *Biochimica et Biophysica Acta*. 1995; 1233: 134–144. https://doi.org/10.1016/0005-2736(94)00235-h.
217. Ross JF, Chaudhuri PK, and Ratnam M. Differential regulation of folate receptor isoforms in normal and malignant tissues in vivo and in established cell lines. Physiologic and clinical implications. *Cancer*. 1994; 73: 2432–2443. https://doi.org/10.1002/1097-0142(19940501)73:9<2432::aid-cncr2820730929>3.0.co;2-s.
218. Kamaly N, Kalber T, Thanou M, Bell JD, and Miller AD. Folate receptor targeted bimodal liposomes for tumor magnetic resonance imaging. *Bioconjugate Chemistry*. 2009; 20: 648–655. https://doi.org/10.1021/bc8002259.
219. Mansouri S, Cuie Y, Winnik F, Shi Q, Lavigne P, Benderdour M, Beaumont E, and Fernandes JC. Characterization of folate-chitosan-DNA nanoparticles for gene therapy. *Biomaterials*. 2006; 27: 2060–2065. https://doi.org/10.1016/j.biomaterials.2005.09.020.
220. Zheng Y, Cai Z, Song X, Chen Q, Bi Y, Li Y, and Hou S. Preparation and characterization of folate conjugated N-trimethyl chitosan nanoparticles as protein carrier targeting folate receptor: in vitro studies. *Journal of Drug Targeting*. 2009; 17: 294–303. https://doi.org/10.1080/10611860902737920.
221. Liang B, He M-L, Xiao Z-P, Li Y, Chan C-Y, Kung H-F, Shuai X-T, and Peng Y. Synthesis and characterization of folate-PEG-grafted-hyperbranched-PEI for tumor-targeted gene delivery. *Biochemical and Biophysical Research Communications*. 2008; 367: 874–880. https://doi.org/10.1016/j.bbrc.2008.01.024.
222. Hwa Kim S, Hoon Jeong J, Joe CO, and Gwan Park T. Folate receptor mediated intracellular protein delivery using PLL-PEG-FOL conjugate. *Journal of Controlled Release*. 2005; 103: 625–634. https://doi.org/10.1016/j.jconrel.2005.01.006.
223. Hwa Kim S, Hoon Jeong J, Chul Cho K, Wan Kim S, and Gwan Park T. Target-specific gene silencing by siRNA plasmid DNA complexed with folate-modified poly(ethylenimine). *Journal of Controlled Release*. 2005; 104: 223–232. https://doi.org/10.1016/j.jconrel.2005.02.006.
224. Benns JM, Mahato RI, and SW Kim. Optimization of factors influencing the transfection efficiency of folate-PEG-folate-graft-polyethylenimine. *Journal of Controlled Release*. 2002; 79: 255–269. https://doi.org/10.1016/s0168-3659(01)00513-2.
225. Benns JM, Maheshwari A, Furgeson DY, Mahato RI, and Kim SW. Folate-PEG-folate-graft-polyethylenimine-based gene delivery. *Journal of Drug Targeting*. 2001; 9: 123–139. https://doi.org/10.3109/10611860108997923.
226. Cheng H, Zhu J-L, Zeng X, Jing Y, Zhang X-Z, and Zhuo R-X. Targeted gene delivery mediated by folate-polyethylenimine-block-poly(ethylene glycol) with receptor selectivity. *Bioconjugate Chemistry*. 2009; 20: 481–487. https://doi.org/10.1021/bc8004057.
227. Kim SH, Mok H, Jeong JH, Kim SW, and Park TG. Comparative evaluation of target-specific GFP gene silencing efficiencies for antisense ODN, synthetic siRNA, and siRNA plasmid complexed with PEI-PEG-FOL conjugate. *Bioconjugate Chemistry*. 2006; 17: 241–244. https://doi.org/10.1021/bc050289f.
228. Oupický D, Ogris M, and Seymour LW. Development of long-circulating polyelectrolyte complexes for systemic delivery of genes. *Journal of Drug Targeting*. 2002; 10: 93–98. https://doi.org/10.1080/10611860290016685.
229. Nguyen HK, Lemieux P, Vinogradov SV, Gebhart CL, Guérin N, Paradis G, Bronich TK, Alakhov VY, and AV Kabanov. Evaluation of polyether-polyethyleneimine graft copolymers as gene transfer agents. *Gene Therapy*. 2000; 7: 126–138. https://doi.org/10.1038/sj.gt.3301052.
230. Choi YH, Liu F, Kim JS, Choi YK, Park JS, and Kim SW. Polyethylene glycol-grafted poly-L-lysine as polymeric gene carrier. *Journal of Controlled Release*. 1998; 54: 39–48. https://doi.org/10.1016/s0168-3659(97)00174-0.
231. Singh RP, Sharma G, Sonali SS, Bharti S, Pandey BL, Koch B, and Muthu MS. Chitosan-folate decorated carbon nanotubes for site specific lung cancer delivery. *Materials Science and Engineering C*. 2017; 77: 446–458. https://doi.org/10.1016/j.msec.2017.03.225.
232. Sidaway P. COVID-19 and cancer: what we know so far. *Nature Reviews Clinical Oncology*. 2020; 17: 336–336. https://doi.org/10.1038/s41571-020-0366-2.
233. Milewska A, Kaminski K, Ciejka J, Kosowicz K, Zeglen S, Wojarski J, Nowakowska M, Szczubiałka K, and Pyrc K. HTCC: Broad range inhibitor of coronavirus entry. *PLoS ONE*. 2016; 11: e0156552. https://doi.org/10.1371/journal.pone.0156552.
234. Chokradjaroen C, Theeramunkong S, Yui H, Saito N, and Rujiravanit R. Cytotoxicity against cancer cells of chitosan oligosaccharides prepared from chitosan powder degraded by electrical discharge plasma. *Carbohydrate Polymer*. 2018; 201: 20–30. https://doi.org/10.1016/j.carbpol.2018.08.037.

27

Anti-angiogenic Therapy for Lung Cancer: Focus on Bioassay Development in a Quest for Anti-VEGF Drugs

Urvashi Bhati[1], Colin J. Barrow[2], Jagat R. Kanwar[3,#], Arnab Kapat[1], Venkata Ramana[1], and Rupinder K. Kanwar[3,*]
[1]Reliance Life Sciences, Dhirubhai Ambani Life Sciences Center, Thane-Belapur Road, Maharashtra, India
[2]School of Life and Environmental Sciences, Centre for Chemistry and Biotechnology, Deakin University, Victoria, Australia
[3]School of Medicine, Faculty of Health, Institute for Mental and Physical Health and Clinical Translation (IMPACT), Deakin University, Victoria, Australia
[*]Department of Translational Medicine Centre All India Institute of Medical Sciences (AIIMS), Madhya Pradesh, India
[#]Department Of Biochemistry, All India Institute of Medical Sciences (AIIMS), Madhya Pradesh, India

27.1 Introduction

Lung cancer is the most prevalent malignancy worldwide. According to the World Health Organization's 2018 fact sheet, with 2.09 million cases and 1.76 million deaths globally, lung cancer topped both the lists of the (i) most common cancers, and (ii) most common causes of cancer death (https://www.who.int/news-room/fact-sheets/detail/cancer).

Based on the validation through prognosis, natural history, pathology, management, and response to therapy, there are two main categories of lung cancer (1). These include the most common types: non small cell lung carcinoma (NSCLC) comprising 80–85% of cases, and the fast growing, aggressive, and highly metastatic small cell lung carcinoma. NSCLC has three further subtypes: squamous cell carcinoma (25–30% of all lung cancer cases); adenocarcinoma (~40% of all lung cancer cases); and the large cell (undifferentiated) carcinoma, accounting for about 10% of all NSCLC (2).

Smoking and tobacco consumption are the major risk factors leading to 80% of lung cancer deaths. Other factors for non-smokers include exposure to carcinogens such as radon, asbestos and arsenic, and genetic factors (3,4). Lung cancer develops by stepwise acquisition of genetic changes resulting in uncontrolled growth, resistance to apoptosis, tumor angiogenesis, tissue invasion, and distant metastasis (1). It is widely established that during tumor development many different factors and signaling cascades become aberrant to transform a dormant cell to a malignant cell. The tumor-associated neovasculature generated through angiogenesis plays a fundamental role in sustaining the tumor and its development and metastatic spread. Angiogenesis is an intricate process, regulated by specific pro-angiogenic and anti-angiogenic factors in a coordinated fashion (5). In terms of the tumor targeted therapies, the investigation and development of different anti-angiogenesis and vascular targeting strategies have been the subject of critical interest and progress for oncology research and clinical management. The focus of targeting the tumor vasculature has been shifted from vascular destruction (proposed by Dr. Judah Folkman in 1971) (6) to vascular normalization (7). This chapter briefly reviews the current importance of anti-angiogenics in the treatment landscape of lung cancer with reference to NSCLC and focuses on the bioassay development in a quest for anti-VEGF drugs.

27.2 Anti-angiogenic Therapy in NSCLC and Approved Anti-angiogenic Agents

For patients with NSCLC, anti-angiogenic agents in combination with other standard chemotherapies have been extensively assessed in the clinical studies, some of these showing survival benefits (8). Although in clinical trials, Axitinib, Motesanib, Sorafenib, Cediranib, Vatalanib, and Linifanib did not show promising outcomes (8–13), for many, trials are still ongoing (Table 27.1). The summary report of current clinical data along with the potential future perspectives of immunotherapy in combination with anti-angiogenic agents for advanced NSCLC have been reviewed in detail by Liang and Wang in 2019 (14).

27.2.1 Bevacizumab

Bevacizumab, a VEGF-A neutralizing recombinant humanized monoclonal antibody, reduces vascular permeability and normalizes the tumor vasculature, thereby increasing cytotoxic drug delivery. Bevacizumab is the first agent to be approved by the US FDA (in 2006) as first-line treatment for advanced non-squamous NSCLC patients, in combination with carboplatin/

DOI: 10.1201/9781003046547-27

TABLE 27.1

Anti-angiogenic Agents for NSCLC

Generic Name	Type	Target	Manufacturer
Bevacizumab (15)	Monoclonal antibody	VEGF-A	Genentech/Roche
Ramucirumab (16)	Monoclonal antibody	VEGFR-2	Eli Lilly and Company
Nintedanib (17)	Small molecule multi-TKI	VEGFR, FGFR, PDGFR	Boehringer Ingelheim
Sunitinib (18)	Small molecule multi-TKI	VEGFR, PDGFR, cKit	Pfizer
Vandetanib (19,20)	Small molecule multi-TKI	VEGFR, EGFR, RET	AstraZeneca
Anlotinib (21)	Small molecule multi-TKI	VEGFR, EGFR, FGFR, PDGFR	Chia Tai Tianqing Pharmaceutical Group Co., Ltd.
Apatinib (22)	Small molecule multi-TKI	VEGFR2	Jiangsu Hengrui Medicine Co., Ltd.
Endostar (23,24)	Modified recombinant human endostatin	VEGF	Simcere Pharmaceutical Group
Pazopanib (25,26)	Small molecule multi-TKI	VEGFR, PDGFR, cKit	GlaxoSmithKline/Novartis

paclitaxel. This has modestly prolonged overall survival (OS) and progression-free survival (PFS) (15,27).

27.2.2 Ramucirumab

Ramucirumab is a recombinant monoclonal antibody that blocks VEGFR-2 which gets upregulated in many cancer types, including lung cancer. In 2014, the FDA approved ramucirumab in combination with docetaxel as a second-line treatment for metastatic NSCLC that has progressed after platinum-based chemotherapy (28). In many trials, this treatment has shown clinically meaningful improvements in OS and PFS compared with placebo plus docetaxel (29,30).

27.2.3 Nintedanib

Nintedanib is a tyrosine kinase inhibitor (TKI) that targets three angiogenesis regulating transmembrane receptors, i.e. vascular endothelial growth factor (VEGF) receptors, fibroblast growth factor receptors, and platelet-derived growth factor receptors. It is sold under the brand name Vargatef (31,32). Nintedanib was approved in 2019 by FDA for a rare lung disease, i.e. systemic sclerosis-associated interstitial lung disease (ILD). Also, it was approved in the EU for advanced adenocarcinoma lung cancer after first-line chemotherapy (17).

27.3 Anti-angiogenics in Combination with Immunotherapy

Since angiogenesis inhibition continues to be an attractive target, new association are being evaluated in clinical trials. In the recent past, the combination of immunotherapy with anti-angiogenic agents has become more popular in tumor treatment demonstrating potential benefits (33).

Immunotherapy has revolutionized significantly the treatment landscape of NSCLC patients without targetable oncogenic mutations or who fail to respond to targeted treatment (34). Yet, in a recent review, Popat et al. identified that, despite the improved immunotherapy outcomes as compared with chemotherapy alone, ~50% of patients with advanced NSCLC do not respond, and ultimately, almost all patients relapse (35). Though the mechanisms of immunotherapy resistance are partially understood, the disease progression often drives re-evaluation of the treatment plan. However, the speed at which immunotherapies have reshaped the treatment landscape caused a scarcity of mature clinical data to guide treatment decisions for patients who progress on chemo-immunotherapy. Until the availability of such robust and prospective clinical trial data, the review authors further highlighted the fact that attention must turn to the potential use of currently licensed agents and any available supporting clinical data in this setting. In this context, clinical trials are currently ongoing to address the question of whether anti-angiogenic drugs can stimulate the immune system and if they have additive effects with immune checkpoint inhibitors (36).

An increasing body of data (preclinical and clinical) for the mechanistic arguments and supporting evidence for the use of anti-angiogenic agents as a means of targeting immunosuppression within the tumor microenvironment has been reviewed recently (35–37). The sustained angiogenesis and immunosuppression are found to be interconnected processes with shared regulators such as proangiogenic molecules. The key regulator of angiogenesis, VEGF-A promotes immunosuppression by different mechanisms. VEGF directly inhibits dendritic cell maturation and activation of antigen-specific T regs and reduces immune cell–EC interactions in angiogenic vessels (35). VEGF inhibits lymphocyte adhesion to activated endothelial cells thereby decreasing lymphocyte trafficking from the circulation into tumors and suppresses adaptive T-cell responses by impairing antigen presentation and T-cell function (38,39).

The normalization of the tumor vasculature with anti-angiogenics can improve blood perfusion and oxygenation with complementary immunomodulatory effects (by changing an immunosuppressive microenvironment to an immunosupportive environment). The preclinical evidence reviewed recently supports consideration of anti-angiogenic therapy to target immunos uppression in the tumor microenvironment and trigger an "angio-immunogenic switch" back toward an immunosupportive environment (35).

27.4 Bioassay Development in a Quest for Anti-VEGF Drugs

Today, selection of a drug target and its bioassay design is an extremely important stage in drug discovery and development. Beginning with the discovery to the final stage of quality control (QC), bioassay screenings play a pivotal role in drug development. The crucial role of the VEGF in tumor angiogenesis has directed the development and entry of novel VEGF inhibitors in market, with many more in the pipeline. Biopharmaceutical companies are required to meet internationally accepted guidelines preceding commercial release, which have mandated bioassays to assess a drug's potency. This section intends to provide a concise update on various anti-VEGF/VEGFR2 drugs and the related bioassays that are on the horizon. Available bioassays employ the use of primary cells or in vivo animal models. Although, in combination, such bioassays can assess the drug's safety and potency. However, for routine analysis, QC requires simple, convenient, and robust cell-based bioassay. Engineering immortalized endothelial cell lines having receptors involved in the VEGF signaling pathway is the key for the development of such cell based bioassays.

27.5 The Regulations and Bioassays

The history of regulations governing biologics is almost as old as biologics itself. Emil von Behring received the first Nobel Prize in medicine in 1901. He had worked on serum therapy, particularly against diphtheria, which had revolutionized medical sciences and provided physicians with a reliable weapon against this devastating disease (40). Following this, in the same year, 13 children died in Saint Louis, Missouri, after getting diphtheria antitoxin contaminated with tetanus (41). In response to this, the *Biologics Control Act* was issued in 1902 (42). This was the first legislation to control the quality of drugs (42).

Regulatory bodies such as the United States Food and Drug Administration (US-FDA), International Conference on Harmonization (ICH), United States Pharmacopeia/European Pharmacopeia (USP/EP), and European Medicines Agency (EMA) control the commercial distribution of biopharmaceuticals/drugs in the market. Biopharmaceutical companies are required to follow set internationally accepted guidelines for manufacturing and marketing their products. A biopharmaceutical's complete analytical profile comprises physiochemical tests as well as bioassays (43). Both are equally important and are required in combination to examine the clinical safety and efficacy of the drug. While the physiochemical tests can well characterize the drug in terms of its structure, bioassays are required for assessing the functionality, clinical efficacy, toxicity, and pharmakokinetics of the drug (44).

Bioassay is a technique carried out on some living systems to estimate the potency or biological activity of a drug. Bioassays play a crucial role in therapeutic drug characterization throughout the span of drug development, manufacture, and release in the market. They are required for establishing the potency of new drug candidates and for the stability and lot release testing of the drugs. A combination of *in vitro, ex vivo*, and *in vivo* bioassays can be employed to assess functionality, safety, and potency of the drugs at preclinical levels. In order to fulfill the regulatory requirements, the bioassay needs to be reflective of the clinical mode of action of the drug (45).

Furthermore, bioassays find application not only in the biologics but also in the successful development of innovator's biosimilars (46). The bioassays determine the relative potency of a biosimilar drug by comparing the biological activity of the biosimilar drug with that of the innovator drug preparation (44). In biopharmaceuticals, a QC group routinely performs bioassays required for the successful regulatory submission packages' stability and release testing of drug substances and drug products. The QC laboratory requires a bioassay, which is simple, convenient for high-throughput screening, robust, and reproducible. Hence, bioassay needs to be designed in a manner to reduce variability and bias, i.e. bioassay's validity should be verifiable.

27.6 VEGF: The Fundamental Controller of Angiogenesis

Angiogenesis is an intricate process, regulated by specific pro-angiogenic and anti-angiogenic factors in a coordinated fashion. The pro-angiogenic factors mainly include VEGF, basic fibroblast growth factor (bFGF), platelet derived growth factor (PDGF), FGF, platelet derived endothelial cell growth factor, angiopoietin-1, angiogenin, transforming growth factor beta-1 (TGF-β1), transforming growth factor alpha (TGF-α), and epidermal growth factor (EGF) (as reviewed by D. Bouïs et al. (47)). Among these, VEGF is the fundamental controller of angiogenesis (48) and one of the best-studied factors that stimulates tumor angiogenesis (1).

VEGF, discovered in the 1980s, is unique among other growth regulators of angiogenesis for its specificity to the vascular endothelium (49). Endothelial cells are the main cells lining the blood vessels and capillaries. VEGF stimulates endothelial cells to undergo proliferation, tube formation, and migration, and also acts as an anti-apoptotic factor for them (50).

27.7 VEGF Family and Its Receptors

The VEGF family consists of PLGF (placental growth factor), VEGF-A, VEGF-B, VEGF-C, VEGF-D, VEGF-E, and snake venom VEGF (sv VEGF). Among these, VEGF-A, also known as VEGF or vascular permeability factor (VPF), has been shown to play the key role in angiogenesis (51). These growth factors exert their impacts by binding with high affinity to different transmembrane tyrosine kinase receptors (52). The three main kinds of VEGF receptors include VEGFR-1 (Flt-1), VEGFR-2 (Flk-1or KDR), and VEGFR-3 (Flt-4). VEGF-R1 and VEGF-R2 are mainly expressed on the surface of blood endothelial cells whereas VEGF-R3 is largely confined to lymphatic endothelial cells (53). Besides these receptors, VEGF also interacts with co-receptors, the neuropilins, namely neuropilin 1 and neuropilin 2 (NRP-1 and NRP-2), known to increase the binding affinity of VEGF to its receptors (54). PLGF is selectively expressed in placenta. VEGF-B binds to VEGFR-1 (55,56), VEGF-C and VEGF-D stimulate lymphangiogenesis and angiogenesis by activating VEGF-R3 and VEGF-R2 (57,58). VEGF-E, encoded by parapoxvirus ORF virus, binds and activates VEGF-R2 (59); svVEGF binds to VEGF-R1 and to VEGF-R2 (60).

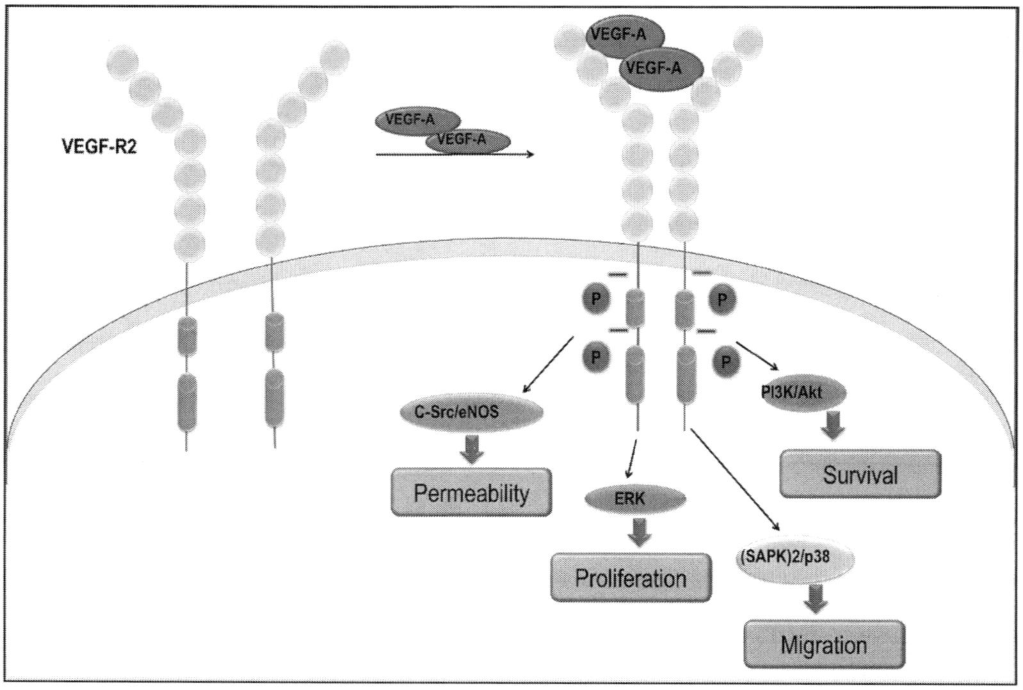

FIGURE 27.1 Schematic Representation of VEGF-R2 Signaling. VEGF-A Binding to VEGF-R2 Causes Dimerization of the Receptor, Followed by Activation of Intrinsic Tyrosine Kinase Activity. VEGF-R2 Signal for Endothelial Cell Proliferation, Migration, Permeability and Survival etc. by Activating Various Pathways

VEGF has nine isoforms that are the different heparin binding, homodimeric molecular species formed by alternative splicing of mRNA (61). All the VEGF isoforms have similar biological activity, but differ in their secretion patterns (62). $VEGF_{121}$, $VEGF_{165}$, $VEGF_{189}$, and $VEGF_{206}$ are more common isoforms, whereas $VEGF_{145}$, $VEGF_{148}$, $VEGF_{162}$, $VEGF_{165b}$, and $VEGF_{183}$ are less frequent forms. Among all the VEGF isoforms, VEGF $_{165}$ is the most important isoform. VEGF binds to both VEGF-R1 and VEGF-R2 and to the co-receptor, neuropilin-1 (63). The binding of VEGF to the VEGFRs causes receptor dimerization, which in turn activates their intrinsic tyrosine kinase activity. Although VEGF-R1 has higher affinity for VEGF as compared to VEGF-R2, still VEGF- R2 is the prime receptor responsible for VEGF stimulated proliferation in the endothelial cells (61). This is because VEGF-R1 exhibits weak receptor tyrosine phosphorylation on VEGF activation (64). VEGF-R2 is responsible for the activation of major signaling pathways, involving the extracellular signal-regulated kinase (ERK) pathway responsible for endothelial cell proliferation. It also activates stress activated protein kinase (SAPK) 2 /p38, which causes endothelial cell migration by cytoskeleton remodeling and actin reorganization (65,66) (Figure 27.1)

27.8 Drugs Targeting VEGF or Its Receptors

VEGF is widely established as the most attractive therapeutic target in tumor angiogenesis for a number of reasons (52). It is the key regulator of angiogenesis and acts mainly on endothelial cells, which are relatively stable. Further, as described earlier here in the chapter, VEGF promotes immunosuppression, thereby leads for the recent consideration of the use of anti-angiogenic therapy as a means of targeting immunosuppression within the tumor microenvironment and trigger an "angio-immunogenic switch" back toward an immunosupportive tumor environment for improved outcome for NSCLC patients (35).

Various approaches to disrupt VEGF mediated signaling in angiogenesis are being investigated (67,68). The main strategies include molecules binding and neutralizing VEGF or VEGF receptors, antisense VEGF mRNA or VEGF receptor mRNA expressing constructs (69,70), toxins conjugated to VEGF for targeting the endothelial cells (69), and the use of ribozymes against VEGF receptors (71).

A number of VEGF targeting drugs are available in the market, which bind to VEGF and prevent its interaction with its receptors, thereby reducing endothelial cell proliferation and formation of new blood vessels. Once the tumor does not get the required vascular support, its growth and expansion ceases. The VEGF inhibitors largely include antibodies, aptamers, and the soluble receptor constructs. US FDA sanctioned drugs for the therapy of various solid tumors are the monoclonal antibodies, fusion proteins, and small molecule TKIs (Table 27.2).

Another approach to block downstream VEGF signaling is to inhibit VEGF receptors. The VEGF receptor inhibitors mainly include the small molecule TKI and monoclonal antibodies. TKI are capable of targeting multiple receptors and can be administered orally. As shown in Table 27.2, the various TKI in the market are sunitinib, sorafenib, vandetanib, axitinib, cabozantinib, pazopanib, and regorafenib.

TABLE 27.2

Drugs Targeting VEGF or VEGF Receptor

Generic Name	Trade Name	Type	FDA Approval date	Target	Indication	Manufacturer
Bevacizumab (52,72)	Avastin	Recombinant humanized monoclonal antibody	February 26, 2004	VEGF-A	Metastatic CRC, Non-squamous NSCLC, metastatic BC, GBM, Metastatic RCC, Advanced CC, Platinum-resistant ovarian cancer	Genentech/ Roche
Pegaptanib sodium (73)	Macugen	Pegylated ribonucleic acid aptamer	December 20, 2004	VEGF 165	Wet macular degeneration, DME, CRC	Eyetech/ Pfizer
Ranibizumab (74)	Lucentis	Recombinant humanized monoclonal Fab fragment	June 30, 2006	VEGF-A	"Wet" type of AMD, DME	Genentech/ Novartis
Aflibercept (75)/ **VEGF Trap**	Eylea	Fusion protein	November 18, 2011	VEGF-A, VEGF-B and PDGF	"Wet"type AMD, metastatic CRC, DME, RVO	Regeneron Pharmaceuticals Inc. (REGN)
Ziv-aflibercept (44)/ **VEGF Trap Eye**	Zaltrap	Fusion protein	August 3, 2012	VEGF-A, VEGF-B and PDGF	Metastatic CRC	Sanofi and Regeneron Pharmaceuticals Inc. (REGN)
Sorafenib (76)	Nexavar	Small molecule multi-TKI	December 20, 2005	VEGFR-2, 3, PDGFR, c-kit, Raf kinase	Advanced RCC, HCC, advanced thyroid cancer	Bayer/Onyx
Sunitinib (77)	Sutent	Small molecule multi-TKI	January 26, 2006	VEGFR-1,2,3, PDGFR, cKit	GIST, advanced RCC, pancreatic neuroendocrine tumor	Pfizer
Pazopanib (78)	Votrient	Small molecule multi-TKI	October 19, 2009	VEGFR-1, 2, 3, PDGFR, cKit	RCC, advanced soft tissue sarcoma	GlaxoSmith Kline
Vandetanib (76)	Caprelsa	Small molecule multi-TKI	April 6, 2011	VEGFR-1, 2, 3,	Medullary thyroid cancer, NSCLC	Astra Zeneca Pharmaceuticals LP
Axitinib (79)	Inlyta	Small molecule multi-TKI	January 27, 2012	VEGFR-1, 2, 3, PDGFR, cKit	Advanced RCC, advanced pancreatic cancer, thyroid cancer and NSCLC	Pfizer
Regorafenib (80)	Stivarga	Small molecule multi-TKI	September 27, 2012	VEGFR-2, PDGFR, TIE-2, FGFR, c-kit	Metastatic CRC, GIST	Bayer
Cabozantinib (81)	Cometriq	Small molecule multi-TKI	November 29, 2012	RET,MET and VEGFR-2	Metastatic MTC, RCC, prostate cancer	Exelixis
Ramucirumab (82)	Cyramza	Humanized Monoclonal antibody	April 21, 2014	VEGFR-2	Advanced gastric or GEJ adenocarcinoma, lung cancer	Eli Lilly and Company
Nintedanib (BIBF1120) (83)	Ofev	Small molecule multi-TKI	October 15, 2014	VEGFR-1,2,3, PDGFR, FGFR	IPC	Boehringer Ingelheim
Lenvatinib (E7080)(84)	Lenvima	Small molecule multi-TKI	February 13, 2015	VEGFR- 2,3	DTC	Eisai Co.
Ponatinib (85)	Iclusig	Small molecule multi-TKI	November 29, 2016	BCR-ABL, VEGFRs,PDGFR,FGFR	CML, ALL	ARIAD pharmaceuticals

Abbreviations: AMD: age related macular degeneration; ALL: Philadelphia chromosome-positive acute lymphoblastic leukemia; BC: breast cancer; CC: cervical cancer; CML: chronic myeloid leukemia; CRC: colorectal cancer; DME: diabetic macular edema; DTC:dDifferentiated thyroid cancer; GBM: glioblastomamultiforme; GEJ: gastroesophageal junction; GIST: gastrointestinal stromal tumor; HCC: hepatocellular carcinoma; IPC: idiopathic pulmonary fibrosis; MTC:mMedullary thyroid cancer; NSCLC: non small cell lung cancer; RCC: renal cell carcinoma; RVO: retinal vein occlusion.

27.9 Global Sales of Anti-VEGF Drugs

Bevacizumab, a VEGF-A neutralizing recombinant humanized monoclonal antibody, reduces vascular permeability and normalizes the tumor vasculature, thereby increasing cytotoxic drug delivery. It became the first anti-angiogenic agent sanctioned by the FDA for the treatment of metastatic colorectal cancer on February 26, 2004. Following the approval of bevacizumab, the anti-VEGF drug market has increased tremendously, considering the implications of angiogenesis in number of diseases. Since then a number of different categories of anti-VEGF drugs have been approved for different pathological conditions, most importantly for eye-related diseases and other cancer types (Table 27.2) while many more are expected to reach the market in near future (Table 27.3). As described earlier here, bevacizumab is the first agent to be approved by the US FDA (in 2006) as first-line treatment for advanced non-squamous NSCLC patients, in combination with carboplatin/paclitaxel. This has modestly prolonged OS and progression-free survival (PFS) (15,27).

27.10 Bioassays Available for Anti-VEGF Drugs

The increasing global market growth of anti-VEGF drugs has created great demand for bioassays for these drugs at all the stages of drug discovery, development, and release. Bioassays available for anti-VEGF drugs are based mainly on animal studies. In vitro cell-based assays are mainly performed using primary cell lines (95).

27.10.1 In Vivo Assays

The in vivo animal study used an animal model to compare changes in biological responses, on application of the drug. The in vivo bioassay facilitates understanding of the full regulation of the angiogenic process, which involves a number of coordinately and spatially regulated steps. The effect of regulators on diverse cell types engaged in the process can be examined. Most commonly used bioassays are mentioned below:

TABLE 27.3

Drugs Under Clinical Development for Inhibiting VEGF or VEGF Receptors

Generic Name	Type	Target	Indication	Manufacturer
Apatinib (86)	Small molecule multi-TKI	VEGFR-2, cKit	Advanced or metastatic gastric cancer, NSCLC, BC, HCC	Advenchen Laboratoraries
Brivanib alaninate (BMS-582664) (87)	Small molecule multi-TKI	VEGFR-1,2,3, FGFR	HCC, Metastatic CRC	Bristol-Myers Squibb
Brolucizumab (RTH258)	Recombinant humanized singlechain monoclonal antibody Fv fragment	VEGF A	n AMD, diabetic macular edema	Novartis
Cediranib (AZD2171) (88)	Small molecule multi-TKI	VEGFR- 1, 2, 3	Metastatic CRPC, recurrent GBM	AstraZeneca
Dovitinib (CHIR 258)	Small molecule multi-TKI	FGFR1,3, FLT3, cKIT, VEGFR 3	Prostate cancer, BC, GBM, RCC	Novartis
Foretinib (GSK1363089)	Multi-kinase inhibitor	cMET, VEGFR2	Ovarian cancer metastatis	GlaxoSmithKline
Fruquintinib (HMPL-013)	Small molecule	VEGFR- 1, 2, 3	NSCLC, CRC, Gastric cancer	Hutchison MediPharma
Icrucumab (IMC-18F1) (89)	Humanized monoclonal antibody	VEGFR- 1	BC	Eli Lilly and Company
JNJ-26483327	Small molecule multi-TKI	VEGFR- 3		Johnson and Johnson Pharmaceutical Research & Development
Linifanib (ABT-869) (13)	Small molecule multi-TKI	VEGFR- 1, 2, 3, PDGFR	HNSCC, Advanced HCC	
Motesanib (AMG 706) (90)	Small molecule multi-TKI	VEGFR-1, 2, 3, PDGFR, cKit	NSCLC, Thyroid cancer and BC	Amgen/Takeda
Orantinib (SU6668) (91)	Small molecule multi-TKI	VEGFR, PDGFR, FGFR	MPM, endometrial cancer	SUGEN/Pharmacia
Telatinib (BAY 57–9352) (92)	Small molecule multi-TKI	VEGFR- 2, 3, PDGFR, c-kit	CRC	ACT Biotech
Tivozanib (AV-951) (93)	Small molecule multi-TKI	VEGFR-1, 2, 3	Advanced RCC	AVEO Pharmaceuticals
Vatalanib (PTK 787/ZK 222584) (94)	Small molecule multi-TKI	VEGFR-1, 2, 3, PDGF, c-kit	NSCLC	Novartis/Bayer Schering

Abbreviations: BC: Breast cancer; CRC: Colorectal cancer; CRPC: Castration –resistant prostate cancer; GBM: Glioblastoma; HCC: Hepatocellular carcinoma; HNSCC: Head and neck squamous cell carcinoma; MPM: Malignant pleural mesothelioma; nAMD: neovascular age-related macular degeneration; NSCLC: non small cell lung cancer; RCC: Renal cell carcinoma.

27.10.1.1 Chick Chorioallantoic Membrane Assay

The chick chorioallantoic assay is used to study tumor angiogenesis. Chorioallantoic membrane (CAM) functions as a respiratory tissue in the chick embryo. It is highly angiogenic until day eleven and is thus considered suitable for studying angiogenesis (96). In this assay, an opening is made in the chick egg shell of seven–nine days pre-incubated eggs, and the test substance is then placed on the extra embryonic membrane. After the re-incubation of eggs, grafts are recovered and the extent of angiogenesis is determined by image analysis (97,98).

CAM assay is relatively inexpensive and easy to carry out. However, it cannot score as a properly validated assay due to the use of CAM which is highly sensitive and undergoes rapid morphological changes. Moreover, only limited reagents including antibodies would be compatible with the non-mammalian, avian system (99).

27.10.1.2 Hindlimb Ischemia Assay

Angiogenesis regulators can be evaluated in a rabbit or mouse model of hindlimb ischemia (100). Hindlimb ischemia assay is based on the hemodynamic changes accountable for the establishment of new blood vessels. In this assay, an incision is made in the skin covering the middle portion of the hind limb of the anesthetized test animal. This is followed by tying of proximal end of femoral artery and the distal end of the saphenous artery and then cutting their side branches. Blood flow can be observed at the ischemic distal location. Vascular outcome is assessed mainly by capillary density, laser Doppler blood perfusion imaging, and Tarlov and ischemia scores (101,102).

27.10.1.3 Chamber Assays

Chamber assays are designed for continuous monitoring of angiogenesis in transparent chambers containing chronic tissues. The various chamber assays are dorsal skinfold chamber, rabbit ear chamber, and the cranial window chamber (103). In these assays, tissue is removed from the anesthetized animal and test preparations are positioned on the exposed surface. It is then covered with glass. Quantification involves microscopy and image analysis (99,104). Such assays allow researchers to study physiological effects like blood flow perturbations but the surgical methods involved make these assays technically difficult to perform.

27.10.1.4 Matrigel Plug Assay

Matrigel is the laminin rich extract excuded by the Engelbreth-Holm-Swarm (EHS) mice sarcoma cells (105). It is liquid at 4 °C and solidifies at 37 °C. The assay is carried out by subcutaneously injecting the matrigel, which contains the test substance into the animal model. The injection of foreign substance in the animal stimulates angiogenesis. Matrigel hardens to form a plug which can be recovered after 7–21 days. Matrigel is avascular when injected and hence, after recovery, newly recruited vessels can be easily distinguished. Angiogenic response can be assessed by measuring hemoglobin in the plug or can be examined histologically. For this assay, mice are the most commonly used animal model (106).

The assay is rapid and easy to perform. Disadvantages include quantification, although there are methods of evaluation using reverse transcription-quantitative PCR (107,108).

27.10.1.5 Corneal Angiogenesis Assay

Originally, this assay was developed for rabbit eyes but now it is more frequently used in mice (109). For carrying out the assay, a pocket is surgically crafted in the cornea. Uniform pellets of the test substance are prepared by mixing them with a stabilizer and a slow release polymer (110). This pellet is then inserted into the micropocket. The new vascular response from limbal vasculature is then monitored and quantified by fluorescence techniques or by measuring the area of vessel penetration (Figure 27.2a). The cornea, being avascular, is an ideal model for analyzing in vivo angiogenesis (111).

Considerable advantages of the assay include the absence of any vascular background in the cornea and its adaptation for use in mice. The assay suffers from the limitations of surgical processes. Moreover, inflammation risks are also common (112).

27.10.2 Ex Vivo (Organ Culture Assays)

The idea behind organ culture assays is to bridge the gap between in vivo and in vitro assays. The in vitro assays cannot recapitulate the full process of angiogenesis, which comprises not only the endothelial cells but also complex interactions with the surrounding cells. The in vivo assays depict the entire regulation of the neo-vascularization. However, sometimes it becomes difficult to distinguish between the autocrine and the paracrine responses.

27.10.2.1 Aortic Ring Assay

The aortic ring assay was first established in rats (113), but now it has also been adapted to mice (114). This assay involves isolation of the mouse thoracic aorta, followed by culturing of rings of aorta in serum free media in three dimensional collagen gels. The injury due to dissection procedure will give rise to a new microvessel network of branched endothelial channels. Test substances are added to analyze their effect on the process. The abundance and lengths of these new vessels is assessed by image analysis software (115) (Figure 27.2b).

The advantage of aortic ring assay is that it is a more physiologically relevant assay and can be performed in serum free media. Nevertheless, aorta is not an ideal choice because angiogenesis is mainly a microvascular event. Other disadvantages include use of non-human tissues and lack of proper quantification method (116).

27.10.2.2 Porcine Carotid Artery Assay

The porcine carotid artery assay is the modification of aortic ring assay. This assay utilizes the porcine carotid artery. Porcine arteries are a better source as they are more closely related to human vasculature, readily available, and can be easily handled. A number of tests can be run from a single carotid artery, thereby decreasing the statistical variations. The drawback is that it utilizes a single donor that would not be a true representative of the population (117).

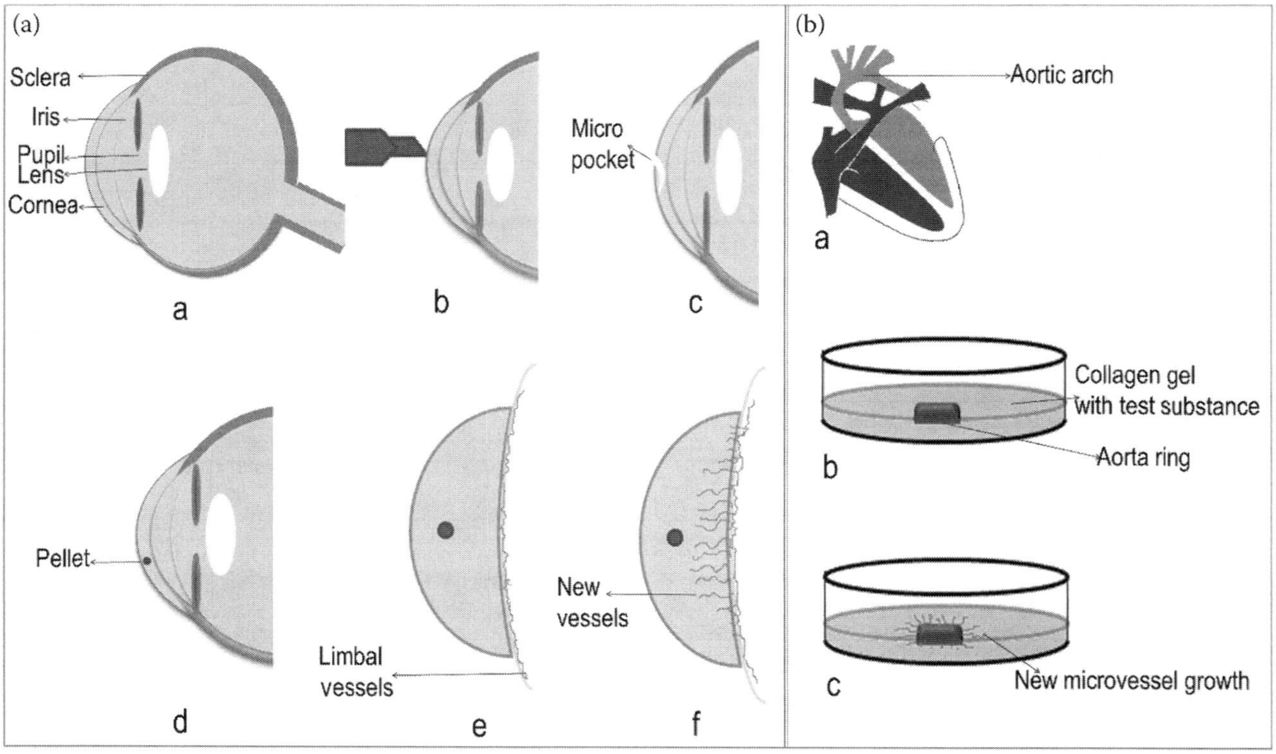

FIGURE 27.2 **(A)** Schematic Representation of Corneal Angiogenesis assay. (a) Schematic Structure of an Eye. (b,c) Creation of an Micropocket by Making an Incision in the Cornea. (d) Insertion of Pellet in Micropocket. (e) Release of Angiogenic Factors from Pellet. f) Leads to New Vessel Growth from the Limbus. **(B)** Schematic Representation of Aortic Ring Assay. (a) Schematic Diagram of Heart with Aortic Arch. (b) Aortic Ring in Collagen Gel with Test Substance in Serum Free Medium. (c) New Vessel Growth from Aortic Ring

27.10.2.3 Chick Aortic Arch Assay

Chick aortic arch assay is also a modification of the aortic ring assay. For this assay, aortic arches are removed from the 12–14 days old chick embryo, cut into rings and then transferred to the matrigel. Substantial microvascular growth is observed within 48 hours. In both the assays, quantification can be performed by using fluorescence labeled lectins or by staining with CD31 labeled antibodies (77).

It is more rapid than the aortic arch assay. The embryonic arch resembles microvascular endothelial cells more closely than the mouse aorta. However, a growing embryo involves rapidly proliferating cell populations (118).

In addition to these, other new ex vivo assays are being reported for testing the anti-angiogenic drugs such as the ex vivo murine retina angiogenesis assay (119).

27.10.3 Endothelial Cell-Based In Vitro Assays

The endothelial cells are the choice of cells for studying VEGF pathways because VEGF is specific to endothelial cells. During the angiogenesis cascade, in the presence of VEGF and other growth regulators, endothelial cells undergo cell proliferation, migration, differentiation, and finally apoptosis (120). These particular cellular events can be assessed in vitro (Figure 27.3.). Researchers have isolated and cultured different endothelial cells for in vitro studies of angiogenesis, including human microvascular endothelial cells (HMVEC) (121), human coronary artery endothelial cells (HCAEC), human aortic endothelial cells, human placental endothelial cells (HPEC) (122), human umbilical vein endothelial cells (HUVEC), (123) etc. Among the various endothelial cells, HUVEC cells are most frequently used because they are relatively easily available from the discarded umbilical cords and can be efficiently isolated (86). HUVECS can be cultured for maximum of ten serial passages.

Furthermore, various commercially available kits are available in the market for endothelial cell proliferation, endothelial tube formation, and endothelial cell migration assays.

27.10.3.1 Endothelial Cell Proliferation Assay

Endothelial cells when set up in culture are capable of undergoing cell division and proliferation. There are numerous endothelial cell proliferation assays. In this assay, a specific number of endothelial cells are seeded, followed by the addition of an angiogenesis regulator. After incubation at 37 °C in CO_2 for a certain time, effect of the test substance on the endothelial cells is determined by measuring the number of cells. Cell number is determined by colorimetric assays like 3-(4, 5-dimethyltetrazolium bromide) (MTT) assay; 2H-Tetrazolium,5-[3-(carboxymethyl) phenyl)-3-(4,5-dimethyl-2-thiazolyl)-2-(4-sulfophenyl)-2-(4-sulfophenyl)-,innersalt (MTS) assay; fluorimetry based assays like Alamar blue (124), and CyQUANT and luminescence assays. Cell cycle analysis can be done by 5-bromo-2'-deoxyuridine (BrdU) colorimetric assay (125).

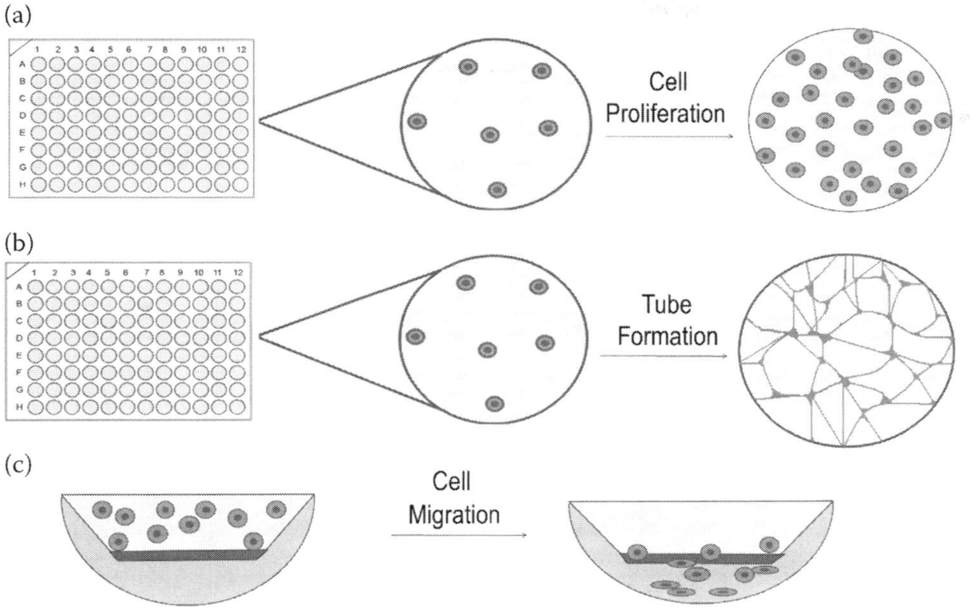

FIGURE 27.3 Schematic Representation of In Vitro Angiogenesis Assays. (a) Endothelial Cell Proliferation. (b) Endothelial Tube Formation. (c) Endothelial Cell Migration

Endothelial cell proliferation assays can be precisely quantitated, and are rapid and easy to perform.

27.10.3.2 Endothelial Cell Migration Assay

During angiogenesis, endothelial cells degrade basement membrane then directionally migrate towards the chemical gradient created by angiogenic factors (126). Cell migration assay is used to study the endothelial cells migratory response (127,128). In the assay, endothelial cells are placed over a layer and allowed to migrate to the attractants kept at lower layer through a filter that allows passage of the cells. Although it is a rapid and highly sensitive assay, it is often technically difficult to carry out.

27.10.3.3 Endothelial Cell Tube Formation Assay

The tube formation assay exploits the capability of the endothelial cells to develop capillary tube-like structures during angiogenesis (129). Endothelial cells specifically form capillary like tubes, when plated on a basement membrane matrix such as collagen or matrigel. The extent of tube formation can be quantified by scanning tube length and branch points, which are analyzed by using various imaging software packages such as ImageJ software. It is one of the most widely used first screen assays for various angiogenic and anti-angiogenic agents. It is quantitative, rapid, easy to perform, and measures multiple steps of angiogenesis comprising cell adhesion, cell migration, alignment, and tube formation (130, 131).

27.11 Requirements for Cell Based Assay

The increasing global market growth of anti-VEGF drugs has created great demand for bioassays for these drugs. A combination of a few of the various bioassays reviewed here can be used to screen the therapeutic drugs. However, for lot releases and stability testing, QC requires robust, reproducible, and convenient bioassays that can be carried out on a routine basis. None of the already existing bioassays can qualify as a QC-lot release bioassay. For example, the in vivo bioassays based on animal studies are expensive, time consuming, labor intensive, and have low throughput. They account for many variations arising due to complexities of living systems and due to species differences. Quantification is a problematic task. Moreover, there are ethical concerns, as well. All these problems limit the use of in vivo bioassays for routine analysis by the QC.

On the other hand, cell-based in vitro bioassays are reliable indicators of the bioactivity of the drug and are easy to use. They are used to evaluate different cellular aspects such as viability, apoptosis, cytotoxicity, signal transduction, and metabolic functions. Cell based assays are consistent with respect to time. This is because the cells in the cell-based bioassays are usually clonally derived (102).

For anti-VEGF drugs, cell-based bioassays use primary cell lines such as HUVEC, which are isolated from the human umbilical vein. Use of primary cell lines for bioassay is not desirable as it involves a tedious method of isolating cells and there are constraints in obtaining the primary source material. They are difficult to propagate and are stable up to few passages only. Also, each time, the cell population obtained would be distinct, resulting in variations (132).

It would be better to have cell-based bioassays with stable cell lines. However, for anti-VEGF drugs, such cell-based bioassays are not developed. This may be because immortalized endothelial cell lines do not show all the required characteristics of endothelial cells, including the expression of all the required receptors (133). Thus, in near future, the challenge is to develop anti-VEGF cell based bioassays, which would be successful in easy technology transfer to QC.

27.12 Engineering Cell Lines with Receptors for the Development of Cell-Based Bioassay

Bioassays are indispensable for drug discovery and throughout their development. Cell-based bioassays for many biopharmaceutical drugs are being commercially developed by engineering the cell lines with appropriate receptors (134). To meet the regulatory requirements, such cell-based bioassays are being efficiently used for high throughput screening of drugs by QC. Moving towards anti-VEGF drugs, transfection of stable cell lines with required receptors would hold the potential for the development of cell-based bioassays. Currently 293\KDR cell line, developed by transfecting VEGFR2 receptor in HEK 293 cell line, is being used for KDR auto phosphorylation assays (135). PAE-VEGF-R2 cell line, developed by overexpressing VEGF-R2 on porcine aortic endothelial cells (PAE), is also being used for binding and receptor phosphorylation bioassays (136). Repoter gene assays are also there, involving HEK293 cells expressing VEGFR2 and luceferase reporter controlled by nuclear factor activated T cells (NFAT) response element (137). Advancement to these could be the transfection of stable human endothelial cell lines with required receptors for the development of functional cell-based assays like cell proliferation assay, tube formation assay, and cell migration assay. These engineered cell lines can be engineered further with reporter genes for easier and more efficient analysis.

Acknowledgments

The authors acknowledge the Financial and in kind support from Deakin India Research Initiative (DIRI) program initiated between Reliance Institute of Life Sciences, India, and Deakin University, Australia.

REFERENCES

1. Korpanty G, et al. Antiangiogenic therapy in lung cancer: Focus on vascular endothelial growth factor pathway. *Experimental Biology and Medicine*. 2010; 235(1): 3–9.
2. Lemjabbar-Alaoui H, et al. Lung cancer: Biology and treatment options. *Biochimica et Biophysica Acta (BBA) - Reviews on Cancer*. 2015; 1856(2): 189–210.
3. Zappa C and Mousa SA. Non-small cell lung cancer: Current treatment and future advances. *Translational Lung Cancer Research*. 2016; **5**(3): 288–300.
4. Dela Cruz CS, Tanoue LT, and Matthay RA. Lung cancer: Epidemiology, etiology, and prevention. *Clinics in Chest Medicine*. 2011; 32(4): 605–644.
5. Kanwar JR, Mahidhara G, and Kanwar RK. Antiangiogenic therapy using nanotechnological-based delivery system. *Drug Discovery Today*. 2011; 16(5-6): 188–202.
6. Folkman, J. Tumor angiogenesis: Therapeutic implications. *The New England Journal of Medicine*. 1971; 285(21): 1182–1186.
7. Jain RK. Normalization of tumor vasculature: An emerging concept in antiangiogenic therapy. *Science*. 2005; 307(5706): 58–62.
8. Alshangiti A, Chandhoke G, and Ellis PM. Antiangiogenic therapies in non-small-cell lung cancer. *Current Oncology*. 2018; 25(Suppl. 1): S45–s58.
9. Kubota K, et al. Phase III, randomized, placebo-controlled, double-blind trial of motesanib (AMG-706) in combination with paclitaxel and carboplatin in East Asian patients with advanced nonsquamous non–small-cell lung cancer. *Journal of Clinical Oncology*. 2017; 35(32): 3662–3670.
10. Metro G, Minotti V, and Crinò L. Years of sorafenib investigation in advanced non-small cell lung cancer: Is there a 'NExUS' linking an unsuccessful treatment and a potentially active one? *Journal of Thoracic Disease*. 2012; 4(6): 635–638.
11. Twelves C., et al. Randomised phase II study of axitinib or bevacizumab combined with paclitaxel/carboplatin as first-line therapy for patients with advanced non-small-cell lung cancer. *Annals of Oncology*. 2014; 25(1): 132–138.
12. Wang F, et al. A phase I study of the vascular endothelial growth factor inhibitor Vatalanib in combination with Pemetrexed disodium in patients with advanced solid tumors. *Investigational New Drugs*. 2019; 37(4): 658–665.
13. Cainap C, et al. Linifanib versus sorafenib in patients with advanced hepatocellular carcinoma: Results of a randomized phase III trial. *Journal of Clinical Oncology*. 2015; 33(2): 172–179.
14. Liang H and Wang M. Prospect of immunotherapy combined with anti-angiogenic agents in patients with advanced non-small cell lung cancer. *Cancer Management and Research*. 2019; 11: 7707–7719.
15. Soria JC, et al. Systematic review and meta-analysis of randomised, phase II/III trials adding bevacizumab to platinum-based chemotherapy as first-line treatment in patients with advanced non-small-cell lung cancer. *Annals of Oncology*. 2013; 24(1): 20–30.
16. Brueckl WM, et al. Efficacy of docetaxel plus ramucirumab as palliative third-line therapy following second-line immune-checkpoint-inhibitor treatment in patients with non-small-cell lung cancer stage IV. *Clinical Medicine Insights. Oncology*. 2020; 14: 1179554920951358.
17. Shiratori T, et al. Effect of nintedanib on non-small cell lung cancer in a patient with idiopathic pulmonary fibrosis: A case report and literature review. *Thoracic Cancer*. 2020; 11(6): 1720–1723.
18. Ming W. and Hui B. Clinical observations of sunitinib combined with docetaxel for advanced non-small cell lung cancer and its effects on serum IL-6 and TNF-alpha. *International Journal of Clinical and Experimental Medicine*. 2019; 12(7): 9030–9039.
19. Kohno T. Promise of vandetanib, a FDA-approved RET kinase inhibitor, for the treatment of RET fusion-positive lung adenocarcinoma. *Translational Cancer Research*. 2016; 5: S237–S239.
20. Zhou Y, et al. The multi-targeted tyrosine kinase inhibitor vandetanib plays a bifunctional role in non-small cell lung cancer cells. *Scientific Reports*. 2015; **5**(1): 8629.
21. Han B, et al. Effect of anlotinib as a third-line or further treatment on overall survival of patients with advanced non-

small cell lung cancer: The alter 0303 phase 3 randomized clinical trial. *JAMA Oncology*. 2018; 4(11): 1569–1575.

22. Zhou C, et al. Efficacy of PD-1 monoclonal antibody SHR-1210 plus apatinib in patients with advanced non-squamous NSCLC with wild-type EGFR and ALK. *Journal of Clinical Oncology*. 2019; 37(15_suppl): 9112.
23. Zhang S-L, et al. Efficacy and safety of recombinant human endostatin combined with radiotherapy or chemoradiotherapy in patients with locally advanced non-small cell lung cancer: A pooled analysis. *Radiation Oncology*. 2020; 15(1): 205.
24. Ji D., et al. Efficacy and safety of endostar (recombinant human endostatin hormone) combined with docetaxel as second-or third-line therapy for patients with non-small-cell lung cancer. *International Journal of Clinical and Experimental Medicine*. 2017; 10: 843–852.
25. Spigel DR, et al. Erlotinib plus either pazopanib or placebo in patients with previously treated advanced non–small cell lung cancer: A randomized, placebo-controlled phase 2 trial with correlated serum proteomic signatures. *Cancer*. 2018; 124(11): 2355–2364.
26. Jo J., et al. Pazopanib for non-small cell lung cancer: The first case report in Korea. *Cancer Research and Treatment: Official Journal of Korean Cancer Association*. 2016; 48(1): 393–397.
27. Zhou C, et al. BEYOND: A randomized, double-blind, placebo-controlled, multicenter, phase III study of first-line carboplatin/paclitaxel plus bevacizumab or placebo in Chinese patients with advanced or recurrent non-squamous non-small-cell lung cancer. *Journal of Clinical Oncology*. 2015; 33(19): 2197–2204.
28. Fala L. Cyramza (ramucirumab) approved for the treatment of advanced gastric cancer and metastatic non-small-cell lung cancer. *American Health & Drug Benefits*. 2015; 8(Spec Feature): 49–53.
29. Uprety D. Clinical utility of ramucirumab in non-small-cell lung cancer. *Biologics: Targets & Therapy*. 2019; 13: 133–137.
30. Yoshimura A, et al. Retrospective analysis of docetaxel in combination with ramucirumab for previously treated non-small cell lung cancer patients. *Translational Lung Cancer Research*. 2019; 8(4): 450–460.
31. Bronte G, et al. Nintedanib in NSCLC: Evidence to date and place in therapy. *Therapeutic Advances in Medical Oncology*. 2016; 8(3): 188–197.
32. Grohé C, et al. Nintedanib plus docetaxel after progression on immune checkpoint inhibitor therapy: Insights from VARGADO, a prospective study in patients with lung adenocarcinoma. *Future Oncology*. 2019; 15(23): 2699–2706.
33. Qu J, et al. Newly developed anti-angiogenic therapy in non-small cell lung cancer. *Oncotarget*. 2017; 9(11): 10147–10163.
34. Horvath L., et al. Overcoming immunotherapy resistance in non-small cell lung cancer (NSCLC) - Novel approaches and future outlook. *Molecular Cancer*. 2020; 19(1): 141.
35. Popat S, et al. Anti-angiogenic agents in the age of resistance to immune checkpoint inhibitors: Do they have a role in non-oncogene-addicted non-small cell lung cancer? *Lung Cancer*. 2020; 144: 76–84.
36. Janning M and Loges S. Anti-angiogenics: Their value in lung cancer therapy. *Oncology Research and Treatment*, 2018; 41(4): 172–180.
37. Lapeyre-Prost A, et al. Immunomodulatory activity of VEGF in cancer. *International Review of Cell and Molecular Biology*. 2017; 330: 295–342.
38. Motz GT and Coukos G. The parallel lives of angiogenesis and immunosuppression: Cancer and other tales. *Nature Reviews Immunology*. 2011; 11(10): 702–711.
39. Fukumura D., et al. Enhancing cancer immunotherapy using antiangiogenics: Opportunities and challenges. *Nature Reviews Clinical Oncology*. 2018; 15(5): 325–340.
40. Hansson N and Enke U. [On the awarding of the first Nobel prize for physiology or medicine to Emil von Behring]. *Deutsche Medizinische Wochenschrift*. 2015; 140(25): 1898–1902.
41. Borchers AT, et al. The history and contemporary challenges of the US Food and Drug Administration. *Clinical Therapeutics*. 2007; 29(1): 1–16.
42. Lilienfeld DE. The first pharmacoepidemiologic investigations - National drug safety policy in the United States, 1901-1902. *Perspectives in Biology and Medicine*. 2008; 51(2): 188–198.
43. Rieder N, Mark Schenerman HGS, Strause R, Fuchs C, Mire-Sluis A, and McLeod LD. The roles of bioactivity assays in lot release and stability testing. *BioProcess International*. 2010. 8(6): 33–42.
44. Food and Drug Administration. Draft guidance for industry from FDA: Quality considerations in demonstrating biosimilarity to a reference protein product. *Biotechnology Law Report*. 2012; 31(2): 185–195.
45. Shah VP, et al. Bioanalytical method validation: a revisit with a decade of progress. *Pharmaceutical Research*. 2000; 17(12): 1551–1557.
46. Bryant K. Biosimilars: the long and winding pathway to approval. *U.S. Pharmacist*. 2013; 38(6): 11–17.
47. Bouïs D, et al. A review on pro- and anti-angiogenic factors as targets of clinical intervention. *Pharmacological Research*. 2006; 53(2): 89–103.
48. Ferrara N. VEGF and the quest for tumour angiogenesis factors. *Nature Reviews Cancer*. 2002; 2(10): 795–803.
49. Korpelainen EI and K. Alitalo. Signaling angiogenesis and lymphangiogenesis. *Current Opinion in Cell Biology*. 1998; 10(2): 159–164.
50. Gerber HP, et al. Vascular endothelial growth factor regulates endothelial cell survival through the phosphatidylinositol 3'-kinase Akt signal transduction pathway - Requirement for Flk-1/KDR activation. *Journal of Biological Chemistry*. 1998; 273(46): 30336–30343.
51. Ferrara N and DavisSmyth T. The biology of vascular endothelial growth factor. *Endocrine Reviews*. 1997; 18(1): 4–25.
52. Ferrara N. Vascular endothelial growth factor: Basic science and clinical progress. *Endocrine Reviews*. 2004; 25(4): 581–611.
53. Karkkainen MJ and TV Petrova. Vascular endothelial growth factor receptors in the regulation of angiogenesis and lymphangiogenesis. *Oncogene*. 2000; 19(49): 5598–5605.

54. Neufeld G, Kessler O, and Herzog Y. The interaction of neuropilin-1 and neuropilin-2 with tyrosine-kinase receptors for VEGF. *Neuropilin: From Nervous System to Vascular and Tumor Biology*. 2002; 515: 81–90.
55. Park JE, et al. Placenta growth-factor - Potentiation of vascular endothelial growth-factor bioactivity, in vitro and in vivo, and high-affinity binding to Flt-1 but not to Flk-1/Kdr. *Journal of Biological Chemistry*. 1994; 269(41): 25646–25654.
56. Olofsson B, et al. Vascular endothelial growth factor B (VEGF-B) binds to VEGF receptor-1 and regulates plasminogen activator activity in endothelial cells. *Proceedings of the National Academy of Sciences of the United States of America*. 1998; 95(20): 11709–11714.
57. Joukov V, et al. A novel vascular endothelial growth factor, VEGF-C, is a ligand for the Flt4 (VEGFR-3) and KDR (VEGFR-2) receptor tyrosine kinases. *EMBO Journal*. 1996; 15(2): 290–298.
58. Achen MG, et al. Vascular endothelial growth factor D (VEGF-D) is a ligand for the tyrosine kinases VEGF receptor 2 (Flk1) and VEGF receptor 3 (Flt4). *Proceedings of the National Academy of Sciences of the United States of America*. 1998; 95(2): 548–553.
59. Meyer M, et al. A novel vascular endothelial growth factor encoded by Orf virus, VEGF-E, mediates angiogenesis via signalling through VEGFR-2 (KDR) but not VEGFR-1 (Flt-1) receptor tyrosine kinases. *Embo Journal*. 1999; 18(2): 363–374.
60. Takahashi H, et al. A novel snake venom vascular endothelial growth factor (VEGF) predominantly induces vascular permeability through preferential signaling via VEGF receptor-1. *Journal of Biological Chemistry*. 2004; 279(44): 46304–46314.
61. Takahashi H and Shibuya M. The vascular endothelial growth factor (VEGF)/VEGF receptor system and its role under physiological and pathological conditions. *Clinical Science*. 2005; 109(3): 227–241.
62. Houck KA, et al. Dual regulation of vascular endothelial growth-factor bioavailability by genetic and proteolytic mechanisms. *Journal of Biological Chemistry*. 1992; 267(36): 26031–26037.
63. Melincovici CS, et al. Vascular endothelial growth factor (VEGF) - Key factor in normal and pathological angiogenesis. *Romanian Journal of Morphology and Embryology*. 2018; 59(2): 455–467.
64. Rahimi N, Dayanir V, and Lashkar K. Receptor chimeras indicate that the vascular endothelial growth factor receptor-1 (VEGFR-1) modulates mitogenic activity of VEGFR-2 in endothelial cells. *Journal of Biological Chemistry*. 2000; 275(22): 16986–16992.
65. Rousseau S., et al. p38 MAP kinase activation by vascular endothelial growth factor mediates actin reorganization and cell migration in human endothelial cells. *Oncogene*. 1997; 15(18): 2169–2177.
66. Lamalice L., et al. Phosphorylation of tyrosine 1214 on VEGFR2 is required for VEGF-induced activation of Cdc42 upstream of SAPK2/p38. *Oncogene*. 2004; 23(2): 434–445.
67. Glade-Bender J, Kandel JJ, and Yamashiro DJ. VEGF blocking therapy in the treatment of cancer. *Expert Opinion on Biological Therapy*. 2003; 3(2): 263–276.
68. Ellis LM and Hicklin DJ. VEGF-targeted therapy: Mechanisms of anti-tumour activity. *Nature Reviews Cancer*. 2008; 8(8): 579–591.
69. Arora N, et al. Vascular endothelial growth factor chimeric toxin is highly active against endothelial cells. *Cancer Research*. 1999; 59(1): 183–188.
70. Becker CM, et al. Gene therapy of prostate cancer with the soluble vascular endothelial growth factor receptor Flk1. *Cancer Biology & Therapy*. 2002; 1(5): 548–553.
71. Weng DE and Usman N. Angiozyme: A novel angiogenesis inhibitor. *Current Oncology Reports*. 2001; 3(2): 141–146.
72. Van Meter MEM and Kim ES. Bevacizumab: Current updates in treatment. *Current Opinion in Oncology*. 2010; 22(6): 586–591.
73. Rinaldi M., et al. Intravitreal pegaptanib sodium (Macugen (R)) for treatment of diabetic macular oedema: A morphologic and functional study. *British Journal of Clinical Pharmacology*. 2012; 74(6): 940–946.
74. Rosenfeld PJ, et al. Ranibizumab for neovascular age-related macular degeneration. *New England Journal of Medicine*. 2006; 355(14): 1419–1431.
75. Sharma T., et al. Aflibercept: A novel VEGF targeted agent to explore the future perspectives of anti-angiogenic therapy for the treatment of multiple tumors. *Mini-Reviews in Medicinal Chemistry*. 2013; 13(4): 530–540.
76. Afonso FJ, et al. Comprehensive overview of the efficacy and safety of sorafenib in advanced or metastatic renal cell carcinoma after a first tyrosine kinase inhibitor. *Clinical & Translational Oncology*. 2013; 15(6): 425–433.
77. Motzer RJ, et al. Overall survival and updated results for sunitinib compared with interferon alfa in patients with metastatic renal cell carcinoma. *Journal of Clinical Oncology*. 2009; 27(22): 3584–3590.
78. Verweij J and Sleijfer S. Pazopanib, a new therapy for metastatic soft tissue sarcoma. *Expert Opinion on Pharmacotherapy*. 2013; 14(7): 929–935.
79. King JW and Lee S-M. Axitinib for the treatment of advanced non-small-cell lung cancer. *Expert Opinion on Investigational Drugs*. 2013; 22(6): 765–773.
80. Strumberg D and Schultheis B. Regorafenib for cancer. *Expert Opinion on Investigational Drugs*. 2012; 21(6): 879–889.
81. Vaishampayan U. Cabozantinib as a novel therapy for renal cell carcinoma. *Current Oncology Reports*. 2013; 15(2): 76–82.
82. Wadhwa R, et al. Ramucirumab: A novel antiangiogenic agent. *Future Oncology (London, England)*. 2013; 9(6): 789–795.
83. Torok S, et al. Nintedanib (BIBF 1120) in the treatment of solid cancers: An overview of biological and clinical aspects. *Magyar onkologia*. 2012; 56(3): 199–208.
84. Cabanillas ME and Habra MA. Lenvatinib: role in thyroid cancer and other solid tumors. *Cancer Treatment Reviews*. 2016; 42: 47–55.
85. Nicolini FE, et al. Overall survival with ponatinib versus allogeneic stem cell transplantation in Philadelphia chromosome-positive leukemias with the T315I mutation. *Cancer*. 2017; 123(15): 2875–2880.
86. Ganguly A., et al. Isolation of human umbilical vein endothelial cells and their use in the study of neutrophil

transmigration under flow conditions. *Journal of visualized experiments: JoVE*, 2012; 66: e4032.

87. Allen E, Walters IB, and Hanahan D. Brivanib, a dual FGF/VEGF inhibitor, is active both first and second line against mouse pancreatic neuroendocrine tumors developing adaptive/evasive resistance to VEGF inhibition. *Clinical Cancer Research*. 2011; 17(16): 5299–5310.
88. Kendrew J, et al. Anti-tumour and anti-vascular effects of cediranib (AZD2171) alone and in combination with other anti-tumour therapies. *Cancer Chemotherapy and Pharmacology*. 2013; 71(4): 1021–1032.
89. Petrylak DP, et al. Randomized phase II study of docetaxel with or without ramucirumab (IMC-1121B) or icrucumab (IMC-18F1) in patients with urothelial transitional cell carcinoma (TCC) following progression on first-line platinum-based therapy. *Journal of Clinical Oncology*. 2012; 30(15).
90. Coxon A, et al. Broad antitumor activity in breast cancer xenografts by motesanib, a highly selective, oral inhibitor of vascular endothelial growth factor, platelet-derived growth factor, and kit receptors. *Clinical Cancer Research*. 2009; 15(1): 110–118.
91. Trung TV, et al. SU6668, a multiple tyrosine kinase inhibitor, inhibits progression of human malignant pleural mesothelioma in an orthotopic model. *Respirology*. 2012; 17(6): 984–990.
92. Langenberg MHG, et al. Phase I evaluation of telatinib, a vascular endothelial growth factor receptor tyrosine kinase inhibitor, in combination with irinotecan and capecitabine in patients with advanced solid tumors. *Clinical Cancer Research*. 2010; 16(7): 2187–2197.
93. Wolpin BM, et al. Multicenter phase II study of tivozanib (AV-951) and everolimus (RAD001) for patients with refractory, metastatic colorectal cancer. *Oncologist*. 2013; 18(4): 377–378.
94. Gauler TC, et al. Phase II trial of PTK787/ZK 222584 (vatalanib) administered orally once-daily or in two divided daily doses as second-line monotherapy in relapsed or progressing patients with stage IIIB/IV non-small-cell lung cancer (NSCLC). *Annals of Oncology*. 2012; 23(3): 678–687.
95. Philip E. Thorpe RAB. Antibody methods for selectively inhibiting VEGF. 2003, Bord of regents, the University of Texas System, Austin TX(US).
96. Lokman NA, et al. Chick chorioallantoic membrane (CAM) assay as an in vivo model to study the effect of newly identified molecules on ovarian cancer invasion and metastasis. *International Journal of Molecular Sciences*. 2012; 13(8): 9959–9970.
97. Brooks PC, AMP Montgomery, and DA Cheresh. Use of the 10-day-old chick embryo model for studying angiogenesis. *Integrin Protocols*, 1999; 129: 257–269.
98. Nguyen M, Shing Y, and Folkman J. Quantitation of angiogenesis and antiangiogenesis in the chick-embryo chorioallantoic membrane. *Microvascular Research*. 1994; 47(1): 31–40.
99. Jain RK, et al. Quantitative angiogenesis assays: Progress and problems. *Nature Medicine*. 1997; 3(11): 1203–1208.
100. Hellingman AA, et al. Variations in surgical procedures for hind limb ischaemia mouse models result in differences in collateral formation. *European Journal of Vascular and Endovascular Surgery*. 2010; 40(6): 796–803.
101. Brenes RA, et al. Toward a mouse model of hind limb ischemia to test therapeutic angiogenesis. *Journal of Vascular Surgery*. 2012; 56(6): 1669–1679.
102. An WF and Tolliday N. Cell-based assays for high-throughput screening. *Molecular Biotechnology*. 2010; 45(2): 180–186.
103. Staton CA, et al. Current methods for assaying angiogenesis in vitro and in vivo. *International Journal of Experimental Pathology*. 2004; 85(5): 233–248.
104. Dellian M., et al. Quantitation and physiological characterization of angiogenic vessels in mice - Effect of basic fibroblast growth factor vascular endothelial growth factor vascular permeability factor, and host microenvironment. *American Journal of Pathology*. 1996; 149(1): 59–71.
105. Hall DM and SA Brooks. In vitro invasion assay using matrigel: A reconstituted basement membrane preparation. *Methods in Molecular Biology*. 2014; 1070: 1–11.
106. Hasan J, et al. Quantitative angiogenesis assays in vivo - A review. *Angiogenesis*. 2004; **7**(1): 1–16.
107. Coltrini D, et al. Matrigel plug assay: Evaluation of the angiogenic response by reverse transcription-quantitative PCR. *Angiogenesis*. 2013; 16(2): 469–477.
108. Kragh M, et al. In vivo chamber angiogenesis assay: An optimized matrigel plug assay for fast assessment of anti-angiogenic activity. *International Journal of Oncology*. 2003; 22(2): 305–311.
109. Gimbrone MA, et al. Tumor dormancy in vivo by prevention of neovascularization. *Journal of Experimental Medicine*. 1972; 136(2): 261–276.
110. Rogers MS, Birsner AE, and D'Amato RJ. The mouse cornea micropocket angiogenesis assay. *Nature Protocols*. 2007; 2(10): 2545–2550.
111. Ziche M. Corneal assay for angiogenesis. *Angiogenesis Protocols*. 2001; 46: 131–142.
112. Auerbach R, et al. Angiogenesis assays: A critical overview. *Clinical Chemistry*. 2003; 49(1): 32–40.
113. Nicosia RF and Ottinetti A. Growth of microvessels in serum-free matrix culture of rat aorta - A quantitative assay of angiogenesis in vitro. *Laboratory Investigation*. 1990; 63(1): 115–122.
114. Aplin AC, et al. The aortic ring model of angiogenesis, in Angiogenesis: In Vitro Systems, Cheresh DA, (Ed.) Elsevier Academic Press Inc, San Diego, 2008. 19–136.
115. Baker M, et al. Use of the mouse aortic ring assay to study angiogenesis. *Nature Protocols*. 2012; 7(1): 89–104.
116. Nowak-Sliwinska P, et al. Consensus guidelines for the use and interpretation of angiogenesis assays. *Angiogenesis*. 2018; 21(3): 425–532.
117. Stiffey-Wilusz J, et al. An ex vivo angiogenesis assay utilizing commercial porcine carotid artery: Modification of the rat aortic ring assay. *Angiogenesis*. 2001; 4(1): 3–9.
118. Muthukkaruppan V, et al. The chick embryo aortic arch assay: A new, rapid, quantifiable in vitro method for testing the efficacy of angiogenic and anti-angiogenic factors in a three-dimensional, serum-free organ culture system. *Proceedings of the American Association for Cancer Research Annual Meeting*, 2000(41): 65.

119. Rezzola S, et al. A novel ex vivo murine retina angiogenesis (EMRA) assay. *Experimental Eye Research*. 2013; 112: 51–56.
120. Marcelo KL, Goldie LC, and Hirschi KK. Regulation of endothelial cell differentiation and specification. *Circulation Research*. 2013; 112(9): 1272–1287.
121. Petzelbauer P, et al. Heterogeneity of dermal microvascular endothelial-cell antigen expression and cytokine responsiveness in situ and in cell culture. *Journal of Immunology*. 1993; 151(9): 5062–5072.
122. Schutz M and Friedl P. Isolation and cultivation of endothelial cells derived from human placenta. *European Journal of Cell Biology*. 1996; 71(4): 395–401.
123. Jaffe EA, et al. Culture of human endothelial cells derived from human umbilical cord veins. *Circulation*. 1972; 46(4): 252-&.
124. O'Brien J, et al. Investigation of the alamar blue (resazurin) fluorescent dye for the assessment of mammalian cell cytotoxicity. *European Journal of Biochemistry*. 2000; 267(17): 5421–5426.
125. Gomez D and Reich NC. Stimulation of primary human endothelial cell proliferation by IFN. *Journal of Immunology*. 2003; 170(11): 5373–5381.
126. Lamalice L, Boeuf FL, and Huot J. Endothelial cell migration during angiogenesis. *Circulation Research*. 2007; 100(6): 782–794.
127. Mastyugin V, et al. A quantitative high-throughput endothelial cell migration assay. *Journal of Biomolecular Screening*. 2004; 9(8): 712–718.
128. Valster A, et al. Cell migration and invasion assays. *Methods*. 2005; 37(2): 208–215.
129. Kubota Y, et al. Role of laminin and basement membrane in the morphological-differentiation of human-endothelial cells into capillary-like structures. *Journal of Cell Biology*. 1988; 107(4): 1589–1598.
130. Arnaoutova I, et al. The endothelial cell tube formation assay on basement membrane turns 20: State of the science and the art. *Angiogenesis*. 2009; 12(3): 267–274.
131. Guidolin D and Albertin G. Tube Formation In Vitro Angiogenesis Assay. In: *Laboratory Methods in Cell Biology: Biochemistry and Cell Culture*, Conn PM (ed.), Elsevier Academic Press Inc, San Diego, 2012. 281–293.
132. Bian C, et al. Immortalization of human umbilical vein endothelial cells with telomerase reverse transcriptase and simian virus 40 large T antigen. *Journal of Zhejiang University Science B*. 2005; 6(7): 631–636.
133. Ebos JM, et al. A naturally occurring soluble form of vascular endothelial growth factor receptor 2 detected in mouse and human plasma. *Molecular Cancer Research*. 2004; 2(6): 315–326.
134. Campbell KS. *Genetically Modified Human Natural Killer Cell Lines*. 2012, US8313943B2.
135. Sisko JT, et al. Potent 2- (pyrimidin-4-yl)amine}-1,3-thiazole-5-carbonitrile-based inhibitors of VEGFR-2 (KDR) kinase. *Bioorganic & Medicinal Chemistry Letters*. 2006; 16(5): 1146–1150.
136. Kendrew J, et al. An antibody targeted to VEGFR-2 Ig domains 4-7 inhibits VEGFR-2 activation and VEGFR-2-dependent angiogenesis without affecting ligand binding. *Molecular Cancer Therapeutics*. 2011; 10(5): 770–783.
137. Wang L, et al. Development of a robust reporter-based assay for the bioactivity determination of anti-VEGF therapeutic antibodies. *Journal of Pharmaceutical and Biomedical Analysis*. 2016; 125: 212–218.

28

Emerging Applications of Nanoparticles for Lung Cancer Diagnosis and Therapy

Raja Reddy Bommareddy[1], Sheshanka Kesani[2], and Yashwant Pathak[2,3]
[1]*Senior Scientists, Tergus Pharma, North Carolina, USA*
[2]*Taneja College of Pharmacy, University of South, Florida, USA*
[3]*Adjunct Professor, Faculty of Pharmacy, Airlangga University, Surabaya, Indonesia*

28.1 Introduction

Cancer is the main health concern throughout the world. As per the 2020 estimation, new cases of cancers diagnosed will be 1,806,590, and 606,520 people die from cancers in the United States. It is expected by 2040, there will be 29.5 million new cases per year and the cancer deaths will be 16.4 million (1). Development of cancer is a complex and multiscale biological process that is correlated with the mutations or molecular changes of vital proteins regulating the cellular functions of the body. Tumors have special physiological properties with abnormal expression of receptors, growth factors, and leaky vasculature. The main objective of any cancer therapy was to improve the patient's survival and quality of life by reducing the systemic toxicity of therapy. The challenges involved in targeting are as follows: setting the goal for the disease, investigating the agent to treat effectively, and how the therapeutic agent is given. Lung cancer has the second major cancer-related mortality in the United States among all the cancer-related deaths. The estimations of new lung cancer cases for 2020 are 228,820 and 135,720 deaths from lung cancer in the United States (1). The patients, when diagnosed with lung cancer, had advanced non small cell lung cancer, because of its asymptomatic nature, thereby necessitating specific and effective treatment after diagnosis. Genetic and epigenetic alterations in the lung epithelium are the causes of lung cancer and make it a heterogenous disease. Therapeutic failures of anti-cancer drugs resulted from the drugs' cytotoxicity and the complexity in treatments. Encapsulation of lung cancer drugs in nanoparticles may facilitate intact drug delivery, avoid first-pass metabolism, and reduce cytotoxicity to normal cells, as well as being attractive to patients. The formulated nanoparticles should facilitate entrance, deposition, retention, and permeability on targeted lung tissues, and escape phagocytosis and mucociliary clearance.

Causes of lung cancer: The major contributing factor in lung cancer is cigarette smoking. Other factors contributing to lung cancer are genetic predisposition, workplace exposure (nickel, asbestos, chromium, arsenic), and environmental exposure (radon, second-hand smoke and air pollution, rapid urbanization, lifestyle change, unhealthy regimen) (2).

28.2 Classification of Lung Cancers

Based on the histological classification, lung cancer is divided into two types: small cell lung cancer and non small cell lung cancer. The small cell lung cancer also called *oat cell cancer* because the cells look like oats under a microscope. The small cell lung cancer is a not very common but aggressive cancer and difficult to treat; it accounts for about 15% of lung cancers. These originate in the inner layer of the wall of bronchi but can migrate to other parts of the body. The earlier symptoms of this cancer are infection, breathing difficulty, and sore throat. Smoking is the main cause of this cancer and other causes are inhalation of radon gas and consuming arsenic in drinking water. The advancement of disease causes difficulty in swallowing, coughing with blood, and fatigue. Non small cell lung cancer is the most common, accounting for 85% of the total lung cancers, and grows slowly when compared to small cell lung cancer. The non small cell lung carcinoma (NSCLC) patients frequently have a poor diagnosis due to diagnostic modalities issues, late detection, increased relapse rate, and rate of metastasis. NSCLC is further classified into three subtypes: adenocarcinoma, large cell carcinoma, and squamous cell carcinoma. Diagnosis of the NSCLC was done by using immunohistochemical (IHC) staining, histological alterations, molecular and mutational genetics analysis, and imaging techniques for the confirmation of the subtype of NSCLC. The adenocarcinoma found centrally in the lung at the joint of trachea and bronchi, the part of the lung which secretes mucus and helps us to breathe. The gene mutations of KRAS, BRAF, EGFR, and ELM4-ALK were exclusive to adenocarcinoma and a complement to the histological analysis to grade accurately the progression of the disease. Squamous-cell carcinoma is the second most common NSCLC subtype and accounts for 20–30% of the all NSCLC cases, and is most common in men. This cancer originates from the central airway. However, there are no specific markers, but the expression of cytokeratin-5, desmocollin-3 and p63 is used to differentiate SCC from other NSCLC variants. Large-cell carcinoma is the third most common NSCLC subtype, accounting for 3–9% of the NSCLC. Inconclusive results were observed in 70% cases

DOI: 10.1201/9781003046547-28

based on the conventional histological methods due to the misrepresentative sampling of the tumor which, indeed, emphasises the need to investigate a more targetable phenotype in relation to which therapeutic approaches to pursue.

28.3 Detection of Lung Cancer

The early detection of lung cancer is very important as it augments the survival rate of the people affected by lung cancer (3). The early detection of cancer lowers the suffering, strain, and overall cost of the treatment for curing the disease, and it is easy to treat the cancer initially. Therefore, there is a dire need to find novel ways to detect the cancer at the early stage (4). The detection of lung cancer is by several histological and biochemical assays, imaging techniques like fluorescence bronchoscopy, chest radiograph, polymerase chain reaction (PCR), bronchial biopsy, computed tomography (CT), and sputum cytology. The above-mentioned methods need specialized equipment and make detection of tumors expensive. The low dose spiral (helical) CT technique is more sensitive than CT in detecting lung cancer, but these techniques need to be cost-effective and sensitive enough to detect the patients affected by lung cancer (5,6). The lung cancer screening with the advent of new technologies should allow accurate detection of cancer in individuals who are at risk of developing the disease (7). The primary role of screening is to detect the presence of disease in asymptomatic patients and the early screening will aid in the survival rate. On cthe other hand, the results may give false positives, leading to more damage to the patient (8). The radiography of the chest is the first procedure to check the individual suspected to have lung cancer and provides the first information, but it is not useful to determine the stage of cancer. Bronchoscopy is the procedure used for imaging and biopsy and it depends on the location of the tumor. Staging of cancer can be determined by whole-body CT scanning or positron emission tomography (PET), and the information gathered is very vital as this provides a prognosis and cancer can be managed efficiently. Many research efforts focused on the predictive biomarkers of lung cancer showed the role of specific genes in lung cancer progression (9). These biomarkers of cancer information allow us to design the therapy and route of therapeutic agent taken. The route of lung cancer treatment is determined by the stage of the cancer. The most common treatments available are surgical resection, radiotherapy, radiofrequency ablation, chemotherapy, radiotherapy, immunotherapy, and palliative therapy (10).

28.4 Treatments Available for Lung Cancer

The most common and effective option available for lung treatment is surgical resection. Patients with poor health and with Stage I cancer may be unable to undergo surgery (11). Radiotherapy is another option if surgery is not viable for the patient. The radiotherapy causes damage to the surrounding cells of the lungs which significantly affects the lung functionality. Radiotherapy is not recommended for patients with compromised pulmonary systems (12). The common side effects of traditional methods were hair loss, lymphedema, blood clot, dental and bone problems, weight loss, vomiting, and blood in urine and stools. Hence, there is a need to develop advanced approaches that effectively kill cancer cells, not harm the healthy cells, and have reduced side effects when compared to surgery and radiotherapy (13).

The first line of treatment options available for late stage (advanced) lung cancer is chemotherapy, which circulates throughout the body, ultimately destroying both cancerous and healthy tissues. The US FDA-approved drugs used for the treatment of lung cancer are abitrexate, avastin, carboplatin (14), docetaxel (15), gefitinib (16), afinitor, doxorubicin (17), folex, tethotrexate (18), and Topotecan hydrochloride (19). Platinum-based drugs such as carboplatin and cisplatin were the standard first-line chemotherapy regimens for lung cancer treatment (20). However, platinum-based regimes pose dose-limiting side effects which include cardio- and neprotoxicity, intestinal injury, anemia, and peripheral neuropathy. To overcome these untoward effects, platinum-based drugs were used in combination with other anti-cancer drugs (20). The anti-cancer drugs used for the chemotherapy lack tumor targeting ability, which also affects normal cells and inconveniences the patients (adverse side effects) (21). To overcome this problem, targeted drug delivery attracted much attention (22). The introduction of the inhibitors targeting receptors like epidermal growth factor receptor (EGFR), anaplastic lymphoma kinase (ALK) (23–25), and other small molecule inhibitors targeting the atypical protein kinase C (aPKC) (2,26–29), which are some of the major causes of lung cancer, yielded some promising results, but their clinical translation has yet to be investigated. This approach delivers the drugs to the tumor and reduces its distribution to other normal tissues and organs (30).

28.5 What Makes Treating Lung Cancer Difficult

There are many fundamental factors present at the lung cancer perplexity: unique tumor heterogenicity of each patient, diagnosis difficulty, the evident gaps in bridging innovative developments, and translating problems to achieve clinical success (31). Chemotherapy is the most widely used treatment strategy for lung cancer among all the available treatments for cancers. The major impediment retarding the clinical success of lung cancer therapies is an adequate amount of drug concentration in the tumor tissues. To overcome this uphill therapeutic challenge, drugs are given at high concentrations, which leads to adverse side effects. The very common side effect originates from the cytotoxic nature of drugs affecting the normal body tissues, apart from the cancerous tissues. These lacunae in the conventional therapies may be attributed to the lack of diagnosis approaches at the early stage of lung cancers (32). The overall survival rate is in question as a majority of patients are diagnosed at the advanced stage (metastatic state), which poses treatment challenges. Of the risk factors associated with the lung cancer, the prominent ones are tobacco smoking, carcinogens, air pollution, and second-hand smoke. The critical determinant of tumorigenesis is the microenvironment of the tumor, which encompasses mesenchymal cells, infiltrating immune cells, close association with vasculature, and extracellular matrix (33). Each patient has unique microenvironment based on the tumor's genetic background, immune architecture, and somatic cells. Furthermore, the main reason that lung cancer is incurable is the development of therapeutic resistance (34). The efficacy of the treatment was low because of the poor biodistribution and non-specific distribution leading to undesirable side effects. The new chemical entities are significantly failing in later stage clinical trials mainly due to lack of safety and efficacy. Thus, there is an emergency in developing novel therapeutic modalities to treat lung cancer effectively.

Lung Characteristics for Drug Delivery: For drug delivery, lung cancer is unique. Fast drug absorption is ensured by the lung's large surface area, rich blood supply, and thin epithelium layer. These special characteristics aid in treating lung diseases, as well as systemic applications of the drugs (35). The metabolic rate was low in lungs when compared to the liver and GI tract. Thus, drugs bypass metabolism when they are delivered straight to the lungs. The most attractive feature of drugs administered to the lungs is its non-invasiveness and chance of self-administration (36).

28.6 Emergence of Nanotechnology in Lung Cancer Treatment

The advent of nanotechnology revolutionized cancer treatment and management. Nanoparticles possess various physical and chemical properties which can be used to investigate their applications in the oncology field.

28.6.1 Advantages of Nanoparticles (37–49)

- Availability of extensive surface area per unit volume
- Tunability of electronic, magnetic, optical, and biological properties
- Ease of engineering to have various shapes, sizes, solid, hollow, and porous structures
- Can be made from diverse materials: metals, silicates, carbon, polymers, metal oxides, lipids, biomolecules
- Existence in many morphologies like cylinders, spheres, platelets
- Ability to carry any nature of drugs (hydrophilic and hydrophobic)
- Nano-drug delivery systems can overcome drug resistance
- Use for specific target drug delivery
- Use in designing novel drug delivery systems
- Improves the stability of drugs
- Can be used as both active and passive drug targeting
- Better image and diagnostic tool for early detection of cancer cells in the biologic system
- Less amount of dosage form is required
- More rapid onset of therapeutic action
- Ability to show multifunctionalities
- Very good biocompatibility and the ability to overcome clearance by the kidney

28.6.2 Disadvantages of Nanoparticles

- Synthesis process is complex
- Subtle changes in the composition of nanoparticles may yield adverse effects
- In vivo clearance and release kinetics of the drug can be complicated by their physicochemical properties
- Induction of immunologic response
- Lack of standards for nanoparticle testing
- From the regulatory perspective, there is a dire need to develop an exhaustive list of tests and a smooth and streamlined approval process which facilitate translation of nano-based drugs to the clinic

In the past two decades, there has been tremendous growth and development of drug delivery systems utilizing nanotechnology. The current research efforts in the nanomedicine are expected to yield safe, efficient, and feasible drug delivery, highly sensitive,disease monitoring, and improved imaging agents for diagnosis. On the other hand, nanomedicine research is obstructed by many challenges in bridging rapidly developing novel ideas and translating them into clinicals.

28.7 Nanoparticles in the Treatment of Lung Cancer

For the treatment of lung cancer, the use of nanoparticles unlocks new avenues to develop novel treatment strategies which are efficient and overcome the shortcomings of the traditional methods. There are many classifications of the nanoparticles used to treat lung cancers. After reviewing the literature, we classify the nanoparticles as follows (50–55) (Figure 28.1):

There are many investigations of different nanoparticles used to treat lung cancer. In this, we will focus on the very recent developments of nanotechnology in lung cancer.

Liposomal nanoparticles have been explored as drug delivery vehicles for their biocompatibility and their safety profile and the ability to carry small and large molecule therapeutics of both hydrophilic and hydrophobic nature. PEG surface modification of liposomes prolongs their half-life in circulation. The liposomes, different from conventional liposomes, are referred to as "Intelligent liposomes" or "smart liposomes." These contain a bilayer of phopsholipids and surface modifiers and these decrease the chances of multi-drug resistance (MDR), apart from precisely targeting the tumor (56). Stealth liposomes, a different type of long-circulating liposome, has more residing and circulation time, which increases the drug delivery of drug at targeted site and improved the interaction of receptors and therapeutic agent in the tumor cells (57). Paclitaxel is insoluble in aqueous solvents and use of this drug is limited due to multi-drug resistance. In the study, the paclitaxel was delivered using polysaccharide nanoparticles while overcoming issues of solubility

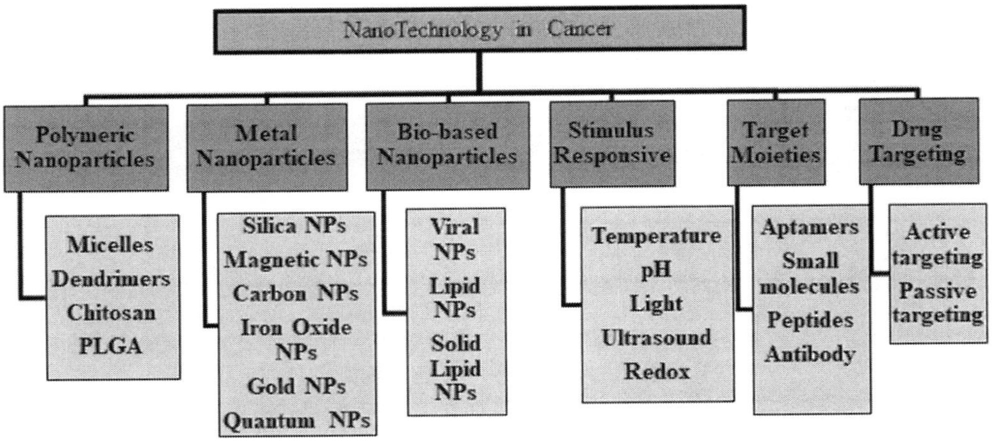

FIGURE 28.1 Classification of the Nanoparticles

and drug resistance. The biocompatible biopolymer galactoxyloglucan used to prepare nanoparticles and the study data provided the evidence that the PST-PTX nanoparticle induces apoptotic cell death and overcomes multi-drug resistance (58).

Table 28.1 presents some of the nanoparticle's advantages and applications in lung cancer treatment.

Nanoparticles are used as drug carriers and there are many research efforts exploiting the advantages of nanoparticles in this therapeutic area (Table 28.2).

28.8 Nanoparticles in the Diagnosis of Lung Cancer

In addition to the applications of nanoparticles in treatment, they can be employed for diagnostic applications. Contrast agents like radionuclides, fluorescent dyes, and gadolinium-based probes have been used for magnetic resonance imaging (MRI), positron emission tomography (PET), optical imaging, computed tomography (CT), single-photon emission computed tomography (SPECT), ultrasound (US) imaging, and photoacoustic imaging (PA) (89). The above molecular imaging methods had greater potential for the detection and diagnosis of the cancer and monitor pathological processes of the body at molecular and cellular levels, including proliferation, apoptosis, signal transduction pathways, angiogenesis, perfusion, and tumor cell metabolism (90).

The imaging technology of the cancerous tissue enables the detection of cancer at an earlier stage. The detection of metastasis in lung cancer is made easy by MRI imaging with the use of immune superparamagnetic iron oxide nanoparticles (SPIONS) (91). The SPIONS that can be delivered directly to the lungs were developed by researchers (92). The application of nanoparticles in the biosensors improves the sensitivity and detection of the test. The gold nanoparticles as biosensors were tested and

TABLE 28.1

Summary of the Different Types of Nanoparticles in the Treatment of Lung Cancer

Type of Nanoparticles	Advantages	Application in Lung Cancer Treatment	References
Liposomes	Improved stability, enhanced circulation of time of drugs, biocompatable, biodegradable	B-amino polymer used in controlled drug delivery, stealth liposomes with PEG in the composition, doxorubicin liposomes, liposomal cisplatin	(59–64)
Solid lipid nanoparticles (SLNs)	Wide range of drug adaptability, suitable for different routes of administration	SLN carrier p53, loaded SLNs with Bcl-2 siRNA	(65,66)
Polymeric nanoparticles	Ease of incorporation of water insoluble drugs, good stability, avoids macrophages phagocytosis, ease of surface functionalization, controlled drug release	Genexol-PM, peptide-Taxol (AEYLR-PNPs),	(67–69)
Dendrimers	Had strong EPR, stable nature, can accommodate multiple functional groups on surface, drug release profile can be customized	Antibody-dendrimer conjugates, PAMAM dendrimers, PEG dendrimers	(68–74)
Bio- nanoparticles	Overcome biological barriers, biocompatable, biodegradable, reduced toxicity and immunogenic response	Human serum albumin modified erlotinib NPs, microRNAs, mesenchymal stem cells (MSC) as drug delivery vehicle with NPs loaded with drugs, albumin NPs	(75–77)
Metal nanoparticles	Simple synthesis process, multifunctional surface modifications	Gold NPs, silver NPs, iron oxide NPs	(78–80)

TABLE 28.2

Summary of Nanoparticles as Drug Carriers to Treat Lung Cancer

Nanocarrier	Carrier	Drug	Vital outcomes	References
Chitosan polyplexes	Mannitol	siRNA	Improved aerosolization and dispersibility by manual grinding	(81)
PBCA NPs	Lactose	DOX	Cytotoxicity of DOX-NPs on both A549 and H460 cells was increased	(82)
SPIONS	Polyrotaxan	5-FU	Improved lung disposition of cubic nanoaggregates by lower PR content	(83,84)
Liposomes	Trehalose	ETP and DTX	Enhanced apoptosis by ETP and DTX ny pre-treatment and co-administration of p53 tumor suppressor genes	(85)
SLNs	Mannitol and leucine	TP5	Increased the bioavailability and activity of TP5	(86)
CUR NS	Mannitol	CUR	Higher cellular uptake and increased cytotoxicity to lung cancer cells	(87)
CS-PLGA NPs	Lactose, leucine, ploxamer	2-ME	Deep lung deposition improved by mucoadhesive properties of chitosan	(88)

able to differentiate lung cancer histologies. The lung cancer detection is efficient with the use of biosensors based on AuNPs and microRNAs (93). Development of biosensors made from the 11-Mercapto-1-undecanol aptamers bound to AuNPs were sensitive and selective to A549 cells (94). The detection of micrometastasis in the peripheral blood of lung cancer was achieved by the development of quantum dots linked to the NSCLC micrometastasis marker lung-specific Xprotein (LUNX) and the surfactant protein-A (SP-A) antibody (95).

Radiomics: The comprehensive method which utilizes data-mining and machine-learning advancements to analyze medical images is referred to as radiomics (96). Radiomics provide quantitative data that may aid in the accuracy of diagnosis and therapy assessments. In lung cancer diagnosis, evaluation of treatment, and prognosis, radiomics is widely used. A deep learning model based on CT images yielded more precise results for the malignant lung nodule when compared to previous methods (97,98). CT radiomic signatures combined with clinical risk factors were used to predict the distant metastasis in a Chinese cohort of 348 lung cancer patients (99)and on a US cohort of 182 pathologically confirmed lung adenocarcinoma patients (100). These studies revealed the good performance of radiomics on distant metastasis (M staging). Radiomics was also used for the prediction of the gene mutation in lung cancers and many studies provide evidence for the detection of EGFR mutation by using CT radiomic features (99). Radiomic signature data from the Chinese cohort study revealed it can serve as a diagnostic factor for the histologic subtype classification of the NSCLC. In addition, radiomics can be used to evaluate the treatment of lung cancer. The delta-radiomic characteristics were used to predict the outcomes in NSCLC stage III patients undergoing radiotherapy (101). The EGFR mutation status in NSCLC before treatment and after gefitinib response, and prediction of progression-free survival after TKI therapy was predicted by CT radiomic features (102).

28.9 Nano-theranostics

The science of integration of both diagnosis and therapeutic applications of nanoparticles is referred to as nano-theranostics and emerged as a propitious paradigm in cancer treatment. It combines the advantages of both therapeutic and diagnostic worlds: nanocarriers to ferry cargo while loading on them both therapeutic and diagnostic agents. The nano-theranostic agents offer many advantages over other theranostic agents because of the sophisticated capabilities in one single platform, which include multimodality therapy/diagnosis or quality performance (e.g. autophagy inhibition, oral delivery) (103–108), stimulus-responsive drug release (e.g. magnetism, temperature, pH, and ultrasound) (109,110), targeted delivery, and synergistic performance (e.g. combination therapy, siRNA delivery) (111–113). There are four crucial aspects in designing efficient therapeutic platforms based on nanoparticles: (a) selection of therapeutic agent, (b) choosing a suitable carrier, (c) adopting a targeting and drug release approach, and (d) carefully isolating the imaging agent. Nano-theranostics was able to monitor drug distribution in the body, drug action site, drug release patterns, and the efficacy of the therapy. There is in vivo data available for either therapeutic or diagnostic agents but not for theranostics. Currently there are only a few studies showing evidence of in vivo results of real theranostic nanomedicines.

Stimulus response	Title of the Theranostic Approach Study	Reference
Heat	Thermosensitive liposomal drug delivery systems: state of the art review	(114)
Light	Near-infrared light-activatable polymeric nanoformulations for combined therapy and imaging of cancer	(115)
Ultrasound	Mechanical force-triggered drug delivery	(116)
Magnetic field	Magnetically triggered nano-composite membranes: a versatile platform for triggered drug release	(117)
Redox	Self-cross-linked polymer nanogels: a versatile nanoscopic drug delivery platform	(118)
pH	Tailor-made dual pH-sensitive polymer–doxorubicin nanoparticles for efficient anti-cancer drug delivery	(119)
Microenvironment	Tumor targeting and microenvironment-responsive nanoparticles for gene delivery	(120)

Other triggers	Multifunctional, stimuli-sensitive nano-particulate systems for drug delivery, smart micro/nanoparticles in stimulus-responsive drug/gene delivery systems	(121,122)

28.10 Innovative Strategies Showed Promising Theranostic Applications in Treating Lung Cancer (35)

The novel strategy to enhance lung cancer treatment with five essential elements is an innovative multi-tier biotechnology treatment approach utilizing the RNA interference mechanism, induction of cell death by anti-cancer agent, local delivery of nanoparticles by inhalation (passive targeting), active targeting of the peptide system to minimize the adverse side effects of traditional agents (harming the normal cells), and constructing the tumor-targeted nanostructured lipid carriers (NLCs) to increase the stability, solubility, and cellular penetration of the drug and siRNA. The most feasible carrier for the inhalatory delivery of siRNA and drug is a lipid-based system selected after screening various nanocarriers like micelles, liposomes, polymers, dendrimers, gold, silica, and NLC nanoparticles. The criteria used for the selection is preferential accumulation and retention of a carrier, siRNA and drug in lungs when compared to other organs. The authors prepared NLC comprising positively charged drug to form complexes with negatively charged siRNAs. The anti-cancer drug, paclitaxel (TAX), which is lipophilic in nature, is used to encapsulate the lipids of NLC. Furthermore, paclitaxel has been used in clinics for the treatment of advanced NSCLC. For improving the treatment efficiency, a dual targeting approach was employed. First, the therapeutic agent was limited to the lungs by its nanoparticle's delivery by inhalation. Secondly, a luteinizing hormone-release hormone (LHRH) peptide was incorporated into the system to shift the preferential accumulation (active targeting) in cancer cells. The data from the study revealed the novel tumor-targeted LHRH-NLC-siRNAs-paclitaxel delivery system is efficient in delivering the active payloads (paclitaxel and siRNA) to cancer cells. The dual active and passive targeting strategies enabled the delivery of the toxic active components specifically to the lungs with tumor and their preferential accumulation in cancer cell, and limiting their adverse effects in the normal (non-cancerous) cells. The data from the study showed all individual components were less effective when compared to the complex tumor-targeted (LHRH-NLC-siRNAs-TAX) system.

Theranostic Applications of Gold Nanorods (123): Accessibility of lung cancer tissues is the limiting factor which makes it difficult to treat lung cancer. Because of the space limitations, a single laser fiber can be used to integrate the diagnostic and therapeutic applications and this is achieved by pulse wave, plasmonic photothermal therapy (PWPPTT technology). The optimization of AuNRs and employing a laser source enabled successful translation of AuNRs for theranostic applications. The size of the AuNR is the most influential characteristic and also exhibits SPR phenomena. The theranostic potential of AuNRs combined with PW lasers is demonstrated for application in the lung cancer.

28.11 Challenges of Nanoparticles in Cancer Treatment

Despite tremendous development and explorational research efforts into the use of nanoparticles to treat cancers, there are vital issues that need to be addressed. Some of the hurdles/challenges to overcome are as follows:

A very good understanding of nanoparticle toxicity, biocompatibility, degradation, and biodistribution is required to exploit their potential in medicinal applications. The nanoparticles' physical and chemical properties, such as their surface size, charge, and shape, will determine the biological response. The rod-shaped nanoparticles are more toxic and harmful when compared to sphere-shaped nanoparticles. New materials used for the applications of nanoparticles and assessment of their nanotoxicity is complicated by the surface modifications, which in turn alter the biologic response (57). There are many possibilities to modify the nanoparticle characteristics that have a significant impact on the biological action (64). Therefore, it is absolute necessity to explore the safety of each material individually. From the perspective of the regulatory agencies and patients, the primary concern is the toxicity of the materials used to make the nanoparticles. There is a possibility of nanoparticles to induce autophagy, which plays an important role in cancer (124). Some of the materials used are not biodegradable, which can cause serious issues that limit their use in nanomedicine. There are no controls or gold standards to validate the functioning of the materials at nanoscale. The productive cost of these nanoparticles is high; to overcome this challenge we must adopt novel production methods, and seek government support and more demand from the consumers.

28.12 Conclusion

Researchers are continuously exploring the infinite potential of nanotechnology and innovative applications in the diagnosing, detecting, imaging, and treating of various cancers. The efforts in this field have already aided in overcoming problems associated with traditional medical methods such as low therapeutic efficiency, undesired side effects, drug resistance, and non-specific targeting. Development of a wide range of applications in the NDDS have shown promising results in treating diseases with more safety, efficacy, and precision. The NDDS approaches aids in targeting the active drugs to a specific site apart from regulating the desired drug level in the blood. Nanotechnology is not ideal or flawless, despite its applications and benefits. The nanoparticles have more surface area, which results in augmented chemical reactivity leading to uncertainty on how these particles behave in different environments. The augmented chemical reactivity produces reactive oxygen radicals, which may produce inflammation, damage to proteins, DNA, and oxidative stress, ultimately leading to toxicity. In lung cancer treatment, NDDS approaches will flourish and unlock a new dimension which replaces traditional dosage approach, which in turn improves health care delivery.

REFERENCES

1. Siegel RL, Miller KD, and Jemal A. Cancer statistics, 2020. *CA: A Cancer Journal for Clinicians*. 2020 ; 70: 7–30.
2. Bommareddy RR, Patel R, Smalley T, et al. Effects of atypical protein kinase C inhibitor (DNDA) on lung cancer proliferation and migration by PKC-ι/FAK ubiquitination through the Cbl-b pathway. *OncoTargets and Therapy*. 2020.
3. Gao X, Guo L, Li J, et al. Nanomedicines guided nanoimaging probes and nanotherapeutics for early detection of lung cancer and abolishing pulmonary metastasis: Critical appraisal of newer developments and challenges to clinical transition. *Journal of Controlled Release*. 2018.
4. Oudkerk M, Devaraj A, Vliegenthart R, et al. European position statement on lung cancer screening. *The Lancet Oncology*. 2017.
5. Ng QS and Goh V. Angiogenesis in non-small cell lung cancer: Imaging with perfusion computed tomography. *Journal of Thoracic Imaging*. 2010.
6. Gibaldi A, Barone D, Gavelli G, et al. Effects of guided random sampling of TCCs on blood flow values in CT perfusion studies of lung tumors. *Academic Radiology*. 2015.
7. Tammemägi MC, Katki HA, Hocking WG, et al. Selection criteria for lung-cancer screening. *New England Journal of Medicine*. 2013.
8. Walther A, Johnstone E, Swanton C, et al. Genetic prognostic and predictive markers in colorectal cancer. *Nature Reviews Cancer*. 2009.
9. Van't Westeinde SC and van Klaveren RJ. Screening and early detection of lung cancer. *Cancer Journal*. 2011; 17(1): 3–10.
10. Report A. Cancer research UK. Annual report. 2017–2018. *FreseniusCom [Internet]*. 2019; 2–2. Available from: https://www.rtda.gov.rw/fileadmin/templates/publications/RWANDA_Annual_Report_2018–2019_SHARING.pdf.
11. Port JL, Parashar B, Osakwe N, et al. A propensity-matched analysis of wedge resection and stereotactic body radiotherapy for early stage lung cancer. *Annals of Thoracic Surgery*. 2014; 98(4): 1152–1159.
12. Hirsch FR, Suda K, Wiens J, et al. New and emerging targeted treatments in advanced non-small-cell lung cancer. *The Lancet*. 2016; 388(10048): 1012–1024.
13. Sharma P, Mehta M, Dhanjal DS, et al. Emerging trends in the novel drug delivery approaches for the treatment of lung cancer. *Chemico-Biological Interactions*. 2019; 309: 108720
14. de Sousa GF, Wlodarczyk SR, and Monteiro G. Carboplatin: Molecular mechanisms of action associated with chemoresistance. *Brazilian Journal of Pharmaceutical Sciences*. 2014; 50(4): 693–701.
15. Herbst RS and Khuri FR. Mode of action of docetaxel - a basis for combination with novel anticancer agents. *Cancer Treatment Reviews*. 2003; 29(5): 407–415.
16. Lenz HJ. Anti-EGFR mechanism of action: Antitumor effect and underlying cause of adverse events. *Oncology (Williston Park, N.Y.)*. 2006; 20(5 Suppl 2): 5–13.
17. Jackson TL. Intracellular accumulation and mechanism of action of doxorubicin in a spatio-temporal tumor model. *Journal of Theoretical Biology*. 2003; 220(2): 201–213.
18. Tian H and Cronstein B. Understanding the mechanisms of action of methotrexate. *Bulletin of the NYU Hospital for Joint Diseases*. 2007; 65(3): 168–173.
19. Palchaudhuri R and Hergenrother PJ. DNA as a target for anticancer compounds: Methods to determine the mode of binding and the mechanism of action. *Current Opinion in Biotechnology*. 2007; 18(6): 497–503.
20. Amarasena IU, Chatterjee S, Walters JAE, et al. Platinum versus non-platinum chemotherapy regimens for small cell lung cancer. *Cochrane Database of Systematic Reviews*. 2015; 2015(8): CD006849.
21. Joo WD, Visintin I, and Mor G. Targeted cancer therapy - are the days of systemic chemotherapy numbered? *Maturitas*. 2013; 76(4): 308–314.
22. Dua K, Malyla V, Singhvi G, et al. Increasing complexity and interactions of oxidative stress in chronic respiratory diseases: An emerging need for novel drug delivery systems. *Chemico-Biological Interactions*. 2019; 299: 168–178.
23. Zhang Z, Lee JC, Lin L, et al. Activation of the AXL kinase causes resistance to EGFR-targeted therapy in lung cancer. *Nature Genetics*. 2012; 44(8):852–860.
24. Paez JG, Jänne PA, Lee JC, et al. EGFR mutations in lung, cancer: Correlation with clinical response to gefitinib therapy. *Science*. 2004; 304(5676): 1497–500.
25. Korpanty GJ, Graham DM, Vincent MD, et al. Biomarkers that currently effect clinical practice in lung cancer: EGFR, ALK, MET, ROS-1 and KRAS. *Frontiers in Oncology*. 2014; 4: 204.
26. Patel R, Islam SA, Bommareddy RR, et al. Simultaneous inhibition of atypical protein kinase-C and mTOR impedes bladder cancer cell progression. *International Journal of Oncology*. 2020; 56(6): 1373–1386.
27. Smalley T, Metcalf R, Patel R, et al. The atypical protein kinase C small molecule inhibitor ζ-stat, and its effects on invasion through decreases in PKC-ζ protein expression. *Frontiers in Oncology*. 2020; 10: 209.
28. Islam SMA, Patel R, and Acevedo-Duncan M. Protein kinase C-ζ stimulates colorectal cancer cell carcinogenesis via PKC-ζ/Rac1/Pak1/β-Catenin signaling cascade. *Biochimica et Biophysica Acta - Molecular Cell Research*. 2018; 1865(4): 650–664.
29. Murray NR, Kalari KR, and Fields AP. Protein kinase Cι expression and oncogenic signaling mechanisms in cancer. *Journal of Cellular Physiology*. 2011; 226(4): 879–887.
30. Badrzadeh F, Rahmati-Yamchi M, Badrzadeh K, et al. Drug delivery and nanodetection in lung cancer. *Artificial Cells, Nanomedicine and Biotechnology*. 2016; 44(2): 618–634.
31. Cryer AM and Thorley AJ. Nanotechnology in the diagnosis and treatment of lung cancer. *Pharmacology and Therapeutics*. 2019; 198: 189–205.
32. Zhang Y, Li M, Gao X, et al. Nanotechnology in cancer diagnosis: Progress, challenges and opportunities. *Journal of Hematology and Oncology*. 2019; 12: 1–13.

33. Kandoth C, McLellan MD, Vandin F, et al. Mutational landscape and significance across 12 major cancer types. *Nature*. 2013; 502(7471): 333–339.
34. Rotow J and Bivona TG. Understanding and targeting resistance mechanisms in NSCLC. *Nature Reviews Cancer*. 2017; 17: 637–658.
35. Abdelaziz HM, Gaber M, Abd-Elwakil MM, et al. Inhalable particulate drug delivery systems for lung cancer therapy: Nanoparticles, microparticles, nanocomposites and nanoaggregates. *Journal of Controlled Release*. 2018; 269: 373–392.
36. Paranjpe M and Müller-Goymann CC. Nanoparticle-mediated pulmonary drug delivery: a review. *International Journal of Molecular Sciences*. 2014; 15(4): 5852–5873.
37. Ferrari M. Cancer nanotechnology: opportunities and challenges. *Nature Reviews Cancer*. 2005; 5: 161–171.
38. Wang X, Yang L, Chen Z, et al. Application of nanotechnology in cancer therapy and imaging. *CA: A Cancer Journal for Clinicians*. 2008; 58(2): 97–110.
39. Bertrand N, Wu J, Xu X, et al. Cancer nanotechnology: The impact of passive and active targeting in the era of modern cancer biology. *Advanced Drug Delivery Reviews*. 2014; 66: 2–25.
40. Alexis F, Rhee JW, Richie JP, et al. New frontiers in nanotechnology for cancer treatment. *Urologic Oncology: Seminars and Original Investigations*. 2008; 26(1): 74–85.
41. Sinha R, Kim GJ, Nie S, et al. Nanotechnology in cancer therapeutics: Bioconjugated nanoparticles for drug delivery. *Molecular Cancer Therapeutics*. 2006; 5(8): 1909–1917.
42. Hassanzadeh P, Fullwood I, Sothi S, et al. Cancer nanotechnology. *Gastroenterology and Hepatology from Bed to Bench*. 2011; 23(10): 2628.
43. Misra R, Acharya S, and Sahoo SK. Cancer nanotechnology: Application of nanotechnology in cancer therapy. *Drug Discovery Today*. 2010; 15(19–20): 842–850.
44. Nie S, Xing Y, Kim GJ, et al. Nanotechnology applications in cancer. *Annual Review of Biomedical Engineering*. 2007; 9: 257–288.
45. Suri SS, Fenniri H, and Singh B. Nanotechnology-based drug delivery systems. *Journal of Occupational Medicine and Toxicology*. 2007; 2: 16.
46. Cuenca AG, Jiang H, Hochwald SN, et al. Emerging implications of nanotechnology on cancer diagnostics and therapeutics. *Cancer*. 2006; 107(3): 459–466.
47. Schroeder A, Heller DA, Winslow MM, et al. Treating metastatic cancer with nanotechnology. *Nature Reviews Cancer*. 2012; 12(1): 39–50.
48. Jabir NR, Tabrez S, Ashraf GM, et al. Nanotechnology-based approaches in anticancer research. *International Journal of Nanomedicine*. 2012; 7: 4391–408.
49. Goldberg MS. Improving cancer immunotherapy through nanotechnology. *Nature Reviews Cancer*. 2019; 19(10): 587–602.
50. Brigger I, Dubernet C, and Couvreur P. Nanoparticles in cancer therapy and diagnosis. *Advanced Drug Delivery Reviews*. 2012; 54(5): 631–651.
51. Haley B and Frenkel E. Nanoparticles for drug delivery in cancer treatment. *Urologic Oncology: Seminars and Original Investigations*. 2008; 26(1): 57–64.
52. Kumar V, Gautam A, and Guleria P. Platinum nanoparticles: Synthesis strategies and application. *Nanoarchitectonics*. 2020; 9(12): 17–19.
53. Woodman C, Vundu G, George A, et al. Applications and strategies in nanodiagnosis and nanotherapy in lung cancer. *Seminars in Cancer Biology*. 2020; 69: 349–364.
54. Hossen S, Hossain MK, Basher MK, et al. Smart nanocarrier-based drug delivery systems for cancer therapy and toxicity studies: A review. *Journal of Advanced Research*. 2019; 15: 1–18.
55. Brigger I, Dubernet C, and Couvreur P. Nanoparticles in cancer therapy and diagnosis. *Advanced Drug Delivery Reviews*. 2002; 54(5): 631–651.
56. Chen J, Guo Z, Tian H, et al. Production and clinical development of nanoparticles for gene delivery. *Molecular Therapy - Methods and Clinical Development*. 2016; 3: 16023.
57. Lombardo D, Kiselev MA, and Caccamo MT. Smart nanoparticles for drug delivery application: Development of versatile nanocarrier platforms in biotechnology and nanomedicine. *Journal of Nanomaterials*. 2019; 2019: 158–164.
58. Reshma PL, Unnikrishnan BS, Preethi GU, et al. Overcoming drug-resistance in lung cancer cells by paclitaxel loaded galactoxyloglucan nanoparticles. *International Journal of Biological Macromolecules*. 2019; 136: 266–274.
59. White SC, Lorigan P, Margison GP, et al. Phase II study of SPI-77 (sterically stabilised liposomal cisplatin) in advanced non-small-cell lung cancer. *British Journal of Cancer*. 2006; 95(7): 822–828.
60. Ansari L, Shiehzadeh F, Taherzadeh Z, et al. The most prevalent side effects of pegylated liposomal doxorubicin monotherapy in women with metastatic breast cancer: A systematic review of clinical trials. *Cancer Gene Therapy*. 2017; 24: 189–193.
61. Zhang CY, Yang YQ, Huang TX, et al. Self-assembled pH-responsive MPEG-b-(PLA-co-PAE) block copolymer micelles for anticancer drug delivery. *Biomaterials*. 2012; 9: 4923–4933.
62. Men W, Zhu P, Dong S, et al. Layer-by-layer pH-sensitive nanoparticles for drug delivery and controlled release with improved therapeutic efficacy in vivo. *Drug Delivery*. 2020; 27(1): 180–190.
63. Immordino ML, Dosio F, and Cattel L. Stealth liposomes: Review of the basic science, rationale, and clinical applications, existing and potential. *International Journal of Nanomedicine*. 2006; 1(3): 297–315.
64. Kedar U, Phutane P, Shidhaye S, et al. Advances in polymeric micelles for drug delivery and tumor targeting. *Nanomedicine: Nanotechnology, Biology, and Medicine*. 2010; 6(6): 714–729.
65. Choi SH, Jin SE, Lee MK, et al. Novel cationic solid lipid nanoparticles enhanced p53 gene transfer to lung cancer cells. *European Journal of Pharmaceutics and Biopharmaceutics*. 2008; 6(3): 696–705.
66. Bae KH, Lee JY, Lee SH, et al. Optically traceable solid lipid nanoparticles loaded with sirna and paclitaxel for synergistic chemotherapy with in situ imaging. *Advanced Healthcare Materials*. 2013; 2(4): 576–584.

67. Kim DW, Kim SY, Kim HK, et al. Multicenter phase II trial of Genexol-PM, a novel Cremophor-free, polymeric micelle formulation of paclitaxel, with cisplatin in patients with advanced non-small-cell lung cancer. *Annals of Oncology*. 2007; 18(12): 2009–2014.
68. Han C, Li Y, Sun M, et al. Small peptide-modified nanostructured lipid carriers distribution and targeting to EGFR-overexpressing tumor in vivo. *Artificial Cells, Nanomedicine and Biotechnology*. 2014; 42(3): 161–166.
69. Han CY, Yue LL, Tai LY, et al. A novel small peptide as an epidermal growth factor receptor targeting ligand for nanodelivery in vitro. *International Journal of Nanomedicine*. 2013; 8: 1541–1549.
70. Liu J, Liu J, Chu L, et al. Novel peptide-dendrimer conjugates as drug carriers for targeting nonsmall cell lung cancer. *International Journal of Nanomedicine*. 2011; 6: 59–69.
71. Wu G, Barth RF, Yang W, et al. Targeted delivery of methotrexate to epidermal growth factor receptor-positive brain tumors by means of cetuximab (IMC-C225) dendrimer bioconjugates. *Molecular Cancer Therapeutics*. 2006; 5(1): 52–59.
72. Ly TU, Tran NQ, Hoang TKD, et al. Pegylated dendrimer and its effect in fluorouracil loading and release for enhancing antitumor activity. *Journal of Biomedical Nanotechnology*. 2013; 9(2): 213–220.
73. Thomas TP, Patri AK, Myc A, et al. In vitro targeting of synthesized antibody-conjucated dendrimer nanoparticles. *Biomacromolecules*. 2004; 5(6): 2269–2274.
74. Dhanikula RS and Hildgen P. Influence of molecular architecture of polyether-co-polyester dendrimers on the encapsulation and release of methotrexate. *Biomaterials*. 2007; 28(20): 3140–3152.
75. Wang X, Chen H, Zeng X, et al. Efficient lung cancer-targeted drug delivery via a nanoparticle/MSC system. *Acta Pharmaceutica Sinica B*. 2019; 9(1): 167–176.
76. Moro M, di Paolo D, Milione M, et al. Coated cationic lipid-nanoparticles entrapping miR-660 inhibit tumor growth in patient-derived xenografts lung cancer models. *Journal of Controlled Release*. 2019; 308: 44–56.
77. Shen Y and Li W. HA/HSA co-modified erlotinib–albumin nanoparticles for lung cancer treatment. *Drug Design, Development and Therapy*. 2018; 12: 2285–2292.
78. Zanganeh S, Hutter G, Spitler R, et al. Iron oxide nanoparticles inhibit tumour growth by inducing pro-inflammatory macrophage polarization in tumour tissues. *Nature Nanotechnology*. 2016.
79. Foldbjerg R, Dang DA, and Autrup H. Cytotoxicity and genotoxicity of silver nanoparticles in the human lung cancer cell line, A549. *Archives of Toxicology*. 2011.
80. Brown SD, Nativo P, Smith JA, et al. Gold nanoparticles for the improved anticancer drug delivery of the active component of oxaliplatin. *Journal of the American Chemical Society*. 2010; 132(13): 4678–4684.
81. Okuda T, Kito D, Oiwa A, et al. Gene silencing in a mouse lung metastasis model by an inhalable dry small interfering RNA powder prepared using the supercritical carbon dioxide technique. *Biological and Pharmaceutical Bulletin*. 2013; 49(1): 112.
82. Azarmi S, Tao X, Chen H, et al. Formulation and cytotoxicity of doxorubicin nanoparticles carried by dry powder aerosol particles. *International Journal of Pharmaceutics*. 2006; 319(1-2): 155–161.
83. Ragab DM, Rohani S, and Consta S. Controlled release of 5-fluorouracil and progesterone from magnetic nanoaggregates. *International Journal of Nanomedicine*. 2012; 7: 3167–3189.
84. Ragab DM and Rohani S. Cubic magnetically guided nanoaggregates for inhalable drug delivery: In vitro magnetic aerosol deposition study. *AAPS PharmSciTech*. 2013.
85. Tomoda K, Ohkoshi T, Hirota K, et al. Preparation and properties of inhalable nanocomposite particles for treatment of lung cancer. *Colloids and Surfaces B: Biointerfaces*. 2009; 71(2): 177–182.
86. Li YZ, Sun X, Gong T, et al. Inhalable microparticles as carriers for pulmonary delivery of thymopentin-loaded solid lipid nanoparticles. *Pharmaceutical Research*. 2010; 27(9): 1977–1986.
87. Taki M, Tagami T, Fukushige K, et al. Fabrication of nanocomposite particles using a two-solution mixing-type spray nozzle for use in an inhaled curcumin formulation. *International Journal of Pharmaceutics*. 2016; 511(1): 104–110.
88. Guo X, Zhang X, Ye L, et al. Inhalable microspheres embedding chitosan-coated PLGA nanoparticles for 2-methoxyestradiol. *Journal of Drug Targeting*. 2014; 22(5): 421–442.
89. Pillai G, Cox A, and Yuen L. The science and technology of cancer theranostic nanomedicines: A primer for clinicians and pharmacists. *SOJ Pharmacy & Pharmaceutical Sciences*. 2018; 5(2): 1–17.
90. Jo SD, Ku SH, Won YY, et al. Targeted nanotheranostics for future personalized medicine: Recent progress in cancer therapy. *Theranostics*. 2016; 5(4): 472–487.
91. Wan X, Song Y, Song N, et al. The preliminary study of immune superparamagnetic iron oxide nanoparticles for the detection of lung cancer in magnetic resonance imaging. *Carbohydrate Research*. 2016; 419: 33–40.
92. Stocke NA, Meenach SA, Arnold SM, et al. Formulation and characterization of inhalable magnetic nanocomposite microparticles (MnMs) for targeted pulmonary delivery via spray drying. International Journal of Pharmaceutics. 2015; 479(2): 320–328.
93. Liu S, Su W, Li Z, et al. Electrochemical detection of lung cancer specific microRNAs using 3D DNA origami nanostructures. *Biosensors and Bioelectronics*. 2015; 71(15): 57–61.
94. Mir TA, Yoon JH, Gurudatt NG, et al. Ultrasensitive cytosensing based on an aptamer modified nanobiosensor with a bioconjugate: Detection of human non-small-cell lung cancer cells. *Biosensors and Bioelectronics*. 2015; 74: 594–600.
95. Wang Y, Zhang Y, Du Z, et al. Detection of micrometastases in lung cancer with magnetic nanoparticles and quantum dots. *International Journal of Nanomedicine*. 2012; 7: 2315–2324.
96. Lambin P, Rios-Velazquez E, Leijenaar R, et al. Radiomics: Extracting more information from medical images using advanced feature analysis. *European Journal of Cancer*. 2012; 48(4): 441–446.
97. Limkin EJ, Sun R, Dercle L, et al. Promises and challenges for the implementation of computational medical

imaging (radiomics) in oncology. *Annals of Oncology*. 2017;28(6): 1191–1206.
98. Hawkins S, Wang H, Liu Y, et al. Predicting malignant nodules from screening CT scans. *Journal of Thoracic Oncology*. 2016; 11: 2120–2128.
99. Zhou H, Dong D, Chen B, et al. Diagnosis of distant metastasis of lung cancer: Based on clinical and radiomic features. *Translational Oncology*. 2018; 11(1): 31–36.
100. Coroller TP, Grossmann P, Hou Y, et al. CT-based radiomic signature predicts distant metastasis in lung adenocarcinoma. *Radiotherapy and Oncology*. 2015; 114(3): 345–350.
101. Huynh E, Coroller TP, Narayan V, et al. CT-based radiomic analysis of stereotactic body radiation therapy patients with lung cancer. *Radiotherapy and Oncology*. 2016; 120(2): 258–266.
102. Wu W, Parmar C, Grossmann P, et al. Exploratory study to identify radiomics classifiers for lung cancer histology. *Frontiers in Oncology*. 2016; 6: 71.
103. Xu C, Mu L, Roes I, et al. Nanoparticle-based monitoring of cell therapy. *Nanotechnology*. 2011; 22(49): 494001.
104. Caldorera-Moore ME, Liechty WB, and Peppas NA. Responsive theranostic systems: Integration of diagnostic imaging agents and responsive controlled release drug delivery carriers. *Accounts of Chemical Research*. 2011; 44(10): 1061–1070.
105. Mei L, Zhang Z, Zhao L, et al. Pharmaceutical nanotechnology for oral delivery of anticancer drugs. *Advanced Drug Delivery Reviews*. 2013; 65(6): 880–890.
106. Ma X, Zhao Y, and Liang XJ. Theranostic nanoparticles engineered for clinic and pharmaceutics. *Accounts of Chemical Research*. 2011; 44(10): 1114–1122.
107. Lammers T, Aime S, Hennink WE, et al. Theranostic nanomedicine. *Accounts of Chemical Research*. 2011; 44(10): 1029–1038.
108. Smith BA and Smith BD. Biomarkers and molecular probes for cell death imaging and targeted therapeutics. *Bioconjugate Chemistry*. 2012; 23(10): 1989–2006.
109. Muthu MS, Rajesh C v, Mishra A, et al. Stimulus-responsive targeted nanomicelles for effective cancer therapy. *Nanomedicine*. 2009; 4(6): 657–667.
110. Muthu MS and Singh S. Targeted nanomedicines: Effective treatment modalities for cancer, AIDS and brain disorders. *Nanomedicine*. 2009; 4(1): 105–118.
111. Ozpolat B, Sood AK, and Lopez-Berestein G. Nanomedicine based approaches for the delivery of siRNA in cancer. *Journal of Internal Medicine*. 2010; 267(1): 44–53.
112. Tokatlian T and Segura T. siRNA applications in nanomedicine. *Wiley Interdisciplinary Reviews: Nanomedicine and Nanobiotechnology*. 2010; 2(3): 305–315.
113. Miele E, Spinelli GP, Miele E, et al. Nanoparticle-based delivery of small interfering RNA: Challenges for cancer therapy. *International Journal of Nanomedicine*. 2012; 7: 3637–3657.
114. Kneidl B, Peller M, Winter G, et al. Thermosensitive liposomal drug delivery systems: State of the art review. *International journal of nanomedicine*. 2014; 9: 4387–4398.
115. Yue X, Zhang Q, and Dai Z. Near-infrared light-activatable polymeric nanoformulations for combined therapy and imaging of cancer. *Advanced Drug Delivery Reviews*. 2017; 13(5): 1607–1616.
116. Zhang Y, Yu J, Bomba HN, et al. Mechanical force-triggered drug delivery. *Chemical Reviews*. 2016; 116(19): 12536–12563.
117. Hoare T, Timko BP, Santamaria J, et al. Magnetically triggered nanocomposite membranes: A versatile platform for triggered drug release. *Nano Letters*. 2011; 11(3): 1395–1400.
118. Ryu JH, Chacko RT, Jiwpanich S, et al. Self-cross-linked polymer nanogels: A versatile nanoscopic drug delivery platform. *Journal of the American Chemical Society*. 2010; 132(48): 17227–17235.
119. Du JZ, Du XJ, Mao CQ, et al. Tailor-Made dual pH-sensitive polymer-doxorubicin nanoparticles for efficient anticancer drug delivery. *Journal of the American Chemical Society*. 2011; 133(44): 17560–17563.
120. Huang S, Shao K, Kuang Y, et al. Tumor targeting and microenvironment-responsive nanoparticles for gene delivery. *Biomaterials*. 2013; 34(21): 5294–5302.
121. Torchilin VP. Multifunctional, stimuli-sensitive nanoparticulate systems for drug delivery. *Nature Reviews Drug Discovery*. 2014; 13: 813–827.
122. Karimi M, Ghasemi A, Sahandi Zangabad P, et al. Smart micro/nanoparticles in stimulus-responsive drug/gene delivery systems. *Chemical Society Reviews*. 2016; 45(5): 1457–1501.
123. Knights OB and McLaughlan JR. Gold nanorods for light-based lung cancer theranostics. *International Journal of Molecular Sciences*. 2018; 19(11): 3318.
124. Mahmud A, Xiong XB, Aliabadi HM, et al. Polymeric micelles for drug targeting. *Journal of Drug Targeting*. 2007; 15(9): 553–584.

29

Metallic Nanoparticles: Technology Overview and Drug Delivery Applications in Lung Cancer

Komal Parmar[1] **and Jayvadan Patel**[2]
[1]*ROFEL, Shri G.M. Bilakhia College of Pharmacy, Gujarat, India*
[1]*Nootan Pharmacy College, Faculty of Pharmacy, Sankalchand Patel University, Gujarat, India*

29.1 Introduction

The term *cancer* describes irrational growth of cells using nutrition from the healthy neighbor living cells. Further, it will lead to secondary growth of tumor in the body which can be malignant or benign. The cancer can be treated if diagnosed at an early stage of ailment. Over 200 types of cancer exist based on various systems. Various factors like lifestyle, age, genetics, and environment are responsible for the cause of cancer. To date, world-wide, cancer is the second major cause of death after heart disease(1) of which lung cancer is the leading cause of cancer-related mortality among both men and women (2). Smoking and air pollution are strongest risk factors for lung cancer (3). Prognosis of lung cancer remains poor with diagnosis rate of 85% of patients at advanced stage (4), and exiguous survival rate of less than 5 years suggests a serious need for efficient diagnosis and therapy. Lung cancer occurs in epithelial cells and is also known as bronchogenic carcinoma. It is broadly classified into small cell lung cancer and non small cell lung cancer, which is comprised of adenocarcinoma, squamous cell cancer, and large cell cancer. Small cell lung carcinoma is the most belligerent form of cancer, caused due to exposure radon gas and substantial intake of tobacco (5).

Early diagnosis and suitable therapy may improve the survival rate in lung cancer patients. Standard therapy includes surgery, chemotherapy, and radiotherapy, depending on the stage of carcinoma. Chemotherapy is the most widely used therapy in the advanced stage of lung cancer. Poor drug therapeutic availability in tumor tissue is the major encumbrance in the successful clinical therapy of lung carcinoma. To overcome this challenge, large doses of neoplastic active agents are given in a repeated manner. This can further seed adverse effects due to exposure to a large concentration of anti-cancer drug in non-cancerous healthy cells. In addition to side effects, resistance to chemotherapy is also observed due to genetic mutations (6). Novel therapeutic tactics should be explored for more effectual medication for lung carcinoma.

The presumption of targeting the carcinoma cells has been proved with the advent of nanotechnology (7,8). Nanoparticles are ultrafine particles with a size range from 1 to 100 nm. Recently extensive research has been carried out with these nanoparticles for the treatment of resilient diseases, such as cancer. An advantage associated with nanoparticles is with the high surface area to volume ratio with desirable surface chemistry that can be manipulated. Nanoparticles have attracted targeted drug delivery of both hydrophilic and hydrophobic drugs on account of their small size and high surface energy (9). Additionally, nanoparticles possess versatile thermal, optical, magnetic, and electrical properties, which can be manipulated further for fabrication of a desired material (10,11). On account of the remarkable properties of nanoparticles, they offer a scope of merits including site specific targeting, stability improvement, dose reduction, controlled drug release, better diagnostic approach, high biocompatibility, and the ability to overcome biological barriers. Site specific targeting by nanoparticles is achieved by either of the mechanisms, namely passive and active targeting (12). Mainly three types of nanoparticles are employed in the treatment of lung cancer, viz. natural and semi-synthetic nanoparticles, organic nanoparticles, and inorganic nanoparticles. Figure 29.1 depicts various types of nanoparticles employed for drug delivery.

Inorganic nanoparticles are also known as metallic nanoparticles, made up of metal or metal oxides. Over the years metal nanoparticles have demonstrated remarkable therapeutic effects in the medical field. However, recent developments have emerged improving the modeling and designing of various medical and biological applications. These metal-based nanoparticles include metals such as gold, silver, copper, iron, zinc, titanium, cerium, platinum, and many more. Compounds of metals such as oxides, sulfides, hydroxides, phosphates, chlorides, and fluorides are also used in the fabrication of metallic nanomaterials. Figure 29.2 illustrates various types of metal nanoparticles. The existence of metal-based nanoparticles in solution form is evident now (13). Metal nanoparticles are widely used in both nano-pharmaceuticals and diagnostic tools. Metal nanoparticles are broadly used as contrast agents in computed tomography scans due to their high X-ray attenuation and high atomic number (14). Furthermore, nanosized metal particles are observed to possess bactericidal properties (15). The probable mechanism involved include the production of reactive oxygen species, cation release, biomolecule damages, ATP depletion, and membrane interaction (16).

DOI: 10.1201/9781003046547-29

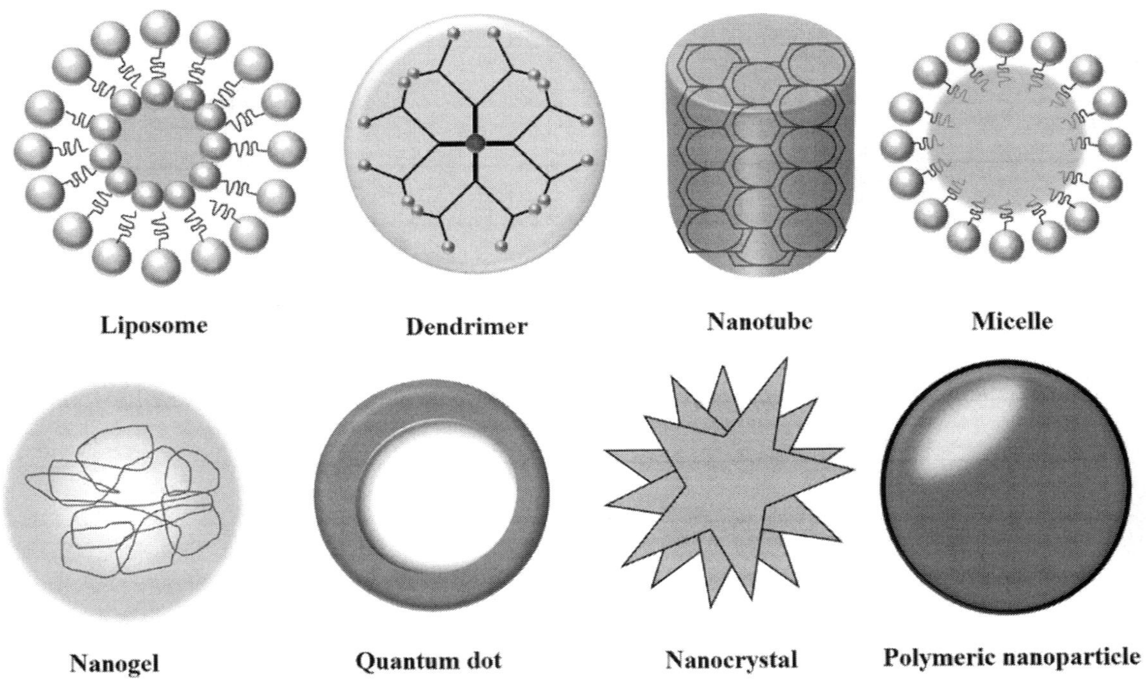

FIGURE 29.1 Various Types of Nanoparticles

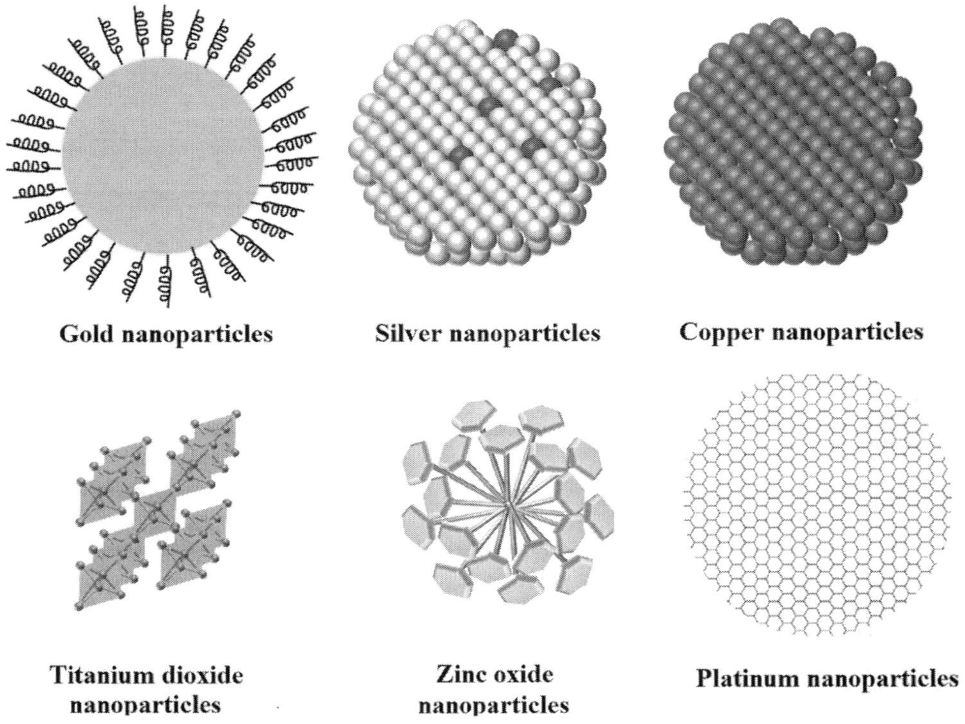

FIGURE 29.2 Different Types of Metal Nanoparticles

29.2 Characteristics of Metal Nanoparticles

From the advent of civilization, metals have been with humankind. Initial use of metal was dependent on the bulk characteristics of metal, for instance, strength, ductility, hardness, melting point, and thermal and electrical conductance. Owing to the beneficial properties of metal, it is contemplated that metal nano systems will change science and technology. Size and shape of metal nanoparticles are the crucial factors influencing most of their characteristics. For instance, metal nanoparticles of different sizes and shapes demonstrate different properti,es like color, melting point, electrical conductance, optical properties, therapeutic efficacy, and so on (17,18).

A nanoparticle consists of thousands of atoms and has a size range of 10–100 nm. Due to the extremely small size of nanoparticles, they are known to influence complicated regions of the biological system (19,20). Nanoparticles are not just tiny parts of a bulk counterpart, but along with their small size, they exhibit extraordinary properties. Since the size of the matter falls to the nanometer scale, quantum effect starts governing and emergence of spatial confinement of electrons, phonons and electrical field of a nanoparticle happens. For example, enhancement of the surface plasmon local field due to electron confinement in the metal nanoparticle resulting in alteration of its spectral properties (21). In addition, nanoparticles have a large surface to volume ratio, which in turn significantly affects nanomaterial properties. For illustration, gold nanoparticles' enhanced drug efficacy towards HeLa cancer cells (22). The shape of nanoparticles is just as important as their nano size. Significant changes in properties of the metallic nanoparticles have been found by a small change in shape. In a study, three different shapes, viz. nanorods, nanostars, and nanospheres of chitosan-gold nanoparticles were synthesized to investigate cellular uptake and cytotoxicity in cancer cells. They observed that nanospheres were taken up more efficiently than nanorods, followed by nanostars. Cytotoxicity was highest in nanorods, then nanostars followed by nanospheres (23,24). Another study involving gold nanoparticles of varying size and shape showed influence of both factors on cellular uptake of siRNA nanoconstructs. It was observed that 50 nm spheres and 40 nm stars showed higher potential for the delivery of siRNA (25). Researchers reported cytotoxicity of gold nanoparticles of various shapes. Results illustrated the highest anti-cancer potential of gold nanostars over gold nanorods and nanospheres. This may be associated with high drug loading due to large surface area of the complex-shaped nanostar (26). In the case of the cage shape of iron oxide, the nanoparticle showed enhanced riluzole cytotoxicity against metastatic cancer cells as compared to solid nanospheres (27). Scholars reported particle shape and size-dependent anti-bacterial activity for a number of silver nanoparticles. For instance, spherical silver nanoparticles demonstrated bactericidal characteristics with the smallest size (15 nm) as compared to triangular and large spherical shaped silver nanoparticles (28).

29.3 Different Types of Metal Nanoparticles, Synthesis and Preparation Methods

Advances in research and development of metallic nanoparticles have introduced numerous new metal nanosystems with varied properties. This has paved the way for scientific research and applications in medicine. Table 29.1 illustrates various metal nanoparticles and their applications. Stability of nanoparticles with desired morphology and properties is the foremost requirement, followed by their diverse applications. Thus, the fabrication technique for the preparation of desired nanoparticles becomes a crucial part of nanotechnology. Method of synthesis of metallic nanoparticles can be broadly classified into two categories, viz. top-down approach (diminution of bulk material) and bottom-up approach (nanomaterial from precursors). Literature abounds with the preparations of metallic nanoparticles using different approaches such as physical, chemical, enzymatic, and biological. Table 29.2 demonstrates metal nanoparticles prepared by different techniques. These fabrication techniques are flexible, technologically affordable, and relatively straightforward to implement. Various additives involved in the method of preparation comprised of surfactants, ligands for surface functionalization, and polymers.

29.4 Application of Metal Nanoparticles in Lung Cancer

Metal nanoparticles are successfully exploited in various types of cancer therapy. Noble metals and others are extensively investigated for clinical applications of lung cancer treatment including imaging diagnosis, detection, and classification of lung cancer. Recently a gold nanoparticle contrast agent was developed to target the epidermal growth factor receptor expressed on lung adenocarcinoma. Results illustrated significant tumor gold accumulation of gold nanoparticles conjugated with cetuximab (66). In another study, cysteamine capped gold nanoparticles were fabricated for early detection of lung cancer specific miRNA-25 in human blood plasma. Under ideal conditions, the projected nanogenosensor proved to have two miR-25 concentration ranges of 1.0×10^{-12} to 1.0×10^{-10} M and 1.0×10^{-10} to 1.0×10^{-6} M and a limit of detection of 2.5×10^{-13} M without using polymerase chain reaction (PCR) or any other augmentation technique. In addition, it exhibited the ability of distinguishing fully matched miRNA strand and miRNA strand having single stage mutation (67). Daraee and colleagues prepared gold nanoparticle-oligonucleotide conjugate to trace the sequence of the hnRNPB1 as a lung cancer biomarker. Uniform GNPs were synthesized by the citrate reduction technique and were found to be better than PCR-based techniques (68). A gold nanoparticle based colorimetric biosensor was developed for detection of lung cancer. The expression of cellular fibronectin is increased mainly due to promotion of fibroblasts proliferation explicitly in non small cell lung carcinoma (NSCLC). In this study A549 lung cancer

TABLE 29.1

Variety of Nanoparticles and Their Medicinal Use

Type of Metal Nanoparticle	Drug/Active Ingredient	Application	Reference
Titanium dioxide nanoparticles	UV-activated titanium dioxide nanoparticles	Antibacterial efficacy against range of Gram-positive (*Staphylococcus aureus, Bacillus cereus, Lactobacillus casei, Lactobacillus bulgaricus, Lactobacillus acidophilus* and *Lactobacillus lactis*) and Gram-negative (*Salmonella enterica* var. *Enteridis* and *Escherichia coli*) bacteria	(29)
Nickel oxide nanoparticles	Extract of Rauvolfia Serpentina	Anti-pesticide efficacy against *Callosobruchus maculatus*	(30)
Platinum nanoparticles	Platinum	Cytotoxic activity on human lung adenocarcinoma (A549), ovarian teratocarcinoma (PA-1), pancreatic cancer (Mia-Pa-Ca-2) cells, and normal peripheral blood mononucleocyte (PBMC) cells	(31)
Platinum nanoparticles	Ononidis radix extract	Efficacy on human non small cell lung carcinoma cells A549	(32)
Iron oxide nanoparticles	Iron oxide	Anti-bacterial activity against *S. aureus*, *K. pneumoniae* and *S. typhi*	(33)
Iron oxide nanoparticles	Iron oxide	Anti-bacterial activity against gram +ve bacteria *Staphylococcus aureus*, gram -ve bacteria *Shigella dysenteriae*, and *Escherichia coli*	(34)
Zinc oxide, iron oxide, and copper nanopowder	Zinc oxide, iron oxide, and copper	Cytotoxicity study	(35)
Zinc oxide nanoparticles	Frizzled-7 antibody	Drug delivery to breast cancer cells	(36)
Zinc oxide nanoparticles	Doxorubicin	Multitarget carrier for cancer	(37)
Zinc oxide nanoparticles functionalized with phenylboronic acid	Curcumin	Breast cancer therapy	(38)
Gadolinium oxide, samarium oxide and erbium oxide nanoparticles	Gadolinium oxide, samarium oxide, and erbium oxide	Anti-microbial activity against *Staphylococcus aureus*, *Enterococcus faecalis*, *Escherichia coli*, *Pseudomonas aeruginosa*	(39)
Ultrasmall silica-based bismuth gadolinium nanoparticles		Image-guided radiation therapy	(40)
Mesoporous silica nanoparticles		Antibacterial activity against both gram-positive and gram-negative	(41)
Mesoporous silica nanoparticles	Doxorubicin and paclitaxel	Anti-cancer therapy for breast cancer	(42)
Titanium dioxide nanoparticle	Doxorubicin	Sonodynamic chemotherapy	(43)
Silver nanoparticles	Transferrin	For chemotherapy	(44)
Polyethylenimine-functionalized silver nanoparticle	Paclitaxel	Chemotherapy against HepG2 cells	(45)
Gold nanoparticles	Doxorubicin	As tumor targeted drug delivery	(46)
Gold nanoparticles	Paclitaxel	Chemotherapy against pancreatic cancer	(47)
Gold nanoparticles	Gold	Anti-bacterial activity against gram-negative (*Pseudomonas aeruginosa* and *Escherichia coli*) and gram-positive bacteria (*Staphylococcus aureus* and *Bacillus* sp.)	(48)

TABLE 29.2

Metal Nanoparticle Fabrication Techniques

Method of Preparation	Metal Nanoparticle	Parameters Assisting the Said Method	Reference
Biosynthesis	Silver nanoparticles	Fungal mediated synthesis	(49)
	Gold nanoparticles	Bacterium mediated synthesis	(45)
	Titanium dioxide nanoparticles	Extract of cola nitida	(50)
	Titanium dioxide nanoparticles	Fruit's peel agro-waste extracts	(51)
Chemical method	Silver nanoparticles	Electrodeposition method	(52)
	Silver and gold nanoparticles	Sol gel synthesis	(53)
	Gold nanoparticles	Turkevich method	(54)
	Tin oxide nanorods	Chemical vapor deposition method	(55)
	Magnetite nanoparticles	In situ precipitation method	(56)
	Silver-gold bi-metallic nanoparticles	Wet-chemical method	(57)
Physical method	Gold nanoparticles	Ultrasonic spray pyrolysis	(58)
	Titanium dioxide nanoparticle	Plasma arcing method	(59)
	Gold nanoparticle	Thermal evaporation	(60)
	Gold-silica nanoparticles	Ablation technique	(61)
	Platinum-copper alloy nanoparticles	Sputter deposition	(62)
	Silver nanoparticles	Layer-by-layer technique	(63)
	Titanium dioxide nanoparticles	Diffusion flame method	(64)
	Silver nanoparticles	Microwave assisted	(65)

cells whereas cultured human skin (AGO-1522) was studied for the expression of fibronectin. Results depicted higher aggregation of gold nanoparticles in the cell line due to a higher amount of fibronectin. The method developed was found to be simple and easy in order to detect the lung cancer mainly using colorimetry techniques (69). Magnetic nanoparticles were developed for the detection of micrometastasis in lung cancer. Magnetic nanoparticles conjugated with the epithelial tumor cell marker pan-cytokeratin competently isolated circulating tumor cells. The cells were further identified using quantum dots coupled to the NSCLC micrometastasis marker lung-specific X protein and surfactant protein-A antibody (70).

Additionally, metal nanoparticles are employed to delivery anti-cancer drugs for enhanced therapeutic effectiveness in lung cancer. Drug resistance in chemotherapy remains a major challenge in cancer therapy. Recently gold nanoparticles were developed for docetaxel drug delivery for lung cancer treatment. Metal nanoparticles were prepared using a chemical reduction method, followed by conjugation of docetaxel and folic acid. The developed nanoparticles showed excellent binding specificity and enhanced cytotoxicity against the lung cancer cell line H520 (71). Gold nanoparticles conjugated with doxorubicin stabilized with polyvinyl pyrrolidone were developed for effective treatment of A549, H460, and H520 human lung cancer cells. The study demonstrated better upregulation of the expression of tumor suppressor genes than free drug and induced intrinsic apoptosis in lung cancer cells (72). Hyaluronic acid-linked selenium nanoparticles were fabricated to enhance the therapeutic efficacy of paclitaxel in lung cancer treatment. The conjugated nanoparticles showed greater uptake in A549 cells and triggered apoptosis (73).

Platinum nanoparticles were synthesized using green technology to investigate the anti-cancer efficiency of cisplatin. Cisplatin conjugated platinum nanoparticles were studied for in vitro cytotoxicity. Results illustrated inhibition of growth of A549 human lung cancer cells successfully (74).

Titanium dioxide nanoparticles was synthesized for targeted delivery of the anti-cancer drug, paclitaxel, by grafting folic acid (FA) onto the PEGylated titanium dioxide nanoparticles (TiO_2–PEG–FA–PAC NPs) by the wet chemical method. Intracellular uptake of paclitaxel conjugated nanoparticles was observed as a result of the receptor-mediated endocytosis. Further enhanced cytotoxic effect was observed by TiO_2–PEG–FA–PAC NPs with folate receptors on HepG2 cells (75).

Magnetic nanoparticles have also demonstrated enhanced action on inhibition of non small cell lung cancer growth. In one such study, paclitaxel loaded core shell magnetic nanoparticles were developed using biodegradable poly(lactic-co-glycolic acid) as the polymer shell through coaxial electrospraying for targeting A549 cells. Nanoparticles inhibited the growth of A549 cells more effectively (76). Recently ultrasmall platinum nanoparticles were synthesized for delivery of gemcitabine for NSCLC tumor. The pH/redox dual stimuli-responsive clustered nanoparticle exhibited enhanced in vivo anti-tumor efficacy of gemcitabine (77).

Metal nanoparticles are found to be potent against cancer cells and themselves act as a therapeutic agent for the treatment. Various studies demonstrate anti-cancer activity of metal nanoparticles in lung cancer. For instance, recently colloidal silver nanoparticles were reported for in vitro cytotoxicity studies against A549 and HOP-62 human lung cancer cells. The silver nanoparticles were prepared using *Momordica charantia* fruit extract. IC50 values of 51.93 μg/ml and 76.92 μg/ml were observed for A549 and HOP-62 human lung cancer cells respectively (78). Gold nanoparticles synthesized using an aqueous extract of *Sesuvium portulacastrum* L. were studied for anti-cancer efficiency of nanoparticles against an A549 lung cancer cell line and the results illustrated that an IC50 dose effectively induces apoptosis and necrosis of A549 cells. Generation of oxidative stress and reactive oxygen species by gold nanoparticles might be the possible mechanism for apoptosis (79). In another study, anti-tumor activity of green-synthesized silver nanoparticles against lung cancer was demonstrated in vitro and in vivo. Cytotoxicity effect was investigated on human lung cancer H1299 cells in vitro by MTT and trypan blue assays. Apoptosis was measured by morphological assessment, and nuclear factor-κB (NF-κB) transcriptional activity was determined by a luciferase reporter gene assay. Silver nanoparticles significantly suppressed the H1299 tumor growth in lung carcinoma (80). Synergistic action of shikonin and silver nanoparticles was investigated on apoptosis of human lung cancer cells A549. The 50% inhibitory concentration (IC50) of shikonin and silver nanoparticles on A549 cells after 24 hours, determined by an MTT assay, was found to be 2.4 ± 0.11 μg/mL, revealing significant inhibited cell viability and proliferation of A549 cells (81). Biosynthesized copper oxide nanoparticles promoted anti-tumor activity in A549 lung cancer cells by inhibition of histone deacetylase. Green synthesized nanoparticles suppressed class I, II, and IV HDACs mRNA expression in A549 cells (82). Platinum nanoparticles were fabricated using phytochemicals as a reducing agent. Platinum nanoparticles showed strong anti-cancer activity against lung cancer cell line A549 (83). Wang and colleagues demonstrated cytotoxicity, DNA damage, and apoptosis were induced by titanium dioxide nanoparticles in human NSCLC-A549 cells. Mechanism primarily involved in the apoptosis was the activation of the intrinsic mitochondrial pathway (84). Zinc oxide nanoparticles were biosynthesized using *Mangifera indica* leaves. Cytotoxicity evaluation represented the noteworthy cytotoxic effect of zinc oxide nanoparticles against the A549 lung cancer cell line (85).

29.5 Conclusion

Metal nanoparticles pose an effective novel drug delivery for lung cancer therapy. They have the ability to solve the conventional problems associated with traditional chemotherapy. In addition to being nanocarriers for efficient drug delivery, metal nanoparticles themselves possess potent anti-cancer properties. Apart from the therapeutic benefits, metal nanoparticles provide a diagnostic tool which will help to revolutionize cancer treatment. Thus, testing therapies that exploit the specific beneficial properties of metallic nanoparticles is a field where cancer immunotherapy can be dedicated to clinically acceptable metal nanoparticles.

REFERENCES

1. Siegel RL, Miller KD, and Jemal A. Cancer statistics, 2020. *CA: A Cancer Journal for Clinicians*. 2020; 70(1): 7–30.
2. Dela Cruz CS, Tanoue LT, and Matthay RA. Lung cancer: Epidemiology, etiology, and prevention. *Clinics in Chest Medicine*. 2011; 32(4): 605–644.
3. Zhou G. Tobacco, air pollution, environmental carcinogenesis, and thoughts on conquering strategies of lung cancer. *Cancer Biology & Medicine*. 2019; 16(4): 700–713.
4. Tas F, Ciftci R, Kilic L, and Karabulut S. Age is a prognostic factor affecting survival in lung cancer patients. *Oncology Letters*. 2013; 6(5): 1507–1513.
5. Torres-Durán M, Ruano-Ravina A, Kelsey KT, Parente-Lamelas I, Provencio M, Leiro-Fernández V, Abal-Arca J, Montero-Martínez C, Vidal-Garcia I, Pena C, Castro-Añón O, Golpe-Gómez A, Martínez C, Guzmán-Taveras R, Mejuto-Martí MJ, Fernández-Villar A, and Barros-Dios JM. Small cell lung cancer in never-smokers. *European Respiratory Journal*. 2016; 47 (3): 947–953.
6. Sosa Iglesias V, Giuranno L, Dubois LJ, Theys J, and Vooijs M. Drug resistance in non-small cell lung cancer: A potential for NOTCH targeting? *Frontiers in Oncology*. 2018; 8: 267. doi: 10.3389/fonc.2018.00267
7. Asghari F, Khademi R, Esmaeili Ranjbar F, Veisi Malekshahi Z, and Faridi Majidi R. Application of nanotechnology in targeting of cancer stem cells: A review. *International Journal of Stem Cells*. 2019; 12(2): 227–239.
8. Wang MD, Shin DM, Simons JW, and Nie S. Nanotechnology for targeted cancer therapy. *Expert Review of Anticancer Therapy*. 2007; 7(6): 833–837.
9. Espanol L, Larrea A, Andreu V, Mendoza G, Arruebo M, Sebastian V, Aurora-Prado MS, Kedor-Hackmann ERM, Santoro MIRM, and Santamaria J. Dual encapsulation of hydrophobic and hydrophilic drugs in PLGA nanoparticles by a single-step method: Drug delivery and cytotoxicity assays. *RSC Advances*. 2016; 6(112): 111060–111069.
10. Kononenko OV, Red'kin AN, Panin GN, Baranov AN, Firsov AA, Levashov VI, and Matveev VN, Vdovin EE. Study of optical, electrical and magnetic properties of composite nanomaterials on the basis of broadband oxide semiconductors. *Nanotechnologies in Russia*. 2009; 4: 822–827.
11. Martin-Gallego M, Verdejo R, Khayet M, Ortiz de Zarate JM, Essalhi M, and Lopez-Manchado MA. Thermal conductivity of carbon nanotubes and graphene in epoxy nanofluids and nanocomposites. *Nanoscale Research Letters*. 2011; 6: 610. https://doi.org/10.1186/1556-276X-6-610
12. Attia MF, Anton N, Wallyn J, Omran Z, and Vandamme TF. An overview of active and passive targeting strategies to improve the nanocarriers efficiency to tumour sites. *Journal of Pharmacy and Pharmacology*. 2019; 71(8): 1185–1198.
13. Edwards PP, and Thomas JM. Gold in a metallic divided state—from faraday to present-day nanoscience. *Angewandte Chemie*. 2007; 46(29): 5480–5486.
14. Aslan N, Ceylan B, Koc MM, and Findik F. Metallic nanoparticles as X-ray computed tomography (CT) contrast agents: A review. *Journal of Molecular Structure*. 2020; 1219: 128599. https://doi.org/10.1016/j.molstruc.2020.128599
15. Sánchez-López E, Gomes D, Esteruelas G, Bonilla L, Lopez-Machado AL, Galindo R, Cano A, Espina M, Ettcheto M, Camins A, Silva AM, Durazzo A, Santini A, Garcia ML, and Souto EB. Metal-based nanoparticles as antimicrobial agents: An overview. *Nanomaterials (Basel)*. 2020; 10(2): 292. doi: 10.3390/nano10020292
16. Slavin YN, Asnis J, Hafeli UO, and Bach H. Metal nanoparticles: Understanding the mechanisms behind antibacterial activity. *Journal of Nanobiotechnology*. 2017; 15: 65. https://doi.org/10.1186/s12951-017-0308-z
17. Dong YC, Hajfathalian M, Maidment PSN, Hsu JC, Naha PC, Si-Mohamed S, Breuilly M, Kim J, Chhour P, Douek P, Litt HI, and Cormode DP. Effect of gold nanoparticle size on their properties as contrast agents for computed tomography. *Scientific Reports*. 2019; 9: 14912. https://doi.org/10.1038/s41598-019-50332-8
18. Sirohi S, Mittal A, Nain R, Jain N, Singh R, Dobhal S, Pani B, and Parida D. Effect of nanoparticle shape on the conductivity of Ag nanoparticle poly(vinyl alcohol) composite films. *Polymer International*. 2019; 68 (12): 1961–1967.
19. Aueviriyavit S, Phummiratch D, and Maniratanachote R. Mechanistic study on the biological effects of silver and gold nanoparticles in Caco-2 cells–induction of the Nrf2/HO-1 pathway by high concentrations of silver nanoparticles. *Toxicology Letters*. 2014; 224(1): 73–83.
20. Vieira LFA, Viana IMMN, Lins MP, Dos Santos JE, Smaniotto S, Reis MDDS. Metallic nanoparticles reduce the migration of human fibroblasts in vitro. *Nanoscale Research Letters*. 2017;12(1): 200. doi: 10.1186/s11671-017-1982-3
21. Pustovit VN and Shahbazyan TV. Quantum-size effects in SERS from noble-metal nanoparticles. *Microelectronics Journal*. 2005; 36 (3-6): 559–563.
22. Farooq MU, Novosad V, Rozhkova EA, Wali H, Ali A, Fateh AA, Neogi PB, Neogi A, and Wang Z. Gold nanoparticles-enabled efficient dual delivery of anticancer therapeutics to HeLa cells. *Scientific Reports*. 2018; 8: 2907. https://doi.org/10.1038/s41598-018-21331-y
23. Lee YJ, Ahn EY, and Park Y. Shape-dependent cytotoxicity and cellular uptake of gold nanoparticles synthesized using green tea extract. *Nanoscale Research Letters*. 2019; 14: 129. https://doi.org/10.1186/s11671-019-2967-1
24. Li J, Li Q, Ma X, Tian B, Li T, Yu J, Dai S, Weng Y, and Hua Y. Biosynthesis of gold nanoparticles by the extreme bacterium *Deinococcus radiodurans* and an evaluation of their antibacterial properties. *International Journal of Nanomedicine*. 2016; 11: 5931–5944. doi: 10.2147/IJN.S119618
25. Yue J, Feliciano TJ, Li W, Lee A, and Odom TW. Gold nanoparticle size and shape effects on cellular uptake and intracellular distribution of siRNA nanoconstructs. *Bioconjugate Chemistry*. 2017; 28 (6): 1791–1800.
26. Steckiewicz KP, Barcinska E, Malankowska A, Zauszkiewicz-Pawlak A, Nowaczyk G, Zaleska-Medynska A, and Inkielewicz-Stepniak I. Impact of gold nanoparticles shape on their cytotoxicity against human osteoblast and osteosarcoma in in vitro model. Evaluation of the safety of use and anti-cancer potential. *The Journal of Materials Science: Materials in Medicine*. 2019; 30(2): 22. doi: 10.1007/s10856-019-6221-2

27. Rampersaud S, Fang J, Wei Z, Fabijanic K, Silver S, Jaikaran T, Ruiz Y, Houssou M, Yin Z, Zheng S, Hashimoto A, Hoshino A, Lyden D, Mahajan S, and Matsui H. The effect of cage shape on nanoparticle-based drug carriers: Anticancer drug release and efficacy via receptor blockade using dextran-coated iron oxide nanocages. *Nano Letters*. 2016; 16(12): 7357–7363.
28. Raza MA, Kanwal Z, Rauf A, Sabri AN, Riaz S, and Naseem S. Size- and shape-dependent antibacterial studies of silver nanoparticles synthesized by wet chemical routes. *Nanomaterials (Basel)*. 2016; 6(4): 74. doi: 10.3390/nano6040074.
29. Ripolles-Avila C, Martinez-Garcia M, Hascoet AS, and Rodriguez-Jerez JJ. Bactericidal efficacy of UV activated TiO2 nanoparticles against Gram-positive and Gram-negative bacteria on suspension. *CyTA Journal of Food*. 2019; 17(1): 408–418.
30. Rahman MA, Parvin A, Khan MSH, War AR, Lingaraju K, Prasad R, Das S, Hussain B, and Bhattacharyya A. Efficacy of the green synthesized nickel-oxide nanoparticles against pulse beetle, *Callosobruchus maculatus* (F.) in black gram (*Vigna mungo* L.). *International Journal of Pest Management*. 2020; https://doi.org/10.1080/09670874.2020.1773572
31. Bendale Y, Bendale V, and Paul S. Evaluation of cytotoxic activity of platinum nanoparticles against normal and cancer cells and its anticancer potential through induction of apoptosis. *Integrative Medicine Research*. 2017; 6(2): 141–148.
32. Dobrucka R, Romaniuk-Drapała A, and Kaczmarek M. Evaluation of biological synthesized platinum nanoparticles using *Ononidis radix* extract on the cell lung carcinoma A549. *Biomedical Microdevices*. 2019; 21(3): 75. doi:10.1007/s10544-019-0424-7
33. Philip S and Kuriakose S. Studies on the antibacterial activity of water-soluble iron oxide nanoparticle – cyclodextrin aggregates against selected human pathogenic bacteria. *Nano-Structures and Nano-Objects*. 2018; 16: 347–353.
34. Saqib S, Munis MFH, Zaman W, Ullah F, Shah SN, Ayaz A, and Farooq M, Bahadur S. Synthesis, characterization and use of iron oxide nano particles for antibacterial activity. *Microscopy Research and Technique*. 2019; 82(4): 415–420.
35. Saranya S, Vijayaranai K, Pavithra S, Raihana N, and Kumanan K. In vitro cytotoxicity of zinc oxide, iron oxide and copper nanopowders prepared by green synthesis. *Toxicology Reports*. 2017; 4: 427–430.
36. Ruenraroengsak P, Kiryushko D, Theodorou IG, Klosowski MM, Taylor ER, Niriella T, Palmieri C, Yague E, Ryan MP, Coombes RC, Xie F, and Porter AE. Frizzled-7-targeted delivery of zinc oxide nanoparticles to drug-resistant breast cancer cells. *Nanoscale*. 2019; 11(27): 12858–12870.
37. Wang J, Lee JS, Kim D, and Zhu L. Exploration of zinc oxide nanoparticles as a multitarget and multifunctional anticancer nanomedicine. *ACS Applied Materials & Interfaces*. 2017; 9(46): 39971–39984.
38. Kundu M, Sadhukhan P, Ghosh N, Chatterjee S, Manna P, Das J, and Sil PC. pH-responsive and targeted delivery of curcumin via phenylboronic acid-functionalized ZnO nanoparticles for breast cancer therapy. *Journal of Advanced Research*. 2019; 18: 161–172.
39. Dedkova K, Kuznikova L, Pavelek L, Matejova K, Kupkova J, Barabaszova KC, Vana R, Burda J, Vlcek J, Cvejn D, and Kukutschova J. Daylight induced antibacterial activity of gadolinium oxide, samarium oxide and erbium oxide nanoparticles and their aquatic toxicity. *Materials Chemistry and Physics*. 2017; 197: 226–235.
40. Detappe A, Thomas E, Tibbitt MW, Kunjachan S, Zavidij O, Parnandi N, Reznichenko E, Lux F, Tillement O, and Berbeco R. Ultrasmall silica-based bismuth gadolinium nanoparticles for dual magnetic resonance-computed tomography image guided radiation therapy. *Nano Letters*. 2017; 17(3): 1733–1740.
41. Liu J, Li S, Fang Y, and Zhu Z. Boosting antibacterial activity with mesoporous silica nanoparticles supported silver nanoclusters. *Journal of Colloid Interface Science*. 2019; 555: 470–479.
42. Yan J, Xu X, Zhou J, Liu C, Zhang L, Wang D, Yang F, and Zhang H. Fabrication of a pH/redox-triggered mesoporous silica-based nanoparticle with microfluidics for anticancer drugs doxorubicin and paclitaxel codelivery. *ACS Applied Bio Materials*. 2020; 3 (2): 1216–1225.
43. Kim S, Im S, Park EY, Lee J, Kim C, Kim T, and Kim WJ. Drug-loaded titanium dioxide nanoparticle coated with tumor targeting polymer as a sonodynamic chemotherapeutic agent for anti-cancer therapy. *Nanomedicine: Nanotechnology, Biology and Medicine*. 2020; 24: 102110. https://doi.org/10.1016/j.nano.2019.102110
44. Montalvo-Quiros S, Aragoneses-Cazorla G, Garcia-Alcalde L, Vallet-Regi M, Gonzalez B, Luque-Garcia JL. Cancer cell targeting and therapeutic delivery of silver nanoparticles by mesoporous silica nanocarriers: insights into the action mechanisms using quantitative proteomics. *Nanoscale*. 2019; 11(10): 4531–4545.
45. Li Y, Guo M, Lin Z, Zhao M, Xiao M, Wang C, Xu T, Chen T, and Zhu B. Polyethylenimine-functionalized silver nanoparticle-based co-delivery of paclitaxel to induce HepG2 cell apoptosis. *International Journal of Nanomedicine*. 2016; 11: 6693–6702.
46. Du Y, Xia L, Jo A, Davis RM, Bissel P, Ehrich MF, and Kingston DGI. Synthesis and evaluation of doxorubicin-loaded gold nanoparticles for tumor-targeted drug delivery. *Bioconjugate Chemistry*. 2018; 29 (2): 420–430.
47. Banstola A, Pham TT, Jeong JH, and Yook S. Polydopamine-tailored paclitaxel-loaded polymeric microspheres with adhered NIR-controllable gold nanoparticles for chemo-phototherapy of pancreatic cancer. *Drug Delivery*. 2019; 26(1): 629–640.
48. Katas H, Lim CS, Azlan AYHN, Buang F, and Busra MF. Antibacterial activity of biosynthesized gold nanoparticles using biomolecules from Lignosus rhinocerotis and chitosan. *Saudi Pharmaceutical Journal*. 2019; 27(2): 283–292.
49. Feroze N, Arshad B, Younas M, Afridi MI, Saqib S, and Ayaz A. Fungal mediated synthesis of silver nanoparticles and evaluation of antibacterial activity. *Microscopy Research and Technique*. 2020; 83(1): 72–80.
50. Akinola PO, Lateef A, Asafa TB, Beukes LS, Hakeem AS, and Irshad HM. Multifunctional titanium dioxide nanoparticles biofabricated via phytosynthetic route using extracts of *Cola nitida*: antimicrobial, dye degradation,

antioxidant and anticoagulant activities. *Heliyon*. 2020; 6(8): e04610. doi: 10.1016/j.heliyon.2020.e04610
51. Ajmal N, Saraswat K, Bakht MA, Riadi Y, Ahsan MJ, and Noushad M. Cost-effective and eco-friendly synthesis of titanium dioxide (TiO2) nanoparticles using fruit's peel agro-waste extracts: Characterization, in vitro antibacterial, antioxidant activities. *Green Chemistry Letters and Reviews*. 2019; 12(3): 244–254.
52. Guo H, Jin H, Gui R, Wang Z, Xia J, and Zhang F. Electrodeposition one-step preparation of silver nanoparticles/carbon dots/reduced graphene oxide ternary dendritic nanocomposites for sensitive detection of doxorubicin. *Sensors and Actuators B: Chemical*. 2017; 253: 50–57.
53. Ramesh S, Kim HS, Lee YJ, Hong GW, and Kim JH. Nanostructured silica/gold-cellulose-bonded amino-POSS hybrid composite via sol-gel process and its properties. *Nanoscale Research Letters*. 2017; 12: 381. https://doi.org/10.1186/s11671-017-2122-9
54. Dong J, Carpinone PL, Pyrgiotakis G, Demokritou P, and Moudgil BM. Synthesis of precision gold nanoparticles using Turkevich method. *KONA Powder and Particle Journal*. 2020; 37: 224
55. Vallejos S, Selina S, Annanouch FE, Gracia I, Llobet E, and Blackman C. Aerosol assisted chemical vapour deposition of gas sensitive SnO2 and Au-functionalised SnO2 nanorods via a non-catalysed vapour solid (VS) mechanism. *Scientific Reports*. 2016; 6: 28464. https://doi.org/10.1038/srep28464
56. Rashid H, Mansoor MA, Haider B, Nasir R, Hamid SBA, and Abdulrahman A. Synthesis and characterization of magnetite nano particles with high selectivity using in-situ precipitation method. *Separation Science and Technology*. 2020; 55(6): 1207–1215.
57. Garcia P.R.A.F., Prymak O, Grasmik V, Pappert K, Wlysses W, Otubo L, Epple M, Oliveira CLP. An in situ SAXS investigation of the formation of silver nanoparticles and bimetallic silver–gold nanoparticles in controlled wet-chemical reduction synthesis. *Nanoscale Advances*. 2020; 2(1): 225–238.
58. Shariq M, Friedrich B, Budic B, Hodnik N, Ruiz-Zepeda F, Majeric P, and Rudolf R. Successful synthesis of gold nanoparticles through ultrasonic spray pyrolysis from a gold(III) nitrate precursor and their interaction with a high electron beam. *Chemistry Open*. 2018; 7(7): 533–542.
59. Wahyudiono, Kondo H, Yamada M, Takada N, Machmudah S, Kanda H, and Goto M. DC-plasma over aqueous solution for the synthesis of titanium dioxide nanoparticles under pressurized argon. *ACS Omega*. 2020; 5(10): 5443–5451.
60. Zhang C and Feng Y. Fabrication of size-controlled gold nanoparticles on silicone oil surface and mechanism for size-controllability. *Journal of Physical Society of Japan*. 2015; 84: 07460. https://doi.org/10.7566/JPSJ.84.074601
61. Riedel R, Mahr N, Yao C, Wu A, Yang F, and Hampp N. Synthesis of gold–silica core–shell nanoparticles by pulsed laser ablation in liquid and their physico-chemical properties towards photothermal cancer therapy. *Nanoscale*. 2020; 12(5): 3007–3018.
62. Deng L, Nguyen MT, Mei S, Tokunaga T, Kudo M, Matsumura S, and Yonezawa T. Preparation and growth mechanism of Pt/Cu alloy nanoparticles by sputter deposition onto a liquid polymer. *Langmuir*. 2019; 35(25): 8418–8427.
63. Detsri E, Kamhom K, and Ruen-ngam D. Layer-by-layer deposition of green synthesised silver nanoparticles on polyester air filters and its antimicrobial activity. *Journal of Experimental Nanoscience*. 2016; 11(12): 930–939.
64. Ismail MA, Memon NK, Hedhili MN, Anjum DH, and Chung SH. Synthesis of TiO2 nanoparticles containing Fe, Si, and V using multiple diffusion flames and catalytic oxidation capability of carbon-coated nanoparticles. *Journal of Nanoparticle Research*. 2016; 18: 22. https://doi.org/10.1007/s11051-016-3332-2
65. Seku K, Gangapuram BR, Pejjai B, Kadimpati KK, and Golla N. Microwave-assisted synthesis of silver nanoparticles and their application in catalytic, antibacterial and antioxidant activities. *Journal of Nanostructure in Chemistry*. 2018; 8: 179–188.
66. Ashton JR, Gottlin EB, Patz EF, West JL, and Badea CT. A comparative analysis of EGFR-targeting antibodies for gold nanoparticle CT imaging of lung cancer. *PLoS ONE*. 2018; 13(11): e0206950. https://doi.org/10.1371/journal.pone.0206950
67. Asadzadeh-Firouzabadi A, and Zare HR. Application of cysteamine-capped gold nanoparticles for early detection of lung cancer-specific miRNA (miR-25) in human blood plasma. *Analytical Methods*. 2017; 9(25): 3852–3861.
68. Daraee H, Pourhassanmoghadam M, Akbarzadeh A, Zarghami N, and Rahmati-Yamchi M. Gold nanoparticle–oligonucleotide conjugate to detect the sequence of lung cancer biomarker. *Artificial Cells, Nanomedicine and Biotechnology*. 2016; 44(6): 1417–1423.
69. Nekouian R, Khalife NJ, and Salehi Z. Development of gold nanoparticle based colorimetric biosensor for detection of fibronectin in lung cancer cell line. *Advanced Techniques in Biology and Medicine*. 2014; 2:1. http://dx.doi.org/10.4172/2379-1764.1000118
70. Wang Y, Zhang Y, Du Z, Wu M, and Zhang G. Detection of micrometastases in lung cancer with magnetic nanoparticles and quantum dots. *International Journal of Nanomedicine*. 2012; 7: 2315–2324.
71. Thambiraj S, Shruthi S, Vijayalakshmi R, and Ravi Shankaran D. Evaluation of cytotoxic activity of docetaxel loaded gold nanoparticles for lung cancer drug delivery. *Cancer Treatment and Research Communications*. 2019; 21: 100157.
72. Vaikundamoorthy R, Krishnamurthy V, Vilwanathan R, and Rajendran R. Target delivery of doxorubicin tethered with PVP stabilized gold nanoparticles for effective treatment of lung cancer. *Scientific Reports*. 2018; 8: 3815. doi:10.1038/s41598-018-22172-5
73. Zou J, Su S, Chen Z, Liang F, Zeng Y, Cen W, Zhang X, Xia Y, and Huang D. Hyaluronic acid-modified selenium nanoparticles for enhancing the therapeutic efficacy of paclitaxel in lung cancer therapy. *Artificial Cells, Nanomedicine, and Biotechnology*. 2019; 47(1): 3456–3464.
74. Bendale Y, Bendale V, Natu R, and Paul S. Biosynthesized platinum nanoparticles inhibit the proliferation of human lung-cancer cells in vitro and delay the growth of a human lung-tumor xenograft in vivo: in vitro and in vivo anticancer activity of bio-Pt NPs. *Journal of Pharmacopuncture*. 2016; 19(2): 114–121.

75. Venkatasubbu GD, Ramasamy S, Ramakrishnan V, and Kumar J. Folate targeted PEGylated titanium dioxide nanoparticles as a nanocarrier for targeted paclitaxel drug delivery. *Advanced Powder Technology*. 2013; 24(6): 947–954.
76. Yu H, Wang Y, Wang S, Li X, Li W, Ding D, Gong X, Keidar M, and Zhang W. Paclitaxel-loaded core–shell magnetic nanoparticles and cold atmospheric plasma inhibit non-small cell lung cancer growth. *ACS Applied Materials & Interfaces*. 2018: 10 (50): 43462–43471.
77. Shi H, Xu M, Zhu J, Li Y, He Z, Zhang Y, Xu Q, Niu Y, and Liu Y. Programmed co-delivery of platinum nanodrugs and gemcitabine by a clustered nanocarrier for precision chemotherapy for NSCLC tumors. *Journal of Materials Chemistry B*. 2020; 8(2): 332–342.
78. Jha M, and Shimpi NG. Green synthesis of zero valent colloidal nanosilver targeting A549 lung cancer cell: in vitro cytotoxicity. *Journal of Genetic Engineering and Biotechnology*. 2018; 16 (1): 115–124.
79. Ramalingam V, Revathidevi S, Shanmuganayagam T, Muthulakshmi L, and Rajaram R. Biogenic gold nanoparticles induce cell cycle arrest through oxidative stress and sensitize mitochondrial membranes in A549 lung cancer cells. *RSC Advances*. 2016; 6(25): 20598–20608.
80. He Y, Du Z, Ma S, Liu Y, Li D, Huang H, Jiang S, Cheng S, Wu W, Zhang K, and Zheng X. Effects of green-synthesized silver nanoparticles on lung cancer cells in vitro and grown as xenograft tumors in vivo. *International Journal of Nanomedicine*. 2016; 11: 1879–1887.
81. Fayez H, El-Motaleb MA, and Selim AA. Synergistic cytotoxicity of shikonin-silver nanoparticles as an opportunity for lung cancer. *Journal of Labelled Compounds and Radiopharmaceuticals*. 2020; 63(1): 25–32.
82. Kalaiarasi A, Sankar R, Anusha C, Saravanan K, Aarthy K, Karthic S, Mathuram TL, and Ravikumar V. Copper oxide nanoparticles induce anticancer activity in A549 lung cancer cells by inhibition of histone deacetylase. *Biotechnology Letters*. 2018; 40(2): 249–256.
83. Ullah S, Ahmad A, Wang A, Raza M, Jan AU, Tahir K, Rahman AU, and Qipeng Y. Bio-fabrication of catalytic platinum nanoparticles and their in vitro efficacy against lungs cancer cells line (A549). *Journal of Photochemistry and Photobiology B: Biology*. 2017; 173: 368–375.
84. Wang Y, Cui H, Zhou J, Li F, Wang J, Chen M, and Liu Q. Cytotoxicity, DNA damage, and apoptosis induced by titanium dioxide nanoparticles in human non-small cell lung cancer A549 cells. *Environmental Science and Pollution Research (International)*. 2015; 22(7): 5519–5530.
85. Rajeshkumar S, Venkatkumar S, Ramaiah A, Agarwal H, Lakshmi T, and Roopan SM. Biosynthesis of zinc oxide nanoparticles using *Mangifera indica* leaves and evaluation of their antioxidant and cytotoxic properties in lung cancer (A549) cells. *Enzyme and Microbial Technology*. 2018; 117: 91–95.

30

Modeling of Pharmaceutical Aerosol Transport in the Targeted Region of Human Lung Airways Due to External Magnetic Field

Anusmriti Ghosh[1], **Mohammad S. Islam**[2], **Mohammad Rahimi-Gorji**[3], **Raj Das**[4], **and Suvash C. Saha**[2]
[1]School of Mechanical, Medical and Process Engineering, Queensland University of Technology, Brisbane, Australia
[2]School of Mechanical and Mechatronic Engineering, University of Technology Sydney, Ultimo, Australia
[3]Faculty of Medicine and Health Sciences, Ghent University, Belgium
[4]School of Engineering, RMIT University, Melbourne, Australia

30.1 Introduction

The lung is a unique organ of the human body that is subjected to oxidative stresses as it flexibly brings down on higher oxygen pressures. Lung cells involve enriched oxygen pressure because of their direct exposure to ambient air by environmental irritations and pollutants (1). Drug delivery in lungs by aerosol inhalation is an authenticated procedure. It has the potential advantage in the treatment of respiratory disorders to deliver drugs via oral and arterial routes. The usage of inhalation aerosols allows direct achievement of high drug concentrations for the selective treatment of the lungs (2). Inhaled aerosol particles transport through the bifurcating airways with tidal air and deposit in the airway wall once the particles' trajectories differ from the original path (3).

Human lung models are often used to demonstrate particle deposition in the targeted region by applying numerical methods. Based on the inlet flow rate conditions, the flow through airways of a human lung can be modeled as laminar, transitional, or turbulent. Particle deposition patterns have a significant effect when turbulence models are applied (4). Among different turbulence models, large eddy simulation (LES), detached eddy simulation (DES) (5), RANS turbulence models (6,7), and k–ω low Reynolds number (LRN) models (8) produced more accurate results compared to experimental data. Turbulent flows affect aerosol deposition mainly in the extrathoracic (upper) respiratory tract and large airways. These types of flows are referred to as an irregular rotation by the fluid in an unstable system, which creates fluid motion. Therefore, the particle trajectories are changed due to the turbulent fluctuation, and eventually, these particles deposit in the airway walls (2,9).

Limited studies have been reported for targeting magnetic drug delivery in a specific region of the lungs. An external magnetic field as a passive technique, which is a potential application tool for drug delivery, was adopted by several researchers (10–13). An aerosol cloud was developed at the beginning of the inspired phase for delivering aerosols to the deepest areas of the lungs by synchronising activation of the magnetic field with the breathing process (11). The authors, however, did not show a deposition pattern for any specific lung model. Dolovich et al. (13) demonstrated therapeutic applications and also did not consider any specific zone deposition pattern for the entire geometry. Goetz et al. (12) studied the particle size for reducing unwanted distribution outside the target area due to the influence of the magnetic force and did not consider specific areas of the lung model for particle deposition. Plank (10) developed a nano-magnetic aerosol drug targeting method for reducing undesired side effects.

This study showed therapeutic applications and did not consider any specific zone deposition pattern for the lung geometry. Ally et al. (14) and Dames et al. (3) developed an in vitro model to investigate the potential of targeted magnetic aerosol deposition for lung cancer. Dames et al. (3) developed nano-magnetosols and demonstrated this for targeting aerosol delivery to the lungs of mice. Pourmehran et al. (15) used Lagrangian magnetic particle tracking, using a discrete phase model (DPM) to investigate the effect of a magnetic field on the behavior of magnetic drug careers. Recently, Pourmehran et al. (16) used a realistic model to investigate the human tracheobronchial airways using computational fluid and particle dynamics. They developed an optimal magnetic drug characteristic coordination for drug delivery to the human lung. Based on several past attempts at studying particle deposition due to the effect of an external magnetic field, it is important to investigate the particle deposition in the specific position of a human lung. There are limited numbers of studies about, numerical or experimental, which provide a full understanding of the effect of an external magnetic field on the particle deposition in a specific targeted portion of human lungs.

The atmospheric airborne particles' size distribution consists of a significant amount of fine and ultrafine particles which are generally smaller than a couple of microns (17). There are billions of airborne particles emitted by a host of man-made

DOI: 10.1201/9781003046547-30

sources, including combustion processes such as industrial, domestic, or vehicle emissions, and from vehicles' brake systems, with a particle size smaller than 10 μm and acquire some magnetic susceptibility (18). These particles are dangerous to human health and can be collected with the help of applied magnetic force; however, this process is still not popular. On the other hand, researchers are conducting research on drug delivery using the susceptibility of magnetic particles in the specific site in the human lung by coating the particles with an appropriate drug.

The present study shows a symmetric third-generation lung model by considering a specific position for targeted drug delivery (see Figure 30.1a). Left lung is chosen as the targeted region for the particle transport direction and deposition, as the left lung is correspondingly longer and narrower than the right lung (19). In this study, the effects of inhalation flow rates, particle sizes, position of magnetic fields, and magnetic field strength on particle deposition in the left lung will be modeled. The outcomes of this work may assist in developing improved and efficient drug delivery systems in targeted regions of lungs.

30.2 Computational Domain and Mesh Generation

The third-generation lung symmetric model is constructed to evaluate the complex flow field in the human lung using the k-ω low Reynolds number turbulence model. This third-generation lung geometry mesh contains 450,429 tetrahedral elements and 179,660 nodes.

An unstructured fine boundary layer mesh is constructed to calculate the complex flow field (Figure 30.1a). Figure 30.1(b) shows the mesh for the first bifurcation of a third-generation lung model. An inflation of 10 boundary layer mesh was constructed near the wall (Figure 30.1c). Figure 30.1(d) shows the inflation layer mesh of a third-generation lung model. Figure 30.1(e) shows the outlet mesh of a third-generation lung model.

30.3 Methodology

A viscous flow solution was obtained using the computational fluid dynamics (CFD) method, implemented in the CFD solver ANSYS (Fluent) 18.0. The continuity and the momentum equations with proper initial and boundary conditions form the governing system. The steady-state flow field was converged when the residuals reduced below 10^{-6}. Air was considered as the working fluid with constant density (ρ), viscosity (μ), and fluid static pressure (p). The governing equations for the continuity and the momentum equations are given respectively as:

$$\frac{\partial \overline{u_i}}{\partial x_i} = 0 \tag{30.1}$$

and

$$\frac{\partial u_i}{\partial t} + u_j \frac{\partial u_i}{\partial x_j} = -\frac{1}{\rho_f}\frac{\partial p}{\partial x_i} + \frac{\partial}{\partial x_j}\left[(\nu_f + \nu_T)\left(\frac{\partial u_i}{\partial x_j} + \frac{\partial u_j}{\partial x_i}\right)\right] \tag{30.2}$$

where u_i and u_j ($i, j = 1, 2, 3$) are the fluid velocity components, and subscripts *i and j* represent velocities in the *x*-, *y*- and *z*-

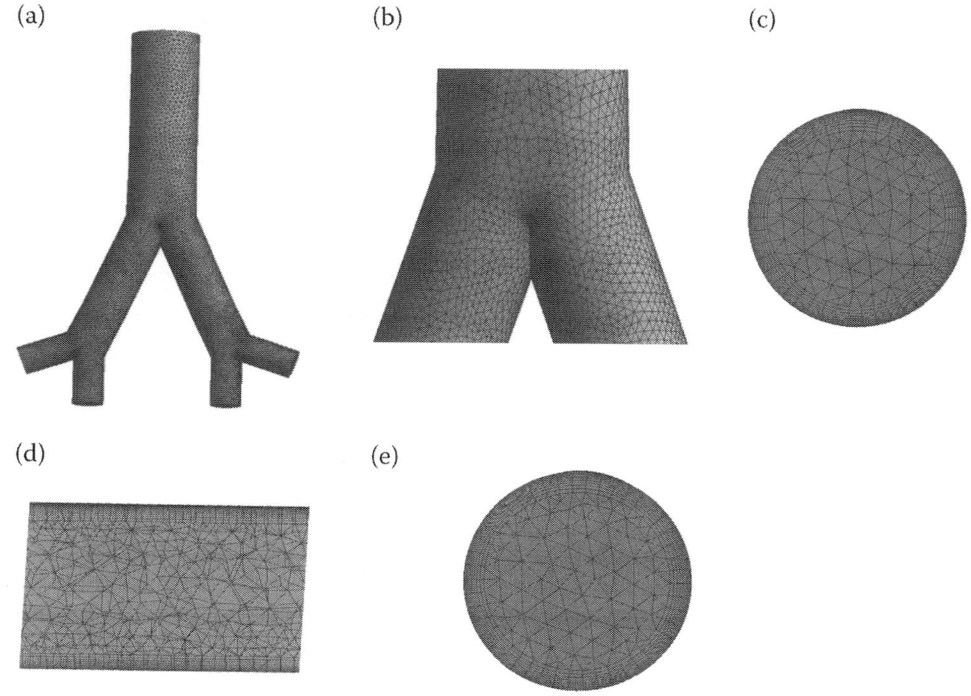

FIGURE 30.1 (a) Anterior View of the Third-Generation Lung Mesh for the Present Model, (b) First Bifurcation, (c) Inlet Mesh, (d) Inflation Layer Mesh Near to the Wall, and (e) Outlet Mesh of the Third-Generation Lung Model

directions. The steady k–ω low Reynolds number turbulence model was adopted to calculate the air flow in the present study. The SIMPLE algorithm (20,21) was used for the pressure–velocity coupling. The second-order upwind numerical scheme was selected to discretize different terms in the transport equations. The k–ω turbulence model governing equations are written as follows:

$$\frac{\partial k}{\partial t} + u_j\frac{\partial k}{\partial x_j} = P - \beta^*\omega k + \frac{\partial}{\partial x_j}\left[\left(v_f + \alpha_k\alpha^*\frac{k}{\omega}\right)\frac{\partial k}{\partial x_j}\right] \quad (30.3)$$

with pseudo vorticity equation:

$$\frac{\partial \omega}{\partial t} + u_j\frac{\partial \omega}{\partial x_j} = \frac{\gamma\omega}{k}P - \beta\omega^2 + \frac{\partial}{\partial x_j}\left[\left(v_f + \alpha_\omega\alpha^*\frac{k}{\omega}\right)\frac{\partial \omega}{\partial x_j}\right] + \frac{\alpha_d}{\omega}\frac{\partial k}{\partial x_j}\frac{\partial \omega}{\partial x_j} \quad (30.4)$$

where the turbulent viscosity, $v_T = C_\mu f_\mu \frac{k}{\omega}$, and the function, f_u is defined as $f_u = \exp\left[-\frac{3.4}{\left(1+\frac{R_T}{50}\right)^2}\right]$ with $R_T = \frac{\rho k}{(\mu\omega)}$. The other coefficients in the above equations are chosen from (22,23) as follows.

$$R_\beta = 8,\ R_\omega = 2.61,\ R_k = 6,\ \alpha_0 = \frac{1}{9},\ \beta_0 = 0.0708,$$

$$\beta_0^* = 0.09,\ \alpha_\infty^* = 1,\ \sigma_\omega = \alpha_k = 0.5$$

For the boundary conditions, wall condition was considered as trap and conducting for the targeted region, outlet condition was considered as a pressure outlet and the turbulence specification method was κ and omega values, and the inlet condition was considered a uniform mass flow.

To simulate the particle trajectories within the flow, the Lagrangian particle tracking approach and the DPM were applied. In this approach, the force–balance equation for the individual particles (particulates) is given as follows:

$$\vec{F} = \vec{F_D} + \vec{F_M} + \vec{F_B} + \vec{F_T} + \vec{F_E} = m_p.\frac{d\vec{U_p}}{dt} \quad (30.5)$$

where $\vec{U_p}$ is the particle velocity and $\vec{F}$ is the force term. $\vec{F_D}$, $\vec{F_M}$, $\vec{F_B}$, $\vec{F_T}$ and $\vec{F_E}$ are the drag, magnetic, Brownian, thermophoretic, and electrical forces, respectively. Brownian force should only be considered for submicron size of particles. Thermoforetic force is related to the temperature gradient. More temperature gradient makes larger thermophoretic force on the particles. Electrical force is created by electrostatic field. In the present work, there is not any electrical field. Consequently, $\vec{F_B}$, $\vec{F_T}$ and $\vec{F_E}$ can be removed from Equation (30.5).

30.3.1. Drag Force

The Stokes drag force acting on a spherical particulate is expressed as:

$$\vec{F_D} = \frac{18\mu}{\rho_p d_p^2}\frac{m_p f Re_P}{24}(u_f - u_p) \quad (30.6)$$

The drag coefficient, C_D, for a smooth particle using spherical drag law can be expressed as:

$$f = a_1 + \frac{a_2}{Re_P} + \frac{a_3}{Re_P^2} \quad (30.7)$$

where Re_P is the particle Reynolds number, which is defined as $Re_P \equiv \frac{\rho_f d_P |u_P - u_f|}{\mu_f}$. u_f, $u_{p,}$, μ_f, ρ_f, ρ_P and d_P are the air velocity, particle velocity, fluid molecular viscosity, fluid density, particle density, and particle diameter. Also in Equation (30.7) a_1, a_2 and a_3 are constants (16).

30.3.2. Magnetic Forces

For the magnetic force, the MHD approach was applied through ANSYS (Fluent) 18.0.

The magnetic force, $\vec{F_M}$ on a small sphere in a non-magnetic fluid, was calculated as

$$\vec{F_M} = \frac{1}{2}\mu_0\chi V_P \nabla(\vec{H^2}) \quad (30.8)$$

Where μ_0 is the magnetic permeability of vacuum, χ is the magnetic susceptibility of the particle, Vp is the particle volume, and $\vec{H}$ is the magnetic field intensity.

The magnetic susceptibility of the particle equation (16) is defined as:

$$\chi = -0.14 d_p.10^6 + 0.98 \quad (30.9)$$

Where, d_p is the particle diameter.

Magnetic number Mn (Tesla) is defined as follows (16):

$$M_n = \mu_0 H_0 \quad (30.10)$$

H_0 is the characteristic magnetic field strength. Magnetic number is dependent on the magnetic field intensity. All parameters that have been used to calculate the desired result are listed in Table 30.1.

30.4 Grid Convergence Study

After completing the meshing, a grid resolution study was performed for selecting the appropriate mesh in terms of convergence and accuracy for the present simulation. Since the fluid flow is complex and the results are sensitive due to regional turbulence effects, it is necessary to use a grid resolution by adequately refining the mesh. To this end, the CFD model

TABLE 30.1

Particle and Fluid (Air) Properties

Parameter	Value(s)
Magnetic number, Mn (Tesla)	0.181, 2.5, 3
Magnetic permeability, χ (h/m)	1.257^{e-06}
Particle diameter, d_P (μm)	$2\mu m$, $4\mu m$, $6\mu m$
Magnetic-drug-density (kg/m^3)	6450
Flow rates, **Q**	15 lpm, 30 lpm, 60 lpm
Air density, ρ (kg/m^3)	1.225
Air viscosity, μ_f (kg/m-s)	1.7894^{e-05}

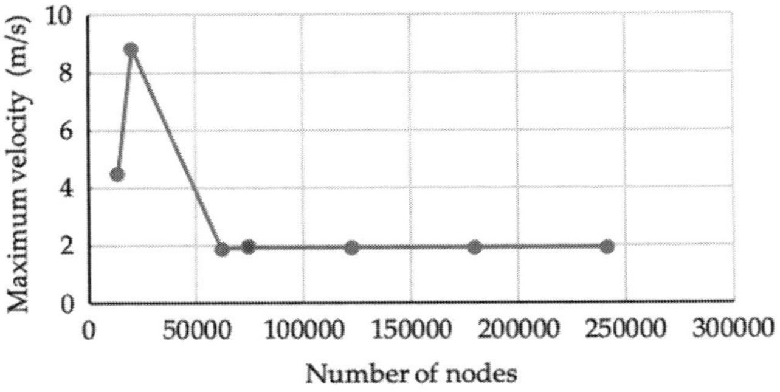

FIGURE 30.2 Maximum Velocity Grid Convergence

was tested for seven grids (meshes) with different cell numbers (see Figure 30.2), and the maximum velocity calculated on the outlet plane was compared among different grids. The flow seems to have converged from the red point, and it is conceivable to use any of the grid numbers after this point. However, the grid (mesh) with 179,660 cells or nodes was adopted for the present simulations. Note that the minimum and the maximum cell sizes are 1×10^{-5} m and 1.9772×10^{-3} m respectively. Also, an inflation of 10 layers was chosen in the boundary layer (near the solid wall).

30.5 Model Validation

A comprehensive model validation was been performed for the present study. The present microparticle simulation results are compared with the experimental data sets of steady laminar flows available in the literature.

For the present airway model, the results are compared with the observations by the Kleinstreuer et al. (24) and Cheng et al. (25) for three inhalation flow rates (Figure 30.3). The overall

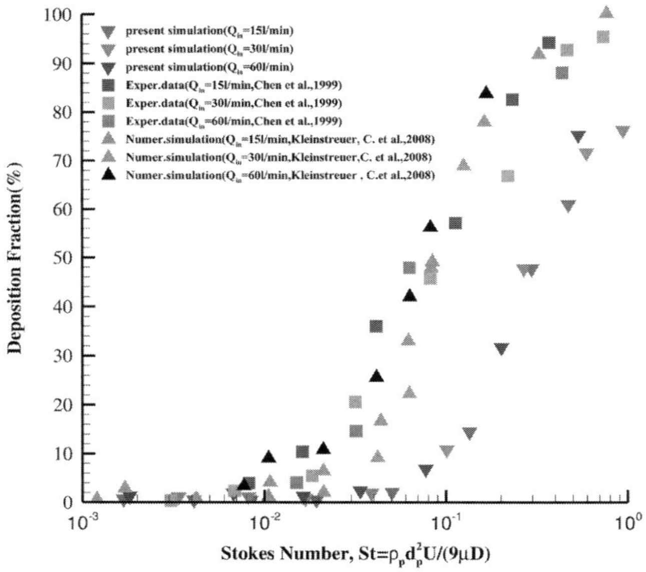

FIGURE 30.3 Comparison of the Present Model Particle Deposition Efficiency With the Experimental Data From Cheng et al. (25) and Numerical Data From Kleinstreuer et al. (24)

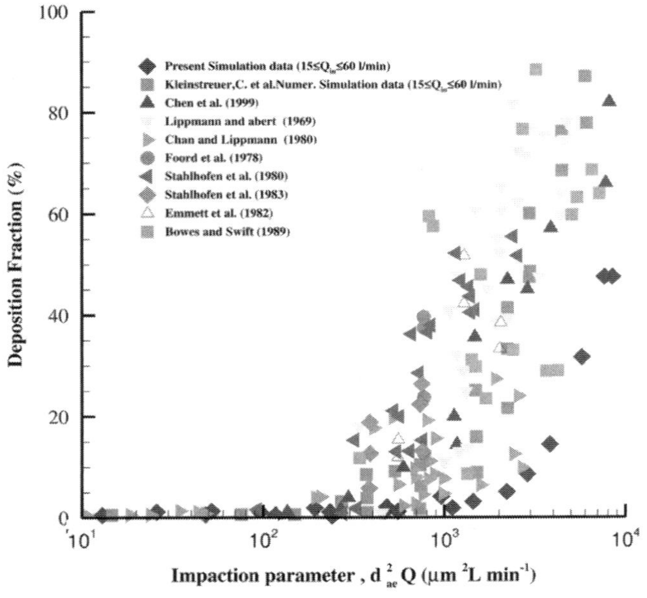

FIGURE 30.4 Comparisons of the Particle Deposition Fraction Between the Present Simulation Model Prediction Data with Different Experimental Data Sets from Kleinstreuer et al. (24), Cheng et al. (25), Lippmann et al. (26), Chan et al. (27), Foord et al. (28), Stahlhofen et al. (29), Stahlhofen et al. (30), Emmett et al. (31) and Bowes et al. (32)

deposition fraction (DF) is compared against the Stokes number. The Stokes number is defined by $st = \rho_p d_p^2 U/(9\mu D)$, where U is the mean velocity and D is the minimum hydraulic diameter. All the experimental results show that the DF is proportional to the Stokes number. The experimental data and the present numerical results show the similar trends against total deposition for the Stokes number. However, the DF of the present model is slightly lower than that of the experimental model, as the present model has considered only three generations instead of the four generations that were considered by the experimental study.

Figure 30.4 shows the present airway model results compared to the results of Kleinstreuer et al. (24), Cheng et al. (25), Lippmann et al. (26), Chan et al. (27), Foord et al. (28), Stahlhofen et al. (29), Stahlhofen et al. (30), Emmett et al. (31) and Bowes et al. (32) for impaction parameter. The impaction parameter is defined as $d_{ae}^2 Q(um^2 Lmin^{-1}$, where d_{ae} is the aerodynamic particle diameter and Q is the flow rate. The results show the comparison of the microparticle DF in the present airway model with the in vivo deposition data as a function of the impaction parameter. All of the experimental results show that the DF is proportional to the impaction parameter. The experimental data and the present numerical results follow similar trends for the impaction parameter. The present model predicted microparticle DF for impaction parameter shows good agreement with the experimental data, but the trend is slightly lower than the experimental data. The potential reason for this is that a fourth-generation lung model was used in the experiments, compared to a third-generation lung model chosen for the present simulation.

The third-generation lung model configuration and magnetic source position are indicated in Figure 30.5. Position 1 and position 2 indicate the magnetic field source locations. The magnetic field positions are selected randomly to demonstrate the particle deposition in the specific position of the lung. Position 1 is set on the left side immediately before the first bifurcation. Position 2 is set on the left lung at the third bifurcation, as indicated in the above Figure 30.5. It is noteworthy position 1 and 2 represents the targeted region in this study. Due to the symmetrical shape of the present model, inlet and outlet are same. Right and left generations of this model are indicated by rg3 and lg3, respectively.

Figure 30.6 shows particle DE comparisons between the present simulation results and the experimental results of Cohen [33], Haverkot [34] 33.,34. and Pourmehran and Gorji-Bandpy (16) for two magnate positions. The total DE under an externally applied magnetics force field for the present model is in the range of the experimental data and agrees sufficiently well with the published literature.

30.6 Results and Discussion

The present third-generation lung airway model was designed to determine the exact deposition in the targeted region. Figure 30.7(a) shows the velocity magnitude for the triple bifurcation lung airways.

The magnitude of velocity at the different outlets of the double bifurcation model is investigated and is shown in Figure 30.7. Figure 30.7(b,c) shows the velocity contours at the outlet planes of the left lung, and Figure 30.7(d,e) shows the same for the right lung. The overall flow contours indicate that a vortex is generated due to the large change in the airway cross-sectional area. However, the turbulence intensity from Figure 30.7 at the left outlet 2 and right outlet 1 appears to be stronger than the other outlets. The highly complicated airway geometry, including the bifurcation structure, the change in the angle, and the curvature along with the centrifugally induced pressure produce the complex velocity field at the selected outlet planes of the airway model.

Figure 30.8(a,b) illustrates the effect of magnetic field intensity at two targeted positions. To identify the magnetic source intensity, the magnitude of magnetic flux density $(\overrightarrow{B})$ is shown for two source locations at position 1 and position 2. It is found that the magnetic field intensity is higher in wall

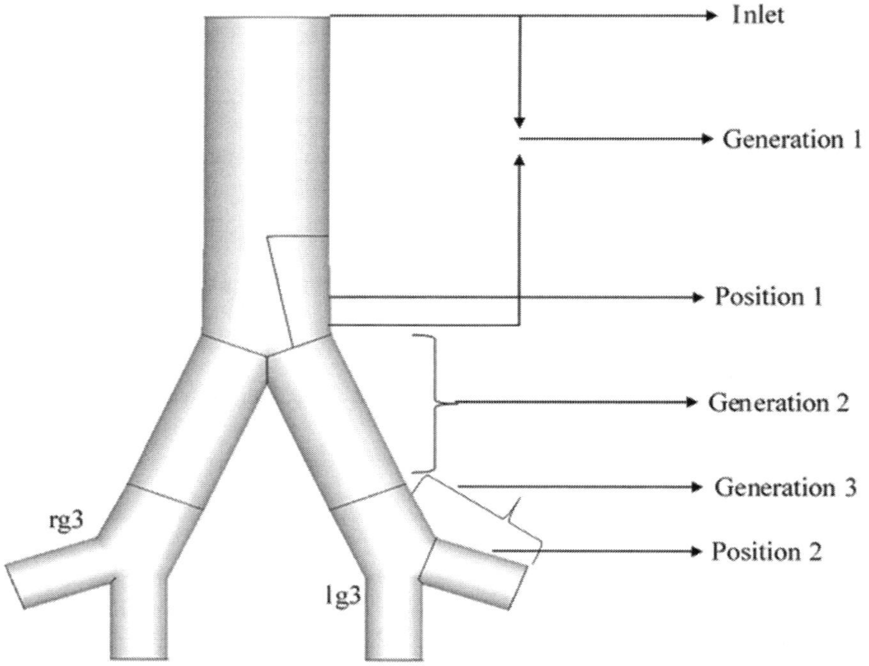

FIGURE 30.5 Geometry Configuration and Specification of the Present Model - Magnet Position is set on the Left Lung. rg3 (Right Generation 3) and lg3 (Left Generation 3) Denote Specific Locations at the Third-generation Bifurcation of the Lung

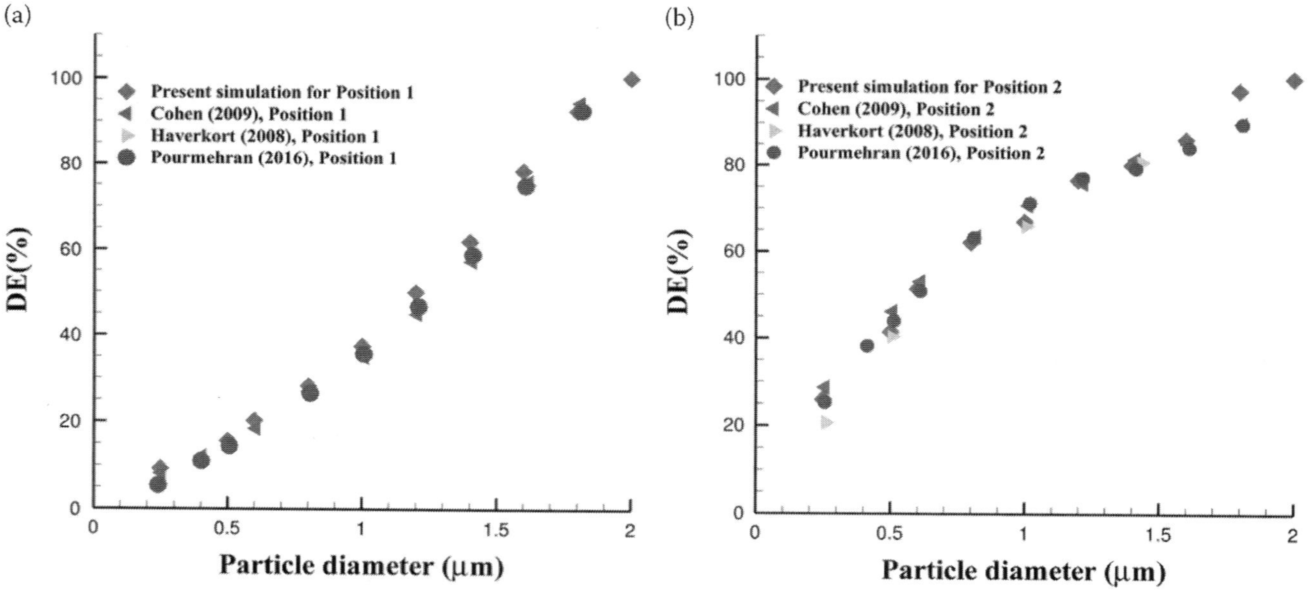

FIGURE 30.6 Comparisons of the Particle Deposition Efficiency (DE) for Two Magnetic (a) Position 1 and (b) Position 2 Between the Present Simulation Model Prediction and the Different Experimental Data Sets (33,34) and (16)

position 1 (targeted position), compared to the other locations (including position 2) of the lung airways, when the magnetic source is located in position 1, as shown in Figure 30.8(a). Likewise, as seen in Figure 30.8(b), the magnetic field intensity is higher in wall at position 2 (targeted position) when the magnetic source is located in position 2. Due to the maximum intensity of magnetic field (flux density) on the two specific targeted positions, the present result shows the particle DE is increased in these two positions. Magnetic flux density ($\vec{B}$) diminishes with increasing distance from the magnetic source position. Figure 30.8(a,b) shows the location where the magnetic field intensity is maximum after applying a magnetic field. To illustrate the interaction of the particles in the presence of this magnetic field, particle traces are shown after producing the magnetic field in two different positions in terms of particle residence time in Figure 30.8(c,d). The particles' trajectory in an electromagnetic field is evaluated by solving the following equation of motion.

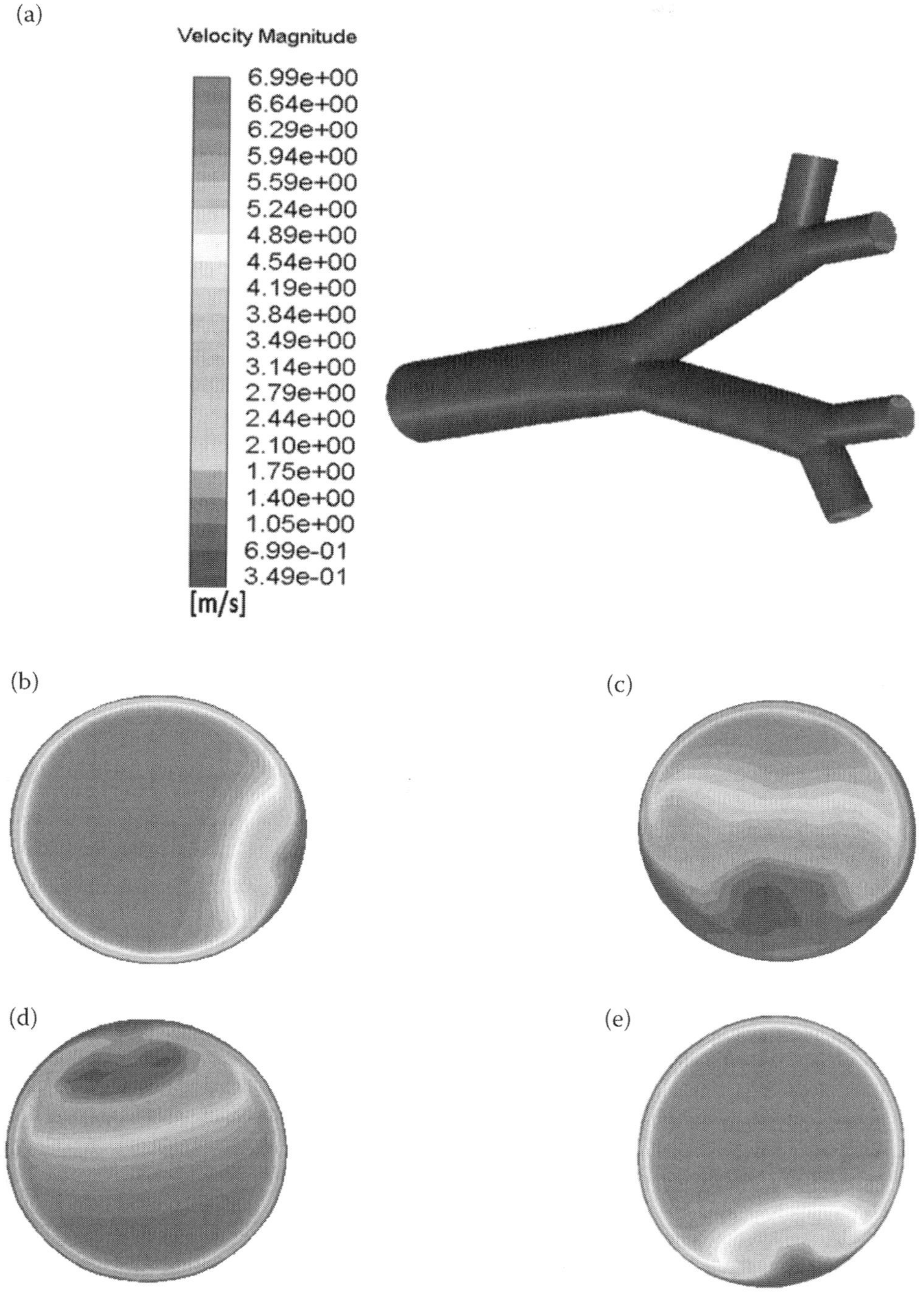

FIGURE 30.7 Velocity Profiles in the Symmetric Bifurcation Airway Model for Steady Inhalation with Q= 60 lpm, Showing Contour of Velocity Magnitude for (a) 3 Generation Lung Model; (b) Left Outlet 1; (c) Left Outlet 2; (d) Right Outlet 1; and (e) Right Outlet 2

$$m\vec{a} = \vec{F} = q(\vec{E} + \vec{V} \times \vec{B}) \tag{30.11}$$

Where m is the mass of the particle, $\vec{F}$ is the Lorentz force vector, q is the charge of the particle, $\vec{E}$ is the electric field vector, and $\vec{V}$ is the velocity vector. Figure 30.8(c,d) depicts that most of the particle trajectories touched the specified targeted location of the airway wall, and hence, the particle deposition is higher on that position.

Figure 30.9 shows the contour of turbulence kinetic energy (TKE) magnitude for two different magnetic source positions in the present model. Figure 30.9(a) shows the magnitude of TKE contour in the targeted magnetic source position 2. Similarly, Figure 30.9(b) shows the TKE contour magnitude for magnetic source position 1. In a turbulent flow, the fluid

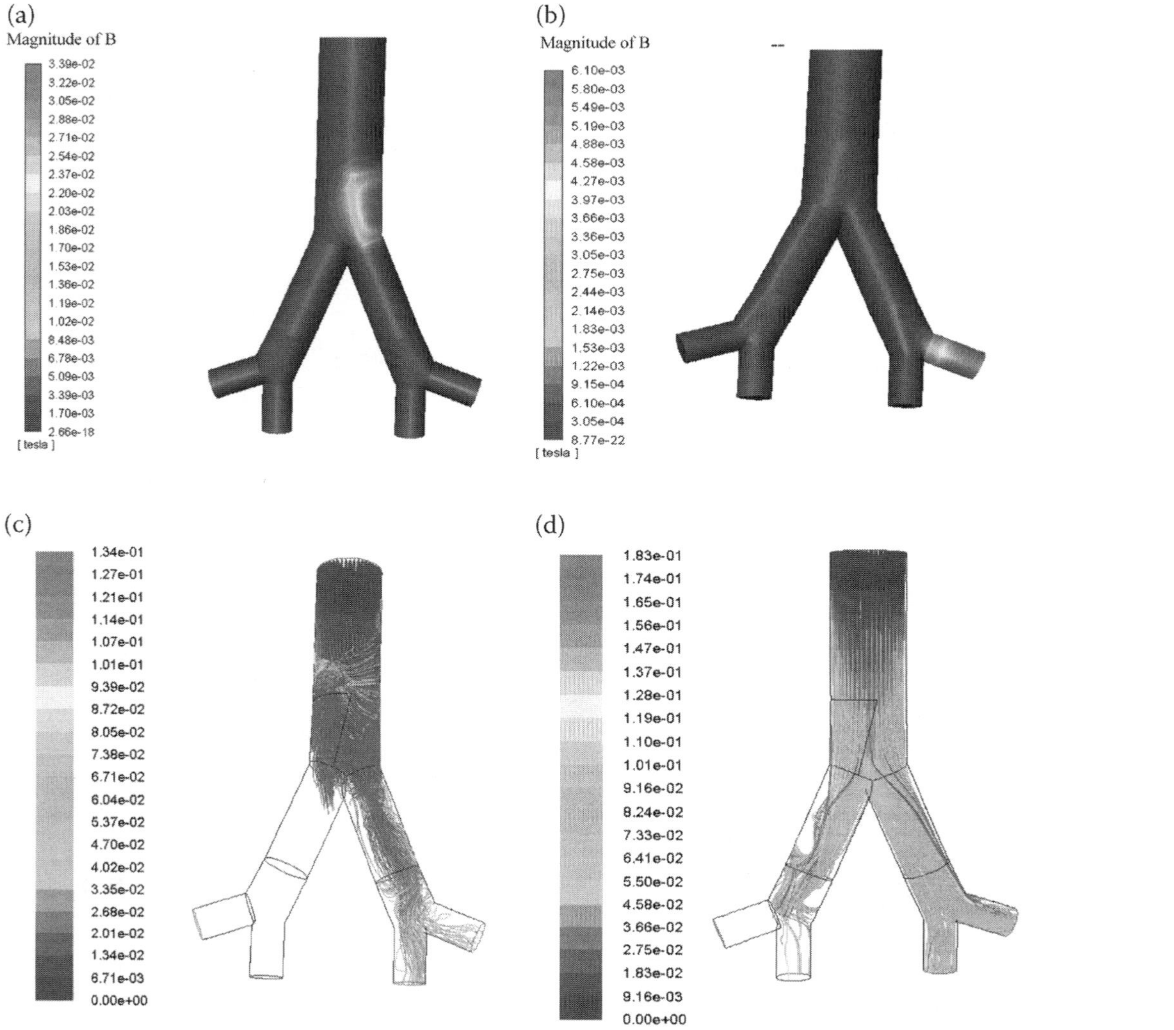

FIGURE 30.8 (a,b) Contours of Magnetic Field for Mn=2.5 for the Magnetic Source Location at Position 1 (a) and Position 2 (b), (c,d) Particle Traces Colored by Particle Residence Time for the Magnetic Source Location at Position 1 (c) and Position 2 (d)

speed at a point is continuously undergoing changes in both direction and magnitude. Turbulent intensity is measured by TKE. It is recognized that with an increase in the turbulent kinetic energy, airflow rapidly becomes faster in the compression region, and as a result, a large number of particles are deposited on that region.

Figure 30.10(a,b,c) represents the deposition efficiency for three different breathing flow rates (slow, medium and fast), i.e. 15 lpm, 30 lpm and 60 lpm, respectively, for the case with magnetic number, Mn = 2.5, magnetic source in position 2 and particle diameter of 4μm. At the slow breathing condition (Q=15 lpm), microparticle deposition at the targeted position is significantly increased more than any other region of lung, as shown in Figure 30.10(a). For the slow breathing condition, the total percentage of deposition is 27.06, and at the targeted position it is 23.84. At the medium breathing pattern (Q=30 lpm), the majority of 4 μm diameter particles are deposited in the left lung, as shown in Figure 30.10(b). The percentage of overall deposition for the medium breathing condition is 42.93, where in the left lung it is 36.56. At the fast-breathing condition represented by a flow rate of 60 lpm, the maximum number of particles deposited in the targeted region, i.e. wall position 2, and the deposition concentration are significantly higher compared to those obtained for the other flow rates considered here, as shown in Figure 30.10(c). The percentage of overall deposition for the fast-breathing condition is 76.70. In the targeted position, the percentage is 36.31, and it is 30.498 in the left lung. During the slow breathing pattern, fewer numbers of particle are deposited. The number of deposited particles increases noticeably with an increase in the flow rate. It is also observed that some particles are deposited at the carinal angle of the first bifurcation. The microparticle inertia has a vital role to deposit particles at the carinal angle. A low number of 4 μm diameter particles are deposited at the first bifurcation for the slow breathing pattern. The number of deposited particles at the carinal angle increases noticeably with the increase in flow rate. Figure 30.10(d) shows the overall deposition efficiency for three different flow rates. The

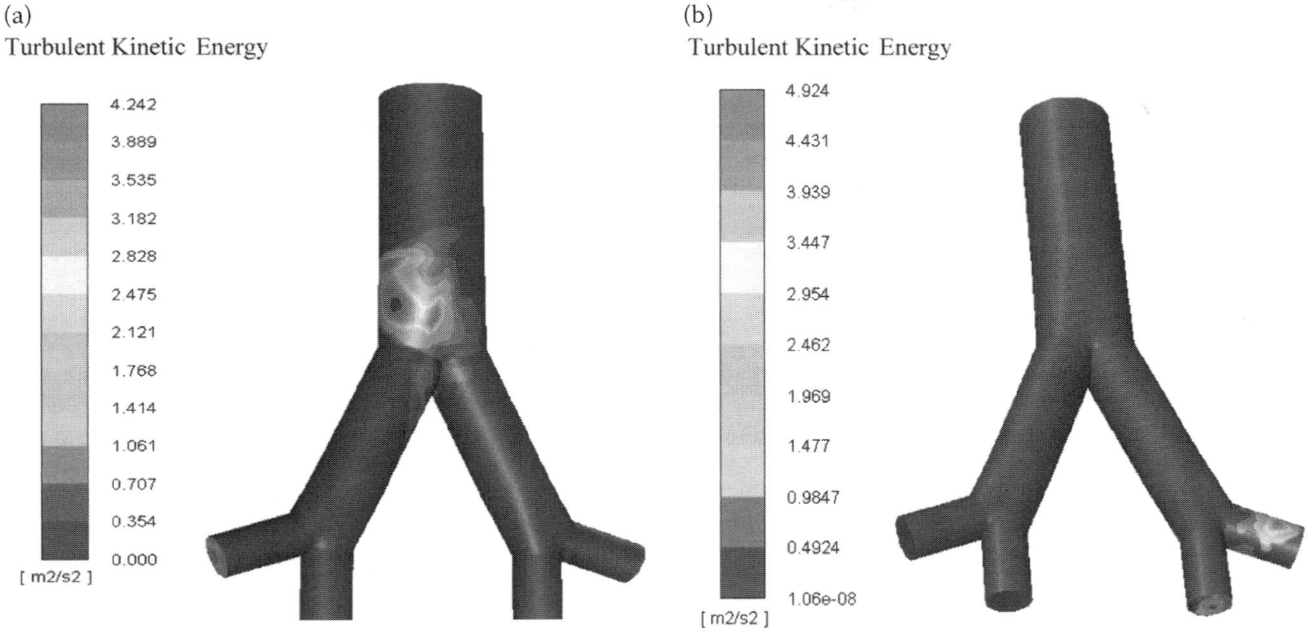

FIGURE 30.9 Turbulent Kinetic Energy Magnitude Contours for (a) Position 1 and (b) Position 1

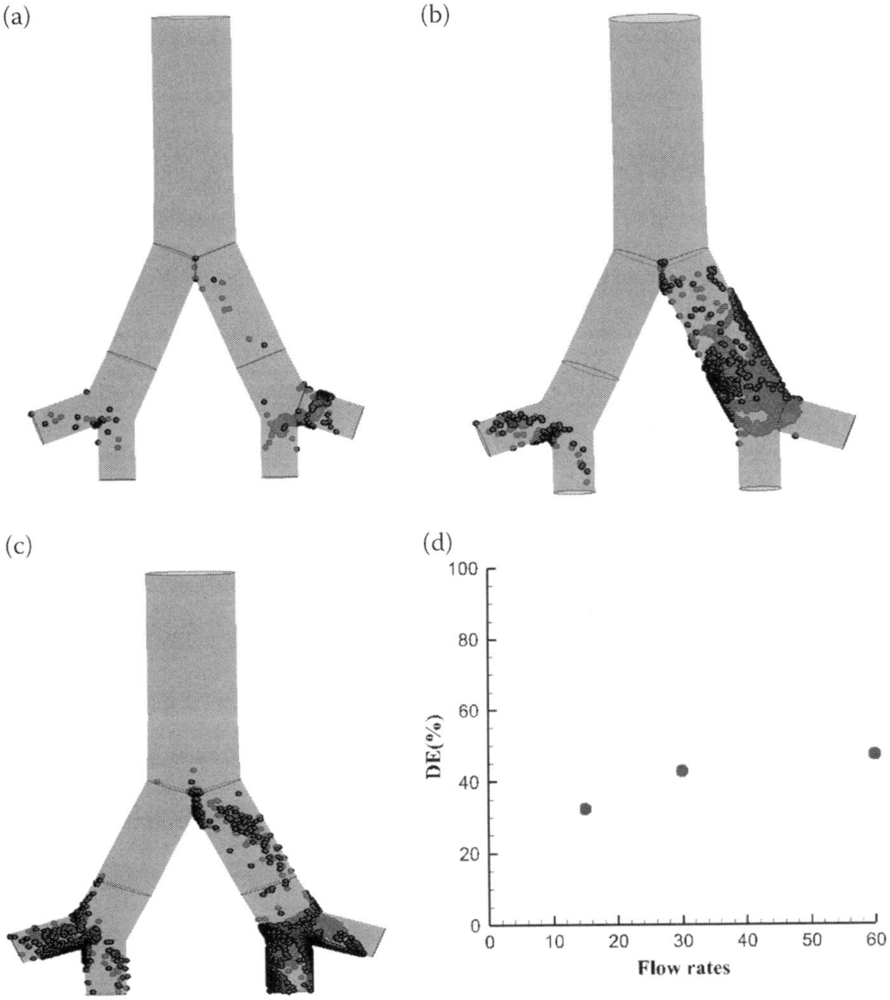

FIGURE 30.10 Effect of Air Flow Rates on the Particle Transport Pattern and DE (%) at Position 2(d_p=4 μm, Mn=2.5) (a) 15 lpm, (b) 30 lpm, (c) 60 lpm, and (d) Total Deposition Efficiency Variation with Flow Rate

deposition scenario also shows that the difference in the DE between the right and left lungs is pointedly higher at higher flow rates. The difference in DE scenario in the targeted region (left lung) is higher than any other region of lung. Hence, the flow rate, inertia of particles, and magnetic field strength play an important role in controlling the particulate deposition in a targeted region of the lung.

To investigate the deposition pattern and efficiency on a targeted region of the lungs for different particle sizes, three different sizes of particles, 2 μm, 4μm, and 6 μm are considered in Figure 30.11 for magnetic source at position 2, and Mn=2.5 and Q=60 lpm. The deposition scenario indicates that a significantly larger number of 2 μm diameter particles are deposited in the targeted lung region than any other region, compared to the cases of particles of diameter 4 μm and 6 μm. Smaller particles reach the targeted region by external magnetic field intensity due to lower inertia, despite fast flow rates. The particles of diameter 2 μm have lower inertia than that of particles of diameter 4 μm and 6 μm. Therefore, the particles reach the targeted region and follow the path line. The total percentage of deposition for the 2 μm particles is 38.45, whereas, at the targeted position, the deposition percentage is 12.34 and for the targeted left lung it is 25.02 as shown in Figure 30.11(a). Figure 30.11(b) shows the deposition scenario for 4 μm diameter particles. For the 4 μm particle diameter, the overall deposition percentage is 47.40, whereas, at the targeted position, it is 36.31. The deposition pattern for 6 μm particle diameter is shown in Figure 30.11(c). The overall deposition

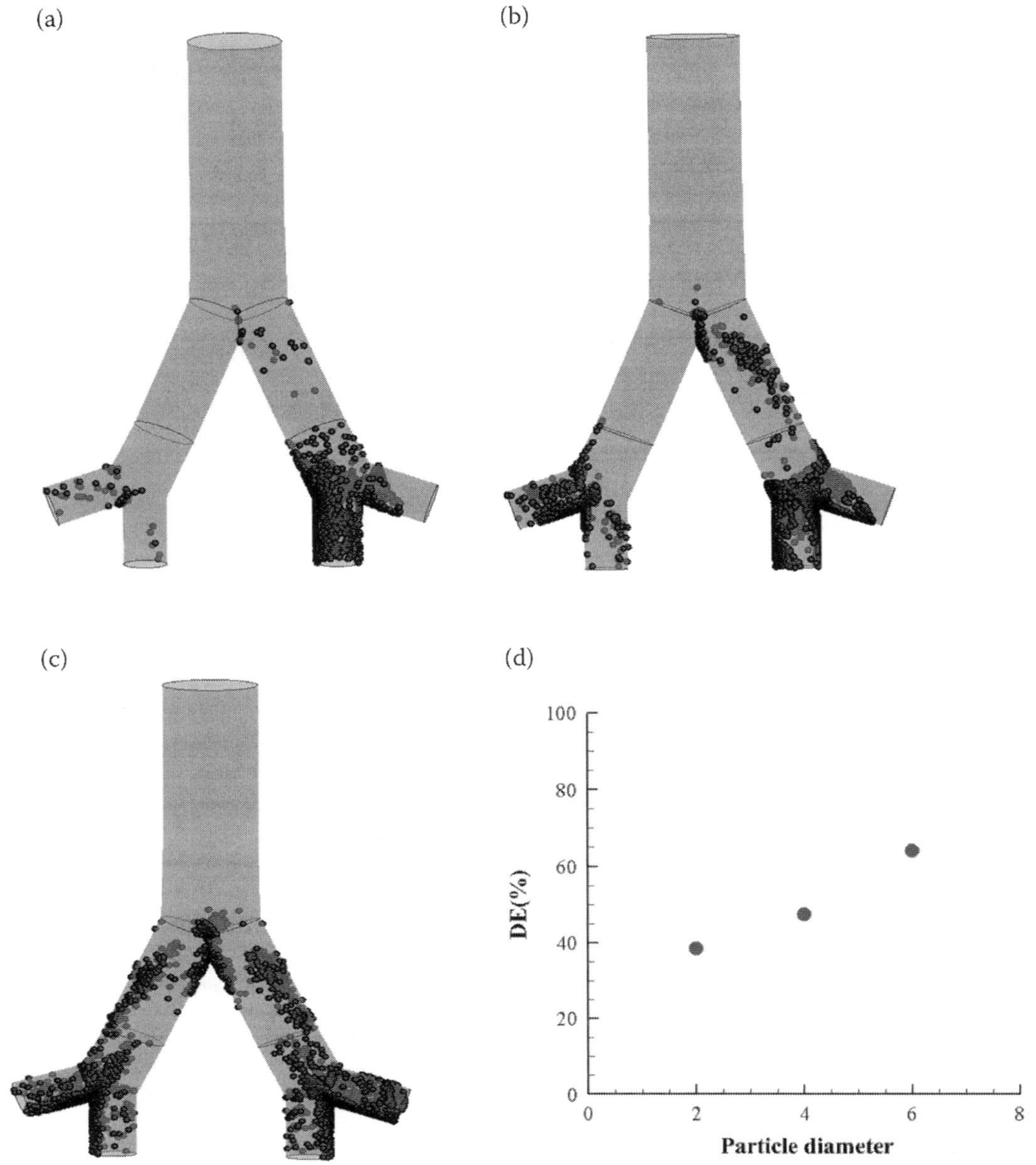

FIGURE 30.11 Effect of Particle Size (Diameter) on the Transport Pattern and DE (%) at Position 2(Q=60 lpm, Mn = 2.5), (a) d_p=2 μm, (b) d_p=4 μm, (c) d_p=6 μm, and (d) Overall Deposition Efficiency Variation with Particle Diameter

percentage for 6 μm particle diameter is 64.03 and for the targeted position, the deposition percentage is 20.11. It is important to note that the influence of the magnetic field on the magnetic drug carrier for d_p = 4 μm is more noticeable in the targeted position than for other particle sizes, as is shown in Figure 30.11(c). Figure 30.11(d) shows that the overall deposition concentration is higher at large particle size. Due to the inertia effect of particles, it is expected that by increasing the particle diameter, the DE will be increased, which is observed in the present study. It is also observed that the DE is higher in targeted position 2 (left branch) compared to the other areas of the lung, which is an expected trend.

Figure 30.12(a,b,c) shows the effect of the magnetic field intensity (given by the magnetic number) for source location in position 2, particle diameter of 4 μm and flow rate of 60 lpm. Figure 30.12(a) illustrates the lung airway deposition pattern at the targeted position 2 for a magnetic number of 0.181. The deposition percentage for magnetic number 0.181 is 12.41. The particle deposition pattern for magnetic number 2.5 is shown in Figure 30.12(b), and the overall deposition percentage is 47.40. The overall deposition percentage for magnetic number 3 is 51.64. It is found that increasing the magnetic number enhances the deposition on the targeted position. Figure 30.12(d) shows the overall deposition variation with the magnetic number intensity. Figure 30.12 shows that with an increase in the magnetic number, more particles are deposited in the target region. Therefore, a large magnetic number, i.e. magnetic field intensity, can play an important role in particulate deposition on the targeted region of a lung.

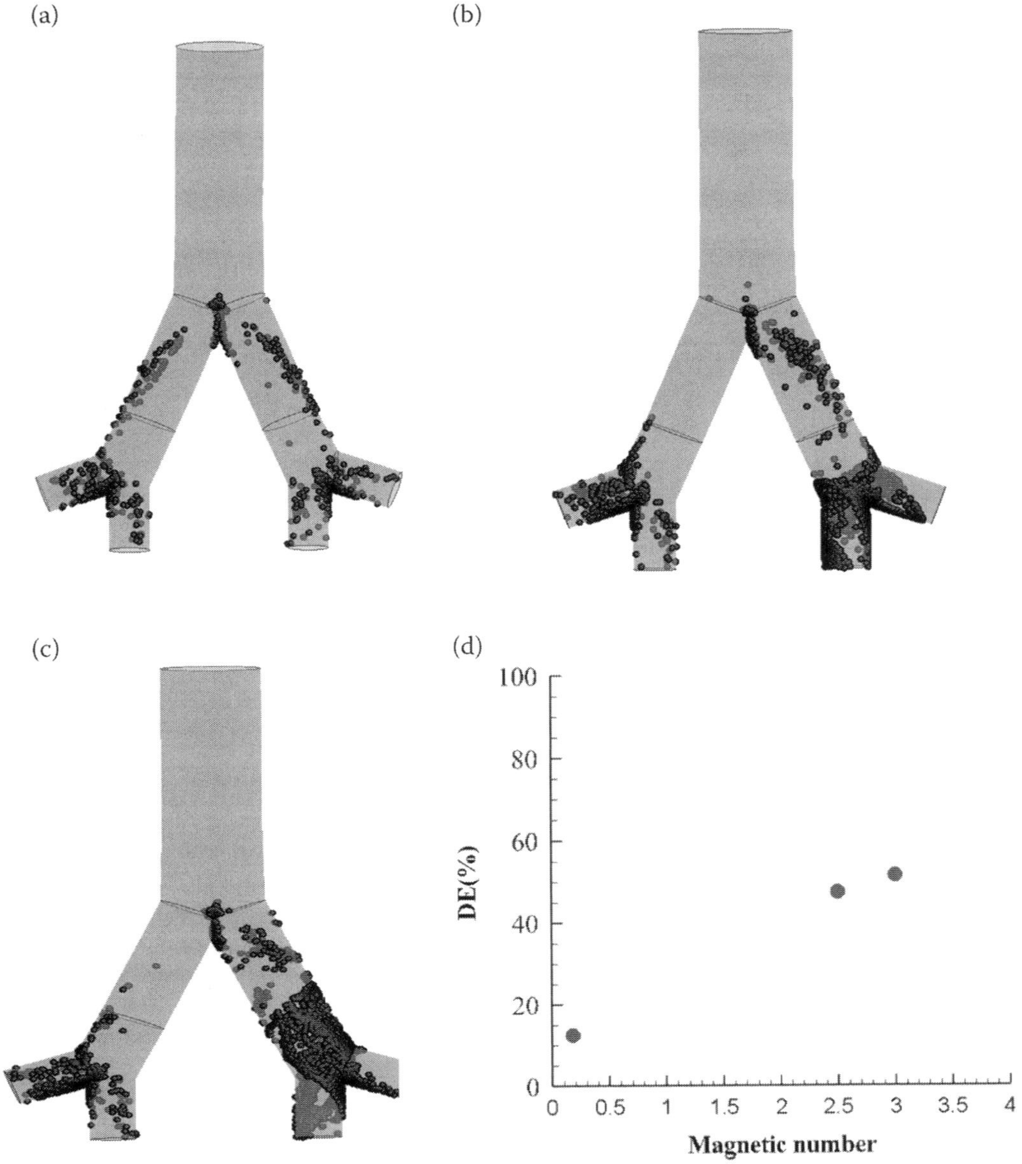

FIGURE 30.12 Effect of Magnetic Number (Flux Value) on the Particle Transport Pattern and DE (%) for Position 2(d_p=4 μm, Q=60 lpm) (a) Mn=0.181, (b) Mn=2.5, (c) Mn=3, and (d) Overall Deposition Efficiency Variation with Magnetic Number

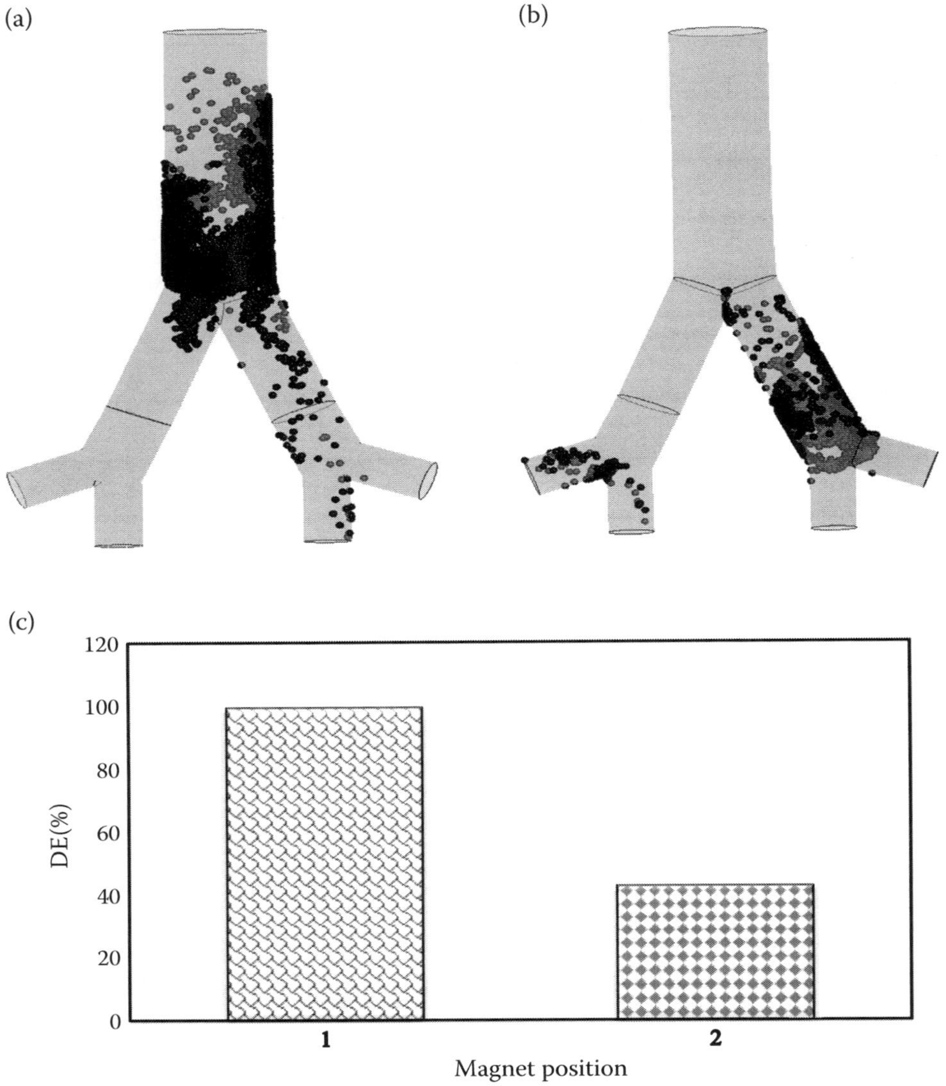

FIGURE 30.13 Effect of the Magnetic Source Position on the Particle Transport Pattern and DE (%) (d_p=4 μm, Q=30 lpm, Mn=2.5) (a) Position 1, (b) Position 2, and (c) Deposition Efficiency Comparison for Magnetic Source Position

Figure 30.13(a,b) represents the effect of external magnetic source in two different positions for the case of 4 μm particle diameter, 30 lpm inhalation flow rate and 2.5 magnetic number. Due to the position of the magnetic source, the drug particles tend to accelerate in the targeted position in the presence of the magnetic force field. Figure 30.13(a) illustrates the deposition scenario for magnetic source (Mn=2.5) in position 1 with the overall deposition percentage being 99.334. Figure 30.13(b) shows the respiratory deposition scenario for 30 lpm inhalation flow rate and magnetic source (Mn=2.5) for position 2. The overall percentage for this magnetic source in position 2 is 42.928. Figure 30.13(c) shows the maximum number of deposited particles in the left lung and the targeted position. It is found that the deposition efficiency is decreased by increasing the distance from the inlet, as shown in Figure 30.13(c). It is noteworthy that an advantage of the current numerical model is identification of the deposition efficiency and concentration at specific regions of interest.

Figure 30.14 shows the local DE for different flow rates, particle diameter, magnetic number, and magnetic source position. Figure 30.14(a) represents the regional deposition scenario for three different flow rates of 15 lpm, 30 lpm and 60 lpm. The deposition concentrations (percentages) in the targeted position (wall position 2) for these three different flow rates are 23.85, 4.03, and 36.31, respectively. The particle deposition concentration in the targeted position is higher compared to the other regions for the particles of diameter 2 μm due to the lower inertia. When the flow rate is high (i.e. 60 lpm) and the magnetic number is 2.5, the maximum number of particles is deposited in wall position 2, which is the targeted position as shown in Figure 30.14(a). Figure 30.14(b) shows the local deposition efficiency for three different particle diameters, i.e. 2 μm, 4 μm and 6 μm.

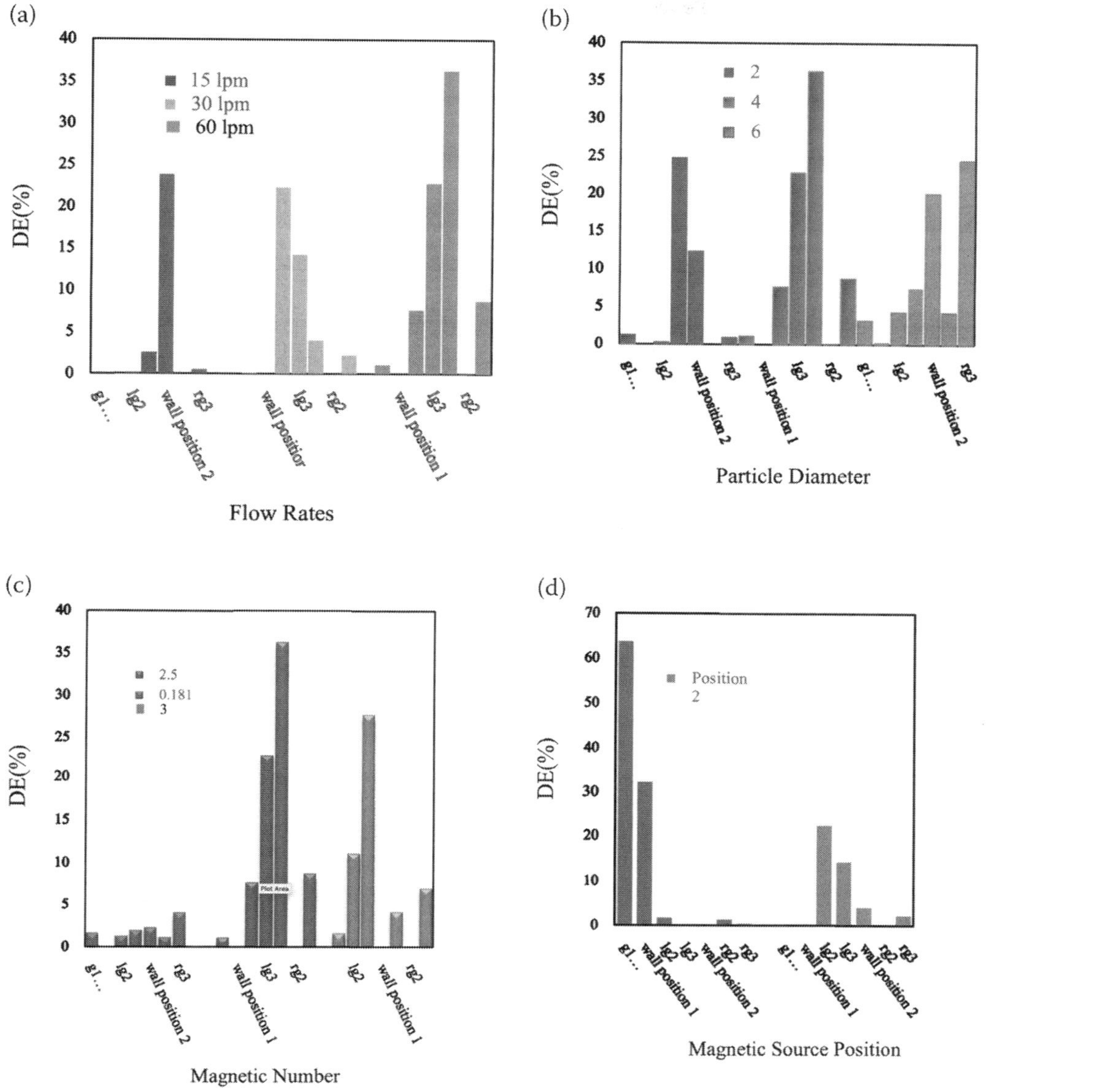

FIGURE 30.14 Local Deposition Efficiency for (a) Flow Rates (b) Particle Diameter, (c) Magnetic Number, and (d) Magnetic Source Position. (Generation 1 (g1…), Left Generation 2 (lg2), Left Generation 3 (lg3), Right Generation 2 (rg2), and Right Generation 3 (rg3))

From Figure 30.14(b), the local particle deposition concentrations (percentages) in the targeted position are 12.34, 36.31 and 20.11 for particle diameters of 2 µm, 4 µm and 6 µm, respectively. For the 4 µm particle diameter, the maximum number of particles is deposited in the targeted position, as shown in Figure 30.14(b). On the other hand, the overall deposition is higher for the 6 µm particle diameter compared to the other particle diameters due to the larger inertia. Figure 30.14(c) illustrates the local deposition scenario for three different magnetic numbers, i.e. 0.181, 2.5, and 3. The corresponding concentrations (percentages) of deposited particles in the targeted position (wall position 2) are 2.28, 36.31, and 4.14, respectively. Due to the large magnetic number, the overall deposition is higher for magnetic number 3. Figure 30.14(d) shows the local depositions for two different magnetic source positions. When the magnetic field source is in position 1, most of the particles are deposited on the wall position 1 and generation 1, for the case with medium flow rate (30 lpm) and particle diameter of 4 µm. Similarly, when the magnetic field source is in position 2, the maximum number of particles is deposited in the left branch and in wall position 2. The deposition scenario shows a higher deposition on the left lung and targeted position in Figure 30.14.

30.7 Conclusions

A numerical model for microparticle transport and deposition in the targeted region of human lungs due to an external magnetic field has been developed for a triple bifurcation lung airway. The present study, together with more specific case studies, would provide inputs in enhancing the biologically and clinically relevant knowledge of the zone-specific targeting drug delivery in human lungs (35). A comprehensive validation has been performed and the following conclusions are drawn from the present study.

- Particle deposition efficiency on the targeted region was investigated as a function of three different breathing conditions, represented by flow rates of 15 lpm, 30 lpm, and 60 lpm. The present third-generation lung model shows that the maximum numbers of particles are deposited in the targeted position for the fast-breathing condition (60 lpm).
- The effect of particle diameter distribution in the targeted position was investigated for 2 μm, 4 μm and 6 μm particle sizes. Most of the particles were deposited on the left lung and in the targeted position for 2 μm particle diameter. For the particle diameter of 4 μm, maximum numbers of particles were deposited in the targeted position (wall position 2).
- The deposition scenario was investigated in the targeted position for three different magnetic field intensities, given by different magnetic numbers, i.e. 0.181, 2.5 and 3. The overall depositions are 12.41%, 47.40% and 51.64%. The overall deposition increases at a higher magnetic number. On the other hand, the maximum number of the deposited particles in the targeted position is noticed for the magnetic number 2.5.
- The microparticle deposition was investigated for two different magnetic source locations, i.e. position 1 (wall position 1) and position 2 (wall position 2). The majority of the particles are deposited on wall position 1 within the generation 1 branch, when the magnetic field source is in position 1 and the particle diameter is 4 μm. Similarly, for the magnetic intensity in position 2, the maximum number of particles deposited is in the left branch and wall position 2.

The present findings would help in designing efficient drug delivery systems for human lungs for the treatment of various lung conditions.

REFERENCES

1. Kinnula VL and Crapo JD. Superoxide dismutases in the lung and human lung diseases. *American Journal of Respiratory and Critical Care Medicine*. 2003; 167(12): 1600–1619.
2. Darquenne C. Aerosol deposition in health and disease. *Journal of Aerosol Medicine and Pulmonary Drug Delivery*. 2012; 25(3): 140–147.
3. Dames P, Gleich B, Flemmer A, Hajek K, Seidl N, Wiekhorst F, Eberbeck D, Bittmann I, Bergemann C, Weyh T, Trahms L, Rosenecker J, and Rudolph C. Targeted delivery of magnetic aerosol droplets to the lung. *Nature Nanotechnology*. 2007; 2: 495.
4. Worth Longest P and Vinchurkar S. Validating CFD predictions of respiratory aerosol deposition: Effects of upstream transition and turbulence. *Journal of Biomechanics*. 2007; 40(2): 305–316.
5. Jayaraju ST Brouns M, Lacor C, Belkassem B, and Verbanck S. Large eddy and detached eddy simulations of fluid flow and particle deposition in a human mouth–throat. *Journal of Aerosol Science*. 2008; 39(10): 862–875.
6. Elcner J, Lizal F, Jedelsky J, Jicha M, and Chovancova M. Numerical investigation of inspiratory airflow in a realistic model of the human tracheobronchial airways and a comparison with experimental results. *Biomechanics and Modeling in Mechanobiology*. 2016. 15(2): 447–469.
7. Singh P, Raghav V, Padhmashali V, Paul G, Islam MS, and Saha SC. Airflow and particle transport prediction through stenosis airways. *International Journal of Environmental Research and Public Health*. 2020; 17(3): 1119.
8. Zhang Z and Kleinstreuer C. Low-Reynolds-number turbulent flows in locally constricted conduits: A comparison study. *AIAA Journal*. 2003; 41(5): 831–840.
9. Islam MS, Paul G, Ong HX, Young PM, Gu Y, and Saha SC. A review of respiratory anatomical development, air flow characterization and particle deposition. *International Journal of Environmental Research and Public Health*. 2020; 17(2): 380.
10. Plank C. Nanomagnetosols: Magnetism opens up new perspectives for targeted aerosol delivery to the lung. *Trends in Biotechnology*. 2008; 26(2): 59–63.
11. Dahmani C, Gotz S, Weyh T, Renner R, Rosenecker M, and Rudolph C. Respiration triggered magnetic drug targeting in the lungs. in Engineering in Medicine and Biology Society, 2009. EMBC 2009. *Annual International Conference of the IEEE*. 2009. IEEE.
12. Goetz SM, Dahmani C, Rudolph C, and Weyh T. First theoretic analysis of magnetic drug targeting in the lung. *IEEE Transactions on Biomedical Engineering*. 2010; 57(9): 2115–2121.
13. Dolovich MB and Dhand R. Aerosol drug delivery: Developments in device design and clinical use. *The Lancet*. 2011; 377(9770): 1032–1045.
14. Ally J, Martin B, Behrad Khamesee M, Roa W, and Amirfazli A. Magnetic targeting of aerosol particles for cancer therapy. *Journal of Magnetism and Magnetic Materials*. 2005; 293(1): 42–449.
15. Pourmehran M, Rahimi-Gorji M, Gorji-Bandpy, and Gorji TB. Simulation of magnetic drug targeting through tracheobronchial airways in the presence of an external non-uniform magnetic field using Lagrangian magnetic particle tracking. *Journal of Magnetism and Magnetic Materials*. 2015; 393: 380–393.
16. Pourmehran TB Gorji and Gorji-Bandpy M. Magnetic drug targeting through a realistic model of human tracheobronchial airways using computational fluid and particle dynamics. *Biomechanics and modeling in mechanobiology*. 2016; 15(5): 1355–1374.
17. Chen J, Li C, Ristovski Z, Milic A, Gu Y, Islam MS, Wang S, Hao J, Zhang H, and He C. A review of biomass burning: Emissions and impacts on air quality, health and climate in China. *Science of the Total Environment*. 2017; 579: 1000–1034.
18. Hoffmann V, Knab M, and Appel. EJ. Magnetic susceptibility mapping of roadside pollution. *Journal of Geochemical Exploration*. 1999; 66(1–2): 313–326.
19. Pityn P, Chamberlain M, King M, and Morgan W. Differences in particle deposition between the two lungs. *Respiratory Medicine*. 1995; 89(1): 15–19.

20. Islam MS, Saha SC, Gemci T, Yang IA, Sauret E, and Gu YT. Polydisperse microparticle transport and deposition to the terminal bronchioles in a heterogeneous vasculature tree. *Scientific Reports*. 2018; 8(1): 16387.
21. Islam MS, Saha SC, Gemci T, Yang IA, Sauret E, Ristovski Z, and Gu YT. Euler-Lagrange prediction of diesel-exhaust polydisperse particle transport and deposition in lung: Anatomy and turbulence effects. *Scientific Reports*. 2019; 9(1): 12423.
22. Pourmehran O, TB Gorji, and Gorji-Bandpy M. Magnetic drug targeting through a realistic model of human tracheobronchial airways using computational fluid and particle dynamics. *Biomechanics and Modeling in Mechanobiology*. 2016; 15(5): 1355–1374.
23. Cheng Y-S, Zhou Y, and Chen BT. Particle deposition in a cast of human oral airways. *Aerosol Science and Technology*. 2010; 31(4): 286–300.
24. Kleinstreuer C, Zhang Z, Li Z, Roberts WL, and Rojas C. A new methodology for targeting drug-aerosols in the human respiratory system. *International Journal of Heat and Mass Transfer*. 2008; 51(23): 5578–5589.
25. Cheng Y-S, Zhou Y, and Chen BT. Particle deposition in a cast of human oral airways. *Aerosol Science & Technology*. 1999; 31(4): 286–300.
26. Lippmann M and Albert RE. The effect of particle size on the regional deposition of inhaled aerosols in the human respiratory tract. *American Industrial Hygiene Association Journal*. 1969; 30(3): 257–275.
27. Chan TL and Lippmann M. Experimental measurements and empirical modelling of the regional deposition of inhaled particles in humans. *American Industrial Hygiene Association Journal*. 1980; 41(6): 399–409.
28. Foord N, Black A, and Walsh M. Regional deposition of 2.5–7.5 μm diameter inhaled particles in healthy male non-smokers. *Journal of Aerosol Science*. 1978; 9(4): 343–357.
29. Stahlhofen W, Gebhart J, and Heyder J. Experimental determination of the regional deposition of aerosol particles in the human respiratory tract. *American Industrial Hygiene Association Journal*. 1980; 41(6): 385–398.
30. Stahlhofen W, Gebhart J, Heyder J, and Scheuch G. New regional deposition data of the human respiratory tract. *Journal of Aerosol Science*. 1983; 14(3): 186–188.
31. Emmett PC, Aitken RJ, and Hannan WJ. Measurements of the total and regional deposition of inhaled particles in the human respiratory tract. *Journal of Aerosol Science*. 1982; **13**(6): 549–560.
32. Bowes SM and Swift DL. Deposition of inhaled particles in the oral airway during oronasal breathing. *Aerosol Science and Technology*. 1989; 11(2): 157–167.
33. Cohen SD. *The development of a discrete particle model for 3D unstructured grids: Application to magnetic drug targeting*. MSc thesis, Delft University of Technology, 2009.
34. Haverkort J. *Analytical and computational analysis of magnetic drug targeting in simplified and realistic arterial geometries.* MSc thesis. Delft University of Technology, 2008
35. Islam MS, Saha SC, Sauret E, Gemci T, Yang IA, and Gu YT. Ultrafine particle transport and deposition in a large scale 17-generation lung model. *Journal of Biomechanics*. 2017; 64: 16–25.

31

Phytochemicals and Biogenic Metallic Nanoparticles as Anti-cancer Agents in Lung Cancer

Nana Ama Mireku-Gyimah[1], Rex Frimpong Anane[2], and Louis Hamenu[3]
[1]*Department of Pharmacognosy and Herbal Medicine, School of Pharmacy, University of Ghana, Accra, Ghana*
[2]*CAS Center for Excellence in Animal Evolution and Genetics, Kunming Institute of Zoology, Chinese Academy of Sciences, Kunming, Yunnan, China*
[3]*Department of Chemistry, School of Physical and Mathematical Sciences, University of Ghana, Accra, Ghana*

31.1 Introduction

Over the years nano-metals with their wide array of chemical properties have been adopted in cancer therapy as diagnostic tools, drug delivery channels, and as drugs themselves. Metals such as silver, gold, and zinc are used in the development of nanoparticles. Nature owns a unique synthetic pathway for the synthesis of metallic nanoparticles. Oxidation/reduction reactions are responsible for the biosynthesis of metallic nanoparticles (MNPs) which are often categorized under the bottom-up approach of synthesis. The biogenic generation of nanoparticles can be achieved using plant phytochemicals and microorganisms such as bacteria and fungi (Figure 31.1). The type of organism involved in the synthesis dictates the type of phytochemicals, proteins, enzymes, and biochemicals that are used in the oxidative or reductive process of metal ion conversion into nanoparticles.

31.1.1 Bacteria as a Biogenic Source for Metallic Nanoparticle Production

The earth has a rich diversity of bacteria populations which makes bacteria easily the most abundant microorganisms on the planet. These prokaryotes have special adaptations that allow them to inhabit even the most extreme environments. Most metal ions are toxic to a variety of bacteria and they try to eliminate them by aggregating them as MNPs both intra- and extracellularly (1). Minute changes in temperature, pH, metal precursor concentrations, and duration of synthesis may affect MNP properties like shape and size. Biochemicals in bacteria that allow for their adaptions in their environment have been exploited by research scientists and mimicked in laboratories for the synthesis of MNPs.

Phytochelatin found in *E. coli* is an example (2). In *E. coli*, phytochelatin is the binding and nucleation site for metal ions. The nanoparticle formed as a result of the nucleation is also stabilized by the same biochemical against continued aggregation. This is useful in the synthesis of CdS nanocrystals with a size distribution of 2–6 nm.

Another biochemical which has been exploited for the synthesis of gold NPs is α-amylase, using a culture supernatant of *B. licheniformis* (3). In bacteria like the *Corynebacterium sp.*, ionized carboxyl groups of amino acid residues and peptide amides trap metal ions on the cell walls. This is followed by bioreduction of the metals by carbonyl groups to complete the bioreduction process to form MNPs. Though slow, the process can be accelerated in alkaline conditions and is quite useful in the synthesis of AgNPs. Other biochemicals responsible for the synthesis of MNPs are the hydrogenase and the nitrate reductase enzymes, which have been used in the synthesis of PdNP and AgNP (4,5). Other nanoparticles which have been synthesized using bacteria are Pt, Hg, Se, Cd/Te, Cr, Co, and Mn.

31.1.2 Plants as a Biogenic Source for Metallic Nanoparticle Production

Plants and their extracts provide some competitive advantages over other biological agents in the synthesis of MNPs. They are easily available, safe and easy to handle, cost-effective, more environmentally friendly, usually require a one-step synthesis process, and have rapid synthesis rates. They are a rich source of multiple biochemicals and metabolites that aid in the synthesis of MNPs. A long list of plants used in the synthesis of MNPs is available in literature. The alfalfa plant is one of the earliest and the most investigated plant species in the synthesis of MNPs (1).

In plants, phytochemicals, mainly carotenoids and polyphenols, are responsible for the relatively quick synthesis of MNP. The polyphenols, especially, are equipped with oxygen rich functional groups like ketones, aldehydes, and carboxylic acids that reduce the metal ions to nanoparticles and cap them in solution to prevent agglomeration and Ostwald ripening. In aqueous systems,

DOI: 10.1201/9781003046547-31

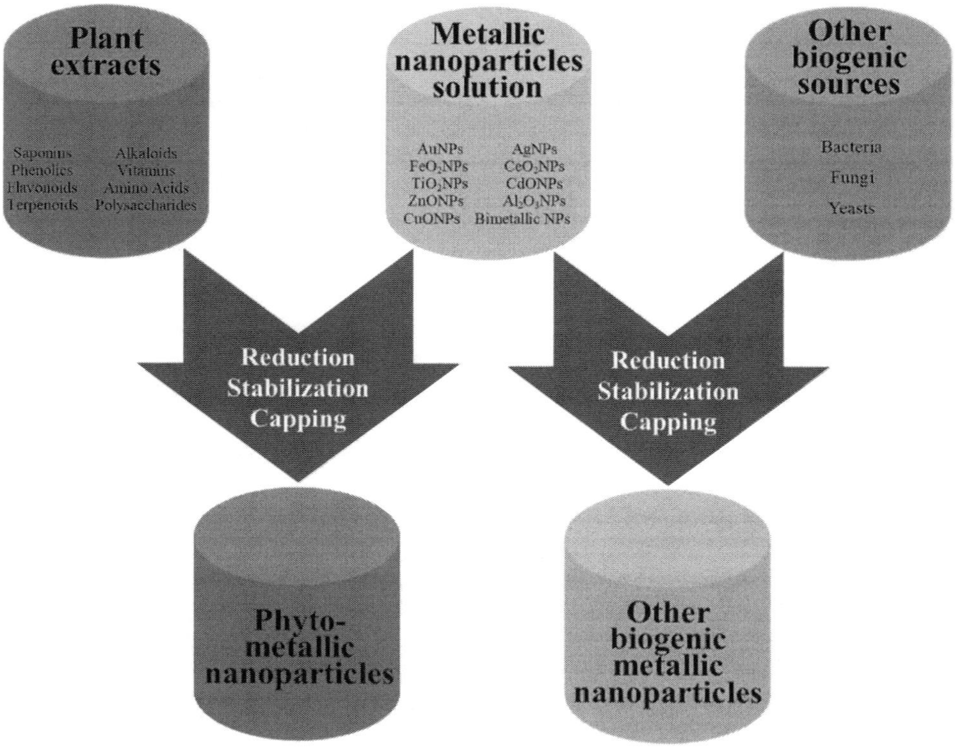

FIGURE 31.1 Generation of Biogenic Metallic Nanoparticles

water soluble phytochemicals are immediately responsible for the reduction process (6). In *Bryophyllum sp.* an anthraquinone type chemical, emodin, was active in the synthesis of AgNP. Cyperoquinone, dietchequinone, and remirin, all benzoquinone derivatives, were the metal-converting agents in *Cyprus sp.,* a mesophyte, and the liberation of hydrogen that accompanied the conversion of catechol and protocatechaldehyde into protocatecheuic acid is responsible for the synthesis of MNPs in hydrophytes like the *Hydrilla sp.*(6). *Azadirachta indica* leaf extract has proven useful also in the synthesis of Au, Ag, and bimetallic Au-Ag nanoparticles (7). In this case, two main terpenoids identified as nimbin and quercetin were responsible for the reduction process, capping and stabilizing effect of the resultant nanoparticle.

31.1.3 Benefits of Biosynthesized Nanoparticles

Biogenic metallic nanoparticles have been explored in recent years because the synthesis process is relatively safe, less toxic, eco-friendly, economically viable, and a single-step method as compared to chemical synthesis (8). Again, the biogenic production of nanoparticles is more advantageous as the therapeutic effect of the secondary metabolites of the biological systems may be conferred on the developed nanoparticle.

Biosynthesized nanoparticles have the ability to modify the smooth function of some mammalian tissues, just like chemically synthesized nanoparticles, therefore, the biocompatibility of these nanoparticles with normal mammalian tissues cannot be ignored. Moulton et al. proved with the synthesis of AgNPs that biosynthesized nanoparticles can be nontoxic to human cells (9). The compatibility of biosynthesized AuNPs with blood is also reported, contrary to chemically synthesized AuNP capped with citrate which agglomerated blood (10).

Nanoparticles which selectively target and destroy some cells or tissues (for example cancer cells) in mammals is just what is needed to fight targeted diseases that affect specific tissues, such as in cancer. The targeted delivery of anti-cancer drugs by noble nanoparticles could minimize the harsh side effects usually associated with conventional chemotherapeutic drug applications, which promote patience non-compliance.

Biosynthesized nanoparticles can also be used for drug delivery to targeted sites such as aberrant cells in cancer chemotherapy. When a nanoparticle is capped or functionalized with different agents, such as the plethora of bioactive compounds present in plants, bacteria, and fungi, the response is different as the different capping agent dictates which cells it will bind to, making the chemotherapy process less harsh.

In this chapter, therefore, we set the tone for the discussion of biogenic nanoparticles from plant phytochemicals for lung cancer therapy. Table 31.1 sumarizes the medicinal plants and their synthesized biogenic metallic nanoparticles as described in the chapter.

31.2 Biogenic Silver Nanoparticles (AgNP), Synthesis and Anti-cancer Activity

Silver nanoparticles are metallic nanoparticles with applications in various fields due to their exceptional physicochemical

TABLE 31.1
Summary of Medicinal Plants and their Synthesized Biogenic Metallic Nanoparticles

Biogenic Source	Size	Shape	Activity in Lung Cancer Cells	Reference
Silver nanoparticles (AgNPs)				
Derris trifoliate seed extract	16.05 ± 5.0 nm		EC_{50} of 86.23 ± 0.22 μg/mL	(11)
Mucuna pruriens seed extract	20–60 nm	Rod-shaped	IC_{50} of 50 μg	(12)
Syzygium aromaticum	5–40 nm	Spherical	IC_{50} of 50 μg/mL	(13)
Dendropanax morbifera leaf extract	100–150 nm	Polygonal	100 μg/mL inhibited 70% cell growth	(14)
Azadirachta indica leaf extract	94 nm (mean diameter)	Spherical		(15)
Acorus calamus rhizome extract	31.83 nm		IC_{50} value of 69.44 μg/mL	(16)
Artemisia princeps leaf extract	20 nm	Spherical	IC_{50} of 30 μg/mL	(17)
Toxicodendron vernicifluum bark extract	2–40 nm	Spherical, oval	320 μg/mL induced 82.5% of cell death	(18)
Scoparia dulcis leaf extract	15–25 nm		At 5 mg/mL concentration 43.4% A549 cells survived	(19)
Rosa damascena petal extract	15–27 nm	Spherical	IC_{50} of 80 g/mL	(20)
Origanum vulgare leaf extract			IC_{50} of 100 g/mL	(21)
Panax ginseng leaf extract				(22)
Gossypium hirsutum	30 nm	Spherical	IC_{50} of 40 g/mL	(23)
Dendropanax morbifera leaf extract	10–20 nm	Hexagonal	25 μM, potentiation of cytotoxic effect of ginsenodise K	(14)
Gold nanoparticles (AuNps)				
Magnolia officinalis leaf extract	128 nm			(24)
Marsdenia tenacissima	30–50 nm	Spherical and oval	IC_{50} of 15 μg/mL	
Musa paradisiaca peel extract	50 nm	Spherical to triangular	IC_{50} of 58 μg/mL	(25)
Rabdosia rubescens leaf extract	130 nm		CC_{50} of 50 μg/mL	(26)
Zinc oxide nanoparticles (ZnONps)				
Mangifera indica leaf extract	60 nm	Nearly spherical and hexagonal		(27)
Deverra tortuosa extract of aerial parts	15.22 nm		IC_{50} of 83.47 μg/mL	(28)
Iron oxide nanoparticles (FeONps)				
Solanum lycopersicum leaf extract	483.8 nm		IC_{50} of 69±0.50 μg/mL	(29)
Phyllanthus emblica fruit extract	98.18 nm		IC_{50} of 104.7 μg/mL	(30)
Cerium oxide nanoparticles (CeONps)				
Musa sapientum fruit peel extract	4–13 nm			(31)
Eucalyptus globulus leaf extract	13.7 nm		IC_{50} value of 45.5 μg/L	(32)

properties such as chemical stability, electrical conductivity, catalytic activity, and large surface area to volume ratio. AgNPs have gained special interest from researchers in various fields, especially in biomedicine, because of the therapeutic effects they offer including anti-infective, anti-inflammatory and cytotoxic activities (33–35). AgNPs are well-known for their broad-spectrum and highly efficient applications in treatments of cancers and microbial infections.

Cyril et al. synthesized biogenic silver nanoparticle AgNP-DTa with an average size of 16.05 ± 5.0 nm using aqueous seed extract of *Derris trifoliata* Lour. of the family Leguminosae (DT) by adding 1 mL of aqueous seed extract to 20 mL of 1 mM solution of $AgNO_3$ at room temperature (11). In their synthesis, they investigated the effect of direct sunlight and absence of same on the nanoparticle synthesis. AgNP-DTa were well dispersed and stabilized in aqueous solution through biological ligands extracted from the seeds of DT. In the formation of AgNP-DTa, the functional groups present in the bioligands extracted from the seeds of DT acted as reducing and stabilizing agents. Per their observations, a longer duration of sunlight exposure was necessary for the synthesis of smaller sized AgNP-DTa. Also sunlight induced and catalyzed the reduction of Ag^+ to nanosilver atoms, which is useful for the synthesis of AgNPs (11). The synthesized AgNP-DTa showed moderate anti-proliferative activity on A549 lung cancer cell lines with an EC_{50} of 86.23 ± 0.22 μg/mL (11). Observation of the cell cultures through an inverted phase contrast tissue culture microscope showed that AgNP-DTa induced apoptotic changes and nuclear condensation on the A549 lung cancer cell lines (11). AgNP-DTa may have caused these changes by penetrating into A549 cells to interfere with the proper functioning of cellular proteins and/or developed reactive oxygen species (ROS) in the cells. This introduces oxidative stress in the cells, which can lead to DNA damage and substantial morphological changes in the A549 cells, resulting in apoptotic cell death.

In a recent study, *Mucuna pruriens* (L.) DC (family Leguminosae) (Mp) aqueous seed extract was employed in the production of silver nanoparticles (Mp-AgNPs) in a one-step procedure. Ten mL of the aqueous seed extract was added to 90 mL of 1 mM silver nitrate solution. The cytotoxicity effects of the produced rod-shaped Mp-AgNPs of size 20–60 nm was then assessed on A549 human lung cancer cell lines by using the MIT assay (12). The Mp-AgNPs exhibited a persistent dose-dependent inhibiting activity on the growth of A549 lung cancer cells with an inhibitory concentration (IC_{50}) value of 50 μg. Although both the standard drug cyclophosphamide and the Mp-AgNPs effectively suppressed the cancerous cell growth, Mp-AgNPs showed better cytotoxic activity as a result of its sustained release in a dose-dependent manner at the same concentration. The cytotoxic effect can be attributed to the interaction of the silver ions with the various intracellular proteins as well as the nitrogenous bases or phosphate groups of the cellular DNA (12).

Similarly, silver nanoparticles were synthesized using *Syzygium aromaticum* (L.) Merr. & L.M.Perry (family Myrtaceae) (Sa) as the reducing and stabilizing agents (13). Synthesis was performed by dissolving 10 mL of the plant extract (prepared by dissolving of 20 g of plant powder in 100 mL of millipore water, followed by heating at 80 °C and filtration) in 90 mL of 1 mM silver nitrate solution to achieve reduction of Ag^+ ions. The synthesized biogenic *Syzygium aromaticum* AgNPs (Sa-AgNPs) were spherical in shape with particle size of 5–40 nm. Examination and analysis of the cytotoxicity impact of Sa-AgNPs on A549 lung cancer cell growth revealed cytotoxicity with IC_{50} of 50 μg/mL. Increasing concentrations of Sa-AgNP resulted in higher growth inhibition of the cancerous cell, thus cell viability decreased with increasing concentrations of Sa-AgNPs as compared to the plant extract only. The mechanism of cell death induced by the biosynthesized AgNPs from *Syzygium aromaticum* (Sa-AgNPs) was through apoptosis as manifested by the characteristic nuclear changes such as chromatin condensation and nuclear fragmentation in the cells. It is also plausible that silver nanoparticles may have stimulated reactive oxygen species (ROS) to effect damage to cellular components which led to cell death (13).

Utilization of the leaf extract of *Dendropanax morbifera* Léveille, a medicinal plant belonging to the *Araliaceae* family, for biosynthesis of silver nanoparticles was studied by Wang et al. (14). In the study, the phytochemicals in *D. morbifera* leaf extract aided in the rapid synthesis of *Dendropanax morbifera* AgNPs (Dm-AgNPs) by acting as the reducing and stabilizing agent. To synthesize Dm-AgNPs, 5 mL of the aqueous leaf extract was mixed with 45 mL of deionized water, and $AgNO^3$ solution was added to a final concentration of 1 mM in the reaction mixture. Bioreduction of Ag^+ ions into Ag^0 in the silver nitrate ($AgNO_3$) solution by the *D. morbifera* leaf extract was indicated by color changes. Characterization of the biosynthesized Dm-AgNPs revealed polygonal shapes with average size of 100–150 nm (14). When analyzed for anti-cancer activity, Dm-AgNPs exhibited potent cytotoxicity in the A549 human lung cancer cells. It was shown that after 48 hours of treatment, 100 μg/mL Dm-AgNPs significantly inhibited A549 cell growth by more than 70%. Additionally, Dm-AgNPs at 50 μg/mL improved the cytotoxicity of ginsenoside compound K (CK), an important secondary metabolite of Korean Ginseng, *Panax ginseng* Meyer, at 25 μM after 48 hours of treatment compared with CK alone. Thus, CK when combined with D-AgNPs decreased the cell viability to 20% at 25 μM after 48 hours, whereas CK alone reduced cell viability to approximately 50% at 25 μM after 48 hours (14). These results suggested that the synthesized Dm-AgNPs did not only inhibit lung cancer cell growth but also enhanced the cytotoxicity of CK in cancer cells. The mechanism by which Dm-AgNPs caused cytotoxicity in the A549 human lung cancer cells was determined to be via changes in the nuclear morphology of the cells by inducing nuclei shrinkage or fragmentation (14). Studies have shown that morphological characteristics such as cell shrinkage, chromatin condensation, and nuclear fragmentation are evidence of cell apoptosis (14,36,37).

The cytotoxicity of AgNPs synthesized by using the leaf extract of *Azadirachta indica* A.Juss. (family Meliaceae) as a reducing agent was assessed by Kummara et al. (15). 50 mL of 2 mM silver nitrate solution was prepared and added in drops into 500 mL of *Azadirachta indica* leaf extract on a magnetic stirrer at 40 °C with 200 rpm for 30 min. After 24 h, a yellowish brown to thick brown color change was observed, indicating that AgNPs were formed from bioreduction of silver ions. The green synthesized *Azadirachta indica* AgNPs (Ai-AgNPs) were spherical with a mean diameter at 94 nm (15). Anti-cancer activity of Ai-AgNPs against NCI-H460 lung cancer cells was determined. It was observed that Ai-AgNPs caused a dose-dependent decrease in cell viability and increase in reactive oxygen species generation. Cell proliferation in NCI-H40 cells was significantly inhibited upon treatment with 160 ppm, 200 ppm, and 240 ppm of *A. indica* mediated AgNPs (Ai-AgNPs) (15). Specifically, *A. indica* mediated AgNPs (Ai-AgNPs) treatment at 240 ppm showed only 2% cell viability in NCI-H40 cells; 240 ppm of Ai-AgNPs generated 78.41% ROS in NCI-H460 cells, suggesting that excessive production of ROS in the NCI-H460 cells might have induced the observed cellular apoptosis in NCI-H460 cells (15).

Silver nanoparticles synthesized from an aqueous rhizome extract of *Acorus calamus* L. (family Acoraceae) (AcAgNPs) and subsequent evaluation of its anti-cancer effects has also been reported (16). AcAgNPs were synthesized by mixing aqueous $AgNO_3$ solution (1 mM) and rhizome extract of *Acorus calamus* in a ratio of 5:1 and incubating the mixture at room temperature for 24 h. The average size of the synthesized AcAgNPs was found to be 31.83 nm (16). Evaluation of the anti-cancer capability of AcAgNPs indicated that 32.1 μg/mL AcAgNPs possesses cytotoxic effect with an IC_{50} value of 69.44 μg/mL on A549 lung cancer cells after 48 h of treatment. Apoptosis, the mechanism of cell death, in ACAgNPs treated cells was indicated by PI and DAPI staining (including dUTP nick end labeling and TUNEL) that confirmed nuclear changes such as condensation and fragmentation of ACAgNPs treated A549 cells (16).

Gurunathan et al. (17) synthesized silver nanoparticles using the leaf extract of *Artemisia princeps* Pamp. (family Asteraceae) as a bio-reductant; 10 mL of 1 mg/mL filtered leaf extract was added to 100 mL of 1 mM $AgNO_3$ in an aqueous solution at room temperature to synthesize AgNPs. The biologically synthesized *Artemisia princeps* AgNPs (ApAgNPs) had size distribution between 10 and 40 nm, with an average size of 20 nm. These ApAgNPs were significantly well separated and spherical in shape. It was shown that ApAgNPs had cytotoxic effect on A549 lung cancer cells. At a concentration of 30 μg/mL, ApAgNPs was able to induce 50% cell death (IC_{50}). Further analysis revealed that the biologically synthesized ApAgNPs induced mitochondrial-mediated apoptosis in A549 cells. Similarly, A549 cells exposed to 30 μg/mL ApAgNPs for 24 h showed a dose-dependent increase in generation of ROS and oxidative stress, and cell membrane damage (17), thereby causing apoptosis in A459 cells.

In another study, the bark extract of *Toxicodendron vernicifluum* (Stokes) F.A. Barkley (Anarcadiadeace) (Tv) plant was used to synthesize silver nanoparticles (18). Tv-AgNPs were synthesized by dissolving 3 mM of $AgNO_3$ in 10 mL of the bark extract at room temperature. These phytogenic Tv-AgNPs were spherical and oval-shaped with a size range of 2–40 nm. The presence of the phytoconstituents amine, amide, phenolic, and alcoholic aromatics derived from the Tv extract was found to be capping and/or reducing agents in the formed Tv-AgNPs. Tv-AgNPs at 320 μg/mL induced 82.5% of cell death in human A549 lung cancer cells, and further 95% of cell death with annexin V FITC/PI based apoptosis. It was indicated that Tv-AgNPs selectively targeted and damaged the A549 cancer cells through ROS generation. The potency of Tv-AgNPs was attributed to the shape and size specific property of Tv-AgNPs that facilitated easy penetration into the cancer cells for targeted therapy (18).

Scoparia dulcis Linn (Plantaginaceae), a plant that has been used extensively as a traditional remedy for various diseases and conditions (38), has also been used to synthesize silver nanoparticles to treat lung cancer (19). 10 mL of *Scoparia dulcis* leaf extract (Sd) was added to 90 mL of aqueous solution of 1 mM $AgNO_3$. The mixture was kept in water bath at 90 °C for 1 hour to reduce silver ions. The resulting change of color from colorless to dark brown was an indication of the development of AgNPs (SdAgNp). The active phytochemicals of *Scoparia dulcis* Linn that include saponin, tannins, amino acids, flavonoids, terpenoids, and catecholamine (39), functioned as both the reducing and stabilizing agents. The SdAgNPs produced was found to be in the size range of 15–25 nm. Cytotoxicity assay indicated that the synthesized phytochemicals-mediated silver nanoparticles SdAgNPs exhibited significant cytotoxicity effects on A549 cells compared to PA-1 cell line in a dose dependent manner. At 5 mg/mL concentration of SdAgNPs, 43.4% of A549 cells survived, whereas at the same concentration, 70% of PA-1 cells survived. The observed apoptosis was as a result of the damage to nuclear DNA of the cells (19).

The petals of the ornamental plant, *Rosa damascena* Mill, of the family Rosaceae, which has been used traditionally for its medicinal and nutritional effects as well as in perfumery (40), has been utilized to synthesize AgNPs. In medicine, various pharmacological properties of *Rosa damascena* have been reported including analgesic, anti-inflammatory, anti-diabetic, antimicrobial, antioxidant, and anti-HIV effects (41). Secondary metabolites such as volatile oils, flavonoids, myrcene, kaempferol, and quercetin are present in *R. damascena* (20,41). In the study (20), AgNPs were synthesized using the aqueous petals extract (Rd) of *Rosa damascena* to further ascertain its anti-cancer effect against A549 lung cancer cells. The process involved the addition of Rd (1 g/mL) to 10 mmol/L silver nitrate solution in a ratio of 2:3 which resulted in a rapid color change from pale yellow to brown yellow, an indication of bioreduction of Ag^+ ions by the phytochemicals in *R. damascena* petals extract. The synthesized biogenic silver nanoparticles were spherical with an average size of 15–27 nm. RdAgNPs were well distributed without any aggregation. Evaluation of the *in vitro* anti-cancer effect of RdAgNPs on human A549 lung cancer cell lines revealed that RdAgNPs in a dose dependent manner decreased the viability of A549 cells. The observed IC_{50} was 80 μg/mL. At this concentration RdAgNPs inhibited the growth of A549 cells by 50% upon treatment (20).

Origanum vulgare L. (family Lamiaceae), commonly known as oregano, is an aromatic herb which is used as a condiment. The essential oil from the dried leaves of oregano, which is rich in terpenes, is said to be chemopreventive and reduces cell viability of A549 lung cancer cells *in vitro*(42). Subsequently, the aqueous extract of *Origanum vulgare* has been used to bioreduce 1 mM $AgNO_3$ solution for the synthesis of silver nanoparticles (OvAgNPs) (spherical shape, average size 136 ± 10.09 nm) (21). In the synthesis, 90 mL of 1 mM silver nitrate solution was mixed with 10 mL of *O. vulgare* leaf extract which was produced by dissolving 10 g of leaf powder in 100 mL of deionized water followed by boiling at 60 °C for 10 min. The formed OvAgNPs mixture was heated gradually in a 60–90 °C water bath for 10 min. The formation of OvAgNPs was indicated by the pale to reddish brown color change of the mixture (21). *In vitro* cytotoxicity studies of the biogenic OvAgNPs against human lung cancer A549 cells at different concentrations (10–500 g/mL) unveiled that there was a direct dose–response relationship between OvAgNPs and tested cells; 100 g/mL OvAgNPs was sufficient to induce 50% of cell mortality, although lower concentrations were able to inhibit the growth of cell line by 6%. At a higher concentration (500 g/mL) OvAgNPs significantly inhibited the cell growth by more than 85%. It was plausible that OvAgNPs may have caused cell death by inducing reactive oxygen species and causing damage to cellular components (21).

The cytotoxic and oxidative effects of silver nanoparticles synthesized from fresh leaves of *Panax ginseng* Meyer (family Araliaceae), a widely used medicinal plant in cancer treatment, were evaluated in human cancer cell lines (22). Fifteen grams of cut fresh leaves of *P. ginseng* were boiled for 20 min in 100 mL of sterile water, and 5 mL of the filtered leaf extract solution was mixed with 25 mL of sterile water. $AgNO_3$ solution was then added to a final concentration of 1 μM in a mixture which was kept at 80 °C for reduction of Ag^+ ions to Ag atoms. The results of the cytotoxic evaluation, as detected by BrdU incorporation assay, revealed that *P. ginseng* mediated synthesis of AgNPs (PgAgNPs) at 5 mg/mL and 10 mg/mL significantly inhibited cell viability by inhibiting DNA synthesis of A549 cells by 75% and 60% respectively (22). Doubling of the dose of PgAgNPs treatment (20 mg/mL) did not significantly decrease the DNA synthesis of A549 when compared to the 10 mg/mL PgAgNPs treated cells. PgAgNPs treatment also inhibited the epidermal growth factor (EGF)-enhanced migration, and also decreased the RNA levels and phosphorylation of EGF receptors in A549 cells (22). Furthermore, PgAgNPs induced oxidative stress in A549 cancer cell lines, altered the morphology of the cell nucleus, and increased apoptosis by stimulating p38MAPK/p53 pathways. These results suggest that PgAgNPs has cytotoxic effects against A549 cells, possibly through the regulation of the EGFR/p38MAPK/p53 pathway (22). *P. gingseng* is a medicinal plant long used in oriental medicine. It is considered a panacea useful for the treatment of a myriad of illnesses including diabetes, hypertension, and cancer. Its anti-neoplastic effect has been reported over the years (43,44). The saponin glycoside phytoconstituents present in the plant, such as the ginsenosides, are responsible for the anti-neoplastic effect of the plant. These may very well contribute to the overall cytotoxic action of PgAgNPs.

Leaf extract of the industrially beneficial crop *Gossypium hirsutum* L. (family Malvaceae) (cotton) has been used as phytoreducer and capping agent to synthesize silver nanoparticles (23). For the reduction of silver ions, 20 mL of aqueous leaf extract was added dropwise into 100 mL of 1 mM silver nitrate with constant stirring. The phyto-synthesized *Gossypium hirsutum* mediated silver nanoparticles GhAgNPs were spherical in shape with an average size of 30 nm. The cytotoxicity effect of GhAgNPs on A549 cell lines was found to be higher with increased concentrations of GhAgNPs. Hence, GhAgNPs caused dose-dependent cytotoxicity in GhAgNPs-treated A549 cells. At 40 g/mL concentration, GhAgNPs caused 50% of cell death (IC_{50}) against A549 cells. The phyto-synthesized GhAgNPs-treated lung cancer cells exhibited loss of membrane integrity, inhibited cell growth, cytoplasmic condensation, and cell clumping that led to cell death (23).

31.3 Biogenic Gold Nanoparticles (AuNP), Synthesis and Anti-cancer Activity

The relatively faster synthesis time for the production of phytomediated gold nanoparticles is much valued (45). Gold nanoparticles top the list of nanoparticles used to prepare drug delivery scaffolds (46). Due to the noble nature of AuNPs, they are often nonimmunogenic and nontoxic in the body making them excellent drug delivery candidates. Gold has high affinity for thiol (-SH) groups which allows the surfaces of AuNPs to be modified with ease (47).

Monodispersed gold nanoparticles of spherical and oval shapes and size 30–50 nm were produced using the extract of *Marsdenia tenacissima* (Roxb.) Wight et Arn. (family Asclepiadaceae), a Chinese medicinal plant with anti-proliferative, anti-angiogenic, and immunomodulation activities. The extract of the plant, which is said to improve the quality of life of cancer patients, is marketed under the name Xiao'aiping injection (48,49). Briefly, $HAuCl_4$ solution (4 mM) was added dropwise under sonication to the extract solubilized in double distilled water to make a final volume of 20 mL and incubated at 25 °C overnight. A mild-yellow to ruby-red solution color change as well as plasmon peak absorbance that ranged between 525 nm and 540 nm with a maximum peak absorbance approximately between 527 nm and 535 nm were indicative of the formation of the gold nanoparticles (MtAuNP). FTIR analysis showed peaks that corresponded to functional groups such as carboxylic acids, amines, aldehydes, and hydroxyl groups. *In vitro* cytotoxicity (IC_{50} = 15 µg/mL) studies showed a dose-dependent cell death action on A549 cells. Further staining of MtAuNP treated A549 cells with acridine orange and ethidium bromide showed morphological changes corresponding to signs of apoptosis. The *M. tenacissima* mediated gold nanoparticle was concluded to induce apoptosis via regulating intrinsic apoptotic pathway as it showed a downregulation of the expression of Bcl-2, Bid, and an upregulation of the expression of Bax, caspase-8, caspase-9, and caspase-3 in lung cancer A549 cell lines in the caspase assay and western blot studies (48). *Marsdenia tenacissima* contains a number of secondary metabolites including flavonoids and is rich in steroidal glycosides such as tenacissoside A–P, marsdenoside A–M, tenacigenoside A–L, and tenacigenin A–D among others, some of which bear the C21 polyoxypregnane glycoside skeletons (50). The presence of these phytoconstituents may help in the synthesis of the gold nanoparticles as well as contribute to the formation of biomolecules with anti-cancer effects on the surface of the AuNP.

Magnolol, a lignin from the *Magnolia* species (family Magnoliaceae) has been reported to be active against human non small cell lung carcinoma (NSCLC), lung squamous carcinoma CH27 cells, and it has a number of therapeutic effects (51). It is thought that magnolol contributes to the overall anticancer activity exhibited by the extract of *Magnolia officinalis* Rehder & E. Wilson. Subsequently, AuNp prepared from the ethanolic leaf extract of *Magnolia officinalis* demonstrated good activity against lung adenocarcinoma A549 cancer cell lines via apoptosis. The AuNP produced was characterized by a red color formation and had a maximum absorbance at 535 nm. The developed AuNP measured 128 nm in size and showed peaks corresponding to aromatic, hydroxyl, carboxyl alkyl, and amide functional groups when subjected to FTIR studies (3908 cm^{-1}, 3744 cm^{-1}, 3500 cm^{-1}, 3255 cm^{-1}, 3010 cm^{-1}, 2925 cm^{-1}, 1705 cm^{-1}, 1612 cm^{-1}, 1479 cm^{-1}, 1295 cm^{-1} respectively) (24).

Dendropanax morbifera Léveille (family Araliaceae), is a traditional medicinal plant native to Korea. The plant has been reported to have anti-inflammatory, antioxidant, anti-microbial, anti-complementary and anti-cancer effects (52,53). Gold metallic nanoparticles (D-AuNPs) were successfully rapidly synthesized with the aqueous leaf extract and gold (III) chloride trihydrate ($HAuCl_4{\cdot}3H_2O$) within 3 minutes at 80 °C giving rise to a dark ruby-red colored solution. The hexagonal shaped D-AuNPs prepared averagely ranged in size from 10–20 nm and were monodisperse in nature. The surface plasmon resonance (SPR) was 540 nm. Reports have further indicated that the leaf extract of *D. morbifera* facilitated the rapid and easy synthesis of the metallic nanoparticles as compared to other biogenic systems. While the gold metallic nanoparticle formed did not exhibit cytotoxicity towards A549 human lung cancer cell line, it showed promising potentiation of the cytotoxic effect of the ginsenoside compound K at 25 μM after 48 hours of treatment against A549 human lung cancer cell line (14).

Banana (*Musa paradisiaca* L., family Musaceae) is a common tropical fruit enjoyed for its delicacy. It is nutritional as well as medicinal. The peels of banana, which are considered an agricultural waste (54), have been effectively used to synthesize gold nanoparticles. The AuNP characterized by a dark wine-red color was synthesized using chloroauric acid ($HAuCl_4$) and the banana peel extract (fruit peel) (saturated in soda water for 5 min, washed twice with distilled water, oven dried at 50 °C, and powdered. Powdered sample is hot extracted (1 g in 10 mL distilled water at 90 °C) for 5 min with continuous mechanical stirring, filtered, centrifuged at 120x*g* for 10 min, and supernatant freeze-dried). The 50 nm sized nanoparticle was found to be spherical to triangular shaped with a zeta potential of +43.5 mV and FTIR peaks corresponding to the functional groups including phenols, carboxylic acids, methoxy groups, amide groups and aromatic compounds. The banana peel gold nanoparticles exhibited a cytotoxic effect against A549 cell lines with an IC_{50} of 58 μg/mL, possibly through cell clumping and loss of membrane stability, maximally observed at 100 μg/mL after 24 h of treatment with the synthesized AuNP (25).

Zhang and colleagues (26) reported on the use of the oriental plant *Rabdosia rubescens* for producing AuNPs which exhibited apoptotic effects on A549 human cancer lines. The plant is a component of the patented herbal supplement PC-SPES which is used for prostate cancer in Complementary and Alternative Medicine (CAM). Fresh leaves of the plant were used to prepare aqueous extracts (leaves shade-dried for 15 days, powdered, 100 g of powder boiled for 30 min at 60 °C in 1 L of distilled water and filtered) of which 10 mL was added to 190 mL of 1 mM of $HAuCl_4 \cdot 3H_2O$ and kept at room temperature. The gold nanoparticles (RR-AuNP) which were polydispersed and of size 130 nm, formed within 15 minutes, evident with a solution color change from yellow to ruby red. The cytotoxicity of RR-AuNP against A549 lung cancer cell lines was ascertained using the MTT test. A CC_{50} (50% cytotoxic concentration) of 50 μg/mL was determined for A549. Apoptosis was observed in cells treated with the synthesized gold nanoparticle as marked by an increase in reactive oxidative species levels and capsases (26,55). This was further established through DAPI staining and TUNEL assay.

31.4 Biogenic Zinc Oxide (ZnONP), Synthesis and Anti-cancer Activity

Aqueous extract of the aerial parts of *Deverra tortuosa* (Desf.) DC. (family Apiaceae) was used in the biogenic synthesis of ZnONPs; 25 mL plant extract, heated (60 °C) on a magnetic stirrer, addition of 2.5 g ($Zn(NO_3)_2.6H_2O$) left for precipitation (1h), mixture oven dried overnight (60 °C), creamy paste formed washed multiple times (distilled H_2O: EtOH, 3:1), creamy paste heated in furnace (400 °C, 2 h) to form white ZnONP powder. The size of the nanoparticles formed averaged 15.22 nm. A UV characteristic absorption peak at 374 nm and a direct band gap (E_g) of 3.32 eV was calculated for the formed nanoparticle. Further FTIR analysis revealed peaks characteristic of primary and secondary amines of alkaloids, proteins/enzymes, and C–O stretching regions of polysaccharides and phenolic groups. *In vitro* MTT colometric assay for the assessment of the anti-cancer activity of the formed nanoparticle was further performed. From the test, the ZnONP was more selectively cytotoxic to A549 human lung cancer cells in a dose dependent manner (IC_{50} 83.47 μg/mL) in relation to WI38 normal human lung fibroblast cell line (IC_{50} 434.60 μg/mL) as compared to the positive control doxorubicin (A549, IC_{50} of 162.86 μg/mL; WI38, IC_{50} of 186.10 μg/mL) (28). One of the major problems with cancer chemotherapy is the destruction of normal body cells, hence this phytomediated nanoparticle holds prospect as a good anti-cancer agent with selectivity benefits. In this study, the aqueous extract of *Deverra tortuosa* alone was also evaluated under same conditions for its cytotoxic effects in A549 cells. Although the extract showed appreciable cytotoxic effects and selectivity towards the normal human lung fibroblast cell line (A549, IC_{50} of 193.12 μg/mL; WI38, IC_{50} of 902.83 μg/mL) its formed ZnONP showed a better activity, further confirming the therapeutic significance of phytomediated metallic nanoparticles.

Mango (*Mangifera indica* L) is a common tropical fruit belonging to the family Anarcadiaceae. Aside from the rich nutritional content of the fruits, various parts of this plant have been employed in ayurvedic and other folk medicines as an anti-tumor, diarrhea, colic, and liver medication, among others. Various phytochemicals have been isolated from different parts of the plant. A xanthone glycoside mangiferin is identified as the major phytoconstituent. Others include volatile oils consisting of humulene, elemene, ocimene, linalool, nerol, etc., from the leaves and flowers. Also extracted are alkyl gallates from the flowers, triterpenoids and flavonoids from the stem, and chromones from the roots (56,57). Pharmacological activities have shown the antioxidant, anthelminthic, anti-infective, anti-inflammatory, and anti-cancer effect of extracts of mango (58). Rajeshkumar et al. (27) biogenically synthesized ZnONP with the aqueous leaf extract of mango and tested it for its antineoplastic effect on human lung cancer A549 cells. From their studies, powdered ZnONPs were produced by adding zinc nitrate (80 mL, 0.1 M) to the aqueous plant extract (20 mL), stirring continuously at room temperature (6h), allowing the resulting solution to settle (2 h), centrifuging (10,000 rpm for 5 min), washing (twice with double distilled water), drying (hot air oven at 80 °C), and calcifying (muffle furnace at 450 °C) formed pellets. XRD analysis showed hexagonal quartzite crystal structure averaging 47.70 nm in size. SEM analysis showed ZnONPs of average size 60 nm with nearly spherical and hexagonal shapes. In vitro anti-cancer tests (MTT colometric assay) of the phytomediated ZnONP formed showed a dose-dependent cytotoxic effect on A549 cells with higher activity with increasing doses (1–100 μg/mL). The Zn nanoparticles also demonstrated free radical scavenging activity in the DPPH antioxidant assay (27).

31.5 Biogenic Iron Oxide Nanoparticles (FeONP), Synthesis and Anti-cancer Activity

The use of iron oxide nanoparticles (FeONPs) in biomedicine, diagnostics, drug delivery, and bioremediation have been reported by various studies. Bharathi et al. (29) demonstrated the cytotoxicity of *Solanum lycopersicum* L. (Solanaceae) FeONPs (Sl-FeONPs) on lung cancer cell line A549. These phytonanoparticles were synthesized by extracting 1 g of dried leaf powder of *S. lycopersicum* with 100 mL of H_2O; 20 mL of this extract was mixed with 0.1 M ferrous sulfate solution in a ratio of 1:1, stirred at 60 °C for 30 min, washed and dried at 60 °C for 18 h. The produced Sl-FeONPs were shown by DLS analysis to be 483.8 nm in size and exhibited anti-cancer activity against A549 cells, with an IC_{50} value of 69±0.50 µg/mL. Bharathi et al. showed that Sl-FeONPs caused A549 cell death by inducing apoptosis, as revealed by cell shrinkage and membrane blebbing (29).

The anti-cancer activity of FeONPs synthesized from the fruit extract of *Phyllanthus emblica* L. (family Phyllanthaceae) (Pe-FeONPs) has been studied by Thoidingjam and Tiku (30). Here, 5 g of crushed fruit of *P. emblica* was extracted with 25 mL of H_2O, and then centrifuged to collect the supernatant, which was added to FeONPs. The obtained solution was then heated at 60 °C while stirring for 15 min. DLS analysis revealed that the synthesized Pe-FeONPs were 98.18 nm in size. Application of Pe-FeONPs and FeONPs to A549 cells showed that Pe-FeONPs was toxic to A549 lung cancer cells with an IC_{50} value of 104.7 µg/mL, while that of FeONPs was 359.8 µg/mL. When the toxicity of Pe-FeONPs on A549 cancer cells and normal lung cell line L132 were compared, Pe-FeONPs was found to be more toxic to cancer cells than to normal lung cells. Proliferation studies revealed that 100 µg/mL Pe-FeONPs significantly inhibited proliferation of A549 cells, as compared to 100 µg/mL of FeONPs. This indicates that *Phyllanthus emblica* fruit extract improved anti-cancer and anti-proliferative actions of iron oxide nanoparticles. The biosynthesized Pe-FeONPs induced apoptosis by generating significant levels of ROS, which consequently triggered DNA damage in A549 cells (30). The generated ROS significantly induced DNA damage to trigger apoptosis in A549 lung cancer cells.

31.6 Biogenic Cerium Oxide (CeONP), Synthesis and Anti-cancer Activity

In a recent study, Miri et al. synthesized cerium oxide nanoparticles (CeONPs) form the fruit peel extract of *Musa sapientum* Linn. (Musaceae) (31). Extraction was carried out by soaking the crushed dried fruit peel in distilled H_2O at a ratio of 1:10. The resulting filtrate was diluted at a ratio of 1:4. A volume of 50 mL of 0.05 M cerium nitrate was then added and the solution was stirred at 70 °C for 4 h and finally dried at 90 °C. The synthesized *Musa sapientum* fruit peel extract CeONPs (Ms-CeONPs) was yellow in color and 4–13 nm in size. MTT assay analysis of Ms-CeONPs on A549 cancer cells showed that 500 µg/mL Ms-CeONPs was cytotoxic to the A549 cancer cells. Ms-CeONPs served as a stabilizing and reducing agent to CeONPs, and may have induced apoptotic cell death via ROS mediated DNA damage (31,59).

Eucalyptus globulus Labill. (family Myrtaceae) leaf extract mediated CeONPs (Eg-CeONPs) has been shown to exhibit cytotoxicity against lung cancer cell line A549 (32). To produce this, 10 g of *E. globulus* leaves dried powder was extracted with 100 mL at 80 °C for 2 h. The resultant *E. globulus* leaf extract filtrate was added to CeONPs to produce Eg-CeONPs. The average size of this green synthesized Eg-CeONPs was shown to be 13.7 nm. Analysis of the cytotoxicity of Eg-CeONPs against A549 cancer cells revealed that Eg-CeONPs were cytotoxic to A549 at an IC_{50} value of 45.5 µg/L. The study also revealed that the observed cytotoxicity was as a result of oxidative stress and apoptotic cell death that was induced by high amounts of ROS released in A549 cells.

31.7 Conclusion

The biogenic generation of metallic nanoparticles has in recent years been explored as drugs and/or drug delivery candidates for lung cancer therapy. Medicinal plants have proved useful in the biosynthesis of these metallic nanoparticles because of the wide array of secondary metabolites they contain. These help promote the development and rapid synthesis of useful metallic nanoparticles. Again, the plethora of therapeutically active compounds they contain contribute to the overall anti-lung-cancer effect of the developed nanoparticles. While there is the danger of destruction of normal cells in cancer chemotherapy, some phytomediated metallic nanoparticles have shown selectivity for cancerous lung cells, thus providing a brighter future for lung cancer chemotherapy. The chapter has outlined 24 medicinal plants, their derived silver, gold, zinc, iron, and cerium metallic nanoparticles, and their anti-cancer effects in lung cancer cells.

REFERENCES

1. Quester K, Avalos-Borja M, and Castro-Longoria E. Biosynthesis and microscopic study of metallic nanoparticles. *Micron*. 2013; 54–55: 1–27. doi: https://doi.org/10.1016/j.micron.2013.07.003.
2. Kang Seung H, Bozhilov KN, Myung NV, Mulchandani A, and Chen W. Microbial synthesis of CdS nanocrystals in genetically engineered *E. coli*. *Angewandte Chemie International Edition*. 2008; 47 (28): 5186–5189. doi: https://doi.org/10.1002/anie.200705806.
3. Kalishwaralal K, Gopalram S, Vaidyanathan R, Deepak V, Babu S, Pandian RK, and Gurunathan S. Optimization of α-amylase production for the green synthesis of gold nanoparticles. *Colloids and Surfaces B: Biointerfaces*. 2010; 77 (2): 174–180. doi: https://doi.org/10.1016/j.colsurfb.2010.01.018.
4. Kalimuthu, K, Pandian RK, Babu S, Venkataraman D, Bilal M, and Gurunathan S. Biosynthesis of silver nanocrystals by Bacillus licheniformis. *Colloids and Surfaces B: Biointerfaces*. 2008; 65 (1): 150–153. doi: https://doi.org/10.1016/j.colsurfb.2008.02.018.

5. Mikheenko IP, Rousset M, Dementin S, and Macaskie LE. Bioaccumulation of palladium by *Desulfovibrio fructosivorans* wild-type and hydrogenase-deficient strains. *Applied and Environmental Microbiology*. 2008; 74 (19): 6144. doi: 10.1128/AEM.02538-07.
6. Nath D and Banerjee P. Green nanotechnology – A new hope for medical biology. *Environmental Toxicology and Pharmacology*. 2013; 36 (3): 997–1014. doi: https://doi.org/10.1016/j.etap.2013.09.002.
7. Tripathi A, Chandrasekaran N, Raichur AM, and Mukherjee A. Antibacterial applications of silver nanoparticles synthesized by aqueous extract of *Azadirachta indica* (Neem) Leaves. *Journal of Biomedical Nanotechnology*. 2009; 5 (1): 93–98. doi: 10.1166/jbn.2009.038.
8. Kang H, Buchman JT, Rodriguez RS, Ring HL, He J, Bantz KC, and Haynes CL. 2019. Stabilization of Silver and gold nanoparticles: Preservation and improvement of plasmonic functionalities. *Chemical Reviews*. 119 (1): 664–699. doi: 10.1021/acs.chemrev.8b00341.
9. Moulton MC, Braydich-Stolle LK, Nadagouda MN, Kunzelman S, Hussain SM, and Varma RS. Synthesis, characterization and biocompatibility of "green" synthesized silver nanoparticles using tea polyphenols. *Nanoscale*. 2010; 2 (5): 763–770. doi: 10.1039/C0NR00046A.
10. Kumar KP, Paul W, and Sharma CP. Green synthesis of gold nanoparticles with Zingiber officinale extract: Characterization and blood compatibility. *Process Biochemistry*. 2011; 46 (10): 2007–2013. doi: https://doi.org/10.1016/j.procbio.2011.07.011.
11. Cyril N, George JB, Joseph L, Raghavamenon AC, and Sylas VP. Assessment of antioxidant, antibacterial and anti-proliferative (lung cancer cell line A549) activities of green synthesized silver nanoparticles from *Derris trifoliata.Toxicology Research*. 2019; 8 (2):297–308. doi: 10.1039/c8tx00323h.
12. Menon S, Agarwal H, Rajeshkumar S, and Kumar S. Anticancer assessment of biosynthesized silver nanoparticles using *Mucuna pruriens* seed extract on lung cancer treatment. *RJET*. 2018; 11(9): 3887–3891.
13. Venugopal K, Rather HA, Rajagopal K, Shanthi MP, Sheriff K, Illiyas M, Rather RA, Manikandan E, Uvarajan S, Bhaskar M, and Maaza M. Synthesis of silver nanoparticles (Ag NPs) for anticancer activities (MCF 7 breast and A549 lung cell lines) of the crude extract of *Syzygium aromaticum. Journal of Photochemistry and Photobiology B: Biology*. 2017; 167:282–289. doi: 10.1016/j.jphotobiol.2016.12.013.
14. Wang C, Mathiyalagan R, Kim YJ, Castro-Aceituno V, Singh P, Ahn S, Wang D, and Yang DC. Rapid green synthesis of silver and gold nanoparticles using *Dendropanax morbifera* leaf extract and their anticancer activities. *International Journal of Nanomedicine*. 2016; 11: 3691–3701. doi: 10.2147/IJN.S97181.
15. Kummara S, Patil MB, and Uriah T. Synthesis, characterization, biocompatible and anticancer activity of green and chemically synthesized silver nanoparticles - A comparative study. *Biomedicine & Pharmacotherapy*. 2016; 84:10–21. doi: 10.1016/j.biopha.2016.09.003.
16. Nakkala JR, Mata R, Gupta AK, and Sadras SR. Biological activities of green silver nanoparticles synthesized with *Acorous calamus* rhizome extract. *European Journal of Medicinal Chemistry*. 2014; 85: 784–794. doi: 10.1016/j.ejmech.2014.08.024.
17. Gurunathan S, Jeong JK, Han JW, Zhang XF, Park JH, and Kim JH. Multidimensional effects of biologically synthesized silver nanoparticles in *Helicobacter pylori, Helicobacter felis*, and human lung (L132) and lung carcinoma A549 cells. *Nanoscale Research Letters*. 2015; 10: 35. doi: 10.1186/s11671-015-0747-0.
18. Saravanakumar K, Chelliah R, MubarakAli D, Oh DH, Kathiresan K, and Wang MH. Unveiling the potentials of biocompatible silver nanoparticles on human lung carcinoma A549 cells and *Helicobacter pylori. Scientific Reports*. 2019; 9 (1):5787. doi: 10.1038/s41598-019-42112-1.
19. Khanra K, Panja S, Choudhuri I, Chakraborty A, and Bhattacharyya N. Evaluation of antibacterial activity and cytotoxicity of green synthesized silver nanoparticles using *Scoparia dulcis. Nano Biomedicine and Engineering*. 2015; 7(3): 128–133.
20. Venkatesan B, Subramanian V, Tumala A, and Vellaichamy E. Rapid synthesis of biocompatible silver nanoparticles using aqueous extract of *Rosa damascena* petals and evaluation of their anticancer activity. *Asian Pacific Journal of Tropical Medicine*. 2014; 7S1: S294–S300. doi: 10.1016/S1995-7645(14)60249-2.
21. Sankar R, Karthik A, Prabu A, Karthik S, Shivashangari KS, and Ravikumar V. *Origanum vulgare* mediated biosynthesis of silver nanoparticles for its antibacterial and anticancer activity. *Colloids and Surfaces B: Biointerfaces*. 2013; 108: 80–84. doi: 10.1016/j.colsurfb.2013.02.033.
22. Castro-Aceituno V, Ahn S, Simu SY, Singh P, Mathiyalagan R, Lee HA, and Yang DC. Anticancer activity of silver nanoparticles from *Panax* ginseng fresh leaves in human cancer cells. *Biomedicine & Pharmacotherapy*. 2016; 84: 158–165. doi: 10.1016/j.biopha.2016.09.016.
23. Kanipandian N and Thirumurugan R. A feasible approach to phyto-mediated synthesis of silver nanoparticles using industrial crop *Gossypium hirsutum* (cotton) extract as stabilizing agent and assessment of its in vitro biomedical potential. *Industrial Crops and Products*. 2014; 55: 1–10.
24. Zheng Y, Zhang J, Zhang R, Luo Z, Wang C, and Shi S. Gold nano particles synthesized from *Magnolia officinalis* and anticancer activity in A549 lung cancer cells. *Artificial Cells, Nanomedicine, and Biotechnology*. 2019; 47 (1): 3101–3109. doi: 10.1080/21691401.2019.1645152.
25. Vijayakumar S, Vaseeharan B, Malaikozhundan B, Gopi N, Ekambaram P, Pachaiappan R, Velusamy P, Murugan K, Benelli G, Kumar RS, and Suriyanarayanamoorthy M. Therapeutic effects of gold nanoparticles synthesized using *Musa paradisiaca* peel extract against multiple antibiotic resistant *Enterococcus faecalis* biofilms and human lung cancer cells (A549). *Microbial Pathogenesis*. 2017; 102: 173–183. doi: https://doi.org/10.1016/j.micpath.2016.11.029.
26. Zhang Xi, Tan Z, Jia K, Zhang W, and Dang M. *Rabdosia rubescens* Linn: Green synthesis of gold nanoparticles and their anticancer effects against human lung cancer cells A549. *Artificial Cells, Nanomedicine, and Biotechnology*. 2019; 47 (1): 2171–2178. doi: 10.1080/21691401.2019.1620249.
27. Rajeshkumar S, Kumar SV, Ramaiah A, Agarwal H, Lakshmi T, and Roopan SM. Biosynthesis of zinc oxide

nanoparticles using *Mangifera indica* leaves and evaluation of their antioxidant and cytotoxic properties in lung cancer (A549) cells. *Enzyme and Microbial Technology*. 2018; 117: 91–95. doi: https://doi.org/10.1016/j.enzmictec.2018.06.009.
28. Selim YA , Azb MA, Ragab I, Mohamed HM, and El-Azim A. Green synthesis of zinc oxide nanoparticles using aqueous extract of *Deverra tortuosa* and their cytotoxic activities. *Scientific Reports*. 2020; 10 (1): 3445. doi: 10.1038/s41598-020-60541-1.
29. Bharathi D, Preethi S, Abarna K, Nithyasri M, Kishore P, and Deepika K. Bio-inspired synthesis of flower shaped iron oxide nanoparticles (FeONPs) using phytochemicals of *Solanum lycopersicum* leaf extract for biomedical applications. *Biocatalysis and Agricultural Biotechnology*. 2020; 27: 101698.
30. Thoidingjam S and Tiku AB. Therapeutic efficacy of Phyllanthus emblica-coated iron oxide nanoparticles in A549 lung cancer cell line. *Nanomedicine (Lond)*. 2019; 14 (17): 2355–2371. doi: 10.2217/nnm-2019-0111.
31. Miri A, Beiki H, and Sarani M. Cerium oxide nanoparticles: Biosynthesis, cytotoxic and UV protection. *Preprints*. 2020; 2020070487.
32. Balaji S, MandalL BK, Reddy VK, and Sen D. Biogenic ceria nanoparticles (CeO2 NPs) for effective photocatalytic and cytotoxic activity. *Bioengineering (Basel)*. 2020; 7 (1): 26. doi: 10.3390/bioengineering7010026.
33. Alexander JW. History of the medical use of silver. *Surgical Infections (Larchmt)*. 2009; 10 (3): 289–292. doi: 10.1089/sur.2008.9941.
34. Kokura S, Handa O, Takagi T, Ishikawa T, Naito Y, and Yoshikawa T. Silver nanoparticles as a safe preservative for use in cosmetics. *Nanomedicine*. 2010; 6 (4): 570–574. doi: 10.1016/j.nano.2009.12.002.
35. Barillo DJ and Marx DE. Silver in medicine: A brief history BC 335 to present. *Burns*. 2014; 40(Suppl 1): S3–S8. doi: 10.1016/j.burns.2014.09.009.
36. Kerr JF, Wyllie AH, and Currie AR. Apoptosis: A basic biological phenomenon with wide-ranging implications in tissue kinetics. *British Journal of Cancer*. 1972; 26 (4): 239–257. doi: 10.1038/bjc.1972.33.
37. Wyllie AH, Kerr JF, and Currie AR. Cell death: The significance of apoptosis. *International Review of Cytology*. 1980; 68: 251–306. doi: 10.1016/s0074-7696(08)62312-8.
38. Das H and Chakraborty U. Anti-hyperglycemic effect of *Scoparia dulcis* in streptozotocin induced diabetes. *Research Journal of Pharmaceutical, Biological and Chemical Sciences*. 2011; 2(2): 334–342.
39. Ratnasooriya WD, Jayakody JR, Premakumara GA, and Ediriweera ER. Antioxidant activity of water extract of *Scoparia dulcis*. *Fitoterapia*. 2005; 76 (2): 220–222. doi: 10.1016/j.fitote.2004.06.012.
40. Nikbakht A and Kafi M. A study on the relationships between Iranian people and damask rose (*Rosa damascena*) and its therapeutic and healing properties. *Acta Horticulturae*. 2008; 790: 251–254.
41. Boskabady MH, Shafei MN, Saberi Z, and Amini S. Pharmacological effects of rosa damascena. *Iranian Journal of Basic Medical Sciences*. 2011; 14 (4): 295–307.
42. Grondona E, Gatti G, López AG, Sánchez LR, Rivero Virginia, Pessah O, Zunino MP, and Ponce AA. Bio-efficacy of the essential oil of oregano (*Origanum vulgare* Lamiaceae. Ssp. Hirtum). *Plant Foods for Human Nutrition*. 2014; 69 (4): 351–357.
43. Helms S. Cancer prevention and therapeutics: *Panax ginseng*. *Alternative Medicine Review*. 2004; 9 (3): 259-274.
44. Shin HR, Kim JY, Yun TK, Morgan G, and Vainio H. The cancer-preventive potential of *Panax ginseng*: A review of human and experimental evidence. *Cancer Causes & Control*. 2000; 11 (6):565–576.
45. Teimuri-Mofrad R, Hadi R, Tahmasebi B, Farhoudian S, Mehravar M, and Nasiri R. Green synthesis of gold nanoparticles using plant extract: Mini-review. *Nanochemistry Research*. 2017; 2 (1): 8–19.
46. Siddiqi KS and Husen A. Recent advances in plant-mediated engineered gold nanoparticles and their application in biological system. *Journal of Trace Elements in Medicine and Biology*. 2017; 40: 10–23. doi: https://doi.org/10.1016/j.jtemb.2016.11.012.
47. Schröfel A, Kratošová G, Šafařík I, Šafaříková M, Raška I, and Shor LM. Applications of biosynthesized metallic nanoparticles – A review. *Acta Biomaterialia*. 2014; 10 (10): 4023–4042. doi: https://doi.org/10.1016/j.actbio.2014.05.022.
48. Sun B, Hu N, Han L, Pi Y, Gao Y, and Chen K. Anticancer activity of green synthesised gold nanoparticles from *Marsdenia tenacissima* inhibits A549 cell proliferation through the apoptotic pathway. *Artificial Cells, Nanomedicine, and Biotechnology*. 2019; 47 (1): 4012–4019. doi: 10.1080/21691401.2019.1575844.
49. Wang X, Yan Y, Chen Xi, Zeng S, Qian L, Ren X, Wei J, Yang X, Zhou Y, Gong Z, and Xu Z. The antitumor activities of *Marsdenia tenacissima*. *Frontiers in Oncology*. 2018; 8: 473-473. doi: 10.3389/fonc.2018.00473.
50. Wang P, Yang J, Zhu Z, and Zhang X. *Marsdenia tenacissima*: A review of traditional uses, phytochemistry and pharmacology. *The American Journal of Chinese Medicine*. 2018; 46(7): 1449–1484. doi: 10.1142/s0192415x18500751.
51. Zhang J, Chen Z, Huang X, Shi W, Zhang R, Chen M, Huang H, and Wu L. Insights on the multifunctional activities of magnolol. *BioMed Research International*. 2019; 2019: 1847130. doi: 10.1155/2019/1847130.
52. Hyun TK, Kim M-o, Lee H, Kim Y, Kim E, and Kim J-S. Evaluation of anti-oxidant and anti-cancer properties of *Dendropanax morbifera* Léveille. *Food Chemistry*. 2013; 141 (3): 1947–1955. doi: https://doi.org/10.1016/j.foodchem.2013.05.021.
53. Im Kyu-Jung, Jang S-B, and Yoo D-Y. Anti-cancer effects of *Dendropanax morbifera* extract in MCF-7 and MDA-MB-231 cells. *The Journal of Korean Obstetrics and Gynecology*. 2015; 28 (2): 26–39.
54. Ranjha MMAN, Irfan S, Nadeem M, and Mahmood S. A comprehensive review on nutritional value, medicinal uses, and processing of banana. *Food Reviews International*. 2020; 1–27.
55. D'Arcy MS. Cell death: A review of the major forms of apoptosis, necrosis and autophagy. *Cell Biology International*. 2019; 43 (6): 582–592.

56. Ediriweera MK, Tennekoon KH, and Samarakoon SR. A review on ethnopharmacological applications, pharmacological activities, and bioactive compounds of *Mangifera indica* (Mango). *Evidence-Based Complementary and Alternative Medicine*. 2017; 2017: 6949835. doi: 10.1155/2017/6949835.
57. Shah KA, Patel MB, Patel RJ, and Parmar PK. *Mangifera indica* (mango) *Pharmacognosy Reviews*. 2010; 4(7): 42–48. doi: 10.4103/0973-7847.65325.
58. Batool N, Ilyas N, Shabir S, Saeed M, and Mazhar R. Mini-review- A mini-review of therapeutic potential of *Mangifera indica* L. *Pakistan Journal of Pharmaceutical Sciences*. 2018; 31 (4): 1441–1448.
59. Miri A, Akbarpour Birjandi S, and Sarani M. Survey of cytotoxic and UV protection effects of biosynthesized cerium oxide nanoparticles. *Journal of Biochemical and Molecular Toxicology*. 2020; 34 (6): e22475. doi: 10.1002/jbt.22475.

32

Pulmonary Applications and Toxicity of Engineered Metallic Nanoparticles

Dr. Deepti Kaushalkumar Jani M. Pharm., Ph.D.
Associate Professor and Head-Department of Pharmacology & Pharmacognosy, Babaria Institute of Pharmacy, Vadodara, India

32.1 Introduction

Nanotechnology has emerged as promising technology in improving the health and quality of life of people worldwide, because of its use in variety of medical applications. Engineered metallic nanoparticles (EMNPs) are at the leading edge of the rapidly developing field of nanotechnology. EMNPs are nanosized metals, developed for biomedical purposes (1). Examples of EMNPs are shown in Figure 32.1.

This chapter focuses on the pulmonary applications and toxicity of the most commonly studied EMNPs such as silver, gold, platinum, copper, cerium oxide, iron oxide, zinc oxide, etc. EMNPs are well known for their unique properties, and are widely studied in the treatment and diagnosis of various pulmonary diseases.

There are several factors that influence the beneficial as well as adverse effects of EMNPs on pulmonary function (2,3). Some important factors, as listed in Table 32.1, most commonly affect the pulmonary function of EMNPs.

EMNPs have unique physicochemical properties. Because of their properties, they have the potential to improve various disease conditions and are also useful in drug delivery, diagnosis, and other applications. However, some properties of EMNPs are also responsible for potential safety concerns, i.e. adverse effects on pulmonary structure and function.

The pulmonary route of nanoparticle delivery is another factor that may result in pulmonary adverse effects of EMNPs due to direct exposure, but at the same time, this route is one of the most important routes in the treatment of respiratory diseases, as it reduces systemic exposure of drugs, avoids first-pass metabolism, and allows local drug delivery to target disease. Therefore, this route of drug administration improves drug availability at target site and facilitates better control of disease. Another advantage of this route is high surface area with rapid absorption due to high vascularization (4). Therefore, route of administration is an important factor, as it may affect desirable and undesirable effects of EMNPs.

32.2 Pulmonary Applications of Engineered Metallic Nanoparticles

EMNPs have numerous therapeutic applications in respiratory diseases. Various studies were performed to study utility of EMNPs for the treatment of pulmonary diseases and beneficial effects were obtained as a study outcome. The majority of the studies focused on the use of EMNPs for the treatment of lung cancer and pulmonary infections.

Most of the recent research focused on the use of EMNPs as vectors for pulmonary delivery of drugs or genes via inhalation or systemic administration. It was observed that drug-loaded EMNPs produce better effects than the drug alone.

Most of the pulmonary diseases are chronic in nature without any effective treatment to completely restore pulmonary function. Drug resistance and undesirable effects of medicines are common because of long term use in the treatment of chronic pulmonary diseases, which supports the need of more specific approach (5). In different studies, it was observed that EMNPs increase drug accumulation in the lungs, thereby increasing the availability of the drug at target site and reducing systemic adverse effects.

Applications of EMNPs are reported in variety of pulmonary diseases like asthma, pulmonary fibrosis, and lung cancer as well as in pulmonary bacterial and viral infections such as tuberculosis, pneumonia, corona virus disease (COVID-19), etc. as shown in Figure 32.2.

Multi-drug therapy is a major problem in the treatment of pulmonary diseases and infections, especially tuberculosis, because it reduces patient compliance. Therefore, nanoparticle-based delivery systems are beneficial in the treatment of pulmonary diseases, as they increase bioavailability, and facilitate controlled release, as well as reducing dosages and the frequency of drug administration.

In a study to target IL4Rα, the receptor for a key pro-inflammatory pathway, Halwani et al. examined the ability of intranasally administered PEGylated dextran-coated anti-IL4Rα-conjugated superparamagnetic iron oxide nanoparticles (SP-IONP)

DOI: 10.1201/9781003046547-32

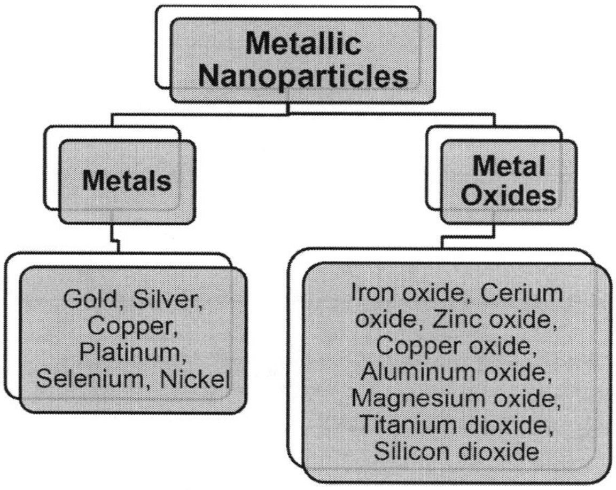

FIGURE 32.1 Types of Engineered Metallic Nanoparticles

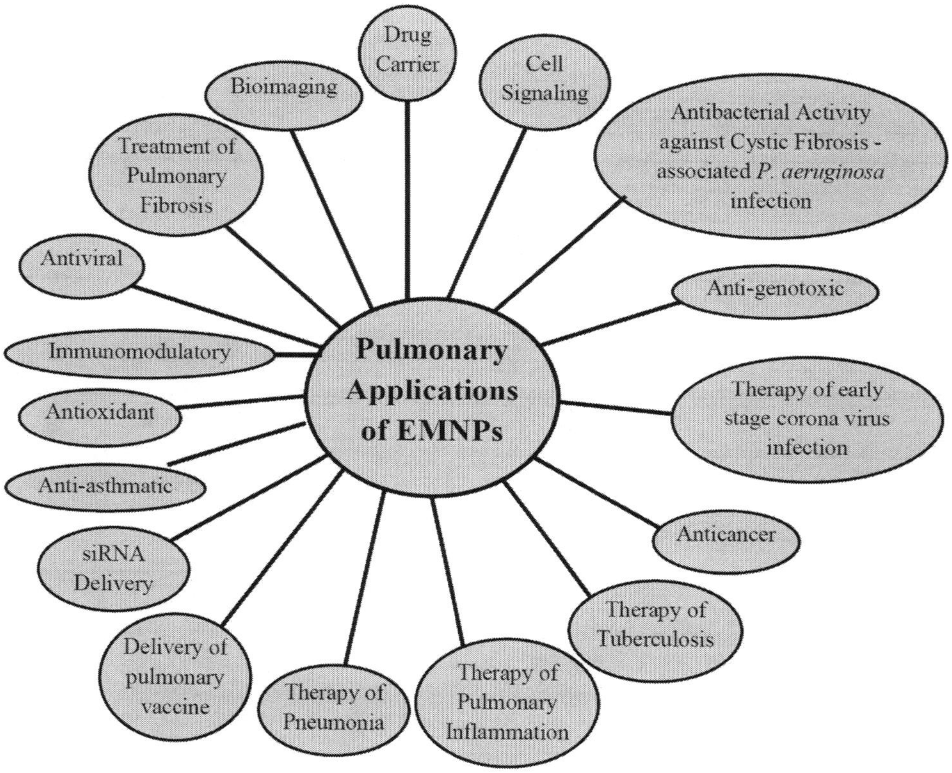

FIGURE 32.2 Pulmonary Applications of Engineered Metallic Nanoparticles (EMNPs) in Various Biomedical Fields

TABLE 32.1
Factors Affecting Pulmonary Function of Engineered Metallic Nanoparticles

I. Physicochemical properties
Size, shape, solubility, chemical composition, physiochemical stability, surface area, surface energy, surface roughness
II. Therapy-related factors
Route of administration: inhalational, intranasal, intra-tracheal, oropharyngeal, systemic; dose of nanoparticles, duration of exposure
III. Host factors
Genetic factors, health status, medical condition

to control lung inflammation in an ovalbumin-sensitized mice model of allergic asthma. They found that the treatment with the anti-IL4Rα nanoparticles significantly reduced pro-inflammatory cytokine expression, lymphocytes, neutrophils, and eosinophils counts in lung tissue. Their study results showed that drug conjugation with metallic nanoparticles increased tissue penetration and duration of action of anti-IL4Rα drug (6).

Smoking is an important risk factor for inflammatory lung diseases, such as chronic obstructive pulmonary disease. Cigarette smoke generates a huge amount of reactive oxygen species that causes pulmonary inflammation. Hence, smoking-induced oxidative stress may be reduced with the use of antioxidant agents. Onizawa et al. investigated the therapeutic effect of platinum nanoparticles (Pt-NP) stabilized with polyacrylate to form a stable colloid solution (PA-Pt-NP) as antioxidants against smoking induced-pulmonary inflammation in mice. The study findings indicated that intranasal administration of PA-Pt-NP prior to exposure to smoke inhibited oxidative stress and inflammation in the lungs of mice. In another in vitro experiment, PA-Pt-NP inhibited smoke-induced cell death of alveolar-type-II-like A549 cells. Thus, study results suggest the potential use of Pt-NPs as an antioxidant to inhibit cigarette smoke-induced pulmonary inflammation (7).

Pulmonary fibrosis is an interstitial lung disease. Drug trials for the treatment of pulmonary fibrosis have shown disappointing results; therefore, Shahabi et al. evaluated the protective effect of selenium nanoparticles (Se-NPs) against early and late phases of pulmonary injury induced by intratracheally administered bleomycin in male rats. The results showed that Se-NPs improved the degree of alveolitis, inflammation, and lung damage in early phase of disease. They concluded that the treatment with the Se-NPs during the early phase of the disease showed protective effects, while no effect was seen in the late phase. Hence, Se-NPs may be used as a new therapeutic agent for the treatment of early phase pulmonary fibrosis (8).

Rubio et al. evaluated the antioxidant and anti-genotoxic activity of cerium oxide nanoparticles (CeO_2-NPs) in the human epithelial lung cell line against $KBrO_3$-induced oxidative stress. They found that pretreatment with CeO_2-NPs significantly reduced the intracellular production of reactive oxygen species and also decreased DNA oxidative damage. Study results support the potential usefulness of CeO_2-NPs as an antioxidant and anti-genotoxic pharmacological agent (9).

Cancer is the leading cause of death worldwide. Lung cancer is one of the most common causes of cancer-related death (10). Chemotherapeutic drugs used for the treatment of cancer are well known for their adverse effects. Recent progress in nanomedicine has encouraged the development of nanoparticle-based delivery systems for targeted delivery of chemotherapeutic drugs to tumor cells and hence reduces their toxicity

Recent advances in cancer therapy involve use of the immune system to fight against cancer, but sometimes immunotherapy also fails to control cancer. It was found that tumors are sensitive to copper oxide nanoparticles (CuO-NPs). These nanoparticles can produce toxic effects, therefore, by creating iron-doped nanoparticles of copper oxide; adverse effects of copper can be eliminated. Naatz et al. synthesized iron-doped CuO-NPs and investigated the anti-cancer effect in combination with immunotherapy in a mice model of lung cancer. Results indicated that the combination of epacadostat and 6% iron-doped CuO-NPs showed increased therapeutic efficacy, complete tumor remission, and absence of tumor relapse in animals with lung adenocarcinoma (KLN-205) cells. Another important finding was that all these animals failed to produce growing tumors after re-engraftment of KLN-205 cells. Therefore, study results indicate that the combination of immunotherapy and iron-doped CuO-NPs is useful as a potent anti-cancer treatment (11).

The platinum-based anti-cancer drugs cisplatin, carboplatin, and oxaliplatin are important chemotherapeutic agents but their use is limited because of the ability of tumors to develop rapid resistance and dose-limiting side effects. These issues may be solved with the use of EMNPs as drug-delivery vehicles to target cancers actively. Gold nanoparticle (Au-NPs)-conjugated drug for targeted drug delivery applications is one of the most promising approaches in the field of cancer research. In a study of anti-cancer drugs based on Au-NPs, the active component of the platinum-based anti-cancer drug oxaliplatin was tethered to Au-NPs for improvement in drug delivery. The oxaliplatin-tethered nanoparticles showed better ability to penetrate the nucleus in the lung cancer cells and significantly better cytotoxicity than oxaliplatin alone in the A549 lung epithelial cancer cell line. The results showed that the use of EMNPs for drug delivery holds a significant potential to improve cancer treatment in the future (12).

In another study, cytotoxicity of Au-NPs conjugated anti-cancer drug docetaxel and folic acid was evaluated using lung cancer cell line (H520). On the basis of the results, it was concluded that Au-NPs can be considered as promising nanocarrier system for chemotherapeutic agents in the treatment of various cancers (13).

In other study, Ullah et al. prepared Pt-NPs by a green route using phytochemicals and studied them for use as anti-cancer agents. They evaluated the in vitro anti-cancer activity of biogenic Pt-NPs using lung cancer cell line A549. Study results revealed that Pt-NPs possess strong anti-cancer activity against lung cancer (14).

In another study, Bendale et al. synthesized Pt-NPs by green technology and investigated their anti-cancer effects using in vitro and in vivo models of human lung cancer. The in vitro study results showed that Pt-NPs inhibited growth of lung cancer cells in a dose-dependent manner, while in vivo study results indicated that Pt-NPs effectively inhibited the growth of lung cancer in SCID mice at mid dose (1000 mg/kg) and high dose (2000 mg/kg). It was concluded that biosynthesized Pt-NPs can be used as a cost-effective alternative for the treatment of patients with lung cancer (15).

Lung cancer is classified into small cell lung cancer (SCLC) and non small cell lung cancer (NSCLC) (16). Majority of lung cancer related deaths has been observed in patients having oncogenic K-RAS driven NSCLC (17). The major limitation of currently existing therapies for K-RAS driven NSCLC is severe toxicity, multi-drug resistance and poor survival outcomes (18,19). Therefore, there is need for safe and effective targeted therapy for K-RAS driven NSCLC. Sulthana et al. synthesized functional CeO_2-NP as drug delivery vehicle with a unique combination of two therapeutic drugs, doxorubicin and ganetespib, for the treatment of K-RAS driven NSCLC. They found that use of ganetespib in combination with doxorubicin synergized the therapeutic efficacy of doxorubicin, while reducing the cardiotoxic potential of doxorubicin. They also found that combination therapy resulted in 80% of NSCLC cell death within 48 hours of incubation and prevention of lung cancer metastasis. It was concluded that the delivery of doxorubicin and ganetespib combination using functional CeO_2-NP offers a robust nanoplatform for the targeted treatment of clinically challenging K-RAS driven NSCLC (20).

SCLC is very responsive to initial treatment. Most of the patients with SCLC experience relapse with relatively resistant disease. Options for second-line chemotherapy against resistant forms of SCLC are limited at present. Thus, Tanino et al. investigated anti-cancer activity of zinc oxide nanoparticles (ZnO-NPs) against SCLC in an orthotopic mouse model in vivo and against lung cancer cell lines in vitro. ZnO-NPs inhibited the proliferation of SCLC cells in vitro by generating reactive oxygen species, but found to be less cytotoxic against normal lung cells. Findings of an in vivo study showed that ZnO-NP showed genotoxic activity against orthotopic small cell lung cancers without producing any adverse effects. Therefore, study results indicate that ZnO-NP has potential for use as novel anti-cancer agent in treatment of SCLC (21).

Ramalingam et al. developed and studied polyvinylpyrrolidone (PVP) stabilized Au-NPs conjugated with doxorubicin (PVP-Au-NPs-DOX). They reported that PVP-Au-NPs-DOX inhibited the proliferation of human lung cancer cells, upregulated the expression of tumor suppressor genes, and induced both early and late apoptosis in lung cancer cells. From the results, they concluded that PVP-Au-NPs-DOX can be used as a potential drug delivery system for effective treatment of human lung cancer (22).

In an anti-cancer study of magnesium oxide nanoparticles (MgO-NPs) against the A-549 cell line, it was observed that MgO-NPs showed apoptotic activity and DNA damage. This study concluded that MgO-NPs showed good anti-cancer effects (23).

In a study to investigate use of intra-tracheally administered siRNA/RGD Au-NPs to target tumor cells in a lung cancer syngeneic orthotopic mouse model, treatment with nanoparticles resulted in successful suppression of tumor cell proliferation, tumor size reduction, and prolongation of survival of lung tumor–bearing mice. These findings support the capability of functional Au-NPs for targeted delivery of siRNA to cancer cells (24).

Computed tomography (CT) imaging is the standard test for the screening of lung cancer. It is challenging to differentiate benign nodules from malignant nodules by this scanning method (25). Therefore, in order to exclude malignancy, invasive procedures are performed on patients with benign nodules (26). Need for invasive diagnostic methods could be reduced by developing non-invasive imaging for cancer discrimination. Epidermal growth factor receptor (EGFR) is highly expressed on the cell surface of most lung adenocarcinoma cells, while benign tumor nodules have a low basal level of EGFR expression, so targeting EGFR-expressing lung tumors can be an effective strategy (27). Ashton et al. developed an EGFR-targeted Au-NPs CT contrast agent and evaluated in vivo CT imaging performance in mice with tumors. They reported that the Au-NPs conjugated with anti-EGFR antibodies (cetuximab) had short blood residence time and significantly high tumor accumulation. This study demonstrated the application of nanoparticle contrast agents to detect overexpression of tumor receptors. They concluded that the targeted EMNPs can be used for the detection of tumor receptor status and to improve tumor discrimination and characterization (28).

Wang et al. reported a new application of folic acid-modified dendrimer-entrapped Au-NPs (Au-DENPs-FA) as nanoprobes for CT imaging of human lung adenocarcinoma. In this study micro-CT images showed that after in vitro incubation with the Au-DENPs-FA, human lung adenocarcinoma cells can be detected under X-ray. With transmission electron microscopy data, it was confirmed that the Au-DENPs-FA is able to be uptaken dominantly in the lysosomes of the cells. It was observed that the Au-DENPs-FA showed good biocompatibility at the given concentration range, as it did not affect cell morphology, cell viability and cell cycle. This study indicated potential application of Au-DENPs-FA as imaging probes for targeted CT imaging of human lung adenocarcinoma (29).

Cell labeling has been widely used for isolation of cells of interest from a heterogeneous population within a magnetic field. Cell isolation has been done with the use of phenotypic markers tagged with magnetic particles (30,31). Cell labeling with magnetic IONPs allows non-invasive monitoring of targeted cells. Thus, cell labeling offers promising new approaches in cell-based therapy with great potential for cancer treatment (32).

In a study of cell labeling by Wang et al., magnetic IONPs were synthesized and surface-modified with cationic poly-l-lysine (PLL) to form the PLL-IONPs. They used synthesized PLL-IONPs to magnetically label human A549 lung cancer cells. The results showed that labeling with PLL-IONPs did not affect cellular viability, proliferation capability, cell cycle, and apoptosis at low concentration. It was also found that at high concentration (400 μg/mL), the PLL-IONPs would slightly disrupt the cytoskeleton, impair cell viability, proliferation, cell cycle, and apoptosis in the treated A549 lung cancer cells. Therefore, adequate concentration of PLL-IONPs can be used efficiently for labeling A549 lung cancer cells. This approach could be considered for targeted anti-cancer drug/gene delivery and targeted diagnosis of lung cancer (33).

Cystic fibrosis (CF) is an inherited disease, caused by mutations in the gene encoding the cystic fibrosis transmembrane conductance regulator. Mutations in this gene cause imbalance of water and ion movement across the airway epithelium, which result in mucus thickening, bacterial infection, airway inflammation, and progressive loss of pulmonary function. Bacterial infection due to *Psuedomonas aeruginosa* is often associated with CF. This infection leads to deterioration of lung function in CF patients (34).

Currently, various studies have performed for the treatment for *P. aeruginosa* infections in CF patients with inhaled antibiotic treatments to directly target the lungs, however, these formulations are absorbed and cleared rapidly from lungs leaving bacteria free to grow (35).

Many bacteria form biofilms during chronic infections, resulting in resistance to antibiotic. Anti-bacterial preparations based on silver nanoparticles (Ag-NPs) can be used against biofilm-forming bacteria. Habash et al. evaluated anti-microbial action of citrate-capped Ag-NPs, alone and in combination with the antibiotic tobramycin, against *Pseudomonas aeruginosa* biofilms. They found that smaller Ag-NPs were more effective in potentiating the activity of tobramycin againt against *P. aeruginosa* and for inhibiting biofilms as compared to larger Ag-NPs. Study results indicate the potential role for Ag-NP/antibiotic combinations for the treatment of patients with *P. aeruginosa* infections in a strain-specific manner (36).

In another study, hybrid silica coated Ag-NPs were prepared using an optimum ratio of chitosan and sodium tripolyphosphate to encapsulate ciprofloxacin and evaluated for anti-microbial activity against *P. aeruginosa* and *P. aeruginosa spp.* biofilm formation. This anti-bacterial preparation based on Ag-NPs was found to be more effective in inhibiting growth and biofilm formation by *P. aeruginosa* as compared to ciprofloxacin alone (37).

In a study for investigation of the anti-bacterial activity and mechanism of Ag-NPs against multi-drug-resistant *P. aeruginosa*, it was found that Ag-NPs produced highly bactericidal effects against drug-resistant or multi-drug-resistant *P. aeruginosa*. This study revealed that the generation of reactive oxygen species is the mechanism for anti-bacterial action of Ag-NPs (38).

Serebrovska et al. designed and evaluated a combination of CeO_2-NP immobilized on the surface of silica nanoparticles in a rat model of pneumonia. Results revealed that the CeO_2-NP treatment in rats with pneumonia produced reduction in tissue injury, reactive oxygen species generation, and pro-inflammatory cytokine expression in lungs. The study showed significant anti-inflammatory and antioxidant effects of CeO_2-NP in rats with pneumonia (39).

Metallic nanoparticles have been reported to display antiviral properties (40), thus they provide a potential opportunity for novel antiviral therapies.

Respiratory syncytial virus (RSV) targets the cells of the bronchial epithelium and alveoli, causing acute lower respiratory infections. This infection can progress to bronchiolitis or pneumonia with an increased chance of significant morbidity and death in the elderly and infants. Currently there is no any effective treatment option for RSV infection (41,42), therefore, investigation of safe and effective antiviral treatments for RSV infection is of great importance.

Effect of curcumin modified silver nanoparticles (cAg-NPs) was evaluated for antiviral activity against RSV infection. It was observed that the cAg-NPs produced a highly efficient inhibition effect against RSV infection, with a decrease of viral titers. Possible mechanism for antiviral activity of cAg-NPs could be prevention of virus from infecting the host cells by inactivating the virus directly, which indicate role of cAg-NPs as a novel promising antiviral drug against RSV infection (43).

In another study, Morris et al. evaluated the antiviral and immunomodulatory effects of Ag-NPs in RSV infection using epithelial cell lines and BALB/c mice. They found Ag-NP-mediated decrease in RSV replication, both in epithelial cell lines and in experimentally infected mice. They also observed reduction in RSV-induced pro-inflammatory cytokines and chemokines, and increased cytokines associated with neutrophil recruitment and activation in the lung tissue. They reported that Ag-NPs produced significant in vivo antiviral activity against RSV infection (44).

In another study, effectiveness of PVP coated Ag-NPs was evaluated for the inhibition of RSV infection. Results showed that PVP-coated Ag-NPs showed low toxicity to cells at low concentrations and inhibited RSV infection significantly as compared to controls. Based on the results, it was concluded that the PVP-coated Ag-NPs have potential as promising candidates for future RSV treatment research (45).

Influenza is a major cause of morbidity and mortality worldwide. Influenza viruses are negative-stranded RNA viruses. They belong to the family Orthomyxoviridae. Influenza A viruses are characterized into different subtypes based on the combinations of their proteins, i.e. hemagglutinin (16 subtypes, H1–H16) and neuraminidase (nine subtypes, N1–N9) molecules, H1-3, and N1 and N2 are commonly represented in humans. Globally, the most common circulating influenza A viruses were H3N2, H1N1, and H1N2 (46).

Ag-NPs have attracted much attention as anti-microbial agents and have demonstrated efficient inhibitory activity against various viruses. Xiang et al. explored the in vitro and in vivo therapeutic antiviral effects of Ag-NPs against H3N2 influenza virus. They observed that Ag-NPs significantly inhibited growth of the influenza virus as well as significantly reduced cell apoptosis induced by H3N2 influenza virus. In mouse model, intranasal Ag-NP administration significantly increased survival, while reducing viral load in lungs. It was concluded that Ag-NPs have beneficial effects against H3N2 influenza virus infection both in vitro and in vivo; therefore, Ag-NPs can be used as potential antiviral agents for the treatment of influenza (47).

In another study, Ag-NP/chitosan composite was evaluated for antiviral activity against the H1N1 influenza A virus. It was observed that antiviral activity was increased with increase in concentration and with decrease in size of Ag-NPs. These results indicate that an Ag-NP/chitosan composite produces significant antiviral activity (48).

Covid-19 is an infectious respiratory disease caused by a newly discovered virus, i.e. severe acute respiratory syndrome corona virus-2 (SARS-CoV-2). Most people infected with the Covid-19 virus experience mild to moderate respiratory illness (49).

Application of Ag-NPs was studied for suppression of viral and bacterial respiratory infections in various respiratory system target locations. In this study, the role of Ag-NPs to treat lung infections in early stage Covid-19 via home treatment and in lowering the risk of ventilator-associated pneumonia was evaluated. It was observed that the effective anti-bacterial minimal inhibitory concentration was achievable, both in the bronchial tree and in the alveoli. It was suggested to use Ag-NPs for the treatment of early stage Covid-19 infection to suppress the progression of the lung infection. It should be used with precautions at home to treat lung infections and in hospital intensive care units for prophylactic treatment for ventilator-acquired pneumonia (50).

In a study of pulmonary vaccine adjuvant-delivery system using aluminum oxide nanoparticles (Al_2O_3-Nps), study results showed that pulmonary vaccination with Al_2O_3-Nps-delivered ovalbumin in mice induced mucosal immune response, as indicated by presence of IgA in saliva, nasal and vaginal washes, and in broncho-alveolar lavage fluid and high levels of anti-antigen IgG in serum. These outcomes suggested that Al_2O_3-Nps may be used as an efficient pulmonary vaccine adjuvant-delivery system against airborne pathogens that often cause pulmonary infection (51).

Thus, the studies summarized in this section highlight the research that is being undertaken to understand the potential pulmonary applications of EMNPs. Table 32.2 gives brief details of studies discussed in this section.

32.3 Pulmonary Toxicity of Engineered Metallic Nanoparticles

Nanomaterials are widely used today, therefore, to understand mechanism of EMNPs-induced pulmonary toxicity as well as to identify their adverse pulmonary effects is of great importance.

In a study, effects of PVP-coated Ag-NPs and silver ions were investigated in the human alveolar cell line A549. The cytotoxicity of silver compounds was greatly decreased by pretreatment with the antioxidant. A strong correlation was noted between the levels of reactive oxygen species and mitochondrial damage or early apoptosis. DNA damage was detected as an increase in bulky DNA adducts after Ag-NPs exposure. The level of bulky DNA adducts was strongly correlated with the cellular reactive oxygen species levels and could be inhibited by antioxidant pre-treatment. Results suggested that Ag-NPs induced cytotoxicity and genotoxicity in the human lung cancer cells by increasing reactive oxygen species (52).

Despite their potential for a variety of applications, copper nanoparticles (Cu-NPs) induce very strong inflammatory responses and cellular toxicity after pulmonary administration. Coating metallic nanoparticles with polysaccharides, such as chitosan, has the potential to decrease this toxicity. Worthington et al. developed a new method to coat Cu-NPs with chitosan and studied the toxicity of chitosan coated Cu-NPs. Results showed that the chitosan coating significantly reduced the toxicity of Cu-NPs and the generation of reactive oxygen species in vitro. Conversely, in mice exposed to chitosan coated CuNPs, increased inflammatory response was observed. The mechanism of this inflammatory response is that coating of metallic nanoparticle with mucoadhesive polysaccharides reduces their ability to be cleared from the lungs, so the prolonged exposure of cells to toxic metal oxides results in acute inflammatory response (53).

Exposure to nickel nanoparticles (Ni-NPs) results in lung inflammation and fibrosis. Mo et al. studied the role of microRNA-21 in lung inflammation induced by Ni-NPs in mice. Exposure to Ni-NPs resulted in upregulation of microRNA-21, severe pulmonary inflammation, and fibrosis in mice. They concluded that that miRNA-21 has important role in Ni-NPs-induced pulmonary toxicity (54).

The molecular mechanism of the toxicity of CeO_2-NPs was studied on lung adenocarcinoma (A549) cells. After internalization, CeO_2-NPs caused significant cytotoxicity and morphological changes in A549 cells. Further, the cell death was found to be apoptotic as shown by loss in mitochondrial membrane potential. A significant increase in oxidative DNA damage was found which was confirmed by phosphorylation of p53 gene and presence of cleaved poly ADP ribose polymerase. This damage could be attributed to increased production of reactive oxygen species with concomitant decrease in antioxidant glutathione level. Our study concludes that reactive oxygen species mediated DNA damage and cell cycle arrest play a major role in CeO_2-NPs induced apoptotic cell death in A549 cells (55).

In an acute inhalational toxicity study of ZnO-NPs, exposure of an aerosol of ZnO-NPs to macrophage depleted mice showed reduction in tidal volume. This study suggested that ZnO-NPs-induced pulmonary toxicity is by a macrophage-independent mechanism. This study also revealed that interaction of ZnO-NPs with lung surfactant layer covering the inside of the alveoli is responsible for inhalational toxicity caused by ZnO-NPs (56).

Most commonly reported mechanisms responsible for pulmonary toxicity of EMNPs are shown in Table 32.3.

The acute inhalation toxicity study of Ag-NPs was performed as per the OECD 403 guideline using seven-week-old Sprague-Dawley rats. Different doses of Ag-NPs, i.e. low-dose (0.94×10^6 particle/cm^3, 76 μg/m^3), middle-dose (1.64×10^6 particle/ cm^3, 135 μg/m^3), and high-dose (3.08×10^6 particle/cm^3, 750 μg/m^3) were selected to study. The animals were exposed to Ag-NPs in a whole-body inhalation chamber for 4 hours. No significant change in body weight and lung function tests was observed in exposed group rats during the 2-week observation period (57).

The subchronic inhalation toxicity of Ag-NPs was studied in eight-week-old Sprague-Dawley rats. Three different doses of Ag-NPs, i.e. low-dose (0.6×10^6 particle/cm^3, 49 μg/m^3), middle-dose (1.4×10^6 particle/cm^3, 133 μg/m^3), and high-dose (3.0×10^6 particle/cm^3, 515 μg/m^3) were evaluated. The rats were exposed to Ag-NPs in a whole-body inhalation chamber for 6 h/day, 5 days/week, for 13 weeks. Results showed a dose-dependent increase in mixed inflammatory cell infiltrate, chronic alveolar inflammation, and small granulomatous lesions in lungs. An increase in bile duct hyperplasia was also observed to be dose dependent. Study results showed that the target organs for Ag-NPs toxicity were lungs and liver in both male and female rats (58).

Particle size and surface chemistry are potential determinants of respiratory toxicity of Ag-NPs. Seiffert et al. studied the impact of size and surface chemistry on the pulmonary toxicity of Ag-NPs with 20 and 110 nm diameters and two coating agents, PVP and citrate. They found that citrate-coated 20 nm and 110 nm Ag-NPs induced greater neutrophilic inflammatory response as compared with the PVP-coated Ag-NPs. Ag-NPs also damaged the blood/alveolar epithelial permeability barrier, induced oxidative stress, and increased release of cytokines. Increased lung resistance and bronchial hyper-responsiveness were noted only with 20 nm Ag-NPs. Thus, Ag-NPs can induce characteristic features of asthma (59).

Wiemann et al. investigated the distribution of silver into remote organs after administration of Ag-NPs to the rat lung. Reversible inflammation was observed at the dose range of 75–150 μg, while DNA damage, increased cell proliferation, and increased numbers of neutrophils were observed at the

TABLE 32.2

Applications of Engineered Metallic Nanoparticles (EMNPs) in Pulmonary Diseases and Infections

EMNPs	Screening Model	Applications in Pulmonary Disease/ Infections	Reference
PEGylated dextran-coated anti-IL4Rα-conjugated superparamagnetic iron oxide nanoparticles	Ovalbumin-sensitized mice model	Allergic asthma	(6)
Polyacrylate stabilized platinum nanoparticles	Mouse model of cigarette smoke-induced pulmonary inflammation, alveolar-type-II-like epithelial cell line (A549 cells)	Pulmonary inflammation	(7)
Selenium nanoparticles	Bleomycin-induced pulmonary injury in male rats	Pulmonary fibrosis	(8)
Cerium oxide nanoparticles	Human bronchial epithelial cell line (BEAS-2B cells)	Antioxidant and anti-genotoxic	(9)
Iron-doped nanoparticles of copper oxide	Mouse model of lung adenocarcinoma (KLN-205 cells)	Lung cancer	(11)
Silver nanoparticles conjugated with oxaliplatin	Human lung cancer cell line (A549 cells)	Lung cancer	(12)
Silver nanoparticles conjugated with docetaxel and folic acid	Human lung cancer cell line (H520 cells)	Lung cancer	(13)
Platinum nanoparticles	Human lung cancer cell line (A549 cells)	Lung cancer	(14)
Platinum nanoparticles	SCID mouse model of lung cancer	Lung cancer	(15)
Cerium oxide nanoparticles conjugated with doxorubicin and ganetespib	Human lung cancer cell line (A549 cells)	K-RAS driven non small-cell lung cancer	(20)
Zinc oxide nanoparticles	Orthotopic mouse model of human small cell lung cancer, lung cancer cell lines (N417 and BEAS-2B cells)	Small cell lung cancer	(21)
Polyvinylpyrrolidone stabilized gold nanoparticles conjugated with doxorubicin	Human lung cancer cell lines (A549, H460, and H520 cells)	Lung cancer	(22)
Magnesium oxide nanoparticles	Human lung cancer cell line (A549 cells)	Lung cancer	(23)
siRNA/RGD gold nanoparticles	Syngenic orthotopic mouse model of lung cancer	Lung cancer	(24)
Epidermal growth factor receptor-targeted gold nanoparticles CT contrast agent	Mouse model of lung cancer	Bioimaging	(28)
Folic acid-modified dendrimer-entrapped gold nanoparticles	Computed tomography imaging of human lung adenocarcinoma	Bioimaging	(29)
Magnetic iron oxide nanoparticles	Human lung cancer cell line (A549)	Cell labeling	(33)
Silver nanoparticles/antibiotic tobramycin combination	Clinical isolates of bacteria *P. aeruginosa* biofilms	Cystic fibrosis-associated *Pseudomonas aeruginosa* biofilms	(36)
Hybrid silica coated silver nanoparticles /ciprofloxacin combination	*P. aeruginosa* PAO1 (wild type) and *P. aeruginosa* NCTC 10662 bacterial strains	Cystic fibrosis-associated *P. aeruginosa* infection and *P. aeruginosa spp.* biofilm formation	(37)
Silver nanoparticles	Clinical isolates of bacteria *P. aeruginosa*	Cystic fibrosis-associated multi-drug-resistant *P. aeruginosa* infection	(38)
Cerium dioxide nanoparticles	Rat model of experimental pneumonia	Pneumonia	(39)
Curcumin modified silver nanoparticles	Antiviral activity assay	Respiratory syncytial virus infection	(43)
Silver nanoparticles	Human alveolar type II-like epithelial cell line (A549 cells), Experimentally infected mouse models	Respiratory syncytial virus infection	(44)
Silver nanoparticles	Antiviral activity assay	Respiratory syncytial virus infection	(45)
Silver nanoparticles	H3N2 influenza virus infected mouse models, in vitro antiviral assay	Influenza (H3N2 virus)	(47)
Silver nanoparticles/chitosan composite	Antiviral activity assay	Influenza (H1N1 virus)	(48)
Silver nanoparticles	Model method and computation to achieve antiviral minimal inhibitory concentration in various respiratory system locations	Early stage Covid-19 infection, ventilator acquired pneumonia	(50)
Aluminum oxide nanoparticles	Immunized female Kunming mouse model	Pulmonary vaccine adjuvant-delivery system	(51)

TABLE 32.3

Mechanisms for Pulmonary Toxicity of Engineered Metallic Nanoparticles (EMNPs)

Mechanisms for Toxicity of EMNPs	Reference
Increased level of reactive oxygen species, DNA damage and mitochondrial damage	(52)
Acute inflammatory response	(53)
Upregulation of microRNA-21	(54)
Cell cycle arrest and apoptosis	(55)
Interaction with lung surfactant layer covering inside the alveoli	(56)

dose of 300 µg in lungs. Exposure of the lungs with the dose of ≥75 µg, resulted in silver accumulation in organs such as liver and kidney (60).

The subchronic inhalational toxicity study of Au-NPs was conducted using seven-week-old Sprague Dawley rats. Three different doses of Au-NPs i.e. low-dose (2.36 × 104 particle/cm^3, 0.04 µg/m^3), middle-dose (2.36×10^5 particle/cm^3, 0.38 µg/m^3), and high-dose (1.85× 106 particle/cm^3, 20.02 µg/m^3) were selected to study. The rats were exposed to Au-NPs for 6 hours/day, 5 days/week, for 90 days. Study results indicated decrease in tidal volume and the presence of mixed inflammatory cell infiltrates. It was also observed that inhaled Au-NPs were accumulated in a dose-dependent manner in lungs and kidneys of both male and female rats (61).

In a study designed to evaluate the effect of Au-NPs on the lung tissue histology of rats, results revealed that Au-NPs produced inflammation, blood vessel congestion, collapse of lung alveoli, and alveolar wall thickening. It was observed that the Au-NPs caused histological changes in the lung tissue (62).

Particle size is an important factor that can influence the adverse effects of EMNPs. In a study designed to investigate the effect of particle size of Au-NPs in lungs of rats, it was observed that no DNA damage was noted with particle size of 2, 20 and 200 nm of Au-NPs. Minor pathological findings were observed in the lungs of the animals exposed to 20 nm and 200 nm-sized nanomaterials. It was concluded that different size of Au-NPs tested in this study were non-genotoxic and showed no systemic or local adverse effects at the given dose (63).

In another study, the size, dose, and time dependent effect of intra-tracheally administered IONPs on pulmonary and coagulation systems were studied. Results showed that IONPs induced inflammatory reaction and immune response after exposure to lungs. The increase in dose caused overloading of phagocytosed nanoparticles in alveolar macrophages. On day 30, IONPs-administered rats showed significantly longer coagulation parameters as compared to controls. It was found that both the size, i.e. 22 nm and 280 nm, of IONPs induced lung injury. The nano-sized IONP particle increased significant lung damage and abnormality in coagulation parameters, as compared to the submicron-sized IONP (64).

Nemmar et al. investigated the acute effects of intra-tracheally administered CeO_2-NPs in mice. Results indicate increased lipid peroxidation and glutathione in the lungs, increased reactive oxygen species in the lung, heart, kidney, and brain, and decreased superoxide dismutase activity in the lung, liver, and kidney. Total nitric oxide was increased in the lung and spleen but it decreased in the heart. DNA damage and increase in tumor necrosis factor-α and interleukin-6 were observed in all organs studied. It was concluded that the pulmonary exposure to CeO_2-NPs induced oxidative stress, inflammation, and DNA damage in multiple organs (65).

In a toxicity study of CuO-NPs using a C57BL/6 mice model, it was observed that CuO-NPs produced dose-dependent increase in pulmonary inflammation and apoptosis of epithelial cells, and promoted collagen accumulation and expression of the progressive fibrosis marker in the lung tissues. Results showed that intranasal delivery of CuO-NPs induces pulmonary toxicity and fibrosis in C57BL/6 mice (66).

Hadrup et al. studied the pulmonary effects of intra-tracheally administered ZnO-NPs in mice. Results showed that ZnO-NPs increased neutrophils in bronchoalveolar lavage fluid and pulmonary inflammation. Weak non-dose-dependent genotoxic effects were also observed. It was concluded that the pulmonary exposure to relatively low doses of ZnO-NPs caused dose-dependent pulmonary acute phase response and pulmonary cytotoxicity (67).

Gelli et al. studied pulmonary toxicity of intra-tracheally administered 1 mg/kg or 5 mg/kg dose of MgO-NPs in rats. Results showed a dose-dependent increase in alkaline phosphatase and lactate dehydrogenase in bronchoalveolar lavage fluid of rat lungs and increased lymphocytic infiltration. In conclusion, exposure to MgO-NPs showed a dose-dependent pulmonary toxicity in rats (68).

Kim et al. performed a 28-day repeated dose inhalational toxicity study of Al_2O_3-Nps in male Sprague-Dawley rats. Results showed increased level of neutrophils, interleukin-6, and tumor necrosis factor-α, and accumulation of alveolar macrophages and high aluminum content in lungs. This study suggested that significant pulmonary toxicity was induced with Al_2O_3-Nps at the dose level of 5 mg/m^3(69).

In a study for investigation and comparison of the toxic effects of four types of metal oxide nanoparticles with similar primary size on human fetal lung fibroblasts in vitro, it was found that the toxic effects were produced in a dose-dependent manner and zinc oxide was found to be most cytotoxic, followed by titanium dioxide, silicon dioxide, and aluminum oxide in descending order (70).

Thus, studies summarized in this section highlight the research that is being undertaken to understand the potential pulmonary toxicity of EMNPs. Figure 32.3 shows common pulmonary toxic effects of EMNPs as discussed in various studies of this section.

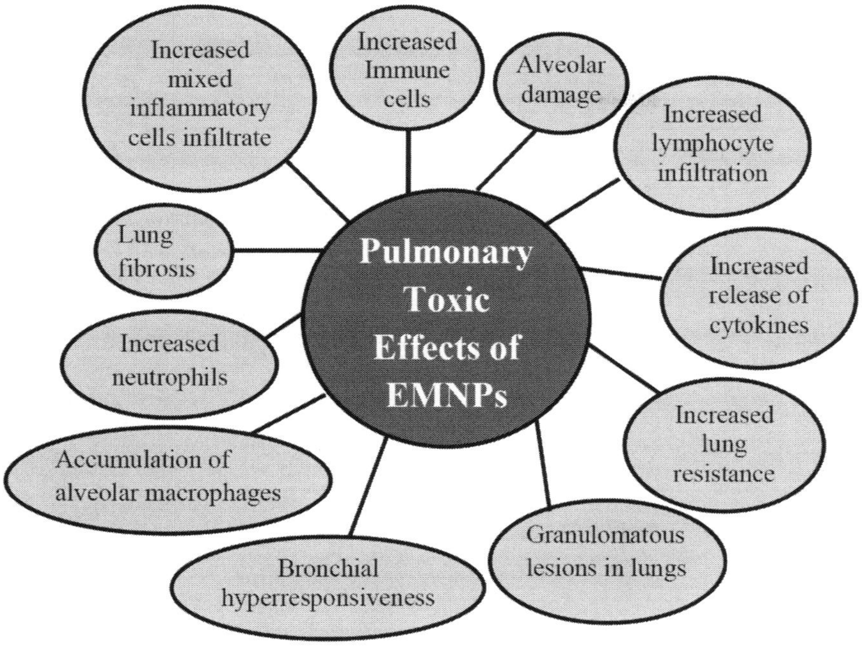

FIGURE 32.3 Pulmonary Toxic Effects of Engineered Metallic Nanoparticles (EMNPs)

32.4 Conclusion

There is an increased demand for EMNPs in pulmonary research and medicine because of their wide range of applications in diagnosis as well as in therapy of pulmonary diseases. EMNPs have unique physicochemical properties, which may render them effective in early diagnosis of pulmonary diseases such as lung cancer and in delivering drugs to the lung. A majority of reported pulmonary applications of metallic nanoparticles include their use in the treatment of pulmonary infections and lung cancer.

With increased demand for metallic nanoparticles, in-depth safety studies are required for the protection of human health. Experimental evidence indicates that there is continuous need to investigate in vitro and in vivo pulmonary adverse effects of EMNPs and to understand mechanisms responsible for their toxic effects in lungs.

Improvement in efficacy and reduction in pulmonary toxic effects of metallic nanoparticles is a challenging task. In future, new treatment strategies can be explored with the use of EMNPs to overcome existing problems and to provide safe and effective therapeutic treatment for a variety of pulmonary diseases.

REFERENCES

1. Li YF and Chen C. 2011. Fate and toxicity of metallic and metal-containing nanoparticles for biomedical applications. *Small*. 7(21): 2965-2980. doi:10.1002/smll.201101059.
2. Gatoo MA, Naseem S, Arfat MY, Dar AM, Qasim K, and Zubair S. 2014. Physicochemical properties of nanomaterials: Implication in associated toxic manifestations. *BioMed Research International*. 2014; 1–8. doi:10.1155/2014/498420.
3. Card JW, Zeldin DC, Bonner JC, and Nestmann ER. 2008. Pulmonary applications and toxicity of engineered nanoparticles. *American Journal of Physiology - Lung Cellular and Molecular Physiology*. 295(3): L400–L411. doi:10.1152/ajplung.00041.2008.
4. Paranjpe M and Müller-Goymann CC. 2014. Nanoparticle-mediated pulmonary drug delivery: A review. *International Journal of Molecular Sciences*. 15(4): 5852–5873. doi:10.3390/ijms15045852.
5. Houglum JE. 2000. Asthma medications: Basic pharmacology and use in the athlete. *Journal of Athletic Training*. 35(2): 179.
6. Halwani R, Shaik AS, Ratemi E, Afzal S, Kenana R, Al-Muhsen S, and Faraj AA. 2016. A novel anti-IL4Rα nanoparticle efficiently controls lung inflammation during asthma. *Experimental & Molecular Medicine*. 48(10): e262–e262. doi:10.1038/emm.2016.89.
7. Onizawa S, Aoshiba K, Kajita M, Miyamoto Y, and Nagai A. 2009. Platinum nanoparticle antioxidants inhibit pulmonary inflammation in mice exposed to cigarette smoke. *Pulmonary Pharmacology and Therapeutics*. 22(4): 340–349. doi:10.1016/j.pupt.2008.12.015.
8. Shahabi R, Anissian A, Javadmoosavi SA, and Nasirinezhad F. 2019. Protective and anti-inflammatory effect of selenium nano-particles against bleomycin-induced pulmonary injury in male rats. *Drug and Chemical Toxicology*. 44(1): 92-100. doi:10.1080/01480545.2018.1560466.
9. Rubio L, Annangi B, Vila L, Hernández A, and Marcos R. 2016. Antioxidant and anti-genotoxic properties of cerium oxide nanoparticles in a pulmonary-like cell system. *Archives of Toxicology*. 90(2): 269–278. doi:10.1007/s00204-015-1468-y.
10. World Health Organisation (WHO). 2018. Global Cancer Data. *International Agency for Research on Cancer*.

11. Naatz H, Manshian BB, Luci CR, Tsikourkitoudi V, Deligiannakis Y, Birkenstock J, Pokhrel S, Mädler L, and Soenen SJ. 2020. Model-based nanoengineered pharmacokinetics of iron-doped copper oxide for nanomedical applications. *Angewandte Chemie - International Edition*. 132(5): 1844–1852. doi:10.1002/anie.201912312.
12. Brown SD, Nativo P, Smith JA, Stirling D, Edwards PR, Venugopal B, Flint DJ, Plumb JA, Graham Duncan, and Wheate Nial J. 2010. Gold nanoparticles for the improved anticancer drug delivery of the active component of oxaliplatin. *Journal of the American Chemical Society*. 132(13): 4678–4684. doi:10.1021/ja908117a
13. Thambiraj S, Shruthi S, Vijayalakshmi R, and Shankaran DR. 2019. Evaluation of cytotoxic activity of docetaxel loaded gold nanoparticles for lung cancer drug delivery. *Cancer Treatment and Research Communications*. 21: 100157. doi:10.1016/j.ctarc.2019.100157.
14. Ullah S, Ahmad A, Wang A, Raza M, Jan AU, Tahir K, Rahman AU, and Qipeng Y. 2017. Bio-fabrication of catalytic platinum nanoparticles and their in vitro efficacy against lungs cancer cells line (A549). *Journal of Photochemistry and Photobiology B: Biology*. 173: 368–375. doi:10.1016/j.jphotobiol.2017.06.018.
15. Bendale Y, Bendale V, Natu R, and Paul S. 2016. Biosynthesized platinum nanoparticles inhibit the proliferation of human lung-cancer cells in vitro and delay the growth of a human lung-tumor xenograft in vivo: In vitro and in vivo anticancer activity of Bio-Pt NPs. *Journal of Pharmacopuncture*. 19(2): 114–121. doi:10.3831/KPI.2016.19.012.
16. American Cancer Society. 2016. Cancer Facts & Figures 2016. *Cancer Facts & Figures 2016*. 2016: 1–66. doi:10.1097/01.NNR.0000289503.22414.79.
17. Acquaviva J, Smith DL, Sang J, Friedland JC, He S, Sequeira M, Zhang C, Wada Y, and Proia DA. 2012. Targeting KRAS-mutant non-small cell lung cancer with the Hsp90 inhibitor ganetespib. *Molecular Cancer Therapeutics*. 11(12): 2633–2643. doi:10.1158/1535-7163.MCT-12-0615.
18. Grossi F, Kubota K, Cappuzzo F, Marinis F, Gridelli C, Aita M, and Douillard J-Y. 2010. Future scenarios for the treatment of advanced non-small cell lung cancer: Focus on taxane-containing regimens. *The Oncologist*. 15(10): 1102–1112. doi:10.1634/theoncologist.2010-0322.
19. Ramalingam S and Belani C. 2008. Systemic chemotherapy for advanced non-small cell lung cancer: Recent advances and future directions. *The Oncologist*. 13: 5–13. doi:10.1634/theoncologist.13-s1-5.
20. Sulthana S, Banerjee T, Kallu J, Vuppala SR, Heckert B, Naz S, Shelby T, Yambem O, and Santra S. 2017. Combination therapy of NSCLC using Hsp90 inhibitor and doxorubicin carrying functional nanoceria. *Molecular Pharmaceutics*. 14(3): 875–884. doi:10.1021/acs.molpharmaceut.6b01076.
21. Tanino R, Amano Y, Tong X, Sun R, Tsubata Y, Harada M, Fujita Y, and Isobe T. 2020. Anticancer activity of ZnO nanoparticles against human small-cell lung cancer in an orthotopic mouse model. *Molecular Cancer Therapeutics*. 19(2): 502–512. doi:10.1158/1535-7163.MCT-19-0018.
22. Ramalingam V, Varunkumar K, Ravikumar V, and Rajaram R. 2018. Target delivery of doxorubicin tethered with PVP stabilized gold nanoparticles for effective treatment of lung cancer. *Scientific Reports*. 8(1): 1–12. doi:10.1038/s41598-018-22172-5.
23. Majeed S, Danish M, and Binti Muhadi NFB. 2018. Genotoxicity and apoptotic activity of biologically synthesized magnesium oxide nanoparticles against human lung cancer A-549 cell line. *Advances in Natural Sciences: Nanoscience and Nanotechnology*. 9(2): 025011. doi:10.1088/2043-6254/aac42c.
24. Conde, J, Tian F, Hernández Y, Bao C, Cui D, Janssen KP, Ibarra MR, Baptista PV, Stoeger T, and de la Fuente JM. 2013. Invivo tumor targeting via nanoparticle-mediated therapeutic sirna coupled to inflammatory response in lung cancer mouse models. *Biomaterials*. 34(31): 7744–7753. doi:10.1016/j.biomaterials.2013.06.041.
25. Markowitz SB, Miller A, Miller J, Manowitz A, Kieding S, Sider L, and Morabia A. 2007. Ability of low-dose helical CT to distinguish between benign and malignant noncalcified lung nodules. *Chest*. 131(4): 1028–1034. doi:10.1378/chest.05-3014.
26. Cronin P, Dwamena BA, Kelly AM, and Carlos RC. 2008. Solitary pulmonary nodules: Meta-analytic comparison of cross-sectional imaging modalities for diagnosis of malignancy.*Radiology*. 246(3): 772–782. doi:10.1148/radiol.2463062148.
27. Chan BA, and Hughes BGM. 2015. Targeted therapy for non-small cell lung cancer: Current standards and the promise of the future. *Translational Lung Cancer Research*. 4(1): 36–54. doi:10.3978/j.issn.2218-6751.2014.05.01.
28. Ashton JR, Gottlin E.B., Patz EF, West JL, and Badea CT. 2018. A comparative analysis of Egfr-targeting antibodies for gold nanoparticle CT imaging of lung cancer. *PLoS ONE*. 13(11): 1–20. doi:10.1371/journal.pone.0206950.
29. Wang H, Zheng L, Peng C, Shen M, Shi X, and Zhang G. 2013. Folic acid-modified dendrimer-entrapped gold nanoparticles as nanoprobes for targeted CT imaging of human lung adencarcinoma. *Biomaterials*. 34(2): 470–480. doi:10.1016/j.biomaterials.2012.09.054.
30. Kamali R, Shekoohi SA, and Binesh A. 2014. Effects of magnetic particles entrance arrangements on mixing efficiency of a magnetic bead micromixer. *Nano-Micro Letters*. 6(1): 30–37. doi: 10.1007/BF03353766.
31. Wu, CH, Huang YY, Chen P, Hoshino K, Liu H, Frenkel EP, Zhang JXJ, and Sokolov KV. 2013. Versatile Immunomagnetic nanocarrier platform for capturing cancer cells. *ACS Nano*. 7(10): 8816–8823. doi:10.1021/nn403281e.
32. Smirnov P. 2009. Cellular magnetic resonance imaging using superparamagnetic anionic iron oxide nanoparticles: Applications to in vivo trafficking of lymphocytes and cell-based anticancer therapy. *Methods in Molecular Biology (Clifton, N.J.)*, 333–353. doi:10.1007/978-1-60327-530-9_19.
33. Wang X, Zhang H, Jing H, and Cui L. 2015. Highly efficient labeling of human lung cancer cells using cationic poly-l-lysine-assisted magnetic iron oxide nanoparticles. *Nano-Micro Letters*. 7(4): 374–384. doi:10.1007/s40820-015-0053-5.
34. Lyczak JB, Cannon CL, and Pier GB. 2002. Lung infections associated with cystic fibrosis. *Clinical Microbiology Reviews*. 15(2): 194–222. doi:10.1128/CMR.15.2.194-222.2002.

35. Chirgwin ME, Dedloff MR, Holban AM, and Gestal MC. 2019. Novel therapeutic strategies applied to pseudomonas aeruginosa infections in cystic fibrosis. *Materials*. 12(24): 4093. doi:10.3390/MA12244093.
36. Habash, MB, Goodyear MC, Park AJ, Surette MD, Vis EC, Harris RJ, and Khursigara CM. 2017. Potentiation of tobramycin by silver nanoparticles against pseudomonas aeruginosa biofilms. *Antimicrobial Agents and Chemotherapy*. 61(11): e00415–e00417. doi:10.1128/AAC.00415-17.
37. Al-Obaidi H, Kalgudi R, and Zariwala MG. 2018. Fabrication of inhaled hybrid silver/ciprofloxacin nanoparticles with synergetic effect against pseudomonas aeruginosa. *European Journal of Pharmaceutics and Biopharmaceutics*. 128: 27–35.
38. Liao S, Zhang Y, Pan X, Zhu F, Jiang C, Liu Q, Cheng Z, et al. 2019. Antibacterial activity and mechanism of silver nanoparticles against multidrug-resistant Pseudomonas Aeruginosa. *International Journal of Nanomedicine*. 14: 1469–1487. doi:10.2147/IJN.S191340.
39. Serebrovska Z, Swanson RJ, Portnichenko V, Shysh A, Pavlovich S, Tumanovska L, Dorovskych A, et al. 2017. Anti-inflammatory and antioxidant effect of cerium dioxide nanoparticles immobilized on the surface of silica nanoparticles in rat experimental pneumonia. *Biomedicine and Pharmacotherapy*. 92: 69–77. doi:10.1016/j.biopha.2017.05.064.
40. Galdiero S, Falanga A, Vitiello M, Cantisani M, Marra V, and Galdiero M. 2011. Silver nanoparticles as potential antiviral agents. *Molecules*. 16(10): 8894–8918. doi:10.3390/molecules16108894.
41. Harris J and Werling D. 2003. Binding and entry of respiratory syncytial virus into host cells and initiation of the innate immune response. *Cellular Microbiology*. 5(10): 671–680. doi:10.1046/j.1462-5822.2003.00313.x.
42. Bueno SM, González PA, Riedel CA, Carreño LJ, Vásquez AE, and Kalergis AM. 2011. Local cytokine response upon respiratory syncytial virus infection. *Immunology Letters*. 136(2): 122–129. doi:10.1016/j.imlet.2010.12.003.
43. Yang XX, Li CM, and Huang CZ. 2016. Curcumin modified silver nanoparticles for highly efficient inhibition of respiratory syncytial virus infection. *Nanoscale*. 8(5): 3040–3048. doi:10.1039/c5nr07918g.
44. Morris D, Ansar M, Speshock J, Ivanciuc T, Qu Y, Casola A, and Garofalo R. 2019. Antiviral and immunomodulatory activity of silver nanoparticles in experimental Rsv infection. *Viruses*. 11(8): 732. doi:10.3390/v11080732.
45. Sun L, Singh AK, Vig K, Pillai SR, and Singh SR. 2008. Silver nanoparticles inhibit replication of respiratory syncytial virus. *Journal of Biomedical Nanotechnology*. 4(2): 149–158. doi:10.1166/jbn.2008.012
46. Clark NM and Lynch JP. 2011. Influenza: epidemiology, clinical features, therapy, and prevention *Seminars in Respiratory and Critical Care Medicine*. 32(4): 373–392. doi:10.1055/s-0031-1283278.
47. Xiang D, Zheng C-L, Zheng Y, Li X, Yin J, Conner MO', Marappan M, et al. 2013. Inhibition of A/Human/Hubei/3/2005 (H3N2) influenza virus infection by silver nanoparticles in vitro and in vivo *International Journal of Nanomedicine*. 8: 4103–4114. doi:10.2147/ijn.s53622.
48. Mori Y, Ono T, Miyahira Y, Nguyen VQ, Matsui T, and Ishihara M. 2013. Antiviral activity of silver nanoparticle/chitosan composites against H1N1 influenza a virus. *Nanoscale Research Letters*. 8 (1): 93.
49. Sohrabi C, Alsafi Z, O'Neill N, Khan M, Kerwan A, Al-Jabir A, Iosifidis C, and Agha R. 2020. World Health Organization declares global emergency: A review of the 2019 novel coronavirus (COVID-19). *International Journal of Surgery*. 76: 71–76. doi:10.1016/j.ijsu.2020.02.034.
50. Zachar O. 2020. Formulations for COVID-19 early stage treatment via silver nanoparticles inhalation delivery at home and hospital. *ScienceOpen Preprints*.
51. Wang N, Wei C, Zhang Z, Liu T, and Wang T. 2020. Aluminum nanoparticles acting as a pulmonary vaccine adjuvant-delivery system (VADS) able to safely elicit robust systemic and mucosal immunity. *Journal of Inorganic and Organometallic Polymers and Materials*. 9: 1–15. doi:10.1007/s10904-020-01572-z.
52. Foldbjerg R, Dang DA, and Autrup H. 2011. Cytotoxicity and genotoxicity of silver nanoparticles in the human lung cancer cell line, A549. *Archives of Toxicology*. 85(7): 743–750. doi:10.1007/s00204-010-0545-5.
53. Worthington, KLS, Adamcakova-Dodd A, Wongrakpanich A, Mudunkotuwa IA, Mapuskar KA, Joshi VB, Guymon CA, et al. 2013. Chitosan coating of copper nanoparticles reduces in vitro toxicity and increases inflammation in the lung. *Nanotechnology*. 24(39): 395101. doi:10.1088/0957-4484/24/39/395101.
54. Mo Y, Zhang Y, Wan R, Jiang M, Xu Y and Zhang Q. 2020. MiR-21 mediates nickel nanoparticle-induced pulmonary injury and fibrosis. *Nanotoxicology*. 14(9): 1175–1197.
55. Mittal S and Pandey AK. 2014. Cerium oxide nanoparticles induced toxicity in human lung cells: Role of ROS mediated DNA damage and apoptosis. *BioMed Research International*. 2014: 1–14. doi:10.1155/2014/891934.
56. Larsen ST, Silva ED, Hansen JS, Jensen ACØ, Koponen IK, and Sørli JB. 2020. Acute inhalation toxicity after inhalation of ZnO nanoparticles: Lung surfactant function inhibition in vitro correlates with reduced tidal volume in mice. *International Journal of Toxicology*. 39(4): 321–327. doi:10.1177/1091581820933146.
57. Sung JH, Ji JH, Song KS, Lee JH, Choi KH, Lee SH, and Yu IJ. 2011. Acute inhalation toxicity of silver nanoparticles. *Toxicology and Industrial Health*. 27(2): 149–154. doi:10.1177/0748233710382540.
58. Sung JH, Ji JH, Park JD, Yoon JU, Kim DS, Jeon KS, Song MY, et al. 2009. Subchronic inhalation toxicity of silver nanoparticles. *Toxicological Sciences*. 108(2): 452–461. doi:10.1093/toxsci/kfn246.
59. Seiffert J, Hussain F, Wiegman C, Li F, Bey L, Baker W, Porter A, et al. 2015. Pulmonary Toxicity of instilled silver nanoparticles: Influence of size, coating and rat strain. *PLoS ONE*. 10(3): e0119726. doi:10.1371/journal.pone.0119726.
60. Wiemann M, Vennemann A, Blaske F, Sperling M, and Karst U. 2017. Silver nanoparticles in the lung: Toxic effects and focal accumulation of silver in remote organs. *Nanomaterials*. 7(12): 441. doi:10.3390/nano7120441.
61. Sung, JH, Ji JH, Park JD, Song MY, Song KS, Ryu HR, Yoon JU, et al. 2011. Subchronic inhalation toxicity of gold nanoparticles. *Particle and Fibre Toxicology*. 8(1): 1–8. doi:10.1186/1743-8977-8-16.
62. Elbakary RH, Okasha EF, Ragab AMH, and Ragab MH. 2018. Histological effects of gold nanoparticles on the lung

tissue of adult male albino rats. *Journal of Microscopy and Ultrastructure*. 6(2): 116–122. doi:10.4103/jmau.jmau_25_18.
63. Schulz, M, Ma-Hock L, Brill S, Strauss V, Treumann S, Gröters S, Ravenzwaay BV, and Landsiedel R. 2012. Investigation on the genotoxicity of different sizes of gold nanoparticles administered to the lungs of rats. *Mutation Research - Genetic Toxicology and Environmental Mutagenesis*. 745(1-2): 51–57. doi:10.1016/j.mrgentox.2011.11.016.
64. Zhu MT, Feng WY, Wang B, Wang TC, Gu YQ, Wang M, Wang Y, Ouyang H, Zhao YL, and Chai ZF. 2008. Comparative study of pulmonary responses to nano- and submicron-sized ferric oxide in rats. *Toxicology*. 247(2–3): 102–111. doi:10.1016/j.tox.2008.02.011.
65. Nemmar A, Yuvaraju P, Beegam S, Fahim MA, and Ali BH. 2017. Cerium oxide nanoparticles in lung acutely induce oxidative stress, inflammation, and dna damage in various organs of mice. *Oxidative Medicine and Cellular Longevity*. 2017: 1–17. doi:10.1155/2017/9639035.
66. Lai X, Zhao H, Zhang Y, Guo K, Xu Y, Chen S, and Zhang J. 2018. Intranasal delivery of copper oxide nanoparticles induces pulmonary toxicity and fibrosis in C57BL/6 Mice. *Scientific Reports*. 8(1): 1–12.
67. Hadrup, N, Rahmani F, Jacobsen NR, Saber AT, Jackson P, Bengtson S, Williams A, Wallin H, Halappanavar S, and Vogel U. 2019. Acute phase response and inflammation following pulmonary exposure to low doses of zinc oxide nanoparticles in mice. *Nanotoxicology*. 13(9): 1275–1292. doi:10.1080/17435390.2019.1654004.
68. Gelli K, Porika M, and Anreddy RNR. 2015. Assessment of pulmonary toxicity of MgO nanoparticles in rats. *Environmental Toxicology*. 30(3): 308-314. doi:10.1002/tox.21908.
69. Kim YS, Chung Y, Seo D, Choi HS, and Lim CH. 2018. Twenty-eight-day repeated inhalation toxicity study of aluminum oxide nanoparticles in male sprague-dawley rats. *Toxicological Research*. 34(4): 343–354. doi:10.5487/TR.2018.34.3.343.
70. Qiang ZX, Hong YL, Meng AT, and Pu PY. 2011. ZnO, TiO2, SiO2, and Al2O3 nanoparticlesinduced toxic effects on human fetal lung fibroblasts. *Biomedical and Environmental Sciences*. 24(6): 661–669. doi:10.3967/0895.

33

Anti-cancer Activity of Eco-friendly Gold Nanoparticles against Lung and Liver Cancer Cells

Anita Patel and Jayvadan Patel
Faculty of Pharmacy, Nootan Pharmacy College, Sankalchand Patel University, Gujarat, India

33.1 Introduction

Cancer is a common term for a kind of genetic disease explained by unrestrained, irregular cell division and invasiveness. The expansion of cancer in the majority of cases resulted from mutations or changes in the expression prototypes of proto-oncogenes, tumor suppressor genes, and genes involved in deoxyribonucleic acid (DNA) repair. The disturbance of pro-apoptotic signaling along with overexpression of several proteins facilitates cell growth and supplementation impedes the expansion of resourceful anti-cancer treatment (1). Most cancers result from the effect of ecological factors, for example, vulnerability to radiation and pollutants, although most prominently, from an unhealthful standard of living, including not enough physical activity, unbalanced diet, tobacco smoke, and trauma. No more than 5–10% of cancer cases are related to hereditary genetics (2,3). The peril of cancer increases notably with age, and many types of this ailment arise increasingly in industrialized countries.

Among all the cancers, lung and liver cancers are the most widespread cancers in the world, and the main cause of cancer fatalities all over the world (4–6). Lung cancer could be categorized into two main subtypes: small cell lung carcinomas (SCLCs) and non small cell lung carcinomas (NSCLCs) as reported by the histological considerations (7,8). The fatality rate of both cancers is too high, and almost every patient dies within a year. The American Cancer Society anticipated that 69410 males and 62470 females would die of lung cancer, while 20300 males and 9930 females would die of liver cancer in the US in 2021 (9).

Contrary to other widespread solid cancers, lung cancer has no well-known techniques for early diagnosis, and the majority of cases are diagnosed at a progressive stage. It accounts for about 14% of all new cancers with a miserable 5-year survival rate of just 15%. On the other hand, most liver cancers are diagnosed during progressive stages, where an invasive strategy is the only therapy with a survival rate of 10–30% (10–12). Latest data indicate that lung cancer is to be expected to exceed breast cancer as the major cause of cancer death among European women by the middle of this decade (13). According to the category of melanoma and stage at the time of diagnosis, lung and liver cancer treatment frequently entails a combination of surgery, chemotherapy, and/or radiations (14–16). Although these techniques have been recognized and practiced in recent decades, they have their downsides and undesirable effects. Surgical retrieval of malignant growth is limited primarily to large, resectable, and easily reached tumors. A chemotherapeutic agent targets quickly dividing cells, and as a result not only destroys cancer cells but kills normal cells such as bone marrow cells as well as immune cells (17). This engenders widespread "collateral damage" in the patient's body. Radiation treatment uses high energy radiation like X-rays and gamma rays to eradicate tumor cells and unavoidably triggers deadly effects on healthy tissues along the radiation path (18).

Regardless of noteworthy progression in its treatment (chemotherapy, surgery, and radiotherapy), cancerous cell suppression remains inadequate, and the rate of survival has not improved to a large extent (12,19).

Taking into account the deficiencies of recent treatment modalities for cancer, a decisive thrust towards upgrading cancer rehabilitation is to explicitly target therapeutic agents to cancer cells while protecting frugal healthy tissues from damage. This is one of the promising focuses of nanotechnology research. Nanotechnology relates to the fabrication of substances having nanoscale dimensions between 1 and 100 nm (20). The smaller size of these nanomaterials empowers their individuality with chemical as well as physical properties that are distinctive from their bulk materials (21). These nanomaterials, attributable to their exclusive physicochemical properties as well as smaller size, have a higher surface to volume ratio, the possibility of surface modification, and discriminatory accessibility to tumor cells, and these novel metal nanoparticles have attracted much attention in the nano-oncology field (22–27). The quick growth in nanomaterials research raises the future perspective of novel diagnostic methods and treatment of diseases in human beings. This division of nanotechnology in the diagnosis of disease, monitoring, and therapy has been referred to as *nanomedicine* by the National Institutes of Health in the USA (20). Presently, cancer nanotechnology has come out as a novel area of medication with the aim to achieve progress in both cancer diagnosis and treatment (28,29). Among numerous nanomaterials

DOI: 10.1201/9781003046547-33

being considered for nanomedicine applications, this chapter will mainly focus on biosynthesized gold nanoparticles for their anti-cancer activity against lung and liver cancer cells.

33.2 Gold Nanoparticles and Their General Properties

At the moment, there is growing attention to nanoparticles of noble metals (30). The consideration of scientists is mainly focused on gold nanoparticles, which have resourceful properties and potential applications in bioimaging, clinical chemistry, and treatment of cancer, in addition to targeted drug delivery persistently being exemplified.

Gold (Au) was one of the first metals found a few thousand years back. Gold in its purest form is a yellow, bright, dense, soft, and malleable metal, solid under ordinary conditions. It is one of the least reactive chemical substances. Since the beginning, gold was treasured because of its rare occurrence, ease of handling and production, and resistance to corrosion as well as other chemical properties, and, obviously, its inimitable color (31). Medicinal applications of gold and its complex form have a long track record, as well. The initial findings on colloidal gold (colloidal suspension of gold nanoparticles in a fluid) can be found in ancient Arabian, Chinese, and Indian papers from the 4th and 5th centuries BCE, which suggested it for the cure of different illnesses, even though the mechanism of action was not properly understood (32). The first research paper on gold nanoparticles was reported in 1857 by Faraday, which attributed the red color to the colloidal nature of gold nanoparticles and described their light scattering features.(33). After 50 years the visible absorption characteristics of gold nanoparticles were described with Maxwell's electromagnetic equations (33). In the year 1971, British investigators Faulk and Taylor designed an innovative technique of antibody–colloidal gold coupling for the direct electron microscopy imaging of surface antigens of Salmonellae (34). This innovation instigated numerous studies over the next 40 years, dedicated to biomedical applications of gold nanoparticles, particularly identifying different biomacromolecules as a result of the surface functionalization and the distinctive features. The latter are generally linked with controlled systems of production, permitting the acquirement of gold nanoparticles with definite sizes and shapes (35–38).

By virtue of much-optimized techniques for production, allowing the control of size and shape, as well as the dimension of gold nanoparticles, they can be specially developed to acquire specific characteristics. For the finding of features of these nanoparticles, interparticle interactions as well as an assemblage of gold nanoparticle networks have a crucial role (39). The size as well as the shape of gold nanoparticles has an insightful influence on their characteristics, affecting compatibility, mobility, stability, etc. (40–46), and needs to be optimized in respect of the specific biomedical applications. For example, nanoparticles fabricated for drug delivery need to be pretty small to cross physiological barriers or go into the target cells, and pretty big to convey an adequate quantity of therapeutic agents to the target site (47,48).

It is worth mentioning that physical and chemical features of materials in nanometer size are noticeably dissimilar than their analogs in larger forms. For gold, the greatest example of this attribute is the yellow color of the larger form of gold and the wine-red color of the gold nanoparticles which is reliant on their shape as well as characteristics (32). Moreover, colloidal gold, unlike bulk gold, is believed to be very reactive, which significantly expands its application perspectives, offering antioxidant, catalytic, and optoelectronic properties, as well as the potential of surface functionalization (32). Because of the smaller size and larger surface, shape, and crystallinity, nanoparticles have turned out to be outstanding therapeutic compounds as they can effortlessly deliver into the target cells and carry a higher drug load (49).

One essential physical attribute of gold nanoparticles is surface plasmon resonance (SPR). This definite miracle takes place when the frequency of the oscillation of free electrons at the exterior part of nanoparticles resonates with the frequency of the arriving light radiation, resulting in a plasmon band. Accordingly, an electromagnetic field emerges at the surface of gold nanoparticles, allowing surface-enhanced optical features. Gold nanoparticles offer absorption as well as scattering effects, the proportions of which are subject to the size and shape as well as the type of solvent, core charge, temperature, surface ligand, and the closeness of other nanoparticles (50,51), affecting the electron charge density on the surface of the particle. In sphere-shaped gold nanoparticles with sizes smaller than 60 nm, the SPR peak absorbance seems near 500–550 nm, creating their red color (52). An added fascinating effect contingent on the form of gold nanoparticles and linked with surface plasmon bands includes the intonation of fluorescence features of proximate fluorophores. This is attributable to the photoinduced electron transfer (PET) method, fluorescence resonance energy transfer (FRET) phenomenon (53–57), and photothermal characteristics, arising from the light absorption and succeeding nonradiative energy dissipation (58,59). As a consequence, modifying the shape of gold nanoparticles presents fascinating optical characteristics that span the broad visible to near-infrared (NIR) spectrum, converting them into excellent tools for bioimaging and theranostic applications (60,61), and permitting the controlling of morphological features of gold nanoparticles during manufacturing. Gold nanoparticles are extensively employed in biomedical science including for tissue or tumor imaging, photothermal therapy, drug delivery, and immunochromatographic detection of pathogens in clinical specimens, thanks to the SPR (62).

In short, the exploitation of gold nanoparticles is increasingly popular in these fields of research for numerous reasons. First of all, gold nanoparticles are believed to be biologically unreactive and so suitable for in vivo appliances when compared with the highly toxic cadmium as well as silver nanoparticles (63). Further beneficial qualities include the strong optical features of gold nanoparticles because of localized SPR (64), effortlessly controllable surface chemistry, which allows flexibility in addition to surface functional groups (65), and finally, the simplicity in control over particle size along with shape during production (66). Gold nanoparticles can be addressed to be completely multifunctional, with the likelihood of combining diverse most-wanted functionalities in one molecular-sized package. All of these points promote a great deal of interest and priorities for the utilization of gold nanoparticles relative to other nanoparticles (67). Moreover, gold nanoparticles have proven to be outstanding therapeutic agents as well as drug transporters.

33.3 Surface Modification of Gold Nanoparticles

Surface modification of gold nanoparticles is one important aspect, together with their size and shape, which decides the fate of particles after administration. Surface functionalization has a marked impact on these two criteria, giving protection against aggregation, improved biocompatibility, specified interactions with cells, and targeted transportation and accumulation in preferred organs. They can exhibit an incredible effect on a gold nanoparticle's blood half-life, averting their elimination by the cells of the mononuclear phagocytic system (MPS), also called the reticulo-endothelial system (RES). Nanoparticles without modification following intravenous administration are sometimes speedily recognized and bound by opsonins in the blood, which permits their phagocytosis and elimination by macrophages. Surface functionalization has the ability to "mask" gold nanoparticles from RES, hence guaranteeing long blood circulation time and permitting them to reach the target site (68).

It must be noted that surface functionalization can alter the optical characteristics of gold nanoparticles (69); this should be taken into consideration while designing gold nanoparticles for a specified purpose, such as in radiofrequency or photothermal therapy.

Functionalization might be carried out either via physical adsorption or covalent linking of ligands on the surface of nanoparticles, generally with thiol linkages. Among the most widely utilized compounds for modification of nanogold is poly (ethylene glycol) (PEG), attached covalently with the surface atoms of gold particles. The PEGylation has been proved to improve the biocompatibility of different nanoparticles, extend their blood half-life (70,71), and avert elimination by RES, growing their hydrophilic nature (72,73). Sphere-shaped gold nanoparticles functionalized with PEG exhibited no cytotoxic effect against in vitro cultured human cell lines (74,75), as well as little take-up by RES, comparatively longer blood circulation time, and superior tumor accumulation during in vivo research on mice (76).

At the same time, different surface functionalization has been revealed to eliminate or reduce the cytotoxicity of gold nanoparticles, as a result enabling their risk-free administration into the living being with no detrimental side effects. An example of that comprises gold nanoparticles functionalized with folic acid (77), polyvinylpyrrolidone (PVP) (78), and polyacrylamide (79).

It's important to mention that the surface charge of nanoparticles has an incredible effect on their cytotoxicity. Particles with positive surface charges are generally more lethal, by reason of non-specific interactions with negatively charged cellular membranes (80). It was proved that cationic gold nanoparticles modified with quaternary amines were seven times more deadly to in vitro cultured cells than their anionic equivalents, achieved by the replacement of amine moiety with a carboxyl group (81).

Considering everything, surface functionalization of gold nanoparticles is meant to modify their biodistribution prototypes, allow targeted delivery, and make possible cellular internalization. Examples of that the exploit of folic acid (77), transferrin (82), carbohydrates (83,84), oligonucleotides (85), and specified antibodies (86–88), attached on the surface of nanoparticles. In addition, PEG molecules have been also employed as linkers for various targeting ligands, such as galactose (89) or tumor necrosis factor α (TNFα) (90,91). Though superficially inert changes may significantly affect the accumulation of nanogold in a variety of organs, which must be taken into consideration. For example, it is known that gold nanoparticles coated with gum arabic or maltose display diverse biodistribution patterns in blood, tissues, and urine. Specifically, the uppermost concentration of gold nanoparticles functionalized with gum arabic was accumulated in the liver, while those coated with maltose were found in the lungs (92). As a result, while designing modified nanogold for anti-cancer applications, it is important to consider not only the targeting features of surface moieties, but also their ability to direct the particles to different body regions.

33.4 Gold Nanoparticle-Mediated Drug Delivery

Targeted drug therapy can be a better choice than conventional drug therapy due to the fact that in a targeted drug delivery system, drugs directly target the site of action, reducing the side effects resulting from conventional drugs (93). Principle objectives of formulating anti-cancer agents are to diminish the different side effects resulting from conventional drugs and to get better efficiency as well as selectivity of drugs (94). Targeted drug delivery is the major field of attention for researchers in recent times, and lots of work has been done regarding manufactured systems for targeted drug delivery, for instance, nanoparticles, polymer gels, and quantum dots (95).

Gold nanoparticles have the aptitude of bio-imaging of the effected malignant cells for treatment (96). For impactful drug therapy, it is essential to explore the biological activities of the nanoparticles (62), as gold nanoparticles have distinctive physical as well as chemical features and have a binding attraction for aptamers (97), carboxylic acids (39), proteins, thiols (95), and disulfides; therefore, they have been widely employed in the biosciences, specifically in drug delivery for cancer treatment. Gold nanoparticles follow three major routes for cellular uptake: phagocytosis, receptor-mediated endocytosis, and fluid-phase endocytosis (98).

The toxic effect of gold nanoparticles is influenced by the size, shape, production method, surface coating, surface charge, and functionalized fragments, however, overall cytotoxicity of gold nanoparticles is tolerable as gold nanoparticles are believed to be nontoxic agents (99). For the resourceful drug delivery system, two factors are most important, and those are drug release and transport. Drugs are loaded on nanocarriers by means of non-covalent bonds or covalent conjugation with pro-drug, which is treated by the cell. Gold nanoparticles have serviceable flexibility, attributable to their monolayers; as a result they give a well-organized system (100).

33.5 Advantages of Gold Nanoparticle-Mediated Drug Delivery

Gold nanoparticle-mediated drug delivery systems have numerous benefits relative to other nanocarriers as well as to conventional medicines. Gold nanoparticles have been extensively utilized as a cancer antigen and in cancer therapies (101). A few benefits are mentioned here: (i) gold nanoparticles have inimitable physical, chemical (102), and optical properties (49), because of their shape and size (103); (ii) gold nanoparticles have higher surface area (62) which enables compact loading of the drug; (iii) gold nanoparticles are biologically compatible (104) and are easily available for attachment with small biomolecules like amino acids, carboxylic acid, DNA, enzymes, and proteins (105); (iv) gold nanoparticles have well-controlled dispersibility (106); (v) as a result of smaller size and homogeneous dispersion they can effortlessly arrive at the targeted site with blood flow (107); (vi) gold nanoparticles are non-cytotoxic to normal cells (95); and (vii) gold nanoparticles are straightforwardly fabricated by different techniques (Figure 33.1) (108).

33.6 Methods for the Synthesis of Gold Nanoparticles

In most cases, the methods of formulation of gold nanoparticles are similar to those of other particles. There are basically two alternatives for the categorization of production methods. The first option is based upon the method of production (bottom-up or top-down) (109), and the second option comprises the methodology-based strategy (biological, chemical, and physical methods).

One of the leading and most popular chemical processes, developed by Turkevich and his team, involves the reduction of chloroauric acid ($HAuCl_4$) with trisodium citrate (plays an additional role of the ligand for newly fabricated gold nanoparticles) at 100 °C. This reaction allowed the acquirement of aqueous solutions of moderately monodisperse sphere-shaped nanoparticles with sizes ranging from 15 to 150 nm, according to the initial concentration of sodium citrate (110). This technique was the base for the progress of further ones, enabling the highly controlled fabrication of gold nanoparticles in water or organic liquids, exploiting different temperatures and pH values but also various reducing agents, similar to sodium borohydride ($NaBH_4$) (111–113), hydroquinone (114), or aspartate (115). The size of gold nanoparticles can be further stabilized with a variety of stabilizing agents, which also function to defend fabricated nanoparticles from aggregation and to manage their characteristics in a specific manner.

Although Turkevich-based techniques produce generally spherical gold nanoparticles, gold nanoparticles can be achieved in different shapes as well, such as rods (116), cages (117), and tubes (118). The most appropriate technique to manufacture different structures of gold nanoparticles is on the basis of seed-mediated growth (119), including the reduction of gold salts with a powerful reducing agent, which leads the fabrication of seed particles, which are afterward added to the solution containing metal salt in the presence of a structure-directing agent as well as a weak reducing agent. Gold nanostructure's geometric shape can be modified by changing the concentration of seeds, structure-directing agents, and reducing agents.

Moreover, physical techniques using microwaves (120), laser ablation (121), ultrasonic waves (122), and photochemical as well as electrochemical reduction (123,124) have been reviewed for fabricating gold nanoparticles. However,

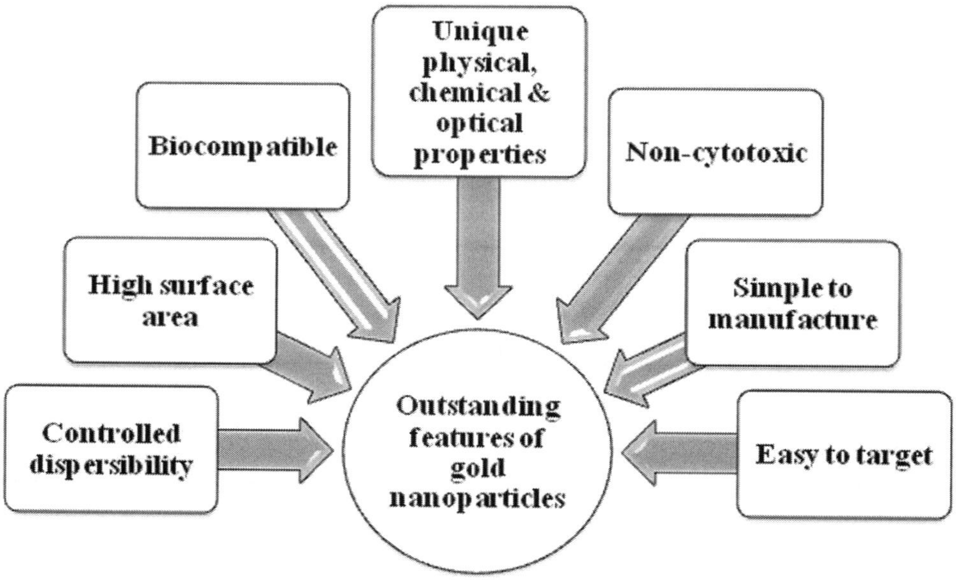

FIGURE 33.1 Outstanding Features of Gold Nanoparticles

seeing as different physical as well as chemical procedures prepared for the production of nanoparticles would be somewhat costly and detrimental to the environment, "green synthesis" techniques have grown to be the main attention of researchers with a view to elaborating an environment-friendly and non-hazardous means of gold nanoparticles fabrication (125). Numerous substrates and reducing compounds have been fruitfully implemented for the safe production of gold nanoparticles, including chitosan (126), extracts from citrus fruit juice (127), eggshell membrane (128), and edible mushrooms (129). In recent times, *Bacillus licheniformis* has been exploited for the fabrication of 10–100 nm gold nanocubes in much smoother conditions relative to classical chemical techniques (130), indicating the probability of the application of different bacterial strains for this process.

33.7 Clinical Applications of Gold Nanoparticles for Lung and Liver Cancer

Gold nanoparticles have distinctive electric and magnetic properties attributable to their size and shape. As a result they have received enormous consideration in research areas, principally in the fields of biological tagging, biomedical imaging, biological and chemical sensing, DNA labeling photothermal therapy (104), photoacoustic imaging and microscopy (131), catalysis, tracking and drug delivery (106), and cancer therapy (103). Gold nanoparticles have been extensively studied for lung as well as liver cancer therapy and diagnosis because of their astonishing and inimitable properties. The synthesized gold nanoparticles induce a dose-dependent inhibition activity against lung and liver cells. A number of the approved chemotherapeutic agents have induced side effects and high cost. For that reason, there is an important need to develop alternative medicines against this lethal ailment. Synthesized gold nanoparticles to fulfill the need for new therapeutic treatments were discovered and some of the potential applications of gold nanoparticles in lung and liver cancer discussed in this section.

Zeng and colleagues formulated gold nanoparticles from *Magnolia officinalis*, which is recognized as an eco-friendly and less toxic technique. The size of nanoparticles is recognized by dynamic light scattering analysis and it shows a value of 128 nm. Besides, energy dispersive X-ray analysis, high-resolution transmission electron microscopy, and atomic force microscopy describe the shape of the gold nanoparticles which are present in the complex materials. The anti-cancer efficiency of gold nanoparticles has been assessed in the human lung cancer cell line (A549). Gold nanoparticles successfully persuade apoptosis and cytotoxicity by intonating intrinsic apoptotic gene expressions in A549 cells. As a result, the gold nanoparticles produced from *Magnolia officinalis* show confirmation of anti-cancer effects (132).

The potential cytotoxicity of mannosylerythritol lipid-gold nanoparticles on human liver cancer cells (HepG2) was studied by the 3-(4,5-dimethylthiazol-2-yl)-2,5 diphenyltetrazolium bromide, a tetrazole (MTT) assay using different concentrations (10, 25, 50, 75, 100, 125, and 150 µg/mL) exposed for 24 and 48 hours. The HepG2 cell population was progressively decreased with an increased mannosylerythritol lipid-gold nanoparticle concentration and treating time. Noticeably, mannosylerythritol lipid-gold nanoparticles established potential cytotoxicity on HepG2 cells, the IC50 values were 75 and 100 µg/mL for 24 and 48 hours, respectively. The highest concentration of mannosylerythritol lipid-gold nanoparticles has inhibited cell growth by about 89% (133).

The anti-cancer effects of gold nanoparticles with *Cordyceps militaris* extract against the hepatocellular carcinoma HepG2 cells was investigated by Ji and colleagues. Gold nanoparticle extract produces reactive oxygen species and induces damage to the mitochondrial membrane potential in the hepatocellular carcinoma HepG2 cells. The gold nanoparticle extracts have a tendency to begin the apoptosis by activating the Bax, Bid, and caspases, and inhibiting the activation of anti-apoptotic Bcl-2 in the HepG2 cells. These results established that the gold nanoparticles with *Cordyceps militaris* would be a capable chemotherapeutic drug against hepatocellular carcinoma cells (134).

Latha and colleagues produced bio-inspired gold nanoparticles using the leaf extract of *Justicia adhatoda* and evaluated the anticancer activity on A549 cells. The bio-synthesized gold nanoparticles were confirmed and characterized by using different spectral studies like UV-Vis spectrum, FTIR analysis, scanning electron microscope with EDAX, transmission electron microscope, and surface-enhanced Raman spectroscopy. The cell viability was verified by the MTT reduction assay. Additionally, cytomorphology and the nuclear morphological study of the A549 cell line were examined under a fluorescence microscope. UV-Vis spectrum verified SPR peak at 547 nm, transmission electron microscopy as well as scanning electron microscopy analysis demonstrated the monodispersed sphere shape, and its average size in the range of 40.1 nm was observed. Fascinatingly, the produced gold nanoparticles confirmed a potent anti-proliferation effect on the A549 lung cancer cell. Cell morphology was monitored and cell death has resulted from apoptosis as revealed by propidium iodide staining. This investigation confirms the anticancer potential of biosynthesized gold nanoparticles. So synthesized gold nanoparticles can be employed for the treatment of A549 cells and it can be exploited for drug delivery in the future (135).

Gold nanoparticles were manufactured using the cetyltrimethylammonium bromide technique and surface properties were enhanced with PVP. The PVP–gold nanoparticles were exploited as a flourishing carrier for doxorubicin delivery for the treatment of lung cancer. Researchers demonstrated the anticancer activity doxorubicin conjugated with PVP coated gold nanoparticles against A549, H460, and H520 lung cancer cells, and outcomes signified that doxorubicin gold nanoparticles efficiently inhibit the propagation of lung cancer cells and promotes p53 mediated mitochondria-dependent apoptosis. Furthermore, doxorubicin gold nanoparticles upregulate the expression of several tumor suppressor genes, proving the capacity of the potential anti-cancer agent in cancer therapy (136).

The steady bioactive gold nanoparticles were produced using *Marsilea quadrifolia* leaf extract. The photochemical found in the extract of *Marsilea quadrifolia* reduces the gold ions into metallic nanoparticles. The biogenic gold nanoparticles show signs of a significant in vitro antioxidant activity and cytotoxicity effect against A549 cancer cells. These biogenic gold nanoparticles are estimated to serve as potent anti-cancer agents and so can be employed in biomedical appliances (137).

Investigators depicted a combination of chemotherapy and improved radiotherapy for in vitro management of lung cancer with gold nanoparticles stabilized by an apigenin-a bioactive gradient that has been explored for cancer treatments. Following interaction with lung cancer cells, the nanoparticles were able to stimulate cell apoptosis, inhibit cell proliferation, and capture cancer cells in G0/G1 phases in a dose-dependent manner. When a treatment combination of X-rays and gold nanoparticles was used, a synergistic anti-cancer effect was observed from chemotherapeutic functions of apigenin and the superior radiation killing effect generated by nanoparticles and X-ray interactions. This study might present a hopeful therapeutic approach for cancer treatment which puts together the benefits of both radiation and chemotherapy (138).

Guo and colleagues formulated PEG-coated gold nanoparticles of two different sizes, 14.4 and 30.5 nm, by a chemical reduction reaction and investigated the radiation enhancing effects in liver cancer cell lines. They investigated cellular uptake, blood stability, cytotoxicity, and radiation therapy. A 3–5 nm red shift of SPR caused by interactions between PEG-coated gold nanoparticles as well as plasma revealed their excellent stability. The in vitro bio-distribution assay indicated PEG-coated gold nanoparticles had high distribution in the cancer cells; almost 10^3 nanoparticles were observed in a single cell. The transmission electron microscopy direct observation demonstrated that PEG-coated gold nanoparticles hybridized with blood proteins constituted a 30–50 nm gold-protein corona. Gold nanoparticles were undergoing endocytosis by cytoplasmic vesicles, located in the intracellular area, and displayed concentration-dependent cell viability. Furthermore, these gold nanoparticles showed little toxicity at the concentration of 10^{-4} M. In vitro radiation therapy indicated that the gold nanoparticles appreciably enhanced the irradiation effect and diminished the endurance of two kinds of liver cancer cells. As a result, PEG-coated gold nanoparticles can be addressed as a prospective agent in liver cancer radiation therapy (139).

Indocyanine green loaded gold nanorod@liposome core-shell nanoparticles showed to be effectiveness in detecting tumor, and surgery guidance in orthotopic liver cancer mouse models by using photoacoustic and fluorescence dual-modality imaging probe. Researchers also investigated their efficacy for tumor detection as well as surgery guidance in orthotopic liver cancer mouse models by using a photoacoustic and fluorescence dual-modality imaging probe. This novel dual-modality nanoprobe gives hope for timely detection, enhances the surgical outcomes of liver cancer, and has immense potential for medical translation (140).

Gold nanoparticles linked with albumin as energetic vectors were utilized to focus liver cells for the development of an alive liver cancer model without any ethical obstacles to evaluate the discriminating features and counteractive capacity of these nanosystems in cancer patients. In order to achieve this goal, samples from cancerous patients were perfused out of the body (ex vivo). The albumin conjugated gold nanoparticles were administered intra-arterially into the model, and delivery of the nanoconjugate to the malignant tissue was established through capillary bedding. Their outcomes demonstrated that albumin conjugated gold nanoparticles build up through receptor-mediated endocytosis triggered to generate a laser-based therapeutic effect at the tumor site, but were not sufficient to have an effect on the healthy parenchyma tissue around it (141).

Aurolase®, made by Nanospectra, is silica-gold nanoshells coated with PEG and created to thermally clear the solid tumors following stimulation with a near infra-r-ed source of light. The absorption of light results in an increase in the local temperature, and it thermally destroys the solid tumors. AuroLase® particles were employed in localized therapy for the management of primary or metastatic lung tumors recently in the clinical trial (142).

The anti-cancer effect of gold nanoparticles against HepG-2 and A549 cell lines was investigated by Rajeshkumar. Results demonstrated that the good cytotoxic activity observed with gold nanoparticles against the cancer cells. The concentration of gold nanoparticles has a significant role in anti-cancer activity. The gold nanoparticles are having the good results against A549 in that 100 μg show fine results followed by 50, 25 and 1 μg. The lowest inhibitory action was observed form the concentration of 10 μg (143).

An in vitro and in vivo research study of gum arabic-conjugated gold nanoparticles and laser combination explained that this approach decreases cell viability as well as the activity of histone deacetylase in HepG2 cells. The findings stated that gum arabic-conjugated gold nanoparticles, with or without laser radiation, may cause apoptosis in cancer cells by activating death receptors (DR5), caspase-3, and in addition suppress pre-neoplastic lesions and primary markers (placental glutathione S-transferase). Moreover, gum arabic–conjugated gold nanoparticle stimulation with laser lessens tumor necrosis factor-α levels. That's why the gum arabic–conjugated gold nanoparticles in combination with laser stimulated the extrinsic pathway of apoptosis and repressed swelling that can stop liver pre-neoplastic lesions (144).

Serum albumin as a simple gold nanoparticle transporter was utilized to augment laser thermal extirpation of HepG2 cells, and therapeutic effects were demonstrated. To reveal the discriminatory internalization of serum albumin gold nanoparticles into HepG2 cells by focusing the Gp60 receptors, darkfield microscopy in addition to immunochemical staining was employed. Their outcomes explained that serum albumin gold nanoparticles resulted in an intracellular uptake rise in liver cancer cells by targeting Gp60 receptors selectively, and it was established that after laser irradiation, gold nanoparticle photo-excitation caused apoptosis by activating caspase-3 (145).

Fibronectin plays a vital role in the extracellular matrix structure and functioning of the normal cells, though, in circumstances like lung carcinoma, its manifestation augments, particularly in NSCLCs. In the present study, researchers linked gold nanoparticles to the human fibronectin antibody (anti-hFN) to form a colorimetric nano biosensor for detecting fibronectin present in the extracellular matrix of cultured cells. For comparison of alterations in color caused by aggregation of gold nanoparticles because of an elevated quantity of fibronectin, three different cell lines, specifically A549 (target cells), AGO-1522 (control cells), and Nalm-6 (negative control cells) were utilized. Their construct was capable of sensing an augmented level of fibronectin, which was identifiable visually by alteration in color and might be established by spectrophotometer, as well (146).

Research in 2014 confirmed that gold nanoparticles fabricated with *Cajanus cajan* phytochemical [3-butoxy-2-hydroxypropyl 2-(2,4-dihydroxyphenyl) acetate] have an ability to provoke apoptosis in liver cancer HepG2 cells (147).

In one more study, a hybrid system made up of gold nanoparticles and the liposomes was exploited to assess the efficacy of paclitaxel in liver cancer treatment. Achieve this goal, two drug delivery systems were analyzed, one by the hybrid system and the other by gold nanoparticles devoid of liposomes. The outcomes of this research confirmed the effectiveness of the hybrid system to be superior to the gold nanoparticles without liposomes in three characteristics of stability, solubility, as well as targeting of liver cells (148). As it was stated, surface modification is imperative to transform a few nanomaterials' distinctiveness. The glycol chitosan layered gold nanoparticles demonstrated a tumor-targeting computed tomography contrast agent in the malignant liver cancer model (149).

The epidermal growth factor receptor (EGFR) is a particular hot topic in cancer therapy, and it can be exclusively targeted with the monoclonal antibody, cetuximab. Cetuximab obstructs signal transduction by coupling to the external arena of EGFR, blocking ligand binding (150–152). In addition to colorectal cancer and head and neck cancer, NSCLC is the third main cancer class for which cetuximab has been assessed. Cetuximab was exploited for the management of EGFR-expressing NSCLC in phase II and III trials, principally in combination with radiotherapy or chemotherapy. Nevertheless, the curative effect of cetuximab in EGFR high-expressing NSCLC is still unsatisfactory. In the present research, investigators stated that the linkage of cetuximab with gold nanoparticles augments the cytotoxicity of cetuximab in NSCLC both in vitro and in vivo. The NSCLC cell lines A549 ($EGFR^{high}$) and H1299 ($EGFR^{low}$) were utilized to explore diverse responses to cetuximab, IgG–gold nanoparticles, and cetuximab–gold nanoparticles. The anti-tumor properties of cetuximab–gold nanoparticles were investigated in vivo by setting up a tumor xenograft model in nude mice. In general, the therapeutic effect of cetuximab–gold nanoparticles was greater in $EGFR^{high}$ A549 cells in comparison to $EGFR^{low}$ H1299 cells. The cytotoxic effect of cetuximab–gold nanoparticles in A549 cells was raised dose-dependently. Cetuximab–gold nanoparticles notably restrained A549 cell propagation and relocation capability, and sped up apoptosis relative to cetuximab, and this effect was most likely attributable to superior EGFR endocytosis and the successive repression of the downstream signaling pathway. Conclusively in the tumor xenograft of nude mice, treatment with cetuximab–gold nanoparticles also resulted in a noteworthy diminution in tumor weight and volume with little toxicity. Their results indicate that a cetuximab–gold nanoparticles conjugate has a hopeful prospect for targeted therapy of EGFR positive NSCLC patients (153).

In recent times, Barash and colleagues recommended a nanodevice rooted in gold nanoparticle sensors that categorize the lung cancer histology by identifying the lung cancer–specific prototypes of volatile organic compound profiles. It aims to distinguish between healthy and lung cancer cells, SCLCs and NSCLCs, and subtypes of NSCLS (154).

There are three common tumor markers of liver cancer, namely alpha-fetoprotein, alpha-fetoprotein variants, and abnormal prothrombin, recognized with three electrochemical redox species with individual voltammetric peaks. In the present research, the electrochemical signals were concurrently attained at different peak potentials, and gold nanoparticle–coated carbon nanotubes were exploited to get better signal response (155). Microorganism-assisted fabricated gold nanoparticles were utilized to explore liver cancer cells by combining them with liver cancer cell surface-specific antibodies. An investigation confirmed that the antibody-conjugated gold nanoparticles attached explicitly to the surface antigens of the cancer cells could productively distinguish normal cell populations from cancerous cells (156).

Arvizo and colleagues coupled the dendrimer-entrapped gold nanoparticles to anti-EGFR antibodies and particularly targeted overexpressed EGFR in NSCLC-type lung cancer cells and facilitated early lung cancer detection (157).

An electrochemical-based immune sensor method has been established to quantitatively test human lung cancer-linked antigens utilizing an (alpha-enolase) ENO1 antibody combined with gold nanoparticles for lung cancer identification (158). In the same way, based on electrochemical and contact angle measurements, a highly susceptible and fast technique has been established for the identification of different cancer cells, including lung and liver cancer (159). Furthermore, Medley and his research team employed the gold nanoparticle–conjugated aptamer for the calorimetric assay for the direct detection of lung cancer cells (160).

The majority of the traditional diagnostic strategies for lung cancer are costly and imprecise. Therefore a novel method has been established for the detection of lung cancer from an exhaled breath sample utilizing an array of sensors anchored in gold nanoparticles. The constitution of volatile organic compounds in exhaled breath is diverse in healthy human beings in comparison to lung cancer patients. Around 42 volatile organic compounds have been discovered, which are employed as lung cancer biomarkers (161). In a similar manner, hollow gold nanospheres have been exploited to fabricate a very susceptible and fast immunoassay system for lung cancer identification which is 100–1000 times more responsive than enzyme-linked immunosorbent assay having a limit of detection 1–10 pg/mL. This surface-enhanced Raman scattering-based immunoassay system makes use of the hollow gold nanospheres for the immunoanalysis of lung cancer markers and carcinoembryonic antigens, whereas magnetic beads are utilized as an immunocomplex-supporting substrate (162). In addition, gold nanoparticles in combination with methotrexate, an analog of folic acid, also generated a cytotoxic effect in Lewis lung carcinoma (163).

Finally, gold nanoparticles delivery systems have shown hopeful outcomes in lung and liver cancer therapy because of their unique features. Together with a growing understanding based on the characteristics and effects of gold nanoparticles, they are at present potential tools for lung and liver cancer therapy.

33.8 Conclusion

Attributable to the swift expansion of nanotechnology over the last several decades, a wide range of particles of different shapes, sizes, and structures are currently available to investigators. Because of distinctive physicochemical and biological properties, gold nanoparticles are of particularly interest in medical appliances, especially for anti-cancer treatment. A great variety of promising production techniques allow for getting gold nanoparticles with definite architecture and characteristics, according to the intended use. Furthermore, their reactivity makes possible further functionalization and modification, additionally enhancing bioavailability and expanding the scope of medical appliances of gold nanoparticles. Plentiful and varied physicochemical properties of gold nanoparticles highlighted in this chapter, differentiating them from other nanoparticles, give us hope for their use, particularly with regard to drug delivery devices and imaging techniques. Gold nanoparticles are employed as sensitive probes in the detection as well as imaging of tumors for diagnostic purposes, delivery agents for the precise targeting of chemotherapeutic medicines to cancer cells, and enhancing agents in plasmonic photothermal therapy and radiation therapy for the eradication of lung and liver cancer cells. Functionalized gold nanoparticles with a variety of biomolecules for example amino acids, carboxylic acids, and proteins, have been employed in cancer treatment and present an outstanding drug delivery system. In the case of lung and liver cancer, there are abundant studies into the utilization of gold nanoparticles. Gold nanoparticle delivery systems have shown hopeful outcomes in lung and liver cancer therapy because of their high surface loading ability of drugs. This inspires hope for the progress of innovative cancer treatment methods, providing an excellent substitute for the most frequently used chemotherapeutics.

REFERENCES

1. National Cancer Institute. https://www.cancer.gov/. Accessed August 21, 2020.
2. Wu S, Zhu W, Thompson P, and Hannun YA. Evaluating intrinsic and non-intrinsic cancer risk factors. *Nature Communications*. 2018; 9(1): 3490.
3. Anand P, Kunnumakara AB, Sundaram C, et al. Cancer is a preventable disease that requires major lifestyle changes. *Pharmaceutical Research*. 2008; 25(9): 2097–2116.
4. Okuda K, Ohtsuki T, Obata H, et al. Natural history of hepatocellular carcinoma and prognosis in relation to treatment study of 850 patients. *Cancer*. 1985; 56: 918–928.
5. Jemal A, Bray F, Center MM, Ferlay J, Ward E, and Forman D. Global cancer statistics. *CA Cancer Journal for Clinicians*. 2011; 61(2):69–90.
6. Siegel RL, Miller KD, and Jemal A. Cancer statistics, 2016. *CA Cancer Journal for Clinicians*. 2016; 66(1): 7–30.
7. Teh E and Belcher E. Lung cancer: Diagnosis, staging, and treatment. *Surgery (Oxford)*. 2014; 32: 242–248.
8. Lemjabbar-Alaoui H, Hassan OU, Yang YW, and Buchanan P. Lung cancer: Biology and treatment options. *Biochimica et Biophysica Acta*. 2015; 1856: 189–210.
9. Siegel RL, Miller KD, Fuchs HE, and Jemal A. Cancer Statistics, 2021. *CA Cancer Journal for Clinicians*. 2021; 71: 7–33.
10. Jemal A, Siegel R, Ward E, Murray T, Xu J, and Thun MJ. Cancer statistics, 2007. *CA Cancer Journal for Clinicians*. 2007; 57(1): 43–66.
11. Ramalingam SS, Owonikoko TK, and Khuri FR. Lung cancer: New biological insights and recent therapeutic advances. *CA Cancer Journal for Clinicians*. 2011; 61(2): 91–112.
12. Abbott A. Cancer research: On the offensive. *Nature*. 2002; 416(6880): 470–474.
13. Malvezzi M, Bertuccio P, Levi F, La Vecchia C, and Negri E. European cancer mortality predictions for the year 2013. *Annals of Oncology*. 2013; 24(3): 792–800.
14. Soren MB, Paul MH, Wolfgang AT, and Minesh PM. Radiation oncology advances: An introduction. *Cancer Treatment & Research*. 2008; 139: 1–4.
15. Yano T, Okamoto T, Fukuyama S, and Maehara Y. Therapeutic strategy for postoperative recurrence in patients with non-small cell lung cancer. *World Journal of Clinical Oncology*. 2014; 5(5): 1048–1054.
16. Sanchez-Santos ME. Therapeutic applications of ionizing radiations. In: Gomez-Tejedor GG and Fuss MC. (eds.) Radiation Damage in Biomolecular Systems. Springer Sciences, Dordrecht, Netherlands. 2012; 397–409.
17. Wagstaff KM and Jans DA. Nuclear drug delivery to target tumour cells. *European Journal of Pharmacology*. 2009; 625:174–180.
18. Boisselier E and Astruc D. Gold nanoparticles in nanomedicine: Preparations, imaging, diagnostics, therapies and toxicity. *Chemical Society Reviews*. 2009; 38: 1759–1782.
19. Ma X, Hui H, Jin Y, et al. Enhanced immunotherapy of SM5-1 in hepatocellular carcinoma by conjugating with gold nanoparticles and its in-vivo bioluminescence tomographic evaluation. *Biomaterials*. 2016; 87: 46–56.
20. Liu Y, Miyoshi H, and Nakamura M. Nanomedicine for drug delivery and imaging: A promising avenue for cancer therapy and diagnosis using targeted functional nanoparticles. *International Journal of Cancer*. 2007; 120: 2527–2537.
21. Lanone S and Boczkowski J. Biomedical applications and potential health risks of nanomaterials: Molecular mechanisms. *Current Molecular Medicine*. 2006; 6: 651–663.
22. Conde J, Doria G, and Baptista PV. Noble metal nanoparticles applications in cancer. *Journal of Drug Delivery*. 2012; 2012(1): 751075.
23. Madni A, Batool A, Noreen S, et al. Novel nanoparticulate systems for lung cancer therapy: An updated review. *Journal of Drug Targeting*. 2017; 25(6): 499–512.
24. Saravanan M, Asmalash T, Gebrekidan A, et al. Nanomedicine as a newly emerging approach to combat human immunodeficiency virus (HIV). *Pharmaceutical Nanotechnology*. 2018; 6: 17–27.
25. Bolukbas DA and Meiners S. Lung cancer nanomedicine: Potentials and pitfalls. *Nanomedicine (London)*. 2015; 10(21): 3203–3212.
26. Cryer AM and Thorley AJ. Nanotechnology in the diagnosis and treatment of lung cancer. *Pharmacology & Therapeutics*. 2019; 198: 189–205.

27. Mottaghitalab F, Farokhi M, Fatahi Y, Atyabi F, and Dinarvand R. New insights into designing hybrid nanoparticles for lung cancer: Diagnosis and treatment. *Journal of Controlled Release*. 2019; 295: 250–267.
28. Ferrari M. Cancer nanotechnology: Opportunities and challenges. *Nature Reviews Cancer*. 2005; 5(3): 161–171.
29. Maddah B. A simple colorimetric kit for determination of ketamine hydrochloride in water samples. *Analytical Methods*. 2015; 7(24): 10364–10370.
30. Arvizo RR, Bhattacharyya S, Kudgus RA, Giri K, Bhattacharya R, and Mukherjee P. Intrinsic therapeutic applications of noble metal nanoparticles: Past, present and future. *Chemical Society Reviews*. 2012; 41(7): 2943–2970.
31. Dykman LA and Khlebtsov NG. Gold nanoparticles in biology and medicine: Recent advances and prospects. *Acta Naturae*. 2011; 3(2): 34–55.
32. Sztandera K, Gorzkiewicz M, and Klajnert-Maculewicz B. Gold nanoparticles in cancer treatment. *Molecular Pharmaceutics*. 2019; 16: 1–23.
33. Mie G. Contributions to the optics of turbid media, particularly of colloidal metal solutions. *Annals of Physics*. 1908; 330(3): 377–445.
34. Faulk W and Taylor G. An immunocolloid method for the electron microscope. *Immunochemistry*. 1971; 8(11): 1081–1083.
35. Boisseau P and Loubaton B. Nanomedicine, nanotechnology in medicine. *Comptes Rendus Physique*. 2011; 12(7):620–636.
36. Magnusson M, Deppert K, Malm J, Bovin J, and Samuelson L. Size-selected gold nanoparticles by aerosol technology. *Nanostructured Materials*. 1999; 12(1–4): 45–48.
37. Mafune F, Kohno JY, Takeda Y, and Kondow T. Full physical preparation of size-selected gold nanoparticles in solution: Laser ablation and laser-induced size control. *Journal of Physical Chemistry B*. 2002; 106(31): 7575–7577.
38. Carnovale C, Bryant G, Shukla R, and Bansal V. Size, shape and surface chemistry of nano-gold dictate its cellular interactions, uptake and toxicity. *Progress in Material Sciences*. 2016; 83: 152–190.
39. Deb S, Patra HK, Lahiri P, Dasgupta AK, Chakrabarti K, and Chaudhuri U. Multistability in platelets and their response to gold nanoparticles. *Nanomedicine*. 2011; 7(4): 376–384.
40. Xia Y, Xiong Y, Lim B, and Skrabalak SE. Shape-controlled synthesis of metal nanocrystals: Simple chemistry meets complex physics? *Angewandte Chemie International Edition in English*. 2009; 48(1): 60–103.
41. Niikura K, Matsunaga T, Suzuki T, et al. Gold nanoparticles as a vaccine platform: Influence of size and shape on immunological responses in-vitro and in-vivo. *ACS Nano*. 2013; 7(5): 3926–3938.
42. Chen H, Kou X, Yang Z, Ni W, and Wang J. Shape- and size dependent refractive index sensitivity of gold nanoparticles. *Langmuir*. 2008; 24(10): 5233–5237.
43. Xie X, Liao J, Shao X, Li Q, and Lin Y. The effect of shape on cellular uptake of gold nanoparticles in the forms of stars, rods, and triangles. *Scientific Reports*. 2017; 7(1): 3827.
44. Jain PK, Lee KS, El-Sayed IH, and El-Sayed MA. Calculated absorption and scattering properties of gold nanoparticles of different size, shape, and composition: Applications in biological imaging and biomedicine. *Journal of Physical Chemistry B*. 2006; 110(14): 7238–7248.
45. Connor EE, Mwamuka J, Gole A, Murphy CJ, and Wyatt MD. Gold nanoparticles are taken up by human cells but do not cause acute cytotoxicity. *Small*. 2005; 1(3): 325–327.
46. Afrooz ARMN, Sivalapalan ST, Murphy CJ, Hussain SM, Schlager JJ, and Saleh NB. Spheres vs. rods: The shape of gold nanoparticles influences aggregation and deposition behavior. *Chemosphere*. 2013; 91(1): 93–98.
47. De Jong WH, Hagens WI, Krystek P, Burger MC, Sips AJAM, and Geertsma RE. Particle size-dependent organ distribution of gold nanoparticles after intravenous administration. *Biomaterials*. 2008; 29(12):1912–1919.
48. Song K, Xu P, Meng Y, et al. Smart gold nanoparticles enhance killing effect on cancer cells. *International Journal of Oncology*. 2013; 42(2): 597–608.
49. Lan M-Y, Hsu Y-B, Hsu C-H, Ho C-Y, Lin J-C, and Lee S-W. Induction of apoptosis by high-dose gold nanoparticles in nasopharyngeal carcinoma cells. *Auris Nasus Larynx*. 2013; 40(6):563–568.
50. Liang A, Liu Q, Wen G, and Jiang Z. The surface-plasmon resonance effect of nanogold/silver and its analytical applications. *TrAC, Trends in Analytical Chemistry*. 2012; 37: 32–47.
51. Toderas F, Baia M, Maniu D, and Astilean S. Tuning the plasmon resonances of gold nanoparticles by controlling their size and shape. *Journal of Optoelectro nics and Advanced Materials*. 2008; 10(9): 2282–2284.
52. Link S and El-Sayed MA. Optical properties and ultrafast dynamics of metallic nanocrystals. *Annual Review of Physical Chemistry*. 2003; 54(1): 331–366.
53. Dulkeith E, Ringler M, Klar TA, Feldmann J, Javier AM, and Parak WJ. Gold nanoparticles quench fluorescence by phase induced radiative rate suppression. *Nano Letters*. 2005; 5(4): 585–589.
54. Anger P, Bharadwaj P, and Novotny L. Enhancement and quenching of single-molecule fluorescence. *Physical Review Letters*. 2006; 96(11): 113002.
55. Sapsford KE, Berti L, and Medintz IL. Materials for fluorescence resonance energy transfer analysis: Beyond traditional donor-acceptor combinations. *Angewandte Chemie International Edition in England*. 2006; 45(28): 4562–4588.
56. Xue C, Kung CC, Gao M, et al. Facile fabrication of 3D layer-by-layer graphene-gold nanorod hybrid architecture for hydrogen peroxide based electrochemical biosensor. *Sensing and Bio-Sensing Research*. 2015; 3(3): 7–11.
57. Same S, Aghanejad A, Nakhjavani SA, Barar J, and Omidi Y. Radiolabeled theranostics: Magnetic and gold nanoparticles. *BioImpacts*. 2016; 6(3): 169–181.
58. El-Sayed MA. Some interesting properties of metals confined in time and nanometer space of different shapes. *Accounts of Chemical Research*. 2001; 34(4): 257–264.
59. Abadeer NS and Murphy CJ. Recent progress in cancer thermal therapy using gold nanoparticles. *Journal of Physical Chemistry C*. 2016; 120(9):4691–4716.
60. Dreifuss T, Barnoy E, Motiei M, and Popovtzer R. Theranostic gold nanoparticles for CT imaging. In: Bulte J and Modo M. (eds.) Design and Applications of Nanoparticles in Biomedical Imaging. Springer International Publishing, Cham. 2016; 403–427.

61. Giljohann DA, Seferos DS, Daniel WL, Massich MD, Patel PC, and Mirkin CA. Gold nanoparticles for biology and medicine. *Angewandte Chemie*. 2010; 49: 3280–3294.
62. Chithrani DB, Dunne M, Stewart J, Allen C, and Jaffray DA. Cellular uptake and transport of gold nanoparticles incorporated in a liposomal carrier. *Nanomedicine*. 2010; 6(1): 161–169.
63. Lewinski N, Colvin V, and Drezek R. Cytotoxicity of nanoparticles. *Small*. 2008; 4: 26–49.
64. Jain KK. The role of nanobiotechnology in drug discovery. *Advances in Experimental Medicine and Biology*. 2009; 655: 37–43.
65. DeLong RK, Reynolds CM, Malcolm Y, Schaeffer A, Severs T, and Wanekaya A. Functionalized gold nanoparticles for the binding, stabilization, and delivery of therapeutic DNA, RNA, and other biological macromolecules. *Nanotechnol ogy, Science and Applications*. 2010; 2010: 53–63.
66. Kim C, Ghosh P, and Rotello V. Multimodal drug delivery using gold nanoparticles. *Nanoscale*. 2009; 1(1): 61–67.
67. Cobley CM, Chen J, Cho EC, Wang LV, and Xia Y. Gold nanostructures: A class of multifunctional materials for biomedical applications. *Chemical Society Reviews*. 2011; 40: 44–56.
68. Owens DE and Peppas NA. Opsonization, biodistribution, and pharmacokinetics of polymeric nanoparticles. *International Journal of Pharmaceutics*. 2006; 307(1): 93–102.
69. Barar J and Omidi Y. Surface modified multifunctional nanomedicines for simultaneous imaging and therapy of cancer. *Bioimpacts*. 2014; 4(1): 3–14.
70. Kobayashi H, Kawamoto S, Saga T, et al. Positive effects of polyethylene glycol conjugation to generation-4 polyamidoamine dendrimers as macromolecular MR contrast agents. *Magnetic Resonance in Medicine*. 2001; 46(4): 781–788.
71. Mishra P, Nayak B, and Dey RK. PEGylation in anticancer therapy: An overview. *Asian Journal of Pharmaceutical Sciences*. 2016; 11(3): 337–348.
72. Peng CA and Hsu YC. Fluoroalkylated polyethylene glycol as potential surfactant for perfluorocarbon emulsion. *Artifical Cells, Blood Substitutes and Immobilization Biotechnology*. 2001; 29(6): 483–492.
73. Zhang X, Wang H, Ma Z, and Wu B. Effects of pharmaceutical pegylation on drug metabolism and its clinical concerns. *Expert Opinion on Drug Metabolism & Toxicology*. 2014; 10(12): 1691–1702.
74. Khan JA, Pillai B, Das TK, Singh Y, and Maiti S. Molecular effects of uptake of gold nanoparticles in HeLa cells. *ChemBioChem*. 2007; 8(11): 1237–1240.
75. Kim D, Park S, Lee JH, Jeong YY, and Jon S. Antibiofouling polymer-coated gold nanoparticles as a contrast agent for in-vivo x-ray computed tomography imaging. *Journal of the American Chemical Society*. 2007; 129(24): 7661–7665.
76. Zhang G, Yang Z, Lu W, et al. Influence of anchoring ligands and particle size on the colloidal stability and in-vivo biodistribution of polyethylene glycol-coated gold nanoparticles in tumor-xenografted mice. *Biomaterials*. 2009; 30(10): 1928–1936.
77. Li G, Li D, Zhang L, Zhai J, and Wang E. One-step synthesis of folic acid protected gold nanoparticles and their receptor mediated intracellular uptake. *Chemistry*. 2009; 15(38): 9868–9873.
78. Zhou M, Wang B, Rozynek Z, et al. Minute synthesis of extremely stable gold nanoparticles. *Nanotechnology*. 2009; 20(50): 505606.
79. Salmaso S, Caliceti P, Amendola V, et al. Cell up-take control of gold nanoparticles functionalized with a thermoresponsive polymer. *Journal of Materials Chemistry*. 2009; 19(11): 1608–1615.
80. Ziemba B, Halets I, Shcharbin D, et al. Influence of fourth generation poly(propyleneimine) dendrimers on blood cells. *Journal of Biomedical Materials Research Part A*. 2012; 100A(11): 2870–2880.
81. Goodman CM, McCusker CD, Yilmaz T, and Rotello VM. Toxicity of gold nanoparticles functionalized with cationic and anionic side chains. *Bioconjugate Chemistry*. 2004; 15(4): 897–900.
82. Li J-L, Wang L, Liu X-Y, et al. In-vitro cancer cell imaging and therapy using transferrin-conjugated gold nanoparticles. *Cancer Letters*. 2009; 274(2): 319–326.
83. Zhao J, Babiuch K, Lu H, Dag A, Gottschaldt M, and Stenzel MH. Fructose-coated nanoparticles: A promising drug nanocarrier for triple-negative breast cancer therapy. *Chemical Communications*. 2014; 50(100): 15928–15931.
84. Porcaro F, Battocchio C, Antoccia A, et al. Synthesis of functionalized gold nanoparticles capped with 3-mercapto-1-propansulfonate and 1-thioglucose mixed thiols and "in-vitro" bioresponse. *Colloids and Surfaces B*. 2016; 142: 408–416.
85. Patel PC, Giljohann DA, Daniel WL, Zheng D, Prigodich AE, and Mirkin CA. Scavenger receptors mediate cellular uptake of polyvalent oligonucleotide-functionalized gold nanoparticles. *Bioconjugate Chemistry*. 2010; 21(12): 2250–2256.
86. Huang X, El-Sayed IH, Qian W, and El-Sayed MA. Cancer cell imaging and photothermal therapy in the near-infrared region by using gold nanorods. *Journal of the American Chemical Society*. 2006; 128(6): 2115–2120.
87. Melancon MP, Lu W, Yang Z, et al. In-vitro and in-vivo targeting of hollow gold nanoshells directed at epidermal growth factor receptor for photothermal ablation therapy. *Molecular Cancer Therapeutics*. 2008; 7(6): 1730–1739.
88. Huang X, Peng X, Wang Y, et al. Reexamination of active and passive tumor targeting by using rod-shaped gold nanocrystals and covalently conjugated peptide ligands. *ACS Nano*. 2010; 4(10): 5887–5896.
89. Bergen JM, Von Recum HA, Goodman TT, Massey AP, and Pun SH. Gold nanoparticles as a versatile platform for optimizing physicochemical parameters for targeted drug delivery. *Macromolecular Bioscience*. 2006; 6(7): 506–516.
90. Paciotti GF, Myer L, Weinreich D, et al. Colloidal gold: A novel nanoparticle vector for tumor directed drug delivery. *Drug Delivery*. 2004; 11(3): 169–183.
91. Goel R, Shah N, Visaria R, Paciotti GF, and Bischof JC. Biodistribution of TNF-α-coated gold nanoparticles in an in-vivo model system. *Nanomedicine*. 2009; 4(4):401–410.

92. Fent GM, Casteel SW, Kim DY, et al. Biodistribution of maltose and gum arabic hybrid gold nanoparticles after intravenous injection in juvenile swine. *Nanomedicine*. 2009; 5(2):128–135.
93. Rezende TS, Andrade GRS, Barreto LS, Costa Jr NB, Gimenez IF, and Almeida LE. Facile preparation of catalytically active gold nanoparticles on a thiolated chitosan. *Materials Letters*. 2010; 64: 882–884.
94. Chen K-S, Hung T-S, Wu H-M, Wu J-Y, Lin M-T, and Feng CK. Preparation of thermosensitive gold nanoparticles by plasma pretreatment and UV grafted polymerization. *Thin Solid Films*. 2010; 518(24): 7557–7562.
95. Lee K, Lee H, Bae KH, and Park TG. Heparin immobilized gold nanoparticles for targeted detection and apoptotic death of metastatic cancer cells. *Biomaterials*. 2010; 31:6530–6536.
96. Pal R, Panigrahi S, Bhattacharyya D, and Chakraborti AS. Characterization of citrate capped gold nanoparticle quercetin complex: Experimental and quantum chemical approach. *Journal of Molecular Structure*. 2013; 1046: 153–163.
97. Tarnawski R and Ulbricht M. Amphiphilic gold nanoparticles: Synthesis, characterization and adsorption to PEGylated polymer surfaces. *Colloids and Surfaces A: Physicochemical and Engineering Aspects*. 2011; 374: 13–21.
98. Nalawade P, Mukherjee T, and Kapoor S. High-yield synthesis of multispiked gold nanoparticles: Characterization and catalytic reactions. *Colloids and Surfaces A: Physicochemical and Engineering Aspects*. 2012; 396: 336–340.
99. Benkovicova M, Vegso K, Siffalovic P, Jergel M, Luby S, and Majkova E. Preparation of gold nanoparticles for plasmonic applications. *Thin Solid Films*. 2013; 543: 138–141.
100. Yu L, Qiu J-J, Cheng H, and Luo Z-H. Facile preparation of gold nanoparticles using the self-assembled ABC non-amphiphilic fluorosilicone triblock copolymer template. *Materials Chemistry and Physics*. 2013; 138: 780–786.
101. Guo Q, Guo Q, Yuan J, and Zeng J. Biosynthesis of gold nanoparticles using a kind of flavonol: Dihydromyricetin. *Colloids and Surfaces A: Physicochemical and Engineering Aspects*. 2014; 441: 127–132.
102. Hartono, D, Hody, Yang KL, and Yung LY. The effect of cholesterol on protein-coated gold nanoparticle binding to liquid crystal-supported models of cell membranes. *Biomaterials*. 2010; 31(11): 3008–3015.
103. Mishra, A, Tripathy, SK, and Yun S-I. Fungus mediated synthesis of gold nanoparticles and their conjugation with genomic DNA isolated from Escherichia coli and Staphylococcus aureus. *Process Biochemistry*. 2012; 47: 701–711.
104. Alanazi FK, Radwan AA, and Alsarra IA. Biopharmaceutical applications of nanogold. *Saudi Pharmaceutical Journal*. 2010; 18: 179–193.
105. Di Guglielmo C, Lopez DR, De Lapuente J, Mallafre JM, and Suarez MB. Embryotoxicity of cobalt ferrite and gold nanoparticles: A first in-vitro approach. *Reproductive Toxicology*. 2010; 30(2): 271–276.
106. Tedesco S, Doyle H, Blasco J, Redmond G, and Sheehan D. Oxidative stress and toxicity of gold nanoparticles in Mytilus edulis. *Aquatic Toxicology*. 2010; 100: 178–186.
107. Kojima C, Umeda Y, Harada A, and Kono K. Preparation of near-infrared light absorbing gold nanoparticles using polyethylene glycol-attached dendrimers. *Colloids and Surfaces B, Biointerfaces*. 2010; 81: 648–651.
108. Gao W, Xu K, Ji L, and Tang B. Effect of gold nanoparticles on glutathione depletion-induced hydrogen peroxide generation and apoptosis in HL7702 cells. *Toxicology Letters*. 2011; 205: 86–95.
109. Daraio C and Jin S. Synthesis and patterning methods for nanostructures useful for biological applications. In: Silva GA and Parpura V. (eds.) Nanotechnology for Biology and Medicine: At the Building Block Level. Fundamental Biomedical Technologies. Springer, New York, NY. 2012, 27–44.
110. Turkevich J, Stevenson P, and Hillier J. A study of the nucleation and growth processes in the synthesis of colloidal gold. *Discussions of the Faraday Society*. 1951; 11: 55–75.
111. Leff DV, Brandt L, and Heath JR. Synthesis and characterization of hydrophobic, organically-soluble gold nanocrystals functionalized with primary amines. *Langmuir*. 1996; 12(20): 4723–4730.
112. Martin MN, Basham JI, Chando P, and Eah SK. Charged gold nanoparticles in non-polar solvents: 10-min synthesis and 2D self-assembly. *Langmuir*. 2010; 26(10): 7410–7417.
113. Brust M, Walker M, Bethell D, Schiffrin DJ, and Whyman R. Synthesis of thiol-derivatised gold nanoparticles in a two-phase liquid-liquid system. *Journal of the Chemical Society Chemical Communication*. 1994; 7: 801–802.
114. Perrault SD and Chan WCW. Synthesis and surface modification of highly monodispersed, spherical gold nanoparticles of 50-200 nm. *Journal of the American Chemical Society*. 2009; 131(47): 17042–17043.
115. Shao Y, Jin Y, and Dong S. Synthesis of gold nanoplates by aspartate reduction of gold chloride. *Chemical Communications*. 2004; 9: 1104–1105.
116. Marangoni V, Cancino-Bernardi J, and Zucolotto V. Synthesis, physico-chemical properties, and biomedical applications of gold nanorods-A review. *J ournal of Biomedical Nanotechnology*. 2016; 12: 1136–1158.
117. Pang B, Yang X, and Xia Y. Putting gold nanocages to work for optical imaging, controlled release and cancer theranostics. *Nanomedicine*. 2016; 11(13): 1715–1728.
118. Bridges CR, DiCarmine PM, Fokina A, Huesmann D, and Seferos DS. Synthesis of gold nanotubes with variable wall thicknesses. *Journal of Mater ial Chemistry A*. 2013; 1(4): 1127–1133.
119. Xu Z-C, Shen C-M, Xiao C-W, et al. Wet chemical synthesis of gold nanoparticles using silver seeds: A shape control from nanorods to hollow spherical nanoparticles. *Nanotechnology*. 2007; 18(11): 115608.
120. Kundu S, Peng L, and Liang HA. New route to obtain highyield multiple-shaped gold nanoparticles in aqueous solution using microwave irradiation. *Inorganic Chemistry*. 2008; 47(14): 6344–6352.
121. Mafune F, Kohno J, Takeda Y, Kondow T, and Sawabe H. Formation of gold nanoparticles by laser ablation in aqueous solution of surfactant. *Journal of Physical Chemistry B*. 2001; 105(22): 5114–5120.

122. Lee JH, Choi SUS, Jang SP, and Lee SY. Production of aqueous spherical gold nanoparticles using conventional ultrasonic bath. *Nanoscale Research Letters*. 2012; 7(1): 420.
123. Fleming DA and Williams ME. Size-controlled synthesis of gold nanoparticles via high-temperature reduction. *Langmuir*. 2004; 20(8): 3021–3023.
124. Ma H, Yin B, Wang S, et al. Synthesis of silver and gold nanoparticles by a novel electrochemical method. *ChemPhysChem*. 2004; 5(1): 68–75.
125. Iravani S. Green synthesis of metal nanoparticles using plants. *Green Chemistry*. 2011; 13: 2638–2650.
126. Huang H and Yang X. Synthesis of chitosan-stabilized gold nanoparticles in the absence/presence of tripolyphosphate. *Biomacromolecules*. 2004; 5(6): 2340–2346.
127. Sujitha M and Kannan S. Green synthesis of gold nanoparticles using citrus fruits (Citrus limon, Citrus reticulata and Citrus sinensis) aqueous extract and its characterization. *Spectrochimica Acta Part A Molecular and Biomolecular Spectroscopy*. 2013; 102: 15–23.
128. Zheng B, Qian L, Yuan H, et al. Preparation of gold nanoparticles on eggshell membrane and their biosensing application. *Talanta*. 2010; 82(1): 177–183.
129. Sen IK, Maity K, and Islam SS. Green synthesis of gold nanoparticles using a glucan of an edible mushroom and study of catalytic activity. *Carbohydrate Polymers*. 2013; 91: 518–528.
130. Kalishwaralal K, Deepak V, Pandian SRK, and Gurunathan S. Biological synthesis of gold nanocubes from Bacillus licheniformis. *Bioresource Technology*. 2009; 100(21): 5356–5358.
131. Kim D and Jon S. Gold nanoparticles in image-guided cancer therapy. *Inorganica Chimica Acta*. 2012; 393: 154–164.
132. Zheng Y, Zhang J, Zhang R, Luo Z, Wang C, and Shi S. Gold nano particles synthesized from Magnolia officinalis and anticancer activity in A549 lung cancer cells. *Artificial Cells Nanomedicine and Biotechnology*. 2019; 47(1): 3101–3109.
133. Bakur A, Niu Y, Kuang H, and Chen Q. Synthesis of gold nanoparticles derived from mannosylerythritol lipid and evaluation of their bioactivities. *AMB Express*. 2019; 9: 62.
134. Ji Y, Cao Y, and Yong S. Green synthesis of gold nanoparticles using a Cordyceps militaris extract and their antiproliferative effect in liver cancer cells (HepG2). *Artificial Cells Nanomedicine and Biotechnology*. 2019; 47(1): 2737–2745.
135. Latha D, Prabu P, Arulvasu C, Manikandan R, Sampurnam S, and Narayanan V. Enhanced cytotoxic effect on human lung carcinoma cell line (A549) by gold nanoparticles synthesized from Justicia adhatoda leaf extract. *Asian Pacific Journal of Tropical Biomedicine*. 2018; 8(11): 540–547.
136. Ramalingam V, Varunkumar K, Ravikumar V, and Rajaram R. Target delivery of doxorubicin tethered with PVP stabilized gold nanoparticles for effective treatment of lung cancer. *Scientific Reports*. 2018; 8: 3815.
137. Balashanmugam P, Mosachristas K, and Kowsalya E. In-vitro cytotoxicity and antioxidant evaluation of biogenic synthesized gold nanoparticles from Marsilea quadrifolia on lung and ovarian cancer cells. *International Journal of Applied Pharmaceutics*. 2018; 10(5): 153–158.
138. Jiang J, Mao Q, Li H, and Lou J. Apigenin stabilized gold nanoparticles increased radiation therapy efficiency in lung cancer cells. *Internationa Journal of Clinical and Experimental Medicine*. 2017; 10(9): 13298–13305.
139. Guo M, Sun Y, and Zhang X-D. Enhanced radiation therapy of gold nanoparticles in liver cancer. *Applied Sciences*. 2017; 7: 232.
140. Guan T, Shang W, Li H, et al. From detection to resection: Photoacoustic tomography and surgery guidance with indocyanine green loaded gold nanorod@ liposome core–shell nanoparticles in liver cancer. *Bioconjugate Chemistry*. 2017; 28(4): 1221–1228.
141. Mocan L, Matea C, Tabaran FA, et al. Selective ex-vivo photothermal nano-therapy of solid liver tumors mediated by albumin conjugated gold nanoparticles. *Biomaterials*. 2017; 119:33–42.
142. Anselmo AC and Samir M. Nanoparticles in the clinic. *Bioengineeing and Translational Med*. 2016; 1(1): 10–29.
143. Rajeshkumar S. Anticancer activity of eco-friendly gold nanoparticles against lung and liver cancer cells. *Journal of Genetic Engineering and Biotechnology*. 2016; 14:195–202.
144. Gamal-Eldeen AM, Moustafa D, El-Daly SM, et al. Photothermal therapy mediated by gum Arabic-conjugated gold nanoparticles suppresses liver preneoplastic lesions in mice. *Journal of Photochemistry and Photobiology B Biology*. 2016; 163: 47–56.
145. Mocan L, Matea C, Tabaran FA, et al. Photothermal treatment of liver cancer with albumin-conjugated gold nanoparticles initiates Golgi Apparatus-ER dysfunction and caspase-3 apoptotic pathway activation by selective targeting of Gp60 receptor. *International Journal of Nanomedicine*. 2015; 10: 5435–5445.
146. Nekouian R, Khalife NJ, and Salehi Z. Development of gold nanoparticle based colorimetric biosensor for detection of fibronectin in lung cancer cell line. *Advanced Techniques in Biology and Medicine*. 2014; 2(1): 118.
147. Ashokkumar T, Prabhu D, Geetha R, et al. Apoptosis in liver cancer (HepG2) cells induced by functionalized gold nanoparticles. *Colloids and Surfaces B Biointerfaces*. 2014; 123: 549–556.
148. Bao Q-Y, Zhang N, Geng D-D, et al. The enhanced longevity and liver targetability of Paclitaxel by hybrid liposomes encapsulating Paclitaxel-conjugated gold nanoparticles. *International Journal of Pharmaceutics*. 2014; 477(1): 408–415.
149. Sun I-C, Na JH, Jeong SY, et al. Biocompatible glycol chitosan-coated gold nanoparticles for tumor-targeting CT imaging. *Pharmaceutical Research*. 2014; 31(6): 1418–1425.
150. Ryu SH, Lee S-W, Yang YJ, Kim JH, Choi EK, and Ahn SD. Intracytoplasmic epidermal growth factor receptor shows poor response to the cetuximab antitumor effect in irradiated non-small cell lung cancer cell lines. *Lung Cancer*. 2012; 77(3): 482–487.
151. Khan JA, Kudgus RA, Szabolcs A, et al. Designing nanoconjugates to effectively target pancreatic cancer cells in-vitro and in-vivo. *PLoS One*. 2011; 6(6): e20347.
152. Khambata-Ford S, Harbison CT, Hart LL, et al. Analysis of potential predictive markers of cetuximab benefit in BMS099, a phase III study of cetuximab and first-line taxane/carboplatin in advanced non-small-cell lung cancer. *Journal of Clinical Oncology*. 2010; 28: 918–927.

153. Qian Y, Qiu M, Wu Q, et al. Enhanced cytotoxic activity of cetuximab in EGFR-positive lung cancer by conjugating with gold nanoparticles. *Scientific Reports*. 2014; 4:7490.
154. Barash O, Peled N, Tisch U, Bunn Jr PA, Hirsch FR, and Haick H. Classification of lung cancer histology by gold nanoparticle sensors. *Nanomedicine*. 2012; 8: 580–589.
155. Li Y, Zhong Z, Chai Y, et al. Simultaneous electrochemical immunoassay of three liver cancer biomarkers using distinguishable redox probes as signal tags and gold nanoparticles coated carbon nanotubes as signal enhancers. *Chemical Communication*. 2012; 48(4): 537–539.
156. Chauhan A, Zubair S, Tufail S, et al. Fungus-mediated biological synthesis of gold nanoparticles: potential in detection of liver cancer. *International Journal of Nanomedicine*. 2011; 6: 2305–2319.
157. Arvizo R, Bhattacharya R, and Mukherjee P. Gold nanoparticles: Opportunities and challenges in nanomedicine. *Expert Opinion in Drug Delivery*. 2010; 7: 753–763.
158. Ho JA, Chang HC, Shih NY, et al. Diagnostic detection of human lung cancer-associated antigen using a gold nanoparticle-based electrochemical immunosensor. *Analytical Chemistry*. 2010; 82(14): 5944–5950.
159. He F, Shen Q, Jiang H, et al. Rapid identification and high sensitive detection of cancer cells on the gold nanoparticles interface by combined contact angle and electrochemical measurements. *Talanta*. 2009; 77(3):1009–1014.
160. Medley CD, Smith JE, Tang Z, Wu Y, Bamrungsap S, and Tan W. Gold nanoparticle-based colorimetric assay for the direct detection of cancerous cells. *Analytical Chemistry*. 2008; 80(4): 1067–1072.
161. Peng G, Tisch U, Adams O, et al. Diagnosing lung cancer in exhaled breath using gold nanoparticles. *Nature Nanotechnology*. 2009; 4: 669–673.
162. Chon H, Lee S, Son SW, Oh CH, and Choo J. Highly sensitive immunoassay of lung cancer marker carcinoembryonic antigen using surface-enhanced Raman scattering of hollow gold nanospheres. *Analytical Chemistry*. 2009; 81: 3029–3034.
163. Chen Y, Tsai C, Huang P, et al. Methotrexate conjugated to gold nanoparticles inhibits tumor growth in a syngeneic lung tumor model. *Mol ecular Pharmacology*. 2007; 4: 713–722.

34

Sub-chronic Inhalational Toxicity Studies for Gold Nanoparticles

Srijan Mishra and Jigna Shah
Department of Pharmacology, Institute of Pharmacy, Nirma University Ahmedabad, Gujarat, India

34.1 Introduction to Gold Nanoparticles

With the increased advancements in the therapeutic field, we have reached the next level of treatment with the help of nanoparticles. Nanoparticles, as the name itself explains, are the particles which relate to the nano (10^{-9}) size range with particle diameter of 1–100 nm. This small size makes it even more magnificent when it comes to targeting individual cells along with their components.

Gold belongs to the 11th group of transition metals; its Latin name is *aurum* and its symbol is Au. Au is the least corrosive metal and is biologically inert in nature, even though it has not been found completely chemically inactive in mammals. Various techniques, such as UV-irradiation (1), bottom-up synthesis (2), γ-irradiation (3), etc., are employed to formulate the gold nanoparticles. The major production of gold nanoparticles is done by the "Turkevich-Frens and Brust-Schiffrin" method (4). Due to their highly reduced size and therefore increased surface area along with its great stability and adaptable properties, gold nanoparticles are one of the best targeted drug delivery systems (5).

34.2 History of Gold Nanoparticles

In the past, gold nanoparticles were not only used as therapeutic agents but for other purposes also. It is reported that gold nanoparticles were used in the staining of glass and were used to make Lycurgus cups (6). Michael Faraday revealed that the properties of colloidal gold were different from those of bulk gold (7). According to the size of gold particles in the colloidal system, they change color according to the light source and location (8). In ancient times, gold was used for the formulation of medicines used for the treatment of various diseases. It was noticed that gold had immense healing capabilities along with immune-boosting activity and had a positive influence upon various organs (9). Gold is reported to offer its role in the therapeutic field in various civilizations. There is evidence of usage of gold in Chinese civilization in very ancient times (2500 BCE). In 17th century Europe, Nicholas Culpepper mentioned the use of gold in hyperthermia, etc. Also, later in the 19th century, it was discovered that when gold chloride was administered along with sodium chloride, which was designated as "muriate of gold and soda," it was effective in the treatment of syphilis (10). Even in Ayurveda, gold ash (swarnabhasma) is described as being used for the treatment of a wide range of diseases, from iinfertility to asthma and various other diseases including arthritis (11,12).

34.3 Studies on Gold Nanoparticles and Their Therapeutic Uses

As seen earlier, use of gold nanoparticles by various civilizations has been practiced as a treatment for various diseases. In 1935, Jacques Forestier studied gold compounds for the treatment of rheumatoid arthritis and found them to be significantly effective. Later, in 1961, a double blind trial funded by the Empire Rheumatism Council reiterated the observations of Forestier's studies (13). Gold nanoparticles are now approved as effective treatment for rheumatoid arthritis under the name of *chrysotherapy*. This therapy is taken into consideration when nonsteroidal anti-inflammatory drugs are found to be ineffective to provide pain relief (14).

A recent clinical trial has shown effectiveness of gold nanoparticles in the treatment of prostate cancer as an adjunct therapy along with surgery (15). Other uses are atherosclerosis (16), gastrointestinal cancer (17), and other diagnostic purposes (18). Further, gold was found to be effective against the *Helicobacter pylori* responsible for stomach ulcers (19). The anti-fungal activity of gold nanoparticles has also been reported (20,21).

According to their shape and features, different types of gold nanoparticles are being made for therapeutic use in various fields like gold nanostars (22,23), nanocages (24), nanorods (25), nanospheres, and nanoshells (26).

34.4 Introduction to Metal Toxicity

It is a well-known fact that if a poison is taken in a balanced and adequate amount, it can act as medicine, and if a medicine

DOI: 10.1201/9781003046547-34

is taken in an excess amount, it can act as a poison. Various metals are essential in our body at different concentrations. Metals such as calcium, sodium, potassium, iron, etc. are required in high concentrations in our body and are known as major minerals. On the other side, metals such as manganese, cobalt, chromium, iodine, etc. are required in minor concentrations and are known as trace minerals. Gold is also required by the body in trace amounts. The presence of gold was found to be approximately 0.2 mg in an adult human of 70 kg body weight. We are very much familiar with the toxicity of heavy metals like lead, arsenic, mercury, cadmium, chromium, etc. (27). Even a small increase in the concentration of these metals can lead to health hazards. Similarly, metals like gold can also cause toxic effects in humans. It has been noticed that gold, when exposed to the skin for a prolonged time, can cause irritation and allergic reactions (28). Similarly, it can also lead to irritation in the eyes. Moreover, it was found that nearly 10% of patients undergo serious adverse effects which leads to the withdrawal of treatment (29). When talking about gold nanoparticles, besides their therapeutic uses, the toxicological aspect of gold nanoparticles must not be left unexplored as it can cause variable toxic effects depending upon the charge on the surface of nanoparticle and according to its shape and size (30). Inhalational toxicity has also been observed in gold nanoparticles (31). Moreover, the portion of skin to which the gold was exposed for a long time (e.g. wearing gold jewelery for long time in the same place) shows traces of solubilized gold (32) in spite of exhibition of intact stratum corneum.

34.5 Types of Inhalational Toxicity Studies

Inhalational toxicity studies are performed to check toxic effects of drugs which are administered via intranasal pathway, like dry powders or aerosols which are formulated generally for respiratory disorders to increase bioavailability or for rapid onset of action. Inhalational toxicity studies confirm the toxic effect of a formulation that is meant to be administered by the inhalatinal method on the respiratory tract and related organs and systems.

Different countries have adopted different guidelines to perform toxicity studies according to their regulatory requirements. In this chapter, a thorough review of inhalational toxicity studies according to the Organisation for Economic Co-Operation and Development (OECD) guidelines, Environmental Protection Agency (EPA) guidelines, and International Council for Harmonisation of Technical Requirements for Pharmaceuticals for Human Use (ICH) guidelines.

The inhalation toxicity studies mainly involve acute inhalational toxicity studies, chronic inhalational toxicity studies, and sub-chronic inhalational toxicity studies. Acute inhalation toxicity studies (33) are performed for a time duration of 24 hours or less. In acute studies, the test substance is exposed to the organism uninterrupted for the specified time duration (not more than 24 hours) and then evaluated for adverse drug reactions. An acute toxicity study is the initial step, which provides information about health hazards which may happen after exposure to the test substance for a short period of time via inhalation. This acute study provides a base on which further studies are planned. It provides information about designing the dose regimen for further long-term studies. It also provides guidance about the mechanisms of toxicity of the test substance.

After the completion of the acute toxicity study, chronic or sub-acute inhalational toxicity studies are performed to evaluate the test drug for long-term exposure. Generally, these studies are performed for a limited time interval of 14–28 days. In this study, the test substance is exposed to the organism for a given time to characterize the toxic effects of test drug when taken for a long time. This type of study suggests the safety index of test substance and thereby usage of the drug for treatment of disease in which repeated doses are to be administered for a long duration. This study also provides data correlating toxic effects and quantity of the drug to be administered.

Sub-chronic inhalation toxicity studies are performed for the longest duration of time (90 days). The longer-duration study is to evaluate the prolonged exposure to the test substance and its related toxic effects. The workers who deal with the hazardous chemicals on daily basis, the patients with chronic disease conditions who are relying on the drug for a longer period of time, or the drugs which are accumulated over time in the body must be evaluated for their toxic effects for prolonged time in order to prevent any hazard. Sub-chronic toxicity studies also contribute to the assessment of risk of unnoticed events, such as the transport of chemicals. Also, they help in the assessment of risks related to the environment.

34.6 Sub-Chronic Inhalational Toxicity Studies According to Various Guidelines

34.6.1 Sub-Chronic Inhalation Toxicity Guidelines According to OECD

The OECD test guideline 413 (TG 413) is the guideline for sub-chronic inhalation toxicity studies. As suggested by the guidelines, repeated doses of the test substance by inhalation route are administered for a time duration of 90 days and the obtained data is evaluated (34). The updated guideline of 2017 also additionally contains the information relevant to conduction of sub-chronic inhalation toxicity studies for testing nano materials and their features.

Test substances used must be clearly declared and provided with the necessary information to the researcher. The formulation used in the study must be clearly specified before use. Test substance concentration in the formulation must be justified and adjusted. For the determination of the concentrations to be administered, the guideline specifies the range-finding studies to be done along with the main study when sufficient data is not available on the dose. The main study is conducted with the doses obtained from data previously available or from the range-finding study. Ideally, the guidelines restrict the use of vehicle and, if necessary, water must be preferred. Other vehicles used must be justified before conducting the study.

As per the guidelines, 10 male and 10 female rodents should be included in each group for 90 days for the conduct of the experiment. Guidelines require exposure of different concentrations of the test substance to not less than three groups.

Purified air should be provided to a single group (negative control) and one group of animals should be subjected to the vehicle used in formulation (vehicle control). The exposure of the test substance should be done for 5 days of each week, but allowance for exposure of the test substance can also be done for all 7 days of the week. Test substance is exposed for 6 hours each day for 90 days, or it can be exposed for 6 hours for 5 days each week for 90 days. As indicated in the grouping of animals, animals of both genders should be selected, but if an established study suggests the effect of a test substance is different for different genders of animals, or either sex being susceptible to the test substance, then different concentrations for different genders must be calculated.

There are certain observations which are mandatory while conducting the study. Body weights of the animals must be measured three times per week after the exposure to the test substance. Food and water measurements must be done every week throughout the study protocol. Clinical pathology includes a wide variety of tests to be done, including hematological parameters and tests, and clinical biochemistry tests. Urine analysis has been included as an optional parameter. Bronchoalveolar lavage, lung burden, ophthalmological parameters, gross pathology of various organs, and organ weights are also included in the observation parameters. Pulmonary function must be observed throughout the study if any kind of increase in temperature or reflex bradypnea is noted in the range-finding study.

According to the guidelines, pliability should be shown for the groups such as satellite groups (for evaluating reversible toxic effects), interim sacrificial groups, bronchoalveolar lavage (BAL) groups, lung burden (LB) groups (for particles), neurological test groups, and clinical histopathological and pathological evaluations for the better characterization of the study for toxic effects of the test substance.

Guidelines specify the data to be reported while the study is conducted. It includes the basic information for individual animals, such as body weight, food consumption, water intake, BALF analysis, and pathological reports of the organs. The data for the range-finding study and the main study must be submitted. Test reports for test animals and animal husbandry, along with data for the test chemical and vehicle used in it, must be reported. Experimental conditions and exposure data must be reported and all these data must be compiled to form a report.

34.6.2 Sub-Chronic Toxicity Guidelines as per ICH

Sub-chronic toxicity guidelines from the ICH are not specifically for inhalation but are more general. Sub-chronic toxicity is most often studied for 90 days but can be extended up to 12 months. Outcome of the study facilitates the determination of proper doses for further chronic studies (35). Moreover, it was also helpful in determining the no observable effect levels (NOELs) in toxicological studies. Sub-chronic toxicity studoes do not deal with resolution of carcinogenicity of the given drug. The guidelines require that all the studies that are to be performed must follow US FDA Good Laboratory Practices (GLP) in the part 58 title 21 code of federal regulations.

Guidelines listed herein are for the conduction of studies including rats and mice (rodents), and if any changes are made with respect to the species, modifications must be made as per the use of animals. Neither gender, i.e. males and females, used in the study is to have been subjected to any preceding study. Selected animals must be healthy and must belong to well-bred colonies. Young animals must be included in studies and, specifically for rodents, animals older than 6 or 8 weeks must not be included. The number of animals of each gender must be the same. Each group must have 20 animals consisting of both genders. Each group including 10 animals of each genders is used for short term studies like finding the drug dose. If sacrifice of animals is to be done, the number of animals in each group should be increased. Caging suggested for the animals is one animal per cage. Diet provided to the animals should be complete in all nutritional aspects for the animals. Any kind of special diet recommendation must be followed.

The animals must be randomly assigned in control and treatment group. Any other parameter taken into consideration for the basis of randomization, it must be justified. Study on the animals must begin on the same day for all the animals and groups. High rate of mortality due to poor care of animals is not acceptable and may lead to repetition of the study. For example, mortality rate in the control group must not be higher than 10%. Similarly, organ loss due to autolysis is less than 10% in a well-managed animal study. If the limit is exceeded, the study must be repeated. Further, as soon as the animal is sacrificed, it must undergo necropsy promptly tto minimize the tissue loss because of necropsy. If necropsy is not carried out promptly, organs must be stored at a temperature where the lowest autolysis rates are observed.

To maintain the composition and percentage purity of the drug, a single lot of drug must be used or different lots with same amount of purity and composition must be taken into consideration. Chemicals must be identified by authenticated agencies and must be provided with Chemical Abstracts Service (CAS) number(s) (numbers for mixture of drugs). Known contaminants must be included in the list of composition. Storage conditions must be appropriate to sustain the stability of the drug.

Exposure of test drug to animals must be 7 days a week continuously for 90 days. Administration of the drug via any route must be justified and must correlate with that of humans. Administration can be done by admixing the drug in the diet of the animals or by dissolving the test drug in water. Also, the drug can be encapsulated and given by oral gavage. Generally, four to five doses are selected based on toxicity of the test drug. Minimum three group of animals for three different dose levels must be undertaken and in addition to that, one control group must be added.

Observations are to be done on the daily basis for all animals in every group for at least one or two times a day until the study ends. Weight of animals and body weight data must be taken at least once a week. Clinical tests include ophthalmological examinations, hematological parameters like hematocrit, hemoglobin concentration, erythrocyte count, total and differential leukocyte counts, mean corpuscular hemoglobin, mean corpuscular volume, mean corpuscular hemoglobin concentration, and a measure of clotting potential. For hepatocellular evaluation, select at least three of the following five: alanine aminotransferase (SGPT, ALT), aspartate aminotransferase (SGOT,

AST), sorbitol dehydrogenase, glutamate dehydrogenase, total bile acids. Hepatobiliary evaluation: select at least three of the following five: alkaline phosphatase, bilirubin (total), gamma-glutamyl transpeptidase (GG transferase), 5' nucleosidase, total bile acids. Other markers of cell changes or cellular function are albumin, calcium, chloride, cholesterol (total), cholinesterase, creatinine, globulin (calculated), glucose (in fasted animals), phosphorous, potassium, protein (total), sodium, triglycerides (fasting), urea nitrogen. Urine analyses must be done. Neurotoxicity tests must be done to determine any kind of neuronal damage. Immunotoxicity is to be tested.

Gross necropsy is to be done for all animals to be sacrificed under supervision of a qualified pathologist. Organs must be properly weighed and carefully dissected. Fat tissues must be removed properly along with other contagious tissues before weighing. The organs to be weighed are adrenals, aorta, bone (femur), bone marrow (sternum), brain (at least three different levels), cecum, colon, corpus and cervix uteri, duodenum, epididymis, esophagus, eyes, gall bladder (if present), Harderian gland (if present), heart, ileum, jejunum, kidneys, liver, lung (with main-stem bronchi), lymph nodes (one related to route of administration and one from a distant location), mammary glands, nasal turbinates, ovaries and fallopian tubes, pancreas, pituitary, prostate, rectum, salivary gland, sciatic nerve, seminal vesicle (if present), skeletal muscle, skin, spinal cord (three locations: cervical, mid-thoracic, lumbar), spleen, stomach, testes, thymus (or thymic region), thyroid/parathyroid, trachea, urinary bladder, vagina, and all tissues showing abnormality. Lesions taken from gross necropsy must be observed microscopically. Lymphoid organ evaluation is given in detail in the immunotoxicity testing section.

34.7 Reported Gold Nanoparticle Inhalation Toxicity Studies

Pernodet et al. studied the adverse effects of gold nanoparticles on fibroblasts derived from human dermal cells. The study resulted in disappearance of actin stress fibers which are vital for cell viability. This affected many vital functions of the cells and thereby altered and damaged many activities in the cells (36). In 2007, Yu et al. investigated gold nanoparticles' translocation along with the effects after exposure to rats via inhalation route. The nanoparticles were in the size range of 30–110 nm. The study also indicated that gold nanoparticles administered by the inhalation pathway were able to make it to the brain as a significant increase in gold was found in lungs and olfactory region. Exposure of gold nanoparticles for 5 days indicated increase of gold concentration in lungs with no significant increase in other organs. But after 15 days of exposure, a high amount of gold was found to be accumulated in other organs (organs were not taken into consideration that came into direct contact with administration of gold nanoparticles) (37).

Sung et al. conducted a sub-chronic 90-day inhalation toxicity study as per Test Guideline 413 from the OECD to detect the probable harmful effects of gold nanoparticles. It was found that the tissue distribution of gold nanoparticles showed a dose-dependent accumulation of gold in only lungs and kidneys with a gender-related difference in gold nanoparticles content in the kidneys (38).

Schulz et al. carried out a study demonstrating the genotoxicity of various sized gold nanoparticles which were administered into the lungs of rats by the intra-tracheal route. The study revealed that gold nanoparticles show neither systemic nor local side effects except slight inflammation in lungs, thus confirming the non-genotoxic nature of the gold nanoparticles (39). In 2013, Coradeghini et al. presented the size-dependent toxicity and cell interaction mechanisms of gold nanoparticles (AuNPs) on Balb/3T3 mouse fibroblasts. The study revealed that cytotoxicity was seen in gold nanoparticles with 5 nm size at concentrations more than 50 μM. Transmission electron microscopy data indicated that gold nanoparticles with 5–15 nm size were found in the endosomal compartments of Balb/3T3 fibroblasts. The uptake and intracellular distribution of the gold nanoparticles revealed that AuNPs stay stable in the culture media and there was a size-dependent cytotoxic effect of AuNPs. Also, autophagosomes were noted after the exposure of the cells to 5 nm AuNPs. This may be due to reduction in clathrin expression and protein cleavage. Thus, a time-dependent increasing Au uptake was seen (40). In the same year, in one more study, the toxicological profiling of gold nanoparticles displayed the effects of gold nanoparticles on the small epithelial cells present in the respiratory tract. The study concluded that gold nanoparticles have the potential to cause cytotoxicity (41).

Han et al. also carried out a study to examine the effect of size of gold nanoparticles on biodistribution and revealed how size of gold nanoparticles plays a critical role in clearance from lungs. Gold nanoparticles of two sizes — small ones measuring 13 nm in diameter with 12.8 ± 2.42 μg/m^3 concentration, and gold nanoparticles measuring 105 nm diameter with 13.7 ± 1.32 μg/m^3 concentration — were administered to Sprague-Dawley rats for a short period by the inhalational route. The study demonstrated that gold nanoparticles with 13 nm size were observed in various organs with significant concentrations shortly after administration, while gold nanoparticles with 105 nm did not have significant concentrations in major organs and were cleared from lungs rapidly. They concluded that bio-distribution and translocation of small gold nanoparticles to the extra pulmonary organs is quite rapid as compared to the large gold nanoparticles (42). Also it is evident that the gold nanoparticles have the capability to cross theblood–brain barrier and reach the olfactory region (43). This highlights the probability of potential toxicity of gold nanoparticles in the brain.

34.8 Conclusion

Scanty reports are available related to the toxicity studies of gold nanoparticles and thus limit the availability of data related to their absorption, distribution, metabolism, and excretion. Gold nanoparticles are circulating in several tissues after their oral, inhalation, or intravenous administration. Thus, there is an obvious need to foster the understanding of their biodistribution and clearance followed by in vivo administration. Further, the toxicity studies on gold nanoparticles needed to be broaden to a wide range of parameters. This will unveil the potential risks of gold nanoparticles and will help in deciding the dose of the nanoparticles safe for organs other than target. This will also facilitate the development of target-specific formulations of gold nanoparticles. Moreover, more toxicity studies targeting the risk associated with gold nanoparticles are warranted to establish safety pillars for gold nanoparticles in future.

REFERENCES

1. Shang Y, Min C, Hu J, Wang T, Liu H, and Hu Y. Synthesis of gold nanoparticles by reduction of HAuCl4 under UV irradiation. *Solid State Sciences*. 2013; 15: 17–23. https://doi.org/10.1016/j.solidstatesciences.2012.09.002
2. Peng Y, Leng W, Dong B, Ge R, Duan H, and Gao Y. Bottom-up preparation of gold nanoparticle-mesoporous silica composite nanotubes as a catalyst for the reduction of 4-nitrophenol. *Chinese Journal of Catalysis*. 2015; 36: 1117–1123. https://doi.org/10.1016/S1872-2067(14)60310-7
3. Misra N, Biswal J, Gupta A, Sainis JK, and Sabharwal S. Gamma radiation induced synthesis of gold nanoparticles in aqueous polyvinyl pyrrolidone solution and its application for hydrogen peroxide estimation. *Radiation Physics and Chemistry*. 2012; 81: 195–200. https://doi.org/10.1016/j.radphyschem.2011.10.014
4. Zhao P, Li N, and Astruc D. State of the art in gold nanoparticle synthesis. *Coordination Chemistry Reviews*. 2013; 257: 638–665. https://doi.org/10.1016/j.ccr.2012.09.002
5. Giljohann DA, Seferos DS, Daniel WL, Massich MD, Patel PC, and Mirkin CA. Gold nanoparticles for biology and medicine. *Angewandte Chemie International Edition*. 2010; 49: 3280–3294. https://doi.org/10.1002/anie.200904359
6. Szunerits S and Boukherroub R. Near-infrared photothermal heating with gold nanostructures, in: Wandelt K (Ed.), *Encyclopedia of Interfacial Chemistry*. Elsevier, Oxford, pp. 500–510. 2018. https://doi.org/10.1016/B978-0-12-409547-2.13228-7
7. Tweney R. Discovering discovery: how faraday found the first metallic colloid. *Perspectives on Science*. 2006; 14: 97–121. https://doi.org/10.1162/posc.2006.14.1.97
8. mengwu4, 2019. What gives gold nanoparticles their color? Sustainable Nano. URL http://sustainable-nano.com/2019/11/12/gold-nanoparticles-color/ (accessed 26.8.2020).
9. Dykman LA, Khlebtsov NG. Gold nanoparticles in biology and medicine: recent advances and prospects. *Acta Naturae*. 2011; 3: 34–55.
10. Pricker SP Medical uses of gold compounds: past, present and future. *Gold Bulletin*. 1996; 29(2): 53–60. https://citeseerx.ist.psu.edu/viewdoc/download?doi=10.1.1.567.2740&rep=rep1&type=pdf
11. Mitra A, Chakraborty S, Auddy B, Tripathi P, Sen S, Saha A, and Mukherjee B. Evaluation of chemical constituents and free radical scavenging activity of Swarn Bhasma (gold ash): an ayurvedic drug. *Journal of Ethnopharmacology*. 2002; 80: 147–153. https://doi.org/10.1016/S0378-8741(02)00008-9
12. Van Riel PLCM, Larsen A, Van De Putte LBA, and Gribnau FWJ. Effects of aurothioglucose and auranofin on radiographic progression in rheumatoid arthritis. *Clinical Rheumatology*. 1986; 5: 359–364. https://doi.org/10.1007/BF02054254
13. Sutton BM. Gold compounds for rheumatoid arthritis. *Gold Bullion*. 1986; 19: 15–16. https://doi.org/10.1007/BF03214639
14. Eisler R. Chrysotherapy: a synoptic review. *Inflammation Research*. 2003; 52: 487–501. https://doi.org/10.1007/s00011-003-1208-2
15. Kim J, Chun SH, Amornkitbamrung L, Song C, Yuk JS, Ahn SY, Kim BW, Lim YT, Oh B-K, and Um SH. Gold nanoparticle clusters for the investigation of therapeutic efficiency against prostate cancer under near-infrared irradiation. *Nano Convergence*. 2020; 7: 5. https://doi.org/10.1186/s40580-019-0216-z
16. Bejarano J, Navarro-Marquez M, Morales-Zavala F, Morales JO, Garcia-Carvajal I, Araya-Fuentes E, Flores Y, Verdejo HE, Castro PF, Lavandero S, and Kogan MJ. Nanoparticles for diagnosis and therapy of atherosclerosis and myocardial infarction: evolution toward prospective theranostic approaches. *Theranostics*. 2018; 8: 4710–4732. https://doi.org/10.7150/thno.26284
17. Singh M, Harris-Birtill DCC, Markar SR, Hanna GB, and Elson DS. Application of gold nanoparticles for gastrointestinal cancer theranostics: a systematic review. *Nanomedicine: Nanotechnology, Biology and Medicine*. 2015; 11: 2083–2098. https://doi.org/10.1016/j.nano.2015.05.010
18. Singh P, Pandit S, Mokkapati VRSS, Garg A, Ravikumar V, and Mijakovic I. Gold nanoparticles in diagnostics and therapeutics for human cancer. *International Journal of Molecular Sciences*. 2018; 19: 1979–1994. https://doi.org/10.3390/ijms19071979.
19. Gopinath V, Priyadarshini S, MubarakAli D, Loke MF, Thajuddin N, Alharbi NS, Yadavalli T, Alagiri M, and Vadivelu J. Anti-Helicobacter pylori, cytotoxicity and catalytic activity of biosynthesized gold nanoparticles: multifaceted application. *Arabian Journal of Chemistry*.2019; 12: 33–40. https://doi.org/10.1016/j.arabjc.2016.02.005
20. Folorunso A, Akintelu S, Oyebamiji AK, Ajayi S, Abiola B, Abdusalam I, and Morakinyo A. Biosynthesis, characterization and antimicrobial activity of gold nanoparticles from leaf extracts of *Annona muricata*. *Journal of Nanostructure in Chemistry*. 2019; 9: 111–117. https://doi.org/10.1007/s40097-019-0301-1
21. Ahmad T, Wani IA, Lone IH, Ganguly A, Manzoor N, Ahmad A, Ahmed J, and Al-Shihri AS. Antifungal activity of gold nanoparticles prepared by solvothermal method. *Materials Research Bulletin*. 2013; 48: 12–20. https://doi.org/10.1016/j.materresbull.2012.09.069
22. Pallavicini P, Cabrini E, Borzenkov M, Sironi L, and Chirico G. Applications of gold nanostars: nanosensing, thermal therapy, delivery systems, in: Chirico G, Borzenkov M, and Pallavicini P (Eds.), *Gold Nanostars: Synthesis, Properties and Biomedical Application, SpringerBriefs in Materials*. Springer International Publishing, Cham, pp. 43–59. 2015. https://doi.org/10.1007/978-3-319-20768-1_3
23. Sironi L, Borzenkov M, Collini M, D'Alfonso L, Bouzin M, and Chirico G. Interactions of gold nanostars with cells, in: Chirico, G, Borzenkov, M, Pallavicini, P (Eds.), *Gold Nanostars: Synthesis, Properties and Biomedical Application, SpringerBriefs in Materials*. Springer

International Publishing, Cham, pp. 61–74. 2015. https://doi.org/10.1007/978-3-319-20768-1_4
24. Xia X and Xia Y. Gold nanocages as multifunctional materials for nanomedicine. *Frontiers in Physics.* 2014; 9: 378–384. https://doi.org/10.1007/s11467-013-0318-8
25. Ma Z, Xia H, Liu Y, Liu B, Chen W, and Zhao Y. Applications of gold nanorods in biomedical imaging and related fields. *Chinese Science Bulletin* 2013; 58: 2530–2536. https://doi.org/10.1007/s11434-013-5720-7
26. Hu J, Sanz-Rodríguez F, Rivero F, Rodríguez EM, Torres RA, Ortgies DH, Solé JG, Alfonso F, and Jaque D. Gold nanoshells: contrast agents for cell imaging by cardiovascular optical coherence tomography. *Nano Research.* 2018; 11: 676–685. https://doi.org/10.1007/s12274-017-1674-4
27. Jaishankar M, Tseten T, Anbalagan N, Mathew BB, and Beeregowda KN. Toxicity, mechanism and health effects of some heavy metals. *Interdisciplinary Toxicology.* 2014; 7: 60–72. https://doi.org/10.2478/intox-2014-0009
28. Rapson WS. Skin contact with gold and gold alloys. *Contact Dermatitis.* 1985; 13: 56–65. https://doi.org/10.1111/j.1600-0536.1985.tb02505.x
29. Kean WF, Lock CJL, Buchanan WW, Howard-Lock H, and Hogan MG. Gold toxicity: chemical, structural, biological and clinical experimental issues, in: Rainsford KDandVelo GP (Eds.), *Side-Effects of Anti-Inflammatory Drugs 3, Inflammation and Drug Therapy Series.* Springer Netherlands, Dordrecht, pp. 321–343. 1992. https://doi.org/10.1007/978-94-011-2982-4_37
30. Senut M-C, Zhang Y, Liu F, Sen A, Ruden DM, and Mao G. Size-dependent toxicity of gold nanoparticles on human embryonic stem cells and their neural derivatives. *Small.* 2016; 12: 631–646. https://doi.org/10.1002/smll.201502346
31. Raftis JB and Miller MR. Nanoparticle translocation and multi-organ toxicity: A particularly small problem. *Nano Today.* 2019; 26: 8–12. https://doi.org/10.1016/j.nantod.2019.03.010
32. Aro T, Kanerva L, Häyrinen-Immonen R, Silvennoinen-Kassinen S, Konttinen YT, Jolanki R, and Estlander T. Long-lasting allergic patch test reaction caused by gold. *Contact Dermatitis.* 1993; 28: 276–281. https://doi.org/10.1111/j.1600-0536.1993.tb03431.x
33. Test No. 403: Acute inhalation toxicity | READ online [WWW Document], n.d. OECD iLibrary. URL https://read.oecd-ilibrary.org/environment/test-no-403-acute-inhalation-toxicity_9789264070608-en (accessed 1.25.21).
34. Test No. 413: Subchronic inhalation toxicity: 90-day study | READ online [WWW Document], n.d. URL https://read.oecd-ilibrary.org/environment/test-no-413-subchronic-inhalation-toxicity-90-day-study_9789264070806-en#page1 (accessed 1.23.21).
35. Nutrition, C. for F.S. and A., 2019. Redbook 2000: IV.C.4.a. Subchronic Toxicity Studies with Rodents [WWW Document]. U.S. Food and Drug Administration. URL https://www.fda.gov/regulatory-information/search-fda-guidance-documents/redbook-2000-ivc4a-subchronic-toxicity-studies-rodents (accessed 1.5.21).
36. Pernodet N, Fang X, Sun Y, Bakhtina A, Ramakrishnan A, Sokolov J, Ulman A, and Rafailovich M. Adverse effects of citrate/gold nanoparticles on human dermal fibroblasts. *Small.* 2006; 2: 766–773. https://doi.org/10.1002/smll.200500492
37. Yu DLE, Yung L-YL, Ong C-N, Tan Y-L, Balasubramaniam KS, Hartono D, Shui G, Wenk MR, and Ong DW-Y. Translocation and effects of gold nanoparticles after inhalation exposure in rats. *Nanotoxicology.* 2007; 1: 235–242. https://doi.org/10.1080/17435390701763108
38. Sung JH, Ji JH, Park JD, Song MY, Song KS, Ryu HR, Yoon JU, Jeon KS, Jeong J, Han BS, Chung YH, Chang HK, Lee JH, Kim DW, Kelman BJ, and Yu, IJ. Subchronic inhalation toxicity of gold nanoparticles. *Particle and Fibre Toxicology.* 2011; 8: 16. https://doi.org/10.1186/1743-8977-8-16
39. Schulz M, Ma-Hock L, Brill S, Strauss V, Treumann S, Gröters S, van Ravenzwaay B, and Landsiedel R. Investigation on the genotoxicity of different sizes of gold nanoparticles administered to the lungs of rats. *Mutation Research/Genetic Toxicology and Environmental Mutagenesis.* 2012; 745: 51–57. https://doi.org/10.1016/j.mrgentox.2011.11.016
40. Coradeghini R, Gioria S, García CP, Nativo P, Franchini F, Gilliland D, Ponti J, and Rossi F. Size-dependent toxicity and cell interaction mechanisms of gold nanoparticles on mouse fibroblasts. *Toxicology Letters.* 2013; 217: 205–216. https://doi.org/10.1016/j.toxlet.2012.11.022
41. Ng C-T, Li JJ, Gurung RL, Hande MP, Ong C-N, Bay B-H, and Yung L-YL. Toxicological profile of small airway epithelial cells exposed to gold nanoparticles. *Experimental Biology and Medicine.* 2013; 238: 1355–1361. https://doi.org/10.1177/1535370213505964
42. Han SG, Lee JS, Ahn K, Kim YS, Kim JK, Lee JH, Shin JH, Jeon KS, Cho WS, Song NW, Gulumian M, Shin BS, and Yu IJ. Size-dependent clearance of gold nanoparticles from lungs of Sprague–Dawley rats after short-term inhalation exposure. *Archives of Toxicology.* 2015; 89: 1083–1094. https://doi.org/10.1007/s00204-014-1292-9
43. Sokolova V, Mekky G, van der Meer SB, Seeds MC, Atala AJ, and Epple M. Transport of ultrasmall gold nanoparticles (2 nm) across the blood–brain barrier in a six-cell brain spheroid model. *Scientific Reports.* 2020; 10: 18033. https://doi.org/10.1038/s41598-020-75125-2

35

Single-Use Dry Powder Inhalers for Pulmonary Drug Delivery

Somchai Sawatdee[1] and Teerapol Srichana[2]
[1]Walailak University, Nakhon Si Thammarat, Thailand
[2]Prince of Songkla University, Hat Yai, Thailand

35.1 Introduction

Pharmaceutical aerosols have been widely used for medical purposes for treatment of local diseases and are particularly beneficial as systemic drug delivery systems. Furthermore, there are many advantages for treatment of a specific disease in the pulmonary system, such as pulmonary infection disease and lung cancer (1). The advantages of pulmonary drug delivery include, for example, drug doses that can generally be reduced and systemic side effects that can be minimized compared to oral administration. The alveolar tissue in the lung provides a large surface area and is effectively perfused with blood and has very little drug metabolization or efflux transporter activities. This represents only a minimal physical barrier between the airspace and the bloodstream. These characteristics enable rapid absorption into systemic circulation without the production of metabolites. It is a convenient, non-invasive mode of administering substances that show no, poor, or less variable systemic bioavailability when applied via the peroral route. In addition, large drug molecules, such as peptides and protein drugs, are degraded in the gastrointestinal conditions and are eliminated by the first-pass metabolism in the liver. These problems can be solved by pulmonary administration.

Delivery of a drug to the pulmonary system must be performed using an inhaler device. Four basic types of inhaler devices have become available in the past few decades. These include the nebulizer (mesh, jet, or ultrasonic nebulizer), pressurized metered dose inhaler (pMDI), soft mist inhaler (SMI), and dry powder inhaler (DPI) (2,3). These systems were primarily developed for the treatment of asthma and COPD. The nature of these diseases generally requires chronic and frequent medication. Nebulizers are advantageously prescribed for patients who are unable to achieve sufficient inspiratory flow, especially for children, elderly patients, and patients with severe airway obstruction or intubation. However, the routine use of nebulizers is limited because of its complexity and the consequent high cost and low drug delivery efficiency, as well as strict cleaning requirements to avoid microbial contamination and infection of patients with pathogens. The Spiriva® Respimat® is the only soft mist inhaler that is available in today's market and is manufactured by Boehringer Ingelheim, Germany. It combines both the advantages of a pMDI and nebulizer. The Respimat® creates aerosols by impingement of two liquid jets. It is a small, portable, hand-held inhaler without propellants. Individual doses are delivered via a precisely engineered nozzle system to produce a slow-moving and long-sustaining aerosol cloud similar to a nebulizer. Thus, it greatly reduces the need to coordinate between actuation and inspiration, like the pMDIs, and bypasses the inspiratory flow limitation of the DPIs (3–13). When the dose-release button is pushed, the energy from a compressed spring forces the drug through a nozzle system (14).

The most commonly used systems are pMDIs and DPIs which can be applied to a broad population of patients for routine medications. This is mainly attributed to their simple press-and-breathe function (simplicity), ease of substitution (similarity), and several cost benefits (1). However, many children and elderly patients have difficulty using a pMDI correctly because it requires high velocity when the dosage is released. Thus, problems may occur when following the inhalation technique recommended in the patient information leaflet. Extensive training is required to achieve correct use of a pMDI. To deliver the drug effectively into the lung, patients must actuate the pMDI as they start to inhale. This requires a high degree of "hand/lung" coordination. Failure to achieve this often results in reduced effectiveness of treatment and poor disease control (15). DPIs avoid the breath coordination and the cold freon issue, but require, at least in the case of passive inhalers, a minimum inspiratory flow rate (1,16). However, DPI products feature some major advantages: (1) combination products can be developed more easily, (2) the formulation prepared as a dry powder generally leads to enhanced stability of the drug product, and (3) high doses of drug can be administered. In recent years, intelligent DPIs have been developed for high delivery efficiencies and convenience for the patient and medical staff.

35.2 Classifications of Dry Powder Inhalers and Single-Use Disposable Devices

DPIs can be categorized into three types, based on their design. The first group is the single unit dose system (usually employing a capsule) and the brands include Spinhaler® (Fisons, Aventis),

DOI: 10.1201/9781003046547-35

Rotahaler® (GlaxoSmithKline), Aerolizer™ (Novartis), Handihaler™ (Boehringer-Ingelheim), Inhalator® (Boehringer-Ingelheim), and Cyclohaler® (Pharmachemie). The second type employs a multiple unit dose dispensing system and examples are the Diskhaler® (GlaxoSmithKline), Aerohaler™ (Boehringer-Ingelheim), and Accuhaler® or Diskus™ (GlaxoSmithKline). The third class of DPI are devices that contain a multiple dose reservoir such as the Turbuhaler® (Astra Zeneca), Autohaler® (Teva Pharmaceuticals), Easyhaler® (Orion Pharma), Chiesi® or Pulvinal™ (Chiesi), Twishaler™ (Schering-Plough), Novolizer™ (ASTA Medica), Clickhaler™ (Innovata Biomed/ML Labs Celltech), and Easyhaler™ (Orion Pharma). In addition, the disposable DPI is a subset of the single unit dose type of DPI such as the Afrezza® Inhaler that uses single-use plastic cartridges and the DirectHaler™ that offers a pre-metered amount of a single dose and is discarded after use (17).

DPIs are not easy to formulate and manufacture, and device development needs considerable expert input. Most of the device systems of DPI use a capsule as a unit dose because it is easy to prepack. The use of a capsule as a unit container, such as the Spinhaler®, Rotahaler®, Inhalator®, and Cyclohaler®, is convenient and often does not lead to a marked loss of the drug through residual formulation that remains in the container after activation. However, moisture retention by the gelatin capsule and drug might cause problems with the stability of the drug and the ease of powder dispersion. Hence, a subsequent generation of multiple unit dose devices has emerged to solve such problems by protecting the drug from moisture until the point of administration. However, in some of these newer devices, the fine particle fraction (FPF) has not shown much improvement. The dose uniformity in some multiple dose systems can be poor. Multiple dose inhalers may therefore require filling with more drug than is needed to maintain a dose uniformity until the last dose is dispensed (17). There is no doubt that modern multiple-use devices are the most cost-effective choice for the treatment of asthma and COPD. However, the treatment of other diseases often requires less frequent, transient, or even a once-only application of the medicament which limits the cost effectiveness of multiple-use devices. Also, occasional use, such as use in hospitals, may favor single-use disposable devices. Thus, these devices can represent a valuable alternative to multiple-use devices. Disposable inhaler devices can have different meanings. It can refer to "single-use" for a single dose, or to "disposable" after a short period of use (e.g. 15 days, as for dry powder insulin, Afrezza) (18). In this chapter, the disposable single dose is introduced. The disposable single unit dose of DPI is easily used for all patients and more stable than multiple dose reservoir systems because the entire unit of formulation is in a sealed container.

35.3 Disposable Dry Powder Inhalers: Advantages

There are many different inhaler device technologies for appropriate patient therapies and treatment regimens. The newer inhaler devices designed for single-use or single-patient inhalers may become safer, more flexible, and/or convenient for the patient and hospital staff. The most important reasons for having a preference for single-use disposable inhaler devices are summarized in Table 35.1 (18).

The disadvantages of single-use devices depend on the situation in which they are used. For example, disposable products impose a heavy burden on waste management and the acceptance of single-use inhalers may be low in societies that have an environmentally friendly lifestyle and the use of durable consumer goods is encouraged. The recycling of waste materials is expensive because it requires that recyclable products be separated from general waste. Single-use inhalers that may contain residues of the drug should not be mixed with other plastic disposables and should be returned to a pharmacy. The collection of used inhalers through pharmacies is relatively easy to organize, and the volume of inhalers is small compared with the volume of other disposable products like packaging materials and PET bottles. The risk of environmental pollution with used inhalers can be minimized by providing new inhalers only upon return of the used ones. Furthermore, if the intent of single-use DPIs is to replace unstable solutions for nebulization or injection solutions, other aspects must be taken into consideration for the total burden imposed on the environment. In addition, cost aspects have to be considered in relation to safety and efficacy of the therapy to decide whether the total balance is for or against disposable inhalers (18). Small single-use inhalers have advantages compared to nebulizers that have many equipment parts, such as the drug solution, and syringes with needles. The empty package and medical device used during treatment are pharmaceutical wastes similar to single-use disposable inhalers. The cost-effectiveness also depends on the situation. For example, in the case of a drug that is administered daily and is not hygroscopic, the cost price per dose will be higher with single-use than with reusable multi-dose MDIs or DPIs. Hygroscopic drug formulations, poor compliance with the cleaning instructions, and/or incorrect operation of the dose

TABLE 35.1

Advantages of Single-Use Disposable Inhaler Devices (Modified from Boer et al. (19)).

- Elimination of the risk of cross-contamination of diseases in mass vaccination programs
- Elimination of the risk of performance failure due to improper cleaning, poor maintenance, or damage
- Avoidance of bacterial resistance development and patient reinfection with drug-resistant strains
- Avoidance of water uptake and particle liquefaction of retained fractions of hygroscopic drug formulations
- Avoidance of time-consuming cleaning and disinfection procedures
- Increased cost-effectiveness, particularly in once-only administrations (e.g. vaccination)
- Simplification of the instructions for use by eliminating the need to operate a dose measuring principle
- Avoidance of desiccant compartments in reusable multi-dose inhalers that protect the drug formulation against moisture absorption
- Greater hygiene to eliminate the spread of inhaler residues in pockets and handbags
- Reduced inhaler size for portability

measuring principle may make the therapy inefficient and increase the frequency of exacerbations and hospitalizations. Also, the costs of cleaning and disinfection of reusable inhaler systems used in hospitals have to be taken into account. In this situation, single-use disposable devices may be cost effective (18).

35.4 Disposable Dry Powder Inhalers: Device Design and Formulation Development

Single-use DPIs should meet the same general requirements as other inhalation drug products regarding clinical efficacy, device reliability, patient compliance, and cost-effectiveness. The disposable design of a single-use inhaler does not need a dose-metering mechanism, a dose counter, or a refillable process.

The design of disposable DPIs is quite different from multi-dose counterparts. The desirable important characteristics of single-use inhalers are summarized in Table 35.2 (1). The specific requirements of disposable DPIs are related to the application used and economic success, for example, single-use devices are not competitive either from an economic or an environmental point of view for continuous and frequent medication treatment of chronic disease states. Single-use DPIs are useful for acute or transient disease state therapy as well as for vaccination. Development and design of a single-use disposable DPI should consider the patients or subjects who are not familiar with inhaled medicine. A good device design has minimal or simple instructions for only one or two operation steps before the drug is inhaled. Furthermore, to facilitate extensive cost-effectiveness through cost-effective development and inexpensive manufacturing and assembly, disposable DPIs should have not more than two to four parts, including the container for the formulation. The disposable DPIs eliminate a great deal of uncertainty on their performance, although delivered doses from passive DPIs may vary more strongly with the patient's inhalation maneuver than those from nebulizers.

The greatest challenge in design and development of disposable DPIs is to keep them small, simple, cheap, and highly effective with a consistent and patient-independent performance. Simplifying the design follows partly from not needing a dose-measuring system, a dose counter, or a desiccant compartment. Simplifying also refers to the operational procedures that must exclude errors as much as possible for keeping the production costs low (18). With regard to economic competitiveness, it is also of particular interest to use a reliable and low-cost mode of powder filling. Although devices that use active powder dispersion principles are increasingly considered as an alternative to passive devices, it is rather questionable whether single-use active devices can compete economically and environmentally against their passive counterparts. Finally, the inhalation system should enable the application of different drugs and doses (1).

Drug formulations for disposable inhalers are not necessarily different from formulations for reusable devices unless they are meant for different applications or therapeutic strategies. The technical design of the main powder formulation for use in a disposable DPI should contain the desired aerodynamic size distribution of the drug for deposition in the target area at the flow rate at which the drug is inhaled. The formulation should be dispersed effectively into the inhaled air stream through the DPI during inhalation. In addition, the drug requires good flow properties of the formulation and control of the interparticulate forces in the powder. Also, the hygroscopic nature of the particles is highly adhesive and cohesive, resulting in poor dispersion performance, and the particles are prone to adhere to the inhaler walls and build up huge inhaler deposits that cannot be resolved. These problems are related to the physicochemical properties for inhalation and are the reasons why particle formulations are engineered with excipients to improve their delivery to the respiratory tract as a dry powder (18). To achieve these requirements, different types of formulations, with and without excipients, and special particle engineering techniques can be applied (20). To avoid inhalation of an excess of powder, the use of excipients in high-dose drugs has to be minimized, but this should not result in more complex and multistep particle engineering processes, as they make the powder formulations expensive. The best techniques to keep disposable DPIs cheap are to use micronized drugs and controlled crystallization and precipitation or spray-drying (18).

35.5 Single-Use Dry Powder Inhalation Devices

35.5.1 MonoHaler

The MonoHaler is a product from Berry Bramlage which is part of Berry Global Company, Germany. The MonoHaler is a breath-actuated, blister-based, single-dose DPI that is available as a reusable or disposable system (Figure 35.1). According to

TABLE 35.2

Important Characteristics of Single-Use Disposable Inhalers (Modified from Friebel and Steckel and Cahndel et al.(1))

- Precise and consistent dose delivery over a wide range of inspiratory flow rates
- Moisture protection of the device as a whole to enable stability of the formulation in the device and uniform dose delivery during the period of application
- Appropriate drug formulation to target the intended site of action
- Simple to use and easy to teach, preferably self-explaining functionality
- Convenient to carry
- Audible and/or visual indication of successful inhalation
- Maximum drug delivery to the target area and minimal drug losses in the device and in the oropharynx
- Suitable for different drugs and doses
- Efficient and reliable mode of powder filling
- Low number of device parts which can inexpensively be produced and assembled

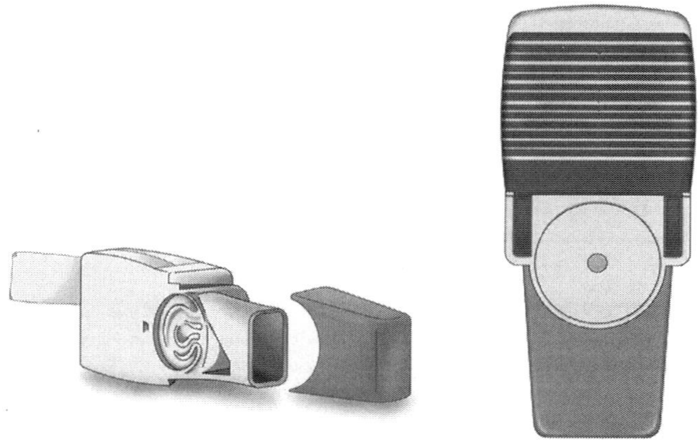

FIGURE 35.1 MonoHaler by Berry Bramlage (Oldenburg, Germany) (Redrawn from https://catalogue.rpc-bramlage.com/en/portals/berrybramlage/assets/12709905/monohaler/)

the company information, important characteristics are easy blister filling and flexibility in blister size. Design of this device is a blister-based single-dose dry powder. The manufacturer claims this device has a high FPF with two filling options. The first is a reusable, blister-replacement system, and the second is a single-use system. The drug is pre-packed in a blister and the dose can vary from 0.5 to 25 mg (www.rpc-bramlage.com). This device is easy to use and the steps of operation are minimal.

35.5.2 TwinCaps®

The TwinCaps® was developed by Hovione Technology Ltd. (Loures, Portugal) and consists of only two plastic components with a movable dose compartment and a non-movable inhaler body (Figure 35.2). The dose compartment contains up to two single doses which can be filled by standard industrial filling machines. After filling, the movable part is inserted into the non-movable inhaler body. In the closed position, the inhaler body serves as a protective housing to avoid environmental contamination of the powder dose. To start inhalation, the patient pushes the dose compartment sideways such that the first dose is in alignment with the mouthpiece of the inhaler. The patient's inspiration creates turbulent air flow inside the dose chamber and the inhalation channel which ensures powder release and de-agglomeration. For inhalation of the second dose, the dose compartment simply needs to be pushed in the other direction until the dose is aligned with the mouthpiece (1,21). The TwinCaps® is marketed for delivery of laninamivir octanoate (Inavir®) for the treatment of influenza. It is different from the short acting predecessor (zanamivir) that required twice daily treatments for 5 days. A single dose of 40 mg inhaled laninamivir can be effective for a week, making it a perfect candidate for a disposable DPI. Since the delivery of Inavir® using TwinCaps® has been approved for children of less than 10 years old (17), this device has sufficient aerosolization performance at lower flow rates.

35.5.3 3M Conix™

The 3M Conix™ DPI (Figure 35.3) is a single-dose system that is both reloadable and disposable. It is also available as multi-unit dose systems. The mechanism of the 3M Conix™ is based on reverse-cyclone technology by Cambridge Consultants Ltd. On inhalation by the patient, inspired air entrains the powder formulation stored in a blister. Air and powder laterally enter a cone-shaped cyclone chamber where a vortex is established. The vortex creates comparably high velocities and de-agglomeration

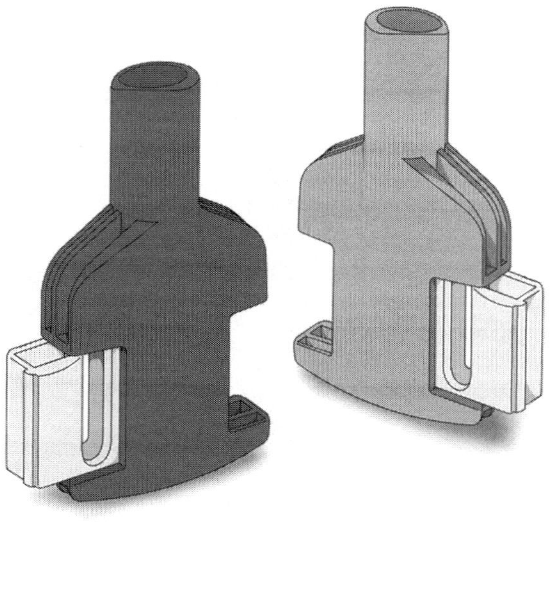

FIGURE 35.2 TwinCaps® by Hovione Technology Ltd (Loures, Portugal) (Redrawn from http://.com/work/cases/hovione-twincaps-inhaler)

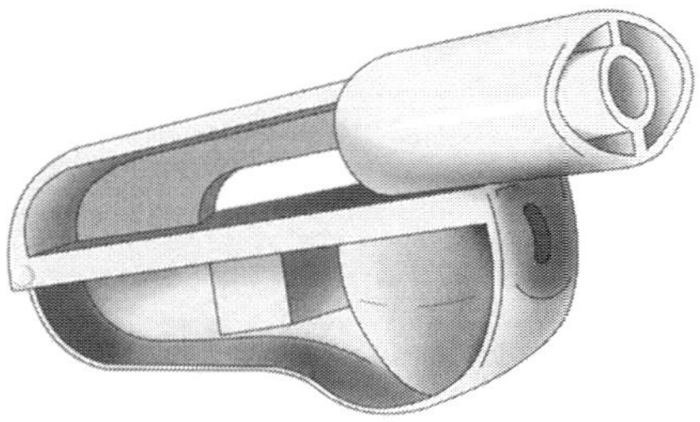

FIGURE 35.3 3M Conix™ 3M (St. Paul, MN, USA) (Redrawn from https://www.medgadget.com/2006/11/conix_one_inhal.html and (22))

is accordingly obtained by collisions of particles with the cyclone wall or other particles, but also by particle shear. Since the chamber is closed at the bottom, the air flow reverses, passes the incoming air flow, and exits the chamber at the top. This mode of operation prevents the release of large particles which would presumably contribute to reduced oropharyngeal drug deposition and higher respirable fractions (1,21).

35.5.4 Twincer™

The Twincer™ is a disposable DPI developed at the University of Groningen in the Netherlands, for the delivery of high drug doses up to 60 mg. This device basically consists of three plate-like plastic parts forming the air flow passages and a blister strip containing the powder formulation (Figure 35.4). The blister has a long cover foil which, by pulling, connects the powder channel and the inlet to the blister chamber. Air passing through the powder channel during inhalation entrains the powder from the blister. The powder flow is then divided between two parallel classifiers, which are circular depressions in the bottom plate of the inhaler where the inertial and shear forces act on the drug agglomerates that result in de-agglomeration and consequent delivery to the patient. The patient opens the blister before inhalation by pulling the folded cover foil. The blister is connected to the powder channel and when the patient inspires, the powder is entrained by air and conveyed to two parallel cylindrical chambers, also called air-classifiers. The tangentially arranged classifier inlets cause a vortex within the chambers ensuring particle breakup, mainly by collisions of the particles with the classifier wall and other particles. By choosing an appropriate layout and dimension of the classifier outlet, the residence time of the powder and the size of particles leaving the classifier toward the mouthpiece can be controlled within certain limits (1,21).

The Twincer™ is used to deliver colistin for the treatment of chronic infections in cystic fibrosis patients (17). In vitro de-agglomeration experiments using micronized colistin sulfomethate with the Twincer™ have shown excellent, flow rate–independent de-agglomeration behavior even at high powder doses of 25 mg. Also, increased respirable fractions were demonstrated by the simple method for controlled agglomeration of micronized substances (1).

35.5.5 DirectHaler™

The DirectHaler™ was developed by Direct-Haler A/S (assets acquired by Trimel Bio-Pharma SRL in 2009). The device resembles a drinking straw and consists of a U-shaped inhaler tube with a corrugated bend. A double cap is used to seal both ends of the inhaler tube. Powders that are not, or are only marginally, susceptible to physical and chemical instabilities during storage and transportation can be filled directly into the tube. Sensitive formulations, however, can be filled into a special cap sealed with a laminate foil strip and thus protected from mechanical stress, light, and moisture until the powder is released into the inhaler tube by removing the sealing. For inhalation, the patient uses one end of the tube as the mouthpiece. The turbulent air flow formed within the tube during inhalation is claimed to generate turbulent whirls at the corrugated surfaces of the tube and to force the powder on the walls of the corrugations from where it is gradually and completely released toward the mouthpiece. Non-peer reviewed in vitro and in vivo data suggest that the performance and safety characteristics are comparable to that of marketed inhalation systems. Due to the very simple design of this device, the cost of materials and goods may be exceptionally competitive (1) (Figure 35.5).

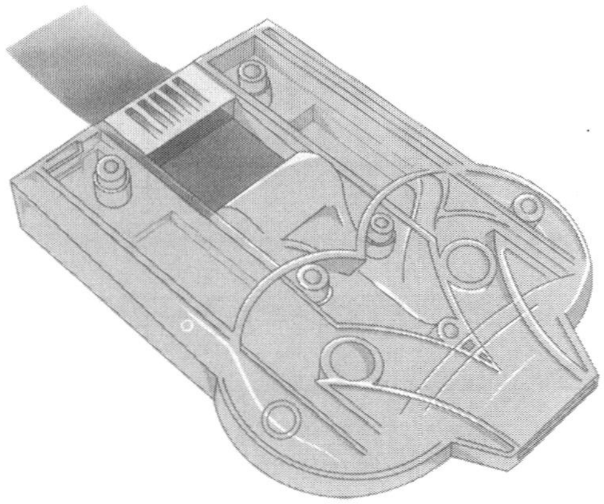

FIGURE 35.4 Twincer™ by the University of Groningen (RuG), Netherlands (Redrawn from http://www.byindes.com/en/product/twincer/)

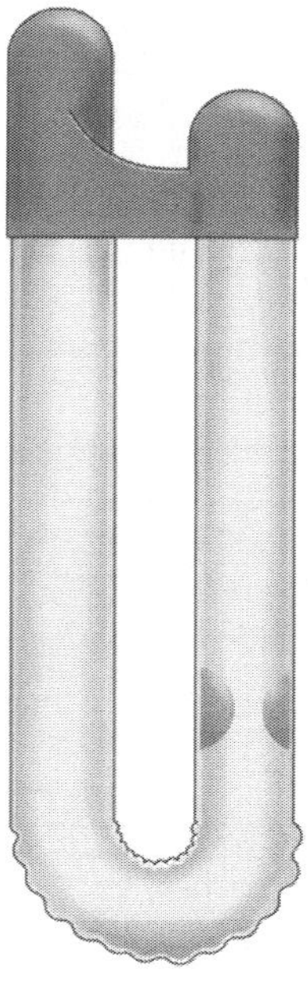

FIGURE 35.5 DirectHaler™ by Direct-Haler A/S (Copenhagen, Denmark) (Redrawn from Keldmann, Lavorini et al. (23,24))

35.5.6 Aespironics DryPod

Aespironics Ltd. is developing a single-use disposable DPI based on its proprietary ActiveMesh™ technology. A unique feature of this technology is the packaging of the powder in a mesh compartment located on a reed or spring, and the beating and vibration of this mesh package on inhalation by the patient. This action forces the powder agglomerates through the mesh holes, thereby releasing and de-agglomerating the powder at the same time. Two different embodiments of this technology are available: a credit card–shaped and a whistle-shaped inhaler. The credit card–shaped inhaler uses a rotor to beat the mesh package, whereas the whistle-shaped embodiment uses a vibrating reed-like rocker. During application, the patient receives audible feedback after successful inhalation from the beating action. Besides other concepts, the technology has already been presented as a potential delivery system for inhaled vaccines. In an in vitro proof-of-concept study, the device prototype has been shown to effectively release and de-agglomerate micronized powders without the need for any additional excipients or special powder engineering. Limited powder release was observed only at low air flow rates. However, appropriate countermeasures for device optimization were proposed. A smaller mesh size increased the number of respirable particles and reduced the fraction of non-respirable agglomerates. The dispersion of powder doses of up to 3 mg resulted in acceptable FPFs. Thus, the Aespironics DryPod DPI platform represents a promising concept for single-use applications (1).

35.5.7 Afrezza® Inhaler (MannKind Corp.)

MannKind Corporation has developed a single-dose device called the Afrezza® Inhaler. The device is available in four unit, eight unit, and 12 unit disposable single dose cartridges of Afrezza® (insulin human) inhalation powder. The patient prepares the Afrezza® Inhaler by loading a cartridge into the open inhaler, closing the inhaler, and then inhaling deeply the powder through the mouthpiece in a single breath. There are two inlet streams; the first stream lifts the powder and delivers it into the second inlet stream to de-agglomerate the fluidized powder. The powder dispersion occurs with low inspiratory flow rate. Therefore, low deposition is expected in the throat while promoting deep lung deposition (Figure 35.6).

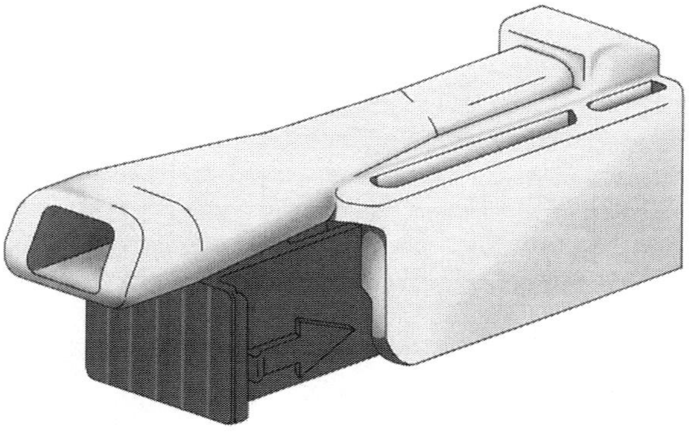

FIGURE 35.6 Afrezza® Inhaler by MannKind Corp., (Danbury, CT, USA) (Redrawn from Leone-Bay (25))

35.5.8 PuffHaler (Aktiv-Dry LLC)

PuffHaler® was used to administer vaccine into cotton rats. The deposition of vaccine in their lungs and successive viral replication was monitored by measles-specific RT-PCR. The immunity in response to dry-powder inhalation was similar to that by injection. One particular study reported three elementary steps for the preparation of dry-powder blends for administration through inhalation. First, the mass median diameter of the particles needs to be reduced to 1–5 μm. The size reduction can be achieved by different procedures, such as spray drying, precipitation from supercritical fluids, and jet milling or micronization. After this, the micronized particles need to be blended with an excipient, usually lactose. These carrier particles facilitate the distribution of the small particles and allow correct filling into the inhaler storage system in a reproducible fashion. The final step is to fill the blend into the storage systems (Figure 35.7).

35.5.9 Perlamed™-BLISTair (Perlen Packaging)

Perlamed™-BLISTair is a smart inhalation device developed by Perlen Packaging. The information from the manufacturer indicates that it is a patented, disposable, single-use DPI made of a high barrier primary packaging film. The BLISTair consists of two thermoformed barrier films to protect the medicinal agent. The two films envelop the powdery agent and form the inhaler through their double funnel shape. The inhaler is extremely intuitive to use: the user simply pulls on a tab to open the powder enclosure. The BLISTair is placed on a flat surface to pull the lid, then the BLISTair is placed horizontally on the lips and the powder is inhaled. After inhalation, the BLISTair is simply disposed of. It is a platform for the pulmonary delivery of a wide range of drug substances. The BlistAir has competitive development and manufacturing costs, adaptable to many powder formulations, is easy to use, and is reliable. There are no moving parts (Figure 35.8).

35.5.10 SOLO™ Inhaler (Manta Devices)

The SOLO™ inhaler was developed by Manta Devices (USA). The SOLO™ inhaler is an easy-to-use low-cost single-use disposable DPI. It is simple in design with high performance dispersion and dose stability. The device automatically removes its foil packaging and opens the powder dispersion for inhalation. The high efficiency dispersion mechanism is able to deliver large or small dose payloads for various applications including respiratory diseases, vaccines, antibiotics, and several medications. This DPI integrates blister opening and metering dose technologies. It is comprised of a standard foil blister and two injection molded parts (www.mantadevices.com) (Figure 35.9).

35.5.11 Cyclops™

The Cyclops™ device is an easy-to-use, preloaded, disposable dry powder inhaler manufactured by PureIMS. Information from the manufacturer declares that the Cyclops™ uses the patient's breath to drive a sophisticated yet simple mechanism. Upon inhalation, the dry powder formulation is circulated and broken into small particles appropriately sized for inhalation. The flow mechanics are engineered to produce collisions and turbulent airstreams for effective de-agglomeration while moving the powder out slowly. This slow-moving powder effectively navigates the airways for delivery into the lung with

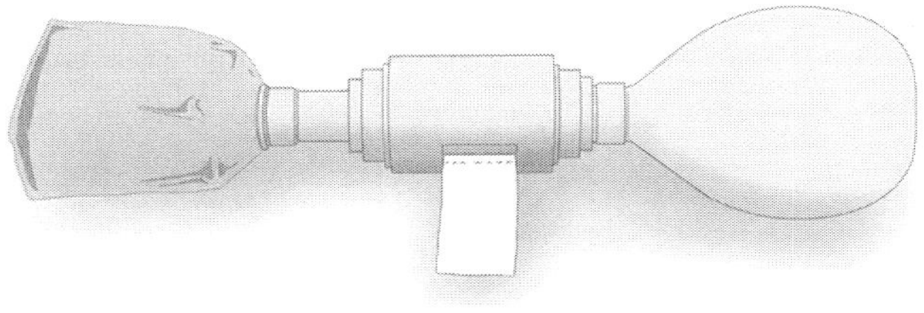

FIGURE 35.7 PuffHaler® by Aktiv-Dry LLC, (Boulder, CO, USA) (Redrawn from https://www.manufacturingchemist.com/news/article_page/AktivDrys_Puffhaler_uses_RPC_knowhow/41574)

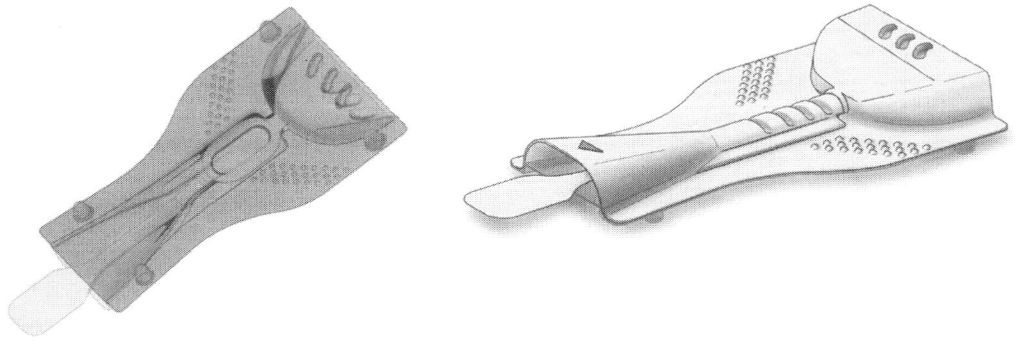

FIGURE 35.8 Perlamed™-BLISTair Perlen Pakaging AG, (Perlen, Switzerland) (Redrawn from https://www.perlenpackaging.com/en/products/perlamedtm-blistair/)

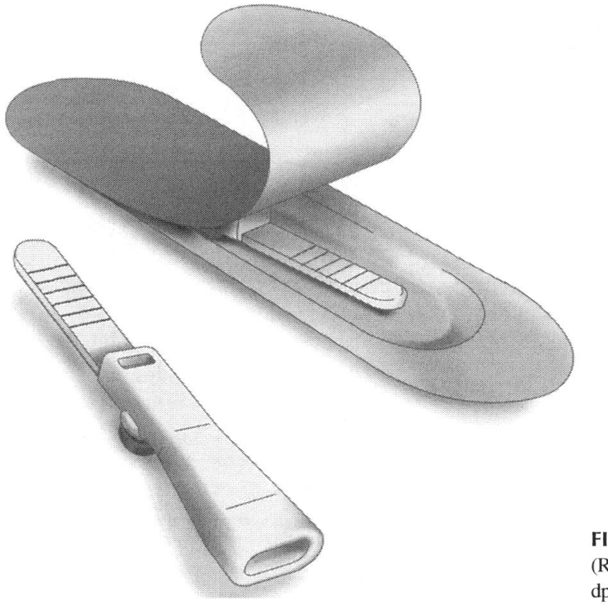

FIGURE 35.9 SOLO™ inhaler by Manta Devices, (Cambridge, MA, USA) (Redrawn from http://mantadevices.com/dry-powder-inhaler-development/disposable-dpi/solo-flyer-image-4-20x3-03/)

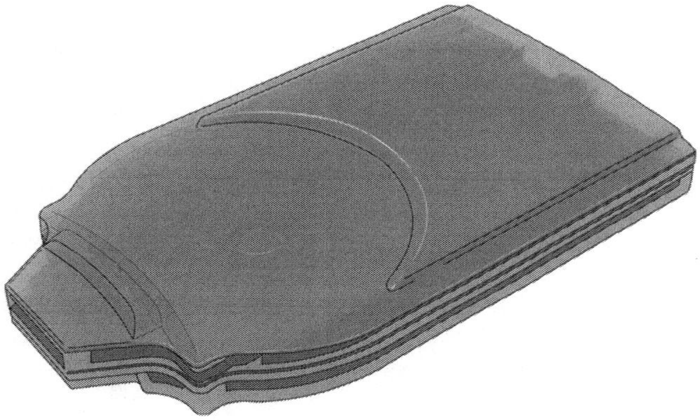

FIGURE 35.10 Cyclops™ by PureIMS, Roden, Netherlands (Redrawn from https://pureims.com)

minimal deposition at the back of the throat. Cyclops™ has several advantages compared to standard-of-care products in multiple clinical indications (www.pureims.com). In addition, the resistance to air flow of the Cyclops™ is minimal (26,27) (Figure 35.10).

35.6 Conclusions

Single-use disposable DPIs are different from the commonly used multiple-use DPIs in device design and instructions for use. The simple design of the device with cost-effective production is the first requirement. The powder technologies of single-use disposable DPI formulations are similar to that of multiple-use DPIs. The advantages of single-use DPIs are rapid applicability, ease of use, hygienic design, and robust. A cost-effective device should be considered in the design. Single-use disposable DPIs seem to cost less than nebulizers or MDIs. Nowadays, several single-use disposable DPIs are currently available in the market and new devices are in the development stages. The performance characteristics must be proved to be safe and effective. The multiple-use systems are more preferrable for chronic diseases than disposable single-use inhaler devices.

REFERENCES

1. Friebel C and Steckel H. Single-use disposable dry powder inhalers for pulmonary drug delivery. *Expert Opinion Drug Delivery*. 2010; 7: 1359–1372.
2. Chandel A, Goyal AK, Ghosh G, and Rath G. Recent advances in aerosolized drug delivery. *Biomedicine & Pharmacotherapy*. 2019; 112: 108601.
3. Zhang W, Xu L, Gao S, Ding N, Shu P, Wang Z, and Li Y. Technical evaluation of Soft Mist Inhaler use in patients with chronic obstructive pulmonary disease: a cross-sectional study. *International Journal of Chronic Obstructive Pulmonary Disease*. 2020; 15: 1471–1479.
4. https://catalogue.rpc-bramlage.com/en/portals/berrybramlage/assets/12709905/monohaler. Accessed on 10 September 2020.
5. http://mantadevices.com/dry-powder-inhaler-development/disposable-dpi/. Accessed on 10 September 2020.

6. www.rpc-bramlage.com. Accessed on 10 September 2020.
7. http://.com/work/cases /hovione-twincaps-inhaler. Accessed on 12 September 2020.
8. http://www.byindes.com/en/product/twincer. Accessed on 11 September 2020.
9. https://www.manufacturingchemist.com/news/article_page/AktivDrys_Puffhaler_uses_RPC_knowhow/41574. Accessed on 17 September 2020.
10. https://www.perlenpackaging.com/en/products/perlamedtm-blistair/. Accessed on 16 September 2020.
11. www.mantadevices.com. Accessed on 13 September 2020.
12. http://mantadevices.com/dry-powder-inhaler-development/disposable-dpi/solo-flyer-image-4-20x3-03/. Accessed on 15 September 2020.
13. https://pureims.com. Accessed on 14 September 2020.
14. Anderson P. Use of Respimat® Soft Mist™ inhaler in COPD patients. *International Journal of COPD*. 2006; 1: 251–259.
15. Chrystyn H. The Diskus™: a review of its position among dry powder inhaler devices. *International journal of Clinical Practice*. 2007; 61: 1022–1036.
16. Dalby R, Spallek M, and Voshaar T. A review of the development of Respimat® Soft Mist™ Inhaler. *International Journal of Pharmaceutics*. 2004; 283: 1–9.
17. Srichana T. *Dry powder inhalers: formulation, device and characterization*. Nova Science Publishers, New York, USA, 2017.
18. Boer AH and Hagedoorn P. The role of disposable inhalers in pulmonary drug delivery. *Expert Opinion Drug Delivery*. 2015; 12: 143–157.
19. Boer AH, Hagedoorn P, Westerman EM, Brun PPH, Heijerman HGM, and Frijlink HW. Design and in vitro performance testing of multiple air classifier technology in a new disposable inhaler concept (Twincer®) for high powder doses. *European Journal of Pharmaceutical Sciences*. 2006; 28: 171–178.
20. Hoppentocht M, Hagedoorn P, Frijlink HW, and Boer AH. Technological and practice challenges of dry powder inhalers and formulations. *Advanced Drug Delivery Reviews*. 2014; 75: 18–31.
21. Berkenfeld K, Lamprecht A, and McConville JT. Devices for dry powder drug delivery to the lung. *AAPS PharmSciTech*. 2015; 16: 479–490.
22. Cipolla D, Chan HK, Schuster J, and Farina D. Personalizing aerosol medicine: development of delivery systems tailored to the individual. *Therapeutic Delivery*. 2010; 1: 667–682.
23. Keldmann T. Simplicity wins – from product conceptualization to drug delivered. *Drug Delivery Report*. 2006. Spring/Summer: 49–52.
24. Lavorini F, Fontana GA, and Usmani O. New inhaler devices – the good, the bad and the ugly. *Thematic Review Series*. 2014; 88: 3–15.
25. Leone-Bay, A. Pulmonary drug delivery – simplified, Pulmonary & Nasal Drug Delivery: could inhalable delivery be appropriate for your small molecule or biotherapeutic? 18–21. Frederick Furness Publishing, East Sussex, UK, 2011.
26. Luinstra M, Rutgers W, Laar T, Grasmeijer F, Begeman A, Isufi V, Steenhuis L, Hagedoorn P, Boer A, and Frijlink HW. Pharmacokinetics and tolerability of inhaled levodopa from a new dry-powder in patients with Parkinson's disease. *Therapeutic Advances in Chronic Disease*. 2019; 10: 1–10.
27. Newman SP and Busse WW. Evolution of dry powder inhaler design, formulation, and performance. *Respiratory Medicine*. 2002; 96: 293–304.

36

Lung Delivery of Nicotine for Smoking Cessation

Hui Wang[1] **and Nazrul Islam**[2]
[1]*Livzon Microsphere Technology Co. Ltd, Zhuhai, Guangdong, China*
[2]*Pharmacy Discipline, School of Clinical Sciences, Queensland University of Technology, Brisbane, QLD, Australia*

36.1 Introduction

Smoking has become one of the leading causes in preventable deaths in the world, accounting for approximately 20% of total deaths and 90% of lung cancer. Cigarette smoking involves the regular and habitual inhalation of smoke owing to a physical addiction to some chemicals, primarily nicotine. It is reported that around 80% of smokers attempt to quit smoking within the first month of abstinence, however, only 3–5% of them remain abstinent after 6 months. "Offer help to quit tobacco use" is one of the proven effective strategies by the World Health Organization (WHO) Framework Convention on Tobacco Control (FCTC) to prevent the expanding epidemic of tobacco dependence (1).

Yet nicotine is only partially responsible for the detrimental effects of smoking, as several constituents from the cigarette smoke result in the development and progression of heart, blood vessel, and cardiovascular damage, particularly atherosclerotic lesions (2). Although there are currently effective FDA-approved therapies which have been rationally designed and tested using a combination of numerous animal models for nicotine addiction (3), few effective drug therapies for nicotine addiction can be effectively applied. Even the currently most effective drugs, varenicline and cytisine, can cause significant side effects, such as nausea, and require substantial dosing in order to bypass the gut and enter the brain. Therefore, a better method of pharmacotherapeutic delivery is highly desirable in order to help improve the efficacy of currently available treatments as well as to open opportunities for the development of novel therapeutics.

36.2 Adverse Health Effects of Smoking

There are approximately 5 million tobacco-related deaths annually and the total is expected to reach 10 million by 2030 (4). The positive effects of cigarette smoking include euphoria, ability to concentrate, weight control, and a decrease in daily tension. Nevertheless, its negative effects (Figure 36.1) lead to some major issues. It not only reduces life expectancy among smokers, it also increases overall costs, both directly and indirectly, and contributes to a loss of productivity during the life span. Neither the deteriorating health of smokers and the aggravating economic impact on society can be ignored. A large number of reports have revealed strong associations between tobacco smoke and various neurological, cardiovascular, and pulmonary diseases (5). What is worse, cigarette smoke also contributes to the health problems of nonsmokers.

36.2.1 Smoking and Cardiovascular Diseases

The cardiovascular system is the first target area to be affected by smoking, and the main cardiovascular diseases associated with cigarette smoking comprise myocardial infarction (6), coronary heart disease (7), stroke (8), type 2 diabetes mellitus (9), hypertension (10), and abnormal cardiopulmonary function and exercise tolerance (11), all of which exhibit a synergistic effect with smoking. Different types of cardiovascular damage from smoking have been identified. The first study by Hammond (12) demonstrated that smoking was related to an increase risk of cardiovascular disease. It stated that smokers would suffer a 70% higher rate of death from coronary artery disease than nonsmokers. The smoker's age also affects the development of coronary heart disease, lung cancer, cerebrovascular disease, and chronic obstructive pulmonary disease (COPD). The greatest risk for developing coronary heart disease occurs in those under the age of 45. Lung cancer is the dominant cause of increased mortality associated with cigarette smoking after age of 50. Risks of death increase steeply, and excess death from COPD is confined largely to the seventh and eighth decades of life. Active smoking leads to clinical and anatomical alterations related to ischemic pathology, whereas passive smoking (second-hand smoking) is more likely to trigger functional disorders chronically which are associated with late atherosclerotic changes (13).

Adverse effects of tobacco smoke on the heart and vessels result from many chemical compounds extracted and concentrated from tobacco mixtures. The tobacco plants and their leaves contain more than 4000 chemical substances which result in increasing risks of cardiovascular effects in humans and animals. Carbon monoxide is one of major components associated with the cardiovascular effects of smoking, which adversely alters the myocardial oxygen supply/demand ratio and produces endothelial injury, leading to the development of atherosclerotic plaque (14). Furthermore, the biochemical or hypoxic injury of tobacco mechanisms also strongly contributes to these pathogenic behaviors.

DOI: 10.1201/9781003046547-36

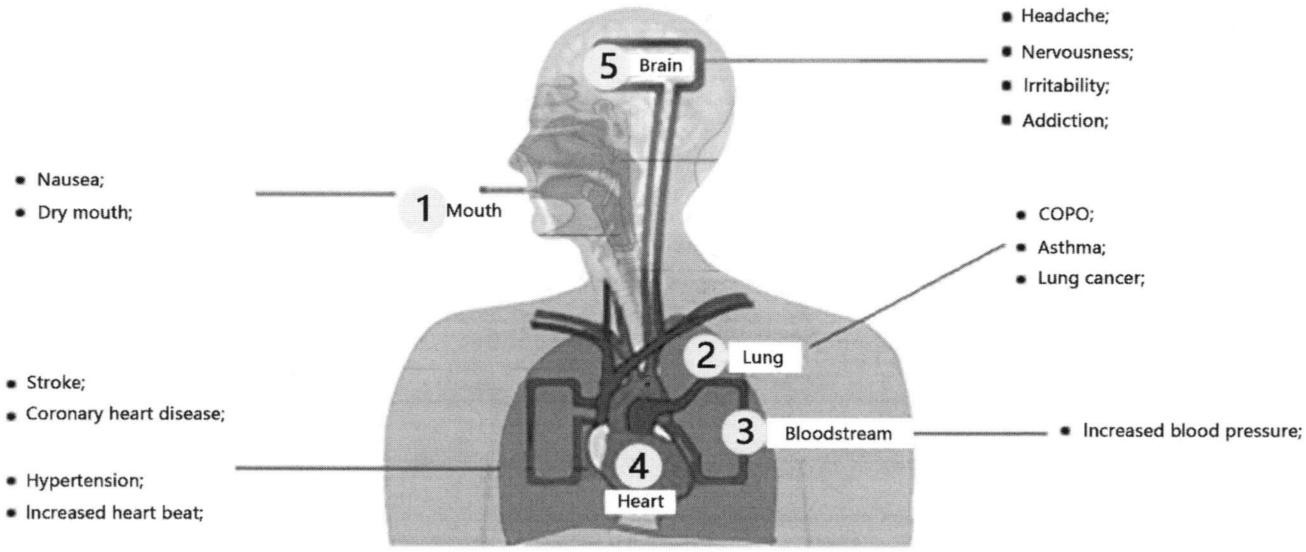

FIGURE 36.1 Smoking and Increased Health Risk

36.2.2 Smoking and Respiratory Diseases

Tobacco smoke is responsible for the pathogenesis of many types of respiratory diseases. Specifically, cigarette smokers have a higher prevalence of respiratory symptoms and lung function abnormalities with a greater mortality rate from COPD compared with nonsmokers (15). The method of consuming tobacco is also important in the risk of developing disease. For example, pipe and cigar smokers have a higher risk of developing COPD and much greater COPD morbidity and mortality rates than cigarette smokers. Tobacco smoke can alter the airway structures and functions, triggering COPD by interfering with the ability of airway epithelial cells to support repair processes (16). There are many other risk factors for the development of COPD, for example, air pollution, occupational exposures, heredity, and infectious and allergic conditions. Thus, not all smokers will develop clinically significant COPD, but mortality rates are higher among cigarette smokers suffering from COPD (17). It has been reported by some researchers that tobacco smoke inhalation results in a decreased level of lung surfactant phosphatidylcholine (18). Nevertheless, other investigators have reported that there is no difference in lung lavage phospholipid concentration between smokers and nonsmokers (19). These findings indicate that the effects of cigarette smoke on surfactant and surfactant-producing cells are complicated, probably due to the heterogeneous nature of the smokers (20).

Asthma is an inflammatory airway disease involving both airway inflammation and impaired airflow. It is well documented that smoking or exposure to passive smoking among asthmatics increases the risk rate of asthma-related morbidity and disease severity because of the decline of lung function. It is reported that approximately 25–35% of individuals suffering from asthma are tobacco smokers, which is much higher compared with nonsmokers (21). Also women are more susceptible to the effects of smoking (22). Asthma is less controlled in asthmatic patients with primary smoking compared with asthmatic nonsmokers (23).

On the other hand, some of the toxic gases in passive or second-hand smoking contribute to the development of asthma (22). Environmental smoking exposure among patients with asthma is not only associated with more severe symptoms, but also with lower quality of life, reducing lung function, increasing health care utilization and hospital admissions. However, young children are the most vulnerable to tobacco smoke, especially maternal smoking or tobacco smoke at home.

36.2.3 Smoking and Lung Cancer

Environmental tobacco smoke, also referred to as second-hand smoke (SHS), has been classified as a known human pulmonary carcinogen (24). So far, a series of epidemiological studies have established a link between SHS exposure and lung cancer development and have shown an association between the exposure to SHS and the risk of lung cancer development. Lung cancer is one of the most commonly diagnosed cancers and the leading cause of cancer death in the world. Some researchers reported that only a small proportion of individuals who were exposed to an environment of extensive tobacco consumption would suffer from lung cancers (25), while epidemiological evidence indicated that cigarette smoking is one of the established risk factors for lung cancer. This controversy has made the mechanisms by which cigarette smoke causes lung cancer unclear. It was estimated that about 90% of male and 75–80% of female deaths from lung cancer in the United States were due to smoking (24). More than 60 chemical compounds in the cigarette smoke have been identified to have a specific carcinogenic possibility, and these carcinogens and their metabolites (e.g. N-nitrosamines and polycyclic aromatic hydrocarbons) can activate multiple pathways, resulting in lung cell transformation in different ways. Nicotine, which is commonly known to be responsible for tobacco addiction, is also involved in tumor promotion and progression with anti-apoptotic and indirect mitogenic properties. Lung nodules are often diagnosed among smokers and can transform into malignant tumors depending on the patients' persistence in smoking.

Any agent, from chemical and physical to viral, may be a carcinogen that causes cancer or increases the incidence, evidenced by experimental animals and human beings. These constituents in tobacco include aromatic amines (e.g. 4-aminobiphenyl, 2-naphthylamine), polycyclic aromatic hydrocarbons, tobacco specific nitrosamines [e.g. 4-(methylnitrosamine)-1-(3-pyridyl)-1-butanone] (26). These carcinogens form the link between nicotine addiction and lung cancer. A genotoxic mode of action for some of these carcinogens generates directly electrophilic species capable of forming covalently bound DNA lesions, known as DNA adducts, which are central to the carcinogenic process (27). Formation of DNA adducts can initiate carcinogenesis, giving rise to the incidence of mutations.

36.3 Mechanism of Nicotine Addiction

Besides a variety of tobacco-related diseases, addiction is one of the most notorious tobacco-related syndromes and is mainly attributed to nicotine. Nicotine withdrawal leads to the many adverse effects of smoking cessation such as nervousness, irritability, anxiety, impaired concentration, and weight gain, which make the process of quitting smoking extremely undesirable for the smoker despite knowing the negative health impact of continued use.

Nicotine, a potent parasympathomimetic alkaloid and a stimulant drug, is a natural ingredient growing in the roots and accumulating in the leaves of tobacco (28). It was first isolated from the tobacco plant in 1828 by physician Wilhelm Heinrich Posselt and chemist Karl Ludwig Reimann of Germany (29). It is a hygroscopic and oily liquid that is miscible with water in its base form, whereas in salt form it is usually solid and water soluble. Nicotine has two enantiomeric forms. Levorotary (S)-isomer is the principal form of nicotine in tobacco, whereas only 0.1–0.6% of total nicotine content is (R)-nicotine (30). The (S)-nicotine is physiologically less active than (R)-nicotine, while (R)-nicotine is more toxic than (S)-nicotine. The salts of (S)-nicotine are usually dextrorotatory (31).

Nicotine addiction is a complex behavioral phenomenon with causes and effects ranging from molecular mechanisms to social interactions, which influences the rate and route of nicotine dosing. The smokers who smoke regularly consume about 17 cigarettes per day, about 0.3 mg/kg daily of nicotine (32). Essentially, nicotine mimics the actions of the hormone epinephrine and the neurotransmitter acetylcholine in the brain. The process of nicotine addiction begins with inhalation of nicotine via cigarette smoking. The lungs distil nicotine from the tobacco smoke and allow absorption of nicotine into the arterial bloodstream rapidly, followed by rapid distribution to body tissues within 8 minutes. This produces a transient exposure of high levels of nicotine to the brain, where neurotransmitters are released, enhancing the smokers' feelings of pleasure. This exposure acts on nicotinic cholinergic receptors, triggering the release of endorphins, dopamine, and other neurotransmitters, e.g. glutamate and γ-aminobutyric acid (GABA) (33).

The release of neurotransmitters is particularly important to the development of nicotine dependence. Neuroadaptation and tolerance involve changes in nicotinic receptors and neural plasticity. High nicotine levels travel through the body to reach the brain within 10 seconds of smoking, followed by a rapid decline only a few minutes after smoking. In human studies, arterial nicotine levels peak at 100 ng/ml, while venous nicotine levels are as high as 10–50 ng/ml (34). Nicotine from sequential cigarette smoking is capable of overcoming tolerance to produce further pharmacological effects. Therefore, smokers have to smoke more and more to maintain the same feelings produced by smoking. In the bloodstream (pH 7.4), 69% of nicotine is ionized and 31% of nicotine is non-ionized where binding to plasma proteins is less than 5%. Liver, brain, kidney, spleen, and lungs have higher tissue affinity compared with the adipose tissue. Nicotine could pass through the blood–brain barrier and bind to brain tissue. Nicotine is metabolized primarily by the enzyme CYP2A6, and variation in the nicotine metabolism rate contributes to differences in vulnerability to tobacco dependence and the response to smoking-cessation treatment (35). An increased understanding of the mechanisms of nicotine addiction (Figure 36.2) has led to the development of novel medications that act on specific nicotinic receptor subtypes in the brain.

36.4 Pharmacokinetic Considerations

Nicotine is extensively metabolized by liver to several metabolites, including cotinine, nicotine-N'-oxide, nicotine glucuronide, and other minor metabolites. Cotinine is the most important metabolite in humans; 70–80% of nicotine is metabolized to cotinine, and about 4% nicotine is converted to nicotine-N'-oxide (36). Cotinine is primarily metabolized to cotinine glucuronide, trans-3′-hydroxycotinine and 3′-hydroxycotinine glucuronide. Subsequently, nicotine and cotinine are further metabolized by N-glucuronidation, while 3′-hydroxycotinine are further metabolized by O-glucuronidation. It is reported that cotinine levels are almost 15 times higher than plasma nicotine in the chronic nicotine exposure subjects (37). Similarly, nicotine in most animal species is primarily metabolized to cotinine. LD_{50} of nicotine for an adult is recorded to be about 60 mg (38). However, nicotine in rats and guinea pigs is converted to nicotine-N′-oxide and 3′-hydroxycotinine as much as cotinine (37). Therefore, they are probably not the ideal animal models when investigating nicotine metabolism in humans.

Metabolites are excreted to a high degree, and only a small number of unchanged metabolites are kept in the kidney. The elimination half-life of plasma nicotine is 2–3 hours during the use of nicotine products. With regular smoking, nicotine levels accumulate over 8 hours and reach a steady state after four half-lives. On the other hand, a long terminal half-life of about 20 hours or more presents in the highly sensitive assays. Although pharmacological implications are unclear about persisting low levels of nicotine in tissues in a long terminal half-life, it reflects the slow release of nicotine from binding sites of body tissues. For example, in a chronic smoker who

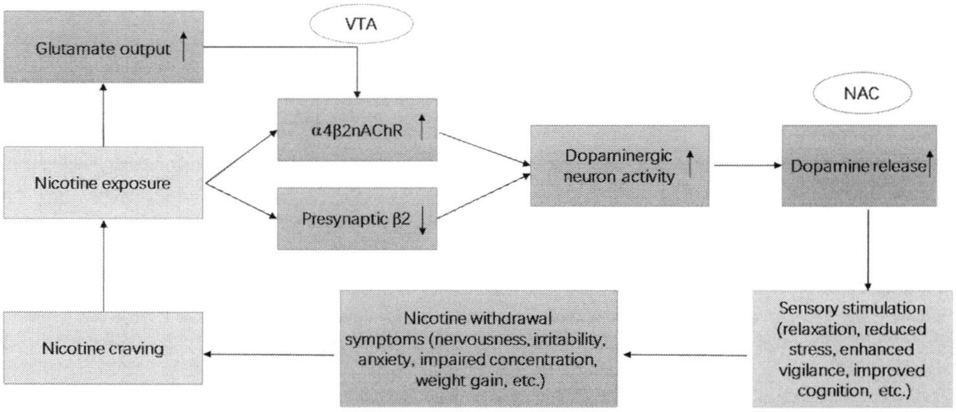

FIGURE 36.2 The Mechanism of Nicotine Addiction

does not smoke overnight, significant nicotine levels (4–5 ng/ml) can still be detected the next morning.

Investigations of pharmacokinetics (PK) and pharmacodynamics (PD) of nicotine provide us with crucial clues to study behavioral, physiological, and biochemical effects of nicotine in different species. Dose selection to study PK/PD of nicotine should consider several factors, including species, route of administration, rate of exposure, uptake, metabolism, and excretion.

36.4.1 Acute Nicotine Treatment

A number of studies have been carried out by researchers to investigate nicotine pharmacokinetics and behavioral response in rodents, providing a wealth of information on physiobiological and neurobiological mechanisms of nicotine, and making translation of preclinical findings to human studies possible. The pharmacokinetics of nicotine were evaluated at the variable doses from 0.08 to 1.0 mg/kg by intravenous (i.v) injection in rats. The maximum plasma concentration was 149 and 297 ng/ml for doses of 0.4mg/kg and 0.8 mg/kg, respectively, and half-life was between 55 and 64 minutes with dose-dependent clearance (39). Studies by Kyerematenl et al. delivered nicotine of 0.1–1.0 mg/kg to rats by intra-arterial injection (i.a) and peak plasma nicotine level was observed at 70 ng/ml; plasma nicotine half-life ranged from 0.9 to 1.1 hours; and total body clearance of nicotine ranged from 2.9 to 3.9 l/hr/kg (40). With advancements of aerosol and inhalation technology in development of inhalation medications, some researchers evaluated pharmacokinetics of nicotine by cigarette smoke inhalation. In one experiment (41), rats were exposed to smoke inhalation and PK was compared with i.v administration. The delivery rate of inhalation chamber is about 68%, which represented the amount of nicotine absorbed by rats, ranging from 0.04 to 0.06 mg/kg. The results indicated absorption rate of inhalation was fast and obtained plasma nicotine levels were parallel with a single dose of I.V. injection. Meanwhile, Lefever et al. compared pharmacokinetic effects between genders in C57BL/6 mice by route of s.c. injection and aerosolization, which indicated both male and female mice were sensitive to nicotine below 0.75 mg/kg (s.c. injection), but stimulus effects of nicotine were decreased in mice when the administration route changed from s.c. injection to aerosolization (42).

As nicotine is capable of altering locomotor activity, locomotor activity of rats and mice is commonly applied to investigate changes in spontaneous activity in response to pharmacological treatments. Nicotine binds to nicotinic acetylcholine receptors (nAChRs) in the brain and different subtypes of nAChRs are responsible for the locomotor stimulant and depressant effects of nicotine (43,44). The depressant effect of nicotine is possibly due to the blockage of neuronal nAChRs (44), but the mechanism by which nicotine acts as an inhibitor remains obscure. Both effects of nicotine can be inhibited by the nicotinic antagonists such as mecamylamine (45). Kumar reported that the dose of 0.4 mg/kg nicotine (s.c.) resulted in depressed activity and induced ataxia in the first 20 minutes, followed by stimulant activity in the later session in non-tolerant rats (46). Locomotion response from mice indicated that 0.5 mg/kg of nicotine caused minimal activity-increasing effects and locomotor activity was decreased with higher concentrations of nicotine because of the blockage of neuronal nAChR (47). However, hypoactive effect was prevented by mecamylamine (0.5, 1.0 mg/kg s.c.). On the other hand, chronic repeated nicotine exposure in rodents would cause nicotine tolerance, as evidenced by a reduced depressant phase and a higher required dose to induce stimulant locomotor response (48). Rats with 0.4 mg/kg s.c. administration every day showed the stimulant activity throughout the session in a dose-dependent manner, while little ataxia was seen in a larger dose of 0.8 mg/kg s.c. nicotine administration (46). The behavior performance induced by nicotine was blocked by mecamylamine (0.1, 0.32, 1.0 mg/kg s.c.) in a dose-related manner.

36.4.2 Chronic Nicotine Treatment

Cigarette smoking is a chronic neurobiological adaptation to nicotine exposure. Smokers are exposed to intensive levels of nicotine, followed by a cycle of desensitization, which causes fluctuations of blood nicotine levels. Therefore, to study chronic nicotine exposure it is extremely important to understand the effects of intermittent nicotine delivery in the long

TABLE 36.1

Pharmacokinetics in Representative Rat and Mouse Studies

Administration Route	Species	Acute/ Chronic	Dosage (mg/kg)	C_{max} (ng/mL)	$t_{1/2}$ (h)	Ref.
i.v	Rats	Acute	0.08–0.8	44-297	0.92–1.10	(39)
i.a.	Rats	Acute	0.1–1.0	30-70	0.9–1.1	(40)
i.v	Rats	Acute	1.0	262	0.86	(52)
inhalation	Rats	Acute	0.04–0.06	79.2	1.4	(41)
Inhalation	Mice	chronic	e-cigarette	21.8	2.1	(50)
inhalation	Rats	chronic	0.1% in solution (nebulizer)	38.8	4.6	(49)

term. Xuesi reported an alveolar region–targeted aerosol exposure system which can produce a clinically relevant animal model from chronic intermittent nicotine exposure (49). Rats were exposed to nicotine aerosol once a day for half an hour and repeated for 10 days in the dark phase of a 12/12 hours circadian cycle by using a whole-body exposure chamber. The particle size of nicotine aerosol is within respirable diameter, with MMAD ranging from 1 to 4 μm. Plasma nicotine levels and cotinine levels were 30–35 and 190–240 ng/ml, respectively. This nicotine levels were clinically relevant with chronic intermittent nicotine exposure and kinetic pattern, and comparable with chronic smokers. A mouse model for chronic nicotine exposure was developed by an e-cigarette aerosol exposure system (50). Mice were exposed to e-cigarettes with three different doses in the 12/12 hours reversed light-dark cycles for 9 consecutive days. The plasma nicotine/cotinine concentration, body weight, food intake, and locomotor activity were observed. The maximum plasma nicotine level was 21.8 ng/ml with exposure dose dependent. Mice body weight and food intake were decreased, whereas locomotor activity was increased after chronic nicotine exposure. Moreover, the mice were vulnerable pulmonary hypertension because both systemic and pulmonary blood pressure were altered with chronic nicotine exposure (51) (Table 36.1).

36.5 Current Therapeutic Options for Smoking Addiction

Present clinical practice guidelines divide pharmacotherapies for the treatment of tobacco addiction into first-line medications, including nicotine replacement therapy (NRT), bupropion, and varenicline and second-line medications, including nortriptyline and clonidine. These medications can also be used in combination. A general strategy of pharmacotherapy is to begin with a single medication and add a second if smokers present severe craving symptoms or difficulties maintaining abstinence.

36.5.1 Nicotine Replacement Therapy (NRT)

NRT is the primary medication used for the treatment of tobacco dependence. Available NRT formulations include nicotine chewing gums, transdermal nicotine patches, sublingual tablets, nasal sprays, and inhalators. All of these are now available on prescription or can be purchased at pharmacies. Each form of drug administration, through randomized, controlled clinical trials, has been proved to double the rate of cessation compared with placebo and the efficacy of each them is similar at week 12 of followup. However, a number of practical problems are associated with each nicotine product.

The basic mechanism of action is the provision of nicotine without other harmful chemicals via cigarette smoke, replacing partially the nicotine formerly obtained from tobacco smoking through stimulation of nicotinic receptors in the ventral tegmental area of the brain and consequent release of dopamine. Thus the severity of withdrawal symptoms and cravings are reduced when smokers try to quit tobacco (53). Concerning the safety issues of NRT, attaining nicotine from NRT is significantly safer than absorbing it from cigarettes, as NRT products avoid the patients being exposed to carbon monoxide or oxidizing gases that are harmful to endothelium. Nicotine gum was the first effective NRT, produced in Sweden, followed by a nicotine patch developed in the US (54). All NRT products appear to be efficacious, leading to a doubling of smokers' quitting chances in comparison with placebo therapy (55). However, NRT could not completely eliminate all symptoms of withdrawal because of its own limitations: (a) no available products could reproduce the rapid and high levels of nicotine in the circulation as achieved by inhaling cigarette smoke; (b) different formulations (nicotine gums, lozenges, patches, nasal sprays, and inhalers) could have different impacts upon withdrawal symptoms and craving, but there is little direct evidence that one nicotine product is more effective than another; and (c) NRT does not fully overcome the issue of side effects after smoking cessation (56). Thus, NRT is less effective for patients who are not motivated to quit or do not expect to experience nicotine withdrawal symptoms. NRT is generally more effective when smokers who have made a previous attempt to quit in conjunction with psychological interventions.

Nasal sprays are a rapid administration route to deliver nicotine to the systemic circulation and the brain. They are absorbed through the nasal mucosa arterial blood, which avoids first-pass effects, and begin to increase nicotine levels within 1–2 minutes. The bioavailability of intranasal administration is about 50–75% (57), and the the rest of unabsorbed nicotine is swallowed, to be absorbed by the gastrointestinal (GI) tract. The efficiency for abstinence of nasal spray is slightly higher than the standard dose of nicotine patch or nicotine gum in the short term according to a US Public Health Service guidelines

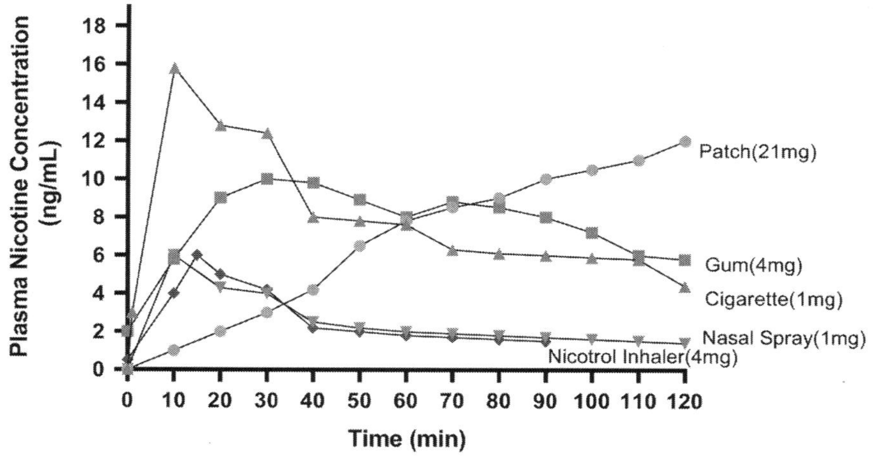

FIGURE 36.3 Plasma Nicotine Levels Administered by Nicotine Nasal Spray, Gum, Patch and Compared with Cigarette Smoking (68,69)

meta-analysis (58). Nicotrol® NS (nicotine nasal spray) is a metered spray absorbed through the nasal mucosa (59). It delivers 1 mg nicotine per dose, with maximum venous concentrations of 2–12 ng/ml in 4–15 minutes; absorption half-life of nicotine is approximately 3 minutes. Patients are encouraged to start with 1–2 doses per hour, up to a maximum 40 doses per day.

Nicotine inhalers are designed to combine pharmacological and behavioral substitutions for treatment of smoking dependence. It stimulates the act of cigarette smoking with delivering nicotine orally through inhalation from a plastic tube. The bioavailability of nicotine inhalers is approximately 40–50% (60). A nicotine inhaler comprises a mouthpiece and a plastic cartridge which contains 10 mg of nicotine, and 4 mg of nicotine can be extracted. Currently, Nicotrol® (Pfizer) and Nicorette® (Johnson & Johnson) inhalers are available as nicotine vapor inhalers, however, these products deliver nicotine into the buccal areas, rather than the deep lungs, lowering plasma maximum concentration and delaying time to reach maximum concentration. According to a randomized, double-blind, placebo-controlled clinical trial with 247 participants who smoked ≥10 cigarettes a day, 123 participants used a nicotine inhaler, while others were given placebo for 1 year. The abstinence rate after 12 months was 28% and 18% ($p = 0.046$) (61). Absorption of nicotine from nicotine inhalers is relatively slow and ineffective. An average maximum arterial nicotine level is 6 ng/ml within 15 minutes after a single inhalation of Nicotrol inhaler, while smokers are given an intensive inhalation of 20 minutes per hour, for 10 hours; a Nicotrol inhaler typically produces peak plasma concentration at 23 ng/ml at steady-state under room temperature (62).

Nicotine patches are a transdermal administration that releases nicotine into the bloodstream through absorption via the skin so that this route avoids first-pass elimination. It usually delivers nicotine continuously over a period of 16 or 24 hours with doses of 14 or 21 mg/patch. The bioavailability of transdermal administration is about 68%. The mean absorption time (MAT) and elimination half-life of nicotine is 4.2 and 2.8 hours, respectively (63). There is a significant difference in effectiveness between genders; transdermal nicotine may be less effective for female compared with male smokers (64). Nicorette Invisi® patch contains nicotine 1.75 mg/sq.cm, releasing nicotine 25 mg for 16 hours, with peak plasma concentrations of 25 ng/ml in 9 hours for 25 mg patch.

Nicotine chewing gum is delivered through buccal mucosa, containing 2 or 4 mg nicotine. Compared with cigarette smoking that reaches maximum nicotine level within 1–2 min about 49 ng/ml, nicotine is absorbed much more slowly from nicotine gum. After a single dose administration of 2 mg nicotine gum, plasma nicotine level was not obtained; however, plasma nicotine levels increase gradually over 15–30 minutes, peaking at 40 ng/ml on the 4 mg gum (65). Heavy smokers (≥25 cigarettes per day) are encouraged to start with 4 mg/piece gum per hour. The users should be instructed properly to chew the gum intermittently for about 30 minutes or until the taste disappears. The bioavailability of nicotine gum is about 50%, means that approximately half of nicotine is absorbed (66). Overall, the slower absorption rate of nicotine gum may limit its therapeutic effects as a substitute for cigarette smoking. However, it was reported that a biphasic nicotine gum with initial rapid delivery and sustained release is in the phase of clinical trials (67). Compared with standard nicotine gums, this product is easier to use, and no parking of gum is required (Figure 36.3, Table 36.2).

36.5.2 Non-nicotine Therapy

36.5.2.1 Bupropion

Bupropion hydrochloride was regarded as an effective smoking cessation aid, with sustained-release oral formulations preferred over immediate release (70). Bupropion SR (Zyban®, GlaxoSmithKline) is taken twice daily, while bupropion XL (Wellbutrin®, GlaxoSmithKline) is taken once daily. Treatment with bupropion doubles the chances of quitting smoking compared with placebo therapy (71). Bupropion in low doses can block brain nicotinic acetylcholine receptor (nAChRs) function, which is associated with dopaminergic and noradrengenergic activity. The blockage of nAChRs function could decrease the positive reinforcement effects of nicotine (72). One randomized, controlled clinical trial compared the efficacy of bupropion alone and in combination with a nicotine patch,

TABLE 36.2

Comparisons mong NRT Products

Administration Route	Dose	Bioavailability	First Pass Metabolism	Common Side Effects	Products
Nasal spray	1–2 dose/h, (maximum 40 mg/day)	50–75%	No	Nasal irritation, cough, sneezing	Nicotrol® NS; Nicotrol®; Nikotugg®;
Nicotine patch	14 mg for 16 h; 16 mg for 24 h	68%	No	Skin irritation, insomnia	Nicorette® Invisi, Nicotine Derm® CQ, Equate™, Habitrol®,
Nicotine gum	1 piece/h (maximum 24 pieces/day)	50%	Yes	Mouth irritation, dyspepsia, sore jaw	Nicorette® Gum; Equate™; Rexall; Nicotinell®; Zonnic®; Rugby®; Zero®; Lucy®; Habitrol®; Nicotex®; Rite Aid; Nicotac®; Nicotrol®; NiQuitin® CQ
Nicotine inhaler	6–16 cartridges/ day, 4 mg/ cartridge)	40–50%	No	Mouth and throat irritation, cough	Nicorette® Inhalator; Nicotrol® Inhaler; Nicosurge®;

which indicated a combination of bupropion and nicotine patch is safe but did not present a significantly higher quitting rate than treatment with bupropion alone. However, physicians tend to use combination therapy for smokers who are heavily dependent on smoking.

Bupropion is safe and effective for the treatment of nicotine dependence in patients who are motivated to quit smoking. At the same time, it is safe for use in patients with cardiovascular disease, but it is limited by an occasional increase in blood pressure which has been reported among smokers with hypertension. In addition, bupropion reduces expired breath CO levels significantly and is not suitable for patients who are at risk of seizure. However, the most common adverse effects with bupropion use are insomnia and dry mouth (73).

36.5.2.2 Varenicline

Varenicline (Chantix/Champix®) was introduced to the market by Pfizer in 2006. It was the first new prescribed drug for smoking cessation in recent years. The drug was developed as a cytisine derivative to increase oral bioavailability and improve brain penetration (74). It is a partial selective agonist for the $\alpha4\beta2$ nAChR subtypes which are responsible for mediating the reinforcing properties of nicotine in the ventral tegmental area of the brain. Coe suggested that $\alpha4\beta2$ nicotinic receptor partial agonists could inhibit activation of dopamine produced by smoking while producing relief from the craving and withdrawal syndrome (75). Varenicline has dual effects, that is, on the one hand, it stimulates nAChRs partially without producing the full effect of nicotine (agonist action), and on the other hand, it blocks nAChRs to prevent the nicotine from tobacco reaching them (75,76). That is, varenicline displays a higher affinity and less functional effect than nicotine, resulting in a smaller release of dopamine. Therefore, the drug can reduce smoking cravings, thereby potentially reducing the risk of relapse.

Varenicline has an acceptable safety and tolerability profile as an option to quit smoking. At the same time, it is also a safe, tolerated, and effective option for treatment of patients with COPD and cardiovascular disease (77). There are more participants quitting cigarettes successfully with varenicline than with bupropion by a meta-analysis from US Public Health Service guidelines (58). However, limited evidence suggests that varenicline may prevent relapse (78). The major adverse effect of varenicline is nausea, but in most patients, nausea tends to subside over time with mild to moderate doses. Varenicline can also produce other side effects, such as headache, insomnia, and gastrointestinal upset. In addition, some patients taking varenicline have experienced depression, suicidal thoughts, suicide, and serious neuropsychiatric symptoms (79).

36.5.3 Electronic Nicotine Delivery Device (E-cigarette)

Electronic nicotine delivery systems (ENDS), also referred to as electronic cigarettes or e-cigarettes, were widely promoted worldwide as a nicotine delivery approach for smoking cessation as using them can reduce urges and withdrawal symptoms and replace many habitual behavioral aspects of tobacco smoking. However, e-cigarettes have not been approved by the FDA and some other countries as a medication for use as a nicotine cessation aid. For example, it is illegal to sell or import them without a license in Australia (80).

E-cigarettes are an aerosolizing delivery device matched with a disposable cartridge. The cartridge contains nicotine in a solution of varying ratios of vegetable glycerin (VG) and propylene glycol (PG) as the humectant, and flavoring agents. The devices generally consist of a battery, a heating element, a power source, and a pressure switch. All these components are embedded in a tube with a mouthpiece (Figure 36.4), and the construction of device may vary. The first generation of e-cigarettes was a look-alike device, but smokers tended to shift from cig-a-likes to second or third generation devices as e-liquid content and physics of aerosolization are more easily controlled. Currently, there are three related products being sold: delivery devices, cartridges, and refill solutions (81). They still have some common characteristics with pipes, multi-dose inhalers and other drug delivery devices.

There is insufficient evidence to indicate the efficacy of e-cigarettes for smoking cessation. Thirty-eight studies were assessed in a random effect meta-analysis that indicated quitting

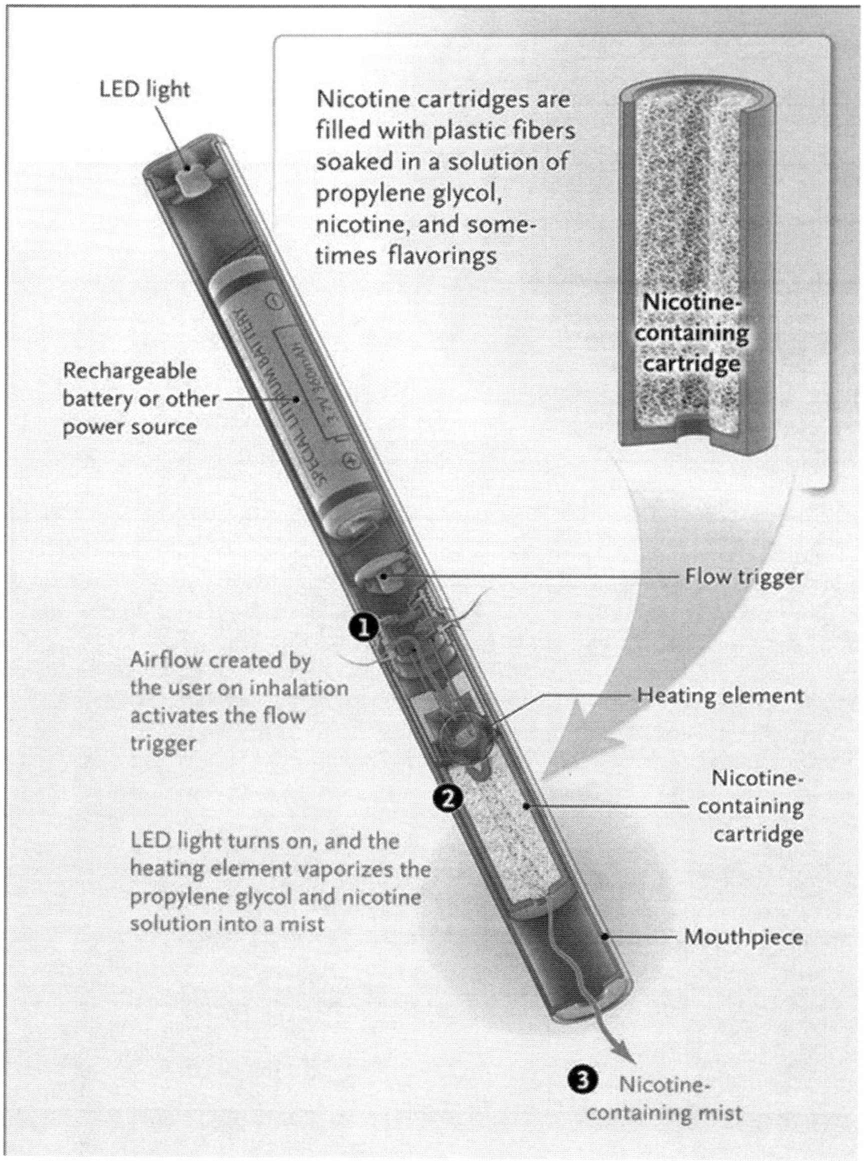

FIGURE 36.4 The Components of an E-cigarette (92)

rate was not significantly different between smokers using e-cigarettes and those not using e-cigarettes (odds ratio [OR] 0.72, 95% CI 0.57–0.91) (80). Moreover, the safety of inhaling e-cigarettes in the long term has not been studied in humans (81). Generally, the contents of cartridge contain coil and wick material such as additional metals and silicate which may cause respiratory tract infection and may also increase the risk of lung cancer. In one study, additional metal material, including chromium, manganese, and nickel, were found in five brands of e-cigarettes (82). Both nickel and chromium are classified as carcinogenic, respiratory toxicant, and reproductive toxicant (83). Silica wicks are also commonly involved in producing e-cigarettes, which is implicated in respiratory disease, silicosis, and autoimmune dysfunction. According to a report from the US Centers for Disease Control (CDC) in 2014, the number of calls to poison control centers due to e-cigarettes increased significantly, rocketing from one call (0.3%) per month in September 2010 to 215 calls (42%) per month in February 2014 (84). In addition, the appropriate dose of nicotine that is delivered also remains unclear (85). The nicotine contents of e-cigarettes varies widely, ranging between 16 and 24 mg/ml in general, while a regular cigarette delivers about 2 mg nicotine to the smoker. PG and VG are common humectants in a typical e-cigarette as both of them are labeled "generally recognized as safe" by the FDA. However, both PG and VG can be converted into toxic aldehydes when they are heated (86). Furthermore, flavor additives could cause potential respiratory disease. There are more than 7,500 flavors of e-liquids on the market (87). Diacetyl is a typical example of an artificial flavoring that can induce lymphocytic bronchitis and bronchiolitis. Still, the safety of inhaling these flavoring additives is a challenge to scientific researchers, manufacturers, and regulatory officers. These uncertainties are worrisome to users. Although no serious adverse effects (e.g. deaths or events requiring hospitalization) have been reported, most smokers who attempted to use e-cigarettes for quitting smoking considered them ineffective (88). In fact, e-cigarettes have become a "bridge product." Smokers tend to use them in places where smoking is forbidden. On April 25, 2011, the FDA had to announce that initial attempts at regulating "electronic cigarettes" as tobacco products had failed (89). In 2015, the U.S. Preventive Services Task Force considered e-cigarettes inappropriate to recommend for smoking cessation in adults because of limited relative risks estimates (90). Since May 2016, no more applications were submitted to FDA to market e-cigarettes for treatment of smoking dependence (91).

36.6 Pulmonary Delivery of Nicotine Inhalers

The pulmonary delivery route was first reported in 1967 by Herxheimer (83), and it was recognized as one of the most efficient treatments for delivering drugs to the targeted area due to the large surface area of the pulmonary alveoli, small airways with rapid action, low metabolism, and high bioavailability. Compared with NRT products administered by other routes, buccal administration of NRT products exhibit a low oral bioavailability, while sublingual tablets and chewing gums can be swallowed before being absorbed (84). Although nasal spray of nicotine is the fastest way for NRT to allow nicotine to enter into the bloodstream, the rate of adsorption still could not be comparable to cigarette smoking (68). As a result, pulmonary delivery of nicotine has the potential to address the limitations of current NRTs and offers great potential therapeutic values.

Inhalation of nicotine is more likely to raise the effectiveness for the treatment of smoking dependence with rapid action, low metabolism, and high bioavailability. On the one hand, small molecules deposited in the lungs are absorbed rapidly into the systemic circulation due to the huge surface area of the lungs, the highly dispersed nature of an aerosol, good epithelial permeability, and small aqueous volume at the absorptive surface (93). On the other hand, it is attributed to the complex physicochemical properties of nicotine. Nicotine is the prototypic agonist at acetylcholine receptor-gated ion channels and is primarily responsible for the abuse of and dependence on cigarette smoking (86), but there is little evidence showing that nicotine is abused in its pure form (87). Dissolution of nicotine products in the fluid of pH 7.4 in the lungs promotes transition across the membrane. It is semi-volatile at vapor pressure of 0.038 mm Hg (25 °C); deposition by inhalation of nicotine is rapid and primarily in the buccal mucosa and larger airways when nicotine is in the vapor phase. Therefore, it is anticipated that the treatment of nicotine addiction will be improved with fewer side effects using the proposed pulmonary drug delivery method, as the potential risks of chronic nicotine inhalation are far lower than the risks of continued smoking.

Even though currently there are some nicotine inhalation products available on the market, most of them cannot achieve the same plasma nicotine levels as cigarettes. In a non-randomized, open-label crossover clinical study, 10 smokers were administered 10 puffs (50 μg/puff or 100 μg/puff nicotine) of 20 inhalations over 5 minutes from a metered dose inhaler (MDI). The plasma concentration reached maximum of about 12.5 ng/ml (~25.9 ng/ml from a cigarette smoking) in 6 minutes (94). Moreover, delivery of a consistent dose that can satisfy the pharmacokinetic requirements in one or two puffs have not been developed. The limitations associated with MDI of nicotine included upper airway irritation and poor deep lung delivery (94). Although Nicotrol® and Nicorette® inhalers are available as nicotine vapor inhalers, these devices deliver nicotine into the buccal areas, rather than the deep lungs, lowering plasma maximum concentration and delaying time to reach maximum concentration (95). Novel products are required for prolonged use that provide a rapid increase in nicotine concentration in brain and plasma, similar to that achieved by cigarette smoking, and a sustained maintenance of an effective brain concentration of nicotine will probably be needed. Therefore, the controlled release of nicotine directly into the deep lung is likely to be more effective to mimick the rapid effects of cigarette smoking on a physiological level, which could eliminate patient craving and allow the tapering of nicotine level over time to alleviate dependence upon nicotine altogether.

The design of novel nicotine formulations with minimized deposition on the oropharynx and upper airways is of great importance, ensuring a reproducible and effective dose to the deep lung. Thus, the performance of the inhaler is a critical issue, and elaborate design is required such that the performance of the inhaler be designed to deliver nicotine aerosol to the deep lung. Philip Morris developed nicotine (2.5% or 5.0%) products as DPI formulations with aerodynamic diameters between 4 and 5 μm using the PreciseInhale™ aerosolization system. The weighted theoretical total deposition fraction was 53% and 50% from 2.5% and 5.0% nicotine, respectively, in in vivo intra-tracheally exposed rats. The PK profile of nicotine levels in plasma and lung were similar, in that C_{max} reached 71.9 (2.5%) and 108 ng/ml (5.0%) within 0.083 hours of the start of aerosol exposure in the plasma, while t_{max} occurred within 0.7 hours, peaking at 100.9 ng/ml (0.25%) and 135.4 ng/ml (5.0%) in the lung (96).

The properly designed nicotine inhaler is essential to improve the performance of therapeutical effects through minimizing the aerosol loss to the oropharynx and upper airways, and ensuring consistent dose to reach the deep lung with better dispersibility of formulations. The optimized nicotine inhalers can be achieved by adjusting the principal parameters of inhalation device properly, such as flow rate, particle size distribution, duration of inhalation exposure, etc. A pure nicotine delivery system of AERx essence (ARD-1600, Aradigm Corporation, US) is comprised of a hand-held aerosol mechanical device and a single-unit discardable dosage strip containing nicotine salt dissolved in water at pH 3.0–3.1 (88). Mass median aerodynamic diameter (MMAD) was between 2.4 and 2.8 μm for the 10–30 mg/ml nicotine concentrations and FPF showed a linear relationship with loaded dose (89). The highest fine particle dose was 0.73 mg from 1.5 mg loaded dose, which equaled about 50% of 1–2 mg from a typical cigarette. Eighteen adult male participants were involved in a phase I trial (69). The maximum arterial blood concentration of nicotine was achieved 1 minute post administration from inhaled nicotine of AERx Essence (89), and the profile of

TABLE 36.3

Comparisons etween Current Nicotine Inhalers and Cigarette Smoking

Brand Name	Dosage	T_{max}	C_{max}	Status	Ref.
Cigarette smoking	0.8–1.9 mg	5 min	49 ng/ml	/	
Nicotrol® Inhaler	4 mg/10 mg	15 min	6 ng/ml	FDA approved	(62)
Nicorette® Inhalator	1.3–6.2 mg/10 mg	40 min	6–8 ng/ml	FDA approved	(96)
Voke® Inhaler	0.45 mg/9 mg	7 min	8 ng/ml	UK MHRA	(69,98)
AERx Essence®	0.76 mg/1.5 mg	2 min	22.5 ng/ml	Phase I	(69)

nicotine levels was comparable to a cigarette smoking. Inhalation of AERx essence did not show significant lesions clinically, though mild to moderate acute adverse effects such as cough and headache were noted (69).

Voke® nicotine inhaler was launched in the UK market in 2019, and is approved by the Medicines and Healthcare Products Regulatory Agency (MHRA) (90). It is designed based on the MDI technology, containing 20 doses of formulated nicotine with propylene glycol, ethanol, saccharin, levomenthol, and HFA134a propellant in an MDI reservoir, which mimics 20 cigarettes per pack (89,91). The delivery of nicotine aerosol is actuated by a plunger within 0.5 seconds with inhalation at a flow rate of at least 2–3 l/min (97), which is different from the typical flow rate of 60 l/min. FPD ranging from 0.10 to 0.30 mg of nicotine was yielded from 0.43 mg delivered nicotine dose with MMAD less than 5 μm. The mean maximum arterial nicotine concentration is achieved around 7 minutes after the start of inhalation at 8 ng/ml, which is only 5–10% blood nicotine level after smoking, from 0.67 mg nicotine dose after 6.5 minutes administration through 6–8 puffs within 2 minutes. It indicates nicotine is absorbed more slowly from the Voke® inhaler compared with cigarettes. Mild or moderate side effects such as dizziness, headache, dry throat, and cough which are associated with cigarette smoking, are reported (91). A clinical study with multiple doses was carried out among subjects aged 18–55 years to determine pharmacokinetics and tolerability of Voke® nicotine inhaler compared to Nicorette 10 mg Inhalator (98). AUC and C_{max} from the Voke inhaler were lower than those of the Nicorette 10 mg Inhalator, while venous nicotine level from Voke inhaler reached maximum concentration in a shorter time (T_{max}: 18.7 vs. 38 minutes). However, the data on product efficacy and human safety are not yet available.

Furthermore, the Staccato® system (Alexza Pharmaceuticals, CA, USA) is a hand-held device that was approved and marketed for delivery of loxapine for acute treatment of schizophrenia (99). The Staccato system for delivery of nicotine is at the early stage of developing a nicotine inhaler, where the nicotine in the form of film layer solubilizing in organic solvents such as ethanol, acetone, hexane, or methanol is applied by a spray processing (100). However, no data from animals or humans are available to date.

Very recently, researchers have developed polymer encapsulated nicotine with a controlled release profile for prevention of smoking dependence and reduction of nicotine craving (47). It is anticipated that pulmonary delivery of nicotine with sustained release properties would not only mimic the effects of tobacco smoking and produce high levels of nicotine in the brain quickly, but also offer a more targeted delivery strategy which reduces the drug dosage required with minimum side effects and better patient compliance. The novel controlled-release chitosan nanoparticles as carriers for delivery of nicotine to the deep lung were developed from the DPI formulation (101). The micro-aggregates of nanoparticles were in the respirable size (<5 μm). Both the amount of chitosan nanoparticles and surface charge can affect the property of aerosolization, of which the fine particle fraction (FPF) was from 33.4% to 38.8% due to reduced ratio of chitosan nanoparticles to nicotine, from 50% to 25%; however, FPF increased significantly with increased surface zeta potential (94,101). The inhaled nano-aerosol was stimulated by software of a Multiple-Path Dosimetry Model (MPPD) to acquire the deposition patterns into the mice lungs, which demonstrated accumulated deposition was about 21.45% in the tracheobronchial tree (47). Interestingly, the degradation behaviors of chitosan nanoparticles in the lung have not been completely understood and more studies are warranted to ensure safety in pulmonary drug delivery (95) (Table 36.3).

36.7 Conclusions

Smoking-related disorders are still the leading causes of preventable deaths worldwide, causing a serious threat and tremendous economic burden to our public health. The issues of smoking are not going to disappear in the near future. Currently, brain nAChRs agonists, such as varenicline, and nicotine replacement therapy (NRT) in the forms of nicotine gums, patches, nasal sprays, and vapor inhalation formulations are the first-line medications for treatment of nicotine addiction. The absorption of nicotine from cigarette smoking is rapid, and none of the currently available pharmacotherapies can deliver high levels of nicotine to the brain as fast as that achieved by tobacco use. Although e-cigarettes were launched on the market in the early stage, uncertainties of safety, proper dosage, and efficacy limit its applications, and it has been banned by the FDA due to limited relative risks estimates. Therefore, the existing products for smoking cessation are not effective to substitute for cigarette smoking to date.

Pulmonary drug delivery is expected to be advantageous over other routes of drug administration. The inhaled nicotine therapy has the potential to address the limitations of current NRTs and offers great potential therapeutic values. The new approaches to develop novel formulations and inhalers are crucial in order to provide a relatively high lung deposition of nicotine producing high levels of nicotine with a better reduction in craving. The inhalable nicotine encapsulated polymer micro/nanoparticles with controlled release profile may be an effective option for the management of smoking cessation.

REFERENCES

1. WHO. Global Report on the Global Tobacco Epidemic: Implementing Smoke-Free Environments. World Health Organization. 2009.
2. Leone A. Biochemical markers of cardiovascular damage from tobacco smoke. *Current Pharmaceutical Design*. 2005; 11: 2199–2208.
3. Pierce RC, O'Brien CP, Kenny PJ, and Vanderschuren LJ. Rational development of addiction pharmacotherapies. Successes, failures, and prospects. *Cold Spring Harbor Perspectives in Medicine*. 2012; 2: a012880.
4. Islam N and Cleary MJ. Developing an efficient and reliable dry powder inhaler for pulmonary drug delivery–A review for multidisciplinary researchers. *Medical Engineering & Physics*. 2012; 34: 409–427.
5. Das SK. Harmful health effects of cigarette smoking. *Molecular and Cellular Biochemistry*. 2003; 253: 159–165.
6. Hay D and Turbott S. Changes in smoking habits in men under 65 years after myocardial infarction and coronary insufficiency. *British Heart Journal*. 1970; 32: 738–740.
7. Baba S, Iso H, Mannami T, Sasaki S, Okada K, Konishi M, and Tsugane S. Cigarette smoking and risk of coronary heart disease incidence among middle-aged Japanese men and women: The JPHC Study Cohort I. *European Journal of Cardiovascular Prevention & Rehabilitation*. 2006; 13: 207–213.
8. Burns DM. Epidemiology of smoking-induced cardiovascular disease. *Progress in Cardiovascular Diseases*. 2003; 46: 11–29.
9. Al-Delaimy WK, Willett WC, Manson JE, Speizer FE, and Hu FB. Smoking and mortality among women with type 2 diabetes. The Nurses' Health Study Cohort. *Diabetes Care*. 2001; 24: 2043–2048.
10. Kannel W and Higgins M. Smoking and hypertension as predictors of cardiovascular risk in population studies. *Journal of Hypertension*. 1990; 8: S3–S8.
11. Louie D. The effects of cigarette smoking on cardiopulmonary function and exercise tolerance in teenagers. *Canadian Respiratory Journal*. 2001; 8: 289–291.
12. Hammond EC and Horn D. Smoking and death rates—Report on forty-four months of follow-up of 187,783 men. *Journal of the American Medical Association*. 1958; 166: 1294–1308.
13. Penn A, Chen L-C, and Snyder CA. Inhalation of steady-state sidestream smoke from one cigarette promotes arteriosclerotic plaque development. *Circulation*. 1994; 90: 1363–1367.
14. Lakier JB. Smoking and cardiovascular disease. *The American Journal of Medicine*. 1992; 93: S8–S12.
15. Vestbo J, Hurd SS, Agustí AG, Jones PW, Vogelmeier C, Anzueto A, Barnes PJ, Fabbri LM, Martinez FJ, Nishimura M, Stockley RA, Sin DD, and Rodriguez-Roisin R. Global strategy for the diagnosis, management, and prevention of chronic obstructive pulmonary disease. *American Journal of Respiratory and Critical Care Medicine*. 2013; 187: 347–365.
16. Wang H, Liu X, Umino T, Skold CM, Zhu Y, Kohyama T, Spurzem JR, Romberger DJ, and Rennard SI. Cigarette smoke inhibits human bronchial epithelial cell repair processes. *American Journal of Respiratory Cell and Molecular Biology*. 2001; 25: 772–779.
17. Matthay MA, Zimmerman GA, Esmon C, Bhattacharya J, Coller B, Doerschuk CM, Floros J, Gimbrone Jr MA, Hoffman E, and Hubmayr RD. Future research directions in acute lung injury: Summary of a National Heart, Lung, and Blood Institute Working Group. *American Journal of Respiratory and Critical Care Medicine*. 2003; 167: 1027–1035.
18. Subramaniam S, Bummer P, and Gairola C. Biochemical and biophysical characterization of pulmonary surfactant in rats exposed chronically to cigarefte smoke. *Toxicological Science*. 1995; 27: 63–69.
19. Mancini N, Bene M, Gerard H, Chabot F, Faure G, Polu J, and Lesur O. Early effects of short-time cigarette smoking on the human lung: A study of bronchoalveolar lavage fluids. *Lung*. 1993; 171: 277–291.
20. Haagsman HP and van Golde LM. Lung surfactant and pulmonary toxicology. *Lung*. 1985; 163: 275–303.
21. Peters JM, Avol E, Navidi W, London SJ, Gauderman WJ, Lurmann F, Linn WS, Margolis H, Rappaport E, and Gong Jr H. A study of twelve southern California communities with differing levels and types of air pollution: I. Prevalence of respiratory morbidity. *American Journal of Respiratory and Critical Care Medicine*. 1999; 159: 760–767.
22. Stapleton M, Howard-Thompson A, George C, Hoover RM, and Self TH. Smoking and asthma. *The Journal of the American Board of Family Medicine*. 2011; 24: 313–322.
23. Schatz M, Zeiger RS, Vollmer WM, Mosen D, and Cook EF. Determinants of future long-term asthma control. *Journal of Allergy and Clinical Immunology*. 2006; 118: 1048–1053.
24. Hecht SS. Cigarette smoking and lung cancer: Chemical mechanisms and approaches to prevention. *The Lancet Oncology*. 2002; 3: 461–469.
25. Duan Y-Z, Zhang L, Liu C-C, Zhu B, Zhuo W-L, and Chen Z-T. CCND1 G870A polymorphism interaction with cigarette smoking increases lung cancer risk: Meta-analyses based on 5008 cases and 5214 controls. *Molecular Biology Reports*. 2013; 40: 4625–4635.
26. Smith CJ, Perfetti TA, Garg R, and Hansch C. Utility of the mouse dermal promotion assay in comparing the tumorigenic potential of cigarette mainstream smoke. *Food and Chemical Toxicology*. 2006; 44: 1699–1706.
27. Tang D, Phillips DH, Stampfer M, Mooney LA, Hsu Y, Cho S, Tsai W-Y, Ma J, Cole KJ, and Shé MN. Association between carcinogen-DNA adducts in white blood cells and lung cancer risk in the Physicians Health Study. *Cancer Research*. 2001; 61: 6708–6712.
28. Benowitz NL, Hukkanen J, and Jacob III P. Nicotine chemistry, metabolism, kinetics and biomarkers. Handbook of Experimental Pharmacology Nicotine Psychopharmacology. Springer. 2009; 129: 29–60.
29. Henningfield JE and Zeller M. Nicotine psychopharmacology research contributions to United States and global

tobacco regulation: A look back and a look forward. *Psychopharmacology (Berlin)*. 2006; 184: 286–291.
30. Carmella SG, McIntee EJ, Chen M, and Hecht SS. Enantiomeric composition of N′-nitrosonornicotine and N′-nitrosoanatabine in tobacco. *Carcinogenesis*. 2000; 21: 839–843.
31. Obaid AL, Koyano T, Lindstrom J, Sakai T, and Salzberg B. Spatiotemporal patterns of activity in an intact mammalian network with single-cell resolution: Optical studies of nicotinic activity in an enteric plexus. *The Journal of Neuroscience*. 1999; 19: 3073–3093.
32. Benowitz NL and Jacob III P. Daily intake of nicotine during cigarette smoking. *Clinical Pharmacology & Therapeutics*. 1984; 35: 499–504.
33. Benowitz NL. Nicotine addiction. *New England Journal of Medicine*. 2010; 362: 2295–2303.
34. Hukkanen J, Jacob P, and Benowitz NL. Metabolism and disposition kinetics of nicotine. *Pharmacological Reviews*. 2005; 57: 79–115.
35. Shiffman S, Ferguson SG, Dunbar MS, and Scholl SM. Tobacco dependence among intermittent smokers. *Nicotine & Tobacco Research*. 2012; 14: 1372–1381.
36. Benowitz NL. Pharmacology of nicotine: Addiction and therapeutics. *Annual Review of Pharmacology and Toxicology*. 1996; 36: 597–613.
37. Benowitz NL and Jacob P. Metabolism, pharmacokinetics and pharmacodynamics of nicotine in man. In: Martin WR, Van Loon GR, Iwamoto ET, and Davis L. (eds.) Tobacco Smoking and Nicotine: A Neurobiological Approach. Springer US, Boston, MA. 1987; 357–373.
38. Mayer B. How much nicotine kills a human? Tracing back the generally accepted lethal dose to dubious self-experiments in the nineteenth century. *Archives of Toxicology*. 2014; 88: 5–7.
39. Miller RP, Rotenberg KS, and Adir J. Effect of dose on the pharmacokinetics of intravenous nicotine in the rat. *Drug Metabolism and Disposition*. 1977; 5: 436–443.
40. Kyerematen GA, Taylor LH, deBethizy JD, and Vesell ES. Pharmacokinetics of nicotine and 12 metabolites in the rat. Application of a new radiometric high performance liquid chromatography assay. *Drug Metabolism and Disposition*. 1988; 16: 125–129.
41. Rotenberg KS, Miller RP, and Adir J. Pharmacokinetics of nicotine in rats after single-cigarette smoke inhalation. *Journal of Pharmaceutical Sciences*. 1980; 69: 1087–1090.
42. Lefever TW, Thomas BF, Kovach AL, Snyder RW, and Wiley JL. Route of administration effects on nicotine discrimination in female and male mice. *Drug and Alcohol Dependence*. 2019; 204: 107504.
43. Wonnacott S, Bermudez I, Millar NS, and Tzartos SJ. Nicotinic acetylcholine receptors. *British Journal of Pharmacology*. 2018; 175: 1785–1788.
44. Licheri V, Lagström O, Lotfi A, Patton MH, Wigström H, Mathur B, and Adermark L. Complex control of striatal neurotransmission by nicotinic acetylcholine receptors via excitatory inputs onto medium spiny neurons. *Journal of Neuroscience*. 2018; 38: 6597–6607.
45. Newman LA and Gold PE. Attenuation in rats of impairments of memory by scopolamine, a muscarinic receptor antagonist, by mecamylamine, a nicotinic receptor antagonist. *Psychopharmacology (Berlin)*. 2016; 233: 925–932.
46. Clarke PBS and Kumar R. The effects of nicotine on locomotor activity in non-tolerant and tolerant rats. *British Journal of Pharmacology*. 1983; 78: 329–337.
47. Wang H, Holgate J, Bartlett S, and Islam N. Assessment of nicotine release from nicotine-loaded chitosan nanoparticles dry powder inhaler formulations via locomotor activity of C57BL/6 mice. *European Journal of Pharmaceutics and Biopharmaceutics*. 2020; 154: 175–185.
48. McMillan DM and Tyndale RF. Inducing rat brain CYP2D with nicotine increases the rate of codeine tolerance; predicting the rate of tolerance from acute analgesic response. *Biochemical Pharmacology*. 2017; 145: 158–168.
49. Shao XM, Liu S, Lee ES, Fung D, Pei H, Liang J, Mudgway R, Zhang J, Feldman JL, Zhu Y, Louie S, and Xie XS. Chronic intermittent nicotine delivery via lung alveolar region-targeted aerosol technology produces circadian pharmacokinetics in rats resembling human smokers. *Journal of Applied Physiology*. 2018; 125: 1555–1562.
50. Shao XM, Lopez B, Nathan D, Wilson J, Bankole E, Tumoyan H, Munoz A, Espinoza-Derout J, Hasan KM, Chang S, Du C, Sinha-Hikim AP, Lutfy K, and Friedman TC. A mouse model for chronic intermittent electronic cigarette exposure exhibits nicotine pharmacokinetics resembling human vapers. *Journal of Neuroscience Methods*. 2019; 326: 108376.
51. Oakes JM, Xu J, Morris TM, Fried ND, Pearson CS, Lobell TD, Gilpin NW, Lazartigues E, Gardner JD, and Yue X. Effects of chronic nicotine inhalation on systemic and pulmonary blood pressure and right ventricular remodeling in mice. *Hypertension*. 2020; 75: 1305–1314.
52. Hwa Jung B, Chul Chung B, Chung S-J, and Shim C-K. Different pharmacokinetics of nicotine following intravenous administration of nicotine base and nicotine hydrogen tartarate in rats. *Journal of Controlled Release*. 2001; 77: 183–190.
53. Stapleton JA, Watson L, Spirling LI, Smith R, Milbrandt A, Ratcliffe M, and Sutherland G. Varenicline in the routine treatment of tobacco dependence: A pre–post comparison with nicotine replacement therapy and an evaluation in those with mental illness. *Addiction*. 2008; 103: 146–154.
54. Fernö O, Lichtneckert SJA, and Lundgren CEG. A substitute for tobacco smoking. *Psychopharmacologia*. 1973; 31: 201–204.
55. Shiffman S. Nicotine replacement therapy for smoking cessation in the "real world". *Thorax*. 2007; 62: 930.
56. Farley AC, Hajek P, Lycett D, and Aveyard P. Interventions for preventing weight gain after smoking cessation. *The Cochrane Library*. 2012.
57. Benowitz NL, Zevin S, and Jacob P. Sources of variability in nicotine and cotinine levels with use of nicotine nasal spray, transdermal nicotine, and cigarette smoking. *British Journal of Clinical Pharmacology*. 1997; 43: 259–267.
58. Panel TUaDG. Treating Tobacco Use and Dependence: 2008 Update. Edited by Services UDoHaH. US Department of Health and Human Services, Rockville, MD. 2008.
59. Drug.com. Nicotrol NS. Drug. com. 2020.
60. Schneider NG, Olmstead RE, Franzon MA, and Lunell E. The nicotine inhaler. *Clinical Pharmacokinetics*. 2001; 40: 661–684.

61. Hjalmarson A, Nilsson F, Sjöström L, and Wiklund O. The nicotine inhaler in smoking cessation. *Archives of Internal Medicine*. 1997; 157: 1721–1728.
62. Inc. P. Nicotrol Inhaler. Drugs.com. 2020.
63. Gupta S, Benowitz N, Jacob 3rd P, Rolf C, and Gorsline J. Bioavailability and absorption kinetics of nicotine following application of a transdermal system. *British Journal of Clinical Pharmacology*.1993; 36: 221–227.
64. Chai SH, Leventhal AM, Kirkpatrick MG, Eisenlohr-Moul TA, Rapkin AJ, D'Orazio L, and Pang RD. Effectiveness of transdermal nicotine patch in premenopausal female smokers is moderated by within-subject severity of negative affect and physical symptoms. *Psychopharmacology (Berlin)*. 2020; 237: 1737–1744.
65. Russell MA, Feyerabend C, and Cole PV. Plasma nicotine levels after cigarette smoking and chewing nicotine gum. *British Medical Journal*. 1976; 1: 1043–1046.
66. Lunell E and Lunell M. Steady-state nicotine plasma levels following use of four different types of Swedish Snus compared with 2-mg nicorette chewing gum: A crossover study. *Nicotine & Tobacco Research*. 2005; 7: 397–403.
67. Shiffman S, Fant RV, Buchhalter AR, Gitchell JG, and Henningfield JE. Nicotine delivery systems. *Expert Opinion on Drug Delivery*. 2005; 2: 563–577.
68. Rigotti NA. Treatment of tobacco use and dependence. *New England Journal of Medicine*. 2002; 346: 506–512.
69. Cipolla D and Gonda I. Inhaled nicotine replacement therapy. *Asian Journal of Pharmaceutical Sciences*. 2015; 10: 472–480.
70. Maccaroni E, Malpezzi L, and Masciocchi N. Structures from powders: Bupropion hydrochloride. *Journal of Pharmaceutical and Biomedical Analysis*. 2009; 50: 257–261.
71. Huibers M, Chavannes N, Wagena E, and Van Schayck C. Antidepressants for smoking cessation: A promising new approach? *European Respiratory Journal*. 2000; 16: 379.
72. Lerman C, Niaura R, Collins BN, Wileyto P, Audrain-McGovern J, Pinto A, Hawk L, and Epstein LH. Effect of bupropion on depression symptoms in a smoking cessation clinical trial. *Psychology of Addictive Behaviors*. 2004; 18: 362.
73. Tonstad S, Farsang C, Klaene G, Lewis K, Manolis A, Perruchoud A, Silagy C, Van Spiegel P, Astbury C, and Hider A. Bupropion SR for smoking cessation in smokers with cardiovascular disease: A multicentre, randomised study. *European Heart Journal*. 2003; 24: 946–955.
74. Tutka P and Zatoński W. Cytisine for the treatment of nicotine addiction: From a molecule to therapeutic efficacy. *Pharmacological Reports*. 2006; 58: 777–798.
75. Coe JW, Brooks PR, Vetelino MG, Wirtz MC, Arnold EP, Huang J, Sands SB, Davis TI, Lebel LA, Fox CB, Shrikhande A, Heym JH, Schaeffer E, Rollema H, Lu Y, Mansbach RS, Chambers LK, Rovetti CC, Schulz DW, Tingley FD, and O'Neill BT. Varenicline: An α4β2 nicotinic receptor partial agonist for smoking cessation. *Journal of Medicinal Chemistry*. 2005; 48: 3474–3477.
76. Vidal C and Changeux JP. Pharmacological profile of nicotinic acetylcholine receptors in the rat prefrontal cortex: An electrophysiological study in a slice preparation. *Neuroscience*. 1989; 29: 261–270.
77. Tashkin DP, Rennard S, Hays JT, Ma W, Lawrence D, and Lee TC. Effects of varenicline on smoking cessation in patients with mild to moderate COPD: A randomized controlled trial. *Chest*. 2011; 139: 591–599.
78. Cahill K, Lindson-Hawley N, Thomas KH, Fanshawe TR, and Lancaster T. Nicotine receptor partial agonists for smoking cessation. *Cochrane Database of Systematic Reviews*. 2016, Art No. CD006103.
79. Kuehn BM. Studies linking smoking-cessation drug with suicide risk spark concerns. *Journal of the American Medical Association*. 2009; 301: 1007–1008.
80. Sutherland R, Sindicich N, Entwistle G, Whittaker E, Peacock A, Matthews A, Bruno R, Alati R, and Burns L. Tobacco and e-cigarette use amongst illicit drug users in Australia. *Drug and Alcohol Dependence*. 2016; 159: 35–41.
81. Adkison SE, O'Connor RJ, Bansal-Travers M, Hyland A, Borland R, Yong H-H, Cummings KM, McNeill A, Thrasher JF, and Hammond D. Electronic nicotine delivery systems: International Tobacco Control Four-country Survey. *American Journal of Preventive Medicine*. 2013; 44: 207–215.
82. Chen I-L. FDA summary of adverse events on electronic cigarettes. *Nicotine & Tobacco Research*. 2013; 15: 615–616.
83. Herxheimer A, Griffiths R, Hamilton B, and Wakefield M. Circulatory effects of nicotine aerosol inhalations and cigarette smoking in man. *The Lancet*. 1967; 290: 754–755.
84. İkinci G, Şenel S, Wilson C, and Şumnu M. Development of a buccal bioadhesive nicotine tablet formulation for smoking cessation. *International Journal of Pharmaceutics*. 2004; 277: 173–178.
85. Patton JS and Byron PR. Inhaling medicines: Delivering drugs to the nody through the lungs. *Nature Reviews Drug Discovery*. 2007; 6: 67–74.
86. Cunningham CS and McMahon LR. The effects of nicotine, varenicline, and cytisine on schedule-controlled responding in mice: Differences in α4β2 nicotinic receptor activation. *European Journal of Pharmacology*. 2011; 654: 47–52.
87. Balfour DJ. The neurobiology of tobacco dependence: A preclinical perspective on the role of the dopamine projections to the nucleus. *Nicotine & Tobacco Research*. 2004; 6: 899–912.
88. Cipolla D, and Gonda I. Inhaled nicotine replacement therapy. *Asian Journal of Pharmaceutical Sciences*. 2015; 10(6): 472–480.
89. Moyses C, Hearn A, and Redfern A. Evaluation of a novel nicotine inhaler device. Part 1: Arterial and venous pharmacokinetics. *Nicotine & Tobacco Research*. 2015; 17: 18–25.
90. Yach D. Transformation of the tobacco industry. *American Journal of Public Health*. 2019; 109: e11.
91. Moyses C, Hearn A, and Redfern A. Evaluation of a novel nicotine inhaler device: Part 2—Effect on craving and smoking urges. *Nicotine & Tobacco Research*. 2015; 17: 26–33.
92. Chen I-L. FDA summary of adverse events on electronic cigarettes. *Nicotine & Tobacco Research*. 2013; 15: 615–616.

93. Patton JS and Byron PR. Inhaling medicines: Delivering drugs to the body through the lungs. *Nature Reviews Drug Discovery*. 2007; 6: 67–74.
94. Wang H, George G, and Islam N. Nicotine-loaded chitosan nanoparticles for dry powder inhaler (DPI) formulations–Impact of nanoparticle surface charge on powder aerosolization. *Advanced Powder Technology*. 2018; 29: 3079–3086.
95. Wang H, Maqbool F, Ferro V, and Islam N. In vitro enzymatic digestibility of glutaraldehyde-crosslinked chitosan nanoparticles in lysozyme solution and their applicability in pulmonary drug delivery. *Molecules*. 2019; 24: 1271.
96. Nicorette Inhalator. 2020. https://www.nicorette.com.au/products: https://www.nicorette.com.au/products.
97. Beard E, Shahab L, Cummings DM, Michie S, and West R. New pharmacological agents to aid smoking cessation and tobacco harm reduction: What has been investigated, and what is in the pipeline? *CNS Drugs*. 2016; 30: 951–983.
98. Romeu ER. Voke® Nicotine Inhaler for Smoking Cessation and Reduction. Regional Drug & Therapeutics Centre Northern Treatment Advisory Group. 2020; 1–18.
99. Dinh K, Myers DJ, Glazer M, Shmidt T, Devereaux C, Simis K, Noymer PD, He M, Choosakul C, Chen Q, and Cassella JV. In Vitro Aerosol Characterization of Staccato® Loxapine. *International Journal of Pharmaceutics*. 2011; 403: 101–108.
100. Simis K, Lei M, Lu AT, Sharma KCV, Hale RL, Timmons R, and Cassella J. Nicotine aerosol generation fromthermally reversible zinc halide complexes using the Staccato® system. *Drug Development and Industrial Pharmacy*. 2008; 34: 936–942.
101. Wang H, George G, Bartlett S, Gao C, and Islam N. Nicotine hydrogen tartrate loaded chitosan nanoparticles: formulation, characterization and in vitro delivery from dry powder inhaler formulation. *European Journal of Pharmaceutics and Biopharmaceutics*. 2017; 113: 118–131.

37

Formulation and Characterization of Dry Powder Inhalers for Pulmonary Drug Delivery

Himangshu Sarma[1,2], Nayanmoni Boruah[2], and Hemanta Kumar Sharma[2]
[1]*Life Sciences Division, Institute of Advanced Study in Science and Technology, Guwahati, Assam, India*
[2]*Department of Pharmaceutical Sciences, Faculty of Science and Engineering, Dibrugarh University, Dibrugarh, Assam, India*

37.1 Introduction

Inhalation drug delivery therapy has been recognized for many years to deliver pharmacologically active substances for treatment of respiratory diseases such as asthma (1). It has tremendous benefits in managing respiratory diseases compared to other formulations such as oral or parenteral dosage forms. In this formulation, a minute therapeutic dose of drug is delivered topically into the airways where the active drug exerts its beneficial effects locally within the lungs. It minimizes the unwanted systemic effects due to the medication effects within the respiratory tract and rapid onset of action (2). The corticosteroids, bronchodilators, mast cell stabilizers, and anti-cholinergic agents are primarily used in modern therapy for asthma or COPD by applying a pressurized metered-dose inhaler (MDI) or aerosol. The MDI is worked by propellants (3). It became rapidly popular owing to the portable size, easy dose administration, comparatively faster dose delivery than other nebulizers, and low cost. However, MDI is creating environmental concerns due to the presence of propellants (4–6). As a result, to minimize the hazard, a range of alternative delivery systems is being formulated. The DPI are being actively developed, formulated, evaluated, and utilized to treat respiratory diseases owing to the uncertain circumstance about MDI. It does not contain any propellants. Therefore, DPI are presently believed to be a practical and significant system for pulmonary drug delivery having potential for improving new therapies. This article addresses the development of dry powder inhalers, including the design, characterization of the formulation, advantages and disadvantages, and feature applications.

37.2 History of Inhalation Therapy

The word *inhaler* was first applied in *A Radical and Expeditious Cure for a recent Catarrhous Cough* by the English physician John Mudgein in 1778 (7). The history of inhalation therapy began more than 4000 years ago in India when powdered formulation containing potent anti-cholinergic alkaloids of *Durata stramonium* and *Durata ferox* were smoked, mixed with pepper and ginger (8). Inhalation therapy acquired importance for managing respiratory diseases, especially after Philip Stern address it in his 1764 book, *Medical Advice to the Consumptive and Asthmatic Peoples of England* (9). In the 18th century, several types of ceramic inhalers were developed to draw the inhaled air through infusions of medicinal product (8). Hand–bulb nebulizers were coming into the market in the early 1900s to administer adrenaline chloride as a bronchodilator, while ultrasonic nebulizers were developed in the 20th century. An MDI was introduced by Riker Laboratories in 1956 (3,8). The first powered or pressurized inhaler was manufactured by Sales-Gironsin France in 1858 (9).

DPI were first patented in the United Kingdom by Vincent Alfred Newton in 1861 (10). This device was not produced in industrial scale to deliver pulverized potassium chloride. Nearly 100 years later, in 1948, Abbott launched a bronchodilator, Aerohalor, and used it to deliver norethisterone and penicillin. It served as a prototype for several capsule inhalers formulated between 1950 and 1980. In 1967 the first inhaler device, FisonsSpinhaler®, was marketed with 20 mg of cromoglycate sodium (9,10).

37.3 Advantages of Dry Powder Inhaler for Drug Delivery

Following are some essential beneficial functions of DPI (11,12).

1. DPI has fewer side effects because the medicament is not exposed to the rest of the body organs.
2. Medicaments are given directly by the patient's pulmonary route to produce a quick onset of action.
3. Medicaments are primarily released in the alveolar surface of the lung, which minimizes the amount of dose.
4. It is a non-invasive drug delivery system.
5. It offers potential and economical treatment for respiratory and other diseases.
6. With the help of DPI the protein molecules, which might otherwise be degraded in the gastrointestinal tract or eliminated by first-pass metabolism, are administered easily through the pulmonary route.

DOI: 10.1201/9781003046547-37

7. Additionally, DPI can be utilized in the management of diabetes mellitus.
8. DPI have been designed to minimize ozone-depleting and greenhouse gases, such as chlorofluorocarbons and hydrofluoroalkanes.
9. The formulations are more stable compared to aqueous drug solutions, and do not necessitate reconstitution of powders into solutions or cold chain storage for nebulization.

37.4 Aspects of DPI Design

Pulmonary drug delivery is a unique way to deliver drugs directly to lung tissues for systemic absorption of drugs from the lung. Further, the application of the right inhaler with good adherence is also needed for effective treatment. Therefore inhaler design, as well as powder formulation, are two essential prerequisites that must be considered during formulation.

In the last few decades, DPI have improved significantly for better drug delivery. A DPI is a pulmonary drug delivery device that delivers medicament or medicaments directly to the alveolar surface of the lungs as well as the respiratory tract in a dry powder form, and where medicament or medicaments are absorbed into systemic blood circulation. The DPIs have directly yielded 1–5 µm particle size medicaments in a mixture with excipients in an aerosol form. The most commonly used carrier is lactose monohydrate (12,13).

DPI are primarily classified into two leading groups, viz. active and passive. The passive breath-actuated DPI devices are primarily available in the market; patients can simply inhale and deeply access the medicament without coordinating breathing. Passive DPI can be subclassed into two categories: pre-metered and device metered. Pre-metered DPI may be a single or multi-dose where the dose is premeasured during the manufacturing process, such as blister, capsule, etc. Device-metered DPI are comprised of a reservoir within the device, which provides a premeasured single dose on actuation (12,14).

37.5 Formulation of Powders for Inhalation

Medicaments delivered by DPI are formulated as either pure medicament or mixed with inactive excipients. Generally, powder blends contain micronized particles of the medicament with an excipient such as lactose, which may act as a carrier molecule. The optimal therapeutic particle size distribution in DPI for an inhaled asthma medication is considered to be 1–5 mm (9) but these small particles are usually not free-flowing.

37.6 The Technical Design of DPI

A typical well-constructed DPI system consists of primary and secondary functional parts. DPI consist of a pre-filled compartment for single dose or multi-dose along with a measuring system for the powder formulation and dispersing system for powder, making it suitable for powder drug delivery. Secondary functional parts involve the activation of correct dose, inhalation of correct dose, and tracking the dose left in the inhaler. Using multi-dose devices, protective measures can be provided against moisture uptake, or exhalation throughthe inhaler can provide - protective measures against moisture uptake during application. One of the most recent developments in DPI is a good signaling process to the patient, which was first introduced in Novolizer® (15).

37.6.1 Dry Powder Inhalation Formulations

The administration as well as accurate measurement of the dose of a microgram-sized medicament are challenges for DPI formulation. It has to be processed into a suitable free-flowing powder formulation for inhalation. The soft spherical agglomerates and adhesive mixtures are two typical powder formulations for low-dose anti-asthmatic and COPD therapy. Soft pellets are more suitable for the administration of milligram-range particle size medicaments. Adhesive mixture type formulations have a limited drug containing capacity (16). For example, Turbuhaler® DPI consists of soft spherical pellets having lactose as a filler agent — or may be without micronized lactose, depending on the patient's dose requirement. This pellet formulation has excellent dispersion behavior. It has improved the accuracy of the dose measuring mechanism, production process, and mechanical stability in the inhaler. So adhesive mixture-type DPIs are formulated typically for low-dose medicament for different respiratory diseases, COPD, and asthma. These types of formulations are formulated by mixing micronized medicaments with carrier crystals, known as ordered mixtures. Carrier crystals act as host particles to carry the micronized medicaments to their absorption site (17). Many variables exist in adhesive mixtures in DPI; this basically occurs as a result of the type and size distribution of the lactose carrier particles and blending conditions during formulation. These variables may influence the size distribution, flow properties, content uniformity of medicament, dispersion performance, and the release of accurate doses as well as fine particle doses. Various other versatile particle engineering techniques have been developed for high dose medicaments.

37.6.2 Dose Measuring Systems

There are several types of dose-measuring systems available, having a mixture of mechanisms. These mechanisms are applied in combination with drug reservoir containers in multi-dose inhalers to release accurate DPI doses from capsules, blisters, and other pre-loaded dose systems (18–20). However, it is observed that drug formulation and dose systems have existing compatibilities that emphasize the necessity for integrated development of the DPI. For example, drug containers with an excess of volume are not preferable for soft spherical pellets or adhesive mixtures in DPI with either high drug content or having a lower mechanical instability in the formulations. Due to violent movements or dropping of powders in such containers, the inhaler may break. Finally, it may lessen the flow properties as well as jeopardize dose consistency. Capsule reservoir-containing DPI may not be suitable for a formulation containing large particle-size carrier crystals. The carrier molecule with medicaments cannot escape from the capsule wall (21). Similarly, moisture-sensitive or hygroscopic medicament-containing formulations are processed in multi-dose reservoir DPIs to protect against water uptake from the environment through ambient air (22,23).

37.7 Principles of Powder Dispersion

Theoretically, different versatile dispersion principles are seen in DPI formulations. The currently available marketed inhalers rely on relatively less efficient and weak drag and lift forces in turbulent air streams. For achieving dispersion of powder, carrier materials should be with relatively smooth surfaces, which exist in relatively fine carriers or are obtained with particle smoothing techniques. Presence of specific amounts of fine lactose particles in the mixture helps in achieving good powder dispersion; however, inconsistencies have been reported (24–26). Dispersion principles based on inertial separation are more effective as they are less dependent on the properties of the mixture (26,27). In contrast with dispersion systems relying on drag and lift forces, better performance of DPIs can be achieved with carriers having relatively high surface rigidity and having less quantity of fine particles (28). Incompatibilities may also occur between the powder formulation and dispersion principle of the system.

37.8 DPI for High Dose Drugs

Inhaler design may vary depending on the dose of the medicament. High dose medicaments including antibiotics and excipients are highly hygroscopic in nature (17,29). Due to their hygroscopic nature and poor flow properties, they are unsuitable for multi-dose DPI with a dosing system. Particle engineering of the formulation is often required to achieve good dispersion of powders. Minimization of inhaler retention is necessary in high dose drug formulations avoiding the use of excipients in order to minimize inhaled powder mass (30). Orbital® disposable DPI consists of a powder-containing puck. It holds high drug doses up to 400 mg. The puck releases the powder by centrifugal force through one or more release rate controlling orifices into the de-agglomeration chamber in several subsequent inhalations (17).

37.9 The Technical Design of DPI Formulations

The aerodynamic size of the particles in powder should be optimum for distribution and deposition of the particles in the target area of the lung at the flow rate while inhaled. The formulation must be capable of achieving reproducibility in dose-measuring and emptying the dose compartment during inhalation. During DPI inhalation, the formulation should be dispersed effectively into the inhaled air stream because incomplete dispersion is observed with most inhalers and the size distribution of the particles in the aerosol may significantly differ from the primary particle size distribution of the particles. The functional requirement of DPI formulation requires control of the inter-particulate forces and good flow properties. To meet these requirements, different development approaches to formulations may be considered, depending on the dose of the drug.

37.9.1 Particle Preparation Techniques

Micronization is one of the particle preparation techniques to get the desired aerodynamic size for medicament particles, especially for anti-asthma and COPD medicaments. The particle size distribution can be controlled to a significant extent by a top-down manufacturing process. However, local distortions of the crystal lattice, generation of flat surfaces, and charging of the particles may affect the cohesiveness of the formulation (31–33). Supercritical fluid drying (SCF) and spray drying are the bottom-up techniques frequently mentioned in the literature (34,35). The particle shape and surface morphology can be controlled to certain extent with these techniques. Powder flow and dispersion of powder can be optimized by controlling morphology, shape, surface porosity, and density of particles (31,36). Although spray drying has been extensively employed to produce inhalable powders, there is a concern that the spray-dried drug particles may exist as amorphous forms and be physically unstable (37).

37.9.1.1 Spray Drying Techniques

Spray drying (SD) is a commonly utilized technique to produce solid particles (38). The primary advantage of the spray-drying technique is the uniformity of the shape, size, and diameter of particles obtained. The particles obtained utilizing other techniques, such as crystallization, do not typically exhibit such uniformity. That is the reason for the widespread use of spray drying when the uniformity of shape and diameter *and* composition of particles has a crucial meaning, especially while designing medication for aerosol therapy. The spray-drying technique may also allow us to get particles with the desired morphology.

Consequently, spherical and hollow particles, or particles with very high porosity or roughness, can be produced. An extensive review, the possibility of obtaining particles with different particle morphologies by spray drying method together with the conditions required to obtain them (temperature of drying gas, the local velocity of dried droplet precursor properties, etc.), is presented by Nandiyanto & Okuyama (39). Parameters that control the formation of particle shape and morphology in a spray-drying method include hosting and templating particle diameter, and the flow rate of droplets through a dryer, as well as droplet evaporation rate (40,41).

37.9.1.2 Spray-Freeze Drying

Spray-freeze drying (SFD) (cryo-spray drying) is another popular formation method of nano- and micro-particulate powders suitable for inhalation drug delivery (42–44). As the process is more complicated and costly than conventional spray drying, it is advised mainly for pharmaceutical products with a high added value. It is a two-step technology consisting of rapid freezing of sprayed API solution and the subsequent sublimation of solvent (lyophilization). Typically the freezing of droplets occurs during liquid spraying in a chamber above the cryogenic medium, typically above liquid nitrogen (–196 °C). Some modifications of the conventional process have been developed, e.g. SFL method (spray freezing into liquids) (45,46) where droplets released from the nozzle are injected instantly beneath the surface of the cryogenic liquid, or SFG process (spray freezing into gas). In this process, droplets are injected into the countercurrent stream of air refrigerated to –60°C (47). At these lower temperatures, droplets rapidly solidify due to the formation of ice of water and the eutectic mixture in some instances. The solidified product is transferred to the second apparatus, where sublimation of ice is completed under reduced pressure and gentle heating. The second step is more time-demanding and requires careful control of the vacuum level and temperature.

37.9.1.3 Supercritical Fluid (SCF) Technologies

The application of supercritical fluids to obtain nano- and microparticles for inhalation has drawn much attention in recent years (44,48,49). An SCF is highly compressible, so it can reach a density similar to that of liquids but still maintain intermediate values of viscosity and diffusivity. These features are essential for mass transfer in a medium, which can be carried out more efficiently than in a liquid phase.

In the medicament, supercritical CO_2 (SC-CO_2) is primarily applied due to relatively mild critical conditions. Other favorable features of SC-CO_2 are minimum cost, non-flammability, low toxicity, and minimum environmental impact. It can also be easily got rid of through the processing stream by merely reducing pressure. However, not many pharmaceutical medicaments are sufficiently soluble in SC-CO_2. Therefore, this medium is often utilized as an anti-solvent to develop the supersaturation needed for crystal nucleation and growth. It is economically justified to use SC-CO_2 as a solvent only when the solubility of the API is low (50). Although some medicament and excipient solubility can be enhanced if SC-CO_2 is mixed with organic solvents such as ethanol and acetone, these co-solvents need to be removed from the final product.

Supercritical fluid (SCF) is used for product development of low dose anti-asthma and COPD drugs. It also has the potential to formulate particles of of relatively high molecular thermo-labile protein-based pharmaceuticals (51). Particles are generated in the form of amorphous state having reduced physical stability and increased moisture. This technique is less suitable for crystalline drugs. Few drugs, such as tobramycin, are hygroscopic in their crystalline modifications.

However, the physical stability of product is not reduced in the spray drying process. Spray drying results in particle compositions which can control release kinetics (52–54), increase bioavailability, increase pulmonary absorption, and improve the dispersion performance (55,56).

To control particle shape and surface morphology, different crystallization or modification techniques have been described. Anti-solvent precipitation is one of the techniques that utilizes ultrasound or stabilizers and growth retarders (57). The controlled spherical agglomeration in liquid, the solid-state modification of crystals, and elongated mannitol particles from confined liquid impinging from jets has been explored (58–60).

Good dispersion behavior of particles can be achieved using spray-freeze technique. However, the product have high porosity and minimal quantity of drug can be measured in single-dose compartments (61,62). A low dose salbutamol sulphate formulation without carrier has been formulated by combining with thermal ink-jet spraying (63). Similar to spray-freeze drying, freeze-drying can be used for formulation of product containing dry thermo-labile compounds. However, the poor particle size distribution and inferior flow properties limit the use of the technique (64). A novel technique, micro-molded PRINT particles, produces particles with narrow size distribution by compression in micro-molds. This technique has gained significant importance in industrial scale production (65).

37.10 Types of DPI Devices

The inhalation device is essential in accomplishing satisfactory delivery of the inhaled drug to the lung. The integration of powder formulation and device performance is necessary for effective drug delivery to deep lungs. The device should be able to deliver high fine particle fraction (FPF) from the delivery system. DPI devices consist of a dose-measuring system, mouthpiece, powder formulation, and powder de-agglomeration principle. Based on a dose metering system or dispersion mechanism, DPI devices are classified into different types. According to the metering system, these devices are mainly classified by dose type into single-dose devices, multiple-unit dose, and multi-dose devices (11). Table 37.1 lists some currently available DPI devices Figure 37.1.

37.10.1 Single-Dose Devices

These are reusable and consist of micronized drug powder with carrier system. Individual doses are delivered in gelatin capsules. It has to be loaded into the inhaler before use. They are inconvenient because an individual dose must be loaded into the device each time it is used (66). The Spinhaler (Aventis) was developed to administer sodium cromoglycate through inhalation therapy in individual gelatin capsules. The patient inserts a single capsule onto a propeller seated inside the inhalation channel, and the capsule is pierced by two needles that are actuated by a sliding cam arrangement. When the patient inhales vigorously through the DPI mouthpiece, it vibrates, dispensing the drug as an aerosol form (67).

37.10.2 Multiple-Unit Dose Devices

These types of inhalers contain several individually packaged doses, either as multiple gelatin capsules or in blisters. These sealed doses are packaged so that the device can hold more than one dose at a time. This pre-packaging ensures control of dose uniformity as well as protection from the environment until use (12,66).

TABLE 37.1

List of the Some Currently Available DPI Devices (66,68)

Type of Device	Marketed Name	Manufacturer
Unit dose device	Cyclohaler®/Aerolizer®	Novartis
	Spinhaler®	Aventis Pharma
	Eclipse	Aventis Pharma
	Breezhaler®	Novartis
	Podhaler™	Novartis
	Turbospin®	PH&T
	Handihaler®	Boehringer Ingelheim
	Eclipse®	Aventis Pharma
Multi-unit dose device	Diskhaler®	GlaxoSmithKline
	Diskus® (Accuhaler)	GlaxoSmithKline
	Prohaler®	AptarPharma
	DreamBoat™	MannKind Corporation
Multi-dose device	Turbuhaler®	AstraZeneca
	Genuair®/Pressair®	Almirall,S.A
	Novolizer®	ASTA Medica
	Easyhaler®	Orion
	Clickhaler®	ML Labs
	Pulvinal®	Chiesi Ltd.
	Certihaler®	SkyePharma

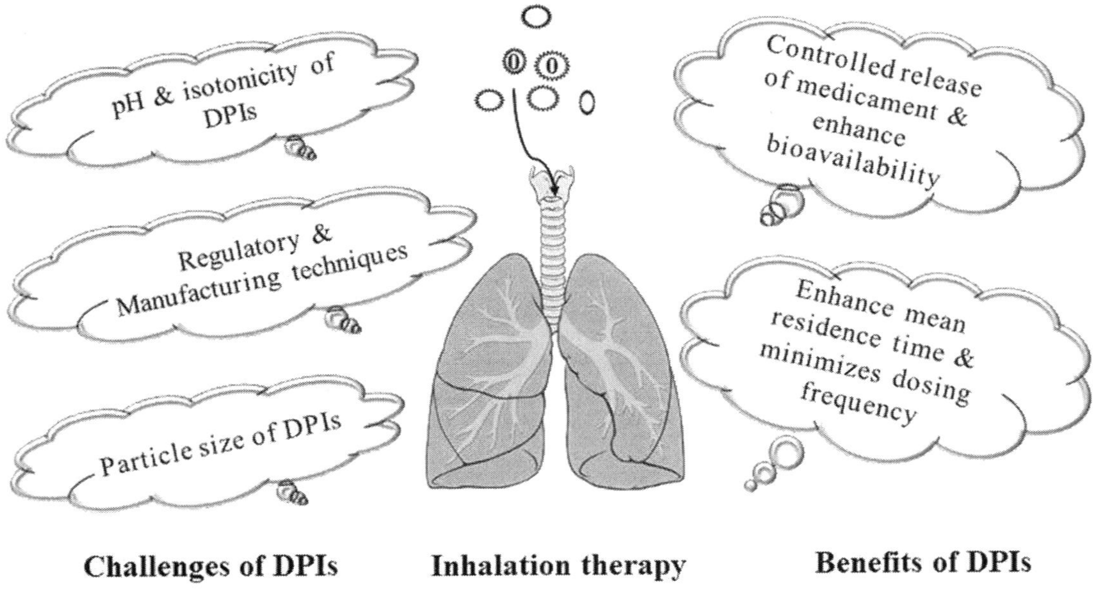

FIGURE 37.1 Some Benefits and Challenges in Novel DPI

37.10.3 Multi-dose Devices

In multi-dose devices, the drug is stored in a bulk powder reservoir from which individual doses are delivered. This device has a powder reservoir, which can deliver more than 100 metered doses. It has an advantage in terms of more accurate metering of individual doses and better protection against moisture ingress, but is generally more expensive to produce (66). Table 37.2 lists some commonly available DPI formulations in the market and their DPI devices.

The device contains the drug product in a multi-dose reservoir and has a built-in measuring system for each individual dose upon actuation. One of the problems associated with this system is moisture ingress into the product reservoir from patient exhalation, which may change particle properties and flow pattern as well as dispersion pattern during the product's life.

TABLE 37.2

List of Some Currently FDA Approve Marketed DPI Product (68)

DPI Product	Active Pharmaceuticals	Device Type	Manufacturer	Dosing
Breo® Ellipta™	Fluticasone furoate and vilanterol	Multi-unit dose	GlaxoSmithKline/Theravance	Once daily
Flovent® Diskus®	Fluticasone propionate	Multi-unit dose	GlaxoSmithKline	Varies
TOBI® Podhaler™	Tobramycin	Multi-unit dose	Novartis	Twice daily (inhale four capsules)
Relenza® Diskhaler®	Zanamivir	Multi-unit dose	GlaxoSmithKline	Once daily
Buventol Easyhaler®	Salbutamol	Multi-unit dose	Orionpharma	N/A
Beclomet Easyhaler®	Beclometasone	Multi-unit dose	Orionpharma	N/A
Giona Easyhaler®	Budesonide	Multi-unit dose	Orionpharma	N/A
Arcapta™ Neohaler™	Indacaterol	Unit dose	Novartis	Once daily
TOBI® Podhaler™	Tobramycin	Unit dose	Novartis	Twice daily
Foradil® Aerolizer®	Formoterol fumarate	Unit dose	Novartis /Merck	Twice daily
Pulmicort® Flexhaler™	Budesonide	Multi-dose	Astra Zeneca	Twice daily
Foradil® Certihaler®	Formoterol fumarate	Multi-dose	Novartis	Twice daily

37.11 Challenges for Future Developments

The therapeutic agent's successful administration to the pulmonary route through a DPI primarily depends on the parameters, such as formulation, the metering system, the inhaler device, and the patient's understanding/training. Various advantages and challenges in DPI are given in Figure 37.1.

37.11.1 DPI Design

Improvement of the efficacy of dry powder inhalation is one of the challenging tasks in the field of DPI development. Although significant improvement in the lung disposition has been made in the past decades, from less than 10% to between 20% and 40%, the dose available at the site of absorption is less than half of the actual dose (17).

A stable balance between three types of forces via inter-particulate forces in the powder formulation, dispersion forces generated by the DPI, and deposition forces in the respiratory tract improve the performance of a DPI system (14,69). The inhaler design is an essential factor for powder dispersion and gaining significant importance in DPI development (70,71). New CFD techniques are playing a crucial role in device performance (72,73). However, most studies have been working on modification of existing device and investigating the performance as well as powder dispersion mechanism thereof (74–77). Most of the pharmaceutical companies are currently working on CFD-assisted new DPI developments.

One of the most important aspects of DPI development is to design optimally tuned combinations of the dispersion principle and powder formulation. Plastic manufacturers have been developing inhaler concepts for the DPI market; however, they initially do not know which type of formulation these concepts will be used to disperse. Low-resistance capsule-based DPIs are still used for new powder formulations (17,78). However, they have disadvantages related to powder dispersion and complete dose delivery. Moreover, the low airflow resistance of these capsule-based DPI causes high flow rates which results in peripheral and central lung deposition. With the aid of CFD, these problems can be avoided to a certain extent.

37.11.2 Dry Powder Formulations

Despite many years of effort to understand powder cohesion and dispersion mechanisms, several challenges remain on the formulation site. Some relevant variables, such as moisture content and roughness, have extensively been described for adhesive mixtures. However, there is a lack of knowledge of the mixture properties during the mixing process that affect the performance of mixture. New measuring techniques should be developed to understand de- and re-agglomeration of particles and distribution over the carrier surface. It is also necessary to know changes in inter-particulate forces caused by inertial and frictional forces during blending. To improve dispersion of particles, particle engineering techniques are widely used. However, these techniques result in increased powder volume when used for low-density particles, thereby increasing the number of inhalations for a single dose (17). It reduces the acceptability of the product by patients (79). Reducing complexity of manufacturing techniques and powder formulations is necessary for large scale manufacturing and reduction of the production cost. Feasibility of inhaler design with physicochemical product properties is necessary for good dispersion of high dose drug formulations with minimum excipient content. Using synergistic combinations of drugs or potent drugs, it is still possible to achieve a suitable dose for effective dose delivery and efficacy. Although few synergistic combinations have been reported in this field, it remains a challenging task for DPI development.

37.11.3 Pulmonary Vaccination

Vaccination involves single administration through the pulmonary route. If the administration of vaccine to the pulmonary route fails, there will be no, or insufficient, protection. Once-only disposable inhalers have been investigated for vaccination through the pulmonary route. Use of add-on devices with measuring equipment can allow monitoring of the amount of dose delivery. The number of disposable inhalers in the field of pulmonary delivery is still small. Moreover, most of these inhalers are still in the development stage (80). To achieveg success in pulmonary vaccination, the development of effective disposable DPIs with robust performance connected to reusable monitoring devices is the future research direction in this field.

37.11.4 Special Patient Groups

There is need of specially designed inhalers for small children and elderly patients. There is a lack of specially designed devices for these age groups. Most currently marketed DPI for pediatric patients are lacking clinical data. Use of DPI in children remains a challenging task and also depends on whether they are capable of inhaling correctly. The operating principle of the DPI is a crucial factor for designing these products for children (81,82). Aspiratory parameters such as peak flow rate must carefully evaluated for successful dose delivery to these age groups (83). DPI development for the pediatric age group must also focus on how these parameters are affected by the airflow resistance of the inhaler (84,85). The dose quantity for the pediatric age group is significantly lower than adult patients (86). Similar problems are observed in elderly patients in using DPI. Lower aspiratory power and inadequate understanding of how to operate an inhaler, and poor manual dexterity and hand strength are the practical problems encountered in elderly patients (87). When designing DPI for the above age group patients, the existing difficulties must be taken into account to develop a easy-to-operate DPI capable of delivering the dose in lower inhaled volumes and at lower flow rates.

37.11.5 Target Sites for Inhaled Drugs

Depending on the nature of the disease, the target sites for inhaled drugs may vary. For example, the target sites for anti-asthmatic and COPD patients depends on the distribution of specific receptors for the drugs (88), the site of inflammation, or smooth muscle (89,90). Formulations administered through the pulmonary route for systemic action are primarily absorbed in the lung's peripheral part and have to be deposited in the most distal airways or even the alveoli (91). In bacterial lung infection, the antibiotic formulation needs to be equally distributed in the lung to achieve the desired concentration. If desired concentration is not achieved by pulmonary delivery due to unequal distribution, it may result in failure of therapy as well as drug resistance. Current studies have demonstrated that approximately one-third of the total pulmonary dose of DPI formulation is deposited in the upper part of airways, one-third in the central airways, and one-third in the peripheral lung (92–94). Thus, the equal distribution of drug concentration varies between the entire regions of lung by more than a factor of 100 in combination with the exponentially increasing surface area of the airways from the trachea to the alveoli (79). Enhancement of peripheral deposition is a future research direction in the field of DPI.

37.11.6 Patient Compliance

Patient compliance is necessary for the product's viability and market success. Leaflet as well as training modules along with a practical demonstration by a healthcare professional can help to achieve this goal to a certain extent. The technical aspects for improving patient compliance already have been discussed.

37.12 Conclusion for Ideal DPI Formulation

There is clear evidence that many major factors in inhaled drug delivery are prioritizing the development of DPI products. However, those formulating DPI face many challenges and must often make compromises. For example, seeking solutions to technical problems associated with optimizing pharmaceutical performance may introduce incompatibilities with patient compliance issues. Multi-dose reservoir devices tend to target the lungs more uniformly than multiple-unit-dose devices; however, they tend to have lower dose uniformity. A range of DPI is already marketed, and many others are in development. Not all new DPI devices and formulations will reach the market, but many of those that do are likely to have successfully addressed perceived limitations in earlier systems. As we go forward into the 21st century, DPI delivery systems are likely to contribute significantly to successful drug delivery by the inhaled route to treat lung cancer, asthma, COPD, and fibrosis and deliver a broader range of drugs intended for local and systemic applications.

Conflict of Interest

Authors have no conflict of interest

REFERENCES

1. Prime D, Atkins PJ, Slater A, and Sumby B. Review of dry powder inhalers. *Advanced Drug Delivery Reviews*. 1997; 26: 51–58. https://doi.org/10.1016/S0169-409X(97)00510-3.
2. Lavorini F, Pistolesi M, and Usmani OS. Recent advances in capsule-based dry powder inhaler technology. *Multidisciplinary Respiratory Medicine*. 2017; 12: 1–7. https://doi.org/10.1186/s40248-017-0092-5.
3. Dessanges JF. A history of nebulization. *Journal of Aerosol Medicine*. 2001; 14: 65–71. https://doi.org/10.1089/08942680152007918.
4. Cutchis P. Stratospheric ozone depletion and solar ultraviolet radiation on earth. *Science*. 1974; 184: 13–19. https://doi.org/10.1126/science.184.4132.13.
5. Lovelock JE. Halogenated hydrocarbons in the atmosphere. *Ecotoxicology and Environmental Safety*. 1977; 1: 399–406. https://doi.org/10.1016/0147-6513(77)90030-6.
6. Noakes TJ. CFCs, their replacements, and the ozone layer. *Journal of Aerosol Medicine*. 1995; 8: S3–S7. https://doi.org/10.1089/jam.1995.8.suppl_1.s-3
7. Mudge J. A Radical and Expeditious Cure for a Recent Catarrhous Cough. Allen, London. 1778.
8. Anderson PJ. History of aerosol therapy: Liquid nebulization to MDIs to DPI. *Respiratory Care*. 2005; 50: 1139–1149.
9. Sanders M. Inhalation therapy: An historical review. *Primary Care Respiratory Journal*. 2007; 16: 71–81. https://doi.org/10.3132/pcrj.2007.00017
10. de Boer AH, Hagedoorn P, Hoppentocht M, Buttini F, Grasmeijer F, and Frijlink HW. Dry powder inhalation: Past, present and future. *Expert Opinion on Drug Delivery*.

2017; 14: 499–512. https://doi.org/10.1080/17425247.2016.1224846
11. Daniher DI and Zhu J. Dry powder platform for pulmonary drug delivery. *Particuology*. 2008; 6: 225–238. https://doi.org/10.1016/j.partic.2008.04.004
12. Kadu P, Kendre P, and Gursal K. Dry powder inhaler: A review. *Journal of Advanced Drug Delivery*. 2016; 3: 42–52.
13. Yadav N and Lohani A. Dry powder inhalers: A review. *Indo Global Journal of Pharmaceutical Sciences*. 2013; 3: 142–155.
14. Telko MJ and Hickey AJ. Dry powder inhaler formulation. *Respiratory Care*. 2005; 50: 1209–1227.
15. Moeller M, Grimmbacher S, and Munzel U. Improvement of asthma therapy by a novel formoterol multidose dry powder inhaler. *Arzneimittel-Forschung/Drug Research*. 2008; 58: 168–173. https://doi.org/10.1055/s-0031-1296488
16. Nakate T, Yoshida H, Ohike A, Tokunaga Y, Ibuki R, and Kawashima Y. Formulation development of inhalation powders for FK888 using the E-haler® to improve the inhalation performance at a high dose, and its absorption in healthy volunteers. *European Journal of Pharmaceutics and Biopharmaceutics*. 2005; 59: 25–33. https://doi.org/10.1016/j.ejpb.2004.08.004.
17. Hoppentocht M, Hagedoorn P, Frijlink HW, and de Boer AH. Technological and practical challenges of dry powder inhalers and formulations. *Advanced Drug Delivery Reviews*. 2014; 75: 18–31. https://doi.org/10.1016/j.addr.2014.04.004.
18. Wetterlin K. Turbuhaler: A new powder inhaler for administration of drugs to the airways. *Pharmaceutical Research, an Official Journal of the American Association of Pharmaceutical Scientists*. 1988; 5: 506–508. https://doi.org/10.1023/A:1015969324799
19. Yang TT, Li S, Wyka B, and Kenyom D. Drug delivery performance of the mometasone furoate dry powder Inhaler. *Journal of Aerosol Medicine*. 2001; 14: 487–494. https://doi.org/10.1089/08942680152744695
20. Zeng XM, O'Leary D, Phelan M, Jones S, and Colledge J. Delivery of salbutamol and of budesonide from a novel multi-dose inhaler Airmax™. *Respiratory Medicine*. 2002; 96: 404–411. https://doi.org/10.1053/rmed.2002.1362
21. Borowski M, Villax P, Eskandar F, and Steckel H. Selecting lactose for a capsule-based dry powder inhaler. *Pharmaceutical Technology Europe*. 2004; 16: 23–35.
22. Fyrnys B, Stang N, and Wolf-Heuss E. Stability and performance characteristics of a budesonide powder for inhalation with a novel dry powder inhaler device . *Current Opinion in Pulmonary Medicine*. 2001; 7: S7–S11.
23. Borgström L, Asking L, and Lipniunas P. An in vivo and in vitro comparison of two powder inhalers following storage at hot/humid conditions. *Journal of Aerosol Medicine*. 2005; 18: 304–310. https://doi.org/10.1089/jam.2005.18.304
24. El-Sabawi D, Edge S, Price R, and Young PM. Continued investigation into the influence of loaded dose on the performance of dry powder inhalers: Surface smoothing effects. *Drug Development and Industrial Pharmacy*. 2006; 32: 1135–1138. https://doi.org/10.1080/03639040600712920
25. Young PM, Cocconi D, Colombo P, Bettini R, Price R, Steele DF, et al. Characterization of a surface modified dry powder inhalation carrier prepared by "particle smoothing." *Journal of Pharmacy and Pharmacology*. 2002; 54: 1339–1344.
26. De Boer AH, Hagedoorn P, Gjaltema D, Goede J, and Frijlink HW. Air classifier technology (ACT) in dry powder inhalation: Part 3. Design and development of an air classifier family for the Novolizer® multi-dose dry powder inhaler. *International Journal of Pharmaceutics*. 2006; 310: 72–80. https://doi.org/10.1016/j.ijpharm.2005.11.030
27. De Boer AH, Hagedoorn P, Gjaltema D, Goede J, and Frijlink HW. Air classifier technology (ACT) in dry powder inhalation: Part 1. Introduction of a novel force distribution concept (FDC) explaining the performance of a basic air classifier on adhesive mixtures. *International Journal of Pharmaceutics*. 2003; 260: 187–200. https://doi.org/10.1016/S0378-5173(03)00250-3
28. De Boer AH, Chan HK, and Price R. A critical view on lactose-based drug formulation and device studies for dry powder inhalation: Which are relevant and what interactions to expect? *Advanced Drug Delivery Reviews*. 2012; 64: 257–274. https://doi.org/10.1016/j.addr.2011.04.004
29. Golshahi L, Tian G, Azimi M, Son YJ, Walenga R, Longest PW, et al. The use of condensational growth methods for efficient drug delivery to the lungs during noninvasive ventilation high flow therapy. *Pharmaceutical Research*. 2013; 30: 2917–2930. https://doi.org/10.1007/s11095-013-1123-3
30. de Boer AH, Hagedoorn P, Westerman EM, Le Brun PPH, Heijerman HGM, and Frijlink HW. Design and in vitro performance testing of multiple air classifier technology in a new disposable inhaler concept (Twincer®) for high powder doses. *European Journal of Pharmaceutical Science*. 2006; 28: 171–178. https://doi.org/10.1016/j.ejps.2005.11.013
31. Geller DE, Weers J, and Heuerding S. Development of an inhaled dry-powder formulation of tobramycin using pulmosphere™ technology. *Journal of Aerosol Medicine and Pulmonary Drug Delivery*. 2011; 24: 175–182. https://doi.org/10.1089/jamp.2010.0855
32. Rasenack N, Steckel H, and Müller BW. Micronization of anti-inflammatory drugs for pulmonary delivery by a controlled crystallization process. *Journal of Pharmaceutical Science*. 2003; 92: 35–44. https://doi.org/10.1002/jps.10274
33. Karner S and Anne Urbanetz N. The impact of electrostatic charge in pharmaceutical powders with specific focus on inhalation-powders. *Journal of Aerosol Science*. 2011; 42: 428–445. https://doi.org/10.1016/j.jaerosci.2011.02.010
34. Mosén K, Bäckström K, Thalberg K, Schaefer T, Kristensen HG, and Axelsson A. Particle formation and capture during spray drying of inhalable particles. *Pharmaceutical Development and Technology*. 2005; 9: 409–417. https://doi.org/10.1081/PDT-200035795
35. Zhou Q (Tony), Morton DAV, Yu HH, Jacob J, Wang J, Li J, et al. Colistin powders with high aerosolisation efficiency for respiratory infection: Preparation and in vitro evaluation. *Journal of Pharmaceutical Science*. 2013; 102: 3736–3747. https://doi.org/10.1002/jps.23685
36. Russo P, Santoro A, Prota L, and Stigliani MPR. Development and investigation of dry powder inhalers for cystic fibrosis. In: Ali DS. (ed.) Recent Advances in Novel Drug Carrier Systems. InTech, Turkey. 2012; 17–38. https://doi.org/10.5772/51408
37. Vehring R. Pharmaceutical particle engineering via spray drying. *Pharmaceutical Research*. 2008; 25: 999–1022. https://doi.org/10.1007/s11095-007-9475-1

38. Gradon L and Sosnowski TR. Formation of particles for dry powder inhalers. *Advanced Powder Technology*. 2014; 25: 43–55. https://doi.org/10.1016/j.apt.2013.09.012
39. Nandiyanto ABD and Okuyama K. Progress in developing spray-drying methods for the production of controlled morphology particles: From the nanometer to submicrometer size ranges. *Advanced Powder Technology*. 2011; 22: 1–19. https://doi.org/10.1016/j.apt.2010.09.011
40. Lee SY, Widiyastuti W, Iskandar F, Okuyama K, and Gradon L. Morphology and particle size distribution controls of droplet-to- macroporous/hollow particles formation in spray drying process of colloidal mixtures precursor. *Aerosol Science Technology*. 2009; 43: 1184–1191. https://doi.org/10.1080/02786820903277553
41. Iskandar F, Nandiyanto ABD, Widiyastuti W, Young LS, Okuyama K, and Gradon L. Production of morphology-controllable porous hyaluronic acid particles using a spray-drying method. *Acta Biomaterialia*. 2009; 5: 1027–1034. https://doi.org/10.1016/j.actbio.2008.11.016
42. Abdelwahed W, Degobert G, Stainmesse S, and Fessi H. Freeze-drying of nanoparticles: Formulation, process and storage considerations. *Advanced Drug Delivery Reviews* 2006; 58: 1688–1713. https://doi.org/10.1016/j.addr.2006.09.017
43. Cheow WS, Ng MLL, Kho K, and Hadinoto K. Spray-freeze-drying production of thermally sensitive polymeric nanoparticle aggregates for inhaled drug delivery: Effect of freeze-drying adjuvants. *International Journal of Pharmaceutics*. 2011; 404: 289–300. https://doi.org/10.1016/j.ijpharm.2010.11.021
44. Shoyele SA and Cawthorne S. Particle engineering techniques for inhaled biopharmaceuticals. *Advanced Drug Delivery Reviews*. 2006; 58: 1009–1029. https://doi.org/10.1016/j.addr.2006.07.010
45. Engstrom JD, Simpson DT, Cloonan C, Lai ES, Williams RO, Barrie Kitto G, et al. Stable high surface area lactate dehydrogenase particles produced by spray freezing into liquid nitrogen. *European Journal of Pharmaceutics and Biopharmaceutics*. 2007; 65: 163–174. https://doi.org/10.1016/j.ejpb.2006.08.002
46. Yu Z, Johnston KP, and Williams RO. Spray freezing into liquid versus spray-freeze drying: Influence of atomization on protein aggregation and biological activity. *European Journal of Pharmaceutical Sciences*. 2006; 27: 9–18. https://doi.org/10.1016/j.ejps.2005.08.010
47. Leuenberger H. Spray freeze-drying - The process of choice for low water soluble drugs? *Journal of Nanoparticle Research*. 2002; 4: 111–119. https://doi.org/10.1023/A:1020135603052
48. Okamoto H and Danjo K. Application of supercritical fluid to preparation of powders of high-molecular weight drugs for inhalation. *Advanced Drug Delivery Reviews*. 2008; 60: 433–446. https://doi.org/10.1016/j.addr.2007.02.002
49. Cape SP, Villa JA, Huang ETS, Yang TH, Carpenter JF, and Sievers RE. Preparation of active proteins, vaccines and pharmaceuticals as fine powders using supercritical or near-critical fluids. *Pharmaceutical Research*. 2008; 25: 1967–1990. https://doi.org/10.1007/s11095-008-9575-6
50. Chow AHL, Tong HHY, Chattopadhyay P, and Shekunov BY. Particle engineering for pulmonary drug delivery. *Pharmaceutical Research*. 2007; 24: 411–437. https://doi.org/10.1007/s11095-006-9174-3
51. Bouchard A, Jovanović N, de Boer AH, Martín Á, Jiskoot W, Crommelin DJA, et al. Effect of the spraying conditions and nozzle design on the shape and size distribution of particles obtained with supercritical fluid drying. *European Journal of Pharmaceutics and Biopharmaceutics*. 2008; 70: 389–401. https://doi.org/10.1016/j.ejpb.2008.03.020
52. Willis L, Hayes D, and Mansour HM. Therapeutic liposomal dry powder inhalation aerosols for targeted lung delivery. *Lung*. 2012; 190: 251–262. https://doi.org/10.1007/s00408-011-9360-x
53. Park C-W, Li X, Vogt FG, Hayes D, Zwischenberger JB, Park E-S, et al. Advanced spray-dried design, physicochemical characterization, and aerosol dispersion performance of vancomycin and clarithromycin multifunctional controlled release particles for targeted respiratory delivery as dry powder inhalation aerosols. *International Journal of Pharmaceutics*. 2013; 455: 374–392. https://doi.org/10.1016/j.ijpharm.2013.06.047
54. Son Y-J and McConville JT. Preparation of sustained release rifampicin microparticles for inhalation. *Journal of Pharmacy and Pharmacology*. 2012; 64: 1291–1302. https://doi.org/10.1111/j.2042-7158.2012.01531.x
55. Sakagami M, Sakon K, Kinoshita W, and Makino Y. Enhanced pulmonary absorption following aerosol administration of mucoadhesive powder microspheres. *Journal of Controlled Release*. 2001; 77: 117–129. https://doi.org/10.1016/S0168-3659(01)00475-8
56. Pilcer G, Sebti T, and Amighi K. Formulation and characterization of lipid-coated tobramycin particles for dry powder inhalation. *Pharmaceutical Research*. 2006; 23: 931–940. https://doi.org/10.1007/s11095-006-9789-4
57. Dhumal RS, Biradar SV, Paradkar AR, and York P. Particle engineering using sonocrystallization: Salbutamol sulphate for pulmonary delivery. *International Journal of Pharmaceutics*. 2009; 368: 129–137. https://doi.org/10.1016/j.ijpharm.2008.10.006
58. Ikegami K, Kawashima Y, Takeuchi H, Yamamoto H, Isshiki N, Momose DI, et al. Improved inhalation behavior of steroid KSR-592 in vitro with Jethaler® by polymorphic transformation to needle-like crystals (β-form). *Pharmaceutical Research*. 2002; 19: 1439–1445. https://doi.org/10.1023/A:1020492213172
59. Tang P, Chan HK, Chiou H, Ogawa K, Jones MD, Adi H, et al. Characterisation and aerosolisation of mannitol particles produced via confined liquid impinging jets. *International Journal of Pharmaceutics*. 2009; 367: 51–57. https://doi.org/10.1016/j.ijpharm.2008.09.024
60. Ikegami K, Kawashima Y, Takeuchi H, Yamamoto H, Mimura K, Momose D, et al. A new agglomerated KSR-592 β-form crystal system for dry powder inhalation formulation to improve inhalation performance in vitro and in vivo. *Journal of Controlle Release*. 2003; 88: 23–33. https://doi.org/10.1016/S0168-3659(02)00460-1
61. Van Drooge DJ, Hinrichs WLJ, Dickhoff BHJ, Elli MNA, Visser MR, Zijlstra GS, et al. Spray freeze drying to produce a stable Δ9- tetrahydrocannabinol containing inulin-based solid dispersion powder suitable for inhalation. *European Journal of Pharmaceutical Sciences*. 2005; 26: 231–240. https://doi.org/10.1016/j.ejps.2005.06.007
62. Saluja V, Amorij JP, Kapteyn JC, de Boer AH, Frijlink

HW, and Hinrichs WLJ. A comparison between spray drying and spray freeze drying to produce an influenza subunit vaccine powder for inhalation. *Journal of Controlled Release*. 2010; 144: 127–133. https://doi.org/10.1016/j.jconrel.2010.02.025
63. Mueannoom W, Srisongphan A, Taylor KMG, Hauschild S, and Gaisford S. Thermal ink-jet spray freeze-drying for preparation of excipient-free salbutamol sulphate for inhalation. *European Journal of Pharmaceutics and Biopharmaceutics*. 2012; 80: 149–155. https://doi.org/10.1016/j.ejpb.2011.09.016
64. Kaialy W and Nokhodchi A. Freeze-dried mannitol for superior pulmonary drug delivery via dry powder inhaler. *Pharmaceutical Research*. 2013; 30: 458–477. https://doi.org/10.1007/s11095-012-0892-4
65. Mark P, Horvath K, Garcia A, Tully J, and Maynor B. Particle engineering for inhalation formulation and delivery of biotherapeutics. *Inhalation*. 2012; 4: 1–5.
66. Newman SP and Busse WW. Evolution of dry powder inhaler design, formulation, and performance. *Respiratory Medicine*. 2002; 96: 293–304. https://doi.org/10.1053/rmed.2001.1276
67. Bell JH, Hartley PS, and Cox JSG. Dry powder aerosols I: A new powder inhalation device. *Journal of Pharmaceutical Sciences*. 1971; 60: 1559–1564. https://doi.org/10.1002/jps.2600601028
68. Muralidharan P, Hayes D, and Mansour HM. Dry powder inhalers in COPD, lung inflammation and pulmonary infections. *Expert Opinion on Drug Delivery*. 2015; 12: 947–962. https://doi.org/10.1517/17425247.2015.977783
69. Hickey AJ, Mansour HM, Telko MJ, Xu Z, Smyth HDC, Mulder T, et al. Physical characterization of component particles included in dry powder inhalers. I. Strategy review and static characteristics. *Journal of Pharmaceutical Sciences*. 2007; 96: 1282–1301. https://doi.org/10.1002/jps.20916
70. Smutney CC, Grant M, and Kinsey PS. Device factors affecting pulmonary delivery of dry powders. *Therapeutic Delivery*. 2013; 4: 939–949. https://doi.org/10.4155/tde.13.77
71. Ngoc NTQ, Chang L, Jia X, and Lau R. Experimental investigation of design parameters on dry powder inhaler performance. *International Journal of Pharmaceutics*. 2013; 457: 92–100. https://doi.org/10.1016/j.ijpharm.2013.08.072
72. Wong W, Fletcher DF, Traini D, Chan H-K, and Young PM. The use of computational approaches in inhaler development ☆. *Advanced Drug Delivery Reviews*. 2012; 64: 312–322. https://doi.org/10.1016/j.addr.2011.10.004
73. Ruzycki CA, Javaheri E, and Finlay WH. The use of computational fluid dynamics in inhaler design. *Expert Opinion on Drug Delivery*. 2013; 10: 307–323. https://doi.org/10.1517/17425247.2013.753053
74. Shur J, Lee S, Adams W, Lionberger R, Tibbatts J, and Price R. Effect of device design on the *in vitro* performance and comparability for capsule-based dry powder inhalers. *AAPS Journal*. 2012; 14: 667–676. https://doi.org/10.1208/s12248-012-9379-9
75. Zhou QT, Tong Z, Tang P, Citterio M, Yang R, and Chan HK. Effect of device design on the aerosolization of a carrier-based dry powder inhaler - A case study on aerolizer® foradile®. *AAPS Journal*. 2013; 15: 511–522. https://doi.org/10.1208/s12248-013-9458-6
76. Longest PW, Son YJ, Holbrook L, and Hindle M. Aerodynamic factors responsible for the deaggregation of carrier-free drug powders to form micrometer and submicrometer aerosols. *Pharmaceutical Research*. 2013; 30: 1608–1627. https://doi.org/10.1007/s11095-013-1001-z
77. Chen L, Heng R-L, Delele MA, Cai J, Du D-Z, and Opara UL. Investigation of dry powder aerosolization mechanisms in different channel designs. *International Journal of Pharmaceutics*. 2013; 457: 143–149. https://doi.org/10.1016/j.ijpharm.2013.09.012
78. Newhouse MT, Hirst PH, Duddu SP, Walter YH, Tarara TE, Clark AR, et al. Inhalation of a dry powder tobramycin pulmosphere formulation in healthy volunteers. *Chest*. 2003; 124: 360–366. https://doi.org/10.1378/chest.124.1.360
79. Hoppentocht M, Hagedoorn P, Frijlink HW, and De Boer AH. Developments and strategies for inhaled antibiotic drugs in tuberculosis therapy: A critical evaluation. *European Journal of Pharmaceutics and Biopharmaceutics*. 2014; 86: 23–30. https://doi.org/10.1016/j.ejpb.2013.10.019
80. Friebel C and Steckel H. Single-use disposable dry powder inhalers for pulmonary drug delivery. *Expert Opinion on Drug Delivery*. 2010; 7: 1359–1372. https://doi.org/10.1517/17425247.2010.538379
81. De Boeck K, Alifier M, and Warnier G. Is the correct use of a dry powder inhaler (Turbohaler) age dependent? *Journal of Allergy and Clinical Immunology*. 1999; 103: 763–767. https://doi.org/10.1016/s0091-6749(99)70417-3
82. Kamps AWA, Van Ewijk B, Roorda RJ, and Brand PLP. Poor inhalation technique, even after inhalation instructions, in children with asthma. *Pediatric Pulmonology*. 2000; 29: 39–42. https://doi.org/10.1002/(SICI)1099-0496(200001)29:1<39::AID-PPUL7>3.0.CO;2-G
83. Kamps AWA, Brand PLP, and Roorda RJ. Variation of peak inspiratory flow through dry powder inhalers in children with stable and unstable asthma. *Pediatric Pulmonology*. 2004; 37: 65–70. https://doi.org/10.1002/ppul.10410
84. Amirav I, Newhouse MT, and Mansour Y. Measurement of peak inspiratory flow with in-check dial device to simulate low-resistance (Diskus) and high-resistance (Turbohaler) dry powder inhalers in children with asthma. *Pediatric Pulmonology*. 2005; 39: 447–451. https://doi.org/10.1002/ppul.20180
85. Desager KN, Geldhof J, Claes R, and De Backer W. Measurement of inspiratory flow through three different dry powder inhalation devices using In-Check ™ in children with asthma. *Pediatric Asthma, Allergy & Immunology*. 2006; 19: 6–13. https://doi.org/10.1089/pai.2006.19.6
86. Kwok PCL and Chan HK. Delivery of inhalation drugs to children for asthma and other respiratory diseases. *Advanced Drug Delivery Reviews*. 2014; 73: 83–88. https://doi.org/10.1016/j.addr.2013.11.007
87. Barrons R, Pegram A, and Borries A. Inhaler device selection: Special considerations in elderly patients with chronic obstructive pulmonary disease. *American Journal of Health-System Pharmacy*. 2011; 68: 1221–1232. https://doi.org/10.2146/ajhp100452
88. Barnes PJ. Distribution of receptor targets in the lung.

Proceedings of the American Thoracic Society. 2004; 1: 345–351. https://doi.org/10.1513/pats.200409-045MS
89. Meurs H, Dekkers BGJ, Maarsingh H, Halayko AJ, Zaagsma J, and Gosens R. Muscarinic receptors on airway mesenchymal cells: Novel findings for an ancient target. *Pulmonary Pharmacology and Therapeutics*. 2013; 26: 145–155. https://doi.org/10.1016/j.pupt.2012.07.003
90. Van Den Berge M, Ten Hacken NHT, Cohen J, Douma WR, and Postma DS. Small airway disease in asthma and COPD: Clinical implications. *Chest*. 2011; 139: 412–423. https://doi.org/10.1378/chest.10-1210
91. Labiris NR and Dolovich MB. Pulmonary drug delivery. Part I: Physiological factors affecting therapeutic effectiveness of aerosolized medications. *British Journal of Clinical Pharmacology*. 2003; 56: 588–599. https://doi.org/10.1046/j.1365-2125.2003.01892.x
92. Cheng YS. Mechanisms of pharmaceutical aerosol deposition in the respiratory tract. *AAPS PharmSciTech*. 2014; 15: 630–640. https://doi.org/10.1208/s12249-014-0092-0
93. Tsuda A, Henry FS, and Butler JP. Particle transport and deposition: Basic physics of particle kinetics. *Comprehensive Physiology*. 2013; 3: 1437–1471. https://doi.org/10.1002/cphy.c100085
94. Newman SP, Pitcairn GR, Hirst PH, and Rankin L. Radionuclide imaging technologies and their use in evaluating asthma drug deposition in the lungs. *Advanced Drug Delivery Reviews*. 2003; 55: 851–867. https://doi.org/10.1016/S0169-409X(03)00081-4

38

Devices for Dry Powder Drug Delivery to the Lung

Vimal Patel[1], Jigar Shah[1], Hiral Shah[2], and Jayvadan Patel[3]
[1]Institute of Pharmacy, Nirma University, Gujarat, India
[2]Arihant School of Pharmacy & BRI, Gujarat, India
[3]Nootan Pharmacy College, Sankalchand Patel University, Gujarat, India

38.1 Pulmonary Drug Delivery System

Pulmonary drug delivery is now gaining the increasing attention of scientists and researchers due to its various advantages. It is the most recommended route for drug administration to directly produce both systemic and local effects. The main purpose of this drug delivery approach is drug formulation being inhaled by mouth and deposited further into the lower respiratory regions of lungs. The determinative aspect is the site of the respiratory tract where the drug particles are being deposited. For systemic administration it is compulsory to reach drug particles to the terminal of the bronchi and absorbed via circulatory blood that exists in the alveolar ducts. While for the treatment of local respiratory diseases, drug particles should not be absorbed into circulatory blood vessels (1,2).

38.1.1 Advantages of Pulmonary Drug Delivery System

The pulmonary system is an interesting approach for drug delivery due to its unique properties. The most important advantages of pulmonary drug delivery are given below (3–6):

1. Direct drug delivery to the site of action. It has several benefits such as reduced drug administration dose resulting in lower side effects and increases in patient compliance.
2. Quick onset of action.
3. Avoids the hepatic metabolic pathway.
4. Minimal enzymatic reaction, which prevents the drug particles' degradation before absorption.
5. Availability of larger surface area due to very thin layer of alveolus epithelial; increases systemic absorption of drug particles.
6. The nanoparticles with diameters not more than 250 nm, exhibit less uptake by phagocytosis of the respiratory tract.
7. Potential in delivering macromolecules, such as proteins and peptides, for both local and systemic targeting.

38.1.2 Mechanism of Drug Particle Deposition

The mechanism of drug deposition mainly depends upon the airflow, drug particle size, and location in the respiratory system (7). Deposition of inhaled drug particles into the different regions is contingent upon several constraints:

- Rate of breathing
- Nasal or mouth breathing
- Respiratory volume
- Lung volume
- Healthiness of the patient
- Distribution of air resulting in continuously fluctuating field of hydrodynamic flow

The mechanism by drug particles being deposited into the lungs are followings (8,9):

1. **Impaction**
   Generally the airflow variates because of bifurcation in the airways, even though the suspended particles prompt to follow their normal path due to inertia effect and may have affect the airway surface. Because of impact on a lung surface, the dispersed particles prompt to travel along their normal path. Aerodynamic diameter of the drug particle is most important factor when roaming distance is low. Mostly impaction happens when the larger particles are very near to the lung walls, close to the initial distribution. Hence, the deposition in bronchial region is higher after impaction. It occurs when most of the particles are deposited based on mass (8,10).
2. **Sedimentation**
   In this case, the particles are getting redistributed into smaller airways to the alveoli, where the dimension and airflow in airways are low. The terminal velocity of particles mostly affects the rate of sedimentation. So it is more important for deposition of particles with larger aerodynamic diameter. Hygroscopic particles develop in size as they go via the warm and humid airway, hence it is possibility to increase deposition by sedimentation (9,10).

DOI: 10.1201/9781003046547-38

3. **Interception**

   This happens while particles directly interact with respiratory surfaces; deposition by interception has shown an insignificant effect on particle airflow. It mostly occurs when the airflow is close to the narrow airway surface. Nanofibers are interacting directly to the smallest region due to their length and small aerodynamic diameter compared to their size (9).
4. **Diffusion**

   The net delivery of particles to the lower respiratory region. It is the principal mechanism of particle deposition, which is mainly driven by geometric diameter less than 0.5 µm instead of aerodynamic diameter. Diffusion of the particles from the higher concentration to the lower concentration region mainly occurs by Brownian motion; a continuous bombardment of particles with air molecules produces casual wiggling movement. It mainly happens with particles just arrived to the nasopharynx region or smaller alveolar region, where air flow is low (11).
5. **Absorption**

   The lungs are naturally permeable to macromolecules such as therapeutic peptides and proteins. The epithelial layer is the important membrane for absorption of inhaled drugs; it is extremely thick in the trachea, 50–60 µm, and thin in the alveoli, about 0.2 µm. The absorption mainly depends upon the type and morphology of cells located in the different parts of respiratory region. The smaller particles are absorbed slowly if they are highly insoluble. Encapsulation of particles also controls the absorption, such as with liposomes and nanoparticles (12).

38.1.3 Drawbacks of Current Inhaler Devices

A serious issue with the utilization of inhaler devices is aerosolized particles that are mostly deposited in the oropharyngeal region and upper respiratory routes due to a lack of coordination between the device actuation and inhalation, generally observed in children and aged patients. With some dry powder inhalers (DPIs), it is necessary for the patient to breath at maximum force to inhale particles, which unless the patient is properly prepared, is infrequently achieved (13). Pressurized metered-dose inhalers (pMDIs) create vapor more quickly than the patient can breathe in. For pMDIs, this issue was addressed by giving a spacer or by using a breath-activated inhaler rather than breath-coordinated devices (14,15).

The breathing pattern of the patient also plays a significant role in achieving effective pulmonary delivery. Fast respiration is not suggested when using nebulizers and pMDIs. It makes dissolute velocity, and turbulence air flow may increase deposition in the upper respiratory tract by impaction (16). However, fast respiration is needed to de-agglomerate drug particles in DPI inhalation. Clinicians must guide the patient when changing the inhalation device for the same medication (17).

38.2 Advancements in Pulmonary Delivery Devices

The lung has been described as a drug administration route for thousands of years. Breathed in treatments were used in India 4000 years ago, when individuals smoked the leaves of the *Atropa belladonna* plant to stifle cough. During the last centuries, asthmatic patients smoked stramonium powder blended into tobacco cigarettes to reduce the side effects (18). The advancement of current inhalation devices could be grouped into three classes: (1) the enhancement of the nebulizer and the development of two minimized compact devices, (2) DPI, and (3) metered-dose inhalers (MDI). The advantages and drawbacks of every device will be addressed below and are summed up by Labiris et al and Ibrahim et al. in their work (19,20).

38.2.1 Nebulizer

Nebulizers have been utilized for a long time in the treatment of asthma and associated pulmonary comorbidities. Nebulizers have been categorized in two types: ultrasonic and jet (21,22). An illustration of the working mechanism of nebulizers is shown in Figure 38.1. In ultrasonic nebulizers (Figure 38.1a), oscillation of a piezoelectric crystal with 1–3 MHz frequency creates a fluid source inside the nebulizer compartment; droplet size decreases with increasing ultrasound frequency (23). The jet nebulizer (Figure 38.1b) works on the Bernoulli principle, in which packed gas (air or oxygen) drives through a thin hole of low pressure at the outlet of the fluid feed tube. This results in drug solution being strained up from the liquid repository and broken into droplets in the gas stream (24,25).

The particle size and nebulization rate are dependent upon the physical characteristics of drug loaded formulation. The pH, osmolarity, viscosity, surface tension, and ionic strength might circumvent the nebulization. If the solution is hypo- or hyperosmolar, or pH is very low, the vapor could prompt coughing, bronchoconstriction, and irritation of the lung mucosa. Also, high drug dose could reduce the drug release with certain nebulizers (26).

Jet nebulizers with continuous outflow could aerosolize most of drug and deliver bulk dose to the patient with little coordination or aptitude. Utilizing these nebulizers is tedious process and also chances of loss of medications is very high during the operation. These single-use nebulizers are cheap, except those required compressed air or oxygen. After nebulization the most of drug particles couldn't reach to the lung, rather they either retained inside the device or disseminated in the air during exhalation. Finally, only 10% of drug particles presented in the nebulizer can truly dispersed inside the lungs (24).

38.2.2 Inhalers

The drug kept inside an inhaler drives directly into the nasal passage, so it reduces the drug dose compared with an orally

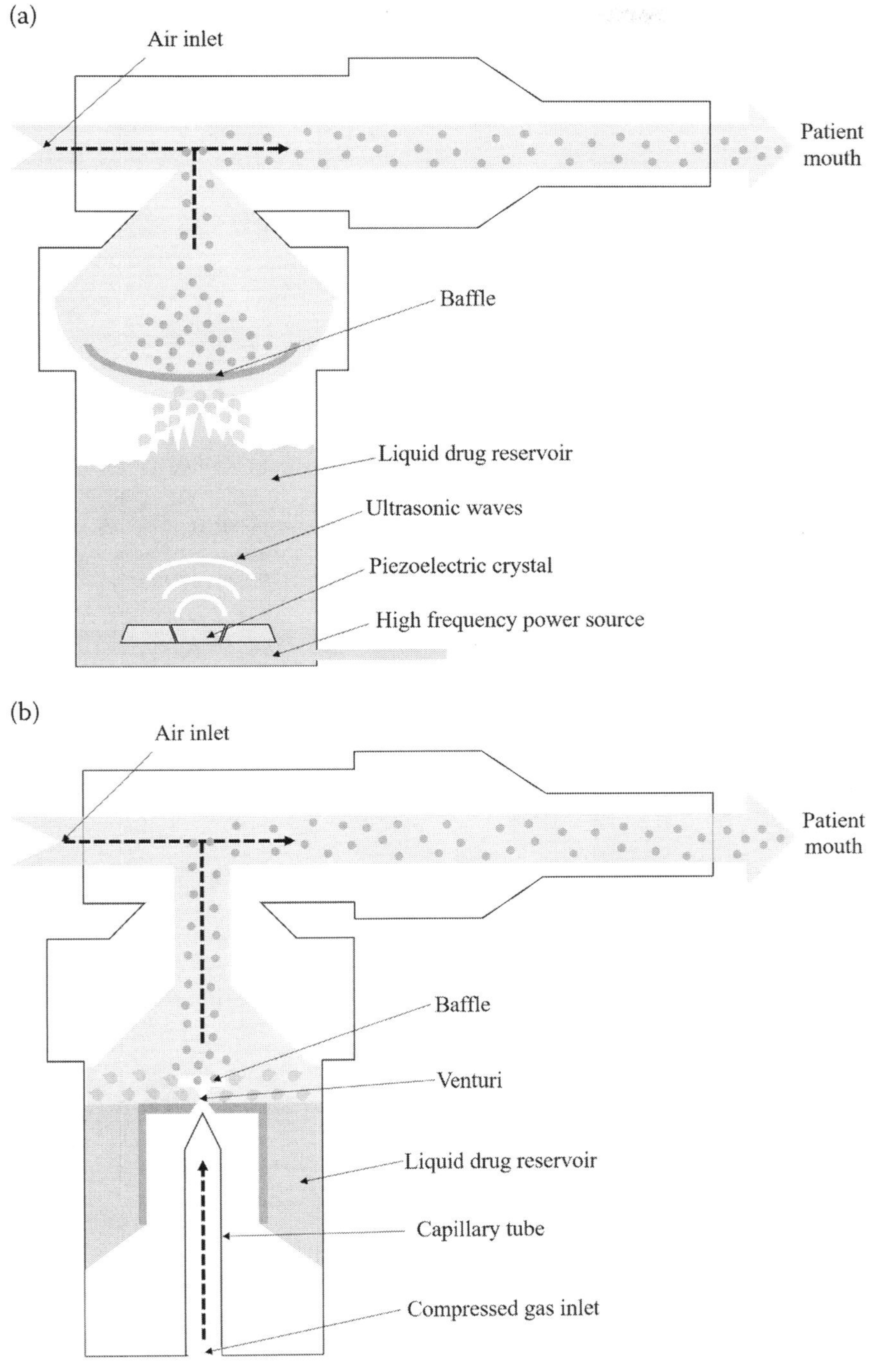

FIGURE 38.1 (a) Working Mechanism of Ultrasonic Nebulizer. (b) Working Mechanism of Jet Nebulizer

administered drug. Inhalation speaks to an alluring, quick, and patient-cooperative delivery route of foundationally active drugs, which are intended to effect locally on the lungs. For the delivery of macromolecules, the major need is the ability to control the consolidated powder and device properties (27). By maintaining these abilities, It is possible to improve the functioning of DPI items, although DPI is having a good scope by looking at below properties, still there is need to handle many obstacles during its performance (28–30):

- Quick onset of action
- Increasing patient compliance by avoiding systemic metabolism
- Accuracy and uniformity in dose delivery
- Ideal sizes of drug particles deposited in the lungs
- Exploration of regulatory endorsement over enhanced reliability of delivery and product stability
- Nominal adhesion between drug particle and device

- Product lifespan improvement
- New systems of inhaled medication frequently need greater and/or higher dose accuracy and efficiency

38.2.2.1 Dry Powder Inhaler

Enthusiasm for DPIs as a viable, productive, and environmentally friendly route of drug delivery to the lung has grown in the last few years. A difficuulty with creating dry powder vaporizers is providing the normal and temporary power source, limited to the powder bed. To be sure, overseeing such particulate strength, for instance through molecule engineering procedures, is presently viewed as integral to effective DPI formulation and development. As a result, much consideration is at present centered around delivering "smart" medications, where it might be conceivable to accomplish great powder flow and low robust forces. Notwithstanding, having a proficient and strong medication innovation in the research facility is just the beginning of making progress toward creating an effective DPI product (31).

Drug researchers very regularly meet significant obstructions when they take part in DPI design — not least due to the plethora of DPI device strategies. There is gigantic variety in the strategies used to store and meter powders and to produce the vaporized cloud. In the DPI aerosol group, there is a lot of variety between various kinds of devices, in the electrostatic condition and liquid dynamic that the powder medication experiences (32).

The medicated vaporizers for DPIs are made through synchronizing air with dry powder. Maximum particles are larger in size, which are taken out into the lungs because the presence of bigger transporters and bulk powder agglomerates (33). Consequently, powder dispersion into inhalable particles relies upon the formation of the turbulent flow of air inside the device. The ferocious air flow makes the dispersion of particles trivial to the lower airways and chances of leakage of drugs from the nanoparticles. Every DPI has substitute air flow obstruction which influences the necessary inspiratory force. The higher the obstruction of the device, the more difficult it is to produce an inspiratory flow sufficient to accomplish the highest dose from the inhaler. Nonetheless, accumulation in the lung has been extended while utilizing high-resistance inhalers (34,35).

38.2.2.2 Metered-Dose Inhaler

The MDI was a progressive development that defeated an issue related to conventional nebulizer, as a versatile inhalation device and maximum utilized for aerosol drug delivery. The device produces a vaporized drug particle by using propellants, for example hydrofluoroalkanes (HFAs) and chlorofluorocarbons (CFC) are atomized at rate (>30 m/s) (36). An illustration of the working mechanism of MDI is shown in Figure 38.2. MDIs release just a little dose of the drug in which only 20% of inhaled drug is accumulated in the lung. High aerosolization rate and larger particle size gives accumulation around 60–80% of the drug particles in the oropharyngeal surface. Discoordination between hand and mouth is one more complication for optimized usage of MDIs (37).

The MDI proficient delivery relies upon a patient's hand–mouth coordination, pattern of breathing, and inspiratory flow rate (IFR). Rises in IFR bring about declines in total lung penetration and deposition into the marginal routes. Quick inhalation (>60 l/min) brings about a decreased peripheral deposition because the aerosol is again promptly accumulated by collision in the oropharyngeal area. When drugs are breathed in gradually, gravitational sedimentation through deposition in peripheral areas of the lung is improved. The decrease in respiratory rate and increasing tidal volume progressively increase peripheral deposition. As an increasing

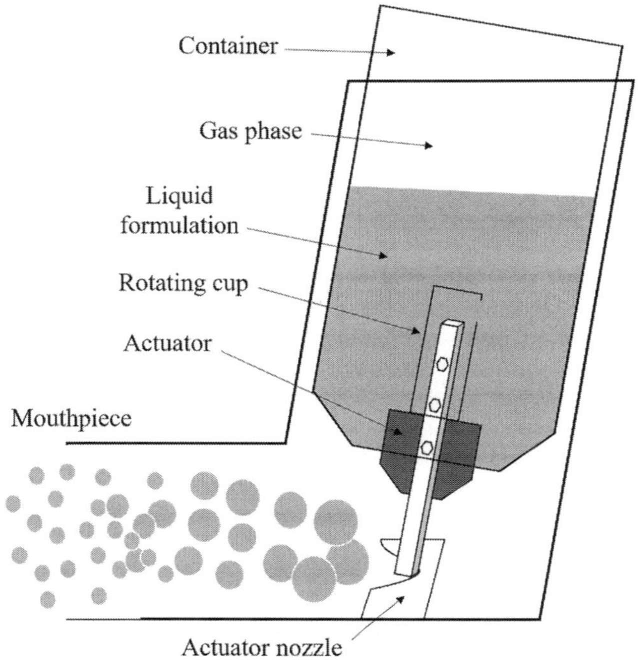

FIGURE 38.2 Working Mechanism of Metered-Dose Inhaler

volume of inhalation, peripheral deposition of drugs is more into the lungs. A time of respiratory hold after completion of inhalation empowers particles which are to be deposited to the periphery region, rather than being breathed out during the expiratory stage (38,39). Therefore, the ideal conditions of MDI inhalation are after a beginning volume corresponding to the tidal retaining volume, atomization of the aerosol at the beginning of inhalation, IFR of <60 l/min be produced by a 10-second breath-hold to the finish of inspiration (40).

38.2.2.3 Pressurized Metered-Dose Inhaler

The pMDI is not used in all drug dosages; it is hard for a physician to prescribe a similar kind of device for different drug inhalations. This is even more difficult when the pharmaceutical industry does not offer more recent drugs to be adminsitered in pMDIs. The strategy requires preliminary and consecutive design of the CFC-propellant for the pMDI. Inefficiency in a device brings about administration of a significantly lessened dosage than that expected. Sadly, continuous inhalation causes a waste of drugs to air (36).

The only problem of pMDI is the conflicting dosing that happens with improper usage. It comprises the effects of nose-breathing, extreme inspiratory rate, hand–breath asynchrony, and the cold freon effect (41,42). For effective drug delivery to the lower respiratory region, most vaporized drug particles must be in a size to inhale and deposited in the lung, largely aerodynamic diameter is about of 1–5 μm (43,44). Generally, the pMDI gives predictable dosing which labelled as number of actuations on the product label, however, due to many factors mentioned above, dosage may change insignificant sometimes (45).

38.3 General Requirements for Pulmonary Delivery Devices

Devices of pulmonary drug delivery are projected to offer reproducible delivery of specific doses to the alveolar regions. It is very much revealed that particles with a mass median aerodynamic diameter (MMAD) of 1–5 μm are effectually accumulated at this location (46). The MMAD is mainly subjected to particle density, morphology, and geometrical diameter, and these characteristics are controlled during the formulation development (47). Due to the existence of inter-particulate forces (i.e. Van der Waal, electrostatic and capillary forces, mechanical interlocking, cohesion/adhesion) (48), micronized powders are automatically producing agglomerates. Degree of the fractional and subsequently of the consolidated forces is subject to major particle characteristics, for example, shape and size of particle, surface morphology and its components (49), and environmental variables, e.g. relative humidity (50). Degree of agglomeration stubbornly influences the fragment of the inhaled powder within the respiration limits (51), so de-agglomeration of those aggregated particles is necessary before or during inhalation and aerosolization (52).

There are mainly two type of DPI devices: (a) passive devices that use a patient's inspiratory flow to defeat the previously mentioned inter-particulate forces, and (b) active devices that utilize additional energy sources. One preferred method is to use a patient's inspiratory flow as a primary energy source for a breath actuation device; this intrinsically evades the necessity of synchronization between patient's inspiration and actuation. The drawback such devices currently face is an airflow resistance which is device specific, and this frequently requires a moderately high inspiratory exertion (53) that is an obstacle for patients with obstructive respiratory complaints, for example, chronic obstructive pulmonary disease (COPD) and/or asthma at young age (54). The degree of deposition is additionally related to the individual patient's inspiratory rate, producing an expected dissimilarity in the consistent delivery of effective drug dose (55). Dose metering is another basic aspect influencing the reproducibility of multi-dosage delivery inhalers.

Single and multiple-dose inhalers can use capsule/blister of pre-metered powder, while bulk powder bed multi-dose devices contain a dose reservoir that separates a single powder dose before actuation (56).

Appropriate flow properties of powder are fundamental, either for precise delivery or discharging of the single-dose vessel completely. Yet micronized powder flow properties are frequently poor, and comprise a physical mixture of drug and larger size of transporter particle (lactose, 30–90 μm), resulting in powder de-agglomeration (33). Considering the above-mentioned reviews, the ideal devices must reproduce accurate dose delivery, irrespective of a patient's complaint. Clinically, it might likewise be invaluable if the devices are safe, simple, breath actuated, and comprise the exact dose control input system associated with efficiency. The sentence can be rephrased as 'The validation and mechanism of dose calculation helps patients to know whether dosage has been given properly and also informed when to replace the device (57).

38.4 Marketed Inhaler Technologies

38.4.1 Technosphere® Insulin

Technosphere® insulin is a gelatin capsule encapsulating a lyophilized dry powder insulin. Mechanism of delivery of insulin utilizes a resistive device incorporated bulk de-agglomeration. Pharmacodynamics and pharmacokinetic outcomes have demonstrated an extremely rapid deposition (insulin tmax: 12–14 min; rate of metabolism: 20–40 min) and a quick onset of action (2–3 hours). The administered dose is directly associated to the bioavailability and 15% of biopotency (58,59).

Particles of dry powder insulin are enhanced for lung delivery. They are breathed in utilizing the MedTone™ inhaler, dry powder delivery at low air-flow rate, a detached inhaler. Dry powder insulin, about 2.5-10 mg, is packed into disposable cartridges which are fixed into the device (60). The powdered insulin is released through inhalation by mouthpiece device directly to the oral cavity. The inhaler does not need manual actuation. Since it is actuated by inhalation of patient, it is not important to coordinate the context of device activation time. Also, the MedTone™ inhaler is a compact device that is subtle, and simple to carry and utilize (61).

38.4.2 Exubera®

In 2006, government authorities (FDA in the US and EMEA in Europe) approved the marketing of Exubera® for the treatment of types 1 and 2 diabetes. This is a device containing powdered insulin, which comprises 60% recombinant human insulin (62); 3 IU and 9 IU of subcutaneous insuin equivalent to 1 and 3 mg of insulin is filled in sachets or in blister respectively. The blister is kept in the base of the device; after actuation by prime, the blister is punctured and releases drug at high speed. Diameter of insulin particles around 5 μm are vaporized into a chamber, and the generated respirable areosolized cloud is inhaled by the patient with a slow breath. Pharmacokinetic studies demonstrated the peak concentration of insulin is reached around 55 min after inhalation, and quickly decreases to normal levels faster than conventional subcutaneous insulin (63).

38.4.3 AERx®

This device releases a single dose of liquid aerosol particles though a nozzle on a dosage strip. AERx is a battery-controlled inhaler device using a chip which electronically regulates the patient's breathing pattern, including flow-rate and acuity of breathing (64,65). The device permits delivery of a metered dose and single unit increases. It is totally non-invasive treatment for proteins and other small molecules that require consistent self delivery. The AERx model comprises an expendable previously filled AERx strip through a built-in nozzle, in which drug is aerosolized through several choices of delivery device. The device works based on an electro-mechanical signal mechanism with accurate dose determination and every mechanical interpretation (AERx Essence®) that convey a pre-determined dose (66).

38.4.4 GyroHaler®

Gyrohaler is minimized and simple to operate, and with few parts it permits minimum maneuver arrangement and lower assembly costs (67). The maneuver is proposed to be expendable after 1 month and is intended to have aerosolization attributes consistent with existing advertised products. Likewise, the GyroHaler appliance uses aluminum foil packed drug for protection from oxygen, humidity, and light. The GyroHaler innovation could be utilized to convey a scope of locally acting items in a proficient, practical, and easy-to-use way (68). The primary objective of GyroHaler is to make improvement in functioning of these items by keeping some room for outsiders to allow innovations in other driving respiratory items. With respect to that, Vectura's researchers have joined their hands with other research organizations to execute the new planning and new items with typical establishments in core innovations, and licensed it (69).

38.4.5 Pressair®

Pressair is a multi-dose dispensable inhaler that was affirmed in 2012 for aclidinium bromide (an anti-muscarinic drug) delivery for the treatment of COPD. It offers audio-video response control mixture for preparation and proper using of device for inhalation. So as to prepare, the patient must be press and delivery a clasp on backside of the device, influx of a single dose from a static container is turned off. A display changes from red to green signal, indication for the device is prepared. An audio sound specifies right inhalation and display signal turned back green to red. When the device is set, a trigger limit instrument forestalls unintentional arrival of a more dose. De-agglomeration of drug powder is accomplished by a cyclone separator (70). In vitro studies utilizing a five-phase multistage fluid impinger (MSLI) at a flow rate of 90 l/min (for 2.7 s) demonstrated a fine particle fraction (FPF) of 40.3±5.6% (n=4) of the delivered dose. The Pressair device has many feedback components to guarantee patient consistence, which is significant since it is intended to be utilized consistently for COPD treatment. The genuinely slow airflow hindrance addresses the issues of this patient populace, and the FPF of about 40% is best in class (71).

38.4.6 Aspirair

Aspirair is a DPI device that efficiently delivers a drug directly to the systemic circulation via the pulmonary route. Prepared powder formulation delivers around 70% or more fine particles to the lungs. Unique among other DPIs, Aspirair is able to deliver extremely fine-particle dosages (<3μm) with nominal deposition in the upper respiratory organs. It is a "functioning" DPI contains a small cyclone diffusion chamber which precisely pressurized air that de-aggregates the particles (72). For the dose administration, a blister containing dry powder is inserted into the device. Then air pressure is compressed by a corkscrew-like manual pump. Then drug dose is circulating into a twister outlet at this point turbulence and shear forces diffuses the powder and delay the air flow, so a vaporized powder rises up out from the mouthpiece that is coordinated with the patient's breathing pattern (73,74). These qualities along with different properties, such as inexpensive and robust design, an easy operating mechanism, suitable size, high drug loading capacity for small and macromolecules, dose reproducibility, optimal atomization rate, straightforward activity, and respiration flow independence mEANS Aspirair is a conceivable candidate for the systemic delivery of macromolecular drug for the treatment of lung disorders (75).

38.4.7 Neohaler®

The Neohaler device was developed for the delivery of glycopyrronium bromide and indacaterol for COPD treatment, designed by Novartis, Switzerland. With the help of two knobs, situated on the edges of the device, the capsules have been punctured and inhalation is carried out using mouthpiece (76). Puncturing the capsule creates a ticking noise showing that the device is prepared. During inhalation, the capsule produces a humming sound, which helps as a feedback control and demonstrates that the powder is being inhaled. Because of its low flow rate hindrance of 0.02 kPa, 0.5 min/l, the Neohaler produces automatic aerosolization that reduces the patient's effort. This makes it a successful inhalation device for the treatment of age-related aspects of COPD (77). In vitro experiments carried out in a cascade impactor with a flow and volume regulator, mimicking a breathing pattern, determined a normal FPF of 26.8±5.8% of the 150 µg indacaterol metered dose. In contrast with Pressair, the Neohaler has a lesser flow-rate hindrance, which increases patient compliance in COPD treatment. A clinical experiment assessing performance and fulfillment of patients regarding the usability of the two inhalers has been carried out, but no results have been published yet (78).

38.4.8 Respimat®

The advantages of nebulizers and pMDIs are combined in the Respimat® inhaler. It is a little, compact inhaler without a power source (like pMDIs) that gradually aerosolizes a soft mist of propellant-free drug formulation (like nebulizers) that decreases the deposition in the oropharyngeal region (79). In asthmatic patients, one-half of total dose administration of fenoterol hydrobromide and ipratropium bromide by Respimat® accomplished a similar therapeutic efficacy as compared to the total dose given by pMDI (80,81). Dalby et al. revealed that this device significantly decreases the fenoterol oropharyngeal deposition and accomplishes maximum absorption in peripheral and central lung areas than do pMDIs, while a spacer containing pMDI had the least oropharyngeal deposition (82).

38.5 Future Perspectives of Inhaler Devices for Lung Delivery

The last decade has seen the development of innovative inhaler devices that enhance targeted delivery of drugs to the lungs and are widely accepted by patients. These advanced technology devices provided clinical improvements in adult patients who are tech-savvy, confident in connecting with technologies, and aware about managing these treatments. But still, these newer approaches have not been successful with highly affected populations like pediatric patients and geriatric patients. The current mobile applications can only help in signaling the time for a dose, but neither track whether the right dose has been taken nor educate at the time of any breathing difficulties felt. Hence, there is need of strong collaborative efforts between various industries and industry and research organizations to share their innovative research ideas and work together in a constructive manner to develop devices for all types of patient populations and develop the concept of inhalers for individuals. This concept will help to motivate and involve patients in self-care management. Beside this, the very important issue which needs to be considered is the customization of devices to accommodate diverse molecules (20). The various different types of molecules administered by pulmonary route for lung delivery include delivery of peptides and proteins, hormones, vaccines, interferons, cytokines, microparticles, antibodies, etc. (83). Therefore, innovative technology needs to develop in such a way that it is flexible enough to accommodate these distinct entities and provide a perfect balance of efficacious treatment, safety to patients, and stability of all these diversified and sensitive compounds. Finally, the future requirement for inhalers is the effective designing of devices and their structures to ensure proper deposition of accurate doses of drugs at the specific sites in the lungs while being simple to operate and economically viable for all segments of the population.

REFERENCES

1. Shaji J and Shaikh M. Current development in the evaluation methods of pulmonary drug delivery system. *Indian Journal of Pharmaceutical Sciences*. 2016; 78(3): 294–306.
2. Mishra B and Singh J. Novel drug delivery systems and significance in respiratory diseases. In: Dua K, Hansbro P, Wadhma R, Haghi M, Pont L, and Williams K, eds. *Targeting Chronic Inflammatory Lung Diseases Using Advanced Drug Delivery Systems*. Australia, PA: Academic Press. 2020; 57–95
3. AboulFotouh K, Zhang Y,Maniruzzamman M, Williams III RO, and Cui Z. Amorphous solid dispersion dry powder for pulmonary drug delivery: Advantages and challenges. International Journal of Pharmaceutics. 2020; 587: 119711.
4. Azarmi S, Roa WH, and Löbenberg R. Targeted delivery of nanoparticles for the treatment of lung diseases. *Advanced Drug Delivery Reviews* 2008; 60: 863–875.
5. Rana H and Patani P. Recent advanced approaches in pulmonary drug delivery system: A review. *International Journal of Pharma and Bio Sciences* 2019; 14(4): 23–27.
6. Fang M, Zhang H, and Cao M. Research progress of pulmonary drug delivery system. *Herald of Medicine*. 2018; 37(3): 302–305.
7. Ou C, Jian H, and Deng Q. Particle deposition in human lung airways: effects of airflow, particle size, and mechanisms. *Aerosol and Air Quality Research*. 2020; 20: 2846–2858.
8. Darquenne C. Deposition mechanisms. *Journal of Aerosol Medicine and Pulmonary Drug Delivery*. 2020; 33(4): 181–185.
9. Gangurde H, Chordiya M, Baste N, Tamizharasi S, and Upasani C. Approaches and devices used in pulmonary drug delivery system: A review. *Asian Journal of Pharmaceutical Research and Health Care*. 2012; 4(1): 11–27.
10. Chen X, Feng Y, Zhong W, Sun B, and Tao F. Numerical investigation of particle deposition in a triple bifurcation airway due to gravitational sedimentation and inertial impaction. *Powder Technology*. 2018; 323: 284–293.
11. Hofemeier P and Sznitman J. Revisiting pulmonary acinar particle transport: convection, sedimentation, diffusion, and their interplay. *Journal of Applied Physiology*. 2015; 118(11): 1375–1385.
12. Ghadiri M, Young PM, and Traini D. Strategies to enhance drug absorption via nasal and pulmonary routes. *Pharmaceutics*. 2019; 11(3): 113.
13. Moral VP and Donaire JG. Inhaled therapy in asthma. *Medicina Clínica*. 2016; 146(7): 316–323.
14. Berkenfeld K, Lamprecht A, and McConville JT. Devices for dry powder drug delivery to the lung. *AAPS PharmSciTech*. 2015;16(3): 479–490.
15. Hampson NB and Mueller MP. Reduction in patient timing errors using a breath-activated metered dose inhaler. *Chest*. 1994; 106(2): 462–465.

16. Roth AP, Lange CF, and Finlay WH. The effect of breathing pattern on nebulizer drug delivery. *Journal of Aerosol Medicine and Pulmonary Drug Delivery*. 2003; 16(3): 325–339.
17. Lærum BN, Telg G, and Stratelis G. Need of education for dry powder inhaler storage and retention–a patient-reported survey. *Multidisciplinary Respiratory Medicine*. 2016; 11(1): 21.
18. Newman SP and Newhouse MT. Effect of add-on devices for aerosol drug delivery: deposition studies and clinical aspects. *Journal of Aerosol Medicine and Pulmonary Drug Delivery*. 1996; 9(1): 55–70.
19. Labiris NR and Dolovich MB. Pulmonary drug delivery. Part II: the role of inhalant delivery devices and drug formulations in therapeutic effectiveness of aerosolized medications. *British Journal of Clinical Pharmacology*. 2003; 56(6): 600–612.
20. Ibrahim M, Verma R, Garcia-Contreras L. Inhalation drug delivery devices: technology update. *The Journal of Medical Devices*. 2015; 8: 131–139.
21. Hess DR. Nebulizers: principles and performance. *Respiratory Care*. 2000; 45(6): 609–622.
22. Ari A. Jet, ultrasonic, and mesh nebulizers: an evaluation of nebulizers for better clinical outcomes. *Eurasian Journal of Pulmonology*. 2014; 16:1–7.
23. Yeo LY, Friend JR, McIntosh MP, Meeusen EN, and Morton DA. Ultrasonic nebulization platforms for pulmonary drug delivery. *Expert Opinion on Drug Delivery*. 2010; 7(6): 663–679.
24. Dennis JH, Stenton SC, Beach JR, Avery AJ, Walters EH, and Hendrick DJ. Jet and ultrasonic nebuliser output: use of a new method for direct measurement of aerosol output. *Thorax*. 1990; 45(10): 728–732.
25. Harvey CJ, O'Doherty MJ, Page CJ, Thomas SH, Nunan TO, and Treacher DF. Comparison of jet and ultrasonic nebulizer pulmonary aerosol deposition during mechanical ventilation. *European Respiratory Journal*. 1997; 10(4): 905–909.
26. Bhattacharyya S and Sogali BS. Inhalation therapy-approaches and challenges. *Asian Journal of Pharmaceutical and Clinical Research*. 2018; 11(4): 9–16.
27. Virchow JC, Crompton GK, Dal Negro R, Pedersen S, Magnan A, Seidenberg J, and Barnes PJ. Importance of inhaler devices in the management of airway disease. *Respiratory Medicine*. 2008; 102(1): 10–19.
28. O'Connor BJ. The ideal inhaler: design and characteristics to improve outcomes. Respiratory Medicine. 2004; 98(Suppl A):S10–S16.
29. Islam N, Gladki E. Dry powder inhalers (DPIs) - a review of device reliability and innovation. *International Journal of Pharmaceutics*. 2008; 360(1-2): 1–11.
30. Molimard M and Colthorpe P. Inhaler devices for chronic obstructive pulmonary disease: insights from patients and healthcare practitioners. *Journal of Aerosol Medicine and Pulmonary Drug Delivery*. 2015; 28(3): 219–228.
31. Newman SP, Busse WW. Evolution of dry powder inhaler design, formulation, and performance. *Respiratory Medicine*. 2002; 96(5): 293–304.
32. Ashurst II, Malton A, Prime D, Sumby B. Latest advances in the development of dry powder inhalers. *Pharmaceutical Science & Technology Today*. 2000; 3(7): 246–256.
33. Timsina MP, Martin GP, Marriott C, Ganderton D, Yianneskis M. Drug delivery to the respiratory tract using dry powder inhalers. *International Journal of Pharmaceutics*. 1994; 101(1-2): 1–13.
34. Ambrus R, Benke E, Farkas Á, Balásházy I, and Szabó-Révész P. Novel dry powder inhaler formulation containing antibiotic using combined technology to improve aerodynamic properties. *European Journal of Pharmaceutical Sciences*. 2018; 123: 20–27.
35. Shetty N, Cipolla D, Park H, and Zhou QT. Physical stability of dry powder inhaler formulations. *Expert Opinion on Drug Delivery*. 2020; 17(1): 77–96.
36. Newman SP. Principles of metered-dose inhaler design. *Respiratory Care*. 2005; 50(9): 1177–1190.
37. Sheth P, Sandell D, Conti DS, Holt JT, Hickey AJ, and Saluja B. Influence of formulation factors on the aerosol performance of suspension and solution metered dose inhalers: a systematic approach. *AAPS Journal*. 2017; 19(5): 1396–1410.
38. Mahler DA. Peak inspiratory flow rate as a criterion for dry powder inhaler use in chronic obstructive pulmonary disease. *Annals of the American Thoracic Society*. 2017; 14(7): 1103–1107.
39. Sharma G, Mahler DA, Mayorga VM, Deering KL, Harshaw O, and Ganapathy V. Prevalence of low peak inspiratory flow rate at discharge in patients hospitalized for COPD exacerbation. *Chronic Obstructive Pulmonary Disease*. 2017; 4(3): 217–224.
40. Fröhlich E. Biological obstacles for identifying in vitro-in vivo correlations of orally inhaled formulations. *Pharmaceutics*. 2019; 11(7): 316.
41. Brocklebank D, Wright J, and Cates C. Systematic review of clinical effectiveness of pressurized metered dose inhalers versus other hand held inhaler devices for delivering corticosteroids in asthma. *The BMJ*. 2001; 323(7318): 896–900.
42. Rubin BK and Fink JB. Optimizing aerosol delivery by pressurized metered-dose inhalers. *Respiratory Care*. 2005; 50(9): 1191–1200.
43. Ali M, Reddy RN, and Mazumder MK. Electrostatic charge effect on respirable aerosol particle deposition in a cadaver-based throat cast replica. *Journal of Electrostatics*. 2008; 66: 401–406.
44. Berkenfeld K, Lamprecht A, and McConville JT. Devices for dry powder drug delivery to the lung. *AAPS PharmSciTech*. 2015; 16(3): 479–490.
45. Pudi K, Feldman G, Fakih F, Mack P, Maes A, Siddiqui S, St Rose E, and Reisner C. An Open-Label Study Evaluating the Performance of the Dose Indicator in a Metered Dose Inhaler Delivering Glycopyrrolate and Formoterol Fumarate in Patients with Moderate-to-Very Severe Chronic Obstructive Pulmonary Disease. *Journal of Aerosol Medicine and Pulmonary Drug Delivery*. 2019; 32(1): 40–46.
46. Labiris NR and Dolovich MB. Pulmonary drug delivery. Part I: physiological factors affecting therapeutic effectiveness of aerosolized medications. *British Journal of Clinical Pharmacology*. 2003; 56(6): 588–599.
47. Vanbever R, Mintzes JD, Wang J, Nice J, Chen D, Batycky R, Langer R, and Edwards DA. Formulation and physical characterization of large porous particles for inhalation. *Pharmaceutical Research*. 1999; 16(11): 1735–1742

48. Telko MJ and Hickey AJ. Dry powder inhaler formulation. *Respiratory Care*. 2005; 50(9): 1209–1227.
49. Dunbar CA, Hickey AJ, and Holzner P. Dispersion and characterization of pharmaceutical dry powder aerosols. *Kona Powder and Particle Journal*. 1998; 16: 7–45.
50. Young PM, Sung A, Traini D, Kwok P, Chiou H, and Chan HK. Influence of humidity on the electrostatic charge and aerosol performance of dry powder inhaler carrier-based systems. *Pharmaceutical Research*. 2007; 24(5): 963–970.
51. Chow AH, Tong HH, Chattopadhyay P, and Shekunov BY. Particle engineering for pulmonary drug delivery. *Pharmaceutical Research*. 2007; 24(3): 411–437.
52. Voss A, Finlay WH. Deagglomeration of dry powder pharmaceutical aerosols. *International Journal of Pharmaceutics*. 2002; 248(1-2): 39–50.
53. Clark AR and Hollingworth AM. The relationship between powder inhaler resistance and peak inspiratory conditions in healthy volunteers - implications for in vitro testing. *Journal of Aerosol Medicine*. 1993; 6(2): 99–110.
54. Tiddens HA, Geller DE, Challoner P, Speirs RJ, Kesser KC, Overbeek SE, Humble D, Shrewsbury SB, and Standaert TA. Effect of dry powder inhaler resistance on the inspiratory flow rates and volumes of cystic fibrosis patients of six years and older. *Journal of Aerosol Medicine*. 2006; 19(4): 456–465.
55. Feddah MR, Brown KF, Gipps EM, and Davies NM. In-vitro characterization of metered dose inhaler versus dry powder inhaler glucocorticoid products: influence of inspiratory flow rates. *Journal of Pharmacy and Pharmaceutical Sciences*. 2000; 3(3): 318–324.
56. Chan JGY, Wong J, Zhou QT, Leung SSY, and Chan HK. Advances in device and formulation technologies for pulmonary drug delivery. *AAPS PharmSciTech*. 2014; 15(4): 882–897.
57. Melani AS. Inhalatory therapy training: a priority challenge for the physician. *Acta Biomedica*. 2007; 78(3): 233–245.
58. Pfützner A, Mann AE, and Steiner SS. Technosphere/Insulina new approach for effective delivery of human insulin via the pulmonary route. *Diabetes Technology & Therapeutics*. 2002; 4(5): 589–594.
59. Mikhail N. Safety of Technosphere Inhaled Insulin. *Current Drug Safety*. 2017; 12(1): 27–31.
60. McGill JB, Peters A, Buse JB, Steiner S, Tran T, Pompilio FM, Kendall DM. Comprehensive Pulmonary Safety Review of Inhaled Technosphere® Insulin in Patients with Diabetes Mellitus. *Clinical Drug Investigation*. 2020;40(10):973–983.
61. Mikhail N. Place of technosphere inhaled insulin in treatment of diabetes. *World Journal of Diabetes*. 2016; 7(20): 599–604.
62. Mohanty RR and Das S. Inhaled Insulin - current direction of insulin research. *Journal of Clinical and Diagnostic Research for Doctors*. 2017; 11(4): OE01–OE02.
63. Harper NJ, Gray S, De Groot J, Parker JM, Sadrzadeh N, Schuler C, Schumacher JD, Seshadri S, Smith AE, Steeno GS, Stevenson CL, Taniere R, Wang M, and Bennett DB. The design and performance of the exubera pulmonary insulin delivery system. *Diabetes Technology & Therapeutics*. 2007; 9 (Suppl 1): S16–S27.
64. Hermansen K, Rönnemaa T, Petersen AH, Bellaire S, andAdamson U. Intensive therapy with inhaled insulin via the AERx insulin diabetes management system: a 12-week proof-of-concept trial in patients with type 2 diabetes. *Diabetes Care*. 2004; 27(1): 162–167.
65. Henry RR, Mudaliar SR, Howland III WC, Chu N, Kim D, An B, and Reinhardt RR. Inhaled insulin using the AERx Insulin Diabetes Management System in healthy and asthmatic subjects. *Diabetes Care*. 2003; 26(3): 764–769.
66. Longest W, Spence B, and Hindle M. Devices for improved delivery of nebulized pharmaceutical aerosols to the lungs. *Journal of Aerosol Medicine and Pulmonary Drug Delivery*. 2019; 32(5): 317–339.
67. Murthy GL, Machale MU, Balireddy N, and Chandrika C. Out lines of aerosols in pharmaceutical technology. *Asian Journal of Research in Pharmaceutical Science*. 2019; 9(3): 215–225.
68. Dadrwal D. Dry powder inhaler; special emphasis to formulation, devices, characterization & process validation protocol: A review. *Journal of Drug Delivery and Therapeutics*. 2017; 7(3): 50–54.
69. Wachtel H. Respiratory drug delivery. In: Dietzel A, ed. *Microsystems for Pharmatechnology. Germany*, PA: Springer, Cham 2016; 257–274.
70. Chrystyn H and Niederlaender C. The Genuair inhaler: a novel, multidose dry powder inhaler. *International Journal of Clinical Practice*. 2012; 66(3): 309–317.
71. Magnussen H, Watz H, Zimmermann I, Macht S, Greguletz R, Falques M, Jarreta D, and Garcia Gil E. Peak inspiratory flow through the Genuair inhaler in patients with moderate or severe COPD. *Respiratory Medicine*. 2009; 103(12): 1832–1837.
72. Cazzola M, Cavalli F, Usmani OS, and Rogliani P. Advances in pulmonary drug delivery devices for the treatment of chronic obstructive pulmonary disease. *Expert Opinion on Drug Delivery*. 2020; 17(5): 635–646.
73. Tobyn M, Staniforth JN, Morton D, Harmer Q, and Newton ME. Active and intelligent inhaler device development. *International Journal of Pharmaceutics*. 2004; 277(1-2): 31–37.
74. Son YJ and McConville JT. Advancements in dry powder delivery to the lung. *Drug Development and Industrial Pharmacy*. 2008; 34(9): 948–959.
75. Atkins PJ. Dry powder inhalers: an overview. *RespirCare*. 2005; 50(10): 1304–1312.
76. Buhl R and Banerij D. Profile of glycopyrronium for once-daily treatment of moderate-to-severe COPD. *International Journal of Chronic Obstructive Pulmonary Disease*. 2012; 7: 729–741.
77. Pavkov R, Mueller S, Fiebich K, Singh D, Stowasser F, Pignatelli G, Walter B, Ziegler D, Dalvi M, Dederichs J, and Rietveld I. Characteristics of a new capsule based dry powder inhaler for the effective delivery of indacaterol. *Current Medical Research and Opinion*. 2010; 26(11): 2527–2533.
78. Chapman KR, Fogarty CM, Peckitt C, Lassen C, Jadayel D, Dederichs J, Dalvi M, and Kramer B. Delivery characteristics and patients' handling of two single-dose dry-powder inhalers used in COPD. *International Journal of Chronic Obstructive Pulmonary Disease*. 2011; 6: 353–363.
79. Hochrainer D, Hölz H, Kreher C, Scaffidi L, Spallek M, and Wachtel H. Comparison of the aerosol velocity and spray duration of Respimat® Soft Mist™ inhaler and pressurized

metered dose inhalers. *Journal of Aerosol Medicine and Pulmonary Drug Deliver*. 2005; 18(3): 273–282.
80. Perriello EAandSobieraj DM. The Respimat Soft Mist Inhaler, a Novel Inhaled Drug Delivery Device. *Connecticut Medicine*. 2016; 80(6): 359–364.
81. Kunkel G, Magnussen H, Bergmann K.,et al. Respimat (a new soft mist inhaler) delivering fenoterol plus ipratropium bromide provides equivalent bronchodilation at half the cumulative dose compared with a conventional metered dose inhaler in asthmatic patients. *Respiration*. 2000; 67(3): 306–314.
82. Dalby RN, Eicher J, and Zierenberg B. Development of Respimat® Soft Mist Inhaler and its clinical utility in respiratory disorders. *Medical Devices*. 2011; 4: 145–155.
83. Zhou QT, Tang P, Leung SS, Chan JG, and Chan HK. Emerging inhalation aerosol devices and strategies: where are we headed? *Advanced Drug Delivery Reviews*. 2014; 75C: 3–17.

39

Numerical Modeling of Agglomeration and De-agglomeration in Dry Powder Inhalers

Tan Suwandecha[1] **and Teerapol Srichana**[1,2]
[1]*Prince of Songkla University, Hat Yai, Songkla, Thailand*
[2]*Department of Pharmaceutical Technology, Prince of Songkla University*

39.1 Introduction

Delivery of a dry powder aerosol into the human airway system for treatment of lung-related disorders or systemic treatment is a major challenge. The aerosolized powder must be small enough to reach the lower airways or the targeted part of the lungs. The aerosolization process begins with transportation of dry powder from the reservoir and the dispersed powder is then de-agglomerated. These processes require a considerably large magnitude of energy that relies on the inhalation force of the patient, which is sometimes limited. Therefore, several factors affect the delivery efficiency of dry powder inhalers (DPIs). The first factor is a dry powder formulation that is needed to optimize the adhesive force between the drug and carrier and the cohesive force between the drug particles themselves. The second important factor is the DPI device. The devices should generate sufficient de-agglomeration of the powder with reasonable inhalation force by the patient. The other factor is the patient's inhalation pattern that affects the de-agglomeration and deposition in the airways of the dry powder aerosol. Due to the complexity of a DPI system, it is quite difficult to identify the influence of a single factor on DPI performance even though many studies have been conducted to explore the effects of particle morphology, physical properties of the particles, dosage containers, device design, and patient factors.

To overcome these limitations, numerical modeling has drawn increasing attention for investigation of the physical properties of the particles, fluid flow, particle behavior, and the deposition pattern in the human airway system. Computational fluid dynamics (CFD) is numerical modeling for analyzing and solving the problem of fluid flow. The flow patterns in both the inhalers and human airway systems can be explored using CFD. The basics of CFD are the Navier–Stokes equations and the Euler equation derived from mass, energy, and momentum conservation laws (1,2). The conservation of mass is described in terms of the density change rate between the inlet flow and outlet flow, and the summation of the mass change rate must be equal to zero (3).

$$\frac{\partial \rho}{\partial t} + \frac{\partial(\rho u)}{\partial x} + \frac{\partial(\rho u)}{\partial y} = 0 \qquad (39.1)$$

ρ = fluid density
t = time
u = fluid velocity

Momentum is the multiplication of mass and velocity which must be conserved in a finite volume system. In a three-dimensional (3D) system, there are two momentum vectors, u and v, that belong to the velocity of the x and y axes (39.1). Furthermore, the first law of thermodynamics states that the total (i.e. internal + kinetic) energy increases in the system and is the summation of the entrance energy, work of surface and body forces, heat from conduction, and heat from any chemical reaction. Finally, the basic form of Navier–Stokes equation is shown in (39.2)(3).

$$\begin{aligned} &\text{umomentum } \frac{\partial(\rho u)}{\partial t} + \frac{\partial(\rho u^2 + P)}{\partial x} + \frac{\partial(\rho u v)}{\partial y} = \frac{\partial \tau_{xx}}{\partial x} + \frac{\partial \tau_{xy}}{\partial y} \\ &\text{vmomentum } \frac{\partial(\rho u)}{\partial t} + \frac{\partial(\rho u^2 + P)}{\partial y} + \frac{\partial(\rho u v)}{\partial x} = \frac{\partial \tau_{xy}}{\partial x} + \frac{\partial \tau_{xx}}{\partial y} \end{aligned} \qquad (39.2)$$

u, v = Cartesian components of velocity along x, y axes

Solving the turbulent flow problem with an appropriate turbulent model, such as the k-epsilon (ε) model, needs to be coupled with the Navier–Stokes equations for laminar flow problems. The k-ε turbulence model was developed from the Reynolds-averaged Navier–Stokes equations (RANS) (4). The RANS equations decompose turbulent flow into time-averaged flow velocity and velocity fluctuation. The k-ε turbulence model includes two extra transport equations to represent the turbulent properties of the flow. This allows some variables such as convection and diffusion of turbulent energy to be traced back effectively. The first transport variable of the k-ε turbulence model is k, which determines the energy in the turbulence. The second variable is ε which determines the scale of the turbulence or turbulent dissipation (2). Despite the robustness, the k-ε turbulence model still has some limitations on complex geometry. Some researchers suggested using large eddy simulation to improve the accuracy of modeling turbulent flow (5).

A particle model, such as the discrete element method (DEM) or discrete particle model (DPM), is coupled to CFD to simulate fluid–particle and particle–particle interactions which is a key point to understanding the dynamics of a DPI system (4).

DOI: 10.1201/9781003046547-39

Particle motion simulation in fluid flow is able to calculate the force balance on the particle (39.3)

$$\frac{d\overrightarrow{u_p}}{dt} = F_D(\overrightarrow{u} - \overrightarrow{u_p}) + \frac{\overrightarrow{g}\,(\rho_p - \rho)}{\rho_p} + \overrightarrow{F} \quad (39.3)$$

where $\overrightarrow{F}$ is the acceleration term and F_D is the drag force.

The particle model can provide in-depth information at the single-particle level, such as drag and lift forces acting on the particles, which are challenging to investigate in conventional experiments.

In addition to CFD and particle modeling, numerical modeling is also applied to simulate the physical interactions between the particles in a DPI formulation such as adhesive-cohesive forces between drug–drug and carrier–drug particles and the effect of electrostatic charges on the surfaces of the particles. These simulations allow us to understand particle dynamics from mixing through handling, packaging, and storage that are beneficial for manufacturing a better DPI system.

Another perspective of numerical modeling in a DPI system is the simulation of particle deposition patterns in the human airway system that can be coupled to pharmacokinetic modeling. Using an accurate airway geometry from magnetic resonance imaging (MRI) scans and computed tomography (CT) in conjunction with numerical modeling should give far more accurate and complete information than that available from the conventional Anderson cascade impactor or the next-generation impactor.

This chapter will comprehensively review all aspects of numerical modeling of DPIs (Figure 39.1).

39.2 Numerical Modeling of Dry Powder Agglomeration and De-agglomeration

The performance of a DPI is usually measured as the fine particle fraction (FPF) which is the fraction of particles smaller than about 5 µm. It is the result of the de-agglomeration of the drug particles from the particle agglomerates. The mechanism of the detachment of drug powder from fluid flow has been studied thoroughly using CFD during the last 10 years and is increasingly used for inhaler development and optimization. The overall de-agglomeration process, including the particle–wall impaction, particle–particle impaction, and aerodynamic force, is shown in Figure 39.2.

DPI formulations usually consist of two major types: drug only formulation and carrier-based formulation. The agglomeration of particles in these formulations can be represented by mathematical differential equations. Thornton et al. published two-dimensional numerical experiments using DEM on monodispersed agglomerated particles that represent the drug-only formulation (6). The simulation started with 1000 particles with a radius of 100 µm and surface energy up to 3.0 J/m^2. The agglomeration was induced by a centripetal gravity field until a satisfactory dense agglomerate was formed. Agglomerate–wall collisions were simulated at 1.0 m/s, 0.1 m/s, and 0.04 m/s. They found that the plastic deformation of an agglomerate was a critical step in high-velocity impaction (1.0 m/s) followed by tensile separation of individual primary particles. The threshold velocity increased exponentially with the bond strength of the primary particles.

The centripetal gravity field agglomeration scheme was also used by Yang et al. (7). Ten thousand monodispersed spherical particles in the size range of 1–1000 µm were agglomerated. It was found that the packing density exponentially decreased as the particle number increased. A stable value was attained when the particle number exceeded 5000. Furthermore, the tensile strength had an inverse relationship with the particle size.

Besides drug-only formulations, several research studies observed agglomeration of carrier-based formulations (8). A carrier particle with a radius in the range of 26.25–52.5 µm and 2.5 µm drug particles were randomly seeded into the simulation. The drug particles settled to the carrier particles until they reached a stable state. The work of adhesion between the drug and the carrier was varied from 0.1 to 0.6 mJ/m^2. The dispersion ratio increased proportionally with increasing air velocity and decreased with increasing work of adhesion. It confirmed that

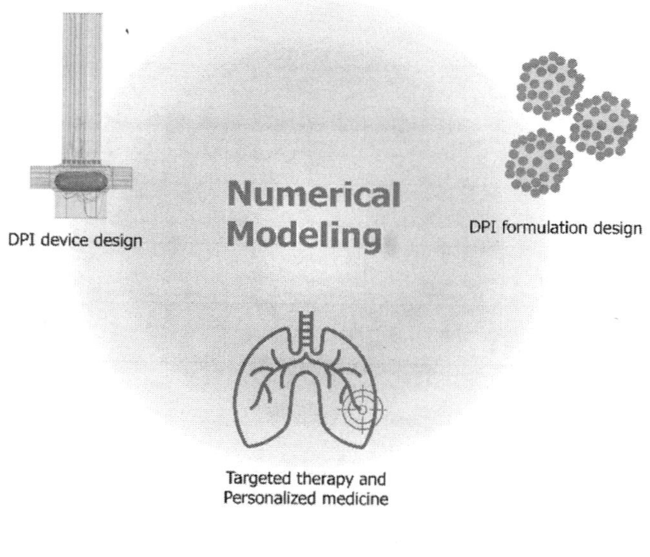

FIGURE 39.1 Overall Aspects of Numerical Modeling in Dry Powder Inhalers

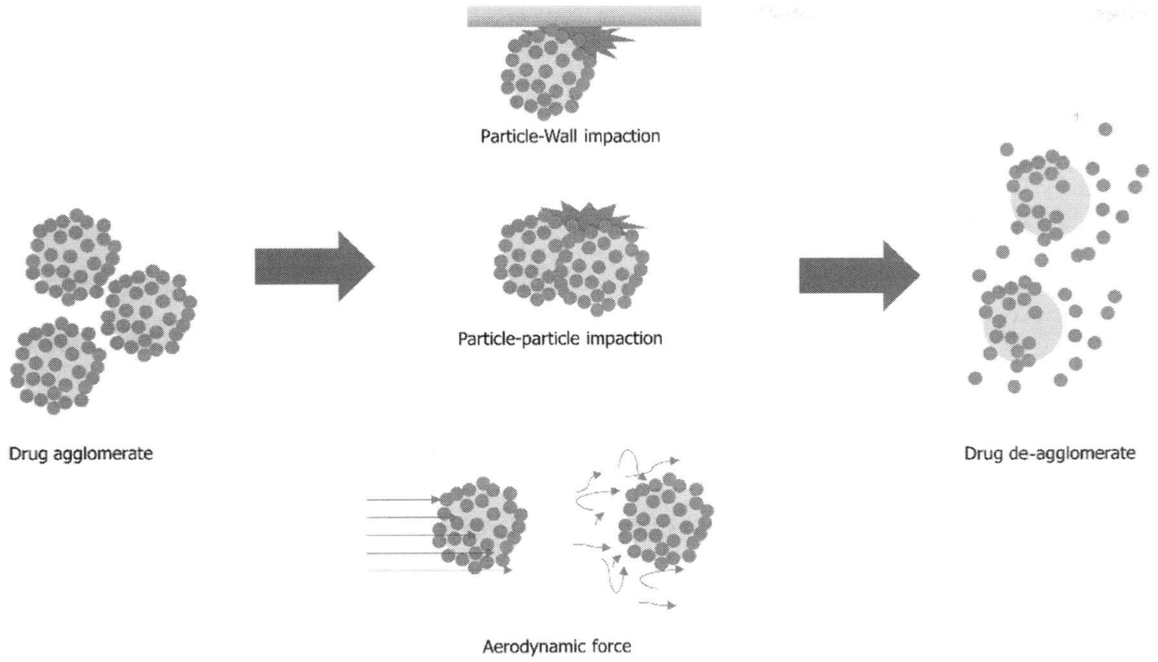

FIGURE 39.2 Dry Powder De-agglomeration Process

the balance between the fluid drag force and the particle–particle interaction force is important in the DPI design. Interestingly, the drug particles in the downstream region were easier to detach than in the upstream region. Thus, the locations of air–agglomerate interaction need to be considered while designing DPI systems to improve overall efficiency. Besides the laminar fluid drag force, de-agglomeration by turbulent kinetic energy (TKE) is also important. Research on the effect of TKE on the de-agglomeration of drug-carrier particles was conducted. The drug-carrier agglomerates were evaluated using three sizes of lactose carrier: large, medium, and small. The TKE was highly correlated to the FPF and mass median aerodynamic diameter (MMAD), especially with smaller carrier particles that had a stronger adhesive force than the large carrier particles (9).

In addition to the fluid drag force acting on agglomerates, the particle–particle and particle–wall impactions are also important in DPI de-agglomeration dynamics. The CFD modeling of a simplified entrainment tube was conducted to evaluate Venturi-driven turbulence effects on agglomerate breakup. The particle system was mannitol particles with a median diameter of about 700 μm. An interesting finding was that the fluid turbulence force and particle de-agglomeration had a poor relationship even though the entrainment tube was designed to maximize turbulence and minimize the impaction. Thus, particle impaction seems to be the important driving force for particle de-agglomeration (10). Impaction de-agglomeration was further investigated by designing a series of entrainment tubes containing different impaction plates. The entrainment tube was designed to minimize TKE and de-agglomeration relied only on impaction. The impaction angle and velocity were varied to investigate their effects. The mannitol particle with a median diameter of about 700 μm was used as the particle model. The model was simulated using commercial CFD software. Particle–wall impacts resulted in powder fragmentation followed by re-entrainment in the airstream. The aerosolization performance was highly dependent on impact velocity and air velocity. The impaction angle was an independent factor for aerosolization. However, it was a crucial factor for wall losses. A shallow impaction angle resulted in less wall mass deposited than in perpendicular impaction (11).

Numerical simulation on real DPI models was performed on the Turbuhaler® device. Two flow conditions were used: steady flow and dynamic flow. The steady flow was modeled using steady airflow rates of 20–70 l/min. For dynamic flows, the outlet pressure varied to create two different peak inspiratory flow rates. The dynamic breakage and particle size distribution were observed in a simplified compartment model of the Turbuhaler® device. It was found that the FPF was directly related to the peak inspiratory flow rates. However, particle deposition in the inhaler device also increased. The compartment model revealed that the de-agglomeration rate of the powder agglomerates had a high impact on the outflowing particle size distribution. A low de-agglomeration rate resulted in a broad particle size distribution, while a high de-agglomeration rate gave a narrower particle size distribution and smaller median size. Thus, rapid de-agglomeration was essential for good DPI performance (12).

39.3 Numerical Modeling for DPI Device Design and Evaluation

The mechanisms of powder de-agglomeration in DPIs rely mainly on aerodynamic drag and impaction as discussed in the previous section. The DPI device is an undeniably important factor. A well-designed DPI device can provide sufficient aerodynamic drag and impaction force for aerosolization without requiring much inhalation effort or drug loss in the device. The geometry of the inhaler device affects many aerodynamic parameters, such as pressure drop, turbulence, kinetic energy, impaction events, and the integral scale strain rate. An interesting design uses a grid which is used in many commercially available DPI devices. The effects of a grid structure were extensively studied by Coates et al. (13). The grid structure of the Aerolizer® device, which is a capsule-based inhaler device, was modified by a reduced density of the grid mesh. Spray-dried mannitol with a median size of 3.2 μm was used as the model drug. The experimental results showed that as the density of the grid mesh decreased, there was a reduction in the %FPF. Interestingly a greater void space in the grid mesh led to increased powder retention in the mouthpiece of the Aerolizer® device. In CFD simulation, it was proven that the grid mesh plays an important role in the reduction of swirly airflow and reduced powder retention due to aerosolized particles impacting the device wall (13). Further investigation of the grid design on aerosolization of a carrier-based DPI formulation (Foradile®) was conducted using a modified Aerolizer® device with a full grid mesh and a cross-grid mesh. The mouthpiece length and air inlet port were also modified to a shorter and narrower geometry, respectively. Numerical simulation showed that the velocity patterns and %FPF were not much different between the full grid mesh and the cross-grid mesh. In contrast, narrowing the air inlet port resulted in a greater swirling flow and significantly increased the %FPF, mostly because high-velocity particle–wall impactions occurred in the inhaler chamber. Interestingly, the amount of drug deposited in the induction port that mimicked the oropharynx was double in the case of the cross-grid mesh. This was due to turbulent flow that was not properly regulated by the grid mesh before emission from the inhaler device (14).

Not only the grid mesh density and grid mesh position but also the surface area affected the aerosolization performance. The Aerolizer® device was modified with different grid positions where (1) the grid was near the mouthpiece opening orifice and (2) the grid was near the inhaler base. The models were evaluated with numerical simulation and verified in vitro using a carrier-based DPI system. The impaction events and impaction energy were recorded and calculated to the probability of breakup of the agglomerated particles by impaction. It was found that the impaction kinetic energy and particle–wall impaction events were higher when the grid was placed near the inhaler base. This was due to a whirlpool-like flow inherited from the design of the Aerolizer® device. This emphasized the importance of the grid in the inhaler design as previously discussed by Coates et al. (13). The probability of de-agglomeration by impaction was higher when the grid was placed near the inhaler base. However, it failed to reflect the % FPF from the in vitro experiment because the researchers did not include the effect of the capsule in the computational simulation (9).

The effect of the grid was also studied on a carrier-based DPI system. Lactose and salbutamol sulfate in four different ratios were properly mixed and used as the test formulations. The grid structure of the Aerolizer® device was modified from the original Aerolizer® grid mesh and cross-grid mesh. It was shown that a low-density grid mesh generated a more cyclonic flow pattern that led to drug loss via mouthpiece deposition. However, a low-density grid mesh enhanced drug dispersion at a low drug-to-carrier ratio due to increased impaction (15).

Numerical simulation of two different inhaler devices, the Aerolizer® and Inhalator®, was conducted by Suwandecha et al. (9). The key differences between the two devices were that the Aerolizer® has the ability to create a whirlpool-like flow and slightly low pressure drop while the Inhalator® has a high velocity flow through a narrow chamber and high pressure drop. The carrier-based salbutamol sulfate with different lactose sizes was used as the model formulation. Turbulent kinetics energy and the probability of de-agglomeration data were extracted from numerical simulation. The high level of TKE clustered mostly around the inlet and inhaler base part of the Aerolizer® and around the capsule chamber and the grid of the Inhalater®. The impaction count of the Inhalater® was slightly higher than the Aerolizer® at a low flow rate because the impaction event was created by different mechanisms that included impacting the grid and impaction by the swirly flow of the Aerolizer® and Inhalater®, respectively. The calculated probability of particle de-agglomeration revealed that de-agglomeration in the Inhalater® was higher than the Aerolizer® because of higher impaction energy. The %FPF emitted from both devices from the experiment was not much different, but the Inhalator® gave a slightly lower MMAD value than the Aerolizer®. Interestingly, the %FPF was highly correlated to the TKE in the Aerolizer®. The %FPF–TKE correlation was poor in the Inhalator® due to the higher mouthpiece outlet velocity that created drug loss in the induction port. It may be possible to design a high-performance DPI device by combining the strengths of the Aerolizer® and Inhalator® since there is a balance between the pressure drop and the dispersion performance of the de-agglomerated particles (9).

The circulation chamber of the Aerolizer® device is also a critical feature. A study adopted modified circulation chambers of the Aerolizer® to make it two times greater, 1.5 times greater, 1.5 times smaller, and two times smaller than the original chamber of the device. Numerical simulation revealed that the maximal velocities were at the grid area and were generally smaller when the circulation chamber had a smaller height. In the DPM, particle tracking showed that tracks with greater height provided more turbulent flow in the circulation chamber. Particle deposition in the device was reduced as the circulation chamber height was decreased. This provided some clues about the circulation chamber geometry for a DPI device, even though the effect of the capsule was not included in the study (16).

The effect of the capsule has been thoroughly studied. The capsule motion in the Aerolizer® was observed using numerical simulation and a high-speed camera at different flow rates of 30 to 100 l/min. Several significant findings were obtained from this study. The high-speed camera revealed the fundamental motion of the capsule from CFD simulation. The capsule predominantly rolled around its longitudinal axis in one direction from 500 rpm to 3700 rpm. The capsule in the swirl chamber also frequently collided with the inhaler wall up to 3 m/s shaking the powder off the walls. The capsule affected the airflow field around the capsule by inducing a swirl of airflow around the capsule. The airflow entered the pierced holes of the capsule at a velocity up to 10 m/s that depended on the capsule angle. This allowed for calculation of the forces acting on the particles inside the capsule by combining gravity and centrifugal force from the capsule rotation. The effect of these forces varied by particle size and position of the particles in the capsule. For example, the aerodynamic force and centrifugal force were roughly equal for a spherical carrier particle (175 μm) at the capsule hole where the air velocity reached to about 30 m/s. The fluid drag was the dominating force in the inner part of capsule for particle sizes below 100 μm. However, the aerodynamic force was insignificant for large carrier particles in the inner part of the capsule (17).

Longest et al. (18) used numerical simulation to design a new high-performance DPI by adapting a previously described methodology. They designed a capsule-based unit-dose DPI device. The inhaler was designed to operate by airflow through a longitudinal axis of the capsule with two pierced holes on each of the size 0 capsule tips. The CFD simulation model was a low Reynolds number k–ω turbulence model in conjunction with particle tracking. Salbutamol sulfate was used as the model drug. Several factors were addressed and included in the design process (Table 39.1).

All in all, the flow and particle parameters, such as $\omega^*_{part} \times k^*_{part}$, $k^*_{field} \times \omega^*_{field}$, and wall shear stress from numerical simulation, were very useful for optimizing inhaler devices to create high-efficiency DPIs (18).

TABLE 39.1

Factors Expected to Influence Dispersion of the Aerosol (18)

Factor	Equation	Findings
Volume-averaged turbulent kinetic energy (TKE) of the flow field (k_{field})	$k_{field} = \frac{1}{Vol}\int_{Vol} kdVol$ Where Vol is the volume of the device and k is the TKE at each location in the volume. (m^2/s^2)	Increasing k_{field} resulted in increasing the MMAD.
Volume-averaged specific dissipation rate of the flow field (ω_{field})	$\omega_{field} = \frac{1}{Vol}\int_{Vol} \omega dVol$ Where Vol is the volume of the DAC unit and ω is the specific dissipation rate at each location in the volume.(1/s)	High values of ω indicated elevated k together with a small turbulent length scale.
Nondimensional TKE of the flow field (k^*_{field})	$k^*_{field} = k_{field}\frac{1}{V^2_{inlet}}\cdot 10^4$ Where V_{inlet} is the mean velocity of the inlet air jet.	Increasing k_{field} resulted in increasing the MMAD-Direct correlation with emitted dose
Nondimensional specific dissipation rate of the flow field (ω^*_{field})	$\omega^*_{field} = \omega_{field}\frac{Vol^{\frac{1}{3}}}{V_{inlet}}$ Provides a representative exposure time of particles to the flow field. $\frac{Vol^{\frac{1}{3}}}{V_{inlet}}$ abbreviated as t_{inlet} (s)	No correlation with emitted dose and MMAD
Nondimensional eddy viscosity (k^*_{field} / ω^*_{field})	$\frac{k^*_{field}}{\omega^*_{field}} = \frac{k_{field}}{\omega_{field}}\frac{1}{V_{inlet}Vol^{\frac{1}{3}}}$	Direct correlation with emitted dose
k^*_{field} x ω^*_{field}	$k^*_{field}\omega^*_{field} = k_{field}\omega_{field}Vol^{\frac{1}{3}}V_{inlet}$ ω field with k as multiplying factor.	k^*_{field} x ω^*_{field} had a strong correlation with the MMAD and emitted dose.
Wall shear stress (WSS)	$WSS = \int_A \mu_{total}\frac{du}{dn}dA$ Where A is the surface area, μ_{total} is the total viscosity at the wall including both laminar and turbulent components, u is the local velocity parallel to the wall, and n is the local wall-normal coordinate.(N/m^2)	WSS correlated well with the emitted dose by removing particles from the wall.
Average particle residence time (t_{part})	$t_{part} = \frac{1}{n}\sum_{i=1}^{n}\int tracjectory\ dt$ Where n particles are considered and the integral is performed for each particle. (s)	
Average particle TKE (k_{part})	$k_{part} = \frac{1}{n}\sum_{i=1}^{n}\int tracjectory^k\ dt$ Where k is the local TKE experienced by each particle along its trajectory (J)	Direct correlation with MMAD, higher values gave larger MMAD values
Nondimensional article TKE (k^*_{part})	$k^*_{part} = k_{part}\frac{1}{V_{inlet}Vol^{\frac{1}{3}}}$	Direct correlation with MMAD and emitted dose, higher values gave a larger MMAD values and higher emitted doses.
Average particle specific dissipation rate (ω_{part} or ω^*_{part})	$\omega^*_{part} = \frac{1}{n}\sum_{i=1}^{n}\int tracjectory^{\omega}\ dt$	Direct correlation with MMAD, higher values gave larger MMAD values- $\omega^*_{part} \times k^*_{part}$ together predicted quite well the MMAD
Nondimensional ω_{part} / t_{part}	$\left(\frac{\omega_{part}}{t_{part}}\right)^* = \frac{\omega_{part}}{t_{part}}\frac{Vol^{\frac{1}{3}}}{V_{inlet}}$	Direct correlation with MMAD, higher values gave larger MMAD values

39.4 Numerical Modeling of Particle Deposition in the Human Airway System

The in vitro aerosol testing apparatus, such as the Andersen cascade impinger, multi-stage liquid impinger, and next-generation impactor, can provide limited aerosol characterization. However, they cannot provide an in-depth deposition profile of a deposition pattern that is highly variable in the geometry of the human lungs. Several attempts have been carried out to create useful human lung models for aerosol deposition studies. Early models of the human lungs were based on a fully symmetrical tree structure of straight cylindrical tubes (19–21) or models that resembled a trumpet-shaped airway tree. Real human airways are far more complicated than those devices. There are variations within the parameters of the tubes of a given airway generation and they are not symmetrical. In addition, inter-subject variation needs to be considered.

The stochastic model was used to describe the asymmetry and randomness of the airway system. Hofmann and Koblinger (22) proposed the IDEAL-2 (Inhalation, Deposition and Exhalation of Aerosols in/from the Lungs-2nd version) which is a stochastic model that allows for variations in the airway geometries and the number of bifurcations to the end of each bronchial pathway. The particle trajectory profile in human airways can be expressed by randomness in a statistical model. Monte Carlo modeling well represented the millions of possible pathways of a particle traveling in the human airway system. The individual particle dynamics, such as Brownian motion, can be calculated based on the geometry of a particle. Spherical particles in the size range of 0.01–10 µm were used to verify the IDEAL-2 model with experimental and reference data. The results from the model had fairly good agreement with the experimental and reference data with some limitations such as inter-subject morphometric variability and differences in the definitions of bronchial and acinar regions (23).

Advancements in medical imaging, such as CT and computational modeling, have opened new opportunities to create more realistic 3D human lung models as opposed to the earlier simplified 2D models. Early efforts at 3D modeling used idealized 3D geometry by averaging CT models, such as the idealized mouth model. Numerical simulation using the k–ω turbulence model showed reasonable fluid dynamics and particle trajectory characteristics. The inlet flow profile to the mouth cavity could create an impinging jet that impacted particle deposition in the mouth (24). The effects of breath pattern on particle behavior and deposition in the CT derived airway models were observed by Kadota et al. (25). Asymmetry of the human airway system results in unequal distribution of airflow into different lobes of the lungs. Most of the airflow is nearly always directed to the right lung. The overall airflow is laminar flow except in the laryngopharynx region where the magnitude of the TKE is considerably higher than in the other parts of the airways. The effect of breath holding was also investigated because DPI manufacturers recommend holding one's breath after inhaling a powder formulation. Breath holding for 10 seconds increased the turbulence in the airways, which improved particle deposition in the airways. It also found that the number of particles that passed through the right trachea was usually higher than in the left trachea. Furthermore, the ratio of inhaled particles that entered the upper lobe was less than the lower lobe.

The airflow entering the human lungs is not a constant flow. A previous study used constant flow at a fixed flowrate. The airflow pattern during inhalation in either normal or unhealthy lungs would have some effects on the deposition of particles in the lungs. The flow pattern can be expressed as a transient inhalation flow rate Q(t) ((39.4))

$$Q(t) = \begin{cases} \frac{PIFR}{T_{PIFR}} \cdot t, & 0 \le t \le T_{PIFR} \\ PIFR \cdot \cos \frac{\pi (t - T_{PIFR})}{2(T - T_{PIFR})}, & T_{PIFR} \le t \le T \end{cases} \tag{39.4}$$

where PIFR is the peak inspiratory flow rate, T is the total inhalation time, and T_{PIFR} is the time to PIFR.

The airflow distributions in each lobe of the lungs were markedly different using either constant flow or flow pattern. Under the flow pattern model, the airflow to the left upper lobe increased while the airflow in the right lower lobe and left lower lobe decreased. Particle deposition simulation revealed that the ratio of the deposited particles deeper into the lungs was greater for the pattern flow while a fairly small number of particles were deposited in the bronchi in the constant flow model (25).

Numerical simulation of particle deposition in the human airway system can be applied to targeting an inhaled aerosol to a certain part of the lungs. Das et al. (26) applied CFD to target aerosol to the upper airways in children. They used a simplified morphometrically realistic upper airway model of the human lungs in different age groups. The inhalation parameters such as tidal volume, PIFR, and total inspiration time as a function of age (5, 10, and 25 years old) were selected as the variable parameters. The CFD simulation showed that inlet velocity and flow velocity around the larynx were larger in the younger ages. Turbulent flow also occurred around the larynx in all age groups with different intensities. The particle deposition efficiency was characterized by the Stokes number (Stk). Particle deposition was found to fit on a single curve as a function of Stk for all age groups. To target the upper airways, particles with Stk ≅ 0.06 gave nearly 80% deposition in the conducting airways (26).

In the design of a DPI with the desired Stk, the DPI formulation design needs to be highly optimized. Kadota et al. (27) demonstrated that milling, spray drying, and spray freezing techniques were able to prepare a DPI powder with the appropriate characteristics to target a certain part of the lungs. Numerical simulation of particle deposition of DPI formulations in the human airways was also conducted in an 82-year-old male patient with COPD and a 69-year-old male patient with COPD. The results showed markedly different airflow distribution and particle deposition patterns (27).

Particle deposition profiles from numerical simulation are also able to evaluate pharmacokinetic profiles. The prediction of an inhaled amiloride plasma concentration-time profile was calculated using pulmonary absorption parameters and drug deposition patterns (28). The physiologically based pharmacokinetics

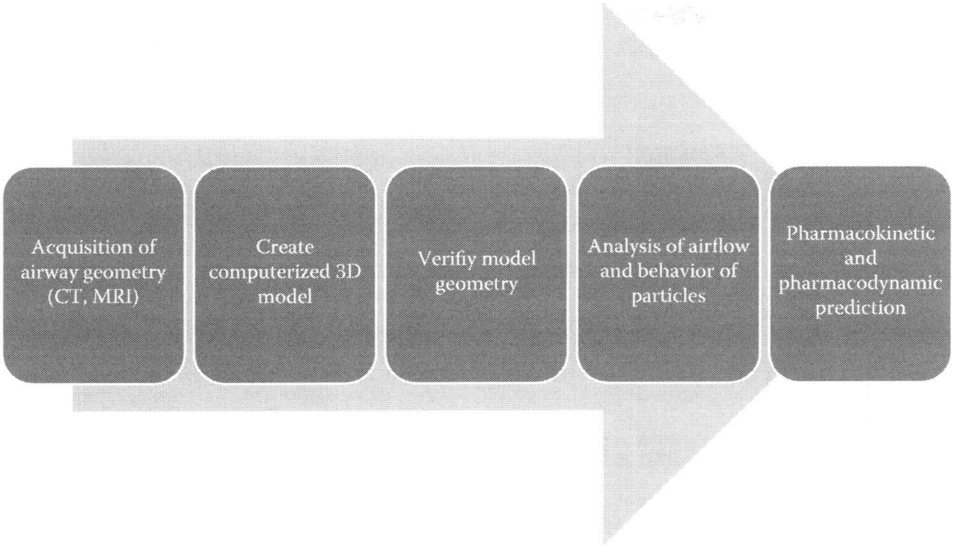

FIGURE 39.3 Flow Diagram for Numerical Modeling of Particle Deposition in the Human Lung Airway System

(PK) model was validated using an oral dosage form and previously studied data. For inhalation dosage, extra-thoracic deposition was assumed according to the oral dosage form. The pulmonary absorption coefficient was estimated from the lung volume, blood flow, drug tissue to plasma partition coefficient, and the blood to plasma concentration ratio. The PK simulation indicated that about 50% of inhaled amiloride was absorbed through the lungs, while the rest of the dose was absorbed in the gastrointestinal tract. The predicted time to peak concentration, peak concentration, and area under the concentration curve were comparable to the in vivo data (29).

In summary, modern techniques for numerical modeling of particle deposition in human airways can be obtained using the following steps: medical imaging of airway geometries, 3D modeling of airways, airflow analysis, behavior of particles, and pharmacokinetic and pharmacodynamic predictions (Figure 39.3).

Combining various numerical modeling techniques with in vitro experiments is expected to improve the ability to design DPI formulations, DPI devices, and to target the site of drug deposition in the lungs of individual patients.

REFERENCES

1. ANSYS Inc. Chapter 1: Basic Fluid Flow. in: ANSYS Inc., (ed.), ANSYS Fluent Theory Guide. Cannonsburg, PA, USA: ANSYS Inc.; 2013.
2. Tu J, Yeoh GH, and Liu C. Chapter 3 - Governing Equations for CFD—Fundamentals. In: Tu J, Yeoh GH, and Liu C, eds. *Computational Fluid Dynamics*. Burlington: Butterworth-Heinemann; 2008. pp. 65–125.
3. Tu J, and Liu C. Appendix A - Full Derivation of Conservation Equations. In: Tu J, Yeoh GH, and Liu C, eds. *Computational Fluid Dynamics*. Burlington: Butterworth-Heinemann. 2008. pp. 410–413.
4. ANSYS Inc. Chapter 16: Discrete Phase. In: ANSYS Inc., editor. ANSYS Fluent Theory Guide. Cannonsburg, PA, USA: ANSYS Inc.; 2013.
5. Stapleton KW, Guentsch E, Hoskinson MK, and Finlay WH. On the suitability of k-epsilon turbulence modeling for aerosol deposition in the mouth and throat: A comparison with experiment. *Journal of Aerosol Science*. 2000; 31(6): 739–749.
6. Thornton C, Yin KK, and Adams MJ. Numerical simulation of the impact fracture and fragmentation of agglomerates. *Journal of Physics D: Applied Physics*. 1996; 29(2): 424–435.
7. Yang RY, Yu AB, Choi SK, Coates MS, and Chan HK. Agglomeration of fine particles subjected to centripetal compaction. *Powder Technology*. 2008; 184(1): 122–129.
8. Yang J, Wu C-Y, and Adams M. Three-dimensional DEM–CFD analysis of air-flow-induced detachment of API particles from carrier particles in dry powder inhalers. *Acta Pharmaceutica Sinica B*. 2014; 4(1): 52–59.
9. Suwandecha T, Wongpoowarak W, Maliwan K, and Srichana T. Effect of turbulent kinetic energy on dry powder inhaler performance. *Powder Technology*. 2014; 267: 381–391.
10. Wong W, Fletcher DF, Traini D, Chan HK, Crapper J, and Young PM. Particle aerosolisation and break-up in dry powder inhalers 1: Evaluation and modelling of venturi effects for agglomerated systems. *Pharmaceutical Research*. 2010; 27(7): 1367–1376.
11. Wong W, Fletcher DF, Traini D, Chan HK, Crapper J, and Young PM. Particle aerosolisation and break-up in dry powder inhalers: Evaluation and modelling of impaction effects for agglomerated systems. *Journal of Pharmaceutical Sciences*. 2011; 100(7): 2744–2754.
12. Alexopoulos AH, Milenkovic J, and Kiparissides C. An integrated computational model of powder release, dispersion, breakage, and deposition in a dry powder inhaler. *Computer Aided Chemical Engineering*. 2013; 32: 139–144.
13. Coates MS, Fletcher DF, Chan HK, and Raper JA. Effect of design on the performance of a dry powder inhaler using computational fluid dynamics. Part 1: Grid structure and

mouthpiece length. *Journal of Pharmaceutical Sciences*. 2004; 93(11): 2863–2876.
14. Zhou QT, Tong Z, Tang P, Citterio M, Yang R, and Chan HK. Effect of device design on the aerosolization of a carrier-based dry powder inhaler - A case study on aerolizer® foradile®. *AAPS Journal*. 2013; 15(2): 511–522.
15. Leung CMS, Tong Z, Zhou QT, Chan JGY, Tang P, Sun S., et al. Understanding the different effects of inhaler design on the aerosol performance of drug-only and carrier-based DPI formulations. Part 1: Grid structure. *AAPS Journal*. 2016; 18(5): 1159–1167.
16. Tuteric T, Vulovic A, Cvijic S, Ibric S, and Filipovic N, eds. Effect of Circulation Chamber Dimensions on Aerosol Delivery Efficiency of a Commercial Dry Powder Inhaler Aerolizer®. 2017 IEEE 17th International Conference on Bioinformatics and Bioengineering (BIBE); 2017 23-25 Oct. 2017.
17. Benque B and Khinast JG. Understanding the motion of hard-shell capsules in dry powder inhalers. *International Journal of Pharmaceutics*. 2019; 567: 118481.
18. Longest W, Farkas D, Bass K, and Hindle M. Use of computational fluid dynamics (CFD) dispersion parameters in the development of a new DPI actuated with low air volumes. *Pharmaceutical Research*. 2019; 36(8): 110.
19. Weibel ER. *Morphometry of the Human Lung*. Elsevier Science; 2013.
20. Horsfield K, Dart G, Olson DE, Filley GF, and Cumming G. Models of the human bronchial tree. *Journal of Applied Physiology*. 1971; 31(2): 207–217.
21. Yeh H-C and Schum GM. Models of human lung airways and their application to inhaled particle deposition. *Bulletin of Mathematical Biology*. 1980; 42(3): 461–480.
22. Hofmann W and Koblinger L . Monte Carlo modeling of aerosol deposition in human lungs. Part II: Deposition fractions and their sensitivity to parameter variations. *Journal of Aerosol Science*. 1990;21(5): 675–688 10.1016/0021-8502(90)90122-e.
23. Hofmann W and Koblinger L. Monte Carlo modeling of aerosol deposition in human lungs. Part III: Comparison with experimental data. *Journal of Aerosol Science*. 1992; 23(1): 51–63.
24. Matida EA, DeHaan WH, Finlay WH, and Lange CF. Simulation of particle deposition in an idealized mouth with different small diameter inlets. *Aerosol Science and Technology*. 2003; 37(11): 924–932.
25. Kadota K, Nishimura T, Nakatsuka Y, Kubo K, and Tozuka Y. Assistance for predicting deposition of tranilast dry powder in pulmonary airways by computational fluid dynamics. *Journal of Pharmaceutical Innovation*. 2017; 12(3): 249–259.
26. Das P, Nof E, Amirav I, Kassinos SC, and Sznitman J. Targeting inhaled aerosol delivery to upper airways in children: Insight from computational fluid dynamics (CFD). *PLoS ONE*. 2018; 13(11): e0207711.
27. Kadota K, Sosnowski TR, Tobita S, Tachibana I, Tse JY, Uchiyama H., et al. A particle technology approach toward designing dry-powder inhaler formulations for personalized medicine in respiratory diseases. *Advanced Powder Technology*. 2020; 31(1): 219–226.
28. DeHaan WH and Finlay WH. Predicting extrathoracic deposition from dry powder inhalers. *Journal of Aerosol Science*. 2004; 35(3): 309–331.
29. Vulović A, Šušteršič T, Cvijić S, Ibrić S, and Filipović N. Coupled in silico platform: Computational fluid dynamics (CFD) and physiologically-based pharmacokinetic (PBPK) modelling. *European Journal of Pharmaceutical Sciences*. 2018; 113: 171–184.

40

Oronasal and Tracheostomy Delivery of Soft Mist and Pressurized Metered-Dose Inhalers with Valved Holding Chamber

Dasharath M. Patel, Unnati D. Chanpura, and Veer J. Patel
Graduate School of Pharmacy, Gujarat Technological University, Gandhinagar Campus, Gujarat, India

40.1 Introduction

Aerosols have been in demand for the treatment of upper respiratory tract diseases. There are no specific guidelines for optimization of aerosol drug delivery in patients, but physicians opt to choose either oronasal or tracheostomy delivery of aerosol formulations. Many pediatric patients receiving inhalation therapy are mostly undergoing tracheostomy. Also, infants requiring aerosol therapy utilize an oronasal mask for delivery of aerosol. Patients on long-term aerosol therapy are mostly adapting oronasal and tracheostomy technique of aerosol delivery. Valved holding chambers (VHCs) are used with aerosols in order to prevent the need to coordinate actuation and inhalation (1).

There are many types of devices available for aerosol drug delivery. Nebulizers are to be used for in-home; otherwise they are bulky and non-portable. The patients are advised to use more portable and relatively easy-to-use devices for everyday use, such as small volume nebulizers (SVNs), metered dose inhalers (MDIs), dry powder inhalers (DPIs), pressurized metered dose inhalers (pMDIs), breath actuated metered dose inhalers (BA-MDIs), and also the latest devices, such as soft mist inhalers (SMI). This chapter focuses on oronasal and tracheostomy delivery of SMI and pMDI and related topics, such as the accessories used with them along with their methodologies for use, mechanisms, advantages, and how to select the right device for the patient. There are two types of pMDIs currently available in the market: conventional pMDIs and liquid MDIs, commonly known as SMIs. The inhaled drugs should be deposited into the lungs to produce the beneficiary effect (2,3). Non-invasive ventilation systems (NIVs) can also be useful after extubation, in order to shorten the duration of invasive ventilation when possible. Since the treatment for patients suffering from chronic obstructive pulmonary disease (COPD) is based on the use of bronchodilators and corticosteroids via the inhalation route, they usually receive a prescription of medication to be delivered via nebulizers or pMDIs. Ventilatory support can either be invasive or non-invasive. The invasive method employs tracheostomic delivery, while non-invasive methods include ventilatory support through nebulizers and a facemask or via inhalers such as MDIs (4). Unlike oral or intravenous (IV) routes, aerosolized therapy delivers the drugs directly to the therapeutic sites, which is the internal lumen of the airways. Hence, the systemic dose of most aerosolized drugs is comparatively lower than their oral or IV treatment counterparts. Direct pulmonary delivery also facilitates rapid bronchodilation in response to β2-adrenergic agonists and anti-cholinergics, and the duration of effect can be longer with some long-acting beta agonists (LABAs) compared to oral treatments (5). Delivery of drug via respiratory tract is more complex than via oral therapy, as the successful therapy demands a delivery system that can generate drug particles having particle size suitable for penetration beyond oropharynx and larynx for deposition into the lungs (6). Patients with diseases other than COPD and asthma can also use inhaled medication during treatment at home, such as patients suffering from pulmonary arterial hypertension (PAH), human immunodeficiency virus (HIV) infection, and patients with cystic fibrosis (CF). Many pediatric patients who have undergone tracheostomy receive long-term treatment with inhalable aerosols, as they have been decannulated, but they still continue to require inhalation treatments. A survey of aerosol delivery practices in tracheotomized children disclosed that 92% of the responders used pMDIs. Additionally, 83% of those using unassisted methods reported using VHCs (air and soul).

40.1.1 Oronasal Delivery

Oronasal delivery is a kind of non-invasive mechanical ventilation technique which involves the use of a facial or oronasal mask. In oronasal delivery, the airways are not directly accessible by the ventilation devices, instead there is an external interface, i.e. face mask or oronasal mask, which utilizes the positive pressure in addition to this mask. These oronasal masks are comfortably sealed, thus providing improvised air-sealing capacity and hence it is useful for patients with profuse air-leaking tendency. In unprecedented circumstances like ventilator failure, in order to avoid rebreathing, these oronasal masks are incorporated with quick-release straps and anti-asphyxia valves. Oronasal delivery of aerosols is utilized for

DOI: 10.1201/9781003046547-40

enhancing the capacity of pulmonary delivery and reducing the deposition in the oropharynx region, thereby improving treatment efficacy. Its flexible feature serves as a valuable tool for effective patient management. Oronasal delivery mainly aids in preventing the risks associated with use of artificial airways in case of tracheal intubation (4).

40.1.1.1 Factors Affecting Oronasal Delivery

40.1.1.1.1 Ineffective Breathing

The physiological condition of a patient can affect the oronasal delivery of aerosol formulations. Patients with conditions like dynamic hyperinflation, weak inspiratory muscles, and a low respiratory drive are prone to have ineffective breathing patterns which ultimately affects the oronasal delivery.

40.1.1.1.2 Impaired Mask Sealing

Impaired mask sealing can occur as a result of expiratory air leak and inspiratory air leak which may lead to gastric distention and loss of positive end expiratory pressure (PEEP).

40.1.1.1.3 Type of Pressure Support

The amount of pressure support, i.e. high or less, directly affects the expiratory cycle, causing delay in the cycle.

40.1.1.1.4 Humidity

If there is low humidity it can increase the chances of nasal airway resistance (NAWR) and mouth leak leading to inaccurate delivery of the aerosol (7,8).

40.1.1.1.5 Physiological Conditions of Patients

The physiological condition of a patient can affect the oronasal delivery of aerosol formulations. Conditions like hypercapnia, thyroid dysfunction, and increased dead space ventilation can affect the delivery of aerosol medication (International Symposium.pdf).

40.1.2 Tracheostomy Delivery

Tracheostomy delivery involves the tracheostomy made in front of the neck and into the trachea, which facilitates the passage of air directly into the lungs. A small tracheal tube (cannula) is used to keep it open. This tube is used to keep the airway open and to remove any lung secretions. Here the aerosol delivery is administered to the patient via tracheostomy tube which enables the patient to breath via the tube and not through nose and mouth in case of severe breathing problems.

A tracheostomy is done for one of several reasons, i.e. for bypassing an obstructed upper airway, for cleaning up the airway and removing secretions of the lungs, and for delivering oxygen to the lungs in an effective and easy way.

40.1.2.1 Conditions Requiring Tracheostomy

40.1.2.2 Airway Complications

These include airway burns resulting from exposure to corrosive material or smoke, infections like epiglottis or croup, laryngeal injury or spasms, obstructive sleep apnea, severe neck or mouth injuries, tracheomalacia, and tumors such as cystic hygroma.

40.1.2.3 Lung Problems

These include bronchopulmonary dysplasia, chronic pulmonary disease resulting reduction in anatomic dead space, chest wall injury, and diaphragm dysfunction.

40.1.2.4 Other Complications

These include neuromuscular diseases that result in weakening of chest muscles and diaphragm, spinal cord injury associated with cervical vertebrae fracture, long term coma, facial burns or surgery, and severe allergic reactions (3).

40.1.2.2 Factors Affecting Tracheostomy Delivery

40.1.2.2.1 The Internal Diameter of a Tracheostomy Tube

The measurement of internal diameter of a tracheostomy tube directly affects the deposition of aerosol content into the airway. Many experimental studies having different inner diameters of tracheostomy tube inferred a result that showed a direct relationship. The smaller the inner diameter of the tube, the less the aerosol deposition in the lungs.

40.1.2.2.2 The Inner Cannula of a Tracheostomy Tube

Before beginning the aerosol therapy, upon removal of inner cannula from the tracheostomy tube there is a noticeable increase in the amount of aerosol deposition in the lungs.

40.1.2.2.3 Fenestration of a Tracheostomy Tube

Tracheostomy tubes with fenestration generally consist of open windows for easy passage of gas from lungs to the upper airway. This reduces the amount of breathing work in tracheostomized patients. In vitro studies done to evaluate the effect of fenestration on aerosol delivery revealed that closing the fenestration (generally by the inner cannula) increases the aerosol delivery in tracheostomized patients.

40.1.2.2.4 The Material Used to Manufacture a Tracheostomy Tube

The tracheostomy tubes are made up of different materials like polyvinylchloride, silicone, silver, etc. In the case of material like polyvinylchloride, which has a potential electrostatic charge associated with it, there may be chances of this electrostatic charge attracting particles to the inner walls of spacers in pressurized MDIs. This may affect the aerosol delivery in tracheostomized patients. However, no substantial findings have been discovered to date that explain the effect of the material of the tracheostomy tube and electrostatic charge on aerosol drug delivery by tracheostomy.

40.1.2.2.5 The Type of Aerosol Device

Various types of aerosol drug delivery devices are utilized for tracheostomy. Devices like nebulizers, pressurized metered dose inhalers, pMDIs with VHC, and soft mist inhalers (SMI) are utilized for tracheostomy delivery. Studies have shown that different aerosol delivery devices have different aerosol drug deposition capacities and it completely depends upon the type of aerosol device utilized for tracheostomy delivery.

40.1.2.2.6 Aerosol Drug Delivery Techniques

Drug delivery techniques used for aerosol devices are generally categorized as assisted and unassisted. Assisted delivery techniques include the utilization of a conjugated manual recovery bag with the aerosol device, whereas the unassisted delivery technique involves direct utilization of aerosols with an aerosol device. No significant observations have been reported stating the effect of delivery technique on the aerosol deposition in the tracheostomized patients (9).

40.2 Pressurized Metered Dose Inhalers (pMDIs)

40.2.1 What is pMDI?

The first reported use of the word *inhaler* dates back to 1778 and the English physician John Mudge (10). An inhalation device is a piece of equipment that holds and dispenses the drug to be delivered into lungs inside the bronchial tree by breathing (2). The pMDI was first introduced by Dr. George Maison (Riker laboratories) in 1956, (marketed under the brand name of Medihaler). It was developed as a handy, multi-dose delivery system for bronchodilators (2,11–13). Since then, inhalation systems have been used as first-line treatments for COPD and asthma, and for delivery of bronchodilators and corticosteroids (13). The pMDI remains the most used inhalation device prescribed to patients of asthma and COPD due to its compact, portable, patient friendly, multiple dose, and reproducible design (2,12,13). At present, more than 250 inhaler devices are available in the market, capable of delivering drugs used in chronic respiratory diseases (14).

A pMDI can contain monotherapy as well as some combination therapy drugs, if needed. It uses pressurized propellants to deliver the drug via inhalation route. They are reliable in terms of delivering a specific amount of dose with each actuation, which is why it's called a metered dose (13).

40.2.2 Drugs Used

The most common drugs include bronchodilators and corticosteroids. The drugs, such as short-acting β2-adrenergic agonists and long-acting β2-adrenergic agonists (LABA), inhaled corticosteroids (ICSs), nonsteroidal anti-inflammatories, antibiotics, and mucolytics are common for use via inhalation route (15). Drugs such as tiotropium, budesonide, salbutamol, beclomethasone diproprion, levosalbutamol, formoterol fumarate, terbutaline, mometasone furoate, beclomethasone diproprionate, umeclidinium bromide, fluticasone furoate, and vilanterol are some examples of the drugs currently present for inhalers in the US and Europe markets (13).

Other drugs such as insulin for treatment of diabetes, gene therapy vectors for treatment of cystic fibrosis (CF), vaccines for measles and papilloma virus, chemotherapy agents for treatment of lung cancer, new formulations for antibiotics, anti-proteases for the treatment of CF and α1-antitrypsin deficiency, morphine for pain relief, and ergotamine for headache relief are some examples under development (15).

40.2.3 Mechanism of Device

The basic components of a pMDI device (Figure 40.1) include the following: a canister, propellant, the drug, metering valve, mouthpiece, and actuator (2). Generally, an aluminum canister is lodged into a plastic support; the canister contains a pressurized solution or suspension of micronized drug particles dispersed in suitable propellants. Typically, a surfactant (such as sorbitan trioleate or lecithin) is also added to the formulation to reduce the particle accumulation. The choice of surfactant is normally the main factor behind the characteristic taste of specific inhaler brands. In suspension formulations, the drug remains as solid powder within the container, rather than solubilizing in the vehicle (16–18).

The metering valve is an essential component of the device, which prepares a precise volume of drug–propellant mixture for each actuation (13). It is designed such that it produces only a specific amount of aerosol upon each actuation (20–100 μl). The conventional pMDI has a press-and-breathe design. On activation, the canister gets depressed into actuator, releasing the drug–propellant mixture, that later expands and vaporizes into aerosol (12). When the bottom of the canister is pressed into the actuator seating, it causes decompression of the formulation within the metering valve. It results to an explosive generation of heterodispersed aerosol droplets consisting of tiny drug particles surrounded by a layer of propellant. The propellant then evaporates with time and distance, resulting in reduction in the size of particles using propellant under pressure to generate an accurate (metered) dose of aerosol through the nozzle (13). The hole in the metering valve is aligned with the metering chamber when the canister is pressed down. The aerosol suspension cools down during initial vaporization of propellant. Then, the high vapor pressure of propellant forces the pre-measured medicated aerosol out of the end hole through the nozzle of the actuator. At last, upon release, the metering valve refills the chamber with another dose (12).

The propellant makes up about 80% of the contents of the pMDI, while the drug is only 1–2%, which is either suspended or dissolved in the mixture that comes out via actuation (12). The pMDI produces a rapid moving plume of aerosol for a short duration of time, typically 0.1–0.4 s; with velocity of about 8 $m{\cdot}s^{-1}$ at a 10 cm distance from the actuator, and it may be even higher in shorter distances to the nozzle (2).

40.2.4 Propellants Used in pMDI

Much innovation in the field of development of pMDIs has been inspired by the Montreal Protocol declaration that banned the use of chlorofluorocarbons (CFCs) because of their ozone-depleting properties. This resulted in substitution of CFCs with hydrofluorocarbons (HFCs) and hydrochlorofluorocarbons (HCFCs). These compounds did not catalyze the ozone layer, but they are potent greenhouse gases and might promote climate change. The propellants currently being used for respiratory medications, HFC-134a and HFC-227ea, are also potent greenhouse gases with carbon dioxide equivalents of 1320 and 3660, respectively. Then the pharmaceutical industries developed hydrofluoroalkanes (HFAs) as propellants.

The drug is solubilized in the HFA propellant, so it is uniformly distributed in the canister, and thus it does not require

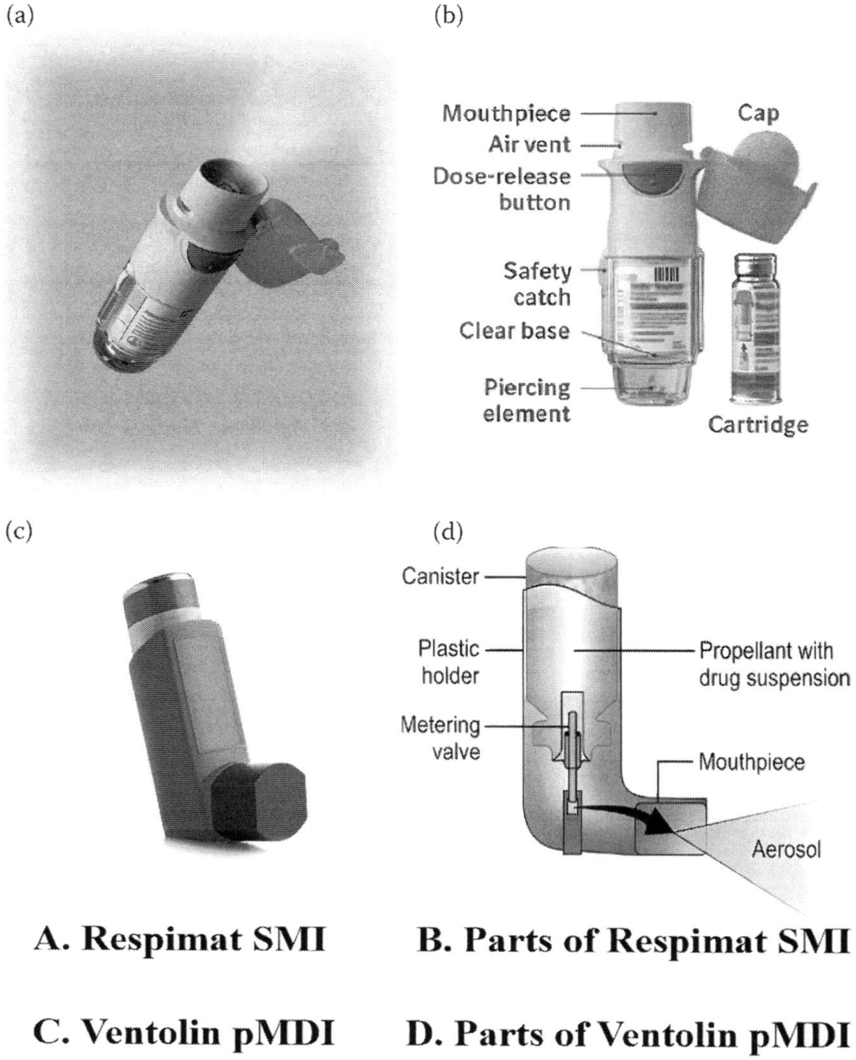

FIGURE 40.1 Respimat and Ventolin Devices with Their Parts (Prepared from Online Images)

shaking prior to usage. The new pMDIs produce aerosols that consist of extra-fine particles providing equivalent efficacy to other preparations with reduced doses. Additionally, they produce a longer plume and also the inhalation technique is less dependent on the inspiratory flow and coordination compared to other MDIs. This property helps to partially overcome the cold freon effect (2,13). The plume released from many HFA-pMDIs tends to have slower velocity and it is warmer. These characteristics help to partially overcome the cold freon effect that causes some patients to stop inhaling their CFC-pMDIs (2,5).

Several HFA-pMDI formulations incorporate a small amount of ethanol, which affects the taste as well as further increasing the temperature and decreasing the velocity of aerosol (5).

40.2.5 Methodology of Use

All pMDI devices are activated by the patient. Contrary to the SVN, the effectiveness of drug delivery with pMDI devices depends upon the patient's ability to apply pressure to the base of canister and simultaneously take a paced, deep breath (12). The instructions for use of the pMDI are described in patient information leaflet (PIL) provided by the manufacturer (2).

The inhaler must be shaken prior to use to ensure a uniform particle distribution (13). Standard practice is to place the mouthpiece between closed lips, but it is also possible to hold the inhaler between open lips or even a few centimeters from an open mouth (19).

The correct practice to use the pMDI involves firing one actuation, while deep breathing in slowly, followed by breath-holding for about 10 seconds. This process requires a good hand-eye coordination by the patient (2). The breath-holding allows particles to sediment on the airway surfaces (2). Table 40.1 shows correct technique to use pMDI.

An adequate inhalation technique is vital to improve pulmonary delivery, reduce oropharynx deposition, and accordingly improve treatment effectiveness (4). The guidelines from authorities suggest that the inhaler technique should always be taught and assessed by competent healthcare professionals (5). The patient's aerosol administration technique should be

TABLE 40.1

Instructions to Use Various Aerosol Delivery Devices (Data From (5))

Type of Device	Instructions
pMDI (for patients having good coordination of actuation-inhalation)	1. Shake 4 to 5 times.
	2. Remove the cap.
	3. Prime the inhaler into the surrounding air for 2-4 doses before use.
	4. Exhale slowly.
	5. Hold the device in upright position.
	6. Without time delay, immediately place the device in mouth between the teeth, keeping the tongue flat under the mouthpiece and ensuring proper seal.
	7. Start to inhale gradually and simultaneously press the canister to actuate a dose.
	8. Maintain the inhalation through the mouth.
	9. At the end of inhalation, remove the device from the mouth.
	10. Hold the breath for up to 10 seconds before breathing out.
	11. Then breathe normally.
	12. If another dose is required, repeat steps 4 to 11.
BA-pMDI (for patients above 6 years of age)	1-6. Repeat the above steps 1 to 6 of pMDI.
	7. Start inhaling slowly through the mouth. The patient is expected to sense the release of the dose either through taste or sound.
	8. Keep maintaining the slow and deep inhalation.
	9. Towards end of inhalation, remove the device from mouth.
	10. Hold the breath for 10 seconds followed by exhalation.
	11. Breathe normally.
	12. If another dose is required, repeat steps 4 to11.
pMDI + spacer with facemask (for patients of 3 years or below 3 years of age and for patients unable to breath consciously through mouth)	1-3. Repeat steps 1 to 3 of directions for use of pMDI.
	4. Inset the mouthpiece of the device into the open end of spacer and ensure that the fitting is tight.
	5. Place the facemask over the mouth and nose with a proper fitting.
	6. Actuate the dose into the spacer chamber.
	7. Inhale and exhale into the spacer at least 10 times.
	8. Take the facemask off the face and repeat steps 1 to 7 if another dose is required.
SMI	Assemble the parts of the device and use as per the directions given by manufacturer.

pMDI, pressurized metered dose inhaler; BA-pMDI, breath actuated pressurized metered dose inhaler; SMI, soft mist inhaler

continuously monitored by their clinicians and should be corrected if necessary (1,12).

40.2.6 Drug Deposition through pMDI

The deposition of drug molecules into the lung and its reach to the small airways may be influenced by the factors related to the specific device characteristics, such as its aerosol generating system, speed of aerosol plume, internal resistance of device, type of propellant or carrier used, and inhalation via nose/mouth; or formulation specifications such as the particle size of drug, charge of particles, lipophilicity and hygroscopicity of the particles; and the inhalation pattern, such as flow rate, volume of each actuation, and breath-holding time after inhalation (20). The chief patient related factors for drug deposition include the morphology of oropharynx and larynx, and the patient's inspiratory volume and flow rate. The flow rate is generally the determining factor of the velocity of the airborne particle, which further affects its probability of impaction in the oropharynx and larynx. To minimize deposition into the upper respiratory region and enhance the delivery in the lungs using a pMDI with or without a using spacer device or BA-pMDI, the patient needs to practice slow inhalation. The inhalation should last over 2–3 s in pediatric patient and minimum of 4–5 s in adults, after a deep exhalation to ensure the ideal flow of around 30 $l{\cdot}min^{-1}$ for a pMDI (21).

The drug particles get deposited in the lungs via inhalation route by impaction, sedimentation, and diffusion processes, depending on their size (5,13). Figure 40.2 shows how particles having different diameters get deposited in the respiratory system. The most important particle-related factor for aerosol deposition is thought to be its aerodynamic diameter. Figure 40.2 displays the relationship between the drug's aerodynamic diameter and lung deposition. Particles greater than 5 μm usually get deposited in the oropharynx by impaction and later swallowed. This is the result of inertia associated with the mass of particle, which diminishes the ability to follow airstream when it changes direction towards the lower

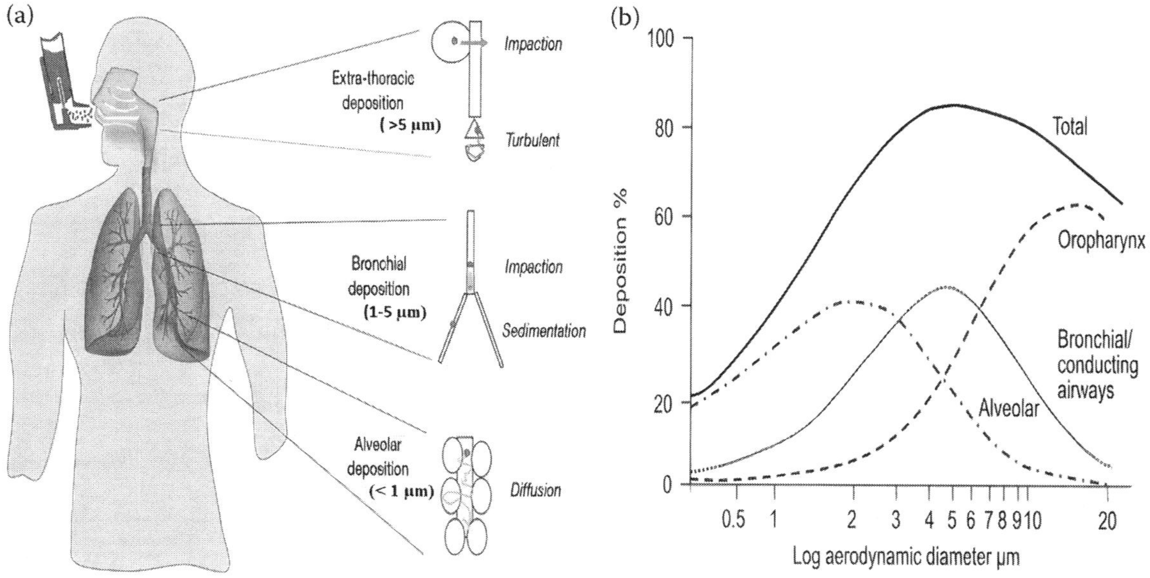

A. Principle of aerosol deposition

B. Relationship between aerodynamic diameter and lung deposition

FIGURE 40.2 Principle and Relationship between Aerodynamic Diameter and Lung Deposition (Modified based on Pablo et al. 2017; (5))

airways. Corticosteroid deposition in the oropharynx is not desirable as it can produce local side effects such as hoarseness and oral candidiasis with ICS (19).

When inhaled via mouth, the airstream instantly takes a bend of 90° at retropharynx level and a majority of large particles fails to flow over this anatomical area. Thus, larger particles first get deposited in the mouth and throat, and then are swallowed (13). Particle size plays a major role in the deposition of drug in the respiratory tract. The particles smaller than 5µm in diameter may flow in airstream past retropharynx to reach the trachea (13). Particles having 2–5 µm diameter are most liable to be deposited in the upper respiratory tract, at the level of trachea and tracheal bifurcation. After that, the airflow slows down and the particles sediment under force of gravity (13). The airflow rate is reduced in the periphery of the lungs and the particles get deposited primarily by sedimentation, as gravity causes them to "rain out" (5). The particles having diameter smaller than 2 µm can get deposited in lower areas of respiratory tract by sedimentation; this is where breath-holding after inhalation plays a significant role in the process. The sedimentation process of the suspended particles having diameters smaller than 0.5 µm is relatively very slow, as these smaller particles demonstrate Brownian motion, and if they come into contact with airway lining, they will be deposited or they will be exhaled (13). Most particles having diameters 0.1–1 µm diffuse by Brownian motion and get deposited when they happen to collide with the airway walls. As the residence time of particles in the smaller, peripheral airways is increased, their possibility of deposition via sedimentation and Brownian motion also increases. The patients are advised to hold their breath after inhalation of dose in order to increase the residence time of aerosol in order to increase the deposition into peripheral airways. The inhaled particles which do not get deposited tend to leave the system during exhalation (5).

The proportion of particles in an aerosol having particle size smaller than 5 µm are often referred to as FPF, or fine particle dose (FPD) if expressed in absolute mass of drug in particles <5 µm. Aerosols with high FPFs have a higher probability of penetrating beyond the upper respiratory tract and deposition in the lungs. The current devices generate aerosols having a substantial proportion of particles of 1–5 µm range. The optimum particle size for pediatric patients is still unknown, but it is likely that it is smaller than the ideal particle size for adults, considering the narrower airway diameters and higher intraluminal flows (5).

The pulmonary deposition fractions of "new generation" devices are about 40–50% of the nominal dose, which is significantly higher than the 10–20% reached by the older devices (13).

While prescribing, the physician should particularly be aware that some countries label the pMDIs with their nominal dose (the dose that is metered), while others mention the emitted dose (i.e. the dose available for inhalation at mouth). For instance, beclomethasone pMDI (HFA formulation) of the same manufacturing brand can be labeled either as 100 mg (nominal dose) or 80 mg (emitted dose), while in both cases, the dose that patient receives remains the same (6).

The intensity of lung disease at the time of inhalation is an important factor for pulmonary deposition. Several studies have demonstrated that central airway deposition enhances with increase in mucus plugging, turbulent airflow and airway obstruction. This concludes that little to no drug deposition in the lungs may occur in case of severe lung diseases. This study may not be clinically significant for bronchodilators, but is important for corticosteroids (5).

40.2.7 Factors Affecting the Performance of pMDI

40.2.7.1 Shaking the Canister

Not shaking the canister which has been standing overnight can lead to reduction in the total and respirable dose by approximately 25–35%. This is due to the separation of drugs from propellant in the solution while standing. Thus, the pMDI has to be shaken several times before the first actuation in order to refill the metering valve with adequately mixed suspension from the canister.

40.2.7.2 Storage Temperature

Outdoor use of pMDI or storage in a cold weather may substantially reduce the aerosol drug delivery.

40.2.7.3 Nozzle Size and Cleanliness

The nozzle size, cleanliness, and lack of moisture is known to affect the amount of medication that is delivered to the patient. The nozzle of the actuator is specific to the device, and the coordination of the nozzle with the medication can influence both the inhaled dose and particle size of the aerosol. White crusty residue produced as a result of crystallization of medication is known to obstruct drug delivery. Thus, the nozzle should be cleaned intermittently based on the manufacturer recommendations.

40.2.7.4 Timing of Actuation Intervals

Rapid actuation of more than one puff using a pMDI may reduce drug delivery as a result of turbulence and the coalescence of particles. A pause between two subsequent actuations may improve bronchodilation, particularly during asthma exacerbations with episodes of wheezing and poor symptom control. In contrast to that, in the case of day-to-day management of preadolescents with β agonist (terbutaline) and a corticosteroid (budesonide), a pause between two actuations has not been found to be beneficial. While early research showed importance of a pause between two actuations, recent studies recommend that there should be a pause of at least 30–60 s between two actuations for effective aerosol therapy.

40.2.7.5 Priming

The action of releasing one or more sprays into the air or VHC is called *priming*. Initial and periodical priming of pMDIs is required to provide an adequate dose. When the pMDI is new or has not been used for a while, the drug may get separated from the propellant and other ingredients in the canister. As the shaking of pMDI will mix the suspension in the canister but not in the metering chamber, priming of the pMDI is important.

40.2.7.6 Characteristics of the Patient

Patient characteristics such as anatomy as well as physical and cognitive abilities, age, etc. can cause variability of aerosol deposition by a pMDI.

40.2.7.7 Breathing Techniques

A closed-mouth technique is universally recommended for pMDIs without using spacers. This method requires the mouthpiece to be placed between the patient's sealed lips during drug administration (12). The nose is a more efficient filter than the mouth. Hence, inhalation through mouth is the preferable route for inhalation. This factor is even more important during treatment of infants and toddlers. While the absolute efficiency in terms of pulmonary dose is low with nose breathing compared to mouth breathing, the total inhaled dose per kg of body weight is relatively higher in nose breathing children compared to adults using a mouthpiece. Thus, the dose to the lungs per kg body weight in nose breathing infants is similar to that achieved by mouth breathing adults (22).

40.2.8 Advantages of pMDIs

Rapid onset can be observed using inhalation route as the drug is delivered directly to the site of action. It also results into low occurrences of side effects (2). pMDIs are compact and portable devices (13). The design of pMDI was to develop a drug–device combination that can deliver a precise dose of specific drug formulations (12). They ensure high dose reproducibility (6). They are convenient to use at home by the patients themselves, or in the case of a young child, with help from their parents (2). They offer consistent dosing and rapid delivery (13). Drug preparation and handling are not needed with pMDI as opposed to nebulizers (12). The pMDIs can contain a minimum of 100 doses and are still compact and portable (2). They may include a dose counter to signal when the device is going to run out of drug (13). The internal components of a pMDI are arduous to contaminate (2). For many patients with CPOD, it would be feasible to achieve the slow inhalation flow required with a pMDI with training (13). They are relatively cheaper than BA-pMDIs or pMDIs incorporated with spacers (2).

40.2.9 Drawbacks of pMDIs

Poeple who use a pMDI need to develop a specific breathing technique that comprises coordination between inspiration and actuation time, slow and steady inspiration followed by breath-holding. Thus, it is difficult to use by patients with poor dexterity or weak grip strength, such as the elderly, who may find it difficult to actuate a pMDI device and, at the same time, coordinate it with their breathing (13). They are not breath actuated and hence, require actuation–inhalation coordination (13). They contain propellants (13). Patients with weak grip strength or poor dexterity like geriatrics may find it difficult to actuate a pMDI (13). There are two main problems associated with the pMDIs for patients: one is failure to accurately time the firing of a dose from the device and inhaling at the same time. This is termed *poor coordination*(2). Failure in proper coordination may result in greatly reduced doses of drug reaching the lungs (13). The second major issue is something called the *cold freon effect* which is experienced as a cold blast of propellant felt at the backside of the throat upon evaporation of the propellant. It either causes the patient to stop inhalation or inhalation via nose instead of mouth. This effect is less observed while using HFA formulations (2,12,13).

Most patients inhale faster than the desired pace (5). Even with a good inhalation technique, most pMDIs only deposit 10–20% of dose into lungs, while the rest gets deposited in the oropharynx (2,12). However, higher pulmonary deposition is marked in some recent formulations using HFA propellants rather than as suspension of micronized particles (2). High solubility and good stability are two requirements for a solution system. Unfortunately, many of the microcrystal-based HFA suspensions demonstrate poor colloidal stability and high dosing variability. Undesirable outcomes of poor colloidal stability are amplified in combination products because of differences in the particle density and size distribution of multiple drug microcrystals (23).

While the guidance for selection of appropriate drugs for inhalation therapy is easily available, the selection of the right inhalation device for a particular patient is difficult (2) due to the high number of options are available. However, the selection should be dependent on the patient's breathing pattern and ability to use a device correctly (13). While there are several types of devices available, not all the devices can be suitable for all patients because of the difference in the way a device performs and the need to master specific inhalation techniques, which require varying degrees of cognitive ability depending on the device. Thus, the Global Initiative for Asthma (GINA) and British Thoracic Society (BTS) guidelines urge that the inhaler technique, as well as degree of adherence with dose regimens, should be assessed before making changes in a patient's inhalation therapy (5). Studies have revealed that a very high proportion of patients are incompetent to use their devices effectively, either because they have forgotten the instructions they were given during training or they have never been properly guided. Many patients tend to soon forget the correct technique to use the device after training. The most common issue is spacer disuse (i.e. patient fails to use a spacer at home) (5). This results in low pulmonary deposition and higher oropharyngeal deposition compared to pMDIs used with spacers or SMIs (5). A common complaint by pMDI users is that it is difficult to determine when their pMDI device will run out of doses without using a dose counter (5,13). Priming of the device is required when the device is new or has been unused for some time (5). Further innovation in technology for pMDIs is represented by "intelligent inhalers" that incorporate a small microprocessor that controls inhalation and adherence, but it also results in more expensive devices (13).

40.2.10 BA-pMDI

In BA-pMDIs, the patient's inhalation through the device activates a mechanism which fires the pMDI, to facilitate automatic coordination of firing and inhalation; this allows them to produce clinical efficiency similar to patients having good coordination (2,13).

BA-pMDI incorporates a flow triggered system driven by a spring that releases the drug dose during inhalation, which results in automatic coordination between firing and inhalation (13). These devices are able to achieve good pulmonary deposition and clinical efficiency in patients facing problems with using a standard "press-and-breathe" pMDI due to coordination difficulties (2,5,12,13).

The preparation of a BA-pMDI is essential (i.e. raising the priming lever, removal of mouthpiece cover etc.); similarly, the inhalation must be strong enough to activate the firing mechanism (i.e. triggering flow of about 30 $l{\cdot}min^{-1}$ for currently available devices) (24). The correct technique to use a BA-pMDI is shown in Table 40.1.

While BA-pMDIs help to solve the coordination problems, they fail to be useful to patients facing the cold freon effect as they do not affect this. Nevertheless, the errors while using BA-pMDI are less common than using a standard pMDI (2). In recent times, the principle limitation with pMDI happens to be the range of drugs/formulations available for use with a BA-pMDI and their adjunctive economic costs (13).

40.2.10.1 Advantages of BA-pMDIs

They are compact, portable, and quick to use (13). They are competent for multiple dose delivery (13) and convenient for poorly coordinating patients (2). The patients suffering from COPD can generate the required flow rate for BA-pMDIs (about 27 $l{\cdot}min^{-1}$) (13). The geriatric population find using BA-pMDI easier than a pMDI or DPI (13). It is hard to contaminate the contents of a BA-pMDI (13).

40.2.10.2 Disadvantages of BA-pMDIs

The cold freon effect needs moderate inspiratory flow to be triggered (13). Patients sometimes stop inhalation once actuation occurs (5). It contains propellants (13). The availability of BA-pMDI device is limited to fewer drugs (13).

40.2.11 Spacers and Valved Holding Chambers

It is essential to coordinate between inhalation and activation of the device while using pMDIs and nearly one-third of the patients using a pMDI face this problem (2,25). High medication loss can be reported in the oropharynx with this type of coordination difficulty using pMDIs (4,5,12,25). This limitation can be overcome by use of a spacer device (Figure 40.3). A spacer is an accessory to be attached to pMDIs. It is simply a tube or an extension device that increases the distance and volume (about 20–750 ml) between the pMDI and oropharynx, in order to reduce oropharyngeal deposition. The spacer device is nothing but a holding chamber for a single metered dose, which results in better lung deposition of drugs and reduction of the oropharyngeal deposition. Furthermore, use of a spacer device helps to eliminate the cold freon effect (2,4,5,25).

A VHC (Figure 40.3) is a spacer device having a low resistance one-way valve(s) to contain the aerosol dose until it is inhaled, and it also directs the exhalation away from the aerosol in the chamber in order to reduce the aerosol losses from poor actuation–inhalation coordination. The aerosol particles can be held inside the VHC for a short period of time until the valve is opened by an inspiratory flow (5,12), hence allowing the patient to breathe tidally (13). In other words, they enable patients to

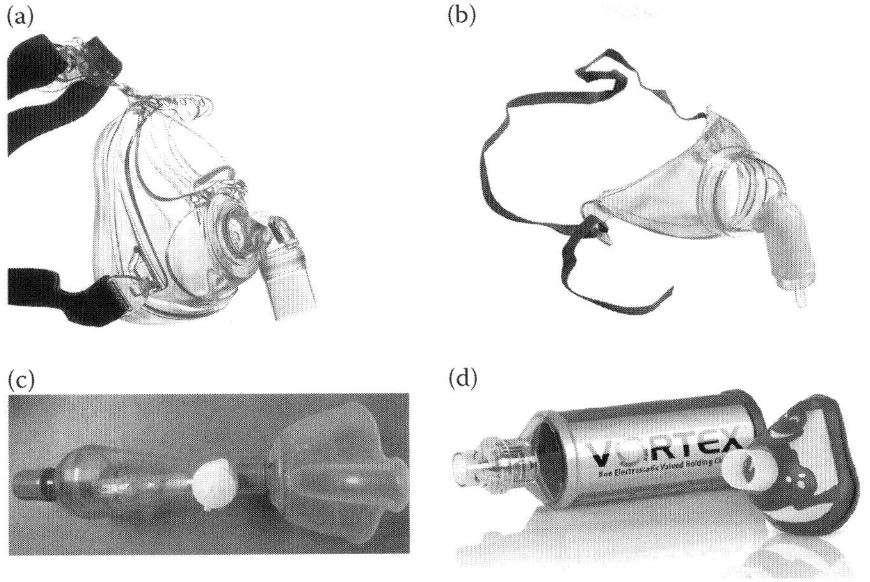

A. Oronasal Mask
B. Tracheostomy Mask
C. Valved Holding Chamber
D. Stainless Steel VHC

FIGURE 40.3 Different Types of Masks and Valved Holding Chambers (Prepared from Online Images)

inhale a static cloud (5). When used with the pMIDs, they decrease the velocity of emitted aerosol, reduce the aerosol dispersion when inhalation is delayed, and considerably increase the pulmonary deposition (5,13). The presence of valves between the chamber and patient also acts as a barrier that further reduces the oropharyngeal deposition (12).

Although both the terms *spacer* and *VHC* are often used interchangeably, the regular spacer device is liable to dispersal of the aerosol within, due to uncoordinated exhalation into the holding chamber as it lacks a one-way valve like a VHC device (25). Therefore, VHCs are preferred over simple spacers (25).

Spacer devices are often prescribed to elderly patients suffering from frequent problems of dexterity and coordination. Nonetheless, it can be difficult to assemble spacers with pMDI for these patients (13). Spacers can be an integral part of the pMDI, but in other cases, the patient needs to remove the canister from the manufacturer's actuator and place it in a special opening in the spacer device. Improper technique for using a spacer/VHC may result in reduction in drug delivery due to drug loss (12).

A collaboration of multidisciplinary experts on the delivery of pharmaceutical aerosols was facilitated by the European Respiratory Society (ERS) and the International Society for Aerosols in Medicine (ISAM), in order to draw up a consensus statement with clear, up-to-date recommendations that enable the pulmonary physician to choose the type of aerosol delivery device that is most suitable for their patient. The focus of the consensus statement is the patient use aspect of the aerosol delivery devices that are currently available. While using a spacer, the patients need to coordinate that their inhalation occurs within 1–2 s after actuation of inhaler to produce an optimal dose (12).

The chamber length helps to increase the distance for drug particles to travel from the device's mouthpiece to patient's mouth, promoting better drug deposition (12). The one-way valve prevents the patient from blowing the dose away after firing. Although the inhalation has to be strong enough to activate the one-way valve, or else no dose would be delivered (2). A spacer device provides additional volume which slows the aerosol velocity from the pMDI, allowing reduction in particle size. The aerosol retention and discharge dose are dependent upon the size and shape of the spacer and also on the electrostatic charge present on the inner wall of the plastic spacers (12).

Spacers and VHCs both can slow down the plume of aerosol. VHCs can even trap the aerosol cloud until inhalation by the patient. This may reduce the oropharyngeal drug deposition by nearly 80–90% through retention of large particles within the holding chamber (25). The spacer nullifies the cold freon effect as the point of aerosol generation is farther from the mouth compared to a standard pMDI (2). In comparison to pMDIs used alone, adding a VHC combination has been proved to increase airway hyper-responsiveness, lung function, and asthma control with decreased local and systematic side effects of inhaled corticosteroids (25).

In selection of a VHC/spacer, a crucial clinical consideration is that the VHC should not reduce the pulmonary delivery of drug compared to the pMDI used alone, and the oropharyngeal dose will be substantially reduced, irrespective of the VHC selection. In spite of this, some studies have shown significant reductions in the respirable dose of certain drugs with certain VHCs. The in vitro and in vivo studies have shown that using a VHC may increase the lung dose by two to three times compared with using a pMDI alone. The pMDI + VHC combination helps to remove the coarse particle fraction that would be otherwise deposited in the oropharynx if only pMDI was used, with no significant difference in particle size distribution or pulmonary deposition. The latter is an important observation as there has been much focus upon optimum particle size in aerosol, which may differ according to the therapeutic class of the drug, and it is also relevant to deposition efficacy. The clinical outcomes are expected to be improved with VHCs in patients with poor pMDI technique and whose disease control is accordingly less likely to be adequate (25).

While tidal breathing may be acceptable after firing with a spacer device, multiple actuations, long delays between firing and inhalation, accumulation of static charge on some plastic spacer devices, etc. are likely to reduce the dose to be inhaled (2). Several factors influence the aerosol therapy, such as selection of spacer device adaptable to the system, device settings, position of leak ports, humidification, synchrony with inspiratory phase, type of used interface and patient-related factors such as type of airway obstruction and support tolerability (4).

The downside of using spacers is that the walls of a spacer retain some amount of drug, which makes it unpredictable to know exactly how much of drug will be delivered in the airways. Furthermore, the spacers can reduce the portability of pMDI by increasing the overall size of the device, which sometimes results in patients using pMDI alone. Using a spacer or VHC can also add to the medical expense (13).

Spacers also tend to reduce the dose output from a pMDI to a variable extent. The accumulation of an electrostatic charge on the plastic walls of the spacer acts as an important factor affecting reduction of dose delivered. Studies suggest that these electrostatic charges reduce the performance of the spacer in such a way that the dose of aerosol available for inhalation is reduced. This effect is more significant in newly purchased devices (5). A significant proportion of the respirable dose can get potentially lost within a non-conducting VHC (i.e. made of polycarbonate or polyester) as a result of electrostatic interactions between aerosol particles and the internal walls of VHC. This happens as a result of charge acquisition by VHC during the manufacturing process, packaging, and storage, or as a result of triboelectrification (frictional charging) of aerosol formulations during their migration out of the canister via a metering valve. While washing the VHC with detergent, a fine coating of the detergent forms on the VHC walls, which helps to reduce this electrostatic charge. About 50–60% patients are reported to wash their VHCs with plain water, while 25–70% towel-dry their VHCs after washing. The aerosol half-life is marked to be superior while using anti-static chambers than that of unconditioned non-conducting VHCs, and in contrast to non-conducting VHCs, it remains unchanged when it is washed or primed with repeated aerosol actuations. Even though scientists have failed several times to show significant differences in using both types of devices in ideal conditions, the data gained from those studies may imply that there are potential benefits of using anti-static VHCs in an emergency room setting for acute asthma and warrant further evaluation (25).

The shape and volume of spacer devices may vary greatly, but they are generally grouped into two categories: small volume (130–300 ml) and large volume (600–800 ml) (5). There are several types of VHCs and spacers available to be used with or without masks (12). There have been several modifications in the basic design of VHCs to improve product performance and usability. These include introduction of a range of patient/age specific face masks, the addition of an "inspiratory flow indicator" indicating inspiration and expiration movements in the presence of an adequate mouth or facemask seal, an alert whistle to indicate attainment of excessive inspiratory flow rate, and charge dissipative capacity by the addition of an anti-static resin display. Some spacers are specially designed to incorporate a reverse flow in order to facilitate small particle delivery. In these spacers, the flow of aerosolized drug is directed away from the patient's mouth on actuation and then it gets directed back toward the mouth during inhalation (5,25). There are currently two variants of VHCs available with respect to their charge dissipative nature: non-conducting VHCs and charge dissipative anti-static VHCs. The latter is considerably newer technology (25). Studies have marked performance differences with a change in non-electrostatic materials used in VHCs (11). There are two types of non-electrostatic materials used in the VHCs: polymer and metal, as shown in Figure 40.3.

There has been much argument to determine the ideal dimensions for a spacer device. In studies, larger volume spacing chambers have performed better in in vitro studies as opposed to their smaller volume counterparts. However, the focus has shifted to other important factors affecting the performance of the device. A small study involving several different spacer devices washed with only water also showed that their performance related to drug delivery was affected by the formulation to be delivered (26).

The non-conducting variant of spacer requires frequently washing with a detergent solution (to be done as per the manufacturer's instructions) to rinse off the particles deposited on the inside surface of the device, while the anti-static versions do not have such requirements as it restricts the particle deposition. When used correctly, both these devices have not yet showed many performance differences in the in vitro studies (25). As spacers are used for several months, they need periodic cleaning in order to prevent deterioration in the function of the valve and also for hygienic reasons. The general recommendation for both the newly purchased and previously purchased device is to wash them with a low concentration of dishwashing liquid and then allowing them to drip-dry. In this manner, the plastic gets coated with detergent, which helps to reduce the electrostatic charge, thus reducing drug losses on spacer walls and promoting lung deposition. For proper cleaning guidelines, the PIL from manufacturer of the device is to be followed, as they are not consistent between different spacers. The recommendation usually ranges from once a week to once a month.

Spacers and VHCs should not be used together with BA-pMDIs. Some pMDIs are licensed to be used along with only a specific spacer (5). Use of spacers/VHCs are recommended by many national and international guidelines on obstructive lung diseases, for patient populations with poor coordination. GINA notes that there may be performance differences with different spacers. This possibly explains why only a few clinical and regulatory guidelines advocate that VHCs are not interchangeable, despite the presence of a large volume of literature principally for in vitro studies (25). The clinical significance of the electrostatic charge on aerosolized medication is still less clear and it may affect some formulations more than others. A change of spacer device represents a change in the delivery system. Thus, while making a change in spacer device, regular monitoring and titration of dose of the ICS dose to the lowest effective dose is recommended.

Since oropharyngeal deposition may still happen, patients receiving ICS via inhalation are advised to rinse their mouths after inhalation. When using a spacer device, only one dose should be actuated at a time as multiple inhalations increase drug losses within the spacer as a result of increased turbulence. Furthermore, the dose should be inhaled immediately after its introduction into the spacer as delays may enable the emitted dose particles to have more time to deposit within the spacer (5). Although the one-way valve in a VHC prevents the aerosol particles from exiting the chamber until inspired by inhalation, an optimal aerosol dosing is still dependent upon the time duration between actuation of dose into chamber and inhalation, as time delays can substantially reduce the available dose for inhalation. One-way valves must be designed to have a low resistance which opens easily with minimal inspiratory effort. Preferably, there should be a feedback mechanism, such as an inspiratory flow indicator, to signal if the inspiratory flow is too high. Pediatric patients having lower tidal volumes may need to take a few breaths from a VHC for a single pMDI actuation. In such cases, VHCs should include one-way valves for inhalation and exhalation, both in order to reduce the need for rebreathing and avoid exhalation of aerosol from the chamber (12).

40.2.11.1 Advantages

Spacer devices make it easier to use for the patients having coordination issues with pMDIs. They help by reducing the oropharyngeal deposition of drugs and elimination of the cold freon effect. VHCs are helpful with tidal breathing. They can be good for inhalation of corticosteroids compared to other methods of administration.

40.2.11.2 Disadvantages

Spacers defeat a potential pMDI benefit of being portable and compact, and to some extent, also its instant readiness for inhalation. Elder patients may face difficulty in assembling the pMDI with spacer. Dosing may be reduced and inconsistent due to electrostatic effects. Requires special washing instructions to be followed. If proper steps are not followed, the result may be a lower amount or no drug being inhaled. It contains propellants. Only a few corticosteroid products are available in the market. It adds to the cost of the medication system (5,13).

40.2.12 Facemasks

Facemasks are possibly the most important components of VHC systems alongside the measures to minimize the static-related aerosol losses. As per the experts, an ideal facemask should be able to facilitate a tight seal, incorporate minimal dead-space, be made out of soft polymer having a countered rolled edge, reflect the facial contours of the anticipated patient subgroups, and contain a low resistance exhalation valve directing the exhaled air away from the VHC (Figure 40.3) (25). The patients who face difficulty sealing their lips to the spacer's nozzle can insert a mask for convenience (4).

Several studies have established that a tight facemask seal is crucial. A 0.4 cm^2 leak in the facemask near the nose has the potential to reduce pulmonary delivery from 10% of the labeled dose to 0%. The gold standard in facemasks for tight seal is the Hans Rudolph anesthetic mask. The force that must be applied to the mask to acquire the intended seal is also an important factor, especially for pediatric or geriatric patient populations. There are masks available in the market that, in the presence of a good seal, demonstrate a respiratory low indicator. This indicator consists of an enclosed flap atop the VHC that moves toward and away from the patient in synchrony with their breathing pattern. The mask dead space is a rather important consideration for infants, as their tidal volumes tend to be very low. For example, an average 6-month-old infant displays tidal volume of only 55 ml (25).

40.3 Soft Mist Inhalers (SMIs)

40.3.1 What is an SMI?

The SMIs are new generation, propellant free, multi-dose inhalers (13). It is a pocket-sized device generating a single-breath inhalable aerosol and designed to enhance pulmonary drug delivery. It reduces the requirement for patient coordination for actuation–inhalation and inspiratory effort and offers improved patient experience and ease of use. The first SMI device (Respimat® Soft Mist™ Inhaler, marketed by Boehringer Ingelheim Pharmaceuticals, Ridgefield, CT, U.S.) was developed to meet the characteristics of an ideal inhaler (27). It is a propellant-free soft mist inhaler that utilizes mechanical energy in the form of a tensioned spring to produce a soft aerosol plume (12). It was first developed in response to the need for a pocket-sized device which can generate a single-breath, inhalable aerosol from a drug solution that is patient-independent, reproducible, and uses an environmentally friendly energy supply (28). SMIs were designed not only to improve drug delivery without the use of propellants, but also to permit reliable drug delivery via a device that was easy for patients to use. For example, the internal mechanism of SMI propels the medication into the lungs without patients having to produce a high respiratory flow, unlike some pMDIs. The slow velocity of the aerosol cloud and longer duration provide the patient with more time to breathe, so that they do not need to coordinate actuation and inhalation as closely, or produce as much inspiratory effort, in order to achieve drug deposition (27). It nebulizes a liquid drug solution through a finely angled nozzle that permits the drug solution to generate aerosol in a slow-moving cloud with a spray time of 1.0–1.5 s. SMIs are designed for patients who breath spontaneously (29,30). The Respimat is designed to deliver a monthly supply of drug, and can be configured to deliver 60 or 120 metered doses, depending on the daily dosing frequency (28). The design of

Respimat® SMI is patient centered, and may help to encourage long-term adherence and improve clinical outcomes with asthma and COPD (27).

40.3.2 Why SMI?

The SMIs are designed to generate a slow-moving and fine mist, be easy to use, and effectively deliver drugs to the lungs (27). The slow velocity and longer duration of the spray of SMIs also aid patient coordination (27). Patients with limited manual dexterity or, cognitive impairment and those with decreased inspiratory flowrate, may be unable to use a pMDI or DPI effectively (27). The SMIs follow a "press-and-breathe" process similar to pMDIs. While coordination between firing and inhalation is required, the low spray velocity and long duration of aerosol mist (about 1–1.5 s) enables patients to coordinate firing and inhalation more easily than with a pMDI (2). As the aerosol generated by SMI has a high fine-particle fraction (FFP) delivered at a slow velocity, pulmonary deposition is maximized and oropharyngeal deposition is minimized, even at low inhalation flows (27). Clinical trials in adults and children with obstructive lung diseases have shown that the efficacy and safety of a pMDI formulation of a combination bronchodilator can be matched by a an SMI formulation containing only one half or one quarter of the dose delivered by a pMDI (28). SMIs show high patient satisfaction, and the long duration of spray is of potential benefit to patients who have difficulty in coordinating inhalation with drug release (28).

40.3.3 Drugs Delivered via SMI

The Respimat® Soft Mist™ Inhaler is a popular SMI available in the market, with various versions available. This inhaler is available in Germany and some other countries for the delivery of tiotropium bromide (2.5 mg per puff), and specifically in Germany for delivery of a combination of fenoterol and ipratropium bromide (50 mg and 20 mg per puff, respectively) (5). Drugs and their combinations approved in US and European markets for administration via SMIs are olodaterol, tiotropium bromide, albuterol/ipratropium bromide, and tiotropium bromide/olodaterol. The pMDI for albuterol/ipratropium bromide has been replaced in the US market with an SMI device (11).

40.3.4 Mechanism of SMI

There are several possible methods for producing a pocket-sized device capable of aerosolizing a drug solution to produce a transient mist that could contain a full dose. Three of the known methods include piezoelectric vibration, extrusion through micron-sized holes, and transformation of mechanical energy into droplet-generating energy in an adequately efficient manner. In the case of the Respimat inhaler, the technical breakthrough was based on the approach of forcing drug solution through a two-channel nozzle by means of mechanical power (Figure 40.1). During this process, the solution is accelerated and split into two converging jets that collide at a sensibly structured angle, causing the drug solution to disintegrate into inhalable droplets. This procedure for aerosolizing a liquid requires only a small amount of mechanical energy, which is easily generated via the twisting action of a patient's hand. The mechanical energy in the SMI comes from a spring which is tensioned by the twisting action of the base of the device before use (28). When the spring is released, the solution containing drug is forced through an extremely fine nozzle system, which produces a fine mist that is slow moving, leading to lower deposition of drug into mouth and throat and relatively higher pulmonary deposition (i.e. 39%) (5).

When a new cartridge is inserted for the first time, the device needs to be primed to expel air from the drug solution flow path. After this one-time setup, the cartridge is permanently attached by a capillary tube containing a non-return valve to the fixed volume dosing chamber, and the device is ready for regular use (28). When the transparent base of the device is given a one-half turn to the right, it generates energy that draws a predetermined metered volume of solution from the cartridge through a capillary tube into a micro-pump. As the dose release button is pressed, the energy from the spring forces the solution to the mouthpiece, creating a soft plume of aerosol lasting for approximately 1.5 s (12). SMI produces an inhalable cloud of aerosol with improved characteristics compared to those produced by pMDIs and DPIs. Therefore, SMIs may be considered to be an innovative device for patient-oriented inhalation therapy (13). SMI uses an extremely fine nozzle system to aerosolize a metered dose of drug solution into tiny particles adequate for inhalation (27). The particle size of the aerosol is less than 5 μm (12). The mechanics of SMI are designed to optimize aerosol velocity, particle size, and internal resistance in order to augment drug delivery into the airways (27). SMIs incorporate use of liquid formulation similar to those used in nebulizers, but these are usually multi-dose devices having the potential to compete with pMDIs and DPIs in the portable inhaler market (2). The SMI can be designed to aerosolized most of the metered volume in the form of droplets with a diameter of >1 μm (in order to avoid loss of small droplets during subsequent exhalation) and < 5 μm (in order to facilitate efficient lung deposition) (27).

Additional features of SMIs may include a dose indicator and a lockout mechanism to avoid tailing-off of the size of dose, a problem sometimes seen with standard pMDIs. The dose indicator shows patients approximately how many doses are remaining and remind the patient to refill the prescription in good time. The dose indicator enters into the red zone a week before the last dose is due to be inhaled. A locking mechanism automatically prevents the use of the device past a specified number of doses actuated. This confirms that there is no detectable "tail-off," which is a common problem with most pMDIs that results in emitted doses becoming ever smaller as the canister is near exhaustion (28). Importantly, SMIs actively generate an aerosol independently of the patient's inhalation effort, with a slow velocity and extended duration, that facilitates the coordination of actuation and inhalation (27). The low aerosol velocity tends to decrease the oropharyngeal deposition and offers the benefits of producing more than 60% of the FPP independent of inspiratory effort for the delivered bronchodilator (29). Higher pulmonary deposition compared with pMDIs can be achieved through the high particle fraction allied with the low velocity and long generation time of the aerosol produced by SMI (28).

40.3.5 Methodology to Use SMI

Since the SMI does not use a propellant base, it does not need to be shaken before use. Just like other pMDIs, the SMI also needs to be primed before use and at times when the device has not been in use for some time. If kept without using for more than 3 days, a single actuation from the device should be made (12).

First, the patient needs to remove the transparent base, insert the cartilage containing the drug solution, and replace the base. For loading a dose, the base of the device needs to be half turned (180°) until it clicks. The helical cam gear transforms the rotation into a linear movement that tenses the spring and moves the capillary with the non-return valve to a defined lower position. During this movement, the drug solution is drawn through the capillary tube into the dosing chamber. When the patient presses the dose-release button to actuate the device, the mechanical power stored in the spring pushes the capillary with the non-return valve. This process drives the metered volume of drug solution (15 µl) through the twin nozzles of the uniblock, so that the two fine jets of liquid converge at a controlled angle. The two jets generate a slow-moving aerosol cloud from which the term *soft mist* is derived (28). The final step in using an SMI consists of breath-hold, same as pMDI. Table 40.2 shows the correct method to use an SMI device. Studies indicate that pulmonary deposition is enhanced after a breath-hold for 10 seconds, rather than for 4 s, as the extra time allows particles to sediment in the small airways of the lung, which increases the amount of inhaled drug that is deposited (2). SMIs usually have a dose indicator that locks once all the medication is used (12).

40.3.6 Statistics

Compared to pMDIs and DPIs, an SMI has a favorable pulmonary deposition profile in patients with asthma or COPD as well as in healthy volunteers compared to pMDIs and DPIs (27). Studies have shown that the pulmonary deposition of SMIs is several times higher than the CFC-pMDIs and clinical trials have confirmed that the drugs delivered by SMI are effective in comparatively lower doses in COPD patients (2). More than 60% of drug dose delivered by SMI falls within a fine-particle dose of <5.0 µm that increases the drug delivery to the smaller bronchi and bronchioles (27). The Respimat® SMI demonstrates fine particle fraction (FPF) of approximately 75% with most formulations, which is nearly double the value reported for aerosols generated by a pMDIs and DPIs (28).

Studies have demonstrated that the aerosol spray velocity of an SMI (0.84 m/s at from the end of nozzle and 0.72 m/s from the nozzle at 100 mm distance) is lower than that of several pMDIs (slowest being 2.47 m/s at 80 mm and 1.71 m/s at 100 mm from the end of nozzle) (27). The SMI exhibits velocity of 0.8 m/s, which is about three to 10 times slower than the speed of release of an aerosol cloud from a pMDI (28). The spray duration of Respimat® SMI (nearly 1.2 s) is considerably longer than for pMDI (typically 0.15–0.36 s). The longer duration of soft mist permits the patient a better opportunity of coordinating the inhalation operation with the drug release (27,28). The pulmonary and oropharyngeal deposition of flunisolide, as investigated by Newman et al., showed Respimat® SMI (39.7%) was higher than using only pMDI (15.3%) and pMDI plus spacer combination (28%). This study suggests that smaller nominal doses could be used with the SMI device to obtain the same pharmacodynamic effect with reduced side effects (28). Clinical trials have marked that using lower doses of tiotropium (5 µm) are not less safe than higher doses (18 µm) (13).

A survey performed on 503 patients having COPD asked them to rank the attributes of an inhaler that are most important to them, based on the connivance characteristics on the Patient Satisfaction adn Preference Questionnaire (PASAPQ). Three main attributes in this study were "feeling that your medicine gets into your lungs," "inhaler works reliably," and "the inhaler makes drug administration easy." In this study, SMIs outperformed other inhalers. Numerous studies in patients with asthma and COPD have reported that SMIs were easy for the patients to use to inhale a dose, and that the inhaled dose was felt by the patients to be going to the lungs. In some comparative studies, patients preferred SMIs over pMDIs and DPIs as they were easy to use and felt the inhaled dose was delivered. Old age appeared to be associated with worsening inhaler technique. In this case, the SMIs have many features that are useful for older patients, taking into consideration the simplicity of preparing a dose, the ability to obtain drug delivery at low inspiratory flowrates, and no requirement for actuation–inhalation coordination. The ease of use is a particularly important feature for pediatric populations, as it might be difficult to use the inhalers correctly due to their lower inspiratory force (27). The physician's choice of inhaler device for a particular patient could include factors like drug products, clinical benefit, economics, ease of use, dosing schedule, portability, taste, adverse effects, and sociocultural factors such as belief, knowledge, and education (28).

40.3.7 Advantages of SMIs

SMIs reduce oropharyngeal deposition and enhance pulmonary deposition of drug. Slower velocity and longer duration of the aerosol cloud delivered by SMIs simplify the coordination for

TABLE 40.2

Selection of Aerosol Delivery Device For Patients (Data from (5))

Type of Patient	Inspiratory Flow	
	Greater than 30 l/min	**Less than 30 l/min**
Patients with good actuation–inhalation coordination	pMDI or BA-pMDI	pMDI
Patients with poor actuation–inhalation coordination	pMDI + spacer or BA-pMDI	pMDI + spacer

pMDI, pressurized metered dose inhaler; BA-pMDI, breath actuated pressurized metered dose inhaler patients. SMIs are easy for patients to use correctly and are associated with a high rate of patient preference and adherence, and a low rate of discontinuation. Several studies concluded that SMIs are easy to use with correct inhalation techniques for patients immediately after training. This may be specifically important for geriatric patients who have more difficulty learning the correct technique for using a new inhaler (27). In general, it is easier to achieve dose-to-dose reproducibility by delivering a small volume of drug solution from a reservoir than a small quantity of suspension or a powder as in pMDI and DPI. The SMI brands incorporate an internal color coding system to identify the specific drug class contained in the device and a transparent base to allow easy identification of the drug inside (28).

40.3.8 Disadvantages of SMIs

The metered volume of 15μl is limiting for the dose delivery capacity of the SMIs, which allows only drugs having high solubility with respect to the required dose. However, this limitation can be overcome by increasing the volume or number of puffs administered; it would have to be balanced versus the risk of reduced patient compliance (13). Probable disadvantages of SMI include the need for basic assembly and priming before first use, and cost compared to other inhaler devices, which can be a consideration for patients (27).

40.4 Selection of Delivery Device

The choice of aerosol drug delivery device depends upon the availability of the device for the drug and the probability of effective utilization of the same device for drug administration by the patient. For optimal aerosol deposition in the lungs by a pMDI there must be a consistent coordination between actuation and inhalation. There are few differences between pMDI with CFC propellants and pMDI with HFA propellants. Out of those, one difference is that the HFA-pMDI formulations have a minute amount of alcohol which affects the velocity of the aerosol. Another difference is that the plume released from the HFA-pMDI aerosols is slower in velocity and warmer in temperature. The physician must inform the patient about these differences while prescribing HFA-pMDI aerosol. The patients must also be instructed to prime the first two to four doses (discharging doses into surrounding air) of the aerosol during first use or when it is to be used after several days or weeks of disuse.

Spacers available with incorporated one-way valve are VHCs. VHCs are used for infants, children, and patients with poor coordination of actuation and inhalation technique for optimization of pulmonary deposition using a pMDI. The patients must be instructed not to actuate the VHC before each inhalation and to inhale the dose immediately, as soon as it is actuated, without any kind of time delay.

In the category of SMIs, only one SMI is commercially available, i.e. Respimat. This inhaler produces a fine mist which is slow moving and thus leads to proper deposition in the lungs and can be used in patients with improper ventilation systems or poor breathing patterns.

40.4.1 Choice of Drug Device Combinations for Home Use

For treatment of asthma and COPD, drugs like bronchodilators, corticosteroids, and combination formulations are used. These are administered using either a pMDI or SMI depending upon the class of aerosol medicament and the capacity of the patient. The same device options are also for patients affected with diseases other than asthma and COPD, i.e. PAH, HIV, and cystic fibrosis (5).

40.4.2 Factors Affecting Choice of Aerosol Drug Delivery Device

40.4.2.1 Patient-Related Factors

Patient-related factors, like diagnosis, age, and the physical and cognitive ability of the patient affect the selection of a medication for inhalation. With ageing there are changes in anatomical and physiological characteristics, like size of airways, and change in respiratory rate and lung volume, which directly affect the selection of aerosol device. Patient's preference for aerosol device is also a crucial factor involved in selection of a device which is optimum for effective aerosol therapy and at the same time not against patient's dislikes. Table 40.2 suggests ideal inhalers according to patients' inspiratory flow.

40.4.2.2 Drug-Related Factors

Certain medications are available with only one kind of aerosol device leading to restricted choice of device for the patient. In cases where multiple devices for a single medication are available, the physician should prescribe the optimum device considering the healthcare insurance and the patient's preference.

40.4.2.3 Device-Related Factors

The ease of using the aerosol delivery device, the duration of treatment with the device, and the portability affect the choice of the delivery device. Also, the amount of maintenance required for the device is an important factor which is considered by the patients. Small devices requiring less cleaning and maintenance are preferred by patients over complex, large-sized aerosol devices.

40.4.2.4 Environmental and Clinical Factors

The time and the location where the aerosol formulation is used also has an impact on the device selection. Noisy compressors are not a good option in cases of late-night treatments in small homes, where the doses are administered at night. In environments where the patient is in close proximity to other people, the secondary exposure to aerosol formulations is a factor that affects the device selection. Here, devices incorporating filter exhaled aerosol should be considered for use.

40.4.2.5 Cost and Reimbursement of Aerosol Devices

Patients tend to choose a device having minimum or affordable out-of-pocket expense. The costs of devices covered by the health insurance of the patient also play a vital role in determining the aerosol device for the patient (12).

40.5 Observations and Considerations

The pMDI can be attached to a tracheostomy tube, but it may be a little difficult to adapt and seal due to difficulty of communication in certain cases, as tracheostomy is responsible for anatomical and physiological alterations of the respiratory system, essential for vocal production. The asynchronicity of the jet shot and the beginning of the inspiratory phase is a common difficulty for these patient groups (4). Patients displayed certain insecurities transitioning from non-invasive ventilation (NIV) to inhalers, about withdrawing the support and carrying out drug administration via inhalers, in addition to having difficulty in inspiring forcibly with a final pause, and many prefer nebulization instead of administration by pMDI (2).

It has been commonly observed that patients face difficulties when they are first prescribed inhalers, even after they are given proper guidelines on the inhalation technique (5). The reports from meta-analysis indicate that if patients use the inhalation technique exactly as recommended by the manufacturer, all the inhaler devices can achieve the same therapeutic effect, although different doses may be required. Nevertheless, many patients fail to use the correct technique when using their inhalers, either because of lack of proper guidance or they have modified the technique after receiving instructions. Poor adherence with the optimal treatment regime is the most common issue reported (5). It has been reported that 76% of patients that use pMDIs and 49–54% of patients using a BA-pMDI make at least one error when using their inhaler. The most common problems with using a pMDI were marked to be lack of actuation–inhalation coordination and stopping inhalation due to the cold freon effect (5).

In vitro studies have marked that the type of VHC, breathing patterns, and other factors affect the drug delivery of pMDI through tracheostomies. Cooper et al. compared oronasal and tracheostomy delivery of albuterol using pMDI with two different VHCs and a continuous output jet nebulizer. They observed that the type of VHC and breathing pattern affected pulmonary dose obtained via oronasal and tracheostomy routes (1).

A study has reported that SMI deliver more albuterol than a pMDI (31,32). In general, albuterol delivery via oronasal and tracheostomy routes using a metallic VHC results in a higher lung deposition with an SMI than in pMDI. Also, generally the tracheostomy route provides a higher pulmonary dose than the oronasal route for both pMDIs and SMIs (5). Ultimately, the finding was that an SMI provides a greater pulmonary dose than pMDI regardless of the characteristics of aerosols produced by both inhalers (5). The aerosol cloud produced by SMIs is 10 times slower than the plume produced by pMDIs and shows a 3-fold longer duration of spray than pMDIs (30).

Studies also suggest that the pulmonary dose increases with increasing tidal volume via oronasal route using pMDIs (5).

The recommendations made based on data obtained by with specific device/formulation should not be assumed to be equal to other combinations until proven. In the absence of correct dosing equivalency, clinicians should consider starting at a lower dose and titrate it to effect while switching from a less efficient to a more efficient delivery system (5).

While a dose increase might be necessary for patients of low tidal volume switching from SMIs to an oronasal or tracheostomy delivery route, the opposite action might be necessary for patients with higher tidal volumes when using SMIs (5). The main limitation of most of the inhaler studies is their in vitro nature, although these are well-established methods in aerosol research (5).

40.6 Conclusion

For effective and optimum delivery of aerosol formulations, oronasal and tracheostomy delivery is used. Oronasal delivery is a non-invasive ventilation technique which uses an oronasal mask for aerosol delivery while tracheostomy includes the delivery of aerosol via tracheal tube (4). Studies have shown that a normal dose delivery via tracheostomy shows a greater amount of pulmonary deposition than the delivery via oronasal mask (11).

To achieve the best administration and therapeutic effects of inhalation drugs, it is recommended to pay attention to the factors that provide greatest comfort and safety to the patient. pMDIs are multiple-dose devices whose indiscriminate use can lead to serious adverse effects or toxicity. Thus, the inhalation technique is important (4). The choice of an inhalation device is known to have a strong scientific basis to affect the dose of drug that is deposited in the lungs. It is of great importance that once a patient is familiar with one type of inhaler, they should not be switched to a new device without their involvement and especially not without follow-up education on how to use the device properly (5). There are a number of variables affecting the drug delivery, such as use of VHCs or spacer devices, masks, aerosol generators, breathing patterns, etc., hence, practitioners should consider prescribing the products that are cost efficient and have been studied in conditions similar to the ones in which they are going to be utilized. Caution should be practiced while switching between pMDIs and SMIs. The pMDI and SMI are not interchangeable and the magnitude of difference in drug delivery may vary considerably with difference in tidal volumes (4). Table 40.3 shows a comparision between pMDIs and SMIs.

40.7 Summary

Oronasal and tracheostomy delivery of pressurized metered dosed inhalers with VHC and soft mist inhalers shows a greater pulmonary deposition into the lungs with a high dose. Certain factors associated with patient, like disease conditions and poor actuation–inhalation coordination, can affect the delivery of aerosol in tracheostomized patients. A variety of portable inhaler devices are now available for treating patients with COPD and more new designs are in development; each type of device has advantages and disadvantages. The plethora

TABLE 40.3

Comparison of pMDI with SMI (Data from (2); (12))

Comparison Parameter	pMDI	SMI
Full form	Pressurized metered-dose inhaler	Soft mist inhaler
Brands available in market	Ventolin, Atrovent, Combivent	Respimat® Soft Mist™ Inhaler
Propellants	Chlorofluorocarbons and hydrofluoroalkanes	No propellants used
Basic mechanism	Use of propellants to generate pressurized aerosol plume	Use of spring to generate mechanical energy for producing aerosol cloud
Drugs administered	Bronchodilators, corticosteroids, combination drugs, insulin, gene therapy vectors, chemotherapy agents, etc.	Bronchodilators and corticosteroids such as tiotropium bromide, fenoterol, combination drugs, etc.
Countries where the device is available	Europe, USA, Japan, India, Africa, Asia-Pacific countries, etc.	Germany, United Kingdom, Croatia, Slovakia, Denmark, Norway, Ireland, Austria, Romania, India, and Spain
Routes it can be used for	Oronasal and tracheostomy	Oronasal and tracheostomy
Velocity of aerosol plume	Slowest is 2.47 m/s at 80 mm and 1.71 m/s at 100 mm from the end of nozzle	0.84 m/s at from the end of nozzle and 0.72 m/s from the nozzle at 100 mm distance (three to 10 times slower than pMDI)
Duration of aerosol cloud	Typically, 0.15–0.36 s	About 1.2 s
Pulmonary deposition	About 40–50% of nominal dose	About 39% of nominal dose
Advantages	Compact and portable, multiple dose, high dose reproducibility, low cost	Compact and portable, multiple dose, high dose reproducibility, locking mechanism prevents actuation after drug runs out, can be easy to learn
Disadvantages	Bulky when used with spacer devices, need for actuation–inhalation coordination, cold freon effect, patients may still actuate propellant after drug has run out	Higher cost compared to pMDIs, limited drugs available

of inhalers with differing instructions may confuse patients and healthcare providers alike. Any inhaler can be misused so that little or no drug is deposited in the lungs. Crucial errors in inhaler technique, resulting in no drug deposition in the lungs, must be avoided. Inhaler choice in COPD should take into account the likelihood of the patient using the inhaler correctly, patient preference, and likely compliance.

REFERENCES

1. Cooper B and A Berlinski. Albuterol delivery via facial and tracheostomy route in a model of a spontaneously breathing child. *Respiratory Care*. 2015; 60(12): 1749–1758. https://doi.org/10.4187/respcare.04142.
2. Newman SP. Inhaler treatment options in COPD. *European Respiratory Review*. 2005; 14(96): 102–108. https://doi.org/10.1183/09059180.05.00009605.
3. Reasons for a Tracheostomy. n.d. Medical and health. John Hopkins Medicine. https://www.hopkinsmedicine.org/tracheostomy/about/reasons.html.
4. Asturian K and Ferreira MA. Administration of inhaled medications during noninvasive ventilation and/or tracheostomy. *Revista Brasileira de Farmácia Hospitalar e Serviços de Saúde*. 2020; 11(1): 0372. https://doi.org/10.30968/rbfhss.2020.111.0372.
5. Laube BL, Janssens HM, de Jongh FHC, Devadason SG, Dhand R, Diot P, Everard ML, et al. What the pulmonary specialist should know about the new inhalation therapies. *European Respiratory Journal*. 2011; 37(6): 1308–1417. https://doi.org/10.1183/09031936.00166410.
6. Labiris NR and Dolovich MB. Pulmonary drug delivery. Part I: Physiological factors affecting therapeutic effectiveness of aerosolized medications: Physiological factors affecting the effectiveness of inhaled drugs. *British Journal of Clinical Pharmacology*. 2003; 56(6): 588–599. https://doi.org/10.1046/j.1365-2125.2003.01892.x.
7. Rubin BK. Air and soul: The science and application of aerosol therapy. *Respiratory Care*. 2010; 55(7): 911–921. PMID: 20587104.
8. Al Otair HA and BaHammam AS. Ventilator- and interface-related factors influencing patient-ventilator asynchrony during noninvasive ventilation. *Annals of Thoracic Medicine*. 2020; 15(1): 1. https://doi.org/10.4103/atm.ATM_24_19.
9. Ari A. Aerosol therapy in pulmonary critical care. *Respiratory Care*. 2015; 60(6): 858–879. https://doi.org/10.4187/respcare.03790.
10. Sanders M. Inhalation therapy: An historical review. *Primary Care Respiratory Journal*. 2007; 16(2): 71–81. https://doi.org/10.3132/pcrj.2007.00017.
11. Berlinski A and Cooper B. Oronasal and tracheostomy delivery of soft mist and pressurized metered-dose inhalers with valved holding chamber. *Respiratory Care*. 2016; 61(7): 913–919. https://doi.org/10.4187/respcare.04575.
12. Gregory KL, Wilken L, and Hart MK. A guide for physicians, nurses, pharmacists, and other health care professionals. 2017; 68.
13. Rogliani P, Calzetta L, Coppola A, Cavalli F, Ora J, Puxeddu E, Matera MG, and Cazzola M. Optimizing drug delivery in COPD: The role of inhaler devices. *Respiratory Medicine*. 2017; 124 (March): 6–14. https://doi.org/10.1016/j.rmed.2017.01.006.
14. Usmani OS. Treating the small airways. *Respiration*. 2012; 84(6): 441–453. https://doi.org/10.1159/000343629.

15. Global Initiative for Asthma. Global Strategy for Asthma Management and Prevention, 2019. Available from: www.ginasthma.org
16. Lavorini F, Fontana GA, and Usmani OS. New inhaler devices - the good, the bad and the ugly. *Respiration*. 2014; 88(1): 3–15. https://doi.org/10.1159/000363390.
17. Coppolo DP, Mitchell JP, Jolyon P, and Mark W. Levalbuterol aerosol delivery with a nonelectrostatic versus a nonconducting valved holding chamber. May 2016.
18. Nebulised antibiotherapy: Conventional versus - nanotechnology-based approaches, is targeting at a nano scale a difficult subject? 2020. *Annals of Translation Medicine*. September 2020. http://atm.amegroups.com/article/view/16842/17767.
19. Connolly CK. Method of using pressurized aerosols. *BMJ*. 1975; 3(5974): 21–21. https://doi.org/10.1136/bmj.3.5974.21.
20. Chorão P, Pereira AM, and Fonseca JA. Inhaler devices in asthma and COPD – An assessment of inhaler technique and patient preferences. *Respiratory Medicine*. 2014; 108(7): 968–975. https://doi.org/10.1016/j.rmed.2014.04.019.
21. Usmani OS, Biddiscombe MF, and Barnes PJ. Regional lung deposition and bronchodilator response as a function of β 2 -agonist particle size. *American Journal of Respiratory and Critical Care Medicine*. 2005; 172(12): 1497–1504. https://doi.org/10.1164/rccm.200410-1414OC.
22. Chua HL, Collis GG, Newbury AM, Chan K, Bower GD, Sly PD, and Le Souef PN. The influence of age on aerosol deposition in children with cystic fibrosis. *European Respiratory Journal*. 1994; 7(12): 2185–2191. https://doi.org/10.1183/09031936.94.07122185.
23. Vehring R, Lechuga-Ballesteros D, Vidya J, Noga B, and Dwivedi SK. Cosuspensions of microcrystals and engineered microparticles for uniform and efficient delivery of respiratory therapeutics from pressurized metered dose inhalers. *Langmuir*. 2012; 28(42): 15015–15023. https://doi.org/10.1021/la302281n.
24. Colthorpe P, VoshaarT, Kieckbusch T, Cuoghi E, and Jauernig J. Delivery characteristics of a low-resistance dry-powder inhaler used to deliver the long-acting muscarinic *Antagonist glycopyrronium. Journal of Drug Assessment*. 2013; 2(1): 11–16. https://doi.org/10.3109/21556660.2013.766197.
25. Dissanayake S and Suggett J. A review of the in vitro and in vivo valved holding chamber (VHC) literature with a focus on the aerochamber plus flow-vu anti-static VHC. *Therapeutic Advances in Respiratory Disease*. 2018; 12 (January): 175346581775134. https://doi.org/10.1177/1753465817751346.
26. Anhøj J, Bisgaard H, and Lipworth BJ. Effect of electrostatic charge in plastic spacers on the lung delivery of hfa-salbutamol in children. *British Journal of Clinical Pharmacology*. 1999; 47(3): 333–336. https://doi.org/10.1046/j.1365-2125.1999.00893.x.
27. Iwanaga T, Tohda Y, Nakamura S, and Suga Y. The Respimat® soft mist inhaler: Implications of drug delivery characteristics for patients. *Clinical Drug Investigation*. 2019; 39(11): 1021–1030. https://doi.org/10.1007/s40261-019-00835-z.
28. Eicher J, Eicher J, Zierenberg B, and Dalby R. Development of respimat® soft mist™ inhaler and its clinical utility in respiratory disorders. *Medical Devices: Evidence and Research*. 2011 September, 145. https://doi.org/10.2147/MDER.S7409.
29. Fang T-P, Chen Y-J, Yang T-M, Wang S-H, Hung M-S, Chiu S-H, Li HH, Fink JB, and Lin H-L. Optimal connection for tiotropium smi delivery through mechanical ventilation: An in vitro study. *Pharmaceutics*. 2020; 12(3): 291. https://doi.org/10.3390/pharmaceutics12030291.
30. National Institute of Health. Revised 2002. Global strategy for asthma management and prevention.
31. Hodder R. Asthma patients prefer respimat® soft mist™ inhaler to turbuhaler® *International Journal of Chronic Obstructive Pulmonary Disease*. 2009 May, 225. https://doi.org/10.2147/COPD.S3452.
32. Critical Care. March 2014. 34th International Symposium on Intensive Care and Emergency Medicine.

41

Modeling for Biopharmaceutical Performance in Lung Drug Discovery

Amy Le[1], **Kianna Samuel**[1], **Truong Tran**[1], **and Yashwant Pathak**[1,2]
[1]*Taneja College of Pharmacy, University of South Florida, Tempa, Florida, USA*
[2]*Adjunct Professor, faculty of Pharmacy, Airlangga University, Surabaya, Indonesia*

41.1 Introduction

With the rise in pharmaceutical medicine production, the pulmonary system is starting to be the route of choice for distributing a range of drugs throughout the body. Benefits of lung drug delivery includes a faster onset, avoidance of degradation in the GI tract, more convenience, less intrusiveness, and relative comfort. In the past, use of inhaled biopharmaceuticals was inhibited by high costs, drug requirements, and stability problems; however, technological advances are actively addressing these problems.

Currently, there are more than 40 different types of inhaled pharmaceuticals in the public domain that are used to combat many diseases: asthma, diabetes, cystic fibrosis, cancer, hypertension, influenza, sarcoma, osteoporosis, growth deficiency, tuberculosis, hepatitis, etc. Overall, inhaled biopharmaceuticals are on a rising trajectory and are expected to become more common as time passes.

41.2 Applications of Modeling and Simulation of Biopharmaceutical Performance Drug Delivery

Modeling and simulation play an essential role in drug discovery. It allows researchers to test drug effects and gather information without using human subjects. In lung drug discovery, modeling and simulation play important roles as the lungs are one of the most complex organs in the body. Researchers can use several different methods, such as cellular and molecular models, to conduct rigorous testing of medications.

41.2.1 Modeling for Lung Studies or Lung Treatment

Lung diseases are unique because of the types of environments that vastly affect them. For many years, drugs have been administered through inhalation and nebulization to reach the lungs; using lung models and simulations allows researchers to mirror the effects of these drugs in the lungs. Breathing patterns, disease progression, and genetics all play a role in how medicine reacts in the body. Recently lung modeling has progressed to allow scientists to "accurately reflect the interactions between cells or within tissue interfaces in the macroenvironment of the lung" (1). By using modeling and simulation, drug developers can predict drug effectiveness. It also helps to know how a drug will interact with the lungs when delivered as different delivery systems or dosage forms.

Using models and simulations could also contribute to the decline of animal testing. In 2015 the Wyss Institute at Harvard created the Lung on a Chip. These micro devices that "reconstitute tissue–tissue interfaces critical to organ function may, therefore, expand the capabilities of cell culture models and provide low-cost alternatives to animal and clinical studies…" (2–4). The lung on a chip mimics the air and blood flow; these chips have inspired researchers to develop chips for other organs such as the heart, kidney, and skin, as well. With the creation of these micro devices, researchers are able to accurately test the effects of drugs without testing on animals.

41.2.2 Cellular Models to Study Lung Treatment

Cellular models can imitate healthy and unhealthy tissues in the body. Scientists have developed a new lung model by using embryonic stem cells to discover drugs that will aid with small cell lung cancer research and treatment (5–8). Researchers can use computational fluid dynamic simulations to mimic the structure of the lungs. Computational fluid dynamic simulations are made by using "3D lung airway models that were reconstructed from CT [computed tomography] scans" (9). These models are accurate when studying certain diseases because they can be modeled after the CT scans of patients with the desired disease. The models and simulations based on CT scans work well to understand the complexities in lifelong diseases such as asthma and chronic obstructive pulmonary disease (COPD).

Stem cells have the ability to become any cell in the body, which aids in the process of cellular modeling. "These cells can be expanded and differentiated to produce a potentially limitless supply of the affected cell type, which can then be used as a tool to improve understanding of disease mechanisms and test therapeutic interventions" (10–12). With the ability to become any type of cell and be replicated endlessly, researchers can

DOI: 10.1201/9781003046547-41

explore the effects of drugs with the diseased cells. Researchers are then able to reproduce the results that they find through computational methods.

41.2.3 Molecular Models for Lung Studies

Molecular modeling and simulation allow researchers to replicate and understand the interactions between molecules with the use of physics. Scientists are able to examine the structures of the molecules and see how they react to the delivery of medications. "Molecular dynamics (MD) is an important computational tool for understanding the physical basis of the structure, the dynamic evolution of the system, and the function of biological macromolecules" (13). MD simulations show meticulously replicated membrane proteins and let researchers see how the proteins work together when introduced to medications.

41.3 The Benefits of Modeling and Simulation

Modeling and simulations of the lungs are a crucial way for scientists to understand the physiology of the organ. By studying simulations, it is possible to understand the effects of certain drugs on the organ.

41.3.1 In Vitro Models for Lung Studies

In vitro models are important in the study of lung drug delivery because of the advantage that exists where the cells are completely isolated from the body's environment. This allows for a more controllable model to test with. They have been a popular model to understand lung cell biology and have given scientists a way to study these tissues for patients.

With past studies of the lungs, it has been seen that much of the information we know about human lungs has been inferred from studies performed on rodents. Although humans and rodents have most of the same cells, differences still exist in the lungs. Human lungs contain basal cells that are located in the trachea and the bronchi, while rodents only have basal cells in the trachea (14). The major differences in both organisms have definitely shone a light on the need for human lung models.

The benefits with in vitro models are that scientists have been able to understand the way that human lungs maintain homeostasis and how the lack of some cellular processes can have detrimental effects on the organ. Using models has also allowed the study of regions in the lungs that are normally too hard to study, such as the alveoli. It has allowed for the discovery of certain therapeutic targets, given scientists a way to test pharmaceutical drugs, and assisted in clinical trials (14). Current models allow for a more controlled cellular environment, which can be compared to studying lung cells in real time.

41.3.2 In Vivo Models for Lung Performance

In vivo models have allowed scientists to see the effects of drugs on lung cells directly in humans. These models are important because studying the cells while they are still in the organism will allow scientists to see the response of the body as a whole.

As mentioned before, many studies done in the lungs have been performed on rodents, which had in vivo environments of the human body (15). This supports the idea again that there is a need for human lung models. While some experiments are successful as in vitro, using the in vivo method allows scientists to create appropriate conclusions about lung cells. A human body environment will allow the lung cells to create real time effects that can be learned from to improve drugs.

Even from these studies, scientists have learned how human lung stem cells divide and renew themselves. This is just one of the benefits of in vivo models. By using these models, scientists have also been able to see remodeling of the extracellular matrix of mouse lungs, which takes part in the development of the lungs. Although not much is known about the lungs due to the lack of human lung models, using rodents as test subjects has given scientists information on which to base their predictions for the effects on humans (14,16,17).

41.3.3 Biopharmaceutical Simulations

Simulations of the lungs assist in designing and developing drug products. Creating simulations has given scientists an easier way to visualize how drugs will affect the environment of the lungs, thus, it has given insight into the best way to deliver drugs into the organ.

Simulations are a very popular method in which it is possible to show a visual of the lungs which has helped in the advancement of insulin delivery, cancer treatments, and nano therapeutics. The ability for scientists to view the effects of drugs on lung cells has created more efficient drug–aerosol drug delivery. This has been crucial for the condition of asthma. By using simulations, inhalers have advanced in targeted delivery methodologies. The medicine moves to the specific sites and has been created to treat those affected sites (18,19).

41.4 Future Usage

Below is a quick outline of future usages of this method, its models, and the challenges that scientists are currently facing.

41.4.1 Lung Cancer Models

Lung cancer is a disease that leads to 30% of all cancer deaths according to research by Min-chul Kwon and Anton Berns (19). This disease is most common with smokers, but also seen in nonsmokers. Annotation of the cancer genes has revealed many common mutations. The advancements of targeted therapy provide scientists with a spark of hope that one day personalized therapy will be used to combat this fatal disease.

On the topic of combating lung cancer, two main models are used to assess the effectiveness of biopharmaceuticals: computerized and mouse models. Computer modeling, or computer process modeling, was first done to diagnose lung cancer and staging the different stages. These models begin with analytical properties that include the Markov chain model and formulas.

The computerized models have both positives and negatives. Currently, computerized models are very time consuming to generate and heavily rely on the environmental factors that are preprogrammed (which does not account for every real-world factor). However, it can provide detailed analysis, graphics, and animations. These models are used to estimate the treatment time for patients and predict the virtual outcome of a drug, before animal testing.

Animal models are very useful in many new drug discovery. The main model for lung drug discovery is with a mouse. Using the mouse as a model is great because they are cost-effective and mimic many similar qualities that human's exhibit. Since the mouse and humans share similar characteristics, this method is useful for studying the response to drug therapy.

41.4.2 Lung Modeling Methods

With limitations in treatments that are controversial in the scientific field, targeted drug delivery methods are gaining popularity. Having drugs that can be delivered through the pulmonary system enables delivery directly to the lung, for both local and system-wide treatments. This is the method that allows for optimization in the number of drugs that the body absorbs. The low efficiency of many drugs in the market is raising a need for a multidisciplinary approach (20,21) which includes many interdisciplinary sciences: polymer science, pharmaceutical technology, biochemistry, and molecular biology. This approach to finding treatments is already in the making. Many drugs have not yet been made for delivery through the lungs but the possibilities include growth hormones and even insulin.

There are two main routes for drugs to be distributed to the body by the lungs — the conducting airways and respiratory region. However, the use of transport through the (upper) airways is limited based on the surface area and the ability of the body to filter out 90% of drug particles. Secondly, the mucus layer that forces the drug out of the body also limits this method of delivery. The current, most common way for similar treatments is through the use of aerosols/inhalation, which provides the best distribution.

Even when working with efficient methods of distribution, there are many techniques to create different sized particles: jet milling, spray drying, spray-freeze drying, supercritical fluid technology, solvent precipitation, solvent evaporation, and particle replication. The first method of creating these nanoparticles is by specially jet milling a drug under nitrogen gas. Spray drying and spray freezing (discovered around 1980) are similar techniques that involve a solution at room temperature being atomized and dried. The freeze-drying method is a restricted and expensive process that adds an extra step by spraying the drug through liquid nitrogen, this improves the performance of the drug and creates close to a 100% yield. When a drug involves proteins and peptides, scientists use the supercritical fluid technology to create controlled sizes of particles by suspending the solution in carbon dioxide. Lastly, Dr. Joseph DeSimone and his team created particle replication (PRINT). The method allows for making uniform-sized, organic microparticles. Being able to control the particle sizes of various solutions will enable a more efficient drug targeting.

41.4.3 Growing Trend

As of 2018, the inhalable drug market, which has a span of five continents, was estimated to be worth about US$25 billion dollars. Due to their convenience, non-invasiveness, efficiency, and minimal toxicities, a projection from Grandview Research predicted a steady increase in market growth (of aerosols, dry powder formulations, and sprays) until 2026, in which the market will almost double in size. The projection also indicated that dry powder formulation is the most popular choice out of the three substances. Now there are more than 220 projects and companies are focusing on developing inhalable drugs and enhanced versions of already existing molecules.

As the discovery of new drugs continues, there are many challenges that must be overcome, such as complying with governmental standards, developing the appropriate physical structures, and developing suitable models. When testing a drug, it is impossible to predict the outcome with so many variables existing; therefore, scientists usually start by using animals to create models to slowly gain an understanding of how a drug reacts. This method is useful, but cannot be fully used to predict the reaction when it is given to humans.

41.5 Conclusion

The lung is a very complex organ and relies heavily on the organs that surround it. Knowing this, it can be hard to research and test the lung in a way that will only affect the lung cells themselves. There have been various ways to model and create simulations of the lungs to test different drugs' effects on the cells. The results of these experiments have helped improve target drug delivery. Scientists have been able to create drugs that treat specific sites in the lungs for conditions such as asthma, diabetes, cystic fibrosis, cancer, etc.

However, although lung drug delivery research has improved with the use of models and simulations, there is still an information gap in that models and simulations cannot directly show the effects on the cells that scientists are less familiar with. For example, some models have yet to be performed on humans, which limits our knowledge of some diseases. Thus, scientists are left to predict outcomes of drug delivery, rather than having the results of an experiment on human lung cells.

REFERENCES

1. Nichols JE, Niles JA, Vega SP, Argueta LB, Eastaway A, and Cortiella J. Modeling the lung: Design and development of tissue engineered macro- and micro-physiologic lung models for research use. *Experimental Biology and Medicine*. 2014; *239*(9): 1135–1169. doi: 10.1177/1535370214536679
2. Huh DD. A human breathing lung-on-a-chip. *Annals of the American Thoracic Society*. 2015; *12*(Suppl 1): S42–S44. https://doi.org/10.1513/AnnalsATS.201410-442MG
3. Jiang W, Kim S, Zhang X, Lionberger RA, Davit BM, Conner DP, and Yu LX. The role of predictive biopharmaceutical modeling and simulation in drug development and regulatory evaluation. *International Journal of Pharmaceutics*. 2011; *418*(2): 151–160. https://doi.org/10.1016/j.ijpharm.2011.07.024

4. Ju Feng, et al. Computer modeling of lung cancer diagnosis-to-treatment process. *Translational Lung Cancer Research.* AME Publishing Company, Aug. 2015, www.ncbi.nlm.nih.gov/pmc/articles/PMC4549484/.
5. Chen H, Elemento O, Poran A, Snoeck H, Unni A, Xuelian S, and Varmus H. Generation of pulmonary neuroendocrine cells and SCLC-like tumors from human embryonic stem cells. *Journal of Experimental Medicine.* 4 March 2019; *216*(3): 674–687. doi: https://doi.org/10.1084/jem.20181155
6. Darquenne C, Fleming JS, Katz I, Martin AR, Schroeter J, Usmani OS, and ... Schmid O. Bridging the gap between science and clinical efficacy: Physiology, imaging, and modeling of aerosols in the lung. *Journal of Aerosol Medicine and Pulmonary Drug Delivery.* 2016; *29*(2): 107–126. doi: 10.1089/jamp.2015.1270
7. Forbes B, et al. Challenges for inhaled drug discovery and development: Induced alveolar macrophage responses. *Advanced Drug Delivery Reviews.* Elsevier, 13 Feb. 2014, www.sciencedirect.com/science/article/pii/S0169409X14000155.
8. Huang S-M, Abernethy DR, Wang Y, Zhao P, and Zineh I. The utility of modeling and simulation in drug development and regulatory review. *Journal of Pharmaceutical Sciences.* 2013; *102*(9): 2912–2923. doi: 10.1002/jps.23570
9. Aarons L, Karlsson MO, Mentré F, Rombout F, Steimer J-L, and Peer AV. Role of modelling and simulation in phase I drug development. *European Journal of Pharmaceutical Sciences.* 2001; *13*(2): 115–122. doi: 10.1016/s0928-0987(01)00096-3
10. Siller R, Greenhough S, Park IH, and Sullivan GJ. Modelling human disease with pluripotent stem cells. *Current Gene Therapy.* 2013; *13*(2): 99–110. https://doi.org/10.2174/1566523211313020004
11. Teuscher N and Patel N (2017, July 3). Modeling & simulation for drug development & formulation. Retrieved from https://www.contractpharma.com/issues/2017-03-01/view_features/modeling-simulation-for-drug-development-formulation/
12. Inhalable Drugs Market Size, Share: Global Industry Report, 2022. Oct 2019. www.grandviewresearch.com/industry-analysis/inhalable-drugs-market.
13. Aminpour M, Montemagno C, and Tuszynski JA. An overview of molecular modeling for drug discovery with specific illustrative examples of applications. *Molecules (Basel, Switzerland).* 2019; *24*(9): 1693. https://doi.org/10.3390/molecules24091693
14. Miller AJ and Spence JR. In vitro models to study human lung development, disease and homeostasis. *Physiology (Bethesda, MD.).* 2017; *32*(3): 246–260. https://doi.org/10.1152/physiol.00041.2016
15. Kajstura J, Rota M, Hall SR, Hosoda T, D'Amario D, Sanada F, Zheng H, Ogórek B, Rondon-Clavo C, Ferreira-Martins J, Matsuda A, Arranto C, Goichberg P, Giordano G, Haley KJ, Bardelli S, Rayatzadeh H, Liu X, Quaini F, Liao R, and ... Anversa P. Evidence for human lung stem cells. *The New England Journal of Medicine.* 2011; *364*(19), 1795–1806. https://doi.org/10.1056/NEJMoa1101324 (Retraction published *The New England Journal of Medicine.* 2018 Nov 8; 379(19): 1870)
16. Moore S, et al. Discovery and Development of Inhaled Biopharmaceuticals. Drug Discovery World. www.ddw-online.com/drug-discovery/p321760-discovery-and-development-of-inhaled-biopharmaceuticals.html.
17. Newman SP. Drug delivery to the lungs: Challenges and opportunities. *Therapeutic Delivery.* 2017; 8(8): 647–661. doi: 10.4155/tde-2017-0037
18. Kleinstreuer C, Zhang Z, and Donohue J. Targeted drug-aerosol delivery in the human respiratory system. *Annual Review of Biomedical Engineering.* 2008; *10*(1): 195–220. doi: 10.1146/annurev.bioeng.10.061807.160544
19. Kwon M-C and Berns A. Mouse Models for Lung Cancer *Molecular Oncology.* John Wiley and Sons Inc., Apr. 2013, www.ncbi.nlm.nih.gov/pmc/articles/PMC5528410/.
20. Patil JS and Sarasija S. Pulmonary drug delivery strategies: A concise, systematic review. *Lung India: Official Organ of Indian Chest Society.* Medknow Publications & Media Pvt Ltd, Jan. 2012, www.ncbi.nlm.nih.gov/pmc/articles/PMC3276033/.
21. Russell V. Four challenges for pulmonary drug delivery. *PharmTech Home.* 11 May 2018, www.pharmtech.com/four-challenges-pulmonary-drug-delivery.

42

Regulatory Consideration for Approval of Generic Inhalation Drug Products in the US, EU, Brazil, China, and India

Urvashi K Parmar[1] and Jayvadan K. Patel[2]
[1]SVKM's Dr. Bhanuben Nanavati College of Pharmacy, Mumbai, India
[2]Nootan Pharmacy College, Sankalchand Patel University, Gujarat, India

42.1 Introduction

Respiratory diseases like asthma and chronic obstructive pulmonary disease (COPD) are results of the prevailing air pollution over the globe. The American Lung Association, percieved from the data of 2016–2018, that the air quality was declining over the period. A paper published in 2020 by researchers at MIT claims that the west to east course of the wind carry half of the polluted air from the other side of the state boundaries (1). The presence of particulate matter floating in the air for longer duration exhibit harmful effects on the public health. The deeper locations of the infection in the lungs are indicative of severe health condition. Young children are prone to the ill effects of air pollution as their breathing rate (20–40 breaths per minute) is much higher than that of an average adult (16–20 breaths per minute), as a consequence to this babies are likely to inhale more pollutants than adults. Particulate Matter 2.5 ($PM_{2.5}$) is one of the causative agent contributing to respiratory and cardiovascular diseases. For the same reason mortality rate is much higher in urban populations. The major sources of air pollution include pollutants from stationary facilities (2–17%), ROG (precursor of ozone) (16–17%), as well as mobile sources including vehicles, and natural mishaps like wildfires. Mobile sources include NOx and CO whereas natural sources are $PM_{2.5}$ and ROG. SOx emissions (40%) come from stationary pollution. Electricity, oil, and gas production increase the SOx level. However, SOx emissions have been reduced in some parts of California lately (2).

The mean life expectancy of people in Europe is shortened by 2.2 years owing to air pollution, which accounts for per-capita mortality of about 133 in 100,000 deaths per year. The yearly mortality rate due to ambient air pollution in Europe exceeds 790,000 (95% confidence interval, CI) in the 27 countries of the EU (3). The fine particulate matter impairs vascular functions which can cause myocardial infarction, heart failure, and respiratory disorders. The burning of fossil fuels and biomass combustion leads to emission of particulate matters. Air pollution in the urban region of Radvanice is generated by industrial pollution (43% for PM_{10} and 27% for PM_1) due to burning of large metallurgical complex, and biomass (25% for PM_{10} and 36% for PM_1) (4). A significant proportion of Europe's population lives in cities where the decreasing air standards are posing grave health threats for the citizens. Particulate matter, nitrogen dioxide (NO_2), and ground level ozone (O_3) are recognized as the main pollutants affecting human health. Around 90% of urban residents in Europe are exposed to high concentration of pollutants.

The particulate matter $PM_{2.5}$ (particulates less than 2.5 microns in diameters) are considered hazardous to public health as they are capable of penetrating deep into the lungs. The WHO sets an average limit of PM_2 to 10 micrograms per cubic meter of air (0.01μg/m^3) and $PM_{2.5}$ to 25 micrograms per cubic meter of air (0.025 μg/m^3). Every year air pollution in Brazil causes deaths, particularly in the urban areas. Household air pollution (HAP) is also a contributory factor of fatal respiratory disorders. One in 26 deaths is ascribed to the air pollution, which makes air pollution the ninth largest mortality threat in the country. Breathing in smoke can cause tiny particles to enter the lungs resulting in chronic lung disease, acute respiratory infections, lung cancer, heart disease, and strokes. About 40% of the total Brazilian population is apparently breathing polluted air. A Farmácia Popular Program (FPP) launched by the Brazilian government for the welfare of citizens targeted elimination of asthma disease by offering free supplies of anti-asthmatic drugs. This attracted manufacturers into the generic sector and encouraged production of generic orally inhaled drug products (OIDPs). Drugs endorsed in Brazil for treatment of respiratory diseases include beclomethasone, fenoterol, fluticasone, fuorate, salbutamol, etc.

Constant development in the economy of China has led to air pollution adversely affecting the society. The Central and Western zones of China are worst affected by the ambient air pollution. In the year 2016, the Central Committee of the Communist Party of China and State Council released the Healthy China 2030 Plan, which emphasized on improving management of health-related environmental problems (5). However, in 2016 it was revealed that the number of cities attaining the national standards of air quality accounted only 84 out of 338 cities. The increased pollution has contributed to the pervasiveness of respiratory diseases of asthma, and COPD amongst the citizen, as a consequence of which has led to the

DOI: 10.1201/9781003046547-42

increase in usage of pressurized metered dose inhalers (pMDIs) in China. The essential national drugs list of China includes pMDIs of salbutamol and ipratropium bromide.

As per the study carried out in 2019, 21 cities out of the 30 most polluted cities in the world were in India. The pollution contributes to 2 million premature deaths every year. Emissions in urban areas are contributed by the automobiles and industries. Biomass burning for cooking (chulha), crop residue burning, tobacco smoke, and smog form the particulate pollution in rural areas. Along with these, there is wide usage of burning mosquito coils as insect repellent. Burning of biomass and firewood as cheap alternatives for light, and heat energy are widely adopted in rural India. The Asian brown cloud is a repercussion of air pollution, that is known to delay the monsoon. Adverse effects of air pollution causes acute respiratory infections, asthma, COPD, and exacerbation of pre-existing respiratory diseases such TB and lung cancer. Nearly three billion people around the globe contribute to almost one third (approximately 33.33%) of environmental air pollution by combustion of solid cooking fuels (6). HAP is cumulative effect of domestic combustion of biomass (fossil) fuels which include wood, coal, and dry cattle dung. Combustion leads to the formation of suspended particulate matter (SPM) and noxious substances like hydrocarbons, carbon monoxide, and oxides of nitrogen (NOx) (7). It may cause respiratory infections, chronic lung diseases, and respiratory cancers. Adverse effects of HAP in pregnant women include risk of premature delivery and post-neonatal infant mortality. The tracheobronchial trunk and lungs, being exposed to external environmental conditions, bear the major brunt of HAP burden. In Tirupur, South India, $PM_{2.5}$ concentrations of 3.8 mg/m^3 have caused respiratory illness among the low income household groups (Figures 42.1 and 42.2).

Generics are copies of (substitutes for) the innovator's product and are termed *second entry*, *test product*, or *subsequent product*. Drugs marketed by their chemical name rather than any brand name are referred as generic drugs (10), whereas the original products are termed *innovator's product*, *ethical product*, *reference listed drug* (RLD), and *reference product*. The patent holder or manufacturer, enjoys monopoly of the patented formulation until the expiry of the exculsivity conferred upon him, thereafter the generic industries can leap in. Generics are captivating attention of small scale manufacturers in the market. Multi-competitors in the market conduces to the substantial decrease in the prices of generic drugs. Generics sold at lower prices provide affordable medication that benefits a large number of patients all over the world. The regulatory systems demand therapeutic equivalency studies for the approval of generics. It is believed that if two drugs are therapeutically equal, then they have similar clinical efficacy and safety profiles, therefore, the test product and original or innovator's product must posses therapeutic similarities. The regulatory procedures required for ensuring the safety, efficiency, and quality of drug product might vary across different countries. The systems for approving generic drugs to enter a market for sale are similar all over the five regions, although there are some variations due to differences in culture and economic circumstances of each region. This chapter summarizes regulatory requirements in the US, EU, Brazil, China, and India for acquiring government approval to market generic inhalation drug products (Figure 42.3).

42.2 Regulatory Considerations in the US (FDA)

The US Food and Drug Administration (FDA) and Office of Generic Drugs (OGD) demands submission of an Abbreviated New Drug Application (ANDA) for acquiring approval for generic medicine. The ANDA submitted to the FDA's Center for Drug Evaluation and Research (CDER) OGD must contain the relevant data of bioequivalence and pharmaceutical equivalence studies [per 21 CFR 320.1 (e)] and [per 21 CFR 320.1 (c)] respectively (12). It must contain all the data required for FDA review and approval. After the enactment of the *Drug Price Competition and Patent Restoration Act* of 1984, also known as the *Hatch-Waxman Act*, the generic market developed greatly in the US. It has created opportunities for developing and marketing generics through the Abbreviated New Drug Application (ANDA). The FDA takes a minimum 18 months to approve an ANDA. Pharmaceutical companies should submit ANDAs to the

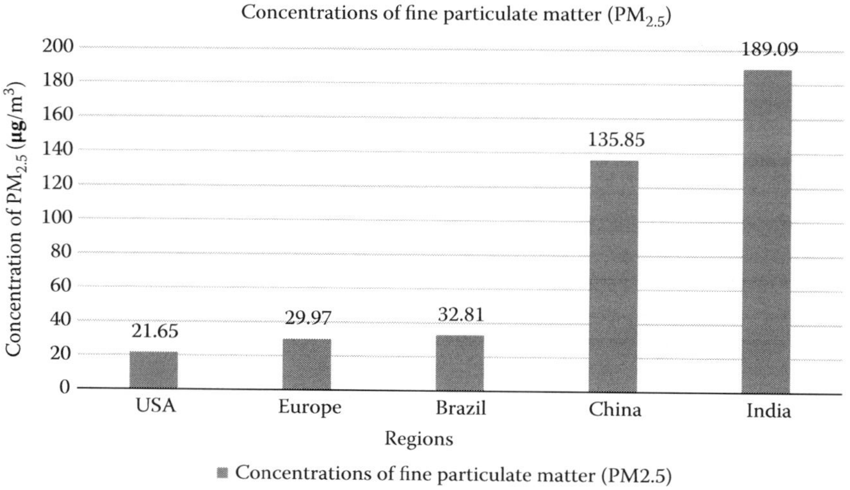

FIGURE 42.1 Concentrations of Fine Particulate Matter ($PM_{2.5}$) (8)

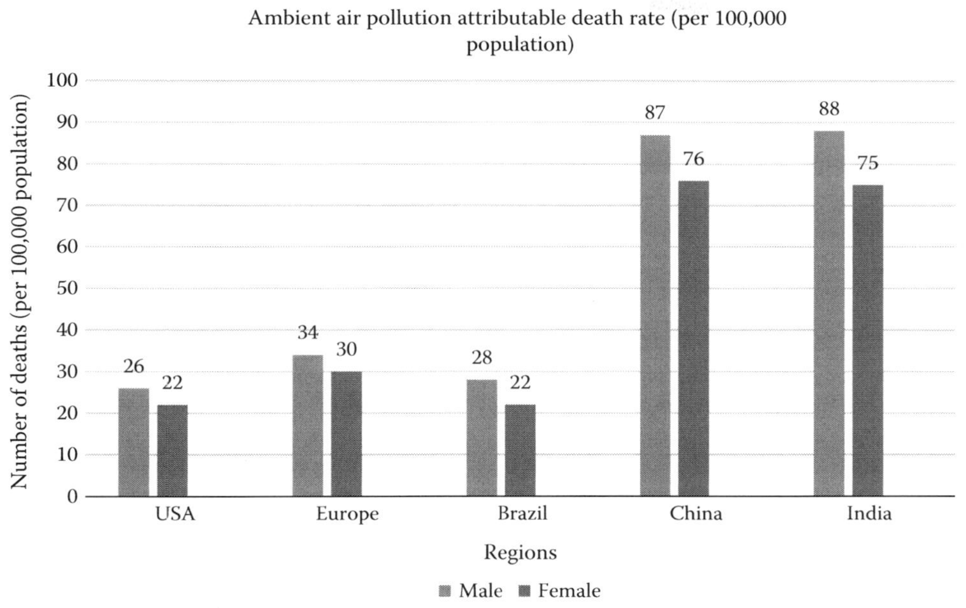

FIGURE 42.2 Ambient Air Pollution–Attributable Death Rate (per 100,000 Population) (9)

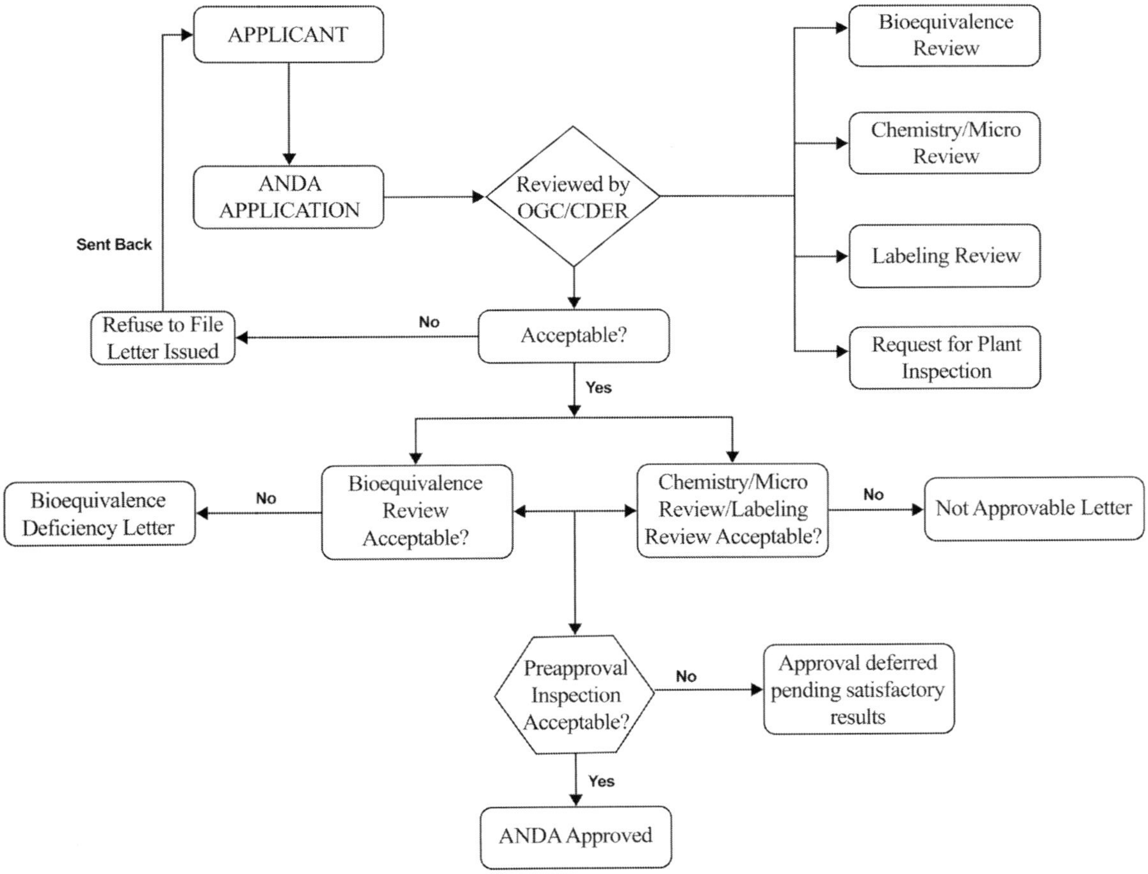

FIGURE 42.3 Generic Drug (ANDA) Review Process (11)

FDA to acquire approval before marketing new generic drugs [per 21 CFR 314.105(d)]. The FDA's Orange Book lists both innovator and generic products. After acquiring the approval, a pharmaceutical manufacturer can market a generic version of an innovator drug at a lower price. An eCTD is required along with the submission of the ANDA for acquiring approval. The dossier for the drug approval application is prepared by following the US FDA guidance (CFR) documents and FDA section 505(j) for ANDA. FDA recommends the applicant to provide patent information along with the expiration date for each patent, whether the RLD is protected by any exclusivity, and expiry date of the exclusivity. Exclusivity is a designed statutory provision for maintaining balance between innovator and generic drug manufacturers (Table 42.1).

Applicant submitting ANDA with a Paragraph IV statement gets a 180-day period of market exclusivity. If an infringement suit is filed against the generic applicant by the patent holder within 45 days of the ANDA notification, FDA postpones market approval of generic drug for 30 months, unless, before that time, the patent expires or is proven to be invalid or not infringed. This 30-month postponement gives time to the patent holder to assert its patent rights in court before a generic competitor is permitted to enter the market. An RLD company enjoys a period of monopoly only as long as the patent survives, which upon expiration removes the monopoly of patent holder (13) (Table 42.2).

The application is submitted by the applicant or any agent certified by the *Generic Drug Enforcement Act*. Administrative information along with cover letter, forms (356h), information of application, a field copy certification, debarment certification, financial certification, patent information, and exclusivity must be submitted. OGD encourages application in electronic format for submissions of bioequivalence, chemistry, and labeling data.

The FDA is aware about the obstacles and difficulties faced by the manufacturers to develop hard-to-copy generic drugs (specifically, inhalation devices). As an initiative to support development of complex generic, the USFDA is coming up with new and revised product specific-guidances (PSGs). Complex generic drug products are inclusive of complex active ingredient, complex formulation, complex route of delivery, complex dosage form and complex drug-device (metered dose inhalers). As per data updated on May 19, 2021, the planned PSGs might include inhalation drug products such as 1) aclidinium bromide; formoterol fumarate, (powder) metered dose 2) albuterol sulfate; ipratropium bromide, (spray) metered dose 3) formoterol fumarate; glycopyrorolate, (aerosol) metered 4) zanamivir (powder) 5) tobramycin (powder) 6) olodaterol hydrochloride, (spray) metered dose 7) olodaterol hydrochloride; tiotropium bromide (spray) metered dose. In August 2017, the generic drug user fee amendments (GDUFA II) were renewed. This committee works to solve the issues faced by the complex generics to establish BE (14).

On (January 30, 2019), first generic of Advair Diskus (fluticasone propionate and salmeterol inhalataion powder, two times a day) for the treatment of asthama in children above 4 years of age was approved. Mylan acquired the approval to manufacture and market the inhaler in three different dose strengths, such as 100 mcg, 250 mcg, and 500mcg of fluticasone, with dose of salmeterol as 50mcg in all three variations (15). Recently, (February 24, 2020), the USFDA approved ProAir HFA (albuterol sulfate inhalation aerosol, 90mcg/actuation). Perrigo Pharmaceuticals Company obtained the approval to market generic version of the said inhalation aerosol for treatment of COPD in children of age 4 years and above (16).

42.2.1 For Generic Inhalation Products

The demonstration of bioequivalence (BE) for inhalation products is quite challenging, so the FDA came up with a "weight of evidence" approach to establish BE for inhalation products and device considerations. FDA includes a step-wise approach listed as follows: (1) in vitro study, (2) pharmacokinetic study, and (3) pharmacodynamic or clinical endpoint studies for proving bioequivalence in OIDPs (Figure 42.4).

TABLE 42.1
Types of Exclusivity (11,13)

Exclusivity	Time for Exclusivity
Orphan drug exclusivity (ODE)	7 years for drugs intended to treat rare disease
Orphan drug exclusivity (ODE)	5 years for first time approval of new chemical entity
Other exclusivity (significant change)	3 years, changes may include new salt, new dosage form, new route, new indication, new strength, new dosing schedule
Pediatric exclusivity (PED)	6 months added to existing patents for exclusivity
Patent challenge	180 days

TABLE 42.2
Paragraphs for Patent Certification (13)

Paragraphs	Certification
Paragraph I	That the patent information has not been submitted to FDA
Paragraph II	That the patent information has expired
Paragraph III	The date on which the patent will expire
Paragraph IV	That the patent is invalid, unenforceable, or will not be infringed by the manufacture, use, or sale of the drug product for which the ANDA is submitted

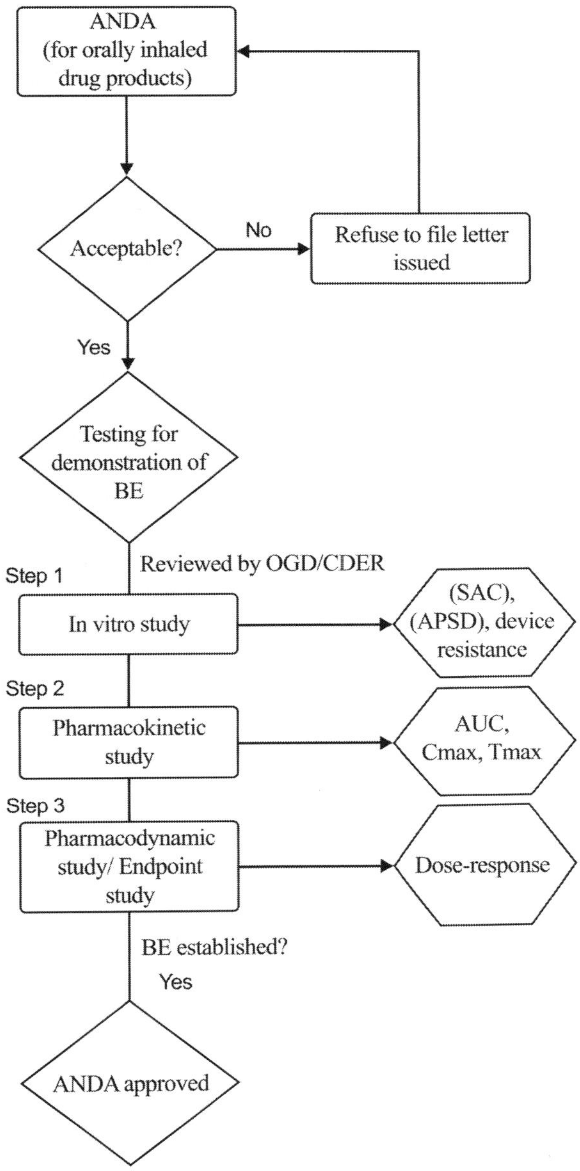

FIGURE 42.4 ANDA Approval for Orally Inhaled Drug Products

42.2.2 In Vitro Studies

The generic product and the innovator's product must be therapeutically equivalent. The in vitro studies for DPI involves single actuation content (SAC) and aerodynamic particle size distribution (APSD) tests. In vitro testing must involve flow rates resembling to the inspiratory flow rates of patient population for determining the SAC and APSD. The results obtained from the in vitro studies are more sensitive and less variable when compared to in vivo studies. Dry powder inhalers (DPIs) consist of a reservoir compartment containing formulation of the drug products, available as multi-dose products or pre-metered single dose blisters. The material used for the device container may not necessarily be the equivalent to that of the inovator's product, which might affect the DPI's in vitro performance owing to the repeated use. Hence it is mandatory to test SAC and APSD for equivalence at various stages of product life span (start, mid, and end stages). The air flow resistance of test and reference DPI must be comparable. Passive DPIs are susceptible to variable flow rates as they are usually breath-actuated. The inflow rate depends upon the range of patient's inspiratory extent. Therefore, it is mandatory to test SAC and APSD equivalence at various flow rates (at least three flow rates). These three flow rates should resemble the flow rates of the patient population. The targeted patients must be able to operate the test device without any major variation in their inspiratory effort.

42.2.3 Pharmacokinetic Study

The pharmacokinetic (PK) studies for solid oral dosage forms and inhalation products are quite similar. OIDPs are used for lung-targeted drug delivery. However, some proportion of the drug swallowed can reach systemic circulation through the gastro-intestinal tract. Measuring the drug concentration in blood or plasma can give the relevant BE of DPIs. The study is based on administering a dose in healthy volunteers/subjects because the data obtained from them are less variable and give more sensitive data for detection of minute differences between the characteristics of drug products, whereas the data obtained from the patients can be variable due to their diseased condition. The dose is selected such that the actuation/inhalation is minimum ensuring the assay sensitivity. The bioequivalence between the test product and the innovator's product are compared by measuring the area under curve (AUC) and Cmax. If 90% CI of the geometric mean ratio for the observed AUC and Cmax falls within a range of 80–125%, then the two inhalation devices are considered to be PK-equivalent. The data of the peak concentration can be used for measurement. Pharmacokinetic BE study is mandatory for OIDPs having multiple strength formulations. The in vitro and PK studies do give considerable weight of evidence to determine the equivalency of the product performance, although the concentration of drug in the local sites remains unknown. Hence, limitations of in vitro and PK studies for determining bioequivalence of the inhalation products are perceptible.

42.2.4 Pharmacodynamic study

Pharmacodynamic (PD) studies are deemed necessary to support the BE of locally acting drugs, it provides greater sensitivity in detecting differences between two drug products. The differences in the design factors of the innovator's and test devices must be minimum to ensure switchability from the patient's point of view. The design factors include shape, size, breath actuation, premetered multidose or device-metered multi-dose, operating principle, dimensions. However, interior design, external appearance, and operating principles may differ considerably. Inactive ingredients can also affect the product performance and product safety, hence, it is necessary for the test drug product to be qualitatively equivalent to the reference drug product. In circumstances where the reference devices are proprietary to the innovator or patent protected, the drug-to-excipient ratio can prove to be one of the most important design variables for adjusting the equivalency of the test product to that of the reference product. Achieving a proper drug-to-excipient ratio helps in establishing equivalence. The device differences should be rationalized by suitable in vitro and in vivo BE studies to confirm safety and efficacy.

For dry powder inhalations the formulation of a test DPI might differ quantitatively but must not differ qualitatively. The DPI performance can be affected by both formulation and device characteristics.

In order to facilitate development of high-quality generic drug products, the FDA sponsored a research study for investigating the capability of PK studies to determine the fate of inhaled drug products as a part of the Generic Drug User Fee Amendment (GDUFA) (10,19).

42.2.5 Other Recommendations

According to FDA's 21st code of federal regulation (21 CFR 200.51), aqueous oral inhalation solutions and suspensions should be sterile and must be used with a specific nebulizer (23). The application must include the details of the drug product, including its active ingredients, excipients, and purity of the drug substance. The drug product must be free from any impurities like residual solvent, organic impurities such as degradation product, inorganic impurities such as heavy metals, and reagents. Monograph must be attached with data of particle size distribution, nature of the drug (crystal or amorphous form), and spray content uniformity. The reproducibility of an in vitro study can be ensured by controlled actuation parameters like length of the stroke and force applied for actuation. Evaluation of the pump performance is inclusive of tests like spray pattern and plume geometry. Pump design, size of metering chamber, size and shape of the nozzle are some of the factors affecting spray pattern. During evaluation of spray pattern, parameters such as the gap between the nozzle, surface to be sprayed on and position of surface to that of the nozzle should be noted. The acceptable shape of the spray pattern is ellipsoid. Spray pattern must be tested at various stages of the pump product's life, whereas plume geometry evaluation is carried out only once during the characterization of product.

A cascade impactor (CI) is a device that collects droplets/particles through series of multi-stage impaction. This device helps to measure the particle size, droplet size fraction of the dose, mass balance of total dose, and size distribution throughout the drug product. The device requires calibration of the flow rate, duration of flow, size and shape of expansion chamber or inlet stem, and adaptor for incorporation of inhalation spray into the specified impactor. The number of sprays should be kept minimum to justify the analytical procedure used to measure the deposited drug substance. The amount of drug substance collected on the surface of the collector is a sufficient assay.

There is strong need to reduce the time duration and cost for the development of the generic OIDPs. The computational fluid dynamics is an in silico-based modeling that helps to determine parameters of fluid and particle in the drug product. The CFD is capable to determine device related characteristics rapidly. It is capable of predicting the influence of the OIPD device on regional lung deposition. The CFD can be coupled with the PBPK modeling to predict drug absorption from the site of action. The in silico modeling of OIPDs has received funding from the OGD in the CDER before 2018 by utilization of funds acquired from the GDUFA regulatory program.

42.3 Regulatory Consideration in EU

The Committee for Medicinal Products for Human Use (CHMP) of the European Medicines Agency (EMA) is the regulatory body for the 27 countries of EU. The legislation of the European Union states that a product application can be declined if it reckons to be a potent risk for the patient population. The generic drug products are considered harmful if they fail to demonstrate equivalent studies. The swapability or switchability of innovator products with test products is completely decided by the national authority of individual European countries, although the failure in proving switchability cannot be the reason of rejection for authorization. The EU works in a two-step process for dealing with the approval of generic inhalations.

In the first step, an EU-level decision is made whether the product is capable for a prescription, after considering the benefit–risk balance. In the second step, the EU decides if the product is interchangeable with the innovator's products in market. Nevertheless, irrespective of their approval as generics, they cannot be considered as switchable. The generic products if not equal to the innovator's product in every aspect, should be bioequivalent to the innovator's product on the grounds of pharmacokinetic (PK) and bioavailability (BA) studies. The products which are proven equivalent via PD or clinical endpoint studies are classified as hybrids. The submission of an application for approval of locally acting orally inhaled drugs involves the PD and endpoint studies for proving equivalency, hence, they are termed as hybrid applications. The patient and the prescriber are not informed whether a product is a generic or a hybrid. The PK studies for locally acting products have been proven to be sufficient, therefore in vitro data for PK studies are accepted in place of PD studies. Hence, the term *generic OIDPs* does not exist in EU regulations/definition. For OIDPs, the approach is similar to that of systemically acting products. In vitro approaches may be considered adequate for demonstration of therapeutic equivalence along with the second-step of PK studies to ensure the safety and deposition of the drug in the lungs. If these two steps are insufficient to prove bioequivalence then the third step of PD or clinical end studies can be conducted for demonstrating therapeutic equivalence in locally acting drugs. Although systemic exposure cannot be considered for determining bioequivalence of locally acting products because drug plasma level can be a result of gastric absorption of drugs due to subsequent swallowing. We cannot deny the fact that the drug concentration in plasma can be a result of downstream of the drug from the site of action. In the events where swallowed fractions are completely stopped or blocked by charcoal blockade, the plasma drug level can be indicative of systemic absorption from the site of action. Therefore, the Committee for Medicinal Products for Human Use (CHMP) European Medicines Agency (EMA) guidelines accept plasma concentration as an indicator of concentration of drug at the site of action. The plasma level can be more sensitive than the PD and clinical endpoint studies.

42.3.1 Step 1: In Vitro Studies

For minor variations between the test and the reference products, in vitro studies are considered enough for proving equivalence. A deviation of range of ±15% is stringent for analyzing the APSD which is based on both individual and group levels. APSD can be assessed by multistage impactor/impingers. Andersen Cascade Impactor is viable currently, however the EU Guidelines does not specify usage of any particular impactor. The comparison study must be carried out in four groups of stages. Grouping of the stages accord less sensitive data, but is acceptable in some EU countries. The use of multi-stage liquid impinger is also described in the European pharmacopoeia. The particle size larger than 5μm are not inhalabale and hence incapable of reaching the lower respiratory tract. The particle size of 3mm are found to well desposited in the lungs. The difference between the in vitro values of the test and reference products should not exceed ±15% (18). The occurrence of inter-batch variations is inevitable since there is possibility of apparent variability even in the same batch of the same reference product. This can be decompensated by increasing the number of sample tests and considering the mean value for the reference and the test product batches. EU regulators suggest three or more batches to compensate for the inter-batch variability.

The check for similarity can be conducted on the basis of viscosity and pH for solutions with identical composition for nebulization. If the viscosity and pH of the test product differed from the innovator's product then, in vitro testing is becomes mandatory. Solutions of the nebulizer of the two drug products if found to be qualitatively as well as quantitatively equal, then the approval may be granted soon.

The inhaled volume and resistance of the device is not crucial for the pMDIs. The handling of the devices must be similar even if there is difference in shape and size of the device. The important test for pMDIs are the plume dynamic, plume geometry, and APSD of the droplets. In Spain, approval has been granted for solution or suspension for nebulization of drugs such as salbutamol and budesonide on the basis of in vitro study solely. However, variation in the shape of the mouth adaptor might lead to greater deposition of nebulized particles in the throat and lesser deposition in the actuator (18). In vitro studies must include droplet size distribution study by considering differences in hygroscopicity and nebulizer efficiency, that might vary due to surfactants or preservatives. For suspensions in nebulization, the similarity in crystallography and particle size distribution is necessary. Particle size distribution in suspension is similar to the aerodynamic particle size. The pMIDs in solution or suspension can be compared by in vitro tests for nasal and oral products as per FDA guidance for industry in some countries like Spain. The EU accepts FDA guidelines for nasal products, however it requires average bioequivalence in lieu of population bioequivalence (PBE). The range of ±15% is considered acceptable.

42.3.2 Step 2: Pharmacokinetic Study

The AUC reflects the dose reaching the site of action (lungs). The only fraction removed is the fraction removed by mucociliary clearance. This fraction is however, not available for absorption, hence, shows no effect on plasma drug concentration. Determination of pattern of deposition is also important. The pattern of lung deposition contributes to the efficacy but cannot be considered sufficient to prove equivalent local efficacy and safety. Cmax and Tmax are capable of providing data for the pattern of lung deposition. The systemic absorption of drugs from peripheral lung deposition site is quick and has short Tmax, whereas the central part of the lungs requires more time for absorption. Nature and physical state of the drug administered (solid or solution, drug lipophilicity, dissolution rate) has impact on Cmax and Tmax.

The time taken by a drug to reach systemic circulation via gastric route is longer compared to the pulmonary route, hence, the drug concentrations can be differentiated by carefully observing the drug–plasma profile. PK studies can be considered sufficient as the lung acceptance of the drug is faster than the gastric absorption. Although charcoal blockade is validated to demonstrate negligible oral contribution. The study is carried out with use of charcoal blockages which obstructs the swallowed fraction of the drug from reaching systemic circulation. The active charcoal blockade must be able to block the GI absorption up to <5% of the drug intake. The efficacy profile of reference and test products shows close similarity if the lung depositions in both cases are equivalent. For bronchodilators, suitable bronchodilation and broncho-provocation models are used for conducting BE studies.

42.3.3 Step 3: Pharmacodynamic Studies

In contrast to the weight-of-evidence approach in the US, the EU considers in vitro data and PK studies as sufficient to establish bioequivalence. EU reviewers place low emphasis on dose scale and response scale analysis. In cases where combination of two bronchodilators are used, the endpoint cannot be ensured by PD study, a suitable endpoint study should is a must. For drugs that lack assay sensitivity, PD studies fail to detect variation between the drug products (e.g. salmeterol and some corticosteroids), however, the approval of such drugs is done by step 1 and 2. The bronchodilation and the broncho-provocation models are available for testing the air flow. Broncho-provocation protocols are presently available for testing short acting beta agonists (SABAs).

Most countries prefer duplicating the entire product development for pediatric drugs intended for children. Other countries are satisfied with PK studies, while some countries test equivalence on adults. Some EU countries consider assay sensitivity to be of great importance while EMA guidelines consider both dose scale and response scale analysis, however, it might be limited to response-scale analysis for some of the EU countries for new chemical entities (Figure 42.5).

42.4 Regulatory Consideration in Anvisa (Brazil)

Brazil's pharmaceutical regulatory government agency for regulatory affairs is known as the Agência Nacional de Vigilância Sanitária (Anvisa).

In 2009, Anvisa specified that demonstration of bioequivalence requires in vitro tests and PK studies that can be supported by clinical studies. In 2011, Anvisa again came up with a set of rules. The law issued under 13.411 in 2016 mandates a one-year frame time for reviewing nonproprietary drugs and it also contained the requirements of the technical note of 2013 (20). They developed a Resolution and Normative Instruction for public consultation in 2018, which was later approved in 2019 (20). Anvisa recommendations are published in the Technical Note, which is available on their website. The standards for OIPDs can be found in the US Pharmacopeia (USP), and European Pharmacopoeia (EP). This is assessed by the Therapeutic Equivalence Committee. It is necessary to submit protocols to Anvisa for evaluation, prior to the conduct of study. The International Pharmaceutical Aerosol Consortium on Regulation and Science (IPAC-RS) is an association seeking advancement in the regulatory science of orally inhaled and nasal drug products (OINDPs) by collection and analysis of data.

A pharmaceutical equivalence study is defined as a physical and chemical test proving the two products as pharmaceutical equivalents. Anvisa requires Population Bioequivalence (PBE) method for determining pharmaceutical equivalence (PE). . Regulatory review is achieved only after the PE test done by the Anvisa-certified lab. The PE of an inhalation product is demonstrated by meeting the compendial quality requirements by the test product; identification, impurities, and microbial content must meet acceptance criteria.

42.4.1 In Vitro Studies

The generic product must contain the same API and similar dose, dosage form, excipients, polymorphic profile, and device handling characteristics as that of the reference product. For metered-dose inhalers (MDIs) automatic actuations are mostly preferred; for dry powder inhalers (DPIs) the mouthpiece shape must be examined. The humidity consideration must be included in APSD evaluation. The standard of identity, strength, quality, and purity must be equal to that of the reference product. Generic OIPDs must contain the same dose, excipients, polymorphic form, dosage form, and device characteristics. Companies seeking approval for a generic must submit protocols prior to the in vitro and in vivo studies.

A complete description of tests for activation of a device must be submitted to Anvisa before conducting any test. The documentation must include device activation instructions. Information regarding the actuator must be given to Anvisa

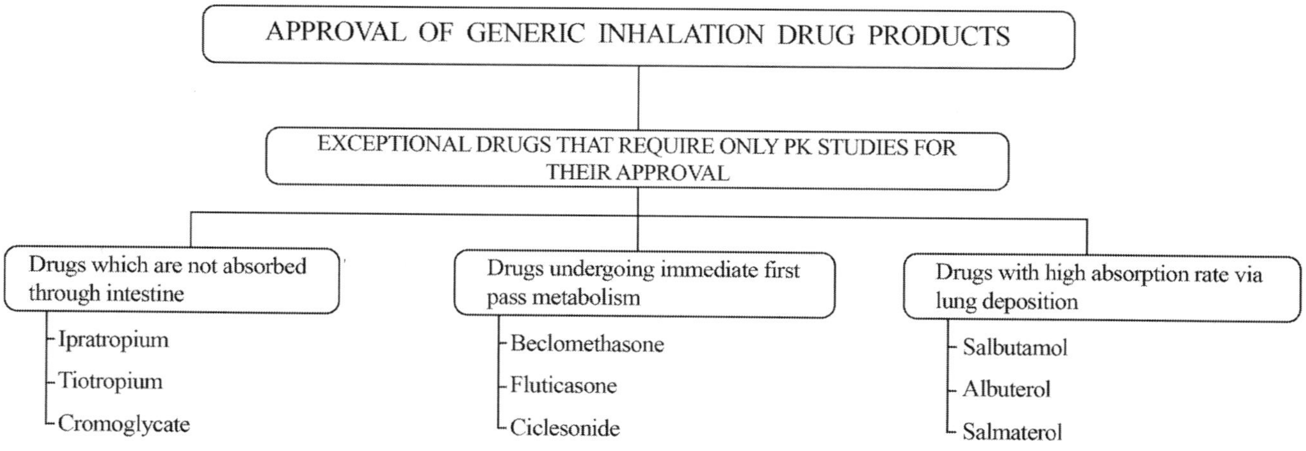

FIGURE 42.5 List of Exceptional Drugs

and must include the description of automatic operation system, performance parameters controlled by it, and acceptance criteria, if any. Automatic actuated systems are prohibited from conducting tests of priming, pre-priming, dose content uniformity, and spray pattern.

As we know, the differences in handling the device by patients can give rise to varied therapeutic performance of the device. Variation depends mainly upon characteristics such as age, gender, disease, and breathing cycle of different patients. Anvisa states that the application must include answers to the questions regarding operation mechanism of the device and contain information about the difference in patients' efforts and handling due to its shape and dimension.

42.4.2 In Vivo Studies

Along with the in vitro studies, the submission requires in vivo studies that demonstrate the pulmonary deposition. This can be achieved by inhibiting the gastrointestinal absorption of drug via activated charcoal. The PK parameter of plasma concentration versus time curve relates to the site of pulmonary deposition and absorption. The legibility of the test, can be confirmed by a similar PK study test without the inhibition of gastrointestinal absorption (via activated charcoal).

APSD for OIPDs can be assessed by CI. The apparatus differs from the in vivo condition as flow rate in a CI test is constant, in contrast to the actual physiological environment where occasions of deposition in oropharyngeal regions as a result of humidity and anatomy might be possible. Generic OIDPs in Anvisa previously required specification of APSD in terms of fine particle dose, fine particle fraction, fine particle mass, mass median aerodynamic diameter, and geometric standard deviation for generic approval. The new normative instruction demands fine particle dose, fine particle fraction, and evaluation of dose content uniformity throughout the product life.

42.4.3 Pharmacokinetic Study

Pharmacokinetic studies provide data of dose available at the site of action (fate of drug) and its residence time (Cmax). AUC parameters are considered adequate to detect variation between lung deposition of the two drug products. Pulmonary deposition can also be detected by gamma scintigraphy, positron emission tomography (PET). The deposition pattern and the time for which the drug remains at the site must not differ between the two drug products that are being compared. The 90% CI of the ratio of geometric mean of the area under the curve ($AUC_{0\text{-}t}$) and maximum concentration (Cmax) of both test product and the reference product should fall within the interval of 0.8–1.25, to confirm the safety and efficacy of the test product.

PK studies are not possible for substances with low assay sensitivity. In this case, a suitable PD/endpoint study is preferable. Biomarkers are directly related to the efficacy of the product but, a dose response should be checked for the marker, only then the study will be sensitive enough to determine differences between the two drug products. If the generic drug product does not meet the similarity criteria, it cannot be registered as a generic and has to follow a full clinical trial for acquiring government approval.

Dissolution rate helps to determine the mucociliary removal of drug from lungs, which makes it necessary for the reference and test products to have similar dissolution rates for demonstration of bioequivalence. Dissolution rate determination faces technical difficulties due to the sub-micron size of the drug particle, density of the particles, and floating sedimentation. As a solution to these problems, a suitable simulated lung fluid (dipalmitoylphosphatidylcholine) is used during dissolution testing. A Transwell membrane apparatus solves floating issues. Characteristics such as dissolution rate, lung deposition pattern, and mucociliary clearance may affect the PK plasma parameters.

42.4.4 Pharmacodynamic Study

Pharmacodynamic studies in Brazil involve conducting tests on healthy groups and determining physiological effects. Anvisa recommends that an administered dose must be in steep dose-response curve in order to observe differences between the two drug products. For evaluation of broncho-protection, the bronchoconstrictor drug metacholine or heavy exercise can be implemented for bronchoconstriction in asthmatic patients. The bronchodilation capacity of both drug products after bronchoconstriction should be similar. Although, the results obtained are highly variable due to differences in broncho-constriction stimulation. Other methods for PD evaluation such as sputum eosinophilia and exhaled nitric oxide (eNO) for testing inhaled corticosteroids is controversial because of their limitations. Limitation of sputum eosinophilia is that enough sputum must be produced for the analysis, which may not be possible for all patients.

There are various advantages and disadvantages of biomarkers and endpoints studies for determining product efficacy. Dose–response relationships in SABAs and long acting beta agonists (LABAs) can be established by spirometry although highly variable responses might be obtained.

42.5 Regulatory Consideration in China

In China, the China Food and Drug Administration (CFDA) previously known as SFDA, is responsible for the regulation and approval of the drug for market. The CFDA comprises 19 departments and bureaus and 18 affiliates. Among these, a few are directly related to drug regulation and approval, including the Center for Drug Evaluation (CDE), National Institute for Food and Drug Control (NIFDC), Center for Certification of Drugs (CCD), Chinese Pharmacopoeia Commission (CPC), and Center for Medical Device Evaluation (CDME) (21,22).

The registration process and regulatory requirements for approval of orally inhaled drug product is similar to that of any other chemical entity. CFDA issues the principles and technical guidelines for the research of chemicals with existing natural standards. The chemical drugs from other countries entering into the Chinese market are classified into six categories. Generic drugs already been sold in other countries are treated as new drugs. They are included in category III drugs. A Clinical Trial Authorization (CTA) for conducting clinical trials (in the

Chinese population) is obtained from the CFDA. In case of an emergency requirement of certain drug, the Chinese population (patients/healthy volunteers) takes part in the clinical trial of a global development process for NDA/ANDA. Generic drugs having national standard in China are considered category VI drugs. The requirement for approval of a generic drug application involves owning a GMP facilitated commercial industrial area that is capable of manufacturing and handling generic drug product. Before the CFDA initiates review of the dossier, three batches of products are manufactured in the presence of local FDA and undergo lab testing in official labs. The CDE is responsible for reviewing the dossier. The waiting period for a generic drug approval in China can be as long as 6 years. However, for drugs having significant clinical value, a fast-track mechanism has been declared by CFDA (19) (Table 42.3).

The CFDA and CDE requirements for bio-equivalency between the reference and test products is inclusive of measuring the deposition of drug at the pulmonary site and systemic drug concentration via pharmacokinetic (PK), pharmacodynamic (PD), and clinical trials (CT).

42.5.1 In Vitro Study

In vitro studies might be sufficient for proving therapeutic equivalence. Biowaiver can be applied by the applicant although the agency has not granted any biowaivers to date. APSD and content uniformity are two important aspects of the study.

42.5.2 In Vivo Study

For cases where in vitro study is insufficient for proving the equivalence between the test and reference products, an in vivo study is expected. In vivo studies must follow the method described below.

42.5.2.1 Method 1: PK Study

Lung deposition comparative study using active charcoal as blockade for downstream of drug from the local site of action can be performed. The drug concentration in the systemic circulation can determine the lung deposition. A single-dose PK study is sufficient.

42.5.2.2 Method 2: PD + PK/EP

If the PK alone isn't enough, a PD study must be conducted for the approval of OIDPs. For bronchodilators, such as short-acting and long-acting beta2 adrenergic receptor agonists (SABA, LABA), a suitable broncho-provocation and single dose bronchodilation might be performed to demonstrate therapeutic equivalence (similar systemic exposure). For inhaled corticosteroids a double blind controlled clinical trial should be performed. Although there is no recommendation available. Hormones can be monitored during trial to understand the effect of a test drug (e.g. cortisol on hypothalamic-pituitary-adrenal). In cases of combination drugs, therapeutic equivalence of all drugs is required. Generic drug intended for two diseases together, such as asthma and COPD, receives approval simultaneously once the equivalency is established for asthma.

42.5.3 For Category VI

Contains generic drugs with reference product approved in other countries but not in China. A drug that has been introduced for the first time in China is considered a new drug and must follow the NDA process, as it is not possible for the new generic drug to be compared with the innovator's product. The clinical trials are performed to determine the drug safety and efficacy. Thetests must be conducted on at least 100 patients from China.

42.6 Regulatory Consideration in India

Drugs and Cosmetic Act of 1940 and 1945 established back in the days of British rule over India are the main legislations governing drug registration, manufacturing, distribution and sale in India. A legislative body called Central Drugs Standard Control Organization (CDSCO) comprises of six zonal, four sub-zonal and 11 airport offices with six laboratories to carry out the test and researches. Drug Controller India (DCI) is designated as the office controller. The CDSCO and Ethical Guidelines for biomedical research on human have issued guidelines for bioavailability, bioequivalence studies, good clinical practice (GCP). The applicant of the Generic drugs must prove the bioequivalency and bioavailability in order to obtain approval from the CDSCO. The clinical trials in India are regulated by the Rule 122A to E of Drug and Cosmetic Act. A duly registered Contract Research Organization (CRO) allows the manufacturer to carry out clinical practices in the country. The registration should be renewed within every 5 years.

In India, OIDPs (nonsolution) that are intended for local action in lungs, a bioequivalence based PK study is not adequate. In such cases, clinical trials or PD studies are required for the establishment of equivalence. A good correlation established

TABLE 42.3
Category of Registration for Imported Drugs in China

Category	Category Details
I	Drugs not marketed in Country of Origin or in China
II	Drug preparations with changed route of administration not marketed in China or Country of Origin
III	Drugs marketed in Country of Origin but not in China
IV	Drug preparations with changed acid or alkaline radicals (or metallic elements),with no pharmacological change, where original drug entity is already approved in China
V	Drug preparations with changed formulation but no change in route of administration, where the original preparation is already approved in China
VI	Drug substance or preparation following China's national standard

between AUC and Cmax in PK studies paves a path for establishment of BE. The measurement of $AUC_{0\text{-}t}$ denotes the concentration of available drug in pulmonary site must be equal for the two drug products being compared.

42.6.1 For Reference Drug Approval in India

Pharmacokinetic study is not considered sufficient, therefore a supporting PD is necessary. If both the drug products contain the same active ingredient with the same concentration and same excipients, a PK bioequivalence study is sufficient. If the test product has same active substance with the same concentration as that of the reference product, then there lies no necessity for a clinical/in vivo study. If the test device is not similar to the reference device, a suitable in vitro study is mandatory for demonstration of bioequivalence.

42.6.2. For Reference Drug not Approved in India

The drug product is considered new under the following circumstances:

1. Drugs which are not used in India or not used for more than 4 years post approval.
2. Drugs with different indication.
3. Combination drugs that were approved as individual drugs.

42.6.3 Process Consideration

The Subject Expert Committee (SEC) and the Technical Review Committee (TRC) are authorized for approval of the study designs. The process usually takes 180 days and is completed in two stages.

Stage 1: The sponsors or designee are informed about any discrepancy in the dossier at the time of screening.

After being accepted, the dossier is sent to the SEC for review. The SEC comprises 10 experts from recognized college and government hospitals having keen knowledge of pulmonary diseases. The meetings of the SEC are held quarterly on a predetermined date chosen by the regulators. The sponsor must present their protocol to the SEC and their recommendations are uploaded as minutes of a meeting on the official CDSCO website within two weeks. The clinical trial applications are then sent to the second stage of TRC.

Stage 2: The function of the TRC is to check whether the sponsor qualifies on the basis of CV, medical registration certificate, registration of ethics committee, active involvement of sponsor in trial, and insurance policy for compensating death or injury in the course of clinical trial. The trial designs are then sent to DCGI for approval in the form of a letter. An investigator can only be involved in three studies at a time, be it a pharmaceutical company sponsored or academic purpose clinical trials. The sponsor must be associated with a hospital having facilities of at least 50 beds. They must register their trials on the Clinical Trials Registry of India.

The responses are to be measured and recorded under double-blind conditions availing objective evaluation. The design baseline must be reproducible and a cross-over or parallel study shall be performed. Maximal responses should not be produced by any of the products during the study, as it renders complications for recognizing variations between the two formulations. A dose–response relationship then becomes necessary. The non-responding class of volunteers should be removed from the study via proper screening. The method for identification of responders and non-responders must be included in the protocol. The acceptance range for PK studies and BE are distinct to that of the PD studies. The protocol must include the acceptance range on the case basis. Only if there is absence of suitable PD measures, the clinical trials must be performed.

42.6.4 Summary of Regulatory Requirements for Approval of Generic Inhaled Drug Products

42.6.4.1 In Vitro Study

An in vitro study for demonstration of bioequivalence is considered necessary for the first step in establishment of equivalence in the US, Europe, and Brazil. APSD parameters in the study are sufficient for BE in China. APSD study parameters are sufficient for BE in China. In vitro studies for aqueous solutions with similar drug and device considerations are sufficient for establishing bioequivalence in India.

42.6.4.2 Pharmacokinetic Study

A pharmacokinetic study involves BE demonstration in healthy volunteers; 90% confidence interval of the geometric mean ratio must fall within 80.00–125.00% in the US and EU. A PK study includes usage of a charcoal blockade in the EU and Brazil. In China, the PK study must be conducted in the Chinese population. Granting of Approval on basis of PK study is not viable in India.

42.6.4.3 Pharmacodynamic Study

A pharmacodynamic study is preferred over an endpoint study in the US. Failure of a PK study to demonstrate BE requires a PD or endpoint study for product approval in the EU and Brazil. Dose–response is the requirement in the US and Brazil. Cross-over or parallel design is acceptable in the EU and India. A broncho-provocation test is required unless a biowaiver is obtained in China.

42.6.4.4 Qualitative and Quantitative Equivalence

The similitude between inactive ingredients of both drug products is necessary in the US and Brazil, but it is not necessary in EU and China. Device similarity is recommended in the US, EU, and Brazil, although not necessary in China and India.

42.6.4.5 Lung Imaging

Lung imaging is considered inadequate for determining BE in the US and Brazil, whereas it is used for demonstrating lung deposition in the EU and China.

42.6.4.6 *Clinical End-Point Study*

A clinical end-point study is carried out if a pharmacokinetic study (PK) and pharmacodynamic study (PD) fail to demonstrate BE in all the five regions.

42.6.4.7 *Statistical Methods*

Statistical methods require PBE for in vitro studies in countries like the US, EU, Brazil, and India. Data for the statistical method is not available in China.

42.7 Conclusion

Regulatory requirements and scientific recommendations for approval of generic inhalation drug products in the five regions were discussed. The overall steps involved in the process for demonstration of bioequivalence are similar over the five regions (US, EU, Brazil, China, and India). Variations can be seen for some parameters. More recommendations are still required for establishing a specific regulation for generic inhalation drug products.

Abbreviations

OIDPs, Orally Inhaled Drug Products; pMDIs, Pressurized Metered Dose Inhalers; RLD, reference listed drug; DPI, dry powder inhalation; SAC, Single Actuation Content;APSD, Aerodynamic Particle Size Distribution; PBE, Population Bioequivalence; MDI, Metered Dose Inhaler; SABA, Short Acting Beta Agonist; LABA, Long acting beta agonist.

REFERENCES

1. Half of U.S. deaths related to air pollution are linked to out-of-state emissions. *MIT News | Massachusetts Institute of Technology.* https://news.mit.edu/2020/half-us-deaths-air-pollution-out-state-0212.
2. Anderson CM, Kissel KA, Field CB, and Mach KJ. Climate Change Mitigation. Air Pollution, and Environmental Justice in California. *Environmental Science & Technology.* 2018; 52: 10829–10838.
3. Lelieveld J., *et al.* Cardiovascular disease burden from ambient air pollution in Europe reassessed using novel hazard ratio functions. *European Heart Journal.* 2019; 40: 1590–1596.
4. Kozáková J. *et al.* The influence of local emissions and regional air pollution transport on a European air pollution hot spot. *Environmental Science and Pollution Research.* 2019; 26: 1675–1692.
5. Liu W, Xu Z, and Yang T. Health effects of air pollution in China. *International Journal of Environmental Research and Public Health.* 2018; 15: 1471.
6. Khilnani GC and Tiwari P. Air pollution in India and related adverse respiratory health effects: Past, present, and future directions. *Current Opinion in Pulmonary Medicine* 2018; 24: 108–116.
7. Jindal SK, Aggarwal AN, and Jindal A. Household air pollution in India and respiratory diseases: Current status and future directions. *Current Opinion in Pulmonary Medicine.* 2020; 26: 128–134.
8. Concentrations of fine particulate matter (PM2.5). https://www.who.int/data/maternal-newborn-child-adolescent/monitor.
9. GHO. By category. Deaths - by country. *WHO.* https://apps.who.int/gho/data/node.main.BODAMBIENTAIRDTHS?lang=en.
10. Rafi N, Ds S, and Narayanan V. Regulatory requirements and registration procedure for generic drugs in USA. *Indian Journal of Pharmaceutical Education and Research.* 2018; 52: 544–549.
11. Shargel L and Kanfer I. Generic drug product development: Solid oral dosage forms. *Drugs and the Pharmaceutical Sciences - Book Series...https://www.routledge.com › IHCDRUPHASCI.* 2005; 389.
12. Guarino RA. New Drug Approval Process. 581. 2009. CRC Press, 552 Pages. ISBN 9781138044937
13. ANDA Submissions — Content and Format of Abbreviated New Drug Applications. 32. https://www.fda.gov/regulatory-information/search-fda-guidance-documents/anda-submissions-content-and-format-abbreviated-new-drug-applications
14. https://www.fda.gov/drugs/guidances-drugs/product-specific-guidances-generic-drug-development
15. https://www.fda.gov/news-events/press-announcements/fda-approves-first-generic-commonly-used-albuterol-inhaler-treat-and-prevent-bronchospasm
16. https://www.fda.gov/news-events/press-announcements/fda-approves-first-generic-advair-diskus
17. A European Perspective on Orally Inhaled Products: In Vitro Requirements for a Biowaiver. DOI: 10.1089/jamp.2014.1130
18. In Silico Methods for Development of Generic Drug–Device Combination Orally Inhaled Drug Products. Doi:10.1002/psp4.12413
19. Lee SL., *et al.* Regulatory considerations for approval of generic inhalation drug products in the US, EU, Brazil, China, and India. *AAPS Journal.* 2015; 17: 1285–1304.
20. Silva MC., *et al.* Overview of brazilian requirements for therapeutic equivalence of orally inhaled and nasal drug products. *AAPS PharmSciTech.* 2019; 20: 235.
21. Zou C., *et al.* Bioequivalence studies of inhaled indacaterol maleate in healthy Chinese volunteers under gastrointestinal non-blocking or blocking with concomitant charcoal administration. *Pulmonary Pharmacology & Therapeutics.* 2020; 61: 101902.
22. Newman B and Witzmann K. Addressing the regulatory and scientific challenges with generic orally inhaled drug products. *Pharmaceutical Medicine.* 2020; 34: 93–102.
23. Nasal Spray and Inhalation Solution. Suspension, and Spray Drug Products--Chemistry, Manufacturing, and Controls Documentation https://www.fda.gov/regulatory-information/search-fda-guidance-documents/nasal-spray-and-inhalation-solution-suspension-and-spray-drug-products-chemistry-manufacturing-and

43

Regulatory Perspectives and Concerns Related to Nanoparticle-Based Lung Delivery

Johirul Islam[5], Hemanga Hazarika[5], Probin Kr Roy[2], Pompy Patowary[1], Pronobesh Chattopadhyay[1], Yashwant V Pathak[3,4], and Kamaruz Zaman[5]
[1]Pharmaceutical Technology Division Defence Research Laboratory, Assam, India
[2]Regional Institute of Paramedical and Nursing Sciences, Mizoram, India
[3]College of Pharmacy, University of South Florida, Tampa, Florida, USA
[4]Adjunct Professor, faculty of Pharmacy, Airlangga University, Surabaya, Indonesia
[5]Department of Pharmaceutical Sciences, Dibrugarh University, Assam, India

43.1 Introduction

Frequent inhalation exposures to airborne pollutants, mineral dusts, cigarette smoke, insecticides and repellent vaporizers, along with anti-cancer treatments and radiotherapy injure the lungs. The most common problems of the respiratory system in the modern society are acute lung injury (ALI), acute respiratory distress syndrome (ARDS), bronchiolitis, common cold, cough, cystic fibrosis (CF), lung cancer, pneumonia, pulmonary hypertension, respiratory diseases of newborns, and acute and chronic inflammatory lung diseases including asthma, lung fibrosis, and chronic obstructive pulmonary disease (COPD)/emphysema (1). The chronic inflammation of lungs causes an increase in airway hyper-responsiveness that leads to frequent breathlessness, chest tightness, coughing, and wheezing, mostly at night or in the early morning (2). These types of lung disease pose a significant health risk to humans and are associated with high morbidity and mortality (1).

The pulmonary route has been regarded an attractive route of drug administration due to its large target surface area, extensive vasculature, good membrane permeability, low enzyme activity, and non-invasive approach for drug delivery (3,4). For asthma and COPD, inhalation therapies are widely acceptable as first-line therapy as an optimal route of drug administration (5). In recent years, the pulmonary route has been explored as a possible route of drug administration for diabetes mellitus (5). Researchers, formulation scientists, and healthcare workers are working on different novel drug delivery nanocarriers and devices to deliver larger drug doses to enhance the bioavailability to and greater deposition efficiency of the drug on the target organ (6).

43.1.1 Pulmonary Drug Delivery

The lung is the major organ of the respiratory system and prevents undesirable airborne particles from entering into the body. It has a large surface area with a comparatively low enzymatic and controlled environment for drug absorption. However, the airway geometry, alveolar macrophages, mucociliary clearance, and humidity of the lung hinders the bioavailability of inhaled medications (5).

43.1.1.1 Lung Anatomy

The respiratory system, composed of the upper respiratory tract (nose, nasopharynx, oropharynx and laryngeal pharynx) and the lower respiratory tract (larynx, trachea, bronchi, and the lungs). The trachea or windpipe (10 cm long) extends from the larynx and divides into two bronchi and projects into the lungs (7). It is lined by a mucous membrane composed of ciliated epithelium and goblet cells, and inhaled dust particles, pollens, etc. are expelled due to the upward movement of cilia. The bronchi formed by the bifurcation of the trachea are similar in structure to the trachea which divides into smaller bronchioles and branches in the lungs, forming a passage for air (8). Alveoli are the terminal parts of the bronchi and are the functional units of the lungs, and they form the site of gaseous exchange. There are approximately 300 million alveoli in each lung (7,9). The nose and oropharynx in the upper airways prevent entry of particles to the lungs (10,11).

Between the alveolar space and the pulmonary capillaries there is a very thin blood barrier present, which is involved in rapid gas exchange. Oxygen diffuses through the alveoli walls and the interstitial area, into the blood stream during inspiration. During expiration/exhalation, carbon dioxide diffuses in the opposite direction (12–14). Drug particles having an aerodynamic diameter of <0.3 μm, are required for delivery to the alveolar epithelium (15,16).

43.1.1.2 Barriers in Pulmonary Drug Delivery

Drug delivery to the lung is relatively complicated due to the defense mechanisms, as the respiratory tract expels the inhaled drug particles out of the lungs (9). The combined effect of mechanical, chemical, and immunological barriers restricts the bioavailability of locally acting as well as systemically acting

DOI: 10.1201/9781003046547-43

drugs in the pulmonary system. Poor adherence and poor inhaler technique also adversely affect the bioavailability of inhaled drugs. Hence, the development of novel, more efficient inhaler systems are required to mitigate the effects of these barriers (17–19).

43.1.1.2.1 Mechanical Barrier

During disease conditions, airways are affected by bronchoconstriction, inflammation, and mucus hypersecretion, or might be blocked by mucus plugs (9). The lung has a natural defense mechanism under mucociliary clearance to remove deposited particles to be delivered to the oropharynx, where they are swallowed or expectorated (20). Mucociliary clearance could be a barrier to drug delivery if it removes/expels drug from target sites (21,22). However, it could be beneficial if it moves deposited drug toward target sites from less favorable areas.

43.1.1.2.2 Chemical and Immunological Barriers

Deposited drug particles are required to dissolve in lung fluids (23). Undissolved drugs may face alveolar macrophages, which are the prevailing phagocytic cells safeguarding against inhaled foreign matters (5). Sometimes the administered drug cannot be eliminated by mucociliary clearance and may be exposed to the actions of chemicals including proteolytic enzymes and surfactants (5,23). Alveolar macrophages constitute an immunological barrier that makes no distinction between potentially harmful substances and potentially beneficial ones (9,24). The macrophages could engulf inhaled drug particles and remove them from the lungs, via the lymphatic system or by assigning them to the mucociliary escalator (23). Surfactant may resist adhesion of inhaled particles to the lung surfaces, making them easily accessible to macrophages (25).

43.1.1.2.3 Behavioral Barriers

Pulmonary drug delivery is crucially influenced by what patients do, or fail to do, with their inhaler devices (26). Major drawbacks in inhaler technique for metered dose inhaler (MDI)s include not actuating the inhaler while breathing in and failing to inhale deeply and slowly. Regarding dry powder inhalers (DPIs), the problems include not inhaling with sufficient force, incorrect device handling pattern, and incorrect device orientation (27). Inadequate training in inhaler use predisposes patients toward poor inhaler technique.

43.1.1.3 Pharmacodynamic Factors in Lung Delivery

After accurate dose adjustment of a drug having a lower receptor binding affinity, the same degree of pulmonary targeting coud be shown as that of a drug having higher receptor binding affinity. This relationship could be supported by pharmacokinetic/pharmacodynamic (PK/PD) studies (28–30).

43.1.1.4 Pharmacokinetic (PK) Factors in Pulmonary Delivery

43.1.1.4.1 Oral Bioavailabilty

The residue of drug available in the systemic circulation after pulmonary delivery is dependent on which amount is delivered to the oropharynx and subsequently swallowed/oral absorption and how much drug in the lungs has been cleared by the mucociliary clearance escalator, as this fraction will also be reaching the GI tract (5,31,32). Therefore, an ideal drug candidate for pulmonary delivery should maintain the lowest possible oral bioavailability (33,34).

43.1.1.4.2 Systemic Clearance and Volume of Distribution

Systemic clearance addresses the capacity of the body to eradicate drug that has been absorbed into the systemic circulation. An ideal drug candidate for pulmonary delivery should possess a high systemic clearance to avoid any side effects (35,36).

Volume of distribution (VD) is the extent of distribution of the drug into tissue compartments. The higher the tissue binding, the larger the VD and, therefore, the more drugs will be in the peripheral compartment. Inhaled drugs with a prolonged half-life do not hamper safety if the half-life is due to extensive tissue binding and sufficient clearance. Moreover, a drug with a higher degree of binding efficiency will show decrease in side effects (28,37).

43.1.1.4.3 Pulmonary Residence Time

Sink condition in the lungs due to the high rate of pulmonary blood flow, and higher number of pores, means the absorption of inhaled drugs is relatively fast across the pulmonary membrane (15,38). Therefore, the inhaled drug will quickly exit the lungs, making a very short pulmonary residence time. The rate of dissolution of inhaled drug is a primary determinant of residence time in the lungs (39,40). Novel drug delivery strategies to prolong the residence time of drug in the lung by modifying the physicochemical properties includes the use of liposomes, microspheres and ultrathin coatings, or lipophilic drugs with slow dissolution characteristics (28,41,42).

43.1.2 Nano-formulations in Pulmonary Delivery

Nano-formulations are the most promising approach of colloidal drug delivery system for enhancement of bioavailability of poorly soluble drugs due to their pharmaceutical advantages and pharmacoeconomic value. Decreases in particle size offers a large surface to volume ratio for which there will be an increase in bioavailability of poorly soluble hydrophobic drugs (43,44). Nanoparticles offer the advantage of sustained release of drug in the target tissue and thus the systemic circulation, which results in a decrease in dosing frequency and increase the patient compliance (45,46).

Over the last two decades, nanoparticles have received immense considerations in drug delivery. They can be delivered via different routes of administration such as parenteral, oral, intraocular, transdermal, and pulmonary. Drug particles conjugating with different carrier systems to be delivered by aerosol therapy are now becoming a popular method to achieve local or systemic effects (47). However, numerous studies reveal that nanoparticles usually accumulate in the organs of the reticulo-endothelial system (RES) such as lungs, liver, and spleen (48).

43.1.2.1 Nanoparticle Delivery for Pulmonary Application

43.1.2.1.1 Solid Lipid Nanoparticles

Solid lipid nanoparticles (SLNs) are the novel discovery in drug delivery, which are the lipid nanocarriers with a solid core, suitable for both hydrophilic and hydrophobic drugs. They can be produced from biocompatible or biodegradible materials and therefore are one of the preferred options for drug delivery. Unique features like mucoadhesiveness or targeting capability of SLNs could be managed by surface modification processes (49,50). An invention related to folic acid targeting of silymarin using solid lipid nanosphere and its preparation method has been disclosed in several reports (49,50). Many of these reports claimed that a drug can be targeted on a tumor cell and thus achieve lung tumor targeting effects, with improved bioavailability and reduced toxic or side effects on cells. The inventors also claimed the method as very easy, economic, and a convenient process with assured environmental safety.

43.1.2.1.2 Polymeric Nanoparticles

Polymers have diverse advantages, like surface modification, high encapsulation efficiency, prolonged drug delivery properties, resistance to rapid degradation of the active drug, and prolonged shelf life (51). For therapeutic purposes, the most commonly used polymers include alginate, chitosan, poly-lactic acid (PLA), poly-lactic-co-glycolic acid (PLGA), polycaprolactone (PCL), and gelatin (50,52). However, the polymeric particles are biodegradable in nature; their degradation rate must be studied. Additionally, the efficacy, safety, and toxicity profiles must be evaluated in various *in vivo, in vitro* and *ex vivo* models.

43.1.2.1.3 Liposomes and Niosomes

As liposomes are mainly prepared from phospholipids, which are inherent in lungs, they could be considered as an attractive drug delivery carrier, especially for the lung. They possess sustained release properties which result in maximum efficacy over a prolonged period of time. The first liposomal product (Alveofact®) was introduced in the 1990s, developed for the treatment of acute respiratory distress syndrome in infants (18,53). Taetz et al. in 2009 designed cationic hyaluronic acid (HA)-modified DOTAP/DOPE liposomes for the targeted delivery of anti-telomerase siRNA to CD44 receptor-expressing lung cancer cells. Their study reveals that their developed formulation has improved stability in cell culture medium with a reduced cytotoxicity (54).

Niosomes are processed from neutral, single-chain surfactants with the incorporation of cholesterol or other amphiphilic moieties (55–57). During pulmonary disease, bronchial mucus hypersecretion remarkably retards the lipophilic drugs such as corticosteroids, from reaching their target receptors. Niosomes have been used to circumvent this problem (58,59). Moazeni et al. in 2010 developed a niosomal formulation containing ciprofloxacin for pulmonary delivery (55). The entrapment of ciprofloxacin in the vesicle showed satisfactory effects. Their study demonstrated that the composition of the vesicle could affect the stability, nebulization capabilities, and cytotoxcicity of the niosome-entrapped ciprofloxacin.

43.1.2.1.4 Nano-emulsion

Nano-emulsion shows the potential to enhance drug bioavailability and has been regarded as an ideal carrier for the hydrophobic drugs. Aerosolized nano-emulsion gains the attraction of formulation scientists to deliver anti-cancer drugs to the lung due to the improved solubility, stability, dissolution profile, and uniform drug distribution in the alveoler region (60–62).

Asmawi et al. in 2020 developed a curcumin and docetaxel encapsulated aerosolized nano-emulsion system using D-optimal mixture (63). The optimized formulation exhibited desirable aspects of physicochemical and aerodynamic properties for inhalation therapy. In another study carried out by Li et al. in 2016, tea tree oil nano-emulsion was developed for bacterial and fungal pneumonia (64). The nano tea tree oil exhibited strong in vitro anti-microbial activities against *Acinetobacter baumannii, Candida albicans, Escherichia coli, Klebsiella pneumoniae*, and *Staphylococcus aureus*. After inhalation, the developed formulation showed higher anti-fungal activity than fluconazole on the fungal pneumonia induced rat models with little lung injury, increased microbial clearance, decrease in pro-inflammatory mediators, and blocking of leukocyte recruitment.

43.1.2.1.5 Dendrimers

Dendrimers are an ideal carrier for drug delivery, synthesized in a layer-by-layer pattern, having precise structures which contributes to the enhancement of solubility and bioavailability of poorly soluble drugs (48,65–67). Lactoferrin-conjugated dendritic nanoconstructs have been developed by Kurmi et al. for lung targeting of methotrexate (68). Their study reveals that the lactoferrin-conjugated dendrimers could be a promising drug targeting carrier for lung cancer. Encapsulation of methotrexate in the developed formulation enhanced the drug residence time, which can reduce the dosing frequency.

43.1.2.1.6 Hydrogels

Hydrogel nanoparticles are basically three-dimentional polymeric matrix networks, known as polymeric nanogels or macromolecular micelles, and are promising drug carriers (69). They have mucoadhesive and swelling characteristics, which could help the drug particles in bypassing the pulmonary barriers (70). Incorporation of drug into respirable nano- or micro-particles is a promising strategy to avoid rapid clearance and to attain controlled release of the target drug. In a research study, ciprofloxacin nanoparticles have been encapsulated into dry, swellable nano-in-micro hydrogel particles (71). The developed hydrogel formulation showed a suitable aerodynamic characteristic with a promising sustained release profile. Their study also reveals higher concentrations of ciprofloxacin in the lung for more than 7 hours.

43.1.2.1.7 Nanosuspension

Drugs having poor aqueous solubility are difficult to formulate using conventional techniques; the nanosuspension approach is useful to solve this issue due to their reduced particle size, large surface area, and longer stability. Nanosuspensions are generally stabilized by surfactants and contain 100% pure drug without any carriers (72). Jacobs et al. 2002 developed a budesonide nanosuspension for pulmonary delivery (73). Their study results open a new insight to formulate budesonide nanosuspension for long-term stability which could be administered by conventional nebulizer or a portable inhaler.

43.1.3 Current Trends

Currently available inhalation products are shorter acting and undergo rapid release, which requires frequent dosing. Controlled drug delivery for pulmonary application is a promising system, but challenges like rapid clearance of airways and macrophages, as well as systemic absorption and metabolic degradation need to be managed. An obvious drawback associated with the available product is that drug concentration peaks at the beginning and then declines rapidly, which can cause undesirable effects of the therapy (74,75).

Nanoparticles can cross the cell membranes and minimize the risk of premature drug clearance. They have outstanding interest in biomedical utilizations due to their well-defined particle size and shape (67). They can minimize the systemic toxicity of drugs by active targeting with enhanced permeation and retention (EPR) effect to the tumor site (76,77). To achieve controlled pulmonary delivery, novel carriers are developed to overcome the airway clearance mechanisms and to retain in the physiological state for longer duration with improved therapeutic efficacy (78,79). Porous and swellable micro- and nanoparticles are the most promising carriers for controlled drug delivery to the lung (80).

Mesenchymal stem cells (MSC) play an important role in the treatment of tissue injury, degenerative disease, and in immune disorder therapy (81). They have anti-tumor effects and show low immunogenicity, for which MSC can be an ideal vehicle for anti-tumor drug delivery (82). However, after intravenous administration, most MSC are trapped in the lung during cell therapy. Wang et al. in 2019 utilized MSC as a drug delivery vehicle by loading nanoparticles (NP) with drug particles in targeted drug delivery to the lung to treat cancer cells (83). Their study results showed that MSC have a higher drug intake capacity than fibroblasts, which established that MSC have predominant lung trapping activity in both rabbits and monkeys.

Costabile et al., in 2015 developed an inhalable antiviral nanosuspension against *Pseudomonas aeruginosa* for lung infections. As their study reports, niclosamide, an anthelmintic drug, has strong inhibiting activity against *P. aeruginosa* and could be repurposed as an antiviral drug (84). They formulated dry powders containing niclosamide nanoparticles which can be reconstituted to produce inhalable nanosuspensions. Overall, their study report provides the rationale for further development of niclosamide, an antiviral drug, as an alternative cystic fibrosis therapy. Repurposing a known drug may also result in considerable reduction in the cost of product development compared to developing new chemical entities.

However, tremendous research has been going on to mitigate and counter the drawbacks of conventional drug delivery; the developed technology should drag the attention of the industrialist to utilize the product for social benefit by production and marketization of the formulation. Development of a novel formulation and filing a patent is a very time-consuming process. Therefore, the introduction of a newer product to the market has to undergo some regulatory formalities which should not be a barrier for the researchers, scientists, and healthcare workers.

43.2 Molecular Mechanism and Associated Pathways for Nano-particulate Lung Delivery

The concept of efficient and targeted drug delivery to the target tissue or organ has historically attracted attention, especially for potent drugs with toxicity concerns. Gene targeting or organ-specific drugs are improvised by numerous orders of magnitude with the aid of targeting ligands connected to the nano-particulate carriers for specifically recognizing the corresponding receptors overexpressed at the target tissues or cells (85). Moreover, drug delivery carriers that target cells by recognition of surface biomarkers is a promising approach with the following advantages.

- Cellular specificity may reduce toxicity and keep adverse reactions to the minimum.
- Introduction of the cellular specificity may overcome the limitations of some previous therapies (86).

There are many types of nanoparticle systems being explored for drug delivery to lungs, especially cancer therapeutics. Nanoparticles attached with drug molecules and conjugated with receptor ligands have been diversely termed as nanoconjugates, nano-formulations, nano-carriers, etc. Receptors have particularly played a key role as molecular targets present novel opportunities for targeting of drug delivery systems intracellularly by not only improving the efficacy of the drug but also reducing the overall systemic cytotoxicity. Furthermore, active targeting can also be achieved by binding targeting ligands to the surface of nano-carriers that bind to molecules specific for the diseased cell. It is worth mentioning that the endothelium has proven to be an important therapeutic target for controlling various anomalies like oxidative stress, inflammation, and thrombosis involved in pulmonary disorders (4). It has been revealed that antibodies conjugated with nanoparticles against constitutive endothelial cell adhesion molecules (CAMs) can be favorably targeted to endothelial cells. For example, the beta2-adrenergic receptor (β2-AR) is the most studied, and abundantly found receptor in lung tissue and can be exploited for targeting therapeutics to lung-related disorders (87). Here, we focus on overexpressed receptors exploited for targeting drugs to treat some common lung disorders.

43.2.1 Cancer

Current research in targeted therapeutics for lung cancer includes inhalable and injectable systems, monoclonal antibodies, folate, growth factors, lung-surfactant proteins, integrin-binding motifs, etc (88). The *recepteur d'origine nantais* (RON) receptor tyrosine kinase is a member of the MET proto-oncogene family associated with the pathogenesis of non small cell lung cancer (NSCLC). Established data stated that novel anti-RON antibody-drug maytansinoid conjugate Zt/g4-DM1 proved to be very much effective in RON-directed drug delivery for targeted inhibition of NSCLC cell-derived tumor growth in mouse xenograft models (89). Again, immune checkpoint inhibitors (ICIs) have emerged in the last decade and have attracted considerable interest in small cell lung cancer (SCLC) (90), since SCLC is characterized by high tumor mutational burden (91), which is promising for the use of ICIs. Yet again, immunotherapy with pembrolizumab, nivolumab, and atezolizumab targeting the programmed death 1 (PD-1) receptors showed a striking response and encouraging improvements in survival in patients with pretreated SCLC.

Several studies had described the role of molecular targets such as epidermal growth factor receptor (EGFR) and anaplastic lymphoma kinase (ALK) in the development and progression of lung cancer, especially in the cases of NSCLC. Furthermore, there are other oncogenic alterations in genes identified in NSCLC that have been reported, viz. vascular endothelial growth factor receptor (VEGF) and mesenchymal epithelial transition factor (MET). It is noteworthy to mention that angiogenesis is regulated through a multifaceted set of mediators and recent evidence demonstrated that integrin $\alpha v\beta 3$ and VEGFs play important regulator roles. Therefore, selective targeting of $\alpha v\beta 3$ integrin and VEGFs can prove to be a novel anti-angiogenesis strategy for treating a wide variety of solid lung tumors (92). It has also been reported that interaction of leukocyte function–associated antigen-1 (LFA-1) and intercellular cell adhesion molecule-1 (ICAM-1) can play crucial roles in tumor metastasis and progression (93).

Studies have demonstrated that nanoparticles can be targeted to lung epithelial cells by means of ICAM-1 and LFA-1 and can be internalized into the cells rapidly (94). Folate receptors expressed at apical membranes of the lung is another receptor investigated in lung malignancies. Cancer cells are overexpressed by folate receptors on cell membranes of lungs. Anticancer drugs like methotrexate target the cancer cells by folate receptor, killing the cells by deactivating the metabolism of diseased cells by inhibition of the enzyme dihydrofolate reductase (DHFR) (95). Furthermore, constitutively activated signal transducer and activator of transcription 3 (STAT3) is reported to be frequently found in lung cancer (96). Cytokines and growth factors possibly bind to their receptors, activating the tyrosine residue of STAT3. The constant upregulation of STAT3 contributes to oncogenesis by the expression of transmembrane molecues like B-cell lymphoma-extra large (bclxl), B-cell lymphoma 2 (bcl2), c-myc, cyclin D1, mcl-1 (97), and VEGF (98), enhancing and promoting cell proliferation and angiogenesis respectively. Investigations also suggest that there is a pragmatic possibility of using nano STAT3 siRNA as a strategy for combating human lung cancer (99).

Among the potential cellular surface targets suitable for use in drug targeting, folate receptors stand out as one of the most investigated and most promising lung cancer markers, constituting a useful target for tumor-specific drug delivery (100). Bombesin receptors (BnRs), also known as gastrin-releasing peptide (GRP) receptors and somatostatin receptors (SSTRs) belonging to the G-protein coupled receptor (GPCR) superfamily have gained attention in cancer prognosis and are widely expressed (101,102). Researchers have also gained interest in exploring sigma receptor ligands for tumor imaging and targeted therapy, since sigma receptors are overexpressed in a variety of human tumors, including NSCLC (103).

Targeting drugs to cancer cells and the tumor microenvironment is a major challenge in the field of anti-cancer therapy. However, nanotechnology has enabled numerous options in manipulating nanocarriers based on lipids, proteins, and metals that can be loaded with anti-cancer drugs and directed mostly to cancer cells by the attachment of moieties that can recognize cancer phenotypes (Figure 43.1).

43.2.2 Infectious Disease

Because they are a first-line defense mechanism against microorganisms invading the lungs via the airways and, consequently, reservoirs of many pathogens, alveolar macrophages have drawn particular attention as a relevant clinical target. Macrophages or neutrophils can be targeted by using nanoparticles conjugated to antibodies against macrophage or neutrophil markers. Various nano-drug delivery systems of different drugs have been developed to promote targeting and uptake of particles by macrophages via receptor-mediated endocytosis, viz. mannose receptors and other lectin-like receptors. In vivo biodistribution studies confirmed that ligand-modified liposomes appeared to be a possible strategy for the treatment of bacterial and fungal infections, since they are accumulated in the lungs for more prolonged time and cleared more slowly through the blood stream (105). Intracellular *Mycobacterium tuberculosis* (Mtb) in macrophages downregulates pro-inflammatory modulators such as interleukin (IL)-12, tumor necrosis factor (TNF), IL-6 and evokes anti-inflammatory cytokines such as IL-10, IL-13, IL-4 and IL-10 (106). Furthermore, virulent strains seem to manipulate the activation of the eukaryotic superfamily of microtubule-associated protein kinase (MAPK) to impair cytokine production or to stimulate anti-inflammatory response. Moreover, Mtb inhibits interferon (IFN)-signaling pathways in human macrophages and it directly or indirectly interrupts the association of STAT-1 with the transcription co-activator cAMP response element-binding (CREB) protein that is essential for the transcriptional response to IFN-g. Several other reports indicate that Mtb infection induces assembly of inflammasomes, and macrophages infected in vitro with mycobacteria continue to release IL-1b through an exocytosis pathway involving the pathogen-encoded proteins of the ESAT secretion system–mycobacterial early secreted antigenic target (ESX-ESAT) system (107). Targeting the above-mentioned cascades, there is scope to choose host responses in the management of pulmonary infections, through the pulmonary drug delivery route.

43.2.3 Asthma

Asthma is an exceedingly complex chronic airway inflammatory disease, involving a wide range of cells and cellular components. Asthma has numerous impending molecular

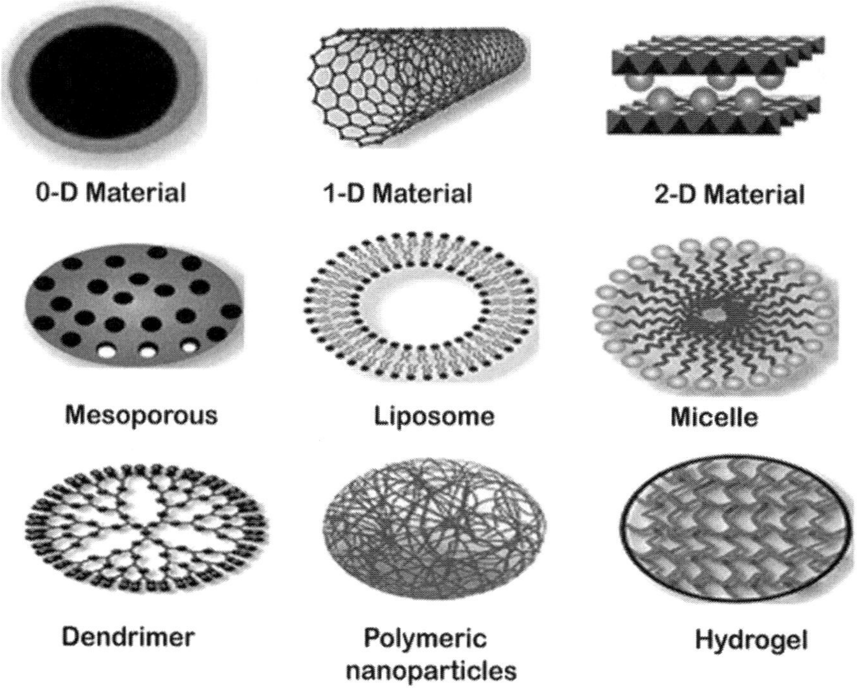

FIGURE 43.1 Different Types of Nanocarriers Used as Controlled Delivery Vehicles for Therapy. Reproduced with permission from Springer Nature (104)

targets including chemokines, cytokines, transcription factors, tyrosine kinases, and their associated receptors which can be delivered together with drugs through nanoparticles (108). The atrial natriuretic peptide (ANP)-natriuretic peptide receptor A (NPRA) signaling axis is highly expressed in airway epithelium and has been implicated in causing persistent pro-inflammatory effects and promoting a Th2-like phenotype. Studies have reported that inhibition of the ANP-NPRA pathway in the lung reduced inflammation in mice model of asthma (109). Therefore, pharmacological intervention to reduce NPRA activity through the inflammatory natriuretic peptide axis in the lung may be a possible and useful adjunct therapy in the treatment of asthma. Again, studies confirmed that the IL-5 antibody substantially reduced asthma exacerbations in human patients with eosinophilic asthma (110).

CXC chemokine receptor (CXCR)-2 antagonist also showed promising results in preventing inflammation by inhibiting the inflammatory mediator IL-8 (111). Again, asthmatic airway inflammation involves the dominance of IL-4 and IL-13 pro-inflammatory cytokines having multiple biological functions that are vital for asthma development (112). Moreover, IL-4 and IL-13 share the IL4Rα subunit in their cognate receptors (113), and studies have confirmed that immunosuppressive effects of anti-IL4Rα inhibit the signaling of both cytokines (114).

43.2.4 Chronic Obstructive Pulmonary Disease

Oxidative stress with increased reactive oxygen species (ROS) and reactive nitrogen intermediates (RNI), is one of the emerging hallmark features in chronic respiratory diseases (CRDs) like asthma, chronic obstructive pulmonary disease (COPD), respiratory infections, and lung cancer (115). Moreover, ROS also initiates inflammatory cascades via multiple mechanisms, such as transcription factors, protein kinase pathways, and genomic expression of pro-inflammatory regulators (116). Furthermore, low levels of oxidative stress initiate the release of antioxidant and anti-inflammatory genes via the activation of nuclear erythroid 2 p45-related factor 2 (Nrf2). Targeting oxidative stress pathways with nanotechnology can help in designing potential and better therapy for CRDs. Moreover, antagonists for muscarinic receptor have been approved and used for treatment of COPD with the aid of inhalation technologies (117).

In addition to anti-cholinergics, neutrophil elastase inhibitors have been suggested as new therapeutic agents to prevent further damage in the lungs of patients with COPD. Elastase increases matrix metallopeptidases (MMP) activity by directly activating MMP-9 and by inactivating the endogenous MMP inhibitor, termed as a tissue inhibitor of matrix metalloprochinateinases (TIMP-1). Moreover, nuclear factor kappa-light-chain-enhancer of activated B cells (NF-κB) regulates the expression of IL-8 and other chemokines, as well as TNF-α and some MMPs, therefore there are several approaches to the inhibition of NF-κB in the treatment of COPD (118). In another study, phosphoinositide 3-kinase (PI3K) inhibitors, which are involved in the differentiation of alveolar epithelial stem cells, were employed to repair pulmonary alveoli in COPD animal models (119). These receptors provide unique opportunities to understand disease biology and its treatment. In attempts to treat a particular disease, overexpressed receptors are directly modulated/inhibited by agents such as antibodies or antibody fragments, and also by other small chemicals that directly bind these receptors and block their activities.

43.3 Regulation and Guidelines

Nanosciences and nanotechnologies open up new avenues of research and lead to new, useful, and sometimes unexpected applications. Novel materials and new engineered surfaces allow the making of products that perform better (120). Although the concept of nanotechnology began couple of decades ago, unprecedented exploration into research and development of nanotechnology and nanomaterials including engineered nanoparticles has been seen since 1995 (121). But assessment of risk associated with the use of nanomaterials and respective regulations have been introduced recently, and these are popping up like mushrooms internationally (122). Current regulations on nanomaterials largely depend on the available toxicological and eco-toxicological data to support and implement regulations. However, it has been argued that methodologies and equipments employed for the risk assessment are not suitable for the determination of hazards associated with nanoparticles (122). Therefore, uniform regulations and guidelines are required for several reasons, including (a) nanoparticles do not appear in nature and living organisms may not have appropriate means to deal with nanoparticles; (b) some of the engineered nanomaterials are dangerous because of their size and reactivity as they have high specific surface area; and (c) as they are tiny, they can float in the air and might easily penetrate into human, animal, and plant cells, causing adverse environmental and health impacts (122,123).

Efforts have been made in the past to create "nano-regulations," however, unfortunately, it has not become possible so far due to disagreement on the existing regulations among "nanonations" on the information relating to the environment and safety concerns of nanomaterials (122). To have a common regulation among nanonations, it is important to have a balance between safety, health, and environment, R and D activity, economic growth, and social benefits. It is worthy to note that nanonations face multiple challenges while implementing authoritative and prescriptive regulatory decisions to ensure that nanotechnology and nanomaterials are safe without hindering innovation, including (1) whether to adapt existing legislation or develop a new regulatory framework, (2) how to define nanotechnology and nanomaterials, (3) whether nanomaterials should be considered as different from bulk, and (4) how to deal with the profound limitations of risk assessment when it comes to nanomaterials (122).

43.3.1 The European Union (EU)

The regulatory framework on nanotechnology had initiated in 2005 in the first public meeting for nanoscale materials stewardship program of the United States Environmental Protection Agency (US EPA) and followed up by the first Implementation Report on the Action Plan (2007). In these commission communications, clear instruction were given on the development of nanotechnology-based products which said that nanotechnology must be developed in a responsible way. A recommendation on the code of conduct on research in nanotechnology (2008) had been provided which specifies voluntary rules for European scientists and researchers active in nanotechnology (124).

In the EU, Registration, Evaluation, Authorisation and Restriction of Chemicals (REACH) regulates the use of chemical substances including nanomaterials (122). Importantly, in 2008, the European Chemicals Agency (ECHA) established the Competent Authorities Sub Group on Nanomaterials (CASG Nano) (125). Two technical guidelines were published in 2010, and three projects were launched devoted to the application of REACH to nanomaterials, especially for 24 nanomaterials pertaining to the application of the related CLP regulation (classification, labeling, and packaging of substances and mixtures) (126). In REACH, special attention has been given to nanomaterials and it is expected to strongly influence regulatory actions, and in fact, at present, provisions of strong legislation have been introduced in many countries including Austria, Belgium, France, Germany, and Italy for introducing notification and registration mechanisms for nanomaterials (127).

43.3.2 United Kingdom (UK)

In 2004, the United Kingdom's Royal Society in its seminal report *Nanoscience and Nanotechnologies: Opportunities and Uncertainties* concluded that: "Many nanotechnologies pose no new risks to health and almost all the concerns relate to the potential impacts of deliberately manufactured nanoparticles and nanotubes that are free rather than fixed to or within a material" (128–130). At the same time, they recommended that nanomaterials should be regulated as new chemicals, and research laboratories and factories that are producing nanomaterials should consider them "as if they were hazardous," and their release to the environment should be minimized as far as possible. In the UK, the Health and Safety Executive (HSE), with others, is responsible for the negotiation, agreement, and enforcement of regulations on nanomaterials by reviewing the available information on the physicochemical and toxicological hazards of relevance in the workplace and the occupational exposure situation (131,132). It is important to note that the EU has standardized legislation to regulate the health and safety hazards of industrial chemicals and these regulations have been accommodated in the UK under the umbrella of the *Health and Safety at Work Act 1974* (the HSW Act) (133). However, the EU chemicals policy is on the verge of a major shift to a different regulatory framework based on REACH and this is likely to influence UK regulation in future.

43.3.3 United States of America (USA)

In the USA, the credit for establishment of nano-regulation may be conferred on different agencies, including the Environmental Protection Agency (EPA), the Food and Drug Administration (FDA), and the Consumer Product Safety Commission (CPSC) (134). The FDA is one of the agencies responsible for regulating policies on nanomaterials, particularly with representatives of FDA centers who performed different assessments and regulate different substances and products. Further, they ensure coordination and regular communications. In 2013, the USA published proposed significant new rules for manufacturers for 14 nanomaterials that were the subject of pre-manufacture notice (PMN). Importantly, the policies of the EPA on nanomaterials are primarily controlled by the existing regulations of the *Toxic Substances Control Act* (TSCA) and *Federal Insecticide, Fungicide, and Rodenticide Act* (FIFRA) (135–137). It is pertinent to mention here that under the section 5 of the TSCA, a manufacturer must notify EPA through a PMN at least 90 days in advance of a new chemical's commercialization, to provide the EPA an opportunity for review (137). In 2014, the FDA issued one draft and three final guidance documents for industry regarding the use of nanotechnology in FDA regulated products. In draft guidance on nanotechnology for industry, the FDA has emphasized assessing the effects of significant manufacturing process changes, including emerging technologies, on the safety and regulatory status of food ingredients and food contact substances, including food ingredients such as color additives, food for animals, etc. (138,139).

43.3.4 Canada

In Canada, regulatory issues on environmental health and safety (EHS) have become of increasing importance for nanotechnologies and nanomaterials, and there has been growing involvement of authorities in different sectors that are working to develop regulatory, product-specific guidance documents for nanomaterials (122,140,141). Nanomaterials are regulated under existing legislation, including the Canadian *Environmental Protection Act*, 1999, the *Pest Control Products Act*, the *Fertilizers Act*, the *Feeds Act,* and the *Food and Drugs Act* (141). Nanomaterials, for instance quantum dots, nanoscale colloid/emulsion/liposome, nano silver, nano gold, etc., fall within the health regulatory mandate of Canada.

43.3.5 Australia

Like Canada, an active authoritative approach has been observed in Australia where industrial nanomaterials are being regulated within the framework for conventional (bulk) chemicals through the *Industrial Chemicals Notification and Assessment Act 1989*(137,142). Any nanotechnology-based product undergoes rigorous and comprehensive scientific safety assessment before it can be legally supplied in Australia and New Zealand. Food Standards Australia New Zealand (FSANZ) is a statutory authority in the Australian Government Health portfolio which basically develops the standards regulations for food and other chemicals, including nanomaterials, with due consultation with experts, other government agencies, and stakeholders (143). The recommendations made by these bodies are open and accountable, and based upon a rigorous scientific assessment of risk to public health and safety.

43.3.6 Japan

In Japan, in 2008, the Ministry of Economy, Trade and Industry (METI) created a committee on Safety Management for Nanomaterials which initiated collecting date related to nanomaterials at industry level (144). In addition, the National Institute of Occupational Safety and Health Japan (JNIOSH) periodically publishes reports on occupational health and safety (OHS) issues of nanomaterials (145).

43.3.7 The Republic of South Korea

The Republic of South Korea has developed an inter-ministerial National Nano-safety Strategic Plan (2012–2016) to investigate the assessment of nanomaterials, including in-depth hazard assessment, exposure analysis, and safety studies. The country has specific guidelines on Safety Management of Nano-based Products, constituted in 2011 (137,146).

43.3.8 Taiwan

In Taiwan, the Industrial Technology Research Institute (ITRI) framed guidelines for regulation of nanomaterials and developed the Nanomark Certification system which has been active since 2004. The Nanomark Certification system is a voluntary reporting and certification scheme that aims to increase public confidence in nanotechnology products (147–150).

43.3.9 Thailand

In Thailand, nanosafety is among the priorities of the national policy on nanotechnologies. The National Nanotechnology Center (Nanotec) manages an industrial standards certification system (NanoQ) for the regulation of nanotechnologies-related products (124,151,152).

43.3.10 China

In China, the State Food and Drug Administration (SFDA) regulates all the aspects related to nanomaterials (153,154). However, the Standard on the Health and Safety Practices in Occupational Settings Relevant to Nanotechnologies (22 August 2011, Standardization Administration of China) are of primary concern to the SFDA.

43.3.11 India

In India, regulations on nanotechnology in terms of the concerns related to human health safety, environmental pollution, and toxicity hves been introduced recently and continuous efforts have been made by different agencies such as the Department of Science and Technology (DST) and Council of Scientific and Industrial Research (CSIR) to establish a specific regulatory system for nanomaterials (127,155,156). For instance, DST has constituted a task force for the regulatory framework for nanotechnology to organize national dialogs to promote research and development (R&D) in developing standards for nanotechnology and laying down a National Regulatory Framework Road-Map for Nanotechnology (NRFR-Nanotech) (PIB 2014) (127). Similarly, the CSIR has initiated a major project, Nano-SHE in 12th Five Year plan to evaluate and create a database on various toxicological aspects of nanostructured materials. It is worth noting that although different laws/regulations are available to regulate nanomaterials in India, their timely interventions and amendments are essential to encompass nanotechnology-related consequences. In this regard, the Energy and Resources Institute (TERI) has classified nanotechnology related consequences into the following categories (TERI 2010) (127):

- Production and Marketing: *Drugs and Cosmetics Act*, 1940, National Pharmacovigilance Protocol, Medical Devices Regulation Bill, and Insecticides Act, 1968.
- Occupational Health and Safety: *Factories Act*, 1948 and OHS under other legislation.
- Environmental Risk Management: Pollution control laws, *Environment Protection Act*, and *Public Liability Insurance* Act.27
- Waste Disposal: *Factories Act*, Hazardous Material (Management, Handling and Transboundary Movement) Rules 2007, Bio-Medical Waste (Management and Handling) Rules 1998, Municipal Solid Wastes Rules 2000.

Importantly, ample efforts have been made to streamline these acts/rules in order to address peculiarities that nanotechnology carries within itself.

43.4 Safety and Toxicity

Inhalation therapy is considered one of the oldest delivery systems to treat respiratory diseases and has a history of delivering medicated aerosol more than 2000 years ago in Ayurvedic medicine in India (157). Considering the current scenario, lungs are considered an attractive target for the delivery of drugs due to their advantages for localized therapy in diseases like lung cancer, cystic fibrosis, asthma, and COPD. It also provides advantages over conventional oral or other delivery methods, such as intravenous injection, because it is a non-invasive route that provides increased bioavailability (158). Unlike the gastrointestinal tract and liver, lungs have limited intracellular and extracellular drug metabolizing enzymes, therefore, pulmonary drug delivery offers enhanced bioavailability with limited first-pass metabolism (4,159). Pulmonary delivery offers selectivity to deliver the drugs directly to the target side therefore reduces side effects (160). Apart from the above-mentioned advantages, delivery through this route offers systemic or local effects. The route also offers high vascularization with extensive surface area leading to high absorption of drugs with limited metabolic activity compared to the oral route (161).

Nanoparticles are the more promising proposed drug delivery route to deliver active chemical entities because of their improved therapeutic outcomes due to their ability to deliver or release the drug to the target, thus minimizing toxicity to other tissues/organs (162). The nanoparticle popularity have grown very fast from the past few decades due to their unique properties such as high surface to mass ratio, better ability to carry active compound and high absorption rate (163). It is obvious that nanoparticles are developed to deliver the chemical entity to the target site or uptake by target cells or to reduce the toxicity or side effects of the free drugs to the non-target cells or organs, but the carrier system itself may possess hazard to the patients. The types of hazard produced by the nanoparticle for drug delivery are far more intense than the conventional hazards of the classical compounds (164). Accumulating evidence suggests that nanoparticles may exert adverse effect on pulmonary structure and function.

In nanoparticle toxicity with respect to pulmonary drug delivery, toxicologists are mainly interested in the mechanisms and the particle size responsible for toxicity (165). Researchers have unveiled that very small sized nanoparticles with larger surface areas may have possible interaction with the biological systems that lead to negative effects compared to surface area of larger particles (165). *In vitro* and *in vivo* studies have shown that the smaller sized particles have higher inflammatory potential per unit mass compared to larger particles within a rat's lungs (166,167). The primary mechanism of nanoparticle-induced toxicity is induction of oxidative stress through the production of ROS (165). To understand the mechanism, the hierarchical oxidative stress model has been proposed by Nel et al. ROS is usually produced by cells during cellular respiration (165). Cells contain built-in antioxidants such as glutathione and antioxidant enzymes which regulate or neutralize the ROS. When there is severe injury in lungs due to inspiration of nanoparticles or delivered nanoparticles, the amount of ROS produced is much higher, therefore the antioxidant contained in the cells is not enough to neutralize the ROS species (165). Under these circumstances, accumulation of reactive oxidized glutathione occurs within the cells. In the presence of nanoparticles, production of these ROS increased in the mitochondria. As a result, accumulation of oxide ion (O^{2-}) and consequently particles are installed within the mitochondria, thereby hindering the electron transport chain and energy production (168). In such condition, cells detect the low ratio of glutathione to oxidized glutathione, consequently inducing inflammation. The interaction of nanoparticles with biological systems and toxicity mechanisms in additional ways are reviewed by Nel et al. (165). Therefore, in order to avoid toxicity of delivered nanoparticles, the agent or the drugs should be targeted carefully.

43.4.1 Preclinical Safety and Toxicity Assessment

When drugs are usually delivered by inhalation, especially highly toxic chemotherapeutic agent/s, lungs are exposed to these agents, therefore chances of side effects are high. In this case, injury to lungs or healthy tissues may occur during the exposure of the agent/s (169). Health workers may experience prolonged exposure to the nanoparticles, especially while administering nebulized drugs to patients, and the exhalation of the remaining nano-particulate or aerosol can be hazardous for the health workers. Various physicochemical factors are responsible for biological interaction and toxicity of nanoparticles, therefore, it is very important to characterize the nanoparticles for the particular formulation. The characterization includes the method of synthesis, particle size and distribution, particle shape, crystal structure, dissolution, aggregation and agglomeration, surface area, and other surface characteristics (164). Without proper characterization of nanoparticles, it is impossible to identify which factors are responsible for the toxicity and very challenging to interpret results of individual studies as well as compare the results with different studies virtually (164). Therefore, identifying the factor responsible for the toxicity is hindered. Numerous nanoformulations have been developed for pulmonary delivery; to ensure the safety of the formulation, many modifications with different types of excipients have been made to reduce the associated toxicity and improve target specific efficacy.

Many in vivo animal studies and cell culture studies were conducted to assess the toxicity of the nano-formulations for pulmonary delivery. The preclinical toxicity study possesses a number of challenges due to the physiological species variability between the species. For preclinical safety studies, an inhalation toxicity test in two animals is mandatory and often starts with rodents (170). Using nebulization in the air, generally inhalation of only a few particles can be achieved; therefore, proper estimation may not be achieved. Since instillation via intra-tracheal route is invasive and skips the upper respiratory defense mechanisms, oropharyngeal aspiration or intra-tracheal intubation are more convenient methods as they cause minimum harm to the animals. The latter method can be considered more physiological than intra-tracheal instillation.

Oropharyngeal aspiration or intra-tracheal intubation can partly be cleared by mucociliary clearance due to the application of solution at the beginning of the trachea (171). Application of the particle doses also varies depending on the application of devices which may alter the lung's physiology. A commonly used device delivers volume of 50–100 μl, which is much higher than the epithelial lining fluid in rodents (45–55 μl in rats and 5–15 μl in mice), and this may indirectly interfere with the result due to differences in normal lung physiology (172). Species variability between rodents and humans also needs to be considered while interpreting the results; it varies species to species and between individuals.

Humans possess dichotomous branching airways whereas rodents possess monopodial airway branching, therefore, particle deposition may differ from humans to rodents (173). Rodents possess lower viscous mucus usually produced by serous glands rather than mucus produced by the submucosal glands and intraepithelial glands, present in humans and which may influence lung physiology. There is a report that higher mucus clearance was observed in humans than rodents (172). Other differences are cellular composition of the lower respiratory tract, and metabolizing enzymes present in the rat lungs and human lungs. Another difference includes specialized cells, such as club cells which are present in the terminal bronchioles of the rat, but absent in humans (174). Similarly, the presence of serous cells in rat lungs differentiates their metabolism when compared to humans, where serous cells are absent (175). To estimate the active entity delivered to the lungs at the respiratory barrier, in vitro tests could be more effective for the most promising formulation, followed by animal experiments. Dissolution testing may provide ideas regarding the non-dissolved particles present in the alveolar surface that may be ingested by alveolar macrophages.

Inhalation toxicity of SLNs for pulmonary delivery had been assessed to be a therapeutic window for nanoparticles (176). A549 cells and murine precision-cut lung slices (PCLS) were selected for the toxicity testing by exposing them to different increased concentrations of SLNs. MTT and NRU assays were conducted to find the cytotoxic effect of SLN in A549 cells. Repeated inhalation toxicity tests of the SLNs used different concentrations in female BALB/c mice (1–200 μg deposit dose). In vivo and in vitro experiments showed toxic effects at concentrations above 500μ/ml. However, a dose of 1–200 μg deposit doses/animal showed no sign of toxicity (176).

Lung cancer often leads to death and the traditional treatment methods include surgical resection. On the other hand, in most of the advanced cancer cases, chemotherapeutic agents are via oral or intravenous administration, which, in most of the cases, possess severe toxic effects (177). To reduce the toxicity of the chemotherapeutic agents, transferrin-conjugated liposomes have been developed to deliver the agents to the lungs via inhalation with the intent to reduce the toxicity (178). The results indicated an increased uptake of transferrin conjugated liposomes by tumor cells via transferrin-receptor mediated endocytosis. It is evident that most of the studied nanoparticles designed for pulmonary delivery either do not show toxicity or are less characterized to assess the toxicity of the nanoparticles. There is a great deal of literature regarding the evaluation of nano-formulations for pulmonary delivery of the agents. The literature itself says how much effort researchers have put into exploring the delivery of therapeutic agents through this route. Some information about the in vitro, in vivo, and ex vivo toxicity of the different nano-formulations for lung delivery are given in Table 43.1.

43.4.2 Clinical Safety and Toxicity Assessment

Different sized nanoparticles exhibit different fractional deposition characteristics within the respiratory tract. Ultrafine nanoparticles <100 nm are found to be deposited in all regions. Particle size <10nm shows maximum tracheobronchial deposition. Similarly, particle size 10–20 nm shows highest alveolar deposition (194). There are limited sources of nanoparticle-associated toxicity to show adverse pulmonary effects involving humans. However, many investigations are done involving healthy volunteers to investigate the deposition pattern of nanoparticles in the lungs (195). Several clinical trials have been conducted to ensure the safety and efficacy of the delivered nano-formulation, either for treatment or diagnosis purposes. Anderson et al., conducted a study with human subjects to find out whether granulocyte macrophage–colony stimulating factor (GM-CSF) can be delivered effectively, incorporated in aerosol, to the lungs for therapeutic efficacy in lung metastases (196). In the phase I dose escalation, three dose levels such as 60 μg, 120 μg, and 240 μg were give twice a day for a week. Pulmonary function and blood count were recorded at the beginning and end of the week. If there was no toxicity, a second level of dose, i.e. 120 μg, was administered. Similarly third level doses were given and all the blood parameters and cytokines levels were investigated. It was reported that aerosol delivery of GM-CSF is safe, feasible, and may be effective for the treatment of lung metastases. In another study, an inhalation of interleukin-2 (IL-2) was administered in a phase I clinical trial (197). They concluded the IL-2 liposome's delivery as aerosol is well tolerated by the patients and further investigations were demanded. So it is obvious that research on nano-formulation for pulmonary delivery in clinical trials is still on and may give a bright future for the effective treatment of lung diseases. Numbers of clinical trials have been conducted or are ongoing regarding the safety and efficacy of the nano-formulation. Therefore, we provide a summary of listed clinical trials that have been conducted for improved efficacy or for better toxicity profiles from the nano-formulated medicine for treatment or diagnosis purposes in the Table 43.2.

TABLE 43.1

Preclinical Toxicity Studies of the Nano-Formulations

Nano-formulation Type	Active Compound and Administration Mode	Animal Model/in vitro cell line/ex vivo model/in vivo	Toxicity Endpoints
Poly(lactic-co-glycolic acid) (PLGA) magnetic core-shell nanoparticles (179)	Quercetin (nebulization)	A549 cells, mice	Cytotoxicity, Glutathione and IL-6 secretion
Solid lipid nanoparticles (SLNs) (176)	Inhalation	A549 cells, BALB/c mice	Cytotoxicity by MTT and SRU assay, lung inflammation, inflammatory cytokines
SLNs (180)	Sildenafil	In vitro human alveolar epithelial cell line (A459) and mouse heart endothelium cell line (MHEC5-T), Ex vivo rat precision cut lung slices (PCLS) and rat heart slices (PCHS) were used	3-(4,5-dimethylthiazol-2-yl)-2,5-diphenyltetrazolium bromide (MTT)
SLNs (176)		Human alveolar epithelial cell line (A549),ex vivo slices of A549 lungs cells	Cytotoxicity by lactate dehydrogenase (LDH) and MTT assay
SLNs (181)	Rifampicin	NR8383 and A549	Cytotoxicity by MTT
SLNs (182)	Paclitaxel (pulmonary delivery)	HeLa, M109-HiFR cells, M109 tumors	Biodistribution, pharmacokinetic proflie
SLNs (181)	Rifampicin	NR8383 and A549	Cytotoxicity by MTT
Surface modified PLGA nanoparticles (183)		A549 human lung epithelial cells	In vitro mitochondrial cytotoxicity, inflammatory cytokines
Liposomes (184)	Amikacin (nebulization)	Sheep	Dynamic compliance (Cdyn), lung resistance (RL), arterial blood pH, pO_2 and pCO_2.
Liposomes (185)	(Intra-tracheal instillations)	Male Balb/c mice	LDH assay and differential cell counts, pulmonary inflammation
Nanoparticles in microgel (186)	Trypsin and neutrophil elastase (Nebulization)	RAW 264.7 macrophage cells and C57BL/6 mice	In vitro degradation, macrophage uptake, organ clearance
Multifunctional dual drug-loaded nanoparticle (187)	NU7441	A549 and H460 lung cancer cells, mice	Biocompatibility, histopathology
PEG-PLA nanoparticles and immunonanoparticles (188)	Paclitaxel palmitate (endotracheally)	Mice	Inflammation, neutrophil count, mortality and hemorrhage
PLGA nanoparticles (189)	Tobramycin		Biodistribution, accumulation
Nanoparticles (190)	Doxorubicin and cisplatin (pulmonary administration)	B16F10 cells, mice	Cytotoxicity, cellular accumulation
Albumin nanoparticles (191)	Doxorubicin and octyl aldehyde (Inhalation)	H226 cell, BALB/c *nu/nu* mice	Cytotoxicity, apoptotic cells
Gold nanoparticles (192)	(Simulation)	Simulated lungs model	Dose enhancement ratio, physicochemical interactions
Nanoparticles (193)	DNA (intranasal)	Mice	BAL neutrophil and cytokine responses, hematologic parameters, serum complement, IL-6, or MIP-2 levels or in the activity, growth, and grooming of the mice

TABLE 43.2

Clinical Trials/Patents of the Nano-Formulation for Pulmonary Drug Delivery

Formulation Type	Drugs/Agents Incorporated and Route of Administration	Clinical Trials Phase	Application
Liposomes (197)	Interleukin-2 (nebulization)	Phase I	Pulmonary metastasis
Aerosol (198)	Sargramostim (nebulization)	Phase I	Metastatic melanoma, anti-tumor immunity
Aerosol (199)	Granulocyte macrophage colony-stimulating factor (nebulization)	Phase I	Lungs malignant metastases
Nanoparticle (200)	Mucin (inhalation)	Phase I	Dispersal of thick mucus produced during lung diseases, diseases at other mucosal surfaces

(Continued)

TABLE 43.2 (Continued)

Clinical Trials/Patents of the Nano-Formulation for Pulmonary Drug Delivery

Formulation Type	Drugs/Agents Incorporated and Route of Administration	Clinical Trials Phase	Application
Recombinant nanoparticles (201)	Cystic fibrosis transmembrane regulator (CFTR) mRNA compositions (inhalation, nebulization, intranasal administration, or aerosolization)		Correction of CFTR gene function in cystic fibrosis patients
Liposomes (202)	PGE2- and nucleic acid (inhalation)		Pulmonary fibrosis
Recombinant nanoparticles (203–205)	Interleukin-2 (inhalation)	Phase I	Pulmonary metastases of renal cell carcinoma
Recombinant nanoparticles (206)	Interleukin-2 (inhalation)	Phase I	Pulmonary metastases in melanoma stage IV
Aerosolized nanoparticles (207)	Carboplatin (nebulization)	Phase II	Non small cell lung acncer
Lipid complexes (208)	Cisplatin (inhalation)	PhaseIb/IIa	Relapsed/progressive osteosarcoma metastatic to the lung
Liposomes (209)	9-nitrocamptothecin (nebulization)	Phase I/ Phase II	Primary or metastatic lung cancer
Liposomes (209,210)	9-Nitro-20(S)-Camptothecin (nebulization)	Phase I	Advanced malignancies in the lungs
Liposomes (211)	Cisplatin (sustained release lipid inhalation targeting) (inhalation)	Phase I	Lung carcinoma
Aerosolized recombinant adeno-associated virus (AAV) vector encoding the complete human cystic fibrosis transmembrane regulator (CFTR) cDNA (tgAAVCF) (212)	Human cystic fibrosis transmembrane regulator (CFTR) cDNA (tgAAVCF) (inhalation)	Phase IIB	Cystic fibrosis
Liposomes (213)	Plasmid DNA (nasal epithelium)		Cystic fibrosis
Liposomes (214)	CF transmembrane conductance regulator (CFTR) iIntranasal)	Phase IIB	Cystic fibrosis

43.5 Conclusion

Nanosciences and nanotechnologies open up new avenues in pulmonary drug delivery for the treatment of diseases like lung cancer, cystic fibrosis, asthma, tuberculosis, and COPD. However, NPs have been implicated in causing damage to both pulmonary structure and function, and this may be attributed to the size of NPs and other issues. Since the hazards concerning to the use of NPs are sometimes considered to be serious, authoritative and prescriptive regulatory measures are essential for them. Although regulatory agencies including the US FDA and EPA have been active in addressing the concerns of NPs, the authoritative and prescriptive regulatory measures are most likely to be limited due to such factors as disagreement on the existing regulations among "nanonations" on the information relating to the environmental and safety concerns of NPs. Therefore, it is important to have consistent regulations among nanonations, considering a balance between safety, health, and the environment; R and D activity; economic growth; and social benefits.

REFERENCES

1. Venkatesan N, Punithavathi D, and Babu M. *The Molecular Targets and Therapeutic Uses of Curcumin in Health and Disease*, Springer, Switzerland. 2007, pp. 379–405.
2. Winkler J, Hochhaus G, and Derendorf H. How the lung handles drugs: pharmacokinetics and pharmacodynamics of inhaled corticosteroids. *Proceedings of the American Thoracic Society*. 2004; 1: 356–363.
3. Scheuch G, Kohlhaeufl MJ, Brand P, and Siekmeier R. Clinical perspectives on pulmonary systemic and macromolecular delivery. *Advanced Drug Delivery Reviews*. 2006; 58: 996–1008.
4. Azarmi S, Roa WH, and Löbenberg R. Targeted delivery of nanoparticles for the treatment of lung diseases. *Advanced Drug Delivery Reviews*. 2008; 60: 863–875.
5. Labiris N and Dolovich M. Pulmonary drug delivery. Part I: physiological factors affecting therapeutic effectiveness of aerosolized medications. *British Journal of Clinical Pharmacology*. 2003; 56: 588–599.
6. Dolovich M. New propellant-free technologies under investigation. *Journal of Aerosol Medicine*. 1999; 12: S-9–S-17.
7. Marieb EN and Hoehn K. *Human Anatomy & Physiology*. Pearson Education, London. 2007.
8. Pearce EC. *Anatomy and Physiology for Nurses: Including notes on Their Clinical Application*. Jaypee Brothers, Guwahati, Assam. 1997.
9. Newman SP. Drug delivery to the lungs: challenges and opportunities. *Therapeutic Delivery*. 2017; 8: 647–661.
10. Shang Y, Dong J, Tian L, Inthavong K, and Tu J. Detailed computational analysis of flow dynamics in an extended respiratory airway model. *Clinical Biomechanics*. 2019; 61: 105–111.
11. Ahookhosh K, Pourmehran O, Aminfar H, Mohammadpourfard M, Sarafraz MM, and Hamishehkar H.

Development of human respiratory airway models: A review. *European Journal of Pharmaceutical Sciences*. 2020; 145: 105233.
12. Hickey AJ and Mansour HM.*Inhalation Aerosols: Physical and Biological Basis for Therapy*. CRC Press, Boca Raton, Florida. 2019.
13. Lemmer HJ and Hamman JH. Paracellular drug absorption enhancement through tight junction modulation. *Expert Opinion on Drug Delivery*. 2013; 10: 103–114.
14. Schwaiblmair M, Behr W, Haeckel T, Märkl B, Foerg W, and Berghaus T. Drug induced interstitial lung disease. *The Open Respiratory Medicine Journal*. 2012; 6: 63.
15. Pilcer G and Amighi K. Formulation strategy and use of excipients in pulmonary drug delivery *International Journal of Pharmaceutics*. 2010; 392: 1–19.
16. Sinha B, Mukherjee B, and Pattnaik G. Poly-lactide-co-glycolide nanoparticles containing voriconazole for pulmonary delivery: in vitro and in vivo study. *Nanomedicine: Nanotechnology, Biology and Medicine*. 2013; 9: 94–104.
17. Respaud R, Vecellio L, Diot P, and Heuzé-Vourc'h N. Nebulization as a delivery method for mAbs in respiratory diseases. *Expert Opinion on Drug Delivery*. 2015; 12: 1027–1039.
18. Cipolla D, Wu H, Eastman S, Redelmeier T, Gonda I, and Chan HK. Modifying the release properties of liposomes toward personalized medicine. *Journal of Pharmaceutical Sciences*. 2014; 103: 314–327.
19. Rajpoot P, Pathak K, and Bali V. Therapeutic applications of nanoemulsion based drug delivery systems: a review of patents in last two decades. *Recent Patents on Drug Delivery & Formulation*. 2011; 5: 163–172.
20. Ganesan S, AT Comstock, and US Sajjan. Barrier function of airway tract epithelium. *Tissue Barriers*. 2013; 1: e24997.
21. Marttin E, Schipper NG, Verhoef JC, and Merkus FW. Nasal mucociliary clearance as a factor in nasal drug delivery. *Advanced Drug Delivery Reviews*. 1998; 29: 13–38.
22. Ruge CA, Kirch J, and Lehr C-M. Pulmonary drug delivery: from generating aerosols to overcoming biological barriers-therapeutic possibilities and technological challenges. *The Lancet Respiratory Medicine*. 2013; 1: 402–413.
23. Patton JS, Fishburn CS, and Weers JG. The lungs as a portal of entry for systemic drug delivery. *Proceedings of the American Thoracic Society*. 2004; 1: 338–344.
24. Cryan S-A, Sivadas N, and Garcia-Contreras L. In vivo animal models for drug delivery across the lung mucosal barrier. *Advanced Drug Delivery Reviews*. 2007; 59: 1133–1151.
25. Green GM, Jakab GJ, Low RB, and Davis GS. Defense mechanisms of the respiratory membrane. *American Review of Respiratory Disease*. 1977; 115: 479–514.
26. Kikidis D, Konstantinos V, Tzovaras D, and Usmani OS. The digital asthma patient: The history and future of inhaler based health monitoring devices. *Journal of Aerosol Medicine and Pulmonary Drug Delivery*.2016; 29: 219–232.
27. Malmberg LP, Rytilä P, Happonen P, and Haahtela T. Inspiratory flows through dry powder inhaler in chronic obstructive pulmonary disease: age and gender rather than severity matters.*International Journal of Chronic Obstructive Pulmonary Disease*. 2010; 5: 257.
28. Tayab ZR and Hochhaus G. Pharmacokinetic/pharmacodynamic evaluation of inhalation drugs: application to targeted pulmonary delivery systems. *Expert Opinion on Drug Delivery*. 2005; 2: 519–532.
29. Hochhaus G. Pharmacokinetics and pharmacodynamics of drugs delivered to the lungs. *Pharmaceutical Inhalation Aerosol Technology*. 2003; 215.
30. Mukker JK, Singh RSP, and Derendorf H. Pharmacokinetic and pharmacodynamic implications in inhalable antimicrobial therapy. *Advanced Drug Delivery Reviews*. 2015; 85: 57–64.
31. Hochhaus G and Möllmann H. Pharmacokinetic/pharmacodynamic characteristics of the beta-2-agonists terbutaline, salbutamol and fenoterol. *International Journal of Clinical Pharmacology, Therapy, and Toxicology*. 1992; 30: 342–362.
32. Taburet A-M and Schmit B. Pharmacokinetic optimisation of asthma treatment. *Clinical Pharmacokinetics*. 1994; 26: 396–418.
33. Derendorf H. Pharmacokinetic and pharmacodynamic properties of inhaled corticosteroids in relation to efficacy and safety. *Respiratory Medicine*. 1997; 91: 22–28.
34. Daley-Yates PT, Price AC, Sisson JR, Pereira A, and Dallow N. Beclomethasone dipropionate: absolute bioavailability, pharmacokinetics and metabolism following intravenous, oral, intranasal and inhaled administration in man. *British Journal of Clinical Pharmacology*. 2001; 51: 400–409.
35. Hochhaus GN. New developments in corticosteroids. *Proceedings of the American Thoracic Society*. 2004; 1: 269–274.
36. Bouhuys A, Lichtneckert S, Lundgren C, and G Lundin. Voluntary changes in breathing pattern and N2 clearance from lungs. *Journal of Applied Physiology*. 1961; 16: 1039–1042.
37. Siekmeier R and Scheuch G. Systemic treatment by inhalation of macromolecules--principles, problems, and examples. *Journal of Physiology and Pharmacology*. 2008; 59: 53–79.
38. Patton JS, Brain JD, Davies LA, Fiegel J, Gumbleton M, Kim K-J, Sakagami M, Vanbever R and Ehrhardt C. The particle has landed--characterizing the fate of inhaled pharmaceuticals. *Journal of Aerosol Medicine and Pulmonary Drug Delivery*. 2010; 23: S-71–S-87.
39. Edwards DA, Hanes J, Caponetti G, Hrkach J, Ben-Jebria A, Eskew ML, Mintzes J, Deaver D, Lotan N, and Langer R. Large porous particles for pulmonary drug delivery. *Science*. 1997; 276: 1868–1872.
40. Yu J and Chien YW. Pulmonary drug delivery: physiologic and mechanistic aspects. *Critical Reviews™ in Therapeutic Drug Carrier Systems*. 1997; 14: 14.
41. Rytting E, Nguyen J, Wang X, and Kissel T. Biodegradable polymeric nanocarriers for pulmonary drug delivery. *Expert Opinion on Drug Delivery*. 2008; 5: 629–639.
42. Tang BC, M Dawson, SK Lai, Y-Y Wang, JS Suk, M Yang, P Zeitlin, MP Boyle, J Fu, and J Hanes. Biodegradable polymer nanoparticles that rapidly penetrate

the human mucus barrier. *Proceedings of the National Academy of Sciences*. 2009; 106: 19268–19273.

43. Sung JC, Pulliam BL, and Edwards DA. Nanoparticles for drug delivery to the lungs. *Trends in Biotechnology*. 2007; 25: 563–570.
44. Sahu BP, Hazarika H, Bharadwaj R, Loying P, Baishya R, Dash S, and Das MK. Curcumin-docetaxel co-loaded nanosuspension for enhanced anti-breast cancer activity. *Expert Opinion on Drug Delivery*. 2016; 13: 1065–1074.
45. Bailey MM and Berkland CJ. Nanoparticle formulations in pulmonary drug delivery. *Medicinal Research Reviews*. 2009; 29: 196–212.
46. Mansour HM, Rhee Y-S, and Wu X. Nanomedicine in pulmonary delivery. *International Journal of Nanomedicine*. 2009; 4: 299.
47. Ely L, Roa W, Finlay WH, and Löbenberg R. Effervescent dry powder for respiratory drug delivery. *European Journal of Pharmaceutics and Biopharmaceutics*. 2007; 65: 346–353.
48. Cheng Y, Xu Z, Ma M, and Xu T. Dendrimers as drug carriers: applications in different routes of drug administration. *Journal of Pharmaceutical Sciences*. 2008; 97: 123–143.
49. Paliwal R, Paliwal SR, Kenwat R, Kurmi BD, and Sahu MK. Solid lipid nanoparticles: a review on recent perspectives and patents. *Expert Opinion on Therapeutic Patents*. 2020; 30: 179–194.
50. Paranjpe M and Müller-Goymann CC. Nanoparticle-mediated pulmonary drug delivery: a review. *International Journal of Molecular Sciences*. 2014; 15: 5852–5873.
51. Shiehzadeh F and Tafaghodi M. Dry powder form of polymeric nanoparticles for pulmonary drug delivery. *Current Pharmaceutical Design*. 2016; 22: 2549–2560.
52. Menon JU, Ravikumar P, Pise A, Gyawali D, Hsia CC, and Nguyen KT. Polymeric nanoparticles for pulmonary protein and DNA delivery. *Acta Biomaterialia*. 2014; 10: 2643–2652.
53. Gaspar MM, Bakowsky U, and Ehrhardt C. Inhaled liposomes–current strategies and future challenges. *Journal of Biomedical Nanotechnology*. 2008; 4: 245–257.
54. Taetz S, Bochot A, Surace C, Arpicco S, Renoir J-M, Schaefer UF, Marsaud V, Kerdine-Roemer S, Lehr C-M, and Fattal E. Hyaluronic acid-modified DOTAP/DOPE liposomes for the targeted delivery of anti-telomerase siRNA to CD44-expressing lung cancer cells. *Oligonucleotides*. 2009; 19: 103–116.
55. Moazeni E, Gilani K, Sotoudegan F, Pardakhty A, Najafabadi AR, Ghalandari R, Fazeli MR, and Jamalifar H. Formulation and in vitro evaluation of ciprofloxacin containing niosomes for pulmonary delivery. *Journal of Microencapsulation*. 2010; 27: 618–627.
56. KhoeeS and MYaghoobianNiosomes: A novel approach in modern drug delivery systems. In *Nanostructures for Drug Delivery*, Elsevier, Amsterdam, Netherlands. 2017, pp. 207–237.
57. Sharma R, Dua JS, Prasad D, and Hira S. Advancement in novel drug delivery system: niosomes. *Journal of Drug Delivery and Therapeutics*. 2019; 9: 995–1001.
58. Muzzalupo R and Mazzotta E. Do niosomes have a place in the field of drug delivery? 2019; 16: 1145–1147.
59. Rinaldi F, del Favero E, Moeller J, Hanieh PN, Passeri D, Rossi M, Angeloni L, Venditti I, Marianecci C, and Carafa M, Hydrophilic silver nanoparticles loaded into niosomes: Physical-chemical characterization in view of biological applications. *Nanomaterials*.2019; 9: 1177.
60. Arbain NH, Salim N, Masoumi HRF, Wong TW, Basri M, and Rahman MBA. In vitro evaluation of the inhalable quercetin loaded nanoemulsion for pulmonary delivery. *Drug Delivery and Translational Research*. 2019; 9: 497–507.
61. Shi L, Qu Y, Li Z, Fan B, Xu H, and J Tang. In vitro permeability and bioavailability enhancement of curcumin by nanoemulsion via pulmonary administration. *Current Drug Delivery*. 2019; 16: 751–758.
62. Karami Z, Zanjani MRS, and Hamidi M. Nanoemulsions in CNS drug delivery: recent developments, impacts and challenges. *Drug Discovery Today*. 2019; 24: 1104–1115.
63. Asmawi AA, Salim N, Abdulmalek E, and Abdul Rahman MB. Modeling the effect of composition on formation of aerosolized nanoemulsion system encapsulating docetaxel and curcumin using d-optimal mixture experimental design. *International Journal of Molecular Sciences*. 2020; 21: 4357.
64. Li M, Zhu L, Liu B, Du L, Jia X, Han L, and Jin Y. Tea tree oil nanoemulsions for inhalation therapies of bacterial and fungal pneumonia. *Colloids and Surfaces B: Biointerfaces*. 2016; 141: 408–416.
65. Abbasi E, Aval SF, Akbarzadeh A, Milani M, Nasrabadi HT, Joo SW, Hanifehpour Y, Nejati-Koshki K, and Pashaei-Asl R. Dendrimers: synthesis, applications, and properties. *Nanoscale Research Letters*. 2014; 9: 247.
66. Tomalia DA. Birth of a new macromolecular architecture: dendrimers as quantized building blocks for nanoscale synthetic polymer chemistry.*Progress in Polymer Science*. 2005; 30: 294–324.
67. Svenson S. Dendrimers as versatile platform in drug delivery applications. *European Journal of Pharmaceutics and Biopharmaceutics*. 2009; 71; 445–462.
68. Kurmi BD, Gajbhiye V, Kayat J, and Jain NK. Lactoferrin-conjugated dendritic nanoconstructs for lung targeting of methotrexate. *Journal of Pharmaceutical Sciences*. 2011; 100: 2311–2320.
69. Gonçalves C, Pereira P, and Gama M. Self-assembled hydrogel nanoparticles for drug delivery applications. *Materials*. 2010; 3: 1420–1460.
70. Du J, Du P, and Smyth HD. Hydrogels for controlled pulmonary delivery.*Therapeutic Delivery*. 2013; 4: 1293–1305.
71. Du J, El-Sherbiny IM and Smyth HD. Swellable ciprofloxacin-loaded nano-in-micro hydrogel particles for local lung drug delivery. *Aaps Pharmscitech*. 2014; 15: 1535–1544.
72. Wang L, Du J, Zhou Y, and Wang Y. Safety of nanosuspensions in drug delivery. *Nanomedicine: Nanotechnology, Biology and Medicine*. 2017; 13: 455–469.
73. Jacobs C and Müller RH. Production and characterization of a budesonide nanosuspension for pulmonary administration. *Pharmaceutical Research*. 2002; 19: 189–194.
74. Hickey AJ. Nanotechnology for multimodal synergistic cancer therapy. *Advanced Drug Delivery Reviews*. 2020; 117: 13566-13638.

75. Fan W, Yung B, Huang P, and X Chen. Nanotechnology for multimodal synergistic cancer therapy. *Chemical Reviews*. 2017; 117: 13566–13638.
76. Malam Y, Loizidou M, and Seifalian AM. Liposomes and nanoparticles: nanosized vehicles for drug delivery in cancer. *Trends in Pharmacological Sciences*. 2009; 30: 592–599.
77. Maeda H, Nakamura H, and Fang J. The EPR effect for macromolecular drug delivery to solid tumors: Improvement of tumor uptake, lowering of systemic toxicity, and distinct tumor imaging in vivo. *Advanced Drug Delivery Reviews*. 2013; 65: 71–79.
78. Liang Z, Ni R, Zhou J, and Mao S. Recent advances in controlled pulmonary drug delivery. *Drug Discovery Today*. 2015; 20: 380–389.
79. Beck-Broichsitter M, Merkel OM, and Kissel T. Controlled pulmonary drug and gene delivery using polymeric nano-carriers. *Journal of Controlled Release*. 2012; 161: 214–224.
80. El-Sherbiny IM and Smyth HD. Biodegradable nano-micro carrier systems for sustained pulmonary drug delivery: (I) self-assembled nanoparticles encapsulated in respirable/swellable semi-IPN microspheres. *International Journal of Pharmaceutics*. 2010; 395: 132–141.
81. Islam MN, Das SR, Emin MT, Wei M, Sun L, Westphalen K, Rowlands DJ, Quadri SK, Bhattacharya S, and Bhattacharya J. Mitochondrial transfer from bone-marrow-derived stromal cells to pulmonary alveoli protects against acute lung injury. *Nature Medicine*. 2012; 18: 759–765.
82. Qiao L, Xu Z, Zhao T, Zhao Z, Shi M, Zhao RC, Ye L, and Zhang X. Suppression of tumorigenesis by human mesenchymal stem cells in a hepatoma model. *Cell Research*. 2008; 18: 500–507.
83. Wang X, Chen H, Zeng X, Guo W, Jin Y, Wang S, Tian R, Han Y, Guo L, and Han J. Efficient lung cancer-targeted drug delivery via a nanoparticle/MSC system. *Acta Pharmaceutica Sinica B*. 2019; 9: 167–176.
84. Costabile G, d'Angelo I, Rampioni G, Bondì R, Pompili B, Ascenzioni F, Mitidieri E, d'Emmanuele di Villa Bianca R, Sorrentino R, and Miro A. Toward re-positioning niclosamide for antivirulence therapy of pseudomonas aeruginosa lung infections: Development of inhalable formulations through nanosuspension technology. *Molecular Pharmaceutics*. 2015; 12: 2604–2617.
85. Roy I and N Vij. Nanodelivery in airway diseases: challenges and therapeutic applications. *Nanomedicine: Nanotechnology, Biology and Medicine*. 2010; 6: 237–244.
86. Xu S, Olenyuk BZ, Okamoto CT, and Hamm-Alvarez SF. Targeting receptor-mediated endocytotic pathways with nanoparticles: rationale and advances.*Advanced Drug Delivery Reviews*. 2013; 65: 121–138.
87. Vhora I, Patil S, Bhatt P, Gandhi R, Baradia D, and Misra A. Receptor-targeted drug delivery: current perspective and challenges. *Therapeutic Delivery*. 2014; 5: 1007–1024.
88. Hood JD, Bednarski M, Frausto R, Guccione S, Reisfeld RA, Xiang R, and Cheresh DA. Tumor regression by targeted gene delivery to the neovasculature. *Science*. 2002; 296: 2404–2407.
89. Feng L, Yao H-P, Sharma S, Y-Q Zhou, Zhou J, Zhang R, and Wang M-H. Erratum to: Biological evaluation of antibody-maytansinoid conjugates as a strategy of RON targeted drug delivery for treatment of non-small cell lung cancer. *Journal of Experimental & Clinical Cancer Research*. 2016; 35: 70.
90. Deneka AY, Boumber Y, Beck T, and Golemis EA.Tumor-targeted drug conjugates as an emerging novel therapeutic approach in small cell lung cancer (SCLC). *Cancers*. 2019, 11, 1297.
91. George J, Lim JS, Jang SJ, Cun Y, Ozretić L, Kong G, Leenders F, Lu X, Fernández-Cuesta L, and Bosco G. Comprehensive genomic profiles of small cell lung cancer. *Nature*. 2015; 524: 47–53.
92. Suri SS, Fenniri H, and Singh B. Nanotechnology-based drug delivery systems. *Journal of Occupational Medicine and Toxicology*. 2007; 2: 16.
93. Skelding KA, Barry RD, and Shafren DR. Systemic targeting of metastatic human breast tumor xenografts by Coxsackievirus A21.*Breast Cancer Research and Treatment*. 2009; 113: 21–30.
94. Kurmi BD, Kayat J, Gajbhiye V, Tekade RK, and Jain NK. Micro- and nanocarrier-mediated lung targeting. *Expert Opinion on Drug Delivery*. 2010; 7: 781–794.
95. Kohler N, Sun C, Wang J, and Zhang M. Methotrexate-modified superparamagnetic nanoparticles and their intracellular uptake into human cancer cells. *Langmuir*. 2005; 21: 8858–8864.
96. Weerasinghe P, Garcia GE, Zhu Q, Yuan P, Feng L, Mao L, and Jing N. Inhibition of Stat3 activation and tumor growth suppression of non-small cell lung cancer by G-quartet oligonucleotides. *International Journal of Oncology*. 2007; 31: 129–136.
97. Epling-Burnette P, Liu JH, Catlett-Falcone R, Turkson J, Oshiro M, Kothapalli R, Li Y, Wang J-M, Yang-Yen H-F, and Karras J. Inhibition of STAT3 signaling leads to apoptosis of leukemic large granular lymphocytes and decreased Mcl-1 expression. *The Journal of Clinical Investigation*. 2001; 107: 351–362.
98. Adachi Y, Aoki C, Yoshio-Hoshino N, Takayama K, Curiel DT, and Nishimoto N. Interleukin-6 induces both cell growth and VEGF production in malignant mesotheliomas. *International Journal of Cancer*. 2006; 119: 1303–1311.
99. Das J, Das S, Paul A, Samadder A, Bhattacharyya SS, and Khuda-Bukhsh AR. Assessment of drug delivery and anticancer potentials of nanoparticles-loaded siRNA targeting STAT3 in lung cancer, in vitro and in vivo. *Toxicology Letters*. 2014; 225: 454–466.
100. Akhtar MJ, Ahamed M, Alhadlaq HA, Alrokayan SA, and Kumar S. Targeted anticancer therapy: overexpressed receptors and nanotechnology. *Clinica Chimica Acta*. 2014; 436: 78–92.
101. Sancho V, Di Florio A, Moody TW, and Jensen RT. Bombesin receptor-mediated imaging and cytotoxicity: review and current status. *Current Drug Delivery*. 2011; 8: 79–134.
102. Herlin G, Kölbeck K-G, Menzel P, Svensson L, Aspelin P, Capitanio A, and Axelsson R. Quantitative assessment of 99mTc-depreotide uptake in patients with non-small-cell lung cancer: immunohistochemical correlations. *Acta Radiologica*. 2009; 50: 902–908.
103. Maurice T and Su T-P. The pharmacology of sigma-1 receptors. *Pharmacology & Therapeutics*. 2009; 124: 195–206.

104. Senapati S, Mahanta AK, Kumar S and Maiti P. Controlled drug delivery vehicles for cancer treatment and their performance. *Signal Transduction and Targeted therapy*. 2018; 3: 1–19.
105. Vyas S, Kannan M, Jain S, Mishra V, and Singh P. Design of liposomal aerosols for improved delivery of rifampicin to alveolar macrophages. *International Journal of Pharmaceutics*. 2004; 269: 37–49.
106. Friedland J, Hartley J, Hartley C, Shattock R, and Griffin G. Inhibition of ex vivo proinflammatory cytokine secretion in fatal Mycobacterium tuberculosis infection. *Clinical & Experimental Immunology*. 1995; 100: 233–238.
107. Mishra BB, Moura-Alves P, Sonawane A, Hacohen N, Griffiths G, Moita LF, and Anes E. Mycobacterium tuberculosis protein ESAT-6 is a potent activator of the NLRP3/ASC inflammasome. *Cellular Microbiology*. 2010;12: 1046–1063.
108. Wang L, Feng M, Li Q, Qiu C, and Chen R. Advances in nanotechnology and asthma. *Annals of Translational Medicine*. 2019; 7: 7.
109. Kandasamy R, Park S, Boyapalle S, Mohapatra S, Hellermann G, Lockey R, and Mohapatra S. Isatin down-regulates expression of atrial natriuretic peptide receptor A and inhibits airway inflammation in a mouse model of allergic asthma. *International Immunopharmacology*. 2010; 10: 218–225.
110. Pavord ID, Korn S, Howarth P, Bleecker ER, Buhl R, Keene ON, Ortega H, and Chanez P. Mepolizumab for severe eosinophilic asthma (DREAM): a multicentre, double-blind, placebo-controlled trial. *The Lancet*. 2012; 380: 651–659.
111. Boppana NB, Devarajan A, Gopal K, Barathan M, Bakar SA, Shankar EM, Ebrahim AS, and Farooq SM. Blockade of CXCR2 signalling: a potential therapeutic target for preventing neutrophil-mediated inflammatory diseases. *Experimental Biology and Medicine*. 2014; 239: 509–518.
112. Kopf M, Le Gros G, Bachmann M, Lamers MC, Bluethmann H and Köhler G. Disruption of the murine IL-4 gene blocks Th2 cytokine responses. *Nature*. 1993; 362: 245–248.
113. Borish L. IL-4 and IL-13 dual antagonism: a promising approach to the dilemma of generating effective asthma biotherapeutics. *Journal*. 2010; 181: 769–770.
114. Halwani R, Shaik AS, Ratemi E, Afzal S, Kenana R, Al-Muhsen S, and Al Faraj A. A novel anti-IL4Rα nanoparticle efficiently controls lung inflammation during asthma. *Experimental & Molecular Medicine*. 2016; 48: e262–e262.
115. Qu J, Li Y, Zhong W, Gao P, and Hu C. Recent developments in the role of reactive oxygen species in allergic asthma. *Journal of Thoracic Disease*. 2017; 9: E32.
116. Dua K, Malyla V, Singhvi G, Wadhwa R, Krishna RV, Shukla SD, Shastri MD, Chellappan DK, Maurya PK, and Satija S. Increasing complexity and interactions of oxidative stress in chronic respiratory diseases: An emerging need for novel drug delivery systems. *Chemico-Biological Interactions*. 2019; 299: 168–178.
117. Barnes P and Stockley R. COPD: current therapeutic interventions and future approaches. *European Respiratory Journal*. 2005; 25: 1084–1106.
118. Onoue S, Misaka S, Kawabata Y, and S Yamada. New treatments for chronic obstructive pulmonary disease and viable formulation/device options for inhalation therapy. *Expert Opinion on Drug Delivery*. 2009; 6: 793–811.
119. Horiguchi M, Oiso Y, Sakai H, Motomura T, and Yamashita C. Pulmonary administration of phosphoinositide 3-kinase inhibitor is a curative treatment for chronic obstructive pulmonary disease by alveolar regeneration. *Journal of Controlled Release*. 2015; 213: 112–119.
120. Logothetidis SNanotechnology: Principles and applications. In *Nanostructured Materials and Their Applications*, Springer, Switzerland. 2012, pp. 1–22.
121. Ramachandraiah K, Han SG, and Chin KB. Nanotechnology in meat processing and packaging: potential applications - a review. *Asian-Australasian Journal of Animal Sciences*. 2015; 28: 290.
122. Hansen SF. A global view of regulations affecting nanomaterials. *Wiley Interdisciplinary Reviews: Nanomedicine and Nanobiotechnology*. 2010. 2: 441–449.
123. Devasahayam S. Nanotechnology and nanomedicine in market: a global perspective on regulatory issues. In*Characterization and Biology of Nanomaterials for Drug Delivery*. Elsevier, Amsterdam, Netherlands. 2019, pp. 477–522.
124. Mantovani E, Porcari A, Morrison M, and Geertsma R. *Brussels, Observatory Nano from.* http://www.observatorynano.eu/project/filesystem/files/ObservatoryNano_Nanotechnologies_RegulationAndStandards_2012.pdf, 2012.
125. Austin T, Denoyelle M, Chaudry A, Stradling S, and Eadsforth C. European chemicals agency dossier submissions as an experimental data source: Refinement of a fish toxicity model for predicting acute LC50 values. *Environmental Toxicology and Chemistry*. 2015; 34: 369–378.
126. Gellert R, Mantovani E, and De Hert P.The EU regulation of nanomaterials: Smoother or harder? The precautionary tool chest as the basis for better regulating nanomaterials. In *Nanoengineering*, Elsevier, Amsterdam, Netherlands. 2015, pp. 339–373.
127. Kumar A. *Nanotechnology development in India: an overview*. Research and Information System for Developing Countries. 2014.
128. Bowman DM. More than a decade on: mapping today's regulatory and policy landscapes following the publication of nanoscience and nanotechnologies: opportunities and uncertainties.*NanoEthics*. 2017; 11: 169–186.
129. H Government. *Journal*. 2005.
130. Pidgeon N, Porritt J, Ryan J, Seaton A, Tendler S, Welland M, and Whatmore R. Nanoscience and nanotechnologies: opportunities and uncertainties. *The Royal Society, The Royal Academy of Engineering*. 2004; 29: 2004.
131. Maynard AD and Pui DY. *Nanoparticles and Occupational Health*. Springer, Switzerland. 2007.
132. Das M, Saxena N, and Dwivedi PD. Emerging trends of nanoparticles application in food technology: Safety paradigms. *Nanotoxicology*. 2009; 3: 10–18.
133. Kerr R, McHugh M, and McCrory M. HSE management standards and stress-related work outcomes. *Occupational Medicine*. 2009; 59: 574–579.

134. Suh J. Study of the Introduction of a nanomaterials regulatory policy for product safety. *Journal of the Korea Academia-Industrial cooperation Society*. 2014, 15: 4987–4998.
135. Hanson N, Harris J, Joseph LA, Ramakrishnan K, and Thompson T. EPA Needs to Manage Nanomaterial Risks More Effectively. 2011; 20: 115–116.
136. Birnbaum LS. Out of the frying pan and out of the fire: the indispensable role of exposure science in avoiding risks from replacement chemicals. 2010.
137. Park H-G and Yeo M-K. Nanomaterial regulatory policy for human health and environment. *Molecular & Cellular Toxicology*. 2016; 12: 223–236.
138. Hamburg MA. Science and regulation. FDA's approach to regulation of products of nanotechnology. *Science*. 2012; 336: 299–300.
139. Kimbrell GA. Nanomaterial consumer products and FDA regulatory challenges and necessary amendments. *Nanotechnology Law & Business*. 2006; 3: 329.
140. Saner MA, Heafey E, and Bowman DM. Mapping the regulatory environment for nanomaterials in Canada. *Nanotechnology Law & Business*. 2012; 9: 343.
141. Lövestam G, Rauscher H, Roebben G, Klüttgen BS, Gibson N, Putaud J-P, and Stamm H. Considerations on a definition of nanomaterial for regulatory purposes. *Joint Research Centre (JRC) Reference Reports*. 2010; 80: 00–41.
142. Batley G and McLaughlin MJ. *Fate of manufactured nanomaterials in the Australian environment*. CSIRO Land and Water, Clayton, Australia. 2010.
143. Fletcher N and Bartholomaeus A. Regulation of nanotechnologies in food in Australia and New Zealand. *International Food Risk Analysis Journal*. 2011; 1: 33–40.
144. Katao K. Nanomaterials may call for a reconsideration of the present Japanese chemical regulatory system. *Clean Technologies and Environmental Policy*. 2006; 8: 251–259.
145. Garduño-Balderas LG, Urrutia-Ortega IM, Medina-Reyes EI, and Chirino YI. Difficulties in establishing regulations for engineered nanomaterials and considerations for policy makers: avoiding an unbalance between benefits and risks. *Journal of Applied Toxicology*. 2015; 35: 1073–1085.
146. Park D. Nanomaterials in Existing and Emerging Chemical Regulation of China, Japan, Korea, US, and the European Union. *Japan, Korea, US, and the European Union (September 17, 2012)*, 2012.
147. Boverhof DR, Bramante CM, Butala JH, Clancy SF, Lafranconi M, West J, and Gordon SC. Comparative assessment of nanomaterial definitions and safety evaluation considerations. *Regulatory Toxicology and Pharmacology*. 2015; 73: 137–150.
148. Guo J-W, Lee Y-H, Huang H-W, Tzou M-C, Wang Y-J, and Tsai J-C. Development of Taiwan's strategies for regulating nanotechnology-based pharmaceuticals harmonized with international considerations. *International Journal of Nanomedicine*. 2014; 9: 4773.
149. Song T-T, Pan E-Y, and Tseng H-C. Nanorisk governance in Taiwan: Studying potential impacts on the environment, health, and safety. *IEEE Nanotechnology Magazine*. 2012; 6: 15–19.
150. Song TT, Pan EY, and Tseng HC. Nanorisk governance in Taiwan: Studying potential impacts on the environment, health, and safety. *IEEE Nanotechnology Magazine*.2012; 11(2): 15–19.
151. Tanthapanichakoon W, Pornsinsirirak T, Pakawech W, Teeratananon M, and Teparkum S. in *Emerging Nanotechnology Power: Nanotechnology R&D and Business Trends in the Asia Pacific Rim*. World Scientific, Singapore. 2009, pp. 359–390.
152. Karlaganis G, Liechti R, Teparkum S, Aungkavattana P, and Indaraprasirt R. Nanoregulation along the product life cycle in the EU, Switzerland, Thailand, the USA, and intergovernmental organisations, and its compatibility with WTO law. *Toxicological & Environmental Chemistry*. 2019; 101: 339–368.
153. Jarvis DS and Richmond N. Regulation and governance of nanotechnology in China: Regulatory challenges and effectiveness. *European Journal of Law and Technology*. 2011; 2(2): 2.
154. Gao Y, Jin B, Shen W, Sinko PJ, Xie X, Zhang H, and Jia L. China and the United States--Global partners, competitors and collaborators in nanotechnology development. *Nanomedicine: Nanotechnology, Biology and Medicine*. 2016; 12: 13–19.
155. Ahmed A, Kumar A, and Desai PN. Mapping the Indian nanotechnology innovation system. *World Journal of Science, Technology and Sustainable Development*. 2014.
156. Fautz C, Fleischer T, Ma Y, Liao M, and Kumar A.Discourses on Nanotechnology in Europe, China and India. In: *Science and Technology Governance and Ethics*, Springer, Cham, 2015, pp. 125–143.
157. Lavorini F, Buttini F, and Usmani OS. 100 years of drug delivery to the lungs. *Concepts and Principles of Pharmacology: 100 Years of the Handbook of Experimental Pharmacology*. Springer Nature, Switzerland. 2019; 143–159.
158. Karathanasis E, Ayyagari AL, Bhavane R, Bellamkonda RV, and Annapragada AV. Preparation of in vivo cleavable agglomerated liposomes suitable for modulated pulmonary drug delivery. *Journal of Controlled Release*. 2005; 103: 159–175.
159. Loira-Pastoriza C, Todoroff J, and Vanbever R. Delivery strategies for sustained drug release in the lungs. *Advanced Drug Delivery Reviews*. 2014; 75: 81–91.
160. Beck-Broichsitter M, Gauss J, Packhaeuser CB, Lahnstein K, Schmehl T, Seeger W, Kissel T, and Gessler T. Pulmonary drug delivery with aerosolizable nanoparticles in an ex vivo lung model. *International Journal of Pharmaceutics*, 2009; 367: 169–178.
161. Hillery AM and Park K. *Drug delivery: Fundamentals and Applications*. CRC Press, 2016.
162. Pontes JF and Grenha A. Multifunctional nanocarriers for lung drug delivery. *Nanomaterials*. 2020; 10: 183.
163. De Jong WH and Borm PJ. Drug delivery and nanoparticles: applications and hazards. *International Journal of Nanomedicine*. 2008; 3: 133.
164. Card JW, Zeldin DC, Bonner JC, and Nestmann ER. Pulmonary applications and toxicity of engineered nanoparticles. *American Journal of Physiology-Lung Cellular and Molecular Physiology*. 2008; 295: L400–L411.

165. Nel A, Xia T, Mädler L, and Li N. Toxic potential of materials at the nanolevel. *Science*. 2006; 311: 622–627.
166. Steimer A, Haltner E, and Lehr C-M. Cell culture models of the respiratory tract relevant to pulmonary drug delivery. *Journal of Aerosol Medicine*. 2005;18: 137–182.
167. Gill S, Löbenberg R, Ku T, Azarmi S, Roa W, and Prenner EJ. Nanoparticles: characteristics, mechanisms of action, and toxicity in pulmonary drug delivery—a review. *Journal of Biomedical Nanotechnology*. 2007; 3: 107–119.
168. Li N, Sioutas C, Cho A, Schmitz D, Misra C, Sempf J, Wang M, Oberley T, Froines J, and Nel A. Ultrafine particulate pollutants induce oxidative stress and mitochondrial damage. *Environmental Health Perspectives*. 2003; 111: 455–460.
169. Kuzmov A and Minko T. Nanotechnology approaches for inhalation treatment of lung diseases. *Journal of Controlled Release*. 2015. 219: 500–518.
170. Fröhlich E. Toxicity of orally inhaled drug formulations at the alveolar barrier: Parameters for initial biological screening. *Drug Delivery*. 2017; 24: 891–905.
171. Ribeiro RS, Ferreira IM, Figueiredo I Jr., and Verícimo MA. Access to the tracheal pulmonary pathway in small rodents. *Journal Brasileiro de Patologia e Medicina Laboratorial*. 2015; 51: 183–188.
172. Fernandes CA and Vanbever R. Preclinical models for pulmonary drug delivery. *Expert Opinion on Drug Delivery*. 2009; 6: 1231–1245.
173. Hofmann W, Koblinger L, and Martonen T. Structural differences between human and rat lungs: implications for Monte Carlo modeling of aerosol deposition. *Health Physics*. 1989; 57: 41–46; discussion 46-47.
174. Harkema JR, Plopper CG, and Pinkerton KE, in *Pulmonary Immunotoxicology*. Springer, 2000, pp. 1–59.
175. Hruban Z. Pulmonary and generalized lysosomal storage induced by amphiphilic drugs. *Environmental Health Perspectives*. 1984; 55: 53–76.
176. Nassimi M, Schleh C, Lauenstein H, Hussein R, Hoymann H, Koch W, Pohlmann G, Krug N, Sewald K, and Rittinghausen S. A toxicological evaluation of inhaled solid lipid nanoparticles used as a potential drug delivery system for the lung. *European Journal of Pharmaceutics and Biopharmaceutics*. 2010; 75: 107–116.
177. Giaccone G, Ruiz MG, Le Chevalier T, Thatcher N, Smit E, Rodriguez JA, Janne P, Oulid-Aissa D, and Soria J-C. Erlotinib for frontline treatment of advanced non-small cell lung cancer: a phase II study. *Clinical Cancer Research*. 2006; 12: 6049–6055.
178. Anabousi S, Bakowsky U, Schneider M, Huwer H, Lehr C-M, and Ehrhardt C. In vitro assessment of transferrin-conjugated liposomes as drug delivery systems for inhalation therapy of lung cancer. *European Journal of Pharmaceutical Sciences*. 2006; 29: 367–374.
179. Verma NK, Crosbie-Staunton K, Satti A, Gallagher S, Ryan KB, Doody T, McAtamney C, MacLoughlin R, Galvin P, and Burke CS. Magnetic core-shell nanoparticles for drug delivery by nebulization. *Journal of Nanobiotechnology*. 2013; 11: 1.
180. Paranjpe M, Neuhaus V, Finke J, Richter C, Gothsch T, Kwade A, Büttgenbach S, Braun A, and Müller-Goymann C. In vitro and ex vivo toxicological testing of sildenafil-loaded solid lipid nanoparticles. *Inhalation Toxicology*. 2013; 25: 536–543.
181. Chuan J, Li Y, Yang L, Sun X, Zhang Q, Gong T, and Zhang Z. Enhanced rifampicin delivery to alveolar macrophages by solid lipid nanoparticles.*Journal of Nanoparticle Research*. 2013; 15: 1634.
182. Rosiere R, Van Woensel M, Gelbcke M, Mathieu V, Hecq J, Mathivet T, Vermeersch M, Van Antwerpen P, Amighi K, and Wauthoz N. New folate-grafted Chitosan derivative To improve delivery of Paclitaxel-loaded solid lipid nanoparticles for lung tumor therapy by inhalation. *Molecular Pharmaceutics*. 2018; 15: 899–910.
183. Grabowski N, Hillaireau H, Vergnaud J, Santiago LA, Kerdine-Romer S, Pallardy M, Tsapis N, and Fattal E. Toxicity of surface-modified PLGA nanoparticles toward lung alveolar epithelial cells. *International Journal of Pharmaceutics*. 2013; 454: 686–694.
184. Schreier H, McNicol KJ, Ausborn M, Soucy DM, Derendorf H, Stecenko AA, and Gonzalez-Rothi RJ. Pulmonary delivery of amikacin liposomes and acute liposome toxicity in the sheep. *International Journal of Pharmaceutics*. 1992; 87: 183–193.
185. Dokka S, Toledo D, X Shi, V Castranova, and Y Rojanasakul. Oxygen radical-mediated pulmonary toxicity induced by some cationic liposomes. *Pharmaceutical Research*. 2000; 17: 521–525.
186. Mejías JC and Roy K. In-vitro and in-vivo characterization of a multi-stage enzyme-responsive nanoparticle-in-microgel pulmonary drug delivery system. *Journal of Controlled Release*. 2019; 316: 393–403.
187. Menon JU, Kuriakose A, Iyer R, Hernandez E, Gandee L, Zhang S, Takahashi M, Zhang Z, Saha D, and Nguyen KT. Dual-drug containing core-shell nanoparticles for lung cancer therapy. *Scientific Reports*. 2017; 7: 1–13.
188. Karra N, Nassar T, Laenger F, Benita S, and Borlak J. Safety and proof-of-concept efficacy of inhaled drug loaded nano- and immunonanoparticles in a c-Raf transgenic lung cancer model. *Current Cancer Drug Targets*. 2013; 13: 11–29.
189. Ungaro F, d'Angelo I, Coletta C, di Villa Bianca RdE, Sorrentino R, Perfetto B, Tufano MA, Miro A, Rotonda MIL, and Quaglia F. Dry powders based on PLGA nanoparticles for pulmonary delivery of antibiotics: modulation of encapsulation efficiency, release rate and lung deposition pattern by hydrophilic polymers. *Journal of Controlled Release*. 2012; 157:149–159.
190. Xu C, Wang Y, Guo Z, Chen J, Lin L, Wu J, Tian H, and Chen X. Pulmonary delivery by exploiting doxorubicin and cisplatin co-loaded nanoparticles for metastatic lung cancer therapy. *Journal of Controlled Release*. 2019; 295: 153–163.
191. Choi SH, Byeon HJ, Choi JS, Thao L, Kim I, ES Lee, BS Shin, KC Lee, and YS Youn. Inhalable self-assembled albumin nanoparticles for treating drug-resistant lung cancer. *Journal of Controlled Release*. 2015; 197: 199–207.
192. Gadoue SM and Toomeh D. Radio-sensitization efficacy of gold nanoparticles in inhalational nanomedicine and the adverse effect of nano-detachment due to coating inactivation. *Physica Medica*. 2019; 60: 7–13.

193. Ziady A-G, Gedeon CR, Muhammad O, Stillwell V, Oette SM, Fink TL, Quan W, Kowalczyk TH, Hyatt SL, and Payne J. Minimal toxicity of stabilized compacted DNA nanoparticles in the murine lung. *Molecular Therapy*. 2003; 8; 948–956.
194. Asgharian B and OT Price. Deposition of ultrafine (nano) particles in the human lung. *Inhalation Toxicology*. 2007; 19. 1045–1054.
195. Moller W, Felten K, Sommerer K, Scheuch G, Meyer G, Meyer P, Haussinger K, and Kreyling WG. Deposition, retention, and translocation of ultrafine particles from the central airways and lung periphery. *American Journal of Respiratory and Critical Care Medicine*. 2008; 177. 426–432.
196. Anderson PM, Markovic SN, Sloan JA, Clawson ML, Wylam M, Arndt CA, Smithson WA, Burch P, Gornet M, and Rahman E. Aerosol granulocyte macrophage-colony stimulating factor: a low toxicity, lung-specific biological therapy in patients with lung metastases. *Clinical Cancer Research*. 1999; 5. 2316–2323.
197. Skubitz KM and Anderson PM. Inhalational interleukin-2 liposomes for pulmonary metastases: a phase I clinical trial. *Anti-Cancer Drugs*. 2000; 11: 555–563.
198. Markovic SN, Suman VJ, Nevala WK, Geeraerts L, Creagan ET, Erickson LA, Rowland KM Jr, Morton RF, Horvath WL, and Pittelkow MR. A dose-escalation study of aerosolized sargramostim in the treatment of metastatic melanoma: an NCCTG study. *American Journal of Clinical Oncology*. 2008; 31: 573.
199. Rao RD, Anderson PM, Arndt CA, Wettstein PJ, and SN Markovic. Aerosolized granulocyte macrophage colony-stimulating factor (GM-CSF) therapy in metastatic cancer. *American Journal of Clinical Oncology*. 2003; 26: 493–498.
200. Chen EY, Garnica M, Wang YC, Mintz AJ, Chen CS, and Chin WC. A mixture of anatase and rutile TiO 2 nanoparticles induces histamine secretion in mast cells. *Particle and Fibre Toxicology*. 2012; 9(1): 1–10.
201. Heartlein M, DeRosa F, and Smith L. mRNA therapy for argininosuccinate synthetase deficiency. World Intellectual Property Organization patent application publication WO 2015/061500 Al, filed October 22, 2014, and published April 30. 2015.
202. Garbuzenko OB, MinkoT, Ivanova V, Kholodovych V, Reimer DC, Reuhl KR, Yurkow E, and Adler D. Combinatorial treatment of idiopathic pulmonary fibrosis using nanoparticles with prostaglandin E and siRNA (s). *Nanomedicine: Nanotechnology, Biology and Medicine*. 2016; 13(6): 1983–1992.
203. Esteban-González E, Carballido J, Navas V, Torregrosa Z, Muñoz A, and de Mon MÁ. Retrospective review in patients with pulmonary metastases of renal cell carcinoma receiving inhaled recombinant interleukin-2. *Anti-Cancer Drugs*. 2007; 18: 291–296.
204. Huland E, Burger A, Fleischer J, Fornara P, Hatzmann E, Heidenreich A, Heinzer H, Heynemann H, Hoffmann L, and R Hofmann. Efficacy and safety of inhaled recombinant interleukin-2 in high-risk renal cell cancer patients compared with systemic interleukin-2: an outcome study. *Folia Biologica-Praha*. 2003; 49: 183–190.
205. Merimsky O, Gez E, Weitzen R, Peretz T, Rubinov R, Ben-Shahar M, Hayat H, Katsenelson R, Mermershtein V, and Loven D. Targeting pulmonary metastases of renal cell carcinoma by inhalation of interleukin-2. *Journal of Clinical Oncology*. 2004; 22: 4662–4662.
206. Posch C, Weihsengruber F, Bartsch K, Feichtenschlager V, Sanlorenzo M, Vujic I, Monshi B, Ortiz-Urda S, and Rappersberger K. Low-dose inhalation of interleukin-2 bio-chemotherapy for the treatment of pulmonary metastases in melanoma patients. *British Journal of Cancer*. 2014; 110: 1427–1432.
207. Zarogoulidis P, Eleftheriadou E, Sapardanis I, Zarogoulidou V, Lithoxopoulou H, Kontakiotis T, Karamanos N, Zachariadis G, Mabroudi M, and Zisimopoulos A. Feasibility and effectiveness of inhaled carboplatin in NSCLC patients. *Investigational New Drugs*. 2012; 30: 1628–1640.
208. Chou AJ, Gupta R, Bell MD, Riewe KOD, Meyers PA, and Gorlick R. Inhaled lipid cisplatin (ILC) in the treatment of patients with relapsed/progressive osteosarcoma metastatic to the lung. *Pediatric Blood & Cancer*. 2013; 60: 580–586.
209. Verschraegen CF, Gilbert BE, Loyer E, Huaringa A, Walsh G, Newman RA, and Knight V. Clinical evaluation of the delivery and safety of aerosolized liposomal 9-nitro-20(s)-camptothecin in patients with advanced pulmonary malignancies. *Clinical Cancer Research*. 2004; 10: 2319–2326.
210. Verschraegen C. Aerosolized liposomal 9-Nitro-20 (S)-Camptothecin in patients with advanced malignancies in the lungs. *Pneumologie*, 2006; 60, A2.
211. Wittgen BP, Kunst PW, Van Der Born K, Van Wijk AW, Perkins W, Pilkiewicz FG, Perez-Soler R, Nicholson S, Peters GJ, and Postmus PE. Phase I study of aerosolized SLIT cisplatin in the treatment of patients with carcinoma of the lung. *Clinical Cancer Research*. 2007; 13: 2414–2421.
212. Moss RB, Milla C, Colombo J, Accurso F, Zeitlin PL, Clancy JP, Spencer LT, Pilewski J, Waltz DA, and Dorkin HL. Repeated aerosolized AAV-CFTR for treatment of cystic fibrosis: a randomized placebo-controlled phase 2B trial. *Human Gene Therapy*. 2007; 18: 726–732.
213. Hyde S, Southern K, Gileadi U, Fitzjohn E, Mofford K, Waddell B, Gooi H, Goddard C, Hannavy K, and Smyth S. Repeat administration of DNA/liposomes to the nasal epithelium of patients with cystic fibrosis. *Gene Therapy*. 2000; 7: 1156–1165.
214. EW Alton, AC Boyd, SH Cheng, S Cunningham, JC Davies, DR Gill, U Griesenbach, T Higgins, SC Hyde, and JA Innes. A randomised, double-blind, placebo-controlled phase IIB clinical trial of repeated application of gene therapy in patients with cystic fibrosis. *Thorax*. 2013; 68: 1075–1077.

44

Clinical Controversies of Pediatric Aerosol Therapy

Jeffrey Cruz[1] **and Charles Preuss**[2]
[1]*Graduate & Postdoctoral Affairs, University of South Florida Morsani College of Medicine, Tampa, Florida, USA*
[2]*University of South Florida Morsani College of Medicine, Department of Molecular Pharmacology & Physiology, Tampa, Florida, USA*

44.1 Introduction

The usage of aerosolized drugs represents a pathway to tackling challenges related to a number of illnesses. Aerosolized drugs are utilized in an effort to alleviate complications related to bronchospasms, enhance pulmonary blood flow, and treatment or prevention of infections, and can reduce inflammation in the airway (1). As it relates to infants and children, however, the usage of aerosolized drugs is typically provided in a manner that is difficult to assess in terms of efficacy. Most aerosolized drugs are studied in great detail, but in adult populations predominantly. As a result, the confidence in treatment therapy when used in pediatric population cannot be without some hesitation related to the ability to determine whether the therapy is actually providing some relief.

Aerosolized therapy can be provided to a patient through a variety of platforms depending on the frequency, convenience, and efficacy that a particular type of therapy would provide to the patient. When deciding the manner in which therapy is provided, the respiratory problem the patient is facing is considered first and foremost. The major forms of aerosol delivery are inhalers and nebulizers, each containing variations in delivery dependent on the goal, the specific type of which will be mentioned in detail later in this chapter. In short, however, inhalers work by providing pressurized medicine through a small container and provide each dose in a pre-determined amount with each activation of the device. A nebulizer works by aerosolizing liquid medication through the usage of a large machine which is then inhaled by the patient (2). Through the utilization of each delivery platform, aerosolized therapy is able to sufficiently ensure that pulmonary care is given to a variety of patients without the need for them to travel to a hospital in order to seek therapy from a respiratory therapist or other allied healthcare providers.

In this chapter we will discuss the specificities of each major delivery platform as it relates to aerosolized care, the reasons these various platforms exist and are effective, the challenges related to pediatric care with respect to aerosolized therapy, the specific controversies related to aerosol care in pediatric populations, and what the data suggest about variations made in an effort to provide a more effective form of care that is specific to pediatric aerosol therapy.

44.2 Differences in Aerosol Drug Delivery — Adult versus Pediatric Patient Groups

Aerosol drug delivery varies widely between adult and pediatric patient populations. Namely, airway morphology can been seen as the most apparent, as the natural progression of physiological development is what allows for adults to receive larger doses of the drug more effectively. When considering adult airway morphology, it can be noted that the entirety of both the conducting and respiratory zones of the airways are fully developed. With these zones fully developed, airway resistance moving from the environment and through the conducting zone will experience fewer challenges related to resistance. Additionally, a fully developed respiratory zone in the airway ensures that there is the maximum amount of surface area possible to ensure gas exchange between the lung alveoli and the pulmonary capillary system. Lung capacities and volumes are also taken into consideration when understanding the reasons that allow for aerosol therapy to be so effective in adults. Specifically, tidal volume and inspiratory reserve volumes are fully developed in adults. On average, the tidal volume (V_T) in an adult is just about 0.5 l, and the inspiratory reserve volume is 3.0 l (3). With these larger amounts of volume as compared to a developing infant or a child, the pulmonary deposition of aerosol particles during treatment is also larger. By having a larger pulmonary deposition of aerosol particles under the context of treatment, it ensures that the patient is exposed to the greatest possible therapeutic dose to ensure the highest level of efficacy in treatment. In doing so, the ailment that is being treated with the aerosol therapy can be alleviated so that the patient can continue about their daily routine generally uninterrupted.

When considering pediatric patients, morphology definitely comes into consideration because of the clear lack of mature development in pulmonary airways. When compared to adults, younger patients tend to have higher levels of airway resistance due to a small cross-sectional diameter of the airway. Additionally, younger patients have a lower tidal volume, which ensures that in general there is less air moving in and out of the airway (1). As a

DOI: 10.1201/9781003046547-44

result of these major morphological differences in the pediatric population, it is easy to note why the pulmonary deposition of any particular aerosolized therapeutic agent would be significantly less than would be the case if the patient been a fully grown adult. In addition to morphological differences seen in the pulmonary development of young patients, there are also mechanical differences in how air is moved into and out of the lungs. Notably, there are variations in the ratio between the inspiratory–expiratory volumes of air that are moving in the lungs of an infant or child — considerably less than that of an adult (1). With a small ratio in young patients, there is concern related to the amount of drug lost in the time of an entire respiratory cycle. These differences in airway morphology make it difficult for pulmonary therapies to yield positive results when utilizing aerosolized methods of delivery in pediatric patients. Even more important to be considered, there are unique challenges which directly apply to patients who are developing infants. Patients in this demographic of pediatrics are especially vulnerable to spontaneously developing obstacles due to the fact that their lungs are underdeveloped as compared to patients who are older. These challenges include nasal obstruction, inflammation, and airway interference. Because of these issues, which are directly related to this age demographic, it should be noted that infants are at a particularly unique disadvantage when considering how to pursue a successful course of treatment related to the pulmonary system.

There is a wide array of differences between adults and pediatric patients when it comes to the challenges with providing an effective course of treatment related to pulmonary pathologies that require aerosolized mechanisms of therapy. It is because of these challenges in simply providing pulmonary care with aerosols that many studies have been done to determine the best course of action when providing treatment. As it has likely been determined by now, adults — considering development of airway morphology and efficiency of treatment — are the population of study subjects that make up the majority of clinically relevant data correlated to how care should be provided. This, in and of itself, provides stark challenges when considering how to utilize these standards of care when attempting to provide care to a pediatric patient. As mentioned, there are clear anatomical differences from a developmental standpoint that clearly show that infants and children are not at all the same as adults from the standpoint of potentially providing aerosol care. But if there are differences, then why are adults used as the comparative model for how to provide pulmonary aerosol care? A large reason for this is that we lack a significant amount of data done in the field to provide solid proof that particular forms of therapy could work over the others. In fact, general studies related to observation of how aerosols distribute through infant or child lung morphology are limited (4). There are, of course, other reasons adults are used as the model, which we will further investigate in the next section.

44.3 Considerations for Age-Appropriate Therapy

44.3.1 Patient–Physician Factors

As mentioned previously, there are physiologic and morphologic differences when considering the pulmonary system for adults versus children or infants. Because of these differences, there must be considerations taken that acknowledge the differences so that a more age-appropriate treatment protocol can be achieved to alleviate ailments related to pulmonary illness. Implementing these protocols, however, can be difficult when there are limitations–which we can fully acknowledge– regarding how we should be treating pulmonary illnesses in adults versus pediatric populations.

When it comes to treatment of pulmonary illnesses, the consistency of knowledge between various members of the medical team is crucial. This knowledge, however, is variable from provider to provider. Interestingly enough, when it comes to knowledge about how to properly use an inhaler, there is a wide discrepancy between typical primary care physicians, nurses, and respiratory therapists (1). Because of this inconsistency in the usage of appropriate knowledge, it is important to see why it is vital to ensure that appropriate knowledge is being shared and implemented in every pulmonary-related case. It would be crucial to enhance the reliability of treatment protocol by first tackling the challenges related to how the information on care is first given to the patient. Increasing the competency of each level of the medical team would ensure a decreased likelihood of worsening impact on the patient and could even ensure that things related to pulmonary care, such as secondary infections related to device usage, could be reduced.

Another factor that should be considered when attempting to treat a patient with a pulmonary illness is selecting a therapy that would be sustainable. When it comes to therapy selection, often what may be the easiest choice may not be the most effective. By this, the idea of *Occam's razor* is not always the correct choice with therapy. Take, for example, a respiratory ailment that may require treatment with a respiratory therapist at a local hospital. While this therapy may be the easiest route to tackle the ailment because of the direct guidance of care by the respiratory therapy, this route may not be effective because of the effort it may take for the patient to receive treatment. Therefore, the chances of the patient being compliant to this treatment protocol over a period of time may not be as high as one would hope. However, a different course of therapy that could be taken at home or remotely–even one that may not be as effective as one that would be given at a hospital–may be more likely to be sustainable over a longer period of time because of the convenience of the treatment.

These are just some of the administrative factors that should be considered when developing a protocol for a patient seeking care due to challenges related to pulmonary functioning. However, at the root of the protocol being used for therapy, with the patient the actual treatment aspect is the predominating aspect of care.

44.4 Treatment Device Selection

44.4.1 Metered-Dose Inhalers

When it comes to the selection of devices that could be utilized to treat the wide variety of pulmonary diseases there are, the selection is almost endless. Each device is utilized for specific reasons which we will discuss further. Despite the array of devices that have been developed and have come to be used at our disposal, the one that most frequently is known about is the inhaler. However, the main goal of each device is to aerosolize the medication that is needed to be taken by the patient so that the intake of the treatment is done via the lungs. To do this effectively, there are three main types of aerosolized medication generators: metered-dose inhalers (MDI), dry powder inhalers (DPI), and small-volume nebulizers (SVN).

MDIs are perhaps the most commonly known aerosolized medical device when considering what is used to treat pulmonary ailments. MDIs are often used in conjunction with a holding chamber or a spacer. The holding chamber or spacer is used to provide an additional chamber to hold the aerosolized medication after the ejected medication is out of the device so that the user is maximizing the ability to take the medication into their lungs. Interestingly enough, when MDIs were initially utilized, the aerosolized medication was propelled through the device using chlorofluorocarbons (CFC), however, as environmental science research developed studies to prove that CFCs were not good when interacting with the ozone, companies developed MDIs to use non-CFCs instead. Additionally, variations to the MDI have been made to assist in the facilitation of medication into the lung via actuation instead of through the use of a propellant, by way of the pressurized metered-dose inhaler (pMDI). Because of the ease with which aerosolized medication can be taken with this device, there are a number of drugs that can be utilized with this method of therapy (2). What this means is that, depending the severity of the respiratory disease in question, a provider would have a variety of drugs to choose from when treating the patient. Figure 44.1 provides a look at the various types of devices that are currently available to use for the treatment of respiratory disease (2).

While the MDI provides a large amount of utility in treating respiratory illnesses, it does not come without its faults — notably, challenges related to dose synchronization. The main challenge with using an MDI is the fact that it can be difficult to utilize appropriately. MDIs require patients to utilieg the device to synchronize actuation with inhalation. This sort of synchronization is also known as hand–breathing coordination. What makes this coordination difficult is that it requires the patient to alter their tidal breathing pattern in order to account for the fact that they'll be receiving medication during a period of time that will require slow inspiration followed by a prolonged breath-hold. What makes this so challenging for so many patients is that inspiration after the administration of the medication may be too quick or the breath-hold may not be long enough for the medication to settle into the lungs of the patient. Thus, the main issues with metered-dose inhalers is that poor coordination when it comes to its usage means that the patient is not receiving as much of the medication as required by dose (2).

Another challenge with the MDI is that there is no actual way to readily measure the amount of doses that could be

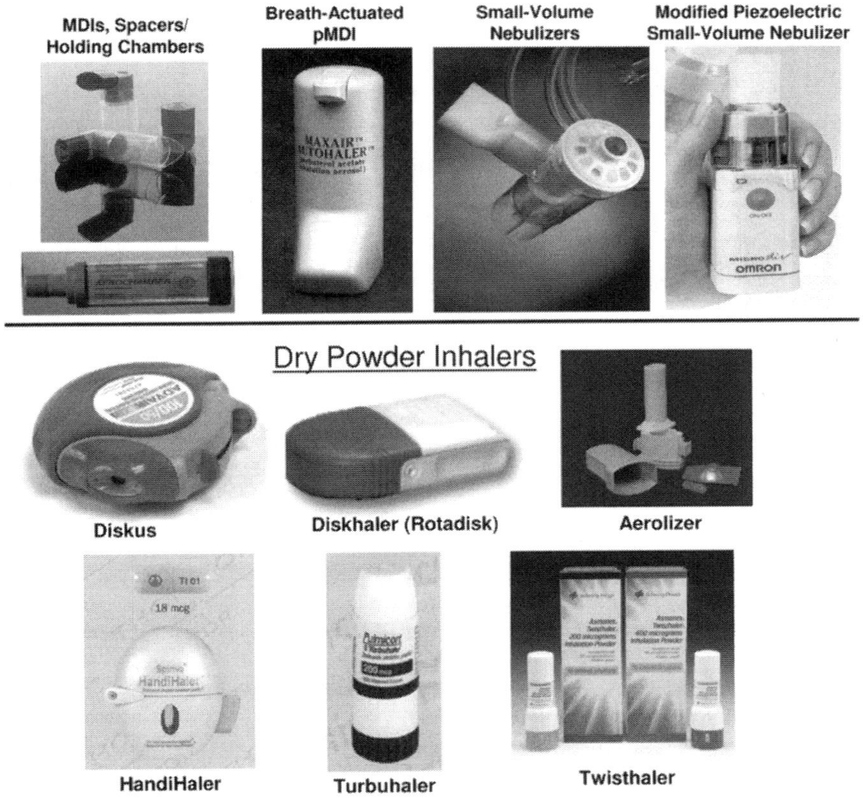

FIGURE 44.1 A Variety of Inhaler Devices Currently Being Used by Patients by the Mode of Aerosolized Medication. Here, Metered-Dose Inhalers Can Be Utilized with or Without a Holding Chamber or Spacer. Additionally, Dry Powder Inhalers Can Be Used, but Not the Variability in Disease. Lastly, Small-Volume Nebulizers Can Be Utilized to Treat Patients, but at the Cost of Portability

provided with any given canister of medication. While the manufacturer of the canister filled with aerosolized medication will often know how many doses per canister could be provided, most users do not know how many doses could be provided. There are ways in which a patient could search for the information needed; however, the fact that additional steps are necessary for the information to be disclosed is inherently the basis of the issue. Interestingly, 54% of patients are unaware of how many doses their MDI could provide them (2). As a result, patients are left to their own devices to decipher whether their MDI is actually out of the aerosolized medication or not. Some patients often utilize auditory clues to determine if they need to refill their prescription at their pharmacy. In fact, 72% of patients in a study done reported that they believed their MDI cannister was empty when they were no longer able to hear a sound when it was shaken (5).

MDI misuse is also another challenge related to why the MDI can make it difficult for patients to utilize aerosolized medication to treat respiratory illnesses. Many of the issues related to MDI misuse are related to the holding chamber or spacer. Recall that the holding chamber or spacer is utilized as a connecting device to the mouthpiece of the MDI. When the canister is activated and the aerosolized medication is expelled, it is first suspended in the holding chamber or spacer. Per proper methodology in MDI usage (i.e. hand–breathing coordination), the aerosolized medicine is then taken from the holding chamber or spacer via inhalation as the patient slowly inspires the medication into their lungs. While many users often utilize their MDI without the holding chamber or spacer, this device is extremely useful in enhancing in the percentage of medication taken into the lungs. Some common problems related to the holding chamber or spacer include incorrect assembly, rapid inhalation, and multiple firings before appropriate. With each problem, the main issue inevitably is that the amount of aerosolized medication that actually gets deposited deep into the patient's lungs decreases. The decrease in aerosolized medication being deposited will affect how much medication is wasted and how many doses it may take for the patient to acquire the adequate amount of effect. Table 44.1 comprises a list of common errors related to holding-chambers and spacers (2).

In addition to errors in technique, there are other features that are less intuitive when it comes to the effectiveness of a MDI. Namely, an electrostatic charge formed within the holding chamber or spacer can alter the ability of the MDI to work appropriately. Aerosolized particles that are discharged upon activation of the inhaler can be lost when the particles collide with the walls of the holding chamber or spacer. Additionally, gravitational forces acting on the aerosolized particles as well as electrostatic attraction between the aerosolized particles and the walls of the holding chamber or spacer can also have an effect on the ability of the particles to then be inhaled by the patient with deposition deep into the lungs (2). The reason these relationships must be considered is because of the technique required in order for an MDI to work properly. Recall, MDIs do not work effectively when the patient activates the inhaler followed by immediate and forceful inhalation into the lungs. In order for proper deposition of the aerosolized medication to occur deep within the lungs, the patient must be able to first correctly coordinate hand–breathing synchronization. Afterward, the patient must slowly inspire so the aerosolized medication can be deposited deep into the lung. To ensure the medication that is deposited in the lungs remains long enough to act on the pulmonary system, the patient must then hold their breath for approximately 10 s followed by slow expiration. Overall, the process of correctly utilizing a MDI can take from 30–40 s, with the slow inspiration of aerosolized medicine occurring for approximately one-third of that time. Figure 44.2 shows the correct breathing required to successfully utilize an MDI compared to what incorrect breathing patterns would look like when utilizing the device (2). During that time the medication that remains in the holding chamber or spacer is exposed to the environment, which can alter the actual amount of medication that inevitably gets inhaled. A study done showed that aerosolized medication delivered via a spacer decreased the actual dose taken in by the patient by almost 66% if there was no delay in inspiration; however, proper delayed inhalation of the medication showed a reduction in the dose by almost 90%. Conversely, the same study also found that reducing electrostatic forces within the spacer with a specialized ionic detergent could increase the amount of aerosolized medication delivered into the patient by roughly 60% (6).

Lastly, multiple MDI actuations also cause issues with the amount of aerosolized medication that is deposited deep into the patient's lungs. When multiple actuations of the MDI are done, there is a reduction in the dose availability when compared to a single actuation of the device. While it is not exactly certain as to why this occurs, observations suggest that the decrease in dose availability is likely due to the prolonged interactions the aerosolized medication would have with the holding chamber or spacer as a result of the proper slow inhalation that is required of the medication in order for it to properly be deposited into the lungs. Research done on this observation showed that the dosage of CFC albuterol delivered per actuation declined by up to 20% with multiple actuations (2).

TABLE 44.1

Common Errors with Holding Chambers and Spacers in MDI Utilization

1. Incorrect device assembly
2. Rapid inhalation
3. Delays related to actuation and breathing
4. Presence of an electrostatic charge in the holding chamber or spacer
5. Lack of knowledge due to limited provider knowledge

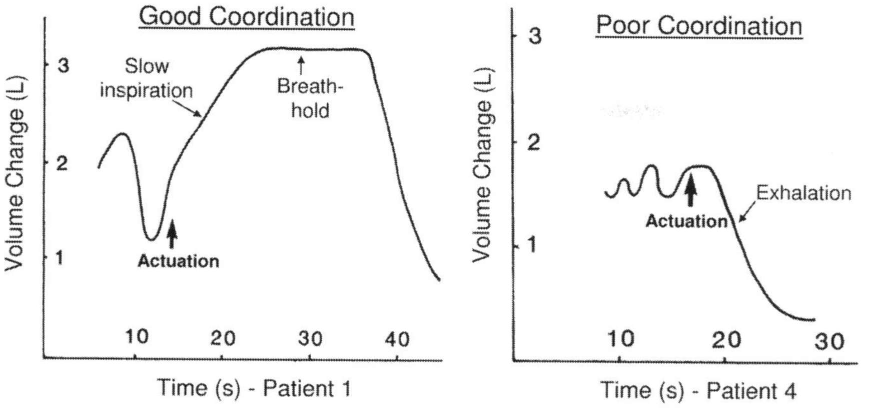

FIGURE 44.2 Note the Comparison between Good and Poor Coordination When Utilizing a Metered-Dose Inhaler. With Proper Hand–Breathing Coordination and Breathing Technique the Concentration of Deep Lung Deposition of Aerosolized Medication is Far GreaterWhen Compared to Those Using Poor Coordination. Variability in Coordination Could D due to Lack of Proper Knowledge Being Transmitted from Medical Team to the Patient.

44.4.2 Dry Powder Inhalers

Dry powder inhalers (DPI) are the second major type of device that provide treatment for those that suffer from pulmonary illnesses. What makes a DPI different from an MDI is that the device is activated via breath-actuation by the individual utilizing the device. Activating the device via breath-actuation means that the inhaler won't release the dry powdered medication until the device is acted upon by the user with a deep enough inhalation to facilitate release. By manually activating the device by breath-actuation only, dry powder inhalers are able to combat some of the challenges related to hand–breath coordination involved with MDI. In a study done, researchers were able to conclude that, compared to MDI users, participants who were utilizing a DPI were able to be categorized into an "ideal" inhaler group based on how effectively users were able to acquire the dosage needed from their device (7).

While dry powder inhalers do have their advantages when it comes to utilization of the actual device, it does not come without a number of challenges that make DPI usage quite difficult outside of a clinical setting. With dry powder inhalers, many of the challenges with successfully utilizing the device stem from issues in technique. In a study done on three major DPIs, Turbohaler, Aerolizer, and Diskus, about one-third of patients utilizing each brand of inhaler had challenges with holding the device correctly. Additionally, approximately 20% of patients also had errors with exhaling through the mouthpiece, not exhaling enough residual lung volume prior to inhalation, not inhaling forcefully enough to engage the breath-actuation processes related to the inhaler, inadequate breath-hold after inhalation, or exhaling into the mouthpiece immediately after the forceful inhalation (8). Each of these errors in technique inevitably decrease the amount of dry powder medication that is able to deposit deep into the lungs of the patient using the inhaler. As a result, the medication being used may not be able to maximize its effect. This would mean that the patient may not be receiving adequate care when attempting to resolve respiratory issues. Additionally, the patient may need to take more doses over a shorter period of time, which could decrease the overall amount of medication being utilized, or could potentially succumb to adverse effects of the drug they are taking more readily as a result of taking the medication more frequently, i.e. more drug side effects. The results from the study did seem to have a correlation to age and frequency of device usage. Most of the patients that reported having errors in technique when it came to utilizing the various dry powder inhalers were younger compared to the overall range of ages of the patients; younger patients on the scale were 18 years of age. Patients that reported errors in technique also utilized their dry powder inhalers less frequently on average (8).

Interestingly, humidity plays a role in the ability of dry powder inhalers to successful administer a dose to the patient when they are utilizing the device. This is because humidity alters the ability of the medication to remain in its powdered form. In fact, humidity causes the powdered medication to clump into variably sized masses. Clumping reduces the ability of the medication to be dispersed into fine particle masses for efficient inhalation. This does beg the question, then: How is the dry powdered medication found within the inhaler exposed to variably humidity such that it compromises the ability of the dry powder inhaler to function normally? An observational study done on humidity and its effects on fine particle dispersion of dry powder medication found that humidity can be formed from the ambient air the DPI is exposed to or through the humid conditions formed in the inhaler when a patient exhales into the mouthpiece erroneously. In fact, in this study about 17–23% of patients being examined were noted to exhale through the mouthpiece, which is consistent with findings from the prior study on errors related to technique (9). Other findings were that 70% ambient humidity was able to reduce the amount of dry powdered medication released by 2 hours, with this reduction lasting up to 4 days post initial dry powder inhaler activation (10). Because humidity can play such a large role when it comes to the altered ability of a DPI to be effective for the patient, the type of drug that is being formulated into a powdered form should be considered.

Another challenge related to dry powder inhalers is the ability of the patient to physically utilize the device. As mentioned previously, a DPI requires that the patient provide enough force during inhalation that the actuation of the device is able to occur. Because of this force-driven breath actuation, the patient who may already have compromised respiratory functioning may not be able to apply the force required to

TABLE 44.2

Common Technique Errors in Dry Powder Inhaler Utilization

1. Holding device incorrectly
2. Exhalation through the mouthpiece
3. Not enough exhalation of residual volume prior to actuation
4. Inhalation not forceful enough to activate
5. Inadequate breath-holding after device activation

receive their dose. Table 44.2 lists common challenges related to technique in patients using dry powder inhalers (2).

44.4.3 Small-Volume Nebulizers

SVNs are the main counter to the inhaler in that SVNs operate differently from a mechanical and usage standpoint with the patient utilizing the device. Most notable is how a SVN operates. Unlike an MDI or a dry powder inhaler, an SVN requires the device to be plugged into a source of electricity. This is because an SVN works to aerosolize the medication being used to treat the patient by mechanically generating enough force by an air compressor to transition the pure liquid form of drug into aerosolized drug droplets that are then inhaled. This transitioning of the medicine via air compressor does take time, however, so it is required for the patient to remain in place anywhere from 60 min to an hour and a half in order to fully acquire the aerosolized medication. Additionally, unlike an MDI or a dry powder inhaler, an SVN requires the patient to utilize a mask or mouthpiece without the need of a holding chamber or spacer. Perhaps the most patient-friendly of the differences between an SVN andeither a metered-dose or dry powder inhaler is that the patient is able to inhale the aerosolized medication with tidal breathing.

Challenges related to SVNs, ironically enough, are not primarily related to the usage of the device itself, but rather in the variability of the manufacturing of the devices and the components that are required to use the device. For example, there are variabilities that exist that alter the ability for a particular SVN to nebulize one solution type compared to another. This inevitably means that a particular SVN may be able to more readily provide aerosolized medication to a patient over another. This does play into account when selecting the type of device to treat a patient that requires a certain medication.

The last major challenge with utilizing an SVN is related to patient adherence to the device itself. A study on compliance with SVNs found that, on average, roughly 60% of all patients using an SVN were able to remain compliant with their treatment protocol. Results also showed that of those who were using an SVN, breathing techniques to inhale the aerosolized medication were relatively effective (11). This would suggest that with patients utilizing a SVN, the issues related to compliance may not be due to difficulties of device utilization as with patients using MDIs or dry powder inhalers.

44.5 Device Comparison

Interestingly, when it comes to the selection of a particular device for treatment of a specific disease, there does not exist at this time a clinical consensus on which device to use. This would suggest that, as it relates to each device earlier discussed, the selection of the device is based upon other factors that are not necessarily related to efficacy of the treatment tools used to combat a particular disease. Each of the three main forms of aerosolized medication devices used to treat pulmonary diseases has its own advantages and disadvantages with respect to treatment.

44.5.1 Metered-Dose Inhaler

The advantages of utilizing the MDI revolves primarily around the portability of the device. The MDI is typically light and compact, meaning that the ability for the patient to easily take the device with them where they need to go is high. Additionally, the device is also easy to use — activation is through the application of pressure on the cannister containing the medication and propellent. These two factors allow for the usage of the device remotely without having to worry about how to methodically utilize the device. The next factors that makes the MDI compelling to utilize are the short treatment times and consistent dosage delivery of the device. By ensuring the treatment times are short, the compliance aspect of the aerosolized medication is more likely to be kept regardless of the age or experience of the patient. Overall, portability and ease of utilization make the MDI a device that could be used anywhere without any issues related to receiving treatment.

The MDI, additionally, is to be used with a holding chamber or spacer which comes with a specific set of advantages. The advantages to using a holding chamber or spacer revolve around maximizing the deposition of the aerosolized medication deep into the lungs of the patient to ensure the efficacy of the treatment plan is sustained. Using a holding chamber or spacer also provides an increased level of comfort for the patient involving the taste of the medication of the mouth. The usage of these auxiliary devices to the MDI reduces the amount of medication in direct contact with the mouth while also ensuring the medication directly passes into the respiratory airways. Additionally, the holding chamber or spacer also allows for an ease of comfort related to the hand–breathing coordination that must be done with all MDIs (12).

The concept of hand–breathing coordination brings up the topic of the possible disadvantages of utilizing an MDI. As mentioned through the chapter, the utilization of an MDI requires the patient to have a proficient level of hand–breathing coordination. Recall that hand–breathing coordination involves the patient's ability to synchronize their inhalation with the time at which they activation their MDI. This coordination maximizes the efficiency with which the aerosolized medication is deposited deep into the lungs without being impeded by anatomical blockages such as the mouth or throat. While the concept of hand–breathing coordination may not seem difficult, it can be quite cumbersome to those patients that are new to the device and for children. It does require a bit of finesse to ensure that the dosage is accurately given. This brings the next disadvantage into the discussion — manual activation by the patient. While the ability to easily carry the device is quite appealing, the fact that the patient must then activate the device on their own can be a formidable challenge. This primarily revolves around the fact that the patient must have enough knowledge to properly prepare the device and then activate the device properly, so the dosage is given appropriately without any issue. Another challenge with utilizing an MDI is the fact that improper care of the device could lead to build up of residue in the mouthpiece which could then be inhaled during a future activation. This would then suggest that maintenance is extremely important to ensure the device is clear of any unwanted debris, which would require additional training and knowledge that must be done for the duration of the time using the device. Lastly, improper usage of the MDI would also mean that the aerosolized medication could end up in an unwanted area, such as the mouth or throat, instead of deep in the lungs where the medication is needed (12). This would then bring about difficulties with unwanted taste or possible taste aversion related to poor usage technique.

Disadvantages with the holding chamber or spacer are also quite persistent. The main disadvantage with a holding chamber or spacer is the fact that its size is counterintuitive to the portability of the MDI by itself. A larger piece to attach to the compact MDI may affect compliance because of the inconvenience it brings. The concept of an additional piece would also suggest that the patient must know how to apply the attachment to the MDI. If the holding chamber or spacer is not used properly, then it can affect the dose availability of the aerosolized medication, which was mentioned earlier. The next major disadvantage is the fact that the holding chamber or spacer can make utilizing the device itself increasingly more difficult. The MDI requires the patient to have good hand–breathing coordination. Using a holding chamber or spacer would also require the patient to time the hand–breathing action more accurately to consider the fact that the medication will first be introduced to the space, prior to being inhaled. This means that if the patient does not have proper coordination, potential loses of medication could be accrued. Lastly, the holding chamber or spacer must be routinely cleaned. This disadvantage primarily revolves around inconvenience, since the MDI provides a sense of portability that is challenged by the utilization of an additional modification (12).

44.5.2 Dry Powder Inhaler

The dry powder inhaler is an alternative to the MDI. Like the MDI, the dry powder inhaler is small and portable to carry around in a pocket or bag. The advantage of this pocket-sized device is that it is easy to transport between locations without having to give much thought to mobility. Additionally, treatments using a dry powder inhaler are also quick, like the treatments related to an MDI. Portability and quickness of treatment provides the most convenient form of respiratory treatment because these devices allow the patient to determine when to facilitate their own therapy. This allows for a more independent approach to care while also ensuring a relatively high form of efficacy for care. Most interestingly, the dry powder inhaler has a built-in dose counter. This is extremely important to consider when discussing the advantages of this particular device, because not having a dosage counter is a huge source for error within patient groups that utilize the MDI. Having a dosage counter is useful to accurately determine how much medication remains in the inhaler. This is important is because it provides the patient utilizing the inhaler with confident knowledge about the medication they actually have on hand. By understanding how much medication remains in the dry powder inhaler, the patient utilizing the device should have a better grasp of what remains and when they should begin considering refilling their current prescription. Ideally, the patient should not have to worry about whether they may run out of medication during a particularly stressful event, such as a respiratory attack related to asthma. Lastly, as mentioned earlier, the dry powder inhaler does not require any form of hand–breathing coordination to properly activate and utilize the respiratory device (12).

While the advantages of the dry powder inhaler revolve around the device being user friendly and portable, it does come with its disadvantages. To start, the activation of the device can be challenging for patients. The reason activation can be difficult to achieve is because the dry powder inhaler requires a forceful enough inhale in order to achieve manual breath actuation of the device. Activation of the device does not necessarily suggest that it is impossible; however, if the patient has difficulties with generating forceful inhalation or has a respiratory disease that impedes their ability to fully exhale followed by complete inhalation, this device would be challenging to use. Additionally, because the device does require forceful inhalation and because the medication found within the device itself is in a powdered form, it is hard to discern with confidence whether the powdered medication has fully achieved deep lung deposition. This unreliability in discerning the amount of deposition achieved is related to the fact the patients utilizing a dry powder inhaler are sometimes unable to feel whether the medication is traveling deep within the lungs or not.

Humidity also plays a factor in the efficacy of the device, as mentioned earlier. Because the medication that is being used with a dry powder inhaler is technically a solid, variability in the environment due to humidity can cause the medication to clump up within the mouthpiece of the device. This clumping inevitably reduces the dose availability to the patient using the device. Therefore, the clumping renders the dry powder inhaler less effective for treating the patient's respiratory illness. Humidity within the dry powder inhaler can develop through a number of avenues, but the more frequent causes are related to improper breathing technique when utilizing the device. If the patient exhales earlier then they are supposed to when utilizing the device, the air from their breath is what causes the environmental changes within the chamber of the dry powder inhaler. Lastly, there is manufacturing variability with dry powder inhalers, which makes it difficult to provide a level of consistency when providing care. Because of these variabilities, not all dry powder inhalers can be utilized to provide care with a specific medication. As a result, the patient may require multiple forms of medication in a powdered form, i.e. multiple dry powder inhalers that all require different levels of knowledge in order to properly utilize and care for each device. The accumulation of multiple devices for treatment may cause a bit of confusion on the part of the patient (12).

44.5.3 Small-Volume Nebulizers

Unlike both the MDI and the dry powder inhaler, the small-volume nebulizer uses a completely different interface to provide treatments for patients. As a result, the SVN comes with unique advantages and disadvantages that trade off portability for enhance efficacy of the form of therapy. Most notable of the advantages to using an SVN is that it requires minimal patient cooperation or coordination for treatments. This is important because, unlike an MDI that requires hand–breath coordination, or a dry powder inhaler that requires a strong enough inhalation to trigger breath actuation, an SVN can be utilized with normal tidal breathing. With only tidal breathing required to receive treatment, the small-volume nebulizer adds a level of comfort when utilizing because there is no requirement for complex coordination to be done for the device to be properly used in therapy. Because of this minimal level of coordination required, the SVN is an excellent option for patients who are not able to successfully receive treatment from an MDI or a dry powder inhaler. Further, the small volume nebulizer is able to deliver a number of medications in aerosolized form with variable concentrations simultaneously. This allows for treatment styles of greater complexity because both the MDI and PDI devices allow for a therapy of only one drug at a time. It is clear to see that the SVN provides care as a sort of safety net option when both the metered-dose and dry powder inhalers are not options. There are, however, a number of disadvantages to the SVN that are detractors when considering for therapy using aerosolized medication (12).

One is the size of an SVN. Compared to both the metered-dose and dry powder inhalers, the SVN is significantly larger and is not nearly as portable as either of the other devices. Additionally, a power source is required in order for the SVN to aerosolize the medication to provide the treatment required by the patient. These two aspects of the SVN greatly reduce the amount of portability and ease with which treatment can be administered. Additionally, the SVN does require a patient receiving treatment to remain in place for anywhere from 15 to 25 min on average for the aerosolized medication to fully be inhaled by the patient. Because of this lengthy duration of time in which the patient must remain attached to the apparatus, compliance with the therapy is also in question because using an SVN does require a serious commitment. Lastly, because there are multiple compartments that are required to put the full nebulizing device together, substantial cleaning is required in order for the device to operate appropriately. As a result of cleaning being required, there is a chance that poor cleaning could lead to pathogenic buildup of microbes, e.g. bacteria. In short, the SVN is not a convenient device to utilize. However, that tradeoff is a level of treatment that is easy to perform by the patient (12)¾

44.5.4 Difficulties with Aerosolized Device Treatment

A review of each of the main devices utilized in therapies related to pulmonary illnesses by means of aerosolized medication has shown that each of the three primary devices utilized has a varying set of knowledge that must be known for the patient to use the device effectively. Without the proper knowledge on behalf of the patient, the treatment received has the potential to be rendered less effective and could put the patient at risk for additional pulmonary complications to develop. Herein lies the challenge related to why respiratory treatment by way of aerosolized medication often is not as consistent as it should be with respect to quantifying data points to suggest which form of care may be more effective. Because the effective use of any given device is dependent on how knowledgeable the patient is at ensuring the device is being used appropriately, it begs the question as to how the knowledge or lack thereof is transmitted from healthcare professional to patient. By understanding this connection, it can help us better understand why devices using aerosolized medication have varying levels of efficacy outside of the physical limitations with each device (12).

When it comes to usage of an MDI, the knowledge that is required in order for maximum efficacy to be achieved can be variable. Shockingly, there is a significant lack of knowledge at multiple levels of the healthcare team as it pertains to the correct use of an MDI. A study done by Interiano and Guntupalli utilized surveys to determine how effectively various members of the healthcare team could usen a MDI with the instructions they had to teach a patient. Results showed that 85% of respiratory therapists could effectively utilize a MDI with good performance. Interestingly though, only 4% of nurses could effectively use the MDI such that they could receive good performance levels (13). Another study with similar methodology, done by Guidry et al., showed that only 65% of internal medicine physicians and 57% of non-pulmonary medical personal could properly utilize an MDI such that they could achieve a minimum of four required steps in order to properly utilize the device (14). These results demonstrate that there is a significant amount of variability in the confidence of different healthcare professionals to utilize a MDI. This is deeply concerning, because if there is such a staggering amount of variability within healthcare personnel, this could easily translate to just as much variability in patients on pulmonary medicines. If this variability does exist in patients, it would then translate to poorer health outcomes due to inappropriate aerosolized medication dosages that could hinder the ability of the patient to improve from a respiratory health standpoint.

Similar issues could be seen when it comes to utilization of a dry powder inhaler. A study done by Hanania et al. showed that when it came to the appropriate use of DPIs, respiratory therapists, physicians, and nurses had only 60%, 21%, and 12%, respectively, of the inhaler knowledge that would enable them to confidently teach appropriate device usage. Interestingly, however, this same study showed that the percentage of those that could prove a sufficient level of device usage was roughly proportional to the length of time the device had been in use. Meaning that the longer the device had been available for use in healthcare, the more likely the healthcare professional was to be able to demonstrate appropriate technique in usage (15). This could be explained by the fact that the longer a device is on the market for usage, the more likely it is for a healthcare professional to be knowledgeable about its correct use. This does not, however, condone the fact that there is such large variability in the knowledge of healthcare professionals when it comes to utilization of the dry powder inhaler.

The only device not mentioned yet is perhaps the device that has the fewest issues related to a lack in knowledge of the provider or other members of the healthcare team. The SVN does not have too many issues related to inconsistencies in the ability to use or educate others to use the device. This is perhaps due to the fact that of the three main devices used to provide patients with aerosolized medication, the SVN is arguably the easiest device to use, albeit at the cost of portability and efficiency.

44.5.5 The "Universal" Design

Each of the three primary devices used in respiratory care when treating patients — MDI, DPI, SVN — has a number of issues related to its usage and how it can provide effective care to patients dealing with illnesses that require aerosolized medication. It has been stated that "management of chronic airway disease is 10% medication and 90% education" (16). What this would suggest is that proper education of the patient is required in order for any treatment to be successful. The issue is that that providing a sufficient level of education that could warrant such success is difficult to acquire. The reasoning for this is that each device requires a certain level of knowledge on behalf of the healthcare professional in order for them to correctly transmit this information to the patient. The likelihood is that the same healthcare provider must also know information for each device that could be utilized. This could create a level of confusion that could then be transmitted to the patient and cause them to inaccurately utilize the device, preventing appropriate treatment. Additionally, if the patient uses more than one inhalation device to support care, there is risk for more confusion on their end, as well. A study done by Serra-Batlles et al. (17) compared the features of each device and calculated which features were most important to create the most ideal inhaler. The features that ranked the highest were ease of use during a respiratory event, having knowledge of the doses remaining, and overall ease of use (17). As mentioned previously, each device has its own set of advantages and disadvantages, making no particular device more ideal than another. Because of all of the different aspects of each device's design that must be considered when selecting a device, scientists and clinicians cannot help but wonder if there is a simpler way to provide respiratory treatment without having to deal with all of the confusion. Instead, perhaps creating a device that essentially combines the best features related to each of the primary pulmonary devices — a sort of "universal" inhaler. Table 44.3 shows a list of ideal features participants believed would allow for easier utilization of aerosol medicine devices (17).

This universal inhaler would be one general type of device that could be utilized to treat any respiratory illness by being able to use all of the available aerosolized drugs on the market. By standardizing the formation of the device and all aspects involved, there is a likelihood of reduction in the amount of confusion related to using each type of device. This reduction in confusion would be a benefit to both the provider and, more importantly, the patient. With a more simplified design, any kind of patient, regardless of cognitive or physical limitations, would be able to utilize the device. Figure 44.3 shows a concept of the idea behind a universal inhaler(2).

44.6 Controversies with Aerosol Therapy in Pediatrics

Considering the various devices that could be utilized to provide aerosolized medicine to respiratory patients, it can be understood why there are difficulties with creating therapies that require these devices to treat patients from the pediatric demographic. Recall that the morphology of the airway in a child or infant is much different than an adult airway. Because of this, there must be considerations taken that account for these morphological differences. Additionally, there is also an aspect of comfort and temperament that must be accounted for when considering what types of aerosolized care that can be provided

TABLE 44.3

Ideal Features for the Optimal Inhaler

Feature	Percentage of Participants Interested in Feature
1. Easiness of usage during respiratory attack	82.8
2. Knowledge of remaining doses	62.1
3. Overall easiness of use	60.9
4. Simplicity in learning how to utilize	59.2
5. Hygiene	58
6. Comfort of pieces	53.3
7. Portability	40.8

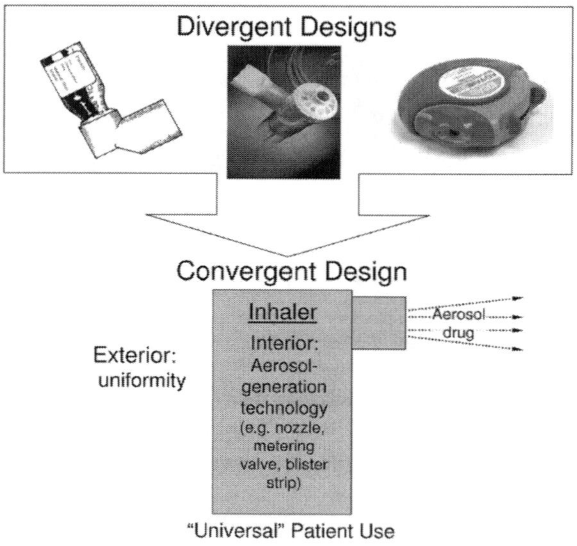

FIGURE 44.3 General Concept behind the Idea of the "Universal Inhaler" To Be Used by All Patients with Respiratory Illnesses. The Convergent Design Allows for Simplicity of Usage and Reduction in Confusion Related to the Educational Aspect of Therapy Design.

to maximize the treatment being provided. All of these will be considered as this section is discussed. Earlier, it was discussed how each device can provide a level of treatment that can be effective depending on portability and efficacy of the inhalation device. These varying advantages and disadvantages for each device primarily apply to adult patients. In order to better understand the totality of this topic, a discussion on how each device is utilized in pediatric populations and why they may not be as ideal as originally thought will be next.

When providing aerosolized medicine to children, comfort and efficiency of the drug dosage form are key items that are considered. Common devices used in pediatric care for respiratory illnesses include the MDI, dry powder inhaler, and SVN, all of which have been covered to a great extent. However, when it comes to treating pediatric patients, other variations of nebulizers can be encountered due to the fact that pediatric patients vary in age and the physical ability to conduct certain actions that are required to activate devices such as an inhaler. The variations of the nebulizer include ultrasonic nebulizers, breath-actuated nebulizers, and vibrating mesh nebulizers. The ultrasonic nebulizer utilizes drug crystals in an apparatus that vibrates them at high frequency in order to generate the force required to aerosolize the medication. Mesh nebulizers utilize a pump to force the liquid medication through a series of apertures within multiple pieces of mesh that vibrate quickly to generate the aerosol used in treatments (18). While both of these devices are effective at providing aerosolized medication that has efficient lung deposition, the largest deterrent is that these devices are costly. Therefore, the utilization of these devices would likely be reserved for patients who are in dire need of treatment and cannot tolerate other devices.

The reason these variations of the SVN were created is due to the fact that the most effective form of inhaled aerosolized drug treatment in children included a mouthpiece. The mouthpiece allows for enhanced aerosolized drug deposition with inhalation. However, infants do not have the ability to use their facial muscles to form the necessary placement required to use a mouthpiece. As a result, there will often be leaks, allowing medication to escape, which renders the treatment less effective (1). The nebulizers ensure that treatment can be done without having to worry about too much waste related to poor technique.

With treatments facilitated with a mouthpiece having been shown to be effective, clinicians must now consider what possible interfaces could be selected to deliver aerosolized medication. Unfortunately, a mouthpiece may not work for all pediatric patients because of the difficulties it could create in infant patient cohorts. To combat this, mask variations have been made to provide treatment to those that cannot utilize a mouthpiece. Because the utilization of a mask could prove to be beneficial, the ability for an infant to tolerate a mask should be considered. For the mask to be tolerable, the material will play a huge factor. Ideally, the mask would be soft, warm, and pliable, since an infant will likely move during treatment. Another reason a mask may be effective is due to the fact that if the infant has to undergo treatment with a nebulizer, the time it takes to receive treatment could also be a factor that could agitate the infant. Should neither a mouthpiece nor mask work, a hood could also be utilized as a tool to assist in drug delivery (1).

In addition to the tools being used to assist with aerosolized drug delivery, there comes the question of how treatment should be done if the patient is crying. While this may not be as much of an issue with an older child, in infants, crying could make treatment more difficult because the distress of the infant would make it more difficult for aerosol medicine to reach the lungs from the device being used. Additionally, the crying of an infant could occur due to a number of reasons related to discomfort of the mask, sound coming from a device such as a nebulizer, requirement of patient to be in a certain position for a prolonged period of time, or other physiological needs due to things like hunger. If the patient is uncomfortable, tolerance often decreases, making treatment more difficult. Which then begs the question of whether or not the clinician should continue pursuing treatment if the patient is already in distress. For moral reasons it is clear to see why it would be appropriate to stop treatment; furthermore, studies have also shown that when an infant is in distress they are four times less likely to receive adequate treatment versus when the infant is calm and receiving their medication (19). Additionally, when considering lung deposition with aerosolized medication treatment, a study showed that when an infant receives treatment while they are calm or asleep that drug deposition into the more distal airways of the lungs wasincreased by as much as five times the amount versus an infant who was in distress (20). Considering these studies, it is clear that the preferable method of treatment would be to perform them when an infant is sleeping or calm.

Because it is preferred that an infant be calm or sleeping when treatment is given, the method in which the treatment inevitably is provided becomes an area of focus. Multiple methods have been developed that could potentially provide successful ways to provide treatments to pediatric patients, including infants, to enhance the efficacy of aerosol therapy. One such method is *blow-by aerosol therapy*. This method is performed with a gas-powered nebulizer with a face mask that is typically funnel shaped and not molded to the patient's face; unlike other nebulizer masks, it is placed at a reasonable distance away from the patient such that the gas-powered nebulizer generates a stream of air that is directed at either the oral or nasal airways. Because there is no direct contact with the patient, this method is quite attractive for those who may not be able to tolerate a mouthpiece or mask. The idea behind this method is that it provides a steady flow that is directed quickly enough at the patient that they receive aerosolized medication without its being too invasive. It is because of this that proponents have gone as far as believing that the minimum distance required for successful treatment to be acquired is less than half an inch in distance from the face. However, those opposed to this treatment style believe that the distance between the source of medication and the patient provides too much space to truly be a successful form of care, and therefore is a waste of resources and money (21). Studies have suggested that, because the blow-by method does not require the use of a properly fitting mask, as much as 85% of the dose could be lost; however, not enough studies have been done in great detail to determine how effective this method actually is. In theory, however, this method could prove to be beneficial for those, such as infants, that are easily distressed to receive treatment with aerosolized medication.

The next treatment considered is *continuous nebulization*. As its name suggests, in continuous nebulization uninterrupted aerosol medication is provided to the patient in an effort to provide a large dose of medication deep into the patient's lungs. To do this, the nebulizer produces a large volume of liquid medication which is continuously aerosolized to deliver as much medication as possible. Interestingly, this form of therapy is used in actual clinical settings, specifically, the pediatric ICU. Because of the continuous delivery of medication, patients are able to receive large dosages without having to receive additional treatments for a large segment of time. The challenges with this form of therapy are related to the fact that there is continuous treatment being administered over a long period of time. As a result of this long period of treatment, a child or infant could be disturbed because of the noisy machine or the temperature of the mask. Because of this, the tolerability of the form of treatment could be questionable based on age (1).

The last form of therapy that has been considered is through primarily noninvasive pathways — humidified high-flow cannula, nasal CPAP, and positive-pressure devices. The reason these devices have stirred significant interest in the scientific and medical communities is because they provide a potential form of treatment that has been proven to work for adults and could be easier for pediatric patients to tolerate. With the expansion of the multiple variations that have developed from the nebulizer, such as the ultrasonic or vibrating mesh, a lot of the non-invasive routes have become of significant interest. Unfortunately, initial studies on these more non-invasive devices have proven to be less effective than other means to provide respiratory care that have already proven to be effective. A lot of the challenges related to these devices is due to the fact that without a certain level of invasiveness, there is more potential for aerosolized medication to be lost, therefore decreasing the actual dosage that is being utilized in providing care to the patient (1).

44.7 Pediatric Models of Study

There are a number of devices being developed that may be able to successfully provide pulmonary care for patients who must use aerosolized medication. The issue related to the devices that have been mentioned thus far is that they are not nearly as effective in pediatric populations, especially infants, because they require a certain level of coordination and tolerance in order to successfully provide treatment. In infants, however, the tolerability of a particular device can vary depending on how comfortable the device is and how much of a disturbance the device can potentially produce as a result of operating. Interestingly, when devices are created that deliver aerosolized medication to patients, the model of the potential patient that could use the device is, by an overwhelming majority, an adult (1). Utilizing data from studies on various respiratory devices from adult study subjects creates difficulties with applying the same knowledge to pediatric populations. The primary reason for this is, as mentioned earlier in the chapter, the fact that children do not have the same pulmonary morphology as adults. Therefore, pediatric patients will respond differently to devices primarily driven by adult-focused processes. By creating devices that are used in studies where the models of the study are those from a pediatric patient demographic, it is hoped we can find a way to enhance pulmonary care for those using aerosol medication.

44.8 Conclusion

Throughout this chapter, it has been discussed which devices are currently available for providing aerosolized medicine treatment in respiratory diseases and how one inhalation dosage form compares to another. Additionally, it has been discussed how these devices apply to pediatric populations and why the same devices may not be as effective in pediatric patients. The reason these differences exist is the airway morphology that differs between a child or infant and an adult. While these differences can reduce the efficacy of the devices that are primarily used, there are developments being made that could potentially result in enhanced care for pediatric patients. It is important to note that many of the challenges related to treating pediatric patients with the devices that are used clinically occur because the research subjects that are used for clinical studies are adults. Because the emphasis is on adult patients, there is an inherent disconnection between how effective these devices could be for children or infants. By creating models of study that use devices focused on the anatomical and physiologic differences that growing infants and children have, we hope to find a way to enhance the efficacy of aerosol treatment with deep lung airway drug deposition as well as the ability to provide this treatment such that it is comfortable for the patient. We hope that doing so will encourage continued development in inhalation dosage forms that can help combat more chronic respiratory illnesses with a high degree of efficacy earlier in life so that quality of care can provide the patient a more ideal control of the challenges they face in dealing with respiratory illnesses.

REFERENCES

1. DiBlasi RM. Clinical controversies in aerosol therapy for infants and children. *Respiratory Care*. 2015; 60(6): 894–916.
2. Rau J. Practical problems with aerosol therapy in COPD. *Respiratory Care*. 2006; 51(2): 158–172.
3. Costanzo L. Respiratory Physiology. In: *Physiology* (5th Ed., pp. 186–236). Philadelphia, PA: Saunders. 2014.
4. Salmon B, Wilson NM, and Silverman M. How much aerosol reaches the lungs of wheezy infants and toddlers? *Archives of Disease in Childhood*. 1990; 65(4): 401–403.
5. Rubin BK and Durotoye L. How do patients determine that their metered-dose inhaler is empty? *Chest*. 2004; 126(4): 1134–1137.
6. Wildhaber JH, Devadason SG, Hayden MJ, James R, Dufty AP, Fox RA, et al. Electrostatic charge on a plastic spacer device influences the delivery of salbutamol. *European Respiratory Journal*. 1996; 9(9): 1943–1946.
7. Brocklebank D, Ram F, Wright J, Barry P, Cates C, Davies L, et al. Comparison of the effectiveness of inhaler devices in asthma and chronic obstructive airways disease: A systematic review of the literature. *Health Technology Assessment*. 2001; 5(26): 1–149.
8. van der Palen J, Klein JJ, van Herwaarden CLA, Zielhuis GA, and Seydel ER. Multiple inhalers confuse asthma patients. *European Respiratory Journal*. 1999; 14(5):1034–1037.
9. Melani AS, Zanchetta D, Barbato N, Sestini P, Cinti C, Canessa PA, et al. Inhalation technique and variables associated with misuse of conventional metered-dose inhalers and newer dry powder inhalers in experienced adults. *Annals of Allergy, Asthma, & Immunology*. 2004; 93(5): 439–446.
10. Meakin BJ, Cainey J, and Woodcock PM. Effect of exposure to humidity on terbutaline delivery from turbuhaler dry powder inhalation devices. *European Respiratory Journal*. 1993; 6(5): 760–761.
11. Corden ZM, Bosley CM, Rees PJ, and Cochrane GM. Home nebulized therapy for patients with COPD: Patient compliance with treatment and its relation to quality of life. *Chest*. 1997; 112(5): 1278–1282.
12. American Association for Respiratory Care (AARC). 2017. A patient's guide to aerosol medication delivery. https://www.aarc.org/wp-content/uploads/2018/01/aerosol-guides-for-patients-3rd.pdf
13. Interiano B and Guntupalli KK. Metered-dose inhalers. Do health care providers know what to teach? *Achieves of Internal Medicine*. 1993; 153(1): 81–85.
14. Guidry GG, Brown WD, Stogner SW, and George RB. Incorrect use of metered dose inhalers by medical personnel. *Chest*. 1992; 101(1), 31–33.
15. Hanania NA, Wittman R, Kesten S, and Chapman KR. Medical personnel's knowledge of and ability to use inhaling devices: Metered-dose inhalers, spacing chambers, and breath-actuated dry powder inhalers. *Chest*. 1994; 105(1): 111–116.
16. Fink JB. Inhalers in asthma management: Is demonstration the key to compliance? *Respiratory Care*. 2005; 50(5): 598–600.
17. Serra-Batlles J, Plaza V, Badiola C, and Morejon E. Inhalation Devices Study Group. Patient perception and acceptability of multidose dry powder inhalers: A randomized crossover comparison of Diskus/Accuhaler with Turbuhaler. *Journal of Aerosol Medicine and Pulmonary Drug Delivery*. 2002; 15(1): 59–64.
18. Ari A. Jet, ultrasonic, and mesh nebulizers: An evaluation of nebulizers for better clinical outcomes. *Eurasian Journal of Pulmonology*. 2014; 16(1): 1–7.
19. Iles R, Lister P, and Edmunds AT. Crying significantly reduces absorption of aerosolized drug in infants. *Archives of Disease in Children*. 1999; 81(2): 163–165.
20. Murakami G, Igarashi T, Adachi Y, Matsuno M, Adachi Y, Sawai M, et al. Measurement of bronchial hyperreactivity in infants and preschool children using a new method. *Annals of Allergy, Asthma & Immunology*. 1990; 64(4): 383–387.
21. Baty R. Blow-by revisited. *Respiratory Care*. 2008; 53(7): 921–922.

45

Drug Delivery Using Aerosols: Challenges and Advances in Neonatal Pediatric Subgroup

Surovi Saikia[1], Aparoop Das[2], and Yashwant Pathak[3,4]
[1]Natural Products Chemistry Division, CSIR North East Institute of Science & Technology, Assam, India
[2]Department of Pharmaceutical Sciences, Dibrugarh University, Dibrugarh, India
[3]USF Health Taneja College of Pharmacy, University of South Florida, Tampa, Florida, USA
[4]Adjunct Professor, faculty of Pharmacy, Airlangga University, Tampa, Surabaya, Indonesia

45.1 Introduction

Frequent administration of aerosolized drugs to pediatric patients is often for facilitating mucus clearance, preventing and treating infections, enhancing pulmonary blood flow, etc., to lessen airway inflammation, and increases bronchospasm etc. Generally, off-label aerosolized drugs are used, which include those that have not been passed to be used as aerosols in neonatals, but in adults, and are being cleared by the US Food and Drug Administration (FDA) for use in different delivery devices. Efficacy and dose ranging studies are seldom performed by drug manufacturers and the overall proposition shows that drug delivery to patients seems rather safe and easy as the patient keeps inhaling the plume produced by the nebulizer. But it is likely that very small amounts of drug reaches the targeted sites and ways to measure this amount are highly challenging and confusing. In younger children, this estimation of efficacy is tedious for clinicians due to a lack of objective measurement techniques (1). Also, the challenges and difficulties associated with infants for aerosol delivery are huge as compared to adults.

A successful aerosolization procedure depends on the aerosol system used, which includes the drug, the target site, aerosol device, and respiratory system of the patient (for mechanically ventilated patients, the ventilator is also added to this list). The performance evaluation of the aerosol system is estimated using the overall emitted dose from the device (ED), the delivered dose to the lungs (the fine particulate factor, or FPF), and bioavailability of the lungs (2,3). ED and FPF are determined in vitro and mostly depend on the design of the device and particulate properties, and patient factors such as anatomy of lung and airway, membrane permeability to drugs, drug metabolism, and clearance of lungs (phagocytic) including FPF (3).

45.2 Issues with Neonates: Different in Many Ways

Neonates can neither be classified as small children nor small adults when viewed from a drug development perspective which encompasses term, preterm, and post-term babies. Adding 27 days to the day of birth defines the neonatal period for term and post-term babies, and for preterm babies this period is 27 days plus the expected day of birth (4). They have small intestinal transit time and surface area, slow gastric emptying and immaturity in transporting, lack of well-formed skin barrier, poor respiratory function, etc., which are critical for oral drug delivery. Other physiological factors, such as surface-to-body volume ratio, pH of of the gastrointestinal tract (GI), ratio of body fat to lean tissue, etc., are also dynamic with time (5).

With a relatively large tongue size in a small oral volume, infants are categorized as obligate nasal breathers. Also, the larynx and epiglottis are in close proximity to the base of the tongue. Added to these, babies have a shorter turbinate region, small nostrils, narrow pharynx–larynx and nasopharynx, and a large anatomical dead space (4). During early development, the fetus has a fully performing airway which changes significantly in the first few years of life. Growth and development lead to changes in breathing patterns. As tidal volume (V_T) and minute ventilation increases with age, resting respiratory rate decreases, wherein V_T increases by 300% from 7 ml/kg as during the initial year of life. A short residence time is accounted for aerosol particles, which disturbs the intended pulmonary deposition due to low vital capacity, low V_T, short respiratory cycle, low capacity for residual function, etc. (6).

45.3 Considerations for Neonatal Aerosol Therapy: What Do We Need to Know?

Various issues regarding neonatal therapy needs to be accounted for in aerosol therapy, otherwise infants may receive a considerably greater amount of aerosolized drug per kilogram of body weight. Factors discussed below make clinicians avoid giving aerosol therapy to infants.

During the first few years of life, the size of the airways changes dramatically, and with growth and development the volumes, flows, and breathing patterns change. With age, the

DOI: 10.1201/9781003046547-45

resting respiratory rate decreases as there is an increase in tidal volume (V_T) and minute ventilation. V_T is around 7 ml/kg during the first year of life and increases to 300% along with increase in inspiratory flow. Pulmonary deposition is hampered in infants as a short residence time is exhibited by aerosol particles due to lower factors like V_T, functional residual capacity, vital capacity, short respiratory cycle, etc. Variable breathing frequencies, lower V_T, and high resistance to small cross sectional airway diameter also lead to poor drug delivery.

Infants have a proportionally larger tongue as compared to a small oral volume, along with the proximity of larynx and epiglottis to the base of the tongue, making them obligate nasal breathers. Further, small nostrils, short turbinate region, narrow nasopharynx and pharynx–larynx, and a large anatomical dead space also need to be taken into consideration for aerosol delivery (7), as do crying and screaming, since in babies they add to these considerations.

45.4 Challenges in Using Aerosol Medicines for Babies: In Baby Size

A report by the European Medicines Agency (EMA) to the European Commission shows that neonates represent a neglected pediatric subpopulation in medicine development (8). Due to lack of medicine specifically developed for pediatric patients, many off-label medicines are used and this remains to be a greater problem for neonatal population due to difficulties of conducting clinical trials owing to lesser number of patients. Also the low incentives provided to pharmaceutical companies for formulating medicines for neonatal populations is another factor.

Mixed outcomes were observed from the use of inhaled pharmaceutical aerosols in infants and neonates receiving mechanical ventilation. Reduced need for improved oxygenation, systematic glucocorticoids, and increased fluid resorption were observed in clinical trials of in ventilated infants (9–11). Delivery efficiency of aerosol to infant lungs is usually low and variable which occurs due to interface device, aerosol generator, and ventilation circuit, including endotracheal tube of around 3 mm internal diameter (in ventilated infants). Additional factors that add to this complexity include an increase in inertial short inhalation period due to low volumes of tidal air and short inhalation periods with an added increase in frequencies of high and small ratios of inspiratoion–expiration (6,12). Another challenge to delivery efficiency is the delivery of around 1% drugs to lungs by conventional metered-dose inhalers and jet nebulizers, which has remained consistent throughout in vitro studies, animal studies, and in humans (13). Some of the major clinical choices that have to be made in order to meet the existing challenges for neonatal aerosol therapies are discussed below.

45.4.1 Unique Cognitive Challenges

Behavioral and emotional development of infants is one of the largest challenges which are unique to pediatric patient population. They are devoid of the physical capabilities to produce coordinate breathing. In many cases, a mouthpiece or a tight-fitting aerosol mask is tolerated poorly by infants, resulting in thrashing, crying, and squirming. There exist fewer choices for children when it comes to ways to promote drug delivery and pulmonary conditions. During the time of hospital stay for a critically ill child, it is quite obvious for the patient to receive multiple drugs (inhaled) with varying delivery options. Thus, many clinical and age-related factors need to be considered in choosing the best strategy for therapy.

45.4.2 Challenges during Aerosol Device Selection

The present situation does not offer any clinical information that can be used as a guide for device selection. Nebulizers used for adults are also used in infants, however, caution has to be maintained for aerosol devices producing particle sizes >5 μm as this can promote tachycardia (1). The effect of jet nebulizers and pressurized metered-dose inhalers (MDIs) were as low as 0.33% and 0.13% in infants (14). The most common delivery devices used for infants include inhalers (dry powder), nebulizers (powered with gas or jet, ultrasonic, breath-actuated, vibrating mesh), valve holding chambers (VHCs) containing pMDIs, and aerosol generators producing small particles. Among them pMDI/VHC and vibrating-mesh nebulizers are the two most preferred devices for infants due to ease of handling, ability to integrate with ventilator systems, and above all, acceptance rate by patients, although they can be categorized a- high cost devices.

45.4.3 Challenges in Face Mask Design for Interface Selection

A mouthpiece is capable of delivering two times the drug amount compared to a simple face mask (aerosol), which also remains to be demonstrated effective in pediatric patients (15,16). The challenge that lies here is that most infants are not able, on command, to open and close their mouths to adequately seal the nebulizer mouthpiece. Further, another challenge is have the infant breathe deeply and subsequently hold their breath. Hence, the face mask design remains a critical challenge for small children in drug delivery as proper sealing of mask is a must to maximize aerosol delivery. Also, any leakage leads to the drug to entering the surrounding air resulting in lung deposition and prevents irritation in patients due to getting the drug to spread into eyes and face (17).

The design for face masks remains to a major challenge in smaller children regardless of the kind of nebulizer used (18). An important aspect to be considered in design is that the design should be warm, soft, flexible, and small, as small masks are required by small patients (19). As with lesser dead space, the probability of more dosage from the device is very likely to reach the lungs, which is true when pMDIs/VHCs or vibrating-mesh nebulizers are used, or another type which does not supply gas (20). In vitro studies done on this front show that a greater aerosol mass-to-lung ratio is provided by a front-loaded mask with low facial and eye deposition when compared to a bottom-loaded mask (21–24). However, contradictory results were obtained for the lung model in a pediatric population where greater drug delivery was observed using a bottom-loaded mask design compared to a front-loaded one (25). As a common nebulizer equipped with a face mask was not utilized for each study, a general conclusion cannot be assumed, but a front-loaded mask directs the aerosol to the oronasal area, while a bottom-loaded mask directs the same toward the upper portion of the mask.

45.4.4 Challenges in Dealing with Patient Preferences

The preference of patients and caregivers in device selection remains to be the major challenge for effective aerosol therapies. Generally, devices that are regularly used are more preferred by patients as compared to devices they rarely use (26,27). Also, the variable impact of nasal breathing by infants and the small caliber of airway should also be considered while implementing preferences. Regular use of a device when device treatment options are many, continues to be a deciding factor in patient preference. Many factors drive these preferences, which also determine the preferences of parents toward their child's treatment.

Preference for a certain type of aerosol device encompasses aspects such as ease of use, portability, time taken for treatment, maintenance etc. Another point considered by patients for preference is the shape and size of the device. One or two time therapy could be easily given by battery operated nebulizers while small battery operated ones requires cleaning between treatments (27). Further treatment time is also another factor for preference as infant rejects each time the treatment is given the therapy is abandoned gradually by the care giver or parents unless and until the child has acute symptoms.

Infants show a poor adherence to aerosol therapy due to their inability to use the device, which remains a major challenge for patient preference. This has to be explained by the therapists to parents regarding dosage and device uses (26).

45.4.5 Challenges in Neonates for Device Synchronizing

A significant reduction of drug loss during exhalation can be achieved by synchronizing the aerosol device to the breathing pattern of the patient (28). This adjustment poses a significant challenge for infants (preterm) due to high respiratory rates and short inspiratory times (29). Aerosols can be damaged by the flow, and pressure-based sensors and their positioning within the vicinity of the aerosol and patient is not advisable. But technologies like the Graseby pneumatic capsule, neurally adjusted ventilators, etc., are used for synchronization of breath which does not require sensors to be placed within the ventilator tubes without causing risks related to aerosol impaction (30,31). These types of breath dependent synchronization methods improve drug delivery but much more data (in vivo) is required from ventilated infants (32).

45.5 Considerations for Neonatal Aerosol Applications

In the last decade, medication requirements for neonates has been largely neglected including various considerations in the areas of formulation, dosage, regulatory challenges, etc. Neonates admitted to Neonatal Intensive Care Units (NICU) require specialized incubators and overall environment care (33). The various influences on the efficiency of aerosol delivery by physical, anatomical, and physiological factors in neonates can be modeled using 3D technologies such as PrINT (34). Even though aerosol delivery has the highest degree of concern with respect to the route of administration, with high likelihood of interaction from packaging component. The dosage criteria still requires the critical aspects to be considered for using aerosol in neonates which needs monitoring in the following factors.

45.5.1 Air Flow Parameters

The aerosol in jet nebulizers is generated by airflow, and commercial nebulizers come with a variety of air-flow parameters for optimal performance, while mesh or ultrasonic vibrating nebulizers require gas flow to supply aerosol particles (35). Lung model studies on a 4 kg infant setting showed that airflow is indirectly related to lung delivery. The study has also showed that smashing of aerosol particles results in higher aerosol deposition within the ventilator circuit in the inspiratory arm (36,37). Similar results were also observed using a nose and throat model of a premature infant (34). However, these model studies have considered a continuous flow of air from the upper respiratory tract rather than considering a patient's breathing pattern. Thus, airflow remains a major factor which needs monitoring in neonatal applications.

45.5.2 Formulation

An aerosol generator with synchronized nebulizer, when placed close to the endotracheal tube (ET) in the inspiratory arm, causes a significant increase in dose emission at the ET as compared to continuous nebulization within the neonatal ventilator circuit. Emitted dose comparisons showed that terbutaline solution, as compared to budesonide suspension, was better, irrespective of synchronized nebulizer or placement (38). Also, improved pulmonary distribution was observed for aerosolized surfactants as compared to a standard liquid instillation (intra-tracheal). Animal model studies showed that large aerosolized surfactant particles, when moved through the vocal chords, form a film at the air–liquid interface, which then spreads to alveoli with better distribution (39). Surfactants have the ability to spread over mucosal surfaces, indicating they are potential carriers for aerosolized agents leading to uniform drug distribution and improved lung deposition. However, some other studies show just the opposite picture of suboptimal emission of dosage due to the highly viscous nature of the surfactant (40).

45.5.3 Aerosol Particle Size

Studies on aerosol lung deposition in both term and preterm infants showed that a marker compound called sodium cromoglycate measured in urine is used heavily. Comparison studies for drug delivery in non-intubated breathing infants (spontaneous) in three different nebulizers [Projet®; Artsana, Grandate, Italy (ultrasonic), LC Star®; Pari, Starnberg, Germany (jet nebulizer) and LS 290®; Systam, Villeneuve sur Lot, France (ultrasonic)] showed that sleeping infants during aerosol delivery carried out breathing through a mask covering both nose and mouth, and the highest lung deposition was observed in the case of a jet nebulizer as compared to ultrasonic nebulizers (41). However, the jet nebulizer exhibited the highest rate of flow and the highest mass for droplets below 2 μm. The amount of retained sodium cromoglycate was tested for each nebulizer, of which the ultrasonic nebulizer was found to be higher as compared to others. Hence, this study concluded that aerosol particles (<2 μm) are more likely to find ways toward the lower airways in spontaneously breathing infants, and the higher deposition rate of the jet nebulizers may be due to the low residual volume of the nebulizer and is not related to particle size. This study has also rejected the concept of using sodium cromoglycate as marker for lung deposition (41).

Even though the jet nebulizers had the highest lung deposition, only around 0.89% of dosage was deposited after inhalation in the lungs. This study was at par with other studies suggesting that for both spontaneous breathing and mechanically ventilated infants, a nominal deposition of <1% is achieved (42). These studies indicate that effective delivery can be expected for ventilated patients thorugh those fine particles that can pass through artificial airways and upper airway tracts, and deposition rates can be achieved by the residual volume differences between nebulizers when expressed as a percentage of nominal dose. It has been observed that particle size between 2 and 6 μm is accumulated in central airways, and those of 6 μm are deposited in the oropharynx (43). However, there are limited studies dealing with particle size evaluation of the upper airways for preterm infants.

The majority of deposition occurs during the exhalation process during the breathing cycle in the tracheostomy tube, indicating that a significant amount of inhaled aerosol is lost during exhalation. Another study carried out on ventilated adults showed that around 53% of the total inhaled dose was found deposited in the lungs and the remaining part was deposited in the expiratory arm (44). The results of this study were of high importance for infants due to their short inspiratory times, which increases the extent of losses in exhalation. Around 20–30% of the aerosol passing via the mainstream bronchi was actually stored in the lungs (45).

The most critical aspect is particle size, which influences patient interface for intubated and non-intubated infants. The particles should be small enough to cross the interface with minimal impact, but at the same time the size should not be so small that they are lost in exhalation. Hence, particle size remains a major factor influencing pulmonary deposition of drugs in infants.

45.5.4 Gastric Deposition and Impact on Upper Airways

Deposition of aerosol in the upper respiratory tract remains a continuous challenge for clinical practitioners. Conventional treatment of infants using a nebulizer with a face mask is as good as hood nebulization with respect to lung deposition fraction, distribution of aerosol, clinical response toward hood nebulizers, etc. Salbutamol, a radiolabeled aerosol, was studied on wheezing infants of 1–19 months and spontaneously breathing infants with an MMAD (mass median aerodynamic diameter) of 4.2 μm. Lung deposition values of around 2.4–2.6% were observed and an improved saturation of oxygen with reduction in respiratory rate was also observed. In addition to local deposition of aerosol around the oropharyngeal region, around 7.6–8.4% was deposited in the gastrointestinal tract due to swallowing (46).

More aerosol deposition occurs in the upper respiratory tract in more distressed infants, which is due to the prolonged expiration on screaming or crying. This is followed by velocity gasps of high inspiratory flow, short in nature, which leads to aerosol impaction inside the throat and swallowing. Clinical impact for this kind of deposition does not hold true for asthmatic infants, but this may create concerns during nebulizer treatments with corticosteroids due to increased side effects and absorption (47).

45.6 Aerosol Therapies Side Effects

Inhaled corticosteroids during aerosol therapy have both systematic and local side effects, while, local effects include ocular and cutaneous effects, oral thrush, hoarseness etc., while systematic ones include absorption in gut via lung absorption, swallowing from deposits in oro- and hypopharynx (48,49). Around 80% of the emitted drugs deposited in the pharynx are due to DPIs and mDPIs, however using a VHC in combination with mDPIs may reduce the local effects of the inhaled aerosols like oral thrush, hoarseness, and dysphonia (48,49). Bruising and skin thinning may result when corticosteroids are administered through a face mask, and caution is needed to avoid eye and face deposition (50).

Decreased growth rate was observed in infants during the first year of inhaled corticosteroid therapy, but in budesonide treatment children did not showed significant changes in growth for a time period of 10 years (51,52). Delayed reduction in growth has been observed (>1 cm) due to poor asthma control. As growth suppression is drug and dose dependent, it is advisable that use of corticosteroids in young children be kept to a minimum. Further, post corticosteroid treatment, the face should be washed to prevent facial deposition. Also, fewer side effects are observed in pMDIs with VHCs, which are just as efficient as jet nebulizers with the same output and wider acceptance rate (1). Higher systematic side effects are observed in oral administration of anti-asthma medicines as compared to topical administration of drugs.

45.7 Improvements in Aerosol Therapies for Infants

A systematic approach to optimize aerosol delivery in infants and children involves the identification of specific determinants and challenges. Inherent limitations are related to pathophysiologic, anatomical, physiologic, and technical aspects of aerosol delivery in infants and young children. Aerosol efficiency can be enhanced by applying sound principles of aerosol delivery and using control over factors which are changeable to intervention. Enhancing factors such as formulation and delivery system increases the efficiency of aerosol delivery with reduced risk, waste, and cost. Other factors, such as close attention to aerosol particle size (1–3 μm aerodynamic diameter and <2 μm geometric standard deviation) and the concentration of this particle produced by an aerosol system may increase the delivery via endotracheal tubes reaching the lower respiratory tracts in infants. Choice of delivery and proper MDI techniques such as priming, shaking prior to inhalation, avoiding multiple actuations and immediate actuations, choice of patient interface (mouthpiece, face mask type, endotracheal) and aerosol spacer, cleaning the spacer, and the medicine considered for aerosol (viscosity, suspension, solution) also contribute to optimal aerosol delivery (52).

Delivery efficacy and therapeutic response can also be improved by considering patient-related, system-related, and operator-dependent issues combined together. Further, motivation for and education to caregivers, parents, and medical professionals also influences efficiency and prioritizing the teaching of proper techniques improves delivery of aerosol therapy. Clear understanding of differences and efficiencies of drugs and devices makes the use of aerosol therapy well suited to infants and children (53). This has been substantially maintained by aerosol scientists and clinicians during the last decade, and must be the first priority for device and drug manufacturers and also regulatory agencies.

The choice of drug dose is determined empirically and the variability in lung doses between inter- and intra-subject is considerable (54). However, this does not have much to contribute toward therapeutic failures in conditions such as asthma because of the use of high doses to fulfill the variability of a wide therapeutic index. Thus, the therapeutic index determines the choice of dose. The delivered dose to an infant can be calculated from the drug deposited on the filter placed between the infant and delivery system as shown by many studies. It is considered to be proportional to lung dose, and doubling this will eventually double the lung dose only if other parameters remain unchanged. Any distressing element, however, may change the breathing pattern, leading to a different lung dose than the inhaled dose. Thus, these studies using filters are essential in product development — but more is required to obtain the lung dose (55).

45.8 Limitations and Advances for Aerosol Delivery

A systematic approach for better aerosol delivery to infants requires the identification of determining factors and challenges associated with it, which include anatomical, technical limitations, pathophysiologic, and physiologic related to infants and young children. Aerosol delivery can be enhanced through sound principles and by executing control over amendable factors. One critical factor on this line is the impact of the particle size (1–3 μm mass median diameter and standard deviation (geometric) < 2 μm) whose concentration enhances the delivery via endotracheal tubes with low V_T and inspiration in children. All issues related to patient, system, and operator-dependent factors greatly influences the aerosol delivery and efficacy improving therapeutic responses. A better understanding of the functional differences and efficiencies of the various drugs and devices will make aerosol delivery well suited for infants and children (53).

Limited factors are available in infants for choosing the right aerosol device. While pulmonary deposition of aerosol in neonates and infants is reduced as compared to toddlers or teenagers, and studies have shown that a similar lung dose/kg body weight was observed in children <3 years, this remains unclear in infants. Thus, due to the uncertainties associated with dire lack of clinical trials, inhaled doses for neonates etc., a careful and cautious approach toward the product used is important, including the therapeutic effects and associated toxicities (56,57).

Recent development of aerosolized vaccine has many advantages, but its clinical feasibility is still in its early stages of discovery. Further work is required for use of such vaccines to ascertain their effects on infants and children for mass use. Infants are highly vulnerable to inhalable medications, even though the aerosol deposition percentage is quite low. This is due to the fact that they receive more drug/kg body weight as compared to their adult counterparts (58). Also, the upper airway deposition in infants is more than in the geriatric population due to the nasal breathing pattern which determines the therapeutic index of drugs (26).

45.9 Conclusions

Thus we have seen that the pediatric population differs on many frontiers as compared to adult populations. Also, development of a drug itself is quite challenging and time consuming, and that, too, for neonates and preterm infants alone is even more difficult due to their rapidly changing physiology and biopharmaceutical conditions. They still fall under the category of therapeutic orphans owing to their access to drugs and formulations backed by studies with regulatory approvals. Certain factors, such as development, physiological studies, care environment, etc., determine the path of formulation preparation for neonates. Also, they constitute a small number of the population for specific formulations, so development remains to be limited in this regard.

A jet nebulizer is a better option for infants as compared to blow-by therapy, which is ineffective. Vibrating-mesh nebulizers can be used together with a total face mask in pediatric NIV (non-invasive ventilation). Hence, a lot of studies in this field are currently needed for aerosol therapy issues relating to neonates.

Conflict of Interest

The authors declare no conflict of interest, financial or otherwise.

Acknowledgments

The authors thank MoTA-New Delhi, India as NFHE fellowship (NFST-2015-17-ST-ASS-3740) for Surovi Saikia.

REFERENCES

1. DiBlasi R. Clinical controversies in aerosol therapy for infants and children. *Respir Care*. 2015; 60(6): 894–914.
2. Shekunov BY, Chattopadhyay P, Tong HH, and Chow AH. Particle size analysis in pharmaceutics: Principles, methods and applications. *Pharm Res*. 2007; 24(2): 203–227.
3. Koushik K and Kompella UB. Particle and device engineering for inhalation drug delivery. *Drug Del Technol*. 2004; 4: 40–50.
4. European Medicines Agency. Ich e11(rl) Guideline on Clinical Investigation of Medicinal Products in the Pediatric Population. Available online: https://www.ema.europa.eu/documents/scientific-guideline/iche11r1-guideline-clinical-investigation-medicinal-products-pediatric-population-revision-1_en.pdf (accessed on 12 February 2020).
5. Allegaert K, van de Velde M , and van den Anker J. Neonatal clinical pharmacology. *Paediatr Anaesth*. 2014; 24: 30–38.
6. Fink JB. Aerosol delivery to ventilated infant and pediatric patients. *Respir Care*. 2004; 49(6): 653–665.
7. Xi J, Si X, Zhou Y, Kim J, and Berlinski A. Growth of nasal and laryngeal airways in children: Implications in breathing and inhaled aerosol dynamics. *Respir Care*. 2014; 59(2): 263–273.
8. Report from the Commission to the European Pparliament and the Council. Available online: https://ec.europa.eu/health/sites/health/files/files/paediatrics/docs/2017_childrensmedicines_report_en.pdf (accessed on 19 March 2019).
9. Cole CH, Colton T, Shah BL, Abbasi S, MacKinnon BL, Demissie S, and Frantz ID. Early inhaled glucocorticoid therapy to prevent broncho pulmonary dysplasia. *N Engl J Med*. 1999; 340: 1005–1010.
10. Sood BG, Delaney-Black V, Aranda JV, and Shankaran S. Aerosolized PGE1: A selective pulmonary vasodilator in neonatal hypoxemic respiratory failure results of a phase I/II open label clinical. *Pediatr Res*. 2004; 56: 579–585.
11. Armangil D, Yurdakok M, Korkmaz A, Yigit S, and Tekinalp G. Inhaled beta-2 agonist salbutamol for the treatment of transient tachypnea of the newborn. *J Pediatr*. 2011; 159: 398–403.
12. Rubin BK and Fink JB. Aerosol therapy for children. *Respir Care Clin N Am*. 2001; 7: 175–213.
13. Dubus JC, Vecellio L, De Monte M, *et al.* Aerosol deposition in neonatal ventilation. *Pediatr Res*. 2005; 58: 10–14.
14. Salmon B, Wilson NM, and Silverman M. How much aerosol reaches the lungs of wheezy infants and toddlers? *Arch Dis Child*. 1990; 65(4): 401–403.
15. Ari A, Dornelas de Andrade A, Sheard M, Al Hamad B, and Fink JB. Performance comparisons of jet and mesh nebulizers using different interfaces in simulated spontaneously breathing adults and children. *J Aerosol Med Pulm Drug Deliv*. 2015; 28(4): 281–289.
16. Ditcham W, Murdzoska J, Zhang G, *et al*. Lung deposition of 99mTc-radiolabeled albuterol delivered through a pressurized metered dose inhaler and spacer with face-mask or mouthpiece in children with asthma. *J Aerosol Med Pulm Drug Deliv*. 2014; 27(Suppl 1): S63–S75.
17. Erzinger S, Schueepp KG, Brooks-Wildhaber J, Devadason SG, and Wildhaber JH. Facemasks and aerosol delivery in vivo. *J Aerosol Med*. 2007; 20(Suppl 1): S78–S83; discussion S83–S84.
18. Janssens HM and Tiddens HA. Facemasks and aerosol delivery by metered dose inhaler-valved holding chamber in young children: A tight seal makes the difference. *J Aerosol Med*. 2007; 20(Suppl 1): S59–S63; discussion S63–S65.
19. Ari A and Fink JB. Effective bronchodilator resuscitation of children in the emergency room: Device or interface? *Respir Care*. 2011; 56(6): 882–885.
20. Amirav I, Mansour Y, Mandelberg A, Bar-Ilan I, and Newhouse MT. Redesigned face mask improves "real life" aerosol delivery for Nebuchamber. *Pediatr Pulmonol*. 2004; 37(2): 172–177.
21. Lin HL, Restrepo RD, Gardenhire DS, and Rau JL. Effect of face mask design on inhaled mass of nebulized albuterol, using a pediatric breathing model. *Respir Care*. 2007; 52(8): 1021–1026.
22. Smaldone GC, Berg E, and Nikander K. Variation in pediatric aerosol delivery: Importance of facemask. *J Aerosol Med*. 2005; 18(3): 354–363.
23. Sangwan S, Gurses BK, and Smaldone GC. Facemasks and facial deposition of aerosols. *Pediatr Pulmonol*. 2004; 37(5): 447–452.
24. Mansour MM and Smaldone GC. Blow-by as potential therapy for uncooperative children: An in-vitro study. *Respir Care*. 2012; 57(12): 2004–2011.
25. Nikander K, Berg E, and Smaldone GC. Jet nebulizers versus pressurized metered-dose inhalers with valved holding chambers: Effects of the facemask on aerosol delivery. *J Aerosol Med*. 2007; 20(Suppl 1): S46–S55; discussion S55–S58.
26. Fink JB and Rubin BK. Problems with inhaler use: A call for improved clinician and patient education. *Respir Care*. 2005; 50(10): 1360–1374.
27. Rau JL. The inhalation of drugs: Advantages and problems. *Respir Care*. 2005; 50(3): 367–382.
28. Miller D, Amin M, Palmer L, Shah A, and Smaldone G. Aerosol delivery and modern mechanical ventilation: In vitro/in vivo evaluation. *Am J Respir Crit Care Med*. 2003; 168(10): 1205–1209.
29. Cole C. Special problems in aerosol delivery: Neonatal and pediatric considerations. *Respir Care*. 2000; 45(6): 646–651.
30. John J, Bjorklund L, Svenningsen N, and Jonson B. Airway and body surface sensors for triggering in neonatal ventilation. *Acta Paediatr*. 1994; 83(9): 903–909.
31. Sinderby C, Navalesi P, Beck J, Skrobic J, Comtois N, Friberg S, *et al*. Neural control of mechanical ventilation. *Nat Med*. 1999; 5: 1433–1436.
32. Nikander K, Turpeinen M, and Wollmer P. Evaluation of pulsed and breath-synchronized nebulization of budesonide as a means of reducing nebulizer wastage of drug. *Pediatr Pulmonol*. 2000; 29: 120–126.
33. Stanford Children's Health. The Neonatal Intensive Care Unit (NICU). Available online: https://www.stanfordchildrens.org/en/topic/default?id=the-neonatal-intensive-care-unit-nicu-90-P02389 (accessed on 21 March 2020).
34. Minocchieri S, Burren JM, Bachmann MA, *et al.* Development of the premature infant nose throat-model

(print-model): An upper airway replica of a premature neonate for the study of aerosol delivery. *Pediatr Res.* 2008; 64: 141–146.
35. Hess D. Aerosol devices in the treatment of asthma. *Respir Care.* 2008; 53(6): 699–723.
36. Coleman D, Kelly H, and McWilliams B. Determinants of aerosolized albuterol delivery to mechanically ventilated infants. *Chest.* 1996; 109(6): 1607–1613.
37. Ari A, Atalay O, Harwood R, Sheard M, Aljamhan E, and Fink J. Influence of nebulizer type, position, and bias flow on aerosol drug delivery in simulated pediatric and adult lung models during mechanical ventilation. *Respir Care.* 2010; 55(7): 845–851.
38. Turpeinen M and Nikander K. Nebulization of a suspension of budesonide and a solution of terbutaline into a neonatal ventilator circuit. *Respir Care.* 2001; 46(1): 43–48.
39. Wagner M, Amthauer H, Sonntag J, Drenk F, Eichstadt H, and Obladen M. Endotracheal surfactant atomization: An alternative to bolus instillation? *Crit Care Med.* 2000; 28(7): 2540–2544.
40. Finer N, Merritt T, Bernstein G, Job L, Mazela J, and Segal R. An open label, pilot study of Aerosurf combined with nCPAP to prevent RDS in preterm neonates. *J Aerosol Med Pulm Drug Del.* 2010; 23: 1–7.
41. Kohler E, Jilg G, Avenarius S, and Jorch G. Lung deposition after inhalation with various nebulizers in preterm infants. *Arch Dis Child Fetal Neonatal.* 2008; 93: 275–279.
42. Grigg J, Arnon S, Jones T, Clarke A, and Silverman M. Delivery of therapeutic aerosols to intubated babies. *Arch Dis Child.* 1992; 67(1): 25–30.
43. Dolovich M. Influence of inspiratory flow rate, particle size, and airway caliber on aerosolized drug delivery to the lung. *Respir Care.* 2000; 45(6): 597–608.
44. O'Riordan T, Palmer L, and Smaldone G. Aerosol deposition in mechanically ventilated patients. Optimizing nebulizer delivery. *Am J Respir Crit Care Med.* 2009; 149(1): 214–219.
45. Heyder J, Gebhart J, and Stahlhofen W. Inhalation of aerosols: Particle deposition and retention. In: Willeke K (ed) Generation of Aerosols. Ann Arbor Science, Ann Arbor. 1980; 65–103.
46. Amirav I, Balanov I, Gorenberg M, Groshar D, and Luder A. Nebulizer hood compared to mask in wheezy infants: Aerosol therapy without tears! *Arch Dis Child.* 2003; 88: 719–723.
47. Mellon M, Leflein J, Walton-Bowen K, *et al.* Comparable efficacy of administration with face mask or mouthpiece of nebulized budesonide inhalation suspension for infants and young children with persistent asthma. *Am J Respir Crit Care Med.* 2000; 162(2 pt 1): 593–598.
48. Everard ML. Guidelines for devices and choices. *J Aerosol Med.* 2001; 14(Suppl 1): S59–S64.
49. Pedersen S. Inhalers and nebulizers: Which to choose and why? *Eur Respir J.* 1996; 12: 1340–1345.
50. Geller DE. Clinical side effects during aerosol therapy: Cutaneous and ocular effects. *J Aerosol Med.* 2007; 20(Suppl 1): S100–S108.
51. Gulliver T, Morton R, and Eid N. Inhaled corticosteroids in children with asthma: Pharmacologic determinants of safety and efficacy and other clinical considerations. *Paediatr Drugs.* 2007; 9(3): 185–194.
52. Bisgaard H, Anhoj J, Klug B, and Berg E. A non-electrostatic spacer for aerosol delivery. *Arch Dis Child.* 1995; 73(3): 226–230.
53. Dolovich M. Aerosol delivery to children: What to use, how to choose. *Pediatr Pulmonol Suppl.* 1999; 18: 79–82.
54. Aswania O, Iqbal SM, Ritson S, Everard ML, Bhatt J, and Rigby AS. Intra-subject variability in lung dose in healthy volunteers using five conventional portable inhalers. *J Aerosol Med.* 2004; 17(3): 231–238.
55. Everard ML. Inhalation therapy for infants. *Adv Drug Deliv Rev.* 2003; 55: 869–878.
56. Rubin B and Fink J. The delivery of inhaled medication to the young children. *Pediatr Clin North Am.* 2003; 50(3): 717–731.
57. Everard ML. Inhaler devices in infants and children: Challenges and solutions. *J Aerosol Med.* 2004; 17: 186–195.
58. Pirozynski M and Sosnowski TR. Inhalation devices: From basic science to practical use, innovative vs generic products. *Expert Opin Drug Deliv.* 2016; 13(11): 1559–1571.

46

Safety and Toxicological Concerns Related to Nanoparticle-Based Lung Delivery

Humera Memon, Vandit Shah, Tejal Mehta, and Jigna Shah
Institute of Pharmacy, Nirma University, Ahmedabad, India

46.1 Properties of Nanoparticles

Nanoparticles are usually utilized as drug carriers where active therapeutically ingredients can get dissolved, encapsulated, entrapped, attached or adsorbed within the particles with the help of distinct fabrication methods (1). These drugs retain within the nanoparticles through either covalent or electrostatic interactions (2). These are prepared using biocompatible and biodegradable materials. These solid colloidal nano-sized particles eventually decompose in the body over a predefined duration and their degradation process can be controlled to modify the release of some physicochemical properties. Within the human body, the absorption, circulation, and elimination phenomena of these particles are different, based on the targeted tissues as well as on the properties of the particles (3).

The nanoparticles having size less than 1 nm are efficient to pass due to the blood–brain barrier, whereas those of around 6 nm can enter the capillaries in lungs, muscles, or skin tissue. Those particles ranging from 40 to 60 nm bear the potential to get rid of fenestrated capillaries that are present within the intestine, kidney, or endocrine/exocrine glands (4). Agglomerations of larger-sized nanoparticles of 600 nm or more are found in bone marrow, spleen, and liver (5). Negatively charged nanoparticles easily attach to the cells having a positively charged surface. To ease or hamper the particle endocytosis, the electrostatic properties of nanoparticles are used. For some time, these nanoparticles have been manipulated as imaging agents and drug carriers, owing to the wide spectrum of attributes like (1) drug protection counter to loss or degradation, (2) reduction of dose frequency, (3) enhanced tissue penetration, (4) large surface–volume ratio, (5) enhanced level of drug on target site, (6) biological mobility, (7) enhanced patient compliance, and (8) sustained and controlled release of drug (6).

46.2 Clearance Mechanism of Nanoparticles from Lungs

Rapid retention time and the non-specific cargo delivery type of barriers are still faced by a majority of the inhalable therapeutics, imputed primarily to 3 methods of clearance (6).

46.2.1 Enzymatic Degradation

Though pulmonary enzymatic degradation is quite low, several isoforms of cytochrome P450 are expressed in lungs representating the primary detoxification enzyme capable of degrading a lot of inhaled therapeutics as theophylline (7). Furthermore, proteases and lung peptidases are potent enough to degrade the peptide and protein drugs like insulin. Hence, further research is required on the process of inhalable therapeutics enzymatic degradation, apart from tracing a new way of escaping lung metabolism (8).

46.2.2 Pulmonary Clearance

Discarding of particles occurs as a part of a natural defense mechanism against the foreign inhaled particles in order to prohibit them from associating with the tissues found in the lungs, causing the therapy to fail. The clearance of all the bits that are inhaled depends primarily on the deposition site (9). Pulmonary clearance is primarily divided into 2 parts. The foremost one is the mucociliary clearance, showcasing major defense through the epithelial cells that are aligned with the mucociliary escalator or the upper portion of the respiratory system and discarding the ones that are 6 μm or more. Cleared by entrapment, these particles are eliminated from the body along with mucus exiting through the trachea by means of swallowing or coughing. Moreover, the mucociliary escalator is evaded by these particles since the alveolar region possesses their preferential deposition (10).

The mucociliary escalator boosts are due to the lung inflammation or infection that ultimately results in therapeutic failure. Owing to the rats' intra-tracheal instillation of 56 kDa PEGylated poly lysine dendrimer–DOX conjugates, a quick 60% clearance of the dendrimer was noticed in just a single day by systemic absorption and mucociliary escalator. Likewise, clearance of albumin NPs from the lungs was done with the help of a translocation technique that effectively covers the mucosal barrier in the lungs as well as the mucociliary clearance (11). Moreover, alveolar macrophages are the monocytes derived phagocytic cells which result in the xenobiotics' clearance from each alveolus, and such clearance leads to reduced reservation duration of the therapeutics that are

DOI: 10.1201/9781003046547-46

being inhaled along with improved frequency of the suggested dose. Since such alveolar macrophages engulf the particles that possess the dimension of 1.5–3 μm, a proposal on the basis of dimension of inhalable therapeutics is utilized to inhibit the consumption by macrophages and prolonged drug retention deep within the lungs (12).

46.2.3 Rapid Systemic Absorption

A noteworthy hurdle is felt by the quick systemic absorption in the way of designing some impressively inhalable formulations. Because the lungs possess visibly more vascularity as well as vast surface area, the majority of the inhalable therapeutics are perfectly sucked up by the lungs. By stopping the inhalable therapeutics within the lung, we can achieve noteworthy local therapeutic effects; otherwise the systemic absorption can result in unwanted side effects (8). Due to the contact with lung surfactants, peptide drugs were aggregated. The alveolar macrophages cleaned them thoroughly, whereas the lung surfactant contact with lipophilic drugs improved its solubility characteristics, leading to better chances of absorption. Inhalable therapeutics are either eliminated or absorbed depending on the type of drug; along with the drugs' reaction with the lung mucosa's surfactant, that ultimately leads to the inhalable therapeutics conveyance, specifically air-to-blood transfer (13).

46.3 Surface-Modified Nanoparticles

A recent development shows that the dimensions as well as the exposed area of the nanoparticles are changed so as to improve the circulation time, reduce the clearance of particles, penetrate through physical barriers, escape the biological protective mechanisms, and lengthen the stay of the drug at the target site (14). Hence, vivid moieties are used for functionalization/variations of nanoparticle surfaces as per various stimuli. Certain exogenous and endogenous factors can stimulate the nanoparticles. The endogenous stimuli include ultrasound, enzyme, redox, and pH, while magnetic fields and light are deployed as the exogenous factors for controlling nanoparticle behavior. Surface alteration along with the polymeric coating of nanocarriers permit the altering of half-life, biodistribution, stimuli reactivity, and circulation duration as well as the therapeutic application (15).

The basic categories of nanoparticles are

- Lipid-based nanoparticles (solid-lipid nanoparticles, multifunctional SLNs, liposomes, nanostructured lipid nanocarriers)
- Polymeric nanoparticles (using natural and synthetic polymers)
- Inorganic nanoparticles (silver nanoparticles, gold nanoparticles, iron oxide nanoparticles)
- Dendrimers (polymeric, PEGylated, and peptide dendrimers)
- Carbon based nanoparticles (graphene oxide, graphene, carbon nanotubes, nano diamonds)
- Mesoporous nanoparticles

Further, particles' binding rate on the components of the blood was determined by the hydrophilicity of nanoparticles. Surface functionalization is usually done using various ligands such as small molecules, surfactants, dendrimers, polymers, and biomolecules. Hydrophobic nanoparticles without surface functionalization are found to be discarded quickly while the circulation is boosted notably since such particles were hydrophilic polymers coated or surfactant to enhance the hydrophilicity (16).

The drugs from nanoparticles release gradually in controlled fashion. Such release of drug is triggered due to the stimuli; otherwise occurs in consistent mode over a fixed duration. Drug accumulation in certain categorized tissues might be triggered due to the stimuli-responsive release of drugs. Such stimuli are deployed through some variations in biological environments, such as alteration of pH, cell environment, and disease-related enzymes or some external forces like electrical, ultrasound, light, magnetic fields, or heat (17).

Ultrasound-sensitive microbubbles are categorized to discard certain therapeutic agents over locally defined targets (18). The microbubbles, which are fabricated from lung surfactants, resulted in a noteworthy boost in drug targeted deposition in comparison with the lipidic microbubbles. Surprisingly, control as well as application of magnetic fields are capable of driving the aerosol droplets enclosing the superparamagnetic iron oxide to the designated target within the lungs of the mice in vivo (19).

46.4 Pulmonary Drug Delivery

The lungs are extremely attractive for targeted drug delivery. The research has shown promising results in treating the respiratory diseases by targeting drug to lungs, to the vast alveolar surface area, extensive blood circulation, a fine epithelial barrier layer, low activity of proteolytic in the alveolar space, and evading of first-pass hepatic metabolism; the lung offers quick as well as effective delivery of drug (20). The metabolic activities in the liver and the gastrointestinal tract are much higher than in the lung. This boosts the systemic drug delivery as well as enhancing the efficacy of treating several lung diseases. Moreover, owing to the scope of self-administration, along with it being non-invasive, drug administration via lungs is preferred. Despite such impressive advantages, limited products are available on the market to deliver drug to the lungs. So far on the market, the sole product of therapeutic inhalable protein is Pulmozyme® (dornase alfa, Genentech Inc., San Francisco, CA). In 2006, the US Food and Drug Administration approved Exubera (inhalable insulin, Pfizer, New York, NY), but in 2007 it was taken off the US market owing to insufficient market demand (2).

Lungs are built to effectively clear foreign agents, thereby making it difficult to deposit drugs within the lungs. Two core mechanisms are available to eliminate the inhalable particles from the lung. Primarily, the particles in the conducting zone of lungs are cleared by patches that are moving in the mucus layer. Ideally, such a procedure works better in the lung diseases with a noteworthy improvement in the mucus production and thickness. Further, macrophages engulf and assimilate to quickly eliminate the insoluble particles from the deep lung (2). A huge number of nanoparticles are being developed right now for pulmonary disorder applications, with an aim to addressing the shortcomings of

TABLE 46.1

Nanoparticles used for respiratory applications.

Nanoparticles	Method of Administration	Use	Animal models
Poly (L -aspartic acid co-lactic acid)/DPPE co-polymer NPs	Intraperitoneal injection	Lung cancer	Mouse xenograft model
LPH (liposome polycation hyaluronic acid) nanoparticles	Intravenous injection	Cancer lung metastasis	Mouse xenograft model
pDNA nanoparticles (NPs)	Intranasal	Allergic asthma	OVA-exposed mice
Poly (beta-amino ester) (PBAE) polymers	Intratumoral injection	Lung cancer	Mouse xenograft model
PEG-dendritic block telodendrimer	Intravenous injection	Allergic asthma	OVA-exposed mice
Poly (DL -lactide coglycolide) NPs	Inhalation	Tuberculosis	*M. tuberculosis* infected guinea pigs
Poly butyl cyanoacrylate NPs	Intravenous injection	Lung cancer	Mouse xenograft model

the conventional drugs (see Table 46.1). These nanoparticles are crucial for treating various lung diseases like tuberculosis, cancer, emphysema, cystic fibrosis, and asthma (21).

46.4.1 Pulmonary Delivery of Nanoparticles Using Dry Powder Carriers

Dry powder carriers are formulated as loose agglomerates by various techniques. The drug with suitable carriers like lactose and dispersing agents are mixed together for formulation development of DPI. Appropriate formulations of the dry powder along with the carrier systems are designed in such a way that nanoparticles accumulate in the alveolar regions for improving the effectiveness of pulmonary delivery (22). For supporting the inhalation characteristics of the pranlukast hydrate dry powder, Kawashima et al. proposed the use of hydroxypropylmethyl cellulose phthalate (HPMCP) nanospheres (hydrophilic nanoparticles). They mixed drug powder and surface nanospheres with the lactose. In the test of in vitro inhalation, the dispersibility as well as the emission of surface modified drug powder was noted to be improved. Also, the deep lungs were successfully provided with the powder. Later, insulin loaded nanospheres of PLGA were fabricated by Kawashima et al. Further, for delivering the nanospheres to the trachea of guinea pigs, an ultrasonically assisted nebulizer was used. A noteworthy reduction in blood glucose was recorded along with a consistent hypoglycemic effect lasting for 48 hours. These outcomes were attributed to the supervised release of insulin from the nanospheres as well as its deposition in various parts of lungs (23). On the other hand, a spray drying method was recorded by Tsapis and colleagues for manufacturing large porous particles. With the use of 1,2-dipalmitoyl-sn-glycero-3-phosphocholine, 1,2-dimyristoyl-sn-glycero-3-phophoethanolamine (surfactants), and lactose, researchers came up with some physicochemical characteristics of the spray-dried powder for pulmonary delivery (24).

Moreover, an interesting fact about these nanoparticles is that they can be efficiently carried in a carrier matrix. After dissolving lactose in nanoparticle suspension prior to spray drying by Sham and colleagues so as to get the nanoparticles-incorporated carrier powder, they noted a significant alteration in the dimension of the particles, thereby reporting scopes of manipulating the delivery as well as release of nanoparticles (25). Other research work by Grenha et al. mixed mannitol with lactose using a spray-drying method for developing microspheres, having insulin incorporated NPs for pulmonary drug delivery. Releasing of the drug from NPs was observed to be unaffected by microencapsulation. Moreover, it permits pulmonary administration for the successful transport of the macromolecules (26).

Ely and colleagues came up with a technology to dry spray for deploying the effervescent carrier particles that could incorporate the nanoparticles which are loaded with ciprofloxacin. The team also successfully altered the dimensions of particles for improving the particle lay-down deep into the lung. Utilizing effervescent carrier particles led to a significant boost in the elimination of drugs. Moreover, while the nanoparticles were released from the effervescent carrier particles, a strange variation was found in their size. When the nanoparticles carrying dry powder are delivered, the most suitable platform is pulmonary administration (27,28). The nebulization parameters have been manipulated so as to reduce the particle collection and facilitating the delivery of the drug deep into the lungs. Table 46.2 demonstrates a brief of dry powder carriers in the pulmonary delivery of nanoparticles.

46.4.2 Use of Nebulization for Pulmonary Delivery of Nanoparticle

The nebulization is another approach explored by researchers for lung targeting. Here the delivery of nanoparticles is done via spraying or nebulizing the nanoparticle suspension. To deliver the pulmonary drug, a surfactant-free nanoparticle suspension was identified by Dailey and colleagues. It boosted the encapsulation efficiency owing to the electrostatic interactions of the nanoparticles with the drug molecules. It was found that use of anionic diethylaminopropylamine-poly (vinyl alcohol)-grafted-poly(lactide-co-glycolide)-contained formulation, as well as boosting the proportion of carboxy methyl cellulose, resulted in reducing the aggregation of the particles (29). Moreover, Yamamoto et al. came up with a modification in the PLGA nanospheres' surface with the help of chitosan for improving the delivery efficacy of calcitonin to the lungs. A significant reduction by 80% was noted in the blood calcium, post administering the chitosan-modified PLGA nanoparticles into guinea pigs' trachea (30).

TABLE 46.2

Using Dry Powder Carriers for Pulmonary Delivery of Nanoparticles

Nanoparticle	Nanoparticle Size (nm)	Active Ingredient	References
Iso-butyl cyanoacrylate	244s	Ciprofloxacin	(27)
Gelatin and iso-butyl cyanoacrylate	173, 242		(25)
Hydroxypropylmethyl cellulose phthalate (HPMCP)	51.6	Pranlukast	(23)
Carboxylate-modified polystyrene (PS) and Nyacol 9950 colloidal silica	PS: 25, 170, 1000 Nyacol: 100		(24)
Chitosan	388, 419	Insulin	(26)

TABLE 46.3

Pulmonary Delivery of Nanoparticle Suspension Using Nebulization

Nanoparticle	Nanoparticle Size (nm)	Active Ingredient	References
Itraconazole nanocrystals	300–800	Itraconazole	(31)
PLGb	186–290	Rifampin, isoniazide, pyrazinamide	(32)
Surface modified PLGAc with chitosan	650	Calcitonin	(30)
Solid-lipid nanoparticles (SLN)	Not provided	Isoniazide, pyrazinamide, and rifampin	(33)
DEAPA-PVAL-g-PLGAa	76.2–213.6		(29)
Solid-lipid nanoparticles (SLN)	200	99mTc	(34)

Moreover, therapeutic levels of the drug lasted for a day and they happened to be larger than those of the unmodified particles. It is justified by the mucoadhesion of the NPs to the local tissue as well as bronchial mucus in the lung, along with a lasting release of drugs from the particles. Further, due to the loosening of intercellular tight junctions, the chitosan improves the drug permeability. One more study was oriented toward fabricating the itraconazole-loaded NPs, dispersed into the aqueous media, and later on nebulized in vivo to murine lungs. It resulted in improved drug concentration in lungs, and a reduced scope of any ill effects (31). Hence, nebulizing of nanoparticle suspensions is a crucial technique to deliver the therapeutic agents to the lung. For clinical success, the physicochemical stability of the suspension needs to be controlled. Table 46.3 depicts the use of nebulization for the pulmonary delivery of nanoparticles (22).

46.4.3 Magnetic Nanoparticles

Lung targeting can also be done by magnetic nanoparticles. Magnetic nanoparticles are used for diagnostics as well as for treatment. Magnetic nanoparticles have been widely employed in the field of pulmonary drug and gene delivery due to their unique properties including their biocompatibility, biodegradability, ease of surface modification, and superparamagnetic and hyperthermia effects (22). During the initial phase of the study, pharmacokinetics of doxorubicin magnetic conjugate (DOX-M) NPs were tested by Mykhaylyk et al. in a mouse model. The team came up with the efficacy of a non-uniform magnetic field on the magnetic DOX-M clearance. Meanwhile, the adult male mice's eye sinus vein was injected with DOX-M suspensions, along with the application of a magnetic field in the left lung.

This study demonstrated non-uniform magnetic fields to be an essential parameter for the modification of the DOX-M conjugate pharmacokinetics. This proved to be a major boost to the DOX-M in the lungs, along with the consumption of the magnetic carrier within the liver as compared to a reference devoid of the magnetic field. This proved that this proposal ultimately improves the bioavailability of DOX-M in the lungs. Though these results were impressive, the results are not verified for application of the magnetic fields in humans so as to establish the drug localization in the lungs having magnetic NPs (35,36). Contradicting the previous studies, Wu et al. demonstrated the effects of magnetic field on rats post intravenous injection of dextran coated Fe_3O_4 and showed that it did not alter the nanoparticle accumulation within the lungs. Ideally, magnetic nanocarriers possesses the biggest advantage for diagnostic as well as treatment purposes. Delivering magnetic nanoparticles to the lungs can lead to advanced studies for utilizing it as an effective drug delivery system, as well as a safe tool used for diagnostic purposes (37).

46.4.4 Delivering Nanoparticles Locally into the Lung

At the target location, enhancement, sustenance, and manipulation of the drug level is allowed via local delivery of nanoparticles for treating various respiratory diseases (1). Such types of deliveries can prevent the degradation of the drugs in the gastrointestinal tract (oral administration), lessen the required dose, as well as reduce the toxicity of the entire system. The itraconazole-loaded nanoparticles were introduced by Vaughn and colleagues for dealing with the fungal infections of *Aspergillus fumigatus*. Pulmonary delivery of the itraconazole NPs to the mice in vivo reported an impressively high concentration of drug in lung tissues; to improve the treatment efficiently and minimize the effective systemic toxicity, the drug level in the serum was managed (38). Additionally, for maintaining sustained drug concentrations, targeted delivery to various tissues is beneficial. Therapeutic compounds like isoniazid, rifampicin, and pyrazinamide were deployed for targeting the delivery of the drugs to the alveolar macrophages for improving the therapeutic efficacy in treatment of pulmonary tuberculosis (33).

Some research works fabricated the drug-encapsulated nanoparticles and characterized them for in vivo tests on the guinea pigs. Lungs are administered with PLGA NPs to control the therapeutic levels of drugs in plasma and they are maintained within the lungs for a remarkable duration of time. This method is also useful in reducing the overall dose as well as minimizing the systemic exposure (32,39). Moreover, the PLGA surface nanoparticles are characterized with the wheat germ agglutinin whose bioadhesive characteristics enable its bonding with the lectin receptors within alveolar epithelium for managing concentration of drugs in lung cells. Such an arrangement can lead to a sustained drug level in plasma for a constant 2 weeks and in the lung for 15 days (40). Likewise, a sustained in vivo drug delivery in guinea pigs to the lungs was provided by PLGA nanoparticles that were formulated for eliminating the tubercle bacilli from *M. tuberculosis* infected guinea pigs (41).

46.4.5 Nanoparticle-Based Gene Delivery to Lungs

Owing to the specific scopes for direct or permanent modification of the functions of cell and organ, gene delivery is a crucial area of drug delivery. Because they possess uniform sizes compared to other viruses with a natural yet pathogenic system of delivering the genes, nanoparticles happen to be the correct option for the said purpose. Owing to the security criteria, the gene delivery systems based on the natural virus possess a rare opportunity for being widely utilized for delivery of genes (42). Nevertheless, the artificial virus technology is yet unsatisfactory. For a therapeutic study of lung cancer, DOTAP cholesterol NPs were considered by Gopalan et al. as a substitutive non-immunogenic gene delivery vector. Their study verified numerous signaling molecules both in vivo and in vitro that may be induced by the systemic administration of the DNA nanoparticles, which are further linked with the inflammation. Small molecule inhibitors are being used against signaling molecules like naproxen, which suggests that such molecules suppress the nanoparticle-mediated inflammation without imparting any impacts on transgene expression. Outcomes are of clinical value both in terms of reducing the toxicity and improving the therapeutic window (43).

Kaul et al. investigated separately to find the chances of gelatin NPs to be used in plasmid DNA delivery systems on the Lewis lung carcinoma (LLC)-bearing mice models. Plasmid DNA encoded reporter for β-galactosidase (pCMV-β) found in the nanoparticles of gelatin and the PEGylated gelatin were encapsulated by them. This ultimately showcased PEGylated gelatin nanoparticles to be prime transfection reagents against the gelatin NPs and lipofectin. Additionally, the PEGylated gelatin NPs are transfected by 61% productivity post I.V. administration as compared to intratumoral administration, which can be verified by in vivo expression of the β-galactosidase in the tumor mass. This led to high transfection efficacy of PEGylated gelatin NPs to get biodegradable, biocompatible and long circulating nature of the carrier systems. Although nanoparticles never tangles the molecules of DNA, sometimes their supercoiled structure is preserved, and it intensifies the cruciality of efficient transfection and nuclear uptake (44).

The nanoparticles having DNA as a single molecule and condensed with the polyethylene glycol substituted lysine-30-mers were demonstrated to effectively transfect the lung epithelium of mice as proved by Fink et al. (45). Later, for the targeted delivery of antisense oligodeoxynucleotides as well as the small interference RNA into the lung cancer cells, Li et al. developed a ligand targeted and sterically stabilized NP formulation. The outcome of this study proved the fact that such NPs have the potential to selectively deliver the antisense oligodeoxynucleotides and the siRNA into the lung cancer cells; that can eventually be utilized for cancer therapy. Such works demonstrate the potential of nanoparticles to be further utilized as a carrier for secured and impressive gene delivery while dealing with various diseases related to lungs (46).

46.5 Advantages of Nanoparticles in Lung Delivery

Alveoli are the functional unit of lungs, with the large surface area allowing the exchange of materials. Also, the thin epithelial layer and highly perfused blood vessels facilitate the delivery of different types of therapeutic agents targeted through this route (26). The application of nanoparticles provides the advantage of targeted delivery, high retention time at the target site, and sustained release of drug (2). Anti-asthmatic (47), anti-tubercular, (41) and anti-cancer drugs (48) based polymeric nanoparticles and liposomal nanoparticles have been widely studied for pulmonary delivery. Liposomes provide a unique benefit of being formulated from the materials that are compatible and endogenous to lungs, reducing the adverse effects, thus it is a preferred platform for pulmonary drug delivery systems. Furthermore, various liposomal based formulations are approved by the US FDA, while many products are under clinical trials (49). For example, colfoscerilpalmitate (Exosurf®) and bovactant (Alveofact®) are the liposomal products used for treating acute respiratory distress syndrome (50). Budesonide encapsulated liposomes are being used to achieve a manageable release rate and uphold the desired therapeutic concentration in the lungs, also reducing the systemic exposure and toxicity (51). Using nanocarrier systems, a combination of drugs can be delivered to the lungs at the same time. Nanoparticles, by virtue of their size and properties, are able to more deeply penetrate the lungs and enter the alveolar areas, also effectively evading macrophage clearance and permeating the lungs' epithelial layer. Surface modification of the nanoparticles enables the enhancement of pharmacokinetic properties of the drugs like bioavailability and penetration to the tissue. Mucoadhesive agents like biodegradable polysaccharide chitosan help in prolonging the deposition of nanocarriers (2). Yamamoto et al. fabricated peptide elcatonin-encapsulated PLGA nanoparticles with chitosan-based surface modification. Hence, the nanoparticles provide a wide array of opportunities for fabricating the delivery of the drugs to the desired location using nanocarrier-based platforms (30).

46.6 Regulatory Concerns Related to Nanoparticles

Nanoparticles have unique high tissue and organ penetration ability, leading to increased drug bioavailability, making them superior as compared to natural chemicals in various biomedical applications. Conversely, there are various potential toxicities related to nanoparticles due to their small size. This makes it essential to frame and implement regulations via rules, law, and legislations drafted by several regulatory bodies in order to reduce the nanoparticle-associated risk factors. However, no specific international regulation or legal definitions have been framed for nanoparticle production, handling, or toxicity testing. However, different government bodies and other agencies are working together to create a regulatory framework in accordance with international harmonization.

In the same context, various modifications in the standards of ethics and medical governance have been made to incorporate nanomaterials (NMs) in the biomedical field (52,53). In the USA, the USNCL, FDA, and EPA are working in close relation to set up a protocol for safety of the products including development, manufacturing, and biological risk assessment in the USA. Similarly, in the European Union,, EUNCL, EMA and others have framed EU legislation as well as technical guidance for NMs. It ensures the usage of NMs across the legislative areas, i.e. NMs used in any sector must be treated as the same. The definition of nanomaterials by the EU is "a natural, incidental or manufactured material containing particles, in an unbound state or as an aggregate or as an agglomerate, and where for 50% or more of the particles in the number size distribution, one or more external dimensions is in the size range of 1 to 100 nm" (36,54). In 2013, European Union cosmetic regulation 1223/2009 was replaced by directive 76/768/EEC which explains NMs as "an insoluble or biopersistent and intentionally manufactured material with one or more external dimensions, or an internal structure in the range of 1 to 100 nm which include man-made fullerene, single-walled carbon nanotubes, and grapheme flakes." In the United Kingdom, DEFRA, FSA, OECD, and ISO are regulating the use and assessment of nanomaterials (55).

The OECD Working Party on Manufactured Nanomaterial (WPMN) is developing new OECD test guidelines and guidance documents in context with nanomaterials. It addresses various nanomaterial-related issues like -physicochemical properties, and effects on health and biotic systems. In this regard, three new test guidelines for nanomaterials have been framed, such as TG318, TG412 (Sub-acute Inhalation Toxicity: 28 days) (56) and TG413 (Sub-acute Inhalation Toxicity: 90 days) (57). Similarly, Inhalation Toxicity Testing like GD39 (Guidance Document on Acute Inhalation Toxicity Testing) (28) has been revised for the concerned issues related to nanomaterials.

Similarly, the International Organization for Standardization (ISO) has developed new technical committees 229 and ISO/TC24. ISO may provide an analytical method for nanomaterials like ISO 18827:2017 (70) and Standard Operating Procedures like ISO 19007:2018 (71). The government of India in 2019 developed the Guidelines for Evaluation of Nanopharmaceuticals in India; it defines nanopharmaceuticals as "A pharmaceutical preparation containing nanomaterials intended for internal or external application on the body for the purpose of therapeutics, diagnostics and any health benefits." It states that any new nanopharmaceutical preparation will be treated as an investigational new drug which needs to establish its quality, safety, and efficacy. It is now widely accepted that NPs are not fundamentally hazardous and many of them are found to be nontoxic, in addition to their incremental health effects. However, various nanoparticle-based studies in the future will elucidate the NPs and NP-based product safety.

46.7 Safety Concerns Related to Nanoparticles

Nanoparticles offer several properties that are not assessable by the currently available chemical assays, making the safety assessment of inhalable NPs of high importance. Accumulation of insoluble nanoparticles at various sites can lead to local responses like increased macrophages, inflammation, and oxidative stress (13). For instance, the carbon nanotubes, on pulmonary administration, caused granulomatous response of leukocytes and increased levels of cytokines in bronchoalveolar lavage fluid. Single dose PEGylated poly- L -lysine dendrimer–DOX conjugates, in intra-tracheal administration to animals, caused substantial lung toxicity ultimately leading to death. This was thought to be because of the lung's exposure for a prolonged period to low levels of DOX, whereas in bolus single time dumping of free DOX on lung tissues is seen. In order to establish the safety of the nanoparticles it is necessary to develop representative models and determine toxicity of inhaled NPs. Also, the applicability of the standardized tests like membrane integrity, inflammatory mediator release, and overall cell metabolic activity should be elucidated. As *in vitro* models for safety assessment, A549 (alveolar) and BEAS-2B (airways) cell lines provide a promising platform for NP uptake-based studies as they lack functional tight junctions (58).

46.7.1 Particle Size and Shape

The NPs' distribution and deposition in the lungs show a discrepancy depending on several factors like rate of breathing, particle size, air flow, and lung volume. Large particles (>10 μm size) are deposited in the oropharyngeal region, while the small particles (1–5 μm size) are accumulated in the deep alveolar regions. Density of the particles, their dimension, and the mass median aerodynamic diameter regulate and facilitate the deposition and distribution of the inhaled particles in the deep lung tissues. In the alveolar region, particle size primarily affects the rate of clearance. Large particles are exhaled without being phagocytized, whereas microparticles are efficiently taken up by the macrophages; nanoparticles (<200 nm size) are capable of penetrating the cellular barrier, further reducing the alveolar macrophage mediated phagocytosis (13).

Smaller sized particles need a higher energy level for assembly and de-aggregation, resulting in the deposition of inhaled particles in lungs by the inertial impaction, diffusion, and sedimentation. The accumulation of particles corresponds to the fixing of spherical particles under the gravitational force via air. Non-deposited NPs are exhaled out of lungs, resulting in loss of the delivered dose. NPs' large-surface-area ultrafine particles of a size less than 100 nm showed more toxic effects such as inflammation, allergic reaction, or cytotoxicity; also leakage to other organs was observed. Similar compositions containing larger particles show lower toxicity (59,60).

NPs' structure and shape could also affect inhalable NP toxicity. Differences in shapes can lead to different patterns of toxicity (61). For instance, carbon black and graphite; when compared to the same chemical composition, carbon nanotubes show different pulmonary effects. Furthermore, NPs of fibrous shape when administered in vivo showed inflammation induction and granuloma formation (59).

46.7.2 Biodegradability

Biodegradability of NPs has a direct correlation with lung toxicity, posing a serious concern regarding the NPs' safety. The use of biodegradable polymers for formulating a safe, inhalable, drug-loaded nanocarrier has gained interest, replacing the non-degradable polymers which promote lung inflammation (62). Furthermore, the use of DPPC-like chemicals that are endogenous to the pulmonary route in preparation of inhalable formulations has shown substantial decrease in the NPs' associated toxicity (63). Dailey et al. assessed the use of linear PLGA and amine-modified branched polyester biodegradable NPs for inflammatory potential in comparison to non-biodegradable NPs like poly(styrene). This study revealed significantly reduced inflammation in the use of biodegradable NPs. Also, the slowly degrading NPs like PLGA are less recommendable when compared to branched polyesters like NPs that are rapidly degradable are promising for pulmonary delivery (64). In another study, albumin NPs, when administered in single doses of 2 µg/mouse and 20 µg/mouse, did not exhibit any inflammatory response; on the other hand, a 390 µg/mouse high dose led to an increase in the mono nucleocytes count and inflammatory response (65). Furthermore, polyalkyl cyanoacrylate and porcine gelatin NPs on pulmonary administration for gene and drug delivery showed no inflammatory and cytotoxicity effects on 16HBE14o-cells and primary airway epithelium cells. Thus, biodegradable or endogenous-based nanocarriers help in reducing the safety concerns related to NP usage (66).

46.7.3 Surface Charge

The role of NPs' surface charge is enormous in determining the NPs' associated lung toxicity (67). Poly(styrene) NPs have a positive surface charge, which lead to higher pulmonary reactions that are evident by increased total protein, cell recruitment and lactate dehydrogenase. Anionic biodegradable NPs showed optimal tolerability, on the other hand cationic NPs lead to severe toxic effects after pulmonary administration (68).

46.7.4 Disease State and Concentration at the Target Site

The small size of nanoparticles tends to result in deposition of inhalable NPs in the lungs. The particles like NPs and MPs tend to concentrate more in asthmatic lungs compared to healthy lungs. This higher level of particle concentration in the cells may induce release of the inflammatory response locally, aggravating preexisting inflammation disease conditions. Hence, the optimum clearance of the NPs from the target site has a significant impact on the therapeutic applicability of the NPs (69–71).

46.7.5 Nanoparticle Interaction with Pulmonary Environment

The pulmonary environment and interactions between the inhalable nanoparticles are largely unknown (11). The pulmonary surfactants are affected by several types of polymeric inhalable NPs which might result in life-threatening consequences. Diminishing activity of the pulmonary surfactants leads to a decrease in the surface tension during expansion/compression cycles, decreasing the gaseous exchange (72).

46.8 Toxicity of Inhaled Nanoparticles

Even with multiple advantages of NPs, there are still various safety concerns. The toxicity of ultrafine particles with aerodynamic diameter <100 nm and the toxicity of polymeric nanoparticles have a direct impact on the decrease of pulmonary function and augmented vulnerability to infection (22).

46.8.1 Toxicity of Inhaled Ultrafine Particles

The translocation of NPs from the respiratory tract to the distinct tissues or organs is recorded via different routes and leads to direct adverse effects on different organs. Endocytosis of NPs by different target cells can interact with subcellular structure and interfere with the normal function of the cellular bodies. Therefore, these effects are particularly considered, thereby posing some prime consequences in a compromised organ. Currently available pulmonary toxicological information is based upon the idea of environmental ultrafine particles (diameter of ~100 nm) like diesel particulates, carbon black, silica, and titanium oxide, inhaled during daily life (73). However, it has been shown that as the particle size decreases, the nanoparticle-associated toxicity increases. This was observed with ultrafine carbon black particles, which have pulmonary toxicity to a much greater extent in rats, as compared with carbon black particles of large size. Warheit et al. carried out in vivo toxicity research on rats using single wall carbon nanotubes (SWCNT). Approximately 15% mortality among the animals who were given a high dose (5 mg/kg) of SWCNT was observed 24 hours after the administration. Their study led to the conclusion that the upper respiratory airway's mechanical blockage induced by SWCNT leads to animal mortality. Their study also showed a non-dose-dependent series of multifocal granulomas that were the result of a foreign tissue body reaction. Nevertheless, these were observed to be non-progressive beyond 1 month post SWCNT exposure and non-uniform in distribution (74).

Shvedova et al. showed a rare inflammatory response in the SWCNT-exposed mouse lungs. The probable reason is significant damage to the pulmonary epithelial cells (Type-1), leading to strong neutrophilic pneumonia-like symptoms followed by macrophage activation and recruitment. This initial acute phase can progress to a fibrogenic phase, ultimately leading to significant pulmonary deposition of elastin and collagen. Furthermore, changing the release and production of pro-inflammatory factors, like interleukin-1β and TNF-α, to anti-inflammatory pro-fibrogenic cytokines like interleukin-10 and TGF-β, ultimately causing a damaging reduction in the pulmonary function and greater vulnerability to various infections (75).

Wiebert et al. assessed 35-nm 99mTc labeled carbonaceous particles' pulmonary retention on asthmatic and healthy volunteers. Their study revealed a minimal or no significant amount of deposited NP translocation into the systemic circulation from lungs of healthy volunteers. Only a trivial fraction (less than 1%) of 35-nm combustion particles diffused out from peripheral lungs into the extrapulmonary organs and even to systemic circulation, which might not be sufficient to cause abnormal conditions (76). Inoue et al. assessed the role of bacterial endotoxin lipopolysaccharides NP size on lung inflammation in mice. Carbon black nanoparticles of two different sizes were administered in mice, for evaluation of the lung coagulation and inflammation. Their study indicated that smaller NPs can worsen the bacterial endotoxin-related lung inflammation (69).

In a study, Xia et al. evaluated the NPs manufactured from titanium dioxide (TiO_2), fullerol, polystyrene, and carbon black on RAW 264.7 (phagocytic cell line), which illustrates a lung target for nanoparticles. These cationic polystyrene nanospheres induce GSH depletion, toxic oxidative stress, and cellular ROS production. Also, cellular structural organelle damage like mitochondrial injury caused by increased calcium uptake was observed. Furthermore, ultrafine particles induced oxidant injury increased TNF-α production, while cationic polystyrene nanospheres caused mitochondrial damage and death of the cell with no inflammation. However, fullerol and TiO_2 based NPs did not lead to toxic oxidative stress (77).

Several studies evaluating the effect of ultrafine nanoparticles on alveolar macrophage functions have been carried out, such as Möller et al. using flow cytometry and cytomagnetometry, to examine the intracellular effects of ultrafine particles on alveolar macrophages, revealing the increased cytoskeletal stiffness and impairment of phagosome transport at high concentrations of NP (78). Takenaka et al. inspected the alveolar macrophages' role by exposing animals to gold NPs and then examining lung tissue and lavage cells for a period of a week. About 29% gold NPs on the first day and 6% gold NPs on seventh day were identified in the rat's lungs. Furthermore, traces of gold were detected in cytoplasm of the alveolar macrophages, indicating that the inhaled ultrafine gold particles might be processed by endocytosis pathways (79). Lin et al. assessed the oxidative stress and cytotoxicity in vitro produced by cerium oxide (CeO_2) nanoparticles on the lung cancer cell line. Their study reported a dose and time dependent toxic effect caused by free radicals generated from CeO_2 nanoparticles, leading to noticeable oxidative stress in the cells, as demonstrated by the decremented glutathione and α-tocopherol levels. Also, the elevated oxidative stress increased malondialdehyde and lactate dehydrogenase levels, which suggests lipid peroxidation and cell membrane damage, respectively (80). The pharmaceutical field needs to decipher this information and address the serious safety issues posed by the use of NPs for delivery to the lungs.

46.8.2 Toxicity of Polymeric Nanoparticles Used in Drug Delivery

A wealth of knowledge related to the toxicity profile of the inhaled, environmentally occurring dust NPs is available, but typically of the ultrafine particles. In order to get a deep lung delivery, various facets are considered, primarily the acute toxicity of the drug delivery system on the lungs' epithelial cells, and secondarily the interplay of drug delivery systems with the alveolar environment. Generally, synthetic or natural polymeric nanoparticles are needed in pharmaceutical sciences for delivering the drugs. Their efficacy and biocompatibility from intravenous routes of administration has been established; however, their acute toxicity profile and action on the pulmonary epithelial cells and alveolar environment needs to be established. To assess such effects, in vitro-based cell culture models are the best way to move ahead (81).

Standard toxicity assays such as release of pro-inflammatory factors, inflammatory mediators, cellular metabolic activity, and membrane integrity can be performed with BEAS-2B (airways) and A549 (alveolar) cell lines after nanoparticle uptake. However, no standardized cell culture model is currently available which mimics the alveolar epithelium permeability except for pneumocyte monolayers in primary culture. For bronchial epithelium, suitability of various cell models like calu-3 and 16HBE14o-cells are being established (82). Brzoska et al., by the usage of 16HBE14o-cells, investigated the nanoparticles formulated from human serum albumin, polyalkyl cyanoacrylate, and porcine gelatin as gene and drug carriers intended for pulmonary administration. Their study showed very little or no inflammation and cytotoxicity was associated with the use of these nanoparticles. This indicated that selection of a suitable carrier is essential for lung targeting (66).

The capability of biodegradable polymeric nanoparticles of a new PLGA derivative like diethylaminopropylamine polyvinyl alcohol grafted-poly (lactic-co-glycolic acid) and PLGA to initiate an inflammatory response in the lungs of animals was studied by Dailey et al. They also considered 2 different dimensions of polystyrene nanospheres (75 nm and 220 nm) as controls in their study and, evaluated the bronchoalveolar lavage fluid for inflammatory parameters like macrophage inflammatory protein-2 mRNA induction, lactate dehydrogenase (LDH) release, and polymorphonucleocyte recruitment. Their study suggests that the non-biodegradable polystyrene particles of corresponding dimension to biodegradable polymeric NPs for pulmonary delivery induce higher inflammatory response (83).

Stuart et al. examined gelatin nanoparticle interactions with the artificial lung surfactants in vitro using biophysical methods. They used dipalmitoyl phosphatidylcholine (DPPC) as an ideal system and assessed pressure acting on the surface of lungs' DPPC monolayer in the presence of gelatin NPs with the help of the Langmuir trough. This study revealed minimal interplay between the lung surfactant film and nanoparticles, without destabilizing the DPPC monolayer, demonstrating the inertness and safety of pulmonary nanoparticle delivery (84).

Further, the nanoparticles–alveolar environment interaction is a secondary safety perspective of deep lung deposition. A fine surfactant film is present on the alveolar space, which accelerates the exchange of various gases and lowers the tension on the surface of alveolar space. Ignorance of such functions by inhalable nanoparticles can lead to some life-threatening outcomes. Hence, the compatibility of a delivery system with the alveolar environment must be given utmost importance (22).

In summary, various studies carried out by different groups with ultrafine inhalable nanoparticles inculcate concerns regarding safety of their usage. However, the use of inhalable biocompatible polymeric NPs of particle size less than 100 nm for gene and drug delivery may cause little or no toxicity. The use of inhalable biodegradable NPs shall soon be considered in clinical studies.

46.9 Conclusion and Future Perspectives

Nanoparticle-based medicine for lung disorders possesses immense possibilities with novel uses such as disease treatment, detection, diagnosis, and imaging. Currently, various models with traditional drugs are being used to address problems like low therapeutic efficacy, drug resistance, non-tissue-specific targeting, and unwanted side effects. The NPs' agile nature makes them an ideal choice for tailored treatment of lung disorders and achieves the goal of personalized medicine. Nowadays, novel combinatorial strategies using tracking and targeting moieties are being utilized for nanoparticle-based lung therapeutics. Such multicomponent structure-based NPs improve water solubility of poorly soluble drugs, diffuse through biological barriers, and potently target the desired site of action. However, the path from bench to bedside is not straightforward; it offers many challenges which need to be addressed to hasten the translation of NP-based therapies (85).

Nanomedicines contain multiple compartments in 3-dimensional structures in a specific spatial arrangement in order to implement their desired action. Minor modifications in the synthesis process or composition of the complex can have serious implications for the chemical and/or physical properties leading to various unwanted effects ranging from pharmacokinetic, pharmacological, and immunological challenges. Additionally, nanoparticles designed for novel routes of administration, like nasal, require higher bioavailability as compared to traditional small molecules. The renal system effectively clears the small molecules, whereas the particles having bigger dimensions are cleared by macrophages and Kupffer cells in the spleen and liver (86). Nanoparticles and/or nanocarriers also pose the challenge of unwanted immunological response. To assure significant reproducibility during the entire formulation process and reduce the unwanted pharmacokinetic, pharmacological, and immunological effects, it is of utmost importance to identify the prime properties of each component along with developing a clearer idea of the NPs' physicochemical characteristics.

Besides the challenges offered by the NPs in their design and discovery, the missing standards in the testing of nanomedicines pose a serious safety concern for NPs usage in clinics. Theoretically, nanoparticle-based therapy needs to deal with the same obstacles faced by any new drug: reproducible manufacturing processes, suitable design of components and characteristics, institution of analysis methods for enough characterization, toxicity profiles, favorable pharmacology, and depiction of safety and efficacy during all the rounds of clinical trials. The complex nature and properties of nanoparticles, with multiple active components, affects numerous pharmacological behaviors, unlike small drug molecules that are composed of single active pharmaceutical ingredients. Thus, new regulatory guidelines and modification of currently available standard examinations of safety, bioequivalence, and pharmacokinetic aspects are of utmost importance. There is a powerful need for the regulatory agencies to establish a comprehensive list of trials and a systematic approval procedure to actively meet the urgency of upcoming nanoparticle-based products and to take nanomedicine delivery to the clinics (87).

REFERENCES

1. Nguyen HX. Targeted delivery of surface-modified nanoparticles: Modulation of inflammation for acute lung injury. 2019; 331–353. doi:10.1007/978-3-030-06115-9_17
2. Rijt SHV, Bein T, and Meiners S. Medical nanoparticles for next generation drug delivery to the lungs. *European Respiratory Journal*. 2014; 44: 765–774. doi:10.1183/09031936.00212813
3. Panyam J, Dali MM, Sahoo SK, … Ma W. Polymer degradation and in vitro release of a model protein from poly (D, L-lactide-co-glycolide) nano- and microparticles. *Journal of Controlled Release*. 2003; 92(1–2): 173–187. doi:10.1016/S0168-3659(03)00328-6.
4. Siafaka PI, Okur NÜ, Karavas E, and Bikiaris DN. Molecular sciences surface modified multifunctional and stimuli responsive nanoparticles for drug targeting: Current status and uses. *International Journal of Molecular Sciences*. 2016; 17(9): 1440. doi:10.3390/ijms17091440.
5. Arruebo M, Fernández-Pacheco R, and Ibarra MR . Magnetic nanoparticles for drug delivery. *Nano Today*. 2007; 2(3): 22–32. doi:10.1016/S1748-0132(07)70084-1.
6. Schütz CA, Juillerat-Jeanneret L, Mueller H, Lynch I, and Riediker M. Therapeutic nanoparticles in clinics and under clinical evaluation. *Nanomedicine*. 2013; 8(3): 449-467. doi:10.2217/nnm.13.8
7. Smola M … Vandamme T. Nanocarriers as pulmonary drug delivery systems to treat and to diagnose respiratory and non respiratory diseases . *International Journal of Nanomedicine*. 2008; 3(1): 1–19.
8. Olsson B, Bondesson E, Borgström L, Edsbäcker S, Eirefelt S, Ekelund K, Gustavsson L, and Hegelund-Myrbäck T. Pulmonary drug metabolism, clearance, and absorption. *Controlled Pulmonary Drug Delivery*. 2011; 21–50. doi:10.1007/978-1-4419-9745-6_2
9. El-Sherbiny IM, Villanueva DG, Herrera D, and Smyth HDC. Overcoming lung clearance mechanisms for controlled release drug delivery . *Controlled Pulmonary Drug Delivery*. 2011; 101–126. doi:10.1007/978-1-4419-9745-6_5

10. Patton JS, Brain JD, Davies LA, Fiegel J, Gumbleton M, Jin Kim K, Sakagami M, Vanbever R, and Ehrhardt C. The particle has landed - Characterizing the fate of inhaled pharmaceuticals. *Journal of Aerosol Medicine and Pulmonary Drug Delivery*. 2010; 23(Suppl. 2): S-71. doi:10.1089/jamp.2010.0836
11. Gill S, Löbenberg R, Ku T, ... Azarmi S. Nanoparticles: Characteristics, mechanisms of action, and toxicity in pulmonary drug delivery—A review. *Journal of Biomedical Nanotechnology*. 2007; 3(2): 107–119. doi:10.1166/jbn.2007.015
12. Oberdörster G. Lung clearance of inhaled insoluble and soluble particles. *Journal of Aerosol Medicine: Deposition, Clearance, and Effects in the Lung*. 1988; 1(4): 289–330. doi:10.1089/jam.1988.1.289
13. Abdelaziz HM, Gaber M, Abd-elwakil MM, Mabrouk MT, Elgohary MM, Kamel NM, Kabary DM, Freag MS, and Elzoghby AO. Inhalable particulate drug delivery systems for lung cancer therapy: Nanoparticles, microparticles, nanocomposites and nanoaggregates. *Journal of Controlled Release*. 2018; 269(October 2017): 374–392. doi:10.1016/j.jconrel.2017.11.036
14. Sung JC, Pulliam BL, and Edwards DA. Nanoparticles for drug delivery to the lungs. *Trends in Technology*. 2007; 25(12): 563–570. doi:10.1016/j.tibtech.2007.09.005
15. Karimi M, Mirshekari H, ... Aliakbari M. Smart mesoporous silica nanoparticles for controlled-release drug delivery. *Nanotechnology Reviews*. 2016; 5(2): 195–207.
16. Yoo JW, Doshi N, and Mitragotri S. Adaptive micro and nanoparticles: Temporal control over carrier properties to facilitate drug delivery. *Advanced Drug Delivery Reviews*. 2011; 63(14-15): 1247–1256. doi:10.1016/j.addr.j2011.05.004
17. De Smedt SC. Crucial factors and emerging concepts in ultrasound-triggered drug delivery. *Journal of Controlled Release*. 2012; 164(3): 248–255. doi:10.1016/j.jconrel.2012.08.014
18. Sirsi SR, Fung C, Garg S, and Tianning MY. Lung surfactant microbubbles increase lipophilic drug payload for ultrasound-targeted delivery. *Theranostics*. 2013; 3(6): 409–419. doi:10.7150/thno.5616
19. Hasenpusch G, Geiger J, Wagner K, Mykhaylyk O, Wiekhorst F, Trahms L, Heidsieck A, et al. Magnetized aerosols comprising superparamagnetic iron oxide nanoparticles improve targeted drug and gene delivery to the lung. *Pharmaceutical Research*. 2012; 29(5): 1308–1318. doi:10.1007/s11095-012-0682-z
20. Courrier HM, ... Butz N. Pulmonary drug delivery systems: Recent developments and prospects. *Critical Reviews in Therapeutic Drug Carrier Systems*. 2002; 19(4–5): 64. doi:101615/critrevtherdrugcarriersyst.v19.i45.40
21. Lu X, Zhu T, Chen C, and Liu Y. Right or left: The role of nanoparticles in pulmonary diseases . *International Journal of Molecular Sciences*. 2014; 15(10): 17577–17600. doi:10.3390/ijms151017577
22. Azarmi S, Roa WH, and Löbenberg R. Targeted delivery of nanoparticles for the treatment of lung diseases. *Advanced Drug Delivery Reviews*. 2008; 60(8): 863–875. doi:10.1016/j.addr.2007.11.006
23. Kawashima Y, Serigano T, Hino T, Yamamoto H, and Takeuchi H. A new powder design method to improve inhalation efficiency of pranlukast hydrate dry powder aerosols by surface modification with hydroxypropylmethylcellulose phthalate nanospheres. *Pharmaceutical Research*. 1998; 15(11): 1748–1752. doi:10.1023/A:1011916930655
24. Tsapis N, Bennett D, Jackson B, Weitz DA, and Edwards DA. Trojan particles: Large porous carriers of nanoparticles for drug delivery. *PNAS*. 2002; 99(19): 12001–12005. doi:10.1073/pnas.182233999
25. Sham JOH, Zhang Y, Finlay WH, ... Roa WH. Formulation and characterization of spray-dried powders containing nanoparticles for aerosol delivery to the lung. *International Journal of Pharmaceutics*. 2004; 269(2): 457–467.
26. Grenha A, Seijo B, and Remunán-López C. Microencapsulated chitosan nanoparticles for lung protein delivery. 2005; 25(4-5): 427–437. doi:10.1016/j.ejps.2005.04.009
27. Ely L, Roa W, Finlay WH, and Löbenberg R. Effervescent dry powder for respiratory drug delivery. *European Journal of Pharmaceuticals and Biopharmaceutics*. 2007; 65(3): 346–353. doi:10.1016/j.ejpb.2006.10.021
28. Environment Directorate Joint Meeting of the Chemicals Committee and the Working Party on Chemicals, Pesticides and Biotechnology Guidance Document on Inhalation Toxicity Studies Series on Testing and Assessment No. 39 (Second Edition). 2018.
29. Dailey LA, Kleemann E, Wittmar M, Gessler T, Schmehl T, Roberts C, Seeger W, and Kissel T. 2003. Surfactant-free, biodegradable nanoparticles for aerosol therapy based on the branched polyesters, DEAPA-PVAL-g-PLGA. *Pharmaceutical Research*. 20(12): 2011–2020. doi:10.1023/B:PHAM.0000008051.94834.10
30. Yamamoto H, Kuno Y, Sugimoto S, ... Takeuchi H. Surface-modified PLGA nanosphere with chitosan improved pulmonary delivery of calcitonin by mucoadhesion and opening of the intercellular tight junctions. *Journal of Controlled Release*. 2005; 102(2): 373–381. doi:10.1016/j.jconrel.2004.10.010
31. McConville JT, Overhoff KA, Sinswat P, Vaughn JM, Frei BL, Burgess DS, Talbert RL, Peters JI, Johnston KP, and Williams RO. Targeted high lung concentrations of itraconazole using nebulized dispersions in a murine model. *Pharmaceutical Research*. 2006; 23(5): 901–911. doi:10.1007/s11095-006-9904-6
32. Pandey R, Sharma A, ... Zahoor A. Poly (Dl-lactide-co-glycolide) nanoparticle-based inhalable sustained drug delivery system for experimental tuberculosis. 2003; 52(6): 981–986. doi:10.1093/jac/dkg477
33. Pandey R and Khuller GK. Antitubercular inhaled therapy: Opportunities, progress and challenges. *Journal of Antimicrobial Chemotherapy*. 2005; 55(4): 430–435. doi:10.1093/jac/dki027
34. Videira MA, Botelho MF, Santos AC, Gouveia LF, Pedroso De Lima JJ, and Almeida AJ. Lymphatic uptake of pulmonary delivered radiolabelled solid lipid nanoparticles. *Journal of Drug Targeting*. 2002; 10(8): 607–613. doi: 10.1080/1061186021000054933
35. Mykhaylyk O, Dudchenko N, and Dudchenko A. Doxorubicin magnetic conjugate targeting upon intravenous injection into mice: High gradient magnetic

field inhibits the clearance of nanoparticles from the blood. *Journal of Magnetism and Magnetic Materials*. 2005; 293: 473–482. doi:10.1016/j.jmmm.2005.01.063
36. Nanomaterials - ECHA. 2020. Accessed April 20. https://echa.europa.eu/regulations/nanomaterials
37. Wu T, Hua MY, Chen J, Wei KC, ... Jung SM. Effects of external magnetic field on biodistribution of nanoparticles: A histological study. *Journal of Magnetism and Magnetic Materials*. 2007; 311(1): 372–375. doi:10.1016/j.ejpb.2006.01.006
38. Vaughn JM, McConville JT, Burgess D, ... Peters JI. Single dose and multiple dose studies of itraconazole nanoparticles. 2006; 63(2): 95–102. doi:10.1016/j.jmmm.2006.10.1202
39. Pandey R and Khuller GK. Solid lipid particle-based inhalable sustained drug delivery system against experimental tuberculosis. *Tuberculosis*. 2005; 85(4): 227–234. doi:10.1016/j.tube.2004.11.003
40. Sharma A, Sharma S, and Khuller GK. Lectin-functionalized poly (lactide-co-glycolide) nanoparticles as oral/aerosolized antitubercular drug carriers for treatment of tuberculosis. *Journal of Antimicrobial Chemotherapy*. 2004; 54(4): 761–766. doi:10.1093/jac/dkh411
41. Zahoor A, Sharma S, and Khuller GK. Inhalable alginate nanoparticles as antitubercular drug carriers against experimental tuberculosis. *International Journal of Antimicrobial Agents*. 2005; 26(4): 298–303. doi:10.1016/j.ijantimicag.2005.07.012
42. Woodle MC and Lu PY. Nanoparticles deliver RNAi therapy. *Materials Today*. 2005; 8(8): 34–41. doi:10.1016/s1369-7021(05)71035-x
43. Gopalan B, Ito I, Branch CD, Stephens C, Roth JA, and Ramesh R. Nanoparticle based systemic gene therapy for lung cancer: Molecular mechanisms and strategies to suppress nanoparticle-mediated inflammatory response. *Technology in Cancer Research and Treatment*. 2004 ; 3(6): 647–657. doi:10.1177/153303460400300615
44. Kaul G and Amiji M . Tumor-targeted gene delivery using poly(ethylene glycol)-modified gelatin nanoparticles: In vitro and in vivo studies. *Pharmaceutical Research*. 2005; 22(6): 951–961. doi:10.1007/s11095-005-4590-3
45. Fink TL, Klepcyk PJ, Oette SM, Gedeon CR, and Hyatt SL. Plasmid size up to 20 Kbp does not limit effective in vivo lung gene transfer using compacted DNA nanoparticles. *Gene Therapy*. 2006; 13(13): 1048–1051. doi:10.1038/sj.gt.3302761
46. Li S-D and Huang L. Targeted delivery of antisense oligodeoxynucleotide and small interference RNA into lung cancer cells. *Molecular Pharmaceutics*. 2006; 3(5): 579–588. doi:10.1021/mp060039w
47. Oh YJ, Lee J, Seo JY, Rhim T, Kim SH, ... Yoon HJ. Preparation of budesonide-loaded porous PLGA microparticles and their therapeutic efficacy in a murine asthma model. 2011; 150(1): 56–62. doi:10.1016/j.jconrel.2010.11.001
48. Azarmi S, Tao X, Chen H, Wang Z, ... Finlay WH. Formulation and cytotoxicity of doxorubicin nanoparticles carried by dry powder aerosol particles. *International Journal of Pharmaceutics*. 2006; 319(1-2): 155–161. doi:10.1016/j.i.jpharm.2006.03.052
49. Zhang L, Gu FX, Chan JM, Wang AZ, Langer RS, and Farokhzad OC. Nanoparticles in medicine: Therapeutic applications and developments. *Clinical Pharmacology and Therapeutics*. 2008 ; 83(5): 761–769. doi: 10.1038/sj.clpt.6100400
50. Müller RH, Mäder K, and Gohla S. Solid lipid nanoparticles (SLN) for controlled drug delivery – A review of the state of the art. *European Journal of Pharmaceutics and Biopharmaceutics*. 2000; 50(1): 161–177.
51. Joshi M and Misra A. Pulmonary disposition of budesonide from liposomal dry powder inhaler. *Methods and Findings in Experimental and clinical pharmacology*. 2001; 23(10): 531–536. doi:10.1358/mf.2001.23.10.677118
52. D'Silva J and Calster GV. Taking temperature – A review of European Union regulation in nanomedicine. *European Journal of Health Law*. 2009; 16(3): 249–259. doi: 10.1163/157180909X453071
53. Marchant GE, Douglas JS, Abbott KW, and Danforth TL. International harmonization of regulation of nanomedicine. *Studies in Ethics, Law, and Technology*. 2009; 3(3). doi:10.2202/1941-6008.1120
54. Siegrist S, Cörek E, Detampel P, Sandström J, Wick P, and Huwyler J. Preclinical hazard evaluation strategy for nanomedicines. *Nanotoxicology*. 2019; 13(1): 73–99. doi:10.1080/17435390.2018.1505000
55. Rasmussen K, Rauscher H, Kearns P, González M, and Sintes JR. Developing OECD test guidelines for regulatory testing of nanomaterials to ensure mutual acceptance of test data. *Regulatory Toxicology and Pharmacology*. 2019; 104(June): 74–83. doi:10.1016/j.yrtph.2019.02.008
56. OECD. 2009. Test No. 412: Subacute Inhalation Toxicity: 28-Day Study.
57. Videira MA, Llop J, Sousa C, Kreutzer B, and Cossío U. Pulmonary administration: Strengthening the value of therapeutic proximity. 2020; 7(50): 1–11. doi:10.3389/fmed.2020.00050
58. Kaminskas LM, McLeod VM, Ryan GM, ... Kelly BD. Pulmonary administration of a doxorubicin-conjugated dendrimer enhances drug exposure to lung metastases and improves cancer therapy. *Journal of Controlled Release*. 2014; 183: 18–26.
59. Badea I, Wettig S , Verrall R , and Foldvari M . Topical non-invasive gene delivery using Gemini nanoparticles in interferon-γ-deficient mice. *European Journal of Pharmaceuticals and Biopharmaceutics*. 2007; 65(3): 414–422. doi:10.1016/j.ejpb.2007.01.002
60. Lacerda L, Bianco A, Prato M, and Kostarelos K. Carbon nanotubes as nanomedicines: From toxicology to pharmacology. *Advanced Drug Delivery Reviews*. 2006; 58(14): 1460–1470. doi:10.1016/j.addr.2006.09.015
61. Oberdörster G, Sharp Z, Atudorei V, Elder A, Gelein R, Kreyling W, and Cox C. Translocation of inhaled ultrafine particles to the brain. *Inhalation Toxicology*. 2004; 16(6-7): 437–445. doi:10.1080/08958370490439597
62. Elgindy N, Elkhodairy K, ... Molokhia A. Biopolymeric microparticles combined with lyophilized monophase dispersions for controlled flutamide release.

International Journal of Pharmaceutics. 2011; 411(1-2): 113–120. doi:10.1016/j.ijpharm.2011.03.047
63. Wittgen BPH, Kunst PWA, Born KVD, Van Wijk AW, Perkins W, Pilkiewicz FG, Perez-Soler R, Nicholson S, Peters GJ, and Postmus PE. Phase I study of aerosolized SLIT cisplatin in the treatment of patients with carcinoma of the lung. *Clinical Cancer Research*. 2007; 13(8): 2414–2421. doi:10.1158/1078-0432.CCR-06-1480
64. Dailey LA and Kissel T. New poly (lactic-co-glycolic acid) derivatives: Modular polymers with tailored properties. *Drug Discovery Today: Technologies*. 2005; 2(1): 7–13.
65. Woods A, Patel A, Spina D, ... Riffo-Vasquez Y. In vivo biocompatibility, clearance, and biodistribution of albumin vehicles for pulmonary drug delivery. *Journal of Controlled Release*. 2015; 210: 1–9. doi: 10.1016/j.jconrel.2015.05.269
66. Brzoska M, Langer K, Coester C, ... Loitsch S. Incorporation of biodegradable nanoparticles into human airway epithelium cells—In vitro study of the suitability as a vehicle for drug or gene delivery in pulmonary disease. 2004; 318(2): 562–570. doi:10.1016/j.bbrc.2004.04.067
67. Harush-Frenkel O, Bivas-Benita M, and Nassar T. A safety and tolerability study of differently-charged nanoparticles for local pulmonary drug delivery. *Toxicology and Applied Pharmacology*. 2010; 246(1-2): 83–90. doi:10.1016/j.taap.2010.04.011
68. Nemmar A, Hoylaerts MF, ... Hoet PHM. Size effect of intratracheally instilled particles on pulmonary inflammation and vascular thrombosis. *Toxicology and Applied Pharmacology*. 2003; 186(1): 38–45. doi:10.1016/s0041-008x(02)00024-8
69. Inoue KI, Takano H, Yanagisawa R, Hirano S, Sakurai M, Shimada A, and Yoshikawa T. Effects of airway exposure to nanoparticles on lung inflammation induced by bacterial endotoxin in mice. *Environmental Health Perspectives*. 2006; 114(9): 1325–1330. doi:10.1289/ehp.8903
70. ISO - ISO/TS 18827:2017 - Nanotechnologies — Electron Spin Resonance (ESR) as a Method for Measuring Reactive Oxygen Species (ROS) Generated by Metal Oxide Nanomaterials. 2020. Accessed April 20. https://www.iso.org/standard/63502.html
71. ISO - ISO 19007:2018 - Nanotechnologies — In Vitro MTS Assay for Measuring the Cytotoxic Effect of Nanoparticles. 2020. Accessed April 20. https://www.iso.org/standard/63698.html
72. Sarlo K, Blackburn KL, Clark ED, Grothaus J, and Chaney J. Tissue distribution of 20 Nm, 100 Nm and 1000 Nm fluorescent polystyrene latex nanospheres following acute systemic or acute and repeat airway exposure in the rat. *Toxicology*. 2009; 263(2-3): 117–126. doi:10.1016/j.tox.2009.07.002
73. Oberdörster G, Maynard A, Donaldson K, Castranova V, Fitzpatrick J, Ausman K, Carter J, et al. Principles for characterizing the potential human health effects from exposure to nanomaterials: Elements of a screening strategy. *Particle and Fibre Toxicology*. 2005 ; 2(1). doi: 10.1186/1743-8977-2-8
74. Driscoll KE, Carter JM, ... Howard BW. Pulmonary inflammatory, chemokine, and mutagenic responses in rats after subchronic inhalation of carbon black. 1996; 136(2): 372–380. doi:10.1006/taap.1996.0045
75. Barlow PG, Clouter-Baker A, Donaldson K, MacCallum J, and Stone V. Carbon black nanoparticles induce type II epithelial cells to release chemotaxins for alveolar macrophages. *Particle and Fibre Toxicology*. 2005; 2(1):1–4. doi:10.1186/1743-8977-2-11
76. Wiebert P, Sanchez-Crespo A, Falk R, Philipson K, Lundin A, Larsson S, Möller W, Kreyling W, and Svartengren M. No significant translocation of inhaled 35-Nm carbon particles to the circulation in humans. *Inhalation Toxicology*. 2006; 18(10): 741–747. doi:10.1080/08958370600748455
77. Xia T, Kovochich M, Brant J, Hotze M, Sempf J, Oberley T, Sioutas C, Yeh JI, Wiesner MR, and Nel AE. Comparison of the abilities of ambient and manufactured nanoparticles to induce cellular toxicity according to an oxidative stress paradigm. *Nano Letters*. 2006; 6(8): 1794–1807. doi: 10.1021/nl061025k
78. Yhee JY, Im J, and Nho RS. Advanced therapeutic strategies for chronic lung disease using nanoparticle-based drug delivery. 2016. doi:10.3390/jcm5090082
79. Takenaka S, Karg E, Kreyling W, Lentner B, Möller W, Behnke-Semmler M, Jennen L, et al. Distribution pattern of inhaled ultrafine gold particles in the rat lung. *Inhalation Toxicology*. 2006; 18: 733–740. doi:10.1080/08958370600748281
80. Lin W, Huang YW, Zhou XD, and Ma Y. Toxicity of cerium oxide nanoparticles in human lung cancer cells. *International Journal of Toxicology*. 2006; 25(6): 451–457. doi:10.1080/10915810600959543
81. Forbes B and Ehrhardt C. Human respiratory epithelial cell culture for drug delivery applications. *European Journal of Pharmaceutics and Biopharmaceutics*. 2005; 60(2): 193–205.
82. Kwang-Jin K, Borok Z, and Crandall ED. A useful in vitro model for transport studies of alveolar epithelial barrier. *Pharmaceutical Research*. 2001; 18(3): 253–255.
83. Dailey LA, Jekel N , Fink L, Gessler T, ... Schmehl T. Investigation of the proinflammatory potential of biodegradable nanoparticle drug delivery systems in the lung. *Toxicology and Applied Pharmacology*. 2006; 215(1): 100–108.
84. Stuart D, Löbenberg R, Ku T, ... Azarmi S. Biophysical investigation of nanoparticle interactions with lung surfactant model systems. *Journal of Biomedical Nanotechnology*. 2006; 2(3-4): 245–252.
85. Babu A, Templeton AK, Munshi A, and Ramesh R. Nanoparticle-based drug delivery for therapy of lung cancer: Progress and challenges. 2013.
86. Moghimi SM, Hunter AC , and Murray JC. Long-circulating and target-specific nanoparticles: Theory to practice. *Pharmacological Reviews*. 2001; 53(2): 283–318.
87. Dobrovolskaia MA, Aggarwal P, Hall JB, and McNeil SE. Preclinical studies to understand nanoparticle interaction with the immune system and its potential effects on nanoparticle biodistribution. *Molecular Pharmaceutics*. 2008; 5: 487–495. doi:10.1021/mp800032f

47

Nanoparticle-Based Lung Drug Delivery: A Clinical Perspective

Vandit Shah[1] **and Jigna Shah**[2]
[1]*Post Graduate Scholar, Department of Pharmacology, L. M. College of Pharmacy, Ahmedabad, Gujarat, India*
[2]*Professor, Department of Pharmacology, Institute of Pharmacy, Nirma University, Ahmedabad, Gujarat, India*

47.1 Introduction

Lung diseases consist of wide variety of deadly and obstinate diseases, such as chronic obstructive pulmonary disease (COPD), asthma, lung cancer, idiopathic pulmonary fibrosis (IPF), and cystic fibrosis (CF) (1,2). Epidemiological data showcases the wide spread of these diseases, affecting a large group of people. About 300 million people globally are currently suffering from asthma, and 210 million people from COPD. Often the diseases are fatal and it is not possible to restore the lungs' capacity to their fullest (3,4). Lung drug delivery is challenging, primarily due to the lungs' defense mechanism against foreign particles. This mechanism prevents the drug particles from entering the respiratory tract also, removing and inactivating the settled particles out of the respiratory tracts (5). In addition, the mechanical, immunological, and chemical barriers play a major role in preventing drug particles from reaching their site of action. Mucus-producing and ciliated columnar epithelial cells, via mucociliary escalator system, mechanically remove the deposited insoluble particles (6). Immunological response is mounted by the macrophages in orchestra with the epithelial cells, T-cells, dendritic cells, and lymphocytes present in the alveoli (6). Chemically proteolytic enzymes and lung surfactants lead to inactivation and less adherence of the drug particles. Lastly, treatment success depends on the patient's adherence to the therapy and correct use of the therapy. These are some of the prominent problems faced by pharmaceutical companies; overcoming them will aid in achieving unmet clinical needs and leveraging the true potential of the pulmonary route (7–10). However, for chronic diseases, such as asthma and COPD, current pharmacotherapy only helps in symptomatic relief. Also, each patient needs to persevere with these drugs for a lifelong period (11). Furthermore, for lung cancer treatment, chemotherapy, radiotherapy, and surgery remain the preferred choices. The efficacy of new drugs depends on various aspects like deep lung deposition, ability to bypass mechanical (mucociliary escalation) and immunological barriers (macrophage clearance), and ultimately whether they are taken up in ample amounts by the target cells (12).

Traditionally, to treat chronic lung diseases, a variety of chemical entities, antibodies, peptides, and genetic molecules such as siRNA, miRNA, and shRNA have been used. Lung drug delivery to the site of action for the topically acting drugs and to the site of absorption for the systemically acting drug is itself a type of targeted drug delivery. The topically acting drugs offers several advantages, like rapid onset of effect, relatively low dose, and fewer side effects. Furthermore, for the systemically acting drugs, lungs provide an attractive target by virtue of the epithelial cells having >100 m^2 of area and <1 µm thickness (13).

In recent times, nano-sized carriers have shown promising results for lung disease pharmacotherapy by the virtue of their inherent physical properties. This platform technology is being used to improve pharmacokinetic properties of the currently available drugs, as well as in targeted therapy by using diverse targeting motifs. This approach helps to bypass limitations offered by the traditional therapy, including low diffusion and absorption, and inappropriate pharmacokinetic profiles (14,15). In gene therapy, the viral vectors have been shown to cause several side effects; on the other hand, nano-particulate based carriers have shown efficient delivery of the genetic materials with significant fewer adverse effects (16–19).

47.2 In Vivo Behavior of Nanoparticles

Nanoparticles, by virtue of their size, have a unique characteristic feature that is being leveraged for achieving disease-specific drug delivery. Their large surface area helps them interact with cells and surrounding tissue, enhancing their efficacy to a greater extent as compared to conventional dosage forms (20,21). In the conditions of tumor, chronic inflammation, and trauma there is an increase in blood flow, abnormal blood vessels, and high endothelial permeability, which helps the nanoparticles in accumulating at the disease-specific site (22–24). Based on these characteristics of cancer, nano-formulations such as Abraxane®, an FDA-approved anti-cancer agent for non small cell lung cancer, was developed (25). Similarly, liposomes of paclitaxel and cisplatin are in phase II clinical trials for lung cancer (26,27).

For other chronic lung diseases, like COPD and cystic fibrosis, in which low permeable blood vessels make it difficult for nanoparticles to accumulate and provide therapeutic effect, studies investigating the nanoparticles accumulating in

DOI: 10.1201/9781003046547-47

different organ systems have shown low accumulation in the lungs as compared to other highly perfused organs systems. Therefore, for lung drug delivery, certain nanocarriers are not capable of improving the pharmacokinetic profile of the drug, but rather lead to toxicity. A specialized approach depending on the properties and type of nanoparticles is necessary, for instance, doxorubicin-loaded solid lipid nanoparticles and methotrexate-loaded albumin nanoparticles have high pulmonary distribution and efficacy (28,29). Solid lipid nanoparticles and liposomes are the most favored forms, considering the fact that they are less liable to aggregation (30). Surface modification with hydrophilic polymers and structural modifications of nanoparticles help in proper absorption and accumulation of drug particles in lungs (31). For instance, surface modified dendrimers with polyethylene glycol (PEG) lead to higher accumulation in the blood as compared to unmodified dendrimers, which showed minimal lung accumulation and were highly absorbed through the bloodstream (31). Various other determinants like size, patient-related factors, breathing problems, and methods of drug delivery need to be addressed for designing a disease-specific delivery method (30,32).

47.3 Drug Delivery System and Its Related Clinical Trials

Many clinical trials are filed on the ClinicalTrials.gov, a service of the United States Institute of Health. The database consists of various nanoparticle-based lung drug delivery clinical trials (Table 47.1). Here we discuss various nanocarrier-based drug delivery systems like polymeric, lipidic, hybrid lipid–polymer nanoparticles, dendrimeri, and Inorganic (Figure 47.1).

47.3.1 Polymer-Based Pulmonary Delivery

Polymers from natural or synthetic origin are widely used for nanoparticles formulations, and are being used to deliver chemotherapeutic agents, genes, or a combination of both to lungs. Polymers are repeated units of monomer, macromolecular in structure. They possess a key property of biodegradability, allowing their efficient and adverse-effect-free application. For instance, PEG polymer is extensively used for NP surface modification. Also, PEGylation of NPs, because of their biologically inert nature has led to a decrease in immune cell-mediated opsonization and deeper penetration in respiratory mucus cells (33,34). Gelatin-based NPs (GNPs) in combination with cisplatin for lung adenocarcinoma cells also enables deeper penetration into the lungs, thereby leading to higher accumulation and therapeutic efficacy (35). Furthermore, polymeric NPs have shown promising results as compared to viral vectors for delivery of genetic materials. The most widely used and studied polymer is polyethyleneimine (PEI), because of its electrostatic affinity it easily binds to nucleotides. Additionally, negatively charged phospholipids help in fusion of this positively charged polymer, facilitating the uptake and escape from endolysosomes (36). However, there have been reports of PEI-based toxicity, which can be overcome by the chitosan-PEI copolymer (37,38). To avoid drug resistance, combined delivery of genes and chemotherapy agents is being investigated, for example, doxorubicin-conjugated PEI linked by cis-aconitic anhydride with Bcl2 siRNA (39). Such characteristic properties of polymeric nanocarriers help in developing targeted and less toxic drug delivery for pulmonary diseases like asthma and tuberculosis.

47.3.2 Lipid Based Pulmonary Delivery

Lipid-based nanocarriers have been used from the dawn of nanobiotechnology and are composed of cholesterol and phosphotidylcholine (40). Liposomes are the key lipid-based NPs because of their ability to carry hydrophobic (lipophilic bilayer core) and hydrophilic drugs (aqueous core) (41). Liposomes with antibiotics, anti-cancer, antioxidant and anti-asthma drugs are being used for pulmonary diseases (42–45). For instance, liposomal 9-nitrocamptothecin (9NC) formulated with dilauroyl phospathidylcholine against lung cancer showed reduced drug load with the same effectiveness, which might lead to decrease in adverse effects (46). Furthermore, increased local lung deposition (310 ng/g) can be observed with liposomal camptothecin (CPT) as compared to conventional therapy (2 ng/g) (47). However, the systemic delivery of liposomal formulations tends to be eliminated by the reticuloendothelial system, limiting their application. On the other hand there is growing evidence of higher retention of liposomes when administered through inhalation.

Solid lipid nanoparticles (SLNs) have a solid lipid core in place of the liquid core of the droplet. A variety of materials are being used to formulate SLNs, like cholesterol (steroid), decanoic acid (fatty acid), tripalmitin (triglycerides), cetyl palmitate (waxes), and glyceryl behenate (partial glycerides) (48). However, the SLNs have several drawbacks, such as drug expulsion on storage and low drug loading capacity (49). Such challenges can be bypassed by the usage of nanostructured lipid carriers (NLCs), which consist of an unstructured solid lipid matrix and an aqueous phase comprising surfactant (50). Using sodium taurocholate as surfactant, and miglyol and compritol as unstructured solid lipid matrix, celecoxib-encapsulated NLCs were formulated for lung cancer treatment (51). The data showed a controlled drug release and higher therapeutic dosage reaching the target site.

47.3.3 Hybrid Lipid–Polymer Nanoparticles

Lipid-polymer hybrid nanoparticles (LPHNs) have a biodegradable polymer core and liposomal shell, with high loading capacity for both hydrophobic and hydrophilic drugs in polymeric and lipid core (52,53). A polymeric shell and lipid core has been used to decrease the lungs' clearance of the drugs via phagocytosis by macrophages. PEG5000–1,2-distearoyl-phosphatidylethanolamine (PEG5000-DSPE) loaded with paclitaxel was able to enhance the lung residence time, with 45-fold higher AUC as compared to I.V. administered drug (54). Hybrid nanoparticles can also be formulated using a hydrophilic polymer core and hydrophobic lipid shell. A 5-FU/poly-glutamic acid core and tripalmitin/cetylalcohol shell based lipid-coated NP showed prolonged release and higher lung accumulation of 5-FU as compared to polymeric microsphere and liposomes formulation (55).

TABLE 47.1
Various Nanoparticle-Based Clinical Trials for Lung Diseases (ClinicalTrials.gov)

Carrier Type	Drug	Clinical Trials ID	Description	Status
Liposome	9-Nitrocamptothecin	NCT00250068	To determine the overall response rate to liposomal 9-nitro-20(S)-camptothecin (L9NC) administered by aerosolization in patients with non small cell lung cancer (NSCLC).	Completed
Liposome	Amikacin	NCT03038178	To study efficacy, safety, and tolerability of once-daily dose of liposomal-amikacin for inhalation (LAI) for *Mycobacterium abscessus* lung disease.	Completed
Liposome	Cyclosporine	NCT01650545	To study efficacy and safety of aerosolized liposomal cyclosporine A in chronic rejection in lung transplant recipient with bronchiolitis obliterans syndrome (BOS).	Completed
Liposome	Cyclosporine A	NCT04107675	Safety in treatment of BOS1 in adult recipient of an allogeneic hematopoietic stem cell transplant.	Recruiting
Liposome	Paclitaxel	NCT02996214	Efficacy and safety of paclitaxel liposomes and cisplatin compared with gemcitabine and cisplatin for squamous NSCLC.	Active, not recruiting
Liposome	Gene therapy	NCT00004806	Efficacy and safety of lipid-mediated transfer of the cystic fibrosis transmembrane conductance regulator gene to nasal epithelium in patient with cystic fibrosis.	Completed
Lipid Nanoparticle	Biological: Quaratusugene ozeplasmid	NCT04486833	Safety and efficacy of GPX-001(a TUSC2, tumor suppressor gene, encapsulate by nonviral lipid nanoparticles) added to osimertinib in NSCLC patients with activating EGFR mutations.	Not yet recruiting
Micelles	Paclitaxel	NCT02667743	Paclitaxel micelles for injection and paclitaxel injection in combination with cisplatin for advanced NSCLC.	Active, not recruiting
Polymeric micelles	Paclitaxel	NCT01023347	Paclitaxel (Genexol®) and cisplatin versus paclitaxel loaded polymeric micelle (Genexol-PM®) and cisplatin in advanced NSCLC.	Completed
Albumin bound	Paclitaxel	NCT02016209	Platinum based albumin bound paclitaxel regimen in advanced NSCLC.	Unknown
Albumin bound	Carboplatin–nanoparticle albumin bound (Nab) paclitaxel	NCT04033354	HLX10 (recombinant anti-PD-1 humanized monoclonal antibody injection) + chemotherapy (carboplatin–nanoparticle albumin bound (Nab)-paclitaxel) for local or metastatic NSCLC.	Recruiting
Nanoparticles	Remdesivir and Neurosivir	NCT04480333	Safety, tolerability, and pharmacokinetics of inhaled nanoparticle formulation of remdesivir (GS-5734) and NA-831 (Neurosivir).	Recruiting
Nanoparticles	Docetaxel	NCT02283320	Docetaxel nanoparticles for injectable suspension for patients with KRAS positive or squamous cell NSCLC.	Completed
Nanoparticles	Docetaxel	NCT01792479	Docetaxel nanoparticles for injectable suspension for patients with NSCLC.	Completed
Nanoparticles	Hafnium oxide	NCT04505267	NBTXR3 and radiation therapy for treatment of inoperable recurrent NSCLC.	Not yet recruiting

47.3.4 Dendrimers-Based Pulmonary Delivery

Dendrimers are highly monodispersed nanoparticles with repeatedly branched molecules. Their surface and size can be controlled in a defined manner (56). Their added advantages are the ability to carry a high amount of drugs and high pulmonary absorption. For instance, PEG-modified dendrimers showed high absorption post inhalation (57). Various agents like antibiotics, steroids, and anti-cancer drugs are being delivered by dendrimers, to obtain enhanced and targeted therapeutics for chronic lung diseases (58–60).

47.3.5 Inorganic Nanocarrier-Based Pulmonary Delivery

Inorganic materials such as gold, silica, and iron oxide are widely used to make the nanoparticles, by the virtue of their plasmonic and magnetic properties. In particular, gold nanoparticles are being extensively investigated for gene therapy because their cationic metal ions are able to bind to anionic nucleic acids (61). Despite such advantages offered by inorganic nanoparticles, their application in pulmonary disease remains limited, as they tend to cause toxicity. Such metal NP-induced

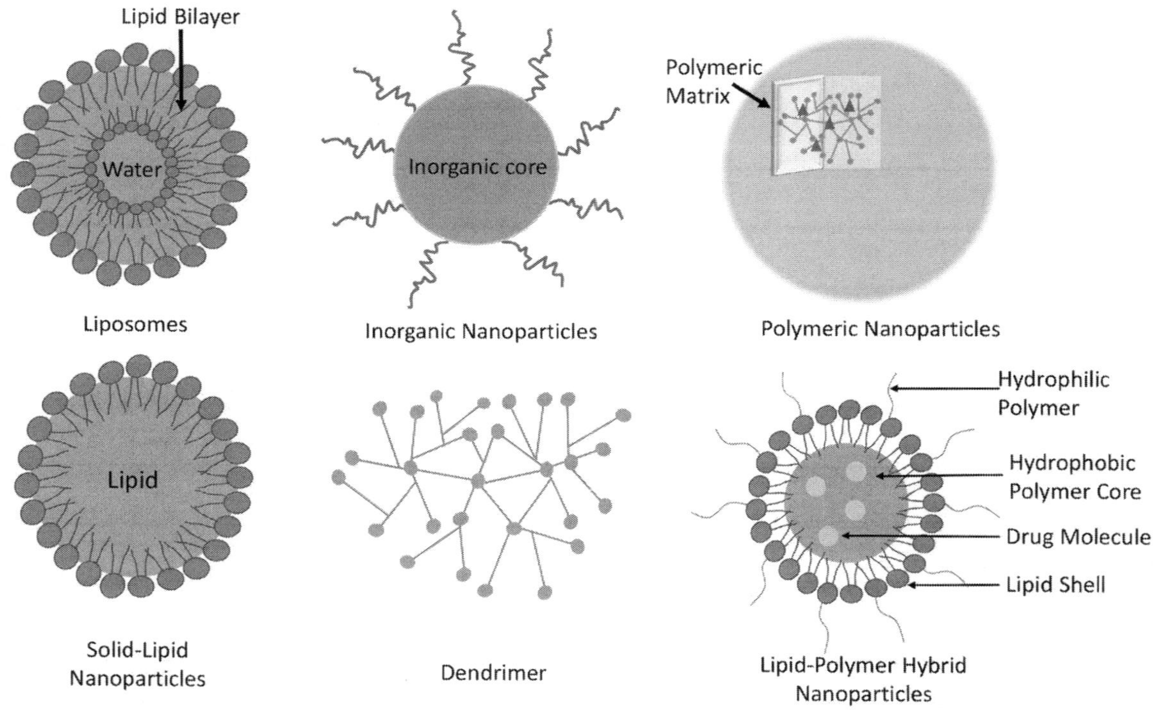

FIGURE 47.1 Sch ematic Representation of Various Nanoparticle-Based Drug Delivery System

toxicity can be controlled by regulating certain attributes like size, time, and concentration of metal ions used in formulating them (62). Additionally, they tend to form aggregates with serum proteins when given I.V. surface modification of these particles with PEG has been shown to reduce the aggregates' in formation (63). Magnetic NPs (MNPs) have emerged as a promising candidate for targeted drug delivery. Tumor ablation can be achieved through magnetic hyperthermia-generated heat. This can be achieved when an interchangeable magnetic field is applied to superparamagnetic iron oxide NPs (SPIONs) (64). Furthermore, mesoporous silica NPs (MSNs) are also used for target specific drug delivery, because of their large surface area and pore volume. The hydrophobic anti-cancer drugs can be entrapped into the pores or conjugated with electrostatic or covalent interaction with silicon groups, thereby protecting them from degradation (65,66). Quantum dots are semiconductors on a nonoscale, made up of group III-V or II-IV elements as a core and a polymer coating as a shell. Their lung retention time is higher, but can also cause cytotoxicity and oxidative stress in lung cells (53,67). Therefore, it is necessary to overcome their limitations and prove their safety before moving toward clinical trials of inorganic nanoparticles.

47.4 Factors Affecting the Toxicological Potential of Nanoparticles

Local deposition and accumulation of the insoluble nanocarriers and the response, such as increased inflammation, macrophages, and oxidative stress, has raised alarming safety concerns (68–70). A single dose of PEGylated poly-1-lysine dendrimer–DOX conjugate when given intra-tracheally caused severe toxicity and death in rats. It is thought to be due to dendrimer-led prolonged doxorubicin release in lungs. Also, the standardized tests need to be developed for membrane integrity, inflammatory mediator's release, and cell metabolic activity.

47.4.1 Shape and Structure

Shape and structure of the nanoparticles have severe implications for the toxicity profile of the nanoparticles. Different structures lead to varying toxicity profiles, for example, carbon black and graphite have different pulmonary effects as compared to carbon nanotubes (smaller structures with same composition). Likewise, fibrous nanoparticles have a tendency to cause granuloma formation and inflammation (71,72).

47.4.2 Particle Size

Inhaled nanoparticles tend to deposit in different areas of the respiratory pathway depending on their size. Larger particles get deposited on the upper respiratory tract, while smaller sized particles tend to get through the distal areas of the respiratory tract. For example, 20 nm diameter nanoparticles had 50% accumulation in alveolar areas, 15% in the nasopharyngeal area, and 15% in the tracheobronchial area (73,74). On the other hand, 1 nm diameter nanoparticles had 90% accumulation in the nasopharyngeal area and 10% in the tracheobronchial area (75). Nanoparticles, because of their large surface area teds to cause severe allergic reactions and cytotoxicity. Larger particles show relatively lower toxicity as compared to particles of 100 nm in size or less, which tend to leak into organs and cause toxicity (71,76).

47.4.3 Surface Charge

Surface charge of the nanoparticles plays a key role in lung toxicity. Nanoparticles with positive surface charge lead to high toxicity as compared to negatively charged particles (77). For instance, the poly(styrene) nanoparticles with positive surface charge induced pulmonary inflammation and high total protein, cell recruitment, and lactate dehydrogenase release. However, no such reactions were observed with negatively charged counterparts of these formulations (78).

47.4.4 Biodegradability of the Nanoparticles

Biodegradable polymers are highly considered for the development of novel drug loaded nanoparticles, overcoming the limitations of inflammation caused by non-biodegradable nanoparticles (79,80). Dipalmitoylphosphatidylcholine (DPPC), a lung surfactant, is used to formulate nanoparticles, decreasing the toxicity by a significant amount. Amine modified branched polyester and linear PLGA biodegradable NPs, when compared to poly(styrene) nanoparticles, showed significantly lower inflammatory response. Furthermore, branched NPs, compared to PLGA, are preferable for frequently administered nanoformulations owing to their biodegradability profile (81).

47.5 Future Perspective and Concluding Remarks

In the recent past, various types of nano-formulation-based lung drug delivery have been developed based on different platform technologies. Novel therapies thus developed are promising candidates for their application in chronic lung diseases with antibodies, nucleic acids, chemotherapeutics, and combinatorial therapies. Alveofact® was the first FDA-approved liposomal formulation for respiratory disease (82). Liposomes are easy to scale up, devoid of irritation, and are biodegradable, making them a prominent candidate for lung drug delivery. Other liposome-based products like Pulmaquin® and Arikace® are under clinical investigation (83). A clinical trial of 9-nitro-20-camphothecin (9NC) with dilauroyl phosphatidylcholine (DLPC) liposomes was performed in 24 patients. Grade 1 or 2 based side effects were observed, with nausea, fatigue, and cough being the most frequent. Thus, concluded the study, safety of the liposomal formulations needs to be checked (84). Similar toxicity was observed in cholesterol- and DPPC-based liposomes loaded with cisplatin in 17 patients with pulmonary carcinoma (85). In the case of cystic fibrosis, highly viscous sputum is secreted that prevents the deposition of nanoparticles on the pulmonary epithelial cells. To tackle this situation, strategies such as surface shielding by PEG, encapsulation of mucolytic agents, and mannitol-based osmotic modification have resulted in increased drug penetration (86–88). Nanotoxicology and especially inorganic nanoparticle-related toxicity, inflammation, and immunogenicity need to be addressed (89). Although in vivo toxicity profile prediction of nanoparticles is hard, optimization and appropriate selection of nanoparticles might help bridge the gap between safety and attractive therapeutic function. Despite the fact that current clinical application of nanocarriers is limited in pulmonary disease, innovative nanomaterials are warranted.

REFERENCES

1. World Health Organization. Global Surveillance, Prevention and Control of Chronic Respiratory Diseases: A Comprehensive Approach, 2007, vii–146.
2. Global Tuberculosis Report 2013 - World Health Organization. Geneva, Switzerland, 2013.
3. Halbert RJ, Natoli JL, Gano A, Badamgarav E, Buist AS, and Mannino DM. Global burden of COPD: Systematic review and meta-analysis. *European Respiratory Journal.* 2006; 28(3): 523–532. 10.1183/09031936.06.00124605
4. Mosali M, Fabian D, Holt S, Beasley R, and Global Initiative for Asthma (GINA) program. The global burden of asthma: Executive summary of the GINA Dissemination Committee Report. *Allergy.* 2004; 59(5): 469–478.
5. Labiris NR and Dolovich MB. Pulmonary drug delivery. Part I: Physiological factors affecting therapeutic effectiveness of aerosolized medications. *British Journal of Clinical Pharmacology.* 2003; 56(6): 588–599. doi:10.1046/j.1365-2125.2003.01892.x
6. Patton JS, Brain JD, Davies LA, Fiegel J, Gumbleton M, Kim KJ, Sakagami M, Vanbever R, and Ehrhardt C. The particle has landed – Characterizing the fate of inhaled pharmaceuticals. *Journal of Aerosol Medicine and Pulmonary Drug Delivery.* 2010; 23(Suppl 2): S71–S87. doi:10.1089/jamp.2010.0836
7. Baker KE, Bonvini SJ, Donovan C, Foong RE, Han B, Jha A, Shaifta Y, Smit M, Johnson JR, and Moir LM. Novel drug targets for asthma and COPD: Lessons learned from in vitro and in vivo models. *Pulmonary Pharmacology and Therapeutics.* 2014; 29(2): 181–198. doi:10.1016/j.pupt.2014.05.008
8. Durham AL, Caramori G, Chung KF, and Adcock IM. Targeted anti-inflammatory therapeutics in asthma and chronic obstructive lung disease. *Translational Research.* 2016; 167(1): 192–203. doi: 10.1016/j.trsl.2015.08.004
9. Fujita Y, Takeshita F, Kuwano K, and Ochiya T. RNAi therapeutic platforms for lung diseases. *Pharmaceuticals.* 2013; 6(2): 223–250. doi: 10.3390/ph6020223
10. Ruppert C, Schmidt R, Grimminger F, Suzuki Y, Seeger W, Lehr CM, and Günther A. Chemical coupling of a monoclonal antisurfactant protein-B antibody to human urokinase for targeting surfactant-incorporating alveolar fibrin. *Bioconjugate Chemistry.* 2002; 13(4): 804–811. doi: 10.1021/bc0255081
11. Meyer KC. Diagnosis and management of interstitial lung disease. *Translational Respiratory Medicine.* 2014; 2(1): 1–13. doi: 10.1186/2213-0802-2-4
12. Patton JS and Byron PR. Inhaling medicines: Delivering drugs to the body through the lungs. *Nature Reviews Drug Discovery.* 2007; 6(1): 67–74. doi:10.1038/nrd2153
13. Brain JD. Inhalation, deposition, and fate of insulin and other therapeutic proteins. *Diabetes Technology & Therapeutics.* 2007; 9(Suppl 1): S-4–S-15. doi:10.1089/dia.2007.0228

14. Burhan E, Ruesen C, Ruslami R, Ginanjar A, Mangunnegoro H, Ascobat P, Donders R, Crevel RV, and Aarnoutse R. Isoniazid, rifampin, and pyrazinamide plasma concentrations in relation to treatment response in Indonesian pulmonary tuberculosis patients. *Antimicrobial Agents and Chemotherapy*. 2013; 57(8): 3614–3619. doi: 10.1128/AAC.02468-12
15. Williams RO, Carvalho TC, and Peters JI. Influence of particle size on regional lung deposition - What evidence is there? *International Journal of Pharmaceutics*. 2011; 406(1-2): 1–10. doi:10.1016/j.ijpharm.2010.12.040
16. Bahadori M and Mohammadi F. Nanomedicine for respiratory diseases. *Tanaffos*. 2012; 11(4): 18–22. https://www.ncbi.nlm.nih.gov/pmc/articles/PMC4153217/
17. Gioia SD, Trapani A, Castellani S, Carbone A, Belgiovine G, Craparo EF, Puglisi G, Cavallaro G, Trapani G, and Conese M. Nanocomplexes for gene therapy of respiratory dseases: Targeting and overcoming the mucus barrier. *Pulmonary Pharmacology and Therapeutics*. 2015; 34(October): 8–24. doi:10.1016/j.pupt.2015.07.003
18. Ratemi E, Shaik AS, Al Faraj A, and Halwani R. Alternative approaches for the treatment of airway diseases: Focus on nanoparticle medicine. *Clinical & Experimental Allergy*. 2016; 46(8): 1033–1042. doi:10.1111/cea.12771
19. Smola M, Vandamme T, and Sokolowski A. Nanocarriers as pulmonary drug delivery systems to treat and to diagnose respiratory and non respiratory diseases. *International Journal of Nanomedicine*. 2008; 3(1): 1–19. doi:10.2147/ijn.s1045
20. Buzea C, Pacheco II, and Robbie K. Nanomaterials and nanoparticles: Sources and toxicity. *Biointerphases*. 2007; 2(4): MR17–MR71. doi:10.1116/1.2815690
21. Oberdörster G, Oberdörster E, and Oberdörster J. Nanotoxicology: An emerging discipline evolving from studies of ultrafine particles. *Environmental Health Perspectives*. 2005; 113(7): 823–839. doi:10.1289/ehp.7339
22. Chung CY, Yang JT, and Kuo YC. Polybutylcyanoacrylate nanoparticles for delivering hormone response element-conjugated neurotrophin-3 to the brain of intracerebral hemorrhagic rats. *Biomaterials*. 2013; 34(37): 9717–9727. doi:10.1016/j.biomaterials.2013.08.083
23. Lee SJ, Lee A, Hwang SR, Park JS, Jang J, Huh MS, Jo DG, et al. TNF-α gene silencing using polymerized SiRNA/thiolated glycol chitosan nanoparticles for rheumatoid arthritis. *Molecular Therapy*. 2014; 22(2): 397–408. doi:10.1038/mt.2013.245
24. Sharma HS, Ali SF, Dong W, Tian ZR, Patnaik R, Patnaik S, Sharma A, et al. Drug delivery to the spinal cord tagged with nanowire enhances neuroprotective efficacy and functional recovery following trauma to the rat spinal cord. *Annals of the New York Academy of Sciences*. 2007; 1122(1): 197–218. doi:10.1196/annals.1403.014
25. Gupta N, Hatoum H, and Dy GK. First line treatment of advanced non-small-cell lung cancer - Specific focus on albumin bound paclitaxel. *International Journal of Nanomedicine*. 2013; 9(1): 209–221. doi:10.2147/IJN.S41770
26. Ait-Oudhia S, Mager D, and Straubinger R. Application of pharmacokinetic and pharmacodynamic analysis to the development of liposomal formulations for oncology. *Pharmaceutics*. 2014; 6(1): 137–174. doi:10.3390/pharmaceutics6010137
27. Chang HI and Yeh MK. Clinical development of liposome-based drugs: Formulation, characterization, and therapeutic efficacy. *International Journal of Nanomedicine*. 2012; 7: 49–60. doi:10.2147/ijn.s26766
28. Santhi K, Dhanaraj SA, Koshy M, Ponnusankar S, and Suresh B. Study of biodistribution of methotrexate-loaded bovine serum albumin nanospheres in mice. *Drug Development and Industrial Pharmacy*. 2000; 26(12): 1293–1296. doi:10.1081/DDC-100102311
29. Zara GP, Cavalli R, Fundarò A, Bargoni A, Caputo O, and Gasco MR. Pharmacokinetics of doxorubicin incorporated in solid lipid nanospheres (SLN). *Pharmacological Research*. 1999; 40(3): 281–286. doi:10.1006/phrs.1999.0509
30. Kuzmov A and Minko T. Nanotechnology approaches for inhalation treatment of lung diseases. *Journal of Controlled Release*. 2015; 219(December): 500–518. doi:10.1016/j.jconrel.2015.07.024
31. Ryan GM, Kaminskas LM, Kelly BD, Owen DJ, McIntosh MP, and Porter CJH. Pulmonary administration of PEGylated polylysine dendrimers: Absorption from the lung versus retention within the lung is highly size-dependent. *Molecular Pharmaceutics*. 2013; 10(8): 2986–2995. doi:10.1021/mp400091n
32. Kumar A, Chen F, Mozhi A, Zhang X, Zhao Y, Xue X, Hao Y, Zhang X, Wang PC, and Liang XJ. Innovative pharmaceutical development based on unique properties of nanoscale delivery formulation. *Nanoscale*. 2013; 5(18): 8307–8325. doi:10.1039/c3nr01525d
33. Jokerst JV, Lobovkina T, Zare RN, and Gambhir SS. Nanoparticle PEGylation for imaging and therapy. *Nanomedicine*. 2011; 6(4): 715–728. doi:10.2217/nnm.11.19
34. Schuster BS, Suk JS, Woodworth GF, and Hanes J. Nanoparticle diffusion in respiratory mucus from humans without lung disease. *Biomaterials*. 2013; 34(13): 3439–3446. doi:10.1016/j.biomaterials.2013.01.064
35. Tseng CL, Su WY, Yen KC, Yang KC, and Lin FH. The use of biotinylated-EGF-modified gelatin nanoparticle carrier to enhance cisplatin accumulation in cancerous lungs via inhalation. *Biomaterials*. 2009; 30(20): 3476–3485. doi:10.1016/j.biomaterials.2009.03.010
36. Hong SH, Park SJ, Lee S, Cho CS, and Cho MH. Aerosol gene delivery using viral vectors and cationic carriers for in vivo lung cancer therapy. *Expert Opinion on Drug Delivery*. 2015; 12(6): 977–991. doi:10.1517/17425247.2015.986454
37. Jiang HL, Kim YK, Arote R, Nah JW, Cho MH, Choi YJ, Akaike T, and Cho CS. Chitosan-graft-polyethylenimine as a gene carrier. *Journal of Controlled Release*. 2007; 117(2): 273–280. doi:10.1016/j.jconrel.2006.10.025
38. Kafil V and Omidi Y. Cytotoxic impacts of linear and branched polyethylenimine nanostructures in A431 cells. *BioImpacts*. 2011; 1(1): 23–30. doi:10.5681/bi.2011.004
39. Xu C, Wang P, Zhang J, Tian H, Park K, and Chen X. Pulmonary codelivery of doxorubicin and SiRNA by pH-sensitive nanoparticles for therapy of metastatic lung cancer. *Small*. 2015; 11(34): 4321–4333. doi:10.1002/smll.201501034
40. Mendoza AE HD, Campanero MA, Mollinedo F, and Blanco-Prieto MJ. Lipid nanomedicines for anticancer drug

therapy. *Journal of Biomedical Nanotechnology.* 2009; 5(4): 323–343. doi:10.1166/jbn.2009.1042

41. Gaber M, Medhat W, Hany M, Saher N, Fang JY, and Elzoghby A. Protein-lipid nanohybrids as emerging platforms for drug and gene delivery: Challenges and outcomes. *Journal of Controlled Release.* 2017; 254: 75–91. doi:10.1016/j.jconrel.2017.03.392
42. Elhissi AMA, Islam MA, Arafat B, Taylor M, and Ahmed W. Development and characterisation of freeze-dried liposomes containing two anti-asthma drugs. *Micro and Nano Letters.* 2010; 5(3): 184–188. doi:10.1049/mnl.2010.0032
43. Hoesel LM, Michael AF, Niederbichler AD, Rittirsch D, McClintock SD, Reuben JS, Pianko MJ, et al. Ability of antioxidant liposomes to prevent acute and progressive pulmonary injury. *Antioxidants & Redox Signaling.* 2008; 10(5): 963–972. doi:10.1089/ars.2007.1878
44. Liu Chunmei, Shi J, Dai Q, Yin X, Zhang X, and Zheng A. In-vitro and in-vivo evaluation of ciprofloxacin liposomes for pulmonary administration. *Drug Development and Industrial Pharmacy.* 2015; 41(2): 272–278. doi:10.3109/03639045.2013.858740
45. Patel AR, Chougule MB, Townley I, Patlolla R, Wang G, and Singh M. Efficacy of aerosolized celecoxib encapsulated nanostructured lipid carrier in non-small cell lung cancer in combination with docetaxel. *Pharmaceutical Research.* 2013; 30(5): 1435–1446. doi:10.1007/s11095-013-0984-9
46. Knight V, Kleinerman ES, Waldrep JC, Giovanella BC, Gilbert BE, and Koshkina NV. 9-Nitrocamptothecin liposome aerosol treatment of human cancer subcutaneous xenografts and pulmonary cancer metastases in mice. *Annals of the New York Academy of Sciences.* 2006; 922(1): 151–163. doi:10.1111/j.1749-6632.2000.tb07033.x
47. Koshkina NV, Waldrep JC, Seryshev A, Knight V, and Gilbert BE. Distribution of camptothecin after delivery as a liposome aerosol or following intramuscular injection in mice. *Cancer Chemotherapy and Pharmacology.* 1999; 44(3): 187–192. doi:10.1007/s002800050966
48. Martins S, Sarmento B, Ferreira DC, and Souto EB. Lipid-based colloidal carriers for peptide and protein delivery – Liposomes versus lipid nanoparticles. *International Journal of Nanomedicine.* 2007; 2(4): 595–607. /pmc/articles/PMC2676808/?report=abstract
49. Mehnert W and Mäder K. Solid lipid nanoparticles: Production, characterization and applications. *Advanced Drug Delivery Reviews.* 2012; 64 (Suppl): 83–101. doi:10.1016/j.addr.2012.09.021
50. Naseri N, Valizadeh H, and Zakeri-Milani P. Solid lipid nanoparticles and nanostructured lipid carriers: Structure preparation and application. *Advanced Pharmaceutical Bulletin.* 2015;5(3): 303–313. doi:10.15171/apb.2015.043
51. Patlolla RR, Chougule M, Patel AR, Jackson T, Tata PNV, and Singh M. Formulation, characterization and pulmonary deposition of nebulized celecoxib encapsulated nanostructured lipid carriers. *Journal of Controlled Release.* 2010; 144(2): 233–241. doi:10.1016/j.jconrel.2010.02.006
52. Elzoghby AO, Mostafa SK, Helmy MW, ElDemellawy MA, and Sheweita SA. Multi-reservoir phospholipid shell encapsulating protamine nanocapsules for co-delivery of letrozole and celecoxib in breast cancer therapy. *Pharmaceutical Research.* 2017; 34(9): 1956–1969. doi:10.1007/s11095-017-2207-2
53. Garbuzenko OB, Mainelis G, Taratula O, and Minko T. Inhalation treatment of lung cancer: The influence of composition, size and shape of nanocarriers on their lung accumulation and retention. *Cancer Biology and Medicine.* 2014; 11(1): 44–55. doi:10.7497/j.issn.2095-3941.2014.01.004
54. Gill KK, Nazzal S, and Kaddoumi A. Paclitaxel loaded PEG5000-DSPE micelles as pulmonary delivery platform: Formulation characterization, tissue distribution, plasma pharmacokinetics, and toxicological evaluation. *European Journal of Pharmaceutics and Biopharmaceutics.* 2011; 79(2): 276–284. doi:10.1016/j.ejpb.2011.04.017
55. Hitzman CJ, Elmquist WF, and Wiedmann TS. Development of a respirable, sustained release microcarrier for 5-fluorouracil II: In vitro and in vivo optimization of lipid coated nanoparticles. *Journal of Pharmaceutical Sciences.* 2006; 95(5): 1127–1143. doi:10.1002/jps.20590
56. Kesharwani P, Jain K, and Jain NK. Dendrimer as nanocarrier for drug delivery. *Progress in Polymer Science.* 2014; 39(2): 268–307. doi:10.1016/j.progpolymsci.2013.07.005
57. Bharatwaj B, Mohammad AK, Dimovski R, Cassio FL, Bazito RC, Conti D, Fu Q, Reineke J, and Da Rocha SRP. Dendrimer nanocarriers for transport modulation across models of the pulmonary epithelium. *Molecular Pharmaceutics.* 2015; 12(3): 826–838. doi:10.1021/mp500662z
58. Bellini RG, Guimarães AP, Pacheco MAC, Dias DM, Furtado VR, De Alencastro RB, and Horta BAC. Association of the anti-tuberculosis drug rifampicin with a PAMAM dendrimer. *Journal of Molecular Graphics and Modelling.* 2015; 60(June): 34–42. doi:10.1016/j.jmgm.2015.05.012
59. Inapagolla R, Raja Guru B, Kurtoglu YE, Gao X, Lieh-Lai M, Bassett DJP, and Kannan RM. In vivo efficacy of dendrimer-methylprednisolone conjugate formulation for the treatment of lung inflammation. *International Journal of Pharmaceutics.* 2010; 399(1–2): 140–147. doi:10.1016/j.ijpharm.2010.07.030
60. Kaminskas LM, McLeod VM, Ryan GM, Kelly BD, Haynes JM, Williamson M, Thienthong N, Owen DJ, and Porter CJH. Pulmonary administration of a doxorubicin-conjugated dendrimer enhances drug exposure to lung metastases and improves cancer therapy. *Journal of Controlled Release.* 2014; 183(1): 18–26. doi:10.1016/j.jconrel.2014.03.012
61. Ding Y, Jiang Z, Saha K, Kim CS, Kim ST, Landis RF, and Rotello VM. Gold nanoparticles for nucleic acid delivery. *Molecular Therapy.* 2014; 22(6): 1075–1083. doi:10.1038/mt.2014.30
62. Elzoghby AO, Hemasa AL, and May SF. Hybrid protein-inorganic nanoparticles: From tumor-targeted drug delivery to cancer imaging. *Journal of Controlled Release.* 2016; 243: 303–322. doi:10.1016/j.jconrel.2016.10.023
63. Alkilany AM and Murphy CJ. Toxicity and cellular uptake of gold nanoparticles: What we have learned so far?

Journal of Nanoparticle Research. 2010; 12(7): 2313–2333. doi:10.1007/s11051-010-9911-8
64. Ahmad J, Akhter S, Rizwanullah M, Amin S, Rahman M, Ahmad MZ, Rizvi MA, Kamal MA, and Jalees Ahmad F. Nanotechnology-based inhalation treatments for lung cancer: State of the art. *Nanotechnology, Science and Applications.* 2015; 8: 55–66. doi:10.2147/NSA.S49052
65. Liang R, Wei M, Evans DG, and Duan X. Inorganic nanomaterials for bioimaging, targeted drug delivery and therapeutics. *Chemical Communications.* 2014; 50(91): 14071–14081. doi:10.1039/c4cc03118k
66. Vivero-Escoto JL, Slowing II, Lin VSY, and Trewyn BG. Mesoporous silica nanoparticles for intracellular controlled drug delivery. *Small.* 2010; 6(18): 1952–1967. doi:10.1002/smll.200901789
67. Madni A, Batool A, Noreen S, Maqbool I, Rehman F, Kashif PM, Tahir N, and Raza A. Novel nanoparticulate systems for lung cancer therapy: An updated review. *Journal of Drug Targeting.* 2017; 25(6): 499–512. doi:10.1080/1061186X.2017.1289540
68. Bermudez E, Mangum JB, Wong BA, Asgharian B, Hext PM, Warheit DB, and Everitt JI. Pulmonary responses of mice, rats, and hamsters to subchronic inhalation of ultrafine titanium dioxide particles. *Toxicological Sciences.* 2004; 77(2): 347–357. doi:10.1093/toxsci/kfh019
69. Warheit DB, Webb TR, and Reed KL. Pulmonary toxicity screening studies in male rats with TiO2 particulates substantially encapsulated with pyrogenically deposited, amorphous silica. *Particle and Fibre Toxicology.* 2006; 3(1): 3. doi:10.1186/1743-8977-3-3
70. Warheit DB, Webb TR, Reed KL, Frerichs S, and Sayes CM. Pulmonary toxicity study in rats with three forms of ultrafine-TiO2 particles: Differential responses related to surface properties. *Toxicology.* 2007; 230(1): 90–104. doi:10.1016/j.tox.2006.11.002
71. Badea I, Shawn Wettig RV, and Foldvari M. Topical non-invasive gene delivery using Gemini nanoparticles in interferon-γ-deficient mice. *European Journal of Pharmaceutics and Biopharmaceutics.* 2007; 65(3): 414–422. doi:10.1016/j.ejpb.2007.01.002
72. Oberdörster G, Sharp Z, Atudorei V, Elder A, Gelein R, Kreyling W, and Cox C. Translocation of inhaled ultrafine particles to the brain. *Inhalation Toxicology.* 2004; 16: 437–445. doi:10.1080/08958370490439597
73. Hagens WI, Oomen AG, de Jong WH, Cassee FR, and Sips AJ AM. What do we (need to) know about the kinetic properties of nanoparticles in the body? *Regulatory Toxicology and Pharmacology.* 2007; 49(3): 217–229. doi:10.1016/j.yrtph.2007.07.006
74. Moghimi SM, Hunter AC, and Murray JC. Nanomedicine: Current status and future prospects. *The FASEB Journal.* 2005; 19(3): 311–330. doi:10.1096/fj.04-2747rev
75. Curtis J, Greenberg M, Kester J, Phillips S, and Krieger G. Nanotechnology and nanotoxicology: A primer for clinicians. *Toxicological Reviews.* 2006; 25(4): 245–260. doi:10.2165/00139709-200625040-00005
76. Lacerda L, Bianco A, Prato M, and Kostarelos K. Carbon nanotubes as nanomedicines: From toxicology to pharmacology. *Advanced Drug Delivery Reviews.* 2006; 58(14): 1460–1470. doi:10.1016/j.addr.2006.09.015
77. Harush-Frenkel O, Bivas-Benita M, Nassar T, Springer C, Sherman Y, Avital A, Altschuler Y, Borlak J, and Benita S. A safety and tolerability study of differently-charged nanoparticles for local pulmonary drug delivery. *Toxicology and Applied Pharmacology.* 2010; 246(1–2): 83–90. doi:10.1016/j.taap.2010.04.011
78. Nemmar A, Hoylaerts MF, Hoet PHM, Vermylen J, and Nemery B. Size effect of intratracheally instilled particles on pulmonary inflammation and vascular thrombosis. *Toxicology and Applied Pharmacology.* 2003; 186(1): 38–45. doi:10.1016/S0041-008X(02)00024-8
79. Khattab SN, Abdel Naim SE, El-Sayed M, Bardan AAE, Elzoghby AO, Bekhit AA, and El-Faham A. Design and synthesis of new S-triazine polymers and their application as nanoparticulate drug delivery systems. *New Journal of Chemistry.* 2016; 40(11): 9565–9578. doi:10.1039/c6nj02539k
80. Elzoghby A, Freag M, Mamdouh H, and Elkhodairy K. Zein-based nanocarriers as potential natural alternatives for drug and gene delivery: Focus on cancer therapy. *Current Pharmaceutical Design.* 2018; 23(35): 5261–5271. doi:10.2174/1381612823666170622111250
81. Dailey LA, Wittmar M, and Kissel T. The role of branched polyesters and their modifications in the development of modern drug delivery vehicles. *Journal of Controlled Release.* 2005; 101: 137–149. doi:10.1016/j.jconrel.2004.09.003
82. El-Sherbiny IM, El-Baz NM, and Yacoub MH. Inhaled nano- and microparticles for drug delivery. *Global Cardiology Science and Practice.* 2015; 2015 (1): 2. doi:10.5339/gcsp.2015.2
83. Paranjpe M and Müller-Goymann C. Nanoparticle-mediated pulmonary drug delivery: A review. *International Journal of Molecular Sciences.* 2014; 15(4): 5852–5873. doi:10.3390/ijms15045852
84. Verschraegen CF, Gilbert BE, Loyer E, Huaringa A, Walsh G, Newman RA, and Knight V. Clinical evaluation of the delivery and safety of aerosolized liposomal 9-nitro-20(S)-camptothecin in patients with advanced pulmonary malignancies. *Clinical Cancer Research.* 2004; 10(7): 2319–2326. doi:10.1158/1078-0432.CCR-0929-3
85. Wittgen BPH, Kunst PWA, Born KVD, Van Wijk AW, Perkins W, Pilkiewicz FG, Perez-Soler R, Nicholson S, Peters GJ, and Postmus PE. Phase I study of aerosolized SLIT cisplatin in the treatment of patients with carcinoma of the lung. *Clinical Cancer Research.* 2007; 13(8): 2414–2421. doi:10.1158/1078-0432.CCR-06-1480
86. Broughton-Head VJ, Smith JR, Shur J, and Shute JK. Actin limits enhancement of nanoparticle diffusion through cystic fibrosis sputum by mucolytics. *Pulmonary*

Pharmacology and Therapeutics. 2007; 20(6): 708–717. doi:10.1016/j.pupt.2006.08.008
87. Lai SK, O'Hanlon DE, Harrold S, Man ST, Wang YY, Cone R, and Hanes J. Rapid transport of large polymeric nanoparticles in fresh undiluted human mucus. *Proceedings of the National Academy of Sciences of the United States of America*. 2007; 104(5): 1482–1487. doi:10.1073/pnas.0608611104
88. Yang Y, Tsifansky MD, Shin S, Lin Q, and Yeo Y. Mannitol-guided delivery of ciprofloxacin in artificial cystic fibrosis mucus model. *Biotechnology and Bioengineering*. 2011; 108(6): 1441–1449. doi:10.1002/bit.23046
89. Bonner JC. Nanoparticles as a potential cause of pleural and interstitial lung disease. *Proceedings of the American Thoracic Society*. 2010; 7: 138–141. doi:10.1513/pats.200907-061RM

48

European Perspective on Orally Inhaled Products: In Vitro Requirements for a Biowaiver

Sunita Chaudhary[1], Dhaval Patel[2], Dasharath M. Patel[3], and Jayvadan K. Patel[4]
[1]Arihant School of Pharmacy and Bio Research Institute, Gujarat, India
[2]Saraswati Institute of Pharmaceutical Sciences, Gujarat, India
[3]Graduate School of Pharmacy, Gujarat Technological University, Gandhinagar Campus, Gujarat, India
[4]Nootan Pharmacy College, Sankalchand Patel University, Gujarat, India

48.1 Introduction

The respiratory route is widely acceptable currently because it has a maximum absorptive area of up to 100 m^2 with good blood supply. Many diseases like asthma, asthmatic rhinitis, and many other respiratory diseases have their own adverse impact on the health of normal human beings (1). Globally, the occurrence of asthma is growing and chronic obstructive pulmonary disease (COPD) is widespread in 65 million people (2). In the European Union (EU), asthma is detected in 7% of the population and COPD in 6% is reported (3). Several kinds of medication are used to treat symptoms of COPD. Currently, inhaled products for the treatment of asthma are being developed (Trends in COPD pdf June 2013, accessed on Sept 2020) (4). Orally inhaled products have gained attention with increased interest in drug deposition in the lungs and alveoli. Devices which can give better targeting of drug in the lungs are largely favored, such as nebulizers, dry powder inhalers (DPIs), and pressurized metered dose inhalers (pMDIs). Use of orally inhaled products for asthma and pulmonary disease has created interest in the development of different combinations of available products and their generic versions.

Orally inhaled products (OIPs) have been successful in the development of some generic products, like Seretide®, Spiriva®, and Symbicort®, with annual revenues of US$10 billion. In the EU, there is an 8% death rate due to diseases of the respiratory system, and this creates a need for the development of safe and effective treatments. Repeatability in the workings of OIP and representing bioequivalence (BE) is difficult because the success of OIPs is mostly dependant on the mutual correlation of patient, device, and formulation. For the development of safe and effective generic products, it is important to develop appropriate approaches to demonstrate BE. One more complication observed at international level with regard to OIP is the testing and registration of new as well as generic products. In 2014, the IPAC-RS/UF conference was organized to provide information regarding the latest advancements regarding OIPs, such as metered dose inhalers (MDIs) and DPIs, and their market registration. At that conference many issues about OIPs were discussed, including the pharmacokinetic and pharmacodynamics aspects of generics and the criteria for acceptance after a demonstration of bioequivalence) (5). For complete and efficient development of an OIP, it is essential to have a through knowledge of formulation variables and their role in product performance for establishment of an *in vitro in vivo correlation* (IVIVC). IVIVC can be established successfully when in vitro and in vivo drug release prove similar in a scalable manner. IVIVC is generally proven by comparison with plasma concentrations. An IVIVC works well on oral formulations and there is much evidence that an in vitro test will not predict the result of a pharmacokinetics study, especially for orally inhaled formulations (6).

48.2 Regulatory Pathways in Europe

In the EU, the authorities responsible for regulation and approval of products follow a systematic methodology for granting of approval where equivalence is tested using in vitro equivalence, followed by pharmaceutical equivalence, and pharmacokinetic equivalence, which includes lung deposition studies, and further pharmacodynamic study that includes efficacy and safety study data. Thus in the EU, an in vitro study to establish equivalence is considered sufficient for approval of a product, but it may not be considered so in the United States (US).

In recent years, the EU as well as some other European countries have been advancing the regulations related to OIPs. Some substantial developments have materialized with regard to medical devices, bioequivalence, quality, and combination products (7).

The guidelines CPMP/EWP/4151/00 Rev.1 and EMEA/CHP/QWP/49313/ 2005 cover the overall necessities and specify the need for clinical and in vitro requirements for generics (Guideline on the requirements for clinical documentation for orally inhaled products pdf. February 2009, Accessed on December 2019) (8). In the EU, it is not compulsory to have complete equivalency between the generic product and reference product in every aspect, but a demonstration of bioequivalency with the reference product can be

DOI: 10.1201/9781003046547-48

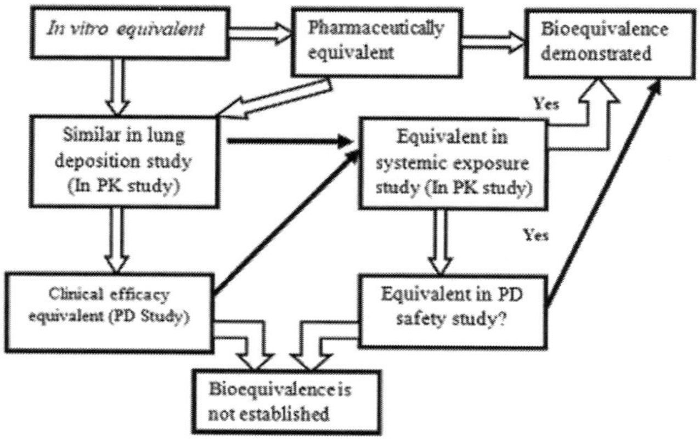

FIGURE 48.1 Diagram Representing EMA's Step-wise Approach for OIPs (10)

made by using pharmacokinetic and bioavailability studies (Guideline on the pharmaceutical quality of inhalation and nasal products pdf, March 2006, accessed on April 2015) (9). There is also specification for hybrid products which demonstrate equivalence by means of pharmacodynamic (PD) studies that can be waived from such kinetic and bioavailability studies. This classification focuses on the methods used to show therapeutic equivalence. During the study, the prescriber and patient are not informed about the approval of a generic product by pharmacokinetic study or as a hybrid which failed pharmacokinetic equivalence (1).

48.2.1 Bioequivalence Requirements in the EU

In the EU guidelines, there are systematic strategies for demonstration of therapeutic equivalence which use in vitro and in vivo data. In short, a study uses initially in vitro data intially for comparison, which is followed by pharmacokinetic data if required, and finally by any requirement for pharmacodynamics data. Figure 48.1 shows the systematic methodology to establish therapeutic equivalence for OIPs.

48.2.1.1 In Vitro Equivalence Testing

A European guideline highlights the main purpose of in vitro equivalence testing as being only for the establishment of therapeutic equivalence of OIPs. Use of in-vitro testing method is very effective for establishing equivalence of OIPs compared to the other strategies due to ease of operation and low variability between products. Generic products are approved depending upon in vitro data where it is found there is little variation with respect to the reference product, because changes from the reference product ire approved without the requirement of in vivo studies. The total aerolized dose administered to the lung and its aerodynamic particle size distribution (APSD) strongly affects the safety and efficacy of orally inhaled drug products (OINDPs). Consequently, the delivered dose and APSD are generally viewed as crucial quality attributes of inhaled products, and corresponding testing is required by regulatory guidance for characterization and quality control purposes (11).

48.2.1.1.1 Aerodynamic Particle Size Distribution

For in vitro characterization of OINDPs, ASPD and delivered dose are considered crucial attributes for quality. Depending upon the aerodynamic particle size of an aerosol, the location of deposition of particles in the lungs is decided. Specifically, the most effective range of particle size is 1 to 5 μm and particles larger than 5 μm have more impact on the oropharynx, if any particle smaller than 1 μm remains present in the air system. Larger particles will deposit at the back of the throat. Extremely small particles are directly exhaled instead of being deposited (European Pharmacopoeia 5.0: Chapter 2.9.18) (12). The cascade impaction method is used to determine APSD and the effect of different particle sizes on drug deposition in the lungs. Generally, for comparison of two OIPs, several methods are used, including inertial impaction methods, Marple-Miller impactor (MMI), Andersen cascade impactor (ACI), and next generation impactor (NGI). In Europe, USP, ACI, and MMI are recommended and NGI is specified in the *European Pharmacopoeia* (EP) (10).

For the in vitro comparison, selection of the stage is based on fine particle size and upper stage of the impactor which is relevant for measuring the safety and efficacy of the medicinal product when administered in vivo. The comparison should be done on at least four stages, which are expected. However, the amount of drug is limited and the quality of information may be degraded by division of the dose in to 4–5 stages. Depending upon the individual stage there is an in vitro acceptance range of ±15% for the aerodynamic particle size distribution (13).

However, the normally used impactor systems are not able to simulate sufficiently the anatomy of the the human respiratory system. When patients use the inhaler, the constant flow rate of the aerosol is not the same as that used in the impactor with respect to time (14). Another difference with the air pathways of the lungs is that the inlets of the impactor are of various diameters and lengths. This problem can be resolved by selecting a device with the proper mouth–throat geometry that closely resembles oropharyngeal deposition 15. If the composition is different qualitatively or quantitatively, then more broad in vitro testing is required with respect to the reference product (16). This study includes particle size distribution by laser diffraction (European Medicines Agency. Guideline on the requirements for clinical documentation for orally inhaled products (OIP) pdf 2009, accessed on March 2015) (8).

In the EU, if suspensions have been evaluated by in vitro testing using crystallographic structure, drug particle size distribution in suspension, and nebulized droplets, and have a qualitative and quantitative composition similar to the reference product, then the suspension for nebulization can be waived from in vivo studies (1).

Approval of pressurized metered dose inhalers (pMDIs) in solution is based on in vitro data in EU. In OIDP guidelines, there is a specific mention of in vitro requirements for pMDIs in suspension, but a description of a detailed test is not given. Hence, it is appropriate to accept in vitro tests for OIDP as per the US Food and Drug Administration (FDA) draft guidance for the industry (1).

48.2.1.1.2 Dose Content Uniformity

This is referred to as the amount of drug substance that is available to deliver the dose to a patient on a per-dose basis. Applicants can justify variation in the amount of dose delivered and also can justify the effects of the dose with regard to safety and efficacy of the product. The European Medicines Agency (EMA) specifies that there should be not more than a 15% difference in the targeted delivered doses of the test and reference products (17).

48.2.1.1.3 Dissolution, Permeation, Particle Clearance, and Tissue Exposure

Orally inhaled products deliver the inhaled drugs to the airways by local action and this generates a high localization of drug with improvement in the potency of the drug and the therapeutic value. When a drug has high water solubility, its molecular properties can influence the extent of tissue binding which ultimately it affects the local tissue concentration and therapeutic effects. For a compound with less solubility, material properties affect solubility and/or dissolution rate and this ultimately affects therapeutic value (18). When two formulations have similar aerodynamic particle size distributions with variation in systemic exposure, this gives a constant predicted difference in the dissolution rates of the two formulations. Even though there is a more significant effect of dissolution rate on therapeutic performance, there are no currently available regulatory guidelines on orally inhaled products and their in vitro testing. Forbes and colleagues (19) have discussed the problem of correlating in vitro dissolution data with the therapeutic and clinical performance of a product. The amount of drug dissolved is affected not only by the properties of the drug, such as surface area and solubility, which can be controlled externally, but also by physiological factors, like composition of the airway lining fluid, particle clearance rate, and permeability of the airway epithelium, all of which can vary between different regions of the lung (20).

48.2.1.1.4 In Vitro Dissolution

All pharmacopoeias specify dissolution testing as a standardized test of solid and semi-solid dosage forms, but due to the unavailability of a standardized method for measuring the dissolution behavior of OIPs, it is challenging to dissolve drug substance with poor solubility in a very limited volume of solvent, which results in poor dissolution behavior in physiological media. Thus, for systemic availability of drug from a DPI, the dissolution rate is a rate limiting step. Currently there is no available standardized in vitro test to explore the in vitro performance and characterization of DPIs (19). The FDA and the EMA recommend measuring the delivered dose and the APSD to determine the quality of the product, and this waives the dissolution testing.

Different techniques, like a modified twin stage impinger, flow through the cell, horizontal diffusion cell, Franz diffusion cell, and ethrough cell USP type IV and paddle type USP type II apparatuses with or without membrane holders, for substances using specified dissolution media are found in the literature. Selection of dissolution media is usually in the range of pH 6.8–7.4 for testing of inhaled products, and the media used can be saline phosphate buffer as well as simulated lung fluid (SLF). SLF is an aqueous solution containing mineral salts and sometimes surfactant, but no protein component or mucus. When orally inhaled products contain poorly soluble drug then it requires the presence of surfactant in the dissolution media. A surfactant that can be used in pulmonary formulations is phospholipid disaturated dipalmitoyl phosphotidyl choline (DPPC) in concentration of 0.02%, which is found in epithelial lung fluid (19). Preparation of DPPC solution is very difficult due to the unclear and lengthy method of preparation, which leads to the use of synthetic surfactants for comparison of in vitro dissolutions. In the pharma field, the FDA and EMA emphasize aerodynamic particle size distribution and delivered dose while there is no need of dissolution testing. As per the *European Pharmacopoeia*, dissolution is performed on particles collected from a cascade impactor in dissolution experiments (6).

SLF is more advantageous because it is a physiologically relevant media, with more complex composition, even with a low buffering effect sometimes, which makes it unsuitable for formulations that show a pH dependent drug release profile as well as a sustained release profile (21). It was observed that in SLF there is increase in pH within 24 hours from 7.4 to 8.8 (22). Son et al. studied the three dissolution media for the drug release profile of budesonide: phosphate buffered saline (PBS), phosphate buffer pH 7.4, and SLF. A similar drug release profile was shown in all three media (23).

48.2.1.2 Pharmacokinetics Approach

Previously in the EU, many inhaled products were approved mainly on the basis of in vitro data and pharmacokinetic data as shown in Table 48.1. EU considers pharmacokinetic bioequivalence studies as an acceptable methodology for analyzing the extent of lung deposition and the pattern of deposition after oral inhalation of products.

In EMEA guidelines, a pharmacokinetics study is performed for two main purposes: (1) to evaluate deposition of drug in the lungs in which there is an exclusion of absorption of active moiety from the gastrointestinal (GI) tract, and (2) to examine the safety of a molecule in a systemic route, where total systemic exposure in the lungs and GI tract should be investigated (1).

According to the EMA guidelines regarding orally inhaled drug products, demonstration of pharmacokinetic parameters such as maximum plasma concentration (Cmax) and area under curve (AUC) for each drug must be carried out to discover the

TABLE 48.1

EU Product Approvals Based on the EMA OIP Guidelines (24)

Products	Company	Year of Approval
Salmeterol HFA pMDI	Neolab Ltd	2011
Fluticasone/Salmeterol (FP/SM) DPI (Elpenhaler)	Pharos Ltd	2011
Fluticasone pMDI	Cipla Ltd.	2013
Ipratropium pMDI	Cipla Ltd.	2013
Fluticasone/Salmeterol pMDI	Cipla Ltd.	2014
Budesonide/Formoterol DPI (DuoResp Spiromax)	Teva Pharmaceuticals	2014
Lifsar (Fluticasone/Salmeterol DPI)	Winthrop Pharmaceuticals	2015
Sirdupla (Fluticasone/Salmeterol pMDI)	Mylan Ltd	2015
Bufomix (Budesonide/Formoterol DPI)	Orien Corporation	2014
Tiotropium Bromide DPI (Braltus or Gregal)	Teva Pharmaceuticals	2016

pMDI: pressurized metered dose inhaler; DPI: dry powder inhaler

bioequivalence of test and reference products (8). The AUC, or amount of drug available in blood from the lungs, determines the amount of dose targeted to lungs. In addition to AUC, the distribution pattern of a drug can be analyzed by Cmax and the time to achieve Cmax (Tmax). Depending upon the area of localisation of drug in the lungs, Tmax can be predicted. If drug is deposited in a peripheral area of the lung then it has shorter Tmax, and if drug from a DPI or solution in pMDI is deposited in the central lungs, then it has a longer Tmax. Sometimes a pharmacokinetic study can be used for analysis of both safety and efficacy of a drug because there is no absorption of drug in the intestine. In the European regulations, there is clear instruction about the consideration of AUC0-30 for measuring the efficacy of drug and AUC0-t for safety (1). The EMA recommends that there is no need to consider absorption of the active moiety from the GI tract in the case of orally inhaled products due to the pulmonary deposition and measurement of that amount via the lung and GI tract (safety) during a pharmacokinetic study. The EMA recommends an adult patient population, for safety purposes, in the selection of study subjects. For drugs with a narrow therapeutic window, the EMA recommends a range of 80–125% along with 90% confidence interval (CI) for ratio of the geometric means of AUC for test to reference. EMA has also increase Cmax ratio range 75–133% with 90% CI for highly variable drugs (24). Generally, pharmacokinetic parameters, including in vitro and in vivo as well as pharmacodynamics studies, are used to prove the bioequivalence of OIDPs. However, when in vitro data and pharmaceutical data are conflicting, then pharmacokinetic (PK) studies are required to prove the bioequivalence of inhaled drugs to ensure the substitutability of generics. There are some issues arising with regard to conducing PK studies to prove bioequivalence (25).

48.2.1.2.1 Dose Selection

A nasal drug delivery system delivers the drug locally; hence the systemic concentration of drug is minimum. Sometimes it is observed that a very low concentration is not detected by available bioanalytical techniques. This type of issue requires an increase in the dose of a drug or the need to develop a more sensitive bioanalytical method for drug assay. Selection of a higher dose is challenging for the safety of human volunteers. It may result in side effects like hypokalemia, tremors, and palpitations, such as in the case of salbutamol (26).

48.2.1.2.2 Subject Selection

To conduct a PK study, selection of human volunteers will be done on the basis of physiological condition and habits. Volunteers who are healthy nonsmokers are selected. There is special emphasis on selection of nonsmokers for the study for several reasons: (1) Regular smokers have chronic respiratory illnesses, which affect comparisons of PK parameters. (2) Regular smoking may induce some metabolic enzymes like CYP 1A1 and 1B1, which decrease the effectiveness of the drug. (3) Smokers have variations in mucocilliary clearance and variations in the pH of the local microenvironment (26). All these factors introduce variations intra-subject in crossover studies because the cumulative effects of these factors may create changes in the same individual at different time intervals. Screening of volunteers for respiratory diseases can be done by reviewing medical histories, checking X-rays, and doing pulmonary function testing (PFT). A spirometer is preferred to conduct PFT and peak flow meters (PFM) are used especially in non-hospitalized patients (27).

48.2.1.2.3 Subject Training

The use of PFM testing is one of the key parameters for measurement of proper performance of OIPs in terms of correct and continuous inhalation technique (26). The important points while providing training for inhalation are as follows: (1) Before starting inhalation, carry out complete exhalation. (2) Ensure complete seal of lip and mouthpiece. (3) Ensure complete coordination of actuation and inhalation as required for metered dose inhaler. (4) Instruct subject to breath slowly and deeply for 5–10 s. During inhalation device should be actuated. Then the breath should be held for 5–10 s followed by exhalation through the nose. (5) For DPIs, the energy required for inhaling and expelling drug is provided by the individual when the device is actuated. So initially there would be rapid and deep inhalation for 4–5 s, then breath-holding for 5–10 s, followed by exhalation through the nose (28).

48.2.1.2.4 Other Factors

i. ***Use of a spacer with an MDI***: By using a spacer with an MDI, we can avoid the need for coordination between inhalation and actuation, and this also avoids oropharynx deposition of active pharmaceutical ingredients (API). As per the EU guidelines, it is mandatory for the product to perform the PK studies with the spacer in order to receive approval of and licencing for the product. If the product can be used both ways, with or without the spacer, then the PK studies need to be performed in both cases (26).

ii. ***Use of charcoal block***: Generally the charcoal block method is used to eliminate drug oral absorption of a drug after a product has been orally inhaled. Administering a charcoal suspension at different time intervals can avoid the oral absorption from the inhaled drug, so the amount of drug available in systemic circulation gives an idea about the amount that was absorbed from the respiratory route (29). For all regulatory submissions, the use of a charcoal block is not compulsory for PK studies, but the EMA recommends two studies, one with the carbon block and one without. It also recommends validation of the method required for the administration of charcoal (30).

48.2.1.3 Pharmacodynamics Approach

The EMA specifies pharmacodynamics (PD) studies for safety purpose. According to the EU, if in vitro and PK studies fail to show equivalenc,e then PD studies are necessary for approvals. When there is a low concentration of drug in the plasma then the EMA also recommends a PD study. Advancement in analytical techniques, however, enables the detection of small amounts of drug in plasma and blood.

EMA highlights various recommendations for a variety of age groups. In adults, they recommend of assessment of the effect on the hypothalamic–pituitary–adrenal (HPA) axis. For children, safety data cannot be exactly predicted by extrapolating from the adult data in asthma. Finding PD equivalence uses two different tests to demonstrate systemic safety in children; for example, the systemic effect of inhaled corticosteroids can be assessed using the HPA axis, and the lower-leg bone growth rate is used as a surrogate marker for growth (8).

For the treatment of asthma and COPD, the main two class of locally acting inhalers are recommended like bronchodilators and corticosteroids. The bronchodilators are long-acting β2-agonists, short-acting β2-agonists, and anti-cholinergics, which are covered under the guidance of the EMA. The EMA also recommends bronchodilation and bronchoprotection studies. The EMA has not covered and specified a dose or criteria for acceptance of locally acting OIPs. PD studies give quantification information about the biological and physiological impacts of drug products and are used to scrutinize both safety and efficacy. Actually, this is very difficult to implement due to high levels of variability and sensitivity (17). Methacholine PD20 and an unspecific forced expiratory volume are accepted methods in the EU for locally acting OIPs. Variation in results can be reduced by crossover design, but as per European guidelines, parallel design is preferable. The time required for study and choice of primary and secondary endpoints are dependent on the therapeutic category of the test product (1).

Therefore, for European regulators, it is crucially important to demonstrate assay sensitivity. As per EMA guidelines, this requires analysis of relative potency and the response per dose. For new chemical entities, some EU reviewers preferred to use a response-scale analysis. By use of clinically related biomarkers, the pharmacological actions of the drug can be linked for successful identification of measurable results in a PD study. Using biomarkers enables a successful demonstration of the dose–response relationship that can be carried out with a representative dose. If the test and reference products show similarity clinically, that indicates that both have similar efficacy if a change in dose is rigorously associated with a measurable response. It is challenging to demonstrate the bioequivalence of orally inhaled corticosteroids based on clinical data because it gives a narrow dose response with a long time to show effect for local action. The EMA recommends different PD studies at different intervals (23).

To study the bioequivalence (BE) of inhaled corticosteroids for the treatment of asthma, the EMA recommends that quality of life be validated by using questionnaires and also by examining patient variables like exhaled nitric oxide (eNO) and sputum eosinophil. Moreover, different studies like parallel double blind or randomized studies are recommended, and a crossover design can also be used as an alternative method of measurement if needed. The EMA recommends a comparison of two different doses of test and reference substances, depending upon their dose response curve, and the selection of dose is from the sharp portion of the dose–response curve (17).

48.3 Biowaiver for OIDP in the European Union

In the European Union, PK studies are used to demonstrate bioequivalence, but under the following criteria, these studies can be waived.

48.3.1 Criteria 1

If two formulations are identical in quantity and quality, in the case of nebulized solution, the approval of a product can be granted without in vitro testing. However, if the composition of both formulations is different, then in vitro testing is necessary. Bioequivalence studies of nebulizers containing budesonide are based on in vitro data, which ensures similarity in particle size distribution in suspension and aerodynamic particle size distribution of the nebulized droplets (6).

48.3.2 Criteria 2

Products that remain in solution at the time of administration like oral solutions, effervescent tablets, and injectable solutions, if some terms and conditions are met, like absence of excipients which change the motility of the GI tract and stability of the preparation.

48.3.3 Criteria 3

As per BCS some classes of drug products can be biowaived by comparing drug release profiles, if there are differences in some key excipients that can affect absorption which can be avoided. However, the drugs that can be biowaived by the EMA for class I and III drugs are slightly different (Guideline on the Investigation of Bioequivalence. 20-1-2010, accessed in June 2020) (31).

48.3.4 Criteria 4

If bioequivalence is proved with the most effective dose, then additional strengths do not require testing to show bioequivalence. In such cases, the strength of the test product must fulfill all requirements of qualitative and quantitative composition, having used similar procedures for manufacturing, and the same dissolution profile is obtained.

When a comparison of test and reference products is carried out using data from in vitro studies, then in vivo studies can be waived, but when a product is available in different strengths and BE is proven with one strength in an in vivo study, the product is also considered proven for equivalence (Community Code Relating to Medicinal Products for Human Use pdf. 2001, accessed in August 2020) (32).

In the case of orally inhaled products available in multiple strengths, the EMA emphasizes in vitro testing for both test and reference products in all possible strengths. When the formulation shows in vitro dose linearity, then an individual dose of drug substance is sufficient to establish bioequivalence clinically, and it is generally preferable to select the minimum dose at different levels, which will enhance the sensitivity of the study. Selection of the dose for study also depends upon the type of study. For PD studies, dose selection should be the minimum in order to obtain maximum sensitivity of drug assay as much possible, as per guideline specification. Conversely, for PK studies, the selected dose for study is highest to get the maximum plasma levels because it gets low systemic exposure (Legal basis for products for local use pdf. December 2012, accessed in January 2020) (33).

Suppose testing reveals that the test and reference product and linearity relationship does not exist, then bioequivalence of the test product to the reference product will have to be recognized with a different product strength and maybe by all possible product strengths. Because it will be necessary to modify the test product so that it becomes similar to the reference product for comparison of their therapeutic equivalence. If, by using different strengths of the test and reference products, it is not possible to show proportionality in vitro, then a bracketing approach is used to establish equivalence. In the bracketing approach, dose strengths are selected from the most similar to the most different for in vitro testing (8).

Therefore, in the design and development of a generic product which acts systemically, the manufacturer should always consider a biowaiver if possible, otherwise it will be necessary to perform a PK bioequivalence study. However, in the legislation of the EU, the focus is on product approvability, whereas interchangeability is not covered in the regulations. Hence PK differences between products are not considered to be clinically significant, so products are considered as therapeutic equivalents based on the PD, and the product can be approved as a hybrid. On the other hand, those generics having bioequivalence based on PK and that waived PD to demonstrate are considered as generic.

48.4 Criteria for the In Vitro Comparison

Based on Committee for Medicinal Products for Human Use (CHMP) guidelines, comparative in vitro data obtained with an above-mentioned method may be acceptable when the test product is compared with the reference product. The following criteria should be satisfied:

i. The product must have similar API in its different forms, like hydrates, solvates, salts, esters.
ii. Product should be pharmaceutically similar like inhaler, metered dose inhaler, dry powder inhaler.
iii. Performance of product, behavior of aerosol particles, and dissolution characteristics should not be influenced by differences in crystalline or amorphous forms.
iv. Product performance, like delivered dose uniformity, aerosol particle characteristics, and inhalation compatibility of the patient should not be influenced by any difference in excipients, and they must not affect change safety of the product (25).
v. The inhaled volume coming through the device must enable a sufficient amount of drug into the lungs, and should be similar (within 15%) to the reference.
vi. The same amount of active substance should be released from both test and reference product.
vii. The inhalation device has the same resistance to airflow (within 15%) compared to reference (8).

48.5 Case Studies of Orally Inhaled Drug Products

48.5.1 Case Study of Glycopyrronium and Formoterol Fumarate

Bevespi Aerosphere was the first medicine in its category to be approved by the European Commission (EC); it contains glycopyrronium and formoterol fumarate in a pressurized metered-dose inhaler (pMDI) for treatment of COPD. It also acts as bronchodilator. This product offers a new choice of device in the inhaler category. Bevespi Aerosphere is to be taken two times a day, a fixed dose, which offers dual action as a bronchodilator containing glycopyrronium which acts as an antagonist of a long-acting muscarinic and formoterol fumarate, which act as agonists of β2. In a phase III trial program, approval of this product was done based on the efficacy and safety of Bevespi Aerosphere and it involved more than 5000 patients in a study (Bevespi Aerosphere approved in China for patients with COPD, May 2020, accessed on August 2020) (34).

48.5.2 Case Study of Fluticasone Furoate/ Vilanterol DPI

This is a combination of fluticasone furoate and vilanterol, available in a fixed dose that contains corticosteroids and a long-acting β2 adrenoreceptor agonist (LABA), which is branded as Relvar or Revinty, and its once-daily dose is delivered via a dry powder inhaler (DPI). The EU approves fluticasone furoate/vilanterol DPI for the treatment of asthma in adults and adolescents. It was observed in this study that a once-daily dose of fluticasone furoate/vilanterol improved pulmonary function and is more effective that a twice-daily dose. It was also observed that fluticasone furoate/vilanterol is more effective compared to fluticasone propionate. With the additional consideration of the frequency of dose administration, fluticasone furoate/vilanterol provides more patient compliance due to the one daily dose where other combinations available for asthma require doses twice a day (35).

48.6 Summary

Approval of an orally inhaled drug product can be allowed in the European Union based on its in vitro data if it follows the step-wise approach of the CHMP plus fulfillment of certain conditions. This step-wise approach can be successfully applied to solutions for nebulization as well as pMDI and suspension for nebulization because variations in the dissolution profile of the suspension depends upon its particle size distribution and surface properties, which can be compared by crystallography. For OIDP, it is essential to demonstrate a comparison of in vitro data as a part of the establishment of equivalence with more than one strength of the same reference product, and it is also necessary to demonstrate similar flow rates from the devices for the test and reference products for the pharmacokinetic data to be accepted.

REFERENCES

1. Lee L, Bhawana S, Alfredo G, Gustavo M, Ying L, Sarah L, Shuguang H, Juliet R, Abhijit V, Jaideep G, Shrinivas P, and Svetlana L. Regulatory considerations for approval of generic inhalation drug products in the US, EU, Brazil, China, and India. *The AAPS Journal*. 2015; 17(5): 1285–1304. doi: 10.1208/s12248-015-9787-8.
2. Global Initiative for Asthma (GINA). "Global burden of asthma." http://www.ginasthma.org/local/uploads/files/GINABurdenReport_1.pdf. Accessed 26 Nov 2014.
3. World Health Organization. Burden of COPD. http://www.who.int/respiratory/copd/burden/en/. Accessed 26 Nov 2014. 3. "Asthma and Allergy Foundation of America. Asthma facts and figures." https://www.aafa.org/display.cfm?sub=42&id=8. Accessed 26 Nov 2014.
4. American Lung Association. "Trends in COPD (Chronic bronchitis and emphysema): morbidity and mortality." 2013. http://www.lung.org/finding-cures/our-research/trend-reports/copd-trendreport.pdf. Accessed 26 Nov 2014.
5. Hochhaus G, Craig D, Martin O, Sau L, and Svetlana L. Current scientific and regulatory approaches for development of orally inhaled and nasal drug products: overview of the IPAC-RS/University of Florida Orlando Inhalation Conference. *The AAPS Journal*. 2015; 17(5): 1305–1310. doi: 10.1208/s12248-015-9791-z.
6. Frohlich E, et al. Biological obstacles for identifying in vitro-in vivo correlations of orally inhaled formulations. *Pharmaceutics*. 2019; 11(7): 316, 1–19. 10.3390/pharmaceutics11070316
7. Dina Al-Numani D, et al. Rethinking bioequivalence and equivalence requirements of orally inhaled drug products. *Asian Journal of Pharmaceutical Sciences*. 2015; 10(6): 461–471. 10.1016/j.ajps.2015.08.006.
8. Committee for Human Medicinal Products (CHMP). "European Medicines Agency: Guideline on the requirements for clinical documentation for orally inhaled products (OIP) including the requirements for demonstration of therapeutic equivalence between two inhaled products for use in the treatment of asthma and chronic obstructive pulmonary disease (COPD) in adults and for use in the treatment of asthma in children and adolescents."22-1-2009. London, European Medicines Agency (EMA). Available at www.ema.europa.eu/docs/en_GB/document_library/Scientific_guideline/2009/09/WC500003504.pdf. Accessed on January 20, 2019.
9. European Medicines Agency. "Guideline on the pharmaceutical quality of inhalation and nasalproducts." http://www.ema.europa.eu/docs/en_GB/document_library/Scientific_guideline/2009/09/WC500003568.pdf, 2006. Accessed on 08.07.15
10. Evans C, David C, Tim C, Eva A, Richard A, Dale C, Sanjeeva D, Myrna D, Eng P, William D, Anders F, Afredo G, Michael G, Robert H, Gu¨ nther H, Susan H, Paul L, Svetlana L, Parameswaran N, Dennis OC, David P, Ilse P, Colin R, Dennis S, Gur J, Marjolein W, and Patricia W. Equivalence considerations for orally Inhaled products for local action—ISAM/IPAC-RS European Workshop Report. *Journal of Aerosol Medicine and Pulmonary Drug Delivery*. 2012; 25(3): 118–123. 10.1089/jamp.2011.0968.
11. Sandell D. "Review of the EMEA guidelines' in-vitro equivalence criteria for cascade impaction data." [internet]. 2010 [cited 2019 Apr 30]. 2010. Available from: https://ipacrs.org/assets/uploads/outputs/Sandell.pdf. Accessed 11 June 2019.
12. Preparations for inhalation: Aerodynamic assessment of fine particles. European Pharmacopoeia 5.0: Chapter 2.9.18.
13. Andersen AA. New sampler for the collection, sizing, and enumeration of viable airborne particles. *Journal of Bacteriology*. 1958; 76(5): 471–484. PMID: 13598704.
14. Khoubnasabjafari M, Elaheh R, Morteza S, Vahid J, Lan C, Donghao C, Hak-Kim C, and Abolghasem J. A new hypothesis to investigate bioequivalence of pharmaceutical inhalation products. *DARU Journal of Pharmaceutical Sciences*. 2019; 27(1): 517–524. doi: 10.1007/s40199-019-00250-x.
15. Newman SP and Chan HK. In vitro/in vivo comparisons in pulmonary drug delivery. *Journal of Aerosol Medicine and Pulmonary Drug Delivery*. 2008; 21(1): 77–84.
16. MacNeish CF, Meisner D, Thibert R, Kelemen S, Vadas EB, and Coates AL. A comparison of pulmonary availability between ventolin (albuterol) nebules and ventolin (albuterol) respirator solution. *Chest*. 1997; 111(1): 204–208. 10.1378/chest.111.1.204

17. Lu D, Lee SL, Lionberger RA, Choi S, Adams W, Caramenico HN, Chowdhury BA, Conner DP, Katial R, Limb S, Peters JR, Yu L, Seymour S, and Li BV. International guidelines for bioequivalence of locally acting orally inhaled drug products: similarities and differences. *The AAPS Journal*. 2015; 17(3): 546–557. 10.1208/s12248-015-9733-9.
18. Backman P, Adelman H, Petersson G, and Jones CB. Advances in inhaled technologies: understanding the therapeutic challenge, predicting clinical performance, and designing the optimal inhaled product. *Clinical Pharmacology & Therapeutics*. 2014; 95(5): 509–520. 10.1038/clpt.2014.27.
19. Forbes B, Per B, David C, Myrna D, Bing V, and Beth M. In vitro testing for orally inhaled products: developments in science-based regulatory approaches. *The AAPS Journal*. 2015; 17(4): 837–852. 10.1208/s12248-015-9763-3.
20. Riley T, David C, Jan A, Andrea C, Agnes C, Andrew C, Monisha D, Janet M, Jolyon M, Maria R, Nastaran S, Terrence T, and Svetlana L. Challenges with developing in vitro dissolution tests for orally inhaled products (OIPs). *AAPS PharmSciTech*. 2012; 13(3): 978–989. doi: 10.1208/s12249-012-9822-3.
21. Radivojev S, Zellnitz S, Paudel A, and Fröhlich E. Searching for physiologically relevant in vitro dissolution techniques for orally inhaled drugs. *International Journal of Pharmaceutics*. 2019; 556: 45–56. 10.1016/j.ijpharm.2018.11.072.
22. Gunther A, Schmidt R, Nix F, Yabut-Perez M, Guth C, Rosseau S, Siebert C, Grimminger F, Morr H, Velcovsky HG, et al. Surfactant abnormalities in idiopathic pulmonary fibrosis, hypersensitivity pneumonitis and sarcoidosis. *European Respiratory Journal*. 1999; 14: 565–573.
23. Son YJ, Horng M, Copley M, and McConville JT. 2010. Optimization of an in vitro dissolution test method for inhalation formulations. *Dissolution*, pp. 6–13.
24. Usmani OS, Molimard M, Gaur V, Gogtay J, Pal Singh GJ, Malhotra Gngfh and Derom E. Scientific rationale for determining the bioequivalence of inhaled drugs. *Clinical Pharmacokinetics*. 2017; 56(10): 1139–1154. 10.1007/s40262-017-0524-6
25. Busse WW, Brazinsky S, Jacobson K, Stricker W, Schmitt K, Vanden BJ, et al. Efficacy response of inhaled beclomethasone dipropionate in asthma is proportional to dose and is improved by formulation with a new propellant. *Journal of Allergy and Clinical Immunology*. 1999; 104(6): 1215–1222. 10.1016/S0091-6749(99)70016-3.
26. Thakkar K, Mhatre S, Jadhav M, Goswami S, and Shah R. Pharmacokinetic studies for proving bioequivalence of orally inhaled drug products—critical issues and concepts. *Frontiers in Pharmacology*. 2015; 6(117): 1–4. 10.3389/fphar.2015.00117.
27. Silvestro L, Savu SR, Savu SN, Tudoroniu A, and Tarcomnicu I. Development of a sensitive method for simultaneous determination of fluticasone propionate and salmeterol in plasma samples by liquid chromatography-tandem mass spectrometry. *Biomedical Chromatography*. 2012; 26: 627–635. 10.1002/bmc.1708.
28. Loke YK and Singh S. Risk of acute urinary retention associated with inhaled anticholinergics in patients with chronic obstructive lung disease: systematic review. *Therapeutic Advances in Drug Safety*. 2013; 4: 19–26. 10.1177/2042098612472928.
29. Fowler SJ and Lipworth BJ. Pharmacokinetics and systemic β2-adrenoceptor-mediated responses to inhaled salbutamol. *British Journal of Clinical Pharmacology*. 2001; 51(4): 359–362.
30. Scott JE. The pulmonary surfactant: impact of tobacco smoke and related compounds on surfactant and lung development. *Tobacco Induced Diseases*. 2004; 2: 3. 10.1186/1617-9625-2-1-3
31. Committee for Medicinal Products for Human Use (CHMP). "Guideline on the Investigation of Bioequivalence."20-1-2010. London, European Medicines Agency (EMA). Available at www.ema.europa.eu/docs/en_GB/document_library/Scientific_guideline/2010/01/WC500070039.pdf. Accessed on June 25, 2020.
32. Directive 2001/83/EC of the European Parliament and of the Council of 6 April 2001 on the "Community Code Relating to Medicinal Products for Human Use." *Official Journal of the European Communities L*. 2004; 311: 67–128.
33. Co-ordination Group for Mutual Recognition and Decentralised Procedures—Human. CMDh Questions & Answers. Generic Applications. Legal basis for products for local use." 12/2012. Available at www.hma.eu/fileadmin/dateien/Human_Medicines/CMD_h_/Questions_Answers/CMDh-272-2012-Rev0-2012_10.pdf. Accessed January 2, 2020.
34. https://www.astrazeneca.com/media-centre/press-releases/2020/bevespi-aerosphere-approved-in-china-for-patients-with-copd.html https://www.astrazeneca.com/media-centre/press-releases/2020/bevespi-aerosphere-approved-in-china-for-patients-with-copd.html. Accessed on 20 June 2020.
35. Deeks ED, Lyseng-Williamson KA, and Syed YY. Fluticasone furoate/vilanterol dry-powder inhaler in asthma: a guide to its use in the EU. *Drugs & Therapy Perspectives*. 2017; 33: 153–159.

Index

Note: *Italicized* page numbers refer to figures, **bold** page numbers refer to tables

E

F

G

H

I

J

K

L

M

N

O

P

Q

R